Volkswagen Jetta, Golf, GTI, Cabrio

Service Manual

Jetta, Golf, GTI: 1993 - 1999
Cabrio: 1995 - 2002
Including 1.9L TDI, 2.0L, and 2.8L VR6

A3 Platform

[B]

Bentley Publishers
Cambridge, Massachusetts

BENTLEY PUBLISHERS® | Automotive Reference™

Bentley Publishers, a division of Robert Bentley, Inc.
1734 Massachusetts Avenue
Cambridge, MA 02138 USA
800-423-4595 / 617-547-4170

Information that makes the difference®

BentleyPublishers®.com

Technical contact information

We welcome your feedback. Please submit corrections and additions to our Volkswagen technical discussion forum at:

`http://www.BentleyPublishers.com`

Updates and Corrections

We will evaluate submissions and post appropriate editorial changes online as updates or tech discussion. Appropriate updates and corrections will be added to the book in future printings. Check for updates and corrections for this book before beginning work on your vehicle. See the following web address for additional information:

`http://www.BentleyPublishers.com/updates/`

WARNING—Important Safety Notice

Do not use this manual for repairs unless you are familiar with automotive repair procedures and safe workshop practices. This manual illustrates the workshop procedures for some maintenance and service work. It is not a substitute for full and up-to-date information from the vehicle manufacturer or for proper training as an automotive technician. Note that it is not possible for us to anticipate all of the ways or conditions under which vehicles may be serviced or to provide cautions as to all of the possible hazards that may result.

We have endeavored to ensure the accuracy of the information in this manual. Please note, however, that considering the vast quantity and the complexity of the service information involved, we cannot warrant the accuracy or completeness of the information contained in this manual.

FOR THESE REASONS, NEITHER THE PUBLISHER NOR THE AUTHOR MAKES ANY WARRANTIES, EXPRESS OR IMPLIED, THAT THE INFORMATION IN THIS MANUAL IS FREE OF ERRORS OR OMISSIONS, AND WE EXPRESSLY DISCLAIM THE IMPLIED WARRANTIES OF MERCHANTABILITY AND OF FITNESS FOR A PARTICULAR PURPOSE, EVEN IF THE PUBLISHER OR AUTHOR HAVE BEEN ADVISED OF A PARTICULAR PURPOSE, AND EVEN IF A PARTICULAR PURPOSE IS INDICATED IN THE MANUAL. THE PUBLISHER AND AUTHOR ALSO DISCLAIM ALL LIABILITY FOR DIRECT, INDIRECT, INCIDENTAL OR CONSEQUENTIAL DAMAGES THAT RESULT FROM ANY USE OF THE EXAMPLES, INSTRUCTIONS OR OTHER INFORMATION IN THIS MANUAL. IN NO EVENT SHALL OUR LIABILITY WHETHER IN TORT, CONTRACT OR OTHERWISE EXCEED THE COST OF THIS MANUAL.

Before attempting any work on your Volkswagen, read the **Warnings and Cautions** on page vii, and any **WARNING** or **CAUTION** that accompanies a procedure in the manual. Review the **WARNINGS** and **CAUTIONS** each time you prepare to work on your Volkswagen.

Your common sense and good judgment are crucial to safe and successful service work. Read procedures through before starting them. Think about whether the condition of your car, your level of mechanical skill, or your level of reading comprehension might result in or contribute in some way to an occurrence which might cause you injury, damage your car, or result in an unsafe repair. If you have doubts for these or other reasons about your ability to perform safe repair work on your car, have the work done at an authorized Volkswagen dealer or other qualified shop.

Part numbers listed in this manual are for identification purposes only, not for ordering. Always check with your authorized Volkswagen dealer to verify part numbers and availability before beginning service work that may require new parts.

Special tools required to perform certain service operations are identified in the manual and are recommended for use. Use of improper tools may be detrimental to the car's safe operation as well as the safety of the person servicing the car.

The vehicle manufacturer will continue to issue service information updates and parts retrofits after the editorial closing of this manual. Some of these updates and retrofits will apply to procedures and specifications in this manual. We regret that we cannot supply updates to purchasers of this manual.

This manual is prepared, published and distributed by Bentley Publishers, 1734 Massachusetts Avenue, Cambridge, Massachusetts 02138. All information contained in this manual is based on the information available to the publisher from Volkswagen at the time of editorial closing. Volkswagen has not reviewed and does not vouch for the accuracy or completeness of the technical specifications and work procedures described and given in this manual.

ISBN 978-0-8376-1660-5 Job code: VG99-10, editorial closing 07 / 2003.

VWoA Inc. Part No. LPV 800 116

Library of Congress Cataloging-in-Publication Data for 1999 edition

Volkswagen Jetta, Golf, GTI, Cabrio service manual : Jetta, Golf, GTI: 1993-1999, Cabrio, 1995-2002, including 1.9L TDI, 2.0L, and 2.8L VR6 : A3 platform.

p. cm.

Includes index.

ISBN 0-978-8376-0366-7 (alk. paper)

1. Jetta automobile--Maintenance and repair--Handbooks, manuals, etc. 2. Golf automobile--Maintenance and repair--Handbooks, manuals, etc. 3. GTI automobile--Maintenance and repair--Handbooks, manuals, etc. 4. Cabrio automobile--Maintenance and repair--Handbooks, manuals, etc. I. Robert Bentley, Inc.

TL215.J49V65 1999
629.28'72--dc21

99-014513

 The paper used in this publication is acid free and meets the requirements of the National Standard for Information Sciences–Permanence of Paper for Printed Library Materials. ∞

Manufactured in the United States of America
202501R003

Selected Books and Repair Information From Bentley Publishers

Motorsports

Alex Zanardi: My Sweetest Victory
Alex Zanardi and Gianluca Gasparini
ISBN 978-0-8376-1249-2

The Unfair Advantage
Mark Donohue and Paul van Valkenburgh
ISBN 978-0-8376-0069-7

Equations of Motion - Adventure, Risk and Innovation *William F. Milliken*
ISBN 978-0-8376-1570-7

Engineering

Mechanical Ignition Handbook: The Hack Mechanic Guide to Vintage Ignition Systems
Rob Siegel ISBN 978-0-8376-1767-1

Hack Mechanic Guide to European Automotive Electrical Systems
Rob Siegel ISBN 978-0-8376-1751-0

Bosch Automotive Handbook *Robert Bosch GmbH* ISBN 978-0-8376-1732-9

Bosch Fuel Injection and Engine Management *Charles O. Probst, SAE*
ISBN 978-0-8376-0300-1

Maximum Boost: Designing, Testing, and Installing Turbocharger Systems
Corky Bell ISBN 978-0-8376-0160-1

Supercharged! Design, Testing and Installation of Supercharger Systems
Corky Bell ISBN 978-0-8376-0168-7

Scientific Design of Exhaust and Intake Systems *Phillip H. Smith & John C. Morrison* ISBN 978-0-8376-0309-4

Physics for Gearheads
Randy Beikmann ISBN 978-0-8376-1615-5

Audi

Audi A4 Service Manual: 2002-2008, 1.8L Turbo, 2.0L Turbo, 3.0L, 3.2L
Bentley Publishers
ISBN 978-0-8376-1574-5

Audi TT Service Manual: 2000-2006, 1.8L turbo, 3.2 L *Bentley Publishers*
ISBN 978-0-8376-1625-4

Audi A6 (C5 platform) Service Manual: 1998-2004 *Bentley Publishers*
ISBN 978-0-8376-1670-4

BMW

Memoirs of a Hack Mechanic
Rob Siegel ISBN 978-0-8376-1720-6

BMW 4 Series (F32, F33, F36) Service Manual: 2014-2016 *Bentley Publishers*
ISBN 978-0-8376-1765-7

BMW 3 Series (F30, F31, F34) Service Manual: 2012-2015 *Bentley Publishers*
ISBN 978-0-8376-1752-7

BMW X3 (E83) Service Manual: 2004-2010 *Bentley Publishers*
ISBN 978-0-8376-1731-2

BMW X5 (53) Service Manual: 2000-2006 *Bentley Publishers*
ISBN 978-0-8376-1643-8

BMW 3 Series (E90, E91, E92, E93) Service Manual: 2006-2011
Bentley Publishers ISBN 978-0-8376-1723-7

BMW 3 Series (E46) Service Manual: 1999-2005 *Bentley Publishers*
ISBN 978-0-8376-1657-5

BMW Z3 (E36/7) Service Manual: 1996-2002 *Bentley Publishers*
ISBN 978-0-8376-1617-9

BMW 5 Series (E60, E61) Service Manual: 2004-2010 *Bentley Publishers*
ISBN 978-0-8376-1689-6

BMW 5 Series (E39) Service Manual: 1997-2003 *Bentley Publishers*
ISBN 978-0-8376-1672-8

Corvette

Corvette: America's Star-Spangled Sports Car *Karl Ludvigsen*
ISBN 978-0-8376-1659-9

Corvette by the Numbers: The Essential Corvette Parts Reference 1955-1982 *Alan Covin*
ISBN 978-0-8376-0228-2

Zora Arkus-Duntov: The Legend Behind Corvette *Jerry Burton*
ISBN 978-08376-0858-7

Porsche

Porsche 911 (996) Service Manual: 1999-2005 *Bentley Publishers*
ISBN 978-0-8376-1710-7

Porsche 911 (993) Service Manual: 1995-1998 *Bentley Publishers*
ISBN 978-0-8376-1719-0

Porsche Boxster Service Manual: 1997-2004 *Bentley Publishers*
ISBN 978-0-8376-1645-2

Porsche 911 Carrera Service Manual: 1984-1989 *Bentley Publishers*
ISBN 978-0-8376-1696-4

Porsche 911 SC Service Manual: 1978-1983 *Bentley Publishers*
ISBN 978-0-8376-1705-3

Porsche: Excellence Was Expected
Karl Ludvigsen
ISBN 978-0-8376-0235-6

Porsche — Origin of the Species
Karl Ludvigsen
ISBN 978-0-8376-1331-4

Volkswagen

Volkswagen Rabbit, GTI Service Manual: 2006-2009 *Bentley Publishers*
ISBN 978-0-8376-1664-3

Volkswagen Jetta Service Manual: 2005-2010 *Bentley Publishers*
ISBN 978-0-8376-1616-2

Volkswagen Jetta, Golf, GTI Service Manual: 1999-2005 *Bentley Publishers*
ISBN 978-0-8376-1678-0

Volkswagen Jetta, Golf, GTI: 1993-1999, Cabrio: 1995-2002 Service Manual *Bentley Publishers*
ISBN 978-0-8376-1660-5

Volkswagen GTI, Golf, Jetta Service Manual: 1985-1992 *Bentley Publishers*
ISBN 978-0-8376-1637-7

Volkswagen Corrado Repair Manual: 1990-1994 *Bentley Publishers*
ISBN 978-0-8376-1699-5

Volkswagen Passat, Passat Wagon Service Manual: 1998-2005
Bentley Publishers
ISBN 978-0-8376-1669-8

MINI Repair Manuals

MINI Cooper Service Manual: 2007-2013 *Bentley Publishers*
ISBN 978-0-8376-1730-5

MINI Cooper Service Manual: 2002-2006 *Bentley Publishers*
ISBN 978-0-8376-1639-1

Mercedes-Benz

Mercedes-Benz C-Class (W202) Service Manual 1994-2000
Bentley Publishers
ISBN 978-0-8376-1692-6

Mercedes Benz E-Class (W124) Owner's Bible: 1986-1995
Bentley Publishers
ISBN 978-0-8376-0230-1

Mercedes-Benz Technical Companion
Staff of The Star and members of Mercedes-Benz Club of America
ISBN 978-0-8376-1033-7

BentleyPublishers®
Automotive Reference

Bentley Publishers has published service manuals and automobile books since 1950. For more information, please contact Bentley Publishers at 1734 Massachusetts Avenue, Cambridge, MA 02138 USA, or visit our web site at **BentleyPublishers.com**

Foreword

For the Volkswagen owner with automotive mechanical skills, and for independent service professionals, this manual includes maintenance and repair procedures likely to be needed during the service life of the vehicle, including many procedures and specifications that were available in an authorized Volkswagen dealer service department at the time this manual was prepared. The Volkswagen owner with no intention of working on his or her car will find that owning and referring to this manual will make it possible to be better informed and to more knowledgeably discuss repairs with a professional automotive technician.

For those intending to do maintenance and repair on their Volkswagen, it is essential that safety equipment be used and safety precautions observed when working on your vehicle. A minimum safety equipment list includes hand protection, eye protection, and a fire extinguisher. A selection of good quality hand tools is also needed. This includes a torque wrench to ensure that fasteners are tightened in accordance with specifications. In some cases, the text refers to special tools that are recommended or required to accomplish adjustments or repairs. These tools are often identified by their Volkswagen special tool number and illustrated.

Disclaimer

We have endeavored to ensure the accuracy of the information in this manual. When the vast array of data presented in the manual is taken into account, however, no claim to infallibility can be made. We therefore cannot be responsible for the result of any errors that may have crept into the text. Please also read **WARNING–Important Safety Notice** on the copyright page.

Prior to starting a repair procedure, read the procedure, **Warnings and Cautions** on page vii and the **warnings** and **cautions** that accompany the procedure. Reading a procedure before beginning work helps you determine in advance the need for specific skills, identify hazards, prepare for appropriate capture and handling of hazardous materials, and the need for particular tools and replacement parts such as gaskets.

Bentley Publishers encourages comments from the readers of this manual with regard to errors, and/or suggestions for improvement of our product. These communications have been and will be carefully considered in the preparation of this and other manuals. If you identify inconsistencies in the manual, you may have found an error. Please contact the publisher and we will endeavor to post applicable corrections on our website. Posted corrections (errata) should be reviewed before beginning work. Please see the following web address:

`http://www.BentleyPublishers.com/updates/`

Volkswagen continues to issue service information and parts retrofits after the editorial closing of this manual. Some of this updated information may apply to procedures and specifications in this manual. For the latest information, please see the following web address:

`http://www.ebahn.com/vw`

Volkswagen offers extensive warranties, especially on components of the fuel delivery and emission control systems. Therefore, before deciding to repair a Volkswagen that may be covered wholly or in part by any warranties issued by Volkswagen of America, Inc., consult your authorized Volkswagen dealer. You may find that the dealer can make the repair either free or at minimum cost. Regardless of its age, or whether it is under warranty, your Volkswagen is both an easy car to service and an easy car to get serviced. So if at any time a repair is needed that you feel is too difficult to do yourself, a trained Volkswagen technician is ready to do the job for you.

Volkswagen and most aftermarket parts suppliers specializing in Volkswagens refer to the Jettas, Golfs, GTIs and Cabrios covered by this manual as the A3 models. The code "A3" is Volkswagen's internal code for the basic platform shared by all of the cars covered by this manual. When ordering parts, especially body parts, knowing the A3 code may be helpful.

Some of the information in this manual applies only to cars of a particular model year or range of years. For example, "1993 m.y." refers to the 1993 model year. The model year does not necessarily match the calendar year in which the car was manufactured or sold. To be sure of the model year of a particular car, check the vehicle identification number (VIN) on the car. Technical changes made in production within a model year are identified in this manual by listing the VIN for the first car produced with this change.

The VIN is a unique sequence of 17 characters assigned by Volkswagen to identify each individual car. 3VWBA21HXPM005678 is an example. Each of the 17 letters and numbers indicates certain facts about the car and its manufacture. VINs used to distinguish information in this manual may refer only to the last eleven digits—the characters 1HXPM005678 in the example above.

Your Volkswagen's VIN can be found on a plate mounted on the top of the instrument panel, on the driver's side where the number can be seen through the windshield. The 10th character is the model year code. The letters "I", "O", "Q" and "U" are not used for model year designation, for example, "P" for 1993 m.y., "R" for 1994 m.y., etc. Beginning with model year 2001, numbers are used to designate model year rather than letters. The table below explains some of the various codes in the VIN numbers of models covered by this manual.

3 V W	R	A	8	1 H	X	R	M	1 2 3 4 5 6
Manufacturer Information	Body	Engine Type	Restraint Type	Body Type	Check Digit	Model Year	Factory	Sequential Chassis No.
3VW-Mexican Passenger car WVW-German passenger car	Depends on model year		Depends on type of seatbelts and airbags installed	1H-Golf or Jetta 1E-Cabrio		P-1993 R-1994 S-1995 T-1996 V-1997 W-1998 X-1999 Y-2000 1-2001 2-2002	M- Puebla, Mexico K-Osnabruk, Germany W-Wolfsburg, Germany	000001-Begin production 999999-End production

Please read these warnings and cautions before proceeding with maintenance and repair work.

WARNING—

- Some repairs may be beyond your capability. If you lack the skills, tools and equipment, or a suitable workplace for any procedure described in this manual, we suggest you leave such repairs to an authorized Volkswagen dealer service department, or other qualified shop.

- Volkswagen is constantly improving its cars. Sometimes these changes, both in parts and specifications, are made applicable to earlier models. Therefore, before starting any major jobs or repairs to components on which passenger safety may depend, consult your authorized Volkswagen dealer about Technical Bulletins that may have been issued since the editorial closing of this manual.

- Do not re-use any fasteners that are worn or deformed in normal use. Many fasteners are designed to be used only once and become unreliable and may fail when used a second time. This includes, but is not limited to, nuts, bolts, washers, self-locking nuts or bolts, circlips and cotter pins. Always replace these fasteners with new parts.

- Never work under a lifted car unless it is solidly supported on stands designed for the purpose. Do not support a car on cinder blocks, hollow tiles or other props that may crumble under continuous load. Never work under a car that is supported solely by a jack. Never work under the car while the engine is running.

- If you are going to work under a car on the ground, make sure that the ground is level. Block the wheels to keep the car from rolling. Disconnect the battery negative (–) terminal (Ground strap) to prevent others from starting the car while you are under it.

- Never run the engine unless the work area is well ventilated. Carbon monoxide kills.

- Finger rings, bracelets and other jewelry should be removed so that they cannot cause electrical shorts, get caught in running machinery, or be crushed by heavy parts.

- Tie long hair behind your head. Do not wear a necktie, a scarf, loose clothing, or a necklace when you work near machine tools or running engines. If your hair, clothing, or jewelry were to get caught in the machinery, severe injury could result.

- Do not attempt to work on your car if you do not feel well. You increase the danger of injury to yourself and others if you are tired, upset or have taken medication or any other substance that may keep you from being fully alert.

- Illuminate your work area adequately but safely. Use a portable safety light for working inside or under the car. Make sure the bulb is enclosed by a wire cage. The hot filament of an accidentally broken bulb can ignite spilled fuel or oil.

- Catch draining fuel, oil, or brake fluid in suitable containers. Do not use food or beverage containers that might mislead someone into drinking from them. Store flammable fluids away from fire hazards. Wipe up spills at once, but do not store the oily rags, which can ignite and burn spontaneously.

- Always observe good workshop practices. Wear goggles when you operate machine tools or work with battery acid. Gloves or other protective clothing should be worn whenever the job requires working with harmful substances.

- Friction materials such as brake or clutch discs may contain asbestos fibers. Do not create dust by grinding, sanding, or by cleaning with compressed air. Avoid breathing asbestos fibers and asbestos dust. Breathing asbestos can cause serious diseases such as asbestosis or cancer, and may result in death.

- Disconnect the battery negative (–) terminal (ground strap) whenever you work on the fuel system or the electrical system. Do not smoke or work near heaters or other fire hazards. Keep an approved fire extinguisher handy.

- Batteries give off explosive hydrogen gas during charging. Keep sparks, lighted matches and open flame away from the top of the battery. If hydrogen gas escaping from the cap vents is ignited, it will ignite gas trapped in the cells and cause the battery to explode.

- Connect and disconnect battery cables, jumper cables or a battery charger only with the ignition switched off, to prevent sparks. Do not disconnect the battery while the engine is running.

- Do not quick-charge the battery (for boost starting) for longer than one minute. Wait at least one minute before boosting the battery a second time.

- Do not allow battery charging voltage to exceed 16.5 volts. If the battery begins producing gas or boiling violently, reduce the charging rate. Boosting a sulfated battery at a high charging rate can cause an explosion.

- The air-conditioning system is filled with chemical refrigerant, which is hazardous. The A/C system should be serviced only by trained technicians using approved refrigerant recovery/recycling equipment, trained in related safety precautions, and familiar with regulations governing the discharging and disposal of automotive chemical refrigerants.

- Do not expose any part of the A/C system to high temperatures such as open flame. Excessive heat will increase system pressure and may cause the system to burst.

- Some aerosol tire inflators are highly flammable. Be extremely cautious when repairing a tire that may have been inflated using an aerosol tire inflator. Keep sparks, open flame or other sources of ignition away from the tire repair area. Inflate and deflate the tire at least four times before breaking the bead from the rim. Completely remove the tire from the rim before attempting any repair.

- Most cars covered by this manual are equipped with a supplemental restraint system (SRS), that automatically deploys an airbag in the event of a frontal impact. The airbag is inflated by an explosive device. Handled improperly or without adequate safeguards, it can be accidently activated and cause serious injury.

- To prevent personal injury or airbag system failure, **only factory trained Volkswagen service technicians** should test, disassemble or service the airbag system.

- Disconnect the power supply before working on the airbag system, or when doing repairs that require removing airbag system components. Disconnect the battery negative (–) terminal and cover the battery.

continued on next page

Please read these warnings and cautions before proceeding with maintenance and repair work.

WARNING (continued) —

- On airbag-equipped cars, never apply stickers or any other type of covering on the steering wheel. Do not let chemical cleaners, oil or grease come into contact with vinyl covering of the airbag unit.

- Never open or otherwise attempt to repair airbag system parts. Always use new parts. Never leave airbag parts or the partially disassembled airbag system unattended.

- Never use a test light to conduct electrical tests on the airbag system. The system must only be tested by trained Volkswagen Service technicians using the Volkswagen VAG 1551/1552 Scan Tool (ST) or an approved equivalent. The airbag unit must never be electrically tested while it is not installed in the car.

- Do not expose the airbag unit to temperatures above 194°F (90°C), even for brief periods. Keep clear of heat sources such as hot plates, soldering irons, heat lamps and welding equipment.

- When driving or riding in an airbag-equipped vehicle, never hold test equipment in your hands or lap while the vehicle is in motion. Objects between you and the airbag can increase the risk of injury in an accident.

- Greases, lubricants and other automotive chemicals contain toxic substances, many of which are absorbed directly through the skin. Read manufacturer's instructions and warnings carefully. Use hand and eye protection. Avoid direct skin contact.

CAUTION—

- If you lack the skills, tools and equipment, or a suitable workshop for any procedure described in this manual, we suggest you leave such repairs to an authorized Volkswagen dealer or other qualified shop. We especially urge you to consult an authorized Volkswagen dealer before beginning repairs on any car that may still be covered wholly or in part by any of the extensive warranties issued by Volkswagen of America.

- Volkswagen offers extensive warranties, especially on components of fuel delivery and emission control systems. Therefore, before deciding to repair a Volkswagen that may still be covered wholly or in part by any warranties issued by Volkswagen United States, Inc., consult your authorized Volkswagen dealer. You may find that he can make the repair for free, or at minimal cost.

- Volkswagen part numbers listed in this manual are for identification purposes only, not for ordering. Always check with your authorized Volkswagen dealer to verify part numbers and availability before beginning service work that may require new parts.

- Before starting a job, make certain that you have all the necessary tools and parts on hand. Read all the instructions thoroughly, do not attempt shortcuts. Use tools appropriate to the work and use only replacement parts meeting Volkswagen specifications. Makeshift tools, parts and procedures will not make good repairs.

- Use pneumatic and electric tools only to loosen threaded parts and fasteners. Never use these tools to tighten fasteners, especially on light alloy parts. Always use a torque wrench to tighten fasteners to the tightening torque specification listed.

- Be mindful of the environment and ecology. Before you drain the crankcase, find out the proper way to dispose of the oil. Do not pour oil onto the ground, down a drain, or into a stream, pond or lake. Consult local ordinances that govern the disposal of wastes.

- On cars equipped with anti-theft radios, make sure you know the correct radio activation code before disconnecting the battery or removing the radio. If the wrong code is entered into the radio when power is restored, that radio may lock up and be rendered inoperable, even if the correct code is then entered.

- Connect and disconnect a battery charger only with the battery charger switched off.

- Do not quick-charge the battery (for boost starting) for longer than one minute. Wait at least one minute before boosting the battery a second time.

- Sealed or "maintenance free" batteries should be slow-charged only, at an amperage rate that is approximately 10% of the battery's ampere-hour (Ah) rating.

- Do not allow battery charging voltage to exceed 16.5 volts. If the battery begins producing gas or boiling violently, reduce the charging rate. Boosting a sulfated battery at a high charging rate can cause an explosion.

0 Maintenance Program

TABLES

GENERAL

All of the maintenance work described in this repair group is important and should be carried out at the specified time or mileage interval. The Owner's Manual, the Maintenance Record, and the Warranty Booklet originally supplied with the car contain the maintenance schedules that apply to your Volkswagen. Following these schedules will ensure safe and dependable operation. In addition, many of the maintenance procedures are necessary to maintain warranty protection.

> ***NOTE —***
> *Volkswagen is constantly updating their recommended maintenance procedures and requirements. The information contained here may not include updates or revisions made by Volkswagen since publication of the Owner's Manual, the Maintenance Record, and the Warranty Booklet supplied with the car. If there is any doubt about what procedures apply to a specific model or model year, or what intervals should be followed, remember that an authorized Volkswagen dealer has the latest maintenance information.*

How To Use This Manual

The manual is divided into 10 main sections, or partitions:

0 Maintenance
1 Engine
2 Fuel, Ignition, and Exhaust Systems
3 Clutch, Transmission, and Final Drive
4 Suspension, Brakes, and Steering
5 Body–Assembly
6 Body–Components and Accessories
7 Body–Interior Trim
8 Heating and Air Conditioning
9 Electrical System
ST Scan Tool

0 Maintenance Program covers the recommended schedules and service procedures needed to do the Volkswagen-specified scheduled maintenance work.

The remaining nine partitions (1 through 9) are repair oriented and are broken down into individual repair groups. Each main partition begins with a general information group, e.g. **1 General Information.** These general groups are mostly descriptive in nature, covering topics such as theory of operation and troubleshooting. The remainder of the repair groups contain the more involved and more detailed system repair information.

A master listing of the 10 partitions and the corresponding individual repair groups can be found on the inside book cover.

Thumb tabs are used on the first page of each repair group page to help locate the groups quickly. Page numbers throughout the manual are organized according to the repair group system. For example, you can expect to find information on engine removal and installation (Repair Group 10) beginning on page 10-1. A comprehensive index can be found on the last pages of the manual.

Warnings, Cautions and Notes

Throughout this manual are many passages with the headings **WARNING**, **CAUTION**, or **NOTE**. These very important headings have different meanings.

> ***WARNING —***
> *The text under this heading warns of unsafe practices that are very likely to cause injury, either by direct threat to the person(s) doing the work or by increased risk of accident or mechanical failure while driving. Warnings are always contained in a box.*

> ***CAUTION —***
> *A caution calls attention to important precautions to be observed during the repair work that will help prevent accidentally damaging the car or its parts. Cautions are always contained in a box.*

> ***NOTE —***
> *A note contains helpful information, tips that will help in doing a better job and completing it more easily.*

Please read every **WARNING**, **CAUTION**, and **NOTE** at the front of the manual and as they appear in repair procedures. They are very important. Read them before you begin any maintenance or repair job.

Some **WARNING**s and **CAUTION**s are repeated wherever they apply. Read them all. Do not skip any. These messages are important, even to the owner who never intends to work on the car.

Work Safety

Although an automobile presents many hazards, common sense and good equipment can help ensure safety. Many accidents happen because of carelessness. Pay attention and stick to these few important safety rules.

> ***WARNING —***
> - *Never run the engine in the work area unless it is well-ventilated. The exhaust should be vented to the outside. Carbon Monoxide (CO) in the exhaust kills.*
> - *Remove all neckties, scarfs, loose clothing, or jewelry when working near running engines or power tools. Tuck in shirts. Tie long hair and secure it under a cap. Severe injury can result from these things being caught in rotating parts.*
> - *Remove rings, watches, and bracelets. Aside from the dangers of moving parts, metallic jewelry conducts electricity and may cause shorts, sparks, burns, or damage to the electrical system when accidentally contacting the battery or other electrical terminals.*
> - *Disconnect the battery negative (–) cable whenever working on or near the fuel system, airbag system or anything that is electrically powered. Accidental electrical contact may damage the electrical system, cause fire, or result in serious personal injury.*
> - *Never work under a lifted car unless it is solidly supported on jack stands that are intended for that purpose. Do not support a car on cinder blocks, bricks, or other objects that may shift or crumble under continuous load. Never work under a car that is supported only by a jack.*
> - *The fuel system retains pressure even when the ignition is off. Loosen the fuel lines very slowly to allow the residual pressure to dissipate gradually. Avoid spraying fuel.*
> - *Fuel is highly flammable. When working around fuel, do not smoke or work near fire hazards. Keep an approved fire extinguisher handy.*
> - *Illuminate the work area adequately and safely. Use a portable safety light for working inside or under the car. A fluorescent type light is best because it gives off less heat. If using a light with a normal incandescent bulb, use rough service bulbs to avoid breakage. The hot filament of an accidentally broken bulb can ignite spilled fuel or oil.*
> - *Keep sparks, lighted matches, and open flame away from the top of the battery. Hydrogen gas emitted by the battery is highly flammable. Any nearby source of ignition may cause the battery to explode.*
> - *Never lay tools or parts in the engine compartment or on top of the battery. They may fall and be difficult to retrieve, become caught in belts or other rotating parts, or cause electrical shorts and damage to the electrical system.*

IDENTIFICATION PLATES AND LABELS

Vehicle Identification Number (VIN)

The vehicle year and model can be determined from the VIN or the identification plate. Because stickers can fall off, be painted over, or be moved, it is best to rely on numbers stamped into the chassis.

The VIN (Chassis number) is stamped into the rear cross panel of the engine compartment. It is visible through a window in the plenum chamber cover. See Fig. 1. The identification plate is located on the driver's side of the padded dash panel. Additional vehicle identification information is also found on stickers attached to the driver's side door post and in the trunk.

> ***NOTE —***
> *Additional information on the 17 digit VIN is given in the* Foreword *on page vi, at the front of the manual.*

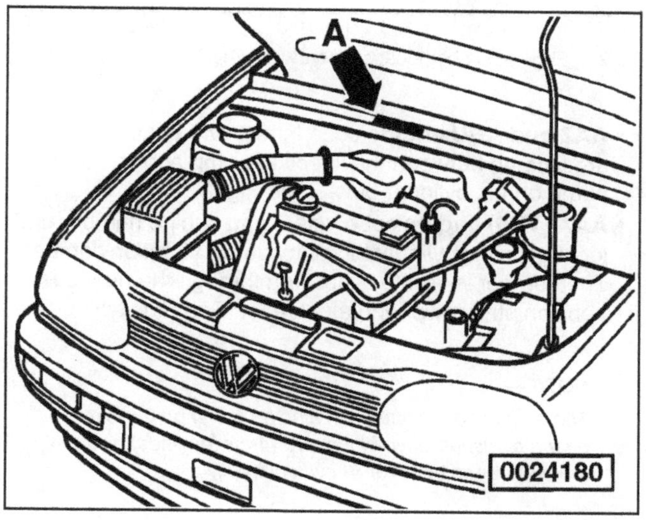

0024180

Fig. 1. Partial vehicle identification number (**A**) stamped into rear cross panel of the engine compartment. Full VIN on plate on driver's side of padded dash panel.

Engine code and engine number

The engines used in the Volkswagens covered by this manual are identified by a three letter code. The engine code and the engine number are located on the engine block. Location depends on engine application.

Engine codes

- ACC .... 1.8 Liter gas
- AAZ, AHU........................ 1.9 Liter diesel
- ABA .... 2.0 Liter gas
- AAA .... 2.8 Liter gas

- **ACC and ABA engine** code letters and engine number are located on the front of the engine block, below the separation point between engine block and cylinder head. See Fig. 2.

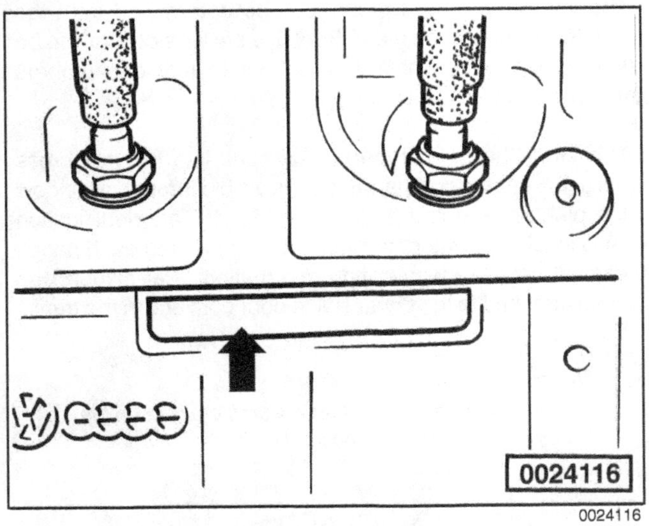

0024116

Fig. 2. Engine number location (**arrow**) on ACC and ABA engines.

- **AAZ and AHU engine** code letters and engine number are located between the injection pump and vacuum pump on the cylinder block. See Fig. 3.
- **AAA (VR6) engine** code letters and engine number are located on the cylinder block above the vibration damper. The number is visible by looking through the right suspension strut tower and air cleaner. See Fig. 4.

NOTE —

The engine codes often can also be found on a sticker on the camshaft drive belt guard (4-cyl. engines) or on the cylinder head cover (6-cyl. engines).

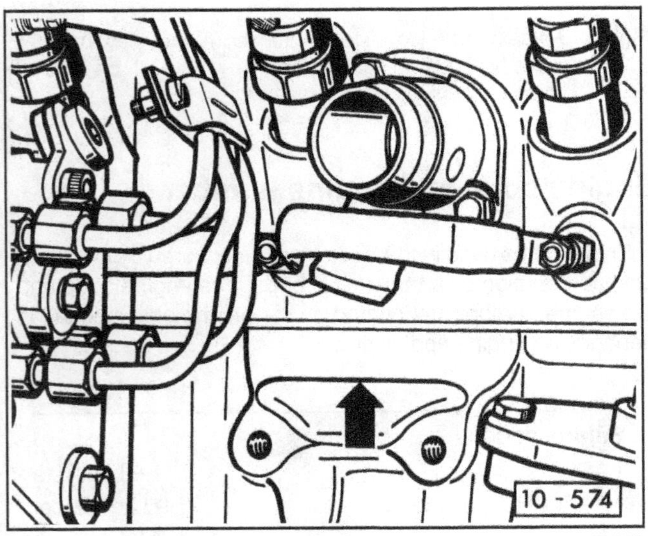

10-574

Fig. 3. Engine number location (**arrow**) on AAZ diesel engine shown. AHU diesel is similar.

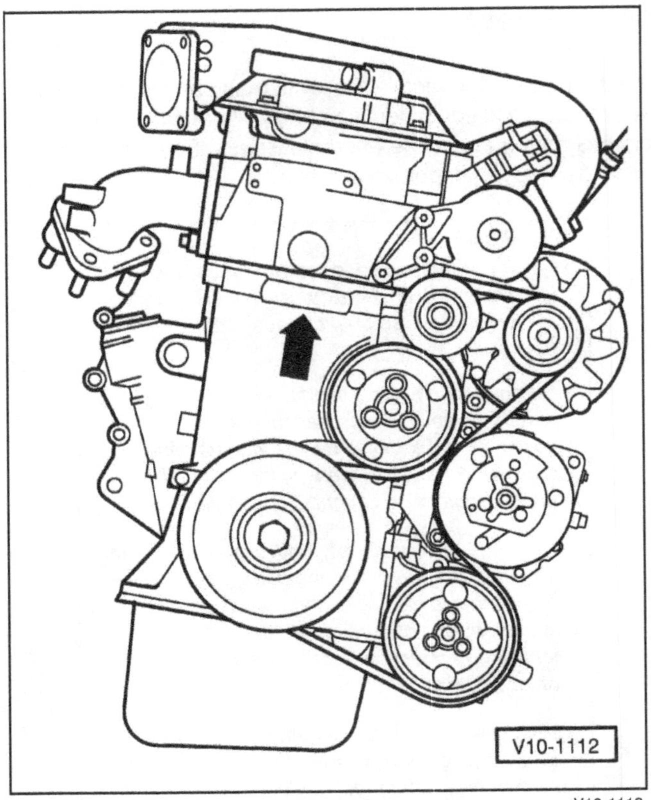

V10-1112

Fig. 4. Engine number location (**arrow**) on VR6 engine.

Transmission identification

Two manual transmissions and two automatic transmissions are used in the cars covered by this manual. Application depends on engine and model year.

020 5-speed manual transmission is used in all cars with 4-cylinder engines and manual transmission, except TDI Diesel. See Fig. 5.

02A 5-speed manual transmission is used in cars with 6-cylinder engine or TDI Diesel, and manual transmission. See Fig. 6.

096 4-speed automatic transmission is used in Golf, Jetta and Cabrio models beginning in 1993. It may be installed in combination with either a 4-cylinder or a 6-cylinder engine. See Fig. 7.

01M 4-speed automatic is installed in some 1995 and later cars with 4- or 6-cylinder engines. See Fig. 8.

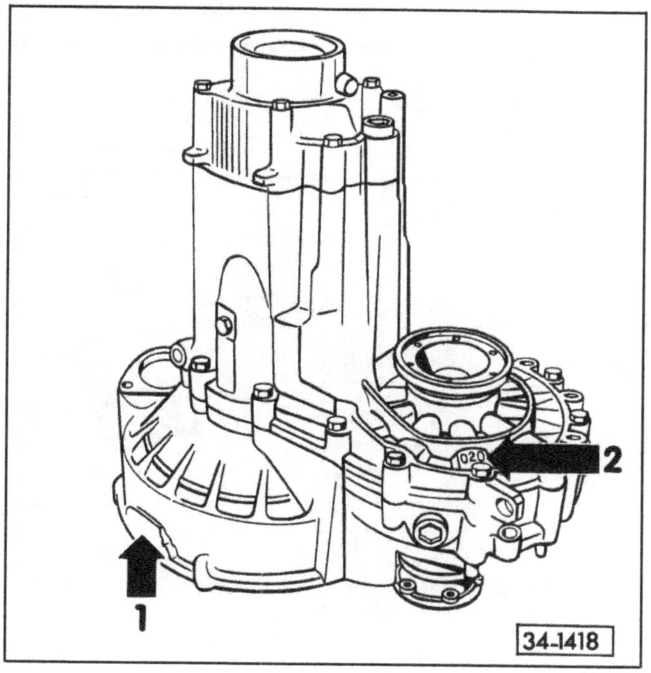

34-1418

Fig. 5. 020 transmission used on cars with 4-cylinder engines. **Arrow 1** indicates location of code letters and date of manufacture. **Arrow 2** indicates location of transmission type.

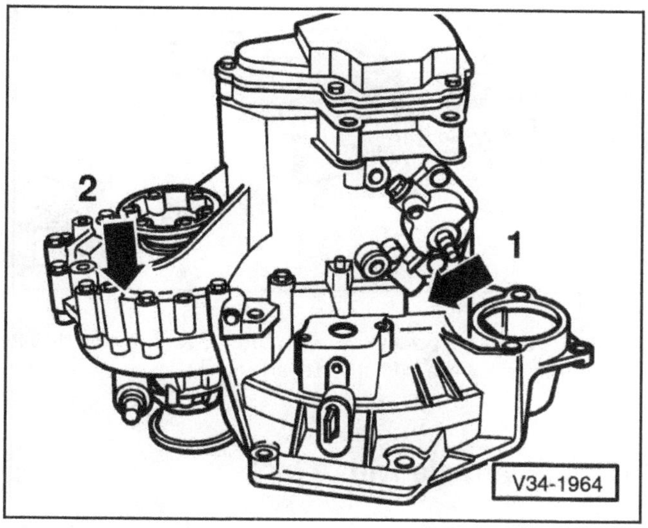

V34-1964

Fig. 6. 02A transmission for 6-cylinder vehicles and 4-cylinder TDI vehicles. **Arrow 1** indicates location of code letters and date of manufacture. **Arrow 2** indicates location of transmission type.

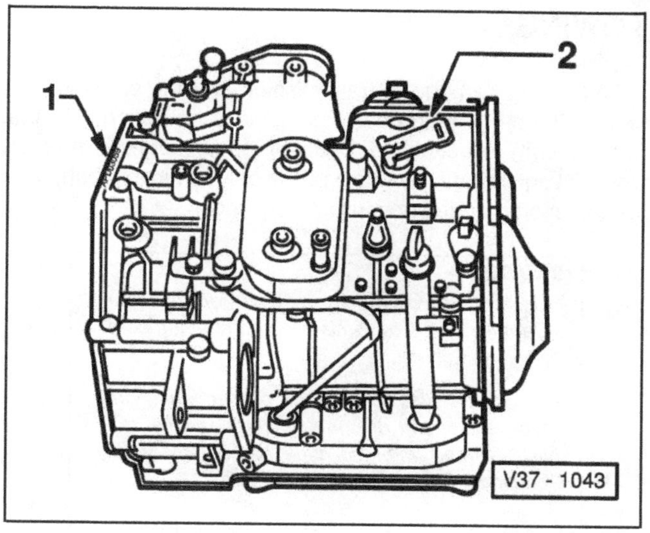

V37-1043

Fig. 7. 096 automatic transmission (with dipstick). **Arrow 1** indicates location of code letters and date of manufacture. **Arrow 2** indicates location of transmission type.

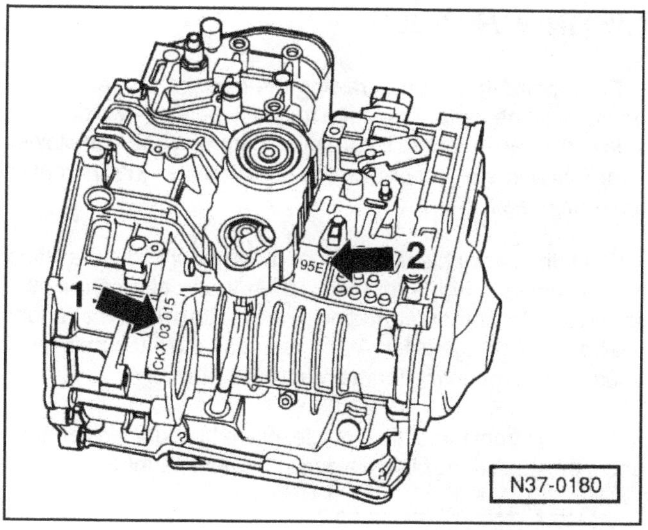

N37-0180

Fig. 8. 01M automatic transmission (without dipstick). **Arrow 1** indicates location of code letters and date of manufacture. **Arrow 2** indicates location of transmission type.

TOWING

To prevent damage to transmission and the steering column anti-theft lock, the car must always be towed with the front (driven) wheels lifted. Use self-loading wheel dollies or wheel lift equipment. Tow the vehicle only within the manufacturer's recommended speeds and distances.

> ***CAUTION —***
> - *Do not tow Volkswagen models from the rear. Damage to the transmission and steering components will result.*
> - *Do not tow the vehicle with conventional sling type equipment or non-self-loading wheel dollies. Towing with this type of equipment will cause bumper and body panel damage.*
> - *If an automatic transmission equipped vehicle cannot be towed with wheel lift equipment, it must be transported with a car carrier (flat bed) to avoid damage to the transmission.*

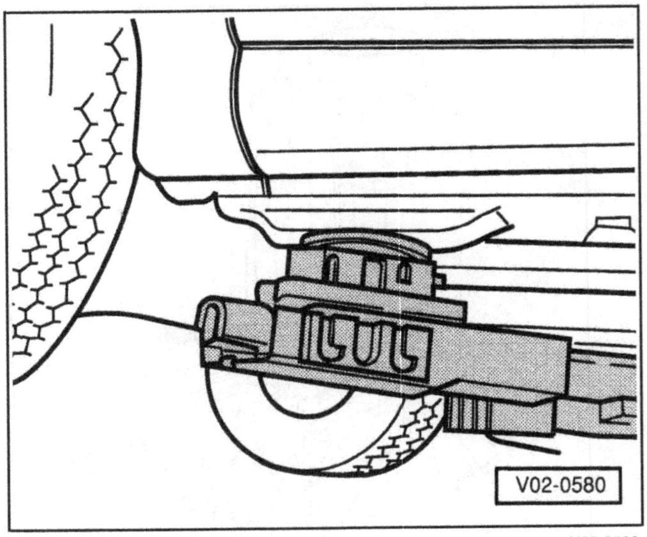

V02-0580

Fig. 9. Front jacking point for use with floor jack or hydraulic lift. Lift on reinforced surface under front door hinge pillar.

LIFTING VEHICLE

For repairs that require raising the car, the proper jacking points must be used. There are four jack points—two on each side of the car—marked by triangular impressions in the lower panel sheet metal just behind the front wheel or just in front of the rear wheel.

To lift the car with a floor jack or hydraulic lift, use the lifting points shown in Fig. 9 and Fig. 10. At the front, use only the small round member protruding at the bottom of the car under the front door hinge pillar. At the rear, use the flat surface just ahead of the mounting for the rear axle assembly.

To lift the front the of the vehicle, place the floor jack or hoist under the reinforcing plate welded to the floor panel. See Fig. 9.

> ***CAUTION —***
> - *Lifting front of vehicle at other than the front jacking points may cause severe body damage.*
> - *Use a rubberized lifting pad between the jack/lift and the body to avoid damaging the underbody.*

To lift the rear of the car place the floor jack or hoist under the rear vertical reinforcing plate of the lower sill in the area of the marking for the vehicle jack. See Fig. 10.

> ***CAUTION —***
> *Be sure to use a lifting pad to avoid damaging vehicle underbody.*

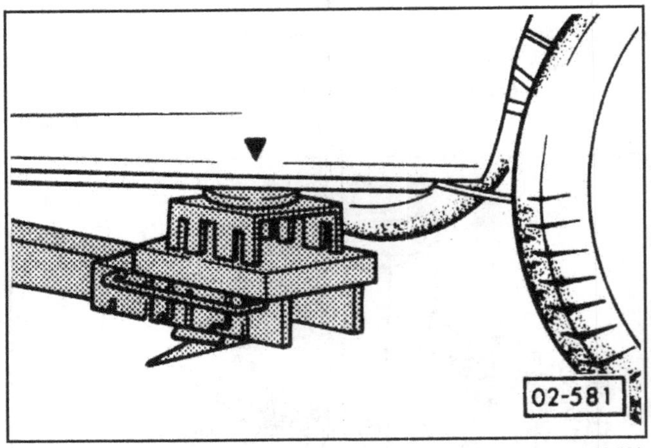

02-581

Fig. 10. Rear jacking point for use with floor jack or hydraulic lift. Lift on flat surface in front of rear axle mount.

To raise vehicle with a jack

> ***WARNING —***
> - *Do not work under a car that is supported only by a jack. If work has to be done under vehicle, the vehicle must be securely supported with jack stands.*
> - *Never start the vehicle and engage a gear when the vehicle is raised.*

1. Place the car on a flat level area with a surface capable of supporting jacks and jackstands, e.g. concrete.
2. Block the wheels to prevent the car from rolling.

3. Place the jack in position as shown earlier in Fig. 9 or 10.

> ***CAUTION —***
> - *Use a rubberized lifting pad between the jack and the body to avoid damaging the underbody.*
> - *The vehicle must never be lifted on the engine oil pan, transmission, rear axle, front axle or bottom sill. Serious damage can occur.*

4. Operate the jack and raise the car slowly. If working under the car, support the weight of the car using jack stands as described below under **To work under car**.

> ***WARNING —***
> *Watch the jack closely and make sure it remains stable and does not shift or tip.*

To work under car

1. Disconnect the negative battery cable so that no one else can start the car. Let others know what you will be doing.

2. Place at least two jack stands under the car. A jack is a temporary lifting device and should not be used alone to support the car while you are under it. Use positively locking jack stands that are designed for the purpose of supporting a car.

> ***WARNING —***
> - *Do not use wood, concrete blocks, or bricks to support a car. Wood may split. Blocks or bricks, while strong, are not designed for that kind of load, and may break or collapse.*
> - *Place jack stands on a hard level surface (e.g. concrete). Ensure car is stable before working under it.*

3. Lower the car slowly until its weight is fully supported by the jack stands. Watch to make sure that the jack stands do not tip or lean as the car settles on them, and that they are placed solidly and will not move.

> ***WARNING —***
> *Check to make sure car is stable before working under car.*

4. Observe all jacking precautions again when raising the car to remove the jack stands.

MAINTENANCE SCHEDULES

The maintenance schedules list all of the routine maintenance specified by Volkswagen, as well as the mileage intervals at which they should be performed. In any instance where the publisher has recommended maintenance or a maintenance interval that differs from that specified by Volkswagen, such recommendations are more conservative, and still meet Volkswagen's maintenance requirements.

In addition to the specified mileage intervals, Volkswagen also recommends that services be carried out on the basis of time. For 1993 to 1996 model years, the time intervals generally occur 6 months from the delivery date and every 6 months thereafter. For 1997 through 1999 model years, the time interval is generally 6 months after delivery, 12 months after delivery, and every 12 months thereafter.

> ***NOTE —***
> *Aside from keeping your Volkswagen in the best possible condition, scheduled maintenance plays a role in maintaining full coverage under Volkswagens extensive warranties. If in doubt about the terms and conditions of your car's warranty, consult an authorized Volkswagen dealer.*

Table a. 7,500/22,500 mile service (1993-1996 model years)

Maintenance item	Tools required	New parts/fluids required	Warm engine required	Dealer service recommended
Engine compartment maintenance				
Change oil and oil filter	*	*	*	
Check engine coolant level, add if necessary		*		
Check and lubricate hood lock (Canada cars only)				
Under car maintenance				
Tires and spare: check condition and pressures	*			
Rotate tires front to rear	*			
Body and interior maintenance				
Service reminder: reset				
Road Test				
Automatic shiftlock: check operation including park/neutral position switch				

Table b. 15,000/45,000 mile service (1993-1996 model years)

Maintenance item	Tools required	New parts/fluids required	Warm engine required	Dealer service recommended
Engine compartment maintenance				
Change oil and oil filter	*	*	*	
Check engine coolant level, add if necessary		*		
Check brake fluid level, add if necessary		*		
Check power steering fluid level, add if necessary		*		
Inspect for fluid leaks				
Check and lubricate hood lock (Canada only)	*			
Dust/pollen filter (where applicable): replace		*		
Battery: check electrolyte level and add if necessary. DO NOT attempt to check or add fluid to a maintenance free battery.	*	*		
Camshaft drive belt (Diesel only): check tension and condition.	*			*
Check engine idle speed	*			*
Under car maintenance				
Tires and spare: check condition and pressures	*			
Rotate tires front to rear	*			
Brake system: check for damage and/or leaks; check pad thickness	*			
Drive axle shafts: check boots for tears, cracks or damage				
Manual transmission: check for leaks				

(continued)

Table b. 15,000/45,000 mile service (1993-1996 model years) (continued)

Maintenance item	Tools required	New parts/fluids required	Warm engine required	Dealer service recommended
Under car maintenance (cont.)				
Automatic transmission final drive: check level, add if necessary[1)]	*	*		*
Automatic transmission: check ATF level, add if necessary. See **37 Automatic Transmission**	*	*		*
Exhaust system: check for damage and/or leaks	*			
Body and interior maintenance				
On Board Diagnostic System: check Diagnostic Trouble Code (DTC) Memory with V.A.G. 1551 Scan tool (ST), and erase stored codes if necessary	*	*		*
Service reminder: reset				
Road Test				
Automatic shift lock: check operation including park/neutral position switch				
Check A/T transmission kickdown, braking, steering, heating, ventilation, air conditioning, power accessories and electrical (including exterior lighting) systems				

1) In the case of severe conditions, such as extremely high temperatures, continuous mountain driving, trailer towing, predominantly stop-and-go traffic conditions, it is necessary to change the ATF every 30,000 miles. During this time the ATF sump should be removed and cleaned and the ATF strainer and sump gasket replaced.

Table c. 30,000/60,000 mile service (1993-1996 model years)

Maintenance item	Tools required	New parts/fluids required	Warm engine required	Dealer service recommended
Engine compartment maintenance				
Change oil and oil filter	*	*	*	
Check engine coolant level, add if necessary		*		
Check brake fluid level, add if necessary		*		
Check power steering fluid level, add if necessary		*		
Engine: inspect for fluid leaks				
Check and lubricate hood lock (Canada only)		*		
Diesel engine only: drain water from fuel filter.	*	*		
Diesel engine only: replace fuel filter	*	*		
Dust/pollen filter (where applicable): replace		*		
Battery: check electrolyte level and add if necessary. DO NOT attempt to check or add fluid to a maintenance free battery.	*	*		
V-belt: check tension and condition, adjust or replace as necessary	*	*		
Poly-ribbed drive belt: check condition, adjust or replace as necessary	*	*		

(continued)

Table c. 30,000/60,000 mile service (1993-1996 model years) (continued)

Maintenance item	Tools required	New parts/fluids required	Warm engine required	Dealer service recommended
Engine compartment maintenance (cont.)				
Diesel engine only: replace camshaft drive belt every 60K miles	*	*		*
4-cyl. gasoline engine only: check camshaft drive belt tension and condition 1)	*	*		*
Spark plugs: replace	*	*		
Air cleaner: replace (30k miles or every 2 years)	*	*		
Under car maintenance				
Tires and spare: check condition and pressures	*			
Rotate tires front to rear	*			
Brake system: check for damage and/or leaks; check pad thickness	*			
Drive axle shafts: check boots for tears, cracks or damage				
Front axle: check dust seals on ball joints and tie rod ends				
Check ball joints and tie rod ends for play/wear	*			
Manual transmission: check for leaks				
Automatic transmission final drive: check level, add if necessary	*	*		*
Automatic transmission: check ATF level, add if necessary 2)	*	*		*
Manual transmission: check fluid, add if necessary	*	*		
Exhaust system: check for damage and/or leaks	*			
Body and interior maintenance				
Check operation of interior and exterior lights				
Service reminder: reset				
On Board Diagnostic system: check Diagnostic Trouble Code (DTC) Memory with VAG1551 Scan tool (ST) and erase if necessary.	*	*		*
Road Test				
Automatic shift lock: check operation including park/neutral position switch				
Check A/T transmission kickdown, braking, steering, heating, ventilation, air conditioning, power accessories and electrical systems				
Check engine idle speed	*			*

1) Volkswagen does not specify a replacement interval for the camshaft drive belt on 4-cyl. gasoline engines. The publisher recommends periodic inspection of the belt and replacement at least every 90k miles or 5 years.

2) In the case of severe conditions, such as extremely high temperatures, continuous mountain driving, trailer towing, predominantly stop-and-go traffic conditions, it is necessary to change the ATF every 30,000 miles. During this time the ATF sump should be removed and cleaned and the ATF strainer and sump gasket replaced.

Table d. First 5,000 mile service (1997-2002 model years)

Maintenance item	Tools required	New parts/fluids required	Warm engine required	Dealer service recommended
Engine compartment maintenance				
Change oil and oil filter	*	*	*	
Water separator: drain water (TDI Diesel)	*			
Timing belt: check condition (TDI Diesel)	*			

Table e. every 10,000 mile service (1997-2002 model years)

Maintenance item	Tools required	New parts/fluids required	Warm engine required	Dealer service recommended
Engine compartment maintenance				
Change oil and oil filter	*	*	*	
Water separator: drain water (TDI Diesel)	*			
Timing belt: check condition (TDI Diesel)	*			
Windshield washer: check fluid level, add as necessary		*		
Dust/pollen filter (where applicable): replace[1)]		*		
Under car maintenance				
Tires and spare: check condition and pressures	*			
Rotate tires front to rear	*			
Brake system: check for damage and/or leaks; check pad thickness and brake fluid level	*	*		
Road Test				
Automatic shift lock (where applicable): check operation including park/neutral position switch				

1)Dust and pollen filter: Volkswagen specifies replacement every 20,000 miles. More frequent intervals may be necessary (i.e. every 10,000 miles) in areas with more dust and pollen.

Table f. every 20,000 mile service (1997-2002 model years)

Maintenance item	Tools required	New parts/fluids required	Warm engine required	Dealer service recommended
Engine compartment maintenance				
Change oil and oil filter	*	*	*	
Water separator: drain water (TDI Diesel)	*			
Timing belt: check condition (TDI Diesel)	*			
Windshield washer: check fluid, add as necessary		*		
Dust/pollen filter (where applicable): replace[1)]		*		
Spark plugs: replace (2.0L gasoline engine only) [2)]	*	*		
Battery: check electrolyte level and add if necessary. DO NOT attempt to check or add fluid to a maintenance free battery.	*	*		
Check engine coolant level, add if necessary		*		
Fuel filter: replace (TDI Diesel)[3)]	*	*		
Brake fluid: check level, add if necessary; replace every 2 years regardless of mileage	*	*		
Under car maintenance				
Brake system: check for damage and/or leaks; check pad thickness	*			
Rotate tires front to rear	*			
Tires and spare: check condition and pressures	*			
Automatic transmission final drive: check level, add if necessary, check for leaks	*	*		
Manual transmission: check fluid, add if necessary, check for leaks	*	*		
Drive axle shafts: check boots for tears, cracks or damage				
Body and interior maintenance				
Door check straps: lubricate		*		
On Board Diagnostic system: check Diagnostic Trouble Code (DTC) Memory with VAG 1551/1552 Scan tool (ST) and erase if necessary.	*	*		*
Road Test				
Automatic shift lock (where applicable): check operation including park/neutral position switch				
Check A/T transmission kickdown, braking, steering, heating, ventilation, air conditioning, power accessories and electrical systems				
After Road Test				
Engine: check for leaks				
Power steering: check fluid level		*		
Exhaust system: check for damage and leaks				
On Board Diagnostic system: check Diagnostic Trouble Code (DTC) Memory with VAG 1551/1552 Scan tool (ST) and erase if necessary.	*		*	*

1)Dust and pollen filter: Volkswagen specifies replacement every 20,000 miles. More frequent intervals may be necessary (i.e. every 10,000 miles) in areas with more dust and pollen.

[2)]Spark plugs: 2.0 Liter: 1997, replace every 30,000 miles; 1998-01, replace every 20,000 miles; 2002, replace every 40,000 miles. 1.8 and 2.8 Liter: 1997-99, replace every 40,000 miles.

[3)]Fuel filter: Gasoline engine fuel filter replacement is not part of Volkswagen's scheduled maintenance.

Table g. every 40,000 mile service (1997-2002 model years)

Maintenance item	Tools required	New parts/fluids required	Warm engine required	Dealer service recommended
Engine compartment maintenance				
Change oil and oil filter	*	*	*	
Water separator: drain water (TDI Diesel)	*			
Timing belt: check tension and condition, adjust if necessary Replace every 60,000 miles (TDI Diesel)[1)]	*	*		*
Timing belt tensioner: replace every 60,000 miles (TDI Diesel)	*	*		*
V-belt: check tension and condition, adjust if necessary	*			
Ribbed belt: check condition	*			
Windshield washer: check fluid, add as necessary		*		
Dust/pollen filter (where applicable): replace[2)]		*		
Spark plugs: replace[3)]	*	*		
Battery: check electrolyte level and add if necessary. DO NOT attempt to check or add fluid to a maintenance free battery.	*	*		
Check engine coolant level, add if necessary		*		
Fuel filter: replace (TDI Diesel)	*	*		
Air filter: replace filter element (required every 40,000 miles and every 2 years)	*	*		
Brake fluid: check level, add if necessary; replace every 2 years regardless of mileage	*	*		
Under car maintenance				
Brake system: check for damage and/or leaks; check pad thickness	*			
Rotate tires front to rear	*			
Tires and spare: check condition and pressures	*			
Automatic transmission final drive: check level, add if necessary, check for leaks	*	*		
Automatic transmission: replace ATF[4)]				
Manual transmission: check fluid, add if necessary, check for leaks	*	*		
Drive axle shafts: check boots for tears, cracks or damage				
Front axle: check dust seals on ball joints and tie rod ends, check tie rods for play/wear	*			

[1)]Volkswagen does not specify a replacement interval for the camshaft drive belt on 4-cyl. gasoline engines. The publisher recommends periodic inspection of the belt and replacement at least every 90k miles or 5 years.

[2)]Dust and pollen filter: Volkswagen specifies replacement every 20,000 miles. More frequent intervals may be necessary (i.e. every 10,000 miles) in areas with more dust and pollen.

[3)]Spark plugs: 2.0 Liter: 1997, replace every 30,000 miles; 1998-01, replace every 20,000 miles; 2002, replace every 40,000 miles. 1.8 and 2.8 Liter: 1997-99, replace every 40,000 miles.

[4)]1997 model year only. 1998 model year requires check level only every 60,000 miles.
In the case of severe conditions, such as extremely high temperatures, continuous mountain driving, trailer towing, predominantly stop-and-go traffic conditions, it is necessary to change the ATF every 30,000 miles. During this time the ATF sump should be removed and cleaned and the ATF strainer and sump gasket replaced.

(continued)

Table g. every 40,000 mile service (1997-2002 model years) (continued)

Maintenance item	Tools required	New parts/fluids required	Warm engine required	Dealer service recommended
Body and interior maintenance				
Door check straps: lubricate		*		
On Board Diagnostic system: check Diagnostic Trouble Code (DTC) Memory with VAG 1551/1552 Scan tool (ST) and erase if necessary.	*	*		*
Road Test				
Automatic shift lock (where applicable): check operation including park/neutral position switch				
Check A/T transmission kickdown, braking, steering, heating, ventilation, air conditioning, power accessories and electrical systems				
After Road Test				*
Engine: check for leaks				
Power steering: check fluid level		*		
Exhaust system: check for damage and leaks				
On Board Diagnostic system: check Diagnostic Trouble Code (DTC) Memory with VAG 1551/1552 Scan tool (ST) and erase if necessary. (Dealer service recommended).	*		*	*

Table h. every 2 year service (all models)

Maintenance item	Tools required	New parts /fluids required	Warm engine required	Dealer service recommended
Engine compartment maintenance				
Brake fluid: replace regardless of mileage	*	*		
Air cleaner: replace regardless of mileage	*	*		

Table i. at 4 years, 8 years, and every 2 years thereafter service (all models)

Maintenance item	Tools required	New parts /fluids required	Warm engine required	Dealer service recommended
Passenger compartment maintenance				
Air bags: To assure continued reliability, the Air Bag System must be inspected by an authorized Volkswagen dealer at 4 and 8 years from the date of manufacture and every two years thereafter.	*	*		*

ENGINE COMPARTMENT MAINTENANCE

The jobs listed under this heading are the engine compartment maintenance items from the maintenance tables. The jobs follow the same sequence in which they appear in the tables.

Engine oil level, checking

It is normal for your engine to consume a small amount of oil. The rate of consumption will depend on the quality and viscosity of the oil and the operating conditions. Make it a habit to check the oil level at every fuel filling.

1. After switching engine OFF, wait at least 3 minutes to allow the oil to flow back into the oil pan.

2. Remove dipstick located in front of intake manifold. Wipe dipstick with a clean cloth and reinsert as far as stop.

3. Remove dipstick again and read oil level. The oil level must be at the top edge of the shaded area but on no account above the maximum level mark. See Fig. 11.

> ***CAUTION —***
> *Risk of damage to catalytic converter if oil level is too high.*

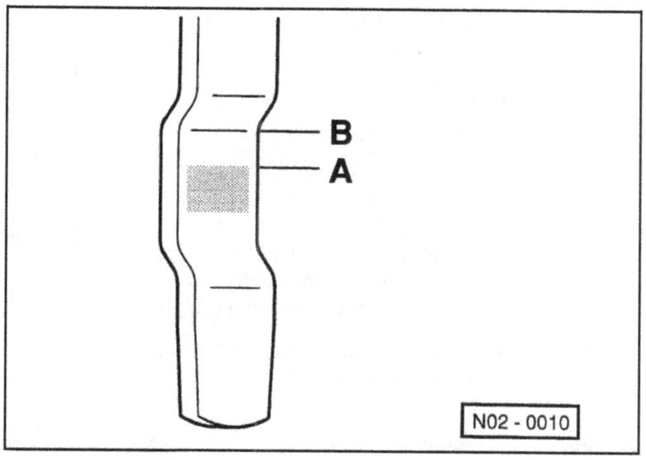

N02-0010

Fig. 11. Oil level must be at the top edge of shaded area **A**. Damage to engine oil seals and/or engine may result if oil is above mark **B**.

4. If oil level is below shaded area, fill to top mark shown in Fig. 11.

Select an oil that conforms to the standards of the American Petroleum Institute (API). A symbol can be found on the oil container with the API rating and viscosity grade.

Recommended Engine Oil
- Gasoline engines . . . . API service rating SJ
- Diesel engine . . . . API service rating CF4 or CG4

Engine oil, changing

1. Warm car to normal operating temperature.

2. Shut off engine and apply parking brake.

3. Raise car and support on jackstands designed for the purpose.

> ***WARNING —***
> *Jack stands should be placed on hard level surface (e.g. concrete).*

4. Remove sound dampening pan (belly pan) from beneath engine compartment. Remove drain plug from oil pan and allow oil to drain. See **17 Engine—Lubrication System** for drain plug location.

> ***WARNING —***
> *Hot oil can scald. Wear protective clothing, gloves and eye protection.*

> ***CAUTION —***
> - *Always replace oil-drain plug sealing ring anytime the drain plug is removed.*
> - *Dispose of oil properly at a facility equipped for recycling or storage.*

5. Replace oil filter. See **Oil filter, replacing** for applicable engine.

6. Install oil drain plug with a new seal. Tighten drain plug to proper torque.

Tightening torques
- Oil drain plug to oil pan
  - 4-cylinder engines . . . . 30 Nm (22 ft-lb)
  - 6-cylinder engines . . . . 50 Nm (37 ft-lb)

7. Fill engine with proper quantity and type of oil.

Oil capacity
- Engine oil capacity (with filter change)
  - ACC, ABA engines . . . . 4.2 qt. (4.0 liters)
  - AAZ, AHU engines . . . . 4.7 qt. (4.5 liters)
  - AAA engine . . . . 5.3 qt. (5.0 liters)

8. Run engine and check for leaks. Shut off engine and reinstall sound dampening pan. Check oil level after about 3 minutes.

Oil filter, replacing (4-cylinder engines)

1. Working at front of engine, loosen oil filter with oil filter strap or suitable wrench. Remove filter.

2. Clean sealing surface on oil filter flange.

3. Lightly lubricate rubber seal with clean oil.

4. Thread on new filter and tighten by hand.

> ***CAUTION —***
> *Do not clean and re-use oil filter.*

Oil filter, replacing (6-cylinder engine)

1. Drain oil from oil filter housing by removing drain screw. See Fig. 12.

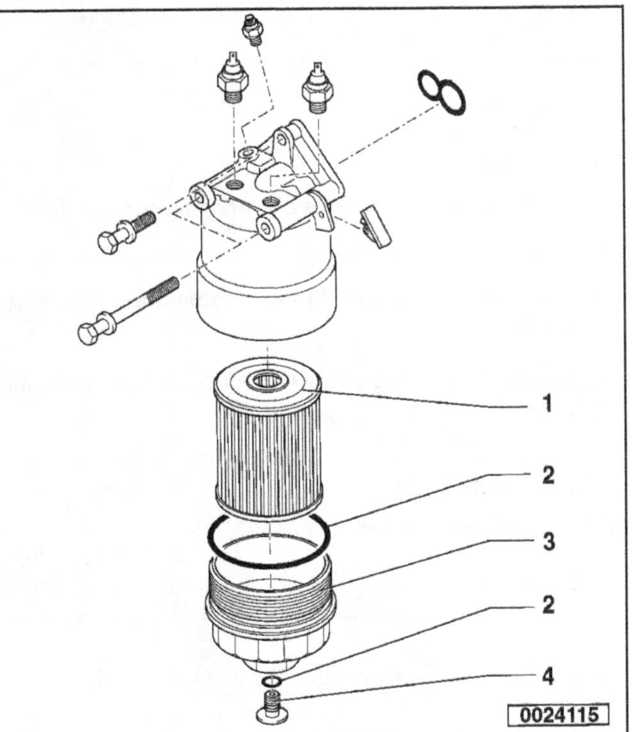

Oil filter (1)
O-rings (2)
Lower filter housing (3)
Drain screw (4)

0024115

Fig. 12. Oil filter assembly used on AAA (VR6) engine.

2. Unscrew lower filter housing. Discard filter and O-ring.

3. Install new filter element and new O-ring to lower filter housing.
   - Lightly lubricate O-ring with clean oil.

> ***CAUTION —***
> *Do not clean and re-use oil filter.*

4. Install and tighten lower filter housing and drain screw.

Tightening torques
- Lower filter housing to upper filter housing . . . . 30 Nm (22 ft-lb)
- Drain screw to lower filter housing (use new O-ring) . . . . 10 Nm (89 in-lb)

Coolant, checking

Cooling system maintenance consists of maintaining the coolant level, checking the coolant freezing point, and inspecting the hoses. Coolant flushing is not part of Volkswagen's scheduled maintenance.

> ***NOTE —***
> *Volkswagen does not require replacing the coolant as part of routine maintenance. The cooling system has been filled at the factory with a permanent coolant. For coolant system draining and refilling procedures, see* **19 Engine-Cooling System**.

> ***WARNING —***
> - *Hot coolant can scald. Do not work on the cooling system until it has fully cooled.*
> - *Use extreme care when draining and disposing of coolant. Coolant is poisonous and lethal. Children and pets are attracted to it because of its sweet smell and taste. See a doctor or veterinarian immediately if any amount is ingested.*

> ***CAUTION —***
> - *Use only Volkswagen original anti-freeze when filling the cooling system. Use of any other anti-freeze may be harmful to the cooling system. Do not use an anti-freeze containing phosphates.*
> - *Do not use tap water in cooling system. Use distilled water only to mix anti-freeze.*

Volkswagen uses two different types of coolant which can be identified by color.

Blue/Green – This coolant/antifreeze is phosphate free and should be used in all water-cooled Volkswagens up to and including 1996 and some early 1997 vehicles. When supplied by Volkswagen, this coolant/antifreeze is generally known as G11.

Red – This coolant/antifreeze is phosphate free and also silicate free. It should be used in all water-cooled Volkswagens from early 1997 production. When supplied by Volkswagen, this coolant/antifreeze is known as G12.

The advantages of G12 over earlier types include improvements in corrosion protection, thermal stability, heat transfer/control, hard water tolerance and environmental protection.

CAUTION —
- *G12 (Red Coolant) must NEVER be mixed with ANY other coolant. Engine damage will result!*
- *Contamination of G12 with other colored coolants is identifiable by discoloration (brown, purple, etc.). This mixture causes a foamy deposit to appear in the expansion tank/radiator and MUST be drained immediately.*
- *The cooling system must be completely free of this mixture before refilling with the correct type of coolant/antifreeze.*

A translucent expansion tank, or overflow reservoir, provides easy monitoring of coolant level without opening the system. The coolant level should always be checked when the engine is cold. The coolant level should be between the maximum and the minimum mark on the expansion tank. See Fig. 13.

0024130

Fig. 13. Coolant level should be between **max** and **min** marks on coolant expansion tank.

Inspect the hose by first checking that all connections are tight and dry. Coolant seepage indicates either that the hose clamp is loose, that the hose is damaged, or that the connection is dirty or corroded. Dried coolant has a chalky appearance. Check the hose condition by pinching them. Hoses should be firm and springy. Replace any hose that is cracked, that has become soft and limp, or has been contaminated by oil or diesel fuel. See Fig. 14.

To check the freezing point of the coolant, use a hydrometer. With the engine cold, remove the cap from the expansion

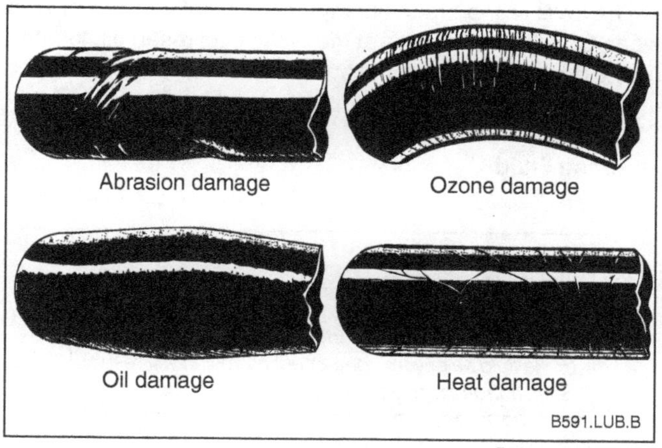

B9110

Fig. 14. Examples of damaged coolant hoses. Any of conditions shown is cause for replacement. Courtesy of Gates Rubber Inc.

tank and draw up some coolant with the hydrometer. Volkswagen recommends using a 50/50 mixture of anti-freeze and water, which has a freezing point of -35°F (-38°C). If the freezing point is too high, drain a small amount of coolant from the system and add new anti-freeze until the proper protection level is obtained.

NOTE —
For protection to approx. -40°C (-40°F) the percentage of anti-freeze may be increased up to 60%. The percentage of anti-freeze must not exceed 60%, as higher amounts will decrease frost protection and cooling capacity.

Brake fluid level, checking

Routine maintenance of the brake system includes maintaining an adequate level of brake fluid in the reservoir, replacing the brake fluid every 2 years, checking the brake pads for wear, checking the hand brake function, and inspecting the system for fluid leaks or other damage. See **Under Car Maintenance** for pad inspection and brake system inspection.

WARNING —
- *Use only new, previously unopened brake fluid conforming to US Standard FMVSS 116 DOT 4.*
- *Brake fluid is poisonous. DO NOT ingest brake fluid. Wash thoroughly with soap and water if brake fluid comes into contact with skin.*

CAUTION —
- *DO NOT let brake fluid come in contact with paint. Wash immediately with soap and water.*
- *Brake fluid absorbs moisture from the air and must be stored in an airtight container.*
- *DO NOT allow brake fluid to exceed the maximum level in the fluid reservoir.*

The level of the brake fluid will drop slightly as the brakes wear. Check the fluid level at the brake fluid reservoir, located on the driver's side, near the bulkhead.

Brake Fluid
- Type . . . . FVMSS 116 SAE DOT 4

WARNING —
Do not mix DOT 5 (silicone) brake fluid with DOT 4 brake fluid as severe component corrosion will result. Such corrosion could lead to brake system failure.

Brake fluid, replacing

Brake fluid readily absorbs moisture from the atmosphere. This moisture can cause brake system corrosion and adversely affect braking performance. Therefore, the old brake should be flushed out of the system and new, fresh fluid added at least every 2 years, regardless of mileage. See **47 Brakes–Hydraulic System** for brake system bleeding procedures and brake fluid replacement.

WARNING —
- *Vehicles equipped with the TEVES 04 ABS/EDL can be bled using conventional methods.*
- *Vehicles equipped with the TEVES 20 GI ABS/EDL can also be bled during maintenance using conventional methods. However, if the reservoir has been completely emptied, the hydraulic unit must be bled using the Volkswagen Scan Tool VAG 1551 or VAG 1552. This is only necessary if the hydraulic unit or control module has been replaced or if there is a leak in the system. This procedure electronically cycles the hydraulic unit and valve body to purge trapped air and is beyond the scope of this manual. See* **45 Anti-Lock Brakes (ABS)** *for application information.*

Power steering fluid level, checking

When checking the power steering fluid level, the engine should be running and at normal operating temperature and the front wheels should be in the straight ahead position. Note that the power steering system uses a special hydraulic oil and not ATF.

To check the fluid level, remove the dipstick from fluid reservoir and check the level on the dipstick. The level should be between MIN and MAX mark. See Fig. 15. If necessary, top up with Volkswagen power steering hydraulic oil (VW part G 002 000).

Power Steering Fluid
- Type . . . . hydraulic oil (VW part no. G 002 000)

CAUTION —
- *Use only Volkswagen hydraulic oil G 002 000 in the power steering system. Do not use ATF or other non-approved types of power steering fluid. If the wrong fluid is used, power steering components may fail.*
- *Part numbers are listed for reference only. Always check with an authorized Volkswagen parts department for the latest parts information.*

NOTE —
If the hydraulic oil level has dropped to the MIN marking the power steering should be checked. It is not sufficient to just add hydraulic oil.

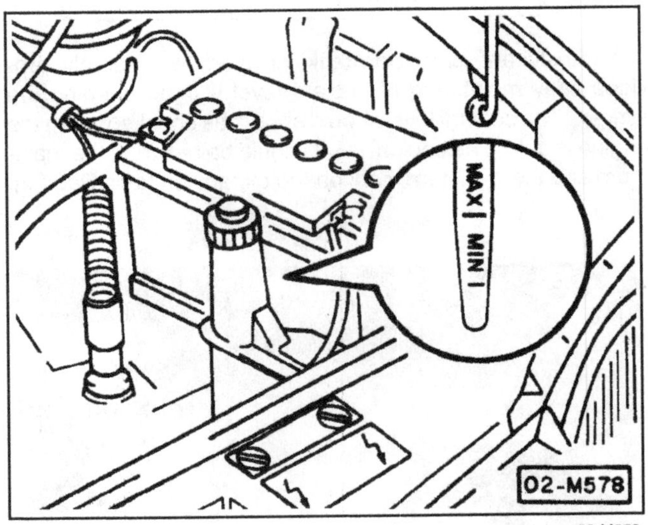

02-M578

Fig. 15. Power steering reservoir and dipstick (inset).

Fluid leaks

Check the engine compartment for signs of fluid leaks. Fluid leaks attract dust making them easier to spot. Many expensive repairs can be avoided by prompt repair of minor fluid leaks.

Inspect for leaks in engine, cooling, fuel, and heating system. Visually inspect hoses and hose connections for leaks, worn areas, porosity and brittleness.

Fuel filter, draining (diesel engines)

The diesel fuel filter functions to trap water. As water is heavier than diesel fuel, it settles to the bottom of the filter and can be drained off.

CAUTION —
Do not allow diesel fuel to contact coolant hoses or other rubber parts. Wipe diesel fuel off hoses immediately and wash with soap and water.

1. Remove control valve retaining clip. See Fig. 16.

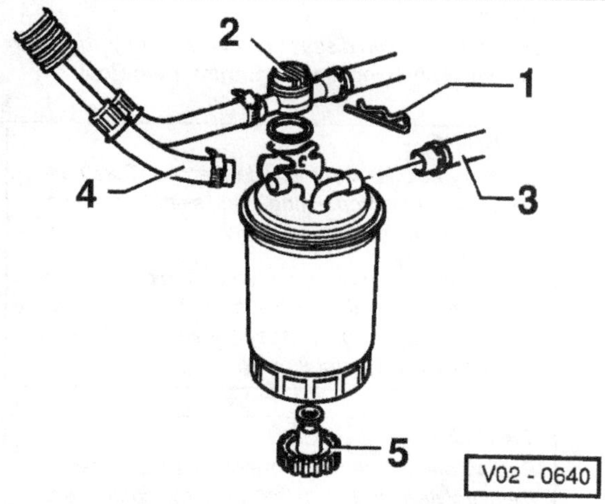

Retaining clip (1)
Control valve (2)
Fuel hose from fuel tank (3)
Fuel hose from fuel pump (4)
Drain screw (5)

V02-0640

Fig. 16. Fuel filter/water separator used on models with diesel engines.

2. Remove control valve with fuel lines connected.

3. Loosen drain screw (on bottom of filter) and drain approx. 100cc (1.7 oz.) of fuel/water.

4. Install control valve using new O-ring.

5. Attach retaining clip and tighten drain screw.

6. Check fuel system for leaks (visual inspection) with the engine running. Raise engine speed several times and let idle. Fuel flow through transparent pipe must be free of bubbles.

Fuel filter, replacing (diesel engines)

NOTE—
Refer to Fig. 16 when replacing the fuel filter.

> ***WARNING—***
> - *Always disconnect the negative battery cable before working on the fuel system.*
> - *Before disconnecting battery be sure to obtain radio anti-theft code.*
> - *Do not smoke or create sparks when working on fuel system. Have an approved fire extinguisher handy.*
> - *Diesel fuel is dangerous to your health. Wear skin, eye and hand protection. Work in a well ventilated area.*

1. Remove control valve retaining clip.

2. Remove control valve with fuel lines connected.

3. Label and disconnect remaining fuel hoses from top of fuel filter.

NOTE—
Do not reverse hoses when installing. Fuel flow direction is indicated by arrows.

4. Remove filter bracket together with filter.

5. Remove filter from the bracket.

6. Installation is the reverse of removal. Be sure to use a new O-ring at the control valve. After installing filter, check for leaks (visual inspection) with the engine running. Raise engine speed several times and let idle. Fuel flow through transparent pipe must be free of bubbles.

> ***CAUTION—***
> *Always replace the fuel filter control valve O-ring.*

Fuel filter, replacing (gasoline engines)

Volkswagen specifies that the fuel filter used on gasoline engines is designed to last the life of the car. Replacement is only necessary if the filter becomes clogged due to contaminated fuel or failure to keep the fuel tank sealed.

Dust and pollen filter, replacing

Some models may be equipped with a ventilation air dust and pollen filter. This filter prevents most dusts and pollens from entering the passenger compartment. Volkswagen specifies replacement of this filter at regular mileage intervals.

1. Working in rear of engine compartment (plenum chamber) remove plastic screws attaching the filter cover. See Fig. 17. See also **80 Heating and Ventillation**.

2. Partially remove rubber seal at bulkhead partition.

3. Push front of cover upward.

4. Push spring catches for filter element inward and remove filter element.

5. Install new filter element. Install rubber seal and cover.

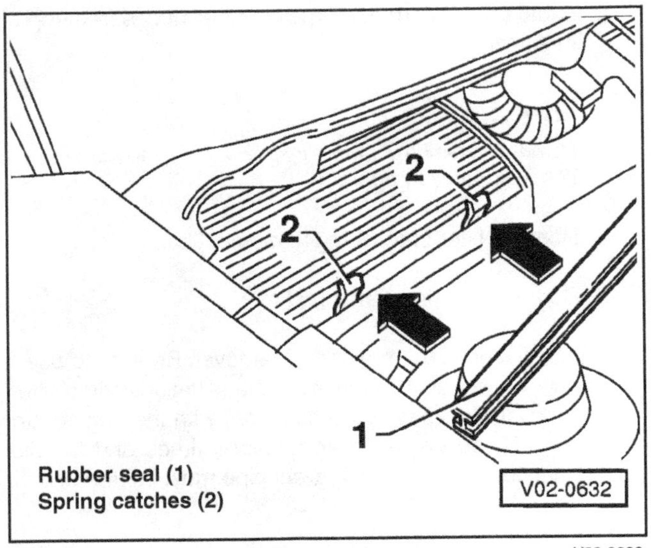

V02-0632

Fig. 17. Dust and pollen filter in rear of engine compartment (right plenum chamber).

Battery Service

Under normal operating conditions, the battery is maintenance free. At high outside temperatures, however, it is advisable to check the electrolyte level through the translucent battery housing. The electrolyte level should be just above the battery plates and their separators. The battery plates can be seen once the filler caps are removed. If the electrolyte level is low, replenish it by adding distilled water only. See Fig. 17.

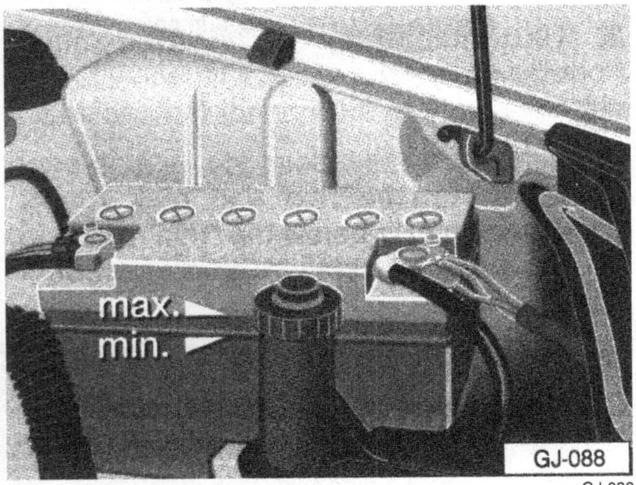

GJ-088

Fig. 18. Battery electrolyte level should be between **min** and **max** area on translucent battery housing.

NOTE —
For a more detailed discussion of the battery, including testing and charging, see **27 Engine Electrical**.

WARNING —
- *Battery acid is extremely dangerous. Take care to keep it from contacting eyes, skin, or clothing. Wear eye protection.*
- *Extinguish all smoking materials and do not work near any open flames.Batteries give off explosive hydrogen gas. Keep sparks and open flames away. Do not smoke.*

CAUTION —
Battery electrolyte can damage painted surfaces. If electrolyte is spilled on the car, immediately clean the area with a solution of baking soda and water.

Power steering V-belt, inspecting/adjusting (4-cylinder engines)

On 4-cylinder engines, a V-belt is used to drive the power steering pump. Check the condition of V-belt for splits in the base material, separation of layers, fraying of cords, flank wear, glazing, traces of oil or grease. If any of these conditions exist, the belt should be replaced. Check V-belt tension by applying firm thumb pressure to the area of the belt between the two pulleys that are furthest apart. Adjust belt tension if necessary.

Power Steering V-belt
- Belt deflection (used belt). . . . approx. 5 mm (3/16 in.)

1. Loosen power steering pump securing bolts and nuts until V-belt is loose. See Fig. 19.

   NOTE —
   Loosen adjustment clamping bolt (Fig. 19, bolt no. 3) at least one full turn.

2. Adjust V-belt tension by torquing tensioning nut to the specification listed below. Tighten clamping bolt when tension is correct. Tighten remaining securing bolts.

   NOTE —
   When the tensioning nut is torqued to the specified setting, belt tension will be correct.

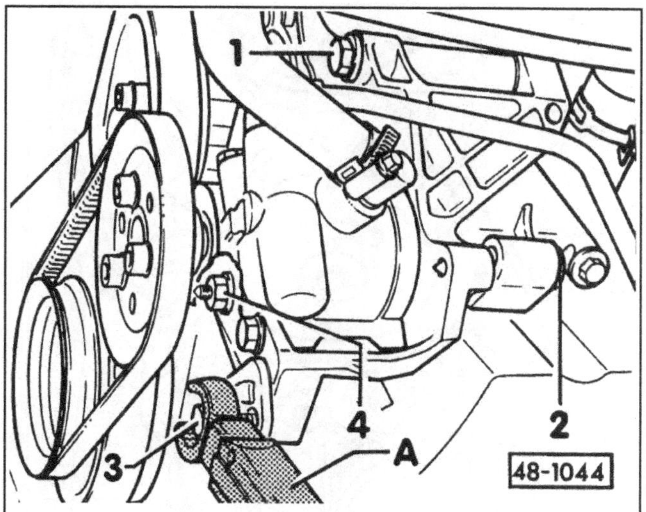

Securing bolts (1 and 2)
Clamping bolt and tensioning nut (3)
Tensioning bar securing nut (4)
Torque wrench with 22 mm crow foot wrench (A)

48-1044

Fig. 19. To loosen V-belt, loosen bolts **1**, **2**, **3**, and nut **4**. Belt tension is adjusted at tensioning nut using torque wrench (**A**).

Tightening torques

- V-belt tensioning nut (3)
  - used belts . . . . 4 Nm (35 in-lb)
  - new belts. . . . . 7 Nm (62 in-lb)
- Clamping bolt (3) . . . . 25 Nm (18 ft-lb)
- Upper securing bolt (1) . . . . 45 Nm (33 ft-lb)
- Lower securing bolt/nut (2/4) . . . . 25 Nm (18 ft-lb)

Poly-ribbed belt, applications

On 4-cylinder engines with a separate tensioner pulley (cars with A/C), and all 6-cylinder engines, the poly-ribbed drive belt is self-adjusting and does not require any routine maintenance aside from periodic inspection for wear.

On 4-cylinder engines without a separate tensioner (cars without A/C), a heavy-duty coil spring is mounted under the alternator. When the alternator mounting bolts are loosened, the coil spring presses up on the alternator and correctly tensions the belt. The alternator mounting bolts are then retightened. This tensioning system does not account for belt wear and/or stretch and must therefore be adjusted at the specified mileage intervals.

NOTE —
It is not necessary to check and adjust the poly-ribbed belt tension on engines with tensioning pulley. Optimal belt tension is assured by means of the spring-loaded tensioning device.

Poly-ribbed belt, adjusting (4-cylinder engines without belt tensioner)

1. Loosen alternator securing bolts.

> ***CAUTION —***
> *Do not remove alternator bolts. Loosen bolts only enough to allow alternator to pivot freely.*

2. Attach tensioning lever (Volkswagen special tool no. 3297) to alternator and secure with locking pin. See Fig. 20.

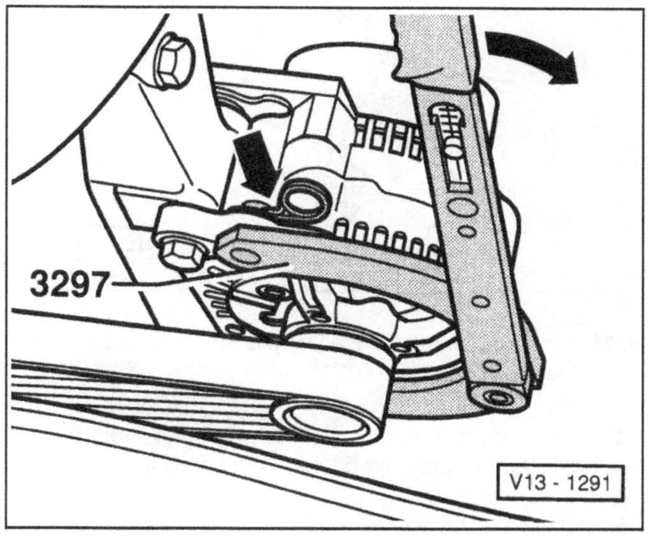

V13-1291

Fig. 20. Alternator belt tension being relieved using Volkswagen special tool no. 3297 and torque wrench. Use locking pin (**left arrow**) to secure tool to alternator.

3. Pivot alternator down to stop three times. Alternator must pivot freely.
4. Remove tensioning lever. Start engine and let idle for at least 10 seconds.
5. Shut off engine and tighten alternator securing bolts.

Tightening torques

- Alternator to mounting bracket . . . . . . 25 Nm (18 ft-lb)

Poly-ribbed belt, removing and installing (4-cylinder engines)

1. Remove power steering V-belt. See **Power Steering V-belt, adjusting (4-cylinder engines)**.
2. On cars without separate belt tensioner, loosen alternator mounting bolts 2 turns. Then press down on the alternator to relieve belt tension. See Fig. 20.

3. On cars with separate belt tensioner, pivot belt tensioner off belt using appropriate spanner wrench (Volkswagen special tool no. 3299). See Fig. 21.

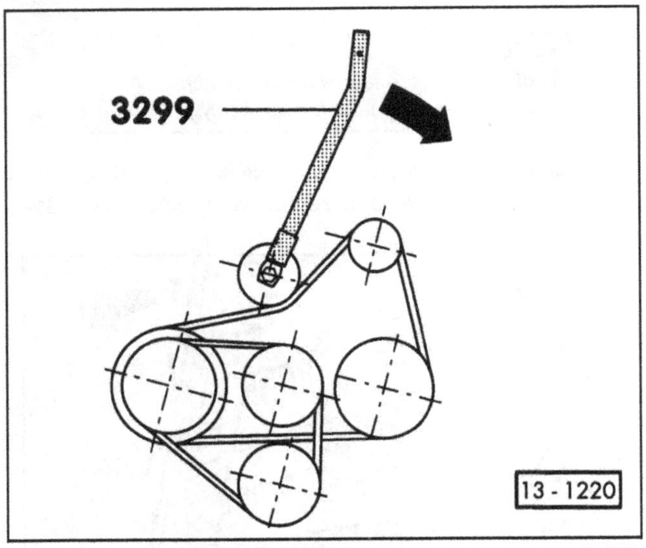

13-1220

Fig. 21. 4-cyl. engine with belt tensioner: Belt tensioner being lifted with spanner wrench (Volkswagen special tool no. 3299).

4. If reinstalling the old belt, mark running direction on belt. Remove poly-ribbed belt from pulleys.

5. Installation is the reverse of removal. Start engine and check that belt runs smoothly.

6. On cars without belt tensioner, adjust the belt tension and tighten alternator mounting bolt as described earlier.

Poly-ribbed belt, removing and installing (6-cylinder engine)

1. Remove evaporative emission (EVAP) cannister purge regulator valve with retaining ring from lower part of air cleaner housing. See **20 Fuel Storage and Supply**.

2. Remove air cleaner housing complete with mass air flow sensor. See **24a Fuel Injection-Motronic**.

3. Temporarily install bolt (M8x80) in threaded hole in tensioner and thread bolt in until poly-ribbed belt is free of tension. See Fig. 22.

> ***CAUTION—***
> *Do not over tighten bolt. Thread bolt in only until poly-ribbed belt can be removed. The tensioner housing may be damaged if the bolt is overtightened.*

4. If reinstalling the old belt, mark running direction on belt. Remove poly-ribbed belt from pulleys.

5. Install poly-ribbed belt in position. See Fig. 23.

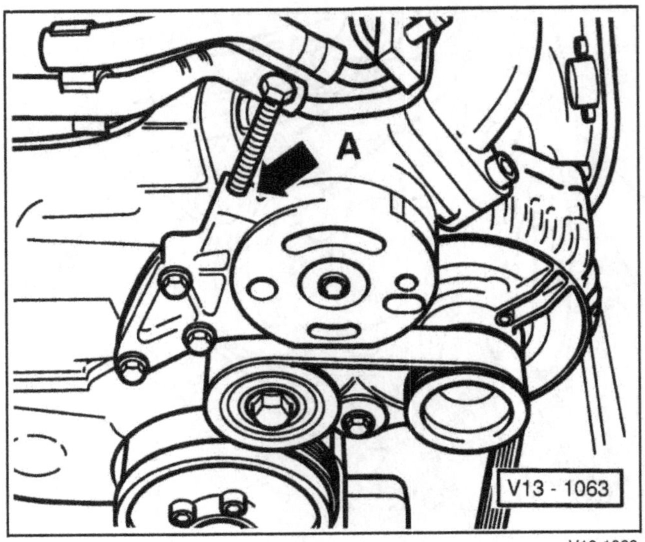

V13-1063

Fig. 22. To relieve poly-ribbed belt tension on 6-cylinder engines, thread M8x80 bolt (**A**) into tensioner until belt is free of tension.

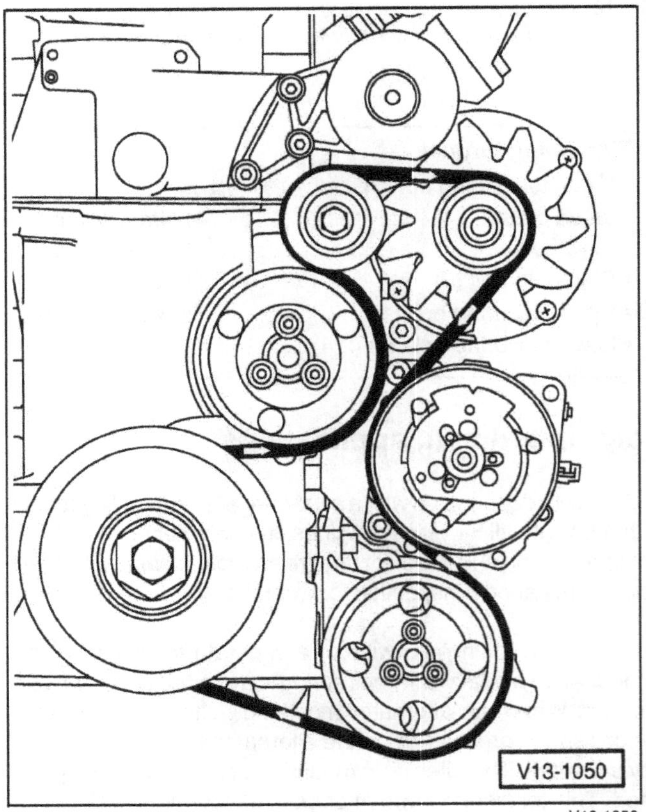

V13-1050

Fig. 23. Poly-ribbed belt routing on 6-cylinder engine. **Arrows** indicate direction of travel.

6. Slowly remove M8 bolt from tensioner, making sure belt is correctly seated in pulley grooves.

7. The remainder of installation is the reverse of removal. Start engine and check that belt runs smoothly.

Camshaft drive belt, checking (4-cylinder engines)

The camshaft drive belt, also called the timing belt, is a toothed rubber belt that drives that camshaft. On diesel engines, the belt also drives the injection pump. The belt is subjected to high temperatures and should be inspected during scheduled maintenance intervals. On diesel engines, Volkswagen specifies replacement of the belt every 60k miles.

> ***NOTE —***
> *Volkswagen does not specify a replacement interval for the camshaft drive belt on gasoline engines. However, the publisher recommends the belt be inspected periodically and replaced at least every 90k miles or 5 years.*

To inspect the drive belt, the upper belt cover must first be removed. Inspect the belt for tears, separation of layers, fraying of belt cords, surface cracks, or traces of oil and grease. Replace the belt if any faults are found. See **15a Cylinder Head and Valvetrain (4-Cylinder Engines)** or **Fuel Injection–Diesel** for replacement procedures.

Camshaft drive belt, adjusting (diesel engine)

A special toothed belt tensioning gauge (Volkswagen special tool no. VW 210) is needed to accurately check and adjust camshaft drive belt tension.

1. Remove upper belt cover.
2. Install belt tensioning gauge (VW 210) between camshaft sprocket and injection pump sprocket and measure belt tension. See Fig. 24.

Camshaft/Injection Pump Drive Belt Tension
- Drive belt tension (using Volkswagen special tool no. VW 210) . . . . 12–13 index marks

3. If the belt tension is incorrect (loose), adjust it as follows:
   - Loosen tensioner locknut and turn tensioner clockwise using a spanner wrench. See Fig. 24.
   - Measure belt tension between camshaft sprocket and injection pump sprocket using special tension gauge (Volkswagen special tool no. VW 210).
   - When tension is correct, tighten tensioner locknut and remove tool.

Tightening Torque
- Drive belt tensioner locknut . . . . 45 Nm (33 ft-lb)

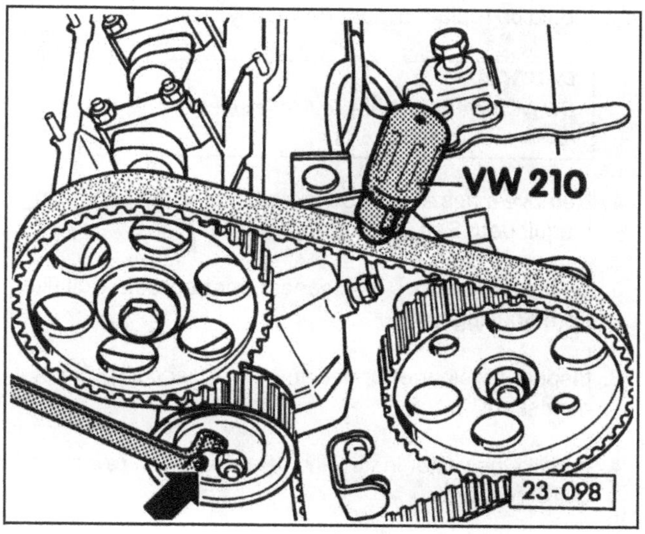

23-098

Fig. 24. Diesel camshaft drive belt tension being measured with Volkswagen special tool no. VW210. Tensioner being adjusted with spanner wrench (special tool Matra V159).

4. Turn the engine by hand, two full revolutions in the direction of normal rotation (clockwise). Recheck drive belt tension, adjust if necessary. Install upper belt cover.

Spark plugs, replacing

The spark plugs should be replaced at the specified mileage interval given in the maintenance tables. Use of the proper spark plugs is important for long engine life and low exhaust emissions. Note that special tools should be used to remove the spark plug connectors from the plugs on 6-cylinder engines. See **28 Ignition System** for spark plug application and replacement information.

> ***NOTE —***
> *Spark plug specifications may change during a model year. Always check with an authorized Volkswagen parts department for the latest information.*

Air cleaner, removing and installing

Air filter elements should be replaced every 2 years or 30,000 mi., whichever comes first. Under severe conditions, dirt roads, desert driving, or dusty conditions, the air filter element should be replaced more frequently.

> ***CAUTION —***
> *DO NOT clean and re-use filter element. Always replace.*

1. Remove mass air flow sensor from air cleaner housing, if equipped. See **24 Fuel Injection**.
2. Unclip and remove air cleaner upper section. Pull filter from housing.
3. Inspect air cleaner upper and lower sections. Clean if necessary.
4. Install a new filter insert and re-assemble in reverse order of removal.

Distributor cap and rotor, inspecting (4-cylinder gasoline engines)

Volkswagen does not specify distributor cap and ignition rotor inspection, although it is a good idea to inspect these components at 60k miles, and every 30k miles thereafter. Distributor cap and ignition rotor inspection and replacement is described in **28a Ignition System–Gasoline**.

> ***NOTE —***
> *When replacing the distributor cap, label the plug wires before removing them from the cap.*

Clutch cable, adjusting

All Volkswagen models covered by this manual have self adjusting clutches. The VR6 engine is equipped with a hydraulically-actuated clutch. 4-cylinder engines use a self-adjusting cable to actuate the clutch.

> ***NOTE —***
> *For more information on the clutch actuating mechanisms, see* **30 Clutch (4-Cylinder Engines)** *or* **30b Clutch (6-Cylinder Engine)**.

UNDER CAR MAINTENANCE

Under car maintenance requires that the car be raised on a lift or properly supported on jackstands. Under car maintenance should not be carried out unless proper (safe) lifting equipment is available. See **Lifting Vehicle** for more information on lifting and working under the car.

> ***WARNING —***
> - *Do not work under a car supported solely by a jack. Jack stands must always be used when working under the car.*
> - *Jack stands must be placed on a hard, level surface (e.g. concrete).*

Tire service

Tire pressure should be checked on a regular basis. It is best to check pressures when the tires are cold. Refer to the data label on the driver's door jamb for proper inflation pressures. Be sure to also check the spare tire pressure. Check that all tires are the same type and tread pattern. Measure tread depth. If the tread wear has exceeded the minimum specification listed, the tire should be replaced.

Tread depth
- minimum height . . . . 1.6 mm (1/16 in.)

> ***CAUTION —***
> *Wheel alignment must be checked after replacing tires to ensure maximum tire life.*

The tires should also be checked for abnormal wear patterns.The tread wear pattern on the front tires is an indication of whether the toe and camber settings need to be checked. "Feathering" on the tread indicates incorrect toe. If the tread is worn on one side, this is usually caused by incorrect camber.

To extended tire life and minimize wear, Volkswagen recommends that the tires be rotated front to rear every 7, 500 miles. For more information on tires, see **44 Wheels–Tires, Wheel Alignment**.

Brake system, visual inspection

Check the brake master cylinder, vacuum brake booster (also hydraulic unit if anti-lock brake system is fitted), brake pressure regulator, and brake calipers for leaks and damage.

- Brake hoses must not be twisted.
- Brake hoses must not touch any part of the vehicle when steering is at full lock.
- Check brake hoses for porosity and deterioration, brake hoses and brake lines for chafing points.
- Check brake connections and attachments for correct seating, leaks and corrosion.
- Any faults found must be repaired.

Brake pads (front and rear), checking

To accurately check the front brake pad thickness, the front wheels should be removed. With the wheel removed, measure the thickness of outer and inner pads. See Fig. 25. If pad thickness (incl. backing plate) is 7 mm (0.28 in.) or less, the brake pads should be replaced.

Disc brakes (front and/or rear)
- brake pad wear limit . . . . 7 mm (0.28 in.)

NOTE —
- *Before removing the front wheels, mark their position in relation to the brake rotor so the wheel can be reinstalled in the same position.*
- *Brake pad wear may be minimally greater on the front passenger's side than on the driver's side. It is therefore recommended that inspection always be performed on passenger's side front wheel.*
- *When tightening wheel bolts, tighten the bolts in stages using a criss-cross pattern.*

Tightening torques
- Wheel bolt to wheel hub . . . . 110 Nm (81 ft-lb)

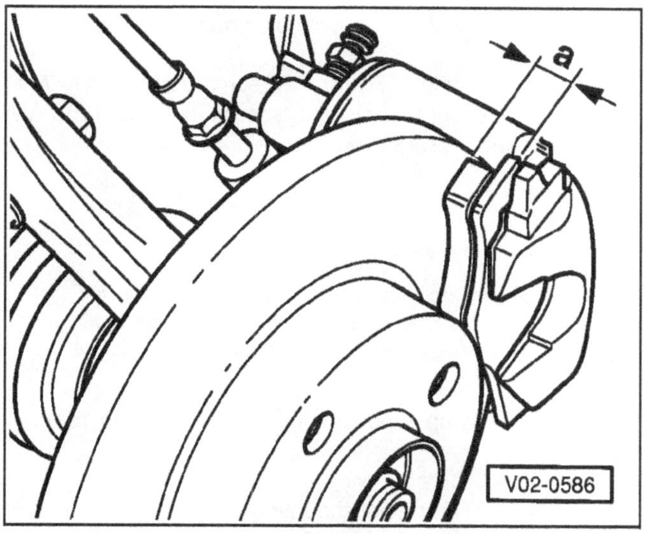

V02-0586

Fig. 25. Brake pad thickness must be at least 7 mm (0.28 in.) as indicated by **dimension a**.

On cars with 4-wheel disc brakes, it is not necessary to remove the wheels for rear brake pad inspection. Visually inspect the thickness of outer and inner brake pads through the opening in wheel using a flashlight and mirror if necessary. If pad thickness (incl. backing plate) is 7 mm (0.28 in.) or less, the brake pads should be replaced.

Brake linings (rear), checking

On cars equipped with rear drum brakes, the rear brake shoe lining thickness can be checked through the inspection holes in the rear of the brake backing plates. If necessary, remove the sealing plugs from the inspection holes. Also check linings for brake fluid or grease contamination.

NOTE —
See **46 Brakes–Mechanical Components** *for brake application information.*

Rear Drum Brakes
- Brake shoe lining wear limit . . . . 2.5 mm (0.10 in.)

Parking brake, adjusting

The parking brake acts on the rear brakes and is self-adjusting to compensate for wear. Adjustment of the parking brake is only necessary if brake components are replaced. See **46 Brakes–Mechanical Components** for parking brake adjustment and service procedures. The parking brake should begin to hold after two clicks of the lever.

Drive axle shafts, checking boots

There is an inner and an outer rubber CV boot on each front axle shaft. The boots should be regularly inspected for tears, cracks, or deterioration. Once the boot is open to the weather, abrasive road debris can enter the joint and quickly destroy it. If boot replacement is carried out promptly, chances are good that the joint can be saved. See **40 Front Suspension and Drive Axles** for drive boot replacement procedures.

Front suspension components, checking

Check ball joint dust boots for damage and correct seating. Check for play in ball joints. Check inner and outer tie rod end boots for damage and correct seating. Check for play by moving tie rods and wheels. Check attachment of ball joints and tie rod ends. See **40 Front Suspension and Drive Axles** for suspension component replacement.

Transmission, checking for leaks

Inspect the transmission for signs of fluid leakage. Pay particular attention to the axle seals. Repair any leaks found, see **34a Manual Transmission (ABA, ACC, AAZ Engines)**, **34b Manual Transmission (AAA, AHU Engines)** or **37 Automatic Transmission**.

Final drive oil level, checking (cars with automatic transmission)

Transmission and final drive assemblies and lubricants used on the cars covered by this manual vary depending on engine installed and on model year. Be sure to identify the transmission installed before checking the final drive oil level. See **Transmission identification**, given earlier.

Engine Code	Manual Transmission Code	Automatic Transmission Code
AAZ, ACC, ABA	**020**	**096** or **01M**
AAA, AHU	**02A**	**096** or **01M**

> ***CAUTION —***
> - *Only genuine VW synthetic ATF (VW part no. part no. G052 162 A1) should be used in the 01M transmission final drive. Use of other ATF types can adversely affect the service life of the transmission.*
> - *Part numbers are listed for reference only. Always check with an authorized Volkswagen parts department for the latest parts information.*

On cars with manual transmission, the final drive lubricant shares a common bath with the transmission lubricant. For final drive and transmission fluid level checking, see **Manual transmission oil, checking**.

On cars with automatic transmission, the final drive oil level is checked by removing the speedometer drive gear from the transmission housing and using the drive gear as a dipstick. The car should be on a level surface when making the check. See **39 Differential and Final Drive** for the procedure on checking the final drive fluid level on cars with automatic transmission.

NOTE —
Volkswagen does not specify a final drive lubricant replacement interval. If the fluid needs to be drained, it can be drawn off using a fluid extraction system.

Automatic transmission fluid (ATF) level, checking

Two automatic transmission are used in the cars covered by this manual: the "096" (also known as Phase I) and the "01M" (also known as Phase II).

The 096 A/T uses a dipstick to check the ATF level and this procedure is described in detail **37 Automatic Transmission**.

NOTE —
Transmission identification and application information is given under **Transmission identification**.

The 01M A/T, on the other hand, does not have a dipstick and the procedure requires measuring and maintaining a specified ATF temperature (which is neither cold nor hot, but somewhere in between) during the ATF level checking. Special Volkswagen tools are required for this operation and it is therefore recommended that ATF level checking on 01M transmission be left to an authorized Volkswagen dealer. See **37 Automatic Transmission** for detailed procedures on checking the 01M A/T fluid level.

ATF, changing (096 automatic transmission)

Absolute cleanliness is necessary when changing the ATF. It is a good idea to clean the transmission sump and surrounding area before beginning work.

1. With engine and transmission cold, raise car and support on jackstands designed for the purpose.

> ***WARNING —***
> - *Jack stands should be placed on a hard, level surface (e.g. concrete).*
> - *Hot ATF can burn and scald. Allow the transmission to cool before beginning work.*

2. Remove ATF filler (dipstick) pipe hold down bolt.

3. Extract the cold ATF from ATF sump using fluid extraction equipment (Volkswagen special tool VAG 1358 A/2 probe and VAG 1358 A).

NOTE —
- *There is no drain plug in the ATF sump.*
- *If extraction equipment is not available, the* ***cold*** *ATF can also be drained by loosening the sump bolts slowly until ATF runs out of one corner and into a drain bucket.*

4. Remove sump mounting bolts and carefully remove sump.

5. Remove ATF strainer mounting bolts and remove ATF strainer (filter). See Fig. 26.

6. Thoroughly clean sump.

> ***CAUTION —***
> *Cleanliness is essential when performing the following installation work.*

7. Fit new gasket (O-ring) into new ATF strainer and install strainer.

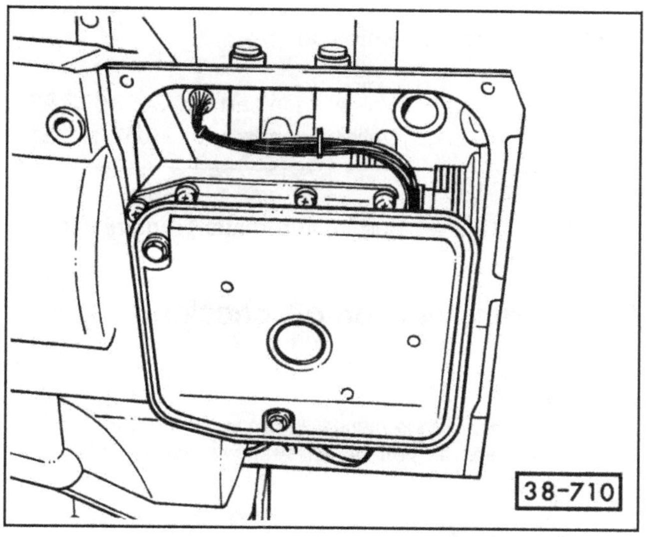

38-710

Fig. 26. 096 automatic transmission fluid strainer.

Tightening torques

- ATF strainer to transmission. . . . . 8 Nm (71 in-lb)

8. Fit spacer bushings into new sump gasket and install sump with gasket. Install filler pipe and pipe hold-down bolt.

Tightening torques

- ATF sump to transmission . . . . 12 Nm (9 ft-lb)

9. Pour in 3 liters (3.2 quarts) of ATF using funnel and extension hose.

> ***CAUTION—***
> - *Only ATF with the designation DEXRON may be used. Do not use any lubricant additives.*
> - *Too much ATF will cause malfunctions and must therefore be removed.*

ATF (096 automatic transmission)

- Lubricant . . . . Dexron II or III, or Genuine VW/Audi ATF G052 162 A1 or G052 162 A2

10. With engine running and vehicle stationary, shift through all selector lever positions. Check ATF level. See **37 Automatic Transmission**.

ATF, changing (01M automatic transmission)

Absolute cleanliness is necessary when changing the ATF. It is a good idea to clean the transmission sump and surrounding area before beginning work.

> ***CAUTION—***
> *01M automatic transmissions requires a special Volkswagen synthetic ATF. Do not use other types of ATF. Do not mix Volkswagen ATF with Dexron ATF.*

1. With engine and transmission cold, raise car and support on jackstands designed for the purpose.

> ***WARNING—***
> - *Jack stands should be placed on a hard, level surface (e.g. concrete).*
> - *Hot ATF can burn and scald. Allow the transmission to cool before beginning work.*

2. Place drain bucket under transmission.

3. Remove level plug and overflow tube and allow ATF to drain. See Fig. 27.

> ***CAUTION—***
> - *When level plug is removed, a small amount of ATF will drain.*
> - *When overflow tube is removed, the ATF will drain rapidly. Do not stand beneath the overflow tube. Wear hand and eye protection.*

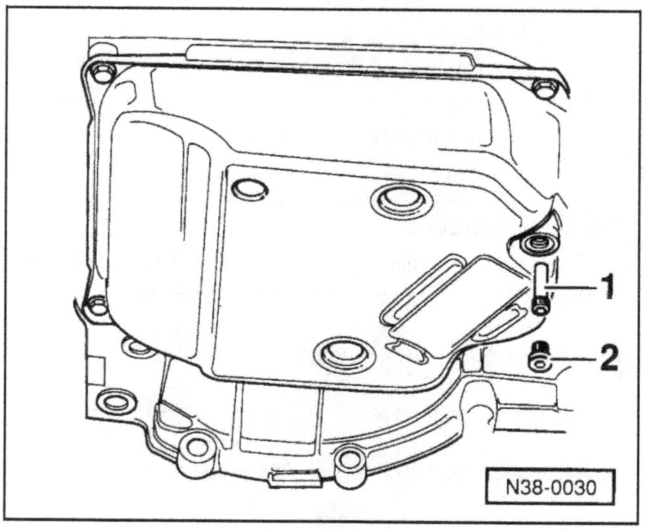

N38-0030

Fig. 27. 01M automatic transmission overflow tube (**1**) and level plug (**2**).

4. Remove sump and ATF strainer.

5. Clean sump and sump magnet.

> *CAUTION —*
> *Cleanliness is essential when performing the following installation work.*

6. Fit new O-ring into the new ATF strainer and install. See Fig. 28.

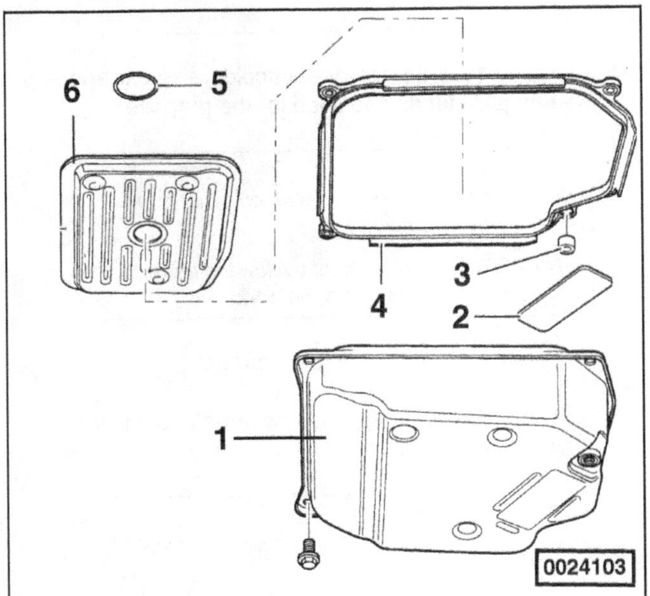

ATF sump (1)	**Gasket (4)**
Magnet (2)	**O-ring (5)**
Spacer bushing (3)	**Strainer (6)**

0024103

Fig. 28. Exploded view of ATF sump on 01M transmission. Be sure to fit spacer bushings into sump gasket.

7. Fit spacer bushings into new gasket. Install magnet into sump. Install oil sump with gasket.

Tightening torques
- Oil sump to transmission . . . . 12 Nm (9 ft-lb)

8. Install overflow tube to stop.

9. Install new seal on level plug and screw in hand tight.

10. Fill with 3 Liters (3.2 qt.) of Volkswagen ATF.

> *CAUTION —*
> *01M automatic transmissions use a special Volkswagen synthetic ATF.*

ATF (01M auto. transmission)
- Lubricant
  - 1 Liter . . . . Volkswagen part No. G052 162 A2
  - 0.5 Liter . . . . Volkswagen part No. G052 162 A1

11. Check ATF level. See **37 Automatic Transmission**.

Manual transmission oil, checking

> *NOTE —*
> *Replacement of the manual transmission lubricant is not specified as part of Volkswagen's maintenance program.*

> *CAUTION —*
> *Car must be level to accurately check transmission oil.*

1. Remove oil filler plug. See Fig. 29 or Fig. 30.

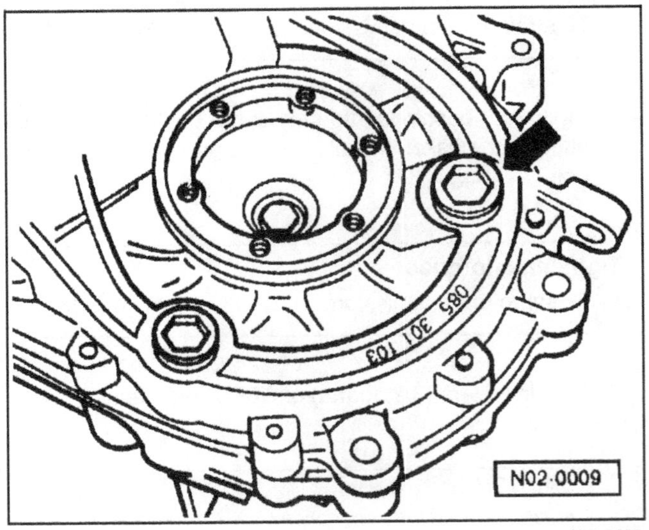

N02-0009

Fig. 29. 020 manual transmission used on cars with 4-cylinder engines. Remove oil filler plug (**arrow**) to check oil level.

2. Check that the oil level is up to lower edge of the filler hole.

3. If necessary, top off by adding oil through the filler hole.

Manual transmission
- Lubricant. . . . . VW G50 synthetic oil, SAE 75W90

4. Re-install oil filler plug.

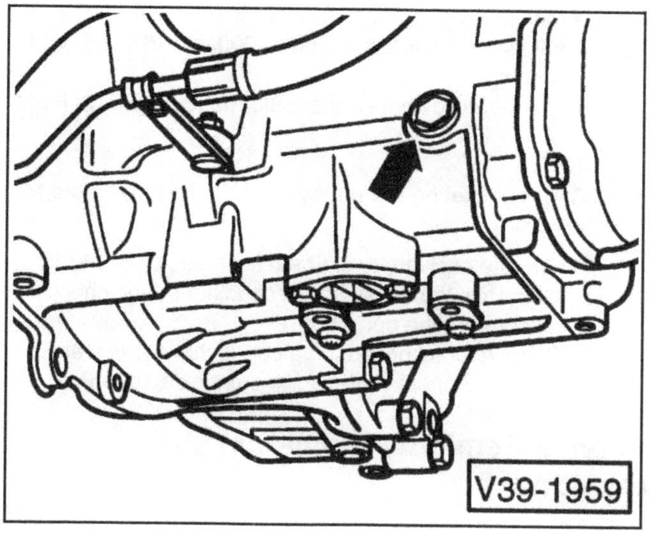

V39-1959

Fig. 30. 02A manual transmission used on cars with AAA and AHU engines. Remove oil filler plug (**arrow**) to check oil level.

Exhaust system, checking

> *WARNING —*
> - *The exhaust system operates at extremely high temperatures. Do not touch the exhaust system while the engine is running.*
> - *Allow exhaust system to cool at least one hour before touching.*

Inspect exhaust system for leaks or damage. Inspect exhaust system mounts. Replace faulty, missing or deteriorated parts.

Underbody sealant, checking

When performing the visual inspection for damage to the underbody sealant, also check the underbody, wheel housings and sill panels. Any faults found should be promptly repaired.

BODY AND INTERIOR MAINTENANCE

Body exterior

Automobile finishes are subjected to abuse from industrial fumes, corrosive road salt, acid rain, and other damaging airborn elements. Regular and correct care will contribute to maintaining and preserving the exterior of your Volkswagen.

> *NOTE —*
> *Proper care may be a condition for upholding the new car warranty, should corrosion damage or paint defects occur.*

The best protection against environmental influences is frequent washing and waxing. How often this is required depends on the environment where the vehicle is used.

Under certain circumstances weekly washing may be necessary. Under other conditions, a monthly washing and waxing may be adequate. Even if a wax solution is used when washing your vehicle, it is advisable to protect the paint with a coat of hard wax at least twice a year. Check the paint for chips and scratches. Paint defects should be touched up soon after they occur to prevent corrosion.

After the winter, the underside of the vehicle should be thoroughly washed.

Exterior plastic and vinyl should be kept clean. Occasionally apply a colorless vinyl or leather preservative. DO NOT wax plastic or vinyl.

Exterior lights, checking

Check operation of headlights (high and low beam), marker lights, taillights, turn signal lights, brake lights, reverse lights and emergency flasher lights. Use a helper when checking lights. See **94 Lights, Accessories–Exterior**.

Interior lights, checking

Check operation of indicator and instrument cluster warning bulbs. Check operation of interior cabin illumination bulbs. See **96 Lights, Accessories-Interior**.

Airbag unit, visual inspection

Models fitted with driver's side have the inscription "AIRBAG" on steering wheel. On cars with passenger side airbags, the inscription is also on the right-hand side of the instrument panel.

> *WARNING —*
> - *The padded airbag covers on the steering wheel and instrument panel must not be covered over or have any objects affixed to them.*
> - *Do not apply any chemical treatment to airbag unit covers. Clean with a dry or water moistened cloth only.*

Inspect padded airbag unit covers on steering wheel and instrument panel for signs of external damage. Check with an authorized Volkswagen dealer if any faults are found.

Door check straps and hinges, greasing

The door checks straps and door hinges should be lubricated periodically. Use a lithium grease for the job. Grease the hinge securing bolt with a lock cylinder spray. It is a good idea to clean the old grease away before applying the new grease.

Windshield wipers and washers

Inspect windshield wiper blades, front and rear. Replace damaged or deteriorated parts. Check spray pattern of windshield washers, front and rear, adjust as necessary. Fill reservoir for windshield cleaning system. Always add a windshield cleaner to the water (windshield washer antifreeze in winter). See **92 Wipers and Washers**.

In cases where cleaning of the windshield and replacement of the wiper blades or inserts fails to eliminate streaking conditions, a more complete cleaning of the windshield may be needed. Thoroughly clean the outside of the windshield with a non-abrasive cleaner such as Bon Ami® or Soft Scrub® using a soft cloth and water. Rub until windshield is completely clean of all foreign material. Rinse windshield thoroughly.

On-board diagnostic (OBD) system

A final part of Volkswagen's maintenance program specifies checking the On-board Diagnostic (OBD) system memory. The ODB system is integrated into the engine management control module and monitors emissions-related and other electrical components on the car, including the adaptive automatic transmission, the ABS system, and the airbag system. Checking the OBD memory for faults requires special tools and training. Therefore, it is recommended that this job be carried out by an authorized Volkswagen dealer.

The following operations are part of the OBD system check and should be performed in the order stated:

1. Visual inspection
   - Make sure all components that influence the exhaust (such as vacuum lines, crankcase ventilation, air cleaner, catalytic converter, oxygen sensor, restricted fuel filler neck) are fitted and connected, and that the exhaust system is free of leaks.
2. Check ignition timing, idle speed and CO content.
3. Check operation of Oxygen Sensor (O2S) control.
   - The oxygen sensor control is operating properly if no malfunction is displayed when the DTC Memory is checked.
4. Check operation of three way catalytic converter.
   - Run engine for at least 1 minute at 2500 ± 200 RPM.
   - Measure CO content at CO tap tube (where applicable, upstream of catalytic converter) at 2500 +200/-0 RPM and note reading.
   - Measure CO content at tail pipe (downstream of catalytic converter) at 2500 +200/-0 RPM and note reading.
   - The catalytic converter should reduce CO by at least 50%.
5. After completing the emission and idling test, note the following:
   - Operate ancillary equipment such as power steering, automatic transmission, air conditioner and check for smooth engine operation.
   - Reconnect all hoses, plug connections that were detached or separated for the check/adjustment.

Service Reminder Indicator (SRI), resetting

On cars equipped with the Service Reminder Indicator (SRI), a graphic will be displayed in the instrument cluster when a service (mileage) interval has been reached. The display appears for a few seconds after switching on the ignition in place of the mileage indicator. If no service is required, the IN 00 will be displayed. The SRI will display one of three maintenance intervals.

- IN 00 = no service required
- OEL = oil service
- IN 01 = inspection I
- IN 02 = inspection II

After performing the specified maintenance, the Service Reminder Indicator (SRI) display should be reset.

1. Switch ignition ON.
2. Press and hold odometer reset button below speedometer. See Fig. 31.

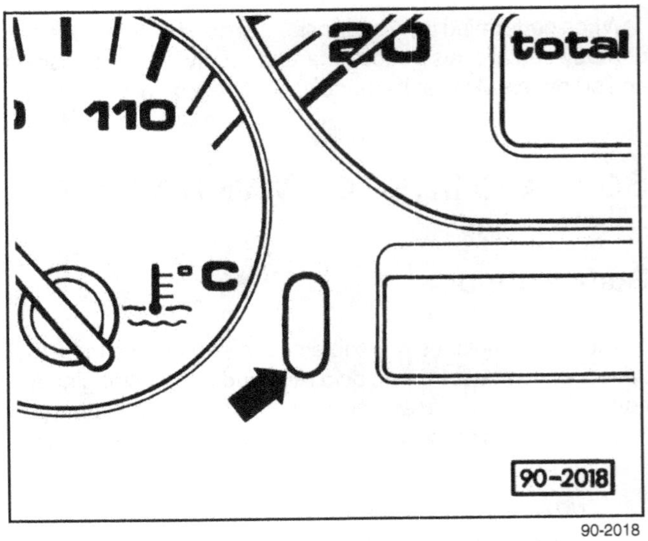

90-2018

Fig. 31. Odometer reset button (**arrow**).

3. Switch ignition OFF and release odometer reset button. OEL will now appear in SRI display.

4. Press and hold the lower digital clock reset button or the analog clock reset button until five dashes (– – – – –) appear in display. See Fig. 32.

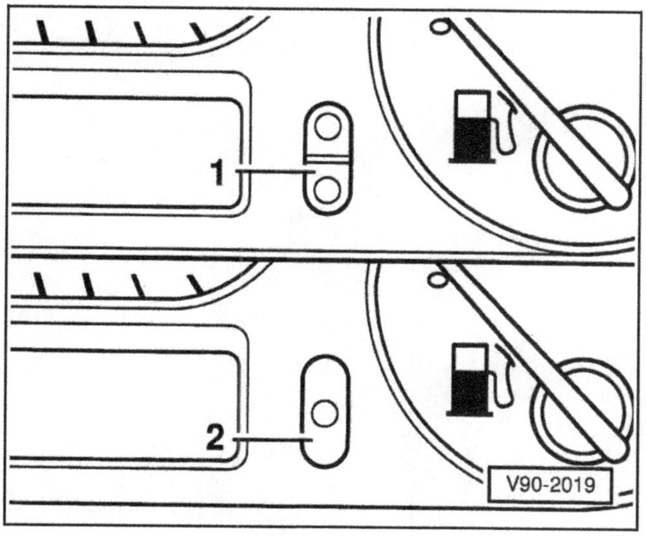

V90-2019

Fig. 32. Top; digital clock reset button (**1**). Bottom; analog clock reset button (**2**).

NOTE —

OEL SRI is now reset for 7,500 miles (12,000 km) or 6 months, whichever comes first. Once OEL is reset, the other displays can be reset in the same manner without switching the ignition on and off again.

5. If necessary, press odometer reset button to proceed to the IN 01 display (15,000 mile maintenance) and reset using the clock reset button.

6. If necessary, repeat procedure to reset IN 02 (30,000 mile maintenance) display.

7. To end reset mode: switch ignition ON; when display IN 01 appears, switch ignition OFF.

ROAD TEST

Upon completion of the maintenance program, the vehicle should be road tested to check for proper operation.

Automatic transmission shift lock

1. Turn the ignition on, but do not start the engine. Apply the parking brake.

2. With the gearshift selector in park attempt to shift into a drive or reverse. The selector lever should not move.

3. Press the brake pedal. The selector lever should move into gear.

4. Place gear selector lever in neutral and release brake pedal.

5. Attempt to shift into gear. The selector lever should not move.

6. Press the brake pedal. The selector lever should move into gear.

7. If shift lock does not operate as described, see **37 Automatic Transmission**.

Automatic park/neutral position switch

Apply the parking brake and step on the brake pedal. Place the gear shift selector in park and attempt to start car. Repeat for all gears. The engine should only start in park or neutral. See **37 Automatic Transmission**.

Brake pedal and parking brake

Check pedal idle travel and operation.

- Idle travel of brake pedal: max. 1/3 of pedal travel
- Idle travel of parking brake lever: 2 notches (clicks)

Manual gearshift

Check for smooth operation of the transmission and its gearshift.

Steering

Check steering play with vehicle weight on the wheels by turning steering wheel back and forth (wheels in straight ahead position).

- Steering play: zero play (engine running if vehicle fitted with power steering).

Kickdown operation

Depress accelerator pedal fully to floor. Depending on vehicle speed and engine speed, upshift is either retarded or the transmission shifts down into the next lower gear.

Air conditioner and heater

Check the full range of the heater and the air conditioner for proper function. Cold or warm air must flow out of the vents.

1 Engine

GENERAL

This general information group gives engine application information and general technical data for the various engines used in the cars covered by this manual.

Much of the engine repair information in **1 Engine** is organized according to engine code. Therefore it is helpful to know the code of the engine installed in your Volkswagen.

Engine Codes
- AAA . . . . . 2.8L, 6-cyl. (VR6), 172 HP
- ABA . . . . . 2.0L, 4-cyl., 115 HP
- ACC . . . . . 1.8L, 4-cyl., 90 HP (Canada)
- AAZ . . . . . 1.9L, 4-cyl. diesel, 75 HP (Canada)
- 1Z/AHU . . . 1.9L, diesel Turbo Direct Injection, 90 HP

AAA ENGINE

The AAA, or VR6 engine, is used in the 6-cylinder GTI and Jetta GLX models. See Fig. 1.

> ***NOTE —***
> *The name VR6 is derived from a combination of Vee (cylinder configuration) and the German word Reihenmotor, which roughly means in-line, or in-line V-6.*

0024131

Fig. 1. Cutaway of VR6 engine.

Technical data–engine code AAA
- Displacement . . . . . 2.8L (2792 cc)
- Bore diameter. . . . . 81.0 mm (3.19 in.)
- Stroke . . . . . 90.3 mm (3.55 in.)
- V-angle . . . . . 15°
- Compression ratio 10:1
- Fuel and ignition systems
  - 1993-1995. . . . . Bosch Motronic M2.9
  - 1996-1999. . . . . Bosch Motronic M5.9
- Horsepower . . . . . 172 @ 5,800 rpm
- Torque . . . . . 173 ft-lb @ 4,200 rpm

The VR6 engine has a unique 15° V-angle between cylinder banks, as compared to the more traditional 60° or 90° angles used in most other V-6 designs. See Fig. 2. This results in a compact engine that can be installed in small spaces, such as in Volkswagen models previously reserved for 4-cylinder engines.

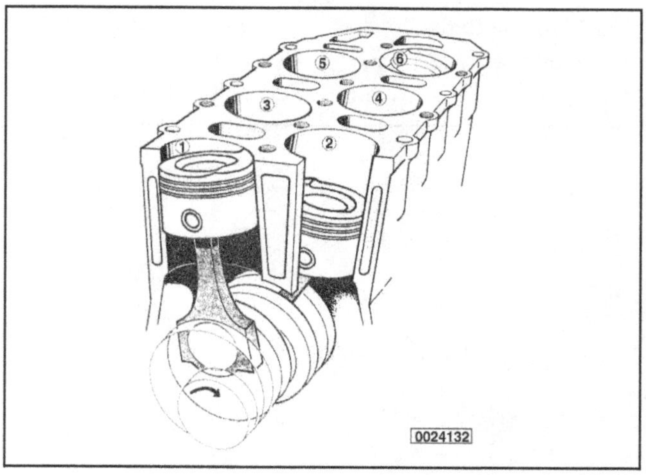

0024132

Fig. 2. VR6 cylinders are staggered along the cylinder block at a15° included angle, resulting in a narrower, more compact engine.

The VR6 engine features a cast iron cylinder block with a one piece light-alloy cylinder head. The overhead camshafts are chain driven and operate two valves per cylinder. The ignition system is distributorless and all fuel and ignition requirements are controlled by the Bosch Motronic Engine Management System.

ABA ENGINE

Fig. 3 shows the ABA engine used in all 4-cylinder Jetta, Golf, GTI and Cabrio models sold in the USA. This 2.0 liter long-stroke engine was manufactured specifically for the North American market. The engine features good fuel economy and a broad torque range.

The ABA engine uses a cross flow cylinder head and a two-piece intake manifold with long intake runners for enhanced low-end torque. The long piston stroke combined with the heightened cylinder block helps depress engine vibrations by reducing the operating angles of the connecting rods.

Other features include a poly-ribbed belt with automatic belt tensioner and a sophisticated Motronic engine management system. A conventional V-belt drives the power steering pump and water pump.

Depending on the date of manufacture, the ABA engine may include any or all of the following: oil spray jets for piston cooling and lubrication; crankshaft/oil pan windage tray, automatic cam/spur belt tensioner and different versions of Bosch Motronic® Engine Management.

Technical data–engine code ABA

• Displacement	2.0L (1984 cc)
• Bore diameter	82.5 mm (3.25 in.)
• Stroke	92.8 mm (3.65 in.)
• Compression ratio	10:1
• Fuel and ignition systems	
1993-1995	Bosch Motronic M2.9
1996-1999	Bosch Motronic M5.9
2000-2002 Cabrio	Bosch Motronic M5.9
• Horsepower	115 @ 5,400 rpm
• Torque	122 ft-lb @ 3,200 rpm

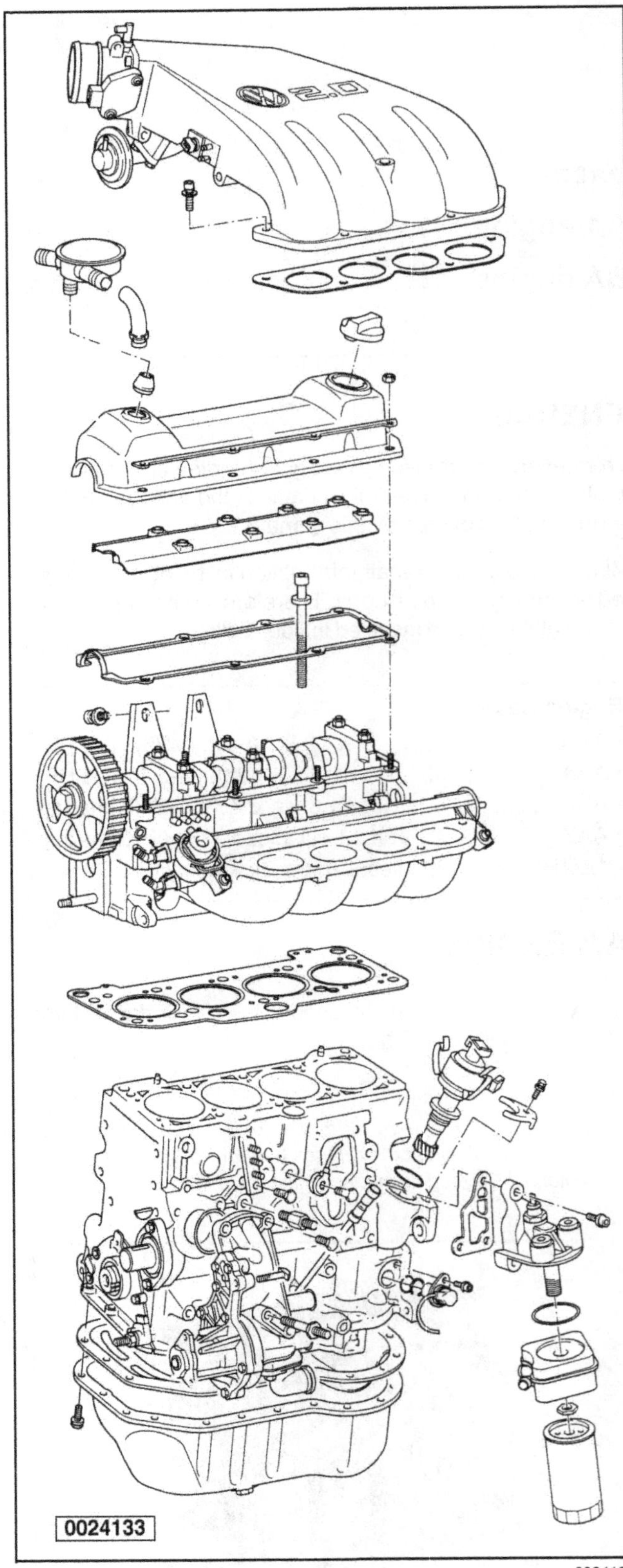

0024133

Fig. 3. Exploded view of 2.0 liter (ABA) engine.

ACC ENGINE

The ACC engine is used in the Golf CL model (Canadian market only). This 1.8 liter, 90 hp engine features 9:1 compression ratio and is designed to provide good mid-range torque. The ACC engine is equipped with throttle-body fuel injection, called Mono-Motronic.

Technical data–engine code ACC

- Displacement .... 1.8L (1781 cc)
- Bore diameter................. 81.0 mm (3.19 in.)
- Stroke....................... 86.4 mm (3.40 in.)
- Compression ratio 9.0:1
- Fuel and ignition systems..... Bosch Mono-Motronic
- Horsepower ....90 @ 5,500 rpm
- Torque .... 107 ft-lb @ 2,500 rpm

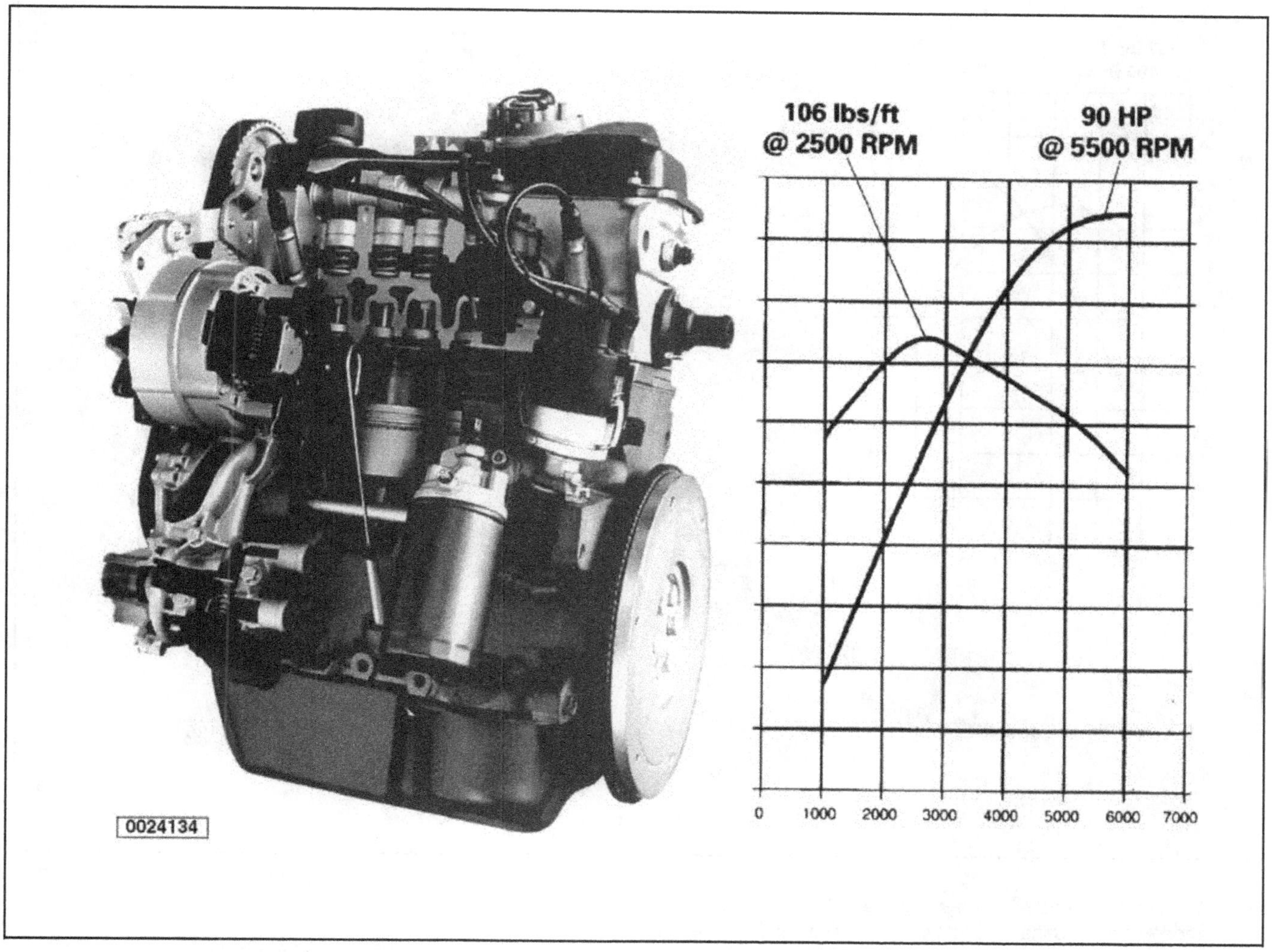

0024134

Fig. 4. 1.8 liter ACC engine used in the Golf CL model (Canadian market only). Chart shows engine performance data.

AAZ ENGINE

The Golf GL, Jetta GL and Jetta GLS models sold in Canada are available with an optional 1.9 liter ECOdiesel engine. The AAZ engine uses a turbocharger for increased performance and reduced particulate emissions. The AAZ engine is equipped with a one-way catalytic converter for reduced hydro-carbon emissions.

Engine data–engine code AAZ

- Displacement . . . . 1.9L (1896 cc)
- Bore diameter. . . . . 79.5 mm (3.19 in.)
- Stroke . . . . 95.5 mm (3.40 in.)
- Compression ratio 22.5:1
- Horsepower .75 @ 4,200 rpm
- Torque . . . . 107 ft-lb @ 2,500 rpm

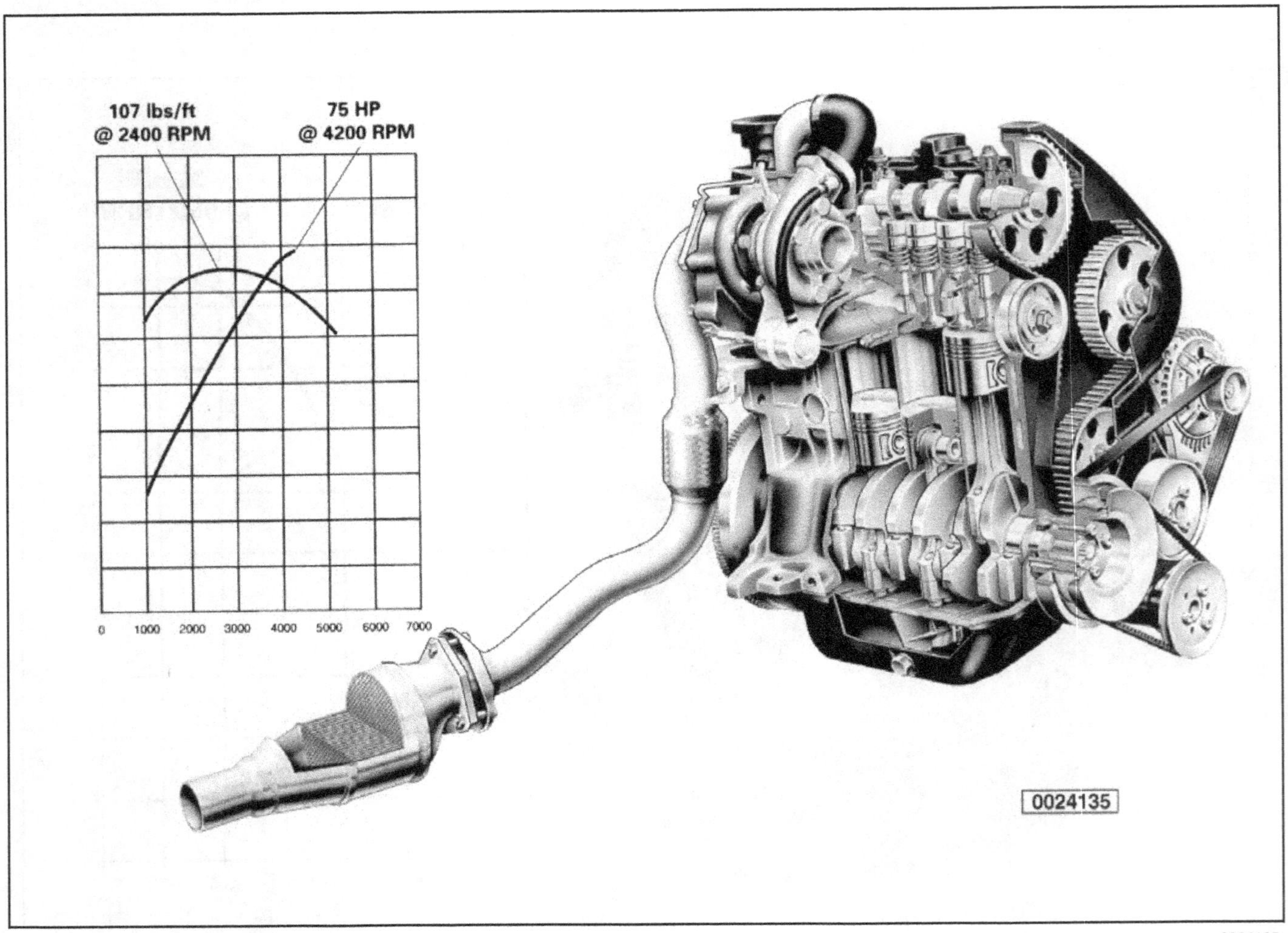

0024135

Fig. 5. 1.9 liter AAZ diesel engine available as an option in Golf and Jetta models sold in Canada. Chart shows engine performance data.

1Z/AHU ENGINE

The Jetta model is available with an optional diesel engine. This engine is known as a Turbocharged direct-injection diesel or TDI. The TDI 1Z/AHU engine features performance on a par with a comparable gasoline engine. It is cleaner than most gas engines and provides significantly better fuel economy. These advantages are achieved through the use of a turbo-charger and an intercooler in addition to a sophisticated electronic engine management system and an oxidation catalytic converter.

Engine data–engine code 1Z/AHU
- Displacement . . . . 1.9L (1896 cc)
- Bore diameter. . . . . 79.5 mm (3.19 in.)
- Stroke . . . . 95.5 mm (3.40 in.)
- Compression ratio 19.5:1
- Horsepower . . . . .90 @ 4,000 rpm
- Torque . . . . .149 @ 1,900 rpm

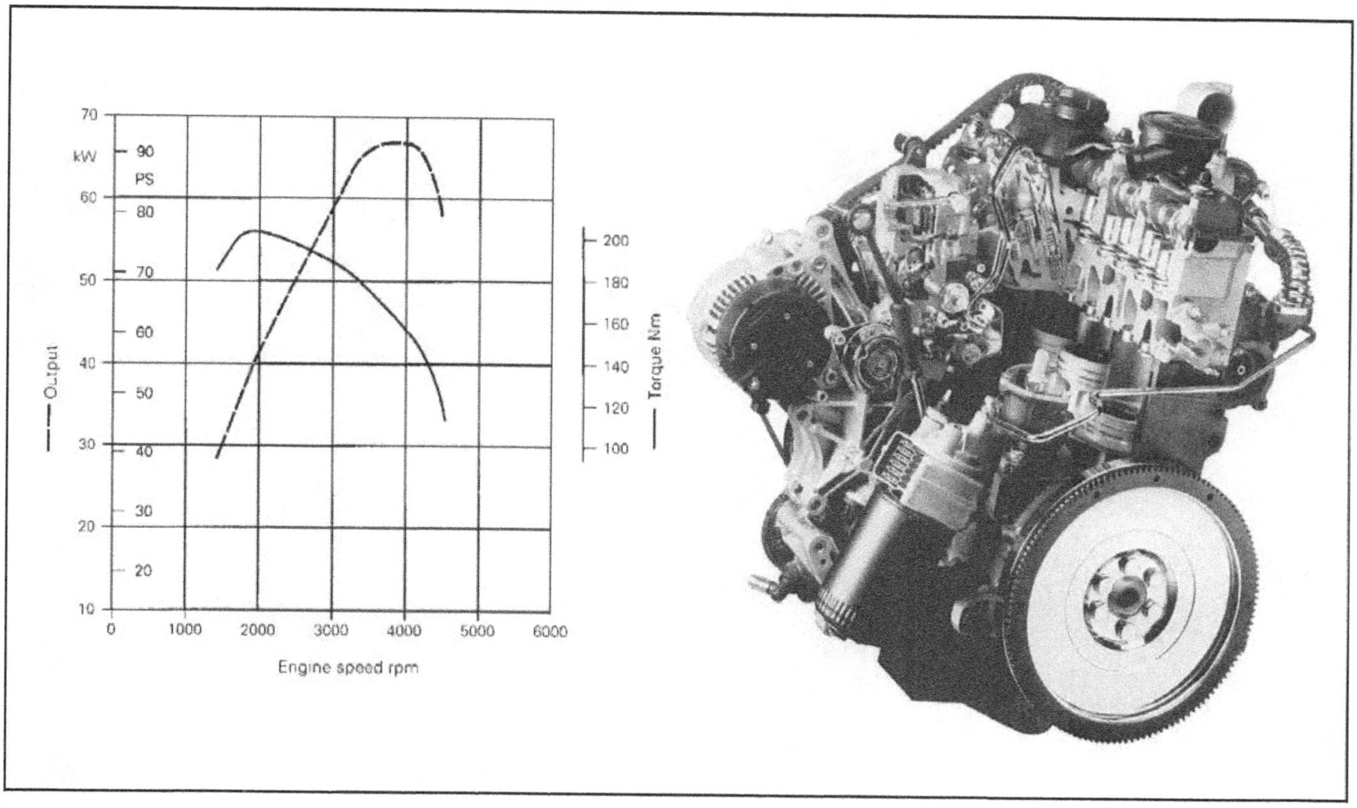

0024181

Fig. 6. 1.9 liter AHU TDI diesel engine available as an option in Jetta models. Chart shows engine performance data.

10 Engine—Removing and Installing

GENERAL

The engine and transmission are removed as a unit and separated from each other once removed from the car. The assembly is lifted out from the front, requiring that the front body assembly (radiator support) be removed. It is not necessary to remove the hood, unless it will interfere with the type of lifting equipment being used.

The operations needed to remove the engine/transmission assembly fall into three areas: jobs done under the hood, jobs done inside the car, and jobs done under the car. The steps below follow this general sequence. Most of the operations are simple and straight forward. References to additional repair groups are provided in bold type where additional information may be helpful to complete a step.

> ***NOTE —***
> *It will be necessary to cut many wire tie wraps when removing the engine. The tie wraps are installed to prevent the wiring harnesses from chaffing or contacting engine parts. Be sure to make note of all tie wraps removed and install new ones during engine installation.*

REMOVING AND INSTALLING ENGINE (4-Cylinder)

To remove engine/transmission assembly (4-cylinder engines)

1. Disconnect the negative (–) battery terminal and then the positive (+) terminal. Remove the battery from car.

> ***NOTE —***
> *Be sure to have the anti-theft radio code on hand before disconnecting the battery.*

2. On AHU diesel engines, remove the large plastic upper engine cover. On AHU and AAZ diesels, remove lower sound dampening (belly) pan.

3. Drain the cooling system as described in **19 Engine–Cooling System**. Remove the coolant hoses connected to the radiator and the heater core.

> ***WARNING —***
> *Hot coolant can scald. Drain the coolant only with engine cold.*

4. Remove the engine drive belts. See **0 Maintenance Program.**

> ***NOTE —***
> *On poly-ribbed drive belts, mark the running direction on the belt before removing it.*

5. On gasoline engines, disconnect and label the hoses, wires, and cables connected to the throttle body, the intake manifold and cylinder head. For more information on fuel injection components, see **24 Fuel Injection**.

> ***WARNING —***
> *Fuel will be expelled when disconnecting fuel hoses. Wrap a cloth around the fuel line fittings before disconnecting them. Do not smoke or work near heaters or other fire hazards. Have a fire extinguisher handy.*

6. Remove the air filter housing together with all intake air ducts. Label and then disconnect the fuel return and fuel supply hoses.

7. On AAZ diesel engines, disconnect the accelerator cable and the cold start accelerator cable from the injection pump. Free the cables from any clamping brackets on the engine. On AAZ and AHU engines, disconnect the fuel supply and return lines from the injection pump. Disconnect the electrical connectors from the pump.

NOTE—
On AHU diesel engines there is no accelerator cable to disconnect from the engine.

WARNING—
Fuel may be expelled when disconnecting hoses. Do not smoke or work near heaters or other fire hazards. Have a fire extinguisher handy.

8. Remove the upper radiator mounting bolts. Disconnect all electrical connectors and coolant hoses from radiator.

9. On cars with air conditioning, remove the A/C hose clamping bracket from the radiator support. Remove the A/C receiver/dryer mounting nuts at the right front corner of the engine compartment and allow it to hang free.

10. On cars with air conditioning, remove the A/C compressor from the engine without disconnecting any refrigerant lines. Support the compressor so that no lines are stressed. See Fig. 1.

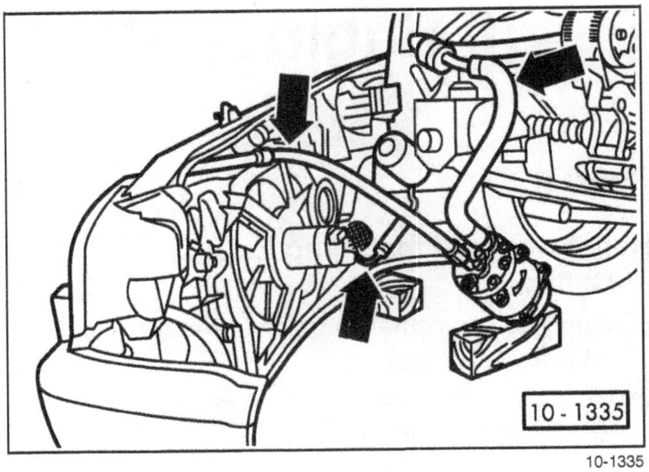

10-1335

Fig. 1. A/C compressor shown removed from engine and supported. Do not stress or kink refrigerant lines (arrows).

CAUTION—
Use care to avoid kinking or bending the refrigerant lines.

11. Remove the front body assembly mounting bolts. Disconnect all electrical connectors and hose brackets from the assembly. Disconnect the hood release cable. Remove the complete assembly with all attachments. See Fig. 2. See also **50 Body–Front**.

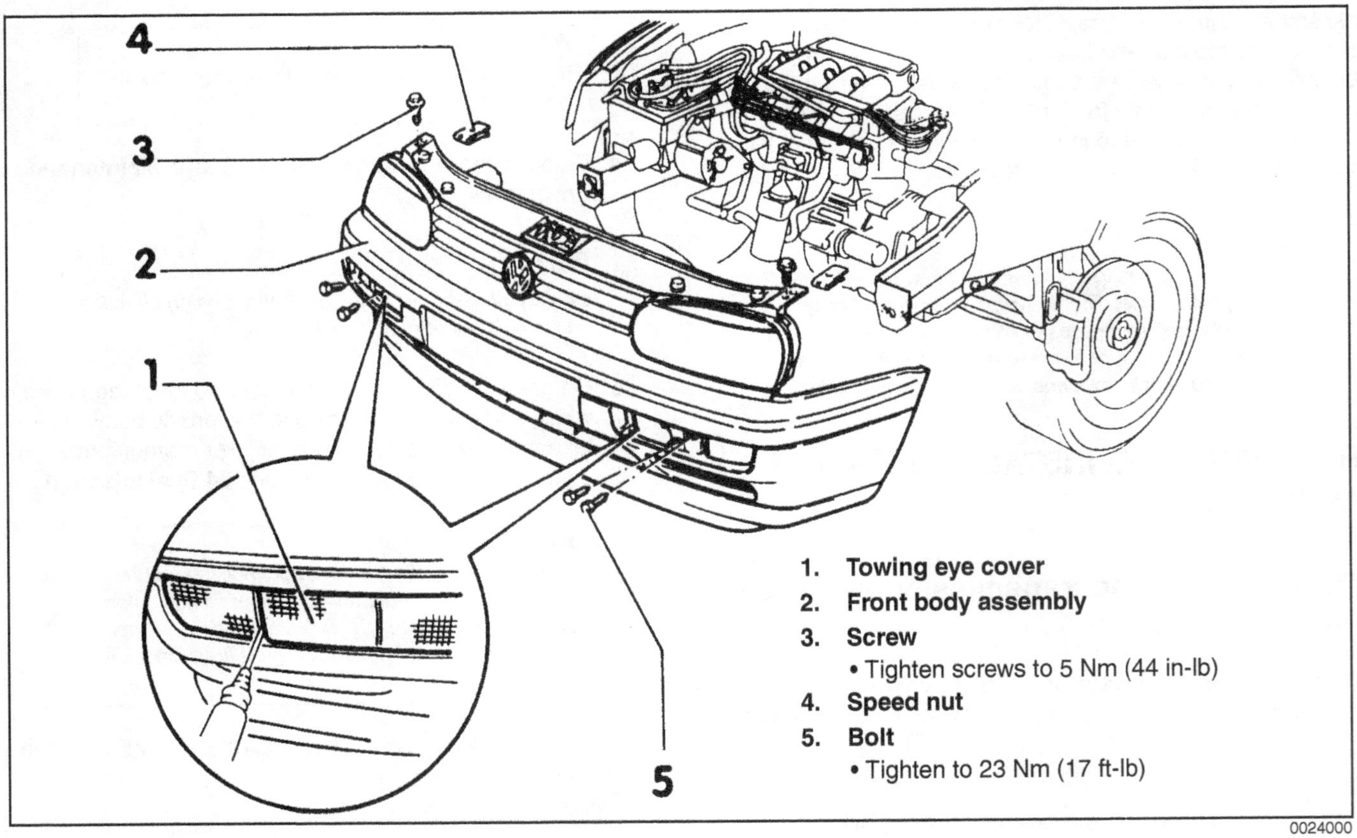

0024000

Fig. 2. Front body assembly. Be sure to remove lower screws from wheel housing trim (not shown).

12. On cars with A/C, swing the radiator together with fan and the A/C condenser out and away from the engine. See Fig. 3.

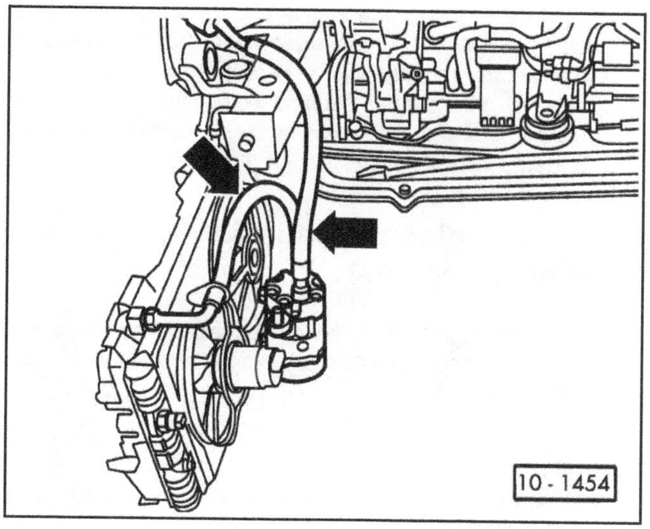

10-1454

Fig. 3. Radiator assembly on cars with A/C shown positioned in front of engine.

13. Unbolt the power steering pump together with its bracket and set the pump aside without disconnecting the power steering fluid lines.

14. Working under the car, disconnect the drive axle inner constant velocity (CV) joints from the transmission drive flanges, as described in **40 Front Suspension and Drive Axles**. Suspend the axles from the body with stiff wire to avoid damaging the CV joints.

15. Disconnect the exhaust pipe from the exhaust manifold. See **26 Exhaust System/Emission Controls**.

16. Label and disconnect the wiring from the starter and the alternator. Disconnect all remaining electrical connections to the engine and transmission, including ground straps.

17. On cars with automatic transmission, remove the engine vibration damper/pulley from the front of the crankshaft. Remove the coolant pump pulley from the pump.

18. On cars with automatic transmission, remove the harness connectors from the transmission housing. Disconnect the selector lever cable from the transmission lever. Free the cable from its bracket. See Fig. 4.

19. On cars with manual transmission, remove any harness connectors from the transmission. On AHU diesel engines, unbolt the clutch slave cylinder from the transmission housing. On all others, detach the clutch cable. Detach the shift linkage for all versions as described in **34 Manual Transmission**.

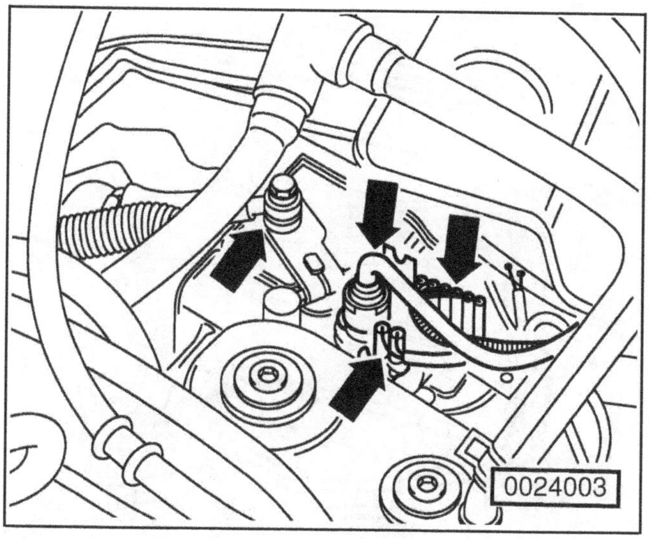

0024003

Fig. 4. Top view of automatic transmission showing harness connectors and selector lever cable (arrows).

20. Install an engine lifting device, using the lifting points illustrated in Fig. 5 or Fig. 6.

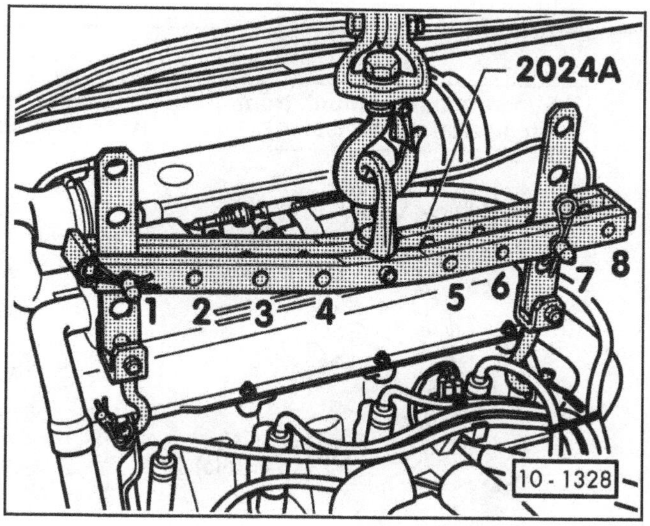

10-1328

Fig. 5. Engine hoist chain properly attached for lifting engine with manual transmission from car. Use of hoist as shown is necessary for proper weight distribution. Note position of left and right lifting hooks in relation to crossbar. Always use securing cotter pins on holding pins and hooks, as shown.

21. Raise the hoist slightly, so that the weight of the engine and transmission assembly is supported by it.

22. Remove the bolts from the rear engine mount and the front and rear transmission mounts. See Fig. 7.
    - Front mount is attached the left side crossmember.
    - Left rear mount is attached the left side subframe.
    - Right rear mount is attached to right side subframe.

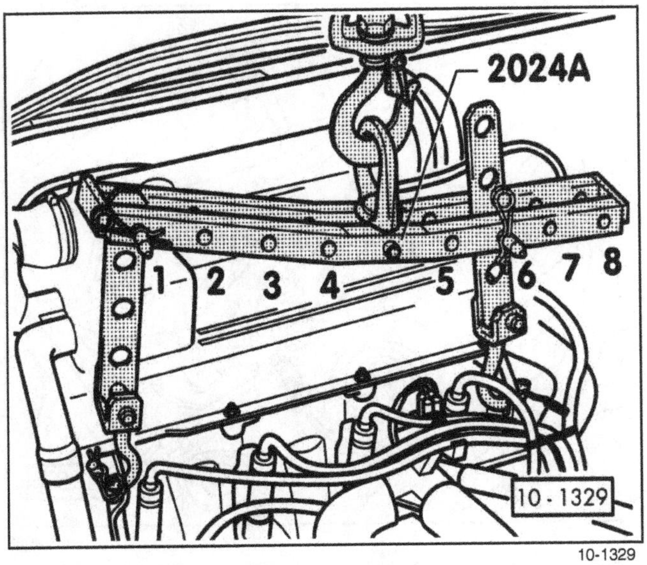

10-1329

Fig. 6. Engine hoist chain properly attached for lifting engine with automatic transmission from car. Use of hoist as shown is necessary for proper weight distribution. Note position of left and right lifting hooks in relation to crossbar. Always use securing cotter pins on holding pins and hooks, as shown.

> ***CAUTION —***
> *The tightening torques given in Fig. 7 apply only to clean and oiled bolts.*

23. Raise and remove the engine and transmission assembly from the car. Proceed slowly. Check frequently to make sure no hoses or wires are interfering with engine removal. If the car has been raised on jack stands, check often to see that it remains stable and adequately supported.

To install engine/transmission assembly (4-cylinder engines)

1. With the engine lifting device connected, lower the engine into position while aligning the front and rear mounts. See Fig. 8.

NOTE —
Be sure to allow for clearance of the drive axles during engine installation.

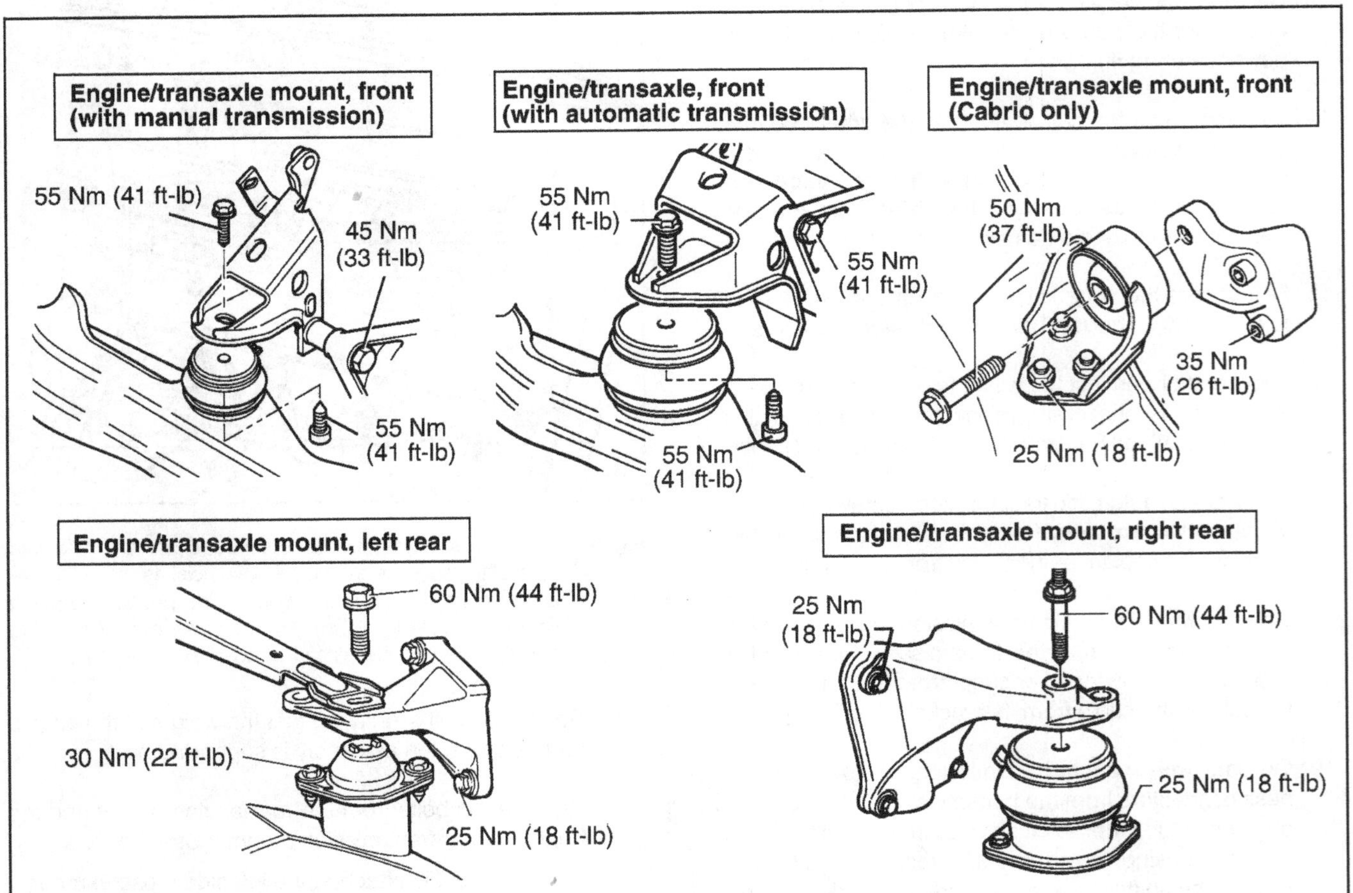

0024001

Fig. 7. 4-cylinder engine/transmission mount tightening torques, which apply only to clean and lightly oiled bolts.

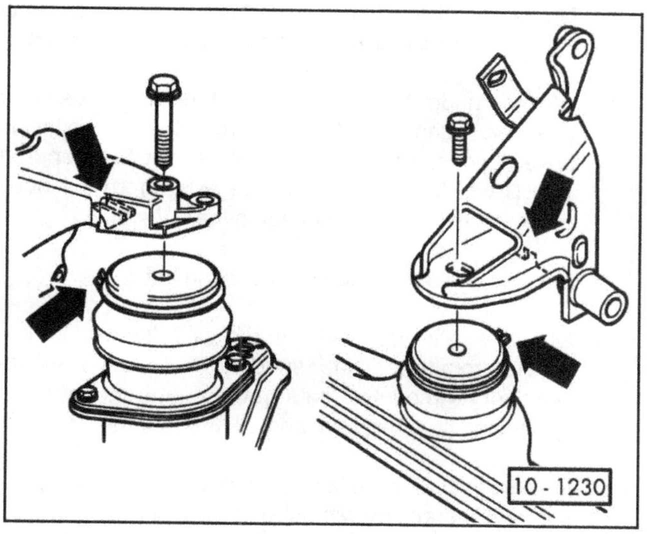

10-1230

Fig. 8. When lowering engine assembly onto mounts, align guide pins on right rear mount (left) and front mount (right) to recesses on brackets (**arrows**).

2. Lightly oil and install the engine/transmission mount bolts. Do not tighten the bolts at this time.

3. Install the front body assembly using Fig. 2 as a guide. Make sure the assembly is stress free before tightening the bolts.

4. Install all components, connectors, hoses and lines previously removed. Install new hose clamps and wire tie wraps as necessary. Inspect the drive belts and replace if worn.

5. With the car on the ground, shake the engine/transmission assembly to allow it to settle on the mounts. Tighten the engine mount fasteners beginning with the front and right rear mounts. Tighten the left rear mount last. See Fig. 7 for mount tightening torques.

> ***CAUTION —***
> - *The tightening torques given in Fig. 7 apply only to lightly oiled bolts.*
> - *Wire tie wraps are installed to prevent wiring harnesses from chaffing or contacting engine parts. Be sure to install new wraps in their original locations.*

Tightening torques
- Coolant pump pulley to coolant pump . . . . 20 Nm (15 ft -lb)
- Drive axle to drive flange. . . . . 45 Nm (33 ft-lb)
- Exhaust pipe to exhaust manifold . . . . 40 Nm (30 ft-lb)
- Vibration damper/pulley to crankshaft hub . . . . 25 Nm (18 ft-lb)

NOTE —
- *After aligning the engine and transmission, check gearshift function. If necessary, adjust the shift mechanism as described in* **3 Clutch, Transmission, and Final Drive.**
- *It may be necessary to adjust headlight aim if a new engine is installed.*

6. Adjust the accelerator cable as described in **24 Fuel Injection**. On 1.8 liter engines, check and adjust ignition timing. See **28 Ignition System**. Check all fluid levels.

7. On diesel engines, reinstall the lower sound dampening (belly) pan and upper engine covers.

REMOVING AND INSTALLING ENGINE (6-Cylinder)

To remove engine/transmission assembly (6-cylinder engine)

1. Disconnect the negative (–) battery terminal and then the positive (+) terminal. Remove battery from the car.

NOTE —
Be sure to have the anti-theft radio code on hand before disconnecting the battery.

2. Remove the engine drive belt. See **0 Maintenance Program.**

NOTE —
On poly-ribbed drive belts, mark the running direction on the belt before removing it.

3. Remove the spark plug connectors from the plugs and the coil. Remove the spark plug wire guides with the ignition wires. See Fig. 9.

NOTE —
A special tool (VW tool no. 3277) is available to remove the spark plug connectors from the plugs without damaging them. A version of this tool is clipped to the front hood support bar.

4. Separate the round 42-pin harness connector near the ignition coil on the rear of the cylinder head.

5. Remove the intake manifold cover from above the intake manifold.

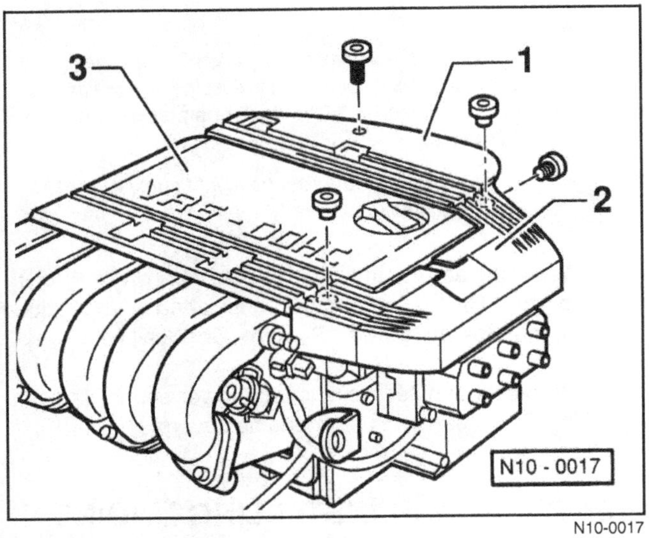

N10-0017

Fig. 9. Spark plug wire guides (**1** and **2**). Intake manifold cover shown at **3**.

6. Disconnect and label the hoses and wires connected to the throttle body, the intake manifold and cylinder head. For more information on fuel injection components, see **24 Fuel Injection**.

7. Remove the air filter housing and intake air ducts (with mass air flow sensor).

8. Disconnect the accelerator cable from the throttle and from its support bracket.

NOTE—

It is not necessary to remove the accelerator cable locating clip.

9. Label and then disconnect the fuel return and fuel supply hoses from the fuel rail. Unclip the fuel lines from the cylinder head and move them out of the way.

> ***WARNING—***
> *Fuel will be expelled when disconnecting fuel hoses. Wrap a cloth around the fuel line fittings before disconnecting them. Do not smoke or work near heaters or other fire hazards. Have a fire extinguisher handy.*

10. Drain the cooling system as described in **19 Engine–Cooling System.** Remove the coolant hoses connected to the radiator and heater core.

> ***WARNING—***
> *Hot coolant can scald. Drain coolant only with the engine cold.*

11. Disconnect the exhaust pipes from the exhaust manifolds. See **26 Exhaust System/Emission Controls**.

12. Working under the car, disconnect the drive axle inner constant velocity (CV) joints from the transmission drive flanges, as described in **40 Front Suspension and Drive Axles**. Suspend the axles from the body with stiff wire to avoid damaging the CV joints.

13. Remove the upper radiator mounting bolts.

14. On cars with A/C, remove the A/C hose clamping bracket from the radiator support. Remove the A/C receiver/dryer mounting nuts at the right front corner of the engine compartment and allow it to hang free.

15. On cars with air conditioning, remove the A/C compressor from the engine without disconnecting any refrigerant lines. Support the compressor so that no lines are stressed. See Fig. 10.

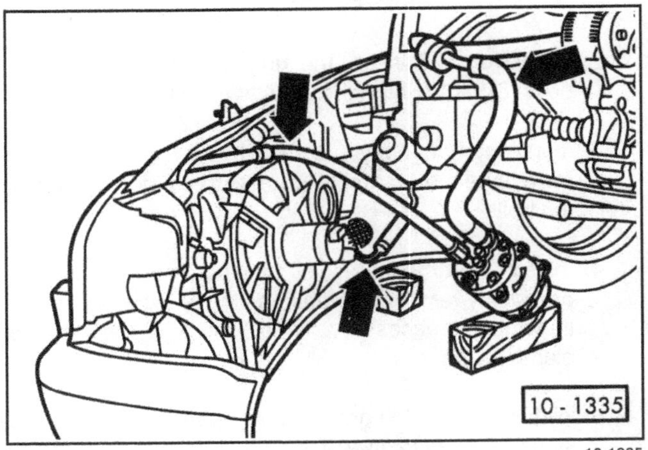

10-1335

Fig. 10. A/C compressor shown removed from engine and supported. Do not stress or kink refrigerant lines (**arrows**).

16. Remove the front body assembly mounting bolts. Disconnect all electrical connectors and hose brackets from the assembly and the radiator. Disconnect the hood release cable. Remove the complete assembly with all attachments. See Fig. 11.

17. On cars with A/C, swing the radiator together with fan and the A/C condenser out and away from the engine. See Fig. 3 given earlier.

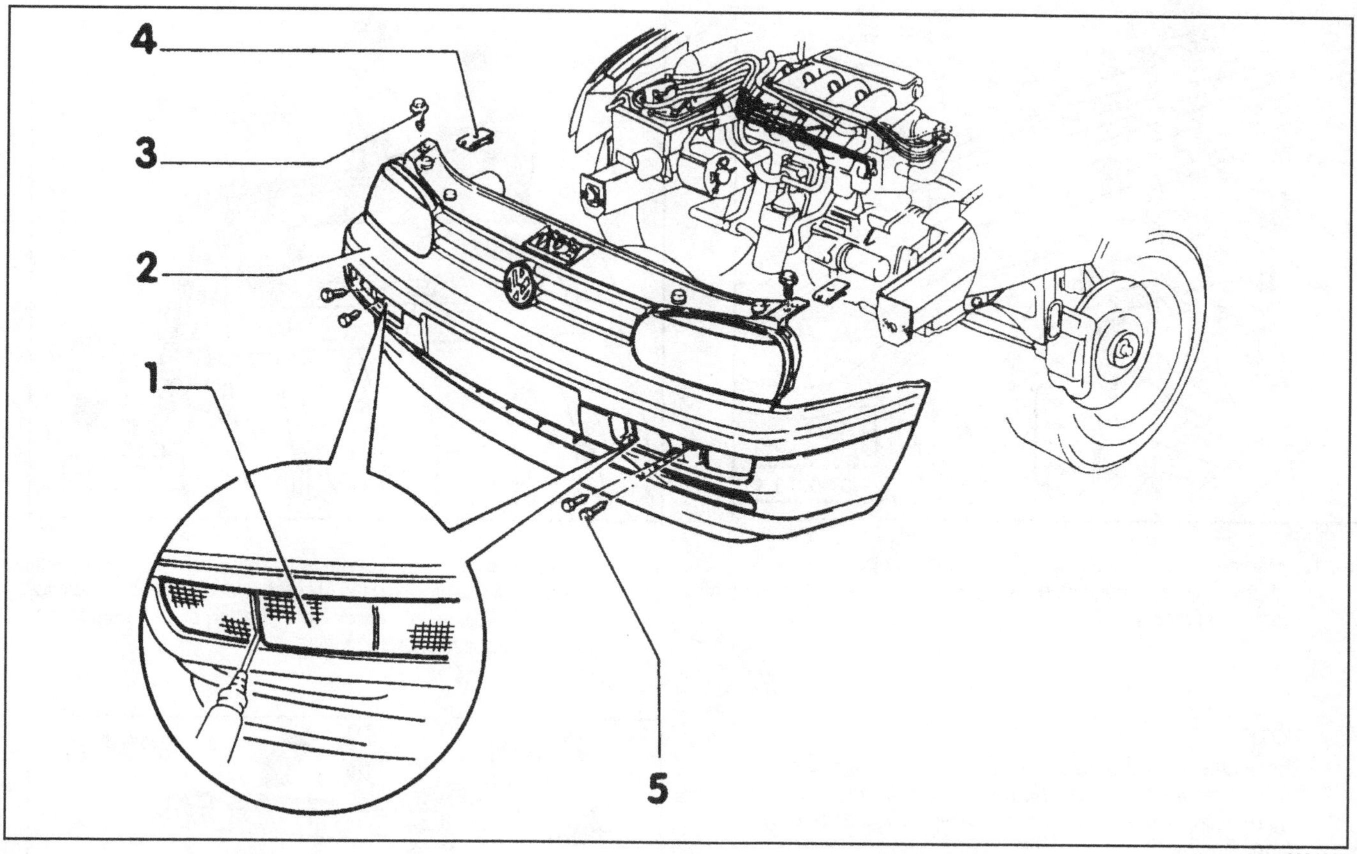

0024000

Fig. 11. Front body assembly. Be sure to remove lower screws from wheel housing trim (not shown).

18. Remove all retainers and mounting brackets holding the power steering lines (including the fluid reservoir lines) to the body, the transmission, and the engine. Unbolt the power steering pump together with its bracket and set the pump aside.

19. On cars with automatic transmission, remove the electrical harness connectors from the transmission housing. Clamp off the ATF cooler lines near the cooler and disconnect the lines. See Fig. 12.

20. On cars with manual transmission, unbolt the clutch slave cylinder from the transmission housing. Disconnect the transmission operating cables from the transmission. Disconnect any harness connectors. See **34 Manual Transmission**.

21. On cars with automatic transmissions, place the selector lever in the "Park" position. Remove the accelerator cable and the selector lever cable as shown in **37 Automatic Transmission**.

22. Label and disconnect the wiring from the starter and the alternator. See **27 Engine Electrical System**. Disconnect all remaining electrical connections to the engine and transmission, including ground straps.

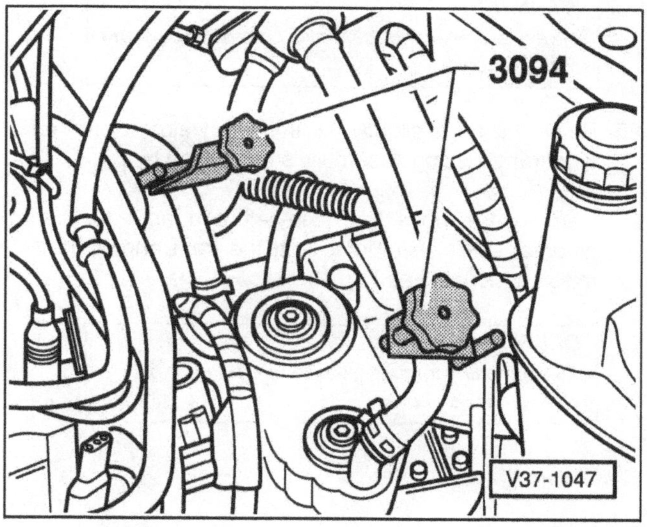

V37-1047

Fig. 12. ATF cooler lines shown clamped off. Number indicates VW special tool.

23. Remove the intake air temperature sensor and disconnect the vacuum hose at the upper intake manifold. See Fig. 13.

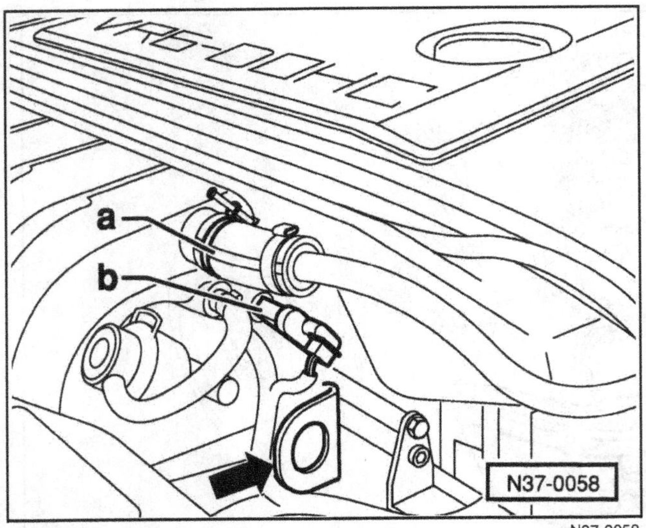

N37-0058

Fig. 13. Remove vacuum hose (**a**) and intake air temperature sensor (**b**) to prevent damage when lifting out engine. Rear lifting eye shown at **arrow**.

NOTE—
The intake air temperature sensor and the vacuum hose at the intake manifold are very close to the rear engine lifting eye and can be damaged by the engine lifting equipment.

24. Install an engine lifting device, using the lifting points illustrated in Fig. 14 or Fig. 15.

> ***WARNING—***
> *Always use securing cotter pins on holding pins and hooks (arrows).*

25. Raise the hoist slightly, so that the weight of the engine/transmission assembly is supported by it.

26. Remove the bolts from the left and right rear engine/transmission mounts and the front engine/transmission mount. See Fig. 16.

> ***CAUTION—***
> *The tightening torques given in Fig. 16 apply only to clean and oiled bolts.*

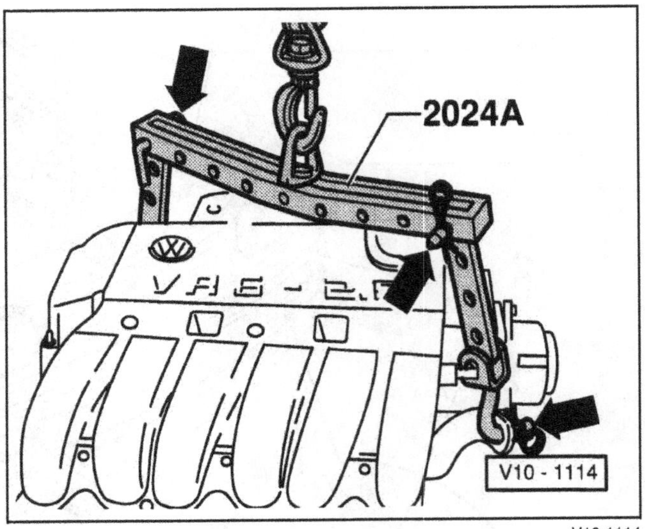

V10-1114

Fig. 14. Engine hoist chain properly attached for lifting engine with manual transmission. Use of hoist as shown is necessary for proper weight distribution. Note position of left and right lifting hooks in relation to crossbar.

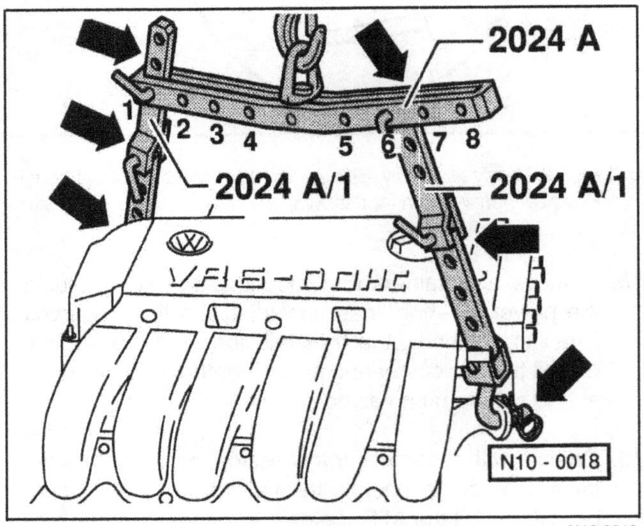

N10-0018

Fig. 15. Engine hoist chain properly attached for lifting engine with automatic transmission. Use of hoist as shown is necessary for proper weight distribution. Note position of left and right lifting hooks in relation to crossbar.

27. Raise and remove the engine/transmission assembly from the car. Proceed slowly. Check frequently to make sure no hoses or wires are interfering with engine removal. If the car has been raised on jack stands, check often to see that it remains stable and adequately supported.

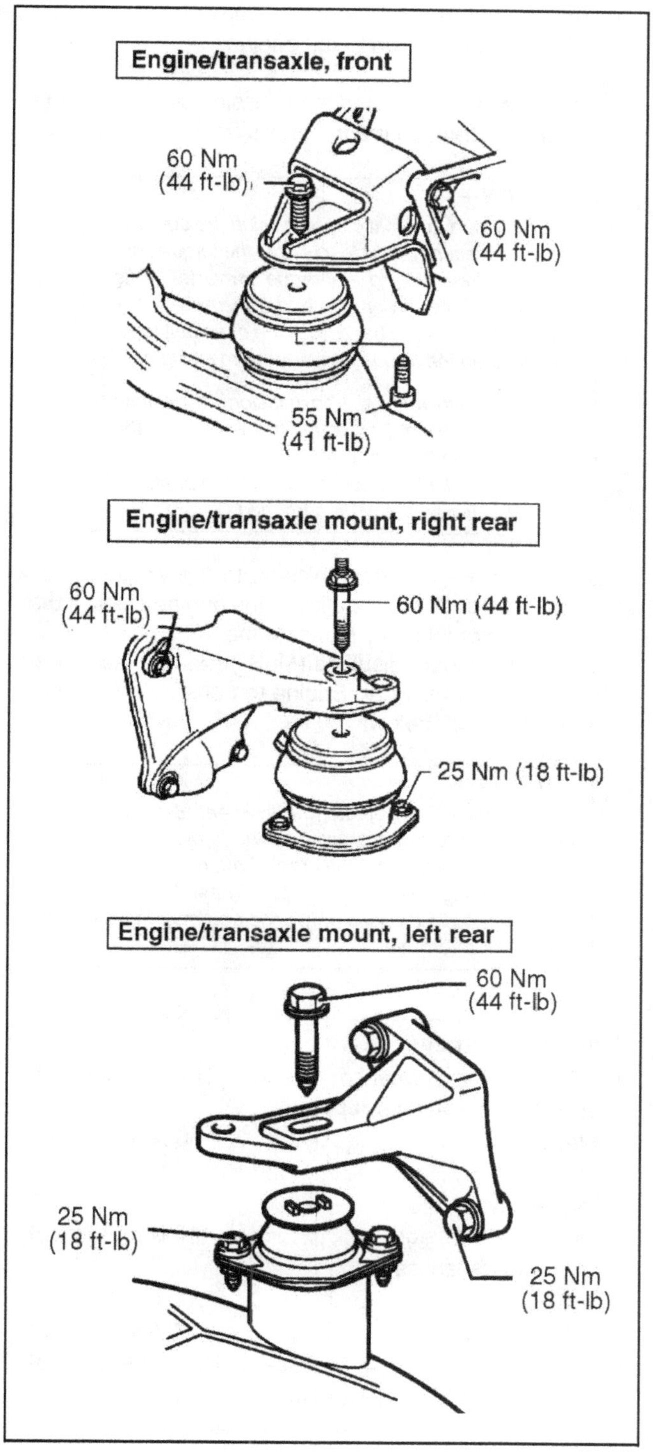

0024002

Fig. 16. 6-cylinder engine/transmission mount tightening torques (bolts oiled). Right rear mount is attached to engine and right side subframe. Left rear mount is attached to transmission and left side subframe. Front mount is attached to transmission and left side crossmember.

To install engine/transmission assembly (6-cylinder engine)

1. With the engine lifting device connected, lower the engine into position while aligning the front and rear mounts. See Fig. 17.

10-1230

Fig. 17. When lowering engine assembly onto mounts, align guide pins on right rear mount (left) and front mount (right) to recesses on brackets (**arrows**).

2. Be sure to allow for clearance of the drive axles during engine installation.
3. Lightly oil and install the engine/transmission mount bolts. Do not tighten the bolts at this time.
4. Install the front body assembly using Fig. 11 as a guide.
5. Install all components, connectors, hoses and lines previously removed. Install new hose clamps and wire tie wraps as necessary. Inspect the drive belt and replace if worn.
6. Additional installation tightening torques are given below. Check all fluids levels as the final step.

> ***CAUTION—***
> *Wire tie wraps are installed to prevent the wiring from chaffing or contacting engine parts. Be sure to install new tie wraps in their original locations.*

Tightening torques

- Drive axle to drive flange. . . . . 45 Nm (33 ft-lb)
- Exhaust pipe to exhaust manifold. . . . 40 Nm (30 ft-lb)

7. If necessary, adjust the accelerator cable. See **24 Fuel Injection**.

8. With the car on the ground, shake the engine/transmission assembly to allow it to settle on the mounts. Tighten the engine mount fasteners beginning with the front and right rear mounts. Tighten the left rear mount last. See Fig. 16 for mount tightening torques.

> ***WARNING —***
> *The tightening torques given in Fig. 16 apply only to oiled bolts.*

NOTE —
- *After the alignment, check gearshift function. If necessary, adjust the shift mechanism as described* **3 CLUTCH, TRANSMISSION, AND FINAL DRIVE.**
- *It may be necessary to adjust headlight aim if a new engine is installed.*

SEPARATING ENGINE AND TRANSMISSION

To separate engine from transmission (all engines)

1. Support the engine and transmission as a unit.

2. On cars with automatic transmission, remove the three nuts holding the torque converter to the drive plate working through the bottom of the transmission. Remove the cover plate. See Fig. 18.

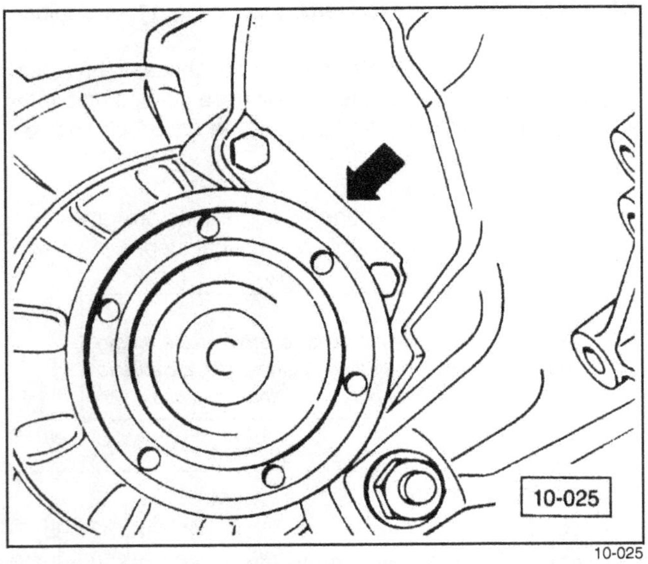

10-025

Fig. 18. Cover plate (**arrow**) to be removed. 4-cylinder engine shown.

3. Remove the starter motor and remove the bolts that hold the transmission to the engine.

4. Separate the engine and transmission using care not to place any strain on the transmission mainshaft.

> ***CAUTION —***
> - *On cars with automatic transmission, be sure the drive plate separates cleanly from the torque converter without pulling the torque converter off its support. Once the engine and transmission are separated, install a bar across the open bellhousing to keep the torque converter from falling out.*
> - *On cars with manual transmission, at no time should the weight of either the engine or the transmission be supported by the transmission mainshaft. If it is, the clutch, the clutch pushrod, or the mainshaft may be damaged.*

When reattaching the transmission to the engine, check that the guide sleeves are installed in the engine block. If they are not, install them. Lightly lubricate the transmission mainshaft with molybdenum disulfide (MoS_2) grease. Inspect the clutch components for wear. Engine to transmission tightening torques are given below.

> ***CAUTION —***
> *If either the A/T driveplate, the flywheel (6-cylinder engines) or the clutch pressure plate (4-cylinder engines) is removed from the crankshaft, new stretch-type mounting bolts must be used. These fasteners are designed to be used only once and may, in service, fail if retorqued.*

Tightening torques
- A/T driveplate, flywheel or pressure plate to crankshaft (stretch bolts–**always replace**)
  - stage I . . . . 60 Nm (30 ft -lb)
  - stage II . . . . .additional 1/4 turn (90°)
- Engine cover plate to engine/transmission . . . . 10 Nm (89 in-lb)
- Engine to transmission
  - M10 . . . . .60 Nm (44 ft-lb)
  - M12 . . . . .80 Nm (59 ft-lb)
- Starter to engine block . . . .See **27 Engine Electrical**
- Torque converter to drive plate. . . . . . 60 Nm (44 ft-lb)

13 Crankshaft/Cylinder Block

TABLES

GENERAL

This repair group provides the special reconditioning information necessary to repair the Volkswagen short block. The information contained here is intended to be used as a reconditioning guide for the professional or experienced automotive technician. Many of the operations and specifications listed here require precision measuring equipment.

> ***NOTE —***
> *The information given in the repair section is organized based on engine codes. A listing of the engine codes for the engines covered by this manual is given at the beginning of this section. For more specific engine information, see* **1 General Information.**

CYLINDER BLOCK OIL SEALS

The front crankshaft and, on 4-cylinder engines, the intermediate shaft oil seals can be replaced with the engine installed. Replacement of the rear crankshaft oil seal requires that the transmission be separated from the engine.

To replace front crankshaft oil seal (4-cylinder engines)

1. Remove the camshaft drive belt as described in **15a Cylinder Head and Valvetrain (4-Cylinder Engines)** for gasoline engines or **23 Fuel Injection–Diesel** for diesel engines.

2. While holding the crankshaft stationary, loosen the crankshaft sprocket (hub) center bolt. Remove the bolt, the hub, and the woodruff key. Discard the bolt.

3. Remove the seal from the its carrier using a seal extractor, or carefully pry it from its housing using a screwdriver. See Fig. 1.

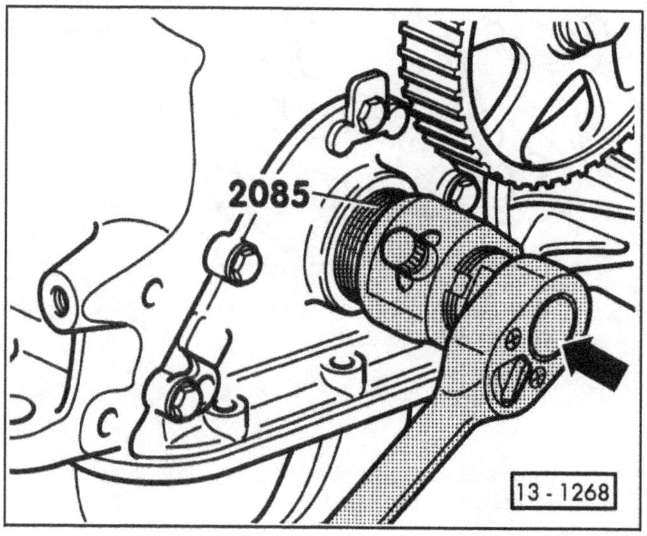

13-1268

Fig. 1. Crankshaft front oil seal being removed using Volkswagen special tool no. 2085. Rotate tool clockwise to remove seal.

> *CAUTION —*
> *On cars with diesel engines, the oil pan to front main seal carrier bolts are self-locking . Self-locking bolts are designed to be used only once. Always replace these fasteners with new parts.*

Tightening torques (4-cylinder engines)
- Front main seal carrier to engine block
  M6 . . . . 10 Nm (89 in-lb)
  M8 . . . . 25 Nm (18 ft-lb)
- Oil pan to front main seal carrier (M6)
  gasoline engines . . . . 20 Nm (15 ft-lb)
  diesel engines (self-locking, always replace) . . . . 25 Nm (18 ft-lb)

4. Install the new seal, lubricated with clean engine oil, with the closed side facing out. Carefully press it into place until it is fully seated in the seal carrier.

> *CAUTION —*
> *Use care not to distort the seal as it is installed. For best results, use a seal press such as Volkswagen special tool no. 3083, shown in Fig. 2.*

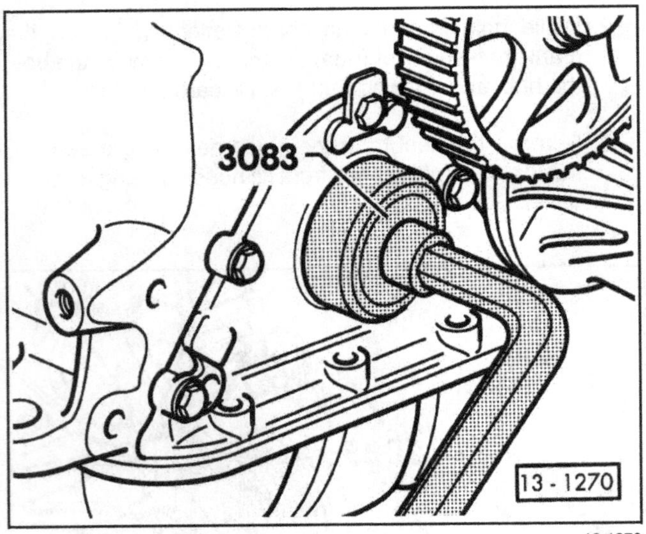

13-1270

Fig. 2. Front crankshaft oil seal being installed using Volkswagen special tool no. 3083.

> *NOTE —*
> *Replacement seals may be pre-coated with lubricant. It is not necessary to lubricate pre-coated seals with engine oil.*

5. Install the woodruff key into the crankshaft keyway. Using a new mounting bolt, install the crankshaft sprocket while aligning the keyway. Hold the crankshaft stationary and tighten the bolt.

> *CAUTION —*
> *Always replace the crankshaft sprocket mounting bolt. It is a stretch bolt designed to be used only once. Lightly oil the threads and bolt shoulder before installing.*

Tightening torques (4-cylinder engines)
- Crankshaft drive belt sprocket to crankshaft (stretch bolt–**always replace**)
  stage I . . . . 90 Nm (66 ft-lb)
  stage II . . . . additional 1/4 turn (90°)

6. Install the camshaft drive belt as described in **15a Cylinder Head and Valvetrain (4-Cylinder Engines)** or **23 Fuel Injection–Diesel**.

To replace intermediate shaft oil seal (4-cylinder engines)

1. Remove the camshaft drive belt, as described in **15a Cylinder Head and Valvetrain (4-Cylinder Engines)** for gasoline engines or **23 Fuel Injection–Diesel** for diesel engines.

2. Lock the intermediate shaft sprocket in place and loosen the sprocket center bolt. Remove the bolt and the sprocket. Remove the woodruff key from the shaft.

> *NOTE —*
> *Hold the intermediate shaft sprocket stationary using a holding fixture (VW special tool 3036, or equivalent) when loosening the sprocket mounting bolt.*

3. Remove the two bolts holding the oil seal carrier to the cylinder block and pull out the seal carrier. Pry or hand-press the seal out of the carrier.

4. Reinstall the seal carrier with a new O-ring.

Tightening torque
- Intermediate shaft oil seal carrier to engine block . . . . 25 Nm (18 ft-lb)

5. Lightly oil the seal and the shaft sealing surface. Fit the seal into position, and carefully press it into place.

> ***CAUTION—***
> *Use care not to distort the seal as it is installed. For best results, use a seal press such as Volkswagen special tool no. 10-203.*

6. Install the woodruff key to the intermediate shaft and install the sprocket while aligning the keyway. Lock the intermediate shaft sprocket in place and tighten the bolt.

Tightening torques (4-cylinder engines)
- Intermediate shaft sprocket to intermediate shaft
  gasoline engines . . . . 80 Nm (59 ft-lb)
  diesel engine . . . . 45 Nm (33 ft-lb)

7. Install the camshaft drive belt, as described in **15a Cylinder Head and Valvetrain (4-Cylinder Engines)** or **23 Fuel Injection–Diesel**.

To replace front crankshaft oil seal (6-cylinder engine)

1. Remove the poly-ribbed drive belt. See **0 Maintenance Program**.

2. While holding the damper stationary using a holding fixture (VW special tool no. 3273 or equivalent) remove the crankshaft vibration damper center bolt. Remove the vibration damper. Discard the bolt.

3. Remove the seal from its carrier using a seal extractor (VW special tool no. 3203, or equivalent), or carefully pry it from its housing using a screwdriver.

4. Install the new seal, lubricated with clean engine oil, with the closed side facing out. Carefully press it into place until it is fully seated. For best results, use a seal press such as Volkswagen special tool no. 3266, shown in Fig. 3.

5. Using a new mounting bolt, install the crankshaft vibration damper. Hold the damper stationary and tighten the bolt.

> ***CAUTION—***
> *Always use a new vibration damper mounting bolt. It is a stretch bolt designed to be used only once.*

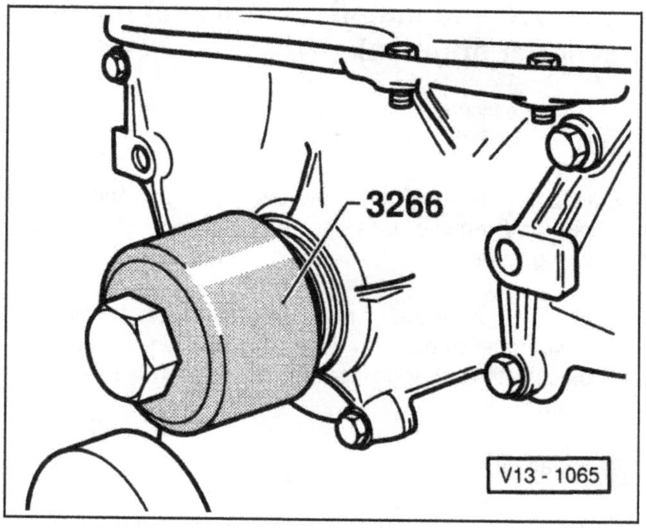

V13-1065

Fig. 3. Front crankshaft oil seal being installed using Volkswagen special tool no. 3266.

Tightening torques (6-cylinder engines)
- Vibration damper to crankshaft
  stage I . . . . 100 Nm (74 ft-lb)
  stage II . . . . .additional 1/4 turn (90°)

6. Install the drive belt and release the drive belt tensioner. See **0 Maintenance Program**.

To replace rear crankshaft oil seal (4 and 6 cylinder engines)

1. Remove flywheel or drive plate as described under **Flywheel or Driveplate**.

2. Remove the rear crankshaft oil seal from the flange by carefully prying it out taking care not to scratch the aluminum housing.

3. Install the new seal, lubricated with clean engine oil, with the closed side facing out using a guide sleeve and a suitable seal installation tool.

4. Install remaining components.

To remove and install rear oil seal flange (4 cylinder engines)

1. Remove flywheel or drive plate as described under **Flywheel or Driveplate**.

2. Unbolt rear flange from the cylinder block and the oil pan and remove. Use care to avoid damaging the oil pan gasket.

3. Use a guide sleeve as necessary to protect the oil seal. Install the flange onto the guide pins on the cylinder block with a new gasket. Torque bolts in a staggered pattern.

4. Install the oil pan.

5. Install the remaining components in reverse order of removal.

Tightening torques

- Rear main oil seal flange to engine block (M6). . . . . 10 Nm (7 ft-lb)

FLYWHEEL OR DRIVEPLATE

Removal of the flywheel or driveplate requires that the engine be separated from the transmission.

On 4-cylinder engines (except AHU diesel) with manual transmission, the flywheel is bolted to the clutch pressure plate. See Fig. 4.

On 6-cylinder and AHU diesel engines, the flywheel is bolted directly to the crankshaft, with the clutch pressure plate mounted to the flywheel.

> ***CAUTION —***
>
> - *On cars with automatic transmission, special mounting procedures are required when installing the driveplate. See the information given below.*
> - *The clutch pressure plate (4-cylinder engines), the flywheel (6-cylinder and AHU diesel engines), and the driveplate (cars with automatic transmission) are mounted to the crankshaft using one-time stretch bolts. These fasteners must always be replaced any time they are removed or loosened.*

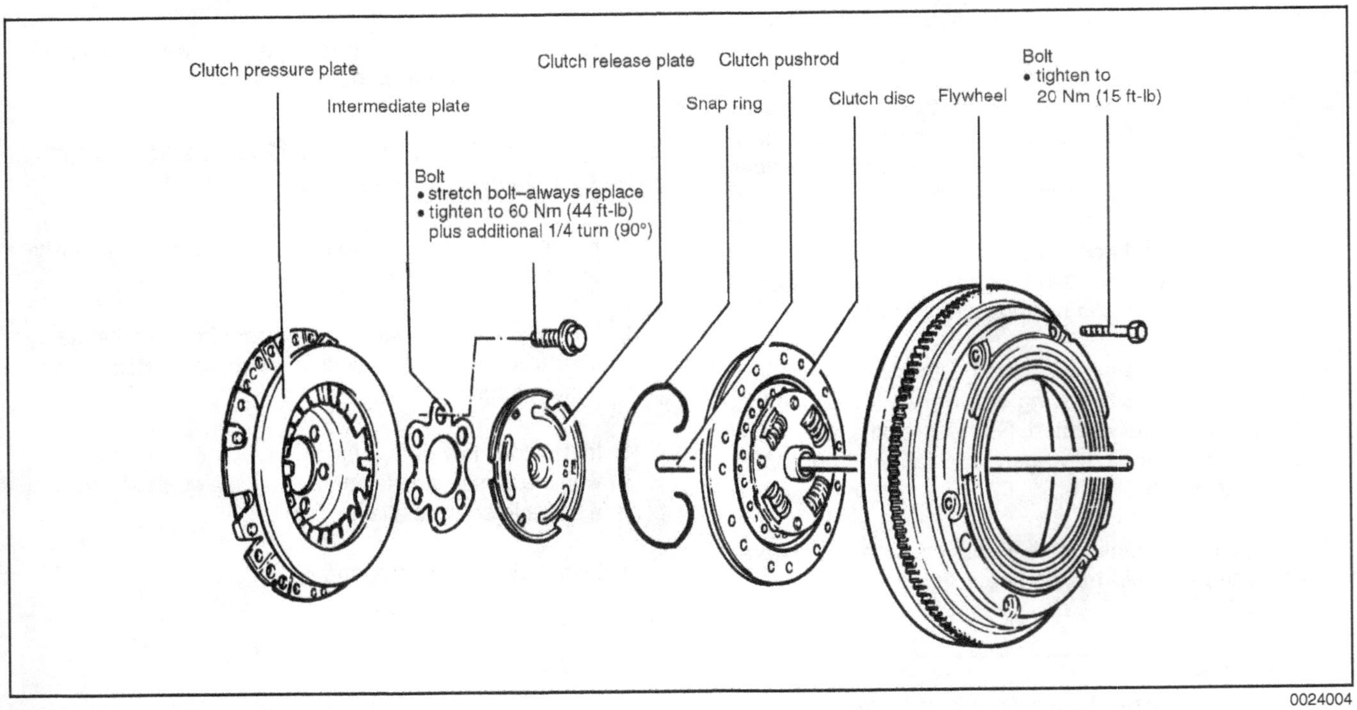

Fig. 4. Flywheel and clutch assembly on cars with ACC, ABA, and AAZ engines with manual transmission.

Fig. 5 illustrates the use of a holding fixture, Volkswagen special tool no. VW 558 (order no. TV0 558 000 13 ZEL) to remove and install the flywheel or driveplate.

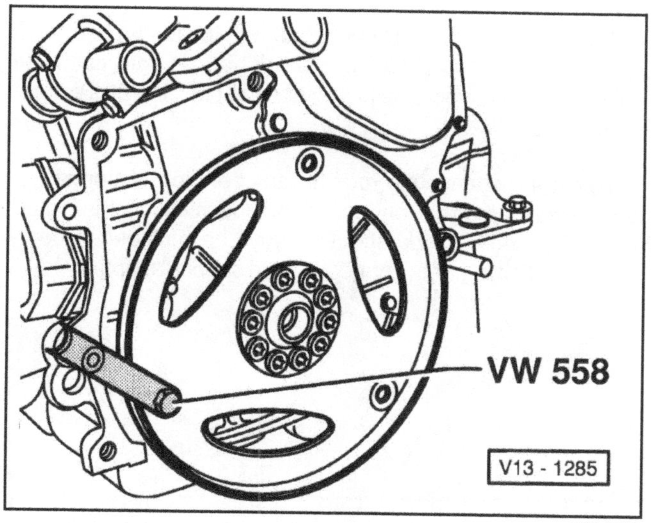

V13-1285

Fig. 5. Driveplate secured with holding fixture VW 558. Flywheel is similar.

Tightening torques

- Flywheel, pressure plate or driveplate to crankshaft (stretch bolts–**always replace**)
  stage I . . . . 60 Nm (44 ft-lb)
  stage II . . . . .additional 1/4 turn (90°)

Driveplate Installation

On cars with automatic transmission, mark the driveplate during disassembly and reinstall it with the same orientation. When reinstalling the driveplate, the clearance between the driveplate and the cylinder block should always be checked and adjusted (shimmed) as necessary.

To check the clearance, install the driveplate using three of the mounting bolts. Tighten the bolts to 30 Nm (22 ft-lb). Then check the clearance as shown in Fig. 6 or Fig. 7. If necessary, remove the driveplate and install an appropriate shim between the driveplate and the crankshaft to achieve the correct dimension.

> ***NOTE —***
> *Shims are available in varying thicknesses. See an authorized Volkswagen dealer for the latest in parts information.*

Tightening torques

- Driveplate to crankshaft (stretch bolts–**always replace**)
  stage I . . . . 60 Nm (44 ft-lb)
  stage II . . . . .additional 1/4 turn (90°)

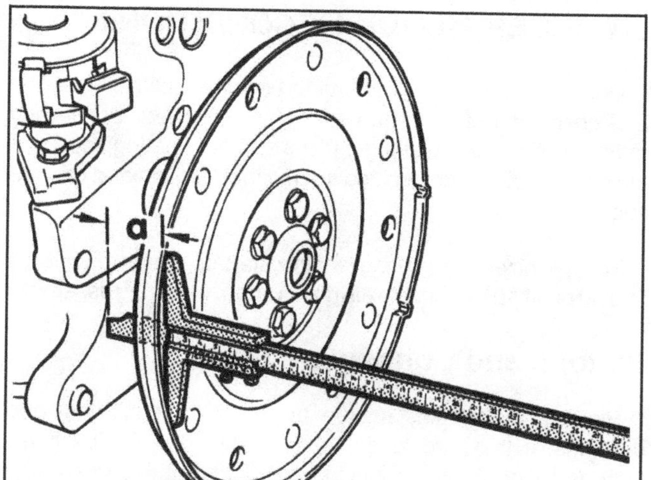

Dimension a:
- 19.5 to 21.1 mm (0.768 to 0.831 in.)
- check dimension at three places and calculate average

0024005

Fig. 6. Driveplate to cylinder block dimension (**a**) being measured on 4-cylinder engine. Measure at three points and calculate average. Adjust distance by adding or removing shim between driveplate and crankshaft.

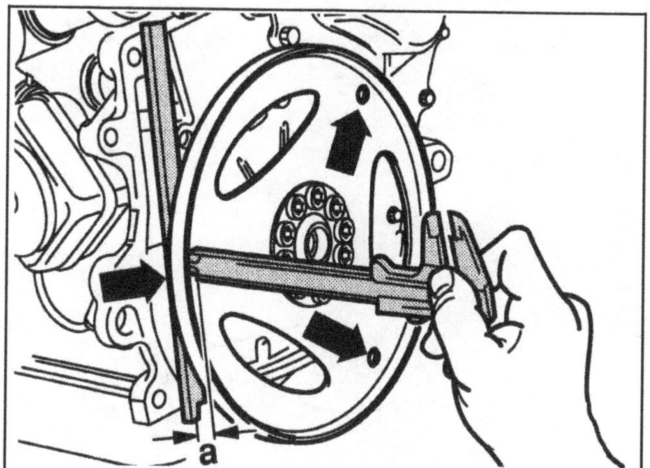

Dimension a:
- 15.7 to 16.5 mm (0.618 to 0.650 in.)
- check dimension at three places and calculate average

0024006

Fig. 7. Driveplate to cylinder block dimension (**a**) being measured on 6-cylinder engine. Use straight edge for reference line. Measure at three points (**arrows**) and calculate average. Adjust distance with shim(s) between driveplate and crankshaft.

CYLINDER BLOCK RECONDITIONING

During engine block disassembly, be sure to mark the position and orientation of all parts as they are removed. Connecting rods, rod caps, rod bearings, pistons, main bearing caps and main bearings are assembled in an exact location and orientation.

To minimize wear during engine start-up, clean engine oil should be used to lubricate all friction surfaces during assembly.

Pistons and Connecting Rods

Pistons, piston pins, piston rings, connecting rods, and bearings, if they are to be reused, should never be interchanged. If necessary, mark the cylinder number and installation orientation on the pistons, connecting rods and connecting rod caps before removal. Components of one piston and connecting rod assembly are shown in Fig. 8.

Pistons from AHU/1Z engines are not all the same. Valve relief pockets machined into the piston crown for cylinders 1 and 2 are the same. Pockets in cylinders 3 and 4 are also the same, but are different from cylinders 1 and 2. See Fig. 9.

The piston pin should fit such that, with the piston heated to approximately 60°C (140°F) in an oil bath, a light push will move the pin. Replace the piston and the pin if the fit is excessively loose.

Inspect the connecting rods for any bending, distortion, or other visual damage. Connecting rod specifications are listed below. Connecting rods should always be replaced in complete sets. **Table a** lists connecting rod specifications.

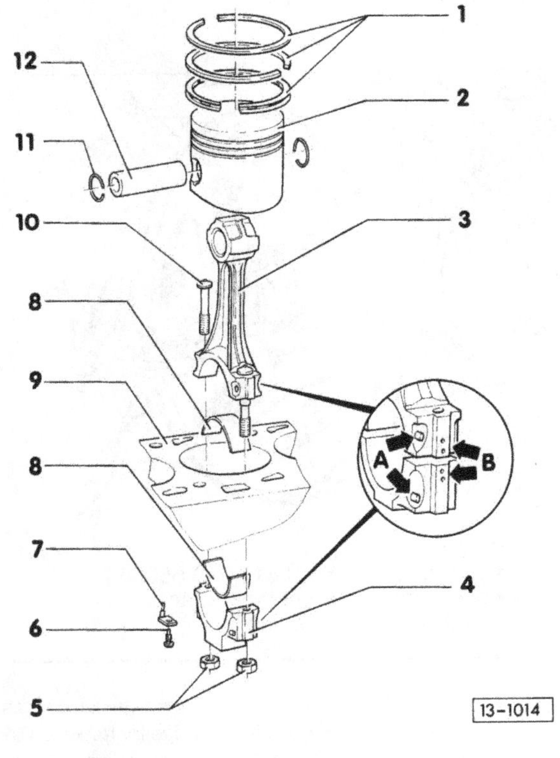

1. **Piston Rings**
   - offset gaps by 120° when installing
   - "TOP" mark faces toward piston crown
2. **Piston**
   - mark installation position and orientation to cylinder block and connecting rod
   - ACC, ABA, AAZ, AHU engines: arrow on piston faces drive belt end
   - AAA engine: high point of piston crown faces toward center of cylinder block
   - AAZ, AHU, 1Z engines: always check piston deck height during assembly. See Fig. 14
3. **Connecting rods**
   - mark cylinder number on rod and cap (B)
   - marks (A) must be above one another and on same side
4. **Connecting rod bearing cap**
   - note installed position
   - AAZ, AHU, 1Z engines: dowel pin must fit tightly in connecting rod
5. **Connecting rod nut (ACC and ABA engines only)**
   - oil threads and nut contact surface during installation
   - tighten to 30 Nm (22 ft-lb) plus an additional 1/4 turn (90°)
6. **Oil spray nozzle bolt**
   - ABA engine: tighten to 25 Nm (18 ft-lb)
   - AAZ, AHU, 1Z engines: install using VW sealant AMV 188 100 02
   - tighten to 10 Nm (89 in-lb)
7. **Oil spray nozzle**
   - not installed on all engines
   - for piston cooling. See **17 Engine-Lubrication System**
8. **Connecting rod bearing shell**
   - note installed position-do not interchange used shells
   - holding tabs must fit tightly in recesses
   - ABA engine: oil hole must face connecting rod
   - AAZ, AHU, 1Z engines: dowel pin must fit tightly in connecting rod
9. **Cylinder block**
10. **Connecting rod bolt**
    - stretch bolt-always replace
    - ACC/ABA shown; on others bolt head is on bearing cap end
    - oil threads and bolt head surface during installation
    - AAZ, 1Z and AAA engines: tighten to 30 Nm (22 ft-lb) plus additional 1/4 turn (90°)
11. **Circlip**
    - always replace
12. **Piston pin**
    - if difficult to remove, heat piston to 140°F (60°C)

13-1014

Fig. 8. Piston and connecting rod assembly for ABA engine. Connecting rod assemblies on other engines are similar.

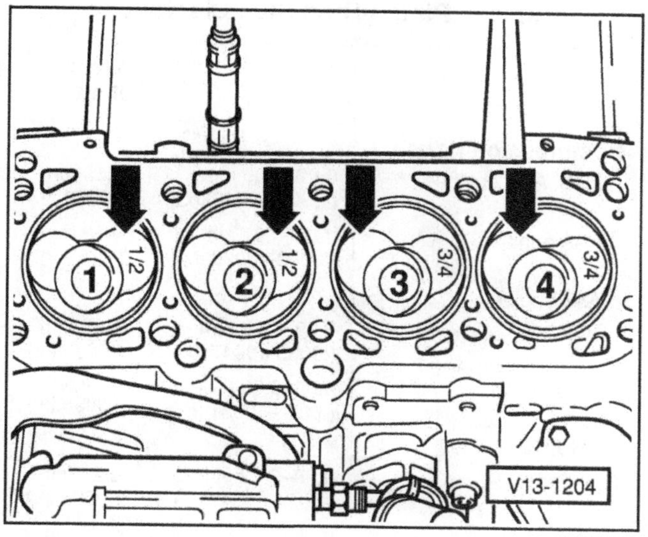

V13-1204

Fig. 9. Piston orientation in AHU/1Z engine. Intake valve relief pocket is larger and points toward flywheel for cylinders 1 and 2 and toward pulley side for cylinders 3 and 4 (**arrows**). Pistons are factory marked for proper location.

NOTE —

When checking radial clearance, lubricate the contact surface of the nut or bolt before tightening. When checking connecting rod radial clearance, tighten the nut or bolt only to 30 Nm (22 ft-lb) and not the additional 1/4 turn.

Table a. Connecting Rods

Radial clearance (Plastigage®)	
new	0.01–0.06 mm (0.0004–0.0024 in.)
wear limit	
ACC, ABA engines	0.12 mm (0.0047 in.)
AAA engine	0.10 mm (0.0039 in.)
AAZ, AHU, 1Z engines	0.08 mm (0.0031 in.)
Axial (side) clearance	
new	0.05–0.31 mm (0.0020–0.0122 in.)
wear limit	
ACC, ABA, AAZ, AHU, 1Z engines	0.37 mm (0.0145 in.)
AAA engine	0.40 mm (0.0157 in.)
Checking torque	30 Nm (22 ft-lb)
Assembly torque	30 Nm (22 ft-lb) plus 1/4 turn (90°)

NOTE —

If connecting rod radial clearance is excessive, the crankshaft connecting rod journals should be checked. If the crankshaft journal diameters are within specifications, recheck radial clearance using new bearing shells.

Piston Rings

Piston ring end gaps are checked with the piston rings inserted approximately 15 mm (5/8 in.) from the bottom of the cylinder. See Fig. 10. **Table b** lists piston ring gap specifications.

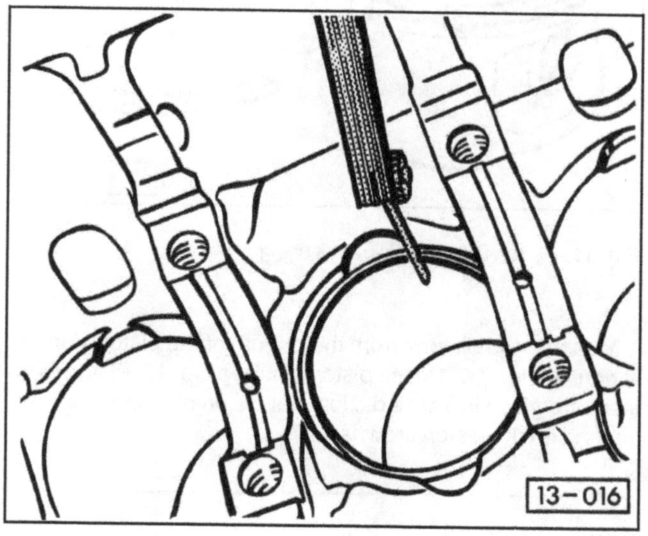

13-016

Fig. 10. Piston ring end gap being measured.

Table b. Piston Ring End Gaps

	New	Wear limit
Top compression ring		
ACC, ABA, AAA, AHU, 1Z engines	0.20–0.40 mm (.0079–.0157 in.)	1.0 mm (.039 in.)
AAZ engine	0.20–0.40 mm (.0079–.0157 in.)	1.2 mm (.0472 in.)
Bottom compression ring		
ACC, ABA, AAA, AHU, 1Z engines	0.20–0.40 mm (.0079–.0157 in.)	1.0 mm (.039 in.)
AAZ engine	0.20–0.40 mm (.0079–.0157 in.)	0.60 mm (.0236 in.)
Oil scraper ring		
ACC, ABA, AAA, AHU, 1Z engines	0.25–0.50 mm (.0098–.0197 in.)	1.0 mm (.039 in.)
AAZ engine	0.25–0.50 mm (.0098–.0197 in.)	1.2 mm (.0472 in.)

Piston ring side clearance is checked using feeler gauges. Measure each ring in its original groove. See Fig. 11. Piston ring side clearance specifications are listed in **Table c**.

NOTE —

Piston ring grooves should be thoroughly cleaned before checking ring side clearance.

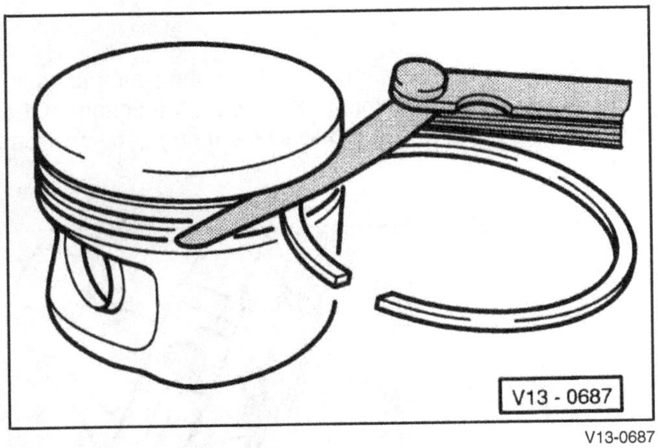

V13-0687

Fig. 11. Piston ring side clearance being measured.

Measure the pistons from the bottom of the piston skirt and at right angles (90°) to the piston pin. See Fig. 12. Piston diameters are given in **Table d**. Nominal piston diameters are also marked on the piston crowns.

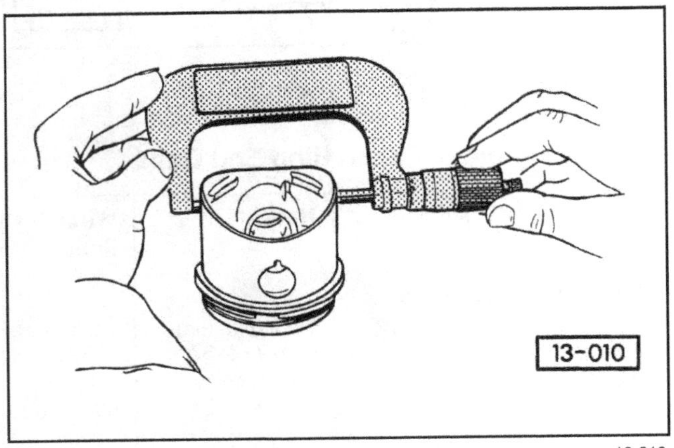

13-010

Fig. 12. Check piston approximately 6–10 mm (0.2–0.4 in.) from bottom of skirt at right angle to piston pin.

Piston Wear Limits

- maximum allowable deviation from nominal piston diameter.................... 0.04 mm (0.0016 in.)

Table c. Piston Ring Side Clearances

	New	Wear limit
Compression rings		
ACC, ABA engines	0.02–0.05 mm (.0008–.0019 in.)	0.15 mm (.0059 in.)
AAA engine	0.02–0.07 mm (.0008–.0028 in.)	0.15 mm (.0059 in.)
AAZ engine		
top ring	0.09–0.12 mm (.0035–.0047 in.)	0.25 mm (.0098 in.)
bottom ring	0.05–0.08 mm (.0020–.0031 in.)	0.25 mm (.0098 in.)
AHU, 1Z engine		
top ring	0.06–0.12 mm (.0024–.0047 in.)	0.25 mm (.0098 in.)
bottom ring	0.05–0.08 mm (.0020–.0031 in.)	0.25 mm (.0098 in.)
Oil scraper ring		
ACC, ABA engines	0.02–0.05 mm (.0008–.0019 in.)	0.15 mm (.0059 in.)
AAA engine	0.02–0.06 mm (.0008–.0024 in.)	0.15 mm (.0059 in.)
AAZ, AHU, 1Z engines	0.03–0.06 mm (.0012–.0024 in.)	0.15 mm (.0059 in.)

Cylinder Block

Measure cylinder bores at approximately the top, the middle, and the bottom of piston travel. Make measurements parallel to the crankshaft and at right angles (90°). See Fig. 13. The top and bottom measurements should be made approximately 10 mm (3/8 in.) from the ends of the cylinder. Nominal piston and cylinder bore diameter specifications are given in **Table d**. Nominal piston diameters are also marked on the piston crowns.

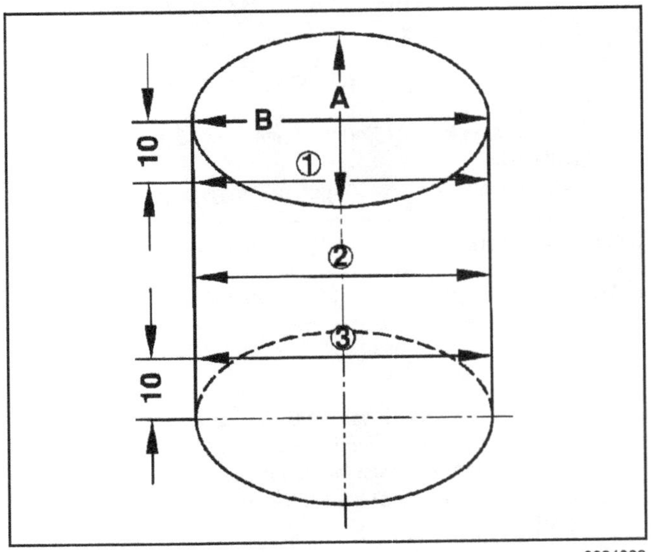

0024009

Fig. 13. Measure cylinder bores in direction perpendicular to the crankshaft (**A**) and parallel to the crankshaft (**B**) at the top, middle, and bottom of cylinder.

Table d. Piston and Cylinder Diameters

Engine code	Piston diameter in mm (in.)	Cylinder bores in mm (in.)
ACC		
standard	80.985 (3.18838)	81.01 (3.1894)
1st oversize	81.235 (3.19822)	81.26 (3.1992)
2nd oversize	81.485 (3.20806)	81.51 (3.2090)
AAZ		
standard	79.48 (3.1291)	79.51 (3.1303)
1st oversize	79.73 (3.1390)	79.76 (3.1402)
2nd oversize	79.98 (3.1488)	80.01 (3.1500)
ABA		
standard	82.485 (3.2474)	82.51 (3.2484)
1st oversize	82.735 (3.2572)	82.76 (3.2583)
2nd oversize	82.985 (3.2671)	83.01 (3.2681)
AAA		
standard	80.985 (3.1884)	81.01 (3.1894)
1st oversize	81.485 (3.2081)	81.51 (3.2091)
2nd oversize	81.985 (3.2278)	82.01 (3.2287)
AHU, 1Z		
standard	79.48 (3.1291)	79.51 (3.1303)
1st oversize	79.72 (3.1385)	79.76 (3.1402)
2nd oversize	79.98 (3.1488)	80.01 (3.1500)

Cylinder Bore Wear Limits

- maximum allowable deviation from nominal diameter
  gasoline engines . . . . 0.08 mm (0.0031 in.)
  diesel engine . . . . 0.10 mm (0.0039 in.)

> ***CAUTION —***
> *Mounting the bare cylinder block to an engine stand can distort its shape and cause inaccurate cylinder bore measurements. Always check cylinder bores with the block resting unstressed on a flat surface.*

NOTE —

On some engines, the cylinder block is fitted with oil spray nozzles for piston cooling. See **17 Engine-Lubrication System** *for more information on spray nozzles.*

Diesel Piston Height

Assembling or replacing a diesel engine short block requires that the height of the pistons be measured in order to select the proper cylinder head gasket. Using a dial indicator, measure the amount that each piston protrudes above the top of the cylinder block at Top Dead Center. See Fig. 14. Cylinder head gaskets are available in three thicknesses. Select the correct gasket from **Table e** based on the largest piston height measurement. The different gaskets are identified by the number of marks, either notches or holes, located near the part number stamped on the gasket face.

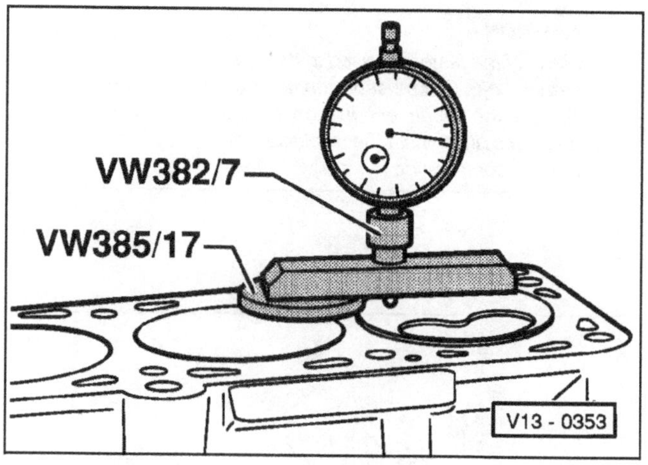

V13-0353

Fig. 14. Piston height being measured on diesel engine.

Table e. Diesel Cylinder Head Gasket Selection

Piston height above cylinder block	Gasket identification or notches
AAZ engine	
0.66–0.86 mm (.0260–.0339 in.)	1
0.87–0.90 mm (.0343–.0354 in.)	2
0.91–1.02 mm (.0358–.0402 in.)	3
AHU, 1Z engines	
0.91–1.00 mm (.0358–.0394 in.)	1
1.01–1.10 mm (.0398–.0433 in.)	2
1.11–1.20 mm (.0437–.0472 in.)	3

Crankshaft and Intermediate Shaft

Fig. 16 and Fig. 17 show the crankshaft assemblies for the engines covered by this manual. Crankshaft main bearing caps must not be interchanged. Crankshaft main bearing shells, if they are to be reused, should only be installed in their original positions. On 4-cylinder engines, the distributor must first be removed to remove the intermediate shaft.

AAA, ABA and AHU crankshafts have sensor wheels for their engine management systems mounted on them. These will only fit on the crankshaft in one position. The sensor wheel on AHU engines uses a dowel pin. See Fig. 15.

> ***CAUTION —***
> *Many of the fasteners used in the cylinder block are stretch bolts that must be replaced once loosened. Review the repair information to identify all bolts and nuts that must be replaced during cylinder block reconditioning.*

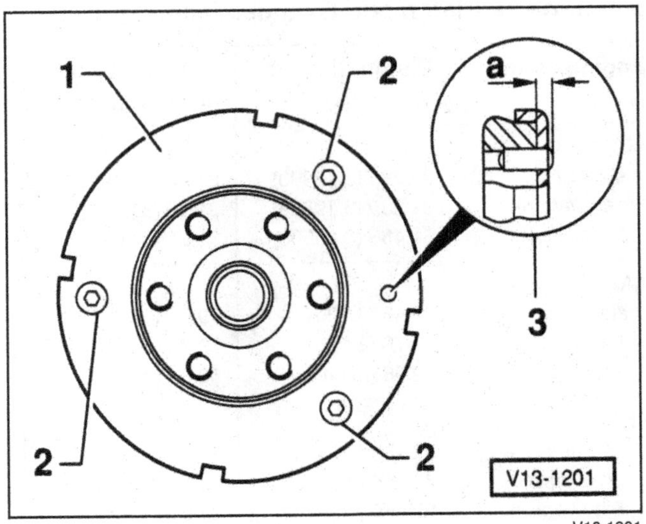

V13-1201

Fig. 15. Sensor wheel (**1**) mounted on AHU crankshaft with screws (**2**). A single dowel pin (**3**) extends 2.5 to 3 mm from the flange end, dimension (**a**).

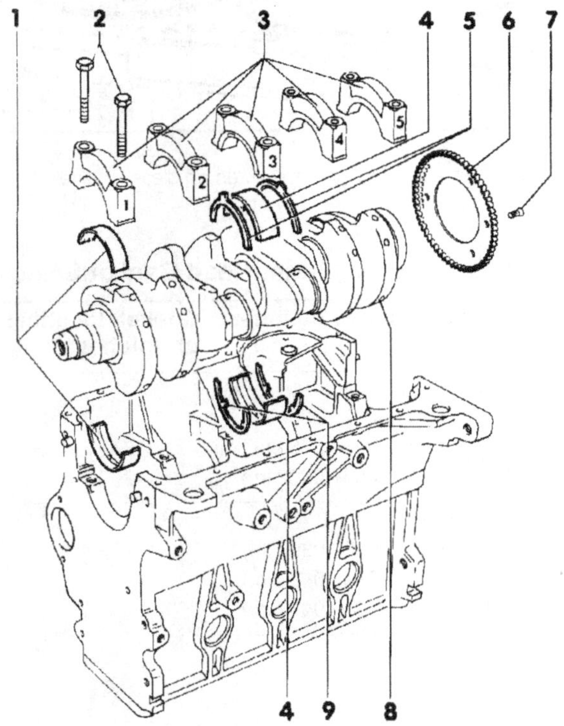

1. **Main bearing shells**
   - without oil groove: for main bearing caps no.1, 2, and 5 (also no. 4 cylinder block shell on diesel engines)
   - with oil groove: for cylinder block and main bearing cap 4 (ex. diesel)
   - do not interchange–mark installed positions
2. **Main bearing cap bolts**
   - new type bolt (replacement) with fully threaded shank tighten to 65 Nm (48 ft-lb) plus 1/4 turn (90°)
   - old type bolt (shown) with partially threaded shank tighten to 65 Nm (48 ft-lb)
   - do not mix old and new type bolts
3. **Main bearing caps**
   - main bearing cap no. 3 with recesses for thrust washers
   - do not interchange–mark installed positions
4. **Main bearing shell no. 3**
   - with oil groove: for cylinder block
   - without oil groove: for bearing cap
5. **Thrust washers, main bearing cap no. 3**
   - do not interchange–mark installed positions
6. **Engine speed sensor wheel (ABA engine only)**
7. **Sensor wheel mounting bolt (ABA engine only)**
   - always replace
   - tighten to 10 Nm (89 in-lb) plus 1/4 turn (90°)
8. **Crankshaft**
9. **Thrust washer, main bearing cap no. 3**
   - do not interchange–mark installed positions

13-1225

Fig. 16. Exploded view of crankshaft and related parts for 4-cylinder engines.

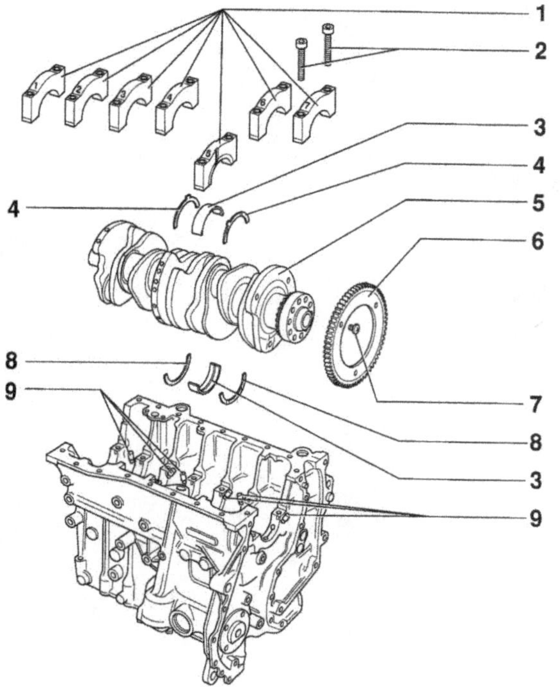

1. **Main bearing caps**
   - main bearing cap no. 5 with recesses for thrust washers
   - do not interchange–mark installed positions
2. **Main bearing cap bolts**
   - always replace
   - tighten to 30 Nm (22 ft-lb) plus an additional 1/2 turn (180°)
3. **Main bearing shells no. 1 through 7**
   - without oil groove: for bearing cap
   - with oil groove: cylinder block
   - do not interchange–mark installed positions
4. **Thrust washers, main bearing cap no. 5**
   - do not interchange–mark installed positions
5. **Crankshaft**
6. **Engine speed sensor wheel**
7. **Sensor wheel mounting bolt**
   - always replace
   - tighten to 10 Nm (89 in-lb) plus 1/4 turn (90°)
8. **Thrust washers, main bearing cap no. 5**
   - do not interchange–mark installed positions
9. **Oil spray nozzle**
   - opening pressure: 2.0 bar (29 psi)

N13-0024

Fig. 17. Exploded view of crankshaft and related parts for 6-cylinder (AAA) engine.

NOTE —

- *On cars with automatic transmission, see* **Flywheel or Driveplate** *given earlier when reinstalling the driveplate. Special installation procedures apply.*
- *On cars with automatic transmission, mark the position of the driveplate on the crankshaft before removing the driveplate.*

Crankshaft journal specifications are listed in **Table f**. Crankshaft and intermediate shaft clearance specifications are listed in **Table g**. If a crankshaft must be replaced, a Volkswagen remanufactured crankshaft is available from an authorized Volkswagen dealer.

NOTE —

The journal diameters given in Table f are nominal diameters and not actual measured dimensions. For example, the first table entry with basic journal diameter of 54.00 mm should actually measure between 53.978 and 53.958, according to the tolerance limits (oil clearance).

NOTE —

On 4-cylinder engines, crankshaft bearings are available in three undersizes to fit reconditioned crankshafts. On the 6-cylinder engine, check with an authorized Volkswagen dealer on the availability of reconditioned crankshafts.

Table f. Crankshaft Journal Diameters

Journal diameters (nominal)	mm (in.)
Basic dimension	
4-cylinder engines	
main	54.00 (2.1260) −0.022 (0.00087) / −0.042 (0.00165)
connecting rod	47.80 (1.8819) −0.022 (0.00087) / −0.042 (0.00165)
6-cylinder engine	
main	60.00 (2.3622) −0.022 (0.00087) / −0.042 (0.00165)
connecting rod	54.00 (2.1260) −0.022 (0.00087) / −0.042 (0.00165)
1st undersize (0.25 mm)	
4-cylinder engines	
main	53.75 (2.1161) −0.022 (0.00087) / −0.042 (0.00165)
connecting rod	47.55 (1.8720) −0.022 (0.00087) / −0.042 (0.00165)
2nd undersize (0.50 mm)	
4-cylinder engines	
main	53.50 (2.1063) −0.022 (0.00087) / −0.042 (0.00165)
connecting rod	47.30 (1.8622) −0.022 (0.00087) / −0.042 (0.00165)
3rd undersize (0.75 mm)	
4-cylinder engines	
main	53.25 (2.0965) −0.022 (0.00087) / −0.042 (0.00165)
connecting rod	47.05 (1.8524) −0.022 (0.00087) / −0.042 (0.00165)

Table g. Crankshaft and Intermediate Shaft Specifications

Main bearing radial clearance (Plastigage®)	
new parts	
ACC, ABA engines	0.02–0.06 mm (.0008–.0024 in.)
AAZ, AHU, 1Z engines	0.03–0.08 mm (.0012–.0031 in.)
AAA engine	0.02–0.06 mm (.0008–.0024 in.)
wear limit	
ACC, ABA engines	0.17 mm (.0067 in.)
AAZ, AHU, 1Z engines	0.17 mm (.0067 in.)
AAA engine	0.10 mm (.0039 in.)
Crankshaft axial play (side clearance)	
new parts	
ACC, ABA engines	0.07–0.17 mm (.0028–.0067 in.)
AAZ, AHU, 1Z engines	0.07–0.17 mm (.0028–.0067 in.)
AAA engine	0.07–0.23 mm (.0028–.0091 in.)
wear limit	
ACC, ABA engines	0.25 mm (.0098 in.)
AAZ, AHU, 1Z engines	0.37 mm (.0146 in.)
AAA engine	0.30 mm (.0118 in.)
Intermediate shaft axial clearance (side clearance)	
wear limit	0.25 mm (.0098 in.)

15a Cylinder Head and Valvetrain– (4-Cylinder Engine)

TABLES

GENERAL

This engine section covers most cylinder head and valvetrain service and repair for four-cylinder engines. For information on short block engine rebuilding and internal engine specifications, see **13 Crankshaft/Cylinder Block**.

Some of the operations described in this repair group require special equipment and experience. If you lack the skills, tools, or a suitable workplace for servicing or repairing the cylinder head, we suggest you leave these repairs to an authorized Volkswagen dealer or other qualified shop.

NOTE —
- *Much of the information given in the repair section is organized according to engine codes. For a listing of the engines and their codes, see* **1 Engine–General**.
- *Most of the operations that pertain to the TDI engine (code AHU, 1Z) require removal of the large plastic sound absorber cover mounted on the top of the engine. Always be sure to reinstall the cover when repairs are completed to insure minimal engine noise levels.*

DIAGNOSTIC TESTING

The tests that follow can be used to help isolate engine problems, to better understand a problem before starting expensive repairs, or just to periodically check engine condition.

Compression Test

A compression test will tell a lot about the condition of the engine without the need for taking it apart. The test is relatively simple. On gasoline engines, it requires only a compression tester and a spark plug wrench. Diesel engines with their higher compression pressures (approximately 500 psi) require a more expensive compression tester. For the most accurate test results, the battery should be fully charged and the engine should be warm.

Low compression is evidence of poorly sealed combustion chambers. Compression pressures which are relatively even but below specifications indicate worn piston rings and/or cylinder walls. Low but erratic values tend to indicate valve leakage. Dramatic differences, such as good values in some cylinders and very low values in one or two cylinders are the sign of a localized failure, such as a burnt valve or a blown head gasket.

To test compression (gasoline engine)

1. With the ignition off, disable the secondary circuit of the ignition system by disconnecting the harness connector at the coil power stage. Disable the fuel pump by removing the fuel pump fuse (no. 18).

   NOTE —
   See **97 Wiring Diagrams, Fuses, and Relays** *for fuse locations.*

2. Remove the spark plug wires from the spark plugs. Use care to pull on both the wire and the boot at the same time. Label the wires for correct reinstallation.

3. Clean the area around the spark plugs using compressed air. Remove the spark plugs and set them aside in order.

4. Thread the compression tester into the first cylinder's spark plug hole, just tight enough to seal the plug hole.

5. With the transmission in neutral and the throttle held wide open, crank the engine with the starter about 4 to 5 revolutions. The gauge reading should increase with each engine revolution. Record the gauge value.

6. Release the pressure in the tester and remove it from the spark plug hole.

7. Repeat the test for each of the other cylinders. Record the data and compare with **Table a**.

Table a. Compression Pressures

Engine code	ACC (1.8 L)	AAZ (1.9 L)	AHU, 1Z (1.9L)	ABA (2.0 L)
new engines	130-174 psi (9-12 bar)	493 psi (34 bar)	363-450 psi (25-31 bar)	145-189 psi (10-13 bar)
wear limit	102 psi (7 bar)	377 psi (26 bar)	276 psi (19 bar)	109 psi (7.5 bar)
maximum difference	44 psi (3 bar)	73 psi (5 bar)	73 psi (5 bar)	44 psi (3 bar)

8. Reinstall the spark plugs and the spark plug wires in their original locations. Reconnect the coil harness connector and the fuel pump fuse.

Tightening torque
- Spark plug to cylinder head
  ABA engine .30 Nm (22 ft-lb)
  ACC engine .25 Nm (18 ft-lb)

To test compression (AAZ diesel engine)

1. Disconnect the wire from the fuel cut-off valve on the injector pump. Tape or insulate the wire end.

2. Unscrew the fuel pipe union nuts from the injection pump and the fuel injectors. Remove the fuel pipes as a unit, taking care not to bend them. Disconnect the return lines from the injectors.

   WARNING —
   Fuel will be expelled as you loosen the unions. Do not smoke or work near heaters or other fire hazards. Have a fire extinguisher handy.

3. Using a 24mm deep socket, remove the injectors, and then remove the heat shields from each injector hole.

   WARNING —
   If the heat shields are not removed, compression pressure will expel them with dangerous force when the engine is turned over.

4. Thread the compression tester with an injector heat shield, into the first cylinder's injector hole. Thread the gauge in tight enough to seal.

5. With the transmission in neutral, crank the engine with the starter. The gauge reading should increase with each engine revolution. Continue until the gauge reading stops increasing (about 4 to 5 revolutions). Record the highest value indicated by the gauge.

6. Release the pressure in the tester and remove it from the injector hole.

7. Repeat the test for each of the other cylinders. Record the data and compare with **Table a**.

8. Reinstall the injectors with new heat shields. Reconnect the fuel lines. Reconnect the wire to the fuel cut-off valve. See **23 Fuel Injection–Diesel**.

Tightening torque
- Diesel injectors to cylinder head . . . . .70 Nm (51 ft-lb)
- Diesel injection line unions. . . . . 25 Nm (18 ft-lb)

To test compression (AHU, 1Z diesel engine)

1. Disconnect the harness connectors for the fuel cut-off valve and quantity adjuster from the Diesel injection pump.
2. Remove the contact bridge wiring from each glow plug by gently pulling up on it. Lay the complete contact bridge aside.
3. Using a 10 mm deep socket and swivel, or Volkswagen special tool 3220, remove the glow plugs.
4. Thread the compression tester with adapter VAG 1323/12 or equivalent into the first glow plug hole just enough to insure a tight seal.
5. With the transmission in neutral, crank the engine with the starter. The gauge reading should increase with each engine revolution. Continue cranking until the gauge reading stops increasing (about 4 to 5 revolutions). Record the highest value indicated by the gauge.
6. Release the pressure from the gauge and remove it from the glow plug hole.
7. Repeat the test for each of the other cylinders. Record the data and compare with Table a.
8. Reinstall the glow plugs and the contact bridge.
9. Reconnect the harness connector for the fuel cut-off valve and quantity adjuster.

Tightening torques
- Diesel glow plug to cylinder head. . . . 15 Nm (11 ft-lb)

Leak-down Test

The most conclusive diagnosis of low compression symptoms requires a leak-down test. Using a special tester and a supply of compressed air, each cylinder is pressurized. The rate at which the air leaks out of the cylinder, as well as the sound of the air escaping, can more accurately pinpoint the magnitude and source of the leakage. Any engine compression diagnosis that will require major disassembly should first be confirmed by a leak-down test. Because the test requires special equipment and experience, it may be desirable to have it performed by a Volkswagen dealer or other qualified repair shop.

CYLINDER HEAD SERVICE

Many cylinder head repairs can be accomplished without removing the cylinder head from the engine. The cylinder head cover gasket, the camshaft, the camshaft oil seal, the valve guide oil seals, and the cam followers are all accessible with the cylinder head installed. This heading describes those repairs that can be done with the cylinder head installed.

To remove and install cylinder head cover (ABA engine)

1. Remove the intake air boot from the throttle housing.
2. Disconnect the accelerator cable from the throttle body.
3. Disconnect all hoses and electrical connectors from the upper intake manifold.
4. On cars with Exhaust Gas Recirculation (EGR), disconnect the pipe from the EGR valve beneath the intake manifold.
5. Remove the upper intake manifold mounting bolts and manifold support bracket. Carefully separate the upper manifold from the lower manifold. Cover the open intake runners using clean shop rags.
6. Disconnect the crankcase breather valve from the cylinder head cover.
7. Remove the nuts and the reinforcing strips from the cover.
8. Unclip and remove the upper camshaft drive belt cover. Remove the small dust shield from the back of the sprocket.
9. Lift off the cylinder head cover and its gasket. Remove the one-piece gasket from the cylinder head or cover.
10. Installation is the reverse of removal. Be sure to use a new gasket at the intake manifold.

Tightening torques
- Cylinder head cover to cylinder head. . . . . 10 Nm (89 in-lb)
- Upper intake manifold to lower manifold (ABA engine) . . . . 20 Nm (15 ft-lb)

To remove and install cylinder head cover (ACC, AAZ, AHU, 1Z engines)

1. On ACC engines, remove the crankcase breather from the cylinder head cover. Disconnect the accelerator cable at the throttle body.
2. Remove the nuts and, if applicable, the reinforcing strips from the cover.
3. On ACC engines, unclip and remove the upper camshaft drive belt cover. Remove the small dust shield from the back side of the cover.
4. Lift off the cylinder head cover and its gasket.
5. On ACC engines, remove the sealing plug from the rear of the cylinder head and remove the semi-circular gasket piece from inside the front of the cover.

> ***NOTE —***
> *On AAZ, AHU and 1Z engines, a one-piece gasket is used.*

6. Installation is the reverse of removal. On ACC engines, be sure the tabs on the semi-circular piece correctly mate the slots on the gasket.

Tightening torque
- Cylinder head cover to cylinder head . . . . 10 Nm (89 in-lb)

To replace camshaft oil seal

1. Set the engine to TDC and remove the camshaft drive belt. For gasoline engines, see **Camshaft Drive Belt (gasoline engines)** given later. For diesel engines, see **23 Fuel Injection–Diesel**.

> ***NOTE —***
> *When using a seal extractor, such as Volkswagen special tool no. 2085, the camshaft oil seal can be removed and installed here, without removing the cylinder head cover. See Fig. 1.*

2. Remove the cylinder head cover as described above.
3. Counterhold the camshaft sprocket with a holding tool (VW special tool 3036 or equivalent). Loosen the center bolt 1/2 turn and tap the sprocket loose. Remove the bolt and sprocket. On gasoline engines, remove the woodruff key from the camshaft.

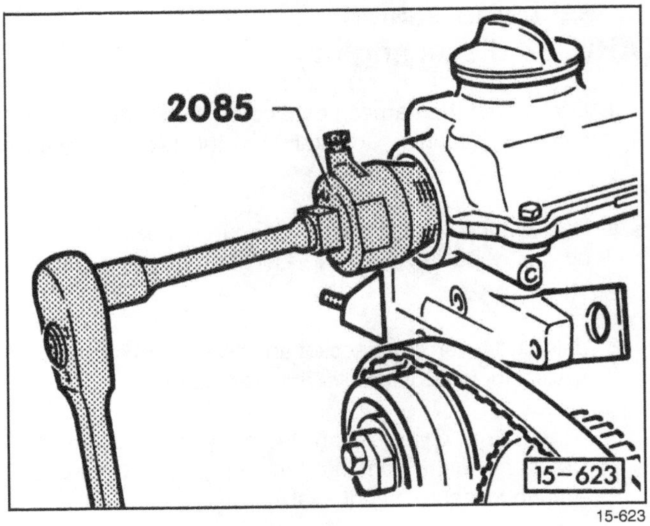

15-623

Fig. 1. Oil seal extractor (VW special tool no. 2085) being used to remove camshaft oil seal.

4. Loosen the camshaft bearing caps in exact order as described in **To remove and install camshaft**. Remove the no. 1 bearing cap (sprocket end).

> ***CAUTION —***
> *Failure to loosen and tighten the bearing caps in the order prescribed may damage the bearing caps or the camshaft.*

5. Lift slightly on the sprocket end of the camshaft and remove the oil seal.
6. Lightly oil the seal lip and camshaft's sealing surface. Fit the seal into position so that it is flush with the front of the cylinder head.
7. Loosely install the no.1 bearing cap. Then tighten the camshaft bearing caps in exact order as described in **To remove and install camshaft**.

> ***CAUTION —***
> *Be sure to install bearing cap correctly. The camshaft bearing caps are bored off-center and go on only one way.*

Tightening torque
- Camshaft bearing caps to cylinder head . . . . 20 Nm (15 ft-lb)

---

8. Install the cylinder head cover, with a new gasket, as described earlier.

9. Reinstall the camshaft drive belt sprocket. Counter hold the sprocket and tighten the bolt.

Tightening torques
- Camshaft sprocket to camshaft
  gasoline engines . . . . 80 Nm (59 ft-lb)
  diesel engine . . . . 45 Nm (33 ft-lb)

10. Reinstall the camshaft drive belt. For gasoline engines, see **Camshaft Drive Belt (gasoline engines)**. For diesel engines, see **23 Fuel Injection–Diesel**.

To remove and install camshaft

The camshaft can be removed with the cylinder head installed. Removing the camshaft allows access to the hydraulic cam followers, the valve springs, and the valve stem oil seals.

1. Remove the camshaft drive belt from the camshaft sprocket. For gasoline engines, see **Camshaft Drive Belt (gasoline engines)**. For diesel engines, see **23a** or **23b Fuel Injection–Diesel**.

NOTE —
On Diesel engines, removal of the drive belt will require that the diesel injection pump be timed during belt installation. This procedure requires special tools.

2. Remove the cylinder head cover and gasket as described above.

3. On ACC engines, remove bearing caps 1 and 3, in that order. Loosen the nuts on each of the two remaining caps, a little at a time so that valve spring tension is relieved evenly.

4. On AAZ, ABA, AHU and 1Z engines, remove bearing caps 5, 1, and 3, in that order. Loosen the nuts on each of the two remaining caps, a little at a time so that valve spring tension is relieved evenly.

CAUTION —
To avoid uneven and accelerated wear, the bearing caps must be reinstalled in their exact original positions.

NOTE —
Once removed, the camshaft should be checked for wear and other visible damage as described later under **Cylinder Head Reconditioning**.

5. Install the camshaft into the head with cylinder No.1 cams lobes pointing up.

6. Install bearing caps 2 and 5 on ACC engines, or 2 and 4 on AAZ, ABA, AHU and 1Z engines. Install the washers and nuts, gradually tightening all nuts until the camshaft is drawn fully and evenly into the bearing saddles. Install the remaining bearing caps with washers and nuts.

CAUTION —
Be sure to install bearing caps correctly. The caps are bored off-center and go on only one way. See Fig. 2.

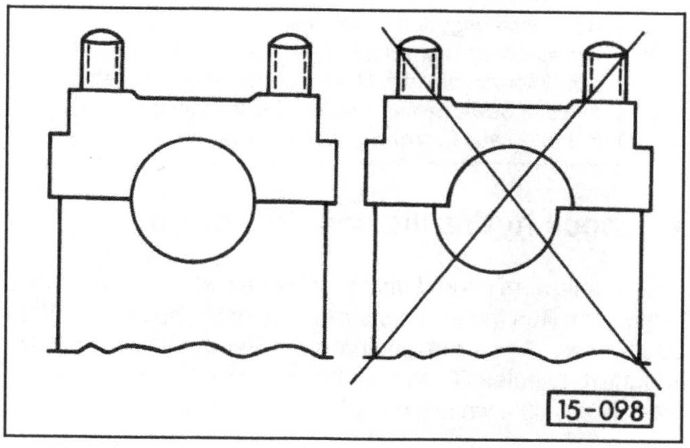

15-098

Fig. 2. Bearing caps are bored off-center. If necessary, install caps before installing camshaft to determine installation direction.

Tightening torque
- Camshaft bearing cap nuts . . . . .20 Nm (15 ft-lb)

7. Reinstall the camshaft drive belt. For gasoline engines, see **Camshaft Drive Belt (gasoline engines)**. For diesel engines, see **24 Fuel Injection**.

8. Reinstall the cylinder head cover with a new gasket. See **To remove and install cylinder head cover**.

NOTE —
- *On ACC engines, be sure to check and, if necessary, adjust ignition timing as described in* **28 Ignition System**.
- *On AAZ, AHU and 1Z engines, be sure to check injection pump timing. See* **23a** *or* **23b Fuel Injection–Diesel**

Hydraulic Cam Followers

All engines covered by this manual are equipped with hydraulic cam followers that automatically maintain proper valve clearance. The cam followers are pumped up by engine oil pressure, expanding as necessary to fill the gap between valve and camshaft lobe.

Some valve noise at start-up is normal at times, due to hydraulic cam followers that have bled down while the engine was not running. Before checking noisy cam followers, check that the engine oil is new and fresh and that the level is correct. Allow a minute or so with the engine running for the lubrication system to properly pump up the cam followers.

> ***CAUTION —***
> *After installing new cam followers, the engine should not be started for at least 30 minutes. New cam followers are at full height and must be allowed to bleed down to their proper height. Failure to do this may cause valve or piston damage.*

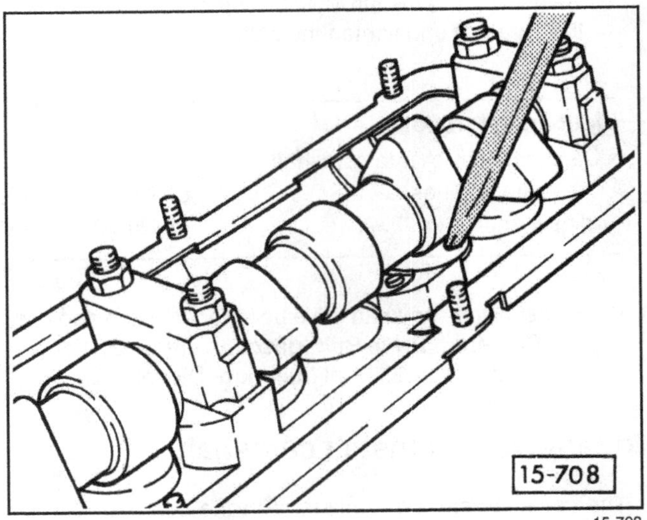

15-708

Fig. 3. Hydraulic cam follower being depressed with a non-metal object.

To check hydraulic cam followers

Cam followers should only be checked when the engine is fully warm. Run the engine, preferably drive the car, for 20 to 30 minutes. Raise the engine speed above 2,500 rpm for about two minutes. Then shut the engine off and proceed as soon as possible while the engine is still warm.

1. Remove the cylinder head cover. See **To remove and install cylinder head cover**.

2. Turn the engine by hand until both the camshaft lobes of one cylinder are pointing approximately up.

3. Using a feeler gauge, check the clearance between the cam follower and the cam lobe. If the clearance exceeds the specified limit, the lifter is faulty and should be replaced.

Hydraulic Cam Follower
- Clearance, maximum allowed . . . . 0.1 mm (0.004 in.)

---

4. If no faults are found in step 3, use a non-metal object such as wood or plastic to apply light pressure to the top of the cam follower. See Fig. 3. If the clearance is excessive, the follower is faulty and should be replaced.

5. Replace a faulty cam follower by removing the camshaft and pulling the follower from the cylinder head. Hydraulic cam followers are replaced only as complete assemblies.

> ***CAUTION —***
> *If new hydraulic cam followers were installed, the engine must not be run for at least 30 minutes. New cam followers are at a maximum height and should be allowed to bleed down or the valve heads could contact the pistons on start up.*

NOTE —
Store removed hydraulic cam followers in order on a clean surface with the camshaft contact surface facing down to prevent bleed down. Cover them with a clean, lint-free rag.

To replace valve stem oil seals

The sign of faulty valve stem oil seals are excessive oil consumption and oil smoke from the exhaust, usually during deceleration. Keep in mind that the valve guides may also be worn and the major cause of oil consumption. See **Cylinder Head Reconditioning** for more information on checking valve guides.

Replacing the valve stem oil seals requires that the camshaft, the cam followers, and the valve springs be removed. In addition, special service tools are required to compress the valve springs.

1. If working with the cylinder head installed, remove the cylinder head cover as described earlier. Otherwise, remove the cylinder head as described later.

2. Remove the camshaft and remove cam followers. See **To remove and install camshaft**. Keep all parts in order so that they can be reinstalled in their original locations.

> *NOTE —*
>
> *Store removed hydraulic cam followers in order on a clean surface with the camshaft contact surface facing down to prevent bleed down. Cover them with a clean, lint-free rag.*

3. If the cylinder head is removed, remove the valve springs. See Fig. 4.

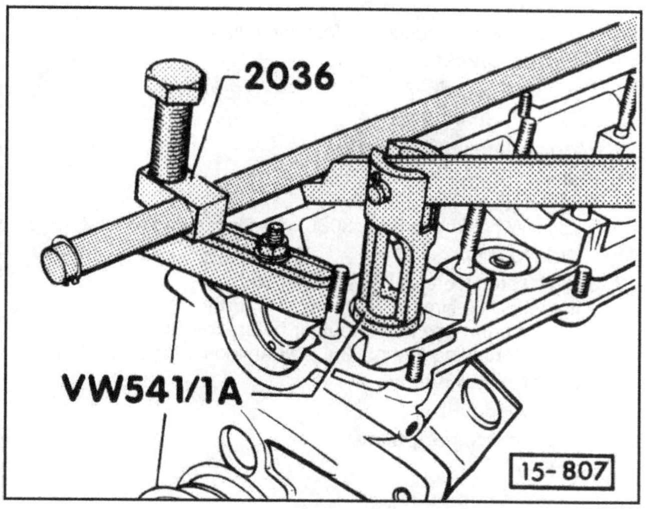

15-807

Fig. 4. Valve spring compressors being used to remove valve springs. Numbers identify Volkswagen special tools.

> *NOTE —*
>
> *The valve spring compressors shown in Fig. 4 are Volkswagen special tools, designed specifically for use on Volkswagen engines.*

4. If the cylinder head is installed, set the cylinder to bottom dead center (BDC). Remove the spark plug and apply compressed air (at least 87 psi) to the cylinder using an appropriate air hose adapter in the spark plug hole. Remove the valve springs as shown in Fig. 4.

> *NOTE —*
>
> - *Lightly tap the top of the valve with a hammer to loosen sticky valve keepers.*
> - *On diesel engines, it is not necessary to use compressed air. Instead, set the piston to TDC before removing the valve springs. The valve can rest on the piston at TDC.*

5. Remove the old oil seals from the valve guides.

6. Install each new oil seal by temporarily fitting a protective plastic cap over the valve stem end. Then hand-press the seal onto the guide as shown in Fig. 5. Remove the protective cap for use on the next valve.

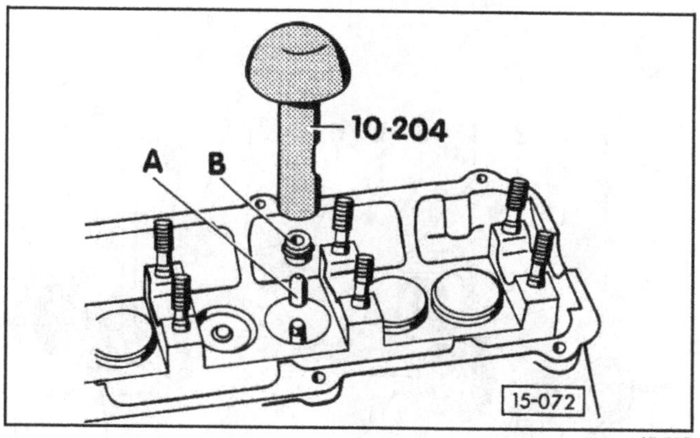

15-072

Fig. 5. Protective cap (**A**), and new valve stem oil seal (**B**).

7. Reinstall the valve springs. The closely spaced coils of the outer valve springs face down, toward the lower spring seats.

8. Repeat the procedure for each pair of valves. Assembly is the reverse of disassembly.

> *CAUTION —*
>
> *To prevent cylinder head or camshaft damage, be sure to follow the camshaft installation procedure when tightening the camshaft bearing caps. See* **To remove and install camshaft**.

Tightening torques

- Spark plug to cylinder head
  - ABA engine . . . . 30 Nm (22 ft-lb)
  - ACC engine . . . . 25 Nm (18 ft-lb)

CAMSHAFT DRIVE BELT (gasoline engines)

The camshaft drive belt and its related parts are shown in Fig. 6. Although no maintenance interval is specified by Volkswagen, the publisher recommends periodic inspection of the belt, and replacement at 60,000 miles, or every 4 years. This will help prevent stripping of the belt due to its stretching and loosening.

> *NOTE —*
>
> *Diesel engine camshaft drive belt replacement can be found in* **23a** *or* **23b Fuel Injection–Diesel**.

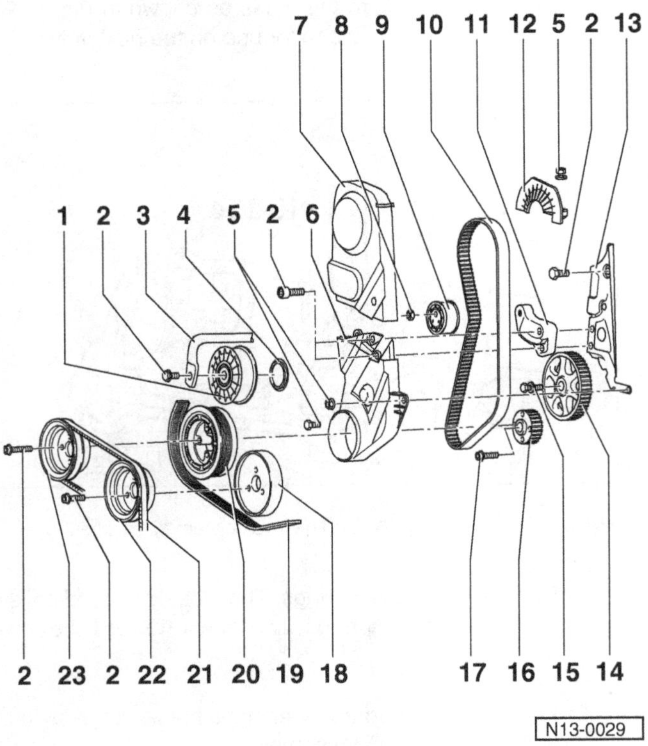

1. **Tensioner**
2. **Bolt**
   - tighten to 20 Nm (15 ft-lb)
3. **Tensioner lever**
4. **Dust sleeve**
5. **Bolt/nut**
   - tighten to 10 Nm (89 in-lb)
6. **Camshaft drive belt cover, lower**
7. **Camshaft drive belt cover, upper**
8. **Nut**
   - tighten to 45 Nm (33 ft-lb)
9. **Tensioner pulley**
   - fixed tension roller shown
   - some vehicles equipped with thermo tension roller, see Fig. 9
10. **Drive belt**
    - mark running direction before removing
11. **Mounting bracket**
12. **Drive belt guard**
13. **Drive belt guard, rear**
14. **Intermediate shaft sprocket**
15. **Bolt**
    - tighten to 80 Nm (59 ft-lb)
    - use VW holding tool no. 3036 when loosening or tightening
16. **Crankshaft sprocket**
17. **Bolt (stretch bolt—always replace)**
    - tighten to 90 Nm (66 ft-lb) plus 1/4 turn
    - lubricate threads and shoulder before installing
    - use VW holding tool no. 3099 when loosening or tightening
18. **Coolant pump pulley**
19. **Poly-ribbed belt**
20. **Vibration damper/pulley**
21. **V-belt**
22. **Coolant pump pulley**
23. **Crankshaft outer pulley**

N13-0029

Fig. 6. Camshaft drive belt and cover assembly.

To remove camshaft drive belt (gasoline engines)

1. Loosen the coolant pump pulley mounting bolts. Mark the engine rotation direction on the poly-ribbed drive belt. Remove the V-belt and the poly-ribbed belt from the engine. See **0 Maintenance Program**.

2. Unclip and remove the camshaft drive belt upper cover.

3. Using a socket wrench on the crankshaft vibration damper/pulley center bolt, rotate the engine clockwise by hand to set the No.1 cylinder at Top Dead Center (TDC). See Fig. 7 or Fig. 8.

> ***CAUTION —***
> *Rotating the crankshaft or the camshaft with the drive belt removed may cause interference which can damage pistons and valves.*

4. Remove the four mounting bolts for the crankshaft outer pulley. Remove outer pulley (item -23- above) and vibration damper/pulley (item -20- above).

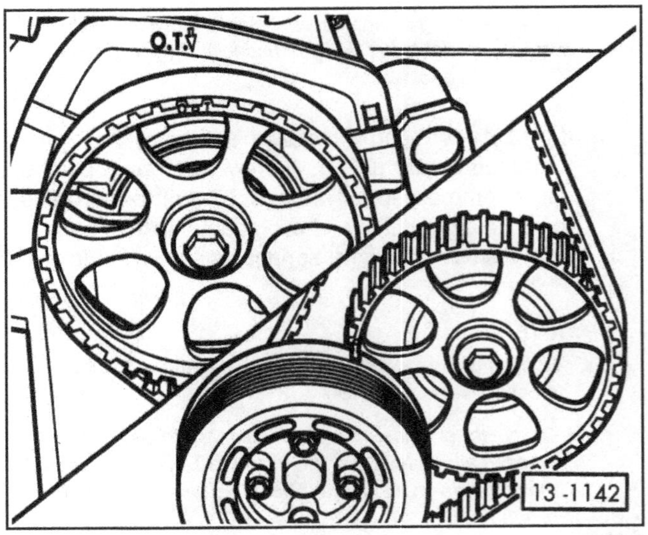

13-1142

Fig. 7. **ACC (1.8 liter) engine**: turn crankshaft so that "OT" mark on sprocket aligns with "OT" mark on rear cover. Check also that mark on crankshaft vibration damper/pulley aligns with mark on intermediate shaft sprocket.

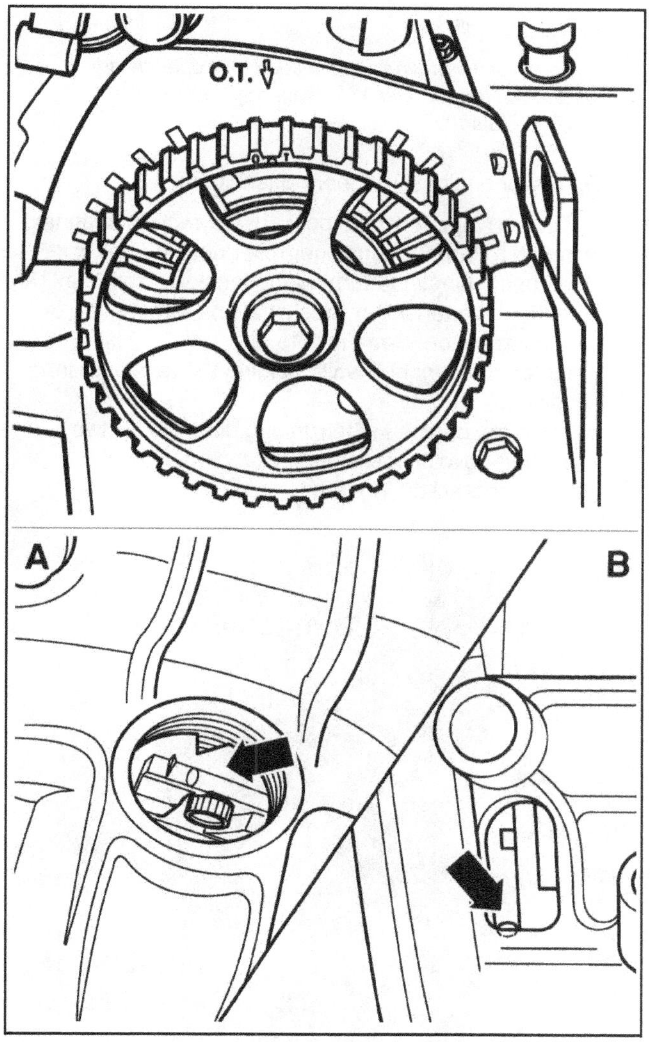

0024011

Fig. 8. **ABA (2.0 liter) engine**: turn crankshaft so that "0T" mark on sprocket aligns with "0T" mark on rear cover (top). Check also that mark on flywheel (**A**) or driveplate (**B**) aligns with mark on bellhousing.

5. Remove the lower camshaft drive belt cover mounting bolts and remove the lower cover.

6. On engines with fixed tensioner pulley (secured with only a nut): loosen pulley nut and turn the pulley counterclockwise to relieve the belt tension. If necessary, use a spanner wrench to turn the pulley.

7. On newer engines with thermo tensioner: loosen Torx bolt on tensioner roller. See Fig. 11. Remove the belt from the sprockets.

NOTE —

Be sure to inspect the tensioner while the belt is removed. Spin the bearing to check that it runs smoothly without play. A damaged or high-mileage tensioner should be replaced.

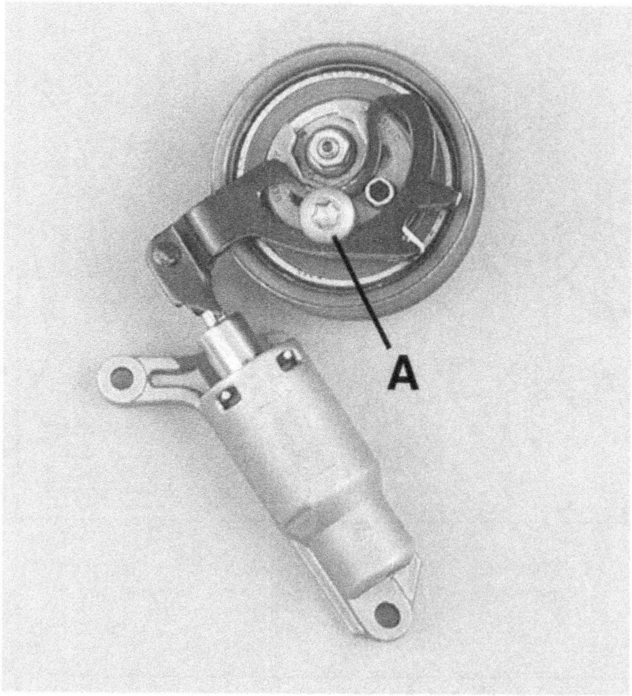

0024182

Fig. 9. Camshaft drive belt thermo tensioner on late models. Torx bolt shown at **A**.

To install camshaft drive belt (gasoline engines)

1. Check that the crankshaft is set at TDC.
   - On ABA engines, check that the timing marks are aligned as shown in Fig. 8.
   - On ACC engines, temporarily install the crankshaft vibration damper/pulley to the crankshaft using at least one mounting bolt. Check that the timing marks are aligned as shown in Fig. 7.

NOTE —

The crankshaft pulley mounting holes are offset, which means that the sprocket/pulley can only go on in one position.

2. Remove the distributor cap and check that the ignition rotor is aligned with the mark on the distributor body. See Fig. 10. If necessary, turn the intermediate shaft to align the mark.

3. Install the camshaft drive belt. Work the belt onto the sprockets so that all belt slack ends up at the tensioner.

4. On engines with fixed tensioner pulley (secured with only a nut): turn the tensioner clockwise using a spanner wrench and tighten the locknut. Check belt tension by twisting the belt. The belt is properly tensioned if it can be twisted by hand no more than 90°. See Fig. 11.

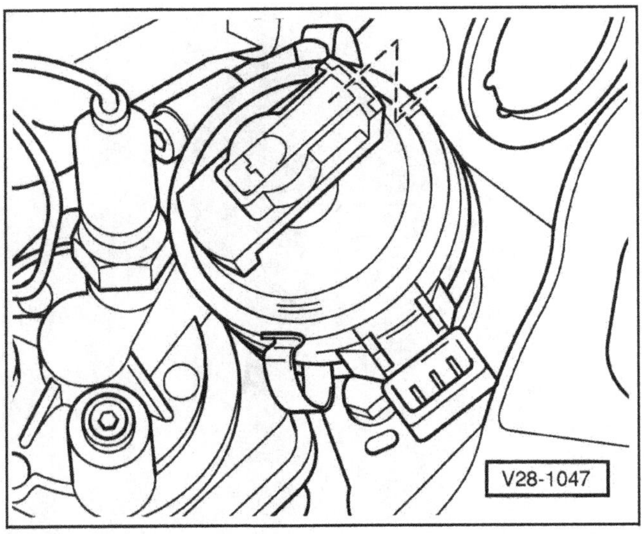

V28-1047

Fig. 10. Ignition rotor aligned with No.1 cylinder TDC mark on distributor housing. ABA engine shown; ACC engine is similar.

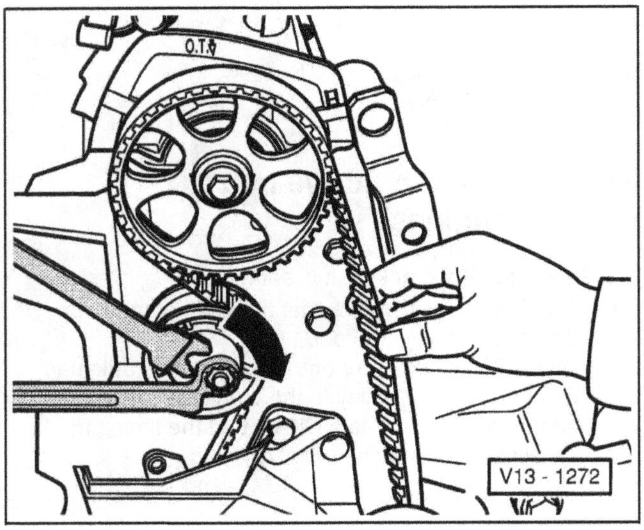

V13-1272

Fig. 11. Camshaft drive belt tension being adjusted. Check tension halfway between the camshaft and intermediate shaft sprockets. VW special tool no. 2587 (spanner wrench) being used to engage holes in tensioner.

Tightening torque

- Drive belt tensioner locknut . . . . 45 Nm (33 ft-lb)

NOTE —

Some movement of the sprockets and their marks is to be expected as belt tension is adjusted. Keep in mind that the smallest possible increment of adjustment is one whole tooth of the belt or sprocket.

NOTE —

A two-pin spanner wrench is needed to accurately adjust the belt tension. Use Volkswagen special tool 2587, 3387, or US 4493.

5. On newer engines with thermo tensioner:
   - With the roller Torx bolt loose, use a two-pin spanner wrench to turn the tensioner roller until the belt is tensioned. Continue turning the tensioner roller until the tip of the tensioner bracket aligns with the top of the outer piston. See Fig. 11.
   - Tighten the Torx bolt while holding the tensioner in position.
   - Rotate the engine in the running direction for two complete revolutions. Re-check the position of the tensioner bracket. Tighten torx bolt to final torque.

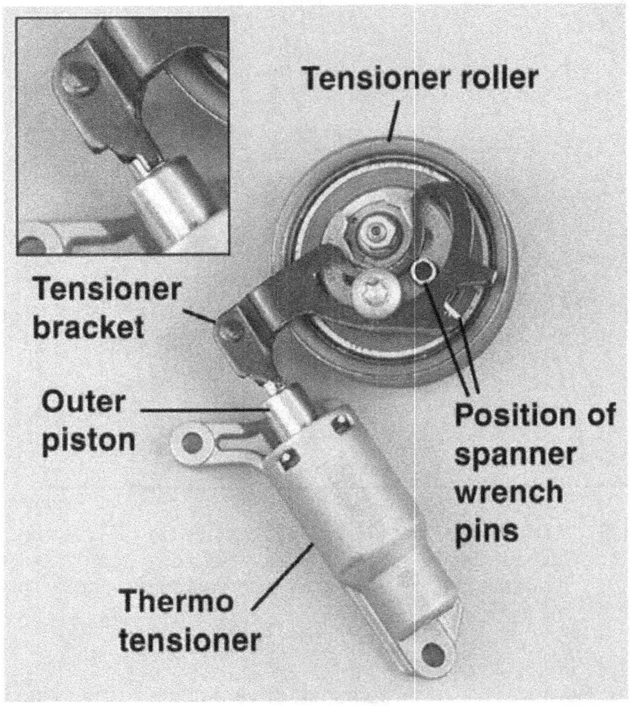

0024182a

Fig. 12. Camshaft drive belt thermo tensioner shown in correctly adjusted position (inset). When installed on the engine, spring tension of inner piston keeps the belt tight. Thermo tensioner moves outer piston to compensate for changes in belt tension due to heat expansion of the engine.

Tightening torques

- Tensioner bracket to tensioner roller (Torx bolt) . . . . .25 Nm (18 ft-lb)
- Tensioner housing to engine. . . . . 7 to 11 Nm (62 to 97 in-lb)

---

6. The remaining installation is the reverse of removal. Use Fig. 6, given earlier, as an installation guide and for bolt tightening torques. Install the engine drive belts as described in **0 Maintenance Program**.

NOTE—

On ACC engines, check and adjust ignition timing as described in **28b Ignition System–Gasoline**.

CYLINDER HEAD REMOVAL AND INSTALLATION

Fig. 13 shows a view of the 2.0 liter cylinder head and related components. Note that the cylinder head bolts are stretch-type fasteners and should not be reused.

NOTE—

When replacing the cylinder head or the cylinder head gasket, it is recommended that the old coolant be flushed out and new coolant added. This will help ensure that any metal shavings and contaminants are removed from the engine and the cooling system. See **19 Engine–Cooling System**.

To remove cylinder head

1. Disconnect the battery negative (–) terminal from the battery.

NOTE—

Be sure to have the anti-theft radio code on hand before disconnecting the battery.

2. On gasoline engines, disconnect the intake air duct from the throttle body on the intake manifold.

3. On diesel engines, remove the air cleaner and the intake air duct. Remove the turbocharger as described in **21 Turbocharger**.

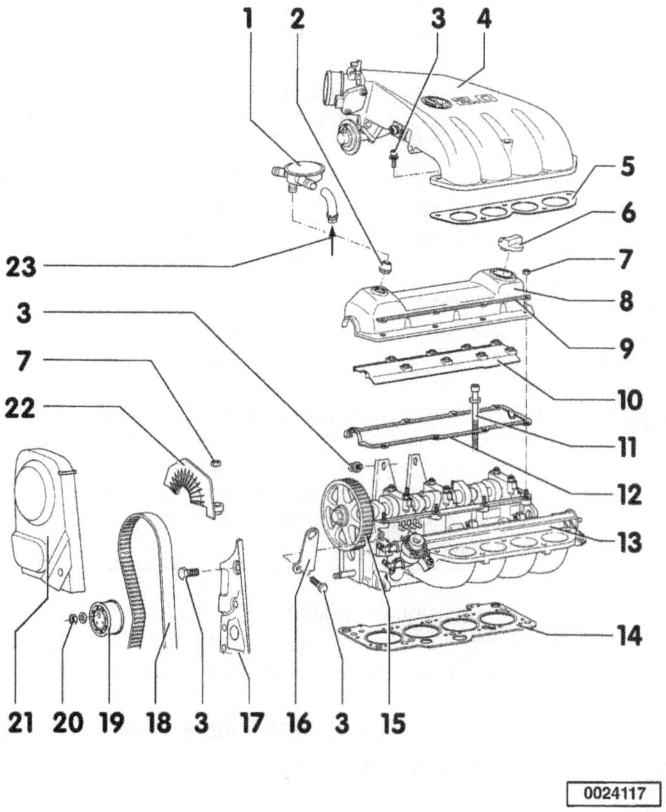

1. **Crankcase pressure regulating valve**
2. **Grommet**
3. **Bolt**
   - tighten to 20 Nm (15 ft-lb)
4. **Intake manifold, upper**
5. **Intake manifold gasket**
   - always replace
6. **Oil filler cap**
7. **Nut**
   - tighten to 10 Nm (89 in-lb)
8. **Cylinder head cover**
9. **Reinforcing strip**
10. **Oil deflector**
11. **Cylinder head bolt**
    - stretch bolt–**always replace**
    - tighten in stages, see Fig. 12
12. **Cylinder head cover gasket**
13. **Cylinder head**
    - check for distortion, see **Cylinder Head Reconditioning**
    - use new coolant when replacing
14. **Cylinder head gasket**
    - use new coolant when replacing
15. **Camshaft sprocket**
16. **Lifting eye**
17. **Drive belt guard, rear**
18. **Drive belt**
19. **Drive belt tensioner**
    - fixed tension roller shown
    - some vehicles equipped with thermo tension roller, see Fig. 9
20. **Drive belt tensioner lock nut**
    - tighten to 45 Nm (33 ft-lb)
21. **Drive belt guard, front**
22. **Drive belt guard, upper**
23. **Hose**

0024117

Fig. 13. Early version ABA (2.0 liter) cylinder head and related components. Other cylinder heads are similar.

4. Drain the coolant as described in **19 Engine–Cooling System**.

> ***WARNING —***
> *Hot coolant can scald. Drain the coolant only with the engine cold.*

5. Remove the camshaft drive belt. For gasoline engines, see **Camshaft Drive Belt (gasoline engines)**. For diesel engines, see **23a** or **23b Fuel Injection–Diesel**.

6. On AAZ Diesel engines, disconnect the accelerator cable and wiring to the glow plug bus. On AHU and 1Z Diesel engines, disconnect the harness connector to the needle lift sensor for cylinder 3 and the harness connector to the glow plug contact bridge. On all Diesel engines, disconnect the fuel injection pipes as described in **23a** or **23b Fuel Injection-Diesel**.

7. On gasoline engines, disconnect the fuel injector harness connectors. Disconnect the throttle cable and harness connectors from the throttle body.

8. On gasoline engines, disconnect and label the hoses and wires connected to the throttle body, the intake manifold and cylinder head. For more information on fuel injection components, see **24 Fuel Injection**.

> ***WARNING —***
> *Fuel will be expelled when disconnecting hoses. Wrap a cloth around the fuel line fittings when disconnecting the lines. Do not smoke or work near heaters or other fire hazards. Have a fire extinguisher handy.*

9. Disconnect the exhaust pipe from the exhaust manifold. See **26 Exhaust System/Emission Controls**.

10. Disconnect all coolant hoses from the outlets on the cylinder head.

11. On gasoline engines, disconnect the spark plug connectors from the spark plugs.

12. Remove the cylinder head cover as described earlier.

13. Gradually and evenly loosen the cylinder head bolts, beginning with the outer bolts and working toward the center. Discard the head bolts.

14. Remove the cylinder head. If the head is stuck, use a soft-faced mallet or pry gently with a wooden stick to loosen it.

To install cylinder head

1. Clean the cylinder head and the gasket surface of the cylinder block. Clean the bolt holes with thread chasers and remove all foreign matter and liquid from the holes.

> ***CAUTION —***
> *Do not use a metal scraper or wire brush to clean the aluminum cylinder head. Instead, use a solvent to soften carbon deposits and old sealing materials. If necessary, use a wooden or plastic scraper.*

NOTE —

- *If installing a VW replacement cylinder head (complete with camshaft), check to see if protective plastic liners are installed for protection of the open valves. If installed, be sure to remove these liners prior to installing the head.*
- *If the camshaft was removed or if installing a new or remanufactured cylinder head, lubricate the contact surfaces between the cam lobes and the hydraulic cam followers prior to start-up.*

2. Check the gasket surfaces of the cylinder head and the cylinder block for warpage. See **Cylinder Head Reconditioning**.

3. On AAZ diesel engines, reinstall new combustion prechamber inserts, if they were removed during cleaning.

4. Place a new cylinder head gasket on the surface of the cylinder block. The word “OBEN”, found imprinted on the gasket near the part number, should face up.

> ***CAUTION —***
> *Three different cylinder head gaskets are used on diesel engines, identified by a number of notches or holes in the gasket. When reassembling the same cylinder head and cylinder block, make sure that the new gasket matches the one it replaces. When assembling a new cylinder head, a new or machined cylinder block, the correct gasket must be selected by measuring piston protrusion. See* **13 Crankshaft/Cylinder Block**.

5. On diesel engines, position the crankshaft to the TDC marking. Then turn the crankshaft opposite engine rotation until all pistons are at nearly equal height.

6. Place the cylinder head in position on the cylinder block. Loosely install all the head bolts.

NOTE —
To ensure proper cylinder head alignment during installation, place two 200 mm long by 11 mm (gasoline engine) or 12 mm (diesel engine) diameter wooden dowels into two of the outermost head bolt holes to hold the gasket and serve as guides as the cylinder head is installed. Install the head and thread in several bolts, then remove the dowels and install the remaining bolts.

7. Tighten the new head bolts in stages using the sequence shown in Fig. 14.

> *CAUTION —*
> *Always replace the cylinder head mounting bolts. They are stretch bolts designed to be used only once.*

Tightening torques
- Cylinder head to cylinder block (stretch bolts–**always replace**)
  - stage I . . . . 40 Nm (30 ft -lb)
  - stage II. . . . . 60 Nm (44 ft -lb)
  - stage III . . . . .additional 1/4 turn (90°)
  - stage IV . . . . .additional 1/4 turn (90°)

8. Installation of the remaining parts is the reverse of removal. See **Camshaft Drive Belt (gasoline engines)**, **23a** or **23b Fuel Injection–Diesel, 24 Fuel Injection, 26 Exhaust System/Emission Controls**, and **19 Engine–Cooling System**.

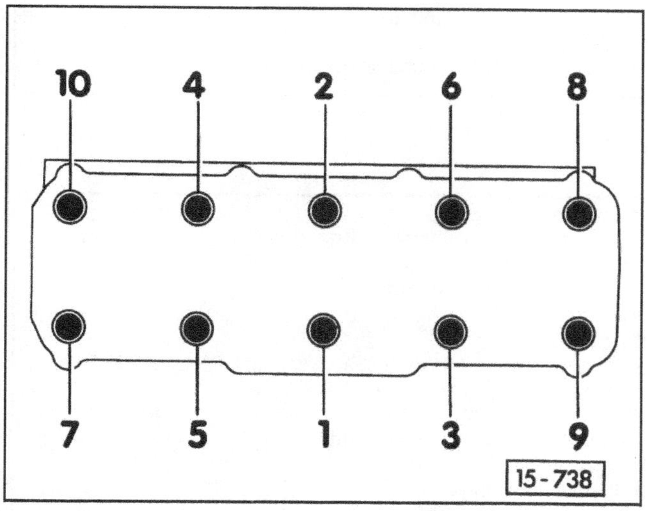

15-738

Fig. 14. Cylinder head bolt tightening sequence.

9. On AAZ Diesel engines, if the old style composition head gasket is used, run the engine until it reaches normal operating temperature. Then retorque the cylinder head bolts in proper sequence an additional 1/4 turn (90°). AAZ, AHU and 1Z Diesel engines that use the newer style metal gasket DO NOT require this step. Retorqueing after the first 1000 miles is NOT needed for either style of head gasket.

Tightening torque
- Cylinder head to cylinder block (AAZ diesel with composition head gasket only)
  - stage V . . . . .additional 1/4 turn (90°)

CYLINDER HEAD RECONDITIONING

This section provides the specifications and special reconditioning information necessary to repair the Volkswagen cylinder head. Special service tools and machine shop services are required for most cylinder head reconditioning.

Fig. 15 through Fig. 17 show the cylinder head and valve train assemblies found on the Golf, Jetta and GTI models. Note that most of the reconditioning specifications are contained within the illustrations.

> *NOTE —*
> *The information given under this reconditioning heading assumes the cylinder head is removed. For cylinder head removal procedures, see* **Cylinder Head Removal and Installation.**

Cylinder Head and Camshaft

Cylinder heads with small, fine cracks between valve seats and plug threads are usable provided the cracks are not more than 0.5 mm (0.02 in.) wide, and do not extend into more than the first few spark plug threads.

If the camshafts were removed or if installing a new cylinder head, lubricate the contact surfaces between the cam lobes and the hydraulic cam followers prior to start-up. When removing and installing the camshaft, be sure to follow the procedure given under **To remove and install camshaft**. Incorrect camshaft removal or installation can result in a damaged camshaft or cylinder head.

> *CAUTION —*
> *Three different cylinder head gaskets are available for diesel engines; identified by a number of notches or holes in the gasket. When reassembling the same cylinder head and cylinder block, make sure the new gasket matches the one it replaces. When assembling a new or machined cylinder block, the correct gasket must be selected by measuring piston protrusion. See* **13 Crankshaft/Cylinder Block**.

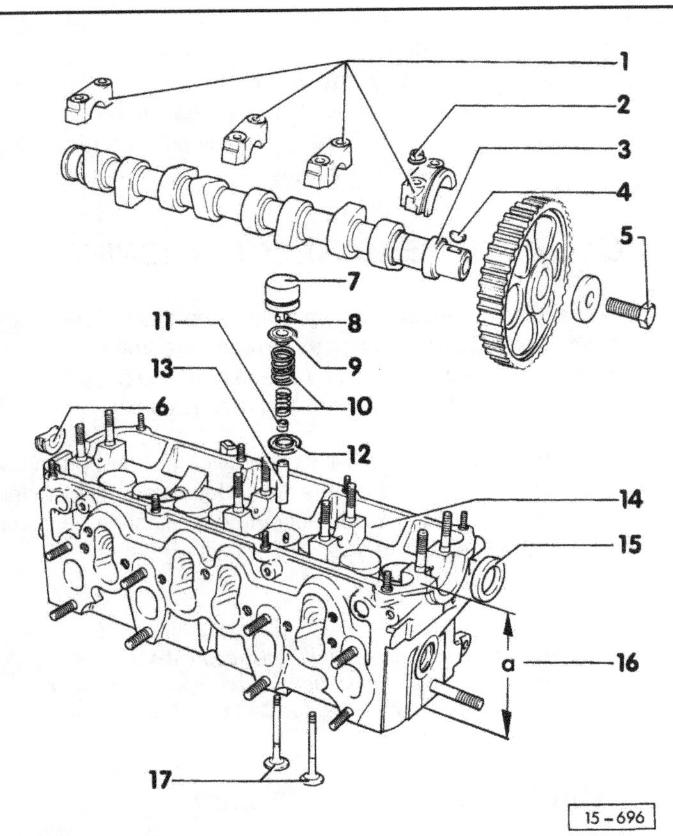

1. **Camshaft bearing cap**
   - note center offset when installing. See Fig. 16.
2. **Camshaft bearing cap nut**
   - 20 Nm (15 ft-lb)
3. **Camshaft**
   - radial clearance (Plastigage®) - wear limit: 0.1 mm (0.004 in.)
   - runout - wear limit: 0.01 mm (0.0004 in.)
   - axial clearance (measure with cam followers removed) wear limit: 0.15 mm (0.0059 in.)
4. **Woodruff key**
   - check for tight fit
5. **Camshaft sprocket mounting bolt**
   - tighten to 80 Nm (60 ft-lb)
6. **Rubber end cap**
7. **Hydraulic cam followers**
   - do not interchange, store with contact surface down
   - oil contact surface when installing
8. **Valve keepers**
9. **Valve spring seat, upper**
10. **Valve springs**
11. **Valve stem oil seal**
12. **Valve spring seat, lower**
13. **Valve guide**
14. **Cylinder head**
    - maximum distortion (warpage): 0.1 mm (0.004 in.)
    - use new coolant when replacing
15. **Camshaft oil seal**
16. **Cylinder head resurfacing height (a)**
    - minimum 132.6 mm (5.220 in.)
17. **Valves**
    - do not reface by grinding, hand-lap only

15-696

Fig. 15. Exploded view of ACC (1.8 liter) cylinder head.

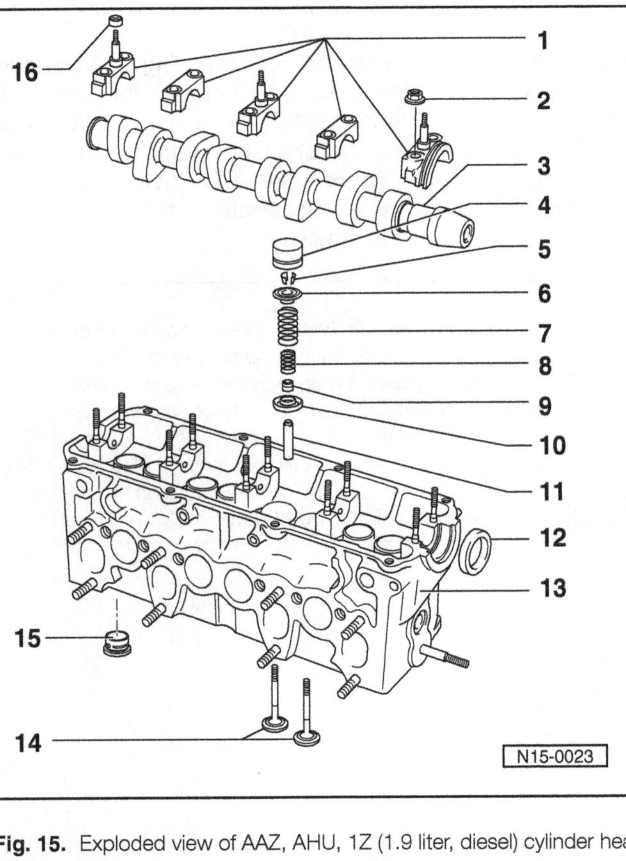

1. **Camshaft bearing cap**
   - note center offset when installing. See Fig. 16.
   - do not interchange
2. **Camshaft bearing cap nut**
   - 20 Nm (15 ft-lb)
3. **Camshaft**
   - radial clearance (Plastigage®) - wear limit: 0.11 mm (0.0043 in.)
   - runout - wear limit: 0.01 mm (0.0004 in.)
   - axial clearance (measure with cam followers removed) wear limit: 0.15 mm (0.0059 in.)
4. **Hydraulic cam followers**
   - do not interchange
   - store with contact surface down
   - oil contact surface when installing
5. **Valve keepers**
6. **Valve spring seat, upper**
7. **Valve spring, outer**
8. **Valve spring, inner**
9. **Valve stem oil seal**
10. **Valve spring seat, lower**
11. **Valve guide**
12. **Camshaft oil seal**
13. **Cylinder head**
    - maximum distortion (warpage): 0.1 mm (0.004 in.)
    - resurfacing diesel cylinder head is not permitted
    - use new coolant when replacing
14. **Valves**
15. **Swirl chamber**
16. **Seal**

N15-0023

Fig. 15. Exploded view of AAZ, AHU, 1Z (1.9 liter, diesel) cylinder head.

CYLINDER HEAD RECONDITIONING

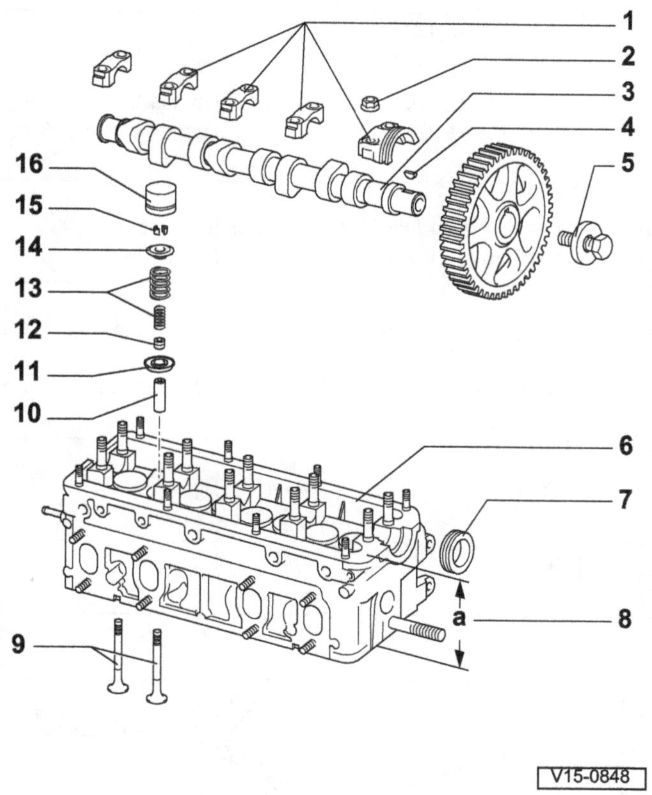

1. **Camshaft bearing cap**
   - note center offset when installing. See Fig. 16.
2. **Camshaft bearing cap nut**
   - 20 Nm (15 ft-lb)
3. **Camshaft**
   - radial clearance (Plastigage®) wear limit 0.1 mm (0.004 in.)
   - runout: wear limit 0.01 mm (0.0004 in.)
   - axial clearance (measure with cam followers removed) wear limit 0.15 mm (0.0059 in.)
4. **Woodruff key**
   - check for tight fit
5. **Camshaft sprocket mounting bolt**
   - tighten to 80 Nm (60 ft-lb)
6. **Cylinder head**
   - maximum distortion (warpage): 0.1 mm (0.004 in.)
   - use new coolant when replacing
7. **Camshaft oil seal**
8. **Cylinder head resurfacing height (a)**
   - minimum 132.6 (5.220 in.)
9. **Valve**
   - do not reface by grinding, hand-lap only
10. **Valve guide**
11. **Valve spring seat, lower**
12. **Valve stem oil seal**
13. **Valve springs**
14. **Valve spring seat, upper**
15. **Valve keepers**
16. **Hydraulic cam follower**
    - do not interchange, store with contact surface down
    - oil contact surface when installing

V15-0848

Fig. 16. Exploded view of ABA (2.0 liter) cylinder head.

> ***CAUTION—***
> *Some later model 4-cylinder engines do not have inner valve springs and lower valve spring seats. Check with an authorized Volkswagen parts department for the latest information.*

Camshaft and Cam Followers

Camshaft bearing caps or cam followers should not be interchanged. Note the bearing cap offset during assembly. See Fig. 16.

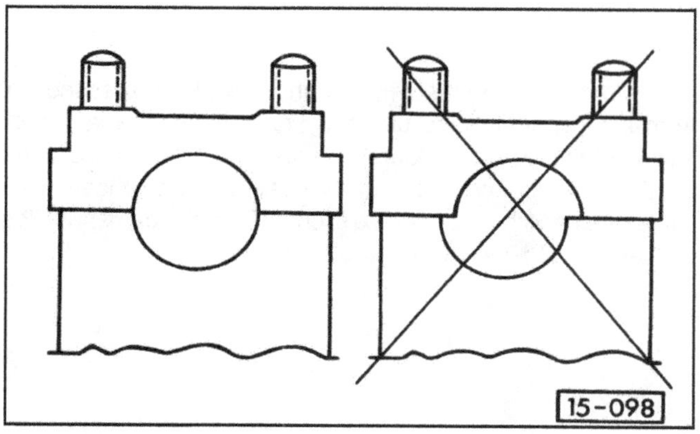

15-098

Fig. 16. Camshaft bearing caps are machined offset and fit only one way.

> ***CAUTION—***
> *After installing new cam followers, the engine should not be started for at least 30 minutes. New cam followers are at full height and must be allowed to bleed down to their proper height. Failure to do this may cause valve or piston damage.*

Valves

Diesel engine exhaust valves and all gasoline engine valves are sodium-filled and should be hand-lapped only. Only intake valves in diesel engines may be refaced by machine. Valve dimensions are given in **Table b**. See Fig. 17.

> ***WARNING—***
> *Sodium filled valves, when discarded, must be disposed of properly to avoid personal injury. Always wear protective goggles or glasses. By hand, cut off the valve stem near the head of each valve. Use only a hack saw. Do not use a power saw. Sodium reacts violently with water. Do not let water contact the valve while cutting. Throw the valve parts (no more than 10 valves at a time) into a bucket of water and stand clear. Discard the valves when the reaction has stopped.*

Table b. Valve Dimensions

Engine Code	ACC	AAZ	AHU, 1Z	ABA
Valve head diameter (**a**) intake exhaust	38.0 mm (1.496 in.) 33.0 mm (1.299 in.)	36 mm (1.417 in.) 31.0 mm (1.220 in.)	35.95 mm (1.415 in.) 31.45 mm (1.238 in.)	39.5±0.15 mm (1.555±0.006 in.) 32.9±0.15 mm (1.295±0.006 in.)
Valve stem diameter (**b**) intake exhaust	7.97 mm (0.3138 in.) 7.95 mm (0.3130 in.)	7.97 mm (0.3138 in.) 7.95 mm (0.3130 in.)	7.97 mm (0.3138 in.) 7.97 mm (0.3138 in.)	6.97 mm (0.2744in.) 6.95 mm (0.2736 in.)
Valve length (**c**) intake exhaust	91.9+0/–0.90 mm (3.6181 0/–0.0354 in.) 91.2+0/–0.40 mm (3.5906+0/–0.0157 in.)	95.0 mm (3.740 in.) 95.0 mm (3.740 in.)	96.85 mm (3.813 in.) 96.85 mm (3.813 in.)	91.85 mm (3.6161 in.) 91.15 mm (3.5886 in.)
Valve face angle (α) intake exhaust	45° 45°	45° 45°	45° 45°	44.4° 45.15°
Valve margin	N/A	0.5 mm (0.020 in.) (intake valves only)	N/A	N/A

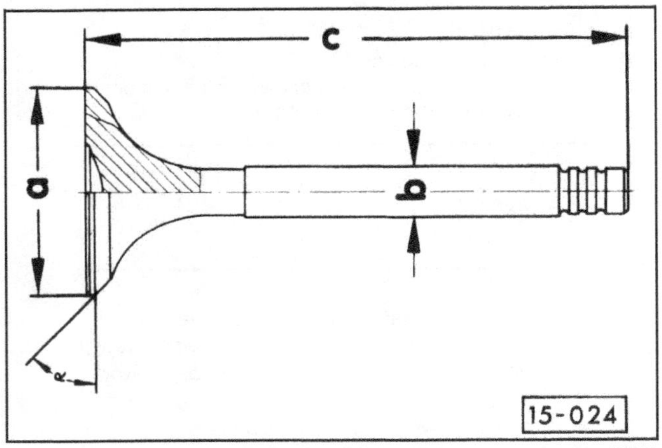

15-024

Fig. 17. Valves dimensions to be used with **Table b**.

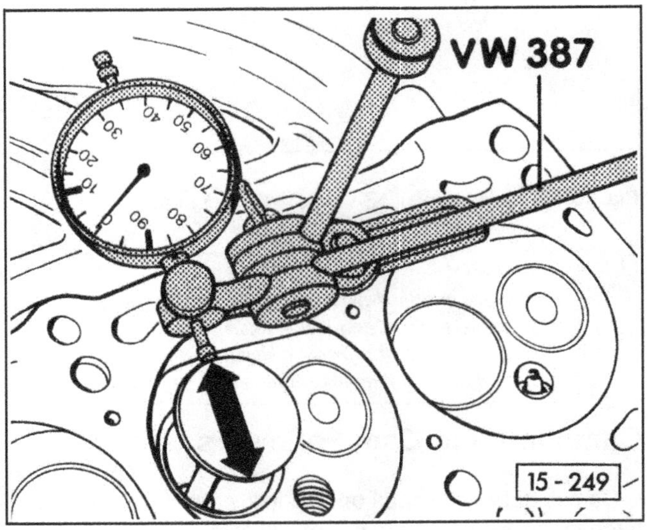

15-249

Fig. 18. Valve guide wear being checked with new valve. Insert valve until stem end is flush with end of guide. Rock valve back and forth to check total travel.

Valve Guides

Special tools and a press are required to replace valve guides. Check valve guide wear with a new valve. See Fig. 18. Inspect the valve seats to ensure that the cylinder head can be reconditioned before installing new valve guides.

Valve Guide (wear limits–maximum play)

- gasoline engines
  - intake valve guide . . . . 1.0 mm (0.039 in.)
  - exhaust valve guide . . . . 1.3 mm (0.051 in.)
- diesel engines
  - intake and exhaust valve guides . . 1.3 mm (0.051 in.)

Original valve guides (without shoulders) are pressed out from the camshaft side of the cylinder head. Replacement valve guides (with shoulders) are pressed out from the combustion chamber side of the cylinder head. Lubricate new valve guides with oil and press them in from the camshaft side until the shoulder is fully seated.

NOTE —

- *Where available, replace valve guides using Volkswagen special tools. Use of other press tools may damage the cylinder head.*
- *On ACC, AAZ, AHU and 1Z engines, use VW tool no. 10-206 for installation and removal. Ream the valves guides using VW special tool 10-215 and appropriate valve cutting fluid.*
- *On ABA engines, use VW tool no. 3121 for installation and removal. Ream the valves guides using VW special tool no. 3120.*

Valve Seats

When resurfacing valve seats, there is a limit to the amount of material that can be removed. If too much material is removed, the final assembly will leave too little space for the hydraulic cam follower to function correctly. The maximum refacing dimension—the maximum amount of material that can be removed from the valve seat—is calculated from the measurement shown. See Fig. 19.

Measure dimension **a** in Fig. 21, and subtract the minimum dimension, as given in **Table c**. The difference is the maximum amount of material that can be removed from the valve seat.

Table c. Minimum Dimensions for Calculating Valve Seat Refacing Dimensions

Engine Code	Intake	Exhaust
ACC, ABA	33.8 mm (1.331 in.)	34.1 mm (1.343 in.)
AAZ, AHU, 1Z	35.8 mm (1.409 in.)	36.1 mm (1.421 in.)

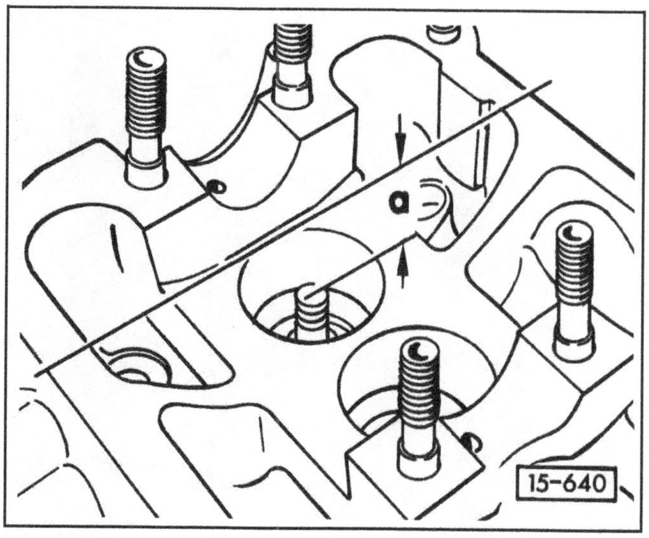

15-640

Fig. 19. Dimension (**a**), distance between top of valve stem and gasket surface of cylinder head, is used to calculate maximum valve seat refacing dimensions.

Valve seat dimensions are given in the tables **Table d** through **Table g.**

Table d. Valve Seat Dimensions: ACC Engine

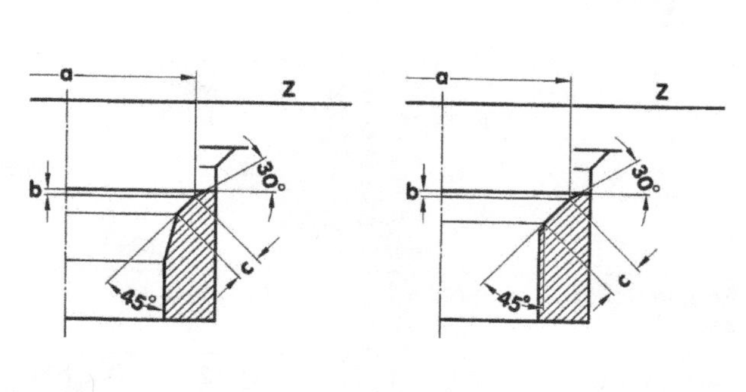

Engine Code: ACC	Intake	Exhaust
Seat diameter (**a**)	37.2 mm (1.465 in.)	32.4 mm (1.276 in.)
Maximum refacing dimension (**b**)	Calculated, See Fig. 19.	Calculated, See Fig. 19.
Seat width (**c**)	approx. 2.0 mm (0.079 in.)	approx. 2.4 mm (0.094 in.)
Valve seat angle	45°	45°
Correction angle, upper	30°	30°

Table e. Valve Seat Dimensions: AAZ Engine

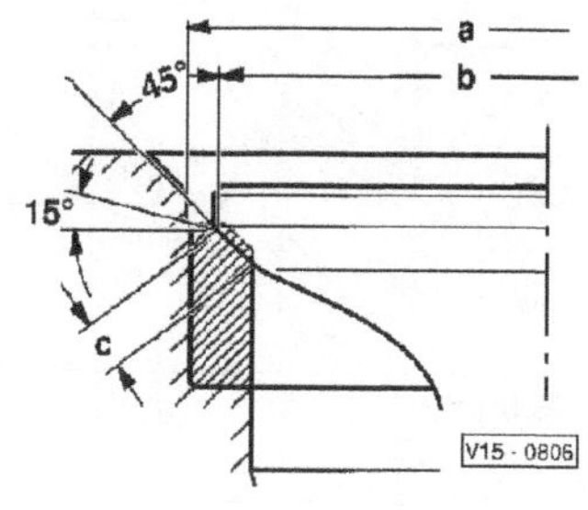

Engine Code: AAZ	Intake	Exhaust
Valve seat insert bore diameter (**a**)	37.20 mm (1.4646 in.)	33.20 mm (1.3071 in.)
Seat diameter (**b**)	34.80 mm (1.3701 in.)	30.40 mm (1.1968 in.)
Seat width (**c**)	2.70 mm (0.1063 in.)	2.05 mm (0.0807 in.)
Valve seat angle	45°	45°
Correction angle, upper	15°	15°

Table f. Valve Seat Dimensions: ABA Engine

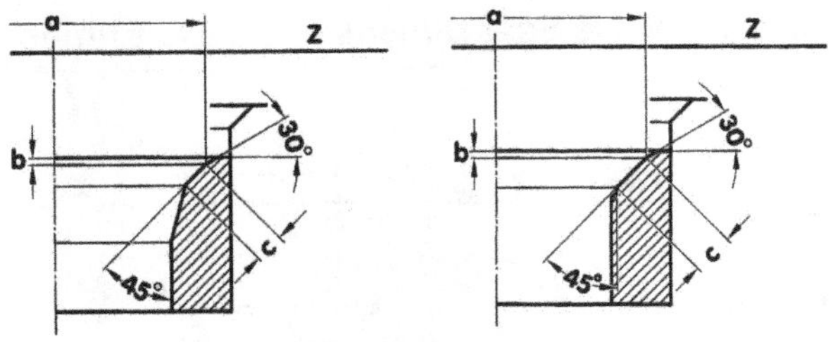

Engine Code: ABA	Intake	Exhaust
Seat diameter (**a**)	39.2 mm (1.543 in.)	32.4 mm (1.276 in.)
Maximum refacing dimension (**b**)	Calculated, see Fig. 19	Calculated, see Fig. 19
Seat width (**c**)	approx. 2.0 mm (0.079 in.)	approx. 2.4 mm (0.094 in.)
Valve seat angle	45°	45°
Correction angle, upper	30°	30°

Table g. Valve Seat Dimensions: AHU, 1Z Engine

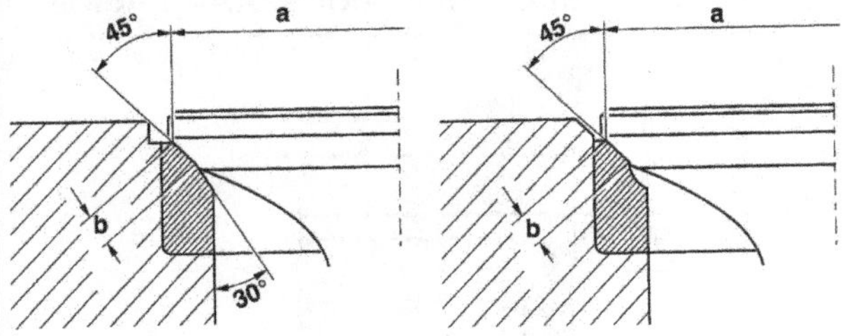

Engine Code: AHU, 1Z	Intake	Exhaust
Seat diameter (**a**)	35.7 mm (1.406 in.)	31.4 mm (1.236 in.)
Seat width (**b**)	1.6 mm (0.063 in.)	2.7 mm (0.106 in.)
Valve seat angle	45°	45°

NOTE —

The 30° chamfer on the intake valve seat is necessary to ensure that the intake channel flow characteristics are maintained.

15b Cylinder Head and Valvetrain–(6-Cylinder Engine)

GENERAL

This engine section covers most cylinder head and valvetrain service and repair for the 6-cylinder VR6 engine. For information on short block engine rebuilding and internal engine specifications, see **13 Crankshaft/Cylinder Block**.

Some of the operations described in this repair group require special equipment and experience. If you lack the skills, tools, or a suitable workplace for servicing or repairing the cylinder head, we suggest you leave these repairs to an authorized Volkswagen dealer or other qualified shop.

DIAGNOSTIC TESTING

The tests that follow can be used to help isolate engine problems, to better understand a problem before starting expensive repairs, or just to periodically check engine condition.

Compression Test

A compression test will tell a lot about the condition of the engine without the need for taking it apart. The test is relatively simple. On gasoline engines, it requires only a compression tester and a spark plug wrench. Diesel engines with their higher compression pressures (approximately 500 psi) require a more expensive compression tester. For the most accurate test results, the battery should be fully charged and the engine should be warm.

> ***NOTE —***
> *Because engine temperature may affect compression, the most accurate results are obtained when the engine is at normal operating temperature.*

To test compression

1. With the ignition off, disable the ignition system by disconnecting the 5-pin harness connector from the coil on the back of the cylinder head.

2. Remove the spark plug wires from the spark plugs.

> ***NOTE —***
> *A special tool (VW tool no. 3277) is needed to remove the spark plug connectors from the plugs without damaging them. A version of this tool may be clipped to the front hood support rod.*

3. Clean the area around the spark plugs using compressed air. Remove the spark plugs and set them aside in order.

> ***NOTE —***
> *Due to the limited clearance around the recessed spark plugs, VW special tool no. 3122B is available to easily remove and install the plugs.*

4. Thread the compression tester into the first cylinder's spark plug hole, just tight enough to seal the plug hole.

5. With the transmission in neutral and the throttle held wide open, crank the engine with the starter about 4 to 5 revolutions. The gauge reading should increase with each engine revolution. Record the gauge value.

6. Release the pressure in the tester and remove it from the spark plug hole.

7. Repeat the test for each of the other cylinders. Record the data and compare with **Table a**.

Table a. Compression Pressure

Engine	**Code: AAA (2.8 L)**
new engine	145–189 psi (10–13 bar)
wear limit	109 psi (7.5 bar)
maximum difference between cylinders	44 psi (3 bar)

8. Reinstall the spark plugs and the spark plug wires in their original locations. Reconnect the coil harness connector.

Tightening torque
- Spark plug to cylinder head
  AAA engine . . . . 25 Nm (18 ft-lb)

Low compression is evidence of poorly sealed combustion chambers. Compression pressures which are relatively even but below specifications indicate worn piston rings and/or cylinder walls. Low but erratic values tend to indicate valve leakage. Dramatic differences, such as good values in some cylinders and very low values in one or two cylinders are the sign of a localized failure, such as a burnt valve or a failed cylinder head gasket.

Leak-down Test

The most conclusive diagnosis of low compression symptoms requires a leak-down test. Using a special tester and a supply of compressed air, each cylinder is pressurized. The rate at which the air leaks out of the cylinder, as well as the location of the air escaping, can more accurately pinpoint the magnitude and source of the leakage. Any engine compression diagnosis that will require major disassembly should first be confirmed by a leak-down test. Because the test requires special equipment and experience, it may be desirable to have it performed by a Volkswagen dealer or other qualified repair shop.

CYLINDER HEAD SERVICE

Many cylinder head repairs can be accomplished without removing the cylinder head from the engine. The cylinder head cover gasket, the camshaft, the valve guide oil seals, and the cam followers are all accessible with the cylinder head installed. This heading describes those repairs that can be done with the cylinder head installed.

> ***NOTE —***
> *A special alignment tool (VW special tool no. 3268, order no. T03 267 000 37 ZEL) is specified for most VR6 cylinder head work. The tool is used to accurately set the camshafts at TDC. This plastic tool is relatively inexpensive and should always be used any time the camshaft timing chains are removed from the camshaft sprockets.*

To remove and install cylinder head cover

1. Remove the spark plug connectors from the plugs. Remove the spark plug wire guides with the ignition wires. See Fig. 1.

> ***NOTE —***
> *A special tool (VW tool no. 3277) is needed to remove the spark plug connectors from the plugs without damaging them. A version of this tool is clipped to the front hood support rod.*

2. Remove the intake manifold cover from above the intake manifold.

3. Disconnect the Positive Crankcase Ventilation (PCV) breather valve from the cylinder head cover.

4. Disconnect the PCV hose from the upper intake manifold. See Fig. 2.

5. Remove the intake air boot from the throttle housing.

6. Disconnect the accelerator cable from the throttle body and from its support bracket.

7. Release the pressure in the cooling system by loosening the cap on the coolant reservoir. Then clamp off the coolant hose at the throttle housing and disconnect the hose from the housing.

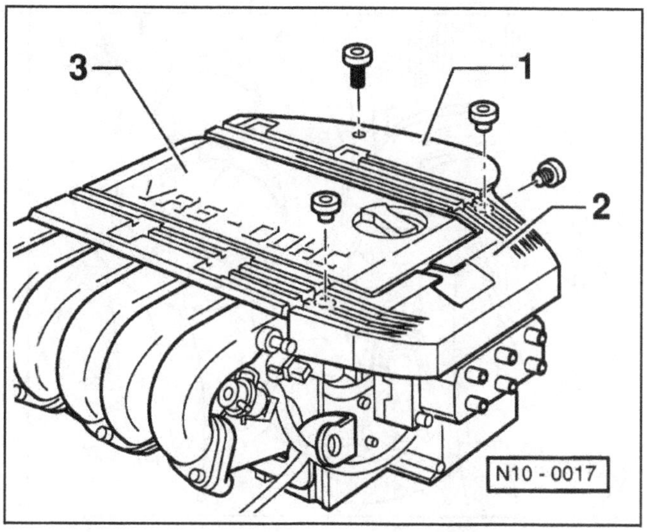

N10-0017

Fig. 1. Spark plug wire guides (**1** and **2**). Intake manifold cover shown at **3**.

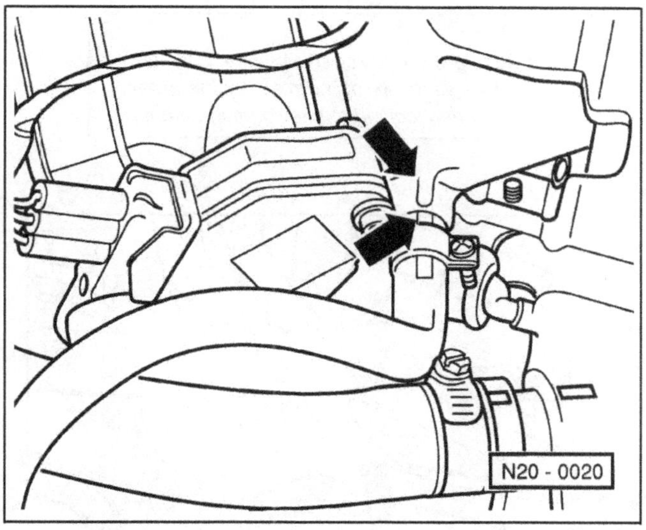

N20-0020

Fig. 2. Charcoal cannister breather hose at upper intake manifold. **Arrows** indicate correct installation orientation: marking on breather hose must align with marking on intake manifold.

8. Disconnect all electrical connectors from the upper intake manifold and the throttle housing. Unclip the fuel lines from the cylinder head cover.

9. Disconnect the vacuum hoses from the fuel pressure regulator and the brake vacuum booster.

10. Remove the oil dipstick guide tube mounting bolt from the lower intake manifold.

11. Remove the upper intake manifold mounting bolts and manifold support brackets. Remove the manifold together with the throttle housing. Cover the open intake runners using clean shop rags.

12. Remove the ten nuts from the perimeter of the cylinder head cover. Lift off the cylinder head cover and its gasket. Remove the one-piece gasket from the cylinder head or cover.

Installation is the reverse of removal. Use a new gasket between the upper intake manifold and the lower intake manifold. Replace the gasket at the cylinder head cover if it is damaged.

Make sure the charcoal cannister breather hose at the intake manifold is installed correctly as shown in Fig. 2. Check also that the hose is routed over the crankcase breather hose and does not contact any other components. Make sure the crankcase breather hose at the cylinder head cover is also installed correctly as shown in Fig. 3.

> ***WARNING —***
> *Evaporative fuel and oil vapors are potentially explosive and flammable. Make sure all evaporative emission and crankcase breather hoses are in good condition and free of cracks or damage. Make sure all hose connections are tight and secure.*

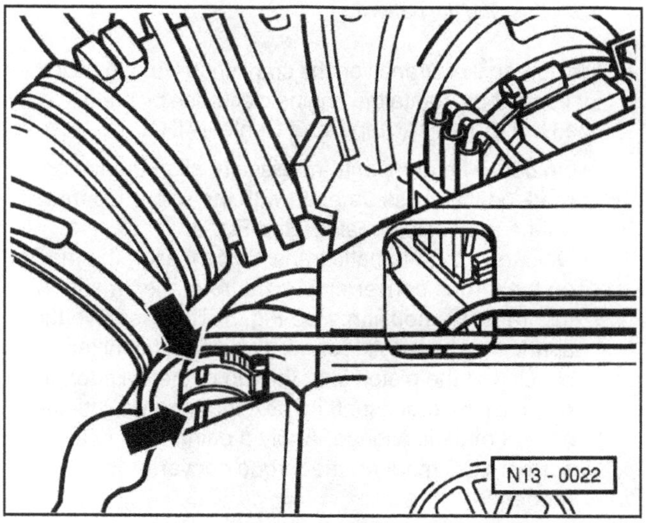

N13-0022

Fig. 3. Charcoal cannister breather hose at cylinder head cover. **Arrows** indicate correct installation orientation: marking on breather hose must align with marking on cover.

Tightening torques
- Cylinder head cover to cylinder head .10 Nm (89 in-lb)
- Intake manifold cover to intake manifold . . . . .5 Nm (44 in-lb)
- Upper intake manifold to lower manifold . . . . 25 Nm (18 ft-lb)

To remove and install camshafts

The camshafts can be removed with the cylinder head installed. Removing the camshafts allows access to the hydraulic cam followers, the valve springs, and the valve stem oil seals.

NOTE —
A special alignment tool (VW special tool no. 3268) is specified for setting up valve timing during camshaft installation. The tool is relatively inexpensive and highly recommended. See Fig. 10 given later.

1. Disconnect the battery negative (–) terminal from the battery.

NOTE —
Be sure to have the anti-theft radio code on hand before disconnecting the battery.

2. Using a socket wrench on the crankshaft vibration damper center bolt, rotate the engine clockwise by hand to set the No.1 cylinder at Top Dead Center (TDC).
   - On cars with manual transmission, align the timing mark on the pressure plate with the reference mark on the transaxle housing. See Fig. 4.
   - On cars with automatic transmission, align the mark on the torque converter with the reference mark on the transaxle housing. See Fig. 5. Then remove the spark plug from cylinder no. 1. Using a flashlight check that the piston is at the top of the cylinder. If not, turn the crankshaft in 120° increments until the correct mark is aligned. Apply a paint mark to the correct TDC mark on the torque converter for future reference.

CAUTION —
On cars with automatic transmission, always verify TDC through the removed spark plug hole for cylinder no.1. The torque converter can be installed to the driveplate in one of three positions. Therefore, three factory TDC marks (120° apart) are made on the torque converter. This insures that a TDC mark will always be present—regardless of where the torque converter is installed on the driveplate.

V15-0842

Fig. 4. Timing mark on clutch pressure plate aligned with mark on transaxle on cars with manual transmission.

CAUTION —
When turning the crankshaft to verify TDC, turn it only in the direction of normal engine rotation (clockwise when looking at the vibration damper).

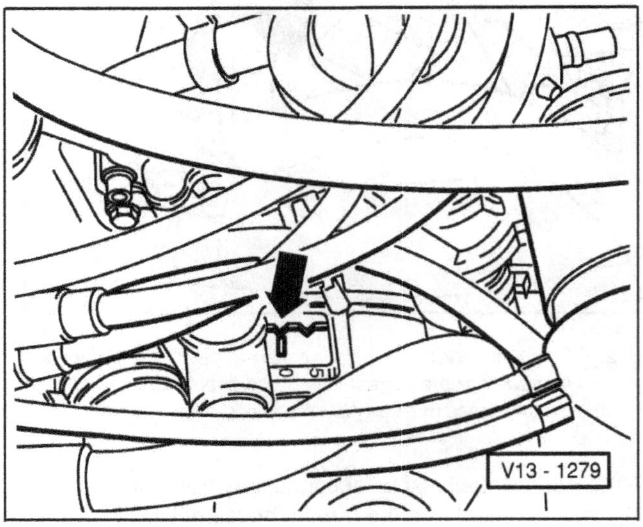

V13-1279

Fig. 5. Timing mark (one of three) on torque converter aligned with mark on transaxle housing on cars with automatic transmission.

3. Remove the cylinder head cover as described earlier.
4. Remove the ignition coil. See **28 Ignition System**.
5. Remove the upper timing chain tensioner from the camshaft sprocket cover.

6. Remove the camshaft sprocket cover from the front of the cylinder head together with the camshaft position sensor.

NOTE —

When removing the cover, note the two M8 socket-head bolts threaded up through the lower cover and the small O-ring on the rear of the cover.

7. Remove both camshaft sprocket mounting bolts. Counterhold the camshafts while loosening the bolts. See Fig. 6. Remove the sprockets from the camshafts. Tie the chain up using stiff wire.

> ***CAUTION —***
>
> *Do not rotate the engine once the camshafts chains are removed from the camshafts. The pistons may contact the open valves.*

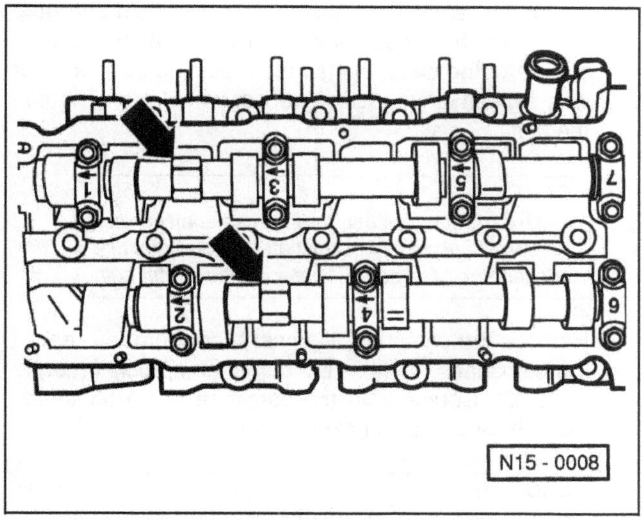

N15-0008

Fig. 6. Counterhold camshaft at 24mm hex flats (**arrows**) when loosening or tightening camshaft sprocket bolt.

8. To remove the camshaft for cylinders 1, 3, and 5, first remove bearing caps 1 and 7. Then loosen the nuts on caps 3 and 5, a little at a time so that valve spring tension is relieved evenly. See Fig. 7.

9. To remove the camshaft for cylinders 2, 4, and 6, first remove bearing cap 4. Then loosen the nuts on caps 2 and 6, a little at a time so that valve spring tension is relieved evenly.

> ***CAUTION —***
>
> *To avoid uneven and accelerated wear, the bearing caps must be reinstalled in their exact original positions.*

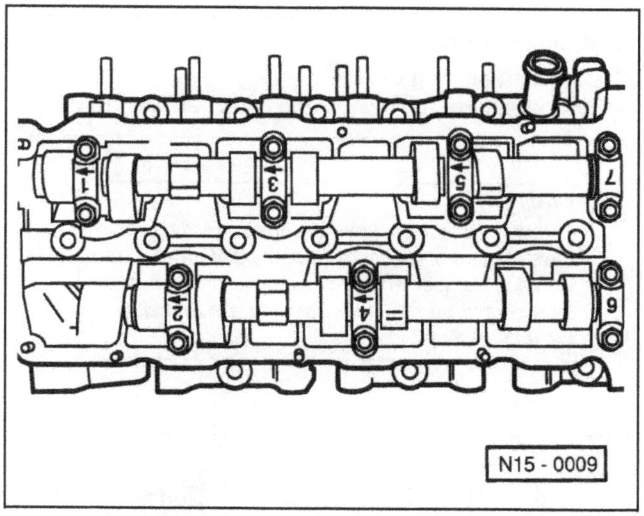

N15-0009

Fig. 7. Cylinder head bearing cap identification.

10. Lubricate the contact surfaces of the camshafts and then install them into the cylinder head so that the camshaft sprocket alignment recesses are at the top. See Fig. 8.

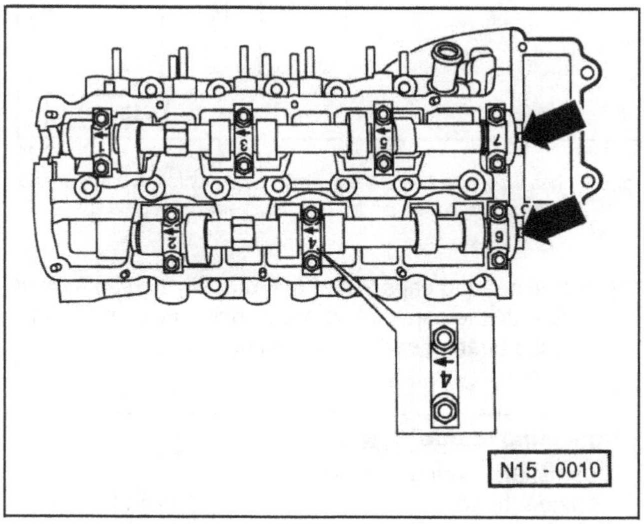

N15-0010

Fig. 8. Install camshafts so that sprocket alignment recesses are facing up (**arrows**).

11. Check that cylinder no. 1 is at Top Dead Center (TDC). See Fig. 4 or FIg. 5 shown earlier.

12. Install bearing caps 3 and 5. Gradually tighten all four nuts until the camshaft is drawn down fully and evenly into the bearing saddles. Install bearing caps 1 and 7.

Tightening torque

Camshaft bearing caps to cylinder head . . . . 20 Nm (15 ft-lb)

> ***CAUTION —***
> *Be sure to install the bearing caps correctly. The arrows marked on the caps should point towards the vibration damper end of the engine and the numbers should be readable from the exhaust side of the engine. See Fig. 9.*

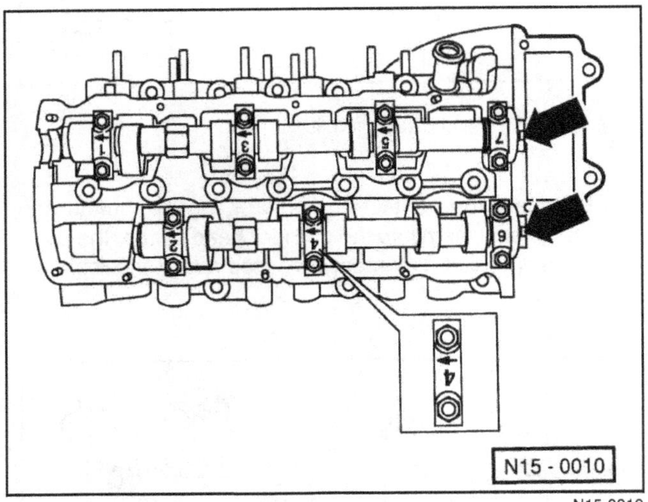

N15-0010

Fig. 9. Install bearing caps so that arrow faces vibration damper and numbers can be read from exhaust side of engine (inset).

13. Install bearing caps 2 and 6. Gradually tighten all four nuts until the camshaft is drawn down fully and evenly into the bearing saddles. Install bearing cap 4.

Tightening torque

Camshaft bearing caps to cylinder head . . . . 20 Nm (15 ft-lb)

14. Position the camshafts so that the slots in the rear of the shafts accept the VW special tool no. 3268. See Fig. 10. Install the special tool and push it forward into the cutouts.

15. Mount the camshaft sprocket to the shorter camshaft (cylinders no. 2, 4, and 6). Install the mounting bolt hand-tight. Install the chain to the sprocket so that there is no slack between the intermediate shaft sprocket and the cam sprocket.

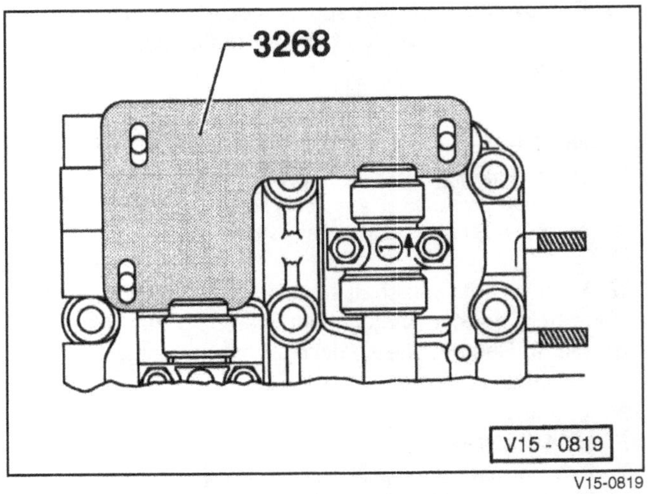

V15-0819

Fig. 10. VW special tool no. 3268 aligns camshafts at TDC. Tool engages slots in rear of camshafts.

16. Install the remaining sprocket into the chain and onto the camshaft so that all the chain slack is at the tensioner side of the chain. Install the mounting bolt together with the camshaft position sensor wheel. Hand tighten the bolt.

> ***CAUTION —***
> *The mating surfaces between the camshaft position sensor wheel and the camshaft sprocket must be dry and free of oil before tightening the center bolt.*

17. Remove the camshaft alignment tool from the rear of the camshafts. Tighten the camshaft sprocket mounting bolts. Counterhold the camshaft when tightening the bolt. See Fig. 6 shown earlier.

> ***CAUTION —***
> *Do not tighten the camshaft bolts with the plastic alignment tool installed. It is easily damaged.*

Tightening torque

- Camshaft sprockets to camshaft (with bolt contact surface oiled) . . . . 100 Nm (74 ft-lb)

18. Clean any old sealer from the camshaft sprocket cover mating surfaces.

19. Install the camshaft sprocket cover (with new O-ring) using Volkswagen sealant AMV 188 001 01 or an equivalent gasket sealer on the gasket surfaces. See Fig. 11. Tighten the larger M8 bolts first.

> ***CAUTION —***
> *Be sure to install a new O-ring to the rear of the cover. Lightly oil the O-ring and install it into the cover before installation.*

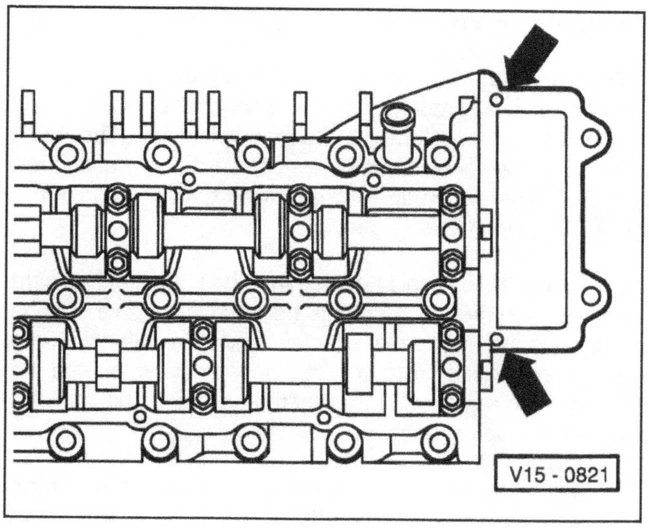

V15-0821

Fig. 11. Clean out sealer from 3mm holes (**arrows**) and re-seal holes and gasket surface using Volkswagen sealant AMV 188 001 02 or equivalent.

Tightening torque

- Camshaft sprocket cover to cylinder head or lower timing chain cover
  M6 . . . . 10 Nm (89 in-lb)
  M8 . . . . 25 Nm (18 ft-lb)

20. Install the chain tensioner using a new O-ring seal. Oil the seal and install it into the cover before installing the tensioner.

> ***CAUTION —***
> *If the tensioner was taken apart, or if the piston has expanded out, it must be bled down. Use a thin wire through the opening in the tensioner piston. See Fig. 12. Push the wire into the piston while apply light pressure. When the piston check valve opens, the tensioner will collapse.*

Tightening torque

- Camshaft chain tensioner to camshaft sprocket cover . . . . 30 Nm (22 ft-lb)

21. Turn the engine over by hand in normal engine rotation direction two full revolutions and realign the timing marks shown earlier in Fig. 4 or Fig. 5.

22. Install VW special tool no. 3268 to the camshafts. If the cam timing is setup correctly, the tool will slide smoothly into the camshaft cutouts.

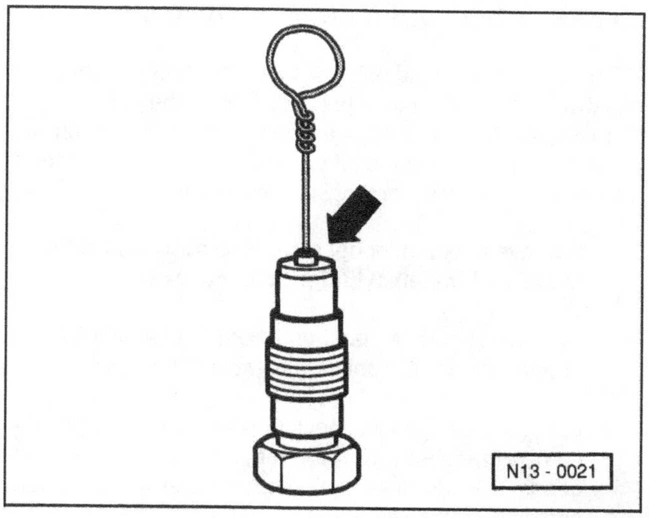

N13-0021

Fig. 12. Bleed tensioner by inserting thin wire (0.8 mm) into piston hole (**arrow**). Apply pressure to tensioner piston while pushing on wire. When piston check valve opens, piston will move into tensioner body.

23. The remainder of installation is the reverse of removal. Install the cylinder head cover and upper intake manifold as described under **To remove and install cylinder head cover**.

> ***CAUTION —***
> *If new hydraulic cam followers were installed, the engine must not be run for at least 30 minutes. New cam followers are at a maximum height and should be allowed to bleed down or the valve heads could contact the pistons on start up.*

Hydraulic Cam Followers

The VR6 engine is equipped with hydraulic cam followers that automatically maintain proper valve clearance. The cam followers are pumped up by engine oil pressure, expanding as necessary to fill the gap between valve and camshaft lobe.

Some valve noise at start-up is normal at times, due to hydraulic cam followers that have bled down while the engine was not running. Before checking noisy cam followers, check to see that the engine oil is new and clean and that the level is correct. Allow a minute or so with the engine running for the lubrication system to properly pump up the cam followers.

> ***CAUTION —***
> *After installing new cam followers, the engine should not be started for at least 30 minutes. New cam followers are at full height and must be allowed to bleed down to their proper height. Failure to do this may cause valve or piston damage.*

To check hydraulic cam followers

Cam followers should only be checked when the engine is fully warm. Run the engine, preferably drive the car, for 20 to 30 minutes. Raise the engine speed above 2,500 rpm for about two minutes. Then shut the engine off and proceed as soon as possible while the engine is still warm.

1. Remove the cylinder head cover as described in **To remove and install cylinder head cover**.

2. Turn the engine by hand until both the camshaft lobes of one cylinder are pointing approximately up.

3. Using a feeler gauge, check the clearance between the cam follower and the cam lobe. If the clearance exceeds the specified limit, the follower is faulty and should be replaced.

Hydraulic Cam Follower
- Wear limit clearance. . . . . 0.1 mm (0.004 in.)

---

4. If no faults are found in step 3, use a non-metal object such as wood or plastic to apply light pressure to the top of the cam follower. See Fig. 13. Recheck the clearance. If the clearance is excessive, replace the follower.

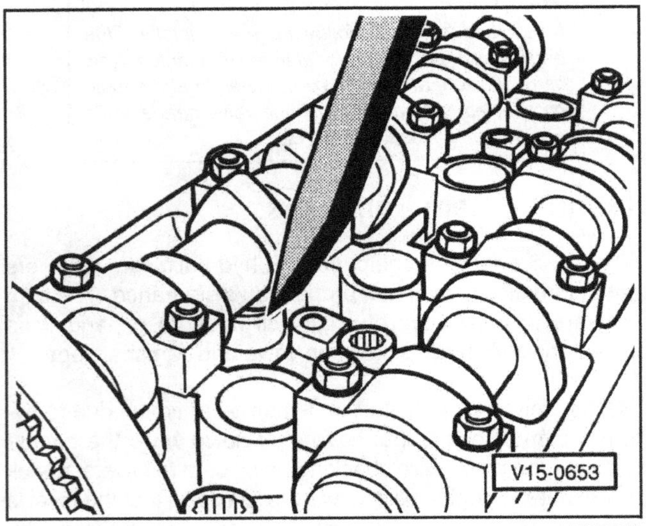

V15-0653

Fig. 13. Hydraulic cam follower being depressed with a non-metal object.

5. Replace a faulty cam follower by removing the camshaft, as described in **To remove and install camshafts** and pulling the follower from the cylinder head.

> ***CAUTION—***
> *Store removed hydraulic cam followers in order on a clean surface with the camshaft contact surface facing down to prevent bleed down. Cover them with a clean, lint-free rag.*

To replace valve stem oil seals

The sign of faulty valve stem oil seals are excessive oil consumption and oil smoke from the exhaust after starting and during sudden deceleration. Keep in mind that the valve guides may also be worn and the major cause of oil consumption. See **Cylinder Head Reconditioning** for more information on checking valve guides.

Replacing the oil seals requires that the camshaft, the cam followers, and the valve springs be removed. Valve springs can be removed while the cylinder head is installed with the use of compressed air to hold the valves closed, or with the cylinder head removed. Either method requires the use of a special valve spring compressor.

1. If working with the cylinder head installed, remove the cylinder head cover, as described in **To remove and install cylinder head cover**.

2. Remove the camshafts, as described in **To remove and install camshafts.**

3. Remove the cam followers.

> ***NOTE—***
> - *Keep all parts in order so that they can be reinstalled in their original locations.*
> - *Store removed hydraulic cam followers in order on a clean surface with the camshaft contact surface facing down to prevent bleed down. Cover them with a clean, lint-free rag.*

4. If cylinder head is removed, remove the valve springs See Fig. 14.

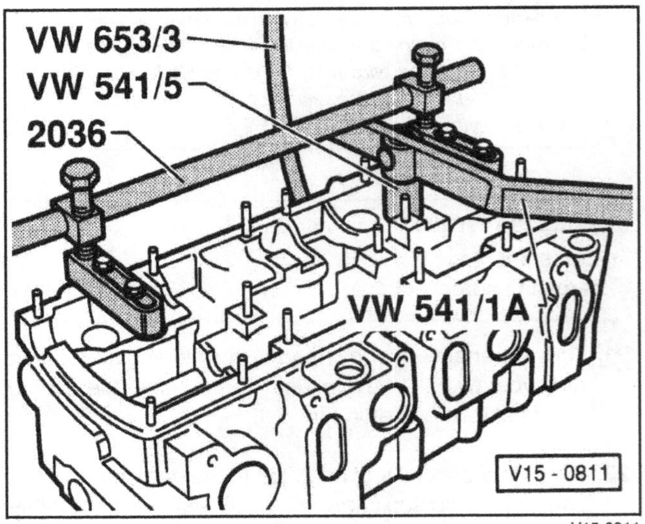

V15-0811

Fig. 14. Valve spring compressors being used to remove valve springs. Numbers identify Volkswagen special tools designed for use on Volkswagen engines. (VW 653/3 is air hose adapter for spark plug hole.)

NOTE —

To prevent the timing chain from jamming when setting the piston to BDC, lift up on and guide the chain while rotating the crankshaft.

5. If cylinder head is installed, set the cylinder to bottom dead center (BDC). Remove the spark plug and apply compressed air (at least 87 psi) to the cylinder using an appropriate air hose adapter in the spark plug hole.

6. Remove the valve springs.

7. Remove the valve stem oil seals from the valve guides.

NOTE —

Volkswagen recommends the use of a slide hammer, VW special tool no. 3047A, to remove the oil seals without damage to the valves or valve guides.

8. Install each new valve stem oil seal by temporarily fitting the protective plastic cap over the valve stem end. Then hand-press the seal onto the valve guide. See Fig. 15. Remove the protective cap for use on the next valve.

NOTE —

The protective cap protects the seal from being damaged by the keeper grooves in the valve stem. In the absence of a protective cap, temporarily wrap the valve stem with plastic tape.

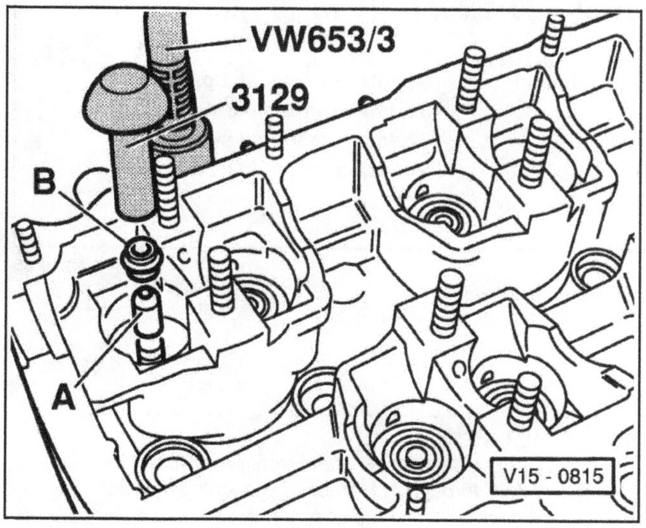

V15-0815

Fig. 15. Protective cap (**A**) and new valve stem oil seal (**B**). VW special tools (with tool numbers) shown.

9. Reinstall the valve springs. Repeat the procedure for each pair of valves.

10. The remainder of assembly is the reverse of disassembly. See **To remove and install camshafts, Cylinder Head Removal and Installation,** and **To remove and install cylinder head cover.**

Tightening torque

- Spark plug to cylinder head
  AAA engine .... 25 Nm (18 ft-lb)

CYLINDER HEAD REMOVAL AND INSTALLATION

Fig. 16 shows an exploded view of the VR6 cylinder head. The cylinder head can be removed without removing the engine from the vehicle and without removing the camshafts. New cylinder head bolts should be used every time the cylinder head is removed and installed.

If a failed head gasket or warped head is suspected, a compression test, as described earlier under **Diagnostic Testing,** may aid diagnosis and should be performed before the cylinder head is removed. A failed head gasket may be caused by a warped cylinder head. When replacing a failed head gasket, always check the cylinder head for straightness. See **Cylinder Head Reconditioning.**

NOTE —

A special alignment tool (VW special tool no. 3268) is specified for removal and installation of the cylinder head. Read the procedure through before beginning the job.

To remove cylinder head

1. Disconnect the battery negative (–) terminal from the battery.

NOTE —

Be sure to have the anti-theft radio code on hand before disconnecting the battery.

2. Remove the cylinder head cover as described in **To remove and install cylinder head cover.**

3. Drain the coolant as described in **19 Engine–Cooling System.** Disconnect all coolant hoses from the outlets on the cylinder head.

WARNING —

Hot coolant can scald. Drain the coolant only with the engine cold.

4. Remove the ignition coil. See **28 Ignition System.**

1. **Bolts**
   - tighten to 5 Nm (44 in-lb)
2. **Ignition wire guide**
3. **Bolt**
   - tighten to 25 Nm (18 ft-lb)
4. **Intake manifold support**
5. **Bracket, for fuel lines**
6. **Flame trap**
7. **Circlip**
8. **Oil filler cap**
9. **Bolt**
   - tighten to 10 Nm (89 in-lb)
10. **Bracket, for wiring harness**
11. **Cylinder head cover**
12. **Cylinder head cover gasket**
    - note installed position
    - replace if damaged
13. **O-ring**
    - always replace
14. **Spacer**
15. **O-ring**
    - always replace
16. **Camshaft position sensor**
17. **Bracket, for wiring harness**
18. **Ignition coil**
19. **Camshaft timing chain tensioner, upper**
    - bleed before installing, See Fig. 5-8
    - tighten to 30 Nm (22 ft-lb)
    - rotate engine only with tensioner installed
20. **Sealing washer**
    - always replace
21. **Camshaft sprocket cover**
    - install using VW sealer AMW 188 001 02 or equivalent
22. **Cylinder head gasket**
    - always replace
    - use new coolant when replacing
23. **Cylinder head**
    - see **Cylinder Head Reconditioning**
    - use new coolant when replacing
24. **Cylinder head bolt**
    - qty. 20, 3 different lenghts
    - stretch bolt–**always replace**

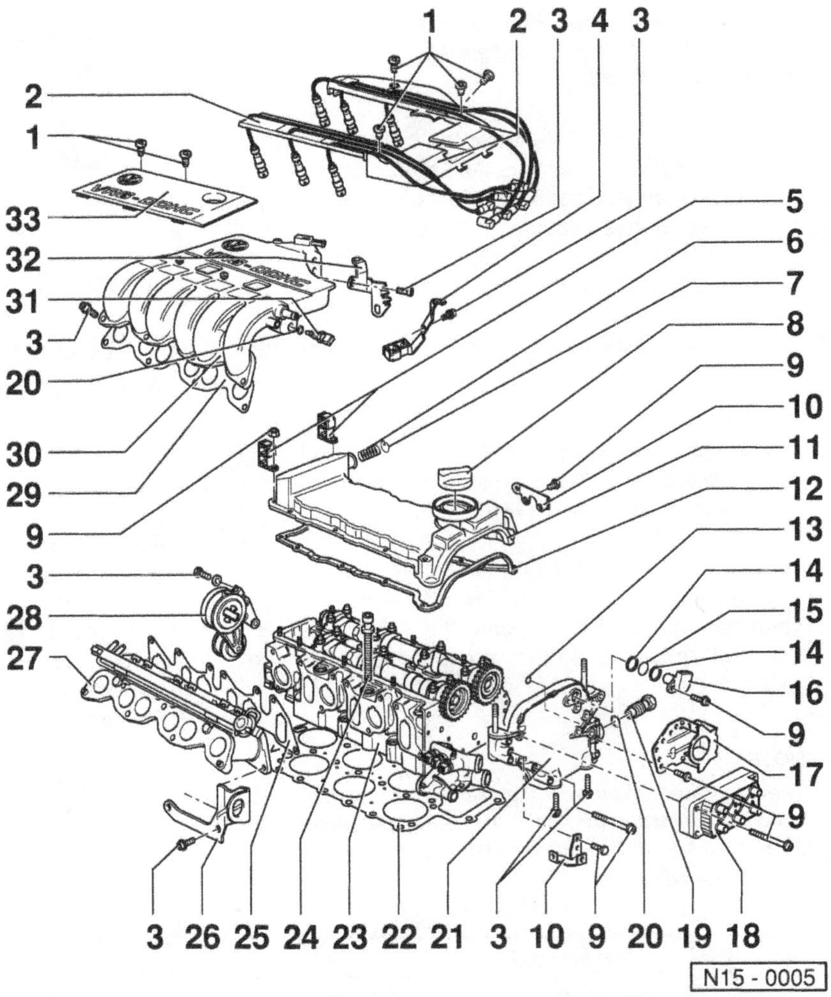

25. **Gasket, lower intake manifold**
26. **Engine lifting eye**
27. **Intake manifold, lower**
28. **Drive belt tensioner**
29. **Gasket**
    - always replace
30. **Intake manifold, upper**
31. **Intake air temperature sensor**
32. **Intake manifold support**
33. **Intake manifold cover**

N15-0005

Fig. 16. Exploded view of AAA (2.8 liter, VR6) cylinder head.

5. Using a socket wrench on the crankshaft vibration damper center bolt, rotate the engine clockwise by hand to set the No.1 cylinder at Top Dead Center (TDC).
   - On cars with manual transmission, align the timing mark on the pressure plate with the reference mark on the transaxle housing. See Fig. 17.
   - On cars with automatic transmission, align the mark on the torque converter with the reference mark on the transaxle housing. See Fig. 18. Then remove the spark plug from cylinder no. 1. Using a flashlight check that the piston is at the top of the cylinder. If not, turn the crankshaft in 120° increments until the correct mark is aligned. Apply a paint mark to the correct TDC mark on the torque converter for future reference.

CAUTION —

- *On cars with automatic transmission, always verify TDC through the removed spark plug hole for cylinder no.1. The torque converter can be installed to the driveplate in one of three positions. Therefore, three factory TDC marks (120° apart) are made on the torque converter. This insures that a TDC mark will always be present—regardless of how the torque converter is installed.*
- *When turning the crankshaft to verify TDC, turn it only in the direction of normal engine rotation (clockwise when looking at the vibration damper).*

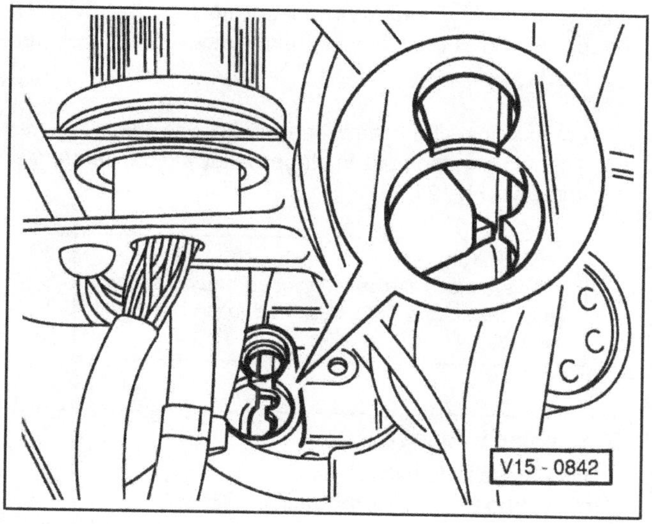

V15-0842

Fig. 17. Timing mark on clutch pressure plate aligned with mark on transaxle on cars with manual transmission.

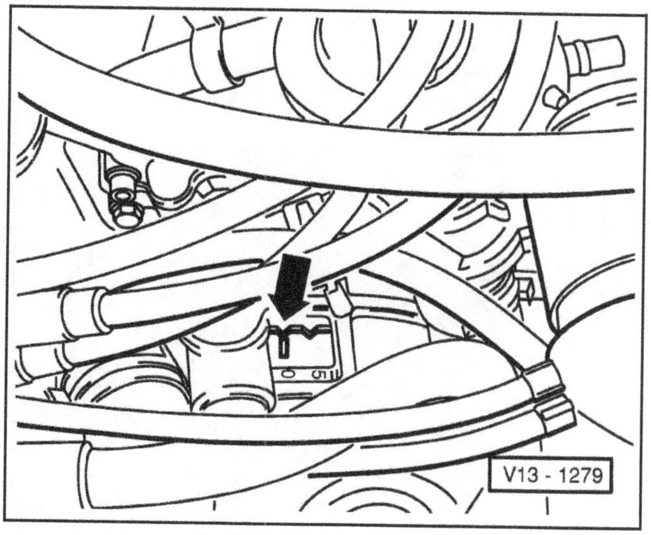

V13-1279

Fig. 18. Timing mark (one of three) on torque converter aligned with mark on transaxle housing on cars with automatic transmission.

6. Remove the upper timing chain tensioner from the camshaft sprocket cover.

7. Remove the camshaft sprocket cover from the front of the cylinder head together with the camshaft position sensor.

NOTE—

When removing the cover, note the two M8 socket-head bolts threaded up through the lower cover and the small O-ring on the rear of the cover.

8. While counterholding the camshafts using a 24mm open end wrench, remove both camshaft sprocket mounting bolts. See Fig. 19. Remove the sprockets from the camshafts. Tie the chain up using stiff wire.

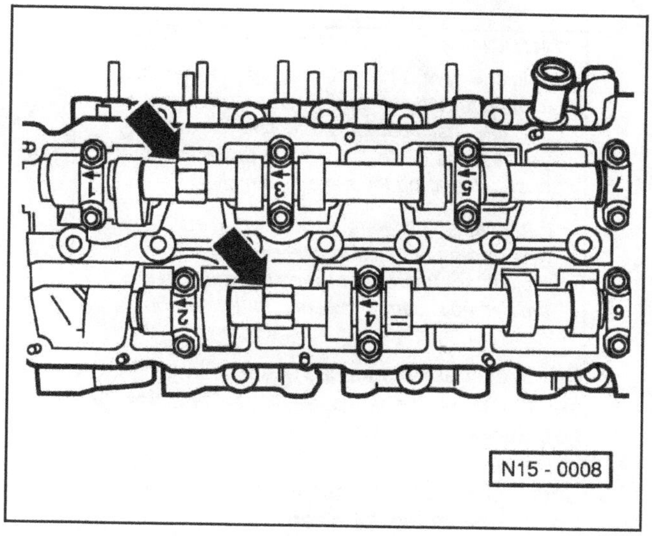

N15-0008

Fig. 19. Counterhold camshaft at 24mm hex flats (**arrows**) when loosening or tightening camshaft sprocket bolt.

9. Disconnect the fuel injector harness connectors. Disconnect and label the hoses, wire connectors and cables connected to cylinder head. For more information on fuel injection components, see **24 Fuel Injection–Gasoline.**

WARNING—

Fuel may be expelled when disconnecting fuel lines. Do not smoke or work near heaters or other fire hazards. Have a fire extinguisher handy.

10. Disconnect the exhaust pipes from the exhaust manifolds, as described in **26 Exhaust System/Emission Controls**.

11. Gradually and evenly loosen the **twenty** cylinder head bolts, beginning with the outer bolts and working toward the center. Discard the bolts.

12. Remove the cylinder head. Proceed slowly, checking to make sure no hoses or wires are interfering with removal.

NOTE—

If the head is stuck, use a soft-faced mallet or pry gently with a wooden stick to loosen it.

To install cylinder head

1. Clean the cylinder head and the gasket surface of the cylinder block. Clean the bolt holes with thread chasers and remove all foreign matter and liquid from the holes.

> ***CAUTION —***
> - *If installing a VW replacement cylinder head (complete with camshafts), protective plastic liners are installed for protection of the open valves. Be sure to remove the protective liners prior to installing the head assembly.*
> - *If the camshafts were removed or if installing a new or remanufactured cylinder head, lubricate the contact surfaces between the cam lobes and the hydraulic cam followers prior to start-up.*

2. Check the gasket surfaces of the cylinder head and the cylinder block for warpage. See **Cylinder Head Reconditioning**.

3. Check that the crankshaft is set to cylinder no. 1 TDC. See Fig. 17 or Fig. 18 given earlier.

4. Place a new cylinder head gasket on the surface of the cylinder block.

> ***CAUTION —***
> *Cylinder head gaskets will give a reliable seal only once. Always use a new gasket that has not previously been compressed by tightening the cylinder head bolts.*

5. Align the camshafts so that the slots in the rear of the shafts are positioned to accept VW special tool no. 3268. See Fig. 20. Install the special tool and push it forward into the cutouts.

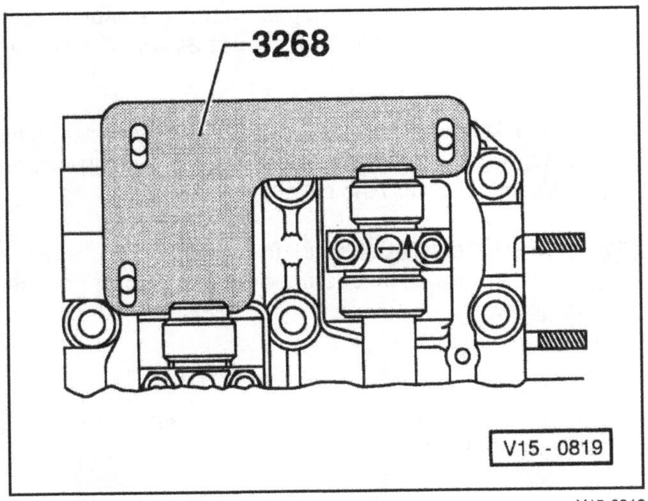

V15-0819

Fig. 20. VW special tool no. 3268 aligns camshafts at TDC. Tool engages slots in rear of camshafts.

6. Place the cylinder head in position on the cylinder block, noting the two alignment sleeves in the cylinder block.

7. Loosely install the new head bolts with the washers. Then tighten the bolts in stages following the sequence shown in Fig. 21.

> ***CAUTION —***
> *Always replace the cylinder head mounting bolts. They are stretch bolts designed to be used only once.*

Tightening torques

- Cylinder head to cylinder block (stretch bolts–**always replace**)
  - stage I . . . . 40 Nm (30 ft -lb)
  - stage II . . . . 60 Nm (44 ft -lb)
  - stage III . . . . .additional 1/4 turn (90°)
  - stage IV . . . . .additional 1/4 turn (90°)

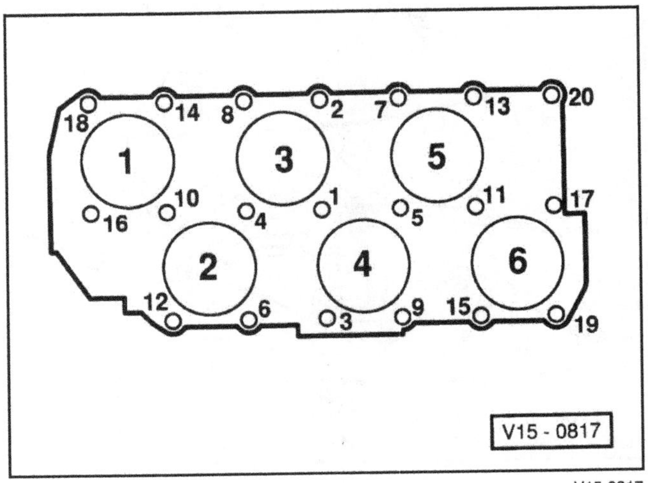

V15-0817

Fig. 21. Cylinder head bolt tightening sequence.

8. Mount the camshaft sprocket to the shorter camshaft (cylinders no. 2, 4, and 6). Install the mounting bolt hand-tight. Install the chain to the sprocket so that there is no slack between the intermediate shaft sprocket and the cam sprocket.

9. Install the remaining sprocket into the chain and onto the camshaft so that all the chain slack is at the tensioner side of the chain. Install the mounting bolt together with the camshaft position sensor wheel. Hand tighten the bolt.

> ***CAUTION —***
> *The mating surfaces between the camshaft position sensor wheel and the camshaft sprocket must be dry and free of oil before tightening the center bolt.*

10. Remove the camshaft alignment tool from the rear of the camshafts. Tighten the camshaft sprocket bolts. Counter hold the camshafts when tightening the bolts.

> ***CAUTION—***
> - *Do not tighten the camshaft bolts with the plastic alignment tool installed. It is easily damaged.*
> - *Always counterhold the camshafts at the 24mm cast hex when tightening the camshaft sprocket bolts. See Fig. 19.*

Tightening torque

- Camshaft sprockets to camshaft (with bolt contact surface oiled) . . . . 100 Nm (74 ft-lb)

11. Clean any old sealer from the camshaft sprocket cover mating surfaces.

12. Install the camshaft sprocket cover (with new O-ring) using Volkswagen sealant AMV 188 001 01 or an equivalent gasket sealer on the gasket mating surfaces. See Fig. 22. Tighten the larger M8 bolts first.

> ***CAUTION—***
> *Be sure to install a new O-ring to the rear of the cover. Lightly oil the O-ring and install it into the cover before installation.*

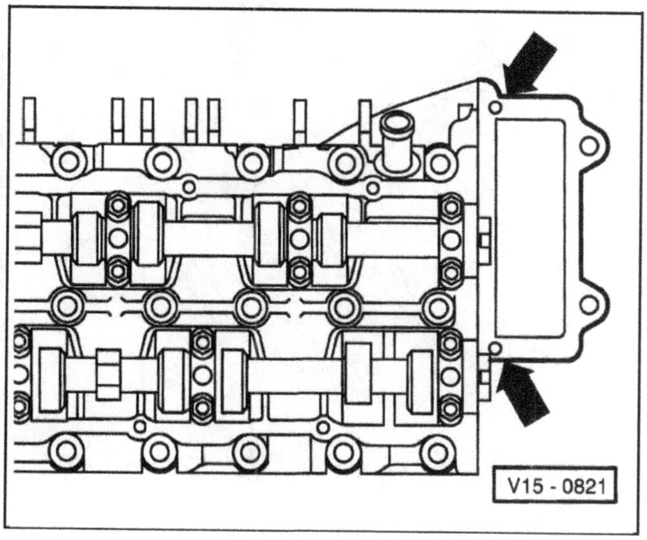

V15-0821

Fig. 22. Clean out sealer from 3 mm holes (**arrows**) and re-seal holes and gasket surface using Volkswagen sealant AMV 188 001 02 or equivalent.

Tightening torque

- Camshaft sprocket cover to cylinder head or lower timing chain cover
  - M6 . . . . 10 Nm (89 in-lb)
  - M8 . . . . 25 Nm (18 ft-lb)

13. Install the chain tensioner using a new O-ring seal. Oil the seal and install it into the cover before installing the tensioner.

> ***CAUTION—***
> *If the tensioner was taken apart or if the tensioner piston has expanded out, it must be bled down. See Fig. 12 given earlier.*

Tightening torque

- Camshaft chain tensioner to camshaft sprocket cover. . . . . 30 Nm (22 ft-lb)

14. Turn the engine over by hand in normal engine rotation direction two full revolutions and realign the TDC timing marks. Reinstall VW special tool no. 3268 to the camshafts. If the cam timing is setup correctly, the tool will slide smoothly into the camshaft cutouts.

> ***CAUTION—***
> *If VW special tool 3268 does not fit into the camshaft cutout, the chain assembly may not be installed correctly. If only very minor readjustment of the cams allows the tool to slide into place, the assembly is probably setup correctly. If any problems are encountered, reinstall the chains from the beginning of this procedure.*

15. Installation of the remaining parts is the reverse of removal. See **To remove and install cylinder head cover**, **24 Fuel Injection–Gasoline**, and **26 Exhaust System/Emission Controls**. Refill the cooling system as described in **19 Engine–Cooling System.**

CAMSHAFT TIMING CHAINS

Removal and installation of the camshaft timing chains requires that the transaxle be separated from the engine. Fig. 23 shows an exploded view of the camshaft timing chain assembly.

> *NOTE —*
> *A special alignment tool (VW special tool no. 3268) is specified for setting up camshaft and valve timing. See Fig. 31.*

To remove camshaft timing chains

1. Remove the engine and transaxle from the car and separate the transaxle from the engine. See **10 Engine–Removing and Installing**.

> *NOTE —*
> *It is also possible to access the chain assembly by removing the transmission from the engine. This way the engine remains in the car. See* **3 CLUTCH, TRANSMISSION, AND FINAL DRIVE** *for transmission removal.*

2. Remove the cylinder head cover. See **To remove and install cylinder head cover.**
3. Remove the ignition coil. See **28 Ignition System**.
4. Using a socket wrench on the crankshaft vibration damper center bolt, rotate the engine clockwise by hand to set the No.1 cylinder at Top Dead Center (TDC).
   - On cars with manual transmission, align the "0/T" mark on the flywheel with the cast reference point mark on the cylinder block. See Fig. 24.
   - On cars with automatic transmission, align the milled mark on the driveplate with the reference point mark on the engine block. See Fig. 25.

1. **Camshaft sprockets**
2. **Not applicable**
3. **Camshaft position sensor (CMP) wheel**
4. **Bolt**
   - tighten to 100 Nm (74 ft-lb)
5. **Tensioner guide pivot pin**
   - tighten to 25 Nm (18 ft-lb)
6. **O-ring**
   - always replace
7. **Upper timing chain tensioner**
   - bleed before installing, See Fig. 5-8
   - tighten to 30 Nm (22 ft-lb)
8. **Tensioner guide**
9. **Intermediate shaft sprocket, inner**
10. **Timing chain, upper (double row)**
    - mark rotation direction before removal
11. **Bolt**
    - tighten to 100 Nm (74 ft-lb)
12. **Intermediate shaft sprocket, outer**
13. **Bolt**
    - tighten to 10 Nm (89 in-lb)
14. **Timing chain tensioner, lower**
15. **Crankshaft with integral sprocket**
    - ground tooth used for TDC timing mark
16. **Timing chain, lower (single row)**
    - mark rotation direction before removal
17. **Guide rail, lower**
18. **Locating pin (w/o collar)**
    - tighten to 25 Nm (18 ft-lb)
19. **Locating pin (with collar)**
    - tighten to 25 Nm (18 ft-lb)
20. **Bolt**
    - install using locking fluid "D6" or equivalent
    - tighten to 20 Nm (15 ft-lb)
21. **Bolt**
    - tighten to 20 Nm (15 ft-lb)

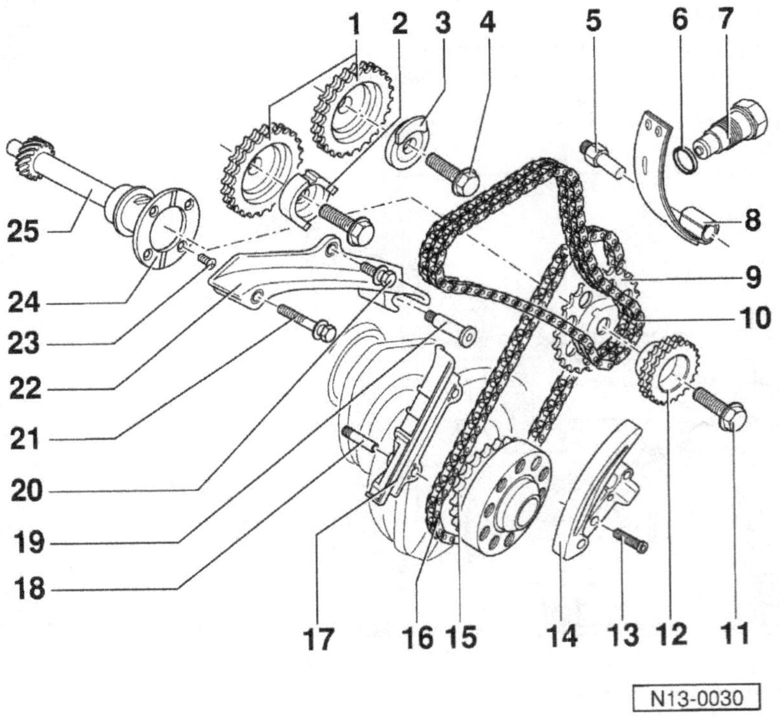

22. **Guide rail, upper**
23. **Bolt**
    - install using locking fluid "D6" or equivalent
    - tighten to 10 Nm (89 in-lb)
24. **24. Thrust washer**
25. **Intermediate shaft**

N13-0030

Fig. 23. Camshaft timing chain assembly.

NOTE —

The camshaft timing can be checked at this time using VW special tool no. 3268. With the cylinder head cover removed, slide the tool into the rear camshaft slots (see Fig. 10 given earlier). The tool should slide into the grooves easily without adjustment of the camshafts.

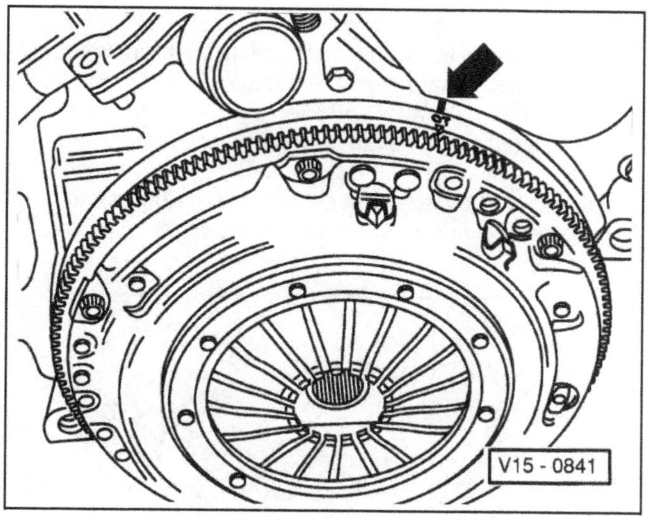

V15-0841

Fig. 24. TDC (0T) mark on flywheel aligned with reference point mark on cylinder block on cars with manual transmission.

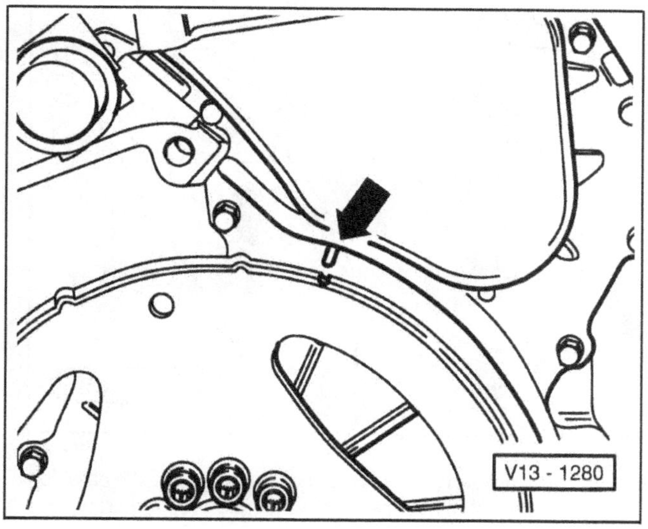

V13-1280

Fig. 25. TDC mark on driveplate aligned with reference point mark on cylinder block on cars with automatic transmission.

5. Remove the upper timing chain tensioner from the camshaft sprocket cover.

6. Remove the camshaft sprocket cover from the cylinder head together with the camshaft position sensor.

NOTE —

When removing the cover, note the two M8 socket-head bolts threaded up through the lower cover and the small O-ring on the rear of the cover.

7. Remove the flywheel or driveplate. Discard the mounting bolts. See **Crankshaft/Cylinder Block**.

CAUTION —

Do not reuse the flywheel or driveplate mounting bolts. They are stretch bolts and designed to be used only once.

8. Remove the lower timing chain cover from the cylinder block and oil pan.

CAUTION —

Use care when removing the lower cover from the oil pan gasket and the cylinder head gasket. If the gaskets are damaged, they will need to be replaced.

9. Mark the engine rotation direction on the timing chains. See Fig. 26.

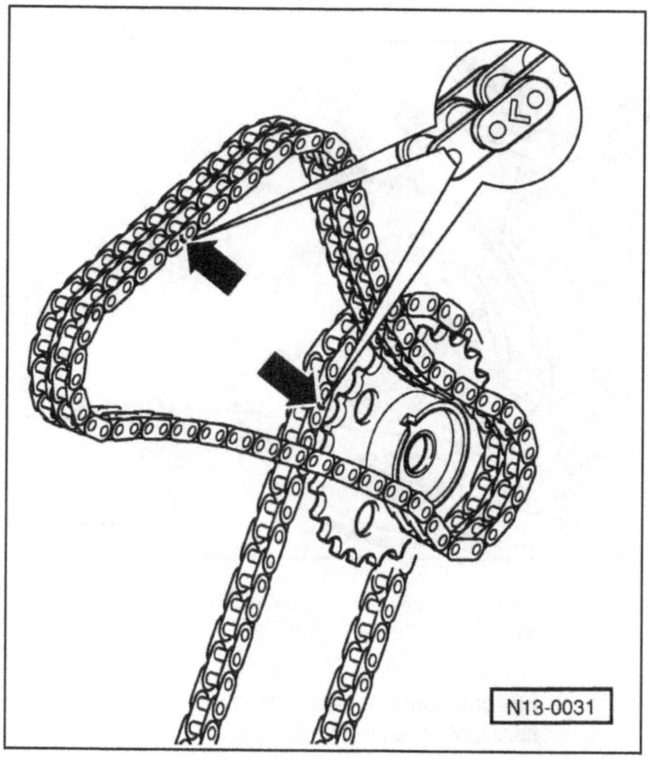

N13-0031

Fig. 26. Before removing timing chains, mark engine rotation direction on chain link (**arrow**).

10. Counterhold the camshafts using a 24mm open end wrench and loosen the camshaft sprocket bolt. See Fig. 27. Repeat the procedure on the other sprocket. Remove the sprockets together with the double row chain.

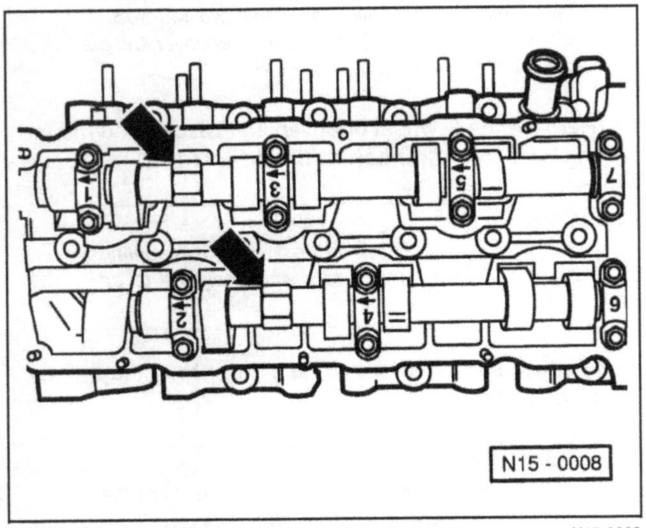

N15-0008

Fig. 27. Counterhold camshaft at 24mm hex flats (**arrows**) when loosening or tightening camshaft sprocket bolt.

11. Temporarily install the flywheel or the driveplate using three of the old bolts. Lock the flywheel or the driveplate using a holding tool. See Fig. 28.

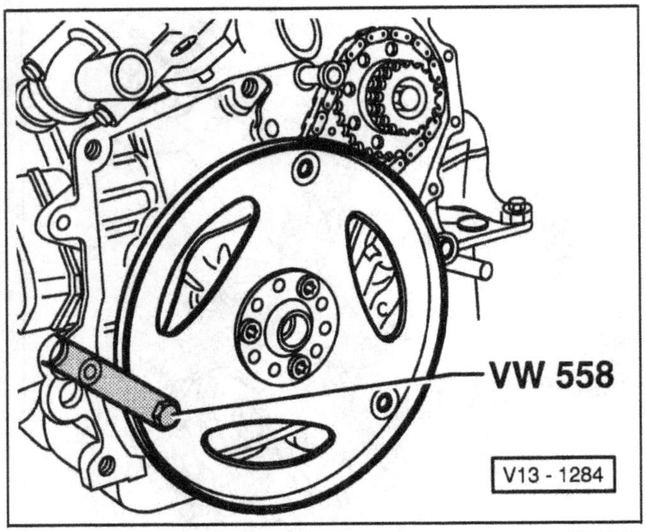

V13-1284

Fig. 28. Installed driveplate shown locked in position using VW special tool no. 558.

12. With the crankshaft locked, remove the intermediate shaft sprocket mounting bolt.

13. Remove the lower timing chain tensioner. Remove the intermediate shaft sprockets together with the lower timing chain and the lower guide rail.

14. Remove the flywheel or driveplate.

15. Inspect the tensioner assemblies and the chain guides for wear or damage. Replace any worn or damaged parts.

To install camshaft timing chains

1. Set the intermediate shaft and the crankshaft to TDC. See Fig. 29.
   - Temporarily install the intermediate shaft single-tooth sprocket. Rotate the intermediate shaft sprocket so that the mark on the sprocket aligns with the notch on the intermediate shaft thrust washer. Remove the sprocket.
   - Position the crankshaft so the ground-down tooth on the sprocket aligns with the main bearing cap joint.

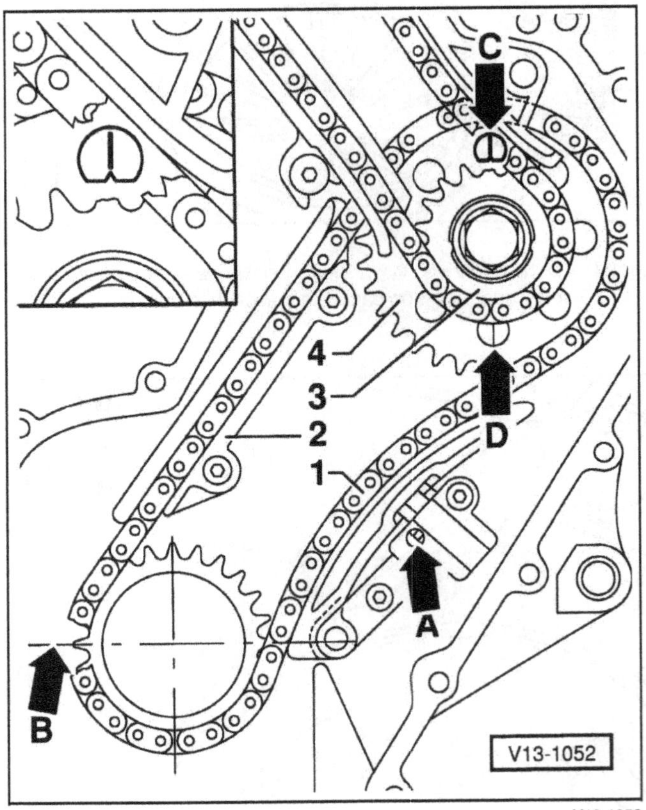

V13-1052

Fig. 29. Align mark on intermediate shaft sprocket (**C**) with notch on thrust washer (see inset). Align ground tooth on crankshaft with main bearing cap joint (**B**).

2. If previously removed, install the lower guide rail locating pin.

Tightening torque

- Lower guide rail locating pin (w/o collar) to engine block . . . . 25 Nm (18 ft-lb)

3. Install the lower guide rail and the intermediate shaft sprocket to the single-row chain, noting rotation direction marks made earlier.

4. Place the chain, guide rail and both intermediate shaft sprockets onto the engine. Install the intermediate shaft sprocket bolt hand tight.

NOTE —

Make sure all chain slack is on the tensioner side of the chain and that the sprocket positions have not been altered. Make sure the intermediate shaft sprockets correctly engage the shaft.

5. Install the lower chain tensioner. Use a small screwdriver to release the locking spline during installation. See Fig. 30. Install and tighten the tensioner mounting bolt, then lightly press the tensioner plate against the chain to remove all chain slack.

Tightening torque

- Lower chain tensioner to engine block . . . . 10 Nm (89 in-lb)

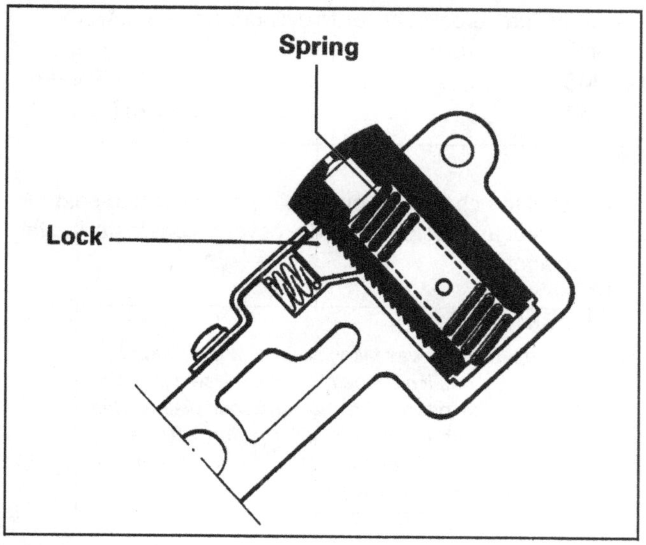

0024010

Fig. 30. Use small screwdriver to release timing chain tensioner lock.

6. Temporarily install the flywheel or driveplate, using at least three of the old bolts.

7. Lock the flywheel or the driveplate using a holding tool. See Fig. 28. Tighten the intermediate shaft sprocket mounting bolt. Recheck the sprocket timing marks.

Tightening torque

- Intermediate shaft sprockets to intermediate shaft . . . . 100 Nm (74 ft-lb)

8. Align the camshafts so that the slots in the rear of the camshafts are positioned to accept the VW special tool no. 3268. Install the special tool and push it forward into the camshaft cutouts. See Fig. 31.

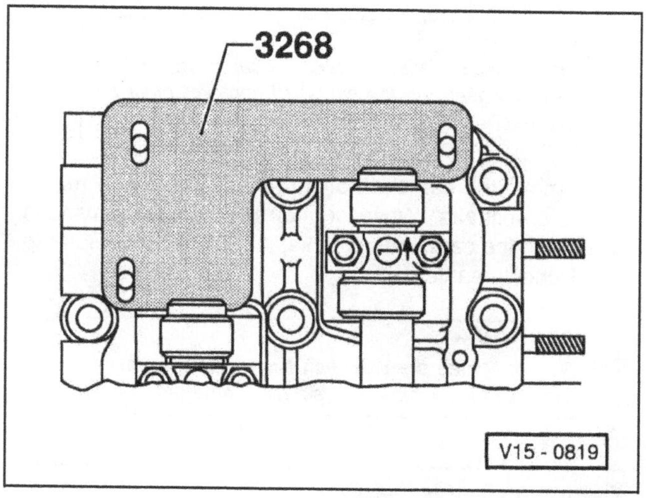

V15-0819

Fig. 31. VW special tool no. 3268 aligns camshafts at TDC. Tool engages slots in rear of camshafts.

9. If previously removed, install the tensioner guide pivot pin and the upper guide rail location pin. Place the upper guide rail into position and install the two mounting bolts. Place the chain tensioner guide into position.

CAUTION —

Use care when working near the cylinder head gasket. If the gasket is damaged, it should be replaced.

NOTE —

Use Volkswagen locking fluid "D6" or equivalent on the shorter guide rail mounting bolt.

Tightening torques

- Tensioner guide pivot pin to engine block . . . . 25 Nm (18 ft-lb)
- Upper guide rail to engine block locating pin (with collar) . . . . 20 Nm (15 ft-lb)
- Hex bolts (use "D6" locking fluid on shorter bolt) . . . . 25 Nm (18 ft-lb)

10. Install the double-row chain onto the intermediate shaft, noting rotation direction marks made earlier.

11. Mount the camshaft sprocket to the shorter camshaft (cylinders no. 2, 4, and 6). Install the sprocket mounting bolt hand-tight. Install the chain to the sprocket so that there is no slack between the sprockets.

12. Install the remaining cam sprocket into the chain and onto the camshaft so that all the chain slack is at the tensioner side of the chain. Install the mounting bolt together with the camshaft position sensor wheel. Hand tighten the bolt.

> ***CAUTION —***
> *The mating surfaces between the camshaft position sensor wheel and the camshaft sprocket must be dry and free of oil before tightening the bolt.*

13. Remove the camshaft alignment tool from the rear of the camshafts. While counterholding the camshafts, tighten the camshaft sprocket mounting bolts. See Fig. 31 shown earlier.

> ***CAUTION —***
> *Do not tighten the camshaft bolts with the plastic alignment tool installed. It is easily damaged.*

Tightening torque
- Camshaft sprocket to camshaft (with bolt contact surface oiled) . . . . 100 Nm (74 ft-lb)

14. Clean off any old sealer from both sides of the head gasket.

15. Remove the flywheel or driveplate.

16. Install the lower timing chain cover using Volkswagen sealant AMV 188 001 01 or an equivalent gasket sealer on the gasket mating surfaces, including the lower side of the head gasket.

> ***CAUTION —***
> *Use care not to damage the crankshaft seal when installing the lower cover. Coat the seal with a light coat of engine oil before installing the cover.*

Tightening torque
- Lower timing chain cover to engine block or oil pan (M6). . . . . 10 Nm (89 in-lb)

17. Install the camshaft sprocket cover (with new O-ring) using Volkswagen sealant AMV 188 001 01 or an equivalent gasket sealer on the gasket mating surfaces. See Fig. 32. Tighten the larger M8 bolts first.

> ***CAUTION —***
> *Be sure to install a new O-ring to the rear of the cover. Lightly oil the O-ring and install it into the cover before installing the cover.*

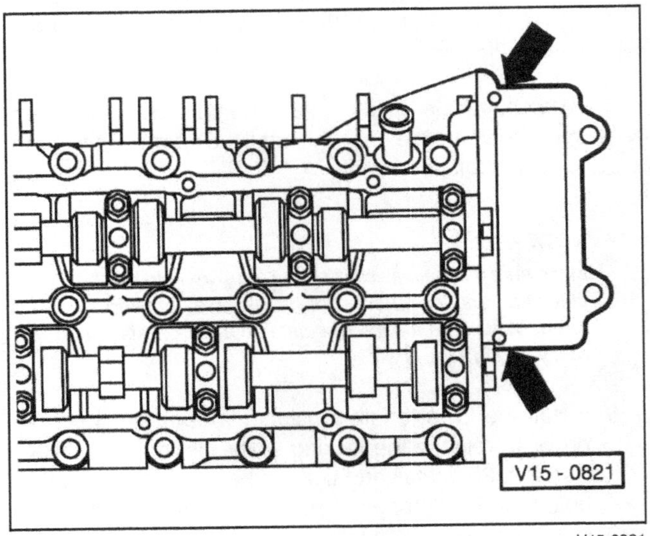

V15-0821

Fig. 32. Clean out old sealer from 3 mm holes (**arrows**) and re-seal holes and gasket surface using Volkswagen sealant AMV 188 001 02 or equivalent.

Tightening torques
- Camshaft sprocket cover to cylinder head or lower timing chain cover
  - M6 . . . . 10 Nm (89 in-lb)
  - M8 . . . . 25 Nm (18 ft-lb)

18. Install the chain tensioner using a new O-ring seal. Oil the seal and install it into the cover before installing the tensioner.

> ***CAUTION —***
> *If the tensioner was taken apart or if the piston has expanded out, it must be bled down. Use a thin wire through the opening in the tensioner piston. See Fig. 33. Push the wire into the piston while apply light pressure. When the piston check valve opens, the tensioner will collapse.*

Tightening torque
- Camshaft chain tensioner to camshaft sprocket cover. . . . . 30 Nm (22 ft-lb)

19. Turn the engine over by hand in normal engine rotation direction two full revolutions and realign the timing marks shown in Fig. 24 or Fig. 25.

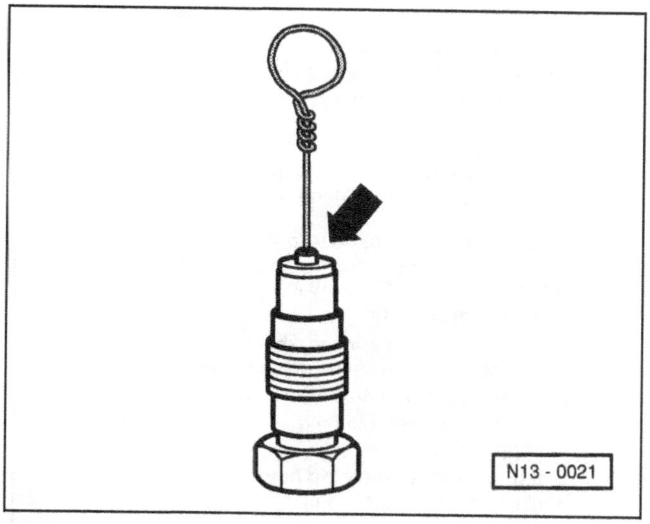

N13-0021

Fig. 33. Bleed tensioner by inserting thin wire (0.8 mm) into piston hole (**arrow**). Apply pressure to tensioner piston while pushing on wire. When piston check valve opens, piston will move into tensioner body.

20. Install VW special tool no. 3268 to the camshafts. If the chains are setup correctly, the tool will slide smoothly into the camshaft cutouts.

> ***CAUTION —***
> *Do not run the engine if VW special tool no. 3268 does not fit into the camshaft cutouts—the chain assembly may not be installed correctly. If only very minor readjustment of the camshafts allows the tool to slide into place, the assembly is probably setup correctly. If any problems are encountered, reinstall the chains from the beginning of this procedure.*

21. The remainder of installation is the reverse of removal. Be sure to use new mounting bolts on the flywheel or driveplate. Install the cylinder head cover and upper intake manifold as described under **To remove and install cylinder head cover**. Install the engine and transaxle assembly as described under **10 Engine–Removing and Installing**.

Tightening torques

- Flywheel or driveplate to crankshaft (stretch bolts–**always replace**)
  - stage I . . . . 60 Nm (44 ft-lb)
  - stage II . . . . additional 1/4 turn (90°)

CYLINDER HEAD RECONDITIONING

Cylinder Head and Camshafts

An exploded view of the cylinder head is given in Fig. 34. Cylinder heads with small, fine cracks between valve seats and plug threads are usable provided the cracks are not more than 0.5 mm (0.02 in.) wide, and do not extend into more than the first few spark plug threads.

If the camshafts were removed or if installing a new or remanufactured cylinder head, lubricate the contact surfaces between the cam lobes and the hydraulic cam followers prior to start-up.

When removing and installing the camshaft, follow the procedure given under **To remove and install camshaft**. Incorrect camshaft removal or installation can result in a damaged camshaft or cylinder head. Lubricate the contact surfaces between the cam lobes and the hydraulic cam followers prior to start-up.

When replacing the cylinder head or the cylinder head gasket, it is recommended that the old coolant be flushed out and new coolant added. The will help ensure that any metal shavings and contaminants are removed from the engine and the cooling system. See **19 Engine–Cooling System** for draining and flushing procedures.

If the camshafts were removed or if installing a new or remanufactured cylinder head, lubricate the contact surfaces between the cam lobes and the hydraulic cam followers prior to start-up.

Follow the camshaft removal and installation procedures given earlier under **To remove and install camshaft**. Specific procedures, and in some cases, special tools are required for camshaft removal. Refer to this section to prevent damaging the camshaft or the cylinder head.

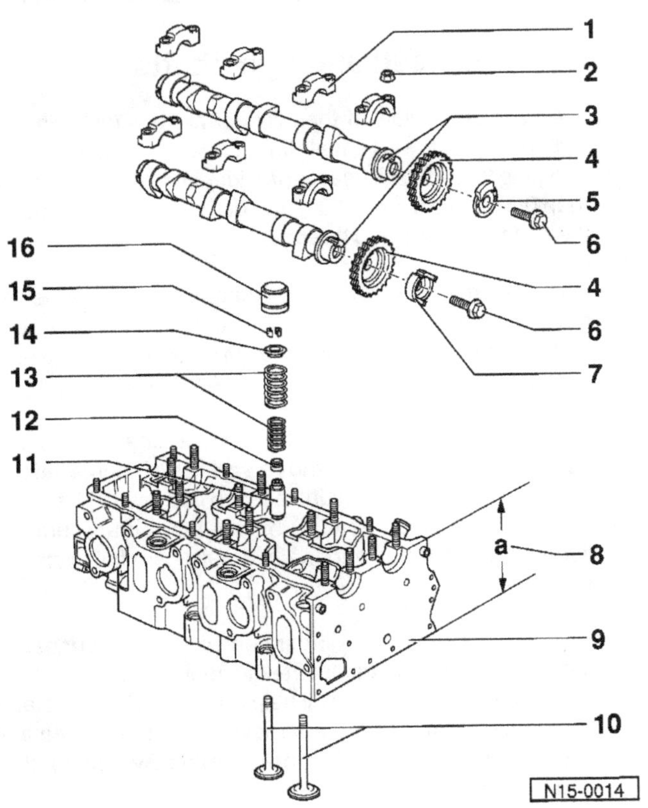

1. **Camshaft bearing cap**
   - note identification markings.
2. **Camshaft bearing cap nut**
   - 20 Nm (15 ft-lb)
3. **Camshaft**
   - radial clearance (Plastigage®) wear limit 0.1 mm (0.004 in.)
   - runout wear limit 0.01 mm (0.0004 in.)
   - axial clearance (measure with cam followers removed) wear limit 0.15 mm (0.0059 in.)
4. **Camshaft sprockets**
5. **Camshaft position sensor wheel**
   - contact surfaces must be dry of oil when installing
6. **Camshaft sprocket mounting bolt**
   - tighten to 100 Nm (74 ft-lb)
7. **not applicable**
8. **Cylinder head resurfacing height (a)**
   - minimum 139.5 (5.492 in.)
9. **Cylinder head**
   - use new coolant when replacing
10. **Valves**
    - do not reface by grinding, hand-lap only
11. **Valve guide**
12. **Valve stem oil seal**
13. **Valve springs**
14. **Valve spring seat, upper**
15. **Valve keepers**
16. **Hydraulic cam follower**
    - do not interchange between cylinders
    - store with contact surface down
    - oil contact surface when installing

N15-0014

Fig. 34. Exploded view of AAA (2.8 liter, VR6) cylinder head.

Camshafts and Cam Followers

Do not interchange camshaft bearing caps or cam followers. The arrows on the numbered bearing caps should point towards the drive belt end of the engine. See Fig. 35.

> ***CAUTION —***
>
> *After installing new cam followers, the engine should not be started for at least 30 minutes. New cam followers are at full height and must be allowed to bleed down to their proper height. Failure to do this may cause valve or piston damage.*

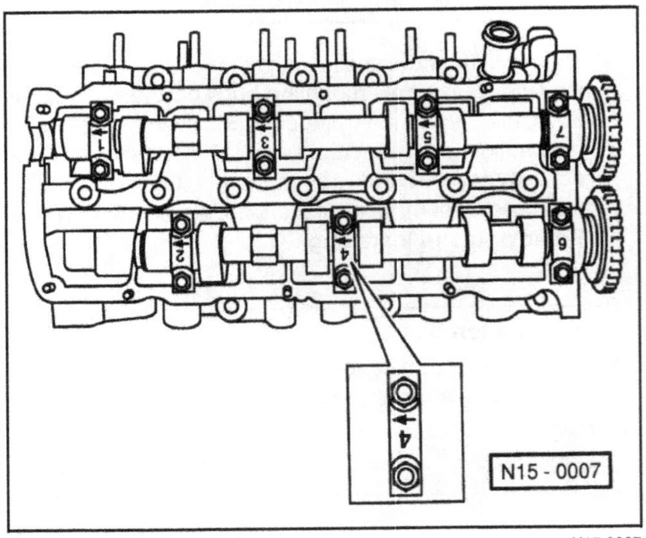

N15-0007

Fig. 35. Arrows on bearing caps face drive belt end of engine. Numbers should be readable from exhaust side of engine.

Valves

Both the intake and exhaust valves are sodium-filled and should be hand-lapped only. Valve and valve spring specifications are given in **Table b**. Refer to Fig. 36 when using the table.

> ***WARNING —***
> *Sodium filled valves, when discarded, must be disposed of properly to avoid personal injury. Always wear protective goggles or glasses. By hand, cut off the valve stem near the head of each valve. Use only a hack saw. Do not use a power saw. Sodium reacts violently with water. Do not let water contact the valve while cutting. Throw the valve parts (no more than 10 valves at a time) into a bucket of water and stand clear. Discard the valves when the reaction has ceased.*

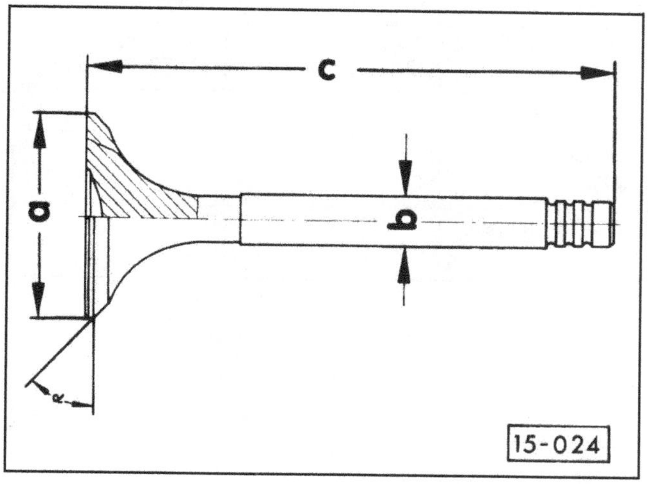

15-024

Fig. 36. Valves dimensions to be used in conjunction with **Table b**.

Table b. Valve Specifications

Engine Code	AAA
Valve head diameter (**a**) intake exhaust	39.00 mm (1.5354 in.) 34.20 mm (1.3465 in.)
Valve stem diameter (**b**) intake exhaust	6.97 mm (0.2744 in.) 6.95 mm (0.2736 in.)
Valve length (**c**) intake exhaust	105.95 mm (4.1713 in.) 106.95 mm (4.2106 in.)
Valve face angle (α) intake exhaust	45° 45°
Valve margin	N/A

Valve Guides

Special tools and a press are required to replace valve guides. Check valve guide wear using new valve. See Fig. 37. Inspect the valve seats to ensure that the cylinder head can be reconditioned before installing new valve guides.

> ***CAUTION —***
> *When checking valves guides, be sure to use an exhaust valve on the exhaust side and an intake valve on the intake side. Exhaust and intake valve stem diameters are slightly different. See **Table b**.*

Valve Guide Wear Limits

- Maximum play
  - intake . . . . 1.0 mm (0.039 in.)
  - exhaust. . . . . 1.3 mm (0.051 in.)

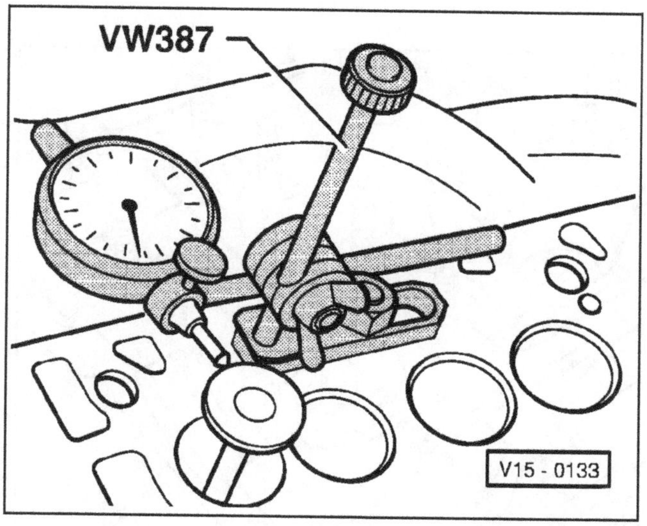

V15-0133

Fig. 37. Valve guide wear being checked with new valve. Insert valve until stem end is flush with end of guide. Rock valve back and forth to check total travel.

Original valve guides (without shoulders) are pressed out from the camshaft side of the cylinder head. Replacement valve guides (with shoulders) are pressed out from the combustion chamber side of the cylinder head. Lubricate new valve guides with oil and press them in from the camshaft side until the shoulder is fully seated.

> ***NOTE —***
> - *Where available, replace valve guides using Volkswagen special tools. Use of other press tools may damage the cylinder head.*
> - *Use VW tool no. 3121 (order no. T03 121 000 15 ZEL) for valve guide installation and removal. Ream the valves guides using VW special tool no. 3120.*

Valve Seats

When resurfacing valve seats, there is a limit to the amount of material that can be removed. If too much material is removed, the final assembly will leave too little space for the hydraulic cam follower to function correctly. The maximum refacing dimension—the maximum amount of material that can be removed from the valve seat—is calculated from the measurement shown in Fig. 38.

Measure dimension (**A**) in Fig. 38, and subtract the minimum dimension, as given in **Table c**. The difference is the maximum amount of material that can be removed from the valve seat. Valve seat dimensions are given in table **Table d**.

Table c. Minimum Dimensions for Calculating Valve Seat Refacing Dimensions

Engine Code	Intake	Exhaust
AAA	33.9 (1.335 in.)	34.1 mm (1.343 in.)

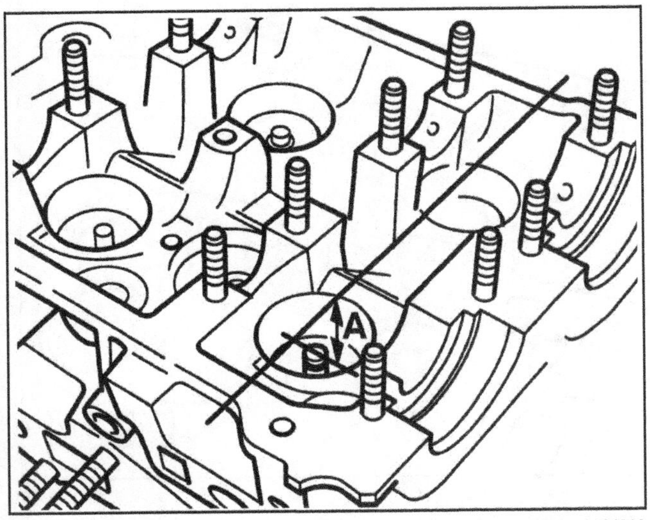

Fig. 38. Dimension **A** (distance between top of valve stem and gasket surface of cylinder head) is used to calculate maximum valve seat refacing dimensions.

Table d. Valve Seat Dimensions: AAA Engine

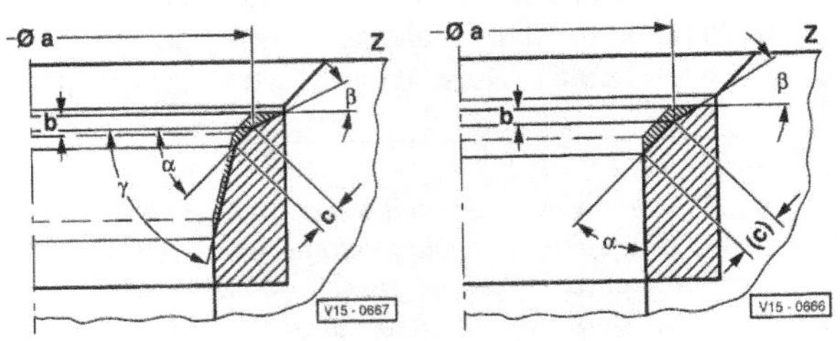

Engine Code: AAA	**Intake** (V15-0667)	**Exhaust** (V15-0666)
Seat diameter (**a**)	38.3 mm (1.508 in.)	33.5 mm (1.319 in.)
Maximum refacing dimension (**b**)	Calculated See Fig. 38.	Calculated See Fig. 38.
Seat width (**c**)	1.4-2.0 mm (0.055-0.079 in.)	2.0-2.5 mm (0.079-0.098 in.)
Valve seat angle (α)	45°	45°
Correction angle, upper (β)	30°	30°
Correction angle, lower (γ)	75°	N/A

17 Engine–Lubrication System

GENERAL

Proper engine lubrication relies on a constant supply of oil, fed to the moving parts under pressure. Pressure is supplied by a gear-type oil pump located inside the oil pan. Engine oil returns to the oil pan where it is stored for pickup by the pump. It is cleaned by circulating through a replaceable filter.

Fig. 1 shows an exploded view of the lubrication system for 4-cylinder engines. Fig. 2 shows an exploded view of the lubrication system for 6-cylinder engines.

NOTE —

Oil change, oil filter changes, and engine oil specifications are covered in greater detail in **0 Maintenance Program**.

1. **Oil filler cap gasket**
2. **Oil filler cap**
3. **Oil dipstick**
4. **Dipstick tube**
5. **Oil pressure switch**
   - brown: 0.3 bar (4.4 psi)
   - blue: 0.2 bar (3.0 psi)
   - tighten to 25 Nm (18 ft-lb)
6. **Oil pressure switch**
   - white: 1.8 bar (26 psi)
   - grey: 0.9 bar (13 psi)
   - tighten to 25 Nm (18 ft-lb)
7. **Check valve**
   - tighten to 5 Nm (44 in-lb)
8. **Plug**
9. **Bolt**
   - tighten to 20 Nm (15 ft-lb)
10. **Oil filter bracket**
11. **O-ring**
    - always replace
12. **Oil cooler (except ACC engines)**
13. **Oil cooler mounting nut**
    - tighten to 25 Nm (18 ft-lb)
14. **Oil filter**
    - hand tighten when installing
15. **Oil pump gears**
16. **Oil pump cover with pressure relief valve**
    - pressure relief opening pressure: 5.7–6.7 bar (82.5–97 psi)
17. **Bolt**
    - tighten to 10 Nm (89 in-lb)
18. **Bolt**
    - tighten to 20 Nm (15 ft-lb)

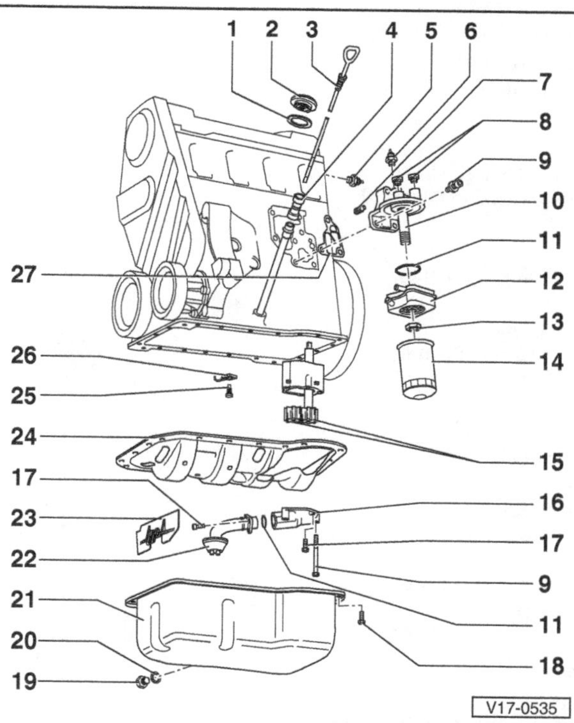

19. **Oil drain plug**
    - tighten to 30 Nm (22 ft-lb)
20. **Drain plug seal**
    - always replace
21. **Oil pan**
22. **Oil pump suction line**
23. **Deflector**
24. **Sound baffle (not installed on all versions)**
25. **Pressure relief valve for oil spray nozzle**
    - See **Oil Spray Nozzles** for tightening torques
26. **Oil spray nozzle (not installed on all versions)**
27. **Oil filter housing gasket**
    - always replace

V17-0535

Fig. 1. Exploded view of lubrication system on ABA (2.0 liter) engine. Lubrication system on other 4-cylinder engines is similar.

OIL PRESSURE WARNING SYSTEM

To prevent serious engine damage, a dynamic oil pressure warning system warns the driver of insufficient oil pressure. Other safety features include a filter by-pass to prevent oil starvation and an oil pump pressure relief valve to prevent excessive system pressure.

The components of the dynamic oil pressure warning system are the two pressure switches, the electronic control unit, the oil pressure warning light in the dash, and the warning buzzer. A low pressure switch provides warning when the oil pressure falls to near zero at any time. A high pressure switch provides earlier warning, at elevated rpm, whenever oil pressure falls below a minimum safe level.

On 4-cylinder engines, the low pressure switch (0.2 bar, blue body) is threaded into the back of the cylinder head and the high pressure switch (1.8 bar, white body) is threaded into the top of the oil filter bracket.

On 6-cylinder engines, both the low pressure switch (0.2 bar, blue body) and the high pressure switch (1.4 bar, black body) are threaded into the top of the oil filter housing. See Fig. 3.

NOTE —
- *On some engines, the low pressure switch may have a brown body and a slightly higher opening pressure (0.3 bar).*
- *On some 4-cylinder engines, the low pressure switch may be threaded into the oil filter bracket, beside the high pressure switch.*

The electronic control unit is integrated into the instrument cluster.

1. **Bolt**
   - tighten to 10 Nm (89 in-lb)
2. **Cover for oil pump drive shaft**
3. **O-ring**
   - always replace
   - lubricate with engine oil before installing
4. **Oil pump drive shaft**
5. **Intermediate shaft**
6. **Thrust washer**
7. **Screw**
   - tighten to 10 Nm (89 in-lb)
   - install using locking fluid "D6"
8. **Oil pump**
9. **Bolt**
   - tighten to 25 Nm (18 ft-lb)
10. **Bolt**
    - always replace
    - tighten to 10 Nm (89 in-lb)
    - install using locking fluid "D6"
11. **Oil drain plug**
    - tighten to 30 Nm (22 ft-lb)
12. **Drain plug seal**
    - always replace
13. **Oil spray nozzle**
    - opening pressure: 2.0 bar (29 psi)
14. **Bolt**
    - tighten to 15 Nm (11 ft-lb)
15. **Oil pan gasket**
16. **Oil pan**
17. **Oil cooler**
18. **Oil cooler cover**
    - tighten to 25 Nm (18 ft-lb)
19. **Oil filter housing gasket**
    - always replace
    - lubricate with engine oil before installing
20. **Oil filter housing**
21. **Dipstick**
22. **Dipstick guide tube**
23. **Non-return valve (check valve)**
    - tighten to 5 Nm (44 in-lb)
    - note installation position before removing

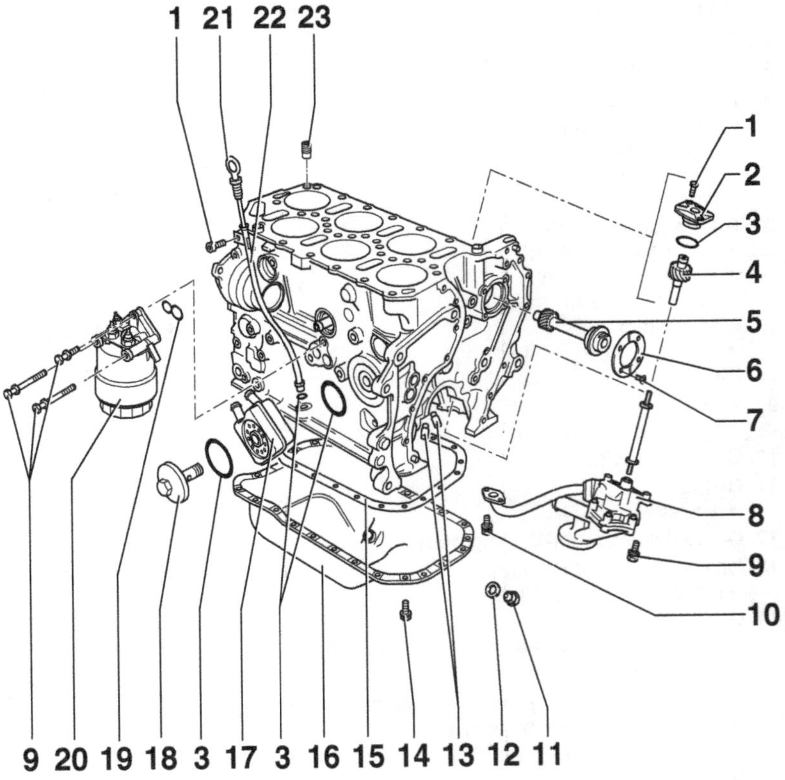

N17 - 0004

N17-0004

Fig. 2. Exploded view of lubrication system on 2.8 liter, VR6 engine.

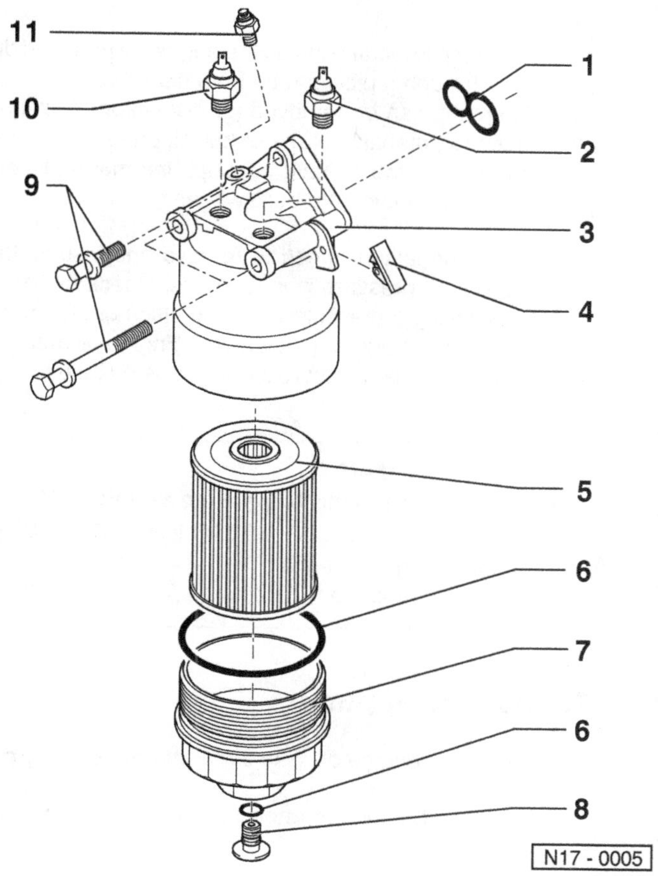

1. **Gasket**
2. **Oil pressure switch: 0.2 bar (3 psi)**
   - blue body, blue/black wire
   - 25 Nm (18 ft-lb)
3. **Oil filter housing, upper**
4. **Stop plate (for oil cooler)**
5. **Oil filter element**
6. **O-ring**
   - always replace
   - lubricate with engine oil before installing
7. **Oil filter housing, lower**
   - tighten to 30 Nm (22 ft-lb)
8. **Filer housing drain screw**
   - tighten to 10 Nm (89 in-lb)
9. **Bolts**
   - tighten to 25 Nm (18 ft-lb)
10. **Oil pressure switch: 1.4 bar (20 psi)**
    - black body, yellow wire
    - tighten to 25 Nm (18 ft-lb)
11. **Engine oil temperature sensor**
    - tighten to 10 Nm (89 in-lb)

N17-0005

Fig. 3. Exploded view of oil filter housing on VR6 engine.

Checking Low Oil Pressure Warning System

With the ignition off, the low pressure switch is closed (complete circuit to ground). When the ignition is turned on, the indicator flashes (no oil pressure). When the engine is started and oil pressure rises, the switch opens and the indicator goes out. With insufficient oil pressure or a stuck switch, the indicator will continue to flash. The switch opens and closes at specific pressures. See **Testing Oil Pressure Switches**.

> ***CAUTION—***
> *If the warning indicator stays on after the engine is started, or flashes on while driving, always assume that there is insufficient oil pressure. Check oil level and test oil pressure before proceeding with tests of the warning system.*

To quick-check the low pressure switch, simulate a closed switch by removing the blue/black wire and grounding it. With the ignition on, the indicator should flash. If not, the problem is in the circuit or the indicator. If the indicator flashes only when tested in this way, replace the switch.

If the indicator flashes with the engine running, and the oil pressure tests OK, remove the blue/black wire from the switch with the engine running. The indicator should go out. If not, the circuit is shorted to ground between the wire and indicator. If the indicator does go out, the switch is stuck closed and should be replaced.

Checking Dynamic Oil Pressure Warning System (high rpm)

The high pressure switch is normally open (no circuit to ground). When the engine is running above 2,000 rpm, oil pressure closes the switch and completes the circuit to ground. If oil pressure is insufficient to close the pressure switch, or the switch is stuck open, the indicator will flash and the buzzer will sound at engine speeds above 2,000 rpm. The pressure switch opens and closes at specific pressures. See **Testing Oil Pressure Switches.**

To quick-check the pressure switch, disconnect the yellow wire from the switch and raise the engine speed to between 2000 to 2500 rpm. The indicator should flash and the buzzer should sound. If the wire is connected to ground, the warning should stop. Remove the wire from ground and reconnect to the switch. If the indicator does not go out, replace the switch.

Testing Oil Pressure Switches

The oil pressure switches can be accurately tested by temporarily installing an oil pressure gauge in parallel with the switch with a T-fitting and monitoring switch operation with a multimeter or test light, as shown in Fig. 3.

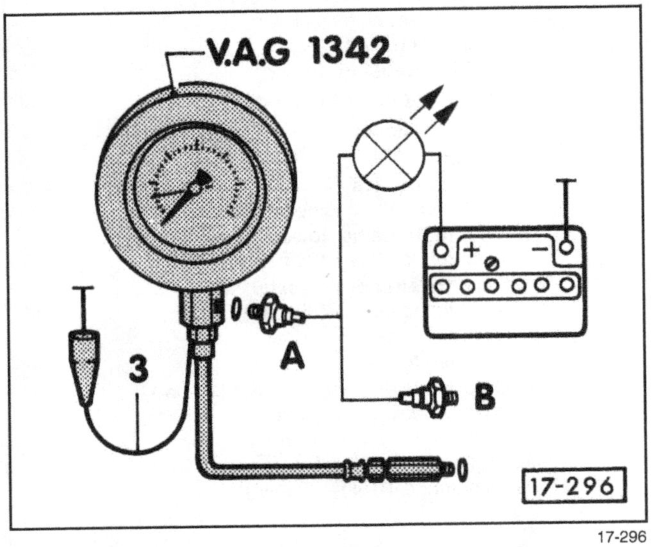

17-296

Fig. 4. Pressure gauge and test light setup for testing oil pressure switches. V.A.G. 1342 is Volkswagen pressure tester. Remove switch from engine (**B**) and install in tester (**A**). Connect wire (**3**) to ground and monitor switch function using an LED or multimeter.

Oil Pressure Switch Test Pressures

- Low pressure switch (engine running, raise idle speed)
  switch opens . . . . 0.15-0.45 bar (2.1-6.5 psi)
- Dynamic pressure switch (idle speed above 2,000 rpm)
  switch closes . . . .
  white switch(1.8 bar) . . . . 1.6-2.0 bar (23-29 psi)
  black switch (1.4 bar). . . . . . . 1.2-1.6 bar (17- 23 psi)
  grey switch (0.9 bar). . . . . 0.7-1.1 bar (10-16 psi)

If the switches are performing properly, the oil pump is producing adequate oil pressure and the system still warns of low oil pressure, then the dynamic oil pressure warning system control unit within the instrument cluster should be replaced.

> ***NOTE —***
> *If an oil pressure switch seal is leaking, a new sealing ring should be installed. The sealing ring is held captive on the switch and needs to be cut off. New sealing rings are available from Volkswagen.*

OIL PUMP

The oil pump is located inside the engine oil pan, and draws engine oil through a pickup tube from near the bottom of the pan. The pump can be removed for inspection of its internal clearances (a potential source of low oil pressure problems) by first removing the oil pan, although internal replacement parts are not available from Volkswagen.

There is normally no need to remove and inspect the oil pump unless oil pressure is inadequate. Check the oil pressure by installing a pressure gauge in place of one of the oil pressure switches and run the engine. Engine oil pressure at normal operating temperature should as listed below.

Engine oil pressure

- engine fully warm (oil temperature @ 80°C (176°F)
  @2,000rpm . . . . at least 2.0 bar (29 psi)
  @higher engine speeds,
  maximum pressure. . . . . 7.0 bar (101 psi)

To inspect oil pump

1. Drain the engine oil. See **0 Maintenance Program.**
2. Remove the oil pan retaining bolts and remove the oil pan.

> ***NOTE —***
> *On 4-cylinder engines, accessing the oil pan mounting bolts on the flywheel end of the engine may be difficult. If available, use VW special tool no. 3185 (10 mm swivel/extension) when loosening or tightening these bolts.*

3. Remove the pump and disassemble it.
   - On 4-cylinder engines, use Fig. 1 as a guide during oil pump disassembly and assembly.
   - On 6-cylinder engines, see Fig. 5 for oil pump disassembly and assembly.
4. Check the oil pump backlash and axial play. See Fig. 6 and Fig. 7. If either clearance exceeds the specifications shown, the pump should be replaced.

> ***NOTE —***
> *On 4-cylinder engines, always replace the pickup tube O-ring if the tube has been removed from the pump cover.*

5. Assembly of the pump and reinstallation of the oil pan are the reverse of removal. Clean the oil pan sealing surfaces thoroughly. Where applicable, install a new pan gasket. Add engine oil, as described in **0 Maintenance Program**. Tightening torques as follows.

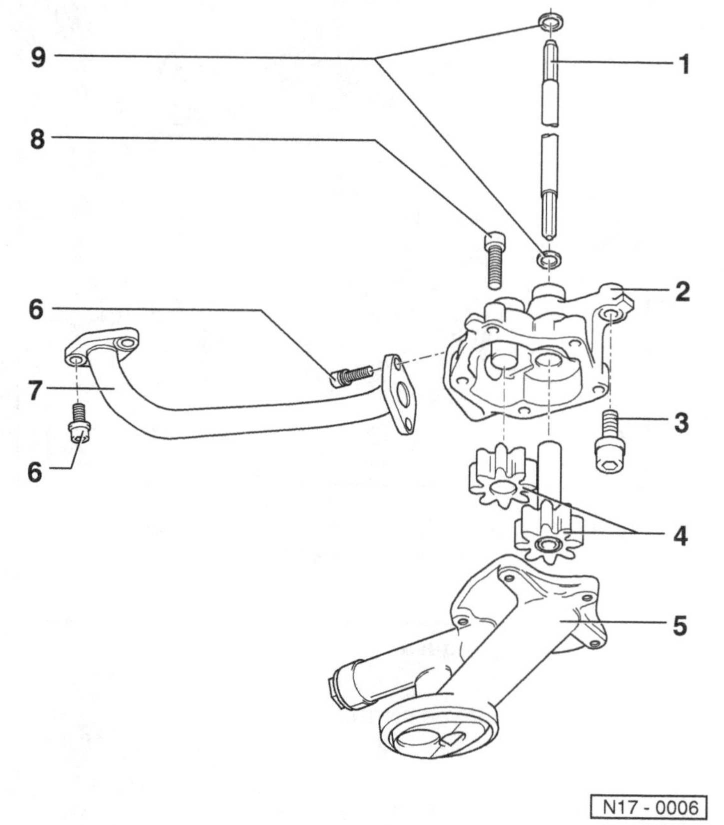

1. **Drive shaft**
2. **Oil pump housing**
3. **Bolts**
   - tighten to 25 Nm (18 ft-lb)
4. **Oil pump gears**
5. **Oil pump cover**
   - clean screen/strainer before installing
6. **Bolt**
   - always replace
   - install using VW "D6" locking fluid
   - tighten to 10 Nm (89 in-lb)
7. **Oil pressure pipe**
   - coat flanges with sealer (VW AMV 188 001 02) during installation
8. **Bolt**
   - tighten to 10 Nm (89 in-lb)
9. **O-ring**
   - replace if damaged

N17-0006

Fig. 5. Exploded view of oil pump on VR6 engine.

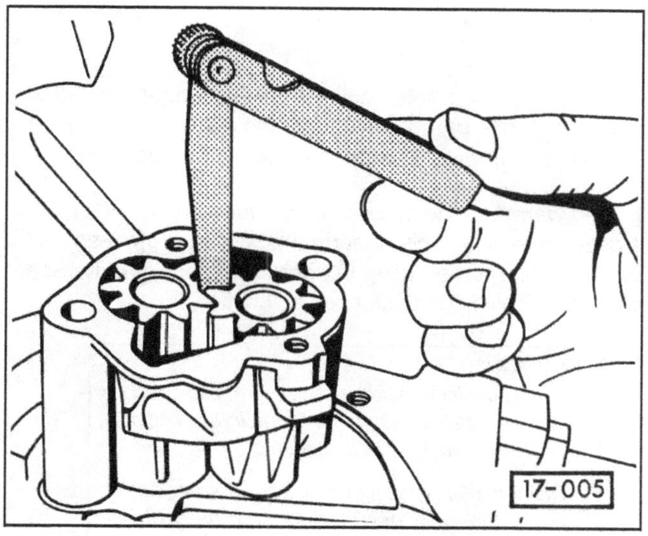

17-005

Fig. 6. Oil pump gear backlash being measured using feeler gauge. Maximum allowable wear limit: 0.20 mm (0.0079 in.)

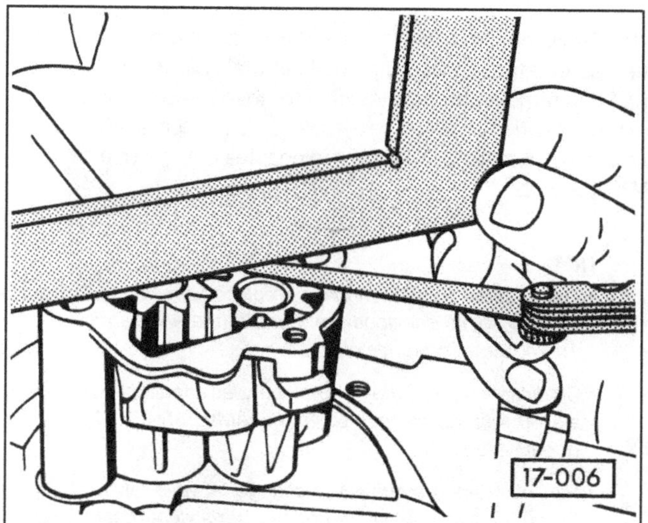

Oil pump gear axial play
- 4-cylinder: wear limit (max. allowable): 0.15 mm (0.0059 in.)
- 6 cylinder: wear limit (max. allowable): 0.10 mm (0.0039 in.)

17-006

Fig. 7. Oil pump gear axial play being measured using straight edge and feeler gauge.

Tightening torques

- Oil pan drain plug to oil pan .... 30 Nm (22 ft-lb)
- Oil pump cover to oil pump .... 10 Nm (89 in-lb)
- Oil pump to cylinder block
  - 4-cylinder...................... 20 Nm (15 ft-lb)
  - 6-cylinder...................... 25 Nm (18 ft-lb)
- Oil pump pickup tube to oil pump cover (4-cylinder) .... 10 Nm (89 in-lb)
- Oil pump pressure pipe to oil pump housing (VR6) use "D6" locking fluid on **new** bolts, coat pipe flanges with VW AMV 188 001 02 sealer)........ 10 Nm (89 in-lb)
- Oil pan to engine block (M6)
  gasoline engines
  - 4-cylinder...................... 20 Nm (15 ft-lb)
  - 6-cylinder .... 15 Nm (11 ft-lb)
  - diesel engines .... 25 Nm (18 ft-lb)

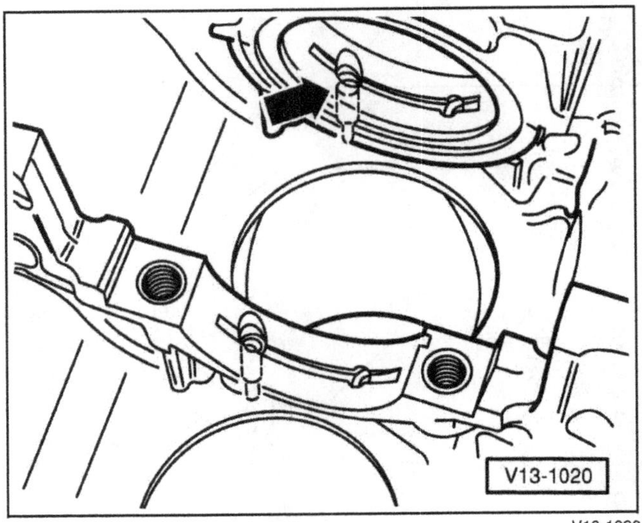

V13-1020

Fig. 8. Oil spray nozzles for each piston are fitted above main bearing caps (**arrow**) on VR6 engine.

OIL SPRAY NOZZLES

AAZ (1.9 liter diesel), 1Z/AHU (1.9 liter diesel), AAA (2.8 liter VR6) and some versions (through 8/95 production) of the ABA (2.0 liter gasoline) engines are equipped with oil spray nozzles. The oil spray nozzles spray oil from the main oil galley against the bottoms of the pistons for added cooling.

On 4-cylinder engines, the oil spray nozzles are bolted to the bottom of the cylinder block. The oil pan must be removed to gain access to the oil spray nozzles. See **13 Engine–Crankshaft/Cylinder Block.**

On 6-cylinder engines, the oil spray nozzles are fitted above the crankshaft main bearing shells in the cylinder block. See Fig. 8. To remove the nozzles, first remove the crankshaft and the main bearing shells. Use a 4mm (5/32 in.) drift from behind to drive the nozzles out. Install the nozzles using a use a 6mm (7/32 in.) drift.

NOTE —

- *The oil spray nozzles should be replaced if metal particles or metal shavings are found in the engine oil. The nozzles cannot be cleaned.*
- *On diesel engines, the oil spray nozzles should be installed with Volkswagen sealer no. AMV 188 100 02, or equivalent.*
- *Part numbers are subject to change. Always rely on an authorized Volkswagen dealer parts department for the latest information.*

Tightening torques

- ABA engine
  Oil spray nozzle pressure relief valve to cylinder block .... 25 Nm (18 ft-lb)
- AAZ engine
  Oil spray nozzle mounting bolt to cylinder block (use VW sealer no. AMV 188 100 02)............ 10 Nm (89 in-lb)

OIL COOLER

The oil cooler is an oil-to-coolant heat exchanger. Engine oil flows through one part of the cooler and gives up heat to the engine coolant flowing through the other part of the cooler.

During warm-up, the process is reversed and engine coolant gets warm first. This speeds the warm-up process and gets the engine to operating temperature more quickly. See Fig.1 or Fig. 2 given earlier for cooler location.

CAUTION —

- *The oil cooler should be replaced if metal particles or metal shavings are found in the engine oil. The cooler cannot be cleaned.*
- *The oil cooler is a potential source of leakage between the lubrication system and the cooling system, and should be considered whenever such leakage is suspected.*

Remove the oil cooler only if it needs to inspected, or replaced. On installation, use sealing paste, Volkswagen sealer no. AMV 188 100 02 on the sealing surfaces and check to see that there is adequate space for the coolant hose connections. Tighten the cooler mounting bolt to 25 Nm (18 ft-lb).

19 Engine–Cooling System

GENERAL

This section covers repairs and troubleshooting information for the engine cooling system only. For heater core and related heating and air conditioning components, see **80 Heating and Ventilation.** For information on the engine oil cooler, see **17 Engine–Lubrication System**. For information on the ATF cooler used on models with automatic transmission, see **37 Automatic Transmission.**

Fig. 1 shows a schematic view of the cooling system and hose routing.

Warnings and Cautions

The following warnings and cautions should be followed when working on the cooling system.

> ***WARNING —***
> - *Hot coolant can scald and result in serious personal injury. Do not work on the cooling system until it has fully cooled.*
> - *At normal operating temperature the cooling system is pressurized. Allow the system to cool as long as possible before opening—a minimum of an hour—then release the cap very slowly to allow safe release of pressure.*

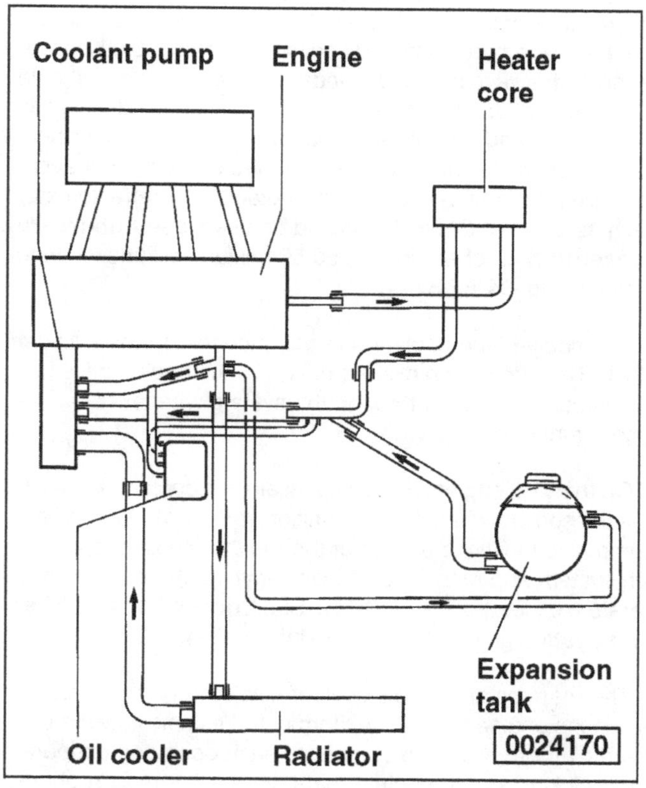

0024170

Fig. 1. Schematic view of typical cooling system components and hose routing for 4-cylinder engine. Arrows indicate coolant flow.

> *WARNING —*
> - *Releasing the cooling system pressure lowers the coolant's boiling point, and the coolant may boil suddenly. Use heavy gloves and wear eye and face protection to guard against scalding.*
> - *Use extreme care when working at or near the cooling fan when the engine is hot. The fan can come on at any time–even if the ignition key is off.*
> - *Use extreme care when draining and disposing of engine coolant. Coolant is poisonous and lethal. Children and pets are attracted to coolant because of its sweet smell and taste. See a doctor or veterinarian immediately if any amount of coolant is ingested.*

> *CAUTION —*
> *Avoid adding cold water to the coolant while the engine is hot or overheated. If it is necessary to add coolant to a hot system, do so only with the engine running and coolant pump turning.*

Coolant, Pump and Thermostat

Volkswagen uses two different types of coolant that are identified by color. The older, blue/green version is known as G11 and the newer red version is G12. Both types have specific characteristics and should not be mixed. **See 0 Maintenance Program** for additional information and cautions.

For best overall performance, the coolant additive proportion must be at least 40%, but not more than 60% to maintain the antifreeze protection and cooling efficiency. Distilled water should always be used. Under no circumstances should 100% coolant/antifreeze be used. This will generally result in overheating and poor heater operation due to the poor heat transfer characteristics of pure coolant/antifreeze. For accurate testing of antifreeze coolant protection, a refractometer such as VW tool CEN5021 should be used, see **Anti-freeze concentration, checking**. A 50/50 mixture will show freeze protection to -34°F (-37°C).

An impeller-type coolant pump is mounted to the cylinder block. The pump is crankshaft-driven by the engine drive belt and circulates coolant through the system whenever the engine is running.

On the VR6 engine, an additional electric coolant pump circulates coolant whenever the ignition key is on. The auxiliary pump also functions as part of the radiator cooling fan after-run system and runs automatically anytime the coolant temperature exceeds 220°F (107°C). See **Cooling Fan and After Run System** given later for more information.

The thermostat controls coolant flow through the radiator. When the engine is cold, the thermostat restricts flow through the radiator to allow the engine to reach operating temperature quicker. As the engine heats up, the thermostat opens and coolant circulates through the whole system, including the radiator.

Radiator and Cooling Fan

A radiator cooling fan assembly provides auxiliary air flow through the radiator. The fan assembly is electrically operated and thermostatically controlled so that it runs only when extra air flow is required to maintain proper coolant temperature.

The radiator is a cross-flow type constructed of an aluminum core and plastic side tanks. A translucent expansion tank, or overflow reservoir, provides for the expansion of the coolant at higher temperatures and easy monitoring of the coolant level.

An electric radiator cooling fan assembly is controlled by a thermoswitch located in the side of the radiator. At high coolant temperatures, the switch closes to start the cooling fans. Any time the coolant temperature is excessive, the fan will start and continue to run until coolant the temperature is in the correct range. On cars with air conditioning, high refrigerant pressure can also activate the cooling fan.

> *WARNING —*
> *The electric cooling fan can come on at any time, even if the key is out of the ignition. To avoid personal injury use extreme caution when working at or near the cooling fan if the engine is hot. As a safety precaution, always disconnect the harness connector from the fan when working in the engine compartment.*

On some models, a radiator cooling fan after-run system, operated by a control unit, runs the cooling fan (and on VR6 engines, an auxiliary coolant pump) after the engine is shut off. This helps to eliminate high under hood temperatures.

TROUBLESHOOTING

When investigating the cause of overheating or coolant loss, begin with a visual inspection of the system. Check the coolant level and for evidence of coolant leaks.

The system becomes pressurized at normal operating temperatures. Leaks may prevent the system from becoming pressurized and allow the coolant to boil at a lower temperature. If visual evidence is inconclusive, a cooling system pressure test will determine whether the system leaks, and may help to indicate the source.

If the cooling system holds pressure, the most probable causes of overheating are an electrical fault with the cooling fan, a faulty thermostat, or poor coolant circulation caused by a worn drive belt or clogged system.

Table a lists overheating and underheating symptoms, their probable causes, and suggested corrective actions. The bold type refers to areas outside of this repair group for suggested repairs.

Table a. Cooling System Troubleshooting

Symptom	Probable cause	Corrective Action
1. Engine overheats	a. Low coolant level	a. Fill the radiator to the **Max.** mark on a cold engine. Check cooling system for leaks with pressure tester.
	b. Radiator hose restricted (lower hose may collapse only at highway speeds)	b. Check hoses for soft, spongy areas. Replace faulty hoses.
	c. Drive belt worn and slipping	c. Replace drive belt. See **0 Maintenance Program**.
	d. Faulty thermostat	d. Remove and test thermostat. Replace as necessary.
	e. Radiator fan not switching on	e. Test thermoswitch and fan. Replace faulty parts.
	f. Faulty radiator cap	f. Test pressure relief valve in cap. Replace faulty caps.
	g. Clogged radiator	g. Clean or replace radiator.
	h. Incorrect ignition timing or valve timing	h. Check camshaft drivebelt/chain installation, See **1 ENGINE**. Adjust ignition timing (if applicable). See **28 Ignition System**.
	i. Coolant pump faulty	i. Replace coolant pump.
	j. Poor air flow through radiator	j. Check for debris buildup on front of radiator. Clean radiator exterior using compressed air.
	k. Incorrect coolant concentration	k. Adjust coolant concentration to a mix of 50% coolant additive and 50% distilled water.
2. Temperature gauge reads low, poor heater output	a. Faulty thermostat	a. Remove and test thermostat.
	b. Electric fan not switching off	b. Replace thermoswitch for fan.
	c. Incorrect coolant concentration	c. Adjust coolant concentration to a mix of 50% coolant additive and 50% distilled water.
3. Temperature gauge reads low, heater output normal	a. Faulty temperature gauge or coolant temperature sensor	a. Test temperature gauge and sensor. Replace instrument cluster if temperature gauge is faulty. See **90 Instruments**.
	b. Coolant old, causing inaccurate gauge readings	b. Replace coolant.
4. Temperature gauge reads normal, inadequate heater output	a. Installed position of heater hoses reversed	a. Install heater hoses. **80 Heating and Ventilation.**
	b. Heater hose restricted	b. Replace hose.
	c. Heat exchanger (heater core) clogged	c. Replace heater core or have core cleaned. See **80 Heating and Ventilation.**
	d. Heater or A/C ventilation controls/flaps not operating correctly	d. Check controls and vent operation. See **80 Heating and Ventilation.**

NOTE —

- *Coolant also circulates through the heater core in the passenger compartment. For problems associated with the heater core and the heating system, see* **80 Heating and Ventilation**.
- *Overheating problems may also be caused by an engine fault that leaks hot combustion gasses into the cooling system, or by ignition timing that is out of specification. See* **1 Engine or 28 Ignition System** *for additional information on these subjects.*

Diagnostic Checks

To pressure test cooling system

A pressure test uses a special tester to pressurize the system and simulate normal running conditions. If the system is unable to hold pressure, the engine will overheat more easily.

NOTE —

A pressure test also checks for internal leaks. Some of the common sources of internal leaks are a faulty cylinder head gasket, a cracked cylinder head, or a cracked cylinder block.

1. With the engine at normal operating temperature, pressurize the system using a cooling system pressure tester.

2. Observe the gauge reading. If the pressure drops, there is a leak in the system. A rapid drop may indicate a faulty cylinder head gasket if no external leaks can be found.

> *CAUTION —*
> *Exceeding the specified test pressure could damage the radiator or other cooling system components.*

Cooling system test pressures

- system test pressure (max.) . . . . 1.25 bar (18 psi)
- cap opening pressure . . . . . . 1.3–1.6 bar (19–22 psi)

3. Using the correct tester adapter, test the coolant expansion tank cap opening pressure. Compare the opening pressure to the specification listed above.

To check temperature gauge circuit

The checks given below should be made at the coolant temperature sensor and its harness connector. Sensor location varies depending on the engine code. See Fig. 2.

0024016

Fig. 2. Engine coolant temperature sensor (4-pin) on rear of cylinder head of ABA (2.0 liter) engine. Other engines are similar.

1. Disconnect the harness connector from the engine coolant temperature sensor.

Component Location

Engine coolant temperature sensor and temperature gauge sensor (4-pin connector)

- AAZ, ACC, AHU, 1Z engines . . side of cylinder head, in coolant outlet
- ABA engine . . . . rear of cylinder head, in coolant outlet
- AAA engine . . . . rear of cylinder head, in thermostat housing

2. Check for ground in the harness connector. On 4-cylinder engines, check between terminal 2 and ground. On the 6-cylinder engine, check between terminal 4 and ground. If ground is not present, check for wiring faults using the appropriate wiring diagram.

3. Turn the ignition on and check for voltage in the harness connector. On 4-cylinder engines, check between terminal 4 and ground. On the 6-cylinder engine, check between terminal 2 and ground. If voltage is not present, check for wiring faults using the appropriate wiring diagram. If approximately 5 volts is not present, check the wiring from the instrument cluster.

NOTE —

- *Wiring diagrams are given in* **97 Wiring Diagrams, Fuses, and Relays.** *The engine coolant temperature sensor is labeled "G2" on the diagrams.*
- *On ACC and ABA engines, the coolant temperature sensor (G2) for the gauge and the coolant temperature sensor (G62) for the Motronic engine management system are combined in the same housing. Disconnecting the harness connector with the engine running may set a fault code in the On-board Diagnostic (OBD) System. See* **24a Fuel Injection–Motronic** *for OBD fault code information.*

4. Connect an ohmmeter across sensor terminals 2 and 4 in the sensor on a cold engine. Note the resistance.

5. Start the engine and allow it to warm up while observing the ohmmeter. The resistance should decrease steadily as the engine warms up. If any faults are found, replace the coolant temperature sensor.

> ***WARNING —***
> *The electric cooling fan can come on at any time, even if the key is out of the ignition. To avoid personal injury use extreme caution when working at or near the cooling fan if the engine is hot. As a safety precaution, always disconnect the harness connector from the fan when working in the engine compartment.*

NOTE —

- *Checking the gauge itself is somewhat more difficult owing to the electronic instrument cluster assembly. The signal from the sensor is processed by the instrument cluster electronic circuit. If no faults are found with the sensor and the wiring to the instrument panel is not faulty, the gauge itself may be faulty.*
- *Individual instruments are not available from Volkswagen. If the coolant temperature gauge is found to be faulty, the complete cluster will need to be replaced.*

To replace a faulty coolant temperature sensor, drain the cooling system down below the level of the sensor. Remove the retaining clip and pull the sensor from the coolant outlet.

To test cooling fan and radiator thermoswitch

Volkswagen cooling systems use 2 speed electric fans on most 4 cylinder engines and 3 speed fans on the VR-6 and diesel engine versions. All vehicles use a thermo-switch located in the lower section of the radiator to control fan speeds II and I. Additionally, vehicles with air-conditioning use a coolant fan control module to link the air-conditioning system to the coolant fans and provide speed III where appropriate.

The cooling system will overheat if the electric cooling fans are not operating correctly. The fans should switch on and off according to engine coolant temperature or whenever the A/C is on. The electric cooling fan operates when the temperature is high enough to close the radiator thermoswitch or, on cars so equipped, the after-run thermoswitch.

> ***WARNING —***
> *The electric cooling fan can come on at any time, even if the key is out of the ignition. To avoid personal injury use extreme caution when working at or near the cooling fan if the engine is hot. As a safety precaution, always disconnect the harness connector from the fan when working in the engine compartment.*

1. Check the fuse (30 amp) in position 19 and replace it if it is blown.

2. Disconnect the harness connector from the fan's thermoswitch on the left side of the radiator.

3. Using a fused jumper wire, short together the harness connector terminals listed in the appropriate wiring diagram. Check that the fan motor runs at the specified speeds.

NOTE —

- *The coolant fan thermo-switch is identified in Volkswagen wiring diagrams as component F18. Most of the wiring diagrams identify which contacts control which speed by printing the numeral I for first speed and the numeral II for the second speed within the F18 symbol. Generally, jumping the red and red/white wires will activate speed I and jumping the red and red/yellow or red and red/black will activate speed II. There can be exceptions for certain vehicles, see* **97 Wiring Diagrams, Fuses and Relays**.
- *On 1996 4-cylinder vehicles without A/C both the high and low speed fans have a red/black wire. See* **97 Wiring Diagrams, Fuses and Relays**.

4. If the fan does not run, check for battery voltage at terminal no. 3 (red wire) in the thermoswitch connector. Disconnect the harness connector from the fan and check for continuity to ground at the brown wire in the connector. If any faults are found, check the wiring as shown in **97 Wiring Diagrams, Fuses, and Relays.**

5. If voltage and ground are present as specified above and no wiring faults can be found, the fan motor is probably faulty and should be replaced.

NOTE —
On VR-6 and diesel engines, fan speed III is a function of air conditioner operation and is controlled via the coolant fan control module. See **97 Wiring Diagrams, Fuses, and Relays** *for the coolant fan wiring schematic for these vehicles.*

6. If the fan runs only when powered directly by the jumpered connector, the radiator thermoswitch may be faulty. Switch opening and closing specifications are given below for more accurate testing of the switch. Drain the coolant before replacing a faulty switch.

Tightening torque
- Thermoswitch to radiator . . . . 35 Nm (26 ft-lb)

Thermoswitch operating temperatures
- Stage 1 (low speed)
  - on. . . . . .197–206°F (92–97°C)
  - off. . . . . .183–195°F (84–91°C)
- Stage II (high speed)
  - on. . . . . .210–221°F (99–105°C)
  - off. . . . . .195–208°F (91–98°C)

Thermostat Quick-Check

A thermostat that is stuck open will cause the engine to warm up slowly and run below normal temperature at highway speed. A thermostat that is stuck closed will restrict coolant flow to the radiator and cause overheating.

To quickly check if the thermostat is opening and if coolant is circulating through the radiator, allow a cold engine to reach operating temperature (temperature gauge needle approximately centered). Feel the radiator hoses. If the hoses are hot to the touch, the coolant is circulating. If the hoses are not hot, coolant flow to the radiator is probably restricted. Check for a faulty thermostat or a plugged radiator.

Anti-freeze concentration, checking

1. Using a refractometer (VW tool CEN5021), take a small amount of coolant and place it onto the measuring prism. See Fig. 4.

2. Close the cover and hold the tester up to a light source.

3. Look into the eyepiece and read the scale for ethylene glycol. The measuring point is where the black and white come together.

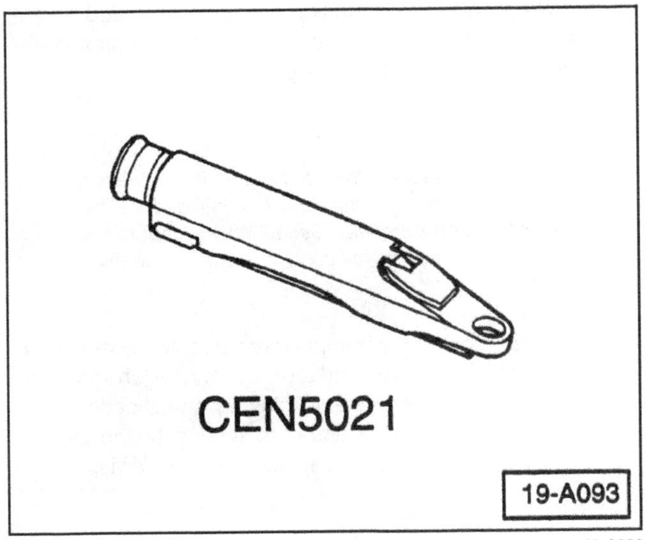

19-A093

Fig. 3. Anti-freeze tester, VW special tool CEN5021.

4. Rinse the tester clean.

5. Periodically check calibration by placing distilled water onto the prism and closing the cover. The measure point should be at 32°F (0°C) which is the freezing point for pure water. If reading is not obtained, follow instructions that came with the tester for adjustment.

DRAINING AND FILLING COOLING SYSTEM

Always use genuine VW coolant and distilled water to avoid the formation of harmful deposits in the cooling system. Use of anti-freeze with phosphate solutions or tap water can be harmful to the cooling system.

Do not reuse the coolant when replacing damaged engine parts. If coolant comes in contact with paint or bodywork, immediately rinse the area with water.

To drain and fill cooling system (4-cylinder engines)

Be sure to have a new coolant pump connector pipe O-ring on hand before draining the coolant.

1. With the engine fully cold, remove the expansion tank cap. Set the temperature control knob in the passenger compartment to full warm.

2. Place a drain pan beneath the coolant pump.

3. Working at the coolant pump, remove the retaining clip from the connector pipe. Pull the connector pipe from the pump housing and disconnect the lower hose from the pump. Allow the coolant to drain. See Fig. 4.

NOTE—

Compressed air may be applied to coolant expansion tank to help blow out remaining coolant.

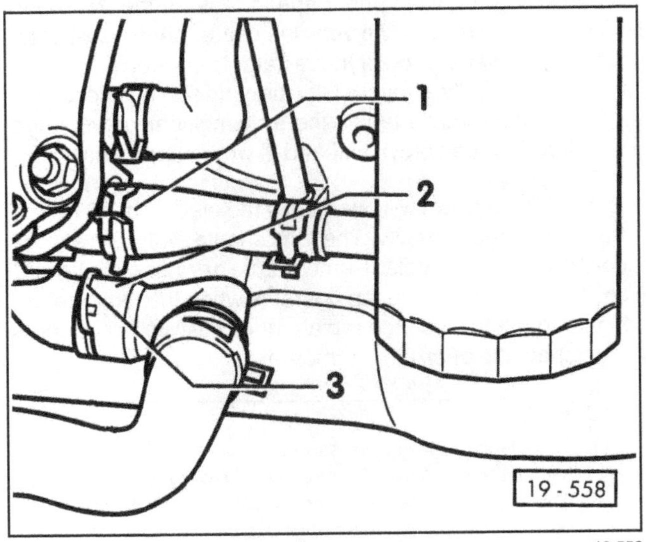

19-558

Fig. 4. Drain coolant by removing retaining clip (**3**) and then connector pipe (**2**). Also disconnect hose (**1**).

4. Install the connector pipe using a new O-ring and install the pipe retaining clip. Install the lower hose to the pump.

5. To flush cooling system, fill with distilled water and run engine a minimum of 2 minutes. Drain system as described above.

6. Using a coolant mixture of 50% anti-freeze and 50% distilled water, fill the expansion tank slowly so that air is allowed to escape. Install the expansion tank cap.

7. Run the engine and check for leaks. Allow the radiator cooling fan to cycle on at least once.

Cooling system capacity (draining and filling)
- 4-cylinder engines
  - ACC, ABA ....5.8 quarts (5.5 liters)
  - AAZ, 1Z, AHU ....6.8 quarts (6.5 liters)

8. Recheck the coolant level when the engine has cooled.

NOTE—

*The coolant level should be at the **MAX** mark at operating temperature and between the **MIN** and **MAX** mark when cold. The final coolant level is best checked when the cooling system is fully cold.*

To drain and fill cooling system (6-cylinder engine)

Be sure to have a new coolant pipe drain plug O-ring on hand before draining the coolant.

1. With the engine fully cold, remove the expansion tank cap. Set the temperature control knob in the passenger compartment to full warm.

2. Place a drain pan beneath the metal coolant pipe that runs between the coolant pump and the thermostat housing.

3. Remove the drain plug from the coolant pipe. See Fig. 5. Allow the coolant to drain while pressing down on the top radiator hose.

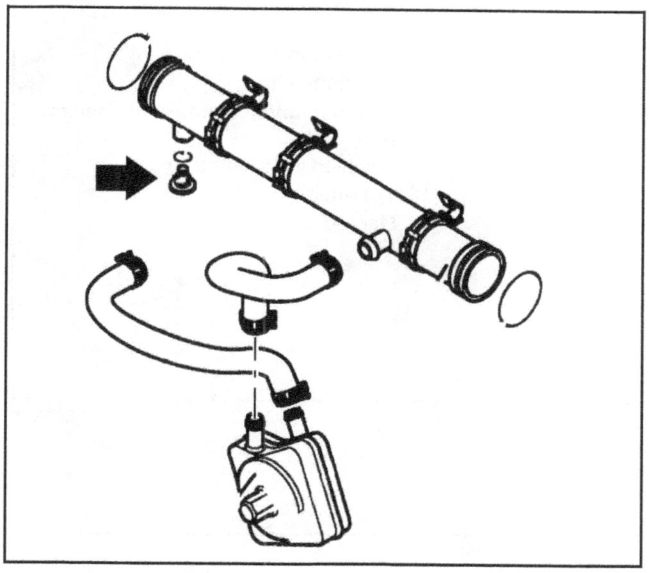

0024020

Fig. 5. Coolant pipe with drain plug (**arrow**) on VR6 engine.

4. Install the drain plug.

Tightening torque
- Coolant drain plug to coolant pipe . . . 10 Nm (89 in-lb)

5. Using a coolant mixture of 50% anti-freeze and 50% distilled water, fill the expansion tank slowly so that air is allowed to escape. Install the expansion tank cap.

6. Run the engine and check for leaks. Allow the radiator cooling fan to cycle on at least once.

Cooling system capacity (draining and filling)
- 6-cylinder engine . . . . 10.6 quarts (10.0 liters)

7. Recheck the coolant level when the engine has cooled.

NOTE —
*The coolant level should be at the **MAX** mark at operating temperature and between the **MIN** and **MAX** mark when cold. The final coolant level is best checked when the cooling system is fully cold.*

COOLANT PUMP

The engine-driven coolant pump is subject to the same wear as any other rotating engine parts. Complete failure of the pump to circulate coolant is unusual, but excessive wear often results in noise or coolant leaks at the pump shaft.

To replace coolant pump (4-cylinder engines)

Fig. 6 is an exploded view of the coolant pump used on 4-cylinder engines. Note the stretch-type studs or bolts that mount the pump to the block. These fasteners should not be reused. Be sure to have new fasteners on hand before beginning the job.

1. Disconnect the negative (–) battery cable.

NOTE —
Be sure to have the anti-theft radio code on hand before disconnecting the battery.

2. Drain the cooling system as described above and remove the remaining hose(s) from the pump.

3. Mark the running direction on the poly-ribbed drive belt. Remove the V-belt and the poly-ribbed belt as described in **0 Maintenance Program**

4. Remove the alternator together with its bracket from the engine. See **27 Engine Electrical Equipment**.

5. Remove the air conditioning compressor from its bracket without disconnecting any refrigerant lines. Place the compressor safely out of the way. See **87 Air Conditioning.**

6. Remove the coolant pump from the engine block. Discard the coolant pump mounting studs or bolts.

7. Installation is the reverse of removal. Clean the pump mating surface before installing. Always use new O-rings and gaskets during installation. Fill the cooling system as described above. Start the engine and check for leaks as the final step.

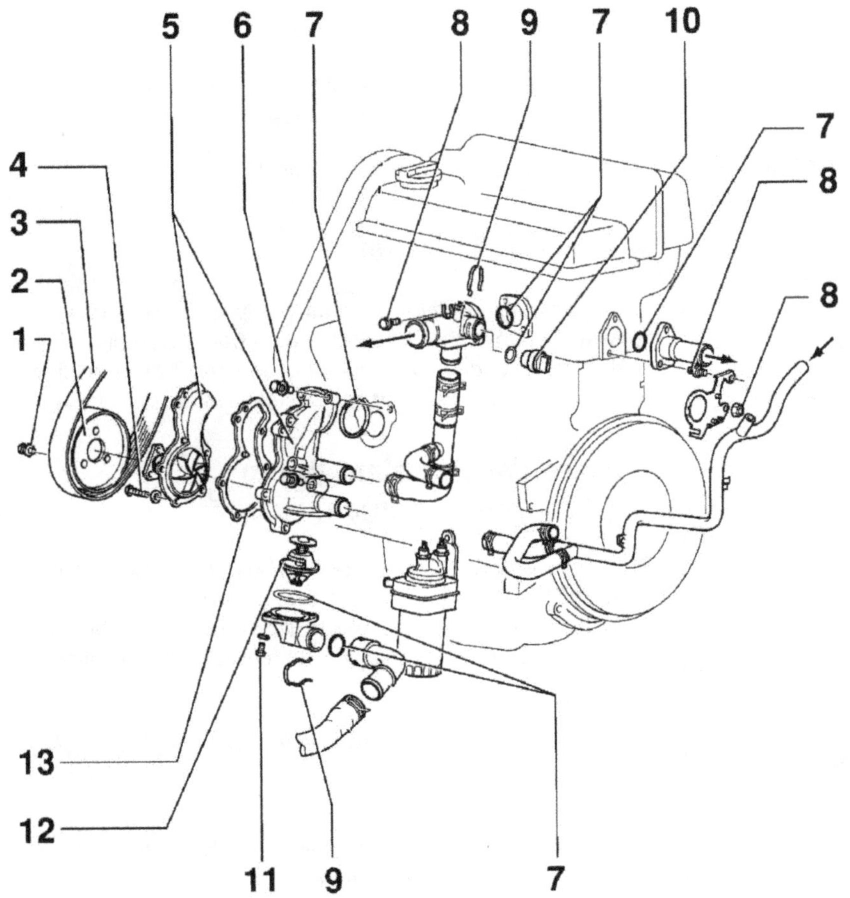

1. **Bolt**
   - tighten to 20 Nm (15 ft-lb)
2. **Pulley**
3. **Poly-ribbed belt**
   - mark running direction before removing
4. **Bolt**
   - tighten to 10 Nm (89 in-lb)
5. **Coolant pump**
6. **Bolt or stud**
   - always replace
   - tighten to 20 Nm (15 ft-lb) plus an additional 1/4 turn (90°)
7. **O-ring**
   - always replace
8. **Bolt**
   - tighten to 10 Nm (89 in-lb)
9. **Retaining clip**
10. **Engine coolant temperature sensor**
11. **Bolt**
    - tighten to 10 Nm (89 in-lb)
12. **Thermostat**
13. **Gasket**

0024018

Fig. 6. Exploded view of coolant pump and related parts on 4-cylinder diesel engine. Other 4-cylinder engines are similar.

Tightening torques (4-cylinder engines)
- Alternator bracket to coolant pump . . . 30 Nm (22 ft-lb)
- Alternator to alternator bracket . . . . . . 25 Nm (18 ft-lb)
- A/C compressor . . . . See **87 Air Conditioning**
- Coolant pump to engine block (stretch bolt–always replace)
  - stage I . . . . 20 Nm (15 ft-lb)
  - stage II . . . . additional 1/4 turn (90°)
- Coolant pump pulley to coolant pump . . . . .20 Nm (15 in-lb)
- Thermostat housing to coolant pump . . . . 10 Nm (89 in-lb)

To replace coolant pump (6-cylinder engine)

Removal of the coolant pump on the VR6 engine requires special engine lifting equipment. Review the procedure to determine what tools, equipment, and supplies will be necessary before beginning the job. Fig. 7 shows an exploded view of the coolant pump assembly.

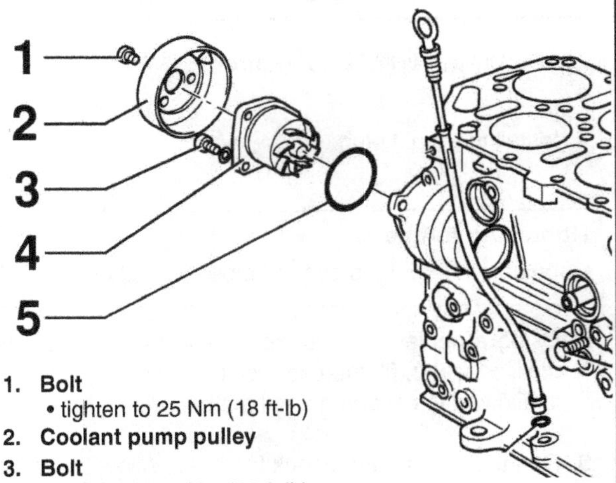

1. **Bolt**
   - tighten to 25 Nm (18 ft-lb)
2. **Coolant pump pulley**
3. **Bolt**
   - tighten to 20 Nm (15 ft-lb)
4. **Coolant pump**
5. **O-ring**

0024019

Fig. 7. Exploded view of coolant pump on 6-cylinder engine.

1. Disconnect the negative (–) battery cable.

NOTE —
Be sure to have the anti-theft radio code on hand before disconnecting the battery.

2. Drain the cooling system as described above.

3. Unclip the evaporative cannister purge valve from the side of the air cleaner housing. Remove the air cleaner housing assembly together with the intake air duct and mass air flow sensor.

4. Disconnect the spark plug wires from the coil. Unclip the wires from the wire guides and place the wire assembly on top of the cylinder head. See Fig. 8.

NOTE —
Depending on type of lifting equipment used, it may also be necessary to disconnect the vacuum hose from the fuel pressure regulator and remove the intake air temperature sensor to access the front engine lifting eye.

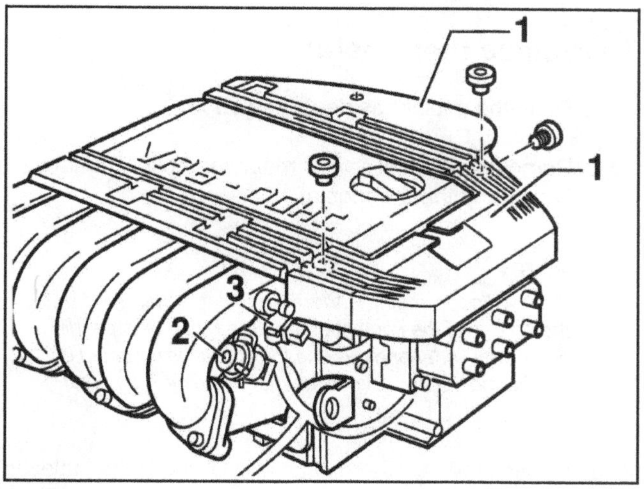

0024021

Fig. 8. Spark plug wires guides (**1**), fuel pressure regulator (**2**) and intake air temperature sensor (**3**).

5. With the drive belt installed, loosen the coolant pump pulley bolts.

NOTE —
If necessary, apply slight pressure to the drive belt to hold the pulley stationary as the bolts are loosened.

6. Mark the running direction on the poly-ribbed drive belt. Release the tension on the belt using an M8X80 bolt. See Fig. 9. Remove the belt from the pulleys.

7. Disconnect the front exhaust pipe from the catalytic converter. See **26 Exhaust System/Emission Controls.**

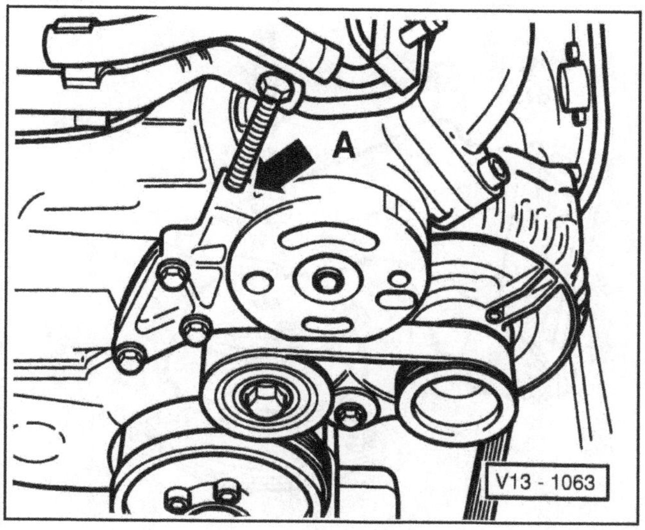

V13-1063

Fig. 9. Thread an 8mm (M8X80) bolt (**A**) into tensioner to release belt tension.

CAUTION —
Thread the M8 bolt in just until the belt can be slipped off the pulleys. Threading the bolt in too far can cause damage to the tensioner assembly.

8. Without disconnecting the coolant hoses, unbolt the coolant reservoir from its mount and place it out of the way.

9. Install an engine lifting hoist, using the left and right lifting points on the cylinder head. Raise the hoist slightly, so that the weight of the engine/transmission assembly is supported.

10. Remove the bolts from the left and right rear engine/transmission mounts and the front engine/transmission mount. See Fig. 10. Also see **10 Engine–Removing and Installing** for engine and transmission mount tightening torques.

11. Remove the coolant pump pulley from the coolant pump. Remove coolant pump mounting bolts.

12. Carefully raise the engine assembly while pushing it towards the driver's side wheel housing. Raise the engine just until the pump can be withdrawn from the block.

13. Clean the pump mating surface before installing the new pump. Always use a new O-ring during installation.

14. Install the engine on its mounts using the engine alignment procedure outlined in **10 Engine–Removing and Installing.**

15. The remainder of installation is the reverse of removal. Be sure to fill the cooling system and check for leaks as the final step.

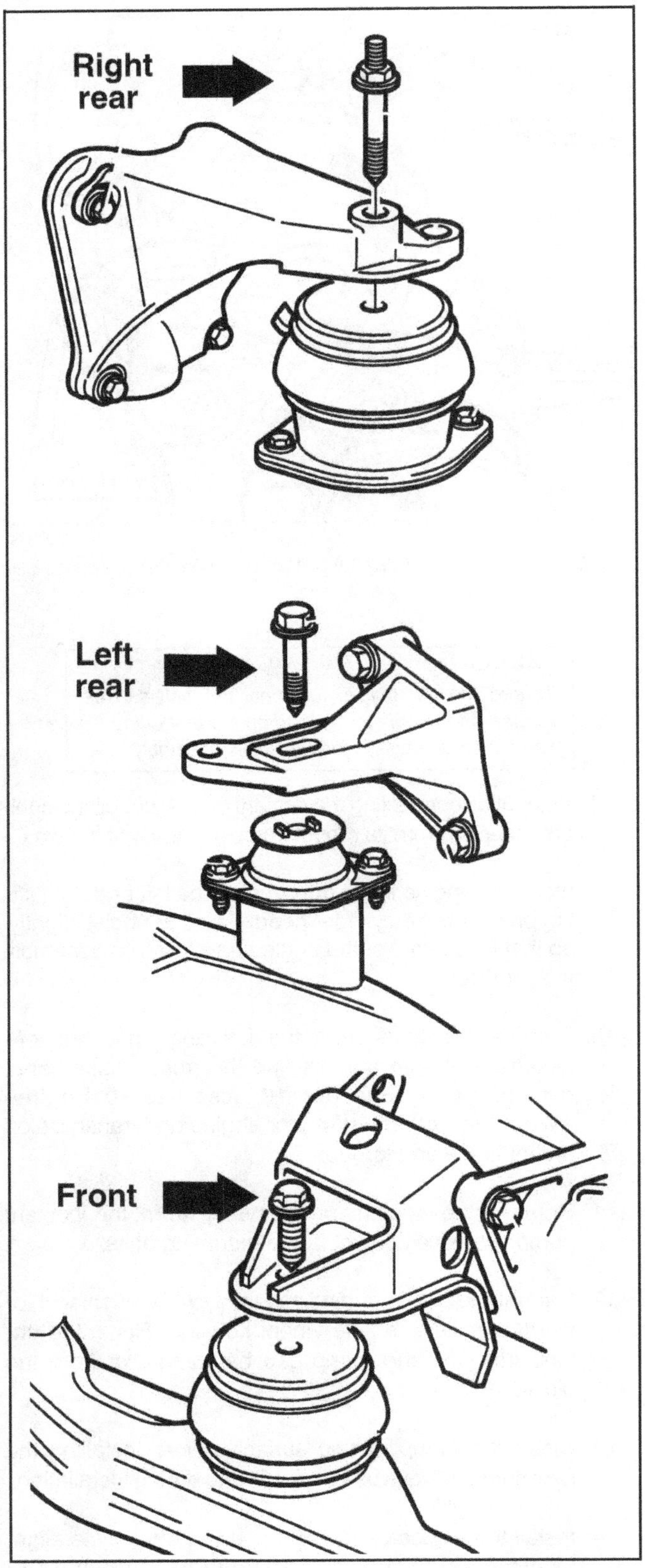

Fig. 10. Engine/transmission mount bolts (**arrows**) to be removed. Right rear mount is attached to engine and right side of subframe. Left rear mount is attached to transmission and left side subframe. Front mount is attached to transmission and left side crossmember.

Tightening torques

- Coolant drain plug to coolant pipe... 10 Nm (89 in-lb)
- Coolant expansion tank to body .... 10 Nm (89 in-lb)
- Coolant pump to engine block.......20 Nm (15 ft-lb)
- Coolant pump pulley to coolant pump....................25 Nm (18 ft-lb)
- Exhaust pipe to catalytic converter (use new fasteners)
  - M8 nuts ....25 Nm (18 ft-lb)
  - M10 nuts ....40 Nm (30 ft-lb)
- Intake air temperature sensor to intake manifold................ 15 Nm (11 ft-lb)

THERMOSTAT

The thermostat controls the coolant temperature by regulating coolant flow to the radiator. A thermostat stuck open will cause the engine to warm up slowly and run below normal temperature at highway speed. A thermostat stuck closed will restrict coolant flow to the radiator and cause overheating.

To replace thermostat

1. Drain the coolant as described above.

2. Remove the outlet flange mounting bolts. Remove the flange and the thermostat.

   NOTE —

   On 4-cylinder engines, the thermostat is mounted in the bottom of the coolant pump. See Fig. 5 given earlier. On 6-cylinder engines, the thermostat is mounted in the thermostat housing on the rear of the cylinder head. See Fig. 11.

3. Position the new thermostat and O-ring in the outlet. Install the thermostat housing.

Tightening torque

- thermostat flange to thermostat housing or coolant pump ....10 Nm (89 in-lb)

4. Refill the cooling system as described earlier.

1. **Thermostat housing**
2. **Bolt**
   • tighten to 10 Nm (89 in-lb)
3. **Thermostat**
4. **Outlet flange**
5. **Bolt**
   • tighten to 10 Nm (89 in-lb)
6. **O-rings**
7. **Outlet flange**
8. **Thermo-switch for A/C overheat cut-out and coolant fan speed III**
   • plug installed for non-A/C equipped vehicles
9. **Thermo-sender with switch for coolant temperature gauge and auxiliary electric coolant pump**
10. **Engine coolant temperature sensor (Motronic)**
11. **Retaining clips**

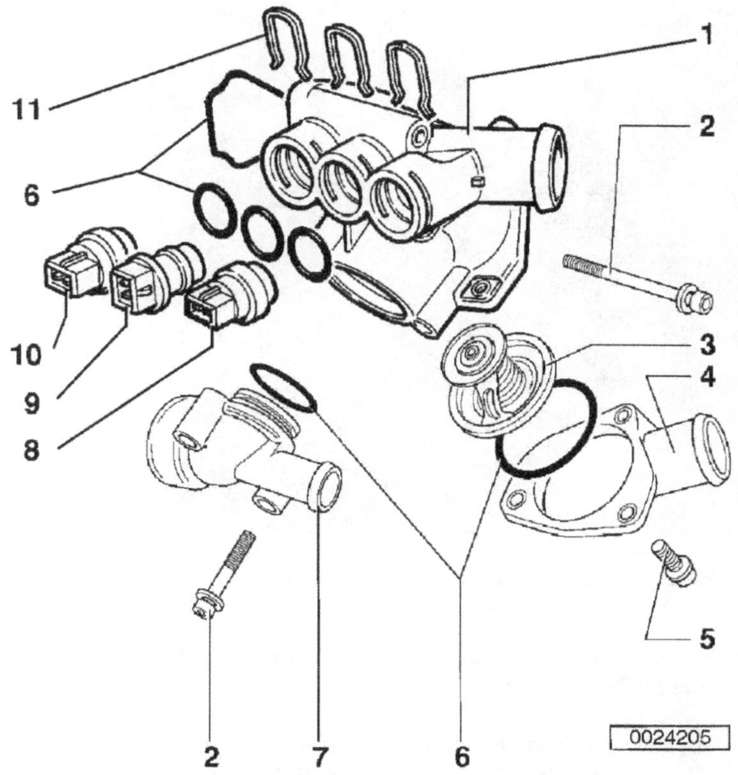

0024205

Fig. 11. Thermostat housing assembly on VR6 engine.

RADIATOR AND COOLING FANS

The cooling fan provides additional air flow through the radiator. A faulty cooling fan motor or thermoswitch may be the cause of insufficient air flow and overheating.

Most 4-cylinder models covered by this manual have a two-speed cooling fan. On 6-cylinder engines, a three speed fan is used. The third speed is used primarily for the cooling fan after-run system.

> ***WARNING —***
> *The electric cooling fan can come on at any time, even if the key is out of the ignition. To avoid personal injury use extreme caution when working at or near the cooling fan if the engine is hot. As a safety precaution, always disconnect the harness connector from the fan when working in the engine compartment.*

To remove radiator and cooling fan assembly

1. Disconnect the negative (–) battery cable. Disconnect the harness connectors from the fan and the thermoswitch.

   NOTE —
   Be sure to have the anti-theft radio code on hand before disconnecting the battery.

2. Remove the tray from above the radiator and disconnect any A/C hose brackets and electrical harness connectors attached to the radiator and fan assembly.

3. Drain the cooling system as described above. Disconnect all coolant hoses from the radiator.

4. Disconnect the evaporative cannister purge valve from the rear of the air cleaner housing. Label and remove all the hoses from the housing. Remove the air cleaner housing retaining bands and remove housing.

5. Remove the upper radiator mounting pins. See Fig. 12.

0024017

Fig. 12. Right radiator pin being removed (**arrow**). Also remove left pin.

6. Remove the front body assembly (radiator support). See **50 Body–Front**.

7. Remove the radiator from the car. On cars with air conditioning, separate the A/C condenser from the radiator and carefully slide the radiator out.

Installation is the reverse of removal. Inspect the lower rubber radiator mountings and replace any that are damaged. Fill the cooling system as described above.

Cooling Fan After-Run System

The VR6 engine uses a radiator cooling fan after-run system to moderate under hood temperatures when the engine is turned off. The system includes a thermoswitch, an auxiliary electric coolant pump, the radiator cooling fan (3rd speed), and a cooling fan control module.

High temperature closes the switch, making a ground connection to the cooling fan control module. The control module then runs the auxiliary coolant pump and the radiator cooling fan on high speed.

The after-run thermoswitch is integrated with the coolant temperature sensor in the thermostat housing. The sensor/switch has a four-pin harness connector and the body of the switch is yellow.

To quickly check the circuit, turn the ignition key on briefly and then turn it off. Remove the cover from above the thermostat housing and disconnect the harness connector from the thermoswitch switch. Jumper terminals 1 and 3 (brown/red and brown wires) in the connector. See Fig. 13. The auxiliary coolant pump should run and the radiator cooling fan should come on.

Switch opening and closing specifications are given below for more accurate testing of the switch. For additional troubleshooting of the circuit, use the wiring diagrams found in **97 Wiring Diagrams, Fuses, and Relays.**

After-Run Coolant Fan Control Thermoswitch

- on (switch closed) . . . . .214–225 °F (101-107°C)
- off (switch open) . . . . 201–212 °F (94-100°C)

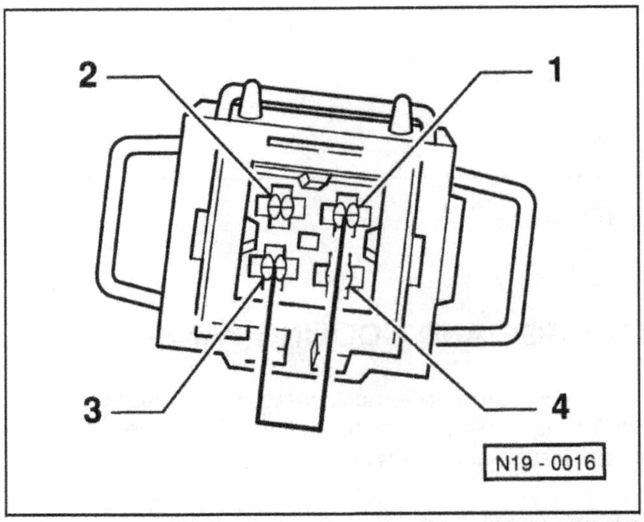

N19-0016

Fig. 13. Terminals 1 and 3 in after-run thermoswitch harness connector shown jumpered.

2 Engine Management, Exhaust, and Engine Electrical Systems

GENERAL

This general information group covers application information and system descriptions for the repair groups listed under **2 Fuel, Ignition, and Exhaust Systems**.

NOTE —

- *For general information on the battery, starter, alternator, see* **27 Engine Electrical**.
- *For emission control system application information, see* **26 Exhaust System/Emission Controls.**

FUEL SUPPLY

The plastic fuel tank is mounted beneath the rear of the car. On gasoline engines, the electric fuel pump with integral fuel level sensor is mounted through the top of the tank. See Fig. 1.

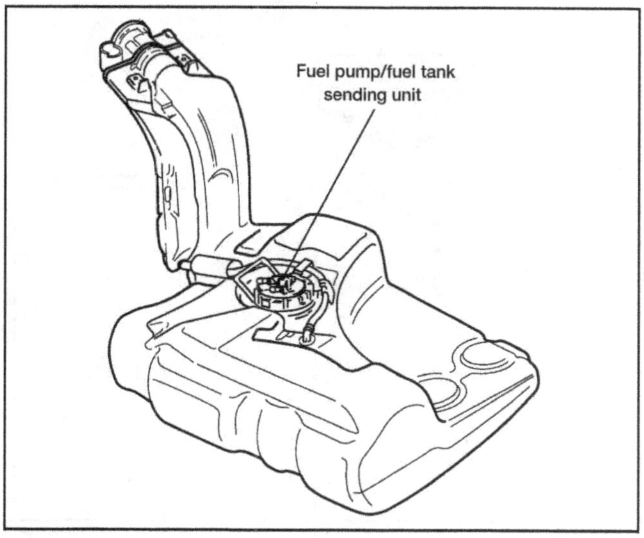

Fig. 1. Plastic molded fuel tank. Diesel engines are not equipped with in-tank pump.

The fuel tank is designed to prevent overfilling and allow for fuel expansion. The filler cap contains a valve to prevent vacuum from occurring in the tank.

On gasoline engines, the fuel pump assembly is submersed in the fuel tank. An inlet strainer helps prevent pump clogging. The fuel pump contains a bypass valve to prevent high pressure from damaging the system. There is a check valve at the pump outlet to hold pressure in the system after the pump is turned off. A roll-over valve is installed in the fuel filler neck to prevent fuel from escaping in the event the car rolls over in an accident.

NOTE —

Diesel engines do not use a fuel tank pump. For more information on Diesel fuel injection and fuel supply, see **23a** *or* **23b Fuel Injection–Diesel**.

FUEL INJECTION

Three basic types of engine management systems are used on the engines covered in this manual depending on engine and model year. Below is brief description each system.

Engine Management System Applications

Engine Code	
• ACC (1.8 liter)	Mono-Motronic
• ABA (2.0 liter)	
1993–1995	Motronic M2.9
1996–2002	Motronic M5.9 (OBD II)
• AAA (2.8 liter)	
1993–2002	Motronic M5.9 (OBD II)
• AAZ (1.9 liter, 1993–1998)	Diesel Fuel Injection
• 1Z, AHU (1.9 liter, 1997–1999)	Diesel Turbo Direct Injection

Mono-Motronic Engine Management (ACC engine)

Fig. 2 shows the Bosch Mono-Motronic engine management system used on the ACC (1.8 liter) engine. The Mono-Motronic system combines the ignition and fuel functions into one system, managed by a single **Engine Control Module (ECM)**. The system features built-in diagnostics that detect and store coded fault information.

Basic fuel metering is determined by engine speed and engine load (throttle position). The ECM receives an engine speed signal from the **Camshaft Position (CMP) Sensor** and an engine load signal throttle from the **Throttle Position (TP) Sensor**. The ECM then meters fuel to the engine by triggering a single injector at a rate proportional to engine speed and throttle position.

The amount of time the injector remains open determines the basic fuel mixture. The **Engine Coolant Temperature (ECT) Sensor**, the **Heated Oxygen Sensor (HO2S)**, and the **Intake Air Temperature (IAT) Sensor** all supply additional input signals to the ECM to further adapt and correct basic fuel metering.

Based on the input signals, the ECM controls the output signal to the following components:

- **Fuel Injector**
- **Ignition Coil**
- **Throttle Position (TP) Actuator** (for idle speed control)
- **Evaporative (EVAP) Canister Purge Regulator Valve**
- **Intake manifold heater**
- **Fuel Pump**

NOTE —
*See **24b Fuel Injection—Mono-Motronic** for troubleshooting and repair information.*

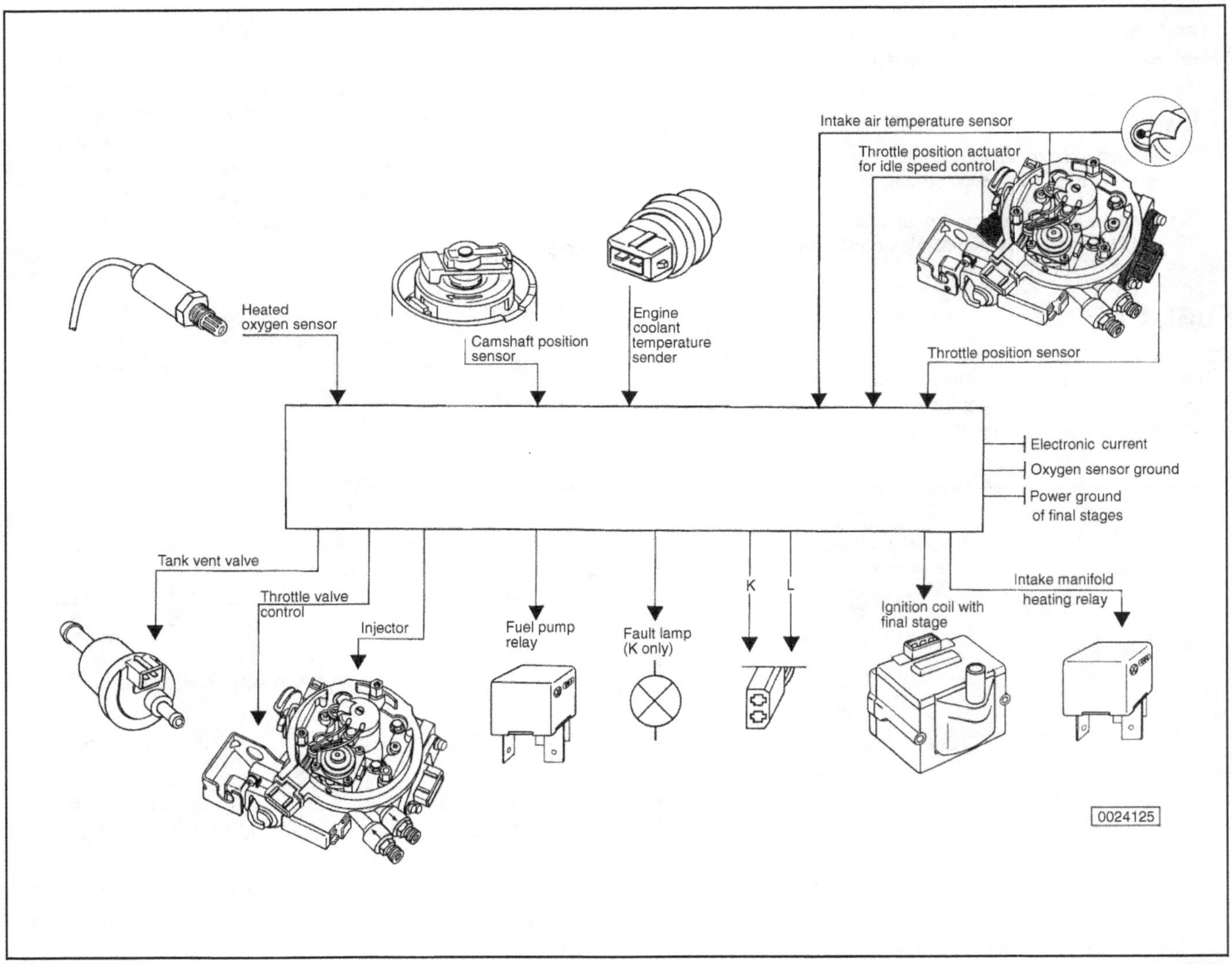

0024125

Fig. 2. Mono-Motronic engine management system with ECM inputs (arrows in) and outputs (arrows out). Inputs to ECM are used to control and adapt output signals to components.

Motronic 2.9 Engine Management (ABA, AAA engines, OBD I, 1993-1995)

The ABA (2.0 liter) and the AAA (2.8 liter) engines are equipped with a sophisticated Bosch Motronic Engine Management System. See Fig. 3. The Motronic system combines the ignition and fuel functions into one system, managed by a single **Engine Control Module (ECM)**. The system is fully adaptive and features built-in diagnostics that detect and store coded fault information. There are no basic adjustments or settings that can be made to the system.

> ***NOTE —***
> *See **24a Fuel Injection—Motronic** for troubleshooting and repair information.*

Basic fuel metering is determined by engine speed and the mass of the air entering the engine. The ECM receives engine rpm and crankshaft position signal from the **Engine Speed (RPM) Sensor** and air mass (engine load) data from the **Mass Air Flow (MAF) Sensor**. The **ECM** then meters fuel to the engine by triggering the injectors at a rate proportional to engine speed and engine load. The amount of time the injectors remain open determines fuel quantity.

The **Camshaft Position (CMP) Sensor** identifies cylinder no. 1 firing position for cylinder-selective injection and ignition knock control. The **Engine Coolant Temperature (ECT) Sensor** supplies engine temperature information to the ECM. The **Throttle Position (TP) Sensor** signals throttle angle. The **Heated Oxygen Sensor (HO2S)** signals combustion efficiency. The **Knock Sensor** supplies engine knock information for ignition timing regulation and the **Intake Air Temperature (IAT) Sensor** is used for idle stabilization and as a correction factor for ignition timing.

Based on the input signals, the ECM precisely controls the following output components:

- **Fuel Injectors**
- **Power Output Stage and Ignition Coil**
- **Idle Air Control (IAC) Valve**
- **Evaporative Canister Purge Valve (EVAP)**
- **Exhaust Gas Recirculation Regulator Valve (EGR)**
- **Fuel Pump and Relay**
- **Heated Oxygen Sensor Relay**

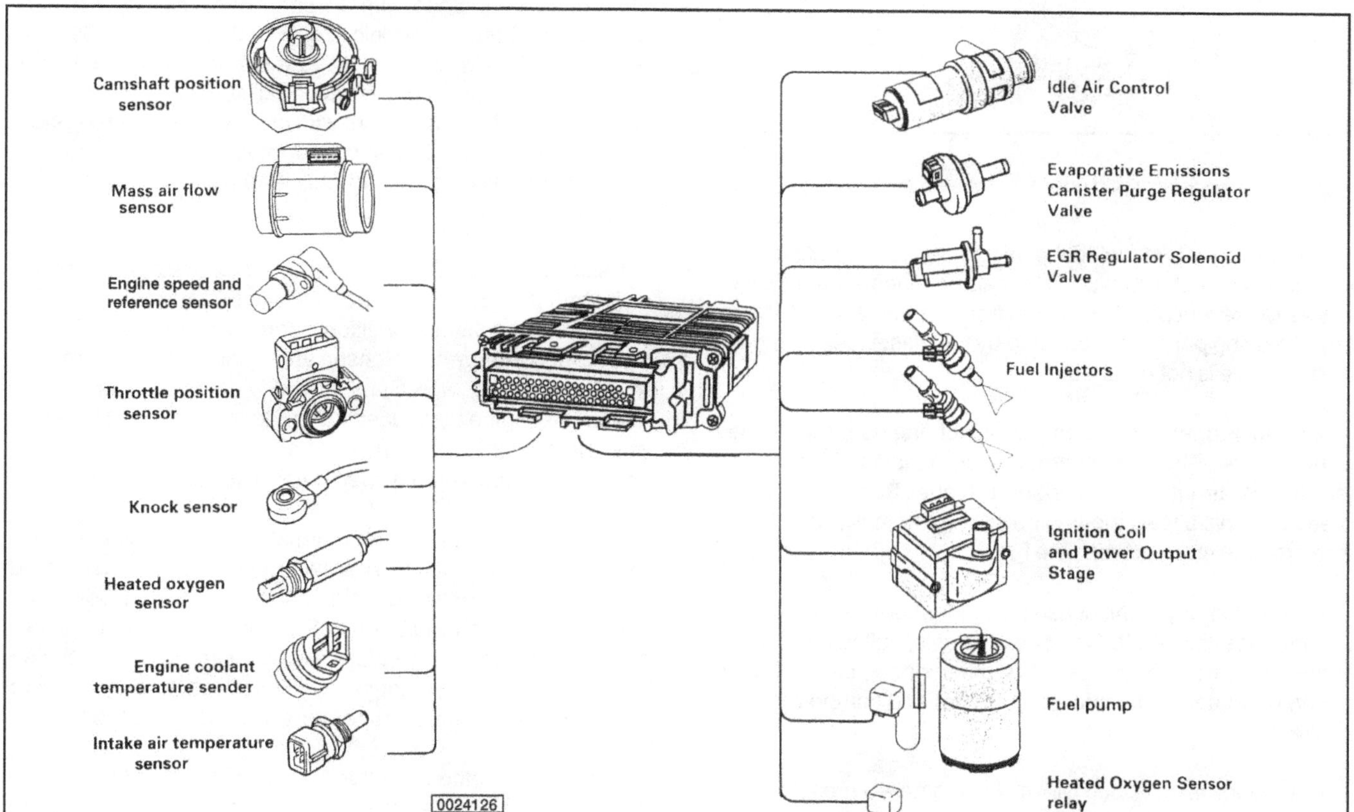

Additional Input Signals: EGR temperature sensor, vehicle speed sensor, A/C system, automatic transmission control module, battery, data link connector

Additional Output Signals: Instrument cluster, automatic transmission control, secondary air pump relay with pump and valve, data link connector

0024126

Fig. 3. Bosch Motronic 2.9 engine management system. Inputs (**left**) to ECM are used to control and adapt output signals (**right**) to the individual components. Components may vary in some verisons.

Diesel Fuel Injection (AAZ engine)

The diesel injection pump with boost pressure enrichment is engine-driven by the same toothed belt that drives the camshaft. It handles virtually all fuel functions including drawing fuel from the tank and controlling the timing and the quantity of fuel injected. A timing advance mechanism changes injection timing in response to engine speed and load and during cold-start conditions. At high boost conditions, the fuel mixture is enriched slightly for increased power. See Fig. 4.

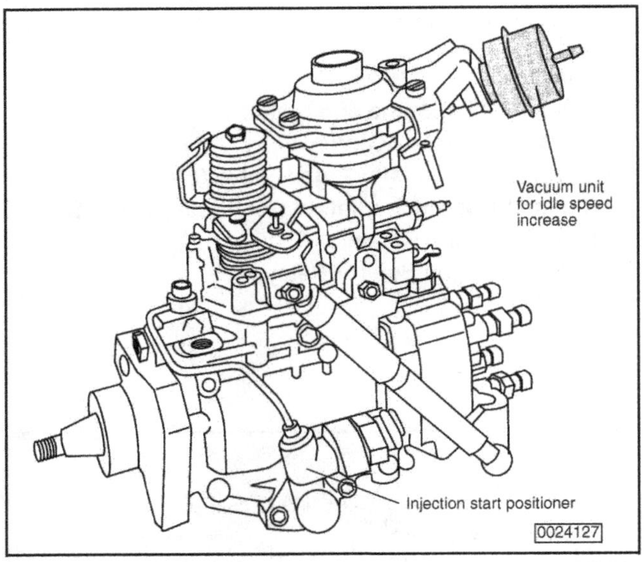

0024127

Fig. 4. Diesel injection pump (engine code AAZ).

The pump meters fuel to the injectors at an operating pressure in excess of 120 bar (1706 psi) and incorporates check valves to prevent backflow of fuel from the injectors. All the internal moving parts are lubricated by the diesel fuel, so routine maintenance is not required.

A turbocharger is used on the diesel engine for increased performance. The turbocharger is an exhaust driven pump that forces air into the cylinders at higher than atmospheric pressure. This allows the engine to burn more fuel and therefore produce more power. See Fig. 5.

The turbocharger unit consists of two wheels mounted on a common rotating shaft. One of the wheels, called the turbine, is driven by the exhaust gas. The other wheel, called the impeller, is located in the intake air stream and compresses the intake air.

The turbine is equipped with a waste gate—a pressure controlled valve that automatically bleeds off any excess pressure and allows exhaust gasses to bypass the system. If the boost pressure exceeds the predetermined value, in spite of the waste gate, a pressure switch will temporarily interrupt the fuel supply to the engine.

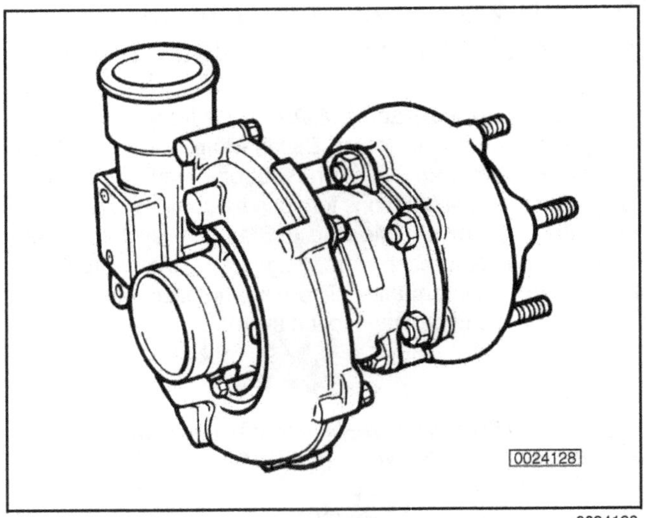

0024128

Fig. 5. Turbocharger used on AAZ diesel engine.

Motronic 5.9 Engine Management (ABA, AAA engines, OBD II, 1996-on)

The ABA (2.0 liter) and the AAA (2.8 liter) engines from 1996-on are equipped with an enhanced version of the sophisticated Bosch Motronic Engine Management System. See Fig. 6. This version complies with the Government mandated OBD II standards. Version 5.9 functions in much the same way as Version 2.9, but with greatly enhanced diagnostics, system monitoring and adaptability. Several components have also been added, and several have been simplified.

Basic fuel metering is determined by engine speed and engine load. The Engine Control Module (ECM) receives engine speed and crankshaft position information from the Engine Speed and Reference Sensor and engine load information from the Mass Air Flow Sensor (MAF). The ECM then meters fuel to the engine by sequentially triggering the Fuel Injectors at a rate proportional to engine speed and load. The length of time the injectors remain open determines fuel quantity.

The ECM uses the same information to determine the correct ignition firing point. A small output signal is generated and sent to the Power Output Stage and Ignition Coil. The Camshaft Position Sensor (CMP), identifies cylinder number 1 firing position for cylinder-selective injection and ignition knock control. The Engine Coolant Temperature Sensor (ECT) supplies engine temperature information to the ECM.

The new Throttle Actuator Control Module (TACM) combines four functions. Three are used on the input side: the Throttle Position Sensor (TPS) signals throttle angle, the Closed Throttle Position Switch (TP Switch) signals fully closed throttle plate, and the Throttle Position Feedback Sensor signals position of the electric motor used to stabilize idle speed.

The Pre-Catalyst Heated Oxygen Sensor (HO2S) signals combustion efficiency, and the Post-Catalyst Heated Oxygen Sensor (HO2S) monitors efficiency of the Three-Way Catalytic Converter (TWC).

The Knock Sensor(s) supply engine knock information for ignition timing regulation and the Intake Air Temperature Sensor (IAT) is used for idle stabilization and as a correction factor for ignition timing.

Based on all of the input signals, the ECM precisely controls the following output components:

- **Fuel Injectors**
- **Power Output Stage and Ignition Coil**
- **Throttle Position Actuator**
- **Evaporative Canister Purge Valve (EVAP)**
- **Fuel Pump and Relay**

In addition, certain versions have additional outputs controlled by the ECM:

- **Secondary Air Pump Relay (AIR), Air Injection Solenoid Control Valve, and Secondary Air Injection Pump**
- **Leak Diagnosis Pump (LDP), Control Relay, and Condensate Drain Valve**

All versions have a warning light known as a Malfunction Indicator Lamp (MIL), in the instrument cluster that signals the vehicle operator when certain systems have failed. In some versions, the light can be switched on if relatively small malfunctions have been recorded such as running very low on fuel or leaving off the gas cap. Most malfunctions that occur will cause the light to stay on steadily but certain very serious situations such as an overheating catalytic converter may cause it to blink.

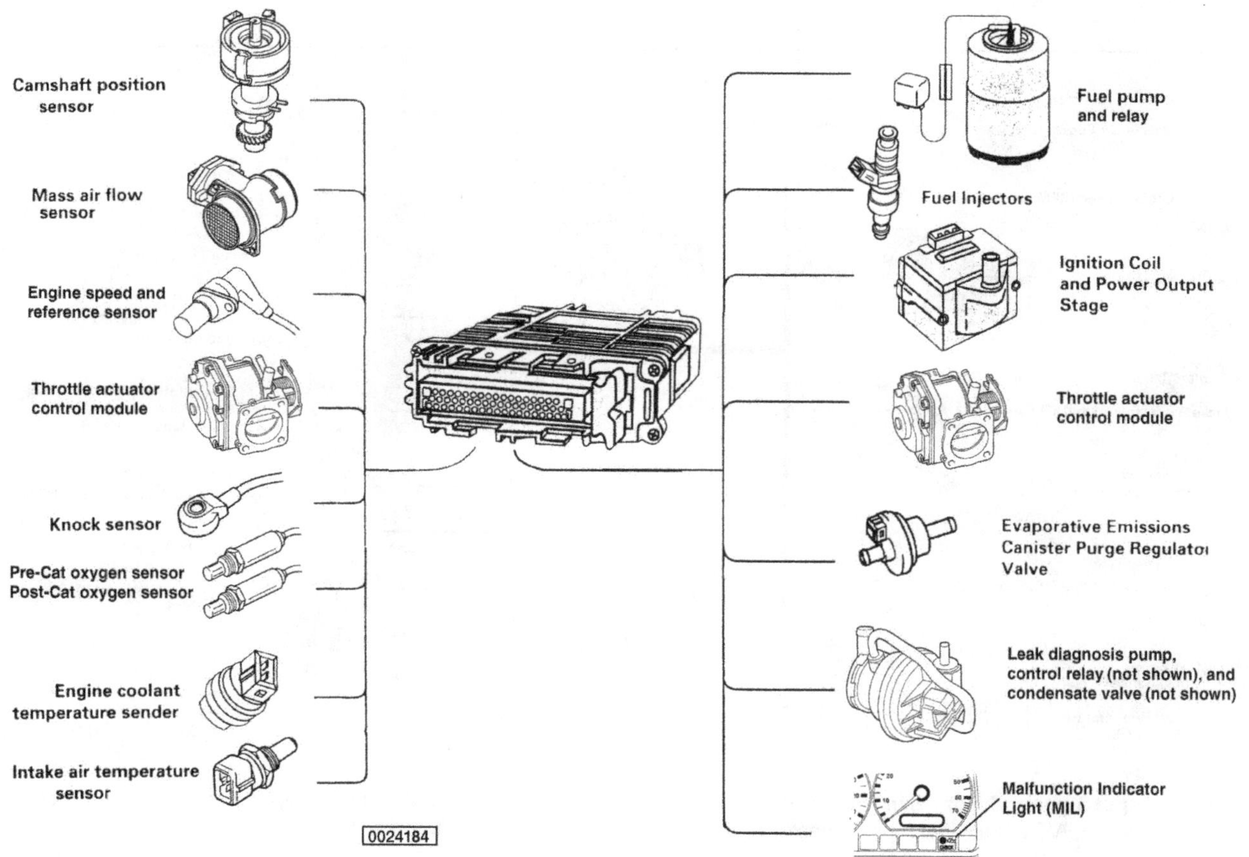

Additional Input Signals: Vehicle speed sensor, A/C system, automatic transmission control module, battery, data link connector

Additional Output Signals: Instrument cluster, automatic transmission control module, data link connector

0024184

Fig. 6. Bosch Motronic 5.9 engine management system. Inputs (**left**) to ECM are used to control and adapt output signals (**right**) to the individual components.

Some of the troubleshooting procedures used for the Motronic 2.9 system will also work on the 5.9 system. However because of its complex nature, proper diagnosis and repair requires the use of a specialized scan tools such as Volkswagen special tool VAG 1551 or VAG 1552.

In addition to diagnostics, Motronic 5.9 systems include the ability to record proper operation of as many as eight monitored functions. These functional checks are called Readiness Codes. Readiness Codes are set when the proper operating parameters are reached at least one time. These codes are important because some areas with emissions testing check these codes first. If the readiness code is NOT set, the vehicle will not pass the test. These codes can be erased if the battery is disconnected or runs low, if the ECM is disconnected, or if faults or DTCs are erased with a scan tool. Generally speaking, the readiness codes in Volkswagen systems will reset themselves after the vehicle has been started and driven under varying conditions several times. They can also be reset using the scan tool.

Diesel Turbo Direct Injection – TDI (1Z, AHU engines)

The AHU (1.9 liter) engine introduced into Jetta models in 1997 is equipped with a management system that is based on the 1Z engine first installed in Passat models for model year 1996. This new Diesel Injection System is different from previous Volkswagen diesel engines and features engine controls that more closely resemble those of a gasoline engine. See Fig. 6. The TDI system combines sophisticated computer control of fuel management and emissions with system monitoring and diagnostics. An exhaust driven turbocharger and an intercooler work with a newly designed combustion chamber and cylinder head to produce more efficient combustion and lower fuel consumption. This also results in reduced engine noise and increased power.

In addition, the accelerator pedal is directly connected to the Engine Control Module (ECM) via a potentiometer. This eliminates the need for an accelerator cable and is known as "drive by wire". The ECM also controls all glow plug functions, all auxiliary coolant glow plug functions and cruise control.

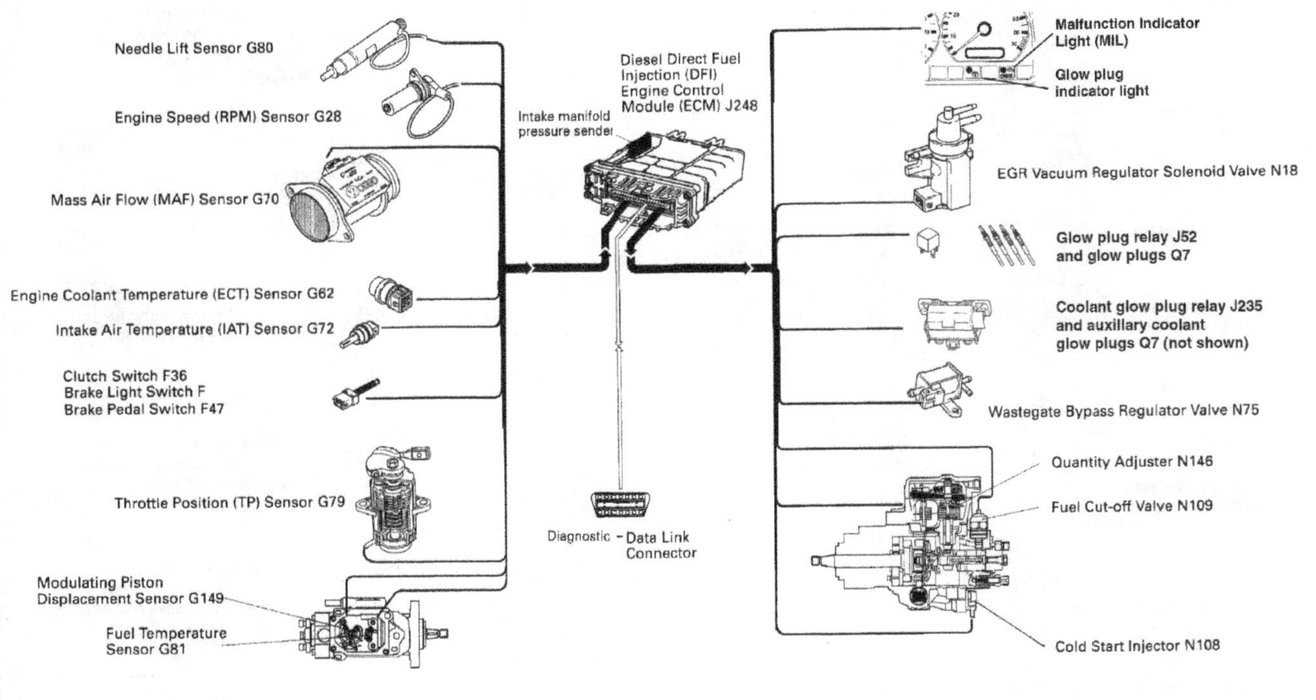

Additional Input Signals: Criuse control switch, vehicle speed sensor, A/C system, data link connector

Additional Output Signals: Instrument cluster, data link connector

0024185

Fig. 7. TDI Engine Management System as installed in 1997 through 1999 Jetta models with engine AHU. Inputs (**left**) to ECM are used to control and adapt output signals (**right**) to the individual components.

Basic fuel metering is determined by engine speed and engine load. When the key is switched on and the engine is started, the fuel cut-off valve opens and fuel flows into the injection pump. The ECM receives engine speed information from the engine speed sensor and engine load information from the Throttle Position Sensor (TPS). These signals are modified by the Engine Coolant Temperature (ECT) sensor, the Mass Air Flow (MAF) sensor, and the fuel temperature sensor. The ECM then signals the quantity adjuster which allows the proper quantity of fuel to be metered sequentially to the mechanical injectors. The ECM also uses the same information to determine the correct moment of injection which is the ignition firing point. An output signal is generated and sent to the cold start valve at the correct time.

The ECM monitors the operation of the quantity adjuster via a signal received from the modulating piston displacement sensor and makes corrections to the fuel quantity as required.

The ECM monitors the operation of the cold start valve via a signal received from the needle lift sensor attached to injector number 3 and again makes corrections as required.

Turbocharger boost pressure is regulated by the ECM based on a signal from the Intake Air Temperature Sensor (IAT), a vacuum connection to the intake manifold, and an ambient pressure signal from an internal Barometric Pressure Sensor (BARO). The wastegate bypass regulator valve receives output from the ECM and controls a vacuum signal sent to the turbocharger wastegate.

In order to prevent possible damage from simultaneous application of the brake, clutch and accelerator pedals, the brake pedal switch, brake light switch, and clutch pedal switch, all signal their respective positions to the ECM. These signals are also used by the ECM during cruise control operation.

Based on all of the input signals, the ECM precisely controls the following output components:

- **Quantity Adjuster**
- **Cold Start Injector**
- **Fuel Cut-off Valve**
- **Wastegate Bypass Regulator Valve**
- **EGR Vacuum Regulator Valve**

In addition, the ECM uses various sensors to control these additional outputs:

- **Glow Plugs**
- **Glow Plug Relay**
- **Auxiliary Heater Coolant Glow Plugs**
- **Coolant Glow Plug Relay**

Tailpipe emissions are further reduced by a two-way oxidation-type catalytic converter.

The TDI equipped cars have two warning lights in the instrument cluster to advise of system status. The glow plug indicator light operates when the key is first turned on to indicate the need to wait for the glow plugs to pre-heat the combustion chambers. The Malfunction Indicator Lamp (MIL) lights if a failure of a monitored component occurs. In some serious failure modes both warning indicators can be lit.

Some troubleshooting procedures used for checking glow plugs are the same as on earlier diesel engines. However because of its complex nature, proper diagnosis and repair requires the use of specialized scan tools such as Volkswagen special tool VAG 1551 or VAG 1552.

IGNITION SYSTEM

The ignition function is handled through the fuel injection/engine management system. On gasoline engines, the Motronic or Mono-Motronic engine control module (ECM) computes ignition timing based on inputs from various sensors. Crankshaft position is the main input to the ECM and is used for starting. The other sensors are used to adapt the basic timing map for varying operating conditions. Motronic engines incorporate adaptive knock control to adjust the ignition timing for individual cylinders. See **28a Ignition System–Gasoline** for more information.

NOTE —

- *On Motronic-equipped engines, crankshaft position is determined by the engine speed (RPM) sensor. The RPM sensor is mounted in the side of the cylinder block and reads a toothed wheel on the end of the crankshaft. See* **24a Fuel Injection–Motronic** *for RPM sensor testing.*
- *On Mono-Motronic cars, crankshaft position is determined by the camshaft position (CMP) sensor, which is located in the ignition distributor. The CMP is an electronic switch and operates based on the Hall effect. See* **24b Fuel Injection–Mono-Motronic** *for testing information.*

EMISSION CONTROLS

Different engines use different emission control systems. Most of these systems are monitored by the On-Board Diagnostic (OBD) system, which is part of the engine management system. See **26 Exhaust System/Emission Controls** for additional information.

20 Fuel Storage and Supply

20

GENERAL

This repair group covers the fuel supply portion of the fuel system—the components that store and supply fuel under pressure to the fuel injection system. For a general system description, see **2 General Information.**

NOTE —
Fuel filter replacement is covered in **0 Maintenance Program**.

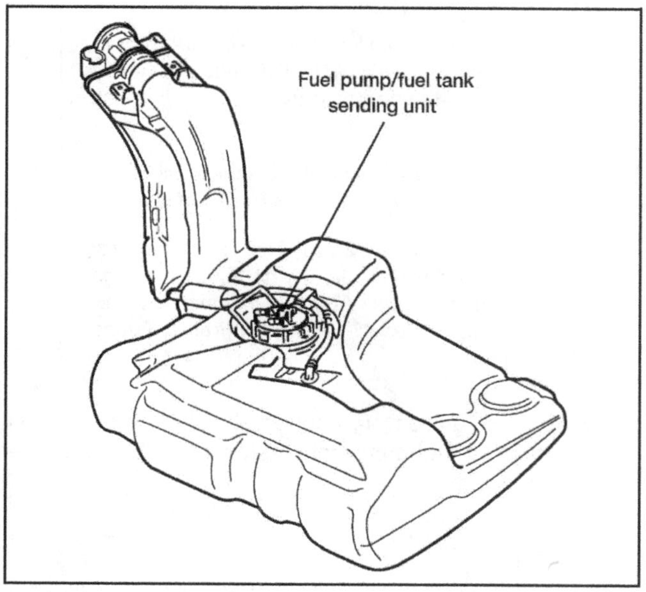

Fig. 1. Molded plastic fuel tank for gasoline engines. Cars with diesel engines are not equipped with in-tank fuel pump.

Safety Precautions

Please read the following warnings and cautions before doing any work on the fuel pump, fuel tank or fuel lines.

WARNING —
- *Always disconnect the negative (–) battery cable and cover the terminal with an insulated material whenever working on any fuel related component.*
- *Gasoline is dangerous to your health. Wear hand and skin protection when working on the fuel system. Do not breathe fuel vapors and always work in a well-ventilated area.*
- *Fuel and fuel vapors will be present during many of the operations described in this repair group. Do not smoke or create sparks. Have an approved fire extinguisher handy.*
- *The fuel supply system is designed to maintain pressure in the system after the engine is turned off. Fuel will be expelled under pressure as fuel lines are disconnected. This can be a fire hazard, especially if the engine is warm. Always wrap a clean shop rag around the fitting before loosening or disconnecting any fuel line.*

CAUTION—
- *Cleanliness is essential when working on the fuel system. Thoroughly clean fuel line unions and hose fittings before disconnecting them. Use only clean tools.*
- *Use only spring-type clamps on fuel hoses. Do not use screw-type hose clamps.*
- *Keep removed components clean, and seal or cover them with plastic or paper, especially if the repair cannot be finished immediately.*
- *When replacing parts, install only new, clean components. Seals and O-rings should always be replaced rather than reused.*

FUEL PUMP TROUBLESHOOTING– GASOLINE ENGINES

The fuel supply system is an integral part of the operation of the fuel injection system. Problems such as a no-start condition, hesitation or stalling may be due to poor fuel delivery. There are some preliminary tests to quickly determine if the fuel pump or its electrical circuit are causing the problem. Some of the tests described below require special test equipment, such as a fuel pressure gauge.

The electrical current that operates the fuel pump is controlled by a relay and protected by a 20-amp fuse. This helps to handle the high current load of the pump and also ensures that the pump will not continue to run in the event of an accident or if the engine stalls. If, for any reason, electric power to run the pump is interrupted, the engine will not run.

Begin troubleshooting with a simple check of the fuel pump electrical circuit. The pump should run while cranking the engine with the starter. If necessary, remove the floor cover and access plate in the luggage compartment and listen or feel to see whether the pump is running. If the pump does not run, see **To check fuel pump electrical circuit**.

CAUTION—
In cold weather, water in the fuel may freeze in the pump, causing the circuit to overload and the fuse to fail. Be sure to check the fuse in position 18. See **97 Wiring Diagrams, Fuses and Relays**.

If the fuel pump runs, begin troubleshooting with a check of the fuel pump delivery rate as described later in this section. The test will indicate whether further tests are necessary. This is especially important on high-mileage cars, where normal pump wear may decrease delivery volume. Also check for correct pump installation and for a clogged pump screen.

The electric fuel pump operates only when the car is running or being started. Because many of the fuel pump and fuel injection tests require that the pump be operated with the engine off, the fuel pump relay can be temporarily bypassed during testing.

To operate fuel pump for tests

The procedure below uses a temporary wiring connection to bypass the fuel pump relay. The preferred method is to use a remote switch, such as the VW special tool VAG 1348/3(3A), but you can accomplish the same thing with a homemade fused jumper wire and an in-line switch. See Fig. 2.

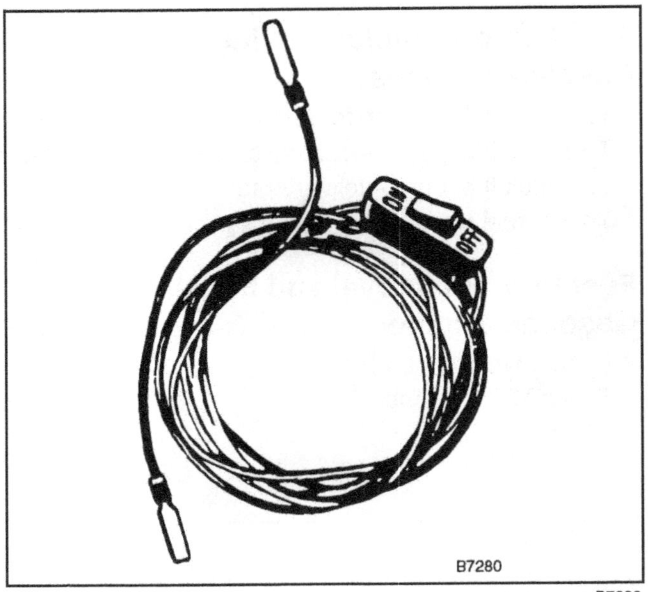

B7280

Fig. 2. Jumper wire with flat-blade connectors and switch for running fuel pump.

1. Remove the fuel pump relay from the fuse/relay panel. See Fig. 3.

CAUTION—
- *A homemade jumper wire with a switch should be at least 1.5 mm metric wire size (14 gauge-AWG) and, for safety, should include a 20 amp in-line fuse.*
- *To avoid damaging the relay sockets, the ends of the jumper wire should be flat-blade connectors that are the same size as the sockets in the relay panel.*
- *Connect and disconnect the remote switch or jumper wire only with the switch in the off position.*
- *Relay locations may vary. The fuel pump relay should have either 67, 80, or 167 stamped on the top of the relay. For more information on relay location, see* **97 Wiring Diagrams, Fuses and Relays**.

2. With the ignition off, connect the battery positive (+) voltage to fuel pump relay terminal 87 using a fused jumper wire.

NOTE—
Terminal numbers should be marked on the fuse/relay panel and on the relay bottom.

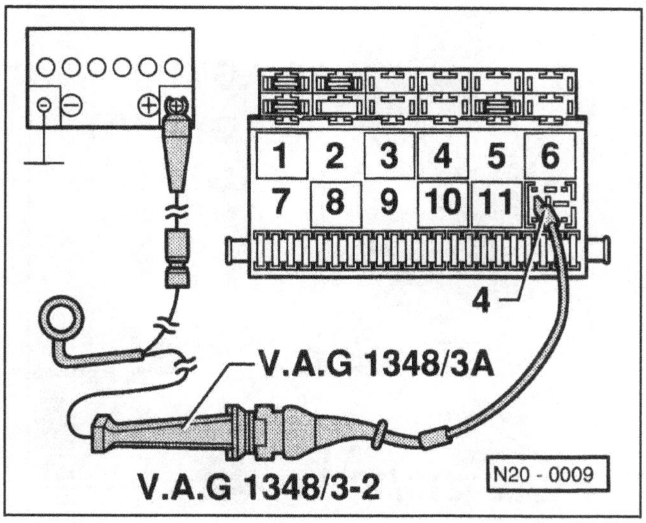

N20-0009

Fig. 3. Fuel pump relay removed from fuse/relay panel. Jumper battery voltage (+) to terminal 87 (**4**) to operate pump.

3. Turn the jumper wire switch on to run the pump.

NOTE —
If the pump runs only when the jumper is connected, the relay, the engine control module (ECM), or the circuit wiring is faulty. If the pump doesn't run with the jumper installed, the problem is most likely in the wiring to the pump or the pump itself is faulty. See **To check fuel pump electrical circuit**.

To check fuel pump electrical circuit

The fuel pump receives power from the fuel pump relay and the relay is energized by the engine control module (ECM). The test given below checks the relay's input (ground) from engine control module (ECM) and the relay's output (B+ power) to the pump.

1. Check that the battery is charged and that Fuse no. 18 is intact.

2. Fold down the rear seat and pull back the floor cover from the luggage compartment. Remove the access panel from above the fuel pump (rear right-hand corner).

3. While a helper listens at the fuel pump, turn the ignition key on. The pump should run briefly. If the pump runs, the circuit is probably functioning correctly. If the pump does not run, proceed to step 4.

NOTE —
Listen carefully for the pump to run—its operation is barely audible.

4. Remove the fuel pump relay and run the fuel pump as described in **Operating Fuel Pump for Tests**.
   - If the pump runs, go to step 5.
   - If the pump does not run, go to step 6.

5. Remove the jumper wire from the relay panel. With the ignition key off, connect an ohmmeter between terminal 3 (corresponds to terminal 85 on the relay) in the relay panel and ground. Turn the key on. There should be a momentary ground signal.
   - If ground is present at terminal 85, the relay is probably faulty.
   - If ground is not present, check the wiring between relay panel terminal 85 and the engine control module (ECM). If no wiring faults can be found, the engine control module (ECM) control unit may be faulty. Check the inputs to engine control module (ECM). See **24a Fuel Injection—Motronic** or **24b fuel Injection—Mono-Motronic**.

NOTE —
See **97 Wiring Diagrams, Fuses and Relays** *for detailed wiring schematics.*

6. If the pump does not run as described in step 4, leave the jumper wire connected and disconnect the harness connector from the fuel pump. Check for voltage at the pump harness connector as shown in Fig. 4.
   - If voltage is not present at the connector, check the wiring between the relay panel and the fuel pump harness connector. Check for power at the relay panel itself. See **97 Wiring Diagrams, Fuses and Relays** for detailed wiring diagrams.
   - If voltage is present, remove the pump and check for internal shorts in the pump wiring. If no faults can be found, the pump is probably faulty.

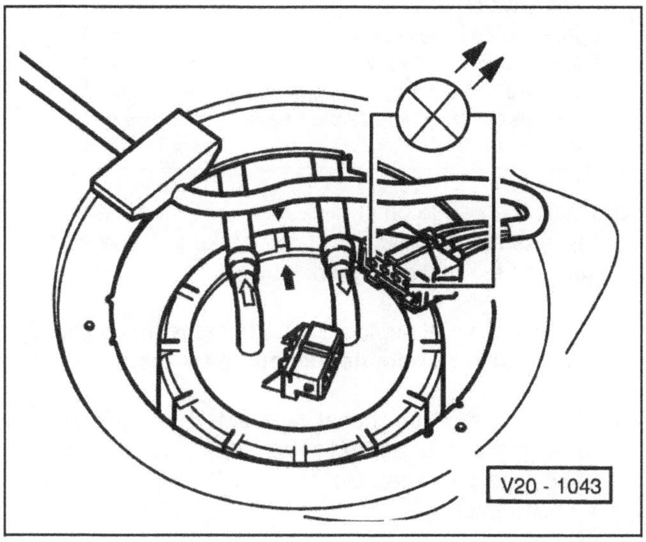

V20-1043

Fig. 4. Voltage supply to fuel pump being checked at outer terminals in connector.

To check fuel pump delivery rate

The test given below requires a fuel pressure gauge (0-100 psi) with a shutoff valve.

1. Remove the filler cap from the fuel tank.

2. Working in the engine compartment, disconnect the fuel supply line and connect a pressure gauge to the line.
   - On Motronic cars (ABA and AAA engines), disconnect the supply line from the fuel rail. See Fig. 5.
   - On Mono-Motronic cars (ACC engines), disconnect the supply line from the throttle body. See Fig. 6.

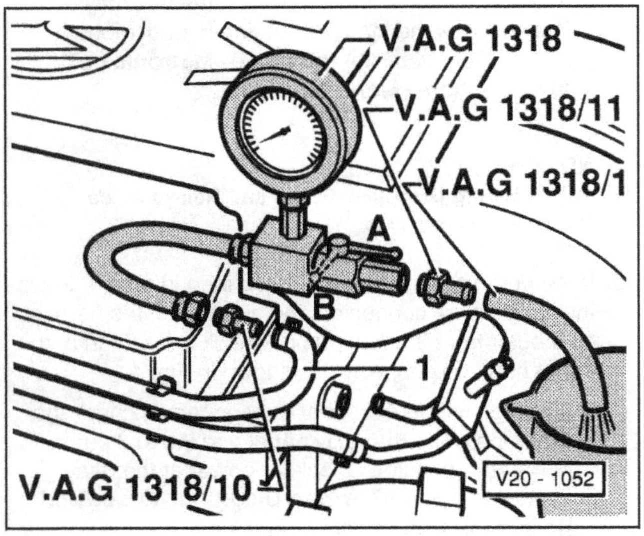

V20-1052

Fig. 5. Fuel gauge hook-up on Motronic cars (ABA, AAA engines). Supply line shown at (**1**). Numbers are for VW special tools.

> ***WARNING —***
> *Fuel will be expelled under pressure as the supply line is disconnected. Wrap a cloth around the line before disconnecting it. Do not smoke or work near heaters or other fire hazards. Keep a fire extinguisher handy.*

3. Route the gauge output line into a measurement container of at least 1000 ml (1 qt.). Make sure the shutoff valve on the gauge is open.

4. Remove the fuel pump relay and run the fuel pump as described in **To operate fuel pump for tests**.

5. Slowly close off the shutoff valve until the gauge reaches the test pressure listed below. Shut the pump off and empty the container. Do not alter the position of the shutoff valve.

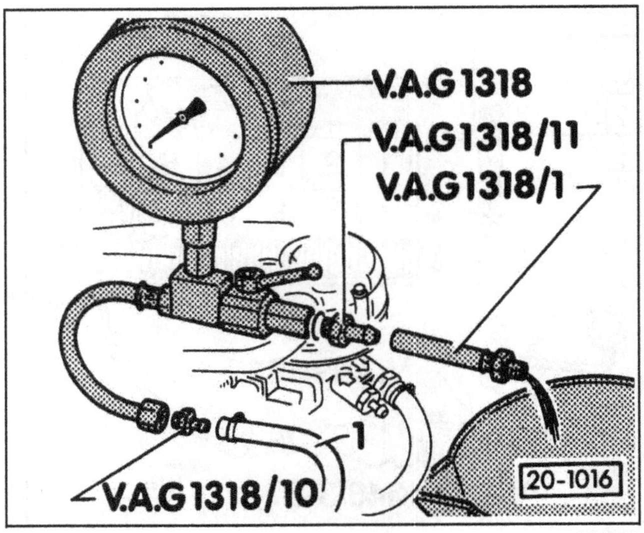

20-1016

Fig. 6. Fuel gauge hook-up on Mono-Motronic cars (ACC engine). Supply line shown at (**1**). Numbers are for VW special tools.

Fuel delivery test pressure

- ABA, ACC engines . . . . 3.0 bar (44 psi)
- AAA engine . . . . 4.0 bar (58 psi)

6. Operate the pump again for exactly 30 seconds. Compare the amount of fuel collected to the chart shown in Fig. 7.

> ***NOTE —***
> *Fuel pump output varies depends on the voltage at the pump. Always measure the voltage at the pump (with the pump running) before using the chart shown in Fig. 7. In general, pump voltage is 1-2 volts less than battery voltage.*

If fuel delivery is below specifications, check for leaks, blocked or kinked lines, a blocked filter, or a blocked pump strainer/filter. If no faults are found, the fuel pump is probably severely worn or faulty and should be replaced.

> ***NOTE —***
> *Fuel filter replacement is covered in* **0 Maintenance Program**.

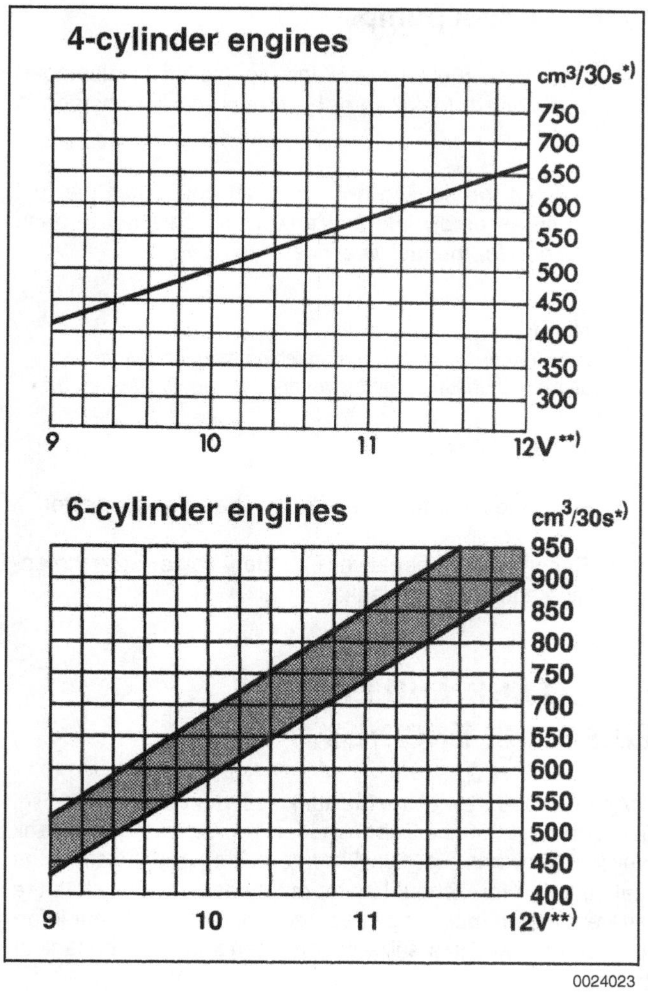

Fig. 7. Minimum amount of fuel to be delivered by fuel pump in 30 seconds of operation. Chart shows fuel delivery vs. voltage at fuel pump with pump running (usually 1-2 volts less than battery voltage).

To test fuel pump check valve

The fuel pump check valve holds residual fuel pressure in the fuel lines after the engine is turned off. This residual pressure aids in quick starting and helps prevent fuel line vapor lock. The fuel pump check valve is part of the pump assembly and not separately replaceable. For additional fuel pressure tests, see **24 Fuel Injection**.

1. Connect the fuel pressure gauge as shown in Fig. 5 or Fig.6.

> ***WARNING —***
> *Fuel will be expelled under pressure when fuel lines are connected and disconnected. Do not smoke or work near heaters or other fire hazards. Keep a fire extinguisher handy.*

2. Close the pressure gauge shutoff valve. Then run the fuel pump in short intervals until the fuel pressure reaches the initial test pressure listed below. Turn the fuel pump off.
   - Pressure will build quickly—operate the pump in short intervals only. If necessary, open the shutoff valve slowly to bleed off excess pressure.
3. After ten minutes, compare the gauge readings to the specifications listed below.

Fuel pump residual test pressure

- initial test pressure
  - ACC engine . . . . 2.0 bar (29 psi)
  - ABA engine . . . . 3.0 bar (44 psi)
  - AAA engine . . . . 4.0 bar (58 psi)
- after ten minutes (minimum pressure)
  - ACC engine . . . . 1.2 bar (17 psi)
  - ABA engine . . . . 2.0 bar (29 psi)
  - AAA engine . . . . 3.0 bar (44 psi)

If the pressure drops too quickly, check for leaks in the fuel lines, fuel injectors, or the throttle body (ACC engine). If no leaks are found, the most likely cause is the fuel pump check valve. To further isolate the problem, repeat the above test and then quickly pinch off the return line to the fuel tank. If the pressure now holds steady, the fault is in the pump check valve. If the pressure still drops off, check for faulty fuel injection components. See **24 Fuel Injection**.

> ***NOTE —***
> *On ABA engines, an in-line check valve is used in the return line near the pump outlet.*

FUEL PUMP REMOVAL AND INSTALLATION–GASOLINE ENGINES

To remove fuel pump

1. Disconnect the negative (–) battery cable.

> ***NOTE —***
> *Be sure to have the anti-theft radio code on hand before disconnecting the battery.*

2. Fold down the rear seat and pull back the floor cover and insulation from the luggage compartment. Remove the access panel from above the fuel pump.
3. Disconnect the harness connector, the supply line, and the return line from the fuel pump. See Fig. 8.

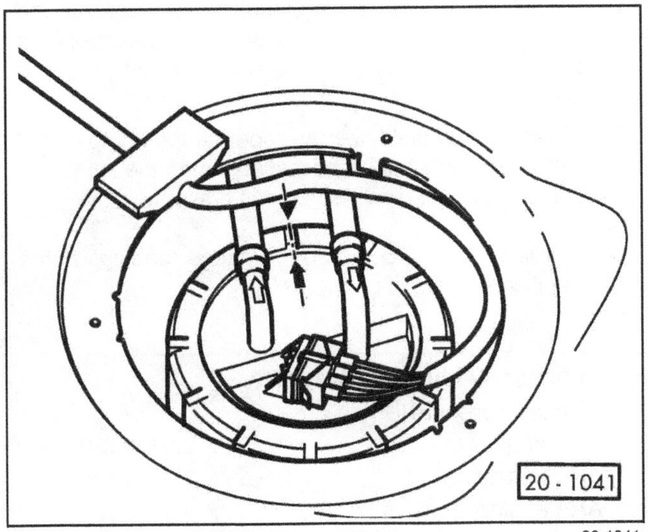

20-1041

Fig. 8. Fuel pump supply line (black), return line (blue) and harness connector. Note matching marks on tank and pump (**arrows**).

> ***WARNING —***
> *Fuel will be discharged and dangerous fuel vapors will be present. Work in a well ventilated area. Do not disconnect wires that could cause sparks. Do not smoke or work near heaters or other fire hazards. Keep a fire extinguisher handy.*

4. Unlock the fuel pump retaining ring by turning/twisting to the left (counter-clockwise) approx. 15°. Once the retaining ring is unlocked, slowly lift the fuel pump assembly from the tank. Empty the fuel pump module into a container once it is removed.

NOTE —
VW special tool no. 3217 is available to loosen and tighten the fuel pump retaining ring.

To install fuel pump

1. Inspect the fuel tank seal and replace it if it is damaged or distorted. Make sure the fuel pump filter screen is clean.

2. Moisten the sealing ring with fuel and install the fuel pump assembly into the tank, using care not to bend the arm for the fuel level float.

NOTE —
Align the pump so that the matching mark on the tank aligns with the mark on the pump. See Fig. 8.

3. Tighten the cap nut.

4. Reconnect the fuel lines and the harness connector.

5. Reconnect the negative (–) battery cable. Start the engine and check for leaks.

FUEL TANK AND LINES– GASOLINE ENGINES

An exploded view of the fuel tank assembly is shown in Fig. 9. From m.y. 2000 the Cabrio models have a revised fuel tank which is shown in Fig. 10. When removing and installing the fuel tank, the tank should be as empty as possible. Note the different length mounting straps and all other pertinent information contained in the illustration when servicing the tank or lines.

Fuel tank assembly–gasoline (1993-1999)

1. **Filler cap**
2. **Circlip**
3. **Rubber boot**
4. **Ground strap**
5. **O-ring**
   - Always replace
6. **Breather valve**
   - See checking below
7. **Expansion tank**
8. **Return line (blue)**
9. **Check valve (ABA engine only)**
   - Arrow points in direction of flow
10. **Fuel pump cap nut**
11. **Hose guide**
12. **Fuel pump flange**
13. **Fuel pump supply hose**
14. **Fuel pump return hose**
15. **Seal**
    - Lubricate with fuel when installing
    - Replace if damaged
16. **Fuel pump**
17. **Filter screen**
18. **Fuel level sensor**
19. **Tank retaining straps**
    - note different lengths
20. **Bracket**
21. **Bolt**
    - Tighten to 25 Nm (18 ft-lb)
22. **Breather hose**
23. **Fuel tank**
24. **Fuel filter clamp**
25. **Fuel filter**
    - Arrow with "AUS" or "AUSLUF" points in direction of fuel flow
26. **Supply line (black)**
27. **Nut**
    - Tighten to 20 Nm (15 ft-lb)
28. **Fuel filter bracket**
29. **Rubber buffer**

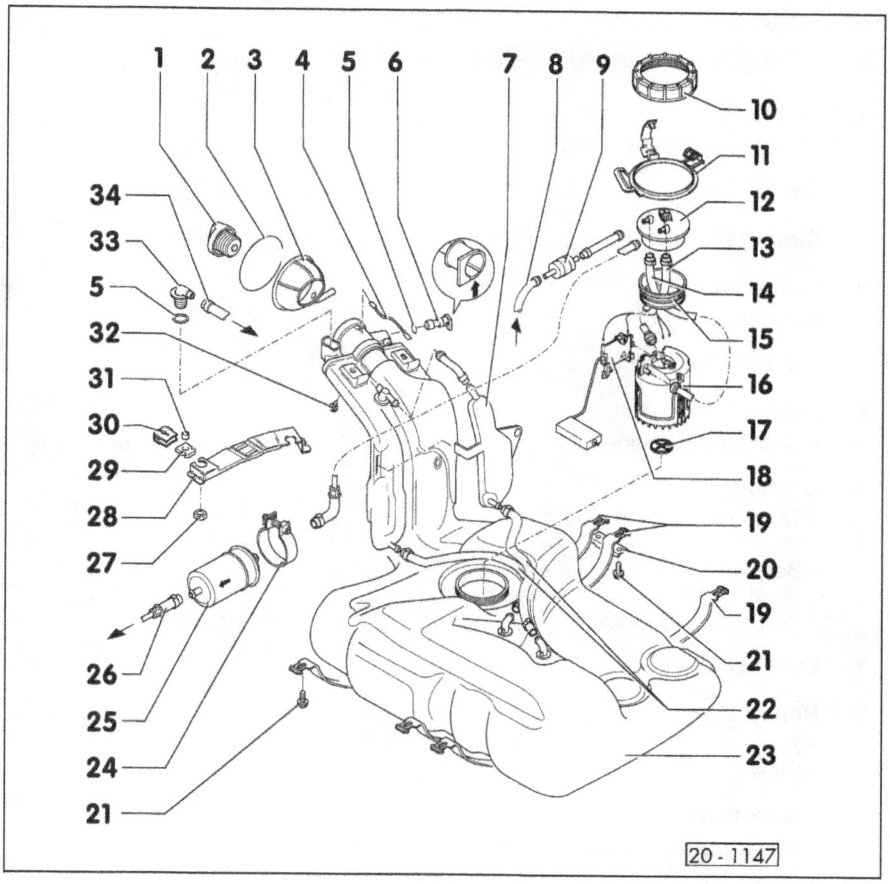

20-1147

Fig. 9. Gasoline fuel tank assembly for all Golf, Jetta, GTI and Cabrio models through 1999.

30. **Spring clip**
31. **Spacer**
32. **Screw**
    - Tighten to 10 Nm (89 in-lb)
33. **Gravity valve**
    - Unclip and pull up and out to remove
34. **Hose**
    - To evaporative emissions (EVAP) canister

Fuel tank breather valve, checking

Remove fuel tank breather valve from the filler neck by depressing the safety catch and pulling it out. With the valve removed, and the lever in the rest position, the valve should be closed. With the lever pressed **(arrow)**, the valve should open. If any faults are found, the breather valve should be replaced.

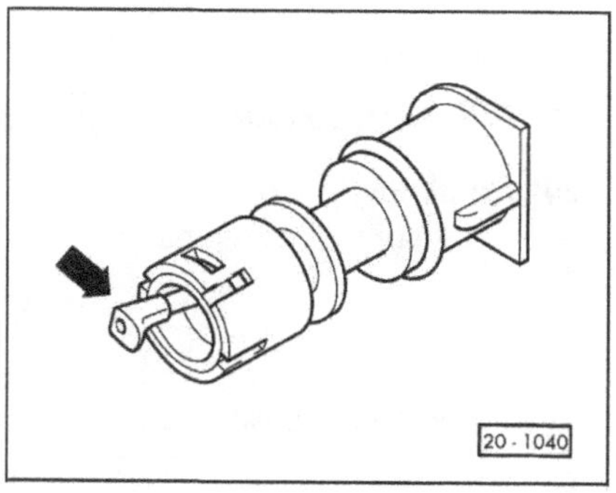

Fuel tank assembly–gasoline (Cabrio 2000-2002)

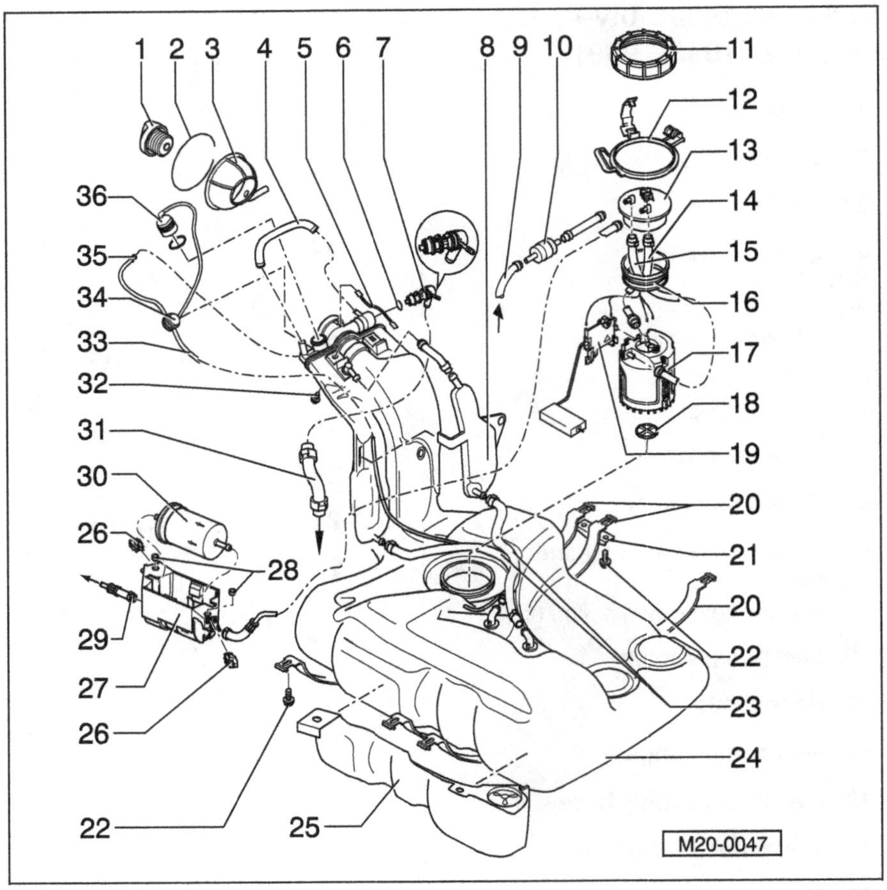

M20-0047

Fig. 10. 2000 m.y. and later Cabrio fuel tank assembly.

1. **Filler cap**
   - Replace seal if damaged
2. **Circlip**
3. **Rubber boot**
4. **Vent pipe**
   - Check for tight fit
5. **Ground connection**
6. **O-ring**
   - Replace if damaged
7. **Change-over valve**
   - To remove: pull to side out of mount
   - Before installing, remove cap (item 1)
   - See checking below
8. **Expansion tank**
9. **Return line**
   - From fuel rail
   - Blue
10. **Check valve**
    - Arrow points in direction of flow
11. **Fuel pump screw cap**
    - Remove/install with 3217 ring nut spanner
12. **Hose guide**
13. **Fuel pump flange**
    - Line up arrows on flange and fuel tank when installing
14. **Supply hose**
15. **Return hose**
16. **Seal**
    - Lubricate with fuel to assemble
    - Replace if damaged
17. **Fuel pump**
18. **Strainer**
19. **Fuel level sensor**
20. **Retaining strap**
    - Note different lengths
21. **Bracket**
22. **Bolt**
    - Tighten to 25 Nm (18 ft-lb)
23. **Breather hose**
24. **Fuel tank**
25. **Heat sheild**
26. **Spring clip**
27. **Fuel filter housing**
28. **Spacer**
29. **Fuel supply line**
    - To fuel rail
    - Black
30. **Fuel filter**
    - Arrow with "AUS" or AUSLAUF" points in direction of flow
31. **Vent line**
    - To EVAP canister
    - Black
32. **Screw**
    - Tighten to 10 Nm (7 ft-lb)
33. **Vent line**
    - From pressure retention valve to EVAP canister purge regulator valve
    - White
34. **Pressure retention valve**
    - For fuel reservoir ventilation
35. **Vent line**
    - From pressure retention valve to change-over valve
    - White
36. **Gavity/oveflow valve**

Change-over valve, checking

- With lever in rest position, valve should be closed.
- With lever depressed (direction of **arrow**) valve should open.

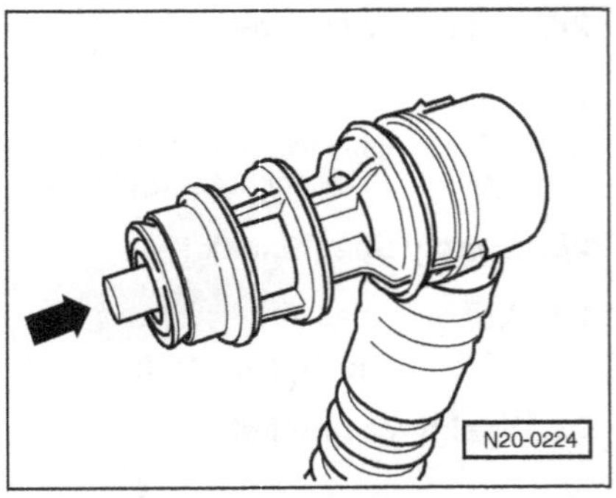

EVAP SYSTEM– GASOLINE ENGINES

Some versions of the Motronic 5.9 Engine Management System are equipped with additional evaporative (EVAP) system monitoring. This system is also referred to as the Leak Detection System. The EVAP system compenents for various model years are shown here and on the following two pages.

The Engine Control Module determines when the appropriate conditions are met and activates a vacuum operated pump in the right front area of the vehicle through an integral solenoid valve. The Leak Detection Pump (LDP) pumps a small amount of air pressure into the fuel tank and lines. At the same time, a timer relay (J335) closes the evaporative canister condensate drain valve and the ECM closes the evaporative canister purge valve. The pump runs until a predetermined pressure or time is reached. If the system does not hold the correct range of pressures, a fault or DTC is stored in the ECM memory.

Basic troubleshooting should begin with a visual inspection of the evaporative system hoses and components with particular attention paid to the fuel cap and seal. Further testing will require use of a scan tool such as Volkswagen special tool VAG 1551 or VAG 1552.

EVAP system components (through 1996)

NOTE —

From 08/95 to 10/95 vehicles were produced with a Leak Detection Pump (LDP) installed, but not activated. Beginning 11/95 the pump was no longer installed. The pump was then reintroduced for the 1997 model year.

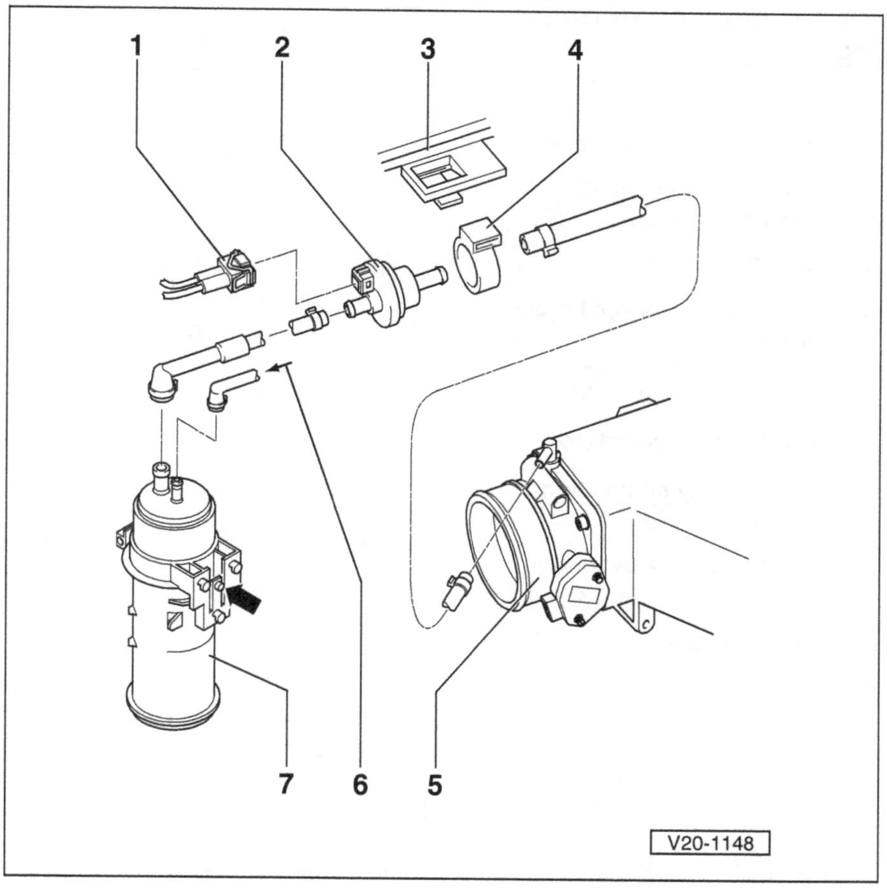

V20-1148

Fig. 11. EVAP system components on vehicles through 1996 m.y.

1. **2-pin harness connector**
2. **EVAP canister purge regulator valve**
   - Blue, normally open
   - Approx. 5 seconds after ignition is switched off, valve is open in direction of arrow
   - Valve is controlled (cycled) by engine control module when engine is warm
3. **Bracket on air cleaner housing, lower**
4. **Retainer**
5. **Throttle valve**
6. **From gravity valve**
7. **EVAP canister**

EVAP system components (1997-1999)

1. **EVAP canister**
   - In right front wheelhousing
2. **Pressure control valve**
   - Flow from fuel tank
3. **Test connection**
4. **EVAP canister purge regulator valve**
   - Black
   - Normally closed
5. **2-pin harness connector**
6. **Throttle body connection**
   - Manifold vacuum
7. **Air filter for leak detection pump**
8. **Leak detection pump (LDP)**
9. **Breather valve**

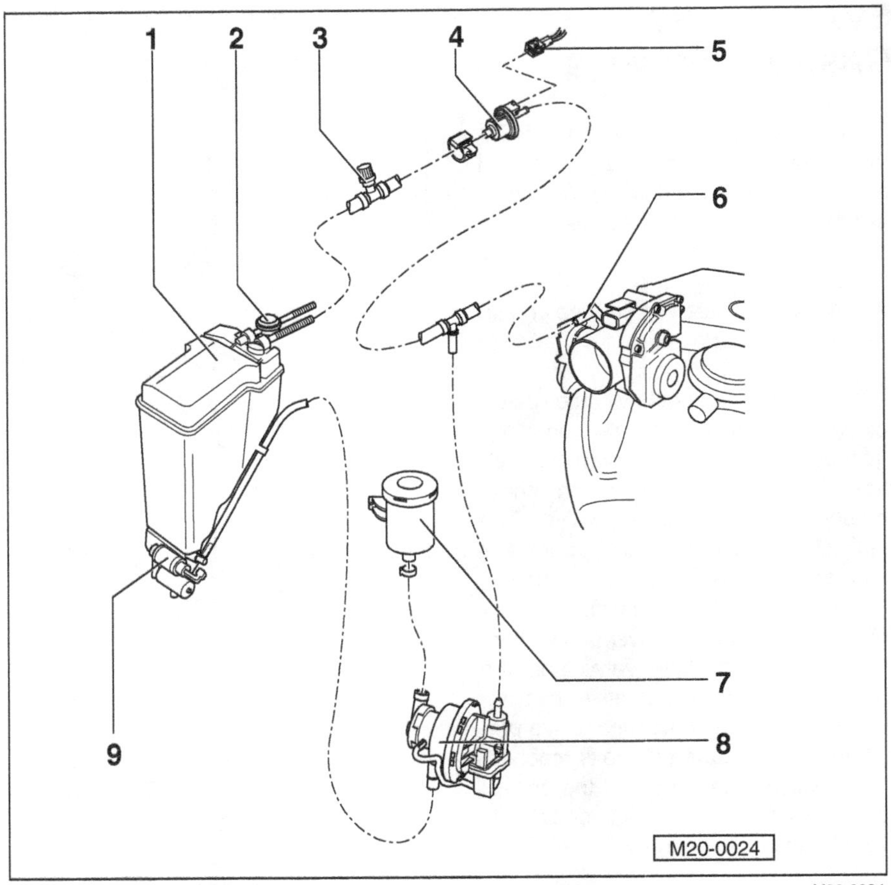

M20-0024

Fig. 12. EVAP system components on vehicles from 1997-1999 m.y.

EVAP system components (Cabrio 2000-2002)

1. **Mounting clip**
2. **EVAP emissions canister**
   - Behind right rear wheelhousing
   - Mounted on body
3. **O-ring**
   - Replace if damaged
4. **Pressure hose**
   - Check for tight fit
5. **3-pin harness connector**
   - Black
6. **Vacuum line**
   - To intake manifold upper section
7. **Leak detection pump (LDP)**
8. **Nuts**
   - Tighten to 25 Nm (18 ft-lb)
9. **Air filter for LDP**
10. **Suction hose**
    - Check for tight fit
11. **Nuts**
    - Tighten to 25 Nm (18 ft-lb)
12. **Vent line**
    - Black
    - Check for tight fit
    - From pressure retention valve
13. **2-pin harness connector**
    - Black
14. **Breather valve**
    - Black
    - To leak detection pump

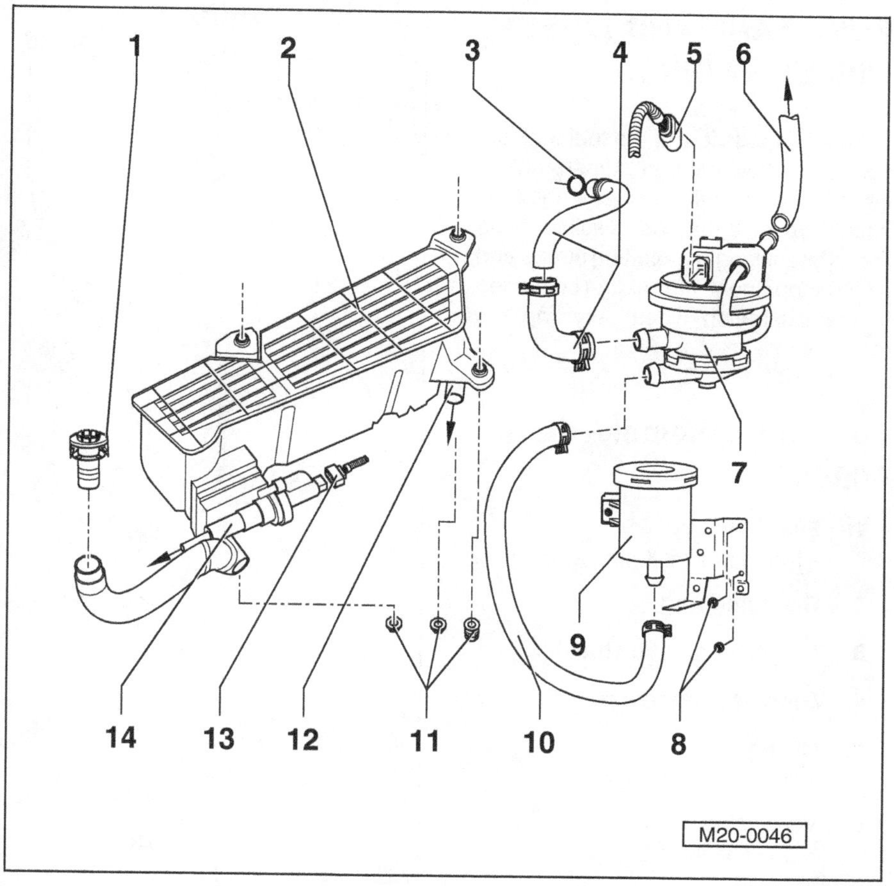

M20-0046

Fig. 13. EVAP system components on 2000-2002 Cabrios.

FUEL TANK AND LINES–DIESEL ENGINES

An exploded view of the fuel tank assembly is shown at right. When removing and installing the fuel tank, the tank should be as empty as possible. Note the different length mounting straps and all other pertinent information contained in the illustration when servicing the tank or lines.

Fuel tank assembly–diesel engines

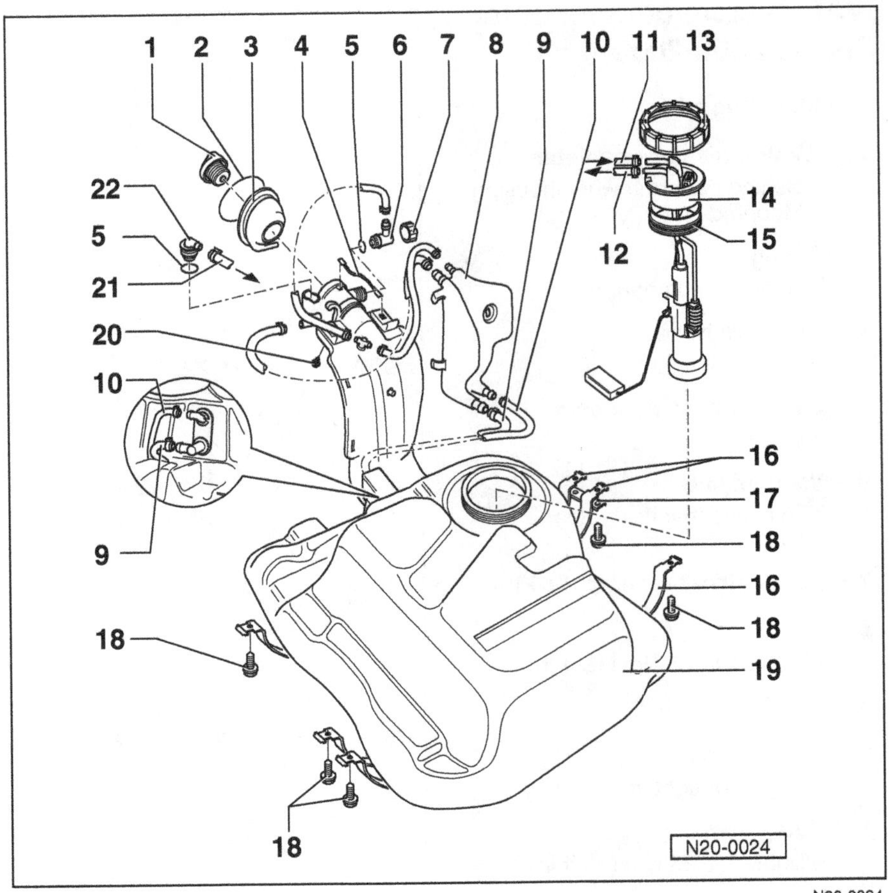

N20-0024

Fig. 14. Diesel fuel tank assembly.

1. **Fuel filler cap**
   - Replace seal if damaged
2. **Retaining ring**
3. **Fuel filler splash shield**
4. **Ground connection**
5. **O-ring**
   - Replace if damaged
6. **Vent valve**
   - See checking below
7. **Union nut**
8. **Overflow tank**
9. **Vent pipe**
   - Filling vent
10. **Vent pipe**
    - Operating vent
11. **Fuel return line**
    - Blue
12. **Fuel supply line**
    - Black
13. **Union nut**
    - Remove and install using Volkswagen special tool, 3218 wrench, or equivalent
14. **Fuel gauge sender**
    - Telescopes with fuel strainer into upper section
    - Line up arrows on flange and fuel tank when installing
15. **Sealing ring**
    - Lubricate with fuel when installing
    - Replace if damaged
16. **Mounting straps**
    - Note different lengths
17. **Bracket**
    - To secure mounting straps
18. **25 Nm (18 ft-lb)**
19. **Fuel tank**
    - Support using VAG 1383A engine and transmission jack or equivalent when removing
20. **10 Nm (7 ft-lb)**
21. **Vent pipe**
    - To fuel filter splash shield (3)
22. **Gravity valve**
    - To remove valve, unclip and pull upward out of filler neck
    - Check valve for flow
    - Valve vertical: open
    - Valve tilted 45°: closed

Vent valve, checking

- With lever in rest position, valve should be closed.
- With lever depressed (direction of **arrow**) valve should open..

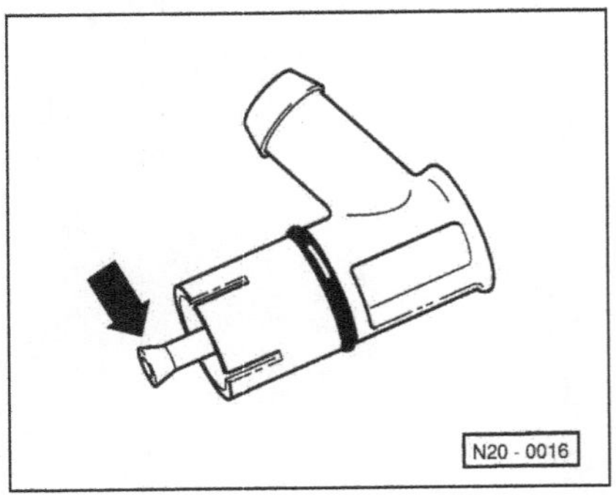

21 Turbocharger and Intercooler

GENERAL

This section covers turbocharger service and repair for the AAZ and AHU diesel engines. Turbocharger boost pressure is not adjustable and separate parts for the turbocharger are not available from VW. If the turbocharger is found to be faulty, the complete unit must be replaced.

Read the following warnings and cautions before servicing the turbocharger or exhaust system.

> ***WARNING —***
> *The turbocharger and related components operate at very high temperature. Always allow the system to cool or use proper protective clothing to prevent severe burns.*

> ***CAUTION —***
> - *Thoroughly clean all joints, pipe unions and connections, before disconnecting or reconnecting the turbocharger or any of its related components.*
> - *Take measures to prevent dust and dirt contamination. Cover all components with dust-free paper or seal them in plastic bags. Do not use cloths. Avoid nearby use of compressed air. Do not move the car or work in dusty conditions while the turbocharger is open or removed.*

A system cutaway view of the turbocharger is shown in Fig. 1. The exhaust-driven turbocharger rotates at very high speeds and is precisely balanced. The turbocharger is cooled and lubricated with engine oil. The boost pressure is limited by a mechanical wastegate. When the boost pressure exceeds a predetermined level, the wastegate opens to bypass some of the exhaust gases around the turbine.

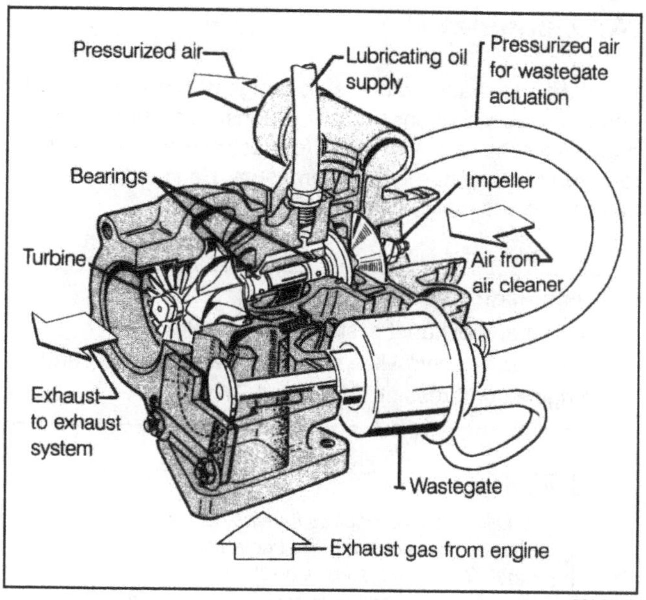

Fig. 1. Cutaway view of turbocharger. The turbocharger unit is not adjustable and must be replaced as a unit if faulty.

TURBOCHARGER DIAGNOSTICS

Due to the extremely high speeds and temperatures, proper lubrication of the turbocharger is critical. Small oil passages are drilled into the turbocharger housing and bearings. Seals are placed around the shaft at each end to prevent oil from entering the compressor and turbine housing. When the turbo bearings and seals become worn, the lubricating oil can slip past the seals and into the combustion chamber. The result is blue-gray oil smoke from the tailpipe. When the bearings become severely worn, the turbocharger compressor and turbine impeller may contact the turbocharger housing, making loud screeching or thumping noises.

The major cause of turbocharger bearing and seal failure is "coking", or baking of the oil in the turbocharger's oil passages. During operation, the temperature of the turbocharger housing can exceed the boiling point of the oil running through it. Because the oil is constantly circulating, it stays cool enough to prevent coking. If the oil pipes are clogged or if the engine is turned off before the turbocharger has had time to cool down, the oil can quickly coke. When these carbon deposits form, oil flow to the turbocharger is reduced. In addition, this substance is abrasive and can wear the seals and bearings prematurely.

If engine performance and turbocharger output is low, first check for any loose or broken hoses or fittings. Make sure the control line to the wastegate is not blocked, loose, or leaking. Make sure there are no restrictions in the intake air system, such as a dirty air filter element. If no faults can be found, check the turbo boost pressure as described below.

To check turbo boost pressure (AAZ engine)

An accurate vacuum/pressure gauge with a long extension hose is required to measure the boost pressure. Hang the gauge from the rear view mirror for ease of viewing. A helper is required to check turbo boost pressure. Be sure the basic requirements are met before making the test.

Basic Requirements
- Intake and exhaust system free of leaks
- Waste gate control line not leaking and in good condition
- Engine compression OK. See **1 Engine**

WARNING —
The following test requires that the car be driven on the road while checking the boost pressure on the gauge. This should always be done using a helper. DO NOT test if a helper is not available to read and record the gauge results.

1. Run the engine until it is fully warmed up.

2. Using a pressure/vacuum gauge with a long hose, connect the gauge between the intake manifold and the injection pump. See Fig. 2. Place the gauge in the passenger compartment.

NOTE —
Use care not to pinch off the hose at the hood or door.

3. While driving the car on a long stretch of road, place the transmission in 2nd gear.

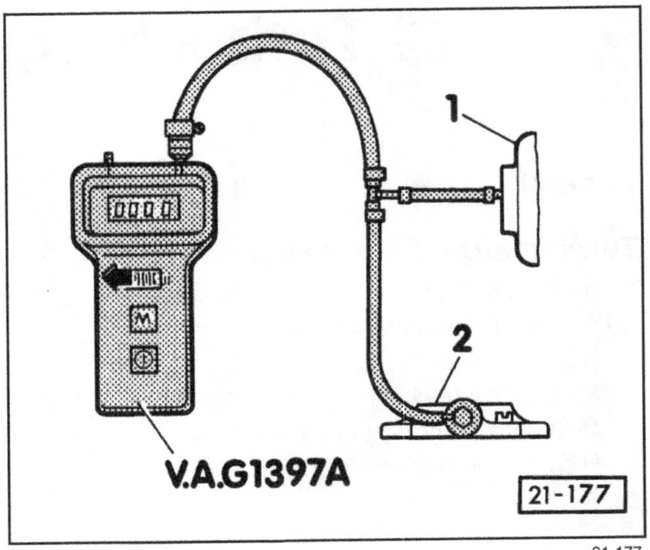

21-177

Fig. 2. Connect vacuum/pressure gauge between intake manifold (**1**) and injection pump (**2**). Use a T-fitting to make the connection. VW special tool VAG 1397A shown.

4. Depress the accelerator pedal fully to the floor until the car speed reaches 38 mph. Apply the brakes as hard as necessary to keep the speed at 38 mph and the accelerator pedal to the floor. Within a short time, the gauge reading will stabilize. Have the helper record the gauge reading at this point.

CAUTION —
Make the test as quickly as possible (preferably within five seconds, and no longer than 10 seconds) to avoid overheating the brakes. If the test is to be made a second time, drive the car about one mile before repeating the test.

Turbocharger boost pressure
- AAZ engine. . . . . 0.60–0.83 bar (8.7–12 psi)

If the boost pressure is incorrect, check that the control line to the wastegate is not damaged, plugged or kinked, If no faults can be found, the turbocharger is faulty.

NOTE —
The boost pressure is not adjustable and individual turbocharger parts are not available from Volkswagen. A faulty turbocharger must be replaced as a complete unit.

To check turbo boost pressure (1Z/AHU engine)

An accurate vacuum/pressure gauge with a long extension hose is required to measure the boost pressure. A helper is required to read and operate turbocharger boost pressure tester while vehicle is being driven. Be sure the following basic requirements are met before doing the test.

Basic Requirements
- Intake and exhaust systems free of leaks
- Wastegate control line is not blocked, loose or leaking
- No engine or fuel injection mechanical malfunctions. See **1 Engine**.
- No malfunctions stored in Diagnostic Trouble Code (DTC) Memory as verified by Volkswagen Scan Tool VAG 1551 or VAG 1552

> ***WARNING —***
> *The following road test requires the car be driven on the road while checking the boost pressure on the gauge. This should always be done using a helper. DO NOT test if a helper is not available to read and record gauge results.*

1. Run the engine until it is fully warmed up.

2. Disconnect harness connector from wastegate bypass regulator valve (N75). See Fig. 3.

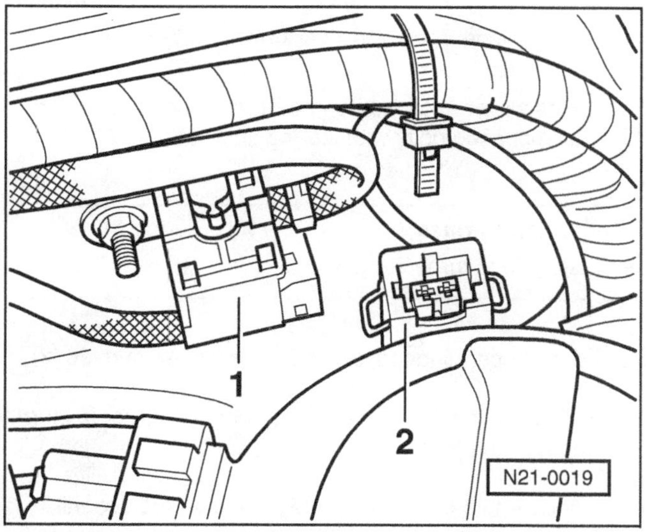

N21-0019

Fig. 3. Disconnect harness connector (**2**) from wastegate bypass regulator valve (**1**).

3. Disconnect the small hose that connects the Engine Control Module (ECM) to the intercooler duct near the connection to the intake manifold. Using a T-fitting, connect Volkswagen Special Tool VAG 1397A turbocharger tester, or equivalent, to the fitting on the duct and the small hose to the ECM. See Fig. 4.

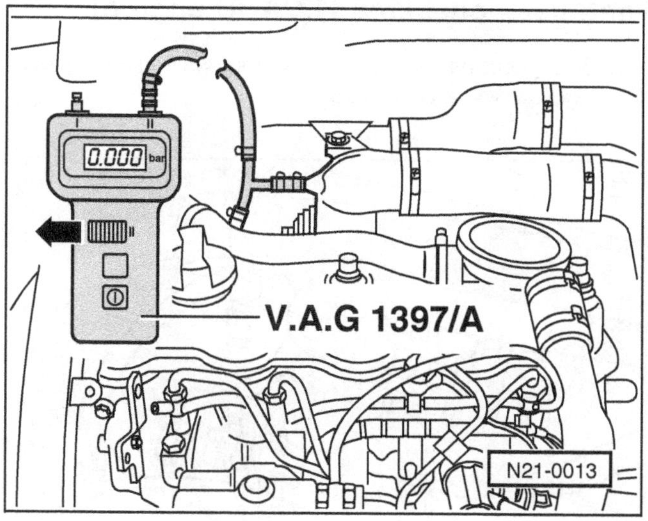

N21-0013

Fig. 4. VAG 1397A turbocharger tester installed to measure boost pressure.

4. Route the tester hose back into the interior of the vehicle in such a way so as to insure that the hood can be fully closed and latched. Also make sure that the hose is not pinched where it passes out of the hood area or into the vehicle interior.

> ***CAUTION —***
> *Be sure that the hood is properly closed and latched before driving vehicle.*

5. Road test the vehicle on open, uncongested roads.
   - In second gear, fully depress the accelerator pedal to the floor.
   - Measure pressure at wide open throttle (WOT) at an engine speed of between 3500 and 4000 RPM as indicated by the vehicle tachometer.
   - Do not apply the brakes while taking boost measurements (unless of course it becomes necessary). The engine control module responds to braking by reducing fuel quantity which can lead to incorrect boost measurements.

Turbocharger boost pressure
- 1Z/AHU engine. . . . . 0.50–0.65 bar (7.4–9.6 psi)

If boost pressure is incorrect, check small diameter hoses connecting wastegate bypass regulator valve, turbocharger and wastegate for routing, kinks and leaks. Repair as necessary. Re-test the system. If boost pressure is still incorrect, further testing will be required using Volkswagen Scan Tool VAG 1551 or VAG 1552 or equivalent.

> ***NOTE —***
> *Boost pressure is not adjustable and individual turbocharger parts are not available from Volkswagen. A faulty turbocharger must be replaced as a complete unit.*

Wastegate, checking (1Z/AHU engine)

1. Disconnect harness connector from wastegate bypass regulator valve. See Fig. 5.

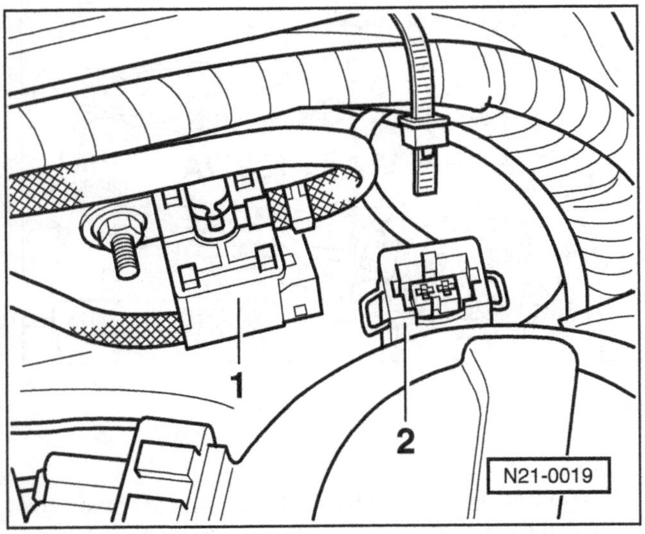

N21-0019

Fig. 5. Disconnect harness connector (**2**) from wastegate bypass regulator valve (**1**).

2. Start engine and raise to maximum speed briefly with throttle.

3. Wastegate operating rod must move. If rod does not move, check wastegate lever for ease of movement. If rod and turbocharger are corroded together, replace turbocharger. If operating rod is free to move but still does not move, replace wastegate. See Fig. 6.

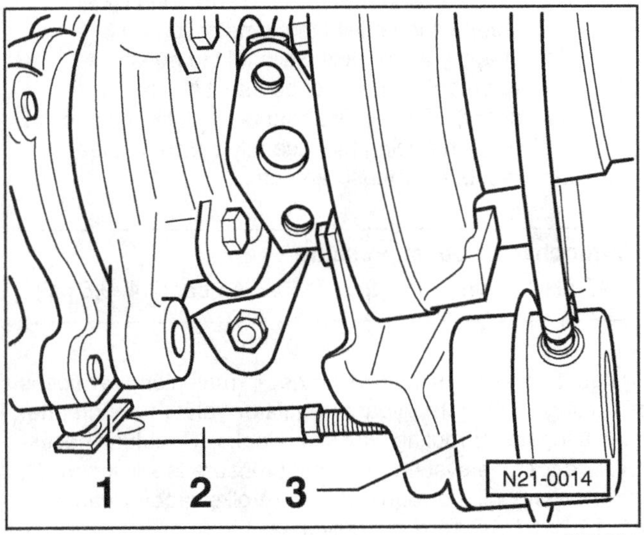

N21-0014

Fig. 6. Check wastegate (**3**) operating rod (**2**) for movement

Wastegate, removing and installing (1Z/AHU engine)

NOTE —

Numbers in parenthesis apply to illustration N21-0015.

1. Remove turbocharger as described later.

2. Unclip wastegate operating rod mounting clip (**4**).

3. Remove wastegate mounting bolts (**2**) and wastegate (**6**).

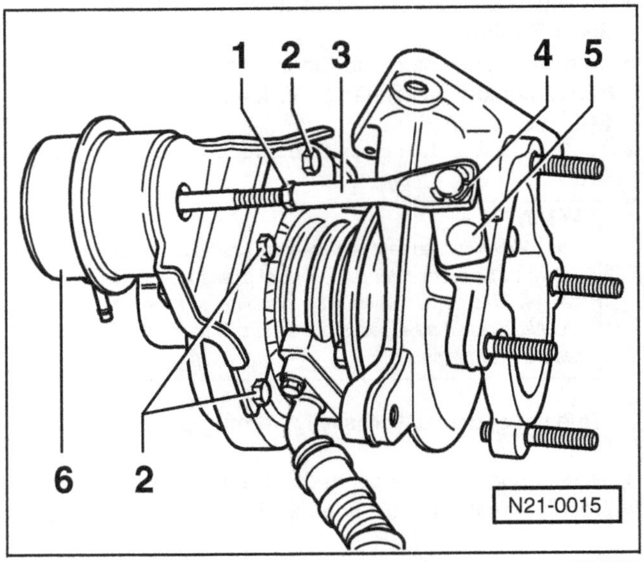

N21-0015

Fig. 7. Turbocharger wastegate components for 1Z/AHU engine.

4. To install, position wastegate on turbocharger and install mounting bolts with locking fluid "D6".

Tightening torques
- Wastegate to turbocharger. . . . . 10 Nm (7 ft-lb)

5. Loosen operating rod lock nut (**1**) and operating rod (**3**).

6. Swing wastegate lever (**5**) toward turbocharger to stop and hold in place.

7. Adjust length of operating rod (**3**) so that rod eye will install easily on lever pin (**5**). Lever should lay against stop with no play.

8. Shorten operating rod 8 full turns from this position and tighten lock nut (**1**).

9. Attach operating rod onto lever and install mounting clip (**4**). Install turbocharger.

TURBOCHARGER SYSTEM COMPONENTS

Turbocharger assembly (AAZ engine)

1. **Exhaust manifold**
2. **Bolt**
   - 25 Nm (18 ft-lb)
3. **Hose**
   - To diesel injection pump
4. **Gasket**
   - Side with coating (beading) faces intake manifold
   - Always replace
5. **Intake manifold**
6. **Bolt**
   - Tighten to 45 Nm (33 ft-lb)
   - Use high temperature anti-seize compound (VW G 000 500) when installing
7. **Gasket**
   - Always replace
   - Note installation position
8. **To Positive Crankcase Ventilation (PCV) valve**
9. **Hose from air cleaner**
10. **Turbocharger**
11. **Wastegate**
12. **Oil return line**
13. **Sealing rings**
    - Always replace
14. **Banjo bolt**
    - 30 Nm (22 ft-lb)
15. **Bolt**
    - 30 Nm (22 ft-lb)
16. **Gasket**
17. **Front exhaust pipe**
18. **Intake air hose**
19. **Mounting bracket**
    - Between exhaust pipe and exhaust manifold

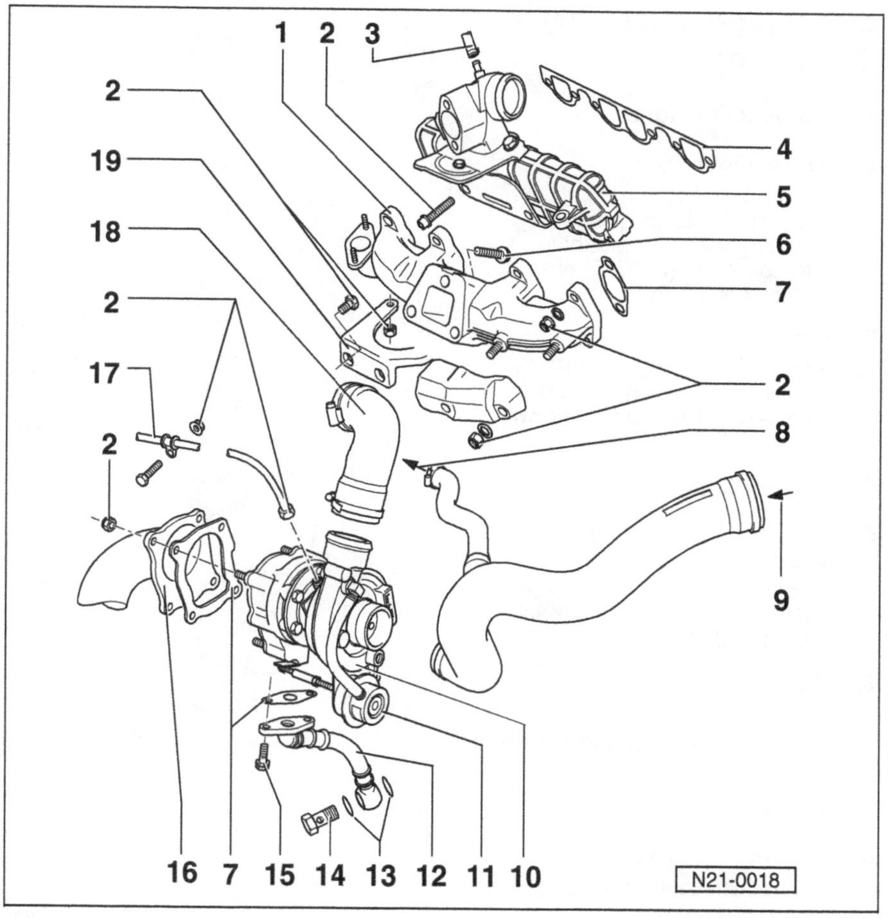

N21-0018

Fig. 8. Exploded view of turbocharger assembly on AAZ diesel engine.

Turbocharger assembly (1Z/AHU engine)

1. **Exhaust manifold**
2. **35 Nm (26 ft-lb)**
   - Always replace
   - Coat threads and bolt head sealing surfaces with high temperature (2000°F) anti-seize compound
3. **Intake manifold**
   - Includes water-cooled EGR valve (not shown), see **26 Exhaust System/Emission Controls**
4. **From charge air cooler**
5. **Hose**
   - Black
   - To Diesel Direct Fuel Injection (DFI) Engine Control Module (ECM) -J248-
6. **Gasket**
   - Always replace
   - Coating (beading) faces intake manifold
7. **25 Nm (18 ft-lb)**
8. **Gasket**
   - Note installation position
9. **25 Nm (18 ft-lb)**
10. **10 Nm (7 ft-lb)**
    - Install using D6
11. **Circlip**
12. **Turbocharger wastegate**
    - See checking below
13. **Hose**
    - Blue
    - To wastegate bypass regulator valve
14. **Oil seal**
    - Always replace
15. **Banjo bolt**
    - 15 Nm (11 ft-lb)
16. **From air cleaner**
17. **O-ring**
    - Replace if damaged
18. **10 Nm (7 ft-lb)**
19. **Banjo bolt**
    - 30 Nm (22 ft-lb)
20. **Oil return line**
    - To cylinder block

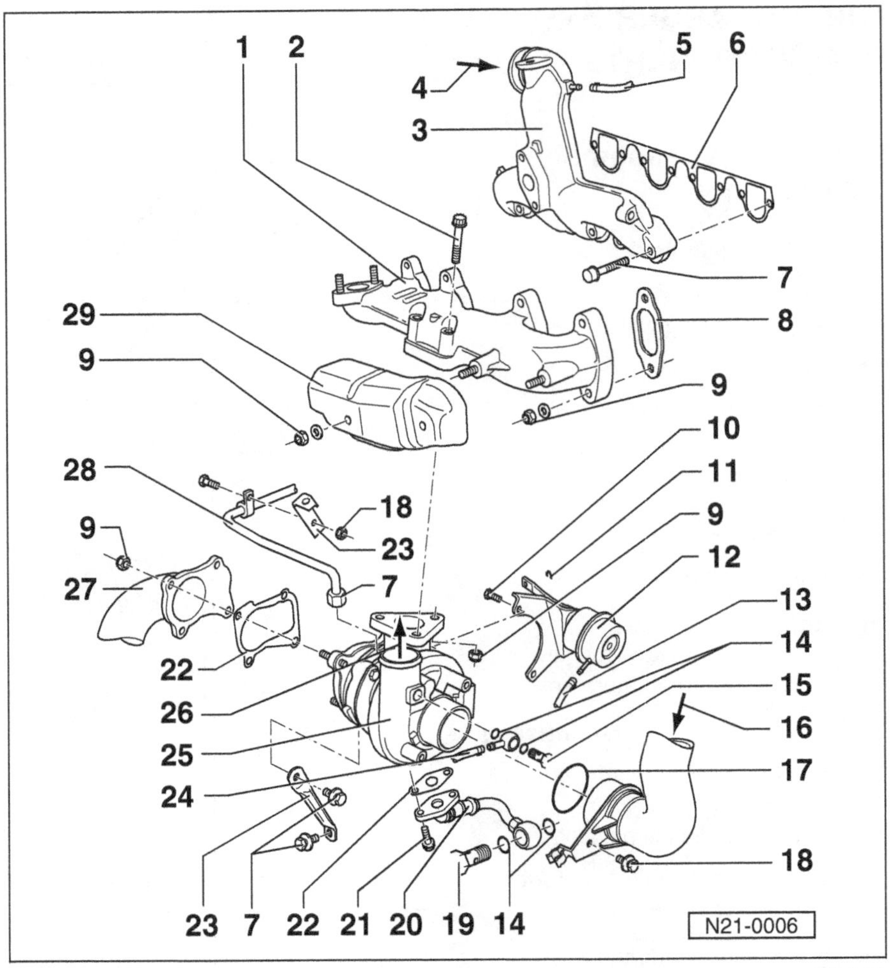

N21-0006

Fig. 9. Exploded view of turbocharger assembly on 1Z/AHU diesel engine.

21. **Bolt**
    - 30 Nm (22 ft-lb)
22. **Gasket**
    - Always replace
23. **Retainer**
24. **Hose**
    - Red
    - To wastegate bypass regulator valve
25. **Turbocharger**
26. **To charge air cooler**
27. **Front exhaust pipe**
28. **Oil supply line**
    - From oil filter bracket
29. **Heat shield**

Turbocharger hoses and lines

AAZ engine

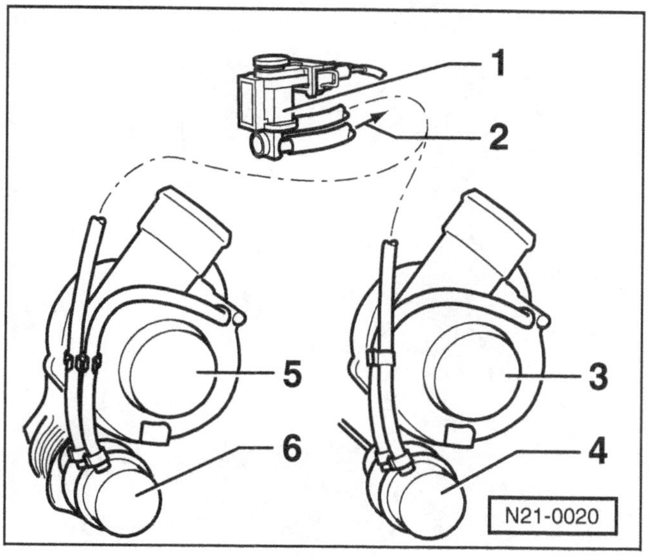

N21-0020

Fig. 10. Turbocharger connections for AAZ engine:
- **1) Two-way valve**
- **2) To vacuum pump**
- **3) Turbocharger (KKK)**
- **4) Wastegate**
- **5) Turbocharger (Garrett)**
- **6) Wastegate**

1Z/AHU engine

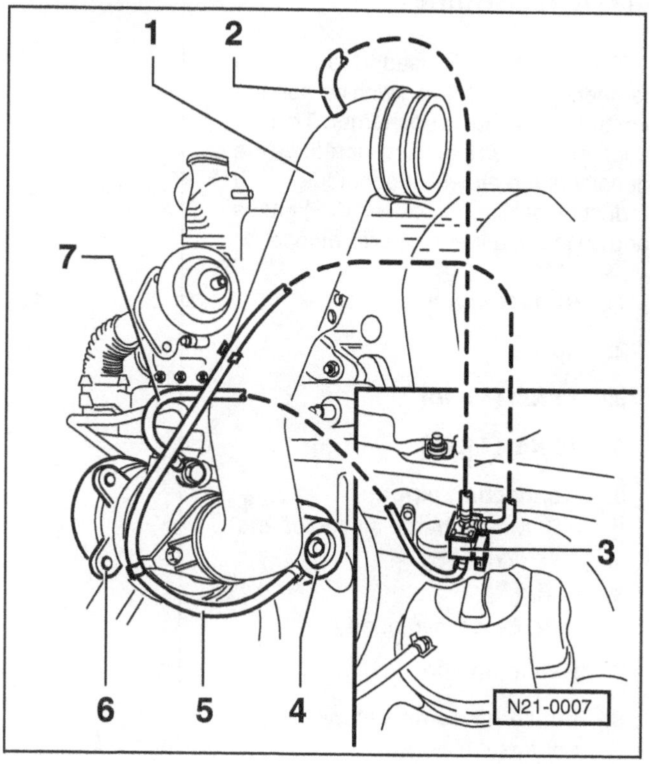

N21-0007

Fig. 11. Turbocharger connections for 1Z/AHU engine:
- **1) Intake air hose**
- **2) Hose, black**
- **3) Wastegate bypass regulator valve**
- **4) Wastegate**
- **5) Hose, blue**
- **6) Turbocharger**
- **7) Hose, red**

Intercooler assembly (1Z/AHU Engine)

The intercooler is used to reduce the temperature associated with compressing the air by the turbocharger. This reduction in air temperature increases the density of the air which contributes to a reduction of Nitrous Oxide (NOS) gases and an improvement in performance.

1. **Rubber mount**
2. **Spacer**
3. **10 Nm (7 ft-lb)**
4. **10 Nm (7 ft-lb)**
5. **Connecting pipe**
   - Between charge air cooler and turbocharger
6. **O-ring**
   - Replace if damaged
7. **Retaining clip**
8. **Intake Air Temperature (IAT) sensor -G72-**
9. **Connecting pipe**
   - Between charge air cooler and intake manifold
10. **Charge air cooler**
11. **Air duct**
12. **Retainers**

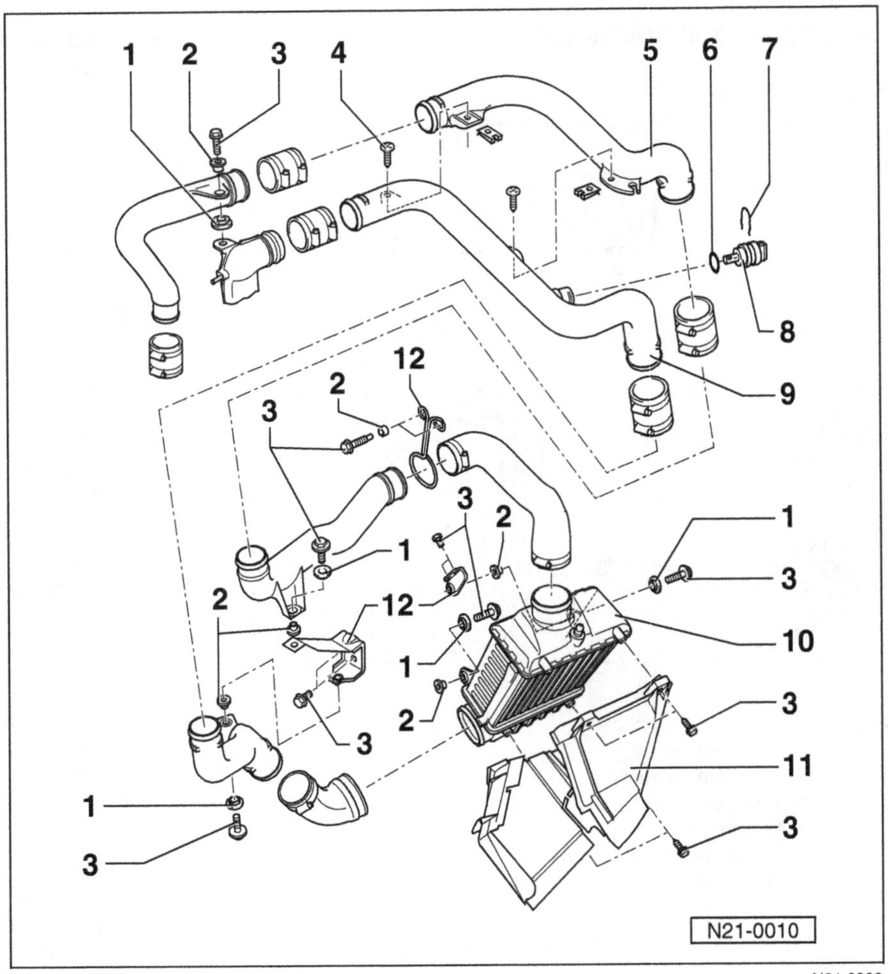

N21-0006

Fig. 12. Intercooler on 1Z/AHU engines.

REMOVING AND INSTALLING TURBOCHARGER

When replacing the turbocharger, the oil return and supply pipes/lines should be removed and thoroughly cleaned using a stiff brush and solvent. If the pipes cannot be completely cleaned, they should be replaced. The engine oil and oil filter should also be changed. Be sure to check the boost pressure once installation is complete. Use the appropriate turbocharger assembly view given earlier to assist with removal and installation.

> ***NOTE —***
> *If the turbocharger is being replaced because of excessive exhaust smoke, there will probably be some residual oil in the exhaust upon start-up. Be sure to test drive the car long enough to clear left over oil from the exhaust system.*

To remove turbocharger

1. Disconnect the battery negative (–) terminal.

> ***NOTE —***
> *Be sure to have the anti-theft radio code on hand before disconnecting the battery.*

2. Loosen the clamps and remove the air ducts that connect the air cleaner to the turbocharger. On AAZ engines, loosen the clamps and remove the air duct that connects the turbocharger to the intake manifold. On 1Z/AHU engines, loosen and remove the air duct that connects the turbocharger to the intercooler. On both engines, remove the 2 small diameter hoses attached to the turbocharger and/or wastegate.

3. Disconnect the oil supply line from the top of the turbocharger. Disconnect the supply line securing clip from the intake manifold. Plug the oil hole in the turbocharger and seal the end of the detached oil line.

4. Disconnect the oil return line from the turbocharger. Plug the end of the disconnected line.

5. Unbolt the exhaust pipe from the turbocharger outlet and remove gasket, if installed. On AAZ engines, remove support bracket between exhaust pipe and manifold. On 1Z/AHU engines, remove support bracket between turbocharger and block.

6. On AAZ engines, working from under the vehicle, remove the mounting bolts that hold the turbocharger to the exhaust manifold. See Fig. 13. Remove the turbocharger.
   On 1Z/AHU engines, remove the turbocharger exhaust manifold nut and bolts. Lift turbocharger out and upward. See Fig. 13.

> ***NOTE —***
> *The 12-point polygon-head mounting bolts are difficult to reach. Use a swivel socket for best access.*

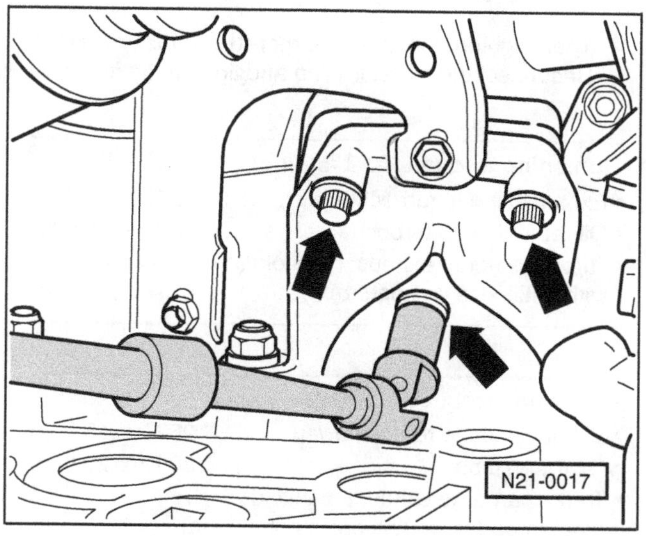

N21-0017

Fig. 13. Turbocharger-to-exhaust manifold bolts (**arrows**) on AAZ engines.

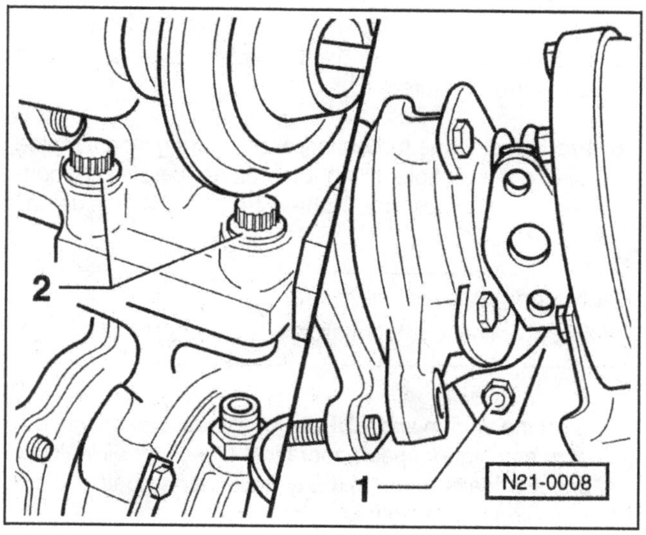

N21-0008

Fig. 14. Turbocharger exhaust manifold nut (**1**), and bolts (**2**) on 1Z/AHU engines.

To install turbocharger

1. Thoroughly clean the mating surfaces between the turbocharger and the exhaust manifold. Make sure surfaces are not damaged, as no gasket is used.

2. Loosely connect the turbocharger to the exhaust pipe. Use a new gasket. Thread the nuts on finger tight.

3. Apply a high temperature anti-seize compound to the threads of the polygon-head bolts. Connect the turbocharger to the exhaust manifold. Evenly tighten the four turbocharger-to-manifold bolts.

4. Then tighten the turbocharger-to-exhaust pipe nuts. Reconnect the oil return line and tighten the fitting.

Tightening torques (AAZ engine)
- Exhaust pipe to turbocharger. . . . . . . 25 Nm (18 ft-lb)
- Oil return line to turbocharger . . . . . . 40 Nm (30 ft-lb)
- Turbocharger to exhaust manifold (coat threads with anti-seize compound) . . . . 45 Nm (33 ft-lb)

Tightening torques (1Z/AHU engine)
- Exhaust pipe to turbocharger. . . . . . . 25 Nm (18 ft-lb)
- Oil return pipe . . . . 30 Nm (22 ft-lb)
- Turbocharger to exhaust manifold (coat threads with anti-seize compound)
  nut. . . . . 20 Nm (15 ft-lb)
  bolts . . . . 35 Nm (26 ft-lb)
- Turbocharger support bracket . . . . . . 25 Nm (18 ft-lb)

---

5. Reinstall the intake air ducts. Replace any ducting that is deformed or cracked.

6. Pre-lubricate the turbocharger by pouring clean engine oil into the oil hole, filling it as full as possible. Reconnect the oil supply line. Tighten the oil supply line union.

Tightening torques
- Oil supply line to turbocharger . . . . . . 25 Nm (18 ft-lb)

---

7. Start the engine and let it idle for at least one minute prior to any higher speed operation. Check for oil leaks or exhaust leaks and make any necessary repairs before returning the car to use.

> ***CAUTION —***
> *Do not increase engine rpm until the oil pressure is stabilized.*

23a Fuel Injection–Diesel (AAZ Engine)

TABLES

GENERAL

This repair group covers the diesel fuel injection system for the AAZ 1.9 liter diesel engine. Testing and repair information on the diesel glow plug ignition system is covered in **28b Ignition System–Diesel**.

> ***CAUTION —***
> - *Repairs and adjustments to the injection pump require specialized equipment, and parts for rebuilding are not generally available. Faulty pumps must be serviced by the pump manufacturer or other qualified diesel fuel injection repair shop. Internal problems usually require replacement of the pump.*
> - *Adjustments and repairs to the fuel injection system should be made carefully. Cleanliness is especially important. Always clean fuel unions before removing lines.*
> - *Diesel fuel is damaging to rubber. Wipe off any fuel that spills on hoses, wiring, and rubber steering and suspension parts, and wash with soap and water. If coolant hoses have been contaminated with diesel fuel, they must be replaced.*

TROUBLESHOOTING

Proper operation of the diesel engine requires a supply of clean fuel under pressure, an unrestricted supply of air, and properly timed fuel delivery. Fuel must be of the correct grade, properly winterized when necessary for cold-start conditions, and clean. Dirt and water can interfere with combustion and cause problems.

Good compression is especially necessary for the engine to run well. Engine mechanical faults leading to poor compression can cause problems that seem to be injection related. Engine lubricating oil that leaks past worn piston rings or valve stem seals may produce exhaust smoke that is mistaken for fuel injection problems.

A faulty glow plug system may contribute to cold starting problems. See **28b Ignition System–Diesel.** The battery and starting system must be able to turn the engine fast enough (at least 150 rpm) to begin compression ignition. See **27 Engine Electrical Systems**. Low power symptoms can also be the result of turbocharger problems. See **21 Turbocharger**.

Because the diesel engine speed is controlled by the amount of fuel injected, and not by the amount of air admitted past a throttle plate, the air intake must be unrestricted. A clogged air filter or faulty turbocharger can restrict the air and reduce power.

The injection pump produces the high pressure necessary to open the injectors and spray the fuel. Fuel leaks or air in the fuel lines may reduce system pressure. Since power is controlled by the amount of fuel injected, low fuel pressure may cause sluggish performance.

For cold starting, when the compression heat necessary for combustion is dissipated quickly by the cold engine, injection timing is advanced to give the fuel more time to burn. At higher rpm, there is less time for fuel to burn, so injection timing is advanced to start the fuel burning sooner. Since the injection pump is driven by the engine and externally mounted, its precise fuel metering and timing can be degraded by such things as a loose timing belt, worn sprocket, or loose mounting bolts.

Table a lists symptoms, their probable causes, and corrective actions. The boldface type refers to other sections in the manual where the repairs are described.

Diagnostic Quick Checks

These tests are used to isolate and diagnose diesel fuel system problems. It is assumed that the glow plug system is functioning correctly and that injection pump timing and valve timing are correct.

> *NOTE —*
> *Testing and repair information on the diesel glow plug ignition system is covered in* **28b Ignition System–Diesel**.

Table a. Diesel Fuel Injection Troubleshooting

Problem	Probable cause	Corrective action
1. Engine does not start	**a.** Cranking speed too low	**a.** Repair starting system or charge or replace battery. See **27 Engine Electrical.**
	b. No fuel in fuel tank	**b.** Check fuel level and add as necessary.
	c. No voltage at fuel cut-off valve on injection pump. Fuel cut-off valve loose or faulty	**c.** Valve should click each time the ignition is turned on and off. If not, check for voltage at the connector. See **97 Wiring Diagrams, Fuses and Relays.** Replace a faulty solenoid.
	d. No voltage at glow plug bus or glow plug(s) faulty.	**d.** The glow plug circuit is activated via the driver's door contact switch and the automatic glow time control module. See **28 Ignition System–Diesel** for glow plug testing and repair information.
	e. Air in fuel system	**e.** Check fuel supply line from fuel tank for cracks and connections for leaks, especially at and around the fuel filter.
	f. Injection pump not delivering fuel	**f.** Check for basic fuel delivery to engine. Check for cracked or damaged lines, clogged fuel filter or other fuel supply problems.
	g. Injection timing incorrect	**g.** Adjust injection timing.
	h. Faulty injectors	**h.** Test and, if necessary, replace injectors.
	i. Engine mechanical faults	**i.** Test compression. See **15a Cylinder Head and Valvetrain.**
	j. Low residual pump pressure	**j.** Check basic settings, adjust as necessary.
	k. Faulty injection pump	**k.** Replace pump.
2. Glow plug warning light not working	**a.** Bulb burned out or trouble in glow plug relay circuit	**a.** Test and repair as described in **28b Ignition System–Diesel.**
3. Idle speed incorrect or idle rough or irregular	**a.** Idle speed incorrectly adjusted	**a.** Check basic settings, adjust as necessary.
	b. Accelerator control binding	**b.** Check that governor lever on pump is not loose, then adjust accelerator cable.
	c. Air in fuel system	**c.** Check fuel supply line from fuel tank for cracks, kinks, or leaks.
	d. Clogged fuel filter, or fuel return line and injection pipes leaking, dirty, kinked, or damaged at connectors	**d.** Inspect and, if necessary, replace lines and hoses, replace fuel filter. See **0 Maintenance Program.**
	e. Faulty injectors	**e.** Check and, if necessary, replace injector.
	f. Injection timing incorrect	**f.** Adjust injection pump static timing.
	g. Engine mechanical faults, described earlier in this heading	**g.** Test compression. See **15a Cylinder Head and Valvetrain.**
	h. High residual pump pressure	**h.** Check basic settings, adjust as necessary.
	i. Faulty injection pump	**i.** Replace pump.

continued on next page

Table a. Diesel Fuel Injection Troubleshooting (continued)

Problem	Probable cause	Corrective action
4. Smoky exhaust (black, blue, or white)	**a.** Maximum rpm incorrectly adjusted	**a.** Check basic settings, adjust as necessary.
	b. Faulty injectors	**b.** Check and, if necessary, replace injectors.
	c. Injection timing incorrect	**c.** Adjust injection timing.
	d. Restricted exhaust system	**d.** Inspect exhaust system for damage and obstructions. See **26 Exhaust System/Emission Controls.**
	e. Engine mechanical faults	**e.** Test compression. See **15a Cylinder Head and Valvetrain.**
	f. Faulty injection pump	**f.** Replace pump.
5. Poor power output, slow acceleration or top speed	**a.** Injection pump throttle lever not reaching maximum rpm stop	**a.** Check basic settings (accelerator cable adjustment and maximum rpm adjustment).
	b. Air filter dirty/restricted	**b.** Clean or replace air filter. See **0 Maintenance Program.**
	c. Clogged fuel filter; or fuel return line and injection pipes leaking, dirty, kinked, or damaged at connections	**c.** Check for basic fuel delivery to engine. Inspect and, if necessary, replace lines and hoses, replace fuel filter.
	d. Air in fuel system	**d.** Check fuel supply line from fuel tank for cracks, kinks, or leaks.
	e. Faulty injectors	**e.** Check and, if necessary, replace injectors.
	f. Injection timing incorrect	**f.** Adjust injection timing.
	g. Engine mechanical faults	**g.** Test compression. See **15a Cylinder Head and Valvetrain.**
	h. High residual pump pressure	**h.** Check basic settings, adjust as necessary.
	i. Faulty injection pump	**i.** Replace pump.
6. Excessive fuel consumption	**a.** Air filter dirty	**a.** Replace air filter element. See **0 Maintenance Program.**
	b. Fuel leaks	**b.** Check and, if necessary, replace or tighten all pipes, hoses and connections.
	c. Return pipe blocked	**c.** Check return line for kinks and dents; replace faulty lines. If line is clogged, blow it out with compressed air, then bleed fuel system.
	d. Faulty injectors	**d.** Check and, if necessary, repair or replace injectors.
	e. Injection timing incorrect	**e.** Adjust injection pump static timing.
	f. Faulty injection pump	**f.** Replace pump.

To check fuel cut-off valve

The diesel engine is stopped by cutting off its fuel supply. This is done by means of an electrical solenoid valve. The valve is normally closed. An electrical signal from the ignition switch (key ON) opens the valve. If the valve is not working, the engine will not start. A faulty valve could also prevent the engine from turning off.

1. Turn the ignition key on and off without operating the starter. Listen for the valve to operate at the pump.
   - The fuel cut-off valve should click open when the key is turned on and click closed when key is turned off.
2. If the valve does not operate, check for voltage at the harness connector with the key on.
   - If voltage is not present, check for wiring faults. See **97 Wiring Diagrams, Fuses and Relays.**
   - If voltage is present, disconnect the battery negative (–) terminal. Clean the area around the valve and remove it. See Fig. 1. Clean and inspect the solenoid plunger and seat. If the plunger does not move freely after cleaning, the valve should be replaced. Install the fuel cut-off valve using a new seal. Reconnect the battery negative (–) terminal.

> ***WARNING —***
> *Fuel will be expelled. Do not smoke or work near heaters or other fire hazards. Have a fire extinguisher handy.*

Tightening Torque
- Fuel cut-off valve to injection pump . . 40 Nm (30 ft-lb)

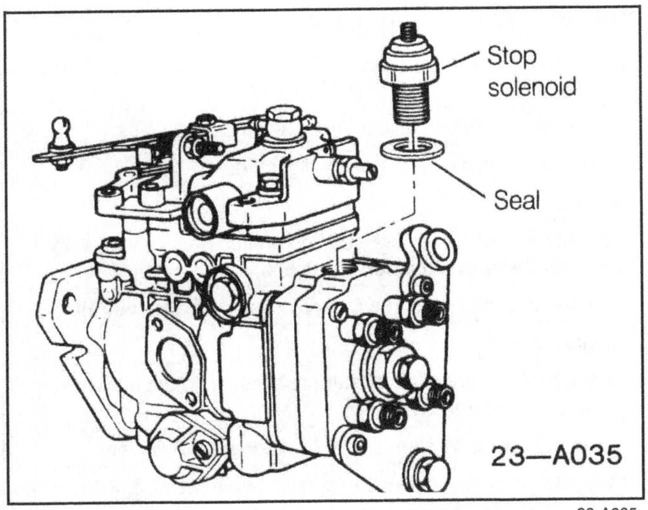

23-A035

Fig. 1. Fuel cut-off valve removed from injection pump.

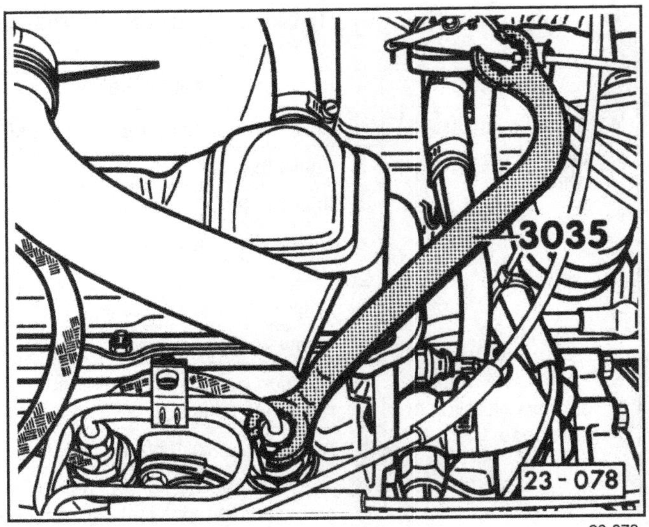

23-078

Fig. 2. Flare-nut wrench (Volkswagen special tool no. 3035) being used to loosen injector union nut at injector.

To check basic fuel delivery

This test checks if fuel is being delivered to the pump and fuel injectors. Make this test when experiencing hard starting or no-start problems.

1. Check visually for fuel leaks at the injector pipe unions or around the injector pump timing plug.

2. If no leaks are found, clean the fuel union nut for fuel injector pipe for No. 1 cylinder either at the pump or at the injector. Then slightly loosen the fuel union nut. See Fig. 2.

> ***WARNING —***
> - *Loosen the nut only about one-half turn to limit the amount of fuel leakage. The fuel injection system operates at very high pressure. Keep hands and eyes clear. Wear heavy gloves and eye protection. Wrap a cloth around the nut when loosening.*
> - *Fuel will be expelled. Do not smoke or work near heaters or other fire hazards. Have a fire extinguisher handy.*

3. Crank the engine and look for fuel leakage at the loosened union.

> ***CAUTION —***
> *Diesel fuel is damaging to rubber. Wipe off any fuel that spills on hoses, wiring, and rubber steering and suspension parts, and wash with soap and water. If coolant hoses are contaminated with diesel fuel, they must be replaced.*

4. If no fuel is observed, the most likely cause is a restriction or air leak in the fuel lines between the fuel tank and the injection pump, or possibly a clogged fuel filter.

5. If fuel is observed, but the engine still fails to start, check the injection pump timing. See **Injection Pump**.

6. When finished testing, tighten the fuel union nut.

Tightening Torque
- Fuel injector union nut to injector or pump . . . . 25 Nm (18 ft-lb)

To check idle speed drop

A defective or clogged injector may be the cause of hard starting, rough idling and running, knocking, or loss of power. An idle speed drop test is used to locate a faulty injector.

1. Run the engine at fast idle. Pull the cold start handle out to the first detent.

2. Clean the fuel union nut for fuel injector pipe for No. 1 cylinder either at the pump or at the injector. Then slightly loosen the fuel union nut. See Fig. 2.
   - Idle speed should roughen and drop, due to loss of pressure at that injector. No change indicates that the injector was already malfunctioning. See **Fuel Injectors**.

> ***WARNING —***
> - *Loosen the nut only about one-half turn to limit the amount of fuel leakage. The fuel injection system operates at very high pressure. Keep hands and eyes clear. Wear heavy gloves and eye protection. Wrap a cloth around the nut when loosening.*
> - *Fuel will be expelled. Do not smoke or work near heaters or other fire hazards. Have a fire extinguisher handy.*

3. Tighten the fuel union nut. Repeat the test for the other three injectors.

Tightening Torque

- Fuel injector union nut to injector. . . . 25 Nm (18 ft-lb)

> *CAUTION —*
> *Diesel fuel is damaging to rubber. Wash or wipe off any fuel that spills on hoses, wiring, and rubber steering and suspension parts.*

To check load-dependant timing cut-off

The injection pump is equipped with load-dependant timing and boost pressure enrichment. The load dependant timing function retards pump timing during low engine loads for better fuel economy and exhaust emissions. In order for the boost pressure enrichment to function, load-dependant timing must be turned off during high boost/load situations. The test below checks this switching-off function.

1. Disconnect the harness connector from the start positioner on the pump. See Fig. 3.

23-241

Fig. 3. Injection start positioner (**1**) on side of injection pump.

2. Connect a test light or voltmeter between the removed connector and ground.

3. Start the engine. Raise the engine speed above 3,000 rpm. The test light should come on or voltage should be present.

4. If voltage is present, remove the start positioner from the pump. Connect positive battery voltage to the start positioner terminal, then ground the positioner body. If the start positioner does not operate, it should be replaced.

5. If voltage is not present, check for wiring faults between the automatic glow time control module and the start positioner. See **97 Wiring Diagrams, Fuses and Relays.**

> *NOTE —*
> *The automatic glow time control module is mounted above the fuse/relay panel.*

BASIC SETTINGS

This section covers basic fuel injection adjustments. Over time, periodic adjustment of the idle speed and maximum rpm may be necessary. Adjustments are also necessary if a new pump is installed.

Many of the tests described below require accurate measurement of the engine rpm. Because the diesel engine has no electrical ignition system, special equipment is required. Two Volkswagen special tools are available: special tool no. VW 1367 measures rpm by detecting timing marks on the engine's flywheel; and special tool VW 1324 converts engine vibrations to electrical signals. Owing to the cost of either of these tools, dealer service may be a more cost-effective alternative.

Fig. 4 should be used as a reference for all basic settings. The tests should be made in the order given. Make sure the following requirements are met before making any checks and adjustments.

Test Requirements

- Engine at normal operating temperature: oil temperature at least 140°F (60°C)
- Cold start handle not pulled out
- All electrical accessories off

> *CAUTION —*
> *If you lack the skills or the correct tools and instruments for adjusting the diesel fuel injection, we suggest you leave these sensitive adjustments to an authorized VW dealer or other qualified and properly equipped shop.*

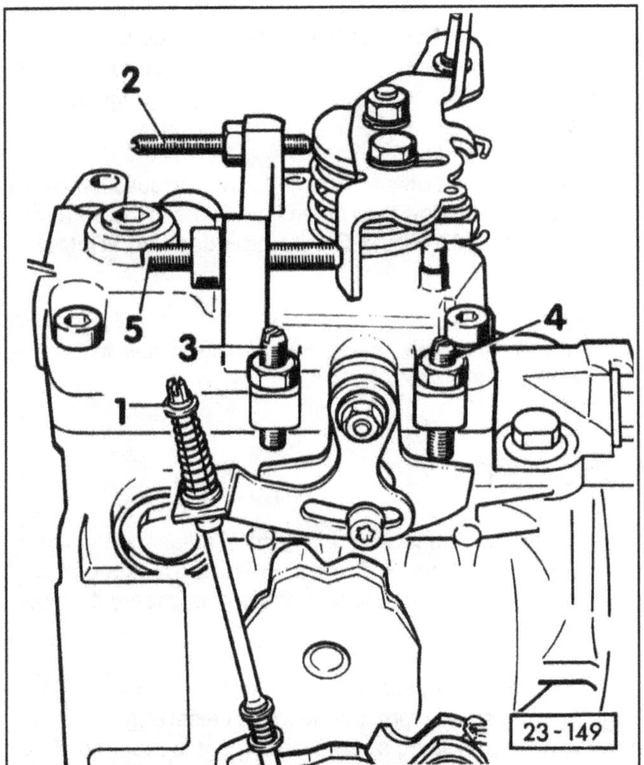

1. **Idle speed adjusting screw**
2. **Max. engine speed adjusting screw**
3. **Minimum idle speed stop screw**
4. **Maximum idle speed stop screw**
5. **Residual fuel volume adjusting screw**

23-149

Fig. 4. Diesel injection pump adjustment screws.

To check and adjust accelerator cable

1. Have a helper press the accelerator pedal fully to the floor (wide open throttle).

2. Check that the full throttle lever just contacts the maximum engine speed adjusting screw (**2**). See Fig. 4.
   - If the lever does not reach the stop or reaches it too soon and stresses the cable, the cable should be adjusted.

3. To adjust the cable, remove the cable retaining clip and reposition the cable in its retaining bracket. Reinstall the clip. See Fig. 5.

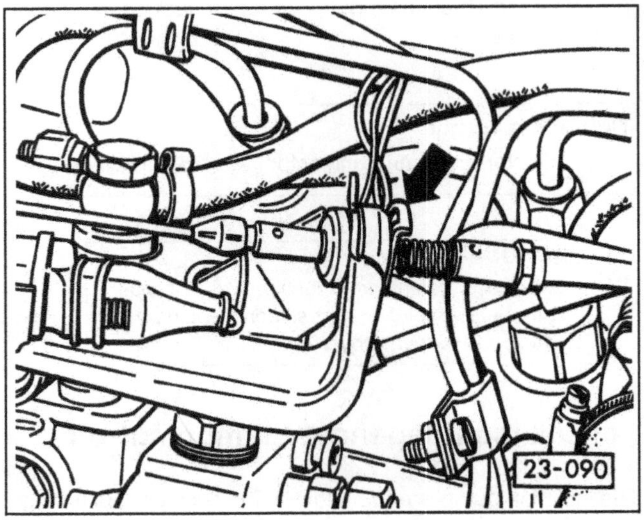

23-090

Fig. 5. Clip (**arrow**) used to adjust accelerator cable.

To check and adjust cold start accelerator cable

1. Check that the cold start accelerator knob in the passenger compartment is pushed fully in.

2. Working at the injection pump, loosen the cable clamping screw and then push the cold start lever in the direction of the arrow as far as it will go. See Fig. 6.

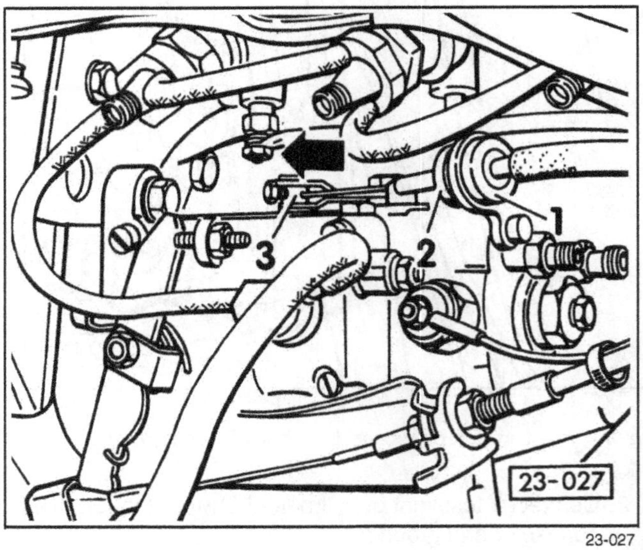

23-027

Fig. 6. Cold start accelerator (CSA) cable attachment to injection pump, top view. Washer (**1**) and clip (**2**) secure cable. Locking screw and pivot pin (**3**) used for adjustment. Push lever in direction of **arrow** to make adjustment.

3. Then without moving the lever, pull taught the end of the cable that comes through the pivot pin, and tighten the clamping screw.

To check and adjust idle speed and idle speed boost

1. Connect a tachometer according to the manufacturer's instructions.

2. With the transmission in neutral and the hand brake set, start the engine and allow it to idle. Check that the idle speed is within the specifications listed.
   - If the basic idle speed is incorrect, turn screw (**1**) until the correct engine speed is obtained. See Fig. 4.
   - If the idle basic speed cannot be adjusted below 930 rpm, check the minimum idle speed stop screw (**3**). If the screw is touching the idle speed throttle lever, loosen the locknut and back the screw out. Then adjust the idle speed at screw (**1**). Turn screw (**3**) back in just until it contacts the throttle lever and tighten the locknut. See Fig. 4.

Basic Idle Speed
- Engine fully warm. . . . . 900 ± 30 rpm

3. Pull the cold start accelerator cable to the first detent. Check that the idle boost speed increases as specified. Pull the cold start accelerator cable all the way out. Check that the idle boost speed increases as specified.
   - If the maximum idle boost speed is incorrect, loosen the locknut and adjust screw (**4**) with the cold start accelerator cable pulled fully out. Tighten the locknut when the correct engine speed is obtained. See Fig. 4.

Idle Boost Speed
- Cold start cable @ 1st detent . . . . . . basic idle speed plus approx. 60 rpm
- Cold start cable @ maximum. . . . . 1050 ± 50 rpm

4. Recheck the adjustment of the acclerator cable and the cold start accelerator cable as described earlier.

To check and adjust maximum engine speed

1. With the transmission in neutral and the hand brake set, start the engine and allow it to idle.

2. Have a helper press the accelerator fully to the floor. Check that the throttle lever contacts the maximum engine speed adjusting screw (**2**). Check that the engine speed is within the specifications listed. See Fig. 4.
   - If the maximum engine speed is incorrect, loosen the locknut and turn screw (**2**) until the correct engine speed is obtained. Tighten the locknut. See Fig. 4.

Maximum Engine Speed
- Wide Open Throttle (WOT). . . . . 5,200 ±100 rpm

3. Recheck the adjustment of the accelerator cable as described earlier.

To adjust residual fuel volume

The residual fuel volume delivery adjusting is set at the factory and sealed with an anti-tamper cap. The screw should never need adjustment unless it has been altered. Incorrect adjustment will result in high residual volume (screw turned in—high engine speed, idle speed adjustment not possible) or low residual volume (screw turned out—long cranking time, delayed acceleration from low rpm or jerking in all gears).

> ***CAUTION —***
> *Do not adjust the residual fuel delivery adjustment screw unless symptoms of high or low residual volume are exhibited and the idle speed can not be corrected by any other method.*

1. Remove the tamper proof cap from the residual fuel volume adjusting screw (**5**) and loosen the locknut. See Fig. 4.

2. With the transmission in neutral and the hand brake set Start the engine and allow it to idle.
   - If residual volume is too high, first turn the screw out until engine speed stops decreasing. Then turn the screw in just until engine speed increases, and back the screw out 1/2 turn. Tighten the locknut and install a new tamper proof cap.
   - If residual volume is too low, turn the screw in just until engine speed increases, then back the screw out 1/2 turn. Tighten the locknut and install a new tamper proof cap.

CAMSHAFT/INJECTION PUMP DRIVE BELT

Any time the camshaft drive belt is removed or installed, camshaft timing and injection pump timing must be checked and adjusted. Special tools are required for setting drive belt tension, camshaft TDC position, and injection pump timing.

To remove camshaft/injection pump drive belt

1. Remove the upper drive belt cover. See Fig. 7.

2. Remove the cylinder head cover and gaskets. See **15a Cylinder Head and Valvetrain (4-Cylinder Engines).**

1. **Bolt**
   - tighten to 45 Nm (33 ft-lb)
2. **Drive belt tensioner**
3. **Spreader clip**
4. **Drive belt cover, upper**
5. **Camshaft sprocket**
6. **Bolt**
   - tighten to 10 Nm (89 in-lb)
7. **Injection pump sprocket**
8. **Drive belt cover, rear**
9. **Injection pump**
10. **Bolt**
    - tighten to 20 Nm (15 ft-lb)
11. **Bracket**
12. **Bracket**
13. **Drive belt**
    - mark running direction on belt before removing
14. **Intermediate shaft sprocket**
15. **Crankshaft sprocket**
16. **Bolt**
    - stretch bolt–**always replace**
    - lightly oil bolt threads and contact shoulder before installation
    - tighten to 90 Nm (66 ft-lb) plus additional 1/4 turn (90°)
17. **Drive belt cover, lower**
18. **Poly-ribbed drive belt**
19. **Vibration damper**
20. **V-belt, power steering**
21. **Pulley**
22. **Bolt**
    - tighten to 25 Nm (18 ft-lb)

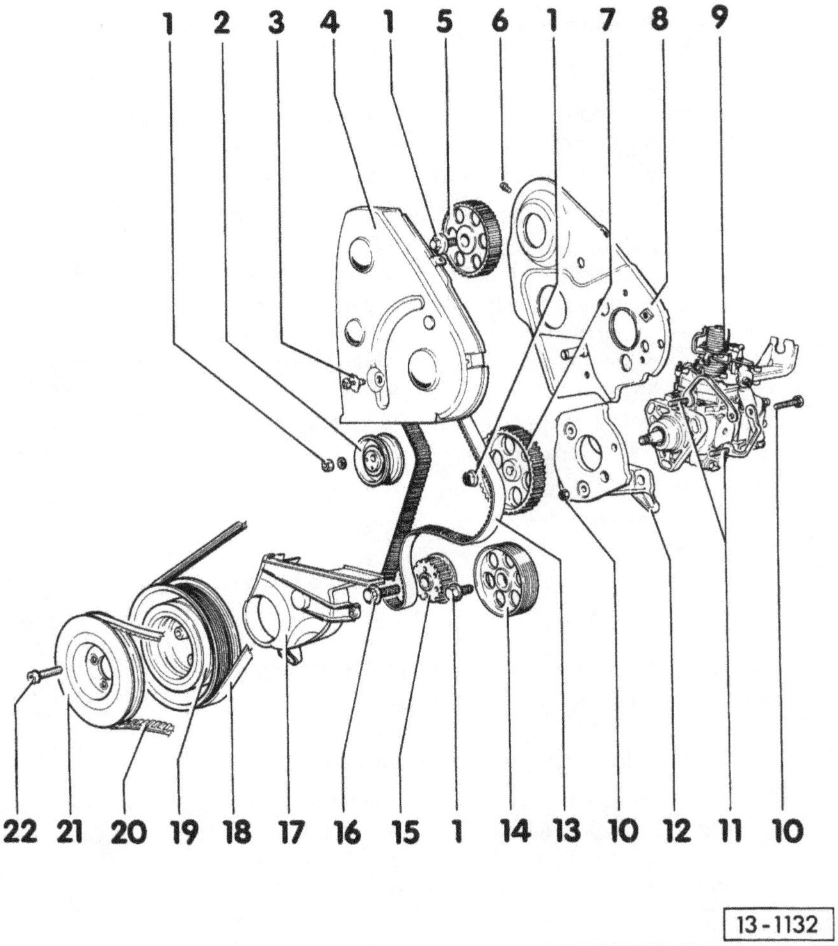

13-1132

Fig. 7. Camshaft drive belt assembly for AAZ diesel engine.

3. Using a socket on the crankshaft pulley bolt, rotate the engine in normal rotation direction (clockwise) to set the No.1 piston at Top Dead Center (TDC). See Fig. 8.

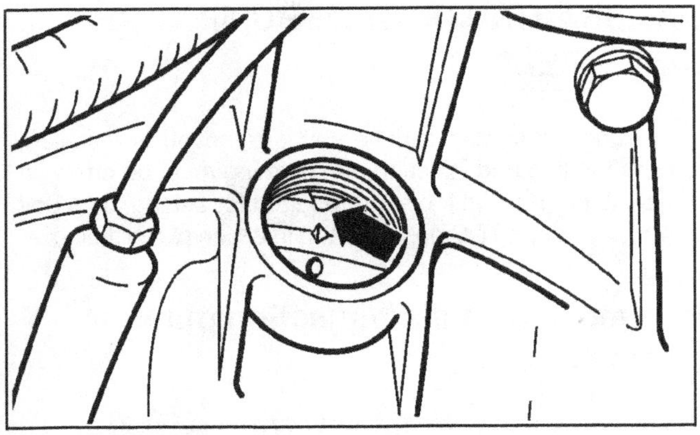

0024025

Fig. 8. TDC mark on flywheel aligned with pointer in transaxle housing.

NOTE—

With the engine at TDC, both cam lobes for No. 1 cylinder will be pointing approximately upward.

4. Lock the camshaft in position using the special setting bar (Volkswagen special tool no. 2065A). See Fig. 9. Then shim tool on either side with feeler gauges of equal thickness so that no play exists.

5. Lock the injection pump in position using the special lock pin (Volkswagen special tool no. 2064). Insert the lock pin through the injection pump sprocket and the rear drive belt cover. See Fig. 10.

CAUTION—

If the drive belt is broken or has jumped time, do not attempt to lock the sprocket in position at this time.

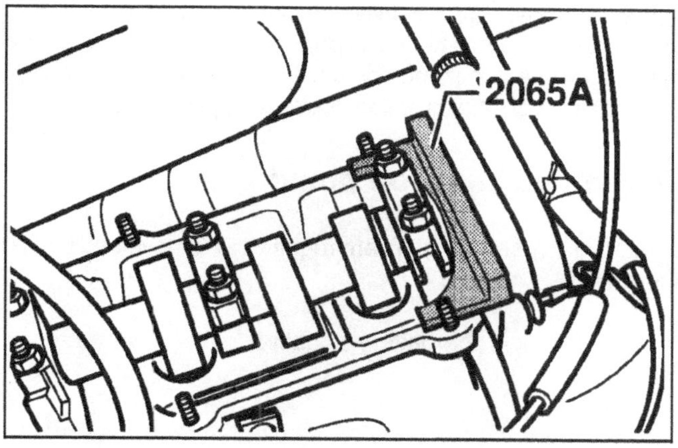

0024026

Fig. 9. Volkswagen special tool no. 2065A used to lock camshaft position. Shim tool on both sides with feeler gauges of equal thickness until there is no play.

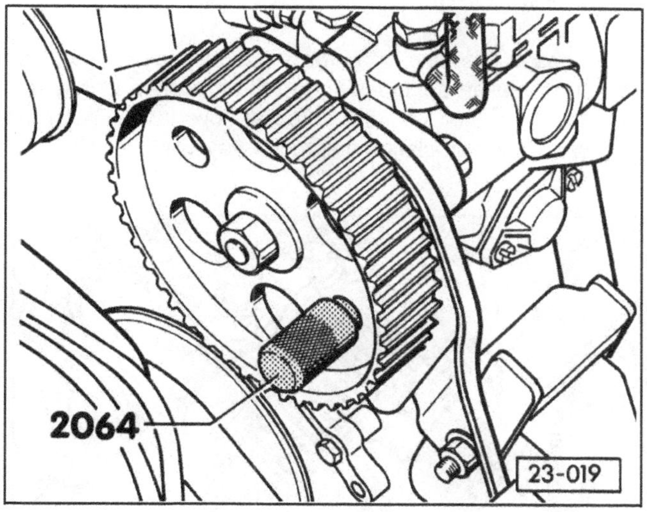

23-019

Fig. 10. Injection pump sprocket locked in position with Volkswagen special tool no. 2064 (order no. T02 064 000 00 ZEL). Snug fit is required, allowing no play.

6. Loosen the coolant pump pulley bolts. Remove the engine drive belts. See **0 Maintenance Program**.

7. Remove the coolant pump pulley. Remove the vibration damper and V-belt pulley.

8. Remove the lower drive belt cover. See Fig. 7 given earlier.

9. Loosen the camshaft drive belt tensioner locknut. Using a spanner wrench (see Fig. 11), turn the tensioner counterclockwise and then remove the belt.

NOTE —
Temporarily tighten the tensioner locknut to inspect the tensioner. Check for smooth rotation and no play. Replace a worn tensioner.

10. Inspect the drive belt for damage such as stretched cords, exposed threads, missing teeth, or other visible damage. Do not reinstall a worn or otherwise faulty belt.

11. Inspect the area around the camshaft oil seal for leaks. If the seal is leaking, it is easiest to replace it before installing the camshaft drive belt. See **15a Cylinder Head and Valvetrain (4-Cylinder Engines)**.

To install camshaft/injection pump drive belt

NOTE —
- *The cold start cable must be completely pushed in during pump installation.*
- *A special belt tensioning tool is needed to correctly install and tension the belt.*

1. Set the crankshaft to TDC (Fig. 8) and lock the camshaft (Fig. 9) and the injection pump (Fig. 10) at their TDC positions using the special tools. See **To remove camshaft/injection pump drive belt.**

> ***CAUTION —***
> *Special care must be taken when rotating the camshaft or the crankshaft with the belt removed. Mistimed rotation could cause the pistons to hit the valves, damaging either or both.*

2. Loosen the camshaft drive belt sprocket bolt 1/2 turn. Loosen the sprocket from the camshaft by tapping the back side of the sprocket with a drift through the hole in the rear drive belt cover.

NOTE —
The camshaft end is tapered. Loosening the sprocket will allow the camshaft sprocket to rotate independently of the camshaft.

3. Install the drive belt to sprockets so that all belt slack is at the tensioner. Temporarily tighten the drive belt tensioner, just enough to keep the belt firmly in place.

4. Remove the lock pin from the injection pump sprocket.

5. Tension the drive belt as follows:
   - Loosen tensioner locknut and turn tensioner clockwise using a spanner wrench. See Fig. 11.
   - Measure belt tension between camshaft sprocket and injection pump sprocket using special tension gauge (Volkswagen special tool no. VW210).
   - When tension is correct, tighten tensioner locknut.

Camshaft/Injection Pump Drive Belt Tension
- Drive belt tension (using Volkswagen special tool no. VW210). . . . . .12–13 index marks

6. Tighten the camshaft sprocket bolt. Remove the belt tension gauge and the camshaft locking tool.

Tightening Torque	
• Drive belt tensioner locknut	45 Nm (33 ft-lb)
• Camshaft sprocket to camshaft	45 Nm (33 ft-lb)

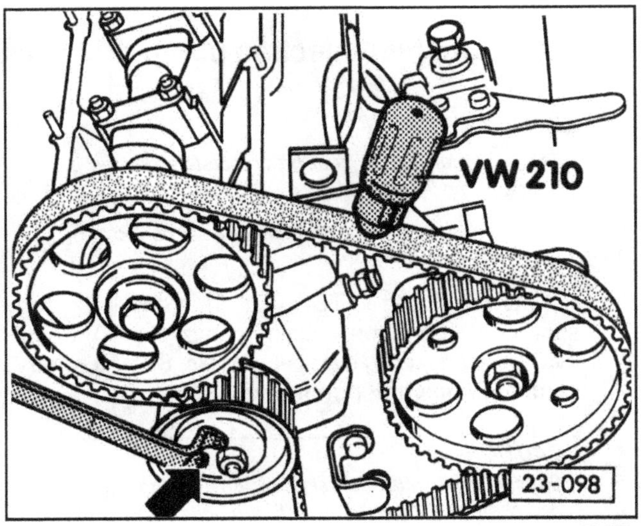

23-098

Fig. 11. Spanner wrench (**arrow**) being used to adjust belt tension. Volkswagen special tool no. VW210 is belt tension gauge.

7. Turn the engine by hand, two full revolutions in the direction of crankshaft rotation (clockwise). Recheck drive belt tension and camshaft alignment. Adjust if necessary.

8. Adjust injection pump static timing as described under **Injection Pump**.

9. The remainder of installation is the reverse of removal. See Fig. 7 for tightening torques. Install and adjust the engine drive belts as described in **0 Maintenance Program.**

INJECTION PUMP

The diesel fuel injection pump performs three functions. It pressurizes the fuel, precisely times the delivery of fuel to the injectors, and meters the amount of fuel delivered. The pump is driven by the camshaft drive belt at 1/2 crankshaft speed. Internal parts are lubricated by the diesel fuel, so no routine maintenance is required.

Most diesel fuel system problems are caused by a problem outside the injection pump. If injection pump problems do occur, it will most likely be due to wear and high mileage. Internal injection pump wear is most often characterized by an unsteady idle, or the inability to precisely adjust pump timing.

Internal repairs to the injection pump require specialized knowledge and equipment, and parts for rebuilding are not generally available. Faulty pumps must be serviced by the pump manufacturer or other qualified diesel fuel injection repair shop.

To remove injection pump

Special VW service tools are required to remove and install the pump.

1. Disconnect the battery negative (–) terminal from the battery

2. Remove the camshaft drive belt. See **Camshaft/Injection Pump Drive Belt.**

NOTE —
The camshaft must be locked at TDC using the special setting bar before removing the drive belt.

3. Loosen the injection pump sprocket retaining nut. Using a puller, apply tension to the sprocket. Lightly tap the puller bolt until the sprocket pops loose. Then remove the sprocket and sprocket key. See Fig. 12.

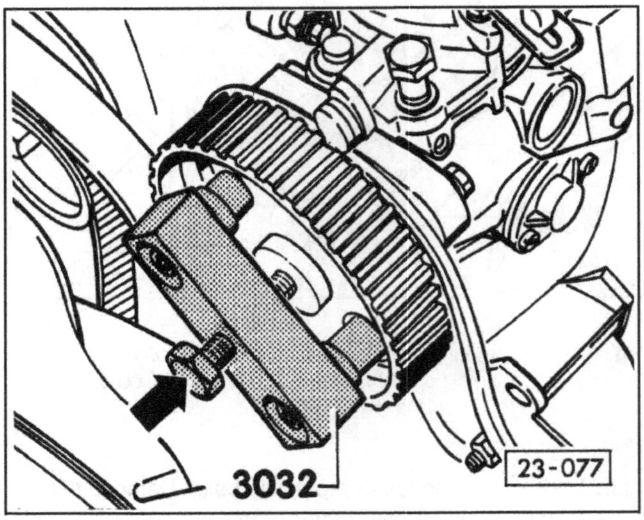

23-077

Fig. 12. Puller (VW special tool no. 3032) installed on injection pump sprocket. Tension puller and tap bolt (**arrow**) to remove sprocket.

4. Label and disconnect all fuel pipes and hoses from the injection pump. Remove the fuel pipes to the injectors as a unit, taking care not to bend them. Cover the unions with a clean, lint-free cloth. Thoroughly clean all unions before disconnecting fuel lines.

WARNING —
Fuel will be expelled. Do not smoke or work near heaters or other fire hazards. Have a fire extinguisher handy.

> ***CAUTION—***
> *Diesel fuel is damaging to rubber. Wipe off any fuel that spills on hoses, wiring, and rubber steering and suspension parts immediately, and wash with soap and water.*

5. Disconnect the electrical wire from the fuel cut-off valve and start positioner. Disconnect the accelerator cable and the cold start cable from the pump.

6. Remove the bolt from the injection pump rear support. Remove the bolts from the front injection pump mounting bracket and remove the pump. See Fig. 13.

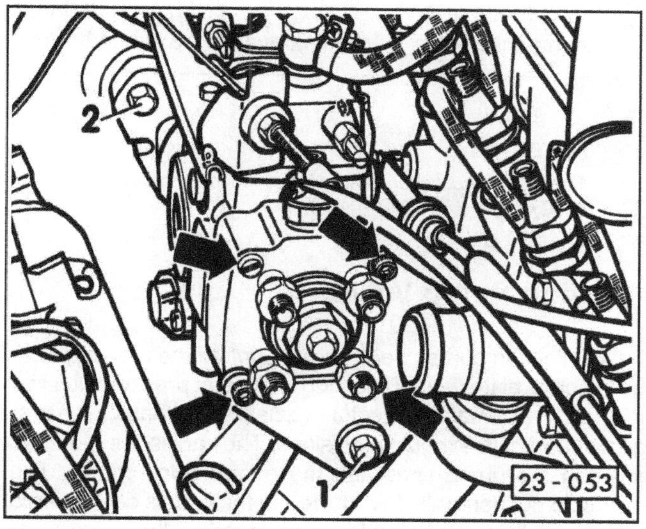

23-053

Fig. 13. Injection pump rear support bolt (**1**) and mounting bracket (**2**). To avoid damage to the distributor plunger, **DO NOT** loosen the bolts (**arrows**) on the fuel distributor head.

To install injection pump

1. Install the pump to the mounting flange and center the pump to the oblong mounting holes. Install and tighten the pump mounting bolts.

Tightening Torque
- Injection pump bracket or support to pump and engine . . . . 25 Nm (18 ft-lb)

> ***CAUTION—***
> *Take care not to interchange the fuel supply union bolt with the return line union bolt. The return bolt has a smaller inside diameter and the bolt head is marked "OUT".*

2. Reconnect the fuel lines and the fuel pipes to the injectors. Tighten the fuel line unions.

Tightening Torque
- Injection fuel lines to pump or injectors . . . . 25 Nm (18 ft-lb)

3. Reconnect the electrical connectors. Reconnect and adjust the accelerator cable and the cold start accelerator cable. See **Basic Settings**.

4. Install the injection pump sprocket while aligning the key to the keyway. Loosely install the sprocket nut. Turn the sprocket so that the lock pin can be installed as shown in Fig. 10. Then tighten the sprocket bolt.

Tightening Torque
- Injection pump sprocket to injection pump . . . . 45 Nm (33 ft-lb)

5. Install the camshaft drive belt as described in **Camshaft/Injection Pump Drive Belt.**

6. Adjust injection pump static timing as described below. Check and adjust idle speed and maximum rpm as described earlier under **Basic Settings**.

To check and adjust injection pump static timing

Precise injection timing is critical, since it also determines when combustion begins. Checking and adjustment of injection timing is necessary any time the injection pump or the timing belt have been loosened or replaced, and may be necessary at other times to maintain precise timing.

Injection timing is checked and adjusted by measuring the injection pump plunger's stroke at TDC with a dial indicator.

> ***CAUTION—***
> *Before adjusting injection pump timing, make sure that the camshaft/valve timing is correct and the drive belt is properly installed and tensioned. See **Camshaft/Injection Pump Drive Belt**.*

1. Check that the cold start accelerator lever is pushed all the way in.

2. Using a socket on the front crankshaft bolt, turn the engine clockwise until No. 1 cylinder is at TDC. Both camshaft lobes should be pointing approximately up and the TDC mark on the flywheel should be aligned. See Fig. 8 given earlier.

3. Unsrew the timing-check plug from the center of the injection pump head.

4. Install a dial indicator with (0-3 mm range) as shown in Fig. 14. The gauge should be installed so that it indicates about 2.5 mm (preload).

> ***CAUTION —***
> *Fuel will be expelled when the timing plug is removed. Do not smoke or work near heaters or other fire hazards. Have a fire extinguisher handy.*

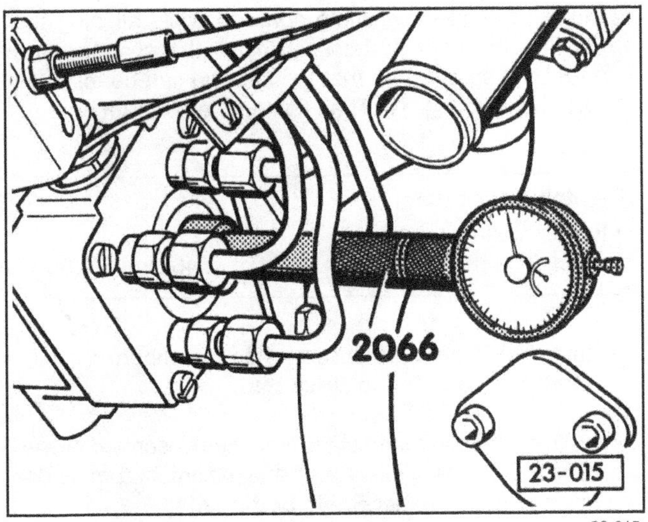

23-015

Fig. 14. Dial indicator with adapter mounted to injection pump. Adapter shown is Volkswagen special tool no. 2066.

5. Slowly turn the crankshaft counterclockwise (opposite normal engine rotation) until the dial indicator stops moving. Zero the dial indicator with about 1 mm of preload.

6. Turn the crankshaft clockwise (normal rotation), stopping precisely at TDC. With the engine at TDC, the dial indicator should read within the checking limits given below.

NOTE —
It is not necessary to adjust the pump if the timing is within the checking range but outside of the actual adjusting value.

Injection Pump Static Timing
- Checking range . . . 0.73-0.87 mm (0.0287-0.0343 in.)
- Adjusting . . . . . 0.80 ± 0.02 mm (0.0315 ± 0.0008 in.)

7. If timing adjustment is necessary, slightly loosen the injection pump rear support bolt and the three front mounting plate bolts. See Fig. 13 given earlier. Then rotate the pump until the dial indicator reading is correct. Tighten all mounting bolts and recheck pump timing. Remove the dial indicator and torque all fasteners.

CAUTION —
Always use a new sealing ring at the timing-check plug.

Tightening Torque
- Injection pump bracket or support to pump and engine . . . . 25 Nm (18 ft-lb)
- Timing-check plug to injection pump . 15 Nm (11 ft-lb)

NOTE —
The timing-check plug can be tightened to a maximum of 25 Nm (18 ft-lb) if leaks persist.

Injection Pump Valves

The injection pump check valves shut off the fuel supply to the injectors at the end of injection to prevent pressure build up and fuel "dribble" at the injector nozzles, which can cause preignition and high exhaust emissions. The valves are mounted between the pump lines and the pump. See Fig. 15. The valves do not generally fail. However, their seals can become loose over time, and should be replaced or the valve body tightened if leakage is seen around them. Leaks around the injection pump valves can also cause a pressure drop in one or more injectors leading to hard starting and rough running.

To correct leaks at pump valves

1. Thoroughly clean the fuel pipes, the unions, and the pump head.

2. Loosen the nut holding the injection pipe to the pump valve housing, then torque the valve to the pump.

> ***WARNING —***
> *Fuel will be expelled. Do not smoke or work near heaters or other fire hazards. Have a fire extinguisher handy.*

> ***CAUTION —***
> *Diesel fuel is damaging to rubber and plastic, and should be cleaned off of hoses and wiring after the leaks are corrected.*

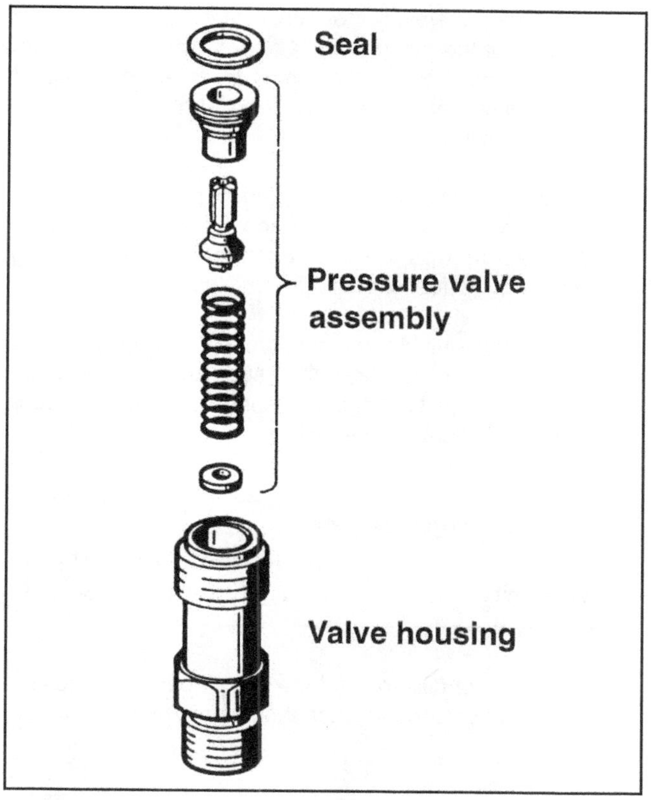

Fig. 15. Injection pump valve and related components.

Tightening torque
- Injection valve to pump body . . . . 45 Nm (33 ft-lb)

---

3. Tighten the injection pipe nut.

Tightening torque
- Injection pipe to injection valve . . . . . 25 Nm (18 ft-lb)

---

4. Run the engine and check for leaks. If there is still leakage, replace the sealing washer and/or the connector.

> ***CAUTION —***
> *Do not mix pressure valve assemblies between cylinders.*

FUEL INJECTORS

The AAZ diesel engine uses double-spring fuel injectors that open at about 150 bar (2175 psi) of pressure. The injectors are not adjustable and should be replaced as complete units if faulty.

> ***NOTE —***
> *To check for a faulty injector, make the idle speed drop test described earlier under* **Diagnostic Quick Checks***.*

The signs of injector trouble usually appears as misfiring and knocking noises from one or more of the injectors or cylinders, engine overheating, loss of power, smoky black exhaust, increased fuel consumption, engine misfire, or excessive blue smoke during cold start.

> ***CAUTION —***
> *When working on the injectors, everything must be kept absolutely clean. Clean all pipe unions before disconnecting.*

To replace injector

1. Unscrew the fuel pipe union nuts at both the injection pump and the injectors. Detach the fuel pipes as a unit.

> ***WARNING —***
> *Fuel will be expelled. Do not smoke or work near heaters or other fire hazards. Have a fire extinguisher handy.*

2. Remove the fuel return hoses from between the injectors.

3. Using a 27 mm deep-well socket, unscrew the injector from the cylinder head.

4. Remove the injector heat shield from the cylinder head and install a new one in its place. See Fig. 16

> ***NOTE —***
> *Be sure to remove the heat shield from the bottom of each injector hole. When reinstalling, always use new heatshields.*

5. Install the injector and fuel lines.

Tightening torque
- Fuel injector to cylinder head . . . . 70 Nm (51 ft-lb)
- Fuel pipe to fuel injector . . . . 25 Nm (18 ft-lb)

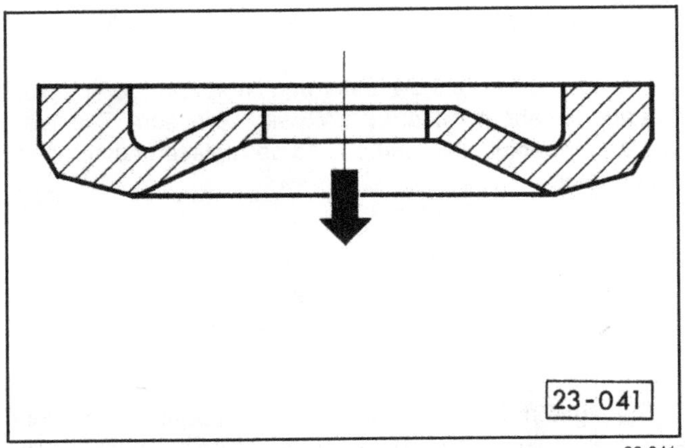

23-041

Fig. 16. Proper installation of the fuel injector heat shield. **Arrow** points down, away from injector.

To pressure test injector

1. Remove the injector as described above.

2. Install the injector to the pressure tester according to the manufacturer's directions. Place the injector in a container to catch the spraying liquid. See Fig. 17.

> ***CAUTION —***
> *Keep hands or other parts of your body clear of the injector spray. Working pressure can cause the fuel oil to penetrate the skin. Work carefully. Do not smoke or work near heaters or other fire hazards. Have a fire extinguisher handy.*

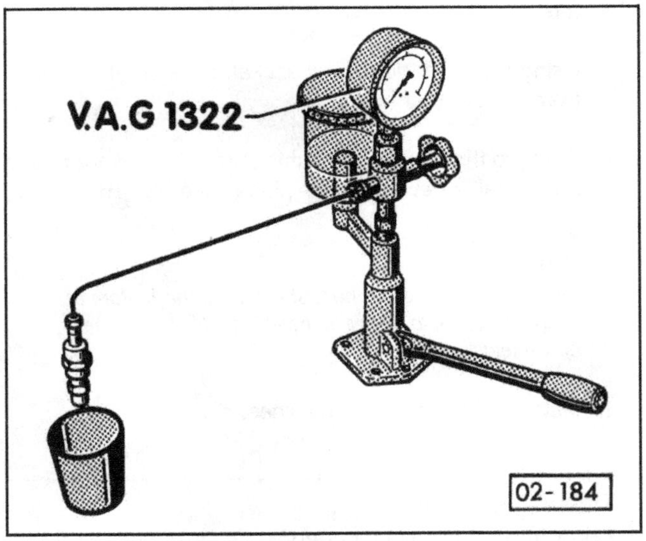

02-184

Fig. 17. Injector installed on hand pressure pump. Volkswagen special tool V.A.G. 1322 shown.

3. To check general injector performance use rapid, short strokes of the pump lever (4 to 6 strokes per second). Check that the injector sprays in an even, compact cone pattern and stops cleanly, and that it does not drip after spraying.

4. To check injectors for leaking, slowly press down on the pump lever to a pressure of about 110 bar (1595 psi). Maintain that pressure for about 10 seconds. During that time, no fuel should drip from the injector nozzle.

5. To measure injector opening pressure, slowly push the pump lever down. Watch the gauge and read the value when the injector begins to spray. Opening pressure specifications are given below.

Injector Opening Pressure
- New .... 150–158 bar (2175–2291 psi)
- Wear limit .... 140 bar (2030 psi)

6. If an injector fails to meet the leak or opening pressure specifications, the injector should be replaced.

23b Fuel Injection–Diesel (1Z/AHU Engine)

TABLES

23b

GENERAL

This repair group covers the diesel fuel injection system for the 1Z/AHU 1.9 liter Turbo Direct Injection (TDI) diesel engine. Testing on the diesel glow plug system is covered in **28b Ignition System–Diesel**.

> ***CAUTION —***
> - *Repairs and adjustments to the injection pump require specialized equipment, and parts for rebuilding are not generally available. Faulty pumps must be serviced by the manufacturer. Internal problems usually require replacement of the pump.*
> - *Adjustments and repairs to the fuel injection system should be made carefully. Cleanliness is especially important. Always clean fuel unions before removing lines.*
> - *Diesel fuel is damaging to rubber. Wipe off any fuel that spills on hoses, wiring, and rubber steering and suspension parts, and wash with soap and water. If coolant hoses have been contaminated with diesel fuel, they must be replaced.*

TROUBLESHOOTING

Proper operation of the diesel engine requires a supply of clean fuel under pressure, an unrestricted supply of air, and properly timed fuel delivery. Fuel must be of the proper grade, and properly winterized for cold-start conditions. Dirt and water can interfere with combustion and will cause problems.

Good compression is especially necessary for the engine to run well. Engine mechanical faults leading to poor compression can cause problems that may seem to be injection related. Engine lubricating oil that leaks past worn piston rings, valve stem seals or turbocharger seals may produce exhaust smoke that also can be mistaken for fuel injection problems.

A faulty glow plug system may contribute to cold starting problems. See **28b Ignition System–Diesel**. The battery must have the correct electrical capacity and the starter must be able to turn the engine fast enough (generally at least 150 rpm) to begin compression ignition. See **27 Engine Electrical Systems**. Insufficient engine power can also be the result of turbocharger or intercooler malfunctions. See **21 Turbocharger and Intercooler**.

Because the diesel engine speed is controlled by the amount of fuel injected, and not by the amount of air admitted past a throttle plate, the air intake must be unrestricted. A clogged air filter, faulty turbocharger, or restricted intercooler can lower the air flow and reduce power output.

The injection pump, under the control of the Engine Control Module (ECM), produces the high pressure necessary to open the injectors and spray the fuel. Fuel leaks or air in the fuel lines may reduce system pressure. Since power is controlled by the amount of fuel injected, low pressure may cause sluggish performance.

For cold starting, when the compression heat necessary for combustion is dissipated quickly by a cold engine, injection timing is automatically advanced by the ECM to give the fuel more time to burn. At higher rpm, there is less time for fuel to burn, so injection timing is advanced to start the fuel burning sooner. Since the injection pump is driven by the engine and externally mounted, its precise fuel metering and timing can be degraded by such things as a loose timing belt, worn sprocket, or loose mounting bolts. Correct Cetane rating as specified by the vehicle's Owner's Manual will profoundly influence starting and overall performance as well.

Most of the in-depth trouble shooting of this electronically controlled engine management system will require the use of a scan tool such as the Volkswagen supplied VAG 1551 or VAG 1552. However, there are still many basic tests that should not be overlooked. **Table a** lists symptoms, their probable causes, and corrective actions. The boldface type refers to other sections in the manual where the repairs are described.

Diagnostic Quick Checks

These tests are used to isolate and diagnose diesel fuel system problems. It is assumed that the glow plug system is functioning within normal parameters and that injection pump timing and valve timing are correct.

> ***NOTE —***
> *Testing and repair information on the diesel glow plug system is covered in* **28b Ignition System-Diesel**.

Table a. TDI Diesel Fuel Injection Troubleshooting

Problem	Probable cause	Corrective action
1. Engine does not start	**a.** Cranking speed too low or will not crank	**a.** Repair starting system or charge or replace battery. See **27 Engine Electrical.**
	b. No fuel or wrong fuel in fuel tank	**b.** Verify sufficient fuel of correct type in tank. Add as necessary.
	c. No voltage at fuel cut-off valve on injection pump. Fuel cut-off valve loose or faulty	**c.** Valve should click each time the ignition is turned on and off. If not, check for voltage at the connector. See **97 Wiring Diagrams, Fuses and Relays.** Replace a faulty solenoid.
	d. No voltage at glow plug bus or glow plug(s) faulty.	**d.** The glow plug circuit is activated by the Engine Control Module. See **28b Ignition System–Diesel** for glow plug testing and repair information.
	e. Excessive air in fuel system	**e.** Check fuel supply lines from fuel tank for cracks and connections for leaks, especially at and around the fuel filter and injection pump inlet.
	f. Injection pump not delivering fuel	**f.** Check for basic fuel delivery to engine. Check for cracked or damaged lines, clogged fuel filter or other fuel supply problems.
	g. Injection timing incorrect	**g.** Adjust injection timing. Requires scan tool VAG 1551/1552.
	h. Faulty injectors	**h.** Test and, if necessary, replace injectors.
	i. Engine mechanical faults	**i.** Test compression. See **15a Cylinder Head and Valvetrain.**
	j. Faulty injection pump	**j.** Replace pump.
	k. Fault in engine speed sensor (G28)	**k.** Test sensor at connector.
	l. Fault in ECM or other engine management system component	**l.** Interrogate permanent fault memory. Repair or replace components as required. Requires scan tool VAG 1551/1552.
2. Glow plug warning light not working	**a.** Bulb burned out or malfunction in glow plug relay circuit	**a.** Test and repair as described in **28b Ignition System–Diesel**.
3. Idle speed incorrect, rough or irregular	**a.** Accelerator pedal or linkage binding	**a.** Remove closeout panels under dash and inspect pedal and linkage for free movement. Repair as required.
	b. Excessive air in fuel system	**b.** Check fuel supply line from fuel tank for cracks, kinks, or leaks.
	c. Clogged fuel filter, or fuel return line and injector pipes leaking, dirty, kinked, or damaged at connectors	**c.** Inspect and, if necessary, replace lines and hoses, replace fuel filter. See **0 Maintenance Program.**
	d. Faulty injectors	**d.** Test and, if necessary, replace injector.
	e. Injection timing incorrect	**e.** Adjust injection timing. Requires scan tool VAG 1551/1552.
	f. Engine mechanical faults	**f.** Test compression. See **15a Cylinder Head and Valvetrain.**
	g. Faulty injection pump	**g.** Replace pump.

continued on next page

Table a. TDI Diesel Fuel Injection Troubleshooting (continued)

Problem	Probable cause	Corrective action
4. Smoky exhaust (black, blue, or white)	**a.** Faulty injectors	**a.** Check and, if necessary, replace injectors.
	b. Injection timing incorrect	**b.** Adjust injection timing. Requires scan tool VAG 1551/1552.
	c. Turbocharger leaking internally	**c.** Inspect and repair, or replace turbocharger.
	d. Engine mechanical faults	**d.** Test compression. See **15a Cylinder Head and Valvetrain.**
	e. Faulty injection pump	**e.** Replace pump.
	f. Fault in ECM or other engine management system component	**f.** Interrogate permanent fault memory. Repair or replace components as required. Requires scan tool VAG 1551/1552.
5. Poor power output, slow acceleration or top speed	**a.** Accelerator pedal or linkage binding	**a.** Remove closeout panels under dash and inspect pedal and linkage for free movement. Repair as required.
	b. Air filter dirty or restricted	**b.** Clean or replace air filter. Inspect housing and ducting for debris. See **0 Maintenance Program**.
	c. Clogged fuel filter; or fuel return line and injection pipes leaking, dirty, kinked, or damaged at connections	**c.** Check for basic fuel delivery to engine. Inspect and, if necessary, replace lines and hoses, replace fuel filter. See **0 Maintenance Program.**
	d. Excessive air in fuel system	**d.** Check fuel supply line from fuel tank for cracks, kinks, or leaks, especially at and around fuel filter and injection pump inlet.
	e. Faulty injectors	**e.** Check and, if necessary, replace injectors.
	f. Injection timing incorrect	**f.** Adjust injection timing. Requires scan tool VAG 1551/1552.
	g. Engine mechanical faults	**g.** Test compression. See **15a Cylinder Head and Valvetrain.**
	h. Faulty injection pump	**h.** Replace pump.
	i. Fault in ECM or other engine management system component	**i.** Interrogate permanent fault memory. Repair or replace components as required. Requires scan tool VAG 1551/1552.
6. Engine runs at a high constant idle and speed does not vary	**a.** Throttle position sensor (G79), malfunctioning	**a.** Interrogate permanent fault memory. Repair or replace components as required. Requires scan tool VAG 1551/1552.
7. Excessive fuel consumption	**a.** Air filter dirty or restricted	**a.** Clean or replace air filter element. Inspect housing and ducting for debris. See **0 Maintenance Program**.
	b. Fuel leaks	**b.** Check and, if necessary, replace or tighten all pipes, hoses and connections.
	c. Fuel return pipe blocked or restricted	**c.** Check return line for kinks and dents; replace faulty lines. If line is clogged, blow it out with compressed air, then bleed fuel system.
	d. Faulty injectors	**d.** Test and, if necessary, repair or replace injectors.
	e. Injection timing incorrect	**e.** Adjust injection timing. Requires scan tool VAG 1551/1552.
	f. Faulty injection pump	**f.** Replace pump.
	g. Fault in ECM or other engine management system component	**g.** Interrogate permanent fault memory. Repair or replace components as required. Requires scan tool VAG 1551/1552.

To check fuel shut-off valve

The diesel engine is stopped by cutting off its fuel supply. This is done by means of an electrical solenoid valve in much the same way as in previous versions. The valve is normally closed. An electrical signal from the Engine Control Module (key ON) opens the valve. If the valve is not working, the engine will not start. Conversely, a faulty valve could also allow the engine to continue running after the key is switched off.

1. Turn the ignition key on and off without operating the starter. Listen for the valve to operate.
   - The fuel cut-off valve should click open when the key is turned on and click closed when the key is turned off. Do not confuse this sound with continuous sounds coming from other electronic components on the pump.

2. If the valve does not operate, check for voltage at the harness connector with the key on.
   - If voltage is not present, check for wiring faults. See **97 Wiring Diagrams, Fuses and Relays.**
   - If voltage is present, disconnect the battery negative (-) terminal. Clean the area around the valve and remove the valve, spring, and plunger. See Fig. 1. Clean and inspect the solenoid plunger and seat. Slide the plunger into the valve without the spring and check for free movement. If the plunger does not move freely, replace the valve assembly. Carefully check for dirt and metal particles in the valve bore in the pump. The slightest amount of debris can cause malfunctions. Install the fuel cut-off valve using a O-ring. Reconnect the battery negative (-) terminal.

> ***WARNING —***
> *Fuel will be expelled. Do not smoke or work near heaters or other fire hazards. Have a fire extinguisher handy.*

> ***CAUTION —***
> *Before disconnecting battery be sure to obtain radio anti-theft code.*

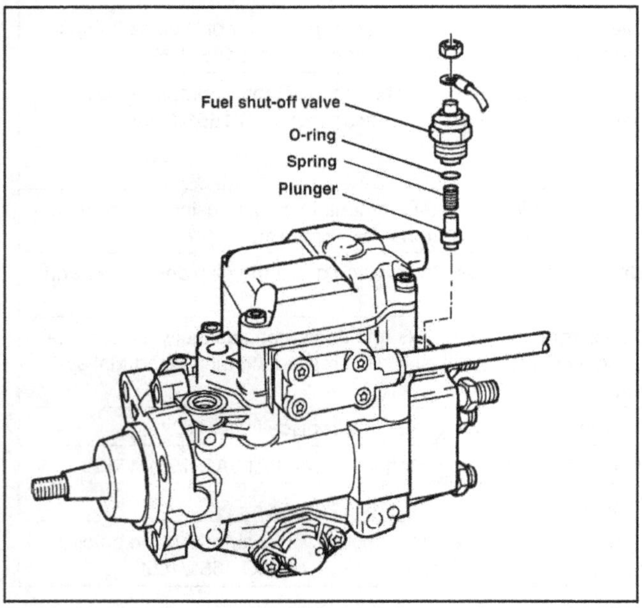

Fig. 1. Fuel shut-off valve removed from injection pump.

Tightening Torque
- Fuel cut-off valve to injection pump . . 40 Nm (30 ft-lb)

To check basic fuel delivery

This test checks if fuel is being delivered to the pump and fuel injectors. Make this test when experiencing hard starting or no-start problems.

1. Check visually for fuel leaks on the injector pump and especially around the injector line unions.

2. If no leaks are found, clean the fuel union nut for cylinder No. 1 fuel injector pipe at the injector. Slightly loosen the fuel union nut. See Fig. 2.

> ***WARNING —***
> - *Loosen the nut only about one-half turn to limit the amount of fuel leakage. The fuel injection system operates at very high pressure. Keep hands and eyes clear. Wear heavy gloves and eye protection. Wrap a cloth around the nut when loosening.*
> - *Fuel will be expelled. Do not smoke or work near heaters or other fire hazards. Have a fire extinguisher handy.*

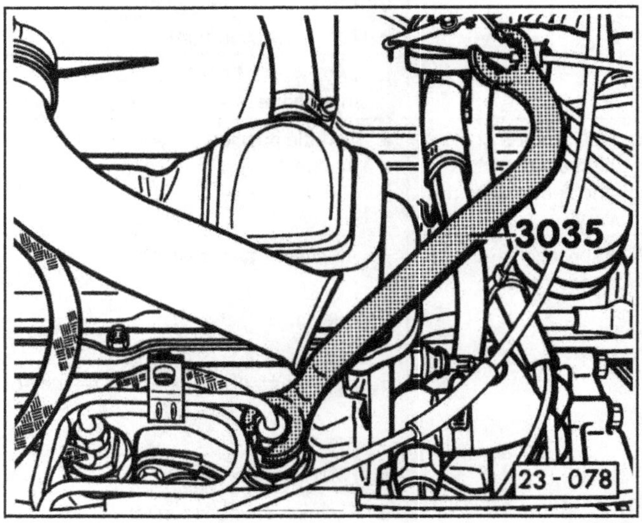

Fig. 2. Flare-nut wrench (VW special tool no. 3035) being used to loosen injector union nut for No. 4 cylinder on an earlier diesel engine version. 1Z/AHU diesel engines are similar.

3. Crank the engine and look for fuel to run out of the loosened union nut at the injector.

> ***CAUTION —***
> *Diesel fuel is damaging to rubber. Wipe off any fuel that spills on hoses, wiring, and rubber steering and suspension parts, and wash with soap and water. If coolant hoses are contaminated with diesel fuel, they must be replaced.*

4. If no fuel is observed, the most likely cause is a restriction or air leak in the fuel lines between the fuel tank and the injection pump, or possibly a clogged fuel filter. If the vehicle has a semi-transparent plastic fuel pipe on the supply side of the diesel injection pump, check for movement of fuel in the line. A stream of tiny air bubbles moving in the line is normal (unlike previous versions of the diesel engine).

5. If fuel is observed, but the engine still fails to start, consult **Table a** for additional troubleshooting checks.

NOTE —

It is not neceessary to test for basic fuel delivery on the remaining three cylinders due to the way the TDI system operates.

6. When finished testing, tighten the fuel union nuts and check for leaks. Clean up any spilled fuel.

Tightening Torque
- Fuel injector union nut to injector or pump . . . . 25 Nm (18 ft-lb)

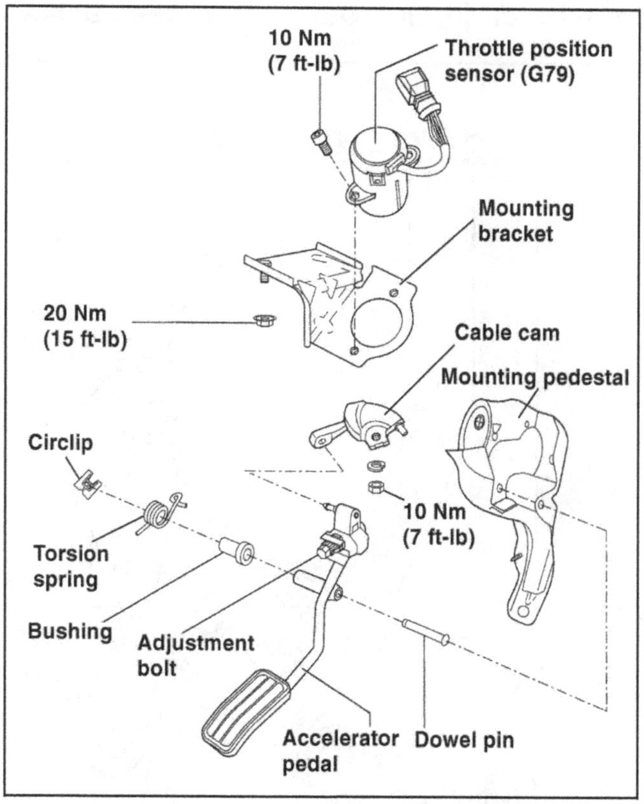

Fig. 3. Throttle position sensor, accelerator pedal and mounting hardware.

THROTTLE POSITION SENSOR

To remove throttle position sensor

Fuel quantity is controlled by the Engine Control Module in TDI equipped cars. There is no accelerator cable connection between the pedal and the engine. This system is described as "drive by wire". A throttle position sensor is mounted on the inside of the firewall near the upper end of the pedal. A very short cable and cam connect the sensor and the pedal.

1. Disconnect the battery negative (-) terminal.

> ***CAUTION —***
> *Before disconnecting battery, be sure to obtain radio anti-theft security code.*

2. Remove the screws attaching the sound deadener panels from under the driver's side of the dash. Carefully pull out the panels with the fiber and foam sound absorber blankets.

3. Remove the instrument cluster. See **90 Instruments**.

4. Remove the pedal from the mount by unclipping the circlip from the pivot pin and sliding the pin out. Unsnap the cable from the pedal and remove the pedal. See Fig. 3.

5. Using a $^1/_4$ inch drive socket set, unbolt the bracket with the sensor still attached from the firewall and pull out from below.

6. Installation is the reverse of removal. Careful attention should be given to installation of the sound deadening panels. Insure that all of the pedals have smooth and unrestricted movement.

Testing and adjustment of the throttle position sensor requires the use of a scan tool such as VAG 1551 or VAG 1552. Faults related to the throttle position sensor will be stored as diagnostic trouble codes (DTCs) in the ECM. If the fault is severe enough, the ECM will ignore the throttle position sensor completely and run the engine a constant high idle speed. This should enable the driver to reach the nearest workshop.

CAMSHAFT/INJECTION PUMP DRIVE BELT

Any time the camshaft drive belt is removed or installed, camshaft timing and injection pump timing must be checked and adjusted. Special tools are required for setting drive belt tension, camshaft TDC position, and injection pump timing. In addition, a scan tool such as VAG 1551 or VAG 1552 is required to properly set injection pump timing on this electronically controlled diesel injection system. Fig. 4 shows an exploded view of the camshaft/injection pump drive belt assembly on the 1Z/AHU Turbo Direct Injection diesel engine.

Camshaft/Injection pump drive belt, assembly

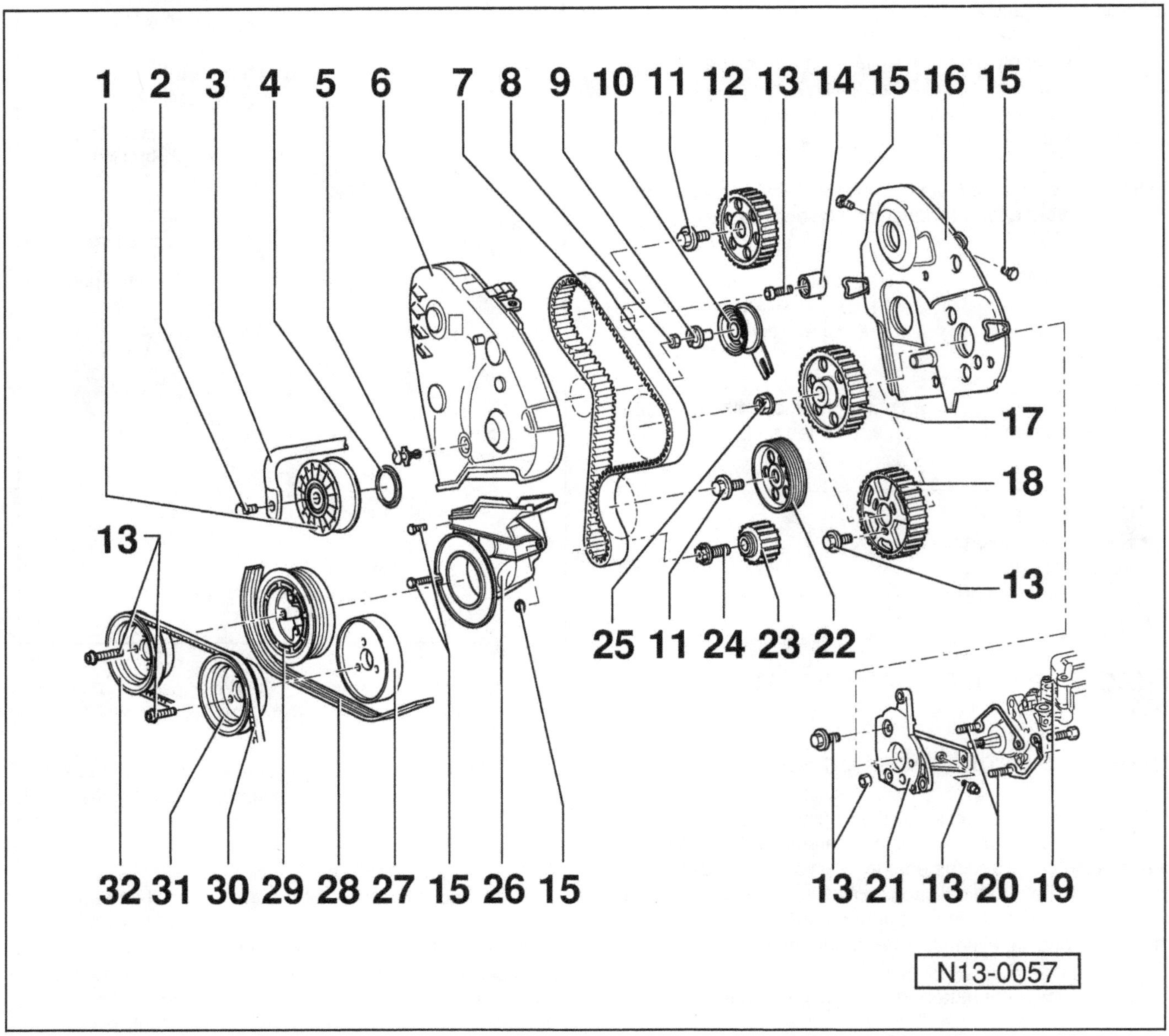

N13-0057

Fig. 4. Camshaft drive belt assembly for 1Z/AHU TDI diesel engine.

1. **Tensioning roller**
2. **20 Nm (15 ft-lb)**
3. **Tensioning lever**
4. **Dust cap**
5. **Expansion clip**
6. **Toothed belt guard, upper**
7. **Toothed belt**
   - Mark running direction before removing
   - Inspect for wear
   - Do not bend or kink
8. **20 Nm (15 ft-lb)**
9. **Cam**
   - For tensioning roller
10. **Tensioning roller**
    - Semi-automatic tension roller
11. **45 Nm (33 ft-lb)**
12. **Camshaft sprocket**
    - Loosen from tapered end of camshaft by tapping with hammer on punch inserted through hole in rear drive belt guard
13. **25 Nm (18 ft-lb)**

(continued from previous page)

14. **Idler pulley**
15. **10 Nm (18 ft-lb)**
16. **Toothed belt guard, rear**
17. **Injection pump sprocket**
18. **Not applicable (AAZ engine only)**
19. **Injection pump**
20. **Bracket**
21. **Mounting bracket**
22. **Intermediate shaft sprocket**
23. **Crankshaft sprocket**
    - For toothed belt
24. **90 Nm (66 ft-lb) + 1/4 turn (90°)**
25. **55 Nm (41 ft-lb)**
26. **Toothed belt guard, lower**
27. **Belt pulley**
    - For engine coolant pump
    - Ribbed belt version
28. **Ribbed belt**
    - Mark running direction before removing
29. **Belt-pulley/vibration damper**
    - Can only be installed in one position
30. **V-belt**
31. **Pulley**
    - For engine coolant pump
    - v-belt version shown
32. **Pulley**

To remove camshaft/injection pump drive belt

1. Remove the sound absorber cover from the engine. Remove the upper toothed belt guard. See Fig. 4. Remove air cleaner and air duct to turbocharger inlet.

2. Remove the cylinder head cover and gaskets. See **15a Cylinder Head and Valvetrain (4-Cylinder Engines).**

3. Using a socket on the crankshaft pulley bolt, rotate the engine in the normal running direction (clockwise) to set the number 1 piston to Top Dead Center (TDC) using the TDC mark on the flywheel and the reference mark on the transmission. See Fig. 5.

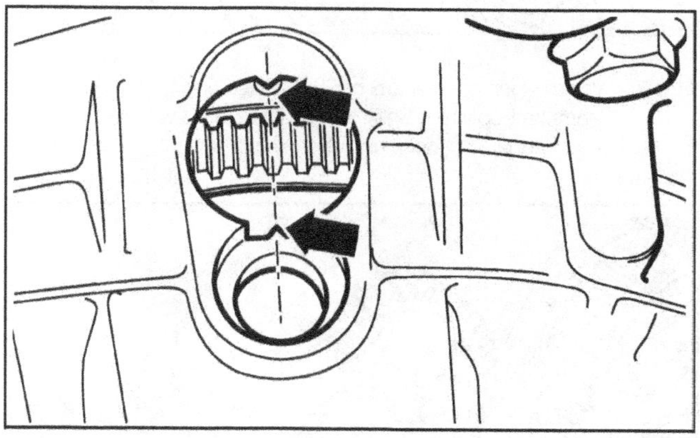

0024188

Fig. 5. TDC mark on flywheel (**top arrow**) lined up with reference mark (**bottom arrow**).

NOTE —
With the engine at TDC, both cam lobes for No.1 cylinder will be pointing approximately upward.

4. Lock the camshaft in position using setting bar (Volkswagen Special tool 2065A, or equivalent). Shim between the gasket surface and both sides of tool 2065A with equal thickness of feeler gauges. This will insure that camshaft is at actual TDC. Remove the idler roller from the support bracket. See Fig. 6.

5. Hold the injection pump in position with the locking pin (Volkswagen special tool 2064, or equivalent) inserted through the single small hole in the pump sprocket and into the support bracket. See Fig. 7.

CAUTION —
If the drive belt is broken or has jumped time, do not attempt to lock the sprocket in position at this time.

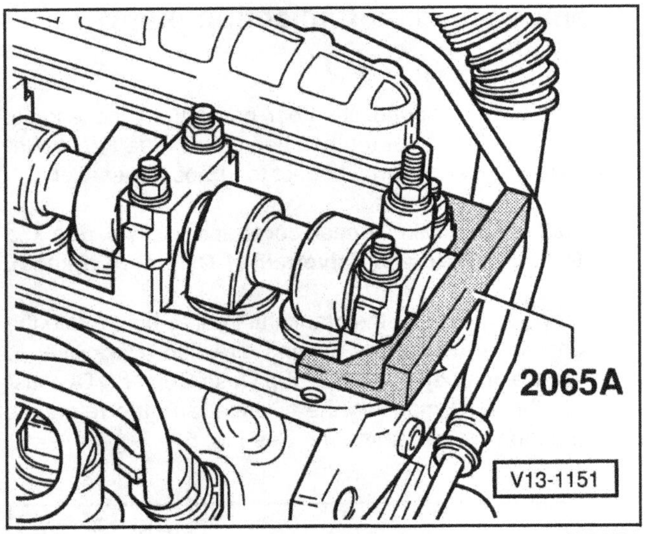

V13-1151

Fig. 6. Volkswagen special tool no. 2065A (setting bar) used to lock camshaft position. Shim tool on both sides with feeler gauges of equal thickness until there is no play.

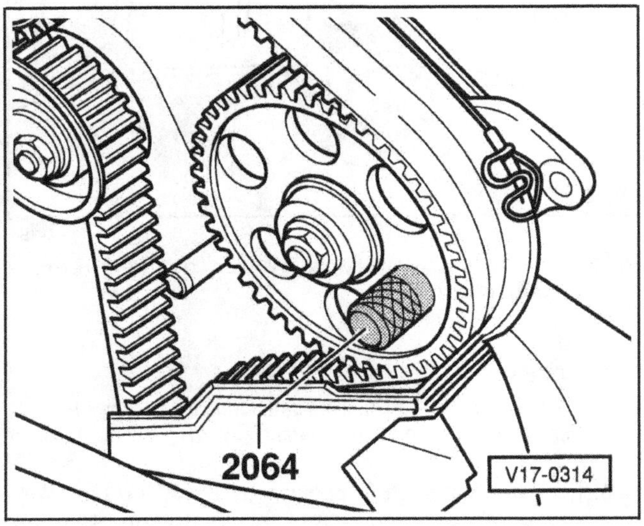

V17-0314

Fig. 7. Injection pump sprocket locked in position with Volkswagen special tool no. 2064. Note that sprocket has only one hole where the tool fits tightly.

6. Relieve the tension from the poly-ribbed belt tensioner and slip the belt off of the pulleys for the generator and air conditioning compressor. Remove the coolant pump pulley bolts and pulley. Mark the running direction of the belts and remove the power steering and coolant pump v-belt and the poly-ribbed belt. See **0 Maintenance Program**.

7. Remove the V-belt drive pulley and vibration dampener.

8. Remove the lower toothed belt guard. See Fig. 4.

9. Remove the idler roller from the support bracket. Remove the nut from the semi-automatic tensioning roller and remove the roller assembly.

10. Mark the running direction of the camshaft/injection pump toothed belt and remove.

11. Inspect the drive belt for damage such as stretched or exposed cords, missing teeth, dry rot, cracks, or evidence of rubbing the covers. Do not reinstall a worn or otherwise faulty belt.

12. Inspect the area around the camshaft, crankshaft, and intermediate shaft oil seal for leaks. If any of these seals are leaking, it is easiest to replace them before installing the camshaft/injection pump toothed belt. See **15a Cylinder Head and Valvetrain (4-Cylinder Engines)**.

NOTE—

If replacing the camshaft/injection pump toothed belt at the factory specified interval of 60,000 miles, remember that the factory also specifies replacement of the semi-automatic tensioner at the same time. If re-using a tensioner, temporarily reinstall the tensioner roller onto the stud and check for smooth rotation with no looseness or play. Always replace a worn tensioner.

To install camshaft/injection pump drive belt

1. Be sure that the crankshaft is still set to TDC for cylinder No. 1. See Fig. 5. It may have moved during disassembly. Also check that the camshaft is still locked with the feeler gauges equally spaced under both sides of the setting bar. See Fig. 6. Also ensure that the locking pin is still installed in the injection pump sprocket. See Fig. 7.

CAUTION—

Special care must be taken when rotating the camshaft or the crankshaft with the belt removed. Mistimed rotation could cause the pistons to hit the valves, damaging either or both.

2. Loosen the camshaft sprocket mounting bolt about 1/2 turn. Release the camshaft sprocket from the camshaft taper by very carefully tapping with a hammer on a soft drift inserted through the hole in the rear of the toothed belt guard.

NOTE—

The camshaft end is tapered. Loosening the sprocket will allow the camshaft sprocket to rotate independently of the camshaft. There may be a keyway cut into the end of the camshaft. However, there is no matching keyway cut into the sprocket nor is there ever a woodruff key.

3. Install the toothed drive belt to the sprockets so that all of the belt slack is at the tensioner. Inspect the idler roller and reinstall the roller to the support. Install the toothed drive belt to the tensioner.

NOTE—

- *Be sure the toothed belt is installed in the original driving direction if belt was reused. New belts can be installed either way.*
- *Be sure that the sprocket is still loose on the camshaft so that the sprocket can move while the camshaft remains locked with the tool.*
- *Be sure that the tool is still locking the injection pump sprocket.*
- *Be sure that the crankshaft is still at TDC. See Fig. 5.*

4. Remove the locking pin from the pump sprocket. Tension the belt by turning the semi-automatic tensioner with a spanner tool (Volkswagen special tool 2587, US4493, 3387 or equivalent) clockwise until the notch and the raised mark are aligned. See Fig. 8.

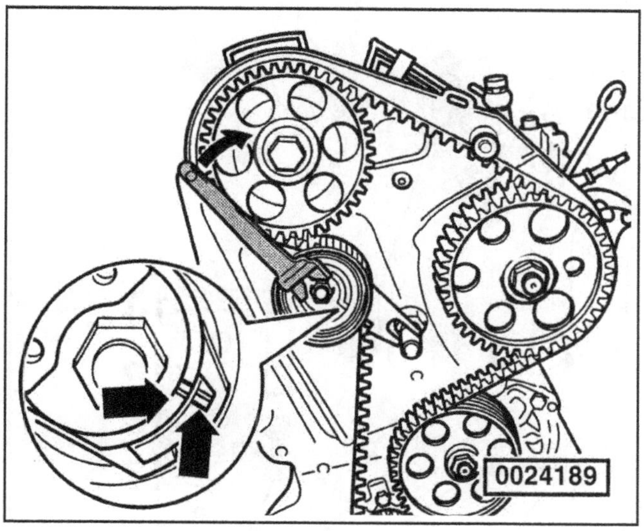

0024189

Fig. 8. Spanner wrench shown inserted into semi-automatic tensioner. Turn tensioner clockwise with spanner wrench until notch and raised mark (**arrows**) are aligned.

5. Tighten semi-automatic tensioner roller clamp nut.

Tightening torques
- Tensioner roller clamp nut . . . . 20 Nm (15 ft-lb)

6. Check to make sure that the TDC mark on the fly-wheel and the reference mark on the transmission case are still aligned. If not, repeat procedure from step 3.
7. Tighten the camshaft sprocket mounting bolt and injection pump sprocket mounting bolt.

Tightening torques
- Camshaft sprocket mounting bolt. . . . 45 Nm (33 ft-lb)
- Injection pump sprocket mounting bolt . . . . 25 Nm (18 ft-lb)

8. Remove the setting bar and feeler gauges from the camshaft.
9. Turn the engine by hand two (2) complete revolutions stopping at TDC as indicated by the marks on the flywheel and the transmission. With the engine at TDC for cylinder No. 1, the setting bar should fit into the back of the camshaft and the locking pin should fit into the injection pump sprocket. If the setting bar fits into the camshaft slot and the crankshaft is at TDC for cylinder No. 1, but the lock pin is slightly out of alignment, loosen the diesel injection pump and move injection pump until the locking pin fits.
10. If all references still do not line up properly, repeat from step 3 until they do.
11. Adjust injection pump timing as described under **To check and adjust injection pump timing**.

CAUTION—
Checking and adjusting the injection pump timing requires the use of Volkswagen scan tool VAG 1551, VAG 1552, or equivalent.

12. The remainder of installation is the reverse of removal. See Fig. 4 for additional tightening torques. Install and adjust the engine drive belts (noting the original direction if reusing the old belts) as described in **0 Maintenance Program.**

INJECTION PUMP

The diesel fuel injection pump is controlled by the ECM and performs three primary functions. It takes the fuel from the tank and pressurizes it to over 3000 psi. It precisely times the delivery of fuel to the injectors, and accurately meters the quantity of fuel delivered to the injectors. The pump is driven by the same toothed drive belt that drives the camshaft at 1/2 crankshaft speed. Internal parts are lubricated and cooled by the diesel fuel. No routine maintenance is required or possible because no internal repair parts are available. Fig. 9 shows an exploded view of the TDI diesel injection pump.

Because the Turbo Direct Injection system is fully electronic, the ECM can detect and store most faults associated with the injection pump. These faults are known as DTCs and are held in the ECM memory. Most DTCs will turn on the malfunction indicator light (MIL) and/or the glow plug light, alerting the driver to this condition. In addition, the ECM has diagnostic capabilities that allow an experienced technician with the scan tool to view certain operating parameters of the injection pump and other parts of the system. These must be investigated before a decision is made to replace an injection pump.

Injection pump, assembly

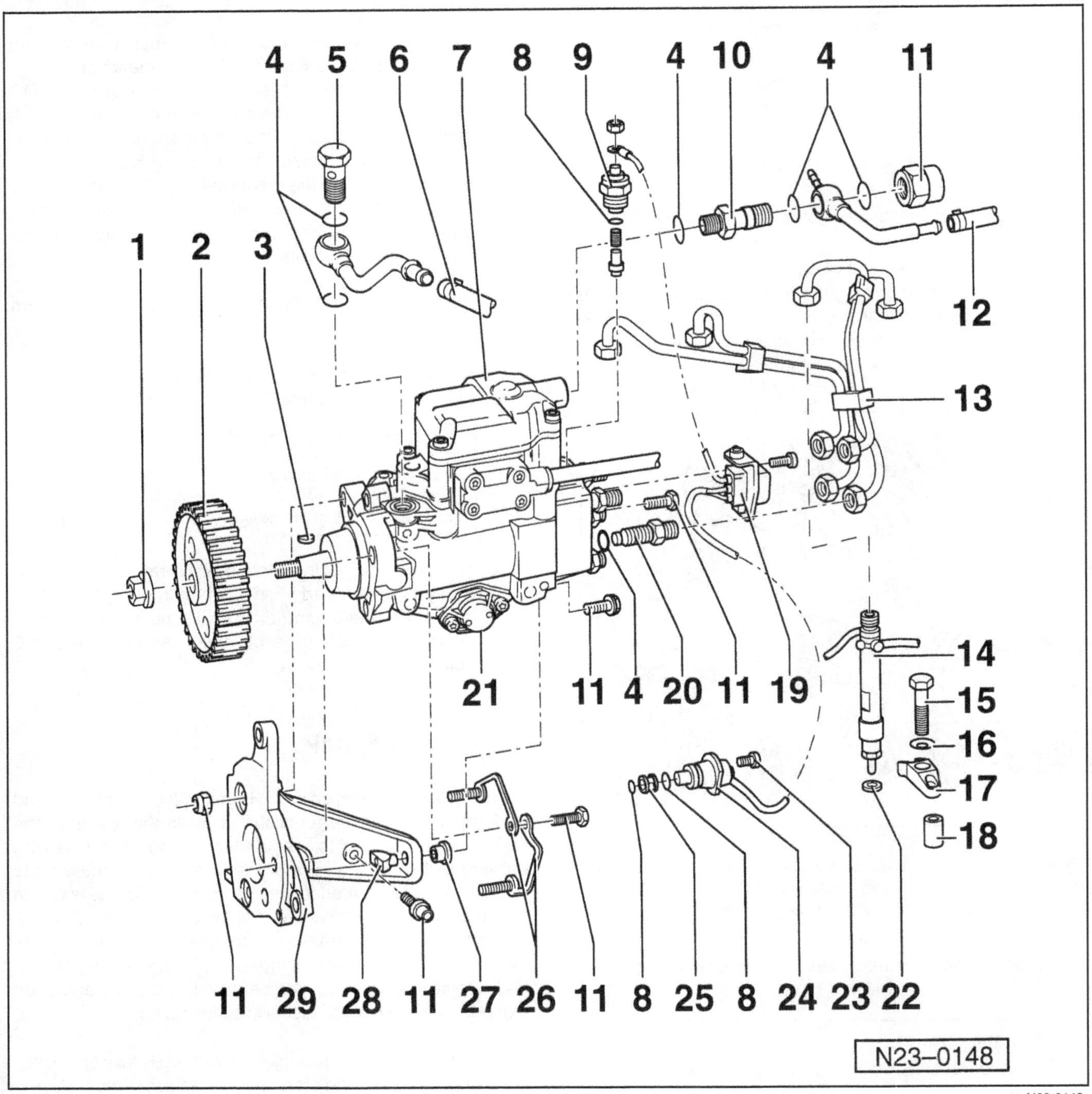

N23-0148

Fig. 9. Exploded view of diesel injection pump with related fuel system components and mounting parts

1. **55 Nm (41 ft-lb)**
2. **Injection pump sprocket**
3. **Woodruff key**
   - Ensure it is seated correctly
4. **Seal**
   - Replace
5. **Banjo bolt**
   - 25 Nm (18 ft-lb)
6. **Supply line**
   - From fuel filter

(continued from previous page)

7. **Diesel injection pump**
   - With quantity adjuster (N146)
   - With modulating piston displacement sensor (G149)
   - With injection start control (N108)
   - With fuel temperature sensor (G81)
8. **O-ring**
   - Replace if damaged
9. **Fuel shut-off valve (N109)**
   - 40 Nm (30 ft-lb)
10. **Union connector**
    - For return line
11. **25 Nm (18 ft-lb)**
12. **Return line**
    - To control valve and fuel filter
13. **Fuel injector lines**
    - Tighten to 25 Nm (18 ft-lb)
    - Remove with 3035 injector line wrench
    - Always remove injector line cluster as an assembly
    - Do not alter shape or bend
14. **Fuel Injector**
    - For third cylinder with needle lift sensor (G80)
15. **20 Nm (15 ft-lb)**
16. **Dished washer**
17. **Clamp**
18. **Bracket**
19. **3-pin harness connector**
    - For fuel cut-off and start of injection control
20. **Union connector**
    - 45 Nm (33 ft-lb)
    - With pressure valve
21. **Timing control cover**
22. **Sealing washer**
    - Always replace
23. **10 Nm (7 ft-lb)**
24. **Solenoid valve (N108)**
    - Controls start of injection (timing)
25. **Screen sieve**
26. **Mounting bracket**
27. **Sleeve bushing**
    - With taper
28. **25 Nm (18 ft-lb)**
29. **Mounting bracket**

To remove injection pump

Special Volkswagen tools including scan tool VAG 1551 or VAG 1552 are required to remove and install the diesel injection pump and properly set it up.

1. Disconnect the battery negative (–) terminal from the battery.

> ***CAUTION —***
> *Before disconnecting battery be sure to obtain radio anti-theft code.*

2. Remove the camshaft drive belt. See **Camshaft/Injection Pump Drive Belt.**

NOTE —
Before removing the toothed drive belt, the crankshaft must be at TDC for cylinder No. 1. The camshaft must also be locked at TDC. See Fig. 5 and 6 given earlier.

3. Insert the locking pin, Volkswagen special tool 2064, into the injection pump sprocket and support bracket. See Fig. 7 given earlier.

4. Loosen and remove the injection pump sprocket retaining nut. Install Volkswagen special tool 3032, US4494, or an equivalent puller onto the injection pump sprocket. Tighten the puller bolt down slightly to apply tension to the sprocket. Lightly tap on the puller bolt to release the sprocket. If the sprocket does not release, tighten the puller bolt further and tap again. See Fig. 10.

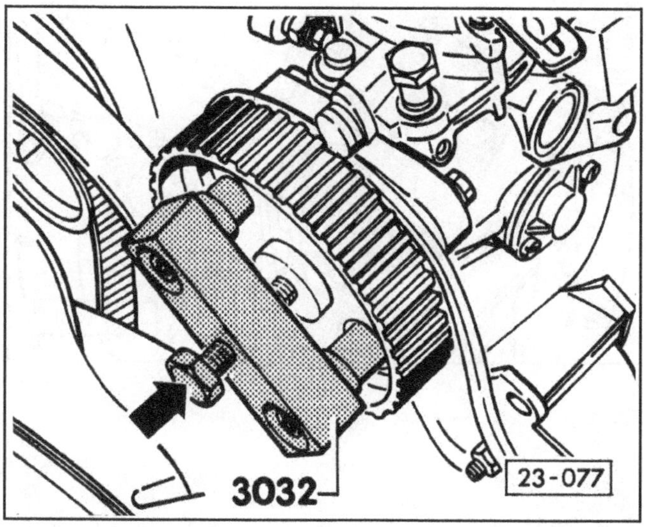

23-077

Fig. 10. Puller (VW special tool no. 3032) installed on injection pump sprocket and shaft. Tension puller and tap bolt (**arrow**) to remove sprocket.

NOTE —
Damage may occur if the puller is simply tightened down all the way without tapping the bolt.

5. Remove the sprocket and woodruff key. Clean the areas around all fuel connections to be removed. Label and disconnect fuel inlet and return hoses. Remove the fuel pipes to the injectors as one assembly taking care not to bend them.

> ***WARNING —***
> *Fuel will be expelled. Do not smoke or work near heaters or other fire hazards. Have a fire extinguisher handy.*

> ***CAUTION —***
> *Diesel fuel is damaging to rubber. Wipe off any fuel that spills on hoses, wiring, and rubber steering and suspension parts immediately, and wash with soap and water. If coolant hoses are contaminated with diesel fuel they must be replaced.*

6. Disconnect the harness connector for the fuel cut-off valve and the solenoid valve from the pump. Disconnect the harness connector for the quantity adjuster, the modulating piston displacement sensor, and the fuel temperature sensor from the pump and slide the terminal housing out of the bracket.

7. Remove the mounting bolts/nuts from the front support. See Fig. 11.

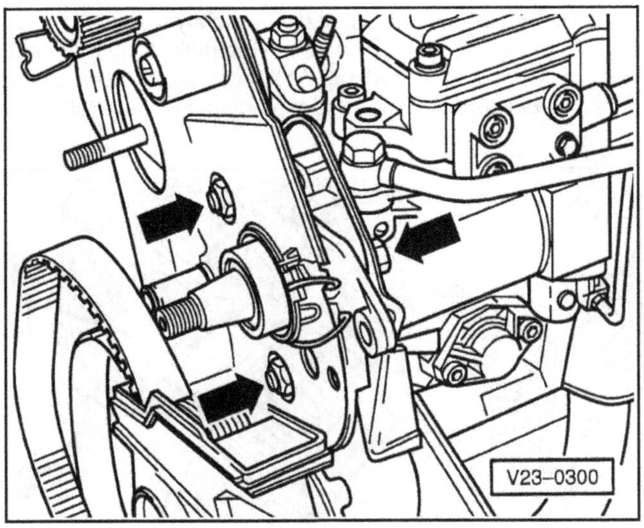

V23-300

Fig. 11. Injection pump front support nuts and bolt (**arrow**).

8. Remove the bolt from the rear pump support. Carefully slide the pump shaft out of the front support and lift the pump out. See Fig. 12.

9. If pump will not be immediately reinstalled, take steps to insure that no dirt can get into any of the pump openings or injector unions.

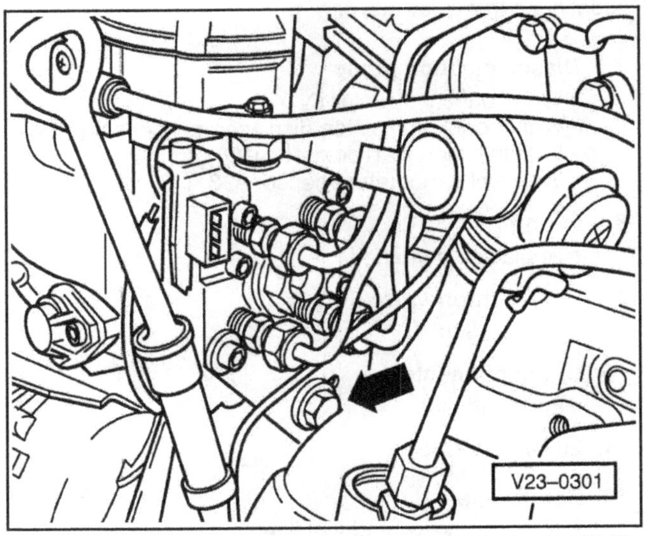

V23-301

Fig. 12. Injection pump rear support bolt (**arrow**). When rear support bolt is removed, watch for conical nut on the opposite side.

To install injection pump

1. Install the injection pump into the rear support and loosely tighten the conical nut.

> ***NOTE —***
> *The conical nut allows the pump to be moved to adjust injection timing and may fall out when the bolt is inserted into the mounting hole. For this reason, it may be necessary to use tape or adhesive to keep the conical nut in place during assembly.*

2. Position the pump in the front mounting bracket so that it is roughly centered in the elongated holes. Hand tighten in this position.

3. Install the injection pump sprocket and woodruff key, lock into position with Volkswagen special tool 2064, and tighten the nut. Remove the locking tool 2064.

Tightening torques
- Injection pump sprocket nut . . . . 55 Nm (41 ft-lb)

4. Camshaft should still be locked with Volkswagen special tool 2065A and shimmed with feeler gauges. If not previously done, remove sprocket from camshaft by loosening the bolt about 1/2 turn and tapping the back of the sprocket with a drift and hammer.

5. Check to make sure that the TDC mark on the fly-wheel and the reference mark on the transmission case are still aligned.

6. Place the toothed belt on the injection pump sprocket and semi-automatic tension roller.

7. Place the camshaft sprocket into the toothed belt and slide the sprocket onto the camshaft. Thread the bolt into the camshaft but leave it sufficiently loose to allow the sprocket to move. Install the idler roller to the front mounting bracket.

8. Tension the semi-automatic toothed belt tensioner as described in **To install camshaft/injection pump toothed drive belt**. Tighten the mounting nut.

Tightening torques
- Semi-automatic tensioner nut ...... 20 Nm (15 ft-lb)

---

9. Check again to make sure that the TDC mark on the flywheel and the reference mark on the transmission case are still aligned.

> ***CAUTION —***
> *Do not proceed until all reference marks are properly aligned. Serious engine damage will result.*

10. Tighten the camshaft sprocket bolt.

Tightening torques
- Camshaft sprocket bolt .... 45 Nm (33 ft-lb)

---

11. Remove the locking tool 2065A from the camshaft.

12. Fill the injection pump with clean diesel fuel through the fuel return line union.

Fill quantity
- Minimum, new pump .... 180 ml (6.1 oz.)

---

13. Re-connect injector pipes, fuel line hoses, and harness connectors. Install belt guards, cylinder head cover, air cleaner and intake air ducts.

14. Check and adjust injection pump timing as described below.

To check and adjust injection pump timing

Diesel injection pump timing must be set using Volkswagen scan tool VAG 1551, VAG 1552, or equivalent in a mode known as Basic Setting. Injection pump timing is correlated to several factors, including fuel temperature. Failure to properly set injection pump timing will result in hard starting, poor performance, poor fuel economy, excessive noise, smoking and increased emissions. The vehicle may not be drivable.

The following instructions assume that the individual technician already has a good working knowledge of the operation and use of the VAG 1551 or VAG 1552 scan tool.

1. Connect the scan tool to the vehicle Data Link Connector (DLC) below the heater controls. The scan tool will be powered up when plugged into the DLC. Start the engine and let it idle. Observe the following sequence in the scan tool display window:
   - 1 – Rapid Data
   - Address Word 01 – Engine Electronics
   - Function 04 – Basic Settings
   - Display Group – 000
   - The display will show 10 fields of information. See Fig. 13.

```
System in Basic Setting                000
XX 53 XX XX XX XX XX XX 90 XX
```

0024190

Fig. 13. Typical VAG 1551/1552 display shows 90 in field 9 and 53 in field 2.

2. Use the table in Fig. 14 to find the timing value. Observe the number in field 9 on the scan tool display. This is a fuel temperature value and is assigned B in the table. The timing value must fall within the shaded area of the table and is assigned the value A. With the engine still running at idle, move the injection pump on its shaft axis until the number in field 2 matches the number from the center of the shaded area A.

Example:
- With the engine running, the fuel temperature value in the scan tool display field 9 is 90. This is value B. Look on the table along the B axis until you see 90. Follow the line from 90 up into the shaded area. The shaded area is the range of timing values. Find the center of the shaded area and look to the left for the value on the A axis. In this case it is 53. Adjust the pump until the value in the scan tool display field 2 reads 53.

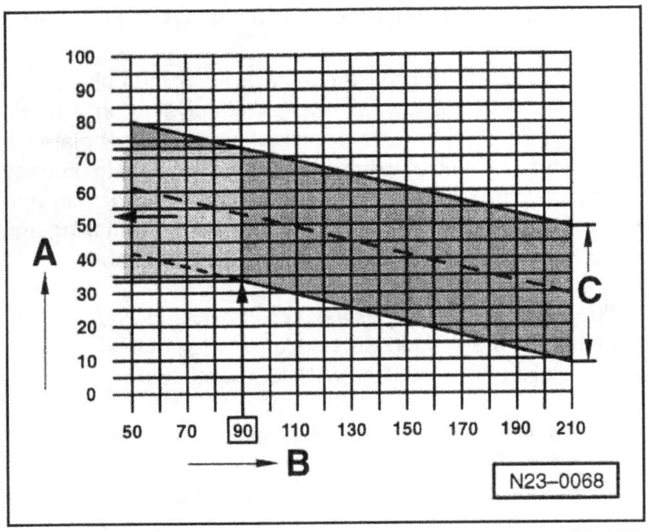

N23-0068

Fig. 14. Graph showing relationship between fuel temperature value **B**, and start on injection timing **A**. Data is only available through use of scan tool VAG 1551 or VAG 1552.

3. Shut off the engine. Tighten all the injection pump mounting bolts using Fig. 4 for torque specifications. Verify that the timing value A still falls within the range of values called for by value B.

4. Loosen all 8 connections of the fuel pipes at the injectors and the pump and then re-tighten all 8. This relieves the strain on the pipes that developed when the injection pump was moved.

> ***CAUTION —***
> *Failure to relieve the strain on the fuel pipes after moving the diesel injection pump may result in damage due to vibration.*

5. Verify that all fasteners and components that were removed have been re-installed and properly tightened.

6. Verify that all electrical connectors have been properly secured and that all seals in the terminal housings are still in place.

7. Verify that all fuel lines are secure and that there are no fuel leaks.

8. With scan tool still connected, check for any Diagnostic Trouble Codes that may have been set. Erase any DTCs, road test and recheck.

9. Re-install any cable binders (wire ties, tie wraps) that were removed. Re-install sound absorber cover on engine.

ENGINE CONTROL MODULE

Diesel engine management is controlled by an Electronic Control Module located in the air plenum under the hood on the passenger's side. See Fig. 15. This system is in many ways similar to the 2.0 liter ABA gasoline engine in other Golf and Jetta models. Inputs are monitored, calibrations are made, outputs are generated and the engine runs. All inputs and all outputs must operate within normal parameters. The ECM itself must also be supplied with correct supply voltage and grounds. In order for the ECM to function correctly, it must also "know" what equipment is installed in the vehicle. This process is known as coding and can only be done with Volkswagen scan tool VAG 1551, VAG 1552, or equivalent. This must be done whenever the ECM is replaced. Coding is, however, retained if the battery is disconnected or if the ECM is unplugged from the harness.

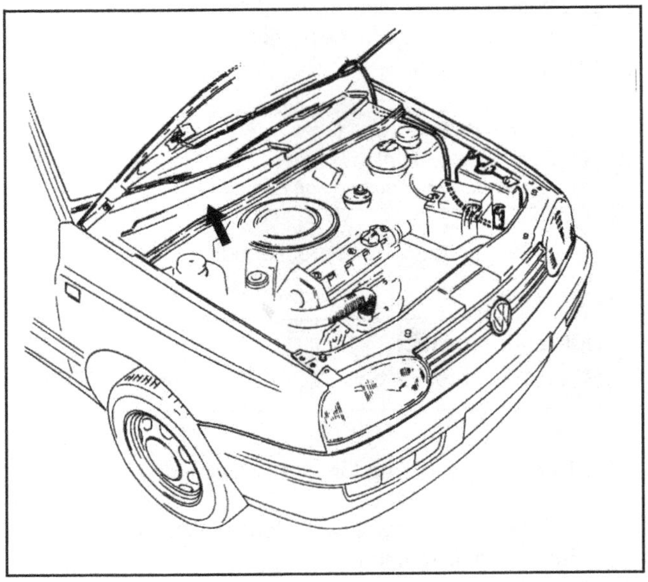

0024191

Fig. 15. Location (**arrow**) of engine control module in Golf and Jetta TDI models.

READINESS CODES

Recent legislation requires that all auto manufacturers build into their engine management systems the ability to check for proper operation of up to 8 functions. This includes both gasoline and diesel engines. For the TDI system on the 1Z/AHU engine, only the exhaust gas recirculation system is relevant. Most of the time, the readiness code will set itself after the appropriate conditions have been met. These may include a cold or a hot start, operation at a certain load or speed for a certain period of time, or operation at different temperatures. After repairs it may be advantageous to confirm repairs rather than wait to see if they set on their own. The setting procedure is simple and requires the scan tool VAG 1551 or VAG 1552 to confirm the setting.

To check TDI readiness codes

1. Connect the scan tool to the vehicle Data Link Connector (DLC) below the heater controls. The scan tool will be powered up when plugged into the DLC. Start the engine and let it idle. Observe the following sequence in the scan tool display window:
    - 1 – Rapid Data
    - Address Word 33 – OBD II
    - Select mode – 41
    - The display will show four fields of information

> ***CAUTION —***
> *These instructions assume that the technician already has a good working knowledge of the operation and use of the VAG 1551 or VAG 1552 scan tool.*

2. Fig. 16 shows the information as it appears on the screen. Disregard the information that is shown here as the letter X. These spaces will have either a 1 or a 0 and do not affect this procedure. The readiness code is the 8 digit number in field 4 on the far right. If the display shows a 1 and the remainder are all zeros, the readiness code needs to be set.

Mode 41		PID 1	Module 08
XXXXXXXX	XXXXXXXX	XXXXXXXX	10000000

0024192

Fig. 16. Display that shows that readiness code is not set and should be. Disregard all Xs.

3. If the screen displays all zeros in the fourth field (far right), the readiness code is set and no further action is required. See Fig. 17.

Mode 41		PID 1	Module 08
00000000	00000101	10000000	00000000

0024193

Fig. 17. Display that shows that readiness code is set and no further action is required.

4. If readiness code is set, disconnect the scan tool and reinstall the appropriate parts.

5. If readiness code is not set, unplug the scan tool and proceed as follows.

To set TDI readiness codes

> ***NOTE —***
> - *Engine temperature must be over 77°F (25°C).*
> - *Do not stop the engine except where directed.*

> ***CAUTION —***
> *This procedure requires that the engine be revved up while stationary; therefore apply the parking brake and place vehicles with automatic transmission into PARK.*

1. Start engine and increase speed to 2200 RPM as observed on the vehicle tachometer.

2. Release accelerator pedal and allow engine to return to idle speed (approx. 900 RPM).

3. Stop engine and wait for 10 seconds.

4. Start engine again and increase speed to at least 2200 RPM as observed on vehicle tachometer.

5. Release accelerator pedal and allow engine to return to idle speed (approx. 900 RPM).

6. Connect scan tool and observe display. See **To check TDI readiness codes**.
    - If display appears as shown in Fig.17, code has been set and no further action is required. Disconnect scan tool and reinstall appropriate parts.
    - If display appears as shown in Fig. 16, code has not been set. Unplug scan tool and repeat from step 1. If code does not set after second attempt, it is likely that there is a malfunction in the system. Consult scan tool directions for additional information on checking for DTCs.

FUEL INJECTORS

The 1Z/AHU TDI diesel engine uses a two spring injector that enables the fuel to be injected in two stages for "softer" combustion. In addition, the injectors are a 5-hole design that spray fuel from 5 ports in a more lateral direction rather than in a single conical pattern as in previous versions. Opening pressures are significantly higher at approximately 190 to 200 bar (2750-2900 psi). All 4 injectors are functionally the same, however, injector for cylinder No. 3 carries the needle lift sensor (G80). This injector must always be installed in cylinder No. 3 to prevent erroneous DTCs or performance problems from occurring.

The signs of injector trouble usually appear as misfiring and knocking noises from one or more cylinders, engine overheating, loss of power, smoky black exhaust, and excessive blue smoke during cold starting.

> ***CAUTION —***
> *When working on the injectors, everything must be kept absolutely clean. Clean all pipe unions before disconnecting.*

To check for a defective fuel injector or weak cylinder

A defective injector can often be identified by simply "turning it off". This can be done by relieving the fuel pressure so that it can not open and inject fuel into the cylinder. Idle speed may not substantially change, however, because the ECM will often be able to compensate for the weak cylinder.

1. Clean the area around the fuel unions where the fuel pipes meet the injectors.

2. Start the engine and let it run at idle.

3. Slightly loosen the union nut for cylinder No. 1 fuel injector at the injector. Cylinder No. 1 is closest to the drive belt. See Fig. 18.

> ***WARNING —***
> - *Loosen the nut only about one half turn to limit the amount of fuel leakage. The fuel injection system operates at very high pressures. Keep hands and eyes clear. Wear heavy duty gloves and eye protection. Wrap a cloth around the nut when loosening. Fuel will be expelled.*
> - *Do not smoke or work near heaters or other fire hazards.*
> - *Have a fire extinguisher handy.*

- If a particular cylinder is defective, the noise or smoke will cease when the union nut is loosened.
- If the problem was roughness, the roughness will be affected very little or not at all because that cylinder was already malfunctioning. Tighten the union nut.

4. Repeat procedure for any suspected cylinders.

> ***CAUTION —***
> *Diesel fuel is damaging to rubber. Wipe off any fuel that spills on hoses, wiring, and rubber steering and suspension parts and wash with soap and water. If coolant hoses are contaminated with diesel fuel, they must be replaced.*

5. When finished, verify that all fuel lines are secure and that there are no fuel leaks.

To replace fuel injector

1. Unscrew the fuel pipe union nuts at both the injection pump and the injectors. Remove the injeector fuel pipes as one assembly taking care not to bend them.

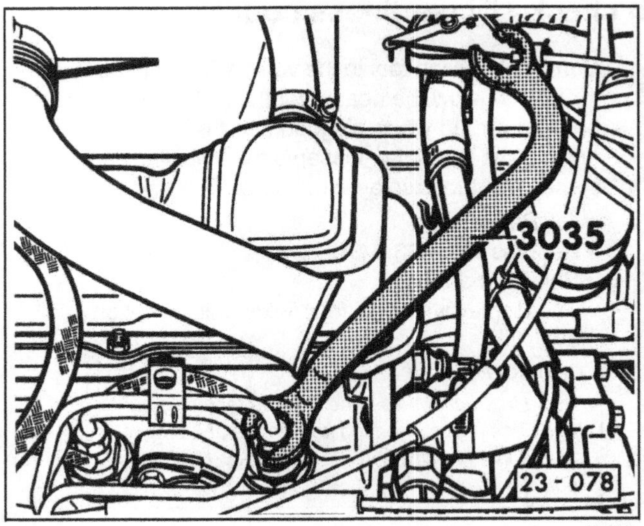

23-078

Fig. 18. Flare-nut wrench (VW special tool no. 3035) being used to loosen injector union nut for No. 4 cylinder on an earlier diesel engine version. 1Z/AHU diesel engines are similar.

2. Remove the fuel return hoses from between the injectors.

3. Unplug the harness connector for the needle lift sensor (G80) on the injector for cylinder No. 3.

4. Loosen and remove the bolt and the special washer and the hold down clamp for each injector. Lift out the injectors.

5. Remove the small copper sealing washer from the tip of the injector or the injector port in the head. Discard.

> ***NOTE —***
> *The crush-type sealing washers must always be replaced to prevent compression leaks.*

6. Install the injector with a new copper sealing washer and torque the bolt to specification.

Tightening torques
- Fuel injector clamp mounting bolt. . . . 20 Nm (15 ft-lb)

7. Install metal fuel pipes and fuel return hoses. Due to the higher pressures involved, the injectors will seldom need to have the air bleed from them. Reconnect harness connector to needle lift sensor.

8. When finished, verify that all fuel lines are secure and that there are no fuel leaks.

To pressure test injector

1. Remove the injector as described previously.

2. Install the injector to the pressure tester according to the manufacturer's directions. Place a suitable container under the injector to catch the spray. See Fig. 19.

> ***WARNING —***
> *When testing fuel injectors, make sure that the high pressure discharge does not contact the hands or any other bare skin. The high pressure spray can be injected directly into the skin causing severe injuries or health complications. Always wear eye protection! Work carefully.*

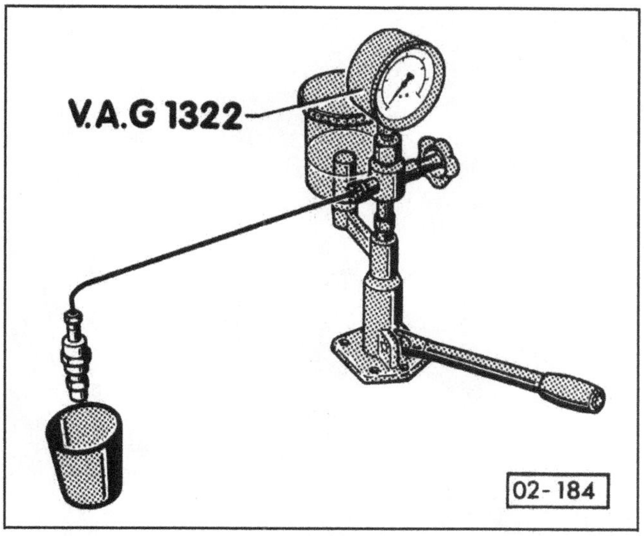

02-184

Fig. 19. Injector installed on hand pressure pump. Volkswagen special tool VAG 1322 shown. This tester is designed to work best with mineral spirits. Older version US1111 will also work with suitable fittings.

3. Pump lever to increase pressure. Read opening pressure when spray begins.

Injector Opening Pressure
- New . . . . 190–200 bar (2750–290 psi)
- Wear limit. 170 bar (2465 psi)

---

4. Spray should discharge evenly from all 5 ports. See Fig. 20.

> ***WARNING —***
> *Fire hazard! No smoking! Do not have anything in the area that can ignite the spray discharged by the injectors. This includes overhead heating units and hidden sources such as water heaters with standing pilot lights.*

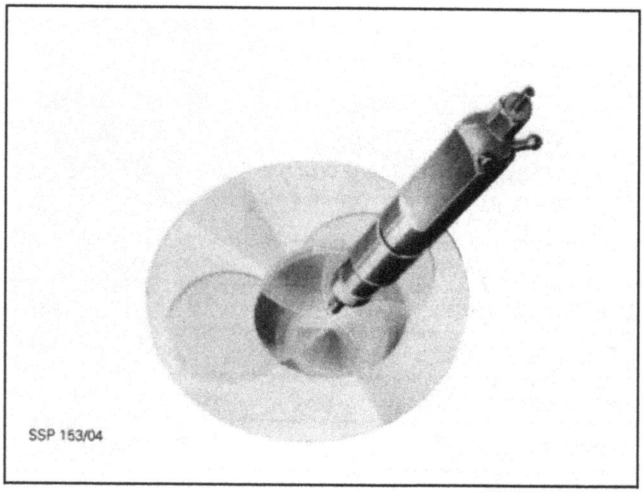

SSP 153/04

Fig. 20. TDI fuel injector spray pattern discharging from all 5 lateral ports.

5. Pump lever slowly to a pressure of approximately 150 bar (2175 psi) for 10 seconds. Check for leaks from injector tip. No fuel should leak.

6. If any injector fails to meet the opening pressure or leak specifications, it should be replaced. No repairs are possible to the fuel injectors.

24a Fuel Injection–Motronic (ABA, AAA engine)

TABLES

GENERAL

This repair group covers Bosch Motronic fuel injection component troubleshooting and repair for the 1993–1999 2.0 liter ABA engine and the 1994–1999 2.8 liter AAA (VR6) engine. Special testing equipment may be necessary for some of the tests and repair procedures given here.

NOTE —
1996-1999 ABA and AAA engines are equipped with Motronic M5.9 (OBD II). It is recommended that fault diagnosis and troubleshooting be carried out using the special Volkswagen scan tool (VAG 1551), or an appropriate "generic" scan tool.

Fuel Injection System Applications

- ACC (1.8 liter) . . . . Mono-Motronic
- ABA (2.0 liter)
  - 1993–1995 models . . . . Motronic M2.9
  - 1996–1999 models . . . . Motronic M5.9
- AAA (2.8 liter)
  - 1993–1995 models . . . . Motronic M2.9
  - 1996–1999 models . . . . Motronic M5.9
- AAZ (1.9 liter) . . . . Diesel Fuel Injection
- 1Z/AHU (1.9 liter) . . . . Diesel TDI

NOTE —
- *Fuel pump and fuel tank testing and repair information is covered in* **20 Fuel Storage and Supply**.
- *Emission systems, such as Evaporative (EVAP) emissions, and Exhaust Gas Recirculation (EGR) are covered separately in* **26 Exhaust System/Emission Controls**.
- *Knock sensor testing and repair information is covered in* **28a Ignition System–Gasoline**.

Fig. 1 shows a schematic view of the Bosch Motronic engine management system. The Motronic system combines the ignition and fuel functions into one system, managed by a single engine control module (ECM). The system is fully adaptive and features built-in diagnostics that detect and store coded fault information. There are no basic adjustments or settings that can be made to the system.

NOTE —
A general description of operation for the Motronic engine management system is given in **2 General Information**.

Safety Precautions

The following warnings and cautions should be adhered to whenever doing work on the fuel injection system.

WARNING —
Fuel will be discharged during many fuel system test procedures. Do not smoke or work near heaters or other fire hazards. Have a fire extinguisher handy. Work only in a well-ventilated area. Wear suitable hand protection when working with gasoline. Prolonged contact with fuel can cause illnesses and skin disorders.

CAUTION —
- *Connect and disconnect wires and test equipment only with the ignition off.*
- *Before making any electrical tests that require the engine to be cranked using the starter, disable the ignition system as described in* **28a Ignition System–Gasoline**.
- *Always wait at least 40 seconds after turning off ignition before removing the ECM connector. If the connector is removed before this time, residual power in the system relay may damage the ECM.*
- *Cleanliness is essential when working on the fuel system. Thoroughly clean fuel line connections and surrounding areas before loosening. Avoid the use of compressed air, and avoid moving the car. Only install clean parts.*

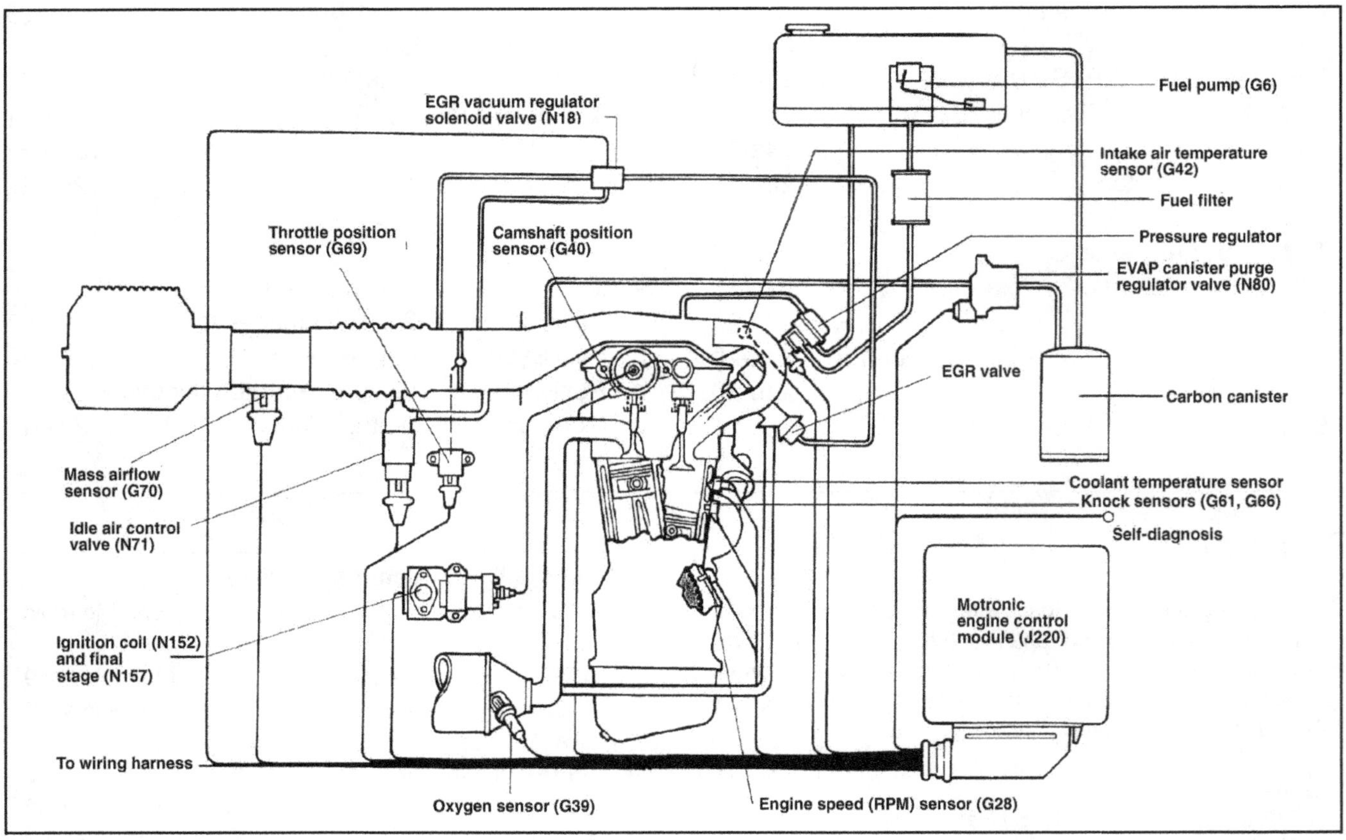

Fig. 1. Schematic view of Motronic engine management system.

Adaptive Circuitry and On-Board Diagnostics (OBD)

The Motronic system is adaptive. Idle speed, ignition timing, injection timing and quantity automatically compensate for and adapt to changes in the engine due to wear or small problems, such as minor vacuum leaks. As a result, idle speed, fuel mixture (%CO) and ignition timing are non-adjustable.

> *NOTE—*
> - *Volkswagen identifies electrical components by a letter and/or a number in the electrical schematics. See* **97 Wiring Diagrams, Fuses and Relays**. *These electrical identifiers are listed as an aid to electrical troubleshooting.*
> - *The Volkswagen Scan Tool V.A.G. 1551 is used to quickly access the Diagnostic Trouble Codes (DTCs) through the data link connector(s) in the dash. Alternatively, the data link connector(s) can be jumpered and the DTC can be read out through the Check Engine light, also called the Malfunction Indicator Lamp (MIL), in the form of a blink code.*
> - *Beginning in 1994 (1993 in California), automakers are required by law to apply uniform terminology, words and terms for certain components. These standardized terms will be used throughout this section.*

The Motronic system features built-in diagnostic circuitry to detect and store Diagnostic Trouble Code (DTC) information. When the system compensates for values that are outside the permitted limits, a fault code is generated and stored in the ECM's permanent memory. For emissions-related faults, the Check Engine or Malfunction Indicator Lamp (MIL) will be turned on. The fault codes can be accessed through the data link connector(s) in the dashboard.

FAULT DIAGNOSIS (1993–1995 models)

Fig. 2 shows Motronic component locations for the 1993–1995 AAA (VR6) engine. Fig. 3 shows Motronic component locations for the 1993–1995 ABA (4-cylinder) engine.

Because the ignition, fuel injection, and emission control functions are inter-related and controlled together by the Motronic engine management system, it is difficult, if not impossible to isolate general driveability problems by examining components of the fuel injection system alone.

For this reason, the first step in locating trouble is to review the basic requirements list shown below. If no faults are found, continue by checking for Diagnostic Trouble Codes (DTCs), especially if the check engine light is illuminated.

1. **Evaporative (EVAP) emissions canister purge regulator valve (N80)**
2. **Mass air flow (MAF) sensor (G70)**
3. **Intake air duct**
4. **Not applicable to vehicles covered by this manual**
5. **Knock sensor 1 (G61)**
   - see **28 Ignition System** for repair information
6. **Motronic engine control module (ECM) (J220)**
7. **Throttle valve housing**
8. **Throttle position (TP) sensor (G69)**
9. **Idle air control (IAC) valve (N71)**
10. **Not applicable to vehicles covered by this manual**
11. **Ignition coil (N152)**
12. **A/C thermal switch (F45)**
13. **Engine coolant temperature (ECT) sensor (G62)**
    - blue, 2-pin connector
14. **After-run coolant fan control thermal switch (F87)/ coolant temperature sensor for temp. gauge (G2)**
    - yellow, 4-pin connector
15. **Intake air temperature (IAT) sensor (G72)**
16. **Fuel pressure regulator**
17. **Engine speed sensor (G28)**
18. **Knock sensor 2 (G66)**
    - see **28 Ignition System** for repair information
19. **Fuel injector, cyl. no.1 (N30)**
20. **Evaporative (EVAP) emission canister**
21. **Air cleaner (ACL)**

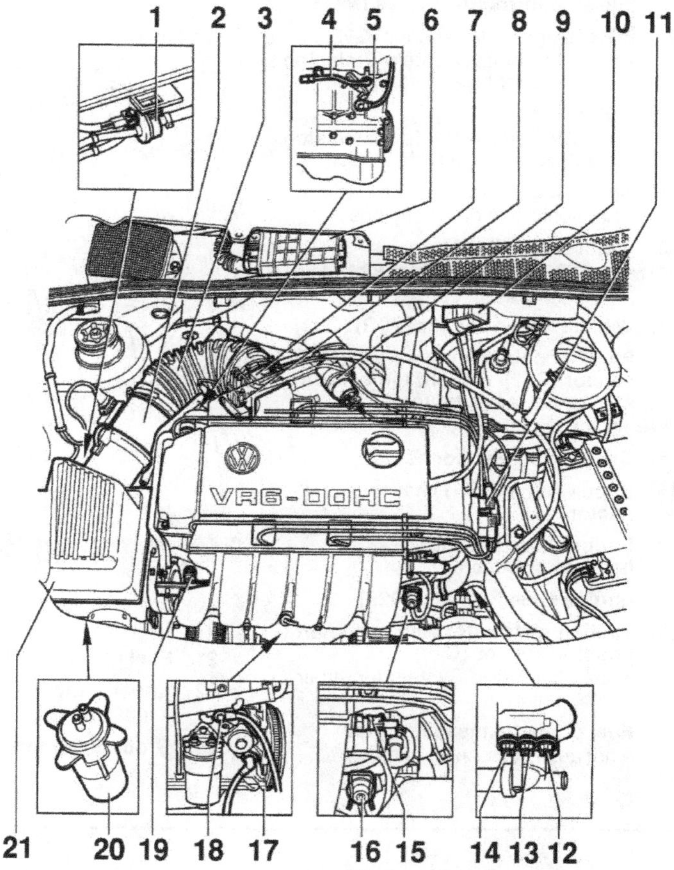

Fig. 2. Motronic component location on 1993–1995 AAA (VR6) engines.

Basic Requirements

Below are a few basic checks that should be made if the Malfunction Indicator Lamp (MIL) is not illuminated, but engine management driveability problems are encountered.

1. Check intake system for leaks. Check for cracked, loose, or disconnected hoses and duct work. Check that all hose clamps are tight. Check small vacuum hoses, especially close to the fittings, and check the fittings themselves.

NOTE —

An air leak allows unmeasured air to enter the engine, often resulting in an overly lean fuel mixture. This can cause driveability problems and may illuminate the MIL.

2. On 4-cylinder engines, check the intake air boot and mass air flow (MAF) sensor for excessive oil deposits. If heavy deposits are found and the engine is experiencing driveability problems, the MAF sensor may be contaminated. Test the air flow sensor as described later in this section.

1. **Evaporative (EVAP) emissions canister purge regulator valve (N80)**
2. **Exhaust gas recirculation (EGR) vacuum regulator solenoid valve (N18)**
   - see **26 Exhaust System /Emission Controls** for repair information
3. **Mass air flow (MAF) sensor (G70)**
4. **Throttle position (TP) sensor (G69)**
5. **Motronic engine control module (ECM) (J220)**
6. **Oxygen sensor (G39) harness connector**
7. **Intake air temperature (IAT) sensor (G72)**
8. **Idle air control (IAC) valve (N71)**
9. **EGR temperature sensor (G98)**
   - see **26 Exhaust System/Emission Controls** for repair information
10. **EGR valve**
    - see **26 Exhaust System/Emission Controls** for repair information
11. **Co tap tube**
    - not applicable on all engines
12. **Ignition coil (N152)**
    - see **28 Ignition System** for repair information
13. **Vehicle speed sensor (VSS) (G22)**
14. **Engine coolant temperature (ECT) sensor (G62)/ coolant temperature sensor for temp. gauge (G2)**
    - black, 4-pin connector
15. **Central harness connector**
16. **Knock sensor (G61) harness connector**
17. **Engine speed sensor (G28) harness connector**
18. **Engine speed sensor (G28)**
19. **Ignition distributor with camshaft position sensor (G40)**
    - see **28 Ignition System** for repair information
20. **Fuel pressure measuring tap**
    - not applicable on all engines

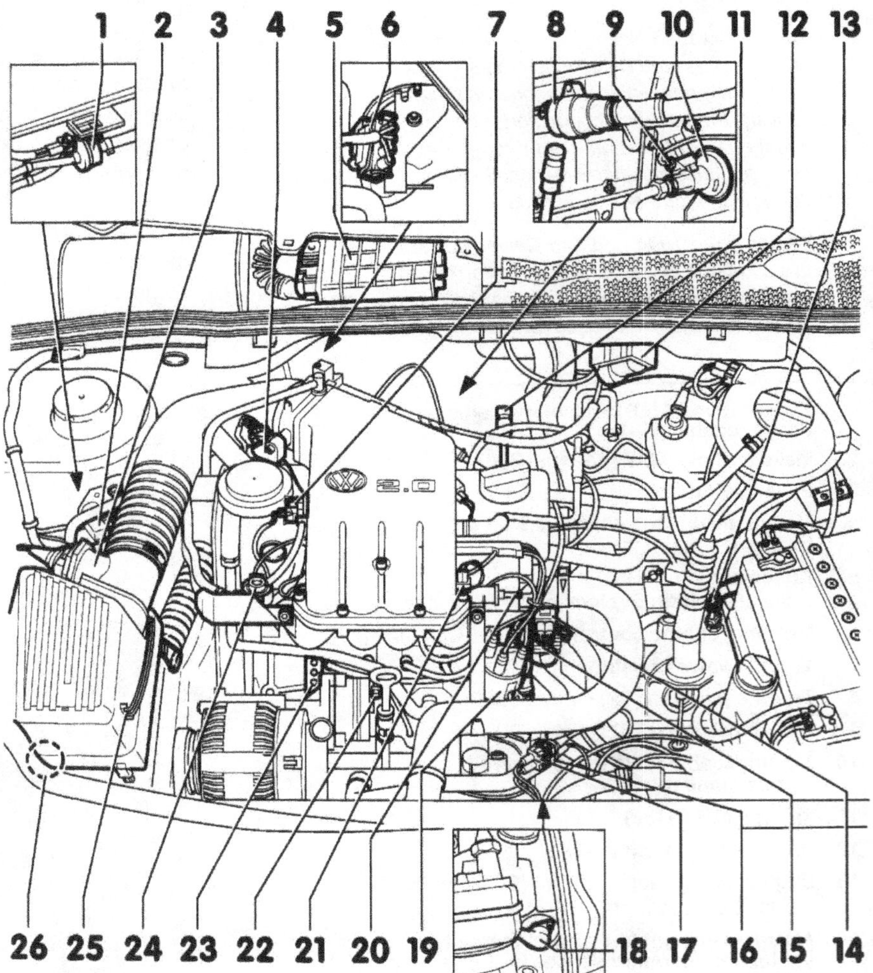

21. **Fuel injectors (N30-N33)**
22. **Knock sensor 1 (G61)**
    - repairing, see **28 Ignition System**
23. **Ground connection**
24. **Fuel pressure regulator**
25. **Thermal vacuum valve (TVV)**
    - for intake air preheating
26. **EVAP canister**
    - located below air cleaner

24-M1047

Fig. 3. Motronic component location on 1993–1995 ABA (4-cylinder) engine.

3. Check that the battery is in good condition. Check that the cables are tight and free of corrosion. Check that all related ground points are firmly connected and in good condition.

4. Check all of the harness connectors for damage and corrosion, especially the connector at the speed sensor (G28), the main 28-pin connector at the rear of the engine, and the knock sensor mounts on the engine block.

NOTE —

4-cylinder engines use one knock sensor. 6-cylinder engines use two knock sensors, one on either side of the engine.

5. If the engine won't start, check for power and ground at the Motronic ECM. See **97 Wiring Diagrams, Fuses and Relays.** Check the fuses.

6. Make sure there is fuel in the tank and fuse no. 22 is OK.

7. As the final step, check for Diagnostic Trouble Codes (DTC) as described below.

Diagnostic Trouble Codes (DTC)

The Motronic systems are equipped with built-in fault diagnostics. Some of the faults will illuminate the check engine light, but many will not.

On 1993-1995 models diagnostic trouble codes (DTC) can be displayed in one of two ways; (1) using the specialized Volkswagen scan tool V.A.G. 1551 or (2) by jumpering the data link connector(s) and interpreting blink codes through the malfunction indicator light (MIL) in the instrument cluster. The procedure below describes the blink code method.

On 1996 and later models the diagnostic trouble codes (DTC) can only be displayed by using the specialized Volkswagen scan tool V.A.G. 1551.

DTCs are accessed through the data link connector(s). The connectors are behind a cover in the center of the dash, below the heating ventilation controls. Two versions of data link connectors are used depending on model year. 1993 and early 1994 cars use two connectors while later model 1994 and later cars use a single connector. See Fig. 4.

NOTE —

On 1995 cars, a special OBD adapter may be stored beneath the rear seat and should be used to access the DTCs. See Fig. 5. On cars with ABS, the connector is stored beneath the ABS coaxial cable next to the ABS control unit. On cars without ABS, the adapter is secured by a wire tie.

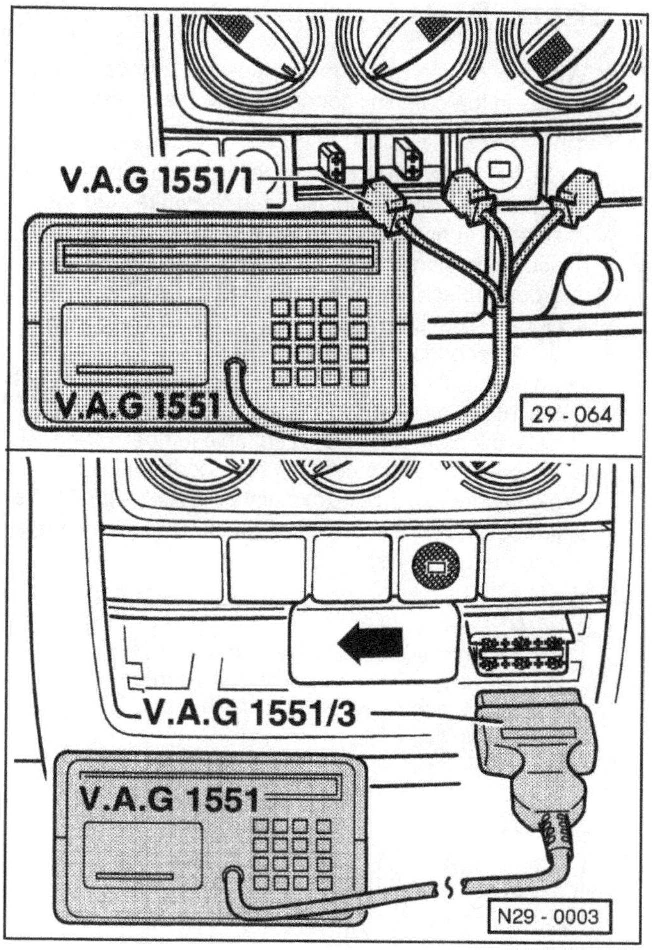

29-064, N29-0003

Fig. 4. Data link connectors in center of dash. 1993 and early 1994 (top) cars use two separate connectors. Later model 1994 (bottom) and later cars use single 16-pin connector. Special VW scan tool V.A.G. 1551 shown.

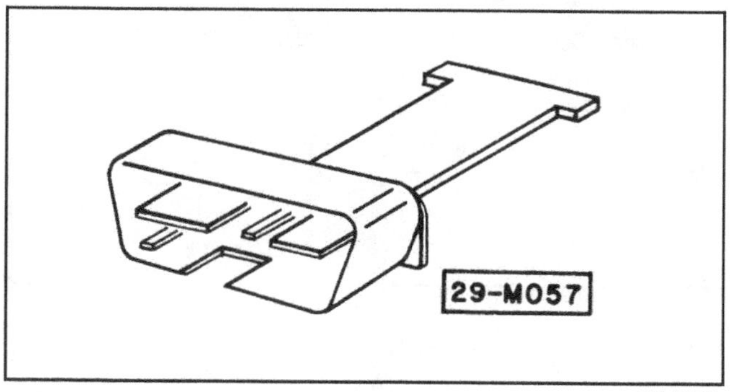

29-M057

Fig. 5. OBD adapter used to access blink codes on later 1994 and 1995 model year vehicles. On some 1995 vehicles, adapter may be attached to the wiring harness under the rear seat.

To display DTCs (models with blink codes)

1. With the ignition key in the on position, jumper the terminals in the data link connector that match the vehicle configuration. See Fig. 6.
   - On 1993 and early 1994 vehicles, jumper position 1 in the black plug to position 1 in the white plug with a suitable jumper wire. This connects the blink code activation signal wire (white plug, position 1) to a ground (black plug, position 1) to initiate the blink signal. The jumper should be connected for about 2-1/2 seconds and then removed.
   - On later 1994 and 1995 vehicles, jumper position 4 of the 16 pin data link connector to position 15 of the data link connector with a suitable jumper wire or adapter. This connects the blink code activation signal wire to a ground to initiate the blink signal. The jumper should be connected for about 2-1/2 seconds and then removed.

> ***CAUTION —***
> *Use care when connecting the jumper wire. Positive (+) battery voltage is present in the connector. A wrong connection can result in damage to ECM or the vehicle's wiring.*

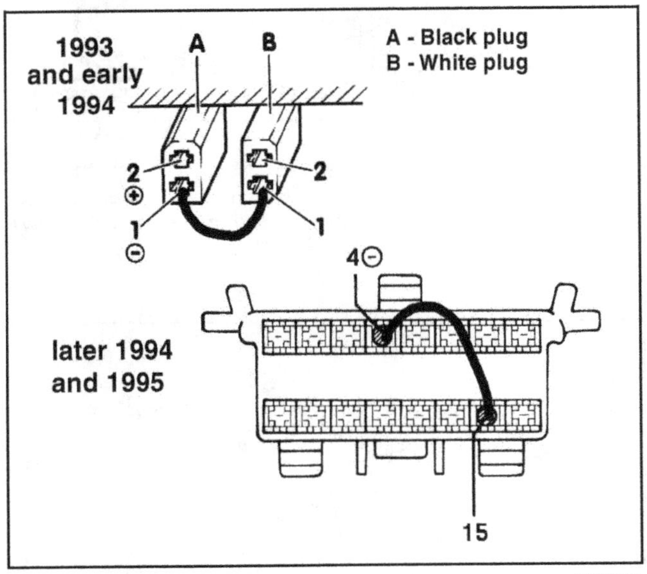

Fig. 6. Jumper wire installed in data link connectors.

2. After one blink of the MIL light, the first 4-digit DTC will be displayed. Approximately one second later, the MIL will blink the first digit, then pause, blink the second digit, then pause, and so on. **Table a** lists DTCs.

> ***NOTE —***
> *A blink code of 2-1-1-1 will always be displayed because the ECM only reads the engine speed (RPM) sensor signal when the engine is running. This code will erase itself once the engine is started.*

3. Once the first DTC has been displayed, reconnect the jumper wire or the adapter for **2.5 seconds** and then remove it. The second DTC code (if present) will be displayed.

> ***NOTE —***
> *Record blink codes on paper for later reference. If an error is made reading blink codes, they can be recalled as many times as needed as long as they have not been erased.*

4. Repeat step 3 until no more codes are displayed.

> ***NOTE —***
> *End of blink code display will be identified by 4 long blinks (0-0-0-0).*

5. When all codes have been displayed and recorded, turn the ignition key off.

Table a. Motronic Diagnostic Trouble Codes (Blink Codes) 1993–1995 model years

DTC	Probable fault	Corrective action, first check
4444	**No DTC recognized**	No corrective action necessary.
1231	**Vehicle speed sensor (VSS) (G22)**–missing or incorrect signal, VSS wiring damaged/open or shorted, VSS faulty. Speedometer faulty.	Check output signal the from speedometer VSS. Check sensor wiring. Check speedometer operation.
1344*	**Secondary air injection system**–general malfunction	Check secondary air injection system components.
2111	**Engine speed (RPM) sensor (G28)**–missing or incorrect rpm signal. (This code will always be present if the test is run with the engine off. It will erase itself when the engine starts. If the engine will not start, the sensor must be checked manually.)	Check the wiring and the resistance of the speed sensor. Check that the sensor is securely mounted and the sensor wheel is not damaged (sensor wheel is mounted to crankshaft).
2113	**Camshaft position (CMP) sensor (G40)**–missing or incorrect signal from sensor, CMP wiring damaged/open or shorted, CMP (Hall) sensor faulty.	Check the wiring and signal from of the CMP (Hall) sensor. On 4-cylinder engines, check that the distributor is correctly installed. On 6-cylinder engines, check that the sensor is securely mounted to the cylinder head. Test the signal from the Hall sensor. Check camshaft/valve timing.
2212	**Throttle position (TP) sensor (G69)**–missing or incorrect signal from TP sensor, TP sensor wiring damaged/open or shorted, TP sensor faulty.	Check the wiring and signal from the TP sensor.
2312	**Engine coolant temperature (ECT) sensor (G62)**–missing or incorrect signal from ECT sensor, ECT sensor wiring damaged/open or shorted, ECT sensor faulty	Check the wiring and signal from the ECT sensor. Check for corrosion at the connector, causing intermittent (sporadic) malfunctions.
2142	**Knock sensor (KS) I (G61)**–missing or incorrect signal from knock sensor to ECM, knock sensor wiring damaged/open or shorted, knock sensor faulty or sensor mounting bolt loose	Check the wiring and signal from knock sensor I. Check the torque of the mounting bolt to the block. Check for corrosion between sensor and engine block. See also DTC 2141. If no faults can be found, the ECM may be faulty.
2144	**Knock sensor (KS) II (G66)**–(6-cylinder engine only) missing or incorrect signal from knock sensor to ECM, knock sensor wiring damaged/open or shorted, knock sensor faulty or sensor mounting bolt loose. ECM faulty.	Check the wiring and signal from knock sensor II. Check the torque of the mounting bolt to the block. Check for corrosion between sensor and engine block. If no faults can be found, the ECM may be faulty.
2342	**Heated oxygen sensor (G39)**–no oxygen sensor signal to ECM. Sensor heater circuit may be open or faulty.	Check the oxygen sensor output signal (voltage fluctuating between 0 and 1.0 volt). Check oxygen sensor heater circuit.
2412	**Intake air temperature (IAT) sensor (G72)** – missing or incorrect signal from IAT sensor, IAT sensor wiring damaged/open or shorted, IAT sensor faulty.	Check the wiring and signal from the IAT sensor.
2234	**Battery voltage**–incorrect system voltage (below 10 volts or over 16 volts with engine running).	Check the charging system, battery connections, and main grounds.
2231	**Idle air control (IAC)** – self-compensating idle circuit problem (incorrect idle speed or idle mixture too lean or too rich).	Check the IAC valve (dirty, sticking, binding). Check for air leak between MAF sensor and intake manifold. Check EGR valve operation.
2141	**Knock sensor (KS) I (G61) Control** – faulty knock sensor control in Motronic ECM.	Check the knock sensor (see DTC 2142 above). If no faults can be found, the ECM is probably faulty and should be replaced.
2341	**Oxygen sensor control**– Incorrect oxygen sensor signal to ECM (fuel mixture too lean or to rich). Sensor heater circuit may be open or faulty.	Check the oxygen sensor heater circuit and sensor output signal first. **Lean**: check exhaust system for leaks between catalytic converter and exhaust manifold, check for intake air leaks (unmeasured air). Check fuel pressure (too low). **Rich**: check ignition system for misfire, check fuel pressure (too high), check for leaking or stuck open fuel injectors, check the Evaporative (EVAP) emission system (purge regulator valve stuck open).
2214	**Maximum engine speed limit reached**–engine over-revved (incorrect downshift) or signal from engine speed signal incorrect	Check the engine speed sensor wiring (shielding) for faults.
2314	**Engine-to-transmission electrical connection**–wire shorted to ground (cars with automatic transmission).	Check the wire between Motronic ECM (pin 18) and transmission control module (TCM) (pin 28).

* This code has no reference in any material supplied by the manufacturer. The publisher has listed it here as a result of field testing.

(continued on next page)

Table a. Motronic Diagnostic Trouble Codes (Blink Codes) 1993–1995 model years (continued)

DTC	Probable fault	Corrective action, first check
2243	**Fuel consumption signal**– signal wire from ECM to instrument cluster shorted to battery positive (B+).	Check wire between ECM (pin 51) and instrument cluster (connector T1e).
2324	**Mass air flow (MAF) sensor (G70)**–MAF sensor faulty, missing or incorrect signal, MAF sensor wiring damaged/open or shorted. Large intake air leak.	Check the MAF sensor connections and wiring. Check for intake air leaks. On VR6 engines, check for sensor burn-off. On 4-cylinder engines, check for excessive oil deposits on sensor element.
2413	**Mixture control adjustment**–fuel mixture too lean or too rich.	Check for conditions causing rich or lean running conditions (see DTC 2342 and 2231, 2341). Check the oxygen sensor heater circuit. Check the EVAP canister purge regulator valve (always open).
2411	**EGR temperature sensor (G98)**–faulty sensor or sensor wire shorted to ground.	Check the wiring and resistance of EGR temperature sensor.
2441*	**EGR system**–control malfunction	Check the EGR components.
3434	**Heated Oxygen Sensor Relay (J278)**–heater relay/circuit malfunctioning (6-cylinder engines).	Check the oxygen sensor heater circuit.
4311	**Secondary air injection solenoid relay (N299)**–faulty relay or wiring to relay damaged/open or shorted.	Check the wiring to relay. Check function of relay.
4313	**Secondary air injection solenoid valve (N112)**–faulty valve or wiring to valve damaged/open or shorted.	Check the wiring to valve. Check function of solenoid valve.
4332	**ECM final output stage**–faulty ECM or mechanical fault/ loose connections to ECM output components.	Check the wiring and test all output component functions (IAC valve, fuel injectors, EVAP canister purge regulator valve, ignition coil, fuel pump). If no faults can be found, the ECM is probably faulty and should be replaced.
4343	**EVAP canister purge regulator valve (N80)**–faulty valve or wiring to valve damaged/open or shorted.	Check the wiring to valve. Check function of solenoid valve.
4411	**Fuel injector no. 1 (N30)**–injector faulty or wiring to valve damaged/open or shorted.	Check the wiring to valve. Check for injector signal using LED test light or injector signal tester. If signal is present, injector is probably faulty and should be replaced.
4412	**Fuel injector no. 2 (N31)**–injector faulty or wiring to valve damaged/open or shorted.	See DTC 4411.
4413	**Fuel injector no. 3 (N32)**–injector faulty or wiring to valve damaged/open or shorted.	See DTC 4411.
4414	**Fuel injector no. 4 (N33)**–injector faulty or wiring to valve damaged/open or shorted.	See DTC 4411.
4421	**Fuel injector no. 5 (N83)**–injector faulty or wiring to valve damaged/open or shorted.	See DTC 4411.
4422	**Fuel injector no. 6 (N84)**–injector faulty or wiring to valve damaged/open or shorted.	See DTC 4411.
4431	**Idle air control (IAC) Valve (N71)**– faulty IAC valve or wiring to valve damaged/open or shorted.	Check the IAC valve (dirty, sticking, binding). Check for air leak between MAF sensor and intake manifold. Check EGR valve operation.
4433	**Fuel pump relay (J17)**–relay not operating or wiring shorted to battery positive (B+).	Check the relay wiring. Test relay circuit. see **20 Fuel Storage and Supply**.
4312	**EGR vacuum regulator solenoid valve (N18)**–faulty valve or wiring to valve damaged/open or shorted.	Check the wiring to valve. Check function of solenoid valve.
1111	**Motronic engine control module (ECM) (J220)**– ECM faulty (electronic component failure in ECM).	If no other faults can be found, the ECM is probably faulty and should be replaced.

* This code has no reference in any material supplied by the manufacturer. The publisher has listed it here as a result of field testing.

To erase DTC memory

1. Connect appropriate jumper or adapter to the DLC as previously described.
2. Switch ignition on, but do not start engine.

   NOTE —
   Fuel pump will switch on without engine running.

3. Leave jumper or adapter connected for at least 5 seconds and then disconnect.
4. Reconnect jumper or adapter for 2.5 seconds and then disconnect.
5. The first blink code will be displayed.
6. Repeat step 4 to display all blink codes stored in memory until 4 long blinks (0-0-0-0) are displayed.

   NOTE —
   All blink codes must be displayed before they can be erased.

7. Reconnect jumper wire or adapter for 2.5 seconds and remove it. The DTC memory is now erased.
8. Switch the ignition off.

   NOTE —
   Confirm that the memory has been erased by displaying blink codes. Only blink code 2-1-1-1 should be displayed.

FUEL PRESSURE TESTS (1993–1999 models)

Checking the fuel pressure is a fundamental part of troubleshooting and diagnosing the Motronic system. Fuel pressure has a direct effect on fuel mixture and driveability. An accurate fuel pressure gauge with a range of at least 0 to 5 bar (approximately 0 to 100 psi) will be needed to make the tests. Fig. 7 shows the connection of the fuel gauge.

There are two significant fuel pressure values: 1) System pressure—created by the main fuel pump and regulated by the fuel pressure regulator, and 2) Residual pressure—the pressure maintained in the closed system after the engine is shut off.

Before making the tests described below, make sure the fuel pump is operating correctly and that the fuel pump delivery rate is within the acceptable limit. See **20 Fuel Storage and Supply**.

> ***WARNING —***
> *Fuel will be discharged when disconnecting fuel lines. Do not smoke or work near heaters or other fire hazards. Have a fire extinguisher handy. Work only in a well- ventilated area. Relieve fuel pressure in the line (as described earlier) or wrap a cloth around the line before disconnecting.*

NOTE —
Some late model cars with the ABA (2.0 liter) engine are not equipped with a fuel rail test port. On these cars, disconnect the fuel supply line and connect the gauge (using a T-fitting) between the fuel rail and the supply line.

To check system line pressure

1. Relieve the fuel pressure in the system and connect a fuel pressure gauge as shown in Fig. 7

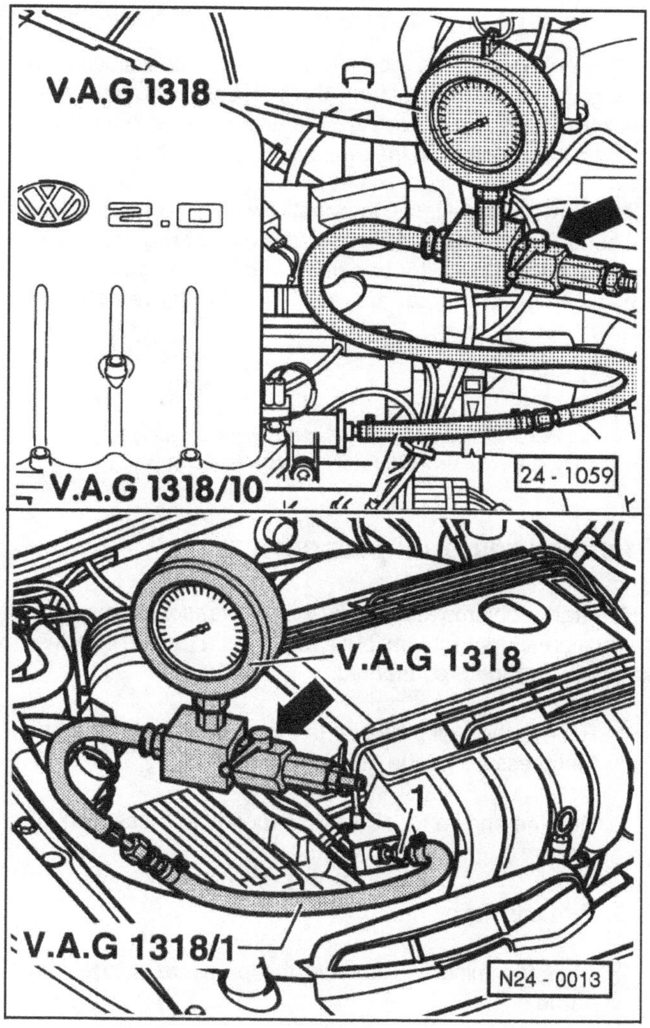

24-1059, N24-0013

Fig. 7. Fuel pressure gauge installation on ABA engine (**top**) and AAA engine (**bottom**). Remove the cap screw from the test port and connect the gauge line.

2. Start the engine and let it idle. See **Table b** for fuel pressure specifications.

NOTE —
If the engine does not start, run the fuel pump by bypassing the fuel pump relay as described in **20 Fuel Storage and Supply.**

3. With the engine idling, disconnect the small vacuum hose from the fuel pressure regulator (FPR) and check that the fuel pressure increases as specified.

Table b. Fuel System Line Pressure

Engine	Fuel Pressure
ABA engine (2.0 liter) vacuum hose @ FPR connected vacuum hose @ FPR disconnected	 2.5 bar (36 psi) 3.0 bar (43 psi)
AAA engine (2.8 liter) vacuum hose @ FPR connected vacuum hose @ FPR disconnected	 3.5 bar (51 psi) 4.0 bar (58 psi)

If system pressure is too high, first check for a blocked or damaged fuel return line from the pressure regulator. Also check the return line check valve on ABA engines. If there are no obstructions, replace the pressure regulator.

If system pressure is too low, first check for fuel leaks. Test fuel pump delivery as described in **20 Fuel Storage and Supply.** A restricted fuel filter or a damaged fuel line are also potential causes. If there are no other faults, replace the pressure regulator.

NOTE —
Additional fuel pump tests, including testing the fuel pump relay, are covered in **20 Fuel Storage and Supply.**

To check residual pressure

Hot-start problems are usually the only reason to suspect a residual pressure problem, but it deserves a routine check as long as the gauge is connected.

1. Relieve the fuel pressure in the system and connect a fuel pressure gauge as shown earlier in Fig. 7.

2. Run the engine for about 30 seconds to pressurize the system.

3. Turn the engine off.

4. After ten minutes, observe the pressure gauge. See **Table c**.

Table c. Residual Fuel Pressure (after ten minutes)

Engine	Fuel Pressure
ABA engine (2.0 liter)	2.0 bar (29 psi)
AAA engine (2.8 liter)	2.5 bar (36 psi)

If the pressure drops beyond specification, check for leaks in the fuel lines, at the fuel pump, and at the fuel injectors. If there are no leaks, either the pressure regulator or the fuel pump check valve is faulty.

To determine which, run the engine briefly to build pressure. Shut the engine off and immediately clamp shut the fuel return line at the pressure regulator. If pressure still drops, the fuel pump check valve is faulty. See **20 Fuel Storage and Supply** for fuel pump replacement. If residual pressure is now OK, the pressure regulator is faulty.

ELECTRICAL CHECKS AND COMPONENT TESTING (1993–1995 models)

The electrical tests under this heading will help to locate electrical component failure. Unless a specific diagnostic trouble code (DTC) has been recorded, make the ECM pin-out test first as listed in **Table d**. The checks listed in the table check the components and wiring to the components.

The ECM is mounted in the rear of the engine compartment. With the ignition off for at least a minute, unplug the connector from the ECM. See Fig. 8. Terminals are marked on the harness connector.

CAUTION —
- *Use a high-quality digital automotive multimeter to make the tests. An analog (swing-needle) meter should not be used as it can permanently damage electronic components.*
- *Always connect and disconnect the ECM connector and meter probes with the ignition off to avoid damage to electronic components. Wait at least 40 seconds after turning off the ignition before removing the ECM connector.*

References to Motronic components and their connector terminals are used throughout the table. Use the appropriate wiring diagram to identify these terminals. See **97 Wiring Diagrams, Fuses and Relays**.

When making tests at harness connectors, probe from the rear of the connector only to avoid damaging the small connector terminals. Separate the connector housing and remove any seals to gain access to the wiring in the connector.

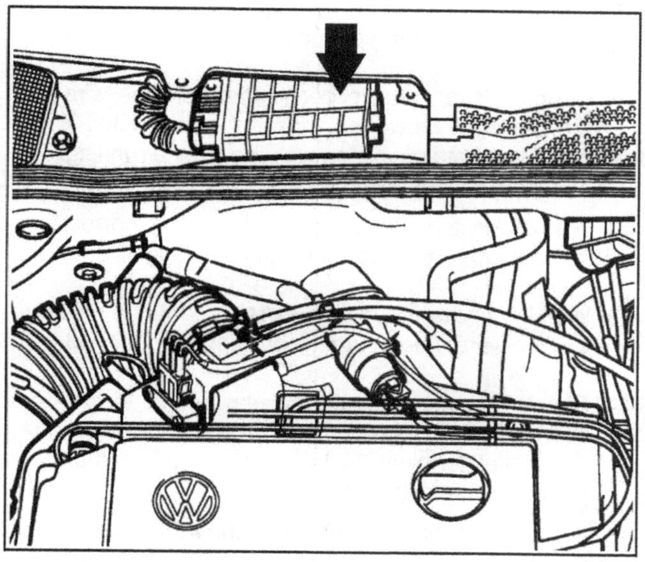

0024029

Fig. 8. Motronic ECM in rear of engine compartment (**arrow**). Separate harness connector and make connections from rear of connector.

Before making any electrical checks or tests, make sure the following conditions are met:

Basic Requirements

- Battery positive (+) voltage OK
- Fuse no. 18 (and no. 21 on AAA engines) OK
- Ground connections on engine and transmission OK, especially the connections at the ground bus below the intake manifold
- Fuel pump and fuel pump relay OK. See **20 Fuel Supply and Storage.**

Power Supply Components

The Motronic engine management system relies on three relays for proper operation; the power supply relay, the oxygen sensor heater relay, and the fuel pump relay. See Fig. 9.

All three relays are energized (grounded) by the Motronic ECM. The power supply relay supplies power to the ECM and many of the Motronic components when the engine is running. If this relay is faulty, the engine will not start. The oxygen sensor heater relay supplies positive (+) voltage to the oxygen sensor heating element to heat the sensor. The fuel pump relay supplies power to the fuel pump. A basic check of these relays is given in **Table d**.

NOTE —

Fuel pump electrical tests are covered in **20 Fuel Storage and Supply**.

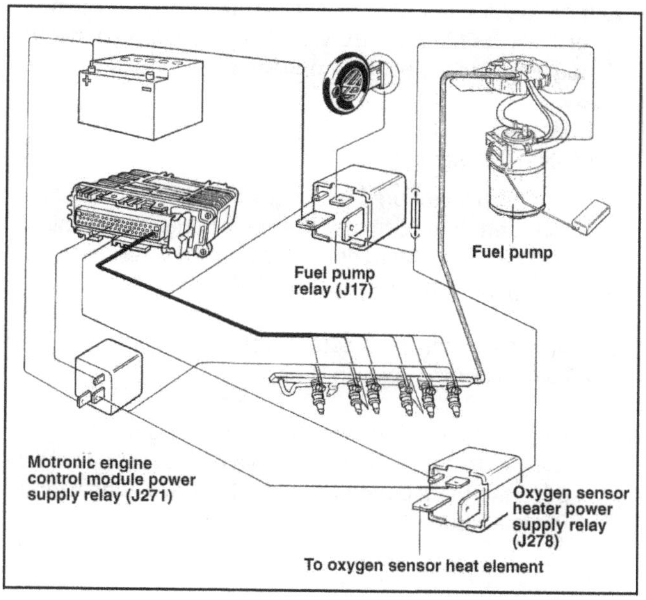

0024030

Fig. 9. Power supply components.

To test power supply relay (J16 or J271)

1. Working at the fuse/relay panel, remove the power supply relay. See Fig. 10.

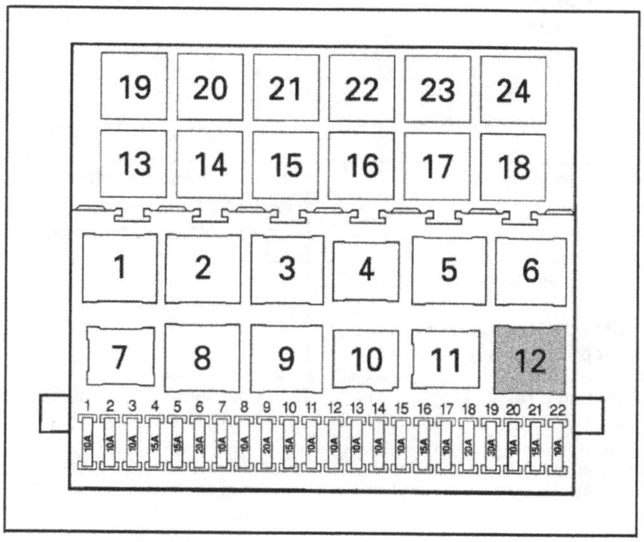

0024031

Fig. 10. Motronic power supply relay in fuse/relay panel (**position 12**).

2. Check for constant (+) battery voltage at relay terminal 30 in the fuse/relay panel.

3. Reinstall the relay and check that the relay clicks each time the ignition key is turned on.

4. If no faults are found, disconnect the harness connector from the Motronic ECM and separate the connector housing. Reconnect the connector to the ECM.

5. Check for ground at ECM terminal 9 each time the ignition is switched on.
   - If ground is not present, check the wiring between relay terminal 85 in the fuse/relay panel and terminal 9 in the ECM connector (black/brown wire). If no wiring faults are found, the ECM is most likely faulty.

6. Check for positive (+) battery voltage at ECM terminal 23 whenever the ignition is on.
   - If power is not present, check the wiring between relay terminal 87 in the fuse/relay panel and terminal 23 in the ECM connector (red/blue wire). If no wiring faults are found, the relay is most likely faulty.

To test oxygen sensor heater relay (J208 or J278)

Be sure both the main relay and the fuel pump relay are functioning correctly before testing the oxygen sensor heater relay.

> ***NOTE —***
> *Fuel pump electrical tests are covered in* **20 Fuel Storage and Supply**.

1. Working at the back of the engine, disconnect the oxygen sensor harness connector.

2. Connect a voltmeter between the red/white wire and the brown or brown/red wire in the connector. There should be battery voltage with the engine running. Reconnect the connector.
   - If voltage is present, the relay is working correctly.
   - If voltage is not present continue testing.

3. Working above the main fuse/relay panel, remove the oxygen sensor heater relay (position 13). See Fig. 10.

4. Check for battery voltage at relay socket terminal 85 with the ignition key on. Check for battery voltage at relay socket terminal 30 (red/yellow wire) in the relay panel with the engine running.
   - If voltage is not present, check the wiring to the relay sockets. Voltage to terminal 85 is supplied via the power supply relay and voltage to terminal 30 is supplied via the fuel pump relay.

5. Reinstall the relay. Then working through the rear of the relay, check for ground at relay terminal 31 (white wire) with the engine running.
   - If ground is not present, check the wire between the relay and the Motronic ECM (terminal 28). If no wiring faults can be found, the ECM may be faulty.
   - If no electrical faults are found up to this point, the relay is probably faulty and should be replaced.

Table d. Motronic ECM Electrical Tests

Component or circuit	Test terminals (at disconnected ECM connector unless otherwise noted)	Test conditions	Test value
ECM power supply (constant)	1 (-) and 54 (+)	—	Approx. battery voltage
ECM power supply (switched, from power supply relay)	1 (-) and 23 (+)	Jumper terminals 9 and 55 at ECM connector	Approx. battery voltage (power supply relay will click)
Fuel pump relay (J17)	Jumper terminals 6 and 55	Turn ignition on	Fuel pump runs (fuel pump relay will click)
Heated oxygen sensor relay (J208 or J278)	**4-cylinder engine only** Jumper terminals 1 and 28 briefly	Jumper terminals 9 and 55 Turn ignition on	Heated oxygen sensor relay closes/opens (clicks) each time wire is connected
Malfunction indicator light (MIL) (K83)	Jumper terminals 5 and 55	Turn ignition on	MIL lights up
Vehicle speed sensor (VSS) (G22)	**4-cylinder engine** 1 and 65	Ignition on Raise and spin front left wheel Cars with **A/T**: selector lever in "D" Cars with **manual transmission**: shift lever in neutral	0-4 volts (VAC), fluctuating
	6-cylinder engine 56 and 65		0-10 volts (VAC), fluctuating

(continued on next page)

Table d. Motronic ECM Electrical Tests (continued)

Component or circuit	Test terminals (at disconnected ECM connector unless otherwise noted)	Test conditions	Test value
Main grounds (GND)	1 and 54	—	Approx. battery voltage
	10 and 54	4-cylinder engine: NY and Calif. cars only 6-cylinder engine: all models	Approx. battery voltage
	33 and 54	—	Approx. battery voltage
	55 and 54	—	Approx. battery voltage
	56 and 54	—	Approx. battery voltage
	with A/T only 58 and 54	—	Approx. battery voltage
	with man. trans. only 7(-) and 54	—	Approx. battery voltage
Park/neutral position relay wiring (A/T)	**4-cylinder engine** 7 and 54	Selector lever in Park or Neutral Ignition on	Approx. battery voltage
	6-cylinder engine 1 and 7	Selector lever in Park. Disconnect connector from ignition coil and operate starter	Approx. 2 volts less than battery voltage
A/C signal On signal	55 (-) and 39 (+)	Turn ignition on Turn fresh air blower on Turn A/C on	Approx. battery voltage
A/C compressor	Jumper terminals 37 and 38 briefly	Ignition on	A/C compressor clutch clicks each time wire is connected
Fuel injector coil resistance **4-cylinder engine** no. 1 (N30) no. 2 (N31) no. 3 (N32) no. 4 (N33)	 24 and 23 2 and 23 25 and 23 26 and 23	 — — — —	 14-21.5 ohms 14-21.5 ohms 14-21.5 ohms 14-21.5 ohms
6-cylinder engine no. 1 (N30) no. 2 (N31) no. 3 (N32) no. 4 (N33) no. 5 (N83) no. 6 (N84)	 23 and 24 3 and 23 23 and 26 4 and 23 23 and 25 2 and 23	 — — — — — —	 14-21.5 ohms 14-21.5 ohms 14-21.5 ohms 14-21.5 ohms 14-21.5 ohms 14-21.5 ohms
Evaporative (EVAP) canister purge regulator valve (N80)	23 and 31	—	40-80 ohms
Engine coolant temperature (ECT) sensor (G62)	14 and 33	—	Resistance varies with temperature. See **Engine Coolant Temperature (ECT) Sensor (G62)**
Data link connector (DLC)–**1993 and early 1994** (See Fig. 6 shown earlier)	21 of ECM and terminal 1 of white data link connector (DLC)	Ignition off	Maximum 1.5 ohms
	43 of ECM and terminal 2 of white data link connector (DLC)	Ignition off	Maximum 1.5 ohms
Data link connector (DLC)–**later 1994 and 1995** (See Fig. 6 shown earlier)	21 of ECM and terminal 4 of 16-pin DLC (L-wire)	Ignition off	Maximum 1.5 ohms
	43 of ECM and terminal 1 of 16-pin DLC (K-wire)	Ignition off	Maximum 1.5 ohms

(continued on next page)

Table d. Motronic ECM Electrical Tests (continued)

Component or circuit	Test terminals (at disconnected ECM connector unless otherwise noted)	Test conditions	Test value
Wiring to heated oxygen sensor (G39)	20 and 42	Disconnect oxygen sensor harness connector (rear of engine) and jumper terminals 3 and 4	Maximum 1.5 ohms
		Reconnect connector	Infinite ohms
Throttle position (TP) sensor (G69) wiring	33 and 41	—	4-cylinder: 1.6 to 2.4 k ohms 6-cylinder: 1.5 to 2.5 k ohms
	33 and 40	Throttle closed	4-cylinder: 1.0 to 2.0 k ohms 6-cylinder: 0.7 to 1.3 k ohms
		Open throttle slowly	Resistance increases steadily
	40 and 41	Throttle closed	4-cylinder: 2.5 to 4.0 k ohms 6-cylinder: 2.5 to 3.5 k ohms
		Open throttle slowly	Resistance decreases steadily
Camshaft position sensor (CMP) (G40) wiring	44 and 56	Disconnect harness connector from camshaft position sensor and jumper terminals 1 (white/red wire) and 2 (brown/black or brown/yellow) in connector	Maximum 1.5 ohms
4-cylinder engine Ignition coil power output stage (N157) wiring		Disconnect harness connector from ignition coil output stage	
	8 and 55	Jumper terminals 1 and 2 in connector	Maximum 1.5 ohms
	8 and 38	Jumper terminals 2 and 3 in connector	Maximum 1.5 ohms
6-cylinder engine Ignition coil (N152) wiring		Disconnect harness connector from ignition coil	
	8 and 55	Jumper terminals 1 and 2 in connector	Maximum 1.5 ohms
	60 and 55	Jumper terminals 1 and 3 in connector	Maximum 1.5 ohms
	52 and 55	Jumper terminals 1 and 4 in connector	Maximum 1.5 ohms
	38 and 55	Jumper terminals 1 and 5 in connector	Maximum 1.5 ohms
Knock sensor I (G61) wiring		Disconnect harness connector to knock sensor I (at rear of engine)	
	33 and 34	Jumper terminals 1 and 2 in connector	Maximum 1.5 ohms
	34 and 56	Jumper terminals 1 and 3 in connector	Maximum 1.5 ohms
Knock sensor II (G66) wiring		Disconnect harness connector to knock sensor II (near starter)	
	33 and 56	Jumper terminals 2 and 3 in connector	Maximum 1.5 ohms
	57 and 56	Jumper terminals 1 and 3 in connector	Maximum 1.5 ohms
Intake air temperature (IAT) sensor (G72)	33 and 36	—	Resistance varies depending on temperature. See **To test intake air temperature sensor**
Idle air control (IAC) valve (N71)	27 and 53	—	4-cylinder: 7 to 10 ohms 6-cylinder: 6 to 12 ohms

(continued on next page)

Table d. Motronic ECM Electrical Tests (continued)

Component or circuit	Test terminals (at disconnected ECM connector unless otherwise noted)	Test conditions	Test value
4-cylinder engine Mass air flow (MAF) sensor (G70) wiring		Disconnect harness connector from MAF sensor	
	1 and 16 1 and 23 1 and 17	Jumper terminals 1 and 2 in connector Jumper terminals 1 and 3 in connector Jumper terminals 1 and 4 in connector	Maximum 1.5 ohms Maximum 1.5 ohms Maximum 1.5 ohms
6-cylinder engine Mass air flow (MAF) sensor (G70) wiring		Disconnect harness connector from MAF sensor	
with 4-pin connector	16 and 17	Jumper terminals 2 and 4 in connector	Maximum 1.5 ohms
with 6-pin connector	1 and 16	Jumper terminals 1 and 2 in connector	Maximum 1.5 ohms
	1 and 17	Jumper terminals 1 and 3 in connector	
	56 and 59	Jumper terminals 1 and 4 in connector	
	1 and 23	Jumper terminals 1 and 5 in connector	
Engine speed/reference sensor (G28)	67 and 68	—	500-700 ohms
Cars with EGR EGR temperature sensor (G98)	15 and 33	Disconnect harness connector from EGR solenoid valve (N18) and jumper terminals 1 and 2 in connector	Maximum 1.5 ohms
Cars with EGR EGR vacuum regulator solenoid valve (N18)	23 and 30	—	27-31 ohm
	23 and 30	Disconnect harness connector from solenoid valve and jumper terminals 1 and 2 in connector	Maximum 1.5 ohms
Secondary air injection (AIR) pump relay (J299)	Jumper terminals 9 and 55	Jumper terminals 1 and 49 briefly	AIR pump relay clicks each time wire is connected
Secondary air injection (AIR) pump motor (V101)	Jumper terminals 9 and 55	Jumper terminals 1 and 49 briefly	AIR pump motor runs each time wire is connected
Secondary air injection (AIR) solenoid valve (N112)	Jumper terminals 9 and 55 Disconnect harness connector from pump motor	Jumper terminals 1 and 50 briefly	AIR solenoid valve clicks each time wire is connected

Engine Coolant Temperature (ECT) Sensor (G62)

The ECT sensor is one of the main inputs to the engine management system. Its signal is used to adapt ignition timing, injection time, and idle speed stabilization. In addition, the ECT signal is used to activate the knock control system, the idle speed control system, oxygen sensor operation, fuel tank venting, and exhaust gas recirculation (EGR).

The ECT sensor sends continuous engine temperature information to the Motronic ECM. If there is a break in the signal from the temperature sensor, the ECU will simulate a fixed signal based on an engine temperature of 176 °F (80°C).

On 4-cylinder engines, the ECT sensor (G62) is mounted in the coolant outlet on the rear of the cylinder head. The ECT sensor is combined with the temperature gauge sensor (G2) in a single housing. The sensor body is black and the sensor has four terminals. See Fig. 11.

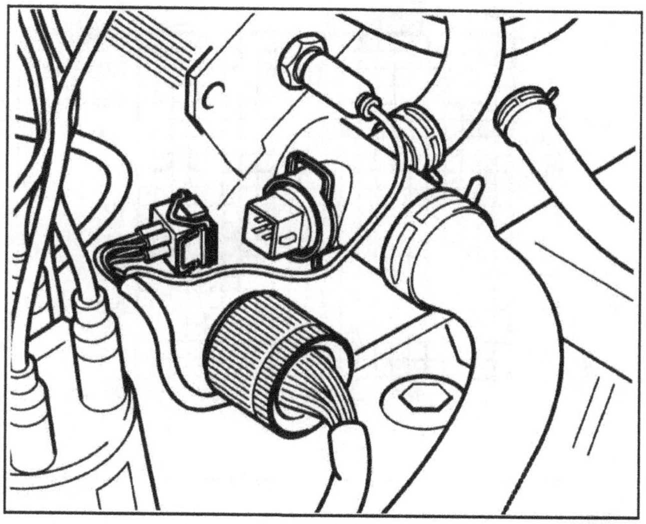

0024043

Fig. 11. ECT sensor on 4-cylinder engine. Check sensor resistance at terminals 1 (blue wire) and 3 (brown/green wire).

On 6-cylinder engines, the ECT sensor is mounted in the thermostat housing. The sensor body is blue and the sensor has two terminals. See Fig 12.

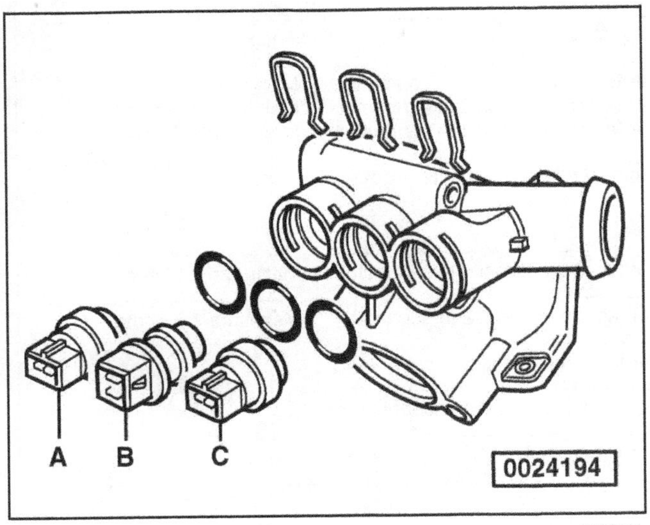

0024194

Fig. 12. Coolant sensor arrangement on VR-6 engines. ECT sensor (G62) (**A**-blue), coolant gauge sensor (G2) with after-run fan switch (F87) (**B**-yellow), and A/C thermal cut-out switch (F163) with third speed coolant fan switch (F165) (**C**-brown).

To test the ECT sensor, disconnect the harness connector and connect an ohmmeter across the sensor terminals with the engine cold. Start the engine and allow it to warm up. Compare the meter readings to the graph shown in Fig. 13. If any faults are found, the sensor should be replaced.

> ***NOTE —***
> *On 4-cylinder engines, test across terminals 1 (blue wire) and 3 (brown/green wire) in the sensor.*

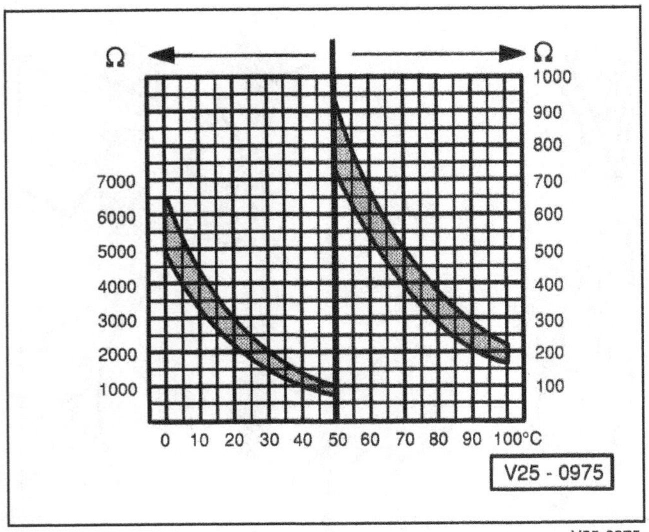

V25-0975

Fig. 13. Graph showing coolant temperature sensor resistance vs. temperature.

Throttle Position (TP) Sensor (G69)

The TP sensor is mounted to the side of the throttle valve housing, directly connected to the throttle valve shaft. See Fig. 14. The ECM sends a 5-volt reference signal into the TP sensor. Depending on throttle position, the output voltage varies in relation to the throttle position.

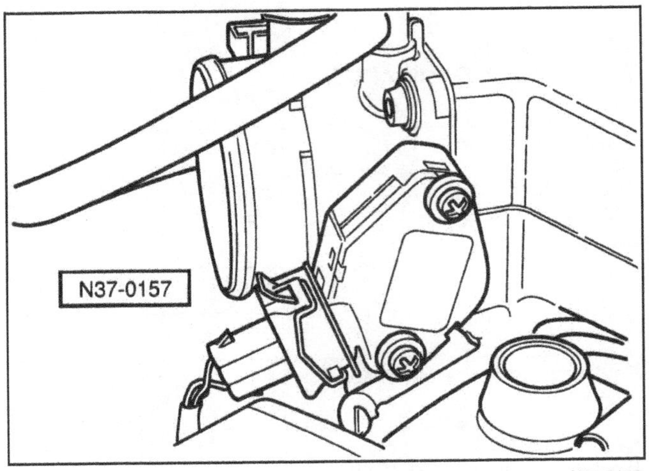

N37-0157

Fig. 14. Throttle position sensor mounted to throttle valve housing.

Test the TP sensor using the information given in **Table d.** If the test results are incorrect, the sensor should be replaced. The sensor is not adjustable. Use a new O-ring at throttle housing when replacing the sensor.

Check the reference voltage between terminals 1 and 3 in the 3-pin harness connector. There should be approximately 5 volts with the ignition on. If any faults are found, check the wiring between the sensor and the ECM. If no wiring faults can be found, the ECM may be faulty.

> ***CAUTION —***
> *The signal from the throttle position (TP) sensor is also used by the automatic transmission control module (TCM). Volkswagen specifies that if the Motronic ECM, the TCM, the TP sensor or the throttle valve housing are replaced, the ECM and TCM "Basic Settings" should be "initiated" using the special V.A.G. 1551 scan tool. If these basic settings are not done, transmission shifting and driveability problems may be encountered.*

Mass Air Flow (MAF) Sensor (G70)

The mass air flow (MAF) sensor provides the engine load signal to the ECM. The MAF sensor is located at the air filter housing and electronically measures the air entering the engine.

When the engine is running, an electrical current is used to heat a thin wire or film in the sensor. The current flowing through the wire/film is regulated to maintain a constant temperature above that of the intake air temperature. The amount of current required to heat the wire is the basis of the engine load signal. The MAF sensor has no internal moving parts and cannot be serviced, adjusted or accurately tested. See Fig. 15.

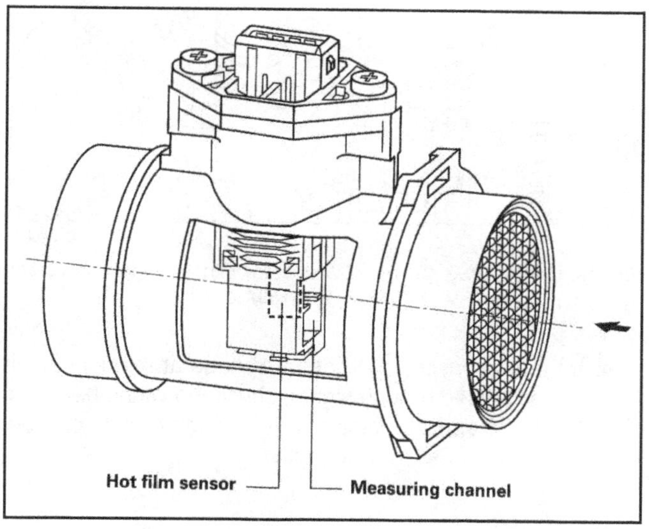

0024032

Fig. 15. Hot-film mass air flow (MAF) sensor used on 4- and 6-cylinder engines.

If there is no signal from the mass air flow (MAF) sensor, the ECM automatically switches to a "limp-home" mode, turns on the Check Engine light, and stores a DTC fault in permanent memory. The engine can still be started and driven in the "limp-home" mode.

To check mass air flow (MAF) sensor voltage and ground

1. Disconnect the harness connector from MAF sensor.

2. Turn the ignition key on and check for voltage between terminals 1 (-) and terminal 3 (+) in the harness connector. See Fig. 16 for terminal identification.
   - If any faults are found, check the wiring to the MAF sensor using the appropriate the diagram found in **97 Wiring Diagrams, Fuses, and Relays**.

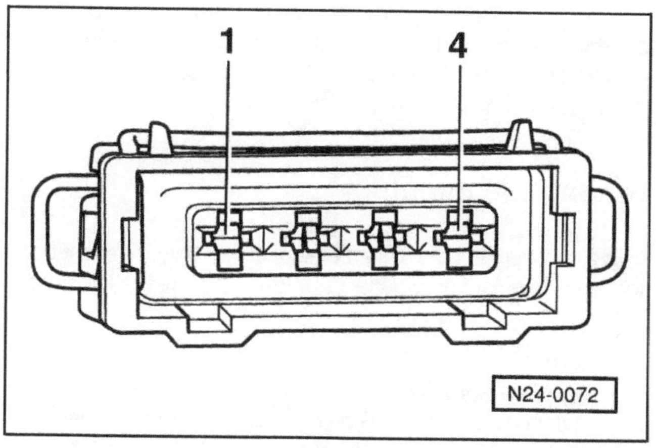

N24-0072

Fig. 16. Mass air flow (MAF) sensor harness connector terminal identification.

> ***NOTE —***
> *On 4-cylinder engines, check for oil deposits on the sensor element. Over time oil vapors can collect on the sensor, owing to its mounted position. If the sensor is oil contaminated and a MAF sensor DTC has been stored, the sensor is most likely faulty and should be replaced.*

Fuel Injectors

The fuel injectors are electrically operated solenoid valves. The injectors are connected to a common fuel supply, called the fuel rail. See Fig. 17.

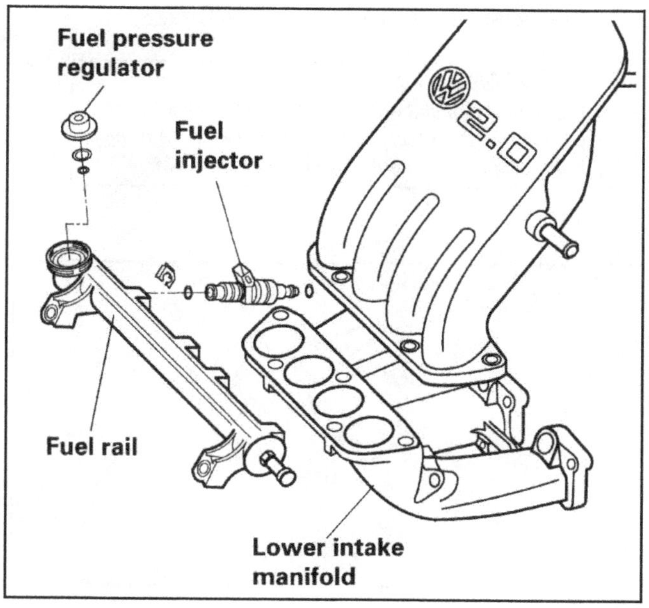

0024034

Fig. 17. Fuel rail is common fuel supply and mounting for fuel injectors.

The injectors are supplied 12 volts by the power supply relay and the ground side of the injectors are sequentially switched on and off by the ECM. Injection quantity is based on the length of time the injectors are open.

To test fuel injectors

The information below is an electrical test only. The test does not check injector spray patterns or fuel supply to the injectors.

> ***NOTE —***
> *If you think an injector is bad, place a screwdriver or an automotive stethoscope on the injector with the engine running. The injector should be buzzing. If not, check for voltage to the injector and check its resistance as described below. If the injector is buzzing, very briefly unplug the harness connector from the injector with the engine running. If the injector is working correctly, there should be a noticeable, but small, rpm drop.*

1. Disconnect a harness connector from an injector and check the resistance across the injector terminals. Replace the injector if the resistance is incorrect.

Injectors	
• resistance	15 to 21.5 ohms

2. Connect an LED test light, an injector tester, or a digital voltmeter across the connector terminals. See Fig. 18. When the engine is cranked, the LED should flash or the meter should indicate some voltage.

> ***CAUTION —***
> *Use only a digital voltmeter, an LED test light or an LED injector tester. Use of an analog VOM or incandescent test light may damage the ECM.*

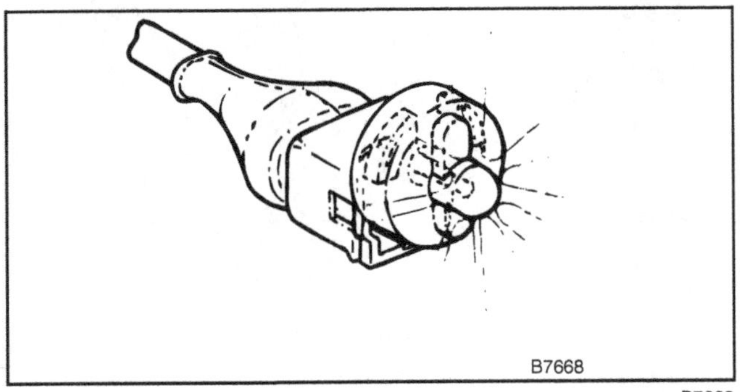

B7668

Fig. 18. Special LED fuel injector tester connected to injector harness.

3. If there is no signal at the connector, check for power to the injector. There should be battery voltage (+) at the red/blue wire of each injector connector with the ignition on. See Fig. 19. If not, check the wiring to the injector.

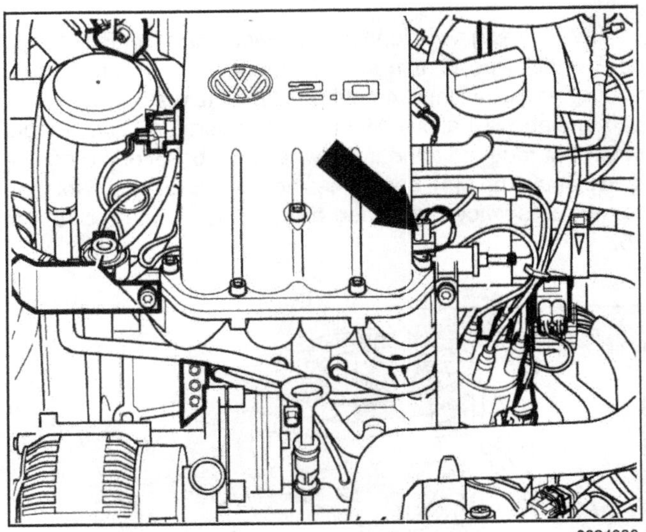

0024036

Fig. 19. Check power to injector connector with ignition on. Check at red-blue wire of connector (**arrow**).

4. If there is positive (+) battery voltage at the connector but no pulsed ground signal, check the wiring between the ECM and the injectors. If no wiring faults can be found, the ECM may be faulty.

Oxygen Sensor (G39)

The heated oxygen sensor monitors the exhaust gas and provides the Motronic ECM with feedback about the air-fuel ratio and combustion efficiency. See Fig. 20. Using this information, the ECM continuously adjusts the air-fuel mixture to ensure optimum driveability and exhaust emissions. The sensor must be at a temperature of at least 600°F to generate voltage, so it is electrically heated to help it reach operating temperature more quickly.

The oxygen sensor produces a small voltage (0-1 volt) based on the oxygen content in the exhaust gas as compared to the oxygen outside the exhaust pipe. The bigger the differential, the greater the output voltage. When the mixture is rich, there is very little oxygen in the exhaust (high output voltage). When the mixture is lean, there is an excess of oxygen in the exhaust (low output voltage).

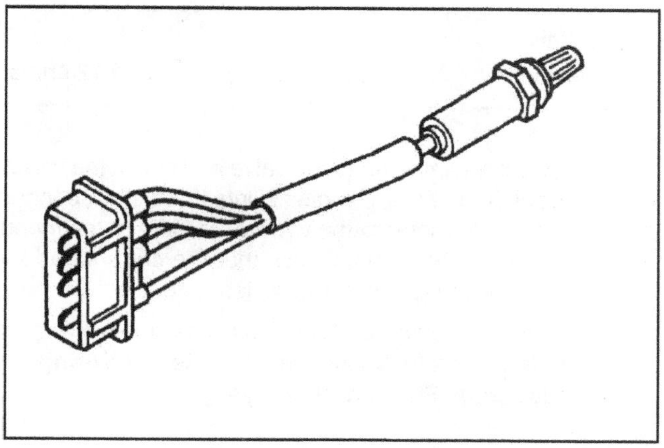

Fig. 20. Heated oxygen sensor monitors exhaust gas.

To test oxygen sensor

1. Working at the back side of the engine, locate the 4-pin oxygen sensor harness connector.

> ***WARNING —***
> *Use care when working near a hot engine. The exhaust pipes and the engine surfaces can cause severe burns.*

2. Without disconnecting the harness connector, connect the positive lead of a voltmeter to terminal 4 (signal wire) of the connector. See Fig. 21. Connect the negative lead to chassis ground.

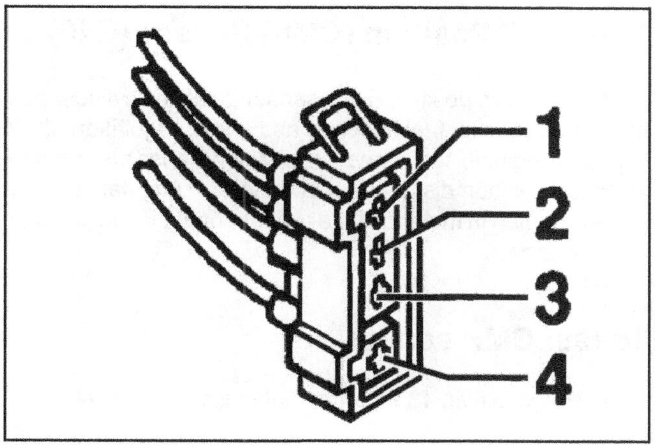

Fig. 21. Oxygen sensor harness connector terminal identification.

3. Start the car and let it idle. The oxygen sensor should start to produce a fluctuating voltage within a short period.

Oxygen sensor output signal
- voltage at idle . . . . 0.3 to 1.0 volt, fluctuating

NOTE —
To further check sensor response to lean and rich mixtures, briefly disconnect an injector connector or pinch off the vacuum hose fuel pressure regulator and observe the sensor response.

4. If the sensor output is incorrect, turn the engine off and check the sensor heater circuit. Disconnect the four-wire connector and check for battery voltage at terminals 1 and 2 with the engine running.
   - If voltage is not present, check the wiring to the heater.
   - If voltage is present, check the heater coil resistance across terminals 1 and 2 of the sensor side of the connector. If any faults are found, replace the sensor.

Oxygen sensor
- heater resistance . . . . approx. 2 ohms

5. If the oxygen sensor signal is not as specified, and the heater circuit is OK, the sensor may be faulty.

NOTE —
If the sensor signal is either very low or very high and fluctuating slightly, the Motronic adaptability limits may have been reached. Look for signs of rich (low output signal) or lean (high output signal) running characteristics before replacing the oxygen sensor.

Tightening Torque
- Oxygen sensor to exhaust pipe . . . . . 50 Nm (37 ft-lb)

NOTE —
Coat the oxygen sensor threads with an anti-seize compound before installation. Do not get the compound on the sensor tip. This can damage the sensor.

Idle Air Control (IAC) System

The Motronic ECM electronically controls the idle speed through the IAC valve (N71). See Fig. 22. The system automatically compensates for engine load and engine operating conditions, so periodic adjustment of the idle speed is unnecessary.

The Motronic ECM monitors engine rpm and other operating conditions such as air conditioning and automatic transmission loads, and then sends an electrical signal to open or close the valve. This changes the amount of air that bypasses the closed throttle valve.

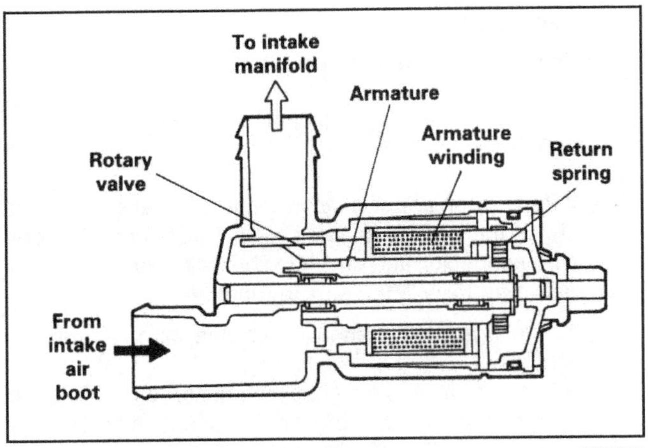

Fig. 22. Cutaway of IAC valve.

CAUTION —
Do not attempt to correct a faulty the idle speed by adjusting the position of the throttle valve stop screw. If the factory setting of this screw is altered, the complete throttle housing will need to be replaced—no adjustment specifications are available from Volkswagen. See **Component Replacement**.

The IAC function is completely adaptive. The ECM remembers how far the valve was open the last time the engine was at idle and reverts to this as a base setting. In addition, the valve has a limp-home feature that maintains a fixed idle speed in the event of an IAC valve failure.

NOTE —
On high mileage cars, enough residue may collect on the throttle valve plate and bore to restrict air flow and affect the IAC system. If heavy residue is found, clean both the IAC valve and the throttle plate and bore before diagnosing IAC problems.

Before proceeding with troubleshooting an idle problem, check that the throttle position sensor is working correctly as described earlier under **Throttle Position Sensor**.

To test IAC system

1. With the engine running, grasp the IAC valve and check that it is buzzing.
2. If not, stop the engine and disconnect the harness connector from the valve. Check the resistance of the valve across its terminals. If resistance is incorrect, replace the valve.

NOTE —
If you suspect an intermittent fault, lightly tap the valve while you are testing resistance.

IAC Valve
- resistance . . . . 6 to 12 ohms

3. Connect an ammeter to the valve and check the valve's current draw with the engine idling. If available, connect a duty cycle meter to the valve according to the manufacturer's instructions. Check that the current and the duty cycle are within the specified ranges.
   - If there is no signal at the valve, check the wiring between the ECM and the valve. See **97 Wiring Diagrams, Fuses and Relays.**

IAC Valve
- current, engine idling . . . . .620 ± 20 mA
- duty cycle, engine idling 30%–50%

NOTE —
- *If the idle speed is high and the current draw is also high or if the idle speed is low and the current draw is low, the problem is most likely a faulty electrical input to the ECM, telling it to maintain an incorrect idle speed. (i.e. engine speed signal, throttle position, A/C-on signals, and A/T drive position).*
- *If the idle speed is low or normal but the current draw is high or if the idle speed is high or normal, but the current draw is low, the problem is most likely a mechanical problem, such as stuck/faulty IAC valve or a maladjusted throttle plate.*

Camshaft Position (CMP) Sensor (G40)

The camshaft position (Hall) sensor identifies cylinder position for sequential fuel injection and knock regulation. On 4-cylinder engines, the camshaft position sensor is mounted within the ignition distributor. On 6-cylinder engines, the sensor is mounted in the end of the cylinder head, in the camshaft sprocket cover.

To test CMP sensor

1. Remove fuse 18 from the main fuse/relay panel.
2. Disconnect the harness connector from the camshaft position sensor.
3. Connect a voltmeter between the two outer terminals in the connector and check for voltage with the ignition key on. If voltage is not present, check for wiring faults using the appropriate wiring diagram.
   - On 4-cylinder engines, there should be a minimum of 4-volts.
   - On 6-cylinder engines, there should be a minimum of 10-volts.

4. If voltage is present, peel back the protective rubber boot on the Hall sensor harness connector and then reconnect it to the sensor. Connect an LED test light between the center terminal and either of the outer terminals.

5. When the starter is actuated, the LED should flicker. If there is no reaction, the Hall sensor is defective and should be replaced. See **Component Replacement**.

NOTE —

On 4-cylinder engines, the Hall sensor is not separately available from Volkswagen. If the sensor is faulty, the complete distributor body will need to be replaced.

Engine Speed (RPM) Sensor (G28)

The engine speed (rpm) sensor supplies engine speed and crankshaft (cylinder no. 1 TDC) position signals to the Motronic ECM. The sensor is mounted on the intake side of the cylinder block and reads a toothed wheel mounted on the end of the crankshaft. See Fig. 23.

If this sensor is faulty, the engine will not start. To check for a faulty sensor, crank the engine for at least six seconds and then check for a stored fault code (DTC) in the On-board Diagnostic memory as described earlier under **Diagnostic Trouble Codes**.

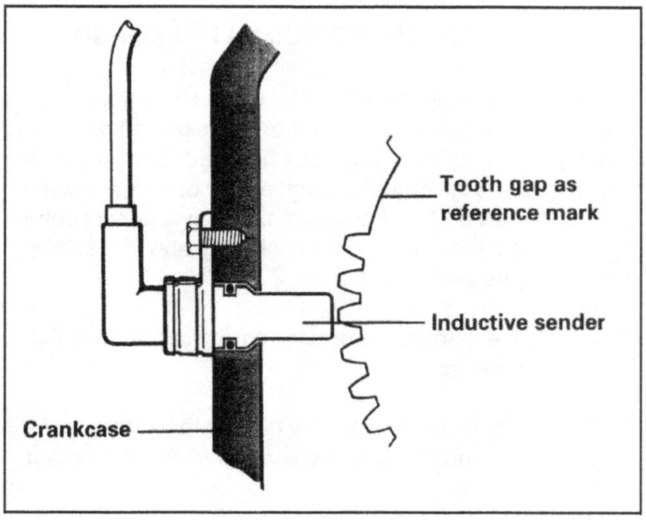

Fig. 23. Engine speed (RPM) sensor (G28) supplies engine speed and crankshaft position. TDC is signaled by two-tooth gap in wheel.

Carefully inspect the toothed wheel on the crankshaft for damage or deformed teeth. This can be accomplished through the sensor mounting hole or by removing the oil pan. There should be 58 teeth plus a 2-tooth gap on the wheel for both 4-cylinder and VR-6 engines. The toothed wheel can be replaced, but requires removal of the crankshaft.

Engine Speed (RPM) Sensor
- coil resistance. 500-700 ohms

Intake Air Temperature (IAT) Sensor (G72)

The intake air temperature (IAT) sensor sends continuous intake air temperature information to the Motronic ECM. This input is used primarily as a correction factor for ignition timing. If there is a break in the signal from the IAT sensor, the ECU will simulate a fixed signal based on an engine temperature of approximately 68°F (20°C).

To test the IAT sensor, disconnect the harness connector and connect an ohmmeter across the sensor terminals. See Fig. 24. Compare the meter reading to the graph shown earlier in Fig. 13. If any faults are found, the sensor should be replaced.

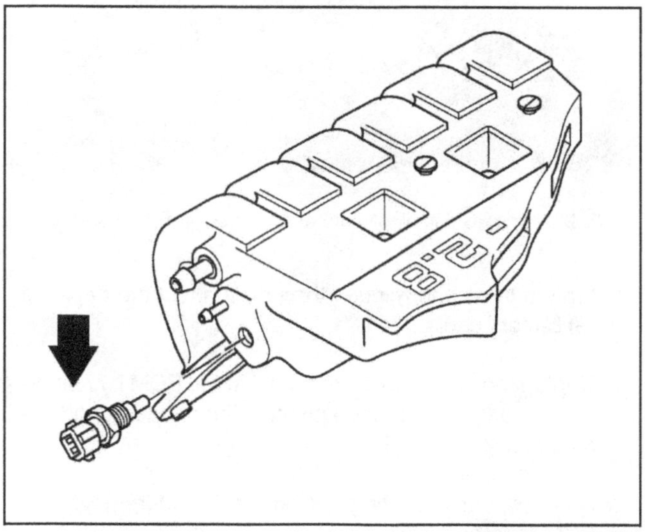

Fig. 24. Intake air temperature sensor (**arrow**) in intake manifold (6-cylinder engine shown).

COMPONENT REPLACEMENT (1993–1995 models)

To remove and install Motronic Engine Control Module (ECM)

The ECM is mounted in the rear of the engine compartment, in the right-hand side of the air plenum. See Fig. 25.

0024052

Fig. 25. Motronic ECM in air plenum at rear of engine compartment.

1. Wait at least one minute, then disconnect the negative (–) battery cable.

2. Unplug the harness connector from the ECM by releasing the fastener and then pivoting the connector off the module.

3. Remove the mounting bracket nut and remove the ECM together with its bracket.

4. Installation is the reverse of removal.

To remove Engine Coolant Temperature (ECT) sensor

For ECT sensor location on 6-cylinder, see Fig. 13 given earlier.

1. Drain approximately one gallon of coolant from the cooling system. This is to prevent coolant from leaking out when the sensor is removed. See **19 Engine–Cooling System**.

2. Disconnect the sensor harness connector.

3. Remove the sensor retaining clip and pull the sensor from its housing. See Fig. 26.

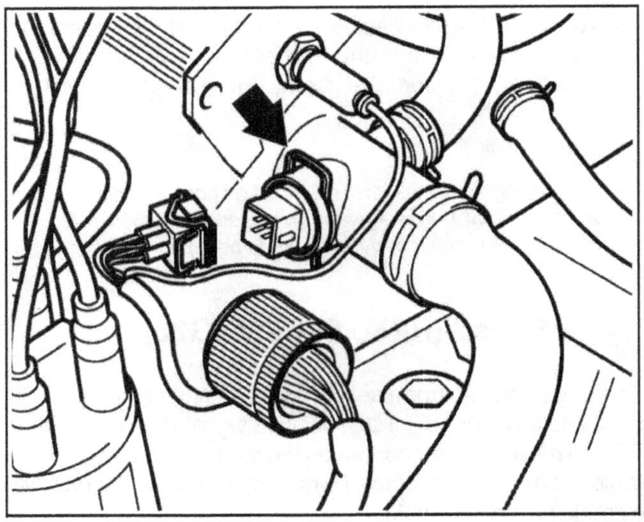

0024047

Fig. 26. ECT sensor retaining clip (**arrow**) on 4-cylinder engine.

Installation is the reverse of removal. Clean the sensor sealing surfaces and use a new O-ring when reinstalling the sensor.

To remove Throttle Position (TP) sensor

Volkswagen specifies that if the TP sensor is replaced, the engine control module (ECM) and transmission control module (TCM) "Basic Settings" should be "initiated" using the special V.A.G. 1551 scan tool. The throttle position signal is used by the ECM and the TCM. If these basic settings are not done using the scan tool, transmission shifting and driveability problems may be encountered.

1. Unplug the harness connector from the sensor. See Fig. 14 given earlier.

2. Remove the mounting screws holding the sensor to the side of the throttle valve housing. Remove the sensor and its O-ring.

Installation is the reverse of removal. Use a new O-ring and tighten the screws to the specified torque.

Tightening torque
- TP sensor to throttle valve housing . . . 3 Nm (27 in-lb)

To remove Mass Air Flow (MAF) sensor

1. Loosen the large hose clamp securing the rubber duct to the mass air flow (MAF) sensor. Remove the duct from the sensor.

2. Disconnect the harness connector from the MAF sensor.

3. On 4-cylinder engines, remove the two mounting screws holding the MAF sensor to the upper air cleaner housing.

4. On 6-cylinder engines, loosen the clamping band holding the MAF sensor to the upper air cleaner housing. See Fig. 27.

1. **Air cleaner housing, upper**
2. **Air cleaner element**
3. **Air cleaner housing, lower**
4. **O-ring**
5. **MAF sensor**
6. **Intake air duct**
7. **Fitting for idle stabilizer valve**
   - not applicable for OBD II

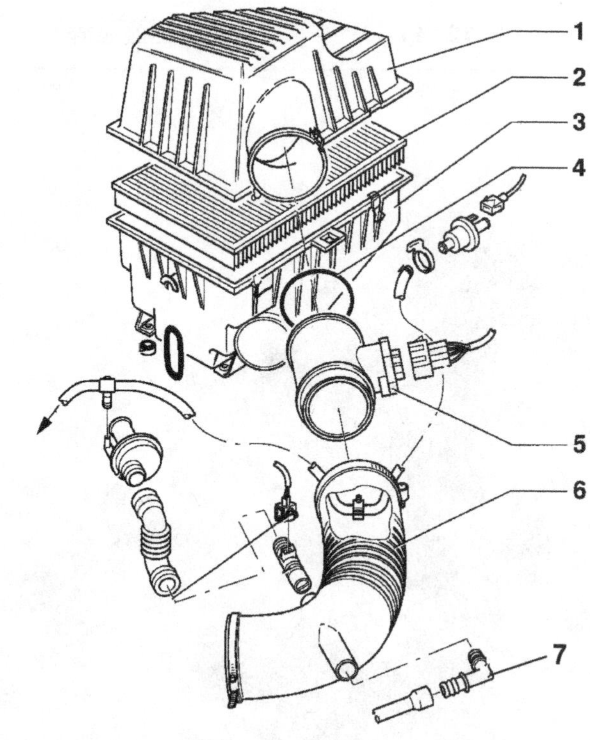

Fig. 27. Mass air flow (MAF) sensor on 6-cylinder engine.

5. Remove the mass air flow (MAF) sensor from the air cleaner housing.

Installation is the reverse of removal. Replace the O-ring if it is damaged.

Tightening torque
- MAF sensor to air cleaner housing (4-cylinder) . . . . 10 Nm (89 in-lb)

To remove fuel injectors

Cleanliness is important when working with fuel circuit components. Thoroughly clean the unions before disconnecting any fuel lines. Use clean tools.

> ***WARNING —***
> *Fuel will be expelled when fuel lines are disconnected. Do not smoke or work near heaters or other fire hazards. Keep a fire extinguisher handy.*

1. Remove the upper intake manifold and related components. Cover the open intake runners using clean shop rags. See **15a** or **15b Cylinder Head and Valvetrain**.

2. Unplug the wiring harness connectors from the injectors.

3. Disconnect the fuel supply and return hoses from the fuel rail. See Fig. 28.

> ***WARNING —***
> *Wrap a clean shop towel around the fuel line fittings before loosening them.*

NOTE —
The fuel supply line is marked with white and the return line is marked with blue.

4. Remove the bolts that hold the fuel rail to the intake manifold and pull the fuel rail away complete with injectors.

5. To remove the injectors from the fuel rail, pry off the retaining clips, then twist the injectors slightly and pull them off.

When installing the injectors, lightly lubricate the O-rings with engine oil. Install the injectors to the fuel rail but do not install the retaining clips. Position the fuel rail and injectors on the intake manifold. Make sure that injectors and the fuel rail are fully seated and then install the retaining clips and mounting bolts. Reconnect the fuel hoses to the fuel rail. Install the upper intake manifold using a new gasket.

Tightening torques
- Upper intake manifold to lower intake manifold
  4-cylinder engine. . . . . 20 Nm (15 ft-lb)
  6-cylinder engine. . . . . 25 Nm (18 ft-lb)

1. **Intake manifold, upper**
2. **Wiring harness**
3. **Test connection plug**
4. **Fuel supply line**
5. **Fuel return line**
6. **Fuel rail**
7. **Bolt**
   • tighten to 10 Nm (89 in-lb)
8. **Bolt**
   • tighten to 25 Nm (18 ft-lb)
9. **Intake manifold, lower**
10. **Gasket**
    • always replace
11. **O-ring**
    • replace
12. **Fuel injector**
13. **Retaining clip**
14. **Fuel pressure regulator**
15. **Injector harness connector**
16. **Intake air temperature (IAT) sensor**
    • tighten to 10 Nm (7 ft-lb)
17. **Vacuum line**

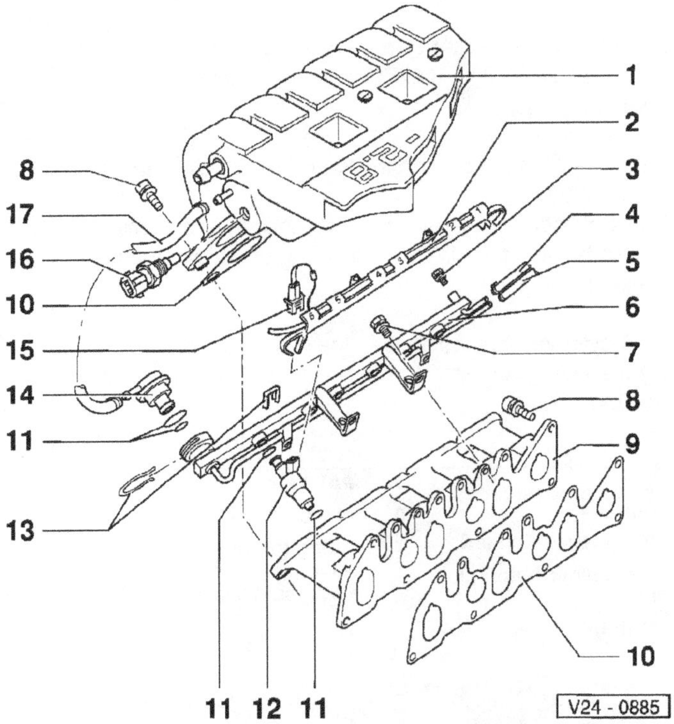

V24-0885

Fig. 28. Fuel rail assembly on 6-cylinder engine.

To remove fuel pressure regulator

> ***WARNING —***
> *Fuel under pressure will be expelled when the fuel pressure regulator is removed. Do not smoke or work near heaters or other fire hazards. Keep a fire extinguisher handy.*

1. Disconnect the vacuum hose from the regulator. See Fig. 29.

2. Remove the fuel pressure regulator retaining clip. Wrap a shop towel around the regulator and then pull it from the fuel rail.

Installation is the reverse of removal. Inspect the O-rings and replace them if damaged.

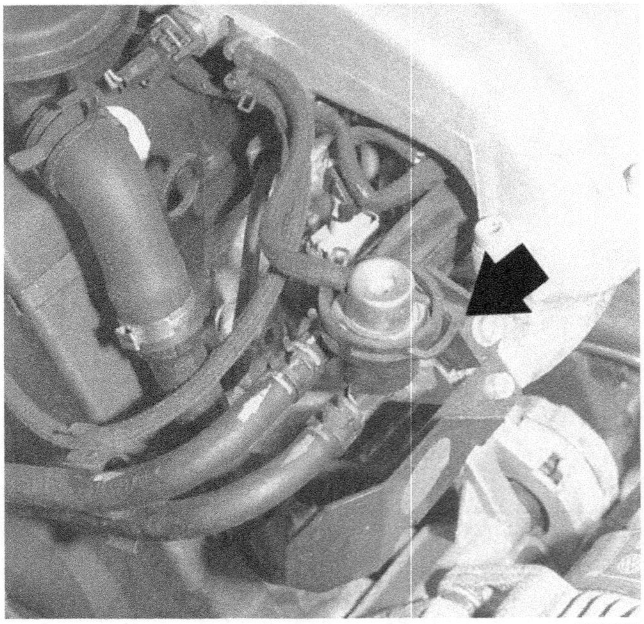

0024053

Fig. 29. Fuel pressure regulator retaining clip (**arrow**).

To remove Idle Air Control (IAC) valve

1. Disconnect the harness connector from the IAC valve.

2. Loosen the valve's hose clamps and disconnect the hoses from the valve.

3. On 4-cylinder engines, loosen the clamping bolt on the valve's bracket. See Fig. 30.

4. Remove the valve from its bracket.

5. Installation is reverse of removal.

0024051

Fig. 30. IAC valve clamping bracket on 4-cylinder engine (**arrow**).

To remove throttle valve housing

Volkswagen specifies that if the throttle valve housing is replaced, the engine control module (ECM) and transmission control module (TCM) "Basic Settings" should be "initiated" using the special V.A.G. 1551 scan tool. If these basic settings are not done using the scan tool, transmission shifting and driveability problems may be encountered.

1. Remove intake air duct from the throttle valve housing.

2. Disconnect the throttle position sensor harness connector and the accelerator cable from the throttle valve housing.

3. On 4-cylinder engines, disconnect the evaporative emissions vent hose from the throttle valve housing.

4. On 6-cylinder engines, relieve the cooling system pressure by removing the cap from the coolant expansion tank. Clamp off the coolant hoses at the throttle valve housing and remove the hoses.

5. Remove the bolts holding the throttle valve housing to the intake manifold and remove the housing.

Installation is the reverse of removal. Use a new gasket during installation. If replacing the throttle valve housing, transfer all parts from the old housing to the new one. On cars with manual transmission, adjust the throttle dashpot as described below.

Tightening torques
- Throttle position sensor to throttle valve housing . . . . 3 Nm (27 in-lb)
- Throttle valve housing to intake manifold
  - 4-cylinder engine. . . . . 25 Nm (18 ft-lb)
  - 6-cylinder engine
    - M6 . . . . 10 Nm (89 in-lb)
    - M8 . . . . 20 Nm (14 ft-lb)

To adjust throttle dashpot

On cars with manual transmission only a dashpot is mounted to the throttle valve housing. The dashpot holds the throttle valve open for a moment when the throttle is suddenly closed. This helps reduce exhaust emission of unburned fuel.

NOTE —

Before making the adjustment below, check that the throttle valve stop screw position has not been altered. If the factory setting of this screw has been tampered with, the throttle valve housing will need to be replaced—no adjustment specifications are available from Volkswagen.

1. Insert a 3 mm drill bit or feeler gauge between the throttle valve stop screw and the throttle lever. See Fig. 31

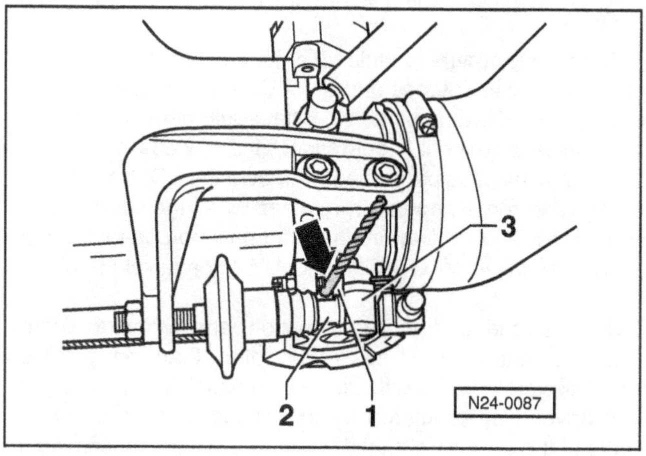

N24-0087

Fig. 31. Drill bit (3 mm) inserted between throttle valve stop screw and throttle lever on 4-cylinder engine.

2. Loosen the dashpot lock nut. Turn the dashpot until its rod just makes contact with the lever roller. Tighten the locknut. Remove the drill bit.

Dashpot Adjustment Specifications
- Checking. . . . . .2.5–3.5 mm (0.98–0.138)
- Adjusting. . . . . 3.0 mm (0.118 in.)

ON BOARD DIAGNOSTICS II (1996–2002 models)

Beginning with the 1996 model year, a second generation On Board Diagnostics (OBD) system known as OBD II was integrated into all Volkswagen engines using the Motronic 5.9 engine management system sold in the United States. The OBD II system refines and enhances previous OBD systems. For a complete description of the system and an overview of operation, see **2 Fuel, Ignition, and Exhaust Systems**.

Because of the large number of operating parameters monitored and the vast quantity of data available, fault diagnosis may no longer be able to be displayed via blink codes as on earlier models. All fault diagnosis must be done using Volkswagen special tool VAG 1551 or VAG 1552. The engine control module or ECM also supports certain limited functions in a government mandated generic scan tool mode. These functions are standardized for all OBD II compliant vehicles but are not as comprehensive as those found by using the VAG 1551/1552 in the original mode of operation.

During repairs where the battery must be disconnected, consideration must now be given to the effect that this will have on the engine management system.

- Diagnostic Trouble Codes (DTCs) will be erased.
- Readiness codes will be erased.
- Fault counters that monitor certain functions will be reset to zero.
- Adaptive leaning values will be set to default.

The engine control module (ECM) must be supplied with appropriate power sources and grounds to function properly. In addition, the ECM must "know" what equipment is installed in the vehicle in which it is operating. This process is known as coding and must be performed whenever the ECM is replaced. Coding memory is, however, retained when the battery is disconnected. This coding process can only be done with Volkswagen scan tool VAG 1551, VAG 1552 or equivalent.

Because the ignition, fuel injection, and emission control functions are even more inter-related than on previous Motronic versions, it is difficult, if not impossible to isolate general driveability problems by examining individual components of the system, although some are tested the same way in both versions.

For this reason, a suitable scan tool must be used for diagnosis if driveability problems occur or if the malfunction indicator light is illuminated. At the back of this book (see **ST Scan Tool**) is a table that lists numerous DTCs and their corresponding P-code. P-codes are what many after-market (non-factory) scan tools will display. The Scan Tool section also has control module coding and other scan tool specific information.

Component locations for the various Motronic fuel injection systems are shown on the next few pages.

Component locations (2.8L AAA engine, 1996–1999)

NOTE —
Volkswagen identifies electrical components by a letter and/or a number in the electrical schematics. See **97 Wiring Diagrams, Fuses and Relays**. *These electrical identifiers are listed as an aid to electrical troubleshooting.*

The list below applies to Fig. 32.

1. **Evaporative Emissions (EVAP) canister purge regulator valve (N80)**
   •EVAP canister system
2. **Mass Air Flow (MAF) sensor (G70)**
3. **Intake air duct**
   •With connections for Positive Crankcase Ventilation (PCV) heating element (N79)
4. **Knock sensor 1 (G61)**
5. **Engine control module (J220)**
6. **Throttle valve control module (J338)**
7. **Ground connection**
8. **Ignition coil (N152)**
9. **A/C cut-out thermal switch (F163) and third speed coolant fan control thermal switch (F165)**
   •Brown, 4-pin
   •On vehicles with A/C
10. **Engine Coolant Temperature (ECT) sensor (G62)**
    •Blue, 2-pin
    •For Motronic ECM
11. **After-run coolant fan control thermal switch (F87) and ECT sensor (G2)**
    •Yellow, 4-pin
12. **Intake Air Temperature (IAT) sensor (G72)**
13. **Fuel pressure regulator**
14. **Engine speed (RPM) sensor (G28)**
15. **Knock sensor 2 (G66)**
16. **Secondary Air Injection (AIR) pump motor (V101)**

(continued on following page)

Component locations
(2.8L AAA engine, 1996–1999)

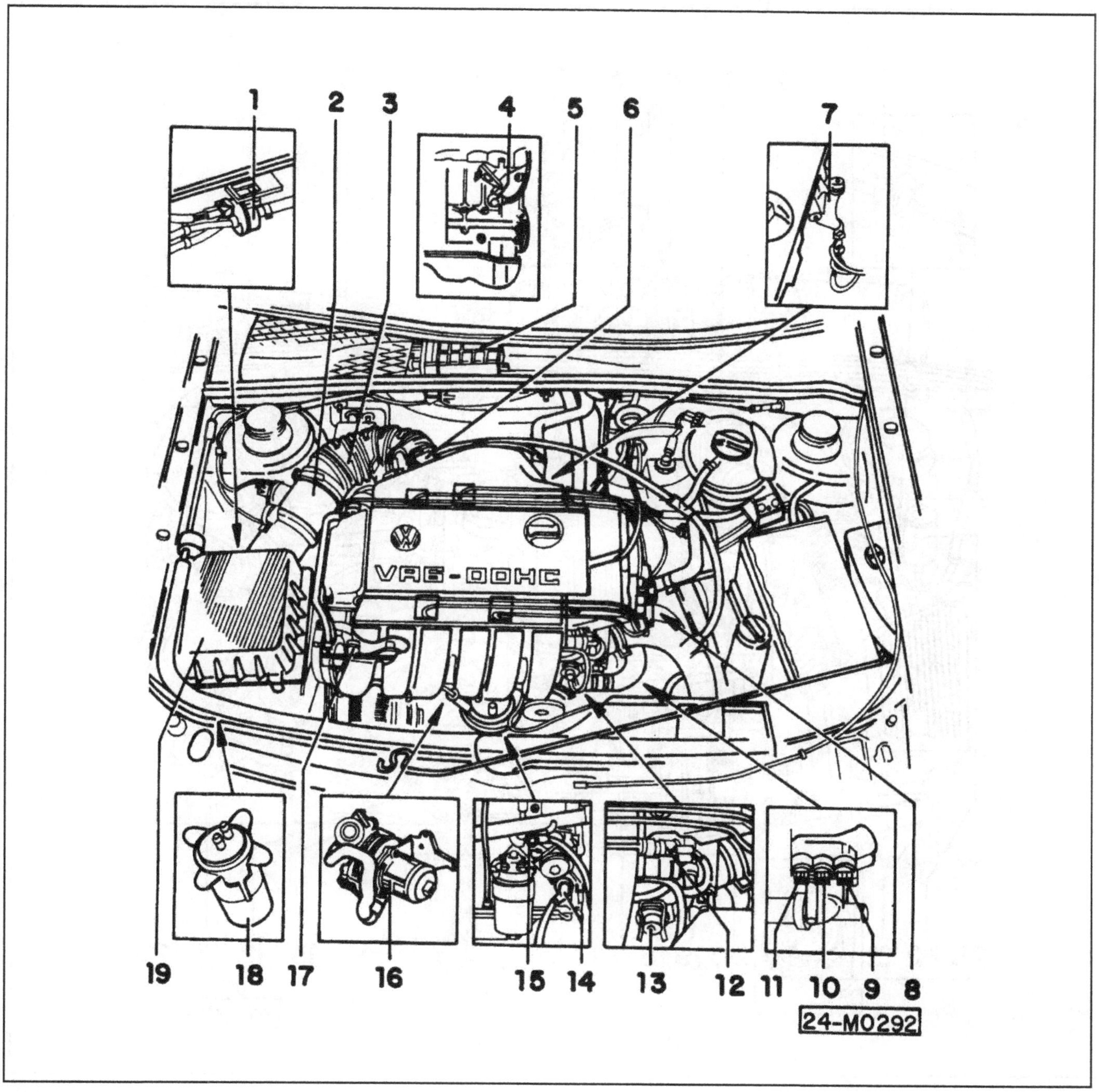

24-M0292

Fig. 32. Motronic fuel injection component locations on AAA (2.8L) OBD II engine, 1996-1999.

17. **Fuel injectors**
    - •Cylinder 1 (N30)
    - •Cylinder 2 (N31)
    - •Cylinder 3 (N32)
    - •Cylinder 4 (N33)
    - •Cylinder 5 (N83)
    - •Cylinder 6 (N84)

18. **EVAP canister**
    - •Below air cleaner
    - •EVAP system

19. **Air cleaner**

Component locations (2.0L ABA engine, 1996–1999)

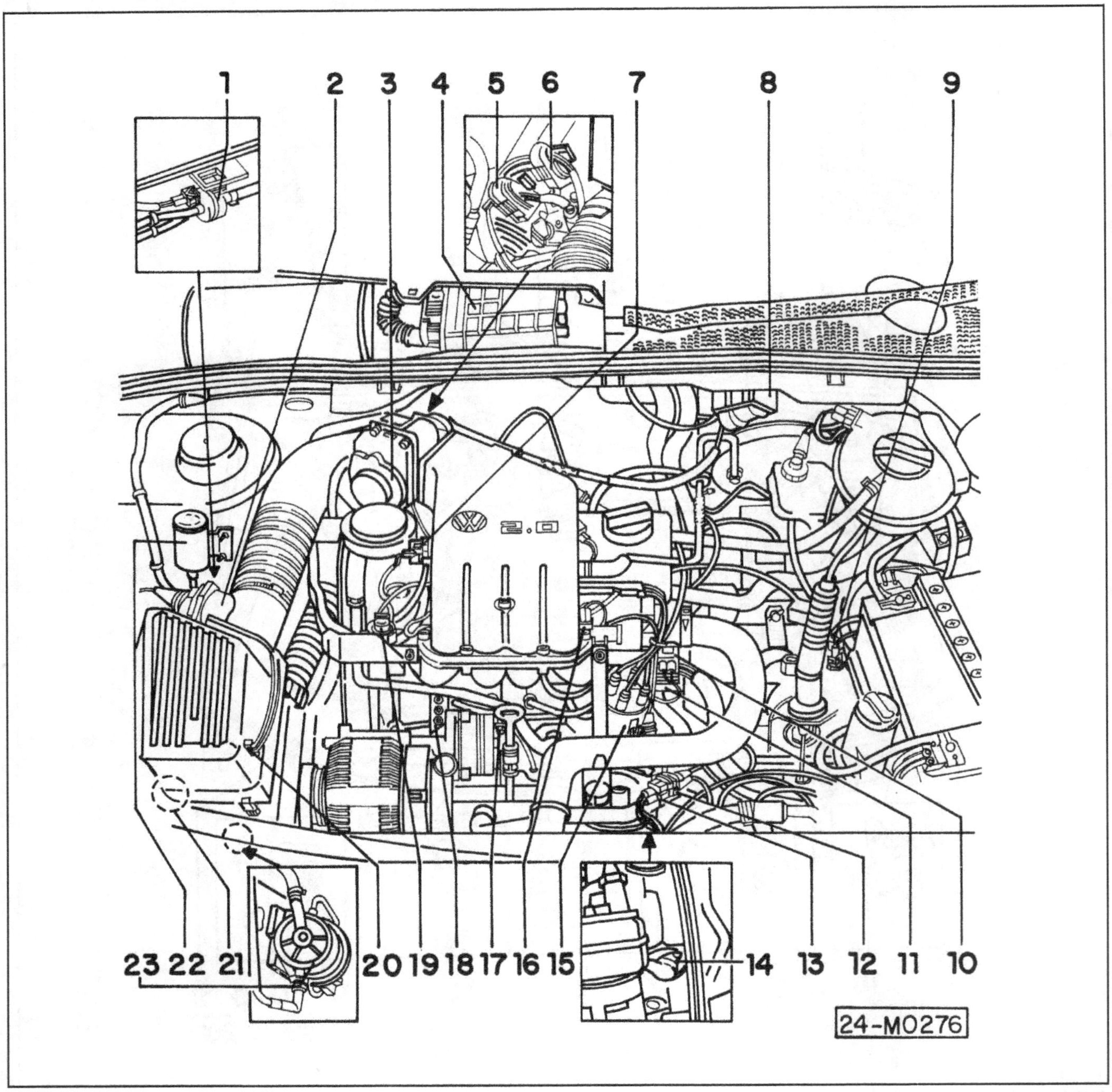

24-M0276

Fig. 33. Motronic fuel injection component locations on ABA (2.0L) OBD II engine, 1996-1999.

1. **Evaporative Emissions (EVAP) canister purge regulator valve (N80)**
   •EVAP canister system
2. **Mass Air Flow (MAF) sensor (G70)**
3. **Throttle valve control module (J338)**
   •With connections for Positive Crankcase Ventilation (PCV) heating element (N79)
4. **Engine Control Module (J220)**
5. **Connector**
   •Brown, 4-pin
   •For Heated Oxygen Sensor (HO2S) (G108) and Oxygen Sensor (O2S) heating (after three way catalytic converter)
   •Installation position: Three Way Catalytic Converter (TWC)

(continued from previous page)

6. **Connector**
   - Black, 4-pin
   - For HO2S (G39) and O2S heating (before three way catalytic converter)
   - Installation position: front exhaust pipe
7. **Intake Air Temperature (IAT) (G72)**
8. **Ignition coil (N152)**
9. **Speedometer Vehicle Speed Sensor (VSS) (G22)**
10. **Engine Coolant Temperature (ECT) sensor (G62)**
    - Black, with ECT sensor (G2)
11. **Center connector**
12. **3-pin connector**
    - For knock sensor
    - Blue
13. **3-pin connector**
    - For engine speed sensor
    - White
14. **Engine speed (RPM) sensor (G28)**
15. **Distributor with Camshaft Position (CMP) sensor (G40)**
16. **Fuel injectors**
    - Cylinder 1 (N30)
    - Cylinder 2 (N31)
    - Cylinder 3 (N32)
    - Cylinder 4 (N33)
17. **Knock sensor (G61)**
18. **Ground connection**
19. **Fuel pressure regulator**
20. **Temperature regulator Thermal Vacuum Valve (TVV)**
    - For intake air preheating
21. **EVAP canister**
    - Below air cleaner (Cabrio)
    - Behind right front wheel housing (Golf, Jetta)
    - EVAP system
22. **Leak Diagnosis Pump (LDP) filter**
    - Golf, Jetta
23. **Leak Diagnosis Pump (LDP)**

NOTE —

Volkswagen identifies electrical components by a letter and/or a number in the electrical schematics. See **97 Wiring Diagrams, Fuses and Relays**. *These electrical identifiers are listed as an aid to electrical troubleshooting.*

OBD II component locations (2.0L ABA engine, Cabrio 2000-2002)

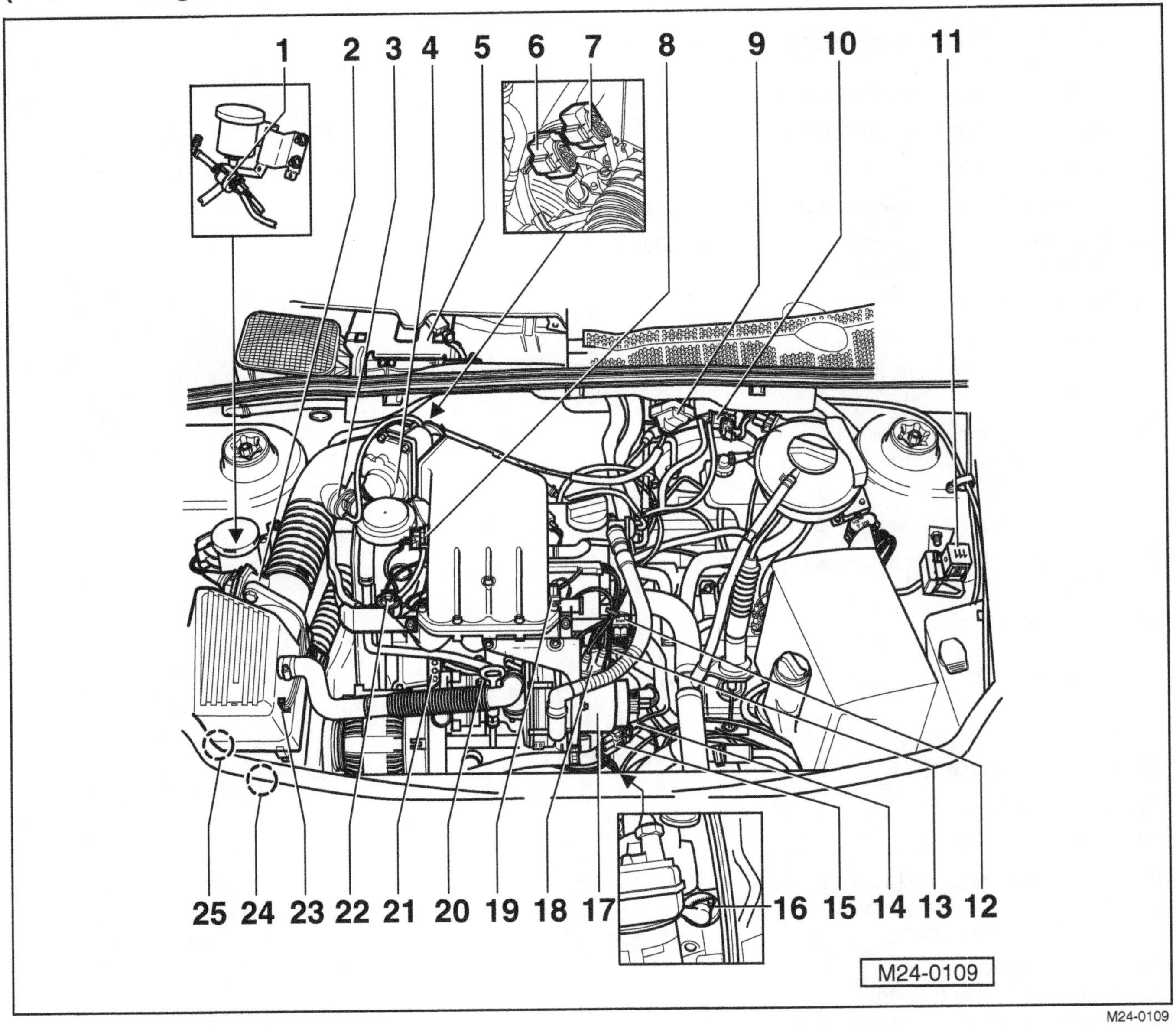

M24-0109

Fig. 34. Motronic fuel injection component locations on ABA (2.0L) OBD II engine, 2000-2002.

1. **Evaporative Emissions (EVAP) canister purge regulator valve (N80)**
   •EVAP canister system
2. **Mass Air Flow (MAF) sensor (G70)**
3. **Positive Crankcase Ventilation (PCV) heating element (N79)**
4. **Throttle valve control module (J338)**
5. **Engine control module (J220)**
6. **4-pin connector**
   • Brown
   • For Oxygen Sensor (O2S) (G130) behind catalytic converter and O2S heater (Z29)
7. **4-pin connector**
   • Black
   • For Oxygen Sensor (HO2S) (G39) before catalytic converter and HO2S heater (Z19)
8. **Intake Air Temperature (IAT) sensor (G42)**
9. **Ignition coil (N152)**
   • With ignition coil output stage (N157)
10. **Secondary Air Injection (AIR) solenoid valve (N112)**
11. **Secondary Air Injection (AIR) pump relay (J299)**
12. **Engine coolant temperature (ECT) sensor (G62)**
13. **Central connector**

(continued from previous page)

14. **3-pin connector**
    - Brown, for knock sensor 1 (G61)
15. **3-pin connector**
    - Gray, for engine speed (RPM) sensor (G28)
16. **Engine speed (RPM) sensor (G28)**
17. **Secondary Air Injection (AIR) pump motor (V101)**
18. **Distributor with Camshaft Position (CMP) sensor (G40)**
19. **Fuel injectors**
    - Cylinder 1 (N30)
    - Cylinder 2 (N31)
    - Cylinder 3 (N32)
    - Cylinder 4 (N33)
20. **Knock sensor 1 (G61)**
21. **Ground connection**
22. **Fuel pressure regulator**
23. **Tempetature regulator**
    - For intake air preheating
24. **Leak Diagnostic Pump (LDP)**
25. **EVAP Canister**

NOTE—
Volkswagen identifies electrical components by a letter and/or a number in the electrical schematics. See **97 Wiring Diagrams, Fuses and Relays**. *These electrical identifiers are listed as an aid to electrical troubleshooting.*

Readiness Codes

Recent legislation requires that all auto manufacturers build into their engine management systems the ability to check for proper operation of up to 8 functions. For the Motronic Engine Management System, the exact number depends on the equipment level of each particular vehicle. Most of the time, the readiness code will set itself after the appropriate conditions have been met. These may include a cold or a hot start, operation at a certain load or speed for a certain period of time, or operation at different temperatures. After repairs, it may be advantageous to confirm repairs rather than wait to see if they set on their own. Other than waiting, there are two methods of setting the readiness codes on the Motronic M5.9 system. The first method requires Volkswagen Scan Tool VAG 1551 or VAG 1552 and extensive knowledge of its operation and capabilities. The second method also requires a scan tool such as VAG 1551 or VAG 1552, or a Generic Scan Tool (GST). The following procedure shows the screens as they appear when the VAG 1551 or VAG 1552 is operated in the GST mode. Generic Scan Tools supplied by other manufacturers should have nearly identical displays due to legislation.

To check readiness codes

1. Connect the scan tool to the vehicle data link connector, (DLC), below the heater controls as shown in Fig. 4, given earlier. The scan tool will be powered up when plugged into DLC. Turn the key to the on position, but do not start the engine. Observe the following sequence in the scan tool display window.
   - 1 – Rapid Data
   - Address Word 33 – OBD II
   - Select mode – 41
   - The display will show four fields of information

CAUTION—
These instructions assume that the technician already has a good working knowledge of the operation and use of the VAG 1551 or VAG 1552 scan tool.

2. Fig. 35. shows the information as it appears on the screen. Disregard the information that is shown here as the letter X. These spaces will have either a "1" or a "0" and do not affect this procedure. The readiness code is the 8 digit number in field 4 on the far right. If the display shows a "1" in any position, the readiness code needs to be set.

Mode 41	PID 1		Module 10
XXXXXXXX	XXXXXXXX	XXXXXXXX	01101101

0024195

Fig. 35. Display that shows that readiness code is not set and should be. Disregard all Xs.

3. If the screen displays all zeros in the fourth field (far right), the readiness code is set and no further action is required. See Fig. 36.

Mode 41	PID 1		Module 10
XXXXXXXX	XXXXXXXX	XXXXXXXX	00000000

0024196

Fig. 36. Display that shows that readiness code is set and no further action is required.

4. If readiness code is set, disconnect the scan tool and reinstall the appropriate parts.

5. If readiness code is not set, unplug the scan tool and proceed as follows.

To set readiness codes

> ***NOTE —***
> - *The engine must not be switched off during the test sequence.*
> - *If a test sequence procedure is interrupted or not adhered to, the readiness code will not be set.*

1. Switch ignition on for approximately 30 seconds but do not start engine.
2. Start engine and let run at idle.
3. Increase engine speed to 3000-3500 RPM for 90 seconds to heat up the Three Way Catalytic Converter (TWC).
4. With the vehicle stationary, run the engine at idle for 40 seconds.
5. Test drive for a minimum of 13 minutes.
   - Automatic transmission: driving range 3
   - Manual transmission: 3rd gear
   - Test drive must be carried out without interruption which means steady pressure on the throttle so that deceleration fuel shut-off is not activated nor is the throttle allowed back to the idle position.
6. Allow the engine to run at idle for a minimum of 13 minutes.
7. After completion of the test sequence, re-connect the scan tool and observe display. See **To check readiness codes**.
   - If display appears as shown in Fig. 36, code has set and no further action is required. Disconnect scan tool and reinstall appropriate parts.
   - If display appears as shown in Fig. 35, code has not been set. Unplug scan tool and repeat from step 1. If code does not set after second attempt, it is likely that there is a malfunction in the system. Consult scan tool directions for additional information on checking for DTCs.

24b Fuel Injection–Mono-Motronic (ACC Engine)

TABLES

GENERAL

This repair group covers Bosch Mono-Motronic fuel injection component troubleshooting and repair. Special testing equipment may be necessary for some of the tests and repair procedures given here.

Fuel System Applications
- ACC (1.8 liter) . . . . . Mono-Motronic
- ABA (2.0 liter) . . . . . Motronic
- AAA (2.8 liter) . . . . . Motronic
- AAZ (1.9 liter) . . . . . Diesel Fuel Injection
- 1Z/AHU (1.9 liter) . . . . . Diesel TDI

Safety Precautions

The following warnings and cautions should be adhered to whenever doing work on the fuel injection system.

WARNING —
- *Fuel will be discharged during many fuel system test procedures. Do not smoke or work near heaters or other fire hazards. Have a fire extinguisher handy. Work only in a well- ventilated area.*
- *Wear suitable hand and eye protection when working with gasoline. Prolonged contact with fuel can cause illnesses and skin disorders.*

CAUTION —
- *Connect and disconnect wires and test equipment only with the ignition off.*
- *Before making any electrical tests that require the engine to be cranked using the starter, disable the ignition system as described in* **28b Ignition System–Diesel**.
- *Always wait at least 40 seconds after turning off the ignition before removing the ECM connector. If the connector is removed before this time, residual power in the system relay may damage the ECM.*

CAUTION —

- *Cleanliness is essential when working on the fuel system. Thoroughly clean fuel line connections and surrounding areas before loosening. Avoid the use of compressed air, and avoid moving the car. Only install clean parts.*
- *Fuel system cleaners and other chemical additives other than those specifically recommended by VW may damage the catalytic converter, the oxygen sensor and the plastic fuel tank.*

NOTE —

The camshaft position (CMP) sensor, located inside the ignition distributor, is also used for ignition timing and dwell angle control. The engine will not operate if the CMP sensor is faulty. CMP sensor testing and repair information is covered in **28 Ignition System.**

Fig. 1 shows the Bosch Mono-Motronic engine management system. The Mono-Motronic system combines the ignition and fuel functions into one system, managed by a single engine control module (ECM). The system features built-in diagnostics that detect and store coded fault information.

NOTE —

A general description of operation for the Mono-Motronic engine management system is given in **2a Fuel, Ignition, and Exhaust Systems—General**.

Adaptive Circuitry and On-Board Diagnostics (OBD)

The Mono-Motronic system is adaptive. Idle speed, ignition timing, and injection quantity automatically compensate for and adapt to changes in the engine due to wear or small problems, such as minor vacuum leaks. As a result, idle speed and fuel mixture (%CO) are non-adjustable. Ignition timing is adjustable, but not a maintenance interval requirement.

The Mono-Motronic system features built-in diagnostic circuitry to detect and store Diagnostic Trouble Code (DTC) information. When the system compensates for values that are outside the permitted limits, a fault code is generated and stored in the ECM's permanent memory. For emissions-related faults, the Check Engine or Malfunction Indicator Lamp (MIL) will be turned on. The fault codes can be accessed through the data link connector(s) in the dashboard.

NOTE —

- *The Volkswagen Scan Tool V.A.G. 1551 is used to quickly access the Diagnostic Trouble Codes (DTCs) through the data link connector(s) in the dash. Alternatively, the data link connector(s) can be jumpered and the DTC can be read out through the Check Engine light, also called the Malfunction Indicator Lamp (MIL), in the form of a blink code.*
- *Volkswagen identifies electrical components only by a number in the electrical schematics. See* **97 Wiring Diagrams, Fuses and Relays**. *These electrical component numbers are listed with the component names on the Table of Contents and under the appropriate repair headings as an aid to electrical troubleshooting.*

FAULT DIAGNOSIS

Because the ignition, fuel injection, and emission control functions are inter-related and controlled together by the Mono-Motronic engine management system, it is difficult if not impossible to isolate general driveability problems by examining components of the fuel injection system alone.

For this reason, the first step in locating trouble is to review the basic requirements list shown below. If no faults are found, continue by checking for Diagnostic Trouble Codes (DTCs), especially if the check engine light is illuminated.

NOTE —

- *Motronic Fuel injection repair information for the 2.0 liter ABA engine and the 2.8 liter AAA engine is covered in* **24a Fuel Injection–Motronic**.
- *Fuel pump testing and repair information is covered in* **20 Fuel Storage and Supply**.
- *Emission systems, such as Evaporative (EVAP) emissions, are covered in* **26 Exhaust System/Emission Controls**.

1. **Harness connector, for oxygen sensor (G39)**
   • on right engine mount
2. **Harness connector, for intake air temperature (IAT) sensor (G42)**
3. **Fuel pressure regulator**
4. **Fuel injector (N30)**
   • with intake air temperature (IAT) sensor (G42)
5. **Throttle position (TP) actuator (V60)**
   • with closed throttle position (CTP) switch (F60)
6. **Throttle position (TP) sensor (G69)**
7. **Mono-Motronic engine control module (ECM) (J257)**
8. **Temperature regulator, for intake air preheating**
9. **Throttle body**
10. **Harness connector, for intake manifold preheater**
11. **CO tap tube**
    • not on all vehicles
12. **Ignition coil with power stage (N152)**
13. **Early fuel evaporation (EFE) relay (J81)**
    • on fuse/relay panel in footwell, driver's side
14. **Ignition distributor**
    • with camshaft position sensor (CMP) (G40)
15. **Engine coolant temperature (ECT) sensor (G62)**
    • with integrated temperature gauge sensor (G2)
16. **Engine ground connection**
17. **Evaporative emission (EVAP) canister purge regulator valve (N80)**
18. **Vacuum unit for intake air preheating**
19. **Air cleaner housing**

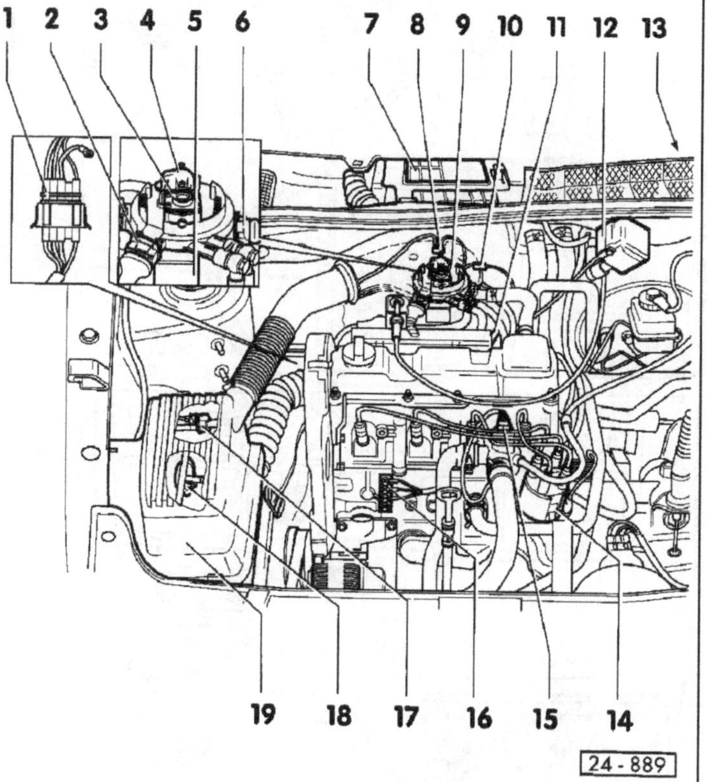

24-889

Fig. 1. View of Mono-Motronic engine management system.

Basic Requirements

Below are a few basic checks that should be made if the Malfunction Indicator Lamp (MIL) is not illuminated, but engine management driveability problems are encountered.

1. Check the intake system for leaks. Check for cracked, loose, or disconnected hoses and duct work. Check that all hose clamps are tight. Check the small vacuum hoses, especially close to the fittings, and check the fittings themselves.

   NOTE —
   An air leak allows unmeasured air to enter the engine, often resulting in an overly lean fuel mixture. This can cause driveability problems and may illuminate the Malfunction Indicator Lamp (MIL).

2. Check that the battery is in good condition. Check that the cables are tight and free of corrosion. Check that all related ground points are firmly connected and in good condition.

3. Check all of the harness connectors for damage and corrosion, especially the connectors at the throttle body housing.

4. If no basic faults are found, check for Diagnostic Trouble Codes (DTC) as described below.

Diagnostic Trouble Codes (DTC)

The Mono-Motronic system is equipped built-in fault diagnostics. Some of the faults will illuminate the check engine light, but some will not.

Diagnostic trouble codes (DTC) can be displayed in one of two ways; (1) using the specialized VW scan tool V.A.G. 1551 or (2) by jumpering the data link connector(s) and interpreting blink codes through the malfunction indicator light (MIL) in the instrument cluster. The procedure below describes the blink code method.

DTCs are accessed through the data link connector(s). The connectors are behind a cover in the center of the dash, below the heating ventilation controls. Two versions of data link connectors are used depending on model year. 1993 and early 1994 cars use two connectors while later model 1994 and later cars use a single connector. See Fig. 2.

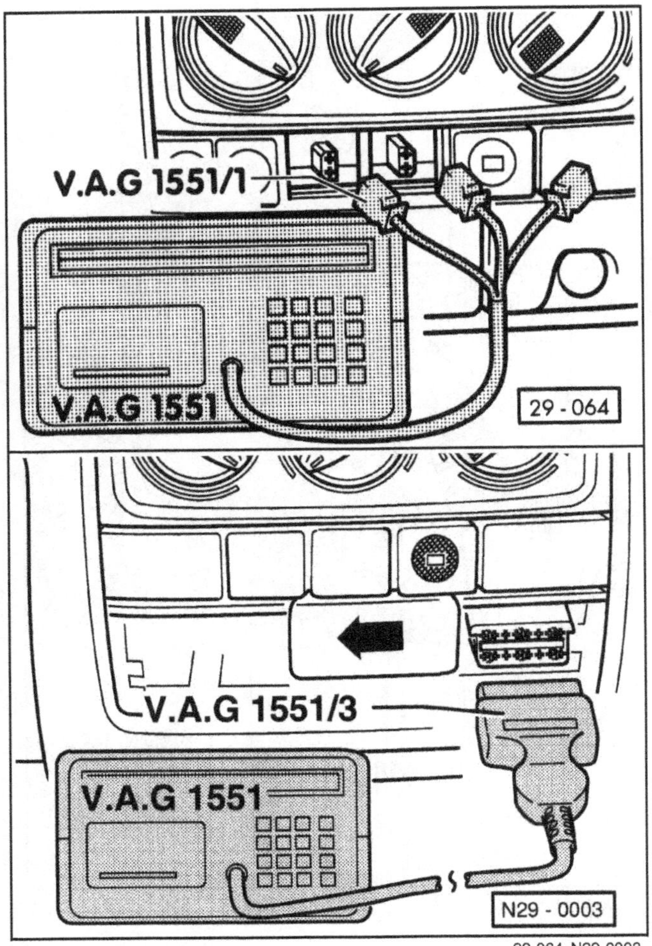

29-064, N29-0003

Fig. 2. Data link connectors in center of dash. 1993 and ealry (top) cars use two separate connectors. Later model 1994 (bottom) and later cars use single 16-pin connector. Special VW scan tool V.A.G. 1551 shown.

To display DTC

1. With the ignition key on, use a jumper wire or the OBD adapter to connect terminal 4 (ground) to terminal 15 (L-wire) for **2.5 seconds** and remove the jumper wire or adapter. See Fig. 3.
   - On 1993 and early 1994 cars jumper socket 1(-) in the black connector to socket 1 in the white connector.
   - On later 1994 and 1995 cars, jumper socket 4 (-) to socket 15.

> ***CAUTION—***
> *Use care when connecting the jumper wire. Positive (+) battery voltage is present in the connector. A wrong connection can result in damage to ECM or the vehicle's wiring.*

2. After one blink of the MIL light, the first 4-digit DTC will be displayed. Approximately one second later, the MIL will blink the first digit, then pause, blink the second digit, then pause, and so on. **Table a** lists DTCs.

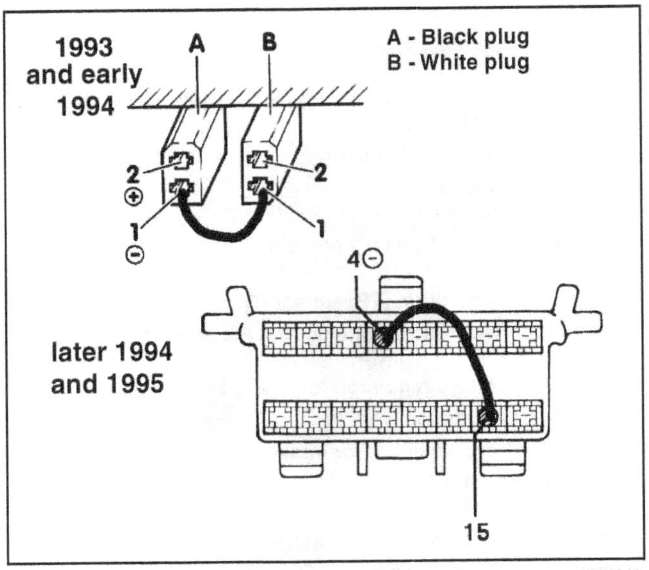

0024041

Fig. 3. Jumper wire installed in data link connectors.

NOTE—
A blink code of 4-4-4-4 indicates that no malfunctions have been stored in permanent memory.

3. Once the first DTC has been displayed, reconnect the jumper wire or the adapter for **2.5 seconds** and then remove it. The second DTC code (if present) will be displayed.

4. Repeat step 3 until no more codes are displayed (DTC 4-4-4-4 displayed).

NOTE—
- *The 4-4-4-4 code will be represented by four long blinks.*
- *On 1995 cars, a special OBD adapter is stored beneath the rear seat and should be used to access the DTCs. See Fig. 4. On cars with ABS, the connector is stored beneath the ABS coaxial cable next to the ABS control unit. On cars without ABS, the adapter is secured by a wire tie.*

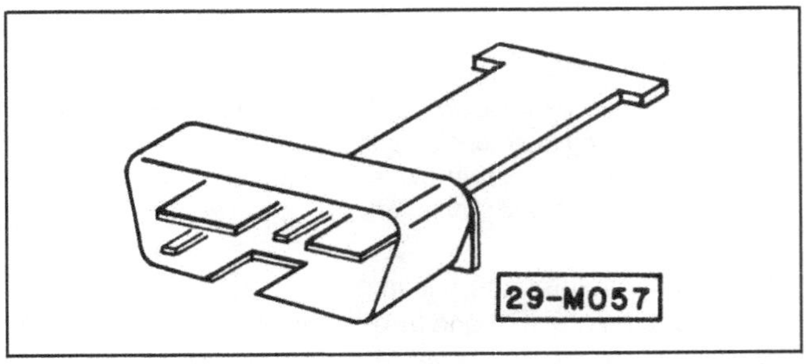

29-M057

Fig. 4. OBD adapter used to access blink codes on 1994 and later cars. On some 1995 cars, adapter may be stored beneath the rear seat.

5. Turn the ignition key off. Erase the DTC codes once all testing and repairs have been made.

To erase the DTC memory

1. With the ignition key off, connect the jumper wire or the adapter to the diagnostic connector(s).
2. Turn the key on, wait at least **5 seconds** and then disconnect the wire or the adapter.
3. Reconnect the jumper wire or the adapter for **2.5 seconds** and then remove it. The first DTC code (if present) will be displayed.
4. Repeat step 3 until no more codes are displayed (DTC 4-4-4-4 displayed).

NOTE —
The 4-4-4-4 code will be represented by continuous long blinks.

5. Reconnect the jumper wire or the adapter for 2.5 seconds and remove it. The DTC memory is now erased. Turn the ignition off.

NOTE —
To confirm that the memory has been erased, repeat step 3. There should be only a 4-4-4-4 code (continuous long blinks) present.

Table a. Mono-Motronic Diagnostic Trouble Codes (DTC)

DTC	Probable fault	Corrective action, first check
4444	**No DTC recognized**	No corrective action necessary.
1231	**Vehicle speed sensor (VSS) (G22)**–missing or incorrect signal, VSS wiring damaged/open or shorted, VSS faulty. Speedometer faulty.	Check output signal the from speedometer VSS. Check sensor wiring. Check speedometer operation. See **97 Wiring Diagrams, Fuses and Relays**
1232	**Throttle Position (TP) Actuator (V60)**—missing or incorrect signal, TP actuator wiring damaged/open or shorted, TP actuator faulty.	Check the wiring and the resistance of the TP actuator. Check that the intake air temperature (IAT) sensor is operating correctly.
2113	**Camshaft position (CMP) sensor (G40)**–missing or incorrect signal from sensor, CMP wiring damaged/open or shorted, CMP (Hall) sensor faulty.	Check the wiring and signal from of the CMP (Hall) sensor. Check that the distributor is correctly installed. Test the signal from the Hall sensor. Check ignition timing.
2121	**Closed throttle position (CTP) switch (F60)**— missing or incorrect signal from CTP switch, CTP switch wiring damaged/open or shorted, CTP switch faulty.	Check the wiring and signal from CTP switch—check the switch opens and closes electrically. Check that the accelerator pedal cable is correctly adjusted. Check that the throttle position actuator is functioning correct and that the throttle lever is contacting the throttle stop.
2212	**Throttle position (TP) sensor (G69)**–missing or incorrect signal from TP sensor, TP sensor wiring damaged/open or shorted, TP sensor faulty.	Check the wiring and signal from of the TP sensor.
2312	**Engine coolant temperature (ECT) sensor (G62)**–missing or incorrect signal from ECT sensor, ECT sensor wiring damaged/open or shorted, ECT sensor faulty	Check the wiring and signal from of the ECT sensor. Check for corrosion at the connector, causing intermittent (sporadic) malfunctions.
2322	**Intake air temperature (IAT) sensor (G42)** – missing or incorrect signal from IAT sensor, IAT sensor wiring damaged/open or shorted, IAT sensor faulty.	Check the wiring and signal from of the IAT sensor.
2341	**Oxygen sensor control**– Incorrect oxygen sensor signal to ECM (fuel mixture too lean or to rich). Sensor heater circuit may be open or faulty.	Check the oxygen sensor heater circuit and sensor output signal first. **Lean**: check exhaust system for leaks between catalytic converter and exhaust manifold, check for intake air leaks (unmeasured air), check fuel pressure (too low). **Rich**: check ignition system for misfire, check fuel pressure (too high), check for leaking or stuck open fuel injector, check the Evaporative (EVAP) emission system (purge regulator valve stuck open).
2342	**Heated oxygen sensor (G39)**–no oxygen sensor signal to ECM. Sensor heater circuit may be open or faulty.	Check the oxygen sensor output signal (voltage fluctuating between 0 and 1.0 volt). Check oxygen sensor heater circuit.
2413	**Mixture control adjustment**–fuel mixture too lean or too rich.	Check for conditions causing rich or lean running conditions (see DTC 2341). Check the fuel pressure, ignition system, and exhaust system.
1111	**Mono-Motronic engine control module (ECM) (J257)**–ECM faulty (electronic component failure in ECM).	If no other faults can be found, the ECM is probably faulty and should be replaced.

FUEL PRESSURE TESTS

Checking the fuel pressure is a fundamental part of troubleshooting and diagnosing the Mono-Motronic system. Fuel pressure has a direct effect on fuel mixture and driveability. An accurate fuel pressure gauge with a range of 0 to 5 bar (approximately 0 to 100 psi) will be needed to make the tests.

There are two significant fuel pressure values: 1) system pressure—created by the main fuel pump and regulated by the fuel pressure regulator, and 2) residual pressure—the pressure maintained in the closed system after the engine is shut off.

Before making the tests described below, make sure the fuel pump is operating correctly and that the fuel pump delivery rate is within the acceptable limit. See **20 Fuel Storage and Supply**.

> ***WARNING —***
> *Fuel will be discharged when disconnecting fuel lines. Do not smoke or work near heaters or other fire hazards. Have a fire extinguisher handy. Work only in a well- ventilated area. Relieve fuel pressure in the line (as described earlier) or wrap a cloth around the line before disconnecting.*

To check system line pressure

1. Relieve the fuel pressure in the system and connect a fuel pressure gauge as shown in Fig. 5.

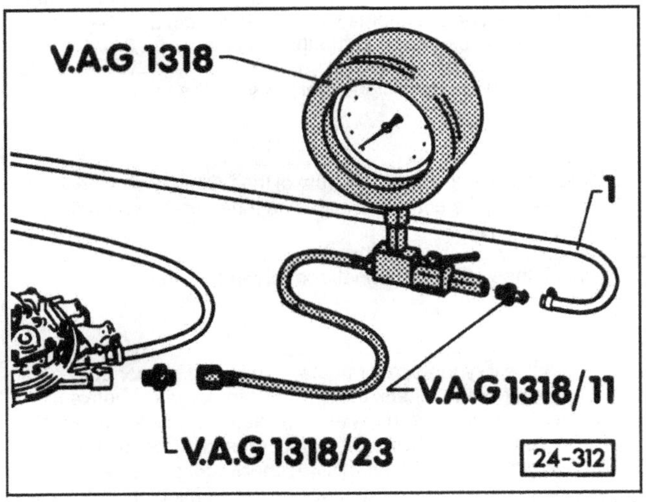

24-312

Fig. 5. Fuel pressure gauge installation. Connect the gauge to the test port on the throttle body using the appropriate adapter. Numbers shown are Volkswagen special tool numbers.

2. Start the engine and let it idle. See fuel pressure specifications given below.

NOTE —
If the engine does not start, run the fuel pump by bypassing the fuel pump relay as described **in 20 Fuel Storage and Supply**.

Fuel System Line Pressure

Engine	Fuel Pressure
ACC engine (1.8 liter)	0.8–1.2 bar (11.6–17.4 psi)

If system pressure is too low, first check for fuel leaks. Test fuel pump delivery as described in **20 Fuel Storage and Supply.** Check for a leaking injector or a stuck open injector. A restricted fuel filter or a damaged fuel line are also potential causes. If there are no other faults, replace the pressure regulator.

If system pressure is too high, first check for a blocked or damaged fuel return line from the pressure regulator. If there are no obstructions, replace the pressure regulator.

NOTE —
Additional fuel pump tests, including testing the fuel pump relay, are covered in **20 Fuel Storage and Supply**.

To check residual pressure

Hot-start problems are usually the only reason to suspect a residual pressure problem, but it deserves a routine check as long as the gauge is connected.

1. Relieve the fuel pressure in the system and connect a fuel pressure gauge as shown in Fig. 7.
2. Run engine for about 30 seconds to pressurize system.
3. Turn the engine off and wit five minutes. Observe the pressure gauge. See residual fuel pressure specifications given below.

Residual Fuel Pressure (after five minutes)

Engine	Fuel Pressure
ACC engine (1.8 liter)	0.5 bar (7.25 psi)

If the pressure drops beyond specification, check for leaks in the fuel lines, at the fuel pump, and at the fuel injector. If there are no leaks, either the pressure regulator or the fuel pump check valve is faulty.

To determine which, run the engine briefly to build pressure. Shut the engine off and immediately clamp fuel return line shut. If pressure still drops, fuel pump check valve is faulty. See **20 Fuel Storage and Supply** for fuel pump replacement. If residual pressure is now OK, the pressure regulator is faulty.

ELECTRICAL CHECKS AND COMPONENT TESTING

The electrical tests under this heading will help to locate electrical component failure. Unless a specific diagnostic trouble code (DTC) has been recorded, make the ECM pin-out test first as listed in **Table b**. The checks listed in the table check the components and wiring to the components.

The ECM is mounted in the rear of the engine compartment. See Fig. 1 given earlier. With the ignition off for about a minute, unplug the connector from the ECM and separate the connector housing.

References to Mono-Motronic components and their connector terminals are used throughout the table. Use the appropriate wiring diagram to identify these terminals. See **97 Wiring Diagrams, Fuses and Relays.**

When making tests at harness connectors, probe from the rear of the connector only to avoid damaging the small connector terminals. Separate the connector housing and remove any seals to gain access to the rear of the connector.

CAUTION—

- *Use a high-quality digital automotive multimeter to make the tests. An analog (swing-needle) meter should not be used as it can permanently damage electronic components.*
- *Always connect and disconnect the ECM connector and meter probes with the ignition off to avoid damage to electronic components. Wait about a minute after turning off the ignition before removing the ECM connector.*

Before making any electrical checks or tests, make sure the following conditions are met:

Basic Requirements

- Battery positive (+) voltage OK
- Fuse no. 15 and no. 18 OK
- Ground connections on cylinder head OK
- Fuel pump and fuel pump relay OK. See **20 Fuel Supply and Storage**.

NOTE—

Fuel pump electrical test are covered in **20 Fuel Storage and Supply.**

Table b. Mono-Motronic ECM Electrical Tests

Component or circuit	Test terminals (at disconnected ECM connector unless otherwise noted)	Test conditions	Test value
ECM power supply (constant)	20 (-) and 21(+)	Ignition off	Approx. battery voltage
ECM power supply (switched from ignition switch, via fuse/relay panel)	20 (-) and 23 (+)	Turn ignition on	Approx. battery voltage
Evaporative emission (EVAP) canister purge regulator valve (N80)	1 and 3	Ignition on	Approx. battery voltage
Fuel pump (G6) wiring	Jumper terminals 20 and 25	—	Fuel pump must run
Fuel injector (N30) wiring	1 and 8	Jumper terminals 20 and 25	Approx. battery voltage (and fuel pump runs)
Early fuel evaporation (EFE) relay (J81)	Jumper terminals 20 and 28	Turn ignition on	EFE relay must click
Signals from A/C	1 and 35	Switch A/C on Ignition on	Approx. battery voltage
	1 and 33	Switch A/C On Ignition on	No voltage
Coding wire for auto. transmission	20 and 40	Ignition on	Approx. battery voltage
Vehicle speed sensor (VSS) (G22)	20 and 36	Ignition on Raise and spin front left wheel Cars with **A/T**: selector lever in "D" Cars with **manual transmission**: shift lever in neutral	0-4 volts (VAC), fluctuating

continued on next page

Table b. Mono-Motronic ECM Electrical Tests (continued)

Component or circuit	Test terminals (at disconnected ECM connector unless otherwise noted)	Test conditions	Test value
Grounds (GND)	1 and 20	Ignition off	Continuity
	1 and ground	Ignition off	Continuity
	20 and ground	Ignition off	Continuity
	with A/T only 12 and 23	Ignition on shift lever in D or N position	Approx. battery voltage
		Ignition on move shift lever in 1, 2, 3, D, or R	No voltage
	with man. trans. only 12 and 23	Ignition on	Approx. battery voltage
Closed throttle position (CTP) switch (NF60)	10 and 20	Throttle closed	Maximum 200 ohms
		Throttle open	no continuity
Throttle position (TP) actuator	2 and 26	—	3 to 200 ohms
Engine coolant temperature (ECT) sensor (G62)	17 and 42	—	Resistance varies with temperature. See **Engine coolant temperature (ECT) sensor**
Intake air temperature (IAT) sensor (G72)	17 and 43	—	Resistance varies depending on temperature. See **Intake air temperature (IAT) sensor**
Data Link Connector (DLC)	11 and 29	**1993 and early 1994 cars** jumper terminals in white Data Link connector (See Fig. 3) **later 1994 - on cars** jumper terminals 7 and 15 in 16-pin Data Link Connector (See Fig. 3)	Maximum 1.5 ohms
Wiring to heated oxygen sensor (G39)	15 and 38	Disconnect oxygen sensor harness connector (rear of engine) and jumper terminals 3 and 4	Maximum 1.5 ohms
		Reconnect connector	Infinite ohms
Throttle position (TP) sensor (G69)	14 and 17	—	520 to 1300 ohms
	17 and 41	Throttle fully closed (throttle positioner not acting on throttle lever) Depress accelerator slowly to about 1/4 throttle	Resistance changes steadily
	17 and 18	Throttle fully closed (throttle positioner not acting on throttle lever) Depress accelerator slowly to about 1/4 throttle	Resistance does not change
		Depress accelerator further past 1/4 throttle	Resistance changes steadily
Camshaft position sensor (CMP) (G40) wiring		Disconnect harness connector CMP at ignition distributor	
	13 and 20	Jumper terminals 1 and 2 in connector	Maximum 1.5 ohms
	8 and 13	Jumper terminals 2 and 3 in connector	Maximum 1.5 ohms
Ignition coil power output stage (N157) wiring		Disconnect harness connector from ignition coil output stage	
	20 and 24	Jumper terminals 1 and 2 in connector	Maximum 1.5 ohms
	23 and 24	Jumper terminals 2 and 3 in connector	Maximum 1.5 ohms

Engine Coolant Temperature (ECT) Sensor (G62)

The ECT sensor is one of the main inputs to the engine management system. Its signal is used to adapt ignition timing, injection time, and idle speed. In addition, the ECT signal is used to activate the oxygen sensor operation and fuel tank venting.

The ECT sensor sends continuous engine temperature information to the Mono-Motronic ECM. If there is a break in the signal from the temperature sensor, the ECU will simulate a fixed signal based on an engine temperature of 90°C (194 °F).

The ECT sensor (G62) is mounted in the coolant outlet on the rear of the cylinder head. See Fig. 6. The ECT sensor is combined with the temperature gauge sensor (G2) in a single housing. The sensor body is black and the sensor has four terminals.

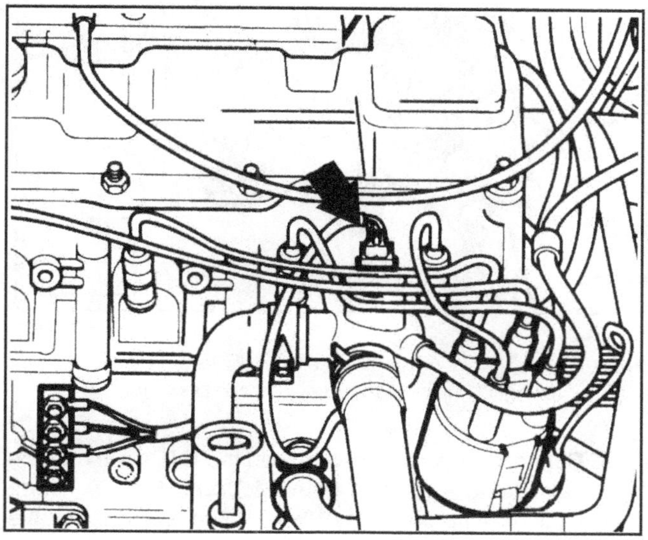
0024044

Fig. 6. Coolant temperature sensor (**arrow**). Check sensor resistance at terminals 1 (brown/white wire) and 3 (grey/green wire).

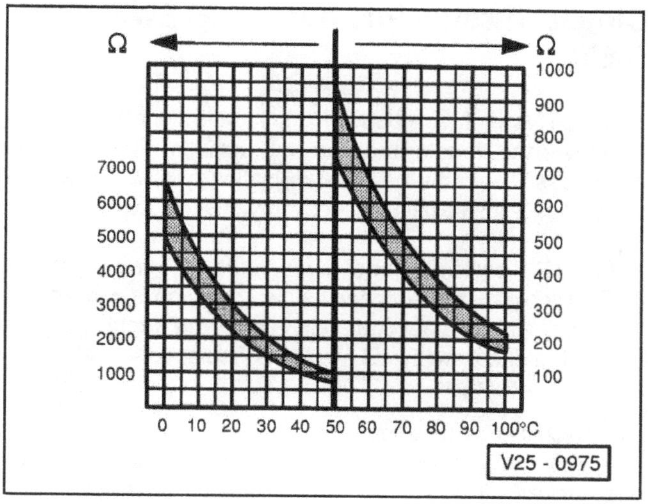

V25-975

Fig. 7. Graph showing coolant temperature sensor resistance vs. temperature.

To test the ECT sensor, disconnect the harness connector and connect an ohmmeter across the sensor terminals with the engine cold. Start the engine and allow it to warm up. Compare the meter readings to the graph shown in Fig. 7. If any faults are found, the sensor should be replaced.

Throttle Position (TP) Sensor (G69)

The TP sensor is mounted to the lower side of the throttle body housing, directly connected to the throttle valve shaft. See Fig. 8. The TP sensor signal provides throttle position signal (engine load) to the Mono-Motronic ECM.

Test the TP sensor signal using the test outlined in **Table b**. If the test results are incorrect, make the same test directly at the sensor terminals using the wiring diagrams shown in **97 Wiring Diagrams, Fuses and Relays.** If the sensor still tests incorrectly, the throttle body lower section should be replaced. The sensor is not adjustable.

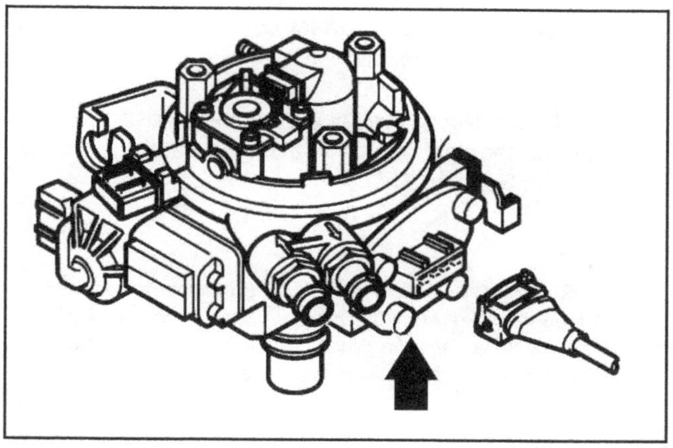
0024046

Fig. 8. Throttle position sensor mounted to side of throttle valve housing (**arrow**).

> ***CAUTION —***
> *The signal from the throttle position (TP) sensor is also used by the automatic transmission control module (TCM). Volkswagen specifies that if the Mono-Motronic ECM, the TCM, or the throttle body housing (lower section) are replaced, the ECM and TCM "Basic Settings" should be "initiated" using the special V.A.G. 1551 scan tool. If these basic settings are not done, transmission shifting and driveability problems may be encountered.*

NOTE —
Volkswagen does not supply the throttle position (TP) sensor as a separate part. If the sensor is faulty, the lower section of the throttle body should be replaced with integral TP sensor.

Closed Throttle Position (CTP) Switch (F60)

The CTP switch provides a closed throttle signal to the ECM whenever the throttle is at the rest position. This signal is mainly used to enable the idle stabilization system. The switch can be easily tested using the test previously listed in **Table b**.

If the CTP switch is not operating correctly, first check the CTP switch gap, as described below. If the gap is correct and the switch still does not operate correctly, the throttle position (TP) actuator assembly will need to be replaced.

> ***NOTE —***
> *The CTP switch is integral with the throttle position (TP) actuator. If the switch is faulty, the TP actuator assembly will need to be replaced.*

To adjust CTP switch gap

1. Disconnect the 4-pin harness connector from the throttle position (TP) actuator.

2. Briefly connect a 6-volt power source across terminal 3 (+) and terminal 4 (-) until the TP actuator is fully retracted. See Fig. 9.

> ***NOTE —***
> *Volkswagen special tool V.A.G. 1655 is a special resistor harness adaptor that reduces battery voltage by about 6 volts.If this harness adaptor is available, battery voltage can be applied directly to the TP actuator.*

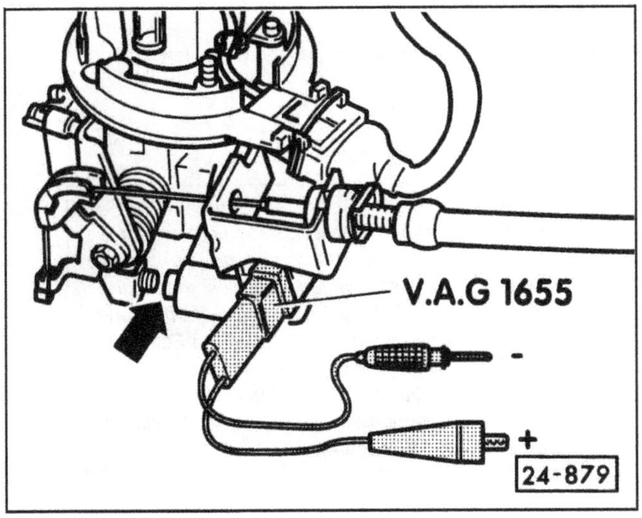

24-879

Fig. 9. Special harness adapter connected to TP actuator (terminals 1 and 2).

3. Push the TP actuator plunger in by hand and then check the gap using a feeler gauge. If necessary, adjust the throttle lever stop screw. See Fig. 10.

Throttle gap
- TP actuator fully retracted
  checking . . . . 0.4–0.5 mm (0.016–0.020 in.)
  adjusting. . . . . .0.5 mm (0.020 in.)

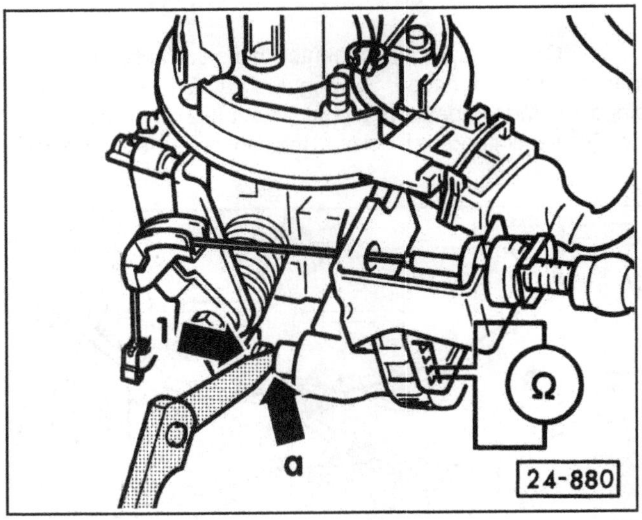

24-880

Fig. 10. Insert feeler gauge between throttle lever and TP actuator plunger (**1**). If necessary, adjust stop screw (**a**).

4. Connect an ohmmeter between the CTP switch terminals 3 and 4 (bottom terminals). See Fig. 10. There should be no continuity (switch open) with the throttle at rest.

5. Insert a 0.5 mm feeler gauge in the throttle gap and check that the switch closes (max. 200 ohms resistance). If the switch does not operate as specified, it is faulty and should be replaced.

Fuel Injector (N30)

The fuel injector is an electrically operated solenoid valve. Fuel pressure to the injector is held constant and regulated by the fuel pressure regulator. See Fig. 11.

> ***NOTE —***
> *Test the fuel pressure regulator by checking fuel pressure as described earlier under* ***Fuel Pressure Tests****.*

The injector is supplied 12 volts (+) via the fuel pump relay. The ground (–) side of the injector is switched on and off by the Mono-Motronic ECM. Injection quantity is based on the length of time the injector is open.

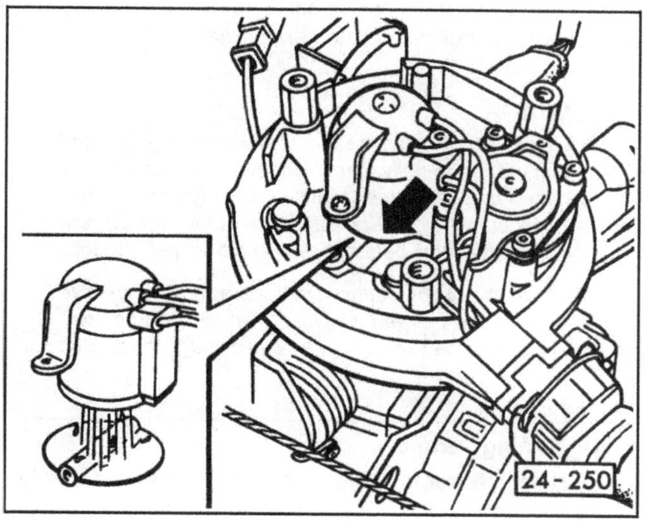

24-250

Fig. 11. Single (mono) fuel injector is positioned directly above throttle plate in throttle body.

To check injector operation, remove the adapter housing from the top of the throttle body. Start the engine and let it idle. If the car does not start, have a helper operate the starter. Look into the throttle body and check that the injector is spraying an even pattern of fuel. Turn the engine off and check the injector for leakage.

Fuel injector leakage
- max. allowable . . . no more than two drops per minute

If no faults are found, disconnect the harness connector from the injector and check the resistance across two center terminals in the injector.

Fuel injector resistance
- resistance @ 15-30°C (60–68°F). . . . 1.2 to 1.6 ohms

Check the signal to the injector by connecting an LED test light across the two center terminals in the harness connector See Fig. 12. When the engine is cranked, the LED should flash.

> ***CAUTION —***
> *Use only an LED test light. Use of an incandescent test light may damage the ECM.*

If the LED does not respond, check for power to the injector. There should be positive (+) battery voltage at terminal 2 in the connector during engine cranking. If not, check the wiring to the injector. If there is no pulsed ground signal, check the wiring between the ECM and the injector. If no wiring faults can be found, the ECM may be faulty.

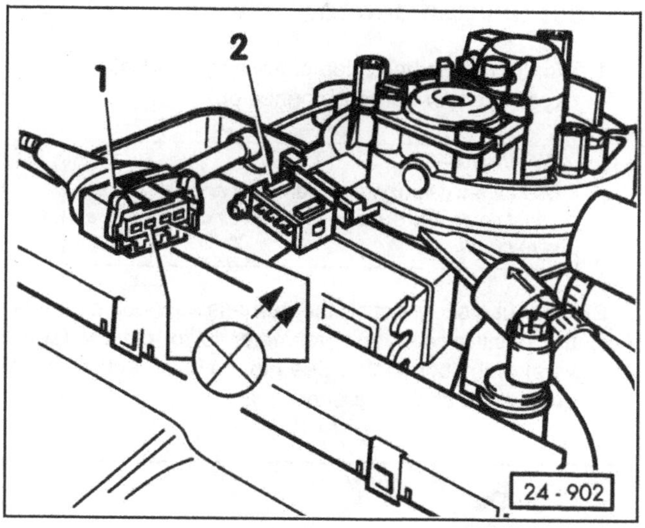

Fig. 12. LED test light (shown schematically) connected across center terminals in harness connector.

Oxygen Sensor (G39)

The heated oxygen sensor is threaded into the rear of the exhaust manifold and provides the Mono-Motronic ECM with feedback about the air-fuel ratio and combustion efficiency. See Fig. 13. Using this information, the ECM continuously adjusts the air-fuel mixture to ensure optimum driveability and exhaust emissions. The sensor must be at a temperature of at least 600°F to generate voltage, so it is electrically heated to help it reach operating temperature more quickly.

The oxygen sensor produces a small voltage (0-1 volt) based on the oxygen content in the exhaust gas as compared to the oxygen outside the exhaust pipe. The bigger the differential, the greater the output voltage. When the mixture is rich, there is very little oxygen in the exhaust (high output voltage). When the mixture is lean, there is an excess of oxygen in the exhaust (low output voltage).

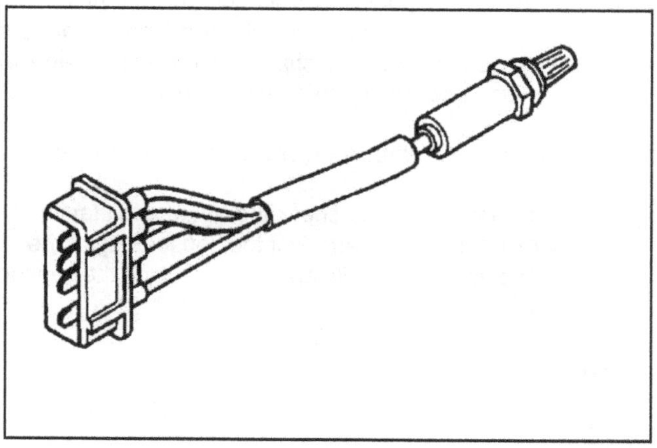

0024033

Fig. 13. Heated oxygen sensor monitors exhaust gas.

To test oxygen sensor

1. Working at the back side of the engine, locate the 4-pin oxygen sensor harness connector.

> ***WARNING —***
> *Use care when working near a hot engine. The exhaust pipes and the engine surfaces can cause severe burns.*

2. Without disconnecting the harness connector, connect the positive lead of a voltmeter to terminal 4 (signal wire) of the connector. See Fig. 14. Connect the negative lead to chassis ground.

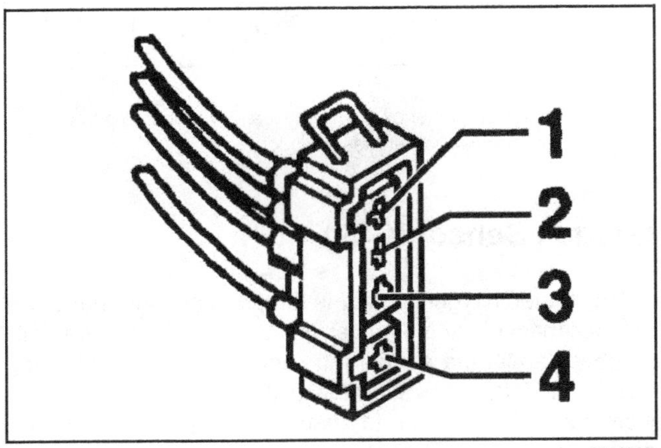

Fig. 14. Oxygen sensor harness connector terminal identification.

3. Start the car and let it idle. The oxygen sensor should start to produce a fluctuating voltage within a short period.

Oxygen sensor output signal
- voltage at idle . . . . 0.3 to 1.0 volt, fluctuating

---

4. If the sensor output is incorrect, turn the engine off and check the sensor heater circuit. Disconnect the 4-pin harness connector and check for battery voltage at terminals1 and 2 with the engine running.
   - If voltage is not present, check the wiring to the heater.
   - If voltage is present, check the heater resistance across terminals 1 and 2 of the sensor side of the connector. If any faults are found, replace the sensor.

Oxygen sensor
- heater resistance . . . . approx. 2 ohms

---

5. If the oxygen sensor signal is not as specified, and the heater circuit is OK, the sensor may be faulty.

NOTE —
- *If the sensor signal is either very low or very high and fluctuating slightly, the Mono-Motronic adaptability limits may have been reached. Look for signs of rich (low output signal) or lean (high output signal) running symptoms before replacing the oxygen sensor.*
- *Coat the oxygen sensor threads with an anti-seize compound before installation. Do not get the compound on the sensor tip. This can damage the sensor.*

Tightening Torque
- Oxygen sensor to exhaust manifold . . 50 Nm (37 ft-lb)

Throttle Position (TP) Actuator (V60)

The Mono-Motronic ECM electronically controls the idle speed through the throttle position (TP) actuator (V60). See Fig. 15. The system compensates for engine load and engine operating conditions, so periodic adjustment of the idle speed is unnecessary.

Idle speed
- engine warm. . . . . .750–1,000 rpm

NOTE —
Mono-Motronic idle speed stabilization also adapts ignition timing as part of the integrated idle speed control system.

The Mono-Motronic ECM monitors engine rpm and other operating conditions such as air conditioning and automatic transmission loads, and then sends an electrical signal to the throttle position (TP) actuator. This changes the position of the throttle plate. See Fig. 15.

> ***CAUTION —***
> *Do not attempt to correct a faulty idle speed by adjusting the position of the throttle valve stop screw. If this screw has been tampered with, check the throttle gap as described earlier under* **To adjust CTP switch gap**.

NOTE —
On high mileage cars, enough residue may collect on the throttle valve plate and bore to restrict air flow and affect the IAC system. If heavy residue is found, clean both the IAC valve and the throttle plate and bore before diagnosing IAC problems.

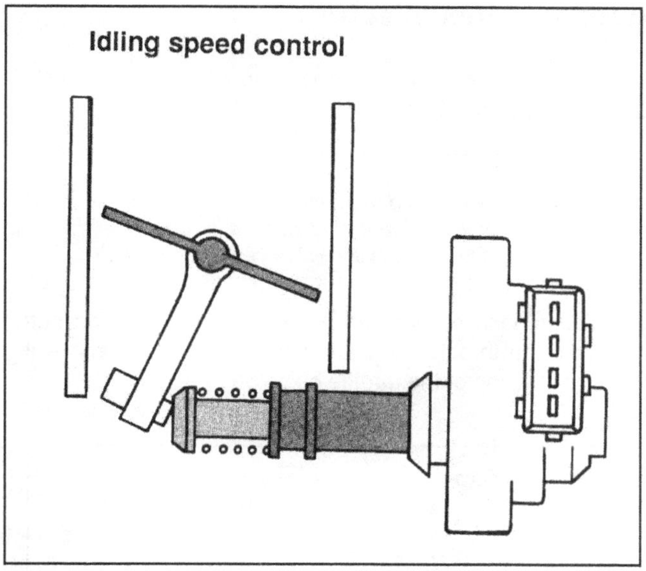

0024048

Fig. 15. Throttle position (TP) actuator acts directly on throttle plate to control idle speed.

Before troubleshooting an idle problem, check that the closed throttle position (CTP) switch is functioning correctly. See **Closed Throttle Position (CTP) Switch**. If the switch is operating correctly, check the resistance across terminals 3 and 4 at the TP actuator. See Fig. 16. If any faults are found, replace the TP actuator.

NOTE —

If you suspect an intermittent TP actuator, lightly tap the unit during the resistance check.

Throttle Position (TP) Actuator
- resistance. . . . . .3 to 200 ohms

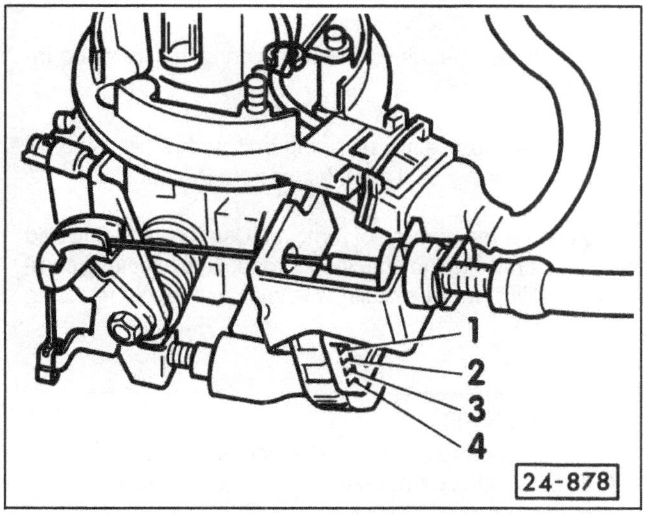

24-878

Fig. 16. Throttle position (TP) actuator terminal identification.

Intake Air Temperature (IAT) Sensor (G42)

The intake air temperature (IAT) sensor sends continuous intake air temperature information to the Mono-Motronic ECM. This input is used primarily as a correction factor for cold running injection duration. If there is a break in the signal from the IAT sensor, the ECU will simulate a fixed signal based on an engine temperature of approximately 68°F (20°C). See Fig. 17.

To test the ECT sensor, disconnect the harness connector and connect an ohmmeter across the two outer terminals in the connector. Compare the resistance reading to the graph shown earlier in Fig. 8. If any faults are found, the sensor should be replaced.

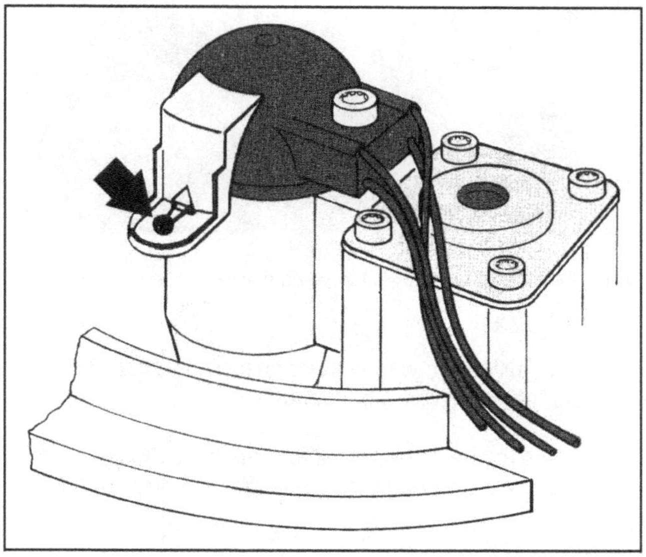

0024049

Fig. 17. Intake air temperature (IAT) sensor (**arrow**) is integrated into harness connector for injector.

COMPONENT REPLACEMENT

The throttle position (TP) sensor and fuel pressure regulator are not available separately from Volkswagen. If either of these parts are faulty, the upper or lower throttle body housing will need to be replaced. In addition, the closed throttle position (CTP) switch is integrated into throttle position (TP) actuator housing and is only available by purchasing the complete TP actuator unit.

To remove Mono-Motronic engine control module (ECM)

The ECM is mounted in the rear of the engine compartment, in the right-hand side of the air plenum. See Fig. 18.

Fig. 18. Mono-Motronic ECM in air plenum at rear of engine compartment.

1. With ignition off, wait at least one minute, then disconnect the negative (–) battery cable.

2. Unplug the harness connector from the ECM by releasing the fastener and then pivoting the connector off the module.

3. Remove the mounting bracket nut and remove the ECM together with its bracket.

Installation is the reverse of removal.

To remove engine coolant temperature (ECT) sensor

For ECT sensor location, see Fig. 6 given earlier.

1. Drain approximately one gallon of coolant from the cooling system. This is to prevent coolant from leaking out when the sensor is removed. See **19 Engine–Cooling System**.

2. Disconnect the sensor harness connector.

3. Remove the sensor retaining clip and pull the sensor from its housing.

Installation is the reverse of removal. Clean the sensor sealing surfaces and use a new O-ring when reinstalling the sensor.

To remove fuel injector

Cleanliness is important when working with fuel circuit components. Use clean tools.

> ***WARNING —***
> *Fuel will be expelled when the injector is removed. Do not smoke or work near heaters or other fire hazards. Keep a fire extinguisher handy.*

1. Disconnect the vacuum hoses and the retaining bolt from the throttle body adapter housing and remove it from the top of the throttle body. See Fig. 19.

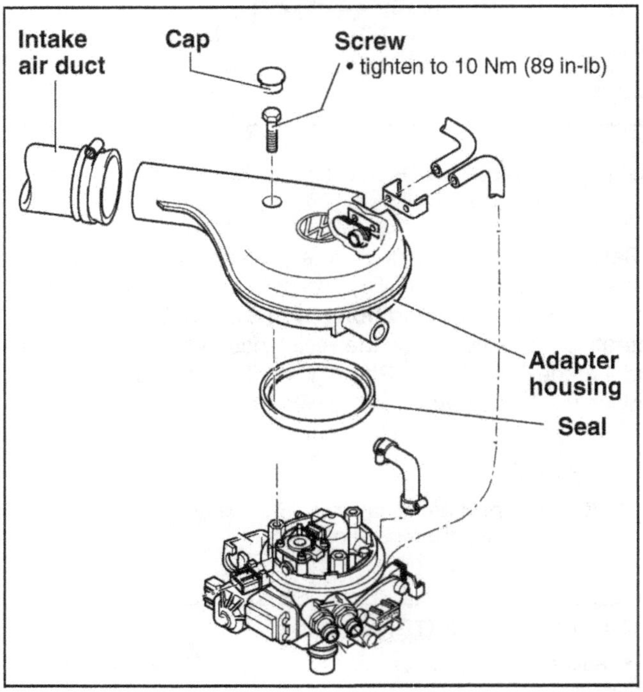

Fig. 19. Throttle body adapter housing and related parts.

2. Disconnect the wiring harness connector from the injector.

3. Remove the injector retaining bolt and remove the injector housing. Pull the injector from the housing.

Installation is the reverse of removal. Use a new O-ring and tighten the screws to the specified torque. Use a locking compound (VW "D6") when installing the injector housing retaining bolt.

Tighting torques

- Fuel injector housing to throttle body housing (use D6 locking compound) . . . . 5 Nm (44 in-lb)
- Throttle body adapter housing to throttle body . . . . 10 Nm (89 in-lb)

To remove throttle position (TP) actuator

1. Disconnect the wiring harness connector from the TP actuator.

2. Remove the TP actuator retaining bolts and remove the actuator. See Fig. 20 given later.

Installation is the reverse of removal. Tighten the screws to the specified torque. Use a locking compound (VW "D6") when installing the injector housing retaining bolt.

Tightening torque

- Throttle position (TP) actuator to throttle body .... 6 Nm (53 in-lb)

To remove and disassemble throttle body housing

Fig. 20 and Fig. 21 show the throttle body lower and upper sections. Use these views for removal, disassembly and assembly of the throttle body. Always replace gaskets and seals when reinstalling components.

> ***CAUTION —***
> *Volkswagen specifies that if the lower throttle body housing is replaced, the engine control module (ECM) and transmission control module (TCM) "Basic Settings" should be "initiated" using the special V.A.G. 1551 scan tool. If these basic settings are not re-established using the scan tool, transmission shifting and driveability problems may be encountered.*

1. **Throttle position (TP) actuator**
2. **Bolt**
   - tighten to 6 Nm (53 in-lb)
3. **Closed throttle position (CTP) switch**
   - to replace, replace complete throttle position actuator
4. **Accelerator cable bracket**
5. **Seal**
6. **Stop screw**
   - for closed throttle position (CTP) switch
7. **Throttle body, lower section**
8. **Stop screw**
   - Do not adjust–Factory sealed
9. **O-ring**
   - always replace
10. **Connector**
    - for EVAP emissions canister
11. **Bracket**
12. **Bolt**
    - tighten to 3Nm (27 in-lb)
13. **Connector**
    - for intake air preheating
14. **Retaining clip**
15. **Throttle position sensor**
    - to replace, replace throttle body lower section
16. **Water trap**
    - for throttle position sensor
17. **Tie wrap**

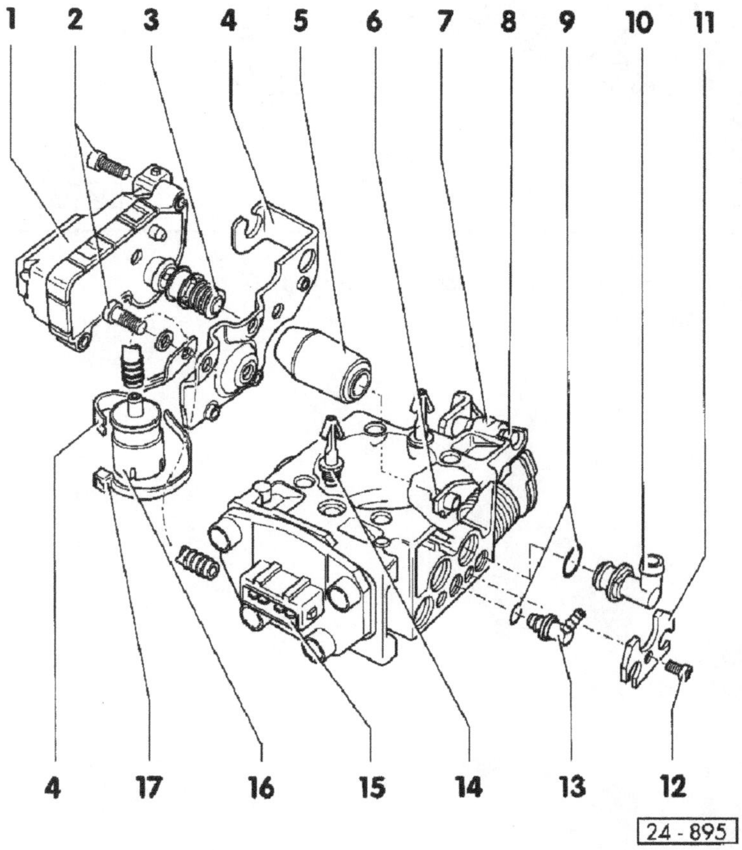

24-895

Fig. 20. Throttle body lower section and related parts.

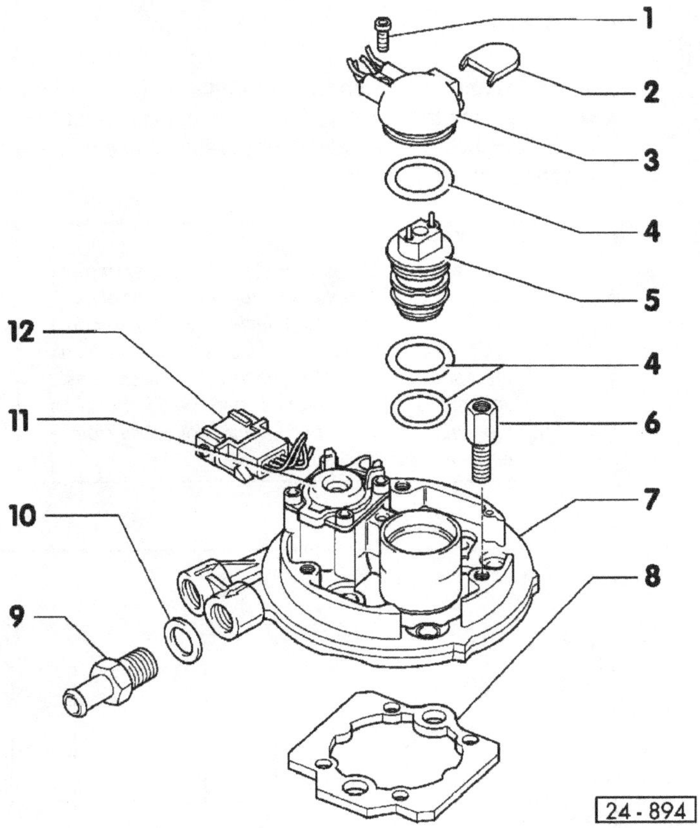

1. **Screw**
   - use "D6" locking compound when installing
   - tighten to 5 Nm (44 in-lb)
2. **Protective cap**
   - only for shipping-discard
3. **Fuel injector housing**
   - with integrated intake air temperature (IAT) sensor
4. **O-ring**
   - always replace
5. **Fuel injector (N30)**
6. **Screw**
   - use "D6" locking compound when installing
   - tighten to 6 Nm (53 in-lb)
7. **Throttle body, upper section**
8. **Gasket**
   - note installation position
   - always replace
9. **Connector**
10. **Seal**
    - always replace
11. **Fuel pressure regulator**
    - to replace, replace throttle body upper section
12. **Harness connector**
    - unlock to remove

24-894

Fig. 21. Throttle body upper section and related parts.

26 Exhaust System/Emission Controls

TABLES

GENERAL

This repair group covers repair and troubleshooting for the emission control systems and removal and installation of exhaust components.

> ***NOTE —***
> - *Oxygen sensor (Lambda) system repair and troubleshooting is covered in* **24a Fuel Injection—Motronic or 24b Fuel Injection—Mono-Motronic**.
> - *All cars have a self-diagnostic program to detect emissions-related problems and store coded fault information in electronic memory. When diagnosing an emissions-related faults, first check for stored fault codes, especially if the Malfunction Indicator (Check Engine) Lamp is illuminated. See* **24a Fuel Injection—Motronic or 24b Fuel Injection—Mono-Motronic**.
> - *The cars covered by this manual have a vacuum hose routing diagram (sticker) located in the engine compartment which can be helpful when working with the vacuum hoses.*

Service Precautions

To help guard against personal injury or damage to car components, the following warnings and cautions apply when servicing the exhaust system.

> ***WARNING —***
> *Exhaust gases are colorless, odorless, and very toxic. Run the engine only in a well-ventilated area. Immediately repair any leaks in the exhaust system or structural damage to the car body that might allow exhaust gases to enter the passenger compartment.*

> ***WARNING —***
> - *The exhaust system, catalytic converter and other emission control systems operate at very high temperatures. Allow these components to cool before servicing, or wear protective clothing to prevent burns. Do not use flammable chemicals near a hot catalytic converter.*
> - *Old corroded exhaust system components crumble easily and often have exposed sharp edges. To avoid injury, wear eye protection and heavy gloves when working with such parts.*
> - *Do not work under a lifted car unless it is solidly supported on jack stands designed for that purpose. Never work under a car that is supported solely by a jack.*

EXHAUST SYSTEM

For safe and proper exhaust system operation, all components must be free of holes and all connections must be airtight. Check the system immediately if it becomes noisy or an exhaust odor is detected inside the car.

> ***NOTE —***
> *Excessive exhaust system back pressure can cause driveability problems such as a rough idle or stalling. Back pressure problems are caused by external damage such as crushed or collapsed parts, or internal damage such as a plugged catalytic converter.*

Exhaust System Replacement

Fig.1 and Fig. 2 show the exhaust system used on the ABA (2.0 liter) 4-cylinder engine. Fig. 3 shows the exhaust system used on the AAA (2.8 liter) 6-cylinder engine. Figs. 4–6 show the exhaust systems used on the AAZ and 1Z/AHU (TDI) diesel engines. Use these illustrations as a guide for removal and installation. Tightening torques and other relevant installation information are given in the illustrations.

A liberal application of penetrating oil to cold exhaust system nuts, bolts and slip joints will make removal easier. New fasteners, clamps and rubber mounts are always recommended when replacing exhaust components. Gaskets should be replaced whenever flange joints are disconnected. Use anti-seize compound on threaded fasteners to extend their service life and make future replacement easier. Some slight smoking and/or odor is normal as new exhaust system parts become hot for the first time.

> ***WARNING —***
> *Inspect the exhaust system heat shields on the car underbody and repair or replace any damaged parts. Heat shields protect the car occupants, undercoating, and various other components from excessive heat. Damaged or missing shields, particularly those above the catalytic converter, will increase interior temperatures and create a fire hazard.*

1. **Bracket**
2. **Nut**
   • tighten to 25 Nm (18 ft-lb)
3. **Heat deflector**
4. **Stud**
   • tighten to 35 Nm (30 ft-lb)
5. **EGR valve (where applicable)**
6. **EGR temperature sensor (where applicable)**
7. **Fitting**
8. **Nut**
   • tighten to 30 Nm (22ft-lb)
9. **Cap, for CO tap tube**
10. **Bolt Nut**
   • tighten to 10 Nm (89 in-lb)
11. **CO tap tube (where applicable)**
12. **Heated oxygen sensor (HO2S)**
   • tighten to 50 Nm (36 ft-lb)
   • use anti-seize when installing
   • testing, see **24 Fuel Injection**
   • from 08/95 HO2S installed in lower portion of front exhaust pipe (pre-cat) and at rear of catalytic converter (post-cat)
13. **Three-way catalytic converter (TWC)**
14. **Sealing ring**
   • always replace
15. **Nut**
   • tighten to 40 Nm (30 ft-lb)
16. **Heat shield**
17. **Front exhaust pipe**
   • when installing pipe to manifold: tighten bolts gradually and evenly, starting from the center and working outward
18. **Gasket**
   • always replace
19. **Exhaust manifold**
20. **Spacer**

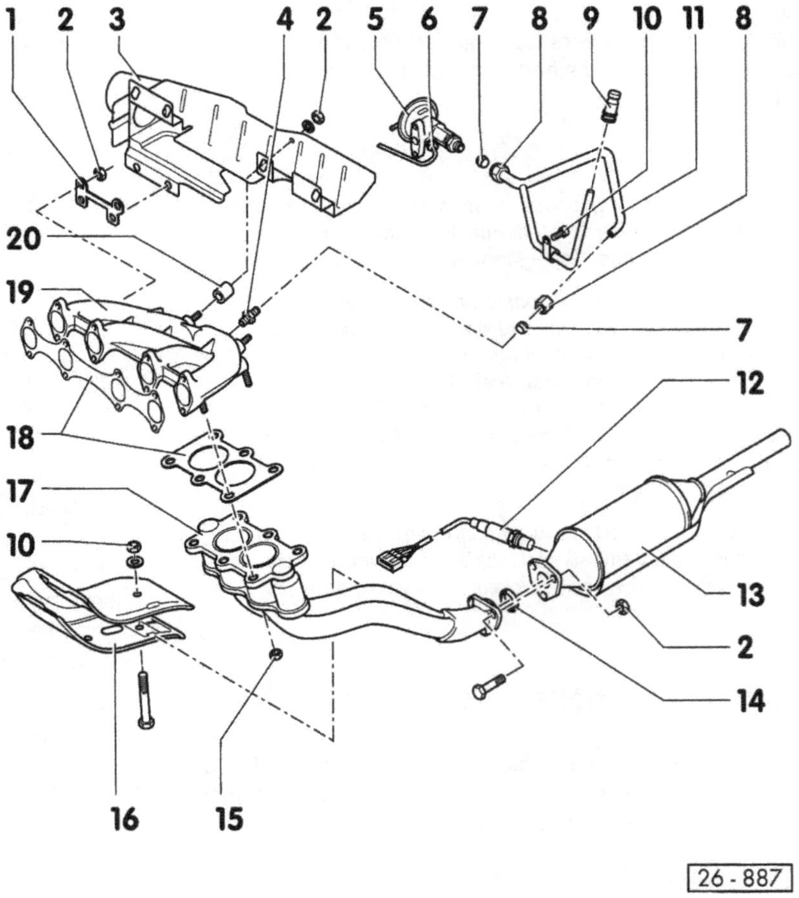

26-887

Fig. 1. Front section of exhaust system for ABA (2.0 liter) 4-cylinder engine.

1. **Nut**
   • tighten to 40 Nm (30 ft-lb)
2. **Locking clamp**
3. **Intermediate pipe**
4. **Bracket**
5. **Bolt**
   • tighten to 25 Nm (18 ft-lb)
6. **Front and rear muffler**
   • available as separate replacement parts
   • when replacing individual muffler: cut connecting pipe at point 9 (**inset B**) and connect using double clamp
7. **Dimension c: approx. 10 mm (0.40 in.)**
8. **Marking for double clamp position**
   • marked 3 times on pipe circumference
9. **Cutting point for replacing individual muffler**
10. **Dimension b: approx. 5 mm (0.20 in.)**
11. **Marking for clamp position**
    • marked 3 times on pipe circumference
12. **Retaining ring**
13. **Bracket**
14. **Nut**
    • tighten to 20 Nm (15 ft-lb)
15. **Bracket**

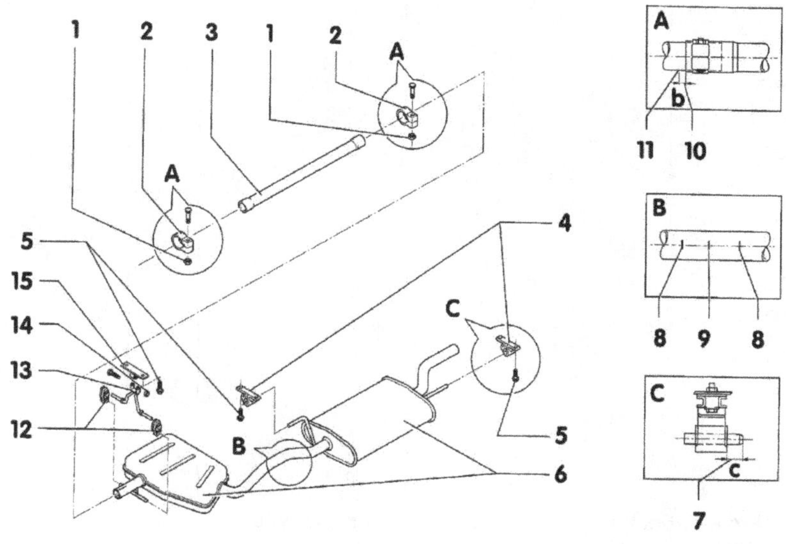

0024058

Fig. 2. Rear section of exhaust system for ABA (2.0 liter) 4-cylinder engine.

1. **Exhaust manifold, two-piece**
2. **Nut**
   • tighten to 25 Nm (18 ft-lb)
3. **CO tap tube (where applicable)**
4. **Harness Connector, 4-pin**
5. **Cap, for CO tap tube**
6. **Heated oxygen sensor (HO2S)**
   • tighten to 50 Nm (36 ft-lb)
   • use anti-seize on threads when installing
   • testing, see **24 Fuel Injection**
7. **Nuts**
   • M8: tighten to 25 Nm (18 ft-lb)
   • M10: tighten to 40 Nm (30 ft-lb)
8. **Double clamp**
9. **Center muffler**
10. **Mounting**
11. **Rear muffler**
12. **Dimension c: approx. 10 mm (0.40 in.)**
13. **Marking for clamp position**
    • marked 3 times on pipe circumference
14. **Cutting point for replacing individual muffler**
15. **Dimension b: approx. 5 mm (0.20 in.)**
16. **Marking for clamp position**
    • marked 3 times on pipe circumference
17. **Marking for clamp position**
    • marked 3 times on pipe circumference
18. **Dimension a: approx. 5 mm (0.20 in.)**
19. **Clamp**
20. **Support**
21. **Mounting**
    • note installed position
22. **Mounting**
23. **Front silencer**
24. **Three-way catalytic converter (TWC)**

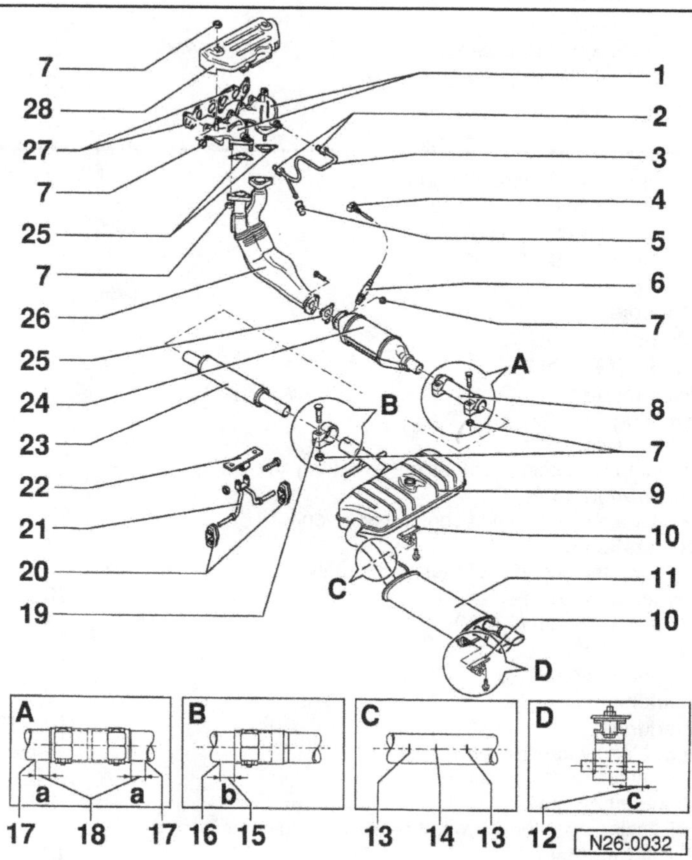

25. **Gasket**
    • always replace
26. **Front exhaust pipe**
27. **Gasket, 2-piece**
    • always replace
28. **Heat shield**

N26-0032

Fig. 3. Exhaust system for AAA (2.8 liter) 6-cylinder engine.

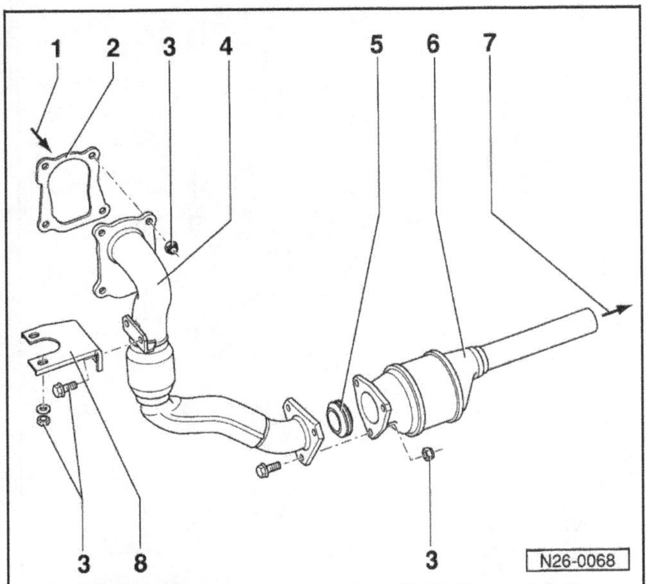

1. **From turbocharger**
2. **Gasket**
   • Always replace
3. **25 Nm (18 ft-lb)**
4. **Front exhaust pipe**
5. **Sealing ring**
   • Replace if damaged
6. **Oxidation catalytic converter**
7. **To intermediate pipe**
8. **Retainer**
   • Between exhaust pipe and exhaust manifold

N26-0068

Fig. 4. Front section of exhaust system for AAZ (1.9 liter) 4-cylinder diesel.

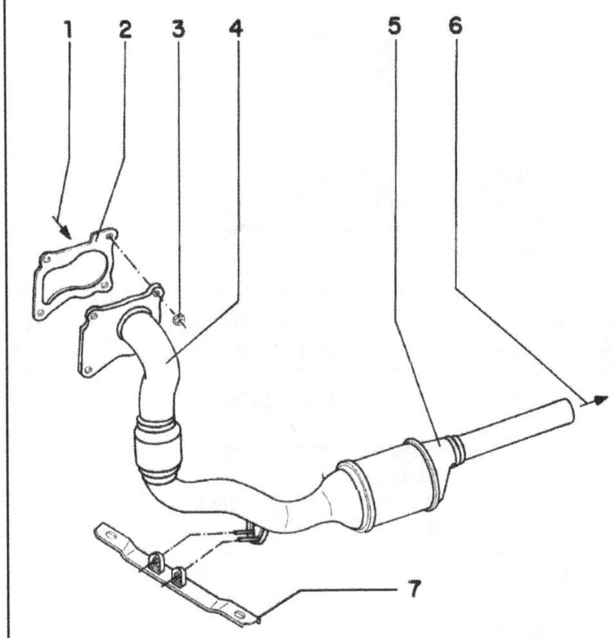

1. **From turbocharger**
2. **Flange gasket**
   • Always replace
3. **25 Nm (18 ft-lb)**
4. **Front exhaust pipe**
5. **Oxidation catalytic converter**
6. **To intermediate tube**
7. **Mounting bracket**

0024197

Fig. 5. Front section of exhaust system for 1Z/AHU (1.9 liter) 4-cylinder TDI diesel.

1. **From oxidation catalytic converter**
2. **40 Nm (30 ft-lb)**
3. **Double clamp**
4. **Intermediate pipe**
5. **Clamp**
6. **Console**
7. **Mount**
8. **25 Nm (18 ft-lb)**
9. **Rear muffler**
10. **Marking**
    • For clamp
    • Clamp to marking (dimension -a- of item 14)
11. **Separation point**
    • Allows replacement of center muffler only
12. **Separation point**
    • Allows replacement of rear muffler only
13. **Dimension -b-**
    • Approx. 10 mm (0.4 in.)
14. **Dimension -a-**
    • Approx. 5mm (0.2 in.)
15. **Marking**
16. **Marking**
    • S: Manual transmission
    • A: Automatic transmission
17. **Connection pipe**
    • Repair pieces for use when replacing single muffler
18. **Center muffler**
    • 1Z/AHU shown, AAZ is similar

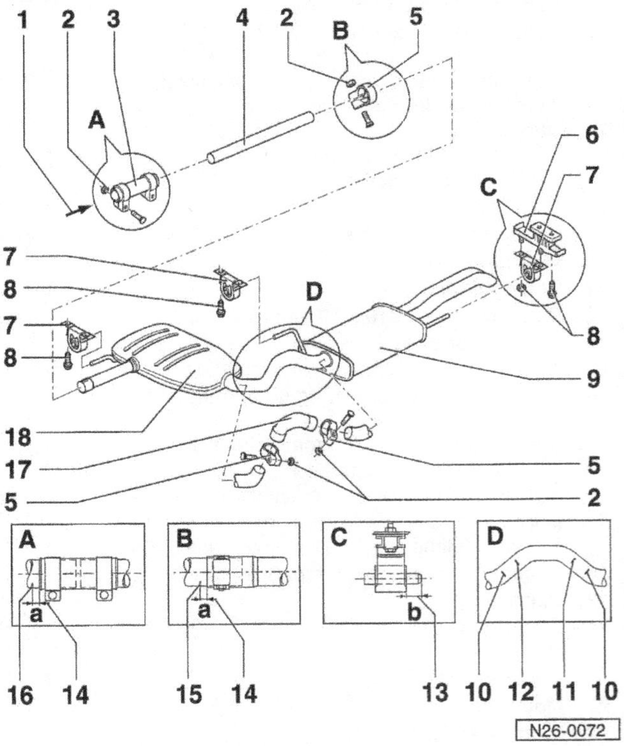

N26-0072

Fig. 6. Rear section of exhaust system for 1Z/AHU (1.9 liter) 4-cylinder TDI diesel. AAZ engine is similar.

EMISSION CONTROLS

Different engines use different emission control systems. See the following application table.

Table a. Emission control system applications

Engine	Emission control systems used
ACC (1.8 liter)	Positive Crankcase Ventilation (PCV) Evaporative Emission (EVAP) System Oxygen Sensor (Lambda) System Three-way catalytic converter
AAZ, 1Z, AHU (1.9 liter)	Positive Crankcase Ventilation (PCV) Oxidation Catalytic Converter Exhaust Gas Recirculation (EGR)
ABA (2.0 liter)	Positive Crankcase Ventilation (PCV) Evaporative Emission (EVAP) System Exhaust Gas Recirculation (EGR) System (where applicable) Oxygen Sensor (Lambda) System Three-way catalytic converter
ABA OBD II (1996-1998)	Leak Detection Pump (LDP) (where applicable) Secondary Air Injection (AIR) (where applicable)
AAA (2.8 liter)	Positive Crankcase Ventilation (PCV) Evaporative Emission (EVAP) System Exhaust Gas Recirculation (EGR) System (where applicable) Oxygen Sensor (Lambda) System Secondary Air Injection (AIR) System Three-way catalytic converter
AAA OBD II (1996-1998)	Leak Detection Pump (LDP) (where applicable)

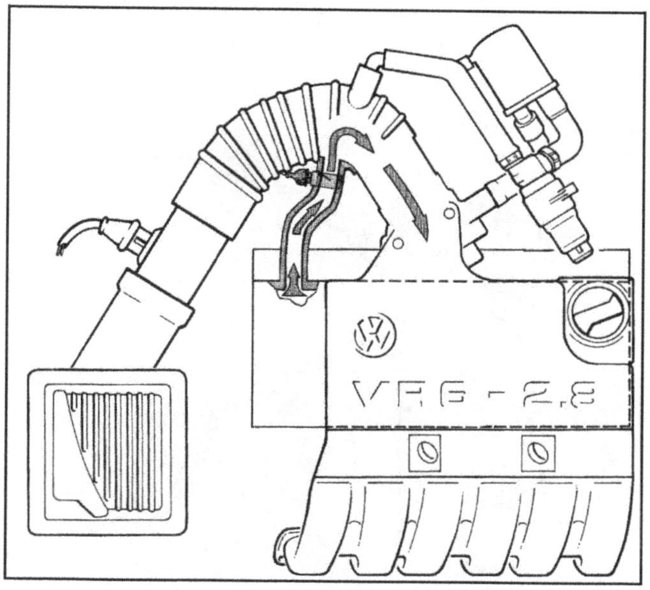

0024054

Fig. 7. PCV system on AAA (2.8 liter) engine. OBD I with idle stabilizer shown.

Most crankcase ventilation problems result when the hoses or valves become clogged with oily residues. Restrictions create excessive crankcase pressure that can eventually cause driveability problems. PCV system service is limited to inspecting and cleaning the breather valves, hoses, and replacing faulty parts.

Positive Crankcase Ventilation (PCV)

The PCV system traps crankcase vapors and routes them back into the intake air stream to be burned.

On 4-cylinder engines, the system consists of a crankcase breather valve on the side of the cylinder block and connecting hoses to the intake air stream. On ABA, AAZ, 1Z and AHU engines, a replaceable pressure regulating valve is used in the top of the valve cover.

On 6-cylinder engines, the system consists simply of a breather hose between the valve cover and the intake air boot. The valve cover contains a flame trap to prevent ignition of the crankcase vapors in the event of a backfire. To prevent icing during cold weather, a heating element is integrated into the breather hose. See Fig. 7.

> ***CAUTION —***
> *Replace PCV hoses only with parts designed for PCV or fuel system service. Conventional vacuum and heater hoses deteriorate rapidly when exposed to oil vapors and combustion gasses.*

Evaporative Emission (EVAP) 1993–1995

The EVAP system uses a charcoal-filter canister to absorb vapors from the fuel tank. The fuel tank is vented to the canister and the canister is in turn vented to the intake manifold via the EVAP canister purge regulator valve.

During certain engine operating conditions, the accumulated vapors in the canister are drawn through the open EVAP regulator valve and into the engine to be burned. The EVAP regulator valve is cycled on and off by the engine control module (ECM).

The charcoal canister is mounted in the driver's side wheel well, behind the headlight. The EVAP canister purge regulator valve is mounted on the side of the air cleaner housing. See Fig. 8.

The EVAP canister purge regulator valve controls the amount of fuel vapors drawn into the engine and therefore can affect the fuel mixture and exhaust emissions. A fault in the EVAP system usually results in overly rich driveability problems, especially at idle. Check the purge regulator valve with the engine idling. The valve should be sporadically clicking on and off. The clicking should vary in intensity depending on engine speed.

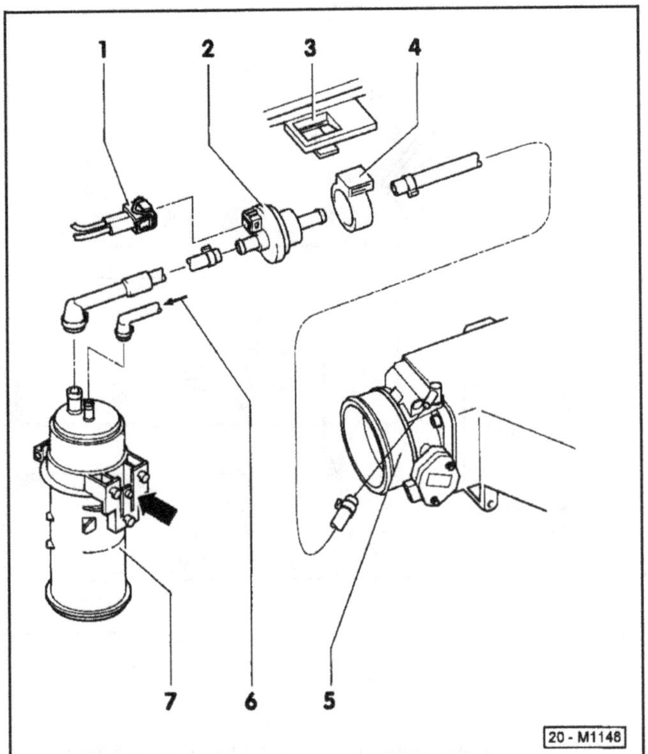

1. **Harness connector**
2. **EVAP canister purge regulator valve**
3. **Bracket, on air cleaner housing**
4. **Retainer**
5. **Throttle valve**
6. **Hose, from fuel tank via gravity valve**
7. **EVAP charcoal canister**

Fig. 8. Evaporative emission (EVAP) system on ABA (2.0 liter) engine.

The electric valve can be tested by disconnecting its harness connector and supplying battery voltage to the valve terminals. The valve should click open when voltage is supplied. If no faults are found, check for voltage to the valve's connector with the engine running. There should be battery voltage between the red/black or red wire and ground. If voltage is not present, check for faults using the appropriate wiring diagram shown in **97 Wiring Diagrams, Fuses and Relays**.

Evaporative Emission (EVAP) 1996–on

The evaporative emissions system found on 1996 and later gasoline engines operates in much the same manner as on previous versions. Additional components required as a result of OBD II standards allow the system to be monitored for leaks. This system can only be properly tested using a scan tool such as VAG 1551 or VAG 1552 and proper operation is included in the system readiness codes. See **20 Fuel Storage and Supply.**

Exhaust Gas Recirculation (EGR) – Gas Engines

The EGR system reduces emissions by directing a small quantity of exhaust gases back into the intake manifold to dilute the air/fuel mixture effectively. This reduces the amount of oxides of nitrogen (NO_x) pollutants in the exhaust.

A schematic view of the EGR system is shown below. See Fig. 9.

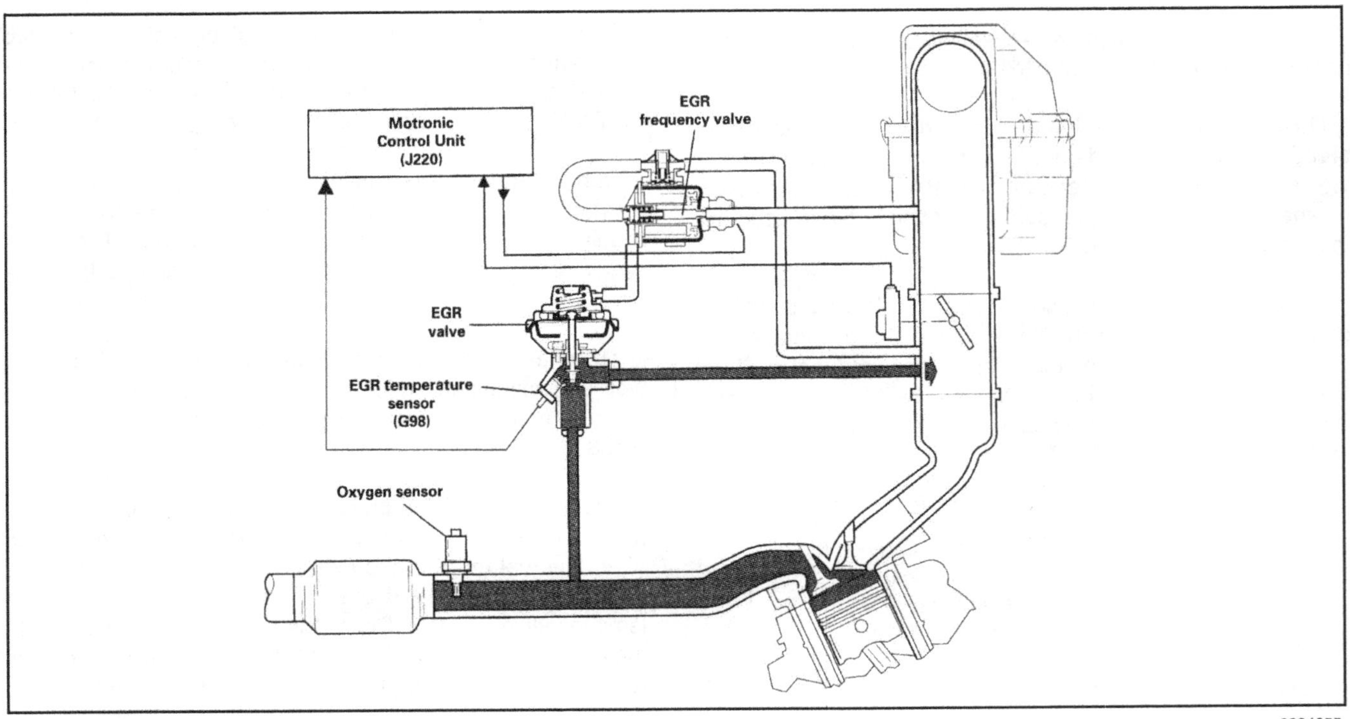

Fig. 9. Schematic view of electronic EGR system.

The system consists of the EGR vacuum regulator solenoid valve, the EGR temperature sensor, and the mechanical EGR valve. The EGR solenoid valve controls the amount of vacuum supplied to the EGR valve based on engine speed and engine load. The EGR temperature sensor monitors the opening and closing of the EGR valve as part of the on-board diagnostic system.

> ***CAUTION —***
> *Use care when handling the temperature sender. The sender is sensitive to shock and impact and can be easily damaged.*

A rough idle may indicate that the EGR valve is sticking open. Stalling or part-throttle surging when the engine is cold may indicate that the EGR valve is incorrectly opening before the engine is fully warmed.

To test EGR system – gas engines

1. Check that the EGR system vacuum hoses are routed correctly and free from obstructions and leaks. See Fig. 10.

> ***WARNING —***
> *Parts of the EGR system valve carry engine exhaust gasses and operate at high temperatures. Exhaust gases are poisonous and they may cause severe burns if handled without proper protection. Wear heavy gloves when working with the EGR valve. Work in a well ventilated area.*

2. With the engine fully warm and idling, disconnect the vacuum hose from the EGR valve. The idle speed should not change.

3. Raise the idle speed and check for varying amounts of vacuum at the disconnected hose. If necessary, connect a vacuum gauge to the hose.

> ***NOTE —***
> *Vacuum at the hose will vary depending on engine load and speed. A vacuum gauge is sensitive enough to pickup the fluctuating vacuum source.*

4. If there is vacuum, attach a hand vacuum pump to the EGR valve vacuum fitting and apply vacuum to the valve. The idle speed should drop noticeably and the engine may even stall.

> ***NOTE —***
> *If there is no change in idle speed, the EGR valve is not opening mechanically and needs to be cleaned or replaced.*

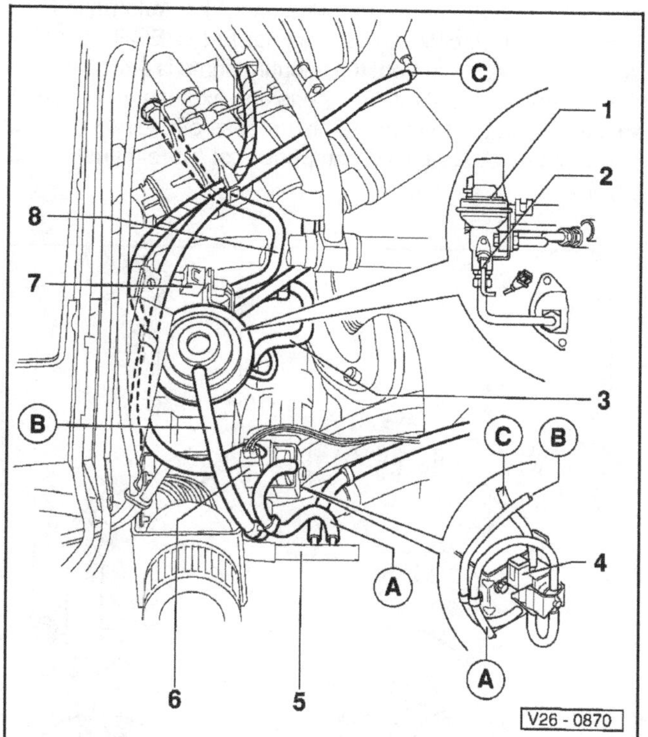

1. **EGR valve**
2. **EGR temperature sensor (G69)**
3. **Exhaust pipe, inlet**
4. **EGR vacuum regulator valve**
5. **Vacuum connection**
6. **Harness connector**
7. **Harness connector**
8. **Exhaust pipe, outlet**

V26-0870

Fig. 10. EGR system layout on AAA engine (2.8 liter).

5. If there is no vacuum, check the EGR vacuum regulator solenoid valve. Check the wiring to the valve and check for a duty cycle (on-off) signal at the valve. If no faults can be found, the signal from the engine control module (ECM) may be missing or the valve or ECM may be faulty.

> ***NOTE —***
> *If the Malfunction Indicator Lamp (Check Engine) light is illuminated, a Diagnostic Trouble Code (DTC) has been stored in the OBD system memory and should be erased. See* **24a Fuel Injection–Motronic or 24b Fuel Injection–Mono-Motronic**.

Exhaust Gas Recirculation (EGR) – AAZ diesel engines

The EGR system on the AAZ Diesel engine reduces the emissions in much the same way as on the gasoline engines. The EGR valve is switched on during certain part-throttle applications. It is switched off at idle and at full throttle, and also during high altitude operation. Control of the EGR system is through the Automatic Glow Time Control Module (J179) by way of input from the EGR Part Throttle Switch (F166) on the injection pump, and Engine Temperature Sensor (G27) on the cylinder head. At the appropriate time, the Automatic Glow Time Control Module sends a signal to the EGR Two-Way Valve where vacuum from the vacuum pump is directed to the EGR Valve. The EGR Valve opens and exhaust gas is directed into the intake manifold. This process effectively reduces the Nitrogen Oxide (NOx) component of the exhaust. See Fig. 10.

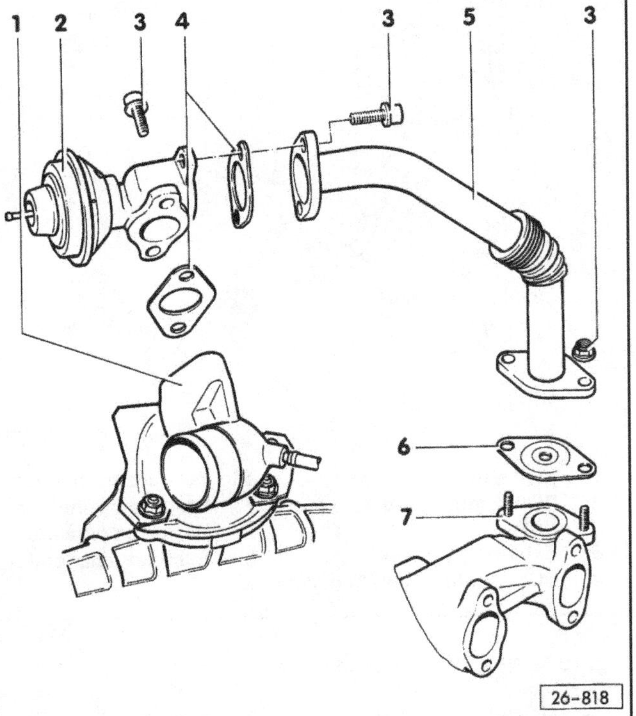

1. **Intake manifold**
2. **EGR valve**
3. **25 Nm (18 ft-lb)**
4. **Gasket**
   • Always replace
5. **Connecting pipe**
6. **Gasket**
   • Always replace
   • Restriction hole 9 mm (0.35 in.) dia.
7. **Exhaust manifold**

V26-0870

Fig. 11. EGR valve and related components fro AAZ diesel engine.

To test EGR valve – AAZ diesel engines

1. Start engine and allow to run until the oil temperature is a minimum of 122°F (50°C).

2. From an idle, slowly raise engine speed to over 1200 RPM. Diaphragm rod in EGR valve should be seen to move towards intake manifold for approximately 2 seconds and then move back. Diaphragm rod is visible through opening in the EGR valve. See Fig. 12.

V26-0821

Fig. 12. EGR valve mounted to intake manifold on AAZ Diesel engines. Diaphragm rod is visible through opening in the valve (**arrow**). Rod should move towards the vacuum line connection.

3. Movement of the rod indicates normal operation. If the rod does not move, visually inspect the vacuum connections to the EGR valve and the Two-Way valve and compare them to the vacuum hose layout sticker under the hood.

4. If no problems are found with the hoses, continue by mechanically checking the EGR valve.

5. Disconnect the vacuum hose from the valve and attach a hand operated vacuum pump such as Volkswagen special tool VAG 1390 or US 8026. Operate the vacuum pump and observe the diaphragm rod. It should move when vacuum is applied.

If the diaphragm rod does not move, there is a mechanical malfunction with the valve and it should be replaced. If the diaphragm rod moves when vacuum is applied, but does not move under the control of the Automatic Glow Time Control Module, there may be an electrical fault in the system. Check the wiring to the Two-Way valve, the Engine Temperature Sensor and the Part Throttle Switch. If no faults can be found, the signal from the Automatic Glow Time Control Module may be missing or faulty, or the Two-Way Valve, Temperature Sensor or Part Throttle Switch may be faulty. See **97 Wiring Diagrams, Fuses, and Relays**.

Exhaust Gas Recirculation (EGR) system – 1Z/AHU TDI diesel engines

The EGR system on the 1Z/AHU TDI Diesel engine reduces the emissions in much the same way as on the AAZ Diesel engine. The EGR valve is functionally controlled by the Engine Control Module (J248) which uses input from the various engine sensors to precisely control the quantity of exhaust gases returned to the intake manifold. The exhaust gas is pulled from a port in the exhaust manifold through large diameter pipes and passed through a cooler. From there, it goes to the EGR valve which is integrated into the intake manifold.

The EGR cooler is connected to the engine cooling system outlet on the cylinder head and the coolant reservoir. See Fig. 13. In operation, the ECM sends a signal to the EGR Vacuum Regulator Solenoid (N18) which controls the vacuum signal sent to the EGR valve. Because the Diesel engine has no throttle plate, exhaust gas recirculation will sometimes occur at idle in addition to part throttle.

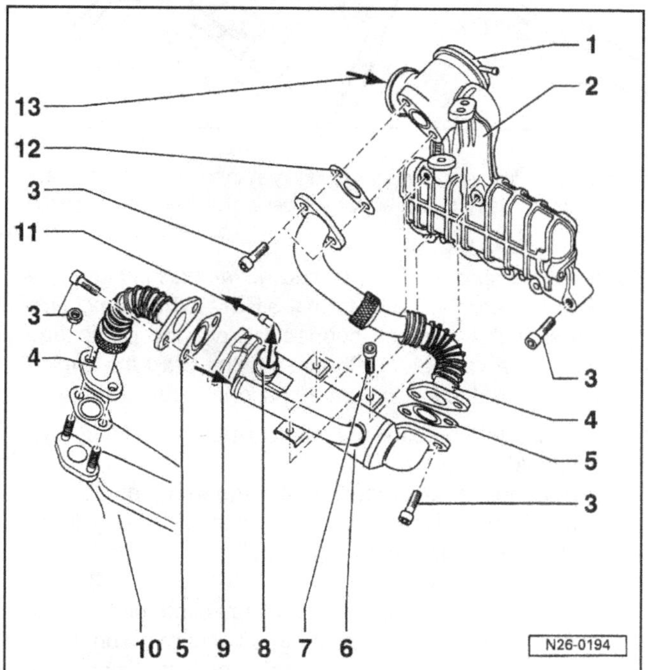

1. **EGR valve**
2. **Intake manifold**
3. **25 Nm (18 ft-lb)**
4. **Connecting pipe**
   - When installing. tighten all bolts finger tight first
5. **Gasket**
   - Always replace
6. **EGR cooler**
7. **10 Nm (7 ft-lb)**
8. **To heater core**
9. **From cylinder head**
10. **Exhaust manifold**
11. **To overflow reservoir**
12. **Flange gasket**
    - Always replace
13. **From intercooler**

N26-0194

Fig. 13. EGR valve, EGR cooler and related components for 1Z/AHU TDI diesel engine.

To test EGR valve – 1Z/AHU TDI diesel engines

NOTE —

Proper diagnosis and testing of the EGR system requires Volkswagen scan tool VAG 1551 or VAG 1552. The EGR valve itself can be mechanically tested.

1. Remove the connecting pipe from the intercooler to the intake manifold.
2. Disconnect the vacuum hose from the EGR valve.
3. Connect a hand operated vacuum pump such as VAG 1390 or US 8026 to the EGR valve.
4. Look into the exposed end of the intake manifold, identify the valve plate, and operate the hand pump. The valve plate must lift from the valve seat when vacuum is applied to the EGR valve. See Fig. 14.

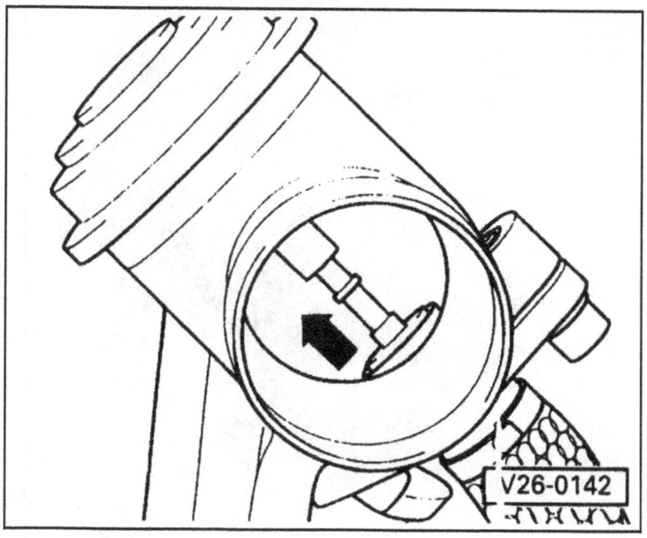

V26-0142

Fig. 14. EGR valve as viewed through the open end of the intake manifold with the pipe from the intercooler disconnected. Valve must move in the direction of the **arrow** when vacuum from a hand operated pump is applied to the valve.

5. Disconnect the vacuum pump and listen for an audible sound as the valve snaps closed against the seat.
6. If valve functions properly, further testing using Volkswagen scan tools VAG 1551 or VAG 1552 must be performed.
7. If the valve fails to open or if it fails to fully close, the intake manifold with the integral EGR valve must be replaced.

Secondary Air Injection (AIR)

The secondary air injection system uses an air pump to inject fresh air behind the exhaust valves for approximately 65 seconds during cold starting (engine coolant between 58°F and 95°F). As part of the on-board diagnostic system, the pump also runs briefly after a restart with the engine warm.

The purpose of the AIR system is to reduce exhaust emissions during engine warm-up when the Motronic engine management system is in open loop.

The system consists of the electric air pump, the vacuum-operated shut-off valve, the inlet (solenoid) valve, and the related duct work. See Fig. 15.

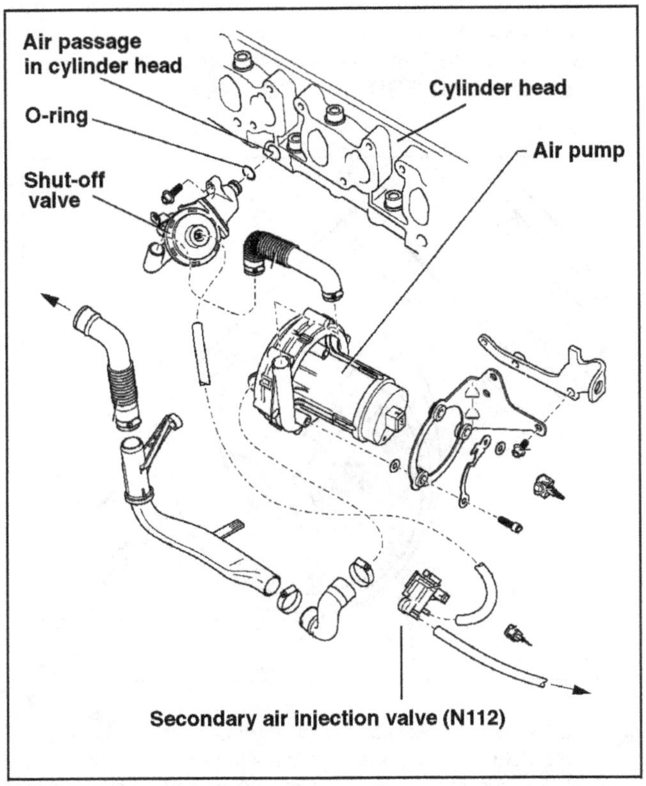

Fig. 15. Secondary air injection (AIR) system used on VR6 engines. 4-cylinder ABA engines are similar.

NOTE—

The AIR system is installed on 4-cylinder ABA engine from 06/96 production.

To test AIR system

1. Disconnect the air hose from the outlet on the top of the air pump.

2. With the engine fully warm, start and allow the engine to idle. Approximately 20 seconds after starting, the air pump should run for about five seconds and secondary air should be felt coming from the pump outlet.
   - If the air pump does not run, check the in-line 30 amp fuse and the secondary air injection (AIR) pump relay (J299). See Fig. 16.

NOTE—

*Wiring diagrams are given in **97 Wiring Diagrams, Fuses and Relays**. If no faults are found, continue testing.*

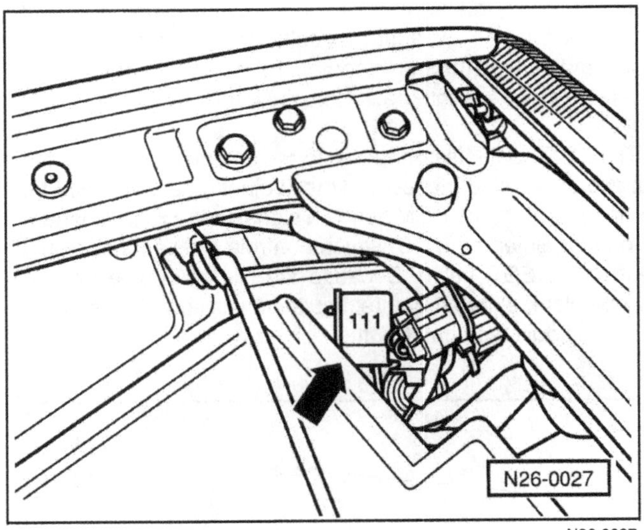

Fig. 16. In-line fuse and relay for secondary air injection (AIR) pump (**arrow**). Fuse and relay are located behind left headlight assembly.

3. If the pump does not run and no wiring faults can be found, check the signal from the ECM to the pump. Disconnect the harness connector from the pump and connect a voltmeter to the connector. Start the engine while checking for voltage 20 seconds after starting.
   - If voltage is present, the pump is most likely faulty and should be replaced.
   - If voltage is not present, check the wiring at the AIR relay and between the ECM and the AIR relay. If no faults are found, the relay or ECM may be faulty.

4. Check the shut-off valve by disconnecting the two hoses from the valve. Attach a hand vacuum pump to the valve's vacuum fitting and apply vacuum. Check that valve opens by blowing air through the large hose fitting. See Fig. 17. Release the vacuum and check that the valve closes (air does not pass through large hose fitting).

CAUTION—

Do not use compressed air to check shut off valve.

5. Check the AIR solenoid valve by disconnecting from the shut-off valve the vacuum hose between the solenoid valve and the shut-off valve. Start the engine while checking for vacuum at the hose 20 seconds after starting. If vacuum is present, the valve is functioning correctly.

6. If vacuum is not present, disconnect the harness connector from the solenoid valve and connect a voltmeter to the harness connector. Start the engine while checking for voltage 20 seconds after starting.
   - If voltage is present, the solenoid valve is most likely faulty and should be replaced.
   - If voltage is not present, check the wire between solenoid valve and the ECM. If no faults are found, the ECM may be faulty.

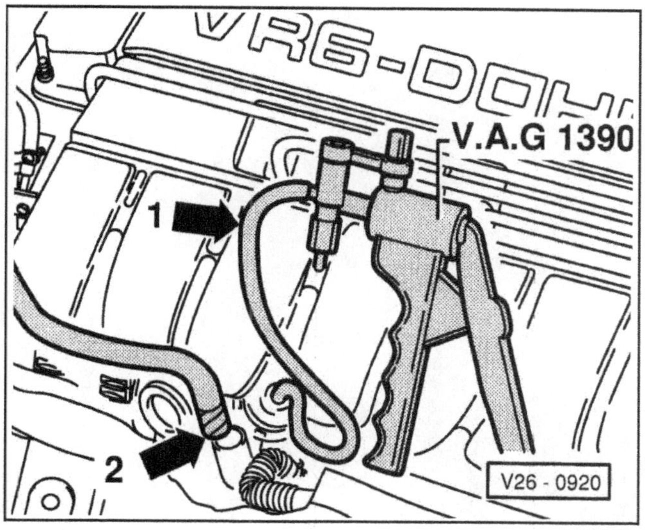

V25-0920

Fig. 17. Hand pump connected to AIR shut-off valve (**1**). Connect pump and apply vacuum while blowing through auxiliary hose (**2**).

To remove AIR system pump

1. With the ignition off, disconnect the negative battery cable from the battery.
2. Remove the trim cover from above the radiator.
3. Remove the air cleaner housing.
4. Remove the clamps holding the A/C hoses to the radiator and the front body assembly.

> ***CAUTION —***
> *Use care not to kink or bend the air A/C hoses.*

5. Remove the lock carrier (radiator support) from the front of the car, complete with headlights. See **50 Body–Front**.
6. Remove the poly-ribbed drive belt. See **0 Maintenance Program**.
7. Remove the alternator. See **27 Engine Electrical Systems**.
8. Remove the air hoses from the air pump and disconnect the vacuum hose from the shut-off valve.
9. Disconnect the dipstick tube from the lower intake manifold.
10. Disconnect the harness connectors from the air pump and the AIR inlet (solenoid) valve. Move the wiring out of the way.
11. Remove the bolts holding the air pump to the engine lifting bracket.

Installation is the reverse of removal.

Tightening torques
- Air pump to air pump bracket . . . . 10 Nm (89 in-lb)
- Air pump bracket to engine lifting eye. . . . . 25 Nm (18 ft-lb)
- Shut-off valve to cylinder head . . . . . . 15 Nm (11 ft-lb)

Catalytic Converter

The three-way catalytic converter chemically reduces pollutants in the engine exhaust. A properly operating converter provides a 90-95 percent reduction of the three major exhaust gas pollutants (nitrogen (NO_X), carbon monoxide (CO), and unburned hydrocarbons (HC).

The catalytic converter works correctly only when the air/fuel ratio is kept within a very narrow range. The oxygen sensor provides the feedback for control of the air/fuel mixture. For more information see **24a Fuel Injection–Motronic or 24b Fuel Injection–Mono-Motronic**.

The catalytic converter itself does not require any routine maintenance. However, any problem that increases converter temperature beyond its normal operating range (incorrect ignition timing or air/fuel mixture, engine misfire, prolonged idling or extended high engine loads) can damage the converter. Reduced power, stalling, exhaust system rattles and excessive emissions are symptoms that may be caused by a faulty catalytic converter.

Catalytic converters found on OBD II equipped vehicles use a second heated oxygen sensor to monitor the operation of the catalyst. If the converter fails to properly oxidize the spent combustion gasses, the heated oxygen sensor will detect this malfunction and a fault or diagnostic trouble code, DTC, will be stored in the engine control module, ECM.

Exhaust leaks anywhere in the system, but especially between the cylinder head and the catalytic converter, can cause running problems, false DTCs or converter failure.

> ***WARNING —***
> *Do not operate the starter for long periods if the engine fails to start. Extended cranking may allow excess fuel to enter the catalytic converter, creating a fire hazard and possibly damaging the converter.*

Accurately testing the function of the catalytic converter requires an infrared exhaust gas analyzer to check the air/fuel mixture adjustment (% CO) at the exhaust system test port upstream of the catalytic converter, and then again at the tailpipe. If the converter is working properly, the tailpipe reading will be lower than the test port reading. If the reading is the same or only slightly less, the converter is probably faulty.

27 Engine Electrical

TABLES

GENERAL

This repair group covers battery, starter, and alternator (generator) troubleshooting and repair.

> ***NOTE —***
> *Beginning with model year 1994, car makers must by law use uniform words and terms (nomenclature) to describe certain components and/or systems. The uniform word for alternator is generator. Therefore, the component labeled "Generator (GEN)" in the wiring diagrams is the alternator.*

> ***NOTE —***
> *Wiring diagrams for the battery, charging system and starter motor are given in* **97 Wiring Diagrams, Fuses and Relays**.

The six-cell, 12-volt lead-acid battery capacity is rated in Ampere/hours (Ah) and cold cranking amps (CCA). The Ah rating is determined by the average amount of current the battery can deliver over time without dropping below a specified voltage. The CCA rating is determined by the battery's ability to deliver starting current at 0°F (–18°C). The battery is installed in the engine compartment, behind the left headlight assembly.

The charging system consists of a belt-driven 14-volt alternator and a voltage regulator. The voltage regulator, which is mounted on the alternator, also serves as the alternator brush holder. The charging system provides the current necessary to keep the battery charged and to operate the vehicle's electrical accessories.

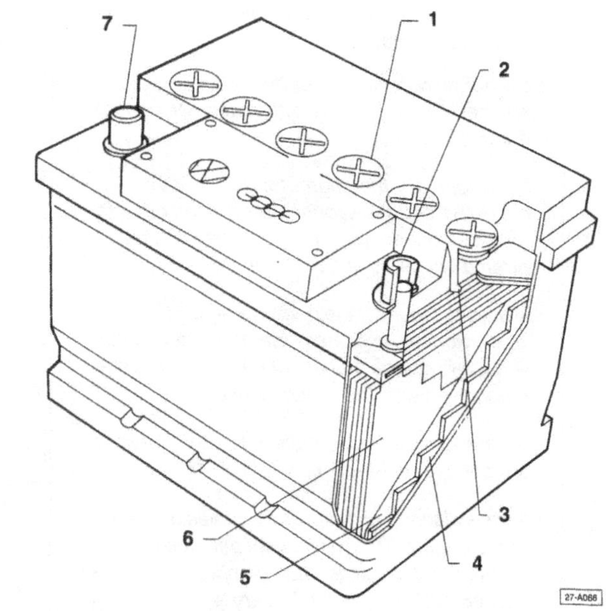

1. **Filler cap**
2. **Positive (B+) terminal**
3. **Electrolyte level indicator**
4. **Negative plate (grey)**
5. **Separator (insulator)**
6. **Positive plate (dark brown)**
7. **Negative (–) terminal**

27-A066

Fig. 1. Battery.

Please read the following warnings and cautions before doing any work on any parts of the engine electrical system.

> *WARNING —*
> - *Wear goggles, rubber gloves, and a rubber apron when working around batteries and battery acid (electrolyte). Battery acid contains sulfuric acid and can cause skin irritation and burning. If acid is spilled on your skin or clothing, flush the area at once with large quantities of water. If electrolyte gets into your eyes, bathe them with large quantities of clean water for several minutes and call a physician.*
> - *Batteries that are being charged or are fully charged give off explosive hydrogen gas. Keep sparks and open flames away. Do not smoke.*

> *CAUTION —*
> - *Disconnecting the negative (-) battery cable may erase fault codes and basic settings in the engine management and automatic transmission control unit memory. Some driveability problems may be noticed until the system re-adapts to operating conditions. OBD II readiness codes, which may be required for emissions testing, may also be erased.*
> - *Do not disconnect the battery cables while the engine is running. The alternator will be damaged.*
> - *Never operate the alternator with its output terminal (B+ or 30) disconnected and the other terminals connected. Never short, bridge, or ground any terminals of the charging system.*
> - *Always disconnect the negative (–) battery cable when working at or near the alternator. Battery voltage is always present at the rear of the alternator, even with the ignition key off.*
> - *Always allow a frozen battery to thaw before attempting to recharge it.*
> - *Always disconnect the battery cables during battery charging. This will prevent damage to the alternator and any solid-state components. Do not exceed 16.5 volts at the battery.*
> - *Never reverse the battery terminals. Even a momentary wrong connection can damage the alternator or electrical components.*
> - *Replace the battery if the case is cracked or leaking. Electrolyte can damage the car. If electrolyte is spilled, clean the area with a solution of baking soda and water.*

BATTERY SERVICE

If the battery quickly discharges, there may be a constant drain or current draw causing it to discharge when the ignition is off. Depending on the draw and the condition of the battery, a full discharge can happen overnight or it may take a few weeks. Although a small static drain on the battery is normal (for example to operate the clock or radio memory), a large drain such as a relay sticking on or a faulty switch will cause the battery to quickly discharge. Make a static current draw test as the first step when experiencing battery discharge.

If the current draw on the battery is not excessive and the battery still discharges, the condition of the battery should be tested. Battery testing determines the state of battery charge. The most common methods are open-circuit and load voltage testing. Batteries with filler caps can also be tested by checking the specific gravity of the electrolyte. Inexpensive specific gravity testers are available at most auto supply stores.

To check static current draw

1. Make sure the ignition and all electrical accessories are switched off.
2. Disconnect the negative (–) cable from the battery.

> *CAUTION —*
> *Be sure to have the anti-theft radio code on hand before disconnecting the battery.*

3. Connect a digital ammeter between the battery negative post and the negative battery cable and measure the current draw. See Fig. 2.

A range of about 0 to 100 milliamps is normal, depending on the number of accessories that need constant power. A current of 500 milliamps (0.5 amp) or more indicates a problem. To determine the circuit or component causing the problem, remove one fuse at a time until the current drops to a normal range. Use the wiring diagrams shown in **97 Wiring Diagrams, Fuses and Relays** to locate wiring or component faults.

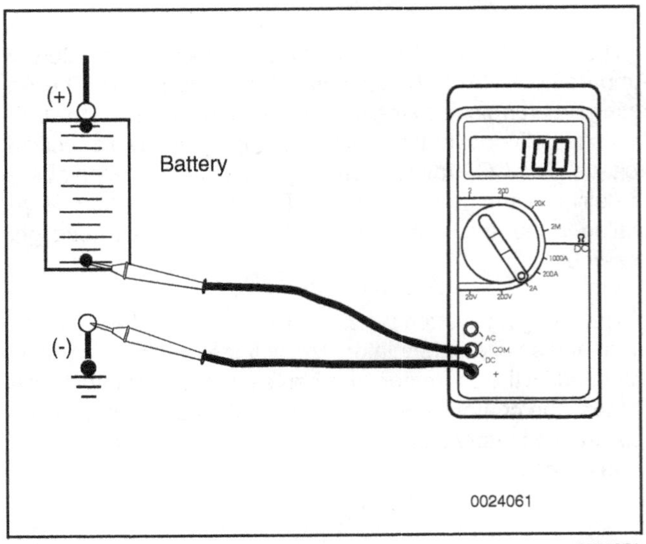

0024061

Fig. 2. Electrical system static current draw being measured.

Open-Circuit Voltage Test

An open-circuit voltage test checks battery voltage by connecting an accurate digital voltmeter to the battery posts after disconnecting the battery ground cable. Before making an open-circuit voltage test on a battery, first load the battery with 15 amps for one minute, for example by turning on the headlights without the engine running. See **Table a** for open-circuit voltage levels and their corresponding percentages of charge.

Table a. Open-Circuit Voltage and Battery Charge

Open-circuit voltage	State of charge
12.6 V or more	Fully charged
12.4 V	75% charged
12.2 V	50% charged
12.0 V	25% charged
11.7 V or less	Fully discharged

The battery is in satisfactory condition if the open-circuit voltage is at least 12.4 volts. If the open-circuit voltage is at this level or above, but the battery still lacks power for starting, make a load voltage test to determine the battery's service condition. If the open-circuit voltage is below 12.4 volts, recharge the battery. If the battery cannot be recharged to at least 75%, it should be replaced.

Load Voltage Testing

A load voltage battery test is made by connecting a specific resistive load to the battery terminals and then measuring the battery's voltage. The test requires a special tester and can generally be performed quickly and inexpensively by an authorized VW dealer or other qualified repair facility.

The battery should be fully charged and at room temperature for the most accurate results. If the equipment is available, disconnect the negative (–) battery cable. Then apply the specified load for 15 seconds and measure the battery's voltage. If the voltage is below that listed, the battery should be replaced. **Table b** lists load current and minimum voltages for original-equipment VW batteries.

> ***WARNING —***
> *Always wear protective goggles and clothing when performing a load test.*

Battery Charging

Discharged batteries can be recharged using a battery charger. Prolonged charging causes gassing that will evaporate the electrolyte to a level that can damage the battery.

Table b. Battery Load Current and Minimum Voltage

Battery Capacity-Ah (Amp-hour)	Cold Cranking Amps (CCA)	Load Current	Minimum voltage
36 Ah	175	100	10.0
40 - 49 Ah	220	200	9.2
50 - 60 Ah	265-280	200	9.4
61 - 80 Ah	300-380	300	9.0
81 - 110 Ah	380-500	300	9.5

> ***WARNING —***
> *The gasses given off by the battery during charging are explosive. Do not smoke. Keep open flames away from the top of the battery, and prevent electrical sparks by turning off the battery charger before connecting or disconnecting it.*

Always read and follow the instructions provided by the battery charger's manufacturer. A slow-charging rate (10% of battery capacity) is best to prevent battery damage caused by overheating.

Battery, removing and installing

1. Disconnect the battery cable at the negative (–) terminal.

> ***NOTE —***
> *Be sure to have the anti-theft radio on hand before disconnecting the battery cables.*

2. Disconnect the battery cable at the positive (+) terminal.
3. Remove the battery hold down nut and clamp. See Fig. 3.

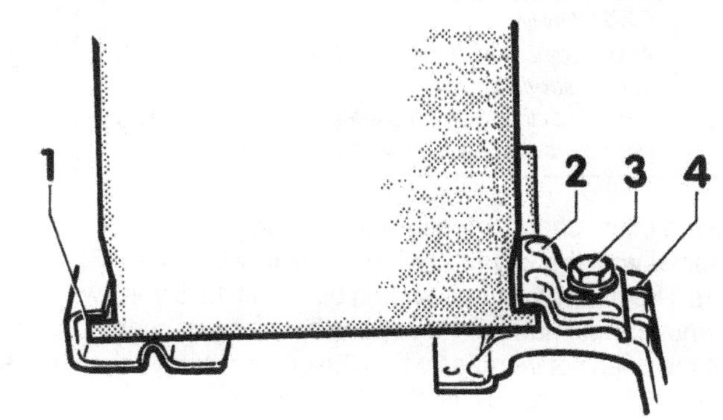

1. **Lower clamping strip**
2. **Hold-down clamp**
3. **Nut**
   • tighten to 20 Nm (15 ft-lb)
4. **Battery securing bracket**

27-595

Fig. 3. Battery hold-down assembly.

4. Position the battery into the lower clamping strip first. Then retain the other side of the battery using the hold-clamp and bolt. Tighten the hold-down bolt and attach the battery negative (-) cable to battery.

> ***CAUTION —***
> *A secure battery hold-down is important. Normal shock and vibration can cause premature damage to a loosely mounted battery.*

5. On cars equipped with airbags, turn on ignition switch before attaching positive (+) battery cable. See **69 Seatbelt, Airbags**.

6. Attach positive (+) battery cable.

Tightening torques
- Battery hold-down bracket . . . . 20 Nm (15 ft-lb)
- Battery cables to battery . . . . 5 Nm (44 in-lb)

7. See "Procedure following interruption of voltage supply" in the **Scan Tool** section at the back of this manual.

CHARGING SYSTEM SERVICE

Charging system trouble is indicated by an illuminated alternator warning light on the instrument panel, or by an under- or overcharged battery.

The alternator generates electrical current by electrical induction. When the engine is running, part of the current it produces energizes its electromagnetic field. When starting, some other current must be provided to initially energize the field and begin the current generating process. This current is provided by the battery through the alternator warning light in the instrument cluster. If the lamp burns out, the alternator will not charge the battery properly.

> ***CAUTION —***
> *Always disconnect the negative (–) battery cable before servicing the charging system. The large red wire at the rear of the alternator comes directly from the battery and is not fuse protected.*

As a quick-check, measure the voltage across the battery terminal with the key off and then again with the engine running. The battery voltage should be about 12.6 volts with key off and approximately 14.0 volts with the engine running. If the voltage does not increase when the engine is running, there is a fault in the charging system.

> ***NOTE —***
> *The regulated voltage (engine running) should be roughly between 13.8 and 14.5 volts, depending on temperature and operating conditions. If voltage is much higher than 14.5 volts, the voltage regulator is most likely faulty.*

To test charging system

1. Inspect the poly-ribbed drive belt for cracking, glazing or wear. Replace the belt if any faults are found. Check that the belt is not slipping and that the belt tension is correct.

2. Make sure the battery is fully charged and capable of holding a charge. Make sure the battery terminals are clean and tight.

3. Check that the charge warning lamp comes on when the key is on. If the light comes on, proceed to step 4. If the light does not come on, repair any wiring or bulb faults before continuing to test.

> ***NOTE —***
> *The charge warning light must come on when the ignition is switched on or the system will not charge. If any faults are found, see* **97 Wiring Diagrams, Fuses and Relays** *for electrical schematics.*

4. Check for battery voltage between ground and terminal **B+** (large red wire) at the back of the generator. Check that the wire is securely fastened.

5. Turn the ignition key on and check for battery voltage between terminal **D+** and ground. If voltage is not present, check the wiring from the warning light in the instrument cluster.

6. If no faults are found up to this point, have the alternator tested with a load tester, such as the Sun VAT40.

> ***NOTE —***
> *If a load test is not possible, an imperfect output test can be done by running the engine at about 2000 rpm and turning on many of the electrical loads such as fans, lights and heated window, wipers, etc. When loaded, the battery should still be 12 volts or higher.*

Alternator, removing and installing (4-cylinder)

1. Disconnect the negative (–) battery cable.

> ***NOTE —***
> *Be sure to have the anti-theft radio on hand before disconnecting the battery cables.*

2. Mark the running direction on the poly-ribbed drive belt and then remove the belt from the alternator pulley. See **0 Maintenance Program**.

3. Disconnect the wiring from the rear of the alternator.

4. On cars with A/C, release pressure on tensioning pulley using VW special tool 3299 or similar (channel-lock pliers can be used).

5. On cars without A/C, loosen the alternator adjustment bolt and push the alternator toward mounting bracket to relieve belt tension. Tighten adjustment bolt and remove poly-ribbed belt from alternator pulley.

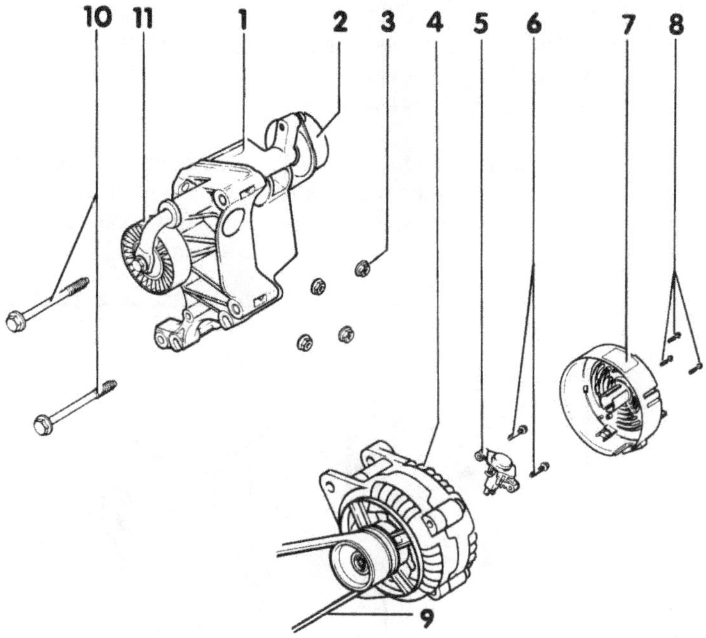

1. **Mounting bracket**
2. **Spring Tensioner**
3. **Nut (M8)**
   • tighten to 30 Nm (22 ft-lb)
4. **Alternator (Generator)**
5. **Voltage regulator/brush assembly**
   • Check carbon brush length, see Fig. 7.
6. **Screw (M4)**
7. **Cover**
8. **Bolts (M3)**
9. **Poly-ribbed drive belt**
10. **Bolt (M8)**
    • tighten to 25 Nm (18 ft-lb)
11. **Tensioner (idler) pulley**

27-682

Fig. 4. Alternator mounting bracket and related parts (4-cylinder gasoline cars with power steering and A/C, diesel cars are similar).

6. Remove the upper and lower alternator through-bolts and remove the alternator from its bracket. See Fig. 4 or Fig. 5.

7. Installation is the reverse of removal. Tightening torques are given in the illustrations.

NOTE —

On cars without A/C, start and run the engine for at least ten seconds before final tightening of the alternator mounting bolts. Tighten the top bolt first.

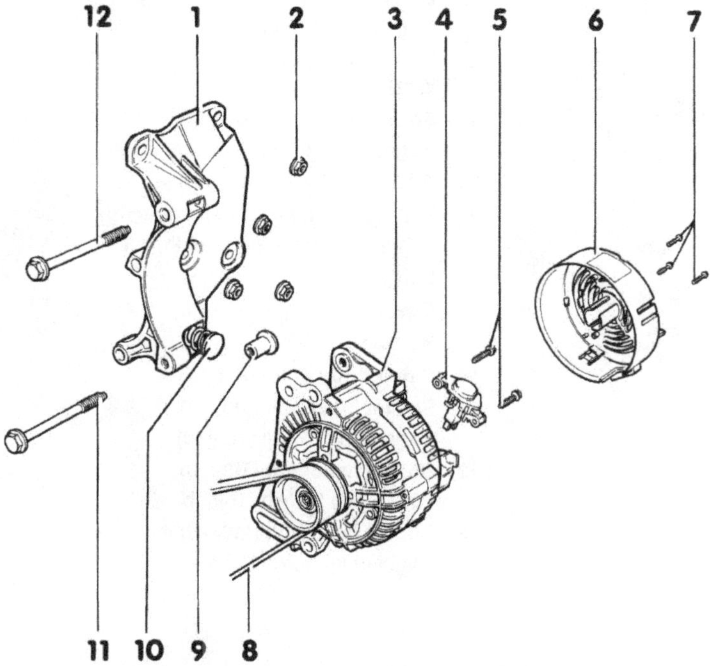

1. **Mounting bracket**
2. **Nut (M8)**
   • tighten to 30 Nm (22 ft-lb)
3. **Alternator (Generator)**
4. **Voltage regulator/brush assembly**
   • Check carbon brush length, see Fig. 7.
5. **Screw (M4)**
6. **Cover**
7. **Bolts (M3)**
8. **Poly-ribbed drive belt**
9. **Sleeve (M8 internal thread)**
10. **Tensioning spring**
11. **Bolt (M8x39)**
    • tighten to 25 Nm (18 ft-lb)
12. **Bolt (M8x85)**
    • tighten to 25 Nm (18 ft-lb)

27-681

Fig. 5. Alternator mounting bracket and related parts (4-cylinder cars with power steering and without A/C).

Alternator, removing and installing (6-cylinder)

1. Disconnect the negative (–) battery cable.

   NOTE —
   Be sure to have the anti-theft radio on hand before disconnecting the battery cables.

2. Mark the running direction on the poly-ribbed drive belt and then remove the belt from the alternator pulley. See **0 Maintenance Program**.
3. Disconnect the wiring from the rear of the alternator.
4. Remove poly-ribbed belt. See **0 Maintenance Program**.
5. Unbolt belt tensioner from cylinder head and remove.
6. Remove alternator mounting bolts. See Fig. 6.
7. In order to remove the alternator, the passenger side of the engine will need to be raised enough to gain clearance to pull the alternator out. This will necessitate unbolting the motor mounts. See **10 Engine-Removing and Installing**. Alternatively, the front body assembly can be moved to the service position to gain access to the alternator. See **50 Body–Front**.
8. Installation is the reverse of removal.

Regulator/brush assembly

The voltage regulator is easily removed from the back of the alternator by removing the two mounting screws. The regulator with brushes is available as a replacement part.

With the regulator removed, measure the brush length. See Fig. 7. Replace the brushes in sets if the length is less than the minimum length specified. Replacing the brushes requires a soldering iron and electric/electronic type solder.

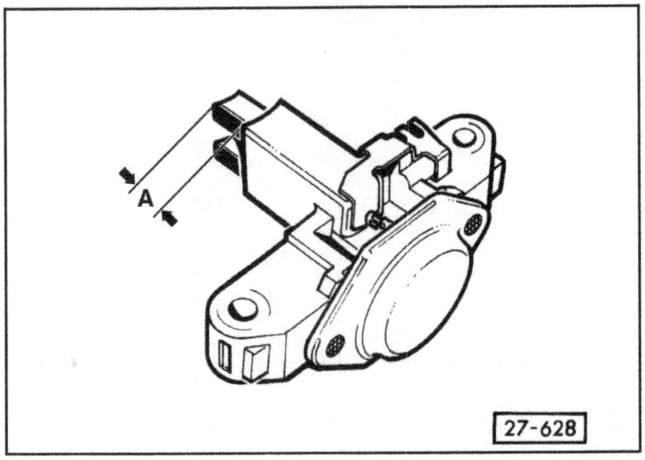

27-628

Fig. 7. Regulator brush length (**A**). Brushes shorter than 5 mm (0.2 in.) should be replaced.

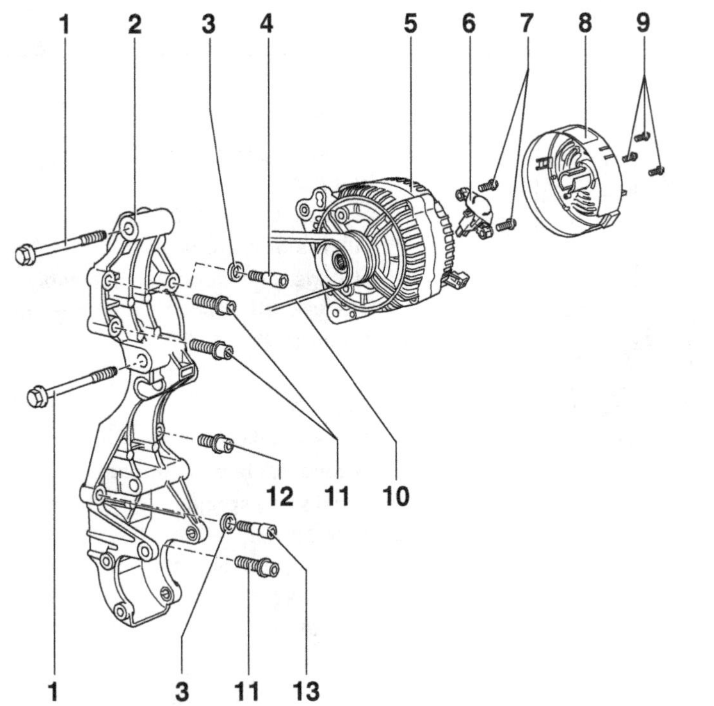

1. **Bolt (M8)**
   - tighten to 25 Nm (18 ft-lb)
2. **Mounting bracket**
3. **Washer**
4. **Socket head bolt (M8x28)**
   - tighten to 25 Nm (18 ft-lb)
5. **Alternator (Generator)**
6. **Voltage regulator/brush assembly**
   - Check carbon brush length, see Fig. 7.
7. **Screws**
8. **Cover**
9. **Screws**
10. **Ploy-ribbed drive belt**
11. **Socket head bolt (M8x30)**
    - tighten to 25 Nm (18 ft-lb)
12. **Socket head bolt (M8x38)**
    - tighten to 25 Nm (18 ft-lb)
13. **Socket head bolt (M8x20)**
    - tighten to 25 Nm (18 ft-lb)

N27-0021

Fig. 6. Alternator mounting bracket and related parts (6-cylinder gasoline cars with power steering and A/C).

STARTER SERVICE

The starter is located at the left-hand (driver's) side of the engine block, just above the oil pan. The starter and solenoid are removed together as an assembly. The solenoid can be separated from the starter motor once the starter has been removed. Although the starter is generally replaced as an exchanged unit, the solenoid and other parts are available from an authorized VW dealer.

Before troubleshooting the starter, make sure the battery is fully charged and the battery cables and ground connections are free of corrosion and in good condition. Troubleshooting information for the starting system appears in **Table c**.

> ***NOTE —***
> *Starter cranking speed is affected by engine oil viscosity. This is especially true in cold weather. Make sure the correct oil is in the engine. See* **0 Maintenance Program.**

Manual transmission vehicles produced from May 1996 have a starter lock-out function requiring the clutch pedal to be depressed before the starter can be activated. This system consists of a pedal switch and a relay on the fuse-relay panel. See Fig. 8.

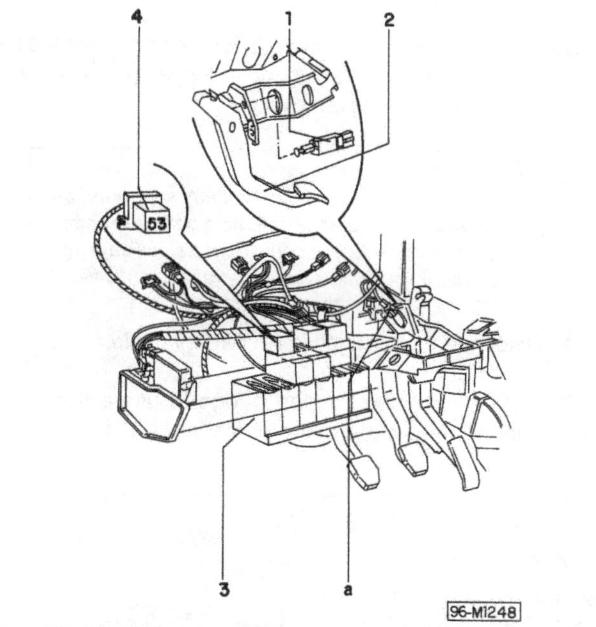

1. **Starter lock-out switch**
   - To remove: disconnect 2-pin harness connector (a). Press and turn switch and pull out of bracket away from clutch pedal (2).
2. **Clutch pedal**
3. **fuse/relay panel**
4. **Relay**
   - Relay position 21

96-M1248

Fig. 8. Starter lock-out switch and relay location found on vehicles with manual transmission starting with May 1996 production.

Table c. Starter System Troubleshooting

Symptom	Probable cause	Corrective action
1. Starter does not operate when ignition switch is turned to START	**a.** Ignition switch or wire leading from ignition switch to solenoid faulty (less than 8 volts to solenoid)	**a.** Test for voltage at terminal 50 of solenoid with ignition at START. If not at least 8 volts, test for voltage at terminal 50 of ignition switch with switch at START. Replace ignition switch or eliminate open circuit between ignition switch and solenoid.
	b. Solenoid faulty (less than 8 volts to starter motor)	**b.** Test for voltage at field-winding strap with ignition at START. If not at least 8 volts, test solenoid.
	c. Starter motor faulty	**c.** Test for voltage at field-winding strap with ignition at START. If 8 volts or more, repair or replace starter motor.
	d. Automatic transmission start inhibitor switch faulty (models with A/T)	**d.** Test switch. See **97 Wiring Diagrams, Fuses and Relays** and **37 Automatic Transmission.**
	e. Starter lock-out switch or relay faulty (models with manual trans)	**e.** Test switch and relay. See **97 Wiring Diagrams, Fuses and Relays.**
2. Starter turns slowly or fails to turn engine	**a.** Dirty, loose, or corroded starter connections	**a.** Clean, and tighten connections. If necessary check voltage drop between the battery and the starter as described in **9 Electrical System—General.**
	b. Dirty, loose, or corroded ground strap between engine and body	**b.** Remove and clean or replace strap
	c. Starter worn or faulty	**c.** Repair or replace starter
3. Starter operates, but does not turn engine	**a.** Flywheel or driveplate teeth missing or damaged	**a.** Replace flywheel or driveplate. See **1 ENGINE.**
	b. Starter drive or armature shaft faulty	**b.** Repair or replace starter
	c. Solenoid mechanism faulty	**c.** Replace starter solenoid

Starter, removing and installing

The procedure given below calls for the use of engine lifting equipment to support the weight of the engine from above while the starter motor is removed.

> ***WARNING —***
> *If the weight of engine is not properly supported before removing the starter mounting bolts, the engine may suddenly fall. This could cause significant injury, damage and make installation/alignment difficult.*

1. Disconnect the negative (–) battery cable.

2. Install an engine support bridge or use a floor jack to support the weight of the engine. See Fig. 9.

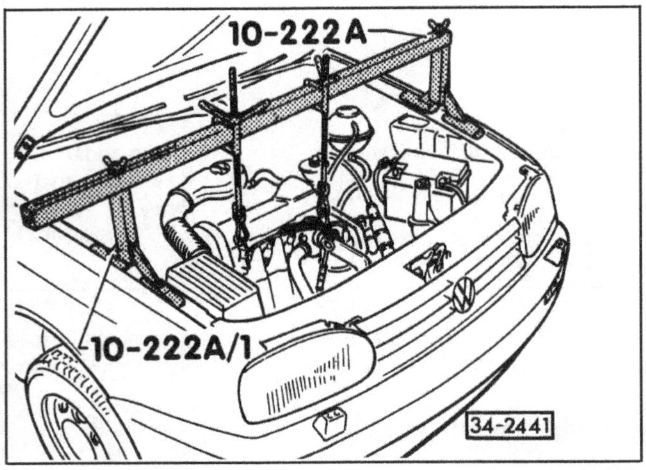

34-2441

Fig. 9. Engine support installed across engine bay. (VW special tools 10-222A and 10-222A/1 shown).

3. Remove the harness connectors, plastic cover and unbolt the battery cable from the starter motor. See Fig. 10.

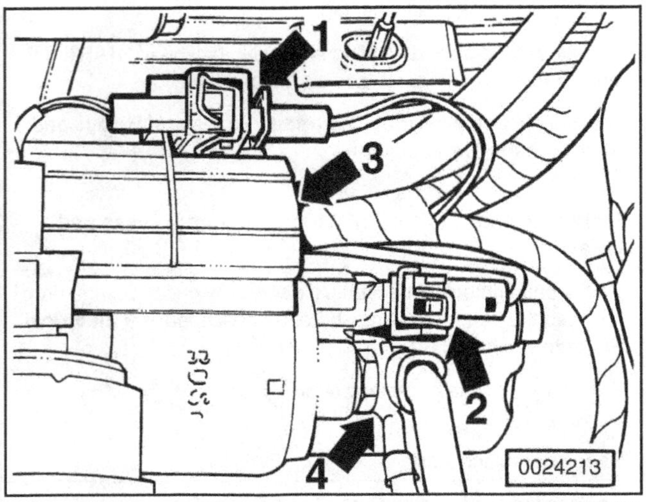

0024213

Fig. 10. Disconnect harness connectors (**1** and **2**), remove shield (**3**) and unbolt battery connection (**4**) from starter.

4. On 6-cylinder and 4-cylinder cars with automatic transmission unbolt the starter mounting bolts/nuts and remove starter. See Fig. 11.

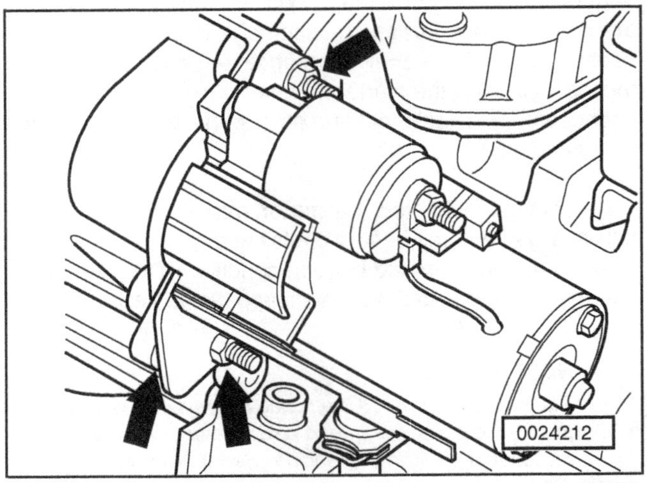

0024212

Fig. 11. Unbolt starter mounting nuts (**arrows**) on cars with automatic transmission. 4-cylinder shown, 6-cylinder is similar.

5. On 4-cylinder cars with manual transmission remove the cable securing bracket from the starter motor. Then remove the upper and lower mounting bolts and remove starter. See Fig. 12.

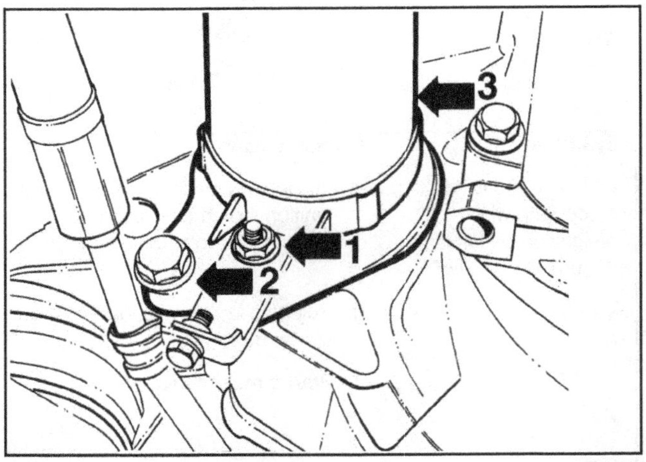

0024060

Fig. 12. Remove nut (**1**) for cable securing bracket and starter mounting nuts (**2** and **3**).

6. Installation is the reverse of removal.

Tightening torques

- Starter motor mounting nuts/bolts (Manual transmission)
  - 6-cylinder. . . . . 47 Nm (44 ft-lb)
  - 4-cylinder (lower). . . . . 45 Nm (33 ft-lb)
  - 4-cylinder (upper) . . . . 60 Nm (44 ft-lb)
  - 4-cylinder cable bracket . . . . 10 Nm (7 ft-lb)
- Starter motor mounting nuts/bolts (Auto. transmission)
  - M10 mounting nuts . . . . 60 Nm (44 ft-lb)
  - M8 nut for protective cover . . . . 10 Nm (7 ft-lb)

28a Ignition System–Gasoline

TABLES

GENERAL

The ignition function is handled through the fuel injection/engine management system. See **Table a**. On gasoline engines, the Motronic or Mono-Motronic engine control module (ECM) computes ignition timing based on inputs from various sensors. See Fig. 1. Crankshaft position is the main input to the ECM and is used for starting. The other sensors are used to adapt the basic timing map for varying operating conditions. Motronic engines incorporate adaptive knock control to adjust the ignition timing for individual cylinders.

Table a. Engine Management System Applications

Engine code	System
ACC (1.8 liter)	Mono-Motronic
AAZ (1.9 liter)	Diesel Fuel Injection
AHU (1.9 liter)	TDI Diesel Fuel Injection
ABA (2.0 liter)	Motronic M2.9/M5.9*
AAA (2.8 liter, VR6)	Motronic M2.9/M5.9*

*1993–1995 models are equipped with Motronic M2.9. 1996 and later models are equipped with Motronic M5.9

NOTE—
The engine management/ ignition system has a built in diagnostic circuit that monitors the ignition system components and detects and stores system faults. Before troubleshooting the ignition system, be sure to check for diagnostic trouble codes (DTCs) as described in **24a Motronic** *or* **24b Mono-Motronic**.

NOTE—
- *On Motronic-equipped engines, crankshaft position is determined by the engine speed (RPM) sensor. The RPM sensor is mounted in the side of the cylinder block and reads a toothed wheel on the end of the crankshaft. See* **24a Fuel Injection–Motronic** *for RPM sensor testing.*
- *On Mono-Motronic cars, crankshaft position is determined by the camshaft position (CMP) sensor, which is located in the ignition distributor. The CMP is an electronic switch and operates based on the Hall effect. See* **24b Fuel Injection–Mono-Motronic** *for testing information.*

Service Precautions

Ignition system service and repair work must be carried out carefully. The ignition system contains sensitive electronic components. To guard against system damage, and for general safety, the following warnings apply to any ignition system troubleshooting, maintenance or repair work.

WARNING—
Ignition systems operate in a dangerous voltage range that could prove to be fatal if exposed terminals or live parts are contacted. Use extreme caution when working on a vehicle with the ignition on or the engine running.

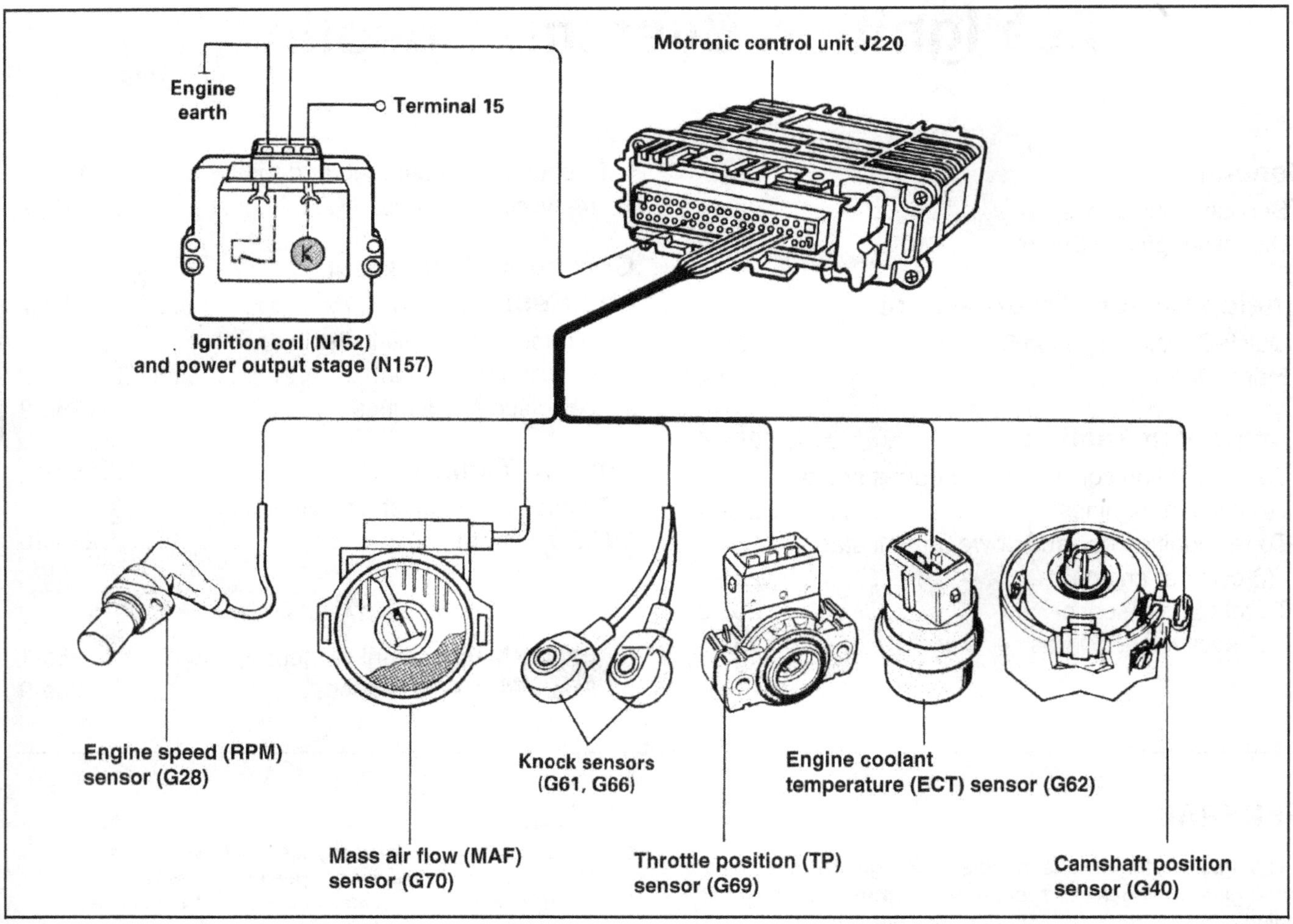

Fig. 1. Ignition system input signals for Motronic-equipped engines.

> ***WARNING —***
> *Connect and disconnect ignition system wires, multiple connectors, and test equipment leads only while the ignition is off. Do not touch or disconnect any of the high voltage wires from the coil, distributor or spark plugs while the engine is running or being cranked by the starter.*

> ***CAUTION —***
> - *Never let the spark gap (between electrode and ground) exceed 5 mm (0.20 in.) when checking the ignition system, as when checking for a spark, as the control unit may be damaged. Always ground the plug on the engine when checking for spark.*
> - *Before operating the starter without starting the engine (as when making a compression test), always disable the ignition as described below.*
> - *A weak spark or no spark can overload the catalytic converter with unburned fuel, leading to converter overheating or plugging. An overheated catalytic converter can be a fire hazard.*

> ***CAUTION —***
> - *Do not conduct ignition system tests with a test lamp that uses a normal incandescent bulb. The high electrical consumption of these test lamps may damage the electronic components. Use only a low-current LED test light.*
> - *Do not connect test instruments with a 12-volt supply to terminal 15(+) of the ignition coil. The voltage backflow can damage the ignition control unit.*
> - *Do not disconnect terminal 4 (center terminal) from the coil or remove the distributor cap to disable the ignition.*

Disabling Ignition System

The ignition system should be disabled to prevent the engine from starting and to prevent the discharge of dangerously high voltage any time work is done on the ignition system. The ignition system should also be disabled any time work is being done in the engine compartment with the ignition key on or if the starter needs to be operated without running the engine, such as when making an engine compression test.

To disable the ignition system, disconnect the harness connector from the ignition coil power output stage.

On 4-cylinder engines, the ignition coil with power output stage is mounted in the rear of the engine compartment, against the firewall. See Fig. 2.

Fig. 2. Harness connector (**arrow**) for ignition coil with integrated power output stage on 4-cylinder engines. Component is mounted in rear of engine compartment.

On 6-cylinder engines, the ignition coil with integrated power output stage is mounted on the rear of the cylinder head. To access the coil harness connector first remove the plastic intake manifold cover from above the intake manifold, then remove the spark plug wire guides with the ignition wires from the top of the cylinder head. See Fig. 3.

> ***NOTE—***
> *If necessary, a special tool (VW tool no. 3277) is needed to remove the spark plug connectors from the plugs without damaging them. A version of this tool is clipped to the front hood support rod.*

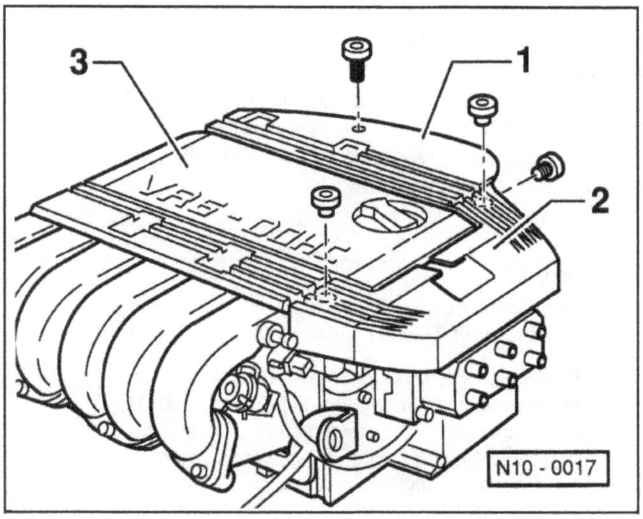

Fig. 3. VR6 spark plug wire guides (**1** and **2**). Intake manifold cover shown at **3**.

A complete failure of the ignition system to produce spark at the spark plugs is self-evident. For other problems such as rough idle, misfiring, or poor starting the cause may not be so clear. The engine management system has a built in diagnostic circuit that monitors the ignition system components and detects and stores system faults. Before troubleshooting the ignition system, always check for diagnostic trouble codes (DTCs) as described in **24a Motronic** or **24b Mono-Motronic**.

IGNITION SYSTEM TROUBLESHOOTING

Poor driveability may have a variety of causes. The fault may lie with the ignition system, the fuel system, parts of the emission control system, or a combination of the three. Because of the interrelated functions of these systems, it is often difficult to know where to begin looking for problems. For this reason, when troubleshooting always consider these systems in unison, as one major system.

Quick-Check of Ignition System

The first step in troubleshooting a no-start condition is to determine whether the problem is caused by the ignition system or some other system, such as a fuel delivery problem. This is done by checking that the spark plugs are firing. If no spark is present, a more detailed testing of the ignition system is necessary.

To make the check, turn the ignition off and remove a spark plug wire from a spark plug. Connect to the plug wire to a known good spark plug. Position the plug so that the outer electrode is grounded on the engine. For accurate test results, the battery should be fully charged.

> ***CAUTION—***
> *Never let the spark plug gap (between electrode and ground) exceed 5 mm (0.20 in.) when checking the ignition system, as when checking for a spark, as the control unit may be damaged. Always ground the plug on the engine when checking for spark.*

While a helper actuates the starter, look and listen for a spark at the plug. A bright blue spark indicates a healthy ignition system. If there is no spark, first make sure that the speed (RPM) sensor (Motronic-equipped cars) or the camshaft position (CMP) sensor (Mono-Motronic-equipped cars) are functioning correctly. See description in **24 Fuel Injection**. Then test the ignition coil output stage and the ignition coil as described below.

> ***WARNING —***
> - *Do not hold the spark plug or its connector during the test, even if using insulated pliers.*
> - *If ignition system failure is not the problem, the engine may start during this test. Be prepared to switch the ignition off immediately. Running the engine with a spark plug wire disconnected may damage the catalytic converter.*

Firing Order

The firing order of the cylinders is given below. Cylinder no. 1 is at the drive belt end of the engine.

Ignition Firing Order
- 4-cylinder engine 1–3–4–2
- 6-cylinder engine 1–5–3–6–2–4

COMPONENT TESTING

Be sure to review the information given in **24a Fuel Injection–Motronic** or **24b Mono-Motronic** for retrieving diagnostic trouble codes (DTCs) and applicable ECM pin-out tests before making the electrical checks and tests given below.

> ***NOTE —***
> *The test given below assumes that the camshaft position (CMP) sensor and the speed (RPM) sensor (where applicable) are functioning correctly. See* **24a Fuel Injection—Motronic** *or* **24b Fuel Injection—Mono-Motronic** *for testing information.*

Basic Requirements
- Battery positive (+) voltage OK
- Fuse no. 15 and no. 18 OK
- Ground connections on cylinder head OK

> ***NOTE —***
> *For engine management and ignition wiring diagrams, see* **97 Wiring Diagrams, Fuses and Relays**.

> ***CAUTION —***
> - *Use a high-quality digital automotive multimeter or an LED test light to make the tests. An analog (swing-needle) meter should not be used as it can permanently damage electronic components.*
> - *Always connect and disconnect the ECM connector and meter probes with the ignition off to avoid damage to electronic components.*
> - *When disconnecting the large harness connector from the engine control module (ECM), wait at least a minute with the ignition off before separating the connector from the ECM.*

To test ignition coil and power output stage (4-cylinder engines)

The ignition coil used on the 4-cylinder engines is mounted in the rear of the engine compartment, against the bulkhead. The coil is integrated with the coil power output stage. Make the test below in the order given, repair any faults before moving on to the next step.

1. Disconnect the 3-pin connector from the power output stage. See Fig. 2, given earlier.

2. Connect a voltmeter across terminals 1(–) and 3(+) in the connector. Turn the ignition on and check for battery voltage. See Fig. 4.

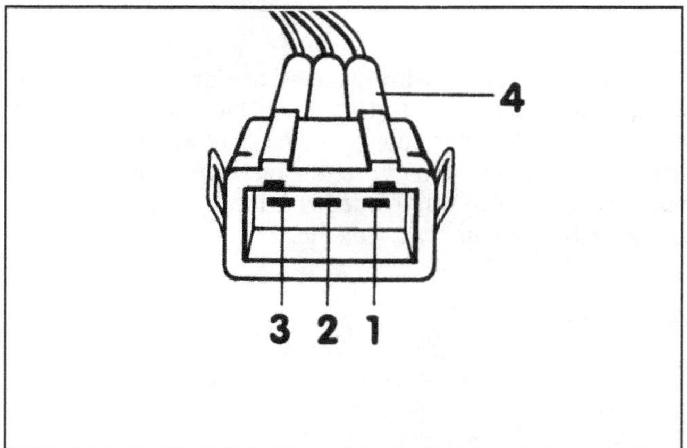

0024199

Fig. 4. 3-pin connector from 4-cylinder coil power output stage.

> ***NOTE —***
> *If voltage is not present, check the wiring to connector. See* **97 Wiring Diagrams, Fuses and Relays.**

3. Connect an LED light across terminals 2 and 3 in the power output stage connector.

4. Remove fuse no. 18 (fuel pump, HO2S control module) from the fuse relay panel.

5. On Motronic cars, disconnect the harness connectors from fuel injectors 1 and 4.

6. On Mono-Motronic cars, disconnect the harness connector from the single fuel injector.

7. Actuate the starter and check that the LED flashes. Reconnect the 3-pin connector to the coil power output stage.

NOTE —

If the LED does not flash, the ignition (ground) signal from the ECM is missing. Check the wire between the ECM and the coil power output stage. If no wiring faults can be found, the ECM may be faulty. See **24 Fuel Injection** *for an ECM pin-out test.*

8. Pry back the cover from the coil terminals. Connect an LED test light across terminals 1 and 15. See Fig. 5.

WARNING —
Do not hold the LED test light during the test. Dangerously high voltage is present.

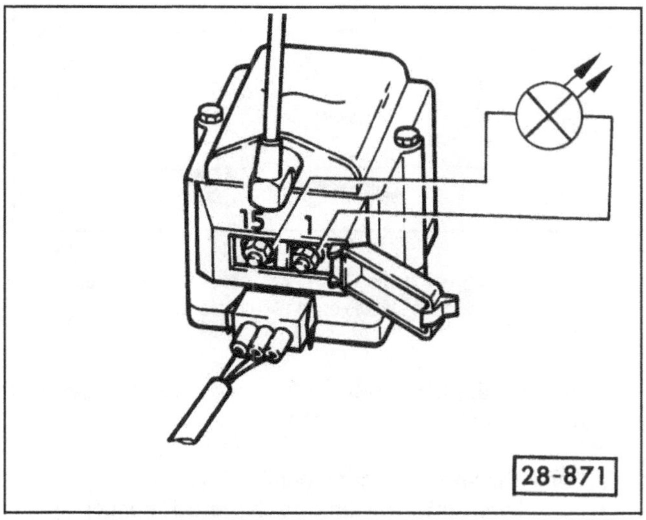

28-871

Fig. 5. LED test light (shown schematically) connected across coil terminals.

9. Switch the ignition on and check that the light illuminates for 1 to 2 seconds. Actuate the starter and check that the LED flashes. Switch the ignition off.

NOTE —

- *If the LED does not light up or does not flash, the coil power output stage is faulty.*
- *The ignition coil and the power output stage are available as individual replacement parts.*

10. If no faults are found up to this point, measure the coil primary and secondary circuit resistance as shown in Fig. 6. If the coil resistance is not as specified, it should be replaced.

Ignition coil resistance
- Primary
  (terminals 1 and 15) . . . . .0.5–0.7 ohms
- Secondary
  (terminals 4 and 15) . . . . .3,000–4,000 ohms

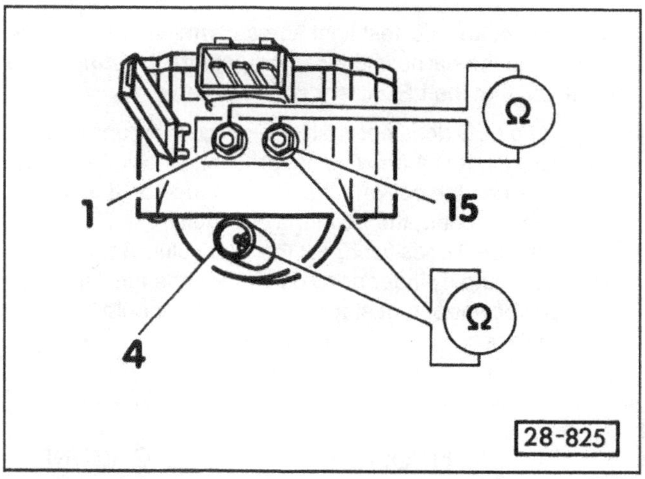

28-825

Fig. 6. Ignition coil primary and secondary resistance being measured (shown schematically).

11. Reconnect the harness connectors to the fuel injector(s) and reinstall fuse no. 18.

To test ignition coil and power output stage (6-cylinder engine)

The ignition coil used on the VR6 engine is comprised of three "double-ended" coils within one housing. When the power output stage triggers one of the three ignition coils, spark is supplied to two spark plugs at once—one plug fires during the compression stroke and the other plug fires during the exhaust stroke. A heat sink and power output stage is mounted to the back of the ignition coil.

Make the test below in the order given, repair any faults before moving on to the next step.

NOTE —

The test given below assumes the speed (RPM) sensor and the camshaft position (CMP) sensor are functioning correctly. See **24a Fuel Injection—Motronic** *for testing information.*

1. Disconnect the 5-pin harness connector from the ignition coil power output stage.

2. Connect a voltmeter across terminals 1(–) and 5(+) in the connector. Turn the ignition on and check that battery voltage is present.

NOTE—

If voltage is not present, check the wiring to the connector. See **97 Wiring Diagrams, Fuses and Relays**.

3. Remove fuse no. 18 (fuel pump, HO2S control module) from the fuse/relay panel.

4. Connect an LED test light across terminal 5 and terminal 2 in the harness connector. Actuate the starter and check that the LED flashes. See Fig. 7.
   - If the LED does not flash, the ignition (ground) signal for cylinders 1 and 6 is not present. Check the wiring between the coil and the ECM connector. If no faults can be found, the ECM may be faulty.
   - If the LED does flash, but the spark plugs for cylinder no. 1 and cylinder no. 6 do not fire, the ignition coil/power output stage is faulty and should be replaced.

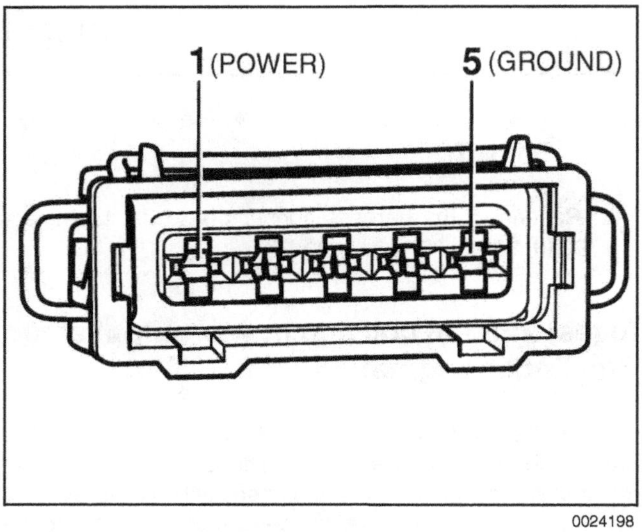

Fig. 7. 5-pin harness connector for 6-cylinder coil power output stage.

5. Repeat above step across terminals 5 and 3 (cylinders 3/4) and terminals 5 and 4 (cylinders 2/5). If any faults are found replace ignition coil/power output stage.

NOTE—

The coils and power output stage are an integrated unit. Individual parts are not separately available for 6-cylinder engines.

6. Turn the ignition off and reinstall fuse no. 18.

To test knock sensor

A knock sensor functions like a microphone and converts mechanical vibration (knock) into electrical signals. The Motronic ECM is programmed to react to the signals whose frequencies are characteristic of combustion chamber noise and change ignition timing accordingly. If the signal from the knock sensor is missing for any reason, the Motronic ECM will retard the timing about 12° to prevent engine damage.

NOTE—

- *4-cylinder engines use a single knock sensor. The sensor is mounted on the front side of the cylinder block, beneath the intake manifold, between cylinders two and three.*
- *6-cylinder engines are equipped with two knock sensors. One sensor monitors cylinders 1, 2, and 3, while the other monitors cylinder 4, 5 and 6. See Fig. 8.*

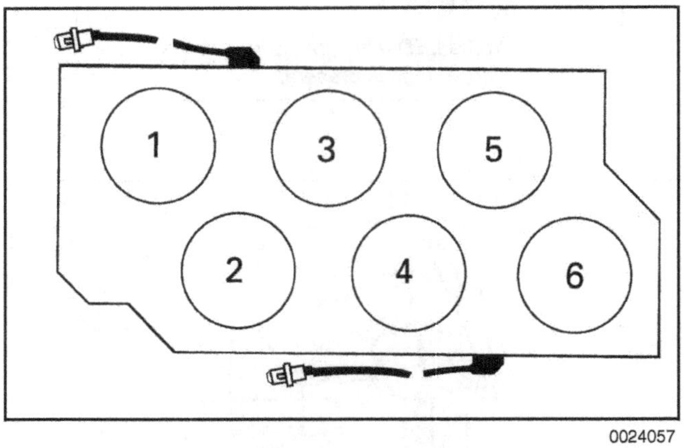

Fig. 8. 6-cylinder engine uses two knock sensors. Knock regulation is done cylinder-selectively on Motronic engines.

1. Remove the knock sensor from the cylinder block and disconnect the 3-pin connector.

2. Using the knock sensor mounting bolt, a washer, and a nut, tighten the bolt through the knock sensor to the specified torque.

NOTE—

Be sure to tighten the knock sensor mounting bolt to the specified torque. Correct mounting bolt torque is critical to knock sensor operation.

Tightening torques
- Knock sensor to cylinder block. . . . . . 20 Nm (15 ft-lb)

3. Connect a voltmeter (DC volts) to terminals 1 and 2 in the connector. See Fig. 9.

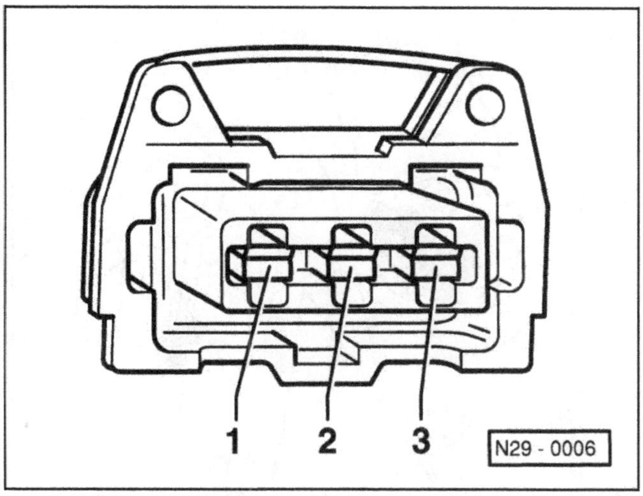

N29-0006

Fig. 9. Terminal identification for knock sensor connector.

4. Using a metal object, lightly tap on the mounting bolt. The voltmeter should register small fluctuating amounts of voltage. Voltage will vary based on the intensity of the tapping.

5. If any faults are found the sensor is faulty and should be replaced.

To check ignition wires

1. Remove one ignition wire and separate the spark plug connector and the interference suppression connector from the wire.

2. Check the wires by gently bending them in several places to expose cracks in the insulation. Wires that are cracked, oil-soaked or dry and brittle should be replaced.

3. Measure the resistance of the individual components. Replace any faulty parts found.

Ignition wire resistances

- Spark plug connector........... 4,000–6,000 ohms
- Interference suppression connector...................... 600–1400 ohms
- Ignition wire (ends removed) .... continuity

4. Assemble and reinstall the wire. Repeat the check for the remaining wires.

To check distributor cap and rotor (4-cylinder engines)

The distributor cap and rotor are subject to wear and electrical breakdown.

1. Unclip and remove the distributor cap.

2. Inspect the contacts inside the distributor cap and at the tip of the rotor for carbon tracks, wear, or pitting. See Fig. 10.

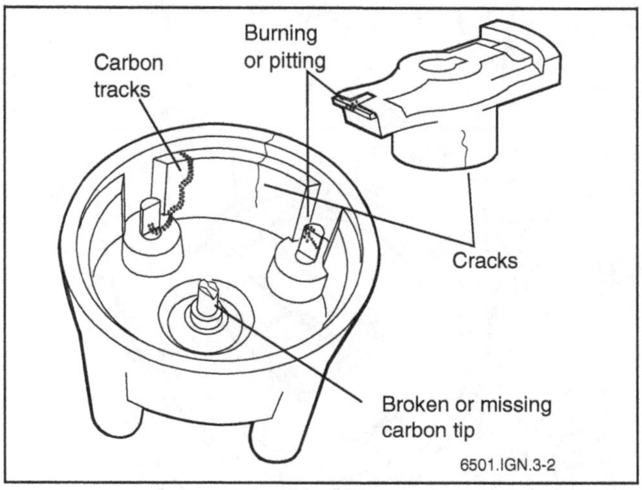

6501.IGN.3-2

Fig. 10. Distributor cap and rotor. Inspect cap and rotor at contact points.

3. If any faults are found, replace the cap and/or rotor.

COMPONENT REMOVAL AND REPLACEMENT

To remove and install distributor (4-cylinder engines)

Individual replacement parts for the distributor are not available from VW. The entire distributor assembly must be replaced if the camshaft position (CMP) sensor is found to be faulty.

> ***NOTE—***
> *On ACC engines, ignition timing must be checked and adjusted any time the distributor is disassembled or removed. See* **To check and adjust ignition timing (ACC engine)** *given later.*

1. Rotate the engine by hand until the distributor rotor tip is aligned with the No. 1 cylinder Top Dead Center (TDC) mark on the distributor housing. See Fig. 10.

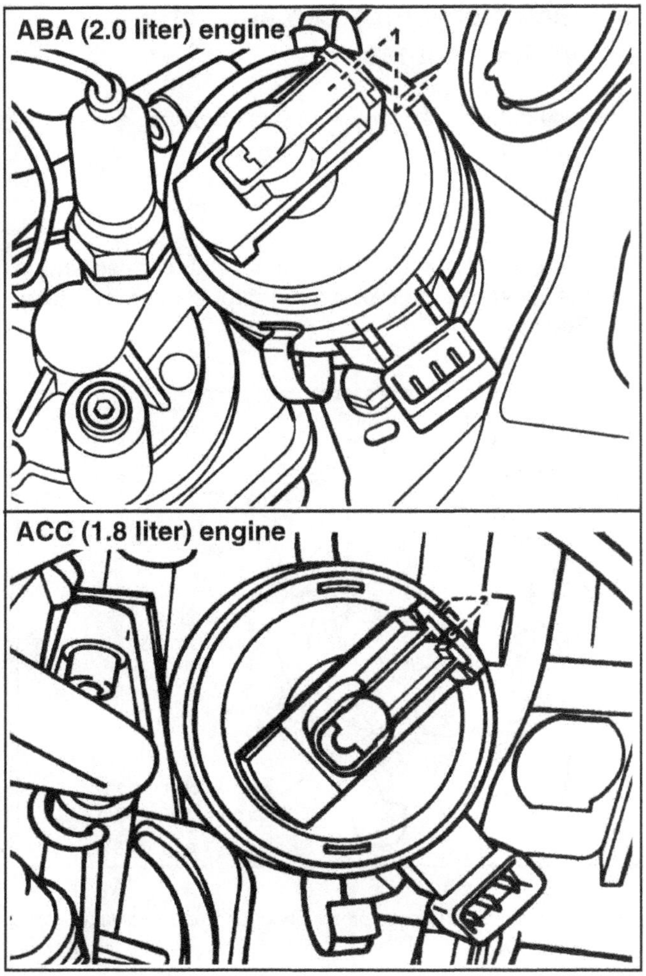

Fig. 11. Ignition rotor aligned with mark on distributor body.

2. Disconnect the electrical connector from the distributor.

3. Remove the distributor hold-down and remove the distributor by pulling it straight out.

4. Retrieve the sealing O-ring from the bottom of the distributor. Cover the opening.

5. If the crankshaft has been turned since the distributor was removed, rotate the crankshaft so that the cylinder no. 1 is at TDC. See Fig. 12.

6. Align the oil pump drive shaft as shown in Fig. 13. Use a screwdriver to turn it if necessary.

7. Using a new O-ring, install the distributor. Align the rotor with the distributor housing, as shown in Fig. 10 earlier.

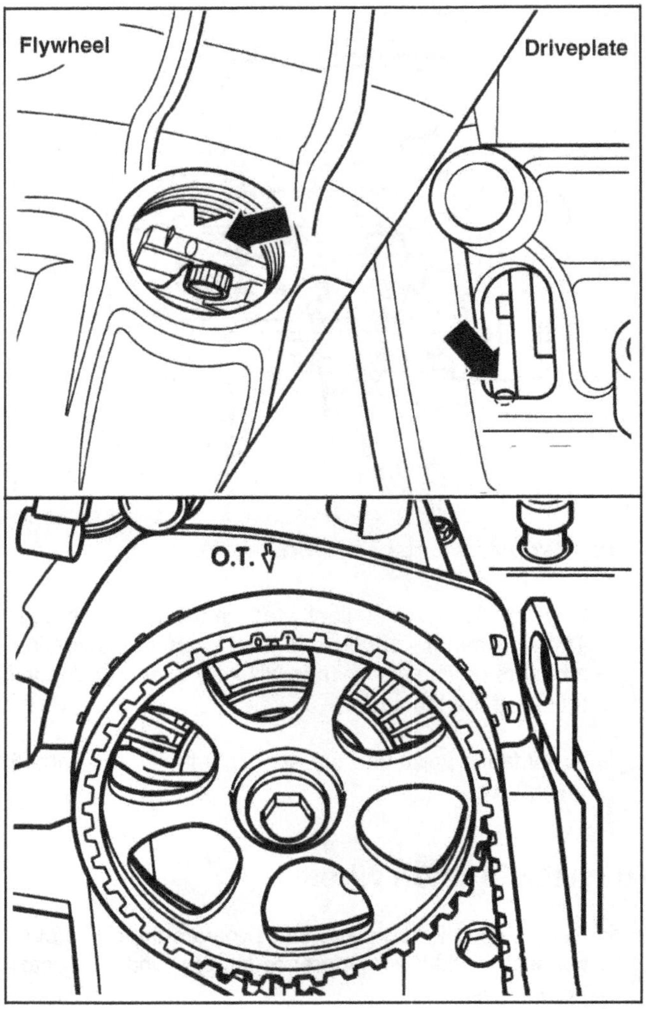

Fig. 12. Engine set to TDC, cylinder no. 1. Position crankshaft (top) and camshaft (bottom) so that TDC marks align as shown.

8. Install the distributor hold-down and the distributor cap. Reconnect the electrical connector.
   - On ABA engines, position the two distributor locating pins so that they are locked in place by the hold down bolt. See Fig. 14.
   - On ACC engines, adjust the ignition timing as described under **Ignition Timing**.

NOTE —
Distributor locating pins may not be present on later production engines or replacement distributors.

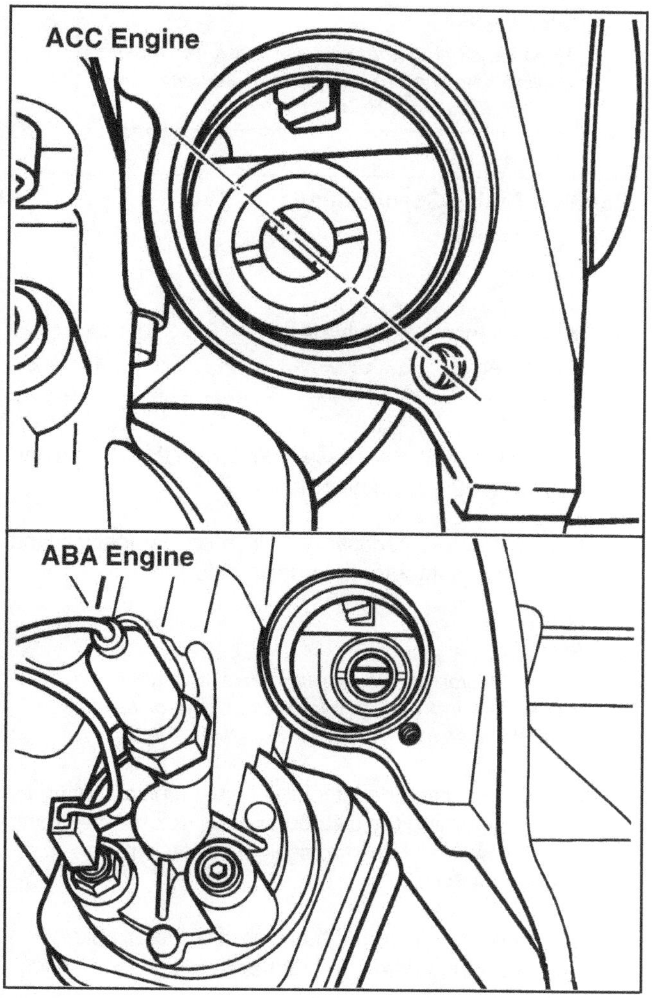

0024065

Fig. 13. **ACC (1.8) engine:** slot on oil pump drive shaft aligned with distributor mounting bolt hole.
ABA (2.0) engine: slot on oil pump drive shaft aligned parallel to crankshaft.

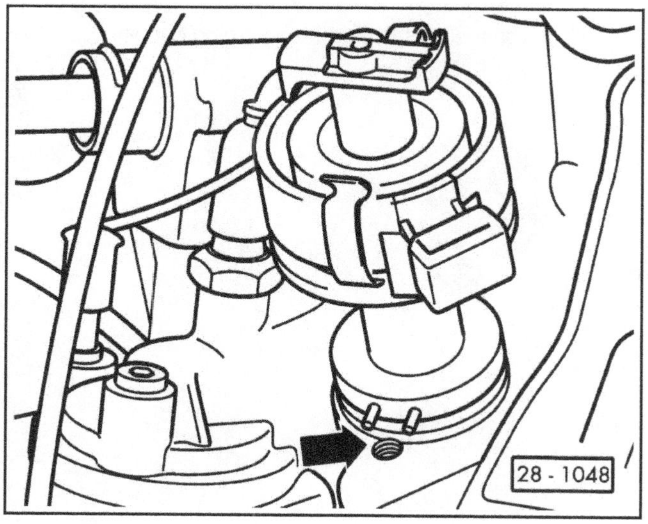

28-1048

Fig. 14. On ABA (2.0 liter) engines, position distributor so that mounting bolt locks distributor in place using alignment pins (if applicable).

To replace spark plugs

1. With the ignition off, disable the ignition system by disconnecting the harness connector from the coil power output stage.

2. Remove the spark plug wires from the spark plugs.

NOTE —

On 6-cylinder engines, a special tool (VW tool no. 3277) is needed to remove the spark plug connectors from the plugs without damaging them. A plastic version of this tool is clipped to the front hood support rod. Do not pull on the wires.

3. Clean the area around the spark plugs using compressed air. Remove the spark plugs and set them aside in order.

NOTE —

On all engines use a 5/8 inch spark plug socket with an extension or VW special tool no. 3122B to remove and install the spark plugs.

4. Install the new spark plugs, making sure the electrode gap is correct before installation. Tighten the plugs to the specified torque.

> *CAUTION —*
> **Table b** *lists general spark applications and specifications. Spark plug designations should be used for reference only. Original spark plugs applications are often changed and updated by the manufacturer over time. Check with an authorized VW dealer for the most up-to-date parts information.*

Table b. Spark Plug Applications

ABA (2.0 liter)	
electrode gap	0.6 mm (0.024 in.)
tightening torque	30 Nm (22 ft-lb)
manufacturer's designation	
Bosch	FR 8 DS
AAA (2.8 liter, VR6)	
electrode gap	0.7 mm (0.028 in.)
tightening torque	25 Nm (18 ft-lb)
manufacturer's designation	
NGK	BKR 5 EKU
ACC (1.8 liter, Canada)	
electrode gap	0.7-0.9 mm (0.028-0.035 in.)
tightening torque	25 Nm (18 ft-lb)
manufacturer's designation	
Bosch	W 7 DTC
NGK	BUR 6 ET
Beru	4-7 DTC

IGNITION TIMING

To check and adjust ignition timing (ACC engine)

1. On cars with manual transmission, remove the plastic plug from the timing check hole at the top of the transaxle bellhousing.

   NOTE —
   The entire plug must be removed. The small center plug (if applicable) is for use with a special timing indicator used by Volkswagen dealer technicians.

2. With the ignition off, connect a timing light according to the instrument manufacturers' instructions.

3. Start the engine and allow it to fully warm up.
   - Oil temperature should be at least 176°F (80°C). The radiator cooling fan should cycle on and off at least once.

4. With the engine idling, aim the timing light at the timing check hole in the bell housing. The timing mark on the flywheel should appear adjacent to the pointer in the hole. See Fig. 15.

   NOTE —
   Watch carefully to be sure that the pointer lines up with the notch in the flywheel and not the "0" which denotes top dead center (TDC).

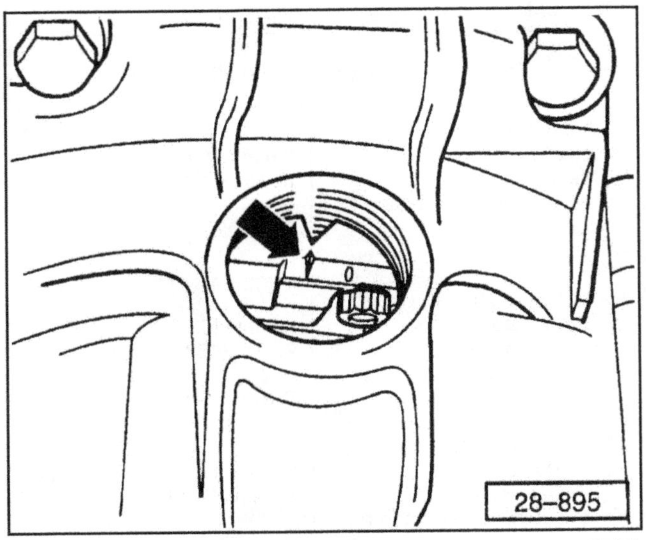

28-895

Fig. 15. Timing mark on flywheel aligned with pointer in timing check hole (**arrow**). Do not align pointer with the 0 mark on the flywheel. This denotes top dead center (TDC).

> *WARNING —*
> *Keep hands and other objects clear of the radiator cooling fan. The fan may start at any time, even when the ignition is switched off.*

Ignition Timing Specifications
- Ignition timing
  - checking . . . . .4-8 ° BTDC
  - adjusting . . . . 6° ± 1° BTDC
- Idle speed (non-adjustable) . . . . .770-1000 rpm
- Firing order. 1-3-4-2

5. If no adjustment is necessary, switch off the ignition and disconnect the test equipment.

6. If adjustment is necessary, switch off the ignition and loosen the distributor hold-down bolt.

   NOTE —
   The distributor should be just loose enough to be moved by hand with deliberate effort. The distributor must not move by itself while the engine is running.

7. Start the engine and let it idle. Adjust ignition timing by gradually turning the distributor housing until the timing mark, viewed with the timing light, meets with the specifications listed above.

8. Stop the engine and tighten the distributor hold-down bolt. Start the engine and recheck the ignition timing.

Tightening torques
- Distributor hold down to distributor . . 25 Nm (18 ft-lb)

9. Turn the ignition off and remove the test equipment. Reinstall the plastic plug in the timing check hole.

28b Ignition System–Diesel

TABLES

GENERAL

The ignition system of a diesel engine is based on the heat generated by the compression of air in the cylinder. Fuel is injected into this hot air at a precise time to obtain ignition and the exact time is varied by the injection pump to obtain ignition advance. Diesel engines are therefore said to be compression-ignition engines. As a result of this design, diesel engines do not require spark plugs and their associated components. They do, however, require a little help to generate sufficient heat during a cold start to burn the fuel. This is because the heat generated is dissipated into the cold engine before it is hot enough to allow the fuel to burn. Volkswagen diesel engines use glow plugs to generate additional heat to aid in starting when cold. **Table a** list the various engine management systems covered by this manual.

Table a. Engine Management System Applications

Engine code	System
AAZ (1.9 liter)	Diesel Fuel Injection
1Z, AHU (1.9 liter)	TDI Diesel Fuel Injection

NOTE —
The TDI engine management system found on 1Z/AHU engines has a built in diagnostic circuit that detects and stores a limited number of system faults related to the glow plug system. These faults are known as Diagnostic Trouble Codes (DTCs). It is recommended that fault diagnosis and troubleshooting be carried out using the special Volkswagen scan tool VAG 1551 or VAG 1552.

Service Precautions

Ignition system service and repair work must be carried out carefully. The ignition system contains sensitive electronic components. To guard against system damage, and for general safety, the following warnings apply to any ignition system troubleshooting, maintenance or repair work.

WARNING —
Glow plug systems operate at high current levels. For this reason, always insure that electrical connections that have been disturbed during testing are properly secured when done.

CAUTION —
Disconnecting engine management components with the key in the "on" position during some procedures will cause diagnostic trouble codes to be stored in the engine control module. Erasing these DTCs and resetting the readiness code will require Volkswagen scan tool VAG 1551 or VAG 1552.

DIESEL GLOW PLUG SYSTEM – AAZ ENGINE

The four glow plugs are small heating elements, one for each cylinder, mounted in the cylinder head with the tips projecting into the combustion pre-chambers. The glow plugs assist combustion during cold starts by providing additional heat for ignition.

The glow plug system is controlled by the automatic glow time control module and the glow plug relay. The system is activated by the driver's door switch. Once the door is opened and then closed, the pre-glow period is initiated. The length of time that the system operates is controlled by an engine temperature sensor. Above a certain temperature, the system does not operate.

NOTE —

- *Some early 1993 models may not have the door-switch activation circuit. On these early models, system operation is activated by the ignition key.*
- *For more information on troubleshooting diesel cold-starting problems, see* **23 Fuel Injection—Diesel**.
- *See the* **97 Wiring Diagrams, Fuses and Relays** *for a complete wiring diagram of the glow plug electrical circuits including the automatic glow time control module, the glow plug relay, fuses, and connector identification.*

Testing and Replacing Glow Plugs

To check the glow plugs, disconnect the wire from the glow plug bus bar, and remove the bus bar from each of the glow plug terminals. Connect one probe of a test light to the battery's positive (+) terminal, and touch the other probe to each of the glow plugs. The test light should come on when it is touched to each glow plug. If not, that plug is faulty and should be replaced. If the test light does not light at all, voltage may not be reaching the bus bar connection due to a problem in the system circuit. See **97 Wiring Diagrams, Fuses and Relays** to check for faulty wiring or components.

Another test of glow plugs is to measure current draw using a special tester, such as the VAT 40. Disconnect the engine temperature sensor and reconnect the bus bar. Connect an ammeter in series between the glow plug bus bar and the wire that connects to it. Activate the pre-glow period (open and close driver's door), start the engine and let it idle.

After an initial surge, the total current draw should drop to a steady 32 amperes (8 amps per glow plug). If the current draw is not steady, or is less than 32 amperes, one or more glow plugs is faulty.

NOTE —
Battery voltage must be greater than 11.5 volts for accurate test results.

Glow Plug Current Draw
- all four glow plugs operating . . . . 32 amps
- three glow plugs operating . . . . 24 amps
- two glow plugs operating . . . . 16 amps
- one glow plug operating . . . . 8 amps

Finally, check the glow plugs by removing and inspecting them. Replace a glow plug if it has been over-torqued, as shown in Fig. 1, or if the tip is burned away, as shown in Fig. 2. Worn away tips can be caused by malfunctioning fuel injectors. See **23 Fuel Injection—Diesel** for fuel injector pressure testing.

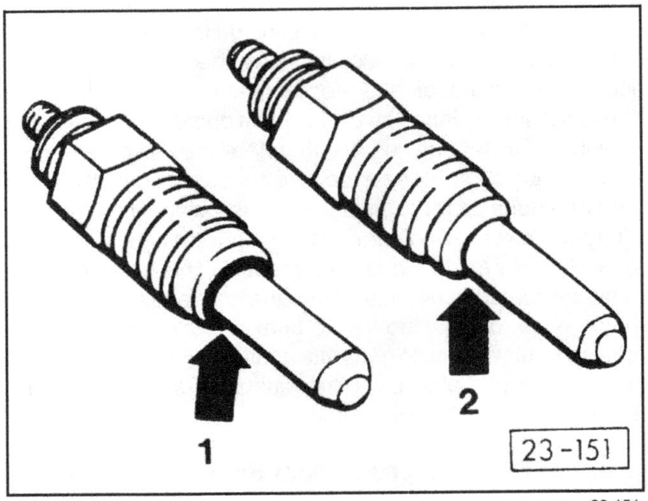

23-151

Fig. 1. Glow plug ring gap closed (**2**) after incorrect torquing. Correct gap (**1**) is approximately 0.5 mm (0.02 in.).

Before installing the glow plugs, clean the glow plug holes and seats in the cylinder head. Clean off carbon deposits before reinstalling used glow plugs. Tighten the plugs to the correct torque and use new self-locking nuts when installing the bus bar.

CAUTION —
Exceeding the recommended tightening torque will damage the glow plugs and lead to premature failure.

Tightening torques
- Glow plug to cylinder head . . . . 25 Nm (18 ft-lb)

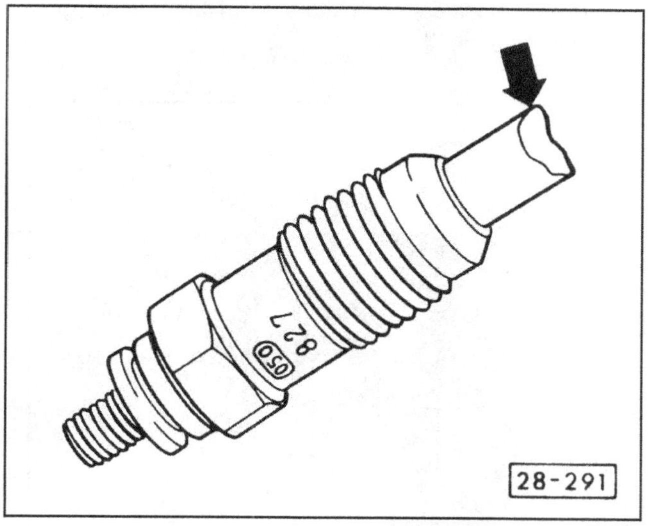

28-291

Fig. 2. Example of glow plug with burned away tip (**arrow**).

DIESEL GLOW PLUG SYSTEM - 1Z/AHU TDI ENGINE

The TDI Diesel engine uses four glow plugs in much the same manner as earlier Diesel versions. Because of the different combustion chamber design, the glow plugs are installed in a vertical manner and are fitted with push-on electrical connectors similar to spark plug connectors. The glow plugs assist combustion during cold starts by providing additional heat for ignition. Due to the efficiency of the TDI engine design, the additional heat that the glow plugs provide is not needed until temperatures drop to approximately 48°F (9°C). This function is known as preglow. In addition to preglowing, the glow plugs will continue to operate after the engine is started. This is known as afterglowing and is used to reduce engine noise, improve idle quality and reduce hydrocarbon emissions. The afterglow function is enabled after every start, both hot and cold until the specified time has expired or the engine speed exceeds 2500 RPM.

The glow plug system is controlled by the Engine Control Module, ECM, and the coolant temperature sensor, ECT. The period of preglow depends on the temperature of the engine coolant as determined by the ECT. A dash mounted Indicator Light signals the driver that preglowing is operating. Preglowing is not linked to operation of the driver's door as in some earlier Diesel systems.

> ***CAUTION —***
>
> *Disconnecting engine management components for this procedure will cause diagnostic trouble codes to be stored in the engine control module. Erasing these DTCs and resetting the readiness code will require Volkswagen scan tool VAG 1551 or 1552.*

NOTE —

- *For more information on troubleshooting Diesel cold starting problems, see* **23b Fuel Injection—Diesel TDI**.
- *See* **97 Wiring Diagrams, Fuses and Relays** *for a complete wiring diagram of the glow plug electrical circuits.*

Testing and replacing glow plugs

Basic checks of the glow plug system are similar to earlier diesel versions.

1. Disconnect the harness connector from the engine coolant temperature sensor, ECT. This simulates a cold engine when the ignition key is switched on. See Fig. 3.

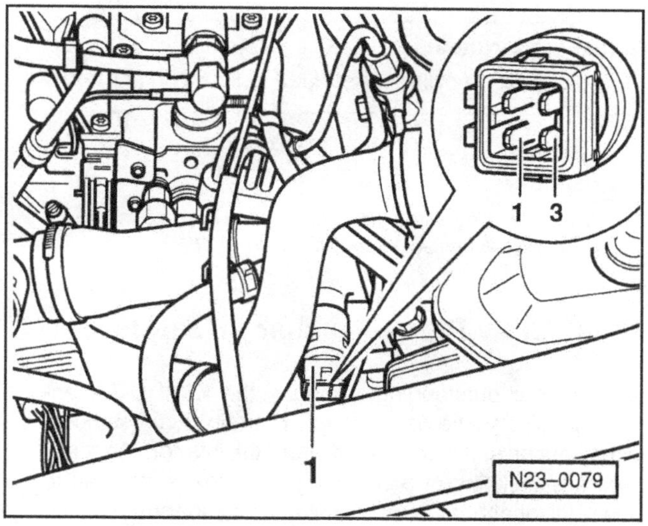

N23-0079

Fig. 3. Engine coolant temperature sensor (1) on coolant outlet pipe on front of engine. Disconnecting the harness connector simulates a cold engine and operates glow plugs for the maximum length of time.

2. Disconnect the bus connector from the glow plugs by gently pulling it off.
3. Connect an accurate voltmeter between the bus connector for cylinder number 1 and ground.
4. Turn on the key and observe battery voltage. Voltage should be present for approximately 20 seconds.
5. Repeat for remaining cylinders.
6. If no voltage is present, check glow plug fuse under dash.

NOTE —

Glow plug fuse is located above the fuse/relay panel inside the vehicle. It is a large plastic-encased design known as a MAXI-FUSE®. Do not confuse the glow plug fuse with the fuse on the firewall in the engine compartment. This old style flat fuse is for the Auxiliary heater.

7. If voltage is present, switch off ignition.
8. Connect a suitable test light to battery positive (+) and touch the other end to the disconnected terminal on the glow plug. The low resistance of the glow plug will cause the test light to light up if the glow plug is good.
9. Replace defective glow plugs using a 10mm deep socket and suitable extension or use Volkswagen special tool 3220. Torque replacement glow plugs to specification.

Tightening torques

- TDI glow plug in cylinder head..............15 Nm (11 ft-lb)

NOTE —

Advanced testing of the glow plug system requires the use of Volkswagen scan tool VAG 1551 or 1552.

Coolant Glow Plugs/Auxiliary Heater

Because of its outstanding efficiency, the 1Z/AHU TDI engine develops very little waste heat. In certain circumstances, there may not be sufficient heat to warm the interior. An auxiliary heater consisting of 3 coolant glow plugs is installed to provide additional heat to the passenger compartment when needed. The 3 coolant glow plugs are operated by the Coolant Glow Plug Relay (J325) under the control of the Engine Control Module. The coolant glow plug relay is mounted under the hood on the driver's side of the firewall and has a fuse for the coolant glow plugs built into the top. The coolant glow plugs can operate in one of three modes depending on the amount of heat required. Coolant glow plug #1 can operate separately, or #2 and #3 together, or all three at the same time.

Testing and Replacing Coolant Glow Plugs

Complete testing of the system requires the use of Volkswagen scan tool VAG 1551 or 1552, however, the glow plugs can be checked in the same manner as the glow plugs used for starting.

1. Visually inspect the fuse on the coolant glow plug relay by lifting the plastic cover. It is the flat style fuse commonly associated with older types of Volkswagen glow plug systems. Replace as required
2. Pull the electrical connectors off of the coolant glow plugs. See Fig. 3.

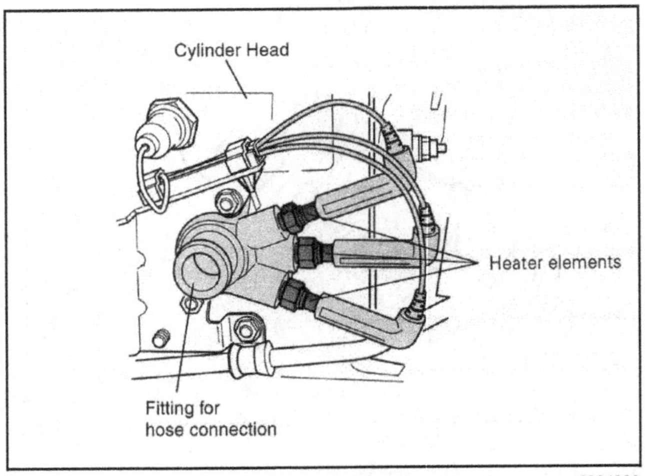

Fig. 4. Coolant glow plugs on the end of the cylinder head with the electrical connectors in place.

3. Connect a suitable test light to the battery and touch the other end to the end of the coolant glow plug. The light will glow due to the low resistance of the glow plug if it is good. Faulty coolant glow plugs will not light the test light.
4. If defective glow plugs are found, unscrew and replace as necessary. Coolant will leak out of the system.

WARNING —

- *Hot coolant can scald and result in serious personal injury. Do not work on the cooling system until it has fully cooled.*
- *At normal operating temperature the cooling system is pressurized. Allow the system to cool as long as possible before opening—a minimum of an hour—then release the cap very slowly to allow safe release of pressure.*
- *Releasing the cooling system pressure lowers the coolant's boiling point, and the coolant may boil suddenly. Use heavy gloves and wear eye and face protection to guard against scalding.*
- *Use extreme care when working at or near the cooling fan when the engine is hot. The fan can come on at any time–even if the ignition key is off.*
- *Use extreme care when draining and disposing of engine coolant. Coolant is poisonous and lethal. Children and pets are attracted to coolant because of its sweet smell and taste. See a doctor or veterinarian immediately if any amount of coolant is ingested.*

5. Re-attach the electrical connectors and replace the lost coolant.

3 Clutch, Transmission, and Final Drive

TABLES

GENERAL

This general information group covers application information and system descriptions for the repair groups listed under **3 Clutch, Transmission, and Final Drive**. This includes the two manual and two automatic transmissions that are available for use in the Jetta, Golf, GTI, and Cabrio models. For transmission identification, see **0 Maintenance Program**.

Transmission applications

- 020 . . . . . . 5-speed manual (ABA, ACC, AAZ engines)
- 02A . . . . 5-speed manual (AAA, 1Z, AHU engines)
- 096 . . . . 4-speed automatic, phase I (ABA, ACC, AAA engines, 1993-1995 models)
- 01M . . . . 4-speed automatic, phase II (ABA, ACC, AAA engines, 1995-1997 models)
- 01M . . . . 4-speed automatic, phase II, "ETA" (ABA, ACC, AAA engines, 1997-1999 models)

MANUAL TRANSMISSIONS

The transmission gears are of the constant-mesh type with balk ring synchronizers. Each gear (1st through 5th) is actually a pair of gears, and constant-mesh simply means the mating gears are always in contact. When shifting, synchronizers cause the moving parts to rotate at the same speed before engaging. This prevents damage to the gears, minimizes wear, and eases shifting. The 3rd, 4th, and 5th gear synchronizers are on the main shaft; the 1st and 2nd gear synchronizers are on the pinion shaft.

The **020 transmission** uses a mechanical clutch release mechanism and a rod operated shift mechanism. See Fig. 1.

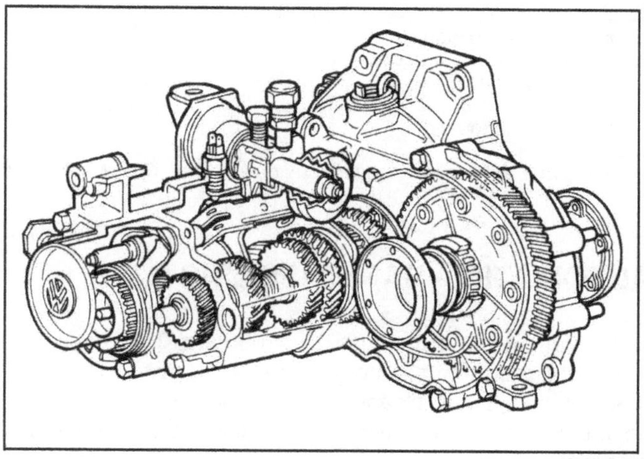

0024157

Fig. 1. Cut-away view of the 020 manual transmission (02A is similar).

The close ratio **02A transmission** uses a hydraulic clutch release mechanism and a cable operated shift mechanism.

Table a lists the engine applications, gear ratios, and other data for the various manual transmissions.

Table a. Manual Transmission Specifications

Designation		020			02A	
Engine (engine code)		1.8L, 90hp (ACC)	1.9L, 75hp (AAZ)	2.0L,115hp (ABA)	1.9L, 90hp (1Z/AHU)	2.8L,172hp (AAA)
Transmission (manuf. dates)*		APW (10.91-8.92) CHB (8.92-12.95) DFN (12.95 on)	ATH (10.91-8.92) CHD (8.92-12.95) DFP (12.95 on)	AMC (10.91-8.92) CHE (8.92-12.95) DFQ (12.95 on)	CTN (8.96 on)	CCM (02.94 on)
Ratio: Z_2:Z_1	Final drive	66:18 = 3.667	66:18 = 3.667	66:18 = 3.667	60:19 = 3.157	61:18 = 3.389
	1st gear	38:11 = 3.455	38:11 = 3.455	38:11 = 3.455	34:9 = 3.777	33:10 = 3.300
	2nd gear	35:18 = 1.944	35:18 = 1.944	35:18 = 1.944	36:17 = 2.117	35:18 = 1.944
	3rd gear	37:27 = 1.370	36:28 = 1.286	36:28 = 1.286	34:25 = 1.360	34:26 = 1.308
	4th gear	32:31 = 1.032	30:33 = 0.909	31:32 = 0.969	34:35 = 0.971	30:29 = 1.034
	5th gear	34:40 = 0.850	38:51 = 0.745	33:41 = 0805	34:45 = 0.755	31:37 = 0.838
	Reverse gear	38:12 = 3.167	38:12 = 3.167	38:12 = 3.167	20:9 x 36:20 = 3.999	17:10 x 36:20 = 3.060
Lubricant	Capacity	1.9 Liters (2.0 qt.)			2.0 L (2.1 qt.)	
	Specification	G50 synthetic oil, SAE 75W/90				
Clutch release mechanism		mechanical			hydraulic	
Clutch disc diameter		210 mm			228 mm	
Ratio, overall in top gear		3.117	2.732	2.952	2.385	2.840

**Note: Other transmission codes may be installed. Check with authorized Volkswagen parts department for the latest information.*

AUTOMATIC TRANSMISSIONS

The two automatic transmissions (096 and 01M) are electro-hydraulically controlled with four forward gears. The 01M does not have a dipstick and has a unique system for checking and changing the transmission fluid. See **0 Maintenance Program**.

The automatic transmission can best be understood by dividing it into five subsystems. These are the torque converter, the ATF pump, the planetary gear system, the hydraulic controls, and the final drive. Fig. 2 shows a cross-section of an automatic transmission.

Torque Converter

The torque converter is a doughnut-shaped assembly located between the engine and the transaxle inside the bellhousing. The torque converter is driven by the engine and in turn drives the transmission. The torque converter provides the desired torque multiplication at low speed to propel the vehicle away from a stop and serves as an efficient fluid coupling at high speed.

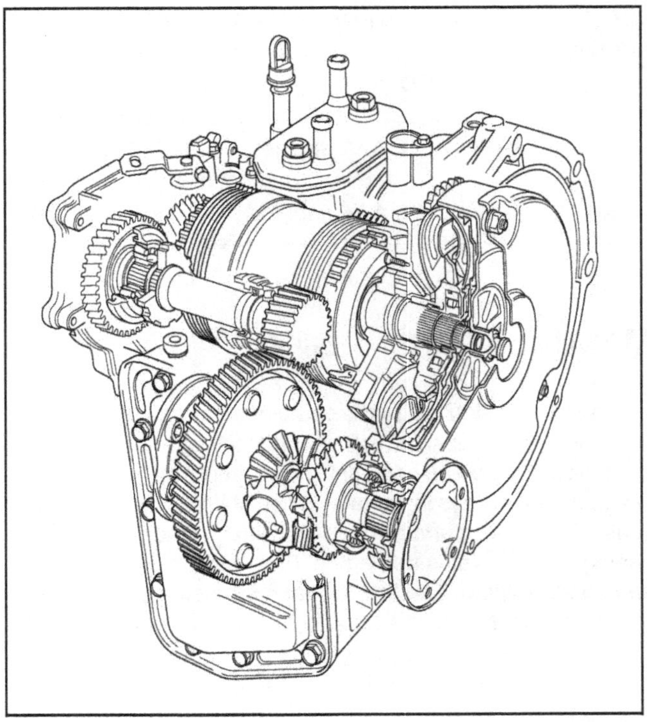

0024158

Fig. 2. Cross-section of Volkswagen automatic transmission.

ATF Pump

The automatic transmission fluid (ATF) circulates through the transmission under pressure whenever the engine is running. The ATF pump that creates this pressure is located behind the torque converter.

Planetary Gear System

A torque converter alone cannot supply the torque multiplication needed for all driving conditions. The torque converter therefore drives through a planetary gear system which can operate at different drive ratios. The planetary gear system consists of three plate clutches, two plate brakes, one planetary gearset, and one roller freewheel.

The plate clutches and plate brakes are operated hydraulically from the valve body. They provide the power flow between converter and planetary gears.

Electro-Hydraulic Controls

The hydraulic control system directs and regulates hydraulic pressure from the ATF pump to control shifting of the planetary gear system. The selector lever cable gives the Transmission Control Module (TCM) information about the position of the selector lever via the multi-function switch. At the same time the selector lever cable and a rod operate the manual valve in the valve body. The manual valve is thus moved into a basic position, in which, in lever position "D", the full automatic program is available.

The TCM switches the solenoid valves in the valve body according to the input signals from engine and transmission sensors. The solenoid valves operate shifting valves in the valve body which supply ATF oil pressure to the shift components (plate clutches and brakes) in the gearbox. Via the shift components the engine torque is transmitted to the planetary gears.

The 096 and 01M transmissions feature adaptive programming and On-Board Diagnostic (OBD) capabilities.

Adaptive programming functions to match driving patterns to one of two shift programs stored in the TCM. When accelerating quickly (throttle at or close to full load), the Sport program is automatically selected and the shift points occur at higher engine speeds. The TCM will automatically shift to the Economy shift program (shift points occur at lower engine speeds) if the throttle is held below the full load range for a predetermined amount of time.

The TCM self-diagnosis system monitors the electrical signals from various sensors and carries out a self check of the TCM. If a fault is detected, it is permanently stored in the TCM, even when the battery is disconnected. This coded fault information can be retrieved using the special Volkswagen scan tool, V.A.G 1551.

Final Drive

The final drive consists of the drive pinion, the intermediate gear, the ring gear, and the differential. The teeth on the intermediate and ring gear, and the drive pinion are helically-cut for quieter running. The differential allows the front wheels to turn at different speeds, as is necessary when turning. **Table b** lists the final drive ratios for the various automatic transmissions.

NOTE —

- *On automatic transmissions, the final drive lubricant is separate from the ATF. Volkswagen does not specify a final drive lubricant replacement interval; it is "filled for life". However, the fluid should be checked regularly.*
- *To check final drive level see* **39 Differential and Final Drive**. *To check ATF level see* **0 Maintenance Program**.

Table b. Automatic Transmission Final Drive

Code letters	Engine application	Final drive ratio
CFA, CFF, CLB, DLL, DMA, DNT	2.8 L (AAA)	3.700:1
CFC, CFH, CLA, CNK, CLK, DKS, DLY, DNS, DLZ	2.0 L (ABA)	4.222:1
CKZ, DLX	1.9 L (AAZ & AHU)	3.273:1
CFD, CFK, CRR, CKX, CKY, DKR, DLV, DLW, DNR	1.8 L (ACC)	4.529:1

Identification and application

The transmission code letters and date of manufacture can be found on the vehicle data plate and stamped on the transmission itself. Fig. 3 and 4 show the location of the code letters for two versions of available automatic transmissions.

Table c. Automatic Transmission Identification

Example:	CFH	03	01	4
	Transmission code letters	Day	Month	Year
		Date (e.g. 3 Jan., 1994)		

There are several versions of the two (096 and 01M) automatic transmissions depending on model and model year. See table **d**.

CAUTION —
Only genuine VW synthetic ATF should be used in the 01M transmission. Do not use other types of ATF. Do not mix Volkswagen ATF with Dexron ATF.

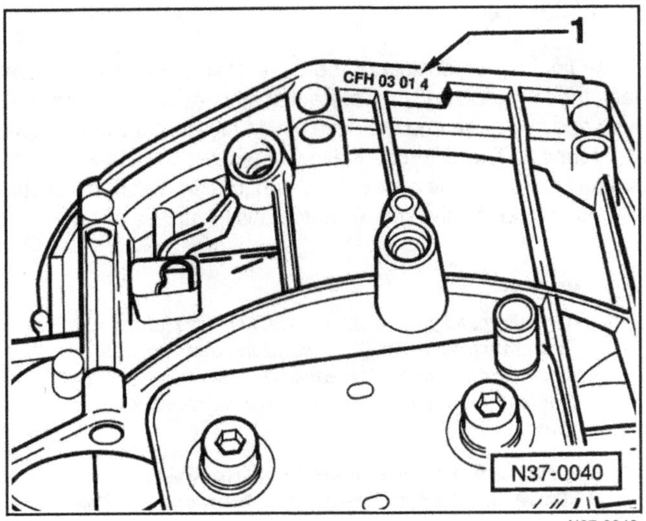

N37-0040

Fig. 3. Code letters (**arrow 1**) for 096 automatic transmission.

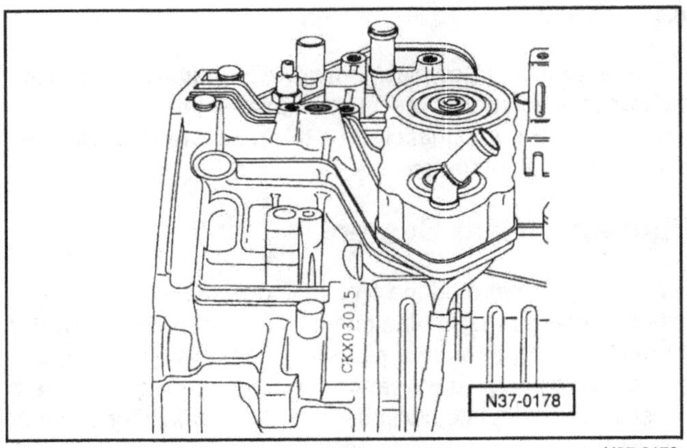

N37-0178

Fig. 4. Code letters for 01M automatic transmission.

Table d. Automatic Transmission Specifications

Designation		096		01M	
Model and model year		Golf from 1993 Jetta from 1993	Golf from 1993 Jetta from 1993 Jetta GLX from 1994	Golf from 1995 Jetta from 1995 Cabrio from 1995	Golf from 1997 Jetta from 1997 Cabrio from 1997
Code letters*		CFA, CFC, CFD	CFF, CFH, CFK, CNK	CLA, CLB, CLK, CKX, CKY, CKZ, DLL, DKR, DKS	DLV, DLW, DLX, DLY, DLZ, DMA, DNR, DNS, DNT
Ratio	1st gear	2.714:1	2.714:1	2.714:1	
	2nd gear	1.551:1	1.441:1	1.441:1	
	3rd gear	1.000:1	1.000:1	1.000:1	
	4th gear	0.679:1	0.743:1	0.742:1	
	Reverse gear	2.111:1	2.884:1	2.884:1	
Transmission lubricant	Initial filling	5.6 Liters (5.9 qt.)		5.3 L (5.6 qt)	
	Oil change	approx. 3.0 Liters (3.2 qt.)		approx. 3.0 L (3.2 qt.)	
	Type	ATF Dexron		Special VW ATF	
Final drive lubricant	Initial filling	0.75 Liter (0.79 qt.)		0.75 Liter (0.79 qt.)	
	Type	G50 synthetic oil, SAE 75W/90		Special VW ATF	

**Note: Other transmission codes may be installed. Check with authorized Volkswagen parts department for the latest information.*

30a Clutch (020 Transmission)

GENERAL

Servicing of the cable operated clutch requires that the transmission be removed from the engine. Special tools and equipment are required to remove the transmission and service the clutch. See **34 Manual Transmission** for transmission removal procedures.

> ***WARNING —***
> - *The cars covered by this manual use an airbag system that automatically deploys an airbag(s) in the event of a frontal impact. The airbag(s) is inflated by an explosive device. Handled improperly or without adequate safeguards, the system can be very dangerous. Special precautions must be observed prior to any work at or near the steering wheel or steering column, including the pedal assembly. See* **69 Seatbelts, Airbags**.
> - *To guard against personal injury or airbag system failure, only trained Volkswagen service technicians should test, disassemble or service the airbag system.*

NOTE —
When servicing the transmission or clutch mechanism, always observe the general repair instructions outlined in **3 General Information**.

CLUTCH RELEASE MECHANISM

Fig. 1 shows an exploded view of the pedal assembly and the clutch release mechanism. Be sure to lubricate all bearings and friction surfaces with MoS2 grease (VW Part No. G 000 602).

> ***WARNING —***
> *Always replace self-locking nuts and circlips during repairs.*

> ***CAUTION —***
> - *Always obtain anti-theft radio code and then disconnect battery ground strap.*
> - *Part numbers are listed for reference only. Always check with an authorized Volkswagen parts department for the latest parts information.*

NOTE —
Before replacing a faulty clutch cable, check the cable adjustment mechanism operation as described under **Faulty cable adjustment mechanism, repairing**.

Clutch release mechanism, assembly

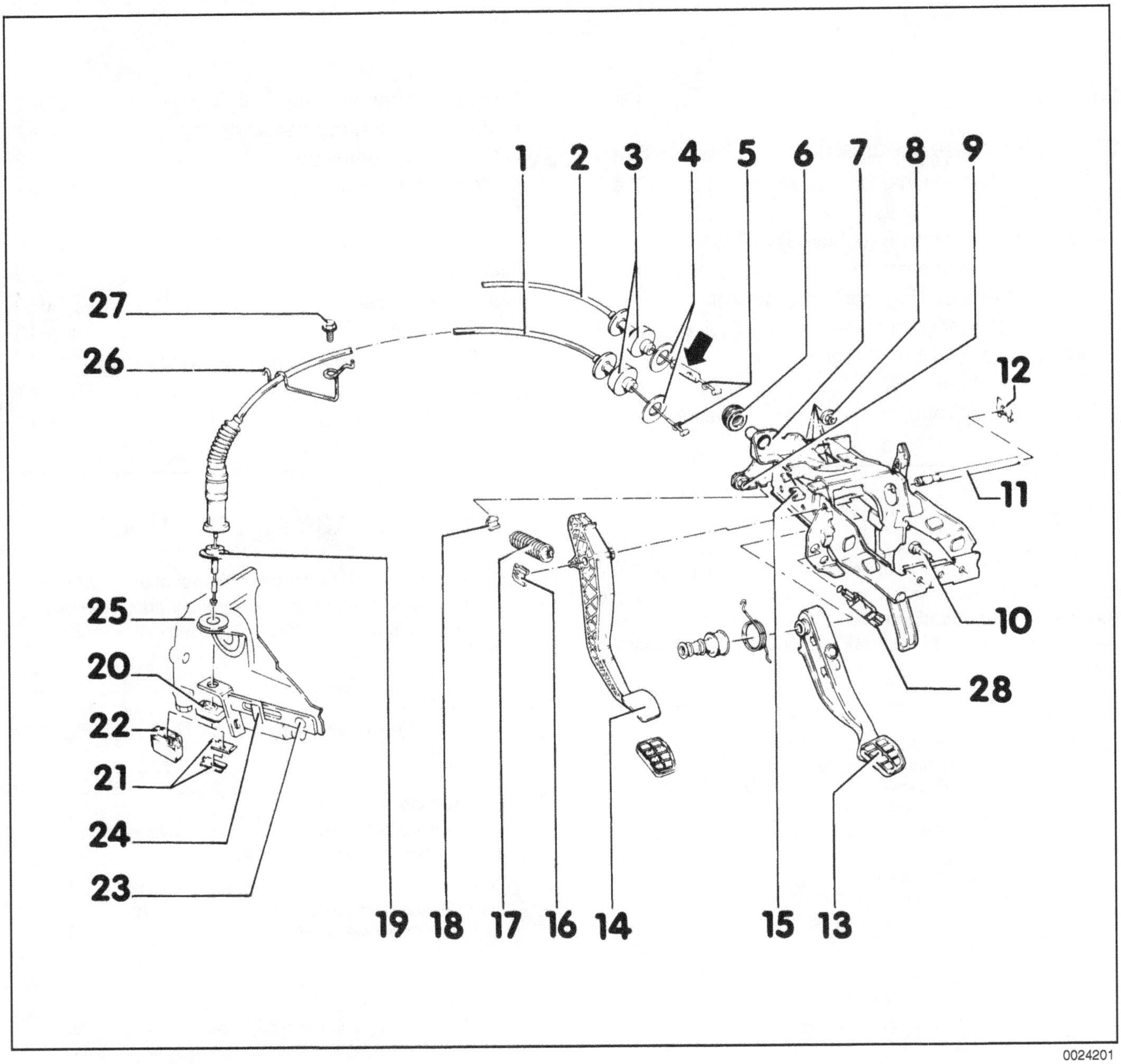

Fig. 1. Exploded view of clutch release mechanism and related parts.

1. **Clutch cable**
   - Vehicles with gasoline spark-ignition engine only
2. **Clutch cable with integral vibration damper -arrow-**
   - For vehicles with Diesel engine, up to August 1993 production only
3. **Rubber stop**
4. **Washer**
5. **Support plate**
6. **Dust boot**
   - Can remain installed during removal of clutch cable
7. **Bracket with accelerator pedal**
   - First remove steering column with sleeve. See **48 Steering.**

> ***WARNING —***
> *Observe airbag precautions as described in* **69 Seatbelts, Airbags**. *Always disconnect the negative (–) battery cable before working at or near the steering column.*

8. **Self-locking nut**
   - Bracket (7) to threaded studs on vacuum brake booster
   - Tightening torque: 25 Nm (18 ft-lb)
   - Always replace
9. **Self-locking nut**
   - Bracket (7) to threaded studs on mounting plate
   - Tightening torque: 25 Nm (18 ft-lb)
   - Always replace
10. **Attachment bolts, bracket to assembly support**
    - Bracket (7) to support
    - Tightening torque: 25 Nm (18 ft-lb)
11. **Pivot pin for clutch and brake pedal**
    - Installation position, **see** (A)
12. **Retaining clip**
13. **Brake pedal**
14. **Clutch pedal**
15. **Stop clip**
16. **Retaining clip**
17. **Over-center spring**
18. **Over-center spring support**
    - Insert in bracket
19. **Rubber guide**
20. **Rubber stop**
21. **Clutch cable attachment components**
22. **Balance weight**
    - With integral clutch cable attachment components
    - For vehicles with Diesel engine, up to August 1993 production only
    - Adding weight to reduce clutch pedal vibration in all vehicles, **see** (B)
23. **Release lever**
24. **Lever stop**
25. **Transmission**
26. **Bracket**
27. **Attachment bolt**
    - Bracket to body
28. **Starter lock-out switch**
    - manual transmission only
    - requires clutch pedal be depressed before starting
    - relay identified as 53 is installed in fuse/relay panel to complete the circuit to the starter

(A) Installation position for brake and clutch pedal pivot pin

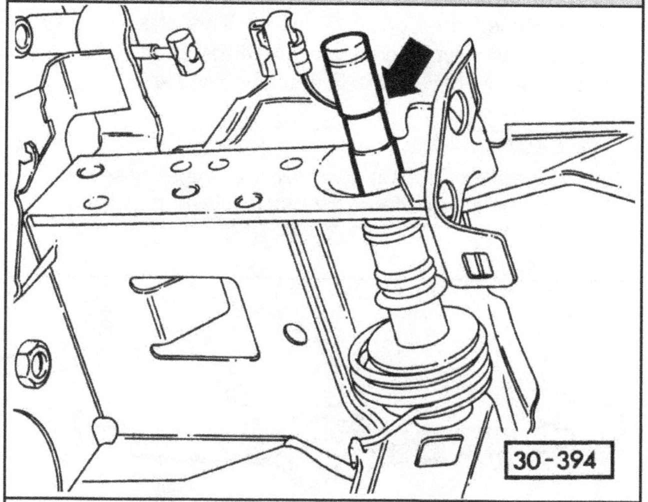

30-394

- **When installing brake pedal/clutch pedal pivot pin, make sure recess (arrow) is on clutch-pedal side.**

(B) Balance weight on clutch release lever (cars built from August 1993)

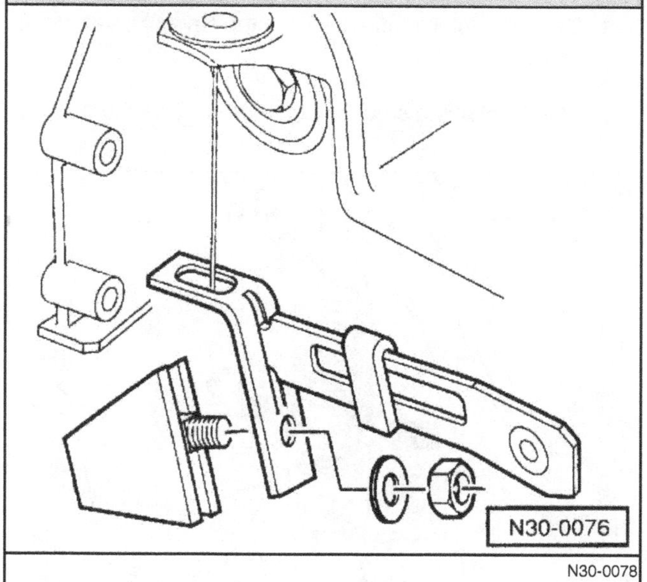

N30-0078

- **The 200 gram balance weight installed on cars with 1.9 liter Diesel engine (up to August 1993) is discontinued.**
- **Beginning August, 1993, a balance weight will be attached to clutch release lever in cases where excessive noise and vibration are transferred to clutch pedal.**
- **The new balance weight can be installed on current vehicles where necessary to address complaints about clutch pedal noise and vibration.**
- **Tightening torque: 25 Nm (18 ft-lb)**

Clutch cable, removing

To remove and install the clutch cable, a special retaining strap is required to hold the cable adjustment mechanism in the compressed position. See Fig. 2. This strap comes installed on a new replacement cable. If the strap is not available, a substitute can be fabricated using stiff wire.

> ***CAUTION —***
> *Avoid sharp bends in clutch cable, as they may affect operation of the adjustment mechanism.*

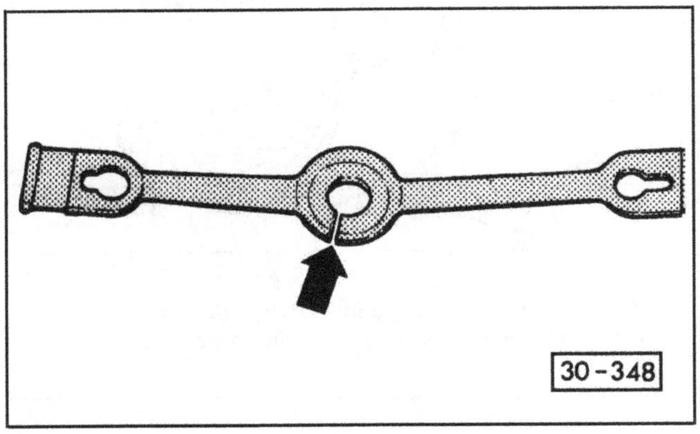

30-348

Fig. 2. Clutch cable retaining strap. When removing strap from new cable, carefully cut open at the perforation (**arrow**).

1. Depress clutch pedal down to stop several times in succession.

2. Attach retaining strap to clutch cable. See Fig. 3.

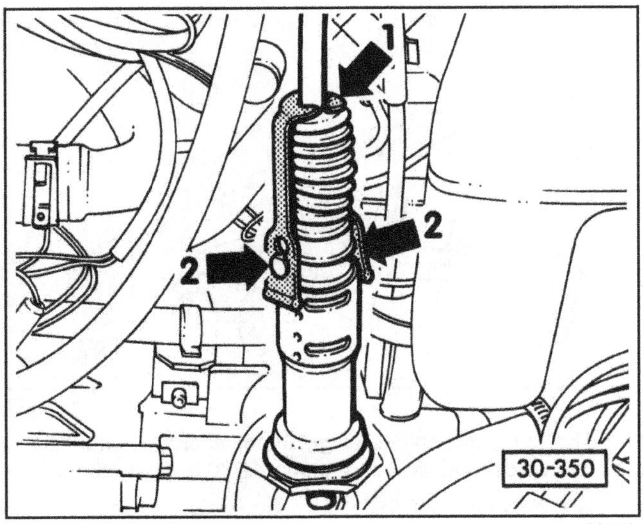

30-350

Fig. 3. Attach retaining strap so that perforation (**arrow 1**) is above protective boot. Then compress adjustment mechanism in area of protective boot and hook tabs on pins (**arrow 2**).

3. Compress adjustment mechanism in area of protective boot.

> ***NOTE —***
> *If the adjustment mechanism cannot be compressed, then it is defective. The clutch cable will be destroyed in the process of being removed.*

4. Hook tabs of retaining straps to pins on adjustment mechanism. See Fig. 3.

5. Disconnect clutch cable attachment components from release lever.

6. Remove over-center spring. See **Clutch over-center spring, removing and installing**.

7. Unhook clutch cable from clutch pedal and remove.

Clutch cable, installing (cable adjustment mechanism compressed)

1. Coat clutch cable suspension point at clutch pedal with MoS2 grease.

2. Press the clutch cable through the bulkhead and suspend in the front recess of the clutch pedal while making sure that the load plate is installed on the cable. See Fig. 4.

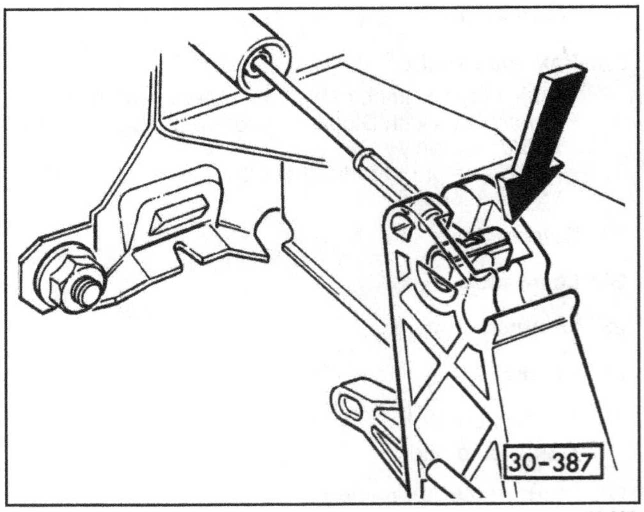

30-387

Fig. 4. When installing clutch cable end to pedal, make sure load plate (**arrow**) is installed on cable end.

3. Route the clutch cable and end piece through the bracket on the transmission and through the release lever while taking care not to damage the plastic sleeve on cable.

4. Hook the attachment parts onto the release lever. See Fig. 5.

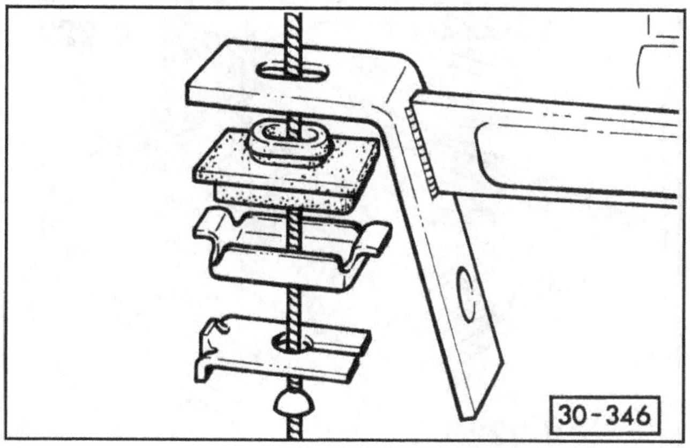

30-346

Fig. 5. Clutch release lever attachment parts (shown here on vehicle without balance weight).

5. Suspend clutch cable in the wire holder (see Fig. 1, item 26).

6. Remove the cable retaining strap so that it can be used again.

7. Install the over-center spring. See **Clutch over-center spring, removing and installing**.

8. Depress the clutch pedal several times in succession.

9. Check the operation of the adjustment mechanism. See **Clutch cable adjustment mechanism, functional check**.

Clutch cable, installing (cable adjustment mechanism not compressed)

1. Mount the cable retaining strap above the dust boot. See Fig. 2 and Fig. 3, given earlier.

2. Press the clutch cable through bulkhead and suspend in the front recess of the clutch pedal while making sure that the load plate is installed on the cable. See Fig. 4, given earlier.

3. Pull the cable end down and hold in place. Then move the adjustment mechanism back and forth several times until it can be pressed toward the bulkhead. See Fig. 6.

4. Hold the adjustment mechanism in this position and hook the retaining strap tabs onto pins on adjustment mechanism (this step requires help from a second technician).

5. Route the clutch cable and end piece through the bracket on the transmission and through the release lever while taking care not to damage the plastic sleeve on the cable.

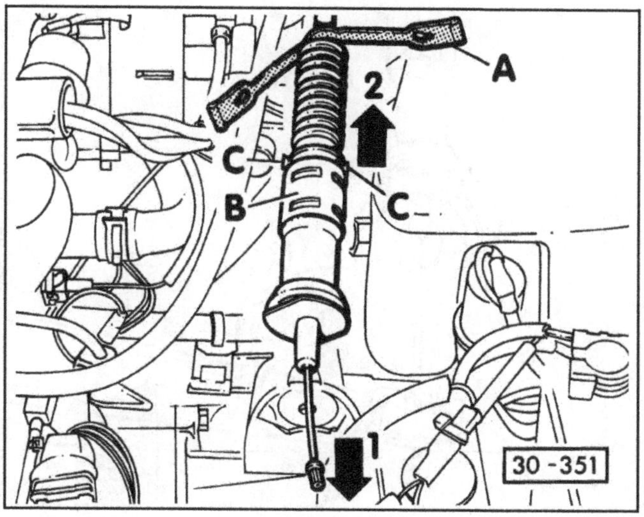

30-351

Fig. 6. Clutch cable retaining strap (**A**). To compress cable adjustment mechanism, pull cable end down (direction of **arrow 1**), while moving adjustment mechanism (**B**) back and forth until mechanism compresses (direction of **arrow 2**). Then hook retaining tabs on pins (**C**).

6. Hook the attachment parts onto the release lever. See Fig. 5, given earlier.

7. Install the over-center spring. See **Clutch cable adjustment mechanism, functional check**.

8. Remove the retaining strap.

9. Check the operation of the adjustment mechanism. See **Clutch cable adjustment mechanism, functional check**.

NOTE —

No other cable (throttle cable, etc.) is to be attached, using wire-ties or other means, to the clutch cable. This could result in the clutch cable failing to operate properly.

Clutch cable adjustment mechanism, functional check

The clutch cable adjustment mechanism must be undamaged.

1. Fully depress the clutch pedal at least 5 times.

2. Move the clutch release lever approx. 10 mm (0.39 in.) in the direction opposite to that of normal travel (down). See Fig. 7. The clutch release lever must move freely.

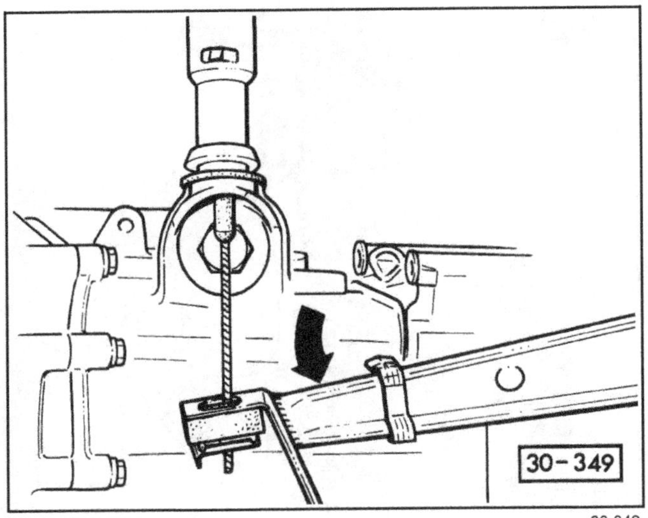

30-349

Fig. 7. Press down on clutch release lever (**arrow**). Lever should move down approximately 10 mm (0.39 in.).

NOTE —
When checking clutch operation, observe the following points in order to avoid premature and thus unnecessary replacement of the clutch cable: (1) Always depress the clutch pedal at least 5 times before repeating the operational check. (2) Replace the cable only if the release lever can not be pushed down as shown in Fig. 7.

Faulty cable adjustment mechanism, repairing

If the functional check of the clutch cable adjusting mechanism cannot be carried out as described above, it is possible that the mechanism has been damaged by mishandling (example: pulling on the cable sleeve). If the clutch release lever cannot move in the return/down direction–even after operating the clutch pedal several times–carry out the adjustment procedure described below.

> *CAUTION —*
> *The malfunctioning adjustment mechanism can cause clutch slippage and premature clutch wear.*

NOTE —
Replacing the cable is to be avoided. If necessary, try repeating the adjustment procedure before installing a new cable.

3. Mount the engine lifting equipment (VW engine sling 10-222A with adapter hooks 10-222 A/2, or equivalent) above the engine. See Fig. 8.

4. Depress the clutch pedal and hold it to the floor (requires a helper).

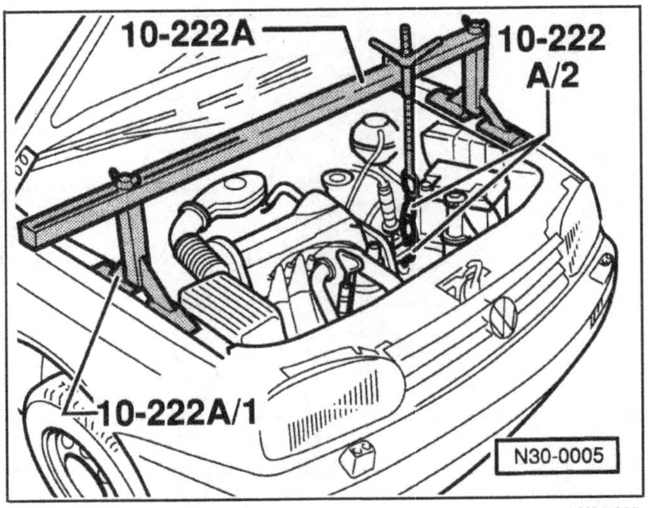

N30-005

Fig. 8. Engine sling shown mounted across engine bay. Sling is used to support clutch release lever in the pedal-depressed position. (Numbers are for VW special tools).

5. With the pedal depressed, lock the release lever in position using appropriate adapter hooks (Volkswagen special tool no. VW 10-222 A/2). See Fig. 9.

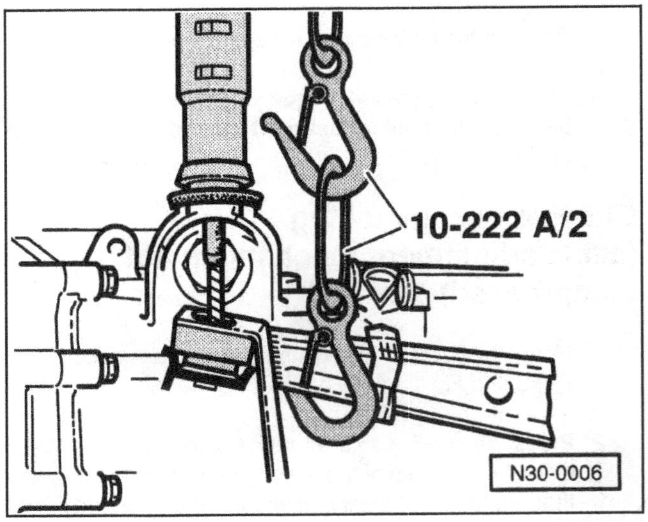

N30-0006

Fig. 9. Clutch release lever held in pedal-depressed position using hooks (VW special tool no. 10-222 A/2).

6. Working in the passenger compartment, pull the clutch pedal fully up and off the floor.

7. Working in the engine compartment, pull the clutch cable down until the adjusting mechanism can be compressed. See Fig. 10.

8. Allow the clutch release lever to return to the rest position (clutch pedal not depressed). Remove the hooks from the clutch release lever.

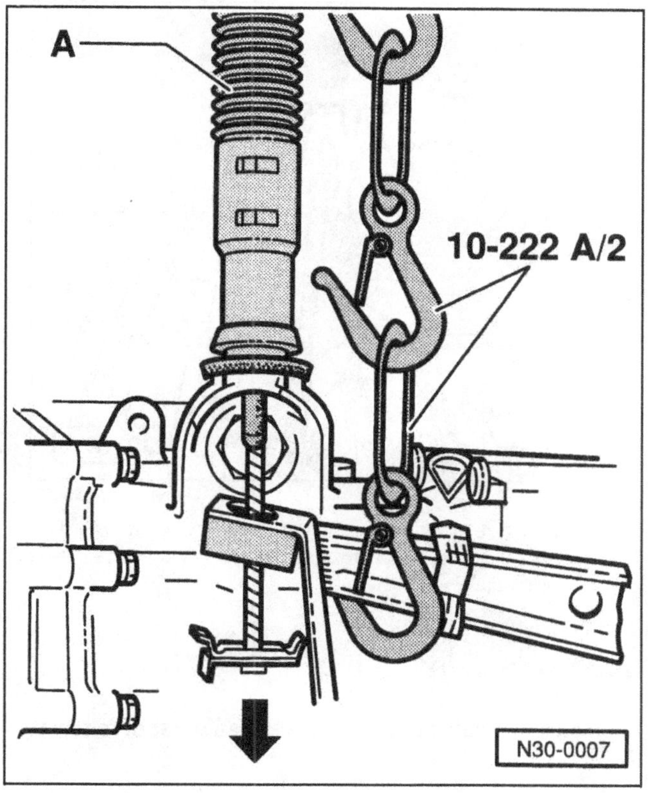

N30-0007

Fig. 10. With clutch pedal fully depressed and lever held stationary, pull cable down (**arrow**).

9. Repeat the functional check as described earlier.

10. Remove the engine lifting equipment.

Clutch over-center spring, removing

1. Disconnect the clutch cable from the transmission. See **Clutch cable, removing and installing**.

2. Remove the interior trim panel at lower left driver's side.

3. Remove the relay panel from its bracket.

4. Remove the guard plate from between the relay panel and the pedal cluster. Then remove the guard plate from below, working carefully to avoid damaging the wiring. See Fig. 11.

5. Insert a screwdriver between the steering column and the clutch pedal and press against the over-center spring. See Fig. 12.

6. Depress the clutch pedal and disconnect the over-center spring from the pedal.

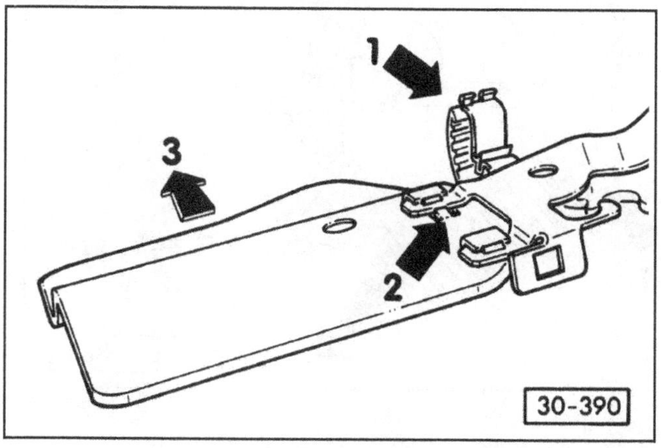

30-390

Fig. 11. Relay panel guard plate. Remove wiring from retainers (**arrow 1**), press detent (**arrow 2**) toward pedal cluster and then press guard plate up toward instrument panel to remove (**arrow 3**).

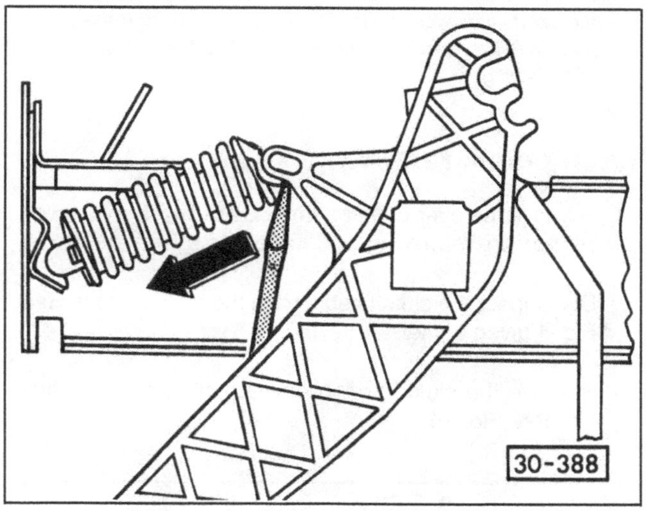

30-388

Fig. 12. Insert screwdriver between steering column and clutch pedal, then depress clutch pedal and disconnect spring by levering off pedal lug (direction of **arrow**).

Clutch over-center spring, installing

1. Install the spring in the reverse order of removal, and note the following:
   - When installing the over-center spring, locate the relay plate in its left mounting first.
   - Working in passenger compartment, pull the clutch pedal into the passenger compartment.
   - Guide the over-center spring between the pedal and the relay panel holder until it is in position in the mounting bracket and located directly below the clutch pedal mounting lug. See Fig. 13.
   - Press the over-center spring onto pedal support, depressing the clutch pedal slightly if necessary.

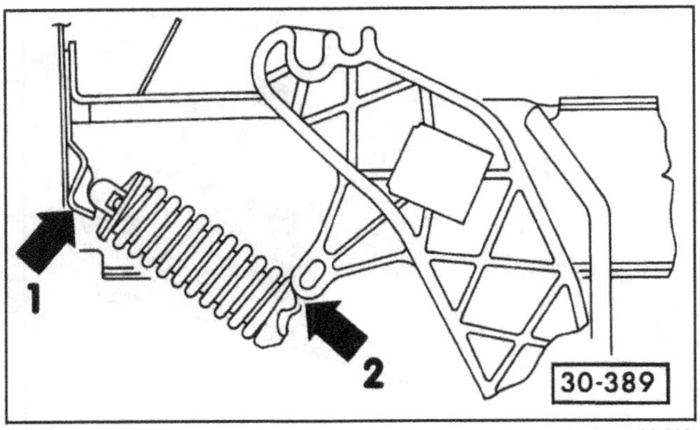

30-389

Fig. 13. Install clutch over-center spring by inserting into bracket (**1**) and then pressing into pedal support (**2**).

NOTE —
If clutch pedal is noisy (squeaks) during operation, lightly lubricate the spring contact points with lithium grease, at the points indicated in Fig. 13 (arrows 1 and 2). If the noise persists, also lubricate the internal sliding surfaces of the spring.

Clutch pedal, removing

1. Remove the over-center spring. See **Clutch over-center spring, removing and installing**, given earlier.

2. Disconnect the clutch cable from the clutch pedal. See Fig. 4 given earlier.

3. Remove the clutch pedal retaining clip from the pivot pin. See Fig. 14.

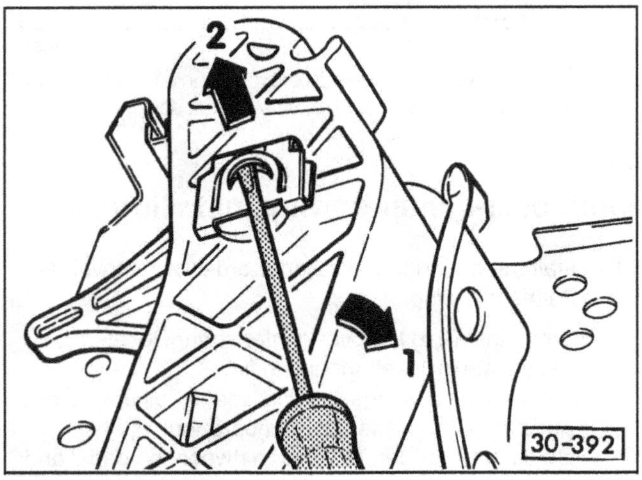

30-392

Fig. 14. Clutch pedal retaining clip being removed. Pry up tab up (direction of **arrow 1**), then slide clip off (direction of **arrow 2**).

4. Working from the brake pedal side, remove the retaining clip from the pivot pin. See Fig. 15.

5. Pull the pivot pin to the right and remove clutch pedal.

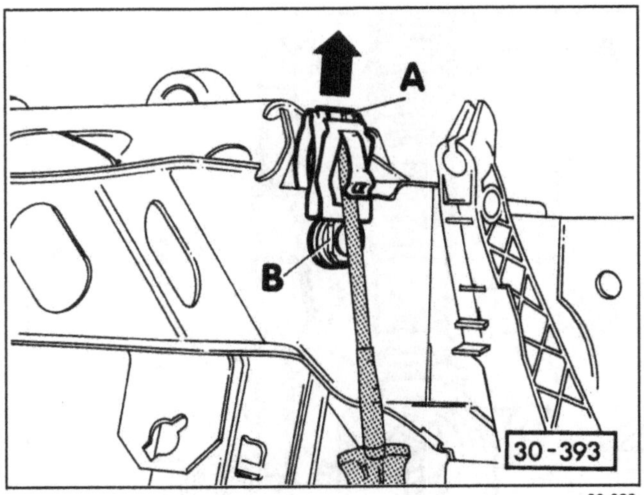

30-393

Fig. 15. Pivot pin retaining clip being removed. Pry up tab (**A**), then press off clip in direction of arrow. Retaining clip recesses shown at (**B**).

Clutch pedal, installing

1. Install the clutch pedal in the reverse order of removal.

NOTE —
Before installing the retaining clip, position the pedal pivot pin so that the recesses in the pin are vertical (See item B in Fig. 15).

2. After installing the clutch pedal, check the operation of the clutch cable adjustment mechanism. See **Clutch cable adjustment mechanism, functional check,** given earlier.

Clutch

An exploded view of the clutch is shown in Fig. 16. Before servicing the clutch assembly, review the general repair instructions given in **3a Clutch, Transmission, and Final Drive—General**. Remove transmission to perform work on clutch. See **34 Manual Transmission**.

CAUTION —
- *Replacement clutch pressure plates are greased and protected against corrosion. Only friction surfaces are to be cleaned as additional cleaning will lead to corrosion and shortened service life.*
- *Assemble the clutch so that the flywheel dowel pin(s) align with the bore(s) or slot(s) in the pressure plate, otherwise the TDC marking on the flywheel will be incorrect.*
- *Friction materials such as clutch linings may contain asbestos fibers. Do not create dust by grinding, sanding or by cleaning with compressed air. Avoid breathing asbestos dust or fibers. Breathing asbestos may result in serious diseases and cause injury or death.*

Clutch assembly

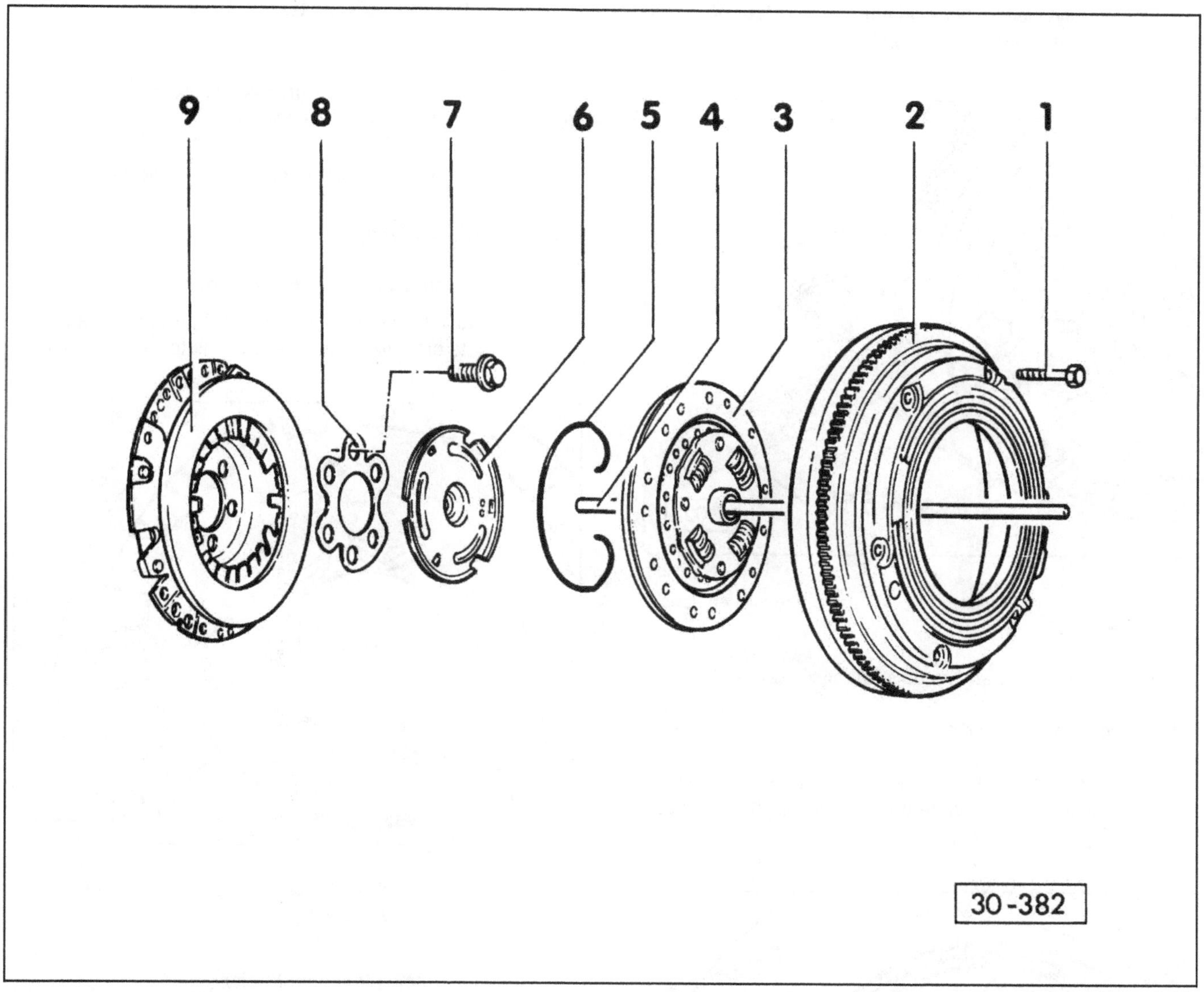

30-382

Fig. 16. Exploded view of clutch and related parts.

1. **Bolt M7 x 40**
   - Tightening torque: 20 Nm (15 ft-lb)
2. **Flywheel**
   - Two versions, see **Flywheel and Pressure plate versions**
   - Make sure that dowel pin(s) seat properly
   - Clutch lining friction surface must be free of grooves, oil and grease
   - Different clutch/flywheel sizes require different clutch housing (timing mark) plugs. See **34 Manual Transmission**
3. **Clutch disc**
4. **Clutch push rod**
   - Grease in area of guide bushing in input shaft
5. **Snap ring**
6. **Release plate**
   - Lightly lubricate pressure plate contact surface and clutch pushrod socket with G 000 100 grease
7. **Attachment bolt, pressure plate to crankshaft**
   - Stretch bolt–always replace
   - See **13 Crankshaft/Cylinder Block** for tightening torque
8. **Intermediate plate**
   - Installation position, see **Clutch, installing**.
9. **Pressure plate**
   - Two versions, see **Flywheel and Pressure plate versions**
   - Maximum distortion of pressure plate inner surface: 0.2 mm (0.008 in.)
   - Replace pressure plate that has damaged or loose rivets.

Clutch, removing

1. Hold the flywheel in place with the locking tool (Volkswagen special tool no. VW 558). See Fig. 17.

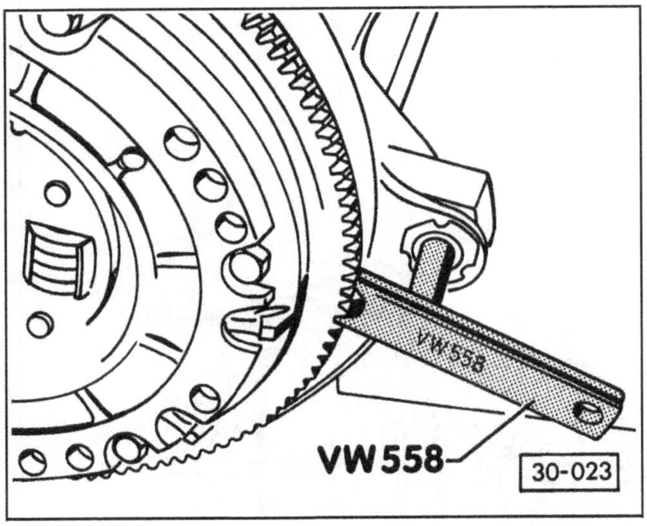

30-023

Fig. 17. Flywheel shown locked in place using Volkswagen special tool VW 558.

2. Loosen the bolts and remove the flywheel.

3. Remove the clutch disc.

4. Use a screwdriver to free the ends of the snap ring from bores in the release plate and remove spring and plate. See Fig. 18.

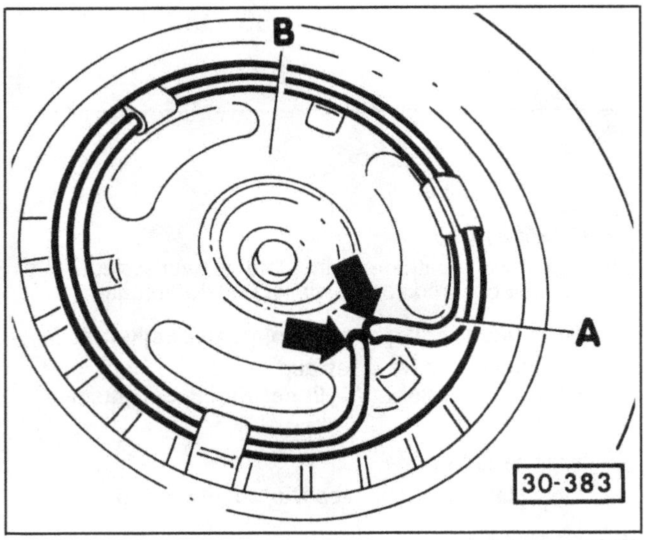

30-383

Fig. 18. Free ends of release plate snap ring (**A**) using a screwdriver, then remove release plate (**B**).

5. Lock pressure plate in place (using Volkswagen special tool no. VW 558) and then remove and discard pressure plate mounting bolts.

> ***CAUTION —***
> *Do not reuse pressure plate mounting bolts. They are one-time only stretch bolts.*

6. Remove pressure plate and intermediate plate.

Clutch, installing

1. Position pressure plate with intermediate plate.
   - Install intermediate plate so that protrusion (arrow) points out away from pressure plate. See Fig. 19.

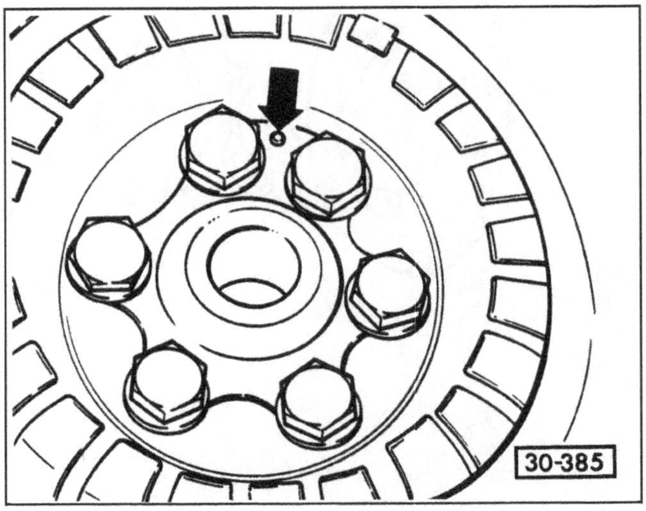

30-385

Fig. 19. Intermediate plate installed with protrusion (**arrow**) facing away from pressure plate.

2. Install the new attachment bolts.

> ***CAUTION —***
> *Always use new pressure plate mounting bolts. They are one-time only stretch bolts.*

3. Lock the pressure plate in position (using special tool VW 558) and tighten the attachment bolts.

> ***NOTE —***
> *For tightening torque of pressure plate attachment bolts, see* **13 Crankshaft/Cylinder Block**.

4. Position the release plate and attach it to the pressure plate with a snap ring. Lock both ends of the snap ring in the release plate. See Fig. 18, given earlier.

5. Coat the splines at center of clutch disc with a thin layer of G 000 100 grease. Then move the clutch disc back and forth along input shaft until disc hub moves on shaft with no resistance.

> ***CAUTION —***
> *Always remove any excess grease.*

6. Mount the clutch and pressure plate to the flywheel.

7. Center the clutch disc using clutch pilot disc (Volkswagen special tool no. VW 547) and hand tighten the flywheel attachment bolts.

8. Install the locking tool (VW 558) to flywheel and tighten the flywheel bolts. Alternate in a diagonal pattern to tighten bolts evenly.

Tightening torques
- Flywheel to pressure plate . . . . 20 Nm (15 ft-lb)

Flywheel and pressure plate versions

Different flywheel and pressure plate versions are installed depending on vehicle date of manufacture. Flywheel and pressure plate versions cannot be interchanged.

> ***CAUTION —***
> *The later style clutch pressure plate can only be used with the later style flywheel.*

Flywheel

On cars built up to December 1992, the flywheel uses two alignment dowel pins, with a dowel pin diameter of 6 mm (0.24 in.). On cars built from December 1992, the flywheel mass is increased, and only one 7 mm (0.28 in.) alignment dowel pin is used. See Fig. 20.

Pressure plate

On cars built up to December 1992, the clutch pressure plate has two recesses to mate with the two-dowel pin flywheel. On cars built from December 1992, the pressure plate has one recess to mate with the single alignment dowel pin. In addition, this pressure plate has circular shoulders to help center the flywheel. See Fig. 21.

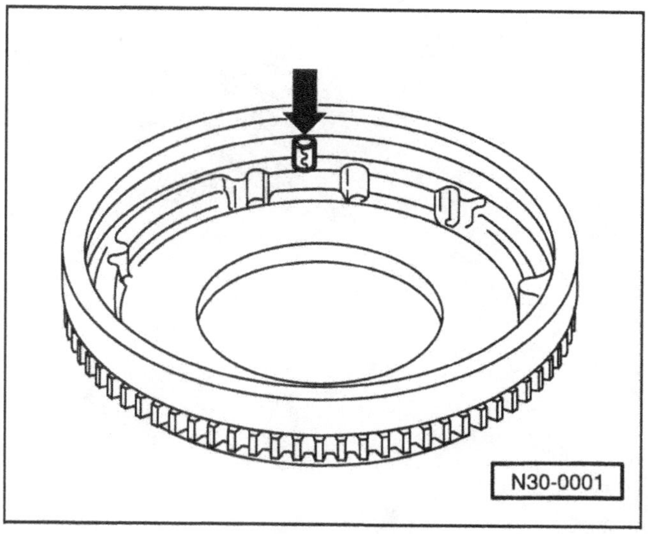

N30-0001

Fig. 20. Flywheel used on cars built from December 1992. Note single dowel pin (**arrow**).

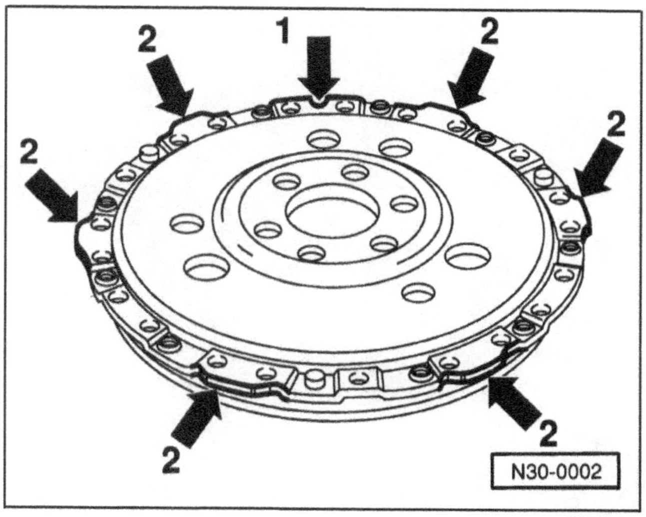

N30-0002

Fig. 21. Clutch pressure plate used on cars built from December 1992. Note single recess for dowel pin (**1**) and circular shoulders (**2**).

30b Clutch (02A Transmission)

GENERAL

Servicing of the hydraulically operated clutch assembly requires that the transmission be removed from the engine. Special tools and equipment are required to remove the transmission and service the clutch. See **34 Manual Transmission** for transmission removal procedures. Read the procedures through to fully understand the scope and nature of the job.

> ***WARNING —***
> - *The cars covered by this manual use an airbag system that automatically deploys an airbag(s) in the event of a frontal impact. The airbag(s) is inflated by an explosive device. Handled improperly or without adequate safeguards, the system can be very dangerous. Special precautions must be observed prior to any work at or near the steering wheel or steering column, including the pedal assembly. See* **69 Seatbelts, Airbags**.
> - *To guard against personal injury or airbag system failure, only trained Volkswagen service technicians should test, disassemble or service the airbag system.*

> ***NOTE —***
> *When servicing the transmission or clutch mechanism, always observe the general repair instructions outlined in* **3a Clutch, Transmission, and Final Drive—General.**

CLUTCH ACTUATING MECHANISM

Fig. 1 shows an exploded view of the pedal assembly. Be sure to lubricate all bearings and friction surfaces with MoS2 grease (VW Part No. G 000 602). Always replace self-locking nuts and circlips during repairs.

> ***CAUTION —***
> *Before working on pedal cluster, always obtain anti-theft radio code and then disconnect battery ground strap.*

Pedal cluster, assembly

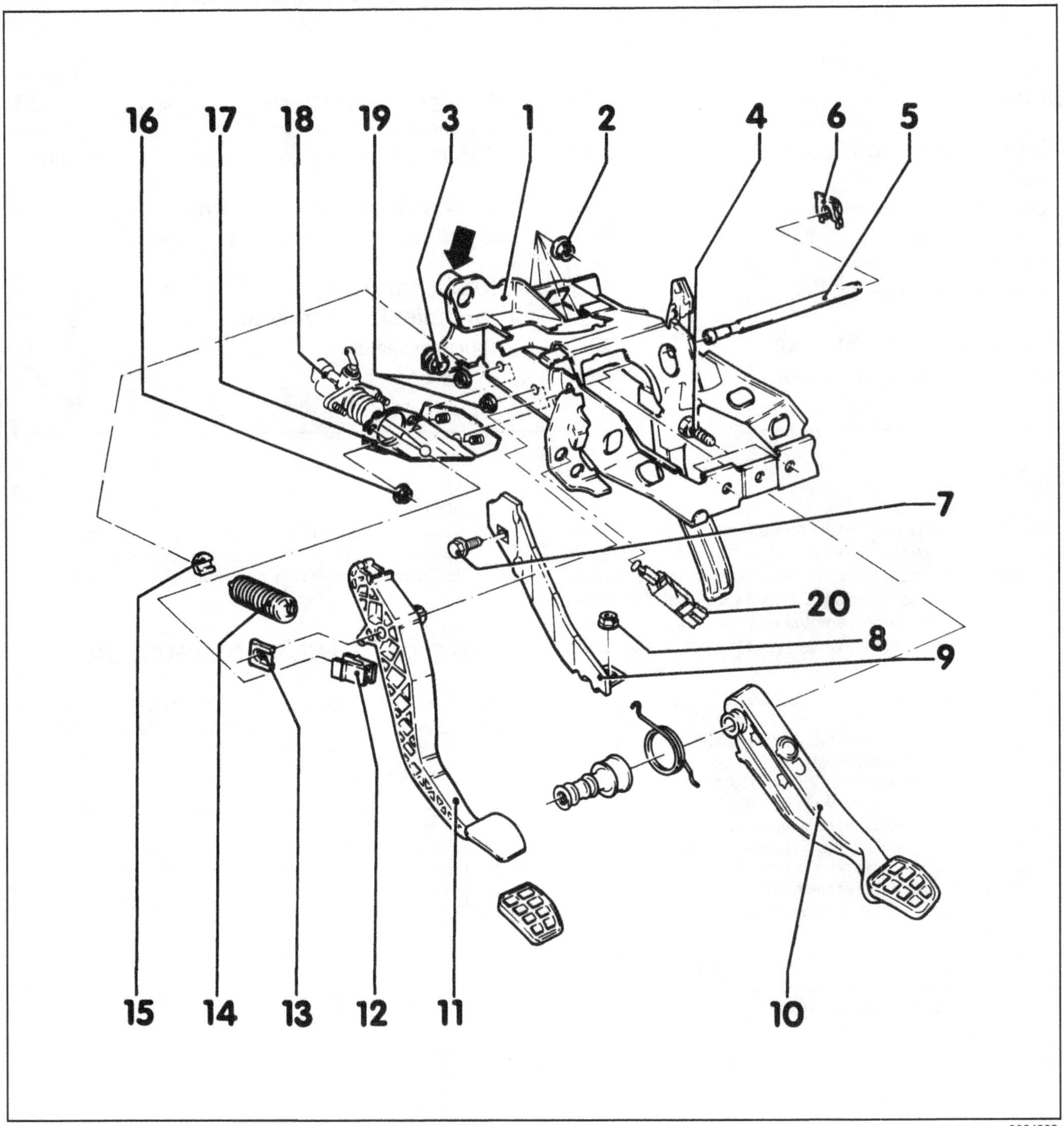

Fig. 1. Exploded view of clutch actuating mechanism and related parts.

1. **Pedal cluster mounting bracket**
   - Removing: first remove steering column with column tube. See **48 Steering**
   - Disconnect brake pedal from push rod of brake master cylinder. See **47 Brakes–Hydraulic System**.

> ***WARNING —***
> *Observe airbag system precautions as described in* **69 Seatbelts, Airbags**. *Always disconnect the negative battery terminal before working at or near steering column.*

> *NOTE —*
> *Before installing the new pedal cluster bracket on a vehicle with hydraulic clutch control, the guide tube must be removed. Saw it off flush with bracket.*

2. **Self-locking nut**
   - Always replace
   - Tightening torque: 25 Nm (18 ft-lb)
   - For attaching bracket to threaded stud on installation plate
3. **Self-locking nut**
   - Always replace
   - Tightening torque: 25 Nm (18 ft-lb)
   - For attaching bracket to threaded studs on vacuum brake booster
4. **Self-locking bolt**
   - Quantity: 2
   - Always replace
   - Tightening torque: 25 Nm (18 ft-lb)
   - For attaching bracket to assembly support,
5. **Pivot pin**
   - For clutch and brake pedals
   - Installation position, **see** (A)
6. **Securing clip**
7. **Bolt**
   - Tightening torque: 25 Nm (18 ft-lb)
   - For attaching additional support (9) to pedal bracket
8. **Self-locking nut**
   - Always replace
   - Tightening torque: 25 Nm (18 ft-lb)
   - For attaching additional support (9) to floor panel
9. **Support**
10. **Brake pedal**
11. **Clutch pedal**
12. **Retainer**
13. **Securing clip**
    - Secures pivot pin (5)
14. **Over-center spring**
15. **Over-center spring mount**
    - Install in mounting bracket
16. **Self-locking nut**
    - Always replace
    - Tightening torque: 25 Nm (18 ft-lb)
    - For attaching master cylinder to mounting bracket
17. **Mounting bracket**
    - Removing: first remove clutch pedal and master cylinder
18. **Master cylinder**
19. **Self-locking nut**
    - Always replace
    - Tightening torque: 25 Nm (18 ft-lb)
20. **Starter lock-out switch**
    - Manual transmission only
    - On cars built from May 1996, **see** (B)

(A) Installation position for brake pedal and clutch pedal pivot pin

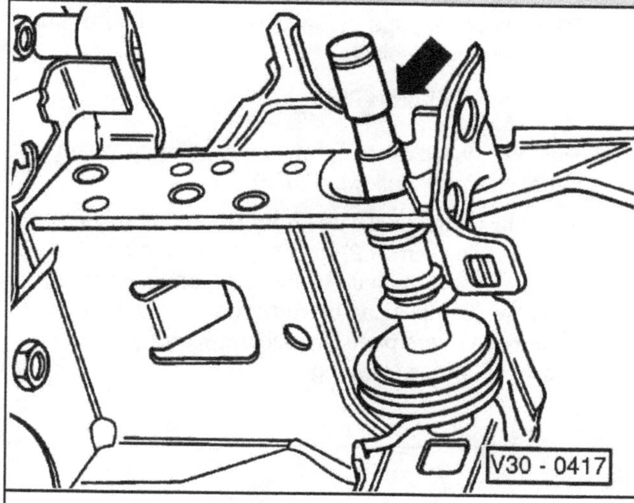

V30-0417

- **Recess (arrow) is on clutch-pedal side.**

(B) Starter lock-out switch on clutch pedal (cars built from May 1996)

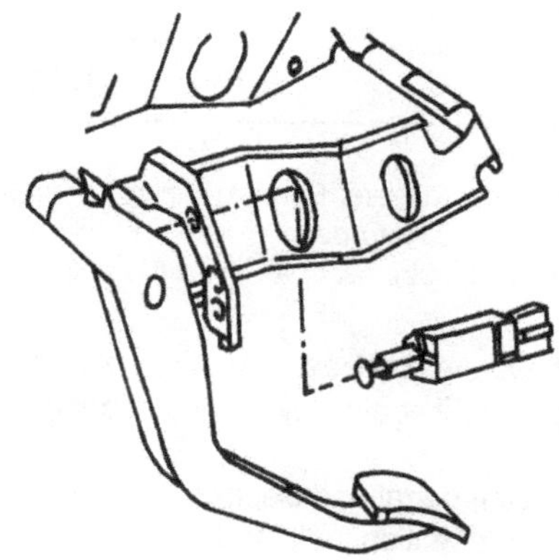

V30-0417

- **Manual transmission vehicles have a starter lock-out function that requires the clutch pedal be depressed before the starter can be activated.**
- **A relay identified as 53 is installed in the fuse/relay panel to complete the circuit to the starter.**

Over-center spring, removing

1. Remove interior trim panel at lower left driver's side.
2. Remove cover plate below pedal cluster (foam rubber liner remains in vehicle.)
3. Unhook relay plate from bracket.
4. Remove guard plate from between relay panel and pedal cluster as follows (Fig. 2):
   - Remove electrical wires from retainers (arrow 1)
   - Press clips (arrow 2) toward pedal cluster.
   - Press guard plate up toward instrument panel (arrow 3) until it can be removed from its mounting.
   - Remove guard plate from below, working carefully to avoid damage to wiring.

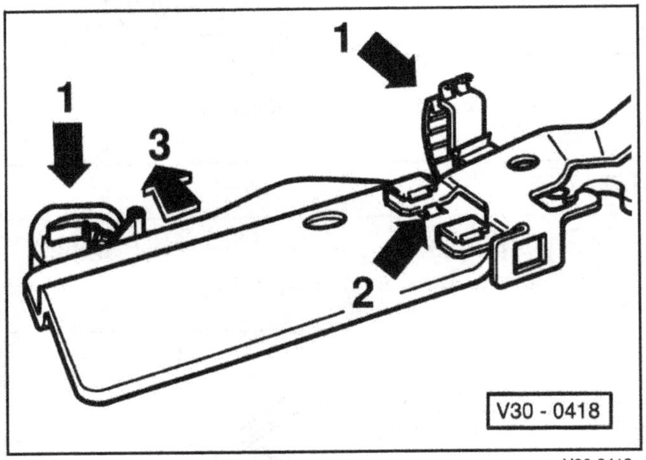

V30-0418

Fig. 2. Relay panel guard plate. Remove wiring from retainers (**1**), press detent (**2**) toward pedal cluster and then press guard plate up toward instrument panel to remove (**3**).

5. Insert screwdriver between steering column and clutch pedal and press it against over-center spring. See Fig. 3.
6. Depress clutch pedal and disconnect over-center spring from pedal in direction indicated by arrow in Fig. 3.

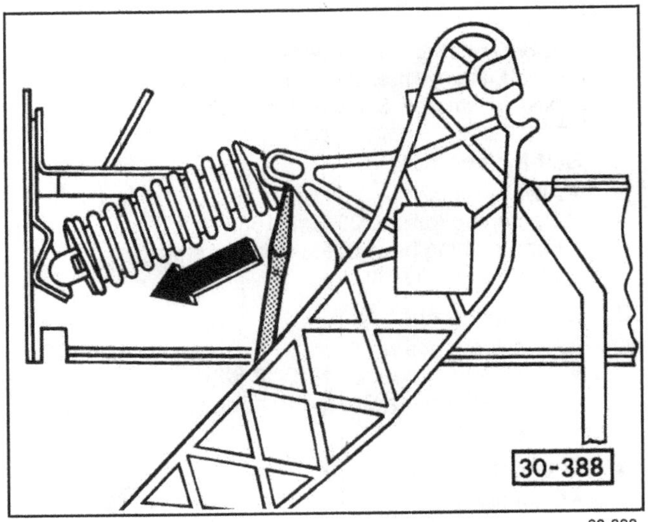

30-388

Fig. 3. Insert screwdriver between steering column and clutch pedal, then depress clutch pedal and disconnect spring by levering off pedal lug (direction of **arrow**).

Over-center spring, installing

Install spring in reverse order of removal, noting the following:

- When installing over-center spring, locate relay plate in its left mounting.
- Working in passenger compartment, pull clutch pedal into passenger compartment.
- Guide over-center spring between pedal and relay panel holder until it is in position in the mounting bracket and located directly below clutch pedal mounting lug. See Fig. 4.
- Press over-center spring onto pedal support, depressing clutch pedal slightly if necessary.

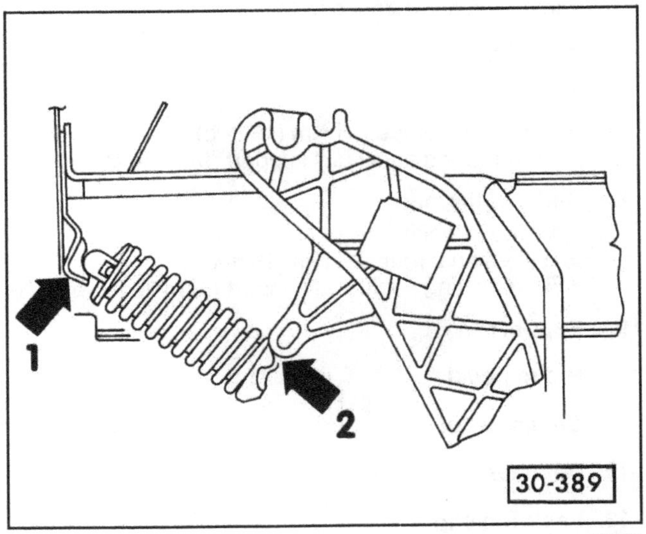

30-389

Fig. 4. Install clutch over-center spring by inserting into bracket (**1**) and then pressing into pedal mounting lug (**2**).

Clutch pedal, removing

1. Open clutch slave cylinder bleeder valve and connect drain hose to bleeder valve. See **Clutch Hydraulic System.**
2. Remove over-center spring. See **Over-center spring, removing and installing.**
3. Disconnect clutch pedal from master cylinder. See **Master cylinder, removing and installing.**
4. Press master cylinder operating rod toward engine compartment until reaching stop.

5. Remove clutch pedal retaining clip from pivot pin. See Fig. 5.

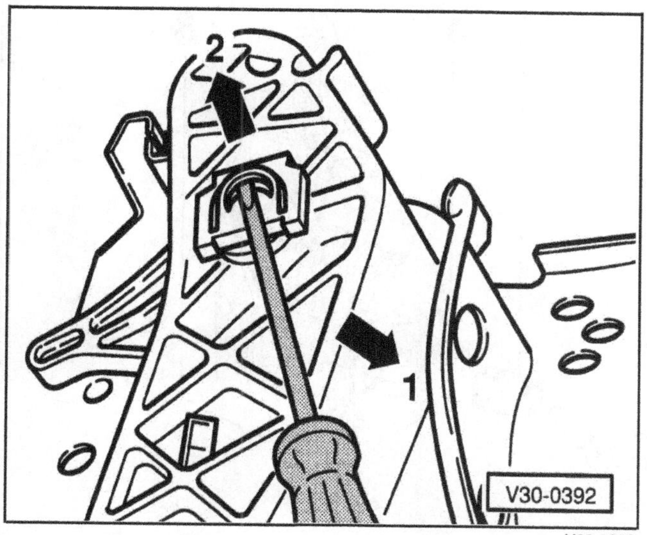

V30-0392

Fig. 5. Clutch pedal retaining clip being removed. Pry up tab (**arrow 1**), then press off clip (**arrow 2**).

6. On brake pedal side, remove retaining clip from pivot pin. See Fig. 6.

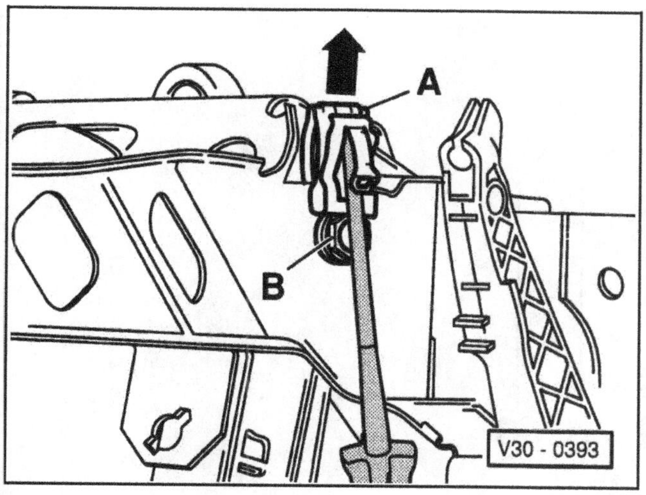

V30-0393

Fig. 6. B rake pedal pivot pin retaining clip being removed. Pry up tab (**A**), then press off clip (**arrow**).

7. Pull pivot pin out toward right side until clutch pedal can be removed.

Clutch pedal, installing

Install in reverse order of removal and note the following:

- Before installing securing clip on brake pedal side, position pivot pin so that flats (B in Fig. 6) are vertical.
- When connecting clutch pedal to master cylinder operating rod, see **Master cylinder, removing and installing**.
- Refill brake fluid reservoir to "max" marking with brake fluid and bleed hydraulic system.

NOTE —

The clutch hydraulic system must be bled if the brake fluid level has dropped below the "min" marking. See **Clutch hydraulic system, bleeding.**

CLUTCH HYDRAULIC SYSTEM

Fig. 7 shows an exploded view of the clutch hydraulic system. Thoroughly clean the area around hydraulic line fittings before removing or loosening fluid lines. Plug all openings to prevent dirt from entering the system. During installation, start hydraulic line fittings by hand to prevent cross-threading.

WARNING —

- *The clutch hydraulic system operates on DOT 4 brake fluid. Brake fluid is poisonous. Wear eye and hand protection when working with brake fluid.*
- *Brake fluid is a hazardous waste. Recycle or dispose of brake fluid in a manner that is mindful of the environment and meets ordinances governing the disposal of such wastes.*

CAUTION —

Brake fluid is very harmful, especially to paint and rubber parts. Spilled brake fluid should be cleaned up immediately.

Clutch hydraulic system, assembly

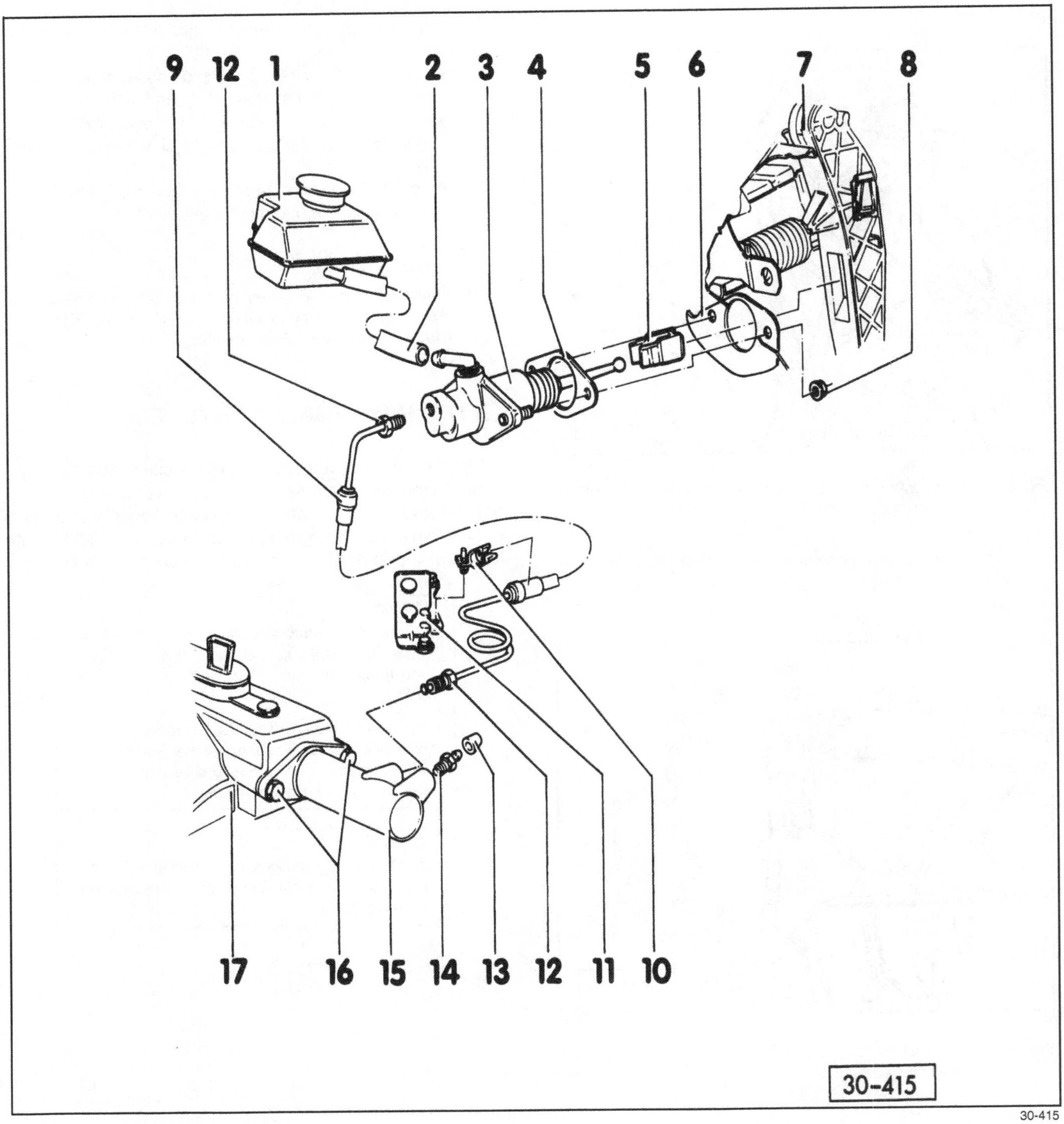

30-415

Fig. 7. Exploded view of clutch hydraulic system.

1. **Brake fluid reservoir**
   - Supplies brake fluid to operate hydraulic clutch
   - Internal baffle prevents hydraulic system from using fluid below approximately 1/2 capacity so that brake system always has a reserve

2. **Supply hose**

3. **Clutch master cylinder**

4. **Gasket**
   - Always replace

5. **Retainer**
   - Replace only with master cylinder removed
   - Removing, **see** (A)
   - Pressing on, **see** (B)
6. **Mounting bracket**
7. **Clutch pedal**
8. **Self-locking nut**
   - Always replace
   - Tightening torque: 25 Nm (18 ft-lb)
9. **Brake line/hose**
10. **Retainer**
    - Mounted to support bracket (11)
11. **Support bracket**
    - For gear shift cable routing
12. **Brake line fitting**
    - Tightening torque: 20 Nm (15 ft-lb)
13. **Dust cap**
14. **Bleeder valve**
15. **Slave cylinder**
    - Before removing: remove gear selector cable from transmission gear selector lever and remove relay lever and actuating arm for gate selector cable. See **34 Manual Transmission**
    - After installing: adjust gear selector mechanism. See **34 Manual Transmission**
16. **Bolt**
    - Tightening torque: 25 Nm (18 ft-lb)
17. **Transmission**

(A) Removing retainer for master cylinder operating rod

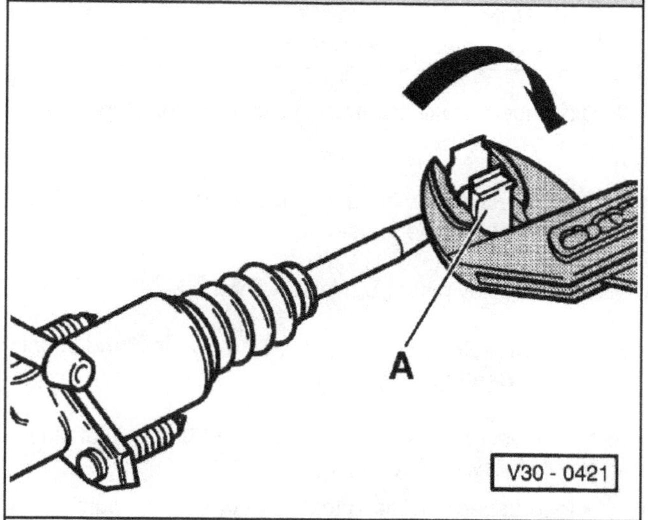

V30-0421

- **Pry off master cylinder operating rod retainer (A) in direction indicated by arrow.**

(B) Pressing on retainer for master cylinder operating rod

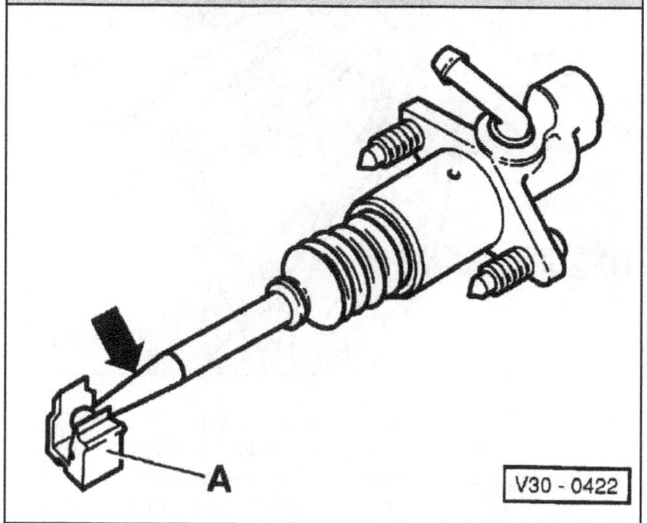

V30-0422

- **Press master cylinder operating rod into guide (A) in direction indicated by arrow.**

Master cylinder, removing

1. Remove coolant expansion tank and lay to one side.

> ***CAUTION—***
> *Do not open the cooling system.*

2. Disconnect brake line from master cylinder. Cap or seal off the open line.
3. Disconnect supply hose from brake fluid reservoir. Cap or seal off the open hose.
4. Unhook relay plate from left retainer.
5. Remove protective plate between relay plate and pedal cluster. See Fig. 2, given earlier.
6. Disconnect master cylinder push rod from pedal cluster, as follows: (See Fig. 8)
   - Insert special release tool into cut-outs on clutch pedal (arrows A). Tool inscription "top/oben" faces toward instrument panel
   - Place special pliers in recesses of release tool (arrows C) and press jaws together.

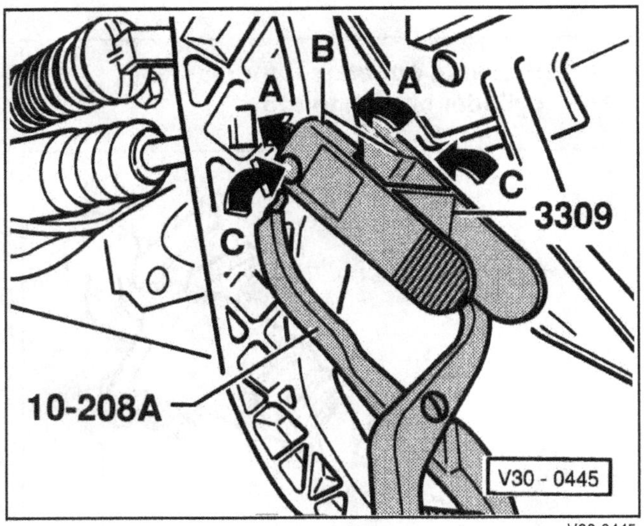

V30-0445

Fig. 8. Special tools being used to remove master cylinder pushrod from clutch pedal. Insert VW special tool no. 3309 (**B**) into cutouts in pedal (**arrows A**). Then insert special pliers (Volkswagen special tool no. 10-208A) at **arrows C** to unlock pushrod from clutch.

7. Remove mounting bracket support by removing bolt and nut shown in Fig. 9

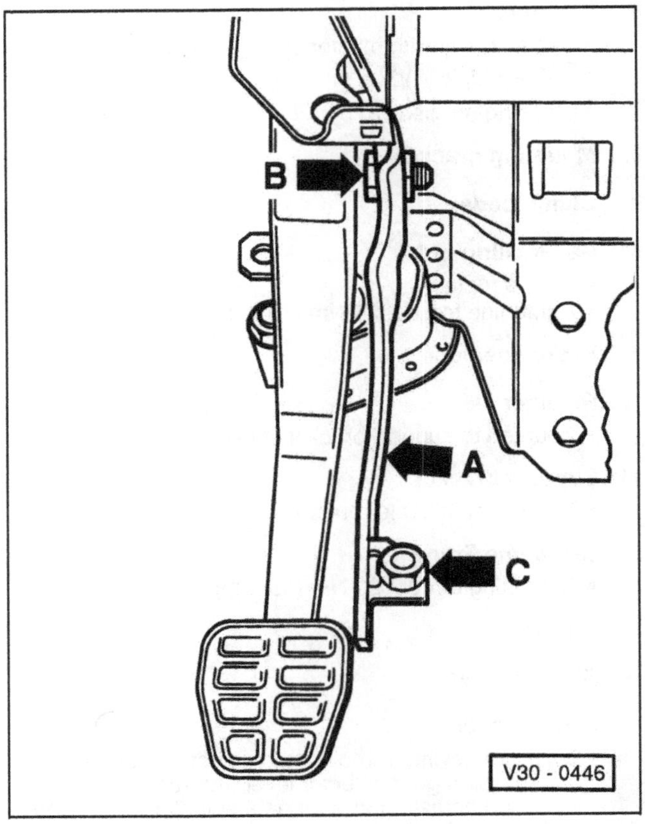

V30-0446

Fig. 9. Remove bolt (**B**) and nut (**C**) to remove support bracket (**A**).

8. Remove master cylinder from passenger compartment by removing mounting nuts on each side of clutch pedal. See Fig. 10.

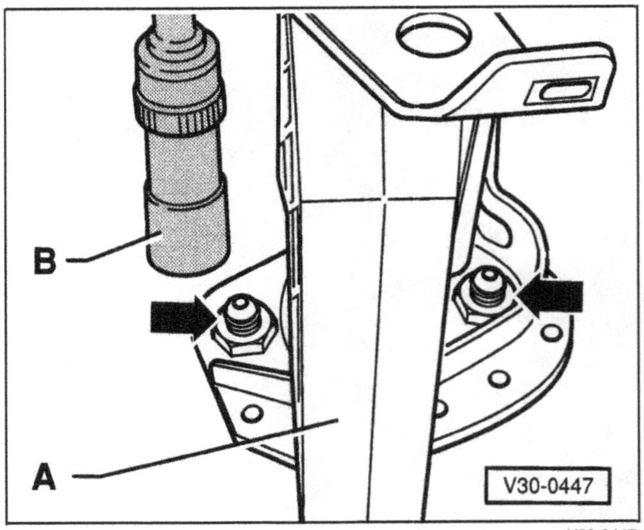

V30-0447

Fig. 10. Master cylinder mounting nuts (**arrows**). Use a 13 mm deep-well socket (**B**) to remove nuts. Clutch pedal shown at (**A**).

Master cylinder, installing

Install in reverse order of removal. While doing so, note the following (Fig. 11):

- Retainer must be installed on master cylinder operating rod.
- To engage retainer, press clutch pedal toward bulkhead (direction indicated by arrow). Make sure retainer is located correctly.
- Bleed clutch hydraulic system after installing master cylinder.

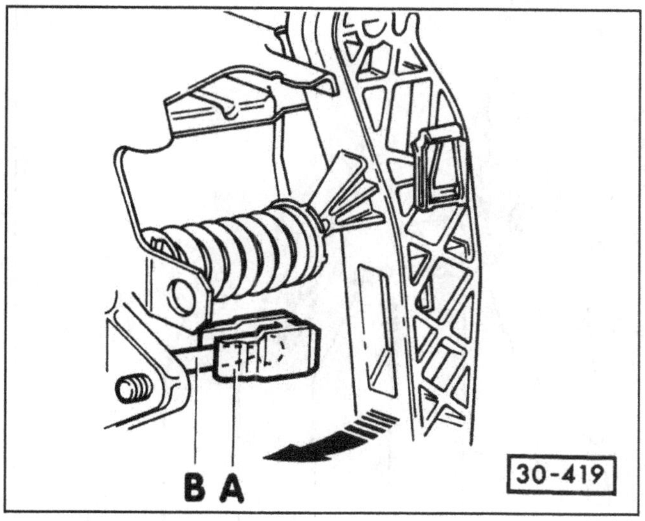

30-419

Fig. 11. When engaging master cylinder pushrod to clutch pedal, make sure retainer (**A**) is installed on pushrod (**B**). To engage retainer to pedal, depress pedal. Make sure retainer locks in pedal.

Clutch hydraulic system, bleeding

Bleed clutch hydraulic system using VAG 1238/1 or VAG 1238B brake filling and bleeding unit or equivalent power bleeding equipment.

1. To bleed system, connect 670 mm (26.38 in.) of bleeder hose to pressure hose fitting on collector bottle. See Fig. 12.

2. Connect bleeder hose to slave cylinder and open bleeder valve to bleed system. See Fig. 13.

3. Depress clutch pedal several times after completing bleeding process.

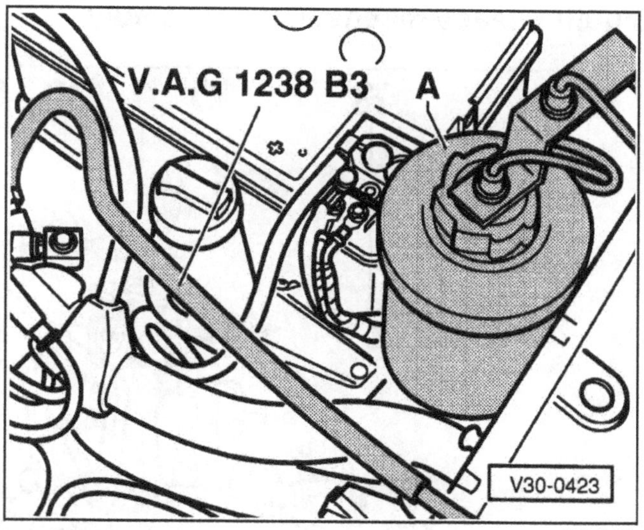

V30-0423

Fig. 12. Brake filling and bleeding unit (**A**). Note long bleeder hose to slave cylinder (V.A.G.1238 B3).

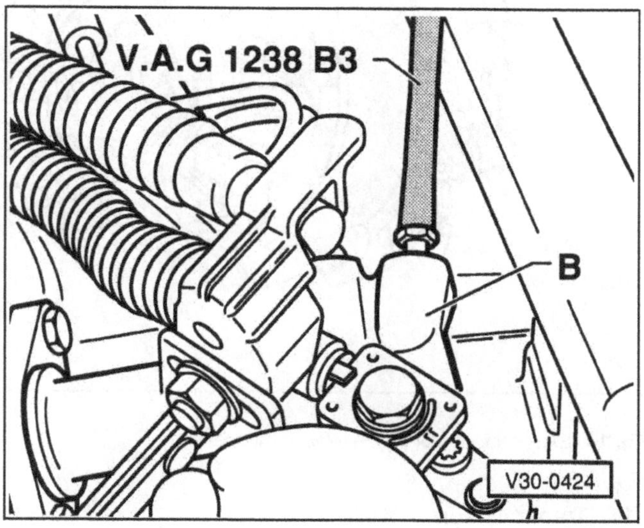

V30-0424

Fig. 13. Bleeder hose shown connected to bleeder valve on slave cylinder (**B**).

CLUTCH RELEASE MECHANISM

Fig. 14 shows an exploded view of clutch release mechanism. The clutch slave cylinder can be replaced with the transmission installed in the car. Replacement of the other clutch release components require that the transmission be removed from the car. See **34 Manual Transmission** for transmission removal and replacement procedures.

Clutch release mechanism, assembly

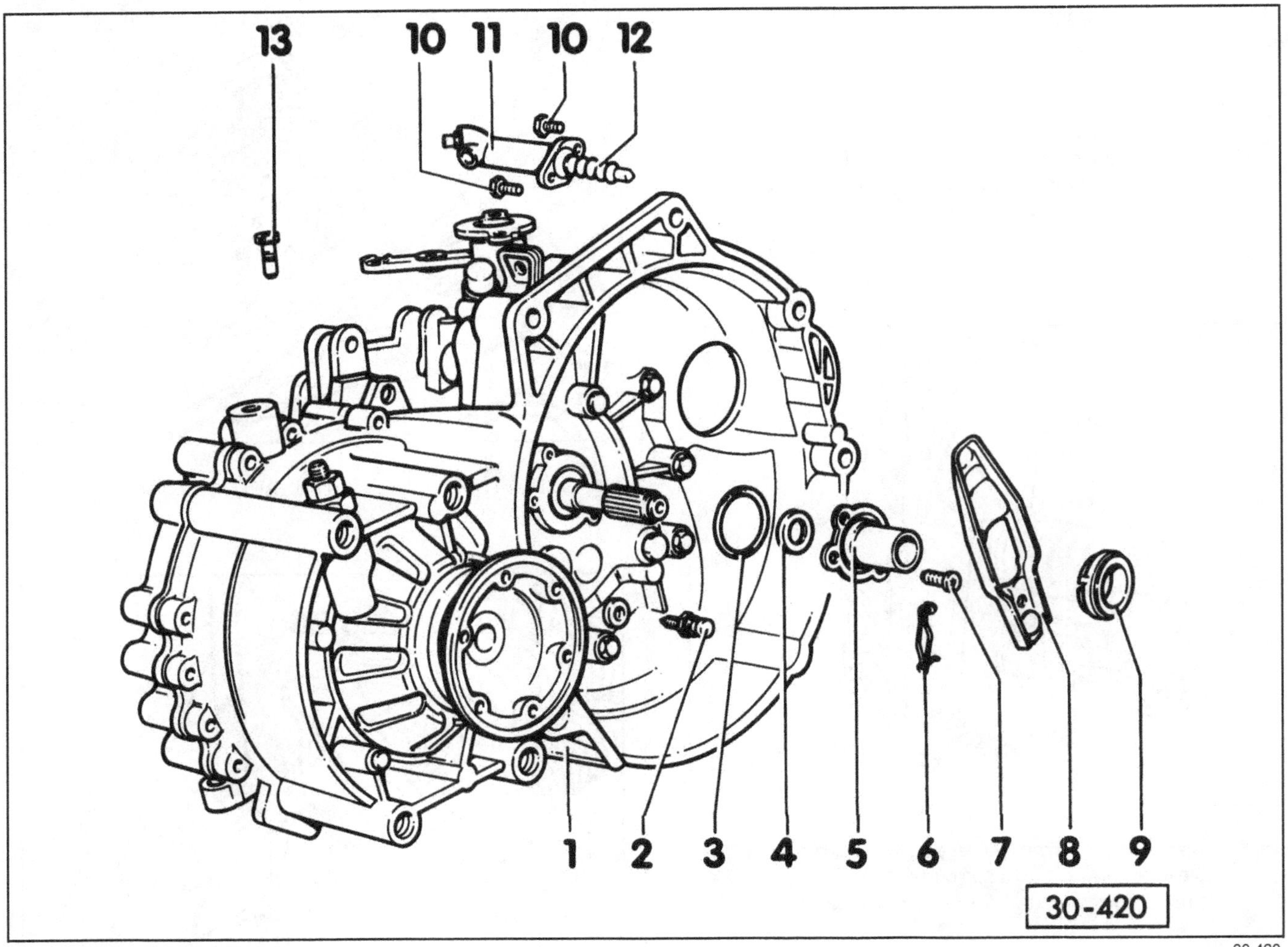

30-420

Fig. 14. Exploded view of clutch release mechanism.

1. **Transmission**
2. **Ball stud**
   - Tightening torque: 25 Nm (18 ft-lb)
   - Lubricate with MoS2 grease
3. **O-ring**
   - Always replace
4. **Input shaft oil seal**
5. **Guide sleeve**
   - Lubricate with MoS2 grease in area of release bearing
6. **Retaining spring**
   - Attach to clutch release lever
7. **Mounting bolt**
   - Tightening torque: 20 Nm (15 ft-lb)
8. **Clutch release lever**
9. **Release bearing**
   - DO NOT wash bearing (wipe off instead)
   - Replace noisy bearings
   - Lubricate surfaces that contact clutch release lever with MoS2 grease
10. **Mounting bolt**
    - Tightening torque: 20 Nm (15 ft-lb)
11. **Slave cylinder**
12. **Push rod**
    - Lubricate end of push rod with MoS2 grease
13. **Assembly pin**
    - Secures clutch release lever while installing transmission.
    - Remove after transmission has been installed.
    - A M8x35 bolt, can be used as a substitute if assembly pin (13) is not available.

CLUTCH, SERVICING (TRANSMISSION REMOVED)

Clutch assembly

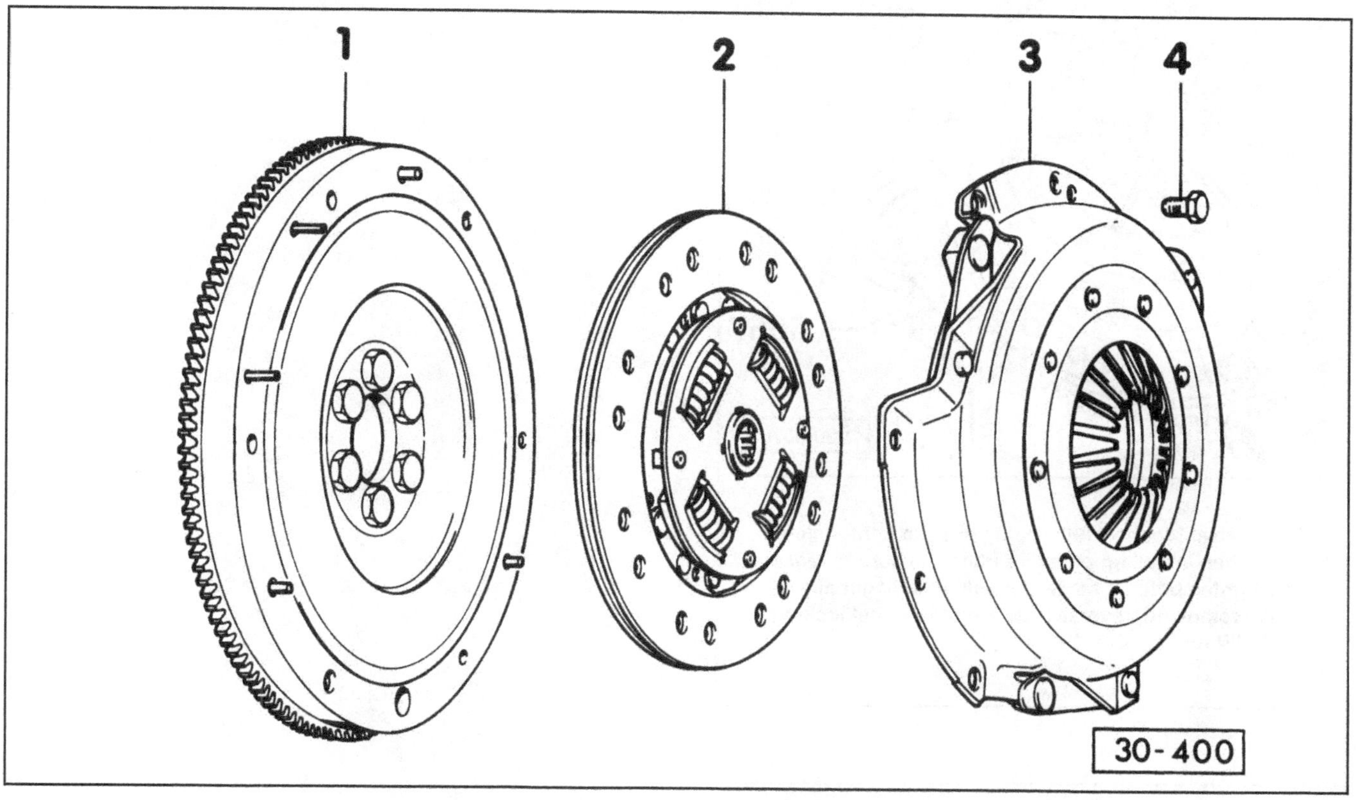

30-400

Fig. 15. Exploded view of clutch assembly.

> *NOTE —*
> *Make sure that the correct replacement clutch disc and pressure plate are selected according to the engine code.*

1. **Flywheel**
   - Make sure centering dowel pins fit tightly
   - Contact surface for clutch disc must be free of grooves, oil and grease
   - Removing/installing. See **13 Crankshaft/Cylinder Block**
2. **Clutch disc**
   - Note installation position: spring cage faces toward pressure plate
   - Centering, **see** (A)
   - Place the clutch disc on the input shaft and make sure it moves back and forth freely.
   - Remove excess grease.

> *NOTE —*
> *Before installing clutch disc remove corrosion from input shaft splines and clutch disc hub splines, clean and apply only a very thin coating of grease, Part No. G 000 100, to the splines of the input shaft only.*

3. **Pressure plate**
   - Removing and installing, **see** (A)
   - Check ends of diaphragm spring, **see** (B)
4. **12-point bolt**
   - Tightening torque: 20 Nm (15 ft-lb)
   - Loosen and tighten in stages, evenly and in a diagonal pattern

A Centering clutch disc and removing and installing pressure plate

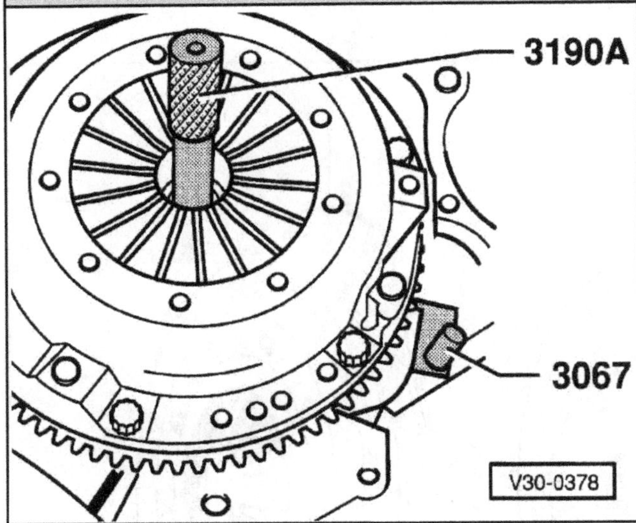

V30-0378

- VW special tool no 3190A being used to center clutch disc when installing pressure plate. Be sure to loosen and tighten bolts in stages, evenly and diagonally. When removing, reverse position of flywheel locking tool (VW tool no. 3067).

B Checking diaphragm spring

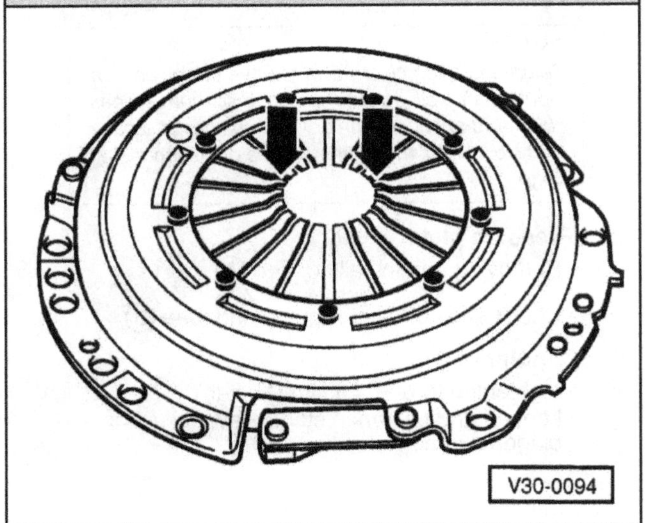

V30-0094

- Check pressure plate diaphragm spring for wear (arrows). Maximum wear–up to half of original diaphragm spring thickness.

34a Manual Transmission (020 Transmission)

GENERAL

This repair group covers shift linkage repair information and removal and installation of the 020 manual transmission for cars with ABA, ACC, 1Z and AAZ engines. This repair group does not cover transaxle or transmission teardown and disassembly.

NOTE—
- *Replacement transmissions may use a modified clutch housing (extra mounting hole in the bellhousing mating surface). If installing a transmission with a modified bellhousing on an early production vehicle (cars built up to September 1993), the casting for the extra hole will interfere with the distributor hold-down clamp. Either use a modified hold-down clamp (available from VW) or saw off enough of the casting (about 4 mm) to allow sufficient clearance for the old hold-down clamp.*
- *To check manual transmission oil, see* **0 Maintenance Program**.

SHIFT LINKAGE

Fig. 1 shows the design and operation of the shift linkage. When assembly or servicing the linkage, lubricate all joints and friction surfaces with MoS2 grease (VW Part No. G 000 602).

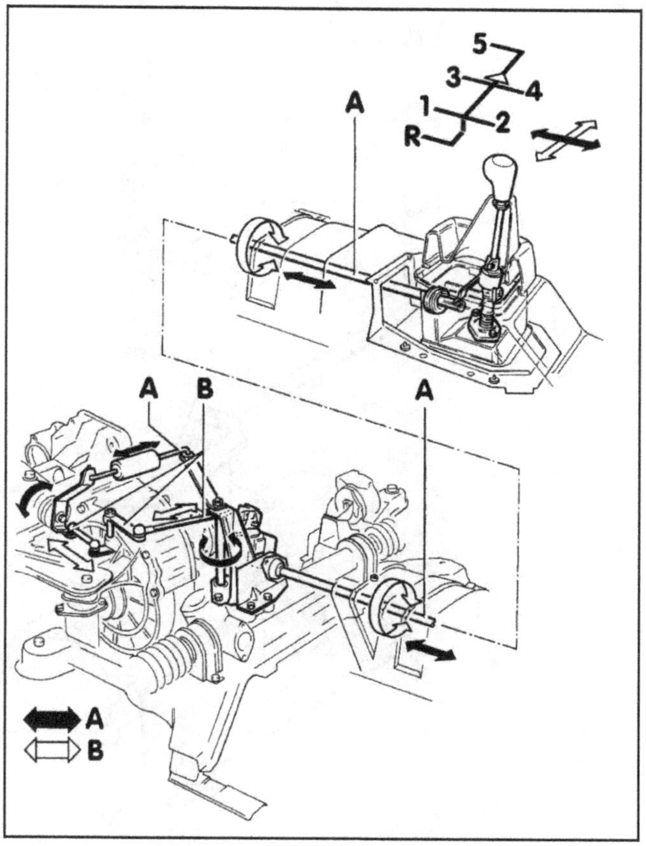

0024067

Fig. 1. Operational view of shift linkage. Selector rods for vertical plane (forward/back travel) are shown at **A**. Selector rods for horizontal plane (side-to-side travel) are shown at **B**.

CAUTION—
Before servicing or repairing the transmission or shift linkage, review the general repair information given in **3 General Information**.

Gearshift Housing and Shift Lever

Fig. 2 shows an exploded view of the gearshift housing and the shift lever assembly. Use this illustration as a guide to servicing the gearshift housing and the shift lever. Applicable tightening torques as well as other pertinent repair information are given with the illustration. Additional shift lever repair information is given in the three detailed views immediately following the illustration.

Gearshift housing and shift lever, assembly

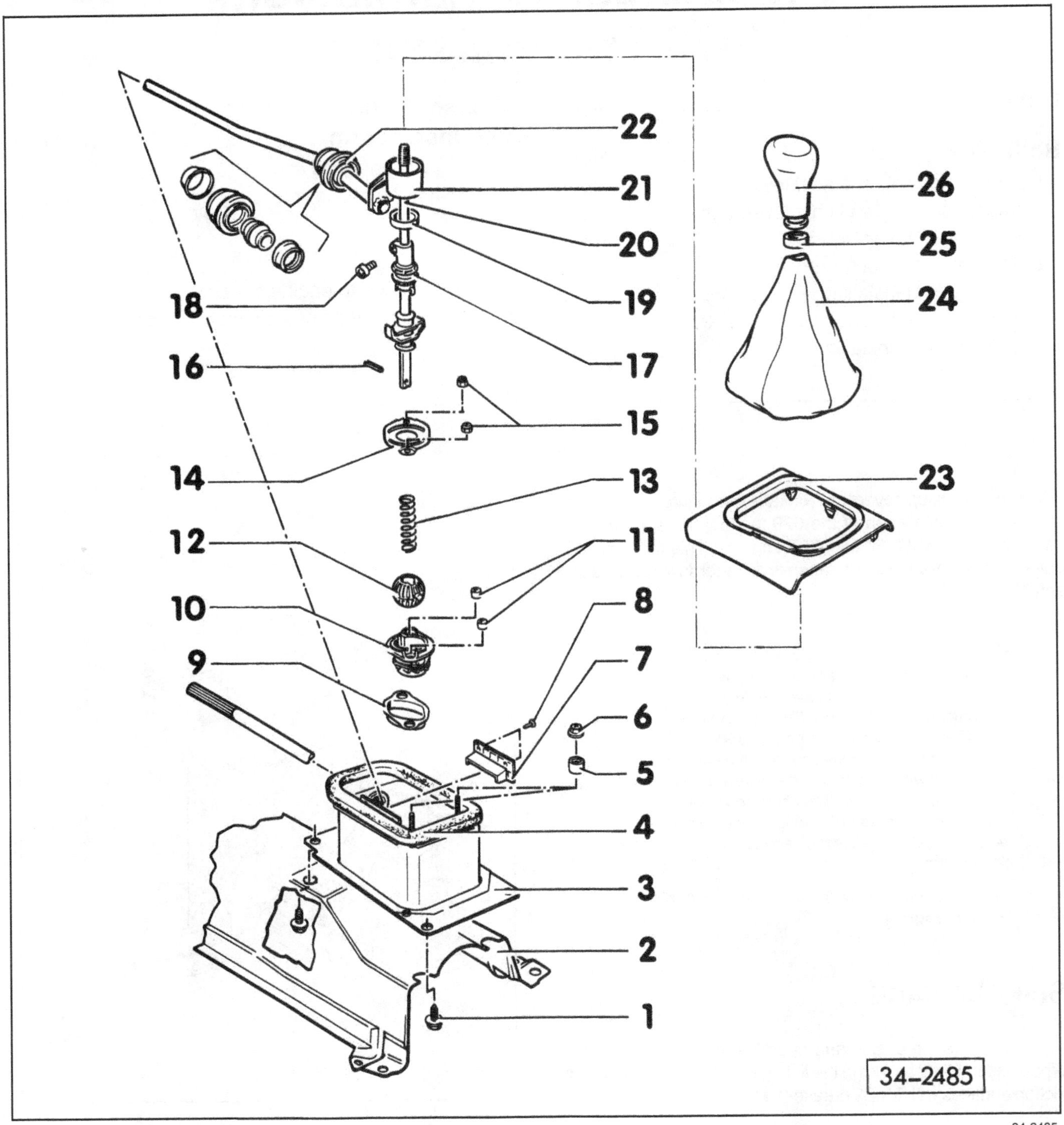

34-2485

Fig. 2. Exploded view of gearshift housing and shift lever.

1. **Attachment bolt, gearshift housing to body**
   - Tightening torque: 20 Nm (15 ft-lb)
2. **Heat shield**
3. **Gearshift housing**
4. **Gasket**
   - Bond to gearshift housing
5. **Spacer**
   - Between center console and gearshift housing

6. **Nut**
   - For center console
7. **Travel stop**
8. **Rivet**
   - For attachment of travel stop
9. **Gasket**
   - For ball housing
10. **Ball housing**
11. **Spacer**
12. **Ball**
13. **Spring**
14. **Retainer plate**
   - Installation position, **see** (A)
15. **Self-locking nuts**
   - Tightening torque: 10 Nm (89 in-lb)
16. **Spring pin**
17. **Eccentric adjuster**
   - After installing shift lever, snap into place on shift lever
   - Removal instructions, **see** (B)
18. **Clamping bolt**
   - Tightening torque: 2 Nm (18 in-lb)
19. **Plastic ring**
   - Attach to eccentric adjuster
20. **Shift lever**
   - Remove eccentric adjuster before removing lever. See Fig. 4
   - Assemble before installation: Slide on retainer plate (14), spring (13) and ball (12) and secure with spring pin (16)
21. **Selector rod**
   - Extract from gearshift housing along with shift lever
   - Removal instructions, **see** (C)
22. **Selector rod support**
   - When installing make sure that the two spherical halves do not separate
23. **Cover plate**
24. **Shift boot**
25. **Rubber ring**
   - Vehicles with leatherette boot only
26. **Shift knob**

(A) Installation position of shift lever retainer plate

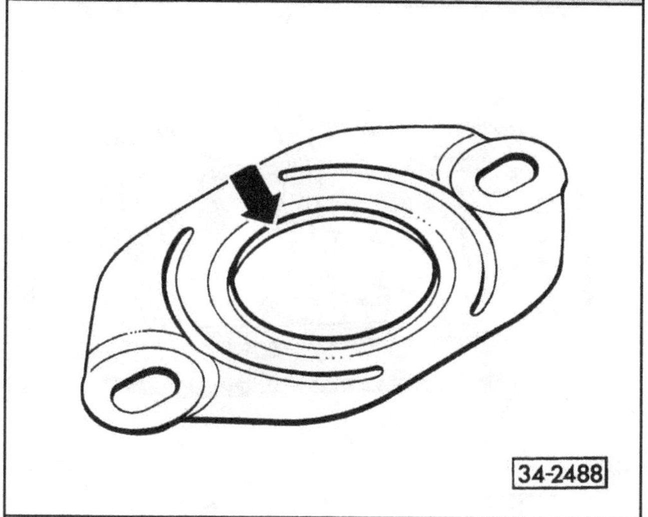

34-2488

- **Install shift lever retainer plate so that shoulder (arrow) is facing away from the shift lever ball.**

(B) Removing shift lever eccentric adjuster

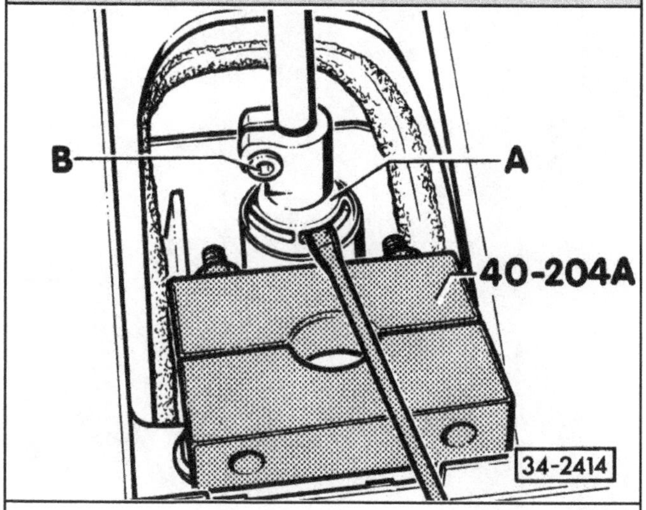

34-2414

- **To remove the shift lever eccentric adjuster, loosen clamping bolt (B) and carefully pry eccentric adjuster (A) from lower section using screwdriver. (VW tool no. 40-204a used as prying block).**

C **Extracting selector rod with shift lever from gearshift housing**

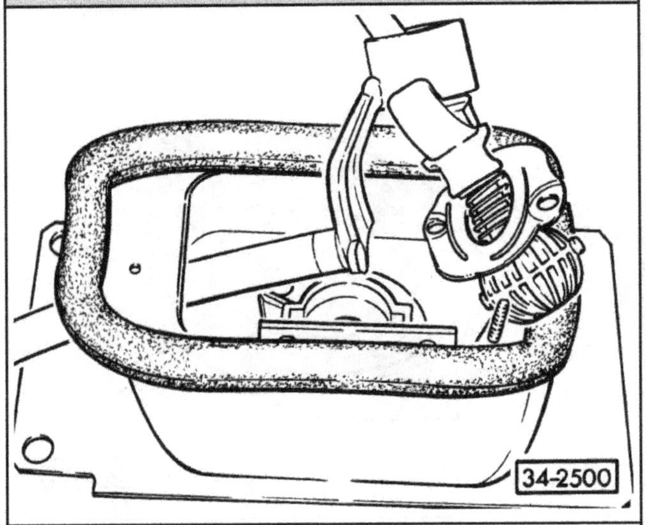

34-2500

- To remove selector rod, pull the shift lever together with selector rod from housing and then remove the shift lever from the rod.

Shift linkage, assembly

Fig. 3 shows an exploded view of the shift linkage mounted to the transmission. Applicable tightening torques as well as other pertinent repair information are given with the illustration.

1. **Selector rod**
2. **Bushing**
3. **Relay lever bracket**
4. **Bolt**
   - Tightening torque: 25 Nm (18 ft-lb)
5. **Bushing**
6. **Relay lever**
7. **Retainer clip**
8. **Selector lever**
9. **Shield**
   - Installation position, **see** (A)
10. **Seal ring**
11. **Clamp**
12. **Bolt**
    - Clamp to selector rod
13. **Nut**
    - Clamp to selector rod
    - Tightening torque: 20 Nm (15 ft-lb)
14. **Intermediate lever**
15. **Nut**
    - Intermediate lever to relay lever
    - Tightening torque: 20 Nm (15 ft-lb)
16. **Relay lever, long**
    - Before removing use a screwdriver to pry back clamp on plastic heads
17. **Bushing**
18. **Selector rod with balance weight**
19. **Split pin**
20. **Lever/gearshift shaft**
    - Fits gearshift shaft in one position only
21. **Attachment nut, lever/gear shift shaft**
    - Tightening torque: 25 Nm (18 ft-lb)
22. **Gearshift shaft on transmission**
23. **Selector rod, short**
24. **Relay lever**

Shift linkage, assembly

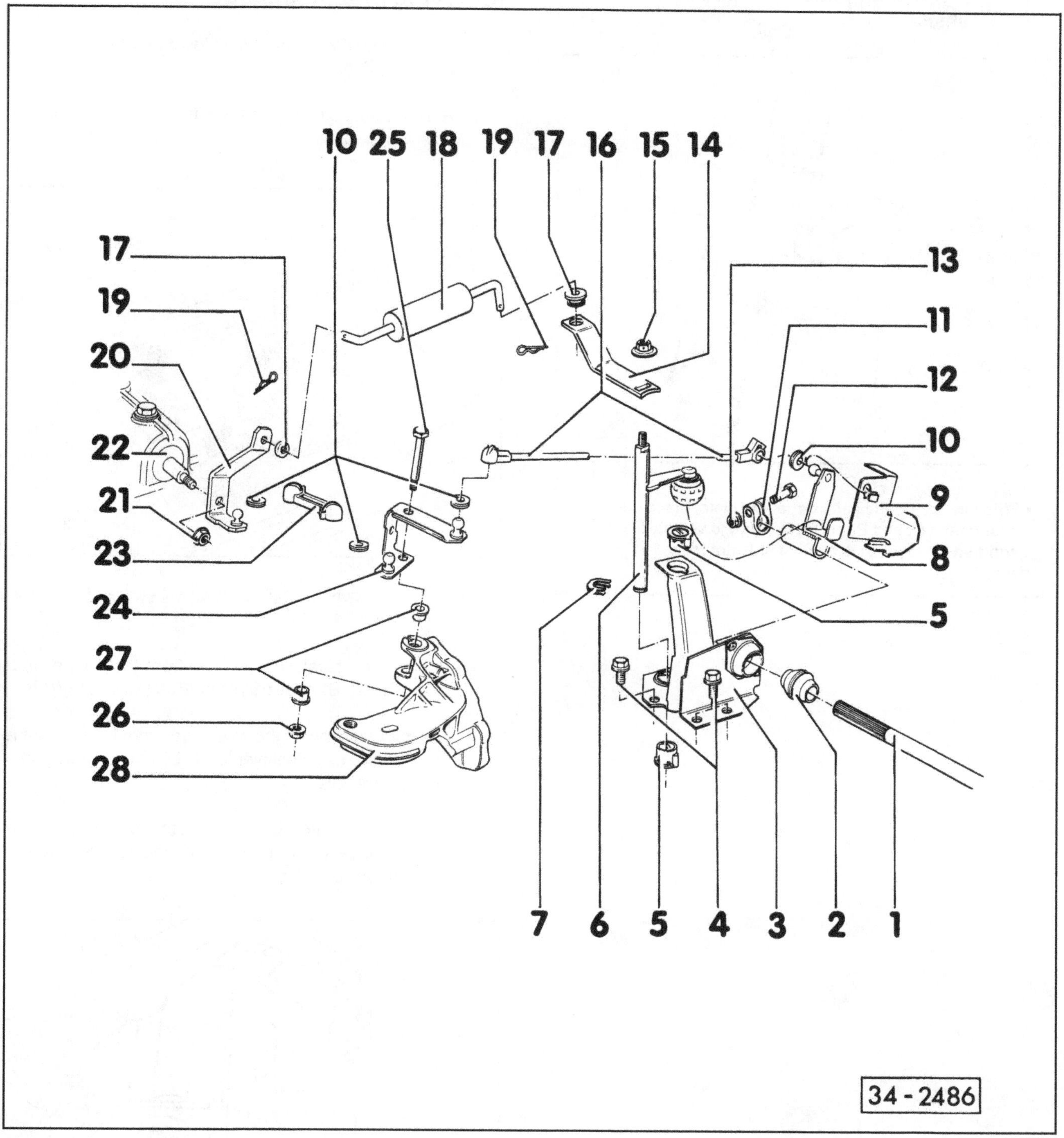

34-2486

Fig. 3. Exploded view of shift linkage mounted to transmission.

25. **Hex-head bolt, relay lever to support bracket**
26. **Attachment nut, relay lever to support bracket**
    - Tightening torque: 15 Nm (11 ft-lb)
27. **Bushing**
28. **Support bracket**
    - Need not be removed to remove relay lever

Installation position of selector lever shield

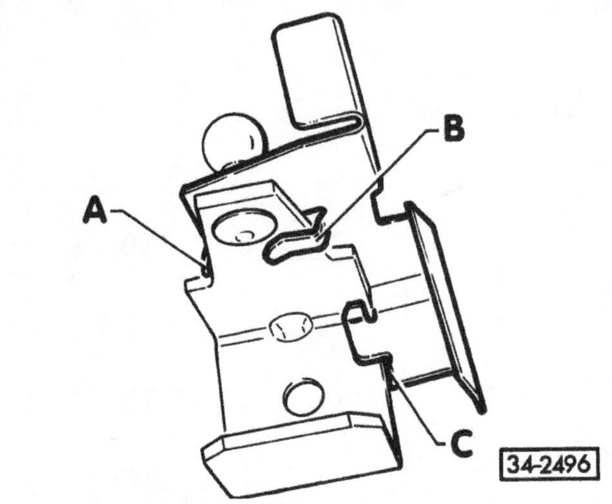

34-2496

- **When installing the selector lever shield, make sure locating tabs (A and B) correctly engage selector lever and that surface (C) is against selector lever.**

Shift lever and selector rod, removing and installing

1. Unscrew shift knob from shift lever and remove together with shift boot.
2. Remove nuts attaching gearshift housing to center console. See Fig. 4.

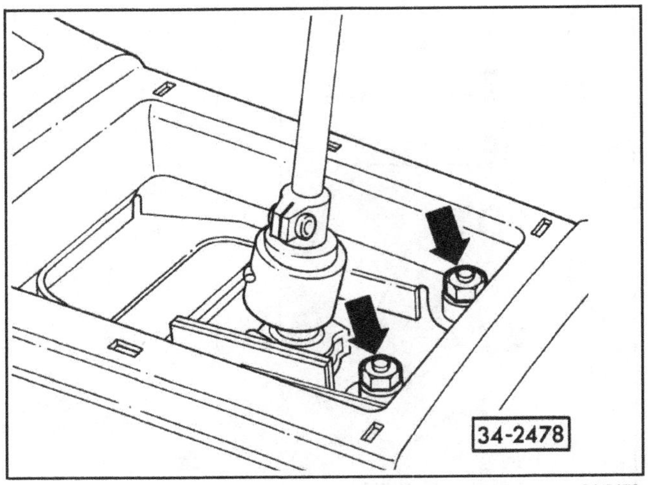

34-2478

Fig. 4. Mounting nuts holding center console to gearshift housing (**arrows**).

3. Remove the catalytic converter from the front exhaust pipe. See **26 Exhaust System/Emission Controls**.
4. Carefully disconnect the relay lever from the selector lever. Remove selector lever shield by pivoting it up and off the selector lever.
5. Remove the clamp for the selector rod/selector lever. Then remove the front mounting bolt from the selector rod support. See Fig. 5.

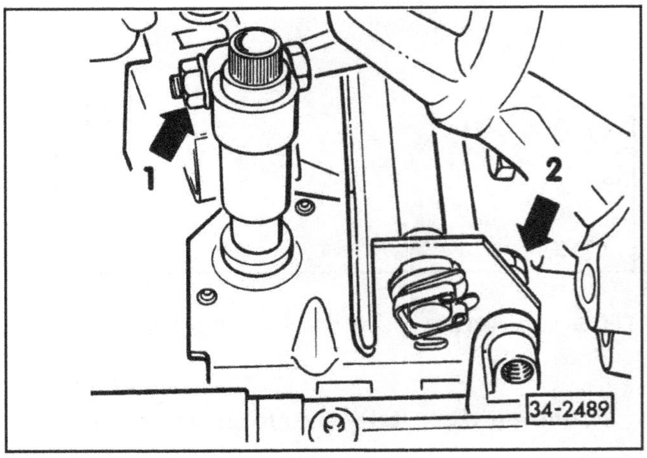

34-2489

Fig. 5. Selector rod/selector lever clamp (**1**) and selector rod support front mounting bolt (**2**).

6. Remove the remaining bolts from the selector rod support. See Fig. 6.

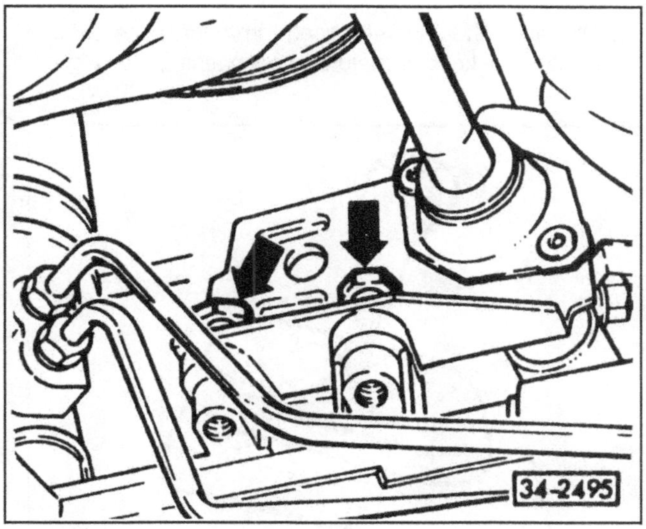

34-2495

Fig. 6. Selector rod support mounting bolts (**arrows**).

7. Working from below, unbolt the gearshift housing from body. See Fig. 2, given earlier.

8. Press the gearshift housing toward the rear. Press up on the selector rod (with bracket) and the selector lever in engine compartment.

9. Disconnect the selector lever from selector rod.

10. Swivel the gearshift housing down together with the selector rod while pressing the housing toward the engine and remove it from the car. See Fig. 7.

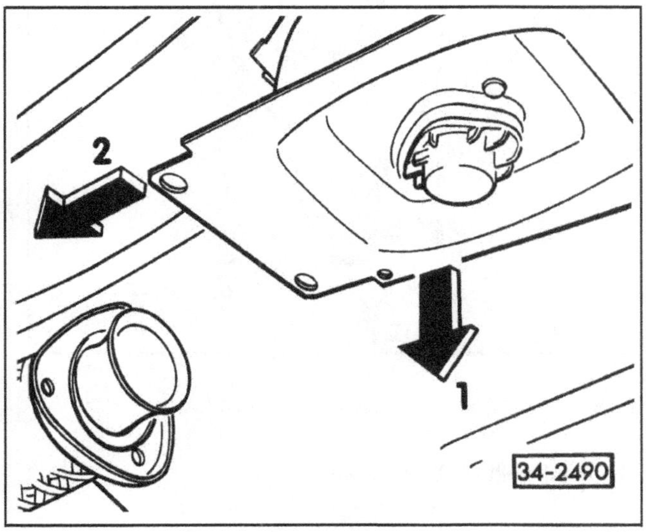

34-2490

Fig. 7. Remove gearshift housing from below by swiveling housing down (**arrow 1**) and then pushing toward engine (**arrow 2**).

Installation is the reverse of removal. For tightening torques, see Fig. 2. Be sure to adjust the shift linkage as described below.

Shift linkage, adjusting

A special tool (alignment gauge) is specified to accurately adjust the shift linkage. The tool (VW tool no. 3104) is used to hold the shift lever in the correct neutral position.

1. Firmly set the emergency brake and then place the transmission in neutral.

2. Loosen the selector rod clamp bolt. See Fig. 8. Check that the selector rod moves freely in the shift lever.

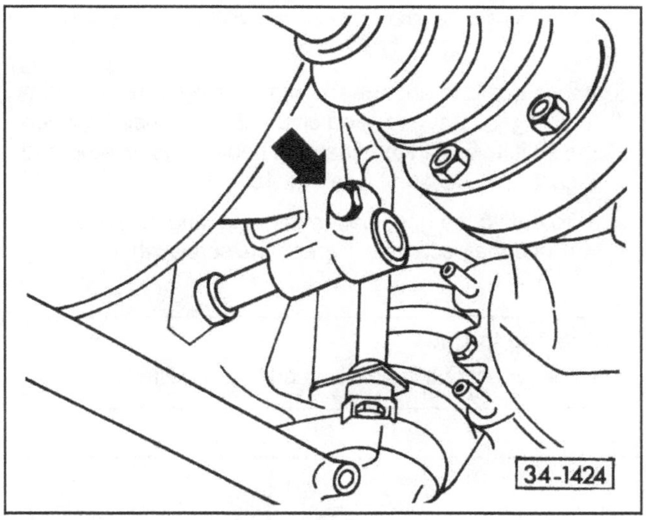

34-1424

Fig. 8. Selector rod clamp bolt (**arrow**). Make sure selector rod moves freely in shift lever once bolt is loosened.

3. Unscrew the shift knob from the shift lever and remove the shift boot with its cover plate.

4. Install the special alignment tool as shown in Fig. 9.

> ***CAUTION —***
> *Check that the gauge does not jam, as the tool can be easily damaged.*

5. Align the selector rod and shift lever (transmission in neutral) and tighten the clamp bolt.

6. Remove the alignment tool and check that the lever now rests in 3rd/4th-gear plane with transmission in neutral.

7. Shift through all the gears and check that they engage smoothly without hanging up. Pay particular attention to the operation of the reverse gear lockout.

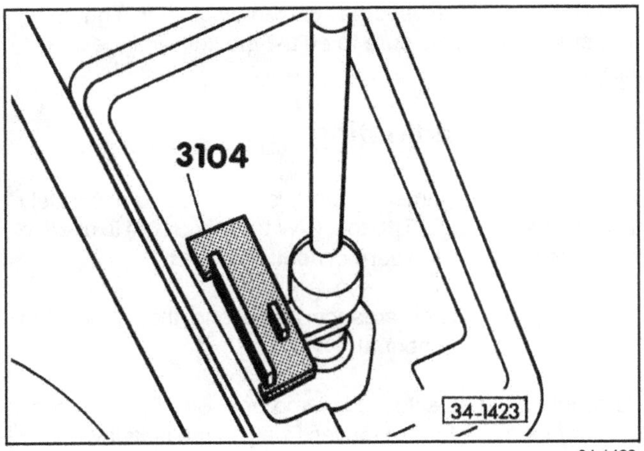

34-1423

Fig. 9. Volkswagen special tool no. 3104 shown installed on alignment lugs on gearshift housing and shift lever.

8. If a gear fails to engage smoothly during repeated shifting, engage 1st gear and check the distance between the shift lever and the gearshift housing travel stop and adjust as necessary. See Fig. 10.
   - To adjust the gap, loosen the clamping screw and rotate the eccentric. Tighten the screw when done.

Tightening torques

- Shift lever eccentric clamping bolt . . . . 2 Nm (18 in lb)

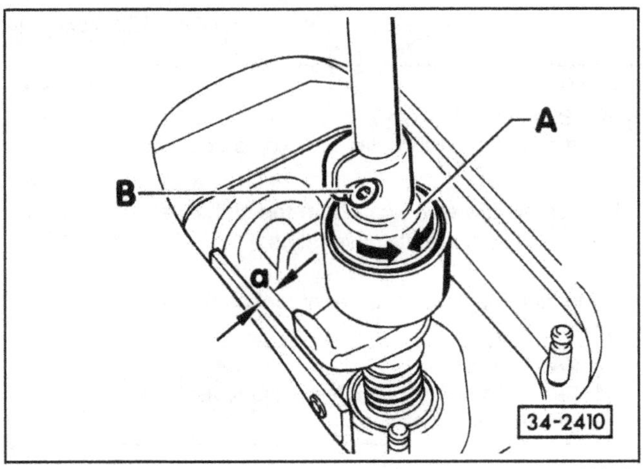

34-2410

Fig. 10. Shift lever adjustment (**a**) should be 1.0–1.5 mm (0.04–0.06 in.) with shift lever pushed to left with light pressure. Clamping screw is shown at (**A**). Rotate eccentric (**B**) as necessary to obtain correct dimension.

> ***CAUTION—***
> *To achieve a satisfactory adjustment:*
> *Shift and linkage components must be in good condition. Shift linkage must operate smoothly without excessive resistance. Transmission, clutch and clutch release mechanism must be in good condition.*

Shift lever boot, installing

1. On cars with a rubber shift boot, snap into shift knob.
2. On cars with leatherette boot, turn boot inside out and position the knob end into the boot opening. See Fig. 11.

34-2482

Fig. 11. Leatherette shift boot shown turned inside out. With boot inside out, install knob into boot (**arrow**).

3. Attach shift knob to boot with rubber ring. See Fig.12

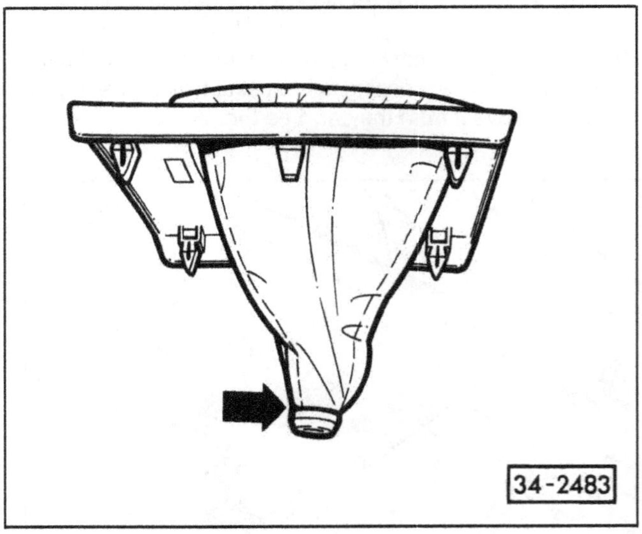

34-2483

Fig. 12. Attach rubber ring (**arrow**) to hold knob to boot.

4. Install shift knob and boot together.

TRANSMISSION REMOVAL AND INSTALLATION

This heading describes removal and installation of the manual transmission used on 4-cylinder engines. Special engine lifting and jacking equipment is needed to support and reposition the engine as the transmission is removed from below. Read the procedure through to determine what tools and equipment are needed for the job.

NOTE—

If the transmission or the clutch housing is being replaced, refer to the note given on the first page of this repair group.

Transmission, removing

1. Disconnect battery ground (GND) strap.

CAUTION—

Be sure to have the anti-theft radio code on hand before disconnecting the battery.

2. Disconnect clutch cable from the transmission. See **30 Clutch**.

3. Disconnect all ground straps and electrical harness connectors from the transmission. See Fig. 13.

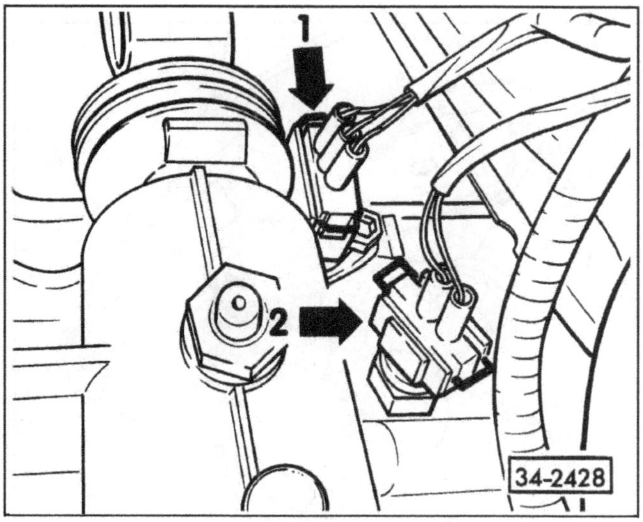

34-2428

Fig. 13. Vehicle speed sensor (**1**) and back-up light switch (**2**) harness connectors to be removed. Also remove any other connectors and the ground strap from transmission.

4. Remove the upper bellhousing bolts connecting the transmission to the engine.

5. Remove the bolt from the right-rear engine mount. bracket. Where applicable, remove connector support plate. See Fig. 14

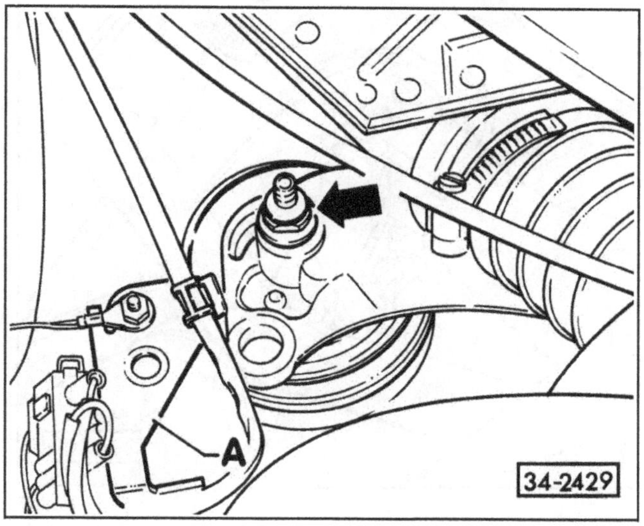

34-2429

Fig. 14. Right rear engine mount bolt (**arrow**) to be removed. Where applicable, also remove the support plate (**A**).

6. On Cabrio models, remove the right engine support from the drive belt side of the engine. See Fig. 15.

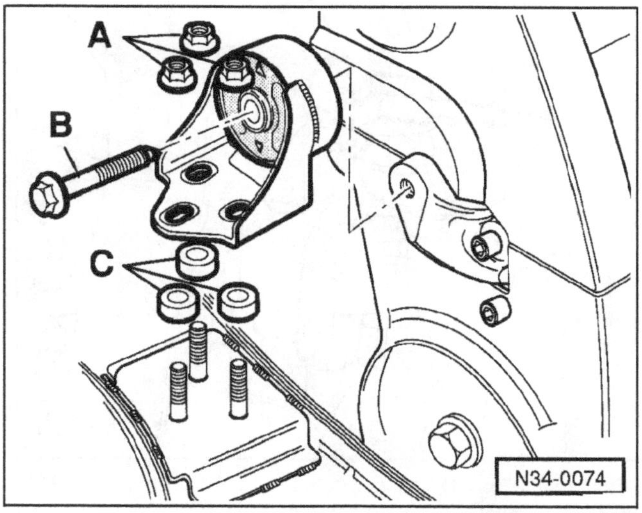

34-0074

Fig. 15. Engine support used on Cabrio models only. Be sure to install spacers (**C**) before installing mount. Tightening torques for bolts (**A** and **B**) are given at the end of the installation procedure.

7. Disconnect the shift linkage selector rods at the points indicated in Fig. 16.

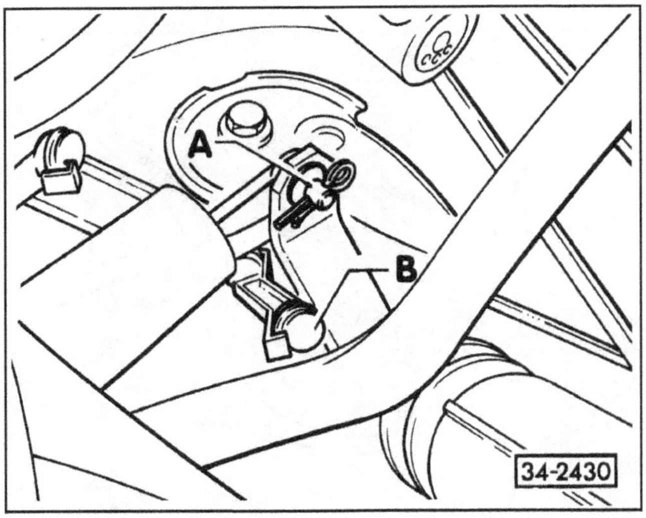

34-2430

Fig. 16. Shift linkage selector rod ends to be disconnected (**A** and **B**).

8. Disconnect the shift linkage long selector rod from relay lever. See Fig. 17.

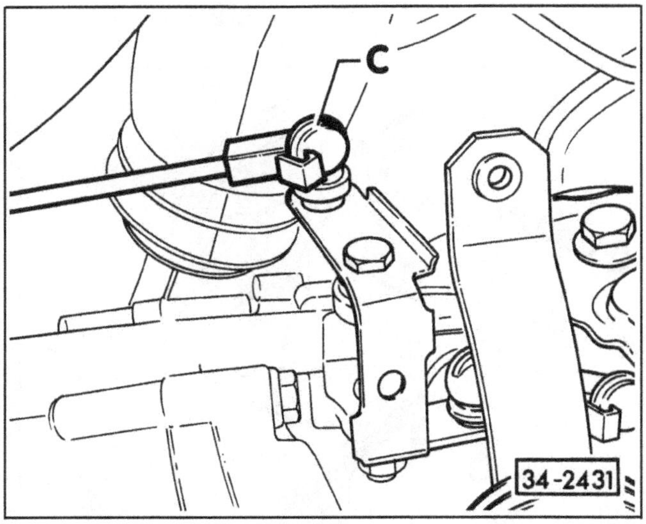

34-2431

Fig. 17. Disconnect long selector rod (**C**) from relay lever by opening clamp and then gently prying off lever.

9. Working from the left side of the transmission, remove the two upper bolts from the transmission bracket. See Fig. 18.

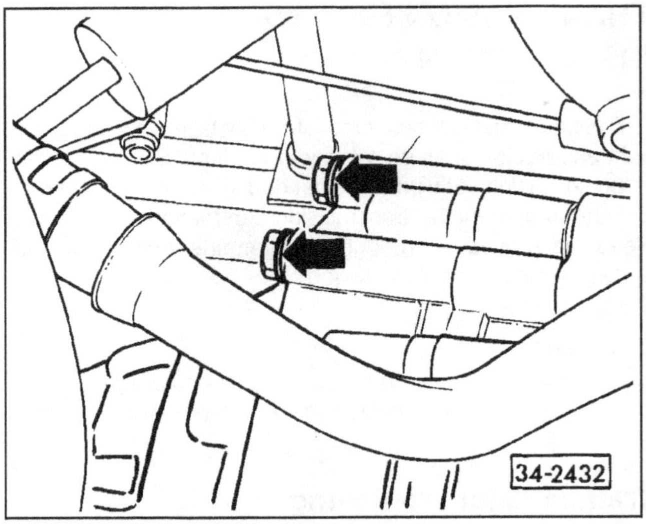

34-2432

Fig. 18. Transmission bracket mounting bolts. Bracket is on left side of transmission.

10. If applicable, remove the transmission support as shown in Fig. 19

NOTE —

- *The transmission support was gradually discontinued after approx. November 1993 production.*
- *When carrying out repairs on a vehicle that is equipped with the support, it should always be re-installed.*

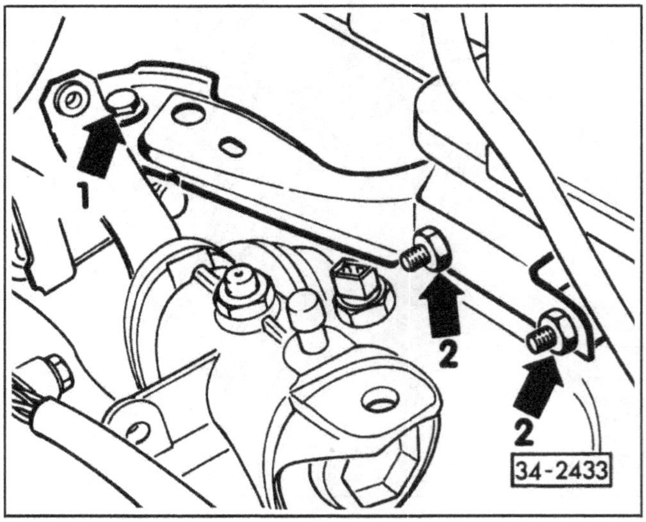

34-2433

Fig. 19. Transmission support used on early production vehicles. To remove bracket remove bolt (**1**) and nuts (**2**).

11. On Cabrio models, remove the transmission support from the left side of the transmission. See Fig. 20.

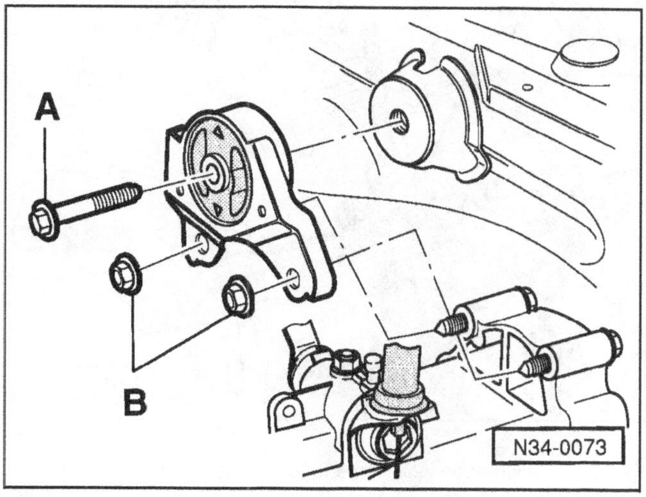

34-0073

Fig. 20. Transmission support used on Cabrio models only. Tightening torques for bolts (**A**) and nuts (**B**) are given at the end of the installation procedure.

12. Remove the bolt from the front engine mount as shown in Fig. 21.

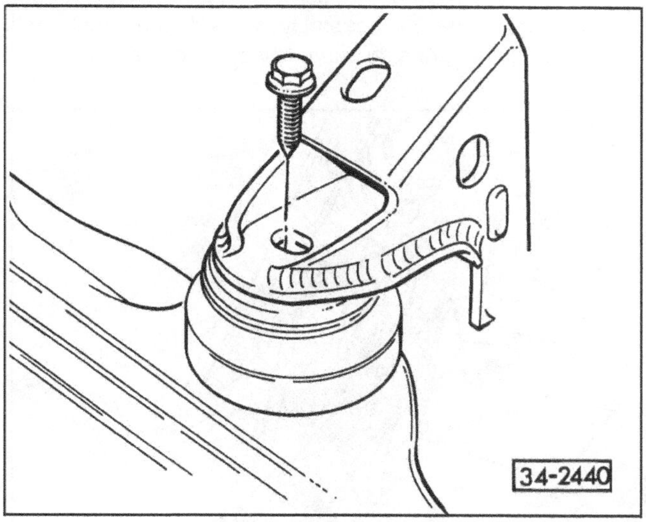

34-2440

Fig. 21. Mounting bolt for front engine mount.

13. Attach engine lifting equipment (VW sling 10-222A and adjust 10-222A/1 adapters, or equivalent) to engine lifting eyes. Adjust the lifting sling until the weight of the engine is supported. See Fig. 22.

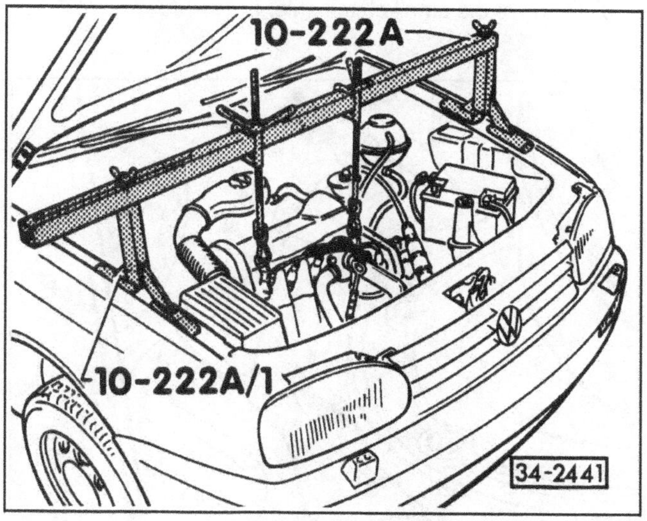

34-2441

Fig. 22. Engine lifting equipment used to support weight of engine. Numbers shown are for VW special tools.

14. If applicable, remove the insulation shield beneath engine/transmission.

15. Remove the starter. See **27 Engine Electrical**.

16. Remove the front engine bracket from the engine block.

17. Working from below, remove the left-side engine mount attachment bolts. See Fig. 23.

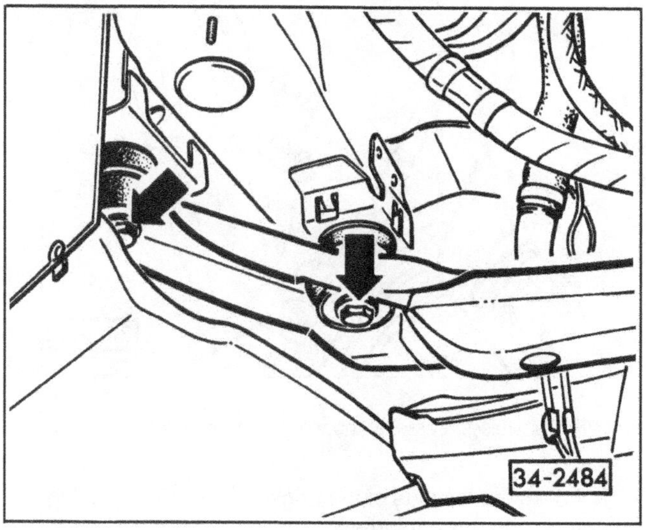

34-2484

Fig. 23. Engine mount attachment bolts (**arrows**).

18. Remove the large clutch shield.

19. Remove the small clutch shield. See Fig. 24.

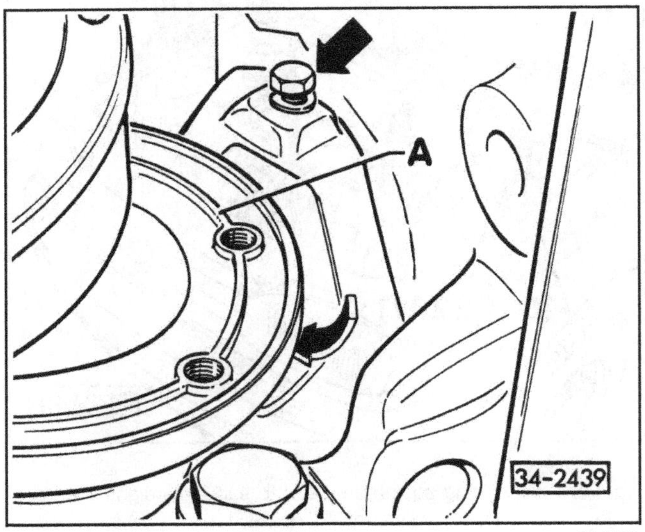

34-2439

Fig. 24. Clutch shield and bolt (**arrows**) located behind drive flange (**A**).

20. Remove the third bolt for the transmission bracket at left side of the transmission and remove the transmission bracket. See Fig. 25.

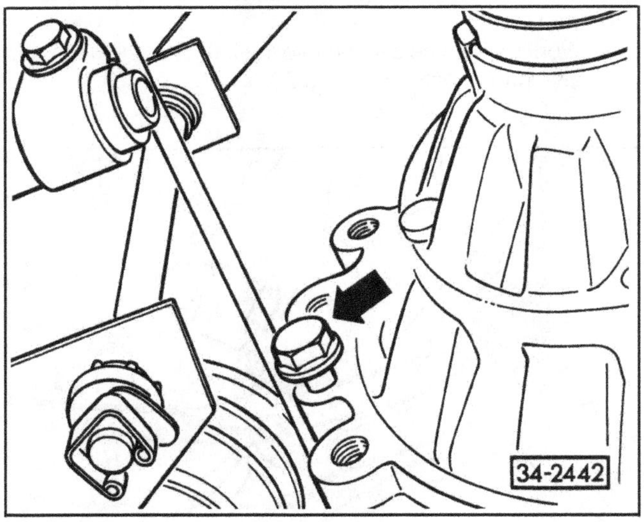

34-2442

Fig. 25. Transmission bracket mounting bolt (**arrow**).

21. Lower the left side of the transmission slightly using engine support equipment.

22. Position transmission jack beneath transmission. See Fig. 26.

23. Remove the bolts for the left-side transmission support and remove it.

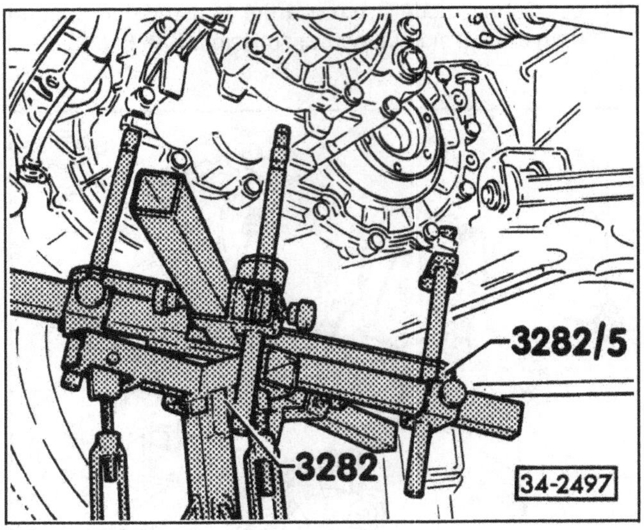

34-2497

Fig. 26. Transmission jack correctly positioned under transmission. Numbers shown are for VW transmission jack and adapters.

24. Disconnect the drive shafts from transmission, removing heat shields as necessary.

25. Turn the steering all the way to the left and suspend drive shafts as high as possible.

26. Using an appropriate scissors jack, push the engine/transmission assembly toward the front and around the right-side assembly support. See Fig. 27.

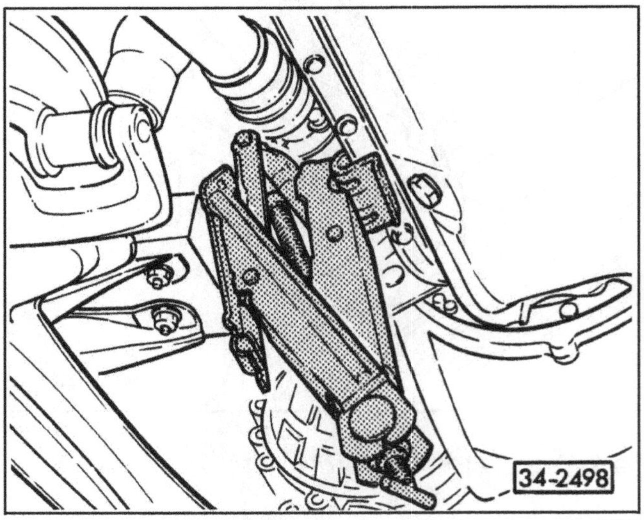

34-2498

Fig. 27. Scissors jack being used to push engine/transmission toward front.

27. Remove the lower bolt attaching the engine to the transmission.

28. Pull the transmission off its alignment dowel pins and carefully lower. See Fig. 28.

> ***CAUTION —***
> *When lowering the transmission, use care to avoid damaging the power steering lines.*

34-2499

Fig. 28. When lowering transmission out of car, guide right drive flange (**A**) past flywheel and left drive flange (**B**) past assembly support.

Transmission, installing

Install the transmission in the reverse order of removal. Check that all engine-to-transmission dowel pins are installed in the engine block. Replace any that are missing. Clean the hub splines of the transmission input shaft and apply a light coat of G 000 100 grease before installing. Install the engine/transmission mounts so that they are free of tension. See **10 Engine–Removing and Installing** for specific engine/transaxle installation procedures.

Make a final check of transmission installation, noting the following:

Final installation checks

- Ensure that the intermediate plate is positioned correctly during transmission installation.
- Check the transmission oil level. See **0 Maintenance Program**.
- Check the operation of the clutch cable (automatic adjustment mechanism). See **30a Clutch (ABA, ACC, AAZ Engines)**.
- Adjust the shift linkage. See **Shift linkage, adjusting**.
- If the clutch housing has been replaced, or if the entire transmission is being replaced, refer to the note given on page 34a-1 for information on modified clutch housings.

Tightening torques for the are given below. Use Fig. 29 together with **Table a** as a guide when installing the transmission to the engine. Use Fig. 30 together with **Table b** when installing the transmission/engine assembly into the vehicle.

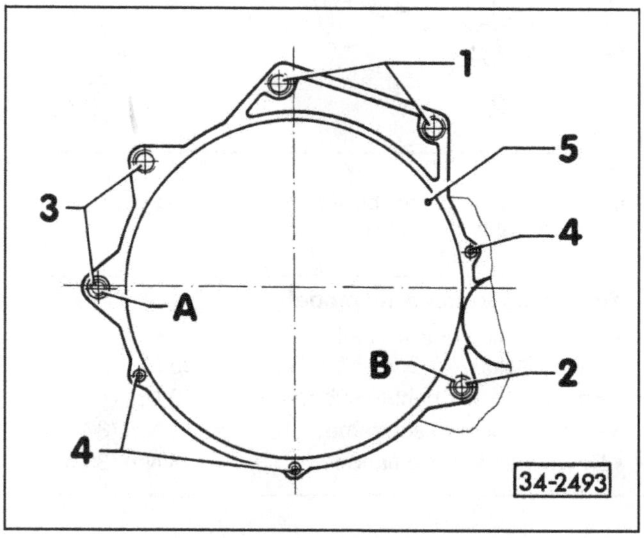

34-2493

Fig. 29. Transmission-to-engine mounting bolt locations (to be used in conjunction with **Table a**). Alignment dowel pins shown at **A** and **B**.

Table a. Transmission to engine fasteners

Fastener	Size	Qty.	Tightening torque
1	M12 X 55	2	80 Nm (59 ft-lb)
2	M12 X 64	1	80 Nm (59 ft-lb)
3*	M10 X 140	2	60 Nm (44 ft-lb)
4**	M7 X 12	3	10 Nm (89 in lb)
5**	M6 X 8	1	10 Nm (89 in lb)

*Also starter and bracket at front of transmission
**Flywheel shield plate

Table b. Transmission/engine assembly mounts

Fastener	Size	Qty.	Tightening torque
1	M10 X 60	1	45 Nm (33 ft-lb)
2	M10 X 30	1	60 Nm (44 ft-lb)
3*	M10 X 45	1	60 Nm (44 ft-lb)
4*	M10 nut	2	40 Nm (30 ft-lb)
*Not applicable on all models			

Tightening torques- Cabrio models only

- Engine support, right (Cabrio models only)
  support to body (**A, Fig. 31**) .... 25 Nm (18 ft-lb)
  support to engine (**B, Fig. 31**)...... 50 Nm (37 ft-lb)
- Transmission support, left (Cabrio models only)
  support to transmission (**A, Fig. 32**) . 50 Nm (37 ft-lb)
  support to body (**B, Fig. 32**) .... 50 Nm (37 ft-lb)

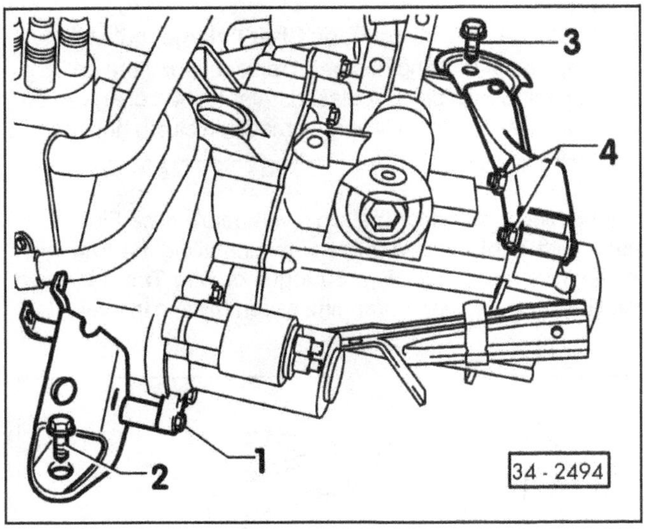

34-2494

Fig. 30. Transmission assembly mounts (to be used in conjunction with **Table b**).

Tightening torques–all models

- Transmission bracket (left) to transmission...................... 35 Nm (26 ft-lb)
- Engine bracket right to subframe.... 60 Nm (44 ft-lb)
- Engine mount to subframe......... 50 Nm (37 ft-lb)
- Drive shaft to drive flange .... 45 Nm (33 ft-lb)

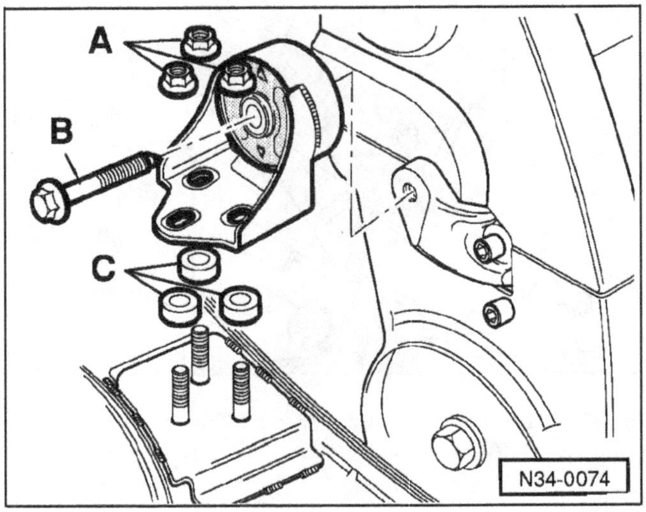

34-0074

Fig. 31. Engine support used on Cabrio models only. Be sure to install spacers (**C**) before installing mount.

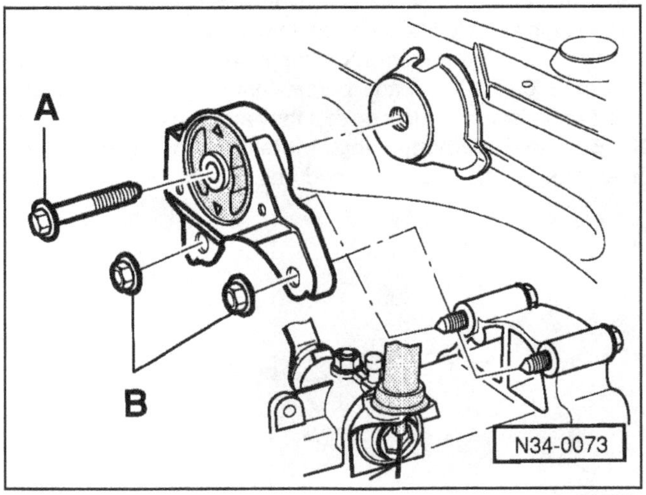

34-0073

Fig. 32. Transmission support used on Cabrio models only.

34b Manual Transmission (02A Transmission)

GENERAL

This repair group covers repair information for the cable-operated gear shift mechanism. Also covered here is the removal and installation of the 02A manual transmission. This repair group does not cover transaxle or transmission teardown and disassembly.

NOTE —
- *For information on drive axles, including drive flange oil seals, see* **39 Differential and Final Drive**.
- *To check manual transmission oil, see* **0 Maintenance Program**.

GEAR SELECTOR MECHANISM

Fig. 1 shows the operation of the gear selector mechanism. Note that the gear selector is cable operated—one cable controls forward/back movement of the shift lever and the other cable controls left/right movement. Access to the gear selector mechanism is difficult without partially removing the exhaust system. For easiest access to the shift mechanism, the catalytic converter and its heat shield should first be removed as described in **26 Exhaust System/Emission Controls**.

NOTE —
- *Before servicing or repairing the manual transmission or shift linkage, review the general repair information given in* **3 General Information.**
- *Always remove the exhaust system heat shield before removing gear selector mechanism.*
- *To replace the selector cables, first remove the complete gear selector mechanism. Be careful not to kink the selector cables during removal.*
- *During reassembly, lubricate all mounting and contact surfaces with MoS2 grease. As the final step, adjust the gear selector mechanism, see* **Gear selector mechanism, adjusting**.

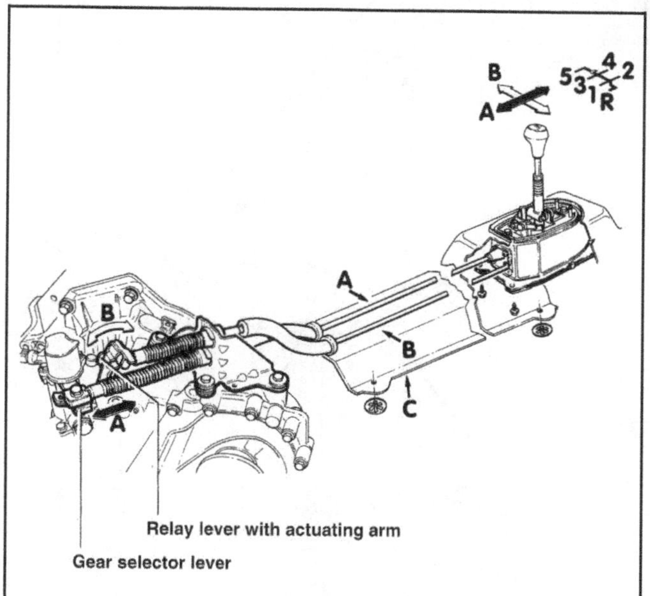

0024068

Fig. 1. Operational view of shift lever control. Gear change cable (**A**) controls forward/back travel. Gear shift cable (**B**) controls side-to-side travel. Heat shield (**C**) holds cables in position.

CAUTION —
- *Before servicing the gear selector mechanism, always disconnect the negative (–) battery cable in the engine compartment.*
- *On vehicles equipped with an anti-theft radio code, make sure you know the correct radio activation code before disconnecting the battery.*

Gear selector control cable, assembly

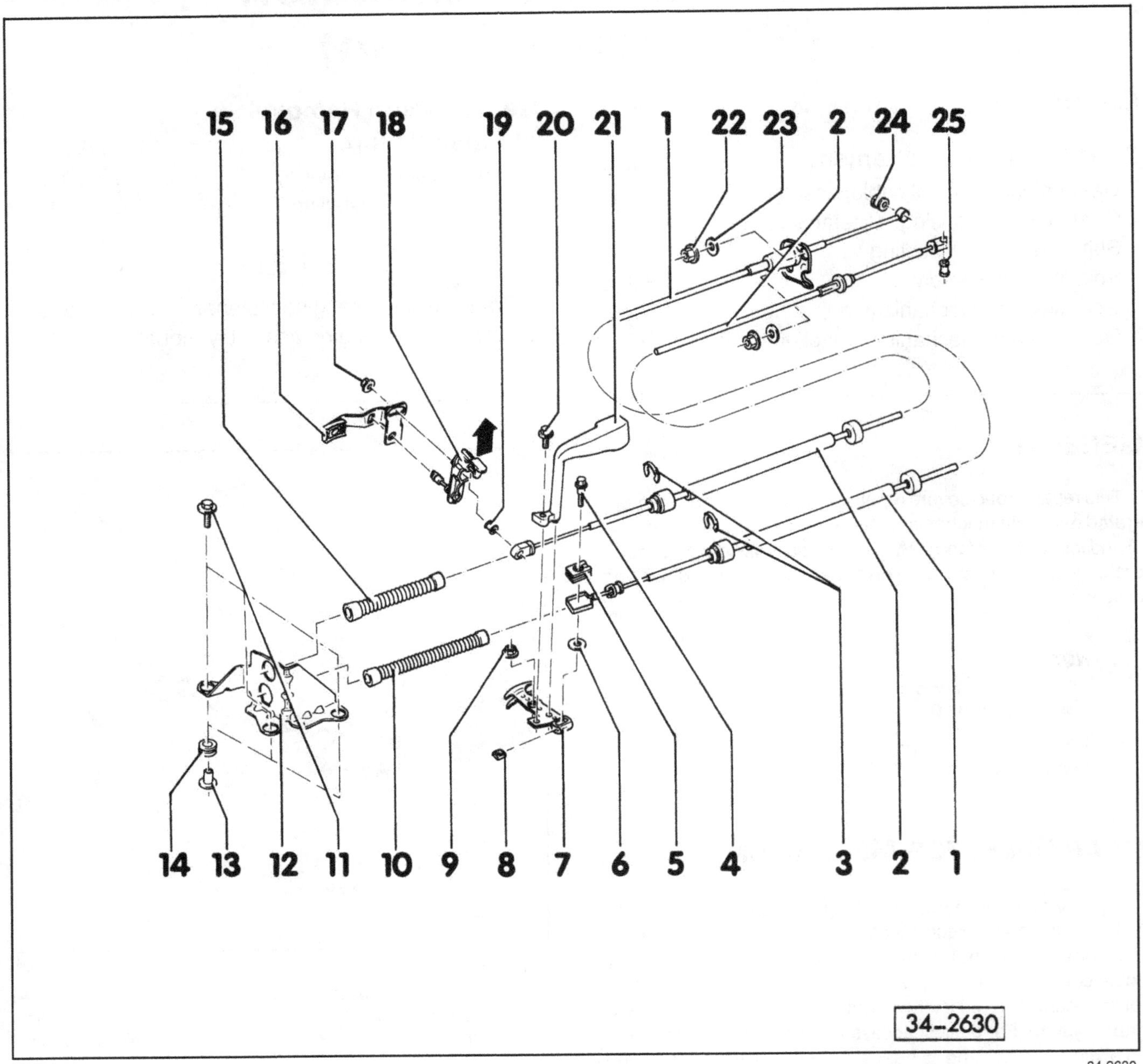

34-2630

Fig. 2. Shift lever control cable and related parts.

1. **Gear selector cable**

> ***NOTE —***
> *Lubricate all mountings and contact surfaces with MoS2 grease.*

2. **Gate selector cable**
   - Pull off actuating arm (18) by pulling lug in direction indicated by arrow
   - Install before installing gear selector cable
3. **Circlip**
   - Be careful to avoid damaging boot when removing
4. **Hex bolt**
   - With shoulder
   - Tightening torque: 25 Nm (18 ft-lb)
   - Attaches gear selector cable to gear selector lever (7)
5. **Rubber bushing**
   - For gear selector cable connection to gear selector lever (7)
   - Press out of cable end piece when replacing boot
6. **Washer**
   - Install between gear selector cable and gear selector lever (7)

7. **Gear selector lever**
   - Fits in one position only
   - After installing, adjust gear selector mechanism, see **Gear selector mechanism, adjusting**
   - Installation position, **see** Ⓐ
8. **Square nut**
   - Insert into gear selector lever
9. **Self-locking hex nut**
   - Always replace
   - Tightening torque: 25 Nm (18 ft-lb)
10. **Boot**
    - Carefully pull over gear selector cable end piece
11. **Hex bolt**
    - For support bracket (12)
    - Tightening torque: 25 Nm (18 ft-lb)
12. **Support bracket**
13. **Spacer**
14. **Bushing**
    - For mounting support bracket to transmission
15. **Boot**
    - Carefully pull over gear selector cable end piece
16. **Relay lever**
    - Remove from transmission before removing clutch controls
    - Installation position, **see** Ⓐ
17. **Self-locking hex nut**
    - Always replace
    - Tightening torque: 15 Nm (11 ft-lb)
    - Loosen only to adjust gear selector mechanism
18. **Actuating arm**
    - With pivot pin to mount relay lever
    - Installation position, **see** Ⓐ
19. **Rubber bushing**
    - For gate selector cable connection to actuating arm
    - Drive out/drive in with drift
20. **Self-locking hex bolt**
    - Always replace
    - Tightening torque: 20 Nm (15 ft-lb)
21. **Balance weight**
    - Remove to adjust gear selector mechanism
22. **Self-locking hex nut**
    - For gear selector cable to selector housing
    - Always replace
    - Tightening torque: 15 Nm (11 ft-lb)
23. **Rubber washer**
24. **Rubber bushing**
    - For gear selector cable connection to shift lever
    - Drive out/drive in with drift
25. **Bushing**
    - For gate selector cable connection to selector bracket
    - Drive out/drive in with drift

Ⓐ Gear selector lever and relay lever, installation position

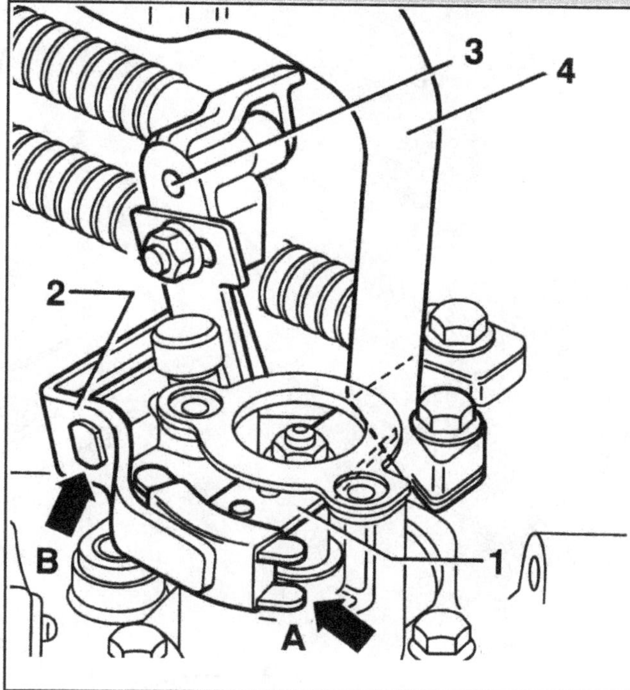

0024069

1. **Gear selector lever**
2. **Relay lever**
   - Engages in guide rail of gear selector lever (A)
   - Lubricate with MoS2 grease
3. **Actuating arm**
   - Flats (B) on pivot pin for relay lever mounting;
   - can be installed in one position only
4. **Balance weight**
   - Fasten to gear selector lever see Fig. 2

NOTE —
Linkage and back-up light switch not shown in illustration.

Gear selector housing, assembly

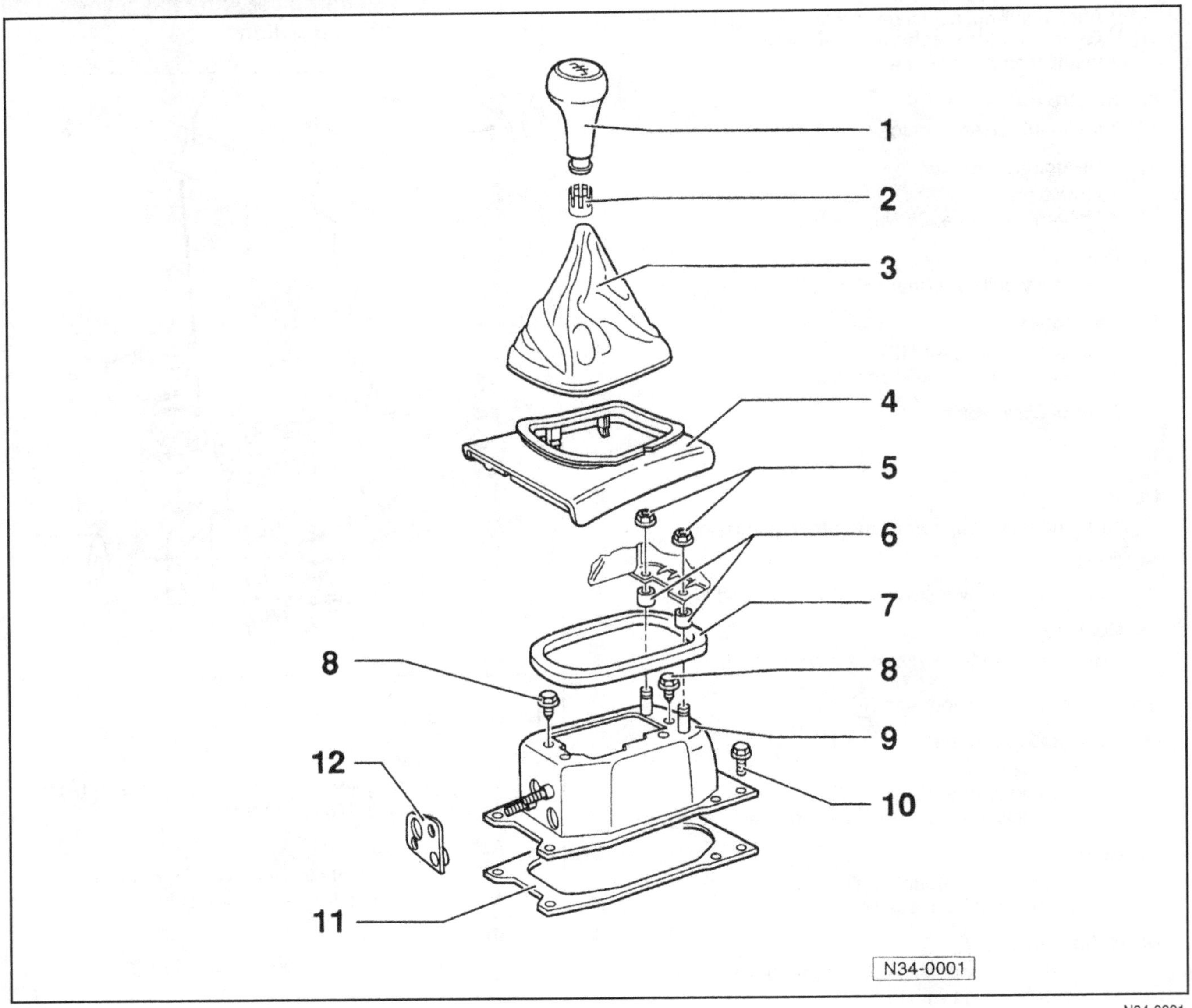

N34-0001

Fig. 3. Gear selector housing and related parts.

1. **Shift knob**
   - Unscrew from shift lever to remove
2. **Sleeve**
3. **Boot**
   - When removing, lift off with frame from center console
4. **Frame**
5. **Hex nut**
   - Attaches console to gearshift housing (9)
   - Remove before removing gear selector mechanism
6. **Spacer**
   - Installed between console and gearshift housing (9)
7. **Gasket**
   - Between gearshift housing and floor
   - Attach to gearshift housing
8. **Hex bolt**
   - For mounting plate to gearshift housing
   - Tightening torque: 15 Nm (11 ft-lb)
9. **Gearshift housing**
10. **Hex bolt**
    - Quantity: 4
    - Tightening torque: 25 Nm (18 ft-lb)
11. **Gasket**
    - Self-adhesive
    - Attach to underside of gearshift housing
12. **Gasket**
    - Install between selector cables and gearshift housing

Shift lever boot, installing

1. Turn boot inside out and position the knob end into the boot opening. See Fig. 4

34-2482

Fig. 4. Turn shift boot inside out and install knob into boot.

2. Attach shift knob to boot with sleeve. See Fig. 5

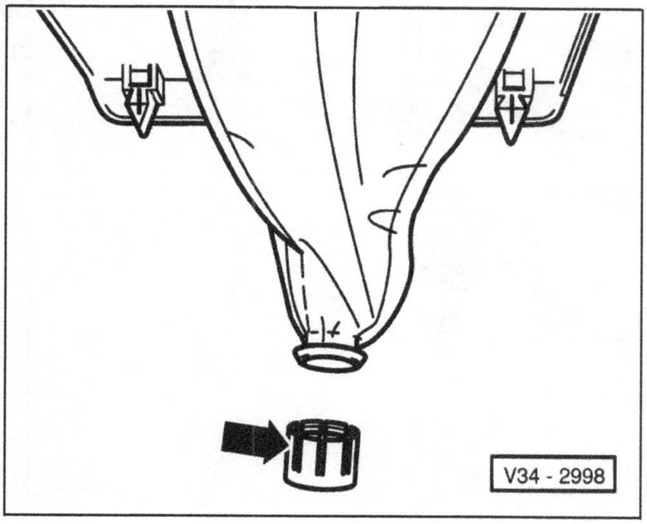

V34-2998

Fig. 5. Shift lever boot retaining sleeve (**arrow**).

3. Install shift knob and boot together.

Shift lever, assembly

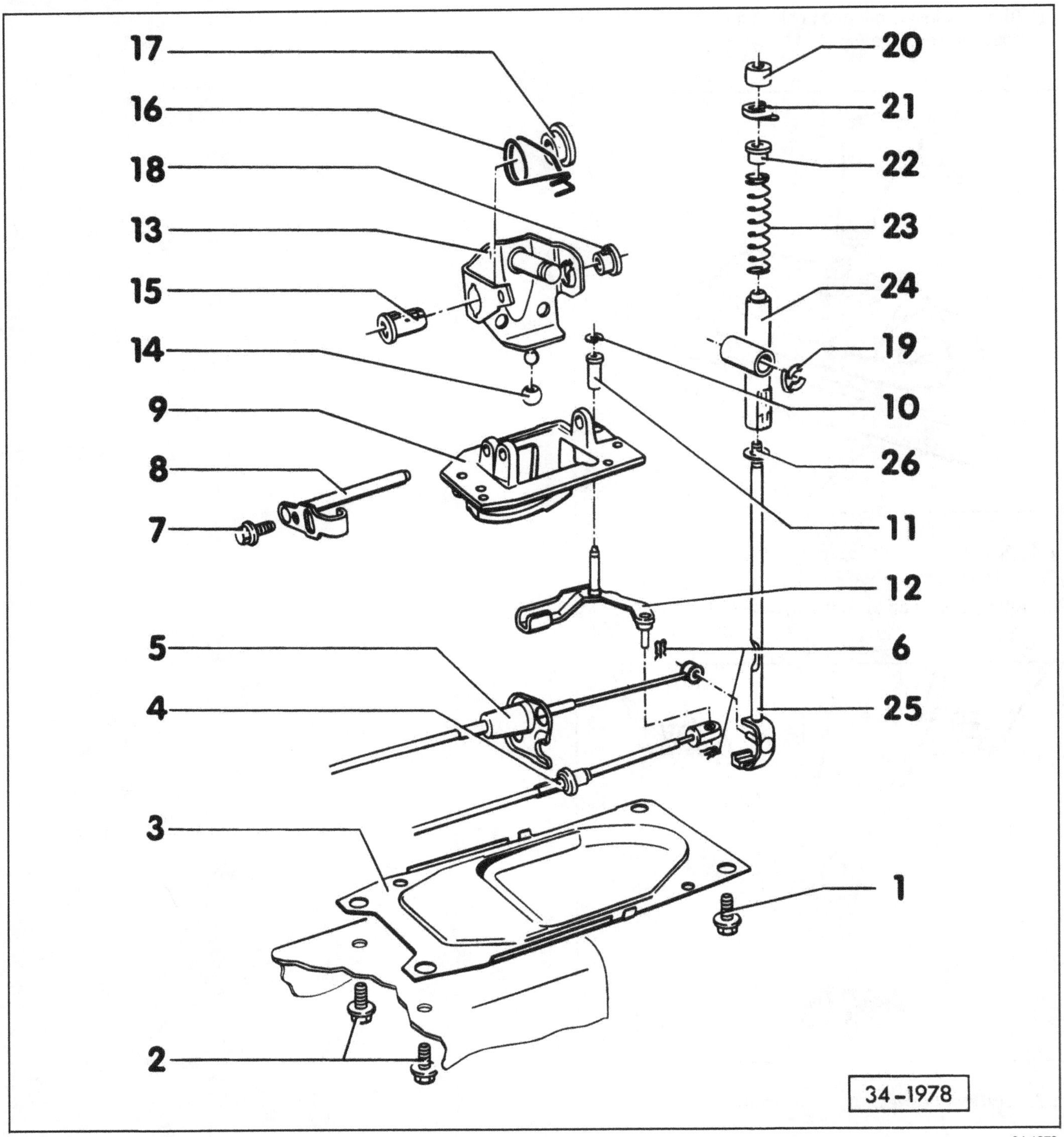

34-1978

Fig. 6. Shift lever assembly.

1. **Hex bolt**
   - Tightening torque: 25 Nm (18 ft-lb)

> ***NOTE —***
> *Lubricate all and contact surfaces with MoS2 grease.*

2. **Hex bolt**
   - Tightening torque: 25 Nm (18 ft-lb)
3. **Cover plate**
4. **Gate selector cable**
   - On gate selector bracket

5. **Gear selector cable**
   • On gear selector lever
6. **Spring clip**
7. **Bolt**
   • Pivot pin (8) to mounting plate
   • Tightening torque: 15 Nm (11 ft-lb)
8. **Pivot pin**
9. **Mounting plate**
10. **Lock plate**
    • Removing and installing, **see** Ⓐ
11. **Bushing**
    • Fits in one position only
12. **Gate selector bracket**
13. **Gate selector housing**
    • Fit into mounting plate after installing complete shift lever
14. **Cap**
    • Assembling, **see** Ⓑ
15. **Bushing**
    • Fits in one position only
16. **Spring**
17. **Guide bushing**
    • Assemble with bushing (15)
18. **Bushing**
19. **Circlip**
20. **Sleeve**
    • Where applicable
21. **Circlip**
    • Removing and installing, **see** Ⓒ
22. **Spacer sleeve**
23. **Spring**
24. **Shift lever guide**
25. **Shift lever**
26. **Damping washer**

Ⓐ Removing lock plate for gate selector bracket

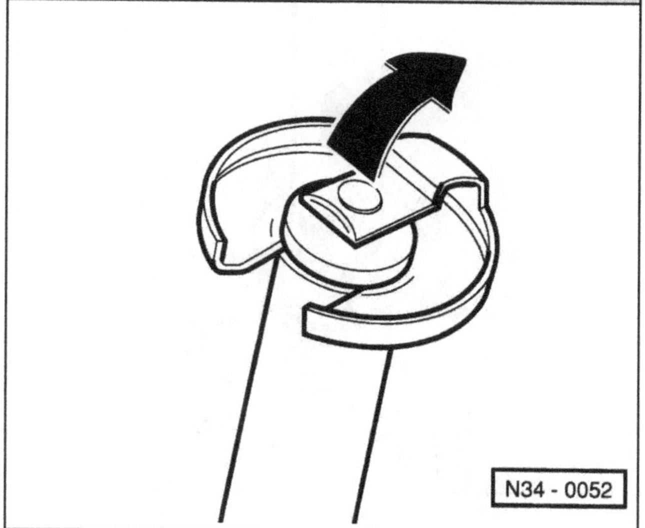

N34-0052

• **Lift clamp in direction of (arrow).**

Ⓑ Installing gate selector housing cap

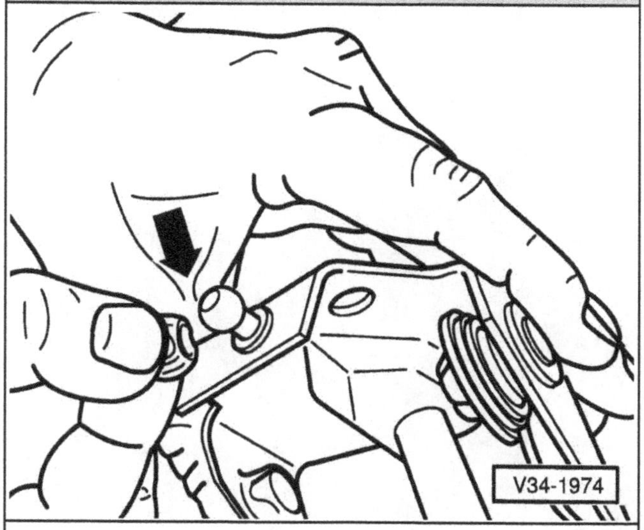

V34-1974

• **Press on from side (arrow).**
• **Press together opening to form a slot.**

NOTE —
If necessary, heat cap in hot water to soften.

Circlip for shift lever guide, removing and installing

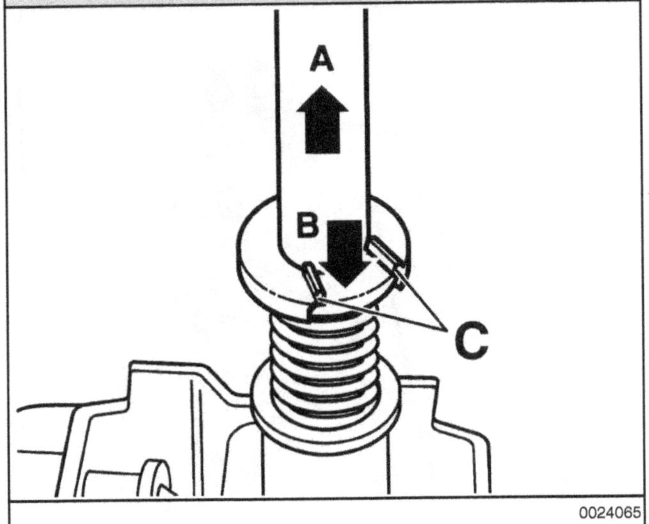

- **Pull shift lever in direction of arrow (A) and simultaneously use screwdriver to push spacer bushing in direction of arrow (B) to stop.**

NOTE —
Do not tilt the spacer bushing when pushing down on it. The holding slot in the shift lever for the lock washer must remain accessible.

Gear selector mechanism, adjusting

NOTE —
Special VW tools are required to accurately adjust the shift mechanism. Read the procedure through to determine what tools will be necessary.

1. Place transmission in neutral.
2. Remove shift knob together with shift boot.
3. Working at the transmission, remove the balance weight from the shift lever. See Fig. 2, given earlier.
4. Loosen bolt and nut for gear selector cables. See Fig. 7. Gear selector cables and actuating arm/gate selector cable should now move freely in elongated holes.

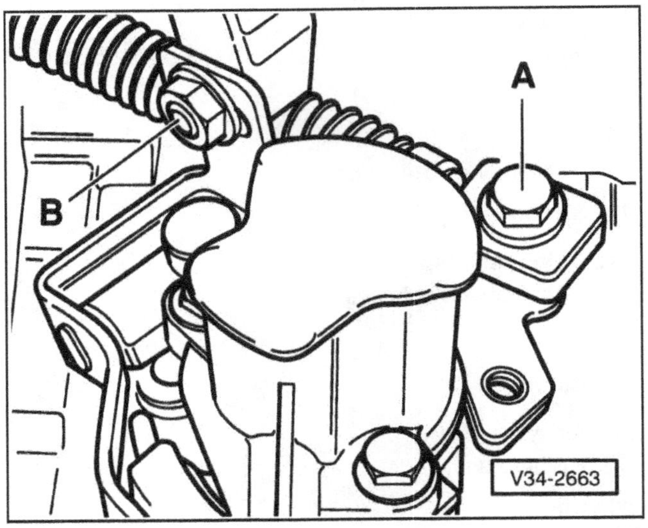

V34-2663

Fig. 7. Gear selector cables mounting bolt (**A**) and nut (**B**).

5. Working in the shift lever housing, loosen bolt shown in Fig. 8.
6. Install shift linkage gauge (VW special tool 3192) over shift lever. Then pivot gauge mounting hook under mounting plate and tighten clamping nut (D in Fig. 8).
7. Press shift lever into left-hand detent of slide. Then press on shift lever and slide toward left stop (direction of arrow in Fig. 8). Tighten clamping bolt (bolt E in Fig. 8).
8. Press shift lever toward right-hand detent and then tighten shift lever housing bolt previously loosened (bolt C in Fig. 8).

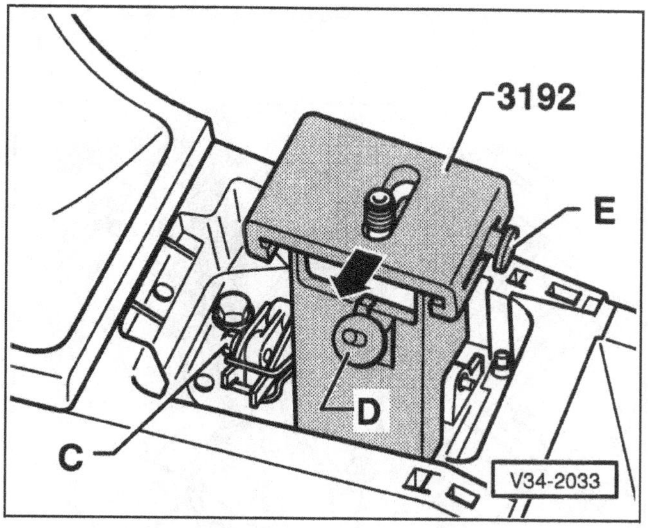

V34-2033

Fig. 8. Shift lever housing mounting bolt to be loosened (**C**). Shift linkage alignment gauge (VW special tool no 3192) shown installed over shift lever.

9. Install special wedge and pin to shift mechanism. See Fig. 9. Slide wedge between gear selector lever and selector cover plate so that there is no play.

NOTE —

When installing the wedge, use care so that the wedge does not cause the shift lever to lift.

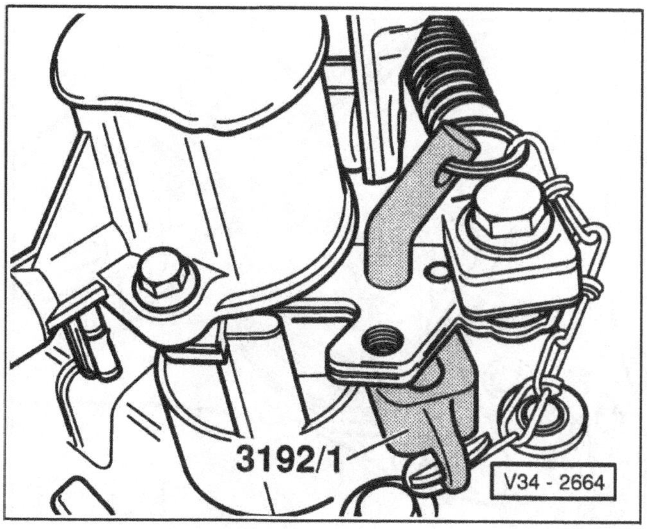

V34-2664

Fig. 9. Shift lever alignment wedge and pin (VW special tool 3192/1) shown installed to shift linkage at transmission.

10. Secure gear selector cable and actuating arm/selector cable by tightening nut and bolt shown in Fig. 7. Check that the wedge is still correctly positioned.

11. Remove shift linkage alignment gauge, wedge, and pin.

12. Install balance weight, shift lever knob and boot.

Gear selector mechanism, functional check

1. With transmission in neutral, make sure shift lever is in 3rd gear/4th gear gate

2. Depress clutch pedal and start engine.

3. Wait approximately 3-6 seconds to allow transmission input shaft to come to a standstill.

4. Select each gear several times, checking all gears. Pay particular attention to operation of reverse gear.

NOTE —

Should any gear fail to engage smoothly after being selected repeatedly, the selector shaft play (lift) should be checked as described below.

5. Have a helper select 1st gear and then press shift lever to left stop, then release. At same time, observe selector shaft on transmission. The selector shaft must move approx. 1 mm (0.04 in.) in the direction shown in Fig. 10.

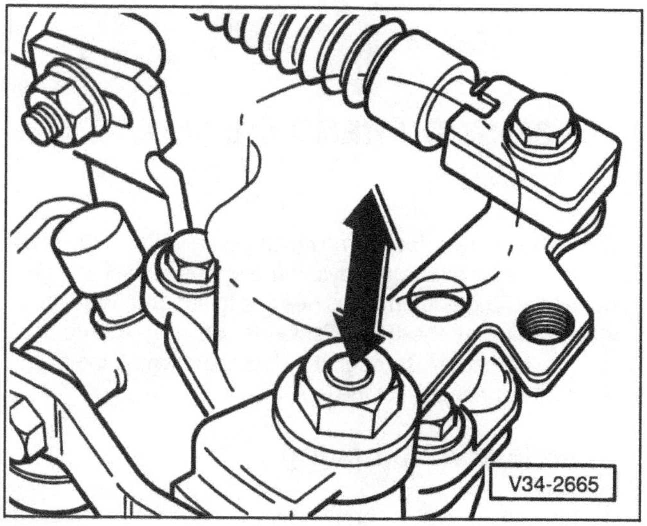

V34-2665

Fig. 10. When moving shift lever to 1st gear stop, transmission selector lever should move in direction of arrow approx. 1.0 mm (0.04 in.).

6. If any faults are found, disengage 1st gear.

7. Then loosen nut at actuating arm/gate selector cable mounting. See Fig. 11. Take up gate selector cable play by lightly pressing on actuating arm and retighten nut.

NOTE —
Gate selector cable has some play at mounting due to transfer elements, i.e. some minimal play at mounting is normal.

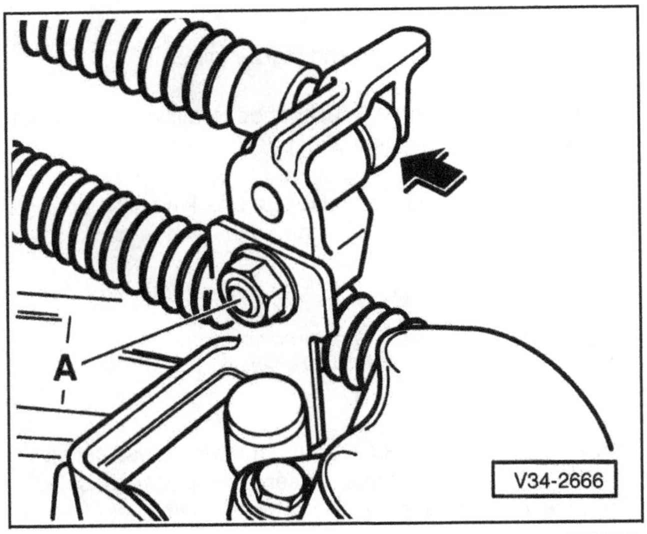

V34-2666

Fig. 11. Actuating arm/gate selector cable mounting nut (**A**). With nut loosened, press lightly on actuating arm (direction of **arrow**) and retighten nut.

TRANSMISSION REMOVAL AND INSTALLATION

This heading describes removal and installation of the manual transmission used on 6-cylinder engines. Special engine lifting and jacking equipment is needed to support and reposition the engine as the transmission is removed from below. Read the procedure through to determine what tools and equipment are needed for the job.

Transmission, removing

1. Disconnect battery ground strap and remove battery.

> ***CAUTION —***
> *On vehicles equipped with an anti-theft radio, make sure you know the correct radio activation code before disconnecting the battery.*

2. Remove power steering fluid reservoir from battery bracket.

3. Remove battery bracket.

4. Disconnect harness connectors from transmission. See Fig. 12.

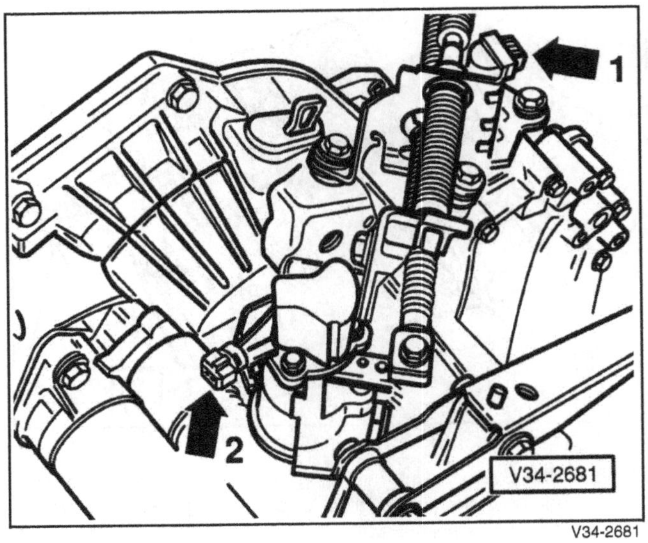

V34-2681

Fig. 12. Vehicle speed sensor (**1**) and backup light switch (**2**) harness connectors to be disconnected from transmission.

5. Disconnect all other electrical connections and ground connections from engine/transmission mounting bolts.

6. Remove mounting bolt and harness connector mounting bracket (if applicable) from right-side engine mount. See Fig. 13.

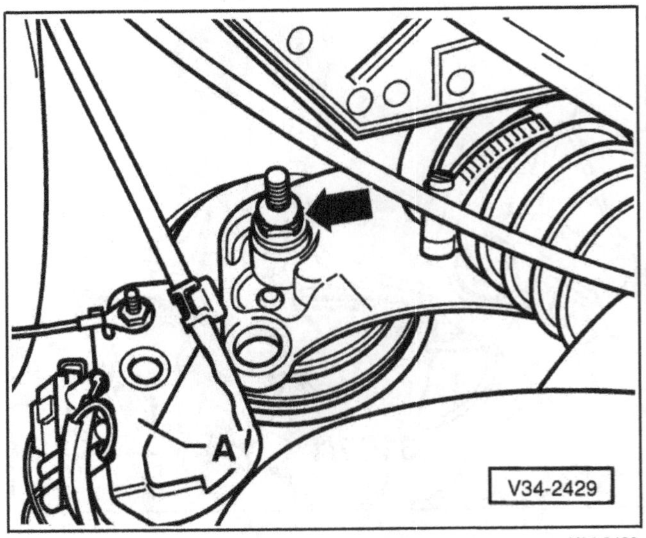

V34-2429

Fig. 13. Right-side engine mount bolt (**arrow**) and harness connector mounting bracket (**A**).

7. Remove balance weight from gear selector lever. Then disconnect gear selector cables from gear shift linkage on transmission. See Fig. 14.

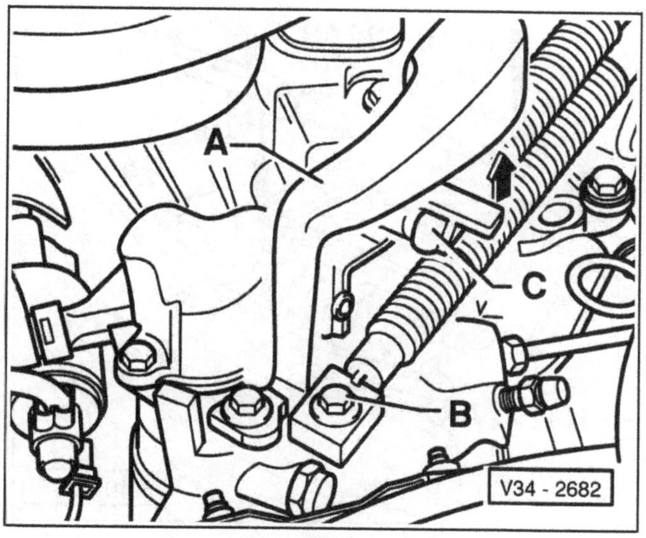

V34-2682

Fig. 14. Gear selector lever balance weight (**A**) on transmission shift lever. Disconnect gear selector cable by removing bolt at (**B**). Disconnect gate selector cable at actuating arm/relay lever (**C**) by pulling tab in direction of **arrow**.

8. Remove cable retaining bracket from transmission. See Fig. 15.

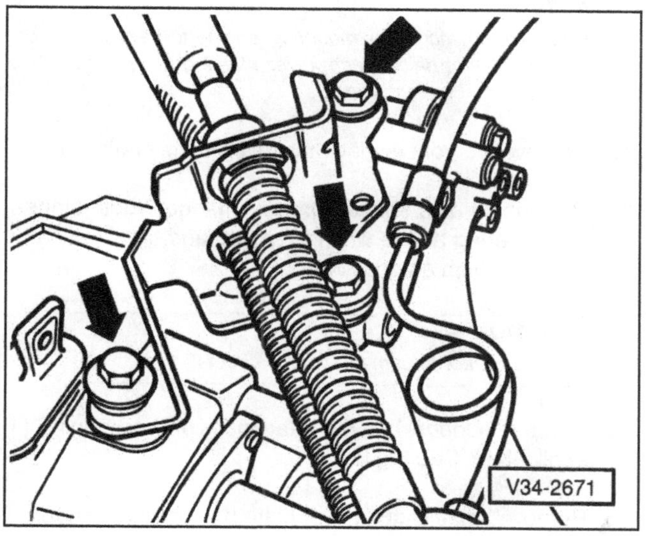

V34-2671

Fig. 15. Cable retaining bracket mounting bolts (**arrows**).

9. Remove clutch slave cylinder from transmission and lay to one side. See Fig. 16.

> ***CAUTION—***
> *Do not open clutch hydraulic system or loosen fluid lines. Secure clutch slave cylinder in place with wire.*

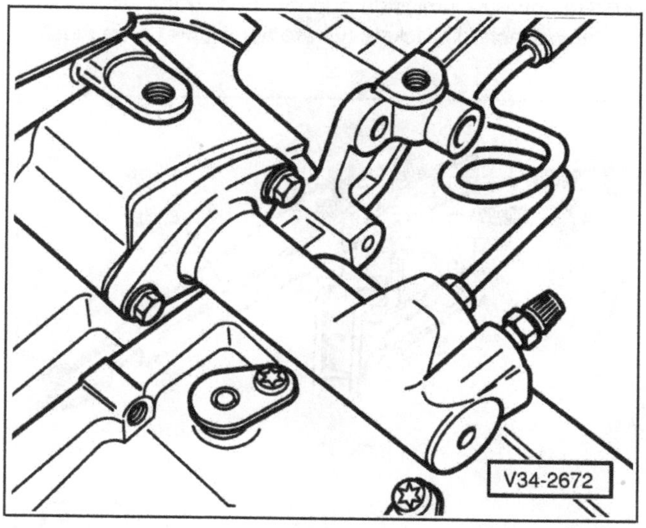

V34-2672

Fig. 16. Clutch slave cylinder on transmission.

10. Disconnect coolant overflow reservoir and lay it aside. See **19 Engine–Cooling System**. Then remove bolt from left transmission mount. See Fig. 17.

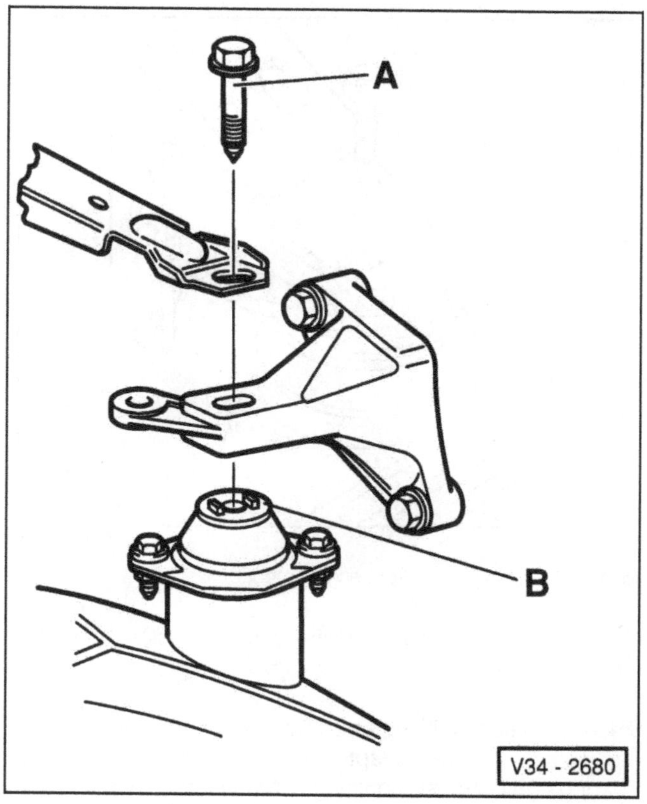

V34-2680

Fig. 17. Mounting bolt (**A**) in left transmission mount (**B**).

11. Remove transmission support from transmission and any attached brackets (where applicable). See Fig. 18.

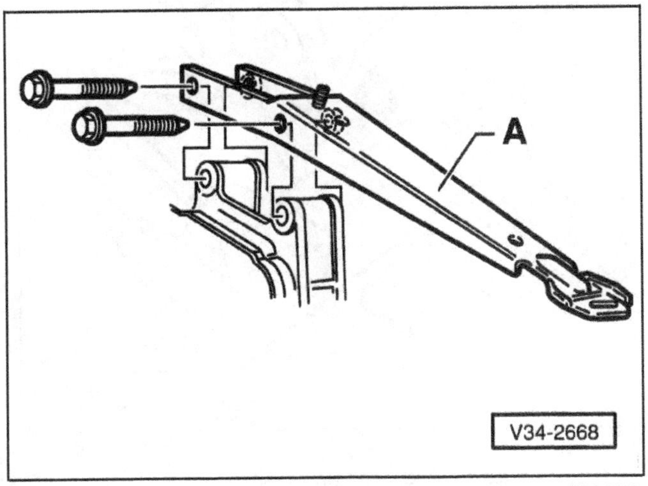

V34-2668

Fig. 18. Transmission support (**A**) used on some models.

12. Remove mounting bolt from front engine mount. See Fig. 19.

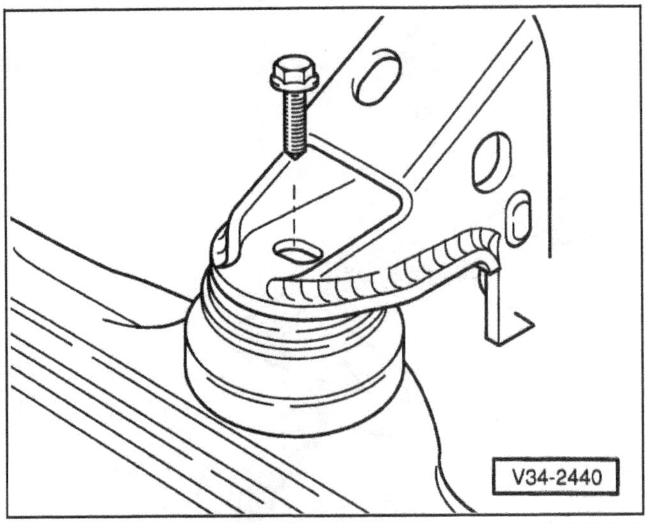

V34-2440

Fig. 19. Mounting bolt for front engine mount.

13. Remove the upper engine-to-transmission mounting bolts.

14. Attach engine lifting equipment (VW sling 10-222A and adjust 10-222A3 adapters, or equivalent) to engine lifting eyes. Adjust the lifting sling until the weight of the engine is fully supported. See Fig. 20.

> ***CAUTION —***
> *Before installing the adapter hooks, disconnect all hoses and wiring in the vicinity of the engine lifting eyes, to prevent damage.*

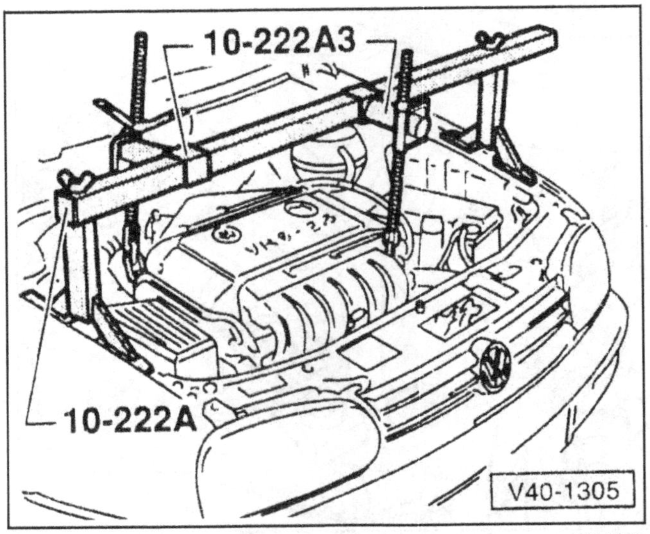

V40-1305

Fig. 20. Engine lifting equipment used to support weight of engine. Numbers shown are for VW special tools.

15. If applicable, remove the insulation shield beneath engine/transmission.

16. Remove the starter. See **27 Engine Electrical**.

> ***NOTE —***
> *Before removing lower mounting engine-to-transmission bolts, first disconnect power steering hose from the bracket at transmission.*

17. Remove the front engine bracket from the engine block.

18. Disconnect axle shafts from axle flanges (both sides), turn steering all the way to the left, and tie axle shafts aside, as high as possible.

> ***CAUTION —***
> *Take care not to damage the axle boots.*

19. Remove support between transmission and engine (if applicable). See Fig. 21.

20. Remove small clutch cover plate behind axle flange.

21. Push engine/transmission assembly as far to the right as possible.

22. Using the engine lifting equipment, raise engine/transmission assembly until exhaust system just contacts exhaust system heat shield.

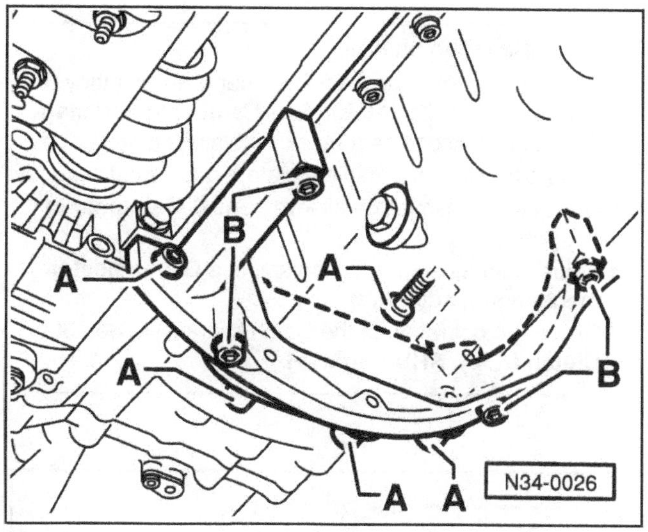

N34-0026

Fig. 21. Engine-to-transmission support mounting bolts (**A** and **B**).

23. Mount scissors jack between engine and body. See Fig. 22. Then press engine/transmission assembly rearward, toward firewall.

> ***CAUTION —***
> *Use care not to damage the power steering lines as the engine/transmission assembly is pressed rearward.*

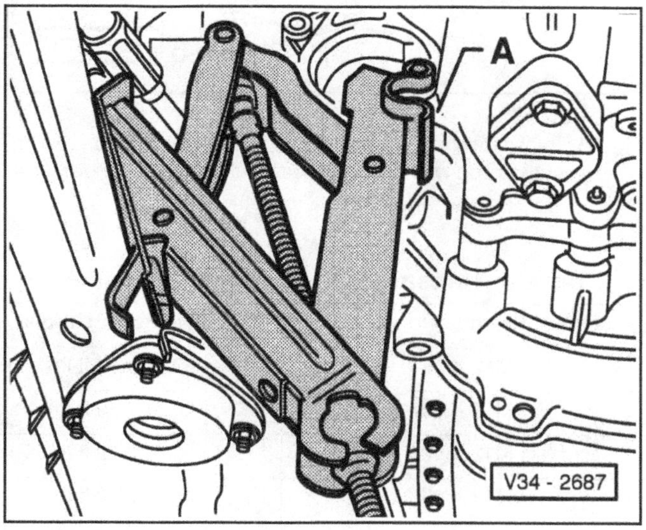

V34-2687

Fig. 22. Scissors jack being used to press engine/transmission assembly rearward. Be sure to mount scissors jack to engine at casting lug (**A**).

24. Remove transmission mount from transmission. See Fig. 23.
25. Lower engine/transmission assembly to its previous position, with the left side being slightly lower than the right side.
26. Remove clutch cover plate.

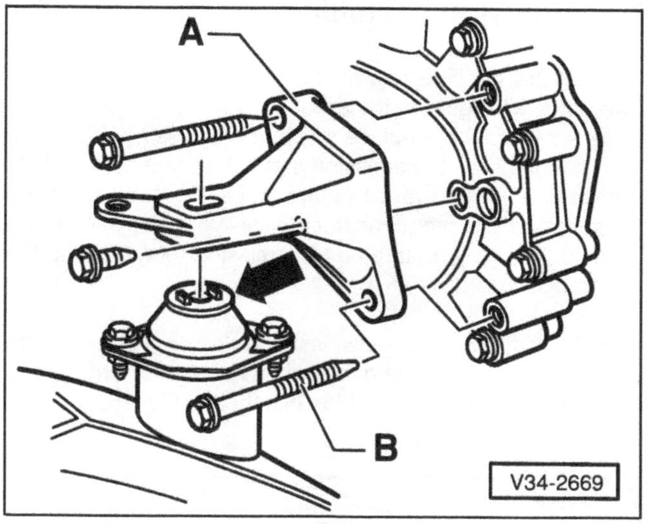

V34-2669

Fig. 23. Mounting bolts (**B**) for transmission mount (**A**).

27. Remove bolt at engine/transmission mount above right axle flange.
28. Place transmission jack under vehicle. See Fig. 24.

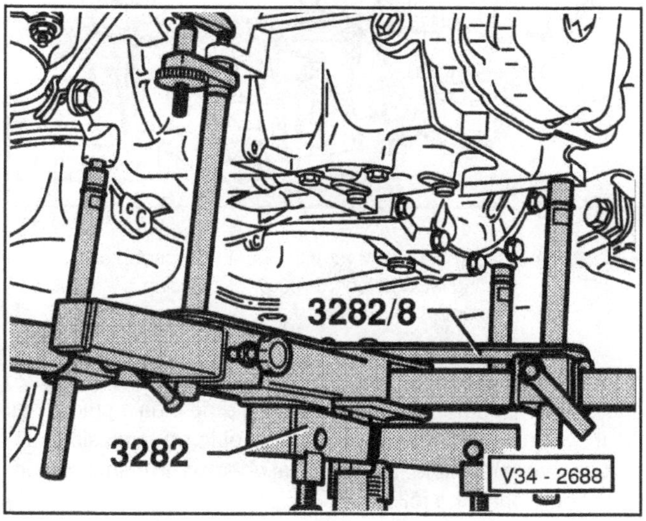

V34-2688

Fig. 24. Transmission jack correctly positioned under transmission. Numbers shown are for VW transmission jack and adapters.

29. Remove bolt for lower engine/transmission attachment.
30. Press transmission off alignment dowel sleeves and lower carefully while pushing forward slightly on engine/transmission assembly.

> ***CAUTION —***
> *While lowering transmission avoid damaging power steering lines.*

Transmission, installing

Install the transmission in the reverse order of removal. Check that all engine-to-transmission dowel pins are installed in the engine block. Replace any that are missing. Clean the hub splines of the transmission input shaft and apply a light coat of G 000 100 grease before installing. If replacing the transmission assembly, transfer the vehicle speed sensor, the back-up light switch, and the transmission relay lever to the new transmission.

Before installing the transmission to the engine, press the clutch release lever toward the transmission case and secure the lever in position using a M8x35 bolt. See Fig. 25.

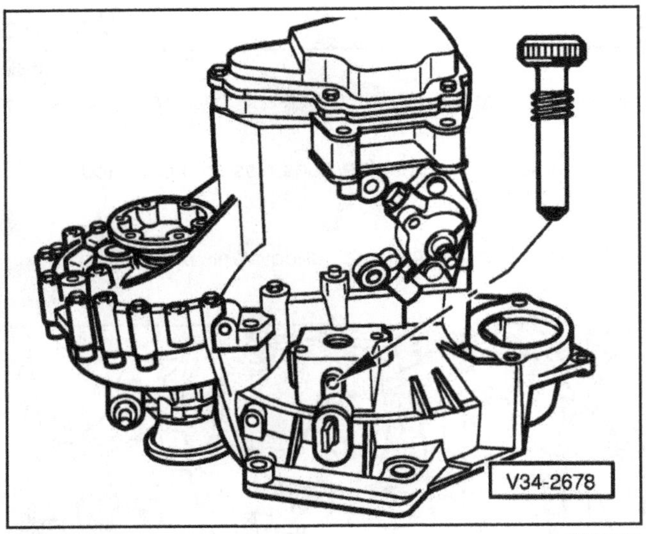

V34-2678

Fig. 25. Press clutch release lever into operating position and secure by inserting M8x35 bolt through bellhousing. Remove bolt once transmission is installed.

Installation tightening torques are given below. Use Fig. 26 and Fig. 27 together with **Table a** and **Table b** as a guide. Fig. 28 shows installation torques for the engine/transmission support, where applicable. Make a final check of transmission installation, noting the following:

Final installation checks

- Install the engine/transmission mounts so that they are free of tension. See **10 Engine–Removing and Installing** for specific engine/transaxle installation procedures.
- Ensure that the intermediate plate is positioned correctly.
- Check the transmission oil level. See **0 Maintenance Program**.
- Adjust gear selector mechanism, see **Gear selector mechanism, adjusting**.
- Check the operation of the clutch hydraulics. See **30b Clutch (AAA, AHU Engines)**.

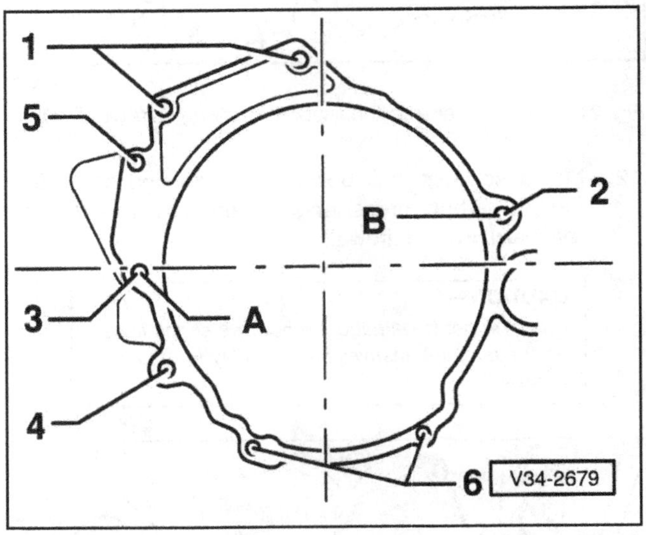

34-2493

Fig. 26. Transmission-to-engine mounting bolt locations to be used in conjunction with **Table a**. Alignment dowel pins shown at **A** and **B**.

Table a. Transmission to engine fasteners

Fastener	Size	Qty.	Tightening torque
1	M12 X 55	2	80 Nm (59 ft-lb)
2	M12 X 62	1	80 Nm (59 ft-lb)
3*	M12 X 62	1	80 Nm (59 ft-lb)
4	M12 X 46	1	80 Nm (59 ft-lb)
5**	M10 X 160	1	60 Nm (44 ft-lb)
6***	M7 X 12	2	10 Nm (89 in lb)

* Also for bracket at front of transmission
** Also for starter and bracket at front of transmission
*** Flywheel shield plate

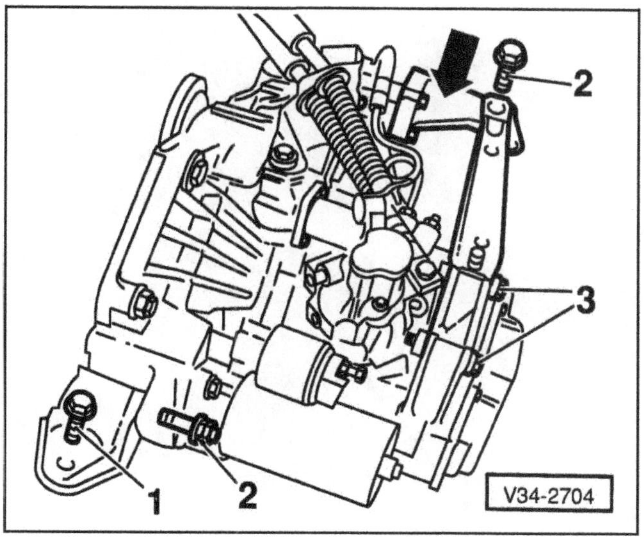

V34-2704

Fig. 27. Transmission assembly mounts to be used in conjunction with **Table b**.

Table b. Transmission/engine assembly mounts

Fastener	Size	Qty.	Tightening torque
1	M10 X 30	1	50 Nm (37 ft-lb)
2	M10 X 45	2	50 Nm (37 ft-lb)
3*	M10 X 60	2	45 Nm (33 ft-lb)
*Not applicable on all models			

When reinstalling the engine to transmission support, tighten the support bolts in the specified sequence shown in Fig. 28.

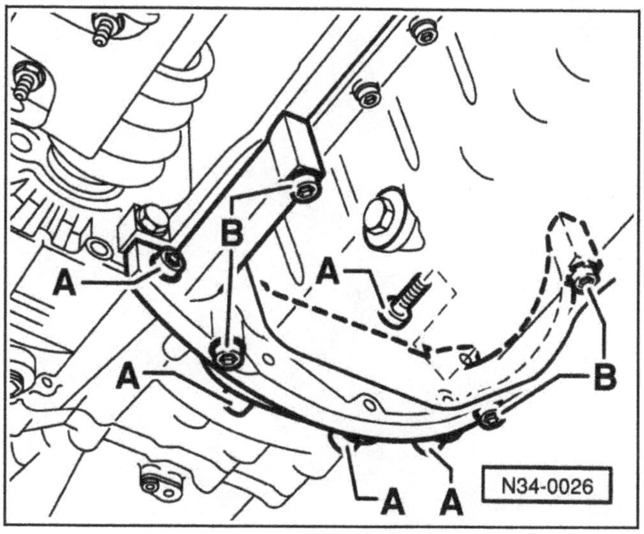

N34-0026

Fig. 28. Tightening sequence for support between engine and transmission (if applicable). First tighten all bolts by hand. Then tighten bolts to torques listed.

Tightening torques

- Lower starter mounting bolt to transmission . . . . 60 Nm (44 ft-lb)
- Left transmission bracket to transmission (**arrow** in Fig. 27) . . . . 25 Nm (18 ft-lb)
- Right engine bracket to subframe. . . . 60 Nm (44 ft-lb)
- Axle shaft to axle flange . . . . 45 Nm (33 ft-lb)
- Support between engine and transmission (**A** and **B** in Fig. 28) . . . . 45 Nm (33 ft-lb)

37 Automatic Transmission

GENERAL

Cars covered by this manual with automatic transmissions are equipped with either the "096" or the "01M" 4-speed automatic transmission. The two transmissions are similar in appearance, with the major differences being mainly internal.

Both the 096 and 01M automatic transmissions are controlled electro-hydraulically and feature adaptive programming and On-Board Diagnostic (OBD) capabilities. For more information on automatic transmission features and application information, see **3 General Information**.

> ***NOTE —***
> *ATF draining and filling procedures, including ATF screen (filter) replacement, is covered in* **0 Maintenance Program**.

Automatic transmission fault diagnosis

Fig. 1 shows the electrical/electronic controls and components used for the automatic transmission. If transmission problems are encountered, review the basic checks given here as well as the information given in the illustration's numbered list.

Basic requirements

1. Check that the ATF level is correct. Check that the fluid is clean and of the correct type. See **Automatic Transmission Fluid (ATF) Service** and **0 Maintenance Program**.

> ***CAUTION —***
> *VW automatic transmission fluids types vary depending on transmission application. The 096 transmission uses Dexron ATF, while the 01M transmission uses a special VW-only ATF. If the fluid was changed, check to make sure that the correct fluid was installed. Dexron is purple/brownish in color. VW ATF is yellowish in color. See* **0 Maintenance Program** *for additional lubricant specifications, including final drive lubricant.*

2. Make a visual check of the components shown in Fig. 1. Check the wiring and harness connectors for loose, damaged or corroded wiring. Check that all related ground points are firmly connected and in good condition. Consult the wiring diagrams shown in **97 Wiring Diagrams, Fuses and Relays**.

3. Check the shift mechanism for proper function. See **Shift Mechanism**.

4. Review the list given below.
   - Engine control module (ECM) replaced
   - New, reconditioned, or used engine installed
   - New or altered/adjusted throttle housing
   - New or altered/adjusted throttle position sensor
   - Transmission control module (TCM) replaced

If any of the conditions are met, the TCM must be reset to its "basic settings" using the VAG. 1551 scan tool. If the TCM basic settings are not re-established, driveability problems may be encountered.

5. If no faults are found up to this point, the next logical step is to have an authorized VW dealer check for faults using the special VW scan tool, VAG 1551. If the transmission problem is electrical/electronic in nature, specific DTCs will most likely be stored in memory.

NOTE—
- *The TCM uses many of the same input signals (engine load, throttle position, etc.) as the Motronic engine control module (ECM). If engine driveability problems are also encountered, check for engine management fault codes first, as a special scan tool is not necessarily required for this check. See* **24a Fuel Injection–Motronic** *for information on displaying DTCs.*
- *Volkswagen identifies electrical components by a letter/number code in the electrical schematics. These component codes are given in the numbered list below as an aid in using the wiring diagrams. See* **97 Wiring Diagrams, Fuses, and Relays**.

Basic settings, restoring

Basic settings can be restored using suitable scan tool and following the manufacturer's instructions or by using the procedure outlined below. In earlier case, the kickdown switch on the accelerator cable must function properly.

1. Switch on the ignition but do not start engine.
2. Push the accelerator pedal all the way to the floor and hold it there for a minimum of 5 seconds.

NOTE—
Be sure the floor mats do not prevent the accelerator pedal from being completely depressed.

3. Release the accelerator pedal and switch off the ignition.

Automatic transmission electronic/electrical components

1. **Transmission control module (TCM) (J217)**
   - Location:
     up to 12/92–in front of right side passenger seat under cover in footwell
     from 01/93–under the rear seat
2. **Engine control module (ECM) (J220)**
   - Location: in the heating/ventilation intake plenum chamber, right side
3. **Data link connector (DLC)**
   - Location:
     up to 7/93–below heating/ventilation controls, next to rear window defogger switch
     from 8/93–behind cover next to ashtray
4. **Valve body with solenoids**
5. **Conductor strip with integrated transmission fluid temperature sensor**
   - On some models conductor strip is integral with valve body
6. **Multifunction transmission range (TR) switch (F125)**
   - Location: on rear of transmission
7. **Vehicle speed sensor (VSS) (G68)**
   - Location: on top of transmission
8. **Transmission vehicle speed sensor (VSS) (G38)**
   - Location: on top of transmission (white connector)
   - Not used on all vehicles

CAUTION—
Do not interchange connectors between sensor G38 and sensor G68. The G38 sensor has a white connector and is not used on all vehicles.

9. **Throttle position sensor (TP) (G69)**
   - Location: on throttle valve housing
   - OBD II vehicles: built into throttle valve control module
10. **Shift lock solenoid (N110)**
    - Location: on selector lever
11. **Cruise control switch (E45)**
    - Location: on the steering column switch
    - Cruise control system is supplied power via TCM
12. **Kick down switch (F8)**
    - Integrated into accelerator cable
    - Location: on bulkhead in engine compartment
13. **Brake light switch (F)**
    - Location: on pedal cluster

Automatic transmission electronic/electrical components

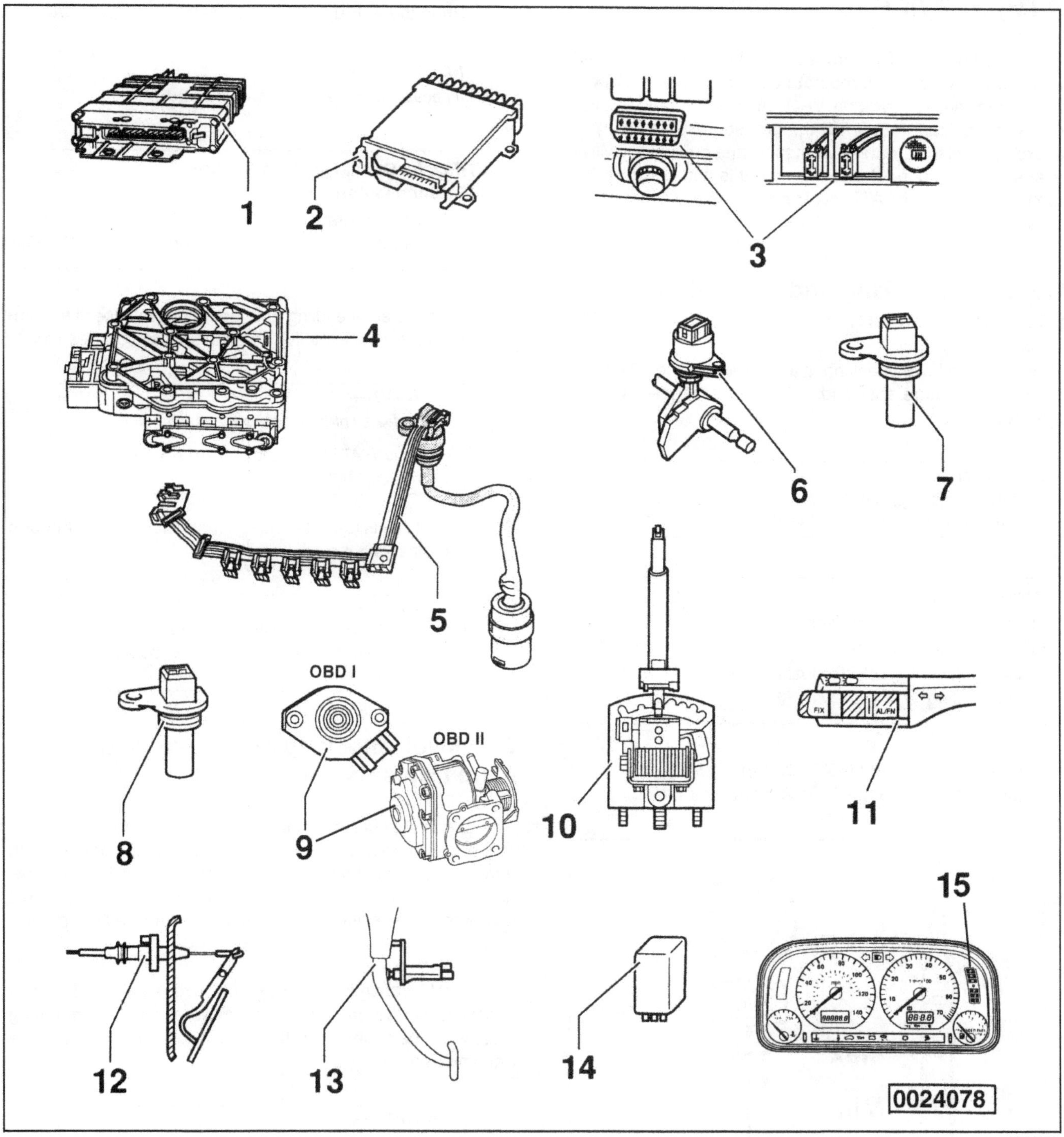

0024078

Fig. 1. Electrical/electronic components for automatic transmission.

14. **Park/neutral position (PNP) relay (J226)**
    - Location: on additional relay panel, under instrument panel left
    - Relay is marked with number "150"
15. **Gear selector display**
    - Location: integral with instrument cluster

AUTOMATIC TRANSMISSION FLUID (ATF) SERVICE

The fluid level checking procedures given under this heading require that the ATF temperature be measured and monitored using the Volkswagen VAG 1551 scan tool (or other equivalent means) throughout the procedure. Read the procedure through to determine what tools and equipment will be necessary. If the correct test equipment is not available, it is recommended that the ATF level be checked by an authorized Volkswagen dealer.

ATF level, checking and correcting (096 transmission)

For accurate level checking, it is critical that the ATF be at the test temperature specified. ATF level changes with temperature.

> *CAUTION —*
> *When checking the ATF level, the transmission must not be in the emergency (limp home) mode.*

Requirements
- Car parked on level surface
- Parking brake applied
- Selector lever in park "P" position
- Engine idling

1. If the engine is cold (68°F (20°C)), check to make sure the fluid level is at the "20°C" mark. See Fig. 2.

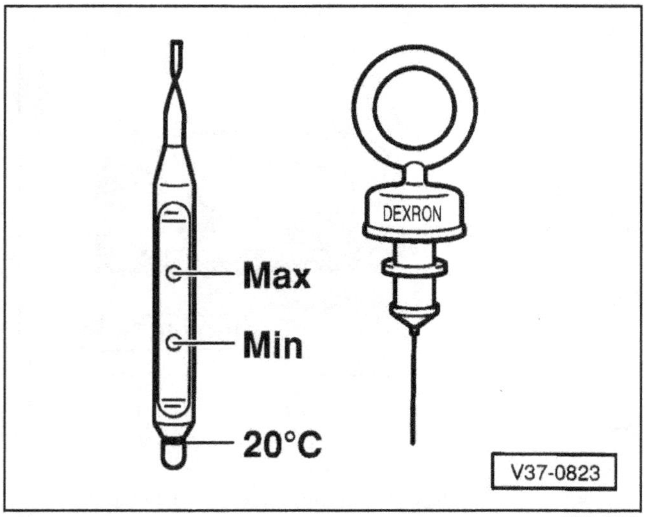

V37- 0823

Fig. 2. ATF dipstick. Use **20°C** mark when checking ATF with transmission cold. Level must be between **Min** and **Max** marks with ATF at 140° F (60° C).

2. Connect VAG 1551 scan tool to the data link connector(s) in the dash and measure the ATF temperature. Bring the ATF up to normal operating temperature.

> *NOTE —*
> *If necessary, drive the car to warm the ATF.*

ATF Level Checking Specification
096 transmission
- ATF normal operating temperature. . . . . approx. 140°F (60°C)

3. With the engine idling and the ATF at the specified temperature, check the fluid level on the dipstick. See Fig. 2.

> *CAUTION —*
> - *Use clean, lint-free rags to wipe off the dipstick.*
> - *Too much ATF will cause transmission malfunctions and must therefore be extracted.*

4. If the level is low, use a clean funnel to add fluid through the dipstick tube.

> *CAUTION —*
> *Only ATF with the designation Dexron may be used in the 096 transmission. Do not use any lubricant additives.*

ATF level, checking and correcting (01M transmission)

For accurate fluid level checking, it is critical that the ATF be at the test temperature specified as ATF level changes with temperature. Also note that the test temperature is well below normal operating temperature, so special Volkswagen test equipment is specified to accurately measure ATF temperature.

The 01M transmission does not have a dipstick for level checking, so the procedure is somewhat entailed, requiring that the car be raised on a lift to access to the underside of the transmission.

> *CAUTION —*
> - *When checking the ATF level, the transmission must not be in the emergency (limp home) mode.*
> - *The 01M transmission uses a specially formulated synthetic ATF. This fluid is available from an authorized VW dealer in one liter and 1/2 liter containers. Use only the special VW-only ATF in the 01M transmission.*

1. Connect VAG 1551 scan tool to the data link connector in the dash and measure the ATF temperature. The ATF temperature must be below 86°F (30°C).

> ***WARNING —***
> *The ATF must initially be below 86°F (30°C). If the ATF is above this temperature, wait until the fluid has cooled.*

2. Place vehicle securely on lift. Apply parking brake and place selector lever in park (P). Raise car evenly so that it is level.

3. Remove ATF level plug from oil pan. See Fig. 3. A small amount of ATF will run out of the overflow tube as the plug is removed. Place a rag under the plug when removing.

> ***WARNING —***
> *DO NOT remove the ATF level plug if the ATF temperature is above 86°F (30°C). At temperatures above 86°F, the ATF may have expanded above the filler tube, resulting in hot ATF spilling from the level plug.*

NOTE —
Anytime the ATF level plug is removed, the sealing ring on the plug must be replaced. Cut the old ring off and install a new ring.

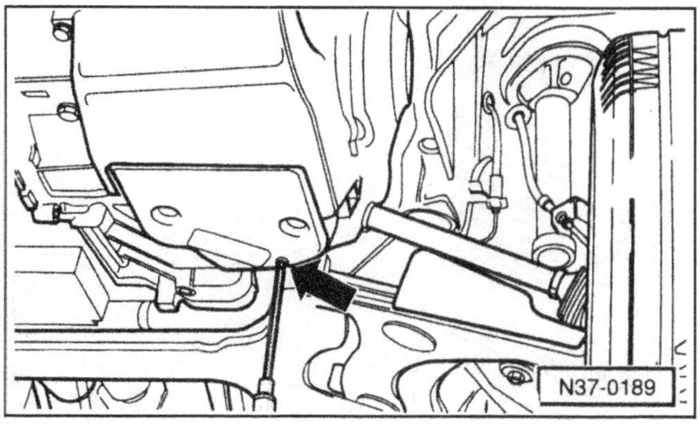

N37-0189

Fig. 3. ATF level plug (**arrow**) being removed from transmission oil pan.

4. Start engine and allow the ATF temperature to reach the specified test temperature.
   - ATF level is correct if ATF drips out of hole with ATF at the correct test temperature.
   - ATF level is low if ATF does not drip out of hole with ATF at the correct test temperature.

ATF Level Checking Specification
01M transmission
- Test temperature . . . . 95–135°F (35–57°C)

5. If ATF needs to be added, pry off filler plug securing cap with a screwdriver. See Fig. 4. Pull filler plug out of filler pipe and slowly add VW ATF through the filler pipe until ATF runs out of the level plug hole.

> ***CAUTION —***
> *If the ATF level is too high or too low it will adversely affect the operation of the transmission.*

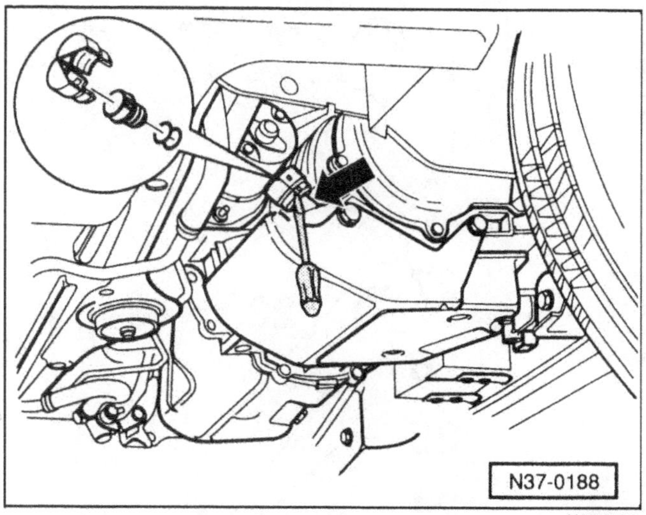

N37-0188

Fig. 4. ATF filler plug is secured with locking cap. Locking cap will be damaged during removal—always replace cap.

NOTE —
Some 01M transmissions may be equipped with a reusable combination cap/lock plug.

6. Turn the engine off and install the level plug with a new sealing ring.

7. If removed, install the filler plug/cap in filler pipe until spring retainer lock locates or secure plug with new cap.

> ***CAUTION —***
> *Always replace the locking cap if it is of the separate cap/plug type. The cap locks the plug in place.*

8. Lower the car to the ground and remove test equipment.

Tightening torques
- ATF level plug to transmission . . . . 15 Nm (11 ft-lb)

SHIFT MECHANISM

Fig. 5 shows an exploded view of the shift mechanism for the 096 and 01M transmissions. The illustrations immediately following the exploded views give additional and more detailed shift mechanism information.

Shift mechanism, assembly

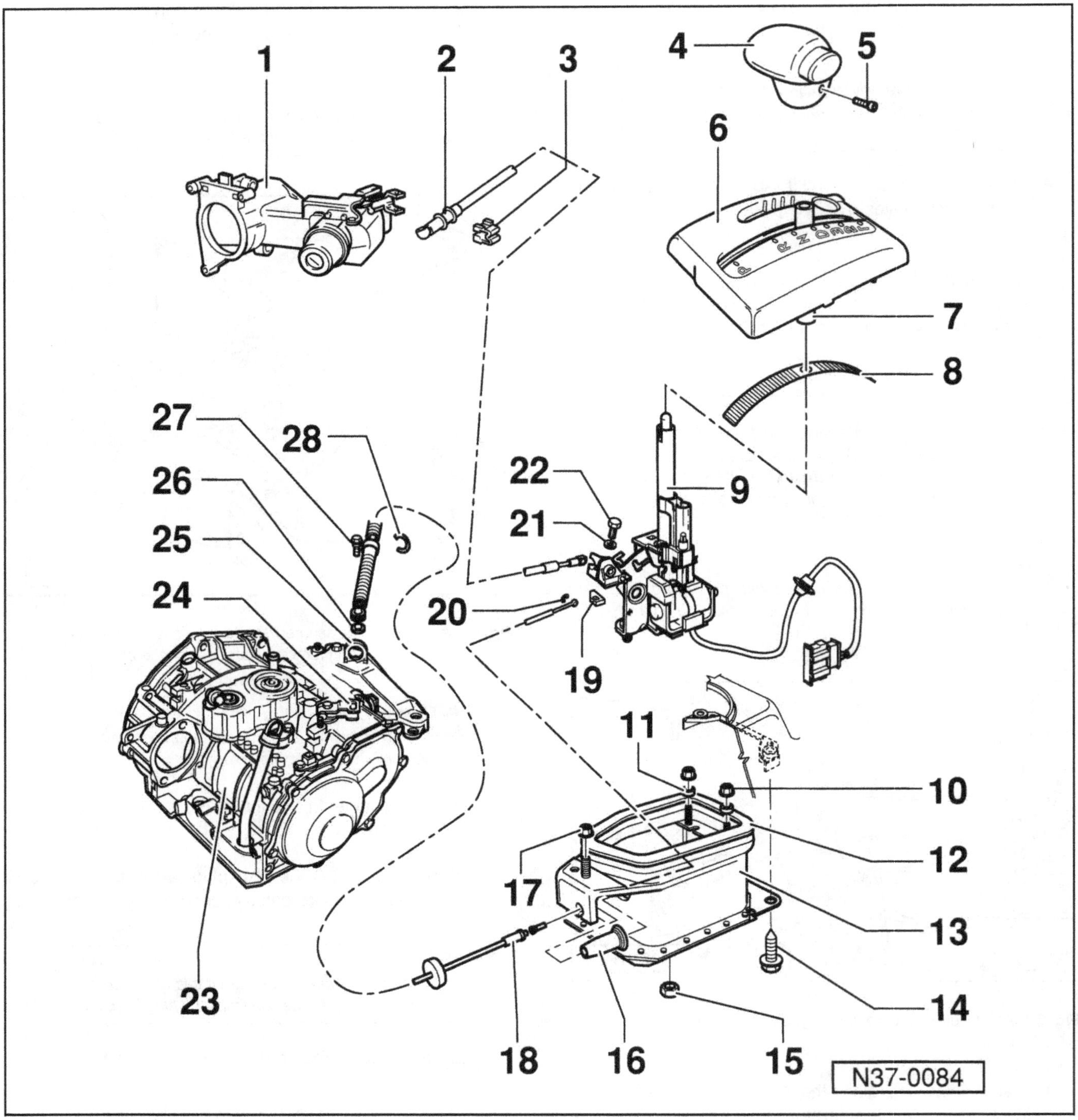

N37-0084

Fig. 5. Exploded view of shift mechanism for automatic transmission.

1. **Ignition/Starter Switch (D)**

> *CAUTION—*
> *Always confirm that the transmission is in the park (P) position and the parking brake is applied before working on the car with the engine running.*

2. **Shift-lock cable**
   - Must not be kinked
   - See **Automatic Shift-Lock**
3. **Clip**
   - For securing shift-lock cable
   - Always replace

4. **Selector lever handle**

5. **Set screw**
   - Tightening torque: 1.5 Nm (13 in-lb)

6. **Cover**
   - With selector indicator

7. **Protective sleeve for selector lever**

8. **Cover strip**
   - Insert in cover (6)

9. **Selector lever**
   - To remove selector lever, lower protective plate is removed and front protective plate removed and pushed forward
   - When installing selector lever, insert shift-lock cable anchor in notch of selector lever housing
   - Securing wiring harness to selector lever housing (shift-lock solenoid with integral wiring harness), **see** (A)
   - Securing wiring harness for shift-lock solenoid and indicator lighting (shift-lock solenoid with separate wiring harness of solenoid), **see** (B)

10. **Nuts**
    - To secure center console

11. **Spacer**

12. **Gasket**
    - Always replace

13. **Selector lever housing**
    - Need not be removed to replace selector lever cable or individual parts of shift mechanism
    - Removing, **see** (C)
    - Installing, **see** (D)

14. **Screw**
    - Tightening torque: 25 Nm (18 ft-lb)

15. **Nut**
    - Always replace
    - Tightening torque: 15 Nm (11 ft-lb)

16. **Protective sleeve**
    - Insert into selector lever housing before attaching selector lever cable

17. **Hex nut**
    - With shoulder
    - Tightening torque: 25 Nm (18 ft-lb)
    - Installing, **see** (D)

18. **Selector lever cable**
    - To remove selector lever, lower protective plate is removed, exhaust system is separated in front of catalytic converter and front protective plate removed (on vehicles with 6-cylinder engine, remove catalytic converter.
    - Do not bend or kink
    - Lightly grease cable eye and cable ends before installing
    - Selector lever cable to cable anchor tightening torque: 20 Nm (15 ft-lb)
    - Check guide tube for tight fit
    - Check protective sleeve between console and lever/gearshift shaft for proper seating. Secure selector lever cable with corrugated locking element (28)
    - Attaching, **see** (E)
    - Adjusting, see **Selector lever cable, adjusting**

19. **Seal**
    - Bonded to cable anchor
    - Replace if damaged

20. **Circlip**
    - Install after installing selector lever cable

21. **Washer**

22. **Bolt**
    - Loosen to adjust shift-lock cable
    - Tightening torque: 6 Nm (53 in-lb)

23. **Transmission**

24. **Lever**
    - For selector shaft

25. **Bracket**
    - With cable anchorage for selector lever cable

26. **Washer**

27. **Combi-bolt**
    - Tightening torque: 25 Nm (18 ft-lb)

28. **Corrugated locking element**
    - For selector lever cable
    - Always replace

A Mounting wiring for shift-lock solenoid (with integral wiring harness)

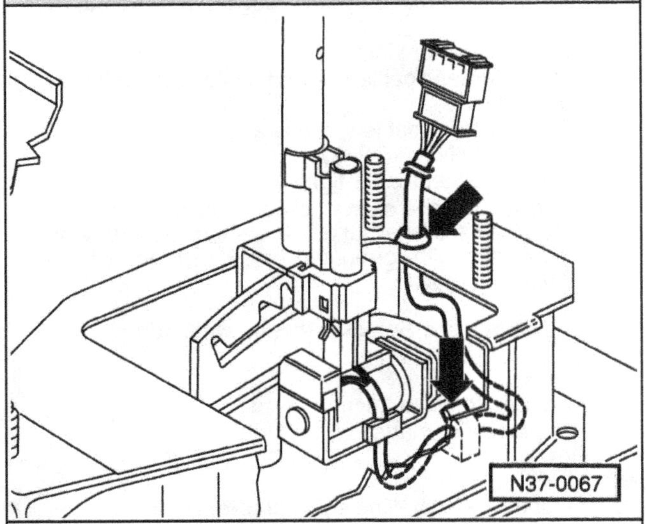

N37-0067

- Shift-lock solenoid with integral wiring harness.
- Press wiring harness into clips and into selector lever housing (arrows).

B Mounting wiring for shift-lock solenoid (with separate wiring harness)

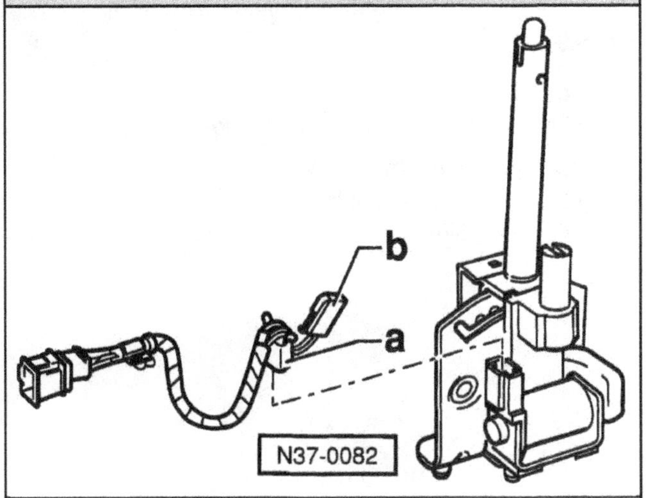

N37-0082

- Shift-lock solenoid with separate wiring harness (a) and indicator lighting (b).

C Removing selector lever housing

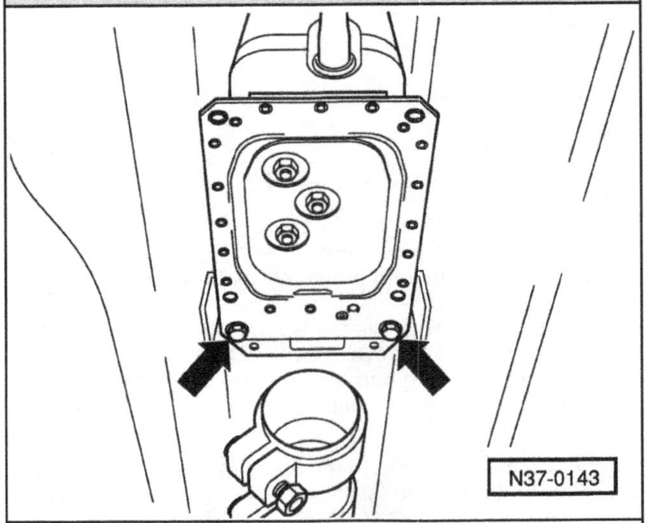

N37-0143

- Remove rear center console.
- Remove hex nut with shoulder.
- Remove lower protective plate.
- Remove catalytic converter.
- Remove front protective plate and rear lower bolts (arrows).

D Installing selector lever housing

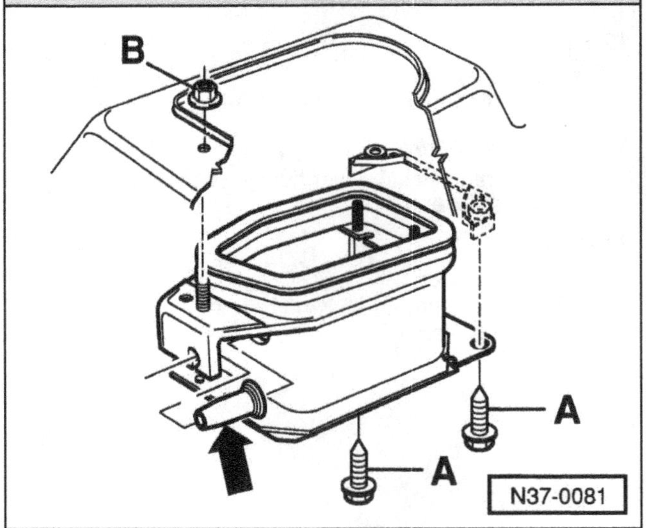

N37-0081

- Before attaching selector lever cable, insert protective sleeve (arrow).
- When securing the selector lever housing with 4 bolts (A), the nut (B) with collar is not used.

(E) Attaching selector lever cable to selector lever

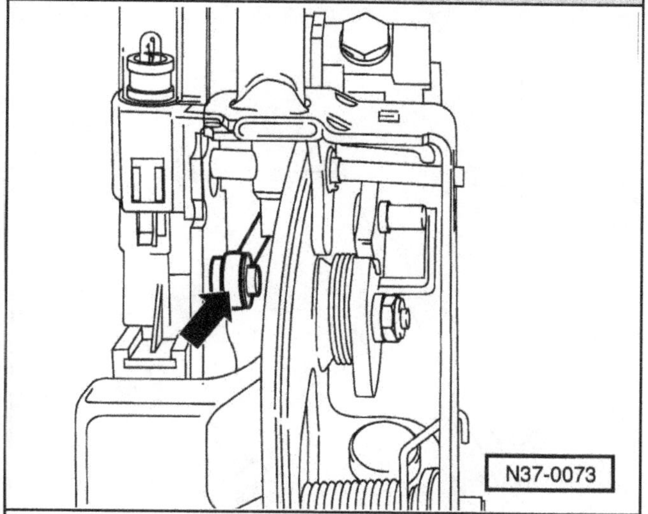

N37-0073

- **Fit selector lever cable eye (arrow) onto selector lever and secure with circlip**

Shift mechanism, functional check

The following procedures apply to both 096 and 01M transmissions.

1. Position selector lever in park (P). Switch ignition on.
   - Without depressing the brake pedal, check that the selector lever is locked and cannot be shifted out of park (P). Shift-lock solenoid blocks selector lever.
   - While depressing the brake pedal, check that the selector lever shift-lock solenoid releases, allowing movement of the selector lever into driving gears.
2. Place selector lever in the "N" position.
   - Without depressing the brake pedal, check that the selector lever is locked and cannot be shifted out of neutral (N). Shift-lock solenoid blocks selector lever.
   - While depressing the brake pedal, check that the selector lever shift-lock solenoid releases, allowing movement of the selector lever into driving gears.
3. Check that the starter does not operate with the selector lever in positions "1", "2", "3", "D" and "R."

NOTE —

- *When travelling at speeds above 5 km/h (3 mph) and shifting into selector lever position "N" the shift-lock solenoid must not engage and block the selector lever. The selector lever can be shifted into a driving gear.*
- *When travelling at speeds below 5 km/h (3 mph) (almost stopped) and shifting into selector lever position "N", the shift-lock solenoid must not engage until after approx. 1 second. The selector lever cannot be shifted out of the "N" position until the brake pedal is depressed.*

Selector lever assembly

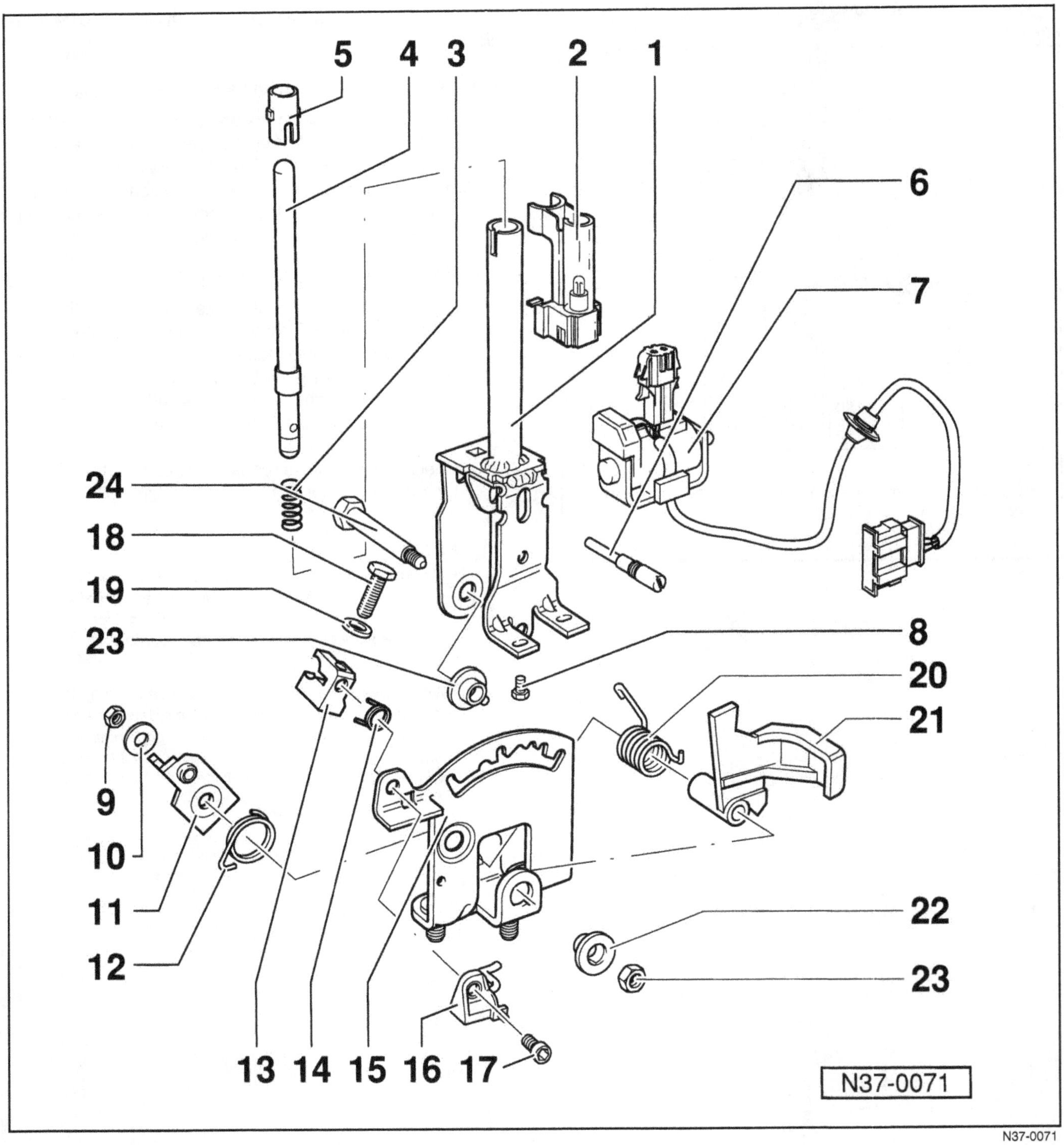

N37-0071

Fig. 6. Exploded view of gear selector lever and related components.

1. **Selector lever**
   - Assembling, **see** Ⓐ

2. **Indicator lighting**

3. **Spring**

4. **Push rod**
   - Grease in area of sliding surfaces

5. **Guide sleeve**
   - Fit after installing push rod
6. **Set screw**
   - Always replace
   - To remove: first remove shift-lock solenoid and indicator lighting (item 2). Clamp set screw in vice and loosen by turning selector lever
   - Install with thread-locking fluid
7. **Shift-lock Solenoid (N110)**
   - Can have integral or two-part wiring harness
   - Adjusting, **see** (B) and (C)
   - Securing wiring harness to selector lever housing (shift-lock solenoid with single wiring harness), see **Shift mechanism assembly**
   - Securing indicator lighting connector, **see** (D)
8. **Screw**
   - Tightening torque: 3 Nm (27 in-lb)
   - Qty. 2
   - Install with thread-locking compound
9. **Nut**
   - Tightening torque: 5 Nm (44 in-lb)
   - Always replace
10. **Washer**
11. **Shift-lock locking lever**
    - For ignition lock
    - Adjust together with shift-lock cable, see **Shift-lock cable, adjusting**
12. **Spring**
    - Hook into hole of detent plate
13. **Retainer**
    - For shift-lock cable
    - See **Shift-lock cable, removing and installing**
    - See **Shift-lock cable, adjusting**
14. **Spring**
    - For locking pin
15. **Detent plate**
    - Assemble to selector lever before fitting in selector lever housing
16. **Locking pin**
    - With bracket
17. **Socket-head screw**
    - Tightening torque: 6 Nm (53 in-lb)
18. **Bolt**
    - Tightening torque: 6 Nm (53 in-lb)
19. **Washer**
20. **Spring**
    - Insert in selector lever together with lever (item 21)
21. **Locking lever**
    - For locking selector lever in position "N" or "P"
    - Check adjustment with shift-lock solenoid when installing locking lever
22. **Bushing**
23. **Nut**
    - Tightening torque: 10 Nm (7 ft-lb)
    - Always replace
24. **Combi-bolt**
    - Grease shank when installing

(A) Assembling selector lever prior to installation

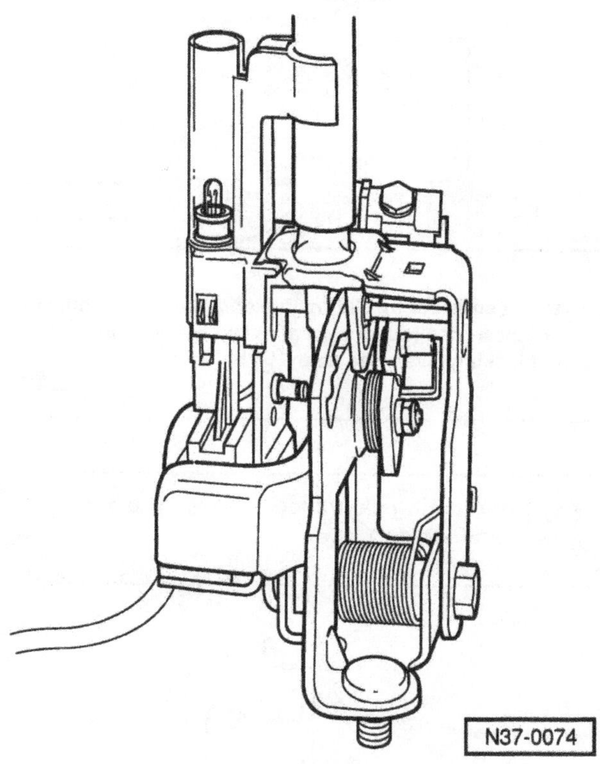

N37-0074

- **Adjust shift-lock solenoid and carry out functional check before installing selector lever in selector lever housing. See (B) and (C).**
- **Install set screw and shift-lock solenoid mounting screws with thread-locking fluid.**

B Adjusting shift-lock solenoid

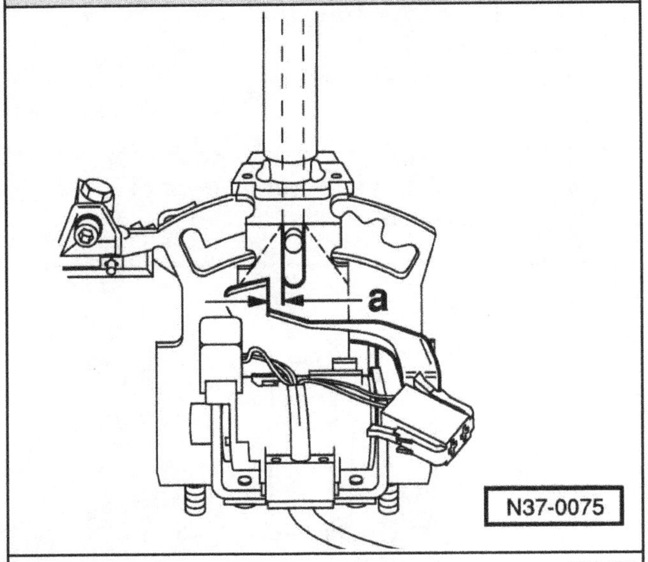

N37-0075

- **Adjust solenoid by sliding in elongated holes until correct clearance (dimension a) exists between push rod and locking lever. See also C below.**

C Adjusting clearance between push rod and locking lever

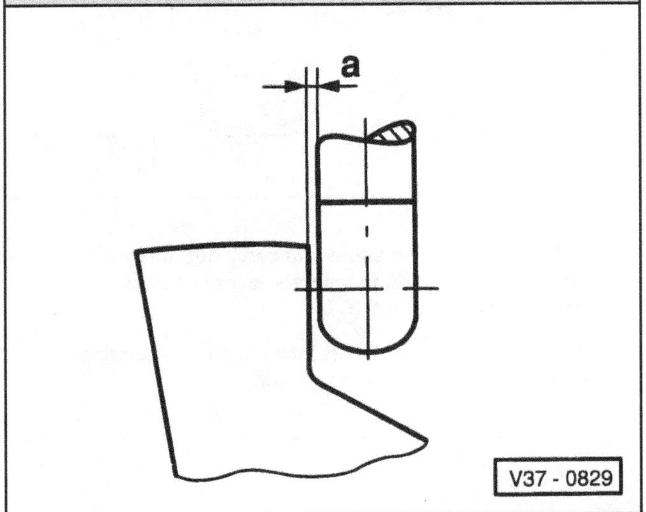

V37-0829

- **Clearance of dimension a = 0.3 mm (0.019 in.)**
- **Check for proper operation and readjust if necessary.**

D Mounting indicator lighting connector (shift-lock solenoid with integral wiring harness)

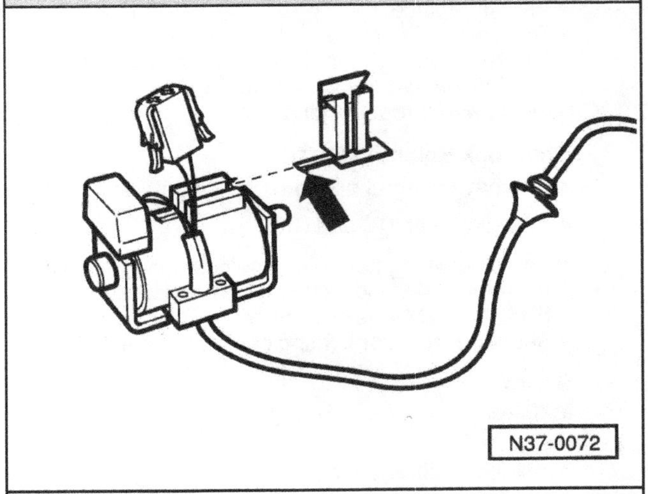

N37-0072

- **Push bracket on until lug (arrow) engages.**
- **Position connector cable in bracket and slide connector onto bracket.**

Selector lever, functional check

1. Shift selector lever into position "P."
2. Supply shift-lock solenoid with 12 Volts. It must not be possible to shift selector lever into a driving gear.
3. Disconnect power from solenoid. It must now be possible to shift into driving gears. Push rod must not catch on lever.
4. Shift selector lever into "N" and repeat steps 2 and 3.

NOTE —
Before installing assembled selector lever, check and adjust clearance between push rod and selector locking lever, as described earlier.

Selector lever cable, checking and adjusting

1. Switch ignition off and set parking brake.
2. Open hood and locate selector lever cable and selector shaft lever on transmission. Use Fig. 5 (given earlier) as a guide.

3. Remove combi-bolt from selector shaft lever on transmission and free cable from lever.

4. Move selector lever from "P" to "1" and check that shift mechanism and selector lever cable move freely. If necessary replace selector lever cable or service shift mechanism.

NOTE—

Always replace locking washer at transmission lever after disassembling.

5. To adjust cable, place selector lever in "P" position and loosely install cable to selector lever using combi-bolt.

6. Place selector shaft lever into position "P" on transmission. Then tighten selector shaft lever combi-bolt. Shift through all gear positions and check for smooth operation.

NOTE—

With selector lever in "P" position, the transmission locking pawl will engage and front wheels will lock.

Tightening torques

• Combi-bolt . . . . 25 Nm (18 ft-lb)

AUTOMATIC SHIFT-LOCK

Automatic shift-lock is used on cars with automatic transmission. Turning the ignition key on and off operates the shift-lock bowden-type cable to lock and unlock the selector lever. See Fig. 7.

When the key is turned to the off position (shift lever in park), the cable pushes the stop lever into the shift lever. This locks the lever button and also allows the ignition key to be withdrawn from the lock cylinder.

When the key is in the ON position, the cable pulls the stop lever away from the shift lever, allowing the shift lever to be moved out of park.

Shift-lock, functional check

1. Shift selector lever to "P" position and turn ignition key OFF.

2. Remove ignition key. Check that selector lever cannot be shifted out of "P" position.

NOTE—

- *It must only be possible to remove ignition key with selector lever in "P" position.*
- *Shift-lock cable must not be kinked.*

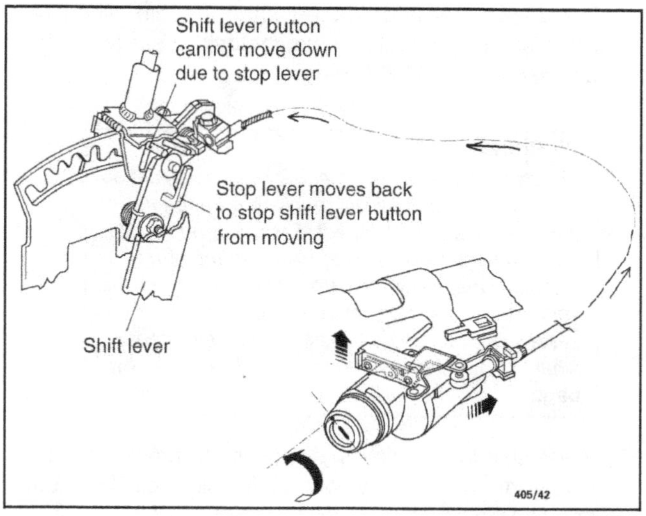

405/42

Fig. 7. Automatic shift-lock shown in the "key off" position.

Shift-lock cable, removing

1. Remove handle from selector lever. See Fig. 5, given earlier.

2. Remove selector lever cover with cover strip.

3. On vehicles equipped with transmission range program switch, disconnect connector from switch.

4. Remove center console from car. See **70 Trim–Interior**.

5. Loosen cable anchor screw. Then separate cable from shift-lock lever and remove cable from its anchor. See Fig. 8.

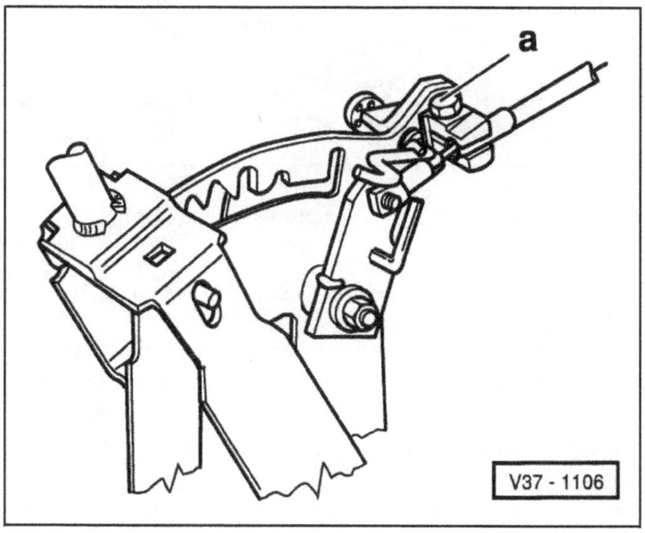

V37-1106

Fig. 8. Loosen cable anchor screw (**a**), remove shift-lock cable from locking lever and pull cable from anchor.

6. Disconnect the negative (–) battery cable. Then remove steering wheel and trim for ignition/starter switch as described in **48 Steering**.

> *WARNING —*
> *The cars covered by this manual may use an airbag system that automatically deploys an airbag(s) in the event of a frontal impact. The airbag(s) is inflated by an explosive device. Handled improperly or without adequate safeguards, the system can be very dangerous. Special precautions must be observed prior to any work at or near the steering wheel or steering column. See* **69 Seatbelts, Airbags**.

7. Press shift-lock cable clip off of support bracket on ignition/starter switch. Remove shift-lock cable from ball head on ignition/starter switch. See Fig. 12, given later.

8. Remove shift-lock cable from car, noting its routing.

Shift-lock cable, installing

1. Install shift-lock cable in vehicle. Be sure the cable is not kinked or stressed. See Fig. 9.

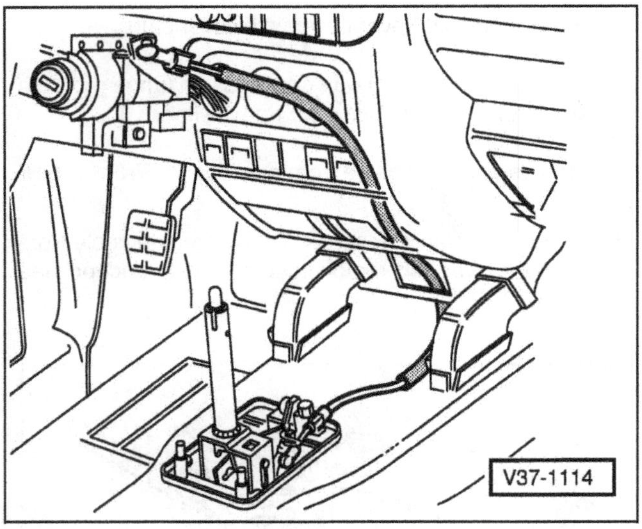

V37-1114

Fig. 9. Locking cable correctly routed through dashboard.

2. Attach shift-lock cable to locking lever. See Fig. 10.

3. Install shift-lock cable in cable anchorage. Hand tighten the clamping screw just enough to hold the cable in position. See Fig. 11.

4. Install cable retaining clip on cable. See Fig. 12.

5. Press cable end onto ball head on ignition/starter switch. Then press cable retaining clip fully into its support bracket.

6. Route shift-lock cable along steering column. See Fig. 13.

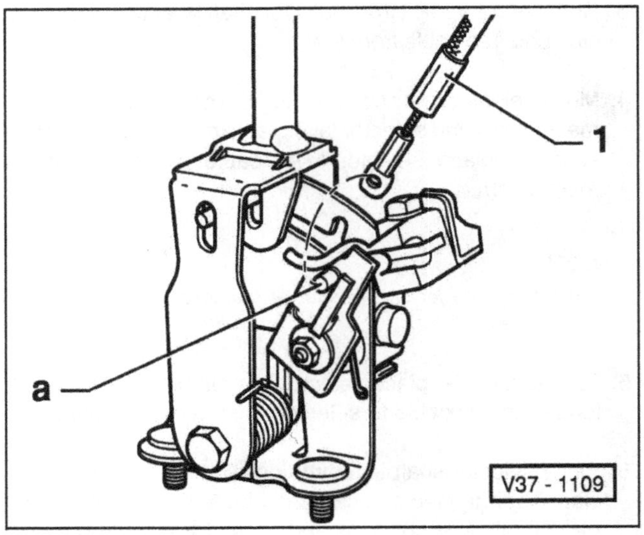

V37-1109

Fig. 10. Attach shift-lock cable (**1**) on locking lever (**a**).

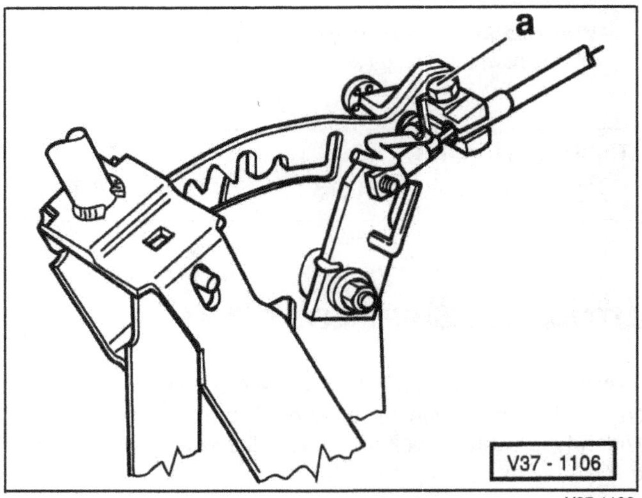

V37-1106

Fig. 11. Locking cable installed at selector lever. Do not tighten clamping screw (**a**).

7. Adjust shift-lock cable as described below. See **Shift-lock cable, adjusting**.

8. Install center console. See **70 Trim–Interior**.

9. Install selector lever cover with indicator and selector lever handle.

10. Attach trim for ignition/starter switch, and install steering wheel.

11. Reinstall airbag unit as described in **69 Seatbelts, Airbags**.

> *WARNING —*
> *Special installation procedures must be followed when reinstalling airbag unit. Be sure to follow the instructions given in* **69 Seatbelts, Airbags**.

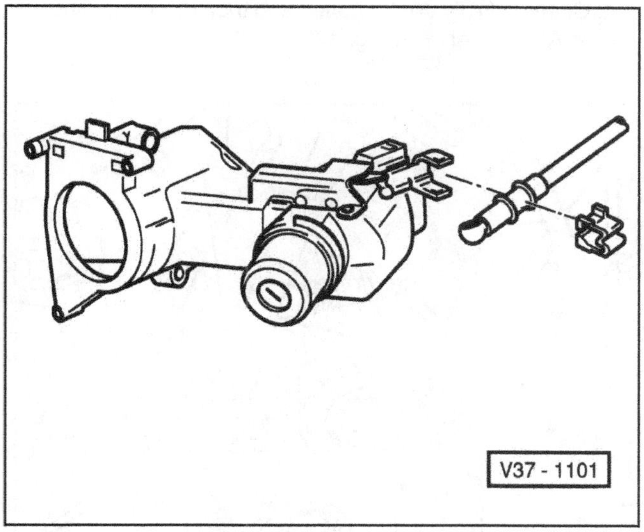

V37-1101

Fig. 12. Shift-lock cable retaining clip should be installed on cable before fitting cable to ignition/starter switch.

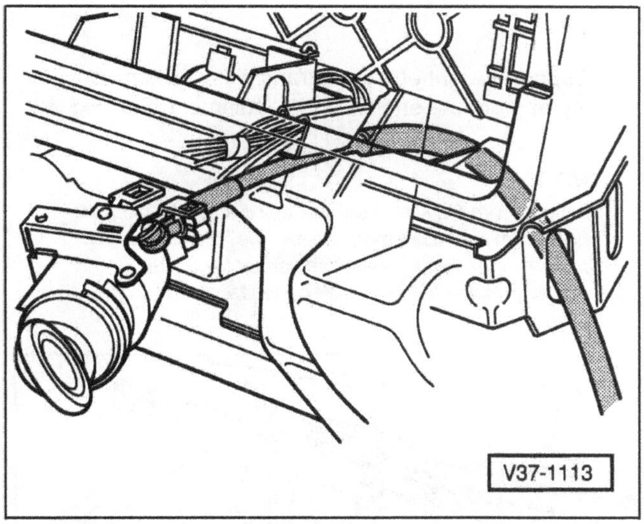

V37-1113

Fig. 13. Shift-lock cable properly routed along steering column.

Shift-lock cable, adjusting

1. Remove handle from selector lever. Remove selector lever cover with cover strip. See Fig. 5, given earlier.

2. Shift selector lever into position "1." Loosen cable anchor screw.

3. Turn ignition key to START position, and then release.

4. Adjust clearance between locking lever and locking pin by repositioning cable at anchor point. See Fig. 14.

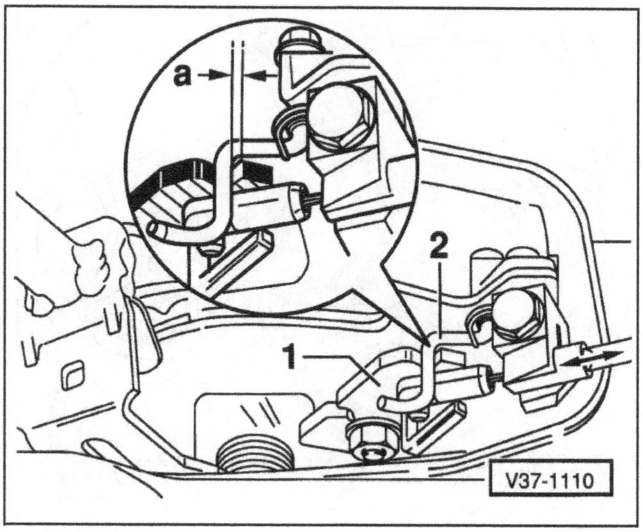

V37-1110

Fig. 14. Clearance (**a**) = 0.7 mm (0.028 in.). Adjust clearance (**a**) between lever (**1**) and locking pin (**2**) and tighten clamping screw.

5. Tighten cable anchor screw.

Tightening torques
- Shift-lock cable anchor screw. . . . . 6 Nm (53 in-lb)

---

6. Shift selector lever into "P" position and turn ignition key to OFF position. Remove ignition key.
   - It must only be possible to remove ignition key with selector lever in "P" position. Check also that selector lever cannot be shifted out of "P" position.

TRANSMISSION, REMOVING AND INSTALLING

This heading describes removal and installation of the automatic transmission for 4- and 6-cylinder engines. Special engine lifting and jacking equipment is needed to support and reposition the engine as the transmission is removed from below. Read the procedure through to determine what tools and equipment are needed for the job.

Transmission, removing

1. Disconnect battery ground strap and remove battery.

> ***CAUTION —***
> *On vehicles equipped with an anti-theft radio, make sure to have the radio code on-hand before disconnecting the battery.*

2. Label and disconnect all electrical harness connections at transmission. See Fig. 15.

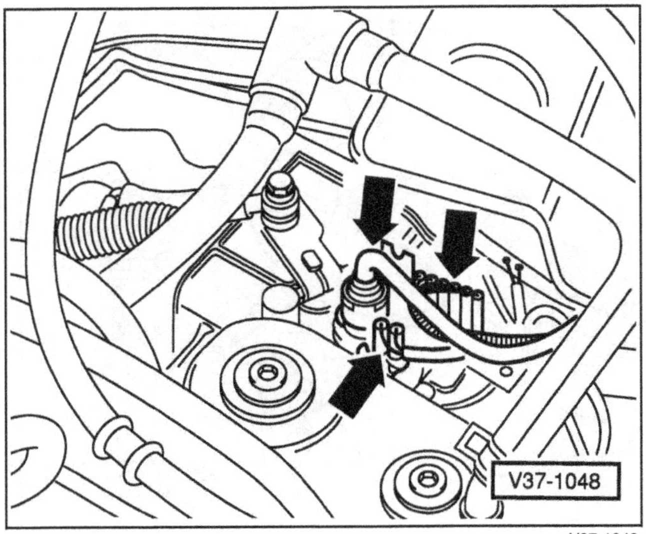

V37-1048

Fig. 15. Electrical connections (**arrows**) on transmission.

3. Remove transmission wiring harness from retaining bracket and set aside.

4. On 6-cylinder engines, remove multi-function transmission range switch from transmission. Cover or plug opening in transmission. See Fig. 16.

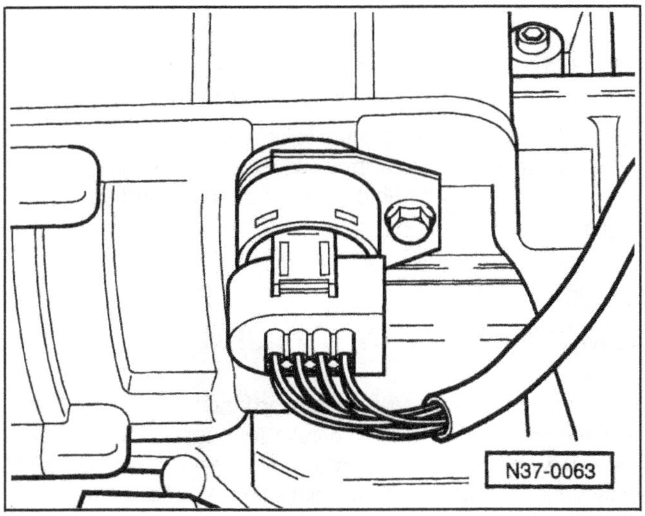

N37-0063

Fig. 16. Multi-function transmission range (TR) switch on rear of transmission.

5. Move selector lever to "P" and disconnect selector lever cable from lever/selector shaft on transmission.

6. Working in passenger compartment, remove selector lever cover/trim and remove selector lever cable circlip. Remove selector lever cable. See Fig. 5, given earlier.

NOTE —
Always replace selector lever circlip anytime it is removed.

7. Clamp off ATF cooler hoses and then disconnect hoses at ATF cooler. See Fig. 17.

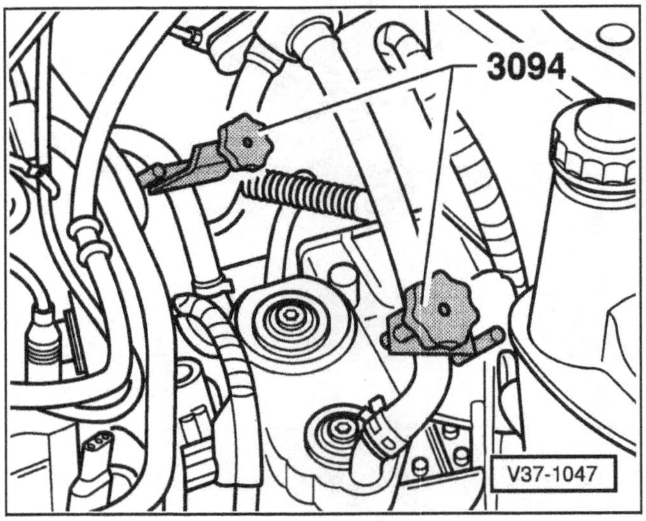

V37-1047

Fig. 17. ATF cooler hoses clamped-off with Volkswagen special tool 3094.

8. Support weight of engine/transmission from above using appropriate engine lifting equipment. See Fig. 18.

NOTE —
On cars with 6-cylinder engine, disconnect the vacuum hose from the intake manifold and the harness connector from intake air temperature sensor to gain access to the rear engine lifting eye. See Fig. 19.

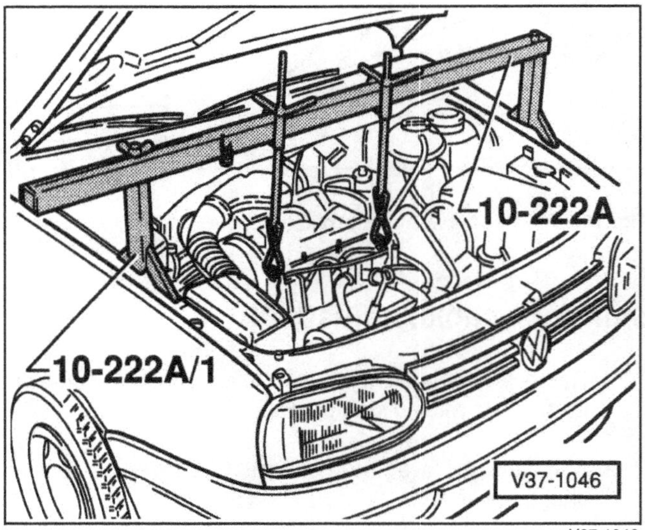

V37-1046

Fig. 18. Engine support installed above engine/transmission assembly. Numbers indicate Volkswagen special tools for 4-cylinder engine. (6-cylinder engines: use VW adapter 10-222A/3).

9. Remove starter. See **27 Engine Electrical**.

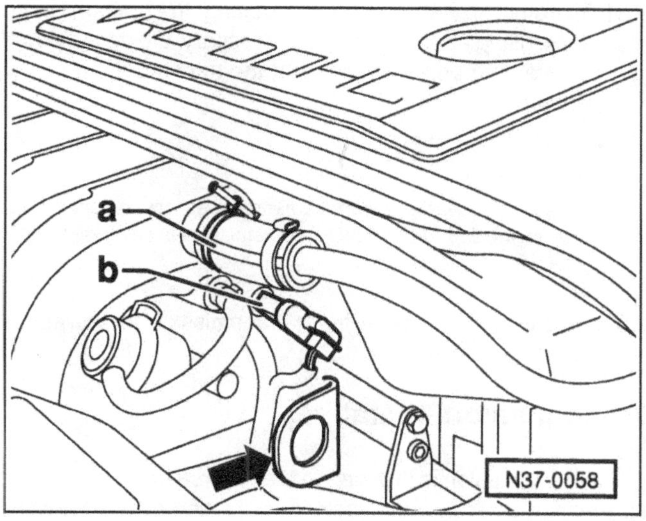

N37-0058

Fig. 19. On cars with 6-cylinder engine, disconnect vacuum hose (**a**) and connector (**b**) for access to rear engine lifting eye (**arrow**).

10. Remove retaining bracket for power steering pressure hose.

11. On cars with 6-cylinder engine, disconnect harness connector from radiator coolant fan.

12. With weight of engine and transmission supported, remove mounting bolt from front engine mount. See Fig. 20.

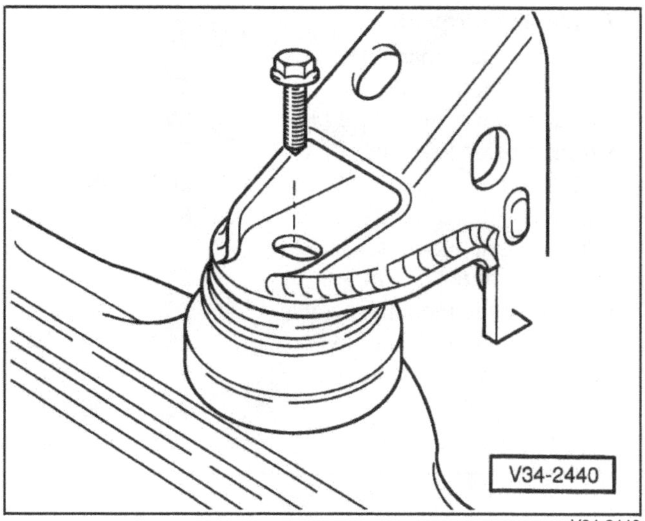

V34-2440

Fig. 20. Front engine mount bolt to be removed.

13. Unbolt front mount from engine block.

14. Remove engine coolant expansion tank and set aside.

15. Remove left rear engine/transmission mount completely.

16. Remove the upper engine-to-transmission mounting bolts.

17. Remove ATF oil pan protective shield.

18. On cars with 6-cylinder engine, remove right-side drive belt cover.

19. Remove torque converter cover plate, and remove nuts from torque converter.

20. On cars with 6-cylinder engine, remove vibration damping weight.

21. Disconnect left and right drive axles from transmission. See **40 Front Suspension and Drive Axles**.

22. Raise right-side drive shaft as far as possible and secure in place with stiff wire.

23. Remove left-front wheel. Turn steering to right lock.

24. Carefully mark installation position of ball joint to left-hand control arm. Remove ball joint mounting bolts. See Fig. 21. Separate ball joint from control arm and place left-side drive axle to one side.

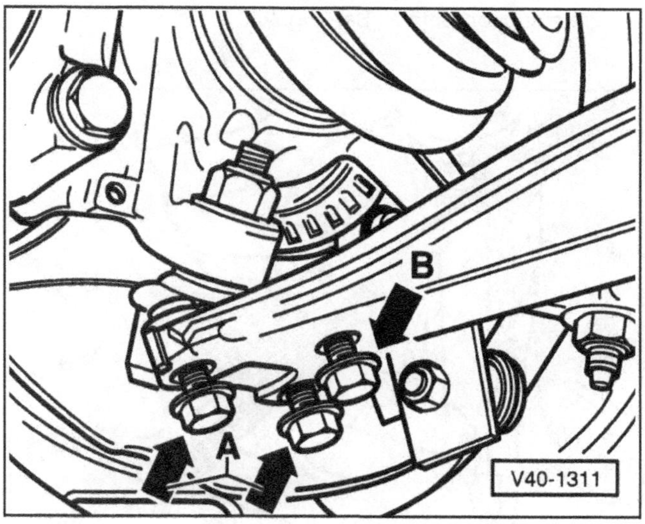

V40-1311

Fig. 21. Ball joint mounting bolts (**A** and **B**).

25. On cars with triple-roller inner CV joints, remove the left-hand drive axle. See **40 Front Suspension and Drive Axles**.

NOTE —

Two variations of drive axles are used on the cars covered by this manual: 4-cylinder engines with automatic transmission use drive axles with triple-roller inner CV joints; all other cars use conventional inner CV joints. See ***40 Front Suspension and Drive Axles*** *for more specific repair information.*

26. Support transmission from below using special transmission jack. Secure transmission to jack. See Fig. 22.

27. Remove lower engine/transmission mounting bolts.

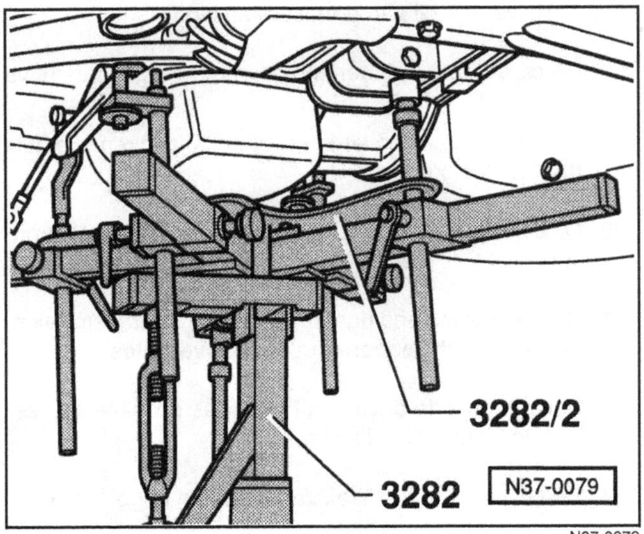

N37-0079

Fig. 22. Transmission jack (Volkswagen special tool no. VAG 1383A) supporting transmission prior to removal. Numbers indicate special VW transmission jack adapter and support.

28. Tilt engine and transmission forward using an appropriate support or jack. See Fig. 23.

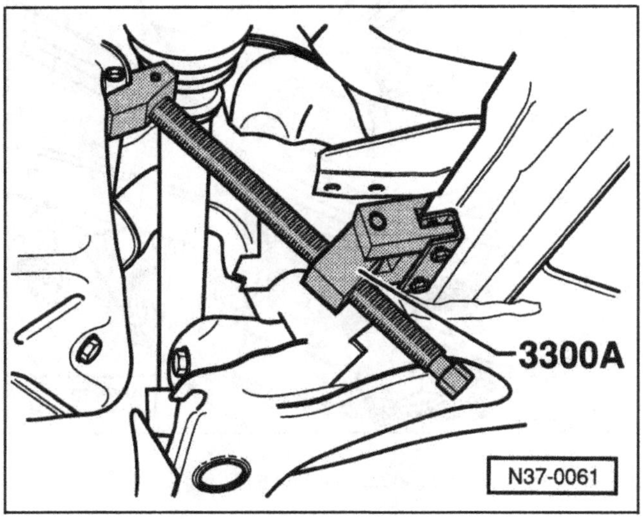

N37-0061

Fig. 23. Engine/transmission assembly supported in forward position with support (Volkswagen special support 3300A shown).

29. Separate engine from transmission. At the same time pry or lever torque converter off of drive plate and press against ATF pump.

30. Lower engine using support or jacking device (shown in Fig. 23).

31. Carefully lower transmission using the transmission jack, tilting and swiveling as necessary. Guide the power steering pressure line past the transmission as it is lowered.

NOTE —
Make sure that the right drive flange does not contact the engine support/jack as the transmission is being removed.

32. Secure torque converter to transmission to prevent it from falling out.

Transmission, installing

The transmission is installed in the reverse order of removal. Replace selector cable locking circlip and adjust selector lever cable. See **Selector lever cable, adjusting**. Have the front wheels professionally aligned once installation is complete.

CAUTION —
- *When installing the torque converter, make sure that both drive pins engage in the ATF pump inner wheel recesses.*
- *Make sure that the dowel sleeves for aligning the engine and transmission are all present and correctly located in the engine block.*

Tightening torques
- Drive axle to transmission drive flange . . . . 45 Nm (33 ft-lb)
- Torque converter to drive plate. . . . . . 60 Nm (44 ft-lb)
- Transmission to engine
  - M10 bolt . . . . 60 Nm (44 ft-lb)
  - M12 bolt . . . . 80 Nm (59 ft-lb)
- Front engine mount to mount assembly . . . . 60 Nm (44 ft-lb)
- Left mount to mount assembly . . . . . . 60 Nm (44 ft-lb)
- Left mount to transmission . . . . 25 Nm (18 ft-lb)
- Cover plate to transmission . . . . 15 Nm (11 ft-lb)
- Ball joint to control arm. . . . . 35 Nm (26 ft-lb)
- Starter to transmission . . . . 60 Nm (44 ft-lb)

TORQUE CONVERTER

When replacing the torque converter, always note the torque converter code number. The code number can be found in one of two places. See Fig. 24 and Fig. 25. The replacement torque converter must have the same code number, as it is matched to the transmission/engine application.

> ***CAUTION—***
> *When installing the torque converter, make sure that both drive pins engage in the recesses in the ATF pump inner wheel.*

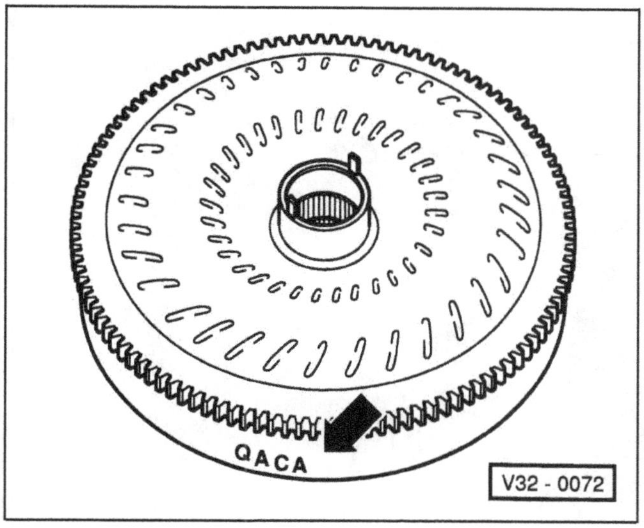

V32-0072

Fig. 24. Torque converter code letter location (**arrow**).

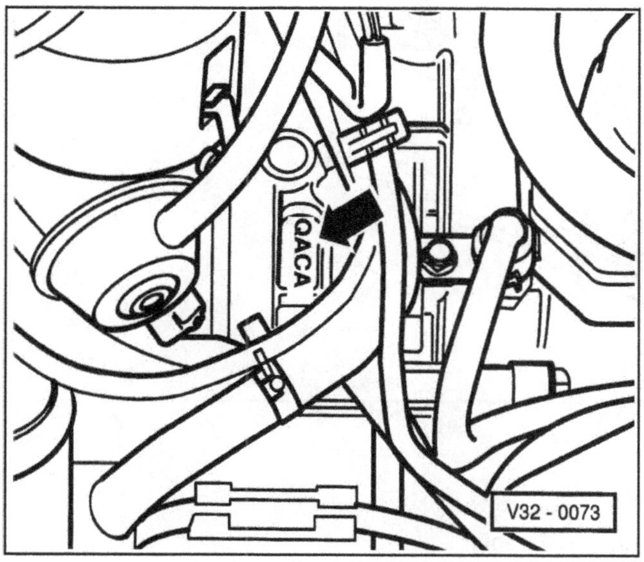

V32-0073

Fig. 25. Torque converter code letters visible through access hole with transmission installed (**arrow**).

If the torque converter has become contaminated by particles resulting from wear or abrasion, or in the course of a major overhaul of the transmission, always flush the torque converter with clean ATF. The torque converter can be flushed with commercially-available extraction equipment, such as EZ1 fluid evacuator, or Volkswagen special tools VAG 1358A and VAG 1358 A/1 probe.

Torque converter oil seal, replacing (transmission removed)

1. Remove the torque converter seal using a seal extractor. See Fig. 26.

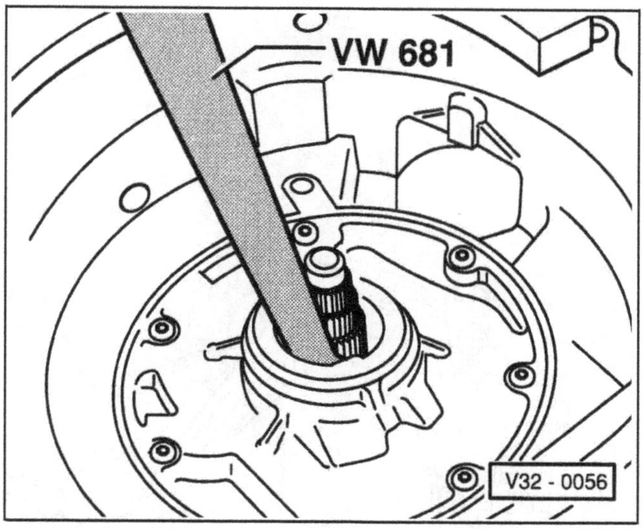

V32-0056

Fig. 26. Torque converter seal being removed. Volkswagen special tool VW 681 shown.

2. Using a seal driver, drive seal in until flush. See Fig. 27.

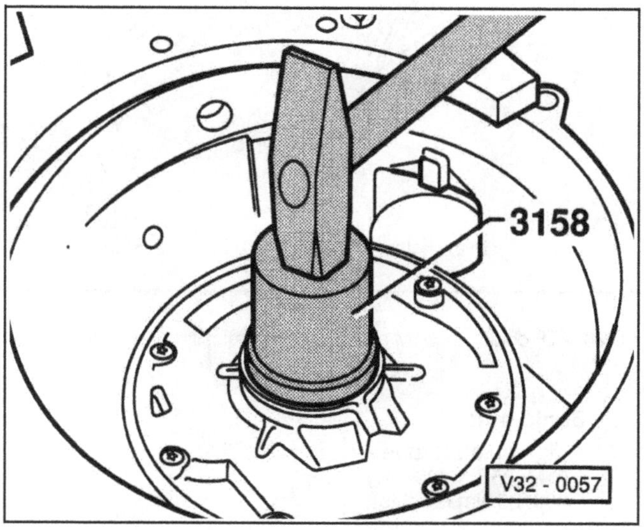

V32-0057

Fig. 27. Installing torque converter seal with seal driver. Volkswagen special tool 3158 shown.

ATF Cooler

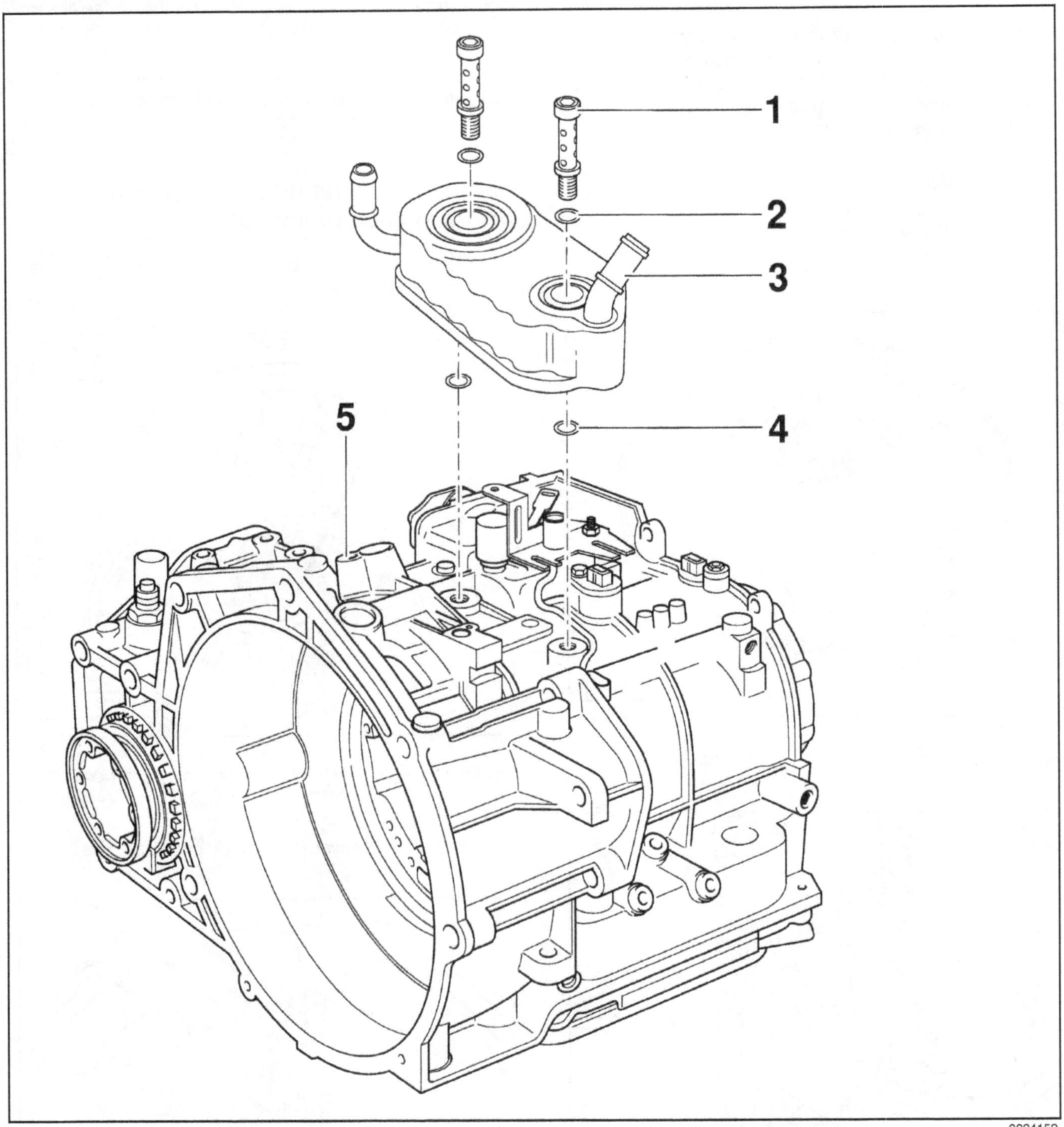

Fig. 28. ATF cooler for 01M automatic transmission (096 is similar).

1. **Banjo bolt**
   - Tightening torque:
     01M: 35 Nm (26 ft-lb)
     096: 35 Nm (26 ft-lb)
2. **O-ring**
   - Always replace
3. **ATF cooler**
4. **O-ring**
   - Always replace
5. **Transmission housing**

39 Differential and Final Drive

39

GENERAL

Transmission and final drive assemblies used on the cars covered by this manual vary depending on engine installed and on model year.

Engine Code	Manual Transmission Code	Automatic Transmission Code
AAZ, ACC, ABA (4-cylinder gasoline)	020	096 or 01M*
AAA (6-cylinder engine)	02A	096 or 01M*
1Z, AHU (4-cylinder TDI diesel)	02A	01M*
*See **3 Clutch, Transmission and Final Drive–General** for specific identification and application information.		

NOTE —

On cars with manual transmission, the differential shares the same gear oil as the transmission. For information on draining and filling the manual transmission and differential oil, see **0 Maintenance Program**.

020 MANUAL TRANSMISSION

Illustrations shown under this heading are for the left-side drive flange assembly only. Removal and installation of the right-side drive flange and oil seal are identical, unless noted otherwise.

Drive flange oil seal, removing (020 transmission)

1. If replacing right drive flange oil seal, disconnect drive axle at right side of transmission, push drive axle upward and suspend with stiff wire.

2. If replacing left drive flange oil seal, remove left drive axle. See **40 Front Suspension and Drive Axles**.

3. Pry off cover in center of drive flange by piercing center of cover with screwdriver.

4. Carefully remove circlip and dished washer from drive flange. See Fig. 1.

WARNING —

The drive flange is under spring force. Once the circlip is removed, the drive flange may release suddenly. Use extreme care when removing the circlip.

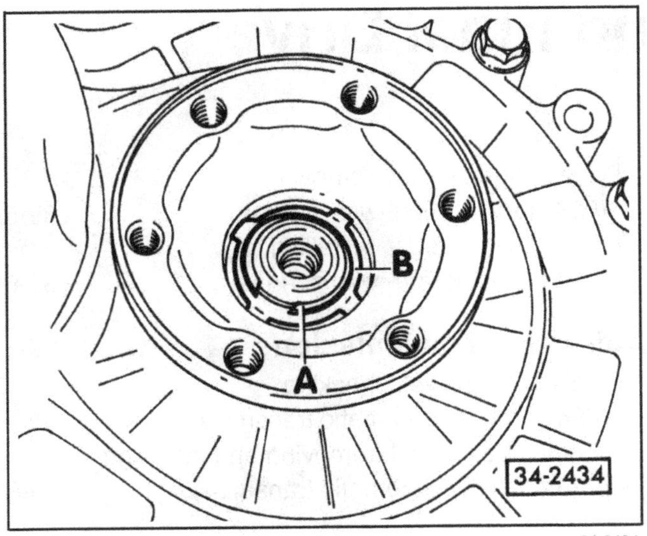

34-2434

Fig. 1. Drive flange retaining circlip (**A**) and dished washer (**B**). Note orientation of dished washer before removing.

5. If necessary, use a press (VW special tool no. VW391 or equivalent) to remove the drive flange together with its spring. See Fig. 2.

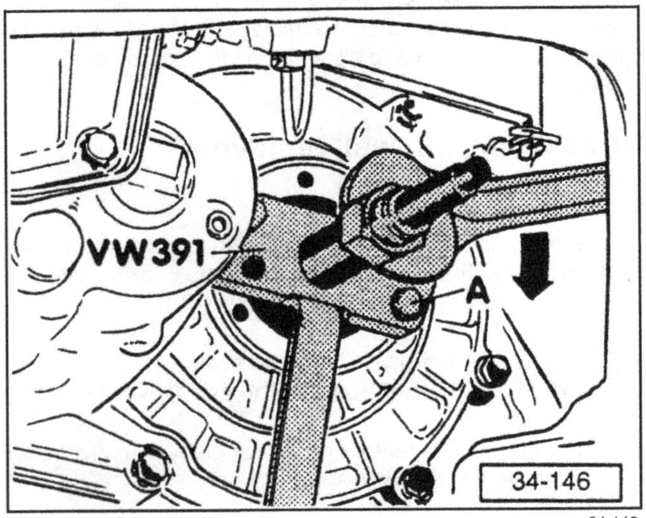

34-146

Fig. 2. Drive flange being pressed off output shaft using Volkswagen special tool no. VW 391.

6. Pry out drive flange oil seal from differential housing. See Fig. 3.

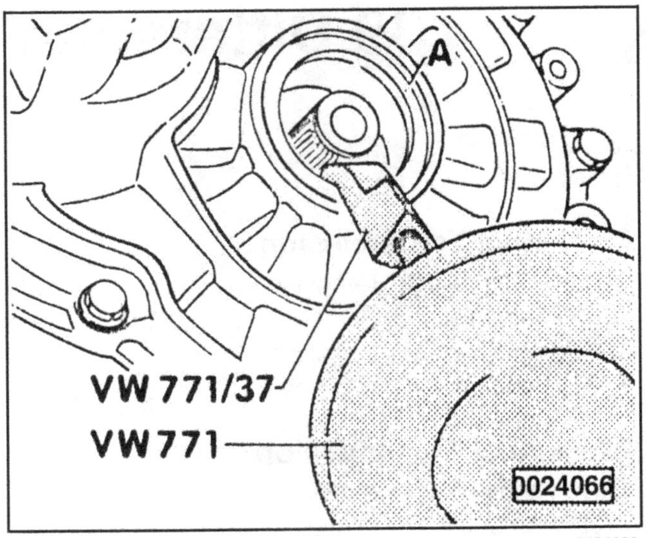

0024066

Fig. 3. Drive flange oil seal (**A**) being removed with seal extractor (VW special tool no. VW 771 and VW 771/37).

Drive flange oil seal, installing (020 transmission)

> ***CAUTION —***
> *Be sure to have the correct transmission code on hand when ordering parts. Drive flange oil seal applications vary depending on transmission application.*

1. Fill space between lips of new oil seal with multi-purpose grease. Then drive seal in against shoulder. See Fig. 4.

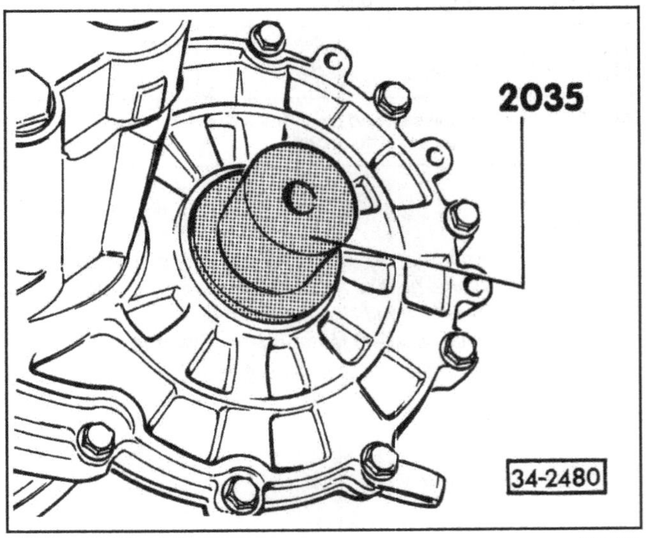

34-2480

Fig. 4. Drive flange oil seal being installed with seal driver (Volkswagen special tool no. 2035).

2. Install spring to drive flange.

3. Carefully heat drive flange on hot-plate to approx. 80°C (176°F). Press the drive flange into position. See Fig. 5.

> ***WARNING —***
> *Use extreme care when handling the heated drive flange. Wear heavy gloves and arm protection to guard against burns and personal injury.*

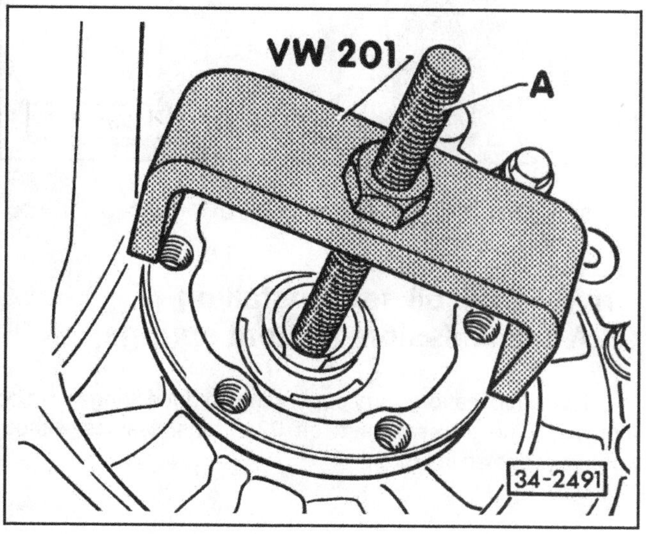

34-2491

Fig. 5. Drive flange being installed using press (VW special tool no. VW 201 with threaded stud and M10 nut (**A**).

4. Install dished washer so that the concave faces in, toward the differential.

5. Fit new circlip to shaft and press down until seated in groove.

6. Press new sealing cover into center of drive flange.

7. Tightening torques are given below. Be sure to check transmission oil level, adding as necessary. See **0 Maintenance Program**.

Tightening torques
- Drive axle to axle flange. . . . . 45 Nm (33 ft-lb)
- Wheel to wheel hub . . . . 110 Nm (81 ft-lb)

02A MANUAL TRANSMISSION

Two versions of drive flange assemblies are used on the 02A manual transmission. One version uses bolts to retain the spring-loaded drive flanges. The other version is not spring-loaded and uses circlips to retain the drive flanges.

To determine which version is installed, the drive axle must first be removed from the drive flange. If a bolt is used to retain the flange, springs are used. If a bolt is not present, the drive flange is retained by a blind circlip.

Drive flange oil seal, removing (02A transmission–without springs)

> ***NOTE —***
> *Illustrations shown under this heading are for the left-side drive flange assembly only. Removal and installation of the right-side drive flange and oil seal are identical.*

1. Turn steering wheel completely to right lock.

2. Disconnect axle shaft from axle flange. Push drive axle upward as far as possible and suspend with stiff wire.

> ***CAUTION —***
> *Avoid damaging the protective surface on the axle shaft.*

3. If replacing left-side oil seal, remove the road wheel. Also remove transmission support, where applicable. See Fig. 6.

> ***NOTE —***
> *For better access to the transmission support, disconnect coolant overflow reservoir and lay it to one side.*

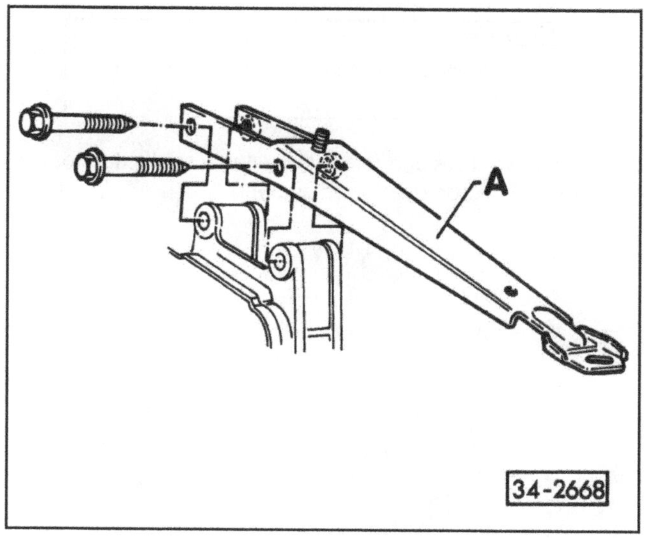

34-2668

Fig. 6. Transmission support (**A**) used on some early models.

4. Pry off cover in center of drive flange by piercing center of cover with screwdriver.

5. Press axle flange from transmission until the circlip releases. See Fig. 7.

NOTE —
If the drive flange fails to come straight off, use an additional spacer and bolt positioned opposite the first bolt and spacer. Tighten both bolts alternately and evenly.

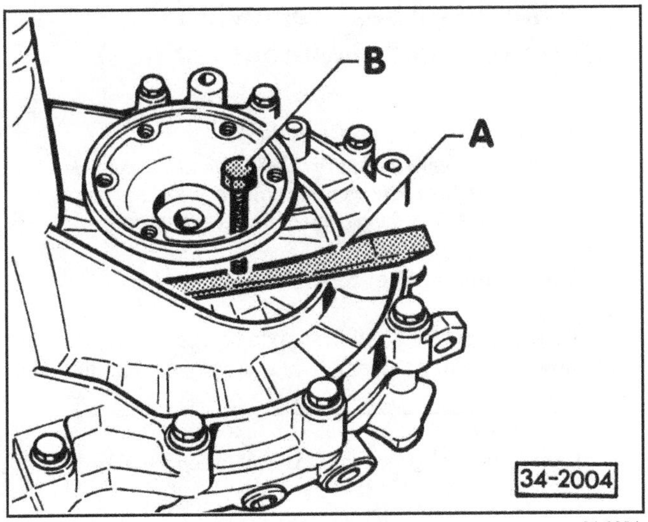

34-2004

Fig. 7. Remove drive flange/axle shaft by placing spacer (**A**) beneath flange and pull flange from differential by threading bolt onto spacer (cold chisel shown).

6. Clamp axle shaft in vise (fitted with protective jaws). Then press old circlip out of shaft groove by installing new circlip. See Fig. 8.

CAUTION —
Always replace drive flange circlip—do not reuse old circlip.

7. Remove drive flange oil seal using a seal extractor (VW 681 extractor lever or equivalent).

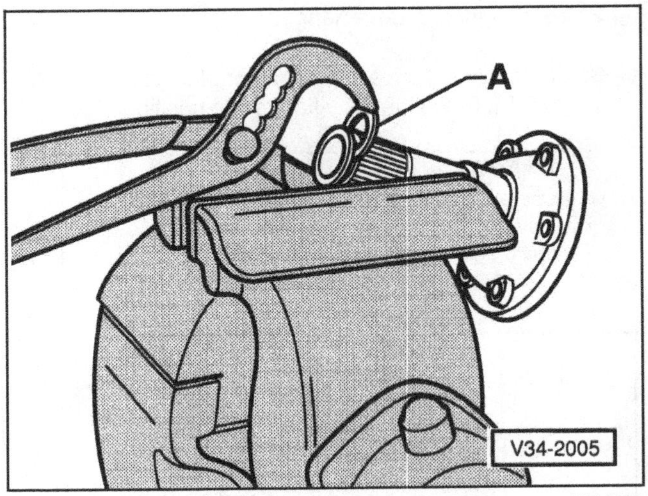

V34-2005

Fig. 8. New circlip being pressed onto axle shaft to remove old circlip.

Drive flange oil seal, installing (02A transmission–without springs)

Oil seal applications vary depending on drive flange version installed. The oil seal used on 02A transmissions without springs is shown in Fig. 9.

Drive Flange Oil Seal Dimensions

• Diameter (**a**)	40.0 mm (1.575 in.)
• Height (**b**)	10.0 mm (0.394 in.)

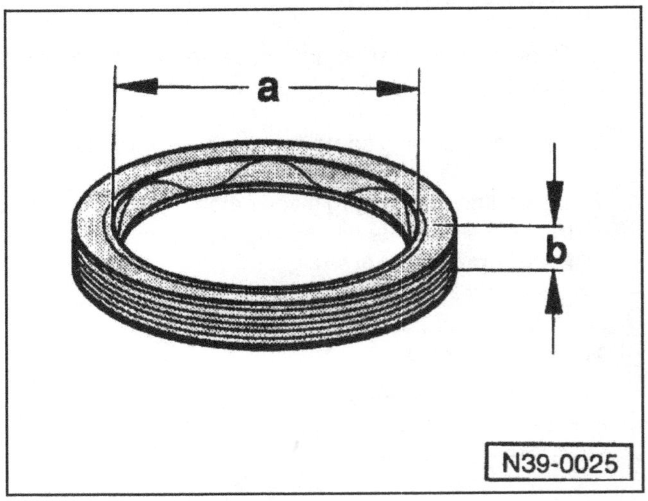

N39-0025

Fig. 9. Drive flange oil seal for 02A transmission without drive flange springs.

1. Fill space between lips of new oil seal with multi-purpose grease. Then drive seal in against shoulder. See Fig. 10.

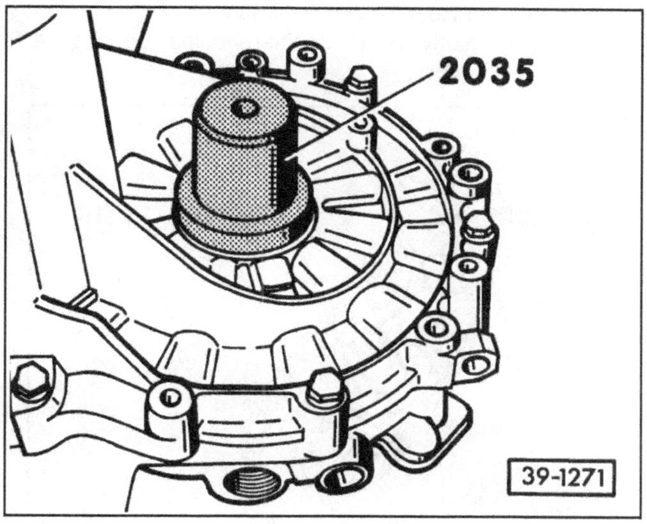

39-1271

Fig. 10. Drive flange oil seal being installed Volkswagen special tool no. VW 2035.

2. Install and drive the drive flange in until the retaining circlip engages circlip groove. See Fig. 11.

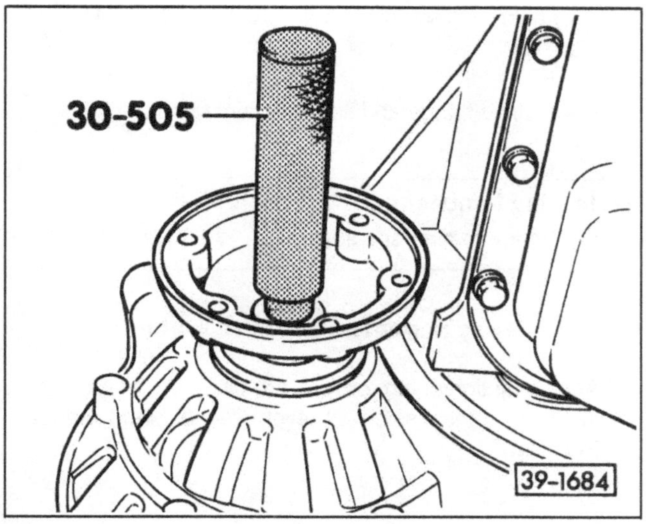

39-1684

Fig. 11. Drive flange being installed using Volkswagen special tool no. 30-505.

3. Press new sealing cover into center of drive flange.

4. Tightening torques are given below. Be sure to check transmission oil level as described in **0 Maintenance Program**.

Tightening torques

- Transmission support to transmission . . . . 45 Nm (33 ft-lb)
- Transmission support to transmission mount . . . . 50 Nm (37 ft-lb)
- Wheel to wheel hub . . . . 110 Nm (81 ft-lb)

Drive flange oil seal, removing (02A transmission–with springs)

NOTE—

Illustrations shown under this heading are for the left-side drive flange assembly only. Removal and installation of the right-side drive flange and oil seal are identical.

1. Turn steering completely to right (full-lock).

2. Disconnect axle shaft from drive flange.

3. Push drive axle upward as far as possible and push drive axle upward and suspend with stiff wire.

CAUTION—

Avoid damaging protective surface on the drive axle.

4. Pry off cover in center of drive flange by piercing center of cover with screwdriver.

5. If replacing left-side oil seal, remove the road wheel. Also remove transmission support, where applicable. See Fig. 6, given earlier.

NOTE—

For better access to the transmission support, disconnect coolant overflow reservoir and lay it to one side.

6. Carefully remove drive flange retaining bolt.

WARNING—

The drive flange is under spring force. Once the bolt is removed, the drive flange may release suddenly. Use extreme care when loosening the bolt.

7. Place oil pan underneath and remove drive flange.

8. Remove oil seal from transmission housing. See Fig. 12.

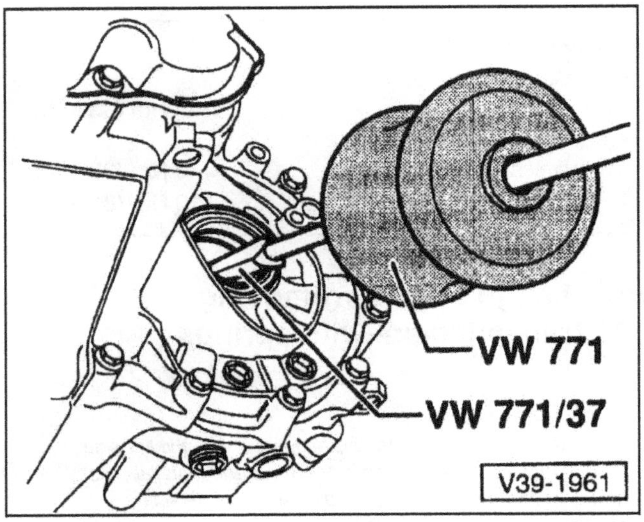

V39-1961

Fig. 12. Drive flange oil seal being removed using slide hammer and hooked adapter (Volkswagen special tool no. VW 771 and VW 771/37).

Drive flange oil seal, installing (02A transmission–with springs)

Oil seal applications vary depending on drive flange version installed. The oil seal used on 02A transmissions with springs is shown in Fig. 13.

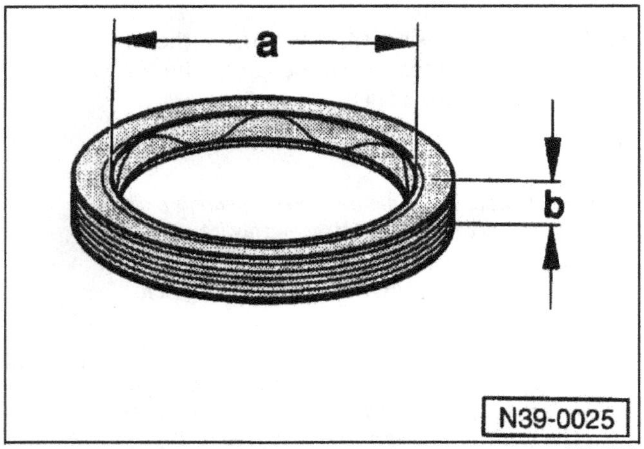

N39-0025

Fig. 13. Drive flange oil seal for 02A transmission with drive flange springs.

Seal dimensions
- Diameter (**a**) . . . . 48.0 mm (1.890 in.)
- Height (**b**) . . . . 6.8 mm (0.268 in.)

1. Fill space between lips of new oil seal with multi-purpose grease. Then drive seal in against shoulder. See Fig. 14.

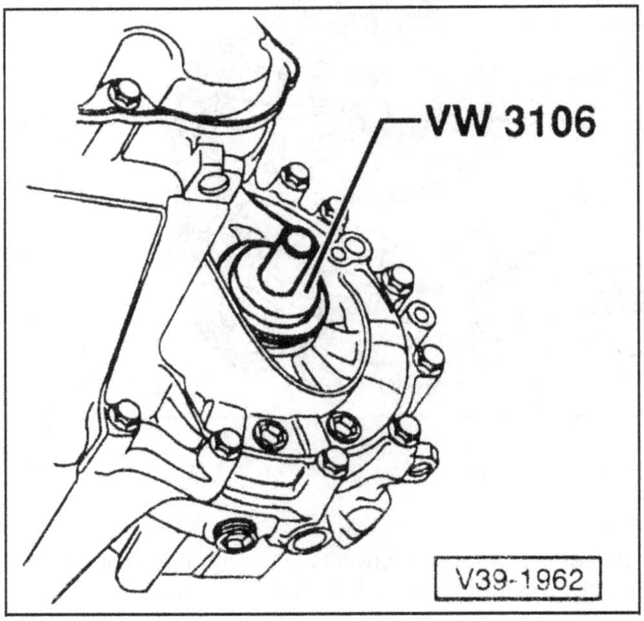

V39-1962

Fig. 14. Drive flange oil seal being installed with seal driver (Volkswagen special tool no. VW 3106).

2. Install drive flange and flange retaining bolt.

Tightening torques
- Drive flange to differential housing . . . 25 Nm (18 ft-lb)

3. Press new sealing cover into center of drive flange.
4. Additional tightening torques are given below. Be sure to check transmission oil level. See **0 Maintenance Program**.

Tightening torques
- Drive axle to axle flange . . . . 45 Nm (33 ft-lb)
- Transmission support to transmission . . . . 45 Nm (33 ft-lb)
- Transmission support to transmission mount . . . . 50 Nm (37 ft-lb)
- Wheel to wheel hub . . . . 110 Nm (81 ft-lb)

AUTOMATIC TRANSMISSION

Fig. 15 shows a cutaway view of the final drive for 096 transmission. Unless otherwise stated, the repair procedures and specifications given under this heading apply to both the 096 and the 01M automatic transmissions.

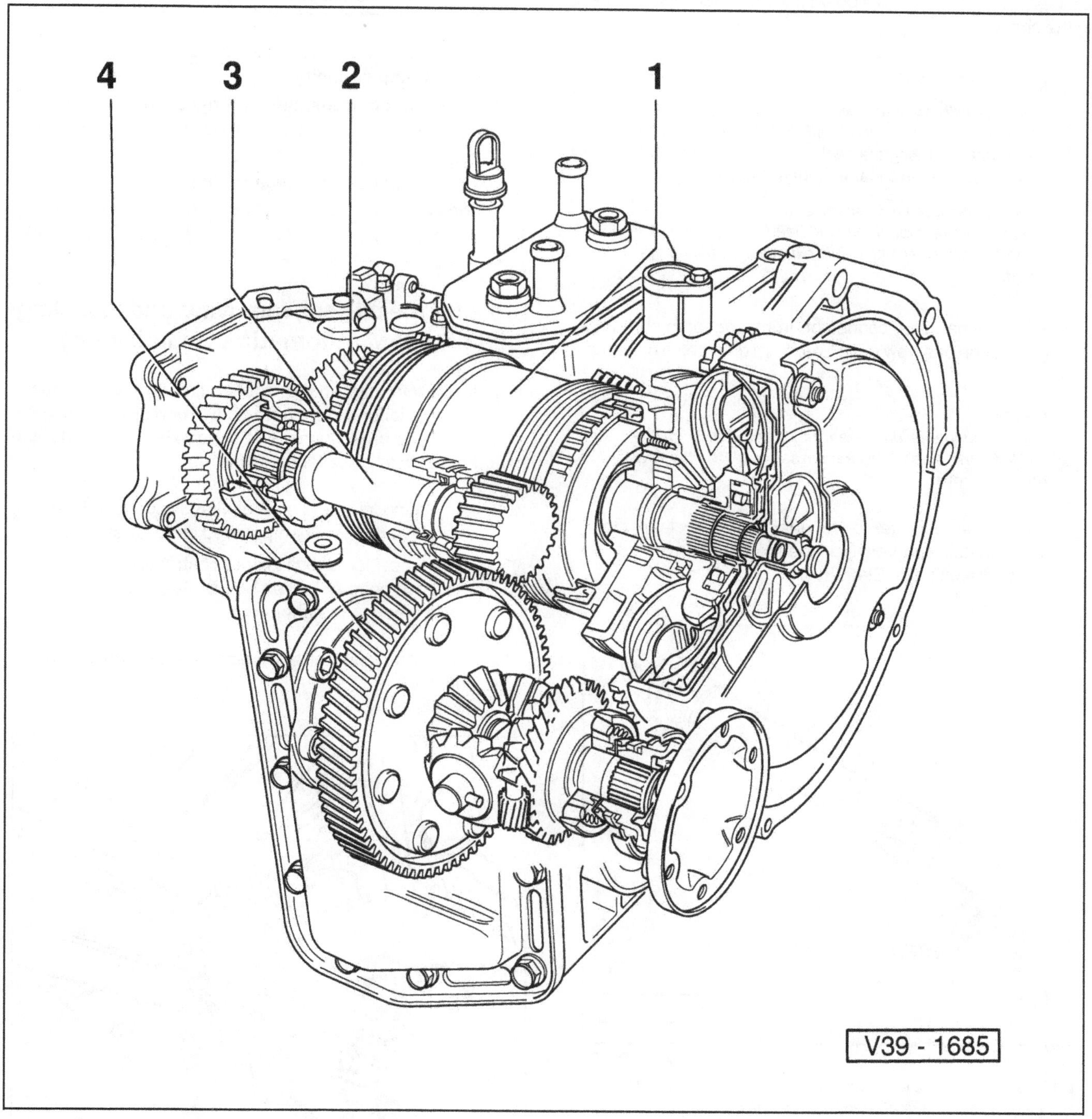

V39-1685

Fig. 15. 096 automatic transmission shown, 01M is similar.

1. **Planetary gear set**
2. **Input gear**
3. **Drive pinion**
4. **Differential**

Final drive oil level, checking (096 and 01M automatic transmission)

On cars with automatic transmission, the final drive oil level should be checked regularly as specified in **0 Maintenance Program**. It is checked by removing the speedometer drive gear from the transmission housing and using the drive gear as a dipstick. The car should be on a level surface when making the check.

> ***NOTE —***
> - *On cars with manual transmission, the final drive lubricant shares a common bath with the transmission lubricant. Checking the final drive/transmission lubricant is covered in* **0 Maintenance Program.**
> - *Volkswagen does not specify a final drive lubricant replacement interval. If the fluid needs to be drained, it can be drawn off using a fluid extraction/pumping system.*

1. Remove harness connector from speedometer drive gear. Then unscrew drive gear from differential housing.

> ***NOTE —***
> *The speedometer drive gear is mounted through the top left-hand side of the transmission, above the final drive end cover.*

2. Clean speedometer drive gear using a clean cloth. Then install and remove speedometer drive gear to check the oil level. See Fig. 16.

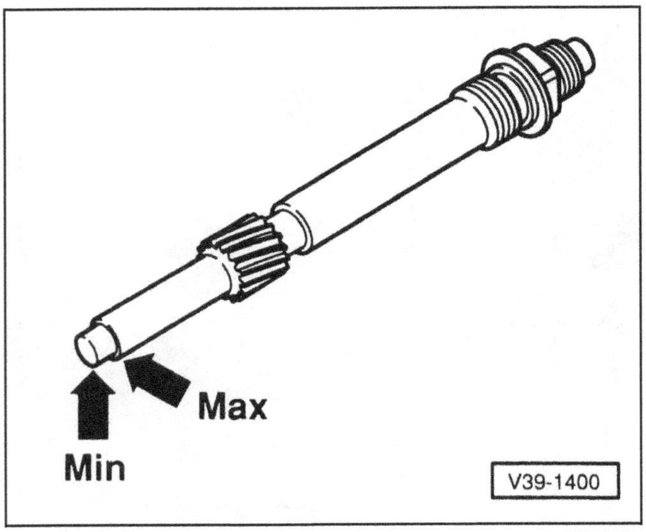

V39-1400

Fig. 16. Speedometer drive gear used to check final drive oil level. Level should be between **Min** and **Max** marks.

3. Add lubricant as necessary. Reinstall drive gear.

> ***NOTE —***
> *The difference between the "Min" and "Max" marks on the speedometer drive gear is approx. 0.1 liter (3.4 fl. oz.).*

> ***CAUTION —***
> *Use only genuine VW ATF in the 01M transmission. Use of other ATF types can adversely affect the service life of the transmission.*

Lubricant and capacity

- **096 automatic transmission final drive**
  lubricant. . . . . SAE 75W/90 synthetic gear oil
  capacity. . . . . 0.75 L (0.79 qt.)
- **01M automatic transmission final drive**
  lubricant. . . . . Genuine VW Synthetic ATF
  capacity. . . . . 0.75 L (0.79 qt.)

Drive flange oil seal, removing and installing (096 and 01M automatic transmission)

1. If replacing right-hand oil seal, first remove the heavy metal balance weight from body. Then disconnect drive axle from drive flange. Set drive axle to one side and hold axle in place using stiff wire.
2. If replacing left-hand oil seal, first disconnect drive axle from drive flange. Then proceed as follows:
   - mark position of ball joint to control arm
   - remove bolts **A**, loosen bolt **B**. See Fig. 17.
   - turn steering to right lock.
3. Swing ball joint out and place left-hand drive shaft to one side.

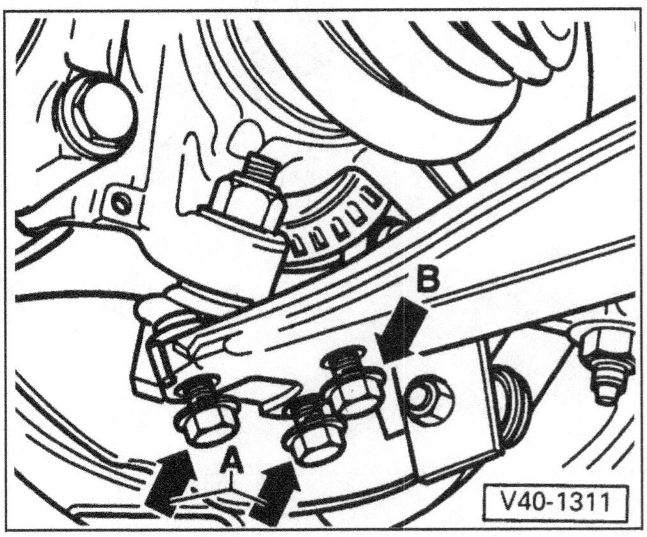

V40-1311

Fig. 17. Ball joint mounting bolts. Loosen bolt (**B**) and remove bolts (**A**).

4. Pry off cover in center of drive flange by piercing center of cover with screwdriver.

5. Mount press tool to drive flange. See Fig. 18. Screw threaded rod (M10) into axle shaft. Then turn nut on tool clockwise to compress spring.

> ***WARNING —***
> *The drive flange is under spring force. Once the circlip is removed, the drive flange may release suddenly. Use extreme care when removing the circlip.*

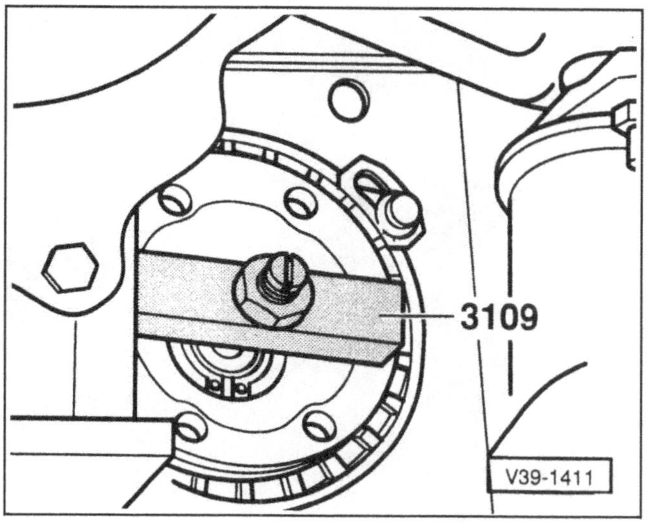

V39-1411

Fig. 18. VW special tool no. 3109 being used to compress spring behind drive flange so that circlip can be removed.

6. Remove circlip and spring (dished) washer. Note orientation of washer before removing.

7. Remove press tool, drive flange, and spring.

NOTE —
If the drive flange does not release easily, use VW special tool no. VW 391 to press the drive flange off the axle shaft. See Fig. 19.

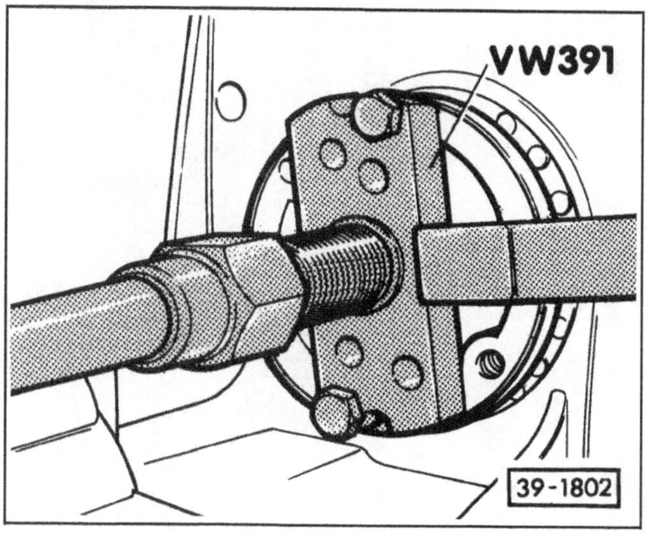

39-1802

Fig. 19. VW special tool no. VW391 being used to press drive flange off axle shaft.

8. Carefully drive new seal into position. Fill space between new oil seal and dust lip with multi-purpose grease.

9. Install drive flange with spring. Then pull drive flange into position using VW special tool no. 3109 or equivalent. See Fig. 18, given earlier.

NOTE —
Install dished washer so that the concave faces in, toward the differential.

10. Seal drive flange using a new sealing cover.

11. Install drive axle. Reconnect ball joint to control arm. Tightening torques are given below.

12. Check final drive oil level as described earlier.

Tightening torques
- Drive axle to drive flange .... 45 Nm (33 ft-lb)
- Ball joint to control arm .... 35 Nm (26 ft-lb)

4 Suspension, Brakes, and Steering

GENERAL

This general information group covers application information and system descriptions for the repair groups listed under **4 Suspension, Brakes, and Steering**.

SUSPENSION

Volkswagen Jetta, Golf, GTI and Cabrio models are equipped with a strut-type independent front suspension and a torsion-type rear axle beam with trailing arms. The front suspension struts are integral spring and shock absorber units, while the rear suspension has separate shocks mounted inside the coil springs. This type of suspension system provides excellent steering and handling in a compact size with comparatively light weight.

Two suspension systems are used on the Jetta, Golf, GTI and Cabrio models. The "base" suspension is used on most 4-cylinder models, while the "plus" suspension is used on the more performance oriented 6-cylinder models. The plus suspension features larger anti-sway bars and brakes, different front struts, and five lug wheels (the base suspension uses four lug wheels).

Front suspension

Fig. 1 shows a system view of the front suspension. The front suspension struts have been designed to provide a negative steering roll radius. That is, the steering axis of each strut intersects the road surface at a point outboard of the wheel's vertical centerline. This geometry tends to steer the car automatically in the direction of a skid caused by unequal front wheel traction. Conventional front suspension geometry, which places the steering axes inboard of the tire centerlines, tends to steer the car in the opposite direction, thereby increasing the severity of the skid.

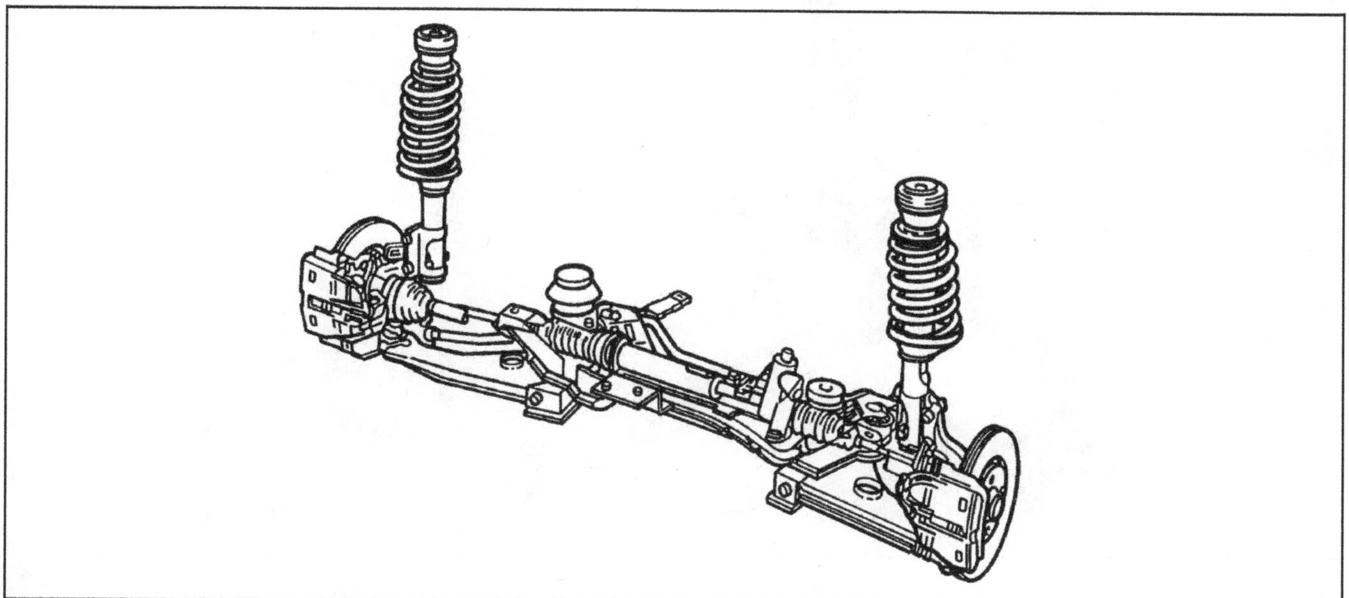

0024138

Fig. 1. Front suspension.

The front suspension struts and control arms can be disassembled for repair, but much of this work requires special tools and experience. If you lack the skills, tools, or a suitable workshop for suspension work, we suggest you leave such repairs to an authorized Volkswagen dealer or other qualified repair shop. Please consult your Volkswagen dealer before beginning any repairs on a car that may be subject to warranty coverage. **Table a** contains suspension technical data

Rear suspension

The rear suspension has been designed for low unsprung weight, easy replacement of springs and shock absorbers, and good ride and handling qualities. A "V" profile independent torsion beam rear axle with integral trailing arms and sway bar is used on both the base and sport suspension models. The advantage of this design is that rear wheel camber and toe remain virtually unaffected by suspension movement. The torsion beam rear axle is mounted to the vehicle with track-correcting bushings. These bushings are asymmetrically shaped and minimize steering effect from the rear wheels during cornering. The rear axle beam is somewhat flexible. Uneven rear suspension loads cause the axle beam to twist slightly, thus acting as a stabilizer bar. This suspension design yields many desired performance benefits.**Table a** contains suspension technical data and Fig. 2 shows the rear suspension components.

Table a. Suspension technical data

Model		**Golf/Jetta/Cabrio with base suspension**	**GTI/Jetta GLX with Plus suspension**
Wheelbase	mm (in.)	2475 (97.44)	
Front track (unladen)	mm (in.)	1478 (58.19) - wheel offset 38mm 1464 (57.64) - wheel offset 45mm	1460 (57.48) - wheel offset 38mm 1450 (57.09) - wheel offset 43mm
Rear track (unladen)	mm (in.)	1460 (57.48) - wheel offset 38mm 1446 (56.93) - wheel offset 45mm	1442 (56.77) - wheel offset 38mm 1432 (56.38) - wheel offset 43mm
Turning circle	meters (feet)	10.7 (35.1)	11.0 (36.1)
Steering roll radius (negative)	mm (in.)	3.2 (0.126)- wheel offset 38mm 10.2 (0.402) - wheel offset 45mm	13.8 (0.543) - wheel offset 38mm 18.8 (0.740) - wheel offset 43mm
Wheel lock angle	inner outer	38° 34°	37° 34°
Ground clearance at permitted total weight	mm (in.)	123 (4.84)	103 (4.06)

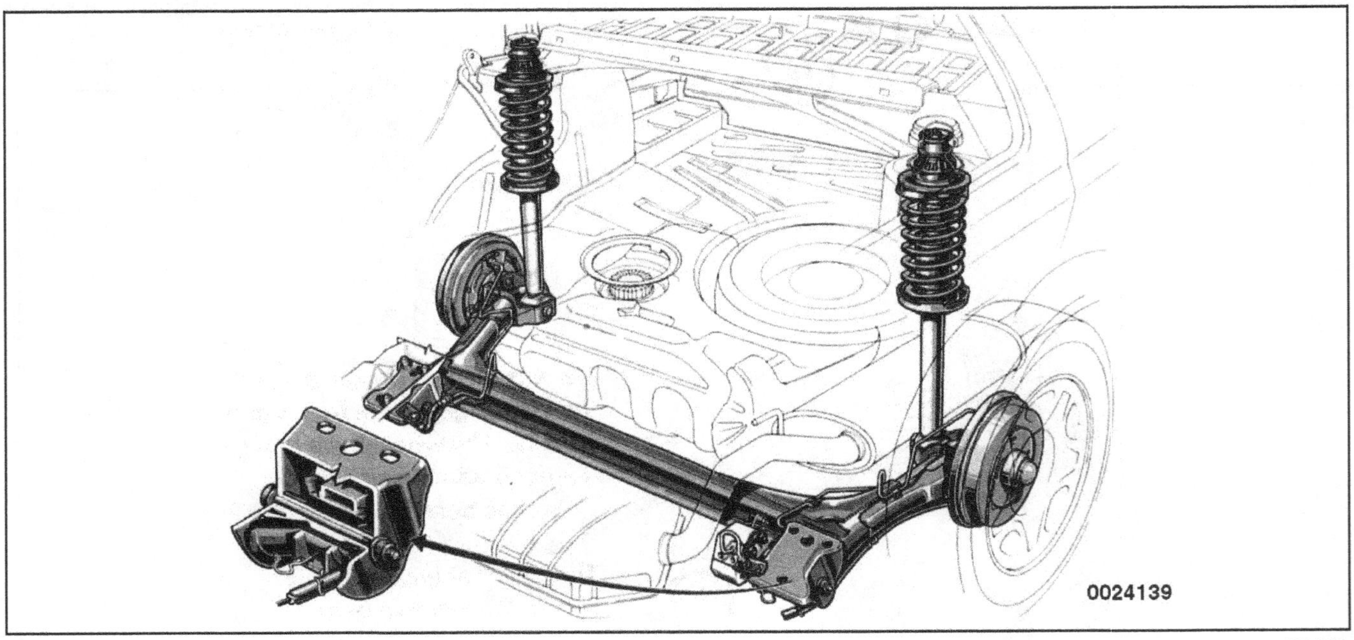

0024139

Fig. 2. Phantom view of rear suspension with detail of track-correcting bushing.

WHEELS

Various tire and wheel combinations are available for use on the models covered by this manual. Different wheel offsets ensure that the tire and wheel combinations are flush with the vehicle body contours. The wheel offset affects front and rear track, and steering roll radius. Some of the tire/wheel combinations available from Volkswagen are listed in **Table b**.

Table b. Tire/wheel Combinations

Model	Wheel	Tire	Offset (mm)
CL	5.5 x 13 steel	175 x 70	38
GL, GLS (early production w/manual transmission)	6J x 14 steel/alloy	185 x 60 R14H	45
GL, GLS, Cabrio, TDI, K2, TREK, Wolfsburg Limited Edition, Sport, GTI (4 cyl.),	6Jx 14 steel/alloy	195 x 60 R14H	45
VR-6 GTI, GLX	6.5J x 15 alloy	205 x 50 R15V	43

BRAKES

The Jetta, Golf, GTI, and Cabrio models are equipped with several different brake systems. All models have a dual diagonal hydraulic brake system with a load dependent brake pressure regulator located on the left side of the rear axle. The brake pressure regulator controls the amount of brake pressure to the rear wheels. **Table c** provides dimensions and technical data for the various brake systems.

VR6 models come standard with an anti-lock brake system (ABS). This is a three channel system which permits individual control of the front wheel and a joint "select low" control of the rear wheels. There is also a vacuum booster that will allow the brakes to operate conventionally with vacuum power assistance if the hydraulic pump should fail.

In addition to ABS, VR6 models also come with a low speed traction control system (Electronic Differential Lock or EDL). The traction control works in conjunction with the ABS system by applying the front brakes to control wheel spin during low speed acceleration. By controlling wheel spin when the vehicle is starting to move from a stop, more power can be transmitted from the vehicle to the road surface. This increases traction and vehicle control when starting off in slippery conditions. Fig. 3 shows the system layout for the anti-lock brake and traction control systems.

Table c. Brake Technical Data

Engine/version		1.8/1.9 Liter CL, GL	1.9/2.0 Liter GL, GLS, Cabrio, TDI, K2, TREK, Wolfsburg Limited Edition, Sport, GTI (4 cyl.)	2.8 Liter VR6 GTI, GLX
Front brake disc	diameter – mm (in.)	239 (9.41)	256 (10.08)	280 (11.02)
Front brake disc thickness	mm (in.)	12 (0.47)	20 (0.79)	22 (0.87) - thru 12/94 25 (0.99) - from 1/95
Front brake disc wear limit	mm (in.)	10 (0.39)	18 (0.71)	20 (0.79) - thru 12/94 23 (0.90) - from 1/95
Front brake pad thickness	mm (in.)	14 (0.55)		
Front brake pad wear limit	mm (in.)	7 (0.28) including backing plate		
Rear brake drum (if applicable)	diameter – mm (in.)	200 (7.87)		(not applicable)
Rear brake lining thickness (if applicable)	mm (in.)	5 (0.20)		(not applicable)
Brake lining width (if applicable)	mm (in.)	40 (1.57)		(not applicable)
Rear brake disc thickness (if applicable)	mm (in.)	–	10 (0.39)	
Rear brake pad thickness (if applicable)	mm (in.)	–	12 (0.47)	

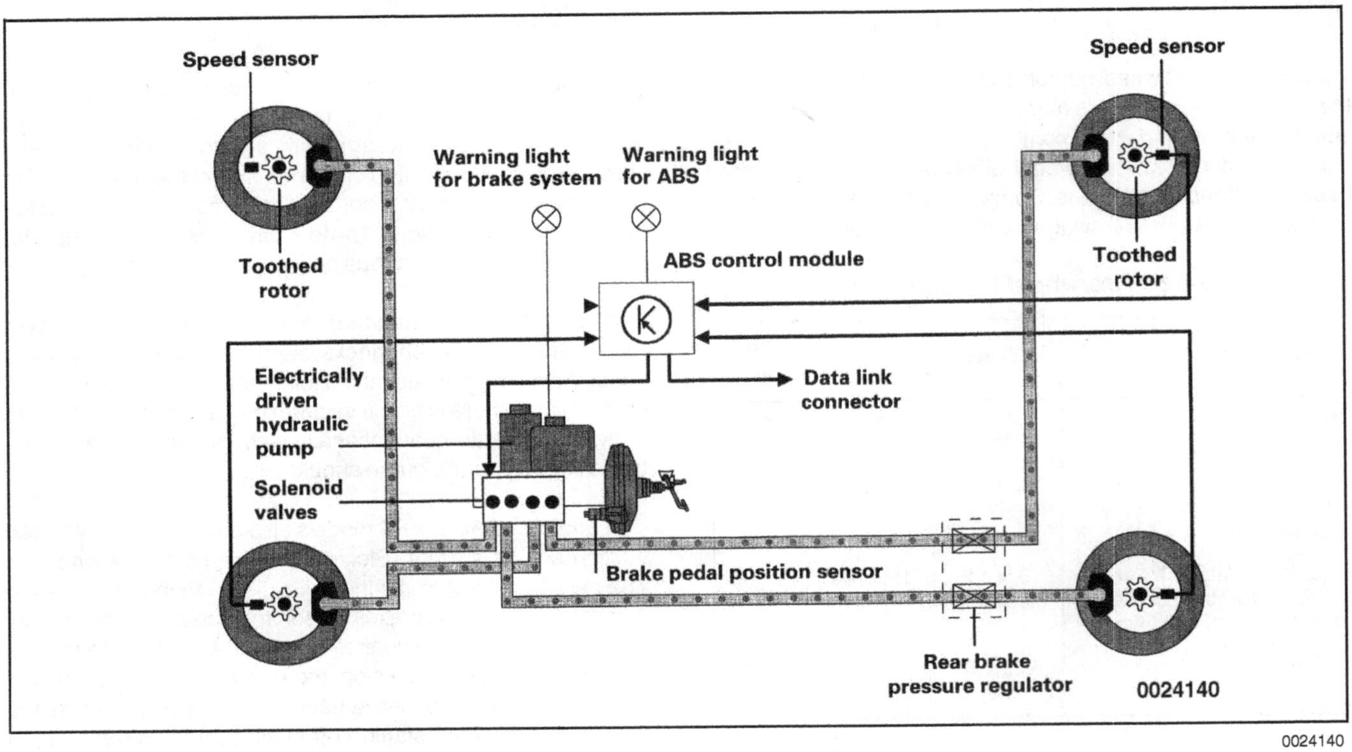

0024140

Fig. 3. System layout for ABS and traction control used on VR6 models.

STEERING

Power assisted rack and pinion steering is standard on all models covered by this manual. The power steering pump is a vane type pump. The hydraulic (power steering fluid) oil is a mineral based oil. The reservoir for the power steering pump has been designed with a long filler neck for easy access.

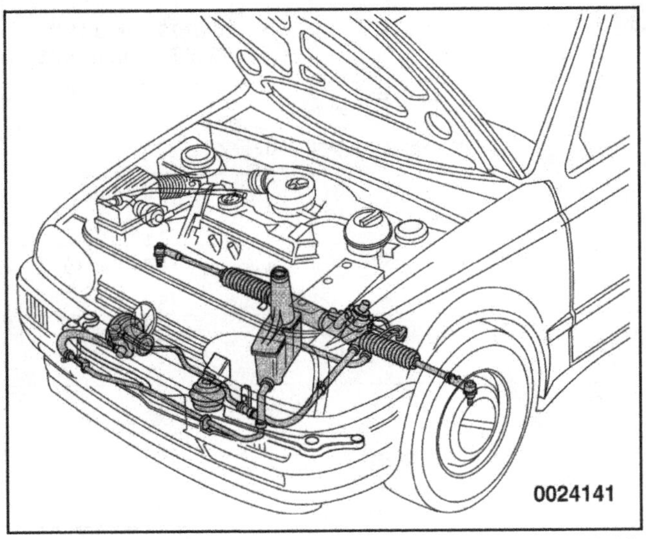

0024141

Fig. 4. Power-assisted, rack and pinion steering system.

NOTE—

For power steering fill capacity, see **0 Maintenance Program**.

The steering column is connected to the steering gear by a universal joint shaft. The universal joint shaft is angled in order to prevent collision impacts from forcing the steering wheel toward the driver. The steering wheel is energy absorbing and acts as a crumple zone. The steering column can retract as much as 170 mm (6.7 in.).

Table d. Steering Specifications

Model	Golf, GTI, Jetta, Cabrio
Steering gear	Power-assisted
Steering wheel turns, lock-to-lock	3.17
Overall steering ratio	17.5:1

NOTE—

For wheel alignment specifications see **44 Wheels–Tires, Wheel Alignment**.

40 Front Suspension and Drive Axles

GENERAL

Special tools and procedures are required for most front suspension repair and component replacement. In addition, front wheel alignment is almost always disturbed when suspension components are removed or replaced.

For general information on front and rear suspension operation, see **4 General Information**. For rear suspension servicing and repair information, see **42 Rear Suspension**. For wheel alignment specifications see **44 Wheels–Tires, Wheel Alignment**.

Most cars covered by this manual are equipped with the "base" front suspension, while the more performance-oriented cars are equipped with the "plus" suspension. Cars with the "plus" suspension system can be identified by the five-lug wheels (base suspension uses four-lug wheels) and ventilated front brake rotors.

> ***WARNING —***
> *Do not re-use any fasteners that are worn or deformed in normal use. Many fasteners are designed to be used only once and become unreliable and may fail when used a second time. This includes, but is not limited to, nuts, bolts, washers, self-locking nuts or bolts, circlips, cotter pins. For replacements, always use new parts.*

> ***WARNING —***
> *Do not reinstall bolts and nuts coated with undercoating wax as correct tightening torque cannot be ensured. Always clean the threads of removed bolts and nuts with a solvent before installation, or replace them with new parts.*

SUBFRAME AND CONTROL ARMS

Fig. 1 shows an exploded view of the front subframe and related parts. The control arm is mounted in replaceable rubber bushings that are subject to wear. The subframe supports the weight of the engine/transmission assembly. Note that the engine assembly must be supported before removing the subframe.

> ***WARNING —***
> - *Always replace self-locking nuts.*
> - *Always replace corroded bolts, nuts, washers.*
> - *DO NOT attempt to straighten or weld suspension strut, wheel bearing housing, control arm or any other wheel locating or load bearing components of the front suspension.*

Subframe/stabilizer bar/control arm, assembly

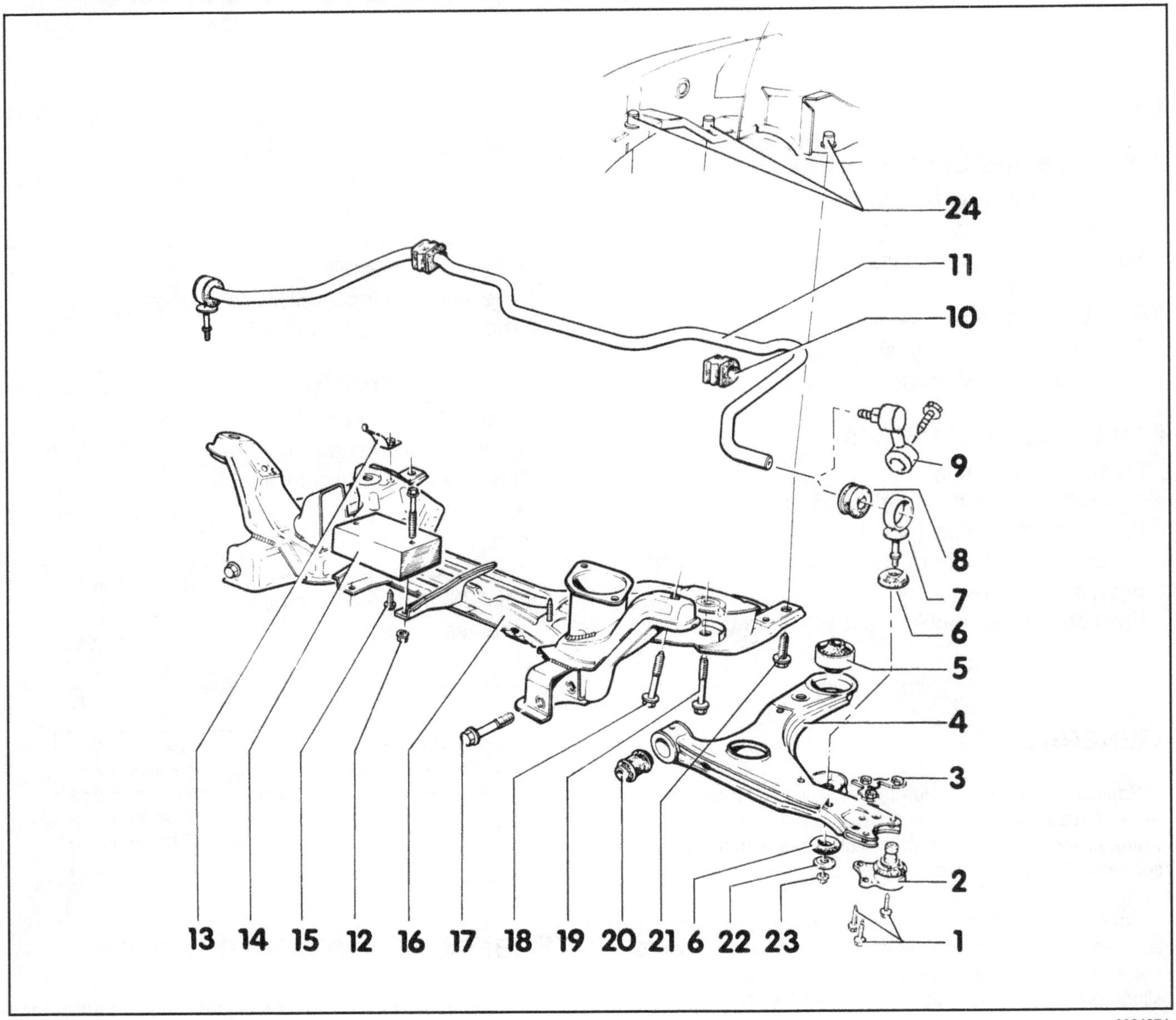

Fig. 1. Exploded view of subframe, stabilizer bar, and control arm.

> ***CAUTION—***
> *If a vehicle has to be moved (rolled) after removing the drive axle, install an outer constant velocity joint into the wheel hub and tighten to 50 Nm (37 ft-lb) to prevent damage to wheel bearing.*

1. **Bolts**
   Tighten to 35 Nm (26 ft-lb)
2. **Ball joint**
   - Checking, see **Ball joint, checking**
   - Mark installation position before removing
   - If replacing, install part centered in elongated hole and align front end
   - Elongated holes are NOT intended for camber adjustment
3. **Mounting plate**
4. **Control arm**
5. **Rear bushing for control arm**
   - Installed position, **see** Ⓐ
   - Pressing out/in, **see** Ⓑ
6. **Link rod bearing (base suspension only)**
   - Conical side faces control arm
7. **Link rod (base suspension only)**
8. **Rubber mounting (base suspension only)**
   - Coat with lubricant (e.g. soft soap) before pressing in
9. **Link rod (plus suspension only)**

(continued from previous page)

10. **Stabilizer bar mounting**

11. **Stabilizer bar**
    - Base suspension diameter:
      ACC, AAZ engines: 18mm
      1Z, AHU engines (Golf models): 18mm
      1Z, AHU engines (Jetta models): 20mm
      ABA engine: 20 mm
    - Plus suspension diameter:
      ABA, AAA engines: 20mm

12. **Nut**
    - Tighten to 25 Nm (18 ft-lb)

13. **Clamp**

14. **Vibration damper**

15. **Bolt**
    - Tighten to 25 Nm (18 ft-lb)

16. **Subframe**
    - Support weight of engine before removing, **see** (C)
    - After installing:
      - align engine and transmission assembly. See **10 Engine-Removing and Installing**.
      - align front wheels if steering gear/subframe attachment has been loosened. See **44 Wheels–Tires, Wheel Alignment**

17. **Bolt**
    - M12 x 1.5 x 82
    - Tighten to 50 Nm (37 ft-lb) plus 1/4 turn (90°)

18. **Bolt**
    - M12 x 1.5 x 65
    - Tighten to 70 Nm (52 ft-lb) plus 1/4 turn (90°)

19. **Bolt**
    - M12 x 1.5 x 78
    - Tighten to 70 Nm (52 ft-lb) plus 1/4 turn (90°)

20. **Front bushing for control arm**
    - Press in/out using hydraulic press
    - **Installing, see** (D)

21. **Bolt**
    - Tighten to 65 Nm (48 ft-lb)

22. **Washer (base suspension only)**
    - Collar faces away from control arm

23. **Nut (base suspension only)**
    - Tighten to 25 Nm (18 ft-lb)

24. **Cap nut**

(A) Rear bushing for control arm, installed position

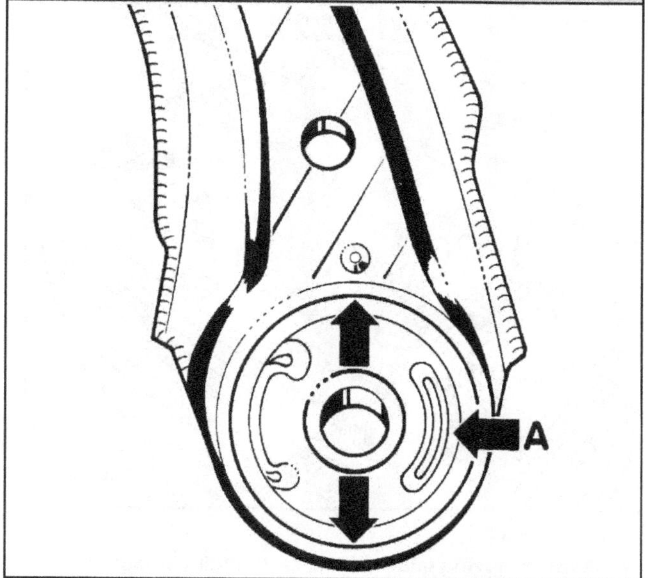

40-1010

- **One of two arrows on bushing must point toward recess in control arm.**
- **Kidney shaped opening (arrow A) must point toward center of vehicle.**

(B) Rear bushing for control arm, pressing out/in

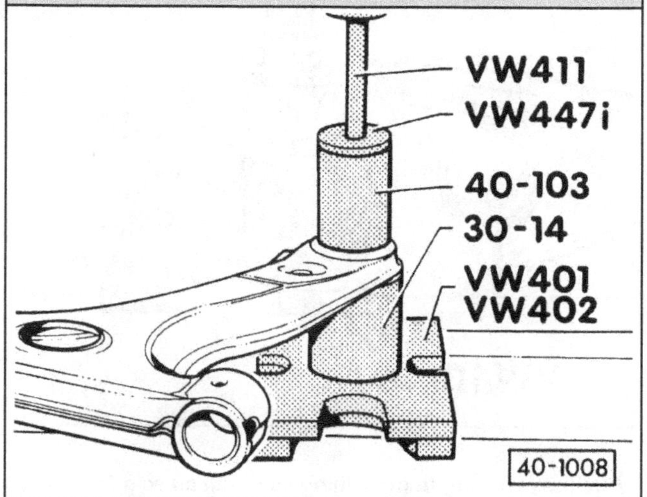

40-1008

- **Press the bushing using Volkswagen special tools as shown, or equivalent.**
- **Installed position, see (A) above**

C Supporting engine/transmission assembly for subframe removal

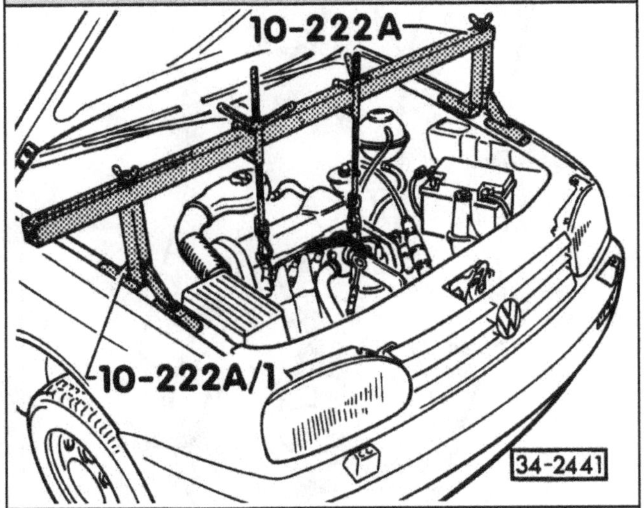

34-2441

- **Before removing subframe support engine/transmission assembly using Volkswagen special tools 10-222A and 10-222A/1, or equivalent.**

D Front control arm bushing, installing

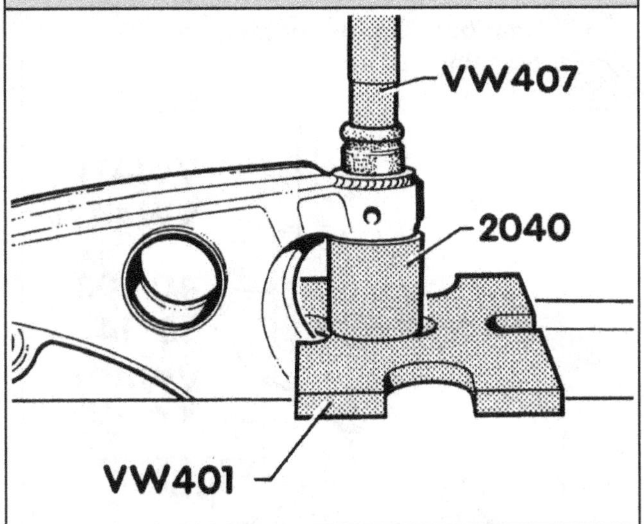

40-1024

- **Before installing front bushing coat with an acid-free lubricant (e.g. soft soap). Install using a hydraulic press and Volkswagen special tools as shown.**

Ball joint, checking

The ball joint can be checked for wear by checking for side-to-side and up-down play in the joint.

1. Check the ball joint axial (up-down) play as shown in Fig. 2. Replace the joint if any play is present.

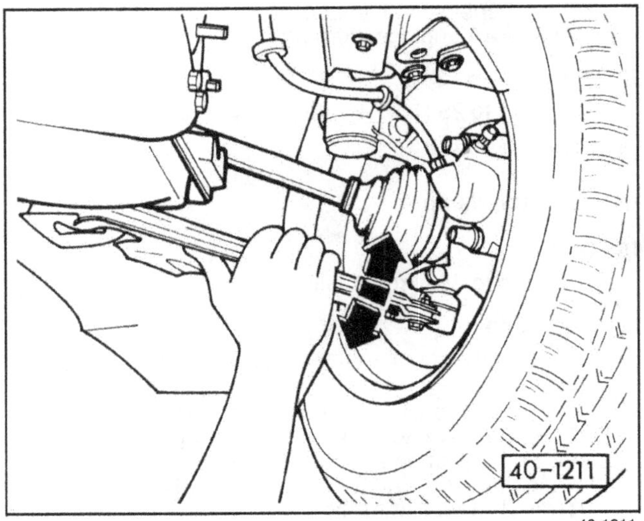

40-1211

Fig. 2. Ball joint axial play being checked. Push/pull control arm (**arrows**) to check for play.

NOTE —

When checking ball joint wear, be sure to consider wheel bearing and upper strut mount play before condemning the ball joint.

2. Check ball joint radial (side-to-side) play as shown in Fig. 3. Replace the joint if any play is present.

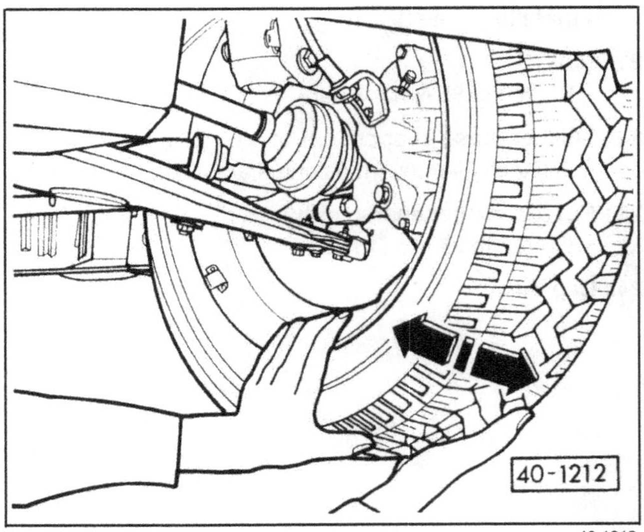

40-1212

Fig. 3. Ball joint radial play being checked. Push/pull lower part of wheel (**arrows**) to check for play.

Ball joint, removing and installing

Replacement of the ball joint affects front wheel alignment. Be sure to have the front wheels aligned as the final step in the job.

1. Disconnect drive axle from transmission drive flange. See **Drive Axles**.

2. Mark the installed position of the ball joint in relation to the control arm. Then remove ball joint bolts. See Fig. 4.

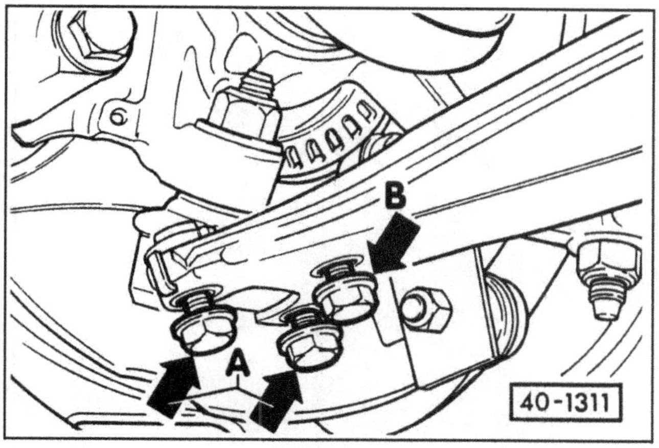

40-1311

Fig. 4. Ball joint mounting bolts (**arrows**). Mark installed position of ball joint to control arm before loosening bolts.

3. Separate ball joint from control arm.

4. Swing wheel and suspension strut outwards and support in position using a block of wood, or equivalent.

5. On cars with base suspension, remove the ball joint clamping nut and bolt. Remove the ball joint from the wheel bearing housing.

6. On cars with plus suspension, loosen the nut on top of the ball joint, but do not fully remove it. Press the ball joint loose. Remove the press and the ball joint. See Fig. 5.

> ***CAUTION —***
> *Use care when pressing the ball joint out of the wheel bearing housing. Once the ball joint breaks loose, the press will want to fall to the ground.*

7. Installation is the reverse of removal. Use a new self-locking nut at the wheel bearing housing. Make sure the matching marks made earlier are exactly aligned before tightening bolts. Check and adjust front wheel alignment.

NOTE —
On cars with plus suspension, use a 6mm hex key to counter hold the ball joint shaft when tightening the self-locking nut.

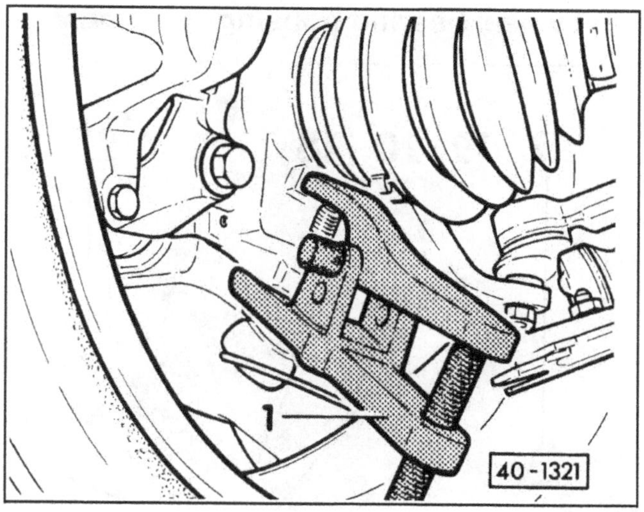

40-1321

Fig. 5. Ball joint being pressed out of wheel bearing housing on cars with plus suspension.

Tightening torques
- Ball joint to wheel bearing housing
  - base suspension .... 50 Nm (37 ft-lb)
  - plus suspension .... 45 Nm (33 ft-lb)
- Ball joint to control arm............ 35 Nm (26 ft-lb)

WHEEL BEARINGS

Removal and/or replacement of the front wheel bearings requires special tools. Note that the wheel is bearing is destroyed any time it is removed. Fig. 6 shows an exploded view of the front wheel bearing housing.

Observe the safety cautions before repairing or servicing the wheel bearing housing or any of its related parts.

> ***WARNING —***
> - *DO NOT attempt to straighten or weld suspension strut, wheel bearing housing, control arm or any other wheel locating or load bearing components of the front suspension.*
> - *Always replace self-locking nuts. Replace any corroded bolts, nuts, washers.*
> - *Wheel alignment must be checked and adjusted if ball joint or front strut assembly is removed.*
> - *If a vehicle has to be moved (rolled) once the drive axle is removed, install the outer CV joint into the wheel bearing housing and tighten the retaining nut to 50 Nm (37 ft-lb) to prevent damage to the wheel bearing.*

Front wheel bearing housing, assembly

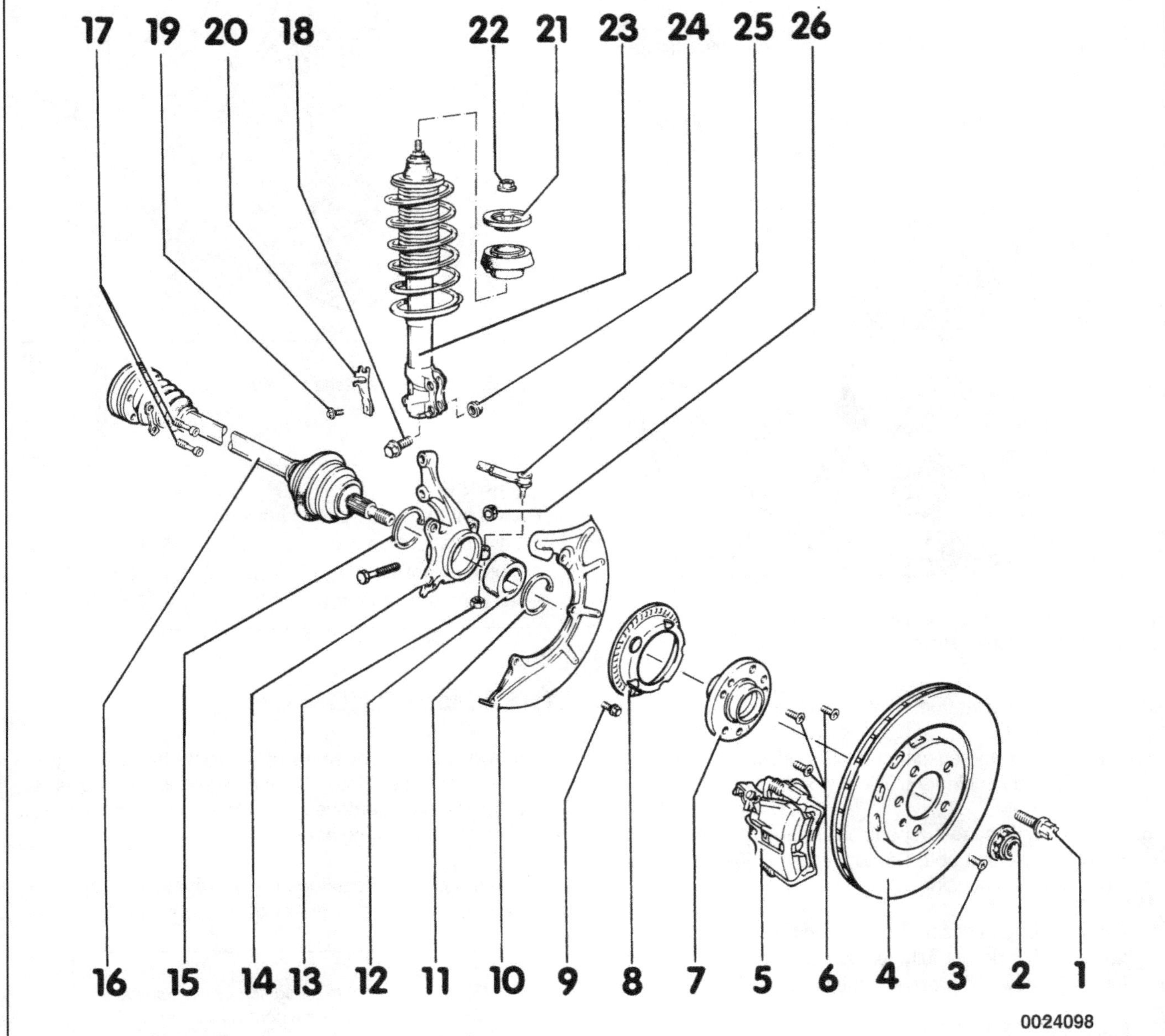

0024098

Fig. 6. Front wheel bearing housing and related components. Note ventilated rotor and ABS wheel speed rotor used on cars with plus suspension.

1. **Lug bolt**
   - Tighten to 110 Nm (81 ft-lb)
2. **Nut**
   - Special tightening procedures apply–See **Drive Axle**
3. **Screw**
4. **Brake Rotor**
   - See **46 Brakes–Mechanical Components**
5. **Brake Caliper**
   - Do not loosen/open hydraulic brake connections when servicing suspension
   - Do not let caliper hang by hose; hang from body using stiff wire.
6. **Screw**
7. **Wheel hub**
   - Removing/installing, see **Wheel bearing, removing and installing**

(continued from previous page)

8. **Rotor for ABS wheel speed sensor**
   - Only for vehicles with ABS
9. **Bolt**
   - Tighten to 10 Nm (89 in-lb)
10. **Splash shield**
11. **Circlip, outer**
    - always replace
12. **Wheel bearing**
    - Always replace (wheel bearing is destroyed when removed)

> ***CAUTION —***
> *Always replace wheel bearing after removing. It is destroyed during removal.*

13. **Nut**
    - Tighten to 35 Nm (26 ft-lb)
14. **Wheel bearing housing**
15. **Circlip, inner (base suspension only)**
    - always replace
16. **Drive axle**
    - Removing, installing, repairing, see **Drive Axle**
    - Clean paint residue and corrosion on threads of outer CV joint before nut (2) is installed
17. **Bolt**
18. **Bolt**
    - Mark installed position before loosening nut (22). See **44 Wheels–Tires, Wheel Alignment**
19. **Bolt**
20. **Bracket**
21. **Stop**
22. **Nut**
    - Loosening/tightening, **see** (A)
    - Tighten to 60 Nm (44 ft-lb)
23. **Suspension strut**
    - Removing, installing, disassembling, see **Front Suspension Struts**

> ***WARNING —***
> DO NOT disassemble strut assembly without spring compressor tools designed for the purpose. See **Front Suspension Struts**.

24. **Nut**
    - Mark position of bolt (18) before loosening
    - Tighten to 95 Nm (70 ft-lb)
25. **Tie rod**
    - Removing/installing, see **48 Steering**
26. **Nut (base suspension only)**
    - For ball joint clamping bolt
    - Tighten to 50 Nm (37 ft-lb)

(A) Front strut, top mount

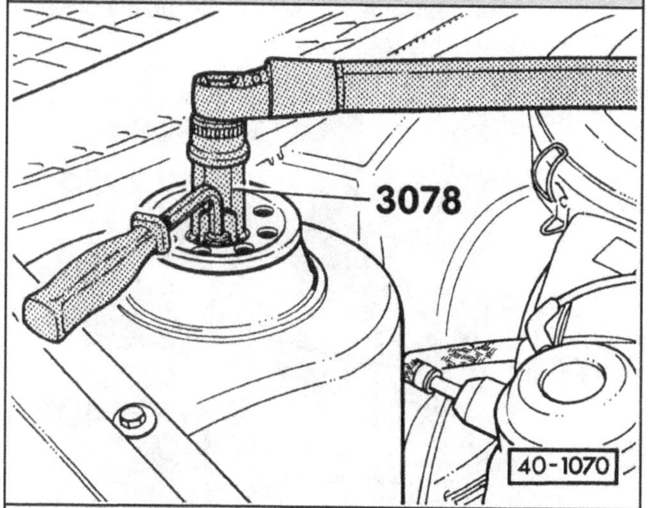

40-1070

- **Loosen strut to body mount using Volkswagen special tool 3078 (22mm special socket), or equivalent.**

Wheel bearing, removing and installing

The front wheel bearing is a permanently lubricated, sealed, double row ball bearing. The wheel bearing is pressed into the wheel bearing housing and held in place with a circlip(s). The wheel hub is in turn pressed into the wheel bearing.

> ***CAUTION —***
> - *Wheel bearing replacement requires the use of a hydraulic press and other specialized tools. Read the procedure through to determine what tools will be necessary.*
> - *Removal of the wheel hub destroys the wheel bearing. If wheel hub is removed, the wheel bearing must be replaced.*
> - *Wheel bearing replacement alters front wheel alignment. The wheels must be professionally aligned any time the wheel bearing housing is removed.*

1. Remove front brake caliper, pads and brake rotor. See **46 Brakes–Mechanical Components**.
2. On cars with ABS, remove ABS wheel speed sensor.
3. Remove drive axle from wheel bearing housing. See **Drive Axles**.

4. Remove suspension strut and steering tie rod end from the wheel bearing housing and remove the housing. See Fig. 6.

5. Using press equipment, press wheel hub out of wheel bearing. See Fig. 7.

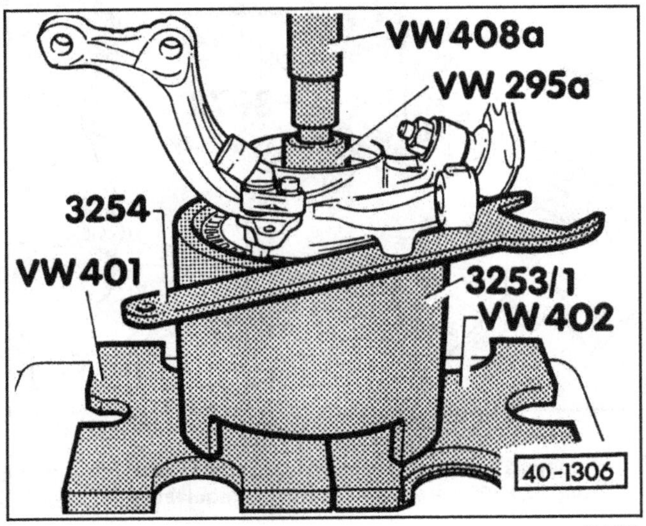

40-1306

Fig. 7. Wheel hub being pressed out of wheel bearing. Numbers indicate Volkswagen special tools. Wheel bearing housing for plus suspension shown.

6. Pull inner wheel bearing race off wheel hub. See Fig. 8.

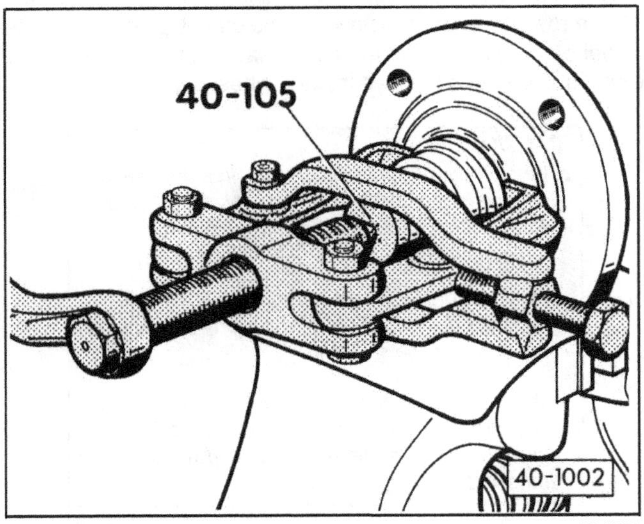

40-1002

Fig. 8. Wheel bearing inner race being removed from wheel hub with puller and adapter.

7. Remove circlip(s) from wheel bearing housing (See Fig. 6, given earlier). Press wheel bearing out of housing.

NOTE —
Cars with base suspension use two circlips; cars with plus suspension use only one circlip.

8. Clean wheel bearing housing bearing surface and oil lightly.

9. Install new inner circlip (if applicable).

CAUTION —
DO NOT re-use circlips.

10. Using a hydraulic press and adapters, press wheel bearing into wheel bearing housing until it contacts inner circlip (base suspension) or shoulder (plus suspension).

11. On car with base suspension, install outer circlip.

12. Install ABS rotor onto wheel hub (if applicable). Then press wheel hub into wheel bearing.

13. The remainder of installation is the reverse of removal. Install brake components as described in **47 Brakes–Mechanical Components**. Once installation is complete check and adjust front wheel alignment.

FRONT SUSPENSION STRUTS

Front suspension strut application varies depending on suspension system installed. Fig. 9 shows the strut assembly used on cars with base suspension through 1995 model year (m.y.). Fig. 10 shows the strut assembly used on cars with plus suspension and cars with base suspension from 1996 m.y.

WARNING —
- *Suspension strut springs are compressed and under pressure when installed on struts.*
- *DO NOT attempt to disassemble or repair suspension without proper tools and experience. Serious injury will result from improper tools or procedures.*

Front strut, assembly (base suspension through 1995 m.y.)

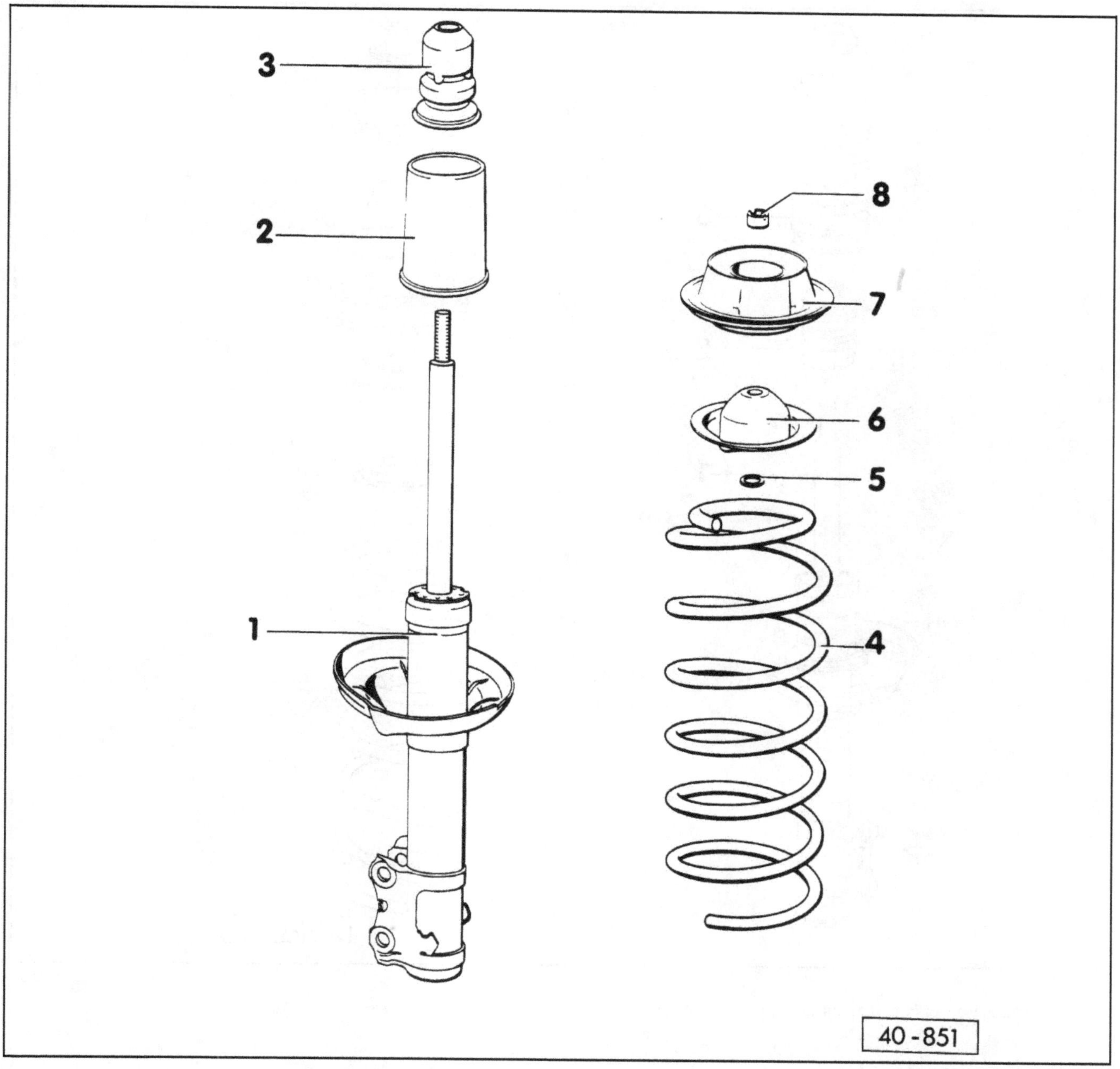

40-851

Fig. 9. Exploded view of front strut assembly used on cars with base suspension through 1995 model year.

1. **Shock absorber insert**
   - Insert may be replaced individually
   - Slight traces of oil do not necessitate replacement
2. **Protective sleeve**
3. **Bump stop**
4. **Coil spring**
   - Note color-code identification when replacing
   - Outer surface of spring must not be damaged
5. **Washer**
6. **Spring retainer**
7. **Strut bearing**
   - Replace complete part
8. **Slotted nut**
   - Tighten to 40 Nm (30 ft-lb)
   - Compress coil spring before loosening nut.

> ***WARNING —***
> ***DO NOT** loosen nut unless coil spring is compressed. The spring must not be exerting force on the nut.*

Front strut, assembly (plus suspension and base suspension from 1996 m.y.)

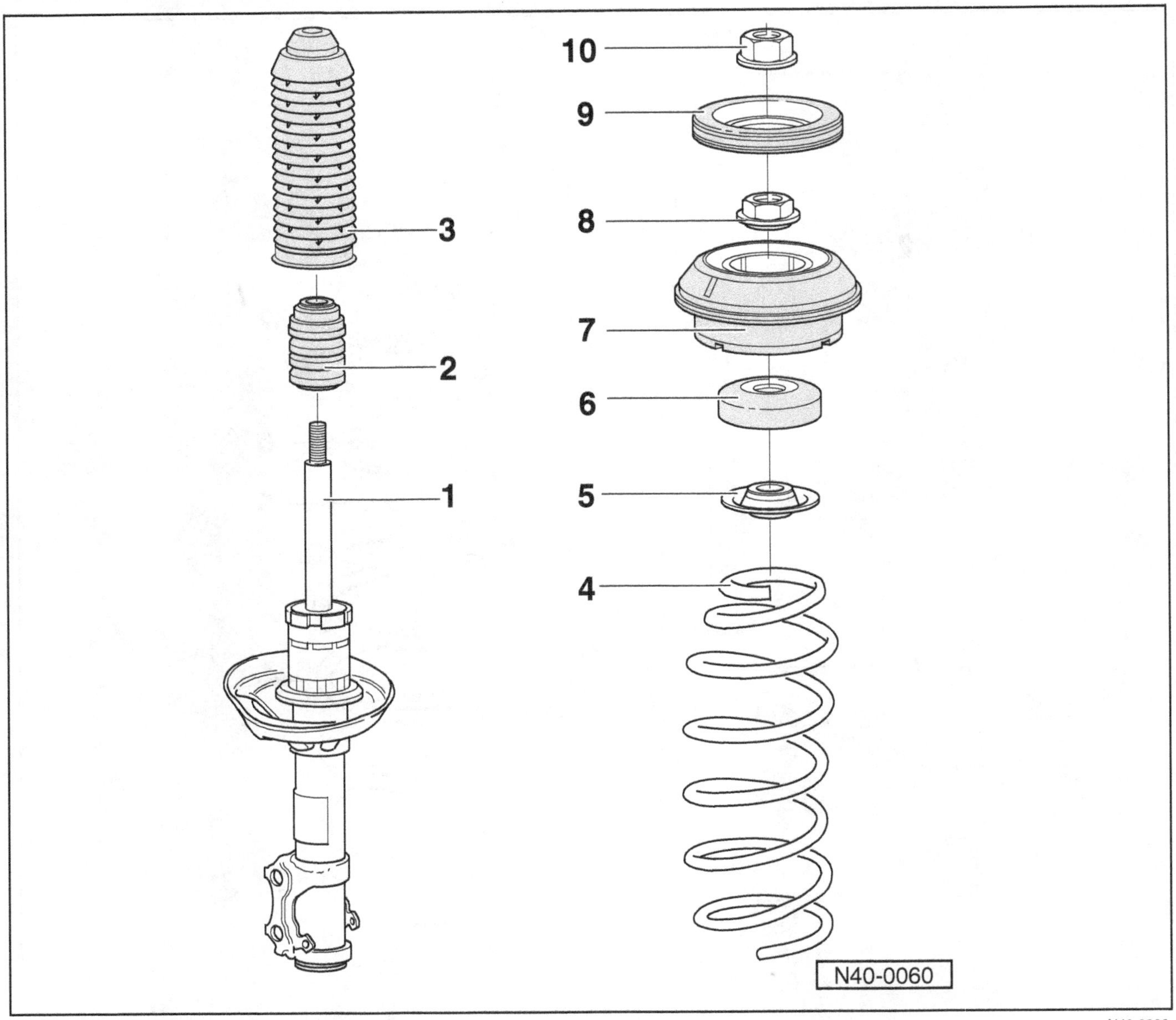

N40-0060

Fig. 10. Exploded view of front strut assembly used on cars with plus suspension and base suspension from 1996 m.y.

1. **Shock absorber insert**
   - Insert can be replaced individually
   - Slight traces of oil do not necessitate replacement
2. **Bump stop**
3. **Protective sleeve**
4. **Coil spring**
   - See **Front strut, disassembling (plus suspension)**
   - Note color-code marking for identification
   - Outer surface of spring must not be damaged
5. **Spring retainer**
6. **Bearing**
7. **Suspension strut mount**
8. **Hex nut**
   - Tighten to 60 Nm (44 ft-lb)
   - Each time nut (item 10) is removed or loosened, nut (item 8) must be tightened to above torque
   - Compress coil spring before loosening

> ***WARNING —***
> *DO NOT loosen nut unless coil spring is compressed. The spring must not be exerting force on the nut.*

9. **Stop**
10. **Self-locking nut**
    - Tighten to 60 Nm (44 ft-lb)

Front strut, removing and installing

> *WARNING —*
> - *Suspension strut springs are compressed and under pressure when installed on struts.*
> - *DO NOT attempt to disassemble or repair suspension without proper tools and experience. Serious injury will result from improper tools or procedures.*
> - *Suspension strut is removed and installed as an assembly. DO NOT disassemble or repair in car.*
> - *Front wheel alignment (camber) is altered when the strut assembly is removed. Wheel alignment must be checked and adjusted any time the strut assembly is removed or it mounting bolts are loosened.*

1. Raise the car and support it on jack stands.

> *CAUTION —*
> *Use jack stands designed for the purpose. The jack stands should be on a level hard surface*

2. Remove the wheel.

3. Remove the nuts from the two bolts that hold the strut to the wheel bearing housing, leaving the bolts in place. See Fig. 11.

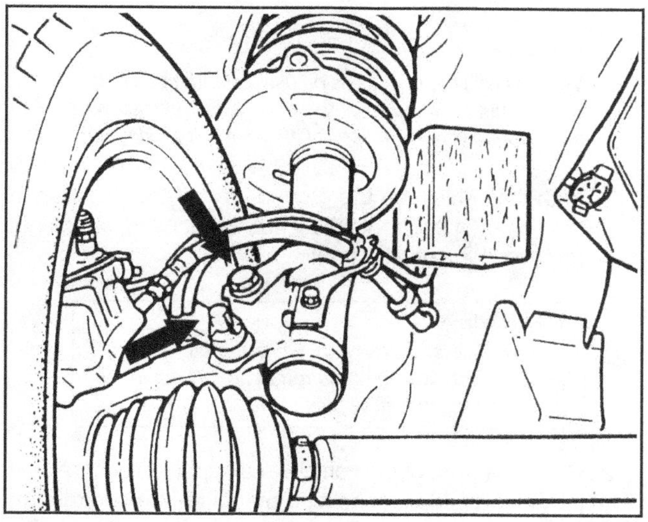

Fig. 11. Strut to wheel bearing housing mounting bolts (**arrow**).

4. Disconnect ABS speed sensor wiring (if applicable) and unhook the brake line from the strut and secure out of the way.

5. Support lower control arm and brake assembly so that it will not fall when the strut is removed.

6. Remove the 21 mm flange nut at the top of the strut. See Fig. 12.

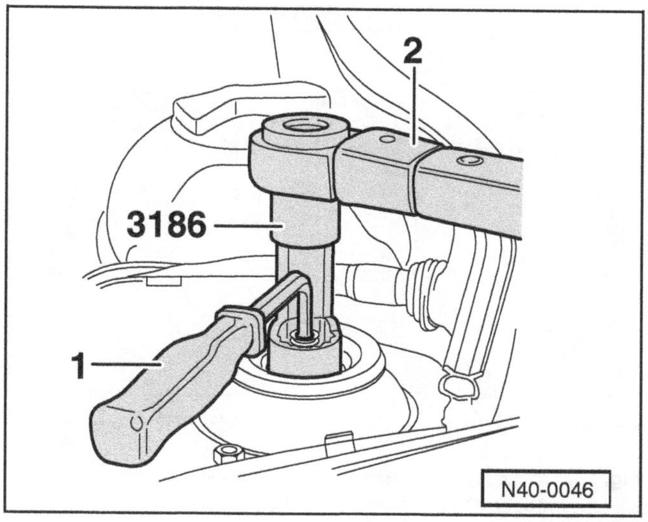

N40-0046

Fig. 12. Flange nut being loosened with Volkswagen special tool 3186, hex wrench holds shock absorber shaft.

7. Remove lower strut bolts and remove the strut assembly from the car.

8. Install the strut in the reverse order of removal. Check and adjust front wheel alignment once installation is complete.

Tightening torques
- Flange nut on top of strut . . . . 60 Nm (44 ft-lb)
- Strut assembly to wheel bearing housing . . . . 95 Nm (70 ft-lb)

Front strut, disassembling and assembling

1. Remove front strut from the car. See **Front strut, removing and installing**.

2. Compress coil spring using a spring compressor designed for the purpose. With the spring compressed, there should be no spring pressure on the strut bearing. See Fig. 13.

> *WARNING —*
> *Suspension coil springs are compressed and under pressure when installed on struts. DO NOT attempt to disassemble or repair suspension struts without proper tools and experience. Serious injury will result from improper tools or procedures.*

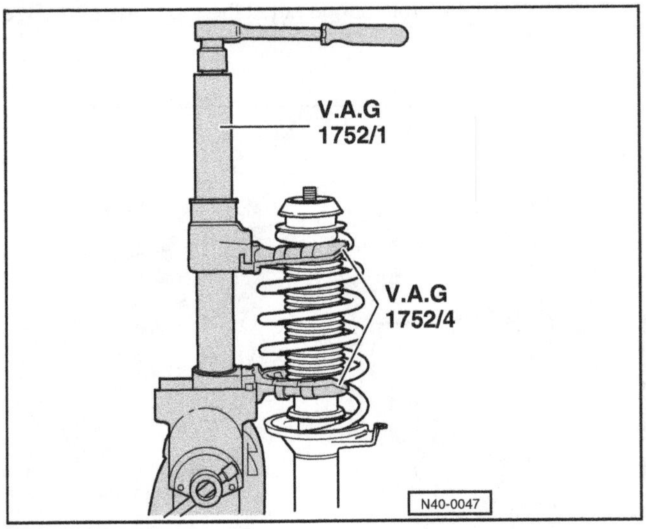

N40-0047

Fig. 13. Front coil spring being compressed. (Numbers indicate Volkswagen special tools.)

3. Remove the nut from the shock absorber rod. Use a wrench to hold the shock absorber shaft.

NOTE —
On cars with base suspension, the retaining nut (Fig. 9, item 8) is a special slotted nut. Use Volkswagen special tool no. VW 524, or equivalent, to loosen and tighten the nut.

4. Assemble the strut by fully extending the shock rod by hand. Install all parts to the strut. With the spring compressed and no pressure on the upper strut bearing, tighten the nut. Align the ends of the coil spring to mate properly with the spring retainers and slowly release the spring compressor.

NOTE —
- *Check the strut bearing for signs of wear and replace it if it does not turn smoothly or if any play is present.*
- *If the coils springs are to be replaced, note the color marking on the spring and always replace in matched pairs.*

Tightening torques
- Slotted nut on top of strut (base suspension) . . . . 40 Nm (30 ft-lb)
- Hex nut on top strut (plus suspension). . . . . 60 Nm (44 ft-lb)

5. Install the strut assembly as described under **Suspension strut, removing and installing**. Check and adjust front wheel alignment.

DRIVE AXLES

The drive axles are equipped at each end with constant velocity (CV) joints which allow power to be continuously delivered to the drive wheels. See Fig. 17. The inner CV joints attach to the transaxle drive flanges, while the outer CV joints are equipped with splined stub axles that engage the splines of the wheel hubs and are secured by the large nut.

Although some inspection of the CV joints is possible while the drive axles are installed, complete inspection and thorough lubrication of the joints require that the drive axles be removed from the car. If the drive axles themselves are damaged, complete remanufactured drive axle assemblies are available from an authorized Volkswagen dealer at lower cost than a new unit.

CAUTION —
- *Drive axles should be removed and installed as a complete unit. DO NOT repair or disassemble drive axles without first removing from vehicle.*
- *Moving (rolling) a car with drive axles removed will damage the wheel bearings. If moving the car is necessary, always secure the bearing races by temporarily installing an outer CV joint in the hub and tighten the retaining nut to 50 Nm (37 ft-lb).*

Drive axle, removing and installing (base suspension)

NOTE —
Vehicles with base suspension use a standard self-locking hex nut on the end of the drive axle; vehicles with plus suspension use a special 12-point self-locking nut.

1. With the vehicle on the ground, pry off the dust cap in the center of the front wheel hub and loosen, but do not remove, the axle shaft nut.

WARNING —
Loosen axle shaft nut only while the car is on the ground. The leverage required to do this could topple the car from a lift or jack stand.

2. Raise the front of the vehicle and support it securely on jack stands. Make matching marks on the control arm and ball joint, then remove the ball joint mounting bolts. See Fig. 14.

3. Clean around the inner CV joint and remove the 12-point socket-head bolts that hold the inner CV joint to the transmission drive flange.

4. Remove the axle shaft nut and pull the drive axle out of the wheel hub. Cover the exposed inner CV joints to prevent contamination.

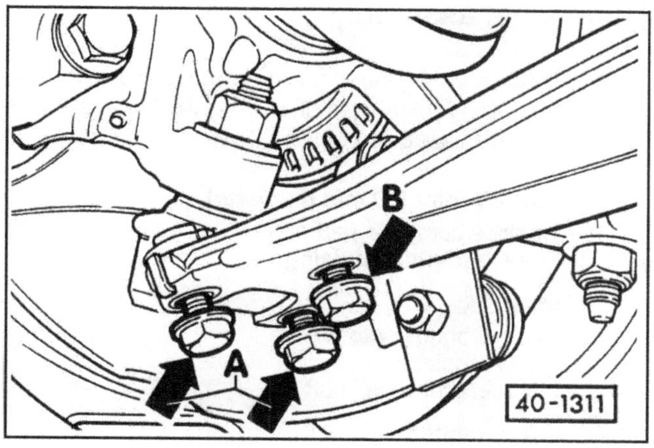

40-1311

Fig. 14. Remove ball joint mounting bolts **A** and **B**.

5. Installation is the reverse of removal, noting the following:
   - Use a new self-locking nut to secure drive axle to wheel hub.
   - Align the ball joint and control arm marks made during removal.
   - Tighten the axle shaft nut as much as possible (to seat CV joint in wheel hub) with the car raised. Make the final tightening torque only with the weight of the car on the ground.
   - Check and adjust front wheel alignment once installation is complete.

> ***WARNING—***
> *Torque the axle shaft nut only while the car is on the ground. The leverage required to do this could topple the car off a lift or jack stand.*

Tightening torques
- Outer CV joint to wheel hub . . . . . . 265 Nm (195 ft-lb)
- Inner CV joint to drive flange . . . . 45 Nm (33 ft-lb)
- Ball joint to control arm . . . . 35 Nm (26 ft-lb)

Drive axle, removing and installing (plus suspension)

> ***NOTE—***
> *Vehicles with base suspension use a standard self-locking hex nut on the end of the drive axle; vehicles with plus suspension use a special 12-point self-locking nut.*

Two types of wheel hubs have been installed on the cars covered by this manual. The early-type hub uses a locking fluid on the CV joint shaft splines. The later type hub uses a modified spline design and no locking fluid. Be sure to note if sealant is present on the drive axle and wheel hub splines when removing the axle. If sealant is present, the drive axle must be reinstalled into the original hub using the specified locking compound.

1. With the vehicle on the ground, pry off the dust cap in the center of the front wheel hub and loosen, but do not remove, the axle shaft nut.

> ***WARNING—***
> *Loosen axle shaft nut only while the car is on the ground. The leverage required to do this could topple the car from a lift or jack stand.*

2. Raise the front of the car and support it on jack stands. Make matching marks on the control arm and ball joint, then remove the ball joint mounting bolts. See Fig. 14.

3. Clean around the inner CV joint and remove the 12-point socket-head bolts that hold the inner CV joint to the transmission drive flange.

4. Remove the axle shaft nut and remove the drive axle from the wheel hub. If necessary, press the axle out of the wheel hub. See Fig. 15. Cover the exposed inner CV joint.

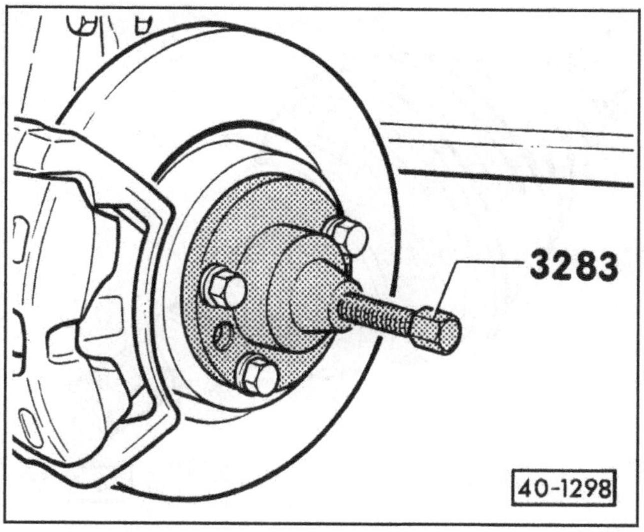

40-1298

Fig. 15. Axle shaft being pressed out of wheel hub. Number indicates VW special tool.

5. Installation is the reverse of removal, noting the following:
   - Use new 12-point self-locking nut to secure drive axle to wheel hub. Note special tightening torques.
   - On cars **without modified wheel hub** (sealant used on axle splines): use special locking sealant (VW part no. D 185 400A2) on the axle splines if reinstalling the drive axle into the same wheel hub. See Fig. 16.
   - On cars **with modified wheel hub** (sealant not used on axle splines): lubricate contact surfaces of nut with oil when installing., tighten axle shaft nut with the car lowered but wheel not touching ground, and brakes applied (requires helper).
   - Align the ball joint and control arm marks made during removal.
   - Check and adjust front wheel alignment once installation is complete.

Tightening torques

- Outer CV joint to wheel hub (w/o modified wheel hub) (use sealant, VW part no. D 185 400 A2)
  - stage 1 . . . . 90 Nm (66 ft-lb)
  - stage 2 . . . . additional 1/8 turn (45°)
- Outer CV joint to wheel hub (with modified wheel hub)
  - stage 1 . . . . 200 Nm (148 ft-lb)
  - stage 2 . . . . loosen one turn
  - stage 3 . . . . 50 Nm (37 ft-lb)
  - stage 4 . . . . tighten 1/12 turn (30°)
- Ball joint to control arm . . . . 35 Nm (26 ft-lb)
- Inner CV joint to drive flange . . . . 45 Nm (33 ft-lb)

CAUTION —
Part numbers are listed for reference only. Always check with an authorized Volkswagen parts department for the latest parts information.

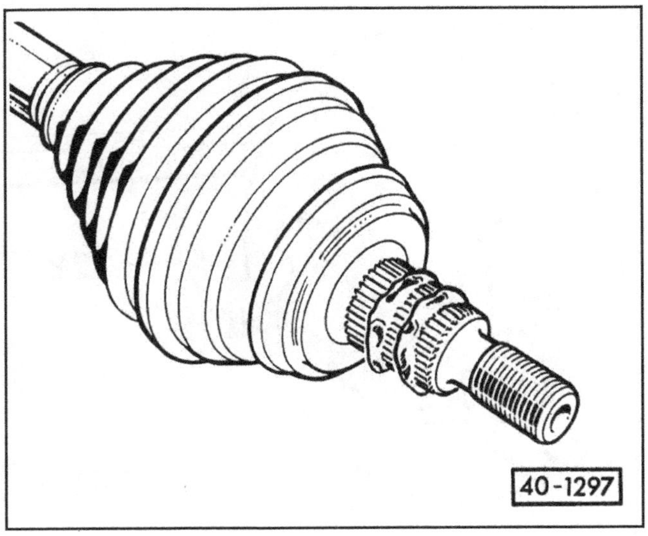

40-1297

Fig. 16. Apply 3 mm (1/8 in.) bead of sealant (VW part no. D 185 400 A2) to CV joint splines on cars without modified wheel hub.

Drive axle, assembly

Fig.17 shows an exploded view of the drive axle components. Applicable tightening torques as well as other pertinent repair information are given with the illustration.

1. **Outer CV joint (90 mm diameter)**
   - Replace complete part only
   - See **CV joints, servicing**
2. **Circlip**
   - Seats (springs) into groove
   - Always replace
   - Insert in groove in shaft
3. **Thrust washer**
4. **Dished washer**
   - Splined on inside diameter
   - Installed position: outer diameter (concave side) faces thrust washer
5. **Outer clamp**
   - Always replace
   - Use pliers designed to install CV boot clamp, such as Snap-On YA 3080
6. **Boot**
   - For 90 mm constant velocity joint (1)
   - Check for tears and chafing
   - Ventilate boot before tensioning small clamp
7. **Inner clamp**
   - Always replace
   - Use pliers designed to install CV boot clamp such as Snap-On YA 3080
8. **Vibration damper**
   - Not applicable to all models
9. **Drive Axle**
10. **Boot**
    - For 100 mm diameter CV joint (12)
    - Installed up to 1/93 with vent hole
    - Check for tears and chafing
11. **Dished washer**
    - Splined on inside diameter
    - Installed position: outer diameter (concave side) faces thrust washer
12. **Inner CV joint (100 mm diameter)**
13. **Gasket**
    - Always replace (self-adhesive backing)
14. **Circlip**
    - Always replace

(continued on following page)

Drive axle, assembly

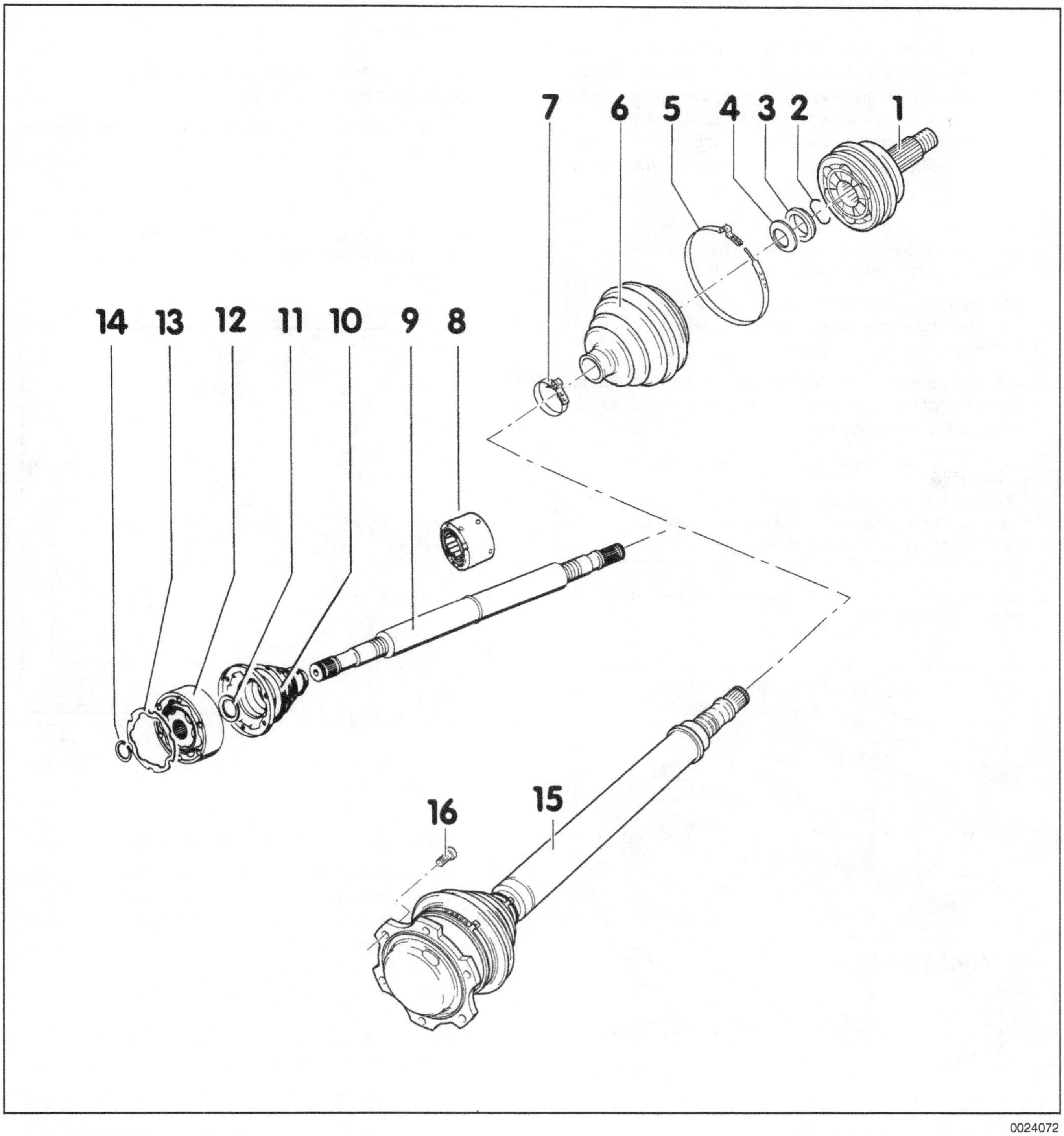

Fig. 17. Drive axle components. Note drive axle with triple-roller inner CV joint used on cars with 4-cylinder engine with automatic transmission.

15. **Drive axle with triple-roller inner CV joint**
    - Installed on 4-cylinder engine with auto. trans.
    - Replace as complete unit (shaft, joint, and boot)

16. **Multi-point socket head bolt**
    - Tighten to 45 Nm (33 ft-lb)

Drive axle with triple-roller joint, removing and installing (left-hand axle only)

Axle shafts with triple-roller joints are installed on vehicles with 4-cylinder engine and automatic transmission. These joints have three offset rollers in a triangular star carrier. If removing the left-hand drive axle, the engine assembly must be supported from above and moved forward off its mounts in order to gain enough clearance to remove the axle.

NOTE —

- *At the time of publication, replacement parts for triple-roller CV joints were not available from Volkswagen.*
- *Removal and installation for the right side drive axle is covered above under* **Drive axle, removing and installing**.

1. With the vehicle on the ground, pry off the dust cap in the center of the front wheel hub and loosen, but do not remove, the axle shaft nut.

WARNING —
Loosen axle shaft nut only while the car is on the ground. The leverage required to do this could topple the car from a lift or jack stand.

2. Install engine support equipment across the top of the engine. See Fig. 18.

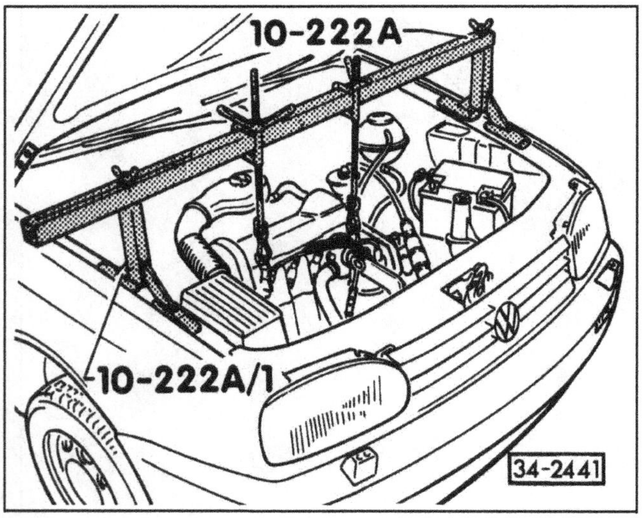

34-2441

Fig. 18. Engine/transmission assembly being supported. (Numbers indicate Volkswagen special tools.)

3. Remove bolt from rear transmission mount and front engine mount. Lift engine/transmission assembly so that engine/transmission weight is fully supported.

4. Raise the front of the vehicle and support it securely on jack stands. Make matching marks on the control arm and ball joint, then remove the ball joint mounting bolts. See Fig. 14.

5. Remove the 12-point socket-head bolts that hold the inner CV joint to the transmission drive flange.

6. Remove the axle shaft nut.

7. Push engine/transmission assembly toward front of car and support in this position.

8. Swing wheel outward and pull drive axle out of wheel hub. Remove axle from car.

NOTE —
If necessary, place block of wood or equivalent between strut and wheel housing to hold strut in position. See Fig. 19.

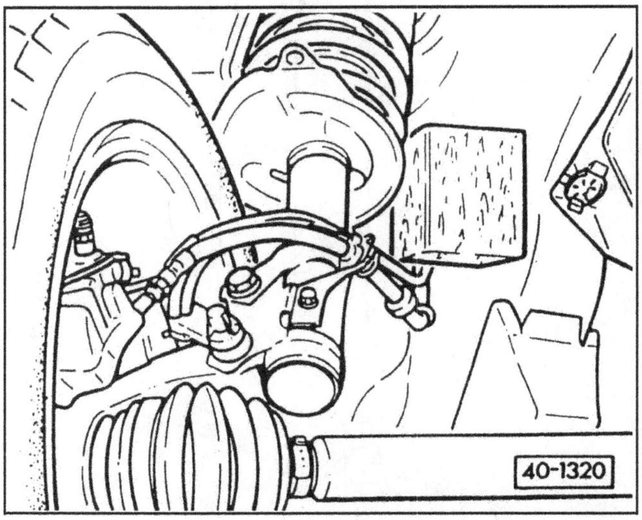

V40-1320

Fig. 19. Wood block placed between strut and wheel housing to facilitate axle removal.

9. Installation is the reverse of removal, noting the following:
   - Use a new self-locking nut to secure drive axle to hub. Tighten the axle shaft nut as much as possible (to seat CV joint in wheel hub) with the car raised. Make the final tightening torque only with the weight of the car on the ground.
   - Align the ball joint and control arm marks made during removal.
   - Check and adjust front wheel alignment once installation is complete.

WARNING —
Torque for axle shaft nut applies only to conventional hex-type self-locking nut. Torque does not apply to special 12-point self-locking nut.

Tightening torques

- Outer CV joint to wheel hub . . . . . . 265 Nm (195 ft-lb)
- Inner CV joint to drive flange . . . . 45 Nm (33 ft-lb)
- Ball joint to control arm. . . . . 35 Nm (26 ft-lb)

CV JOINTS, SERVICING

The components of each CV joint are precisely matched during manufacture and they cannot be serviced individually. The joint can be disassembled to replace the grease or to check the balls and ball tracks for wear and damage.

> ***CAUTION —***
> - *CV joints must be removed from the axle shaft prior to disassembly.*
> - *DO NOT re-use drive axle circlips or CV boot clamps.*
> - *DO NOT interchange parts between CV joints.*

CV Joint Grease Quantity
- Outer CV joint .... 110 grams
- Inner CV joint .... 120 grams

Outer CV joint and boot, removing and installing

1. Remove drive axle from car. See **Drive Axles**.
2. Mark position of CV boot on drive axle.
3. Remove drive axle boot clamps. Disconnect boot from CV joint and cut boot from drive axle.
4. Carefully drive CV joint off drive axle with firm blow from soft mallet. See Fig. 20.

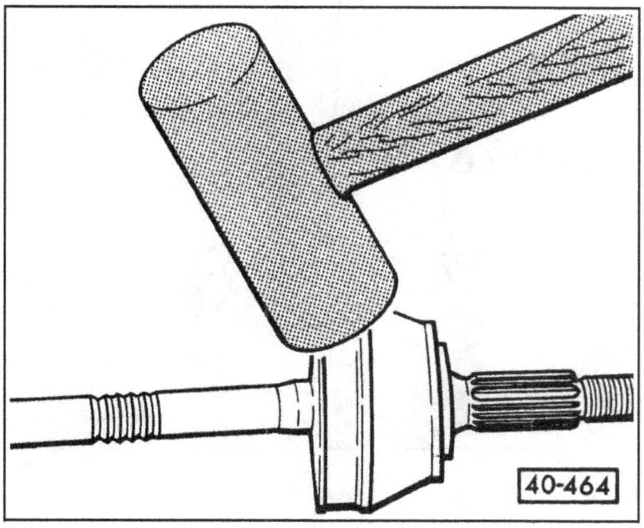

40-464

Fig. 20. CV joint being driven off axle with soft mallet.

5. Remove thrust washer and dished washer.
6. Clean drive axle. Install new circlip on drive axle.
7. Slide inner (small) CV boot clamp onto drive axle.
8. Slide new CV boot onto drive axle. Position CV boot on axle, aligning marks made earlier.
9. Pack CV boot with CV joint grease (55 grams).
10. Install dished washer and thrust washer onto axle. Dished washer outer diameter faces thrust washer.
11. Pack CV joint with 55 grams of CV joint grease.
12. Using a soft mallet drive the CV joint onto the axle until the circlip is seated (springs into groove).
13. Install grease packed boot to CV joint. Boot should fit into groove in CV joint.
14. Install outer clamp and crimp.

> ***NOTE —***
> *Special pliers should be used to crimp CV boot clamps. Use Snap-On YA 3080 pliers, or equivalent.*

15. Lift the boot slightly off the joint to ventilate and equalize pressure. Then slide inner clamp onto boot and crimp.

Outer CV joint, disassembling and assembling

1. Remove CV joint as described above.
2. Mark position of ball hub in relation to ball cage and housing with electric scriber.
3. Swivel ball hub and ball cage so that balls are exposed. See Fig. 21.

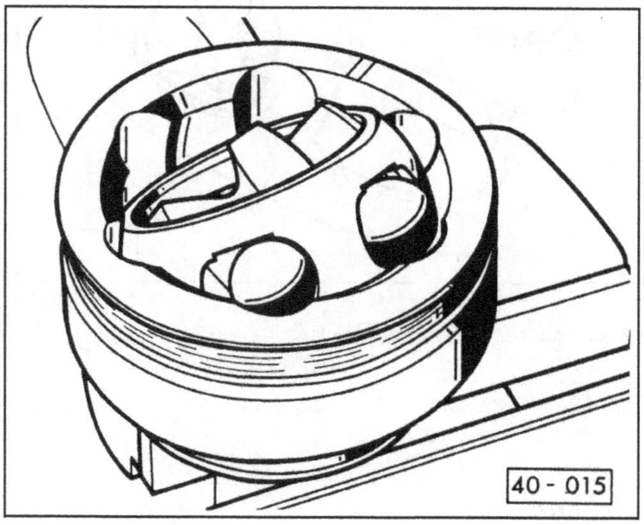

40-015

Fig. 21. Ball track swivelled for ball removal.

4. Remove each ball in turn, swiveling the ball hub and cage as necessary.

5. Turn cage until rectangular openings on either side of ball cage align with joint housing and lift out ball cage together with hub. See Fig. 22.

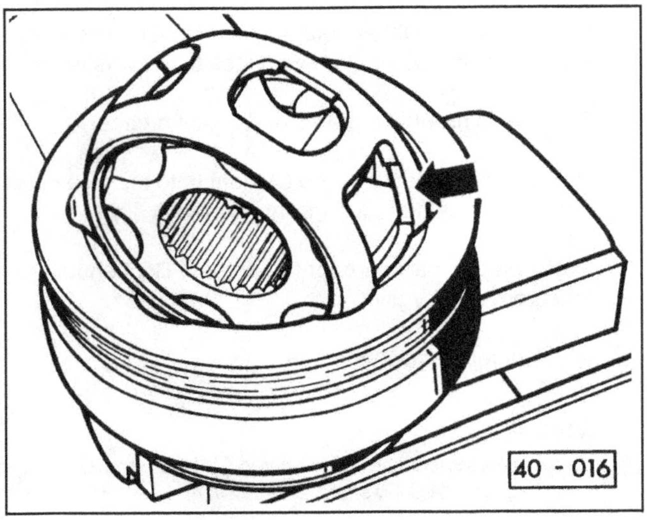

40-016

Fig. 22. Rectangular openings (**arrow**) aligned with housing.

6. Remove ball hub from ball cage by swinging segment of hub into rectangular opening of cage and tilting hub out of cage. See Fig. 23.

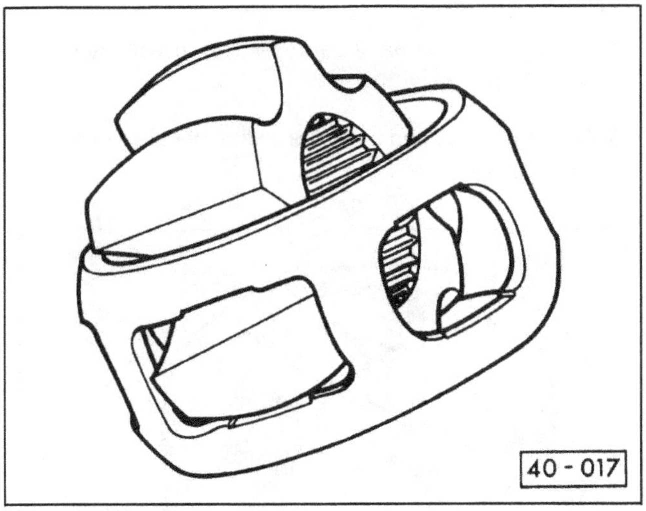

40-017

Fig. 23. Hub being aligned with rectangular opening in cage.

7. Check housing, hub, cage, and balls for pitting and signs of seizure.

NOTE —

Excessive backlash in the joint will be noticed as a knock when changing from acceleration to overrun or vice versa. Replace the joint if necessary. Do not replace the joint because of a polished appearance or because ball tracks are visible.

8. Insert ball hub into ball cage at rectangular opening. See Fig. 23.

9. Pack CV joint housing with 55 grams CV joint grease.

10. Install cage together with hub into joint housing while aligning rectangular openings in ball cage. Make sure marks made earlier align when installing the cage into the housing.

11. Press in balls separately from alternate sides. Ensure original position of hub in relation to cage.

12. Install CV joint to drive axle as described above.

Inner CV joint and boot, removing and installing

1. Remove drive axle from car. See **Drive Axles**.

2. Mark the position of the inner CV boot on the drive axle.

3. Wipe away the grease from the end of the drive axle. Remove the circlip holding the inner CV joint to the end of the drive axle.

4. Using a drift, tap off the CV joint boot protective cap The cap is pressed onto the CV joint.

5. While supporting the ball hub, press the drive axle out of the CV joint. Remove the dished washer from the axle. See Fig. 24.

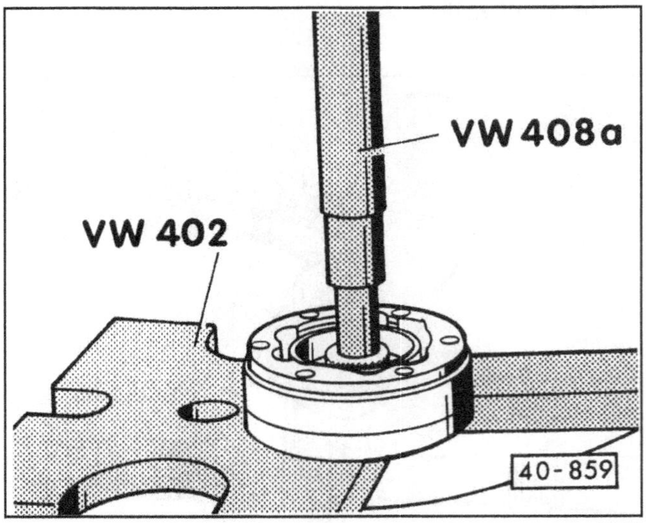

40-859

Fig. 24. Axle shaft being pressed out of inner CV joint. Numbers indicate VW special tools.

6. Remove the CV boot clamp and slide the boot off.

7. Clean drive axle.

8. Slide inner (small) CV boot clamp onto axle.

9. Slide CV boot onto drive axle.

10. Pack boot with 60 grams of CV joint grease.

11. Install dished washer so that dished washer outer diameter faces CV joint.

12. Pack CV joint with 60 grams of CV joint grease.

13. Press CV joint onto axle and install circlip. See Fig. 25.

NOTE —
Chamfer on inner ball hub (splined diameter) must face the contact shoulder on the drive axle.

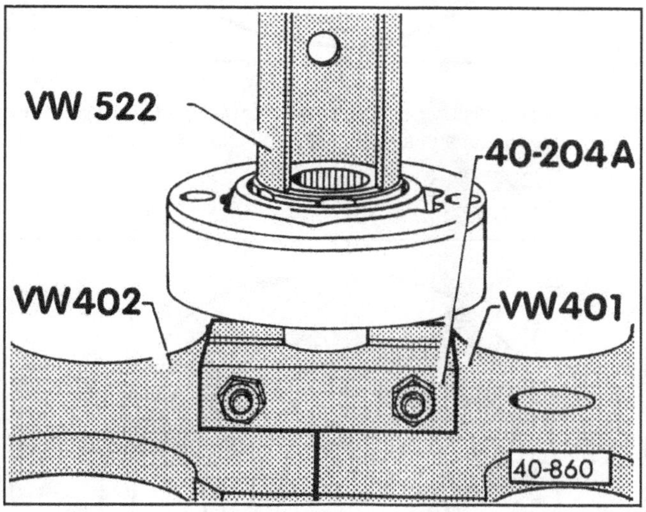

40-860

Fig. 25. Inner CV joint being pressed onto drive axle. Numbers indicate Volkswagen special tools.

14. Slide axle flange bolts through protective cap and into openings in CV joint. Press protective cap onto CV joint.

15. Slide inner clamp onto boot and crimp.

NOTE —
Special pliers should be used to crimp CV boot clamps. Use Snap-On YA 3080 pliers, or equivalent.

Inner CV joint, disassembling and assembling

NOTE —
This section does not cover triple-roller inner CV joints used on cars with 4-cylinder engine and automatic transmission. At the time of this publication, replacement parts for triple-roller CV joints were not separately available from Volkswagen.

1. Remove CV joint from drive axle as described previously.

2. Mark position of ball hub in relation to ball cage and housing with electric scriber.

3. Swivel ball hub and ball cage 90°. Pull ball hub together with cage and balls from housing. See Fig. 26.

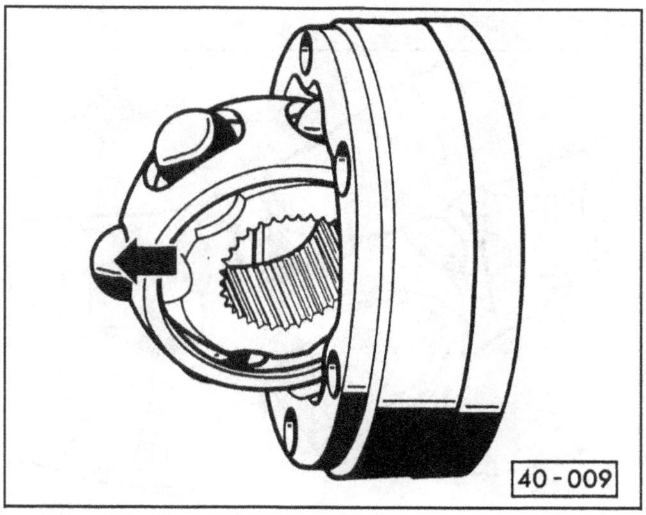

40-009

Fig. 26. Ball hub and cage being removed from housing. Pull in direction of **arrow**.

4. Press balls out of cage. Tilt ball hub so that ball tracks are aligned with edge of ball cage and remove hub from cage. See Fig. 27.

5. Check housing, hub, cage, and balls for pitting and signs of seizure.

NOTE —
Excessive backlash in the joint will be noticed as a knock when changing from acceleration to overrun and vice versa. Replace the joint if necessary. Do not replace a joint because of polished areas or because the ball tracks are visible.

6. Insert ball hub over both chamfers into ball cage.

7. Align ball hub to cage using marks made earlier. Press balls into cage. See Fig. 28.

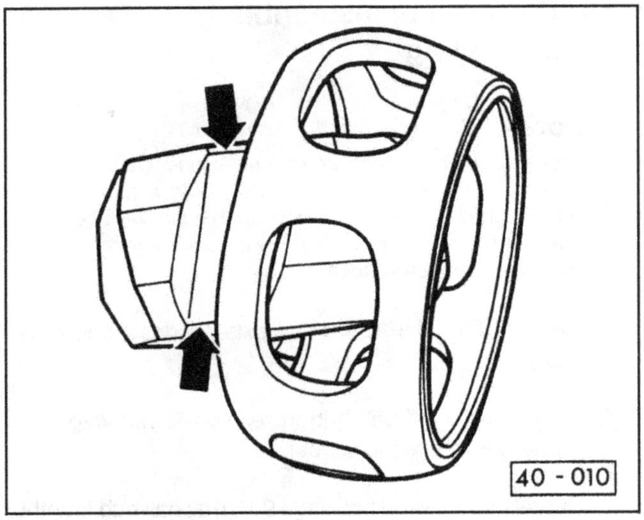

40-010

Fig. 27. Align ball tracks (**arrows**) on hub with ball cage edge when installing hub into cage.

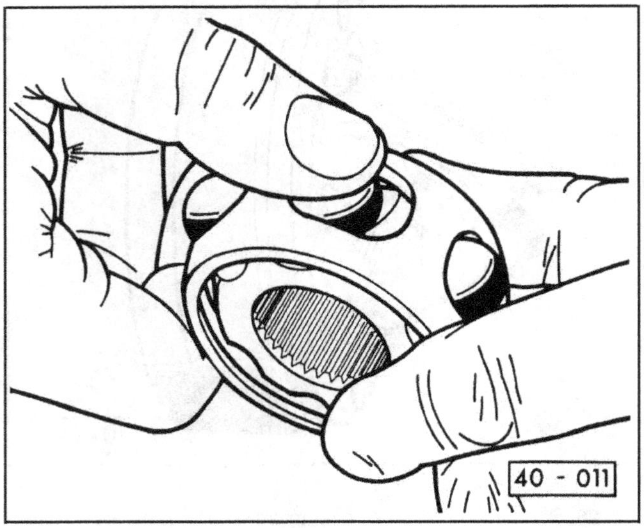

40-011

Fig. 28. Balls being pressed into cage.

8. Insert hub with cage and balls into joint at right angles, ensuring the following (see Fig. 29):
   - Wide ball groove in outer ring (**a**) and narrow groove in hub (**b**) are together on one side when hub is pivoted into housing.
   - Matching marks on hub, cage, and housing are aligned.
   - Chamfer on inside diameter of ball hub faces larger diameter of joint.

9. Swivel ball hub with cage and balls into housing. Pivoting hub and cage until balls align. See Fig. 30.

10. Press cage firmly by hand until hub swings fully into position.

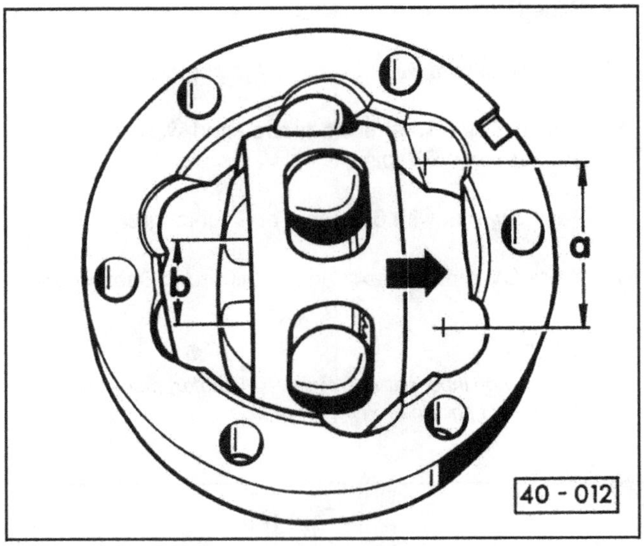

40-012

Fig. 29. Hub assembly correctly in housing with wide ball groove in housing (**a**) and a narrow groove in hub (**b**) together on one side.

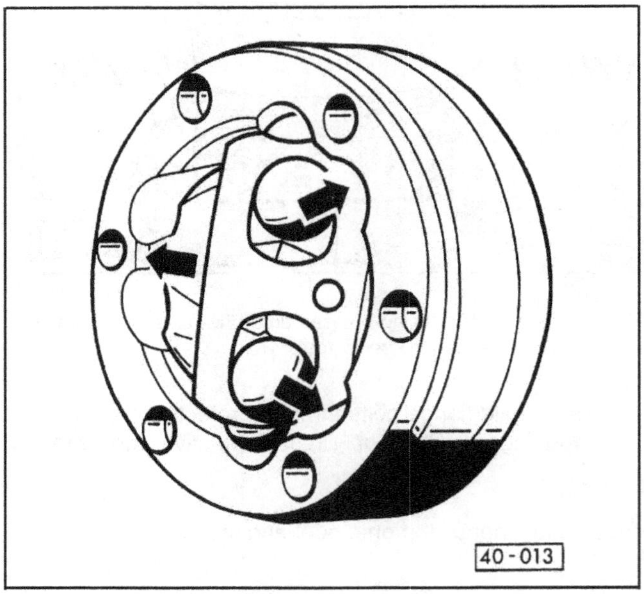

40-013

Fig. 30. Ball hub correctly installed into housing. Pivot hub out of cage (**arrows**) until balls are spaced to fit grooves.

NOTE —

The joint is correctly assembled when the ball hub can be moved in and out over the full range of axial movement by hand.

11. Pack CV joint with 60 grams of CV joint grease. Pack grease from both sides and put an additional 60 grams in the CV boot.

12. Install CV joint to axle using a new circlip.

42 Rear Suspension

GENERAL

The beam-type rear axle is a one-piece welded assembly of axle beam and trailing arms. Rear wheel stub axles are bolted to the trailing arms. The rear strut or shock absorber assemblies are separate coil springs mounted concentrically with shock absorbers. The shocks are easily removed and installed and, unlike the front suspension, the rear springs and shock absorbers can be safely disassembled without the use of a spring compressor. For general information on the suspension system used on the cars covered by this manual, see **4 General Information.**

When performing service or repair of the rear suspension, observe the following warnings and cautions.

> ***WARNING —***
> - *Do not re-use any fasteners that are worn or deformed in normal use. Many fasteners are designed to be used only once and become unreliable and may fail when used a second time. This includes, but is not limited to, nuts, bolts, washers, self-locking nuts or bolts, circlips, cotter pins. For replacements always use new parts.*
> - *Do not reinstall bolts and nuts coated with undercoating wax as correct tightening torque cannot be assured on installation. Always clean the threads of removed bolts and nuts with a solvent before installation, or replace them with new parts.*

> ***WARNING —***
> - *If the axle beam has been removed, or if for any other reason the brake lines have been disconnected, reassembly must include bleeding air from the brake system and readjusting the brakes as described in* **46 Brakes–Mechanical Components** *and* **47 Brakes–Hydraulic System**.
> - *Do not weld or attempt to straighten a damaged rear axle beam or rear stub axles.*

> ***CAUTION —***
> *Use only multipurpose (lithium) grease to lubricate the rear wheel bearings. Other greases will not maintain adequate lubrication and may lead to premature bearing failure.*

REAR SUSPENSION COMPONENTS

Fig. 1 shows the rear suspension components. Special repair information and tightening torques are given in the illustration's numbered list.

Rear suspension, assembly

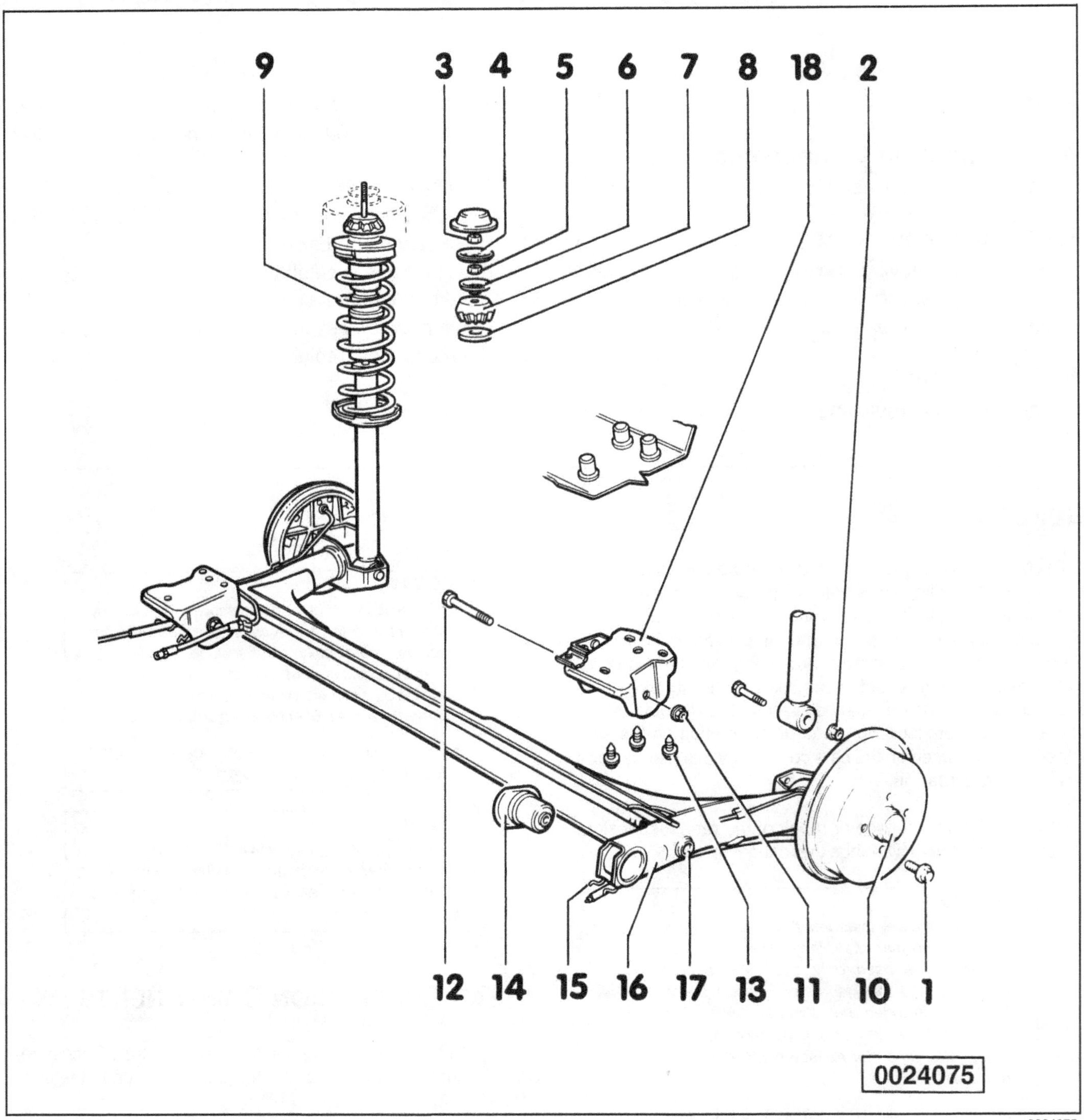

0024075

Fig. 1. Exploded view of rear suspension components. Rear suspension with drum brakes shown.

1. **Wheel lug bolt**
   - Tighten to 110 Nm (81 ft-lb)
2. **Nut**
   - Tighten to 70 Nm (52 ft-lb)
   - Axle beam must be in horizontal position (vehicle on ground, not loaded) when tightening nut
3. **Nut (locknut)**
   - Tighten to 25 Nm (18 ft-lb)
4. **Dished washer**
5. **Nut**
   - Tighten to 15 Nm (11 ft-lb)

6. **Cover (large)**
7. **Upper mounting ring (rubber)**
8. **Cover (small)**
9. **Shock absorber assembly**
10. **Grease cap**
11. **Nut**
    - Tighten to 80 Nm (59 ft-lb)
    - Use only self-locking nut with captive washer
    - Axle beam must be in horizontal position (vehicle on ground, not loaded) when tightening nut
12. **Bolt**
    - Install from center of vehicle
13. **Bolt**
    - Tighten to 70 Nm (52 ft-lb)
    - If threads in cap nut are damaged, drill out and re-tap
14. **Axle beam bushing**
    - See **Rear axle beam bushing, replacing**
15. **Bracket**
    - For parking brake cable
16. **Axle beam**
    - Keep contact surface and threaded holes for stub axles free of paint and dirt
    - See **Rear axle beam**
17. **Stabilizer bar**
18. **Mounting bracket**
    - Positioning on axle beam, see **Rear axle beam, installing**
    - With bracket for mounting brake pressure regulator
    - Do not loosen bracket nut and bolt (for bushing) when removing rear axle beam

REAR SHOCK ABSORBERS

Rear shock, removing and installing

> ***CAUTION —***
> *Remove and install the rear shock absorber assemblies one side at a time in order to prevent the rear axle from hanging unsupported and damaging the brake lines.*

1. Pry off the cap from the upper shock mount. While holding the shock shaft stationary, loosen and remove the lock nut and dished washer. See Fig. 2.

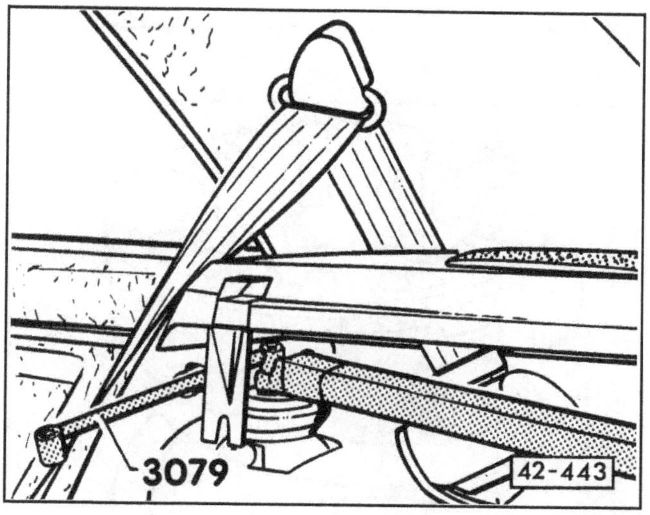

42-443

Fig. 2. Lock nut being removed from shock. Volkswagen special tool no. 3079 is used to hold shock shaft stationary.

2. Remove the second (lower) nut from the shock and remove the upper mounting hardware.

3. Raise the car slowly until the coil spring is unloaded and the top of the shock absorber begins to pull away from the body. Support the car safely.

   NOTE —
   Do not use any part of the suspension as a lifting point. The axle beam must not be supporting the car.

4. Remove the bolt from the lower shock mount and remove the shock from the car.

5. Installation is the reverse of removal. Install the lower shock mounting bolt with a new self-locking nut. Tighten the nut after the car has been lowered to the ground.

Tightening torques
- Upper shock to body. . . . . 15 Nm (11 ft-lb)
- Locknut to shock mounting nut. . . . . . 25 Nm (18 ft-lb)
- Lower shock mount to rear axle . . . . . 70 Nm (52 ft-lb)

Rear shock, disassembling and assembling

Fig. 5 shows an exploded view of the rear shock absorber assembly. Note that the upper mounting hardware is not shown in the illustration.

When assembling the shock, make sure the spring ends fit correctly into the spring seats. See Fig. 3. Make sure the lower spring seat is properly aligned. See Fig. 4

> ***CAUTION —***
> *The lower spring seat must be installed so that the holes in the seat align with the shock bushing eye. If not, the spring seat could interfere with the tire. See Fig. 4.*

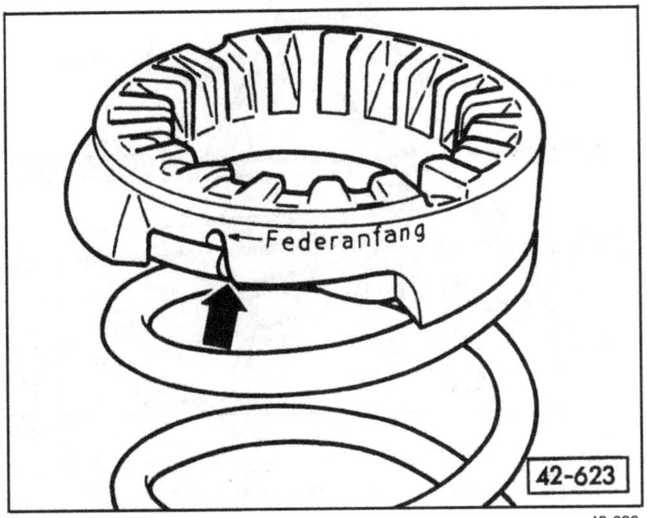

42-623

Fig. 3. End of coil spring aligned with notch in upper spring seat (**arrow**).

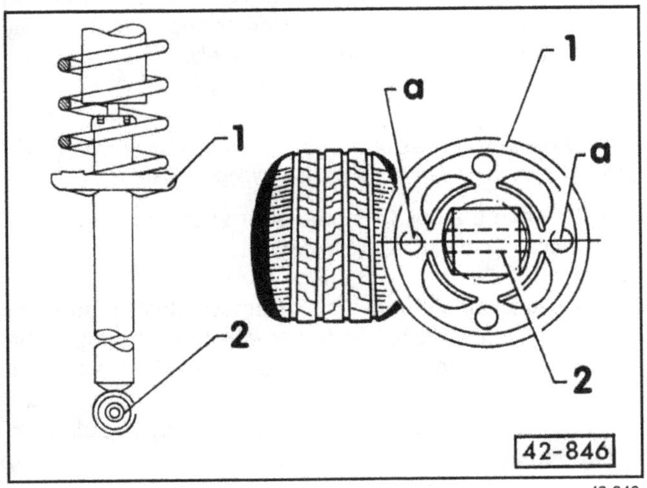

42-846

Fig. 4. Hole (**a**) in spring seat (**1**) must be aligned with bushing eye (**2**) in shock absorber.

Rear shock absorber, assembly

Fig. 5 shows an exploded view of the rear shock assembly. Applicable tightening torques as well as other pertinent repair information are given with the illustration.

1. **Shock absorber**
   - Slight traces of oil do not necessitate replacement
2. **Lower spring seat**
   - Installation position, see **Rear shock, disassembling and assembling**
3. **Insulator**
4. **Circlip**
5. **Coil spring**
6. **Protective cap**
7. **Protective tube**
8. **Bump stop**
9. **Upper spring seat**
   - Installation position, see **Rear shock, disassembling and assembling**
10. **Washer**
11. **Spring support**
12. **Spacer**
13. **Rubber lower mounting**
14. **Nut**
    - Tighten to 15 Nm (11 ft-lb)

Rear shock absorber, assembly

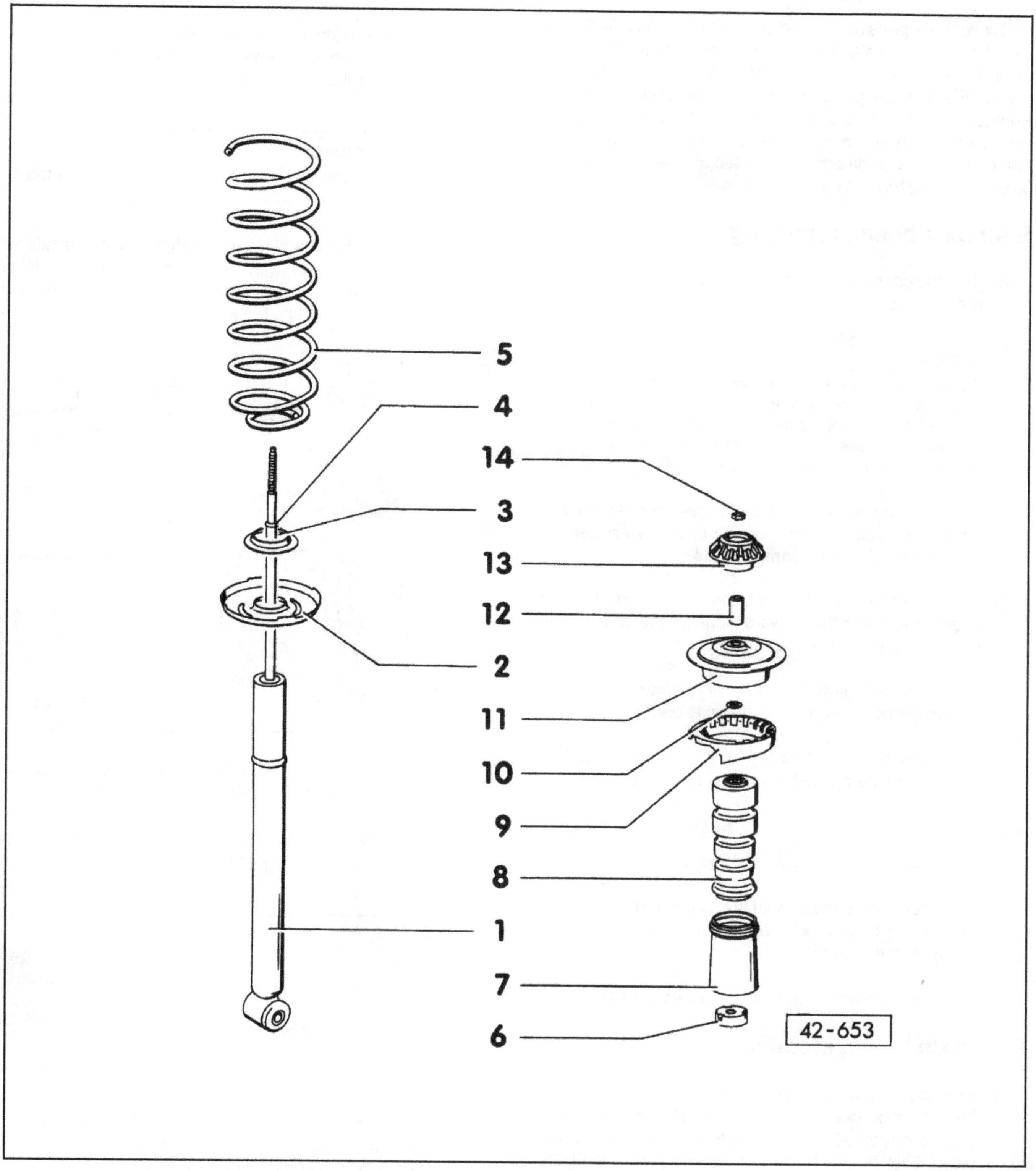

42-653

Fig. 5. Rear shock absorber components.

REAR AXLE BEAM

The rear suspension can be removed and installed as a complete unit for replacement or for facilitating rear body repairs. The axle beam is mounted to the body by brackets at the front of the axle, supported by replaceable rubber bushings. Replacement of the axle beam bushings require removal of the axle assembly and mounting brackets. Note that a special Volkswagen tool (protractor) is available to set up the axle beam mounting brackets prior to installation of the axle.

Rear axle beam, removing

1. With the car raised and properly supported, remove the rear wheels.

NOTE —
The car must be raised high enough to remove the axle assembly, so a lift, or a jack and suitable jack stands are required. Removal of the axle assembly also requires disconnecting brake lines to the rear wheels, and bleeding the brakes when reinstallation is complete.

2. Working between the front seats, disconnect the parking brake cables at the parking brake lever. See **46 Brakes–Mechanical Components**.

3. Disconnect the flexible rear brake hoses from the rigid brake lines near the forward edge of the axle beam. Cap both ends of the open connection.

4. While supporting the axle beam from below, disconnect the rear shock absorbers at the axle beam.

5. If applicable, disconnect the brake pressure regulator spring from the bracket attached to the axle beam.

NOTE —
Not all cars are fitted with a brake pressure regulator.

6. While supporting the front of the axle beam, loosen and remove the bolts which hold the axle beam mounting brackets to the body.

7. Remove the rear suspension by pulling out to the rear.

Rear axle beam, installing

1. If the axle beam mounting brackets have been removed from the axle beam, or if the bracket pivot bolts have been loosened, align the brackets at 12° ± 2° to the axle beam. Then tighten the pivot bolts and nuts. See Fig. 8.

Tighteninig torques
- Mounting bracket to bushing .... 80 Nm (59 ft-lb)

2. Position the rear suspension assembly under the car and loosely install the mounting bolts.

3. Position the right (passenger-side) mounting bracket so that the bolts are centered in the elongated holes. Tighten right bracket bolts.

Tightening torques
- Mounting bracket to body .... 70 Nm (52 ft-lb)

4. Use a set of prybars, lever the left mounting bracket so that only a minimal gap is left between the inside of bushing and mounting bracket. See Fig. 6. Tighten the left bracket mounting bolts.

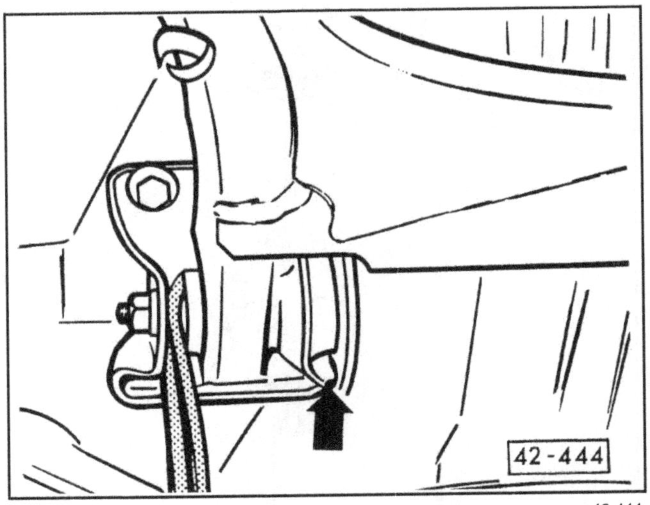

42-444

Fig. 6. Press left axle beam mounting bracket to left so that only small gap exists on right side of bracket (**arrow**).

5. Clean the brake line unions and reconnect the flexible brake hoses to the rigid brake lines.

Tightening torques
- Brake line unions .... 15 Nm (11 ft-lb)

6. Remount the rear shock absorbers.

7. Reconnect the parking brake cables.

8. Install the rear wheels and carefully lower the car. Install and tighten the shock mounts as described under **Rear shock, removing and installing**.

9. Bleed the brakes and adjust the parking brake as described in **47 Brakes–Hydraulic System** and **46 Brakes–Mechanical Components**.

Rear axle beam bushing, replacing

1. Remove rear axle beam as described previously.
2. Remove pivot bolt and nut from mounting bracket and remove bracket from the axle beam.
3. Using an appropriate press tool, press bushing out of axle. See Fig. 7.

NOTE—
Remove any corrosion near bushing before pressing out.

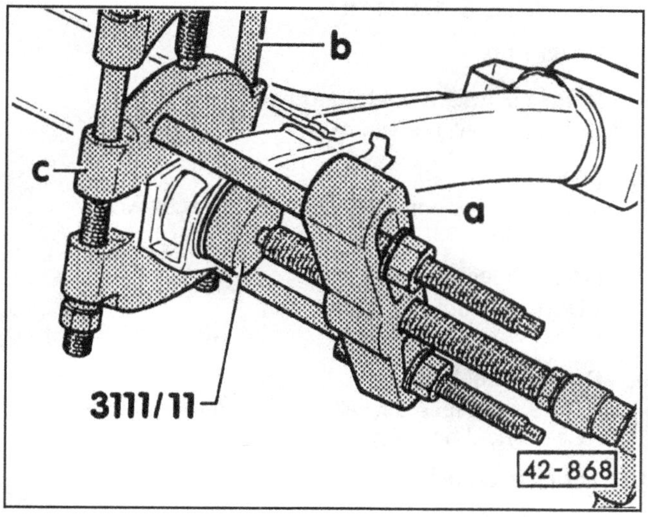

42-868

Fig. 7. Rear axle bushing being pressed out. Numbers indicate Volkswagen special tools.

4. Press new bushing into the axle, noting the following:
   - The new bushing must protrude from axle 8.0 mm ± 1.0 mm (0.32 + 0.04 in.).
   - Prior to installing rear axle, adjust the mounting bracket to 12° ± 2° to the axle and then tighten the bracket mounting nut. See Fig. 8.
5. The remainder of installation is the reverse of removal.

Tightening torques
- Mounting bracket to bushing .... 80 Nm (59 ft-lb)

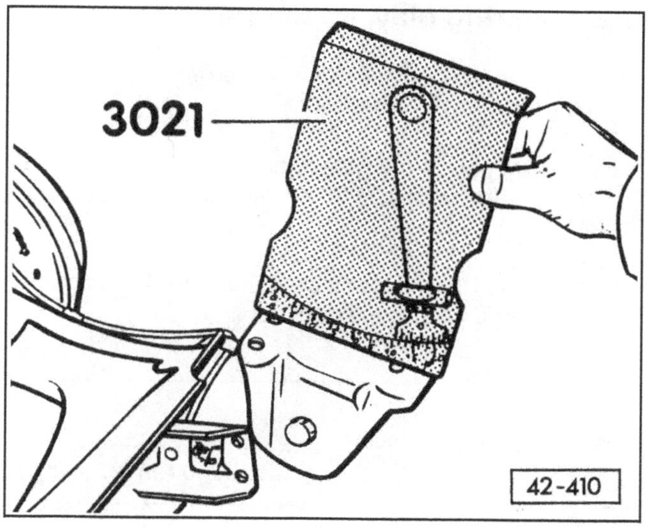

42-410

Fig. 8. Axle beam mounting bracket being aligned at 12° ± 2° angle to axle beam prior to axle beam installation. Protractor shown is Volkswagen special tool no. 3021.

WHEEL BEARINGS

The rear wheel bearings are of the tapered-roller design. Aside from periodically adjusting play, the rear wheel bearings are generally maintenance free. When replacing the wheel bearings, press equipment should be used to install the bearing races into the brake hub.

Fig. 10 shows an exploded view of the wheel bearing and stub axle assembly for cars with 4-wheel disc brakes. Fig. 11 shows an exploded view of the wheel bearing and stub axle assembly for vehicles with rear drum brakes.

CAUTION—
- *Do not use solvent to clean bearings. Wipe with clean rag and repack before installing.*
- *Adjust wheel bearing play carefully.*
- *Proper sealing, cleanliness and precise adjustment are essential to long wheel bearing service life.*

Wheel bearing play, adjusting

1. Raise the car and support it on jack stands. Remove the wheel.

2. Carefully pry off the grease cap in the center of the brake drum or disc. Remove the cotter pin and lock ring.

3. Loosen and then tighten the axle nut to the torque given below while rotating the disc/drum.

Tighting torques

- Wheel bearing axle nut
  stage 1 . . . . 10 Nm (87 in-lb)

4. Loosen the nut slightly.

5. While rotating the disc/drum, tighten the axle nut slowly in small increments until the thrust washer can just barely be moved back and forth with a light push on the tip of a screwdriver (do not pry or twist screwdriver). See Fig. 9.

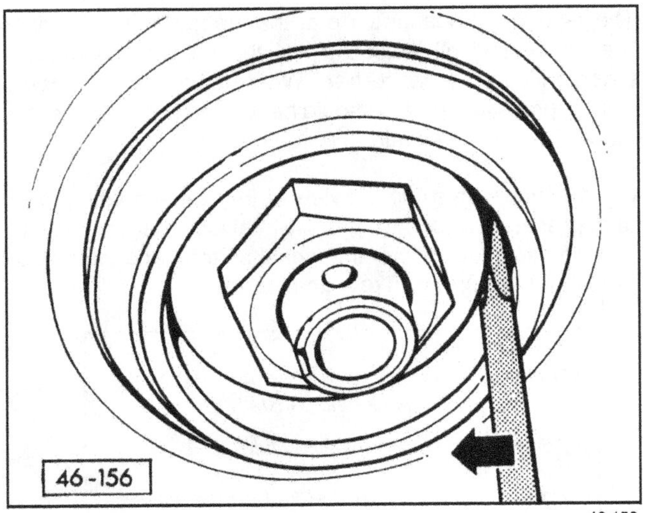

46-156

Fig. 9. Wheel bearing adjustment being checked. Axle nut is tightened correctly when thrust washer moves (**arrow**) only with slight pressure.

6. Secure nut with locking ring and new cotter pin. Install the grease cap. If necessary, tighten the axle nut slightly so that the locking ring aligns with the hole in the stub axle.

> ***CAUTION —***
> *Always use a new cotter pin. Do not reuse the previously installed cotter pin.*

Wheel bearing and axle (disc brakes), assembly

1. **Stub axle**
   - Contact surface against axle beam must be completely clean
   - Do not attempt to straighten
2. **Splash shield**
3. **Dished washer**
   - Large contact surface (concave side) faces splash shield
4. **Bolt**
   - Tighten to 60 Nm (44 ft-lb)
5. **Cover ring**
6. **Rotor for ABS Wheel Speed Sensor**
   - Only for vehicles with ABS
7. **Seal**
   - Fill area between sealing lips with grease when installing
8. **Inner wheel bearing**
   - Drive race out with copper drift
   - See **Rear wheel bearing, installing (disc brakes)**
9. **Outer wheel bearing**
   - Drive outer race out with copper drift
   - See **Wheel bearings, replacing**
10. **Thrust washer**
11. **Hex nut**
    - Tighten to adjust wheel bearing. See **Wheel bearing play, adjusting**.
12. **Lock ring**
13. **Cotter pin**
    - Always replace
14. **Grease cap**
15. **Brake disc**
    - Coat inside of hub with grease before installing
16. **Bolt**
    - Tighten to 56 Nm (41 ft-lb)
17. **Brake pad carrier**
18. **Protective cap**
19. **Guide bolt**
20. **Brake pads**
21. **Bolt**
    - Tighten to 35 Nm (26 ft-lb)
    - Always replace
22. **Brake caliper**
    - See **47 Brakes–Hydraulic System**
23. **Brake hose bracket**

Wheel bearing and axle (disc brakes), assembly

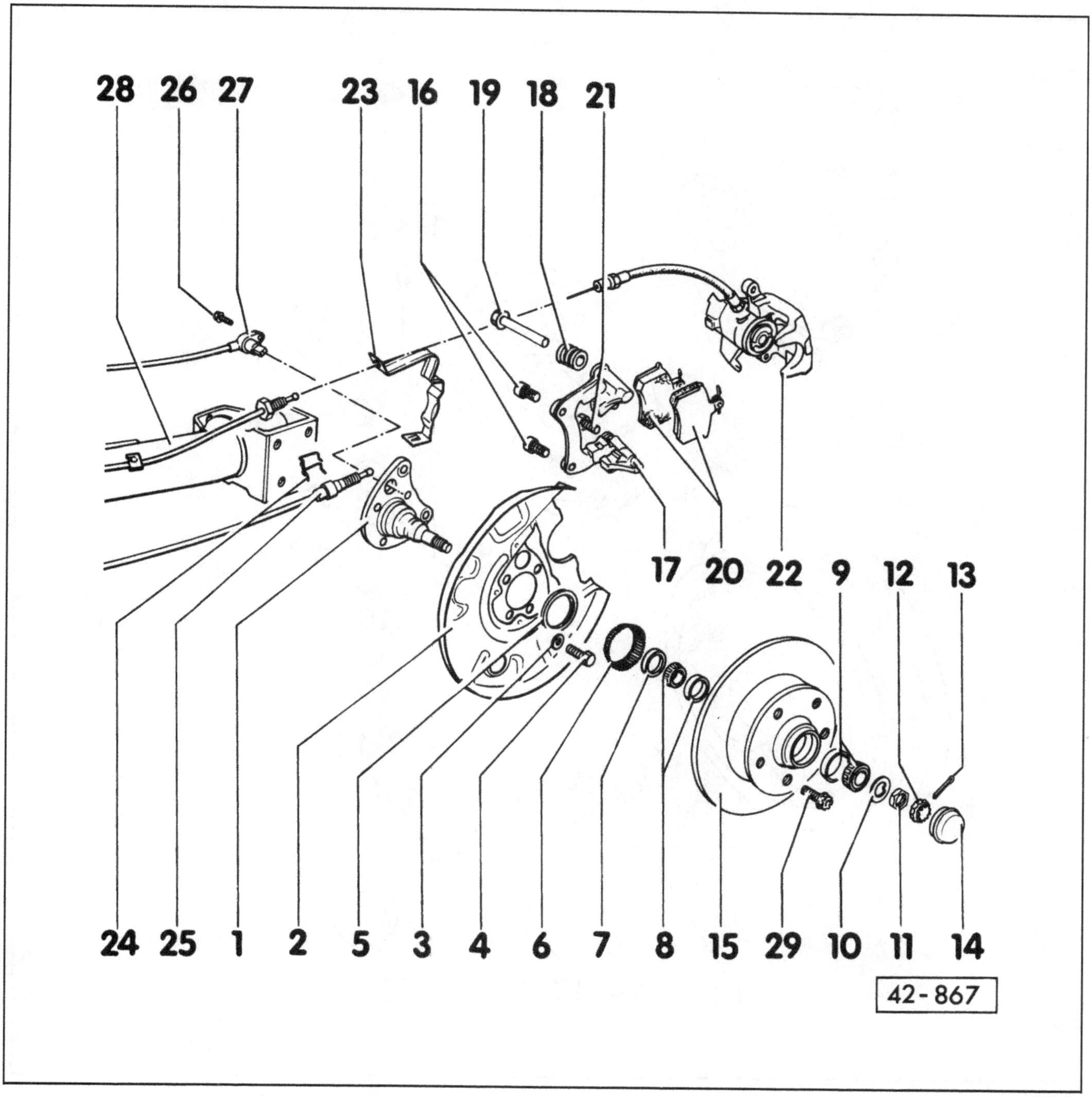

42-867

Fig. 10. Exploded view of rear wheel bearings and stub axle for cars with 4-wheel disc brakes.

24. **Spring clamp**
25. **Parking brake cable**
    - Adjusting See **46 Brakes–Mechanical Components**
26. **Bolt**
    - Tighten to 10 Nm (89 in-lb)
27. **ABS wheel speed sensor**
    - Only for vehicles with ABS
28. **Axle beam**
29. **Wheel lug bolt**
    - Tighten to 110 Nm (81 ft-lb)

Wheel bearing and axle (drum brakes), assembly

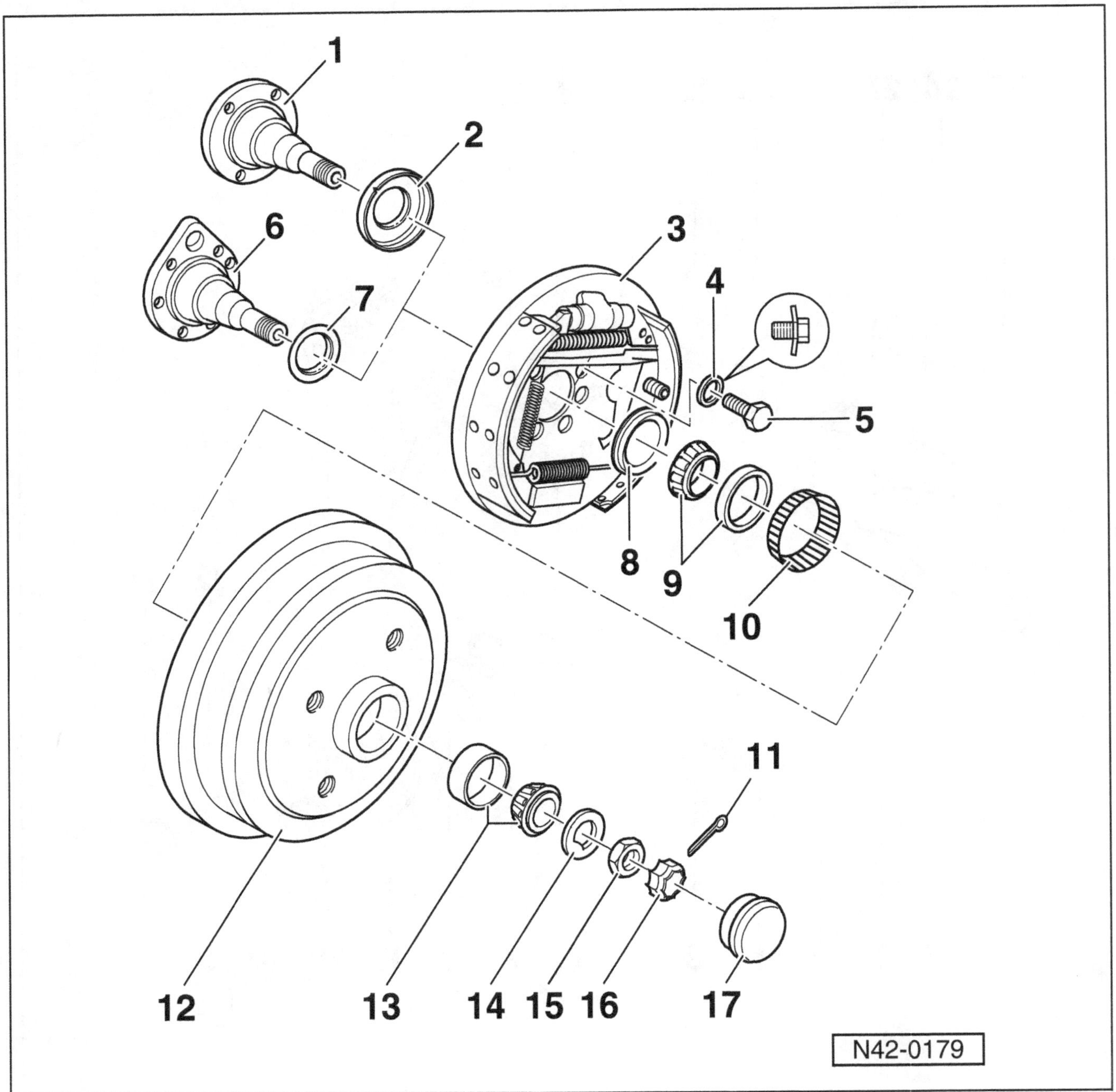

N42-0179

Fig. 11. Exploded view of rear wheel bearings and stub axle for cars with rear drum brakes, with and without ABS.

1. **Stub axle**
   - Contact surface against axle beam must be clean
   - Do not attempt to straighten
2. **Cover ring for vehicles without ABS**
3. **Backing plate**
   - Bolt onto rear axle beam along with stub axle
4. **Dished washer**
   - Large contact surface (concave side) faces brake backing plate
5. **60 Nm (44 ft-lb)**
6. **Stub axle for vehicles with ABS**
   - Do not attempt to straighten

(continued from previous page)

7. **Cover ring for vehicles with ABS**
8. **Seal**
   - Fill gaps between sealing lips with grease
9. **Inner wheel bearing**
   - Drive race out with copper drift
   - See **Wheel bearings, replacing**
10. **Rotor**
    - Only on vehicles with ABS
11. **Cotter pin**
    - Always replace
12. **Brake drum**
    - Release brake shoe adjusting wedge through wheel bolt hole before removing. See **Wheel bearings, replacing**
    - When removing make sure outer wheel bearing does not fall out
    - Coat inside of hub with grease before installing
13. **Outer wheel bearing**
    - Drive race out with copper drift
14. **Thrust washer**
15. **Hex nut**
    - Tighten/loosen to adjust wheel bearing play. See **Wheel bearing play, adjusting**
16. **Lock ring**
17. **Grease cap**

Wheel bearings, replacing

1. Raise the car and support it on jack stands. Remove the wheel.

> ***WARNING —***
> *Jack stands should be placed on a hard level surface (e.g. concrete).*

2. On cars with rear drum brakes, release brake shoes by inserting a screwdriver through front, top wheel bolt hole and pushing up adjusting wedge. See Fig. 12.

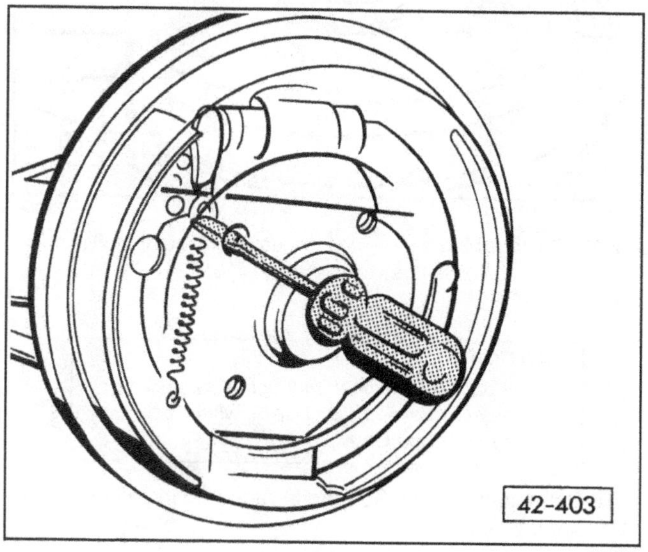

42-403

Fig. 12. To remove brake drum, release brake shoes by inserting screwdriver through wheel bolt hole while pushing up on adjusting wedge.

3. On cars with 4-wheel rear disc brakes, remove brake caliper. See **46 Brakes–Mechanical Components**.

4. Remove grease cap from center of brake drum/rotor.

5. Remove cotter pin and lock ring from stub axle. Remove wheel bearing nut. Slide brake drum/rotor straight off axle.

> ***NOTE —***
> *If necessary, rock the brake drum/rotor side-to-side to aid removal of the washer and outer wheel bearing.*

6. Remove grease seal and inner wheel bearing from brake drum/rotor.

7. Using a drift, drive the wheel bearing races from the brake drum or brake rotor.

8. Press new races into the brake rotor or brake drum. See Fig. 13.

9. Pack wheel bearings with lithium-based wheel bearing grease. Lightly coat bearing races and hub with grease.

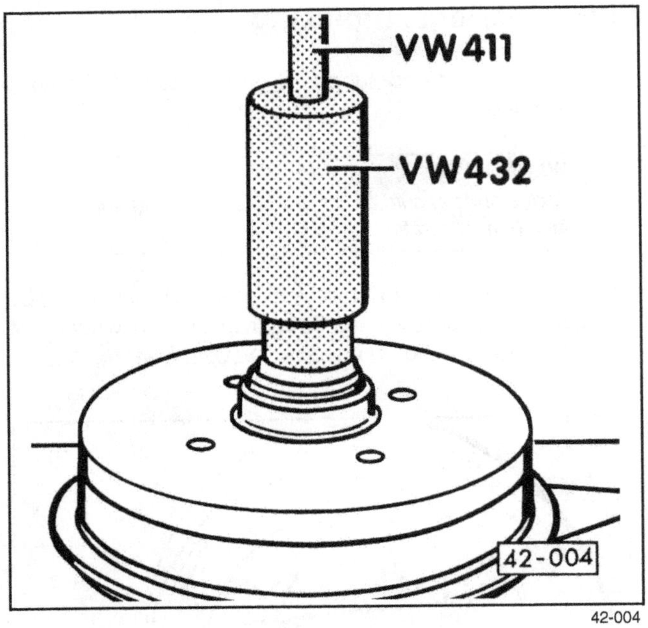

42-004

Fig. 13. Outer bearing race being pressed into brake drum. Numbers indicate Volkswagen special tools.

> ***CAUTION—***
> *Proper packing of wheel bearings is essential for long bearing service life. Use a wheel bearing packing tool if available.*

10. Place inner wheel bearing in race and install new inner seal.

11. Apply grease to gap between sealing lips on inner seal.

12. Slide the brake drum or rotor onto the stub axle. Use caution to avoid damaging the grease seal.

13. Install the outer wheel bearing, thrust washer and axle nut. Tighten the axle nut while rotating the brake drum or rotor.

14. Adjust rear wheel bearing play as described earlier under, **Wheel bearing play, adjusting**. Install lock ring and new cotter pin.

15. Fill grease cap with 10 grams of grease and install.

16. The remainder of installation is the reverse of removal.
    - On cars with disc brakes, install the caliper and adjust the parking brake as described in **46 Brakes–Mechanical**.
    - On cars with rear drum brakes, adjust the rear brake shoes and parking brake as described in **46 Brakes–Mechanical**.

Tightening torques
- Road wheel to wheel hub. . . . . 110 Nm (81 ft-lb)

---

44 Wheels–Tires, Wheel Alignment

TABLES

GENERAL

This repair group covers basic tire, wheel, and wheel alignment information. Also covered here is wheel alignment specifications to be used in conjunction with professional alignment tools and measuring equipment.

Wheels and tires

Wheels and tires approved by the manufacturer have been matched to the vehicle and contribute largely to road handling and driving characteristics. To retain the handling characteristics, it is recommended that the tires be replaced only with tires having the same specifications with regard to size, design, load carrying capacity, speed rating, tread pattern, tread depth, etc. This information can be found on the tire's sidewall. See Fig. 1.

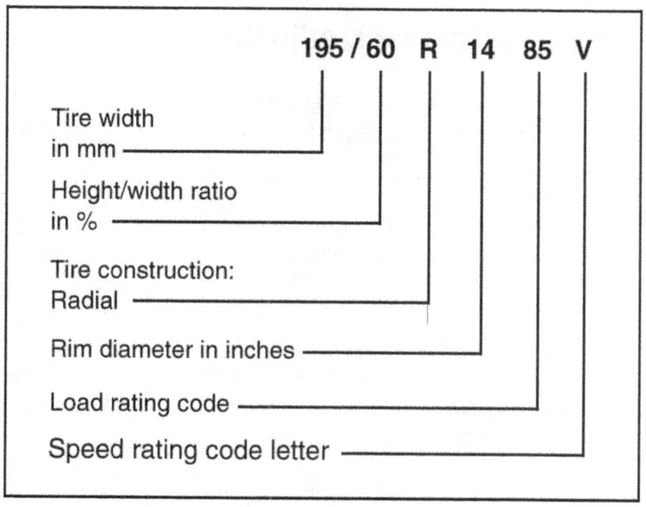

Fig. 1. Tire sidewalls are marked with important tire specifications.

Volkswagen recommends that the tires be rotated front to back, with the tires remaining on the same side of the vehicle. Only when tires show unusual wear should they be rotated diagonally. See **0 Maintenance Program** for maintenance schedules regarding tire rotation.

WHEEL ALIGNMENT

Tire pressures, tire wear, and wheel alignment will all influence how the car feels and responds on the road. For stability and control, all four wheels and tires must be in good condition, balanced, and be properly aligned. Precise wheel alignment can only be accomplished when the tires, the suspension, and the steering are in good condition. Reputable wheel alignment technicians will always inspect the front and rear suspension and the steering for worn parts before an alignment, and will recommend that any necessary repairs be made before proceeding. The important front wheel alignment angles are camber, caster, and toe. In the rear, the important angles are camber, toe and thrust line. Although rear alignment angles are not generally adjustable, they should be checked because of the effect that they have on rear tire wear and straight-line stability of the vehicle.

Camber is the angle that the wheels tilt from vertical when viewed from front or rear, as illustrated in Fig. 2. Wheels which tilt out at the top have positive (+) camber. Wheels that tilt in at the top have negative (–) camber. On the Volkswagens covered by this manual, camber adjustment is not normally required, although there are provisions for making camber adjustments if necessary.

Camber influences cornering, directional stability, and tire wear. Different camber on the two front wheels may cause the car to pull to one side. Incorrectly adjusted camber will cause uneven tire wear.

Caster is the angle at which the steering axis deviates from vertical, as shown in Fig. 3. Most cars are designed with positive caster, which improves directional stability and tends to make the steering more self-centering. Caster angle should be checked as part of wheel alignment, but it is not adjustable on cars covered by this manual.

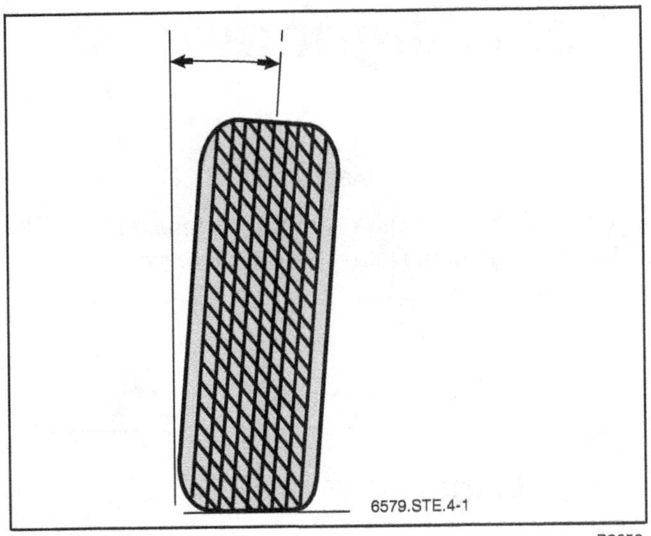

Fig. 2. Camber is the wheel/tire deviation from vertical as viewed from front or rear.

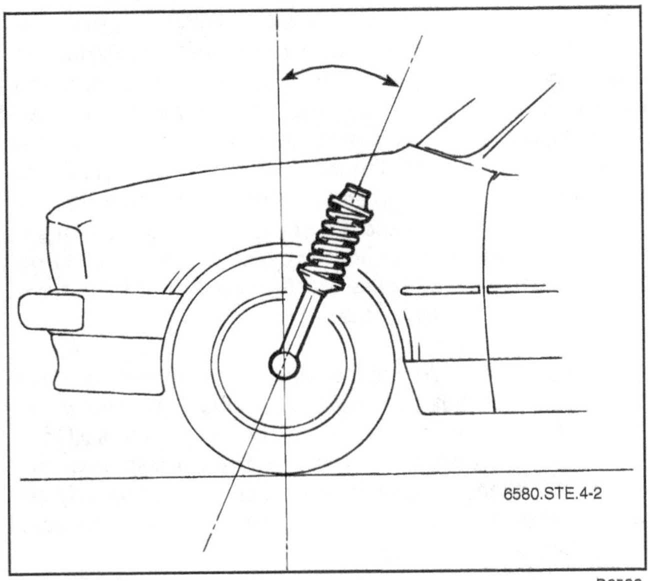

Fig. 3. Caster is the angle of steering axis inclination from vertical.

Toe is a measurement of the amount that two wheels on the same axle point toward each other (toe-in) or away from each other (toe-out). Toe affects directional stability and tire wear. Toe also affects response to steering input. Too much toe will cause tires to scrub and to wear unevenly and quickly. Too little toe may cause the car to be less stable and wander at highway speeds. Toe-in is illustrated in Fig. 4.

Wheel alignment, checking

Wheel alignment specifications are given in **Table a** and **Table b**. Be sure the checking requirements are met before carrying out a wheel alignment.

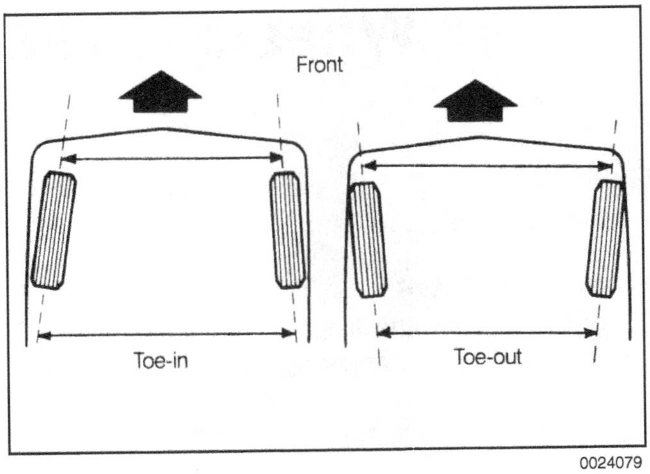

Fig. 4. Front wheel toe.

Checking requirements

- Vehicle at curb weight [1)]
- Tires correctly inflated [2)]
- Suspension bounced several times
- No excessive play or damage to steering or linkage
- No excessive play or damage to suspension

[1)]Curb weight is defined as the weight of vehicle ready for the road:
- Full fuel tank
- Equipped with spare wheel
- Equipped with tool kit, if included
- Equipped with vehicle jack

[2)]Tire sizes and pressures, see specification label on vehicle, located on passenger-side rear door pillar (B-pillar on 2-door vehicles, C-pillar on 4-door vehicles)

Front wheel camber, adjusting

Front wheel camber can be adjusted as part of a front wheel alignment. The procedure requires a special Volkswagen tool.

> *CAUTION —*
> *This adjustment procedure must only be carried out on specialized wheel alignment equipment, as part of a front wheel alignment.*

1. Working in the front wheel housing, install Volkswagen special tool no. 3270 between the front wheel and the body as shown in Fig. 5. Tension the tool slightly so that all slack is removed.

2. Loosen the two bolts connecting the suspension strut to wheel bearing housing.

3. Set camber to specification by turning the threaded spindle on the tool.

4. Once the adjustment is correct, tighten the suspension strut/wheel bearing housing bolts.

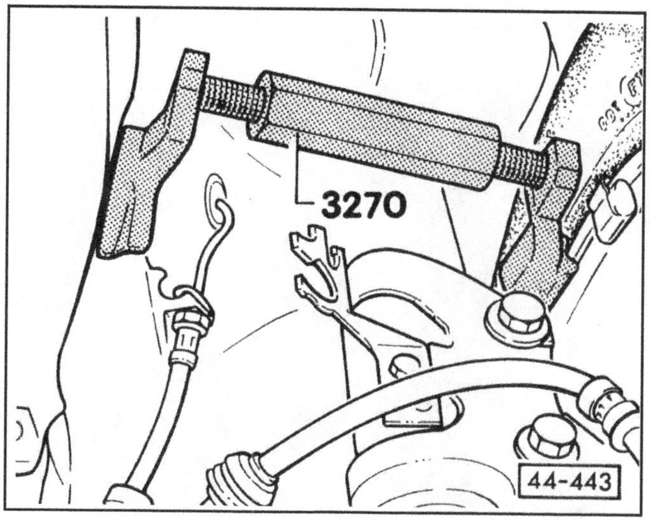

44-443

Fig. 5. Volkswagen special tool no. 3270 shown installed in wheel housing.

Tightening torques

- Strut assembly to wheel bearing housing . . . . 95 Nm (70 ft-lb)

---

5. Check camber and repeat procedure if necessary.

Table a. Wheel Alignment Data—Front Suspension

Suspension	Base Suspension 1993-1994		Base Suspension 1995-on	Plus Suspension 1993-on
Engine	1.8 Liter (ACC) 1.9 Liter (AAZ)	2.0 Liter (ABA)	All 4-cylinder engines	All VR-6 engines
Total toe [1] (wheels not pressed)	0° ± 10'			
Camber [2] (in straight-ahead position)	–30' ± 20'	–35' ± 20'	–36' ± 20'	–40' ± 20'
Maximum permissible difference between left and right	20'			
Toe angle difference at 20° lock from left to right	–1° 5' ± 20'			–1° 30' ± 20'
Caster (not adjustable)	+1° 45' ± 30'	+1° 50' ± 30'		+3° 25' ± 30'
Maximum permissible difference between left and right	30'			

[1]Adjust total toe on right tie rod only, set left tie rod to length. See **48 Steering**
[2]Camber can be corrected only at wheel bearing housing/suspension strut connection

Table b. Wheel Alignment Data—Rear Suspension (non-adjustable)

Camber	–1° 30' ± 10'
Maximum permissible difference between left and right	20'
Total toe (with specified camber)	+20' ± 10'
Maximum permissible deviation from direction of travel (see Fig. 4)	20'

45 Anti-Lock Brakes (ABS)

GENERAL

Various ABS systems with and without traction control are used on the cars covered by this manual, depending on model and date of manufacture. The traction control system, called EDL (electronic differential lock), works in conjunction with the ABS system by alternately applying the front brakes to control wheel spin during low speed acceleration. For more general information on ABS and traction control, see **4 General Information**.

This repair group covers basic troubleshooting and service to the Anti-Lock Braking Systems. System function, component locations, and adjustments are covered in detail. In-depth electrical/electronic troubleshooting and diagnosis are not included as part of in this repair group, as special Volkswagen test equipment is needed to properly diagnose the system.

NOTE —
- *The ABS system features built-in diagnostic circuitry that detects and stores Diagnostic Trouble Code (DTC) information. When the system detects a fault, a DTC is generated and stored in the ABS control module's permanent memory. These fault codes can be accessed through the data link connector(s) in the dashboard using Volkswagen Scan Tool V.A.G. 1551.*
- *If an ABS malfunction is stored, ABS is switched off and the ABS warning light comes on. Malfunctions of the ABS system do not influence the brake system in normal braking conditions.*

Special precautions for ABS cars

The ABS control module processes precise, low-level electrical signals and can be sensitive to changes in the power supply and the environment. The following precautions apply to cars with ABS.

WARNING —
- *ABS is a vehicle safety system; appropriate knowledge and special equipment are necessary to work on the system.*
- *The ABS/EDL system must be bled after work that requires opening the brake hydraulic system. See **47 Brakes-Hydraulic System**.*

CAUTION —
- *On vehicles with an anti-theft radio, obtain the radio code before disconnecting the battery.*
- *Always disconnect the battery terminals before quick-charging the battery. Never disconnect the battery while the engine is running. Never use a battery charger or auxiliary battery producing more than 16 volts when jump-starting.*
- *Remove the ABS control module from the car before repairs that involve high temperatures. The control module should never be exposed to temperatures greater than 176°F (80°C).*
- *Do not work with compressed air or move the vehicle when the hydraulic system is open.*
- *Do not drive car with harness connector disconnected from ABS control module.*

CAUTION—
- *Always disconnect the negative (–) battery terminal before working on the ABS or ABS/EDL system.*
- *Before doing any welding on the vehicle with an electric welder disconnect the ABS control module.*
- *Absolute cleanliness is required when working on the ABS system. DO NOT use mineral oils, grease or aggressive cleaning agents such as brake cleaner, gasoline or thinners.*
- *Use clean lint-free cloths and take care to seal or cover opened components if a repair cannot be completed immediately.*

Applications–identifying features

Most early cars are equipped with the "Teves 04" ABS system (also referred to as "Mark IV"). Late cars are equipped with the "Teves 20 GI" ABS system. In addition, both systems are available with traction control, called EDL.

NOTE—
EDL stands for electronic differential lock, although the ABS/EDL system does not act on the differential.

The Teves 04 and the Teves 20 GI systems are easily distinguishable from each other. Fig. 1 shows the Teves 04 ABS/EDL unit. Fig. 2 shows the Teves 20 GI ABS/EDL unit. Distinguishing features are listed below each illustration.

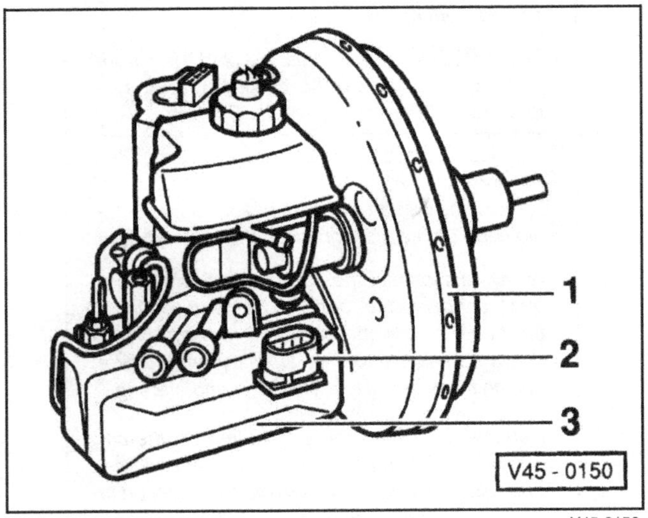

V45-0150

Fig. 1. Teves 04 ABS/EDL unit with vacuum brake servo (**1**), harness connector (**2**), and valve block (**3**).

Distinguishing features–Teves 04 ABS/EDL
- **ABS only**
  - -10" brake servo (**1**)
  - -10 pins used on 15-pin harness connector (**2**)
- **ABS with EDL**
  - -10" brake servo (**1**)
  - -14 pins used on 15-pin harness connector (**2**)
  - -Longer valve block (**3**) for ABS/EDL than for ABS alone
  - -Series resistor for EDL (see Fig. 3, given later)

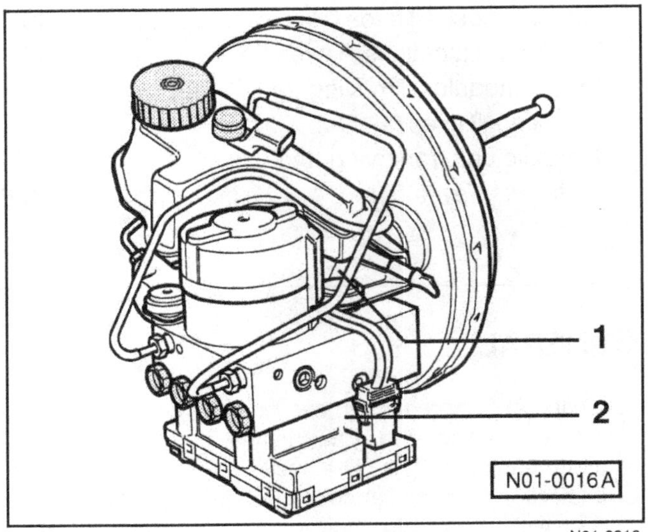

N01-0016

Fig. 2. Teves 20 GI ABS/EDL unit. The ABS hydraulic unit (**1**) and the ABS control module (**2**) together make up one component.

Distinguishing features–Teves 20 GI ABS/EDL
- **ABS only**
  - - 10" brake servo
  - - 8-valve hydraulic unit (valve sleeves are visible when ABS control module is separated from ABS hydraulic unit)
- **ABS with EDL**
  - - 10" brake servo
  - - 10-valve hydraulic unit (valve sleeves are visible when ABS control module is separated from ABS hydraulic unit)
  - -Longer valve block for ABS/EDL than for ABS alone

NOTE—
Volkswagen identifies electrical components alpha-numerically in the electrical schematics. See **97 Wiring Diagrams, Fuses and Relays**. *Where applicable, these alpha-numeric codes will be listed with the component as an aid to electrical troubleshooting.*

ABS, ABS/ EDL–TEVES 04

The Teves 04 ABS /EDL system is a three channel system; the front brakes are controlled independently and the rear wheels are controlled as one circuit. The system utilizes dual diagonal brake circuits. On cars equipped with the VR6 engine, a low speed traction control system, called EDL, is integrated into the ABS system.

Self-diagnosis function–Teves 04

After switching the ignition on, the ABS warning light lights up for approx. 2 seconds. During this period a self-test sequence is carried out by the ABS control module with checks of the following functions:

- Hydraulic pump runs at low speed
- Brief valve activation after switching the ignition on

If an ABS malfunction is stored, ABS is switched off and the ABS warning light comes on. If an EDL malfunction is stored, the EDL system is also switched off, although these malfunctions are not indicated by the ABS warning light. If ABS or ABS/EDL is switched off due to a malfunction, the conventional hydraulic brake system remains fully operational.

NOTE —
- *If the ABS warning light does not light up, but the brake system is not functioning correctly, then the malfunction must be diagnosed in the conventional braking system (e.g. leaks in the brake hydraulic system).*
- *In some cases, certain ABS malfunctions may not be recognized at speeds below 40 km/h (24 mph).*

Fault diagnosis–Teves 04

Fig. 5 shows the electrical/electronic components for the Teves 04 system. When troubleshooting the ABS/EDL system, make sure the requirements listed below are met. Check that the ground connection for the ABS control module is tight and in good condition. Location of the ground point depends on production date. See Fig. 3 and Fig. 4.

NOTE —
If no faults can be found after reviewing the troubleshooting information given under this heading, have the system serviced by an authorized VW dealer who is equipped and trained to properly diagnose and service the Teves 04 ABS/EDL system.

Troubleshooting requirements
- Tires on all four wheels must be of the same (approved) size, inflated to correct pressure.
- Conventional brake system OK, including brake light switch and brake lights.
- Brake hydraulic lines and connections not leaking (visual check of ABS hydraulic unit, brake calipers).

(cont.)

Troubleshooting requirements (cont.)
- Wheel bearings and wheel bearing play OK.
- Connector on ABS control module correctly attached (retainer is engaged).
- Check terminals and sockets on ABS or ABS/EDL components for damage and correct seating.
- ABS relay, ABS hydraulic pump relay and electronic differential lock relay correctly installed/connected
- All fuses according to wiring diagram OK (remove fuse from fuse holder to check).
- Battery (supply) voltage OK (at least 10.5 V).
- ABS control module ground connection OK.

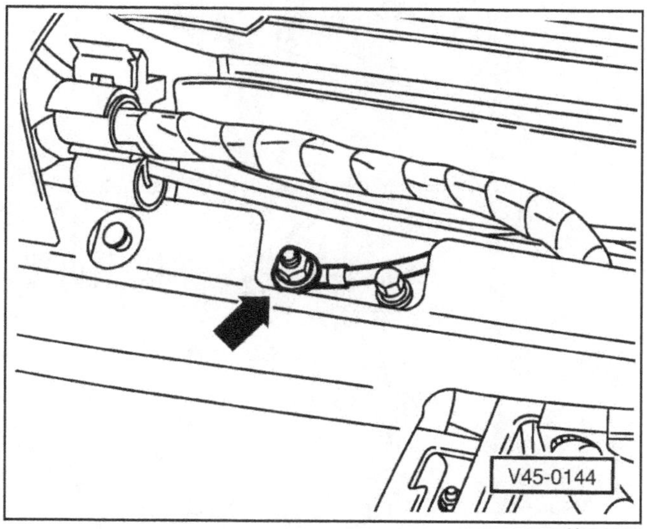

V45-0144

Fig. 3. ABS/EDL control module ground location (**arrow**) on cars built up to 01/93 is under right (passenger side) sill panel trim.

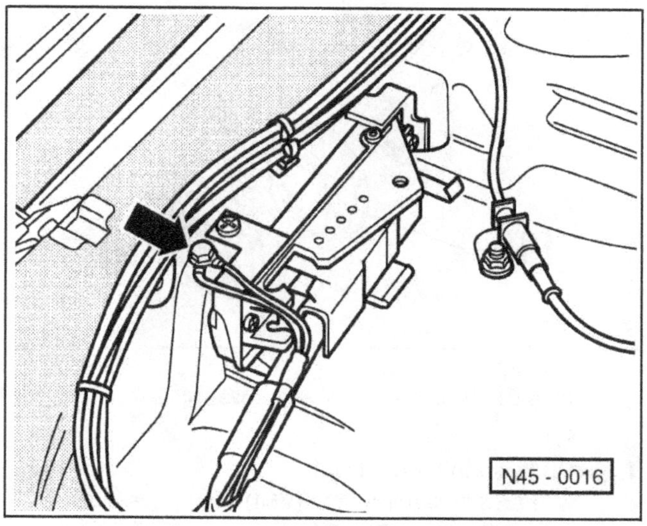

N45-0016

Fig. 4. ABS/EDL control module ground location (**arrow**) on cars built after 01/93 is under rear seat near ABS control module.

Component locations–Teves 04

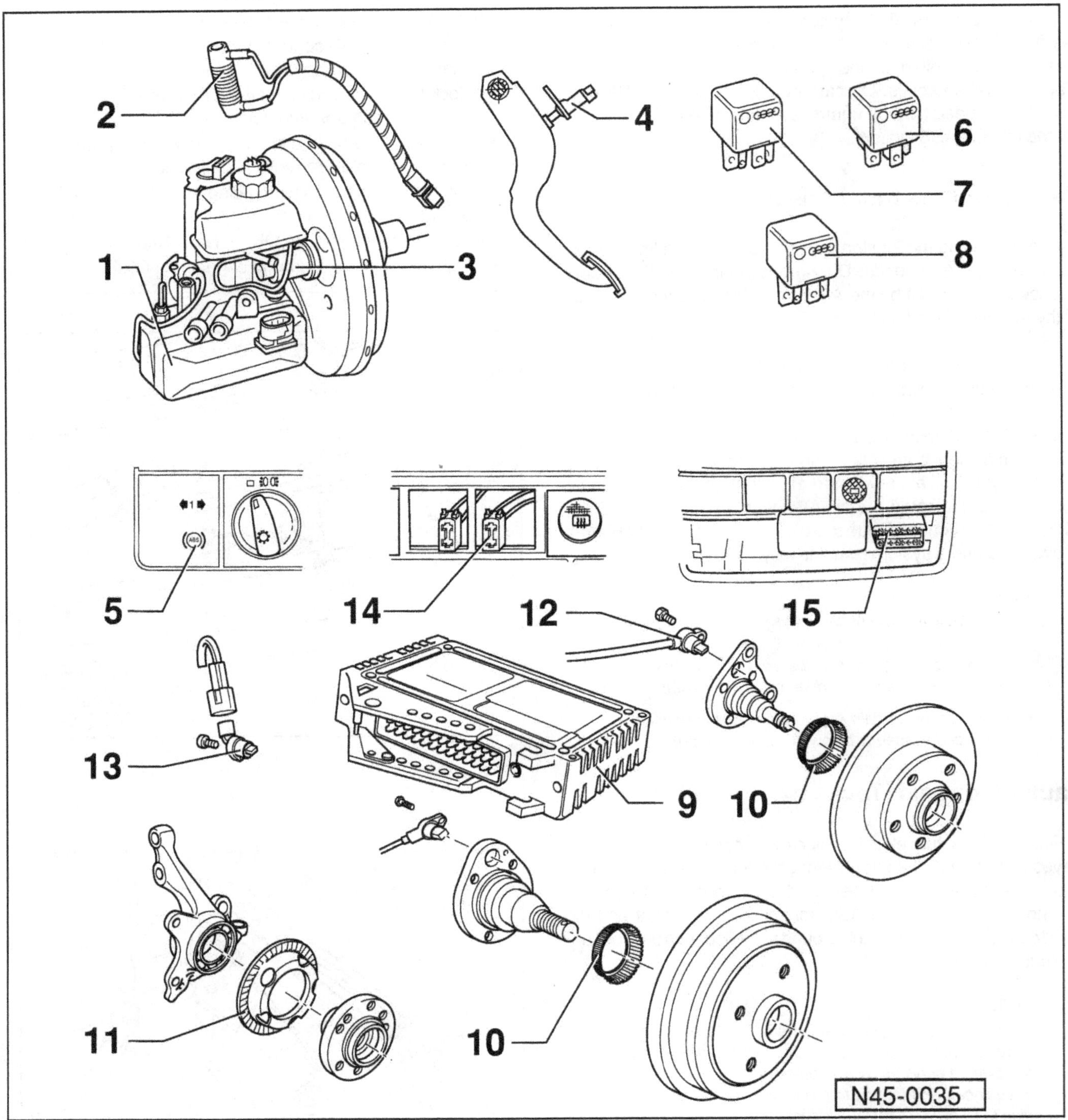

N45-0035

Fig. 5. Teves 04 ABS and ABS/EDL electrical/electronic components.

1. **ABS hydraulic unit (N55)**
   - with ABS hydraulic pump (V64)* and valve block
   - ABS hydraulic pump and valve block must not be separated from one another
2. **EDL series resistance (N159)***
   - Location: on ABS hydraulic unit (N55)

> ***NOTE —***
> *Components marked with an asterisk (*) are monitored by On Board Diagnostic (OBD) system.*

3. **Brake pedal position sensor (G100)***

4. **Brake light switch (F)***
   - Open in the rest position; see **Brake light switch, adjusting**
5. **ABS warning light (K47)**
   - Function: lights for a few seconds after switching ignition on, remains on if one or more malfunctions are recognized
6. **ABS relay (J102)**
   - Location: above relay panel
7. **ABS hydraulic pump relay (J185)**
   - Location: above relay panel
8. **EDL relay (J310)***
   - Location: above relay panel
9. **ABS control module (J104)***
   - Location:
     - Cars built up to 01/93: in footwell, right front below floor mat in front of passenger seat
     - Cars built from 01/93: under rear seat, right side

> *NOTE —*
> *The vehicle production date can be found on the safety compliance sticker on driver's door jamb (B-pillar).*

10. **Impulse rotor**
    - For left-rear and right-rear ABS wheel speed sensors
11. **Impulse rotor**
    - For left-front and right-front ABS wheel speed sensors
12. **Left-rear/right-rear ABS wheel speed sensor (G44)*/(G46)***
    - Before inserting the sensor, clean the inner surface of the sensor mounting and coat with lubricating paste G 000 650
    - When connecting the ABS wheel speed sensor wire, make sure it is not twisted in wheel housing
    - Mounting bolt tightening torque: 10 Nm (89 in-lb)
    - Resistance: 1.0–1.3 k ohms
13. **Left-front/right-front ABS wheel speed sensor (G45)*/(G47)***
    - Before inserting the sensor, clean the inner surface of the sensor mounting and coat with lubricating paste G 000 650
    - When connecting the ABS wheel speed sensor wire, make sure it is not twisted in wheel housing
    - Mounting bolt tightening torque: 10 Nm (89 in-lb)
    - Resistance: 1.0–1.3 k ohms
14. **Data Link Connectors (DLC) for vehicles built up to 07/93**
    - Location: below the heating/ventilation controls, next to heated rear window switch
15. **Data Link Connectors (DLC) for vehicles built from 08/93**
    - Location: behind cover, near ashtray

Control module, removing and installing–Teves 04

Location of the ABS control module depends on the production date. On early cars (built up to 01/93), the ABS control module is in the left front footwell under the floor covering in front of the passenger seat. On later cars (built after 01/93), the ABS control module is under the right-hand side of the rear seat.

> *CAUTION —*
> *Disconnecting the harness connector from the ABS control module may erase the diagnostic trouble code (DTC) memory. If the system is malfunctioning, check for DTCs using the Volkswagen diagnostic scan tool before disconnecting connector.*

1. Make sure ignition is off. Press the ABS control module fastening clips together and pull out.

> *CAUTION —*
> *The ignition must be off when disconnecting the harness connector from the ABS control module.*

2. Release the connector latch (fold latch down) and pull connector off.

3. Installation is the reverse order of the removal.
   - When positioning the harness connector into the control module, make sure the retaining lugs align with the grooves on ABS control module before pressing the connector onto the module. See Fig. 6.

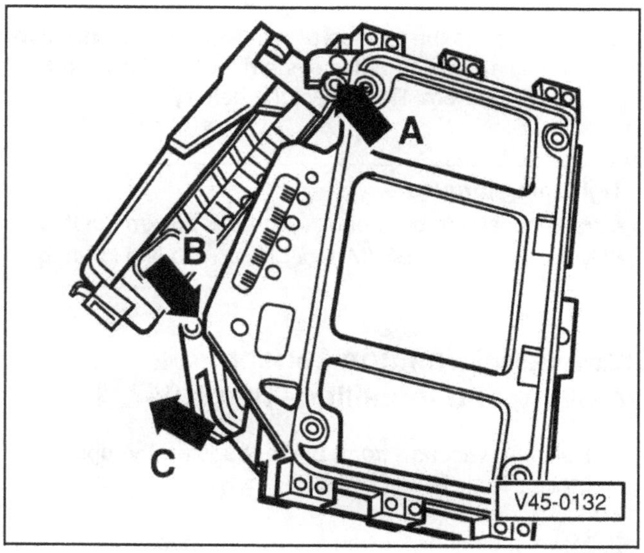

V45-0132

Fig. 6. Connector being installed into ABS control unit. Align lugs with grooves (**arrow A**), press connector into control module (**arrow B**), and lock in place using latch (**arrow C**).

Hydraulic unit, removing and installing–Teves 04

> ***WARNING —***
> - *Review Warnings, Cautions, and Notes on page 45-1 before removing hydraulic unit.*
> - *ABS hydraulic units supplied by Volkswagen as replacement parts may be filled with hydraulic fluid and pre-bled. Check with an authorized Volkswagen dealer parts department for the latest parts information when replacing the hydraulic unit.*

1. Disconnect negative (-) battery cable.
2. Disconnect all electrical connections from ABS hydraulic unit.
3. Remove brake fluid from reservoir (use a clean syringe).
4. Thoroughly clean around all hydraulic lines on ABS hydraulic unit. Then remove lines and seal/plug open connections.
5. Remove shelf below left side instrument panel.
6. Disconnect brake pedal from vacuum brake booster. See **47 Brakes-Hydraulic System**. Remove nuts securing vacuum brake booster to bulkhead.
7. Remove ABS hydraulic unit, complete with brake master cylinder and vacuum brake booster, from engine compartment.
8. Installation is the reverse of removal. Once installation is complete, bleed brake system, see **47 Brakes–Hydraulic System**. Reconnect the battery.

Tightening torques
- Brake booster to bulkhead . . . . 25 Nm (18 ft-lb)
- Hydraulic lines to ABS (M10 or M12) . . 15 Nm (11 ft-lb)

Brake pedal position sensor, removing and installing–Teves 04

1. Release vacuum from brake booster by operating brake pedal several times.
2. Move coolant expansion tank to one side.
3. Remove wiring harness connector from brake pedal position sensor.
4. Pry out circlip from brake pedal position sensor with scribe or similar tool. See Fig. 7.

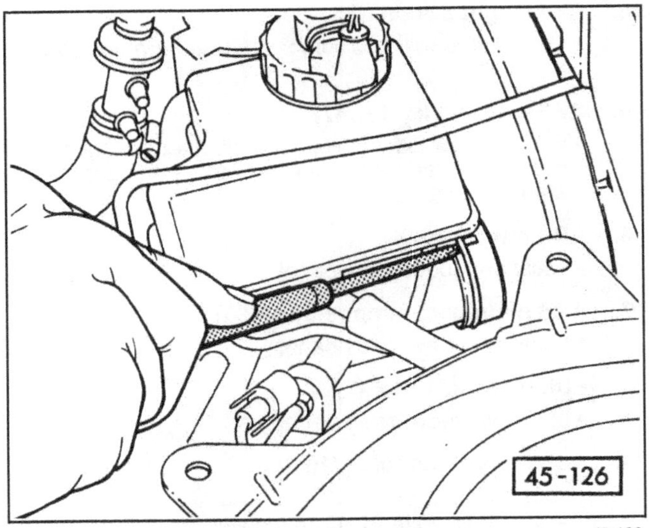

45-126

Fig. 7. Circlip being removed from brake pedal position sensor.

5. Pull brake pedal position sensor from brake booster. See Fig. 8.

> ***CAUTION —***
> *The brake pedal position sensor is supplied from Volkswagen as a replacement part in repair kit form. The kit contains the O-ring, circlip and four different spacer caps of different length and color (white, green, blue, and red). The replacement brake pedal position sensor spacer cap must be the same color as the part removed from the vehicle. The color is also marked on the top of the brake booster housing near the sensor.*

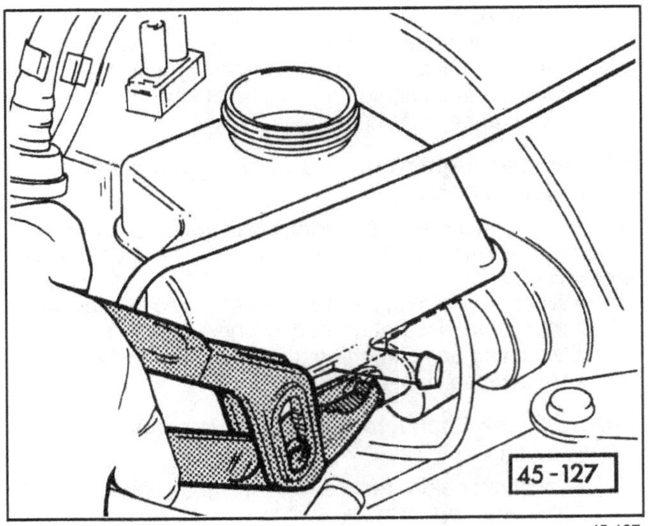

45-127

Fig. 8. Brake pedal position sensor being removed from brake booster.

6. Install new circlip so that gap in circlip points downward. See Fig. 9.

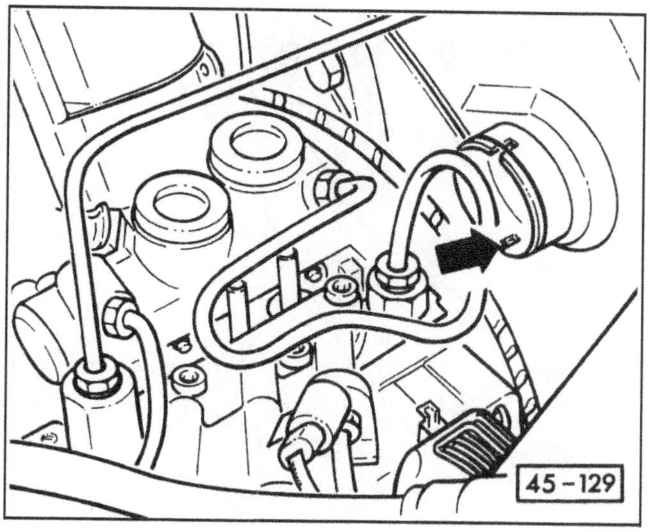

45-129

Fig. 9. Install circlip so that gap in circlip points down (**arrow**).

7. Coat new O-ring seal with lubricant (e.g. soft soap) and install on sensor.

> ***CAUTION —***
> *If the O-ring does not seal properly, insufficient braking power and poor idling might result.*

8. Insert the sensor into the brake booster so that the groove in the sensor is 180° opposed to the circlip gap. See Fig. 10.

NOTE —

- *Listen for a click as circlip locks into groove of sensor. Gently pull on sensor by hand to ensure that it is seated.*
- *Fig. 10 shows the brake fluid reservoir removed for illustration purposes only.*

9. Reconnect wiring harness connector. Move coolant expansion tank back to original position.

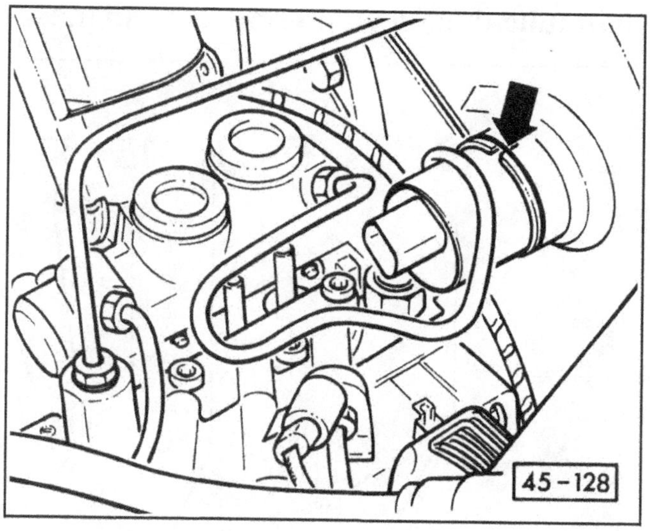

45-128

Fig. 10. Install brake pedal position sensor so that groove (**arrow**) is 180° opposed to circlip gap.

Brake light switch, adjusting–Teves 04

> ***CAUTION —***
> *Shut off engine and apply parking brake before removing brake light switch.*

1. Remove brake light switch by rotating it 90° clockwise.

NOTE —

- *It is necessary to first remove brake light switch to adjust it. Brake light switch is located above brake pedal in pedal bracket.*
- *Once removed, the brake light switch plunger should extend fully out. If not, switch is defective.*

2. Press brake pedal down as far as possible by hand.
3. Guide brake light switch through pedal bracket opening until plunger contacts brake pedal.
4. Lock in position by turning 90° counter-clockwise.
5. Release brake pedal and check function of brake lights.

Hydraulic unit, vacuum booster and master cylinder–Teves 04 ABS/EDL, assembly

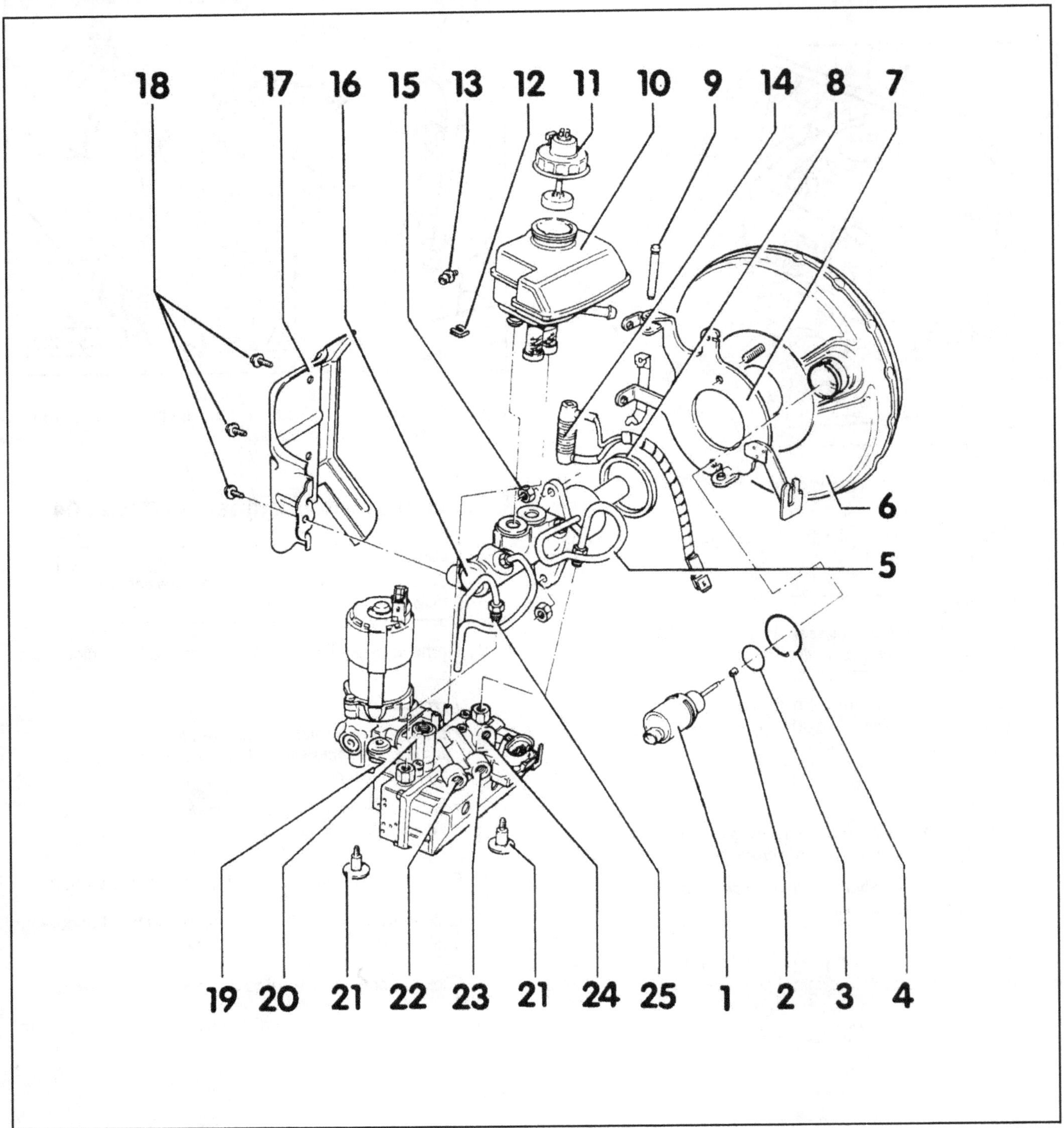

Fig. 11. Exploded view of the Teves 04 ABS/EDL unit.

1. **Brake pedal position sensor (G100)**
   - See **Brake pedal position sensor, removing/installing**.
2. **Spacer cap**
   - Four versions, see **Brake pedal position sensor, installing**.
3. **O-ring**
   - Always replace
4. **Circlip**
   - Always replace

(continued from previous page)

5. **Hydraulic line**
   - Brake master cylinder (primary piston circuit) to ABS hydraulic (N55)
6. **Vacuum brake booster**
7. **Bracket**
8. **Seal**
9. **Pin**
10. **Brake fluid reservoir**
11. **Cap**
12. **Clip**
13. **Screw**
14. **EDL series resistance (N159)**
    - On Board Diagnostic (OBD) component
15. **Self-locking nut**
    - Tightening torque: 25 Nm (18 ft-lb)
16. **Brake master cylinder**
17. **Shield**
18. **Bolt**
19. **ABS/EDL hydraulic unit (N55)**
    - See **Hydraulic unit, removing/installing–Teves 04**
20. **Hydraulic line connection**
    - To right-front brake caliper
21. **Torx bolts (T25)**
22. **Hydraulic line connection**
    - To left-rear brake caliper
23. **Hydraulic line connection**
    - To right-rear brake caliper
24. **Hydraulic line connection**
    - To left-front brake caliper
25. **Hydraulic line**
    - Brake master cylinder (secondary piston circuit) to ABS hydraulic unit (N55)

ABS, ABS/EDL–TEVES 20 GI

The Teves 20 GI ABS system is a compact design that incorporates the ABS control module as an integral part of the ABS hydraulic unit. On cars equipped with the VR6 engine, a low speed traction control system, called EDL, is also integrated into the ABS system. Fig. 12. shows the Teves 20 GI ABS electrical/electronic components. Fig. 13. shows the Teves 20 GI hydraulic unit, vacuum brake booster and master cylinder.

Most service and repair to the Teves 20 GI ABS/EDL brake system requires the use of the Volkswagen V.A.G. 1551 scan tool. This includes "coding" of replacement control modules and brake bleeding. Therefore, it is recommended that if no faults can be found after reviewing the information given in this repair group, the system be serviced by an authorized VW dealer who is equipped and trained to service the Teves 20 GI ABS and ABS/EDL systems.

> ***CAUTION —***
> - *Special Volkswagen test equipment is needed to bleed the brakes on cars with Teves 20 GI ABS/EDL. The test equipment tool is used to electronically cycle the ABS hydraulic unit on and off to purge the hydraulic unit/valve body of trapped air.*
> - *The Teves 20 GI ABS/EDL control module is "coded" for the specific vehicle application. New ABS control modules from Volkswagen are not coded and must be coded by an authorized Volkswagen dealer. DO NOT use an ABS control module from another vehicle as coding may be different.*

Self-diagnosis function–Teves 20 GI

After switching the ignition on, the ABS warning light and the warning light for the brake system light up for approx. 2 seconds. During this period a self-test sequence is carried out by the ABS control module to check the following:

- Battery positive voltage (B+); must be at least 10.0 volts
- ABS control module, including valve windings
- ABS wheel speed sensors (electrical test; fully completed at approx. 12 mph)
- ABS control module coding

Fault diagnosis–Teves 20 GI

If an ABS malfunction is recognized, ABS is switched off and the ABS warning light comes on.

> ***NOTE —***
> *ABS malfunctions are indicated by the ABS warning light coming on; certain malfunctions will only be recognized at speeds above 12 mph; carry out a road test.*

If the ABS warning light does not go out after completion of self-test sequence, the possible malfunctions are:

- Battery voltage (B+) is below 10 volts.
- Sensor malfunction (sporadic) after last starting the vehicle.
- Open circuit between the ABS warning light module and ABS control module or ABS warning light module faulty.
- Malfunction in ABS circuit.

NOTE —
In the case of a sporadic sensor malfunction, the ABS warning light will go out after restarting the engine and reaching a speed in excess of 12 mph.

If ABS warning light goes out, but brake system warning light remains on, the possible malfunctions are:

- The parking brake is pulled on.
- The brake fluid level is too low.
- Malfunction in control of warning light for brake system. See **97 Wiring Diagrams, Fuses and Relays**.

NOTE —
If both the ABS warning light and the brake system warning light stay on, then ABS and electronic brake pressure distribution (EBPD) is non-functional. In this case, there will be a change in braking performance, as the brake pressure to the rear wheels will not be regulated.

Brake light switch, adjusting–Teves 20 GI

CAUTION —
Shut off engine and apply parking brake before removing brake light switch.

NOTE —
It is necessary to first remove brake light switch to adjust it. Brake light switch is located above brake pedal in pedal bracket.

1. Remove brake light switch by rotating it 90° clockwise.

NOTE —
Once removed, the brake light switch plunger should extend fully out. If not, the switch is defective.

2. Press brake pedal down as far as possible by hand.
3. Guide brake light switch through pedal bracket opening until plunger contacts brake pedal.
4. Lock in position by turning 90° counter-clockwise.
5. Release brake pedal.
6. Check function of brake lights.

Numbered list below applies to Fig. 12.

1. **ABS hydraulic unit (N55)**
   - with ABS hydraulic pump (V64)* and valve block
   - ABS hydraulic pump and valve block must not be separated
2. **ABS control module (J104)***
   - Located on ABS hydraulic unit in left side of engine compartment

NOTE —
Components marked with an asterisk (*) are monitored by On Board Diagnostic (OBD).

3. **Brake light switch (F)**
   - The brake light switch is open in the rest position
   - Adjusting see **Brake light switch, adjusting**
4. **Warning light for brake system (K118)**
   - Location: in instrument cluster
   - Function-lights up for approx. 2 seconds after ignition is switched on and/or engine is started. Remains on when parking brake is applied or brake fluid level too low, or when the ABS warning light lights up and there is a loss of electronic brake pressure distribution
5. **ABS warning light (K47)**
   - Location: left next to instrument cluster
   - Function-lights up for approx. 2 seconds after ignition is switched on and/or engine is started. Remains on if one or more ABS malfunctions are detected (e.g. ABS wheel speed sensor open circuit)
6. **Impulse rotor**
   - For right- and left-rear speed sensors
   - Removing and installing see **46 Brakes-Mechanical Components**
7. **Impulse rotor**
   - For right- and left-front speed sensors removing and installing see **40 Front Suspension and Drive Axles**
8. **Right-rear/left-rear ABS wheel speed sensor (G44)*/(G46)***
   - Before installing sensor, clean inner surface of mounting and coat with lubricating paste G 000 650
   - When connecting the ABS wheel speed sensor wire, make sure it is not twisted in the wheel housing
   - Mounting bolt tightening torque: 10 Nm (89 in-lb)
   - Resistance: 1.0–1.3 k ohms

(continued on following page)

Component locations–Teves 20 GI

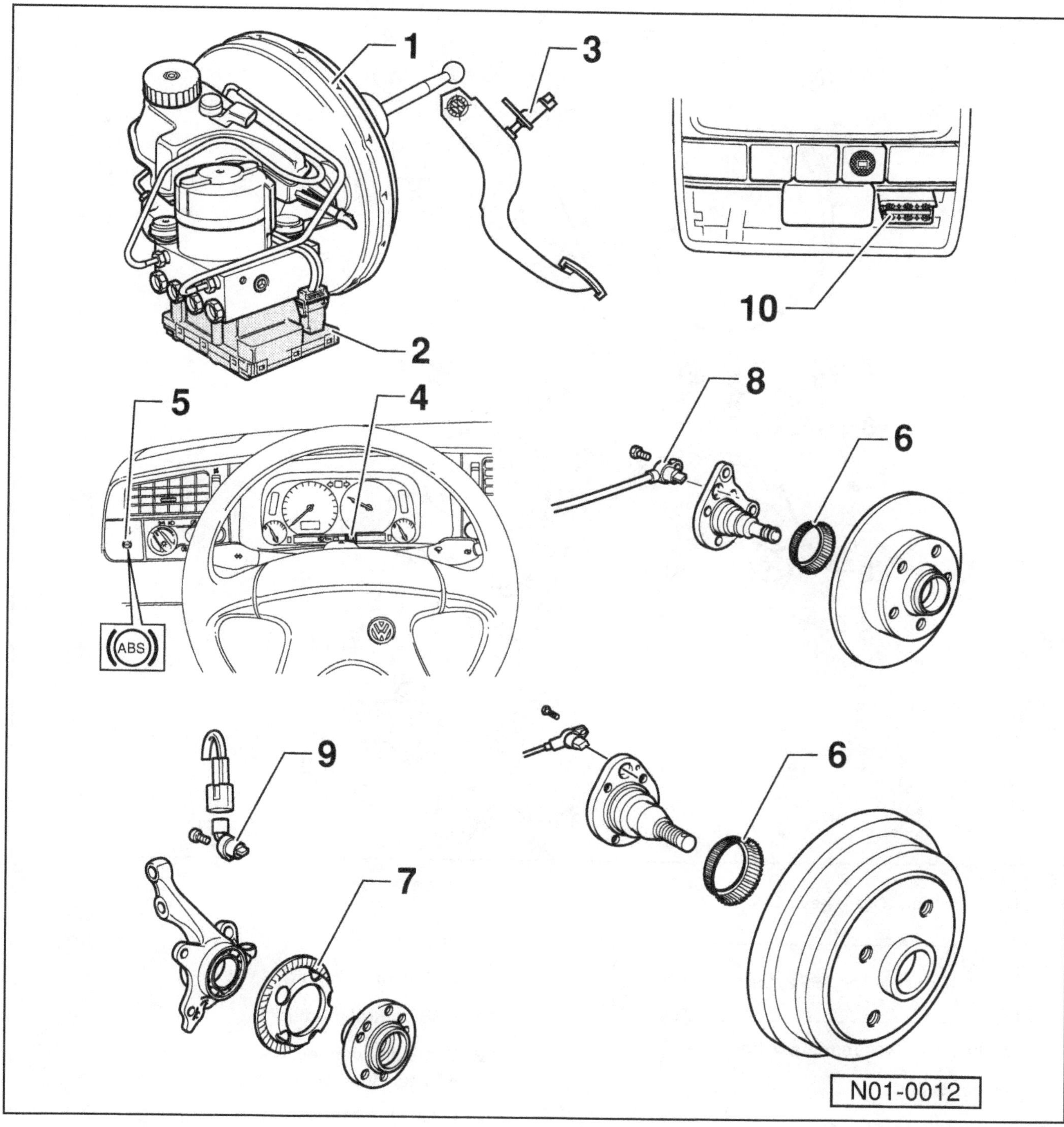

N01-0012

Fig. 12. Teves 20 GI ABS and ABS/EDL electrical/electronic components.

9. **Right-front/left-front ABS wheel speed sensor (G45)*/(G47)***
   - Before installing sensor, clean inner surface of mounting and coat with lubricating paste G 000 650
   - When connecting the ABS wheel speed sensor wire, make sure it is not twisted in the wheel housing
   - Mounting bolt tightening torque: 10 Nm (89 in-lb)
   - Resistance: 1.0–1.3 k ohms

10. **Data Link Connector (DLC)**
    - Location: behind a cover on the right next to the ashtray

Hydraulic unit, vacuum booster and master cylinder–Teves 20 GI ABS/EDL, assembly

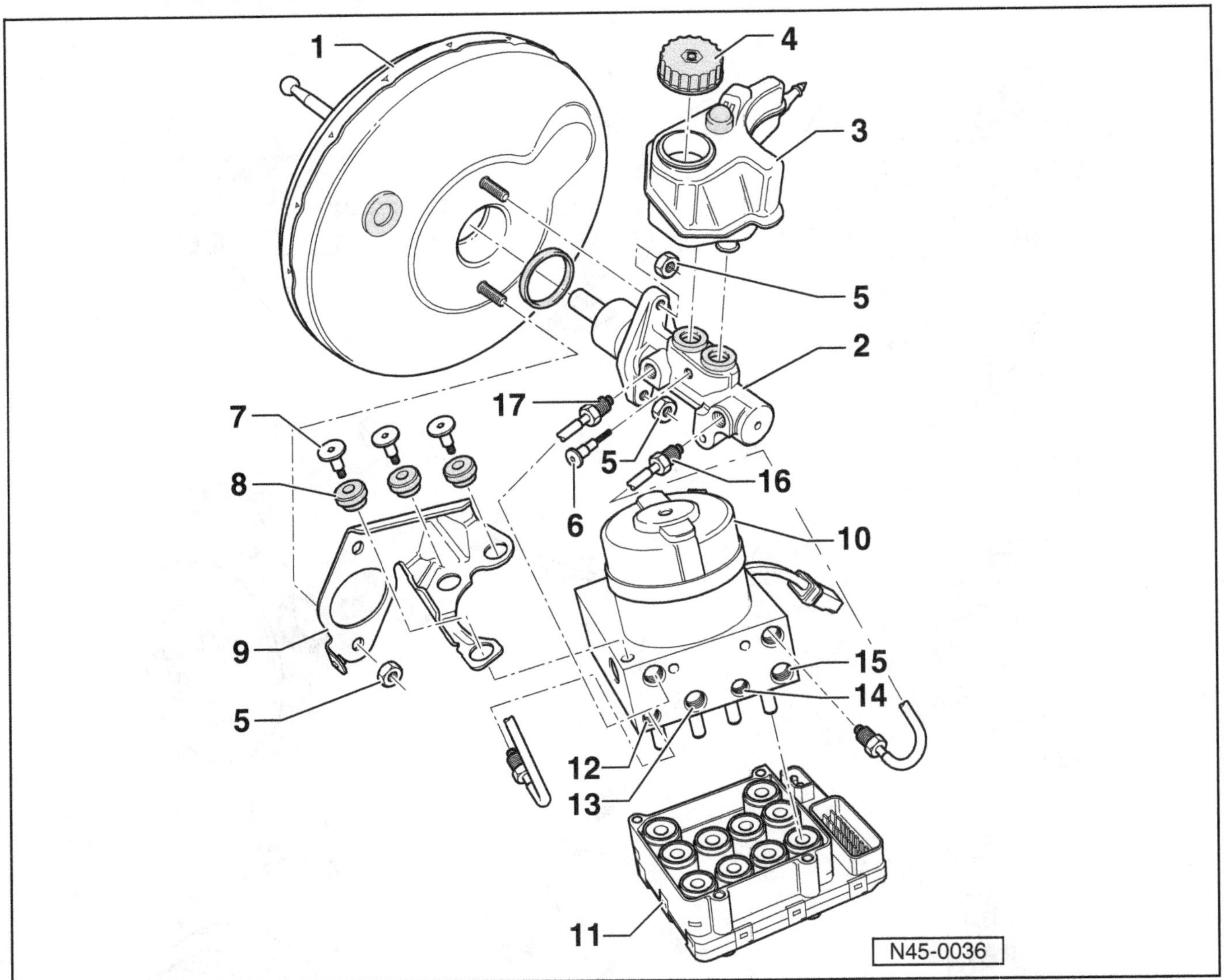

N45-0036

Fig. 13. Exploded view of the Teves 20 GI ABS/EDL unit.

> ***CAUTION—***
> *Do NOT separate the ABS hydraulic pump from the valve block.*

1. **Brake servo (vacuum brake booster)**
2. **Master cylinder**
3. **Brake fluid reservoir**
4. **Cap**
5. **Self-locking nut**
   - Always replace
   - Tightening torque: 25 Nm (18 ft-lb)
6. **Torx (T25) socket head bolt**
7. **Torx (T25) socket head bolt**
8. **Rubber damper**
9. **Retainer**
10. **ABS hydraulic unit**
11. **ABS control module**
12. **Brake line connection**
    - To right front brake caliper
13. **Brake line connection**
    - To left rear brake caliper/wheel cylinder
14. **Brake line connection**
    - To right rear brake caliper/wheel cylinder
15. **Brake line connection**
    - To left front brake caliper
16. **Brake line**
    - Brake master cylinder (secondary piston) to ABS unit
17. **Brake line**
    - Brake master cylinder (primary piston) to ABS unit

46 Brakes–Mechanical Components

GENERAL

This repair groups covers service and repair to brake pads, rotors, shoes, drums and the parking brake components.

Cars covered by this manual are equipped either with front disc brakes and rear drum brakes, or with disc brakes on all four wheels. Some cars with four-wheel disc brakes are also equipped with Anti-Lock Brakes (ABS).

NOTE —
- *For brake caliper and rear brake wheel cylinder repair information, see* **47 Brakes–Hydraulic System**.
- *For information on the ABS hydraulic unit and ABS service, see* **45 Anti-Lock Brakes (ABS)**.
- *Brake fluid should be flushed from the system every two years. See* **47 Brakes–Hydraulic System**.

WARNING —
- *A properly functioning brake system is essential to safe driving. If the red brake/parking warning light or ABS warning light illuminates while driving, it is imperative that the system be given a thorough check, even if braking action still seems satisfactory. The brakes should be inspected regularly.*
- *Brake fluid is poisonous. Wear safety glasses when working with brake fluid, and wear rubber gloves to prevent brake fluid from entering the bloodstream through cuts or scratches. Do not siphon brake fluid by mouth.*
- *New brake pads and shoes require some break-in. Allow for slightly longer stopping distances for the first 100 to 150 miles of city driving, and avoid hard stops.*

CAUTION —
- *All brake work must be done with cleanliness, careful attention to specifications, and proper working procedures. If you lack the skills, the tools, or a clean workplace for servicing the brake system, we suggest you leave these repairs to an authorized VW dealer or other qualified shop.*
- *After replacing brake components, depress the brake pedal firmly several times to seat the brakes in their normal operating position.The pedal should be firm and at its normal height, if not, further work is required before driving vehicle.*
- *Brake fluid is very damaging to paint. Immediately wipe up any brake fluid that spills on the vehicle.*

FRONT BRAKES

Two types of front discs–solid and ventilated–are used on the cars covered by this manual. Cars with ventilated rotors are equipped with a Girling two piece caliper or in the case of later VR6 cars, a Teves/ATE one piece caliper. Cars with solid rotors use a one piece VW II caliper. Repair procedures vary depending on rotor/caliper application. Exploded views of the various front brake assemblies and procedures for pad and rotor replacement are given on the following page.

Rotors should be inspected for cracks, scoring, glazing and warpage. Rotors must be replaced in pairs if any faults are found. Rotors must always be replaced if either disc is worn below the minimum thickness specification.

Front brake assembly (Girling caliper with ventilated rotor)

1. **Screw**
2. **Brake pads**
   - VR6 engine: with pad wear indicators
   - Thickness: 14 mm (0.551 in.)
   - Wear limit: 7 mm (0.276 in.) including backing plate
   - Always replace all pads on one axle at the same time
3. **Ventilated brake rotor**
   - Remove brake caliper before removing
   - 4-cylinder rotor specifications: Diameter: 256 mm (10.08 in.) Thickness: 20 mm (0.787 in.) Wear limit: 18 mm (0.709 in.)
   - 6-cylinder rotor specifications: Diameter: 280 mm (11.02 in.) Thickness: 22 mm (0.866 in.) Wear limit: 20 mm (0.787 in.)
4. **Screw**
5. **Wheel hub**
   - Removing/installing see **40 Front Suspension and Drive Axles**.
6. **ABS rotor**
   - For ABS wheel speed sensor, where applicable
7. **Bolt**
   - Tighten to 10 Nm (89 in-lb)
8. **Splash shield**
9. **Wheel bearing housing**
10. **Brake pad carrier**
11. **Heat deflector**
    - Insert in piston
12. **Brake caliper**
    - Do not loosen hydraulic line when replacing pads
13. **Bracket**
14. **Self-locking nut**
    - Tighten to 35 Nm (26 ft-lb)
    - Always replace
15. **Bolt**
    - Tighten to 125 Nm (92 ft-lb)
16. **Protective cap**
17. **Guide bolt**

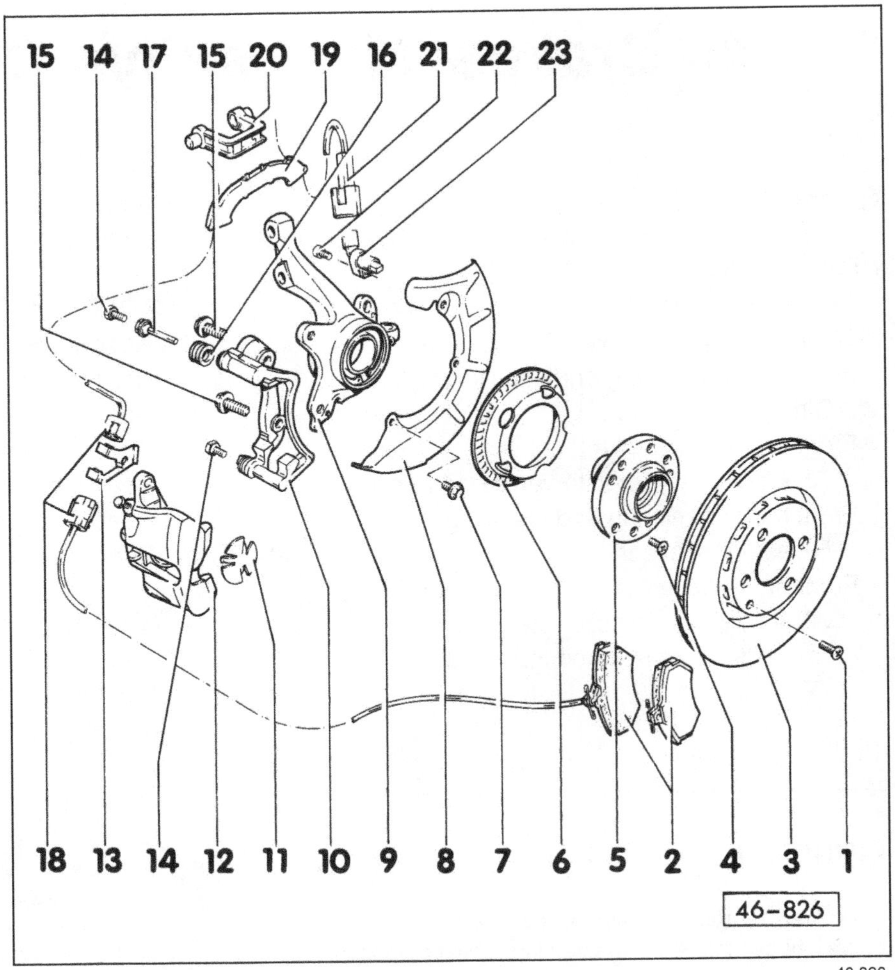

46-826

Fig. 1. Front disk brake assembly for vehicles with Girling two-piece caliper.

18. **Connector**
    - For brake pad wear indicator (where applicable)
19. **Bracket**
20. **Bracket**
21. **Connector**
    - For ABS Wheel Speed Sensor
22. **Bolt**
    - Tighten to 10 Nm (89 in-lb)
23. **ABS wheel speed sensor**
    - Before installing, clean and coat with lubricating paste (VW part no. G 000 650)

Front pads, removing and installing (Girling caliper)

1. Raise the car and support on jack stands.

> *WARNING —*
> *Jack stands should be placed on a hard level surface (e.g. concrete).*

2. Remove the front wheels.

3. Separate connector for brake pad wear indicator, where applicable.

4. Remove lower caliper mounting bolt while holding guide bolt stationary. See Fig. 2.

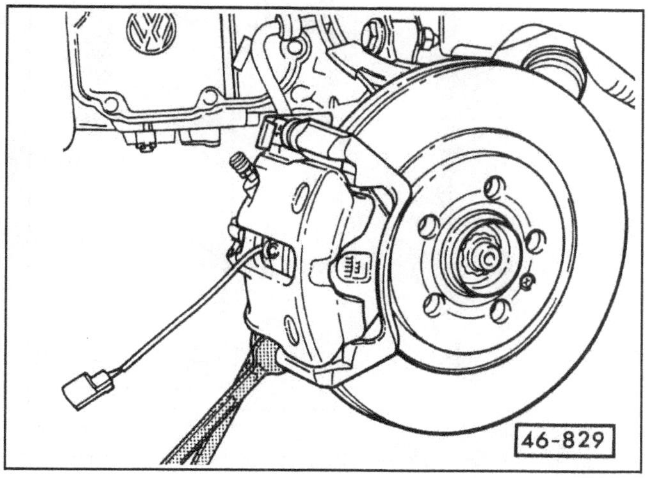

46-829

Fig. 2. Lower caliper mounting bolt being removed. Use wrench to hold guide bolt when removing mounting bolt.

5. Swing brake caliper housing upward and off brake pads. Remove brake pads from pad carrier. See Fig. 3.

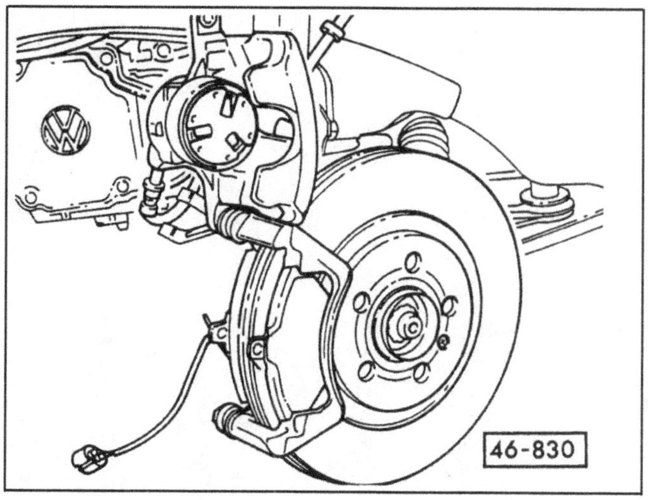

46-830

Fig. 3. Pivot caliper up to access pads.

6. Press piston into caliper housing. See Fig. 4.

> *CAUTION —*
> *Always remove some brake fluid from the reservoir before installing new brake pads. When the caliper piston is pushed in, fluid is forced up into the reservoir.*

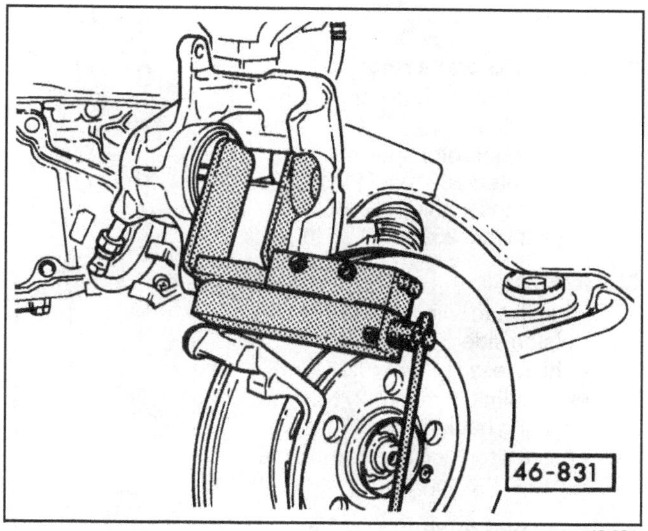

46-831

Fig. 4. Caliper piston being pressed into caliper using VW special tool.

7. Install heat shield and brake pads.

> *NOTE —*
> *On pads with wear indicators (cars with VR6 engine), the pad with the wear indicator is the inner pad.*

8. Swing brake caliper down.

9. Install and tighten new caliper mounting bolt while holding guide bolt stationary. Reconnect wear indicator, where applicable.

> *WARNING —*
> *Always replace the lower caliper bolt. It is a one-time only self-locking bolt.*

Tightening torques

- Caliper mounting bolt to guide bolt (self-locking) . . . . 35 Nm (26 ft-lb)

10. Depress brake pedal firmly several times to seat brake pads. Check brake fluid level and add as necessary. Test drive vehicle and operate brakes several times.

Front brake assembly (Teves/ATE caliper with ventilated rotor)

NOTE —
The Teves/ATE caliper is installed on vehicles with VR6 engine from 01/1995.

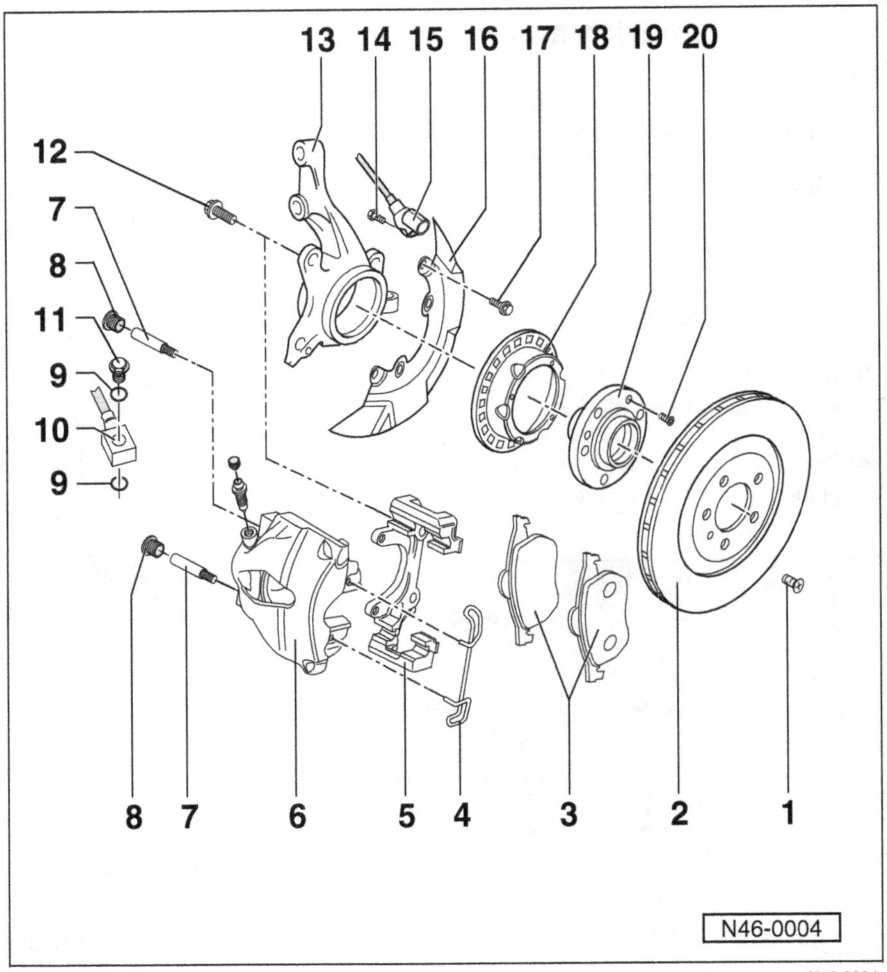

N46-0004

Fig. 5. Front disk brake assembly for VR6 equipped vehicles (from 01/1995) with Teves/ATE caliper.

1. **Screw**
2. **Ventilated brake rotor**
   - Remove brake caliper before removing
   - 6-cylinder rotor specifications: Diameter: 288 mm (11.34 in.) Thickness: 25 mm (0.98 in.) Wear limit: 23 mm (0.90 in.)
3. **Brake pads**
   - Install pad with retaining clip on piston side
   - Thickness: 14 mm (0.551 in.)
   - Wear limit: 7 mm (0.276 in.) including backing plate
   - Always replace all pads on one axle at the same time
4. **Brake pad retaining spring**
5. **Brake carrier**
   - Factory replacement caliper comes with guide pins greased
   - Lubricate guide pins with grease, as necessary
6. **Brake caliper**
   - Do not disconnect brake hydraulic line from caliper when replacing pads
   - Do not allow the caliper to hang by the brake hydraulic line. The unsupported weight can stretch and damage the hose
7. **Guide pin**
   - Tighten to 25 Nm (18 ft-lb)
8. **Protective cap**
9. **Seal**
   - Always replace
10. **Brake hose**
11. **Banjo bolt**
    - Tighten to 30 Nm (22 ft-lb)
12. **Bolt**
    - Tighten to 125 Nm (92 ft-lb)
13. **Wheel bearing housing**
14. **Bolt**
    - Tighten to 10 Nm (89 in-lb)
15. **ABS wheel speed sensor (G45/G47)**
    - Before installing, clean mounting hole inner surface and coat with lubricating paste (part no. G 000 650)
16. **Splash shield**
17. **Bolt**
    - Tighten to 10 Nm (89 in-lb)
18. **ABS rotor**
    - For ABS wheel speed sensor, where applicable
19. **Wheel hub**
    - Removing/installing see **40 Front Suspension and Drive Axles**.
20. **Countersunk bolt**

Front pads, removing and installing (Teves/ATE caliper)

1. Raise the car and support on jack stands.

> ***WARNING —***
> *Jack stands should be placed on a hard level surface (e.g. concrete).*

2. Remove the front wheels.
3. Separate connector for brake pad wear indicator, where applicable.
4. Remove guide pin protective caps from back of caliper. See Fig. 6.

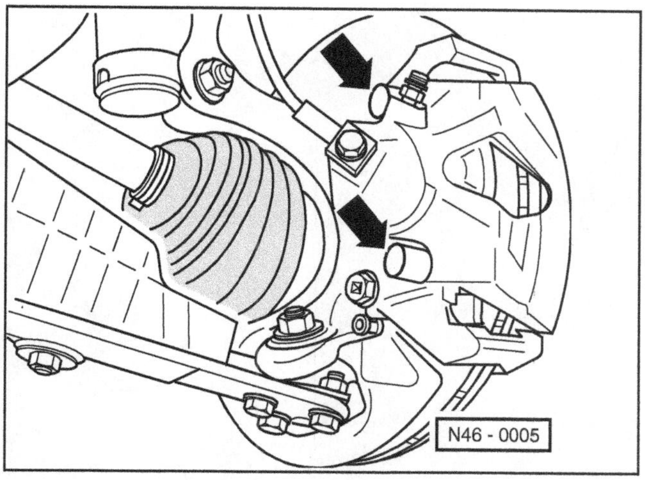

N46-0005

Fig. 6. Remove guide pin protective caps (**arrows**).

5. Use a screwdriver to pry off brake pad outer retaining spring. Make note of installation position. See Fig. 7.

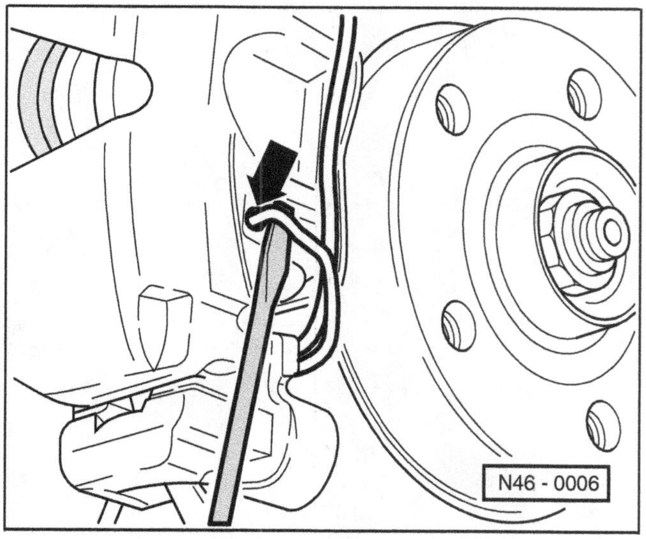

N46-0006

Fig. 7. Pry off brake pad retaining spring.

6. Remove guide pins from back side of caliper.
7. Remove caliper and suspend from vehicle with wire or heavy-duty zip ties.

> ***CAUTION —***
> - *Do not disconnect brake hydraulic line from caliper when replacing pads.*
> - *Do not allow the caliper to hang by the brake hydraulic line. The unsupported weight can stretch and damage the hose.*

8. Remove old brake pads from caliper.
9. Press piston into caliper housing to allow clearance for installation of new pads. See Fig. 8.

> ***CAUTION —***
> *Always remove some brake fluid from the reservoir before installing new brake pads. When the caliper piston is pushed in, fluid is forced up into the reservoir and can overflow.*

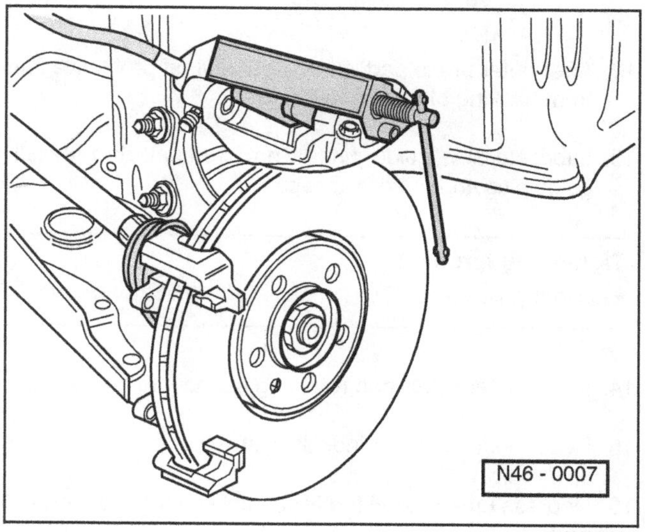

N46-0007

Fig. 8. Caliper piston being pressed into caliper using piston compressing tool.

10. Clean caliper thoroughly before installing new pads.
11. Install brake pads.
    - Install pad with retaining clip (inner pad) into caliper piston.
    - Make sure that arrow on inner pad is facing downward when installed. See Fig. 9.

> ***NOTE —***
> *On pads with wear indicators (cars with VR6 engine), the pad with the wear indicator is on the inner pad.*

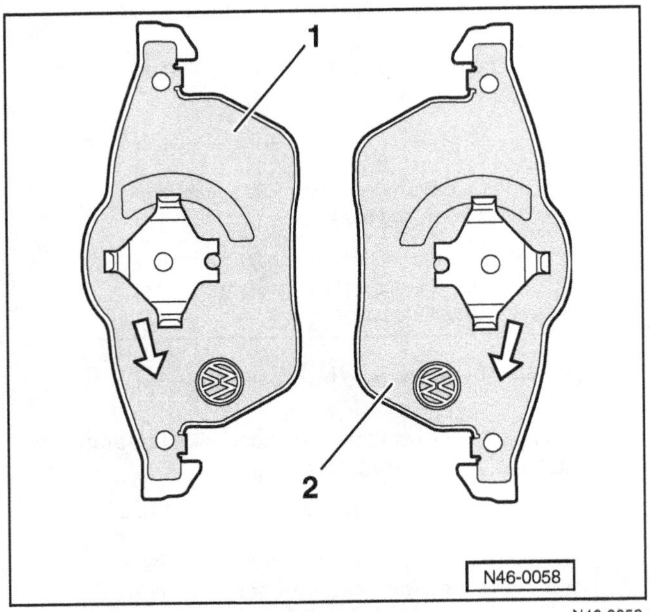

N46-0058

Fig. 9. Make sure **arrow**, on pad backing plate, is pointing down when inner pads are installed in caliper. Right side (**1**) and left side (**2**) pads are shown above.

12. After outer brake pad is installed, pull off protective foil from backing plate, if applicable.

13. Lubricate new guide pins, if not pre-lubricated. Install guide pins and protective caps.

Tightening torques

- Caliper guide pins . . . . 25 Nm (18 ft-lb)

14. Install outer brake pad retaining spring into caliper.

15. Reconnect pad wear indicator, as applicable.

16. Depress brake pedal firmly several times to seat brake pads.

17. Check brake fluid level and add as necessary. Test drive vehicle and operate brakes several times.

Front brake assembly (VW II caliper with solid rotor)

1. **Bolt**
   - Tighten to 10 Nm (89 in-lb)
2. **ABS wheel speed sensor**
   - Before installing, clean mounting hole inner surface and coat with lubricating paste (part no. G 000 650)
3. **ABS wheel speed sensor rotor**
4. **Wheel hub**
   - Must be pressed in and out, see **40 Front Suspension and Drive Axles, Wheel bearing, removing and installing**
5. **Screw**
6. **Splash shield**
7. **Brake rotor**
   - Always replace rotors in pairs
   - Use penetrating oil to loosen rotor from wheel hub if necessary; do not use force to remove rotors
   - Rotor specifications:
     Diameter: 239 mm (9.41 in.)
     Thickness: 12 mm (0.472 in.)
     Wear limit: 10 mm (0.394 in.)
8. **Wheel bolt**
   - Tighten to 110 Nm (81 ft-lb)
9. **Bolt**
   - Tighten to 10 Nm (7 ft-lb)
10. **Brake pads**
    - For 1996 m.y. and later
    - Thickness: 14 mm (0.551 in.)
    - Wear limit: 7 mm (0.276 in.)
    - Pad with larger friction area fits on outside (where applicable)
    - Always replace pads on both sides
11. **Brake caliper**
    - For 1996 m.y. and later
    - Do not disconnect brake hydraulic line from caliper when replacing pads
    - Do not allow the caliper to hang by the brake hydraulic line. The unsupported weight can stretch and damage the hose
12. **Air deflector plate**
13. **Socket-head bolt, lower**
    - Tighten to 25 Nm (18 ft-lb)
14. **Socket-head bolt, upper**
    - Tighten to 25 Nm (18 ft-lb)

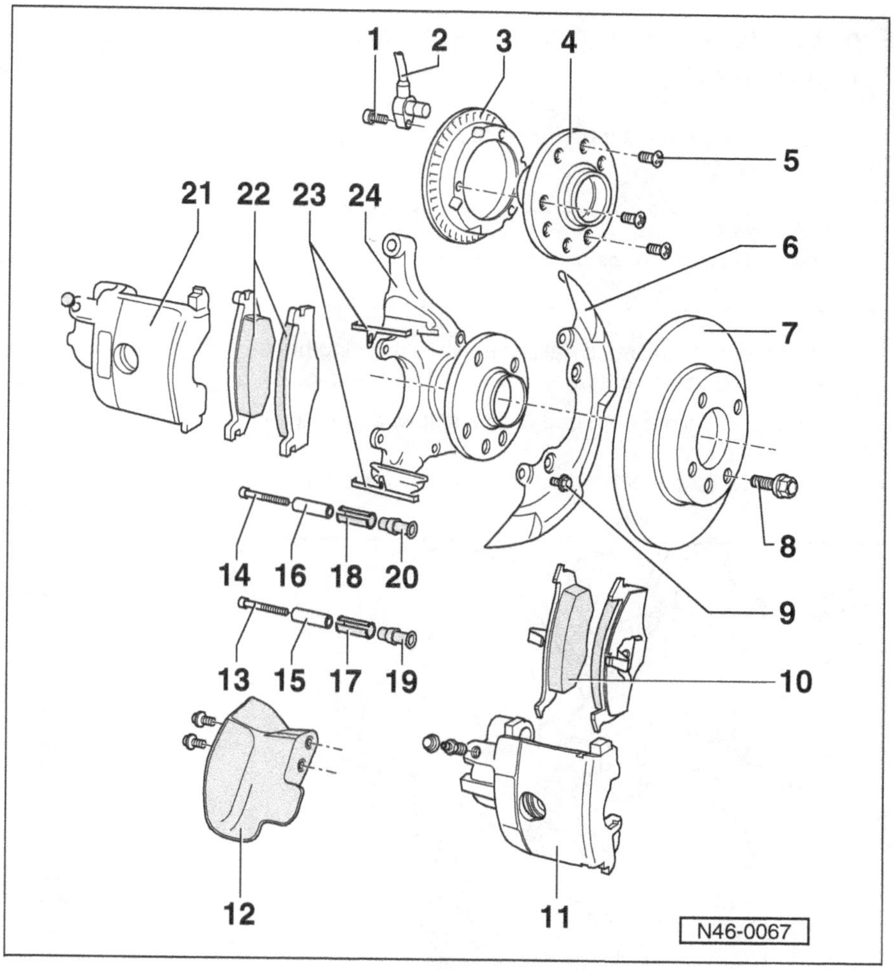

N46-0067

Fig. 10. Front disk brake assembly with non-ventilated solid rotor and VW II caliper.

15. **Lower spacer sleeve**
16. **Upper spacer sleeve**
17. **Upper sleeve**
18. **Upper sleeve**
19. **Lower bushing**
20. **Upper bushing**
21. **Brake caliper**
    - For 1995 m.y. and earlier
22. **Brake pads**
    - For 1995 m.y. and earlier
    - Thickness: 14 mm (0.551 in.)
    - Wear limit: 7 mm (0.276 in.)
    - Pad with larger friction area fits on outside (where applicable)
    - Always replace pads on both sides
23. **Brake pad retaining spring**
24. **Wheel bearing housing**

Front pads, removing and installing (VW II caliper)

1. Raise the car and support on jack stands. Remove the front wheels.

> ***WARNING —***
> *Jack stands should be placed on a hard level surface (e.g. concrete).*

2. Remove bolts securing caliper to wheel bearing housing. Swing caliper out from bottom to top. See Fig. 11.
   - Support caliper from body using stiff wire or heavy-duty zip tie.
   - Do not disconnect hydraulic line from brake caliper.

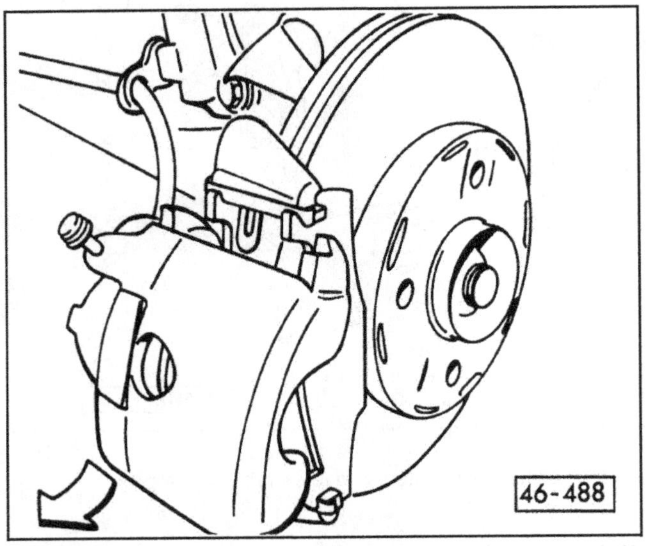

46-488

Fig. 11. Caliper being removed from wheel bearing housing. Swing caliper out from bottom first.

3. Remove brake pad retaining springs and brake pads.
4. Install inner brake pad and pad retaining springs on wheel bearing housing. Install outer brake pad.
   - Pad with larger friction surface is the outer pad.
5. Push piston into caliper housing.

> ***CAUTION —***
> *Always remove some brake fluid from the reservoir before pushing in piston. When the piston is pushed in, fluid is forced up into the reservoir.*

6. Carefully mount caliper just far enough to start caliper mounting bolts. Tighten caliper mounting bolts.

> ***NOTE —***
> *The pad retaining springs could be bent if the caliper is pushed on too far. Bent springs can cause noise when braking.*

Tightening torques
- Caliper to wheel bearing housing. . . . 25 Nm (18 ft-lb)

7. Depress brake pedal firmly several times to seat brake pads. Check brake fluid level and add as necessary. The remainder of installation is the reverse of removal.

Front brake rotor, removing and installing (ventilated and solid rotors)

1. Raise corner of car and support on jack stand. Remove the front wheel.

> ***WARNING —***
> *Jack stands should be placed on a hard level surface (e.g. concrete).*

2. On cars with ventilated rotors, remove bolts securing brake pad carrier to wheel bearing housing.

> ***CAUTION —***
> *On cars with ventilated rotors, do not remove the caliper from the brake pad carrier. The caliper is mounted to the guide bolts using self-locking bolts that must be replaced once removed.*

3. On cars with solid rotors, remove the caliper and brakes pads as described earlier under **Front Brake pads, removing and installing (solid rotors)**.
4. Remove countersunk screw from front of rotor hub and remove rotor from wheel hub.

> ***NOTE —***
> *If necessary, use a rust-penetrating oil to free the brake rotor from the hub. Do not use force to remove the rotor.*

5. Installation is reverse of removal. Depress brake pedal firmly several times to seat brake pads.

Tightening torques
- Brake pad carrier to wheel bearing housing (ventilated rotor, 2-piece caliper) . . . 125 Nm (92 ft-lb)
- Brake caliper to wheel bearing housing (solid rotor, 1-piece caliper) . . . . 25 Nm (18 ft-lb)
- Wheel to wheel hub . . . . 110 Nm (81 ft-lb)

REAR BRAKES

Two types of rear brake systems–drum and disc–are used on the cars covered by this manual. The following pages show assembly views of both systems and brake shoe/pad removal and installation procedures.

Rear drum brake, assembly

1. **Grease cap**
2. **Cotter pin**
   - Always replace
3. **Lock ring**
4. **Nut**
   - Adjusting wheel bearing play, see **42 Rear Suspension**
5. **Thrust washer**
6. **Outer wheel bearing**
7. **ABS wheel speed sensor rotor**
   - Installed on ABS equipped cars
   - Pry out from brake drum using two screwdrivers on either side
8. **Brake drum**
   - Diameter: 200 mm (7.874 in.)
   - Wear limit: 201 mm (7.913 in.)
   - Clean thoroughly, check for wear, damage, damaged wheel bolt threads and check dimensions
9. **Stub axle**
10. **Bolt**
    - Tighten to 10 Nm (7 ft-lb)
11. **ABS wheel speed sensor**
    - Installed on ABS equipped cars
    - Before installing, clean mounting hole inner surface and coat with lubricating paste (part no. G 000 650)
12. **Backing plate brake shoes**
13. **Dished washer**
    - Large contact surface faces backing plate
14. **Bolt**
    - Tighten to 60 Nm (44 ft-lb)

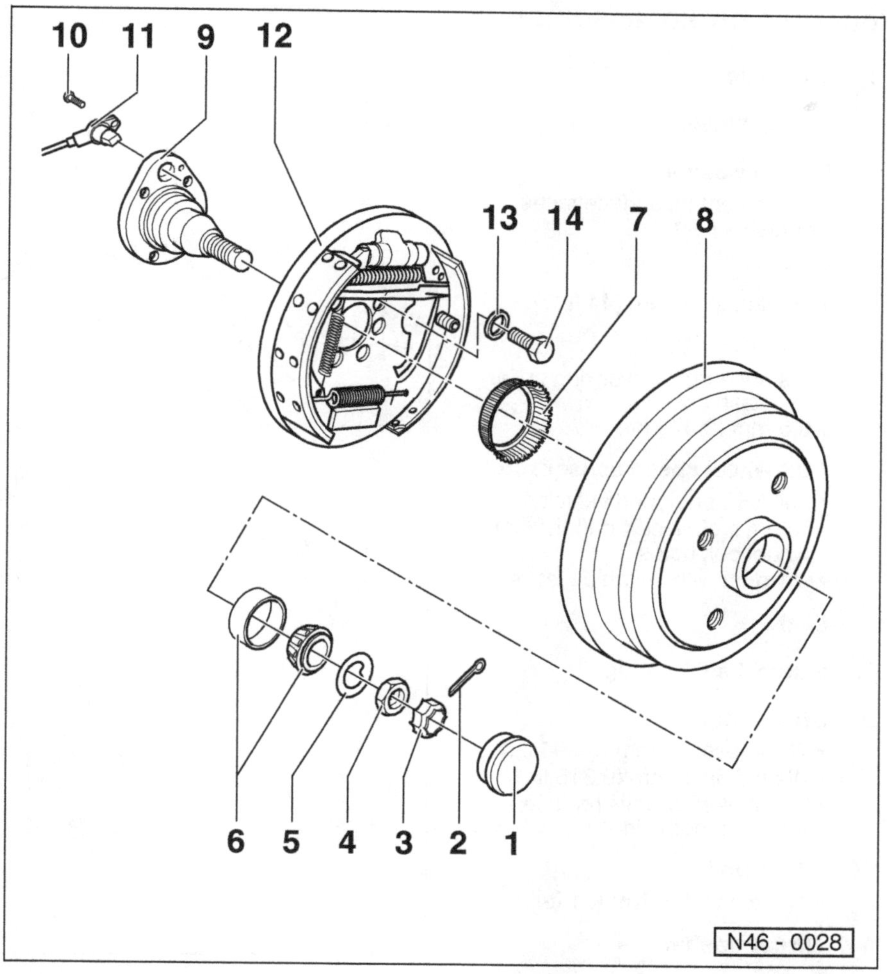

N46-0028

Fig. 12. Rear drum brake assembly.

Rear disc brake, assembly

1. **Stub axle**
2. **Splash shield**
3. **Dished washer**
   - Large contact surface faces splash shield
4. **Bolt**
   - Tighten to 60 Nm (44 ft-lb)
5. **Cover ring**
   - Distance from cover ring collar to splash shield: 9.5 mm (0.374 in.)
6. **ABS wheel speed sensor rotor**
   - For ABS equipped vehicles
   - To drive off, insert drift through wheel bolt holes
   - Drive on with suitable sleeve
7. **Seal**
8. **Inner wheel bearing**
9. **Brake rotor**
   - Thickness: 10 mm (0.394 in.)
   - Wear limit: 8 mm (0.315 in.)
   - When worn always replace rotors on both sides
10. **Wheel bolt**
    - Tighten to 110 Nm (81 ft-lb)
11. **Thrust washer**
12. **Nut**
    - Adjusting wheel bearing play, see **42 Rear Suspension**
    - Secure with lock ring and new cotter pin
13. **Grease cap**
14. **Rear axle beam**
15. **Socket head bolt**
    - Tighten to 10 Nm (7 ft-lb)
16. **ABS wheel speed sensor (G44/G46)**
    - Before installing, clean mounting hole inner surface and coat with lubricating paste (part no. G 000 650)
17. **Brake hose bracket**
18. **Bolt**
    - Tighten to 56Nm (41 ft-lb)
19. **Guide pin**
20. **Protective cap**

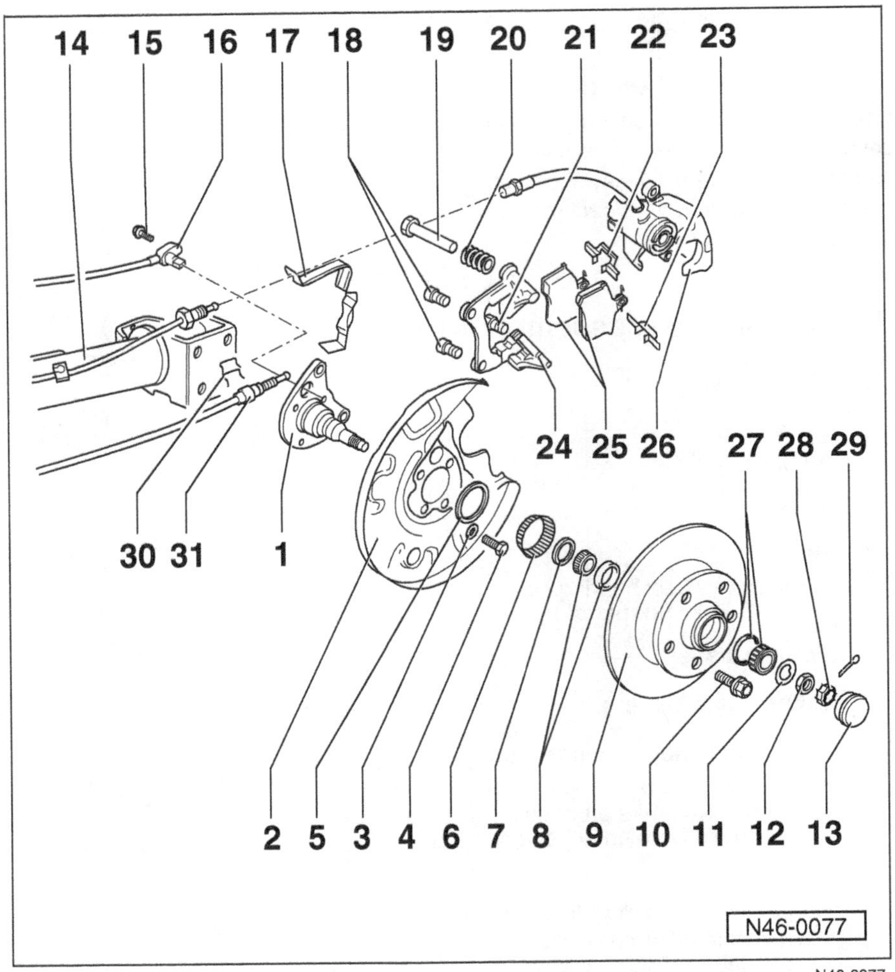

N46-0077

Fig. 13. Rear disc brake assembly. 5-bolt wheel version shown, 4-bolt is similar.

21. **Self-locking bolt**
    - Always replace
    - Hold guide bolt when loosening/tightening
    - Tighten to 35 Nm (26 ft-lb)
22. **Upper pad retaining spring**
    - Always replace
23. **Lower pad retaining spring**
    - Always replace
24. **Brake carrier**
    - Factory replacement caliper comes with guide pins greased
    - Lubricate guide pins with grease, as necessary
25. **Brake pads**
    - Thickness:
      Base suspension: 12 mm (0.472 in.)
      Plus suspension: 14 mm (0.551 in.)
    - Wear limit: 7 mm (0.315 in.) (includes backing plate)
    - Always replace both sides
26. **Brake caliper**
    - Do not disconnect brake hydraulic line from caliper when replacing pads
    - Do not allow the caliper to hang by the brake hydraulic line. The unsupported weight can stretch and damage the hose
27. **Outer wheel bearing**
28. **Lock ring**
29. **Cotter pin**
    - Always replace
30. **Spring clip**
31. **Parking brake cable**

Rear pads, removing and installing

1. Raise the car and support on jack stands.

> ***WARNING —***
> *Jack stands should be placed on a hard level surface (e.g. concrete).*

2. Remove rear wheels.

3. Disconnect parking brake cable from caliper. See Fig. 14.

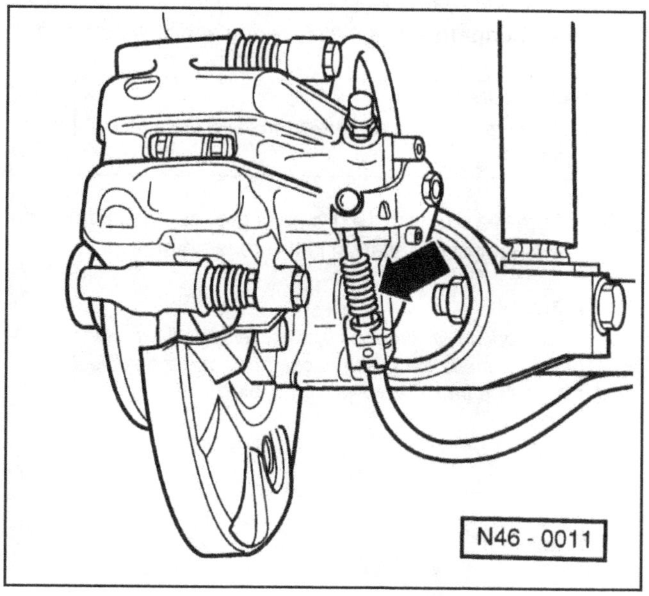

N46-0011

Fig. 14. Parking brake cable (**arrow**) at rear brake caliper.

4. While holding caliper guide bolt stationary, remove caliper mounting bolts. Remove brake caliper and brake pads.

> ***NOTE —***
> *If brake pads are to be reused, mark pad location.*

5. Reset the caliper's automatic adjustment mechanism by turning the piston clockwise while pushing the piston in. See Fig. 15.

> ***CAUTION —***
> - *If the piston is not reset correctly, or if the brake pedal is operated with the caliper removed, the automatic adjustment mechanism will be destroyed.*
> - *Always remove some brake fluid from the reservoir before resetting and pushing the caliper piston in. When the piston is pushed in, fluid is forced up into the reservoir.*

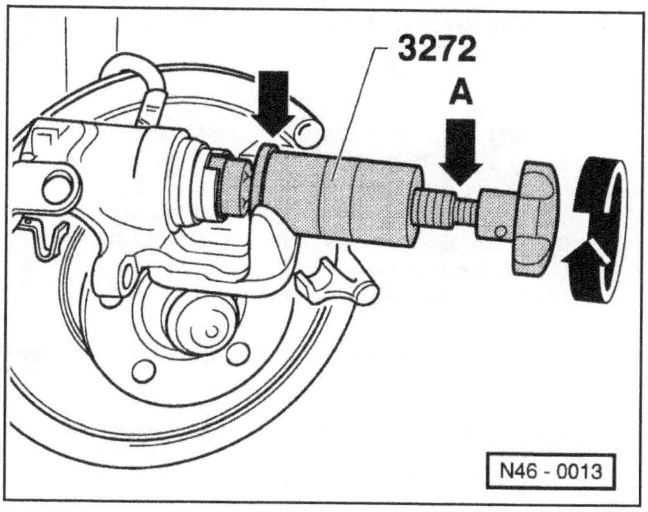

N46-0013

Fig. 15. Automatic adjustment mechanism being reset using Volkswagen special tool 3272. Turn piston clockwise (**rotational arrow**) while pushing piston in.

6. Install brake pads onto pad carrier.

7. Install brake caliper housing using new self-locking bolts.

> ***WARNING —***
> *Always replace the caliper mounting bolts. They are one-time only self-locking bolts.*

Tightening torques
- Caliper mounting bolt to guide bolt (self-locking) . . . . 35 Nm (26 ft-lb)

8. Connect parking brake cable to lever on caliper.

9. Depress brake pedal firmly several times to seat brake pads. Check brake fluid level and add as necessary. The remainder of installation is the reverse of removal.

Rear brake drum, removing and installing

NOTE—
The rear brake drum contains the wheel bearings and bearing races. The procedure given below covers only removal and installation of the brake drum and not wheel bearing service. If the brake drum is being replaced, the wheel bearing assembly must be transferred to the new drum. Wheel bearing replacement and adjustment is covered in **42 Rear Suspension**.

1. Raise the car and support it on jack stands. Remove the wheels.

> *WARNING—*
> *Jack stands should be placed on a hard level surface (e.g. concrete).*

2. Release the brake shoes by inserting a screwdriver through the front, top wheel bolt hole and pushing up the adjusting wedge. See Fig. 16.

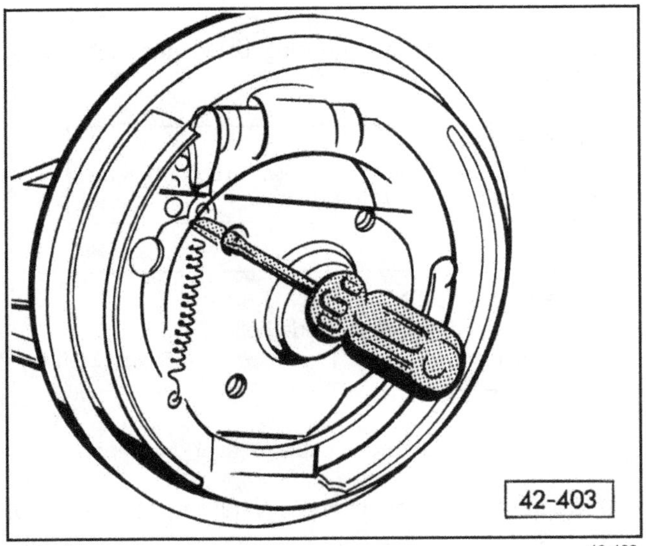

42-403

Fig. 16. Using a screwdriver through wheel bolt hole, push adjusting wedge upward.

3. Pry off grease cap from front of brake drum.

4. Remove cotter pin and lock ring. Remove wheel bearing adjusting nut.

5. Rock brake drum from side to side and remove washer and outer wheel bearing. Slide brake drum straight off axle.

6. Apply grease to gap between sealing lips on inner grease seal in brake drum. Slide the brake drum onto the stub axle.

> *CAUTION—*
> *Use care to avoid damaging the grease seal when installing the brake drum.*

7. Install the outer wheel bearing, thrust washer and axle nut. Adjust rear wheel bearing play as described in **42 Rear Suspension**. Install lock ring and new cotter pin.

> *CAUTION—*
> *Do not reuse cotter pin. Always install a new cotter pin.*

8. Fill grease cap with 10 grams of grease and install. The remainder of installation is the reverse of removal.

NOTE—
After working on the rear brakes, release the parking brake, depress the brake pedal firmly and then adjust the parking brake, see **Parking Brake**.

Rear brake shoes, assembly

1. **Upper return spring**
2. **Push rod**
   - Lubricate contact points with grease
3. **Adjusting wedge**
   - To remove/install brake drum: work through wheel bolt hole to push wedge up first
4. **Locating spring**
5. **Wheel cylinder**
   - If brake fluid is found under dust seal, replace wheel cylinder.
6. **Cap**
7. **Backing plate**
8. **Socket-head bolt**
   - Tighten to 10 Nm (89 in-lb)
9. **Tensioning pin**
10. **Spring retainer**
    - To remove push against spring and turn 90°
11. **Spring**
12. **Wedge spring**
13. **Lower return spring**
    - Lubricate contact points with grease
14. **Brake shoe with lever for parking brake**
    - Minimum lining thickness: 2.5 mm (0.098 in.)
15. **Brake shoe**
    - Minimum lining thickness: 2.5 mm (0.098 in.)

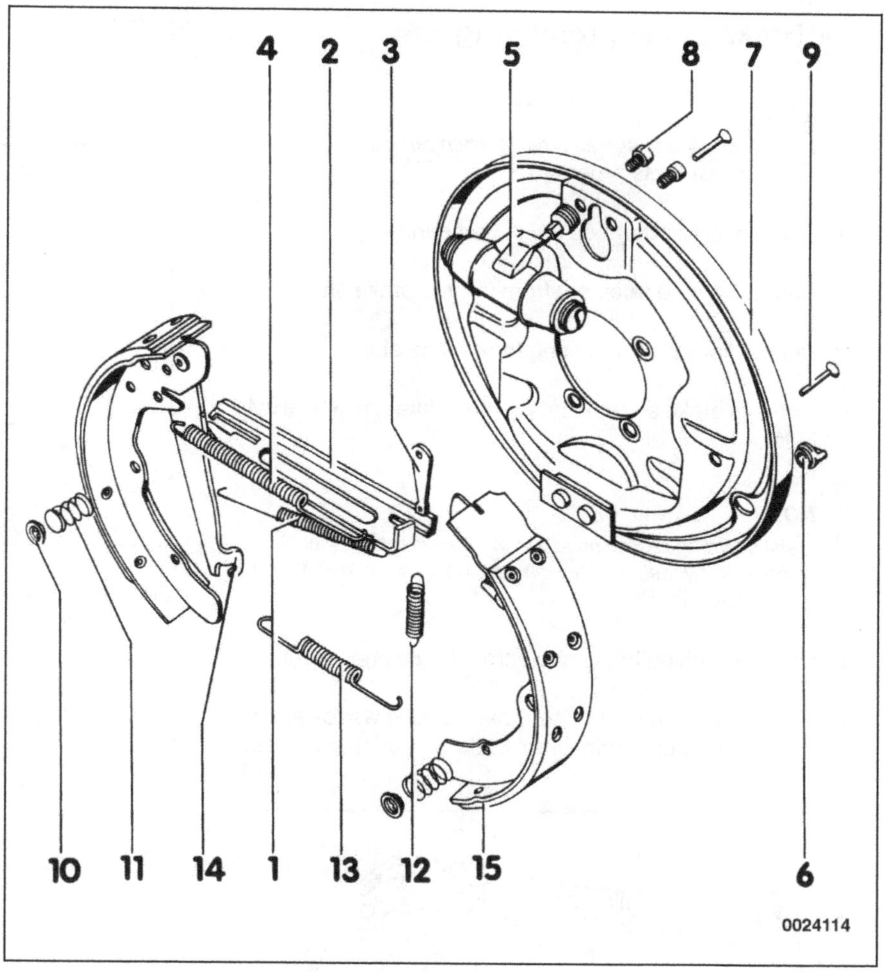

0024114

Fig. 17. Rear brake shoes and related hardware.

Rear brake shoes, removing

NOTE—
Use the previous page as a guide when removing and installing rear brake shoes.

1. Remove brake drum as described earlier.
2. Remove spring retainers from front of brake shoes.
3. Unhook lower return spring from brake shoe.
4. Remove brake shoes from backing plate and wheel cylinder.

NOTE—
Install a strong rubber band around the wheel cylinder to hold the wheel cylinder pistons in place while the brake shoes are removed.

5. Unhook parking brake cable from brake shoe lever.
6. Clamp brake shoes in vice. Then remove wedge spring and upper return spring from brake shoes. See Fig. 18.

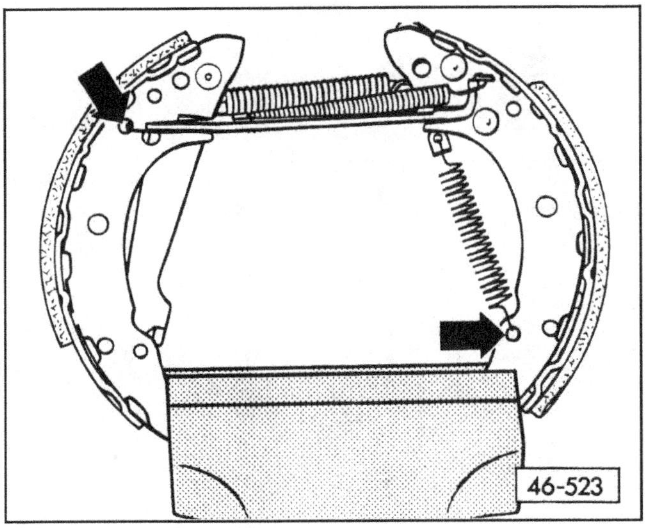

46-523

Fig. 18. Remove wedge spring (**right arrow**) and upper return spring (**left arrow**) with shoes in vice.

7. Unhook locating spring from push rod and remove from brake shoe. See Fig. 19.

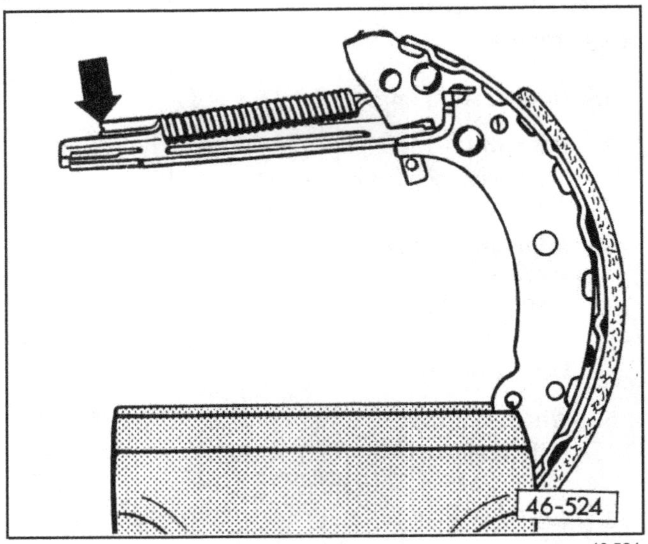

46-524

Fig. 19. Unhook locating spring from push rod (**arrow**) and remove from brake shoe.

Rear brake shoes, installing

1. Connect locating spring and insert brake shoe into slot in push rod. See Fig. 20.

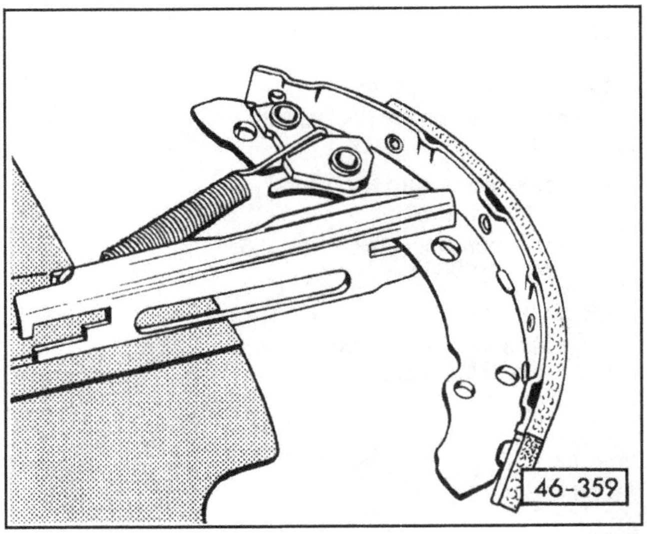

46-359

Fig. 20. Connect locating spring and insert brake shoe into slot in push rod.

2. Insert adjusting wedge into slot.

3. Insert brake shoe with brake lever in push rod. See Fig. 21.

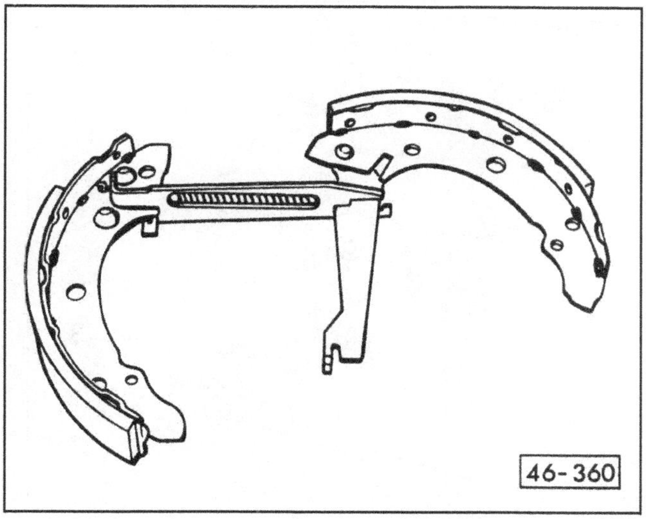

46-360

Fig. 21. Insert brake shoe with brake lever in push rod.

4. Install upper return spring to shoe assembly.
5. Connect parking brake cable to brake lever.
6. Place brake shoes into position on wheel cylinder pistons.
7. Install lower return spring and install brake shoes onto lower support.
8. Install wedge spring and connect to wedge.
9. Install retaining springs and retainers.
10. Install brake drum and adjust wheel bearing play as described in **42 Rear Suspension**.
11. Depress brake pedal firmly to set self-adjusting mechanism. Adjust parking brake as described below.

PARKING BRAKE

The parking brake lever assembly is shown on the following page. The cable-operated parking brake mechanically actuates either the rear brake shoes or the rear caliper pistons, independent of the hydraulic brake system. On cars with rear disc brakes, the caliper contains a self-adjusting mechanism to compensate for brake pad wear.

Adjustment of the parking brake is only necessary if rear brake components or the parking brake cables have been replaced.

Parking brake, adjusting (rear disc brakes)

1. Release parking brake. Remove the trim from parking brake lever. Loosen locknuts at cable ends.
2. Tighten adjusting nuts at base of parking brake lever until levers on brake calipers just begin to lift off of stops on caliper. See Fig. 22.
   - Gap between lever and stop: 1.0 - 3.0 mm (0.04 - 0.12 in.)

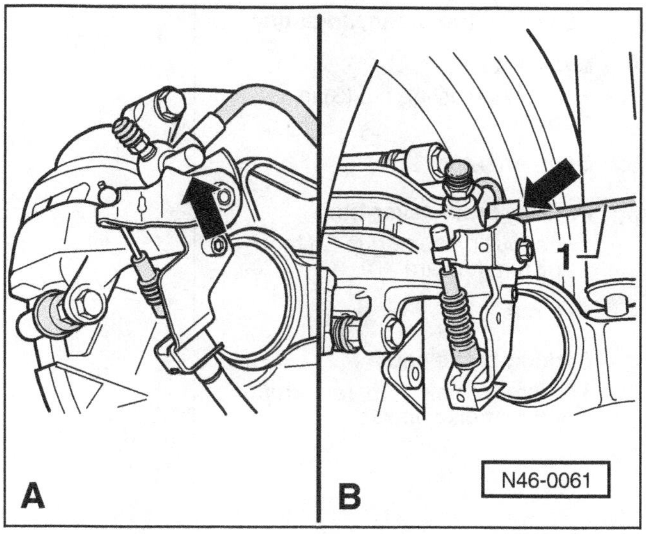

N46-0061

Fig. 22. Adjust parking brake cables until lever on caliper (**arrow**) lifts off stop. Check adjustment at both calipers. 1993-1995 shown at **A**, 1996 and later shown at **B**.

3. Pull up on parking brake and then release. Check that both wheels turn freely. Tighten cable locknuts.

Parking brake, adjusting (rear drum brakes)

1. Release parking brake. Remove trim from parking brake lever. Loosen locknuts at cable ends.
2. Depress brake pedal firmly one time.
3. Pull parking brake lever up until four clicks are heard.
4. Turn adjustment nuts until both rear wheels can just be turned by hand.
5. Release parking brake and check that both rear wheels rotate freely. Tighten cable locknuts.

Parking brake lever, assembly

1. **Parking brake lever**
   - Different version for rear drum and rear disc brakes
   - Before removal, remove center console
2. **Hand grip**
   - Pull down locating tab under hand-grip and pull forward to remove
3. **Adjusting nut**
   - See **Parking brake, adjusting**
4. **Lock nut**
   - To lock adjusting nut (3) in position
5. **Compensator**
6. **Parking brake lever trim**
   - First pull off hand grip (2) then pull trim forward
7. **Hex nut**
8. **Parking brake cable**
   - Different versions for rear drum and rear disc brakes

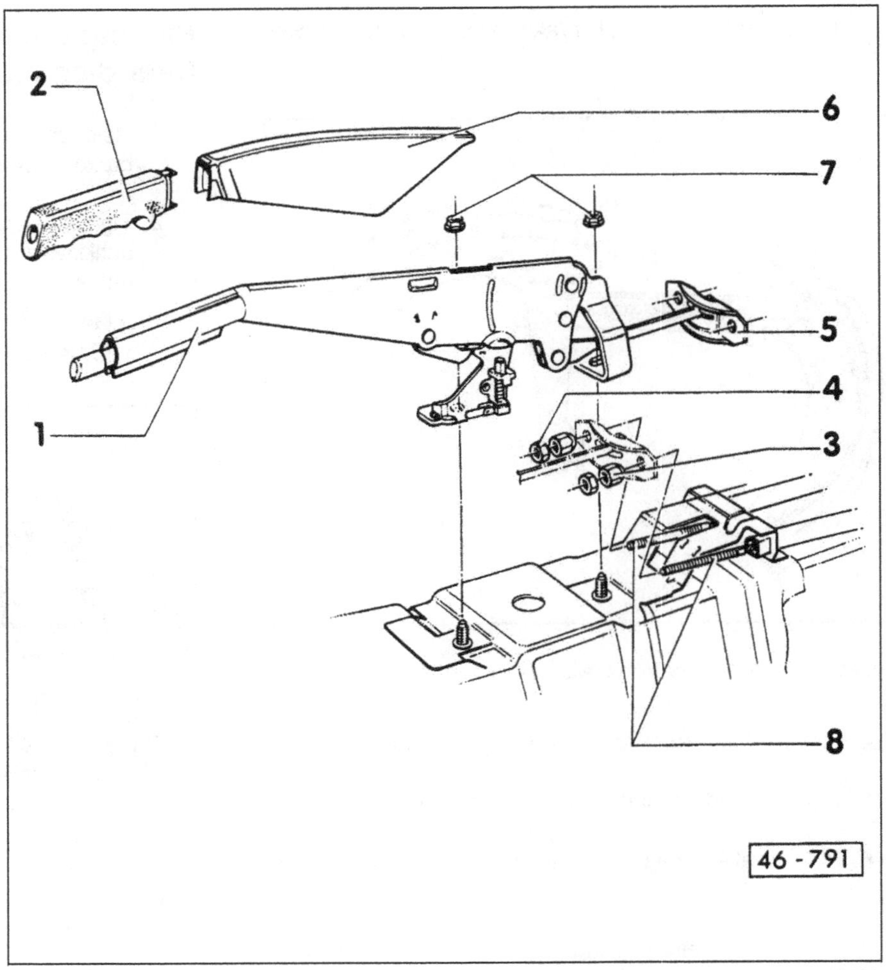

46-791

Fig. 23. Parking brake lever and related components.

47 Brakes–Hydraulic System

GENERAL

This repair group covers service and repair to the hydraulic and vacuum-assist brake system components. Also included here is brake bleeding. Cars covered by this manual are equipped either with front disc brakes and rear drum brakes, or with disc brakes on all four wheels. Some cars with four-wheel disc brakes are also equipped with Anti-Lock Brakes (ABS)

> ***CAUTION —***
> - *This repair group does not cover the anti-lock braking system hydraulic unit (with integral master cylinder). For information on the ABS hydraulic unit and ABS service, see* **45 Anti-Lock Brakes (ABS)**.
> - *Vehicles equipped with the TEVES 04 ABS/EDL can be bled using conventional methods.*
> - *Vehicles equipped with the TEVES 20 GI ABS/EDL can also be bled using conventional methods. However, if the reservoir has been completely emptied, the hydraulic unit must be bled using the Volkswagen Scan Tool VAG 1551 or VAG 1552. This procedure electronically cycles the hydraulic unit and valve body to purge trapped air and is beyond the scope of this manual.*

Brake service precautions

The following warnings, cautions, and notes should be read before servicing the brake hydraulic system:

> ***WARNING —***
> *Brake fluid is poisonous. Wear safety glasses and rubber gloves when working with brake fluid. Prevent brake fluid from entering the bloodstream through cuts or scratches. Do not siphon brake fluid by mouth.*

> ***WARNING —***
> - *DO NOT mix DOT 5 silicone type brake fluid with DOT 4 brake fluid as severe corrosion will result. Such corrosion could lead to complete brake system failure. Use only new, approved brake fluid that complies with MVSS 116 DOT 4.*
> - *Brake pads and brake shoes may contain asbestos fibers. Do not create dust by grinding, sanding, or cleaning pads with compressed air. Avoid breathing any brake material fibers or dust as this may result in serious diseases or death.*
> - *A properly functioning brake system is essential to safe driving. If the red brake/parking warning light or the ABS warning light illuminates while driving, it is imperative that the system be given a thorough check, even if braking action still seems satisfactory. The brakes should be inspected regularly.*

> ***CAUTION —***
> - *Always disconnect battery negative (–) cable when working at or near pedal cluster.*
> - *Before disconnecting the battery be sure to obtain the anti-theft radio code.*
> - *After replacing brake components, depress the brake pedal firmly several times to seat the brakes in their normal operating position. The pedal should be firm and at its normal height, if not, further work is required before driving vehicle.*
> - *Brake fluid is very damaging to paint. Immediately wipe up any brake fluid that spills on the vehicle.*

MASTER CYLINDER, VACUUM BOOSTER, BRAKE PEDAL

Fig. 3 shows the vacuum booster and master cylinder assembly used on cars without anti-lock braking (ABS).

> *NOTE —*
> *This repair group does not cover the master cylinder/ABS hydraulic unit used on cars with anti-lock braking. See* **45 Anti-Lock Brakes (ABS)**.

Master cylinder, removing and installing (cars without ABS)

1. Disconnect the battery negative (–) terminal.
2. Remove the fluid reservoir cap. Remove the fluid from the reservoir using a clean syringe.

> ***WARNING —***
> *Brake fluid is poisonous. Do not siphon brake fluid with your mouth.*

3. Disconnect the brake lines from the master cylinder.
4. Remove the two nuts holding the master cylinder to the vacuum booster and remove the master cylinder.
5. Fill the master cylinder with new, fresh brake fluid and pre-bleed.
6. Using a new O-ring and new locking washers behind the mounting nuts, install the master cylinder. Bleed the brakes as described under **Brake Bleeding**.

Tightening torques
- Master cylinder to vacuum booster . . 10 Nm (89 in-lb)

Brake pedal, separating from vacuum booster

The brake pedal is retained to the vacuum booster pushrod using a ball-and-socket type fitting. The separating procedure described below shows the use of a special Volkswagen tool.

1. Disconnect negative (–) battery cable.
2. Remove trim under instrument panel on left side.
3. Remove brake light switch from its bracket.
4. Push brake pedal in direction of vacuum brake booster, and hold in this position.
5. Install VW special tool no. 3289 to pedal cluster and pull tool toward rear while holding brake pedal stationary (pedal must not be allowed to move back). See Fig. 1.

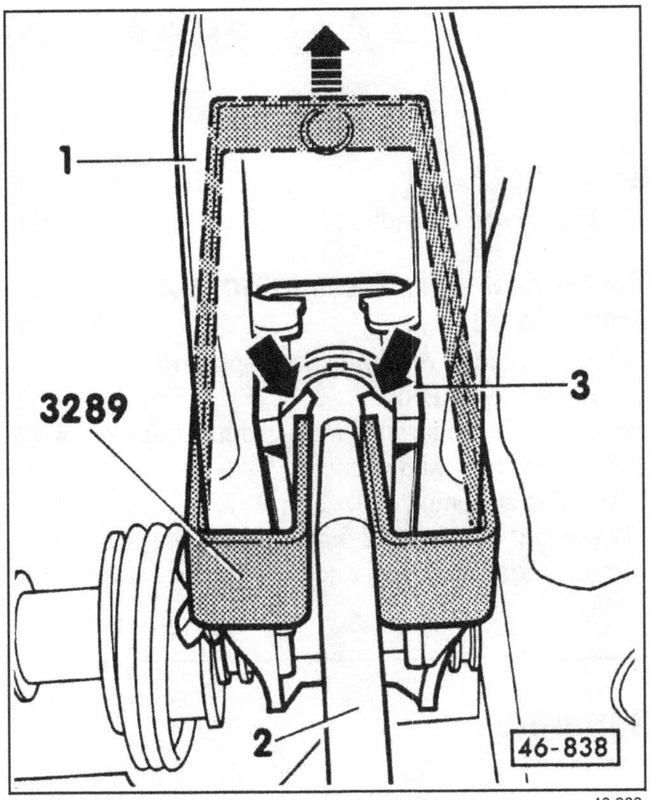

46-838

Fig. 1. VW special tool no. 3289 being used to separate brake pedal (**1**) from brake booster pushrod (**2**). Ball retaining tabs shown at **arrows**.

6. Pull special tool and brake pedal together toward the rear until pedal separates from pushrod.

To reconnect pedal to pushrod, hold ball-head of pushrod and push pedal in until parts lock together with an audible click. See Fig. 2.

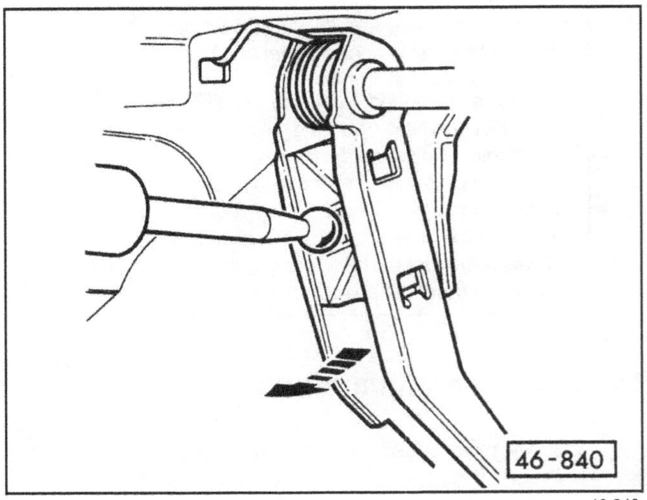

46-840

Fig. 2. Brake pedal being installed to vacuum booster pushrod. Pedal should engage pushrod with an audible click.

Master cylinder and vacuum booster, assembly

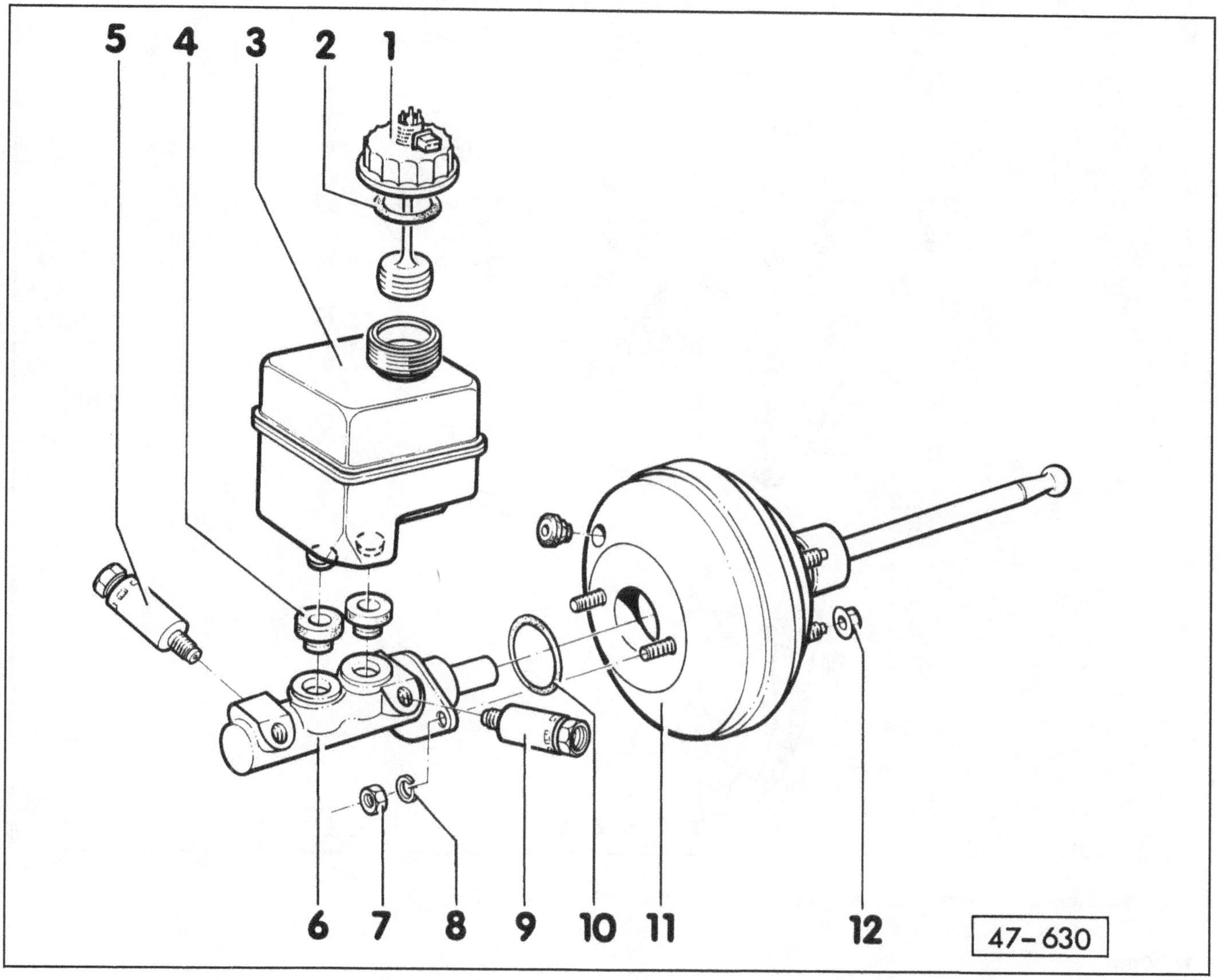

47-630

Fig. 3. Exploded view of master cylinder and vacuum booster for cars without ABS.

1. **Cap (with level sensor)**
2. **Seal**
3. **Brake fluid reservoir**
4. **Plug**
   - Moisten with brake fluid and press into reservoir
5. **Not for USA/Canada models**
6. **Master cylinder**
7. **Nut**
   - Tighten to 20 Nm (15 ft-lb)
8. **Lock washer**
   - Always replace
9. **Not for USA/Canada models**
10. **Seal**
    - Always replace
11. **Vacuum brake booster**
    - Checking function:
      Depress brake pedal several times with engine not running. Depress and hold brake pedal while starting engine. If booster function is OK, pedal will drop slightly as vacuum builds and boost takes effect.
    - Cannot be repaired, replace complete unit if necessary
    - Check valve installed in vacuum line
      It should be possible to blow air through valve in direction of arrow. Blowing air through valve in opposite direction should close valve.
12. **Nut**

Brake and clutch pedal bracket, assembly

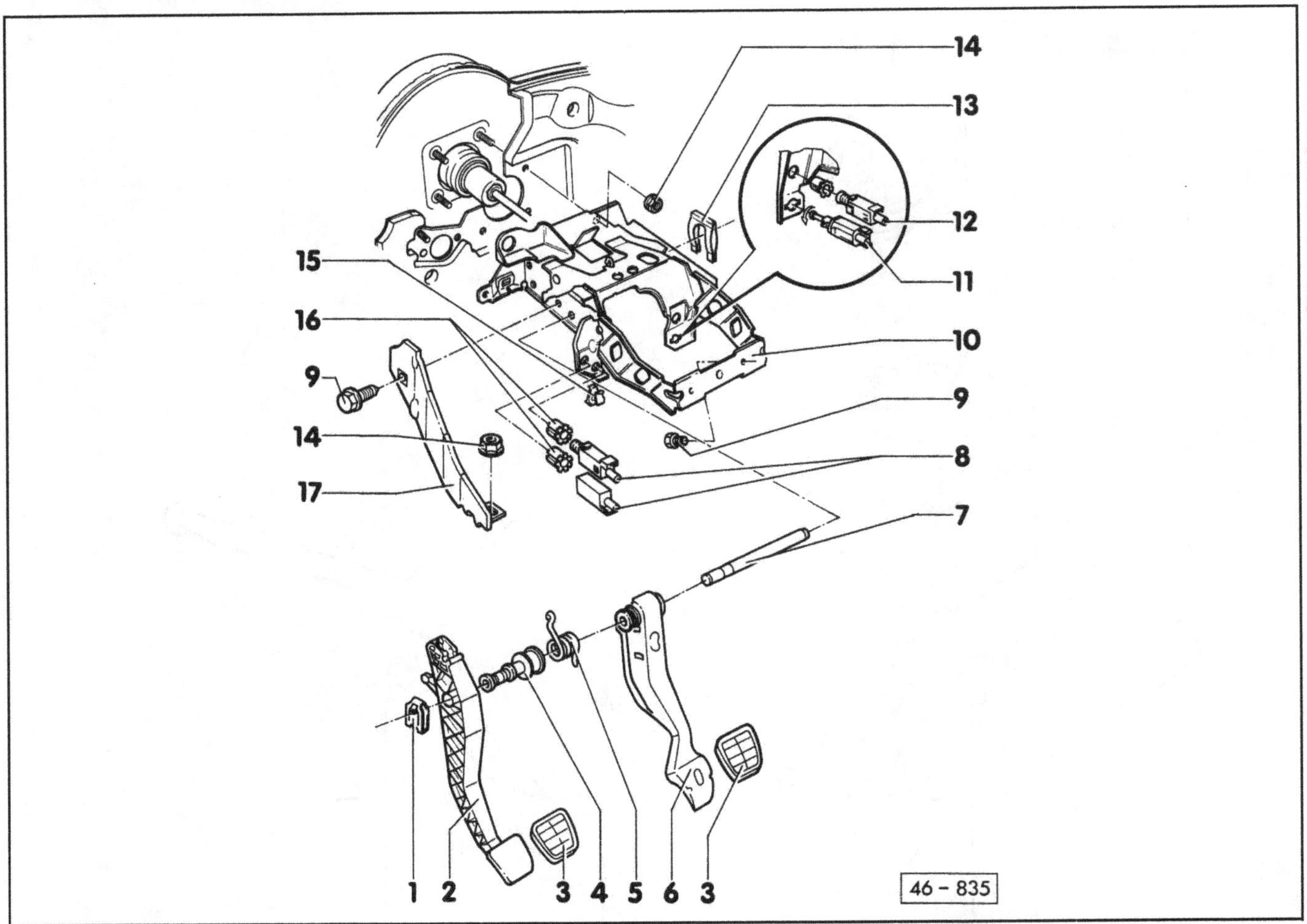

Fig. 4. Brake and clutch pedal bracket and related components.

1. **Clip**
   - Pry out to remove clutch pedal
2. **Clutch pedal**
3. **Pedal cover**
4. **Bushing**
5. **Spring**
6. **Brake pedal**
   - Disconnect from brake booster to remove
7. **Pedal shaft**
8. **Clutch pedal switches**
   - Clutch pedal position switch (F194) upper switch on bracket requires clutch pedal to be fully depressed before starting, see **30 Clutch**
   - Clutch vacuum vent valve switch (F36) lower switch on bracket for cruise control
9. **Bolt**
   - Tighten to 20 Nm (15 ft-lb)
10. **Bracket**
11. **Brake light switch (F)**
    - To install:
      Clip brake pedal to brake booster push rod
      Pull switch plunger out
      Push brake pedal down and hold
      Guide switch into opening and turn 90° counterclockwise
      Release brake pedal
      Verify proper operation of brake lights
12. **Brake vacuum vent valve switch (F47)**
    - For cruise control
    - TDI diesel uses electrical portion only (no vacuum)
13. **Brake pedal clip**
14. **Nut**
    - Tighten to 20 Nm (15 ft-lb)
15. **Clip**
16. **Clip**
17. **Bracket**

Brake pedal, removing and installing

1. Disconnect negative (–) battery cable.
2. Disconnect brake pedal from vacuum booster pushrod as described earlier.
3. Remove over-center spring for clutch pedal, see **30a Clutch (4-Cylinder Engines) or 30b Clutch (6-Cylinder Engine)**.
4. Remove pedal shaft retaining clip. See Fig. 5.

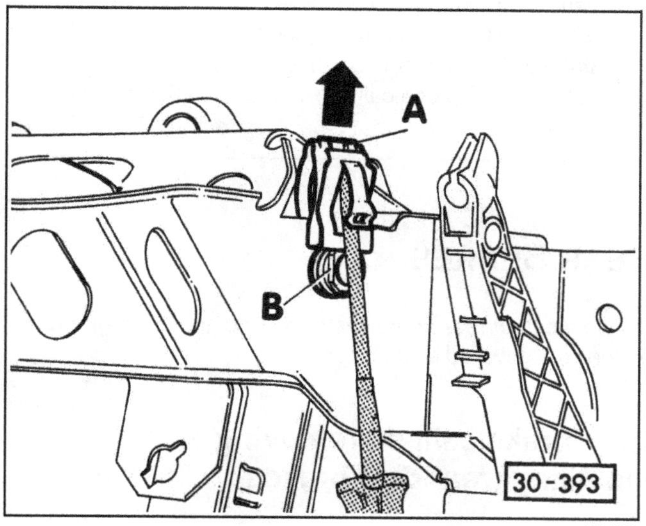

30-393

Fig. 5. Remove pedal shaft clip (**A**) by prying up and off (direction of **arrow**). When installing clip, flats on shaft (**B**) must be vertical.

5. Unhook spring from brake pedal.
6. Pull out pedal shaft to left side and remove brake pedal.
7. Install in reverse order of removal. Make sure the pedal shaft is installed with the narrow section on the left. See Fig. 6.

NOTE —
When installing the retaining clip, make sure the retaining clip flats on the pedal shaft are vertical.

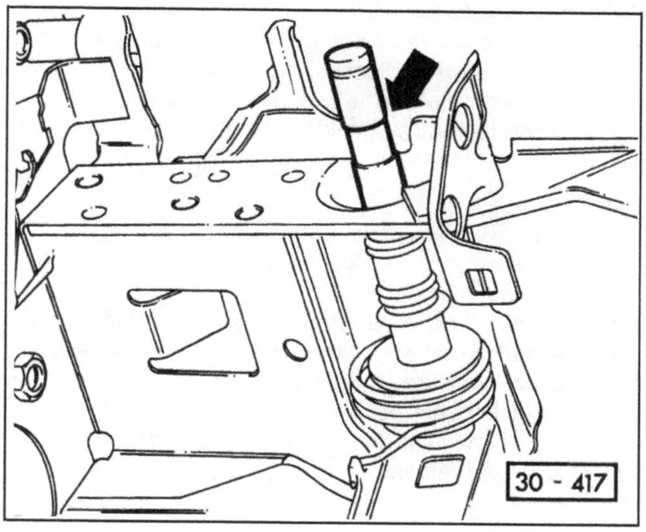

30-417

Fig. 6. Install pedal shaft so that narrow section of shaft (**arrow**) goes in on left side.

Brake pressure regulator, checking

The load-sensing brake pressure regulator is mounted underneath the car with an actuating lever connected to the rear axle beam by a small spring. See Fig. 7. When the axle changes position, due to a heavy load or during hard braking, the pressure regulator varies the pressure to the rear brakes.

NOTE —
The brake pressure regulator should be checked and if necessary adjusted following repairs to the rear suspension, or when there is excessive rear brake wear.

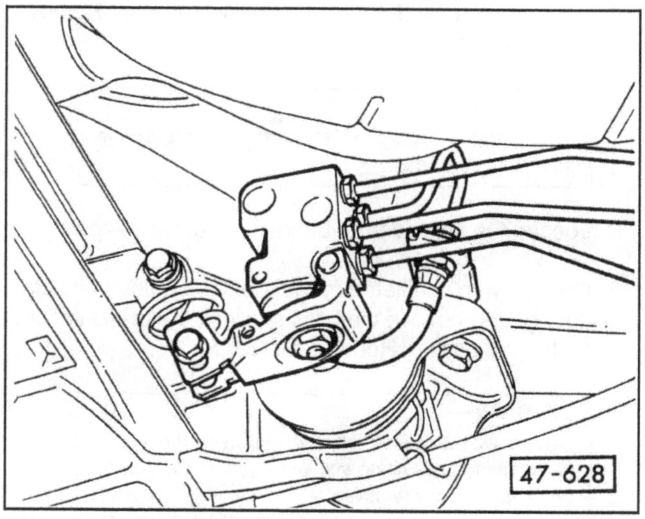

47-628

Fig. 7. Brake pressure regulator mounted in front of rear axle beam.

With the car resting on all four wheels and the fuel tank full, observe the pressure regulator actuating lever while a helper depresses and quickly releases the brake pedal. The lever should move slightly when the brake is quickly released. If not, the regulator is faulty and should be replaced.

> *NOTE —*
> *For this check to be accurate, the car should be emptied of all cargo and occupants, except driver.*

Testing and adjusting the pressure-regulating function requires measuring brake system pressure at each wheel caliper, using two pressure gauges with a range of at least 160 bar (2320 psi). Because of the need for this specialized equipment, we recommend having this test performed by an authorized Volkswagen dealer.

FRONT BRAKE CALIPER

Two types of front brake calipers are used depending on type of brake rotor installed. Cars with ventilated rotors use a two-piece caliper assembly. Cars with solid rotors use a single piece caliper. Repair procedures vary depending on caliper application. See **46 Brakes–Mechanical Components** for more information on front brake assemblies.

> *NOTE —*
> *Complete replacement of the caliper is recommended to remedy a leaking caliper and avoid future problems. A damaged caliper dust seal can be replaced separately.*

Front brake caliper, removing and installing

1. Raise the car and support it on jack stands. Remove the front wheel.

> ***WARNING —***
> *Jack stands should be placed on a hard level surface (e.g. concrete).*

2. Loosen the flexible front brake hose at the caliper.

3. On cars with ventilated rotors, hold the guide bolts stationary and remove the bolts securing the brake caliper to the pad carrier. Remove the caliper.

> ***CAUTION —***
> *On cars with ventilated rotors, the caliper is mounted to the guide bolts using self-locking bolts that must be replaced once removed.*

4. On cars with solid rotors, remove the bolts securing the brake caliper to the wheel bearing housing. Swing the caliper out from bottom to top to remove.

5. Disconnect the pad wear sensor connector, where applicable. Unscrew the caliper from the brake hose.

6. Installation is the reverse of removal. Fill the caliper with brake fluid and pre-bleed before installing, if not supplied as such. Bleed the brakes as described under **Brake Bleeding**.

Tightening torques
- Brake pad carrier to wheel bearing housing (ventilated rotor, 2-piece caliper) . . . 125 Nm (92 ft-lb)
- Brake caliper to guide bolts (ventilated rotor, 2-piece caliper) self-locking bolts (always replace) . . . 35 Nm (26 ft-lb)
- Brake caliper to wheel bearing housing (solid rotor, 1-piece caliper) . . . . 25 Nm (18 ft-lb)
- Wheel to wheel hub . . . . 110 Nm (81 ft-lb)

REAR BRAKES

Two types of rear brake systems–disc and drum–are used on the cars covered by this manual.

Rear brake caliper, removing and installing (rear disc brakes)

Replace the caliper if fluid is leaking at parking brake lever connection.

1. Raise the car and support on jack stands.

> ***CAUTION —***
> *Jack stands should be placed on a hard level surface (e.g. concrete).*

2. Remove the rear wheel.

3. Loosen the flexible brake hose at the caliper.

4. Disconnect the parking brake cable from the caliper.

5. While holding the guide bolts stationary, remove the bolts securing the caliper to the brake pad carrier. Remove the caliper.

> ***CAUTION —***
> *The caliper is mounted to the guide bolts using self-locking bolts that must be replaced once removed.*

6. Unscrew the caliper from the brake hose. Cap the open hose connection to prevent contamination.

7. Install parts in reverse order of disassembly, noting the following installation instructions:
   - Lightly lubricate the guide bolts with grease.
   - Before tightening the bleeder screw, lubricate the threads with brake fluid.
   - Fill the caliper with brake fluid and pre-bleed before installing, if not supplied as such. Bleed the brakes as described under **Brake Bleeding**.

Tightening torques
- Brake pad carrier to stub axle ...... 56 Nm (41 ft-lb)
- Brake caliper to guide bolts (ventilated rotor, 2-piece caliper) self-locking bolts (always replace) . . . 35 Nm (26 ft-lb)
- Wheel to wheel hub . . . . 110 Nm (81 ft-lb)

Rear brake wheel cylinder, removing and installing (rear drum brakes)

Complete replacement of the brake cylinder is recommended to remedy a leaking cylinder and avoid future problems. If brake fluid is present under the dust seal, the wheel cylinder is leaking and should be replaced. See Fig. 8.

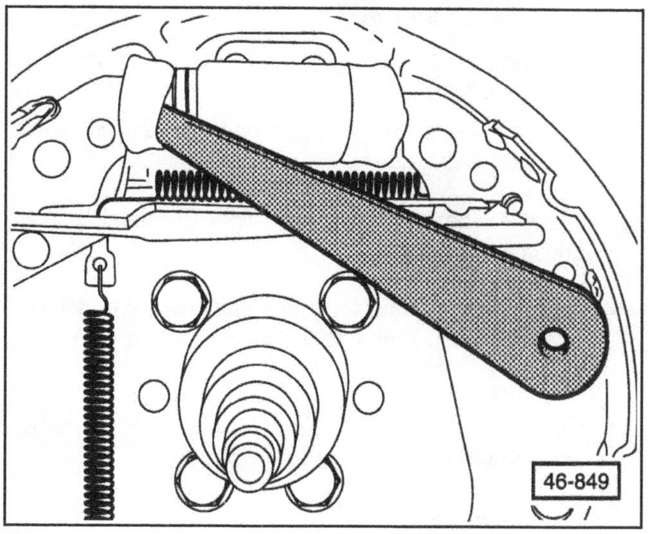

46-849

Fig. 8. Dust seal being removed from wheel cylinder to check for leakage.

1. Remove the rear brake shoes. See **46 Brakes–Mechanical Components**.

2. Working from the rear of the backing plate, disconnect the brake line from the wheel cylinder.

3. Remove the bolts that fasten the wheel cylinder to the backing plate. See Fig. 9.

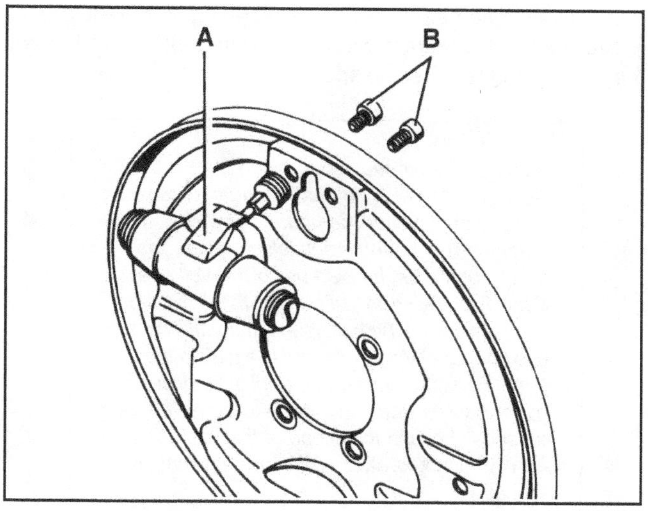

0024081

Fig. 9. Rear brake cylinder (**A**) with allen head mounting bolts (**B**).

4. Remove the wheel cylinder from the backing plate.

NOTE —
Cap the open brake line to prevent contamination while the wheel cylinder is removed.

5. Installation is reverse of removal. Fill the wheel cylinder with brake fluid and pre-bleed before installing. Bleed the brakes as described below under **Brake Bleeding**.

Tightening torques
- Rear brake wheel cylinder to backing plate . . . . 10 Nm (7 ft-lb)

BRAKE BLEEDING

The procedure given here applies to all vehicles covered in this manual with and without ABS.

The brake system can be bled using a pressure bleeder or manually using a helper. Pressure bleeding, if the equipment is available, is the fastest. Manual bleeding requires a helper, but is easy and requires no special tools.

> ***WARNING —***
> - *Vehicles equipped with the TEVES 20 GI ABS/EDL can also be bled during maintenance using conventional methods. However, if the reservoir has been completely emptied, the hydraulic unit must be bled using the Volkswagen Scan Tool VAG 1551 or VAG 1552. This is only necessary if the hydraulic unit or control module has been replaced or if there is a leak in the system. This procedure electronically cycles the hydraulic unit and valve body to purge trapped air and is beyond the scope of this manual. See* **45 Anti-Lock Brakes (ABS)** *for application information.*
> - *DO NOT mix DOT 5 silicone type brake fluid with DOT 4 brake fluid as severe corrosion will result. Such corrosion could lead to complete brake system failure.*
> - *Brake fluid is poisonous. Wear safety glasses and rubber gloves when working with brake fluid. Prevent brake fluid from entering the bloodstream through cuts or scratches. Do not siphon brake fluid by mouth.*

> ***CAUTION —***
> - *Brake fluid is very damaging to paint. Immediately wipe up any brake fluid that spills on the vehicle.*
> - *Use only new, approved brake fluid that complies with MVSS 116 DOT 4*

Bleeding, with pressure bleeder (all systems)

Pressure bleeding using Volkswagen special tool US1116 or equivalent is the preferred method of removing air from all brake systems because it does a more thorough purge.

1. Connect pressure bleeder with regulated compressed air supply to fluid reservoir according to manufacturer's instructions.
2. Connect hose from bleeder bottle to caliper/wheel cylinder bleeder screw and bleed brakes in following sequence:
   - 1 - Right rear caliper or wheel cylinder
   - 2 - Left rear caliper or wheel cylinder
   - 3 - Right front caliper
   - 4 - Left front caliper
3. Brake fluid should be allowed to flow from the bleeder valve/screw until it runs clear.

> ***CAUTION —***
> *The brake fluid level in the reservoir must not fall below the MIN mark during bleeding.*

NOTE —
- *If installed, the brake pressure regulator lever must be pressed toward the rear during rear brake bleeding (see Fig. 6, given earlier).*
- *Depress the brake pedal several times during the bleeding operation.*

Bleeding, manually (all systems)

1. Connect hose from bleeder bottle to brake bleeder screw at the right rear caliper or wheel cylinder.

> ***CAUTION —***
> *The brake fluid level in the reservoir must not fall below the MIN mark during bleeding.*

NOTE —
- *If installed, the brake pressure regulator lever must be pressed toward the rear during rear brake bleeding (see Fig. 6, given earlier).*
- *Depress the brake pedal several times during the bleeding operation.*

2. Have a helper pump the brake pedal several times and then hold pedal down.
3. Open bleeder screw at the caliper or wheel cylinder and collect fluid.
4. Close bleeder screw and then release brake pedal. Repeat operation until brake fluid runs clear and flows without air bubbles.
5. Repeat the above procedure at the remaining wheels, using the following sequence:
   - Left rear caliper/wheel cylinder
   - Right front caliper
   - Left front caliper/wheel cylinder

Vehicles equipped with the TEVES 20 GI ABS/EDL can also be bled during maintenance using conventional methods. However, if the reservoir has been completely emptied, the hydraulic unit must be bled using the Volkswagen Scan Tool VAG 1551 or VAG 1552. This is only necessary if the hydraulic unit or control module has been replaced or if there is a leak in the system. This procedure electronically cycles the hydraulic unit and valve body to purge trapped air and is beyond the scope of this manual. See **45 Anti-Lock Brakes (ABS)** for application information.

48 Steering

GENERAL

Power assisted rack and pinion steering is standard on all models covered in this manual. A belt driven vane type pump provides hydraulic pressure to the rack and pinion steering gear. Service of the steering gear, including repair of housing seal leaks, is by replacement only. The power assisted steering gear is available only as a complete unit.

Special tools and procedures may be required for steering system repair and component replacement. In addition, front wheel alignment is almost always disturbed when steering components are removed or replaced. For wheel alignment specifications, see **44 Wheels–Tires, Wheel Alignment**.

> ***WARNING —***
> - *Most cars covered by this manual are equipped with an airbag system. The airbag is a dangerous device if handled improperly. Before starting any work on the steering components of an airbag equipped car, read the safety precautions listed in* **69–Seatbelts, Airbags**.
> - *Do not re-use any fasteners that are worn or deformed in normal use. Many fasteners are designed to be used only once and become unreliable and may fail when used a second time. This includes, but is not limited to, nuts, bolts, washers, self-locking nuts or bolts, circlips, cotter pins. For replacements, always use new parts.*

> ***WARNING —***
> - *Do not reinstall bolts and nuts coated with undercoating wax as correct tightening torque cannot be assured. Always clean the threads of removed bolts and nuts with a solvent before installation, or replace them with new parts.*
> - *Do not attempt to repair damaged steering components by welding or straightening.*

> ***CAUTION —***
> - *Front wheel alignment should be checked after replacement of steering components.*
> - *Do not hold the steering wheel at full lock for extended periods with the engine running. The constant load may overheat and damage the power steering pump.*

STEERING WHEEL AND STEERING COLUMN

Fig. 1 shows the steering column assembly. Fig. 2 shows the steering column tube assembly for cars with height adjustable steering. Note that most models are equipped with a driver's side airbag unit mounted to the front of the steering wheel. Special service precautions should be followed when servicing vehicles with airbag(s). See **69–Seatbelts, Airbags**.

Steering wheel and column, assembly

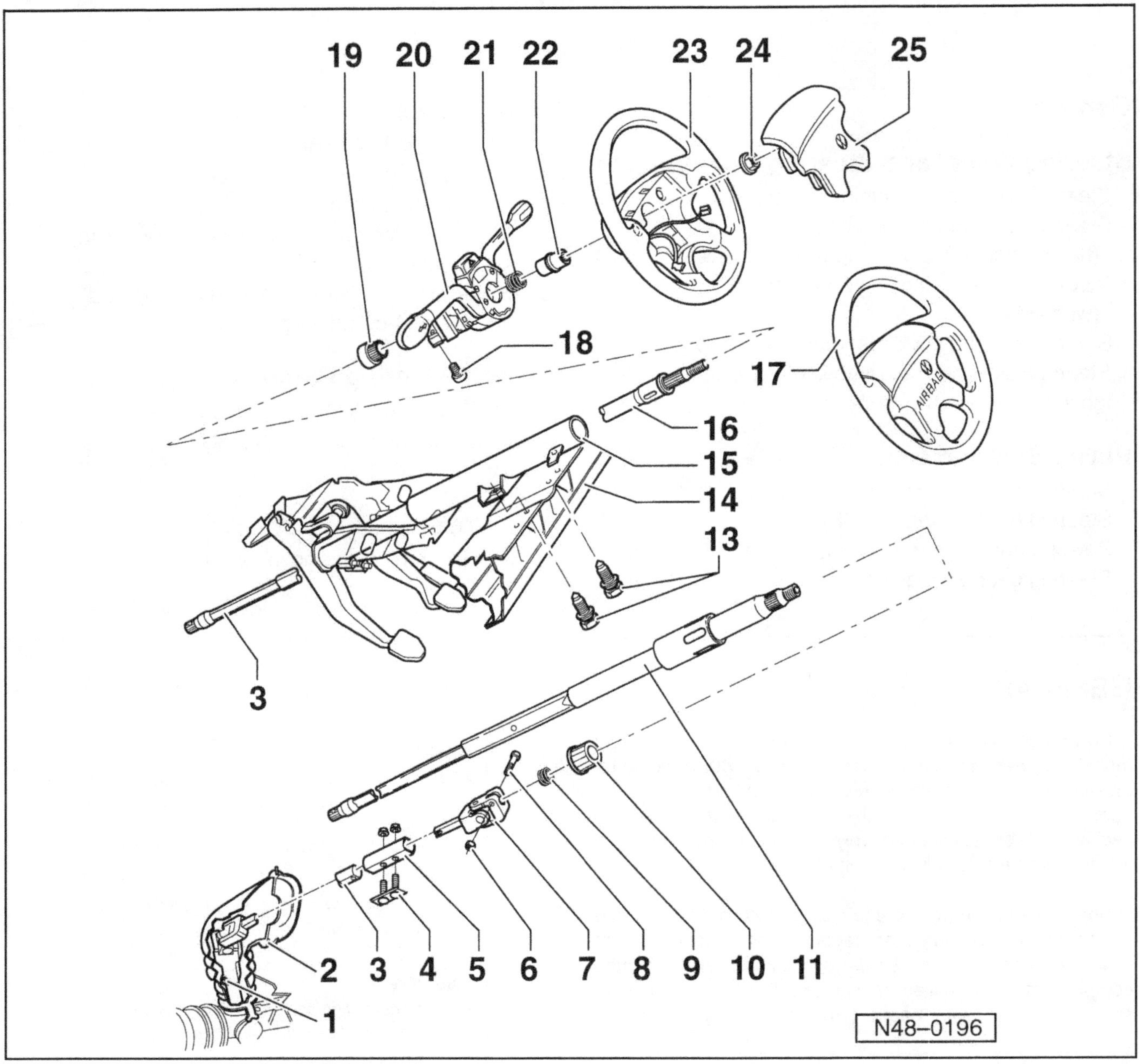

N48-0196

Fig. 1. Steering wheel, column, and column tube exploded view. Steering wheel with and without airbag shown.

1. **Hex bolt**
   - Tighten to 30 Nm (22 ft-lb)
2. **Protective cover**
3. **Lower universal joint**
   - Different number of splines from VIN_1HN 002 450
4. **Connecting piece**
5. **Connecting sleeve**
6. **Hex nut**
   - Tighten 30 Nm (22 ft-lb)
7. **Upper universal joint**
8. **Hex bolt**
9. **Spring**
10. **Lower steering column bearing**
    - Drive on using slide hammer set VW 771 and large washer
    - Drive off downward using a tube

11. **Steering column**
    - With torque overload clutch, cannot be dismantled
    - Standard on Cabrio from m.y. 1997
    - Can be retrofitted to all models
    - Use corresponding steering column tube and steering column switch when retrofitting
    - If a torque of over 100 Nm (74 ft-lb) is exerted on steering wheel, then the steering column will also turn without damaging the locking pin. The straight ahead position will not be altered.
12. **Lower steering column section**
    - Remove by pulling out of steering column tube
    - Supplied as replacement part with upper section
    - Checking length, **see** (A)
13. **Shear bolt**
    - Drill out to remove
    - Tighten new bolt until head shears off
14. **Assembly carrier**
15. **Steering column tube**
    - Lower bracket must lie fully on foot pedal cluster mount
    - Steering column with torque overload clutch has a longer cut-out on underside of tube
16. **Steering column**
    - Remove by pulling upward out of steering column tube
    - Supplied as replacement part with upper section
17. **Steering wheel for airbag**
    - See **Steering wheel, removing/installing (with airbag)**
18. **Shear bolt**
    - Before tightening, check distance between steering wheel and steering lock housing trim
    - Drill out to remove
19. **Support ring**
20. **Steering column switches**
    - Adjusting distance between steering wheel and steering lock housing, **see** (B)
21. **Spring**
22. **Splined adapter sleeve**
    - Remove with two-arm puller (i.e. Kukko 204/2)
    - Install using steering wheel mounting nut and washer to press sleeve firmly on its seat
23. **Steering wheel**
    - Install in centered position
    - For vehicles not equipped with airbag
24. **Hex nut**
    - Tighten to 50 Nm (37 ft-lb)
25. **Cover plate**
    - For vehicles not equipped with airbag

(A) Checking length of steering column

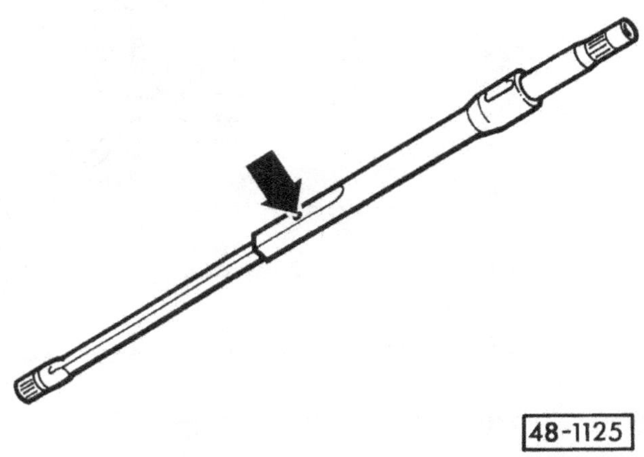

48-1125

- **The small sheet metal lug on the lower part must be visible in the hole in the upper part (arrow). If necessary pull both parts firmly apart onto stop.**

(B) Checking distance between steering wheel and steering lock housing trim

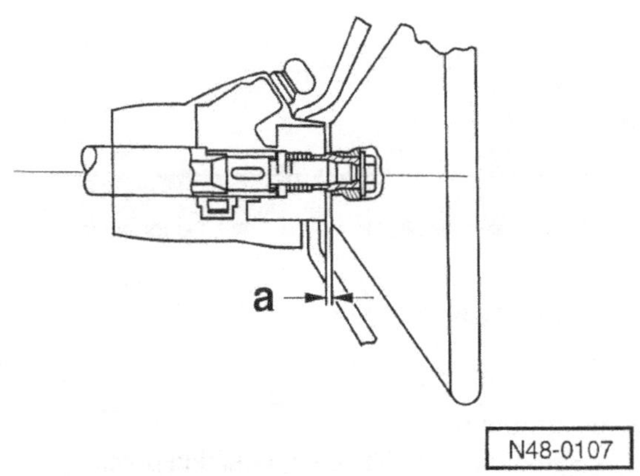

N48-0107

- **Adjust to dimension -a- by moving steering lock housing.**
- **Dimension -a- = 2.0-5.5 mm (0.078-0.216 in.)**
- **Tighten shear-head bolt until head shears off.**

Steering column tube, assembly (height-adjustable steering wheel)

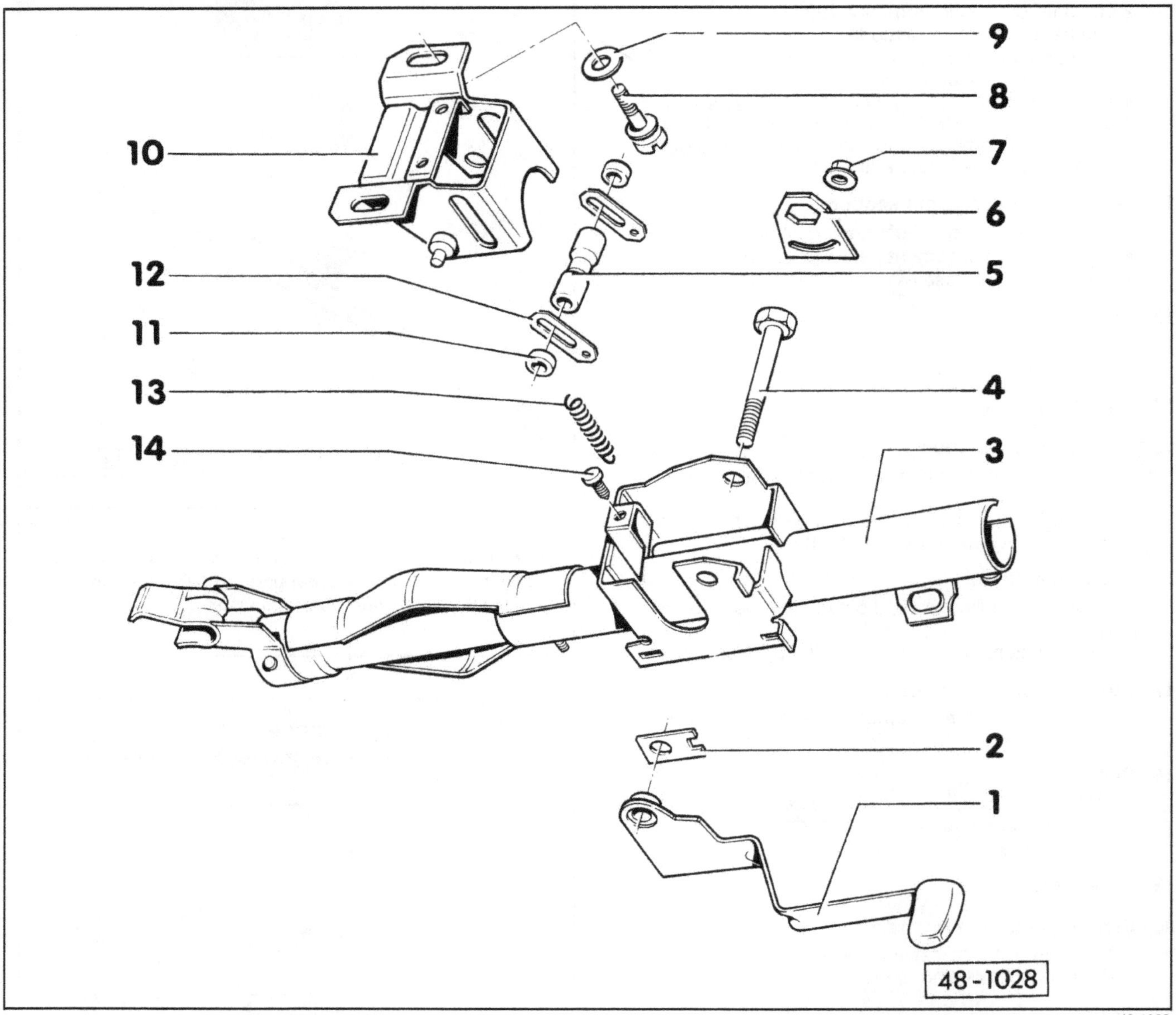

48-1028

Fig. 2. Exploded view of steering column tube for height-adjustable steering wheel.

1. **Operating lever**
2. **Packing plate**
3. **Column tube**
   - Square housing fits over mounting bracket
4. **Bolt with left-hand thread**
   - Tighten to 22 Nm (16 ft-lb) with lever on upper stop
   - After tightening install lock plate
5. **Clamping sleeve**
6. **Lock plate**
7. **Nut**
   - Always replace
   - Tighten to 8 Nm (71 in-lb)
8. **Shear bolt**
   - Tighten until head shears off
9. **Washer**
10. **Mounting bracket**
    - Coat sliding surfaces with MoS2 grease before assembling
11. **Thrust washer**
12. **Plate**
13. **Return spring**
    - Hooked into mounting bracket on column tube
14. **Rubber stop**

Steering wheel, removing and installing (with airbag)

> ***CAUTION —***
> *Cabrio models from 02/98 production use a different style airbag mount (release pins). Please see* **69 Seatbelts, Airbags** *for airbag removal and installation instructions.*

1. Disconnect the negative (-) battery cable.
2. Center steering wheel in straight-ahead position.
3. Remove both socket head screws from rear of steering wheel (airbag unit mounting screws). See Fig. 3.

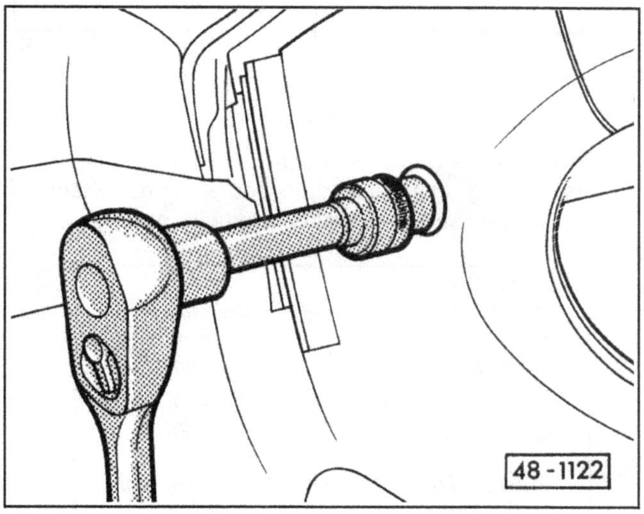

48-1122

Fig. 3. Socket-head (self-locking) screw being removed from rear of steering wheel.

4. Carefully remove airbag unit from steering wheel and disconnect harness connector from rear of unit.

> ***WARNING —***
> *Store the removed airbag in a safe place with the airbag pad facing up.*

5. Remove trim below steering column switch. Disconnect wiring harness connector (remove foam sleeve from connector if fitted). See Fig. 4.
6. Remove hex nut in center of steering wheel. Make matching marks on steering wheel and steering column. Pull steering wheel off steering column.

> ***CAUTION —***
> - *The airbag spiral spring on the rear of the steering wheel must be "locked" in its centered position. To do this, hold the wheel in the normally installed position. Then with wire on the spiral spring facing downward, turn the connecting ring just until the spiral spring locks in position. See Fig. 5.*
> - *Replacement of the airbag spiral spring is covered in* **69 Seatbelts, Airbags**.

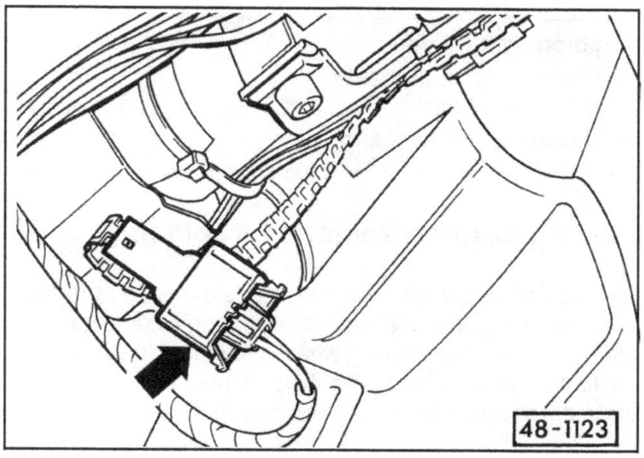

48-1123

Fig. 4. Wiring harness connector (**arrow**) for airbag unit.

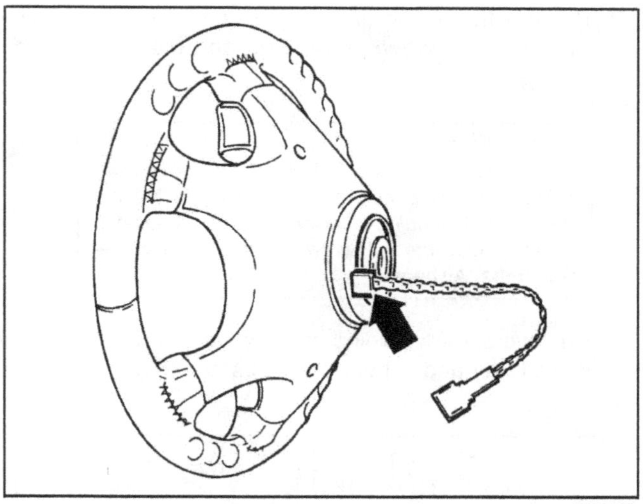

0024082

Fig. 5. Wire (**arrow**) on spiral spring should be in downward position so that spiral spring locks in center position.

7. Reinstall steering wheel in reverse order of removal. Use new self-locking bolts to mount airbag to steering wheel. When installing trim, ensure that airbag harness connector sits in recess in trim.
8. After installing airbag unit and reconnecting airbag electrical connectors, switch ignition on. Then reconnect battery negative (–) cable.

> ***WARNING —***
> - *DO NOT reconnect the battery cable until airbag and steering wheel installation is complete, including reconnection of all harness connectors.*
> - *The ignition switch must be turned on BEFORE reconnecting the negative (–) battery cable.*
> - *Ensure that no one is in the vehicle when reconnecting the negative (–) battery cable.*

Tightening torques
- Airbag to steering wheel (self-locking bolts) . . . . .6.5 Nm (58 in-lb)
- Steering wheel to steering column. . . 50 Nm (37 ft-lb)

Steering column switches, replacing

The turn signal switch and the windshield wiper and washer switch are fastened to the top of the steering column housing by three screws. The steering wheel and airbag (if applicable) must first be removed to access the switches. **Table a** lists terminals identification for the column switches.

1. Disconnect the negative (–) battery cable.
2. Center steering wheel in straight-ahead position.
3. Remove the airbag (if equipped) and steering wheel. See **Steering wheel, removing and installing (with airbag)**.

> ***WARNING —***
> *The airbag is a dangerous device if handled improperly. Always disconnect the negative (–) battery cable before beginning work on an airbag equipped vehicle. Read the safety precautions listed in* **69–Seatbelts, Airbags**.

4. Remove screws holding steering column trim and remove lower and upper trim covers. See Fig. 6.

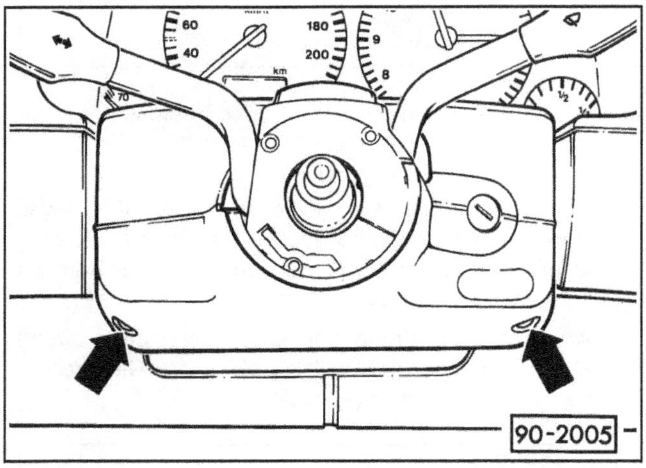

90-2005

Fig. 6. Steering column trim securing screws (**arrows**).

5. Remove the screws holding the switches to the steering column. See Fig. 7.
6. Pull the switches off the steering column and disconnect the wiring harness connectors from the switches.

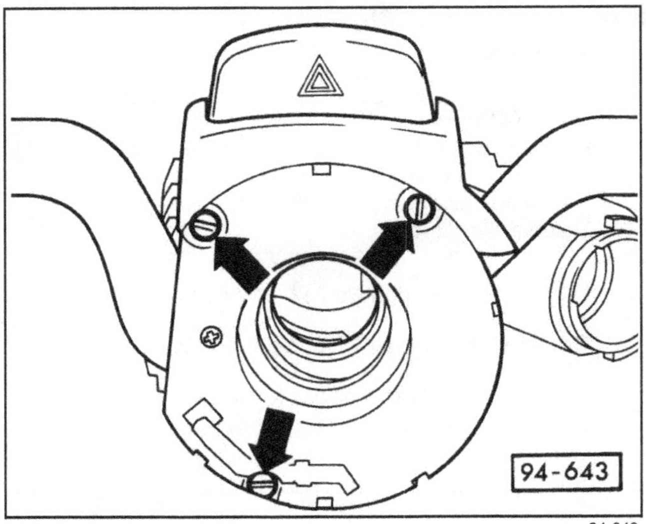

94-643

Fig. 7. Steering column switch securing screws (**arrows**).

7. Installation is the reverse of removal. Be sure to follow the airbag installation procedures. See **Steering wheel, removing and installing (with airbag)**.

> ***WARNING —***
> - *DO NOT reconnect the battery cable until airbag and steering wheel installation is complete, including reconnection of all harness connectors.*
> - *The ignition switch must be turned on BEFORE reconnecting the negative (–) battery cable.*
> - *Ensure that no one is in the vehicle when reconnecting the negative (–) battery cable.*

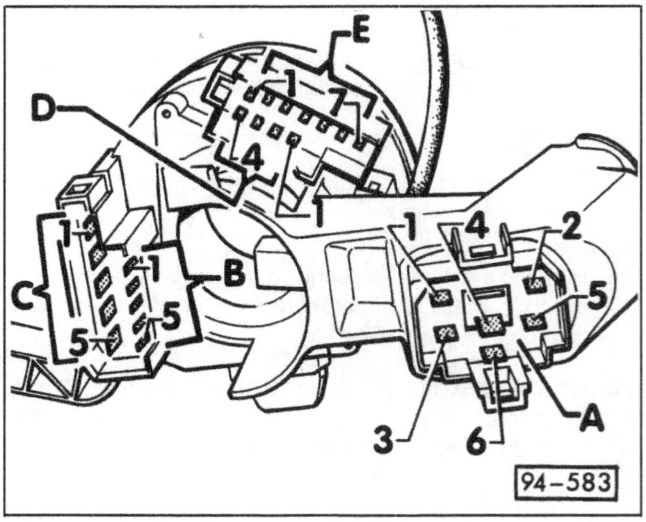

94-583

Fig. 8. Terminal identification for the electrical connectors on rear of steering column switch assembly. Use with **Table a** to make electrical tests.

Table a. Terminal identification, steering column switches

Terminal Number	A) Ignition/starter switch	B) 5-point connector	C) 5-point connector	D) 4-point connector	E) 7-point connector
1	Ignition/starter switch - Terminal 15 (B+ with key on)	Windshield wiper/wash switch- Terminal 53b	Emergency flasher switch - Terminal 49	Windshield wiper/wash switch- Terminal L	Horn button - Terminal 71
2	Ignition/starter switch - Terminal X (B+ with key on)	Windshield wiper/wash switch- Terminal 53a	Headlight dimmer/ flasher switch - Terminal 56b	Windshield wiper/wash switch- Terminal T	Emergency flasher switch - Terminal 49a
3	Ignition/starter switch - Terminal 50	Windshield wiper/wash switch- Terminal 53	Headlight dimmer/ flasher switch - Terminal 56	Windshield wiper/washer switch - Terminal 31 (GND)	Turn signal switch - Terminal L
4	Ignition/starter switch - Terminal 30 (B+)	Windshield wiper/wash switch- Terminal 53e	Headlight dimmer/ flasher switch - Terminal 56a	Emergency flasher switch- Terminal 15 (B+ with key on)	Parking light switch - Terminal PL
5	Ignition/starter switch - Terminal SU	Windshield wiper/wash switch- Terminal J	Headlight dimmer/ flasher switch - Terminal 30 (B+)	–	Parking light switch - Terminal P
6	Ignition/starter switch - Terminal P	–	–	–	Parking light switch - Terminal PR
7	–	–	–	–	Turn signal switch - Terminal R

Steering lock housing, removing and installing

1. Disconnect the negative (–) battery cable.

2. Center steering wheel in straight-ahead position.

3. Remove the airbag (if equipped) and steering wheel. See **Steering wheel, removing and installing (with airbag)**.

> ***WARNING —***
> *The airbag is a dangerous device if handled improperly. Always disconnect the negative (–) battery cable before beginning work on an airbag equipped vehicle. Read the safety precautions listed in* **69–Seatbelts, Airbags**.

4. Remove steering column switches as described earlier.

5. On cars with automatic transmission, disconnect the shift-lock cable from the ignition lock housing. See **37 Automatic Transmission**.

6. Using an appropriate puller, remove splined adapter sleeve from steering column (see Fig. 1).

7. Remove spring and horn contact ring.

8. Remove shear bolt that clamps ignition lock housing to steering column tube. If necessary, drill bolt head and remove using screw extractor.

9. Insert ignition key and unlock lock cylinder. Disconnect wiring connectors from ignition switch. Slide steering lock housing off of steering column.

10. Installation is the reverse of removal. Use a new shear bolt when installing the steering lock housing. Be sure to follow the airbag installation procedures. See **Steering wheel, removing and installing (with airbag)**. On cars with automatic transmission, adjust the shift lock cable. See **37 Automatic Transmission**.

> ***WARNING —***
> - *DO NOT reconnect the battery cable until airbag and steering wheel installation is complete, including reconnection of all harness connectors.*
> - *The ignition switch must be turned on BEFORE reconnecting the negative (–) battery cable.*
> - *Ensure that no one is in the vehicle when reconnecting the negative (–) battery cable.*

Ignition lock cylinder, replacing

1. Remove steering lock housing as described earlier.

2. Mark drilling location on lock housing using Fig. 9 as a guide. Drill 3 mm (1/8 in.) hole into steering lock housing at mark until stop spring is visible

3. Compress stop spring using opposite end of drill and pull lock cylinder out of housing.

4. Insert lock cylinder into steering lock housing.

5. Insert key into lock cylinder.

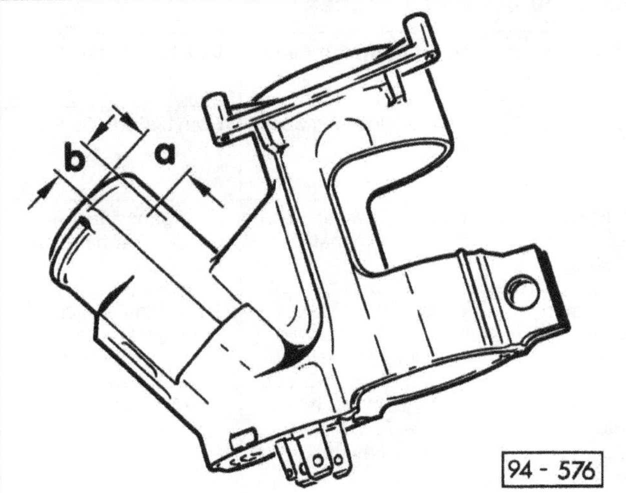

•Dimension a = 12 mm (15/32 in.)
•Dimension b = 10 mm (25/64 in.)

94-576

Fig. 9. Mark location at dimension **a** and **b** and drill 3mm (1/8 in.) hole, approx. 3 mm (1/8 in.) deep.

6. Push lock cylinder fully into housing while gently turning key.

7. The remainder of installation is the reverse of removal. Use a new shear bolt when installing the steering lock housing to the steering column. Be sure to follow the airbag installation procedures. See **Steering wheel, removing and installing (with airbag)**.

> ***WARNING —***
> - *DO NOT reconnect the battery cable until airbag and steering wheel installation is complete, including reconnection of all harness connectors.*
> - *The ignition switch must be turned on BEFORE reconnecting the negative (–) battery cable.*
> - *Ensure that no one is in the vehicle when reconnecting the negative (–) battery cable.*

POWER STEERING GEAR

Two variants of power steering racks are installed on the cars covered by this manual; TRW manufactured and ZF manufactured. See Fig. 10 and Fig. 11. Removal and installation procedures are identical for both racks, however parts between the two systems are not interchangeable.

> ***CAUTION —***
> - *Always replace self-locking nuts and bolts.*
> - *Pay special attention to the rules for cleanliness when working on the power steering. Thoroughly clean all connections and adjacent areas before disconnecting.*

Power steering specifications (TRW or ZF rack)

- Oil type...... Hydraulic oil (VW Part No. G 002 000)
- Oil capacity............ 0.7-0.9 liter (0.74-0.95 qt.)
- Lubricant.................. power steering grease (VW part # AOF 063 000 04)

Steering gear, assembly (TRW)

The numbered list below applies to Fig. 10.

1. **Tie rod end**
   - Pressing off, see **Power steering gear, removing and installing**
   - Checking see **0 Maintenance Program**
2. **Nut**
   - Tighten to 50 Nm (37 ft-lb)
3. **Hose clamp**
4. **Boot**
   - Can be replaced with steering gear installed
   - Must not be twisted after adjusting toe
   - Not interchangeable with boot for ZF steering gear
   - Loosen clamp (item 5) and push boot out of way to check steering rack seals or to lubricate rack
5. **Clamp**
   - Always replace
6. **Tie rod, right**
   - Tighten to 70 Nm (52 ft-lb)
   - Install inner joint with locking compound (Volkswagen "D6" or equivalent). See **Tie rods, removing and installing**
   - Adjusting basic length, see **Tie rods, removing and installing**
   - Seat diameter for boot at tie rod: 12.7 mm (0.50 in.)
7. **Clamp**
   - Always replace
8. **Tie rod, left**
   - Tighten to 70 Nm (52 ft-lb)
   - Install inner joint with locking compound (Volkswagen "D6" or equivalent).
   - Adjusting basic length, see **Tie rods, removing and installing**
   - Seat diameter for boot at tie rod: 12.7 mm (0.50 in.)
9. **Tie rod end, left**
10. **Self-locking hex nut**
    - Always replace
    - Tighten to 30 Nm (22 ft-lb)
11. **Pressure balancing line**
    - Check for damage (dents, kinks)
12. **Self-locking hex nut**
    - Always replace
    - Tighten to 30 Nm (22 ft-lb)
13. **Clamp**

Steering gear, assembly (TRW)

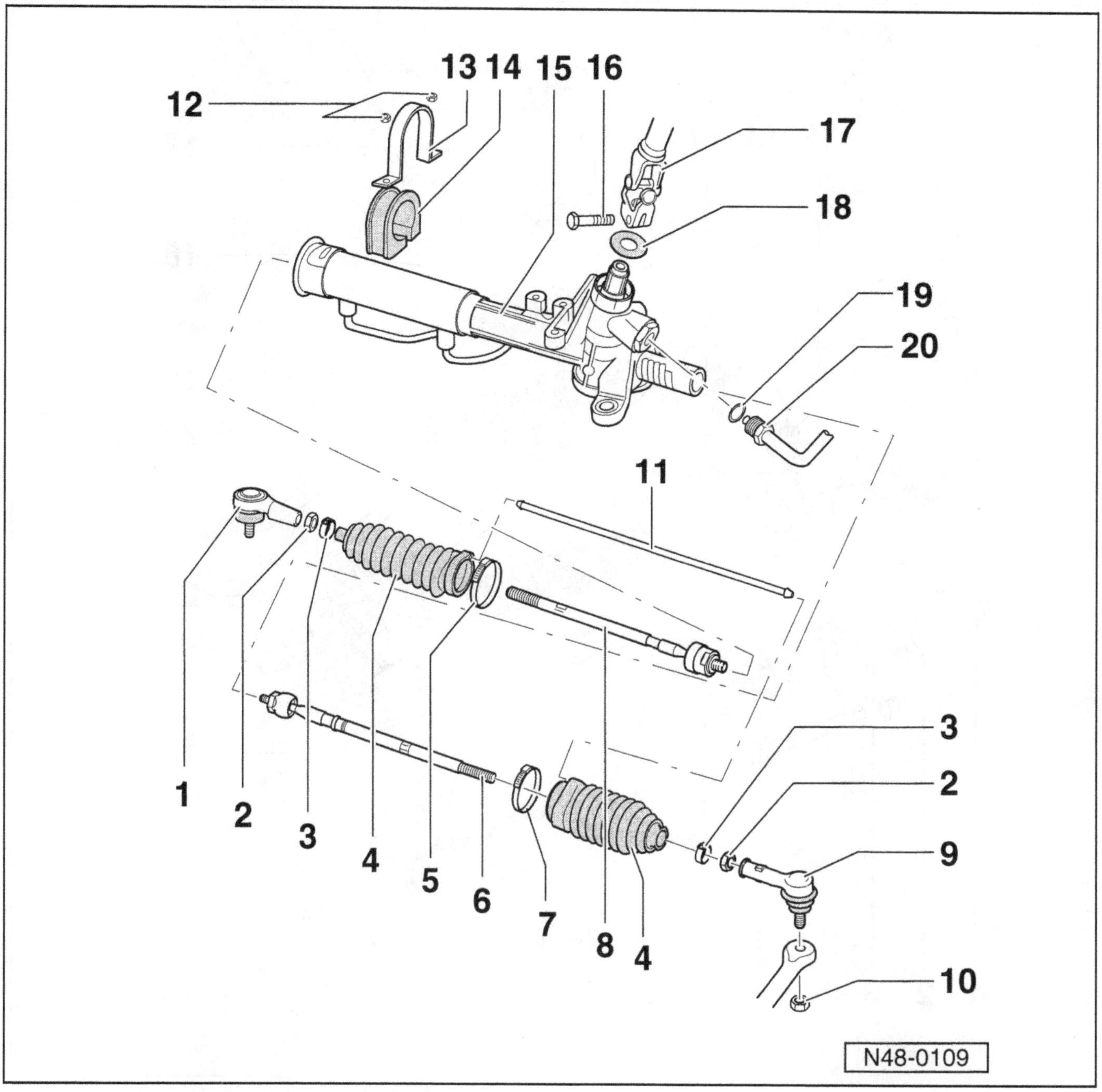

N48-0109

Fig. 10. Exploded view of TRW steering gear assembly.

14. **Rubber mounting**
    - Inside diameter: 44 mm (1.73 in.)
15. **Steering gear**
    - See **Steering gear, removing and installing**
    - See **Steering gear play, adjusting**
16. **Bolt**
    - Tighten to 30 Nm (22 ft-lb)
17. **Lower universal joint**
18. **Cover disc**
19. **O-ring**
    - Always replace
20. **Union nut**
    - Tighten to 30 Nm (22 ft-lb)

Steering gear, assembly (ZF)

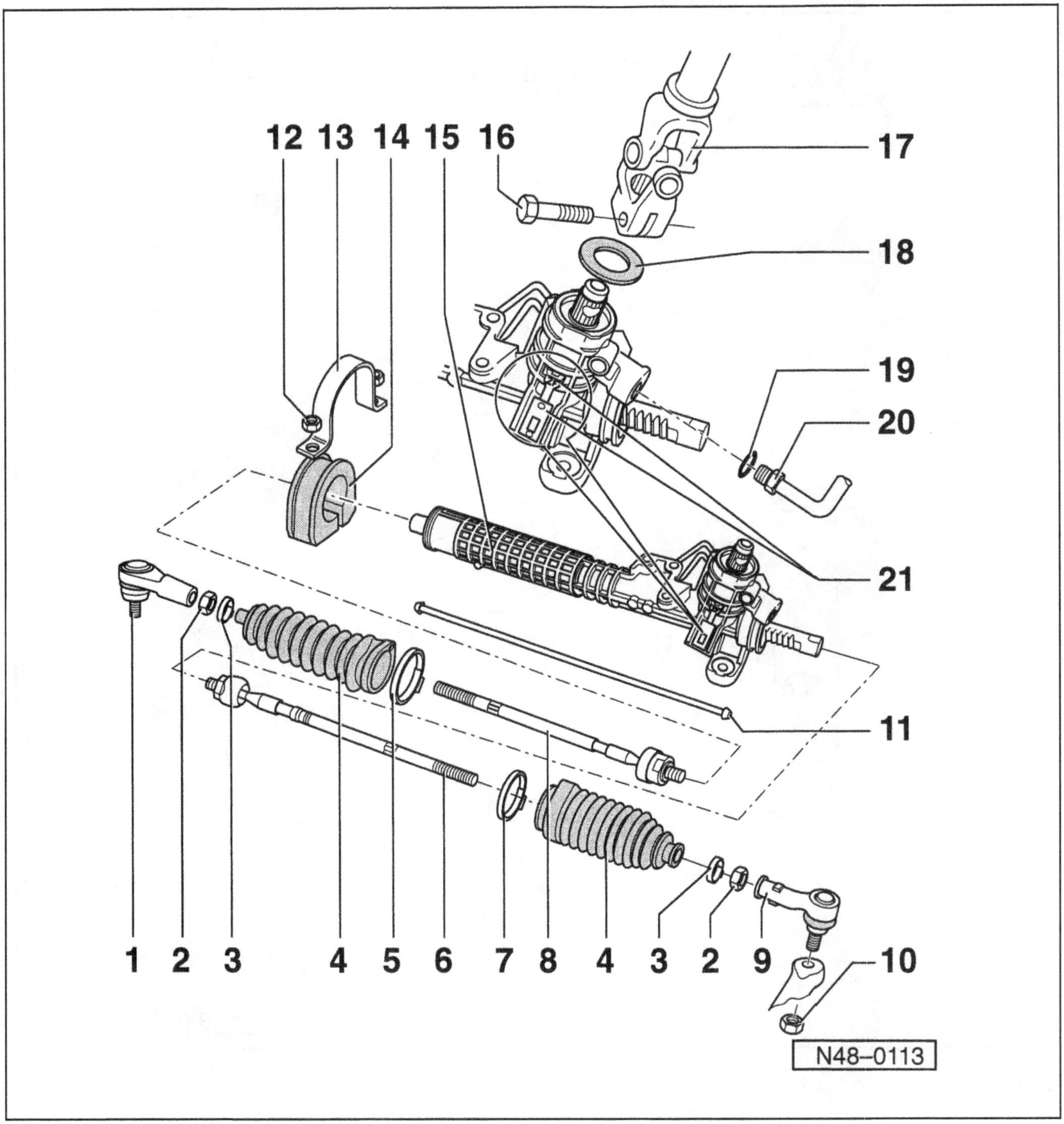

N48-0113

Fig. 11. Exploded view of ZF steering gear assembly.

1. **Tie rod end, right**
   - Pressing off, see **Power steering gear, removing**
   - Checking see **0 Maintenance Program**

2. **Nut**
   - Tighten to 50 Nm (37 ft-lb)

3. **Hose clamp**

4. **Boot**
   - Remove tie rod end before removing
   - Check for wear (cuts, cracks), check sealing surfaces and sealing lips for cleanliness/damage
   - Not interchangeable with boot for TRW steering gear
5. **Clamp**
   - Always replace
   - Tensioning, see **Tie rods, installing**
6. **Tie rod, left**
   - Install inner joint with locking compound (Volkswagen "D6" or equivalent).
   - Adjusting basic length, see **Tie rods, removing and installing**
   - Seat diameter for boot at tie rod: 15 mm (0.59 in.)
7. **Clamp**
   - To remove, cut open with pliers
   - Tensioning, see **Tie rods, installing**
8. **Tie rod, right**
   - Install inner joint with locking compound (Volkswagen "D6" or equivalent). See **Tie rods, removing and installing**
   - Seat diameter for boot at tie rod: 15 mm (0.59 in.)
9. **Tie rod end, left**
   - Pressing off, see **Power steering gear, removing**
   - Checking see **0 Maintenance Program**
10. **Self-locking hex nut**
    - Always replace
    - Tighten to 30 Nm (22 ft-lb)
11. **Pressure compensation line**
12. **Self-locking hex nut**
    - Always replace
    - Tighten to 30 Nm (22 ft-lb)
13. **Clamp**
14. **Rubber bushing**
    - Inside diameter: 50 mm (1.97 in.)
15. **Steering gear**
    - See **Power steering gear, removing and installing**
    - See **Steering gear play, adjusting**
16. **Bolt**
    - Tighten to 30 Nm (22 ft-lb)
17. **Lower universal joint**
18. **Cover disc**
19. **Sealing ring**
    - Always replace
20. **Union nut**
    - Tighten to 30 Nm (22 ft-lb)
21. **Manufacturer data plate and company initials**

Power steering gear, removing and installing

NOTE —

Removal and installation of the steering gear requires the use of engine support equipment. The support equipment is needed to lower and raise the engine and subframe assembly.

1. Raise the car and support on jack stands.

CAUTION —

Jack stands should be placed on a hard level service (e.g. concrete).

2. Separate tie rods ends from wheel bearing housing. See Fig. 12.

CAUTION —

DO NOT use a ball joint fork to separate tie rods. The tie rod end boots or the tie rod could be damaged.

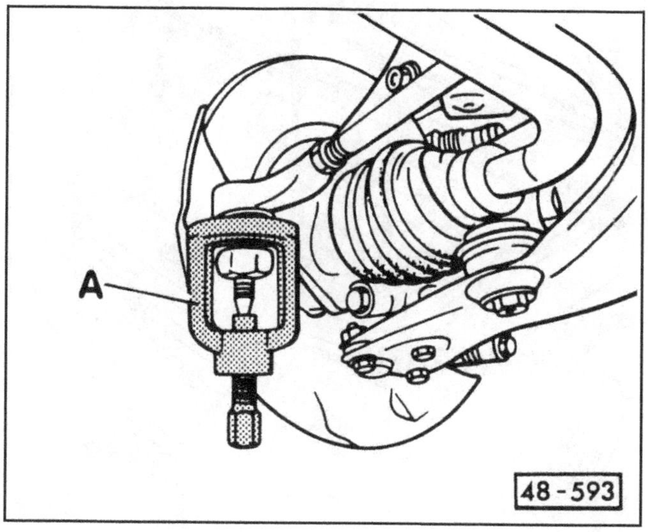

48-593

Fig. 12. Tie rod end being separated using press (**A**).

3. Remove ball joint bolts and separate ball joint from control arm.
4. Remove bolts from universal joint shaft. See Fig. 13.
5. Disconnect universal joint protective boot from its seat on steering gear.
6. Support weight of engine/transmission assembly with appropriate engine lifting equipment. See Fig. 14.
7. Place a drain pan below the power steering pump. Remove lower hose from the pump and drain power steering fluid.

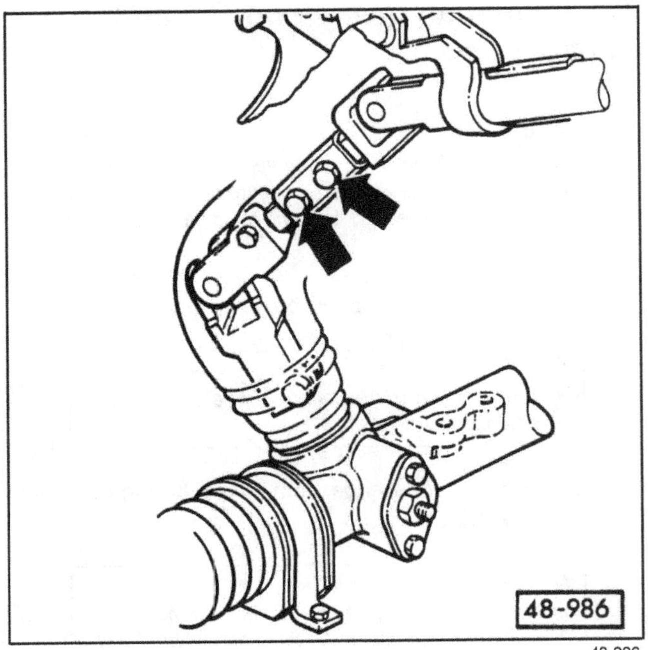

48-986

Fig. 13. Universal joint shaft bolts (**arrows**).

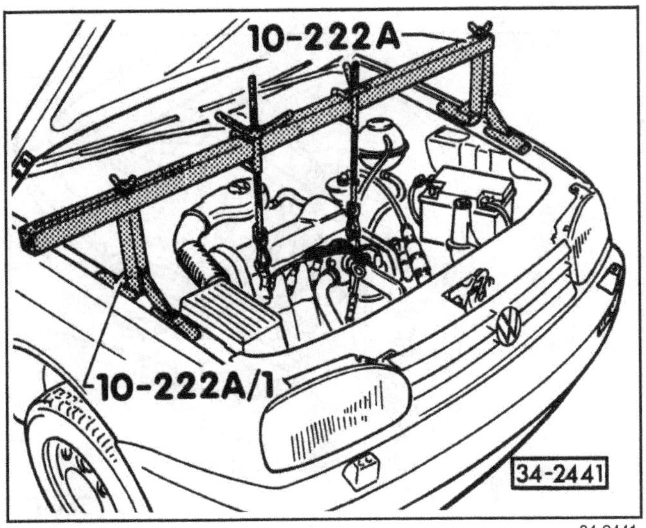

34-2441

Fig. 14. Engine/transmission assembly being supported from above. Volkswagen special tools shown.

8. Unbolt right rear engine/transaxle mount from subframe.

9. With engine/transmission assembly supported, remove subframe bolts.

10. Lower engine/transmission assembly slowly until two halves of universal joint separate.

11. Disconnect hydraulic lines from steering gear and seal lines with plastic bags and tape.

NOTE —

Seal or plug threaded holes on steering gear.

12. Remove steering gear retaining nuts (steering gear bolts remain in subframe). Remove steering gear towards the rear.

13. Install in reverse order of removal. Align front end and check position of steering wheel. Tightening torques are given below. Fill system with specified hydraulic fluid as specified below.

NOTE —

- *When lifting the engine/transmission assembly, the lower half of the U-joint shaft must be guided into the upper half of U-joint shaft (helper required). See Fig. 15.*
- *Coat U-joint protective boot with lubricant, or soap, before pulling onto the steering gear.*

Tightening torques

- Universal joint shaft halves.......... 25 Nm (18 ft-lb)
- Steering gear to subframe .... 30 Nm (22 ft-lb)
- Tie rod end to wheel bearing housing 35 Nm (26 ft-lb)
- Ball joint to control arm............ 35 Nm (26 ft-lb)
- Right rear engine/transaxle mount to subframe .... 60 Nm (44 ft-lb)
- Subframe to body (front and rear)
  - stage I .... 70 Nm (52 ft-lb)
  - stage II .... additional 1/4 turn (90°)
- Rear subframe bracket to body ..... 65 Nm (48 ft-lb)

Power steering specifications

- Oil type ...... Hydraulic oil (VW Part No. G 002 000)
- Oil capacity............ 0.7-0.9 liter (0.74-0.95 qt.)

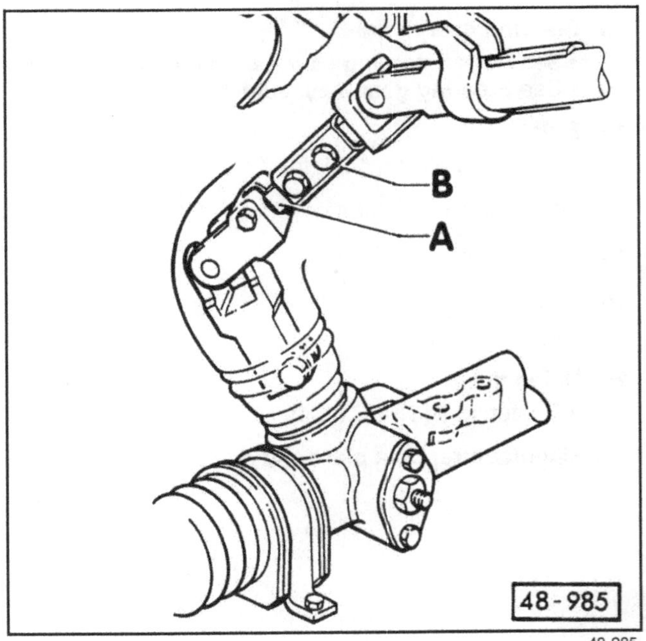

48-985

Fig. 15. When lifting engine/transmission assembly, lower half of joint shaft (**A**) must be guided into upper half of joint shaft (**B**).

Steering gear play, adjusting

Two persons are required to adjust steering gear play. The adjustment should also be made with the car on a lift and the steering and suspension system in its normal position.

NOTE —
On cars with TRW steering gear, a special Volkswagen tool is required to make the adjustment.

1. Raise the vehicle on a lift.
2. Place wheels in straight-ahead position.
3. Turn steering wheel back and forth (about 30° from center position) and listen for knocking noise which will be heard if there is excessive steering gear play.
4. If excessive play is present, adjust steering gear play.
   - On cars with TRW steering gear, use Volkswagen special tool no. 524 to turn adjusting screw in until knocking noise can no longer be heard inside vehicle. See Fig. 16.
   - On cars with ZF steering gear, loosen locknut and turn in adjusting screw in until knocking noise can no longer be heard inside vehicle. Tighten locknut. See Fig. 17.

CAUTION —
Do not overtighten adjusting screw. Only minor adjustments should be necessary. Make adjustments in small increments, just until the rattling noise is no longer heard inside the car.

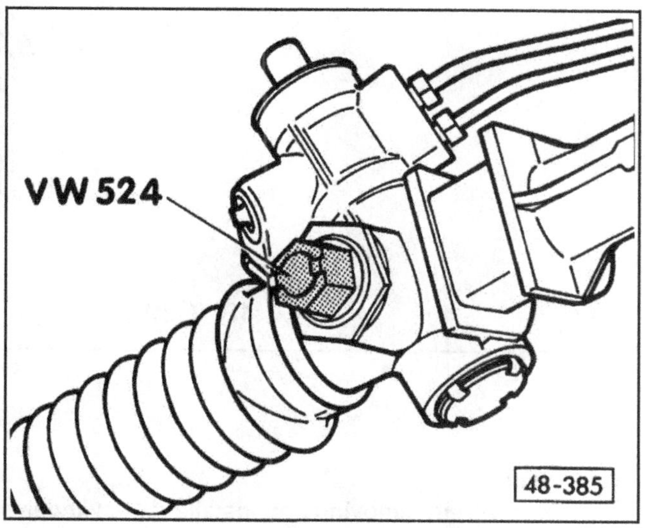

48-385

Fig. 16. TRW steering gear being adjusted. Volkswagen special tool VW 524 shown.

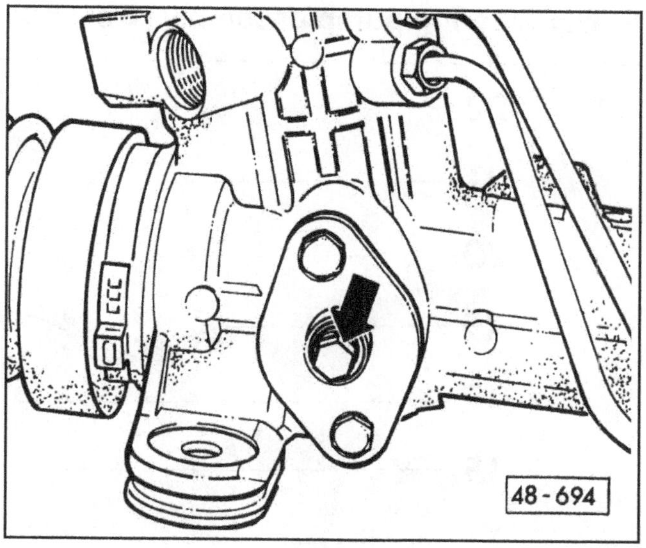

48-694

Fig. 17. ZF steering gear adjusting screw (**arrow**).

5. Road test vehicle and ensure that steering self-centers without binding after negotiating a corner. Repeat adjustment procedure if necessary.

NOTE —
The power steering gear must be replaced as a complete unit if faulty.

POWER STEERING PUMP (4-CYLINDER ENGINES)

CAUTION —
- *Always replace seals.*
- *Do not use hydraulic oil that has been drained.*

Fig. 18 shows an exploded view of the power steering pump and related mounting hardware used on cars with 4-cylinder engines. When servicing hydraulic system or replacing hydraulic pump, be sure to bleed the pump before starting the engine.

Power steering pump assembly (4-cylinder engines)

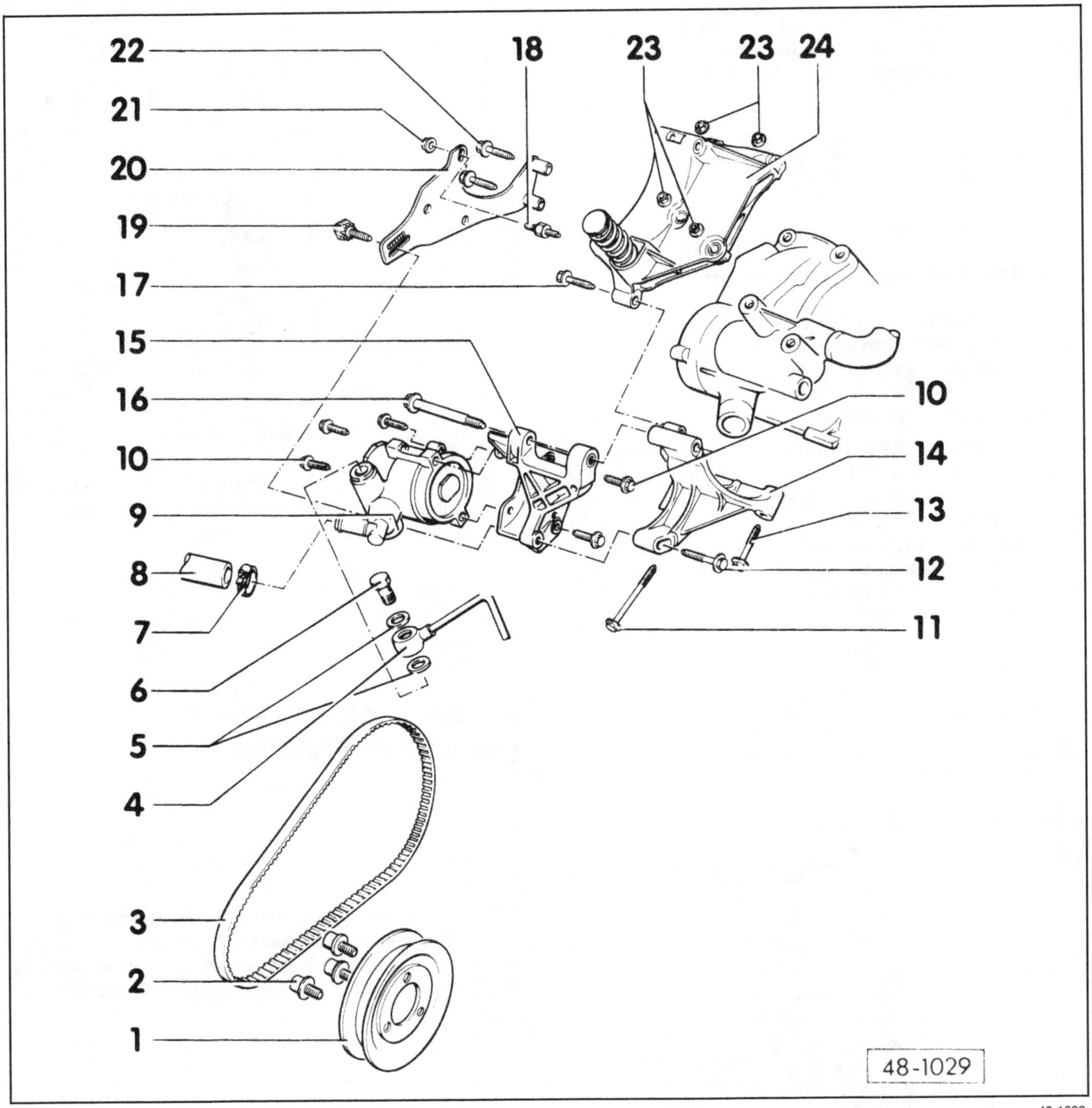

48-1029

Fig. 18. Power steering pump assembly used on cars with 4-cylinder engines.

1. **V-belt pulley**
   - When installing, ensure alignment with crankshaft pulley
2. **Bolts**
   - Tighten to 25 Nm (18 ft-lb)
   - Counterhold with water pump wrench VAG 1590 when loosening and tightening
3. **V-belt**
   - See **V-belt, removing and installing (4-cylinder engines)**
4. **Pressure line**
5. **Seal**
   - always replace

6. **Banjo bolt**
   • Tighten to 30 Nm (22 ft-lb)
7. **Hose clamp**
8. **Suction hose**
9. **Power steering pump**
   • See **Power steering pump, removing and installing (4-cylinder engines)**

> *NOTE —*
> *Replacement pumps are not filled with oil. Always fill with hydraulic oil (Volkswagen part No. G 000 200) and turn by hand before installing.*

10. **Bolt M8 x 32**
    • Tighten to 25 Nm (18 ft-lb)
11. **Bolt M8 x 85**
    • Tighten to 25 Nm (18 ft-lb)
12. **Bolt M8 x 50**
    • Tighten to 25 Nm (18 ft-lb)
13. **Bolt M8 x 50**
    • Tighten to 25 Nm (18 ft-lb)
14. **Pump pivot bracket**
    • Must be installed free of tension
15. **Pump mount bracket**
    • Must be installed free of tension
16. **Bolt M10 x 98**
    • Tighten to 45 Nm (33 ft-lb)
17. **Bolt M8 x 42**
    • Tighten to 25 Nm (18 ft-lb)
18. **Stud**
    • Tighten to 25 Nm (18 ft-lb)
    • Only for gasoline engines
19. **Adjustment bolt**
    • Tighten to 25 Nm (18 ft-lb)
20. **Adjusting bracket**
21. **Nut**
    • Tighten to 25 Nm (18 ft-lb)
    • Only for gasoline engines
22. **Bolt M8 x 42**
    • Tighten to 25 Nm (18 ft-lb)
23. **Nuts**
    • Tighten to 25 Nm (18 ft-lb)
24. **Mounting bracket**

Power steering pump, removing and installing (4-cylinder engines)

1. Remove power steering V-belt. See **Power Steering Pump V-belt, removing and installing (4-cylinder engines)**.

2. Remove hoses from power steering pump and catch draining fluid. Seal hose connections on pump.

3. Remove bolts/nuts from adjusting bracket and remove bracket. See Fig. 19.

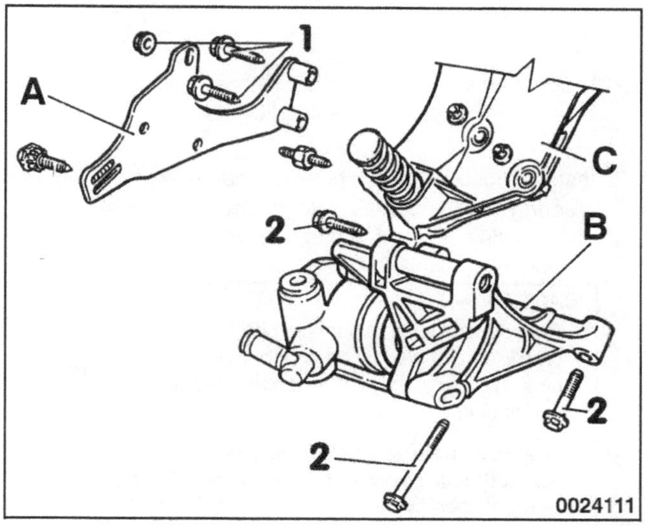

0024111

Fig. 19. Power steering pump adjusting bracket (**A**), pivot bracket (**B**) and mounting bracket (**C**). To remove adjusting bracket remove bolts and nuts at **1**. To remove pump and pivot bracket remove bolts at **2**.

4. Remove bolts from pivot bracket and remove pump together with pivot bracket.

5. Install pump with brackets to the engine block and mounting bracket. Install adjusting bracket to engine block and pump. Do not tighten pivot bracket bolts at this time. See Fig. 20.

6. Install pump adjusting bracket. Use Fig. 18 for tightening torques.

> *NOTE —*
> *If replacing the pump, transfer the pump pulley and brackets to the new pump. Refer to Fig. 18 for tightening torques.*

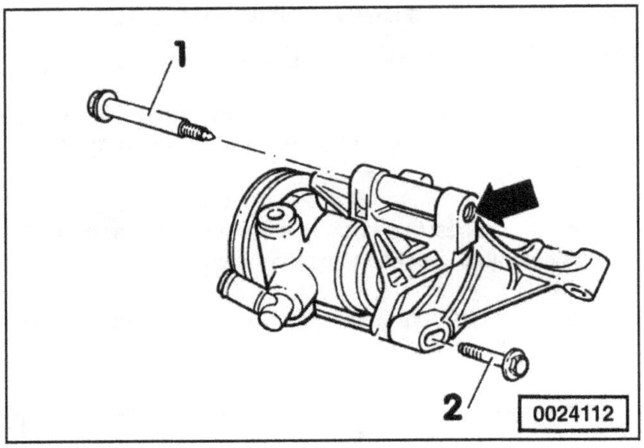

0024112

Fig. 20. Before installing pump, make sure pivot bolts (**1** and **2**) are loose. Threaded bushing shown at **arrow**.

7. Install suction (lower) hose to pump and fill power steering fluid reservoir with hydraulic oil. Rotate the pump pulley by hand to bleed pump.

> ***CAUTION—***
> - *The power steering pump must be filled with hydraulic oil and bled by hand prior to starting engine. The pump may become damaged or noisy during operation if improperly bled.*
> - *Use only hydraulic oil (Volkswagen part No. G 002 000) in the power steering system. Do not reuse drained hydraulic oil.*
> - *Part numbers are listed for reference only. Always check with an authorized Volkswagen parts department for the latest parts information.*

Power steering system hydraulic oil capacity
- 4-cylinder engines . . . . 0.7-0.9 liter (0.74-0.95 qt.)

8. Once the pump is bled, connect the pressure line to the pump using new sealing washers. Install V-belt and adjust tension. Recheck hydraulic oil level.

> ***CAUTION—***
> *Always replace sealing washers.*

Power steering pump V-belt, removing and installing (4-cylinder engines)

1. Loosen power steering pump bolts and nuts shown in Fig. 21. Remove belt from pulleys.

> ***NOTE—***
> *Loosen adjustment clamping bolt (Fig. 21, bolt no. 3) at least one full turn.*

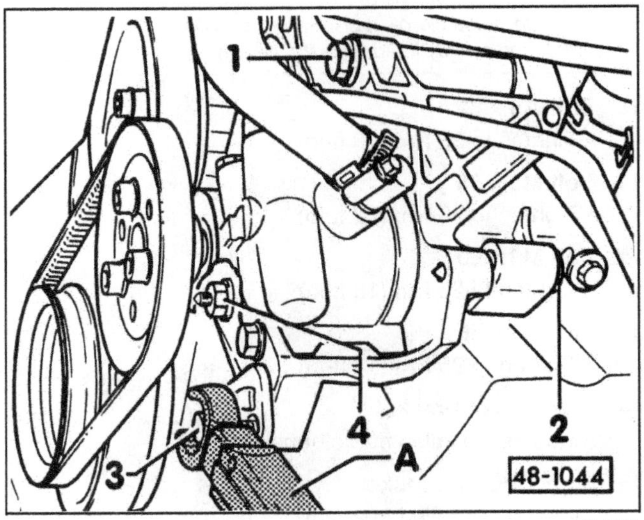

48-1044

Fig. 21. To remove V-belt, loosen bolts **1**, **2**, **3**, and nut **4**. Belt tension is adjusted at tensioning nut using torque wrench (**A**).

2. Install the V-belt. Adjust V-belt tension by tightening tensioning nut with torque wrench. Tighten clamping bolt when tension is correct.

> ***NOTE—***
> - *When the tensioning nut is torqued to the specified setting, belt tension will be correct. See Fig. 21.*
> - *Refer to Fig. 18 for tightening torques.*

Tightening torques
- V-belt tensioning nut
  - used belts . . . . 4 Nm (35 in-lb)
  - new belts . . . . 7 Nm (62 in-lb)
- Clamping bolt . . . . 25 Nm (18 ft-lb)

POWER STEERING PUMP (6-CYLINDER ENGINE)

Fig. 22 shows an exploded view of the power steering pump and related mounting hardware used on cars with 6-cylinder engine. Use Fig. 22 as a guide to remove and install the pump. When servicing hydraulic system or replacing hydraulic pump, be sure to bleed the pump before starting the engine.

Power steering pump, assembly (VR6 engine)

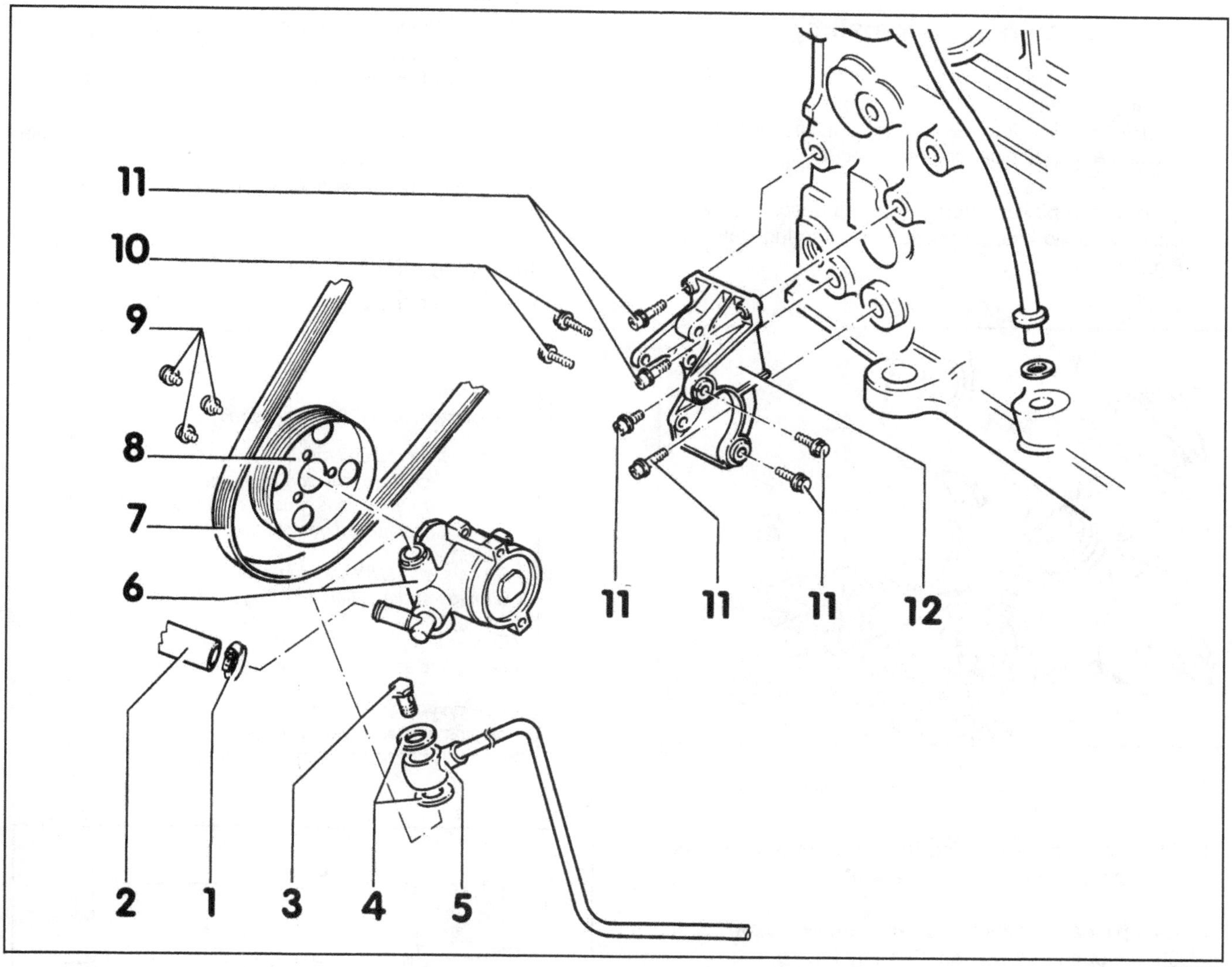

Fig. 22. Exploded view of power steering pump assembly used on cars with VR6 engine.

1. **Hose clamp**
2. **Suction hose**
3. **Banjo bolt**
   - Tighten to 30 Nm (22 ft-lb)
4. **Seal**
5. **Pressure line**
6. **Power steering pump**
   - Delivery pressure: 85-95 bar (1233-1377 psi)
   - Fill with hydraulic fluid before installing

> ***CAUTION —***
> *Replacement pumps are not filled with oil. Always fill with hydraulic oil (Volkswagen part No. G 000 200) and turn by hand before installing.*

7. **Poly-ribbed belt**
   - Removing and checking, see **0 Maintenance Program**
   - Tensioned by spring-operated roller

> ***CAUTION —***
> *On poly-ribbed drive belts, mark the running direction on the belt before removing it.*

8. **Pulley**
9. **Bolts**
   - Tighten to 20 Nm (15 ft-lb)
10. **Bolt M8 x 30**
    - Tighten to 25 Nm (18 ft-lb)
11. **Bolt**
    - Tighten to 25 Nm(18 ft-lb)
12. **Bracket**

TIE RODS

Tie rods, removing and installing

NOTE —
Removal and installation procedure given here applies to both TRW and ZF power steering racks.

1. Remove the power steering gear as described earlier. Seal hydraulic line connections in steering gear. See Fig. 23.

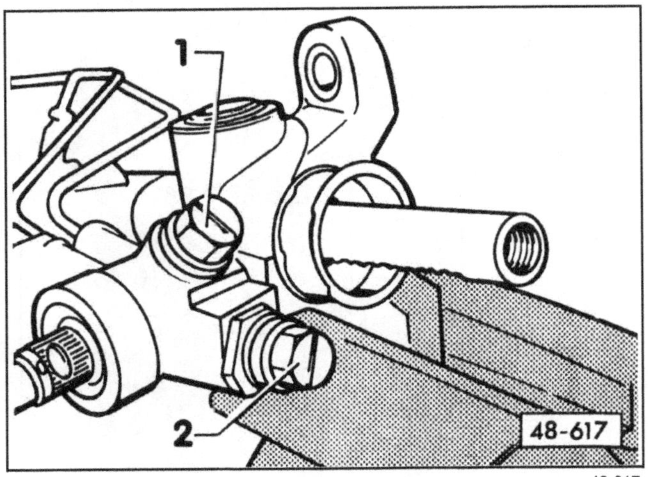

48-617

Fig. 23. Hydraulic line connections sealed with plastic plugs (**1** and **2**).

2. Clean outside of steering gear in area of boot. Then open boot clamp and push back boot.

3. Clamp steering gear in vice and remove tie rod using a 32 mm open end wrench. See Fig. 24.

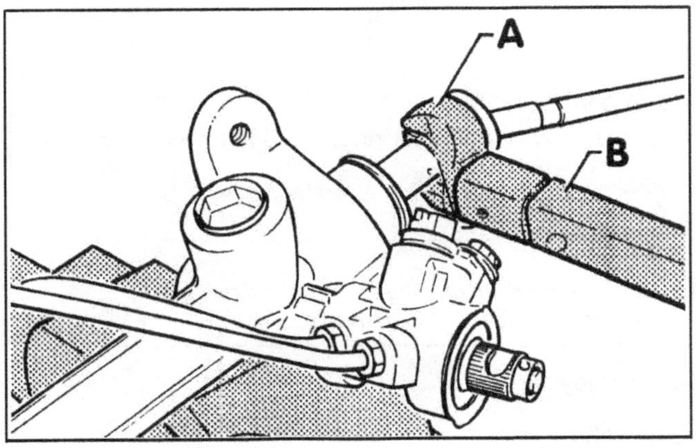

0024083

Fig. 24. Tie rod being removed from steering gear with 32 mm wrench.

4. Install in the reverse order of removal, noting the following:
   - Be sure to lock the tie rod end to the steering gear rack. Use locking compound (Volkswagen "D6" or equivalent) on the inner joint when installing to the rack.
   - When installing the left-hand tie rod, check that the tie rod length is correct. See Fig. 25.
   - Align the front wheels when installation is complete.

Tightening torques
- Tie rod to steering gear . . . . 80 Nm (59 ft-lb)

CAUTION —
- *Only adjust toe on the right tie rod.*
- *If the left tie rod end must be removed, adjust to correct length (dimension **a**) when reinstalling. Also, new left tie rods must be checked and, if necessary, adjusted before installation.*
- *On cars with TRW steering gear, use care not to damage sticker when installing boot.*
- *Ensure that the boots and the pressure compensating line are correctly fitted.*
- *Use only original equipment type replacement clamps on steering gear boots.*

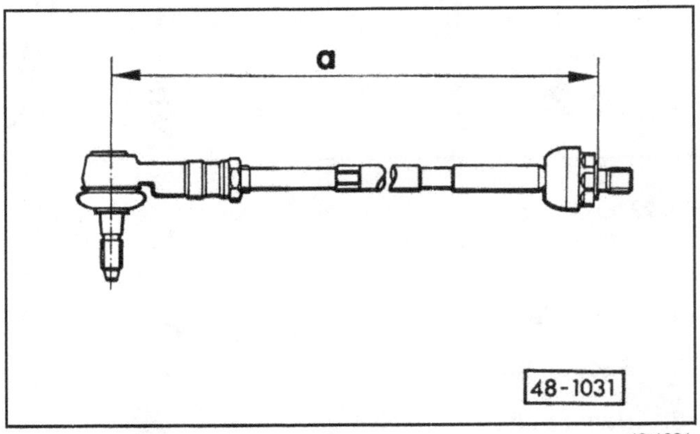

48-1031

Fig. 25. Left tie rod length (dimension **a**) = 406 ± 1 mm (15.98 ± 0.04 in.)

NOTE —
Beginning with model year 1995 (from VIN # 1E_S_000001) all Cabrio vehicles are equipped with a revised tie rod end with a new pin that is 3 mm (1/8 in.) longer than the previous pin. When replacing tie rod ends on earlier vehicles (1994 m.y. and earlier) if one tie rod end is replaced with the newer revised tie rod end then both tie rod ends must be replaced.

5 Body–Assembly

GENERAL

This general information group covers application information and system descriptions for the repair groups listed under **5 Body–Assembly, 6 Body–Components and Accessories,** and **7 Body–Interior Trim**.

BODY ASSEMBLY

The front fenders, radiator support, grille, both bumpers, doors, hood, and rear lid are bolted to the main body structure for easy replacement in the event of damage.

Jetta, Golf, GTI

The body assemblies for the Jetta, Golf, and GTI are designed to form a very rigid passenger cell around the interior passenger compartment. The lower door sills have been significantly changed in profile and the metal thickness increased to provide good lateral protection. Side impact cross-members are bolted into each door to form profiled stiffening sections for additional side impact protection. See Fig. 1.

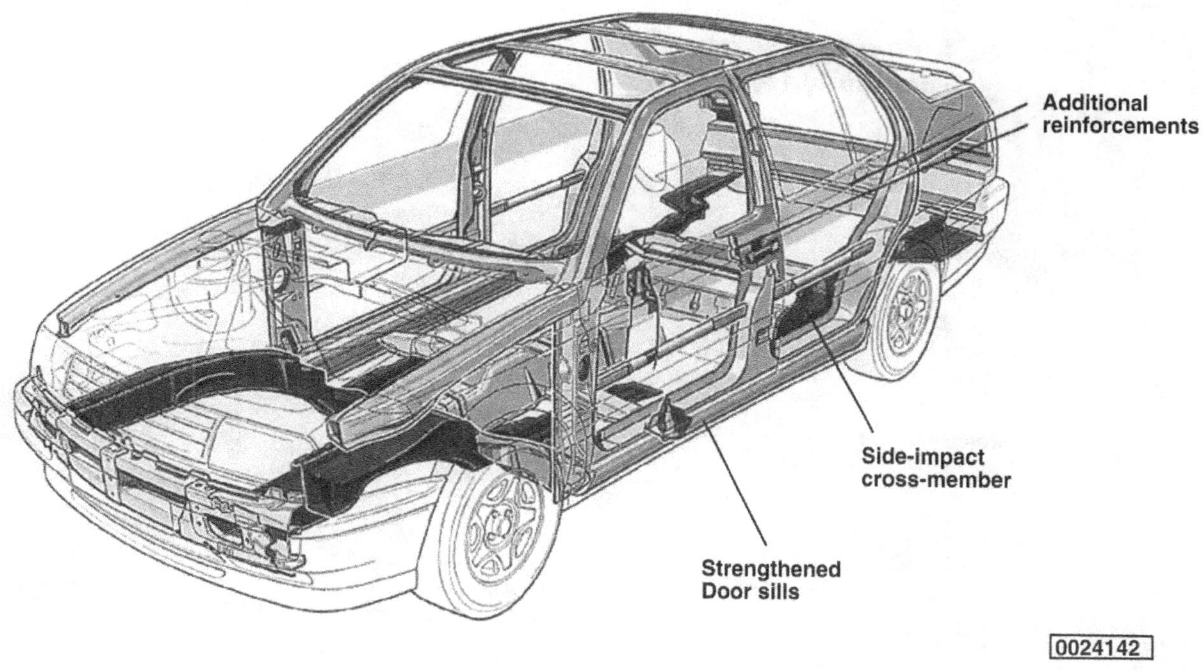

0024142

Fig. 1. Structural body members of Jetta models (Golf and GTI similar).

Cabrio

The Cabrio is a modification of a basic two-door unitized body. When the steel roof is deleted and the convertible top and its mounting frame are added, extra body-strengthening members and a steel roll-bar are also added to maintain body structure.

Two elastically suspended masses are utilized to reduce body shake that is common in convertibles. The mass at the front (engine and transmission) is suspended from specially designed mounts that allow the engine and transmission to dampen front to rear shake. The rear mass is located in the left side of the trunk. It weighs approximately 20 pounds and reduces side to side shake. See Fig. 2.

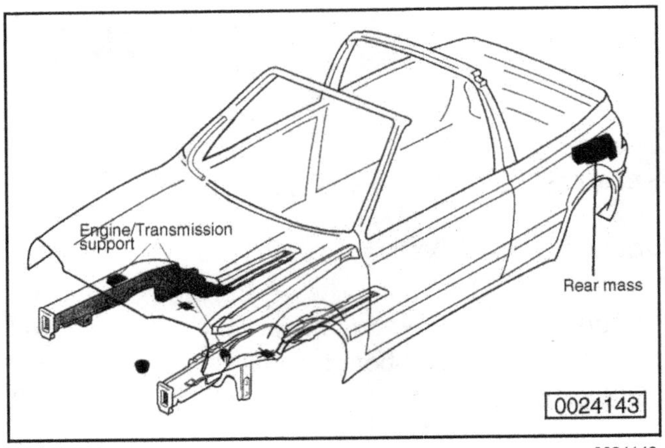

0024143

Fig. 2. Structural components specifically designed for Cabrio model.

Front bulkhead/dashboard module

The dashboard unit, together with the heating and air conditioning unit, steering column, and pedal cluster with master cylinder, are assembled as a module outside the vehicle. These components are attached to the reinforcement cross-member and a bulkhead cross panel. The entire assembly is then installed in the vehicle. For service work, the dashboard can be removed from the reinforcement cross-member by unbolting it.

This modular construction permits many of the more difficult assembly procedures on these units to be performed outside the vehicle where access to the components is easier. This same technique is also used on the front end, door assemblies and sunroof.

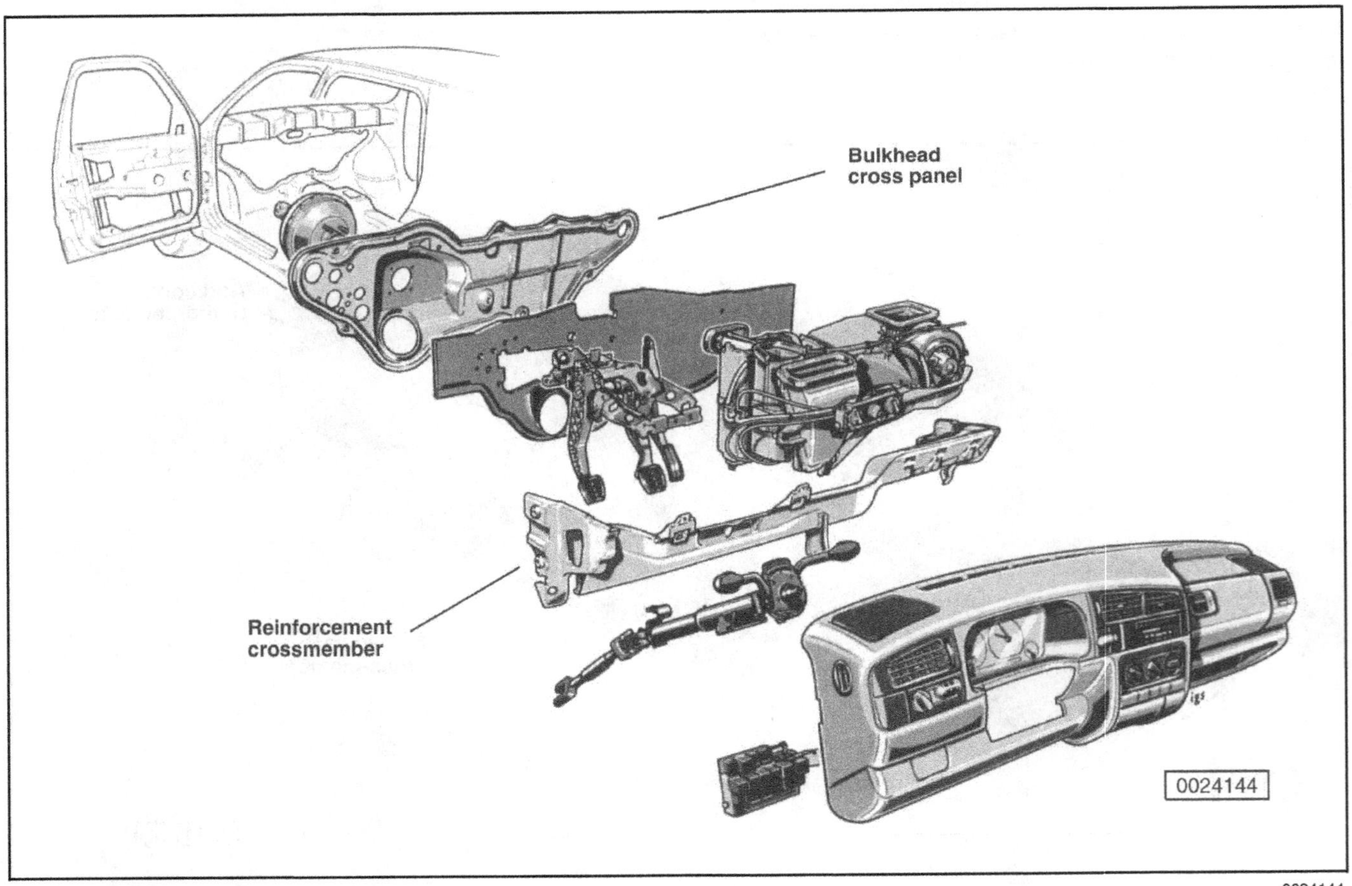

0024144

Fig. 3. Bulkhead/dashboard module.

Bumpers

The bumper assemblies on the all models utilize a cross-member that is bolted directly to the body structure, instead of using brackets. A molded styrofoam insert is installed over this cross-member. The body colored bumper covers are then bolted to the bumper cross-members with the styrofoam insert sandwiched in-between. See Fig. 4.

The bumper covers use special plastics to optimize their ability to absorb minor impacts. The styrofoam insert will compress and absorb major impacts. The bumper cover can be removed to replace this insert. This design spreads the impact over a larger area and limits unnecessary bumper weight.

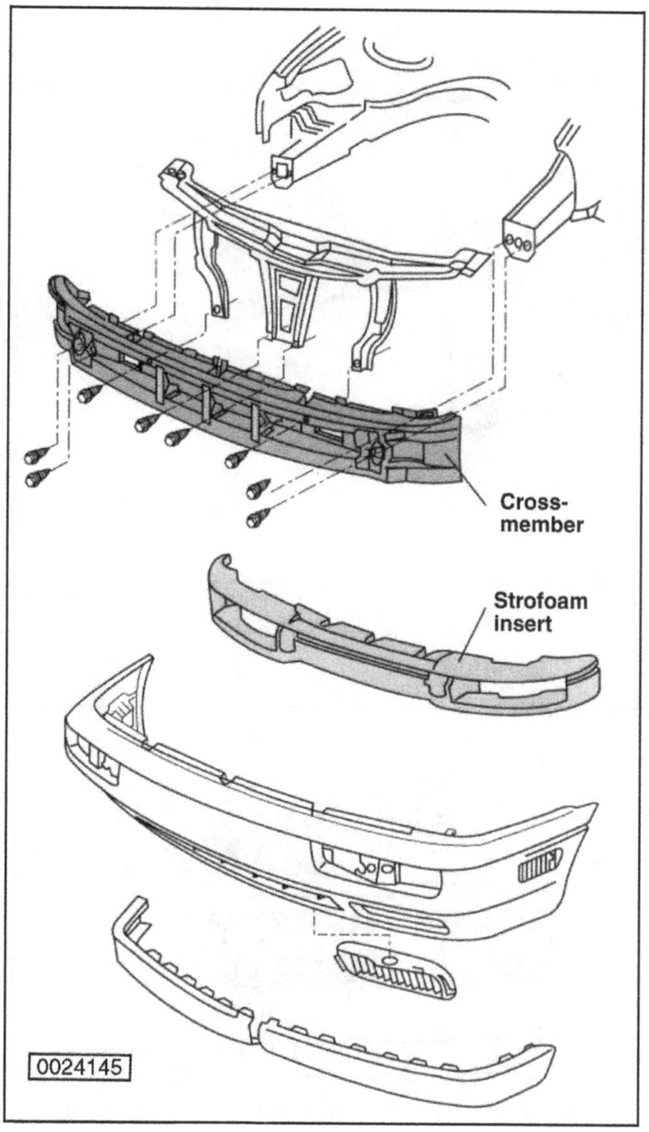

0024145

Fig. 4. Front bumper assembly.

POWER WINDOWS

Models equipped with power windows feature the ability to raise and lower all four windows by using the key in either front door lock.

By holding the key in the locked position for more than 1 second, all windows that are opened will be raised to the closed position at the same time. This feature is known as convenience close. See Fig. 5.

In addition, if the vehicle is equipped with a sunroof, it will close with the power windows in one of two ways. The sunroof will either close while the windows are in motion or it will close after all of the power windows are raised to the fully closed position. The sunroof and the windows will stop as soon as the key is released.

As a ventilation aid in hot weather, all four windows can be lowered. By holding the key in the unlock position for more than 2 seconds, all four windows will be lowered at the same time. The sunroof, however, will not open with this function.

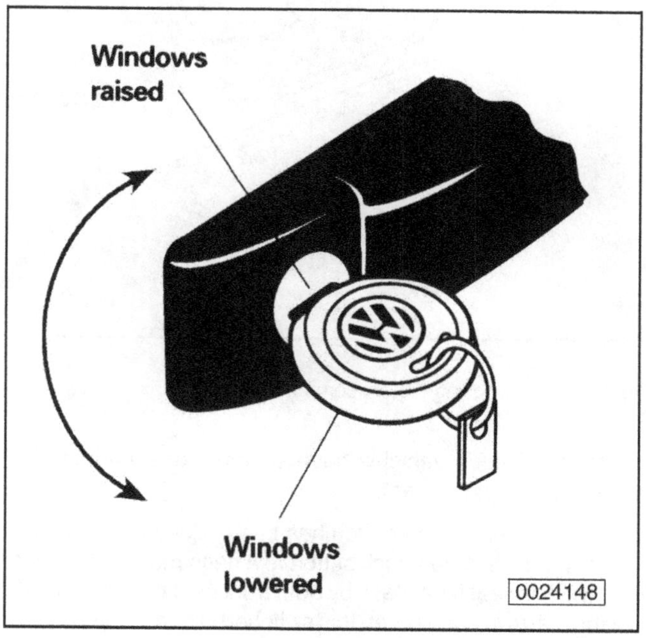

0024148

Fig. 5. On models with power windows, the windows and sunroof can be operated via the door lock.

CENTRAL LOCKING

All models are equipped with central locking for the doors. Some versions also lock the rear lid and the fuel filler door. A vacuum servo is part of each latching assembly. A micro-switch that is operated by the front door lock cylinders activates the central locking system and the alarm system. Fig. 6 shows the location of the various components of the central locking system.

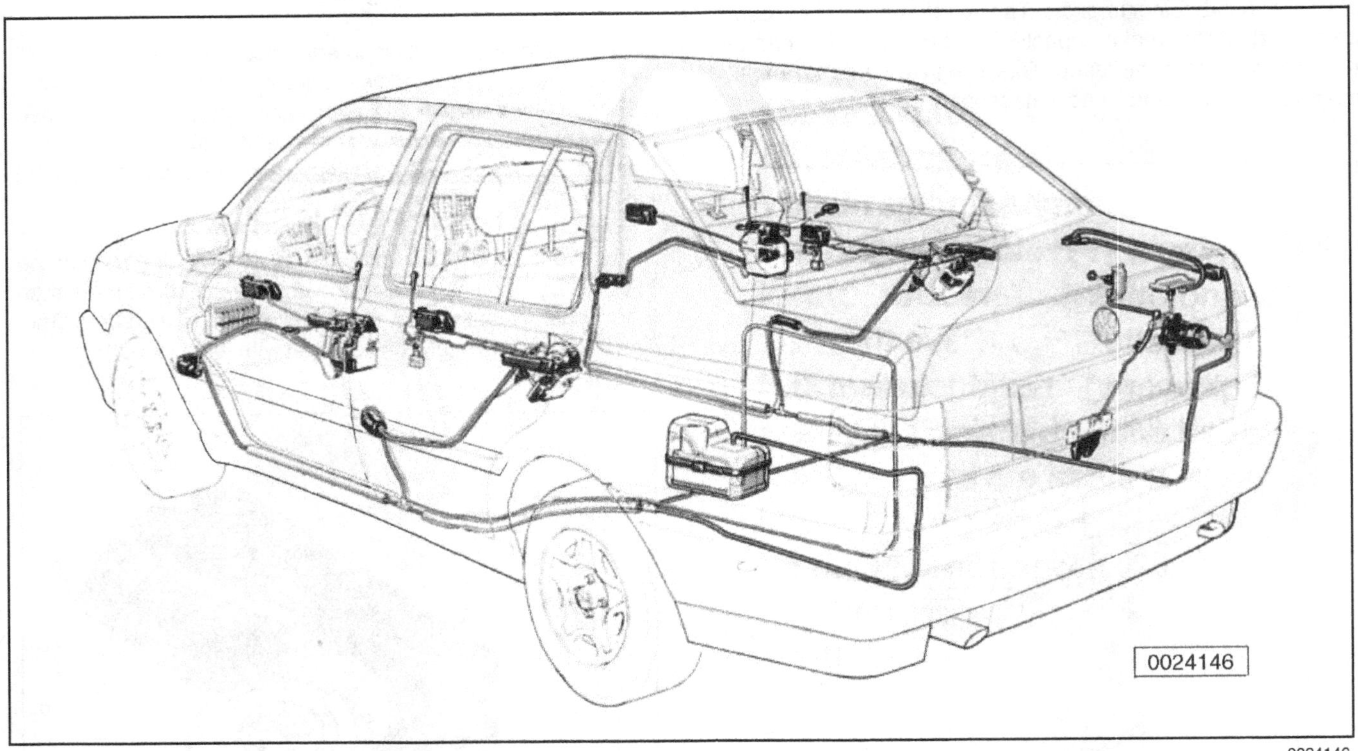

0024146

Fig. 6. Central locking system component locations.

Later production vehicles have several enhancements to the central locking system.

- The master locking switch has been relocated from the inside driver's door lock button to a dash mounted rocker switch for easier access by both front seat passengers. If the alarm system is set, this switched is deactivated. See Fig. 8.
- When opening a locked vehicle from the outside with the key, there are now two options. If the key is turned once, only that particular door will unlock. If the key is quickly turned twice, all of the doors will be unlocked. This function is called selective unlocking.
- Central locking is connected to the factory installed Keyless Remote System and the vehicle can be locked and unlocked with the remote transmitter. Selective unlocking does not function in this mode of operation.

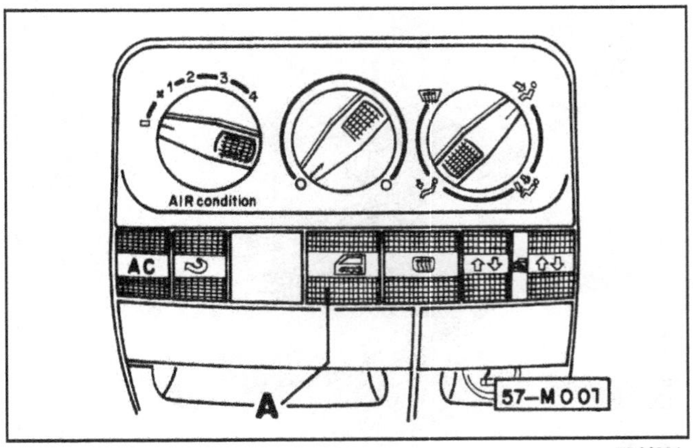

57-M001

Fig. 7. Master door lock switch, **A**, that was previously part of the driver's side door lock and latch assembly is now dash mounted. All door locks can be locked and unlocked with it.

ALARM SYSTEM

All models are equipped with a factory installed alarm system that is linked to the central locking system. The alarm system is armed when either front door is locked with the ignition key, and the trunk lid, hood and all doors are closed. The alarm will not arm if a door or lid is open. See Fig. 8.

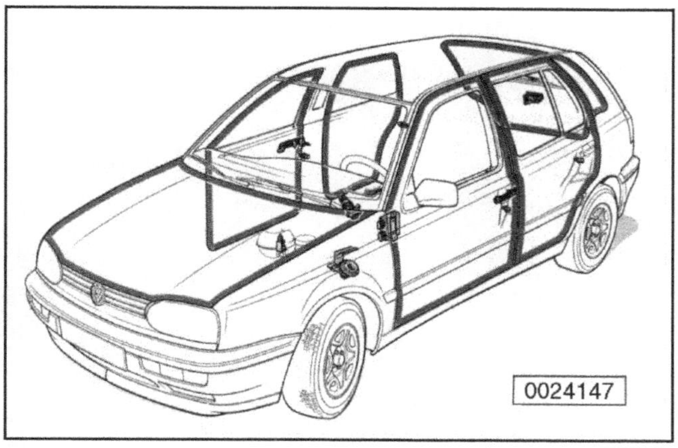

0024147

Fig. 8. Volkswagen central alarm system (Golf shown, others similar).

When set, the alarm system will trigger if the trunk or any door is opened without the key. It will also be triggered if the hood is opened or the factory installed radio is disturbed. It will also trigger if it detects that the ignition has been switched on.

When triggered, the separate alarm horn will sound and the directional signals will flash. The starter motor circuit will also be interrupted. The system can have multiple triggers. If, for example, the trunk is breached, the alarm will trigger and the horn and lights will operate for the prescribed period and then switch off. The alarm system remains armed. Disturbing a door will trigger the system again in the same way. This can happen as often as the system is breached.

The system has a defeat switch in the luggage compartment lock that allows the luggage compartment to be opened using the key, without activating the alarm.

To prevent the alarm system from becoming a nuisance, the alarm horn and directional lights will stop after approximately 30 seconds. The starter disable will, however, still function to deter vehicle theft.

The alarm control module is located behind the upper left side of the dashboard, next to the left side dashboard fresh air vent. An LED is located next to the driver's door lock knob. This LED will flash at 1-second intervals after the doors are locked to indicate that the system will soon be armed. After 30 seconds, the LED will begin to flash faster to indicate that the system is armed. To prevent battery drain, some versions of the system will turn off the flashing LED after approximately 24 hours. Other versions let it continue to flash indefinitely. In all versions, however, the alarm system will remain armed until deactivated with a key in a door.

KEYLESS REMOTE LOCKING

Beginning with the 1998 model year, a factory installed keyless remote locking system is standard equipment. This enhancement to the central locking and alarm systems uses a remote radio frequency transmitter about the size and shape of a regular vehicle key. Four system functions can be enabled through this battery-powered transmitter up to a range of approximately 10 feet. See Fig. 9.

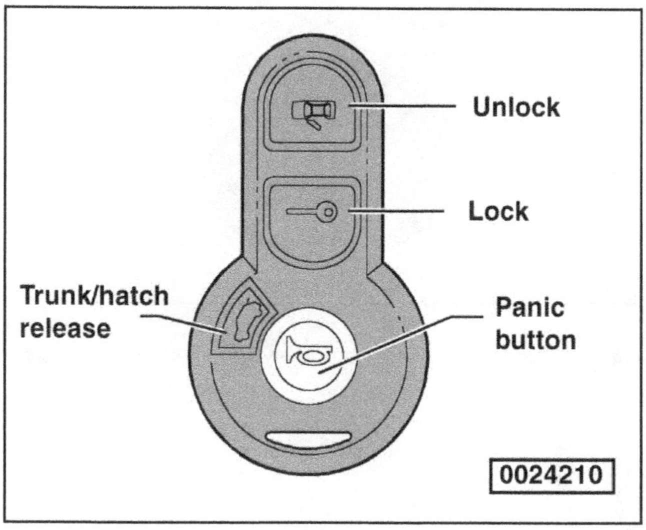

0024210

Fig. 9. Radio frequency remote transmitter used in Golf and Jetta vehicles beginning with 1998 model year.

The four functions of the keyless remote transmitter are:

- Lock the vehicle and arm the alarm.
- Unlock the vehicle and dis-arm the alarm.
- Dis-arm the alarm and open the trunk or hatch. The system will re-arm when the trunk or hatch is closed if the alarm system had previously been armed.
- Red panic button to trigger the alarm at any time.

The selective unlock feature of the central locking and the convenience close feature of the power windows do not function with the keyless remote system.

If the vehicle has been unlocked using the keyless remote transmitter and a door is not opened within approximately 30 seconds, the system will re-lock all of the doors and re-arm the alarm system.

Up to 4 transmitters can be programmed into each vehicle's system. Each transmitter uses two "button cell" batteries that are readily obtainable and easily replaced. See **96 Lights, Accessories-Interior**.

Several different keyless remote systems are available in the aftermarket for earlier vehicles. These include systems offered by Volkswagen through authorized dealers. See Fig. 10.

Because these systems vary considerably in operation and installation, they are outside the scope of this manual.

0024204

Fig. 10. Typical aftermarket 2 function keyless remote transmitter supplied by Volkswagen.

50 Body–Front

GENERAL

The repair group covers the front body section (radiator support) and front fender replacement. The radiator support can be removed from the car as one unit to facilitate repairs, such as engine removal. The front fenders are easily removable panels, although some new parts may be required for fender installation.

FRONT BODY ASSEMBLY

Radiator support, removing and installing

The radiator support assembly is shown in Fig. 1.

1. Disconnect the battery negative (–) cable.

2. Disconnect all wiring harness connectors and other electrical connections from radiator support as required.

3. Disconnect hood lock cable from hood lock on radiator support. See **55 Hood and Lids**.

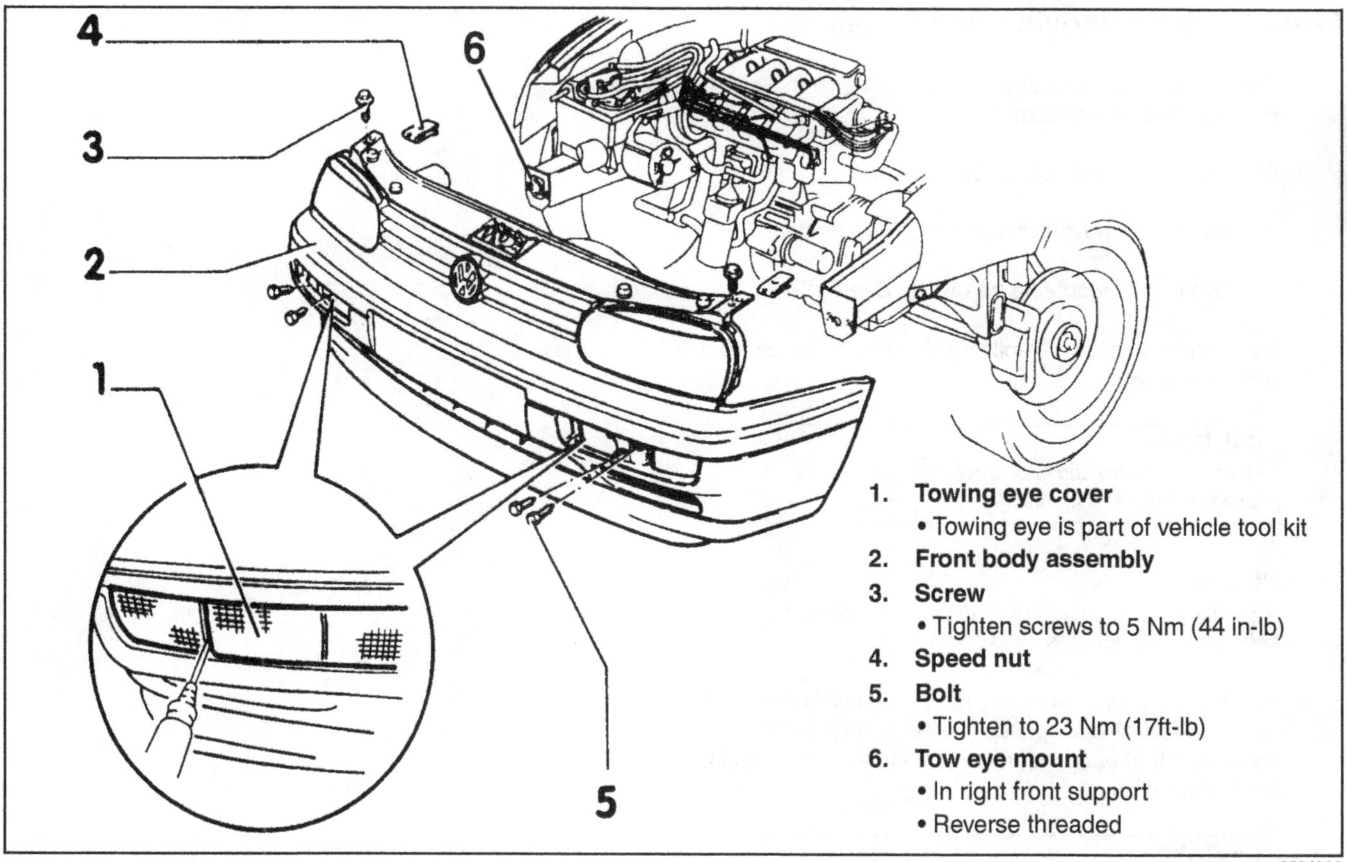

0024000

Fig. 1. Radiator support assembly. Remove upper bolts from fenders and lower bolts from bumper. Also remove screws from wheel housing trim (not shown). Inset shows towing eye covers that must be removed to access lower bolts.

4. Loosen A/C refrigerant lines from mounting brackets and hose clamps on radiator support (cars with air conditioning only). Do not disconnect or loosen any refrigerant lines or fittings.

> ***WARNING —***
> *Loosening or disconnecting A/C refrigerant lines will allow the refrigerant to be discharged under pressure, possibly causing personal injury. In addition, a discharged A/C system will need to be recharged using special equipment.*

5. Remove upper radiator mounting pins. Lift the radiator up and off up the lower mounts. See **19 Engine–Cooling System**.

6. Unbolt A/C condenser from radiator support. See **87 Air Conditioning**.

> ***CAUTION —***
> *Use care not to bend or kink A/C refrigerant lines.*

7. Remove mounting bolts shown in Fig. 1 and lower screws from wheel housing liner and air guides.

8. Remove the radiator support with all attachments.

Front fender, removing and installing

1. Remove radiator grille and wheel housing liner. See **66 Body Accessories–Exterior**.

2. Remove front bumper. See **63 Bumpers**.

3. If applicable, remove radio antenna from fender.

4. Remove fender mounting hardware. See Fig. 2.

5. Heat fender in area of A-pillar with a hot air blower and remove fender.

> ***CAUTION —***
> *Heat PVC slightly and for a short time only. Do not discolor PVC or blister paint.*

NOTE —
If necessary, cut through the PVC with an upholstery knife while a helper pulls out on fender.

6. Installation is the reverse of removal. Before installing fender, install zinc foil plates (Volkswagen part No. AKL 381 035 50) at each mounting point in the wheel housing/A-pillar area.

> ***CAUTION —***
> *Part numbers are listed for reference only. Always check with an authorized Volkswagen parts department for the latest parts information.*

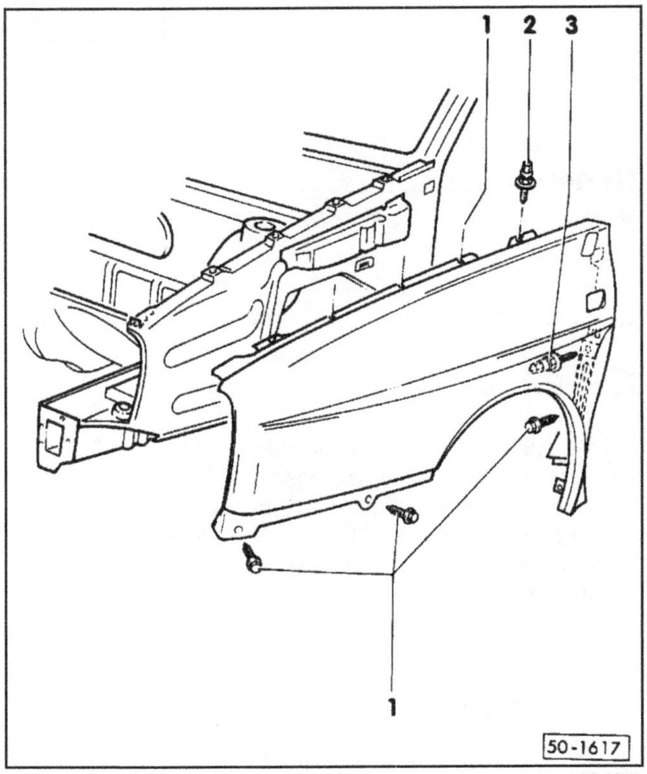

50-1617

Fig. 2. Front fender attachment bolts (**1**,**2**,**3**).

55 Hood and Lids

GENERAL

The hood and rear lid are easily unbolted and removed with the aid of a helper. Where applicable, the support struts should first be disconnected. Be sure to mark the lid location in reference to the hinges for proper alignment during installation.

HOOD LOCK

The hood lock cable assembly is shown in Fig. 1. To replace the cable, remove the operating lever and the hood lock and disconnect the cable ends. Then pull the cable out through the bulkhead. When installing the new cable, be sure to install the sealing grommet on the cable.

Hood lock cable assembly

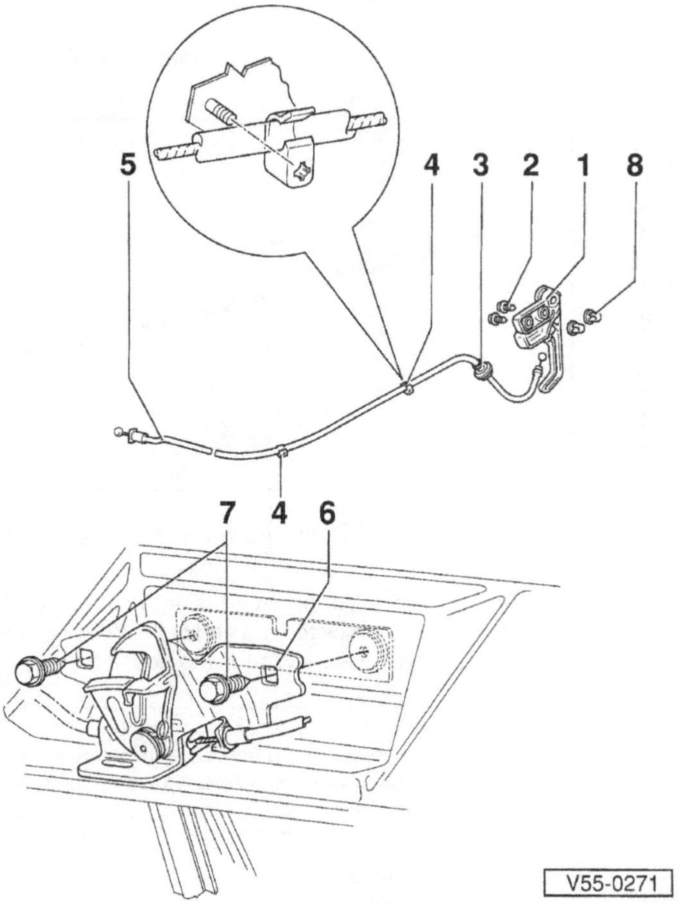

1. **Operating lever**
2. **Self-tapping screw**
3. **Sealing grommet**
4. **Clip**
5. **Cable**
6. **Lock**
   - Cable attached
7. **Self-locking bolt**
   - Always replace fastener
   - Tighten to 12 Nm (9 ft-lb)
8. **Spreader clip**

V55-0271

Fig. 1. Hood lock cable assembly.

REAR LID

NOTE —
For repair information on the power locking system, see **57 Doors–Front**.

Gas-filled strut for rear lid, replacing

Fig. 2 shows the strut used on Golf and GTI models. Fig. 3 shows the strut used on Jetta models. To remove the strut, support the rear lid, then carefully pry out the retainers from the strut ends using a screwdriver. Pry the strut ends off of the ball studs.

> ***WARNING —***
> *Be sure to support rear lid before removing strut.*

NOTE —
If reinstalling the strut, do not remove the retainer from the strut. If the retainer is fully removed, it will be damaged. The retainer is available as a replacement part from Volkswagen.

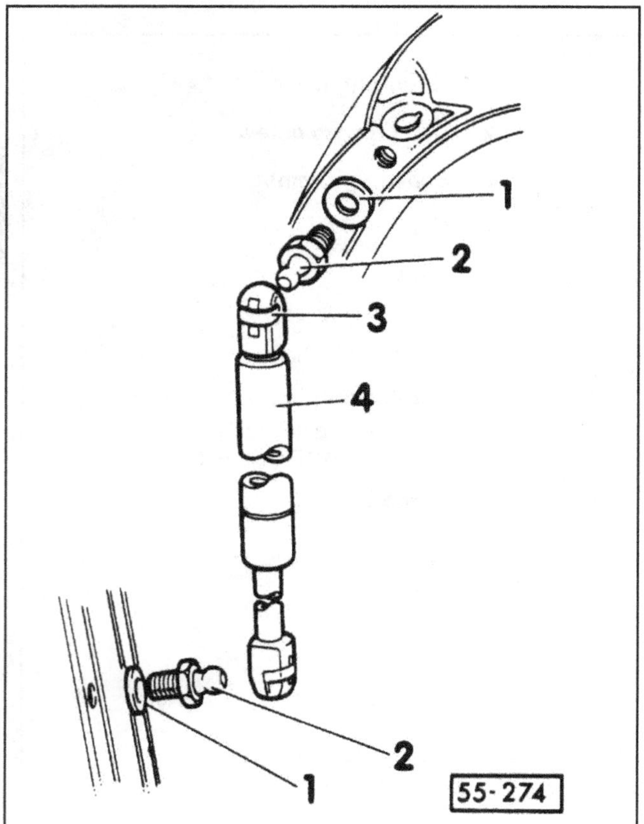

1. **Washer**
2. **Ball stud**
   - Tighten to 15 Nm (11 ft-lb)
3. **Retainer**
4. **Gas-filled strut**

55-274

Fig. 2. Gas-filled strut for Golf and GTI rear lid (Cabrio is similar).

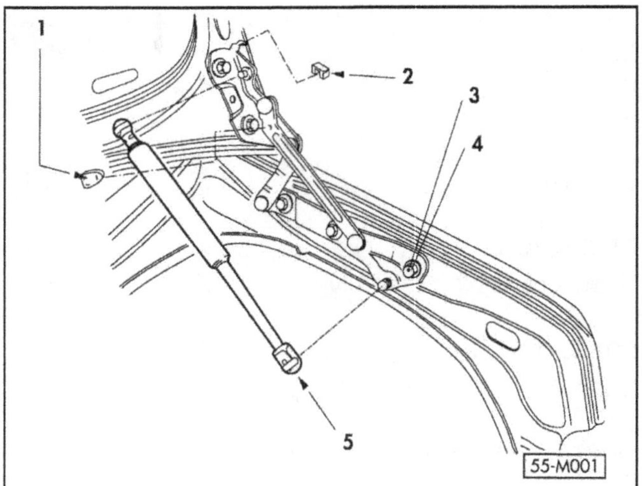

1. **Rubber damper**
2. **Rubber damper**
3. **Screw**
   - Tighten to 7.5 Nm (66 in-lb)
4. **Spring washer**
5. **Gas-filled strut**

55-M001

Fig. 3. Gas-filled strut for Jetta models.

Gas-filled strut, discarding

The gas charge in the strut should always be released before scrapping the strut.

1. Clamp strut in vise in area shown in Fig. 4.

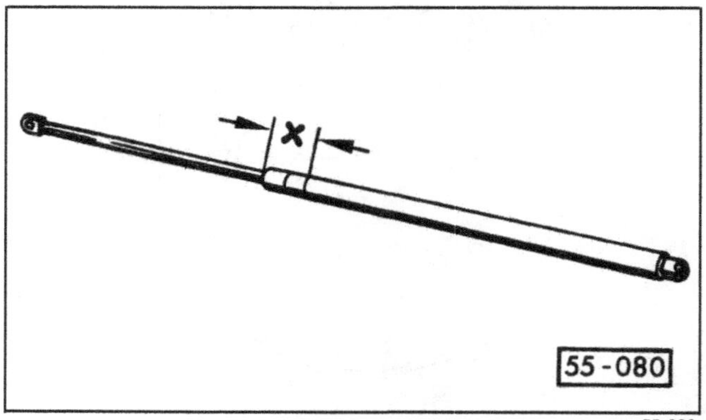

55-080

Fig. 4. Clamp strut in vise at first 2 inches (**x**) of strut housing.

2. Make a mark on the first 1/3 of the strut housing, as measured from piston rod end of the housing. Then using a hacksaw, saw through the housing at the mark.

> ***WARNING —***
> *Protective goggles must be worn during sawing. In addition, it is recommended that the area of the saw cut be covered with a rag in order to catch oil spray*

3. Dispose of strut.

Rear lid lock

The rear lid lock assembly is shown in Fig. 5. To access the lock assembly, remove the trim from the trunk or rear hatch. Fig. 6 shows the remote trunk lid opener, used on some Jetta models.

Rear lid lock assembly

1. **Screw**
   • Tighten to 4 Nm (35 in-lb)
2. **Rear lid lock assembly**
   • Removing:
   remove screws (1), disconnect push rod
   (6) from lock assembly
3. **Key**
   Must be inserted to remove lock
4. **Rear lid latch**
5. **Screw**
   • Tighten to 7 Nm (62 in-lb)
6. **Push rod**

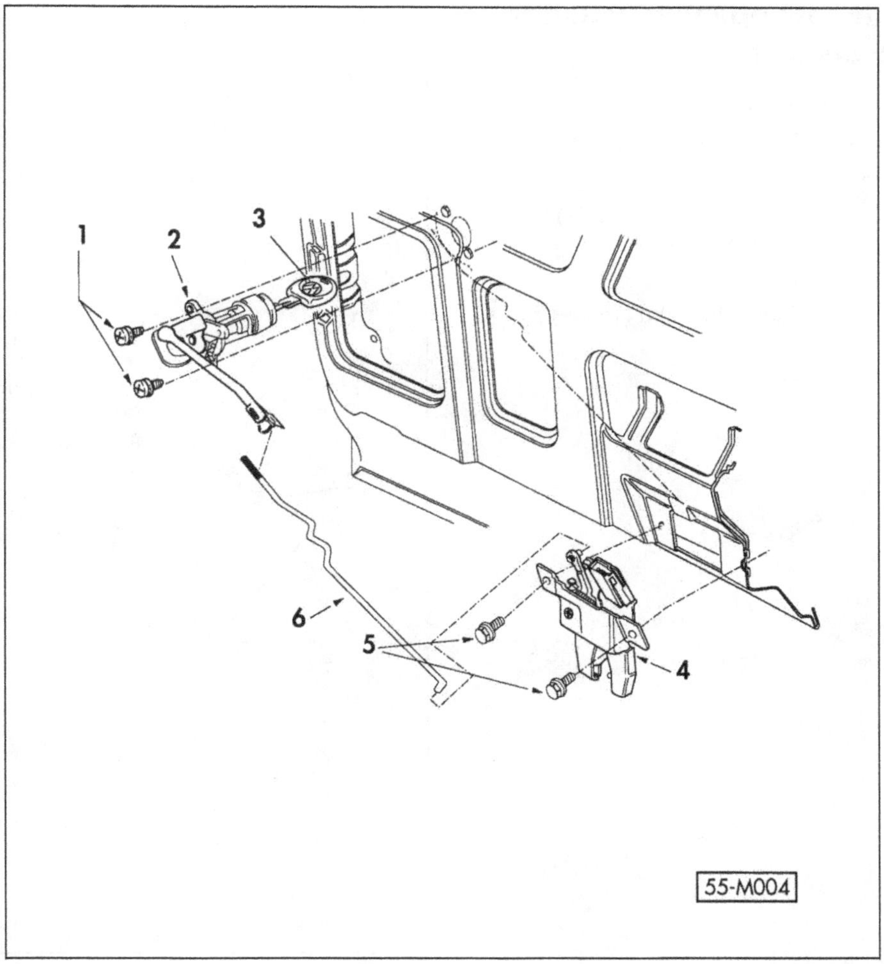

55-M004

Fig. 5. Rear lid lock mechanism. Mark location of latch (**4**) before loosening mounting screws.

Remote opening actuator assembly

1. **Push rod**
   - Color: orange
2. **Actuator**
   - To remove, push rod (1) must be disconnected
3. **Screw**
   - Tighten to 2 Nm (18 in-lb)

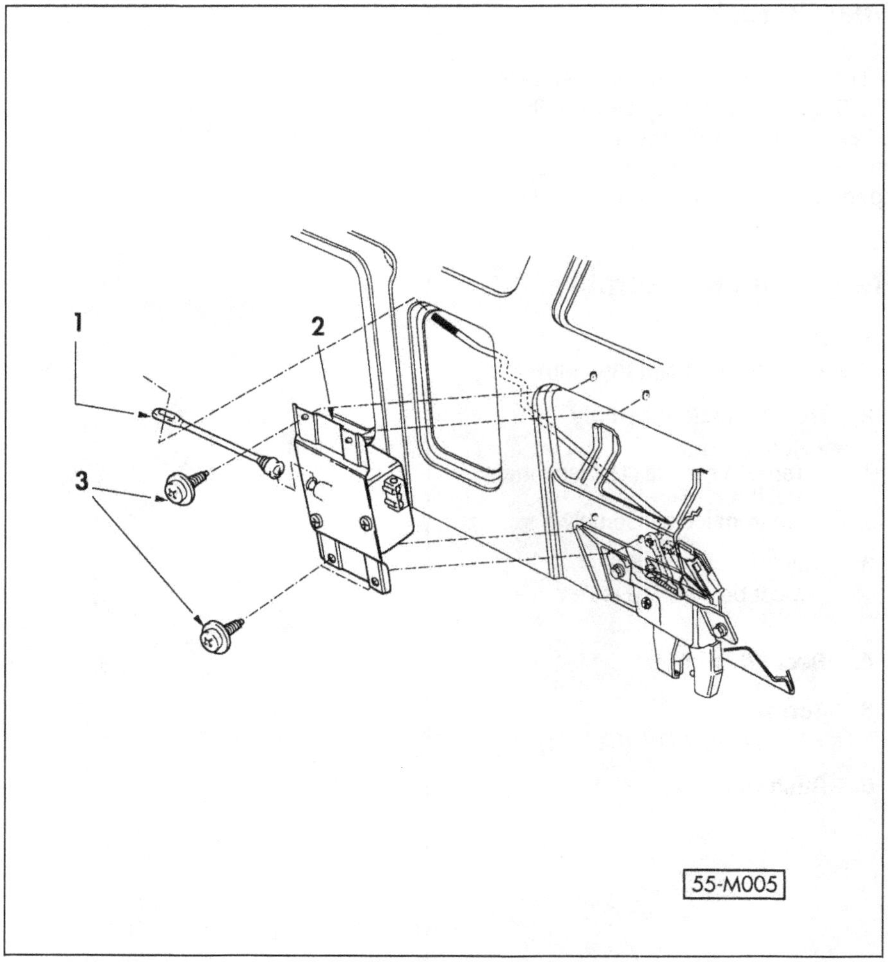

55-M005

Fig. 6. Remote rear lid electric actuator assembly used on some Jetta models.

Rear lid lock assembly (Cabrio through 03/98)

1. **Pull rod**
2. **Push button**
3. **Spreader nut**
4. **Screw**
5. **Handle trim**
6. **Screw**
   - Tighten to 20 Nm (15 ft-lb)
7. **Rear lid lock**
8. **Striker pin plate**
9. **Socket head bolts**
   - Tighten to 20 Nm (15 ft-lb)

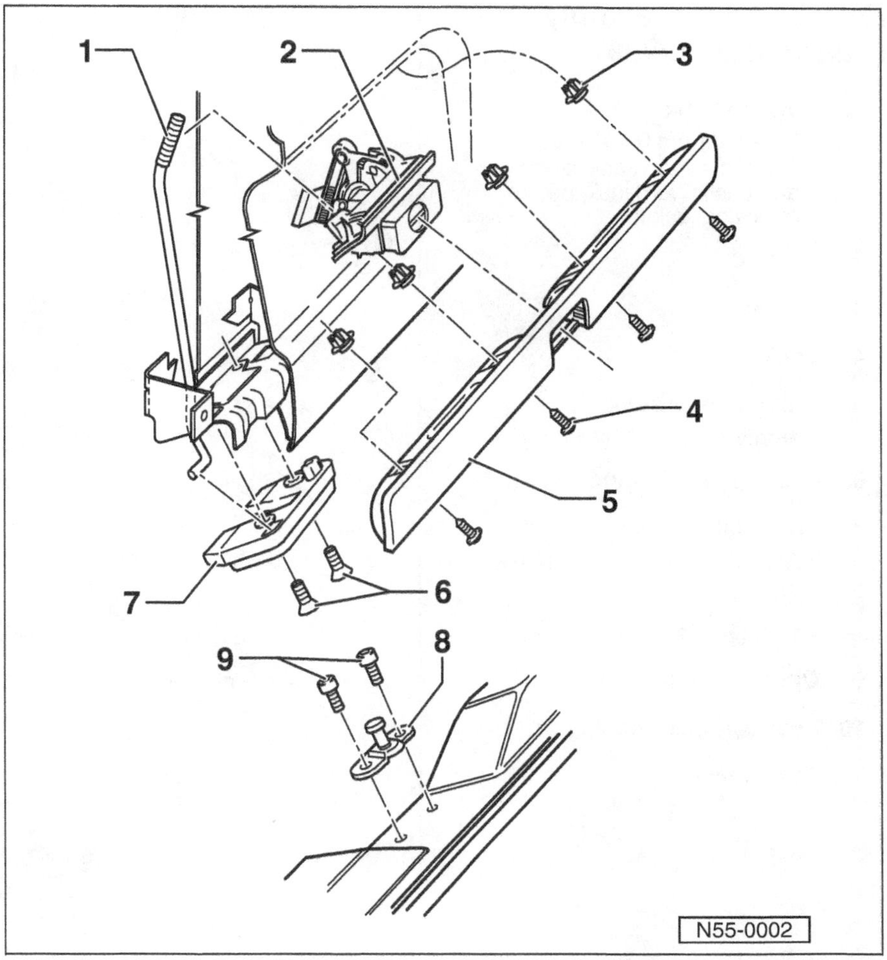

N55-0002

Fig. 7. Rear lid lock assembly for Cabrio models through 03/98 production.

Pull rod, adjusting

NOTE —

Pull rod must be adjusted each time it is removed or installed or when repairs are performed.

1. Secure locking link by inserting 3.2 mm dia. (1/8 in.) drill bit into lock hole.

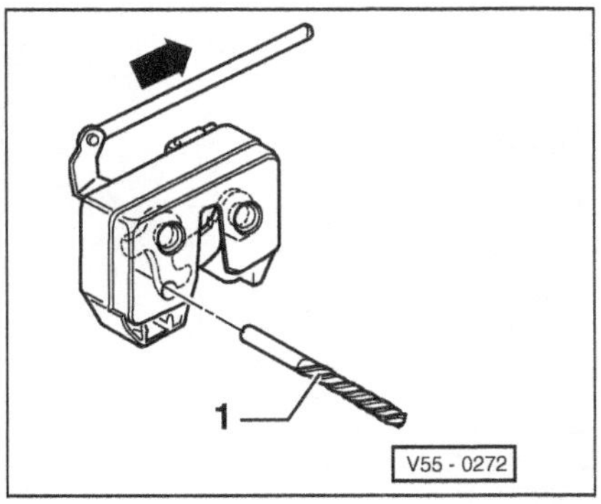

2. Draw pull rod in direction of **arrow**.
3. Press pull rod in and close hinged sleeve.
4. Pull out drill bit.
5. Perform functional check. Locking link must spring back to end position.

Push button, removing

1. Remove exterior screws and trim from handle.
2. Remove rear lid.
3. Unhook pull rod.
4. Press inner locking tabs together (**arrows**) and remove push button from rear lid.

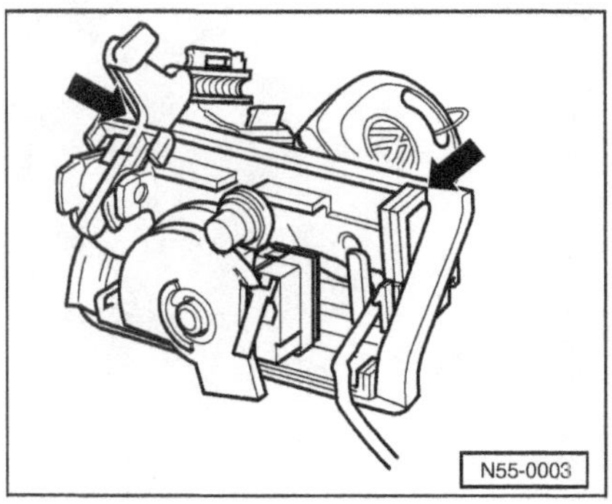

Rear lid lock assembly (Cabrio from 04/98)

1. **Bowden cable**
   - Installed from 06/01 prod.
   - To remove, turn cable clockwise and unclip from lock, before removing lock trim
2. **Rear lid**
3. **Handle**
4. **Lock**
5. **Bolt (counter sunk)**
   - Tighten to 20 Nm (15 ft-lb)
6. **Cross panel (body)**
7. **Lock plate**
   - Adjust within oversized holes
8. **Bolt**
   - Tighten to 20 Nm (15 ft-lb)
9. **Operating rod**
10. **Lock cylinder housing**
11. **Torx screw**

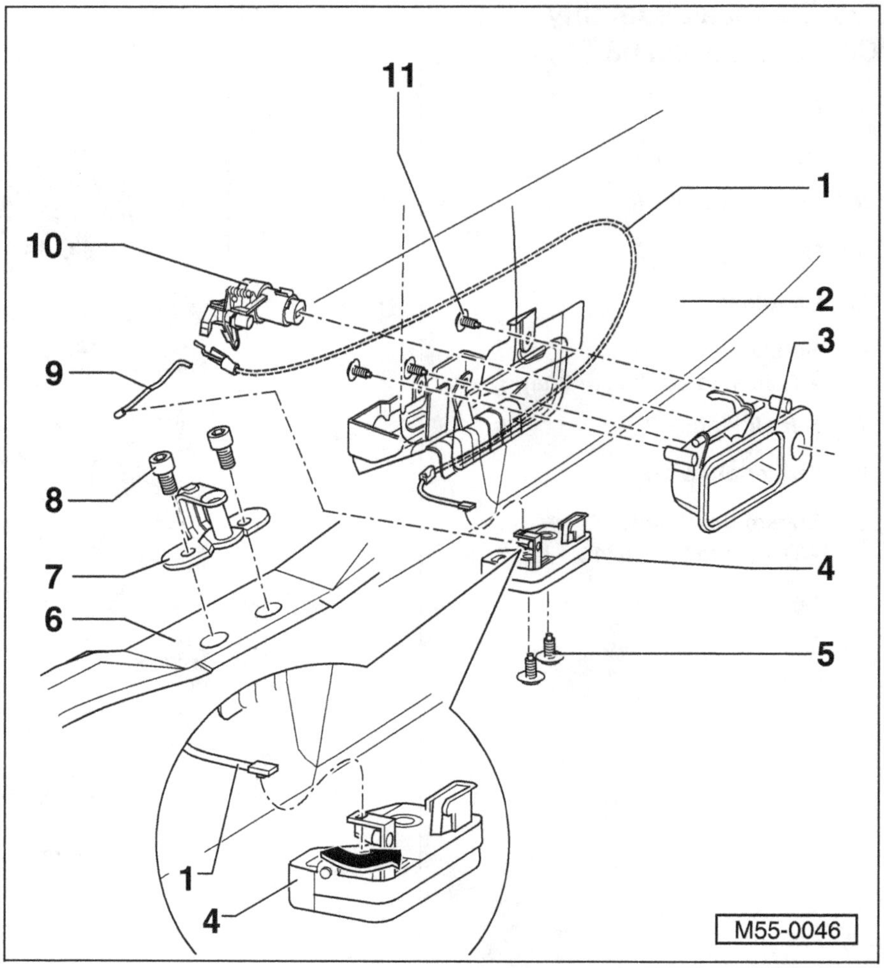

M55-0046

Fig. 8. Rear lid lock assembly for Cabrio models from 04/98 production.

57 Doors–Front

GENERAL

This repair group covers removal and installation of the door assembly, the door glass and window regulators, and service to door lock assemblies, including the central locking system.

> ***NOTE—***
> *Interior door panel removal and installation instructions are given in* **70 Trim–Interior**.

FRONT DOOR, SERVICING

An exploded view of the front door assembly is shown in Fig. 2.

Front door, removing and installing

1. Disconnect harness connector at A-pillar. On vehicles with central locking, disconnect bi-pressure line in harness connector (see Fig. 18, given later).

2. Remove door stop bolt at A-pillar. Remove threaded pin bolt from upper and lower hinges. See Fig. 1.

3. Lift door upward and off of hinge brackets. If necessary, use a helper when lifting off door.

4. Installation is the reverse of removal. Door adjustment should not be necessary unless the hinge mounting bolts have been loosened.

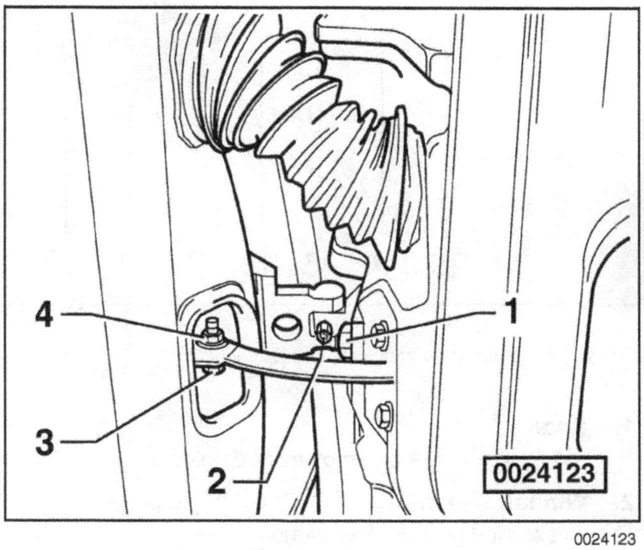

0024123

Fig. 1. Door stop bolt (**3**) and nut (**4**). Threaded pin bolt shown at **2**. Hinge bolt (T45 torx bit) shown at **1**.

Tightening torques
- Hinge to door or body. . . . . 36 Nm (27 ft-lb)
- Hinge screw to hinge . . . . 23 Nm (17 ft-lb)

> ***NOTE—***
> *If the door requires adjustment, the hinge bolts can be loosened and the door can be repositioned in the bolt's elongated holes. A misaligned door can result in wind noise, incorrect latching, and possible paint damage.*

Front door, assembly

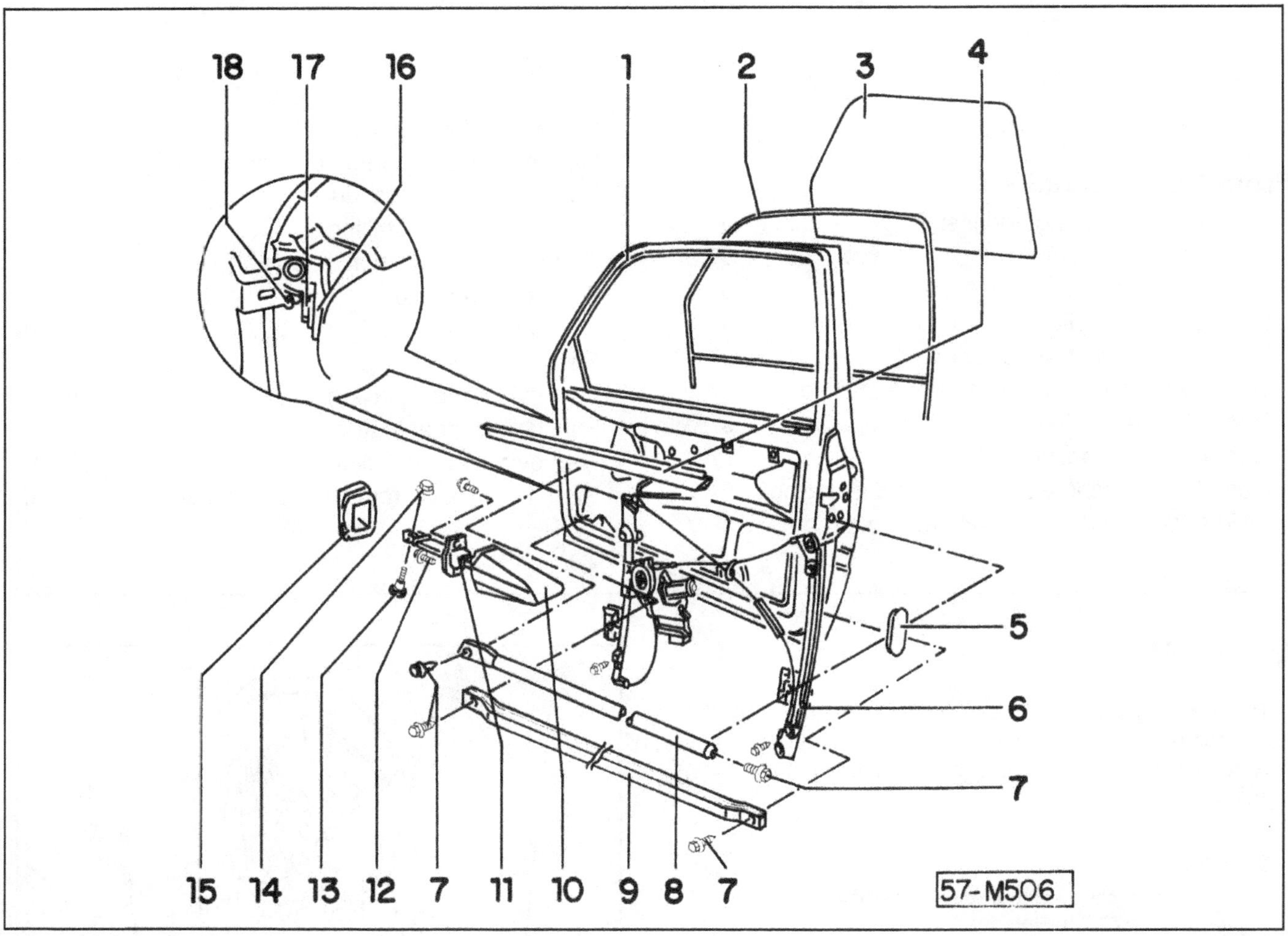

57-M50(

Fig. 2. Exploded view of front door assembly for Golf, GTI, or Jetta (Cabrio is similar).

1. **Door**
   - Adjust at hinge within oversized holes
2. **Window channel**
   - Fastened into window frame
3. **Door window**
   - See **Door glass, removing and installing**
4. **Window slot seal**
   - Pushed onto flange

(continued from previous page)

5. **Cover**
   - To close off impact member installation opening
6. **Window regulator**
   - See **Window regulator, removing and installing**
7. **Hex bolt**
   - Tighten to 23 Nm (17 ft-lb)
8. **Upper impact member**
9. **Lower impact member**
   - From model year 1997
   - Aluminum
10. **Door stop sleeve**
11. **Door stop**
12. **Hex-head screw**
    - Tighten to 7.5 Nm (66 in-lb)
13. **Bolt**
14. **Locking nut**
    - Tighten to 6.5 Nm (58 in-lb)
15. **Door check strap cover**
    - Self-adhesive, attached to A-pillar
16. **Door hinge**
    - Bolted to door and A-pillar
17. **Torx bolt T45**
    - Tighten to 36 Nm (27 ft-lb)
18. **Screw**
    - Tighten to 23 Nm (17 ft-lb)

Door glass, removing and installing

1. Remove interior door panel and vapor barrier. See **70 Trim–Interior**.

> ***CAUTION—***
> *Use care when removing the vapor barrier. Water leaks will occur if the vapor barrier is damaged or wrinkled when installed. Always replace vapor barrier if damaged.*

2. Pry off window slot seal. Lower door glass into door.

3. Remove hex nuts from the front and rear clamping jaws. Press jaws apart. See Fig. 3.

N57-0132

Fig. 3. Rear clamping jaw nut (**1**). Door glass shown lowered into door. Golf, GTI, Jetta shown (Cabrio is similar).

4. Pull door glass upward, tilt toward front and lift out of window slot.

5. Install in the reverse order of removal. Adjust the glass as described below.

Door glass, adjusting (Golf, GTI, Jetta)

1. Remove interior door panel and vapor barrier. See **70 Trim–Interior**.

2. Crank door window up or down until clamping jaws are located in the installation openings. See Fig. 4.

3. Loosen clamping jaw nuts, then push door window into rear window channel and tighten nuts.

Tightening torques
- Clamping jaw hex nuts . . . . 10 Nm (7 ft-lb)

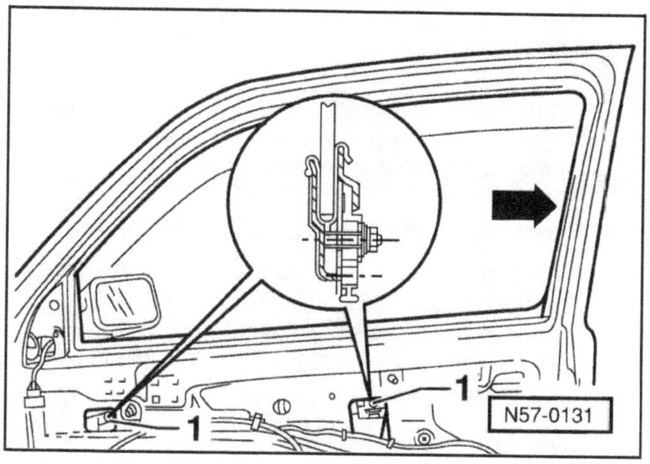

N57-0131

Fig. 4. Clamping jaws positioned in installation openings. Adjust window by pushing into rear window channel (**arrow**). Inset shows clamping jaws.

Door glass, adjusting (Cabrio)

On Cabrio models, the front power window regulators are cable operated. There are two height adjusters per window regulator, one in the front and one in the rear. The height adjustment can be done with the door panel installed by inserting a Torx driver through the openings in the underside of the door frame. The top of the window can also be adjusted inward or outward by repositioning wedges. A wedge is located on each window carrier. See Fig. 5.

> ***NOTE—***
> *Before adjusting the glass, door to body and convertible top to windshield frame adjustments should be correct. See* **61 Convertible Top** *for top adjustments.*

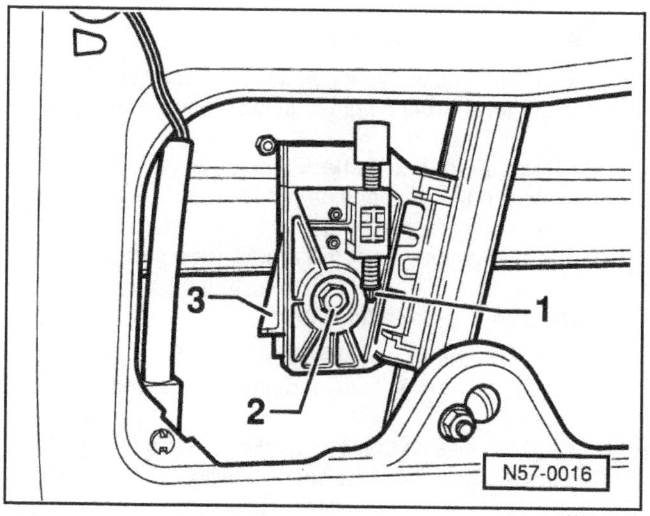

N57-0016

Fig. 5. Cabrio door glass adjustment mechanism as viewed with door panel removed. Height adjustment screw is shown at **1**. Clamping jaw nut shown at **2**. Adjusting wedge shown at **3**.

1. Remove interior door panel and vapor barrier. See **70 Trim–Interior**.

> *CAUTION —*
> *Use care when removing the vapor barrier. Always replace the vapor barrier if damaged.*

2. **Window to front window channel adjustment**:
   - Position window such that clamping jaws are visible in installation opening. Loosen nut on front and rear clamping jaws and loosen jaws. See Fig. 6.
   - Push window firmly into front window channel and tighten clamping nuts.

Tightening torques
- Clamping jaw hex nuts . . . . 10 Nm (7 ft-lb)

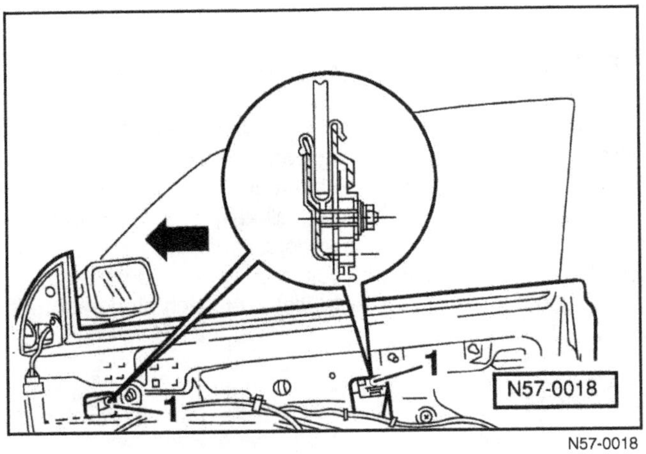

N57-0018

Fig. 6. With window clamping jaws loose, press window into front window channel (direction of **arrow**). Inset shows clamping jaws.

3. **Window height adjustment (rear)**:
   - Lower window. Working through underside of door, fully lower front height adjustment screw using Torx driver. See Fig. 5.
   - Raise window and check that top of glass meets upper edge of door/side window seal. See Fig. 7. If necessary, adjust height by turning rear height adjustment screw at the rear window carrier.

4. **Window B-pillar adjustment**:
   - With clamping jaws visible in installation openings, loosen clamping jaws and height adjustment bolt. See Fig. 8.
   - Adjust glass so that it is parallel to B-pillar seal. Move adjustment wedges to center position (align with lower edge of clamping jaws) and tighten bolt and nuts.
   - With window raised and door closed, check window contour in relation to B-pillar using a strip of paper between glass and glass seal. The paper should be able to be pushed along length of window with the same force. If necessary, adjust window by repositioning adjustment wedge.

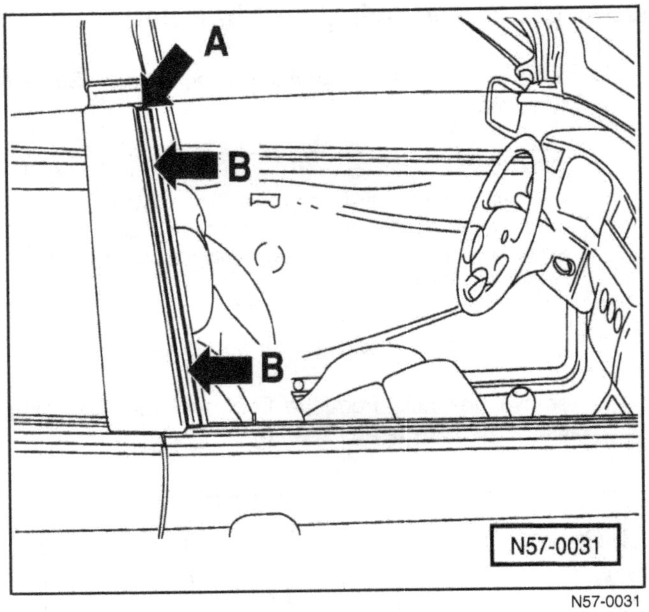

N57-0031

Fig. 7. Adjust rear corner of window so top of glass meets edge of rear door/window seal (**A**). Also adjust gap at B-pillar to be parallel with window seal (**B**).

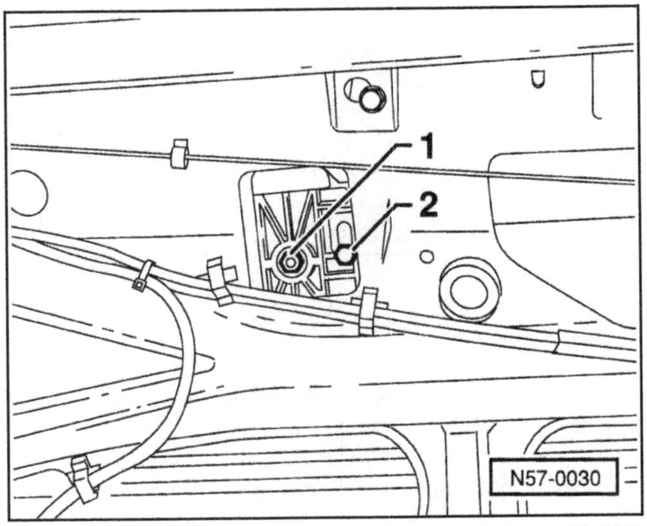

N57-0030

Fig. 8. Rear clamping jaw nut (**1**) and height adjustment bolt (**2**).

5. **Window height adjustment (front)**:
   - Raise door window. Then check for play at front height adjustment/stop screw. Lower window slightly (approx. 2 in.) and turn (raise) front height adjustment screw until no play exists. Raise window and check adjustment. Repeat adjustment as necessary. See Fig. 9.

6. **Window pretension adjustment**:
   - Fully raise window. Loosen tension wedge bolt and press wedge against glass using a screwdriver through opening and tighten bolt. See Fig. 10.

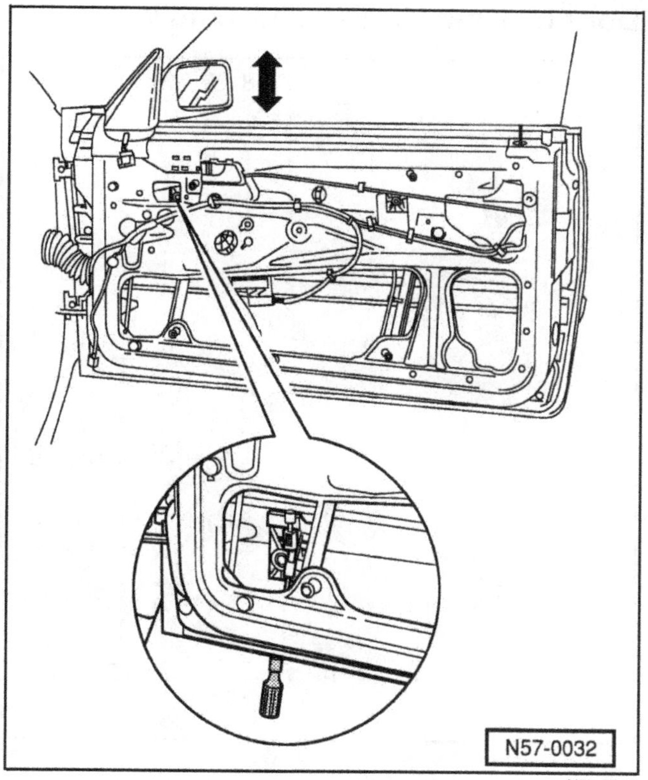

N57-0032

Fig. 9. Adjust front adjustment screw until no play exists when pulling on window (direction of **arrow**). Adjust window height with Torx driver (inset).

Tightening torques

- Hex bolt for pretension wedge . . . . 10 Nm (7 ft-lb)

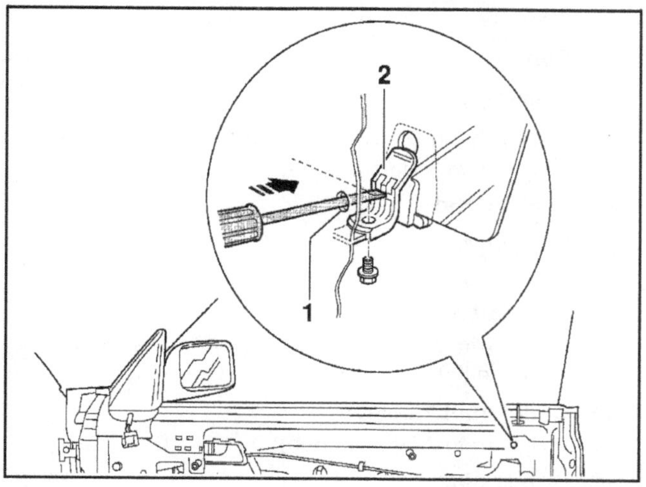

0024097

Fig. 10. Adjust pretension on door glass by pressing wedge (**2**) against glass and tightening bolt.

Window regulator, removing and installing

1. Remove interior door panel and vapor barrier. See **70 Trim–Interior**.

> *CAUTION —*
> *Use care when removing the vapor barrier. Water leaks will occur if the vapor barrier is damaged or wrinkled when installed. Always replace vapor barrier if damaged.*

2. Remove the front window as described earlier. See **Door glass, removing and installing**.
3. Remove regulator mounting screws and nuts. If applicable remove cable clip. See Fig. 11.

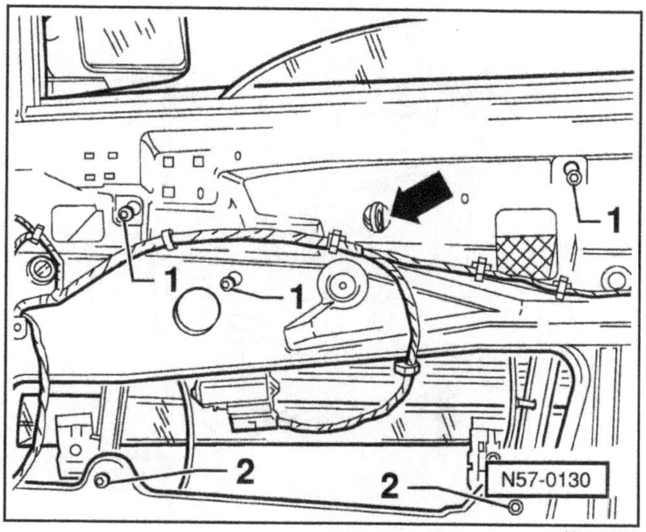

N57-0130

Fig. 11. Remove hex head screws and nuts (**1** and **2**). On cars with electric window regulators, unclip harness retaining clip (**arrow**).

4. Lift window regulator slightly so that screws can be pulled through installation openings.
5. Pull window regulator downward through opening.
6. On electric regulators, disconnect harness connector by sliding connector forward and down off of regulator motor.
7. Install in reverse order of removal.

Tightening torques

- Window regulator to door . . . . 10 Nm (89 in-lb)

Door latch, removing and installing

1. Remove interior door panel and vapor barrier. See **70 Trim–Interior**.

> ***CAUTION —***
> *Use care when removing the vapor barrier. Water leaks will occur if the vapor barrier is damaged or wrinkled when installed. Always replace vapor barrier if damaged.*

2. Close door window. Remove outside door handle as described later. Remove door latch mounting screws. See Fig. 12.

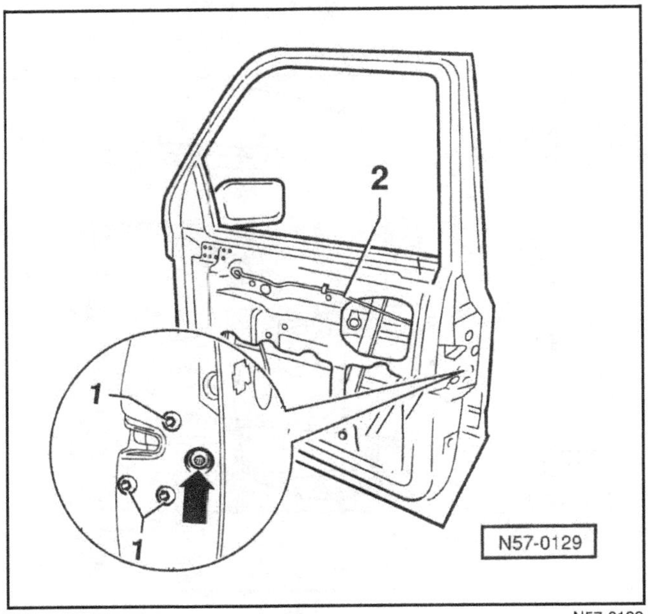

N57-0129

Fig. 12. Door latch screws (**1**), pull rod (**2**). Door latch adjusting screw shown at arrow.

3. Disconnect pull rod from latch and remove latch.

4. Installation is reverse of removal. Loosen latch adjusting screw (see Fig. 12) before installing latch. Tighten adjusting screw after installing door handle.

> ***NOTE —***
> *Screw on left-side latch has right-hand threads. Screw on right-side latch has left-hand threads.*

Tightening torques
- Door latch to body . . . . 8 Nm (71 in-lb)
- Door latch adjusting screw (Torx T20) . . . 3 Nm (27 in-lb

Door lock mechanism, assembly

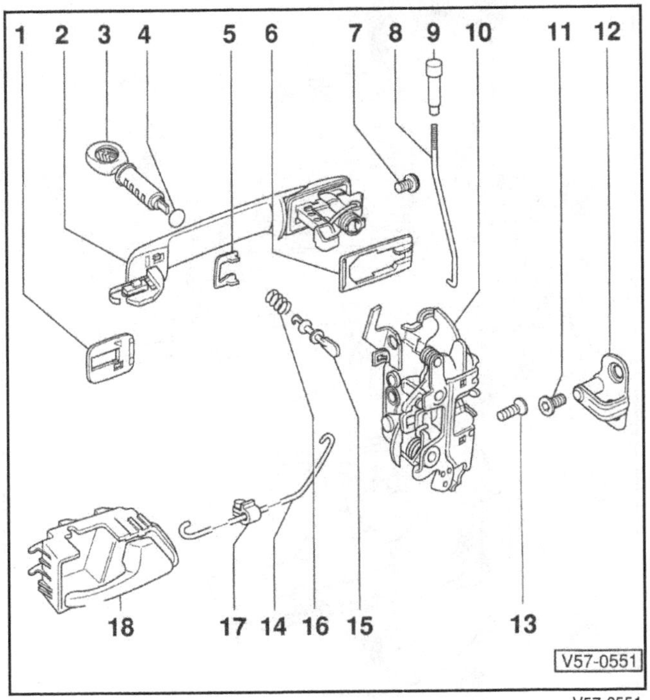

V57-0551

Fig. 13. Exploded view of door lock and operating mechanism.

1. **Backing piece**
2. **Door handle**
3. **Lock cylinder**
   - Only remove with key inserted
   - Lubricate with G 000 400 prior to installation
   - Some 1998 models with factory installed keyless remote may not have passenger side lock cylinder
4. **Seal**
5. **Clip**
6. **Backing piece**
7. **Socket-head screw with collar**
   - Tighten to 8 Nm (71 in-lb)
8. **Locking rod**
9. **Locking button**
   - Threaded on
10. **Door latch**
    - Door window must be closed for removal and installation
11. **Countersunk screw**
    - Tighten to 8 Nm (71 in-lb)
12. **Striker pin**
13. **Socket-head screw with collar**
    - Tighten to 8 Nm (71 in-lb)
14. **Pull rod**
15. **Connecting rod**
    - Attached to lock cylinder housing
16. **Spring**
17. **Clip**
18. **Inside door handle**

Outside door handle, removing and installing

1. Remove socket-head screw from side of door. See Fig. 14.
2. Slide door handle forward and pivot door handle out of door.
   - On driver's door, insert key and turn key 1/4-turn before pivoting handle out of door (when key is turned, door handle carrier releases operating lever.)

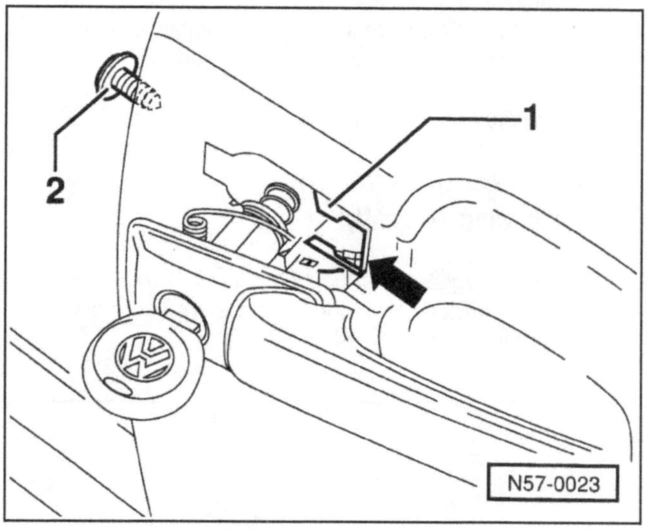

N57-0023

Fig. 14. Door handle mounting screw (**2**). Check clip (**1**) and replace if damaged.

3. Before installing door handle, make sure carrier for connecting rod is in the vertical position. See Fig. 15.
4. Pivot door handle into door and install door handle screw.

Tightening torques
- Door handle socket-head screw. . . . . . 8 Nm (71 in-lb)

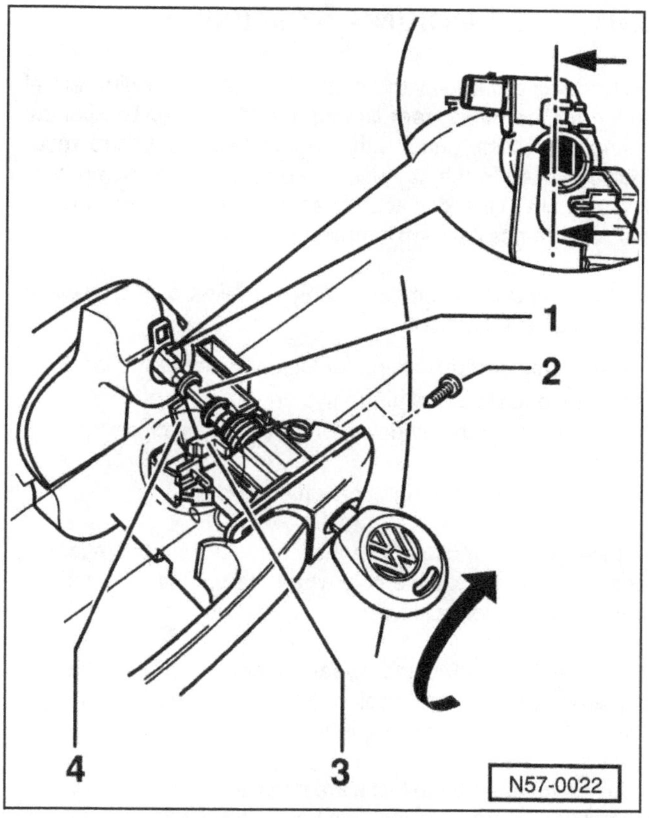

N57-0022

Fig. 15. When installing driver's door handle, connecting rod carrier (**1**) must be in vertical position (inset) before installing handle. Door handle carrier shown at **3** and operating lever shown at **4**.

Inside door handle, removing and installing

1. Depress retaining (locking) clip and then slide handle assembly forward and off door. See Fig. 16.
2. Disconnect link rod from handle and remove from handle.
3. Installation is the reverse order of removal.

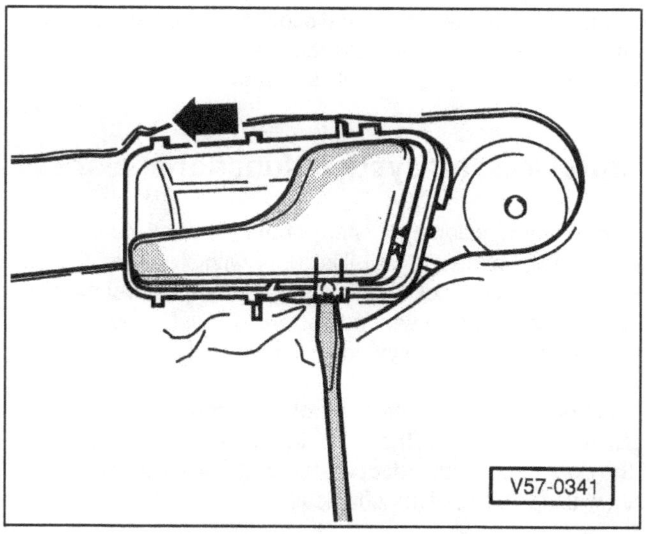

V57-0341

Fig. 16. Pry out retaining clip with screwdriver and slide handle forward (**arrow**) to remove.

CENTRAL LOCKING SYSTEM

All models covered by this manual are equipped with central locking for the doors, rear lid and fuel filler door. The central locking bi-pressure pump with integral electronic control module are located in the luggage compartment. The pump supplies vacuum to the lock actuators for locking and pressure for unlocking, hence the term bi-pressure. See Fig. 18.

The driver's door lock assembly contains a micro-switch that activates 4 functions:

- Activates central locking for lock and unlock
- Arms and dis-arms alarm system
- Activates power windows for convenience close and open
- Closes sunroof during convenience close

These same functions are also activated by the passenger's side door lock assembly when the passenger's side is equipped with a lock cylinder.

Vehicles from 1998 model year are equipped with a factory installed keyless remote system and may not have a lock cylinder installed in the passenger's side outside door handle.

Servos to operate the locks are integrated into the door lock assemblies. Some versions also have servos for the gas flap and the trunk or hatch. The servos are connected to the bi-pressure pump through a series of hard plastic pipes and flexible rubber hoses that are sometimes glued together. Starting with mid-1995 production, all versions use quick disconnect couplers on all components. Most of the parts supplied as replacements also use the quick disconnect couplers. Several adapter fittings are available to mate the newer style quick disconnect servos to the older style systems.

Central locking system, functional test

A properly functioning system should lock all doors, luggage compartment and fuel filler door within approx. 2 seconds. If the central locking system pump and control module runs for longer than 5 seconds, there is most likely a vacuum/pressure leak in the system.

If the pump shuts off after approx. 5 seconds and starts up again after 12 seconds, the most likely fault is that at least one of the door knobs was not operated or the actuator switching point was not reached (mechanically).

1. Begin the functional test with doors, rear lid and fuel tank flap closed and door windows completely open.
2. Push down lock knob on driver door. Check locking on every door, rear lid and fuel tank flap.
   - All lock knobs must go down.
   - All doors and the rear lid should be locked.
   - Fuel tank flap should be locked.
3. To test anti-theft system, lock driver's door with key. then quickly pull all lock knobs up within 12 seconds.
   - 12 seconds after locking, the central locking system pump and control module must run, and all doors should be locked again.
4. After another 15 seconds, try to pull all lock knobs up in succession (wait at least 24 seconds after locking with key).
   - It must not be possible to pull the lock knobs into the unlocked position. The pump starts up and pulls the lock knob back. At the same time, the door lock is deactivated and it is no longer possible to open the door. In this process, there is an audible locking sound as the door locks are deactivated.
5. Repeat the test at the passenger door.

Pump and control module, removing and installing

1. Working in luggage compartment, open wire ties or disconnect retainer straps and remove pump/control module housing covers. See Fig. 17.

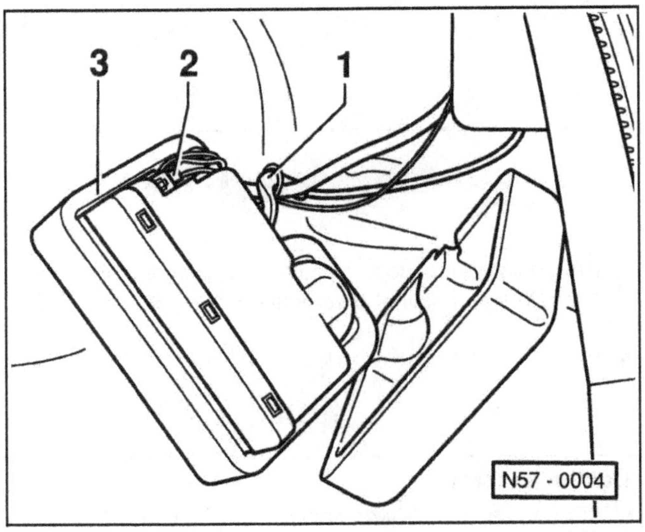

N57-0004

Fig. 17. Central locking system pump and control module (**2**), housing covers (**3**) and multi-pin connector with bi-pressure hose (**1**). Note the bi-pressure hose connection when removing multi-pin connector.

2. Separate multi-pin connector and bi-pressure hose from pump.
3. Installation is the reverse of removal. When connecting multi-pin connector, make sure bi-pressure hose is also properly connected.

Central locking system, assembly

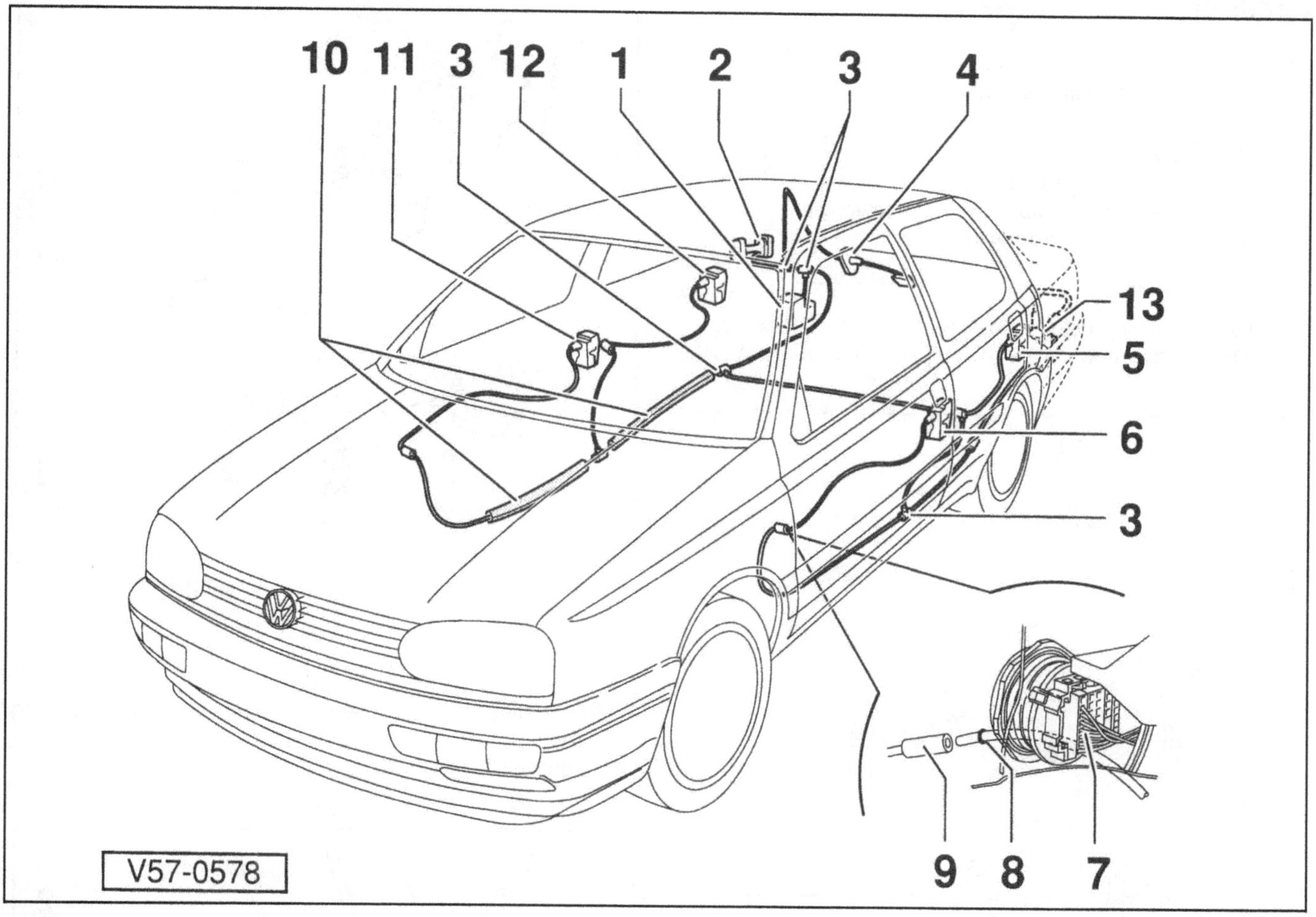

V57-0578

Fig. 18. Central locking system components.

1. **Central locking system pump and control module (V69) (Golf)**
   - Located in rear luggage compartment, right hand side
   - See **Pump and control module, removing and installing**
2. **Actuator, fuel tank door**
   - To remove, loosen luggage compartment trim near wheel housing
   - See **Fuel tank door actuator, removing and installing**
3. **T-connector**
   - Location: near B-pillar/rocker panel, covered by sill molding
4. **Actuator, rear lid**
   - See **Rear lid actuator, removing and installing**
5. **Actuator, rear door lock**
   - See **Rear door actuator, removing and installing**
6. **Actuator, front door lock**
   - Includes driver's door central locking system switch
   - See **Front door actuator, removing and installing**
7. **Multi-pin connector**
   - Location: near A-pillar (also near B-pillar on 4-door vehicles)
8. **Color marking**
   - Bi-pressure hoses must be pushed into bi-pressure hose connectors (9) up to color marking
9. **Bi-pressure hose connector**
   - Two types:
     Integral with multi-pin connector (7), or separate
   - Integral type:
     Multi-pin connector must be separated to allow connection of bi-pressure hose
   - Separate type:
     Bi-pressure hose connector (9) must be pushed back into multi-pin connector (7) after bi-pressure hose is connected
10. **Foam tube**
    - Passenger's side only
11. **Actuator, front door lock**
    - Includes passenger's door central locking switch
12. **Actuator, rear door lock**
13. **Central locking system pump and control module**
    - See **Pump and control module, removing and installing**

Front door actuator, removing and installing

1. Remove interior door panel and partially remove vapor barrier. See **70 Trim–Interior**.

> ***CAUTION—***
> *Use care when removing the vapor barrier. Water leaks will occur if the vapor barrier is damaged or wrinkled when installed. Always replace vapor barrier if damaged.*

2. Disconnect bi-pressure hose and harness connectors from actuator and central locking switch. See Fig. 19.

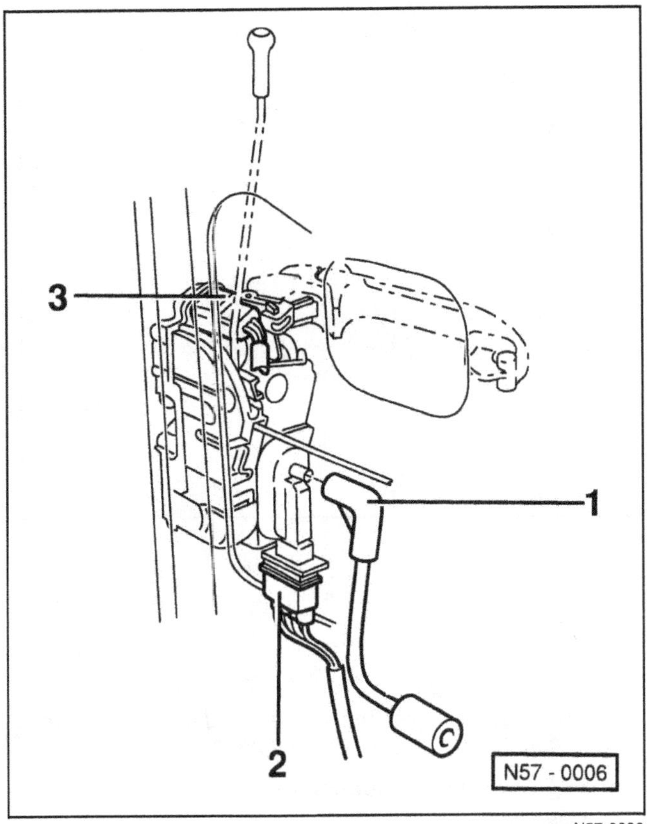

N57-0006

Fig. 19. Front door actuator. Bi-pressure hose shown at **1** and harness connectors shown at **2** and **3**.

3. Remove door latch/lock with actuator, see **Door latch, removing and installing**.

NOTE—
On four-door vehicles the window regulator should be temporarily locsened for better access.

4. Separate actuator from lock assembly;
   - Turn lock rotor to "closed" position. Then remove Torx screw (T15) and pry back retainers while separating actuator from lock. See Fig. 20.

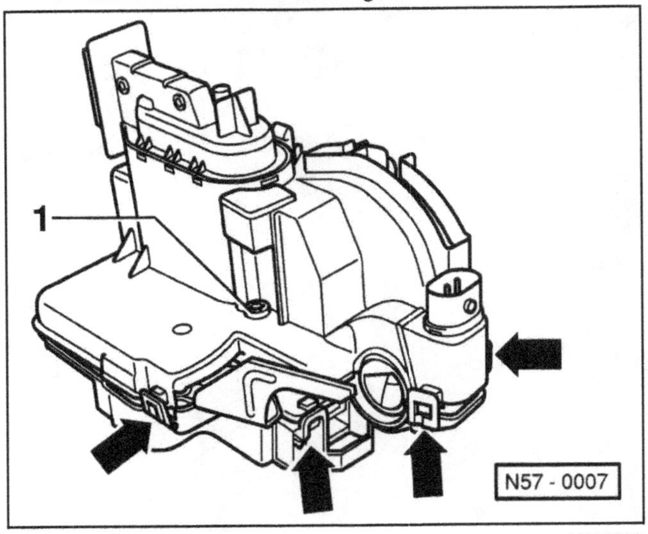

N57-0007

Fig. 20. Separate actuator from door latch/lock by removing Torx screw (**1**) and prying back retainers (**arrows**).

5. Installation is the reverse of removal. When installing actuator to lock assembly, check that locking lever engages slot in striker. See Fig. 21.

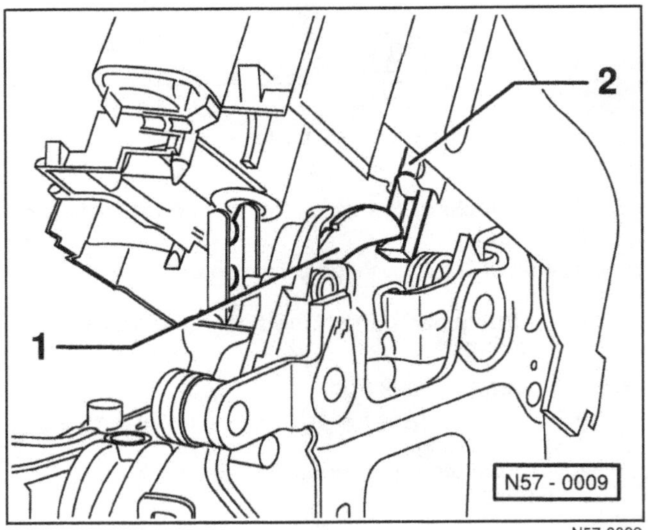

N57-0009

Fig. 21. When installing actuator, slide locking lever (**1**) into striker slot (**2**).

NOTE—
If the door latch/lock does not operate properly after the actuator is replaced, the complete assembly should be replaced as a unit.

Rear door actuator, removing and installing

1. Remove interior door panel and partially remove vapor barrier. See **70 Trim–Interior**.

> *CAUTION —*
> *Use care when removing the vapor barrier. Water leaks will occur if the vapor barrier is damaged or wrinkled when installed. Always replace vapor barrier if damaged.*

2. Disconnect bi-pressure hose from actuator and separate multi-pin connector. Unhook pull rod and lock rod from actuator. See Fig. 22.

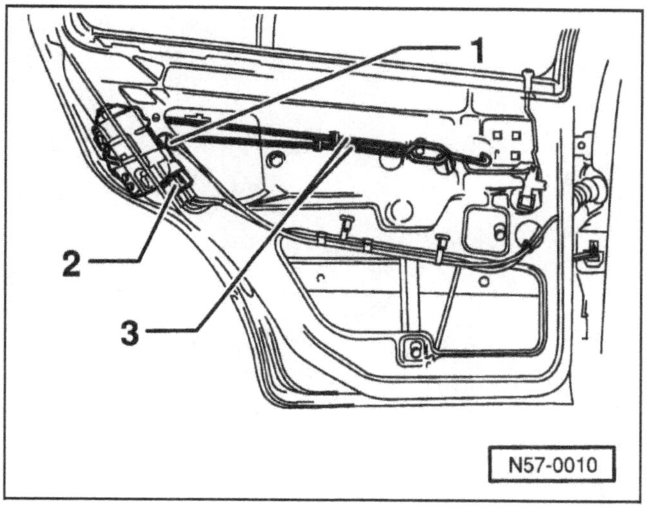

N57-0010

Fig. 22. Disconnect hose (**1**) and electrical connector (**2**) from rear door actuator. Pull and lock rods shown at **3**.

3. Unbolt lock from door and remove rear door lock with actuator.

4. Separate actuator from lock assembly;
   - Turn lock rotor to "closed" position. Then remove Torx screw (T15) and pry back retainers while separating actuator from lock. See Fig. 20.

5. Installation is the reverse of removal. When installing actuator to lock assembly, check that locking lever engages slot in striker.

> *NOTE —*
> *If the door latch/lock does not operate properly after the actuator is replaced, the complete assembly should be replaced as a unit.*

Fuel tank door actuator, removing and installing

1. Remove central locking system pump and control module as described earlier.

2. Partially remove luggage compartment trim on fuel tank door side.

3. Remove actuator screws and pull actuator with locking rod out of stop. See Fig. 23.

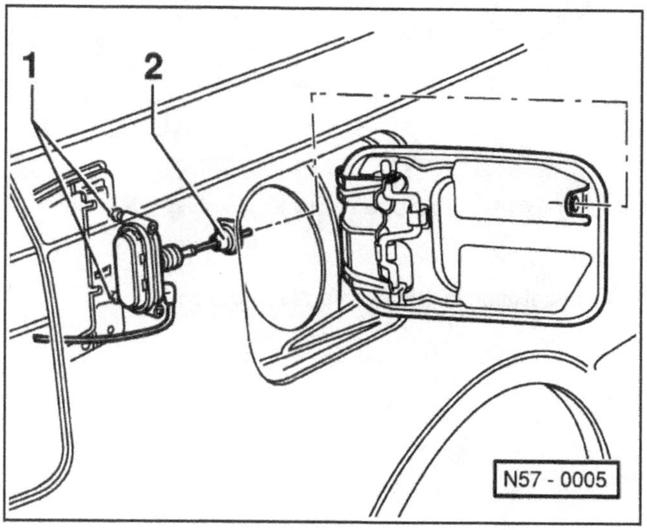

N57-0005

Fig. 23. Fuel tank door actuator mounting screws (**1**) and stop (**2**).

4. Disconnect bi-pressure hose from actuator.

5. Installation is the reverse of removal.

Rear lid actuator, removing and installing

1. Remove handle and trim piece from rear lid.

2. Disconnect locking rod from rear lid latch push button, and disconnect bi-pressure hose.

3. Remove actuator mounting screws and pull actuator with locking rod out of rear lid. See Fig. 24.

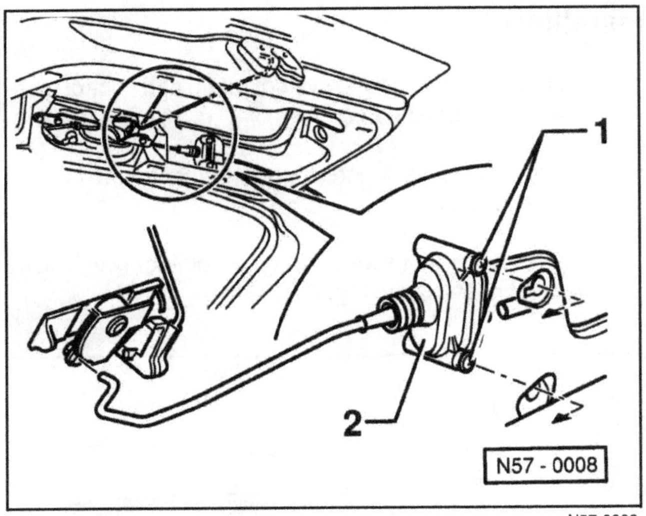

N57-0008

Fig. 24. Rear lid actuator (**2**) and actuator mounting screws (**1**).

4. Installation is the reverse of removal.

58 Doors–Rear

GENERAL

This repair group covers removal and installation of the rear door assembly, and removal and adjustment of the door glass.

> ***NOTE—***
> - *Service to door lock assemblies, including the central locking system is covered in* **57 Doors–Front**.
> - *Interior door panel removal and installation instructions are given in* **70 Trim–Interior**.

REAR DOOR, SERVICING

An exploded view of the rear door assembly is shown in Fig. 2.

Rear door, removing and installing

1. Disconnect harness connector at B-pillar. Disconnect central locking bi-pressure line in harness connector.

> ***NOTE—***
> *Tape or wire tie the removed harness connector to the B-pillar. If the connector falls into the pillar, it will be difficult to retrieve.*

2. Remove door check strap bolt at B-pillar. Remove threaded pin bolt from upper and lower hinges. See Fig. 2.
3. Lift door upward and off of hinge brackets.
4. Installation is the reverse of removal.

Tightening torques
- Hinge to door or body . . . . 40 Nm (30 ft-lb)
- Hinge screw to hinge . . . . 23 Nm (17 ft-lb)

> ***NOTE—***
> *If the door requires adjustment, the hinge bolts can be loosened and the door can be repositioned in the bolt's elongated holes. A misaligned door can result in wind noise, incorrect latching, and possible paint damage.*

Door glass, removing and installing

1. Remove interior door panel and vapor barrier. See **70 Trim–Interior**.

> ***CAUTION—***
> *Use care when removing the vapor barrier. Water leaks will occur if the vapor barrier is damaged or wrinkled when installed. Always replace vapor barrier if damaged.*

2. Remove nuts from glass clamping jaws. Press jaws apart. See Fig. 1.

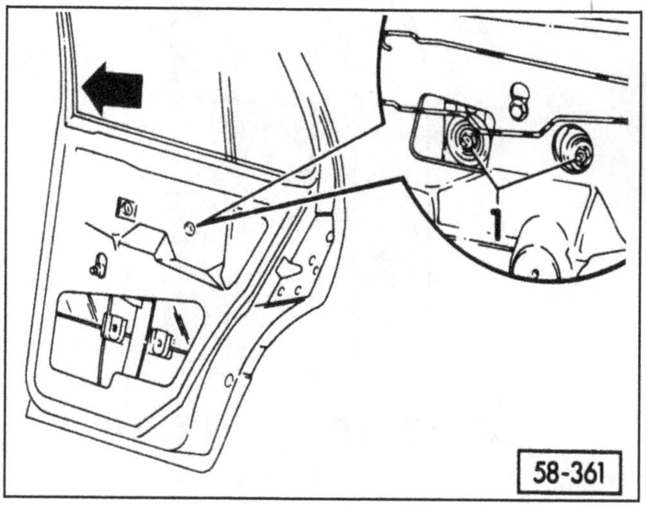

58-361

Fig. 1. Glass clamping jaw nuts (**1**). Door glass shown lowered into door.

3. Lower door glass into door. Pull window channel out of door frame and remove center connector. See Fig. 2.
4. Pull door glass upward, tilt toward front and lift out of window slot.
5. Install in the reverse order of removal. Adjust the glass as described later.

Tightening torques
- Clamping jaw hex nuts . . . . 10 Nm (7 ft-lb)

Rear door, assembly

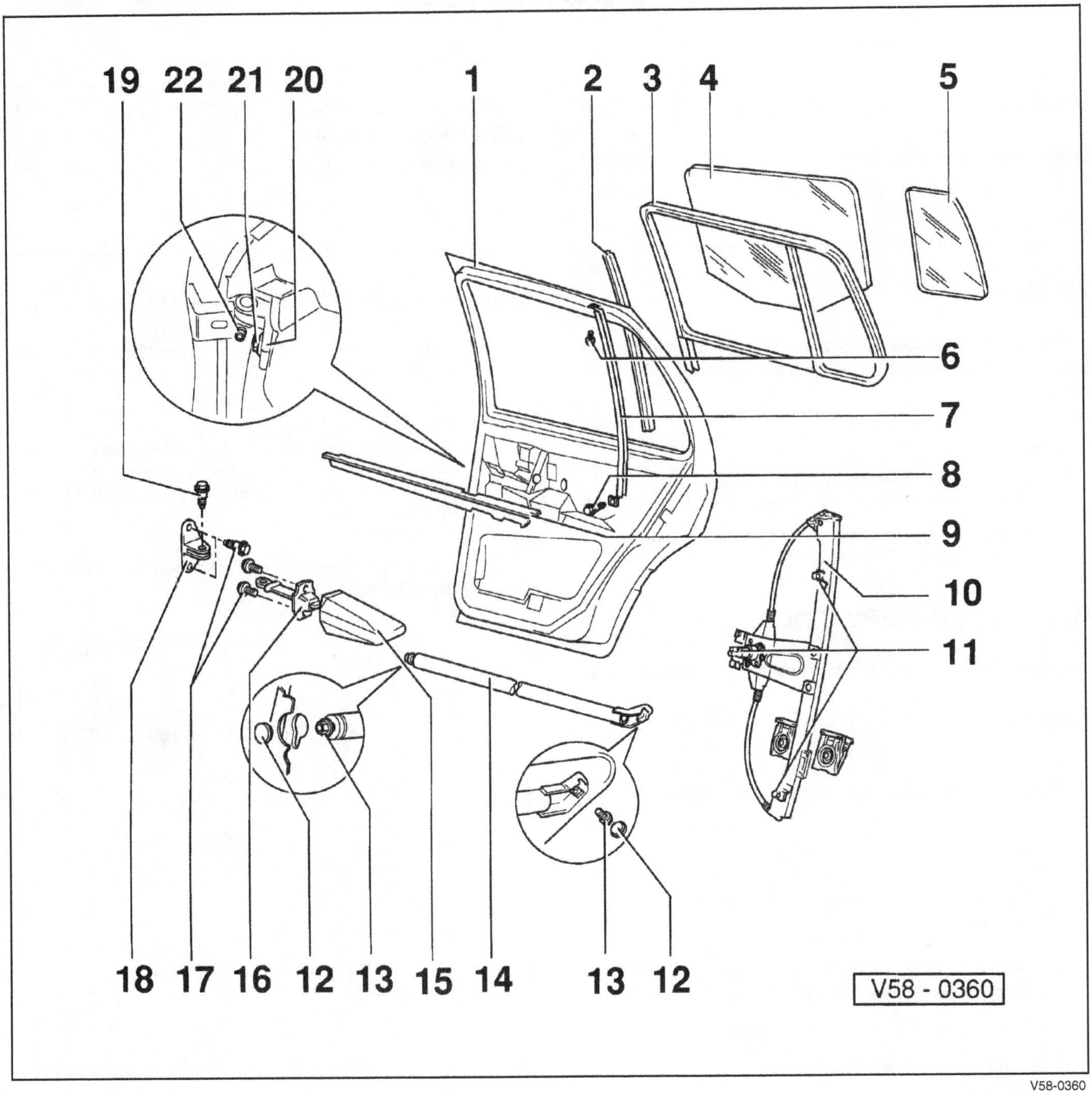

V58-0360

Fig. 2. Exploded view of rear door assembly.

1. **Door**
   - Adjust at hinge within over-sized holes
2. **Window channel connector**
3. **Window channel**
4. **Door window**
5. **Vent window**
6. **Screw**
   - Tighten to 3.5 Nm (31 in-lb)
7. **Center connector**
8. **Hex-head screw**
   - Tighten to 6 Nm (53 in-lb)
9. **Window slot seal**

(continued from previous page)

10. **Window regulator**
11. **Hex-head screws**
    - Only loosened for removal of window regulator
    - Tighten to 10 Nm (7 ft-lb)
12. **Cover**
13. **Bolt**
    - Tighten to 25 Nm (18 ft-lb)
14. **Impact member**
15. **Door check strap sleeve**
16. **Door check strap**
17. **Hex-head screw**
    - Tighten to 7.5 Nm (66 in-lb)
18. **Console**
19. **Bolt**
    - Tighten to 9 Nm (80 in-lb)
20. **Door hinge**
21. **Bolt (Torx T45)**
    - Tighten to 40 Nm (30 ft-lb)
22. **Threaded pin bolt**
    - Tighten to 23 Nm (17 ft-lb)

Door glass, adjusting

1. Remove interior door panel and vapor barrier. See **70 Trim–Interior**.

> ***CAUTION —***
> *Use care when removing the vapor barrier. Water leaks will occur if the vapor barrier is damaged or wrinkled when installed. Always replace vapor barrier if damaged.*

2. Crank door window up or down until glass clamping jaws are located in the installation openings. See Fig. 3.

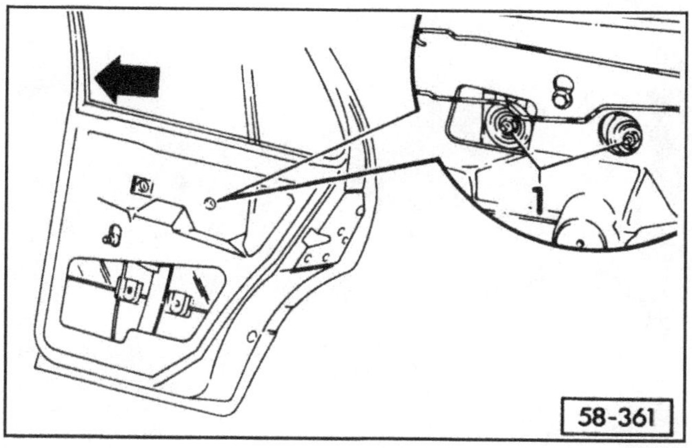

58-361

Fig. 3. Clamping jaws positioned in installation openings. Adjust window by pushing into front window channel (**arrow**). Inset shows clamping jaws.

3. Loosen clamping jaw nuts, then push door window into front window channel and tighten nuts. The remainder of installation is the reverse of removal.

Tightening torques
- Clamping jaw hex nuts . . . . 10 Nm (7 ft-lb)

60 Sunroof

GENERAL

The power sunroof is a two-way design that tilts open or slides back into the roof. The sunroof is a glass panel with an interior sliding shade (headliner). An electronic control unit is combined with the sunroof motor and regulates the motor operation.

SUNROOF, SERVICING

Fig. 3 shows an exploded view of the sunroof assembly.

Sunroof glass panel, removing

1. Tilt sunroof open and slide shade to rear.

2. Unclip and slide trim frame to rear. Trim frame is clipped into front guide. See Fig. 1.

3. Remove screws (Torx T25) and adjusting plates from each corner of sunroof panel. See Fig. 2.

4. Carefully push sunroof panel upward and out.

NOTE —

If necessary, remove trim frame by pulling forward and upward out of roof cutout.

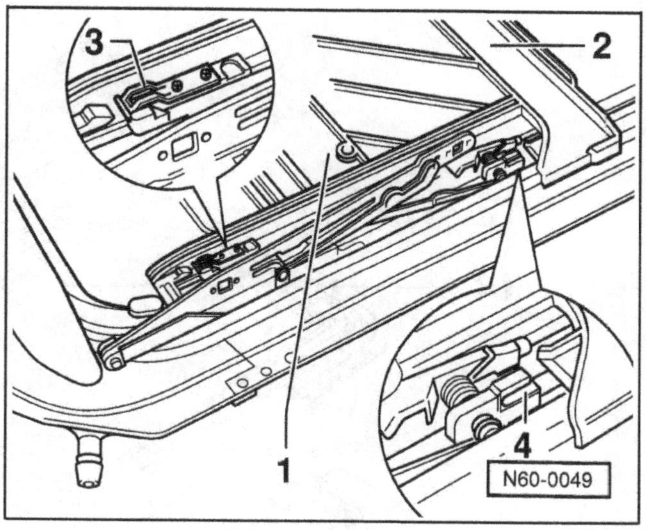

N60-0049

Fig. 1. Slide headliner (**1**) and trim frame (**2**) to rear. Insets show trim frame clip (**3**) and trim frame guide (**4**).

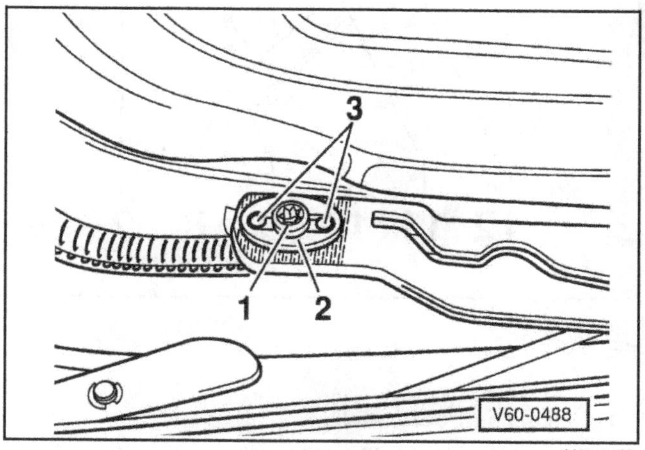

V60-0488

Fig. 2. Sunroof mounting screws (**1**). Note position of mounting plates (**2**) in reference to guide pins (**3**) for installation. Front right mounting screw shown.

60

Sunroof, assembly

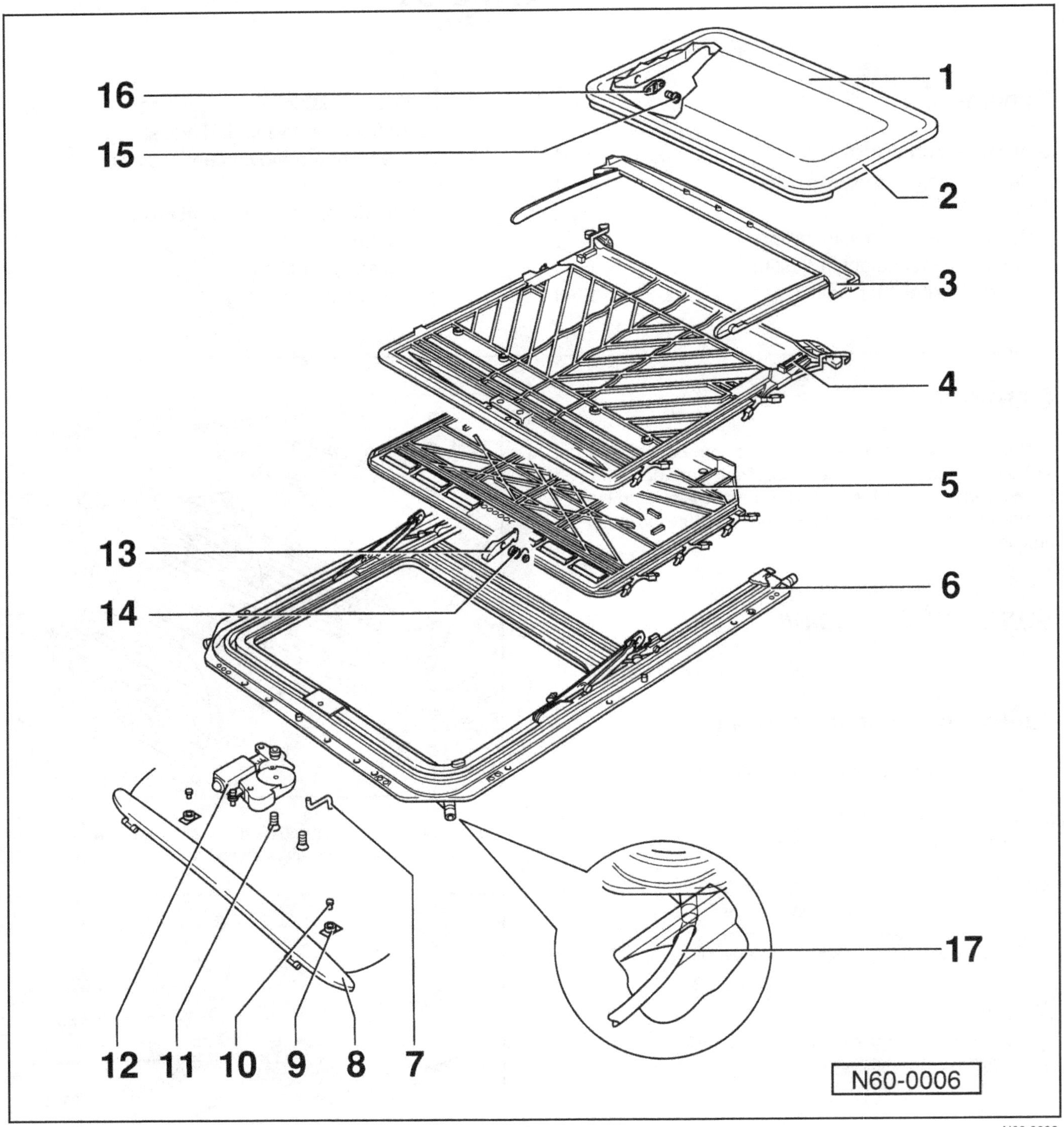

N60-0006

Fig. 3. Exploded view of sunroof assembly.

1. **Glass sunroof panel**
   - Single pane green safety glass

2. **Sunroof panel seal**
   - Seal joint is in center of right side section
   - Use water to lubricate panel seal clamping channel before installing

(continued from previous page)

3. **Trim frame**
   - See **Sunroof glass panel, removing**
4. **Headliner (cars built up to 8/94)**
5. **Headliner (cars built from 9/94)**
   - See **Sunroof headliner, removing and installing**
6. **Carrier unit**

   Carrier unit consists of the following parts:
   - U frame with guide channels
     (If required, guide channels are to be lubricated with special grease (VW part no. G 000 450 02) only. Multipurpose grease must not be used under any circumstances.)
   - Rear guides with cables
   - Front guides

   Some individual parts are available for the carrier unit

> ***CAUTION —***
> *Part numbers are listed for reference only. Always check with an authorized Volkswagen parts department for the latest parts information.*

7. **Hex key for emergency operation**
   - Clipped onto cable driver
8. **Wind deflector**
   - Removing: open panel and remove screws (10)
9. **Threaded fitting**
   - Thread must be at an angle toward the rear
10. **Screw**
    - For wind deflector (8)
11. **Countersunk self-locking screws**
    - Always replace
    - Tightening torque 3 Nm (27 in-lb)
12. **Sunroof motor with cable drive and sunroof control module**
    - See **Cable drive closed position, checking**
13. **Locating hooks**
14. **Locating hook spring**
15. **Torx screw**
    - Torx T25
    - Tighten to 6 Nm (53 in-lb)
16. **Adjusting plate**
    - Quantity: 4
17. **Water drain hose**
    - See **Water drain hoses, cleaning**

Sunroof glass panel, installing

1. Begin with sunroof panel in the closed position.
2. Check that locating hooks and pins on both sides of carrier unit are correctly positioned. See Fig. 4.

> ***NOTE —***
> *If the pins are not correctly aligned, see* **Parallel running, checking and adjusting.**

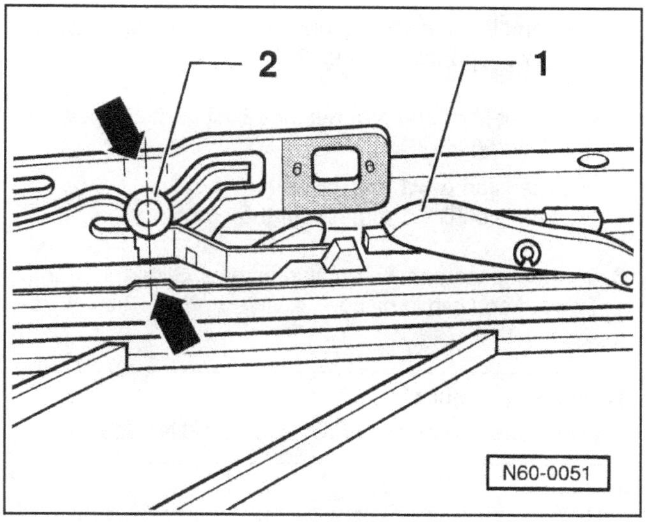

N60-0051

Fig. 4. Before installing sunroof panel, locating hooks (**1**) must be in guide rails and pins (**2**) must be within marks or notches (**arrows**).

3. Place sunroof in position from above and install adjusting plates and securing screws. Lightly tighten screws (Torx T25).

> ***NOTE —***
> *When installing adjusting plates, make sure guide pins are in elongated holes of adjusting plate. See Fig. 2.*

4. Position panel to correct height in opening and tighten sunroof panel securing screws. See **Sunroof panel height, adjusting.**

Tightening torques
- Sunroof panel to carrier unit
  (Torx T25 screws) . . . . 6 Nm (53 in-lb)

Sunroof panel height, adjusting

Before adjusting sunroof panel height, ensure that sunroof runs parallel, see **Parallel running, checking and adjusting**.

1. Tilt sunroof panel open and push back sunroof headliner.
2. Unclip and slide trim frame to rear. Trim frame is clipped into front guide. See Fig. 1, given earlier.
3. Close glass panel.
4. Loosen all four screws (Torx T25) and adjusting plates on sunroof panel. See Fig. 2.
5. Carry out front and rear height adjustment as shown in Fig. 5 and 6.
   - Dimension A = 1 mm lower than roof
   - Dimension B = 1 mm higher than roof
6. Tighten glass panel mounting screws. Slide trim frame forward and clip in place.

Tightening torques
- Glass panel mounting screws .... 6 Nm (53 in-lb)

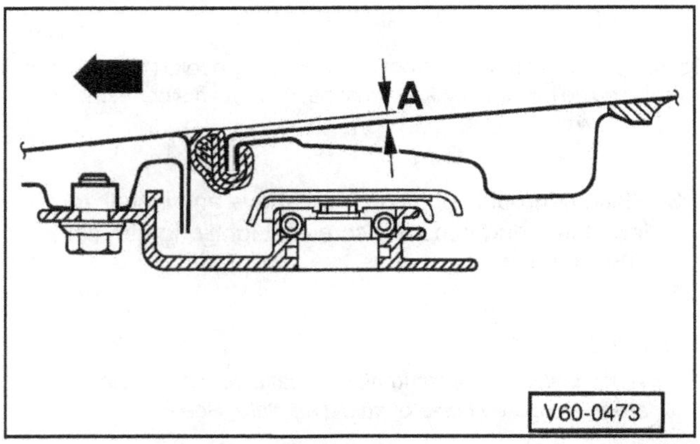

V60-0473

Fig. 5. Front of sunroof panel should be 1mm (dimension **A**) lower than body (roof). Arrow indicates direction of travel.

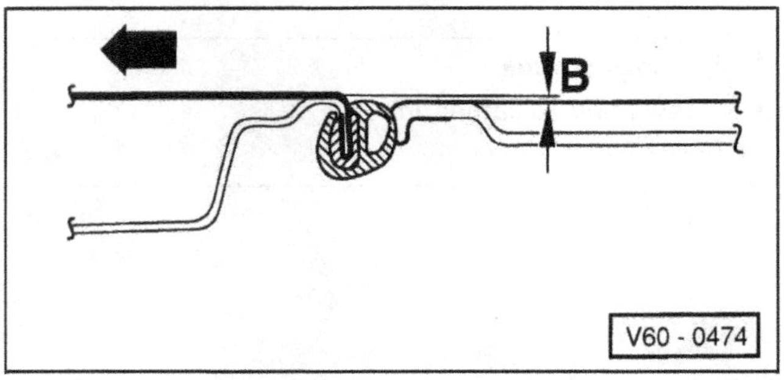

V60-0474

Fig. 6. Rear of sunroof panel should be 1mm (dimension **b**) higher than body (roof). Arrow indicates direction of travel.

Sunroof panel seal, adjusting

The preload between the sunroof panel seal and the body should be uniform on all sides. See Fig. 7.

Panel seal gap
- Dimension a .... 0.3 mm (0.012 in.)

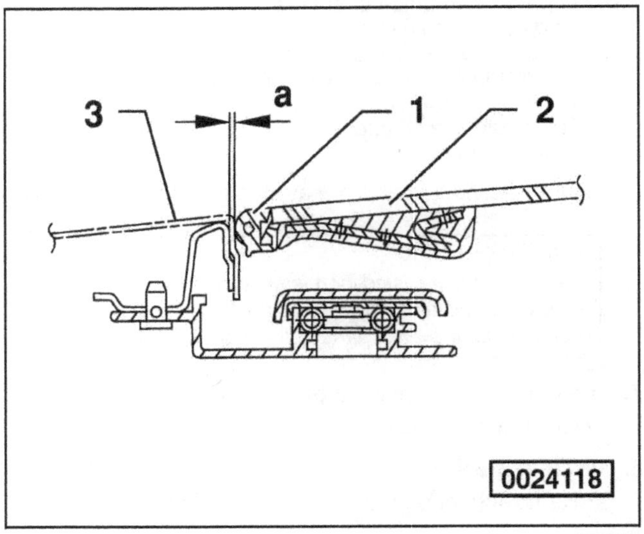

0024118

Fig. 7. Gap (**a**) between panel seal (**1**) and body (**3**) should be 0.3 mm (0.012 in), or about the thickness of a business card.

Check panel seal preload using strip of paper 0.3 mm (0.012 in.) thick, such as a business card. When seal is correctly adjusted, it should be possible to pull paper strip through, between panel seal and body, with some resistance.

If necessary, remove sunroof panel to adjust seal preload. Spread seal apart using a wedge to increase preload or push seal together to reduce preload.

Cable drive closed position, checking

This procedure should be carried out when replacing the sunroof motor/cable drive. The purpose of this procedure is to position the motor in the closed position before installing it to the cable drive. Before making the check, ensure that the sunroof runs parallel, see **Parallel running, checking and adjusting**.

1. Begin with sunroof panel closed and motor removed from sunroof carrier, connect power harness connector to motor.
2. Operate sunroof switch, and count cable drive pinion rotations until end-switch-off position is reached. See Fig. 8.

3. Operate sunroof switch again to reach closed position.
   - The closed position is 2 full turns from the tilting switch-off position (**A** in shown Fig. 8) and 8.25 turns from the open end-switch-off position (**B** shown in Fig. 8).

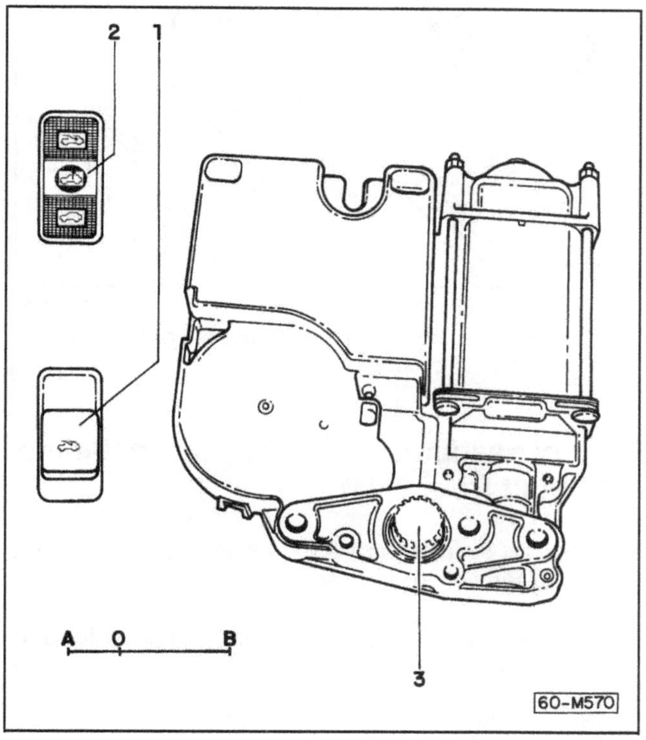

60-M570

Fig. 8. Drive pinion shown at **3**. Sunroof motor switch-off positions represented at **A** (tilt) and **B** (full-open). Closed position represented at **O**.

4. Install sunroof motor to sunroof carrier using new self-locking screws.

> ***CAUTION —***
> *Always use new self-locking screws.*

Parallel running, checking and adjusting

1. Begin with sunroof panel closed and sliding headliner (sunshade) pushed back.

2. Unclip and slide trim frame to rear. Trim frame is clipped into front guide. See Fig. 1, given earlier.

3. Check alignment of left and right locating hooks and pins in relation to carrier. See Fig. 9.
   - Pins must be aligned within marks on carrier and locating hooks must be engaged in guide rails.

4. If cable drive is out of parallel alignment, disconnect the negative (-) battery cable.

5. Remove motor access cover by sliding toward rear. See Fig. 10.

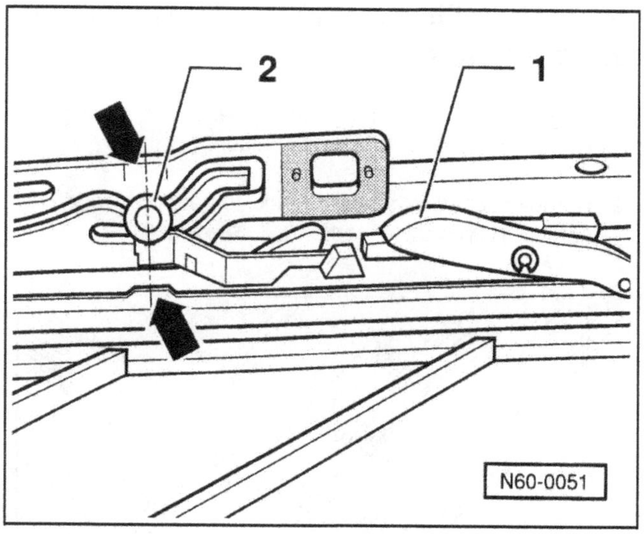

N60-0051

Fig. 9. Locating hooks (**1**) must be in guide rails and pins (**2**) must be within marks or notches (**arrows**) for parallel operation.

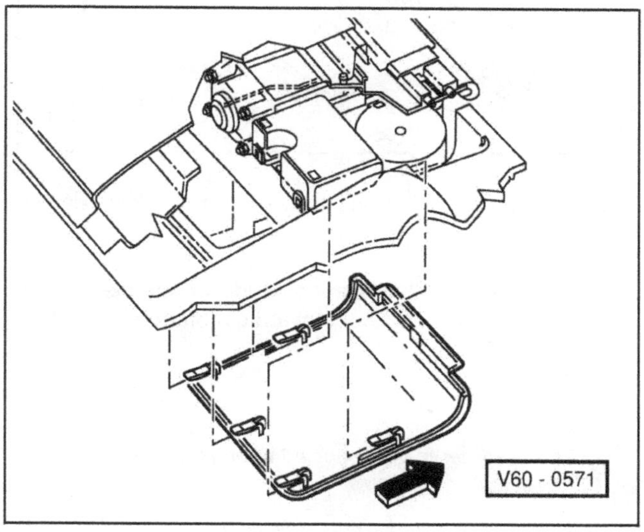

V60-0571

Fig. 10. Remove motor access cover by sliding in direction of **arrow**.

6. Remove sunroof motor mounting screws and motor.

7. Align cable drive for parallel operation. See Fig. 9.
   - Engage locating hooks in guide rails.
   - Slide guide pins from front to rear to align with marks.

8. Install sunroof motor using new self locking screws.

> ***CAUTION —***
> *Always use new self-locking screws.*

Tightening torques
- Sunroof motor mounting screws. . . . . . 3 Nm (27 in-lb)

Sunroof headliner, removing and installing (cars built up to 8/94)

1. Remove sunroof glass panel as described earlier. Position sunroof unit in closed position.
2. Remove screw that mounts slide to headliner (one side only). Pull slide out of carrier unit. See Fig. 11.

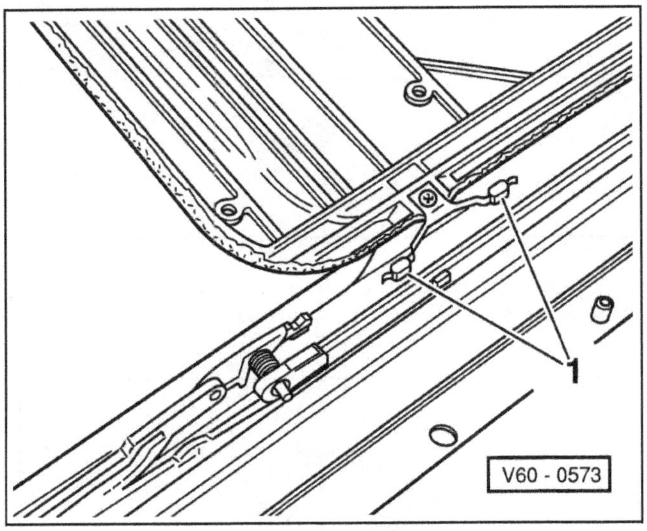

V60-0573

Fig. 11. Remove screw that mounts slide (**1**) (remove from one side only).

3. Remove screws that release headliner springs on both sides and pull headliner forward out of roof cutout. See Fig. 12.

NOTE —
- *The headliner must be angled when removing from roof cutout.*
- *When installed, springs are difficult to see.*

4. Before installing headliner, remove left and right springs from headliner.
5. Insert headliner in roof cutout.
6. Insert headliner slide in guide channel. See Fig. 11.
7. Insert slide on other side into guide channel and secure with new self-locking screw.
   - Always use new self-locking screws.
8. Install left and right springs with new self-locking screws.
9. Slide headliner to rear, until springs engage.

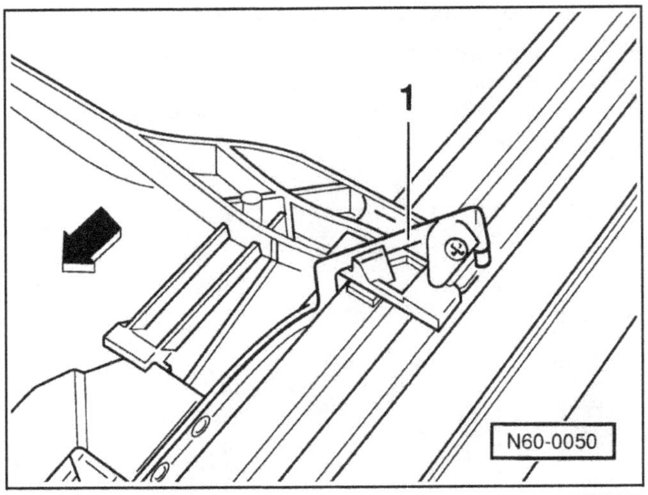

N60-0050

Fig. 12. Release left and right springs (**1**) and pull headliner forward and out of roof cut-out.

Sunroof headliner, removing and installing (cars built from 9/94)

1. Remove sunroof glass panel as described earlier. Position sunroof unit in closed position.
2. Unclip stop on one side and press headliner towards this side. See Fig. 13.

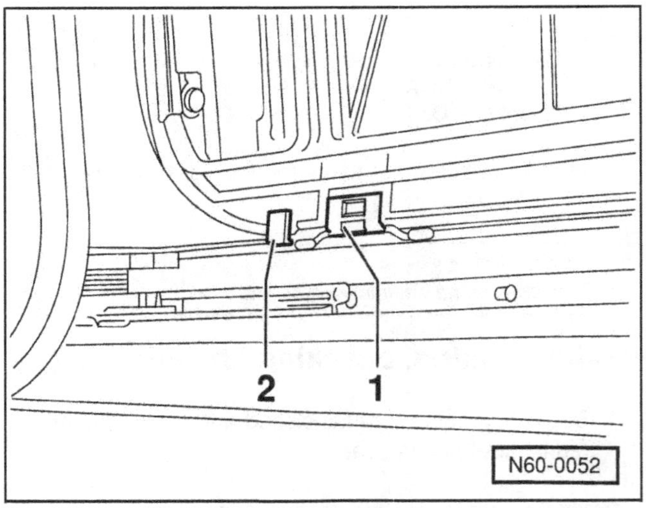

N60-0052

Fig. 13. Unclip stop (**2**) on one side, press headliner toward this side. Slide shown at (**1**).

3. On opposite side, carefully pry slide out of guide channel, working from front to rear.
4. Swing headliner out of assembly frame.

5. To install, insert headliner in the roof cutout.

6. Insert headliner slide into guide channel (slide engaged into headliner).

7. Insert slide on opposite side into guide channel. Fit side stops.

8. The remainder of installation is the reverse of removal.

Sunroof noise, correcting

Noise created while driving with the sunroof open may be caused by some amount of adhesive that has not adhered properly and is hitting the reinforcement crossbar. This problem was corrected in production from 08.94, starting with VIN# 3VWRA81H8SM010425. A pair of foam strips should be installed on each side of the sunroof assembly between the metal roof and the crossbar. See Fig. 14.

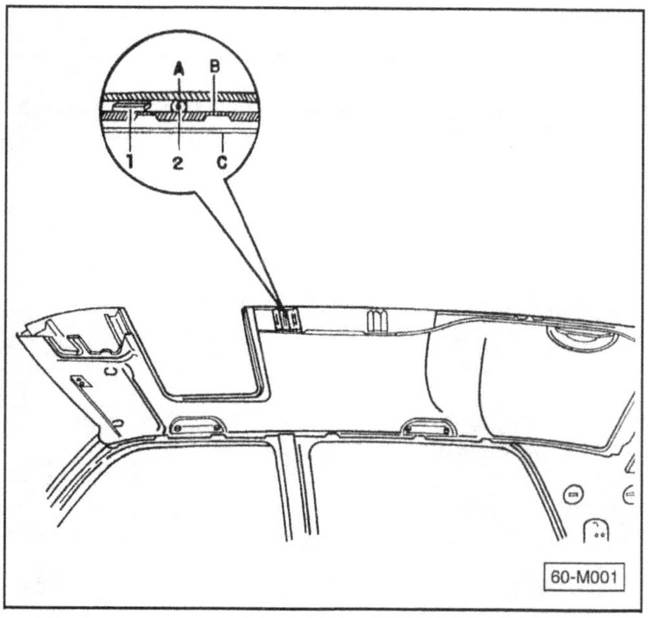

60-M001

Fig. 14. Sunroof noise on vehicles produced up 08.94 could be caused by adhesive (**2**) that is not adhered properly and is hitting the reinforcement crossbar (**B**). Two foam strips (**1**) should be installed on the left and right sides of the sunroof between the metal roof (**A**) and crossbar (**B**).

Foam strip dimensions
- Thickness . . . . 4 mm (5/32 in.)
- Length . . . . 5 cm (2 in.)
- Width . . . . 2 cm (3/4 in.)

NOTE —

*In order to install the foam strips it will be necessary to remove the A-, B-, and C-pillar trim panels and the molded headliner as described in **70 Trim–Interior**.*

For vehicles produced from 08.94 noise from the sunroof heard while driving on rough roads may be caused by vibration between the sunroof trim frame and the headliner. A new sunroof trim frame has been installed in production from 08.94. From 09.94 production, the noise absorbers on the trim frame have been relocated. Vehicles manufactured during 08.94 may still exhibit some noise and vibration problems. If so, the sunroof trim frame must be replaced with the later version. See Fig. 15.

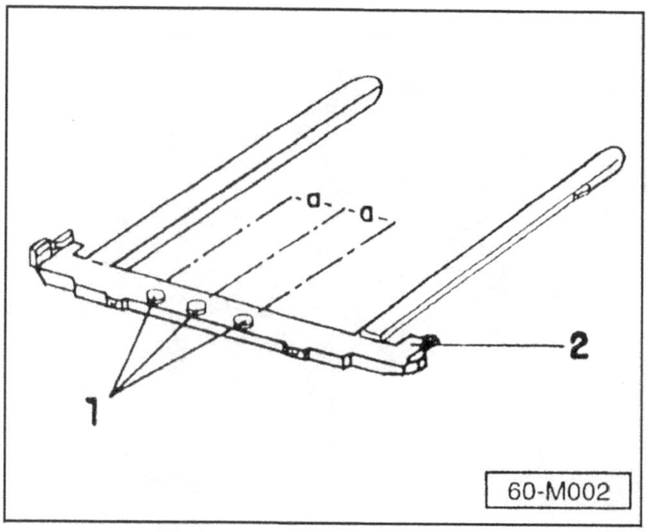

60-M002

Fig. 15. Position of noise absorbers (dimension **a**) on sunroof trim frame was increased as of 09.94 production to further help reduce sunroof noise.

Sunroof noise absorber specifications
- During 08.94 production
  Dimension **a**, Fig. 15. . . . . .106 mm (4.17 in.)
  Part number 357 877 261 01C
- From 09.94 production
  Dimension **a**, Fig. 15. . . . . .140 mm (5.51 in.)
  Part number 6NO 877 261

To remove the sunroof trim frame see Fig. 1. given earlier in this repair group.

Water drain hoses, cleaning

The front water drain hoses are routed through the A-pillars and terminate between the door and A-pillar. Cleaning is done from the sunroof panel opening. See Fig. 16.

> ***NOTE —***
> *A flexible cable, approx. 2300 mm (7.5 ft) long, is best for clearing the drain hoses.*

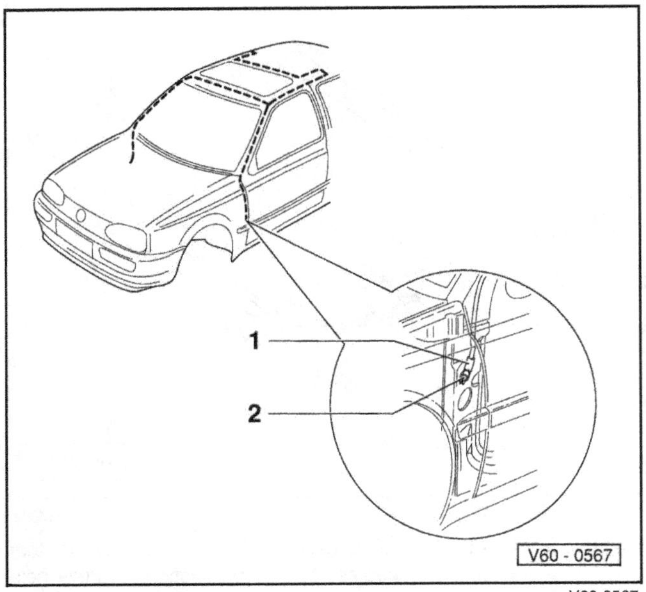

V60-0567

Fig. 16. Front water drain hoses (**1**) and water drain valve (**2**).

The rear water drain hoses are routed in the C-pillars and terminate at the sides behind the bumper cover. See Fig. 17. Cleaning the rear hoses should be done from the lower end, after removing the bumper, as described in **63 Bumpers**.

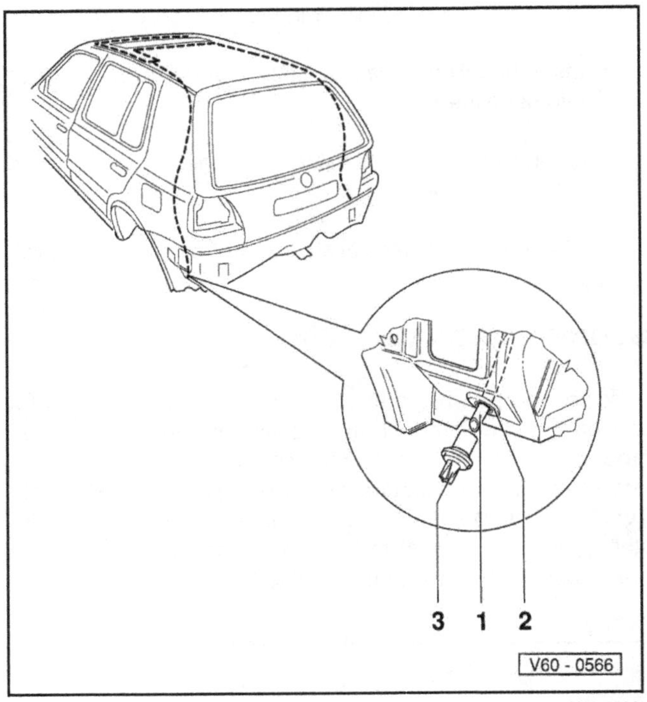

V60-0566

Fig. 17. Rear water drain hoses (**1**) with grommet (**2**) and drain valve (**3**).

61 Convertible Top

GENERAL

This section covers adjustment, removal, and replacement of both the manual and the hydraulically operated convertible top. Some procedures described require specialized tools and adhesives and an assistant.

The multi-layered convertible top consists of the outer top cover, the mounting frame, the top padding, headlining, side wind deflectors, and the heated glass rear window.

On manual versions, a gas strut slows the top movement during storage and assists when unfolding the top. Early versions use a pair, while later versions use a single strut on the left side only. The top is held in the stowed position in its compartment by snap hooks on the frame. Both snap hooks are operated simultaneously from the left side of the vehicle via a Bowden cable.

On power versions, a rear mounted hydraulic pump (V82) operates the top mechanism by way of double-acting hydraulic cylinders on each side. The hydraulic pump is controlled by a dash mounted switch (E137), a relay (J321), and a control module (J256). The control module has no OBD capabilities.

In order to help reduce body shake, also known as "cowl shake", two elastically suspended masses are used; one at the front, and one at the rear.

The front mass is the engine/transmission assembly supported with special mountings; four on vehicles with an automatic transmission, and five on vehicles with a manual transmission. These mounts allow the engine/transmission assembly to dampen fore and aft (longitudinal) shake.

In the rear, a countermass of approximately twenty pounds is located in the left rear area of the trunk. This mass, also known as a vibration counterweight, is also elastically suspended with special mountings and serves to dampen side to side (lateral) shake.

Vehicles with a power top have the hydraulic pump mounted in the left rear as a countermass in place of the inert countermass. See Fig. 1.

Convertible tops are produced in two major versions, both can be manual or power operated. For vehicles produced through 09/2000, See Fig. 2. For production from 10/2000, See Fig. 3.

Vibration counterweight (countermass)

1. **Nut, hex head**
   - Qty. 1
   - Tighten to 46 Nm (34 ft-lb)
2. **Trunk trim panel, left rear**
3. **Screw, Phillips head**
   - Qty. 2
4. **Vibration countermass (inert countermass)**
   - With mounting bracket
5. **Nut, hex head**
   - Qty. 2
   - Tighten to 27 Nm (20 ft-lb)
6. **Shock tower, left rear**

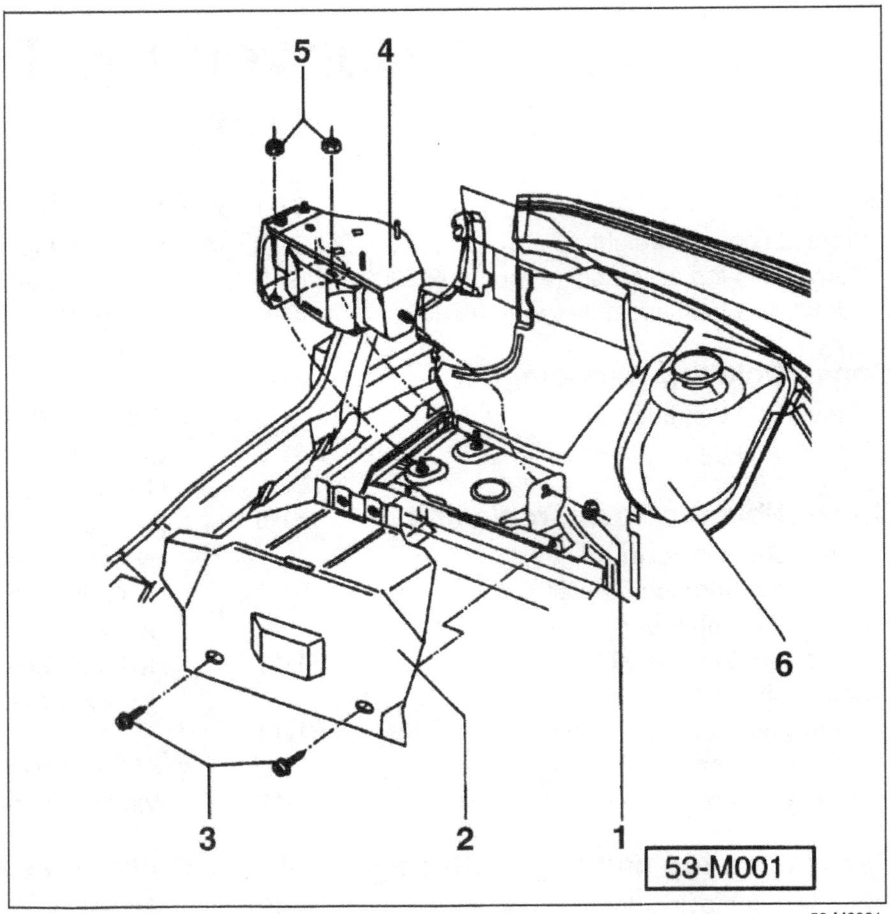

53-M0001

Fig. 1. Left rear body area on vehicles with a manually operated convertible top.

Convertible top assembly (through 09/2000)

1. **Convertible top frame**
2. **Padding**
3. **Convertible top cover**
4. **Rear window frame**
5. **Hinge**
6. **Screw**
   - Tighten to 8 Nm (71 in-lb)
7. **Bolt, hex**
   - Tighten to 8 Nm (71 in-lb)
8. **Seal, convertible top frame**
9. **Bolt**
   - Tighten to 8 Nm (71 in-lb)
10. **Sealing band**
11. **Convertible top fabric mount**
12. **Bolt, hex**
    - Tighten to 50 Nm (37 ft-lb)
13. **Convertible top catch pin**
14. **Convertible top catch**
15. **Lining (headliner)**
16. **Retaining rail**
17. **Roof frame seal**
18. **Nut, speed**
19. **Screw, self tapping**
    - Tighten to 1.5 Nm (13 in-lb)
20. **Screw, self tapping**
    - Tighten to 1.5 Nm (13 in-lb)
21. **Clamping strip**

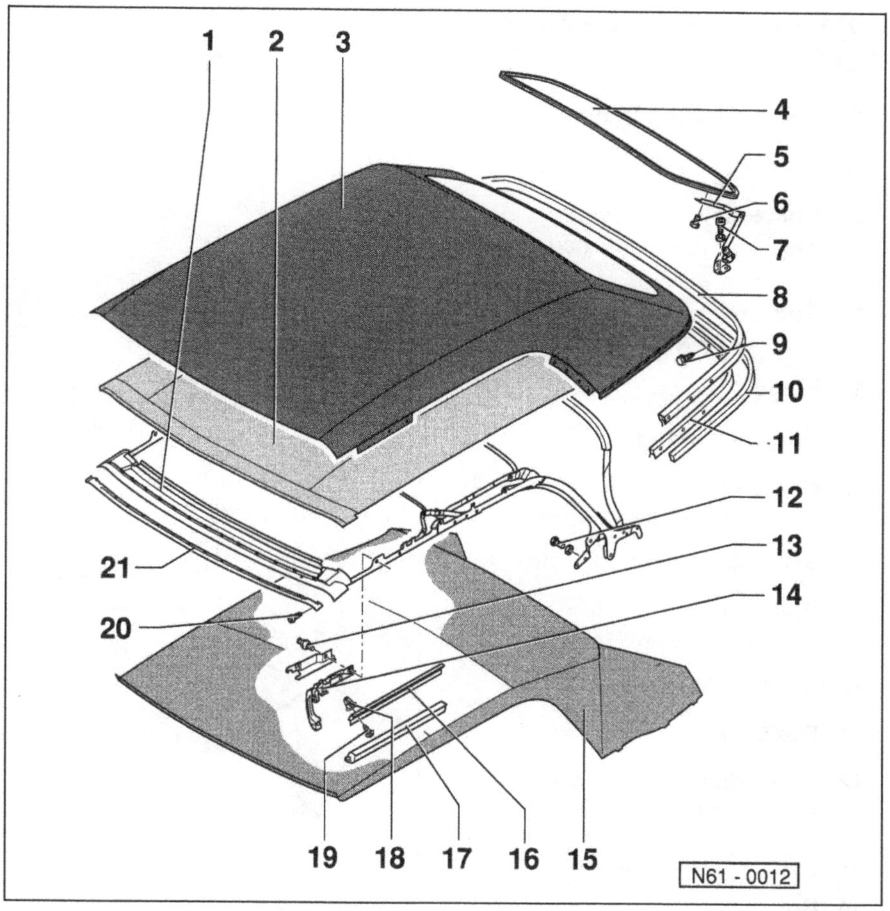

N61-0012

Fig. 2. Exploded view of convertible top components for vehicles produced through 09/2000.

Convertible top assembly (from 10/2000)

1. **Convertible top frame**
2. **Padding**
3. **Convertible top cover**
   - Includes integral heated rear window glass
4. **Bolt, hex**
   - Tighten to 8 Nm (71 in-lb)
5. **Hinge**
6. **Hinge mounting**
7. **Bolt, hex**
   - Tighten to 8 Nm (71 in-lb)
8. **Seal, convertible top frame**
9. **Bolt, hex**
   - Tighten to 8 Nm (71 in-lb)
10. **Connecting plate**
11. **Sealing band**
12. **Convertible top fabric mount, split frame**
13. **Pop rivet**
14. **Bolt, hex**
    - Tighten to 50 Nm (37 ft-lb)
15. **Lining (headliner)**
16. **Retaining rail**
17. **Roof frame seal**
18. **Nut, speed**
19. **Screw, self tapping**
    - Tighten to 1.5 Nm (13 in-lb)
20. **Convertible top catch**
21. **Convertible top catch pin**
22. **Screw, self tapping**
    - Tighten to 1.5 Nm (13 in-lb)
23. **Clamping strip**

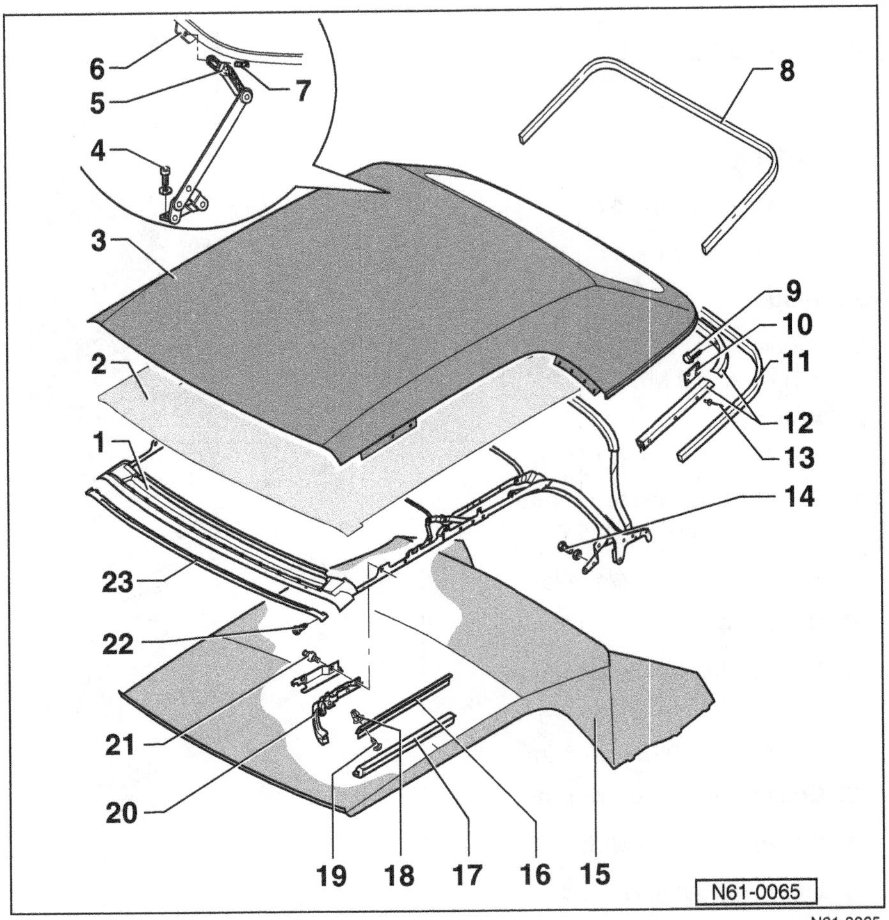

N61-0065

Fig. 3. Exploded view of convertible top components for vehicles produced from 10/2000.

CONVERTIBLE TOP, SERVICING

Convertible top, adjusting

Proper convertible top adjustment can help reduce and/or eliminate water leaks and wind noises.

1. Check leading edge of convertible top in relation to windshield frame. Top should be about 5 mm (0.20 in.) below edge of windshield frame. See Fig. 4.

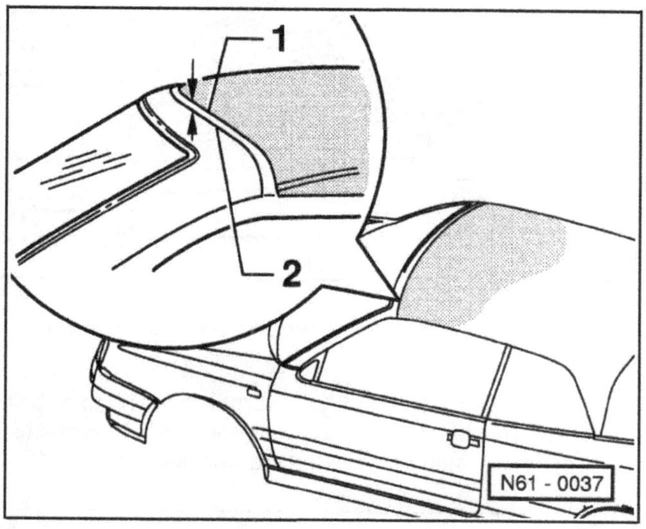

N61-0037

Fig. 4. Leading edge of top (**1**) should be seated about 5 mm (0.20 in.) below (**arrows**) windshield frame (**2**).

2. To adjust top position, open convertible top and loosen self-locking securing hook screws. Adjust positions of securing hooks to obtain correct height and tighten screws. See Fig. 5.

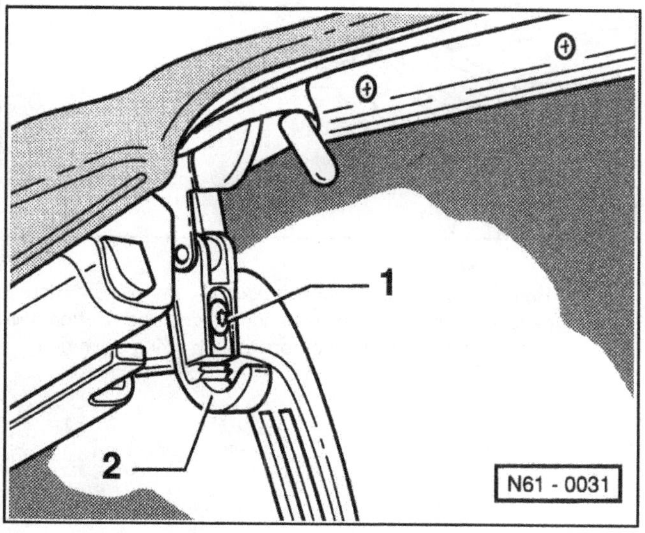

N61-0031

Fig. 5. Loosen Torx (T20) screws (**1**) to adjust securing hook (**2**). Repeat for other side.

Tightening torques

- Securing hook mounting screws . . . . . 5 Nm (44 in-lb)

Check door window sealing by closing window on a strip of paper. The paper should be able to be pushed along the length of the window seal with the same force. If necessary, adjust door glass. See **57 Doors-Front**.

Convertible top, removing

1. With top up, release rear seat backrests (backrest lower section) and fold down.

2. Remove backrest upper part and pull upper part upward, off bulkhead. Place backrest with seat belts in luggage compartment. See Fig. 6.

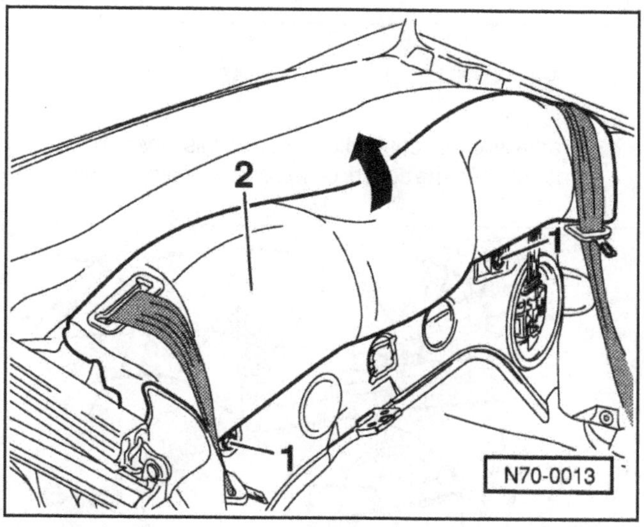

N70-0013

Fig. 6. Remove backrest upper part mounting bolts (**1**) and slide backrest (**2**) upward (**arrow**) and place in luggage compartment. Top shown in down position for clarity of illustration.

3. Release tension in top by unlocking front securing hooks and lifting up slightly.

4. Working inside car, unhook convertible top liner in left rear and right rear corners. See Fig. 7.

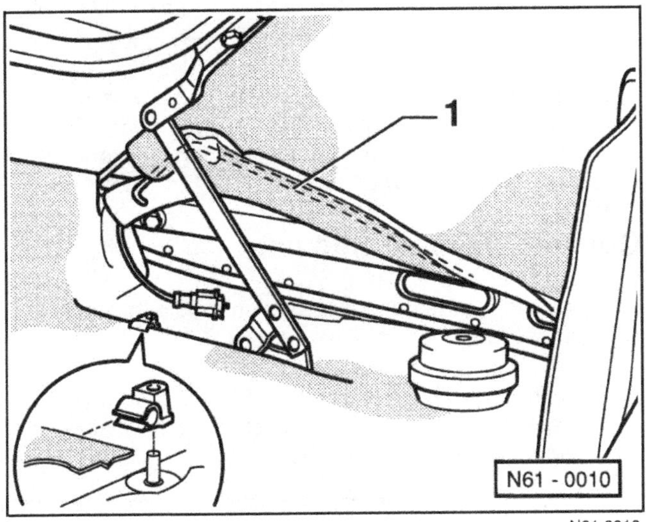

N61-0010

Fig. 7. Unhook top liner (**1**) from left and right corners (**inset**).

5. Partially open convertible top. In this position, remove mounting frame covers on left and right sides. See Fig. 8.

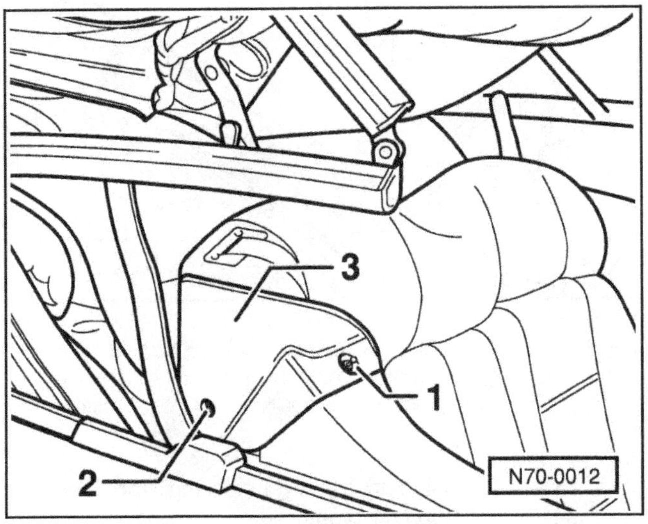

N70-0012

Fig. 8. To remove cover (**3**), unscrew ball pins (**1**) and screws (**2**).

6. Lower convertible top completely. Then remove screws from liner on left and right sides. See Fig. 9.

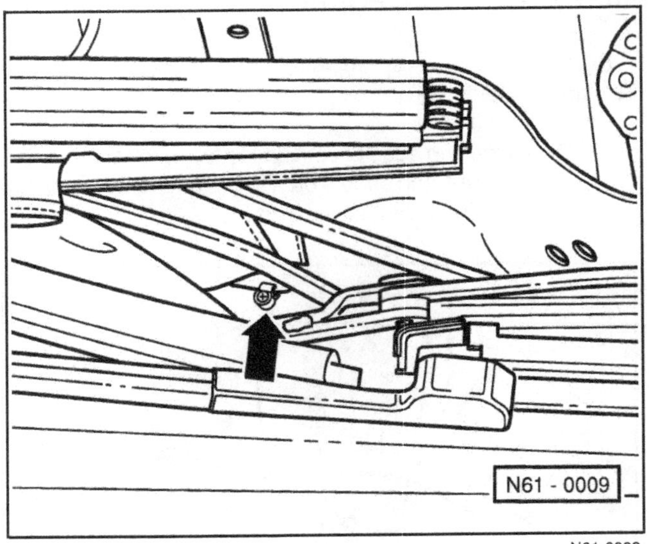

N61-0009

Fig. 9. Remove screw from liner (**arrow**) on left and right sides.

7. For vehicles up to 09/2000 production: remove bolts from left side and right side rear window hinges and disconnect wiring from heated rear window. See Fig. 10.

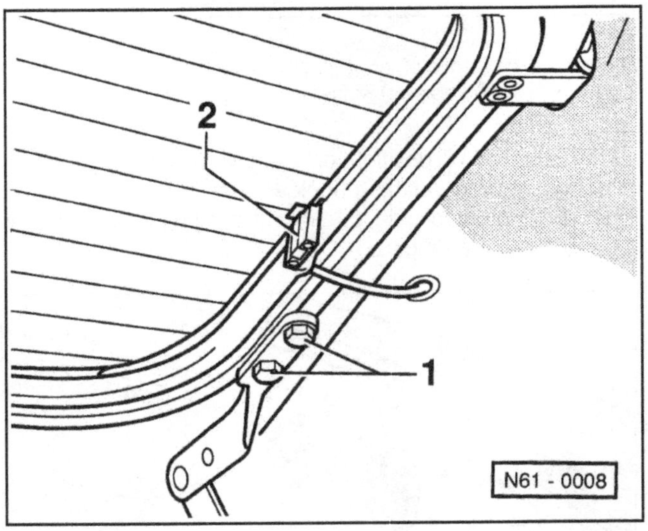

N61-0008

Fig. 10. Rear window hinge mounting bolts (**1**). Pry off rear window defroster electrical harness connector housing (**2**). from wire end. Then pull wiring out of top (though wiring grommet)

8. For vehicles from 10/2000 production: remove hex bolts from left side and right side rear window hinges and disconnect heated rear window wiring. See Fig. 11.

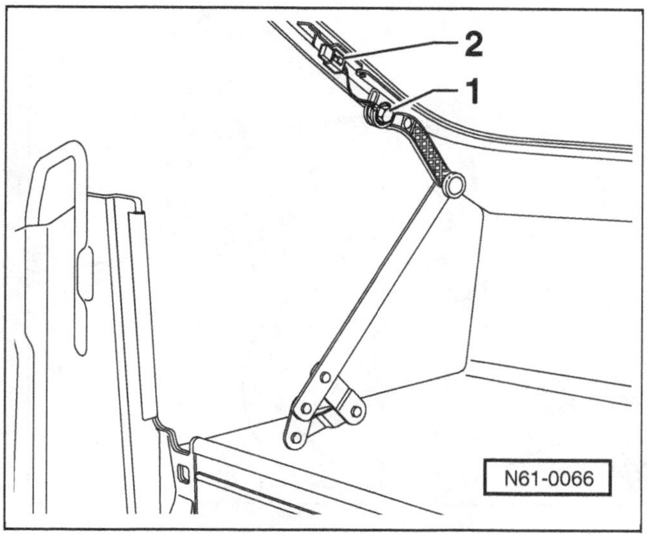

N61-0066

Fig. 11. Remove rear window hinge hex mounting bolts (**1**) from left and right. Separate rear window defroster electrical connector (**2**), pry terminal housing out of rear window seal and then pull out electrical connector.

9. Remove securing clips from trunk liner and remove liner from car.

10. Remove mounting frame bolts (quantity 12). Separate mounting frame from body. See Fig. 12.

NOTE—

The sealing band between the mounting frame and the body may cause the frame to bind slightly when removing.

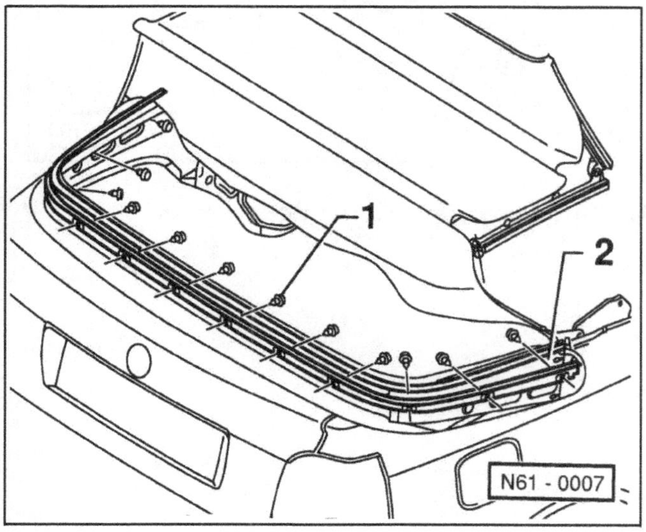

N61-0007

Fig. 12. Remove bolts (**1**) and remove mounting frame and top (**2**).

11. Fold mounting frame and rear window fully forward. Be sure to remove wiring harness for rear window defogger from guides as frame is folded forward. See Fig. 13.

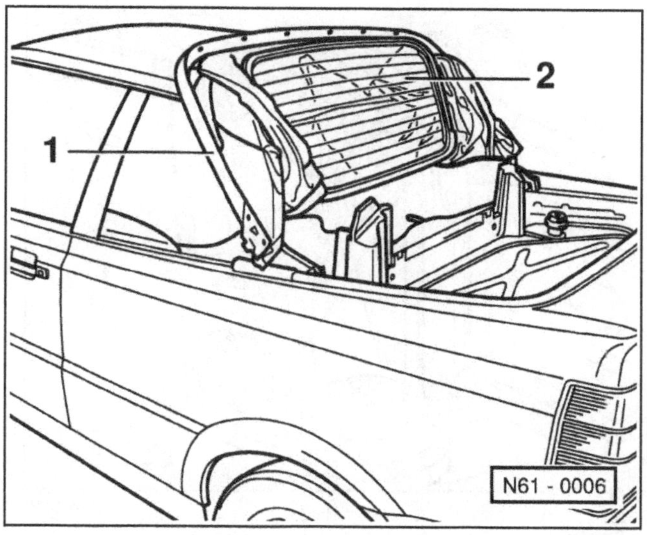

N61-0006

Fig. 13. Fold mounting frame (**1**) and rear window (**2**) forward.

12. Remove circlip from pin securing hydraulic or gas filled strut and slide pin out of bracket. Remove gas-filled strut or hydraulic cylinder with lines. See Fig. 14.

NOTE—

- *From 09/1995 production, there is a gas-filled strut on the left side only.*
- *When removing hydraulic cylinder, note routing of hydraulic lines.*

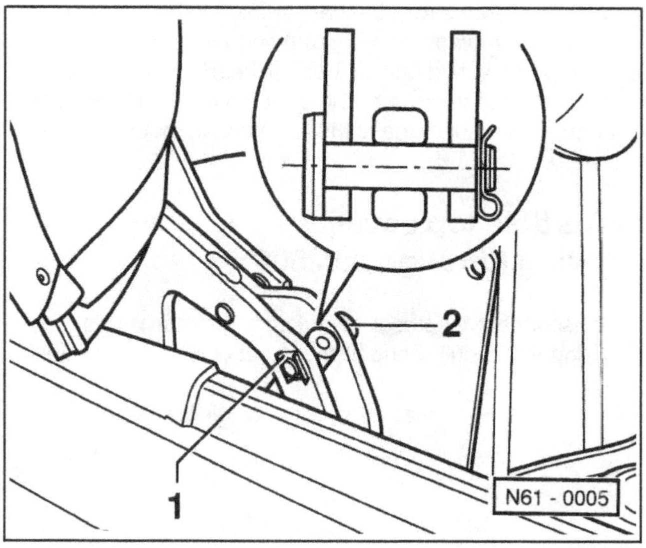

N61-0004

Fig. 14. Remove circlip (**1**) from pin (**2**) of hydraulic cylinder or gas-filled strut. Slide pin out to the inside.

13. Remove bolts from main brackets on left and right sides. See Fig. 15.

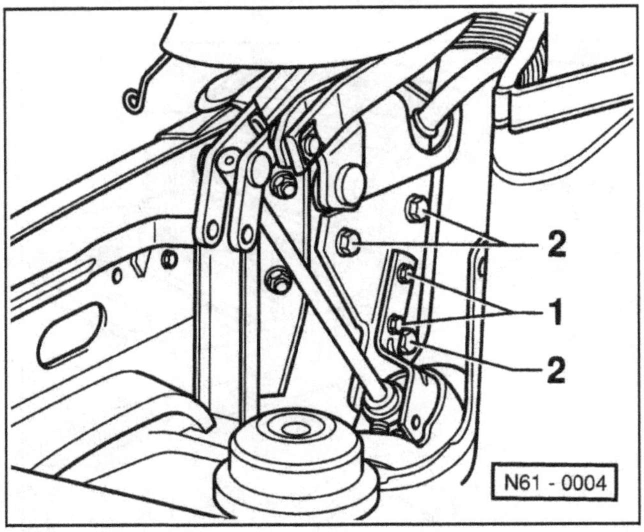

N61-0004

Fig. 15. Remove bolts with washers (**1**) from hydraulic cylinder or gas-filled strut retaining bracket. Remove bolts with spring washers (**2**) from main bracket.

14. Carefully remove convertible top from vehicle. This operation requires a helper to avoid damage due to size and weight of top.

CONVERTIBLE TOP COVER, REPLACING

The convertible top cover is available separately and can be replaced from above with the convertible top frame, padding and headliner in place. When removing rivets from the convertible top frame, first drill off the rivet head and then drift the rivet shank out of the frame. Clean up shavings immediately (vacuum) to avoid damage to the top material, bobywork, and upholstery.

Convertible top cover, removing (through 09/2000)

1. Disconnect electrical connectors for rear window defogger and pull wiring harness out of rear window seal.

2. Carefully push rear window out of window seal.

3. Separate top completely from rear window frame by prying out staples. See Fig. 16.

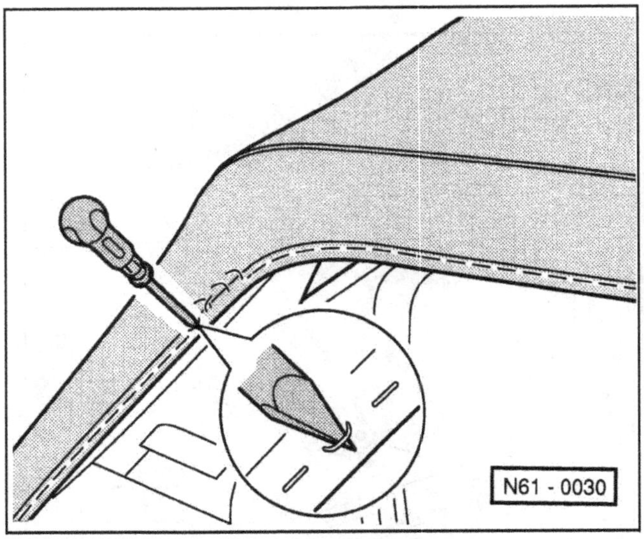

N61-0030

Fig. 16. Pry out staples (**inset**) as shown.

4. Unhook top liner in left rear and right rear corners. See Fig. 17.

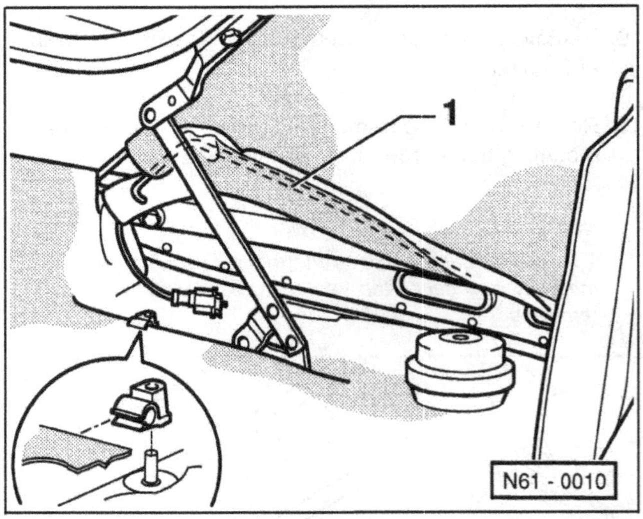

N61-0010

Fig. 17. Unhook top liner (**1**) from left and right corners.

5. Peel front side seal off frame, remove seal retaining rail, and pry out clips. See Fig. 18.

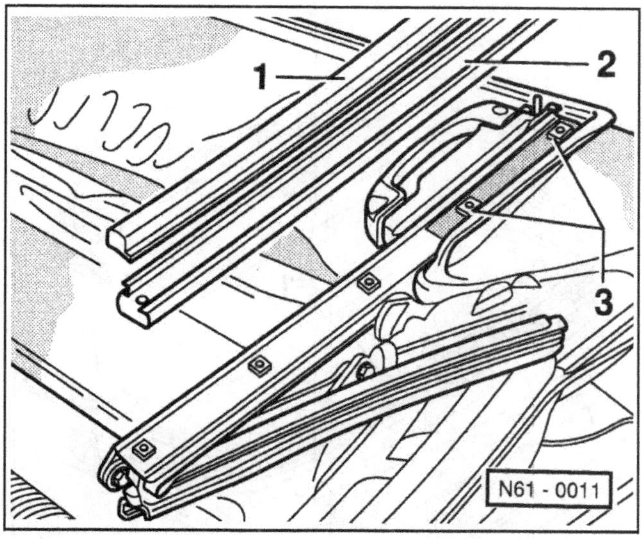

N61-0011

Fig. 18. Remove front seal (**1**), retaining rail screws (T20) and retaining rail (**2**), and pry out clips (**3**).

6. Remove screws from front (leading) edge clamping strip and remove strip. Peel top cover off front of frame (top is adhesive bonded to frame). See Fig. 19.

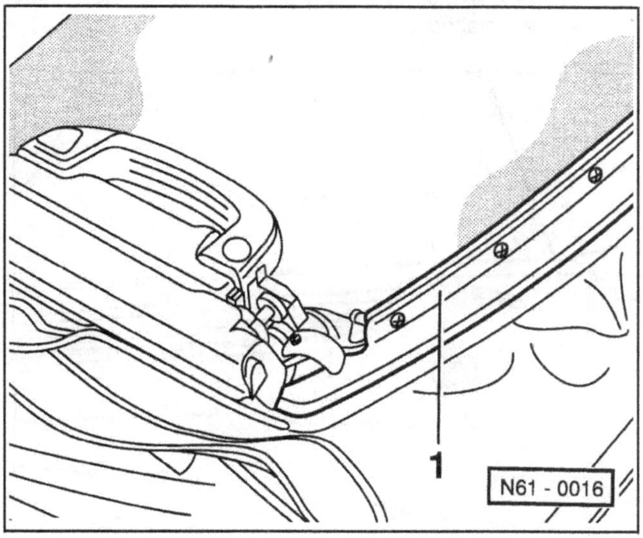

N61-0016

Fig. 19. Remove top cover leading edge clamping strip (**1**), and pull adhesive bonded top cover off of leading edge.

7. Remove rivets holding tensioning cords to left and right front corners of top frame. See Fig. 20.

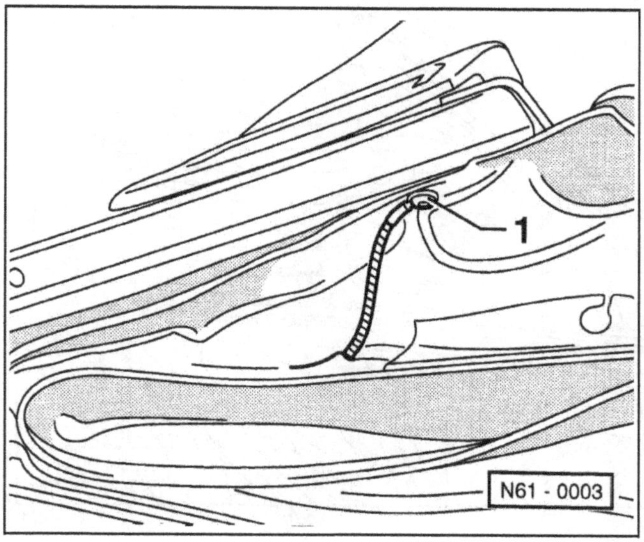

N61-0003

Fig. 20. Drill out tensioning cord mounting rivet (**1**) at front of frame and drive out shank with a drift.

8. Peel seal off rear part of frame, remove seal retaining rail, clips and rivet. See Fig. 21.

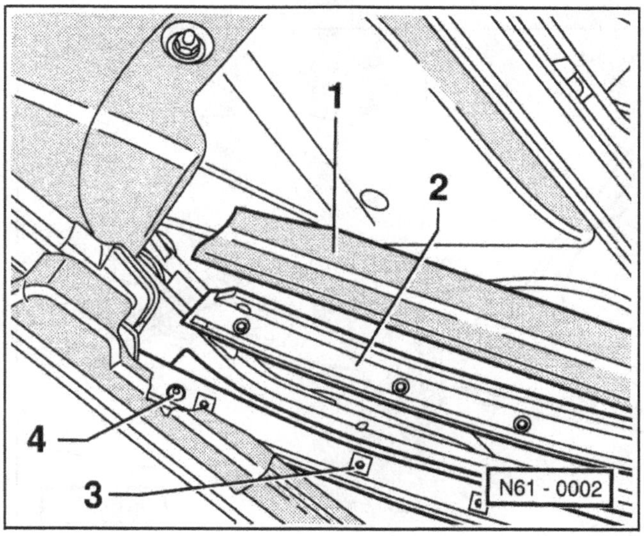

N61-0002

Fig. 21. Pull off rear seal (**1**) and unbolt retaining rail screws (T20) and retaining rail (**2**). Carefully pry out clips (**3**). Drill out rivet (**4**) and drive out shank with a drift.

9. Peel top cover off rear of frame (top is adhesive bonded to frame).

10. Pull tensioning cords out of top and remove rivets from left and right side of frame. See Fig. 22.

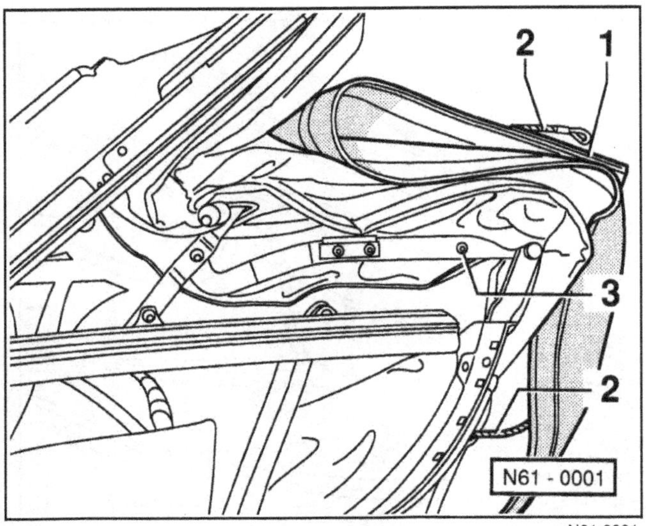

N61-0001

Fig. 22. Pull tensioning cord (**2**) out of top cover (**1**) and drill out rivet (**3**) and drive out shank with a drift.

11. Cut convertible top cover off sides (allows for easier removal). Use care not to damage the tensioning cord and the lining when cutting. See Fig. 23.

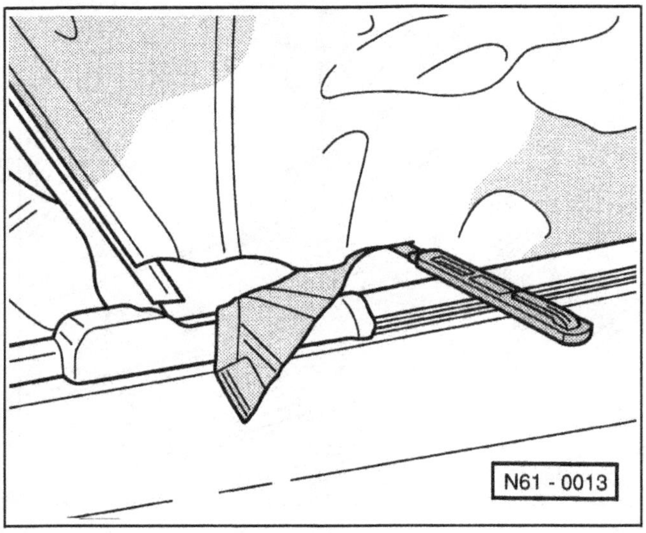

N61-0013

Fig. 23. Cut top with a sharp knife in the areas shown on each side to facilitate removal. Take care to avoid cutting tensioning cord during this operation.

12. Pull out carpet securing clips using special tool 80-200 or equivalent.

13. Remove mounting frame bolts (quantity 12). Separate mounting frame from body. See Fig. 24.

NOTE —

The sealing band between the mounting frame and the body may cause the frame to bind slightly when removing.

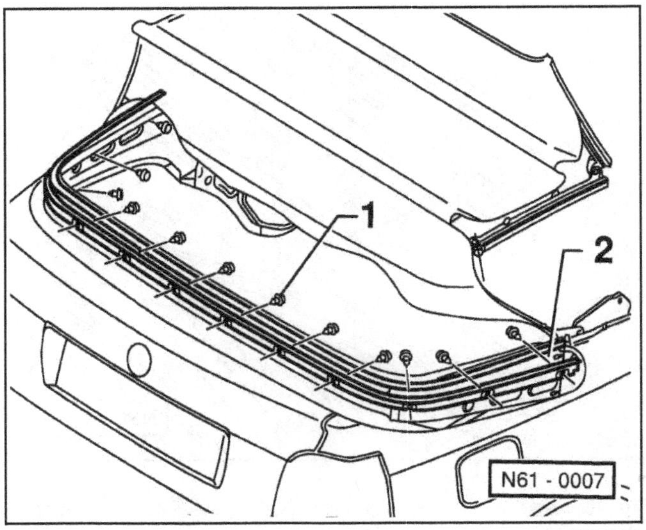

N61-0007

Fig. 24. Remove bolts (**1**) and remove mounting frame and top (**2**).

14. Remove convertible top frame as previously described.

15. Remove rivets from transportation protection plate where installed. See Fig. 25.

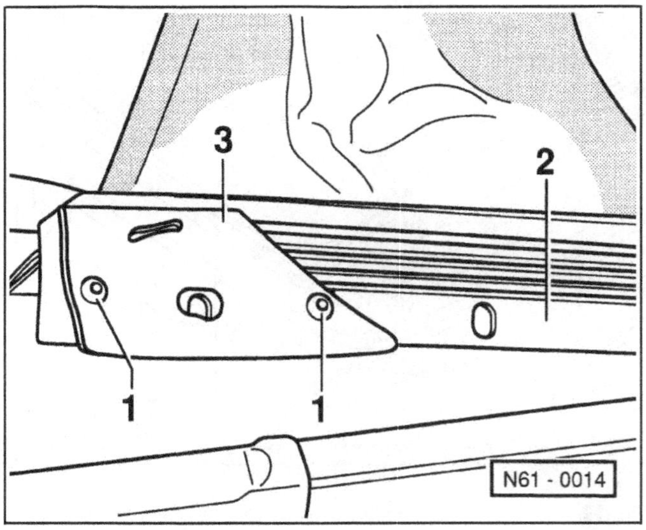

N61-0014

Fig. 25. Drill out rivets (**1**) in frame (**2**) and drive out shank with a drift. Securing plate (**3**) is for new vehicle transportation only and can be discarded.

16. Pull strips out of mounting frame. See Fig. 26.

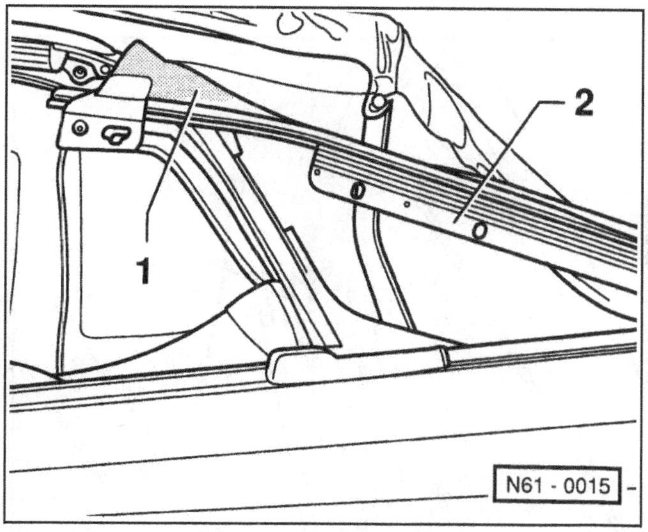

N61-0015

Fig. 26. Pull strips (**1**) out of mounting frame (**2**).

Convertible top cover, installing (through 09/2000)

New rivets, 8mm galvanized (corrosion-resistant) steel staples, and special adhesives are needed to install a new convertible top cover. Read the procedure through to determine what tools and supplies will be needed.

> ***NOTE —***
> *To facilitate installation, special Volkswagen assembly brackets VAG 1887 are highly recommend for top installation. See Fig. 27.*

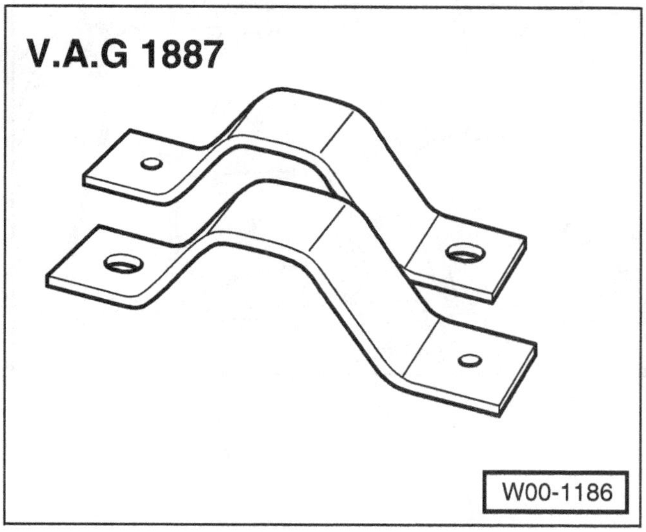

W00-1186

Fig. 27. Assembly brackets VAG 1887. One pair is required.

1. Secure convertible top frame to assembly brackets and then secure to body. See Fig. 28.

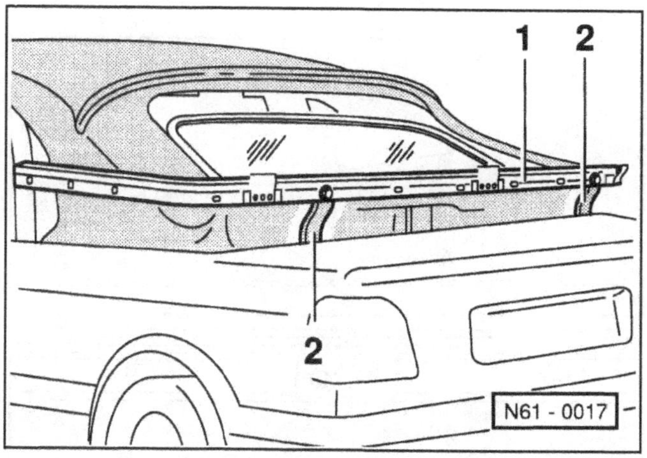

N61-0017

Fig. 28. Convertible top mounting frame (**1**) secured to body with Volkswagen special assembly brackets VAG 1887 (**2**).

2. Slide new top completely into channel in mounting frame (use a lubricant such as silicone lubricant spray for easier installation). See Fig. 29.

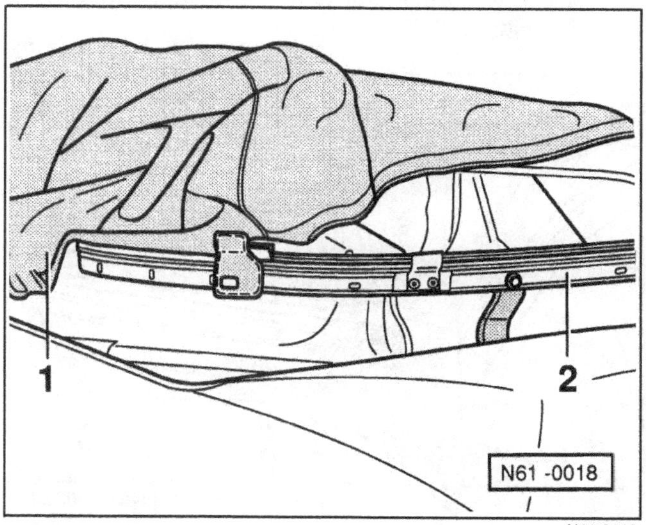

N61-0018

Fig. 29. Pull convertible top cover (**1**) onto mounting frame (**2**). Use a silicone lubricant spray (D 007 000 A2) as needed.

3. Align convertible top center marking (spot weld) with convertible top mounting frame center marking (drilled hole). See Fig. 30.

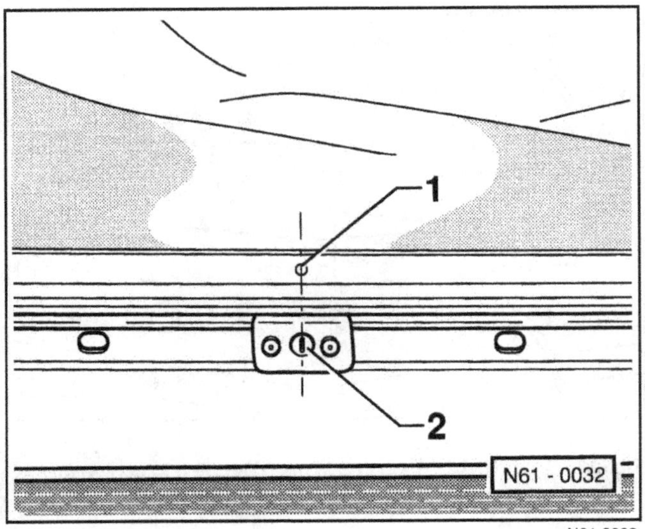

Fig. 30. Align convertible top center mark (spot weld **1**) with frame center marking (hole in convertible top frame **2**).

4. Attach water deflector to mounting frame with rivets. See Fig. 31.

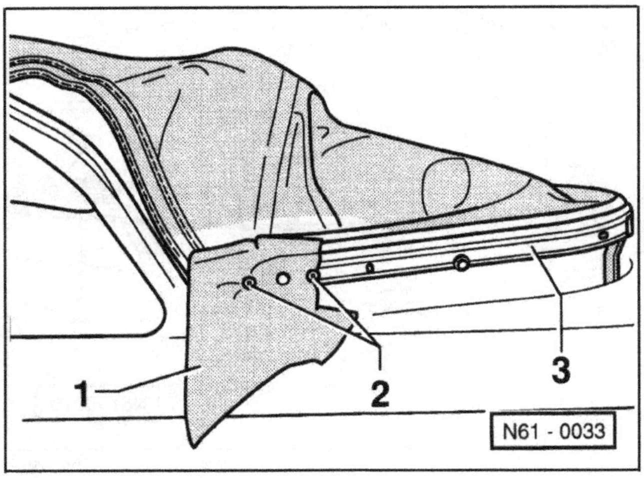

Fig. 31. Install water deflector (**1**) to mounting frame (**3**) with rivets (**2**).

5. Remove assembly brackets VAG 1887 and bolt mounting frame to body.

6. Push left and right tensioning cords through side guides. See Fig. 32.

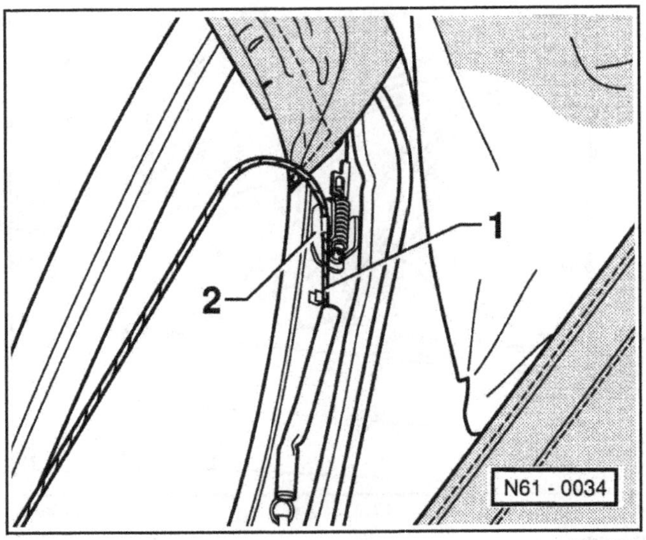

Fig. 32. Push tensioning cord (**1**) into left side and right side guides (**2**).

7. Pull tensioning cables through guide channels in the top. Rivet cord to forward area of frame. See Fig. 33.

NOTE—

Insert a stiff wire through channels in top and connect wire to cable end and pull cable though channels in top.

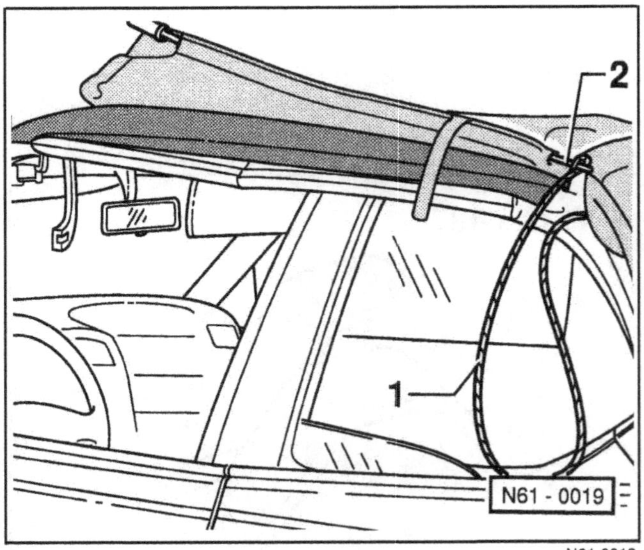

Fig. 33. Pull tensioning cord (**1**) into left side and right side with the help of a guide wire and rivet to forward area.(**2**).

8. Secure tension cord. See Fig. 34.

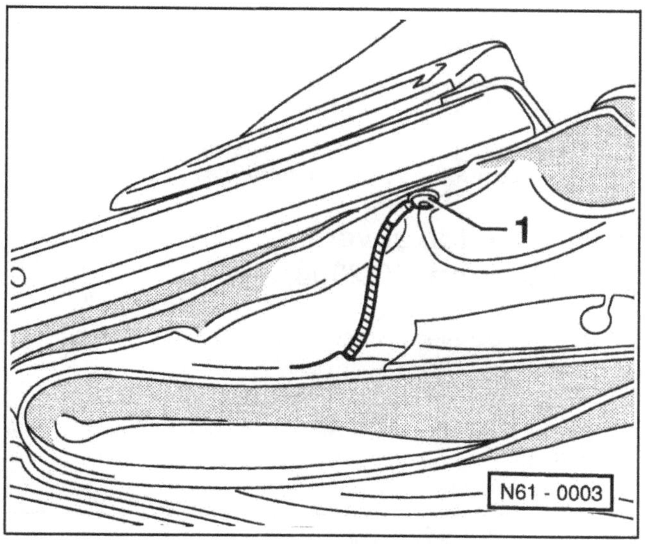

N61-0003

Fig. 34. Install tensioning cord with a new rivet (**1**).

9. Fit top to front of frame, align and bond in place using special adhesive (2-part synthetic leatherette adhesive, VW part no. D 000 801). See Fig. 35.

> ***CAUTION—***
> *Part numbers are listed for reference only. Always check with an authorized Volkswagen parts department for the latest parts information.*

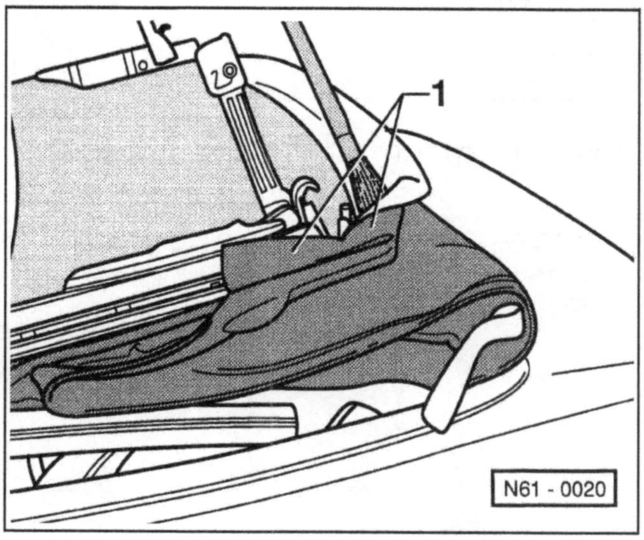

N61-0020

Fig. 35. Fasten top cover (**1**) to leading edge, align and bond.

10. Install leading edge clamping strip.

11. Bond convertible top cover securing strips to frame using special adhesive D 000 801.

12. Install rear seal retaining rail, rivets and clips. Install seal to rail. See Fig. 36.

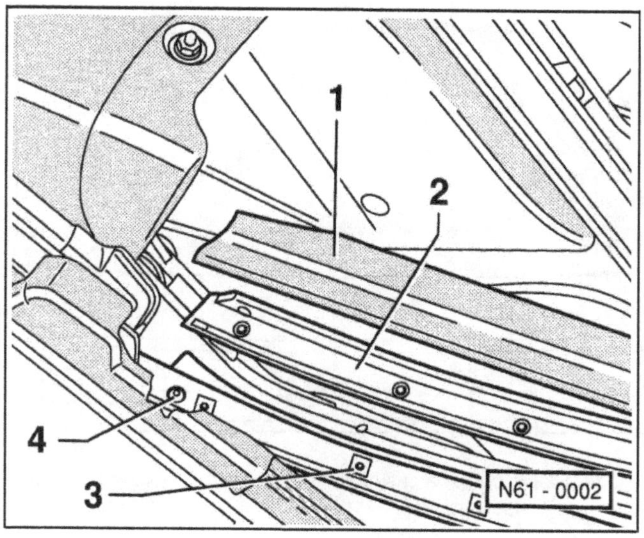

N61-0002

Fig. 36. Install rivet (**4**) and insert clips (**3**). Install retaining rail (**2**) with screws (T20). Install seal (**1**).

13. Align rear window frame parallel to corner bow. See Fig. 37.

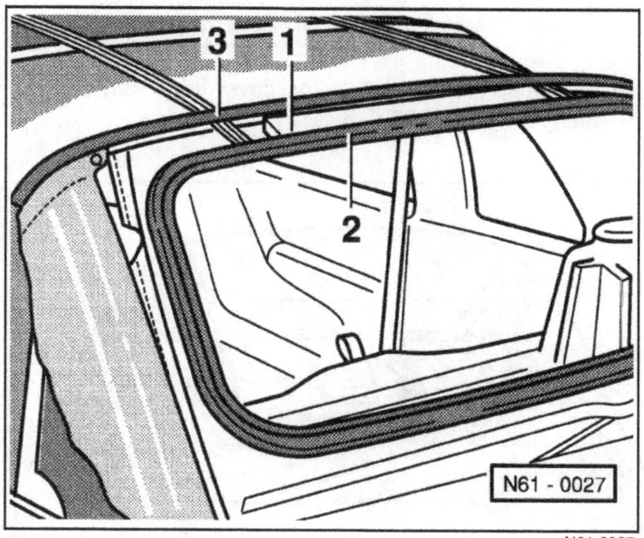

N61-0027

Fig. 37. Clamp wooden block (**1**) between rear window frame (**2**) and corner bow (**3**) as shown. (Wooden block size 60 x 20 x 460 mm (2. 4 x 0.8 x 18 in.).

14. Spread out convertible top on frame.

> ***NOTE—***
> *Convertible top cover must lie against rear window frame over complete perimeter.*

15. Using 8 mm, corrosion-resistant staples, fasten top to window frame, working at the corners first. See Fig. 38.

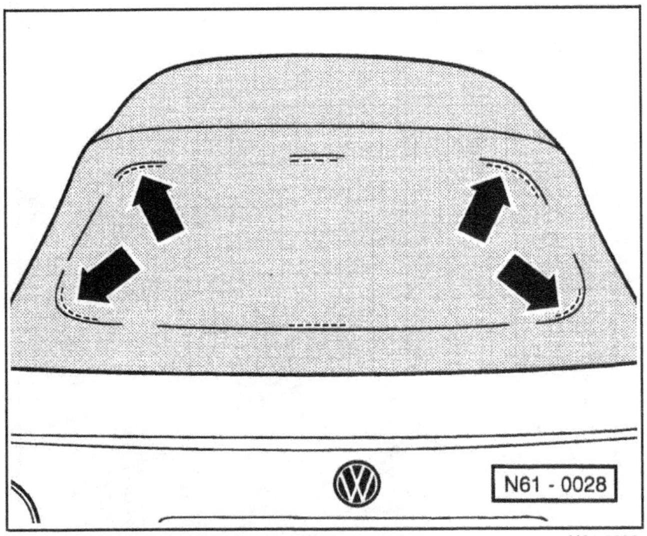

N61-0028

Fig. 38. Staple corner areas (**arrows**) first when fastening top cover to window frame.

16. Check measurement between rear window frame and convertible top transverse seam.

> ***CAUTION —***
> - *Do not cut convertible top cover before stapling to rear window frame.*
> - *Double check fit between rear window frame and convertible top seam before cutting top.*

17. Carefully cut convertible top cover flush with window flange. See Fig. 39.

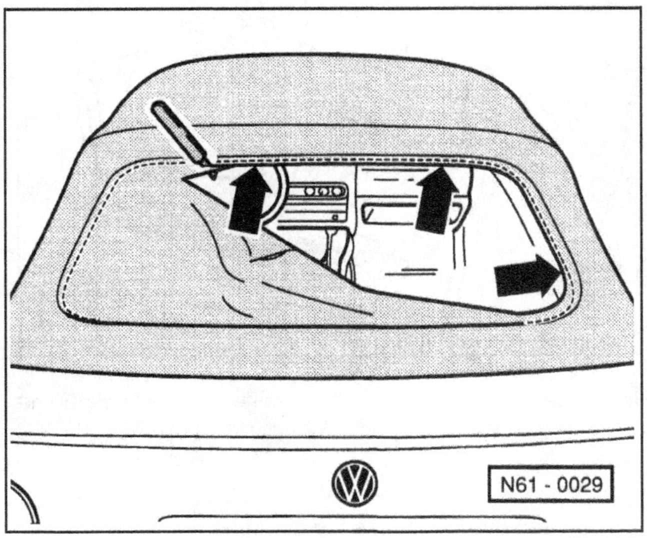

N61-0029

Fig. 39. Carefully cut convertible top cover flush with window flange (**arrows**).

18. Install rear window and window seal. Use adhesive D 469 101 A3 to secure and seal as required.

19. Check top adjustment as described earlier.

> ***NOTE —***
> *Folds that may have formed can be removed by warming the convertible top cover with a hot air blower.*

Convertible top cover, removing (from 10/2000)

1. Disconnect electrical connectors for rear window defroster.

2. Unhook top liner in left rear and right rear corners. See Fig. 40.

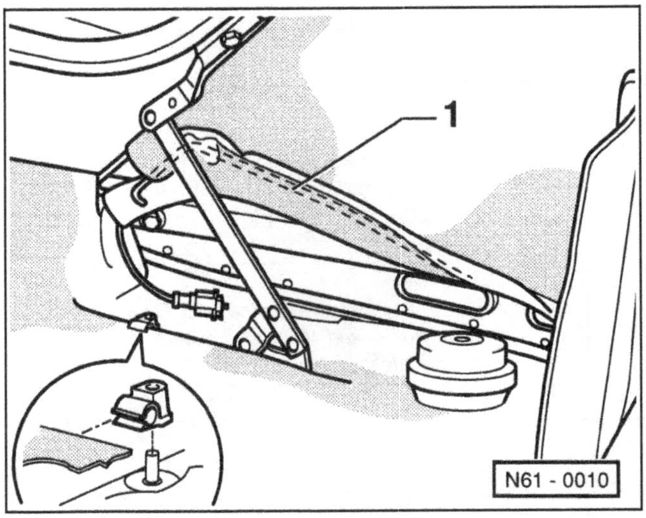

N61-0010

Fig. 40. Unhook top liner (**1**) from left and right corners (**inset**).

> ***NOTE —***
> *Rear window is part of convertible top cover and is **not** removable or available separately.*

3. Remove seal from frame, remove seal retaining rail, and pry out clips. See Fig. 41.

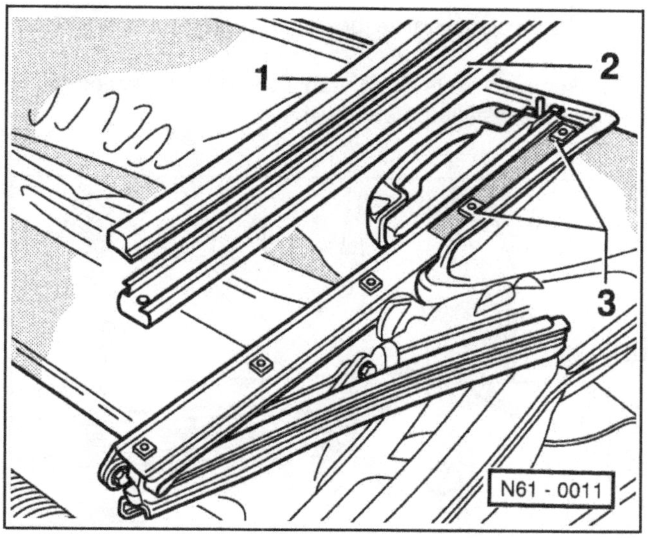

N61-0011

Fig. 41. Remove seal (**1**), retaining rail screws (T20) and retaining rail (**2**), and pry out clips (**3**).

4. Remove screws from front (leading) edge clamping strip and remove strip. Pull top cover off front of frame (top is adhesive bonded to frame). See Fig. 42.

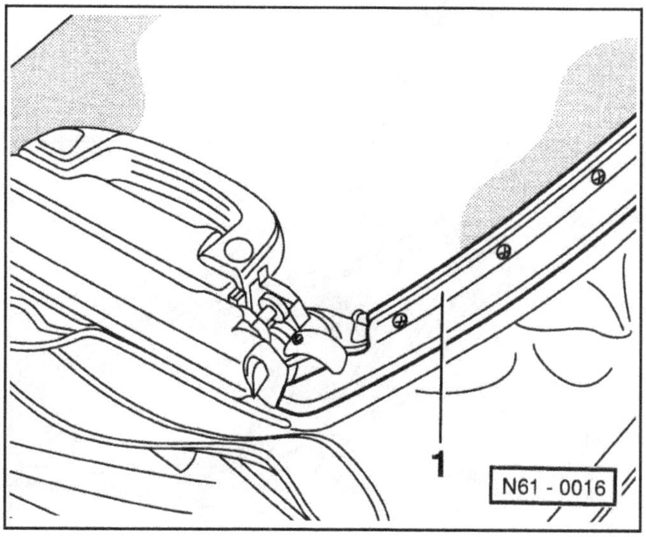

N61-0016

Fig. 42. Remove top cover leading edge clamping strip (**1**), and pull adhesive bopnded top cover off of leading edge.

5. Remove rivets holding tensioning cords to left and right front corners of top frame. See Fig. 43.

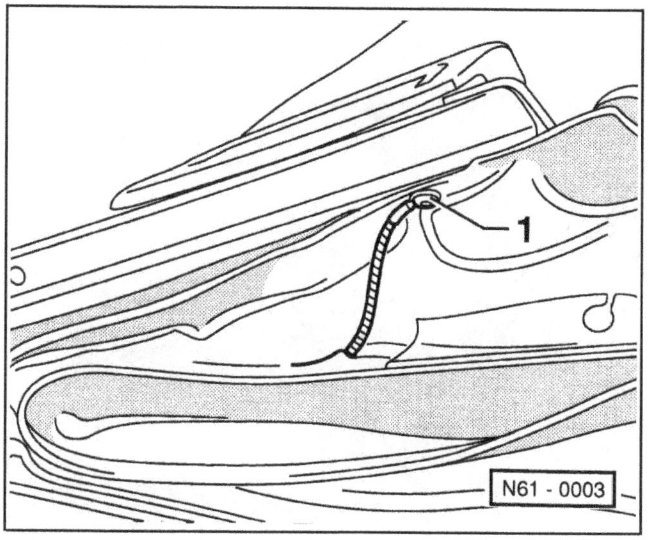

N61-0003

Fig. 43. Drill out tensioning cord mounting rivet (**1**) at front of frame and drive out shank with a drift.

6. Pull seal off rear part of frame, remove seal retaining rail, clips and rivet. See Fig. 44.

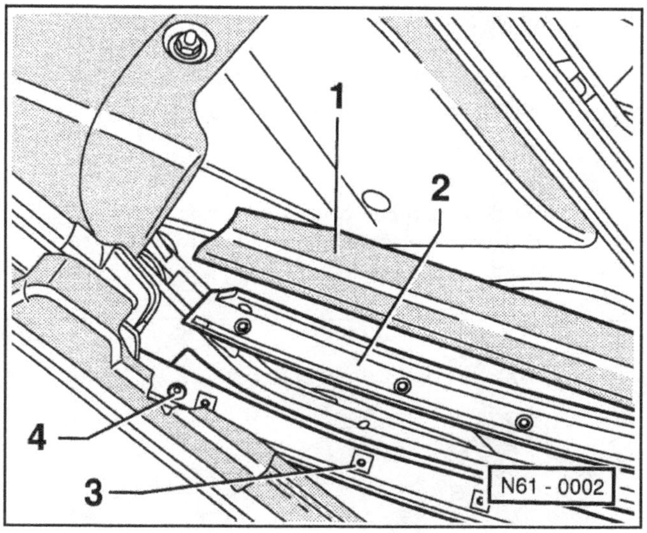

N61-0002

Fig. 44. Pull off rear seal (**1**) and unbolt retaining rail screws (T20) and retaining rail (**2**). Carefully pry out clips (**3**). Drill out rivet (**4**) and drive out shank with a drift.

7. Cut convertible top cover off sides (allows for easier removal). Use care not to damage the tensioning cord and the lining when cutting. See Fig. 45.

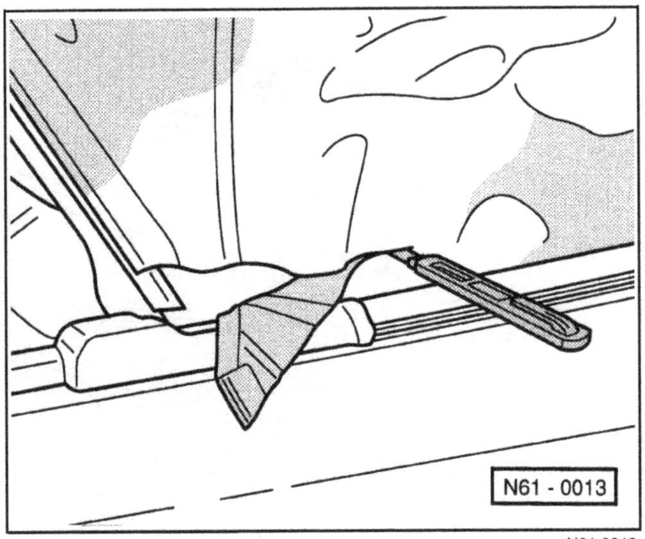

N61-0013

Fig. 45. Cut top with a sharp knife in the areas shown on each side to facilitate removal. Take care to avoid cutting tensioning cord during this operation.

8. Pull out carpet securing clips using special tool 80-200 or equivalent.

9. Remove mounting frame bolts (quantity 12). Separate mounting frame from body. See Fig. 46.

NOTE—
The sealing band between the mounting frame and the body may cause the frame to bind slightly when removing.

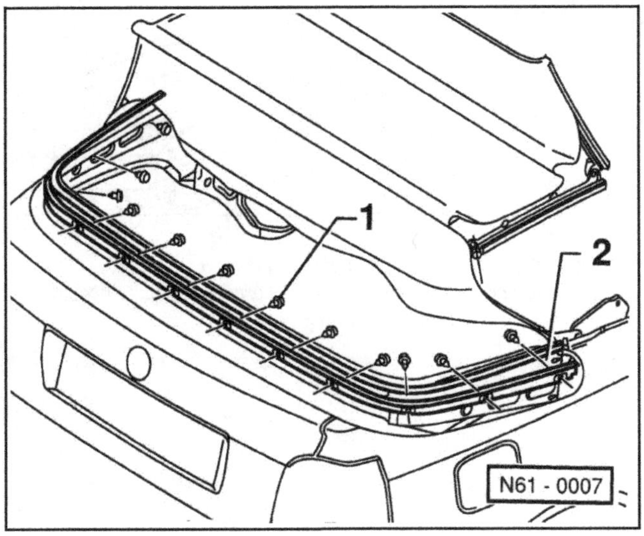

N61-0007

Fig. 46. Remove bolts (**1**) and remove mounting frame and top (**2**).

10. Assembly brackets VAG 1887 can be installed to facilitate removal.

11. Remove convertible top frame as previously described.

12. Remove rivets from transportation protection plate where installed. See Fig. 47.

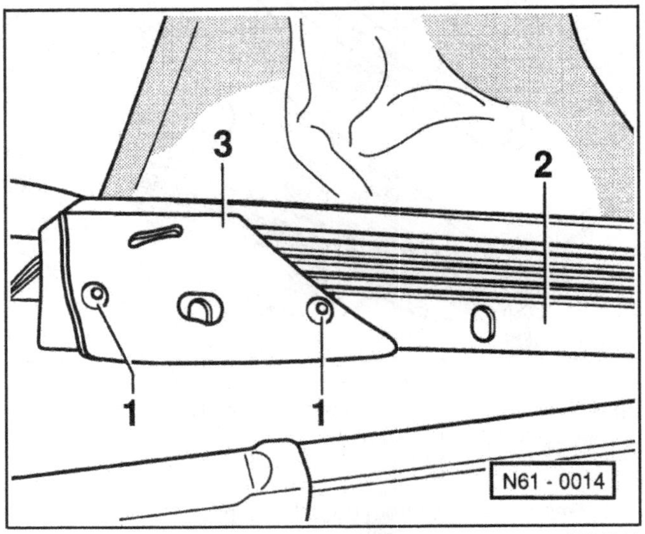

N61-0014

Fig. 47. Drill out rivets (**1**) in frame (**2**) and drive out shank with a drift. Securing plate (**3**) is for new vehicle transportation only and can be discarded.

13. Pull strips out of mounting frame. See Fig. 48.

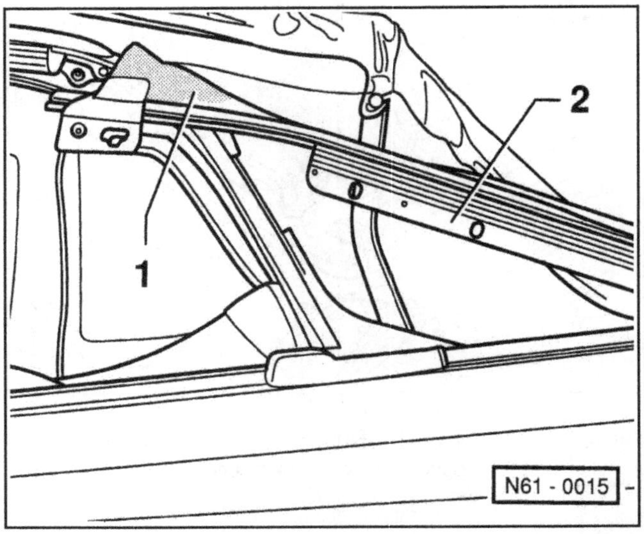

N61-0015

Fig. 48. Pull strips (**1**) out of mounting frame (**2**).

Convertible top cover, installing (from 10/2000)

New rivets, 8mm galvanized (corrosion-resistant) steel staples, and special adhesives are needed to install a new convertible top cover. Read the procedure through to determine what tools and supplies will be needed.

> ***NOTE —***
> *To facilitate installation, special Volkswagen assembly brackets VAG 1887 are highly recommend for top installation. See Fig. 49.*

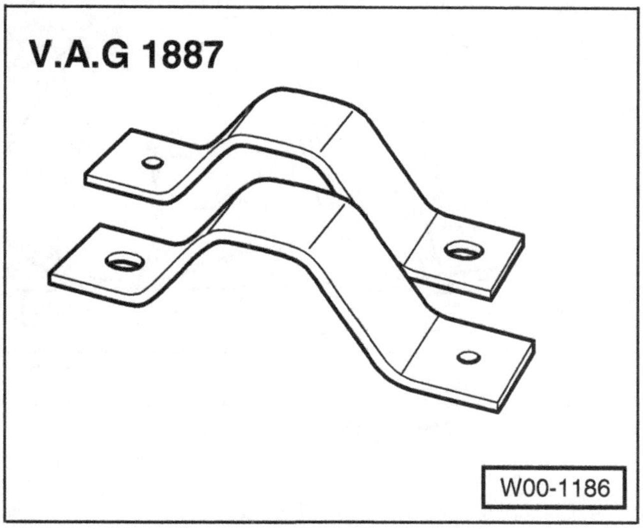

W00-1186

Fig. 49. Assembly brackets VAG 1887. One pair is required.

1. Secure convertible top frame to assembly brackets and then secure to body. See Fig. 50.

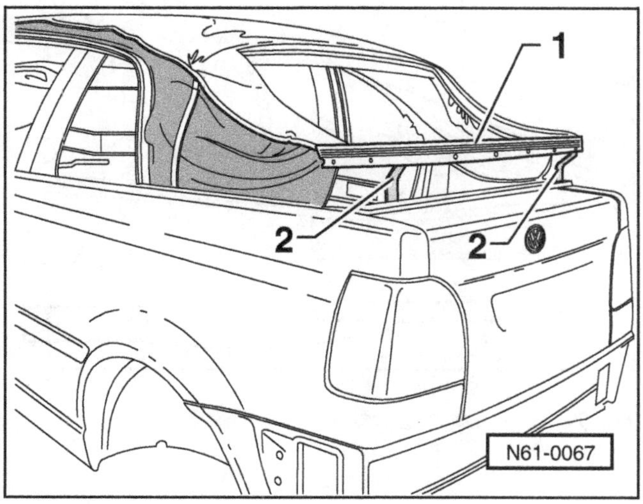

N61-0067

Fig. 50. Convertible top mounting frame (**1**) secured to body with Volkswagen special assembly brackets VAG 1887 (**2**).

2. If installed on the frame, the transportation securing plate(s) should be removed by drilling out the rivets.

3. Pull convertible top cover onto top mounting frame (use a lubricant such as silicone lubricant spray D 007 000 A2 for easier installation). Ensure that rear window is properly positioned. See Fig. 51.

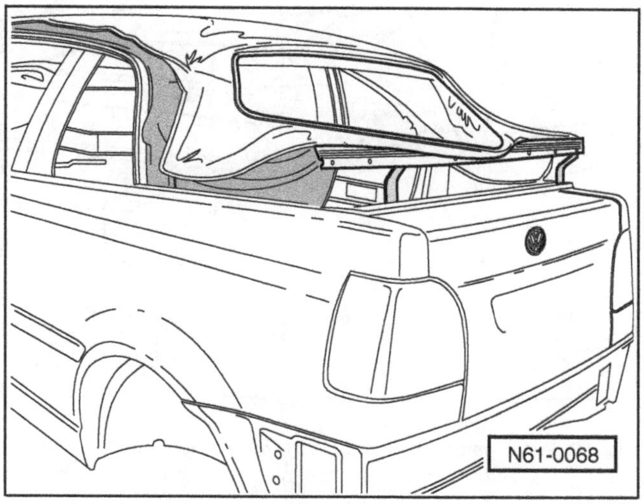

N61-0068

Fig. 51. Pull convertible top cover (**1**) onto top mounting frame (**2**). Ensure that the rear window is properly positioned on the frame.

4. Align convertible top center marking (spot weld) with convertible top mounting frame center marking (drilled hole). See Fig. 52.

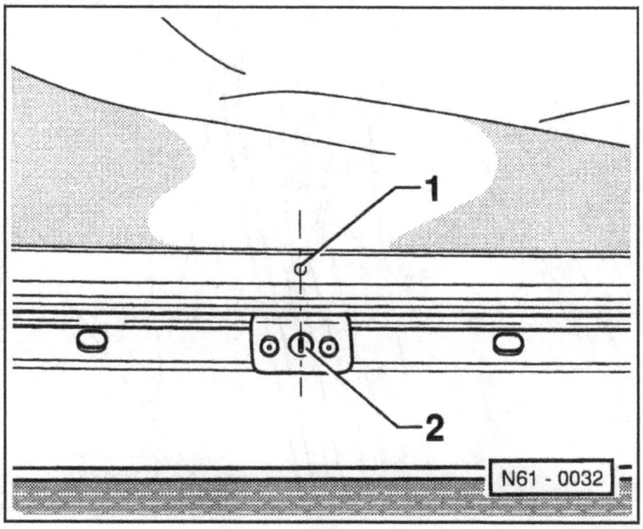

N61-0032

Fig. 52. Align convertible top center mark (spot weld **1**) with frame center marking (hole in convertible top frame **2**).

5. Attach water deflector to mounting frame with rivets. Connect top frame sections with connecting plates. See Fig. 53..

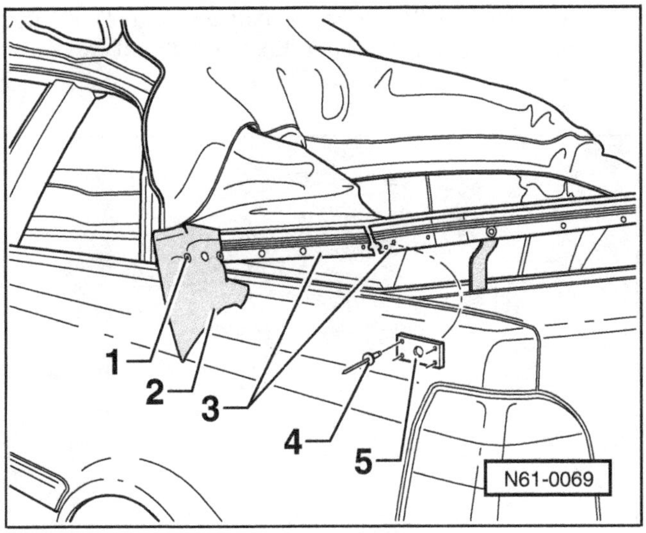

N61-0069

Fig. 53. Install water deflector (**2**) to top mounting frame (**3**) with rivets (**1**). Attach top mounting frame sections with connecting plates (**5**) and rivets (**4**).

6. Remove assembly brackets VAG 1887 and bolt mounting frame to body.

7. Push left and right tensioning cords through side guides. See Fig. 54.

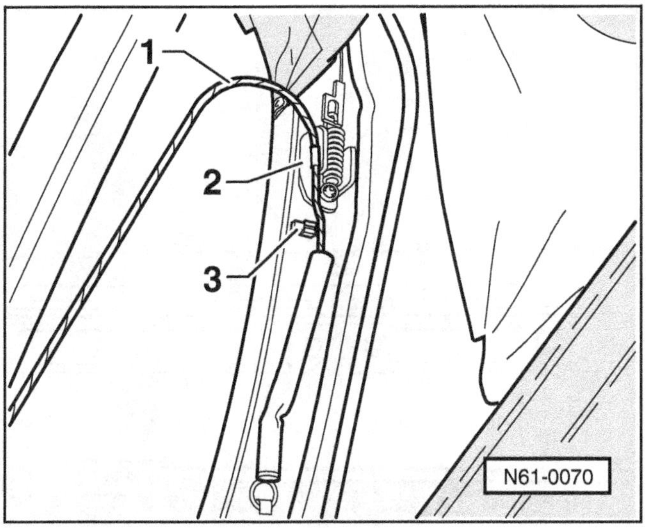

N61-0070

Fig. 54. Push tensioning cord (**1**) into left side and right side guides (**2**). The slot in spreader (**3**) near the tensioning cord guide must be at a right angle (90°) to the direction of travel.

8. Pull tensioning cables through guide channels in the top. Rivet cord to forward area of frame. See Fig. 55.

NOTE—
Insert a stiff wire through channels in top and connect wire to cable end and pull cable though channels in top.

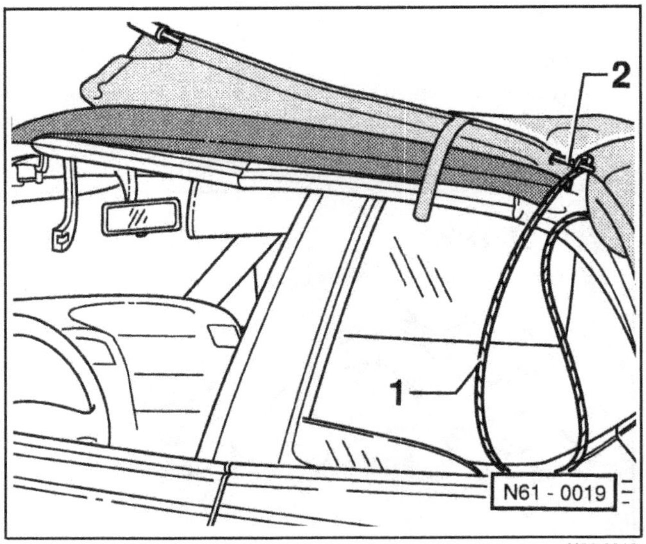

N61-0019

Fig. 55. Pull tensioning cord (**1**) into left side and right side with the help of a guide wire and rivet to forward area.(**2**).

9. Secure tension cord. See Fig. 56.

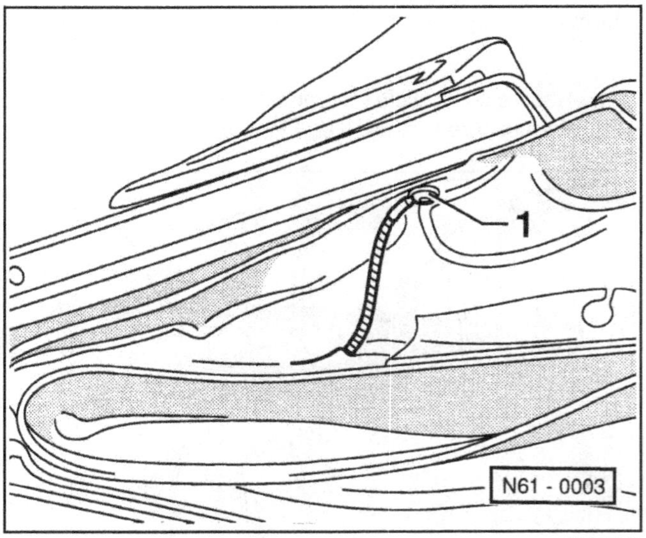

N61-0003

Fig. 56. Install tensioning cord with a new rivet (**1**).

10. Fit top to front of frame, align and bond in place using special adhesive (2-part synthetic leatherette adhesive, VW part no. D 000 801). See Fig. 57.

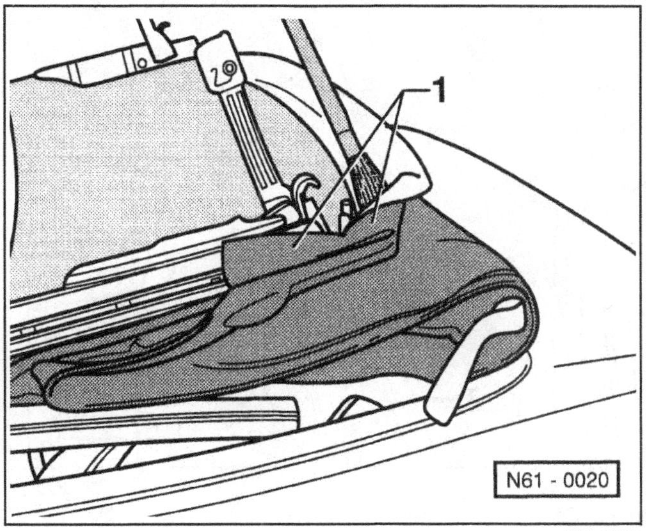

N61-0020

Fig. 57. Fasten top cover (**1**) to leading edge, align and bond.

11. Install leading edge clamping strip.

12. Bond convertible top cover securing strips to frame using special adhesive D 000 801.

13. Install rear seal retaining rail, rivets and clips. Install seal to rail. See Fig. 58.

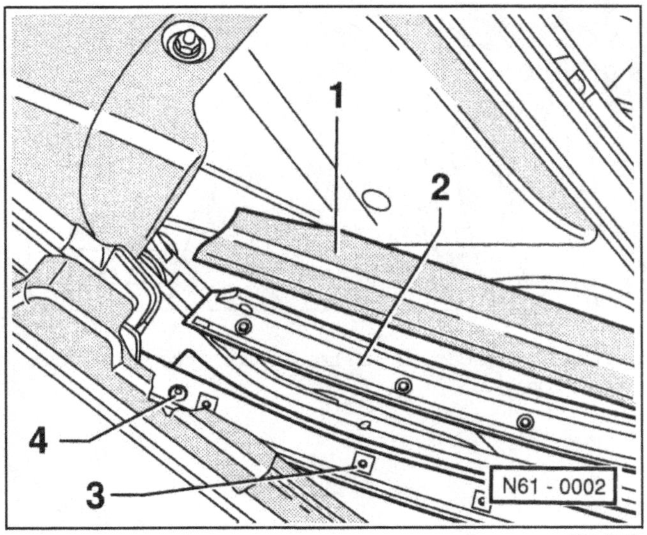

N61-0002

Fig. 58. Install rivet (**4**) and insert clips (**3**). Install retaining rail (**2**) with screws (T20). Install seal (**1**).

14. Check top adjustment as described earlier.

NOTE—
Folds that may have formed can be removed by warming the convertible top cover with a hot air blower.

CONVERTIBLE TOP PADDING, REPLACING

Padding, removing (through 08/1998)

1. Remove leading edge clamping strip, detach convertible top padding tensioning cord near leading edge.

2. Remove roof frame seals and retaining rails. Detach convertible top padding from convertible top leading edge and roof frame.

3. Pull front padding pocket off convertible top leading edge. See Fig. 59.

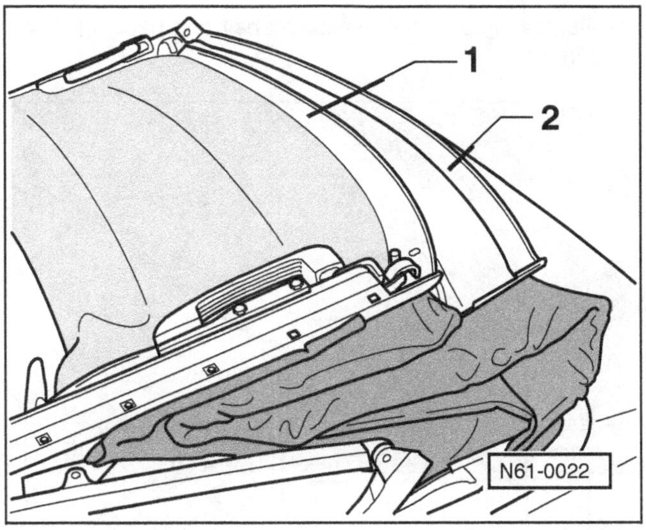

N61-0022

Fig. 59. Pull front padding pocket (**2**) off of convertible top leading edge (**1**).

4. Carefully lay padding to the rear and remove rivets from bows. See Fig. 60.

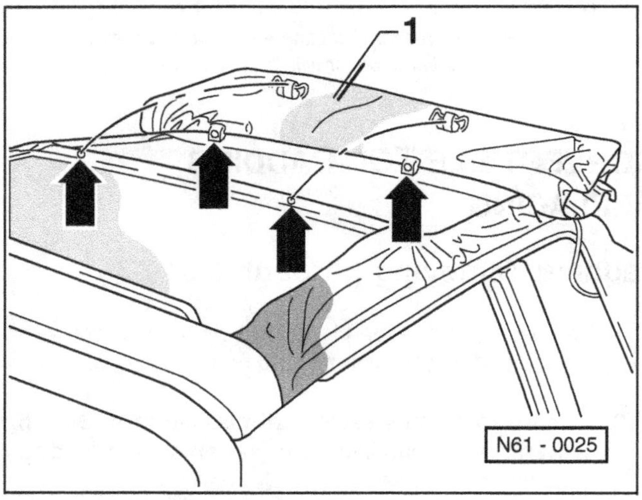

N61-0025

Fig. 60. Fold and lay padding (**1**) to the rear. Remove rivets (**arrows**) from bows.

5. Remove rivets from tensioning belt and padding. See Fig. 61.

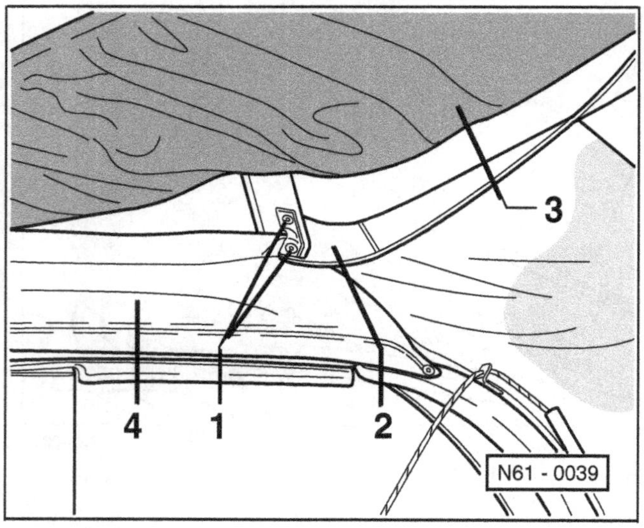

N61-0039

Fig. 61. Remove rivets (**1**) from tensioning belt (**2**) and padding (**3**). Use care to avoid damaging cover (**4**).

6. Remove screws and washers from corner bows. See Fig. 62.

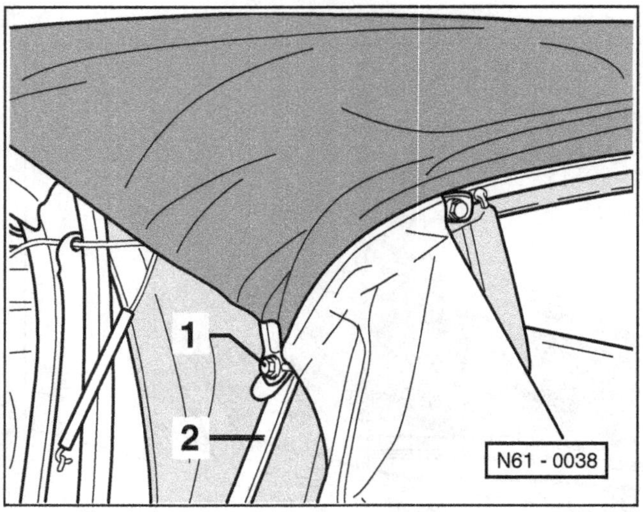

N61-0038

Fig. 62. Remove screws with washers (**1**) from corner bows (**2**).

7. Remove padding and beading strip from corner bow channel. See Fig. 63.

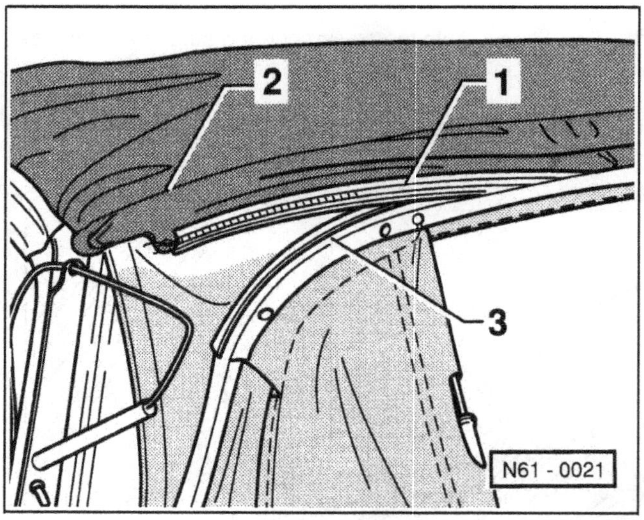

N61-0021

Fig. 63. Remove padding (**2**) and beading strip (**1**) from corner bow channels (**3**).

Padding, installing (through 08/1998)

1. Pull or press beading strip into corner bow channel. See Fig. 64.

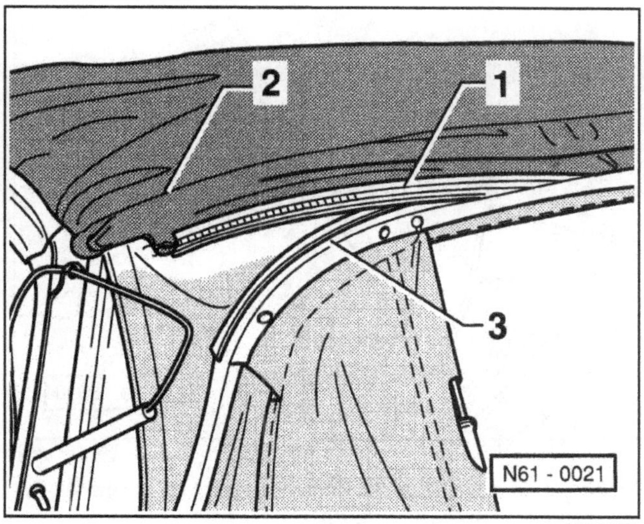

N61-0021

Fig. 64. Pull or press beading strip (**1**) into corner bow channel (**3**) in such a way so as not to be visible when fully installed.

2. Pull padding loops out and place in position by pulling in cords. See Fig. 65.

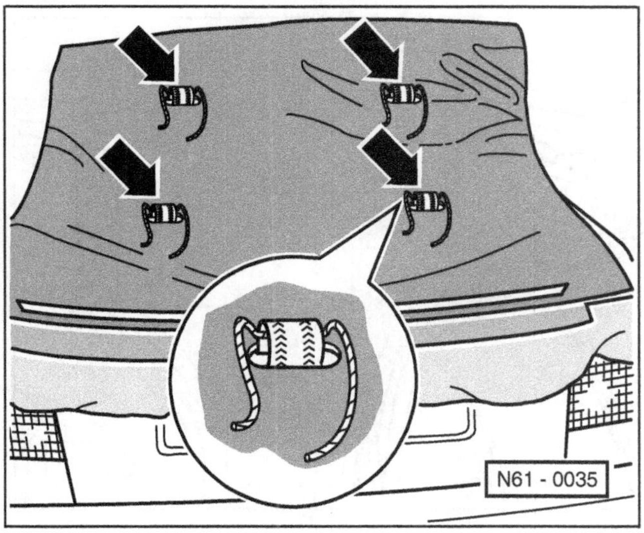

N61-0035

Fig. 65. Pull padding loops (**inset**) out and place in position by pulling in cords (**arrows**).

3. Bolt padding tongues to corner bow. See Fig. 66.

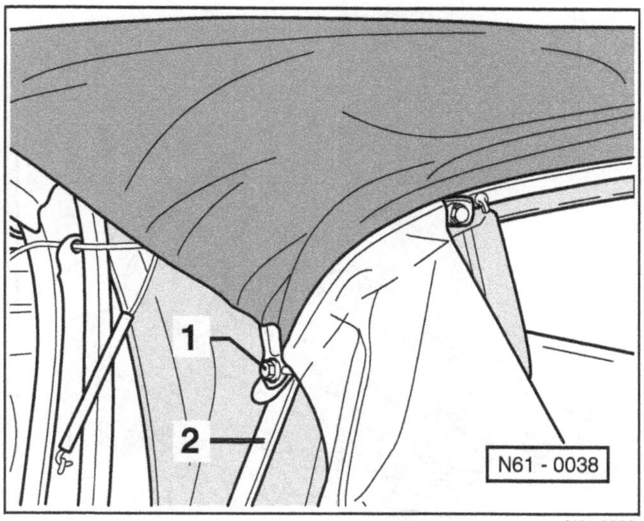

N61-0038

Fig. 66. Bolt padding tongues (**1**) to corner bow (**2**).

4. Rivet padding tongues to main and intermediate roof bows. See Fig. 67.

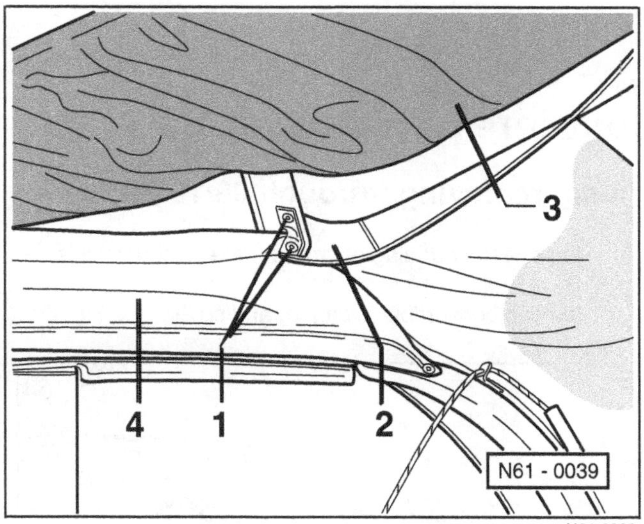

N61-0039

Fig. 67. Rivet padding tongues (**1**) to main and intermediate bows.

5. Install padding pocket over convertible top leading edge. See Fig. 68.

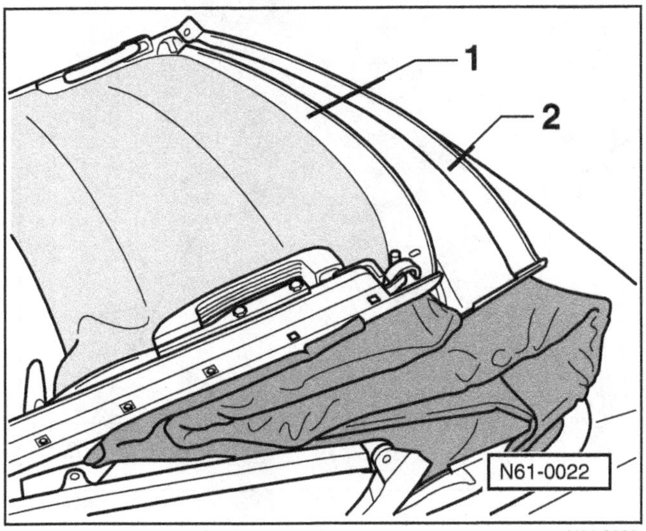

N61-0022

Fig. 68. Install padding pocket (**2**) over convertible top leading edge (**1**).

6. Install convertible top cover.

CONVERTIBLE TOP LINING, REPLACING

Lining, removing (through 08/1998)

1. Remove top padding as appropriate for the vehicle.
2. Remove convertible latch pin, and cover. See Fig. 69.

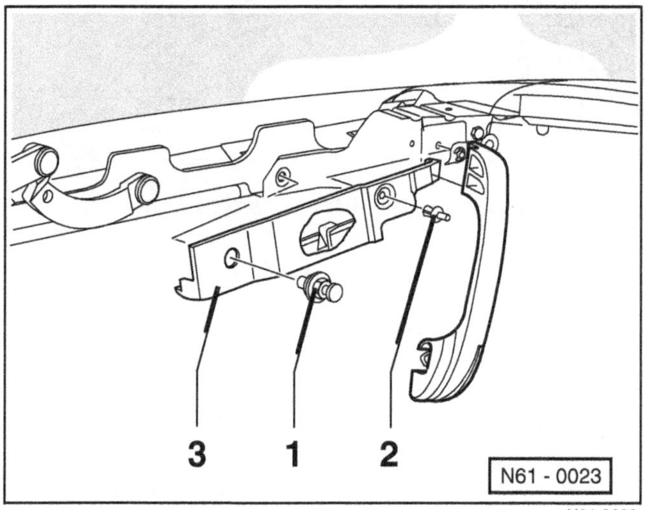

N61-0023

Fig. 69. Remove top latch pin (**1**), spreader rivet (**2**), and cover (**3**).

3. Remove rivet from lining and pull off of guide pins and away from leading edge. See Fig. 70.

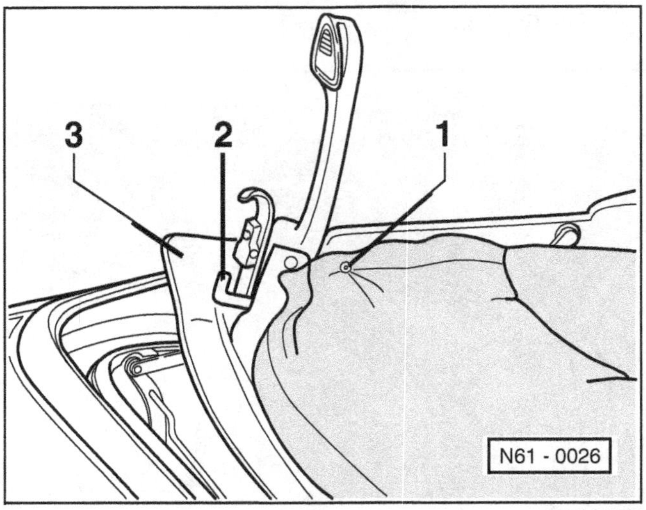

N61-0026

Fig. 70. Remove rivet (**1**) and pull top lining off of guide pins (**2**). Pull away from leading edge (**3**).

4. Remove tensioning belt rivets and wind deflector from intermediate bow.
5. Continue removing lining at inside rear of convertible top. See Fig. 71.

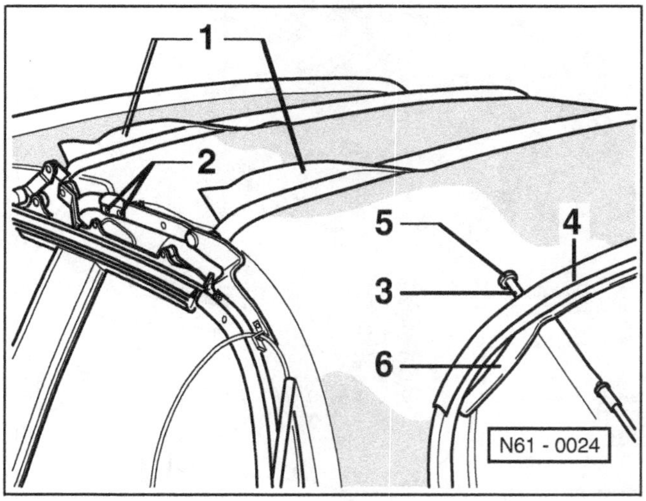

N61-0024

Fig. 71. Remove rivets (**2**) from top lining side securing points.
Pull top lining (**1**) from bows.
Pull clamping sleeve (**3**) out of corner bow (**4**) and remove from tensioning cord (**5**).
Pull tensioning cord (**5**) out rearward.
Pull rear window defroster wiring out of top lining.
Pull clamping strip (**6**) out from corner bow (**4**).

6. Remove convertible top lining screws and pull contour wire out of top. See Fig. 72.

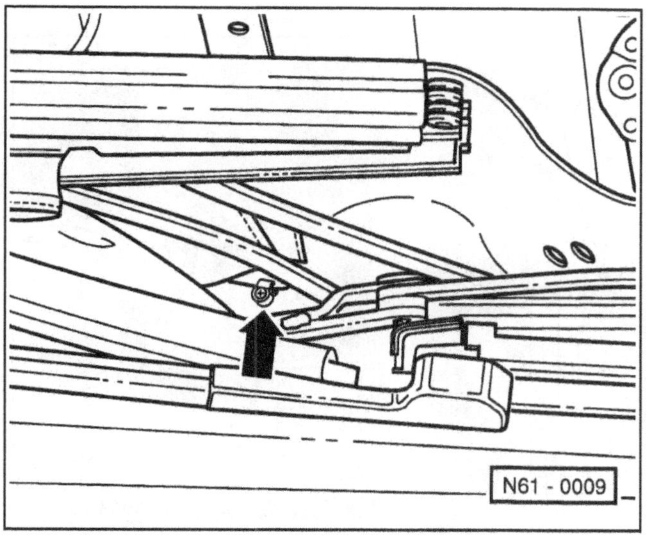

N61-0009

Fig. 72. Remove left and right convertible top lining screws (**arrow**) and pull contour wire out from convertible top lining.

Lining, installing (through 08/1998)

1. Install lining starting from inside rear. See Fig. 73.

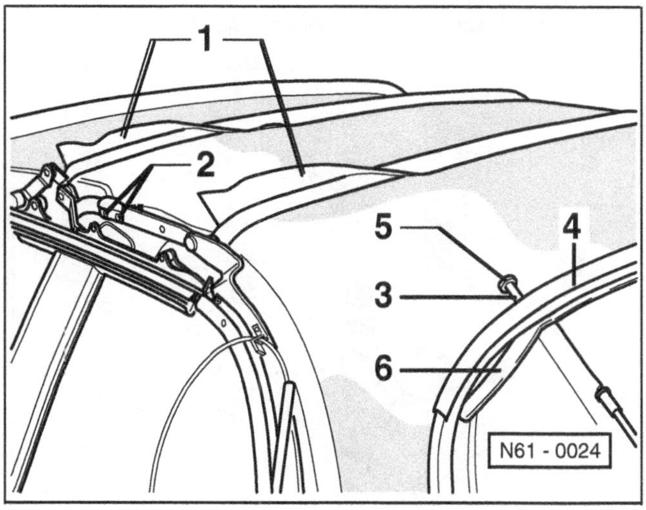

N61-0024

Fig. 73. Press convertible top lining clamp (**6**) into corner bow (**4**). Pull tension cord (**5**) through convertible top lining and corner bow. Secure clamping sleeve (**3**) to corner bow (**4**).

2. Continue installation at front inside of convertible top. See Fig. 74.

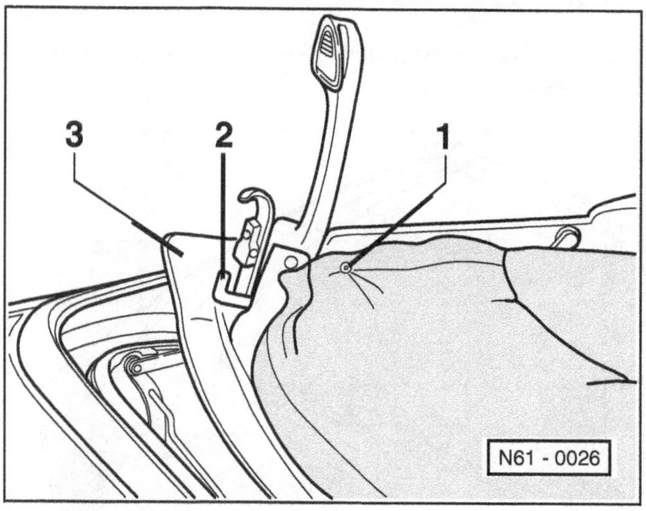

N61-0026

Fig. 74. Place convertible top lining at leading edge over guide pins (**2**) and bond in position with adhesive D 000 801. Install left and right rivets (**1**).

3. Install convertible top latch cover.
4. Rivet side strips to frame.
5. Bond strips to main and intermediate bow with adhesive D 000 801.
6. Rivet tensioning belt and wind deflector to intermediate bow.
7. Install padding as previously described.

Lining, removing (from 09/1998)

1. Remove convertible top cover.

2. Remove upholstery lining from front edge of convertible top. See Fig. 75.

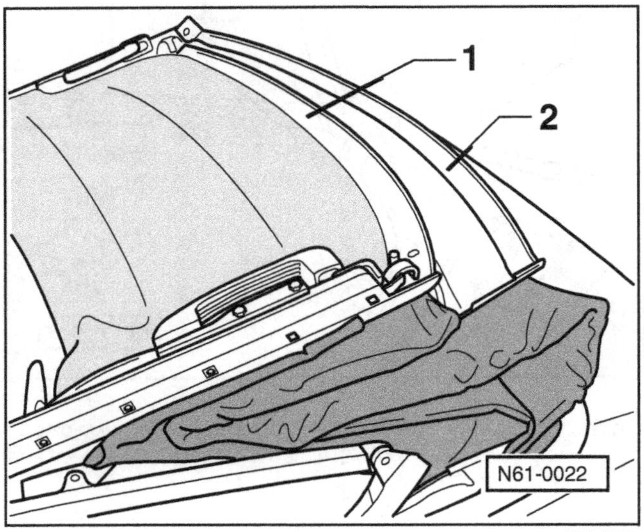

N61-0022

Fig. 75. Remove upholstery lining (**2**) from front edge of convertible top (**1**). Clean off all residual adhesive.

3. Remove screws from cover and remove window belt. This will facilitate threading of the upholstery lining. See Fig. 76.

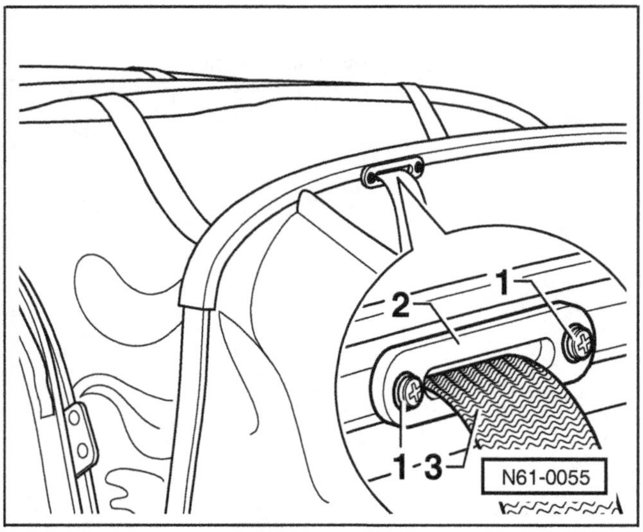

N61-0055

Fig. 76. Working at roof bow (**inset**), remove screws (**1**) from guide cover (**2**) and remove window belt (**3**) by unscrewing. This procedure facilitates threading of the upholstery lining

4. Remove mounting bolts from corner ribs and remove tension straps with mounting bracket. These components are not re-used. See Fig. 77.

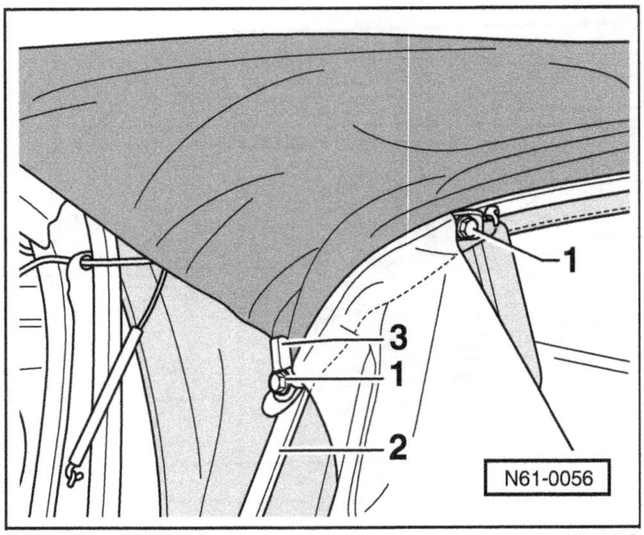

N61-0056

Fig. 77. Remove mounting bolts (**1**) from corner ribs (**2**) and remove tension straps with mounting bracket (**3**). These components are not re-used on re-assembly.

5. Carefully lay padding to the rear and remove rivets from bows. See Fig. 78.

NOTE —
New replacement upholstery lining is no longer riveted.

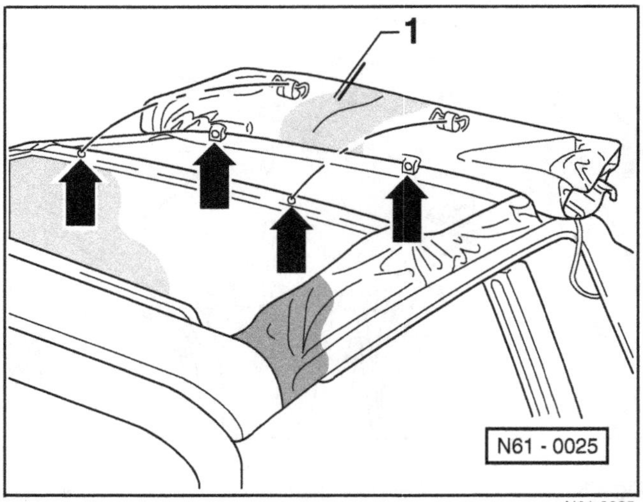

N61-0025

Fig. 78. Fold and lay upholstery padding (**1**) to the rear. Remove rivets (**arrows**) from bows.

6. Remove rivets from tensioning belt and padding. See Fig. 79.

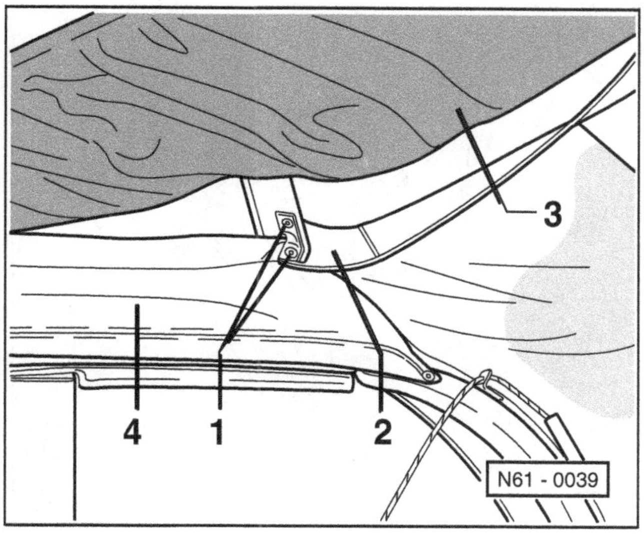

N61-0039

Fig. 79. Remove rivets (**1**) from tensioning belt (**2**) and padding (**3**). Use care to avoid damaging cover (**4**).

7. Remove engagement strip for upholstery lining from groove in corner rib. See Fig. 80.

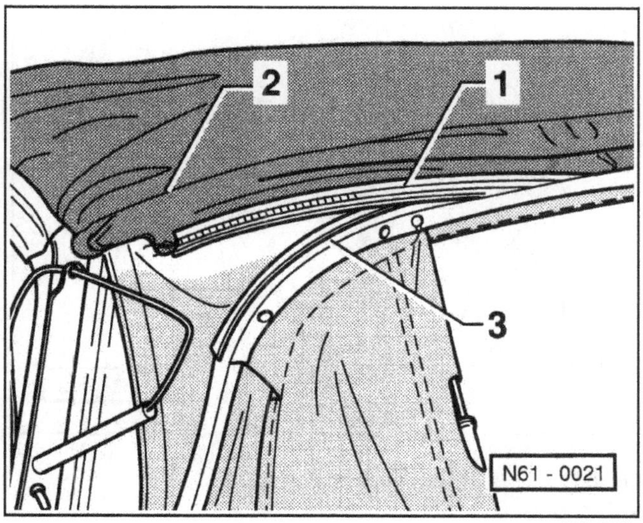

N61-0021

Fig. 80. Remove engagement strip (**1**) for upholstery lining (**2**) from engagement groove of corner rib (**3**).

8. Remove lining.

Lining, installing (from 09/1998)

1. Secure convertible top frame to assembly brackets and then secure to body. See Fig. 81.

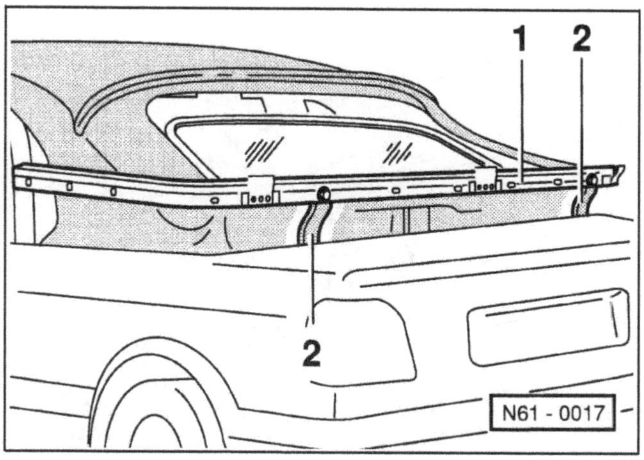

N61-0017

Fig. 81. Convertible top mounting frame (**1**) secured to body with the pair of assembly brackets VAG 1887 (**2**).

2. Remove bolts, mounting bracket and nuts from tension strap anchorage. These components are not re-used. See Fig. 82.

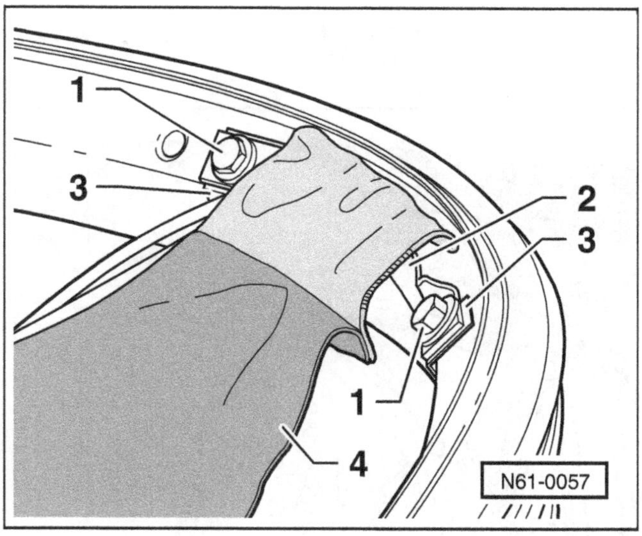

N61-0057

Fig. 82. Remove bolts (**1**), mounting brackets (**2**) and nuts (**3**) from tensioning strap anchorage points. These components are not re-used on re-assembly.

3. Modify top frame (material retaining bracket) with holes spaced as shown for the upholstery liner tension strap. See Fig. 83.

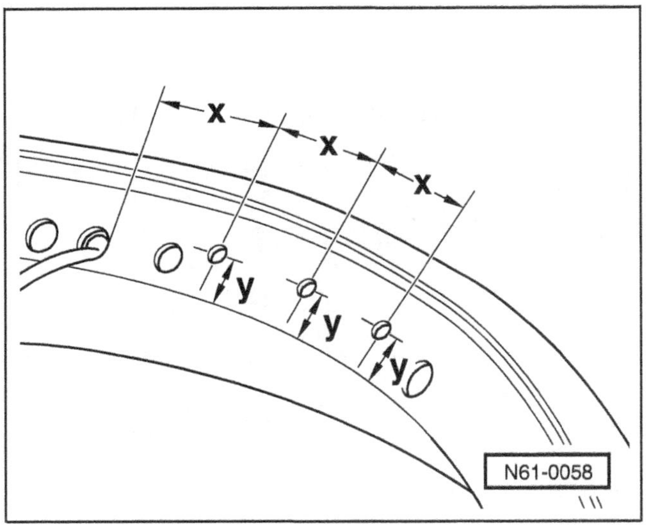

N61-0058

Fig. 83. Drill six holes in the top frame (material retaining bracket) for the upholstery liner tensioning strap.

Quantity: 6
Diameter: 3.2 mm (approximately 1/8 in.)
Distance X: 22 mm
Distance Y: 10 mm

4. Insert engagement strip for upholstery lining in groove in corner rib. See Fig. 84.

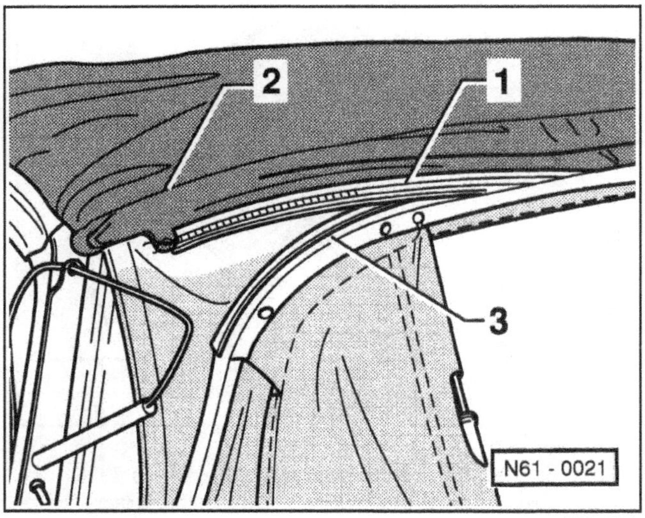

N61-0021

Fig. 84. Insert engagement strip (**1**) for upholstery lining (**2**) into groove of corner rib (**3**) in such a way so as not to be visible when fully installed.

5. Secure tension strap using rivets. See Fig. 85.

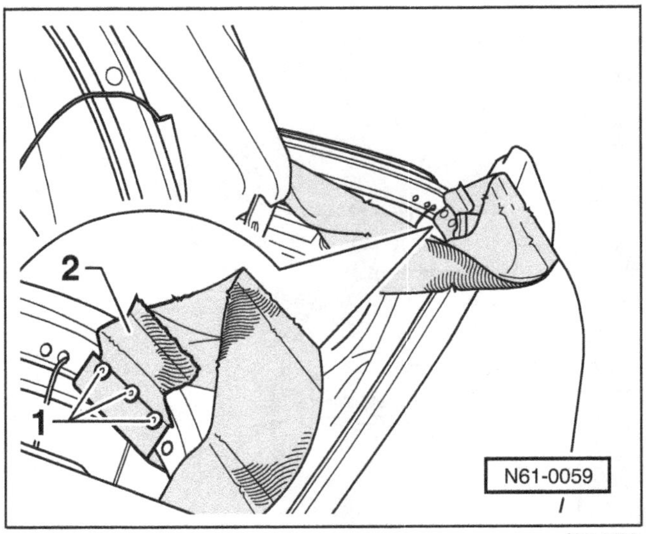

N61-0059

Fig. 85. Secure tension strap (**2**) using rivets (**1**),

6. Secure upholstery liner at rib. See Fig. 86.

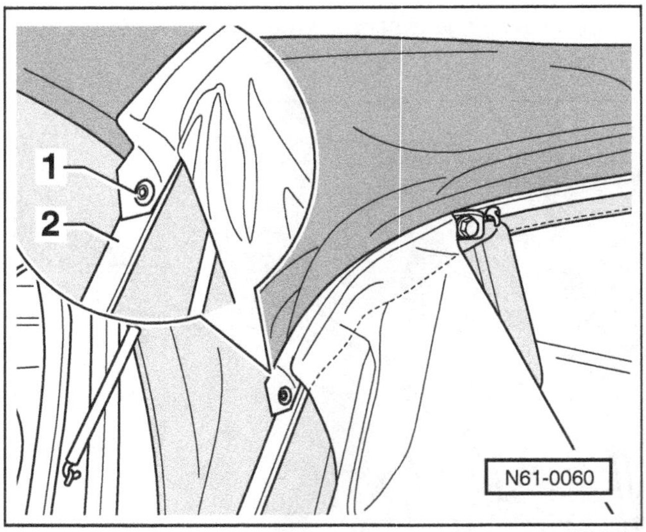

N61-0060

Fig. 86. Secure upholstery liners at rib (**2**) using rivets (**1**),

7. Secure upholstery strap at rib. See Fig. 87.

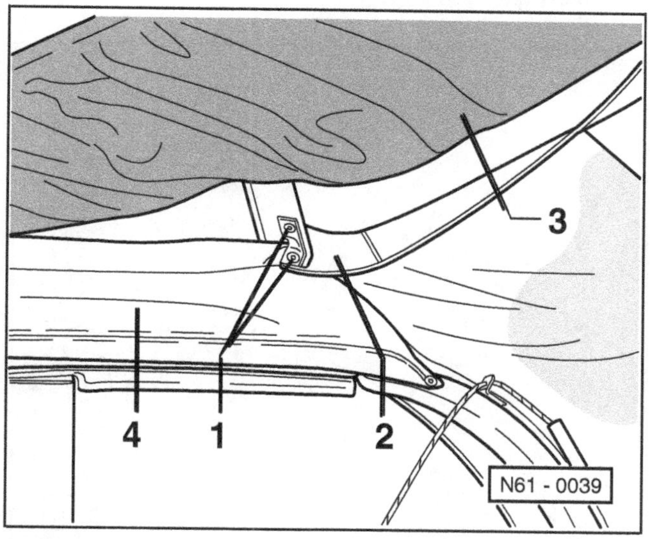

N61-0039

Fig. 87. Secure upholstery strap at rib (**2**) using rivets (**1**),

8. Insert convertible top cover completely into channel in mounting frame (use a lubricant such as silicone lubricant spray for easier installation). See Fig. 88.

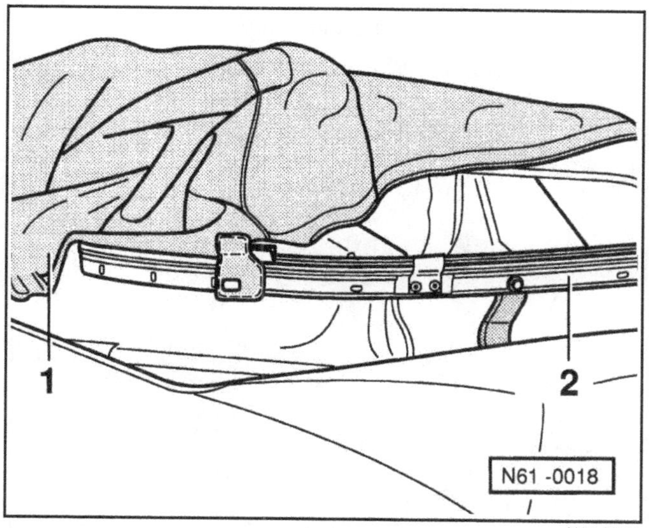

N61-0018

Fig. 88. Insert convertible top cover (**1**) onto mounting frame (**2**). Use a silicone lubricant spray (D 007 000 A2) as needed.

9. Align convertible top center marking (spot weld) with convertible top mounting frame center marking (drilled hole). See Fig. 89.

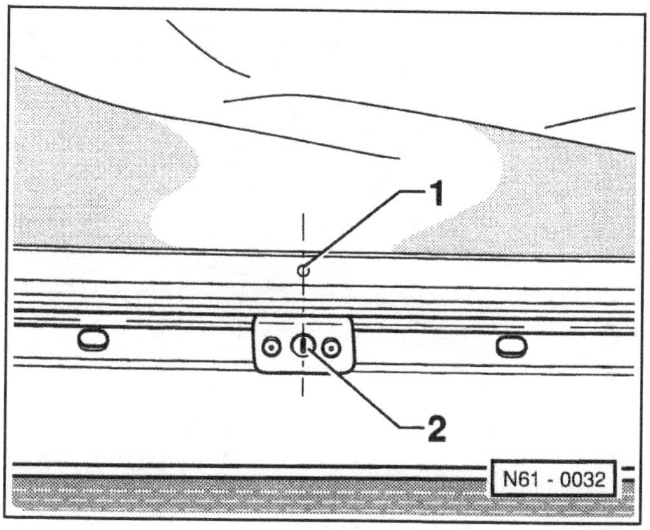

N61-0032

Fig. 89. Align convertible top center mark (spot weld **1**) with frame center marking (hole in convertible top frame **2**).

10. Attach water deflector to mounting frame with rivets. See Fig. 90.

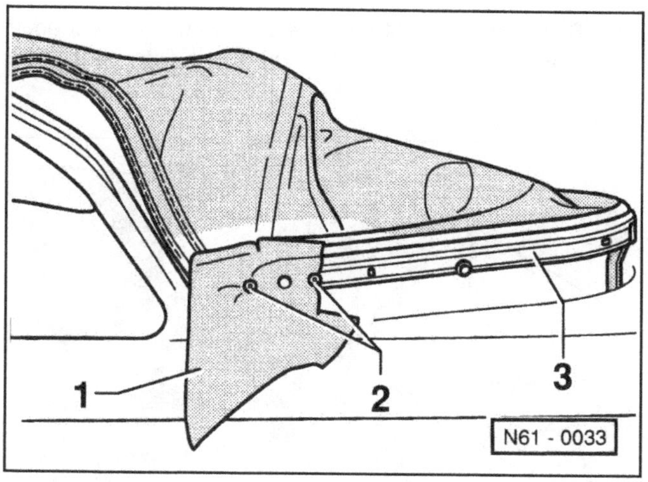

N61-0033

Fig. 90. Install water deflector (**1**) to mounting frame (**3**) with rivets (**2**).

11. Fold edge of convertible top forward and clean residual felt from edge of top. Attach double sided adhesive tape to front edge of top. See Fig. 91.

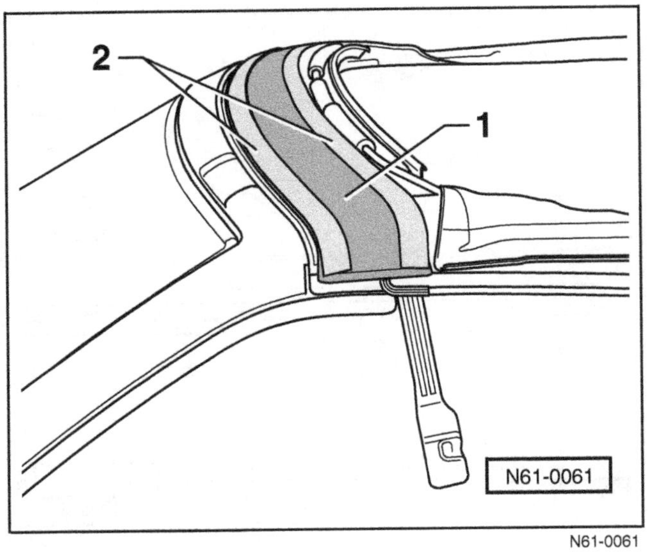

N61-0061

Fig. 91. Attach double sided adhesive tape (**2**) to front edge of convertible top (**1**).

12. Engage convertible top catches into latches on windshield frame, but do not pull down and lock.

13. Bond lower part of upholstery liner flush to front edge of top. Attach double sided tape to upholstery liner. See Fig. 92.

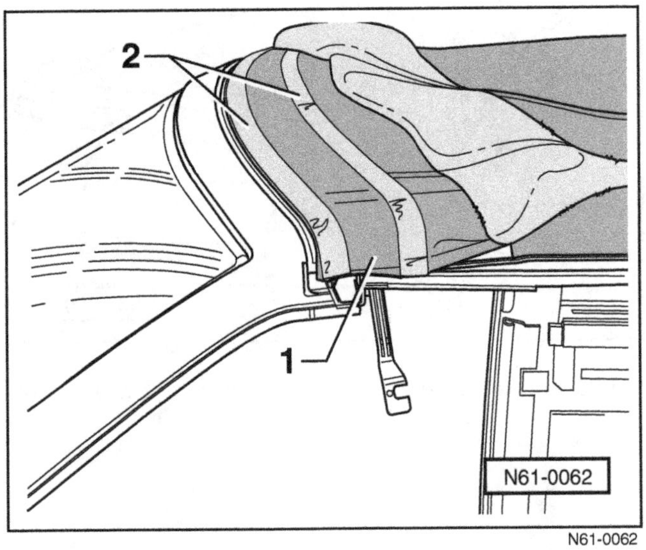

N61-0062

Fig. 92. Attach double sided adhesive tape (**2**) to front edge of convertible top (**1**).

14. Adhere inner section of upholstery flush with upholstery liner and edge of top. See Fig. 93.

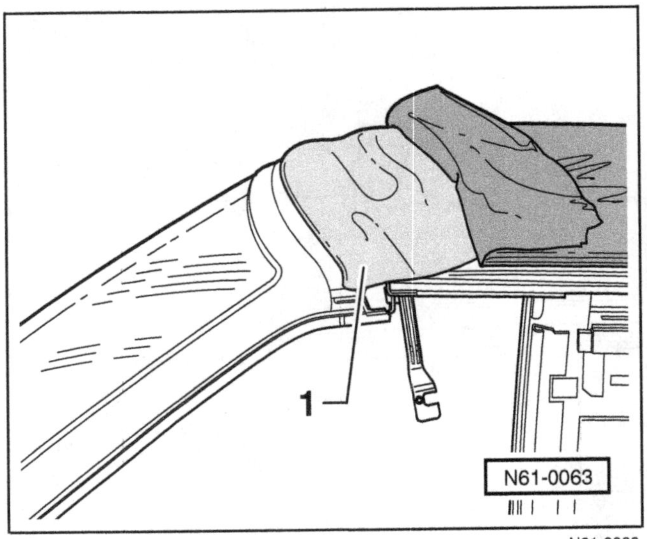

N61-0063

Fig. 93. Adhere inner section of upholstery (**1**) flush with upholstery liner and edge of convertible top.

15. Fold convertible top rearward and securely bond upper section of upholstery liner to lower part of top edge with double sided adhesive tape. See Fig. 94.

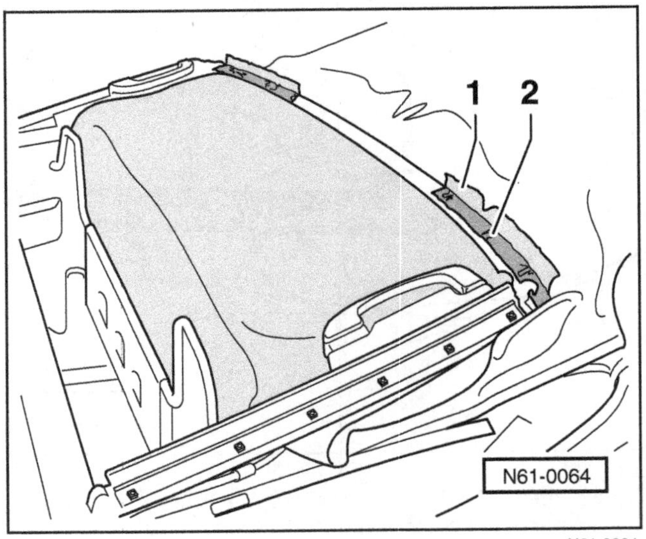

N61-0064

Fig. 94. Fold convertible top rearward and securely bond upper section of upholstery liner (**2**) to lower part of top edge (**1**) with double sided adhesive tape.

16. Install convertible top cover for using instructions for the appropriate version.

POWER CONVERTIBLE TOP

Electro/hydraulic convertible top assembly

1. **Convertible top switch**
   - Located in center dash panel, exact position varies. See **Convertible top switch location**
2. **Convertible top control module (J256)**
   - Located under rear seat
   - Not equipped with self diagnosis
3. **Convertible top frame main bracket**
   - Quantity 2
4. **Hydraulic cylinder**
   - Quantity 2
   - Double action
5. **Hydraulic fluid level indicator**
   - Correct level is obtained with convertible top in open position
   - Correct level is visually verified when between MIN and MAX markings
   - Use only hydraulic fluid G 002 000. See **Hydraulic fluid type**
6. **Oil filler plug**
   - Tighten to 5 Nm (44 in-lb)
7. **Pressure relief screw**
   - Removing pressure relief screw releases internal system pressure allowing manual convertible top operation.
8. **Hydraulic hoses**
9. **Hydraulic pump**
   - Located in left rear of trunk compartment
   - Replaces countermass found in vehicles with manual top
   - If the hydraulic pump is overloaded, a thermo-fuse opens the circuit and the system is switched off for a maximum of 35 seconds.
10. **Reed contact switch**
    - Quantity 2
    - Located in protection trim cover
    - Contact switches on top of rear side trim together with rod magnets prevent power convertible top operation when cover is installed

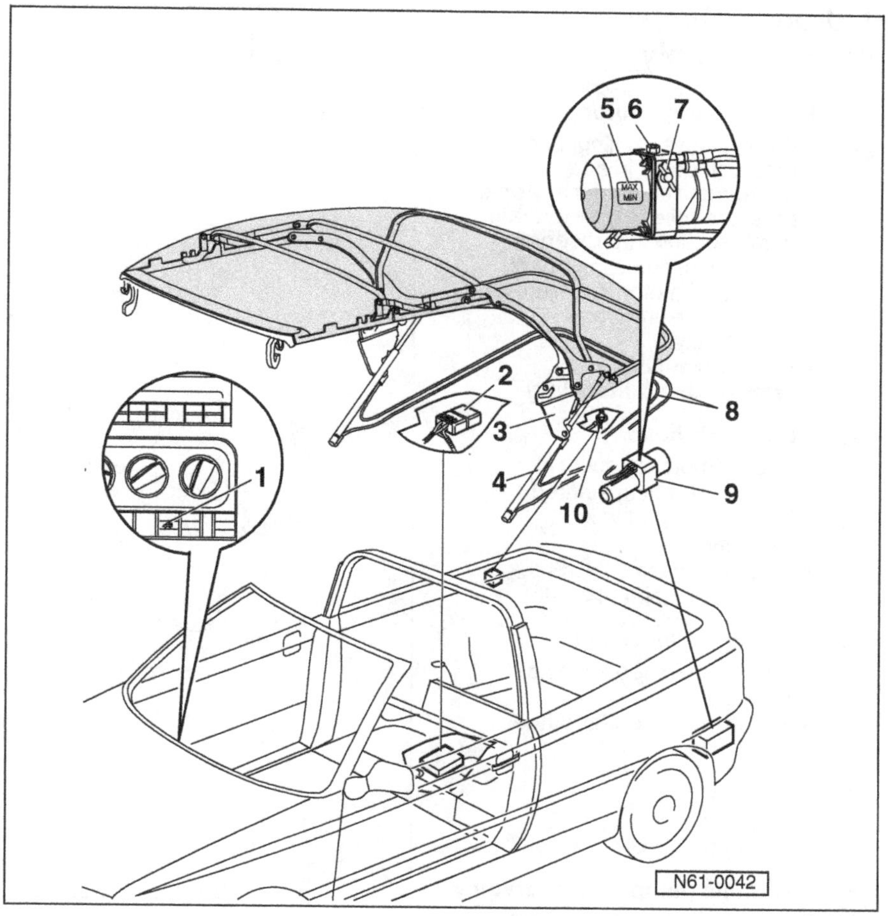

N61-0042

Fig. 95. Exploded view of electro-hydraulic convertible top components.

Convertible top switch location

- Location of switch (**arrow**) within row of switches varies with vehicle model year.

Hydraulic fluid type

- Use only VW specified hydraulic fluid G 002 000.

Hydraulic pump, motor assembly

1. **Electric pump motor**
   - When disassembling, position reservoir down in an appropriate container and open pressure relief screw to drain pump section into reservoir
   - When disassembling, mark pump and motor to ensure correct indexing upon reassembly
   - See **Hydraulic pump, removing and installing**
2. **Bolt, Torx® head**
   - Secure motor to pump
   - With micro-encapsulated locking compound
   - Quantity 2
   - Always replace
   - Tighten to 5 Nm (44 in-lb)
3. **Pump impeller wheel**
   - See **Pump impeller wheel, removing**
   - See **Pump impeller wheel, installing**
4. **O-ring seal**
   - Always replace
   - Part of repair kit
5. **Hydraulic pump and reservoir**
   - Reservoir available as part of repair kit.
   - See **Hydraulic pump, reservoir, assembly**

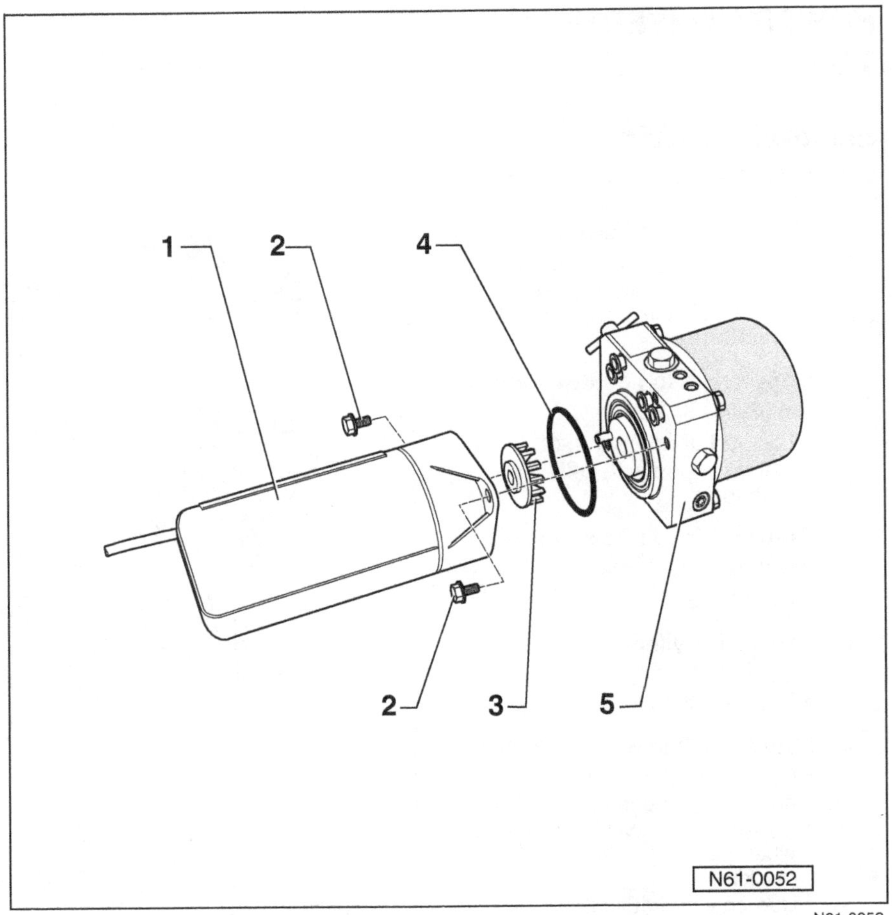

N61-0052

Fig. 96. Exploded view of electric motor and pump components.

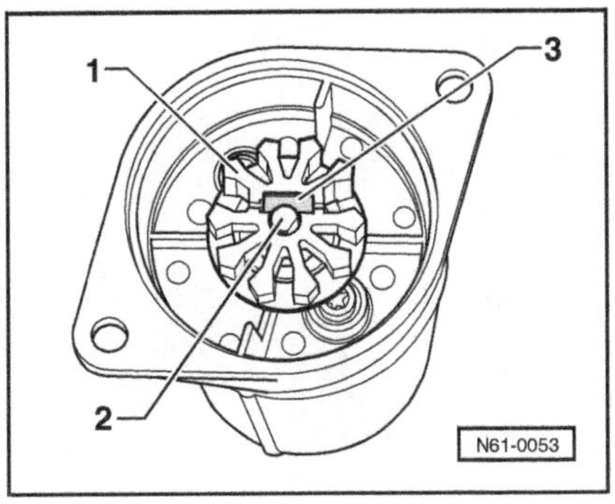

Pump impeller wheel, removing

- Remove pump impeller wheel (**1**) from shaft (**2**).
- Remove and discard metal key (**3**) from pump housing. Replacement impeller wheels do not use metal key (**3**).

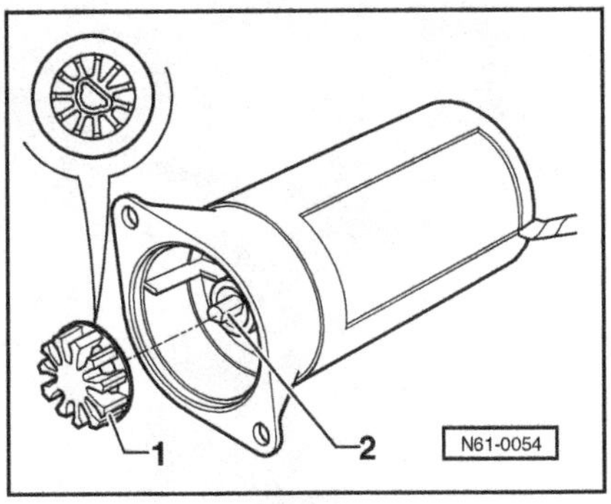

Pump impeller wheel, installing

- Press new impeller wheel (**1**) onto shaft (**2**). Replacement impeller wheels fit pump motor shaft in one position only.

Hydraulic pump, oil reservoir assembly

NOTE —
Hydraulic pump assembly must be removed from vehicle before removal of hydraulic fluid reservoir.

> *CAUTION —*
> *Place a suitable container under pump assembly to contain hydraulic fluid before removing reservoir.*

1. **Bolt, Torx® head**
   - Secure reservoir to pump
   - With micro-encapsulated locking compound
   - Quantity 4
   - Always replace
   - Tighten to 2 Nm (18 in-lb).

> *CAUTION —*
> *Do not over tighten! Over tightening will distort reservoir and prevent proper sealing*

2. **Hydraulic fluid reservoir**
   - Part of repair kit
   - Moisten O-ring seating area with hydraulic fluid prior to assembly
   - Install to hydraulic pump with twisting motion

3. **O-ring seal**
   - Always replace
   - Part of repair kit
   - Moisten O-ring with hydraulic fluid before installing

4. **Hydraulic pump and reservoir**
   - Reservoir available as part of repair kit

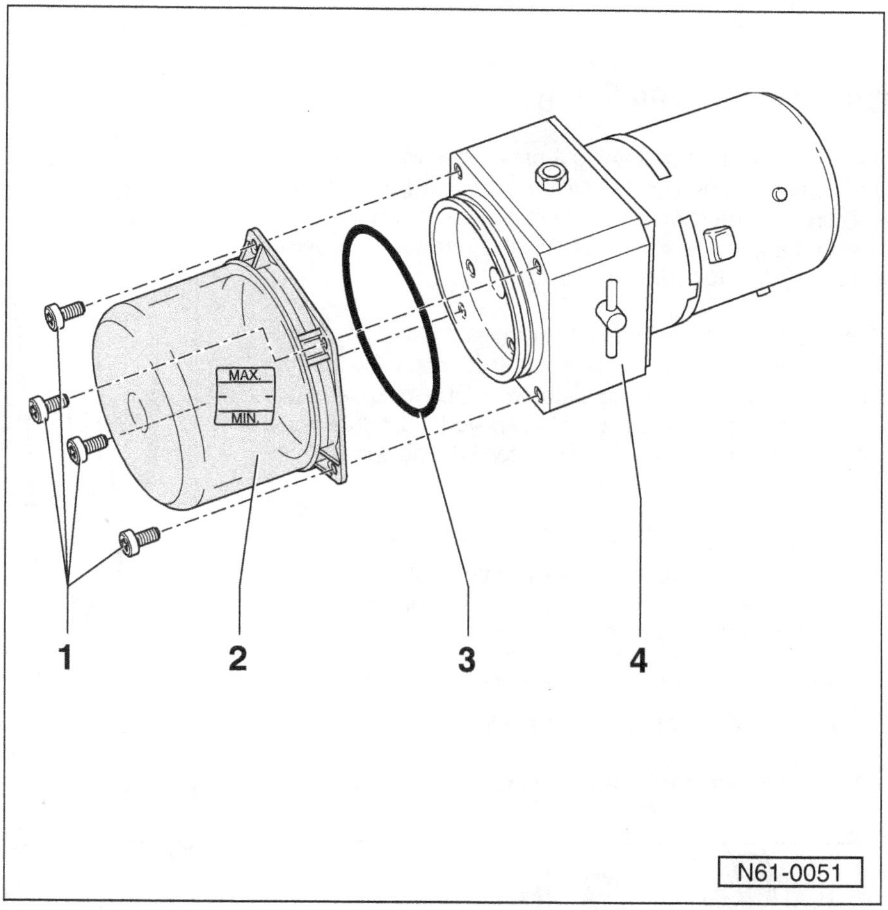

N61-0052

Fig. 97. Exploded view of hydraulic pump reservoir components.

Hydraulic system, filling

The hydraulic unit containing a pressure relief screw is located in the trunk behind the trim cover. Turning the pressure relief screw counterclockwise releases pressure in the system allowing manual operation. Turning the screw clockwise returns the system to hydraulic operation.

The electro/hydraulic convertible top is a closed hydraulic system. Filling is normally only required after opening the system such as after replacing a component. Because of the stroke and the displacing effect of the piston rod, hydraulic oil level must only be checked when the top is in the open (down) position.

> ***CAUTION—***
> *Do not fill the system above the "MAX" mark. Failure to observe the "MAX" level requirement may allow system pressure to rise high enough to cause damage.*

1. Remove CD changer where equipped.

2. Remove left rear trunk trim panel. See Fig. 98.

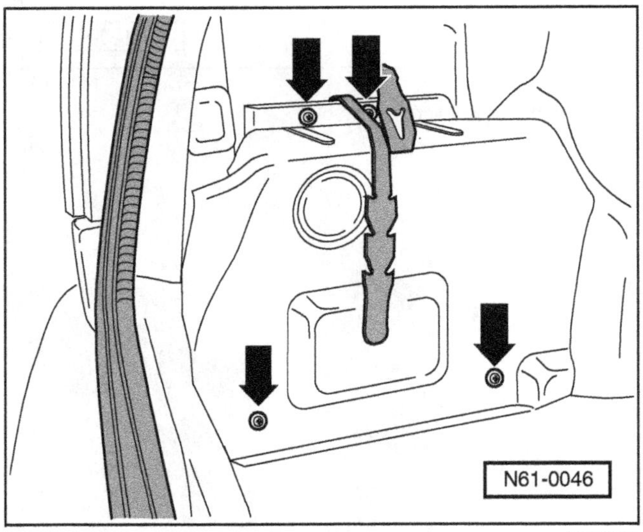

N61-0046

Fig. 98. Remove screws (**arrows**) from left rear trim panel and remove panel.

3. Open (lower) the top to fill system. Correct level is only registered in this position.

4. Remove oil filler plug and add oil until level is between "MIN" and "MAX" marks on reservoir. See Fig. 99.

> ***NOTE—***
> *An clean oil can or squeeze bottle is suitable for this purpose and must be used to prevent spillage and contamination of hydraulic oil.*

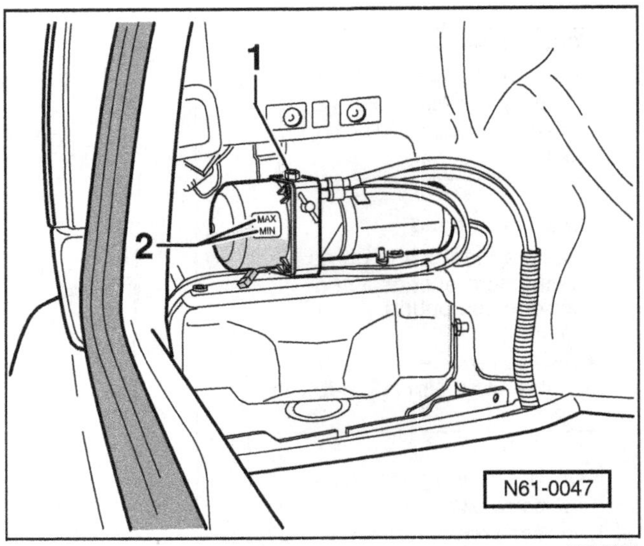

N61-0047

Fig. 99. Remove oil filler plug (**1**). Add specified type of oil (G 002 000) until level is between "MIN" and "MAX" marks (**2**). Tighten filler plug (**1**) to 5 Nm (44 in-lb).

5. Reinstall CD changer (where applicable) and left rear trunk trim panel.

Hydraulic system, bleeding

The hydraulic system is self-bleeding during normal operation, but can also be bled manually. The system is bled by operating it several times with the fill plug loosened.

1. Remove CD changer where equipped.

2. Remove left rear trunk trim panel.

3. Loosen hydraulic fluid filler plug and operate power top via switch on dash.

NOTE —
- *System is bled during opening and closing cycle.*
- *Convertible top must open and close smoothly without shaking and bouncing.*
- *Hydraulic fluid level must be between "MIN" and "MAX" marks when top is open (down). See Fig. 100.*

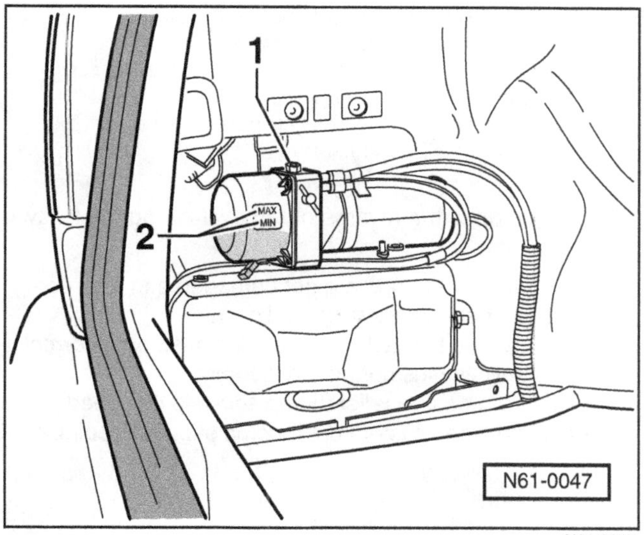

N61-0047

Fig. 100. Loosen oil filler plug (**1**) and operate convertible top switch on dash through 4 or 5 complete cycles. Ensure that hydraulic fluid level is between "MIN" and "MAX" marks (**2**) upon completion. Tighten filler plug (**1**) to 5 Nm (44 in-lb).

4. Reinstall CD changer (where applicable) and left rear trunk trim panel.

Hydraulic pump, removing and installing

The hydraulic pump can be removed for repair or replacement. Use care to avoid hydraulic system contamination.

> ***CAUTION —***
> - *Hydraulic system is sensitive to all types of contamination. Use care when system is open.*
> - *As soon as system is opened, any opened fittings must be capped to prevent contamination ingress.*
> - *Disconnect battery ground (GND) strap before working on the electrical system.*

NOTE —
Before disconnecting battery be sure to obtain radio anti-theft code.

1. Disconnect battery ground strap.

2. Remove CD changer where equipped.

3. Remove left rear trunk trim panel.

4. Continue removal of electrical and hydraulic connections at left rear. See Fig. 101.

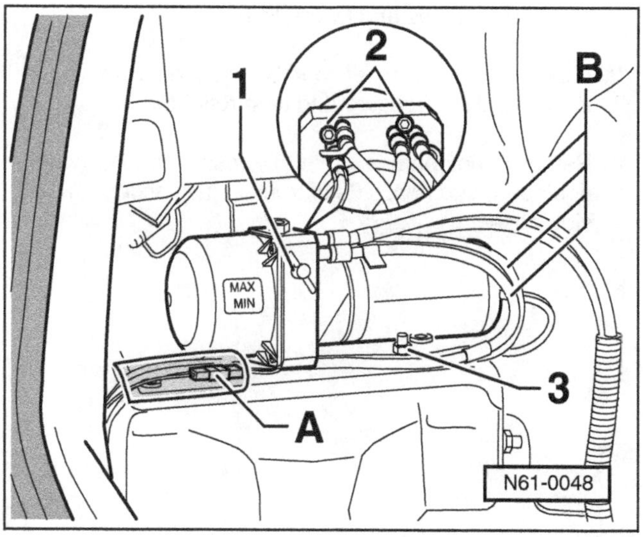

N61-0048

Fig. 101. Disconnect harness connector (**A**).
Loosen pressure relief thumb screw (**1**) to release residual pressure in system and then retighten.
Mark hydraulic lines (**B**) before removing.
Remove screws (**2**).
· Tighten to 7 Nm (62 in-lb)
Pull hydraulic lines (**B**) from holes.
Remove nuts (**3**) on rubber mounts (second nut behind pump)
· Tighten to 5 Nm (44 in-lb)

5. Installation is the reverse of removal noting the following points:
   - Ensure that O-rings are not damaged and seat correctly when installing hydraulic lines.
   - Ensure that hydraulic lines are installed onto correct fittings during reassembly.
   - Fill and bleed system upon completion of repairs.
   - After connecting battery, check vehicle equipment (radio, clock, electric windows) and reset as required.

Hydraulic hoses, removing and installing

The hydraulic hoses can be removed separately from the pump for replacement. Use care to avoid hydraulic system contamination.

> ***CAUTION—***
> - *Hydraulic system is sensitive to all types of contamination. Use care when system is open.*
> - *As soon as system is opened, any opened fittings must be capped to prevent contamination ingress.*

1. Remove CD changer where equipped.
2. Remove trim panels on left side of trunk to gain access to. left side hoses and hydraulic pump.
3. Loosen pressure relief thumb screw on pump to release residual pressure and then retighten.
4. Remove trim panels on rear cross member (where equipped) and right side of trunk to gain access to right side hoses.
5. Unclip right side hoses from cross panel. See Fig. 102.

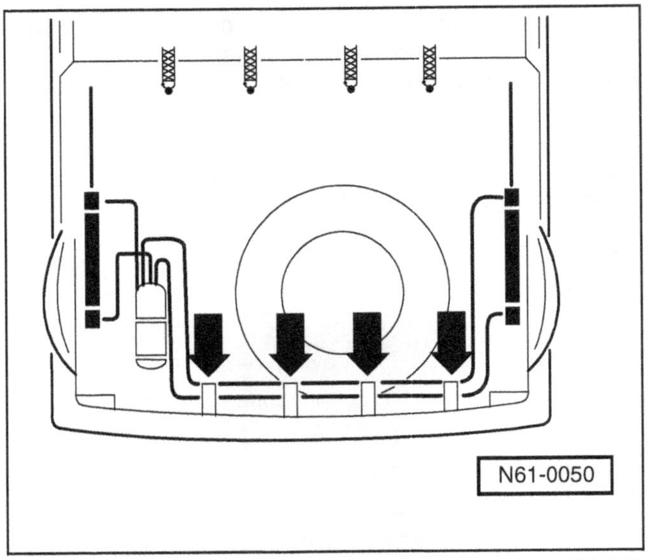

N61-0050

Fig. 102. Unclip hoses from rear crossmenber (**arrows**).
Mark hydraulic hoses before removing.
Disconnect hose connections (Torx®) at hydraulic pump.
- Tighten to 7 Nm (62 in-lb)

Disconnect hose connections at union on hydraulic cylinder.
- Tighten to 15 Nm (11 ft-lb)

6. Installation is the reverse of removal noting the following points:
   - Ensure that O-rings are not damaged and seat correctly when installing hydraulic lines.
   - Ensure that hydraulic lines are installed onto correct hydraulic fittings during reassembly.
   - Ensure pressure relief thumb screw is tightened.
   - Fill and bleed system upon completion of repairs.

Hydraulic cylinders, removing and installing

The hydraulic cylinders can be removed separately from the pump for replacement. Use care to avoid hydraulic system contamination.

> ***CAUTION—***
> - *Hydraulic system is sensitive to all types of contamination. Use care when system is open.*
> - *As soon as system is opened, any opened fittings must be capped to prevent contamination ingress.*

1. Remove CD changer where equipped.
2. Remove trim panels on left side of trunk to gain access to. hydraulic pump, and left side hydraulic cylinder and hose.

3. Loosen pressure relief thumb screw on pump to release residual pressure and then retighten.

4. Remove trim panels on right side of trunk to gain access to right side hydraulic cylinder.

5. Continue removal of electrical and hydraulic connections at left rear. See Fig. 103.

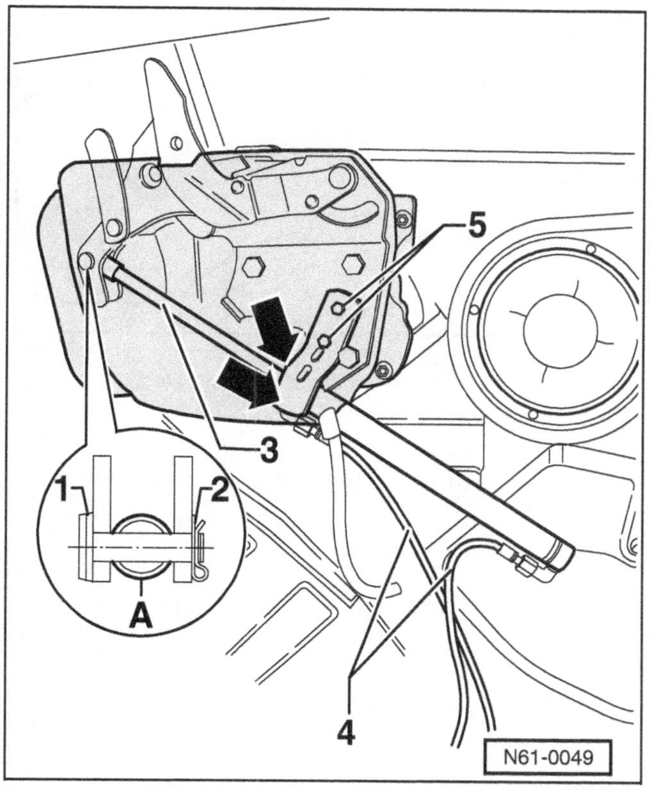

N61-0049

Fig. 103. Pry off circlip (**2**) to press out retaining pins (**1**). Observe position of plastic sleeve (**inset, A**).
Remove hydraulic cylinder piston rod (**3**) from convertible top frame.
Mark and remove hydraulic hoses (**4**) from cylinder union nuts and cover/cap all open lines.
· Tighten to 15 Nm (11 ft-lb)
Remove bolts with washers (**5**) from cylinder bracket.
· Tighten to 8 Nm (70 in-lb)
Observe position of plastic washers (**arrows**) when removing.

6. Installation is the reverse of removal noting the following points:
   - Ensure that plastic sleeve and plastic washers between hydraulic cylinder and top frame are properly positioned when reassembling.
   - Ensure that O-rings are not damaged and seat correctly when installing hydraulic lines.
   - Ensure that hydraulic lines are installed onto correct hydraulic fittings during reassembly.
   - Ensure pressure relief thumb screw is tightened.

7. Fill and bleed system upon completion of repairs.

WINDSHIELD FRAME SEAL

Replacing

1. Carefully remove front interior light assembly. See Fig. 104.

V70-0437

Fig. 104. Carefully pry out front interior light assembly and disconnect harness connector.

2. Remove sun visor and retaining clip. See Fig. 105.

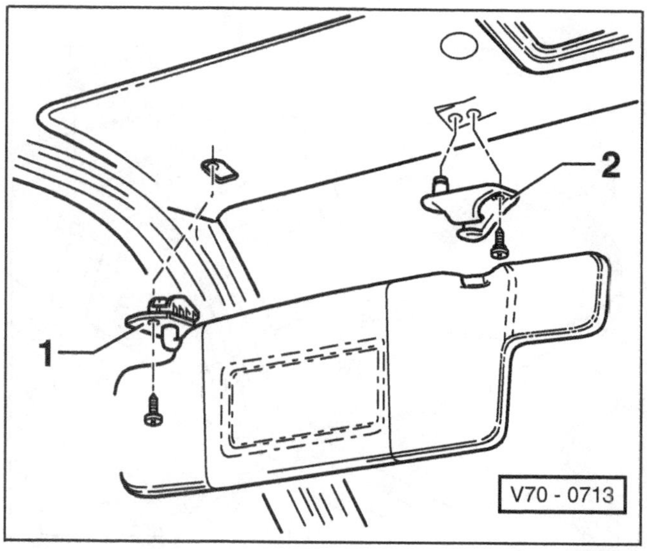

V70-0713

Fig. 105. Remove sun visor screws and pull sun visor (**1**) out of windshield frame. Remove retaining clip screw and turn retaining clip (**2**) ¼ turn (90°) to remove. Disconnect harness connectors as required.

3. Remove windshield trim.

4. Remove mount and windshield seals. See Fig. 106.

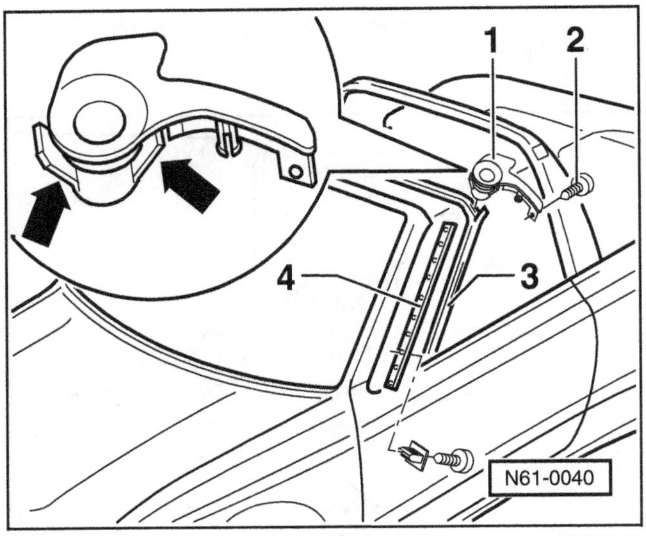

N61-0040

Fig. 106. Remove mount screws (**2**). Press mount locking tabs (**inset, arrows**) together and remove mount (**1**). Pull windshield seal (**3**) out from retainer (**4**). Upper area of windshield of seal is attached and sealed with butyl tape.

5. Begin installation by coating retainer and frame seal with a lubricant such as cleaner D009 401 04. Pull seal into retainer. See Fig. 107.

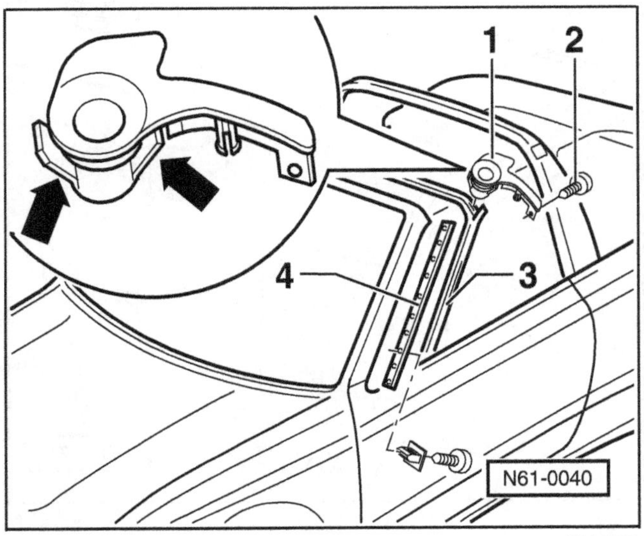

N61-0040

Fig. 107. Coat retainer (**4**) and windshield frame seal (**3**) with a lubricant such as D 009 401 01 and pull seal (**3**) into retainer (**4**).

6. Seal windshield frame. See Fig. 108.

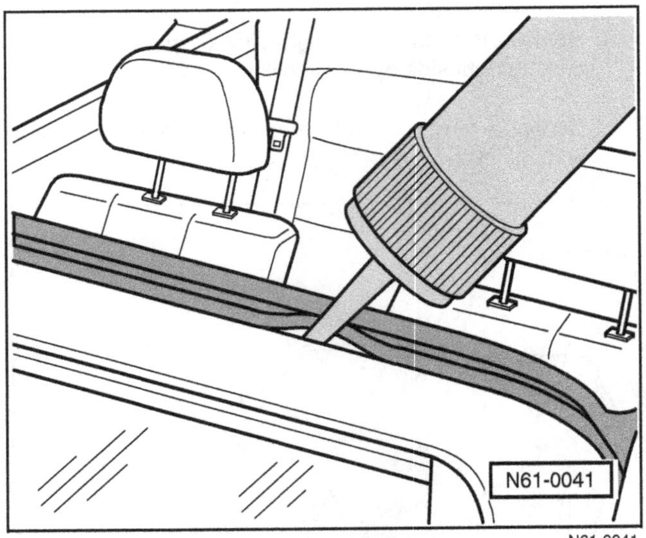

N61-0041

Fig. 108. Seal upper area of windshield frame with sealant D 469 101 A3 applied from a gun as shown.

7. Reinstall trim, clips, sun visor, and interior light.

CONVERTIBLE TOP OPERATION

Thanks to its design, the convertible roof, whether vinyl or fabric, is particularly easy to operate. Its multi-layer construction provides good thermal insulation and makes it wind and waterproof. under normal operating conditions. Only under extreme conditions such as heavy downpours or high pressure car washing, for example, might a small amount of water enter the interior between the top and the bodywork. This situation is normal and should not be cause for any concern.

Factory recommendations include the following:

- Clean the top regularly.
- Do not lower the top when it is dirty. Gritty dirt particles can scratch the rear window and the top itself. This can lead to damage over time. It should probably be mentioned that the same applies to ice and snow.
- Do not lower the fabric top unless it is completely dry (also suggested for vinyl convertible tops). Lowering a damp fabric top may result in unsightly marking, and mildew staining and odor are possible for both.
- Do not leave the top folded down for extended periods (several weeks). This avoids discoloration occurring in the folds and allows creases to stretch out.

> ***CAUTION—***
> *For safety reasons, the top should only be opened and closed when the vehicle is stationary!*

Manual top

1. Remove any objects from the luggage compartment cover area.

2. Release both locking levers. See Fig. 109.

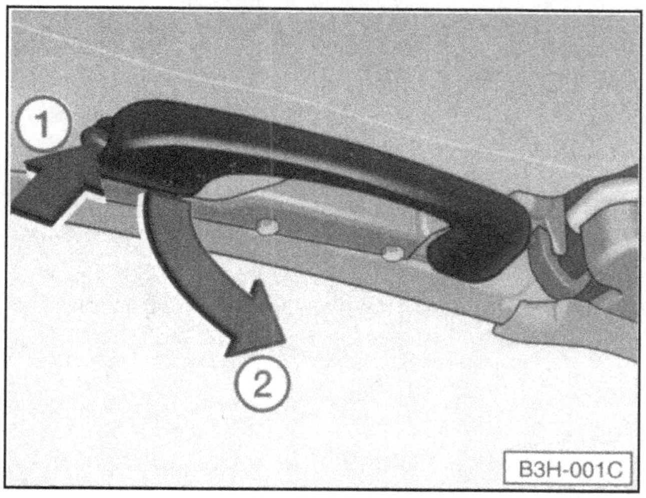

B3H-001C

Fig. 109. Release locking levers by pressing release button (**arrow 1**) and turning the lever downward (**arrow 2**). When released, turn the lever upward again.

3. Fold the top back. See Fig. 110.

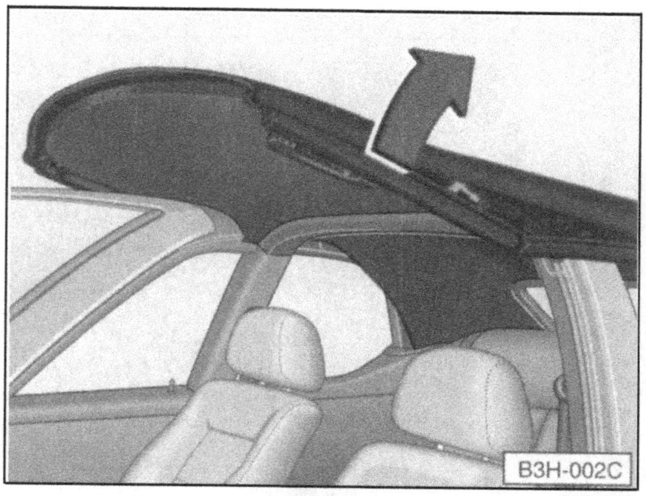

B3H-002C

Fig. 110. Grab the top and fold it back (**arrow**) towards the rear of the vehicle

4. Press the middle of the roof down until the latches on the frame engage on both sides. See Fig. 111.

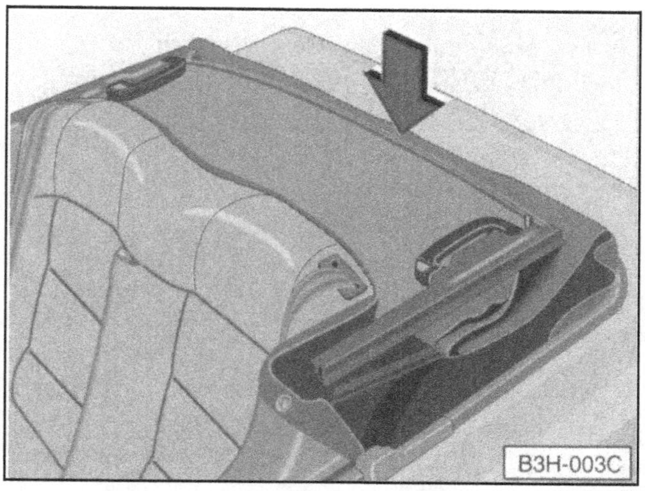

B3H-003C

Fig. 111. Press the middle of the roof (**arrow**) down until the latches on the frame engage on both sides

5. Install the protective cover (boot).

6. To close the top, first remove the protective cover.

7. Unlatch the convertible top with the release button on the left side latch handle. See Fig. 112.

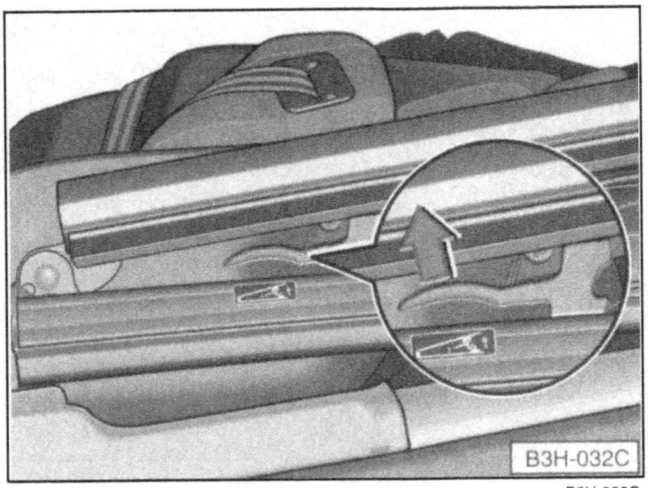

B3H-032C

Fig. 112. Unlatch the top by pressing the release button on the left side handle (**inset arrow**).

8. Pull the top toward the front of the vehicle, and unlatch the locking levers.

9. Pull down on the locking levers, engage the catches and lock the top. See Fig. 113.

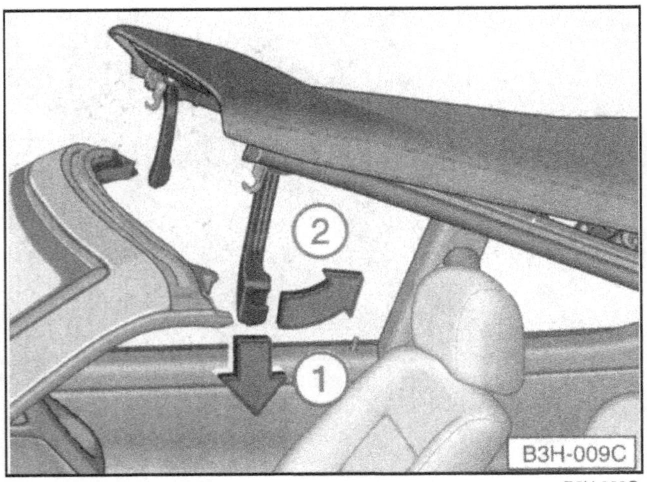

B3H-009C

Fig. 113. Pull the top down by grasping the locking lever (**arrow 1**). Engage the catches on the windshield frame and rotate the levers upward until they stop (**arrow 2**). The roof is now locked.

10. Stow the protective cover (boot).

Protective cover (boot)

1. Secure the front of the protective cover using the attached snaps and tuck the cover in behind the vehicle body. See Fig. 114.

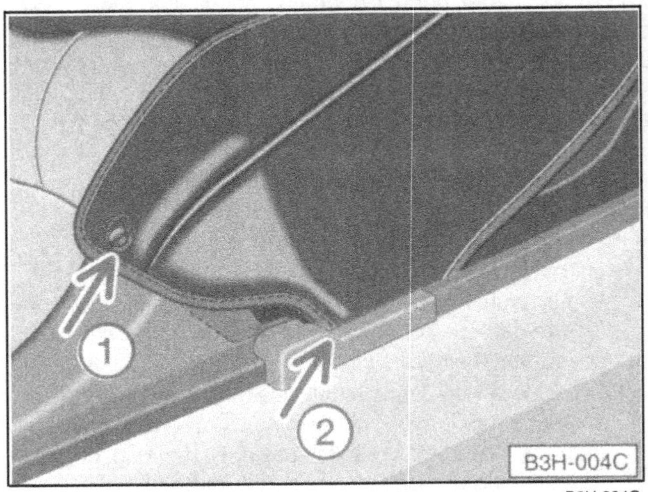

B3H-004C

Fig. 114. Secure the front of the protective cover using the snap buttons (**arrow 1**). Tuck the cover in behind the vehicle body on the sides (**arrow 2**).

2. Tuck the protective cover in at the rear of the body. See Fig. 115.

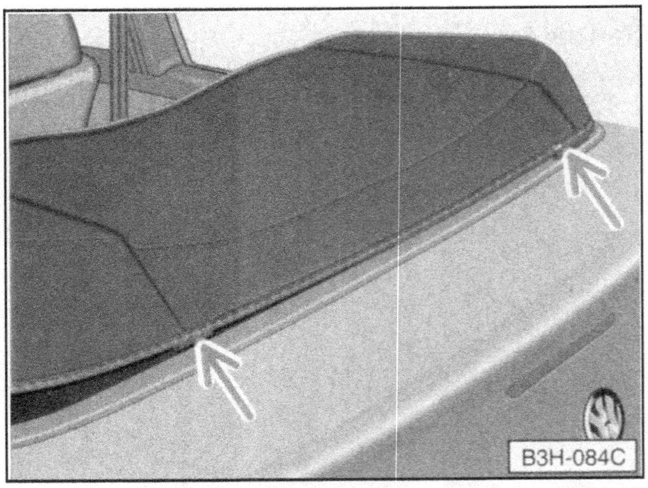

B3H-084C

Fig. 115. Secure the rear of the protective cover into the body at the indicated (**arrows**) locations first. When the first part of the protective cover is properly installed, tuck in the remainder.

3. Tuck the front part of the protective cover in behind the back seat backrest. See Fig. 116.

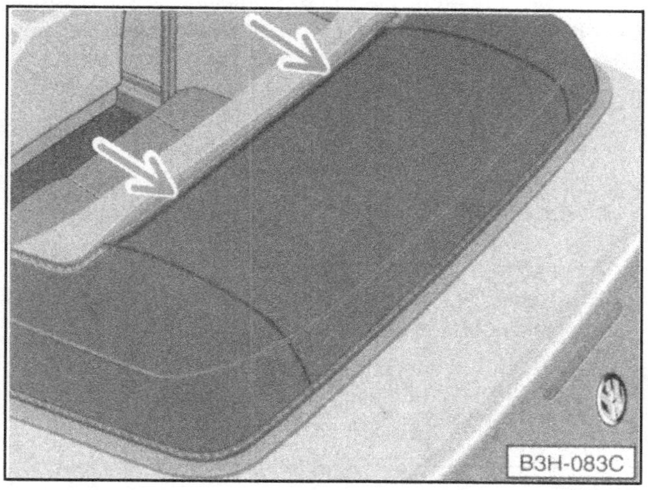

B3H-083C

Fig. 116. Tuck the front part of the protective cover in behind the back seat back rest (**arrows**).

NOTE—

Vehicles with power convertible tops have an electrical connection in the front snaps (figure 112, item 1). This electrical connection prevents the power top from operating when the boot is properly installed.

Power top

1. Remove any objects from the luggage compartment cover area.

2. Release both locking levers. See Fig. 117.

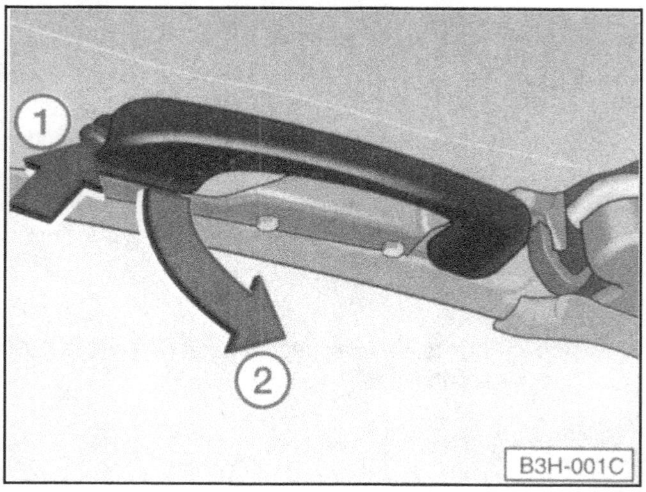

B3H-001C

Fig. 117. Release locking levers by pressing release button (**arrow 1**) and turning the lever downward (**arrow 2**). When released, turn the lever upward again.

3. Pull up on the parking brake handle.

NOTE—

For safety reasons, the power convertible top will only work when the parking brake is set.

4. Insert the key into the steering lock and switch it to the ignition on position (position 2).

5. Push and hold the bottom part of the dash switch until the top is **completely** folded down. See Fig. 118.

> ***CAUTION—***
>
> *Never press the switch unless the roof is unlocked! Pressing the switch with the convertible top still in the locked position could cause substantial damage to the frame due to the powerful hydraulic system needed to operate the top.*

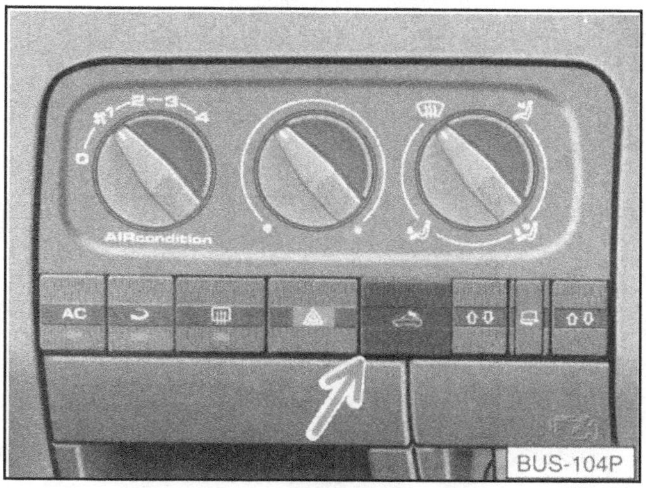

BUS-104P

Fig. 118. Push and hold the bottom part of the dash switch (**arrow**) until the top is completely folded down.

6. Install the protective cover (boot).

NOTE—

- *The power top is protected by a circuit breaker which will automatically switch off the power to the hydraulic system operating the top in the event of an overload. Power is restored automatically a few seconds later.*
- *Vehicles with power convertible tops have an electrical connection in the front snaps (figure 112, item 1). This electrical connection interrupts the electrical supply to the hydraulic system and prevents the power top from operating when the boot is properly installed. This safety system prevents damage to the power top frame and the protective cover.*
- *Never drive the vehicle without the protective cover (boot) installed when the roof is folded. The cover protects the top from dust and dirt, and keeps it in place to prevent lifting or "flying" while driving.*

7. To close the power top, first remove the protective cover.

8. Pull up on the parking brake handle.

NOTE —
Remember, for safety reasons, the power convertible top will only work when the parking brake is set.

9. Insert the key into the steering lock and switch it to the ignition on position (position 2).

10. Push and hold the upper part of the dash switch until the top is **completely** raised to the up position.

> ***CAUTION —***
> *Never press the switch unless the roof is unlocked! Pressing the switch with the convertible top still in the locked position could cause substantial damage to the frame due to the powerful hydraulic system needed to operate the top.*

11. Unlatch the locking levers.

12. Pull down on the locking levers, engage the catches and lock the top. See Fig. 119.

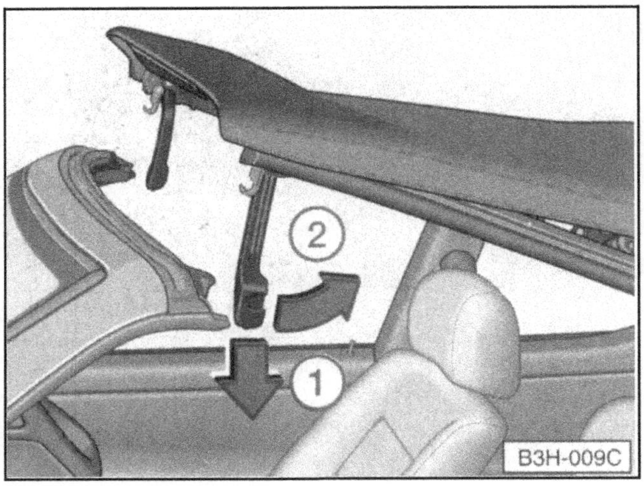

B3H-009C

Fig. 119. Pull the top down by grasping the locking lever (**arrow 1**). Engage the catches on the windshield frame and rotate the levers upward until they stop (**arrow 2**). The roof is now locked.

13. Stow the protective cover (boot).

Power top, manual operation

In the event of an electrical power malfunction, the power convertible top can still be raised and lowered by hand. The following procedure will allow the top to be raised; reverse the procedure to lower the top.

1. Remove the protective cover.

2. Open the trunk.

3. On some vehicles, it may be necessary to remove the CD changer (where equipped) and/or trim panels on left side of trunk to gain access to the hydraulic pump.

4. Turn the pressure relief thumb screw on pump fully counterclockwise to the stop. See Fig. 120.

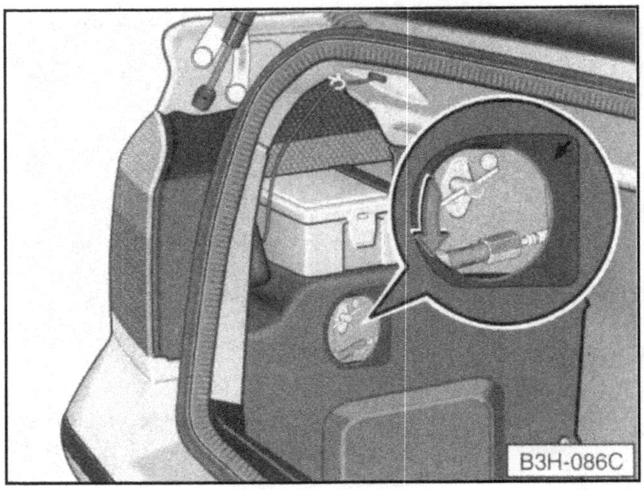

B3H-086C

Fig. 120. Turn the pressure relief thumb screw on pump fully counterclockwise to the stop (**inset**).

5. Close the trunk lid.

6. Swing the top forward by grabbing it with both hands, and lift and push forward.

7. Latch and lock the top in place.

8. Open the trunk.

9. Turn the pressure relief thumb screw on pump fully clockwise to the stop.

10. Stow the protective cover.

11. Close the trunk lid.

63 Bumpers

GENERAL

The bumpers consist of a cross-member that is bolted directly to the body structure. A molded styrofoam insert is installed over this cross-member. The bumper covers are then bolted to the bumper cross-member over the styrofoam insert.

Towing eye cover, removing and installing

A tow hook (eye), which is included in the vehicle tow kit, can be threaded into the front bumper cross-member after removing the plastic access cover.

1. Insert screwdriver between turn signal housing and towing eye cover. Carefully lift cover from plastic retaining clips on turn signal housing. See Fig. 1.

> ***CAUTION —***
> *To avoid damaging the retaining clips, insert the screwdriver between the housing and towing eye cover lens and NOT between the cover and bumper opening or fog light.*

2. Detach spring clips from opposite side or from fog light housing (if equipped) and remove cover from opening.

> ***NOTE —***
> *Before installing cover, inspect plastic retainer clips located on turn signal housing. Clips must not be broken or deformed.*

3. To install, insert cover at 30° angle into turn signal housing clips and make sure metal pin on rear of cover engages fully.

> ***CAUTION —***
> *To avoid damaging the retaining clips, do not insert cover at more than a 45° angle.*

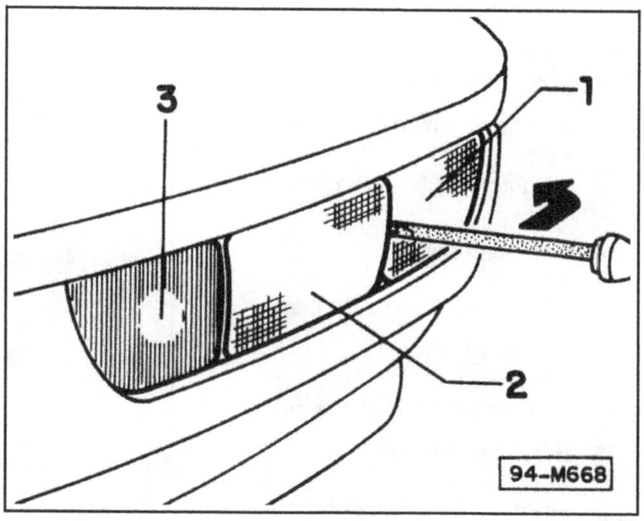

94-M668

Fig. 1. Carefully lift cover (**2**) in direction of arrow from retaining clips and detach from spring clips.

4. Carefully swing cover into opening and engage spring clips.

BUMPERS, SERVICING

The following pages show assembly views of the front and rear bumpers with removal/installation instructions. The bumper covers can be removed for replacement of the styrofoam impact insert.

63

Front bumper assembly

1. **Hex-head screws**
   - Qty: 4
   - Tighten to 10 Nm (7 ft-lb)
2. **Carrier**
3. **Plastic foam insert**
4. **Plastic insulator**
5. **Guide**
   - Remove/install cover (8) by pulling out/pushing into left and right guides in parallel
6. **Front side marker light**
7. **Spreader rivet**
   - Connects cover to wheel housing liner.
8. **Cover**
   - Material: PP/EPDM
9. **Hex-head screw**
   - Qty: 8
   - Tighten to 23-25 Nm (17-18 ft-lb)
10. **Hex-head screw**
    - Qty: 4
    - Tighten to 30 Nm (22 ft-lb)
11. **Hex-head screw**
    - Qty: 2
    - Tighten to 5 Nm (44 in-lb)
12. **Hex-head screw**
    - Qty: 2
    - Tighten to 10 Nm (7 ft-lb)
13. **Cover element**
    - For painted covers only
14. **Hex-head screw**
    - Qty: 5
    - Tighten to 4.5 Nm (40 in-lb)
15. **Spoiler**
    - Fits into cover

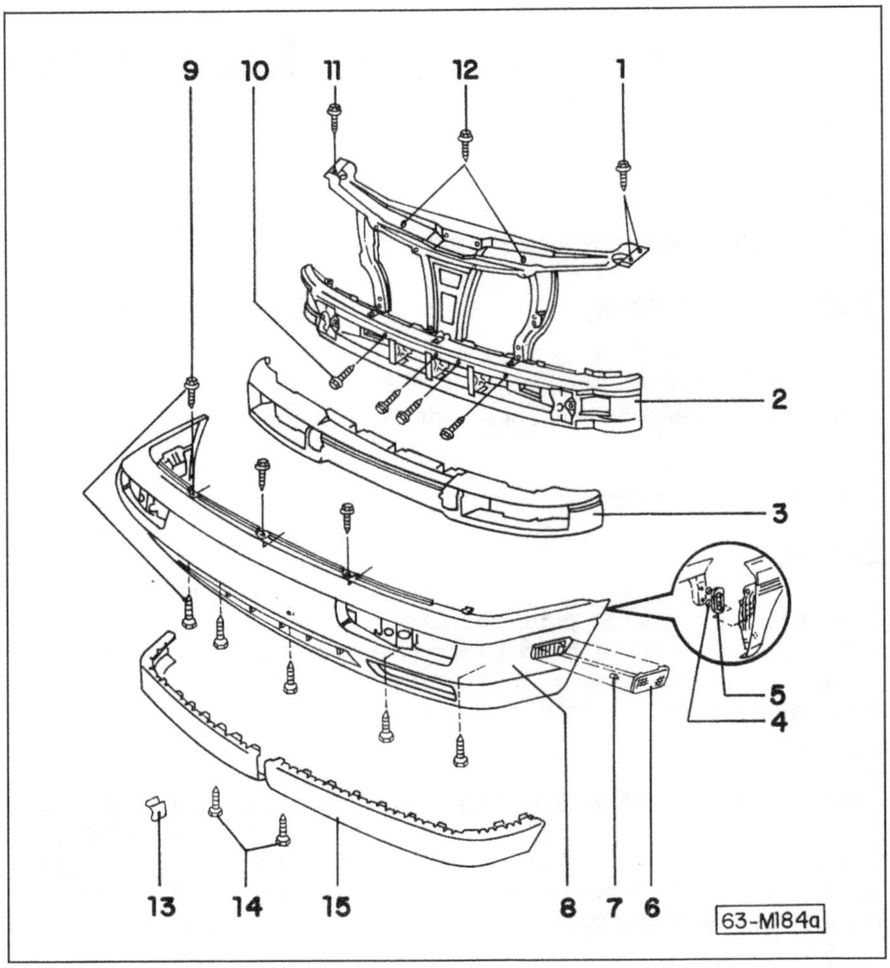

63-M184a

Fig. 2. Exploded view of front bumper assembly.

Front bumper cover, removing and installing

1. Remove radiator grille and spreader clips. See **66 Body Accessories–Exterior**.
2. Remove bumper cover mounting screws.
3. Disconnect spoiler (air guides) from bumper cover.
4. Disconnect harness connectors from lighting.
5. Push down on left/right retainers (**arrow**) with screwdriver and pull bumper cover out of side guides, keeping cover parallel.

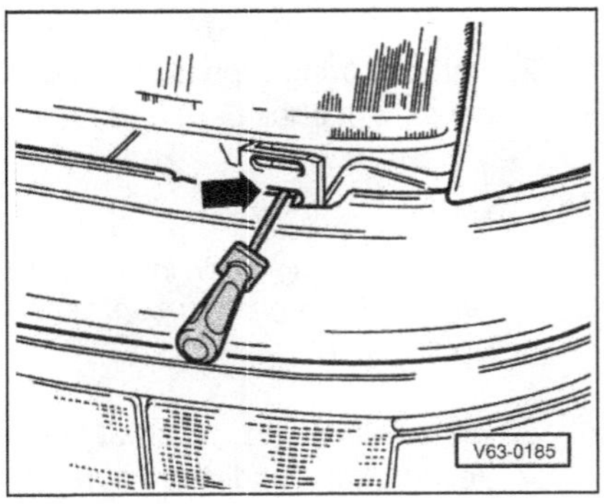

6. Install in reverse order of removal.

Front bumper assembly (Cabrio from 03/98)

1. **Cover**
   - Material: PP/EPDM
2. **Clip**
   - Quantity: 6
3. **Nut**
4. **Bracket**
5. **Guide**
6. **Bolt**
7. **Spreader nut**
8. **Impact absorber**
   - Secured to cover with four clips
9. **Bolt**
10. **Corner piece**
11. **Bumper strip**
    - Clipped to bumper cover
12. **Bolt**
13. **Bolt**
14. **Grille**
15. **Clip**
    - Quantity: 6
16. **Spoiler**
    - Clips in cover
    - Spoiler can stay attached when removing cover

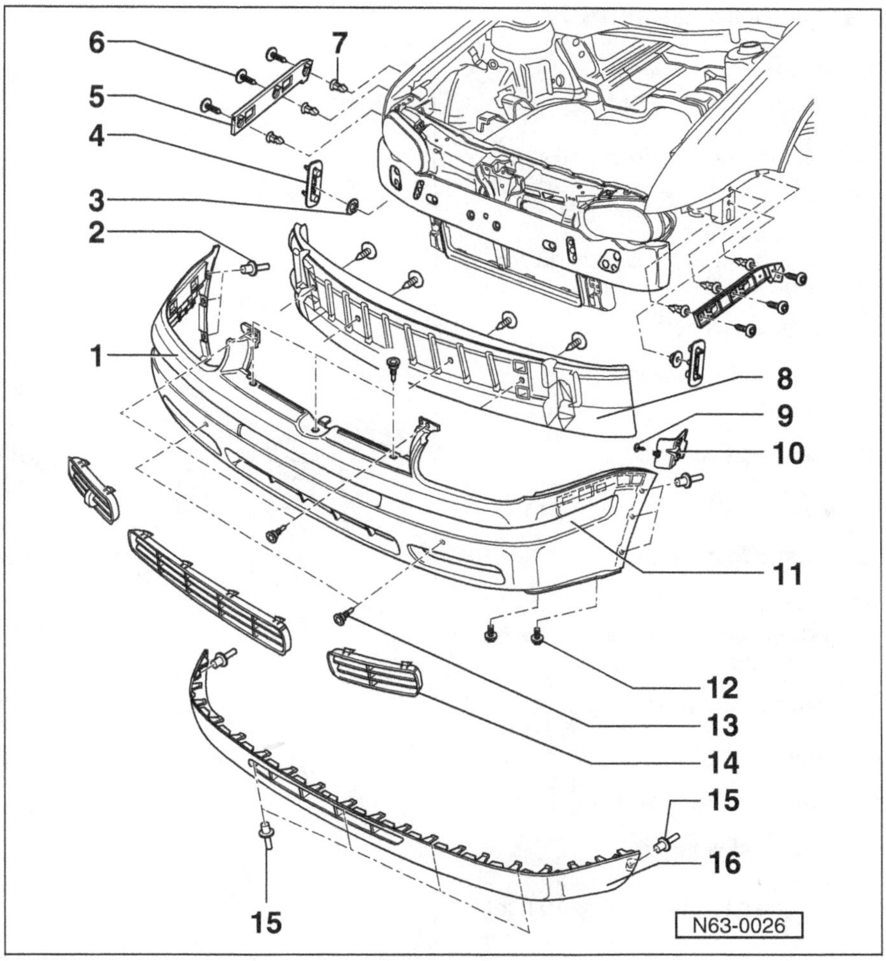

N63-0026

Fig. 3. Exploded view of front bumper assembly on late model Cabrio.

Front bumper cover, removing and installing (Cabrio from 03/98)

1. Remove radiator grille and spreader clips. See **66 Body Accessories–Exterior**.

2. Unclip lower air grilles (item **14** above).

3. Pull out clips (**2** and **15**).

4. Remove corner part (**10**).

5. Remove grille from cover (**1**).

6. Remove bolts (**13**). Pull cover (**1**) evenly off guides (**5**) and brackets (**4**).

7. Install in reverse order of removal.

Rear bumper, assembly

1. **Cover**
   - Material: PP/EPDM
   - To remove:
     Open rear lid and pry off four covers (3)
     Remove seven hex-head screws (4)
     Pull bumper out of side guides (8)
2. **Towing eye cover**
3. **Cover for upper hex-head screws**
4. **Hex-head screws**
   - Qty: 7
   - Tighten to 4.5 Nm (40 in-lb)
5. **Plastic foam insert**
6. **Carrier**
   - Remove four nuts with washers (10) and pull carrier out
7. **Plastic insulator**
8. **Guide**
9. **Carrier reinforcement**
10. **Nuts and washers**
    - Qty: 4
    - Tighten to 11-13 Nm (8-10 ft-lb)

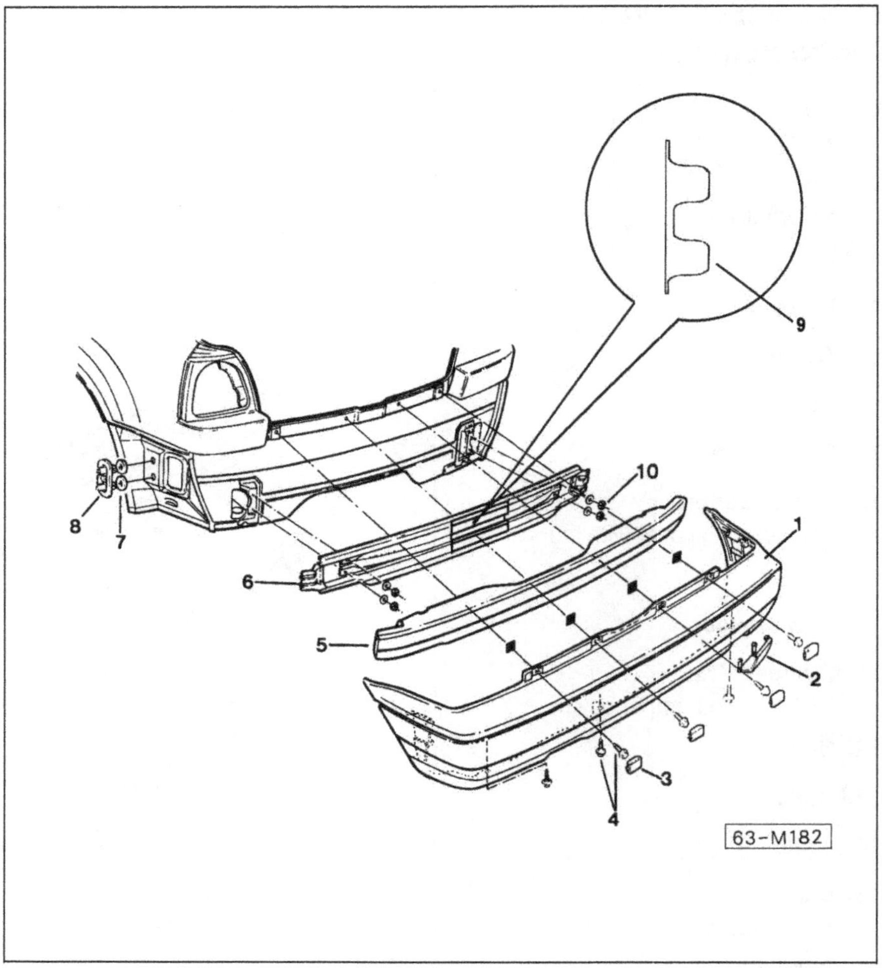

N63-0026

Fig. 4. Exploded view of rear bumper assembly.

Rear bumper, removing and installing

1. Open rear lid.
2. Pry off covers for hex head mounting screws.
3. Remove outer and lower hex head screws (item **4** above).
4. Pull bumper cover evenly out of side guides (**8**).
5. Remove mounting nuts with washers (**10**) and bumper carrier.
6. Install in reverse order of removal.

Rear bumper, assembly (Cabrio from 03/98)

1. **Cover**
   - Material: PP/EPDM
2. **Securing strip**
3. **Impact absorber**
4. **Guide**
5. **Clip**
   - Quantity: 4
6. **Bolt**
   - Quantity: 6
7. **Bolt**
   - Quantity: 8
8. **Plug**
   - Quantity: 4
9. **Bumper strip**
   - Cannot be removed
10. **Spoiler**
    - Clips in cover
    - Spoiler can stay attached when removing cover
11. **Spreader clip**
    - Quantity: 2

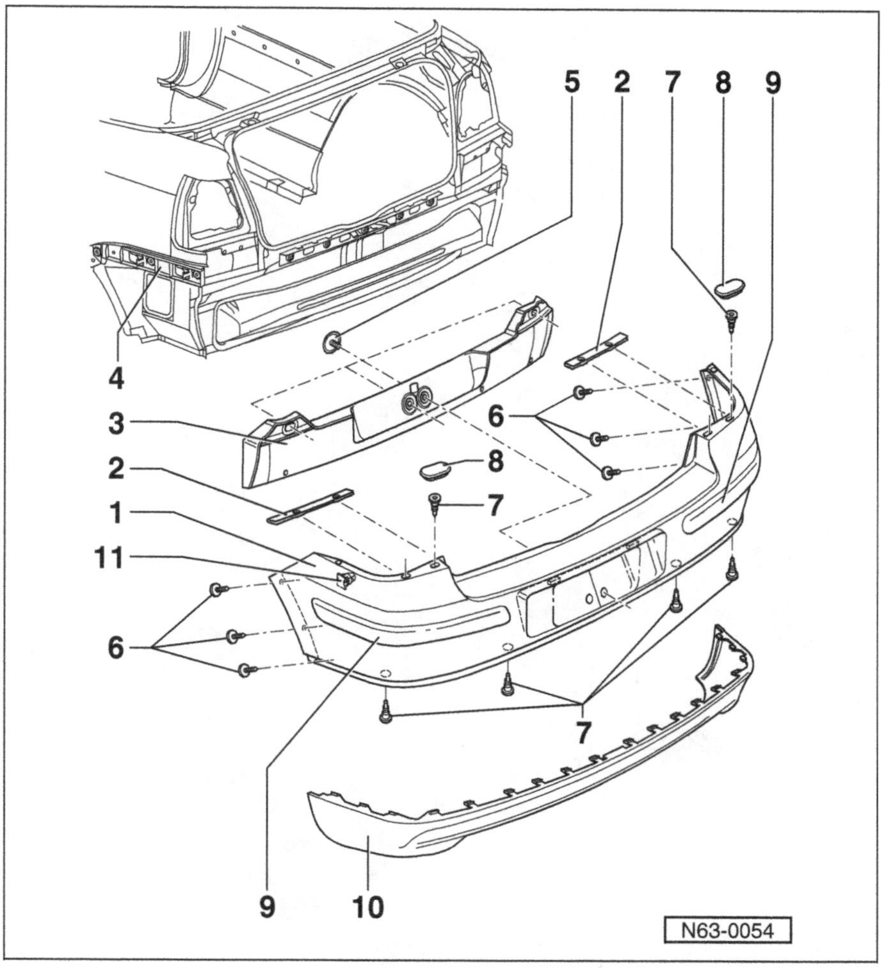

N63-0054

Fig. 5. Exploded view of rear bumper assembly on late model Cabrio.

Rear bumper, removing and installing (Cabrio from 03/98)

1. Remove taillight clusters, see **94 Lights, Accessories-Exterior**.
2. Disconnect bumper cover at wheel housing and unclip from body.
3. Remove bolts (item **7** above).
4. Pull cover evenly off guides (**4**) and securing strip (**2**).
5. Disconnect harness connector for license plate light.
6. Continue removing bumper cover.
7. Install in reverse order of removal.

66 Body Accessories–Exterior

GENERAL

This repair group covers various exterior trim items including the wheel housing liner, the front grille assembly, the rear spoiler, and the side-view mirror.

EXTERIOR COMPONENTS

Wheel housing liner, removing and installing

1. Raise car and support on jack stands. Remove wheel.

> ***CAUTION —***
> *Jack stands should be placed on a hard level surface (e.g. concrete).*

2. Remove seven screws from expander nuts. See Fig. 1.

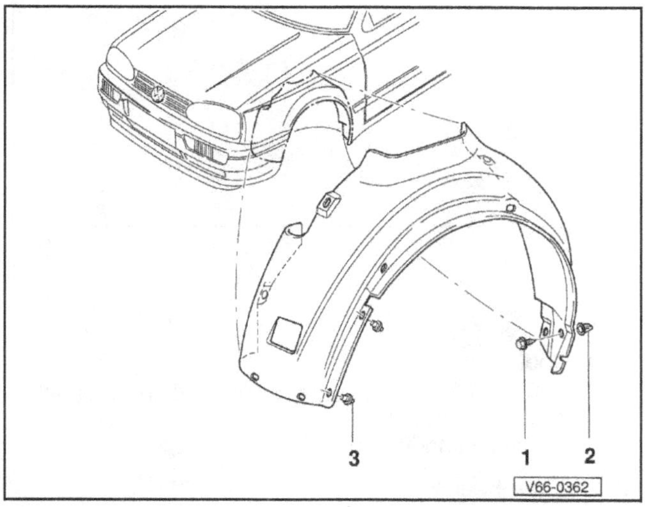

V66-0362

Fig. 1. Wheel housing liner mounting screw (**1**), expander nut (**2**), and expander rivet (**3**).

3. Remove expander rivets that connect wheel housing liner and front bumper panel.

4. Remove wheel housing liner from wheel house.

5. Installation is the reverse of removal. Replace any damaged fasteners.

Tightening torques	
• Wheel to wheel hub (lug bolt)	81 Nm (110 ft-lb)

Radiator grille, removing and installing

Fig. 2 shows an exploded view of the grilles for the Golf, GTI, and early Cabrio models. The grille used on Jetta models is similar. Refer to this illustration when removing and installing the grille.

1. Open hood and remove four screws from top of grille.

2. Using a screwdriver, release/unlock lower catches at each end of grille. Then release/unlock remaining lower catches.

3. Using both hands, pull radiator grille upward off radiator support evenly.

4. Insert small tool (e.g. ice pick) into top mounting holes to release catches and remove, then remove radiator grille.

5. Align radiator grille with center positioner (see arrow in Fig. 2).

6. Place grille into radiator support and fasten lower catches, making sure end catches are fastened last.

7. Fasten top catches.

8. Install and tighten mounting screws.

66

Radiator grille, assembly

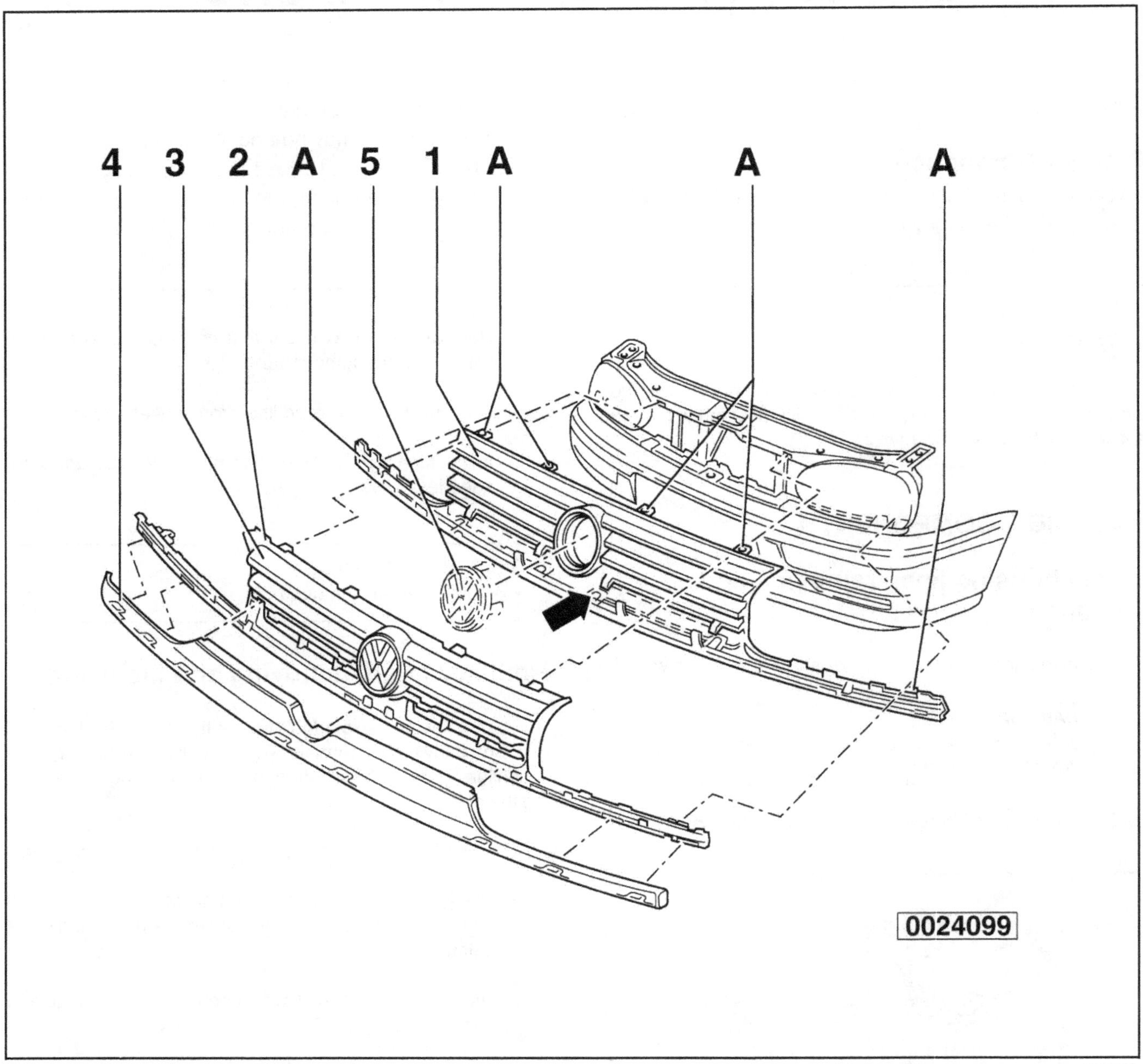

0024099

Fig. 2. Exploded view of radiator grille used Golf, GTI, and early Cabrio models. Grille used on Jetta models is similar.

1. **Radiator grille (Golf)**
   - Material: ASA
   - Screws (A)
   - Center positioner (arrow)
2. **Radiator grille (GTI, Cabrio)**
   - Material: ASA
3. **Radiator grille fins (GTI, Cabrio)**
   - Material: ABS (acrylic-nitrile butadiene-styrene copolymeride)
   - Clipped in radiator grille (2).
   - Painted in body color
4. **Panel**
   - Removing: remove radiator grille and unclip panel from radiator grille
   - Painted in body color
5. **VW emblem**
   - Clipped in radiator grille

> ***CAUTION—***
>
> *Remove and install the VW emblem carefully to avoid damaging the paint finish.*

Radiator grille, removing and installing (Cabrio from 03/98)

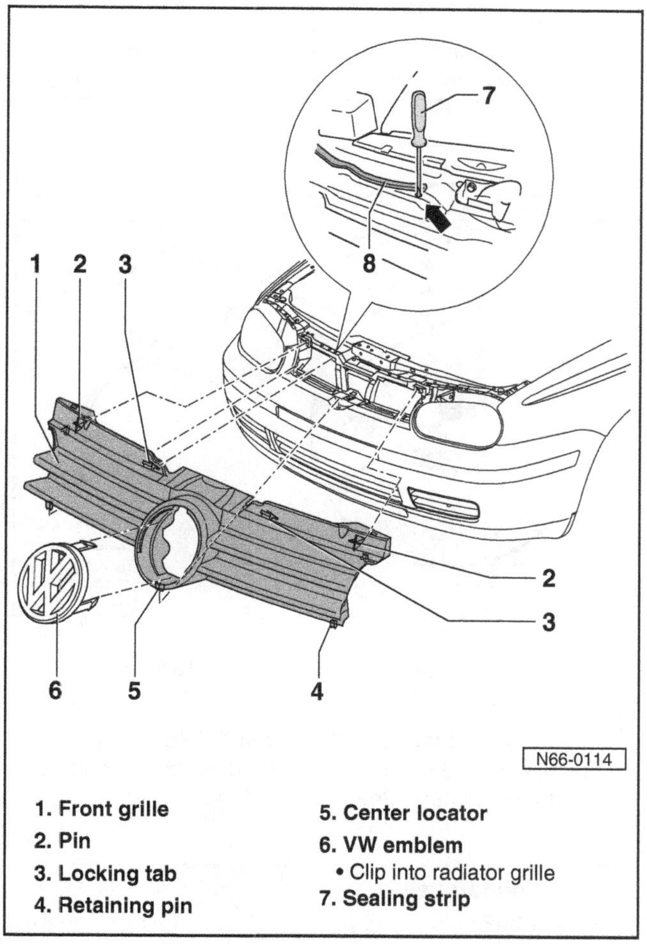

N66-0114

Fig. 3. Front grille for later model Cabrio.

1. Loosen sealing strip (item **8** above).
2. Release locking hook (**3**) with a screwdriver through hole (**arrow**).
3. Pull pins (**2**) out from lock carrier and remove radiator grille by pulling upward.
4. To install, align radiator grille with locator (**5**), insert lower pin (**4**) into bumper cover and clip in upper area.

Rear spoiler, retrofitting (Jetta)

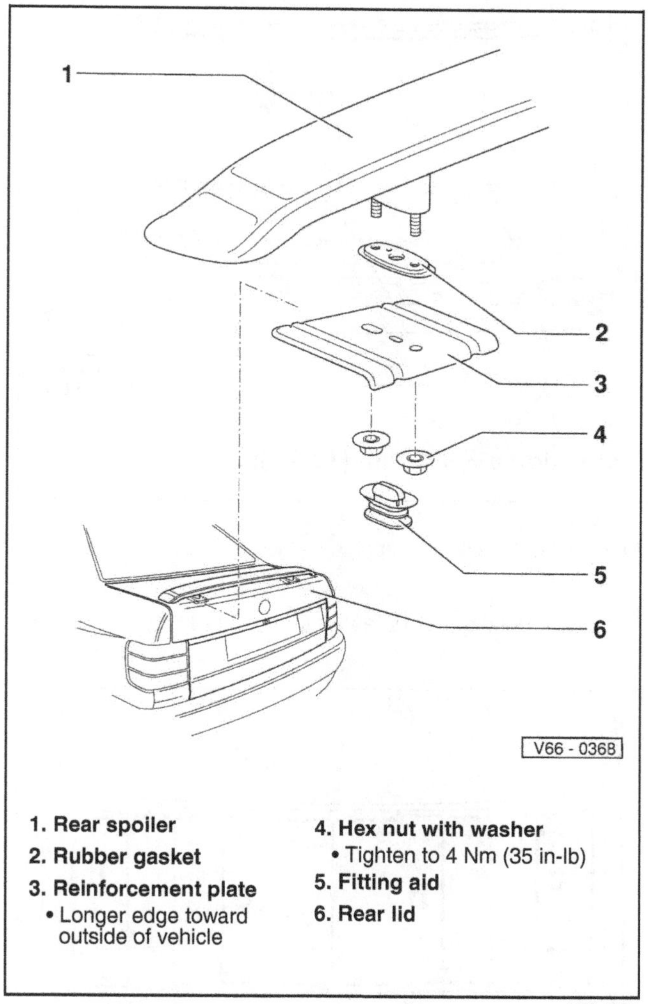

V66-0368

Fig. 4. Rear spoiler assembly for Jetta models. Fitting aid (**5**) is not applicable when retrofitting spoiler.

1. Place a 50 mm (2 in.) wide strip of adhesive tape on center of rear lid.
2. Determine center line of rear lid with a tape measure and mark on adhesive tape.
3. Apply adhesive tape and mark lines as shown in Fig. 5.

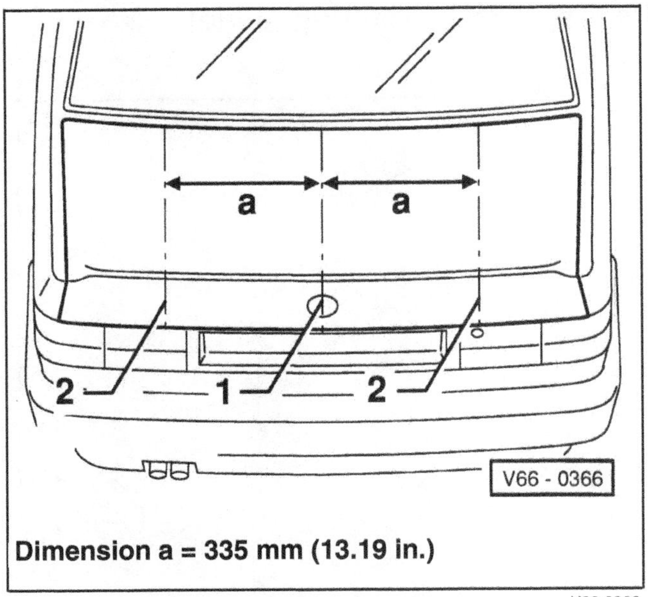

V66-0366

Fig. 5. Draw lines (**2**) after applying adhesive tape to trunk surface.

4. Mark position for holes as shown in Fig. 6.

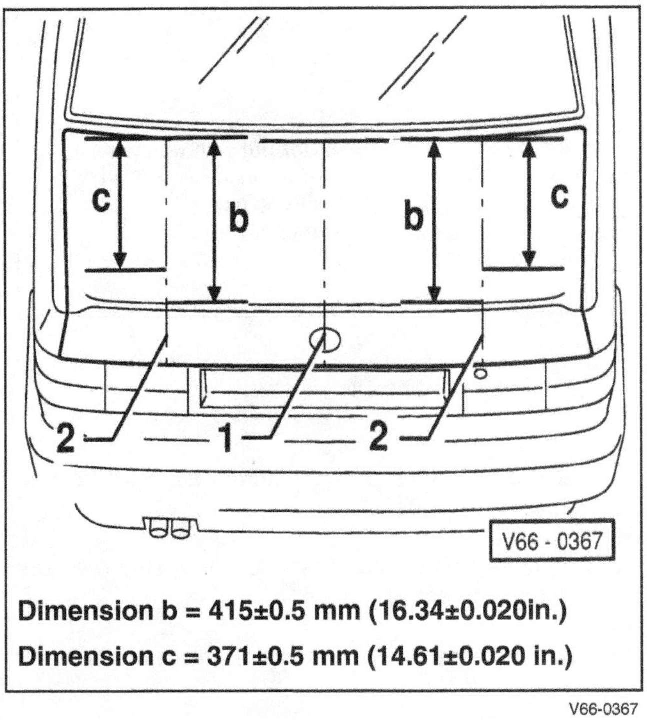

V66-0367

Fig. 6. Mark position of holes on line **2** at dimensions **b** and **c**.

5. Carefully center punch holes and drill with 7.5 mm (9/32 in) drill.

6. Remove burrs from holes.

7. If necessary, rework the holes slightly with a round file to fit the rear spoiler.

8. After completing the painting/corrosion protection measures, place rubber gaskets and spoiler on rear lid. See Fig. 7.

> ***CAUTION —***
> *Corrosion and paint protection at drilled holes must be carried out.*

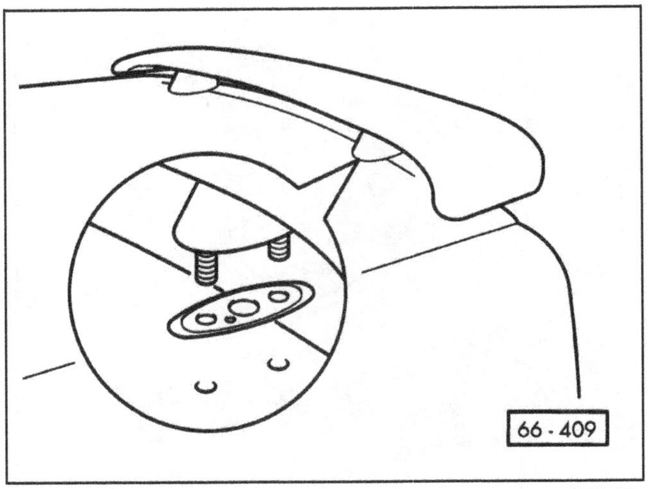

66-409

Fig. 7. Place rubber gaskets and spoiler on rear lid (inset).

9. From inside rear lid, install left and right reinforcement plates, with longer edge toward outside of vehicle. Tighten hex nuts. See Fig. 8.

Tightening torques

- Spoiler hex nuts . . . . 4 Nm (35 in-lb)

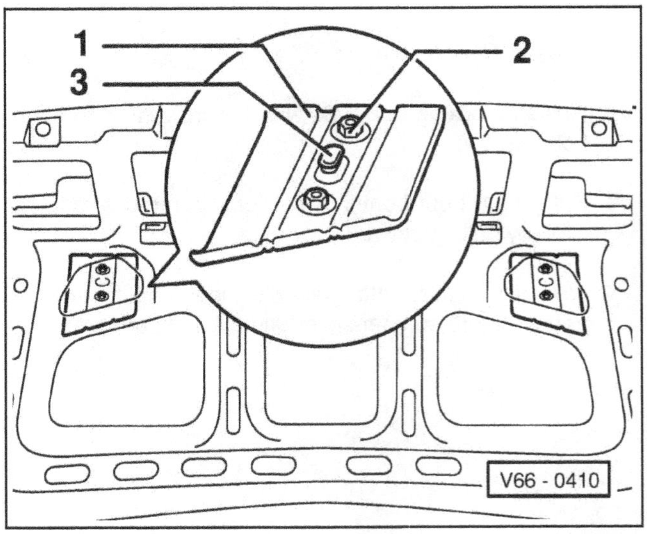

V66-0410

Fig. 8. Install left and right spoiler reinforcement plates (**1**), if applicable, remove fitting aid (**3**), and tighten hex nut (**2**).

Side-view mirror, assembly

Fig. 9. shows an exploded view of the side-view mirror assembly. The mirror glass is clipped into the mirror housing. A special assembly lever (Volkswagen special tool no. 800-200) is available to aid in removal of the glass.

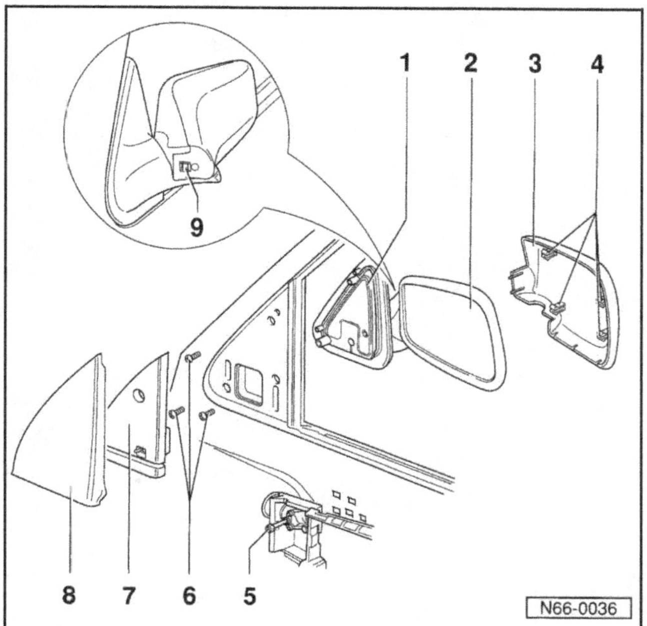

1. **Mirror housing**
2. **Mirror glass**
   - clipped in
3. **Housing cap**
   - Removing: remove mirror glass and squeeze catches (4) and (9) while pulling off cap
4. **Catches**
5. **Mirror actuator**
6. **Oval head screw**
   - Tighten to 2 Nm (17 in-lb)
7. **Insulation**
8. **Cover**
9. **Catches**

N66-0036

Fig. 9. Exploded view of side-view mirror assembly.

69 Seatbelts, Airbags

GENERAL

This repair group covers seat belt assemblies and removal and installation of the airbag system components. This repair group does not cover airbag system fault diagnosis or repair. Service and repair to the airbag system requires special test equipment, knowledge and training and should only be carried out by an authorized Volkswagen dealer.

> ***WARNING —***
> - *The airbag(s) are inflated by an explosive device. Handled improperly or without adequate safeguards, the system can be very dangerous. Special precautions must be observed prior to any work at or near the steering wheel or steering column, including the pedal assembly.*
> - *The airbag is a vehicle safety system. To guard against personal injury or airbag system failure, only trained Volkswagen service technicians should test, disassemble or service the airbag system.*
> - *Always replace any airbag system component that has been mechanically damaged (example: dented or cracked).*
> - *Supplementary child restraint anchorage equipment is available from an authorized VW dealer. Depending on the child restraint system being used, additional anchorage points may be required. Check with your Volkswagen dealer for proper installation and positioning of one or more anchorage points and the required hardware for the attachment of the tether.*

Seat belts, inspecting

> ***WARNING —***
> *After every accident the seat belt system must be inspected systematically. If damage is found when inspecting the seat belts must be replaced.*

Inspect front and rear belt webbing, and replace belts with broken or pulled threads, cut loops at the belt edge, bowed webbing, faded areas, or cuts. Pull the belt out fully and let it retract. If it does not move smoothly in either direction, check for dirt, grease, or gum in the webbing. If the belt cannot be cleaned using only a mild soap solution, replace the belt.

Replace the belt if the buckle cover is cracked, if the push button is loose, or if the buckle does not lock securely. Check all mounting points, and replace any corroded hardware.

On three-point seat belts, check the belt lock-up mechanism by grasping the webbing near the retractor and tugging quickly. Replace any belt that does not lock up.

A road test should also be performed to check belt lock-up mechanism. Drive in an open area away from other cars. At approximately 5 to 15 mph, with the belt securely fastened, quickly stop the car. Replace any belt that does not lock up.

> ***WARNING —***
> *For safety reasons, road test should be carried out in a traffic free area to ensure that other motorists or pedestrians are not endangered.*

69

Check the belt latching mechanism by pushing belt tongue into latch until it engages with an audible "click". Give the belt a firm pull to ensure that mechanism is properly engaged. If belt tongue fails even only once to engage properly in belt latch during at least 5 tests, replace seat belt and belt latch.

Check the release of belt latching mechanism by pressing button on belt latch with finger pressure. When belt is slack, tongue must spring out of belt latch on its own. If it fails to do this only once in 5 tests, replace seat belt and latch.

> ***WARNING —***
> *Never use lubricant to eliminate noise or stiffness at seat belt lock buttons.*

FRONT SEAT BELTS

Fig. 1 shows the front seat belt used on cars without airbags. Fig. 2 shows the front seat belt used on cars with airbags and emergency tensioning seat belt retractors. The illustrations that immediately follow Fig. 1 and Fig. 2 show detailed information on front seat belt mounting points.

Front seat belt, assembly (cars without airbags)

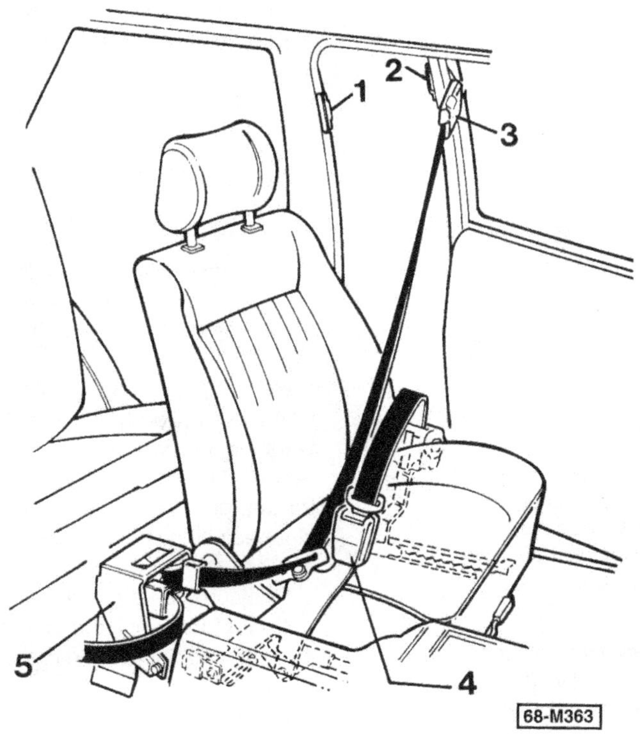

68-M363

Fig. 1. Front seat belt system used on cars without airbag.

1. **Upper pillar catch**
   - Tighten screws to 15 Nm (11 ft-lb)
2. **Upper door catch**
   - Removing, **see** Ⓐ
3. **Upper belt buckle**
4. **Lap belt buckle**
   - Removing, **see** Ⓑ
5. **Lower mounting point**
   - Removing, **see** Ⓒ

A Upper door catch, removing (front seat belts, cars without airbag)

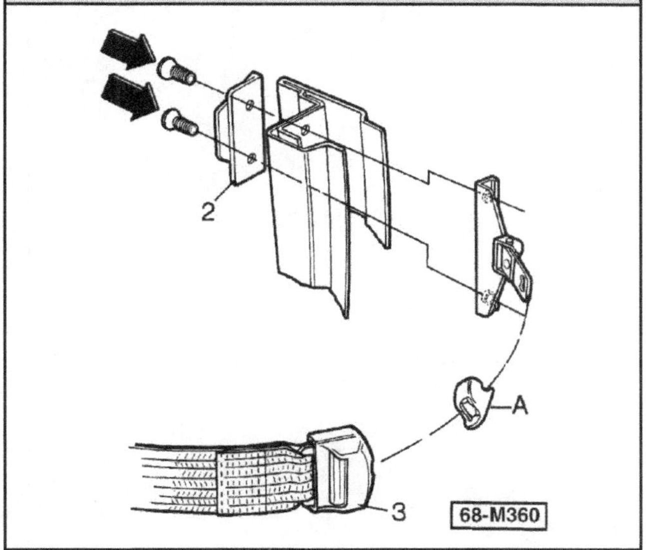

68-M360

- Remove trim (A)
- Pull off inner window seal near latch
- Remove screws (arrows)
  Tighten to 15 Nm (11 ft-lb)

B Lap belt buckle, removing (front seat belts, cars without airbag)

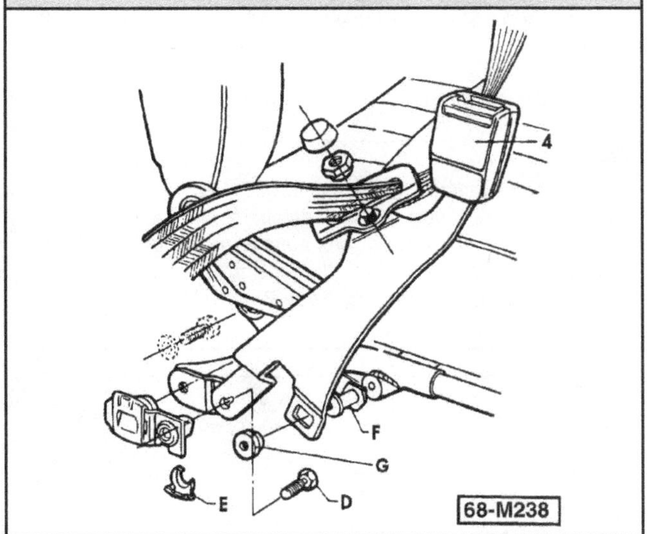

68-M238

- Unbolt nut (G) and bolt (D)
- Do not misplace the spacer bushing (F) or clip (E)
- To install, tighten nut (G) to 25 Nm (18 ft-lb)
  and bolt (D) to 60 Nm (44 ft-lb).

C Lower mounting point, removing (front seat belts, cars without airbag)

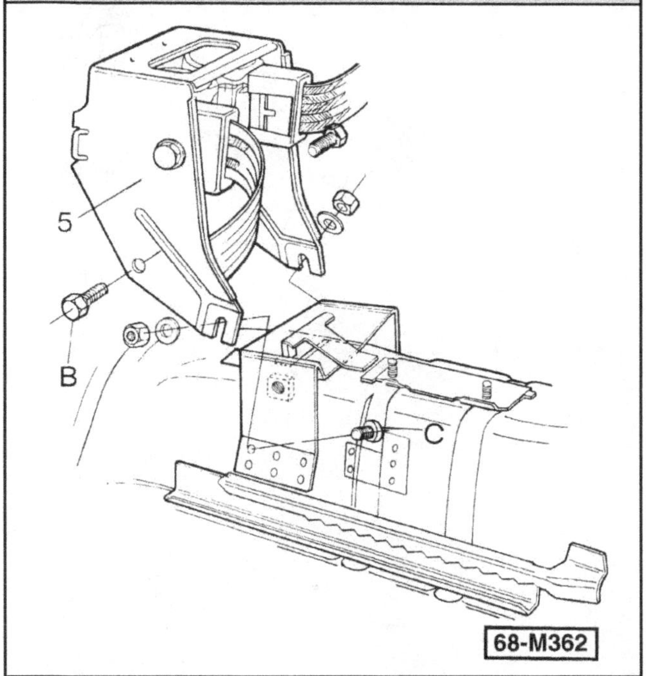

68-M362

- First remove front seat, see 72 Seats
- Remove hex bolts (B) and (C)
- Remove lower mounting point
- Install in reverse order of removal
  Tighten (B) to 40 Nm (30 ft-lb)
  Tighten (C) to 19 Nm (14 ft-lb)

> ***WARNING —***
> *The seat belt retractor tensioning unit shown in Fig. 2 is an explosive device. Handled improperly or without adequate safeguards, it can be very dangerous. Special precautions must be observed prior to any work on the tensioning unit. See* **Emergency tensioning seat belt retractors**.

Front seat belt assembly (cars with airbag and emergency tensioning belt retractors)

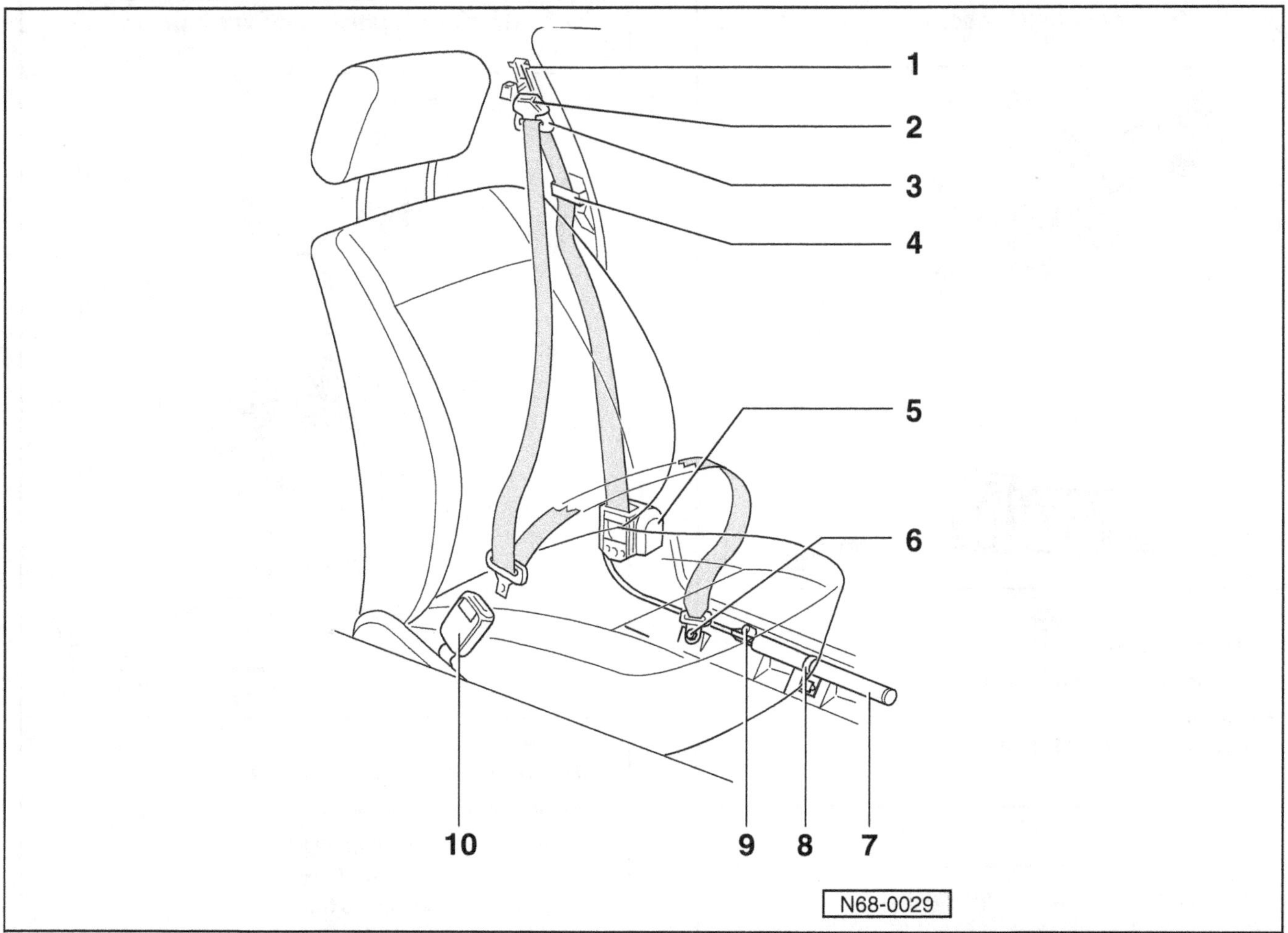

N68-0029

Fig. 2. Front seat belt system used on cars with airbag and emergency tensioning seat belt retractors. Illustration shows 4-door and Cabrio models, 2-door and GTi models are similar with the exception of front anchor (**6**).

1. **Belt height adjustment fitting**
   - See **Belt height adjustment, assembly**
2. **Cap**
3. **Belt relay (upper anchor loop)**
4. **Belt guide**
   - Tighten to 1.5 Nm (13 in-lb)
5. **Belt reel**
   - Tighten (hex bolt) to 40 Nm (30 ft-lb)
6. **Front anchor**

> ***CAUTION —***
> *Very Important! There are three different types of front anchors, depending on model.*

   - 4-door models, **see** Ⓐ
   - 2-door models, **see** Ⓑ
   - Cabrio models, **see** Ⓒ

7. **Tensioning unit**

> ***WARNING —***
> *The tensioning unit is an explosive device. Handled improperly or without adequate safeguards, it can be very dangerous. Special precautions must be observed prior to any work on the tensioning unit. See* **Emergency Tensioning Seat Belt Retractors.**

> ***NOTE —***
> *Some cars with airbags may not be equipped with tensioning unit.*

8. **Bracket**
9. **Locking nut**
10. **Front belt buckle (latch)**
   - Removing, **see** Ⓓ

A Front anchor, 4-door models (front seat belts, cars with airbag)

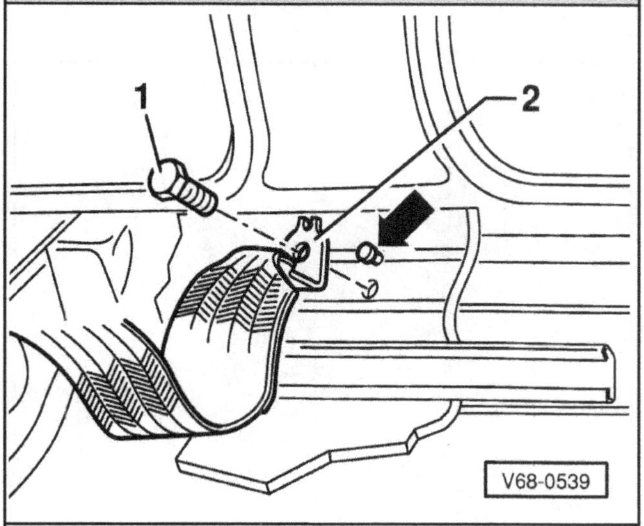

V68-0539

- **Hex bolt (1)**
  Tighten to 40 Nm (30 ft-lb)
- **Seat belt (2)**
  Align metal fitting with retaining lug (arrow) to install.

B Front anchor, 2-door models (front seat belts, cars with airbag)

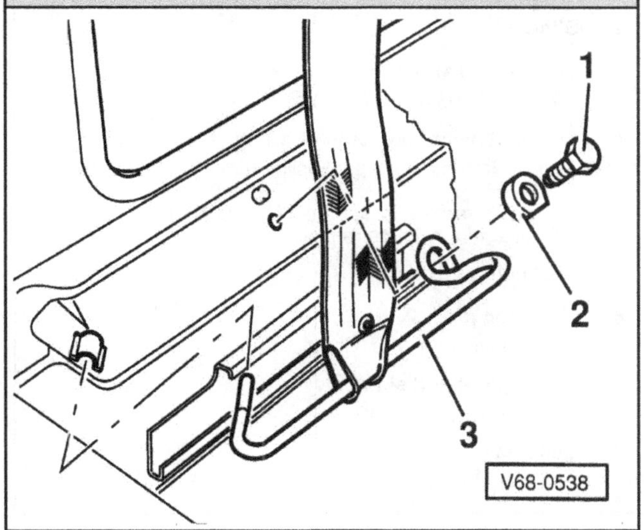

V68-0538

- **Hex bolt (1)**
  Tighten to 40 Nm (30 ft-lb)
- **Cap (2)**
- **Anchor rail (3)**

C Front anchor/belt reel, Cabrio models (front seat belts, cars with airbag)

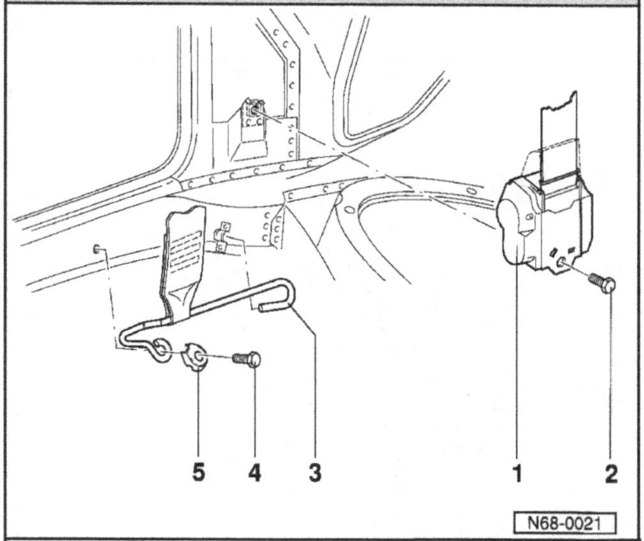

N68-0021

- **Front belt reel (1)**
  Retaining lugs determine mounting position
- **Hex bolt (2): tighten to 40 Nm (30 ft-lb)**
- **Anchor rail (3)**
- **Hex bolt (4): tighten to 40 Nm (30 ft-lb)**
- **Eye cap (5)**

D Belt buckle (front seat belts, cars with airbag)

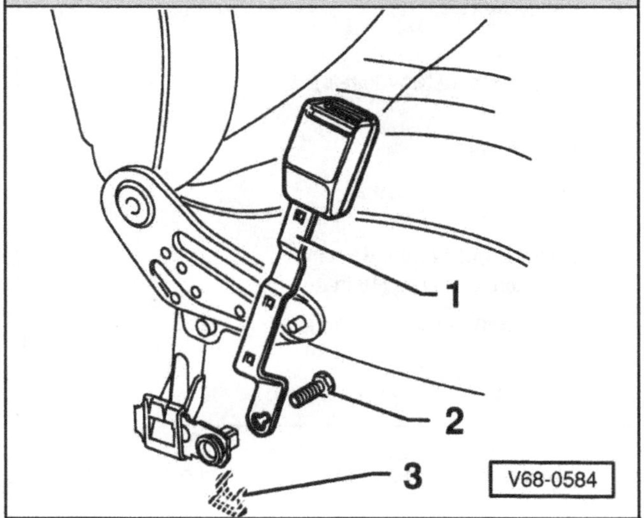

V68-0584

- **Belt buckle (1)**
- **Hex bolt (2): tighten to 40 Nm (30 ft-lb)**
- **Clip (3)**
  slide on/off

Seat belt height adjustment

Fig. 3 and Fig. 4 show seat belt height adjustment mechanisms. The illustrations contain removal and installation information, including tightening torques.

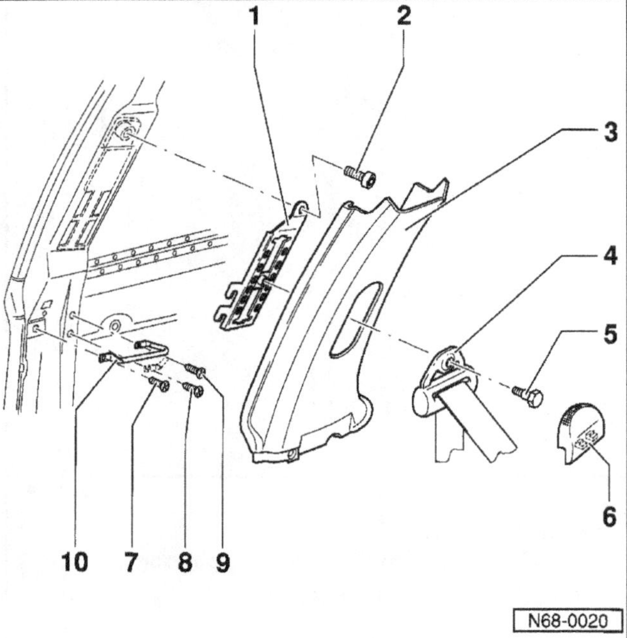

1. **Seat belt height adjustment fitting**
   - Removing: remove socket-head bolt (2) and pull belt relay (4) upward
2. **Sockethead bolt**
   - Tighten to 23 Nm (17 ft-lb)
3. **B-pillar trim**
   - Removing, see **70 Trim–Interior**
4. **Belt relay**
5. **Hex bolt**
   - Tighten to 40 Nm (30 ft-lb)
6. **Cover**
   - Pushed on
7. **Screw**
   - Tighten to 1.5 Nm (13 in-lb)
8. **Screw (right-hand side only)**
   - Tighten to 1.5 Nm (13 in-lb)
9. **Screw (left-hand side only)**
   - Tighten to 1.5 Nm (13 in-lb)
10. **Belt relay**

N68-0020

Fig. 3. Exploded view of seat belt height adjustment for Cabrio only.

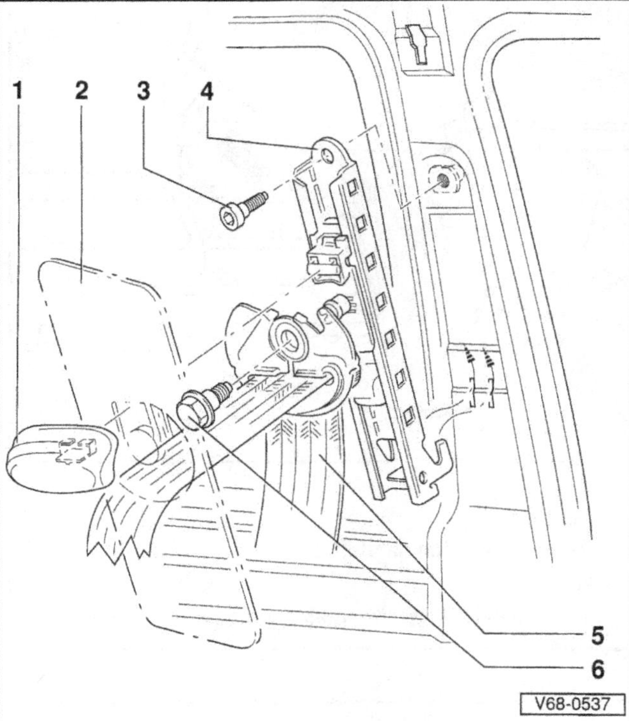

1. **Seat belt height adjustment control**
   - Pull out
2. **Slide**
3. **Sockethead screw**
   - Tighten to 23 Nm (17 ft-lb)
4. **Seat belt height adjustment fitting**
   - Front: To remove, upper B-pillar trim must be removed first
   - Rear: To remove, side parcel shelf support and C-pillar trim must be removed first
5. **Seat belt**
6. **Threaded pin**
   - Tighten to 1.5 Nm (13 in-lb)
7. **Screw (left-hand side only)**
   - Tighten to 1.5 Nm (13 in-lb)
8. **Belt relay**

V68-0537

Fig. 4. Exploded view of seat belt height adjustment fittings for cars with airbag.

EMERGENCY TENSIONING SEAT BELT RETRACTORS

NOTE —
Some early cars with airbags may not be equipped with the emergency seat belt retractors.

The emergency tensioning seat belt retractor uses a pyrotechnic device to tension the front seat belts during frontal impacts. The tensioning unit is a potentially dangerous device that requires special handling and disposal. The tensioning unit has no expiration date and is an integral part of the seat belt retractor (belt reel). Fig. 5 shows the tensioning unit/retractor used on 4-door models. Fig. 6 shows the tensioning unit/retractor used on 2-door and Cabrio models.

Always follow the prescribed work sequence when removing or installing the tensioning unit and retractor. Pay attention to all warnings associated with the replacement of this component.

WARNING —
- *After an accident in which one or both belt tensioners have been activated, both seat belts must be replaced (the belts will no longer retract).*
- *Handle tensioning unit with extreme care. DO NOT drop tensioning unit. Belt tensioner units that have been dropped on the floor must not be installed in a vehicle.*
- *Belt tensioning unit testing, removing, and installing should only be carried out by qualified personnel.*
- *Belt tensioner components may be neither opened or repaired; always use new parts.*
- *Belt tensioner units that are mechanically damaged (dents, cracks) must be replaced.*
- *The belt tensioner unit must not be treated with grease, cleaning solution or similar substances.*
- *Belt tensioner units MUST NOT be exposed to temperatures above 100°C (212°F).*
- *The belt tensioner unit should be installed immediately after removing it from its packaging.*
- *The belt tensioner unit gas generator must be made unusable before disposal. It is a pyrotechnic device that can cause injury if activated improperly (e.g. scrapping with a cutting torch). Consult your authorized Volkswagen dealer for tensioning unit disposal.*
- *Before starting any body disassembly, straightening or other body repair work, the belt tensioning units must be removed.*

Front belt reel with tensioning unit, removing and installing (4-door models)

1. Remove upper and lower B-pillar trim. Remove sill plate trim. See **70 Trim–Interior**.
2. Loosen tensioning unit locknut, this detaches the tensioning unit and makes it inactive. See Fig. 5.
3. Remove belt reel mounting bolts and remove reel.
4. Pull tensioning unit out of guide in mounting bracket.
5. Installation is the reverse order or removal. Fig. 5 lists assembly tightening torques.

Front belt reel with tensioning unit, removing and installing (2-door and Cabrio models)

1. Remove B-pillar trim and sill plate trim. See **70 Trim–Interior**.
2. Loosen tensioning unit locknut. This detaches the tensioning unit and makes it inactive. See Fig. 6.
3. Remove front anchor rail. See Fig. 2 given earlier.
4. Remove cover cap from upper belt relay and remove relay fitting and belt guide. See Fig. 2 given earlier
5. Remove belt reel.
6. Pull tensioning unit out of bracket and remove.
7. Installation is the reverse order or removal. Fig. 6 lists assembly tightening torques.

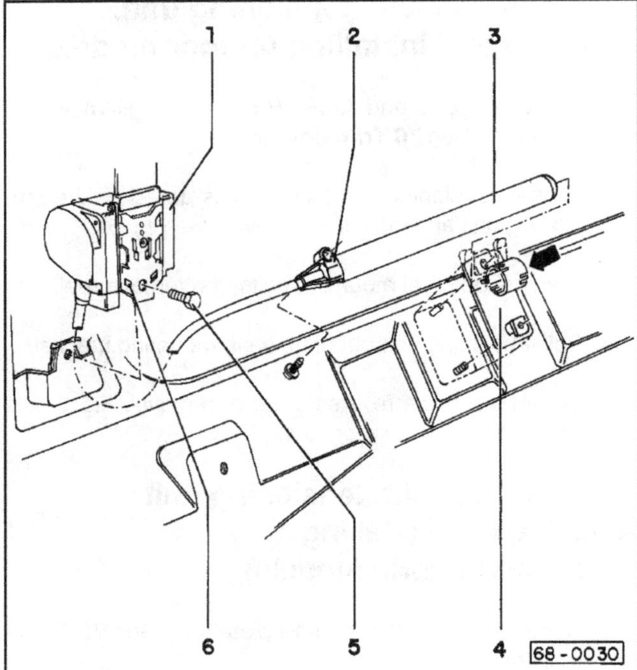

1. **Belt reel**
2. **Lock nut**
   • Tighten to 23 Nm (17 ft-lb)
3. **Tensioning unit**
   • Pyrotechnic device

> ***WARNING —***
> *The tensioning unit is an explosive device. Handled improperly or without adequate safeguards, it can be very dangerous.*

4. **Cable channel/mounting bracket for tensioning unit**
5. **Hex bolt**
   • Tighten to 40 Nm (30 ft-lb)
6. **Retaining slots**
   • Engage with mounting lugs to determine belt reel mounting position

68-0030

Fig. 5. Front belt reel tensioning unit for 4-door models.

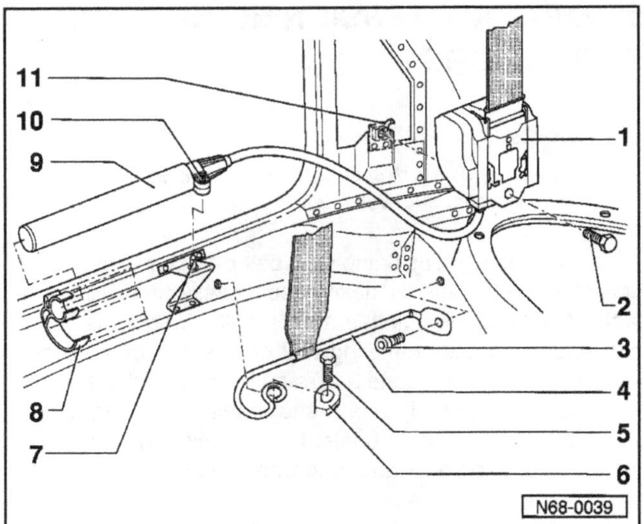

1. **Belt reel**
2. **Hex bolt**
   • Tighten to 40 Nm (30 ft-lb)
3. **Socket head cap screw**
   • Pyrotechnic device
4. **Anchor rail**
5. **Hex bolt**
   • Tighten to 40 Nm (30 ft-lb)
6. **Cap**
7. **Threaded bolt**
8. **Cable channel/mounting for tensioning unit**
9. **Tensioning unit**
   • Pyrotechnic device

> ***WARNING —***
> *The tensioning unit is an explosive device. Handled improperly or without adequate safeguards, it can be very dangerous.*

10. **Lock nut**
    • Tighten to 20 Nm (15 ft-lb)
11. **Retaining lugs**
    • Determines belt reel mounting position

N68-0039

Fig. 6. Front belt reel tensioning unit for 2-door and Cabrio models.

REAR SEAT BELTS

Fig. 7 shows the rear seat belts with inset details for various models.

Rear seat belts, assembly

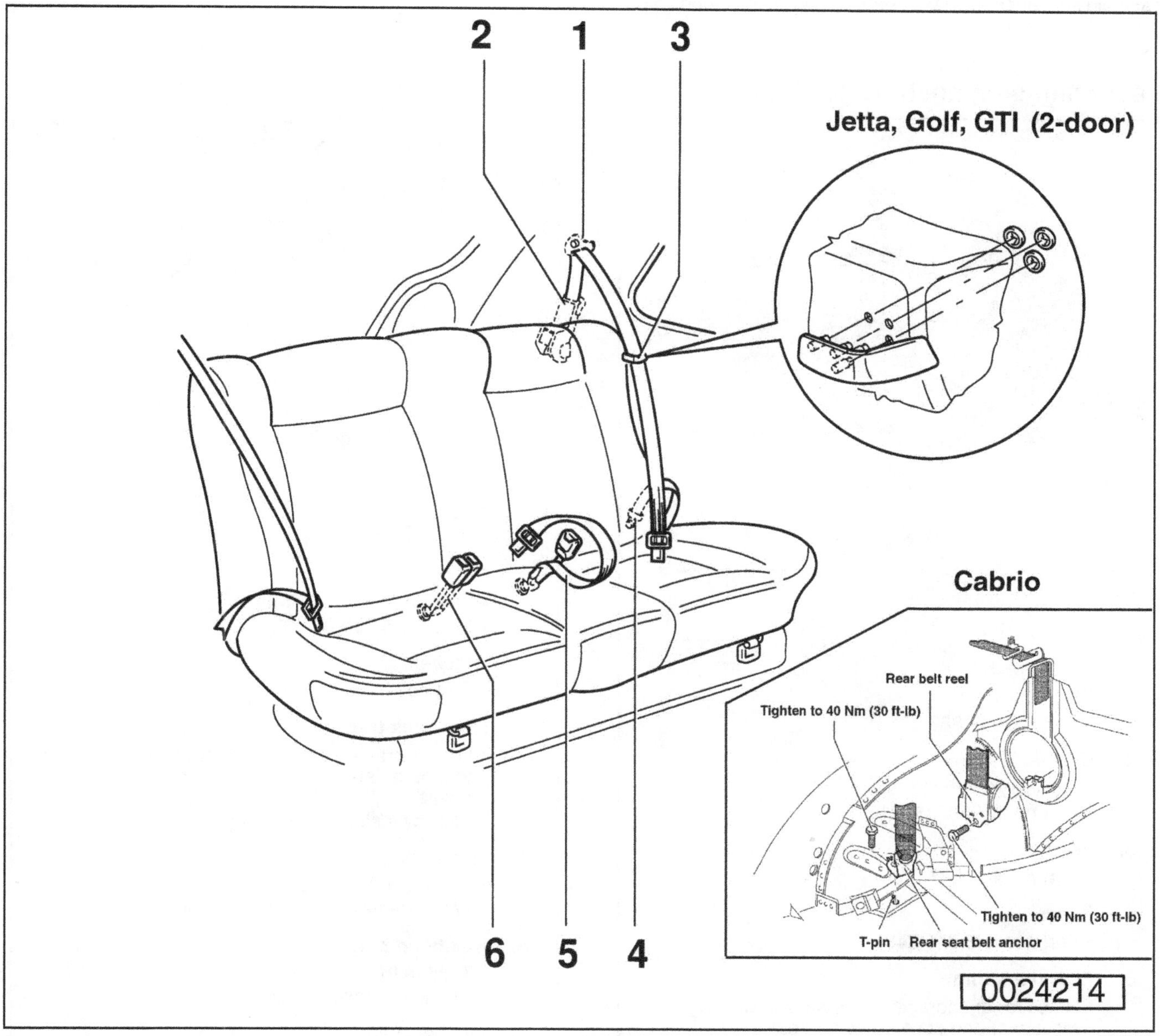

0024214

Fig. 6. Rear seat belt assembly.

1. **Seat belt height adjustment fitting**
2. **Belt reel**
   - Removing and installing
     Golf, GTI, Jetta: remove wheel housing panel, side liner for hat rack, and C-pillar trim to access
     Cabrio: remove seat backrest, fold seat forward, and loosen upper part of rear seat backrest to access. Anchor and belt must be pulled through upper part of rear backrest. When installing, belt anchor must be properly located on T-pin before tightening mounting bolt. See **Cabrio** inset above.
3. **Clasp (2-door models only)**
   - Secured to side trim panel with three spring nuts
4. **Rear belt anchor**
   - Tighten mounting bolt to 40 Nm (30 ft-lb)
5. **Lap belt**
   - Tighten mounting bolt to 40 Nm (30 ft-lb)
6. **Double buckle**
   - Tighten mounting bolt to 40 Nm (30 ft-lb)

Rear child seat anchorage points

Front facing child seats that are installed in the rear seat using the center seatbelt should also be anchored at the top of the child seat. This keeps the top of the child seat securely anchored in the event of a collision. The illustrations shown below show the proper components and mounting details for rear child seat anchors in the Jetta, Golf/GTI and Cabrio models.

Rear child seat anchorage, Jetta

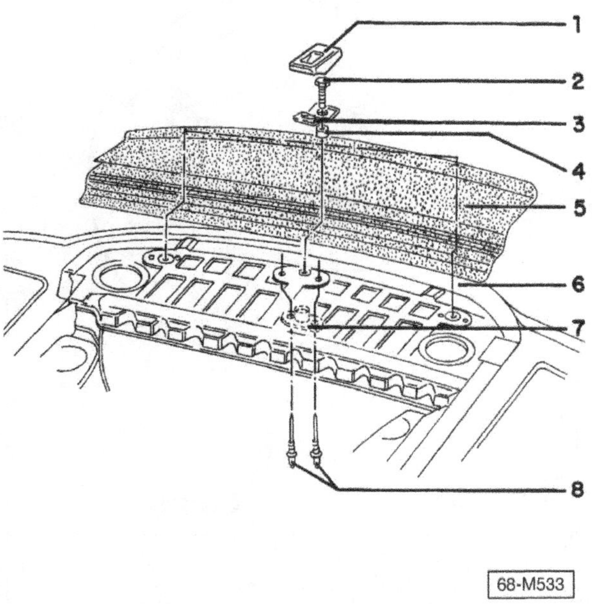

1. **Cap**
   - Part no. 443 887 301 B
2. **Bolt**
   - Tighten 23 Nm (17 ft-lb)
   - Part no. N 969 108.01
3. **Bracket**
   - Part no. 443 887 269
4. **Spacer**
   - Part no. 1H5 858 846
5. **Rear deck trim**
   - Holes for anchorage points are pre-drilled (3 locations) and covered by trim panel
   - Locate holes by hand from top and make small cut in material to allow installation of anchorage components
6. **Rear filler panel sheet metal**
   - With pre-drilled holes for anchorage points
7. **Washer-nut**
   - Part no. N 989 514.01
8. **Rivets**
   - Part no. N 905 763.01
   - Install from below

Rear child seat anchorage, Golf/GTI

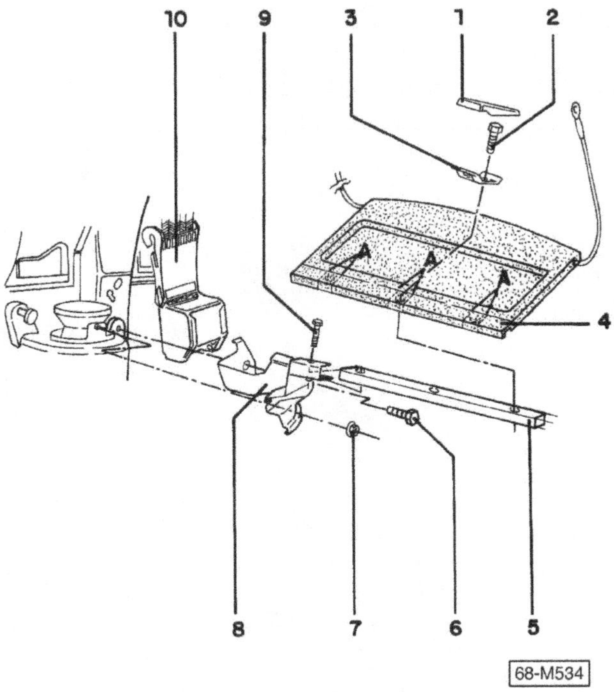

1. **Cap**
   - Part no. 443 887 301 B
2. **Bolt**
   - Tighten 25 Nm (18 ft-lb)
   - Part no. N 989 099.02
3. **Bracket**
   - Part no. 443 887 269
4. **Rear deck trim**
   - Locate pre-cut slots and make two cuts in carpet to create flap at **A**, lift flap to expose anchorage points
   - Already installed in vehicle
5. **Metal bar**
   - Part no. 1H6 857 875
   - With pre-drilled holes for anchor points
6. **Hex-head bolt**
   - Tighten to 40 Nm (30 ft-lb)
   - Already installed in vehicle
7. **Washer-nut**
   - Already installed in vehicle
8. **Bar supports**
   - Left: part no. 1H6 858 877
   - Right: part no. 1H6 858 878
9. **Bolt**
   - Part no. N 010 247.8
   - Tighten to 30 Nm (22 ft-lb)
10. **Belt reel**
    - Already installed in vehicle

Rear seat child anchorage, Cabrio

The rear seat upper tethers in Cabrio models are located in the luggage compartment on the floor. See Fig. 7.

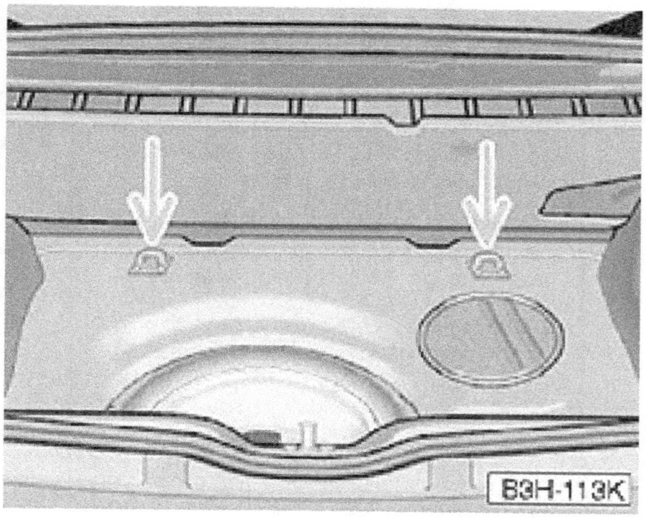

B3H-113K

Fig. 7. Rear child restraint upper anchor points (**arrows**) are located on the floor of the Cabrio luggage compartment.

Guide rear upper tether strap under seat headrest and through guide as shown. Route strap into luggage compartment and attach to tether anchors. See Fig. 8.

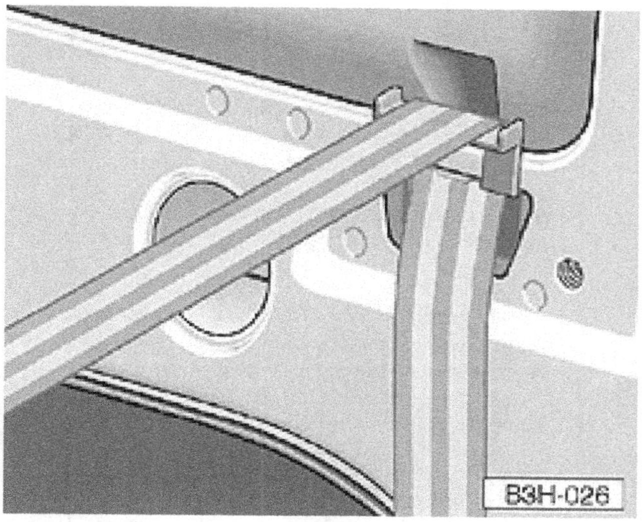

B3H-026

Fig. 8. Rear upper tether strap should be positioned in guide as shown.

The rear lower anchors secure the child restraint in the seat without using the vehicle's safety belts. Anchors provide a secure and easy-to-use attachment and minimize the possibility of improper child restraint installation. The lower anchor attachment points are located between the rear seat back and rear seat cushion for the outboard seating positions. There is no rear middle seat position. See Fig. 9.

B1S-027M

Fig. 9. Rear lower tether anchors are located at **arrows** for Cabrio.

Vehicles manufactured as January 2001 (01/01) are equipped with buttons to identify the rear lower anchor points for securing a child restraint. See Fig. 10. For vehicle production date, see the data label attached to the lower B-pillar.

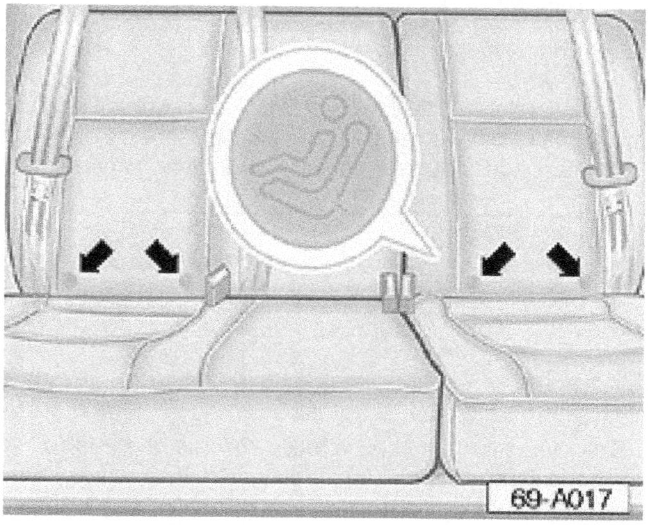

69-A017

Fig. 10. Rear lower anchor points in Cabrio from 01/01 production.

AIRBAGS

The heading covers only the removal and installation of the airbag units from the steering wheel, the passenger side instrument panel and the front seats on cars so equipped. Airbag system repair and fault diagnosis is not covered here. Diagnostics, component testing, and repair should be carried out by properly trained Volkswagen technicians using specialized test equipment.

When the ignition key is turned on, the indicator light will illuminate for approximately 5 seconds. The illuminated light indicates the self-test of the electronic control/monitoring module, the ignitors and sensor circuits, and the system wiring. The airbag system should be inspected by an authorized Volkswagen dealer if the indicator light does not come on, if it does not go out in approximately five seconds after switching on the ignition, or if the light comes on during driving. If the system has detected a fault, the light will come on and stay on.

If the vehicle has been involved in an accident where the airbag was deployed, Volkswagen specifies that the following components must be replaced:

- All airbag units that have been deployed
- Passenger's side airbag retaining frame
- Airbag control module with holder
- Airbag spiral spring
- Backrest padding with triggered side airbag and sensor
- Driver's side seat belt
- Front passenger's side seat belt (if in use at the time of the accident)

In addition, the following must also be replaced if necessary (visual check):

- All damaged components

If the vehicle was in an accident that did not involve the use of the airbag(s) and the warning lamp does not indicate any malfunctions, it is not necessary to replace the airbag components.

> ***WARNING —***
> *The seat belts should be checked separately. See* **Seat belts, inspecting**.

General

The airbag system in all vehicles consists of the following components: an electronic control unit/monitoring module, an indicator light in the instrument panel, an inflatable airbag with integral gas generator in the steering wheel, and an inflatable airbag with integral gas generator in the right side of the padded dash.

In addition some vehicles are equipped with optional side airbags which add the following extra components: inflatable airbags with integral gas generators in both outer front seat backrests and lateral crash sensors mounted on the floor under the front seats. All vehicles are also equipped with knee bars for extra protection. See Fig. 11.

Airbag, safety precautions

Special precautions must be observed prior to any work at or near the airbag units, the steering wheel or steering column, including the pedal assembly.

> ***WARNING —***
> - *Always disconnect the negative (–) battery cable before working on or near any part of the airbag system. Accidental electrical contact may damage the electrical system, cause fire, or result in serious personal injury.*
> - *The airbag(s) is inflated by an explosive device. Handled improperly or without adequate safeguards, the system can be very dangerous.*
> - *The airbag is a vehicle safety system. To guard against personal injury or airbag system failure, only trained Volkswagen service technicians should test, disassemble or service the airbag system.*
> - *Always replace any airbag system component that has been mechanically damaged (example: dented or cracked).*
> - *Always disconnect the negative (–) battery cable before working on or near the airbag(s).*
> - *No waiting time is necessary after disconnecting the battery before working on the airbag system.*
> - *When reconnecting the battery, there must be NO person inside the vehicle.*
> - *DO NOT leave an undeployed airbag unit unattended.*
> - *If the airbag unit or airbag control module has been dropped from a height of 0.5 meter (19.7 in.) or more, DO NOT install.*
> - *Always place an airbag unit that has been removed from its packaging with the padded side facing upward.*
> - *Airbag units that have NOT been activated, which have been removed from vehicles, should be marked and returned to an authorized Volkswagen dealer for proper disposal (use packaging from replacement airbag unit).*
> - *Airbags that have been deployed DO NOT have to be disposed of as hazardous waste and can be discarded in the normal manner. It is recommended to have the unit recycled.*
> - *The storage, transportation and disposal of airbag units are subject to the laws for flammable solids.*

Front airbag components, overview

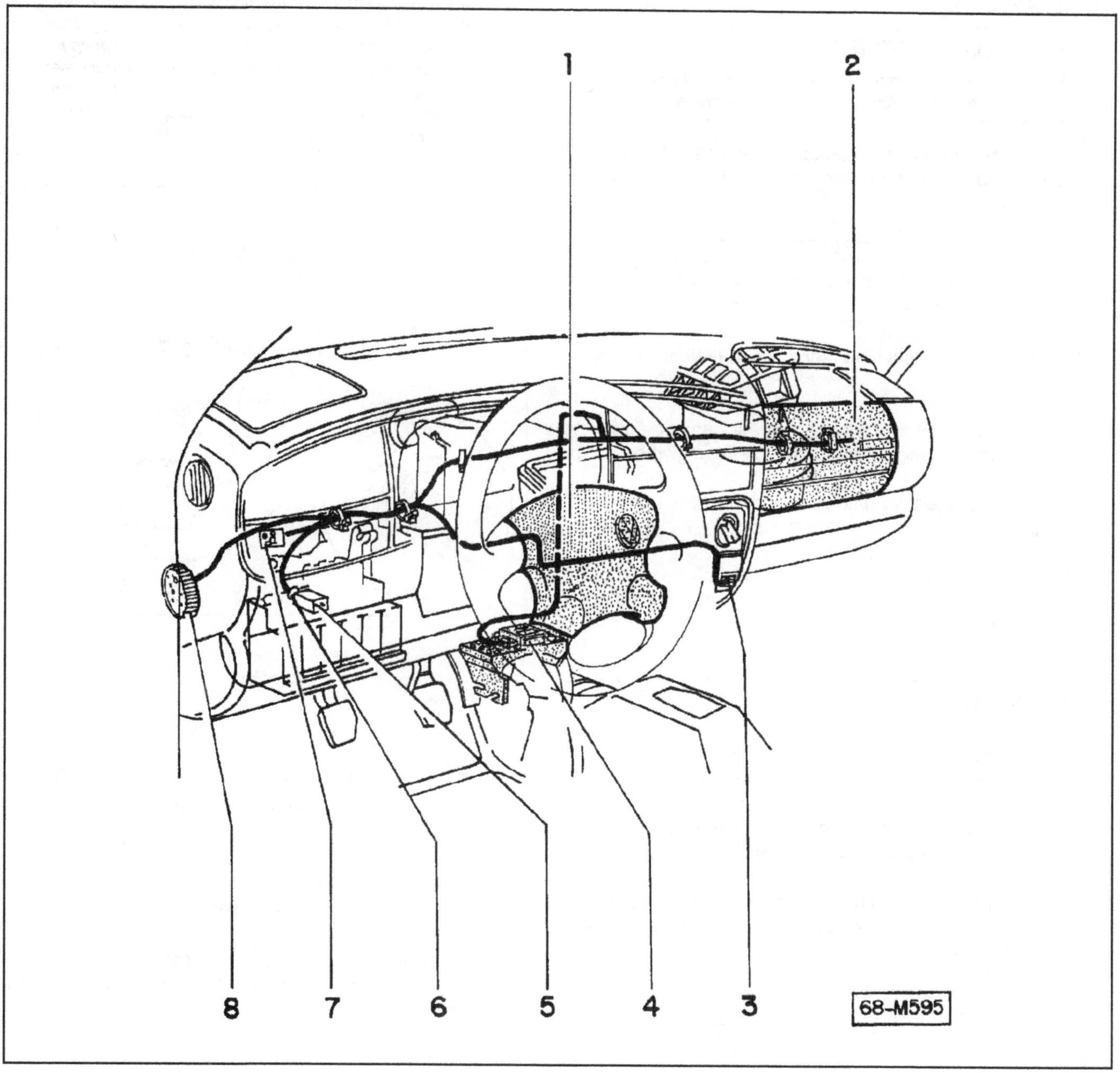

68-M595

Fig. 11. Overview of airbag components. Car with dual airbags shown.

1. **Driver's side airbag unit**
   - See **Driver's side airbag, removing and installing**
2. **Passenger's side airbag unit**
   - See **Passenger's side airbag, removing and installing**
3. **Diagnostic (data link) connector**
4. **Airbag control module**
5. **Horn relay**
6. **Wiring harness distributor**
7. **Airbag Malfunction Indicator Lamp (MIL)**
8. **Central ground terminal**

Driver's side airbag unit, removing and installing

> *WARNING —*
> *Before removing the airbag unit, read the warnings given earlier under* **Airbag, safety precautions**

1. Disconnect negative (–) battery cable. Insulate battery cable end to prevent it from accidentally touching battery post.

> *CAUTION —*
> *Before disconnecting battery be sure to obtain the radio anti-theft code.*

2. Turn steering wheel to center position (wheels straight ahead).

3. Working from rear of steering wheel, remove socket-head bolts that mount the airbag to steering wheel.

4. Carefully pull airbag unit from steering wheel and unplug red connector from airbag unit.

5. With battery negative (–) cable disconnected, reconnect harness connector to airbag unit and reinstall to steering wheel.

Tightening torques
- Airbag to steering wheel (socket-head bolts). . . . . 6.5 Nm (58 in-lb)

6. Once airbag unit installation is complete, switch ignition ON. Leave vehicle and close driver's door.

7. Connect battery negative (–) cable. Check that no malfunctions exist, as indicated by an illuminated or malfunctioning warning light.

> *WARNING —*
> *Ensure that there is no person inside the vehicle before reconnecting battery. Always connect the battery as the very last step.*

> *CAUTION —*
> *If the ignition is not switched on BEFORE connecting the ground strap, the airbag control unit could be damaged.*

NOTE —
Steering wheel removal is described on **48 Steering**. *Be sure the steering wheel is in the center position (wheels straight ahead) when the steering wheel is removed.*

NOTE —
Note that the airbag spiral spring is mounted to the rear of the steering wheel. See Fig. 12. If the steering wheel is removed, be sure to lock the airbag spiral spring in its center position. To lock the spiral spring, hold the steering wheel in the normal operating position. Then turn the spiral spring housing left and/or right just until the spiral spring locks in place.

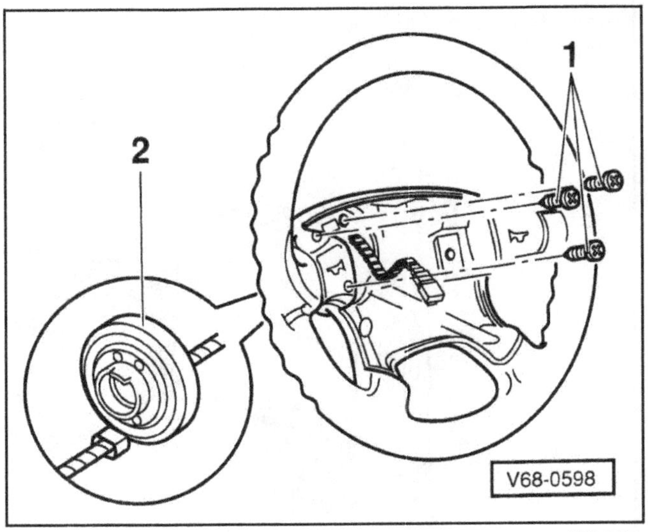

V68-0598

Fig. 12. Airbag spiral spring (**2**) is mounted to rear of steering wheel. Be sure to lock spiral spring in center position once wheel is removed.

Driver's side airbag unit, removing and installing (Cabrio from 02/98)

> *WARNING —*
> *Before removing the airbag unit, read the warnings given earlier under* **Airbag, safety precautions**

1. Disconnect negative (–) battery cable. Insulate battery cable end to prevent it from accidentally touching battery post.

> *CAUTION —*
> *Before disconnecting battery be sure to obtain the radio anti-theft code.*

2. Release steering column adjustment lever.

3. Turn steering wheel until a spoke is vertical. Extend steering column out fully toward driver.

4. Secure steering column adjustment lever.

5. Using a screwdriver approximately 175 mm (7 in.) long, insert approximately 45 mm (1¾ in.) from reverse side into hole on dashboard side of steering wheel hub.

6. Press screwdriver upward to press back spring clip and release locking lug of airbag unit. See Fig. 13.

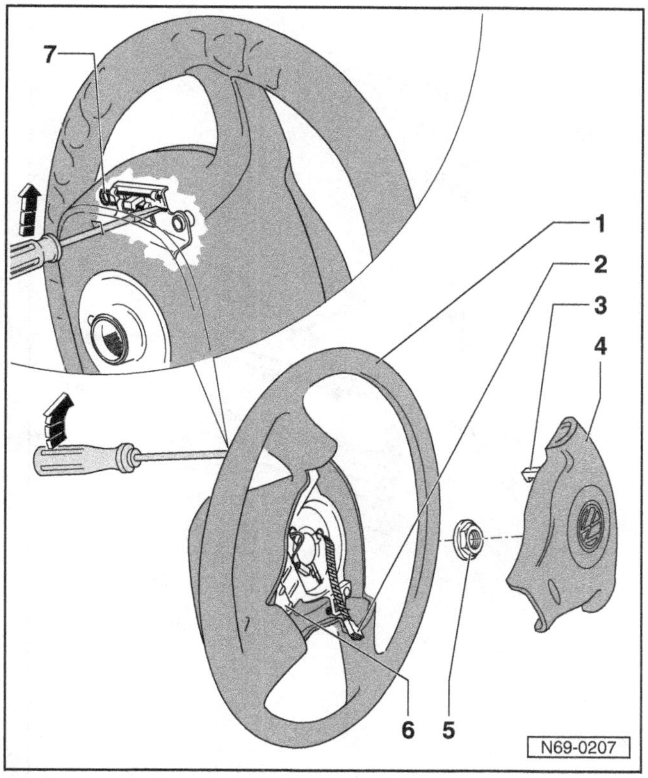

N69-0207

Fig. 13. Steering wheel with airbag release mechanism installed on Cabrio models from 02/98 production.

7. Turn steering wheel back ½ turn (180°) and release second locking lug on opposite side.

8. Turn steering wheel to center position (wheels straight ahead).

9. Disconnect harness connector from airbag unit.

10. Place the removed airbag unit in a secure location with the padding side facing up.

11. Begin installation by connecting harness connector for airbag unit.

12. Position airbag unit on steering wheel and snap into place until retaining tabs are locked.

13. Airbag unit locking lugs must be heard and felt to snap into place.

14. Switch ignition on.

15. Connect battery ground strap.

> ***WARNING —***
> - *Make sure the passenger compartment is not occupied before connecting the battery Ground (GND) strap.*
> - *Observe all cautions, warnings and notes when completing repairs involving airbag systems.*

Passenger's side airbag unit, removing and installing

1. Disconnect negative (–) battery cable.

> ***CAUTION —***
> *Before disconnecting battery be sure to obtain radio anti-theft code.*

2. Remove passenger side knee bar, or on later cars the glovebox with kneebar, see **70 Trim–Interior**. The knee bar is held by five screws on the bottom and three catches on top. See Fig. 14.

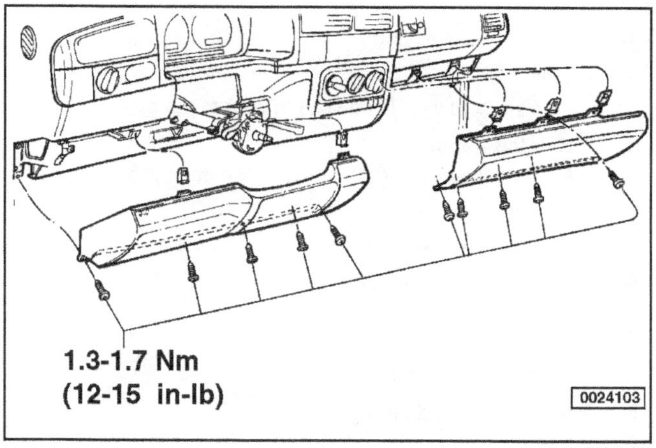

0024103

Fig. 14. Knee bars used on cars with airbags.

3. Remove right-hand air vent from instrument panel by pulling it off its housing.

4. Remove air vent housing from instrument panel by removing two screws. On cars with VR6 engine, unplug harness connector from air vent housing.

5. Remove airbag cover mounting bolts from lower part of instrument panel. Then slide cover to right and remove. See Fig. 15. or Fig. 16.

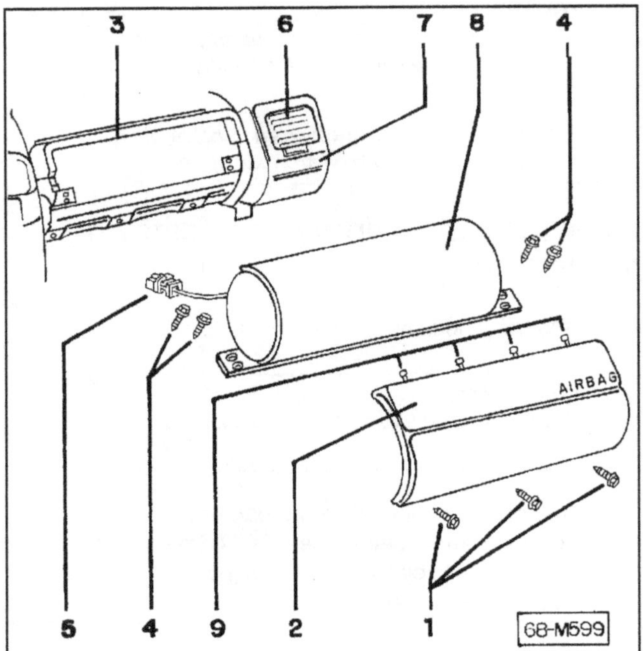

1. **Hex bolt**
2. **Cover**
   • Remove screws (1) and slide to right to remove
3. **Retaining frame**
   • Always replace after airbag has been deployed
4. **Hex bolt**
   • Tighten to 6.5 Nm (58 in-lb)
5. **Airbag harness connector**
6. **Air vent**
7. **Air venting housing**
8. **Passenger's side airbag unit**
9. **Guide pins**

68-M599

Fig. 15. Components of passenger side airbag for vehicles produced before March 1996.

6. Remove retaining frame from instrument panel.

7. Remove airbag mounting bolts. Disconnect airbag harness connector and remove airbag unit.

8. With battery negative (–) cable disconnected, reconnect harness connector to airbag unit and reinstall into instrument panel.

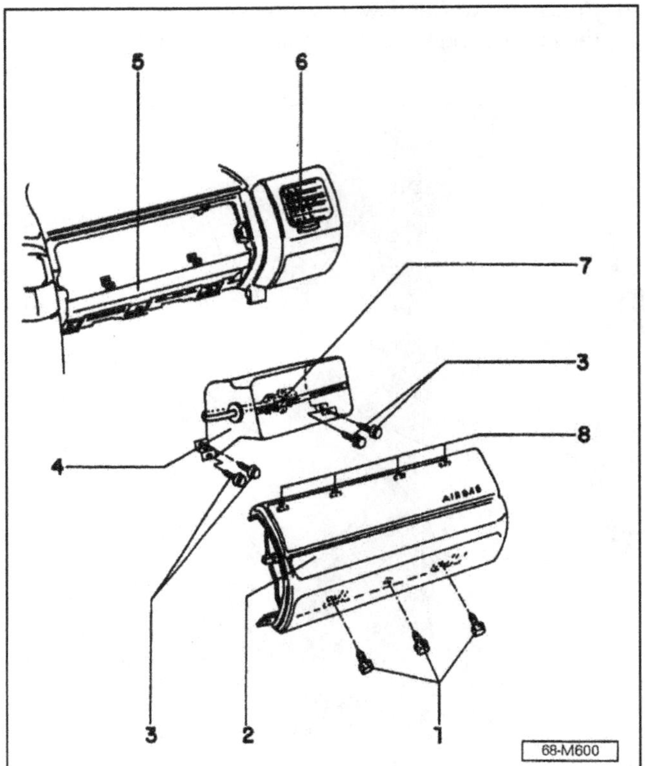

1. **Hex bolt**
2. **Cover**
   • Remove screws (1) and slide to right to remove
3. **Hex bolt**
   • Tighten to 10 Nm (7 ft-lb)
4. **Passenger's side airbag unit**
5. **Crossbar**
6. **Air vent**
7. **Airbag harness connector**
8. **Guide pins**

68-M600

Fig. 16. Components of passenger side airbag for vehicles produced from March 1996.

Tightening torques
- Airbag to instrument panel (up to March 1996 production) .... 6.5 Nm (58 in-lb)
- Airbag to instrument panel (from March 1996 production) .... 10 Nm (7 ft-lb)

9. Reinstall air vent housing, air vent, and airbag retaining frame. On cars with VR6 engine connect harness connector at air vent housing.

10. Install airbag cover.
    - Lock airbag cover in place by aligning its four pins with holes in retaining frame, slide cover to left and install lower mounting bolts.

11. Install knee bar or glove box with knee bar on later cars.

Tightening torques
- Knee bar to instrument panel . . . . 1.3-1.7 Nm (12-15 in-lb)

12. Once airbag unit installation is complete, switch ignition ON. Leave vehicle and close driver's door.

13. Connect battery negative (–) cable. Check that no malfunctions exist, as indicated by an illuminated or malfunctioning warning light.

WARNING —
Ensure that there is no person inside the vehicle before reconnecting battery. Always connect the battery as the very last step.

CAUTION —
If the ignition is not switched on BEFORE connecting the ground strap, the airbag control unit could be damaged.

Side airbag unit, driver's and front passenger side

Side airbags built into the outside sections of both front seat backrests are available as an option beginning with model year 1998.

These airbags are deployed when side deceleration forces reach predetermined levels and can operate independently of the front airbags. Deployment is controlled by the airbag control module and one additional crash sensor per side. The side crash sensors are mounted under each front seat and detect lateral deceleration.

WARNING —
Volkswagen of America has issued a directive specifically warning against the installation of aftermarket upholstery on any vehicle equipped with side airbags. The factory-installed upholstery is designed to separate in specific places at specific rates, and in specific directions. Installation of non-factory upholstery including, but not limited to, "beads" and "sheepskins", may cause seat mounted airbags to deploy when they are not supposed to; fail to deploy when they should; or to deploy in some manner other than designed. This is a safety hazard and could result in serious injury or death to occupants of the vehicle.

Side airbag unit, removing and installing

1. Disconnect negative (–) battery cable.

2. Remove backrest frame. See **72 Seats**.

3. Loosen backrest upholstery and work it up to a point where the side airbag mounting bolts are visible. See Fig. 17.

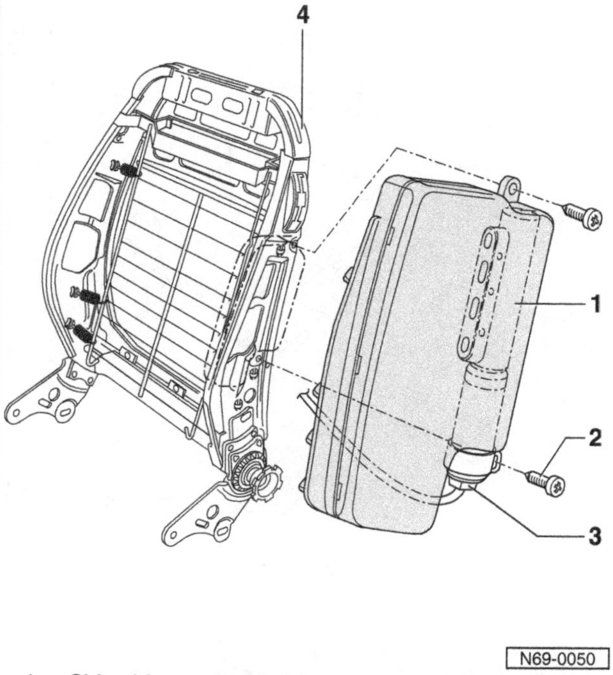

1. **Side airbag unit, driver's and front passenger side**
2. **Bolt**
   - Ground wire secured with bolt -2-
   - Tightening torque: 7 Nm (62 in-lb)
3. **Connector**
4. **Backrest frame**

N69-0050

Fig. 17. Side airbag unit with front seat backrest frame. Driver's side shown, passenger side is symmetrically opposite.

4. Remove both mounting bolts.

5. Disconnect electrical connector from airbag unit and remove airbag from backrest.

6. Installation is the reverse of removal, noting the following:
   - Negative (–) battery cable remains disconnected
   - Switch ignition on
   - Close doors
   - Re-connect negative (–) battery cable

WARNING —
Make sure that no one is in the vehicle when reconnecting the negative (–) battery cable.

Side airbag crash sensor, removing and installing

1. Disconnect negative (–) battery cable.

2. Remove driver's or front passenger seat. See **72 Seats**.

3. Release carpet and insulation mat in area of sensor. See Fig. 18.

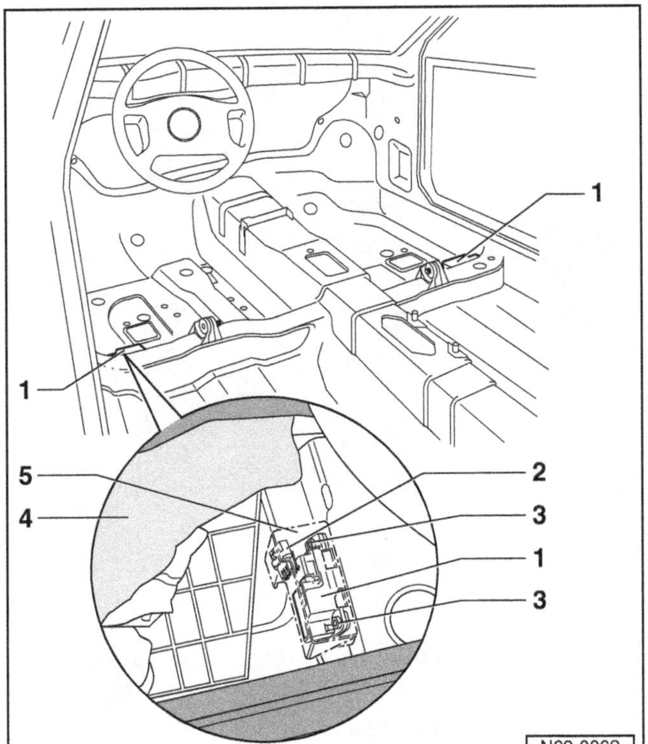

1. **Crash sensor for side airbag, driver's and front passenger side**
2. **Connector**
3. **Bolt**
   • Tightening torque: 10 Nm (7 ft-lb)
4. **Carpet**
5. **Cover**

N69-0062

Fig. 18. Side airbag crash sensors mounted to the seat supports on the floor. One sensor is used for each side airbag.

4. Carefully pry off cover.

5. Disconnect electrical connector and remove mounting screws.

6. Installation is the reverse of removal, noting the following:
   - Negative (–) battery cable remains disconnected
   - Switch ignition on
   - Close doors
   - Re-connect negative (–) battery cable

> ***WARNING —***
> *Make sure that no one is in the vehicle when reconnecting the negative (–) battery cable.*

70 Trim–Interior

GENERAL

This group covers removal and installation of the interior trim, including the headliner.

> *NOTE —*
> • *Instrument cluster removal and installation is covered in* **90 Instruments**.

> *CAUTION —*
> *Before disconnecting battery be sure to obtain radio anti-theft code.*

CENTER CONSOLE AND INSTRUMENT PANEL

Fig. 1 and Fig. 2 show center console assemblies, depending on whether or not the vehicle is equipped with an airbag(s).

Center console, removing and installing (cars with airbag)

Use Fig. 1 as a guide when removing and installing the center console used on cars with airbag.

1. Disconnect negative (–) battery cable. Insulate battery cable end.
2. Remove shift lever boot or selector lever cover from center console. See **3 Clutch, Transmission, and Final Drive**.
3. Carefully pry off screw covers and then remove left and right parking brake lever console mounting screws.
4. Pull out and disconnect connector from remote rear trunk opener (where applicable).
5. Slide parking brake lever console to rear.
6. Remove three mounting nuts from the center console.
7. Remove screws from center console (one on left-side, two on right-side).
8. Lift rear end of console slightly and remove it from instrument panel.
9. Install center console in reverse order of removal. Be sure the spacer bushings and locating pins are positioned correctly when installing console.

Center console, assembly (cars with airbag)

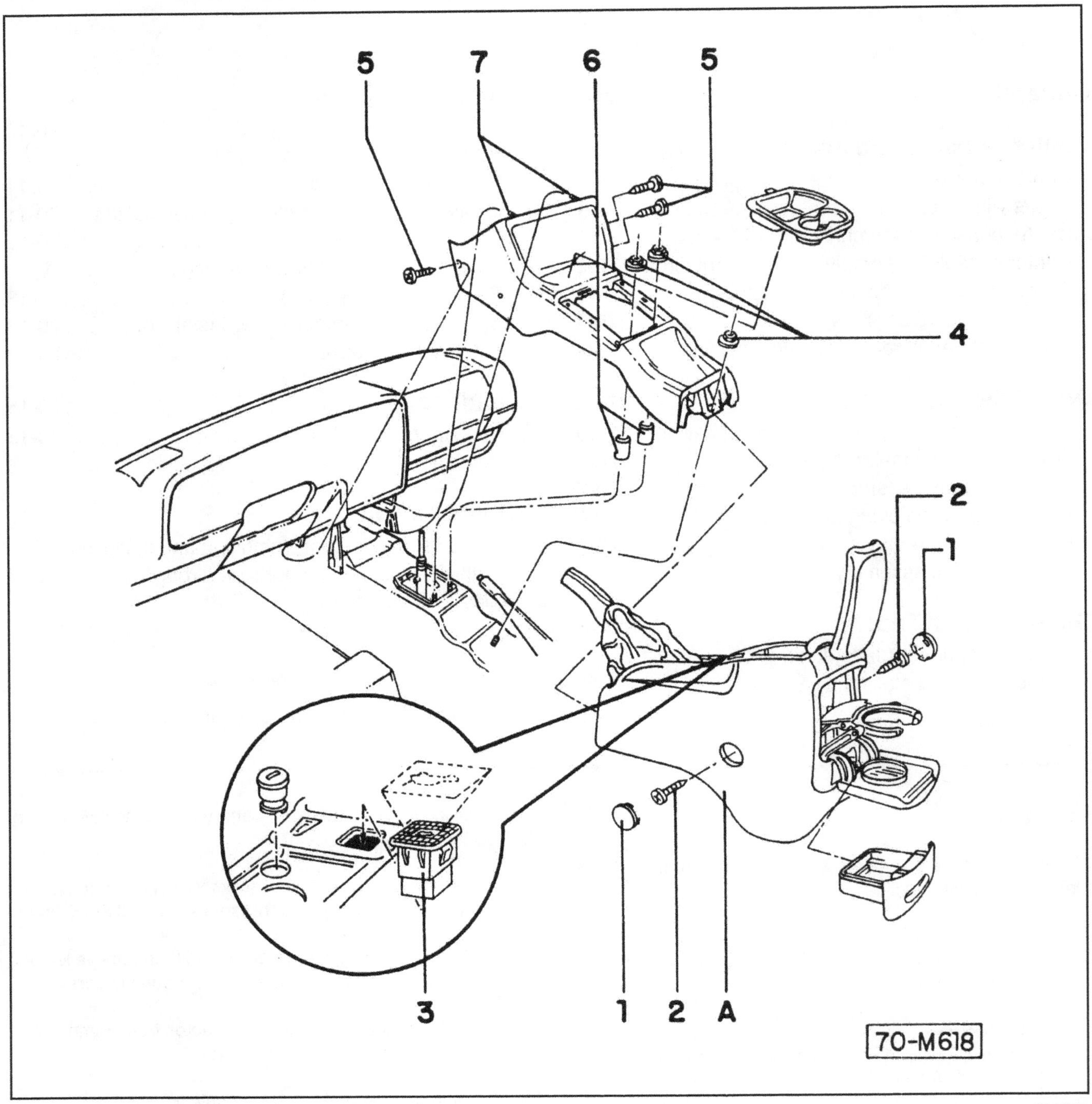

70-M618

Fig. 1. Exploded view of center console for vehicles equipped with airbags.

1. **Screw covers**
2. **Screws**
3. **Remote trunk switch**
   Location varies
   - Under lid in console
   - On driver's seat trim panel
   - In glove box on early cars without airbags
4. **Nuts**
5. **Screws**
   - One on left, and two on right side
6. **Spacers**
7. **Locating/retaining pins**

A. **Parking brake lever console**

Center console assembly (cars without airbag)

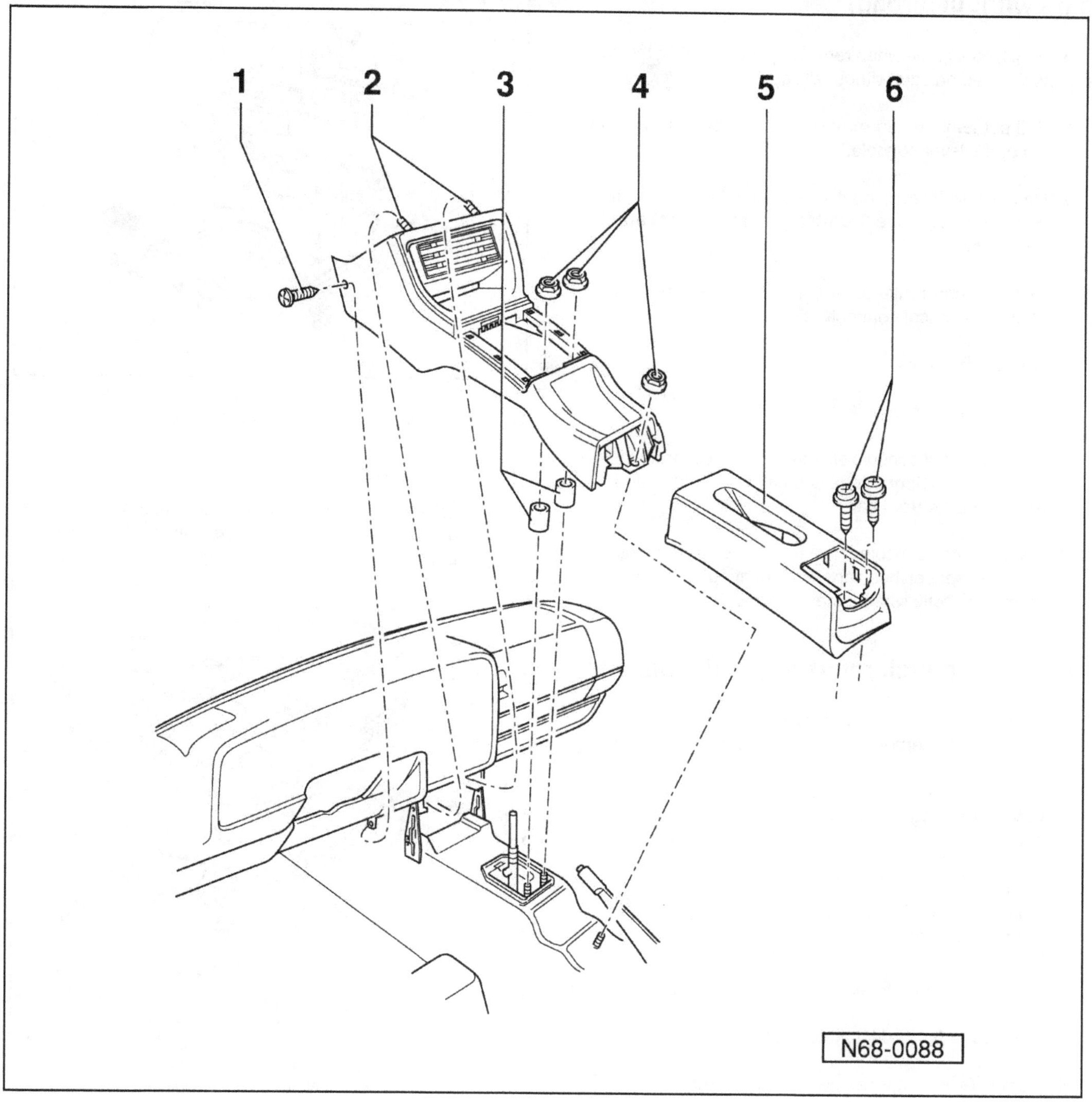

N68-0088

Fig. 2. Exploded view of center console for vehicles without airbags.

1. **Screws**
   • On left and right side
2. **Locating/retaining pins**
3. **Spacers**
4. **Hex nuts**
5. **Parking brake lever console**
6. **Hex head screw**

Center console, removing and installing (cars without airbag)

Use Fig. 2 as a guide when removing and installing the center console used on cars without airbag.

1. Pull out ashtray and remove screws from rear of parking brake lever console.
2. Remove shift lever boot or selector lever cover from center console. See **3 Clutch, Transmission, and Final Drive**.
3. Slide parking brake lever console to rear and disconnect all electrical connectors.
4. Remove hex nuts.
5. Remove screws on left and right sides.
6. Lift rear end of console slightly and remove from instrument panel. Separate electrical connectors and remove console from car.
7. Install center console in reverse order of removal. Be sure the spacer bushings and locating pins are positioned correctly when installing console.

Instrument panel, removing and installing

Center console, radio, steering wheel and steering column switch must first be removed before the instrument panel can be removed.

1. Disconnect negative (–) battery cable. Insulate battery cable end.

> *CAUTION —*
> *Before disconnecting the battery, be sure to obtain the radio anti-theft code.*

2. Remove center console as described previously.
3. Remove radio as described in **91 Radio**.
4. Remove steering wheel. See **48 Steering**.
5. Remove passenger side airbag unit as described in **69 Seat belts, Airbags**.
6. Remove relay panel cover from under left-hand side of instrument panel. See Fig. 3.

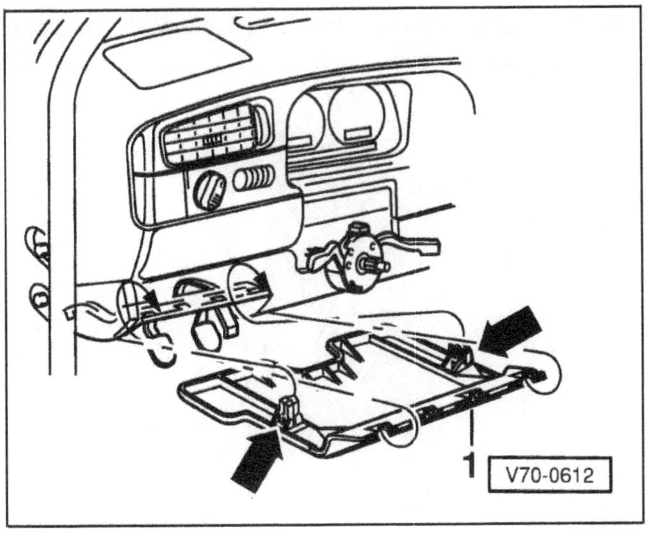

V70-0612

Fig. 3. Press locking buttons (**arrows**) to remove relay cover (**1**).

7. Remove lower instrument panel trim:
   - Pry off center cover on driver's side, remove screws, and pull large tray downward at an angle. Make sure that all guides disengage. See Fig. 4.

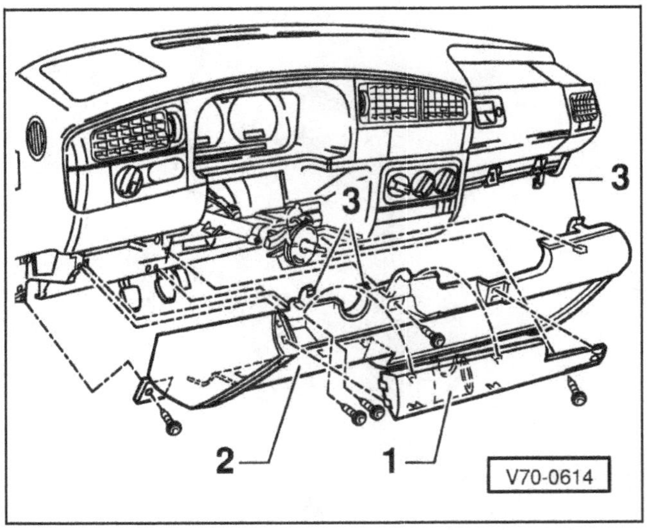

V70-0614

Fig. 4. Pry off center cover (**1**), remove screws, and pull down tray (**2**), making sure that guides (**3**) disengage.

8. Remove screws and pull passenger side storage shelf downward at an angle. Make sure that tabs disengage. See Fig. 5. For vehicles equipped with an airbag and glovebox, remove the glovebox as described later in **Glovebox**.

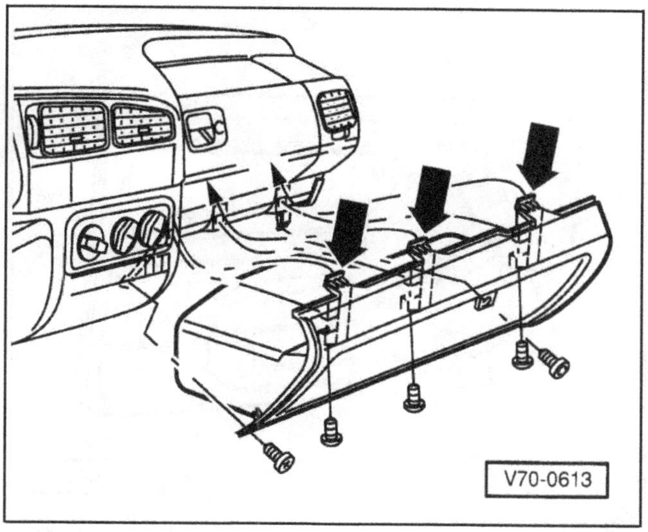

V70-0613

Fig. 5. Remove screws and pull passenger side shelf downward at an angle, making sure tabs (**arrows**) disengage.

9. Pry off cover for heating and ventilating controls. Remove control panel screws and position heating and ventilating controls (with attached cables) below instrument panel. See Fig. 6.

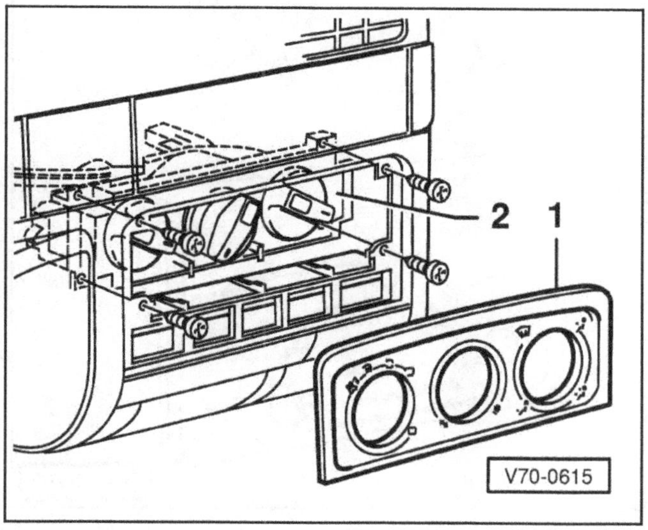

V70-0615

Fig. 6. Pry off cover (**1**) and press controls (**2**) (with cables attached) under instrument panel.

10. Unclip covers and light switch from instrument panel. Remove screws and pull out instrument cluster cover. See Fig. 7.

11. Remove instrument cluster securing screws, tip instrument cluster forward, and disconnect multi-pin connector on rear of instrument cluster. See Fig. 8.

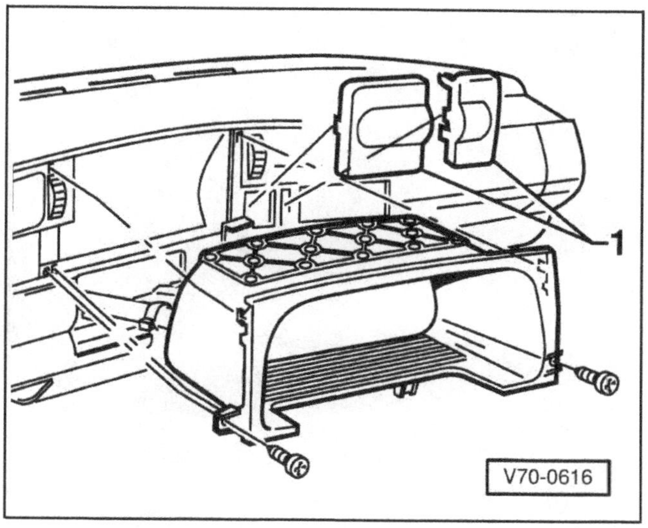

V70-0616

Fig. 7. Unclip covers (**1**), and remove headlight switch. Then remove screws and pull out instrument cluster cover.

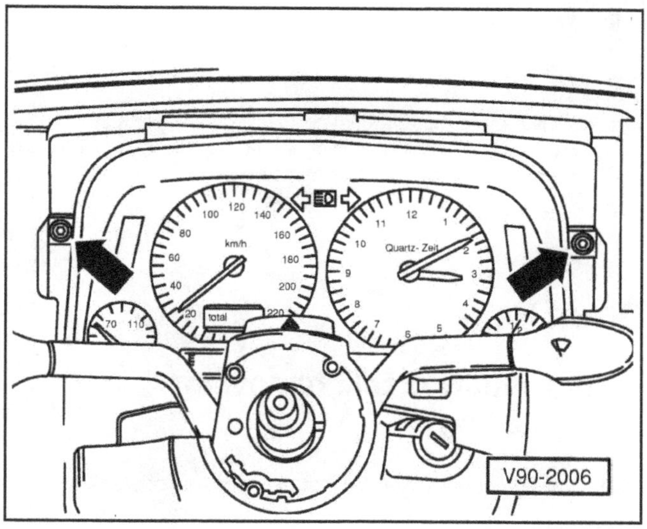

V90-2006

Fig. 8. Instrument cluster securing screws (**arrows**).

12. Remove instrument cluster.

13. Remove instrument panel mounting screws (11 screws, including glove box) and hex nuts (located in plenum chamber). See Fig. 9.

14. Partially remove instrument panel from carrier and disconnect wiring harnesses from instrument panel.

15. Remove instrument panel.

16. Installation is the reverse of removal. Replace any wire-ties cut during removal. Make sure the mounting nut insulating washers are reinstalled during installation (see Fig. 9).

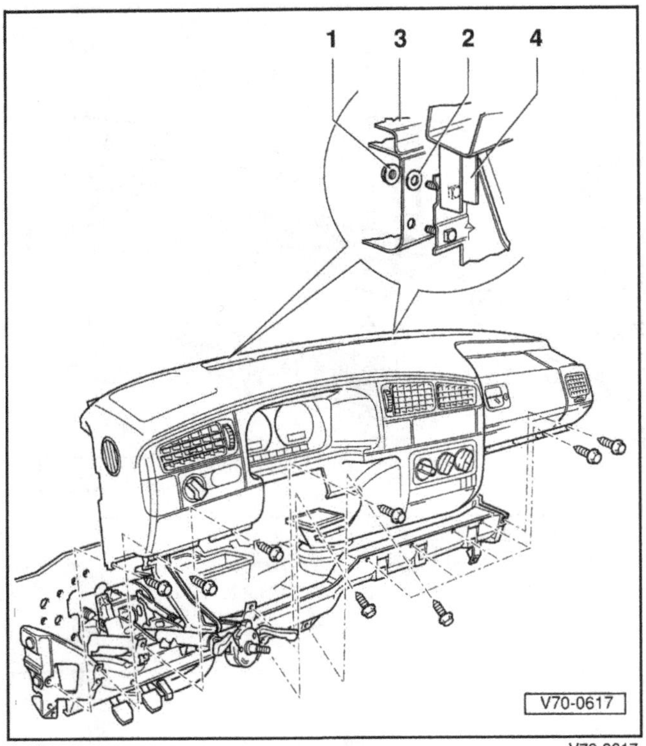

V70-0617

Fig. 9. Instrument panel mounting (**4**) hardware. Nuts (**1**) are accessed through plenum chamber wall (**3**). Note insulating washers (**2**) that must be reinstalled.

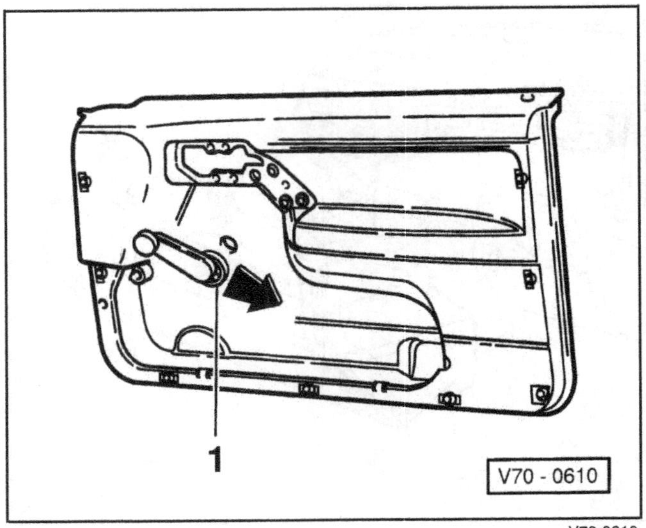

V70-0610

Fig. 10. Slide spacer ring (**1**) in direction of arrow to unlock window crank retaining clip.

DOOR TRIM

Front window crank, removing and installing

1. Raise window fully. Note installed position of crank handle.
2. Slide spacer ring at base of crank to rear to unlock crank retaining clip. Then pull crank off. See Fig. 10.
3. Installation is the reverse of removal. Be sure to reinstall crank in original position.

Front door trim panel, removing

1. Unscrew lock knob.
2. Unclip grab handle trim and remove grab handle attachment screws together with handle. See Fig. 11.
3. Remove window crank as described earlier.
4. Using both hands, place fingers behind outside remote rear-view mirror control knob and pull forward to remove.
5. Pull inside door latch release lever trim piece straight out, bottom first then top.

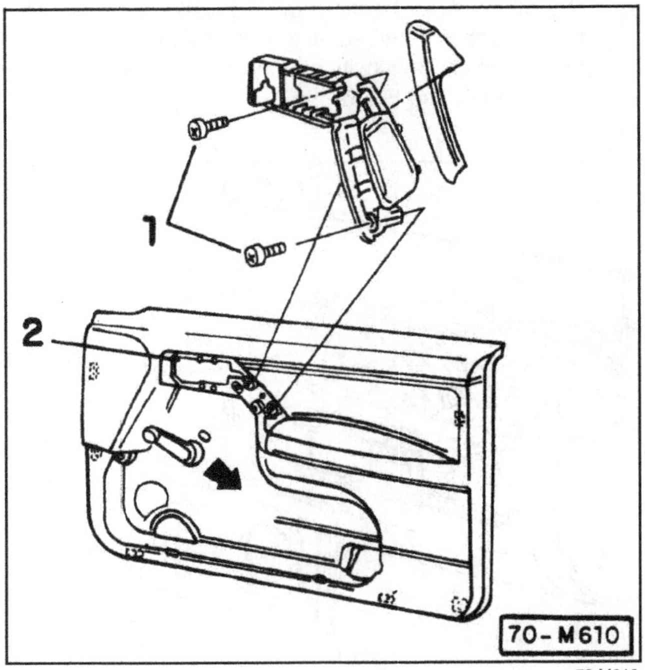

70-M610

Fig. 11. Grab handle retaining screws (**1**).

6. Remove top screw (in front of inside door release lever housing) for outside remote rear-view mirror control housing and door trim panel.
7. Remove six door trim panel attachment screws from perimeter of door panel.
8. Separate door panel from trim piece at top of door, starting from either end, not both. Do not pull hard, this will cause the door panel to break.
9. Disconnect electrical connector for speaker. Disconnect electrical connector for anti-theft alarm system (driver's door only).

Front door trim, assembly

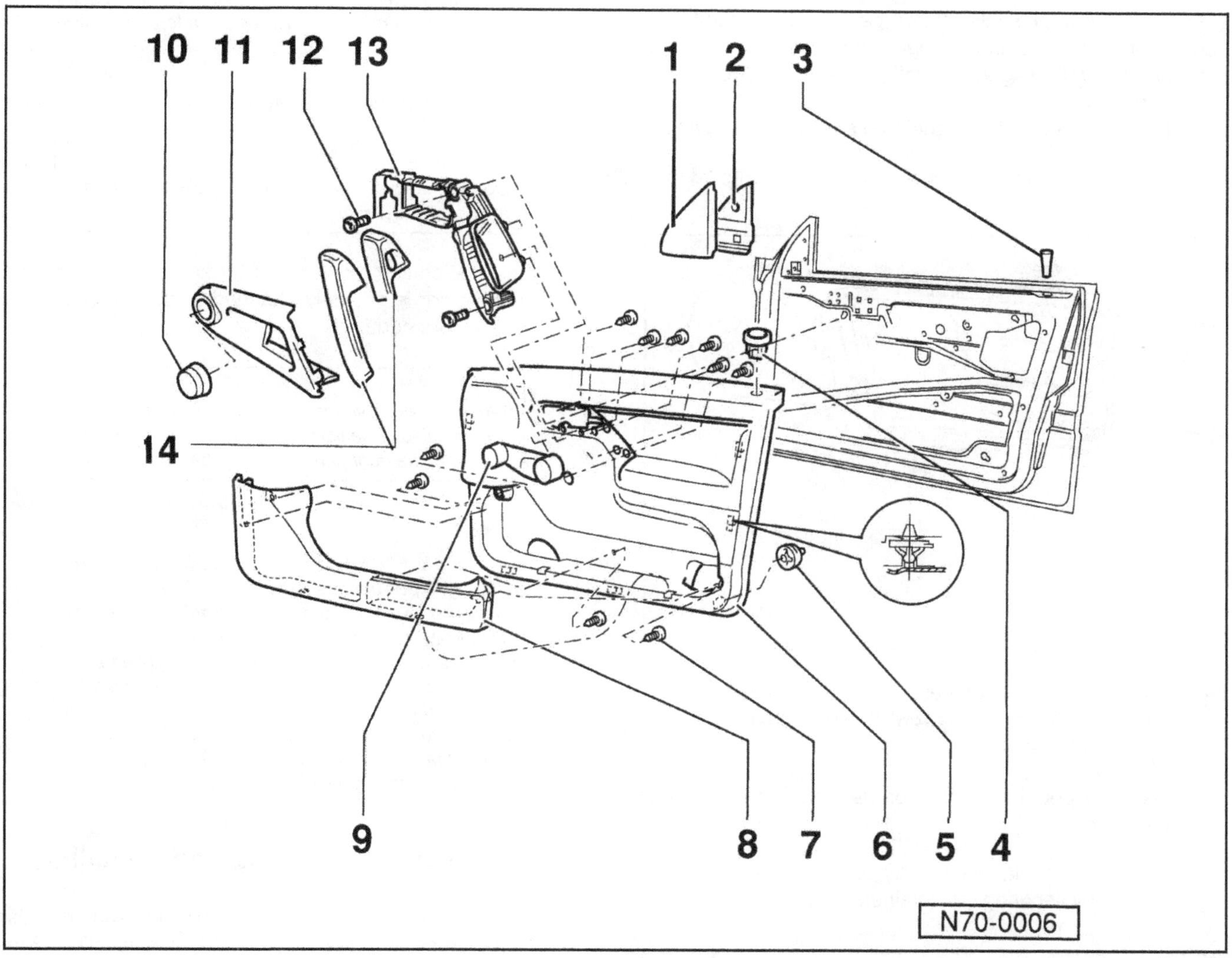

N70-0006

Fig. 12. Exploded view of door trim for Cabrio (other models similar).

1. **Cover plate**
   - clipped to door frame
2. **Insulating plate**
3. **Locking knob**
4. **Trim ring**
5. **Securing clip**
   - Two-piece
6. **Trim panel**
7. **Countersunk screw**
8. **Door tray**
   - Removed only after first removing trim panel
9. **Window crank**
10. **Mirror control knob**
    - Pushed onto control
11. **Inner door release trim panel**
    - Clipped into door/door pull
12. **Countersunk screw**
13. **Door pull**
14. **Cover**
    - Connected on inside opener plate
    - Clipped into door/door pull

Vapor barrier, removing and installing

The vapor barrier seals off the passenger compartment from water, noise, and drafts. Replacement vapor barriers are supplied from Volkswagen with self-adhesive surfaces.

1. Remove interior door panel trim as described earlier.

2. Remove inside door release housing. See Fig. 13.

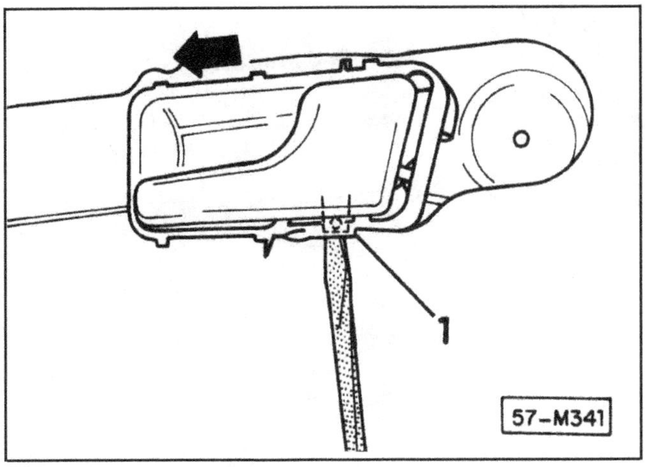

57-M341

Fig. 13. Remove door release housing by releasing locking tab (**1**) and sliding housing forward (**arrow**). If necessary, unhook link rod from release mechanism.

3. Carefully peel back vapor barrier, starting from either top edge.

4. Using a razor blade, carefully cut vapor barrier around trim panel securing clip retainers. See Fig. 14.

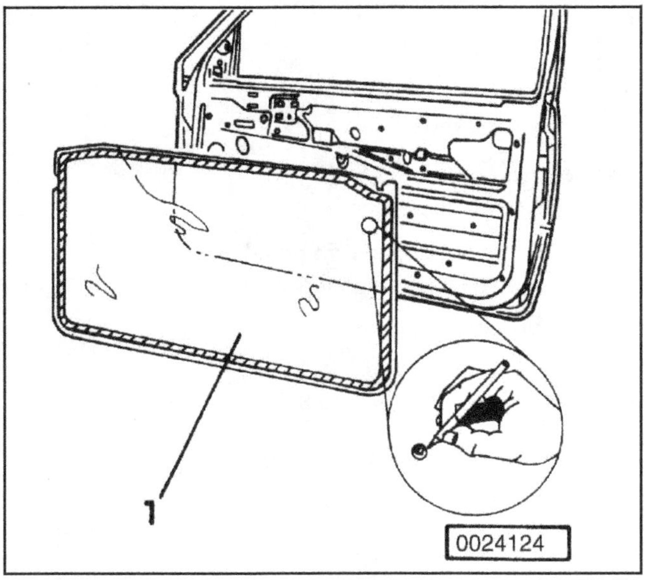

0024124

Fig. 14. Trim around securing clip, then pull vapor barrier (**1**) down only as far as is necessary.

> ***NOTE —***
> *Scratches in paint caused by cutting the vapor barrier should be repaired with touch up paint to prevent corrosion.*

5. Carefully peel vapor barrier from door.

> ***CAUTION —***
> *Do not remove vapor barrier completely unless absolutely necessary.*

6. Installation is the reverse of removal. Be sure that vapor barrier is securely attached to door, especially at the bottom edge.

> ***CAUTION —***
> *When installing vapor barrier, there must not be any wrinkles in the barrier. This will cause water to leak onto the door panel and into the vehicle.*

> ***NOTE —***
> *Beginning with 03.95 production, Volkswagen began installing vapor barriers without a self-adhesive strip. This newer barrier is bonded with an adhesive (VW part # D469 101 A3) that can be removed from the door after the vapor barrier has been carefully pulled away. When replacing vapor barrier, remove remaining adhesive from door by rolling the adhesive into a ball and pulling away from door (self-bonding of adhesive is greater than bonding action to door). Use new adhesive when installing new vapor barrier.*

Rear door trim, removing and installing

Fig. 15 is an exploded view of the rear door interior trim. Use the illustration as a guide for removal and installation of the door trim.

Rear door trim, assembly

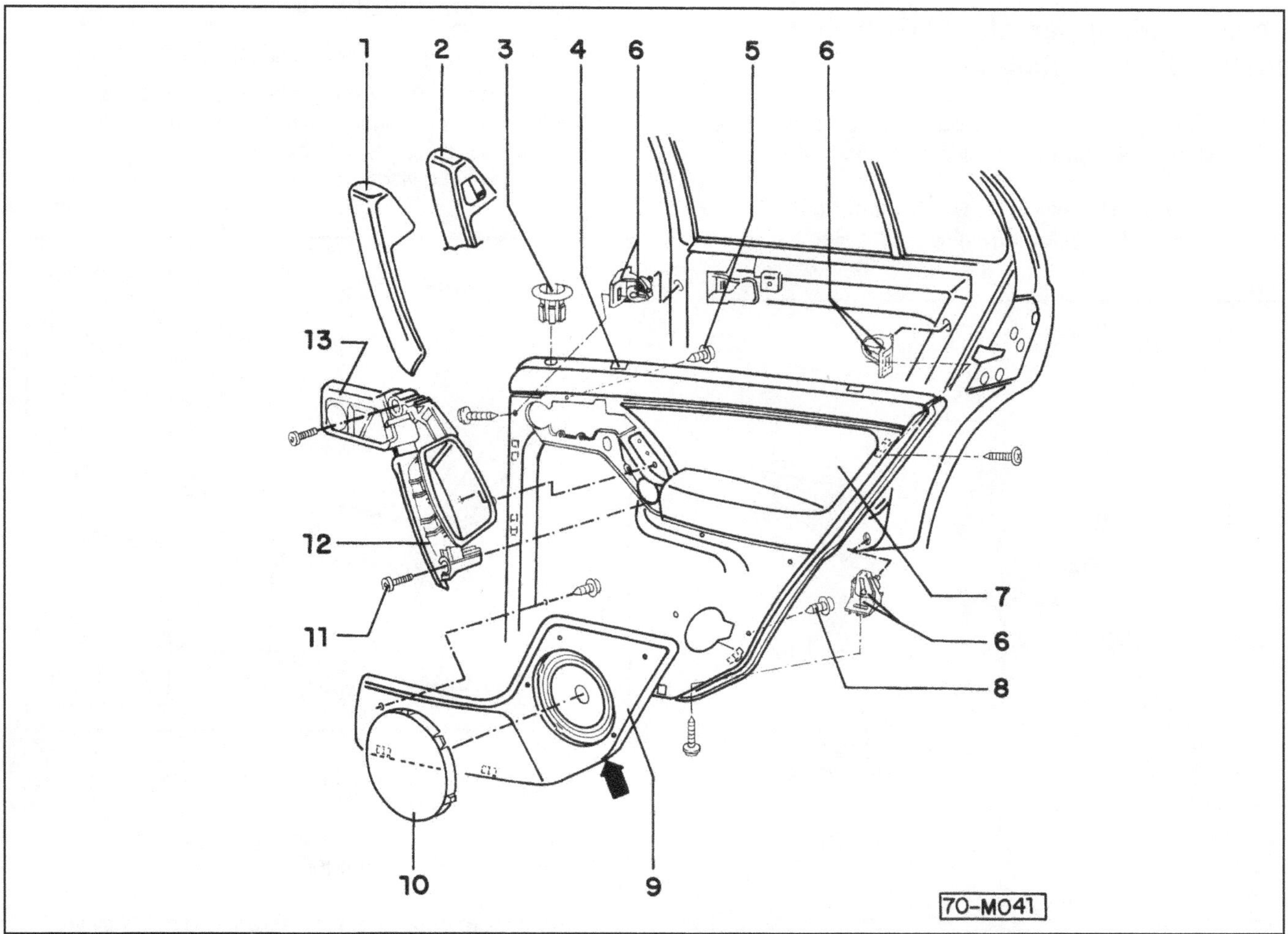

70-M041

Fig. 15. Exploded view of rear door trim for 4-door models for 1993 through 1995 models. Door speakers on 1996 and later models relocated to rear parcel shelf.

1. **Door pull trim**
   - For vehicles without electric windows
   - Clipped into door/door pull
2. **Door pull trim**
   - For vehicles with electric windows
   - Clipped into door/door pull
3. **Trim ring**
4. **Clip**
5. **Screw**
6. **Trim clip**
7. **Trim**
   Removing:
   - Unscrew locking knob
   - Remove window crank (if applicable)
   - Unclip door pull trim (**1**) or (**2**)
   - Remove screws (**11**)
   - Pry off trim
   - Disconnect electric wiring (window switch, speaker)
8. **Screw**
9. **Door tray**
   - Secured by screws through rear of trim
   - Speaker fastened to door tray
10. **Speaker cover**
    - Removing: press down on retaining lugs with screwdriver through opening in door tray (arrow)
11. **Screw**
12. **Door pull**
    - Secured by screws through rear of trim
13. **Cover**
    - Clipped into door pull
    - Mounting location for speaker (tweeter)
    - If speaker is installed, remove only after removing door trim and disassembling door pull (danger of damage)

OTHER INTERIOR TRIM

A-pillar trim, upper, (Jetta, Golf, GTI), removing and installing

1. Remove upper part of B-pillar trim. See **B-pillar trim (4-door Golf, Jetta), removing and installing**.

2. Remove trim retaining screw. Pull trim from clips and then unhook from lower retaining lugs. See Fig. 16.

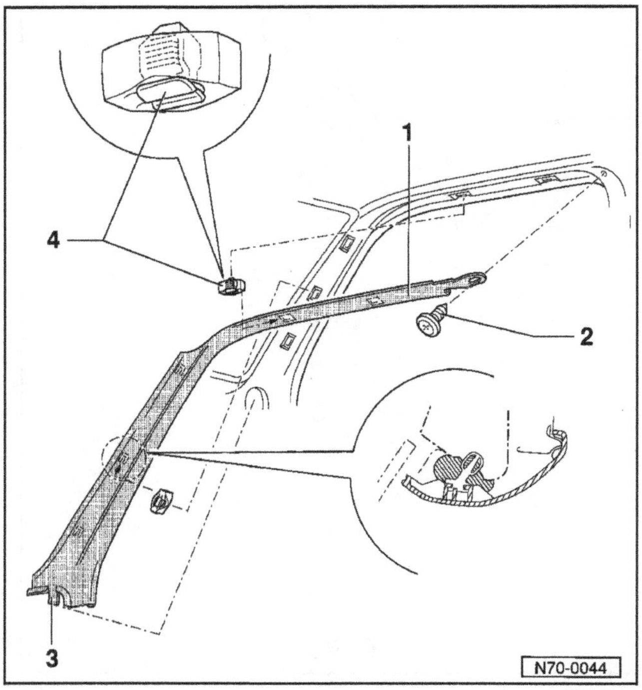

N70-0044

Fig. 16. A-pillar trim (**1**), trim screw (**2**), retaining lugs (**3**), and clips (**4**). Upper inset shows detail of mounting clip with foam padding.

NOTE —

Trim fits over sealing flange of door inner seal, see sectional view.

> *CAUTION —*
>
> *Use care when pulling trim from clips. Tabs on trim are easily damaged.*

3. Installation is the reverse of removal. Before installing trim, ensure that mounting clips are properly seated in slots on trim and not damaged.

A-pillar trim (Cabrio), removing and installing

1. Remove wind protector trim as follows:
   - Disconnect negative (–) battery ground strap.
   - Pry out interior light above rear view mirror and separate electrical connectors.
   - Remove sun visor. See Fig. 17.

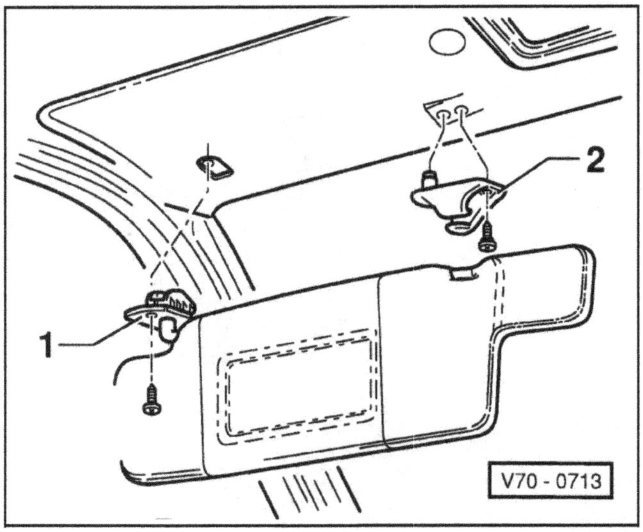

V70-0713

Fig. 17. Sun visor mounts (**1** and **2**) on Cabrio models. Remove sun visor by turning 90°.

2. Remove wind protector trim.

3. Pry off screw cover, remove screw, and pull A-pillar trim upward. See Fig. 18.

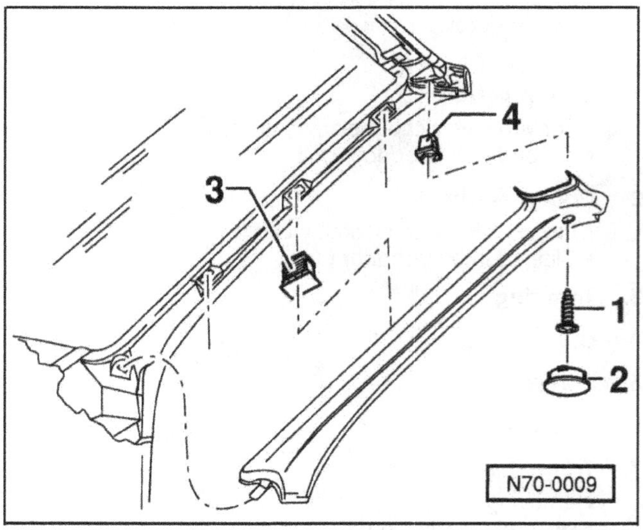

N70-0009

Fig. 18. Pry off screw cover (**2**), remove screw (**1**), and pull A-pillar trim upward to remove.

4. Before installing trim, remove clip from A-pillar (see Fig. 18) and re-insert clip into trim. The remainder of installation is the reverse of removal.

B-pillar trim (4-door Golf, Jetta), removing and installing

1. Remove sill panel trim, see **Sill panel trim assembly**, given later.

2. Remove front seat belt anchor. See **69 Seat belts, Airbags**.

3. Pull upper trim off, starting from bottom. See Fig. 19.

NOTE —

For 2-door model with one piece B-pillar trim, follow steps 1 and 2 (above), remove rear side panel trim, and pull B-pillar trim off starting from the bottom.

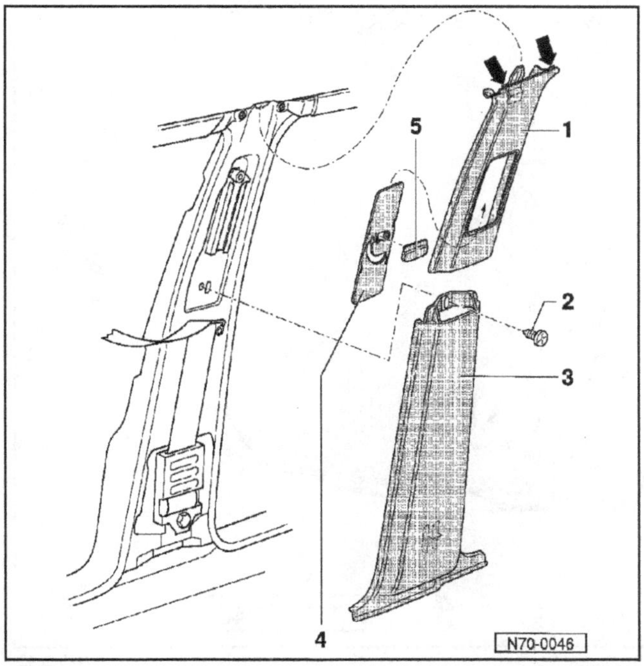

N70-0046

Fig. 19. B-pillar trim for 4-door models (2-door models have one piece B-pillar trim).

4. Remove lower trim retaining screw and pull lower trim off, starting from top.

5. Install in reverse order of removal, noting the following.
   - For upper trim piece, first locate retaining lugs in A-pillar and C-pillar trim.
   - Position seat belt height adjuster cover and adjuster (items 4 and 5 in Fig. 20) behind upper trim.
   - Check function of belt height adjuster.

C-pillar trim (Jetta), removing and installing

The C-pillar trim used on the Jetta models is shown in Fig. 20. C-pillar trim used on other models is similar.

1. Remove sill panel trim, see **Sill panel trim, assembly**.

2. Remove upper B-pillar trim, as described previously.

3. Remove trim retaining screw and pull off trim along roof edge, removing squarely from mounting clips. See Fig. 20.

4. Unhook parcel shelf trim.

5. Remove lower trim piece at C-pillar.

6. Remove rear seat belt anchor.

7. Installation is the reverse of removal. Before installing, check seating of clips and correct if necessary. Be sure to check function of belt height adjuster once installation is complete.

C-pillar trim (Jetta), assembly

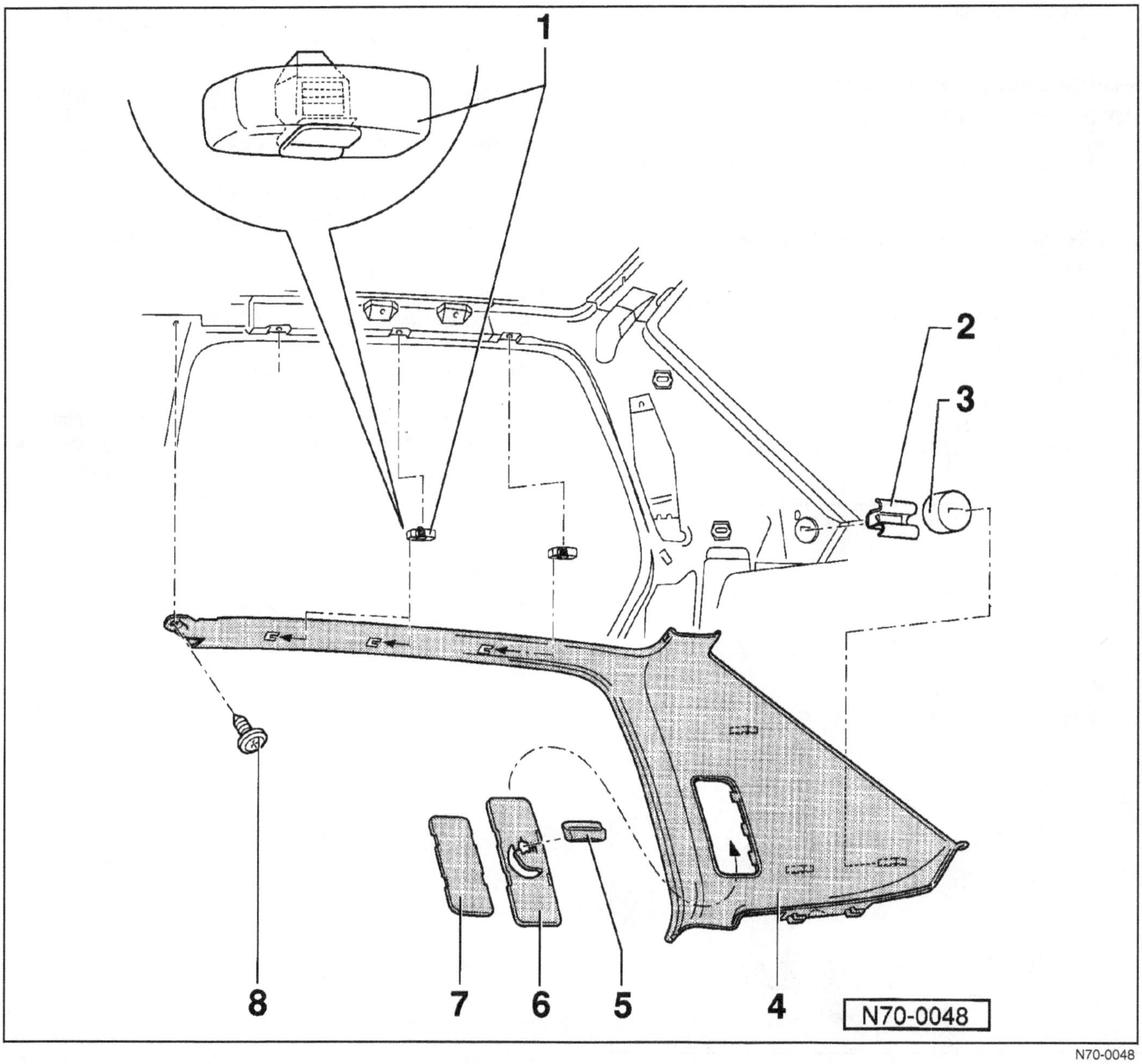

Fig. 20. C-pillar trim for Jetta models (other models similar).

1. **Clip**
   - With foam padding
   - Pushed into trim mounting
2. **Clip**
3. **Foam padding**
4. **Trim**
5. **Adjuster**
6. **Cover**
7. **Cover**
   - Only for vehicles without rear seat belts
8. **Screw**

Parcel shelf trim (Jetta), removing

1. Remove upper B-pillar trim and upper C-pillar trim as described previously.

2. Lift front edge of trim piece and unhook from mounting clips. Pull trim forward off of securing pins. See Fig. 21.

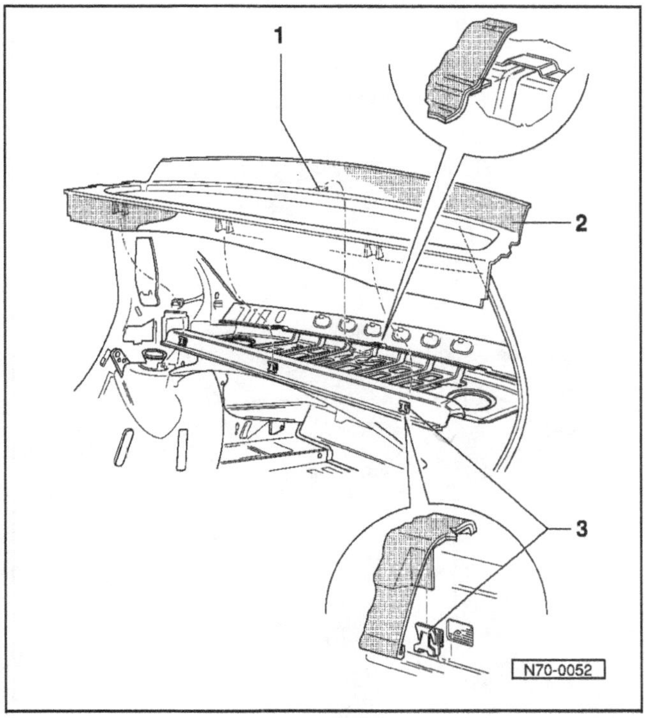

N70-0052

Fig. 21. Parcel shelf trim (**2**) for Jetta models. Upper inset shows securing pin (**1**) and lower inset shows mounting clips (**3**).

3. Installation is the reverse of removal. Replace any damaged mounting clips.

Rear lid trim (Golf, GTI)

Fig. 22 shows the rear lid trim used on Golf and GTI models. Use the illustration as a guide for removal and installation of the rear lid trim.

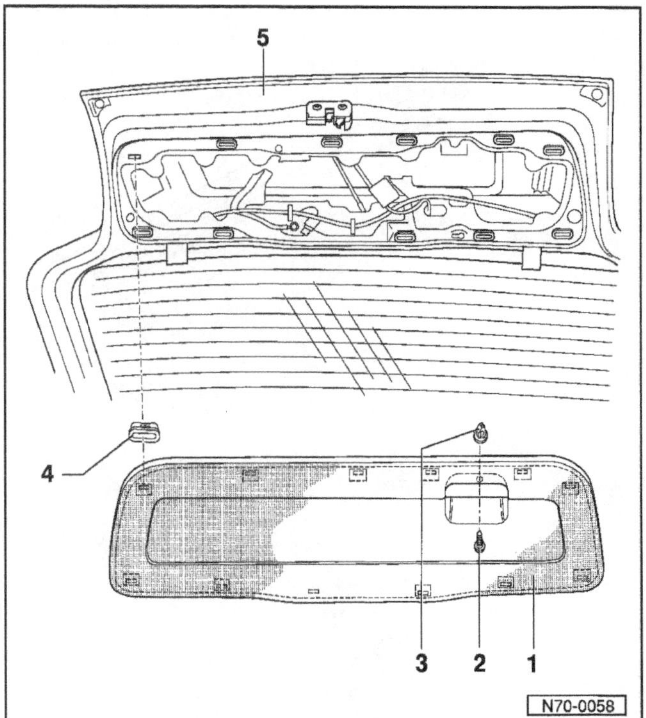

1. **Rear lid trim**
   - Remove screw and unclip trim
2. **Phillips-head screw**
3. **Spreader nut**
4. **Mounting clip**
5. **Rear lid**

N70-0058

Fig. 22. Rear lid trim for Golf and GTI .

Sill panel trim, removing and installing

Fig. 23 shows an exploded view of the sill panel trim used on 4-door Jetta models. Use the illustration as a guide for removal and installation of the sill panel trim.

Sill panel trim, assembly

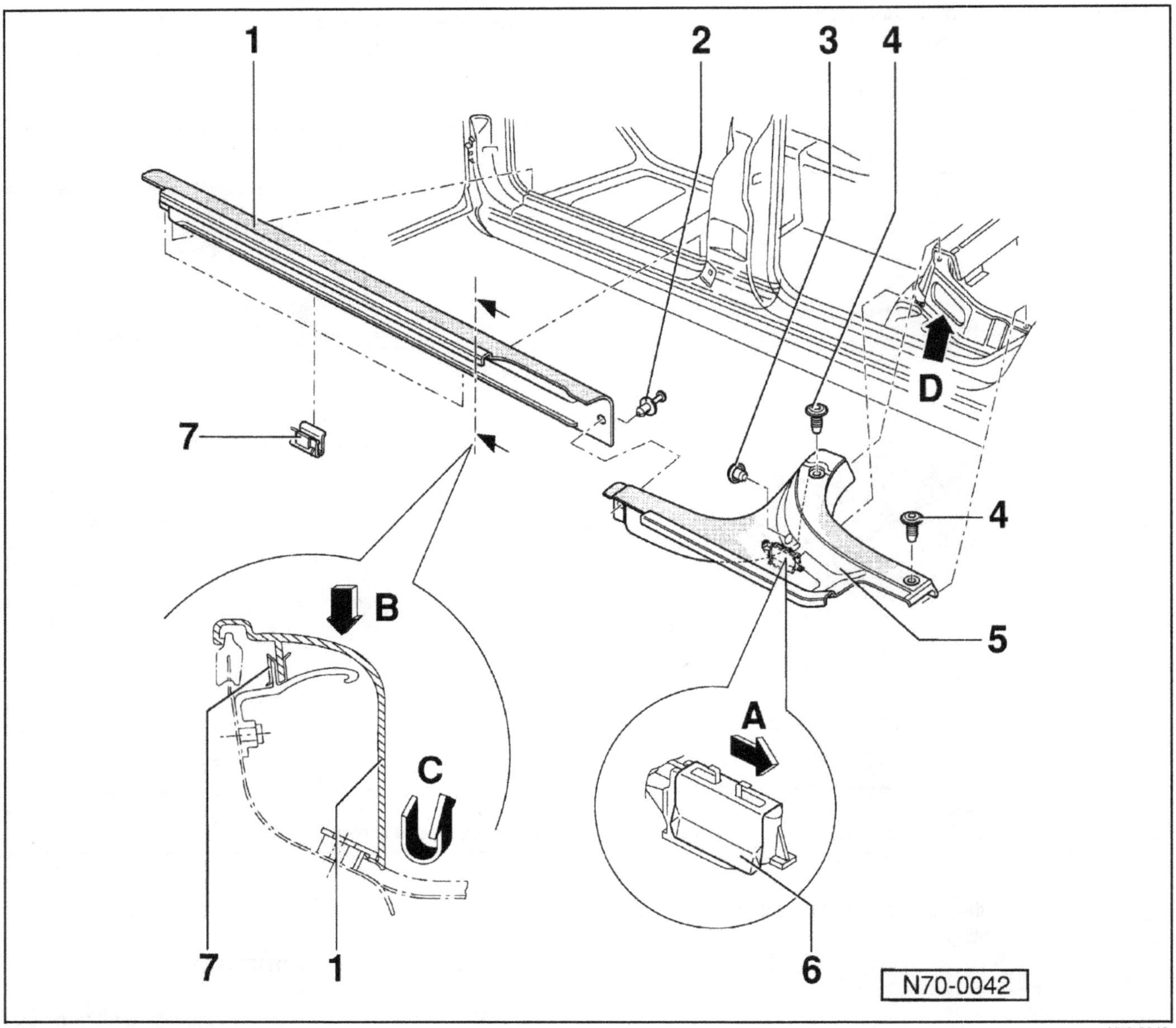

N70-0042

Fig. 23. Exploded view of front and rear door sill panel trim for 4-door Jetta models. Other models are similar.

1. **Sill panel trim**
   - Removing:
     Remove securing clip (**2**)
     Press down sill panel trim (**arrow B**) and simultaneously unhook bottom of sill panel trim (**arrow C**). Pull sill panel trim off clips (**7**).
2. **Securing clip**
   - Unscrew to remove
   - Press outer part into mounting hole to install, then press inner part
3. **Press button**
   - Unscrew or press on
4. **Plug**
   - Carefully pry out
5. **Rear trim**
   - Removing:
     Unscrew securing clip (**2**) and remove sill panel trim (**1**)
     Unscrew press button (**3**) and pry out plugs (**4**)
     Slide clip (**6**) rearward (**arrow A**). To do this, grip from inside of seat frame through opening (**arrow D**)
     Remove trim
   - Installing:
     First slide clip (**6**) back onto its seat. Remainder of installation is reverse of removal.
6. **Clip**
7. **Clip**

HEADLINER

Fig. 24 shows the one-piece headliner assembly. A helper should be used when removing the headliner from the vehicle.

Headliner, assembly

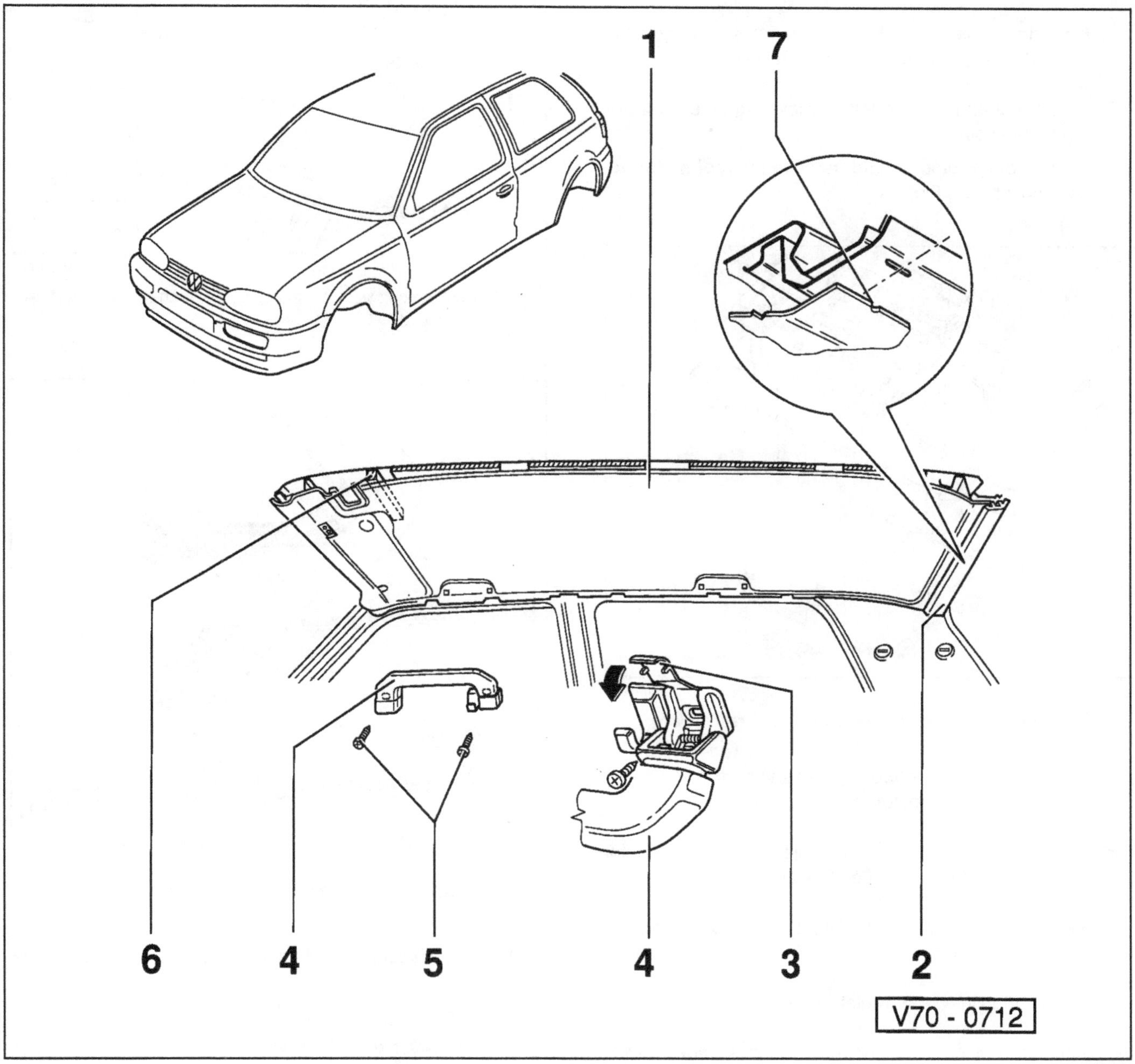

V70-0712

Fig. 24. Components of molded one-piece headliner. (Does not apply to Cabrio models).

1. **Headliner**
   - See **Headliner, removing**
2. **Rear trim strip**
   - Clipped onto roof frame
3. **Cover cap/grab handle**
4. **Grab handle**
5. **Screw**
6. **Spacer/interior light**
   - On vehicles with sunroof, spacer is located in front of cut-out/interior light.
7. **Center marking**
   - Align to center of vehicle (center clip hole)

Headliner, removing

1. Disconnect negative (–) battery cable.

> ***CAUTION —***
> *Before disconnecting battery be sure to obtain radio anti-theft code.*

2. On 2-door Golf and GTI models, remove sill panel trim, and side panel trim as described earlier.

3. On all Golf and GTI models, remove luggage compartment cover supports.
4. On sunroof equipped cars, remove sunroof motor access cover. See Fig. 25.

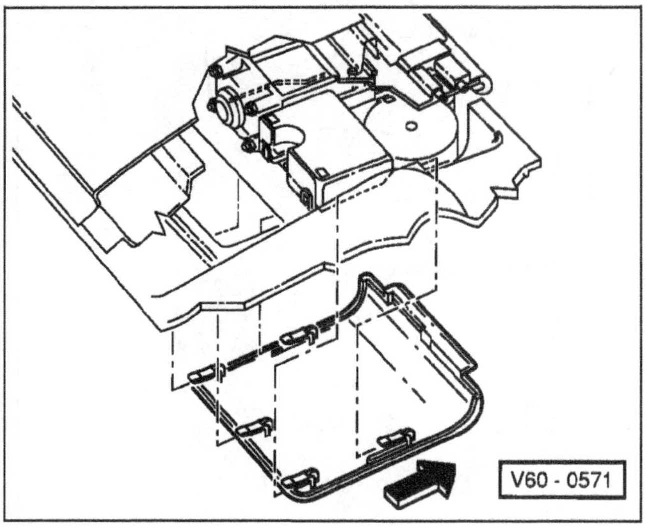

V60-0571

Fig. 25. Remove sunroof motor cover by sliding to rear (**arrow**).

5. Pry out interior light above rear view mirror and separate electrical connectors.

6. Remove sun visors. Separate electrical connector for vanity light, if applicable. See Fig. 26.

7. Remove upper A-pillar, B-pillar, and C-pillar trim as described previously.

8. Pry off headliner rear trim strip. See Fig. 27.

9. Remove roof handles (mounting screws under plastic covers).

NOTE —
The roof handles for sunroof equipped vehicles are marked with a "V" (front) or an "H" (rear).

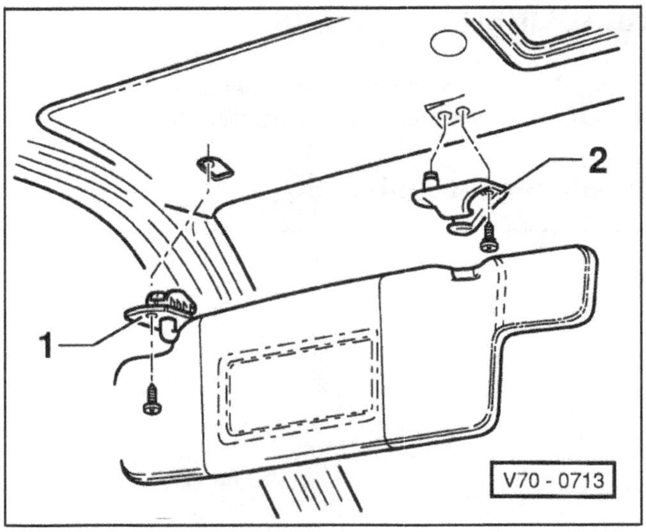

V70-0713

Fig. 26. Remove sun visor mounting screws then turn mount (**1**) 90° to remove.

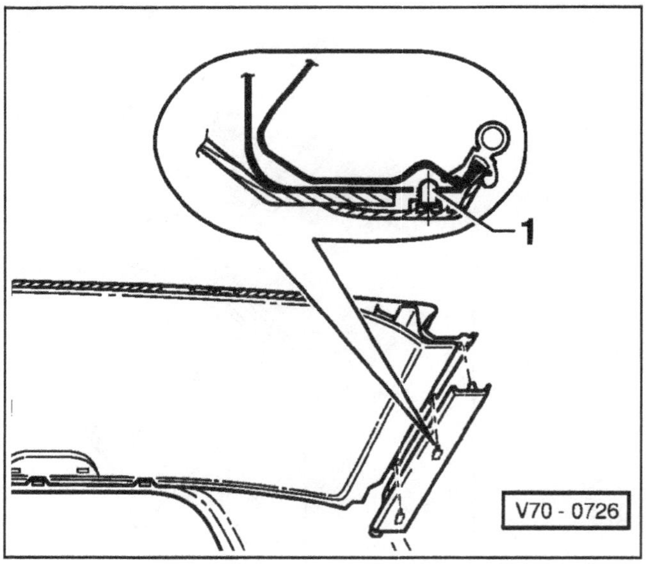

V70-0726

Fig. 27. Headliner rear trim strip. Inset shows detail of mounting clip (**1**).

10. Using a helper, lower headliner and remove from vehicle.

NOTE —
- *On Golfs, the headliner is removed through the rear hatch opening.*
- *On Jettas, the headliner is carefully removed through a rear door opening after fully reclining both front seat backrests.*
- *To avoid staining headliner, wash hands or use clean gloves before handling.*

> ***CAUTION —***
> *Removing the headliner without a helper could result in damage to headliner.*

GLOVEBOX

Several different glove box configurations have been used on Golf, Jetta and Cabrio vehicles depending on the model year, production date and if equipped with airbags. Glove boxes in the upper dash area are part of the padded dash. Lower dash glove boxes are built into the knee bar structure on the passenger side and the door is part of the passenger protection.

Table a. Glove box applications by year

Year	Glove Box	Airbags
1993	Yes - upper dash	No
1994 - early	Yes - upper dash	No
1994 - late	No	Yes
1995	No	Yes
1996 - early	No	Yes
1996 - late	Yes - lower dash	Yes
1997	Yes - lower dash	Yes
1998	Yes - lower dash	Yes
1999	Yes - lower dash	Yes
2000	Yes - lower dash	Yes
2001	Yes - lower dash	Yes
2002	Yes - lower dash	Yes

Vehicles without glove boxes should not have them retro-fitted due to the differing structure of the body in this area. Alterations to the dashboard crossmember and knee bar will change their performance in a collision resulting in a safety hazard.

> ***WARNING —***
> *Volkswagen of America has issued a directive specifically warning against the installation of aftermarket glove boxes into 1994, 1995, and 1996 Golf, Jetta and Cabrio vehicles not originally equipped with them. Installation may change compliance with Federal Motor Vehicle Safety Standards.*

Glove box in knee bar, removing and installing, vehicles with passenger side airbag

1. Remove the lower Phillips head screws. See Fig. 28.
2. Pull the glovebox downward making sure that the tabs are disengaged. Remove glovebox.
3. Installation is the reverse of removal. Make sure that the clips are properly located and that the phillips-head screws are properly installed.

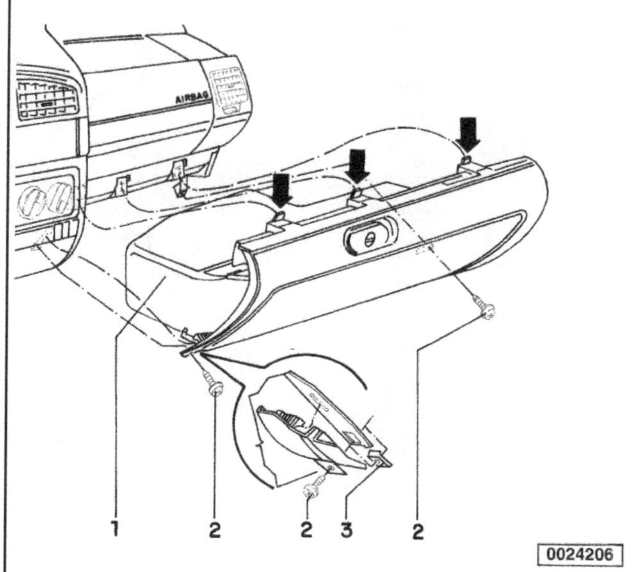

1. **Glovebox**
2. **Phillips-head screws**
3. **Clip**

0024206

Fig. 28. Glovebox in knee bar as installed in later vehicles that have a passenger side airbag. Be sure tabs (**arrows**) are disengaged when removing.

An assembly view of the glovebox installed in earlier vehicles without a passenger side airbag and kneebar is shown on the following page. See Fig. 29.

Glovebox assembly (vehicles without passenger side airbag)

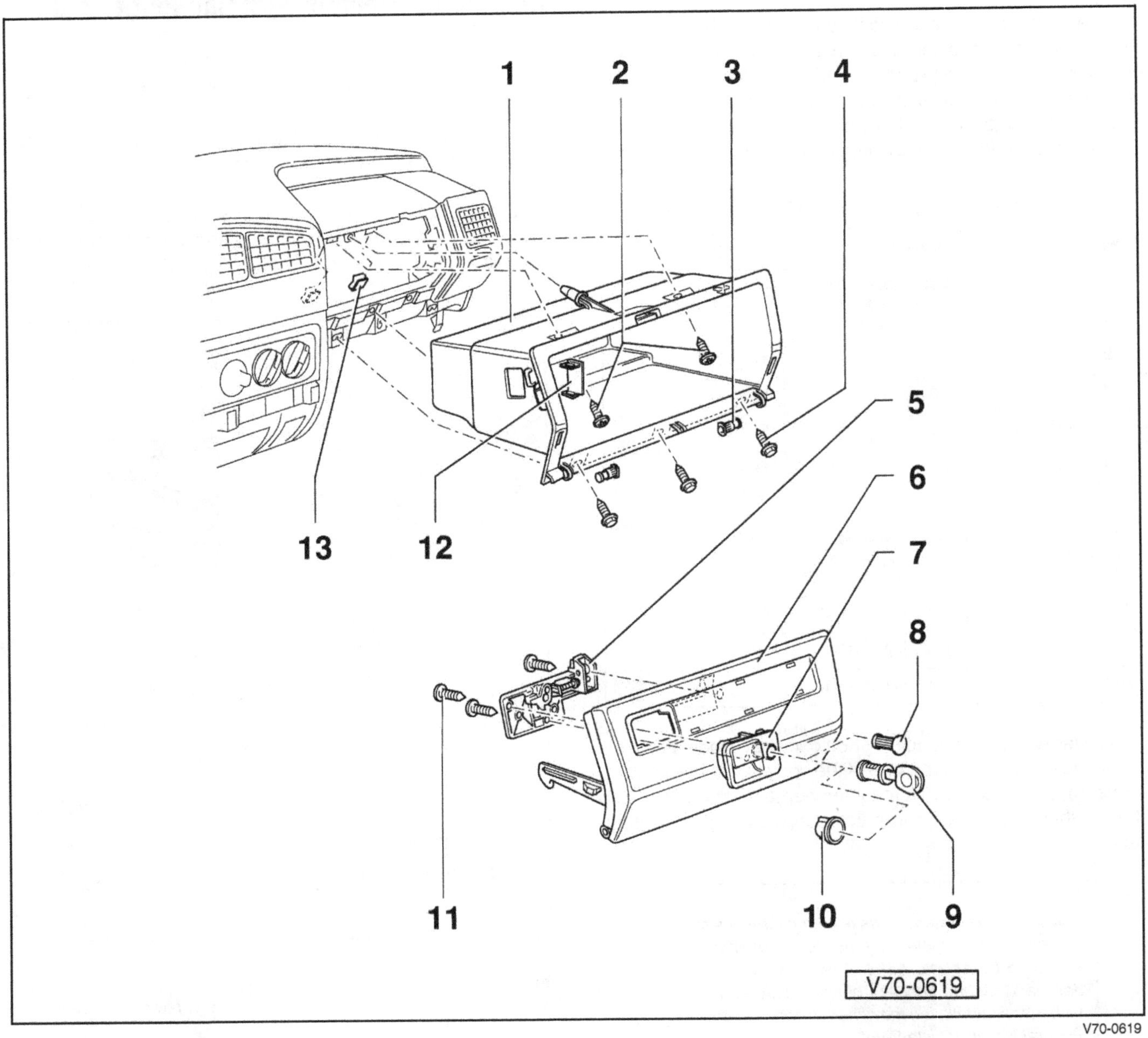

V70-0619

Fig. 29. Glovebox assembly for vehicles without passenger side airbag.

1. **Glovebox**
   - Removing:
     Remove passenger side storage shelf
     Unclip glovebox light
     Remove screws (items 2 and 4)
2. **Phillips-head screw**
3. **Hinge pin**
4. **Hex-head screw**
5. **Latch link**
6. **Glovebox lid**
   - Removing:
     Remove glovebox, hinge pins and take lid off
7. **Latch**
8. **Cap**
9. **Lock cylinder**
10. **Sleeve**
11. **Phillips-head screw**
12. **Lens cover for glovebox light**
13. **Snap nut**

72 Seats

GENERAL

This repair group covers front and rear seat removal, partial seat installation, and seat mounting.

> ***NOTE —***
> *For wiring diagrams of electrically heated/adjustable front seats, see* **97 Wiring Diagrams, Fuses and Relays**.

FRONT SEATS

Front seat, removing and installing

Front seats on some models have side airbags built into the backrests. These are part of the airbag system. Special precautions must be observed prior to any work on or near the front seats on vehicles equipped with side airbags.

> ***WARNING —***
> *Before removing or working on the front seats of side airbag equipped vehicles, read the warnings under* ***Airbag, safety precautions****, see* ***69 Seatbelts, Airbags****.*

In addition to all of the safety precautions noted in **69 Seatbelts, Airbags,** removal of the front seats requires the use of special tool VAS 5061. See Fig. 1.

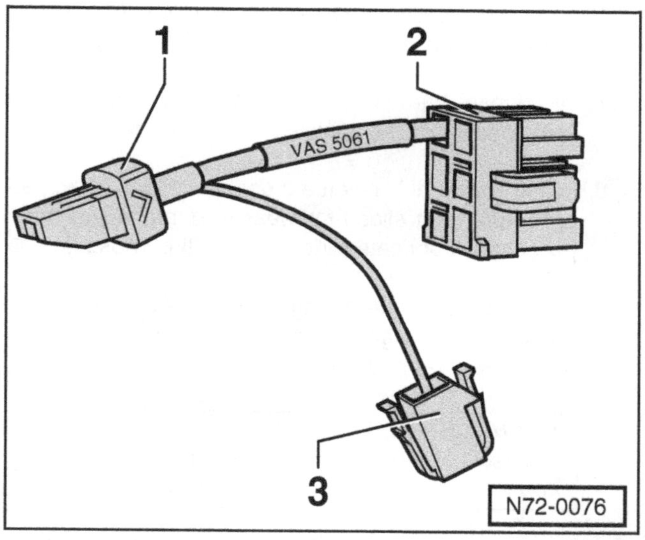

N72-0076

Fig. 1. VAS 5061 is used to safely remove front seats with side airbags. Connector **1** and connector **3** are inserted into seat harness connectors. Connector **2** is not used in this application.

72

Fig. 3 shows the front seat mounting points. Use this illustration as a guide when removing and installing the front seat. Depending on the type of seat and the electrical equipment installed in the vehicle, slight deviations may be required when removing and installing.

1. For seats with side airbags and/or seat heating, disconnect the negative (–) cable from the battery.

2. Working at the seat bottom, remove side cover retaining clip and pull cover off guide rail toward rear.

3. Pull wedge from back of cap using pliers.

4. Slide seat forward, remove cap nut and remove stop screw, or stop pin, if so equipped.

5. Slide seat back out of guide rails.

6. On seats with side airbags a special harness adapter (VAS 5061 or VAS 5094) needs to be installed in order to prevent accidental deployment of the airbag when removing the seat. See Fig. 2.
   - Be sure you are electrostatically discharged by briefly touching a body or chassis ground.
   - Disconnect airbag harness connector under seat.
   - Install airbag adapter as shown in Fig. 2.

7. Disconnect harness connectors for seat heating and seat adjustment, as applicable. Remove seat.

8. Install front seat in reverse order of removal. Check left, right, and front slides for wear or damage; replace if necessary. Lubricate slides with multi-purpose grease.

9. For seats with side airbags, install seat and harness connectors, turn ignition on, close doors, and then connect battery (–) negative cable.

> ***WARNING —***
> *Make sure the passenger compartment is not occupied before connecting the battery ground strap.*

10. If ignition was switched on while seat with side airbag was removed, a Diagnostic Trouble will be stored. Erase with a scan tool or scan tool program.

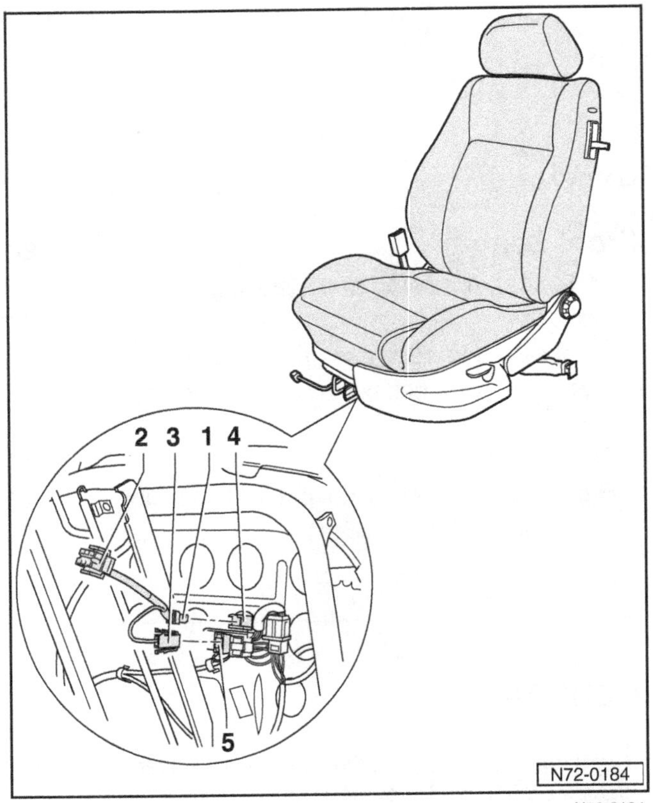

N72-0184

Fig. 2. On seats with side airbags install airbag adapter connector (**1**) into seat harness connector housing (**4**) and adapter connector (**3**) into seat harness connector housing (**5**). Adapter connector (**2**) is not used.

Front seat, mounting

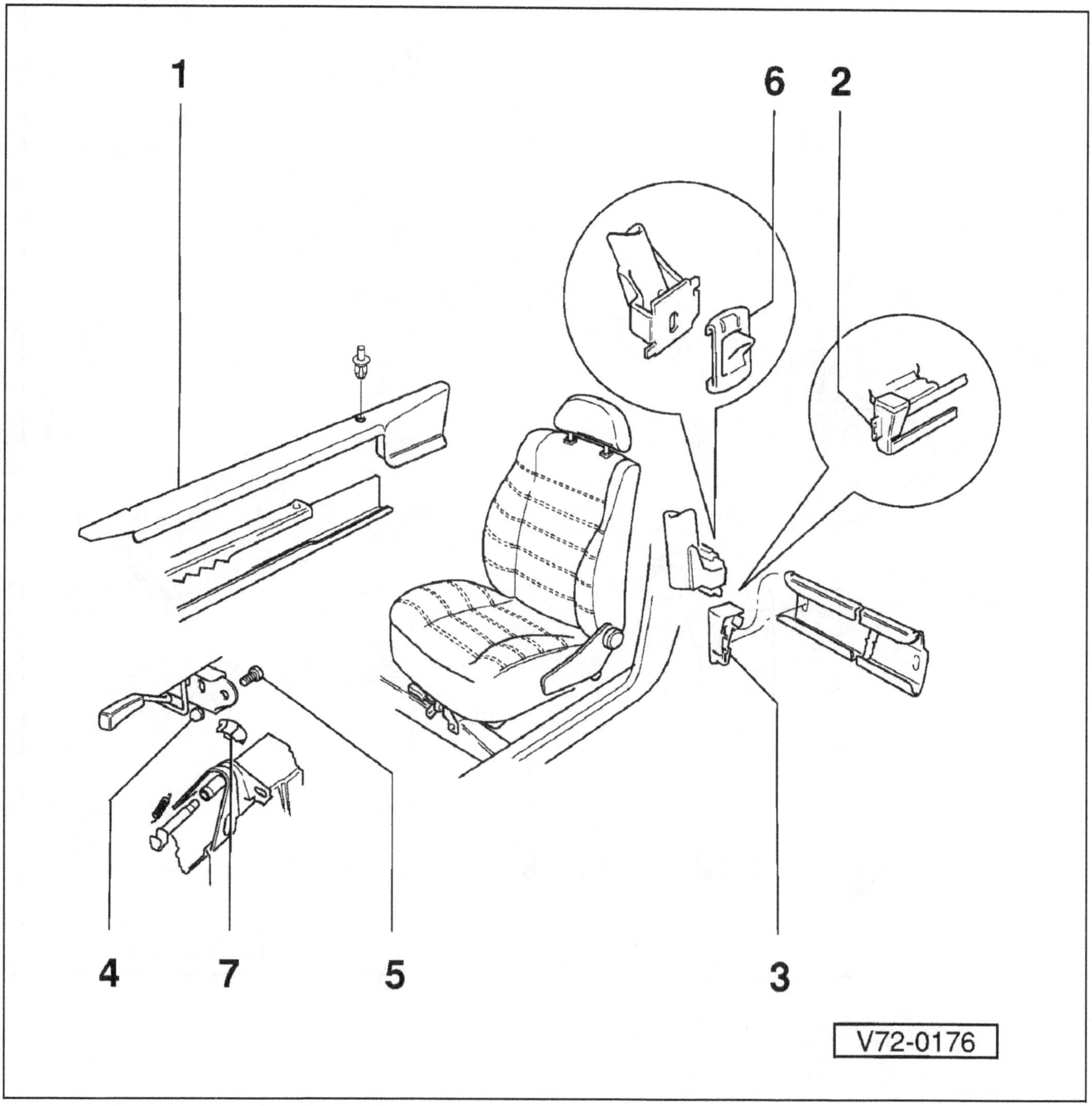

V72-0176

Fig. 3. Front seat securing hardware and related components.

> ***WARNING —***
> *Before removing or working on front seats with side airbags, read the warnings under **Airbag, safety precautions**, see **69 Seatbelts, Airbags**.*

1. **Side cover**
2. **Wedge**
3. **Cap**
4. **Cap nut**
5. **Stop screw**
6. **Slide (left/right)**
   - Check for damage, replace if necessary
7. **Slide (front)**
   - Check for damage, replace if necessary

Front seat backrest, assembly

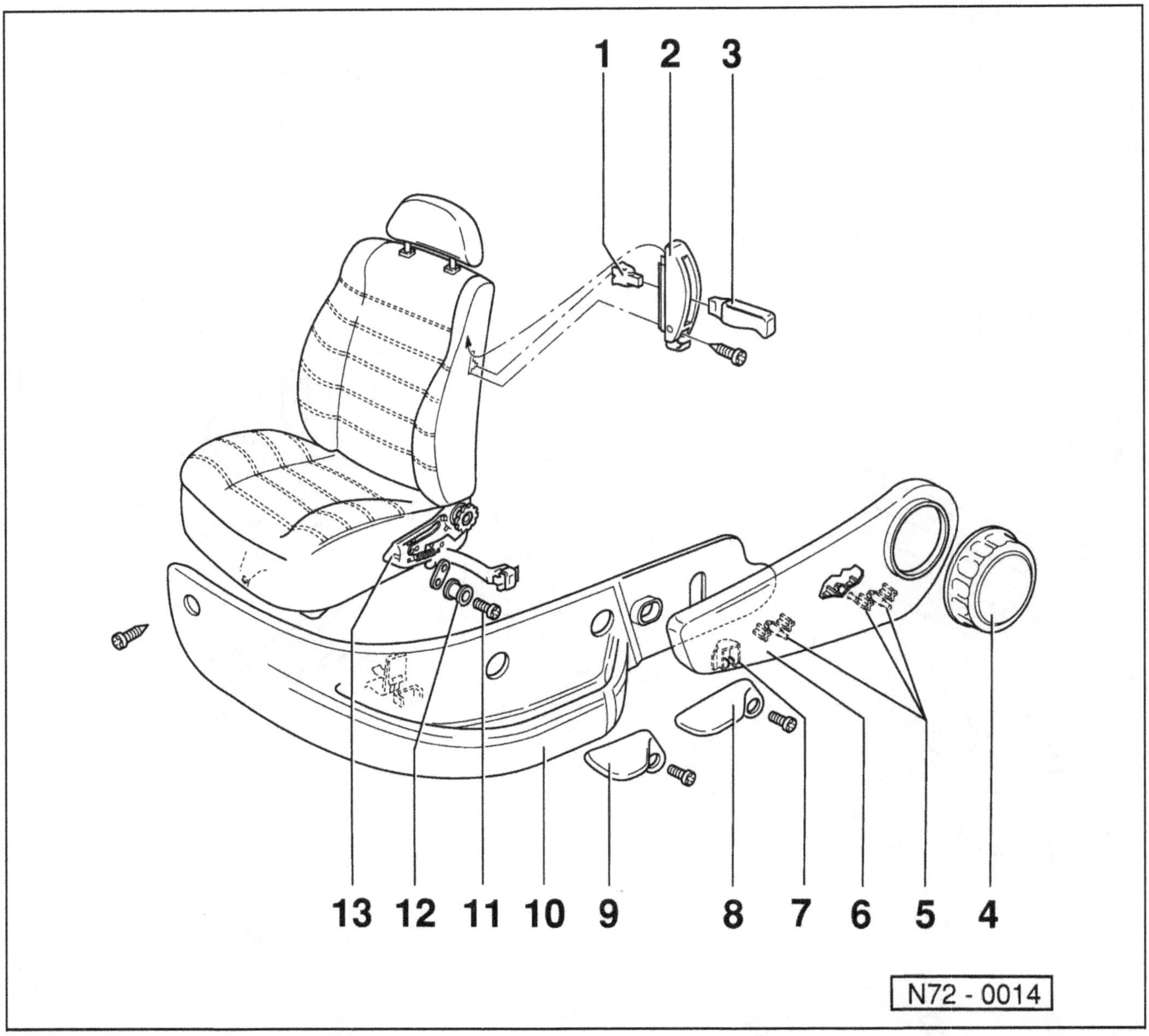

N72-0014

Fig. 4. Exploded view of front seat back rest components.

> ***WARNING —***
> *Before removing or working on front seats with side airbags, read the warnings under* ***Airbag, safety precautions****, see* ***69 Seatbelts, Airbags****.*

1. **Securing clip**
2. **Trim**
3. **Release knob**
4. **Control knob**
   - Seats with height adjustment only
5. **Expander pins**
6. **Cover**
7. **Retainer**
8. **Control handle (height adjustment only)**
9. **Control handle (comfort seat only)**
10. **Trim panel**
    - Seats with height adjustment only
    - Mounts remote trunk switch on some vehicles
11. **Screw**
    - Self-locking, always replace
12. **Bushing**
13. **Mounting pin**

Front seat backrest, removing and installing

Fig. 4 shows the front seat backrest trim and mounting hardware. Use this illustration as a guide when removing and installing the seat backrest.

1. Remove front seat from vehicle as described earlier.

2. Pull off control knob from side of seat.

3. Drive side cover expander pins (left and right sides) in, toward seat center.

   NOTE—
   The front expander pins will contact the seat frame when driving them in. Use a screwdriver from above to press the pin downward to clear the seat frame.

4. Remove side cover from backrest frame.

5. Remove bushing retaining screws.

6. Remove retainers (left and right sides) from backrest frame.

7. Pull backrest frame off mounting pin.

8. Installation is the reverse of removal, noting the following:
   - Check the bushings (left and right sides) for wear or damage; replace if necessary.
   - Use new self-locking screws (item 11 in Fig. 4) at the bushings and remove any locking compound from nut threads with tap or thread chaser.

> ***WARNING—***
> *Before removing or working on front seats with side airbags, read the warnings under **Airbag, safety precautions**, see **69 Seatbelts, Airbags**.*

REAR SEATS

Rear seat cushion, removing and installing (Golf and Jetta)

1. Working at front of rear seat, remove both screws from seat hinges and remove seat cushion using pull strap. See Fig. 5.

   NOTE—
   Access to the rear seat cushion can be substantially improved by sliding both front seats all the way forward and folding (tilting) them.

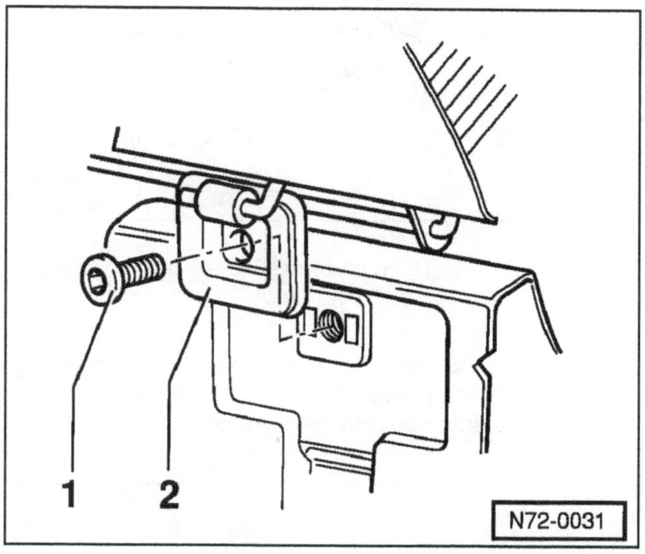

N72-0031

Fig. 5. Rear seat cushion hinge plate (**2**) retaining screw (**1**). Screws are usually socket headed (Allen®) screws.

2. Lift seat cushion up and out.

3. Installation is the reverse. Ensure correct positioning of seat belts as cushion is pushed in place.

Rear seat backrest, removing and installing (Golf and Jetta)

1. Fold down seat backrests(s).
2. On vehicles with 60/40 split rear seat, remove pivot pin locks from center adapters on pivot bracket. See Fig. 6..

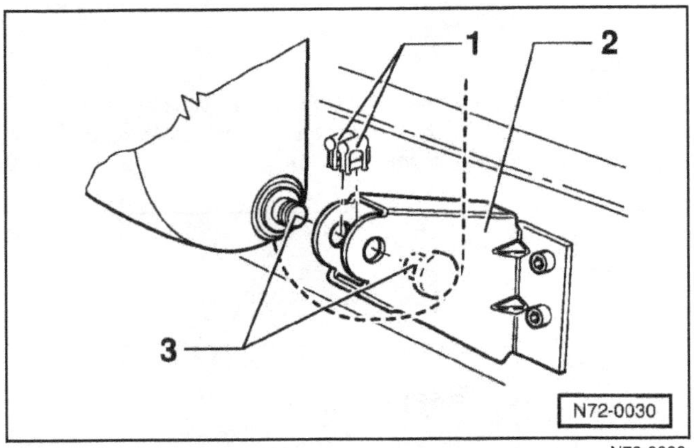

N72-0030

Fig. 6. Remove locks (**1**) on adapters (**3**) at rear seat back center bracket (**2**). (Split rear seat only).

3. Press back latching hook on left and/or right mounting bracket, with screwdriver, and pull seat back upward out of the seat back bracket. See Fig. 7.

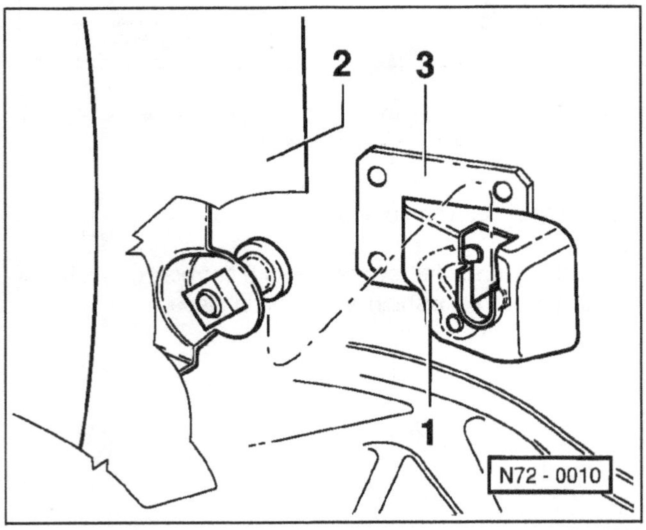

N72-0010

Fig. 7. Release latching hook (**1**) with screwdriver and pull seat backrest (**2**) out of seat back bracket (**3**).

4. Pull seat backrest out of center bracket and remove from vehicle.
5. Installation is the reverse. When installing seat backrest into latching hook, position pivot pin into latching hook and press backrest down to lock.

Rear seat cushion, removing and installing (Cabrio)

1. Working at front of rear seat, remove bolts from seat hinges and remove seat cushion using pull strap. See Fig. 8.

NOTE —

Access to the rear seat cushion can be substantially improved by sliding both front seats all the way forward and folding (tilting) them.

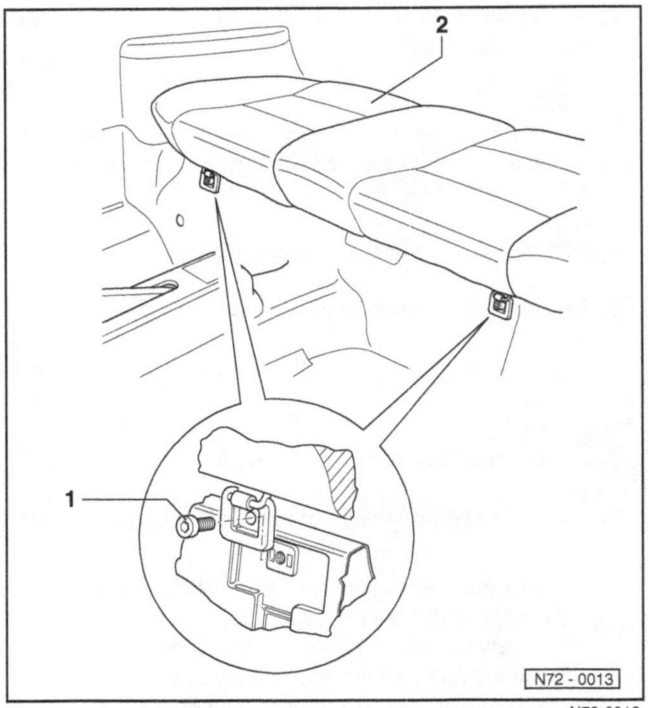

N72-0013

Fig. 8. Rear seat cushion (**2**) hinge plate retaining screw (**1**). Screws are usually socket headed (Allen®) screws.

2. Lift seat cushion up and out.
3. Installation is the reverse. Ensure correct positioning of seat belts as cushion is pushed in place.

Rear seat backrest, removing and installing (Cabrio)

On Cabrios, the rear seat backrest is divided into lower and upper sections with trim bonded to the upper backrest. Use a universal adhesive (VW part number D 001 200) when installing. Remove lower and upper areas and trim as follows:

1. Open (lower) convertible top.
2. Pull backrest release handle (in luggage compartment) and fold lower rear seat backrest.
3. Press back latching hook on left and right mounting bracket with screwdriver and pull lower seat backrest upward out of the seat back bracket. See Fig. 7.

4. Remove seat belt anchor bolts. See Fig. 9.

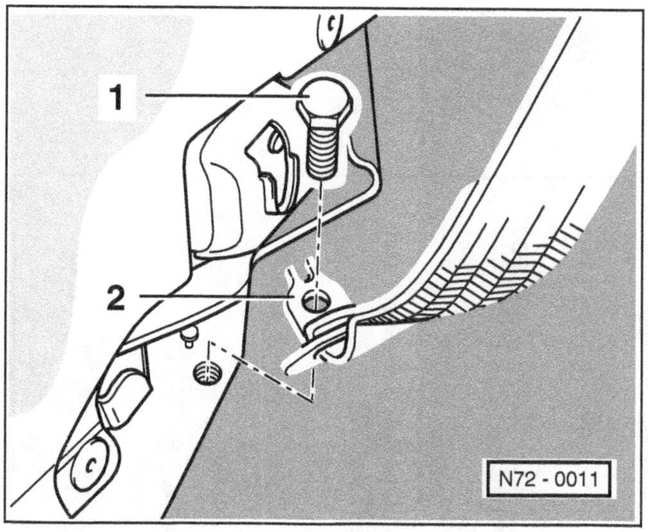

N72-0011

Fig. 9. Remove seat belt anchor bolt (**1**) and pull seat belt anchor (**2**) out of positioning lug.

5. Pry off the rear bulkhead trim from the seat backrest flange (glued in place). Pull off rearward. With the lower backrest removed, remove upper area securing bolts, pull seatbelts out and lift off upper area. See Fig. 10.

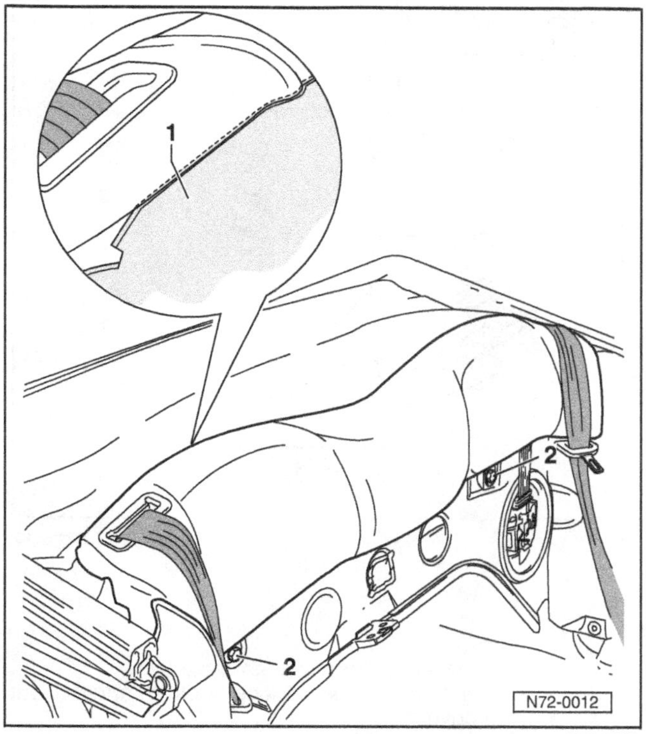

N72-0012

Fig. 10. Pry off rear bulkhead trim (**1**) from the seat backrest flange (glued in place). Pull off rearward. Remove upper area securing bolts (**2**), pull seatbelts out and lift off upper area.

6. Installation is the reverse. When installing seat backrest into latching hook, position pivot pin into latching hook and press backrest down to lock.

Tightening torques
- Rear seat belt anchor to body (Cabrio model only) . . . . 40 Nm (30 ft-lb)

Center arm rest

The center arm rest assembly for Jetta models is shown in Fig. 11. Arm rest removal instructions are contained within the numbered list.

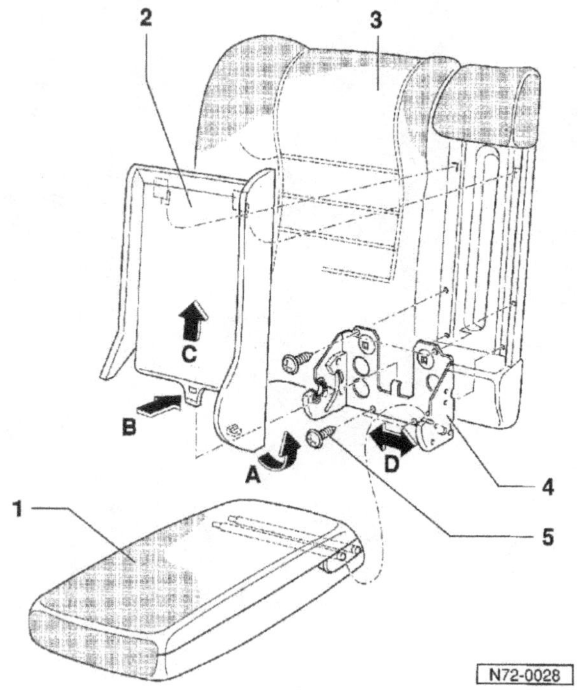

1. **Center arm rest**
   - Removing:
     Remove split seat back rest (**3**)
     Fold down center arm rest
     Remove trim (**2**) and screws (**5**)
     Spread bearing on bracket (**4**) apart (**arrow D**) and disconnect bearing rod for center arm rest. (bearing rod cannot be taken out through the opening on the bearing).
2. **Trim**
   - Disconnect side tab (**arrow A**), center tab (**arrow B**), and push trim upward (**arrow C**) to remove.
3. **Back rest (split seat)**
4. **Arm rest bracket**
5. **Screw**

N72-0028

Fig. 11. Exploded view of center arm rest (rear seat) for Jetta models.

Backrest lock and latch assembly

The rear seat backrest lock and latch assembly is shown in Fig. 12. Some models do not use the key lock. Use the illustration as a guide when removing and installing lock components.

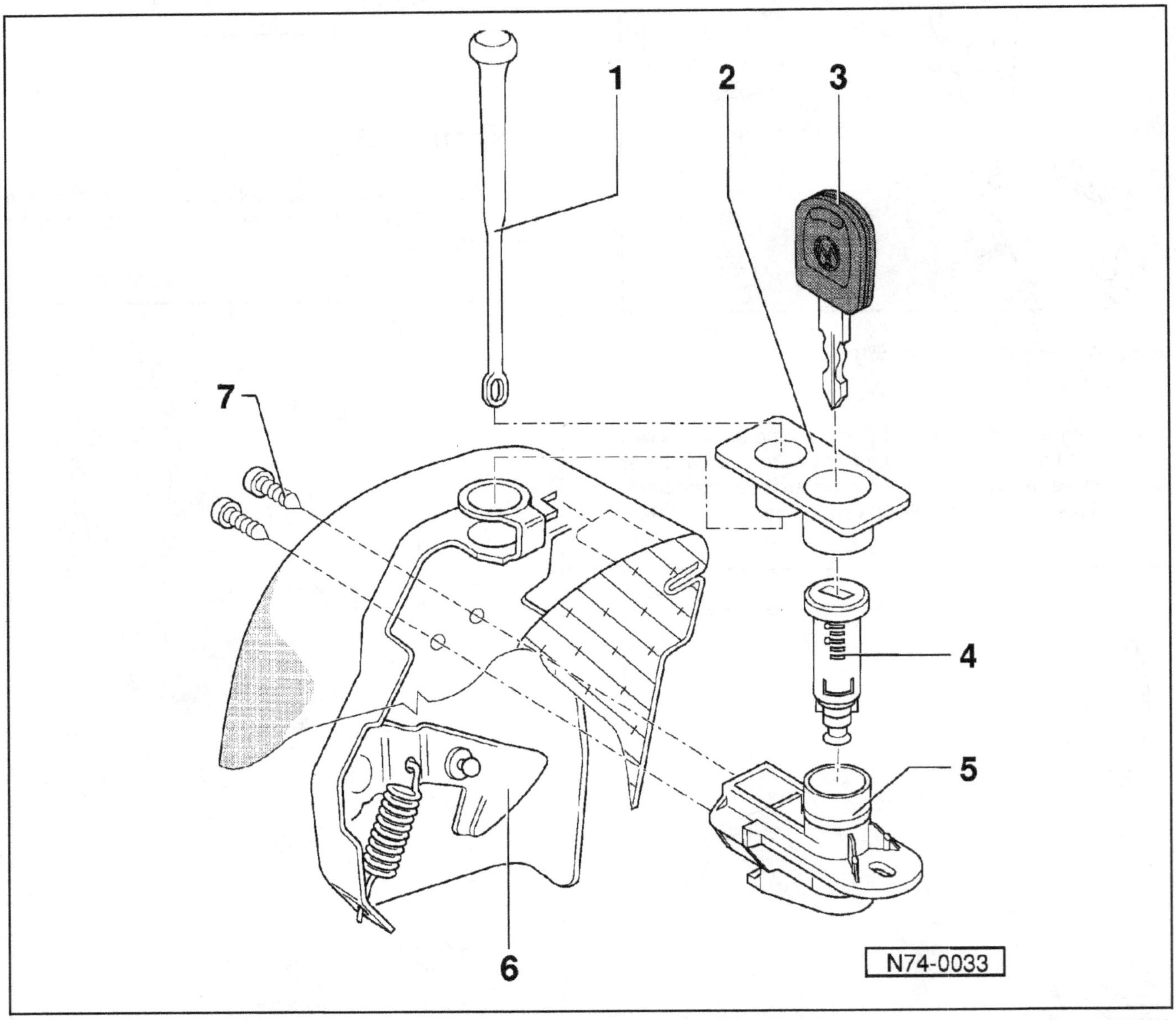

N74-0033

Fig. 12. Exploded view of rear seat back lock for Jetta models.

1. **Pull rod**
2. **Cover**
3. **Key**
4. **Lock cylinder**
   - Press out carefully, only with key inserted
5. **Lock**
   - Removing:
     Loosen cover (**2**) from rear area of lock
     Pry off pull rod (**1**) from locking hook (**6**) and pull out
     Pull off cover (**2**)
     Remove screws (**7**)
6. **Locking hook**
7. **Screw**

8 Heating and Air Conditioning

GENERAL

This general information group gives a basic explanation of the heating and ventilation system. More specific information is provided in **80 Heating and Ventilation** for cars without air conditioning and **87 Air Conditioning** for cars with air conditioning.

AIR FLOW

Ventilation is by a flow-through system. Fresh air enters through the grille below the engine hood and into the passenger compartment via the dashboard and footwell vents. The interior air is extracted through vents that are located in the rear of the vehicle, behind the bumper covers. This is a relatively neutral pressure zone, which results in the flow of air being quiet and less affected by changes in vehicle speed.

The duct work is designed to provide two layers of air flow for good occupant comfort. The air flowing from the center dashboard vents moves at a higher speed than air from the footwell and rear heating vents. This results in an upper layer of air that is slightly cooler than the lower air layer around the feet and legs. See Fig. 1.

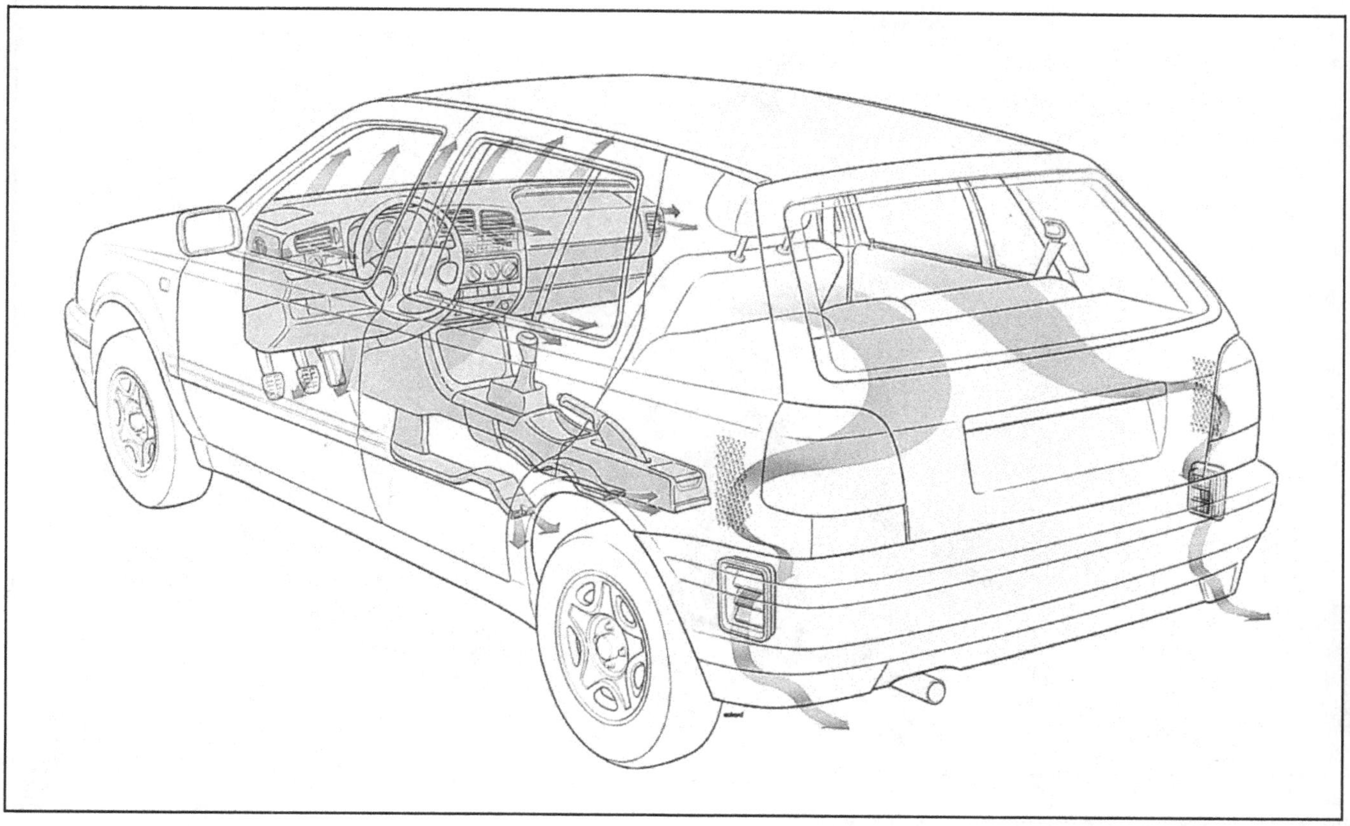

0024088

Fig. 1. Phantom view of Golf showing vehicle airflow.

HEATING AND VENTILATION

The heating and ventilation system is a blend type system. A heater on/off valve is not used. Coolant flows unrestricted through the heat exchanger. Interior temperature is regulated by a blend door that controls the amount of air that is passed over the heater core. This blend door is operated by a cable connected to the temperature control dial. See Fig. 2.

Three round dials are used to control blower speed, air temperature and air distribution. The fresh air blower has four speeds. The series resistance for the blower is located in the blower motor housing.

AIR CONDITIONING (A/C)

Air conditioning is standard equipment on most A3 Golfs and Jettas sold in the United States and Canada. On A/C-equipped vehicles, a blend type system is also used. Fresh air is first drawn in and passed through the evaporator to be cooled and dehumidified when the A/C compressor is switched on. A portion of this air can then be directed over the heat exchanger by the temperature regulation door to maintain the desired temperature. See Fig. 3.

Vehicles with A/C have two additional buttons on the ventilation controls. The A/C button turns the A/C compressor on and off. A second button is used to switch closed the air recirculation door for maximum cooling. The air circulation door is also closed when the blower switch is turned off.

The series resistance for the blower motor is located in the A/C Evaporator case.

The A/C refrigerant system uses a variable displacement A/C compressor. The A/C compressor does not cycle on and off when the A/C is switched on. Instead, the displacement of the A/C compressor will increase or decrease to match the cooling load on the A/C system. This eliminates the surging caused by a cycling A/C compressor clutch and increases the thermal efficiency of the system.

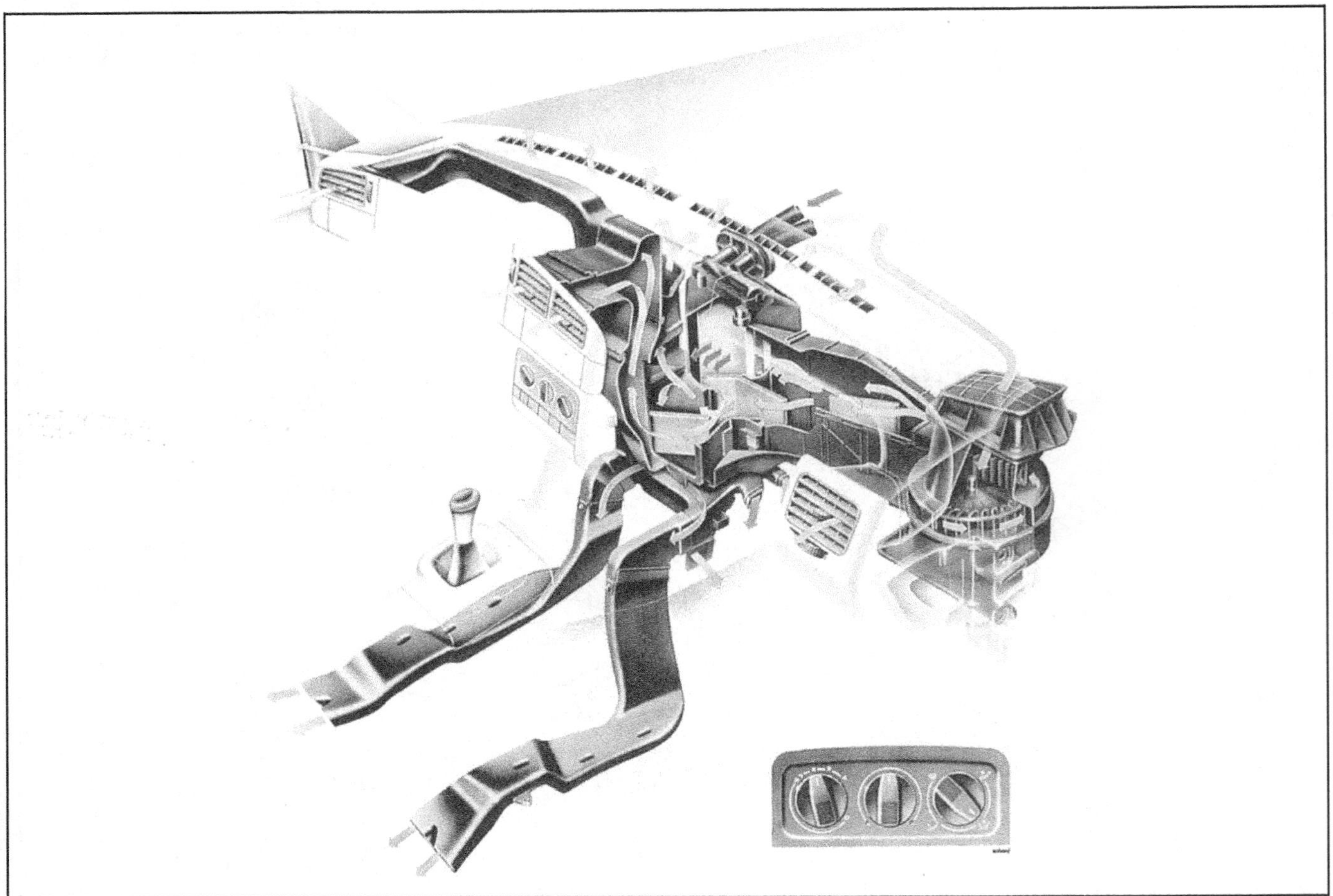

0024089

Fig. 2. Blend-type heating and ventilation system on cars without A/C. Air flows past heater core and blending door before entering passenger compartment.

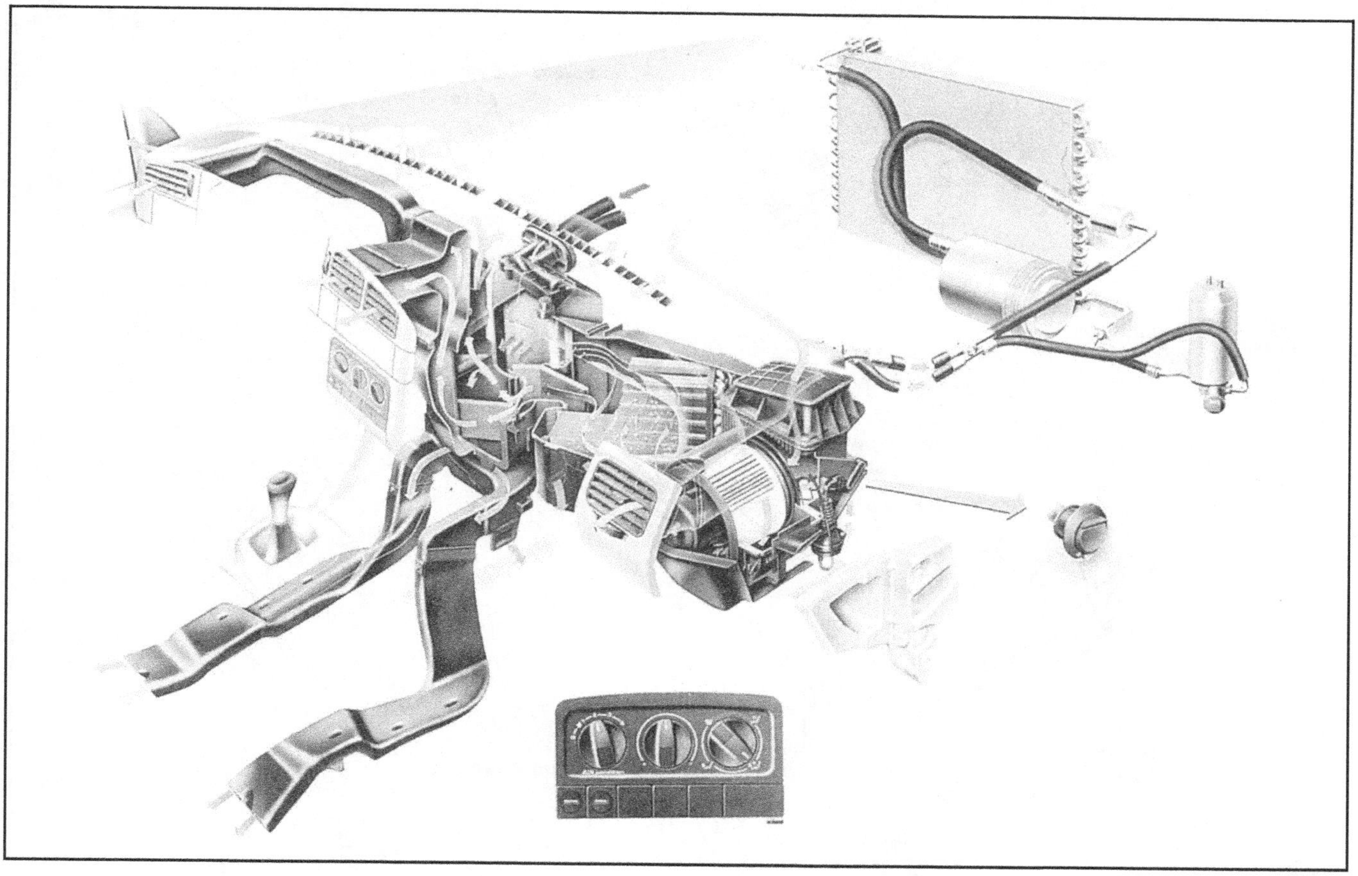

0024090

Fig. 3. Phantom view of A/C system.

R134a refrigerant

1993 and later Volkswagens use R-134a A/C refrigerant. All of the system's components are specially made to be compatible with R-134a. These components must not be interchanged with other systems which use R-12 or other refrigerants. Fig. 4 shows an overall view of the A/C system components.

Due to the harmful effect that R-12 refrigerant has on the Earth's ozone layer, R-12 refrigerant was replaced by R-134a refrigerant. The benefit of using R-134a (hydrofluorocarbon) refrigerant is that it is chlorine free. Chlorinated fluorocarbons (CFCs), present in R-12, destroy the ozone in the stratosphere through chemical reaction.

The R-134a and the R-12 A/C systems are very similar in appearance; don't let this fool you. There are differences in the refrigerant, lubrication, components and servicing.

A/C components used in R-134a systems can be identified by green markings. For example, the A/C hoses have a green stripe on the hose; the identification label for the A/C compressor is green.See Fig. 5.

CAUTION —

- *R-134a components, refrigerants, and refrigerant oils must not be interchanged with other system components or refrigerants which use R-12 or other refrigerants.*
- *The equipment needed to service an R-134a system is different than the equipment needed to service an R-12 system . Do not use R-12 equipment on an R-134a system.*

The refrigerant oil used in R-134a A/C systems is a synthetic oil (polyalkyleneglycol) otherwise known as PAG oil. The R-12 systems use very different mineral based refrigerant oil. The two refrigerants and refrigerant oils must not be mixed under any circumstances or component failure will result. There are also different kinds of PAG oils, so make sure to use the correct oil for the R-134a system.

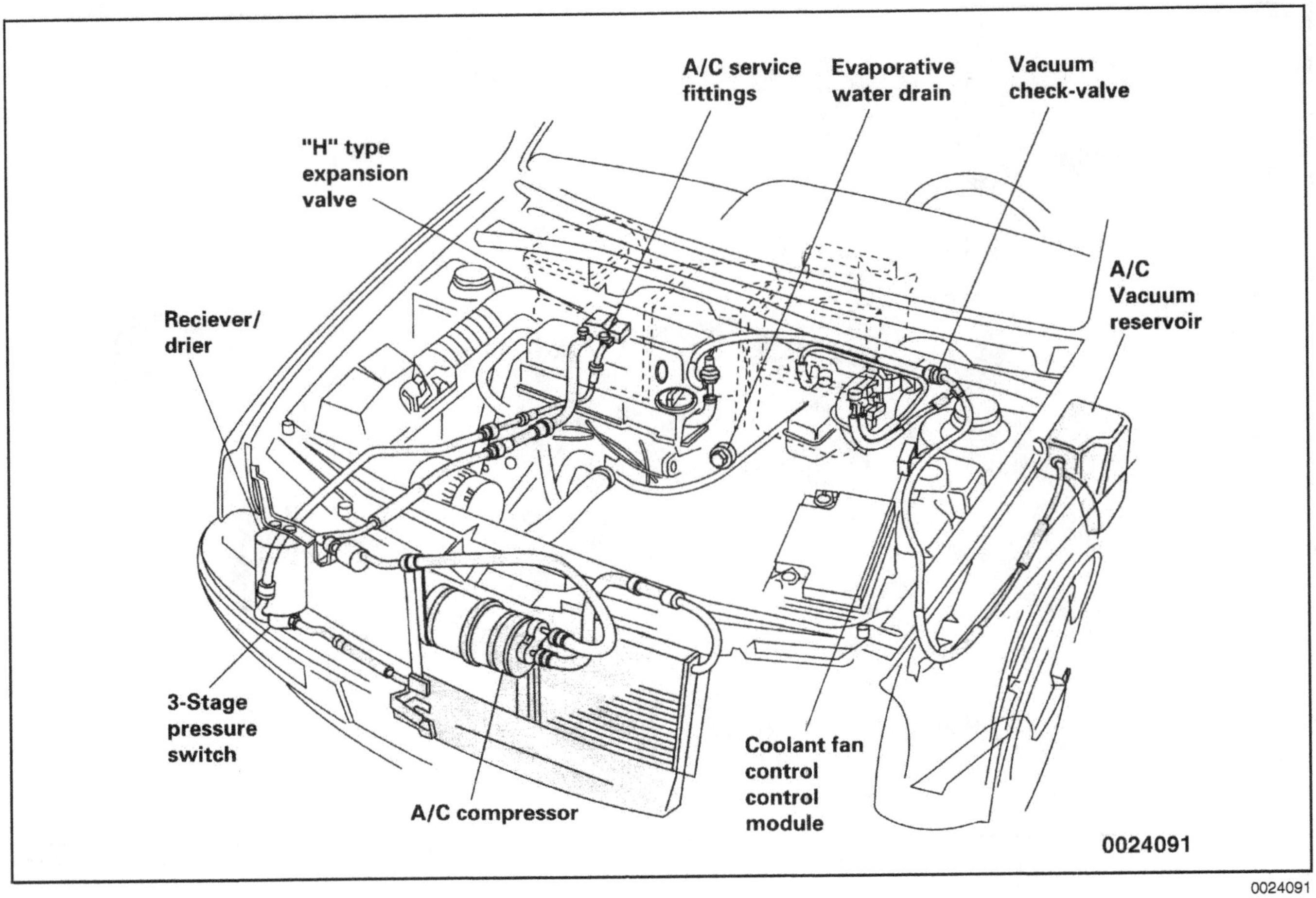

0024091

Fig. 4. Air conditioning components.

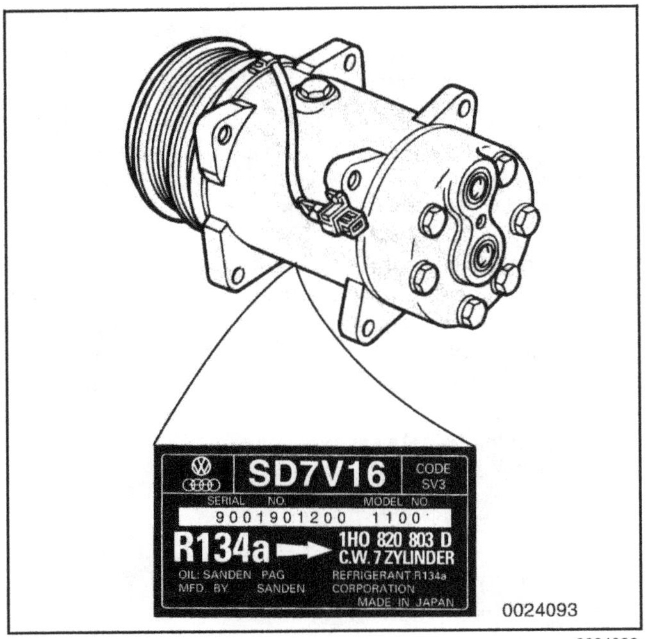

0024093

Fig. 5. A/C components used in the R134a system can be identified by green markings. The ID label on the A/C compressor is green and the A/C hoses have a green stripe.

80 Heating and Ventilation

GENERAL

This section primarily applies to cars without air conditioning (A/C). For cars with A/C, the system must be professionally discharged before working on most of the heating and ventilation components. See **87 Air Conditioning**.

> ***WARNING —***
> *On cars with A/C, replacement of the heater and ventilation components may require discharging/charging the A/C system. This requires special equipment and training. For more information, see* **87 Air Conditioning**.

HEATING AND VENTILATION, SERVICING

> ***WARNING —***
> *Hot coolant is under pressure and can scald. Allow engine to cool before performing any work which requires contact with heating or cooling system.*

> ***CAUTION —***
> - *Before working on the electrical system disconnect negative (–) battery cable.*
> - *Be sure to have radio anti-theft code before disconnecting battery.*

Ventilation control head, removing and installing

1. Carefully remove trim plate. See Fig. 1.

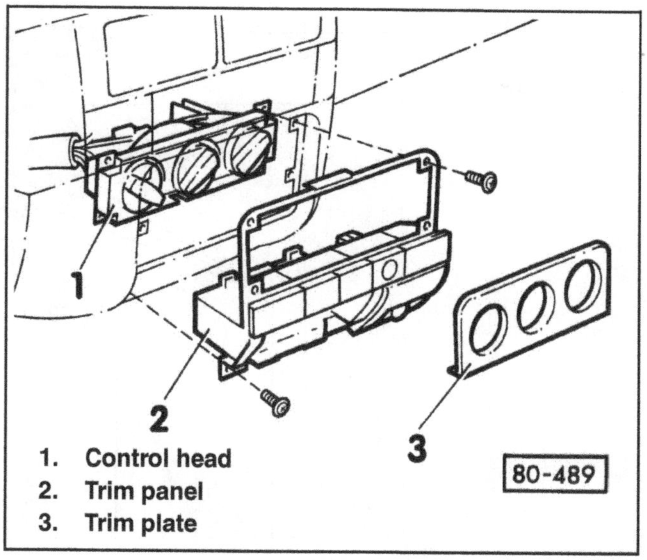

80-489

Fig. 1. Ventilation control head and trim.

2. Remove screws and remove trim panel.
3. Pull ventilation control head from instrument panel.
4. Disconnect control cables from control head and remove control head. See **Control Cables**.
5. Install in reverse order of removal.

> ***NOTE —***
> *Install control cables on the control head before installing the control head into instrument panel. See* **Control Cables**.

80

Heater assembly (cars without A/C)

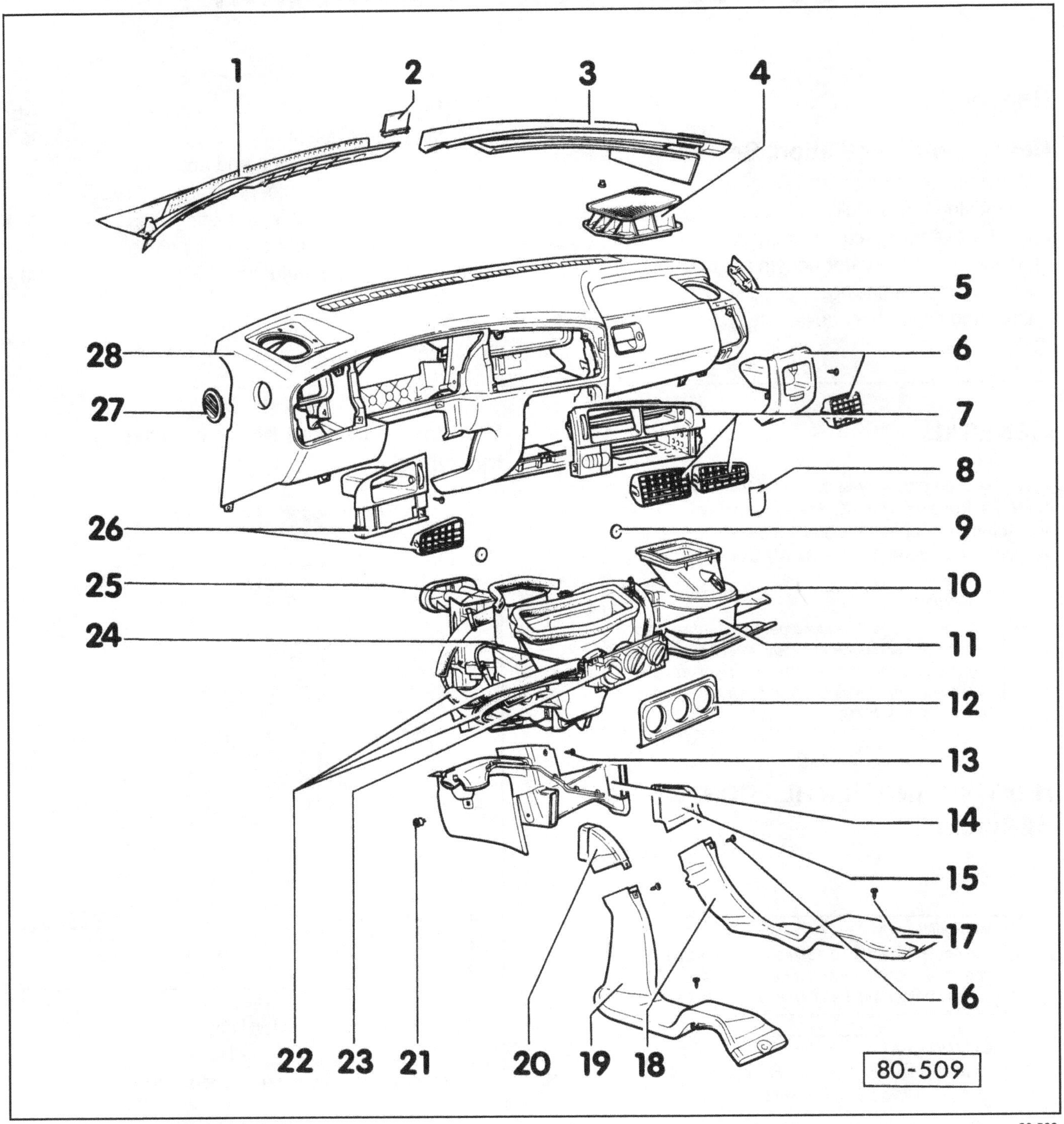

80-509

Fig. 2. Exploded view of heating and ventilation system used on cars without A/C.

1. **Left plenum cover**
2. **Connecting clip**
3. **Right plenum cover**
4. **Air intake adapter**
   - Dust and pollen filter (if applicable), removing/installing, **see 0 Maintenance Program**
5. **Side window air outlet**

6. **Right air outlet**
7. **Center air outlet**
8. **Trim cover**
9. **Nut**
10. **Fresh air blower series resistance with fuse (N24)**

> *NOTE —*
> *Volkswagen identifies electrical components by a letter and/or a number in the electrical schematics. These electrical identifiers are listed as an aid to electrical troubleshooting.*

- Removing/installing (with A/C) see **87 Air Conditioning**
- Removing/installing (without A/C), **see** (A)

11. **Fresh air blower (V2)**
- Removing/installing (without A/C), **see** (A)
- Removing/installing (with A/C), see **87 Air Conditioning**

12. **Trim plate**
13. **Self-tapping screw**
14. **Footwell air outlet**
- To remove, first remove center console

15. **Rear footwell air duct connector, right**
16. **Self-tapping screw**
17. **Self-tapping screw**
18. **Rear footwell air duct, right**
19. **Rear footwell air duct, left**
20. **Rear footwell air duct connector, left**
21. **Clip**
22. **Control cables**
- See **Control cables installing/adjusting**

23. **Ventilation control head**
- See **Control cables, installing and adjusting**

24. **Heater box/evaporator housing**
- To remove, see **Heater box, removing**

25. **Heater core**
26. **Left air outlet**
- Removing/installing, **see** (B)

27. **Side window air outlet**
28. **Instrument panel**
- Do not disassemble further than shown
- Removing/installing see **70 Trim-Interior**

(A) Fresh air blower series resistance, removing/installing (cars without A/C)

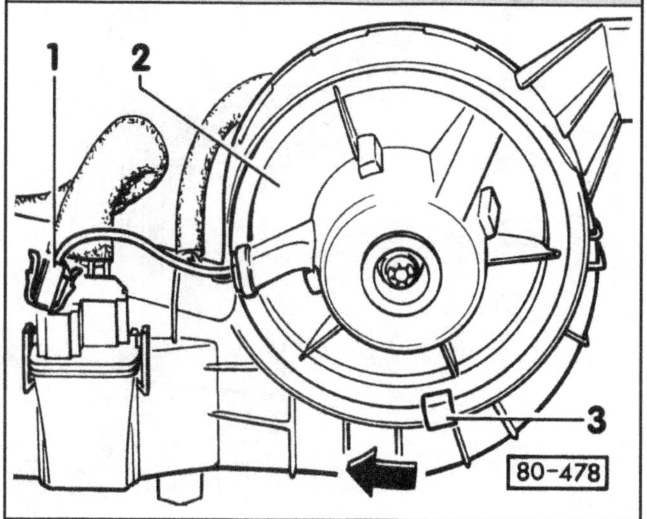

80-478

- **To remove series resistance:** **disconnect wire connector (1) from series resistance unit. Unclip series resistance unit from fresh air blower housing.**
- **To remove blower motor:** **Remove clip (3), rotate fresh air blower (2) in direction of arrow and remove.**

(B) Left air outlet, removing

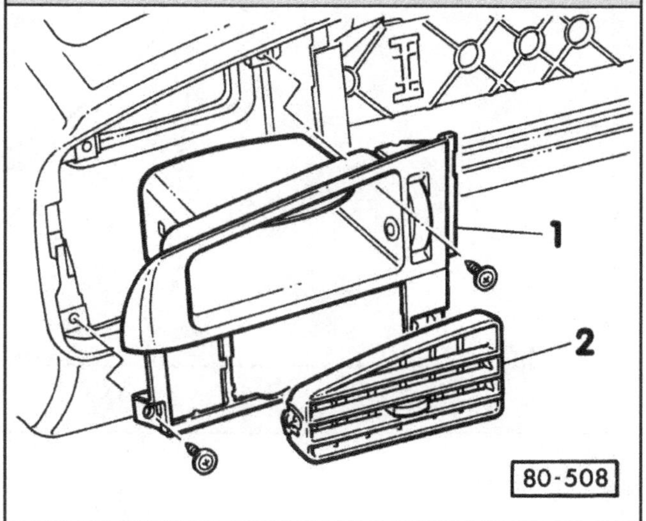

80-508

- **Remove headlight switch, see 96 Lights–Interior**
- **Using long nose pliers, carefully pull out grill (2).**
- **Remove screws and remove air outlet (1).**

Heater box/evaporator housing, assembly

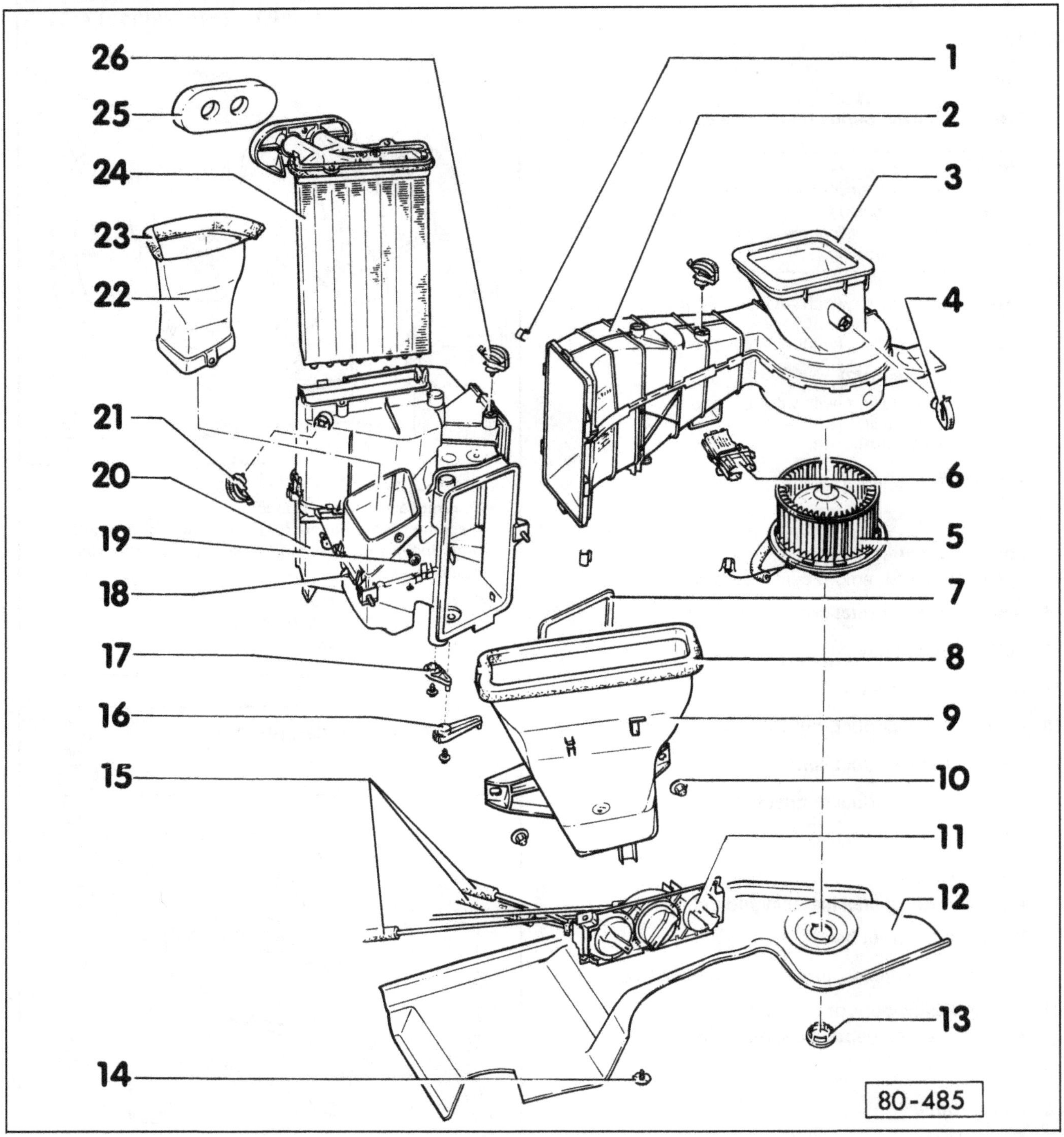

80-485

Fig. 3. Heater box/evaporator housing and related components.

1. **Clip**
2. **Air duct with main shut-off flap**
   - Do not disassemble further
3. **Gasket**
4. **Wire clip**
5. **Fresh air blower motor (V2)**

6. **Fresh air blower series resistance with fuse (N24)**
   - If series resistor fails in open circuit mode (burned out), check fresh air blower motor (V2) for excessive current draw and shaft bearings for binding.
7. **Gasket**
8. **Gasket**
   - Always replace
9. **Air duct**
10. **Nut**
11. **Ventilation control head**
    - Assembly, **see** (A)
12. **Cover**
13. **Plug**
14. **Clip**
15. **Control cables**
    - When replacing, first remove cover (12)
    - Cars without A/C, see **Control Cables**
    - Cars with A/C, see **87 Air Conditioning**
16. **Central flap lever**
17. **Temperature flap lever**
18. **Footwell/defrost flap lever**
19. **Self-tapping screw**
20. **Air distribution housing**
    - Do not disassemble further
21. **Wire clip**
22. **Air duct connector**
23. **Gasket**
    - Always replace
24. **Heater core**
    - Install seals around perimeter of heater core so there are no gaps.
    - If heater core does not lock into place during installation, fasten to heater box using screws on top of heater core.
    - Installing vacuum hose adaptor, **see Heater core, replacing**

> *NOTE—*
> *After replacing heater core, the coolant should be drained and refilled with coolant. See* **19 Engine–Cooling System**

25. **Gasket**
    - Always replace
26. **Wire clip**

(A) Ventilation control head assembly

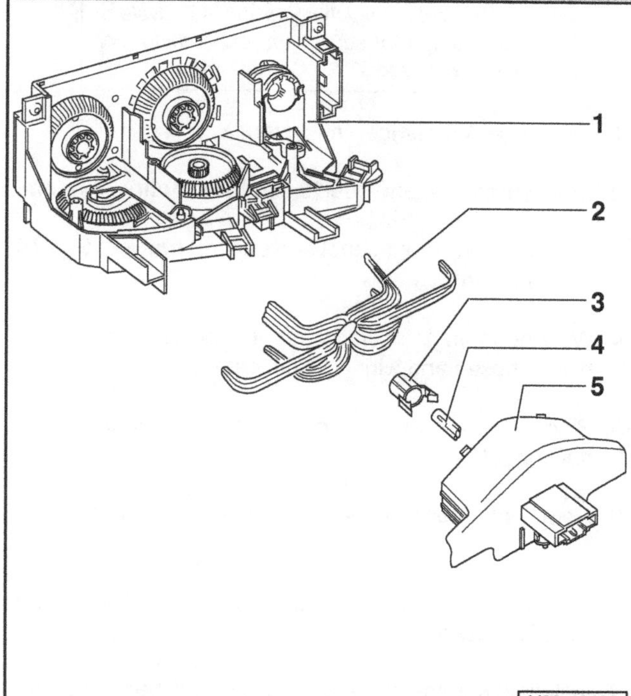

V80-0486

1. **Fresh air/heating controls with integral fresh air blower switch (E9).**
2. **Light transmitting tube (fiber optic).**
3. **Color filter (green).**
4. **Fresh air control lever light (L16).**
5. **Cover with bulb holder.**

Heater core, replacing

> ***WARNING —***
> *On cars with A/C, replacement of the heater core will require discharging/charging the A/C system. This requires special equipment and training. For more information see* **87 Air Conditioning**.

1. Discharge A/C refrigerant system.
2. Drain engine coolant. See **19 Engine-Cooling System**.
3. Remove instrument panel and support bracket. See **70 Trim-Interior**.
4. Working in engine compartment, label and disconnect heater hoses and wiring as necessary.
5. Plug A/C lines and fittings immediately after disconnecting.
6. Remove heater box/evaporator housing from car. See Fig. 3.
7. Unclip heater core housing and pull heater core straight out of heater box.
8. Installation is the reverse of removal. When installing, make sure seals are installed around perimeter of heater core so there are no gaps. Seal grooves and contact surfaces with silicone rubber sealant. On cars with A/C be sure to reinstall vacuum connection. See Fig. 4.

> ***CAUTION —***
> *A/C system must be evacuated for at least 30 minutes before recharging. See* **87 Air Conditioning**.

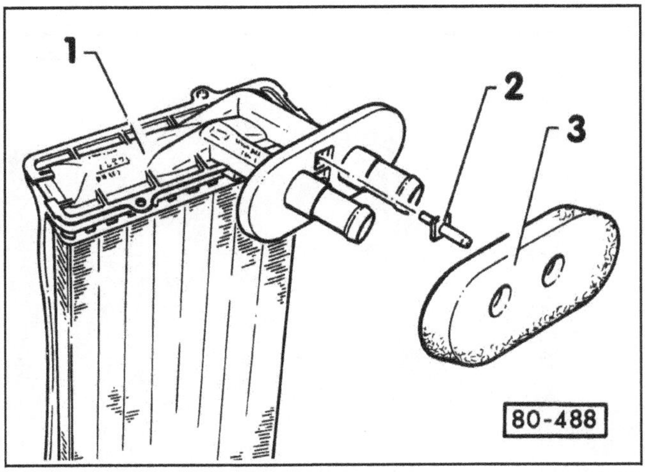

80-488

Fig. 4. Install vacuum hose adapter (**2**) and replace gasket (**3**) when installing new heater core (**1**) on cars with A/C.

Dust and pollen filter, assembly

A dust and pollen filter is installed in some vehicles and can be retro-fitted in most cases. See Fig. 5. See Fig. 6.

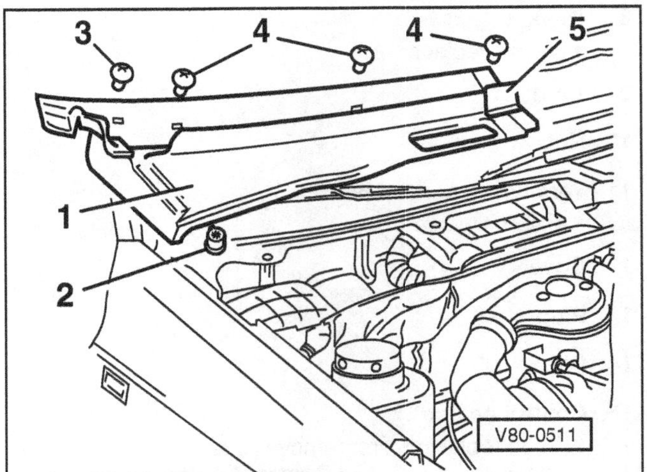

V80-0511

Fig. 5. Plenum cover and attaching components.

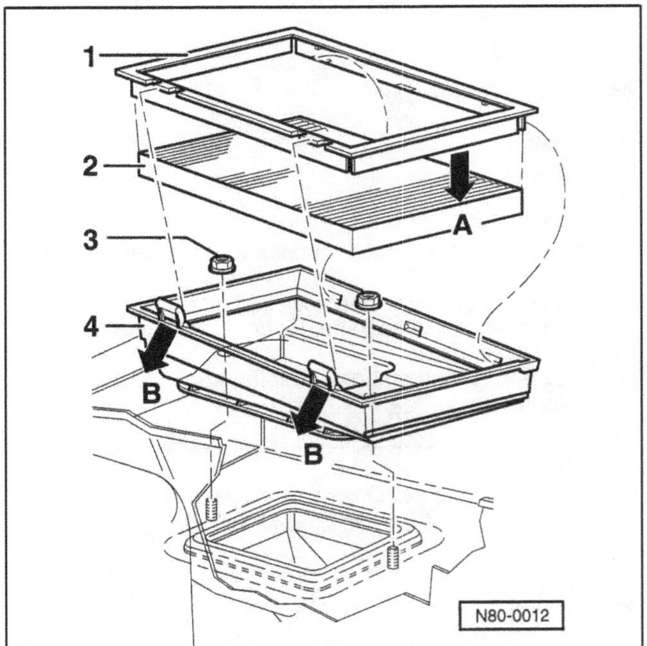

N80-0012

Fig. 6. Release filter frame (**1**) by gently pulling filter base (**4**) (**arrows B**). Install filter frame to filter element (**arrow A**) before installing.

CONTROL CABLES

Use Fig. 5 as a guide when installing and adjusting control cables.

Control cables, assembly

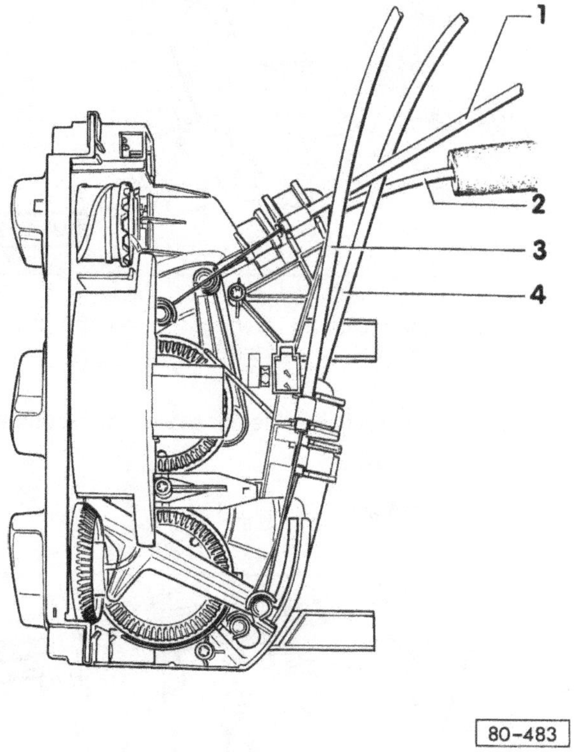

1. **Main shut-off flap cable**
   - Connects blower control knob to main shut-off flap
   - Black sleeve
   - Length: 807 mm (31.8 in.)
   - Not applicable to A/C equipped cars
2. **Temperature flap cable**
   - Connects temp. control knob to temp. flap
   - Blue sleeve
   - Length: 585 mm (23.0 in.)
3. **Footwell/defroster flap cable**
   - Connects upper control lever to footwell/defroster flap
   - Black sleeve
   - Length: 488 mm (19.2 in.)
4. **Central flap cable**
   - Connects lower control lever to central flap
   - Black sleeve
   - Length: 639 mm (25.2 in.)

80-483

Fig. 7. Exploded view of control cables on ventilation control head (control head shown for cars without A/C, others similar).

Control cables, installing and adjusting

1. Install cable ends identified with colored bands to control head levers.
2. Install cables to control head before installing on evaporator/heater box levers.
3. After connecting cables to control head levers, position cable sleeves against stops and fasten to control head using clips.
4. Adjust cables at flap levers with control head installed.
5. When control knobs are turned stop to stop, flaps must move and audibly contact stops.

Main shut-off flap cable, adjusting

NOTE —
Main shut-off flap cable is only installed on cars without A/C.

1. Install ventilation control head into instrument panel. See **Ventilation control head, removing and installing**.
2. Adjust fresh air blower switch to stop at position "0".
3. Connect main shut-off flap cable (black sleeve) to main shut-off flap lever.
4. Push lever to stop. See Fig. 8.

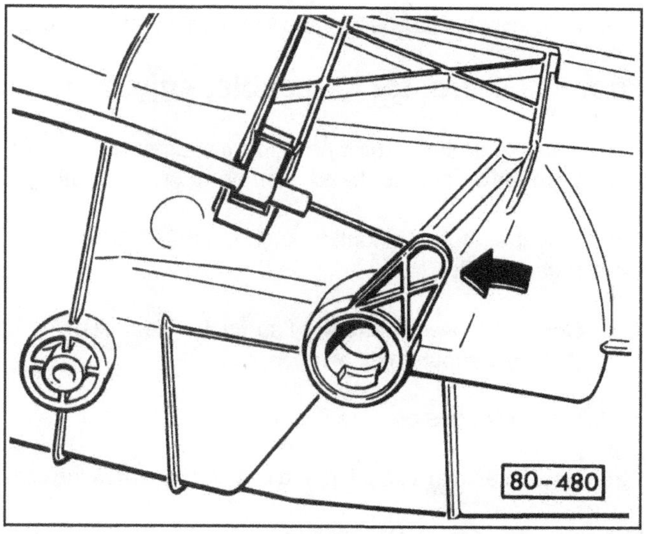

80-480

Fig. 8. Push main shut-off lever against stop (direction of **arrow**), then install cable retaining clip.

5. Hold lever in position and install cable retaining clip.

Temperature flap cable, adjusting

1. Install ventilation control head into instrument panel. See **Ventilation control head, removing and installing**.

2. Adjust temperature control knob to full cold.

3. Connect temperature flap cable (blue sleeve) to temperature flap lever.

4. Push lever to stop. See Fig. 9.

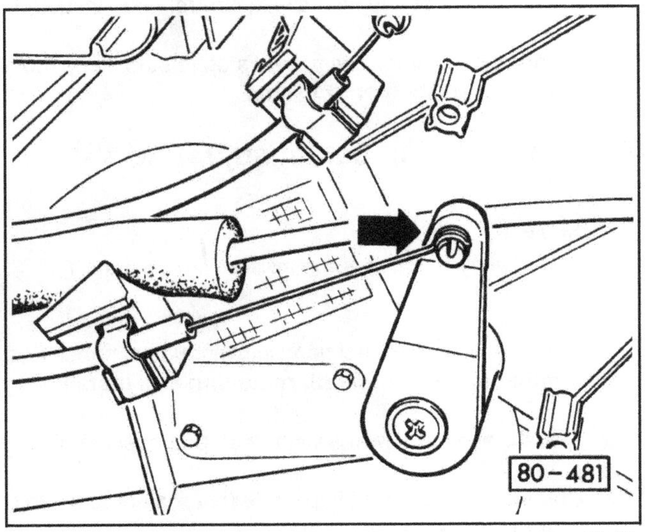

80-481

Fig. 9. Push temperature lever against stop (direction of **arrow**), then install cable retaining clip.

5. Hold lever in this position and install cable retaining clip.

Footwell/defroster flap cable, adjusting

1. Install ventilation control head into instrument panel. See **Ventilation control head, removing and installing**.

2. Adjust air flow distribution knob to DEFROST position (against stop).

3. Connect footwell/defroster flap cable (black sleeve) to footwell/defroster flap lever.

4. Push lever to stop. See Fig. 10.

5. Hold lever in this position and install cable retaining clip.

Central flap cable, adjusting

1. Install ventilation control head into instrument panel. See **Ventilation control head, removing and installing**.

2. Adjust air flow distribution knob to DEFROST position (against stop).

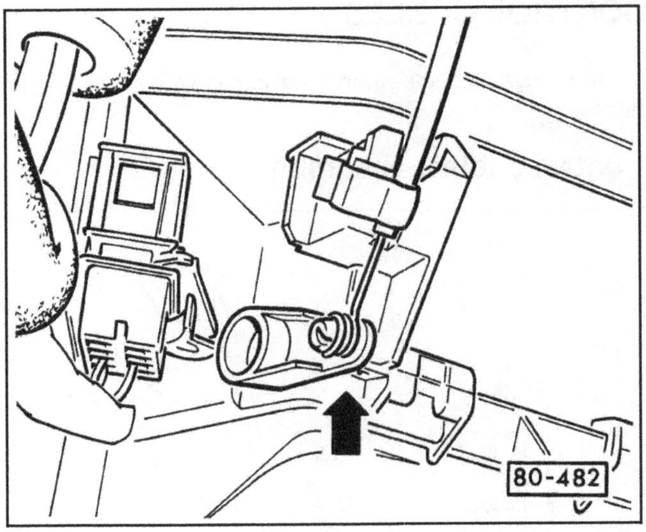

80-482

Fig. 10. Push footwell/defroster lever against stop (direction of **arrow**), then install cable retaining clip.

3. Connect central flap cable (black sleeve) to central flap lever.

4. Push lever to stop. See Fig. 11.

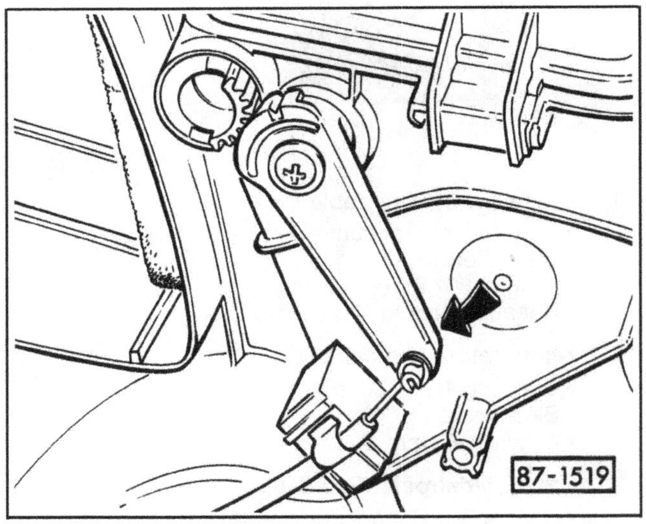

87-1519

Fig. 11. Push central flap lever against stop, then install cable retaining clip.

5. Hold lever in this position and install cable retaining clip.

87 Air Conditioning

TABLES

GENERAL

This section covers the air conditioning refrigerant components and the vacuum distribution assembly. To service the ventilation control head or make adjustments to the ventilation control cables, see **80 Heating and Ventilation**.

> ***CAUTION —***
> *It is recommended that all service to the A/C system refrigerant components be left to an authorized Volkswagen dealer or other qualified A/C service facility. If any of the refrigerant components are disconnected or removed, special equipment, supplies, and new parts will be needed to remove moisture from the system and recharge the refrigerant before it is closed back up.*

The subassemblies and lines of the air conditioner are filled with refrigerant R-134a (Tetrafluoroethane). Refrigerant R-134a replaces refrigerant R-12, which is reported to have harmful effects on the earth's ozone layer. R-134a and R-12 systems are similar in design, but there are very important differences in the refrigerant, lubricating oils, service equipment, and A/C system components.

A/C system, warnings and cautions

> ***WARNING —***
> - *Work in a well ventilated area. Refrigerant gases are heavier than air, displace oxygen and may cause suffocation in areas of poor circulation, for example under a car. Switch on exhaust/ventilation systems when working on refrigerant system.*
> - *Pressurized R-134a refrigerant in the presence of oxygen may form a combustible mixture. Never introduce compressed air into any R-134a container (full or empty), capped off A/C component, or piece of service equipment.*
> - *As of January 1, 1992 any person who services a motor vehicle air conditioner must, by law, be properly trained and certified, and use approved refrigerant recycling equipment. Technicians must complete an EPA-approved recycling course to be certified.*
> - *State and local governments may have additional requirements regarding air conditioning servicing. Always comply with state and local laws.*
> - *The A/C system is filled with refrigerant gas which is under pressure.*
> - *Avoid breathing refrigerant vapors. Exposure may irritate eyes, nose, and throat.*

87

WARNING —
- *Always be careful that refrigerant does not come in contact with your skin. Always wear hand and eye protection (gloves and goggles) when working around the A/C system. If refrigerant does come in contact with your skin or eyes:*
  -Do not rub skin or eyes
  -Immediately flush skin or eyes with cool water for 15 minutes
  -Rush to a doctor or hospital–Do not attempt to treat yourself
- *Do not expose any component of the A/C system to high temperatures (Above 80°C/ 176°F) or open flames. Excessive heat will cause a pressure increase which could burst the system.*
- *Switch on exhaust/ventilation systems when working on the refrigerant system.*
- *Keep refrigerant away from open flames. Poisonous gas will be produced if it burns. Do not smoke near refrigerant gases for the same reason.*
- *Electric welding near refrigerant hoses causes R-134a to decompose. Discharge system before welding.*

CAUTION —
- *Do not top-off a partially charged refrigerant system. Discharge system, evacuate and then recharge system.*
- *R-134a and R-12 refrigerants and refrigerant oils and system components, including hoses and sealing O-rings, are NOT compatible. Component damage and/or system contamination will result.*
- *Service equipment for the R-134a and R-12 systems are not compatible. Use proper service equipment. Damage to both vehicle and service equipment will result from R-12 contamination of R-134a.*
- *The mixture of refrigerant oil (PAG oil) and refrigerant R-134a attacks some metals and alloys (for example, copper) and breaks down certain hose materials. Use only hoses and lines that are identified with a green mark (stripe) or the lettering "R-134a".*
- *Immediately plug open connections on A/C components to prevent dirt and moisture contamination.*
- *R-134a and R-12 systems use different size service fittings. Never use adapters to convert an R-12 fitting to R-134a size.*
- *DO NOT steam clean condensers or evaporators. Use only cold water or compressed air.*

Technical data, capacities

NOTE —
As of publication, new A/C compressors supplied by Volkswagen are filled with 115 cc (3.9 oz.) of refrigerant oil. This is the total A/C system refrigerant oil capacity.

Table a. A/C system capacities

	Quantity
R-134a refrigerant	800g+50g (28.2 oz.+1.8 oz.)
PAG oil SP-10	115 cc ± 15 cc (3.9oz ± 0.5 oz.)

Table b. Refrigerant oil distribution

Component	Approx. amount of refrigerant oil in component / %of total
Compressor	57.5 cc (1.9 oz.) / 50%
Condenser	11.5 cc (0.4 oz.) / 10%
Low pressure hose	11.5 cc (0.4 oz.) / 10%
High pressure hose	– – –
Evaporator	23 cc (0.8 oz.) / 20%
Receiver dryer	11.5 cc (0.4 oz.) / 10%

A/C system, operation

Fig. 1 shows the major components of the A/C system. Refrigerant entering the **evaporator** absorbs heat from air passing through the evaporator fins and cools the air. As the refrigerant absorbs heat it turns to vapor and then is suctioned by the compressor. Moisture is removed at the evaporator in the same way that water drops form on a cold glass. The moisture drips onto the water tray in the evaporator housing and is routed away via a drain hose. The **expansion valve** controls the refrigerant flow to the evaporator thus lowering refrigerant temperature and pressure.

The **compressor** is driven via a belt on the engine when the A/C clutch engages (A/C ON). Low pressure refrigerant gas from the evaporator is compressed by the compressor. After compression, the refrigerant gas (now high pressure) flows to the condenser. The compressor contains refrigerant oil (PAG oil) that is mixable under all temperatures with R-134a refrigerant. The **pressure relief valve** is mounted on the compressor or the receiver drier. At pressures above 40 bar (580 psi), the pressure relief valve opens to vent excessive pressure. When the system pressure is reduced, the valve closes to prevent total refrigerant loss. A cap on the pressure relief valve will pop out if the valve has opened.

The **condenser**, which looks like a small radiator and is located in front of the engine radiator, transfers heat from the compressed refrigerant gas to the outside air which causes the refrigerant to change state from a gas to a liquid. The **receiver drier** acts as a reservoir for the A/C system. Any moisture in the system is absorbed in the drier desiccant.

A/C system, component layout

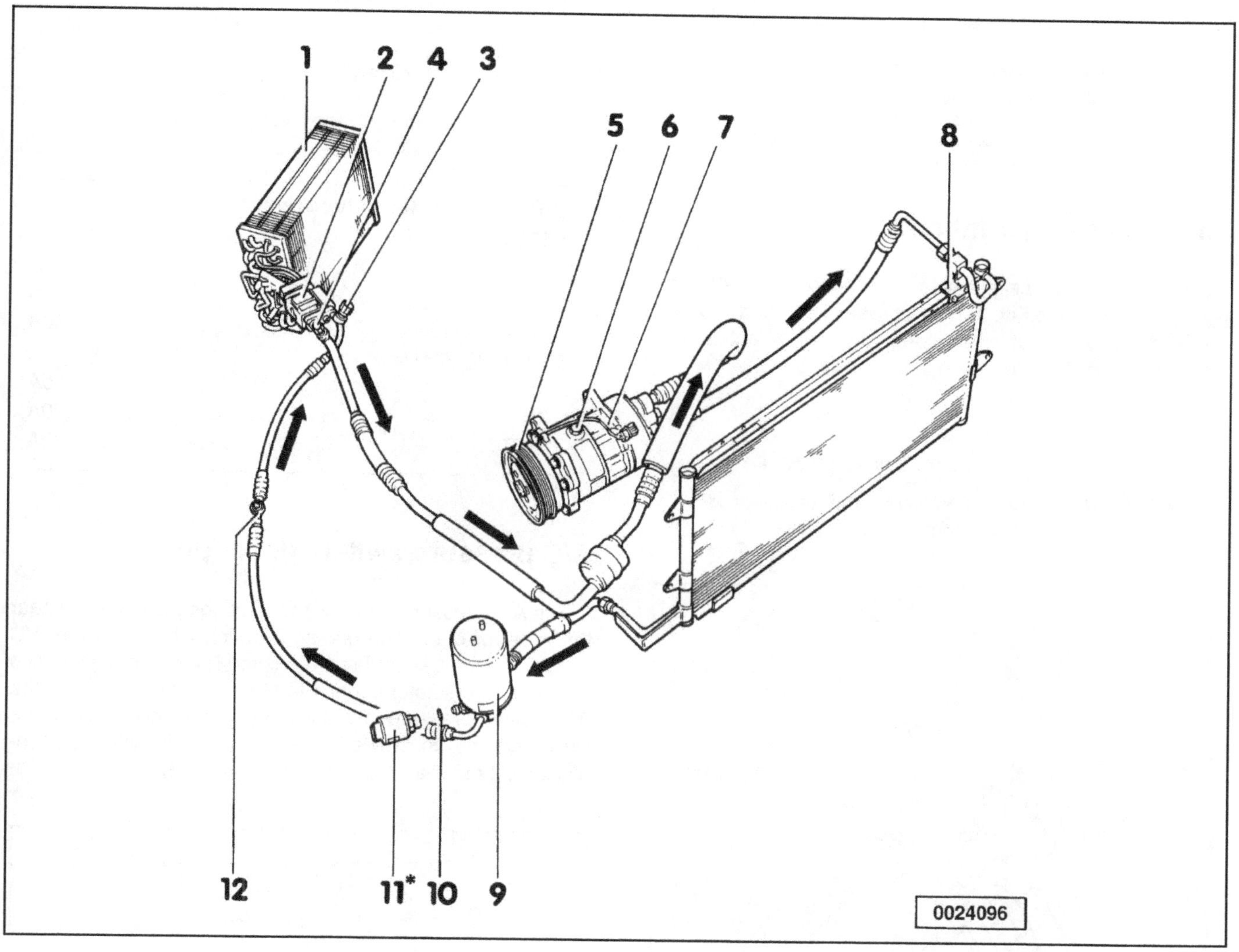

0024096

Fig. 1. R-134a A/C components. Arrows indicate direction of refrigerant flow.

1. **Evaporator**

> *CAUTION —*
> *A/C components specified for use with R-134a are identified by a green label or stripe.*

2. **Expansion valve**
   - On right side of engine compartment
3. **High pressure service valve**
4. **Low pressure service valve**
5. **A/C Clutch (N25)**
6. **Oil fill/drain plug**
7. **Compressor**
   - May be equipped with pressure relief valve
8. **Condenser**
9. **Receiver drier**
   - Replace whenever A/C system is discharged
   - May be equipped with pressure relief valve
10. **O-ring**
    - Use only new red or green O-rings that are compatible with R-134a refrigerant and refrigerant oil.
    - Always use correct size O-rings
    - Do not reuse O-rings, always replace
    - Lubricate O-rings with refrigerant oil before installing
11. **A/C Pressure Switch (F129)**
    - Can remove/install without discharging A/C system
    - Ensure O-ring fits properly in groove
12. **Sight glass**
    - Not applicable to all vehicles
    - See **Checking refrigerant charge**, given later

A/C SYSTEM TROUBLESHOOTING

> *WARNING —*
> *Discharging/charging the A/C system requires special equipment and training. Do not discharge/charge the A/C without proper equipment and training, damage to the vehicle and personal injury may result.*

Checking refrigerant charge

On some models, a sight glass is provided to check the refrigerant charge. See Fig. 2. With a properly charged A/C system and the A/C clutch engaged, the refrigerant should remain clear and bubble-free in the sight glass.

NOTE —
- *Not all vehicles are equipped with a sight glass.*
- *The mixture of R-134a refrigerant and refrigerant oil (PAG oil) may appear milky in the sight glass.*

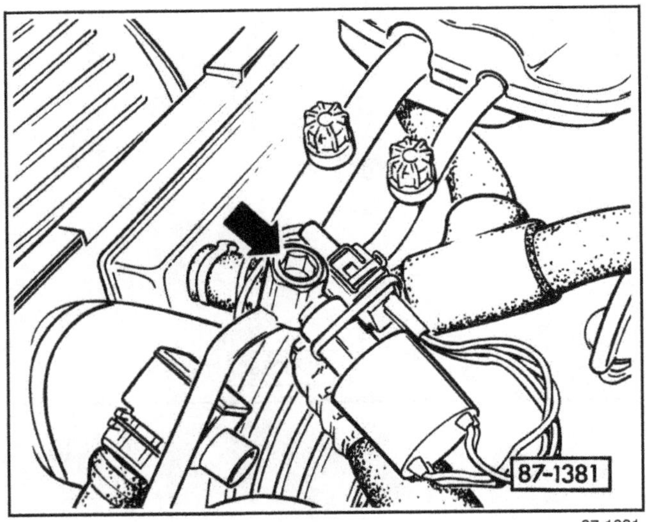

87-1381

Fig. 2. Sight glass (**arrow**) for A/C system. Sight glass is in-line between the receiver drier and the expansion valve. Sight glass may also be located on top of receiver drier.

On cars not equipped with a sight glass, the refrigerant charge cannot be accurately checked. If a problem with the charge is suspected, check the system for leaks. If no leaks are found the A/C system must be discharged, evacuated and recharged.

> *CAUTION —*
> *DO NOT top off a partially filled A/C system, system failure will result.*

A/C system fuses

NOTE —
Coolant fan fuses vary depending on model year of car. For wiring diagrams and exact fuse and wiring information, see **97 Wiring Diagrams, Fuses and Relays**.

A/C system fuses in fuse/relay panel
- 1993 m.y.
  - S6 .... 20A
  - S19 .... 30A
  - S23 .... 30A
- 1994 m.y. and later
  - S4 .... 15A
  - S6 .... 30A
  - S19 .... 30A

A/C pressure switch, description

The A/C pressure switch is mounted the pressure line near the receiver drier. The pressure switch will switch the A/C clutch off if the system pressure is too low or too high. It also switches the radiator coolant fan to second speed when the A/C system pressure increases above a specified level. The switch can be unscrewed from its Schraeder valve and replaced without discharging the A/C system. See Fig. 3.

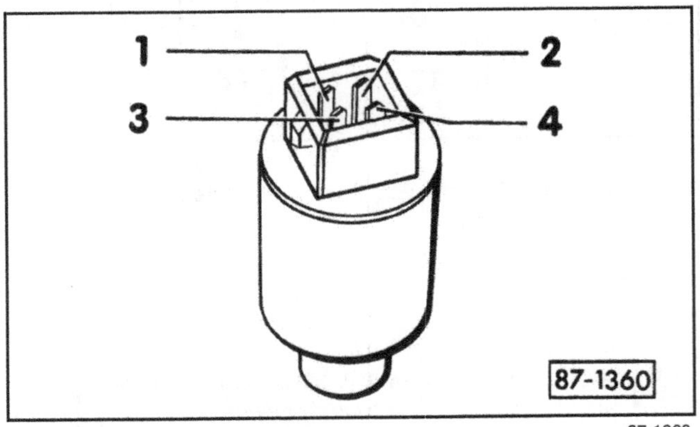

87-1360

Fig. 3. A/C pressure switch terminal identification.

NOTE —
When installing the pressure switch, ensure that the O-ring (color: red, 10.8 mm x 1.8 mm) fits properly in groove.

A/C pressure switch operating limits (low pressure)

Terminals 1 and 2
- Switch opens (compressor OFF) .... below 1.2 bar (17.4 psi)
- Switch closes (compressor ON) .... above 2.4 bar (34.8 psi)

A/C pressure switch operating limits (high pressure)

Terminals 1 and 2
- Switch opens (compressor OFF) . . . . above 32.0 bar (464 psi)
- Switch closes (compressor ON) . . . . below 24.0 bar (348 psi)

A/C pressure switch operating limits (senses increased pressure)

Switching between terminals 3 and 4
- Switch closes (radiator fan on, 2nd speed) . . . . above 16.0 bar (232 psi)
- Switch open (radiator fan off, 2nd speed) . . . . below 12.5 bar (181.3 psi)

NOTE —

A/C pressure switch is identified with a green label or the lettering "R-134a".

Tightening torques
- A/C pressure switch . . . . 8 Nm (71 in-lb)

A/C pressure switch, checking

Perform this test if A/C clutch does not engage when the A/C is turned on with the engine running.

1. Disconnect the harness connector from the pressure switch and use a jumper wire to briefly bridge terminal 1 and terminal 2 of A/C pressure switch connector.

2. If A/C Clutch now engages, either the refrigerant is discharged (empty) or the switch itself is faulty. Check A/C system for leaks, repair as necessary.

NOTE —

Check the switch by first removing it and then operating the A/C system. With the A/C pressure switch removed and connector attached, the A/C clutch should not engage and A/C system should not cool.

A/C compressor pressure relief valve

The pressure relief valve is mounted on the A/C compressor and protects the refrigerant circuit from excessive pressure.

When the valve has been actuated by excessive pressure, an adhesive-attached plate is pushed out. If the plate is pushed, out, repairs should be carried out by an A/C specialist. See Fig. 4.

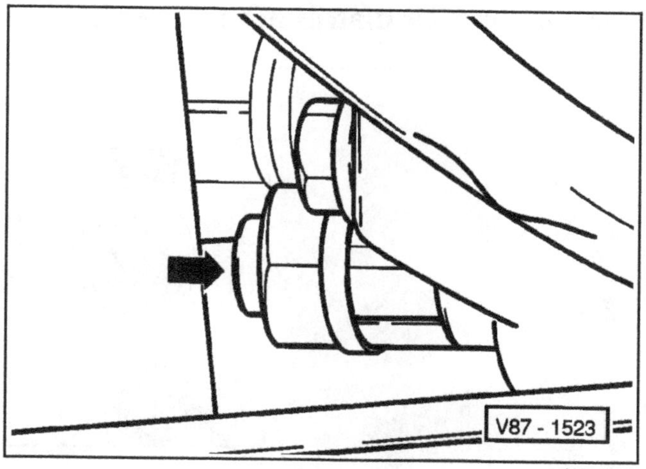

V87-1523

Fig. 4. Pressure relief valve on A/C compressor will indicate excessive pressure if plate (**arrow**) has been pushed out.

NOTE —

Insert vacuum hose into reservoir 30 mm (1 3/16 in.). Insert vacuum hoses onto plastic lines up to stop, approx. 15 mm (19/32 in.).

Vacuum and air distribution, fault tracing

The fresh air blower supplies air flow for heating and air conditioning. Blower speed is determined by the blower switch and blower series resistance. The blower series resistance is mounted to the evaporator housing, underneath the glovebox. The fresh air blower must run in all 4 speeds for the A/C to work correctly. See **80 Heating and Ventilation** and **97 Wiring Diagrams, Fuses and Relays**.

Fig. 5 shows a schematic of the ventilation air flap and control assembly. The footwell/defrost flap, temperature flap and central flap are cable controlled. The control head, which activates the cables, and the cables themselves must be correctly functioning and properly adjusted. See **80 Heating and Ventilation** for cable adjustment and control head replacement procedures.

The fresh air/recirculating flap has 2 positions (fully open or closed) and is controlled by the vacuum servo. The servo is vacuum operated and is controlled by the fresh air recirculating flap switch. Check that the flap is operating correctly. Inspect the vacuum system for routing, leaks, and operation. See Fig. 6. If any electrical faults are found test the switch housing the appropriate diagram shown in **97 Wiring Diagrams, Fuses and Relays**.

NOTE —

The position of the other two ventilation control knobs does not affect fresh air cut-off.

Vacuum and air distribution assembly

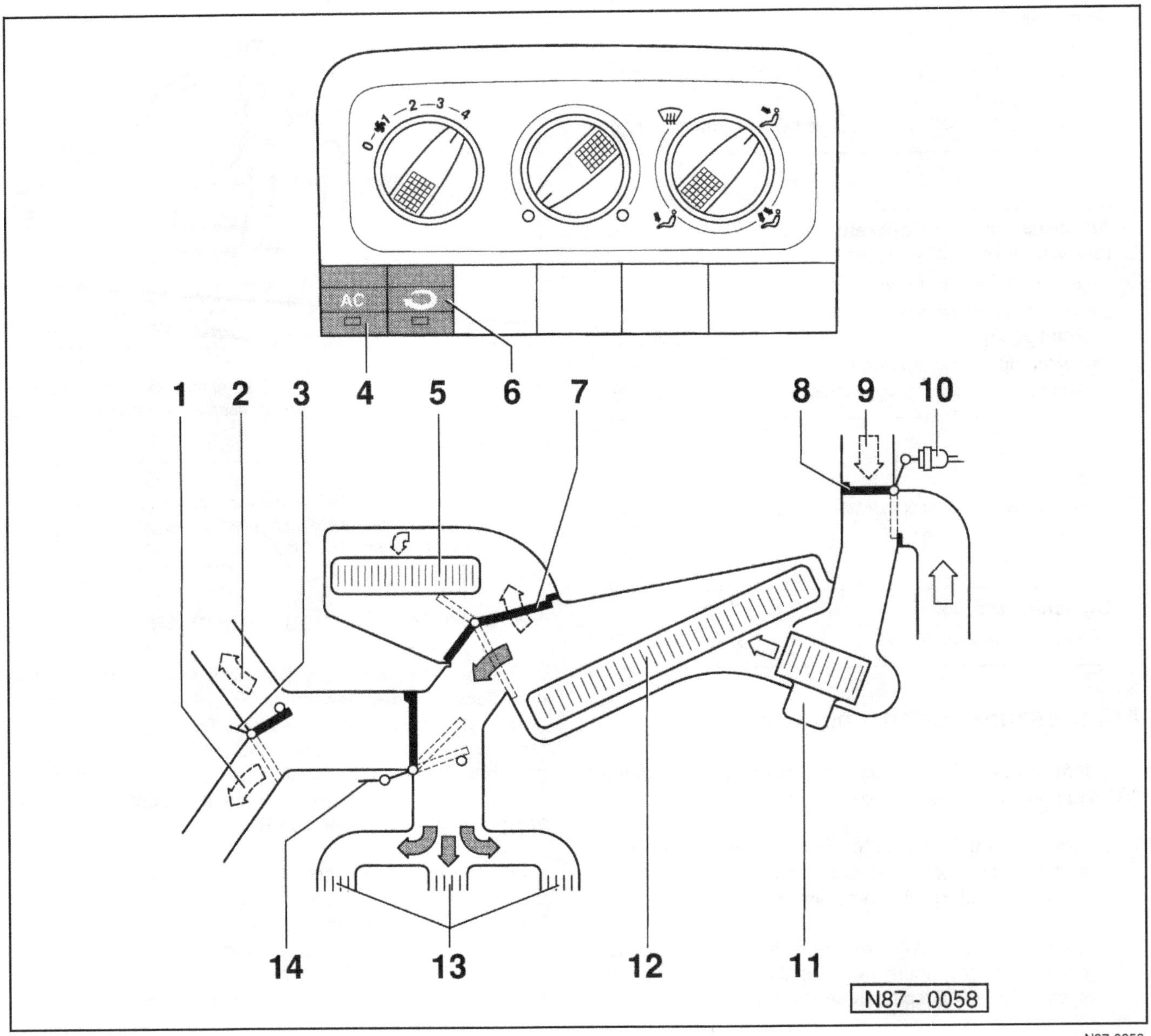

N87-0058

Fig. 5. Air flaps and controls for 1994 and later models (1993 similar).

1. **Footwell vents**
2. **Defroster vents**
3. **Footwell/defrost flap**
4. **A/C switch (E35)**
   - With engine off, ignition on and blower switch on high A/C clutch must audibly click when A/C switch is pressed
5. **Heater core**
6. **Fresh air/recirculating flap switch (E159)**
7. **Temperature flap**
8. **Fresh air/recirculating flap**
9. **Fresh air inlet**
10. **Vacuum servo**
    - Extended no vacuum applied, fresh air inlet open
    - Retracted vacuum applied, fresh air inlet blocked
11. **Fresh air blower (V2)**
    - Must run in all 4 speeds
12. **Evaporator**
13. **Instrument panel vents**
14. **Central flap**

Vacuum hose layout

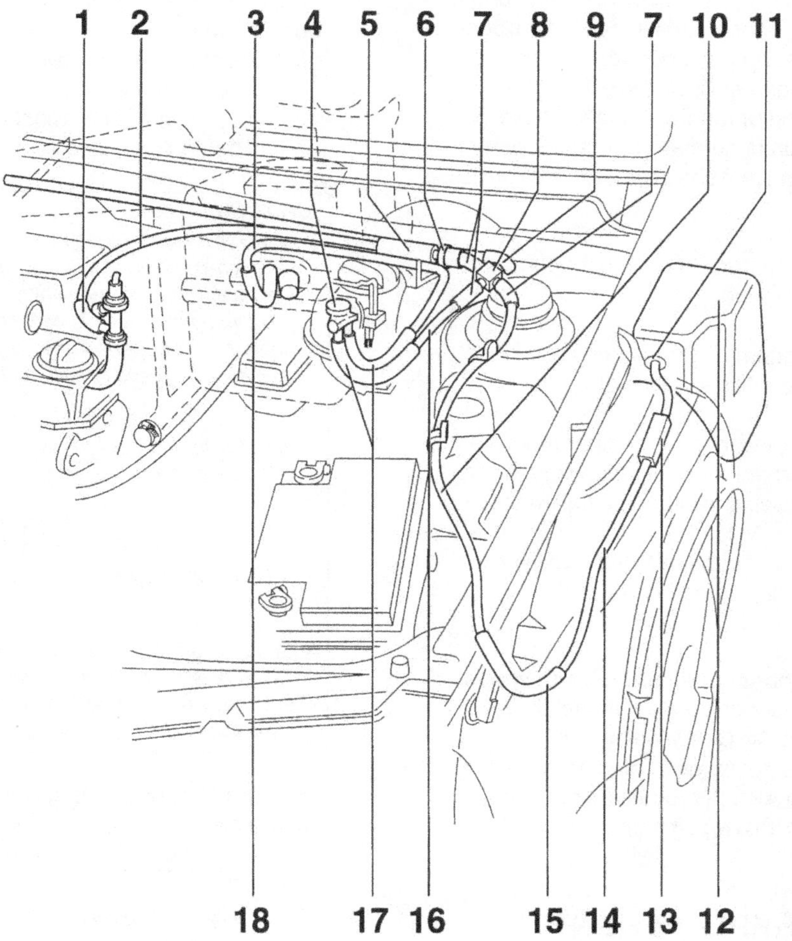

N87-0015

Fig. 6. Vacuum hose routing. Hose routing may vary, depending on level of equipment installed.

1. **Hose (3.5 x 2 x 40 mm)**
2. **Plastic line (4 x 1 x 540 mm)**
3. **Plastic line (4 x 1 x 490 mm)**
4. **Fresh air/recirculating flap two-way valve (N63)**
5. **Foam insulator**
6. **Check valve**
   - Arrow on valve points to vacuum source
7. **Hose (3.5 x 2 x 40 mm)**
8. **Vacuum tee connector**
9. **Hose (3.5 x 2 x 40 mm)**
10. **Plastic line (4 x 1 x 1170 mm)**
11. **Rubber grommet**
12. **Vacuum reservoir**
13. **Foam insulator**
14. **Plastic line**
15. **Hose (3.5 x 2 x 130 mm)**
16. **Plastic line (4 x 1 x 125 mm)**
17. **Curved hose**
18. **Vacuum hose**
    - For fresh air/recirculating flap

A/C system odor, checking

An unpleasant smell or "musty" odor can be emitted from any automotive air conditioning system under certain conditions. Many times this "musty" odor is caused by residual condensate in and around the A/C evaporator mixing with airborne pollutants. This environment is suitable for the formation and growth of bacteria and/or fungi which can cause undesirable odors. Condensation is normally formed as part of the air conditioning process and must be allowed to flow out of the system.

To minimize the growth of bacteria and/or fungi, there are several steps that can be performed.

1. Water drain(s) in the bottom of the plenum chamber(s) under the hood must be open and free of any debris.

2. The bulkhead between the engine compartment and the passenger compartment must be sealed along with any cable grommets passing through the bulkhead.

3. The evaporator water (condensate) drain valve must open freely. This valve is located behind the bulkhead insulation. See Fig. 12.

If the above steps fail to correct an odor problem, professional preparation and treatment using a commercially available product such as Airsept Air Conditioning Treatment^{TM} should be performed. Special tools, equipment and training are required to properly apply Airsept products. These procedures are outside the scope of this repair manual

A/C COMPONENT REPLACEMENT

As a general rule, the receiver drier should always be replaced anytime A/C components are replaced. Be sure to install the receiver drier immediately after opening to prevent moisture contamination of drier desiccant. Moreover, keep refrigerant system and other replacement components sealed for as long as possible to minimize the chance of dirt and moisture contamination.

If the A/C system has been discharged due to a damaged or leaking component (refrigerant hose/line, compressor, evaporator, condenser, etc.), the system should be flushed as described under **Flushing A/C system**.

The A/C system should be flushed if any of the following conditions are met:

- Moisture or dirt have entered the system (i.e. following an accident).
- Refrigerant oil is dark and thick.
- Too much refrigerant oil is in the system following compressor replacement.
- The compressor is replaced due to internal damage (i.e. noises).

After flushing the system:

- If the compressor is not replaced with a new unit, fill compressor with the correct type and total system quantity of refrigerant (PAG) oil, see **Technical data, capacities**.
- If the compressor is replaced new, confirm that the correct quantity and type of oil (PAG) is in the compressor, see **Technical data, capacities** given at the beginning of this section.

CAUTION—
- *A/C compressors, evaporators and condensers supplied by Volkswagen as replacement parts are filled with a Nitrogen charge and sealed. If a gas (Nitrogen) does not escape when the component is first opened, the component may be faulty (leaking) and should not be installed.*
- *Do not top-off a partially charged refrigerant system. Discharge system, evacuate and then recharge system.*
- *Replacement A/C compressors from Volkswagen are filled with the total refrigerant oil (PAG oil) quantity needed for the entire refrigerant system.*
- *Do not flush R-134a refrigerant system with R-12. R-12 is not compatible with R-134a refrigerant and PAG oil and will cause total system contamination.*

Observe the following warnings and cautions when replacing A/C components.

CAUTION—
- *Use only the specified PAG oil from a sealed container and always reseal oil container immediately after use. Do not use oil if it has become contaminated with moisture or the container has been left open.*
- *Dispose of contaminated refrigerant (PAG) oil following laws governing hazardous waste disposal. Do not combine PAG oil with any other old oils such as engine oil or transmission fluid.*
- *Always replace O-rings, do not re-use. Use only the correct size and type of O-rings specified for use with R-134a refrigerant. Lubricate O-ring with refrigerant oil before installing.*
- *Always re-install caps over A/C service valves.*
- *Always wear protective goggles and gloves when working on A/C system.*
- *Switch on shop exhaust/ventilation system when working on the refrigerant system.*
- *Immediately plug open connections on A/C components to prevent dirt and moisture contamination.*

Discharging and charging A/C system

All A/C systems covered by this manual use R-134a refrigerant system. R-134a systems are not compatible with earlier R-12 systems. R-134a systems require specialized equipment and training. The A/C system charge cannot be accurately checked. If a problem with the charge is suspected, first check the system for leaks. If no leaks are found the A/C system must be discharged, evacuated and recharged.

> ***WARNING —***
> - *Do not discharge/charge the A/C without proper equipment and training, damage to the vehicle and personal injury may result.*
> - *Any person who services a motor vehicle air conditioner MUST, by law, be properly trained and certified and use approved refrigerant recycling equipment.*

Following the manufacturer's instructions, connect an approved refrigerant recovery/recycling/recharging unit to the A/C system and discharge system.

> ***CAUTION —***
> *Do not release refrigerant into the air.*

Before recharging, evacuate the system for at least 30 min. Recharge system following the equipment manufacturer's instructions. For capacities, see **Technical data, capacities**.

Flushing A/C system

1. Discharge A/C system as described above.

2. Disconnect refrigerant lines from all A/C components.

3. Blow compressed air then nitrogen through evaporator, condenser, receiver drier and refrigerant lines in opposite direction of refrigerant flow.

> ***WARNING —***
> *Pressurized R-134a refrigerant in the presence of oxygen may form a combustible mixture. Never introduce compressed air into any R-134a container (full or empty), capped off A/C component, or piece of service equipment.*

> ***CAUTION —***
> *Do not blow compressed air or nitrogen through compressor or expansion valve.*

4. Reconnect A/C components and lines.

> ***CAUTION —***
> *Do not re-use O-rings when reconnecting refrigerant lines. Always replace O-rings.*

5. Drain refrigerant oil from compressor. Refill compressor with total amount of oil, see **Technical data, capacities** given earlier.

6. Evacuate system for at least 30 min.

7. Recharge system according to manufacturer's instructions. See **Technical data, capacities** for refrigerant capacity.

A/C thermal switch, removing and installing

1. Allow engine to cool completely and relieve cooling system pressure by removing cap from coolant expansion tank.

> ***WARNING —***
> - *Hot coolant can scald and result in serious personal injury. Do not work on the cooling system until it has fully cooled.*
> - *At normal operating temperature the cooling system is pressurized. Allow the system to cool before opening, then release the cap slowly to allow release of pressure.*

2. Remove harness connector from A/C thermal switch. For 6-cylinder, See Fig. 7. For 4-cylinder, See Fig. 8.

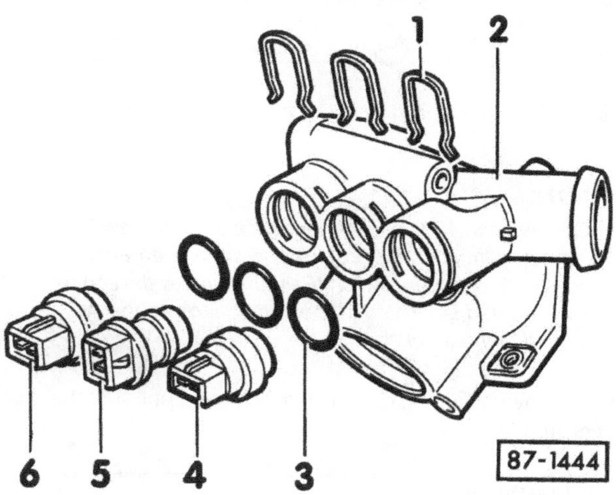

1. **Clip**
2. **Engine coolant regulator housing**
3. **O-ring**
4. **A/C thermal switch (F45)**
   - Brown
5. **Engine coolant temperature sensor (G2)/ after-run coolant fan control thermal switch (F87)**
   - Yellow
6. **Engine coolant temperature sensor (G62)**
   - Blue

87-1444

Fig. 7. A/C thermal switch used on cars with VR6 engine.

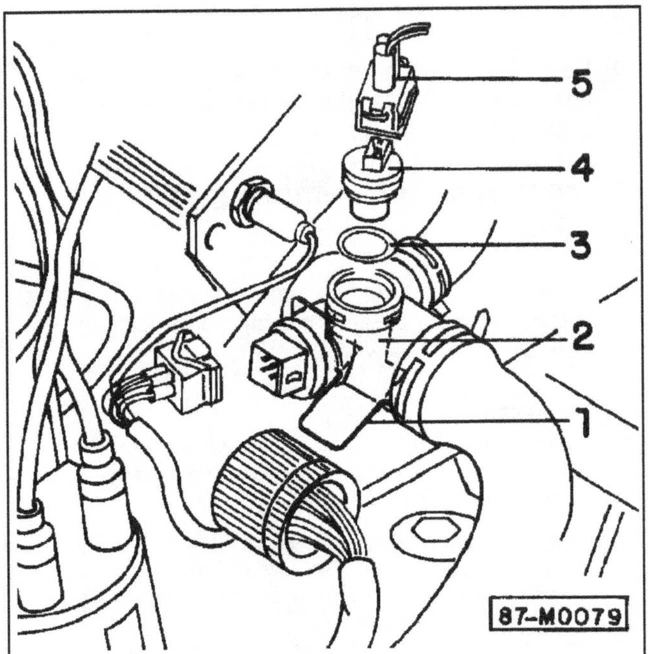

1. **Clip**
2. **Coolant flange**
3. **O-ring**
4. **A/C cut-out thermal switch (F163)**
5. **Harness connector**

87-M0079

Fig. 8. A/C thermal switch used on cars with 4-cylinder engine.

NOTE —
On 1993 and 1994 m.y. 4-cylinder cars there is no A/C cut-out thermal switch, but instead there is an engine coolant temperature switch (F14) that serves the same purpose and is located in the same place as the A/C cut-out thermal switch.

3. Remove retaining clip from switch and pull A/C thermal switch from housing.

WARNING —
Cooling system may be under pressure. Relieve system pressure before removing any engine coolant temperature sensors or switches.

NOTE —
Coolant may spill as the A/C thermal switch is removed. Wrap a shop towel around the switch during removal.

4. Install in reverse order of removal. Refill engine coolant to proper level.

A/C condenser, removing and installing

WARNING —
Replacement of A/C condenser requires discharging/charging the A/C system. Do not discharge/charge the A/C without proper equipment and training, damage to the vehicle and personal injury may result.

1. Discharge A/C refrigerant system.
2. Remove front bumper and bumper cross support, see **63 Bumpers**.
3. Disconnect refrigerant lines from condenser.

CAUTION —
Immediately plug open connections on A/C components to prevent dirt or moisture contamination.

4. Remove condenser mounting screws and condenser. On 1998 and later models, remove grille. See Fig. 9.

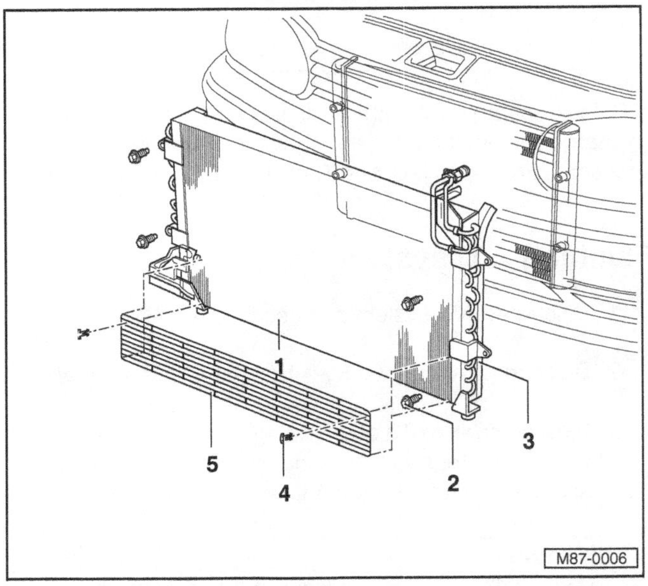

M87-0006

Fig. 9. A/C condenser (**1**), mounting screw (**2**), and side seal (**3**). From m.y. 1998 remove and install plastic grille (**5**) with clips (**4**).

5. Install in reverse order of removal.

Tightening torques
- A/C condenser to body. . . . . 10 Nm (7 ft-lb)
- Upper refrigerant line to condenser . . 35 Nm (26 ft-lb)
- Lower refrigerant line to condenser . . 15 Nm (11 ft-lb)

Fig. 10 shows the air conditioning components behind the instrument panel, including the ventilation ducts. Fig. 11 shows the evaporator housing and the evaporator.

Heating and A/C duct assembly

WARNING —
Removal of the heater box requires discharging/charging the A/C system. Do not discharge/charge the A/C without proper equipment and training, damage to the vehicle and personal injury may result.

CAUTION —
Discharge A/C system before removing components marked with an asterisk (*).

1. **Instrument panel**
   - Removing and installing, see **70 Trim-Interior**.
2. **Retaining strap**
3. **Evaporator housing***
4. **Seal***
5. **Fresh air blower series resistance**
   - Removing and installing, see (A)
6. **Fresh air/recirculating flap vacuum servo***
7. **Vacuum hose**
8. **Seal**
9. **Fresh air blower (V2)**
   - Removing and installing, see (B)
10. **A/C evaporator temperature switch (E33)**
    - Not applicable to all vehicles
    - Prevents evaporator from freezing
    - Located behind cover (16)
    - Insert sensor tube of temperature switch into evaporator 330 mm (13.0 in.)
    - Switch opens (compressor OFF): below -25°C (28°F)
11. **Seal**
    - Always replace
12. **Intermediate connecting duct**
    - With seal
13. **Plug**
14. **A/C switch (E35)**
15. **A/C (ventilation) control head trim**

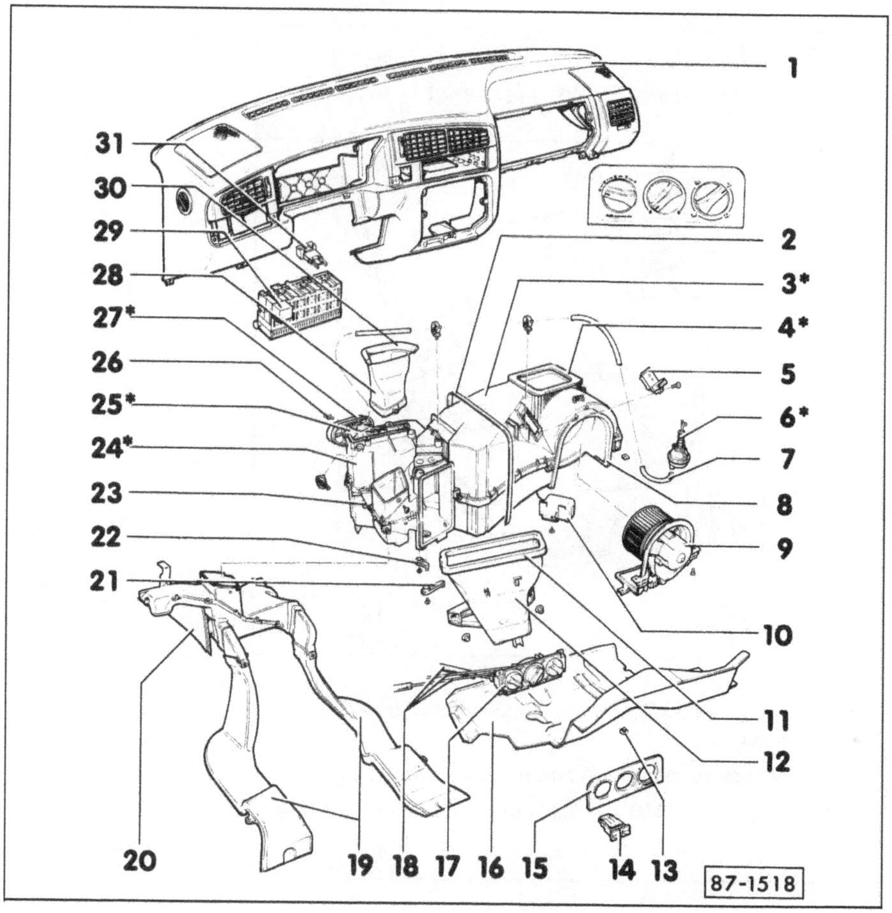

87-1518

Fig. 10. Exploded view of heater and air conditioner housing and ducts.

16. **Cover**
17. **A/C (ventilation) control head**
    - Removing and installing, see **80 Heating and Ventilation**
18. **Control cables**
    - Remove footwell air outlet (20) before removing
    - Installing and adjusting (with A/C), see **80 Heating and Ventilation**
19. **Rear footwell air ducts**
20. **Footwell air outlet**
21. **Central flap lever**
22. **Temperature flap lever**
23. **Footwell/defrost flap lever**
    - Lever and flap are one piece
24. **Heater (air distribution) box***
25. **Heater core***
    - Install seals around perimeter of heater core so there are no gaps
    - If heater core does not lock into place during installation, fasten to heater box at tabs
    - After removing heater core, drain entire cooling system and refill with new coolant
26. **Vacuum hose clip**
27. **Seal***
28. **Intermediate connecting duct**
29. **A/C relay (J32)**
30. **Seal**
    - Always replace
31. **Heater fuse (S23)**
    - Not applicable to all models

(A) Fresh air blower series resistance, removing/installing (cars with A/C)

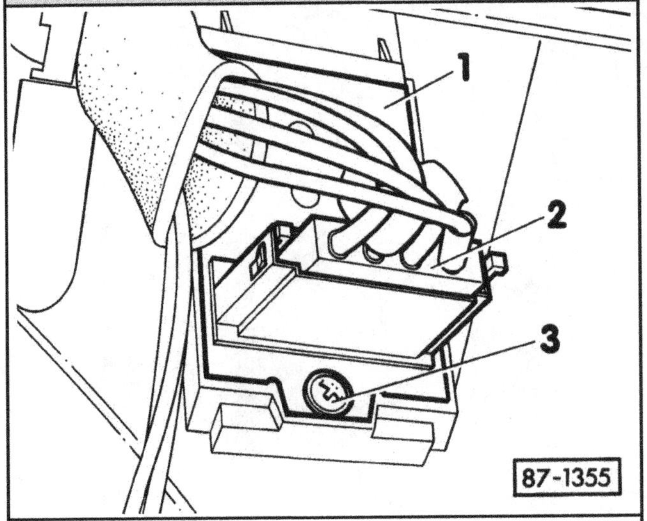

87-1355

- Remove glove compartment to gain access to series resistance (1).
- Disconnect connector (2) and remove screw (3).
- Apply adhesive (AMV 176 000 05) to mating surface before installing.

(B) Fresh air blower, removing/installing (cars with A/C)

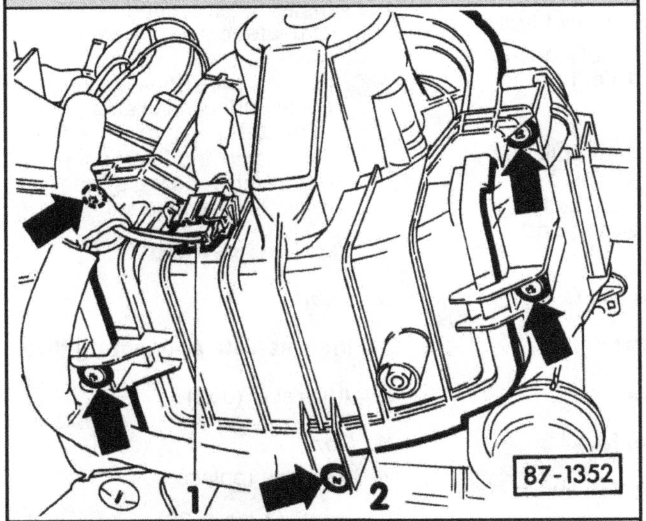

87-1352

- Disconnect connector (1)
- Remove 5 screws (arrows), pull fresh air blower (2) downward.

Evaporator housing, removing and installing

1. Discharge A/C refrigerant system.

> ***WARNING —***
> *Replacement of the evaporator housing requires discharging/charging the A/C system. Do not discharge/charge the A/C without proper equipment and training, damage to the vehicle and personal injury may result.*

2. Drain engine coolant. See **19 Engine-Cooling System**.
3. Working in engine compartment, label and disconnect heater hoses at rear bulkhead.
4. Remove instrument panel and support bracket. See **70 Trim-Interior**.
5. Label and disconnect all harness connectors from the heater box. Disconnect the refrigerant lines from the evaporator. Plug A/C lines and fittings immediately after disconnecting.

> ***CAUTION —***
> *Do not leave any part of the A/C system open, system contamination and expensive failure will result.*

6. Remove heater box from car. See Fig. 10.
7. Separate the evaporator housing from the heater box housing.
8. Remove clips and separate upper and lower evaporator housings and remove evaporator. See Fig. 11.
9. Installation is the reverse of removal. Seal grooves and contact surfaces with silicone rubber sealant.

Evaporator housing, assembly

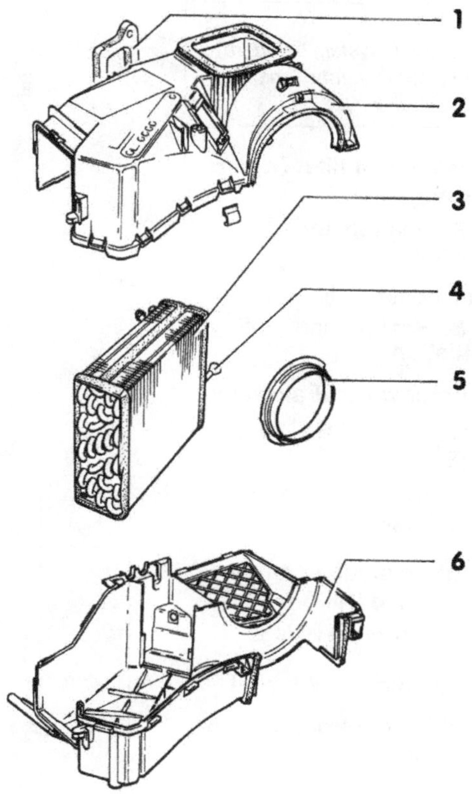

1. **Seal**
2. **Upper evaporator housing**
   - Seal grooves and contact surfaces with silicone rubber sealant
3. **Evaporator**
4. **Funnel**
   - For sensor tube of A/C evaporator temperature switch (E33)
5. **Air intake ring**
6. **Lower evaporator housing**
   - Seal grooves and contact surfaces with silicone rubber sealant

87-1388

Fig. 11. Exploded view of evaporator and evaporator housing with component listing

A/C components in engine compartment

> *CAUTION —*
> *Discharge A/C system before removing components marked with an asterisk (*).*

1. **Dust and pollen filter (where applicable)**
   - See **80 Heating and Ventilation**
2. **Plenum cover**
   - See **80 Heating and Ventilation**
3. **Evaporator water drain valve**
   - Removing and installing, **see** (B)
4. **Heater core connections and vacuum hose**
5. **Fresh air/recirculating flap two-way valve (N63)**
   - Valve open when voltage applied
   - See **Vacuum hose layout**
6. **Coolant fan control (FC) control module (J293)**
7. **Vacuum reservoir**
   - Insert vacuum hose 30 mm (1.2 in.) into reservoir
   - Removing, **see** (A)
8. **Vacuum hose**
   - See **Vacuum hose layout**
9. **A/C Thermal Switch (F45)**
   - Only on vehicles equipped with VR6 engine
   - Switches coolant fan to third speed when engine coolant temperature increases
   - Switch closes (third speed ON): above 112°C (234°F)
   - Switch opens (third speed OFF): below 108°C (226°F)
   - Switches A/C clutch OFF if engine coolant temperature is too high
   - Switch opens (compressor OFF): above 119°C (246°F)
   - Switch closes (compressor ON): below 112°C (234°F)
   - Removing/installing, see **A/C thermal switch, removing and installing**

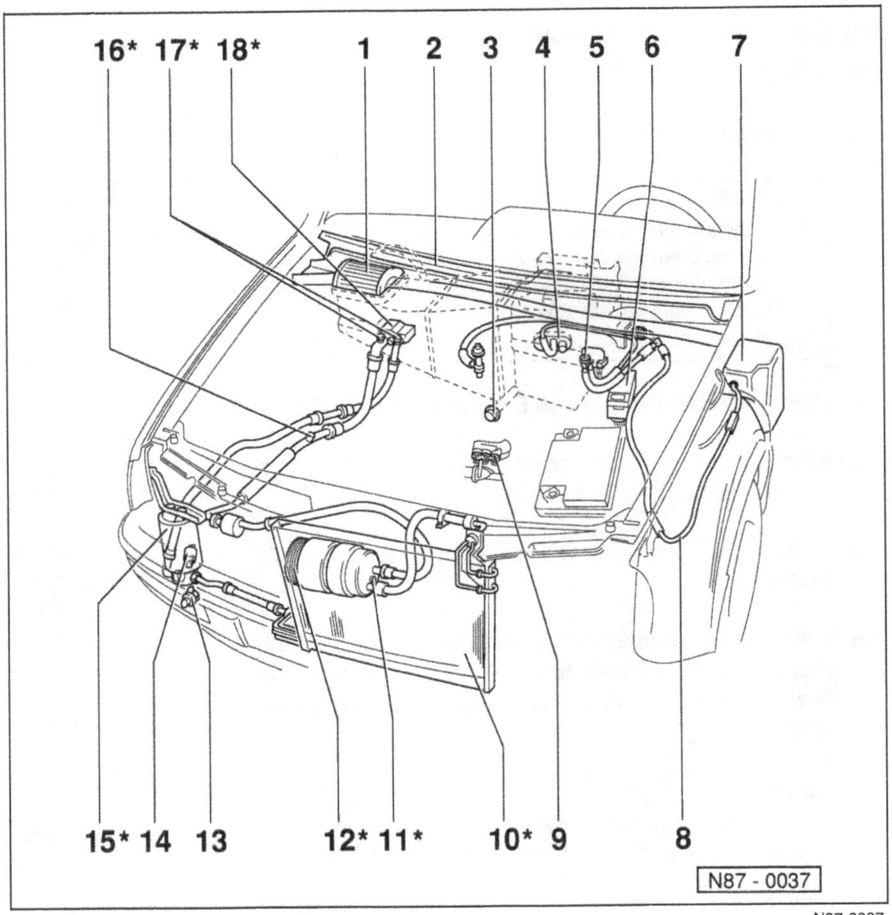

N87-0037

Fig. 12. A/C refrigerant system (R-134a) components in engine compartment.

10. **Condenser***
11. **Pressure relief valve***
12. **A/C clutch (N25)***
13. **Ambient temperature switch (F38)**
    - Switches A/C clutch OFF when outside temperature is too low
    - Switch opens (compressor OFF): below 25°C (36°F)
    - Switch closes (compressor ON): above 75°C (45°F)
    - Location:
      From 08.95 production, in plenum at base of windshield
      Through 07.95 production, near horn, behind front lower grille, **see** (C)
14. **A/C pressure switch (F129)**
    - Tightening torque: 8 Nm (71 in-lb)
    - Always replace O-ring seal (note Part No.)
15. **Receiver drier***
16. **Sight glass***
    - Not applicable to all vehicles
17. **High and low pressure service valves***
18. **Expansion valve***
    - Insulation is used around expansion valve to prevent moisture (condensation) build-up and to protect expansion valve from high engine compartment temperatures and the resulting decrease in A/C system cooling performance
    - Check insulator for damage and proper fit, replace if necessary

A Vacuum reservoir, removing

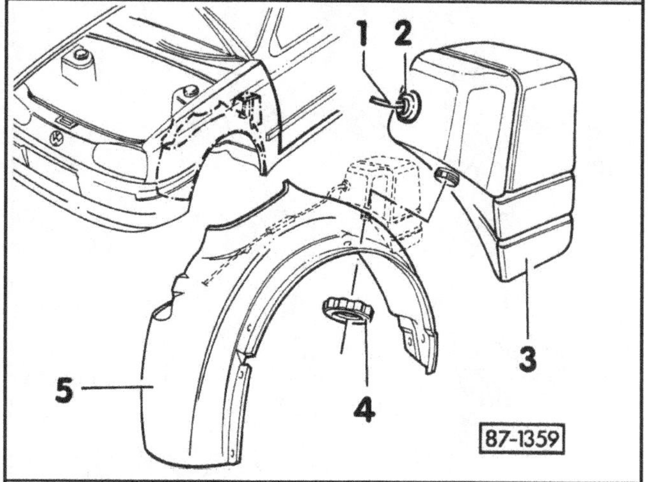

87-1359

- Remove wheel housing liner (5).
- Disconnect vacuum hose (1) from rubber grommet (2).
- Remove nut (4) and reservoir (3).

B Evaporator water drain valve, removing and installing

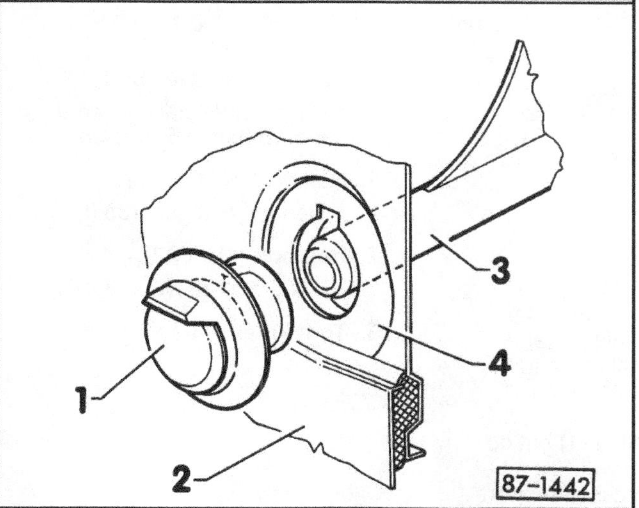

87-1442

- Bulkhead (2) has a mounting recess (4) for drain valve (1).
- To remove, carefully pull water drain valve (1) from drain tube (3).
- Install water drain valve so that opening is at bottom. Valve must be free of dirt, wax, undercoating etc. Clean if necessary.

C Ambient temperature switch (F38), through 07.95 production

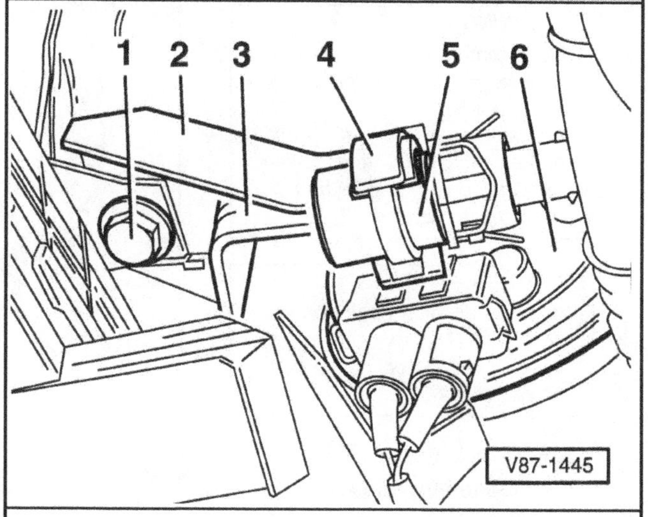

V87-1445

- Ambient temperature switch (5) will slide out of plastic clip (4) to remove.
- Location is driver's side lower front near horn (6).
- Inspect for water damage and replace as necessary.

A/C compressor assembly (4-cylinder gasoline engines)

> ***WARNING —***
> *Replacement of A/C compressor requires discharging/charging the A/C system. Do not discharge/charge the A/C without proper equipment and training, damage to the vehicle and personal injury may result.*

> ***CAUTION —***
> *Immediately plug open connections on A/C components to prevent dirt or moisture contamination.*

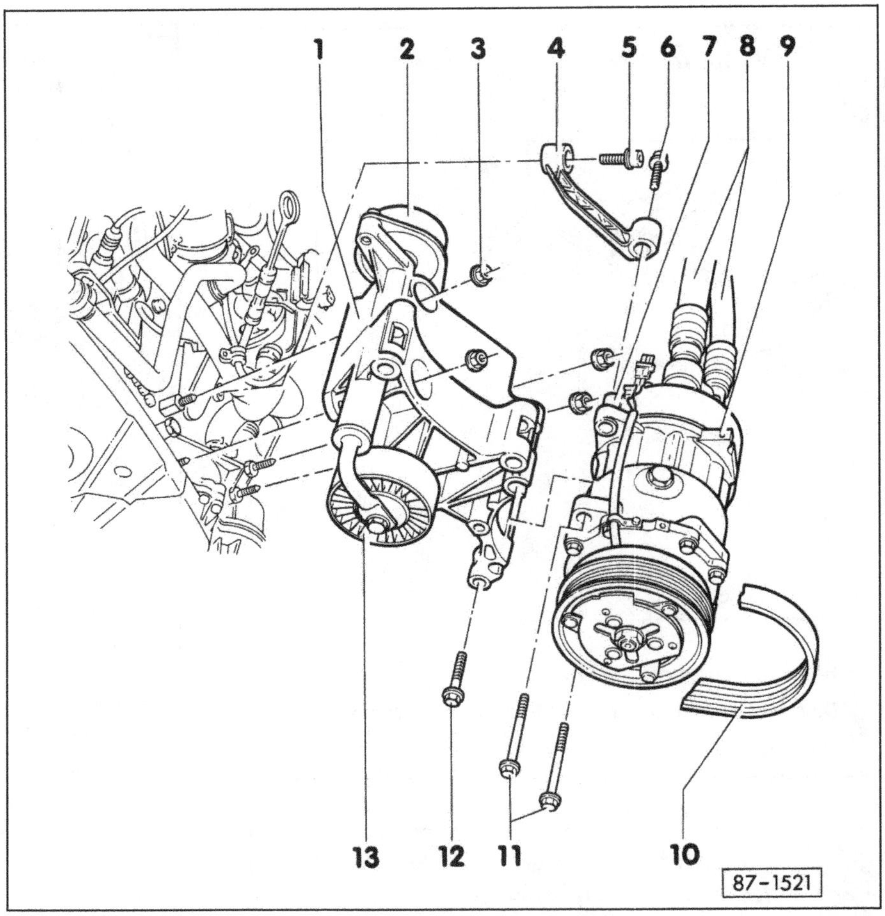

87-1521

Fig. 13. A/C compressor and mounting bracket for 4-cylinder gasoline engine vehicles.

1. **Mounting bracket**
   - For Generator and A/C compressor
   - Generator removing and installing, see **27 Engine Electrical**

> ***NOTE —***
> *The mounting bracket can be removed and installed without discharging the A/C system.*

2. **Spring tensioner**
3. **Nut - M8**
   - Tighten to 30 Nm (22 ft-lb)
4. **Compressor support**
   - Not installed in vehicles from 12.95 production equipped with Sanden A/C compressor
   - Installing:
     Install support using bolts (5) and (6), do not tighten
     Thread bolt (6) into threaded bushing (7) until support contacts bushing
     First tighten bolt (5) to 45 Nm (33 ft-lb) then tighten bolt (6) to 45 Nm (33 ft-lb)
5. **Bolt (M10 x 30)**
   - Tighten to 45 Nm (33 ft-lb)
6. **Bolt (M10 x 30)**
   - Tighten to 45 Nm (33 ft-lb)
7. **Threaded bushing**
8. **Refrigerant hoses**
9. **A/C compressor**
   - Removing:
     Discharge A/C system, see **WARNING** on this page!
     Disconnect A/C lines and harness connector from compressor
     Loosen hex bolts (11) but do not remove
     Remove compressor support (4)
     Press out threaded bushing (7)
     Completely remove hex bolts (11) and remove compressor
10. **Poly-ribbed drive belt**
    - Removing/installing see **0 Maintenance Program**
11. **Hex bolt (M10 x 112)**
    - Tighten to 45 Nm (33 ft-lb)
12. **Hex bolt (M8 x 42)**
    - Tighten to 25 Nm (18 ft-lb)
13. **Tensioner pulley**

A/C compressor, assembly (TDI diesel engine)

> ***WARNING —***
> *Replacement of A/C compressor requires discharging/charging the A/C system. Discharging/charging the A/C system requires special equipment and training. DO NOT discharge/charge the A/C without proper equipment and training, damage to the vehicle and personal injury may result.*

> ***CAUTION —***
> *Immediately plug open connections on A/C components to prevent dirt or moisture contamination.*

1. **Bolt (M8 x 22)**
   - Tighten to 25 Nm (18 ft-lb)
2. **Mounting bracket**
   - For alternator and A/C compressor
3. **Bolt (M8 x 85)**
   - Tighten to 25 Nm (18 ft-lb)
4. **Alternator (generator)**
   - Remove alternator, see **27 Engine Electrical**, before removing A/C compressor
5. **Tensioning element**
6. **Socket-head bolt (M8 x 22)**
   - Quantity: 3
   - Tighten to 25 Nm (18 ft-lb)
7. **Nut (M8)**
   - Quantity: 4
   - Tighten to 25 Nm (18 ft-lb)
8. **Bolt (M8 x 85)**
   - Tighten to 25 Nm (18 ft-lb)
9. **Compressor**
   - Removing:
     Loosen hex-head bolts (item 11) approximately two turns and drive threaded sleeve back on compressor
     Remove bolts
10. **Ribbed belt**
11. **Bolt (M10 x 112)**
    - Quantity: 2
    - Tighten to 45 Nm (33 ft-lb)
12. **Belt tensioner**

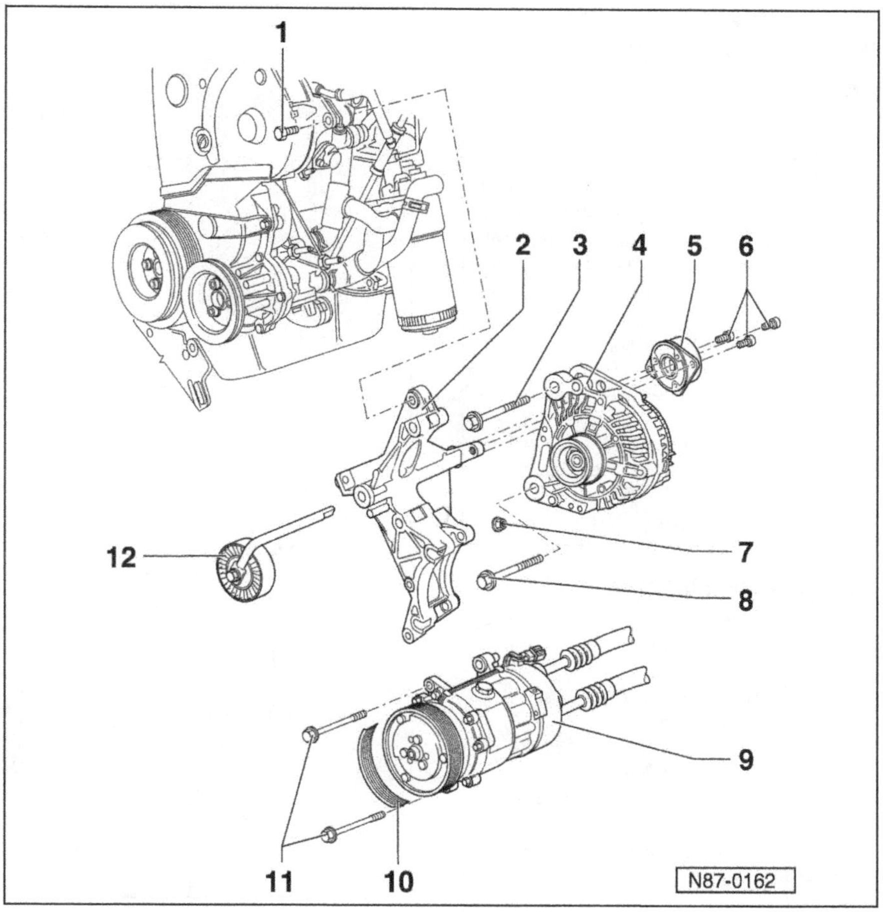

N87-0162

Fig. 14. A/C compressor and mounting bracket for 4-cylinder TDI engine vehicles.

A/C compressor assembly (6-cylinder gasoline engines)

> ***WARNING —***
> *Replacement of A/C compressor requires discharging/charging the A/C system. Do not discharge/charge the A/C without proper equipment and training, damage to the vehicle and personal injury may result.*

> ***CAUTION —***
> *Immediately plug open connections on A/C components to prevent dirt or moisture contamination.*

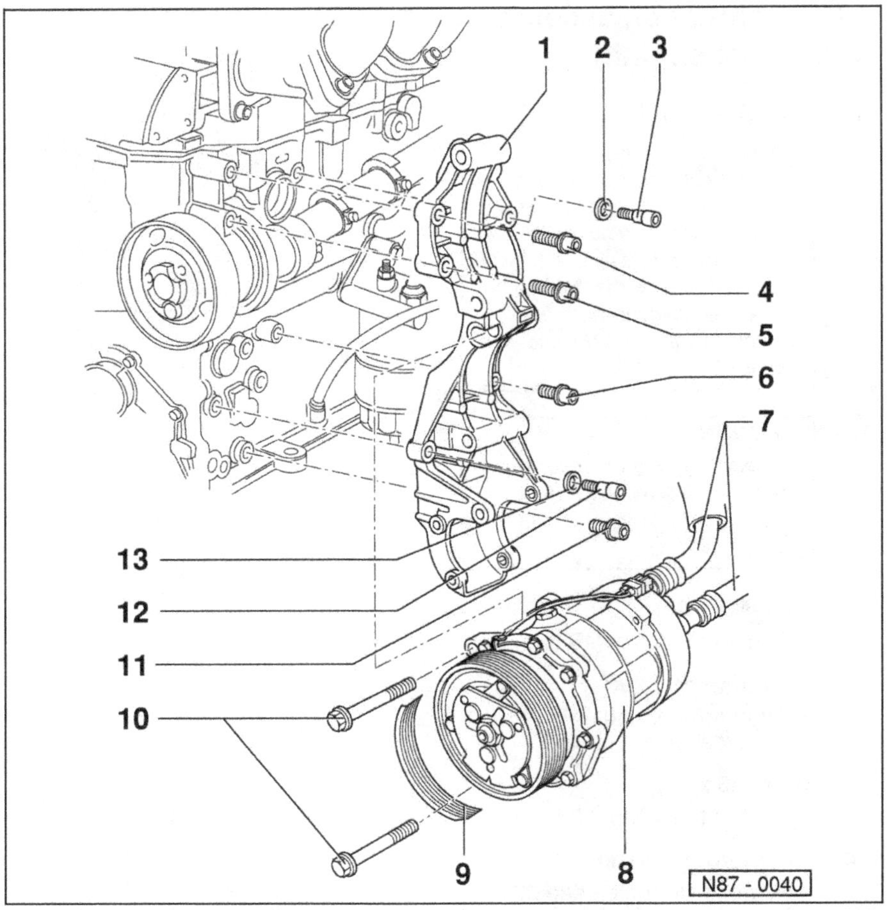

N87-0040

Fig. 15. A/C compressor and mounting bracket for 6-cylinder VR-6 gasoline engine vehicles.

1. **Mounting bracket**
   - For alternator, A/C compressor, and power steering pump
   - Removing:
     Remove A/C compressor (8) from bracket
     Remove alternator, see **27 Engine Electrical**
     Remove power steering pump
   - Installing
     First install fitted bolts (3) and (12)

> ***NOTE —***
> *The mounting bracket can be removed/installed without discharging the A/C system.*

2. **Washer (8.4 x 16 x 1.6 mm)**
3. **Fitted bolt (M8 x 28)**
   - 25 Nm (18 ft-lb)
4. **Socket-head bolt (M8 x 30)**
   - 25 Nm (18 ft-lb)
5. **Socket-head bolt (M8 x 30)**
   - 25 Nm (18 ft-lb)
6. **Socket-head bolt (M8 x 38)**
   - 25 Nm (18 ft-lb)
7. **Refrigerant hoses**
8. **A/C compressor**
   - Removing:
     Discharge the A/C system, see **WARNING** on previous page
     Remove front bumper, see **63 Bumpers**
     Loosen lock carrier (radiator support) and pull to front, see **50 Body–Front**
     Remove poly-ribbed drive belt (9)
     Remove Alternator, see **27 Engine Electrical**
     Disconnect A/C lines and harness connector from compressor
     Loosen A/C compressor bolts (item 10) approx. 2 turns and drive back threaded tensioning sleeves from compressor
     Remove bolts (10) and remove compressor
9. **Poly-ribbed drive belt**
   - Removing and installing, see **0 Maintenance Program**
10. **Hex bolts (M10 x 112)**
    - Tighten to 45 Nm (33 ft-lb)
11. **Socket-head bolt**
    - Tighten to 25 Nm (18 ft-lb)
12. **Fitted bolt (M8 x 20)**
    - Tighten to 25 Nm (18 ft-lb)
13. **Washer (8.4 x 16 x 1.6 mm)**

9 Electrical System

GENERAL

This general information group covers a brief description of the principal parts of the electrical system. Also covered here is general electrical system troubleshooting and instructions on using Volkswagen wiring diagrams.

Voltage and polarity

Volkswagen electrical systems are 12-volt direct current (DC) negative-ground systems. A voltage regulator controls the output of the alternator to approximately 13.5 volts. All circuits are grounded by direct or indirect connection to the negative (–) terminal of the battery. A number of ground connections throughout the car connect the wiring harness to chassis ground. These circuits are completed by the battery cable or ground strap between the body and the battery negative (–) terminal.

Electrical system safety precautions

Please read the following warnings and cautions before doing any work on your electrical system.

> ***WARNING —***
> - *Ignition systems operate in a dangerous voltage range that could prove to be fatal if exposed terminals or live parts are contacted. Use extreme caution when working on a vehicle with the ignition on or the engine running.*
> - *On cars equipped with Airbags, special precautions apply to any electrical system testing or repair. The airbag unit is an explosive device and must be handled with extreme care. Before starting any work on an airbag equipped car, refer to the warnings and cautions in* **69 Seatbelts, Airbags**.

> ***WARNING —***
> *Before operating the starter without starting the engine (as when making a compression test), disable ignition system as described in* **28 Ignition System**.

> ***CAUTION —***
> - *Always switch the ignition off and remove the negative (–) battery cable before removing any electrical components.*
> - *Before disconnecting battery be sure to obtain radio anti-theft code.*
> - *Connect and disconnect ignition system wires, multiple connectors, and ignition test equipment leads only while the ignition is switched off.*
> - *Do not disconnect battery while the engine is running. Never reverse the battery terminal connections. Even a momentary wrong connection can damage the alternator or electrical components. If the polarity markings on the battery are not visible, confirm the polarity of battery using a voltmeter.*
> - *Always remove the battery cables before quick-charging the battery. Never use a quick-charger as a booster for starting the car. Do not exceed 16.5 volts at the battery.*
> - *Many solid-state modules operate on very low current and can be permanently damaged if exposed to static discharge. Always handle the modules using proper static prevention equipment and techniques.*
> - *Always switch a test meter to the appropriate function and range before making test connections.*
> - *Disconnect the battery before doing any electric welding on the car.*
> - *Do not wash the engine while it is running, or anytime the ignition is switched on.*
> - *Do not try to start the engine of a car which has been heated above 176°F (80°C), (for example, in a paint drying booth) until allowing it to cool to normal temperature.*

Electrical test equipment

Many of the electrical tests described in this manual call for measuring voltage, current or resistance using a digital multimeter (DMM). DMMs are preferred for precise measurements and for electronics work because they are generally more accurate than analog meters. The DMM is also safe for most solid state components whereas an analog meter can damage some components.

An LED test light is a safe, inexpensive tool that can be used to perform many simple electrical tests that would otherwise require a multimeter. The LED indicates when voltage is present between any two test-points in a circuit.

> ***CAUTION —***
> - *Choose test equipment carefully. Use a meter with at least 10 megohm input impedance, or an LED test light. An analog meter (swing-needle) or a test light with a normal incandescent bulb may draw enough current to damage sensitive electronic components.*
> - *An analog meter must not be used to measure resistance on solid state components such as control units or time delay relays.*
> - *Always disconnect the battery before making resistance (ohm) measurements on the circuit.*

Wiring diagrams, fuses and relays

Nearly all parts of the wiring harness connect to components of the electrical system with keyed, push-on connectors that lock into place. Notable exceptions are the heavy battery cables and the alternator wiring.

With the exception of the battery charging system, all electrical power is routed from the ignition switch or the battery through the fuse panel, located in the passenger compartment behind the driver's left side knee bar. Fuses prevent excessive current from damaging components and wiring. Fuses are color coded to indicate their different current capacities. Most relays are electro-mechanical switches that operate on low current to switch a high-current circuit on and off.

The wiring diagrams shown in **97 Wiring Diagrams, Fuses and Relays** are organized according to model year and engine type, with complete diagrams for each year.

ELECTRICAL TROUBLESHOOTING

Four things are required for current to flow in any electrical circuit: a voltage source, wires or connections to transport the voltage, a consumer or device that uses the electricity, and a connection to ground or a return to the voltage source. Most problems can be found using only a digital multimeter (volt/ohm/amp meter) to check for voltage supply, for breaks in the wiring (infinite resistance/no continuity), or for a path to ground that completes the circuit.

Electric current is logical in its flow, always moving from the voltage source toward ground. Keeping this in mind, electrical faults can be located through a process of elimination. When troubleshooting a complex circuit, separate the circuit into smaller parts. Be sure to analyze the problem. Use the wiring diagrams to determine the most likely cause of the problem. Get an understanding of how the circuit works by following the circuit from ground back to the power source.

> ***CAUTION —***
> *When making test connections at connectors and components, use care to avoid spreading or damaging the connectors or terminals. Some electrical tests may require jumper wires to bypass components. When connecting jumper wires, use blade connectors at the wire ends that match the size of the terminal being tested. The small internal contacts are easily spread apart, and this can cause intermittent or faulty connections that can lead to more problems.*

Checking for voltage and ground

Checking for the presence of voltage or ground is usually the first step in troubleshooting a problem circuit. For example, if a parking light does not work, a check for voltage at the bulb socket will quickly determine if the circuit is functioning properly or if the bulb itself is faulty. If voltage and ground are found at the socket, then the bulb is most likely faulty.

Another valuable troubleshooting technique is a voltage drop test. This is a good test to make if current is flowing through the circuit, but the circuit is not operating correctly. Sluggish wipers or dim headlights are examples of this. A voltage drop test will help to pinpoint a corroded ground strap or a faulty switch. Normally, there should be less than 1 volt drop across most wires or closed switches. A voltage drop across a connector or short cable should not exceed 0.5 volts.

A voltage drop is caused by higher than normal resistance in a circuit. This additional resistance actually decreases or stops the flow of current. Some common sources of voltage drops are faulty wires or switches, dirty or corroded connections or contacts, and loose or corroded ground wires and ground connections.

A voltage drop can be checked only when current is flowing through the circuit, such as by operating the starter motor or turning on the headlights. Making a voltage drop test requires measuring the voltage in the circuit and comparing it to what the voltage should be. Since these measurements are usually small, a digital voltmeter should be used to ensure accurate readings. If a voltage drop is suspected, turn the circuit on and measure the voltage at the circuit's load.

NOTE —

- *A voltage drop test is generally more accurate than a simple resistance check because the resistances involved are often too small to measure with most ohmmeters. For example, a resistance as small as 0.02 ohms would result in a 3 Volt drop in a typical 150 amp starter circuit (150 amps x 0.02 ohms = 3 volts).*
- *Keep in mind that voltage with the key on and voltage with the engine running are not the same. With the ignition on and the engine off (battery voltage), voltage should be approximately 12.6 volts. With the engine running (charging voltage), voltage should be approximately 14.0 volts. Measure voltage at the battery with the ignition on and then with the engine running to get exact measurements.*

To measure voltage

1. Set voltage meter to 20V DC scale and connect the negative lead to a reliable ground point on the car.
2. Connect the voltmeter positive lead to the point in the circuit you wish to measure. See Fig. 1.
3. If a reading is obtained, there is voltage at that point in the circuit. The voltage reading should not deviate more than 1 volt from the voltage at the battery. If the voltage is less than this, there is probably a fault in the circuit, such as a corroded connector or a loose ground wire.

To check for voltage drop

1. Connect the digital voltmeter positive lead to positive (+) connector of component to be tested.
2. Connect the voltmeter negative lead to the negative (–) connector of the component being tested. See Fig. 2.

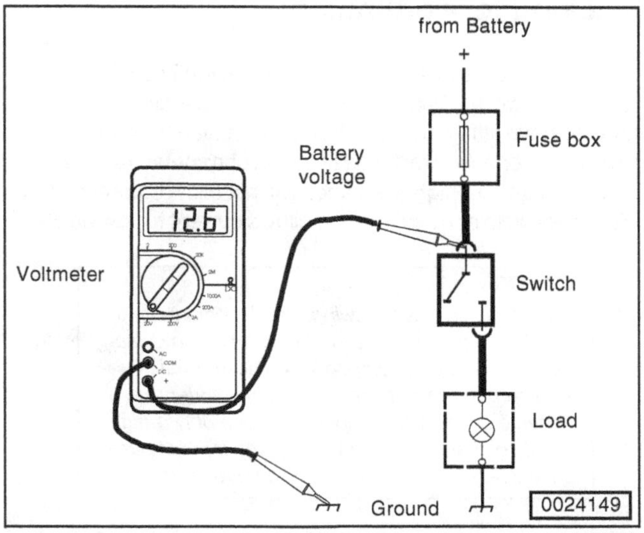

0024149

Fig. 1. Voltmeter being used to check for voltage.

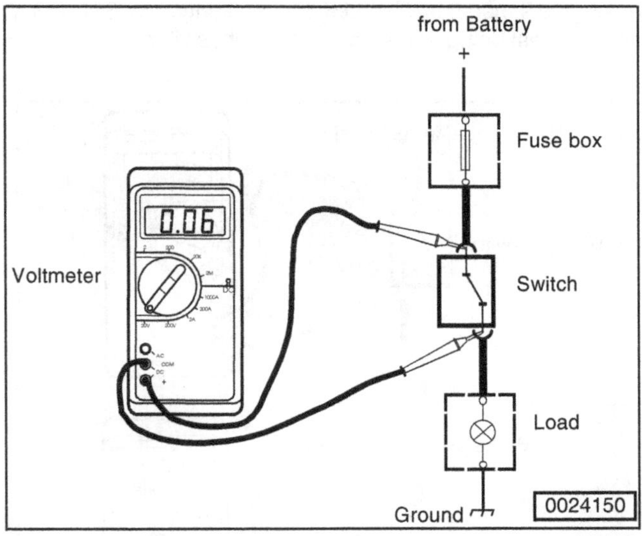

0024150

Fig. 2. Voltmeter being used to check for voltage drop across switch.

3. With the power on and the circuit working, the meter shows the voltage drop (the difference between the two points). This value should not exceed 1 volt.

NOTE —

The maximum voltage drop in an automotive circuit, as recommended by the Society of Automotive Engineers (SAE), is as follows: 0 Volts for small wire connections; 0.1 Volts for high current connections; 0.2 Volts for high current cables; and 0.3 Volts for switch or solenoid contacts. On longer wires or cables, the drop may be slightly higher. In any case, a voltage drop of more than 1.0 Volt usually indicates a problem.

Checking for continuity

The continuity test can be used to check the basic integrity of a circuit or switch. Because most automotive circuits are designed to have little or no resistance, a circuit or part of a circuit can be easily checked for faults using an ohmmeter. An open circuit or a circuit with high resistance will not allow current to flow. A circuit with little or no resistance allows current to flow easily.

> **CAUTION —**
> *Do not use an analog (swing-needle) ohmmeter to check circuit resistance or continuity on any electronic (solid-state) components. The internal power source used in most analog meters can damage solid state components. Use only a high quality digital ohmmeter having high input impedance when checking electronic components.*

When checking continuity, the ignition should be off. On circuits that are powered at all times, the battery should be disconnected. Using the appropriate wiring diagram, a circuit can be easily tested for faulty connections, wires, switches, relays, and engine sensors by checking for continuity. Fig. 3 shows a continuity test being made on a brake light switch.

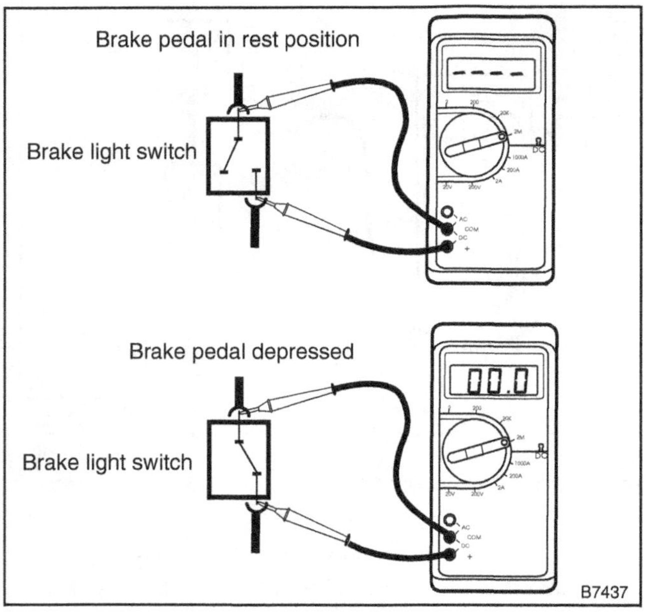

B7437

Fig. 3. Brake light switch being tested for continuity (battery disconnected). With brake pedal in rest position (switch open) there is no continuity (infinite ohms). With the pedal depressed (switch closed) there is continuity (zero ohms).

Checking for short circuits

A short circuit is exactly what the name implies. The circuit takes a shorter path than it was designed to take. The most common short that causes problems is a short to ground where the insulation on a positive (+) wire wears away and the metal wire is exposed. When the wire rubs against a metal part of the car or other ground source, the circuit is shorted to ground. If the exposed wire is live (positive battery voltage), the direct current flow to ground will blow a fuse or damage an unfused circuit.

> **CAUTION —**
> - *On circuits protected with large fuses (25 amp and greater), the wires or circuit components may be damaged before the fuse blows. Always check for damage before replacing fuses of this rating.*
> - *When replacing blown fuses, use only fuses having the correct rating. Always confirm the correct fuse rating printed on the fuse panel cover.*

To check for short circuit with voltmeter

1. Remove the blown fuse from the circuit.

2. Disconnect the harness connector from the circuit's load or consumer.

3. Using a voltmeter, connect the test leads across the fuse terminals. See Fig. 4. Make sure power is present in the circuit. If necessary, turn the key on.

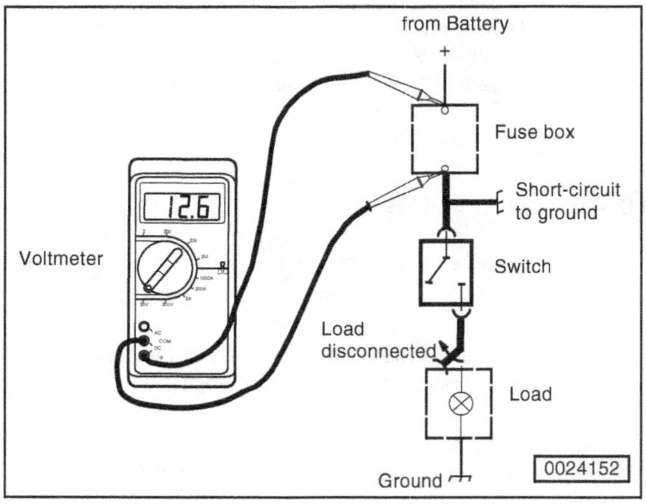

0024152

Fig. 4. Voltmeter being used to find short circuit.

4. If voltage is indicated at the voltmeter, there is a short to ground somewhere in the circuit.

5. If voltage is not indicated, work from the wire harness nearest to the fuse panel and move or wiggle the wires while observing the meter. Continue to move down the harness until the meter displays a reading. This is the location of the short to ground.

6. Visually inspect the wire harness at this point for any faults. If no faults are visible, carefully slice open the harness cover or the wire insulation for further inspection. Repair any faults found.

90 Instruments

GENERAL

An electronic speedometer with an LCD digital odometer display is used on all models. The display will switch between trip mileage or elapsed mileage if the odometer reset button is pushed quickly. Trip mileage is reset when the button is pushed and held for more than 2 seconds. The speedometer receives an electronic signal from Hall sender G22 on the transmission. Mileage is permanently held in memory and will be retained if power to the instrument cluster is interrupted.

Jetta GLX and GTI VR-6 models include the multi-function indicator system. Vehicles with the VR6 engine also have a front brake pad wear indicator system. A light will come on in the instrument cluster indicating brake wear. See Fig. 1.

With the exception of the light bulbs used for indicator lights and for illumination, individual components for the instrument cluster are not available from Volkswagen and therefore disassembly of the cluster is not recommended. Remanufactured instrument cluster assemblies are available through the Volkswagen parts department.

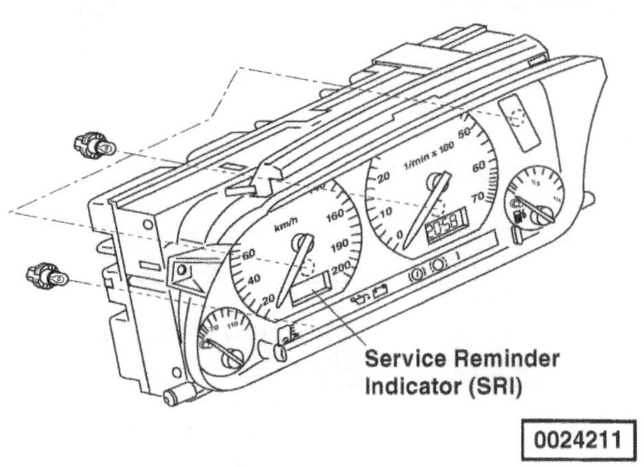

0024211

Fig. 1. Instrument cluster assembly. Service reminder indicator (used on some models) appears in the odometer display at start-up.

INSTRUMENT CLUSTER

Instrument cluster, removing and installing

> ***WARNING —***
> *The illustrations given on the following pages show the steering wheel removed in order to better illustrate the procedure. It is not necessary to remove the steering wheel and not recommended on cars equipped with airbags.*

1. Disconnect (–) negative battery cable.

> ***CAUTION —***
> - *Before working on the electrical system disconnect negative (–) battery cable.*
> - *Before disconnecting the battery, be sure to obtain radio anti-theft code.*

2. Carefully pry off left and right side trim panels. See Fig. 2.

3. Remove light switch by pressing release button on upper left-hand corner of light switch behind trim panel. If necessary, see **96 Lights, Accessories–Interior**.

> ***NOTE —***
> - *Some vehicles may not have left and right trim panels but will have additional switches or warning lamps, depending on vehicle equipment level.*
> - *On vehicles equipped with heated seats, the heater switches are located on the right and left sides of the instrument cluster and must be removed. See* **96 Lights, Accessories–Interior.**

90

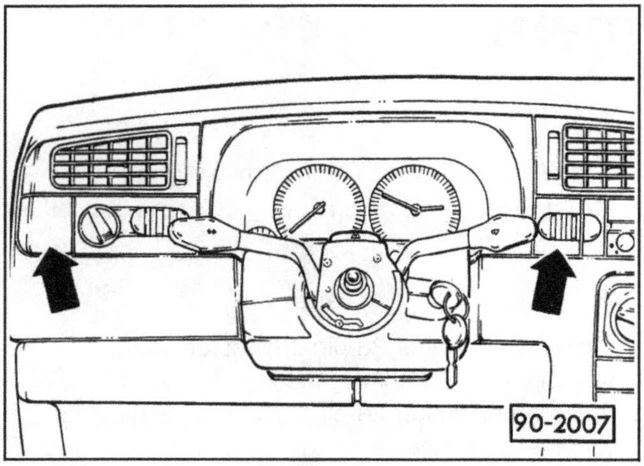

90-2007

Fig. 2. Left and right side trim panels (**arrows**).

4. Remove upper part of steering column trim. See Fig. 3.

> ***WARNING —***
> - *DO NOT remove steering wheel. Steering wheel is shown removed for clarity of illustrations.*
> - *The cars covered by this manual use an airbag which is inflated by an explosive device. Handled improperly or without adequate safeguards, the system can be very dangerous. Special precautions must be observed prior to any work at or near the steering wheel or steering column. See* ***69 Seatbelts, Airbags****.*

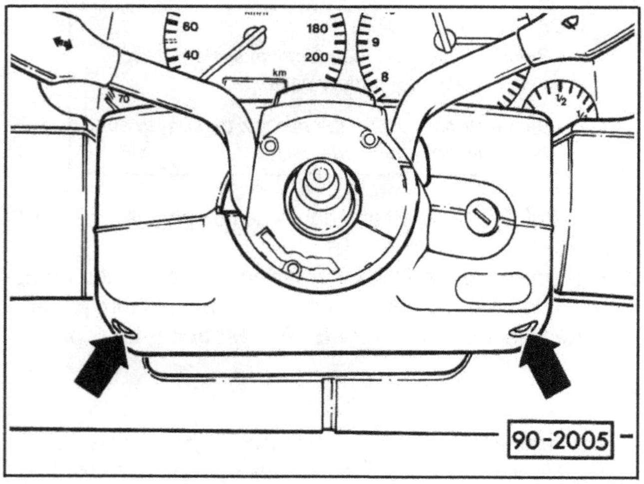

90-2005

Fig. 3. Steering column trim retaining screws (**arrows**).

5. Remove two screws securing instrument panel trim. See Fig. 4.

6. Release instrument cluster trim side retainers by pulling firmly at top and remove instrument cluster trim.

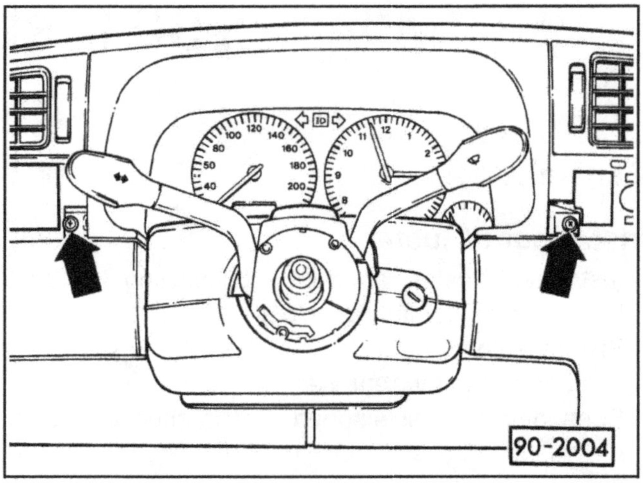

90-2004

Fig. 4. Instrument panel trim retaining screws (**arrows**).

7. Remove mounting screws for instrument cluster. See Fig. 5.

90-2006

Fig. 5. Instrument cluster retaining screws (**arrows**).

8. Tilt instrument cluster downward from the top.

9. Disconnect two multi-point connectors from back of instrument cluster.

10. Remove instrument cluster.

11. Installation is the reverse of removal.

> ***CAUTION —***
> *Be careful to push connectors straight on to avoid bending or damaging pins.*

Service reminder indicator (SRI)

The Service Reminder Indicator (SRI) is used on some vehicles covered by this manual to indicate when vehicle service is necessary. The SRI monitors vehicle mileage and time and appears in the odometer display window below the speedometer. The SRI will only be shown for 3 seconds after the ignition is switched on (engine not running).

When the ignition is switched on, one of the following displays will appear for approximately three seconds indicating the service required. See Fig. 6.

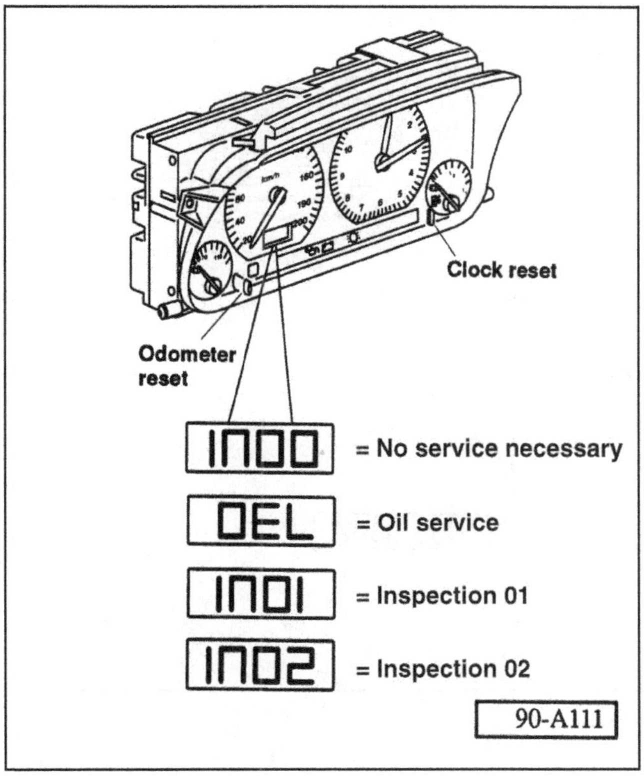

90-A111

Fig. 6. Instrument cluster with service reminder display.

IN00: No service necessary

OEL: 7500 MILE (12,000 km)/6-month engine oil change

IN01: 15,000 mile (24,000 km)/12-month maintenance (inspection service)

IN02: 30,000 mile (48,000 km)/24-month maintenance (inspection service)

After a service has been performed on the vehicle, each affected display must be reset individually. For example, if the 30,000 mile maintenance has just been completed, the OEL, IN 01 and IN 02 displays must be reset.

The SRI uses a hard permanent-type memory and will not be erased if the battery is disconnected.

SRI, resetting

1. Switch ignition ON.
2. Press and hold the odometer reset button located below speedometer.
3. Switch ignition OFF and release odometer reset button. OEL will now appear in SRI display
4. Press and hold the lower digital clock reset button or the analog clock reset button until five dashes (- - - - -) appear in display. OEL SRI is now reset for 7,500 miles (12,000 km) or 6 months, whichever comes first.

Once OEL is reset, the other displays can be reset the same way without switching the ignition on and off again.

5. Press odometer reset button to proceed to IN 01 display (15,000 mile maintenance) and reset using clock reset button.
6. If necessary, repeat procedure to reset IN 02 (30,000 mile maintenance) display.
7. To escape reset mode: Switch ignition ON; when display IN 01 appears, switch ignition OFF.

Instrument cluster, rear view

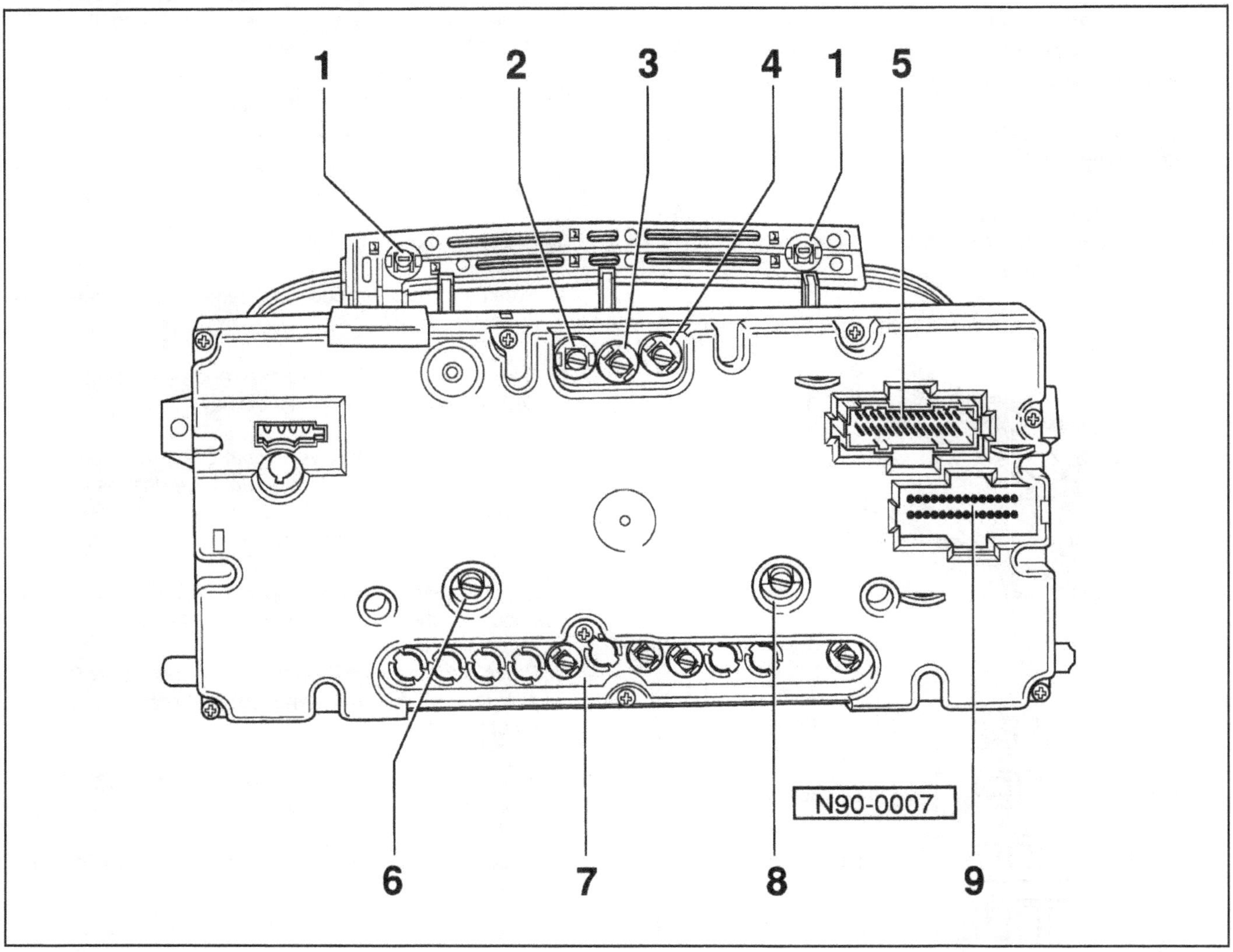

N90-0007

Fig. 7. Rear view of instrument cluster.

1. **Instrument cluster light (L10)**
   - 12V/2W bulb
2. **Right turn signal indicator light (K94)**
   - 12 V/1.1W bulb
3. **Headlight high beam indicator light (K1)**
   - 12V/1.1W bulb
4. **Left turn signal indicator light (K65)**
   - 12 V/1.1W bulb
5. **28-pin connector socket 1 (T28)**
   - Terminal identification, **see table a**
6. **Digital display light (L75)**
   - Illumination for digital clock or multi-function indicator
   - 12V/1.1W bulb
7. **Indicator warning lights**
   - Applications vary depending on vehicle equipment
   - 12V/1.1W bulbs
8. **Illumination for speedometer (G21) and odometer**
   - 12V/1.1W bulb
9. **28-pin connector socket 2 (T28b)**
   - Terminal identification, **see table b**

> ***CAUTION —***
> - *Do not use anything except a digital multi-meter to test instrument cluster or wiring to instrument cluster. Use of test lights, analog meters, etc. may cause expensive damage.*
> - *Individual components for the instrument cluster are not available from Volkswagen and therefore disassembly of the cluster is not recommended.*

Table a. Instrument cluster 28-pin connector (T28) terminal identification

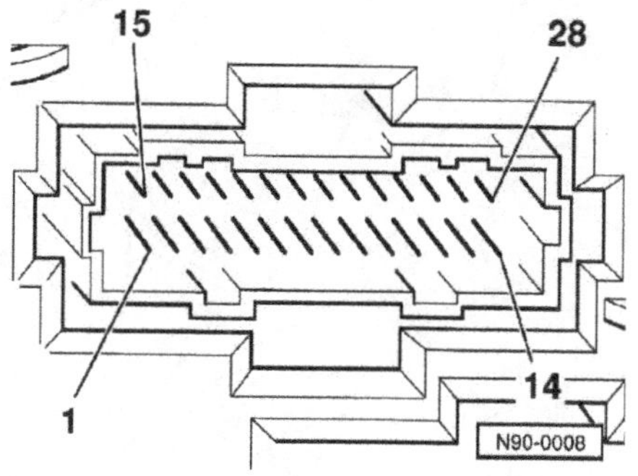

Terminal	Circuit
1	Outside Air Temperature Sensor (G17), ground *
2	Engine Coolant Level (ECL) Sensor (G32)
3	Terminal 31, Ground (GND)
4	Multi-Function Indicator Memory Switch (E109) (reset)*
5	Terminal 31, Ground (GND)
6	Multi-Function Indicator Memory Switch (E109) (memory)*
7	Vehicle Speed Sensor (VSS), output
8	Oil Pressure Switch (F1), 1.8 bar
9	0.3 Bar Oil Pressure Switch (F22)
10	Terminal 1/Terminal W
11	Terminal 30, Battery Positive Voltage (B+)
12	Terminal 58b, lighting
13	Terminal 15 (B+ with ignition switched on)
14	Open
15	Multi-Function Indicator Mode Select Switch (E86)*
16	Generator (GEN) Warning Light (K2), terminal 61
17	Engine Oil Temperature Sensor (G8)*
18	Brake and Parking Brake Warning Light (K7)
19	Outside Air Temperature Sensor (G17), signal*
20	Malfunction Indicator Lamp (MIL) (K83), or Glow Plug Indicator Light (K29) (Diesel)
21	Fuel Gauge (G1)
22	Left Turn Signal Indicator Light (K65)
23	Engine Coolant Temperature (ECT) Gauge (G3)
24	Right Turn Signal Indicator Light (K94)
25	Headlight High Beam Indicator Light (K1)
26	Input signal, fuel consumption*
27	Vehicle speed signal from speedometer Vehicle Speed Sensor (VSS) (G22)
28	Transmission Range (TR) Sel. Lever display (AG4)

(*)Only applicable to cars with multi-function indicator (J119).

Table b. Instrument cluster 28-pin connector (T28b) terminal identification

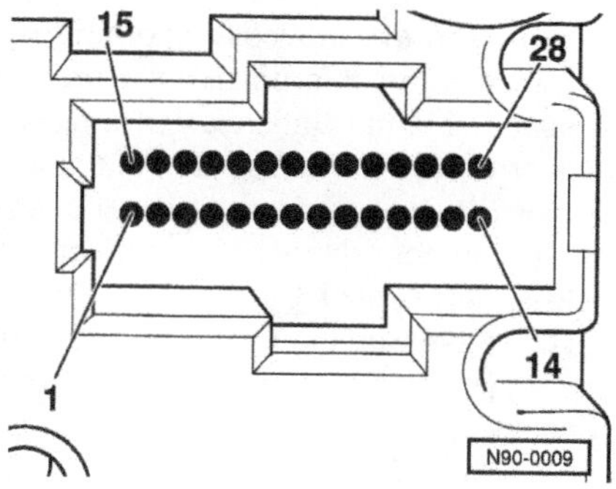

Terminal	Circuit
1	Airbag function check**
2	Airbag control wire **
3	Seat Belt Warning Light (K19)
4	Open
5	Open
6	Rear Hatch/Trunk Lid Ajar Warning Light (K102)
7	Open
8	Open
9	Open
10	Open
11	Open
12	Airbag terminal 15 **
13	Airbag terminal 31 **
14	Open
15	Open
16	Open
17	Open
18	Open
19	Open
20	Open
21	Open
22	Open
23	Open
24	Open
25	Open
26	Open
27	Open
28	Open

(**) Only applicable to vehicles with airbag warning light mounted in dash. Early vehicles have airbag warning light and/or ABS warning light mounted into the left side trim panel. See Fig. 2.

Speedometer vehicle speed sensor, VSS G22, checking

The speedometer is electronically operated by a sensor driven by the transmission differential. This three wire sensor is a Hall-type sensor. Manual transmission vehicles have only the single sensor on the transmission. Automatic transmission vehicles could have as many as three sensors on the transmission. The number depends on the version and may include vehicle speed sensor G68 and transmission speed sensor G38. These two wire inductive sensors send information to the transmission control module, TCM, for use by the automatic transmission only.

1. Disconnect the 3-pin connector from the vehicle speed sensor. See Fig. 8.

2. Switch the ignition on.

3. Connect a multimeter between terminals 1 and 3 (the two outermost terminals). Meter should read between 9 volts and 14.5 volts. If no voltage is present, check appropriate wiring diagram for a break in the wiring.

4. If voltage is present, re-connect the 3-pin connector. Attach one lead of the multimeter to the battery positive terminal. Backprobe the center terminal of the connected 3-pin connector with the other multimeter lead.

5. Raise one front wheel sufficiently to allow wheel to be turned by hand. Slowly turn wheel. Voltage must change between low and high when wheel is rotated. Low voltage is typically around 2.5 volts and high voltage is approximately battery voltage.

6. If voltage is not present, speed sensor is defective. If the low/high voltage signal is present, sensor is good.

7. If sensor is good, but speedometer does not work, test the cruise control (if equipped) as described in **96 Lights, Accessories–Interior.** If the signal for the cruise control is good, speedometer is probably defective.

8. The vehicle speed sensor, VSS G22, can also be tested with the VAG 1551/1552 through the engine control module. These procedures are outside of the scope of this repair manual.

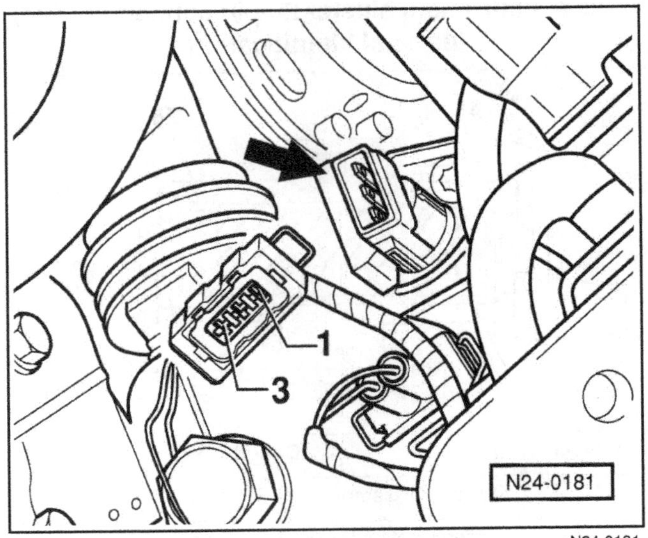

N24-0181

Fig. 8. Disconnected harness connector from the vehicle speed sensor, G22 (**arrow**). Harness connector terminals 1 and 3 supply power and ground.

Modified instrument cluster warning lights (from 01/1996)

The airbag Malfunction Indicator Lamp (MIL) has been relocated from a location near the headlight switch to the instrument cluster with production from 01/1996. Functionality remains the same. See Fig. 9.

A low-fuel indicator warning light has also been added. The light functions as follows:

- ON: Less than1.6 gallons
- OFF: At least 2.4 gallons
- ON or OFF: 1.6 to 2.4 gallons

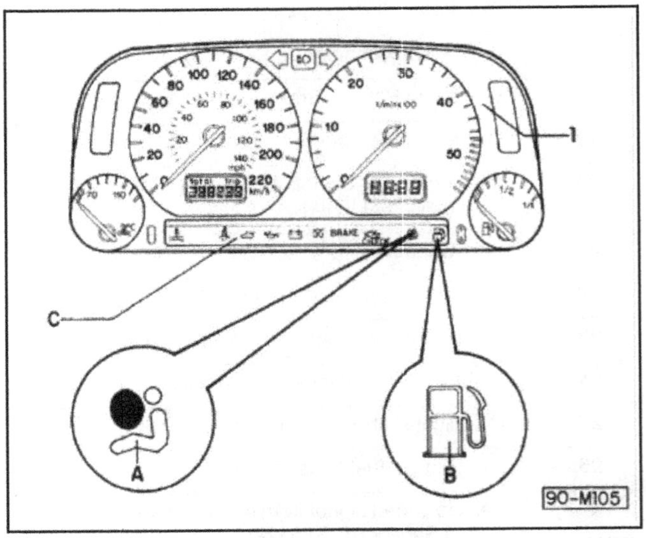

90-M105

Fig. 9. Warning light area (**C**) now contains airbag warning light (**A**) and low fuel warning light (**B**) as shown.

91 Radio

GENERAL

This section covers the Volkswagen installed stereo system, including: radio recoding, radio removal and installation and the antenna assembly. For specific circuit tracing it is necessary to see **97 Wiring Diagrams, Fuses and Relays**. For component layout and location, See Fig. 7.

Volkswagen has offered several different model radios, all featuring ignition switch power shutoff, blinking security LED, and anti-theft coding. See Fig. 1, 2, and 3. Some feature the ability to control an optional CD changer. Four different types of radio antennas are used.

- Roof mounted/amplified/flexible; GLX, GTI, TREK, K2
- Fender mounted/amplified/flexible; Cabrio
- Fender mounted/telescopic mast; Golf, Jetta
- Fender mounted/fixed length mast; Golf, Jetta, Cabrio

Starting with the 1998 model year, Volkswagen radios are capable of communicating with scan tool VAG 1551/1552 through the Data Link Connector (DLC). These radios can store Diagnostic Trouble Codes, DTCs, and must be coded via the scan tool in order to function properly. The correct code varies depending on the type of vehicle, the type of antenna and if equipped with a CD changer.

> ***NOTE —***
> *Radio malfunctions and equipment changes requiring the use of scan tool VAG 1551/1552 should be handled by an authorized Volkswagen dealer or other qualified repair facility. Repairs of this nature fall outside the scope of this repair manual.*

> ***CAUTION —***
> - *The radio is wired to the vehicle alarm system. If the alarm system is armed, removing the radio will activate the alarm even if the proper tools are used. Do not attempt to remove the radio without disarming alarm.*
> - *Do not remove radio or disconnect battery without obtaining anti-theft code.*
> - *The factory installed connectors are designed for Genuine Volkswagen Radios. If installing a different radio, remember that the radio may fit poorly into the space provided, the electrical connections may not be compatible and different terminals may be needed.*
> - *Keep in mind that factory installed radios are electrically linked to the ignition and alarm system circuits in all cars and the data link connector in some cars.*

STEREO SYSTEM

> ***CAUTION —***
> - *Before beginning work on radio or stereo system verify that the radio is an original equipment Volkswagen radio and that the wiring harness has not been modified.*
> - *Aftermarket radios, amplifiers, speakers, etc. are beyond the scope of this manual. Servicing of aftermarket sound equipment should be referred to an authorized agent of the equipment in question.*

Radio, re-entering anti-theft code ("Deluxe/Premium", 1993–1997)

When the power to the radio is interrupted, the radio will electronically lock-up. The word "SAFE" appears in the frequency display when the radio is switched on and it will be necessary to enter the anti-theft code to make the radio operational again.

1. Turn on the radio and press the MODE and SCAN buttons simultaneously and hold them down until "1000" lights up in the display. Release the buttons! See Fig. 1 or Fig. 2.

> ***CAUTION —***
> *Do not continue to hold MODE and SCAN buttons after "1000" appears in display or press them again. This would enter the number "1000" as the input code number.*

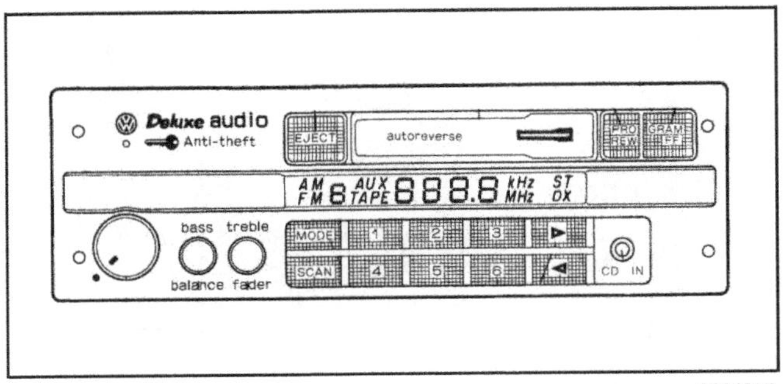

0024207

Fig. 1. Volkswagen installed "Deluxe" sound system. This unit has a front panel input for a portable CD player.

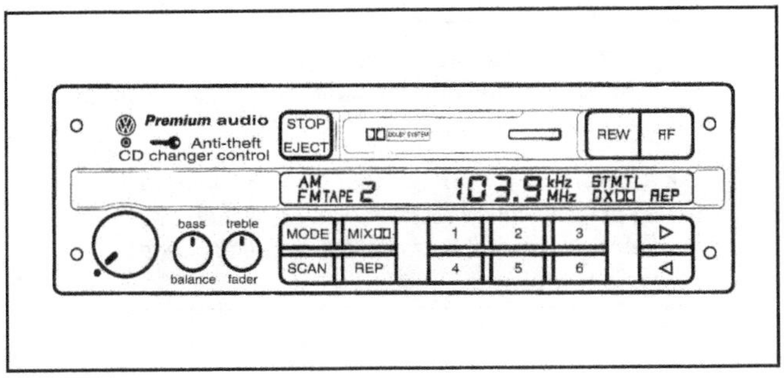

0024208

Fig. 2. Volkswagen installed "Premium" sound system. This unit can control a trunk mounted CD changer. These units are supplied by two different manufacturers (Matsushita and Clarion) and look the identical from the front, but the optional CD changers and cables are not interchangeable.

2. Use station buttons 1 to 4 to input the correct anti-theft code number. Use button 1 to input the first digit, button 2 for the second, and so on.

> ***NOTE —***
> *The first digit of the anti-theft code can only be a 1 or left blank.*

3. Once the anti-theft code is shown in the display, hold the MODE and SCAN buttons down until the word "SAFE" appears in the frequency display. Release buttons! Soon afterward, a radio frequency will be displayed and the radio will be operational.

> ***NOTE —***
> - *Should you inadvertently input the incorrect code number, the word "SAFE" will appear in the display – first blinking and then continuous. It is possible to repeat the coding procedure* **once**.
> - *If the incorrect code is again entered, the radio will be locked up for approximately one hour. Leave the radio switched on. After an hour, the display with the number of attempts goes out and you can again re-enter the anti-theft code as described above. The two attempt, one hour lock-up cycle will continue to apply.*

Radio connector, terminal identification ("Deluxe/Premium", 1993-1997)

The three connectors shown in Fig. 3. supply all of the inputs and outputs used for the Deluxe and Premium radios except for the antenna coaxial input. Connectors II and III are used for all applications, although some pins may not be used. Connector I is only used for CD changer control on Premium units with different pin assignments for Matsushita and Clarion versions.

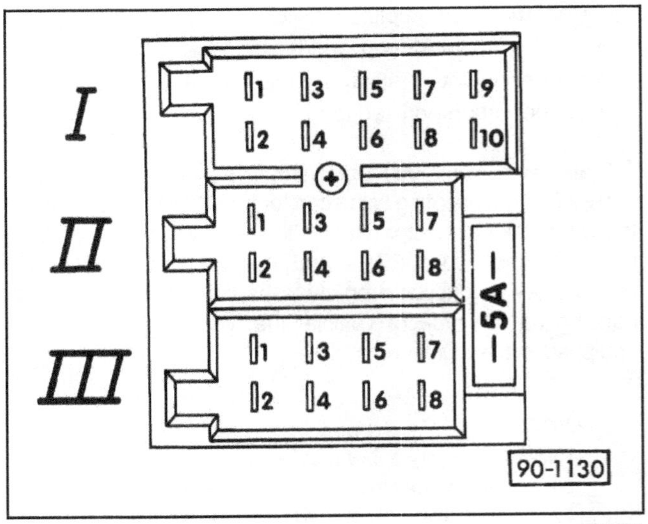

90-1130

Fig. 3. Radio connectors and fuse on rear of Deluxe/Premium radio. Top connector (**I**) is used for optional CD changer control on Premium versions only.

Multi-pin connector II, T8g, 8-pin, brown

1 .... Right rear speaker (+)
2 .... Right rear speaker (-)
3 .... Right front speaker (+)
4 .... Right front speaker (-)
5 .... Left front speaker (+)
6 .... Left front speaker (-)
7 .... Left rear speaker (+)
8 .... Left rear speaker (-)

Multi-pin connector III, T8, 8-pin, black

1 .... to Alarm System Control Module (J284)
2 .... Open
3 .... Switched positive voltage (B+) (for auxiliary amplifier)
4 .... Ignition key on/off connection (from ignition/starter switch, terminals S/SU)
5 .... Switched positive voltage (B+) (for amplified roof antenna)
6 .... Radio illumination, terminal 58b
7 .... Battery Positive Voltage (B+) (terminal 30 from fuse S22)
8 .... Ground (GND), terminal 31

Radio, re-entering anti-theft code, ("Sound System/Monsoon", 1998–2002)

When the power to the radio is interrupted, the radio will electronically lock-up. The word "SAFE" appears in the frequency display when the radio is switched on and it will be necessary to enter the correct anti-theft code to make the radio operational again.

1. Turn on the radio. The word "SAFE" will appear for 3 seconds and then the number "1000" will automatically appear in the display.

2. Use the radio station preset buttons 1 to 4 to enter the correct anti-theft code number. Use button 1 to enter the first digit, button 2 for the second digit, and so on.

NOTE —
The value range for each of the 4 digits is between 0 and 9. The digit will increase by 1 each time you tap the preset button and will return to 0 after 9.

3. After you finish entering the code, press the right side ">" of the SEEK button (Fig. 4., button A) for longer than 2 seconds until you hear an audible signal.

4. If you've entered the code correctly, the display will show "FM1 88.3MHZ" or "Monsoon" (depending on version) and the unit is ready to use again.

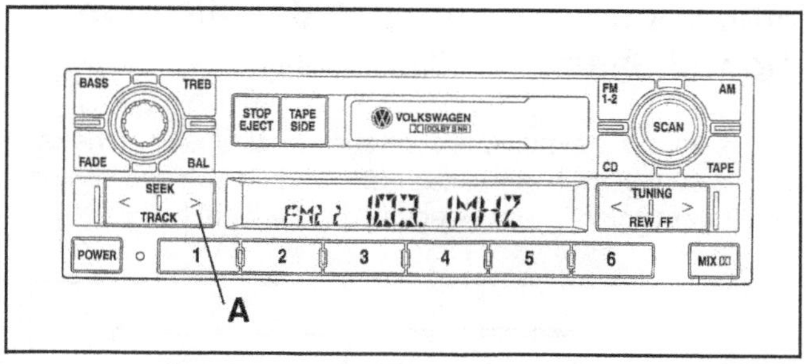

0024209

Fig. 4. Volkswagen installed "Sound System" used from model year 1998. This unit can also control a trunk mounted CD changer, however, the changer is different from those used with other sound system versions. This sound system has built-in diagnostics and is connected to the vehicle Data Link Connector (DLC).

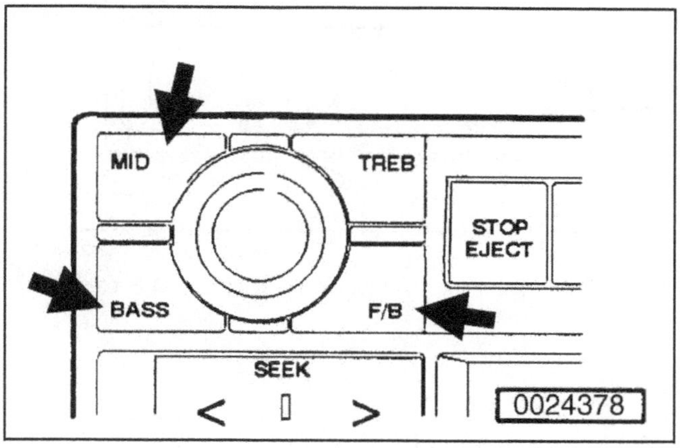

0024378

Fig. 5. Volkswagen installed "Monsoon" system used from model year 1998. This unit is similar to the "Sound System" unit installed as standard and differs primarily in the area of the controls as noted above. It shares most of the same features and functions as "Sound System".

NOTE —

- *Should you inadvertently enter a wrong code number, "SAFE" will appear in the display. Initially it will flash, then stay on. You can repeat the procedure one time. The number of attempts will appear in the display.*
- *If the incorrect code is again entered, the radio will be locked up for approximately one hour. Leave the radio switched on. After an hour, the display with the number of attempts goes out and you can again re-enter the anti-theft code as described above. The two attempt, one hour lock-up cycle will continue to apply.*
- *If you lose your Radio Card, contact your authorized Volkswagen Dealer.*

Radio connector, terminal identification ("Sound System", 1998–2002)

Fig. 6. shows the multi-pin connector on the back of the radio. This connector has five separate parts, each designated by a different color. The connector pins are identified below.

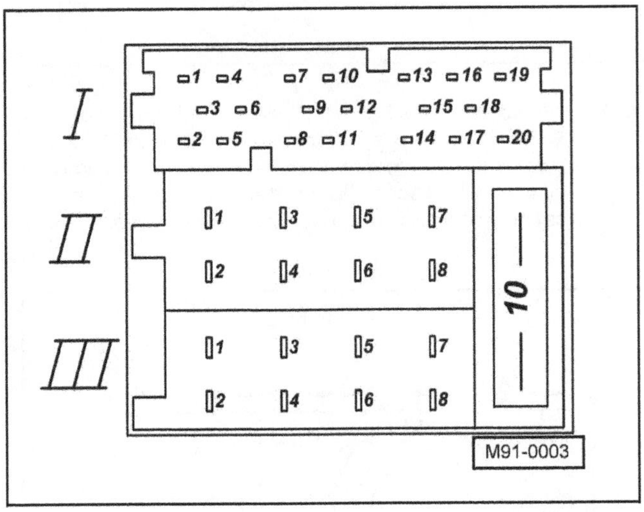

M91-0003

Fig. 6. Radio connectors and fuse on rear of Sound System and Monsoon radios.

Multi-pin connector I, part I, yellow

1	Line out left rear (LR)
2	Line out right rear (RR)
3	Ground (GND)
4	Line out left front (LF)
5	Line out right front (RF)
6	Switched positive (B+ out) for amplifier

Multi-pin connector I, part II, green

7	Telephone input signal (TEL+)
8	Second display (CLOCK)
9	Second display (DATA)
10	Second display (ENA)
11	Remote control (REM)
12	Telephone input signal (TEL)

Multi-pin connector I, part II, blue

13	CD changer (DATA IN)
14	CD changer (DATA OUT)
15	CD changer (CLOCK)
16	CD changer, power supply (+), terminal 30
17	CD changer, control signal
18	CD changer, left and right port, ground (GND)
19	CD changer, left port (CD/L)
20	CD changer, right port (CD/R)

Multi-pin connector II, T8g, 8-pin, brown

1	Right rear speaker (+)
2	Right rear speaker (-)
3	Right front speaker (+)
4	Right front speaker (-)
5	Left front speaker (+)
6	Left front speaker (-)
7	Left rear speaker (+)
8	Left rear speaker (-)

Multi-pin connector III, T8, 8-pin, black

1	Vehicle speed sensor (GALA)
2	Mute (telephone mode)
3	On Board Diagnostic (OBD)/k wire
4	Ignition switch -S- contact
5	Control signal for anti-theft system (SAFE)
6	Radio illumination, terminal 58b
7	Battery Positive Voltage (B+) (terminal 30)
8	Ground (GND), terminal 31

> ***CAUTION —***
> *Do not connect terminal 3 (OBD) of connector III to a power or ground source, or use a Deluxe/Premium radio in a car with a Sound System radio. Damage to the OBD system could result. If installing an aftermarket radio, disconnect the OBD wire at terminal 3 and wrap connector with electrical tape.*

Stereo system, component layout

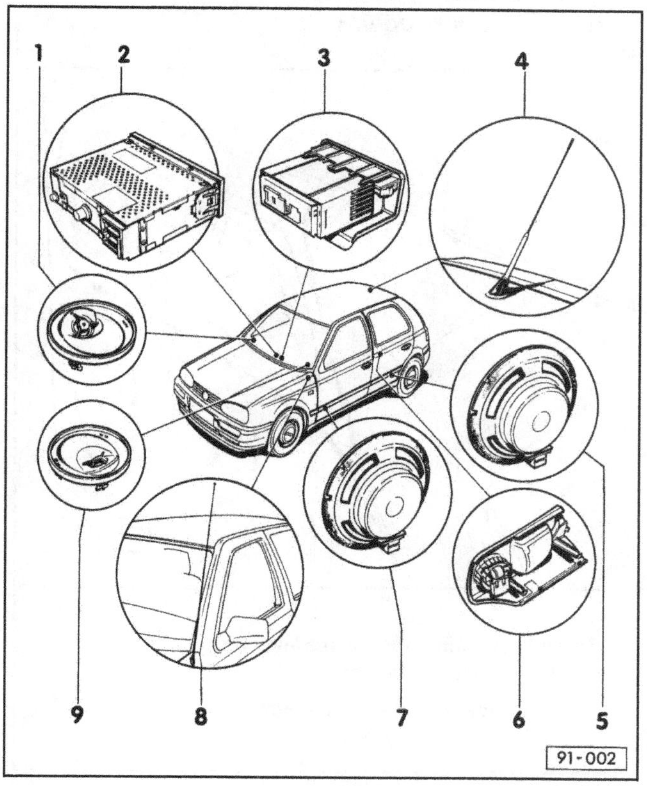

91-002

Fig. 7. Volkswagen stereo components. Speaker locations may vary on 1996 and later Jetta models.

> ***CAUTION —***
> *Before disconnecting battery be sure to obtain radio anti-theft code.*

1. **Right front tweeter (R22)**
   - Dome tweeter
   - Diameter: 100 mm (3.9 in.)
   - Mounted on base in instrument panel
   - Removing and installing, **see** (A)
2. **Stereo radio (Deluxe and Premium)**
   - See **Stereo radio, removing and installing**
3. **Cassette storage (where applicable)**
   - Installed in front of center console
4. **Roof-mounted antenna, amplified**
   - Installed on certain models only; replaces telescoping antenna (8)
   - See **Amplified roof-mounted antenna, assembly**
5. **Left rear woofer (R15)**
   - Diameter: 168 mm (6.6 in.)
   - In rear side (2-door) or door panel (4-door), **see** (B)
   - Relocated to parcel shelf in 1996 m.y. Jettas
6. **Left rear tweeter (R14)**
   - Dome tweeter
   - Connected to rear base speaker
   - Right rear tweeter is identical
   - 2-door: Mounted on rear side panel, **see** (C)
   - 4-door: Mounted near door pull, **see** (D)
   - Relocated to parcel shelf in 1996 m.y. Jettas
7. **Left front woofer (R21)**
   - Diameter: 168 mm (6.6 in.)
   - In front door panel, removing/installing **see** (B)
8. **Telescoping antenna (R11)**
   - Installed on certain models only, where roof-mounted antenna (4) is not used
   - See Telescoping antenna, assembly
9. **Left front tweeter (R20)**
   - Dome tweeter
   - Diameter: 100 mm (3.9 in.)
   - Mounted on base in instrument panel
   - Removing and installing, **see** (A)

(A) Front tweeter, removing and installing

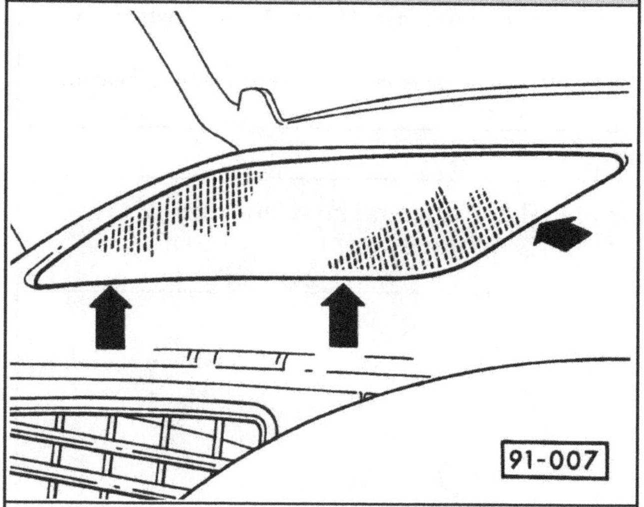

91-007

- **Carefully pry up speaker cover (arrows) with screwdriver.**
- **Pry out tweeter mounting base and remove connector.**
- **Install in reverse order of removal noting installation marks on speaker.**

(B) Removing and installing woofer (except Jetta from 1996 m.y.)

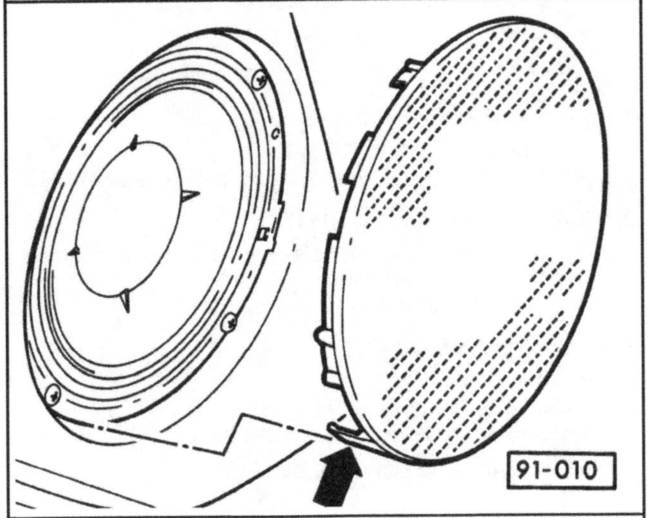

91-010

- Using screwdriver, carefully pry off speaker cover.
- Remove four screws, pull speaker from panel and remove connector.
- Install in reverse order of removal noting installation marks on speaker.
- Install cover; ensure tab (arrow) is properly positioned.

(C) Removing and installing rear tweeter (2-door models)

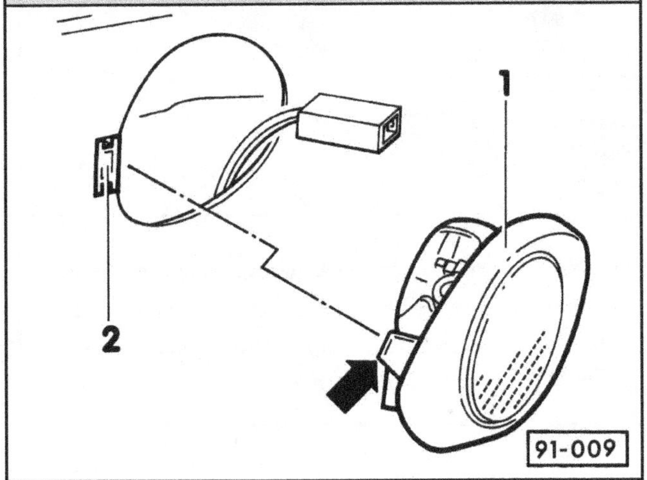

91-009

- Using screwdriver, carefully pry out speaker (1) and remove connector.
- Install in reverse order of removal ensuring clip (2) is positioned properly.
- Install speaker so tab (arrow) fits into clip (2).

(D) Removing and installing rear tweeter (4-door models)

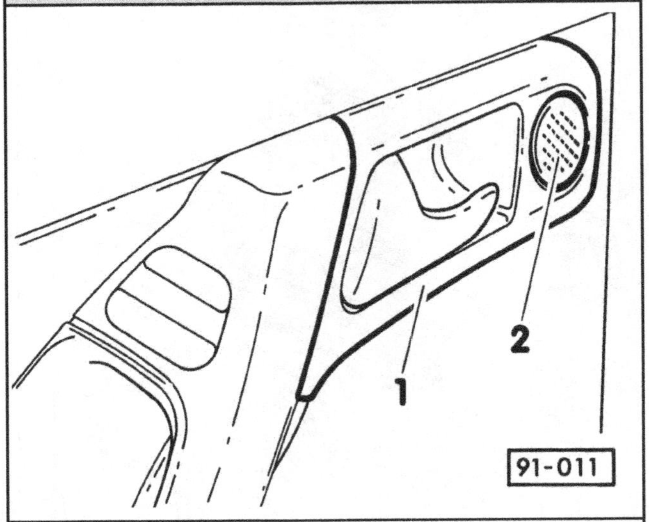

91-011

- Carefully pry out door release lever trim (1) with tweeter (2) and remove connector.
- Install in reverse order of removal.

Stereo radio, removing and installing ("Deluxe/Premium", 1993-1997)

> ***CAUTION —***
> *Before disconnecting battery be sure to obtain radio anti-theft code.*

1. Disconnect negative (–) battery cable.
2. Insert radio removal tools VW 160 into radio. See Fig. 8.

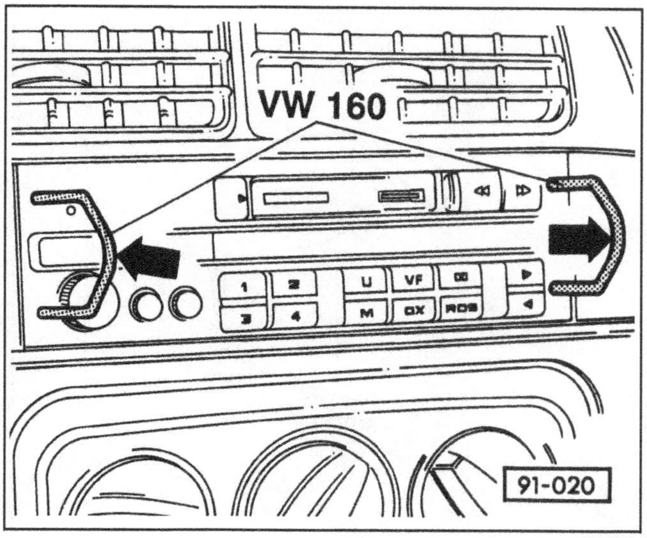

91-020

Fig. 8. Press special removal tools VW160 outward (**arrows**) to remove radio. Take the tools out before reinstalling the radio.

3. Press removal tools outward while carefully pulling the radio from instrument panel.
4. Remove connectors, antenna cable, and remove radio.
5. Install all radio connectors.
6. Install antenna cable and secure to radio. See Fig. 9.
7. With removal clips removed, slide radio into instrument panel until radio locks into place
   - Rubber bushing on rear of radio must contact stop.
   - Radio must lock into place.

> ***CAUTION —***
> *Do not pinch or damage wires when installing radio.*

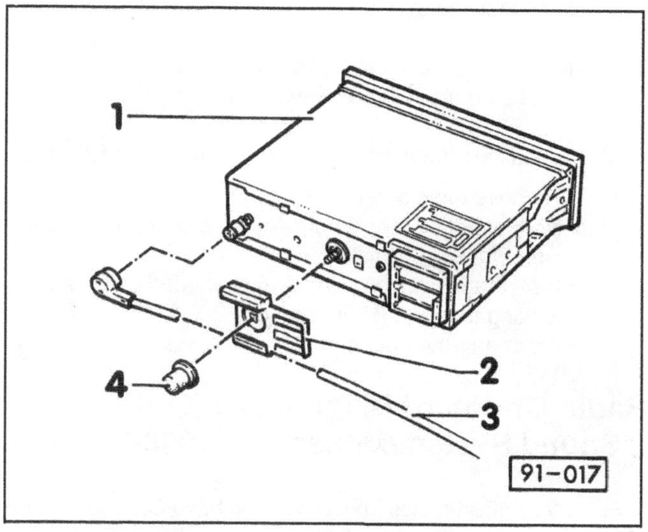

91-017

Fig. 9. Radio (**1**), cable guide (**2**), antenna cable (**3**), rubber bushing (**4**).

Radio, removing and installing ("Sound System/Monsoon", 1998-2002)

> ***CAUTION —***
> *Before disconnecting battery be sure to obtain radio anti-theft code.*

1. Disconnect negative (-) battery cable.
2. Insert radio removal tools (Volkswagen special tool number 3316) into vertical slots in radio face until they are felt to lock into place. See Fig. 10.

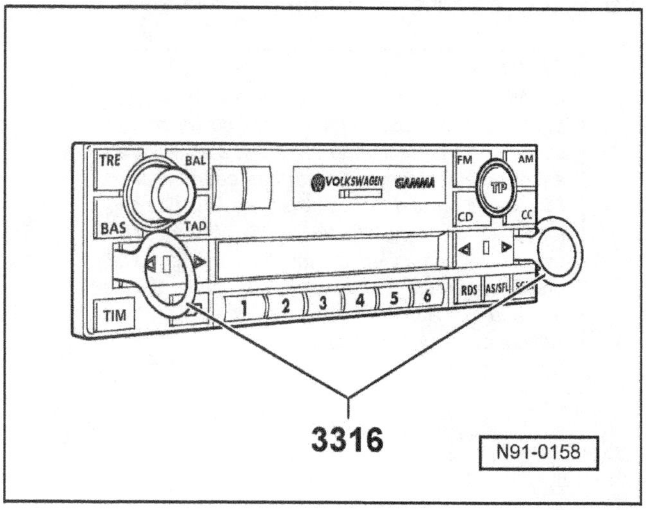

N91-0158

Fig. 10. Radio removal tools 3316 shown in position on Sound System radio. Take the tools out before reinstalling the radio.

3. Gently pull radio straight out from opening in instrument panel using loops in the tools as handles.

4. Disconnect connectors and antenna from rear of radio.

5. Remove tools from radio by pressing the release locking tabs on side of radio before re-installing.

6. Installation is the reverse of removal, noting the following:
   - Remove tools before installing
   - Reconnect electrical connectors and antenna wire. See Fig. 9., given earlier.
   - Insert radio into opening and push straight in until locking tabs are felt to engage
   - Enter anti-theft code

Radio On-Board Diagnostics (OBD) ("Sound System/Monsoon", 1998-2002)

Factory installed radios from m.y. 1998 have On-Board Diagnostics. This includes the ability to store Diagnostic Trouble Codes (DTCs) and the requirement for coding. Diagnostics require a scan tool for access, see **ST Scan Tool**.

Vehicles with OBD radios are wired into the vehicle in the same manner as non-OBD radios with one exception, which is not always shown in the wiring diagrams. If the vehicle in question has one of the above-noted radios, there will be a gray wire with a white tracer inserted into the black T8/3 connector on the back of the radio. This wire connects to the data bus junction located above the pedal cluster near the fuse/relay panel. The sole purpose of this wire is diagnostics. If an alternate radio is installed which does not have this diagnostic capability, this wire must be left disconnected and insulated from contact with all other wires, positive and negative. If this data wire comes into contact with a ground or positive source, vehicle OBD may be affected. In addition, vehicle systems may be influenced to the extent that driveability problems may be encountered, such as engine stalling and hard transmission shfts.

> ***CAUTION—***
> *DO NOT attempt to connect any scan tool, such as the VAG 1551/1552, which is not equipped with over-current protection to any vehicle in which an aftermarket accessory (radio) has been installed. Check beforehand through the Data Link Connector (DLC) by using the VAS 5051/5052 or other over-current protected Diagnostic Scan Tool. Failure to observe this precaution can cause severe damage to V.A.G.1551/1552 and other scan tools without over-current protection!*

Fender-mounted antennas, assembly

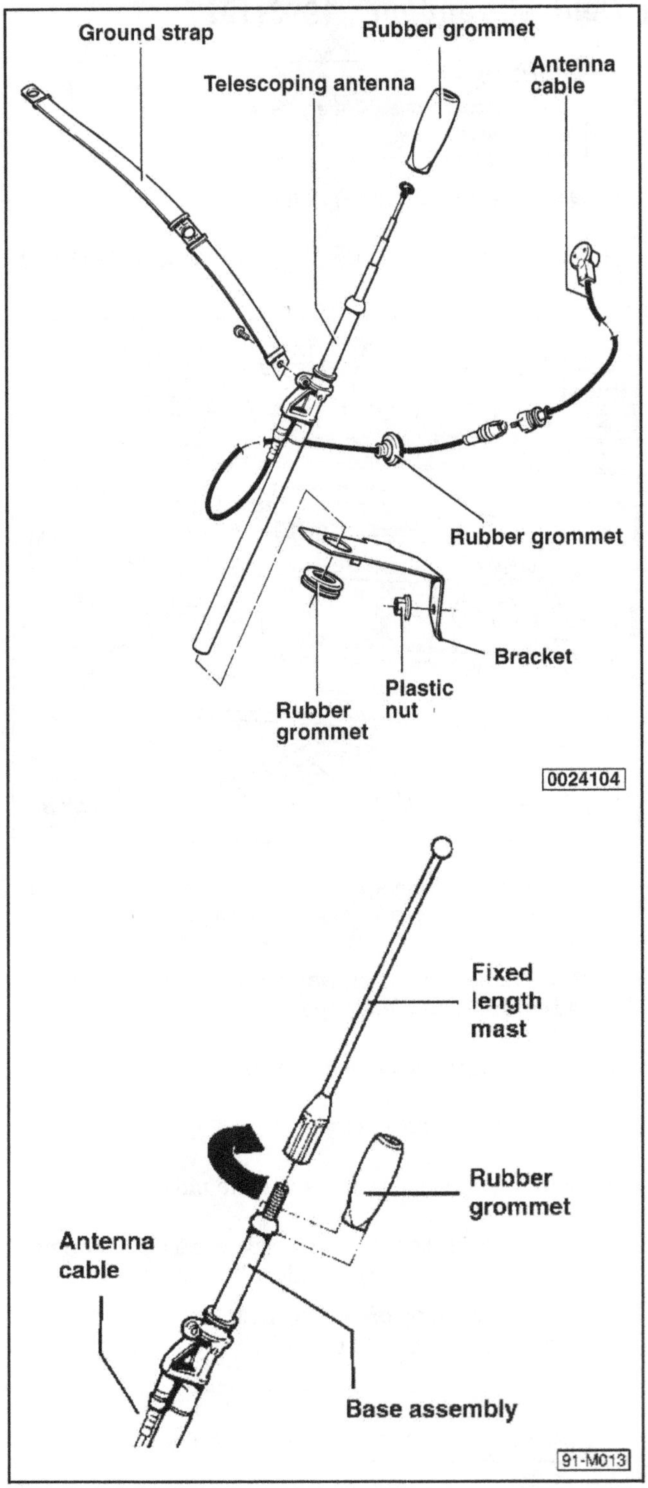

0024104, 91-M013

Fig. 11. Telescoping and fixed length antennas are mounted on the left front fender. Both versions mount in the same manner using the same brackets. Version used depends on model and model year. Fixed length mast is threaded onto base assembly (**arrow**) and in some cases may be a flexible rubber mast.

Fender-mounted antennas, removing and installing

Radio antennas for Golfs and Jettas can be mounted on either the left front fender or the roof. Fender-mounted telescoping antennas are used on Golf and Jetta through 10/1995 production, with fixed length versions installed from 11/1995. Cabrios use a fixed length antenna. Masts can be stainless steel, matte black, or flexible rubber.

All antenna types mounted on the left front fender are removed and installed in a similar manner. See Fig. 11.

1. Remove antenna ground strap from hood hinge and fender. See Fig. 12.

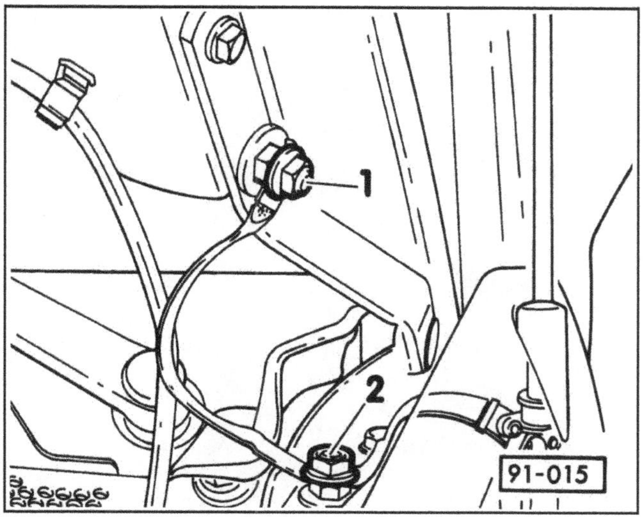

91-015

Fig. 12. Antenna ground strap on hood (**1**) and left wheel housing (**2**).

2. Disconnect antenna cable coaxial connector from behind lower left A-pillar side trim (where equipped) or remove radio and unplug coaxial connector from rear of radio.
3. Pull antenna cable into air intake plenum.
4. Remove left wheel housing liner as described in **66 Body Accessories–Exterior**.
5. Completely collapse telescoping antenna or unscrew fixed antenna mast.
6. Working from underneath vehicle, remove antenna mounting bracket. See Fig. 13.
7. Pull antenna downwards and remove fender grommet.
8. Install in reverse order of removal.
9. Hold fender grommet in position and push antenna into grommet from below.

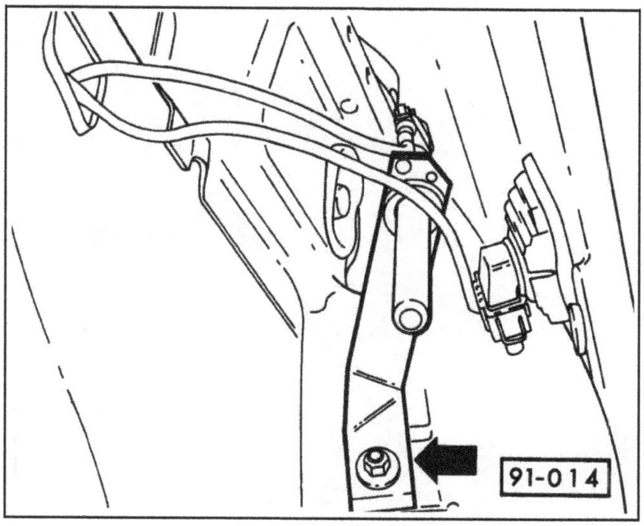

91-014

Fig. 13. Remove bracket (**arrow**) from antenna.

Amplified roof-mounted antenna, assembly

All of the radios listed in this section supply power to the amplified antenna through the center coaxial cable wire that supplies the signal back to the radio. This is known as a "Phantom" power supply. For this reason, a separate 12 volt wire is not needed for the amplifier in the antenna. Fig. 14. shows the amplified roof-mounted antenna assembly.

NOTE —

- *The majority of the electrical consumers/components in the vehicle are noise suppressed to ensure minimal electrical interference with radio reception. On vehicles equipped from the factory with a radio or radio prep wiring, the following components are also suppressed: coolant fan, windshield wiper motor, and rear window wiper motor.*
- *In addition, vehicles with a radio or radio prep wiring will have the following additional ground connections: ground strap from left wheel housing to hood and from ignition coil to engine cylinder block (vehicles with gasoline spark-ignition engine only).*

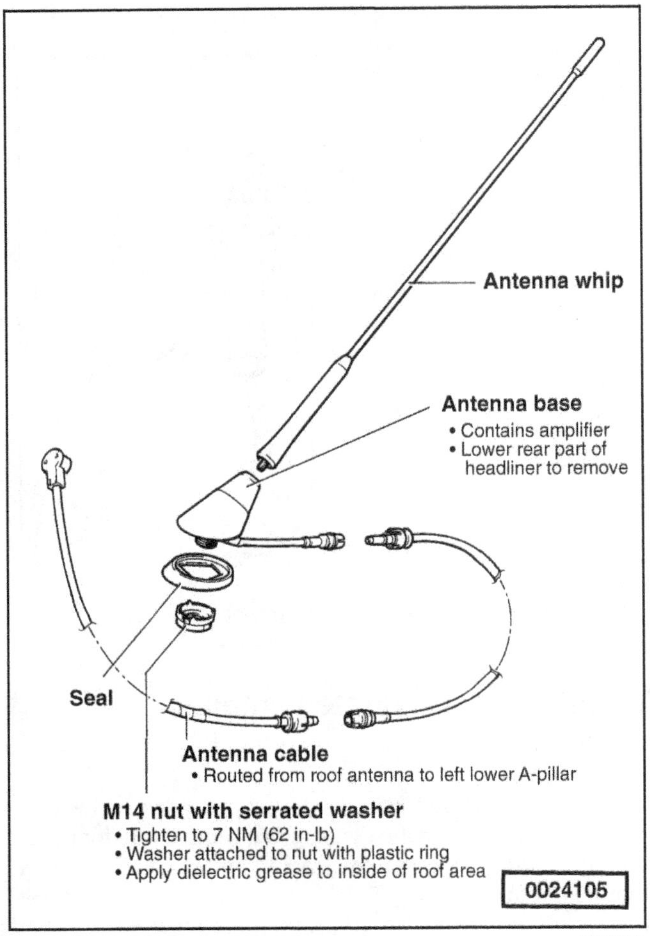

0024105

Fig. 14. Amplified roof-mounted antenna assembly.

Stereo system, electrical interference suppression

The majority of the electrical consumers/components in the vehicle are suppressed to ensure minimal electrical interference with radio reception. On vehicles equipped from the factory with a radio or radio prep wiring, the following components are also suppressed:

- Coolant fan (V7)
- Windshield wiper motor (V)
- Rear window wiper motor (V12)
- Fresh air blower fan motor (V2)

In addition, vehicles with a factory installed radio or radio prep wiring will have the following supplemental ground (GND) connections:

- Ground (GND) strap from left wheel housing to hood (integral with antenna base ground)
- Ground (GND) strap from ignition coil (N152) to engine cylinder block (vehicles with gasoline engines only)

92 Wipers and Washers

GENERAL

The front windshield wipers have 2 speeds and an intermittent position. The rear wiper (Golf, GTI) runs intermittently. The front wipers, rear wipers, and washers are controlled by the wiper switch on the right side of the steering column.

The front and rear washers share a common washer pump and washer fluid reservoir. The front washers are mounted in the hood. The rear washer is mounted in the rear wiper arm pivot.

WIPERS AND WASHERS

CAUTION —

- *Worn or dirty wiper blades will reduce visibility, making driving hazardous. Clean blades regularly to remove road film and car wash wax build-up. Use an alcohol based cleaning solution, a lint free cloth, and wipe along the rubber blade.*
- *Clean all inside and outside glass regularly. Use an alcohol based cleaning solution and wipe dry with a lint free cloth.*
- *Do not use the wiper/washer in freezing weather without first warming the windshield with the defroster, otherwise the washer solution may freeze on the windshield and obscure your vision.*
- *Avoid running the wiper blades over a dry windshield to prevent scratching the glass. A scratched windshield will reduce visibility and increase glare.*

Rubber wiper inserts, replacing

1. Free end of wiper insert from retaining hooks by squeezing metal strip in insert with pliers.

2. Slide rubber insert from retaining hooks.

3. Slide rubber insert into lower retaining hooks of wiper blade.

   NOTE —
   If necessary, transfer metal strips from old insert to new insert. Notches in metal strips must face rubber and engage rubber lugs in groove.

4. Squeeze metal strips at open end of rubber insert using pliers and install insert into hooks so that hook engages rubber retaining slot on insert. See Fig. 1.

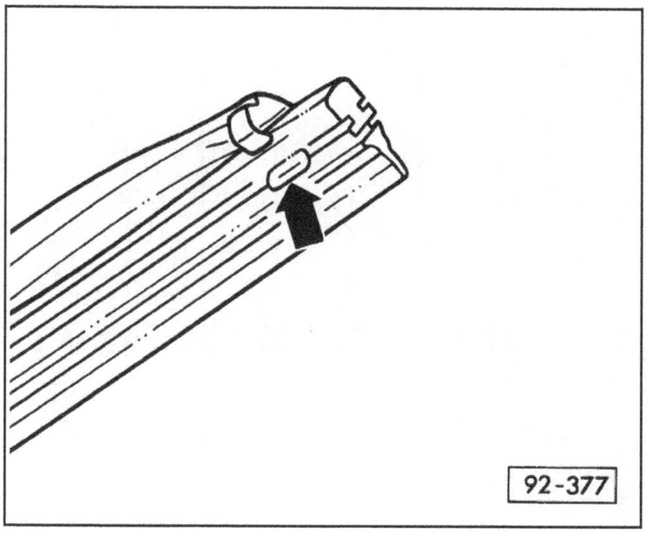

92-377

Fig. 1. Install rubber insert so that hook engages rubber retaining slot (**arrow**).

92

Windshield wiper, assembly

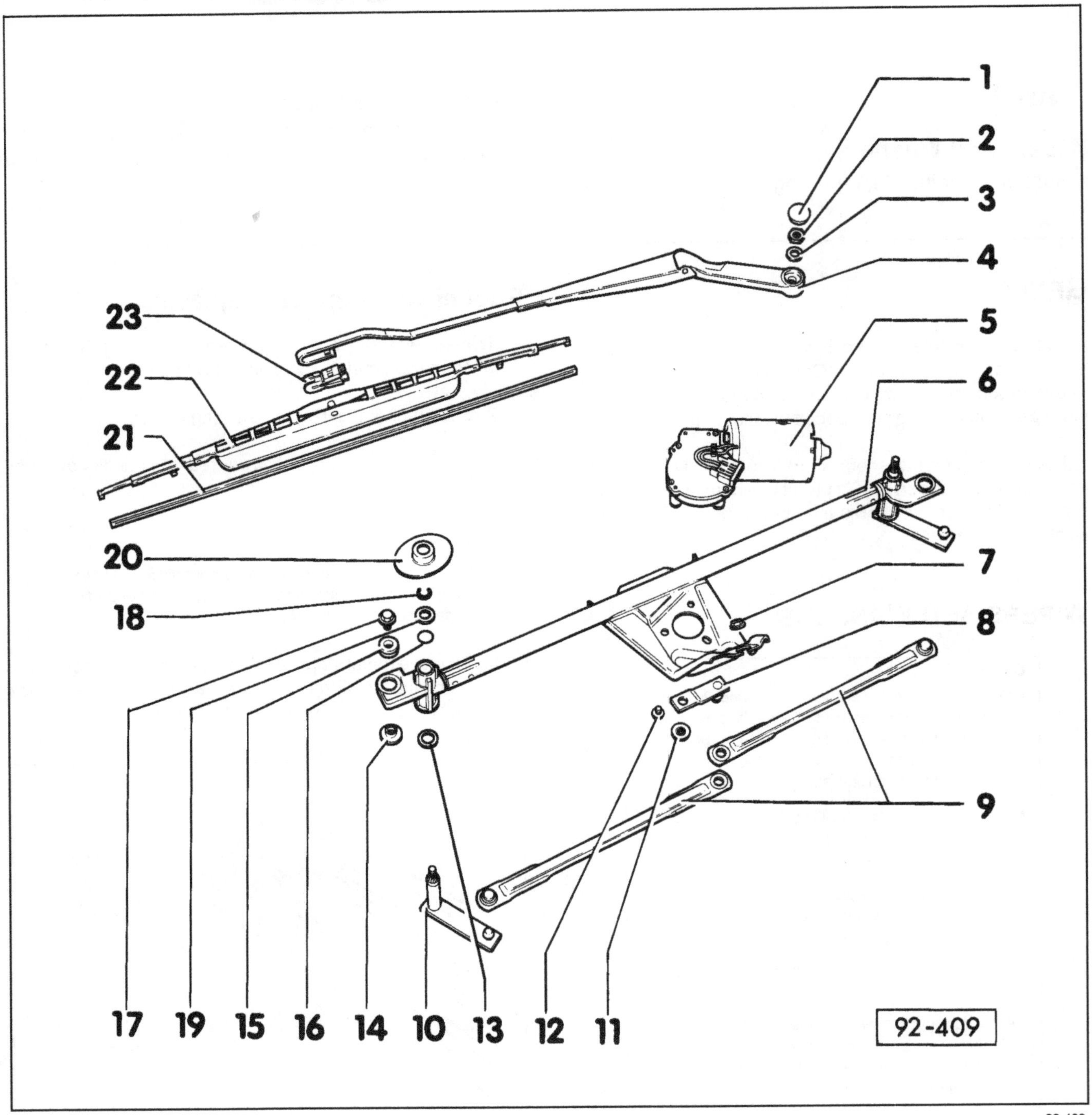

92-409

Fig. 2. Windshield wiper assembly.

1. **Cap**
2. **M8 hex nut**
   - Tighten to 20 Nm (15 ft-lb)
3. **Washer**
4. **Wiper arm**
5. **Windshield wiper motor**
   - Removing: first remove wiper frame (6)
6. **Wiper frame**
7. **M6 hex nut**
8. **Crank**
   - Installing and adjusting, **see** (A)

9. **Relay rods**
10. **Wiper shaft**
11. **M8 hex nut**
    - Tighten to 20 Nm (15 ft-lb)
12. **M6 hex bolt**
    - Tighten to 5 Nm (44 in-lb)
13. **Spring washer**
14. **Washer**
15. **Rubber seal**
16. **O-ring**
17. **M6 hex bolt**
    - Tighten to 5 Nm (44 in-lb)
18. **Snap ring**
19. **Washer**
20. **Cap**
21. **Rubber wiper insert**
22. **Wiper blade**
23. **Retaining clip**

(A) Wiper motor crank, installing and adjusting

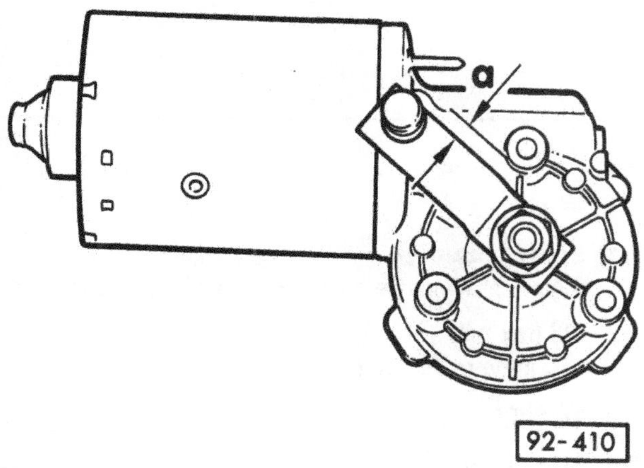

92-410

- **Wiper motor must be in the "park" position before installing the crank.**
- **Run wiper motor to stop.**
- **Install crank and align.**
  Dimension (a) = 10 mm (0.39 in.)

Windshield washer system

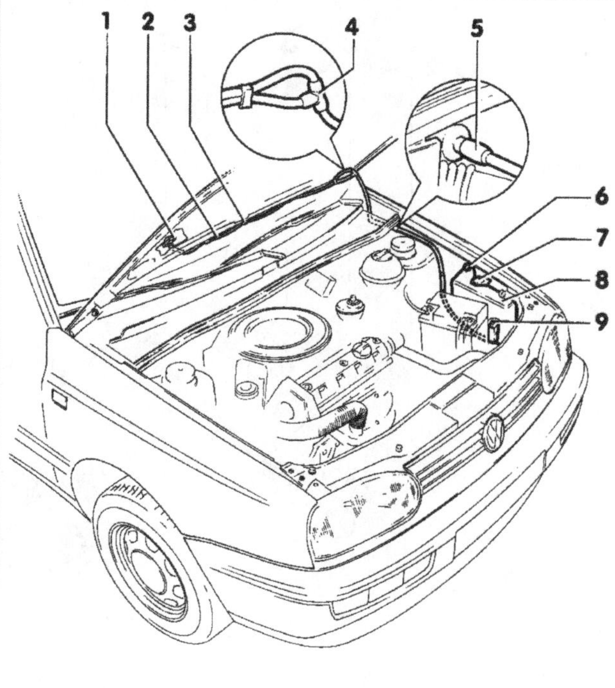

1. **Washer nozzles**
2. **Hose**
3. **Retaining clip**
4. **T-fitting**
5. **Hose connector**
6. **M5 plastic nut**
7. **Washer fluid reservoir cap**
8. **Washer fluid reservoir**
    - Capacity: 4.0 liters (4.2 qt)
9. **Windshield washer pump**
    - Mark hose connections before removing

92-408

92-408

Fig. 3. Windshield washer system components.

Rear window wiper, assembly

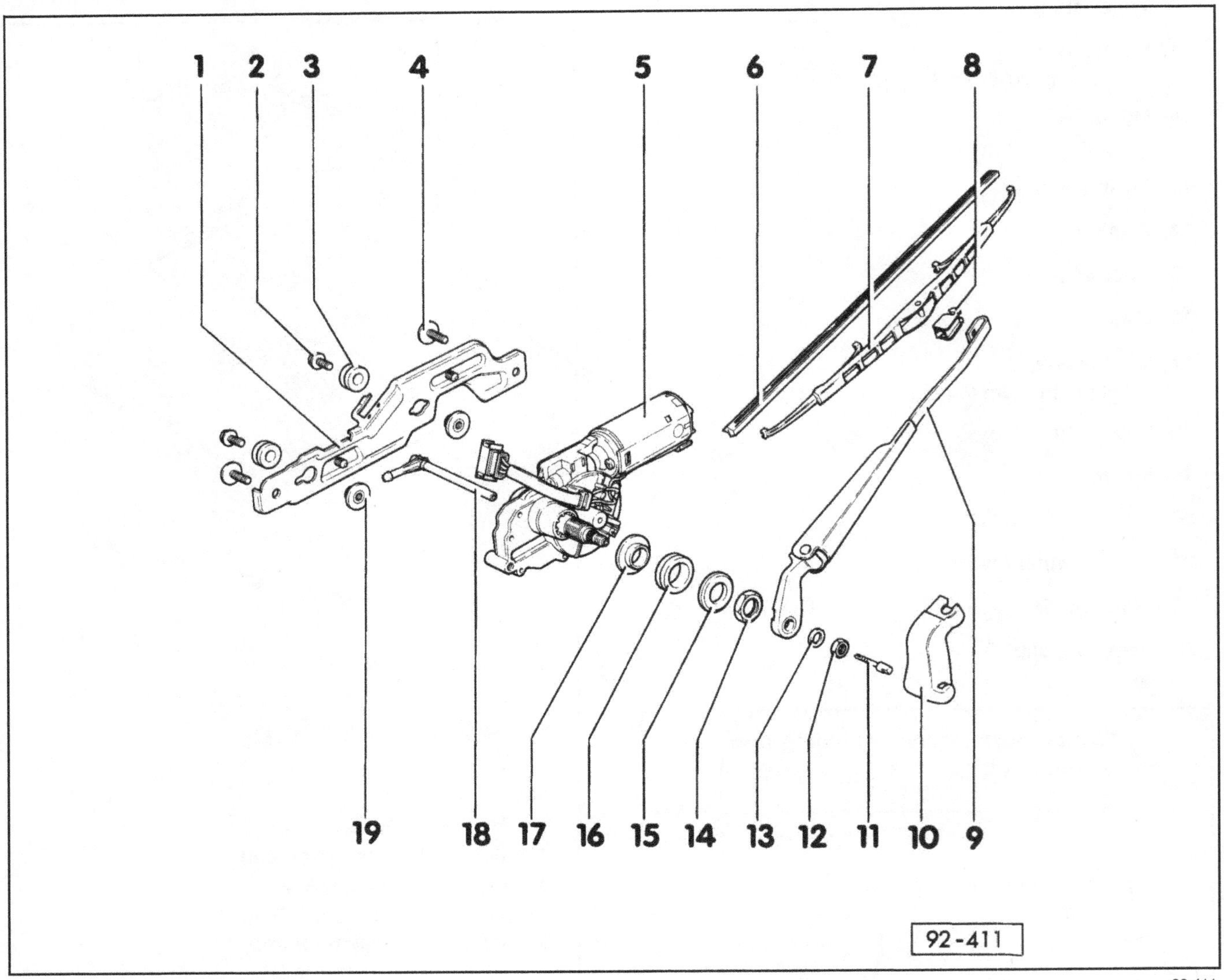

92-411

Fig. 4. Rear window wiper assembly.

1. **Wiper motor bracket**
2. **M6 hex bolt**
   - Tighten to 5 Nm (44 in-lb)
3. **Rubber washer**
4. **M6 hex bolt**
   - Tighten to 5 Nm (44 in-lb)
5. **Rear window wiper motor**
6. **Rubber wiper insert**
7. **Wiper blade**
   - Adjust to one inch from bottom window seal
8. **Retaining clip**
9. **Wiper arm**
10. **Cap**
11. **Spray nozzle**
    - Aim spray nozzle toward center of wiper coverage
12. **M8 hex nut**
    - Tighten to 15 Nm (11 ft-lb)
13. **Washer**
14. **M16 hex nut**
    - Tighten to 7 Nm (62 in-lb)
15. **Plastic washer**
16. **Rubber washer**
17. **Spacer ring**
18. **Connector tube for spray nozzle**
    - Routed through wiper motor shaft to spray nozzle
19. **Washer**

94 Lights, Accessories–Exterior

GENERAL

The models covered by this manual come with different headlight systems depending on the model, country (USA and Canada), and model year. Base models are equipped with single bulb headlights. Some GTI and Cabrio models are equipped with dual bulb headlights. Both of these versions use a glass lens. GTI GLX and Jetta GLX models are equipped with fog lights mounted in the front valence panel under the bumper. Cabrio models from October 1998 production use headlights that are similar to those used on the later Golf (A4 platform) with poly-carbonate lenses. All lights have replaceable bulbs.

Daytime running lights

Daytime running lights (DRL) have been installed on Canadian Golf and Jetta vehicles starting with the 1993 model year and on USA models beginning with the 1995 model year. Different versions are used depending on the model year. All versions turn on the headlights when the key is switched on and the handbrake is released and turn off the headlights when the handbrake is applied. Some versions operate the headlights at a reduced brightness. Some switch on the taillights and dashlights. See **97 Wiring Diagrams, Fuses and Relays** for fuse/relay positions and schematic for the vehicle in question.

NOTE —
If daytime running light (DRL) relay (marked "25") is noisy, install insulator (VW part 1HM 963 571) over relay.

DRL relay position (at fuse/relay panel)

• Gasoline vehicles	position 23
• Diesel vehicles	position 19

Different headlight switches are installed to accommodate the variety of headlight versions. Both 2- and 3-position switches are used and include provisions for front fog light operation. See **96 Lights, Accessories-Interior** for switch removal and installation information.

EXTERIOR LIGHTS

CAUTION —
- *Before working on electrical system, disconnect battery negative (–) cable.*
- *Before disconnecting battery be sure to obtain radio anti-theft code.*

94

Headlight bulbs

Different types of headlight bulbs are used in the vehicles covered by this manual. See Fig. 1.

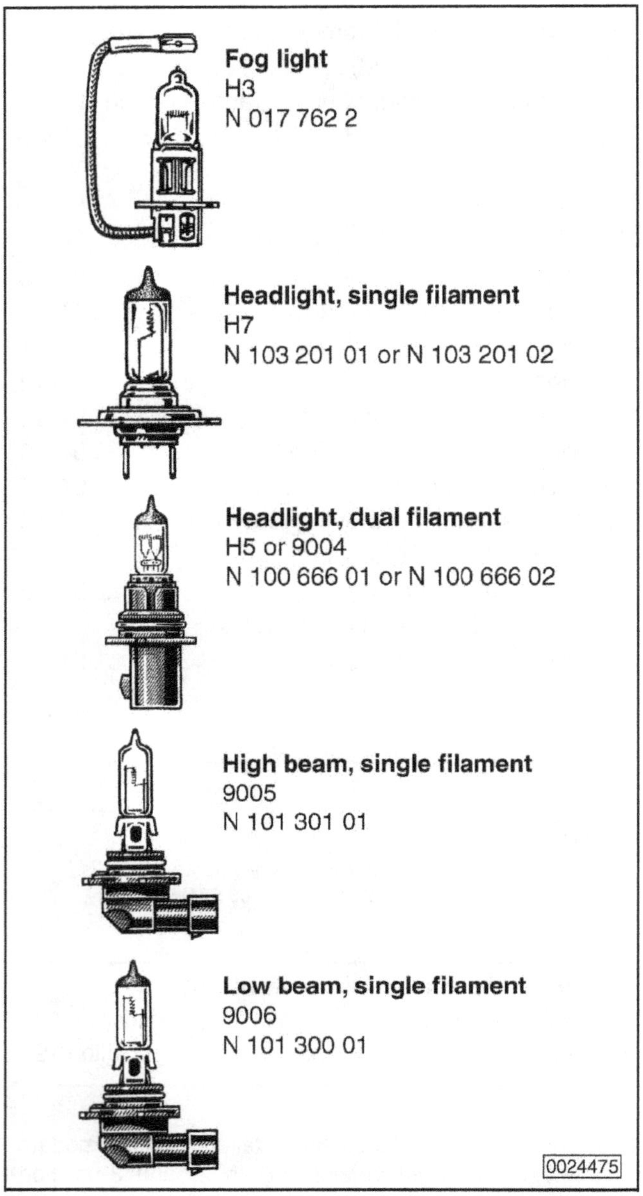

0024475

Fig. 1. Headlight and fog light bulbs used in the vehicles covered by this manual. Information includes industry or trade number on the second line, and Volkswagen/Audi part number on the third line (illustrations not to scale). Note that industry or trade number will sometimes have suffix letters such as HO, BP, LL, etc. to denote additional specifications.

> ***CAUTION —***
> *Do not touch bulb glass with bare fingers:*
> - *Skin oil will burn onto the glass envelope causing hot spots when bulb gets hot, damaging bulb.*
> - *Skin oil will evaporate when bulb gets hot, dulling the reflector.*

Bulb application and usage depends on model and year. Several bulbs are available in long-life versions and in wattages higher than originally specified by Volkswagen. Use of these higher power bulbs may be in contradiction to law and may cause damage to glass and plastic lenses and reflectors due to substantially higher heat generation. Additionally, in some cases, electrical load may be greater than the capacity of the wiring.

> ***CAUTION —***
> *Volkswagen part numbers are given for reference only! Always consult an Authorized Volkswagen Dealer Parts Department or aftermarket Parts Specialist for the latest parts information.*

Headlight bulb, replacing (single bulb system)

NOTE —
Check and adjust headlights after performing any work that may affect headlight aim.

1. Remove headlight harness connector and cut tie-wrap (if installed).

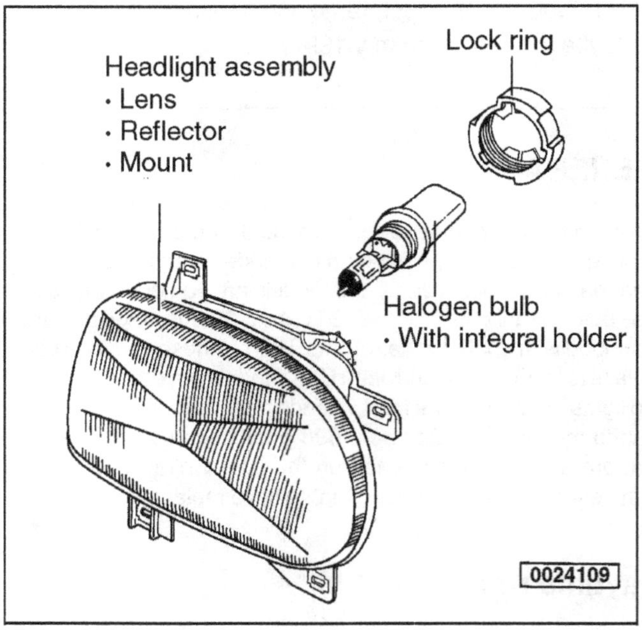

0024109

Fig. 2. Single bulb headlight assembly (Golf version shown).

2. Twist bulb lock ring counter-clockwise approx. 1/4 turn and remove. See Fig. 2.
3. Pull bulb from headlight assembly.
4. Insert new bulb into headlight assembly taking care to align lugs on bulb holder with matching grooves in reflector housing.

> *CAUTION —*
> *Do not touch bulb glass with bare fingers:*
> - *Skin oil will burn onto the glass envelope causing hot spots when bulb gets hot, damaging bulb.*
> - *Skin oil will evaporate when bulb gets hot, dulling the reflector.*

5. Install bulb holder lock ring and twist ¼ turn clockwise to lock.
6. Reconnect headlight connector and new tie-wrap.
7. Check headlight operation.

Headlight bulbs, replacing (dual bulb system)

NOTE —
Check and adjust headlights after performing any work that may affect headlight aim.

1. If equipped with plastic cover on rear of headlight assembly, unsnap and remove.

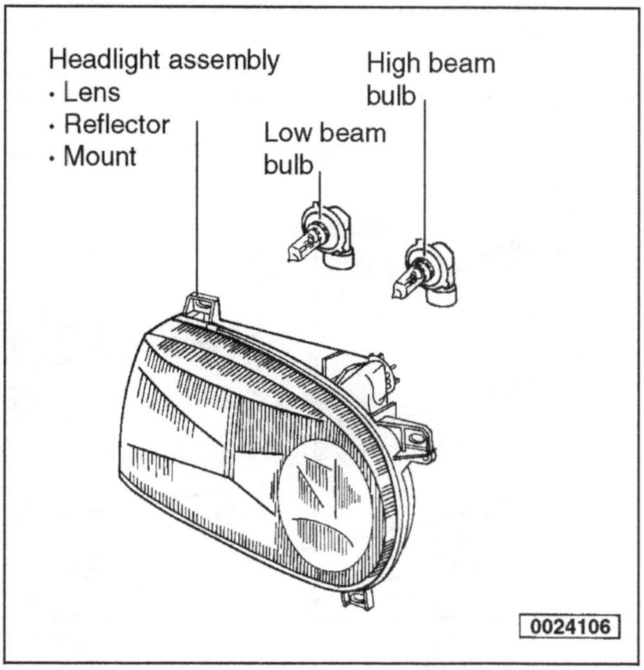

0024106

Fig. 3. Headlight assembly for two bulb system (Golf version shown).

2. Remove headlight harness connectors and cut tie-wrap (if installed).
3. If equipped with wire bulb retaining bail, unclip bail from tabs on back of reflector and move aside.
4. To remove bulb, either twist bulb counter-clockwise approx. ¼ turn and remove it, or pull it straight out. See Fig. 3.
5. Insert new bulb into headlight assembly taking care to align lugs on bulb holder with matching grooves in reflector housing.

> *CAUTION —*
> *Do not touch bulb glass with bare fingers:*
> - *Skin oil will burn onto the glass envelope causing hot spots when bulb gets hot, damaging bulb.*
> - *Skin oil will evaporate when bulb gets hot, dulling the reflector.*

6. Install wire bulb retaining bail (where equipped) or twist bulb clockwise approx. ¼ turn to lock.
7. Reconnect headlight connector and new tie-wrap.
8. Install plastic cover on rear of headlight assembly (where equipped).
9. Check headlight operation.

Headlight assembly, removing and installing

1. Remove radiator grille, **see 66 Body Accessories-Exterior**.
2. Remove headlight harness connector(s) and cut tie-wrap (if installed).
3. Remove securing screws and pull headlight assembly forward to remove. See Fig. 4..

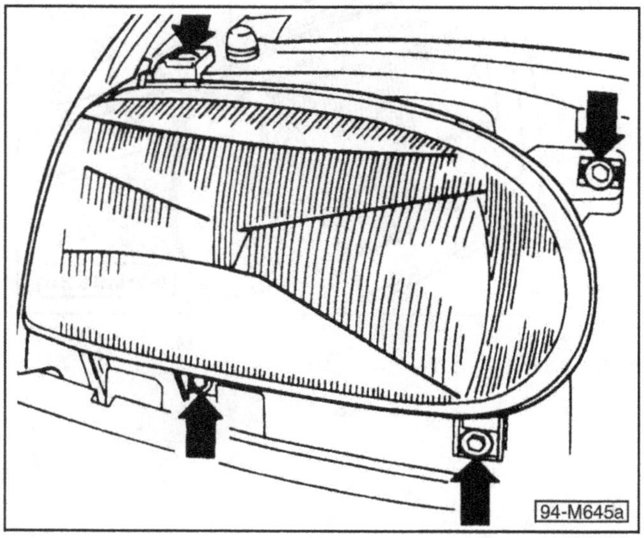

94-M645a

Fig. 4. Remove screws (**arrows**) and pull headlight assembly forward and out (Golf version shown).

4. Installation is the reverse of removal.

5. Check headlight operation.

6. Check headlight aim.

Headlights, adjusting

Accurately aiming the headlights requires special adaptors and equipment. Authorized Volkswagen dealers and most qualified independent automotive repair facilities are equipped to make these adjustments. In addition, headlight aiming may be governed by law in many areas with adjustment parameters varying slightly between the different State and Provincial governments.

To adjust horizontal and vertical headlight aim, turn the respective adjuster with a screwdriver through the slots above or below the headlight. See Fig. 5.

Headlight adjusting requirements:

- Tires inflated to the correct pressure.
- Fuel tank at least half full.
- Lenses and lens reflectors clean and not damaged.
- Driver's seat loaded with approx. 75 kg (165 lbs)
- Vehicle and headlight beam adjusting equipment standing on level ground and properly aligned.

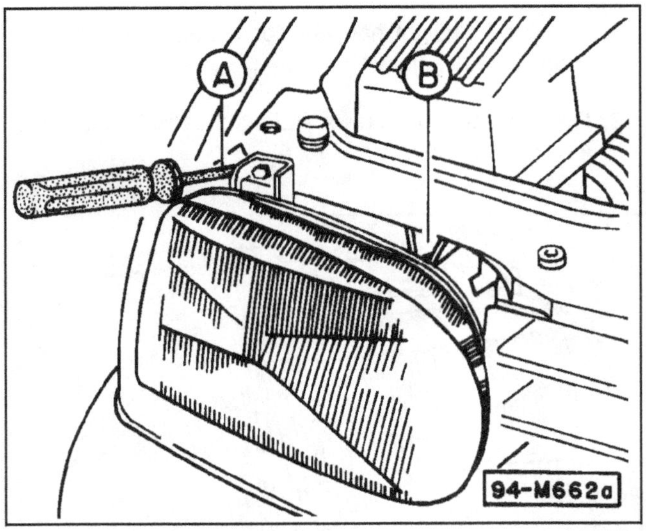

94-M662a

Fig. 5. Vertical adjustment (**A**) and horizontal adjustment (**B**).

Front turn signal housing, removing and installing

1. Remove center reflector lens (towing eye cover). See Fig. 6.

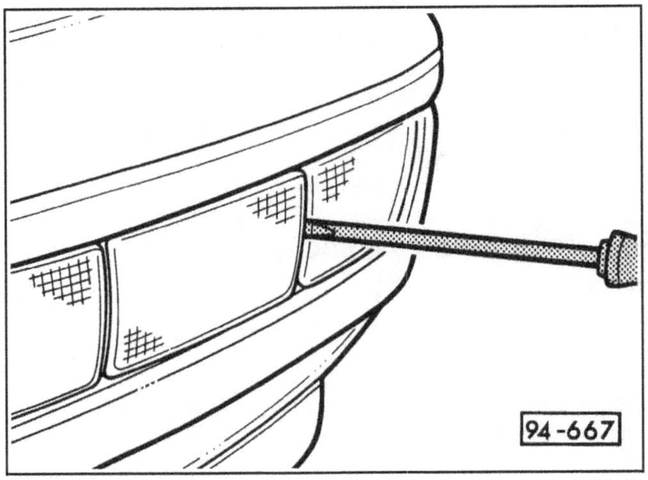

94-667

Fig. 6. Use screwdriver to carefully pry out towing eye cover.

2. Push in locking tab and pull turn signal housing from bumper. See Fig. 7.

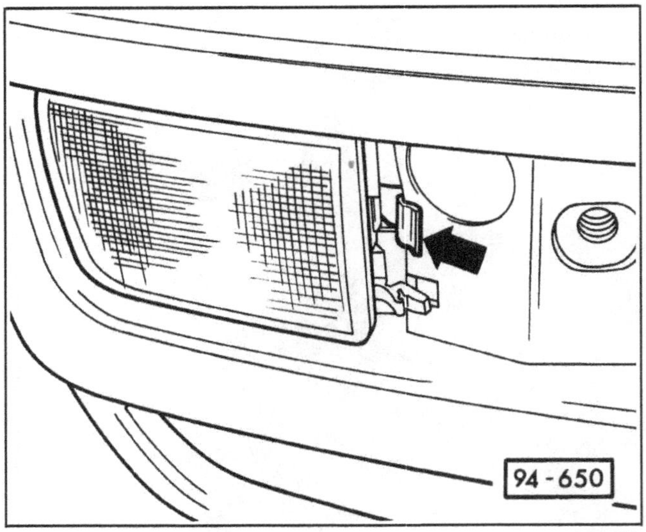

94-650

Fig. 7. Front turn signal locking tab (**arrow**).

3. Disconnect harness connector and remove turn signal housing.

4. Install in reverse order of removal.

5. Check light operation.

Front turn signal bulb, removing and installing

1. Remove turn signal housing as described above.
2. Twist bulb holder from rear of turn signal housing approx. 1/8 turn counter-clockwise and remove from housing.
3. Remove bulb from holder.
4. When installing, twist bulb holder clockwise approx. 1/8 turn until holder locks into turn signal housing.
5. The remainder of the installation is the reverse of the removal.

Fog light housing, removing and installing

1. Remove center reflector lens (towing eye cover). See Fig. 6.
2. Remove fog light housing mounting screws. See Fig. 8.

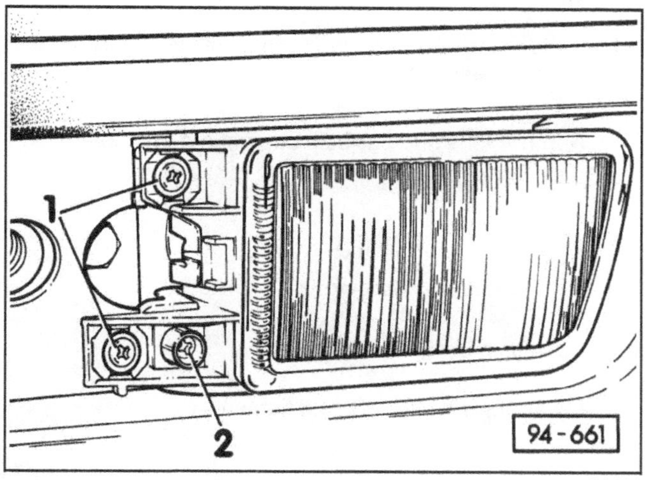

94-661

Fig. 8. Fog light mounting screws (**1**) and adjuster (**2**).

3. Pull fog light housing from bumper and disconnect harness connector.
4. Install in reverse order of removal.
5. Adjust fog lights by turning adjusting screw counter-clockwise to lower beam or clockwise to raise beam.

> ***NOTE —***
> *Horizontal or lateral (side-to-side) fog light aim is fixed and can not be adjusted. Only vertical (up-down) adjustment is possible.*

6. Check light operation.

Fog light bulb, removing and installing

1. Remove fog light housing as described previously.
2. Unplug 2-pin electrical connector from fog light.
3. Twist rear bulb cover counter-clockwise and remove.
4. Disconnect fog light bulb connector (inside of cover).
5. Unclip bulb spring retaining clip (wire bail) and swing clip out of the way. See Fig. 9.

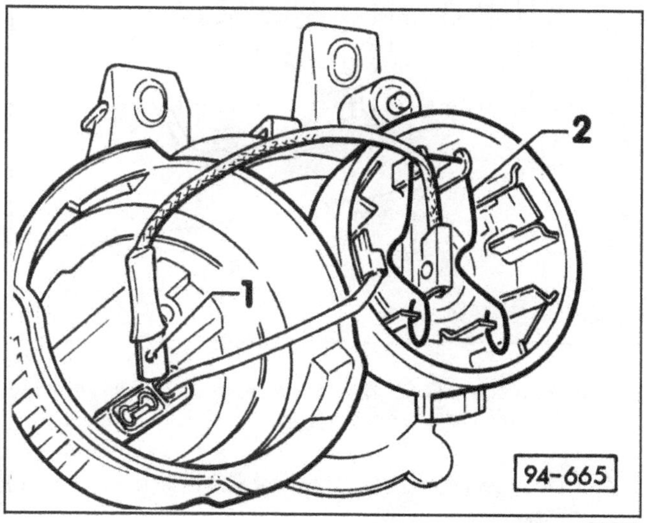

94-665

Fig. 9. Fog light bulb connector (**1**) and spring clip (**2**).

6. Remove bulb from housing.

> ***CAUTION —***
> *Do not touch bulb glass with bare fingers:*
> - *Skin oil will burn onto the glass envelope causing hot spots when bulb gets hot, damaging bulb.*
> - *Skin oil will evaporate when bulb gets hot, dulling the reflector.*

7. Insert bulb so that recess in bulb holder aligns with housing guide lug.
   - Halogen bulb: H3 12 V/55 W. See Fig. 1.
8. Install bulb spring retaining clip (wire bail) into retainers on reflector.
9. Plug bulb "pigtail" wire into connector in cover and ensure that ground wire is securely in place.
10. Install rear bulb cover.
11. The remainder of installation is the reverse of removal.
12. Check light operation.

Front side marker light bulb, removing and installing

1. Using a small screwdriver, carefully pry side marker light lens from bumper cover until retaining lug is released. See Fig. 10.

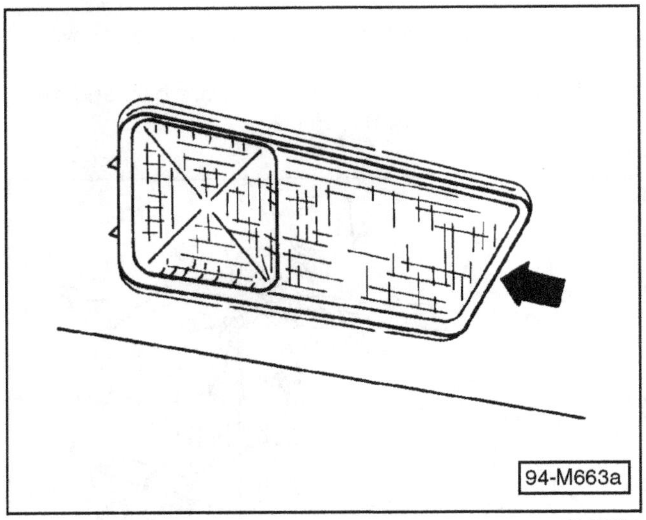

94-M663a

Fig. 10. Push side marker light lens in direction of arrow and carefully pry out to release retaining lug.

2. Disconnect electrical connector and twist bulb holder to release.
3. Remove side marker light bulb from holder.
4. Install bulb into holder and holder into lens.
5. Hook front edge of side marker light into bumper cover and push rear edge until retaining lug engages.
6. Check light operation.

License plate bulb, removing and installing

1. Remove plastic trim from underside of rear hatch/trunk lip (where equipped).
2. Remove license plate light retaining screws. See Fig. 11.

94-655

Fig. 11. License plate light retaining screws (**arrows**). Golf version shown, Jetta and Cabrio are similar.

3. Remove lens and pull bulb straight out of holder.
   - Bulb: 12 V/5 W

   NOTE—
   Do not twist bulb when removing from holder.

4. Push bulb straight into holder.
5. Install screws into light and tighten.

   NOTE—
   Do not over-tighten! Plastic lens can be distorted and will crack.

6. Install plastic trim panels.
7. Check light operation.

Taillight bulbs, removing and installing

1. Remove rear trim as required to access rear taillight.

NOTE—
Jettas have some bulbs in the lens on the trunk lid. Access requires removal of trim panels in most cases, but bulb replacement is otherwise the same.

2. Unclip bulb holder. See Fig. 12.

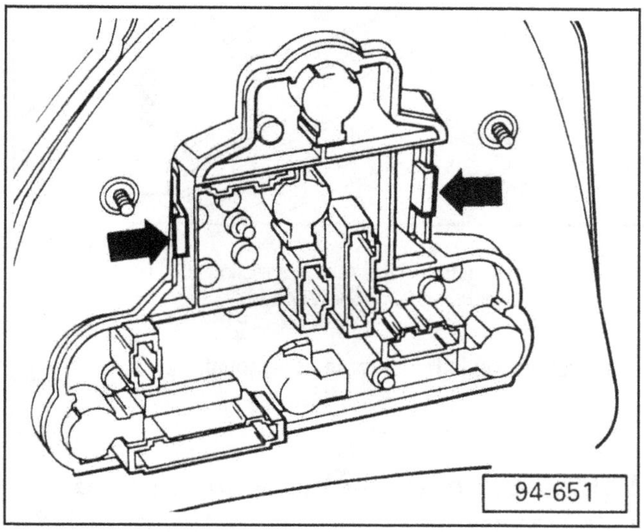

94-651

Fig. 12. Squeeze retailing tabs (**arrows**) to release bulb holder from taillight. Golf shown, others similar.

3. Push bulbs in slightly and twist counter-clockwise to remove.

NOTE—
It is not usually necessary to cut tie wrap on wiring harness. If, however, access requires cutting, always replace tie wrap to prevent harness chafing on body.

4. When bulbs have been replaced, position bulb holder in lens and snap in place.

5. Install trim panels.

6. Check light operation.

Taillight assembly

All taillight bulbs can be accessed from inside the trunk. Some vehicles have bulbs in the trunk lid, but access and removal is similar with similar bulbholders. See Fig. 13.

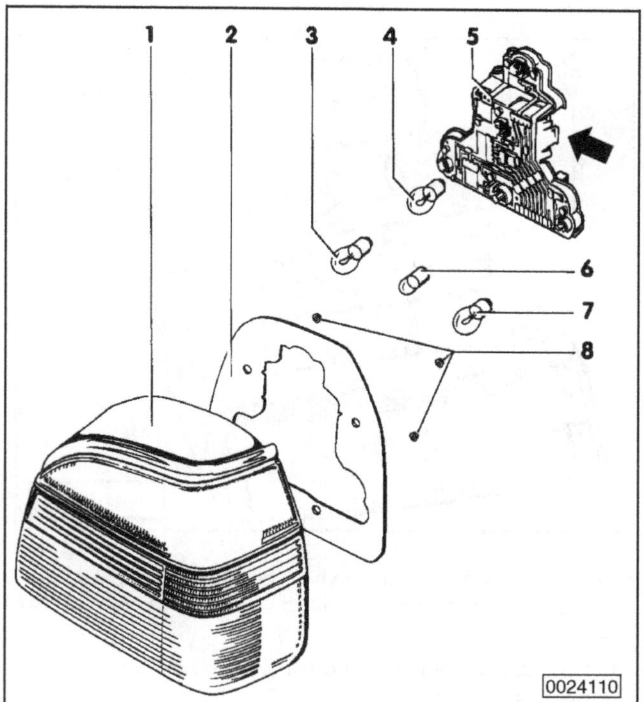

1. **Rear lens unit**
   • With rear and side reflectors
2. **Gasket (seal)**
3. **Back-up light bulb**
   • 12 V/21 W
4. **Turn signal bulb**
   • 12 V/21 W
5. **Bulb holder**
   • Squeeze tab (**arrow**) to remove
6. **Taillight bulb**
   • 12 V/10 W
7. **Brake light bulb**
   • 12 V/21 W
8. **Nut M5**
   • Tighten evenly to avoid distortion and cracking

0024110

Fig. 13. Right rear tail light assembly with rear side marker and reflecters. Golf shown; Jetta and Cabrio are similar. Jetta back-up lights are in trunk lid.

High mount brake light, removing and installing (Jetta, all; Golf from m.y. 1997)

The high mounted brake light is installed in the upper portion of the rear window. It is comprised of four major components. See Fig. 14.

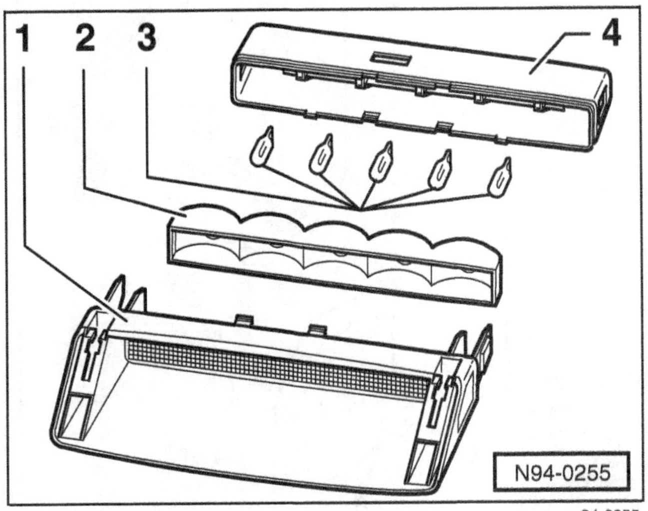

94-0255

Fig. 14. Major components of the high mounted brake light include: lamp housing (**1**), reflector (**2**), bulb(s) (**3**), and bulb holder (**4**).

1. Pull bulb holder from housing assembly. and disconnect harness connector. See Fig. 15.

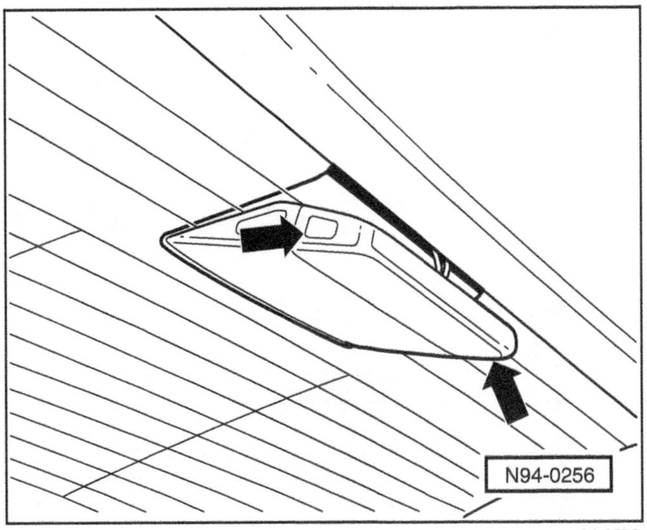

94-0256

Fig. 15. Depress tabs (**arrows**) and slide bulbholder off of housing.

2. Replace bulbs as required.

3. If cleaning of the glass or reflector is required, the housing can be removed. See Fig. 16.

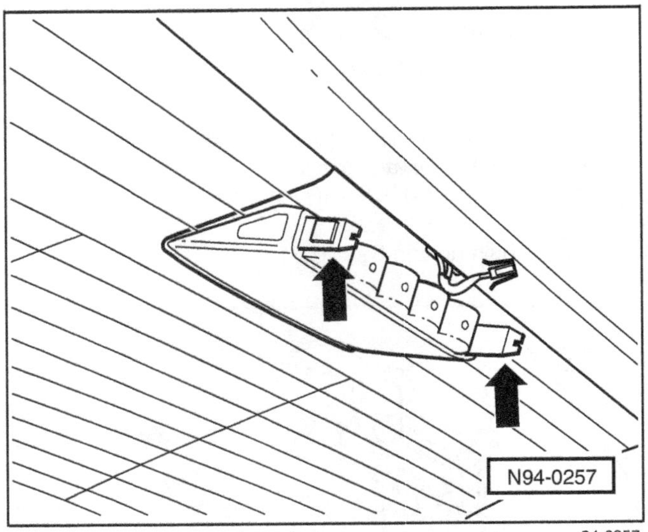

94-0257

Fig. 16. Depress metal retainer tabs in direction shown (**arrows**) and pull housing out from rear window.

4. Installation is the reverse of removal.

5. Check light operation.

High mount brake light, removing and installing (Golf through m.y. 1996)

The high mounted brake light is installed in the upper portion of the rear window in a rubberized spoiler. Access is from the outside by removing two screws. See Fig. 17.

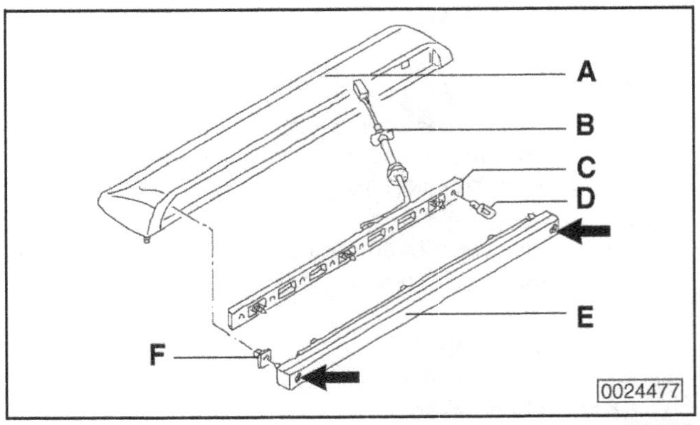

0024477

Fig. 17. Major components of the high mounted brake light on early Golfs include: spoiler (**A**), wiring harness (**B**), bulb holder (**C**), bulb(s) (**D**), lens/reflector assembly (**E**), and retainers (**F**). Removal of screws (**arrows**) removes assembly from spoiler. Removal of 3 small screws on the back of the bulbholder (not shown) allows access to bulbs.

High mount brake light, removing and installing (Cabrio through 09/1998)

The high mounted brake light is located in the upper portion of the rear lid. It is comprised of four major components. See Fig. 18.

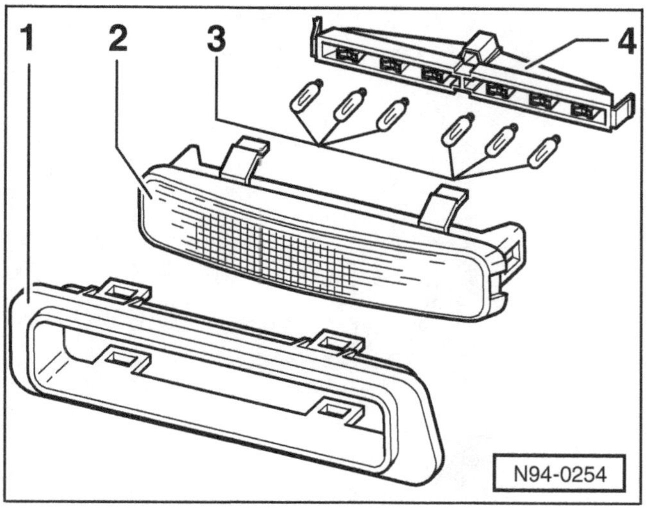

94-0254

Fig. 18. Major components of the Cabrio high mounted brake light include: mounting frame (**1**), lamp housing (lens) (**2**), bulb(s) (**3**), and bulb holder (**4**).

1. Remove rear lid trim.

2. Pull bulb holder from housing assembly. and disconnect harness connector. See Fig. 19.

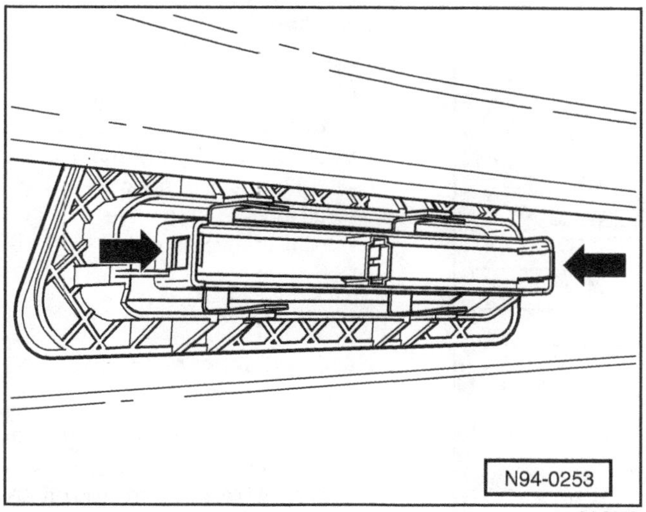

94-0253

Fig. 19. Depress tabs (**arrows**) and slide bulbholder off of housing.

3. Replace bulbs as required.

4. If cleaning of the lens or reflector is required, the housing can be removed. See Fig. 20.

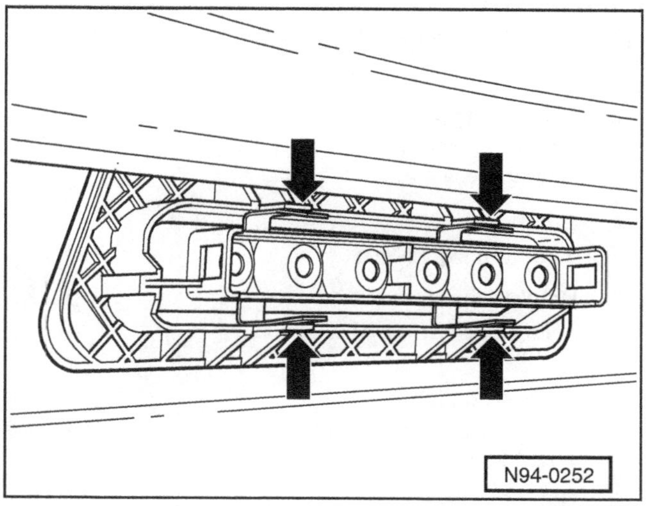

94-0252

Fig. 20. Depress tabs (**arrows**) and unclip housing from frame.

5. Installation is the reverse of removal.

6. Check light operation.

EXTERIOR LIGHTS, (CABRIO FROM 10/1998)

Head light assembly

1. **Rear covers**
2. **Bulb holder**
   - For turn signal bulb
3. **Height adjuster motor**
   - Not installed on USA/Canadian vehicles
4. **Bulb, turn signal**
   - Single filament
   - 12V, 21W
5. **Bulb, fog light**
   - H3, 12V, 55W
6. **Bulb, low beam**
   - H7, 12V, 55W
7. **Retaining clip (wire bail)**
8. **Head light lens, reflector, and mounting assembly**
   - See **Headlights, removing and installing**
   - Mounting tabs repairable if housing is otherwise undamaged, see **Headlight upper securing tabs, servicing**
9. **Retaining clip (wire bail)**
10. **Bulb, high beam**
    - 9005, 12V, 65W
11. **Bulb, parking light**
    - 12V, 5W
12. **Bulb holder**
    - For parking light bulb

A. **Low beam**

B. **Turn signal**

C. **Fog light (where equipped)**
- Blanked out if not equipped

D. **High beam**
- With parking light in lower part of reflector.

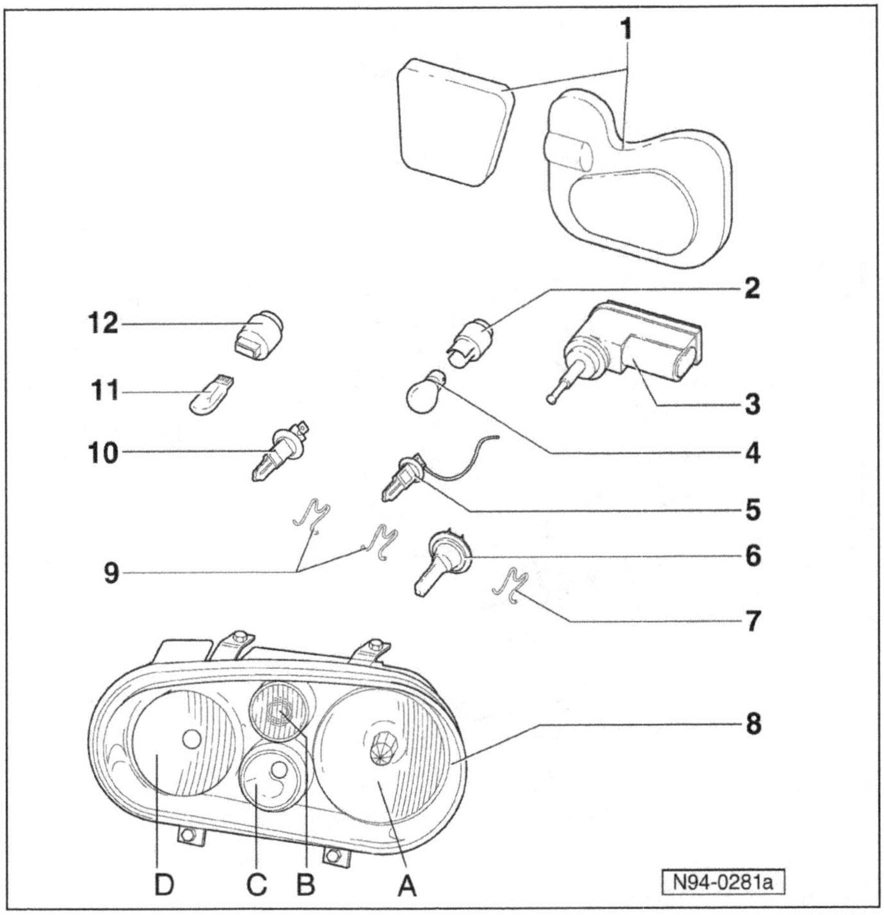

N94-0281a

Fig. 21. Headlight assembly for Cabrio from 10/1998 production.

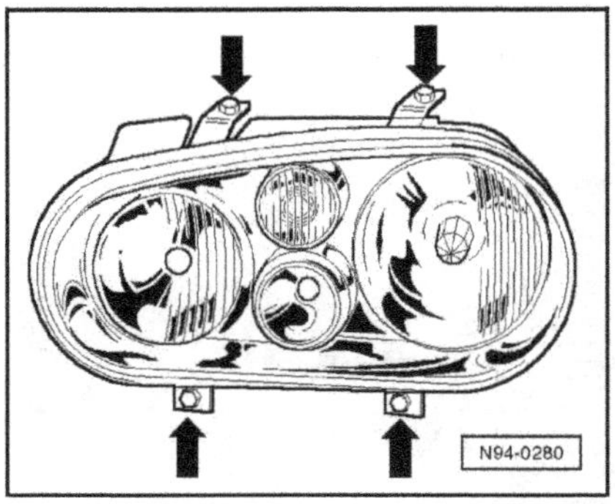

Headlights, removing and installing

- Remove front bumper
- Remove securing screws (**arrows**)
- Pull forward slightly and disconnect harness connector
- Pull forward to remove
- Install in reverse order

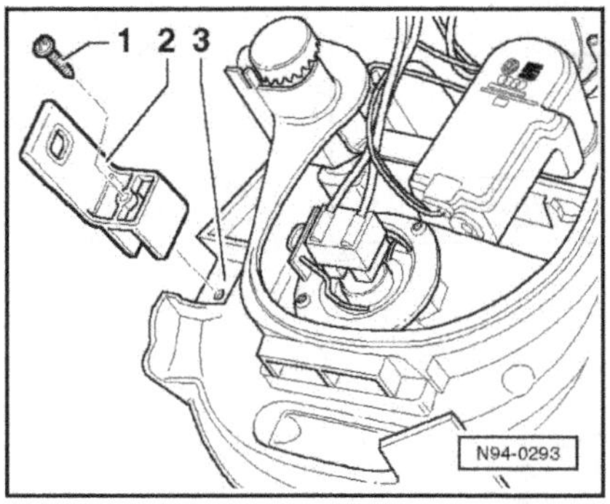

Headlight upper securing tabs, servicing

- If upper headlight securing tabs break off, headlight may still be serviceable if assembly is otherwise undamaged.
- Remove headlight assembly
- Remove remnants of old securing tab (sand or grind)
- Install replacement tab (**2**) onto area of old tab (**3**) and secure with screw (**1**)
- Aim headlights

Headlight bulb, replacing (low beam)

> ***CAUTION—***
> *Do not touch bulb glass with bare fingers:*
> - *Skin oil will burn onto the glass envelope causing hot spots when bulb gets hot, damaging bulb.*
> - *Skin oil will evaporate when bulb gets hot, dulling the reflector.*

1. Unclip spring clip on rear of headlight assembly and remove cover. See Fig. 22.

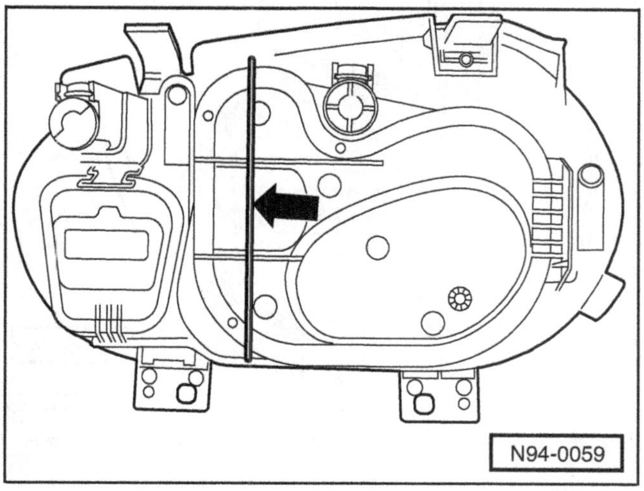

N94-0059

Fig. 22. Release (unclip) spring clip (**arrow**) and remove rear cover.

2. Disconnect harness connector at rear of lamp, unclip wire bail and pull bulb out of reflector. See Fig. 23.

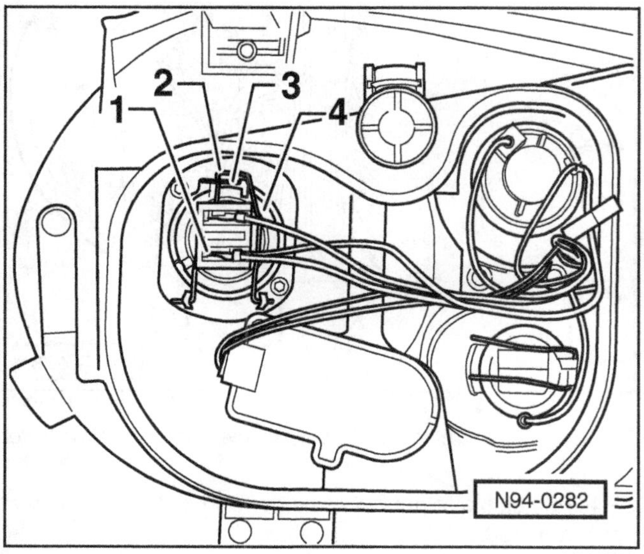

N94-0282

Fig. 23. Disconnect harness connector (**1**) at rear of lamp. Unclip both ends of wire bail (**2**) from tabs (**3**) and pull bulb (**4**) out of reflector.

3. Insert new bulb into reflector and align lugs on bulb holder base with matching slots in reflector.
4. Clip both ends of retaining wire bail into tabs.
5. Connect harness connector to bulb.
6. Install cover.
7. Check headlight operation.
8. Check headlight aim.

Headlight bulb, replacing (high beam)

> ***CAUTION—***
> *Do not touch bulb glass with bare fingers:*
> - *Skin oil will burn onto the glass envelope causing hot spots when bulb gets hot, damaging bulb.*
> - *Skin oil will evaporate when bulb gets hot, dulling the reflector.*

1. Disconnect harness connector at rear of headlight assembly.
2. Unclip spring clip on rear of headlight assembly and remove cover. See Fig. 24.

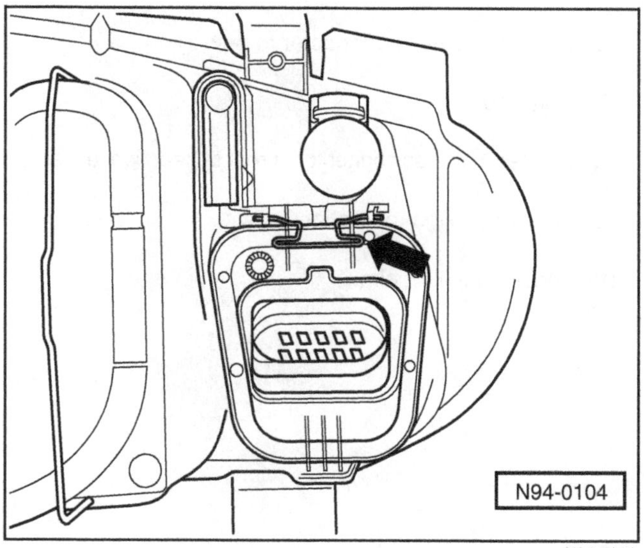

N94-0104

Fig. 24. Release (unclip) spring clip (**arrow**) and remove rear cover.

3. Disconnect harness connector at bulb, unclip wire bail and pull bulb out of reflector. See Fig. 25.

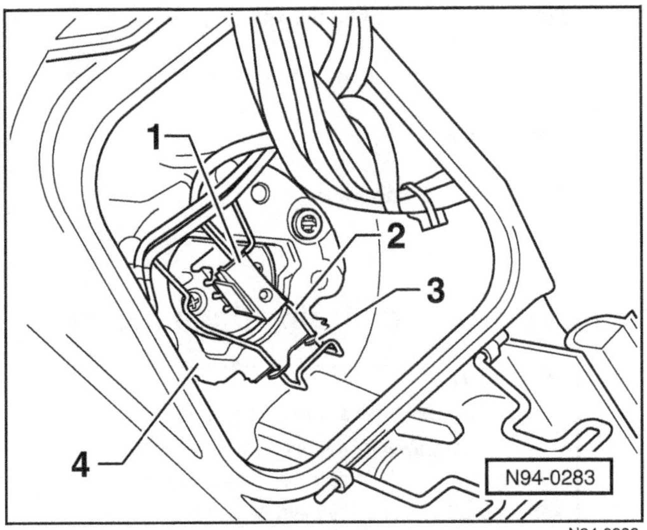

N94-0283

Fig. 25. Disconnect harness connector (**1**) at bulb. Unclip both ends of wire bail (**2**) from tabs (**3**) and pull bulb (**4**) out of reflector.

4. Insert new bulb into reflector and align lugs on bulb holder base with matching slots in reflector.
5. Clip both ends of retaining wire bail into tabs.
6. Connect harness connector to bulb.
7. Install cover.
8. Connect harness connector at rear of headlight assembly.
9. Check headlight operation.
10. Check headlight aim.

Headlight bulb, replacing (parking light)

1. Disconnect harness connector at rear of headlight assembly.
2. Unclip spring clip on rear of headlight assembly and remove cover. See Fig. 26.

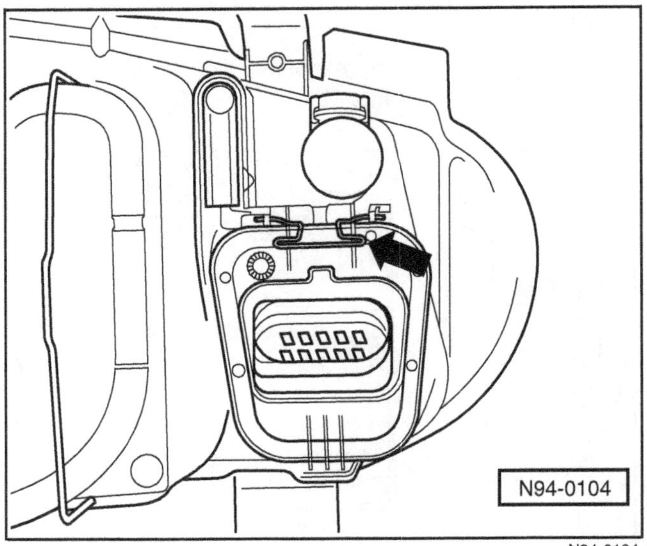

N94-0104

Fig. 26. Release (unclip) spring clip (**arrow**) and remove rear cover.

3. Pull out bulb holder and socket assembly. See Fig. 27.

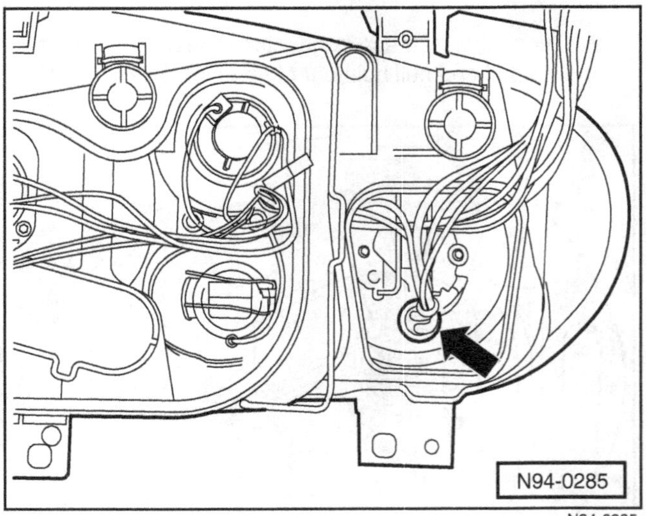

N94-0285

Fig. 27. Pull out bulb holder with bulb (**arrow**) from socket in reflector.

4. Pull bulb straight out of holder and install new bulb as required.
5. Insert bulb holder fully into socket in reflector. Ensure that wire is tightly connected.
6. Install cover.

7. Connect harness connector at rear of headlight assembly.

8. Check parking light operation.

Headlight bulb, replacing (turn signal)

1. Unclip spring clip on rear of headlight assembly and remove cover. See Fig. 28.

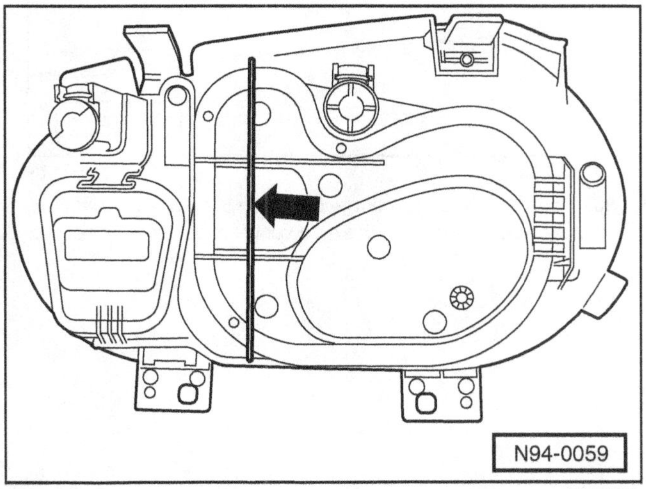

N94-0059

Fig. 28. Release (unclip) spring clip (**arrow**) and remove rear cover.

2. Turn bulbholder counterclockwise and pull out of reflector. See Fig. 29.

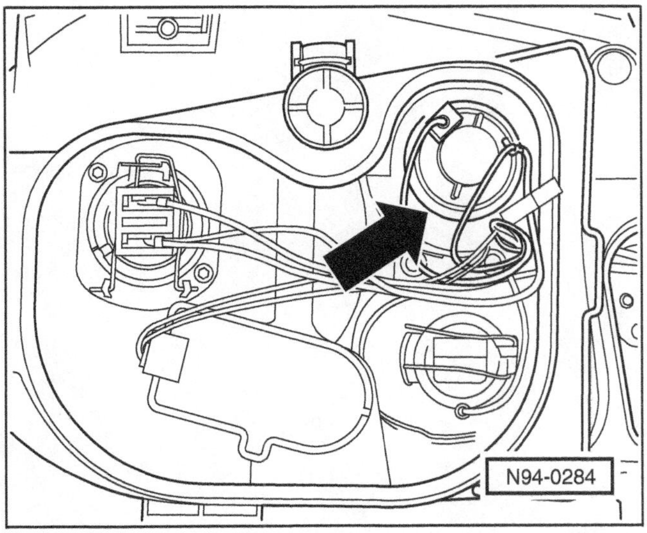

N94-0284

Fig. 29. Turn bulbholder (**arrow**) counter-clockwise and pull out of reflector.

3. Insert new bulb into bulbholder and insert bulbholder into reflector.

4. Install cover.

5. Check turn signal operation.

Headlight bulb, replacing (fog light)

1. Unclip spring clip on rear of headlight assembly and remove cover. See Fig. 30.

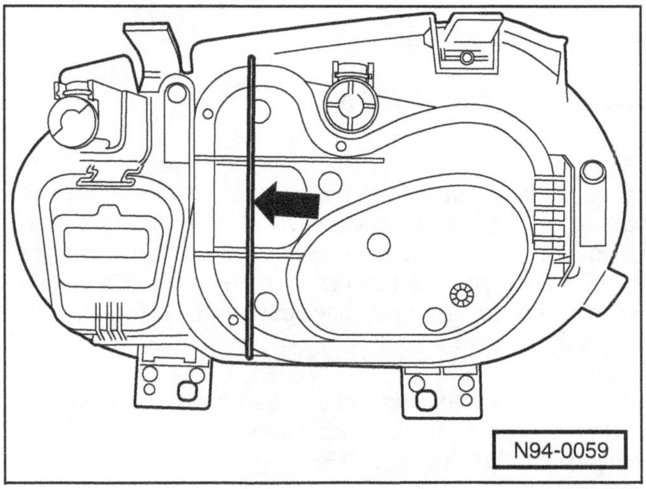

N94-0059

Fig. 30. Release (unclip) spring clip (**arrow**) and remove rear cover.

2. Disconnect harness connector at bulb "pigtail" wire, unclip wire bail and pull bulb out of reflector. See Fig. 31.

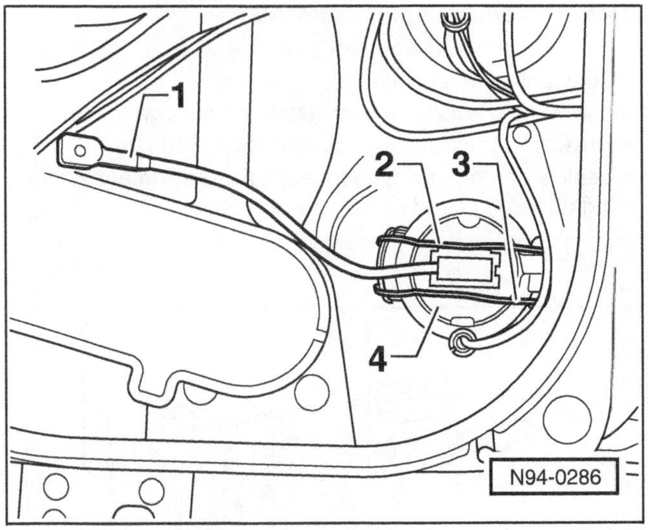

N94-0286

Fig. 31. Disconnect harness connector at bulb "pigtail" wire (**1**), unclip wire bail (**2**) at tabs (**3**) and pull bulb (**4**) out of reflector.

3. Insert new bulb into reflector and align lugs on bulb holder base with matching slots in reflector.

> ***CAUTION —***
> *Do not touch bulb glass with bare fingers:*
> - *Skin oil will burn onto the glass envelope causing hot spots when bulb gets hot, damaging bulb.*
> - *Skin oil will evaporate when bulb gets hot, dulling the reflector.*

4. Clip both ends of retaining wire bail into tabs.

5. Connect harness connector to bulb pigtail.

6. Install cover.

7. Check fog light operation.

Headlights, adjusting

Accurately aiming the headlights requires special adaptors and equipment. Authorized Volkswagen dealers and most qualified independent automotive repair facilities are equipped to make these adjustments. In addition, headlight aiming may be governed by law in many areas with adjustment parameters varying slightly between the different State and Provincial governments.

To adjust horizontal and vertical headlight aim, turn the respective adjuster with a screwdriver through the slots above the headlight. See Fig. 32. and See Fig. 33.

NOTE —
Adjustment of the low beam headlight adjusts fog light and high beam simultaneously.

Headlight adjusting requirements:
- Tires inflated to the correct pressure.
- Fuel tank at least half full.
- Lenses and lens reflectors clean and not damaged.
- Driver's seat loaded with approx. 75 kg (165 lbs)
- Vehicle and headlight beam adjusting equipment standing on level ground and properly aligned.

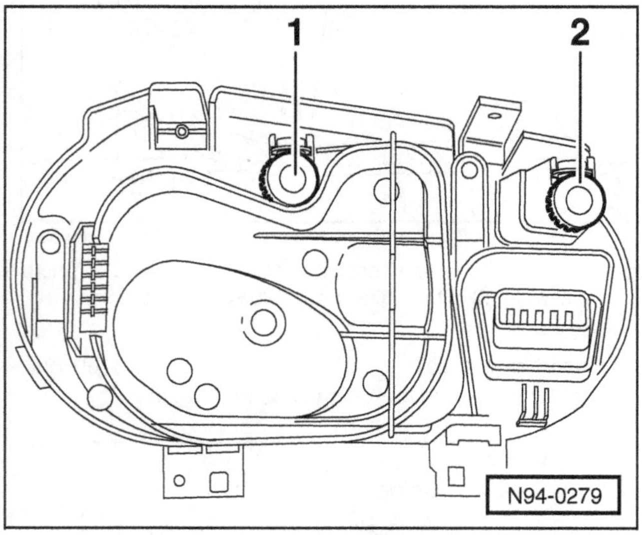

N94-0279

Fig. 32. Horizontal (lateral) adjustment (**1**) and vertical adjustment (**2**) as seen from the rear of the headlight assembly.

N94-0273

Fig. 33. Horizontal and vertical headlight adjustments (**arrows**) are made using a long Phillips screwdriver in guides from above.

License plate bulb, removing and installing

1. Remove screws. See Fig. 34.

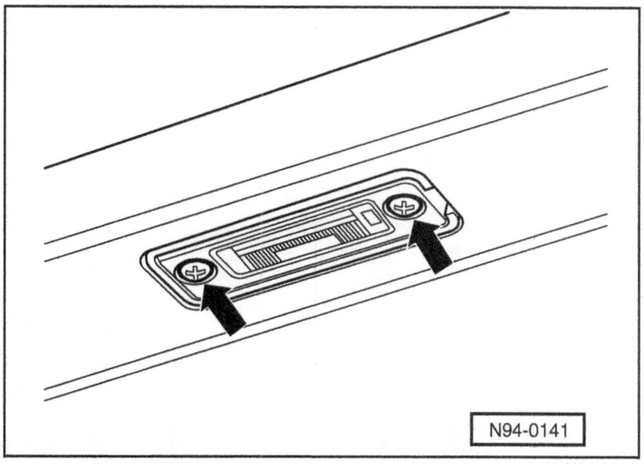

N94-0141

Fig. 34. Remove screws (**arrows**) from assembly.

2. Carefully pull out lens with bulb and bulbholder.

3. Installation is the reverse; note lens orientation. See Fig. 35.

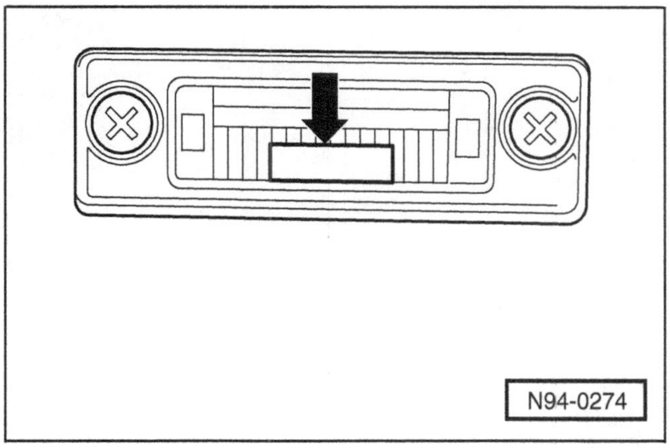

N94-0274

Fig. 35. Install lens so that the small silvered reflector (**arrow**) is towards the bumper.

4. Check light operation.

High mount brake light, removing and installing

The high mounted brake light is located in the upper portion of the rear lid. The design is changed from previous versions. The brake light assembly contains 16 light emitting diodes (LEDs) which are NOT replaceable separately.

1. Loosen the three nuts that secure the retainer clamp to the brake light housing. It is not necessary to remove the nuts completely. See Fig. 36.

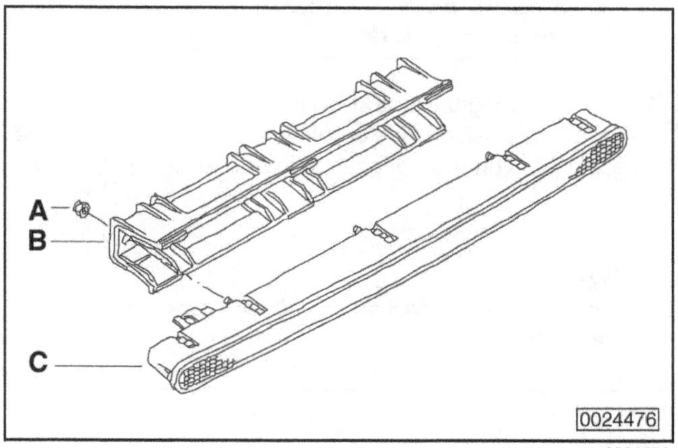

0024476

Fig. 36. Loosen the 3 nuts (**A**) securing the retainer (**B**) to the brake light housing (**C**).

2. Slide retainer away from electrical connection so that nuts pass through the openings in the clamp, remove clamp.

3. Depress both tabs on either side of brake light assembly and push the assembly through the opening in the rear lid.

4. Disconnect electrical connector.

5. Installation is in reverse order of removal.

6. Check light operation.

Side mounted turn signal repeaters, removing and installing

1. Switch off all electrical consumers, switch off ignition and remove ignition key

> ***CAUTION —***
> *Removal of turn signal repeater lens/housing is only possible in one direction. The side of the housing equipped with retainer clip is not obvious in it's installed position. Use care when removing to prevent damage to retainer clip or housing*

2. Push turn signal repeater with a suitable tool on the mounting position side carefully against the force of the spring clip on the opposite side and lever out. See Fig. 37.

NOTE —
Protect painted surfaces with adhesive tape beforehand if necessary.

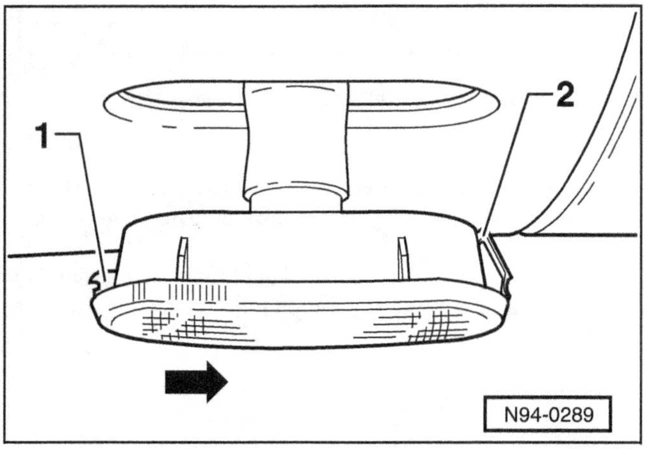

N94-0289

Fig. 37. Carefully push turn signal repeater with a suitable tool on the mounting position side (**1**) in the direction of arrow against the force of the spring clip (**2**) on the opposite side. While pushing pry out entire assembly.

3. Pull rubber holder with bulb out of bulb housing.

NOTE —
Do not twist bulb when removing from holder.

4. Push new bulb straight into holder.
5. Fit rubber holder with new bulb onto bulb housing and insert into fender opening.
6. Check turn signal repeater operation.

96 Lights, Accessories–Interior

TABLES

GENERAL

This section covers electrical equipment located in the passenger compartment; instrument panel switches, door switches, alarm system, cruise control. Items such as radio, heating and A/C that require detailed coverage have their own repair groups. See **97 Wiring Diagrams, Fuses and Relays** for additional information.

> ***CAUTION —***
> - *Before working on electrical system disconnect negative (–) battery cable.*
> - *Before disconnecting battery be sure to obtain radio anti-theft code.*

> ***WARNING —***
> *Volkswagen identifies electrical components by a letter and/or a number in the electrical schematics. These electrical identifiers are listed in this repair group as an aid to electrical troubleshooting.*

NOTE —
If the battery has been disconnected, or if the power window regulator is being connected for the first time, it will be necessary for the electronic control system to establish a reference point for the upper window frame. This is accomplished by the programming procedure that follows.

ELECTRICAL SWITCHES

Power window child safety system, programming (limited production, from March 1995)

A very limited number of vehicles produced from March 1995 had power windows that feature a "one touch up" function. This feature was NOT built into the majority of vehicles equipped with power windows. If equipped, this feature works when the upper part of the power window switch is pushed briefly (approx. 0.3 seconds). The window will be raised automatically.

If the system detects an obstruction of any kind during window closing, the window is lowered again automatically. The safety function is activated if an obstruction is detected in the range from 200mm to 4 mm (7-7/8 to 5/32 in.) below the upper window frame.

1. From outside vehicle, raise windows using key (front door).

2. Turn key two times in "closed" direction, then wait at least three seconds.
   - At this point the safety function is programmed, and objects that interfere with the window being closed will cause it to be lowered again. Once the power window regulator is programmed, it will not be necessary to repeat the procedure unless the battery is disconnected

96

Door, rear lid, and power roof switches

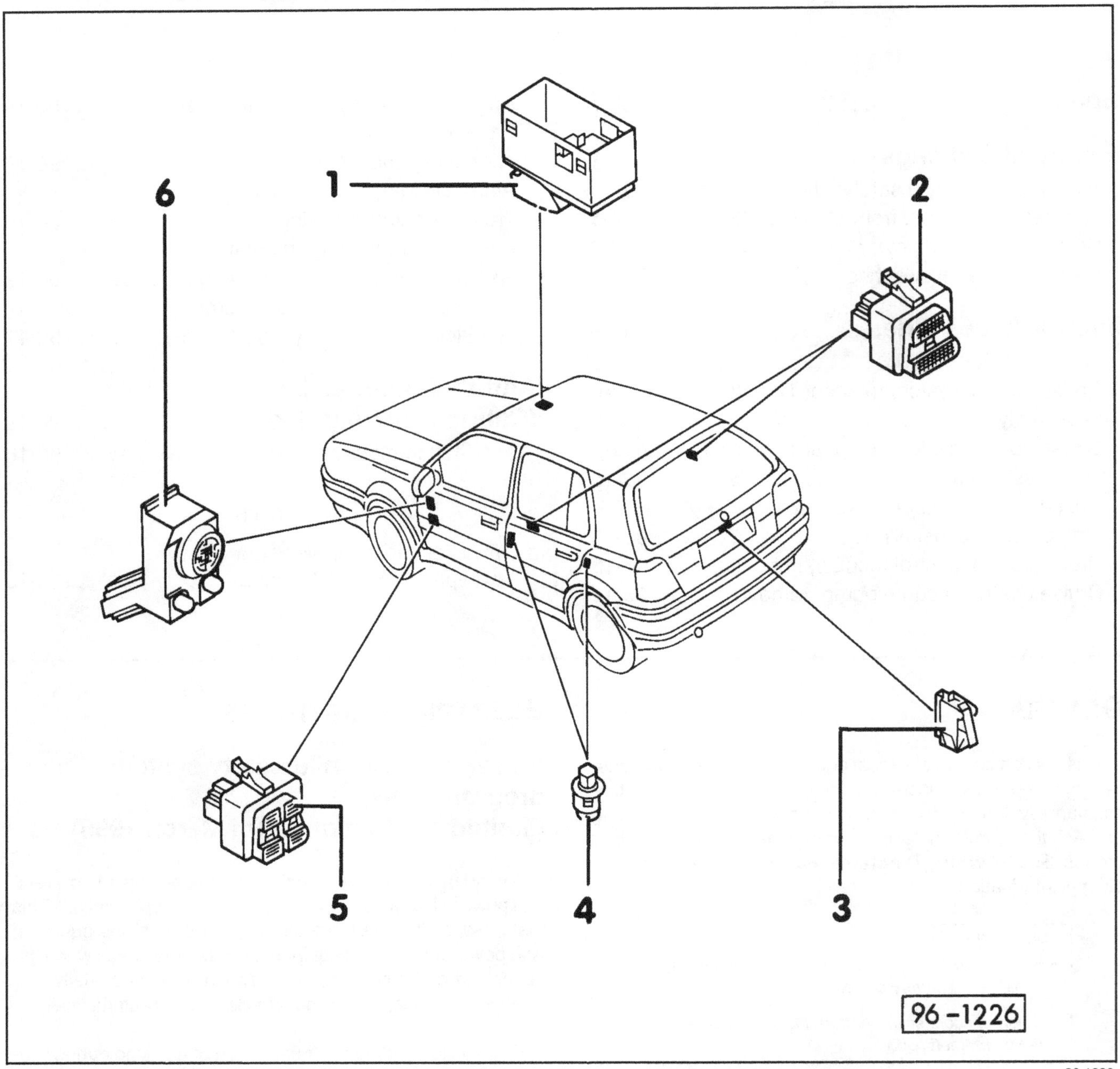

96-1226

Fig. 1. Switch locations for power accessories.

1. **Sunroof switch (E8)**
   - Removing and installing, **see** (A)
2. **Left rear window switch (E52)**
   - See **Power window child safety system, programming**
   - Removing and installing, **see** (B)
3. **Luggage compartment light switch (F5)**
   - Removing and installing, **see** (C)
4. **Left front door contact switch (F2)**
   - Remove rubber boot and pry out with screwdriver
5. **Left front power window switch (E40)**
   - See **Power window child safety system, programming**
   - Removing and installing, **see** (B)
6. **Mirror adjustment switch (E43) with mirror selector switch (E48)**
   - Removing and installing, **see** (D)

(A) Sunroof switch, removing and installing

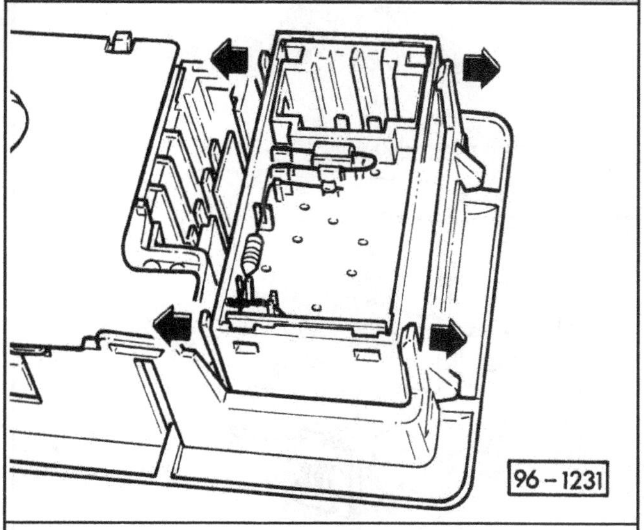

96-1231

- Remove front interior light and sunroof switch connector.
- Spread switch retaining tabs (arrows) and pry switch from interior light assembly.
- Install in reverse order of removal.

(B) Power window switches, removing and installing

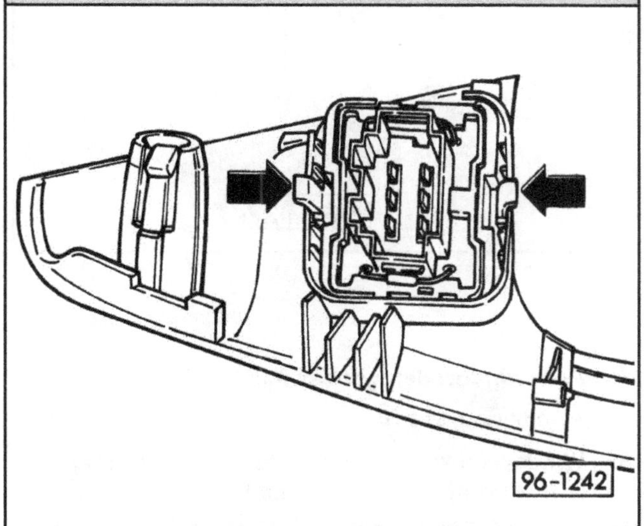

96-1242

- Carefully pry off upper trim of door grab handle and remove window switch connector.
- Squeeze switch retaining tabs (arrows) and remove switch.
- Install in reverse order of removal.

(C) Luggage compartment light switch, removing and installing

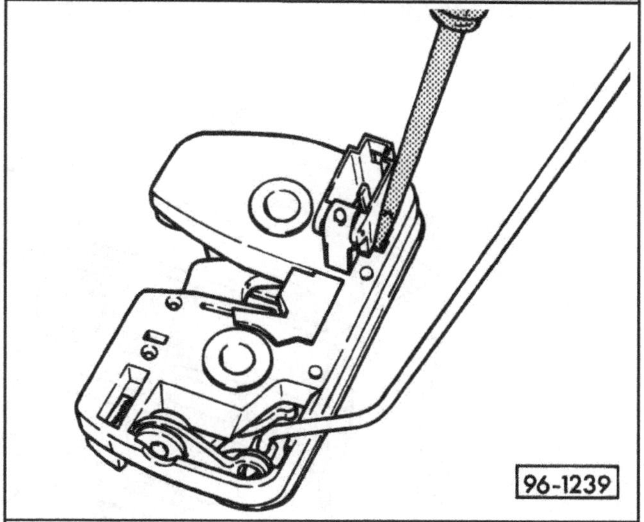

96-1239

- Located on the rear hatch/trunk lid lock assembly.
- Remove rear lid latch, see 55 Hood and Lids.
- Remove switch connector.
- Using a screwdriver, disengage retaining tabs and pull switch from lock assembly.
- Install in reverse order of removal.

(D) Mirror adjustment switch with selector switch, removing and installing

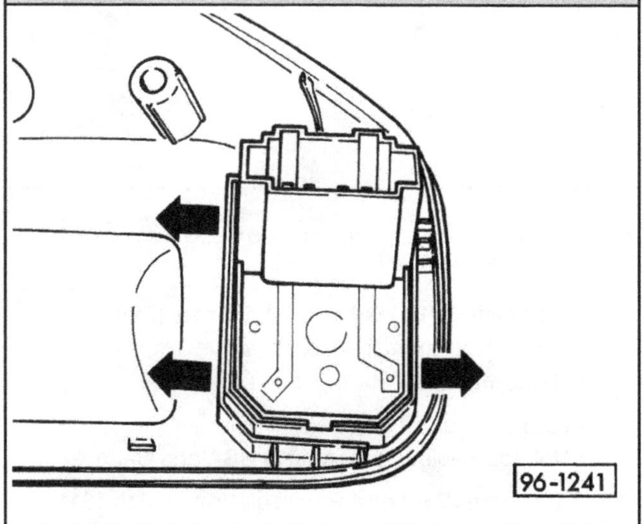

96-1241

- Pry off door handle trim and remove switch connector.
- Disengage switch retaining tabs (arrows) and remove switch from trim.
- Install in reverse order of removal.

Instrument panel switches

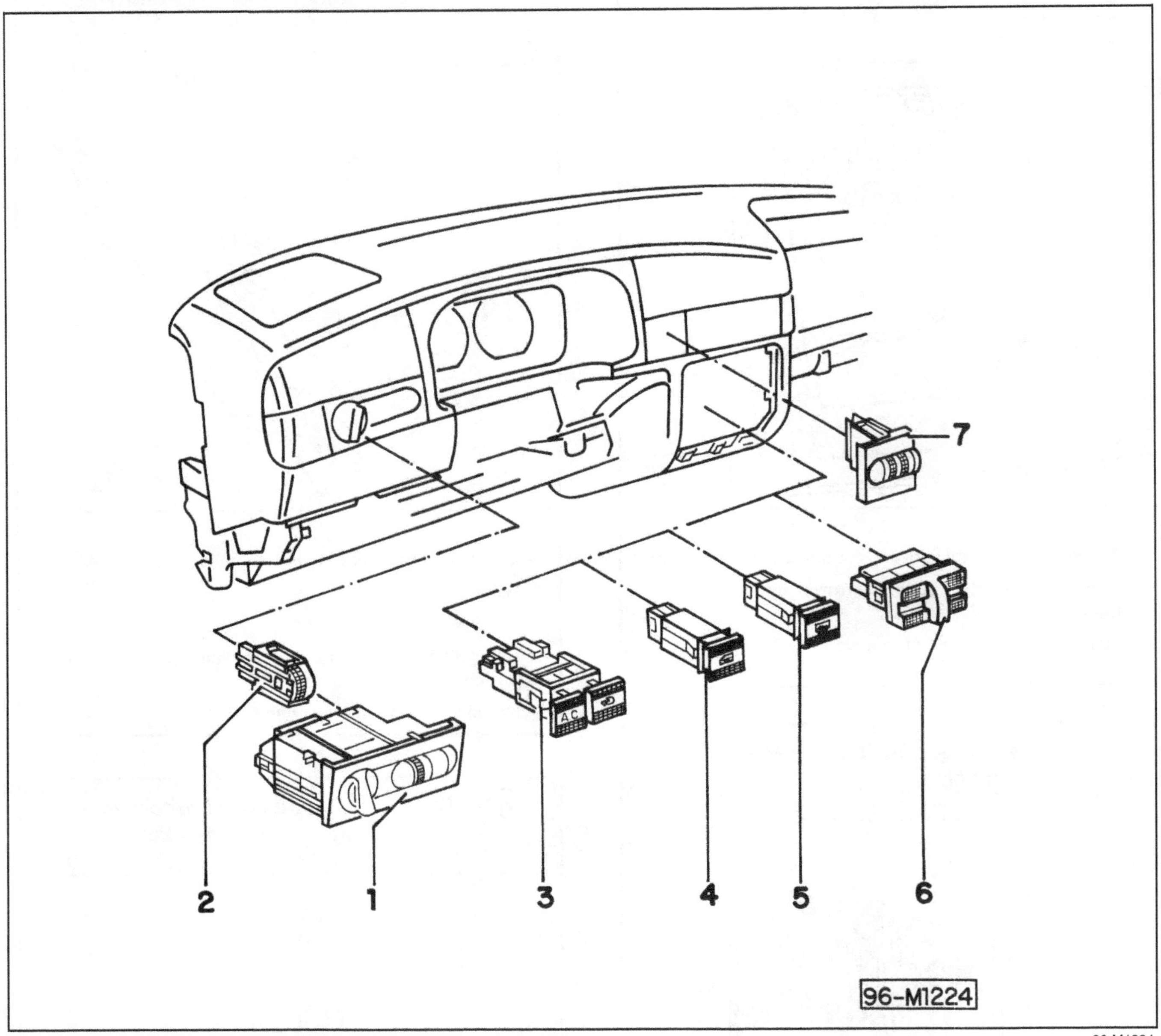

96-M1224

Fig. 2. Instrument panel switches for 1996 model year, other model years are similar.

1. **Light switch (E1) with panel dimmer (E20)**
   - For parking lights, headlights and fog lights
   - Removing and installing, **see** (A)
2. **Headlight height adjuster**
   - Not applicable to cars sold in USA and Canada
3. **A/C switch (E35) and Recirculation switch (E159)**
   - Removing, **see** (B)
4. **Door lock switch (E150)**
   - Not applicable to all models
   - Some models use a double switch located in the driver's side and passenger's side door pockets
   - Removing, **see** (B)
5. **Rear window defogger switch (E15)**
   - Removing, **see** (B)
6. **Window lockout switch (E39) with left and right rear window switch (E53 and E55)**
   - Four door vehicles with power windows.
   - See **Power window child safety system, programming**
   - Removing, **see** (B)
7. **Adjuster for heated driver's and passenger's seat (E94 and E95)**
   - Available as an option on some Golf, Jetta, and Cabrio models.
   - Removing and installing, **see** (C)

(A) Light switch, removing and installing

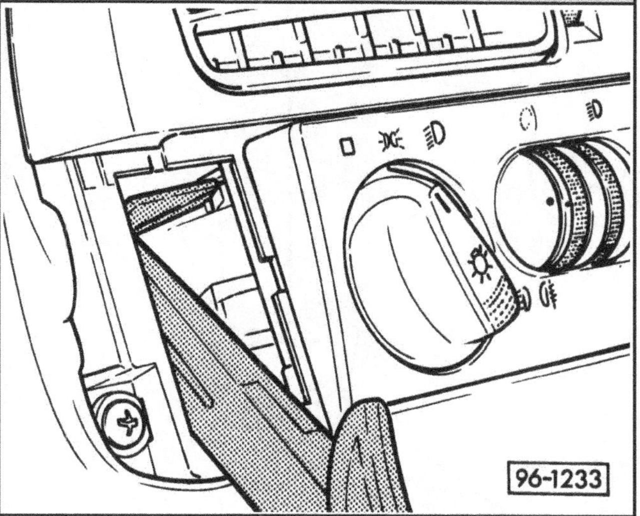

96-1233

- Light switch shown in illustrations differs slightly from light switch installed in vehicle.
- Carefully pry off trim piece on left side of light switch.
- Press switch release button on side of switch as shown in illustration.
- Pull light switch from instrument panel, remove connector and light switch.
- Install in reverse order of removal.

(B) Switches below heater controls, removing

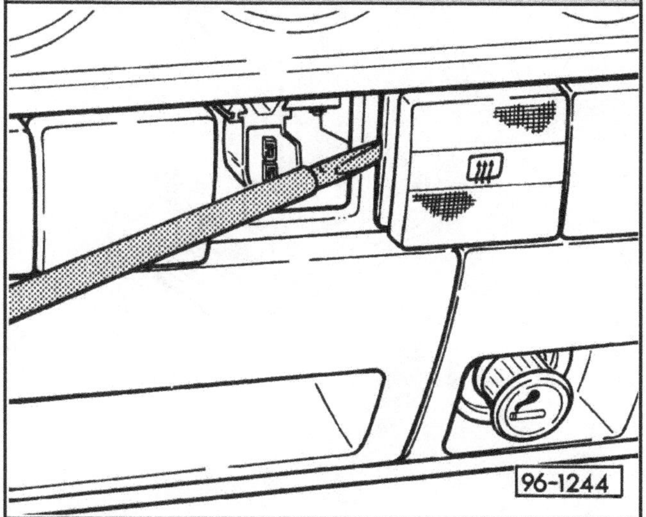

96-1244

- The procedure applies to rear window defogger switch, A/C switch, window lockout switch, left rear window switch, right rear window switch, and door lock switch.
- Carefully pry off cover plate (not all models).
- Insert screwdriver between switch housing and instrument panel, carefully pry out switch and remove connector.

(C) Heated seat switch

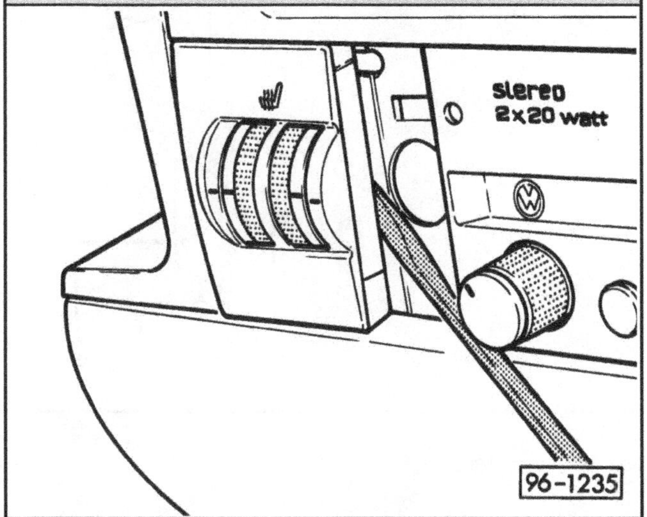

96-1235

- Carefully pry off trim plate located to right of switch.
- Insert screwdriver on right side of switch between switch housing and instrument panel, carefully pry out switch and remove connector.

INTERIOR ACCESSORIES

Interior lights

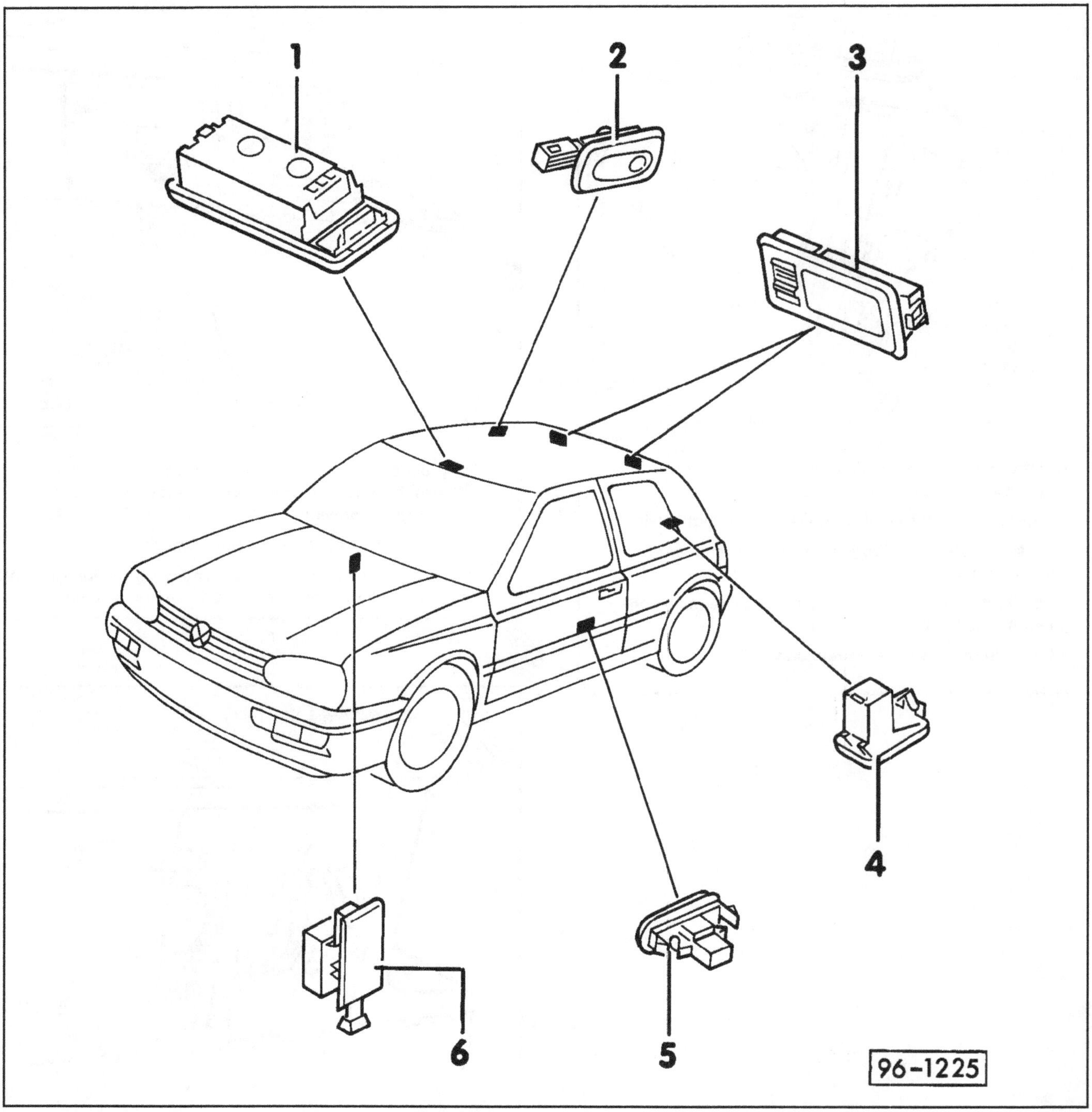

96-1225

Fig. 3. Interior lighting.

1. **Front interior light (W)**
   - Includes right front map/reading light (W13) and sunroof switch (E8)
   - Removing and installing, **see** Ⓐ and Ⓑ
2. **Right rear reading light (W12) (Jetta only)**
   - Left rear reading light (W11) also installed in Jetta
3. **Rear/luggage compartment interior light (W1)**
   - Not applicable to cars sold in USA and Canada
4. **Luggage compartment light (W3)**
   - Replacing, **see** Ⓒ
5. **Door courtesy light (Jetta only)**
6. **Glove compartment light (W6)**
   - Replacing, **see** Ⓓ

A Front interior light, removing

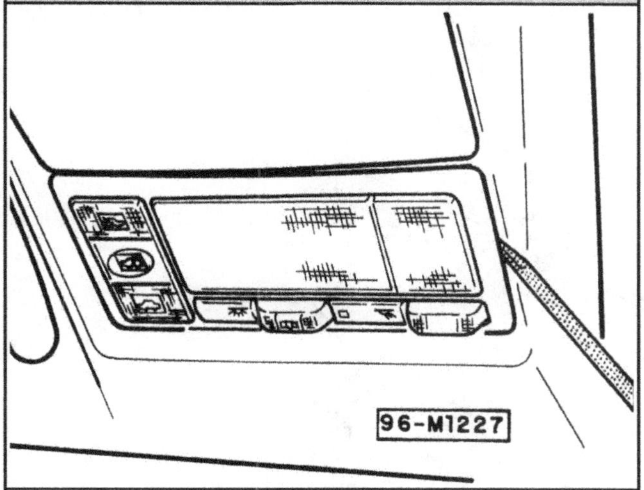

96-M1227

- Using a screwdriver, carefully pry out front interior light assembly from headliner and remove wire connectors.
- When checking or replacing the front interior light bulb, remove only the lens; it is not necessary to remove the entire light assembly.

B Right front map/reading light bulb, removing/replacing

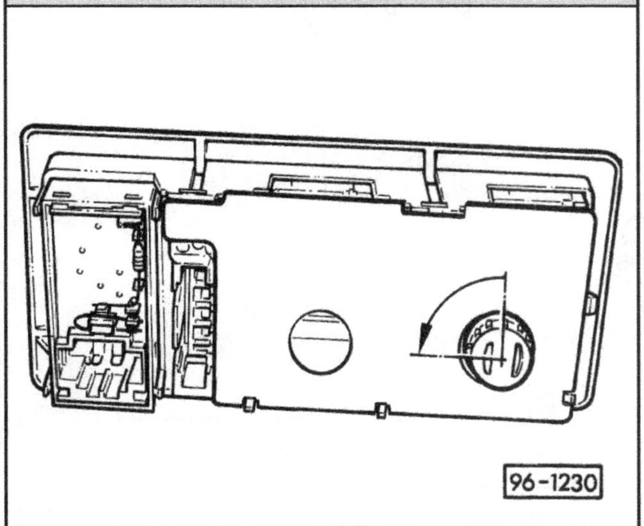

96-1230

- Remove front interior light assembly.
- Twist bulb holder counter-clockwise (arrow) and remove from interior light.
- Pull bulb (12V/5W) straight out of holder (do not twist), replace with new bulb as necessary.

C Luggage compartment light, replacing

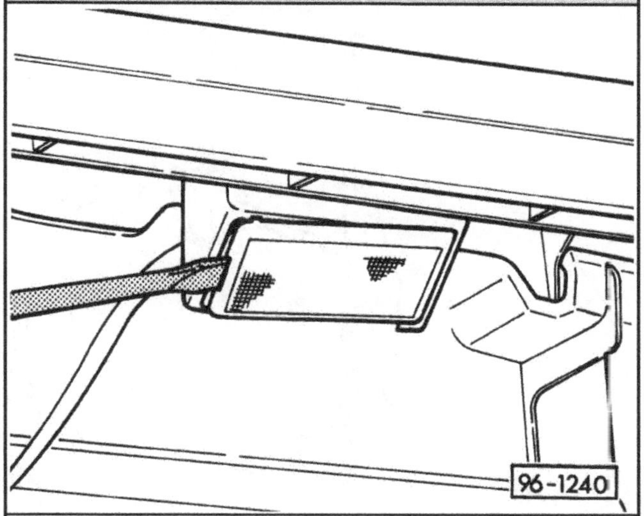

96-1240

- Using a screwdriver, carefully pry out luggage compartment light.
- Remove wire connector and replace bulb (12V/3W) as necessary.

D Glove compartment light, replacing (cars without passenger's side airbag)

96-1229

- Position screwdriver at top portion of glove compartment light as shown and carefully pry out light.
- Remove lens, bulb holder and microswitch and replace bulb (12V/3W) as necessary.

Cruise control system, (except 1Z/AHU TDI engine) assembly

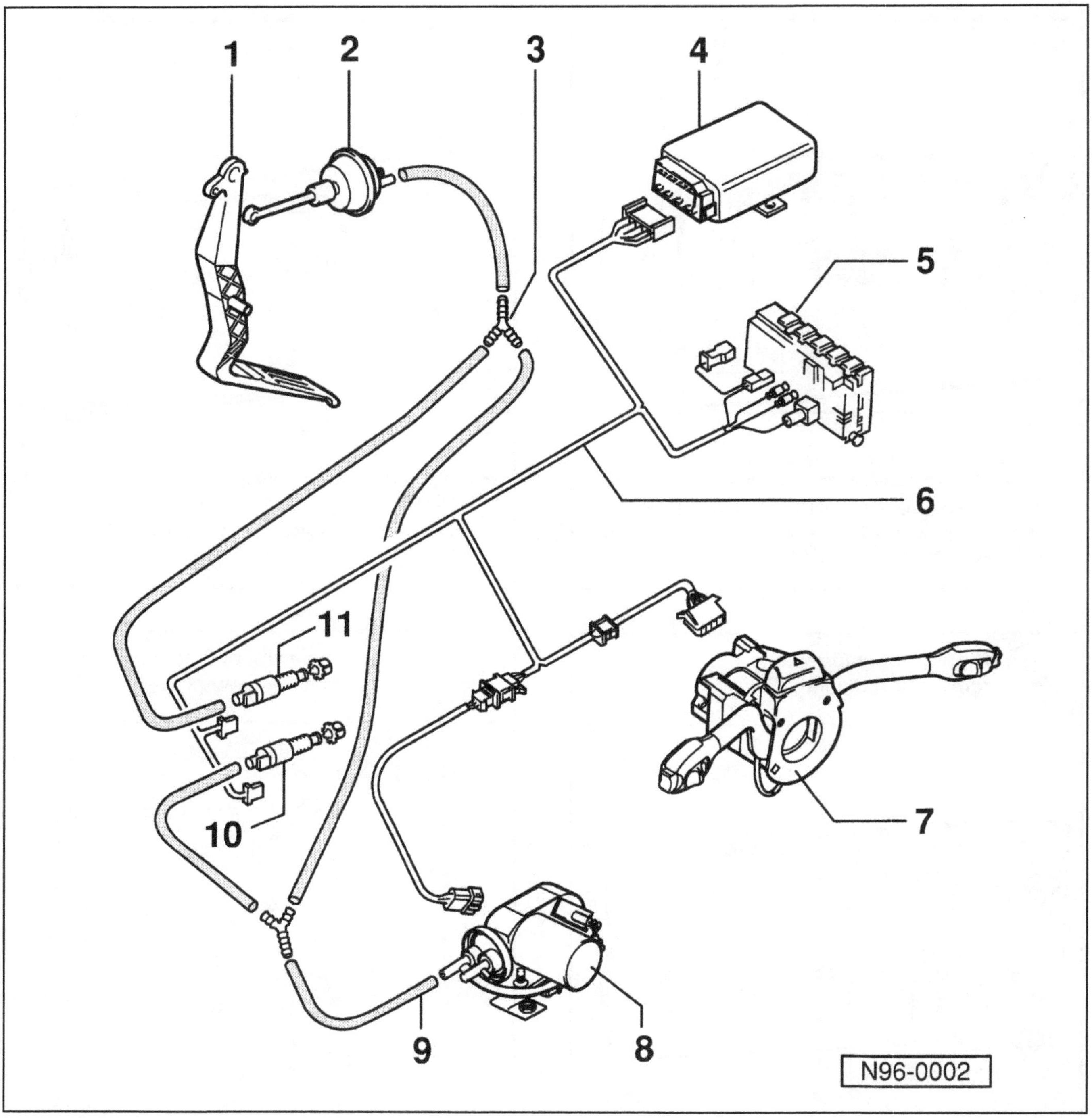

N96-0002

Fig. 4. Cruise control system components.

> ***CAUTION —***
> - *Before working on electrical system disconnect negative (–) battery cable.*
> - *Before disconnecting battery be sure to obtain radio anti-theft code.*

1. **Accelerator Pedal**
2. **Throttle control diaphragm**
   - Check for leaks
   - Removing, **see** (A)
   - Adjusting, **see** (B)

3. **T-connector**
4. **Cruise control control module (J213)**
   - Located in driver's side A-pillar
   - See **Cruise control control module, removing**
5. **Relay panel**
6. **Wiring harness**
7. **Cruise control switch (E45)**
   - Removing and installing, see **48 Steering**
8. **Cruise control vacuum pump (V18)**
   - See **Cruise control vacuum pump, removing and installing**
9. **Vacuum hose**
   - Check for leaks
10. **Vacuum vent valve with cruise control clutch pedal position (CPP) switch (F36)**
    - Vehicles with manual transmission only
11. **Vacuum vent valve with cruise control brake pedal position (BPP) switch (F47)**

(A) Throttle control diaphragm, removing

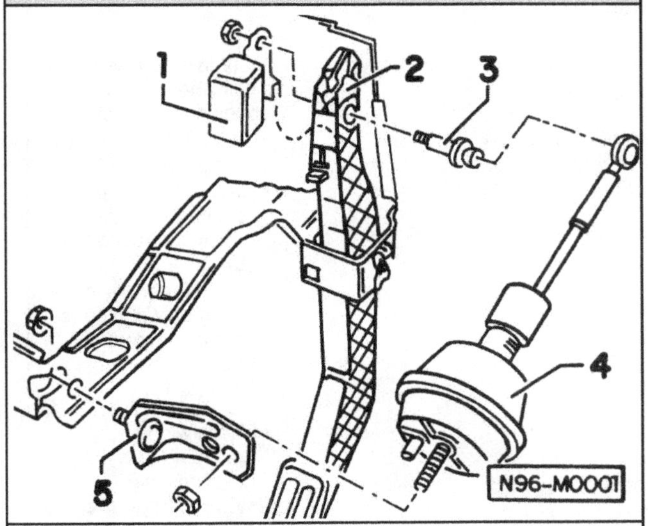

N96-M0001

- **Balance weight (not installed on all vehicles) (1)**
- **Accelerator pedal (2)**
- **Remove M8 securing nut and throttle control element (4) from bracket (5).**
- **Carefully pull throttle control rod off ball socket (3).**

(B) Throttle control diaphragm, adjusting

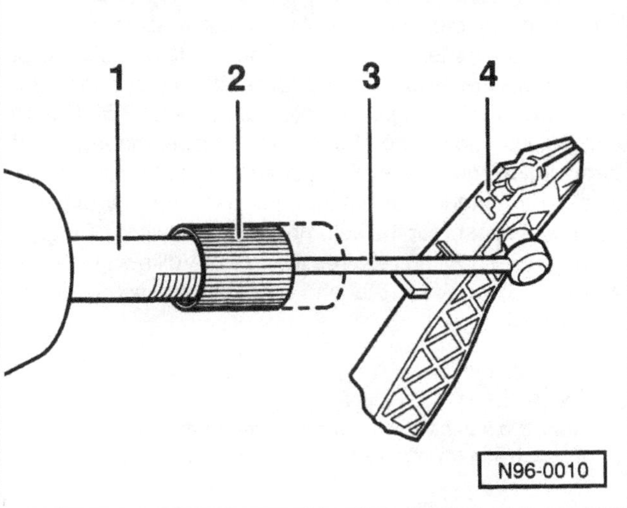

N96-0010

- **Remove adjusting sleeve (2) from throttle control diaphragm(1) and push fully forward.**
- **Adjust accelerator pedal cable (3).**
- **Pull adjusting sleeve to rear until there is maximum play of 1 mm (0.04 in.) and lock by turning clockwise onto stop.**

Cruise control system, (except AHU 1Z/TDI engine) troubleshooting

Factory installed cruise control is available as an option or as standard equipment on most Golf and Jetta models. All of these systems are essentially the same differing only in the location of fuses. The single exception is the Jetta TDI. Due to the fully electronic nature of this diesel engine management system, cruise control is simply added into the function of the engine control module eliminating the need for any external components except for the column mounted switch. Diagnosis and troubleshooting requires the use of Volkswagen scan tool VAG 1551 or VAG 1552 and is beyond the scope of this repair manual.

NOTE —

Always have the correct wiring diagram available before starting any repairs or troubleshooting.

CAUTION —

Use only a multimeter and jumper wires that have the correct terminal ends. Damage to the harness connector terminals could result if incorrect terminals are used.

1. Disconnect the 12 terminal harness connector from the cruise control ECU under the left side of the dash near the fuse/relay panel.

2. Use appropriate adapters and carefully probe the harness connector or jumper it as required, see **table a**. The cruise control ECU terminals are clearly marked on both the housing and the ECU.

3. When troubleshooting is completed, be sure to remove all jumper wires including the power source to fuse S51. Reinstall fuse.

4. If all tests are OK, substitute a known good cruise control ECU and road test.

Table a. Troubleshooting cruise control (except 1Z/AHU TDI engine)

Step	Item to be checked	Connect multimeter to cruise control harness connector terminals	Test procedures and/or extra steps	Expected results	Area to check if result is NOT obtained
1-Automatic Transmission	Voltage supply, terminal 15	**9** and vehicle **GROUND**	**Automatic Transmission:** Using a helper to read multimeter, road test car with transmission in **D**, **3**, or **2** and accelerate to above 25 m.p.h. Switch on cruise control.	Battery voltage	Voltage is supplied from TCM when step 1A conditions are met • Check TCM or break in wiring from TCM • Check for voltage at terminals 1 and 4 of cruise control switch (E45) • Check for break in wiring between cruise control switch (E45) and cruise control ECU (J213)
1A-Automatic Transmission			**Automatic Transmission, to continue testing after above test is completed:** Remove fuse S51. Using a fused jumper, supply 12 volts back into the socket of the removed fuse. Try both terminals until battery voltage is indicated on multimeter. Switch ignition on. Switch on cruise control. Leave jumper in place for all remaining tests.		• Check wiring according to wiring diagrams, see **97 Wiring Diagrams, Fuses and Relays** • Confirm that fused jumper wire works

continued on next page

Table a. Troubleshooting cruise control (except 1Z/AHU TDI engine)

Step	Item to be checked	Connect multimeter to cruise control harness connector terminals	Test procedures and/or extra steps	Expected results	Area to check if result is NOT obtained
1-Manual Transmission	Voltage supply, terminal 15	**9** and vehicle **GROUND**	**Manual Transmission:** Switch on ignition and cruise control	Battery voltage	• Check appropriate fuse in fuse/relay panel • Check for voltage at terminals 1 and 4 of cruise control switch (E45) • Check for break in wiring between cruise control switch (E45) and cruise control ECU (J213)
2	ECU grounds, terminal 31	**9** and **4** **9** and **12** (some cars)	Switch on ignition and cruise control	Battery voltage	• Check for break in wiring between cruise control ECU and ground
3	Cruise control switch (E45), set function	**8** and **4**	Switch on ignition and cruise control. Push and hold set button.	Battery voltage	• Check for voltage at terminal 3 of cruise control switch (E45) • Check for break in wiring between cruise control switch (E45) and cruise control ECU (J213)
4	Cruise control switch (E45), resume function	**5** and **4**	Switch on ignition and cruise control. Push and hold resume button.	Battery voltage	• Check for voltage at terminal 2 of cruise control switch (E45) • Check for break in wiring between cruise control switch (E45) and cruise control ECU (J213)
5	Cruise control brake pedal position switch (F47) and clutch pedal position switch (F36) (as applicable)	**3** and **9**	Switch on ignition and cruise control.	Battery voltage	The cruise control ECU senses continuity to ground through this part of the circuit. This check tests the circuit using the brake lights. • At least one brake light must work properly • Check brake pedal position and/or clutch pedal position switches for adjustment • Check brake and/or clutch adjustment
			Switch on ignition and cruise control. Depress brake pedal. Depress clutch pedal.	No voltage when either pedal is depressed	
6	Vehicle speed sensor (G22)	**7** and **4**	Switch on ignition. Raise one front wheel sufficiently to allow wheel to be turned by hand. Slowly turn wheel.	Voltage must change between low and high as noted below (typical values) when wheel is rotated. • Low voltage: 0-4 volts • High voltage: 7-12 volts	The cruise control ECU uses the same Hall speed sensor as the speedometer. • Speedometer must work properly • Check for break in wiring between speedometer and fuse/relay panel and fuse/relay panel and cruise control ECU

continued on next page

Table a. Troubleshooting cruise control (except 1Z/AHU TDI engine)

Step	Item to be checked	Connect multimeter to cruise control harness connector terminals	Test procedures and/or extra steps	Expected results	Area to check if result is NOT obtained
7	Vacuum pump and valve (V18), servo, connecting hoses, brake and clutch vacuum components, pedal position switches	No multimeter connections required for this step	Use suitable jumper wires. Jumper all of these harness connector terminals: 9 and 11 1 and 4 2 and 4 Switch on ignition Do not start engine! Switch on cruise control	Cruise control vacuum pump (V18) runs and throttle opens fully (goes to floor)	• Check cruise control vacuum pump • Check wiring between cruise control vacuum pump and ECU • Check vacuum servo and all connecting hoses for cracks and leaks
			Depress brake pedal (jumpers remain in place)	Throttle must close fully (return)	• Check brake pedal position switch and adjustment
			Release brake pedal (jumpers remain in place)	Throttle must open fully (goes to floor)	–
			Depress clutch pedal (manual transmission only) (jumpers remain in place)	Throttle must close fully (return)	• Check clutch pedal position switch and adjustment
			Release clutch pedal (manual transmission only) (jumpers remain in place)	Throttle must open fully (goes to floor)	–
			Disconnect jumper from **2** and **4**	Cruise control vacuum pump must shut off and throttle must stay open (on the floor)	• Check vacuum servo and all connecting hoses for cracks and leaks
			Remove remaining jumpers	Throttle must close fully (return). Operation must be smooth.	• Check cruise control vacuum pump • Check servo and linkage for mechanical problems and binding

Cruise control vacuum vent valve, removing and installing

1. Remove driver' side lower trim from below instrument panel, see **70 Trim-Interior**.
2. Remove wiring connector and vacuum hose from vent valve. Unscrew vent valve from bracket. See Fig. 5.

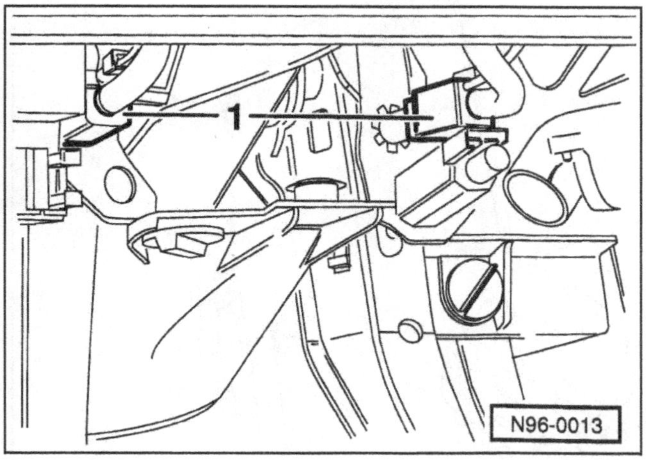

N96-0013

Fig. 5. Cruise control vacuum vent valves (**1**).

3. To install, screw vent valve into bracket to stop.
4. Using pedal, pull vent valve back to pedal rest position. Vacuum vent valve is now adjusted.
5. Reconnect harness connector and vacuum hose.

Cruise control control module, removing

Control module is located behind bracket on left A-pillar.

1. Remove driver's side lower trim from below instrument panel, see **70 Trim-Interior**.
2. Disconnect 12-pin connector from cruise control control module.
3. Disconnect 6-pin and 10-pin connectors from Alarm system control module.

NOTE —

The alarm system control module and cruise control control module are located on the same mounting bracket.

4. Remove assembly carrier by removing retaining nut. See Fig. 6.
5. Pull control module mounting plate locking lugs out from carrier and remove mounting plate with control module.

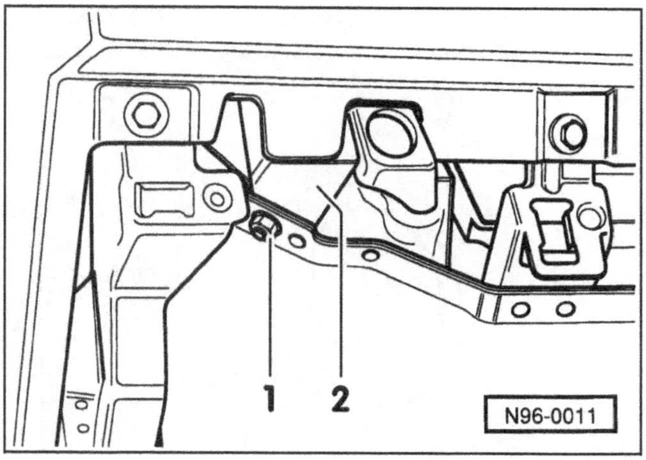

N96-0011

Fig. 6. Cruise control control module (**2**) and assembly carrier retaining nut (**1**).

6. Remove control module mounting bolts from mounting plate and remove control module.

Cruise control vacuum pump, removing

1. Remove battery, see **27 Engine Electrical**.
2. Remove windshield washer fluid reservoir.
3. Separate 3-pin connector.
4. Remove vacuum pump mounting bracket bolt. Then slide bracket with pump from its mount. See Fig. 7.

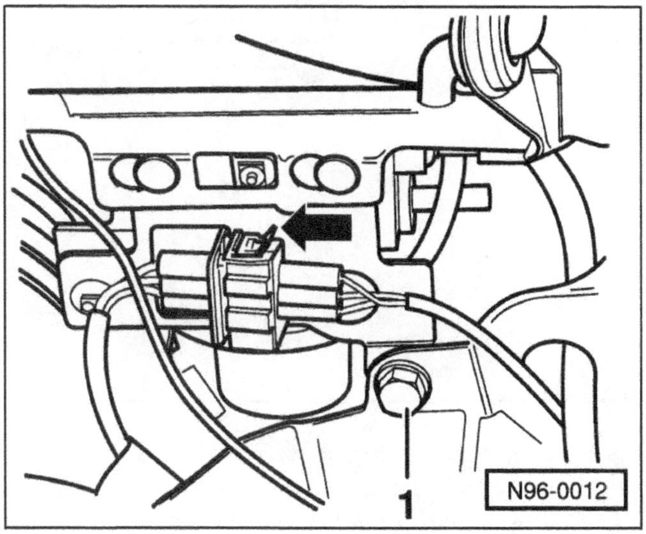

N96-0012

Fig. 7. Remove bolt (**1**) and slide mounting bracket in direction of arrow.

Anti-theft alarm system, components

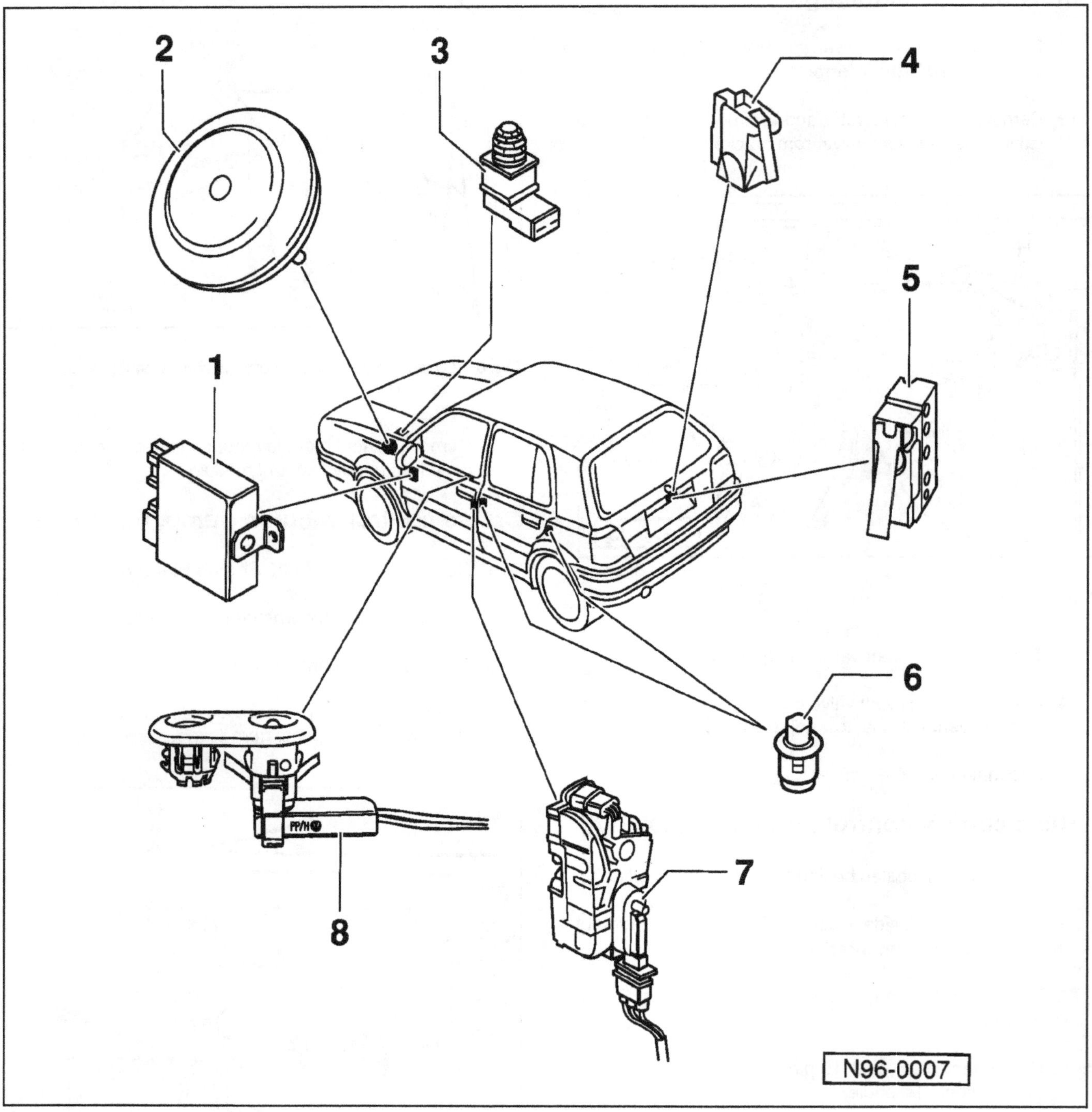

N96-0007

Fig. 8. Anti-theft alarm system, components.

1. **Alarm system control module (J284)**
2. **Alarm horn (H8)**
   - Removing and installing, **see** Ⓐ
3. **Hood alarm switch (F120)**
   - Removing and installing, **see** Ⓑ
4. **Trunk lid alarm switch (F123)**
5. **Trunk lock alarm/central locking switch (F124)**
6. **Door contact switch (F2, F3, F10, F11)**
7. **Driver's door central locking system switch (F59)**
8. **Alarm system indicator light (K95)**
   - Removing and installing, **see** Ⓒ

(A) Alarm horn, removing

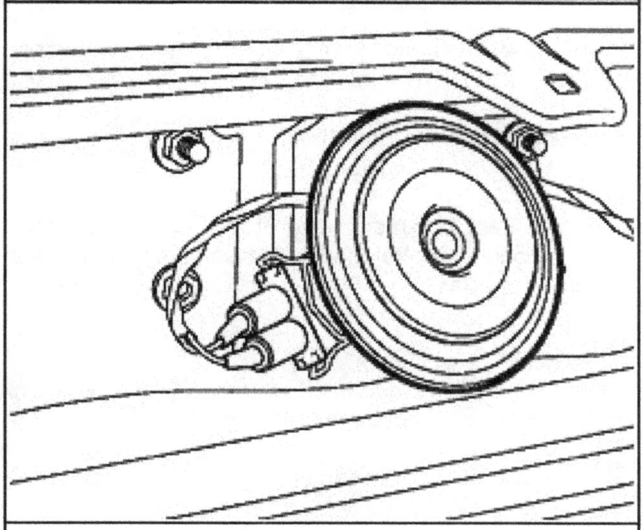

VA3H8

- Located in fresh air plenum, left side, behind windshield wiper motor and linkage.
- Access requires removal of plastic water shield and wiper assembly.
- Some early production vehicles may have alarm horn located behind left front inner fender liner near coil spring.

(B) Hood alarm switch, removing

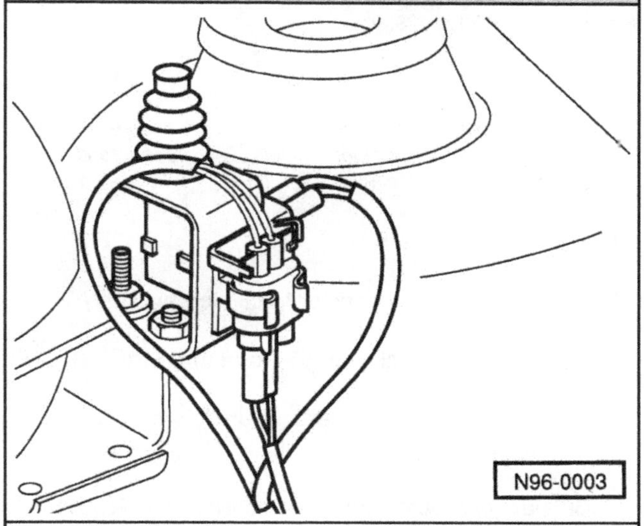

N96-0003

- Located in left side of engine compartment, near shock tower.
- Remove 2-pin connector from switch and remove contact switch from mounting bracket.

(C) Alarm system indicator light, removing

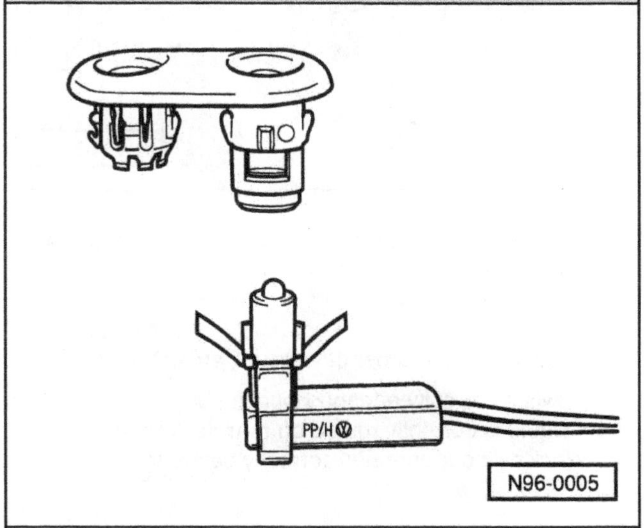

N96-0005

- Remove door trim (see 70 Trim-Interior), and disconnect 2-pin connector.
- Release warning lamp lower section from locating device and pull out. Pull upper section out of door trim.

Trunk lock microswitch, removing and installing

The trunk lock microswitch functions as part of the alarm/central locking system.

NOTE —
The following procedure describes removal and installation of the switch for Golf models. The procedure is similar for Jetta and Cabrio.

1. Remove rear lid trim.
2. Disconnect 2-pin connector from microswitch.
3. Remove rear lid lock, see **55 Hood and Lids**.
4. Insert key into lock cylinder and turn lock cylinder clockwise to stop.
5. Carefully pry microswitch from locating pins. See Fig. 9.

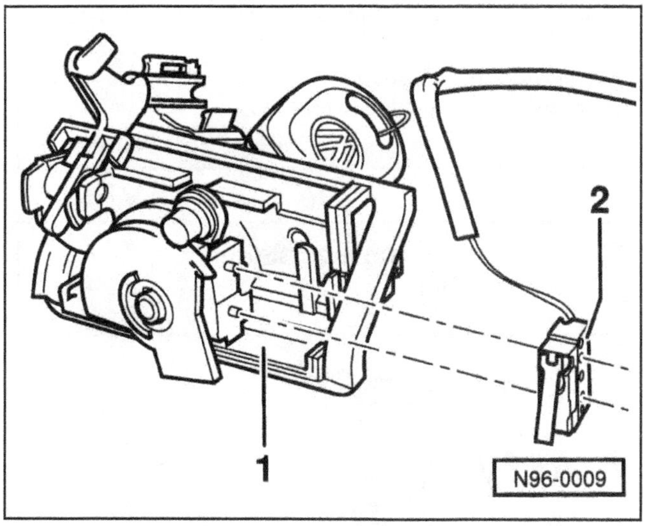

N96-0009

Fig. 9. Trunk lock alarm/central locking mechanism with microswitch (**2**) and installation location (**1**).

6. Install in reverse order of removal, ensuring that:
   - switch correctly engages locating pins
   - wiring is correctly routed on rear deck lid lock
   - lock cylinder operates correctly before installing rear lid lock

NOTE —
Various types of switches have been used in production. Installing the incorrect contact switch will render the alarm inoperable on vehicles with central locking.

Alarm system control module, removing and installing

Control module is located behind bracket on left A-pillar.

1. Remove driver' side lower trim from below instrument panel, see **70 Trim-Interior**.
2. Disconnect 6-pin and 10-pin connectors from alarm system control module.
3. Disconnect 12-pin connector from cruise control control module. Control module is located behind bracket on left A-pillar.

NOTE —
The alarm system control module and cruise control control module are located on the same mounting plate.

4. Remove hex nut from carrier bracket. See Fig. 10.

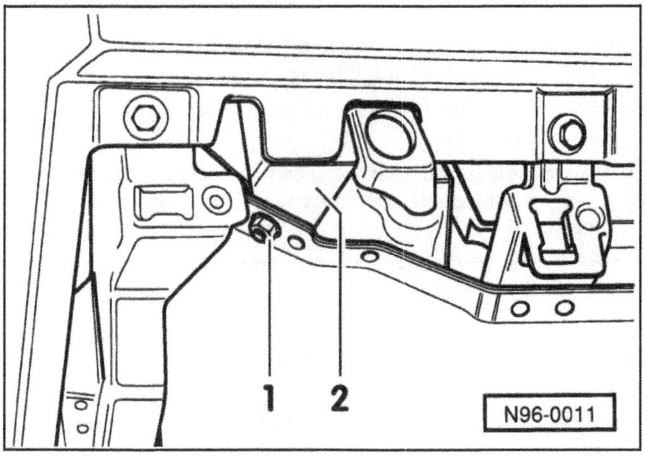

N96-0011

Fig. 10. Alarm system control module (**2**) and assembly carrier retaining nut (**1**).

5. Pull mounting plate locking lugs out from bracket and remove mounting plate with control module.
6. Remove control module mounting bolts from mounting plate and remove control module.
7. Install in reverse order of removal.

Keyless remote entry transmitter, programming (m.y. 1998-1999)

Beginning with the 1998 model year, a factory installed keyless remote locking system is standard equipment. This enhancement to the central locking and alarm systems uses a remote radio frequency transmitter about the size and shape of a regular vehicle key. See **5 Body–Assembly, General Information** for a description of the keyless remote system and its various functions.

If new or additional keyless remote entry transmitters are required, use the following procedure to program the vehicle to accept the new transmitter or reprogram existing transmitters. Up to a maximum of 4 transmitters can be programmed into each vehicle. If one transmitter is activated 200 times consecutively out of range of the vehicle, all transmitters may require reprogramming.

NOTE —

This procedure requires 2 keys and must be completed within one (1) minute.

1. Switch ignition ON.
2. Close all doors and rear hatch/trunk lid.
3. Using a second key in driver's door, turn key to UNLOCK position and hold for approximately 10 seconds or until alarm horn sounds 3 times.
4. Press the UNLOCK/DISARM button on the first transmitter (alarm horn will briefly sound once to confirm programming).
5. Press the UNLOCK/DISARM button on the second transmitter (alarm horn will briefly sound twice to confirm programming).
6. Repeat the procedure for the remaining transmitters up to a maximum of 4.
7. Turn and hold the key in the driver's door to the UNLOCK position for 10 seconds or until the alarm horn sounds 3 times.
8. Switch off ignition and remove all keys.

Keyless remote entry transmitter, replacing battery (m.y. 1998-1999)

The keyless remote entry transmitters are powered by a pair of batteries and are simply snapped together. See Fig. 11.

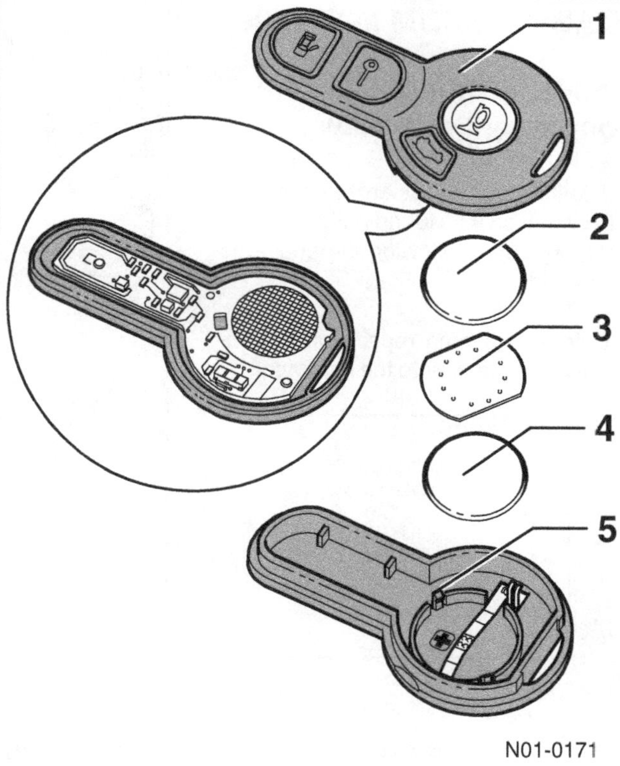

1. **Upper housing**
2. **Battery**
3. **Metal contact**
4. **Battery**
5. **Lower housing**

N01-0171

Fig. 11. Exploded view of radio frequency keyless remote entry transmitter. Observe polarity of batteries when replacing.

1. Using a screwdriver or a coin, separate the upper housing from the lower housing at an appropriate location.
2. Note the polarity of the upper battery and using a screwdriver or small pick, remove the upper battery.
3. Under the upper battery is a metal contact plate. Carefully turn the contact plate until the flat edges line up with the catches and remove the contact plate.
4. Note the polarity of the lower battery and remove.
5. Installation is the reverse. Note the polarity of the batteries. The positive side goes down into the lower housing.
6. When all components are properly positioned, snap both housing sections together.

ANTI-THEFT IMMOBILIZER (CABRIO FROM M.Y. 1999)

Anti-theft immobilizer component overview

The Immobilizer system is a theft-deterrent that checks for an authorized vehicle key before allowing the engine to be started.

Before beginning repairs on the immobilizer system, note the following:

> ***CAUTION—***
> - *Before disconnecting battery, be sure to obtain radio anti-theft code.*
> - *Switch ignition off.*
> - *Disconnect negative (–) battery cable.*

NOTE—
Anti-theft immobilizer has scan tool OBD capabilities.

1. **Immobilizer Control Module (J362)**
   - Located under the left side of the instrument panel near the steering column
2. **Immobilizer (data) reader coil (D2)**
   - Attached to steering lock
   - See **Immobilizer reader coil, removing and installing**
3. **Ignition key with electronic transponder**
   - Transponder is activated by signal from reader coil and sends appropriate signal back to reader coil
   - Transponder is integral with key
   - All keys have different coding

- **Engine Control Module (not shown)**
   - Communicates with Immobilizer control module to increase security

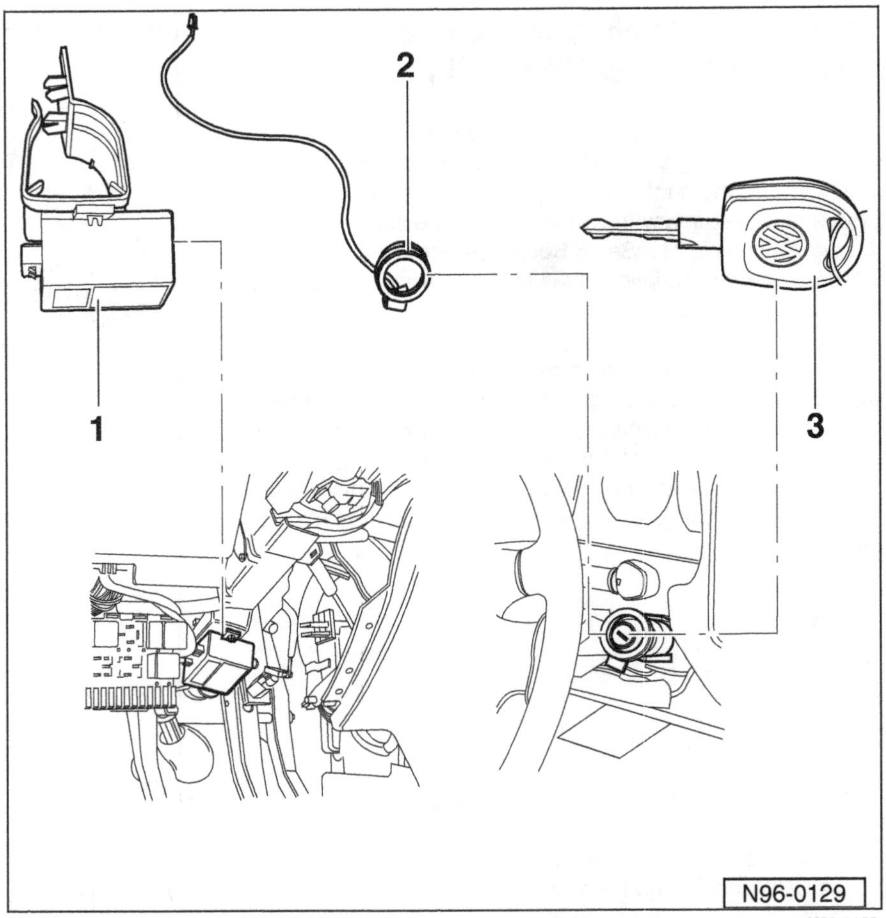

N96-0129

Fig. 12. Headlight assembly for Cabrio from 10/1998 production.

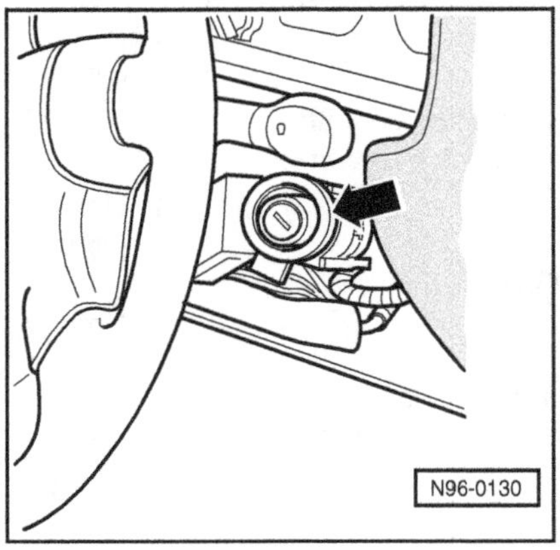

Immobilizer reader coil removing and installing

- Remove steering column switch upper and lower trim
- Pull reader coil (**arrow**) off of steering lock housing
- Disconnect harness connector and remove reader coil
- Install in reverse order

97 Wiring Diagrams, Fuses and Relays

GENERAL

This section contains wiring diagrams for 1993 through 1999 m.y. Golf, Jetta, and Cabrio models. Also included here are the fuse and relay positions.

> ***WARNING —***
> *On cars equipped with Airbags, special precautions apply to any electrical system testing or repair. The airbag unit is an explosive device and must be handled with extreme care. Before starting any work on an airbag equipped car, refer to the warnings and cautions in* **69 Seatbelts, Airbags**.

> ***CAUTION —***
> - *Always switch the ignition off and remove the negative (–) battery cable before removing any electrical components.*
> - *Before disconnecting battery, be sure to obtain radio anti-theft code.Connect and disconnect ignition system wires, multiple connectors, and ignition test equipment leads only while the ignition is switched off.*
> - *Always switch a test meter to the appropriate function and range before making test connections.*

How to find an electric circuit

1. Locate the index page for your Volkswagen by model year, model, and engine in the table of contents above.
2. Go to the applicable index page and find the electrical component or electrical circuit you are interested in diagnosing.
3. Go to the applicable wiring diagram page.
4. Using the list at the bottom of the wiring diagram page, find the component code for the circuit in question. Then find the component on the wiring diagram.

> ***NOTE —***
> *The page listed on the wiring diagram index is where a particular circuit starts.Circuits or systems will frequently run across more than one page. This is indicated by arrows at either end of the fuse/relay panel (gray bar) at the top of the diagrams.*

97

FUSES AND RELAYS

The fuse/relay panel is located under the left side of the dash board. The wiring harness connectors are attached to the back of the panel. A sliding lock is located between two rows of harness connectors. The lock will only slide and lock into place when all the connectors are fully seated.

Fuse positions

NOTE —

- *Fuse locations may not exactly match those listed below depending on vehicle accessories and optional equipment.*
- *Fuse locations for options not listed below can be found in the corresponding wiring diagram.*

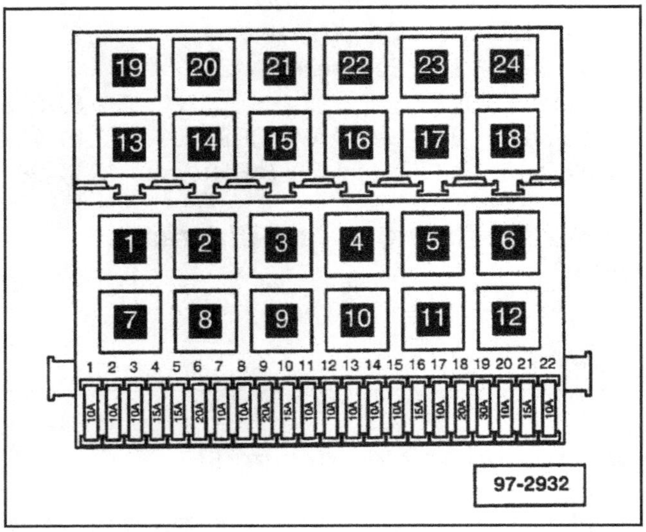

97-2932

Fig. 1. Fuses are located on the bottom of the fuse/relay panel.

Table a. Fuse identification

Fuse position	Description	Rating in Amps
1	Headlight, low beam, left	10
2	Headlight, low beam, right	10
3	License plate lights	10
4	Rear window wiper and washer system	15
5	Windshield wiper/washer	15
6	Fresh air blower, A/C	20
7	Tail lights, side marker lights, right	10
8	Tail lights, side marker lights, left	10
9	Rear window defogger	20
10	Fog lights	15
11	Headlight, high beam, left	10
12	Headlight, high beam, right	10
13	Horn	10
14	Back-up lights	10
15	Engine electronics	10
16	Power sunroof, front map/reading light	15
17	Emergency flashers	10
18	Fuel pump, heated oxygen sensor (HO2S) control module	20
19	Coolant fan, A/C relay	30
20	Brake lights	10
21	Interior lights, clock, luggage compartment light	15
22	Cigarette lighter, radio, luggage compartment release	10

The fuses come in different colors that correspond to different ratings. Each fuse is specifically chosen to protect its circuit against excess current flow that might damage the circuit components. When replacing fuses, it is never appropriate to substitute a fuse of a higher rating.

Fuse colors

- green 30A
- white 25A
- yellow 20A
- blue 15A
- red 10A

CAUTION —
To avoid damaging the electrical system, turn off all lights and accessories and remove the ignition key before replacing a fuse.

Relay positions

NOTE —
Relay locations may not exactly match those listed below depending on vehicle accessories and optional equipment. Relay locations are indicated in wiring diagrams.

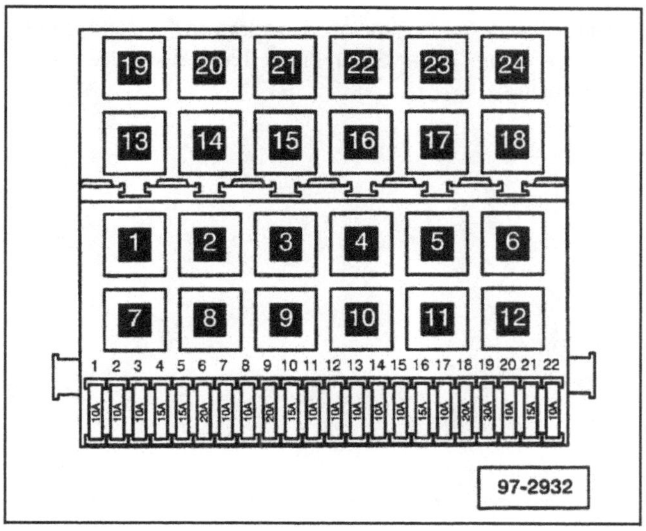

97-2932

Fig. 2. Relay locations are numbered as shown.

NOTE —
A large number is stamped on each relay. This is the production control number and can be used as an aid to identification.

Table b. Relay identification

Relay position	Description	Control number
1	Air conditioning	13
2	Rear window wiper/washer relay	72
3	ECM (engine control module) power supply	30,109, 288
4	Load reduction	18
5	Open	
6	Emergency flasher relay	21
7	Open	
8	Intermittent washer/wiper	19,99
9	Seat belt warning system	4, 29
10	Fog light relay	
11	Dual-horn relay or bridge for horn	53
12	Fuel pump relay – or – glow plug relay	67,80,167 102,104
13	Heated oxygen sensor (HO2S) Control Module	
14	Park/neutral position relay	150
15 thru 24	Available for optional equipment	

Fuse/relay panel, removing and installing

1. Disconnect negative (–) battery cable.

CAUTION —
Before disconnecting battery be sure to obtain radio anti-theft code.

2. Remove knee bar from left side of instrument panel. See **69 Seatbelts, Airbags**.

3. Remove lower instrument panel trim. See **70 Trim–Interior**.

4. Disengage retaining clip lock pins from brackets. See Fig. 3.

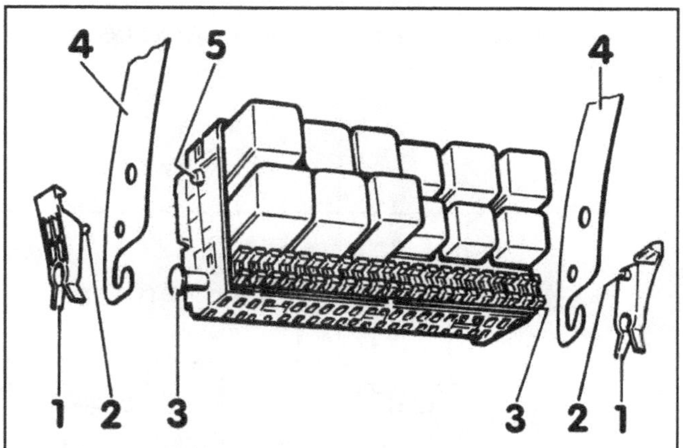

1. **Retaining clips**
2. **Lock pins**
3. **Fuse/relay panel pins**
4. **Brackets**
5. **Alignment pin**

97-3012

Fig. 3. Fuse/relay panel with mounting hardware.

5. Swing retaining clips forward and pull from fuse/relay panel pins.

6. Spread left retaining bracket and disengage from alignment pin.

7. Lift fuse/relay panel from brackets.

8. Pull out sliding lock on back of fuse/relay panel. See Fig. 4.

9. Depress connector locking tabs and pull multi-point connectors from fuse/relay panel.

10. Remove fuse/relay panel.

11. Install in the reverse order of removal, making sure to position alignment pins properly and to engage retaining clips in brackets. See Fig. 3.

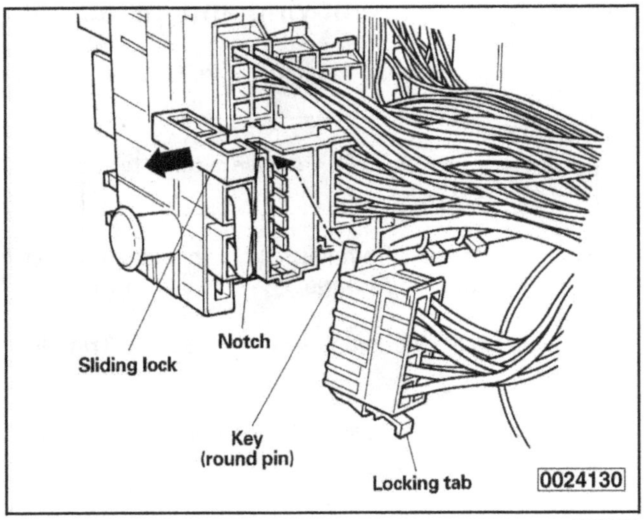

0024130

Fig. 4. Pull sliding lock in direction of arrow and depress locking tab to remove multi-point connectors.

Connections and plugs on fuse/relay panel

Each connector on the fuse/relay panel has a unique combination of pins and placement. The connectors are keyed to align with notches in the appropriate receptacle. This ensures that the connectors will not be incorrectly positioned. See Fig. 4.

The locations of the multi-point connectors on the back side of the fuse/relay panel are shown in Fig. 5. **Table c** identifies connector terminals.

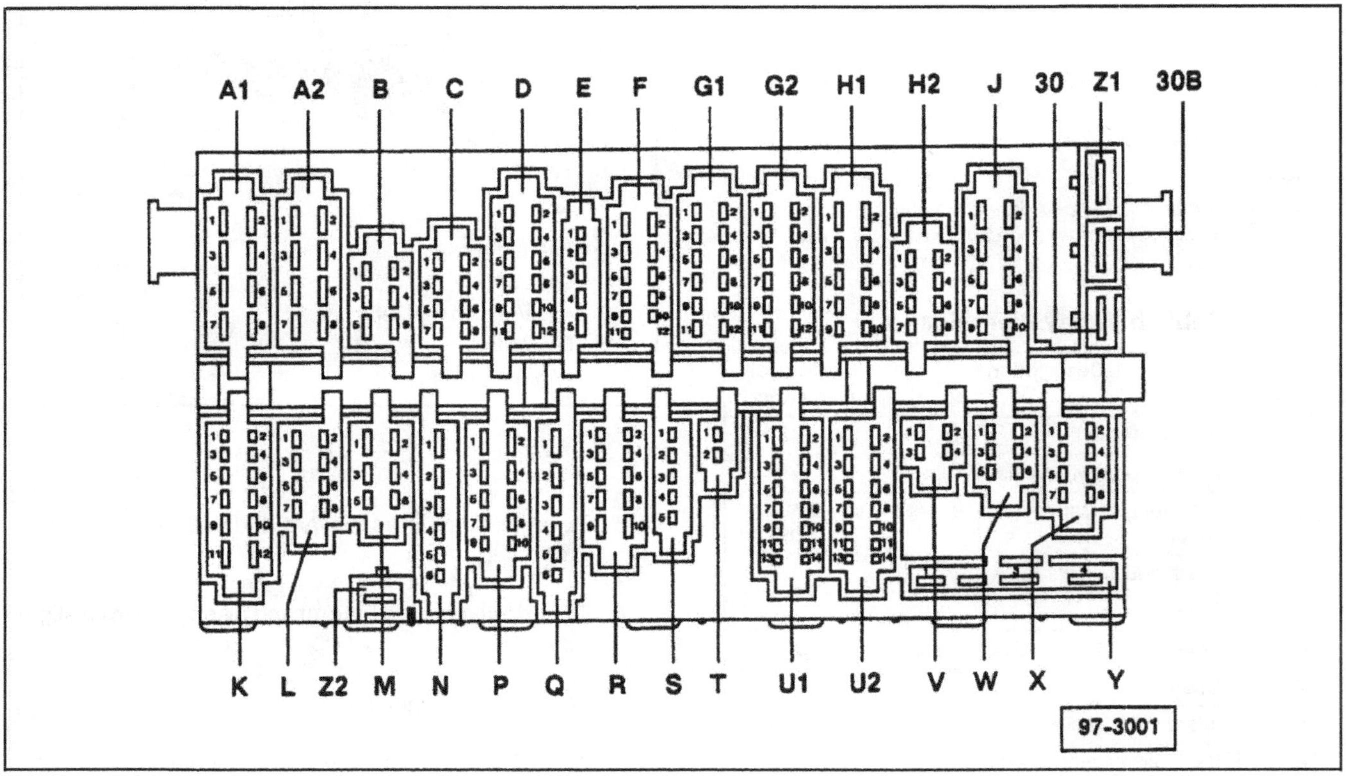

97-3001

Fig. 5. Multi-point connector locations on back of fuse/relay panel.

Table c. Fuse/relay panel connections

Position	No. of pins	Description	Color
A1	8-pin	Headlight wiring harness	yellow
A2	8-pin	Headlight wiring harness	yellow
B	6-pin	Headlight washer system wiring harness	green
C	8-pin	Headlight wiring harness	green
D	12-pin	Optional equipment wiring harness	green
E	5-pin	Instrument wiring harness	green
F	9-pin	Engine compartment wiring harness	white
G1	12-pin	Engine compartment wiring harness	white
G2	12-pin	Engine compartment wiring harness	white
H1	10-pin	Steering column switch harness	red
H2	8-pin	Steering column switch harness	red
J	10-pin	Steering column switch harness	red
K	12-pin	Rear wiring harness	black
L	7-pin	Rear wiring harness	black
M	6-pin	Rear wiring harness to fuel tank	black
N	6-pin	Headlight wiring harness	green
P	9-pin	Rear window defroster/fog light switch wiring harness	blue
Q	6-pin	Instrument wiring harness	blue
R	10-pin	Light switch wiring harness	white
S	5-pin	Engine compartment wiring harness	white
T	2-pin	Connector	green
U1	14-pin	Instrument cluster wiring harness	blue
U2	14-pin	Instrument cluster wiring harness	blue
V	4-pin	Steering column (multi-function indicator) wiring harness	green
W	6-pin	Optional equipment connector	green
X	8-pin	Warning lamp harness	yellow
Y	1-pin	Four single connectors, (terminal 30) (B+)	
Z1	1-pin	Single connector	
Z2	1-pin	Single connector (terminal 31, ground)	
30	1-pin	Single connector (terminal 30, B+)	
30B	1-pin	Single connector	

USING WIRING DIAGRAMS

A great deal of information is included in each wiring diagram. For example, you will notice that all electrical components are identified using a unique number. Each component with it's corresponding number is identified below the diagrams.

Reading the wiring diagrams is detailed under **How to read wiring diagrams**. An explanation of the various symbols used in the diagrams can be found under **Symbols used in wiring diagrams**.

Most terminals are identified by numbers on the components and harness connectors. The terminal numbers for major electrical connections are shown in the diagrams. Though many terminal numbers appear only once, several other numbers appear in numerous places throughout the electrical system and identify certain types of circuits. A listing of the most common standard circuit numbers can be found in **Standard circuits.**

All wire colors in the diagrams are abbreviated. Combined color codes indicate a multi-colored wire. For example the code ws/bl indicates a white (ws) wire with a blue (bl) stripe. Wire color abbreviations are listed in **Terminal designations** as well as on each individual wiring diagram. Immediately preceding the wire color is the nominal cross-sectional area of the wire given in mm^2.

NOTE —

Sometimes the color of an installed wire may be different than the one on the wiring diagram. Don't be concerned, just be sure to confirm that the wire connects to the proper terminals.

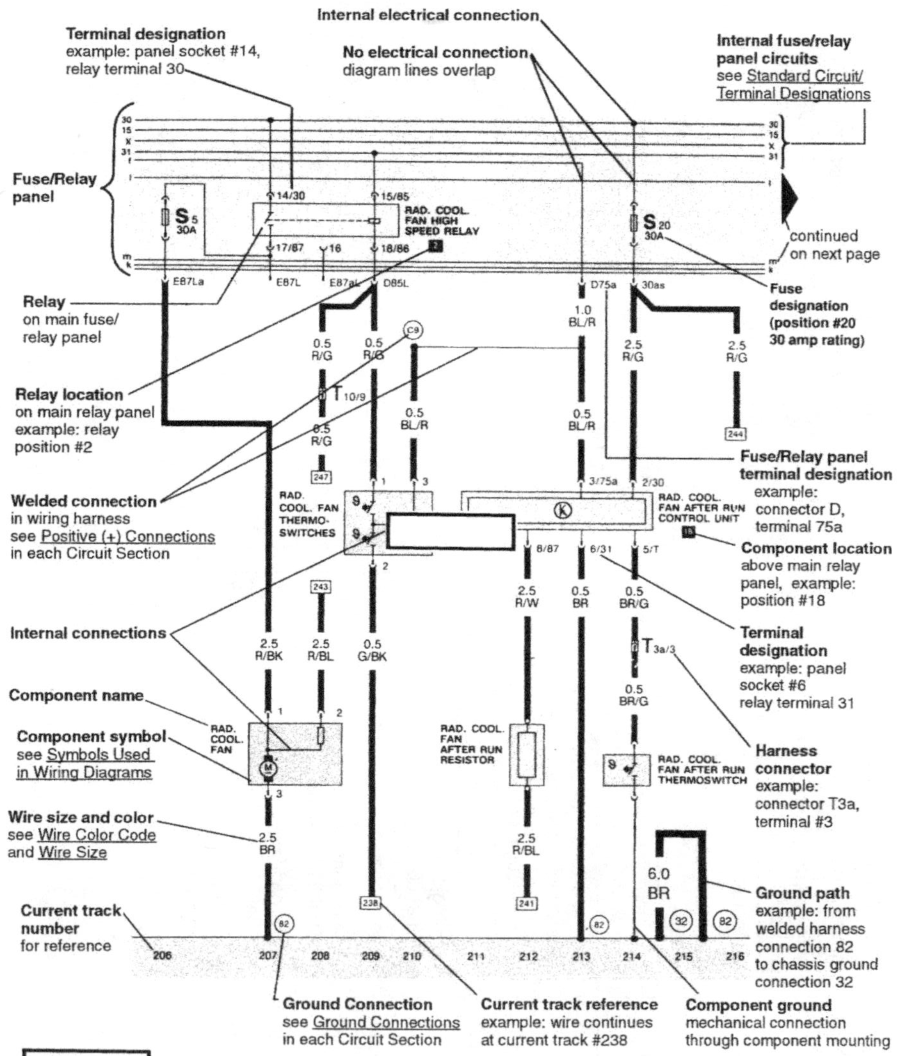

0024119

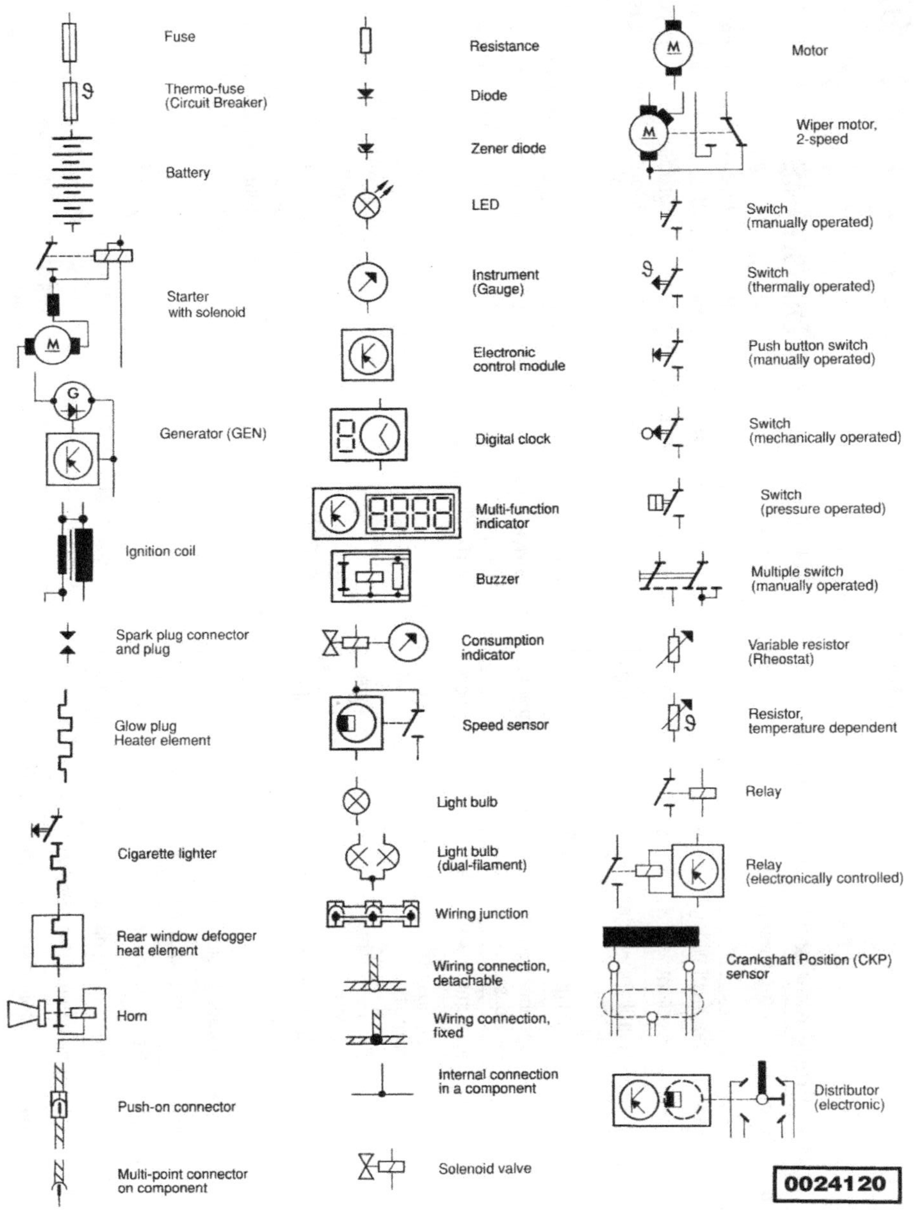

0024120

Standard Circuit/Terminal Designations

Circuit Number	Circuit Description	Most Common Wire Color
30	Battery positive (+) voltage whenever battery is connected	**Red (ro)**
15	Battery positive (+) voltage with ignition On	**Black (bk)**
X	Load-reduction circuit Battery positive (+) voltage via load-reduction relay (with ignition On) Interrupted for starting (ignition in "Start" position)	**Black/yellow (bk/ye)**
31	Ground or battery negative (–)	**Brown (br)**
50	Starter circuit Battery positive (+) voltage (with ignition in "Start" position)	**Red/black (ro/bk)**
B+	Alternator output (charging voltage to battery)	**Red (ro)**
D+	Alternator warning light and field energizing circuit	—
85	Ground (–) for switching relay	**Brown (br)**
86	Power input (+) for switching relay	—
87	Relay change-over contact	—

0024121

Wire Color-Code

ws..................white   bl.....................blue

sw..................black   gr.....................grey

ro......................red   li...................... lilac

br................ brown   ge ....yellow

gn.................green

Combined color codes indicate a multi-colored wire.

- example: The code ro/gn indicates a red wire with a green tracer stripe.

Sometimes the wire color on the car may be different than the one indicated on the wiring diagram. Don't be concerned. Just be sure to confirm that the wires in question connect to the proper terminals.

Wire Size

The wiring diagrams in this manual identify wires by their metric wire size—their cross-sectional area in square millimeters (mm^2). The table below lists metric wire sizes and their approximate equivalents in American Wire Gauge (AWG) sizes.

Metric (mm^2)	American Wire Gauge (AWG)
0.35	22
0.5	20
1.0	16
1.5	14
2.5	12
6.0	8
16.0	4
25.0	2
32.0	2

Connectors

T1..........................Single-point
T2..................................2-point
T3..................................3-point
T4..................................4-point
T5..................................5-point
T6..................................6-point
T8..................................8-point
T10..............................10-point
T12..............................12-point
T19..............................19-point
T21..............................21-point
T28..............................28-point
T30..............................30-point
T32..............................32-point
T38..............................38-point
T45..............................45-point
T68..............................68-point

In any particular diagram, each connector has a unique number.

example: T6, T6a, and T6b are three different 6-point connectors

Wiring diagrams usually identify a particular connector terminal for each wire connection

example: T3a/2 indicates connector T3a, terminal #2

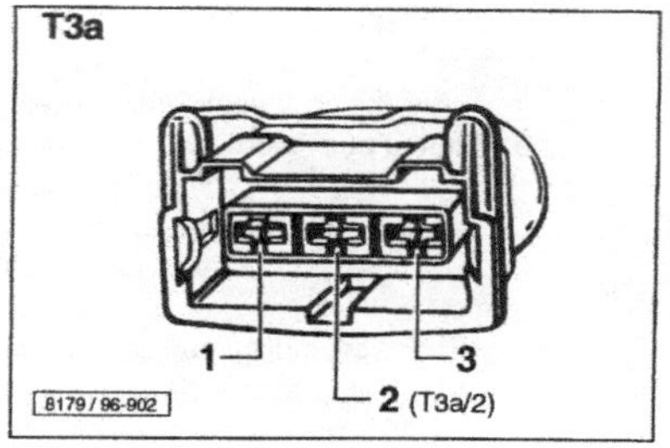

Golf, Jetta, Cabrio — 2.0L gas engine (ABA)

Golf — 1.8L gas engine (ACC), Canada only

Golf, Jetta — 1.9L Eco diesel engine (AAZ), Canada only

Golf, Jetta — 2.8L gas engine (AAA)

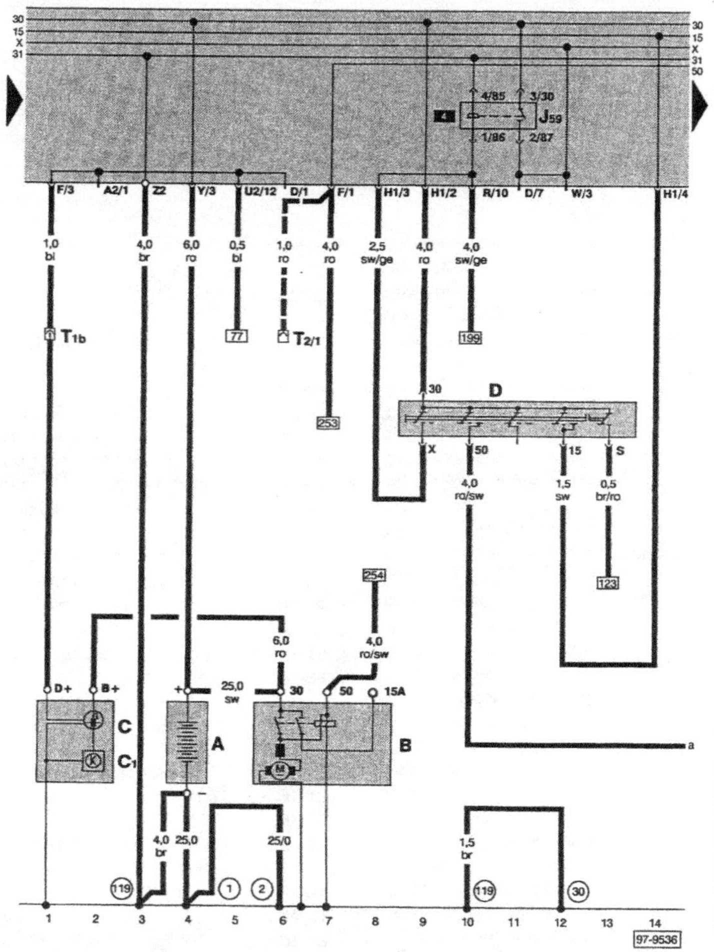

WIRING COLOR CODE

ws	=	white
sw	=	black
ro	=	red
br	=	brown
gn	=	green
bl	=	blue
gr	=	grey
li	=	lilac
ge	=	yellow

A - Battery
B - Starter
C - Generator (Gen)
C1 - Voltage Regulator (VR)
D - Ignition/Starter Switch
J59 - Load Reduction Relay
T1b - Connector, Single, near battery
T2 - Connector, Double, behind fuse/relay panel, only on automatic transmission

(1) - Ground strap, battery to body
(2) - Ground strap, transmission to body
(30) - Ground connection - beside fuse/relay panel
(119) - Ground connection -1-, in headlight wiring harness
- - - Automatic transmission only

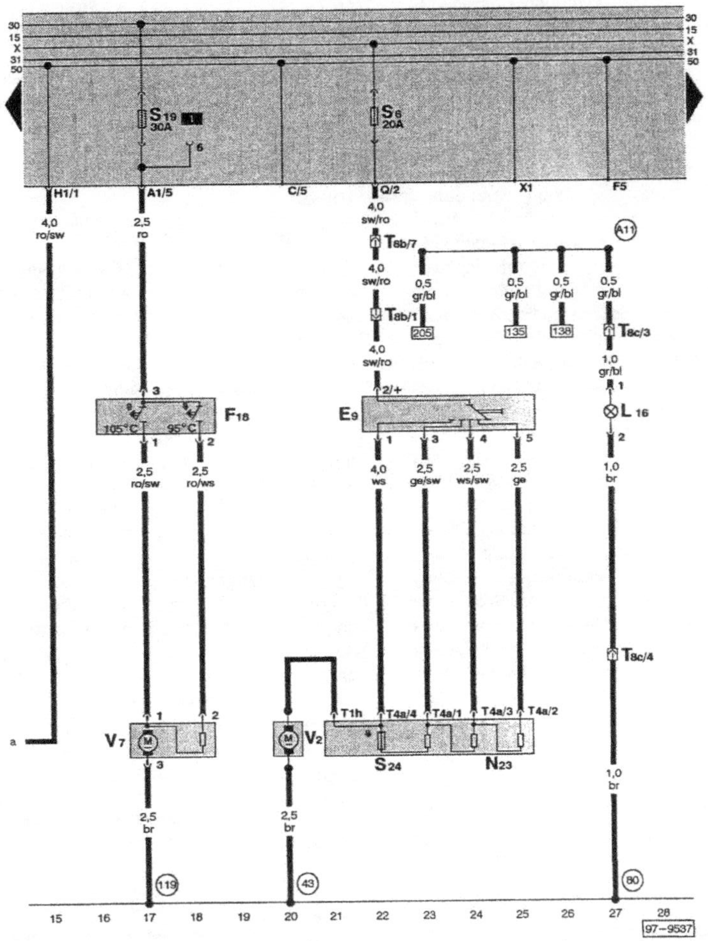

WIRING COLOR CODE

ws	=	white
sw	=	black
ro	=	red
br	=	brown
gn	=	green
bl	=	blue
gr	=	grey
li	=	lilac
ge	=	yellow

E9 - Fresh Air Blower Switch
F18 - Coolant Fan Control (FC) Thermal Switch
L16 - Fresh Air Control Lever Light
N23 - Fresh Air Blower Series Resistance
S24 - Heater, A/C Clutch Fuse
V2 - Fresh Air Blower
V7 - Coolant Fan
T1h - Connector, Single, behind right side of dash
T4a - Connector, 4 Pin, behind right side of dash
T8b - Connector, 8 Pin, behind fuse/relay panel
T8c - Connector, 8 Pin, behind lower center of dash panel

(43) - Ground connection, on right A-pillar, lower part
(80) - Ground connection -1-, in instrument panel wiring harness
(119) - Ground connection -1-, in headlight wiring harness
(A11) - Plus connection (58b), in instrument panel wiring harness

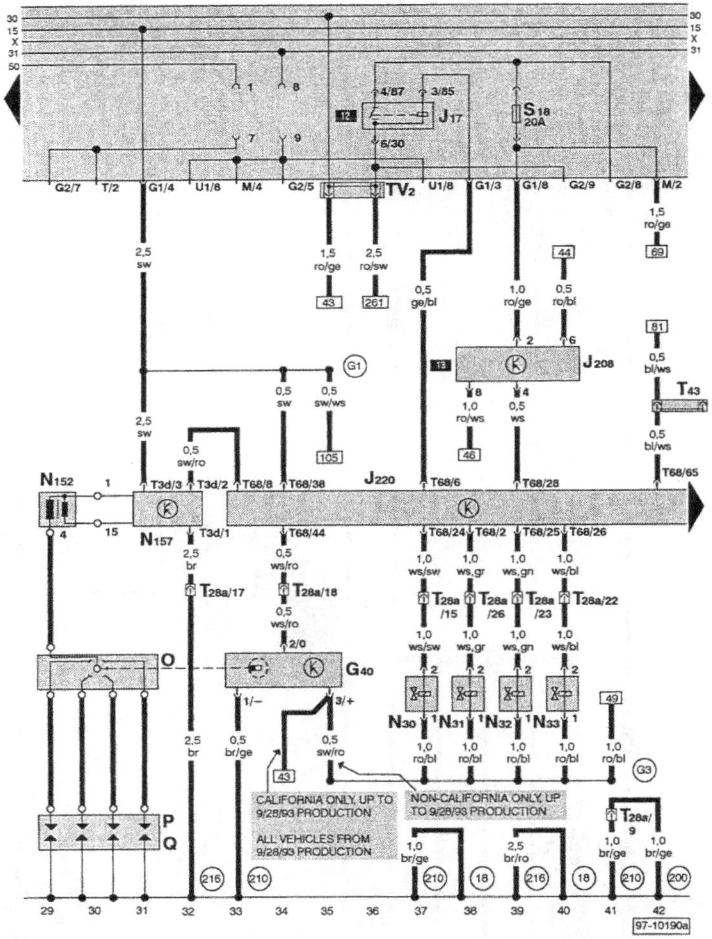

WIRING COLOR CODE		
ws	=	white
sw	=	black
ro	=	red
br	=	brown
gn	=	green
bl	=	blue
gr	=	grey
li	=	lilac
ge	=	yellow

G40 – Camshaft Position (CMP) Sensor
J17 – Fuel Pump (FP) Relay
J208 – Heated Oxygen Sensor (HO2S) Control Module
J220 – Motronic Engine Control Module (ECM)
N30 – Injector, Cyl. 1
N31 – Injector, Cyl. 2
N32 – Injector, Cyl. 3
N33 – Injector, Cyl. 4
N152 – Ignition Coil
N157 – Ignition Coil Power Output Stage
O – Ignition Distributor
Q – Spark Plugs
P – Spark Plug Connector
T3d – Connector, 3 Pin, left rear of engine compartment
T28a – Connector, 28 Pin, left side of engine, near distributor
T43 – Connector, vehicle speed signal

T68 – Connector, 68 Pin, under right side of hood, on Motronic Engine Control Module (ECM)
TV2 – Wire connector, terminal 30 (B+)
(18) – Ground connection, on engine block
(200) – Ground connection, (shielding) in engine compartment wiring harness
(210) – Ground connection –1–, in Motronic Multiport Fuel Injection (MFI) wiring harness
(216) – Ground connection –2–, in Motronic Multiport Fuel Injection (MFI) wiring harness
(G1) – Plus connection, in Multiport Fuel Injection (MFI) wiring harness
(G3) – Plus connection, in injector wiring harness

1993–1994

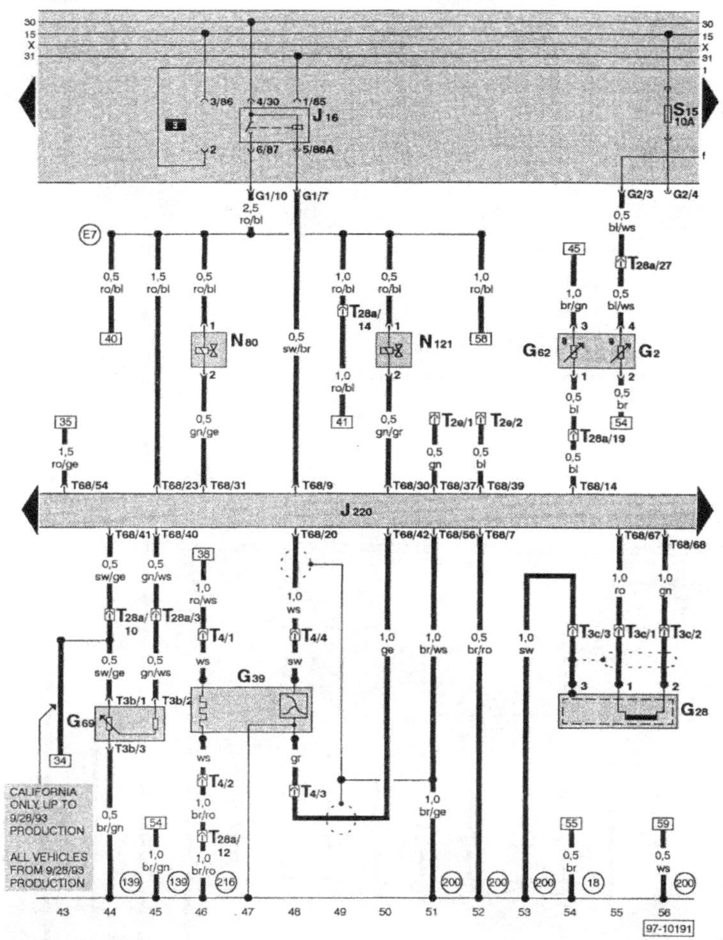

G2 – Engine Coolant Temperature (ECT) Sensor
G28 – Engine Speed (RPM) Sensor
G39 – Heated Oxygen Sensor (HO2S)
G62 – Engine Coolant Temperature (ECT) Sensor
G69 – Throttle Position (TP) Sensor
J16 – Power Supply Relay
J220 – Motronic Engine Control Module (ECM)
N80 – Evaporative Emission (EVAP) Canister Purge Regulator Valve
N121 – EGR Vacuum Regulator Valve (California models only)
T2e – Double Connector, behind fuse/relay panel, A/C connection
T3b – 3–Pin Connector, on throttle position sensor
T3c – 3–Pin Connector, left front of engine
T4 – 4–Pin Connector, bottom right rear of engine compartment

T28a – 28–Pin Connector, left side of engine, near distributor
T68 – 68–Pin Connector, under right side of hood, on Motronic Engine Control Module (ECM)
(18) – Ground connection, on engine block
(139) – Ground connection (sensor ground), in Motronic Multiport Fuel Injection (MFI) wiring harness
(200) – Ground connection, (shielding) in engine compartment wiring harness
(216) – Ground connection –2–, in Motronic Multiport Fuel Injection (MFI) wiring harness
(E7) – Plus connection (87a), in Multiport Fuel Injection (MFI) wiring harness

1993–1994

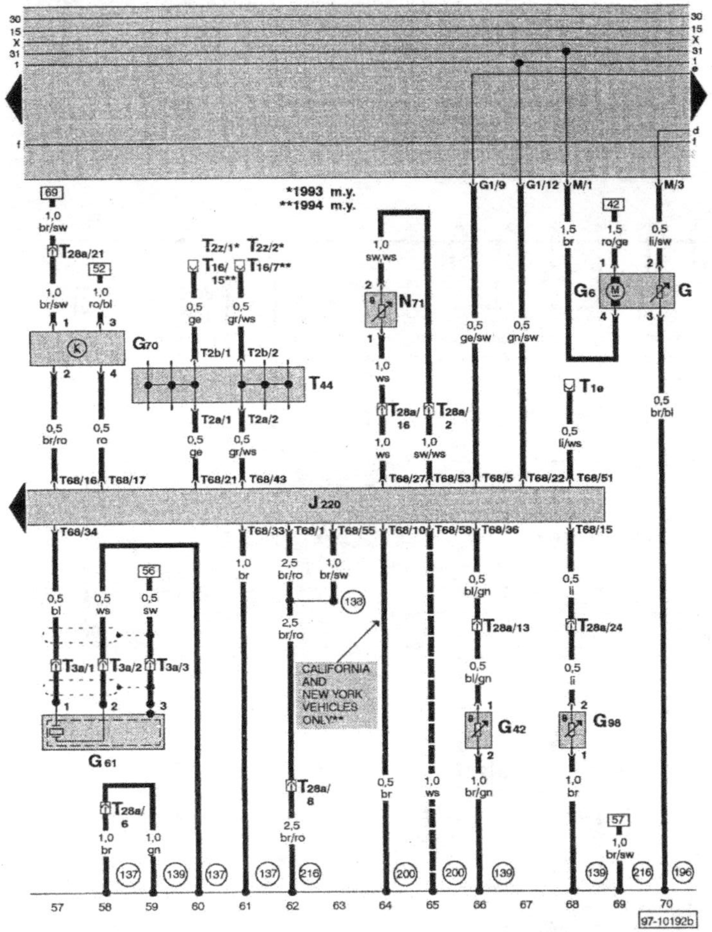

WIRING COLOR CODE		
ws	=	white
sw	=	black
ro	=	red
br	=	brown
gn	=	green
bl	=	blue
gr	=	grey
li	=	lilac
ge	=	yellow

G – Fuel Level Sensor
G6 – Fuel Pump (FP)
G42 – Intake Air Temperature (IAT) Sensor
G61 – Knock Sensor (KS) 1
G70 – Mass Air Flow (MAF) Sensor
G98 – EGR Temperature Sensor (California models only)
J220 – Motronic Engine Control Module (ECM)
N71 – Idle Air Control (IAC) Valve
T1e – Single Connector, behind fuse/relay panel
T2a – Double Connector, on junction connector (T44) above fuse/relay panel
T2b – Double Connector, on junction connector (T44) above fuse/relay panel
T2z – Data Link Connector (DLC), behind center of dash panel
T3a – 3–Pin Connector, on engine
T16 – Data Link Connector (DLC), behind center of dash panel
T28a – 28–Pin Connector, left side of engine, near distributor
T44 – Junction Box for On Board Diagnostic, above fuse/relay panel

T68 – 68–Pin Connector, under right side of hood, on Motronic Engine Control Module (ECM)
(137) – Ground connection (end stage), in Motronic Multiport Fuel Injection (MFI) wiring harness
(138) – Ground connection (control module) in Motronic Multiport Fuel Injection (MFI) wiring harness
(139) – Ground connection (sensor ground), in Motronic Multiport Fuel Injection (MFI) wiring harness
(196) – Ground connection –1–, in rear wiring harness
(200) – Ground connection, (shielding) in engine compartment wiring harness
(216) – Ground connection –2–, in Motronic Multiport Fuel Injection (MFI) wiring harness
– – – Automatic transmission only

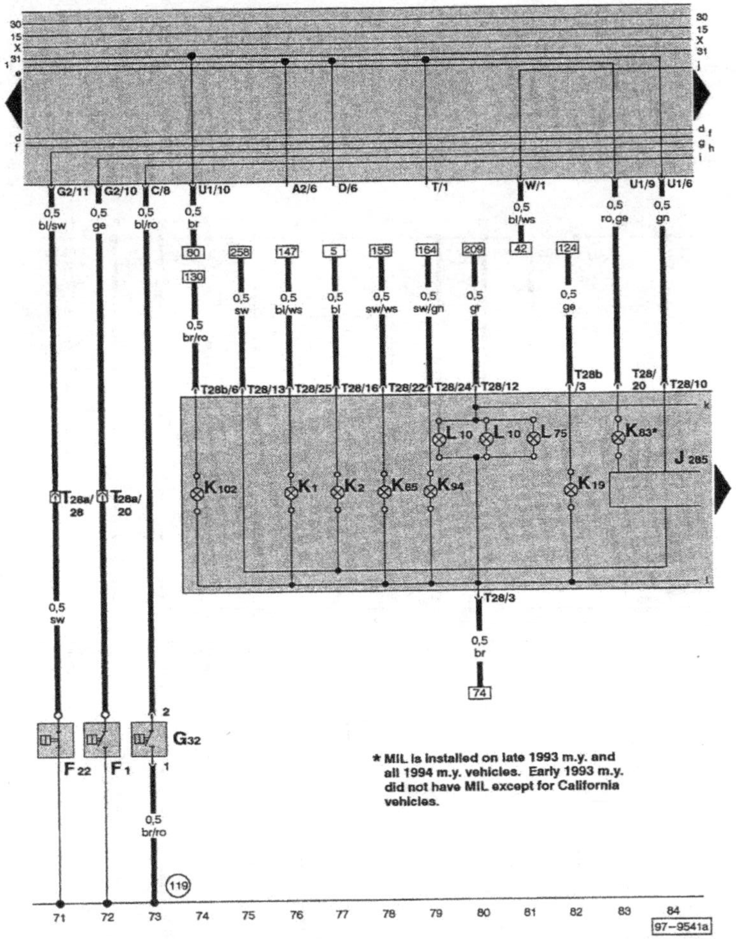

* MIL is installed on late 1993 m.y. and all 1994 m.y. vehicles. Early 1993 m.y. did not have MIL except for California vehicles.

WIRING COLOR CODE		
ws	=	white
sw	=	black
ro	=	red
br	=	brown
gn	=	green
bl	=	blue
gr	=	grey
li	=	lilac
ge	=	yellow

F1 – Engine Oil Pressure Switch
F22 – Engine Oil Pressure Switch (0.3 Bar)
G32 – Engine Coolant Level (ECL) Sensor
J285 – Control Module With Display Unit In Instrument Cluster
K1 – Headlight High Beam Indicator Light
K2 – Generator (GEN) Warning Light
K19 – Seat Belt Warning Light
K65 – Turn Signal Indicator Light, Left
K83 – Malfunction Indicator Lamp (MIL)
K94 – Turn Signal Indicator Light, Right
K102 – Rear Hatch/Trunk Ajar Warning Light
L10 – Instrument Cluster Light
L75 – Digital Display Light
T28 – Connector, 28 Pin, instrument cluster
T28a – Connector, 28 Pin, left side of engine, near distributor
T28b – Connector, 28 Pin, on instrument cluster
(119) – Ground connection –1–, in headlight wiring harness

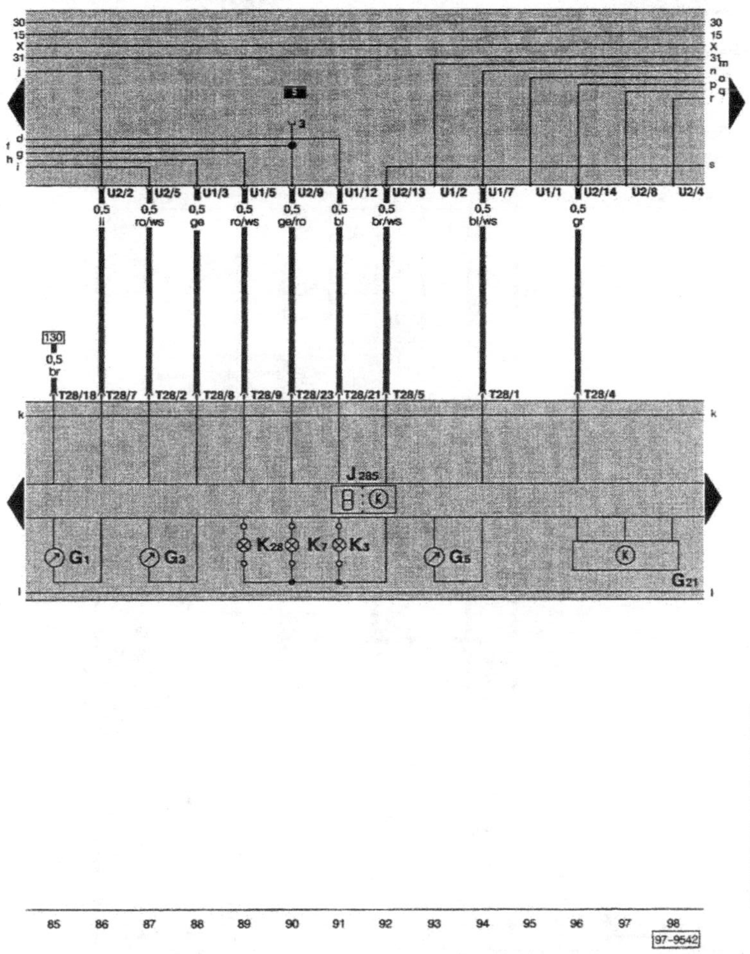

WIRING COLOR CODE

ws	=	white
sw	=	black
ro	=	red
br	=	brown
gn	=	green
bl	=	blue
gr	=	grey
li	=	lilac
ge	=	yellow

G1 – Fuel Gauge
G3 – Engine Coolant Temperature (ECT) Gauge
G5 – Tachometer
G21 – Speedometer
J285 – Control Module With Display Unit In Instrument Cluster
K3 – Oil Pressure Warning Light
K7 – Brake and Parking Brake Indicator Light
K28 – Engine Coolant Level/Temperature (ECL/ECT) Warning Light
T28 – Connector, 28 Pin, instrument cluster

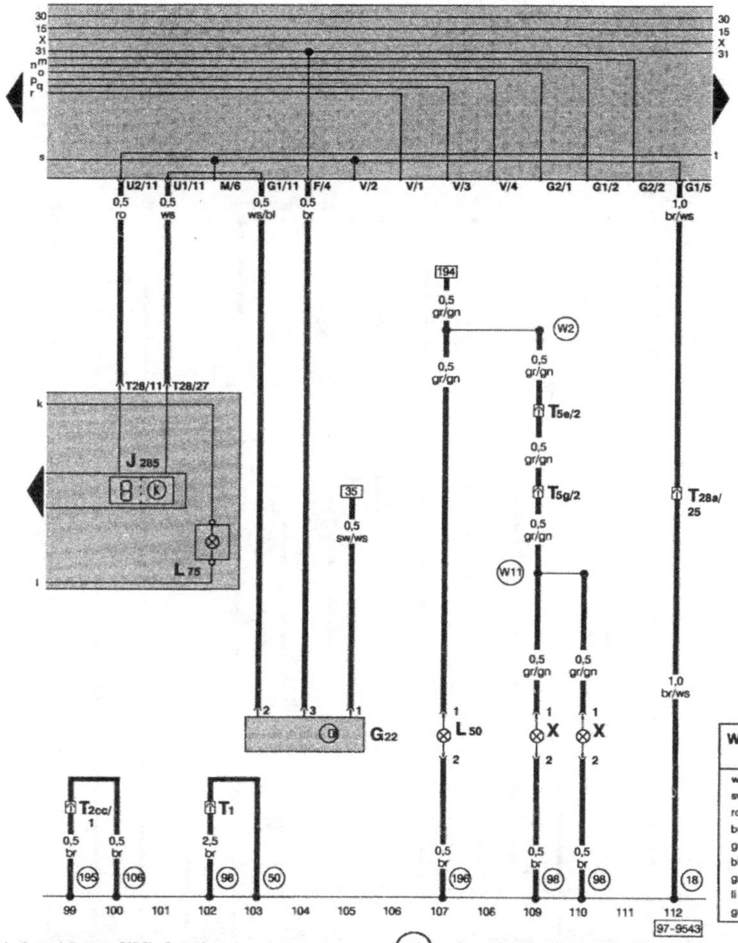

WIRING COLOR CODE

ws	=	white
sw	=	black
ro	=	red
br	=	brown
gn	=	green
bl	=	blue
gr	=	grey
li	=	lilac
ge	=	yellow

G22 – Vehicle Speed Sensor (VSS), Speedometer
J285 – Control Module With Display Unit In Instrument Cluster
L50 – Center Rear Ashtray Light
L75 – Digital Display Light
X – License Plate Light
T1 – Connector, single, in luggage compartment lid
T2cc – Connector, double, on lower right "A" pillar
T5e – Connector, 5 Pin, in luggage compartment, left
T5g – Connector, 5 Pin, in luggage compartment lid
T28 – Connector, 28 Pin, instrument cluster
T28a – Connector, 28 Pin, left side of engine, near distributor
(18) – Ground connection, on engine block
(50) – Ground connection, in luggage compartment, left
(98) – Ground connection, in rear lid wiring harness
(106) – Ground connection, in alarm system wiring harness
(195) – Ground connection, in rear door contact switch wiring harness
(196) – Ground connection – 1 – in rear wiring harness
(W2) – Plus connection (58), in rear wiring harness
(W11) – Wiring connection (58), in rear lid wiring harness

Door switches, interior light. seat belt warning system relay

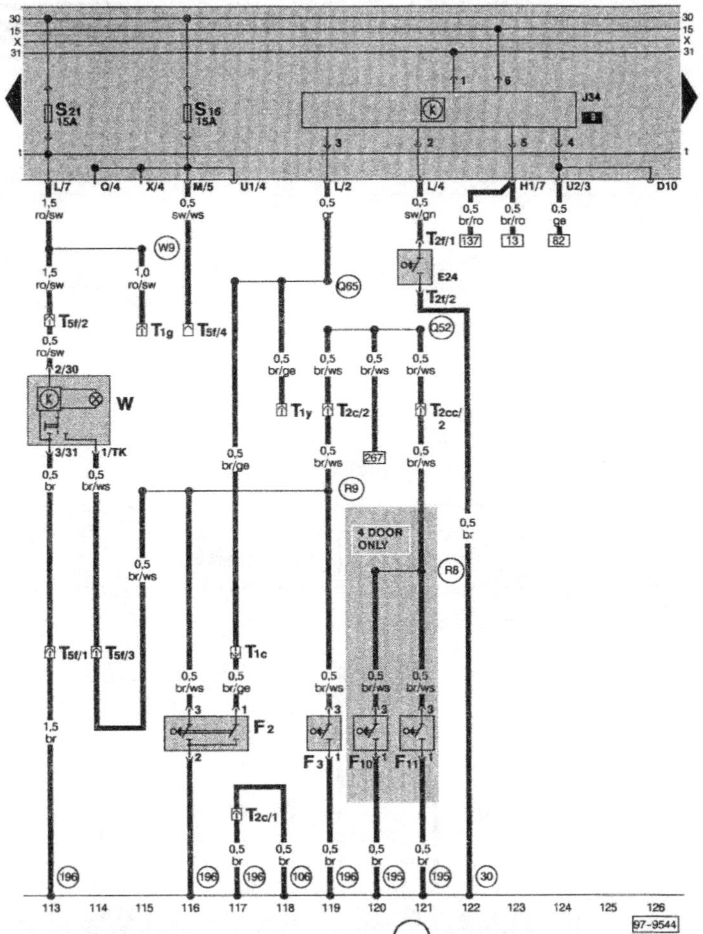

WIRING COLOR CODE		
ws	=	white
sw	=	black
ro	=	red
br	=	brown
gn	=	green
bl	=	blue
gr	=	grey
li	=	lilac
ge	=	yellow

- E24 – Seat Belt Switch, Left
- F2 – Door Contact Switch, Left Front
- F3 – Door Contact Switch, Right Front
- F10 – Door Contact Switch, Left Rear
- F11 – Door Contact Switch, Right Rear
- J34 – Seat Belt Warning System Relay
- W – Interior Light, Front
- T1c – Connector, Single, behind fuse/relay panel
- T1g – Connector, Single, behind fuse/relay panel
- T1y – Connector, Single, behind fuse/relay panel
- T2c – Connector, Double, on lower right "A" pillar
- T2cc – Connector, double, on lower right "A" pillar
- T2f – Connector, Double, behind rear of center console
- T5f – Connector, 5 Pin, above fuse/relay panel
- (30) – Ground connection – beside fuse/relay panel
- (106) – Ground connection, in alarm system wiring harness
- (195) – Ground connection, in rear door contact switch wiring harness
- (Q52) – Wire connection (door contact switch), in alarm system wiring harness
- (Q65) – Wire connection, in seat belt warning system wiring harness
- (R8) – Wire connection (door contact switch) in rear door contact switch wiring harness
- (R9) – Wire connection (30), in rear wiring harness
- (W9) – Plus connection (30), in rear wiring harness

Luggage compartment release solenoid, luggage compartment light, cigarette lighter

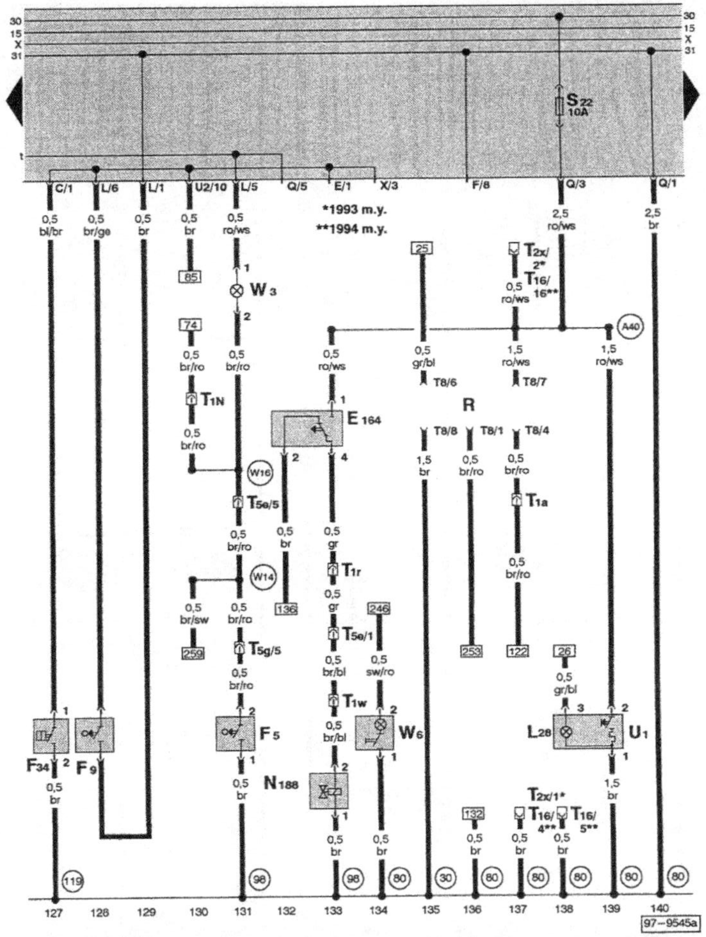

WIRING COLOR CODE		
ws	=	white
sw	=	black
ro	=	red
br	=	brown
gn	=	green
bl	=	blue
gr	=	grey
li	=	lilac
ge	=	yellow

- E164 – Luggage Compartment Release Switch
- F5 – Luggage Compartment Light Switch
- F9 – Parking Brake Warning Light Switch
- F34 – Brake Fluid Level Warning Switch
- L28 – Cigarette Lighter Light
- N188 – Luggage Compartment Release Solenoid
- R – Radio
- U1 – Cigarette Lighter
- W3 – Luggage Compartment Light
- W6 – Glove Compartment Light
- T1a – Connector, Single, behind fuse/relay panel
- T1n – Connector, Single, behind fuse/relay panel
- T1r – Connector, Single, behind fuse/relay panel
- T1w – Connector, Single, on luggage compartment lid
- T2x – Data Link Connector (DLC), behind center of dash panel
- T5e – Connector, 5 Pin, in luggage compartment, left
- T5g – Connector, 5 Pin, in luggage compartment lid
- T8 – Connector, 8 Pin, on radio
- T16 – Data Link Connector (DLC), behind center of dash panel
- (30) – Ground connection – beside fuse/relay panel
- (80) – Ground connection –1–, in instrument panel wiring harness
- (98) – Ground connection, in rear lid wiring harness
- (119) – Ground connection –1–, in headlight wiring harness
- (A40) – Plus connection (30), in instrument panel wiring harness
- (W14) – Wire connection (31s), in rear wiring harness
- (W16) – Wire connection (31s), in rear warning harness

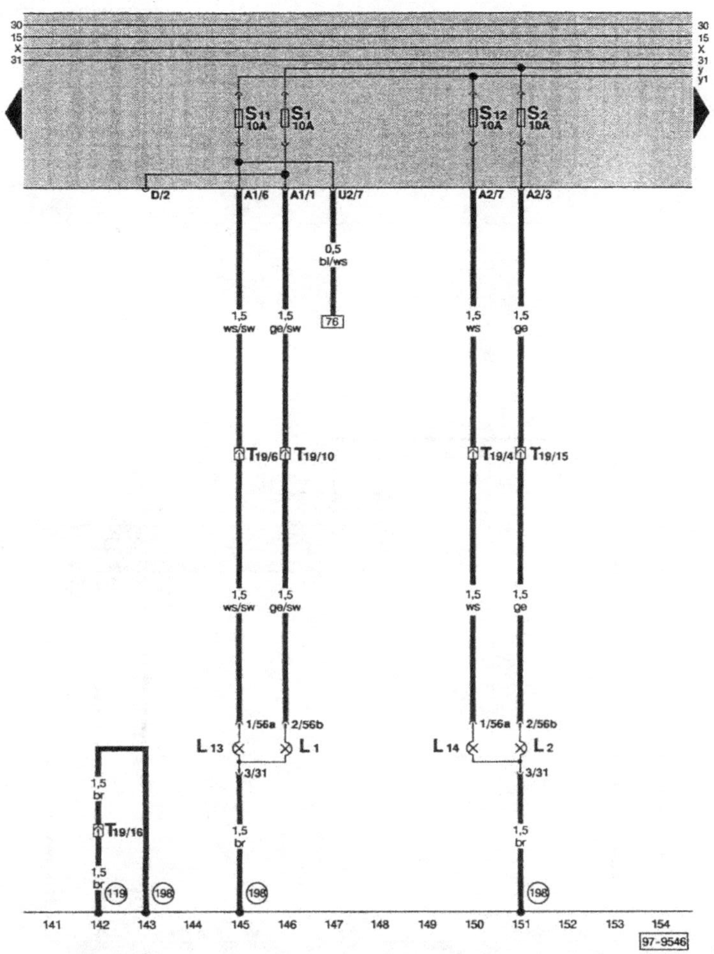

ws	=	white
sw	=	black
ro	=	red
br	=	brown
gn	=	green
bl	=	blue
gr	=	grey
li	=	lilac
ge	=	yellow

WIRING COLOR CODE

L1 – Headlight, Left
L2 – Headlight, Right
L13 – High Beam Headlight, Left
L14 – High Beam Headlight, Right
T19 – Connector, 19 Pin, behind radiator grill, center
(119) – Ground connection –1–, in headlight wiring harness
(198) – Ground connection, in front light wiring harness

1993–1994

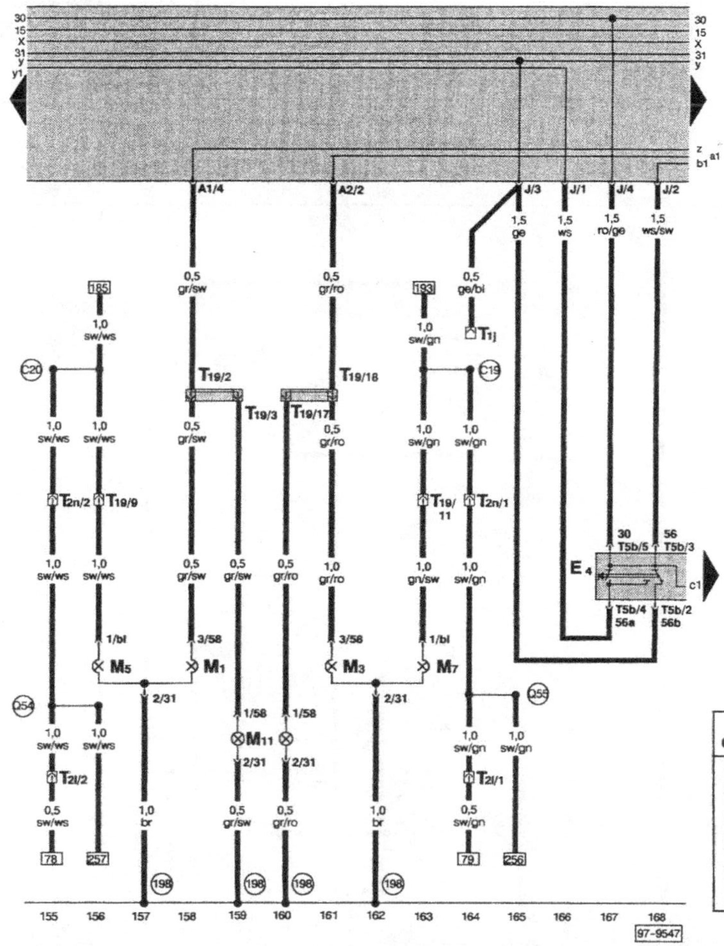

WIRING COLOR CODE

ws	=	white
sw	=	black
ro	=	red
br	=	brown
gn	=	green
bl	=	blue
gr	=	grey
li	=	lilac
ge	=	yellow

E4 – Headlight Dimmer/Flasher Switch
M1 – Parking Light, Left
M3 – Parking Light, Right
M5 – Turn Signal Light, Left Front
M7 – Turn Signal Light, Right Front
M11 – Side Marker Lights, Front
T1j – Connector, Single, behind fuse/relay panel
T2l – Connector, Double, behind fuse/relay panel
T2n – Connector, Double, behind fuse/relay panel
T5b – Connector, 5 Pin, behind steering column switch trim
T19 – Connector, 19 Pin, behind radiator grill, center

(198) – Ground connection, in front light wiring harness
(C19) – Wire connection (right turn signal), in headlight wiring harness
(C20) – Wire connection (left turn signal), in headlight wiring harness
(Q54) – Wire connection (left turn signal), in alarm system wiring harness
(Q55) – Wire connection (right turn signal), in alarm system wiring harness

1993–1994

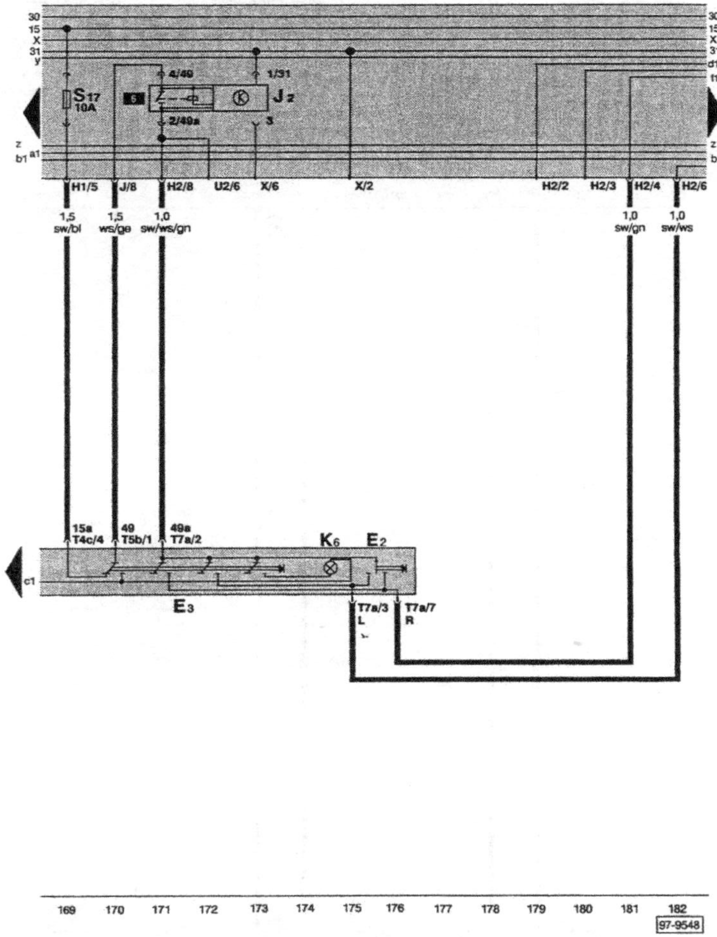

WIRING COLOR CODE		
ws	=	white
sw	=	black
ro	=	red
br	=	brown
gn	=	green
bl	=	blue
gr	=	grey
li	=	lilac
ge	=	yellow

E2 - Turn Signal Switch
E3 - Emergency Flasher Switch
J2 - Emergency Flasher Relay
K6 - Emergency Flasher Warning Light
T4c - Connector, 4 Pin, behind steering column switch trim
T5b - Connector, 5 Pin, behind steering column switch trim
T7a - Connector, 7 Pin, behind steering column switch trim

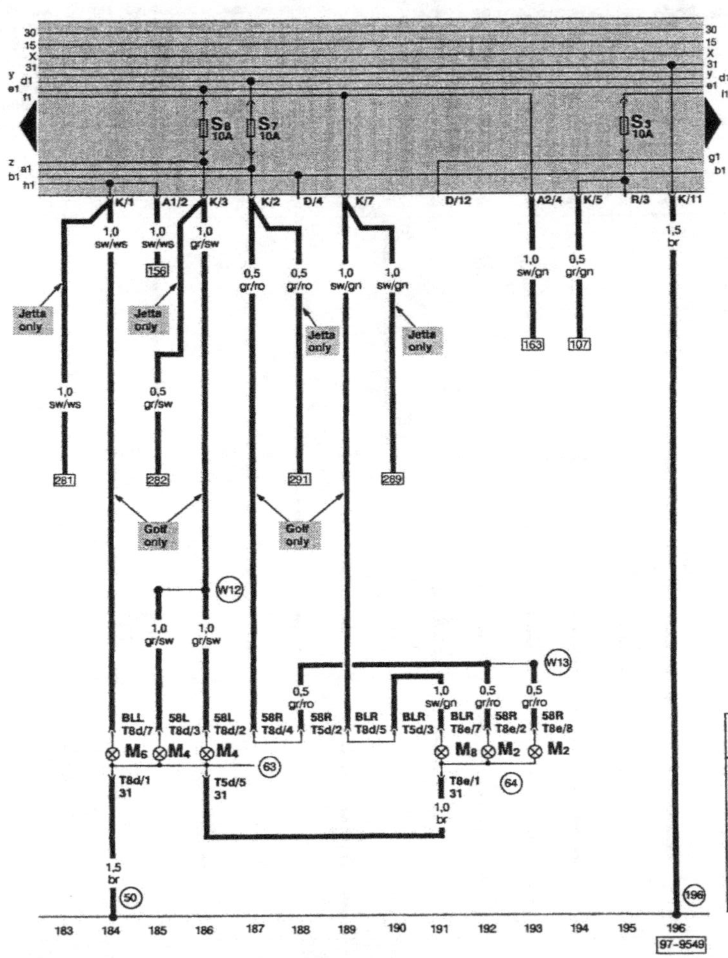

WIRING COLOR CODE		
ws	=	white
sw	=	black
ro	=	red
br	=	brown
gn	=	green
bl	=	blue
gr	=	grey
li	=	lilac
ge	=	yellow

M2 - Tail Light, Right
M4 - Tail Light, Left
M6 - Turn Signal Light, Left Rear
M8 - Turn Signal Light, Right Rear
T5d - Connector, 5 Pin, on taillight, left
T8d - Connector, 8 Pin, on taillight, left
T8e - Connector, 8 Pin, on taillight right
(50) - Ground connection, in luggage compartment, left

(63) - Ground connection, on left taillight bulb holder
(64) - Ground connection, on right taillight bulb holder
(196) - Ground connection -1-, in rear wiring harness
(W12) - Wire connection (58L), in rear wiring harness
(W13) - Wire connection (58R), in rear wiring harness

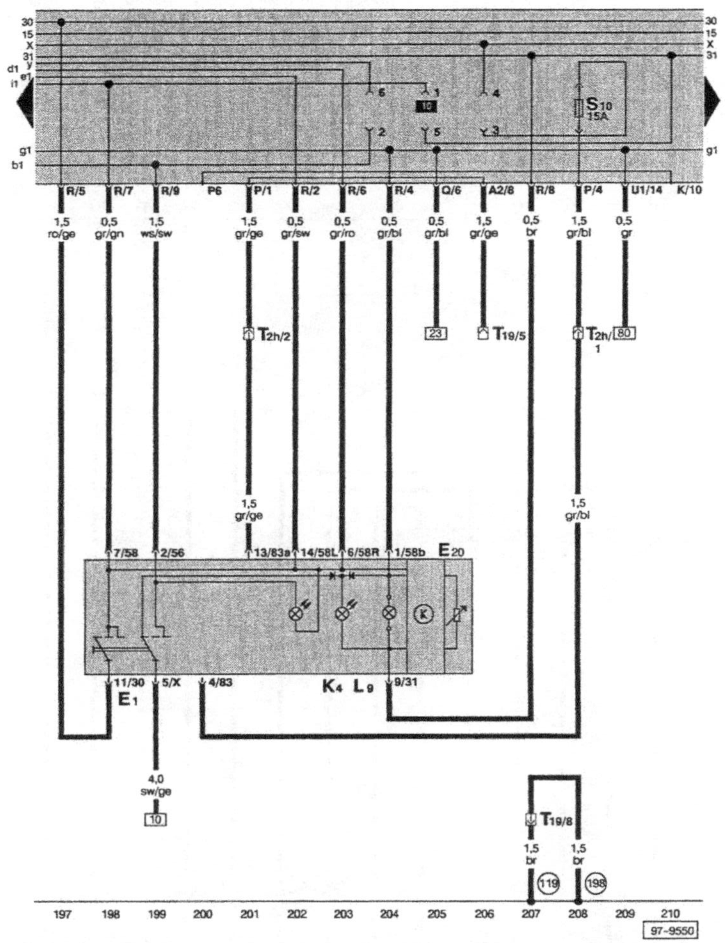

WIRING COLOR CODE		
ws	=	white
sw	=	black
ro	=	red
br	=	brown
gn	=	green
bl	=	blue
gr	=	grey
li	=	lilac
ge	=	yellow

E1 - Light Switch
E20 - Instrument Panel Light Dimmer Switch
K4 - Park Light Indicator Light
L9 - Headlight Switch Light
T2h - Connector, Double, behind fuse/relay panel
T19 - Connector, 19 Pin, behind radiator grill, center
(119) - Ground connection -1-, in headlight wiring harness
(198) - Ground connection, in front light wiring harness

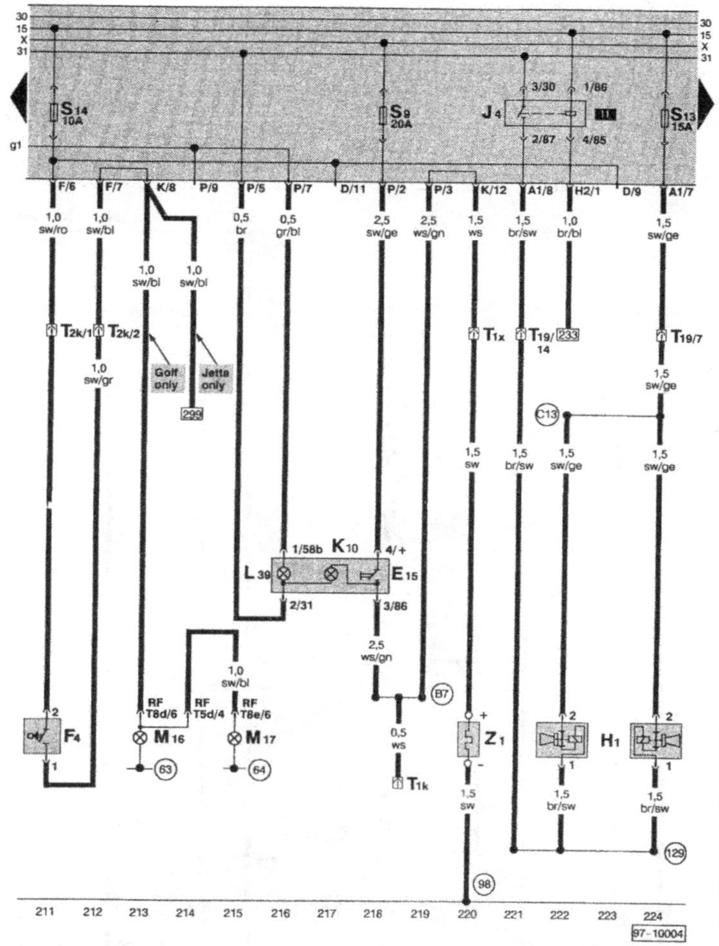

WIRING COLOR CODE		
ws	=	white
sw	=	black
ro	=	red
br	=	brown
gn	=	green
bl	=	blue
gr	=	grey
li	=	lilac
ge	=	yellow

E15 - Rear Window Defogger Switch
F4 - Back-Up Light Switch
H1 - Dual Horns
J4 - Dual Horn Relay
K10 - Rear Window Defogger Indicator Light
L39 - Rear Window Defogger Switch Light
M16 - Back-Up Light, Left
M17 - Back-Up Light, Right
Z1 - Rear Window Defogger (Heat Element)
T1k - Connector, Single, behind fuse/relay panel
T1x - Connector, Single, in luggage compartment, left
T2k - Connector, Double, behind fuse/relay panel
T5d - Connector, 5 Pin, on taillight, left
T8d - Connector, 8 Pin, on taillight, left
T8e - Connector, 8 Pin, on taillight right
T19 - Connector, 19 Pin, behind radiator grill, center
(63) - Ground connection, on left taillight bulb holder
(64) - Ground connection, on right taillight bulb holder
(98) - Ground connection, in rear lid wiring harness
(129) - Ground connection, in dual tone horn wiring harness
(B7) - Wire connection (86), in rear window defogger wiring harness
(C13) - Plus connection, in dual tone horn wiring harness

Windshield wiper motor, horn button

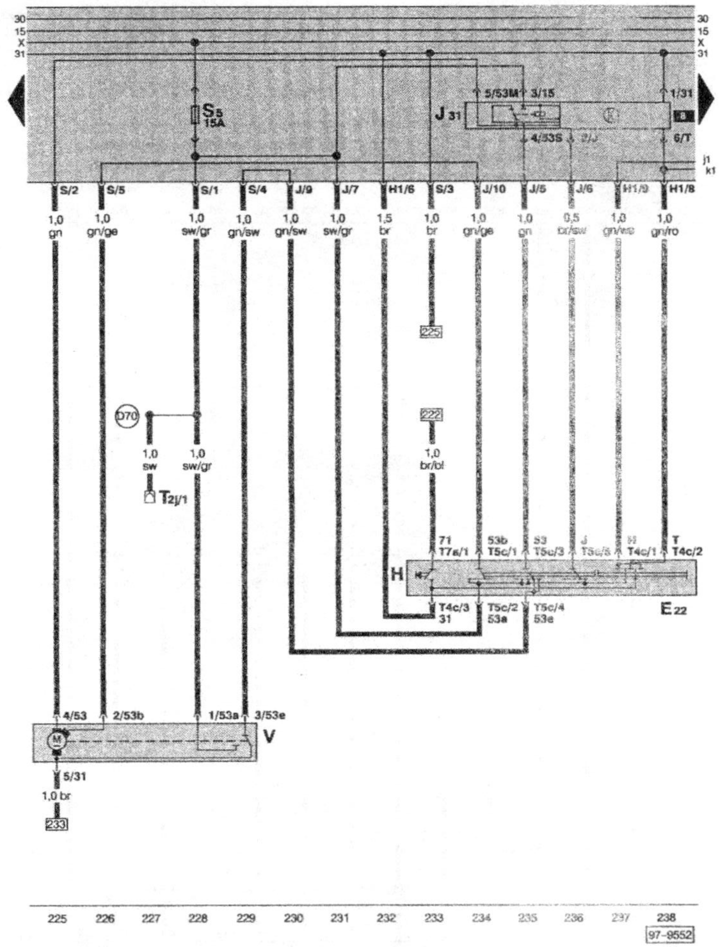

WIRING COLOR CODE		
ws	=	white
sw	=	black
ro	=	red
br	=	brown
gn	=	green
bl	=	blue
gr	=	grey
li	=	lilac
ge	=	yellow

E22 - Windshield Wiper Intermittent Switch
H - Horn Button
J31 - Washer/Wiper Intermittent Relay
V - Windshield Wiper Motor
T2j - Connector, Double, behind fuse/relay panel
T4c - Connector, 4 Pin, behind steering column switch trim
T5c - Connector, 5 Pin, behind steering column switch trim
T7a - Connector, 7 Pin, behind steering column switch trim
(D70) - Wire connection (53a), in engine compartment wiring harness

1993–1994

Rear window wiper motor, brake lights (Golf only)

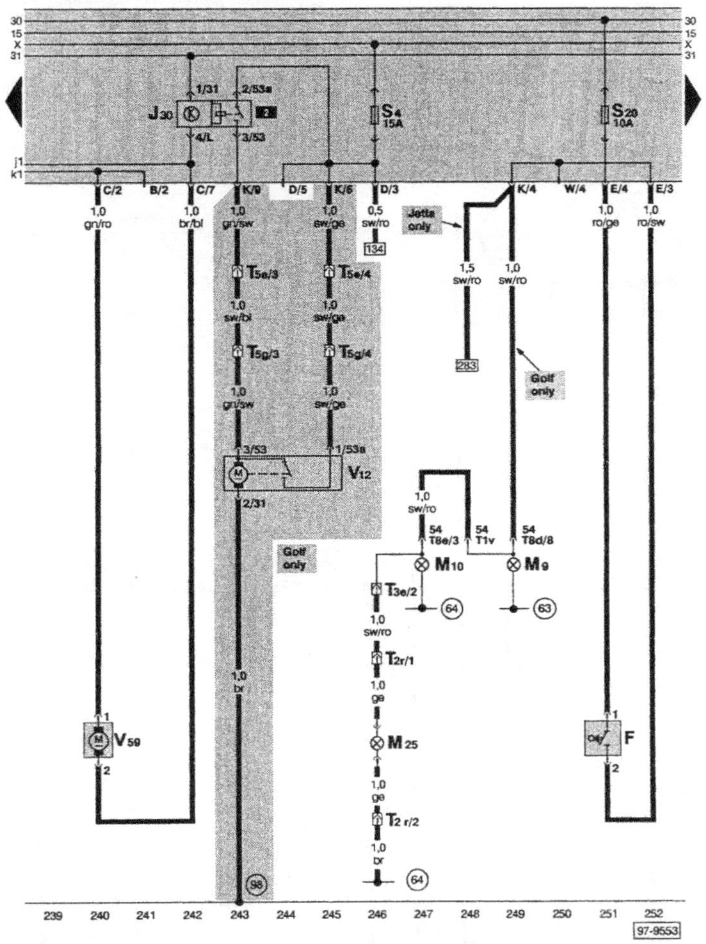

WIRING COLOR CODE		
ws	=	white
sw	=	black
ro	=	red
br	=	brown
gn	=	green
bl	=	blue
gr	=	grey
li	=	lilac
ge	=	yellow

F - Brake Light Switch
J30 - Rear Window Wiper/Washer Relay
M9 - Brake Light, Left
M10 - Brake Light, Right
M25 - High-Mount Brake Light
V12 - Rear Window Wiper Motor
V59 - Windshield and Rear Window Washer Pump
T1v - Connector, Single, on taillight, left
T2r - Connector, Double, behind right grommet, on luggage compartment lid
T3e - Connector, 3 Pin, on tail light, right
T5e - Connector, 5 Pin, in luggage compartment, left
T5g - Connector, 5 Pin, in luggage compartment lid
T8d - Connector, 8 Pin, on taillight, left
T8e - Connector, 8 Pin, on taillight right
(63) - Ground connection, on left taillight bulb holder
(64) - Ground connection, on right taillight bulb holder
(98) - Ground connection, in rear lid wiring harness

1993-1994

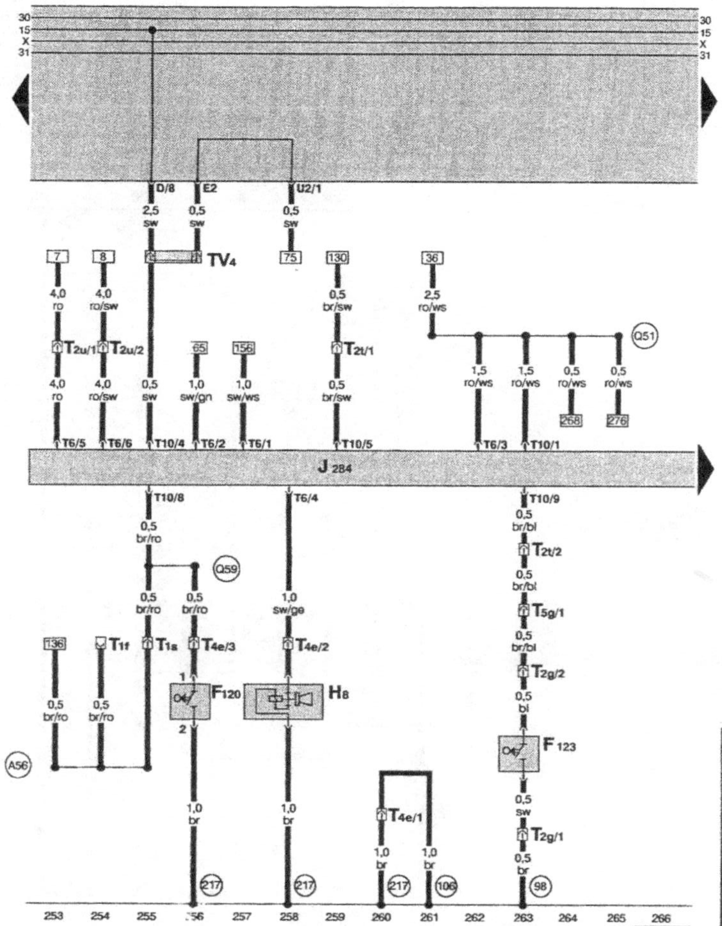

WIRING COLOR CODE		
ws	=	white
sw	=	black
ro	=	red
br	=	brown
gn	=	green
bl	=	blue
gr	=	grey
li	=	lilac
ge	=	yellow

F120 – Alarm Switch, Hood
F123 – Alarm Switch, Luggage Compartment Lid
H8 – Alarm Horn
J284 – Alarm System Control Module
T1f – Connector, Single, behind dash panel, center, behind radio
T1s – Connector, Single, behind fuse/relay panel
T2g – Connector, Double, on luggage compartment lid
T2t – Connector, Double, on luggage compartment left
T2u – Connector, Double, behind fuse/relay panel
T4e – Connector, 4 Pin, behind fuse/relay panel
T5g – Connector, 5 Pin, in luggage compartment lid
T6 – Connector, 6 Pin, behind dash panel, left, on Alarm System Control Module
T10 – Connector, 10 Pin, behind dash panel, left, on Alarm System Control Module
TV4 – Connector, terminal 15

98 – Ground connection, in rear lid wiring harness
106 – Ground connection, in alarm system wiring harness
217 – Ground connection, in alarm horn wiring harness
A56 – Wire connection (31SDS) in instrument panel wiring harness
Q51 – Plus connection (30), in alarm system wiring harness
Q60 – Wire connection (MHKS) in alarm system wiring harness

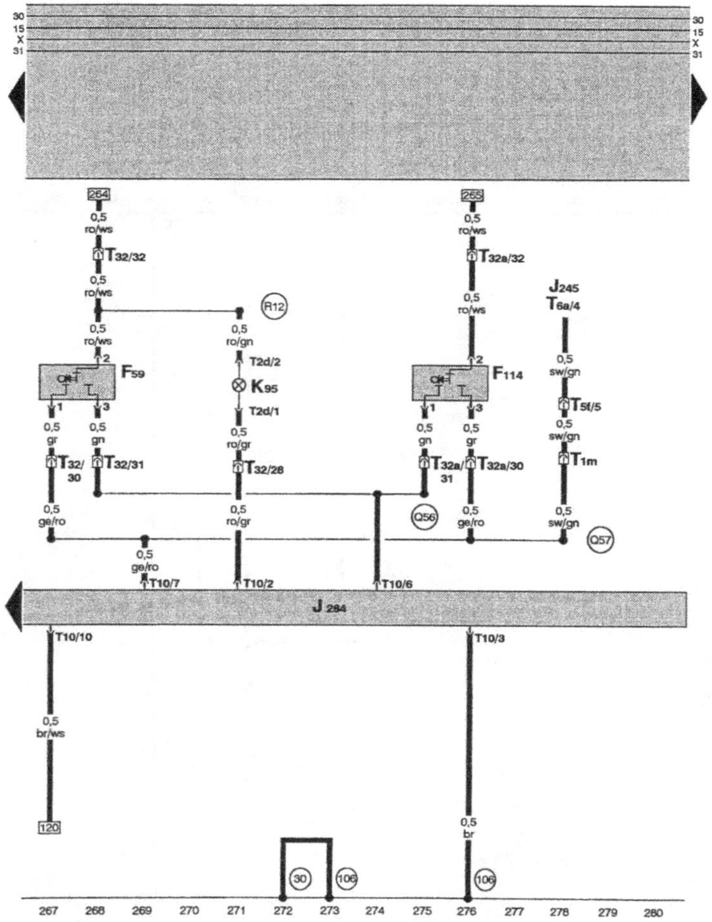

WIRING COLOR CODE		
ws	=	white
sw	=	black
ro	=	red
br	=	brown
gn	=	green
bl	=	blue
gr	=	grey
li	=	lilac
ge	=	yellow

F59 – Central Locking System Door Switch, Driver's Door
F114 – Central Locking System Door Switch, Passenger's Door
J245 – Power Sunroof Control Module
J284 – Alarm System Control Module
K95 – Alarm System Indicator Light
T1m – Connector, Single, behind fuse/relay panel
T2d – Connector, Double, in left front door
T2m – Connector, Double, in right front door
T5f – Connector, 5 Pin, above fuse/relay panel
T6a – Connector, 6 Pin, behind interior light, on Power Sunroof Control Module
T10 – Connector, 10 Pin, behind dash panel, left, on Alarm System Control Module
T32 – Connector, 32 Pin, on A–pillar, lower left
T32a – Connector, 32 Pin, on A–pillar, lower right

30 – Ground connection – beside fuse/relay panel
106 – Ground connection, in alarm system wiring harness
Q56 – Wire connection (open), in alarm system wiring harness
Q57 – Wire connection (closed), in alarm system wiring harness
R12 – Plus connection –3– (30), in driver's door wiring harness

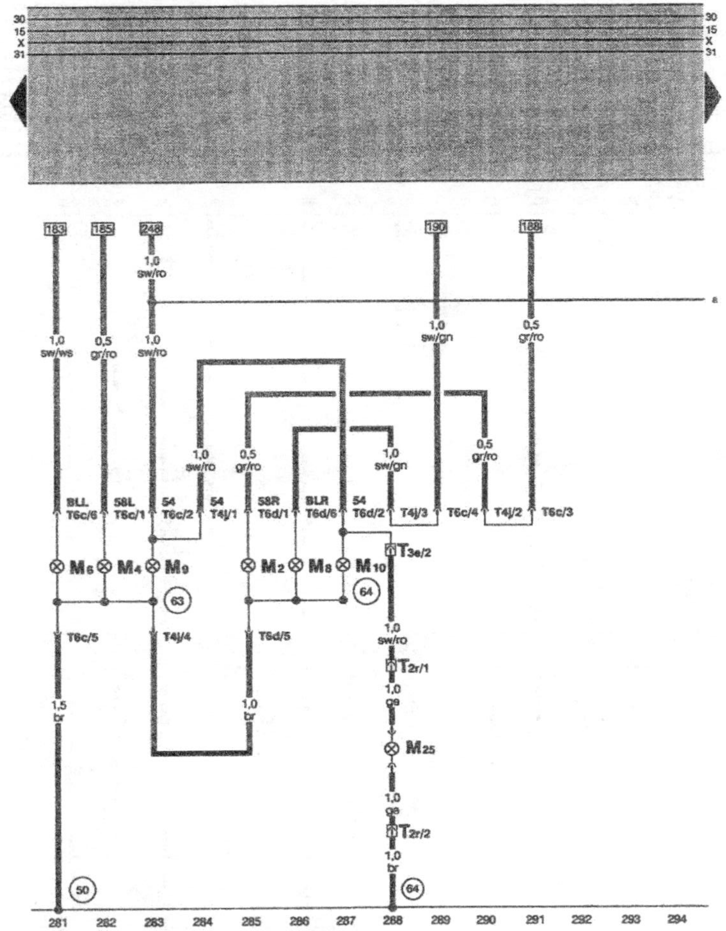

WIRING COLOR CODE

ws	=	white
sw	=	black
ro	=	red
br	=	brown
gn	=	green
bl	=	blue
gr	=	grey
li	=	lilac
ge	=	yellow

- M2 – Tail Light, Right
- M4 – Tail Light, Left
- M6 – Turn Signal Light, Left Rear
- M8 – Turn Signal Light, Right Rear
- M10 – Brake Light, Right
- M25 – High Mount Brake Light
- T2r – Connector, double, behind right grommet, on luggage compartment lid
- T3e – Connector, 3 pin, on tail light, right
- T4j – Connector, 4 pin, on tail light, left
- T6c – Connector, 6 pin, on tail light, left
- T6d – Connector, 6 pin, on tail light, right

- (50) – Ground connection, in luggage compartment lid
- (63) – Ground connection, on left tail light bulb holder
- (64) – Ground connection, on right tail light bulb holder

1993–1994

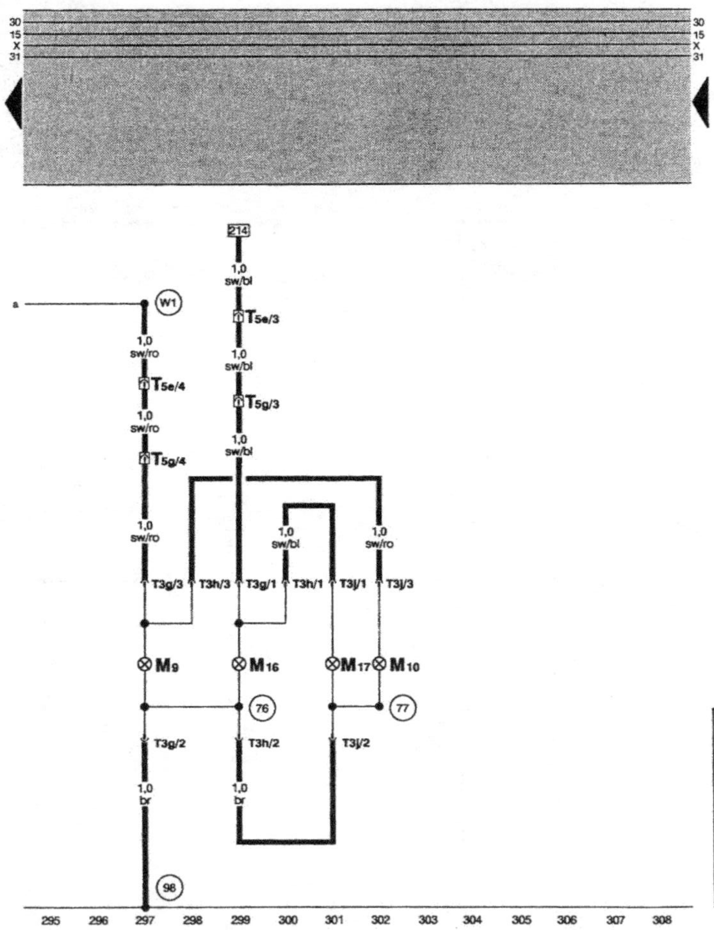

WIRING COLOR CODE

ws	=	white
sw	=	black
ro	=	red
br	=	brown
gn	=	green
bl	=	blue
gr	=	grey
li	=	lilac
ge	=	yellow

- M9 – Brake Light, Left
- M10 – Brake Light, Right
- M16 – Back-up Light, Left
- M17 – Back-up Light, Right
- T3g – Connector, 3 pin, on back-up/brake light, left
- T3h – Connector, 3 pin, on back-up/brake light, left
- T3j – Connector, 3 pin, on back-up/brake light, right
- T5e – Connector, 5 pin, on luggage compartment lid
- T5g – Connector, 5 pin, on luggage compartment lid

- (W1) – Plus connection, (54), in rear wiring harness
- (76) – Ground connection, on left back-up/brake light bulb holder
- (77) – Ground connection, on right back-up/brake light bulb holder

1993-1994

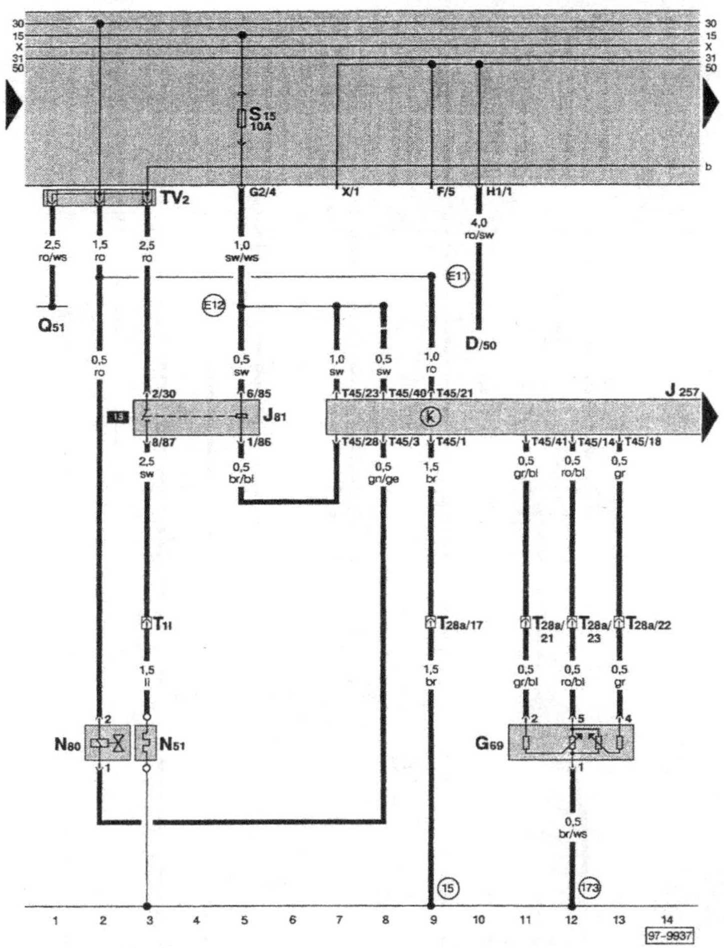

WIRING COLOR CODE		
ws	=	white
sw	=	black
ro	=	red
br	=	brown
gn	=	green
bl	=	blue
gr	=	grey
li	=	lilac
ge	=	yellow

D - Ignition/Starter Switch
G69 - Throttle Position (TP) Sensor
J81 - Early Fuel Evaporation (EFE) Relay
J257 - Mono-Motronic Engine Control Module
N51 - Early Fuel Evaporation (EFE) Heating Element
N80 - Evaporative Emission (EVAP) Canister Purge Regulator Valve
T1l - Connector, Single, left rear of engine
T28a - Connector, 28 Pin, on engine
T45 - Connector, 45 Pin, on Mono-Motronic Engine Control Module (ECM)
TV2 - Wire connector, terminal 30 (B+)

(15) - Ground connection, on cylinder head
(173) - Ground connection -1-, in Mono-Motronic Throttle Body Fuel Injection (TBI) wiring harness
(E11) - Plus connection (30) in Throttle Body Fuel Injection (TBI) wiring harness
(E12) - Plus connection -1-(15) in Throttle Body Fuel Injection (TBI) wiring harness
(Q51) - Plus connection (30), in alarm system wiring harness
- - - Automatic transmission only

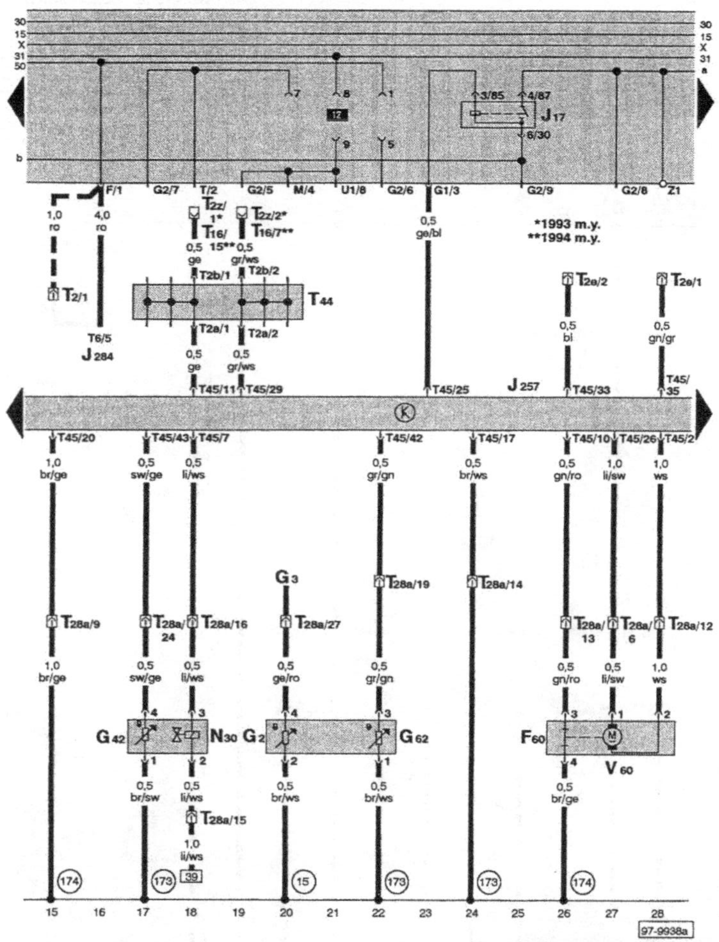

WIRING COLOR CODE		
ws	=	white
sw	=	black
ro	=	red
br	=	brown
gn	=	green
bl	=	blue
gr	=	grey
li	=	lilac
ge	=	yellow

F60 - Closed Throttle Position (CTP) Switch
G2 - Engine Coolant Temperature (ECT) Sensor
G3 - Engine Coolant Temperature (ECT) Gauge
G42 - Intake Air Temperature (IAT) Sensor
G62 - Engine Coolant Temperature (ECT) Sensor
J17 - Fuel Pump (FP) Relay
J257 - Mono-Motronic Engine Control Module
J284 - Alarm System Control Module
N30 - Injector, Cyl. 1
V60 - Throttle Position (TP) Actuator
T2 - Connector, Double, behind fuse/relay panel, only on automatic transmission
T2a - Connector, Double, on junction connector (T44) above fuse/relay panel
T2b - Connector, Double, on junction connector (T44) above fuse/relay panel
T2e - Connector, Double, behind fuse/relay panel, A/C connection
T2z - Data Link Connector (DLC), behind center of dash panel
T6 - Connector, 6 Pin, behind dash panel, left, on Alarm System Control Module
T16 - Data Link Connector (DLC), behind center of dash panel
T28a - Connector, 28 Pin, left side of engine, near distributor
T44 - Junction Box for On Board Diagnostic, above fuse/relay panel
T45 - Connector, 45 Pin, on Mono-Motronic Engine Control Module (ECM)
(15) - Ground connection, on cylinder head
(173) - Ground connection -1-, in Mono-Motronic Throttle Body Fuel Injection (TBI) wiring harness
(174) - Ground connection -2-, in Mono-Motronic Throttle Body Fuel Injection (TBI) wiring harness
- - - Automatic transmission only

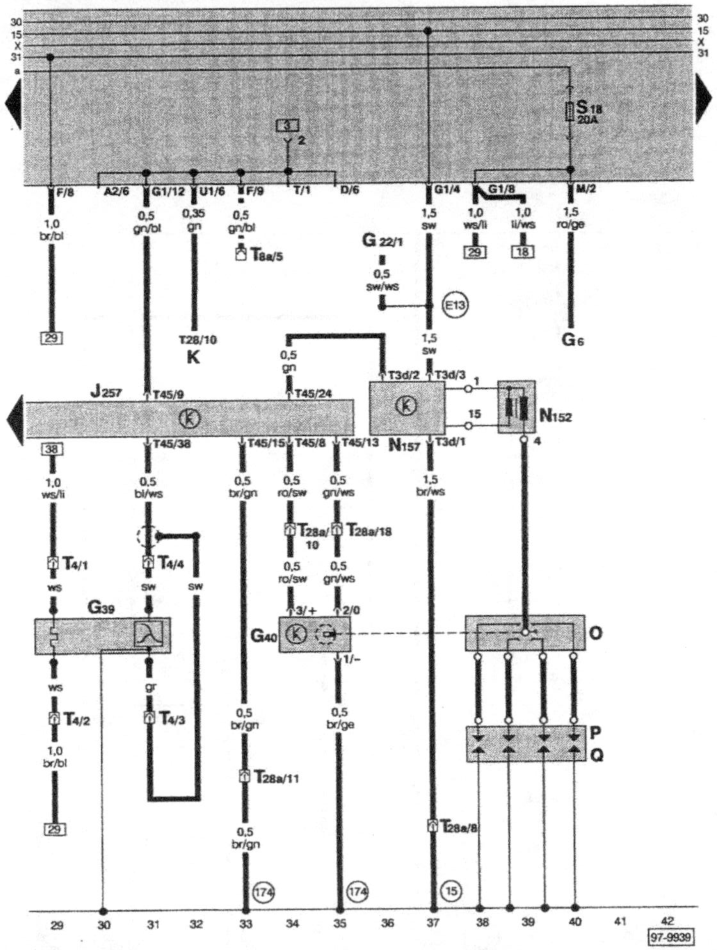

WIRING COLOR CODE		
ws	=	white
sw	=	black
ro	=	red
br	=	brown
gn	=	green
bl	=	blue
gr	=	grey
li	=	lilac
ge	=	yellow

G6 – Fuel Pump (FP)
G22 – Vehicle Speed Sensor (VSS), Speedometer
G39 – Heated Oxygen Sensor (HO2S)
G40 – Camshaft Position (CMP) Sensor
J257 – Mono–Motronic Engine Control Module
K – Instrument Cluster
N152 – Ignition Coil
N157 – Ignition Coil Power Output Stage
O – Ignition Distributor
Q – Spark Plugs
P – Spark Plug Connector
T3d – Connector, 3 Pin, left rear of engine compartment
T4 – Connector, 4 Pin, bottom right rear of engine compartment

T8a – Connector, 8 Pin, behind fuse/relay panel
T28 – Connector, 28 Pin, instrument cluster
T28a – Connector, 28 Pin, left side of engine, near distributor
T45 – Connector, 45 Pin, on Mono–Motronic Engine Control Module (ECM)

(15) – Ground connection, on cylinder head
(174) – Ground connection –2–, in Mono–Motronic Throttle Body Fuel Injection (TBI) wiring harness
(E13) – Plus connection –2– (15) in Throttle Body Fuel Injection (TBI) wiring harness

1993–1994

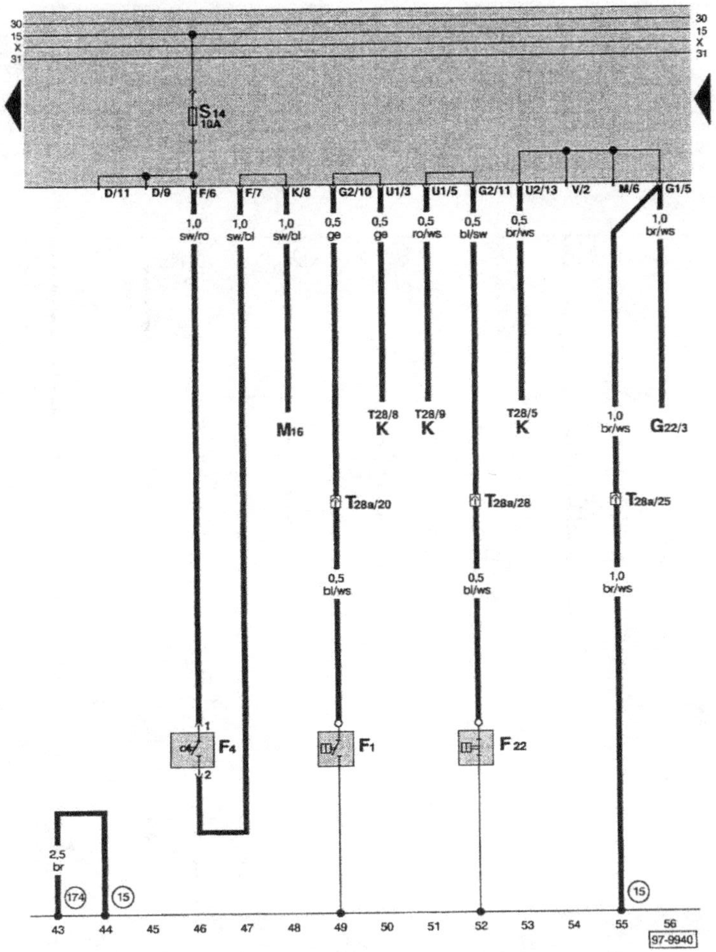

WIRING COLOR CODE		
ws	=	white
sw	=	black
ro	=	red
br	=	brown
gn	=	green
bl	=	blue
gr	=	grey
li	=	lilac
ge	=	yellow

F1 - Oil Pressure Switch
F4 - Back-Up Light Switch
F22 - Oil Pressure Switch (0.3 Bar)
G22 - Vehicle Speed Sensor (VSS), Speedometer
K - Instrument Cluster
M16 - Back-Up Light, Left
T28 - Connector, 28 Pin, instrument cluster
T28a - Connector, 28 Pin, left side of engine, near distributor
(15) - Ground connection, on cylinder head
(174) - Ground connection –2–, in Mono-Motronic Throttle Body Fuel Injection (TBI) wiring harness

1993-1994

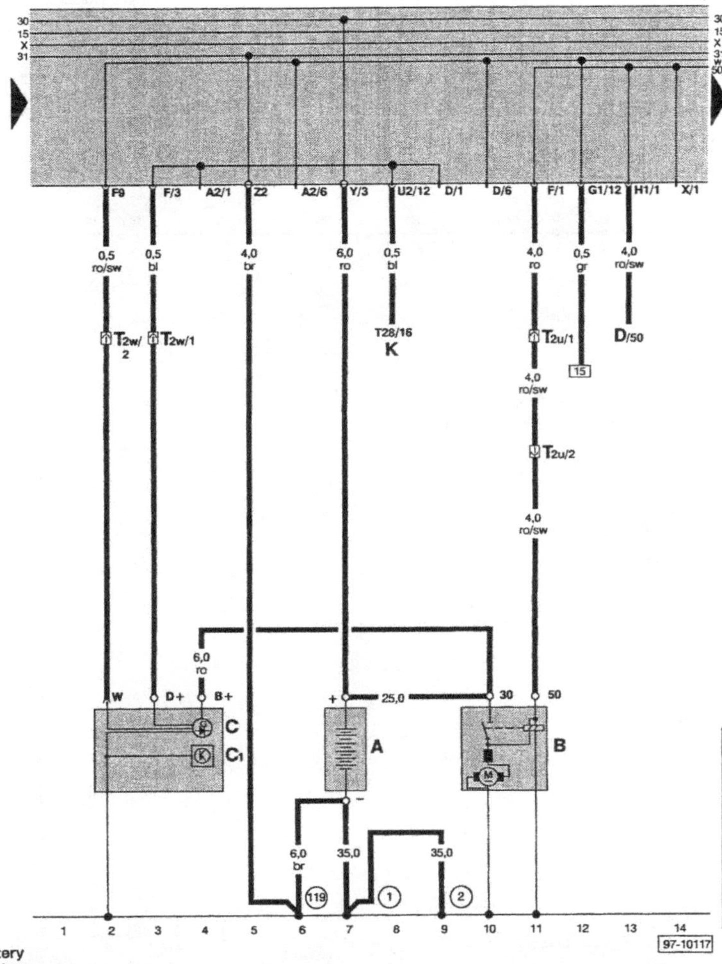

WIRING COLOR CODE		
ws	=	white
sw	=	black
ro	=	red
br	=	brown
gn	=	green
bl	=	blue
gr	=	grey
li	=	lilac
ge	=	yellow

A - Battery
B - Starter
C - Generator (Gen)
C1 - Voltage Regulator (VR)
D - Ignition/Starter Switch
K - Instrument Cluster
T2u - Connector, Double, behind fuse/relay panel
T2w - Connector, Double, near battery
T28 - Connector, 28 Pin, instrument cluster
(1) - Ground strap, battery to body
(2) - Ground strap, transmission to body
(119) - Ground connection -1-, in headlight wiring harness

1993–1994

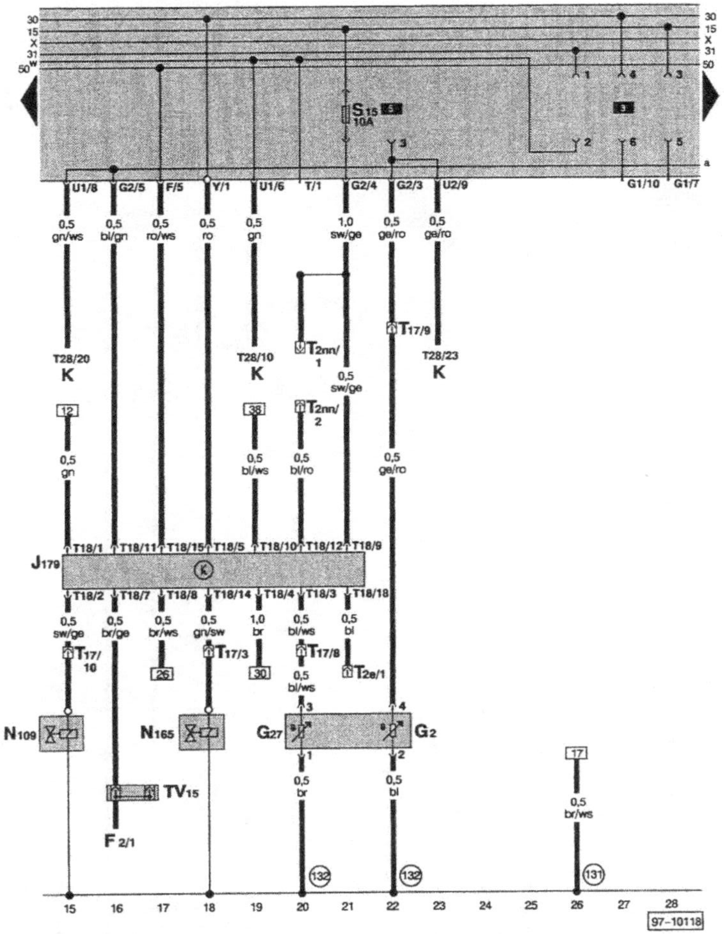

WIRING COLOR CODE		
ws	=	white
sw	=	black
ro	=	red
br	=	brown
gn	=	green
bl	=	blue
gr	=	grey
li	=	lilac
ge	=	yellow

F2 - Door Contact Switch, Left Front
G2 - Engine Coolant Temperature (ECT) Sensor
G27 - Engine Temperature Sensor
J179 - Automatic Glow Time Control Module
K - Instrument Cluster
N109 - Fuel Cut-off Valve
N165 - Injection Start Positioner
T1y - Connector, Single, behind fuse/relay panel
T2nn - Connector, Double
T2e - Connector, Double, behind fuse/relay panel, A/C connection
T17 - Connector, 17 Pin, on engine
T18 - Connector, 18 Pin on Automatic Glow Time Control Module
T28 - Connector, 28 Pin, instrument cluster
TV15 - Wire Connector, door contact switch
(131) - Ground connection -2- in engine compartment wiring harness
(132) - Ground connection -3- in engine compartment wiring harness

1993–1994

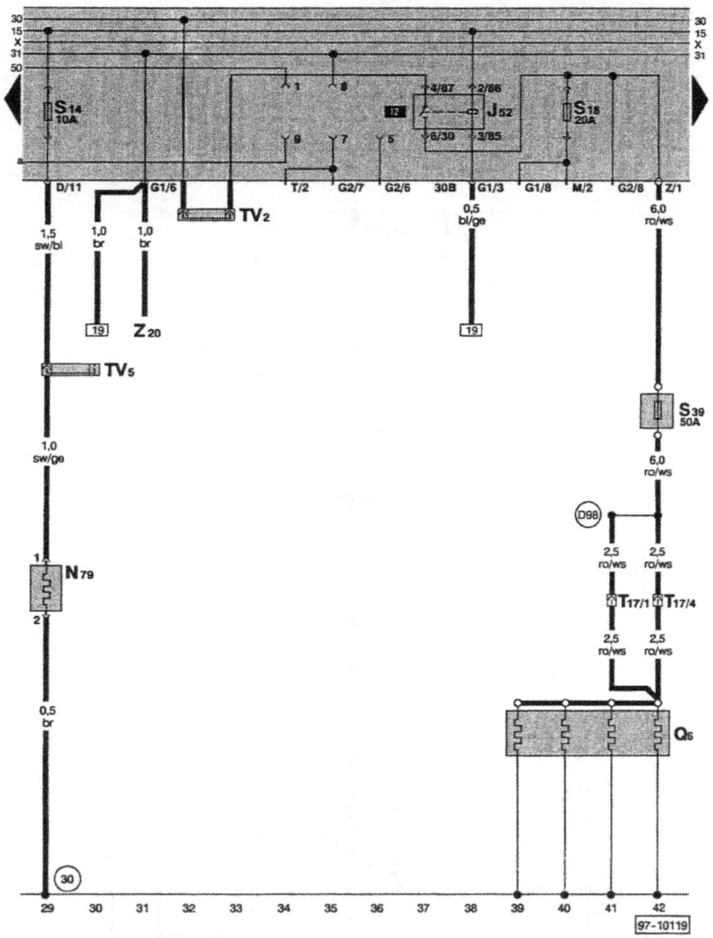

WIRING COLOR CODE		
ws	=	white
sw	=	black
ro	=	red
br	=	brown
gn	=	green
bl	=	blue
gr	=	grey
li	=	lilac
ge	=	yellow

J52 - Glow Plug Relay
N79 - Positive Crankcase Ventilation (PCV) Heating Element
Q6 - Glow Plug
S39 - Glow Plug Fuse Strip, 50A
Z20 - Washer Nozzle Heater, Left
T17 - Connector, 17 Pin, on engine
TV2 - Wire connector, terminal 30 (B+)
TV5 - Connector, terminal 15a

(30) - Ground connection, beside fuse/relay panel

(D98) - Wire connection, (glow plugs) in engine compartment wiring harness, right

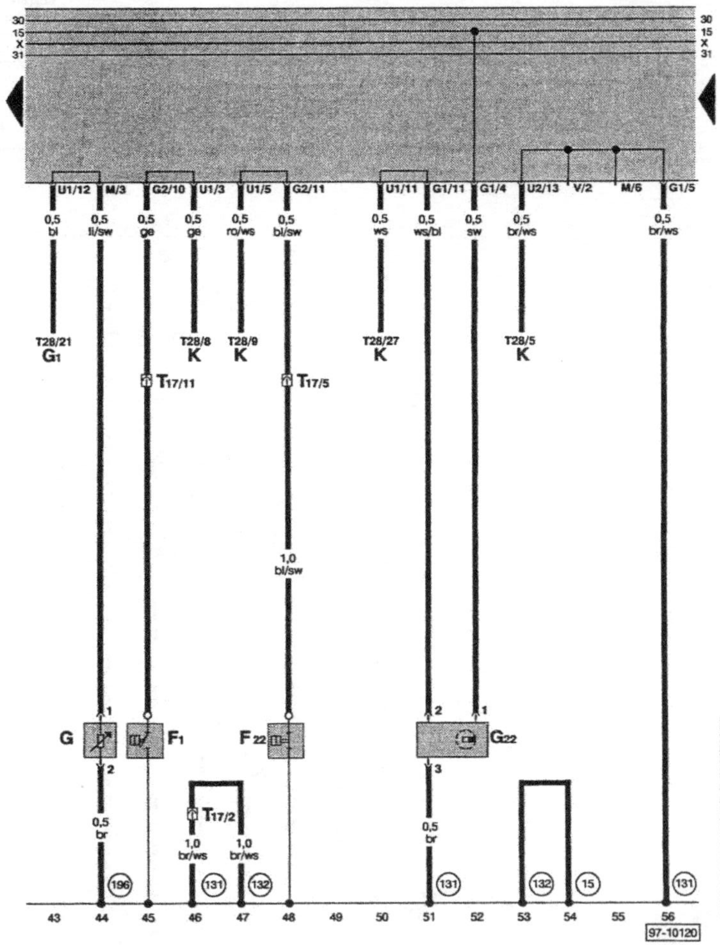

WIRING COLOR CODE		
ws	=	white
sw	=	black
ro	=	red
br	=	brown
gn	=	green
bl	=	blue
gr	=	grey
li	=	lilac
ge	=	yellow

F1 - Oil Pressure Switch
F22 - Oil Pressure Switch (0.3 Bar)
G - Fuel Level Sensor
G1 - Fuel Gauge
G22 - Vehicle Speed Sensor (VSS), Speedometer
K - Instrument Cluster
T17 - Connector, 17 Pin, on engine
T28 - Connector, 28 Pin, instrument cluster

(15) - Ground connection, on cylinder head

(131) - Ground connection -2- in engine compartment wiring harness

(132) - Ground Connection -3- in engine compartment wiring harness

(196) - Ground connection -1-, in rear wiring harness

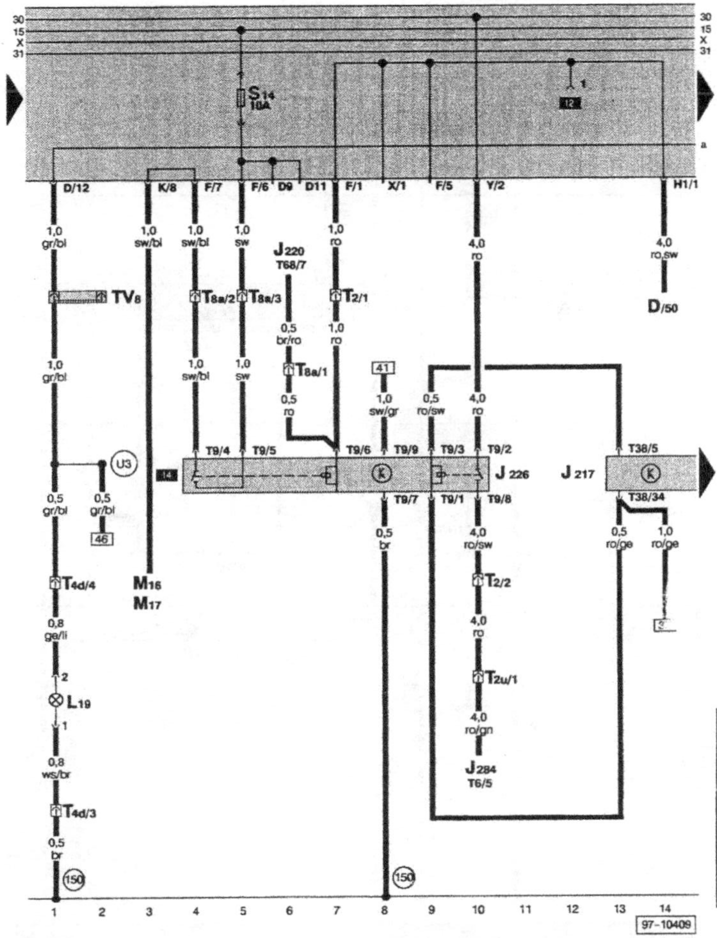

WIRING COLOR CODE

ws	=	white
sw	=	black
ro	=	red
br	=	brown
gn	=	green
bl	=	blue
gr	=	grey
li	=	lilac
ge	=	yellow

D - Ignition/Starter Switch
J217 - Transmission Control Module (TCM)
J220 - Motronic Engine Control Module (ECM)
J226 - Park/Neutral Position (PNP) Relay
J284 - Alarm System Control Module
L19 - Automatic Transmission Console Light
M16 - Back-Up Light, Left
M17 - Back-Up Light, Right
T2 - Connector, Double, behind fuse/relay panel, only on automatic transmission
T2u - Connector, Double, behind fuse/relay panel
T4d - Connector, 4 Pin, below center console
T6 - Connector, 6 Pin, behind dash panel, left, on Alarm System Control Module
T8a - Connector, 8 Pin, behind fuse/relay panel
T9 - Connector, 9 Pin, on Park/Neutral Position (PNP) Relay
T38 - Connector, 38 Pin, below right front seat, on Transmission Control Module (TCM)
T68 - Connector, 68 Pin, under right side of hood, on Motronic Engine Control Module (ECM)
TV8 - Connector, terminal 58b
(150) - Ground connection, in automatic transmission wiring harness
(U3) - Wire connection, in automatic transmission wire harness

1993–1994

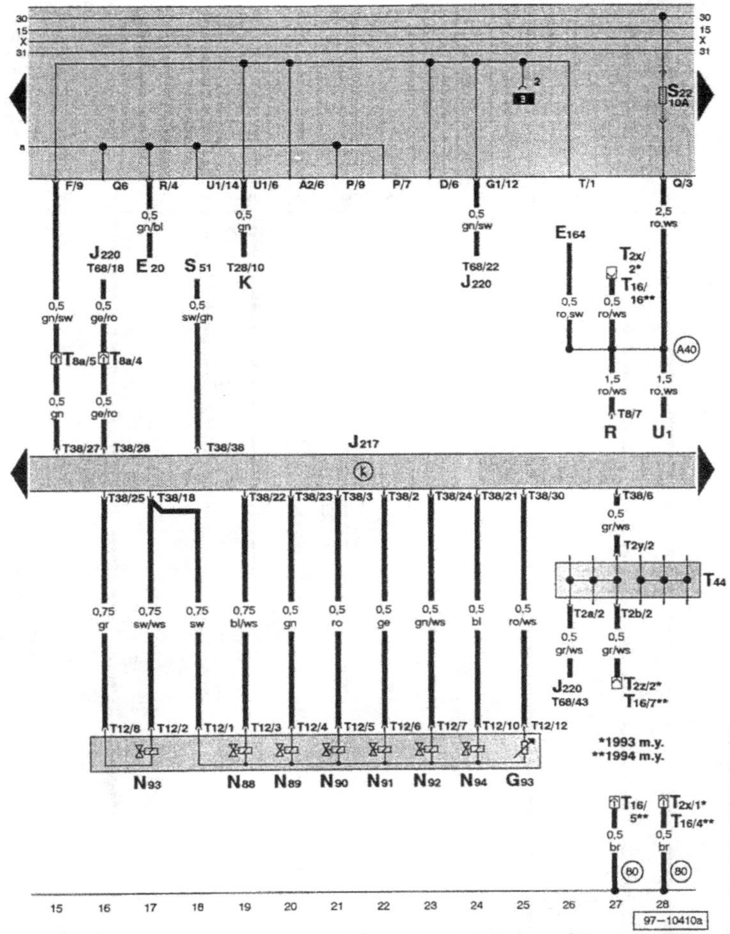

WIRING COLOR CODE

ws	=	white
sw	=	black
ro	=	red
br	=	brown
gn	=	green
bl	=	blue
gr	=	grey
li	=	lilac
ge	=	yellow

E20 - Instrument Panel Light Dimmer Switch
E164 - Luggage Compartment Release Switch
G93 - Transmission Fluid Temperature Sensor
J217 - Transmission Control Module (TCM)
J220 - Motronic Engine Control Module (ECM)
K - Instrument Cluster
N88 - Solenoid Valve 1
N89 - Solenoid Valve 2
N90 - Solenoid Valve 3
N91 - Solenoid Valve 4
N92 - Solenoid Valve 5
N93 - Solenoid Valve 6
N94 - Solenoid Valve 7
R - Radio
S51 - Fuse
U1 - Cigarette Lighter
T2a - Connector, Double, on junction connector (T44) above fuse/relay panel
T2b - Connector, Double, on junction connector (T44) above fuse/relay panel
T2x - Data Link Connector (DLC), behind center of dash panel
T2y - Connector, Double, on junction connector (T44) above fuse/relay panel
T2z - Data Link Connector (DLC), behind center of dash panel
T8 - Connector, 8 Pin, on radio
T8a - Connector, 8 Pin, behind fuse/relay panel
T12 - Connector, 12 Pin, on transmission
T16 - Data Link Connector (DLC), behind center of dash panel
T28 - Connector, 28 Pin, instrument cluster
T38 - Connector, 38 Pin, below right front seat, on Transmission Control Module (TCM)
T44 - Junction Box for On Board Diagnostic, above fuse/relay panel
T68 - Connector, 68 Pin, under right side of hood, on Motronic Engine Control Module (ECM)
(80) - Ground connection -1-, in instrument panel wiring harness
(A40) - Plus connection (80), in instrument panel wiring harness

1993–1994

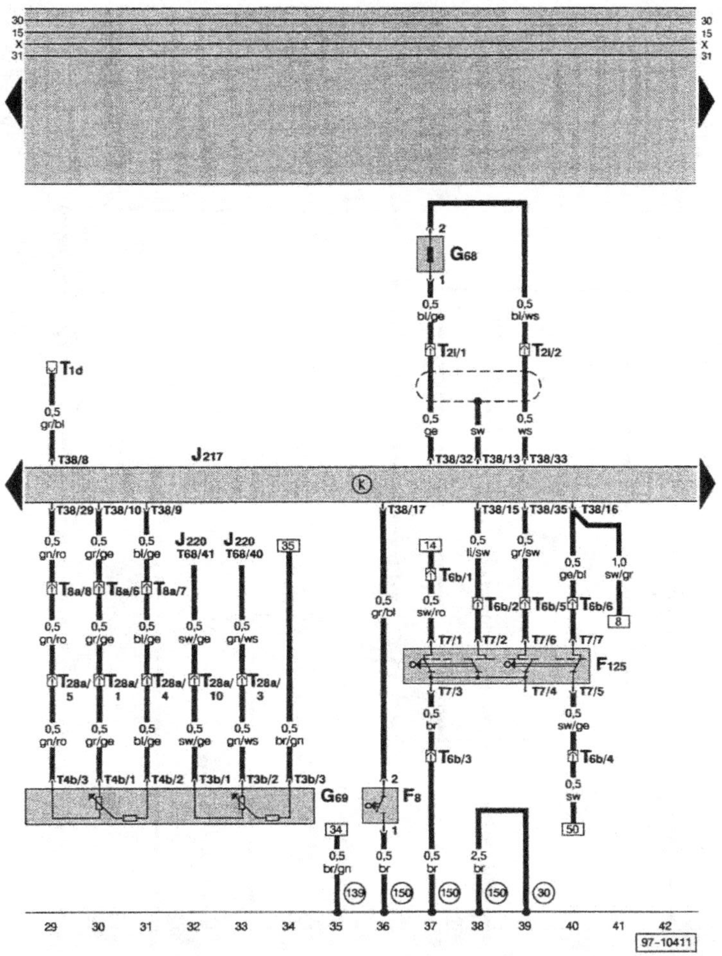

WIRING COLOR CODE		
ws	=	white
sw	=	black
ro	=	red
br	=	brown
gn	=	green
bl	=	blue
gr	=	grey
li	=	lilac
ge	=	yellow

F8 – Kick Down Switch
F125 – Multi-Function Transmission Range (TR) Switch
G68 – Vehicle Speed Sensor (VSS)
G69 – Throttle Position (TP) Sensor
J217 – Transmission Control Module (TCM)
J220 – Motronic Engine Control Module (ECM)
T1d – Connector, Single, behind fuse/relay panel
T2i – Connector, Double, below battery
T3b – Connector, 3 Pin, on throttle position sensor
T4b – Connector, 4 Pin, on throttle position sensor
T6b – Connector, 6 Pin, below battery
T7 – Connector, 7 Pin, on transmission
T8a – Connector, 8 Pin, behind fuse/relay panel

T28a – Connector, 28 Pin, left side of engine, near distributor
T38 – Connector, 38 Pin, below right front seat, on Transmission Control Module (TCM)
T68 – Connector, 68 Pin, under right side of hood, on Motronic Engine Control Module (ECM)
(30) – Ground connection – beside fuse/relay panel
(139) – Ground connection (sensor ground), in Motronic Multiport Fuel Injection (MFI) wiring harness
(150) – Ground connection, in automatic transmission wiring harness

1993–1994

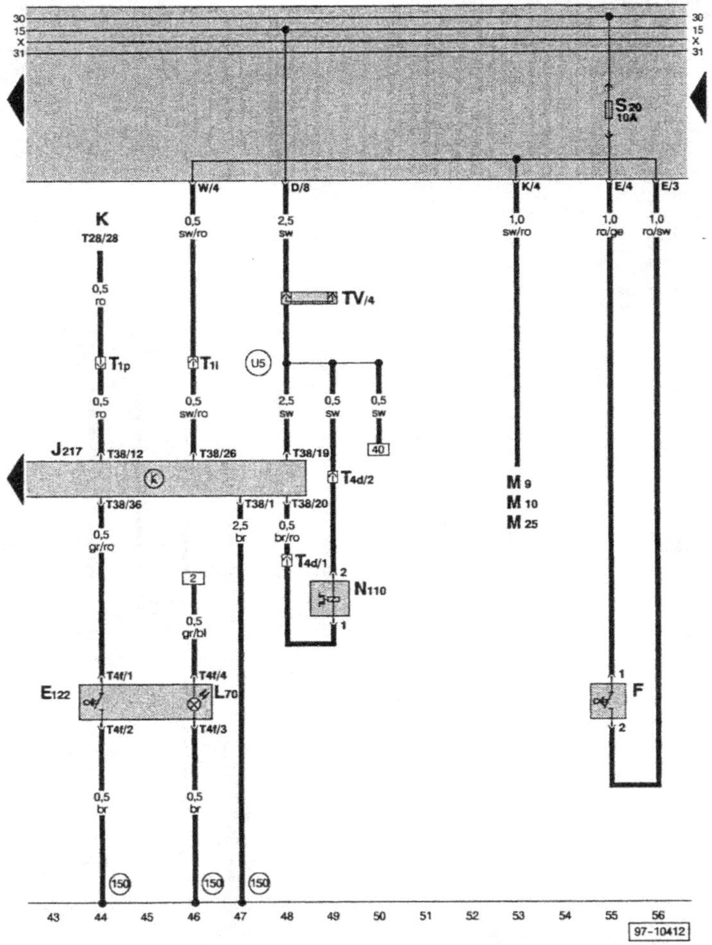

WIRING COLOR CODE		
ws	=	white
sw	=	black
ro	=	red
br	=	brown
gn	=	green
bl	=	blue
gr	=	grey
li	=	lilac
ge	=	yellow

E122 - Transmission Range (TR) Program Switch
F - Brake Light Switch
J217 - Transmission Control Module (TCM)
K - Instrument Cluster
L70 - Program Switch Illumination
M9 - Brake Light, Left
M10 - Brake Light, Right
M25 - High-Mount Brake Light
N110 - Shift Lock Solenoid
T1i - Connector, Single, behind fuse/relay panel
T1p - Connector, Single, behind fuse/relay panel
T1v - Connector, Single, on taillight, left
T4d - Connector, 4 Pin, below center console

T4f - Connector, 4 Pin, on Transmission Range (TR) Program Switch
T8d - Connector, 8 Pin, on taillight, left
T8e - Connector, 8 Pin, on taillight right
T28 - Connector, 28 Pin, instrument cluster
T38 - Connector, 38 Pin, below right front seat, on Transmission Control Module (TCM)
TV4 - Connector, terminal 15
(150) - Ground connection, in automatic transmission wiring harness
(U5) - Wire connection -2- (15), in automatic transmission wiring harness

1993-1994

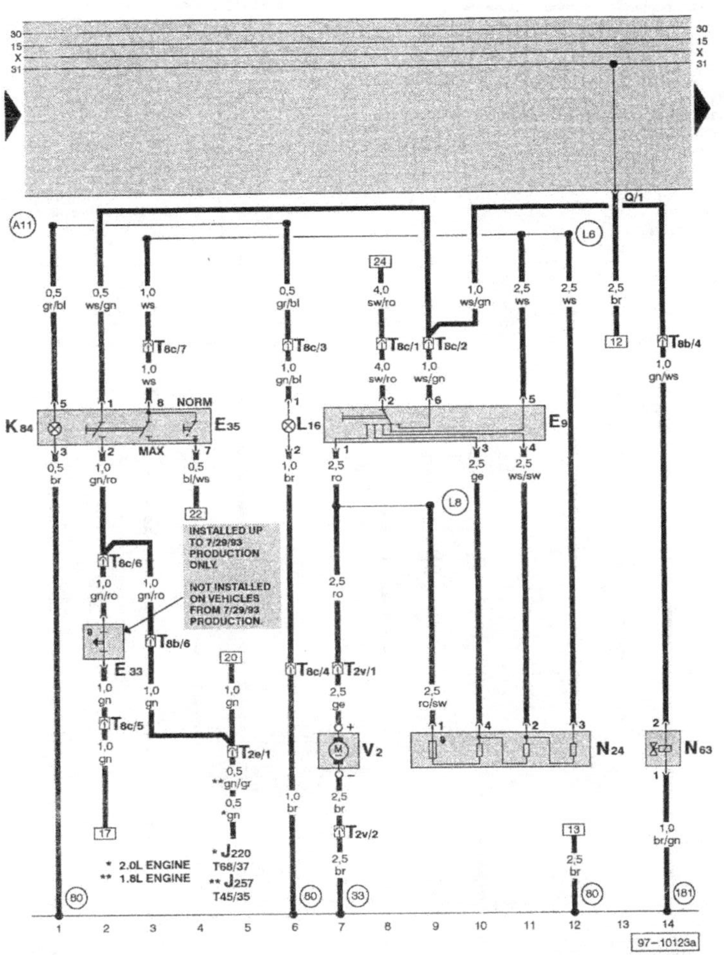

WIRING COLOR CODE

ws	=	white
sw	=	black
ro	=	red
br	=	brown
gn	=	green
bl	=	blue
gr	=	grey
li	=	lilac
ge	=	yellow

- E9 – Fresh Air Blower Switch
- E33 – A/C Evaporator Temperature Switch
- E35 – A/C Switch
- J220 – Motronic Engine Control Module (ECM)
- J257 – Mono–Motronic Engine Control Module (ECM)
- K84 – A/C Indicator Light
- L16 – Fresh Air Control Lever Light
- N24 – Fresh Air Blower Series Resistance with Fuse
- N63 – Fresh Air/Recirculating Flap Two–Way Valve
- V2 – Fresh Air Blower
- T2e – Double Connector, behind fuse/relay panel, A/C Connection
- T2v – Double Connector, below right side of dash at blower motor
- T8b – 8–Pin Connector, behind fuse/relay panel
- T8c – 8–Pin Connector, behind lower center of dash panel
- T45 – 45–Pin Connector, on Mono–Motronic Engine Control Module (ECM)
- T68 – 68–Pin Connector, under right side of hood, on Motronic Engine Control Module
- (33) – Ground connection, behind instrument panel, right
- (80) – Ground connection –1–, in instrument panel wiring harness
- (181) – Ground connection (for A/C), in engine compartment wiring harness
- (A11) – Plus connection (58b) in instrument panel wiring harness
- (L6) – Wire connection –2–, in evaporator housing wiring harness
- (L8) – Wire connection –4–, in evaporator housing wiring harness

1993–1994

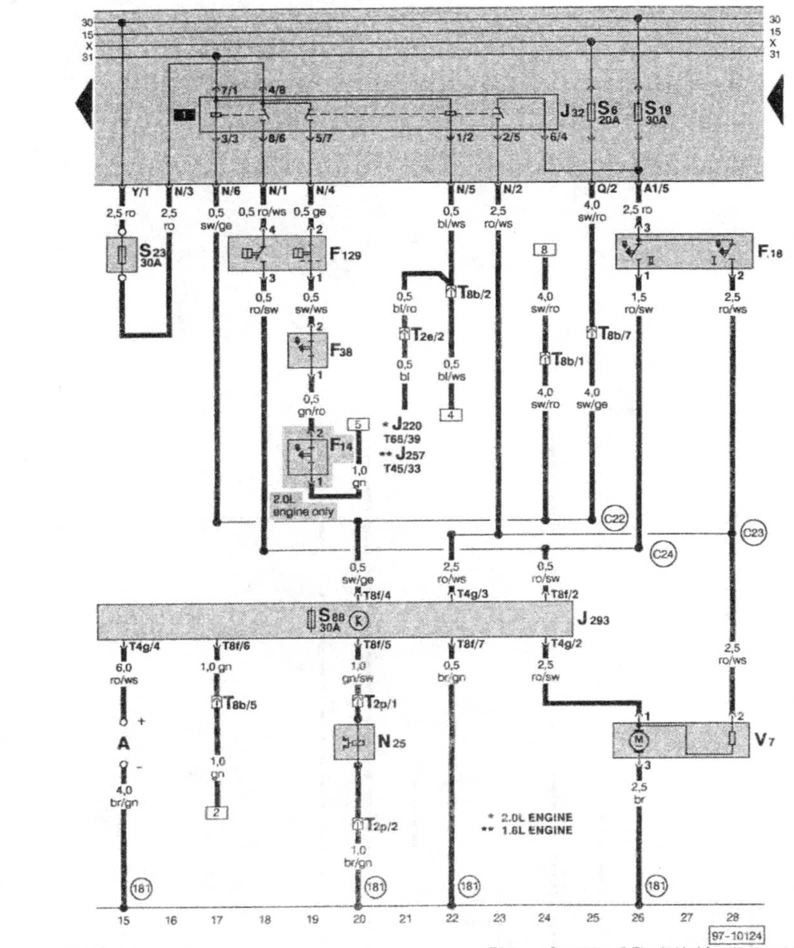

WIRING COLOR CODE

ws	=	white
sw	=	black
ro	=	red
br	=	brown
gn	=	green
bl	=	blue
gr	=	grey
li	=	lilac
ge	=	yellow

- A – Battery
- F14 – Engine Coolant Temperature (ECT) Warning Switch
- F18 – Coolant Fan Control (FC) Thermal Switch
- F38 – Ambient Temperature Switch
- F129 – A/C Pressure Switch
- J32 – A/C Relay
- J220 – Motronic Engine Control Module (ECM)
- J257 – Mono–Motronic Engine Control Module (ECM)
- J293 – Coolant Fan Control (FC) Module
- N25 – A/C Clutch
- S23 – Heater Fuse
- S88 – Fuse Strip
- V7 – Coolant Fan
- T2e – Connector, Double, behind fuse/relay panel, A/C connection
- T2p – Connector, Double, on compressor
- T4g – Connector, 4 Pin, left side of engine, on Coolant Fan Control (FC) Module
- T8b – Connector, 8 Pin, behind fuse/relay panel
- T8f – Connector, 8 Pin, left side of engine, on Coolant Fan Control (FC) Module
- T45 – Connector, 45 Pin, on Mono–Motronic Engine Control Module (ECM)
- T68 – Connector, 68 Pin, under right side of hood, on Motronic Engine Control Module (ECM)
- (181) – Ground connection (for A/C), in engine compartment wiring harness
- (C22) – Wire connection (X), in headlight/air conditioner wiring harness
- (C23) – Wire connection (blower, 1st speed) in headlight/air conditioner wiring harness
- (C24) – Wire connection (blower, 2nd speed) in headlight/air conditioner wiring harness

1993–1994

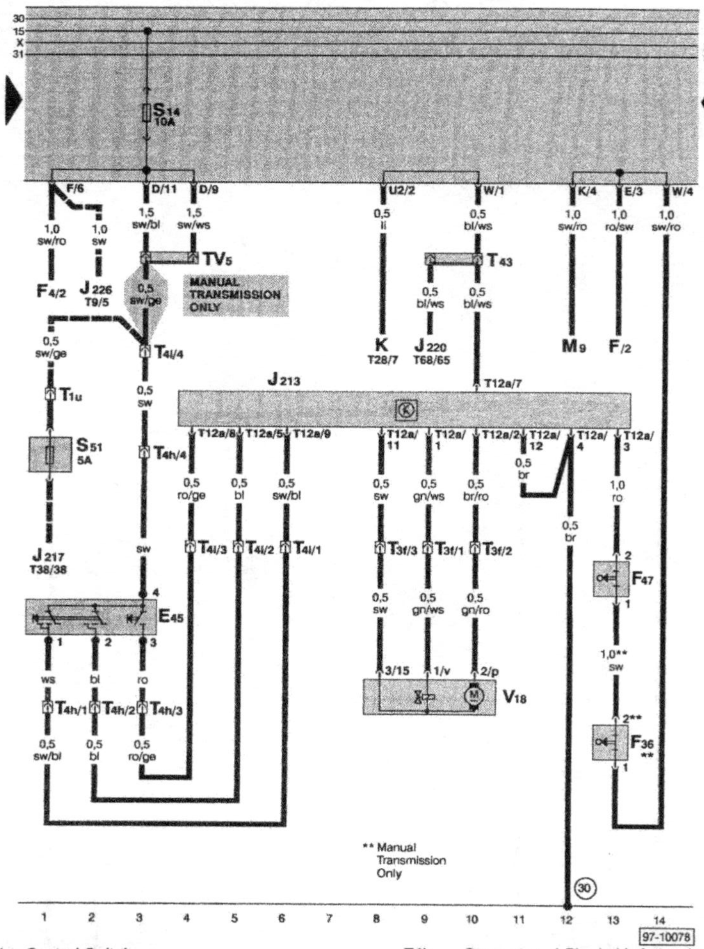

WIRING COLOR CODE

ws	=	white
sw	=	black
ro	=	red
br	=	brown
gn	=	green
bl	=	blue
gr	=	grey
li	=	lilac
ge	=	yellow

E45 – Cruise Control Switch
F – Brake Light Switch
F4 – Back-Up Light Switch
F36 – Cruise Control Clutch Pedal Position Switch
F47 – Cruise Control Brake Pedal Position Switch
J213 – Cruise Control Module
J217 – Transmission Control Module (TCM)
J220 – Motronic Engine Control Module (ECM)
J226 – Park/Neutral Position (PNP) Relay
K – Instrument Cluster
M9 – Brake Light, Left
S51 – Fuse
V18 – Cruise Control Vacuum Pump
T1u – Connector, Single, behind fuse/relay panel
T3f – Connector, 3 Pin, below left side of dash
T4h – Connector, 4 Pin, behind steering column switch trim
T4l – Connector, 4 Pin, behind fuse/relay panel
T9 – Connector, 9 Pin, on Park/Neutral Position (PNP) Relay
T12a – Connector, 12 Pin, on Cruise Control Module
T28 – Connector, 28 Pin, instrument cluster
T38 – Connector, 38 Pin, below right front seat, on Transmission Control Module (TCM)
T43 – Connector, vehicle speed signal
T68 – Connector, 68 Pin, under right side of hood, on Motronic Engine Control Module (ECM)
TV5 – Connector, terminal 15a
(30) – Ground connection – beside fuse/relay panel
– – – Automatic transmission only

1993–1994

Daytime running lights (except GTI, Canada only)

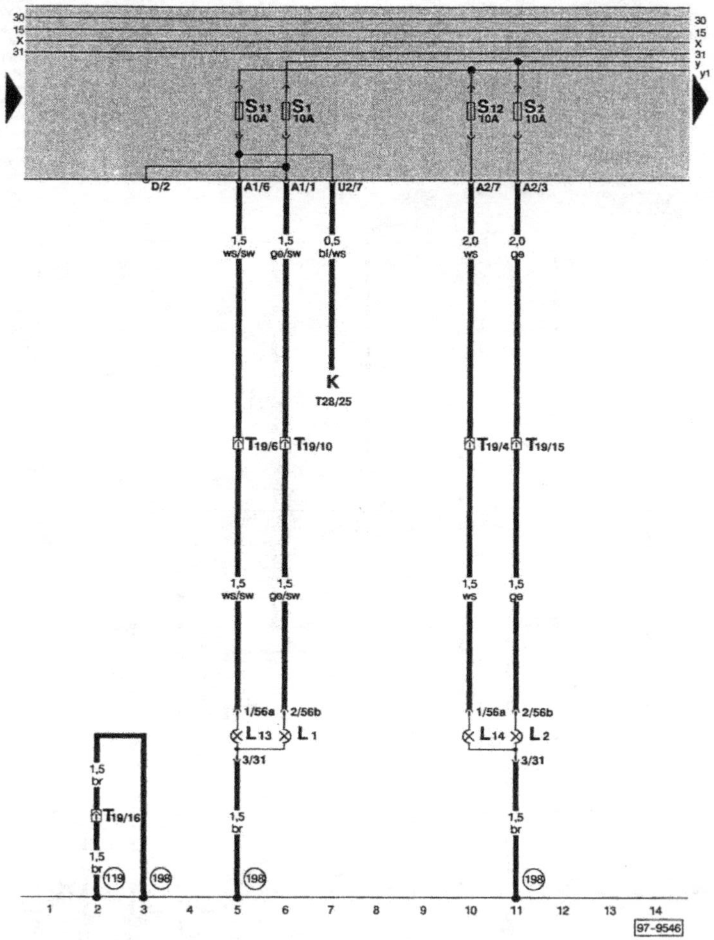

WIRING COLOR CODE

ws	=	white
sw	=	black
ro	=	red
br	=	brown
gn	=	green
bl	=	blue
gr	=	grey
li	=	lilac
ge	=	yellow

K – Instrument Cluster
L1 – Headlight, Left
L2 – Headlight, Right
L13 – High Beam Headlight, Left
L14 – High Beam Headlight, Right
T19 – Connector, 19 Pin, behind radiator grill, center
T28 – Connector, 28 Pin, instrument cluster
(119) – Ground connection -1-, in headlight wiring harness
(198) – Ground connection, in front light wiring harness

1993-1994

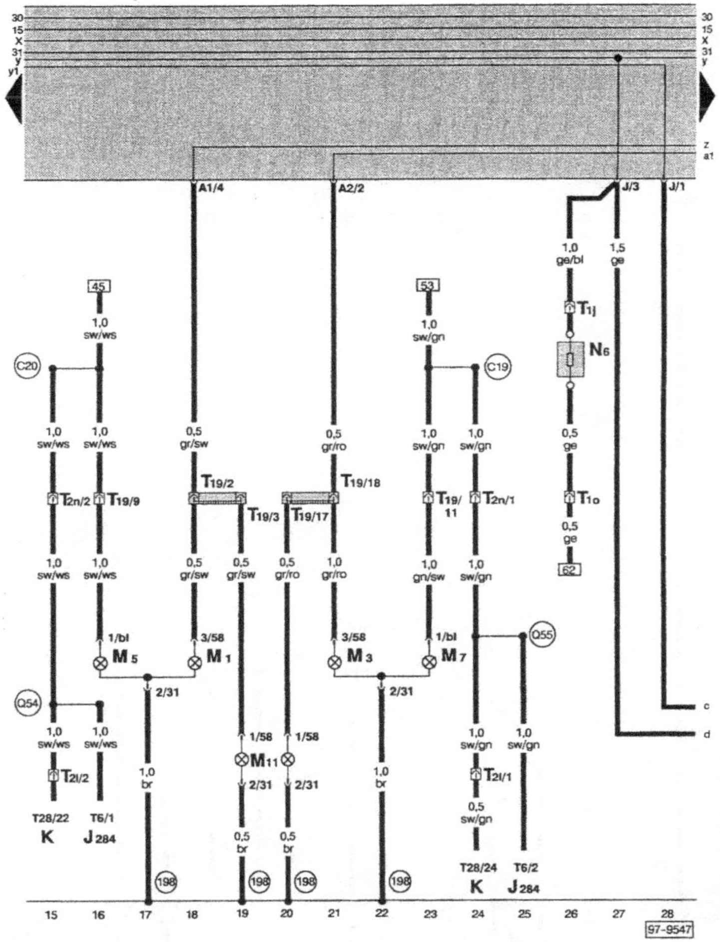

WIRING COLOR CODE

ws	=	white
sw	=	black
ro	=	red
br	=	brown
gn	=	green
bl	=	blue
gr	=	grey
li	=	lilac
ge	=	yellow

- J284 – Alarm System Control Module
- K – Instrument Cluster
- M1 – Parking Light, Left
- M3 – Parking Light, Right
- M5 – Turn Signal Light, Left Front
- M7 – Turn Signal Light, Right Front
- M11 – Side Marker Lights, Front
- N6 – Series Resistance
- T1j – Connector, Single, behind fuse/relay panel
- T1o – Connector, Single, behind fuse/relay panel
- T2l – Connector, Double, behind fuse/relay panel
- T2n – Connector, Double, behind fuse/relay panel
- T6 – Connector, 6 Pin, behind dash panel, left, on Alarm System Control Module
- T19 – Connector, 19 Pin, behind radiator grill, center
- T28 – Connector, 28 Pin, instrument cluster
- (198) – Ground connection, in front light wiring harness
- (C19) – Wire connection (right turn signal), in headlight wiring harness
- (C20) – Wire connection (left turn signal), in headlight wiring harness
- (Q54) – Wire connection (left turn signal), in alarm system wiring harness
- (Q55) – Wire connection (right turn signal), in alarm system wiring harness

1993–1994

Daytime running lights
(except GTI, Canada only)

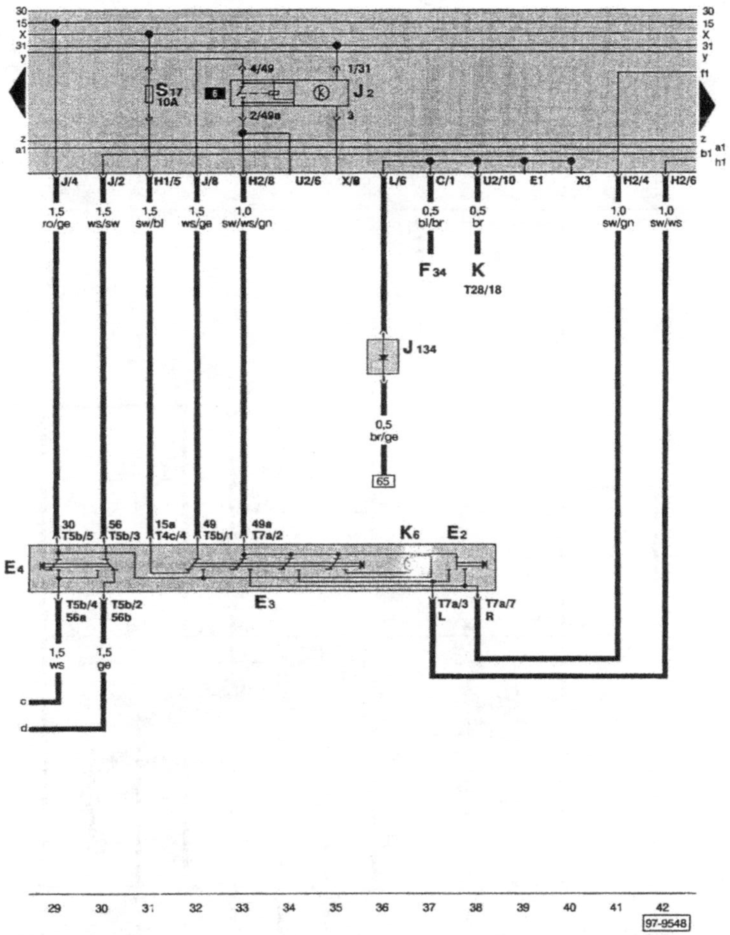

WIRING COLOR CODE

ws	=	white
sw	=	black
ro	=	red
br	=	brown
gn	=	green
bl	=	blue
gr	=	grey
li	=	lilac
ge	=	yellow

- E2 – Turn Signal Switch
- E3 – Emergency Flasher Switch
- E4 – Headlight Dimmer/Flasher Switch
- F34 – Brake Fluid Level Warning Switch
- J2 – Emergency Flasher Relay
- J134 – Daytime Running Light Diode
- K – Instrument Cluster
- K6 – Emergency Flasher Warning Light
- T4c – Connector, 4 Pin, behind steering column switch trim
- T5b – Connector, 5 Pin, behind steering column switch trim
- T7a – Connector, 7 Pin, behind steering column switch trim
- T28 – Connector, 28 Pin, instrument cluster

1993–1994

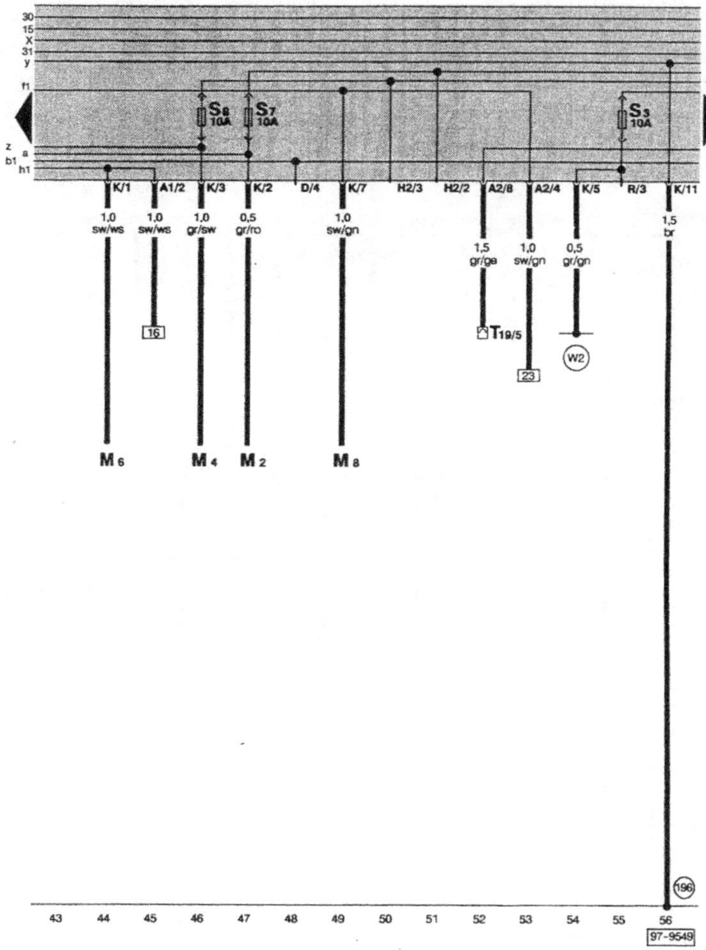

WIRING COLOR CODE		
ws	=	white
sw	=	black
ro	=	red
br	=	brown
gn	=	green
bl	=	blue
gr	=	grey
li	=	lilac
ge	=	yellow

M2 – Tail Light, Right
M4 – Tail Light, Left
M6 – Turn Signal Light, Left Rear
M8 – Turn Signal Light, Right Rear
T19 – Connector, 19 Pin, behind radiator grill, center
(196) – Ground connection -1-, in rear wiring harness
(W2) – Plus connection (58), in rear wiring harness

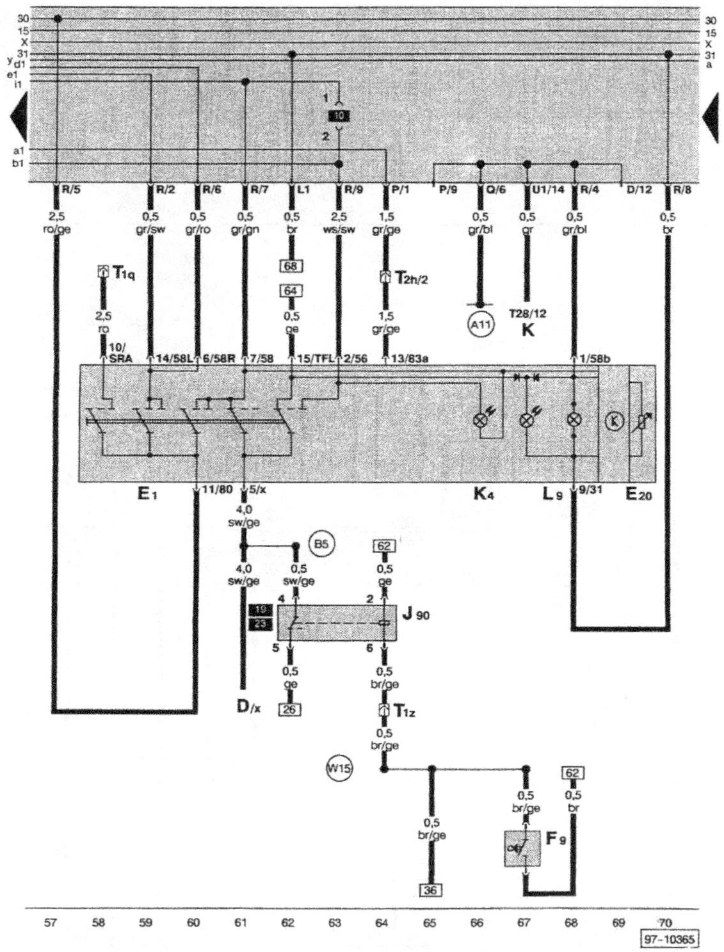

WIRING COLOR CODE		
ws	=	white
sw	=	black
ro	=	red
br	=	brown
gn	=	green
bl	=	blue
gr	=	grey
li	=	lilac
ge	=	yellow

D – Ignition/Starter Switch
E1 – Light Switch
E20 – Instrument Panel Light Dimmer Switch
F9 – Parking Brake Warning Light Switch
J90 – Daytime Running Lights Relay (Switch on)
K – Instrument Cluster
K4 – Park Light Indicator Light
L9 – Headlight Switch Light
T1q – Connector, Single, behind fuse/relay panel
T1z – Connector, Single, behind fuse/relay panel
T2h – Connector, Double, behind fuse/relay panel
T28 – Connector, 28 Pin, instrument cluster
(A11) – Plus connection (58b), in instrument panel wiring harness
(B5) – Wire connection, (X), in headlight switch wiring harness
(W15) – Wiring connection, (HB) in rear wiring harness

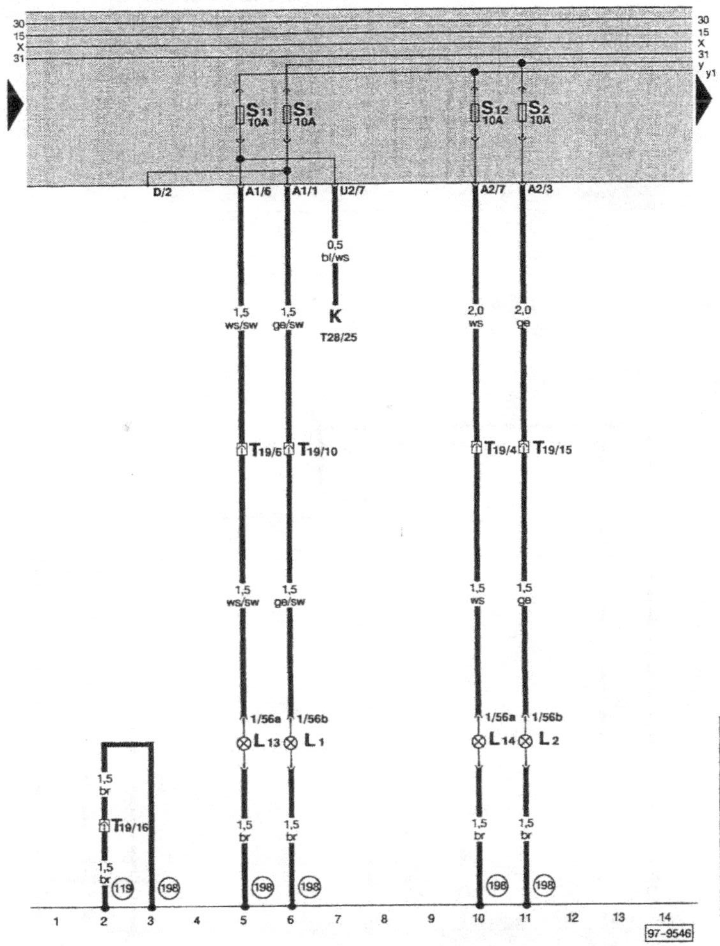

WIRING COLOR CODE		
ws	=	white
sw	=	black
ro	=	red
br	=	brown
gn	=	green
bl	=	blue
gr	=	grey
li	=	lilac
ge	=	yellow

K - Instrument Cluster
L1 - Headlight, Left
L2 - Headlight, Right
L13 - High Beam Headlight, Left
L14 - High Beam Headlight, Right
T19 - Connector, 19 Pin, behind radiator grill, center
T28 - Connector, 28 Pin, instrument cluster
(119) - Ground connection -1-, in headlight wiring harness
(198) - Ground connection, in front light wiring harness

1993–1994

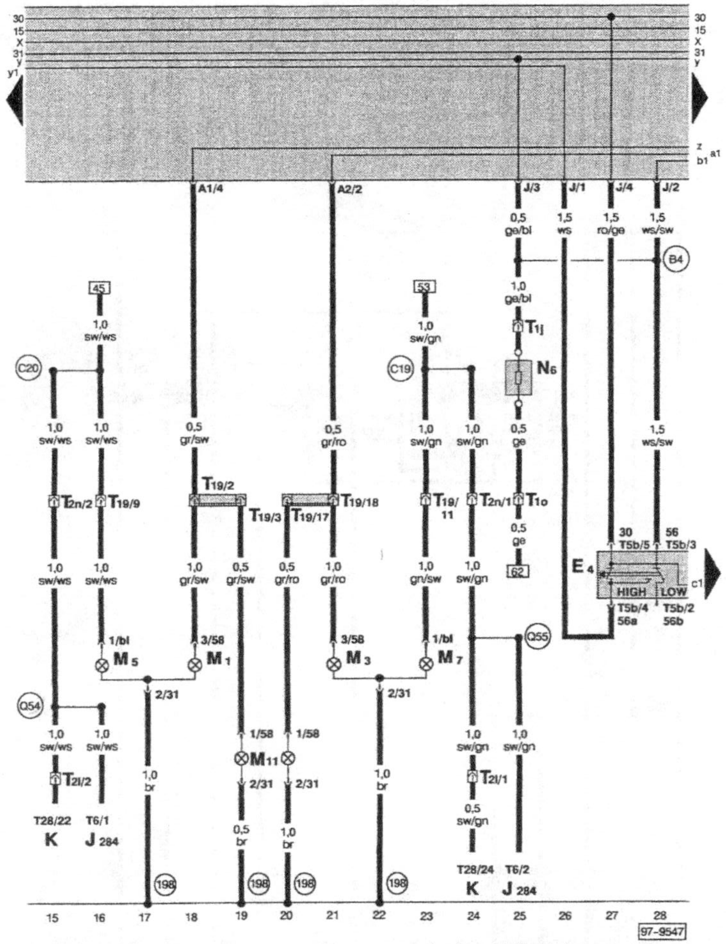

WIRING COLOR CODE		
ws	=	white
sw	=	black
ro	=	red
br	=	brown
gn	=	green
bl	=	blue
gr	=	grey
li	=	lilac
ge	=	yellow

E4 - Headlight Dimmer/Flasher Switch
J284 - Alarm System Control Module
K - Instrument Cluster
M1 - Parking Light, Left
M3 - Parking Light, Right
M5 - Turn Signal Light, Left Front
M7 - Turn Signal Light, Right Front
M11 - Side Marker Lights, Front
N6 - Series Resistance
T1j - Connector, Single, behind fuse/relay panel
T1o - Connector, Single, behind fuse/relay panel
T2l - Connector, Double, behind fuse/relay panel
T2n - Connector, Double, behind fuse/relay panel
T5b - Connector, 5 Pin, behind steering column switch trim
T6 - Connector, 6 Pin, behind dash panel, left, on Alarm System Control Module
T19 - Connector, 19 Pin, behind radiator grill, center
T28 - Connector, 28 Pin, instrument cluster
(198) - Ground connection, in front light wiring harness
(B4) - Plus connection, 58e in light switch wiring harness
(C19) - Wire connection (right turn signal), in headlight wiring harness
(C20) - Wire connection (left turn signal), in headlight wiring harness
(Q54) - Wire connection (left turn signal), in alarm system wiring harness
(Q55) - Wire connection (right turn signal), in alarm system wiring harness

1993–1994

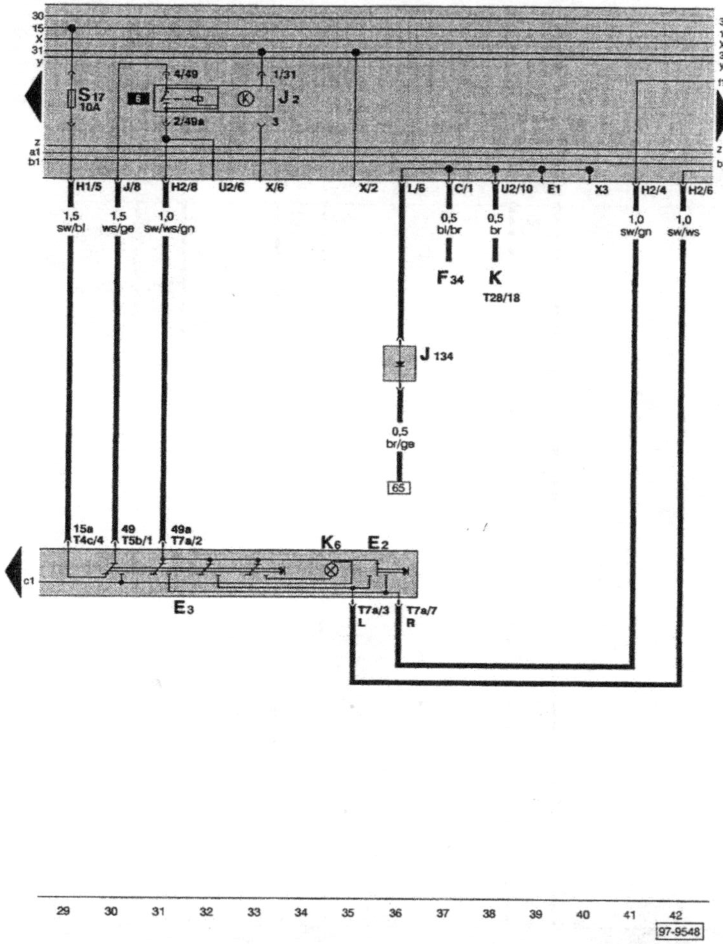

WIRING COLOR CODE

ws	=	white
sw	=	black
ro	=	red
br	=	brown
gn	=	green
bl	=	blue
gr	=	grey
li	=	lilac
ge	=	yellow

E2 - Turn Signal Switch
E3 - Emergency Flasher Switch
F34 - Brake Fluid Level Warning Switch
J2 - Emergency Flasher Relay
J134 - Daytime Running Light Diode
K - Instrument Cluster
K6 - Emergency Flasher Warning Light
T4c - Connector, 4 Pin, behind steering column switch trim
T5b - Connector, 5 Pin, behind steering column switch trim
T7a - Connector, 7 Pin, behind steering column switch trim
T28 - Connector, 28 Pin, instrument cluster

1993–1994

Daytime running lights
(GTI, Canada only)

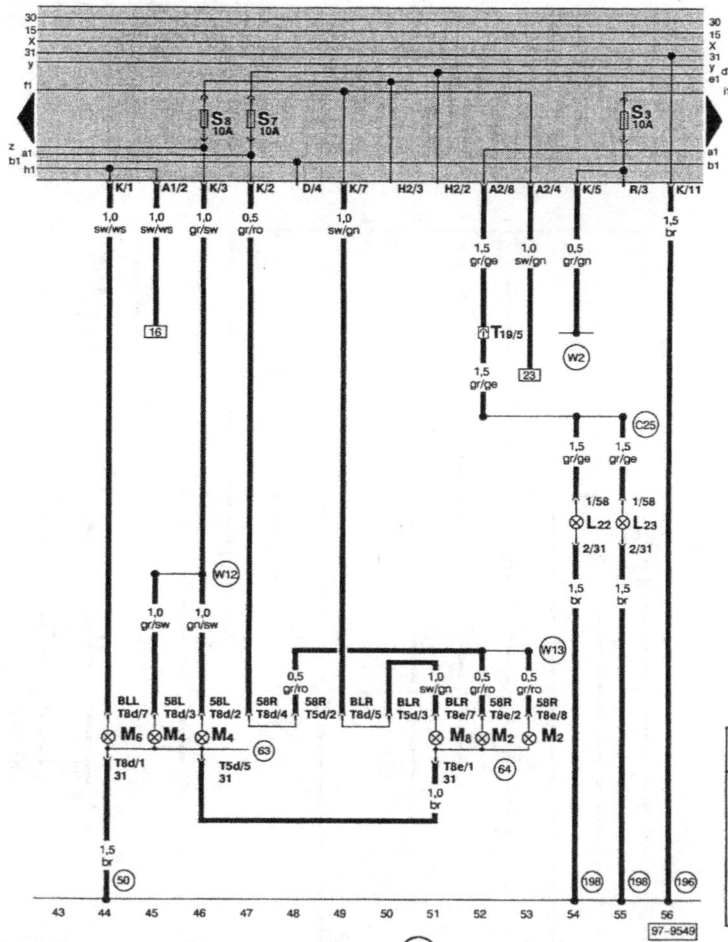

WIRING COLOR CODE

ws	=	white
sw	=	black
ro	=	red
br	=	brown
gn	=	green
bl	=	blue
gr	=	grey
li	=	lilac
ge	=	yellow

L22 - Left Front Fog Light
L23 - Right Front Fog Light
M2 - Tail Light, Right
M4 - Tail Light, Left
M6 - Turn Signal Light, Left Rear
M8 - Turn Signal Light, Right Rear
T5d - Connector, 5 Pin, on taillight, left
T8d - Connector, 8 Pin, on taillight, left
T8e - Connector, 8 Pin, on taillight, right
T19 - Connector, 19 Pin, behind radiator grill, center

(50) - Ground connection, in luggage compartment, left
(63) - Ground connection, on left taillight bulb holder
(64) - Ground connection, on right taillight bulb holder
(196) - Ground connection -1-, in rear wiring harness
(198) - Ground connection, in front light wiring harness
(C25) - Wire connection (fog lights), in front left wiring harness
(W2) - Plus connection (58), in rear wiring harness
(W12) - Wire connection (58L), in rear wiring harness
(W13) - Wire connection (58R), in rear wiring harness

1993-1994

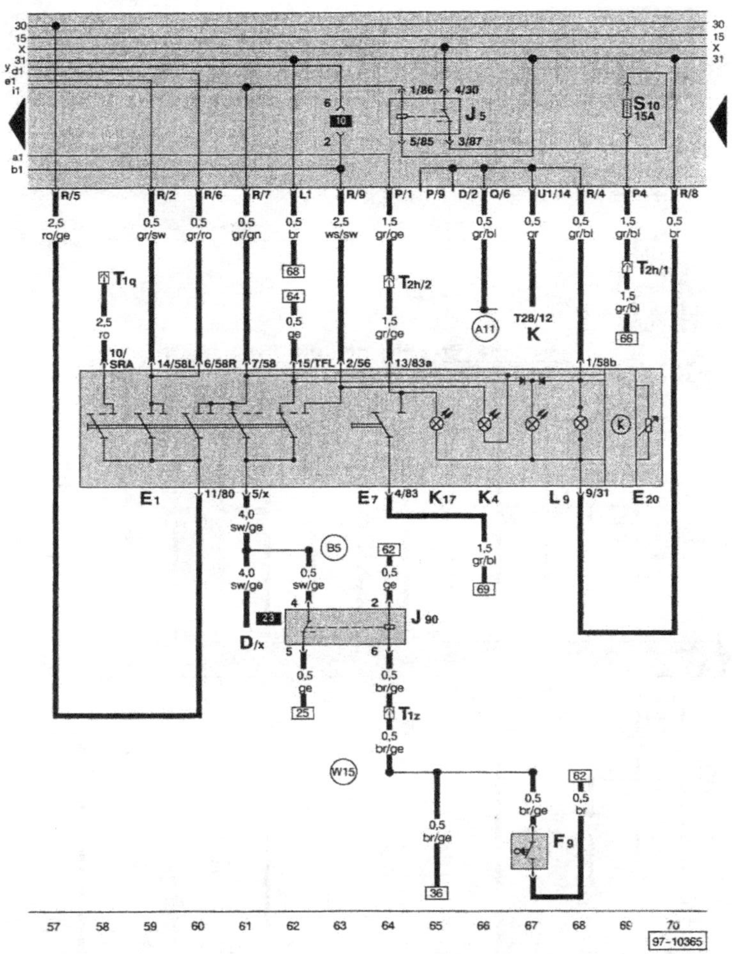

WIRING COLOR CODE		
ws	=	white
sw	=	black
ro	=	red
br	=	brown
gn	=	green
bl	=	blue
gr	=	grey
li	=	lilac
ge	=	yellow

- D – Ignition/Starter Switch
- E1 – Light Switch
- E7 – Fog Light Switch
- E20 – Instrument Panel Light Dimmer Switch
- F9 – Parking Brake Warning Light Switch
- J5 – Fog Light Relay
- J90 – Daytime Running Lights Relay (Switch on)
- K – Instrument Cluster
- K4 – Park Light Indicator Light
- K17 – Fog Light Indicator Light
- L9 – Headlight Switch Light
- T1q – Connector, Single, behind fuse/relay panel
- T1z – Connector, Single, behind fuse/relay panel
- T2h – Connector, Double, behind fuse/relay panel
- T28 – Connector, 28 Pin, instrument cluster
- (A11) – Plus connection (58b), in instrument panel wiring harness
- (B5) – Wire connection, (X), in headlight switch wiring harness
- (W15) – Wiring connection, (HB) in rear wiring harness

1993–1994

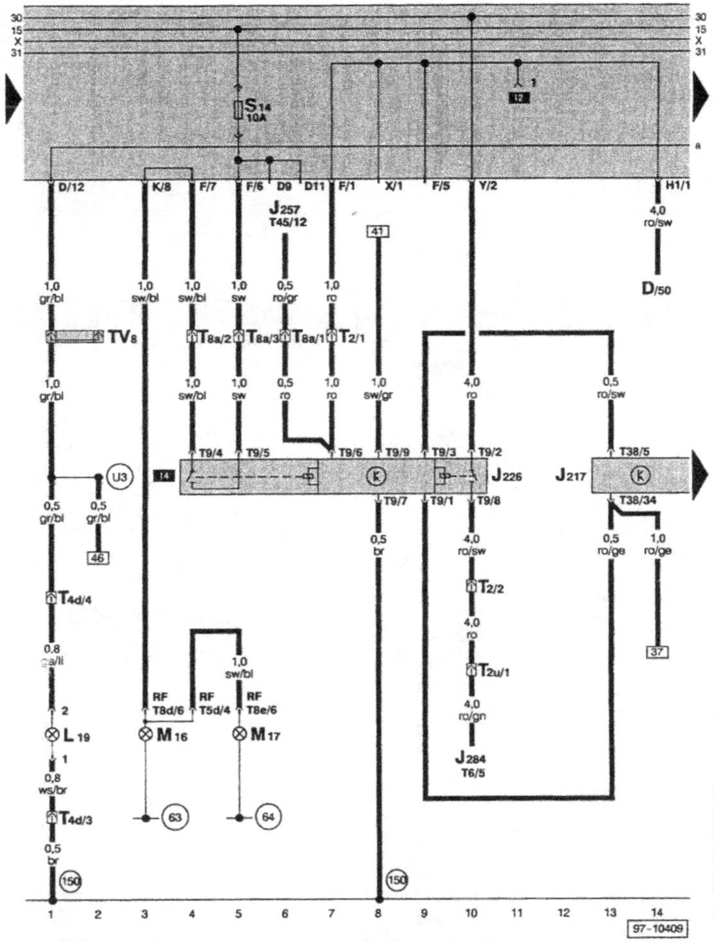

WIRING COLOR CODE		
ws	=	white
sw	=	black
ro	=	red
br	=	brown
gn	=	green
bl	=	blue
gr	=	grey
li	=	lilac
ge	=	yellow

- D – Ignition/Starter Switch
- J217 – Transmission Control Module (TCM)
- J226 – Park/Neutral Position (PNP) Relay
- J257 – Mono-Motronic Engine Control Module
- J284 – Alarm System Control Module
- L19 – Automatic Transmission Console Light
- M16 – Back-Up Light, Left
- M17 – Back-Up Light, Right
- T2 – Connector, Double, behind fuse/relay panel, only on automatic transmission
- T2u – Connector, Double, behind fuse/relay panel
- T4d – Connector, 4 Pin, below center console
- T5d – Connector, 5 Pin, on taillight, left
- T6 – Connector, 6 Pin, behind dash panel, left, on Alarm System Control Module
- T8a – Connector, 8 Pin, behind fuse/relay panel
- T8d – Connector, 8 Pin, on taillight, left
- T8e – Connector, 8 Pin, on taillight, right
- T9 – Connector, 9 Pin, on Park/Neutral Position (PNP) Relay
- T38 – Connector, 38 Pin, below right front seat, on Transmission Control Module (TCM)
- T45 – Connector, 45 Pin, on Mono-Motronic Engine Control Module (ECM)
- TV8 – Connector, terminal 58b
- (63) – Ground connection, on left taillight bulb holder
- (64) – Ground connection, on right taillight bulb holder
- (150) – Ground connection, in automatic transmission wiring harness
- (U3) – Wire connection, in automatic transmission wire harness

1993–1994

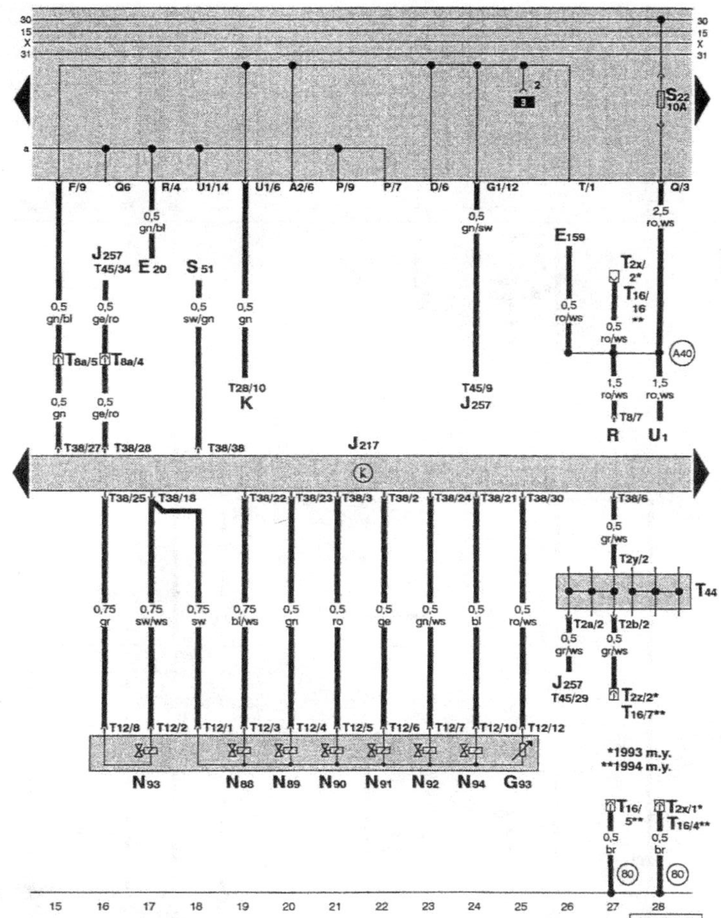

WIRING COLOR CODE		
ws	=	white
sw	=	black
ro	=	red
br	=	brown
gn	=	green
bl	=	blue
gr	=	grey
li	=	lilac
ge	=	yellow

E20 – Instrument Panel Light Dimmer Switch
E159 – Luggage Compartment Release Switch
G93 – Transmission Fluid Temperature Sensor
J217 – Transmission Control Module (TCM)
J257 – Mono-Motronic Engine Control Module
K – Instrument Cluster
N88 – Solenoid Valve 1
N89 – Solenoid Valve 2
N90 – Solenoid Valve 3
N91 – Solenoid Valve 4
N92 – Solenoid Valve 5
N93 – Solenoid Valve 6
N94 – Solenoid Valve 7
R – Radio
S51 – Fuse
U1 – Cigarette Lighter
T2a – Connector, Double, on junction connector (T44) above fuse/relay panel
T2b – Connector, Double, on junction connector (T44) above fuse/relay panel
T2x – Data Link Connector (DLC), behind center of dash panel
T2y – Connector, Double, on junction connector (T44) above fuse/relay panel
T2z – Data Link Connector (DLC), behind center of dash panel
T8 – Connector, 8 Pin, on radio
T8a – Connector, 8 Pin, behind fuse/relay panel
T12 – Connector, 12 Pin, on transmission
T16 – Data Link Connector (DLC), behind center of dash panel
T28 – Connector, 28 Pin, instrument cluster
T38 – Connector, 38 Pin, below right front seat, on Transmission Control Module (TCM)
T44 – Junction Box for On Board Diagnostic, above fuse/relay panel
T45 – Connector, 45 Pin, on Mono-Motronic Engine Control Module (ECM)
(80) – Ground connection -1-, in instrument panel wiring harness
(A40) – Plus connection (30), in instrument panel wiring harness

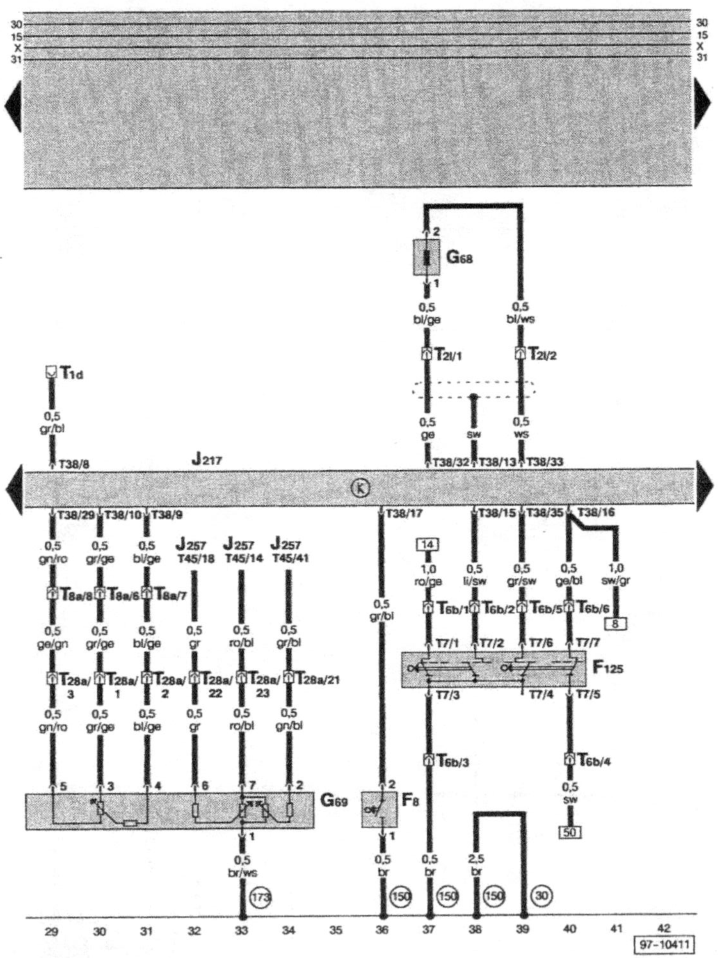

WIRING COLOR CODE		
ws	=	white
sw	=	black
ro	=	red
br	=	brown
gn	=	green
bl	=	blue
gr	=	grey
li	=	lilac
ge	=	yellow

F8 – Kick Down Switch
F125 – Multi-Function Transmission Range (TR) Switch
G68 – Vehicle Speed Sensor (VSS)
G69 – Throttle Position (TP) Sensor
J217 – Transmission Control Module (TCM)
J257 – Mono-Motronic Engine Control Module
T1d – Connector, Single, behind fuse/relay panel
T2i – Connector, Double, below battery
T6b – Connector, 6 Pin, below battery
T7 – Connector, 7 Pin, on transmission
T8a – Connector, 8 Pin, behind fuse/relay panel
T28a – Connector, 28 Pin, left side of engine, near distributor
T38 – Connector, 38 Pin, below right front seat, on Transmission Control Module (TCM)
T45 – Connector, 45 Pin, on Mono-Motronic Engine Control Module (ECM)
(30) – Ground connection – beside fuse/relay panel
(150) – Ground connection, in automatic transmission wiring harness
(173) – Ground connection -1-, in Mono-Motronic Throttle Body Fuel Injection (TBI) wiring harness

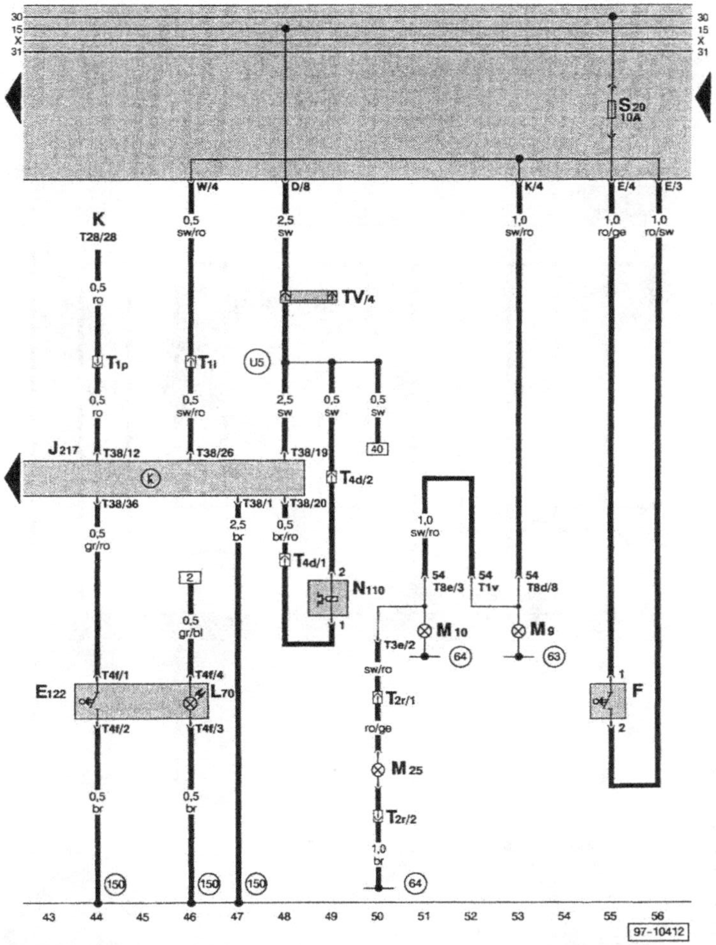

WIRING COLOR CODE		
ws	=	white
sw	=	black
ro	=	red
br	=	brown
gn	=	green
bl	=	blue
gr	=	grey
li	=	lilac
ge	=	yellow

E122 – Transmission Range (TR) Program Switch
F – Brake Light Switch
J217 – Transmission Control Module (TCM)
K – Instrument Cluster
L70 – Program Switch Illumination
M9 – Brake Light, Left
M10 – Brake Light, Right
M25 – High-Mount Brake Light
N110 – Shift Lock Solenoid
T1i – Connector, Single, behind fuse/relay panel
T1p – Connector, Single, behind fuse/relay panel
T1v – Connector, Single, on taillight, left
T2r – Connector, Double, behind right grommet, on luggage compartment lid
T3e – Connector, 3 Pin, on taillight, right
T4d – Connector, 4 Pin, below center console
T4f – Connector, 4 Pin, on Transmission Range (TR) Program Switch
T8d – Connector, 8 Pin, on taillight, left
T8e – Connector, 8 Pin, on taillight right
T28 – Connector, 28 Pin, instrument cluster
T38 – Connector, 38 Pin, below right front seat, on Transmission Control Module (TCM)
TV4 – Connector, terminal 15
(63) – Ground connection, on left taillight bulb holder
(64) – Ground connection, on right taillight bulb holder
(150) – Ground connection, in automatic transmission wiring harness
(U5) – Wire connection -2- (15), in automatic transmission wiring harness

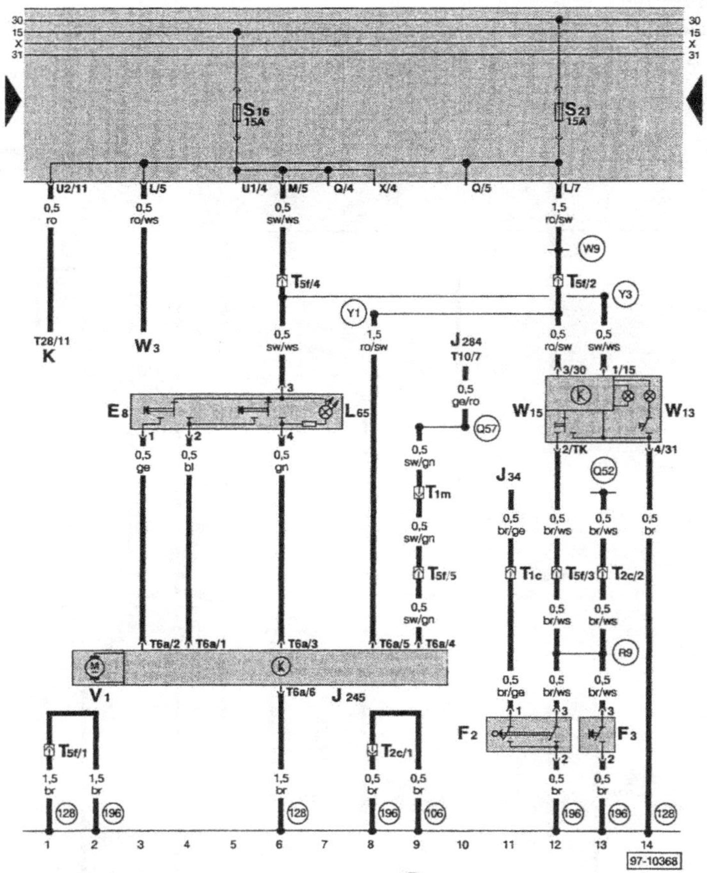

WIRING COLOR CODE		
ws	=	white
sw	=	black
ro	=	red
br	=	brown
gn	=	green
bl	=	blue
gr	=	grey
li	=	lilac
ge	=	yellow

E8 – Sunroof Switch
F2 – Door Contact Switch, Left Front
F3 – Door Contact Switch, Right Front
J34 – Seat Belt Warning System Relay
J245 – Power Sunroof Control Module
J284 – Alarm System Control Module
K – Instrument Cluster
L65 – Sunroof Switch Light
V1 – Sunroof Motor
W3 – Luggage Compartment Light
W13 – Map/Reading Light, Right Front
W15 – Interior Light With Delay Switch
T1c – Connector, Single, behind fuse/relay panel
T1m – Connector, Single, behind fuse/relay panel
T2c – Connector, Double, on lower right "A" pillar
T5f – Connector, 5 Pin, above fuse/relay panel
T6a – Connector, 6 Pin, behind interior light, on Power Sunroof Control Module
T10 – Connector, 10 Pin, behind dash panel, left, on Alarm System Control Module
T28 – Connector, 28 Pin, instrument cluster
(106) – Ground connection, in alarm system wiring harness
(128) – Ground connection -1-, in interior light wiring harness
(196) – Ground connection -1-, in rear wiring harness
(Q52) – Wire connection (door contact switch), in alarm system wiring harness
(Q57) – Wire connection (closed), in alarm system wiring harness
(R9) – Wire connection (30), in rear wiring harness
(W9) – Plus connection (30), in rear wiring harness
(Y1) – Plus connection (30), in interior light wiring harness
(Y3) – Plus connection (15) in interior light/sunroof wiring harness

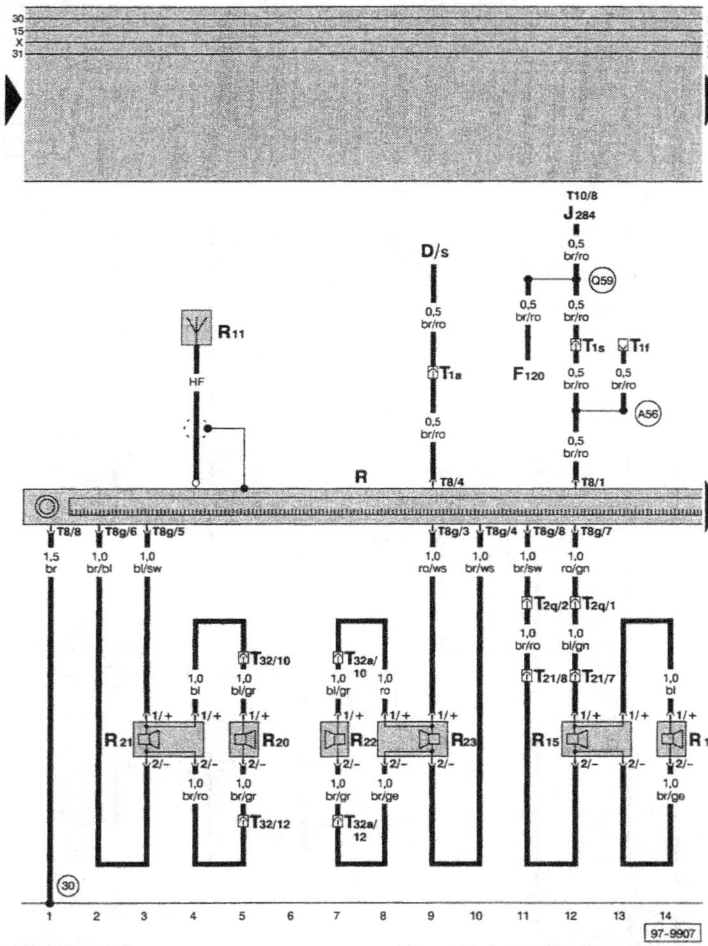

WIRING COLOR CODE

ws	=	white
sw	=	black
ro	=	red
br	=	brown
gn	=	green
bl	=	blue
gr	=	grey
li	=	lilac
ge	=	yellow

D – Ignition/Starter Switch
F120 – Alarm Switch, Hood
J284 – Alarm System Control Module
R – Radio
R11 – Antenna
R14 – Tweeter, Left Rear
R15 – Woofer, Left Rear
R20 – Tweeter, Left Front
R21 – Woofer, Left Front
R22 – Tweeter, Right Front
R23 – Woofer, Right Front
T1a – Connector, Single, behind fuse/relay panel
T1f – Connector, Single, behind dash panel, center, behind radio
T1s – Connector, Single, behind fuse/relay panel

T2q – Connector, Double, on lower left "A" pillar
T8 – Connector, 8 Pin, on radio
T8g – Connector, 8 Pin, on radio
T10 – Connector, 10 Pin, behind dash panel, left, on Alarm System Control Module
T21 – Connector, 21 pin, on left B-pillar
T32 – Connector, 32 Pin, on A-pillar, lower left
T32a – Connector, 32 Pin, on A-pillar, lower right
(30) – Ground connection – beside fuse/relay panel
(A56) – Wire connection (31SDS) in instrument panel wiring harness
(Q59) – Wire connection (MHKS) in alarm system wiring harness

Stereo radio, 4 door

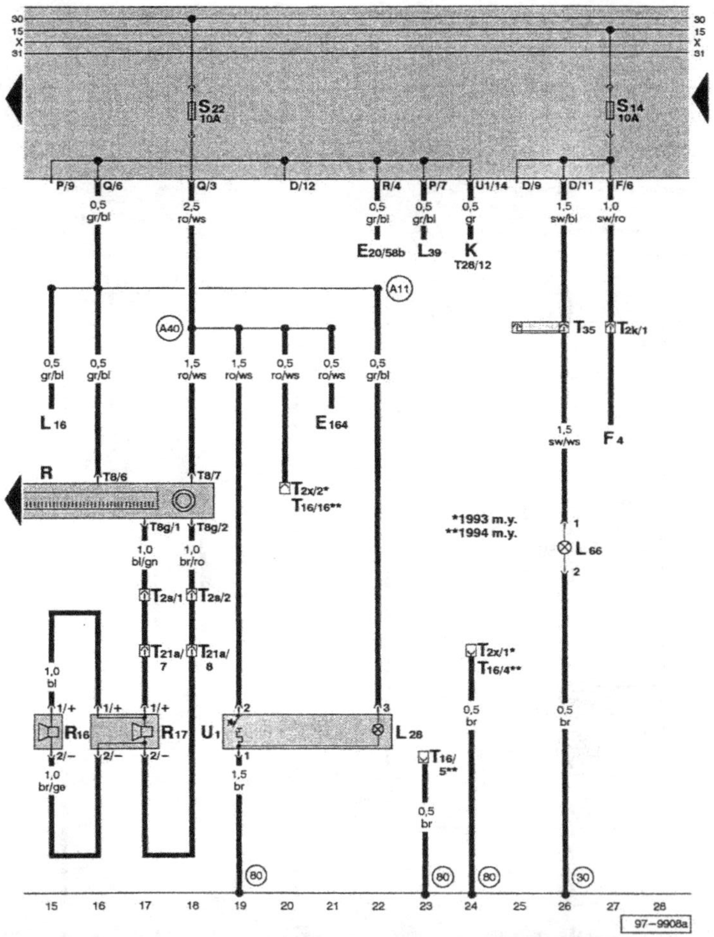

WIRING COLOR CODE

ws	=	white
sw	=	black
ro	=	red
br	=	brown
gn	=	green
bl	=	blue
gr	=	grey
li	=	lilac
ge	=	yellow

E20 – Instrument Panel Light Dimmer Switch
E164 – Luggage Compartment Release Switch
F4 – Back-Up Light Switch
K – Instrument Cluster
L16 – Fresh Air Control Lever Light
L28 – Cigarette Lighter Light
L39 – Rear Window Defogger Switch Light
L66 – Cassette Storage Light
R – Radio
R16 – Tweeter, Right Rear
R17 – Woofer, Right Rear
U1 – Cigarette Lighter
T2k – Connector, Double, behind fuse/relay panel
T2s – Connector, Double, on lower right "A" pillar
T2x – Data Link Connector (DLC), behind center of dash panel

T8 – Connector, 8 Pin, on radio
T8g – Connector, 8 Pin, on radio
T16 – Data Link Connector (DLC), behind center of dash panel
T21a – Connecter, 21 pin, on right B-pillar
T28 – Connector, 28 Pin, instrument cluster
T35 – Connector, terminal 15
(30) – Ground connection – beside fuse/relay panel
(80) – Ground connection -1-, in instrument panel wiring harness
(A11) – Plus connection (58b), in instrument panel wiring harness
(A40) – Plus connection (30), in instrument panel wiring harness

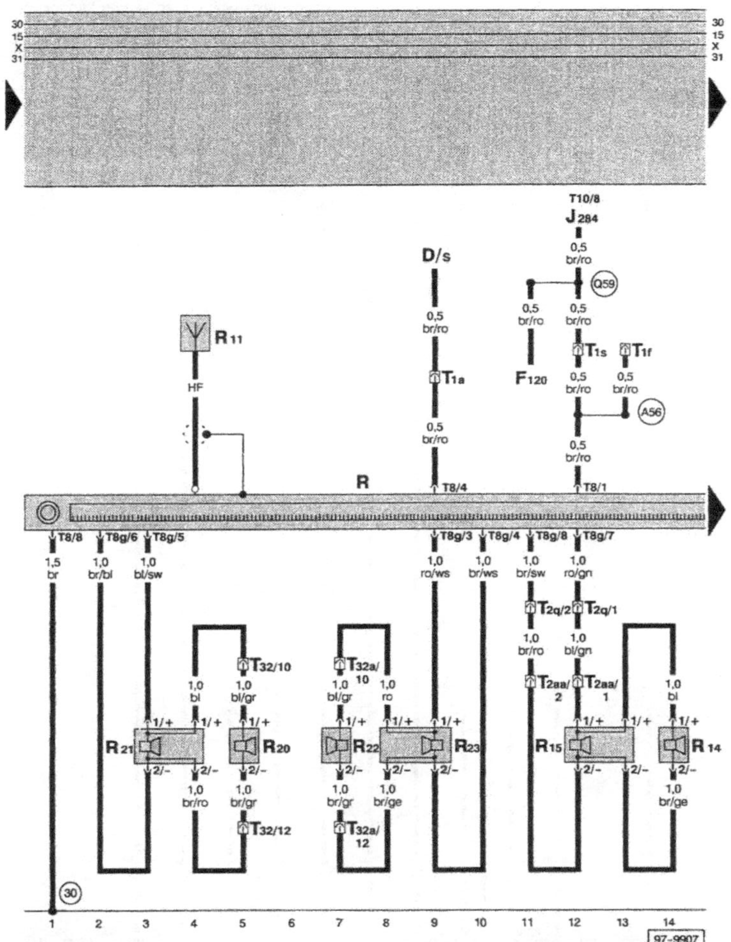

WIRING COLOR CODE	
ws	= white
sw	= black
ro	= red
br	= brown
gn	= green
bl	= blue
gr	= grey
li	= lilac
ge	= yellow

97-9907

D - Ignition/Starter Switch
F120 - Alarm Switch, Hood
J284 - Alarm System Control Module
R - Radio
R11 - Antenna
R14 - Tweeter, Left Rear
R15 - Woofer, Left Rear
R20 - Tweeter, Left Front
R21 - Woofer, Left Front
R22 - Tweeter, Right Front
R23 - Woofer, Right Front
T1a - Connector, Single, behind fuse/relay panel
T1f - Connector, Single, behind dash panel, center, behind radio
T1s - Connector, Single, behind fuse/relay panel
T2aa - Connector, Double
T2q - Connector, Double, on lower left "A" pillar
T8 - Connector, 8 Pin, on radio
T8g - Connector, 8 Pin, on radio
T10 - Connector, 10 Pin, behind dash panel, left, on Alarm System Control Module
T32 - Connector, 32 Pin, on A-pillar, lower left
T32a - Connector, 32 Pin, on A-pillar, lower right
(30) - Ground connection - beside fuse/relay panel
(A56) - Wire connection (31SDS) in instrument panel wiring harness
(Q59) - Wire connection (MHKS) in alarm system wiring harness

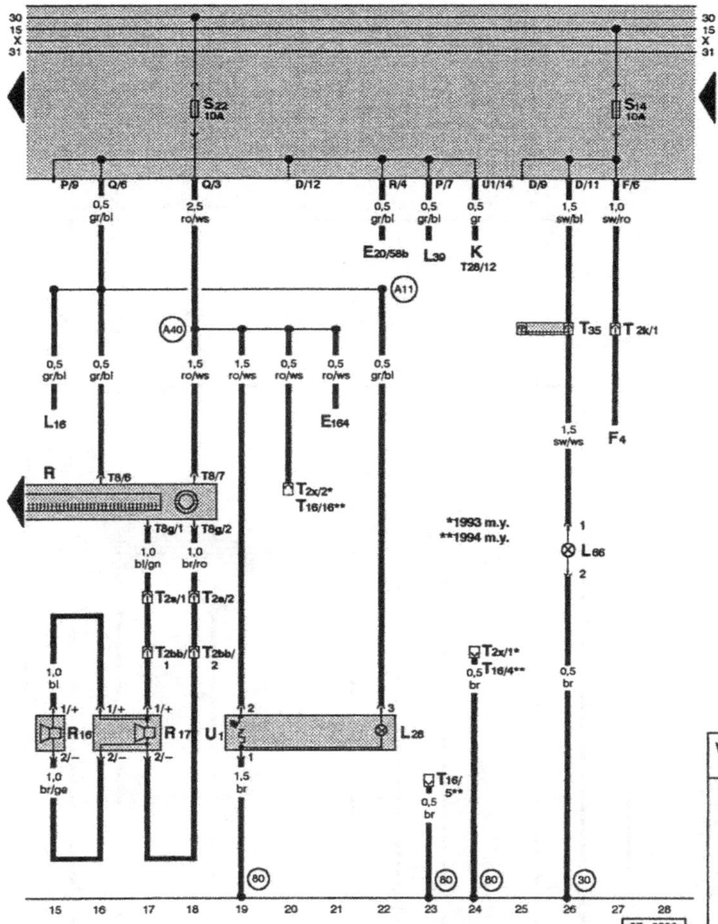

WIRING COLOR CODE	
ws	= white
sw	= black
ro	= red
br	= brown
gn	= green
bl	= blue
gr	= grey
li	= lilac
ge	= yellow

97-9908a

E20 - Instrument Panel Light Dimmer Switch
E164 - Luggage Compartment Release Switch
F4 - Back-Up Light Switch
K - Instrument Cluster
L16 - Fresh Air Control Lever Light
L28 - Cigarette Lighter Light
L39 - Rear Window Defogger Switch Light
L66 - Cassette Storage Light
R - Radio
R16 - Tweeter, Right Rear
R17 - Woofer, Right Rear
U1 - Cigarette Lighter
T2bb - Connector, Double
T2k - Connector, Double, behind fuse/relay panel
T2s - Connector, Double, on lower right "A" pillar
T2x - Data Link Connector (DLC), behind center of dash panel
T8 - Connector, 8 Pin, on radio
T8g - Connector, 8 Pin, on radio
T16 - Data Link Connector (DLC), behind center of dash panel
T28 - Connector, 28 Pin, instrument cluster
T35 - Connector, terminal 15
(30) - Ground connection - beside fuse/relay panel
(80) - Ground connection -1-, in instrument panel wiring harness
(A11) - Plus connection (58b), in instrument panel wiring harness
(A40) - Plus connection (30), in instrument panel wiring harness

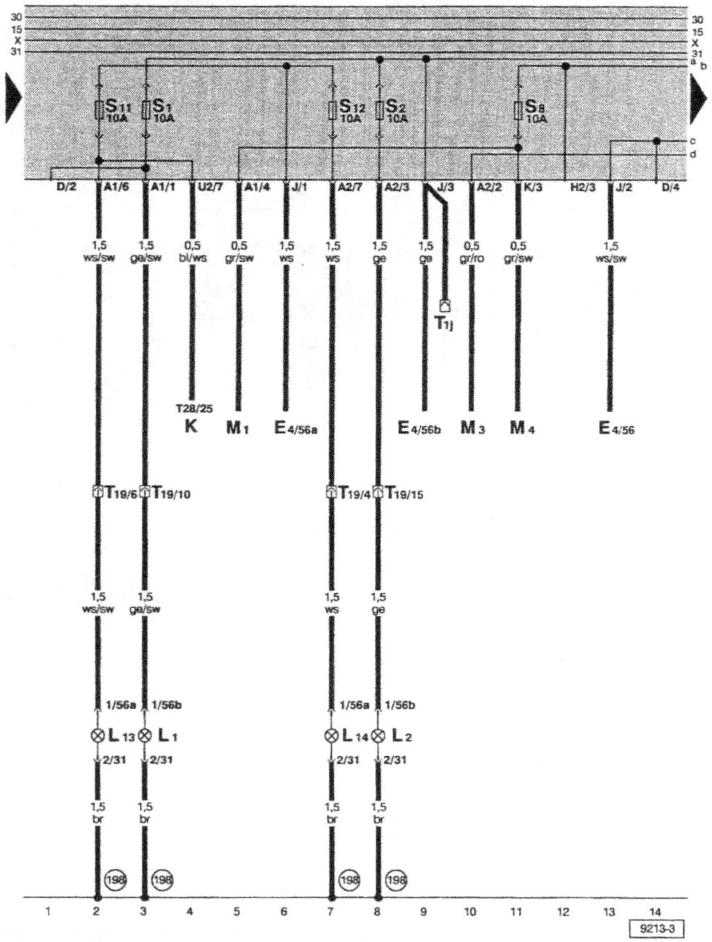

WIRING COLOR CODE		
ws	=	white
sw	=	black
ro	=	red
br	=	brown
gn	=	green
bl	=	blue
gr	=	grey
li	=	lilac
ge	=	yellow

- E4 – Headlight Dimmer/Flasher Switch
- K – Instrument Cluster
- L1 – Headlight, Left
- L2 – Headlight, Right
- L13 – High Beam Headlight, Left
- L14 – High Beam Headlight, Right
- M1 – Parking Light, Left
- M3 – Parking Light, Right
- M4 – Tail Light, Left
- T1j – Connector, Single, behind fuse/relay panel
- T19 – Connector, 19 Pin, behind radiator grill, center
- T28 – Connector, 28 Pin, instrument cluster
- (198) – Ground connection, in front light wiring harness

1993–1994

Headlights and fog lights
(GTI, USA only)

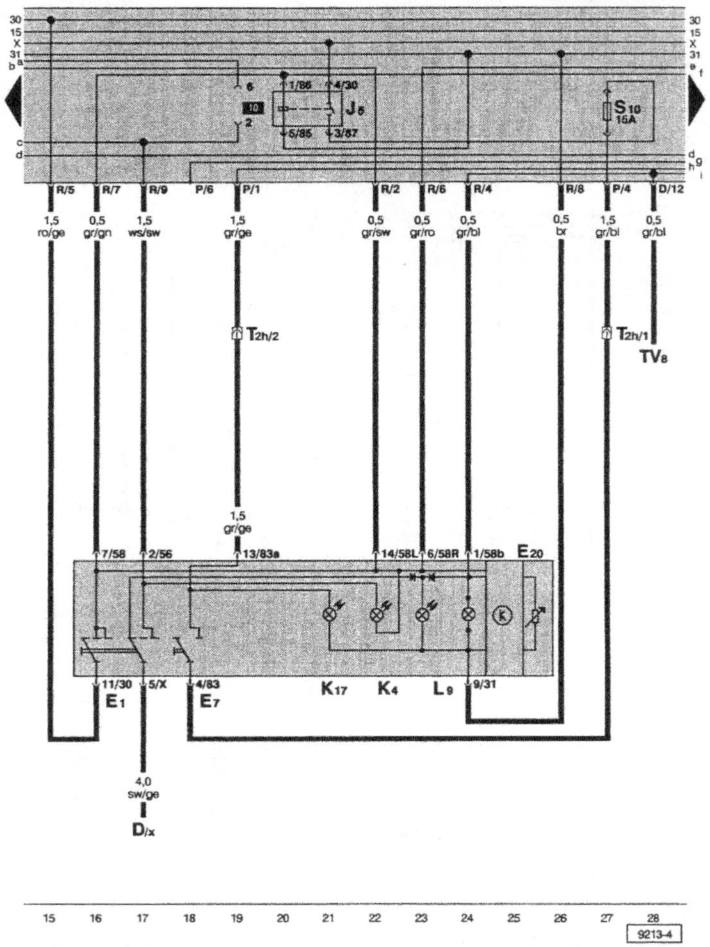

WIRING COLOR CODE		
ws	=	white
sw	=	black
ro	=	red
br	=	brown
gn	=	green
bl	=	blue
gr	=	grey
li	=	lilac
ge	=	yellow

- D – Ignition/Starter Switch
- E1 – Light Switch
- E7 – Fog Light Switch
- E20 – Instrument Panel Light Dimmer Switch
- J5 – Fog Light Relay
- K4 – Park Light Indicator Light
- K17 – Fog Light Indicator Light
- L9 – Headlight Switch Light
- T2h – Connector, Double, behind fuse/relay panel
- TV8 – Connector, terminal 58b

1993-1994

Headlights and fog lights (GTI, USA only)

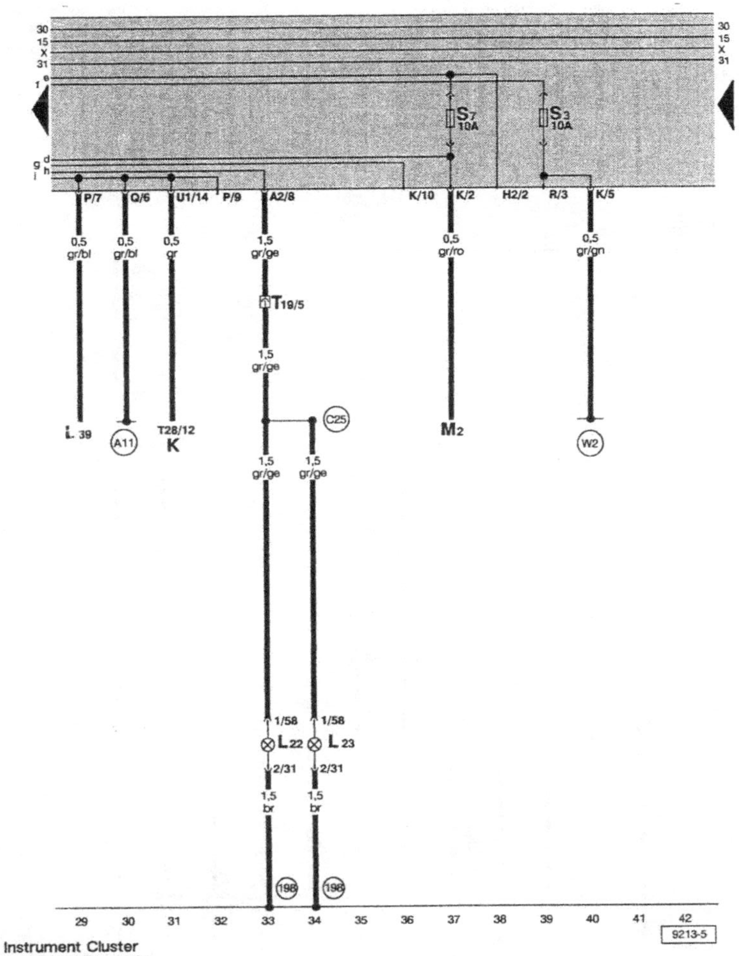

WIRING COLOR CODE		
ws	=	white
sw	=	black
ro	=	red
br	=	brown
gn	=	green
bl	=	blue
gr	=	grey
li	=	lilac
ge	=	yellow

K - Instrument Cluster
L22 - Left Front Fog Light
L23 - Right Front Fog Light
L39 - Rear Window Defogger Switch Light
M2 - Tail Light, Right
T19 - Connector, 19 Pin, behind radiator grill, center
T28 - Connector, 28 Pin, instrument cluster
(198) - Ground connection, in front light wiring harness
(A11) - Plus connection (58b), in instrument panel wiring harness
(C25) - Wire connection (fog lights), in front left wiring harness
(W2) - Plus connection (58), in rear wiring harness

Heated washer nozzles

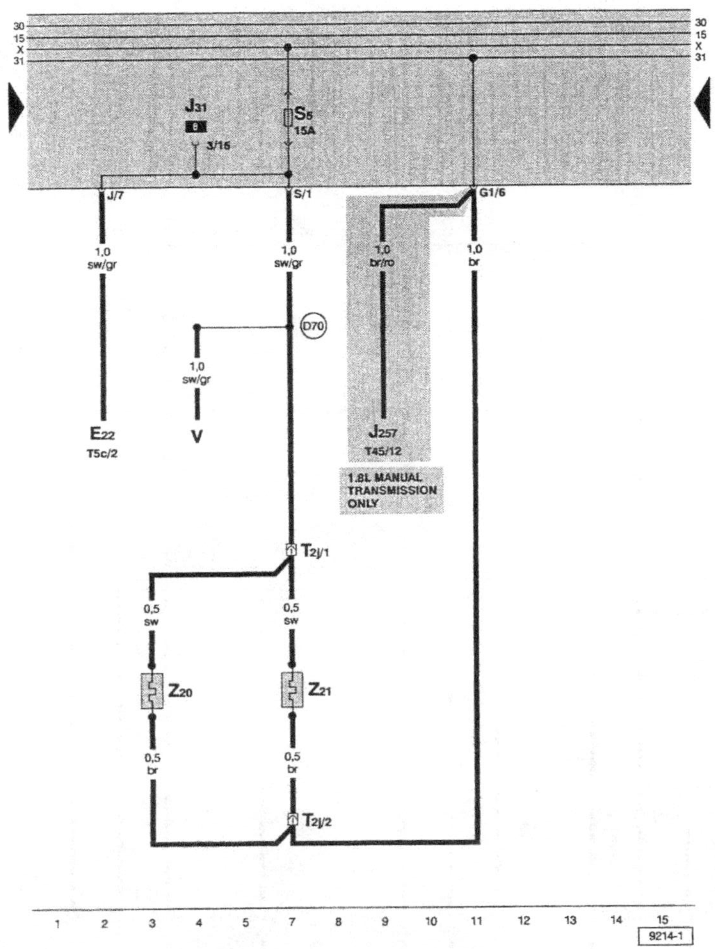

WIRING COLOR CODE		
ws	=	white
sw	=	black
ro	=	red
br	=	brown
gn	=	green
bl	=	blue
gr	=	grey
li	=	lilac
ge	=	yellow

E22 - Windshield Wiper Intermittent Switch
J31 - Washer/Wiper Intermittent Relay
J257 - Mono-Motronic Engine Control Module
V - Windshield Wiper Motor
Z20 - Washer Nozzle Heater, Left
Z21 - Washer Nozzle Heater, Right
T2j - Connector, Double, behind fuse/relay panel
T5c - Connector, 5 Pin, behind steering column switch trim
T45 - Connector, 45 Pin, on Mono-Motronic Engine Control Module (ECM)
(D70) - Wire connection (53a), in engine compartment wiring harness

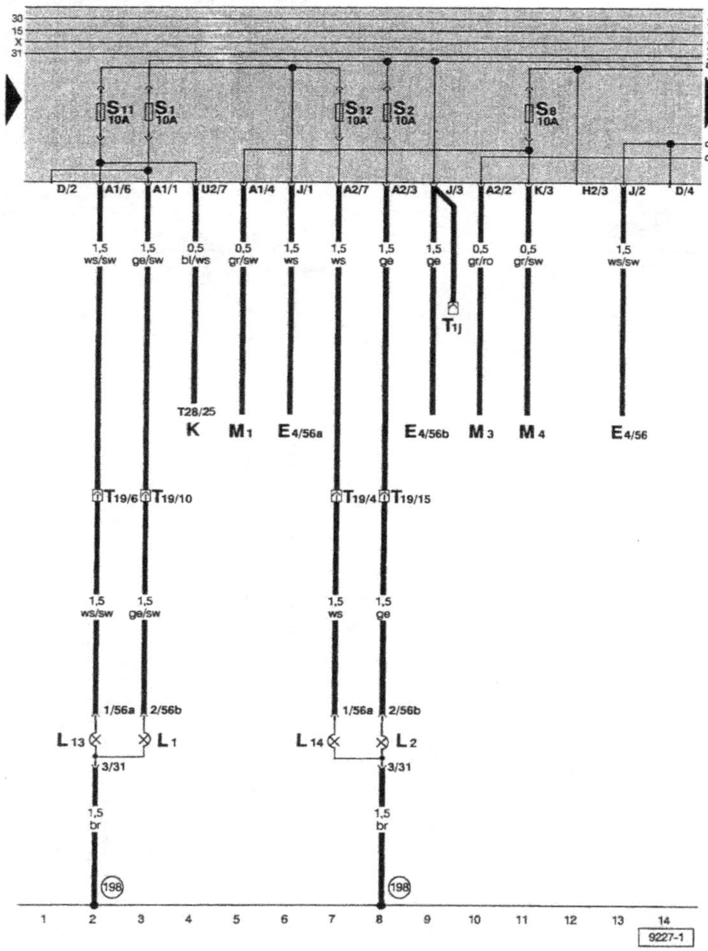

WIRING COLOR CODE		
ws	=	white
sw	=	black
ro	=	red
br	=	brown
gn	=	green
bl	=	blue
gr	=	grey
li	=	lilac
ge	=	yellow

E4 - Headlight Dimmer/Flasher Switch
K - Instrument Cluster
L1 - Headlight, Left
L2 - Headlight, Right
L13 - High Beam Headlight, Left
L14 - High Beam Headlight, Right
M1 - Parking Light, Left

M3 - Parking Light, Right
M4 - Tail Light, Left
T1j - Connector, Single, behind fuse/relay panel
T19 - Connector, 19 Pin, behind radiator grill, center
T28 - Connector, 28 Pin, on instrument cluster
(198) - Ground connection, in front light wiring harness

1993–1994

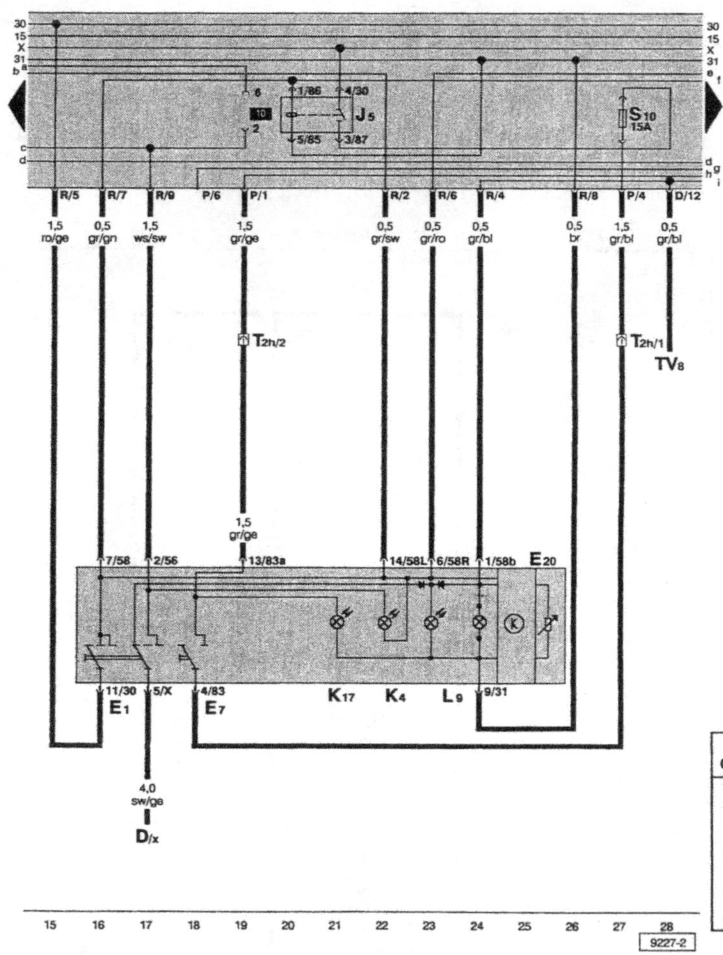

WIRING COLOR CODE		
ws	=	white
sw	=	black
ro	=	red
br	=	brown
gn	=	green
bl	=	blue
gr	=	grey
li	=	lilac
ge	=	yellow

D - Igntion/Starter Switch
E1 - Light Switch
E7 - Fog Light Switch
E20 - Instrument Panel Light Dimmer Switch
J5 - Fog Light Relay
K4 - Park Light Indicator Light
K17 - Fog Light Indicator Light
L9 - Headlight Switch Light
T2h - Connector, Double, behind fuse/relay panel
TV8 - Connector, terminal 58b

1993-1994

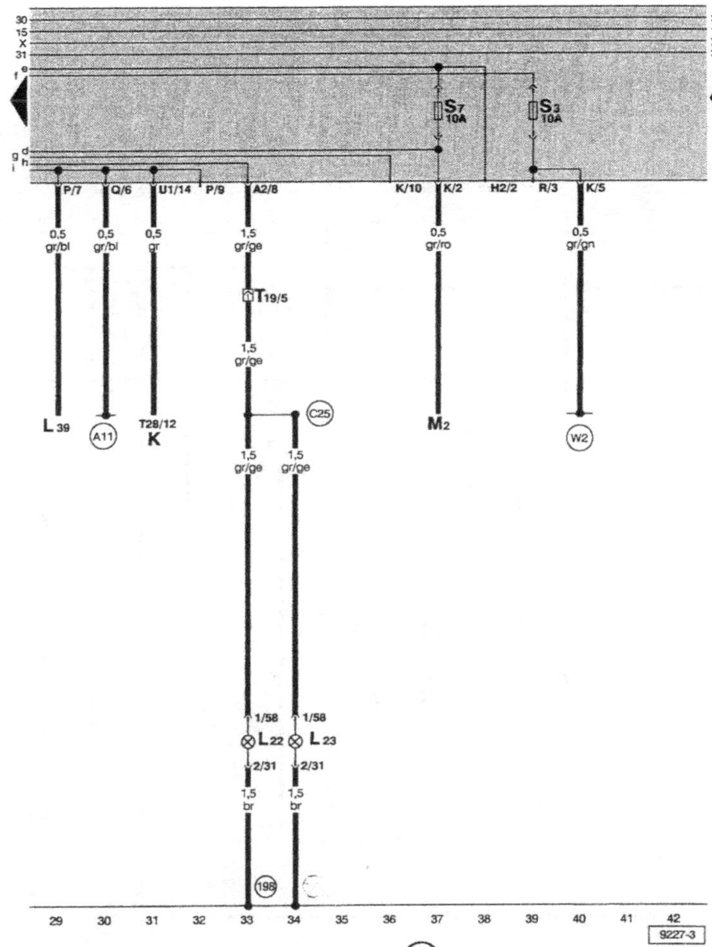

WIRING COLOR CODE

ws	=	white
sw	=	black
ro	=	red
br	=	brown
gn	=	green
bl	=	blue
gr	=	grey
li	=	lilac
ge	=	yellow

K - Instrument Cluster
L22 - Left Front Fog Light
L23 - Right Front Fog Light
L39 - Rear Window Defogger Switch Light
M2 - Tail Light, Right
T19 - Connector, 19 Pin, behind radiator grill, center
T28 - Connector, 28 Pin, on instrument cluster

(198) - Ground connection, in front light wiring harness

(A11) - Plus connection (58b), in instrument panel wiring harness

(C25) - Wire connection (fog lights), in front left wiring harness

(W2) - Plus connection (58), in rear wiring harness

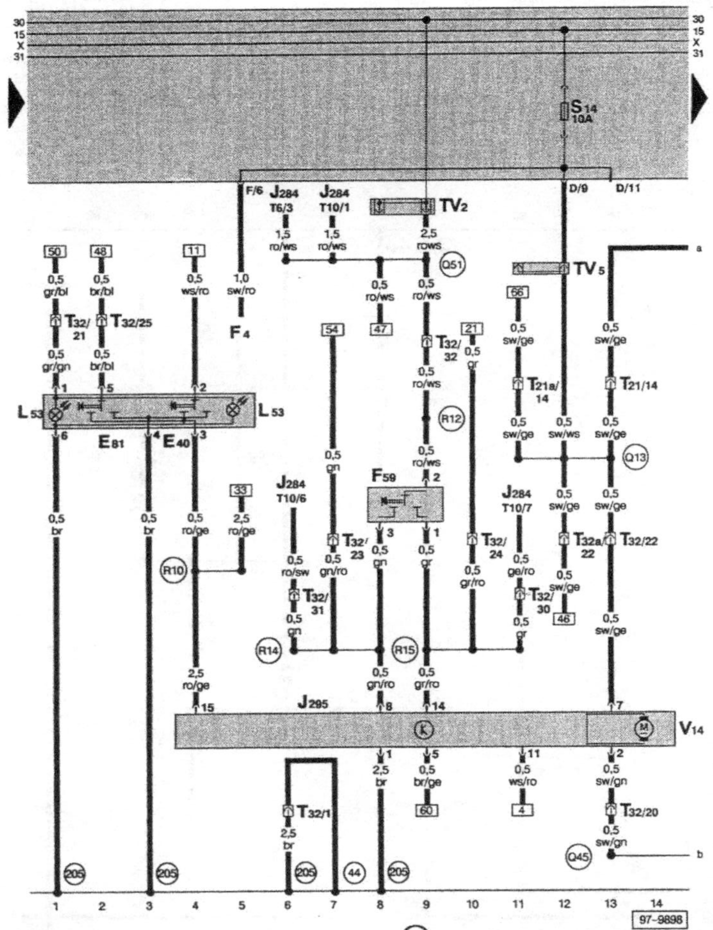

WIRING COLOR CODE

ws	=	white
sw	=	black
ro	=	red
br	=	brown
gn	=	green
bl	=	blue
gr	=	grey
li	=	lilac
ge	=	yellow

E40 - Window Switch, Left Front
E81 - Right Front Window Switch (In Driver's Door)
F4 - Back-Up Light Switch
F59 - Central Locking System Door Switch, Driver's Door
J284 - Alarm System Control Module
J295 - Power Window Control Module, Left Front
L53 - Power Window Switch Light
V14 - Window Motor, Left
T6 - Connector, 6 Pin, behind dash panel, left, on Alarm System Control Module
T10 - Connector, 10 Pin, behind dash panel, left, on Alarm System Control Module
T21 - Connector, 21 pin, on left B-pillar
T21a - Connecter, 21 pin, on right B-pillar
T32 - Connector, 32 Pin, on A-pillar, lower left
T32a - Connector, 32 Pin, on A-pillar, lower right
TV2 - Wire connector, terminal 30 (B+)
TV5 - Connector, terminal 15a

(44) - Ground connection, on left "A" pillar, lower part

(205) - Ground connection, in driver's door wiring harness

(Q13) - Plus connection (75as/15), in power window/central locking system and door contact switch wiring harness

(Q45) - Wire connection (comfort), in power window wiring harness

(Q51) - Plus connection (30), in alarm system wiring harness

(R10) - Plus connection, -1- (30), in driver's door wiring harness

(R12) - Plus connection -3- (30), in driver's door wiring harness

(R14) - Wire connection, -1- (open), in driver's door wiring harness

(R15) - Wire connection, -2- (closed), in driver's door wiring harness

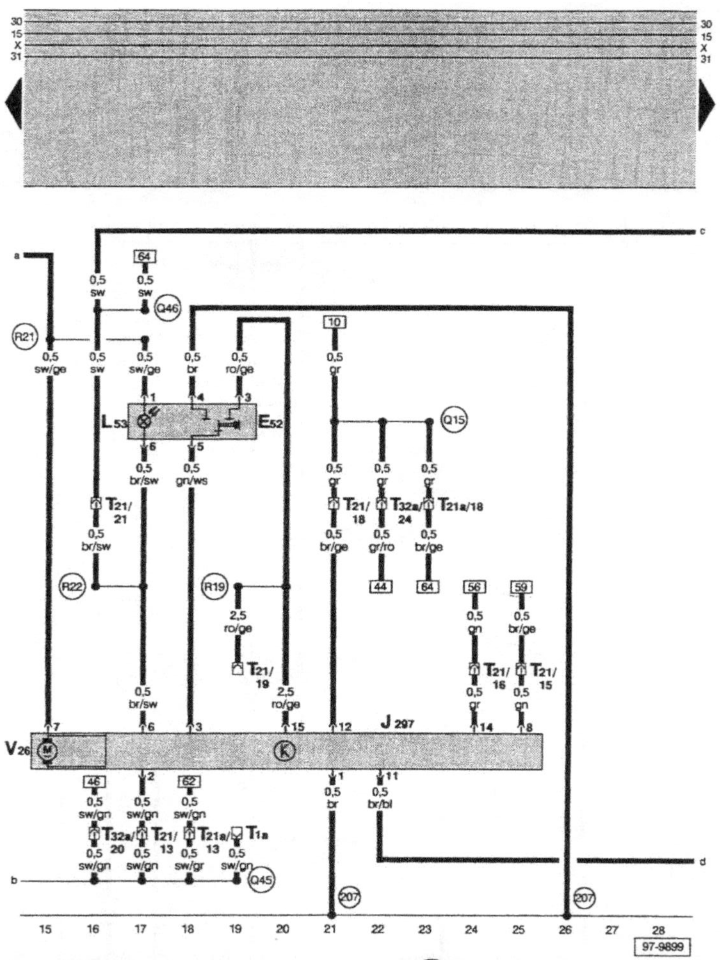

WIRING COLOR CODE		
ws	=	white
sw	=	black
ro	=	red
br	=	brown
gn	=	green
bl	=	blue
gr	=	grey
li	=	lilac
ge	=	yellow

E52 - Left Rear Window Switch, (In Door)
J297 - Power Window Control Module, Left Rear
L53 - Power Window Switch Light
V26 - Window Motor, Left Rear Door
T1a - Connector, Single, behind fuse/relay panel
T21 - Connector, 21 pin, on left B-pillar
T21a - Connector, 21 pin, on right B-pillar
T32a - Connector, 32 Pin, on A-pillar, lower right

(207) - Ground connection, in left rear door wiring harness

(Q15) - Wire connection in power window wiring harness

(Q45) - Wire connection (comfort), in power window wiring harness

(Q46) - Wire connection (child lock) in power window wiring harness

(R19) - Plus connection -1- (30), in left rear door wiring harness

(R21) - Plus connection,-3- (15), in left rear door wiring harness

(R22) - Wire connection (child lock) in left rear door wiring harness

1993–1994

Power windows

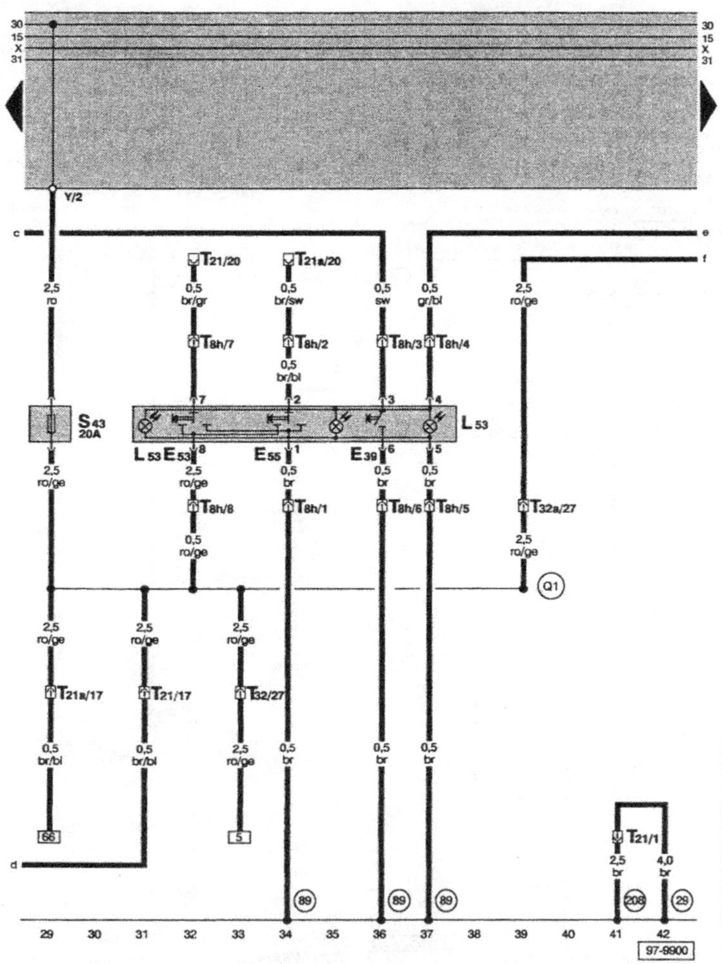

WIRING COLOR CODE		
ws	=	white
sw	=	black
ro	=	red
br	=	brown
gn	=	green
bl	=	blue
gr	=	grey
li	=	lilac
ge	=	yellow

E39 - Window Lockout Switch
E53 - Left Rear Window Switch (In Console)
E55 - Right Rear Window Switch (In Console)
L53 - Power Window Switch Light
S43 - Fuse, power windows
T8h - Connector, 8 pin, behind center of dash panel
T21 - Connector, 21 pin, on left B-pillar
T21a - Connector, 21 pin, on right B-pillar
T32 - Connector, 32 Pin, on A-pillar, lower left
T32a - Connector, 32 Pin, on A-pillar, lower right

(29) - Ground connection, near "B" pillar, right

(89) - Ground connection, in power window wiring harness

(208) - Ground connection, in right rear door wiring harness

(Q1) - Plus connection (30) in power window wiring harness

1993-1994

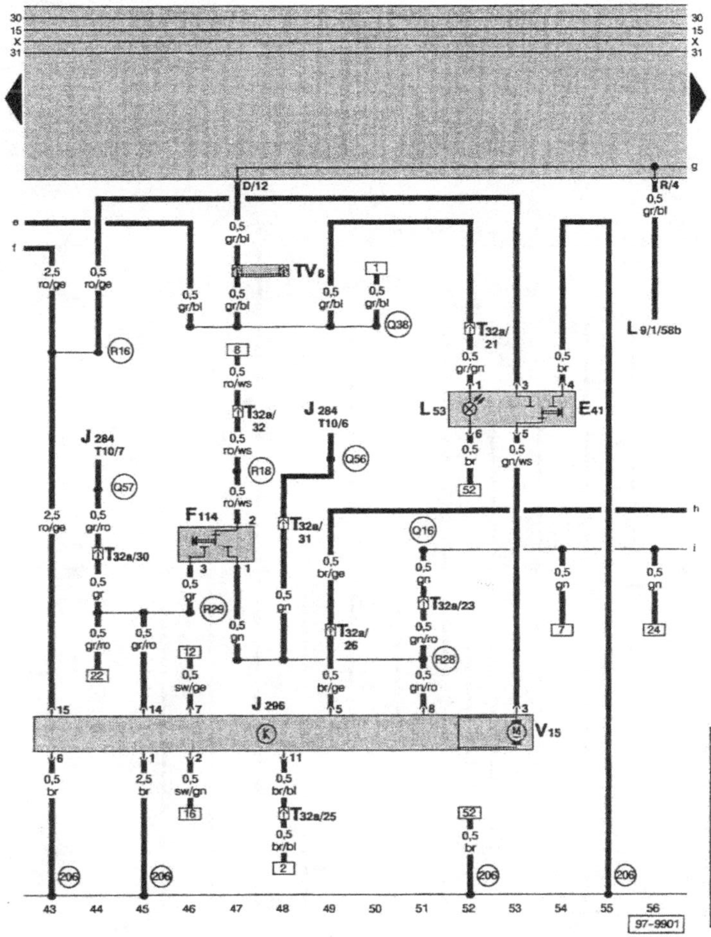

WIRING COLOR CODE

ws	=	white
sw	=	black
ro	=	red
br	=	brown
gn	=	green
bl	=	blue
gr	=	grey
li	=	lilac
ge	=	yellow

E41 - Window Switch, Right Front
F114 - Central Locking System Door Switch, Passenger's Door
J284 - Alarm System Control Module
J296 - Power Window Control Module, Right Front
L9 - Headlight Switch Light
L53 - Power Window Switch Light
V15 - Window Motor, Right
T10 - Connector, 10 Pin, behind dash panel, left, on Alarm System Control Module
T32a - Connector, 32 Pin, on A-pillar, lower right
TV8 - Connector, terminal 58b
(206) - Ground connection, in passenger's door wiring harness

(Q16) - Wire connection, in power window wiring harness (left down)
(Q38) - Plus connection (58b) -1- in power window wiring harness
(Q56) - Wire connection (open), in alarm system wiring harness
(Q57) - Wire connection (closed), in alarm system wiring harness
(R16) - Plus connection -1- (30), in passenger's door wiring harness
(R18) - Plus connection -3- (30), in passenger's door wiring harness
(R28) - Wire connection, (open), in passenger's door wiring haness
(R29) - Wire connection, (closed), in passenger's door wiring harness

1993–1994

Power windows

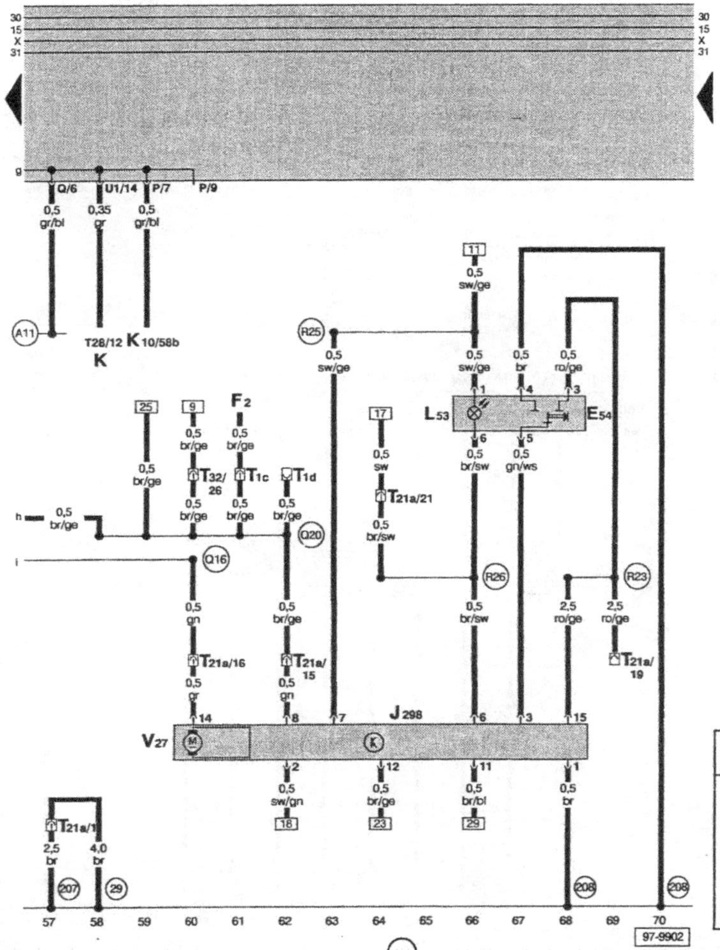

WIRING COLOR CODE

ws	=	white
sw	=	black
ro	=	red
br	=	brown
gn	=	green
bl	=	blue
gr	=	grey
li	=	lilac
ge	=	yellow

E54 - Right Rear Window Switch, (In Door)
F2 - Door Contact Switch, Left Front
J298 - Power Window Control Module, Right Rear
K - Instrument Cluster
K10 - Rear Window Defogger Indicator Light
L53 - Power Window Switch Light
V27 - Window Motor, Right Rear Door
T1c - Connector, Single, behind fuse/relay panel
T1d - Connector, Single, behind fuse/relay panel
T21a - Connector, 21 pin, on right B-pillar
T28 - Connector, 28 Pin, on instrument cluster
T32 - Connector, 32 Pin, on A-pillar, lower left
(29) - Ground connection, near "D" pillar, right

(207) - Ground connection, in left rear door wiring harness
(208) - Ground connection, in right rear door wiring harness
(A11) - Plus connection (58b), in instrument panel wiring harness
(Q16) - Wire connection, in power window wiring harness (left down)
(Q20) - Wire connection -1- (door contact switch), in power window wiring harness
(R23) - Plus connection -1- (30), in right rear door wiring harness
(R25) - Plus connection -3- (15), in right rear door wiring harness
(R26) - Wire connection (child lock), in right rear door wiring harness

1993–1994

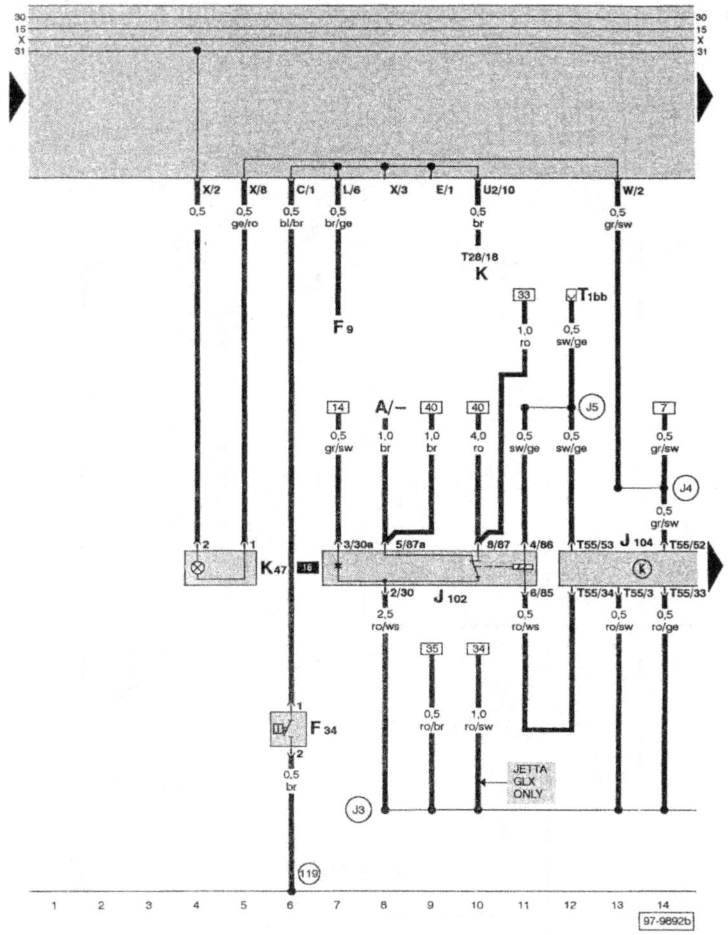

WIRING COLOR CODE

ws	=	white
sw	=	black
ro	=	red
br	=	brown
gn	=	green
bl	=	blue
gr	=	grey
li	=	lilac
ge	=	yellow

A – Battery
F9 – Parking Brake Warning Light Switch
F34 – Brake Fluid Level Warning Switch
J102 – ABS Relay
J104 – ABS (with EDL) Control Module
K – Instrument Cluster
K47 – ABS Warning Light
T1bb – Single Connector, behind fuse/relay panel
T28 – 28–Pin Connector, on instrument cluster
T55 – 55–Pin Connector, on ABS Control Module
(119) – Ground connection –1–, in headlight wiring harness

(J3) – Wire connection –1–, in ABS wiring harness
(J4) – Wire connection (OBD), in ABS wiring harness
(J5) – Wire connection –2–, in ABS wiring harness

Anti-lock brake system (ABS)

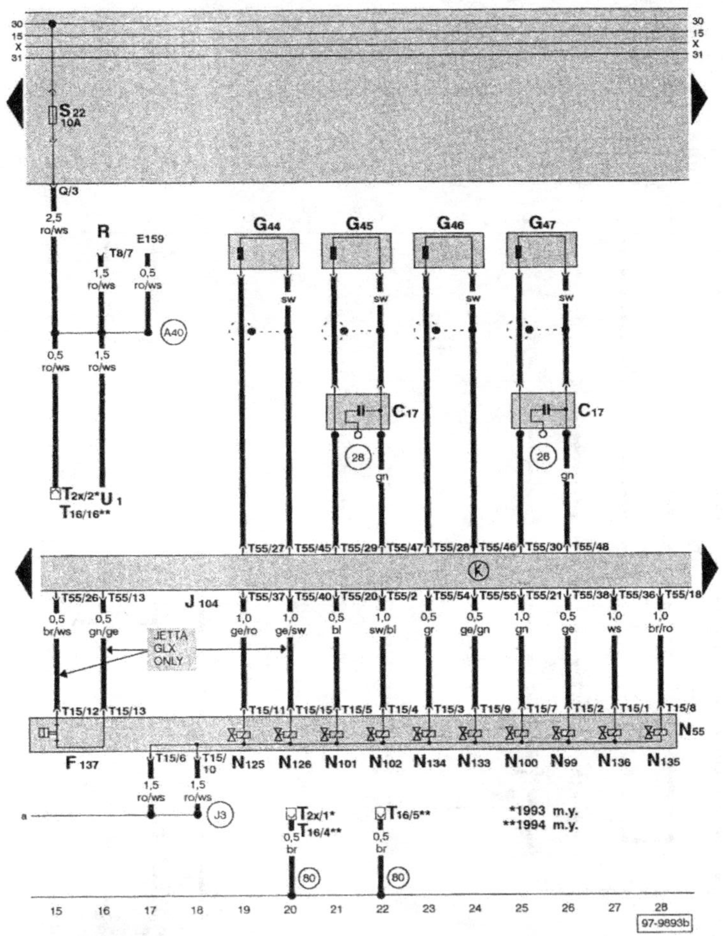

WIRING COLOR CODE

ws	=	white
sw	=	black
ro	=	red
br	=	brown
gn	=	green
bl	=	blue
gr	=	grey
li	=	lilac
ge	=	yellow

C17 – ABS Speed Sensor Capacitor
E159 – Fresh Air/Recirculating Flap Switch
F137 – ABS Pressure Control Switch
G44 – Right Rear ABS Wheel Speed Sensor
G45 – Right Front ABS Wheel Speed Sensor
G46 – Left Rear ABS Wheel Speed Sensor
G47 – Left Front ABS Wheel Speed Sensor
J104 – ABS (with EDL) Control Module
N55 – ABS Hydraulic Unit
N99 – Right Front ABS Inlet Valve
N100 – Right Front ABS Outlet Valve
N101 – Left Front ABS Inlet Valve
N102 – Left Front ABS Outlet Valve
N125 – Differential Lock Valve 1
N126 – Differential Lock Valve 2
N133 – Right Rear ABS Inlet Valve
N134 – Left Rear ABS Inlet Valve
N135 – Right Rear ABS Outlet Valve
N136 – Left Rear ABS Outlet Valve
R – Radio
U1 – Cigarette Lighter
T2x – Data Link Connector (DLC), behind center of dash panel
T8 – 8–Pin Connector, on radio
T15 – 15–Pin Connector, on ABS Hydraulic Unit
T16 – Data Link Connector (DLC), behind center of dash panel
T55 – 55–Pin Connector, on ABS Control Module
(28) – Ground connection, on firewall
(80) – Ground connection –1–, in instrument panel wiring harness
(A40) – Plus connection (30), in instrument panel wiring harness
(J3) – Wire connection –1–, in ABS wiring harness

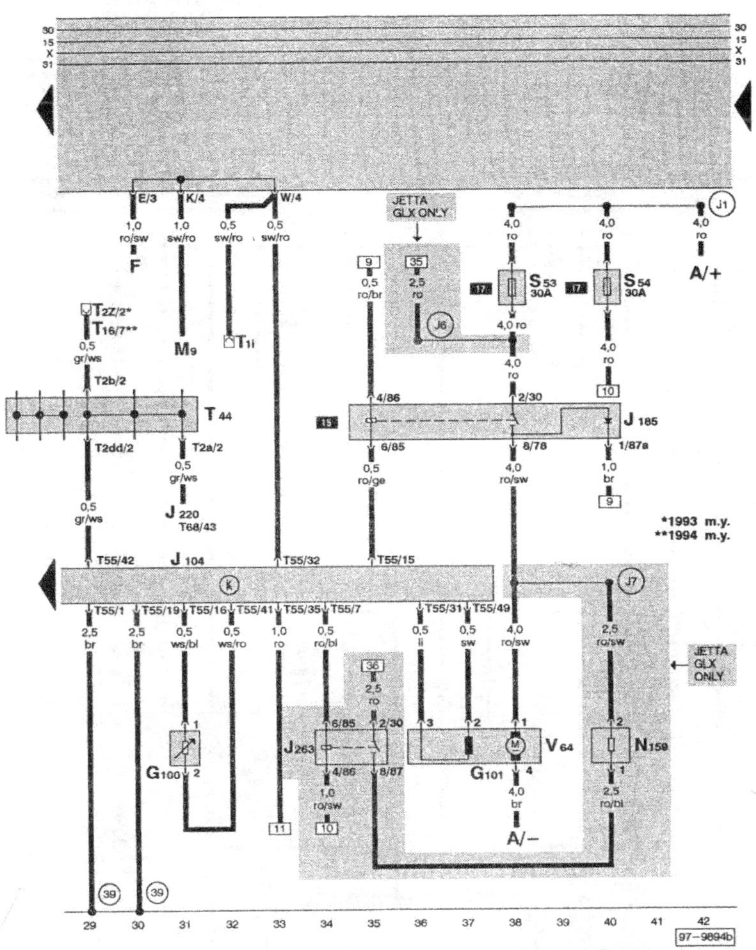

WIRING COLOR CODE		
ws	=	white
sw	=	black
ro	=	red
br	=	brown
gn	=	green
bl	=	blue
gr	=	grey
li	=	lilac
ge	=	yellow

- A – Battery
- F – Brake Light Switch
- G100 – Brake Pedal Position Sensor
- G101 – Hydraulic Pump Sensor
- J104 – ABS (with EDL) Control Module
- J185 – ABS Hydraulic Pump Relay
- J220 – Motronic Engine Control Module (ECM)
- J263 – Electronic Differential Lock Cut–Off Relay
- M9 – Left Brake Light
- N159 – Electronic Differential Lock Series Resistance
- S53 – ABS Hydraulic Pump Fuse
- S54 – ABS Valves Fuse
- V64 – ABS Hydraulic Pump
- T1i – Single Connector, behind fuse/relay panel
- T2a – Double Connector, on junction connector (T44) above fuse/relay panel
- T2b – Double Connector, on junction connector (T44) above fuse/relay panel
- T2dd – Double Connector, on junction connector (T44) behind fuse/relay panel
- T2z – Data Link Connector (DLC), behind center of dash panel
- T16 – Data Link Connector (DLC), behind center of dash panel
- T44 – Junction Box for On Board Diagnostic, above fuse/relay panel
- T55 – 55–Pin Connector, on ABS Control Module
- T68 – 68–Pin Connector, under right side of bonnet, on Motronic Engine Control Module (ECM)
- (39) – Ground connection, below rear seat bench, left
- (J1) – Plus connection (30) in ABS wiring harness
- (J6) – Plus connection –2– (30), in ABS wiring harness
- (J7) – Wire connection –3–, in ABS wiring harness

1993–1994

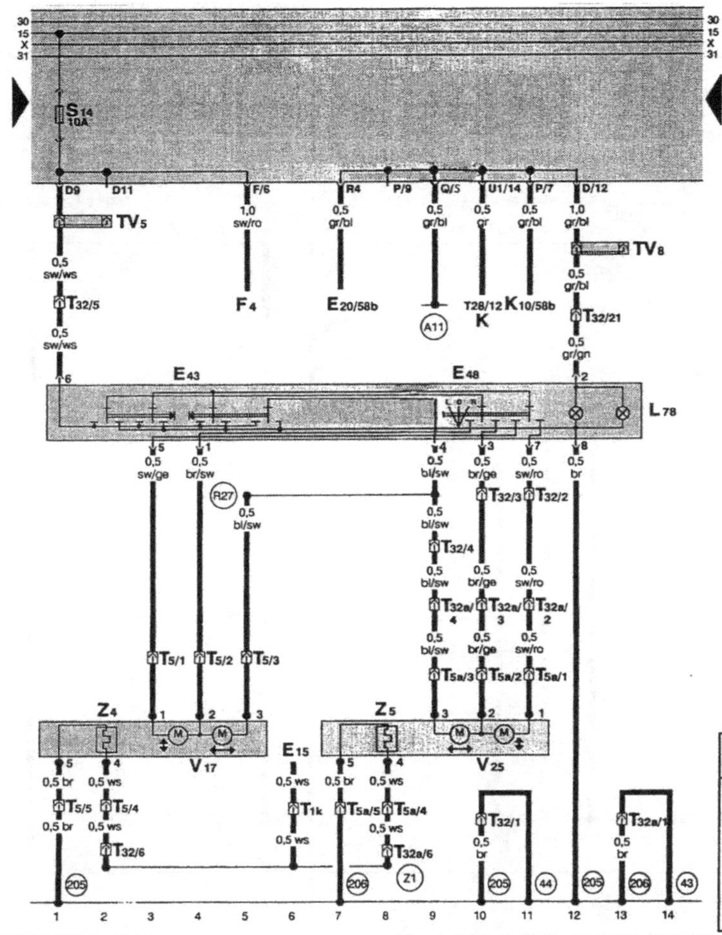

WIRING COLOR CODE		
ws	=	white
sw	=	black
ro	=	red
br	=	brown
gn	=	green
bl	=	blue
gr	=	grey
li	=	lilac
ge	=	yellow

- E15 – Rear Window Defogger Switch
- E20 – Instrument Panel Light Dimmer Switch
- E43 – Mirror Adjustment Switch
- E48 – Mirror Select Switch
- F4 – Back-Up Light Switch
- K – Instrument Cluster
- K10 – Rear Window Defogger Indicator Light
- L78 – Mirror Adjusting Switch Light
- V17 – Driver's Side Mirror Adjustment Motor
- V25 – Passenger's Side Heated Mirror Adjustment Motor
- Z4 – Driver's Side Heated Mirror
- Z5 – Passenger's Side Heated Mirror
- T1k – Connector, Single, behind fuse/relay panel
- T5 – Connector, 5 pin, in driver's door
- T5a – Connector, 5 pin, in passenger's door
- T28 – Connector, 28 Pin, on instrument cluster
- T32 – Connector, 32 Pin, on A-pillar, lower left
- T32a – Connector, 32 Pin, on A-pillar, lower right
- TV5 – Connector, terminal 15a
- TV8 – Connector, terminal 58b
- (43) – Ground connection, on right A-pillar, lower part
- (44) – Ground connection, on left "A" pillar, lower part
- (205) – Ground connection, in driver's door wiring harness
- (206) – Ground connection, in passenger's door wiring harness
- (A11) – Plus connection (58b), in instrument panel wiring harness
- (R27) – Wire connection (power mirror), in driver's door wiring harness
- (Z1) – Wire connection, –1–, in power heated mirror wiring harness

1993–1994

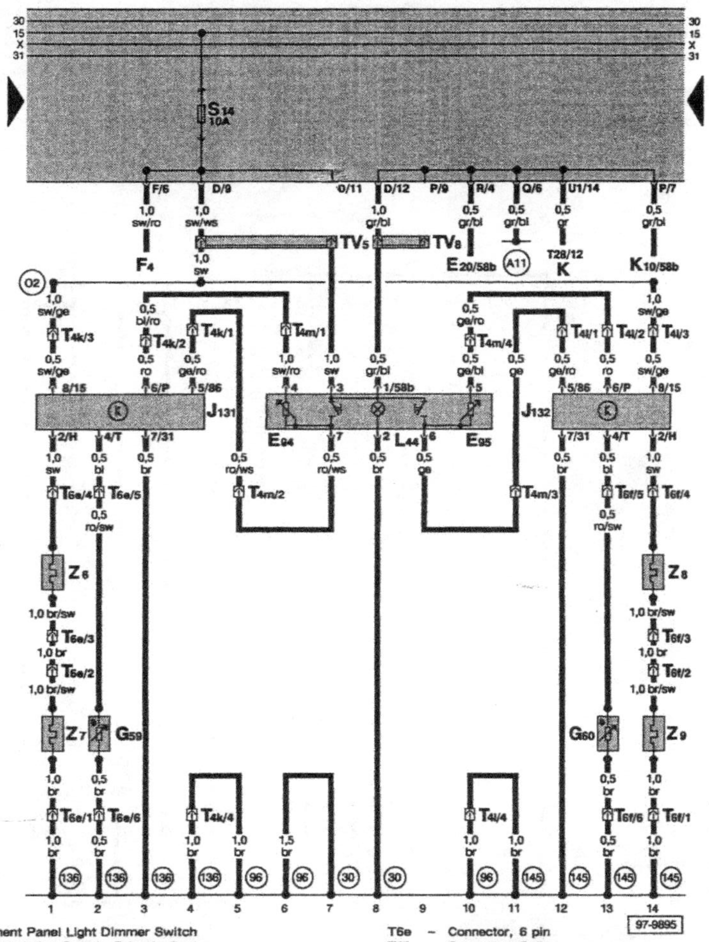

WIRING COLOR CODE		
ws	=	white
sw	=	black
ro	=	red
br	=	brown
gn	=	green
bl	=	blue
gr	=	grey
li	=	lilac
ge	=	yellow

E20 – Instrument Panel Light Dimmer Switch
E94 – Heat Regulating Switch, Driver's Seat
E95 – Heat Regulating Switch, Passenger's Seat
F4 – Back-Up Light Switch
G59 – Driver's Seat Heater Temperature Sensor
G60 – Passenger's Seat Heater Temperature Sensor
J131 – Driver's Seat Heater Control Module
J132 – Passenger's Seat Heater Control Module
K – Instrument Cluster
K10 – Rear Window Defogger Indicator Light
L44 – Seat Heater Switch Illumination Light
Z6 – Driver's Seat Heat Element
Z7 – Driver's Backrest Heat Element
Z8 – Passenger's Seat Heat Element
Z9 – Passenger's Backrest Heat Element
T4k – Connector, 4 pin
T4l – Connector, 4 pin
T4m – Connector, 4 pin
T6e – Connector, 6 pin
T6f – Connector, 6 pin
T28 – Connector, 28 pin, on instrument cluster
TV5 – Connector, terminal 15a
TV8 – Connector, terminal 58b

(30) – Ground connection – beside fuse/relay panel
(96) – Ground connection -1-, in heated seats wiring harness
(136) – Ground connection -2-, in heated seats wiring harness
(145) – Ground connection -3-, in heated seats wiring harness
(A11) – Plus connection (58b), in instrument panel wiring harness
(O2) – Plus connection (15) in heated seats wiring harness

1993–1994

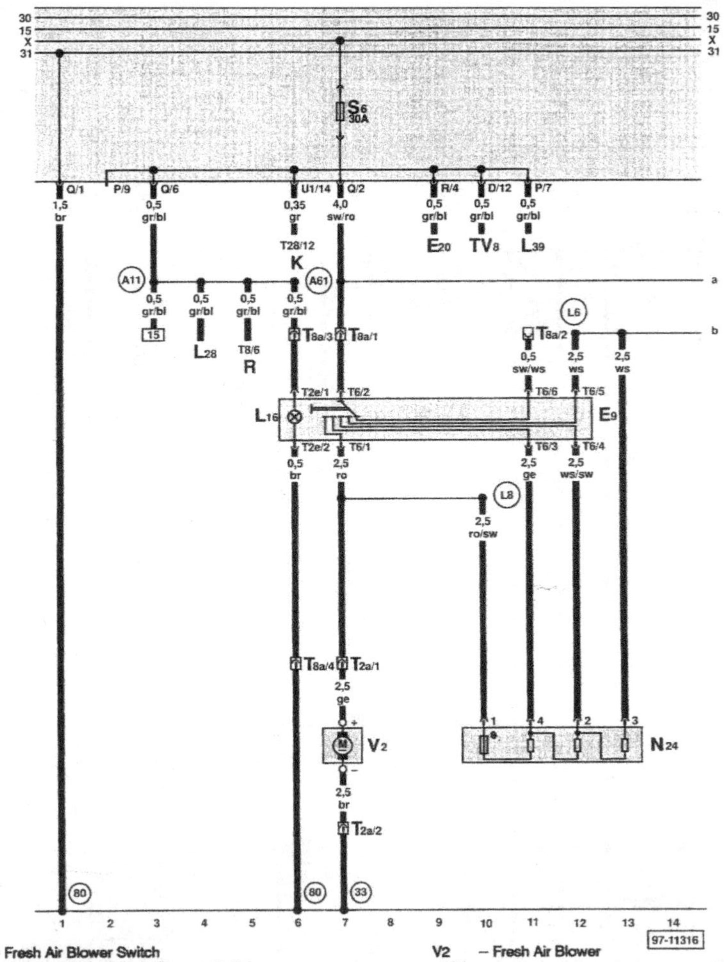

WIRING COLOR CODE		
ws	=	white
sw	=	black
ro	=	red
br	=	brown
gn	=	green
bl	=	blue
gr	=	grey
li	=	lilac
ge	=	yellow

E9 – Fresh Air Blower Switch
E20 – Instrument Panel Light Dimmer Switch
K – Instrument Cluster
L16 – Fresh Air Control Lever Light
L28 – Cigarette Lighter Light
L39 – Rear Window Defogger Switch Light
N24 – Fresh Air Blower Series Resistance with fuse
R – Radio
T2a – Double Connector, behind instrument panel, right
T2e – Double Connector, on E9
T6 – 6-Pin Connector, on E9
T8 – 8-Pin Connector, on radio
T8a – 8-Pin Connector, behind instrument panel, center
T28 – 28-Pin Connector, on instrument cluster
TV8 – Terminal 58b Wire Connector, above fuse/relay panel
V2 – Fresh Air Blower

(33) – Ground connection, behind instrument panel, right
(80) – Ground connector -1-, in instrument panel wiring harness
(A11) – Plus connection (58b), in instrument panel wiring harness
(A61) – Plus connection (X), in instrument cluster wiring harness
(L6) – Wire connection -2-, in evaporator housing wiring harness
(L8) – Wire connection -4-, in evaporator housing wiring harness

1993-1994

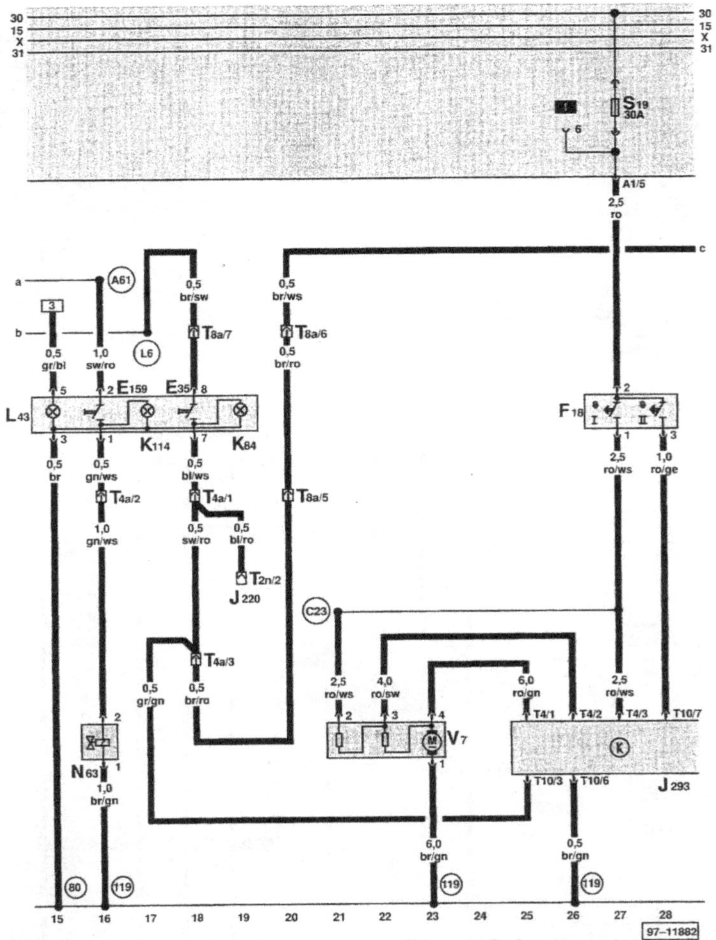

WIRING COLOR CODE		
ws	=	white
sw	=	black
ro	=	red
br	=	brown
gn	=	green
bl	=	blue
gr	=	grey
li	=	lilac
ge	=	yellow

E35 – A/C Switch
E159 – Fresh Air/Recirculation Flap Switch
F18 – Coolant Fan Control (FC) Thermal Switch
J220 – Motronic Engine Control Module (ECM)
J293 – Coolant Fan Control (FC) Control Module, in left side of engine compartment
K84 – A/C Indicator Light
K114 – Fresh Air and Recirculating Air Mode Indicator Light
L43 – A/C Switch Illumination Light
N63 – Fresh Air/Recirculating Flap Two-Way Valve (00605), in rear of engine compartment
T2n – Double Connector, behind fuse/relay panel
T4 – 4-Pin Connector, on J293
T4a – 4-Pin Connector, behind fuse/relay panel
T8a – 8-Pin Connector, behind instrument panel, center
T10 – 10-Pin Connector, on J293
V7 – Coolant Fan
(80) – Ground connection -1-, in instrument panel wiring harness
(119) – Ground connection -1-, in headlight wiring harness
(A61) – Plus connection (X), in instrument cluster wiring harness
(C23) – Wire connection (blower, 1st speed) in headlight/air conditioner wiring harness
(L6) – Wire connection -2-, in evaporator housing wiring harness

1993–1994

Air conditioning

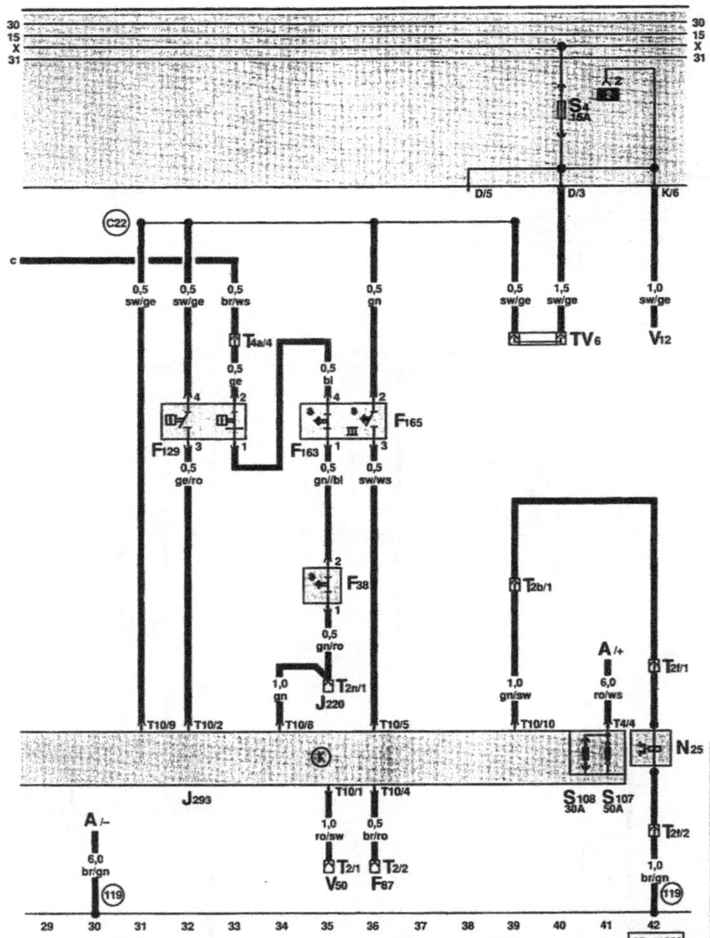

WIRING COLOR CODE		
ws	=	white
sw	=	black
ro	=	red
br	=	brown
gn	=	green
bl	=	blue
gr	=	grey
li	=	lilac
ge	=	yellow

A – Battery
F38 – Ambient Temperature Switch, near horn
F87 – After-Run Coolant Fan Control (FC) Thermal Switch
F129 – A/C Pressure Switch, in right side of engine compartment
F163 – A/C Cut-Out Thermal Switch, in center of engine compartment
F165 – Third Speed Coolant Fan Control (FC) Thermal Switch, in center of engine compartment
J220 – Motronic Engine Control Module (ECM)
J293 – Coolant Fan Control (FC) Control Module
N25 – A/C Clutch
S107 – Coolant Fan Second and Third Speed Fuse
S108 – Coolant Fan First Speed and A/C Clutch Fuse
T2 – Double Connector, in engine compartment, left
T2b – Double Connector, in engine compartment, front
T2f – Double Connector, near compressor
T2n – Double Connector, behind fuse/relay panel
T4a – 4-Pin Connector, behind fuse/relay panel
TV6 – Terminal X Wire Connector, above fuse/relay panel
V12 – Rear Window Wiper Motor (not applicable to Jetta III)
V50 – Coolant Pump
(119) – Ground connection -1-, in headlight wiring harness
(C22) – Wire connection (X), in headlight/air conditioner wiring harness

1993–1994

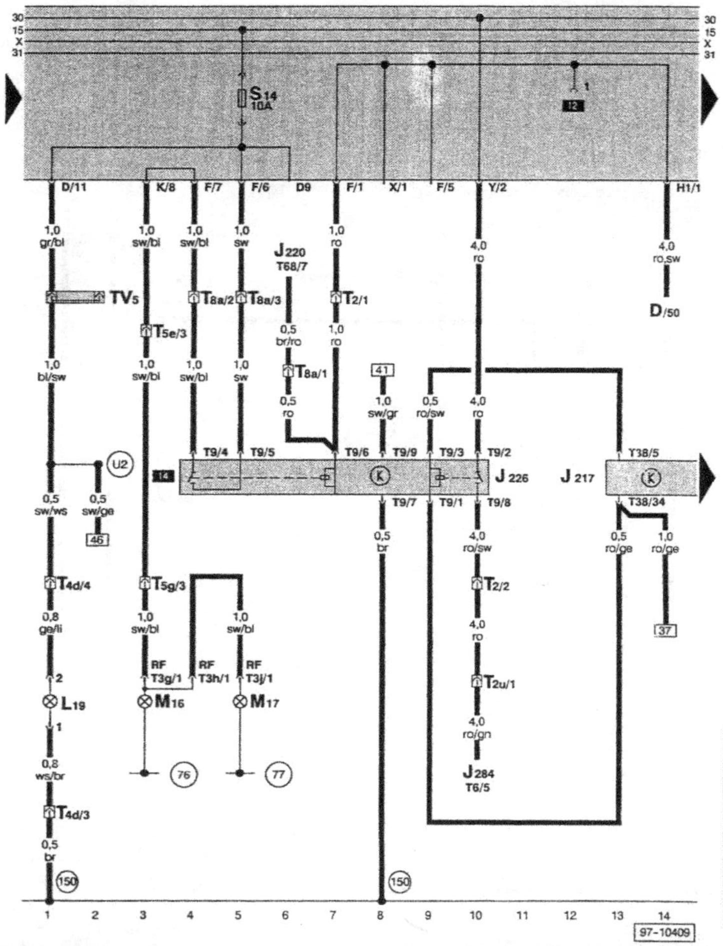

WIRING COLOR CODE		
ws	=	white
sw	=	black
ro	=	red
br	=	brown
gn	=	green
bl	=	blue
gr	=	grey
li	=	lilac
ge	=	yellow

D – Igntion/Starter Switch
J217 – Transmission Control Module (TCM)
J220 – Motronic Engine Control Module (ECM)
J226 – Park/Neutral Position (PNP) Relay
J284 – Alarm System Control Module
L19 – Automatic Transmission Console Light
M16 – Back-Up Light, Left
M17 – Back-Up Light, Right
T2 – Connector, Double, behind fuse/relay panel, only on automatic transmission
T2u – Connector, Double, behind fuse/relay panel
T3g – Connector, 3 pin, on back-up/brake light, left
T3h – Connector, 3 pin, on back-up/brake light, left
T3j – Connector, 3 pin, on back-up/brake light, right
T4d – Connector, 4 Pin, below center console
T5e – Connector, 5 Pin, in luggage compartment, left
T5g – Connector, 5 Pin, in luggage compartment lid
T6 – Connector, 6 Pin, behind dash panel, left, on Alarm System Control Module
T8a – Connector, 8 Pin, behind fuse/relay panel
T9 – Connector, 9 Pin, on Park/Neutral Position (PNP) Relay
T38 – Connector, 38 Pin, below right front seat, on Transmission Control Module (TCM)
T68 – Connector, 68 Pin, under right side of bonnet, on Motronic Engine Control Module (ECM)
TV5 – Connector, terminal 15a
(76) – Ground connection, on left back-up/brake light bulb holder
(77) – Ground connection, on right back-up/brake light bulb holder
(150) – Ground connection, in automatic transmission wiring harness
(U2) – Wire connection -1- (15), in automatic transmission wiring harness

1993–1994

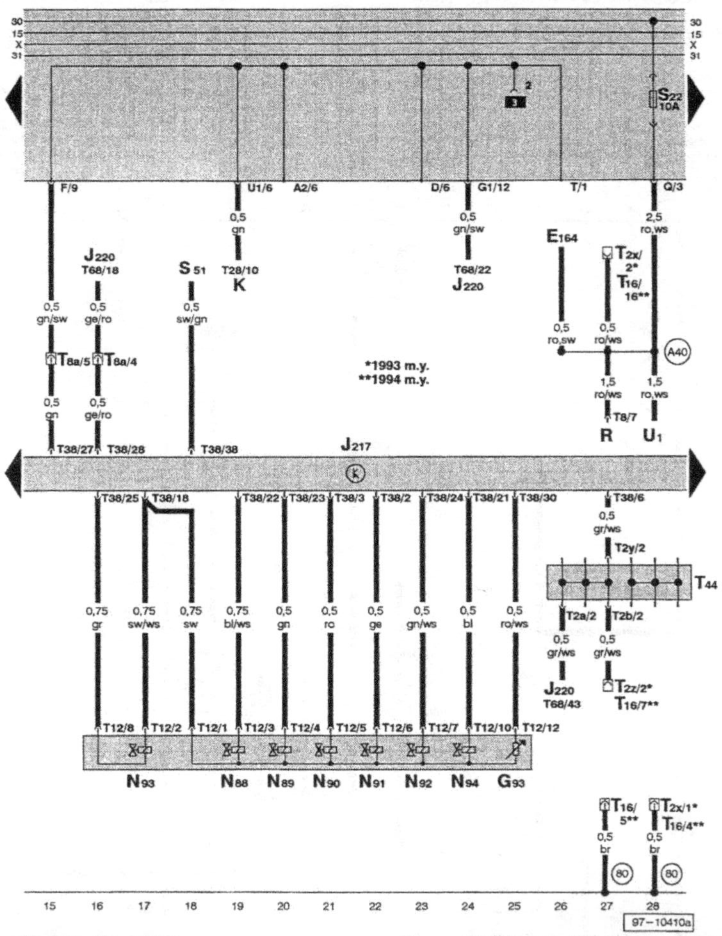

WIRING COLOR CODE		
ws	=	white
sw	=	black
ro	=	red
br	=	brown
gn	=	green
bl	=	blue
gr	=	grey
li	=	lilac
ge	=	yellow

E164 – Luggage Compartment Release Switch
G93 – Transmission Fluid Temperature Sensor
J217 – Transmission Control Module (TCM)
J220 – Motronic Engine Control Module (ECM)
K – Instrument Cluster
N88 – Solenoid Valve 1
N89 – Solenoid Valve 2
N90 – Solenoid Valve 3
N91 – Solenoid Valve 4
N92 – Solenoid Valve 5
N93 – Solenoid Valve 6
N94 – Solenoid Valve 7
R – Radio
S51 – Fuse
U1 – Cigarette Lighter
T2a – Connector, Double, on junction connector (T44) above fuse/relay panel
T2b – Connector, Double, on junction connector (T44) above fuse/relay panel
T2x – Data Link Connector (DLC), behind center of dash panel
T2y – Connector, Double, on junction connector (T44) above fuse/relay panel
T2z – Data Link Connector (DLC), behind center of dash panel
T8 – Connector, 8 Pin, on radio
T8a – Connector, 8 Pin, behind fuse/relay panel
T12 – Connector, 12 Pin, on transmission
T16 – Data Link Connector (DLC), behind center of dash panel
T28 – Connector, 28 Pin, on instrument cluster
T38 – Connector, 38 Pin, below right front seat, on Transmission Control Module (TCM)
T44 – Junction Box for On Board Diagnostic, above fuse/relay panel
T68 – Connector, 68 Pin, under right side of bonnet, on Motronic Engine Control Module (ECM)
(80) – Ground connection -1-, in instrument panel wiring harness
(A40) – Plus connection (30), in instrument panel wiring harness

1993-1994

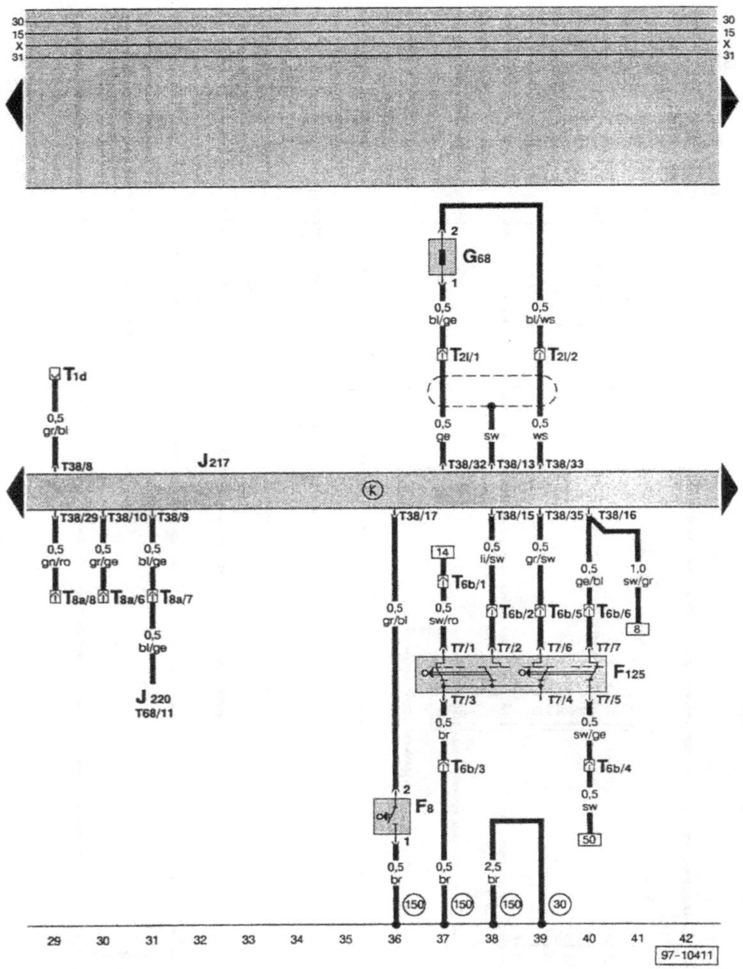

WIRING COLOR CODE

ws	=	white
sw	=	black
ro	=	red
br	=	brown
gn	=	green
bl	=	blue
gr	=	grey
li	=	lilac
ge	=	yellow

F8 – Kick Down Switch
F125 – Multi–Function Transmission Range (TR) Switch
G68 – Vehicle Speed Sensor (VSS)
J217 – Transmission Control Module (TCM)
J220 – Motronic Engine Control Module (ECM)
T1d – Connector, Single, behind fuse/relay panel
T2i – Connector, Double, below battery
T6b – Connector, 6 Pin, below battery
T7 – Connector, 7 Pin, on transmission
T8a – Connector, 8 Pin, behind fuse/relay panel

T38 – Connector, 38 Pin, below right front seat, on Transmission Control Module (TCM)
T68 – Connector, 68 Pin, under right side of bonnet, on Motronic Engine Control Module (ECM)
(30) – Ground connection – beside fuse/relay panel
(150) – Ground connection, in automatic transmission wiring harness

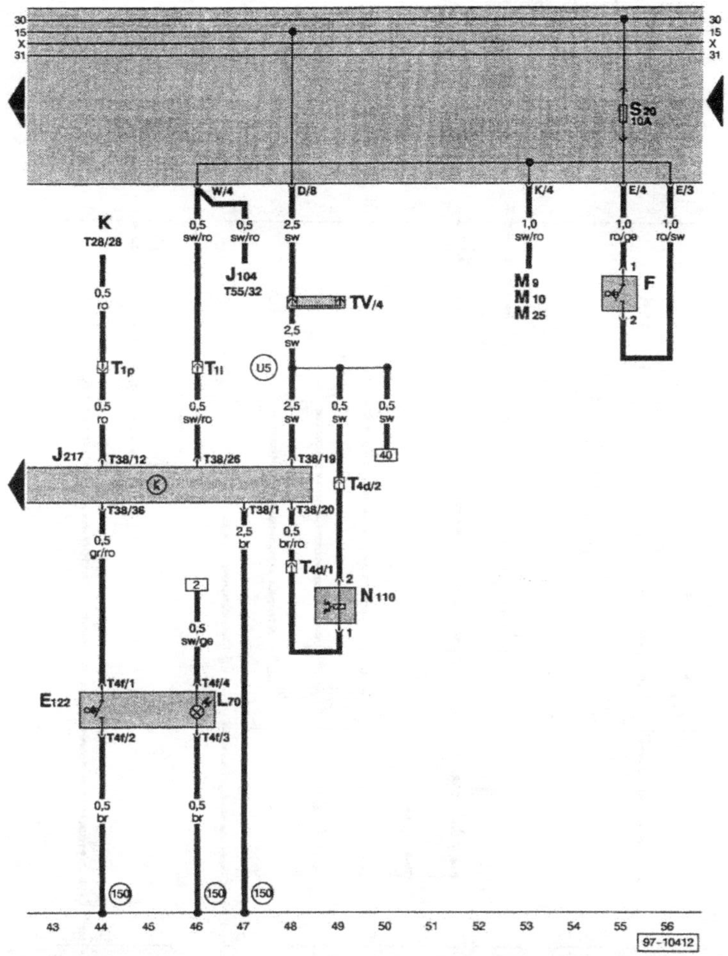

WIRING COLOR CODE

ws	=	white
sw	=	black
ro	=	red
br	=	brown
gn	=	green
bl	=	blue
gr	=	grey
li	=	lilac
ge	=	yellow

E122 – Transmission Range (TR) Program Switch
F – Brake Light Switch
J104 – ABS Control Module (w/EDL)
J217 – Transmission Control Module (TCM)
K – Instrument Cluster
L70 – Program Switch Illumination
M9 – Brake Light, Left
M10 – Brake Light, Right
M25 – High-Mount Brake Light
N110 – Shift Lock Solenoid
T1i – Connector, Single, behind fuse/relay panel
T1p – Connector, Single, behind fuse/relay panel
T4d – Connector, 4 Pin, below center console

T4f – Connector, 4 Pin, on Transmission Range (TR) Program Switch
T28 – Connector, 28 Pin, on instrument cluster
T38 – Connector, 38 Pin, below right front seat, on Transmission Co
T55 – Connector, 55 pins, on ABS Control Module
TV4 – Connector, terminal 15
(150) – Ground connection, in automatic transmission wiring harness
(U5) – Wire connection -2- (15), in automatic transmission wiring harness
(W1) – Plus connection (54), in rear wiring harness

Battery, starter, generator, ignition switch

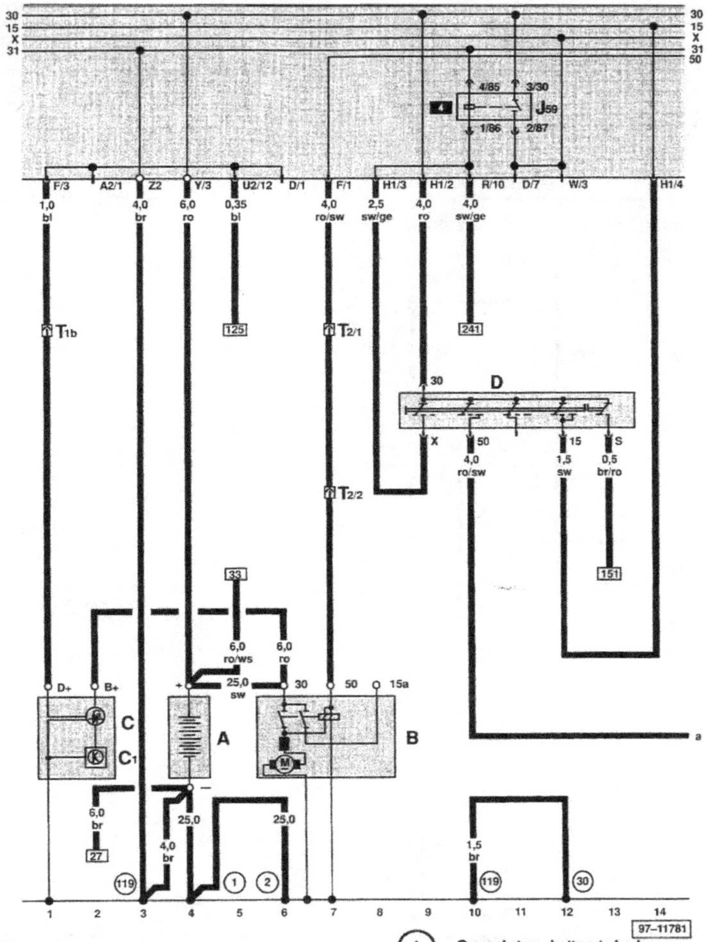

WIRING COLOR CODE		
ws	=	white
sw	=	black
ro	=	red
br	=	brown
gn	=	green
bl	=	blue
gr	=	grey
li	=	lilac
ge	=	yellow

A – Battery
B – Starter
C – Generator (GEN)
C1 – Voltage Regulator (VR)
D – Ignition/Starter Switch
J59 – Load Reduction Relay
T1b – Single Connector, near battery
T2 – Double Connector, behind fuse/relay panel

(1) – Ground strap, battery to body
(2) – Ground strap, transmission to body
(30) – Ground connection -1-, beside fuse/relay panel
(119) – Ground connection -1-, in headlight wiring harness

Coolant fan, fresh air blower

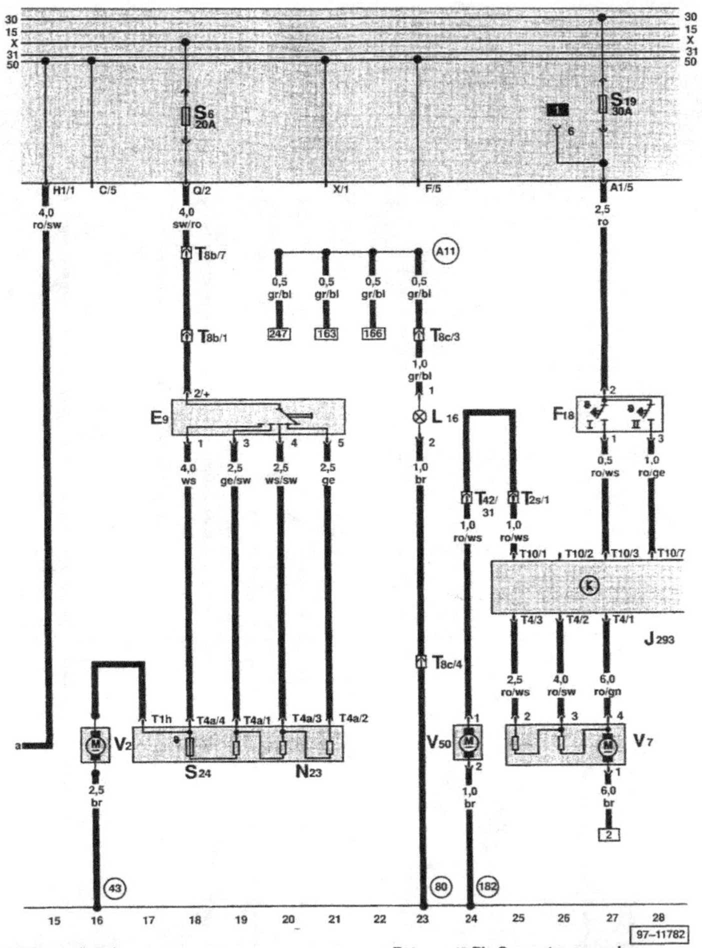

WIRING COLOR CODE		
ws	=	white
sw	=	black
ro	=	red
br	=	brown
gn	=	green
bl	=	blue
gr	=	grey
li	=	lilac
ge	=	yellow

E9 – Fresh Air Blower Switch
F18 – Coolant Fan Control (FC) Thermal Switch
J293 – Coolant Fan Control (FC) Control Module
L16 – Fresh Air Control Lever Light
N23 – Fresh Air Blower Series Resistance
S24 – Heater, A/C Clutch Fuse (8A)
T1h – Single Connector, on N23
T2s – Double Connector, behind fuse/relay panel
T4 – 4-Pin Connector, on J293
T4a – 4-Pin Connector, on N23
T8b – 8-Pin Connector, behind instrument panel, center
T8c – 8-Pin Connector, behind instrument panel, center
T10 – 10-Pin Connector, on J293
T42 – 42-Pin Connector, on engine
V2 – Fresh Air Blower (01273-4124)
V7 – Coolant Fan (01018-3433)
V50 – Coolant Pump (00739)

(43) – Ground connection, on right A-pillar, lower part
(80) – Ground connection -1-, in instrument panel wiring harness
(182) – Ground connection -1-, in engine compartment wiring harness (6 cylinder)
(A11) – Plus connection (58b), in instrument panel wiring harness

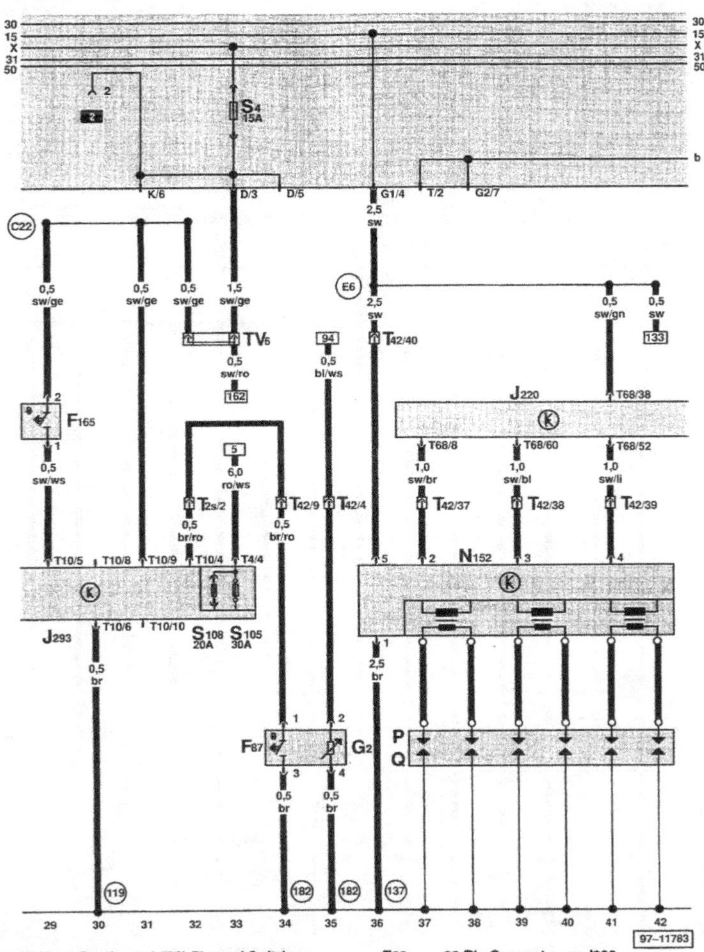

WIRING COLOR CODE		
ws	=	white
sw	=	black
ro	=	red
br	=	brown
gn	=	green
bl	=	blue
gr	=	grey
li	=	lilac
ge	=	yellow

F87 – After-Run Coolant Fan Control (FC) Thermal Switch
F165 – Third Speed Coolant Fan Control (FC) Thermal Switch
G2 – Engine Coolant Temperature (ECT) Sensor (01039)
J220 – Motronic Engine Control Module (ECM)
J293 – Coolant Fan Control (FC) Control Module
N152 – Ignition Coil
P – Spark Plug Connector
Q – Spark Plugs
S105 – First Speed Coolant Fan Fuse
S108 – Coolant Fan First Speed and A/C Clutch Fuse
T2s – Double Connector, behind fuse/relay panel
T4 – 4-Pin Connector, on J293
T10 – 10-Pin Connector, on J293
T42 – 42-Pin Connector, on engine
T68 – 68-Pin Connector, on J220
TV6 – Terminal X Wire Connector
(119) – Ground connection -1-, in headlight wiring harness
(137) – Ground connection (end stage), in Motronic Multiport Fuel Injection (MFI) wiring harness
(182) – Ground connection -1-, in engine compartment wiring harness (6 cylinder)
(C22) – Wire connection (X), in headlight/air conditioner wiring harness
(E6) – Plus connection -1- (15), in Motronic Multiport Fuel Injection (MFI) wiring harness

1993–1994

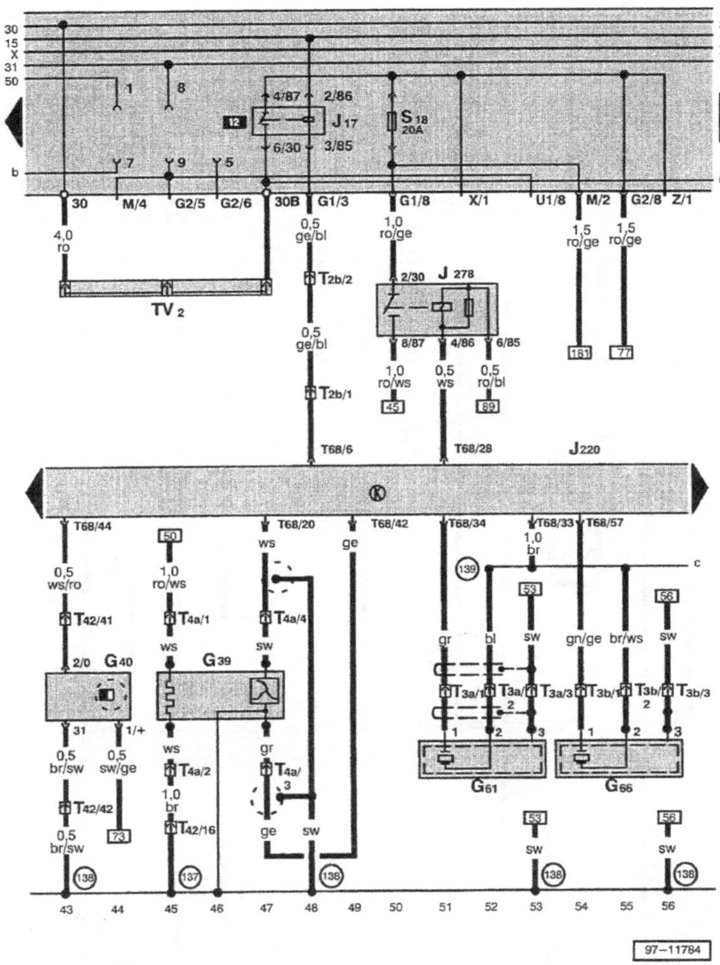

WIRING COLOR CODE		
ws	=	white
sw	=	black
ro	=	red
br	=	brown
gn	=	green
bl	=	blue
gr	=	grey
li	=	lilac
ge	=	yellow

G39 – Heated Oxygen Sensor (H02S) (00525)
G40 – Camshaft Position (CMP) Sensor (00515)
G61 – Knock Sensor (KS) 1 (00524)
G66 – Knock Sensor (KS) 2 (00549)
J17 – Fuel Pump (FP) Relay (01259)
J220 – Motronic Engine Control Module (ECM)
J278 – Heated Oxygen Sensor (HO2S) Relay (00640)
T2b – 2-Pin Connector, behind fuse/relay panel
T3a – 3-Pin Connector, near intake manifold
T3b – 3-Pin Connector, near intake manifold
T4a – 4-Pin Connector, near intake manifold
T42 – 42-Pin Connector, on engine
T68 – 68-Pin Connector, on J220
TV2 – Terminal 30 (B+) Wire Connector
(137) – Ground connection (end stage), in Motronic Multiport Fuel Injection (MFI) wiring harness
(138) – Ground connection (control module), in Motronic Multiport Fuel Injection (MFI) wiring harness
(139) – Ground Connection (sensor ground), in Motronic Multiport Fuel Injection (MFI) wiring harness

1993–1994

Engine control module, sensors, fuel injectors

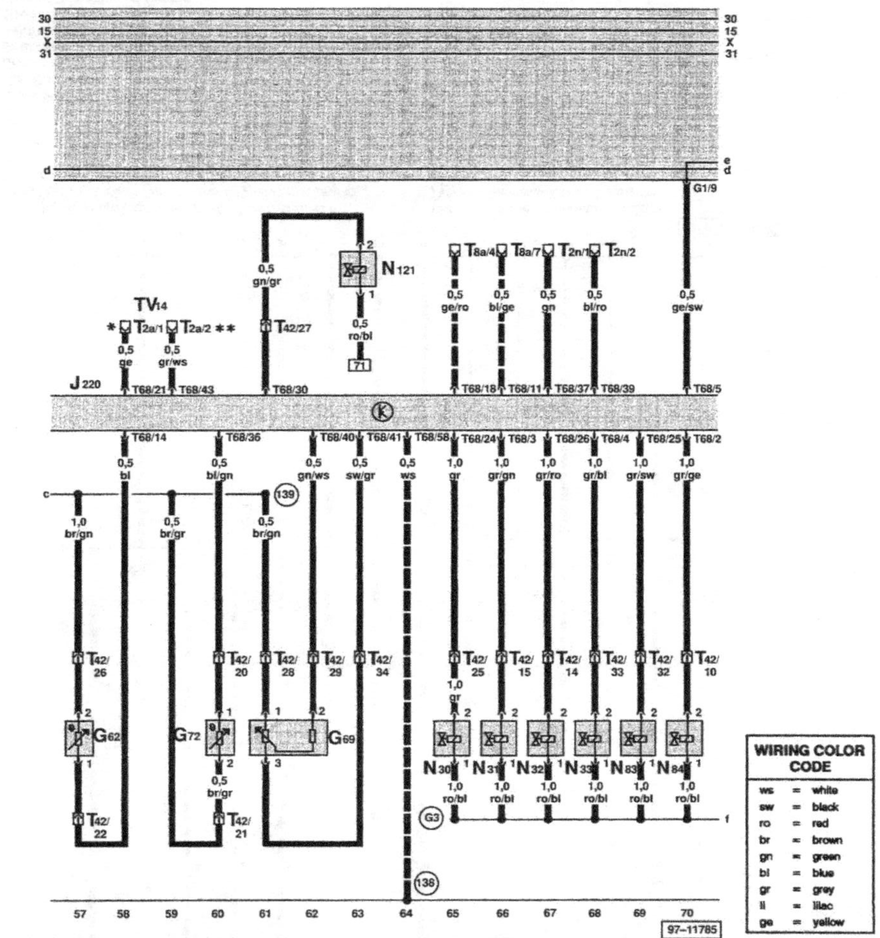

G62 – Engine Coolant Temperature (ECT) Sensor (00522)
G69 – Throttle Position (TP) Sensor (00518)
G72 – Intake Air Temperature (IAT) Sensor (00527)
J220 – Motronic Engine Control Module (ECM)
N30 – Cylinder 1 Fuel Injector (01249)
N31 – Cylinder 2 Fuel Injector (01250)
N32 – Cylinder 3 Fuel Injector (01251)
N33 – Cylinder 4 Fuel Injector (01252)
N83 – Cylinder 5 Fuel Injector (01253)
N84 – Cylinder 6 Fuel Injector (01254)
N121 – EGR Vacuum Regulator Valve
T2a – Double Connector, behind fuse/relay panel
T2n – Double Connector, behind fuse/relay panel
T8a – 8-Pin Connector, behind fuse/relay panel
T16 – Data Link Connector (DLC), behind center of dash panel

T42 – 42-Pin Connector, on engine
T68 – 68-Pin Connector, on J220
TV14 – Data Link Connector (DLC) wire connector

(138) – Ground connection (control module), in Motronic Multiport Fuel Injection (MFI) wiring harness

(139) – Ground connection (sensor ground), in Motronic Multiport Fuel Injection (MFI) wiring harness

(G3) – Plus connection in fuel injector wiring harness

– – – Automatic transmission only
* Connector T2a/1 connected to Data Link Connector T16/15
** Connector T2a/2 connected to Data Link Connector T16/7

Engine control module, idle air control valve, mass air flow sensor

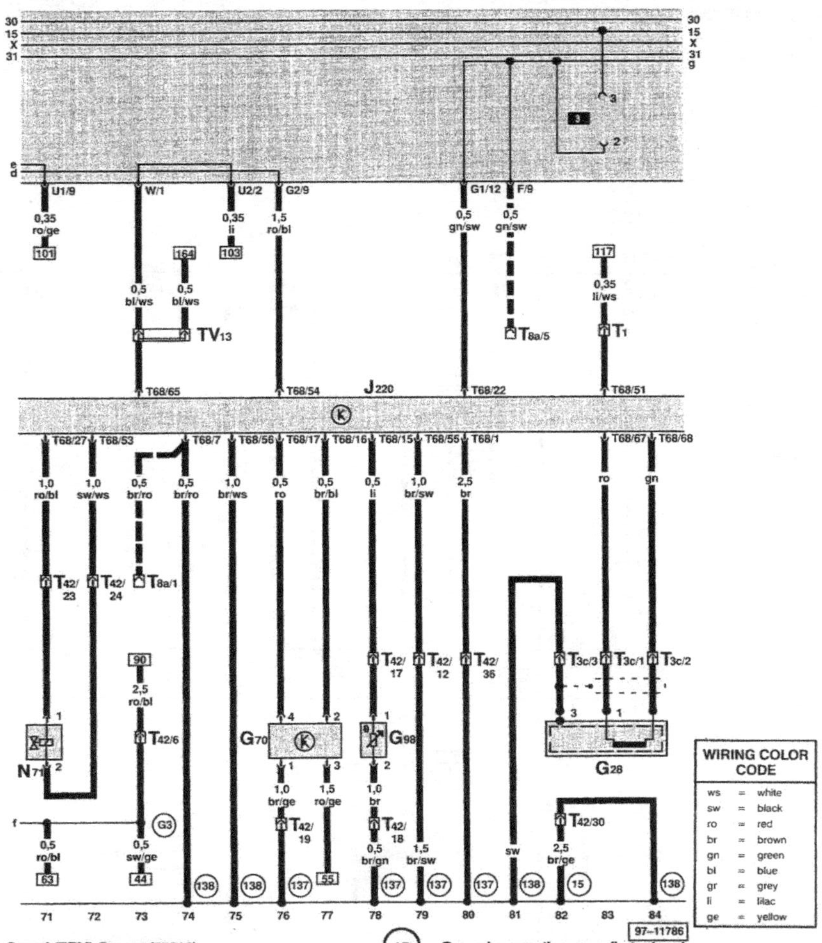

G28 – Engine Speed (RPM) Sensor (00513)
G70 – Mass Air Flow (MAF) Sensor (00553)
G98 – EGR Temperature Sensor (00585)
J220 – Motronic Engine Control Module (ECM)
N71 – Idle Air Control (IAC) Valve (01257)
T1 – Single Connector, behind fuse/relay panel
T3c – 3-Pin Connector, near starter
T8a – 8-Pin Connector, behind fuse/relay panel
T42 – 42-Pin Connector, on engine
T68 – 68-Pin Connector, on J220
TV13 – Vehicle Speed Signal Wire Connector

(15) – Ground connection, on cylinder head

(137) – Ground connection (end stage), in Motronic Multiport Fuel Injection (MFI) wiring harness

(138) – Ground connection (control module), in Motronic Multiport Fuel Injection (MFI) wiring harness

(G3) – Plus connection, in fuel injector wiring harness

– – – Automatic transmission only

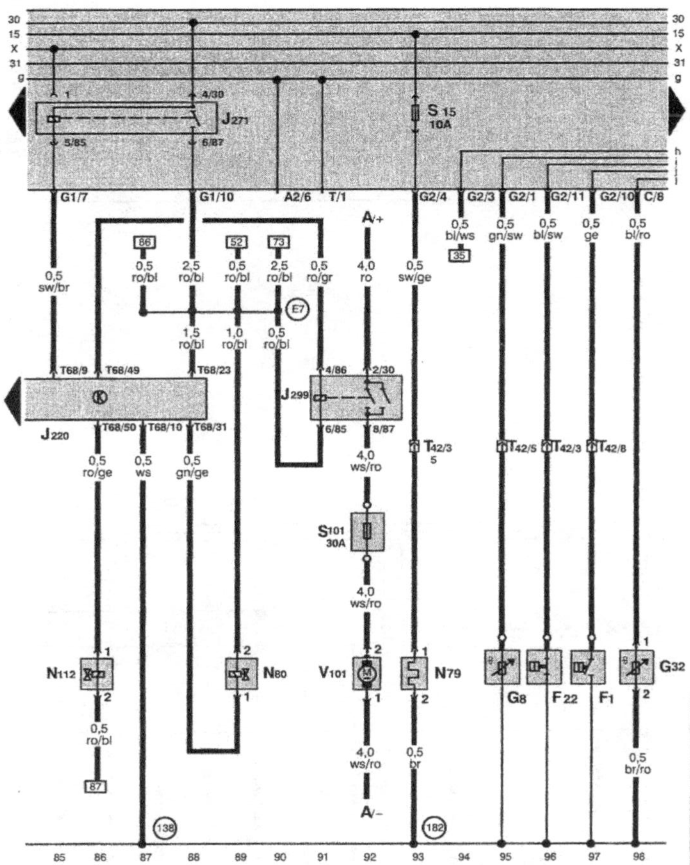

WIRING COLOR CODE		
ws	=	white
sw	=	black
ro	=	red
br	=	brown
gn	=	green
bl	=	blue
gr	=	grey
li	=	lilac
ge	=	yellow

A – Battery
F1 – Engine Oil Pressure Switch
F22 – 0.3 Bar Engine Oil Pressure Switch (00772)
G8 – Engine Oil Temperature Sensor
G32 – Engine Coolant Level (ECL) Sensor
J271 – Motronic Engine Control Module (ECM) Power Supply Relay
J299 – Secondary Air Injection (AIR) Pump Relay (01264)
N79 – Positive Crankcase Ventilation (PCV) Heating Element
N80 – Evaporative Emission (EVAP) Canister Purge Regulator Valve (01247)
N112 – Secondary Air Injection (AIR) Solenoid Valve (01235)
S101 – A/C Climatronic Control Module Fuse
T42 – 42-Pin Connector, on engine
T68 – 68-Pin Connector, on J220
V101 – Secondary Air Injection (AIR) Pump Motor
(138) – Ground connection (Control Module), in Motronic Multiport Fuel Injection (MFI) wiring harness
(182) – Ground connection –1–, in engine compartment wiring harness (6 cylinder)
(E7) – Ground connection (87a), in Motronic Multiport Fuel Injection (MFI) wiring harness

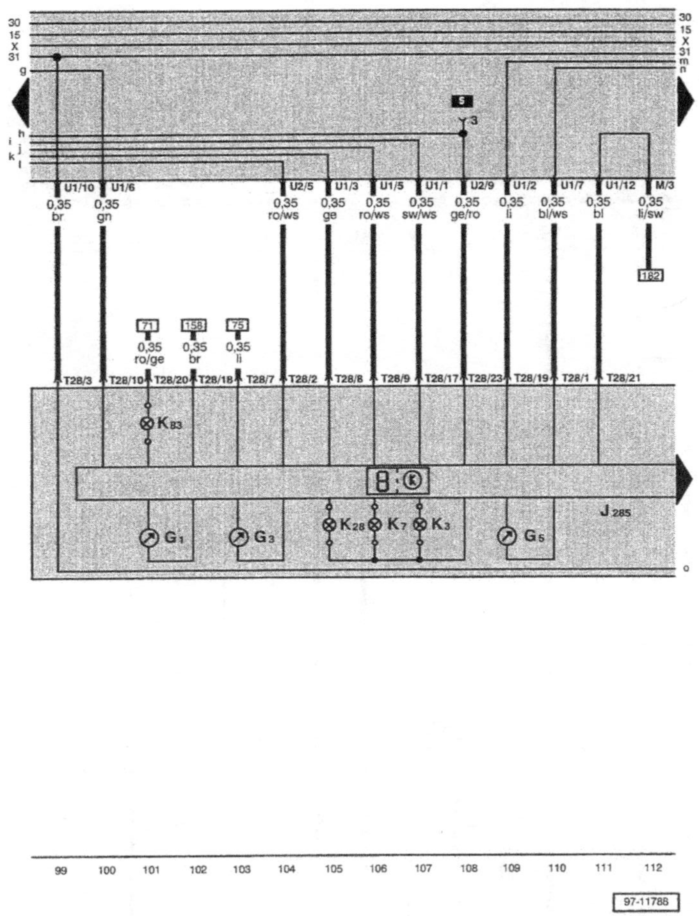

WIRING COLOR CODE		
ws	=	white
sw	=	black
ro	=	red
br	=	brown
gn	=	green
bl	=	blue
gr	=	grey
li	=	lilac
ge	=	yellow

G1 – Fuel Gauge (00664)
G3 – Engine Coolant Temperature (ECT) Gauge
G5 – Tachometer (00662)
J285 – Control Module With Display Unit In Instrument Cluster
K3 – Oil Pressure Warning Light (00755)
K7 – Brake and Parking Brake Warning Light (01047)
K28 – Engine Coolant Level/Temperature (ECL/ECT) Warning Light
K83 – Multi-Function Indicator (MIL)
T28 – 28-Pin Connector, on instrument cluster

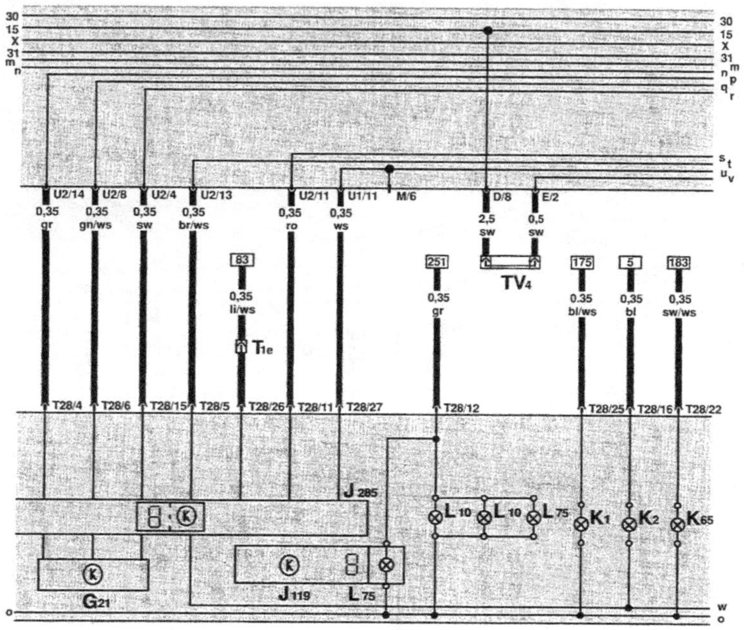

WIRING COLOR CODE

ws	=	white
sw	=	black
ro	=	red
br	=	brown
gn	=	green
bl	=	blue
gr	=	grey
li	=	lilac
ge	=	yellow

113 114 115 116 117 118 119 120 121 122 123 124 125 126

97-11789

- G21 – Speedometer (00663)
- J119 – Multi-Function Indicator
- J285 – Control Module with Display Unit in Instrument Cluster
- K1 – Headlight High Beam Indicator Light
- K2 – Generator (GEN) Warning Light
- K65 – Left Turn Signal Indicator Light
- L10 – Instrument Cluster Light
- L75 – Digital Display Light
- T1e – Single Connector, behind fuse/relay panel
- T28 – 28-Pin Connector, on instrument cluster
- TV4 – Terminal 15 Wire Connector

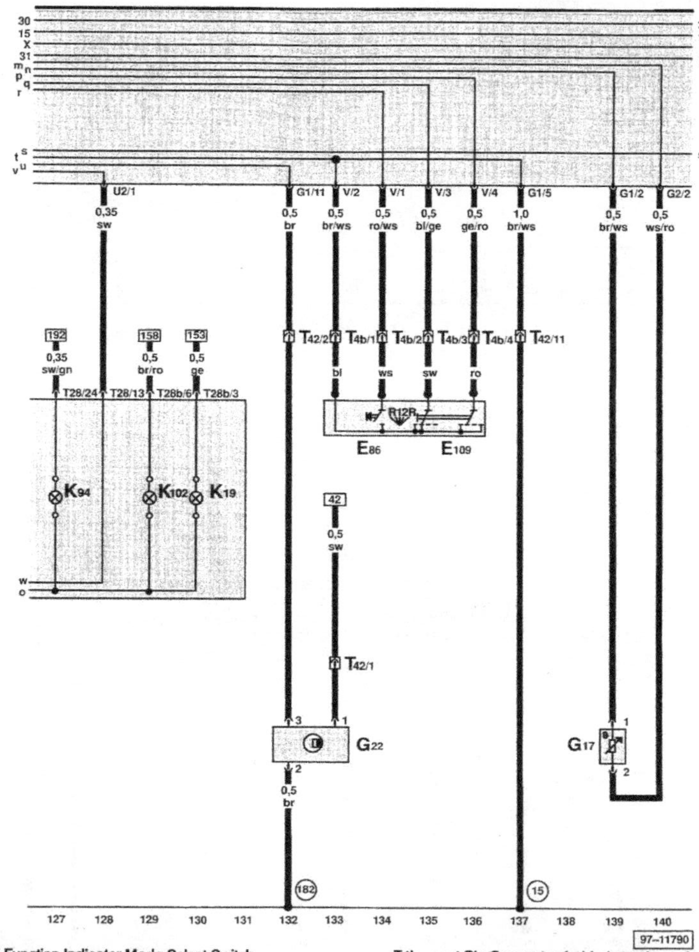

WIRING COLOR CODE

ws	=	white
sw	=	black
ro	=	red
br	=	brown
gn	=	green
bl	=	blue
gr	=	grey
li	=	lilac
ge	=	yellow

- E86 – Multi-Function Indicator Mode Select Switch
- E109 – Multi-Function Indicator Memory Switch
- G17 – Outside Air Temperature Sensor (00779-3133)
- G22 – Speedometer Vehicle Speed Sensor (VSS) (01086), on transmission
- K19 – Seat Belt Warning Light (00665)
- K94 – Right Turn Signal Indicator Light
- K102 – Rear Hatch/Trunk Lid Ajar Warning Light
- T4b – 4-Pin Connector, behind steering column switch cover
- T28b – 28-Pin Connector, on instrument cluster
- T42 – 42-Pin Connector, on engine
- (15) - Ground connection, on cylinder head
- (182) - Ground connection -1-, in engine compartment wiring harness (6 cylinder)

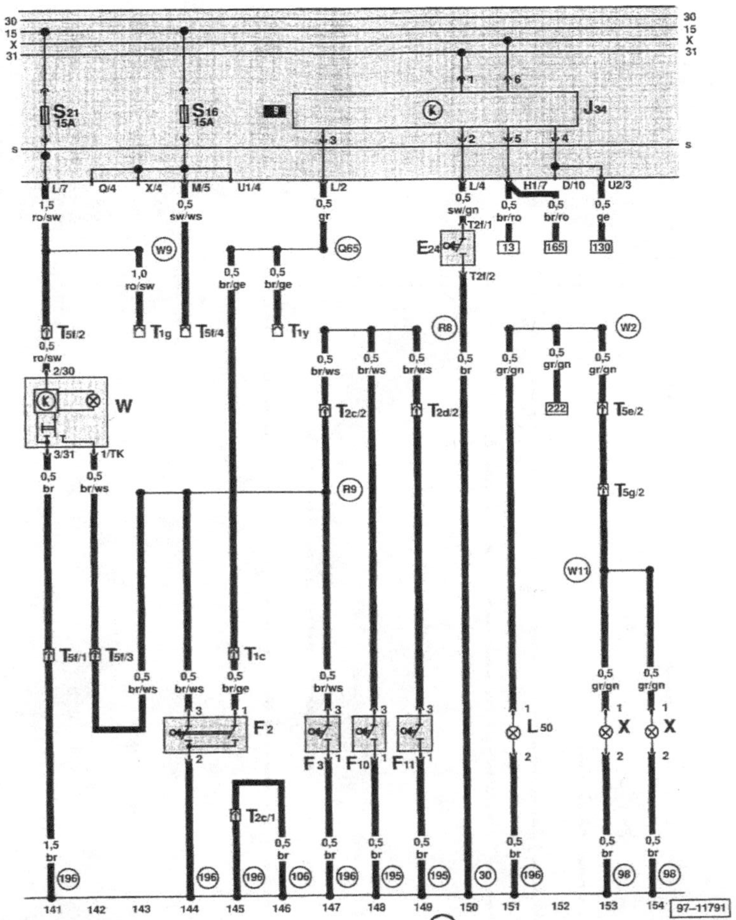

WIRING COLOR CODE

ws	=	white
sw	=	black
ro	=	red
br	=	brown
gn	=	green
bl	=	blue
gr	=	grey
li	=	lilac
ge	=	yellow

- E24 – Left Seat Belt Switch (00591)
- F2 – Left Front Door Contact Switch
- F3 – Right Front Door Contact Switch
- F10 – Left Rear Door Contact Switch
- F11 – Right Rear Door Contact Switch
- J34 – Seat Belt Warning System Relay
- L50 – Center Rear Ashtray Light
- T1c – Single Connector, behind fuse/relay panel
- T1g – Single Connector, behind fuse/relay panel
- T1y – Single Connector, behind fuse/relay panel
- T2c – Double Connector, on right A-pillar, lower part
- T2d – Double Connector, on left A-pillar, lower part
- T2f – Double Connector, behind console
- T5e – 5-Pin Connector, in luggage compartment, left rear
- T5f – 5-Pin Connector, behind fuse/relay panel
- T5g – 5-Pin Connector, on rear apron
- W – Front Interior Light
- X – License Plate Light
- (30) – Ground connection -1-, beside fuse/relay panel
- (98) – Ground connection, in rear lid wiring harness
- (106) – Ground connection -1-, in alarm system wiring harness
- (195) – Ground connection, in rear door contact switch wiring harness
- (196) – Ground connection -1-, in rear wiring harness
- (Q65) – Wire connection, in seat belt warning system wiring harness
- (R8) – Wire connection (door contact switch), in rear door contact switch wiring harness
- (R9) – Wire connection (door contact switch), in rear door wiring harness
- (W2) – Plus connection (58), in rear wiring harness
- (W9) – Plus connection (30), in rear wiring harness
- (W11) – Wire connection (58), in rear lid wiring harness

1993–1994

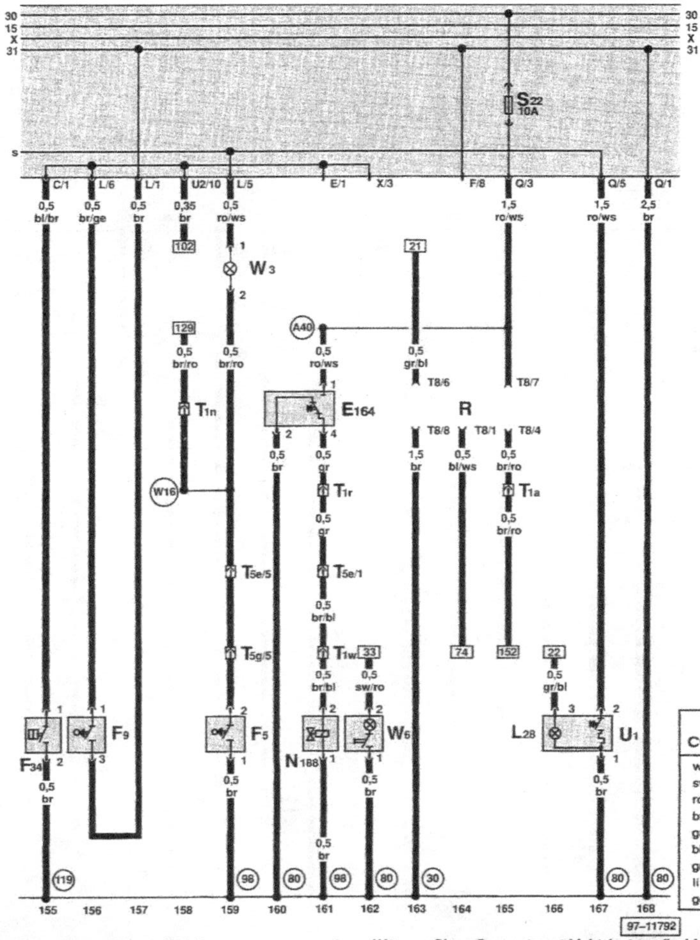

WIRING COLOR CODE

ws	=	white
sw	=	black
ro	=	red
br	=	brown
gn	=	green
bl	=	blue
gr	=	grey
li	=	lilac
ge	=	yellow

- E164 – Luggage Compartment Release Switch
- F5 – Luggage Compartment Light Switch
- F9 – Parking Brake Warning Light Switch (01114)
- F34 – Brake Fluid Level Warning Switch
- L28 – Cigarette Lighter Light
- N188 – Luggage Compartment Release Solenoid
- R – Radio
- T1a – Single Connector, behind fuse/relay panel
- T1r – Single Connector, behind fuse/relay panel
- T1w – Single Connector, on rear apron
- T1h – Single Connector, behind fuse/relay panel
- T5e – 5-Pin Connector, in luggage compartment, left rear
- T5g – 5-Pin Connector, on rear apron
- T8 – 8-Pin Connector, on radio
- U1 – Cigarette Lighter
- W3 – Luggage Compartment Light
- W6 – Glove Compartment Light (not applicable on vehicles with passenger side airbag)
- (30) – Ground connection -1-, beside fuse/relay panel
- (80) – Ground connection -1-, in instrument panel wiring harness
- (98) – Ground connection, in rear lid wiring harness
- (119) – Ground connection -1-, in headlight wiring harness
- (A40) – Plus connection (30), in instrument cluster wiring harness
- (W16) – Wire connection, in rear lid wiring harness

1993–1994

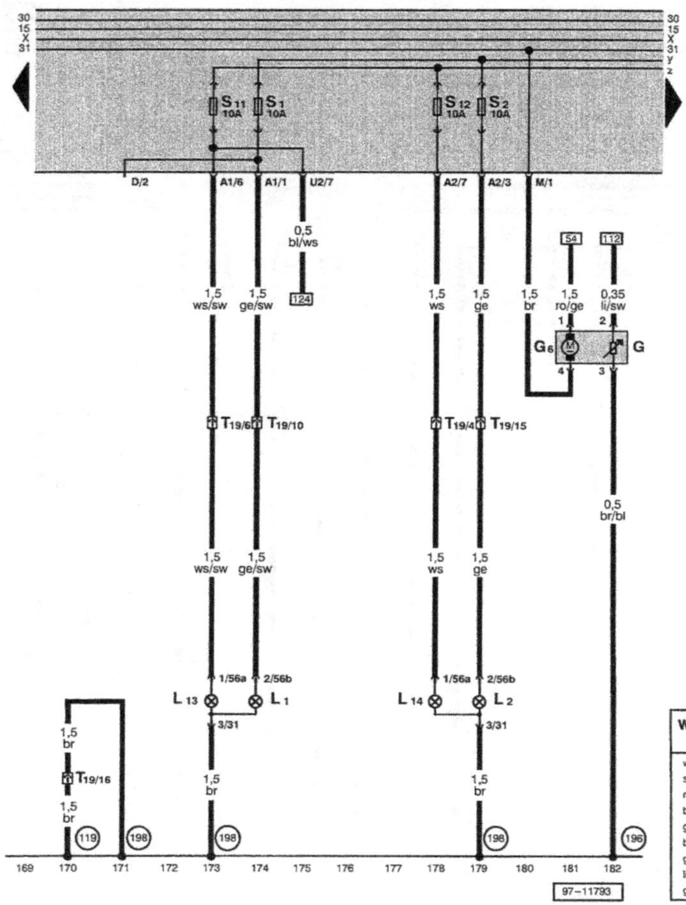

WIRING COLOR CODE		
ws	=	white
sw	=	black
ro	=	red
br	=	brown
gn	=	green
bl	=	blue
gr	=	grey
li	=	lilac
ge	=	yellow

G – Fuel Level Sensor
G6 – Fuel Pump (FP)
L1 – Left Headlight
L2 – Right Headlight
L13 – Left High Beam Headlight
L14 – Right High Beam Headlight
T19 – 19-Pin Connector, behind radiator grill, center
(119) – Ground connection –1–, in headlight wiring harness
(196) – Ground connection –1–, in rear wiring harness
(198) – Ground connection, in front light wiring harness

1993–1994

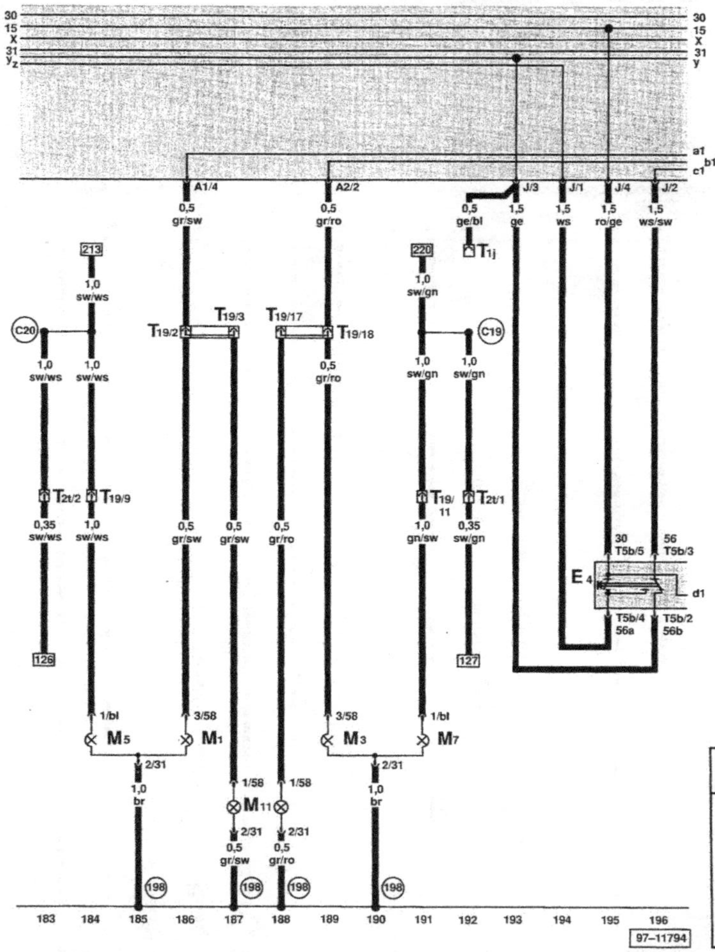

WIRING COLOR CODE		
ws	=	white
sw	=	black
ro	=	red
br	=	brown
gn	=	green
bl	=	blue
gr	=	grey
li	=	lilac
ge	=	yellow

E4 – Headlight Dimmer/Flasher Switch
M1 – Left Parking Light
M3 – Right Parking Light
M5 – Left Front Turn Signal Light
M7 – Right Front Turn Signal Light
M11 – Front Side Marker Lights
T1j – Single Connector, behind fuse/relay panel
T2t – Double Connector, behind fuse/relay panel
T5b – 5-Pin Connector, behind steering column switch cover
T19 – 19-Pin Connector, behind radiator grille, right
(198) – Ground connection, in front light wiring harness
(C19) – Wire connection (right turn signal), in headlight wiring harness
(C20) – Wire connection (left turn signal), in headlight wiring harness

1993-1994

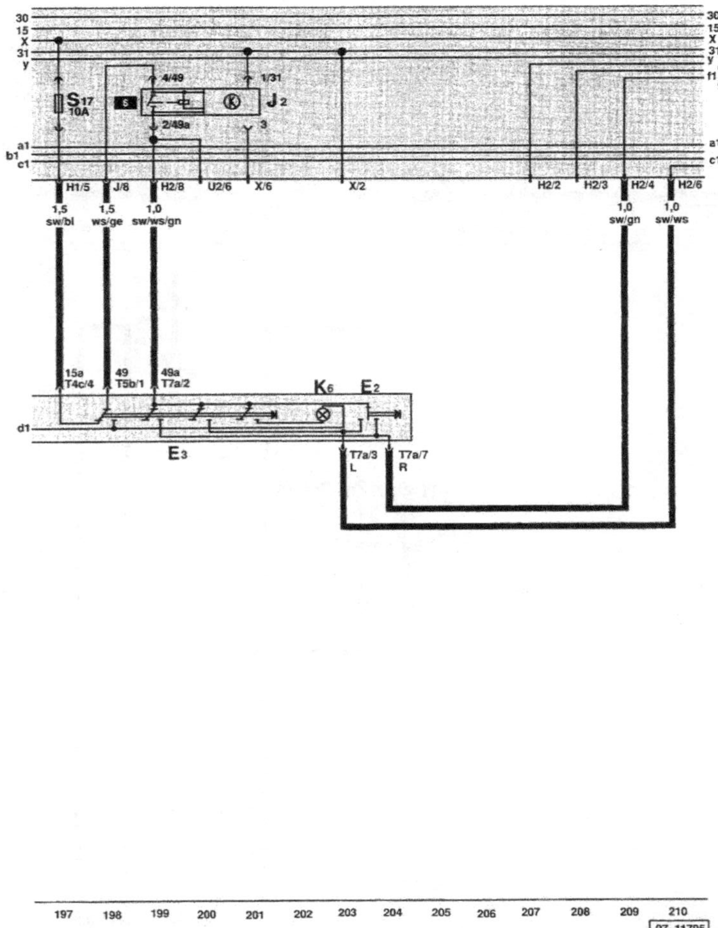

WIRING COLOR CODE

ws	=	white
sw	=	black
ro	=	red
br	=	brown
gn	=	green
bl	=	blue
gr	=	grey
li	=	lilac
ge	=	yellow

E2 – Turn Signal Switch
E3 – Emergency Flasher Switch
J2 – Emergency Flasher Relay
K6 – Emergency Flasher Warning Light

T4c – 4-Pin Connector, behind steering column switch cover
T5b – 5-Pin Connector, behind steering column switch cover
T7a – 7-Pin Connector, behind steering column switch cover

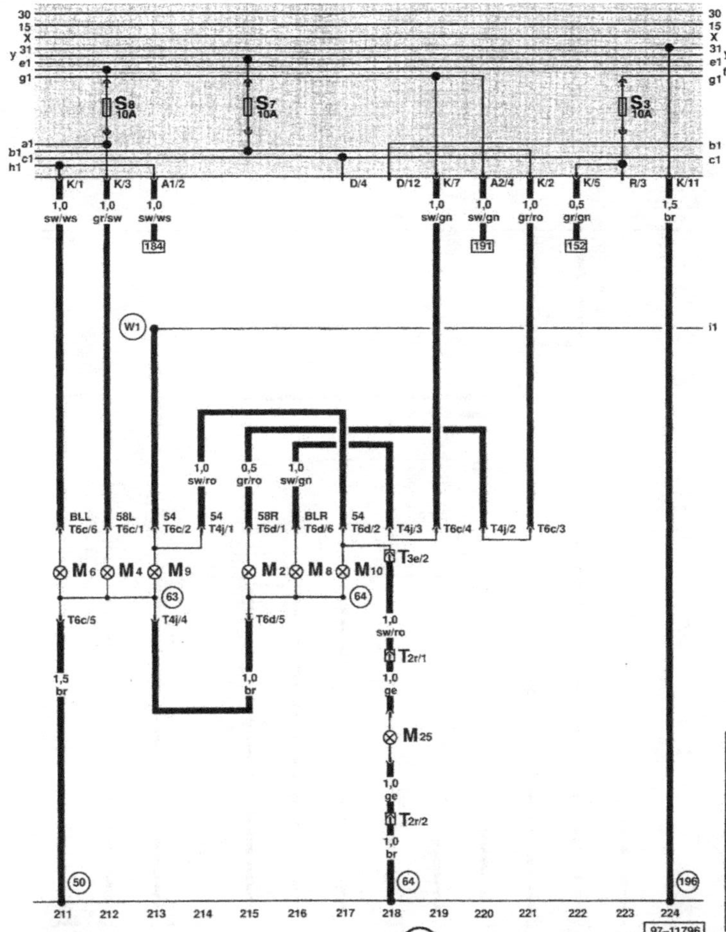

WIRING COLOR CODE

ws	=	white
sw	=	black
ro	=	red
br	=	brown
gn	=	green
bl	=	blue
gr	=	grey
li	=	lilac
ge	=	yellow

M2 – Right Taillight
M4 – Left Taillight
M6 – Left Rear Turn Signal Light
M8 – Right Rear Turn Signal Light
M9 – Left Brake Light
M10 – Right Brake Light
M25 – High-Mount Brake Light
T2r – Double Connector, in luggage compartment, right rear
T3e – 3-Pin Connector, on right taillight
T4j – 4-Pin Connector, on left taillight
T6c – 6-Pin Connector, on left taillight
T6d – 6-Pin Connector, on right taillight

(50) – Ground connection, in luggage compartment, left
(63) – Ground connection, on left taillight bulb holder
(64) – Ground connection, on right taillight bulb holder
(196) – Ground connection -1-, in rear wiring harness
(W1) – Plus connection (54), in rear wiring harness

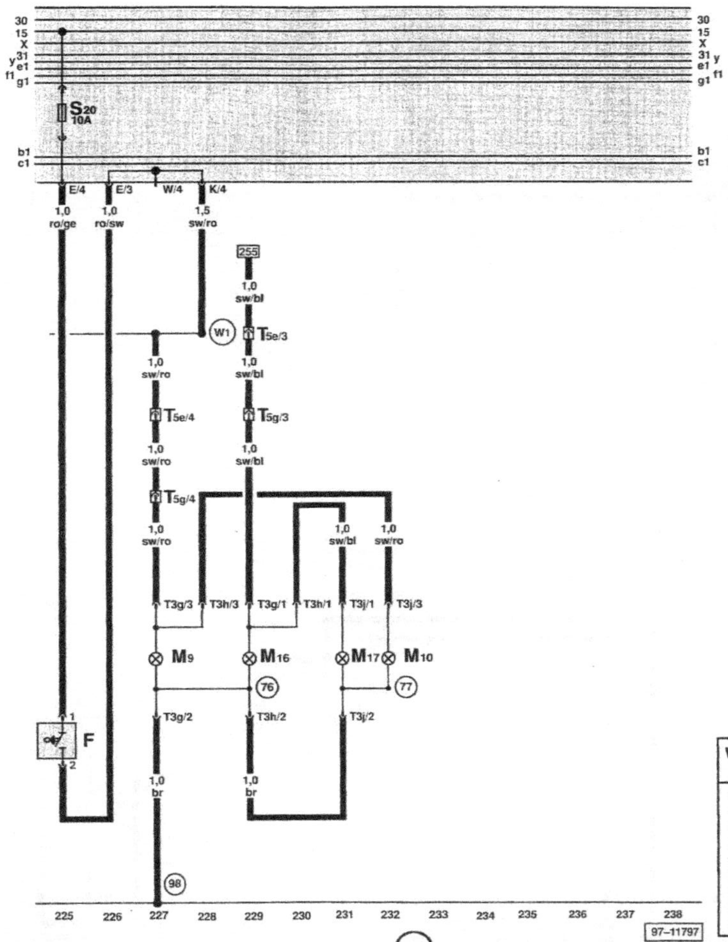

WIRING COLOR CODE

ws	=	white
sw	=	black
ro	=	red
br	=	brown
gn	=	green
bl	=	blue
gr	=	grey
li	=	lilac
ge	=	yellow

F – Brake Light Switch (00526-2131)
M9 – Left Brake Light
M10 – Right Brake Light
M16 – Left Back-Up Light
M17 – Right Back-Up Light
T3j – 3-Pin Connector, on right taillight
T3h – 3-Pin Connector, on left taillight
T3g – 3-Pin Connector, on left taillight
T5e – 5-Pin Connector, on rear apron
T5g – 5-Pin Connector, on rear apron

(76) – Ground connection, on left taillight bulb holder
(77) – Ground connection, on right taillight bulb holder
(98) – Ground connection, in rear lid wiring harness
(W1) – Plus connection (54), in rear wiring harness

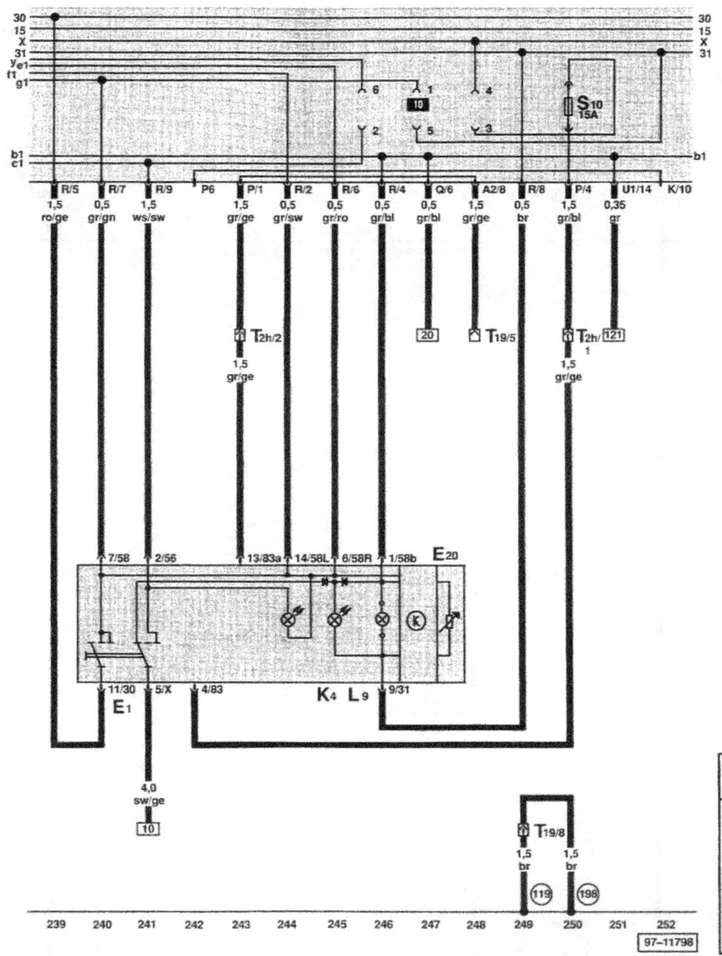

WIRING COLOR CODE

ws	=	white
sw	=	black
ro	=	red
br	=	brown
gn	=	green
bl	=	blue
gr	=	grey
li	=	lilac
ge	=	yellow

E1 – Light Switch
E20 – Instrument Panel Light Dimmer Switch
K4 – Park Light Indicator Light
L9 – Headlight Switch Light
T2h – Double Connector, behind fuse/relay panel

T19 – 19-Pin Connector, behind radiator grille, right
(119) – Ground connection -1-, in headlight wiring harness
(198) – Ground connection, in front light wiring harness

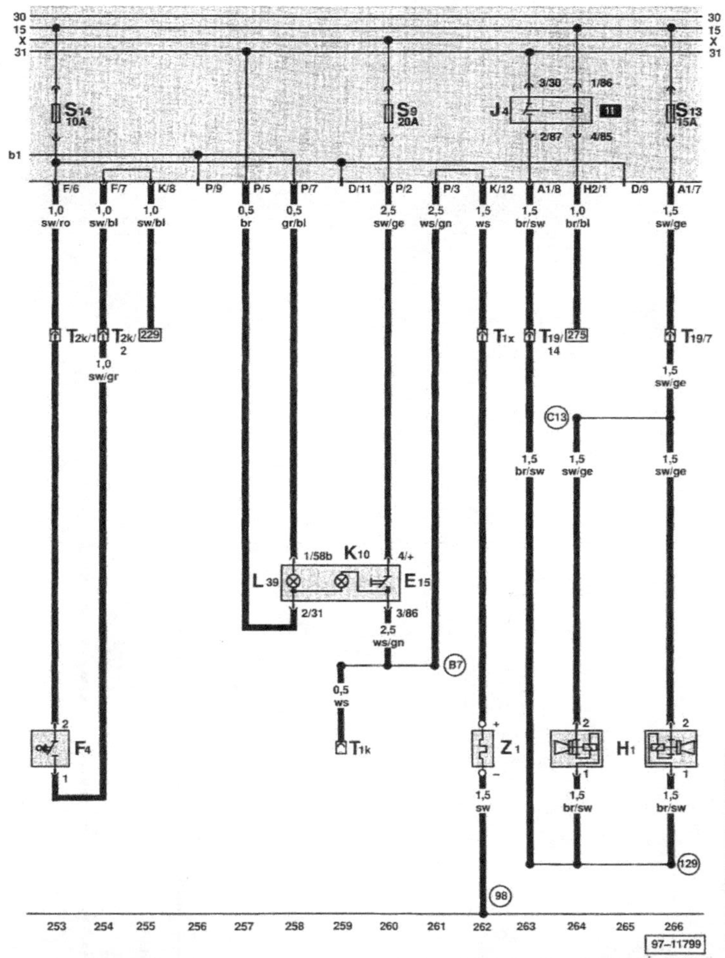

WIRING COLOR CODE

ws	=	white
sw	=	black
ro	=	red
br	=	brown
gn	=	green
bl	=	blue
gr	=	grey
li	=	lilac
ge	=	yellow

E15 – Rear Window Defogger Switch
F4 – Back-Up Light Switch
H1 – Dual Horns
J4 – Dual Horn Relay
K10 – Rear Window Defogger Indicator Light
L39 – Rear Window Defogger Switch Light
T1x – Single Connector, in luggage compartment, left
T1k – Single Connector, behind fuse/relay panel
T2k – Double Connector, behind fuse/relay panel

T19 – 19-Pin Connector, behind radiator grille, right
Z1 – Rear Window Defogger/Heat Element
(129) – Ground connection, in dual-tone horn wiring harness
(B7) – Wire connection (86), in rear window defogger wiring harness
(C13) – Plus connection, in dual-tone horn wiring harness

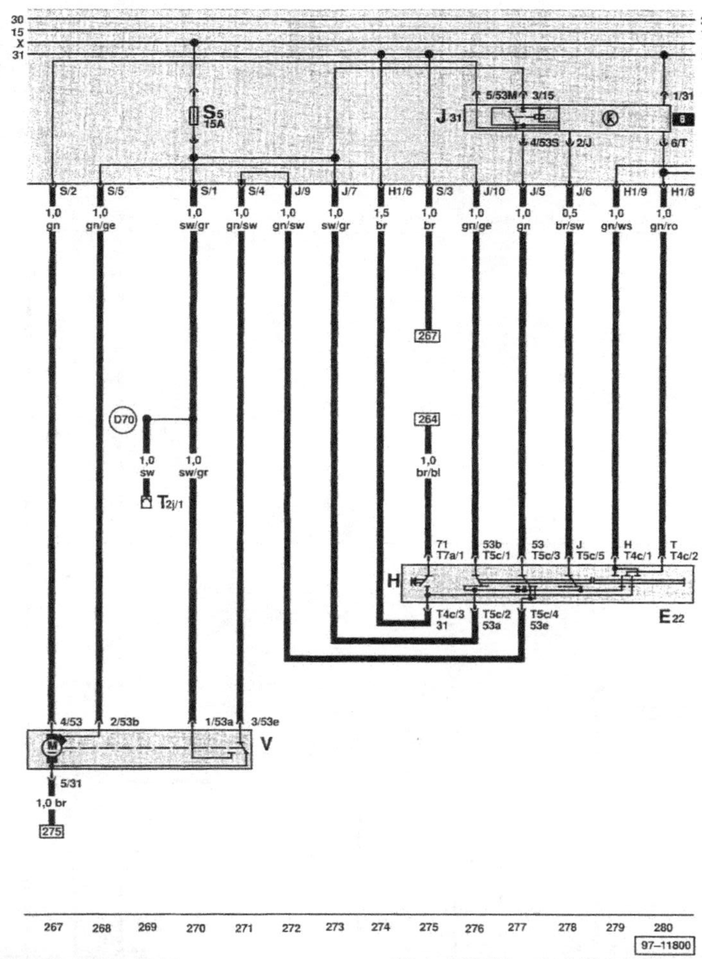

WIRING COLOR CODE

ws	=	white
sw	=	black
ro	=	red
br	=	brown
gn	=	green
bl	=	blue
gr	=	grey
li	=	lilac
ge	=	yellow

E22 – Windshield Wiper/Washer Switch
H – Horn Button
J31 – Wiper/Washer Intermittent Relay
T2j – Double Connector, behind fuse/relay panel
T4c – 4-Pin Connector, on E22
T5c – 5-Pin Connector, behind steering column switch cover

T7a – 7-Pin Connector, behind steering column switch cover
V – Windshield Wiper Motor
(D70) – Wire connection (53a), in engine compartment wiring harness

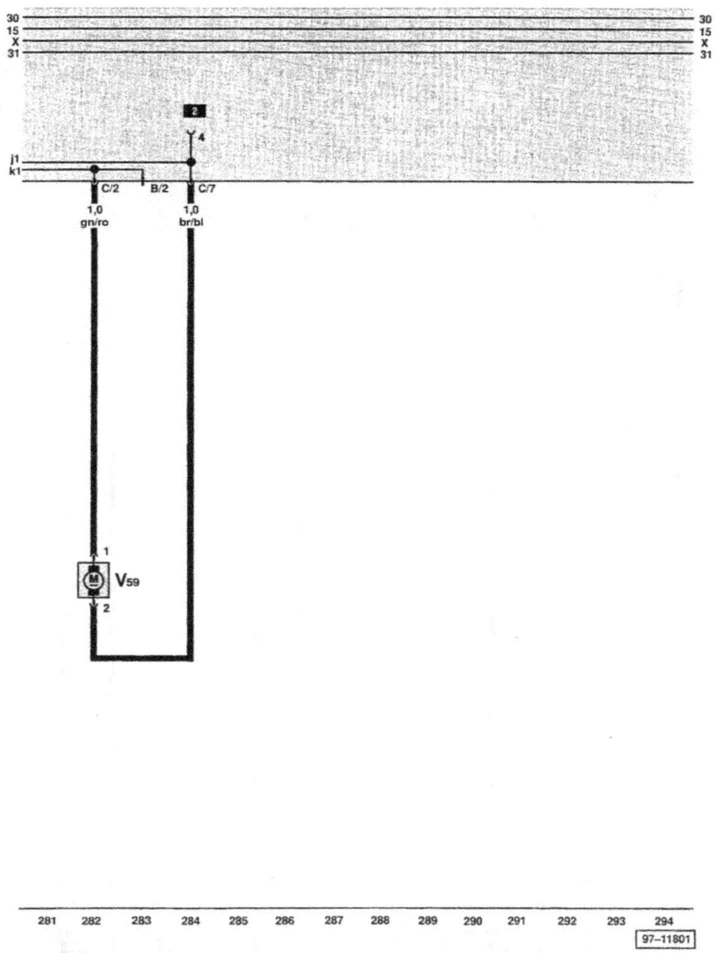

WIRING COLOR CODE		
ws	=	white
sw	=	black
ro	=	red
br	=	brown
gn	=	green
bl	=	blue
gr	=	grey
li	=	lilac
ge	=	yellow

V59 – Windshield and Rear Window Washer Pump

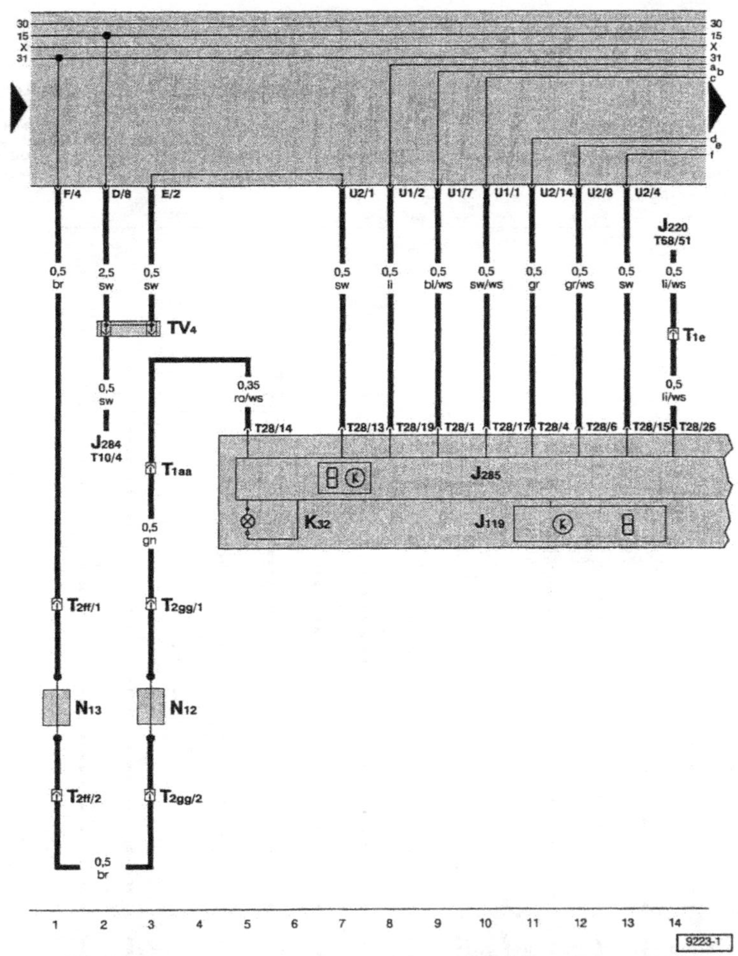

WIRING COLOR CODE		
ws	=	white
sw	=	black
ro	=	red
br	=	brown
gn	=	green
bl	=	blue
gr	=	grey
li	=	lilac
ge	=	yellow

J119 - Multi-Function Indicator
J220 - Motronic Engine Control Module (ECM)
J284 - Alarm System Control Module
J285 - Control Module With Display Unit In Instrument Cluster
K32 - Brake Pad Wear Indicator Light
N12 - Brake Pad Wear Indicator, Right (Sheer Element)
N13 - Brake Pad Wear Indicator, Left (Sheer Element)
T1aa - Connector, Single, behind fuse/relay panel
T1e - Connector, Single, behind fuse/relay panel
T2ff - Connector, Double, behind fuse/relay panel
T2gg - Connector, Double, engine compartment, right
T10 - Connector, 10 Pin, behind dash panel, left, on Alarm System Control Module
T28 - Connector, 28 Pin, on instrument cluster
T68 - Connector, 68 Pin, under right side of bonnet, on Motronic Engine Control Module (ECM)
TV4 - Connector, terminal 15

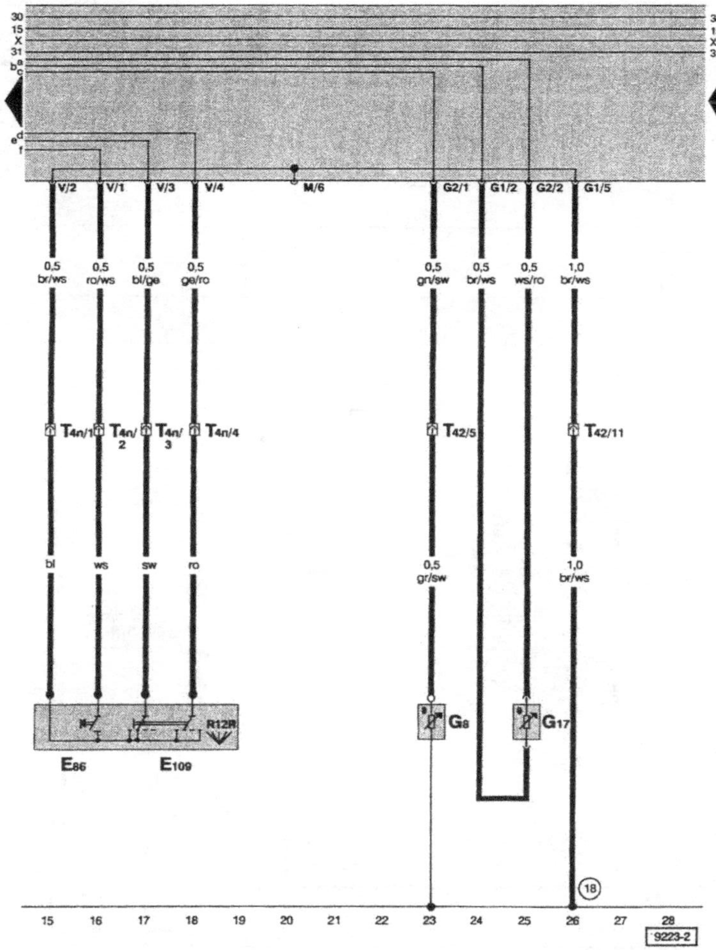

WIRING COLOR CODE		
ws	=	white
sw	=	black
ro	=	red
br	=	brown
gn	=	green
bl	=	blue
gr	=	grey
li	=	lilac
ge	=	yellow

E86 - Multi-Function Indicator Mode Select Switch
E109 - Multi-Function Indicator Memory Switch
G8 - Oil Temperature Sensor
G17 - Outside Air Temperature Sensor
T4n - Connector 4 pin
T42 - Connector, 42 pin, left side of engine
(18) - Ground connection, on engine block

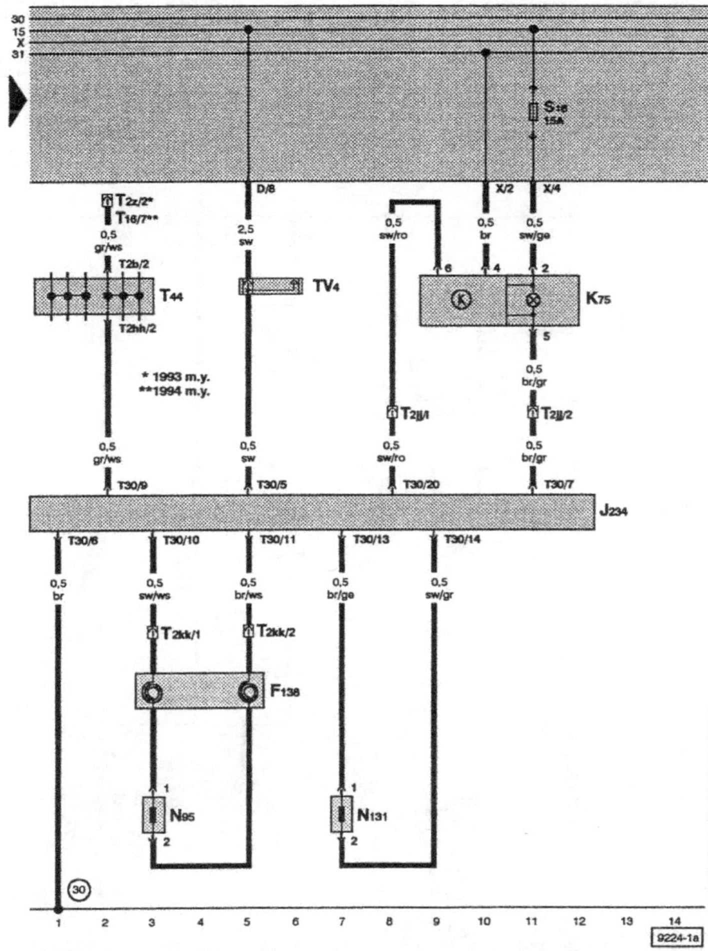

WIRING COLOR CODE		
ws	=	white
sw	=	black
ro	=	red
br	=	brown
gn	=	green
bl	=	blue
gr	=	grey
li	=	lilac
ge	=	yellow

F138 - Airbag Spiral Spring
J234 - Airbag Control Module
K75 - Airbag Malfunction Indicator Lamp (MIL)
N95 - Driver's Side Airbag Ignitor
N131 - Passenger's Side Airbag Ignitor 1
T2b - Connector, Double, on junction connector (T44) above fuse/relay panel
T2hh - Connector, Double, on junction connector (T44) above fuse/relay panel
T2jj - Connector, Double, behind fuse/relay panel
T2kk - Connector, Double, behind fuse/relay panel
T2z - Data Link Connector (DLC), behind center of dash panel
T16 - Data Link Connector (DLC), behind center of dash panel
T30 - Connector, 30 pin, on Airbag Control Module
T44 - Junction Box for On Board Diagnostic, above fuse/relay panel
TV4 - Connector, terminal 15
(30) - Ground connection, beside fuse/relay panel

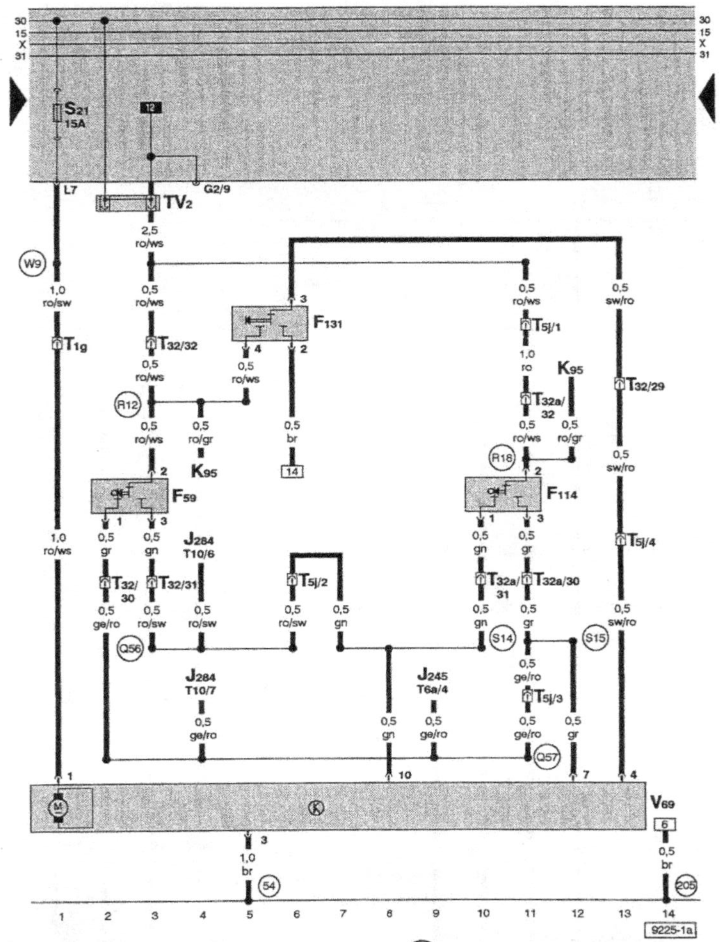

WIRING COLOR CODE

ws	=	white
sw	=	black
ro	=	red
br	=	brown
gn	=	green
bl	=	blue
gr	=	grey
li	=	lilac
ge	=	yellow

F59 - Central Locking System Door Switch, Driver's Door
F114 - Central Locking System Door Switch, Passenger's Door
F131 - Left Front Central Locking System Activator/Switch
J245 - Power Sunroof Control Module
J284 - Alarm System Control Module
K95 - Alarm System Indicator Light
V69 - Central Locking System Pump and Control Module
T1g - Connector, Single, behind fuse/relay panel
T5j - Connector, 5 pin, behind dash, left
T6a - Connector, 6 Pin, behind interior light, on Power Sunroof Control Module
T10 - Connector, 10 Pin, behind dash panel, left, on Alarm System Control Module
T32 - Connector, 32 Pin, on A-pillar, lower left
T32a - Connector, 32 Pin, on A-pillar, lower right
TV2 - Wire connector, terminal 30 (B+)
(54) - Ground connection, on rar apron

(205) - Ground connection, in driver's door wiring harness
(Q56) - Wire connection (open), in alarm system wiring harness
(Q57) - Wire connection (closed), in alarm system wiring harness
(R12) - Plus connection -3- (30), in driver's door wiring harness
(R18) - Plus connection -3- (30), in passenger's door wiring harness
(S14) - Wire connection (open) in central locking system wiring harness
(S15) - Wire connection (closed) in central locking system wiring harness
(W9) - Plus connection (30), in rear wiring harness

1993–1994

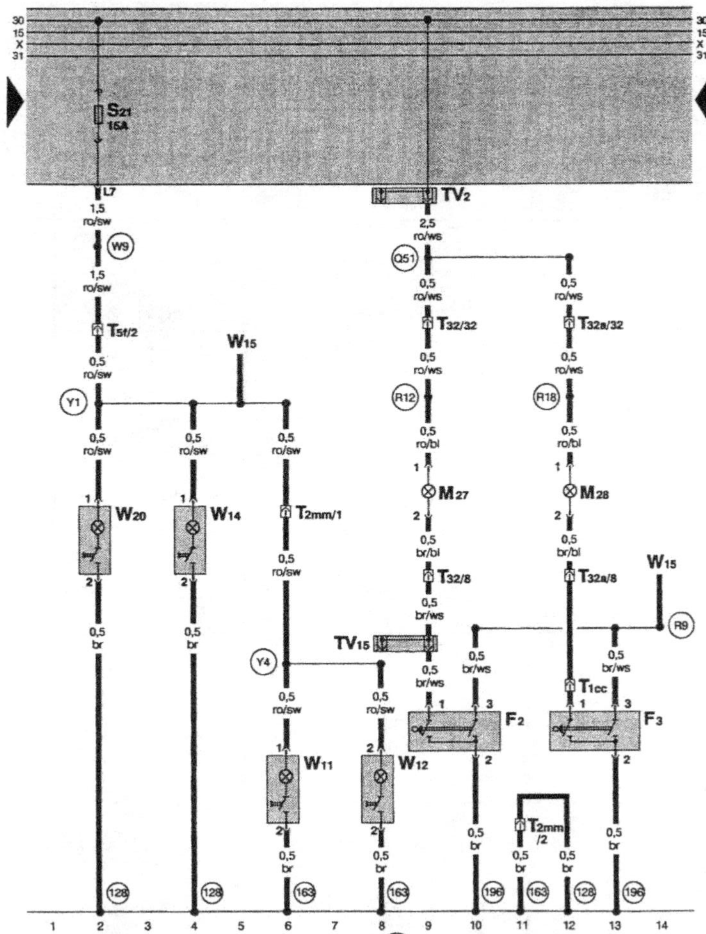

WIRING COLOR CODE

ws	=	white
sw	=	black
ro	=	red
br	=	brown
gn	=	green
bl	=	blue
gr	=	grey
li	=	lilac
ge	=	yellow

F2 - Door Contact Switch, Left Front
F3 - Door Contact Switch, Right Front
M27 - Left Door Warning Light
M28 - Right Door Warning Light
W11 - Reading Light, Left Rear
W12 - Reading Light, Right Rear
W14 - Make-up Mirror Light, Right
W15 - Interior Light with Delay Switch
W20 - Make-up Mirror Light, Left
T1cc - Connector, Single, behind fuse/relay panel
T2mm - Connector, Double
T5f - Connector, 5 Pin, above fuse/relay panel
T32 - Connector, 32 Pin, on A-pillar, lower left
T32a - Connector, 32 Pin, on A-pillar, lower right
TV2 - Wire connector, terminal 30
TV15 - Wire connector, door contact switch
(Q51) - Plus connection (30), in alarm system wiring harness

(R9) - Wire connection (30), in rear wiring harness 9227-4
(R12) - Plus connection -3- (30), in driver's door wiring harness
(R18) - Plus connection -3- (30), in passenger's door wiring harness
(W9) - Plus connection (30), in rear wiring harness
(Y1) - Plus connection (30), in interior light wiring harness
(Y4) - Plus connection (30), in reading light wiring harness
(128) - Ground connection -1-, in interior light wiring harness
(163) - Ground connection -2-, in interior light wiring harness
(196) - Ground connection -1-, in rear wiring harness

1993-1994

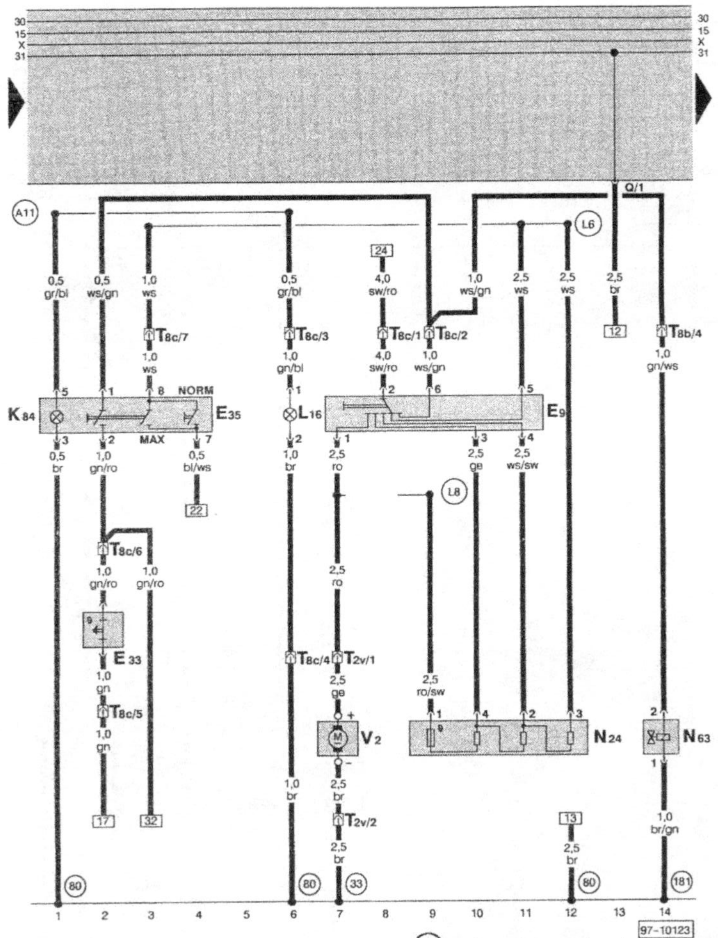

WIRING COLOR CODE		
ws	=	white
sw	=	black
ro	=	red
br	=	brown
gn	=	green
bl	=	blue
gr	=	grey
li	=	lilac
ge	=	yellow

E9 – Fresh Air Blower Switch
E33 – A/C Evaporator Temperature Switch
E35 – A/C Switch
K84 – A/C Indicator Light
L16 – Fresh Air Control Lever Light
N24 – Fresh Air Blower Series Resistance with Fuse
N63 – Fresh Air/Recirculating Flap Two-Way Valve
V2 – Fresh Air Blower
T2v – Connector, Double, below right side of dash at blower motor
T8b – Connector, 8 Pin, behind fuse/relay panel
T8c – Connector, 8 Pin, behind lower center of dash panel
(33) – Ground connection, behind instrument panel, right

(80) – Ground connection -1-, in instrument panel wiring harness
(181) – Ground connection (for A/C), in Engine compartment wiring harness
(A11) – Plus connection (58b), in instrument panel wiring harness
(L6) – Wire connection -2-, in evaporator housing wiring harness
(L8) – Wire connection -4-, in evaporator housing wiring harness

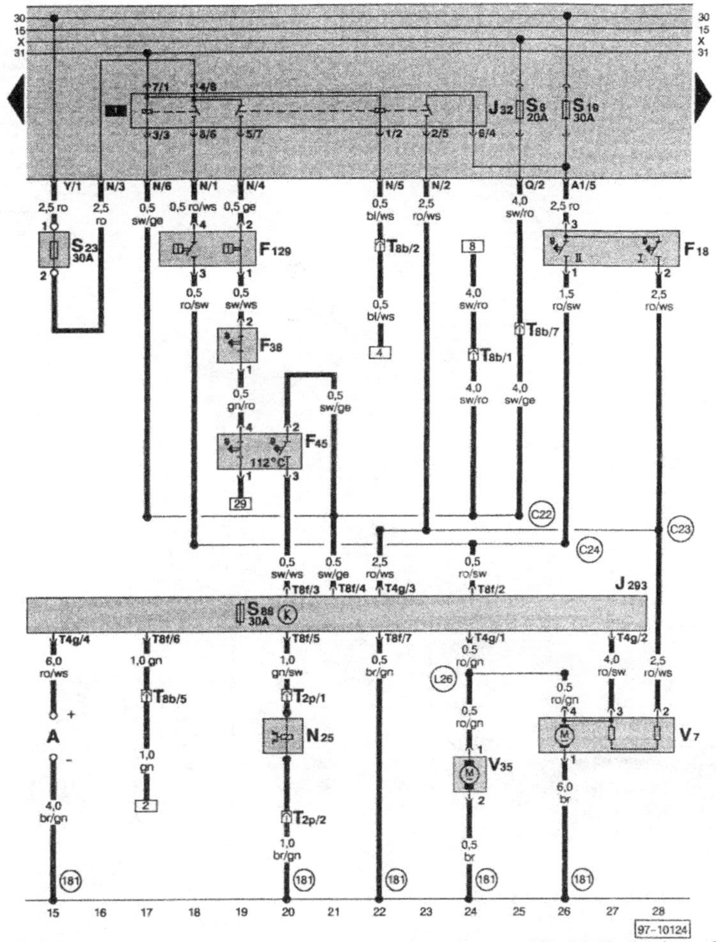

WIRING COLOR CODE		
ws	=	white
sw	=	black
ro	=	red
br	=	brown
gn	=	green
bl	=	blue
gr	=	grey
li	=	lilac
ge	=	yellow

A – Battery
F18 – Coolant Fan Control (FC) Thermal Switch
F38 – Ambient Temperature Switch
F45 – A/C Thermal Switch
F129 – A/C Pressure Switch
J32 – A/C Relay
J293 – Coolant Fan Control (FC) Module
N25 – A/C Clutch
S23 – Heater Fuse
V7 – Coolant Fan
V35 – Coolant Fan, Right
T2p – Connector, Double, on compressor
T4g – Connector, 4 Pin, left side of engine, on Coolant Fan Control (FC) Module
T8b – Connector, 8 Pin, behind fuse/relay panel

T8f – Connector, 8 Pin, left side of engine, on Coolant Fan Control (FC) Module
(181) – Ground connection (for A/C), in Engine compartment wiring harness
(C22) – Wire connection (X), in headlight/air conditioner wiring harness
(C23) – Wire connection (blower, 1st speed) in headlight/air conditioner wiring harness
(C24) – Wire connection (blower, 2nd speed) in headlight/air conditioner wiring harness
(L26) – Wire connection (coolant fan), in Engine compartment wiring harness

1993–1994
1993–1994

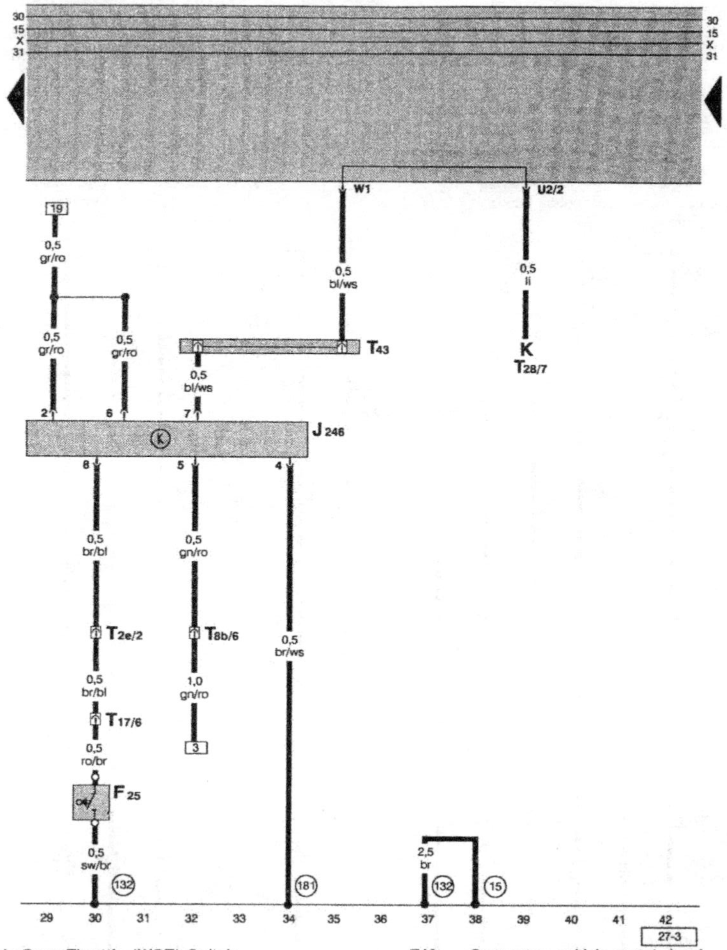

WIRING COLOR CODE		
ws	=	white
sw	=	black
ro	=	red
br	=	brown
gn	=	green
bl	=	blue
gr	=	grey
li	=	lilac
ge	=	yellow

F25 - Wide Open Throttle (WOT) Switch
J246 - A/C Clutch Cut Off Relay
K - Instrument Cluster
T2e - Connector, Double, behind fuse/relay panel, A/C connection
T8b - Connector, 8 pin, behind fuse/relay panel
T17 - Connector, 17 pin, on engine
T28 - Connector, 28 pin on instrument cluster

T43 - Connector, vehicle speed signal
(132) - Ground connection-3-in engine compartment wiring harness
(181) - Ground connection (for A/C), in Engine Compartment Wiring Harness
(15) - Ground connection, on cylinder head

1993–1994

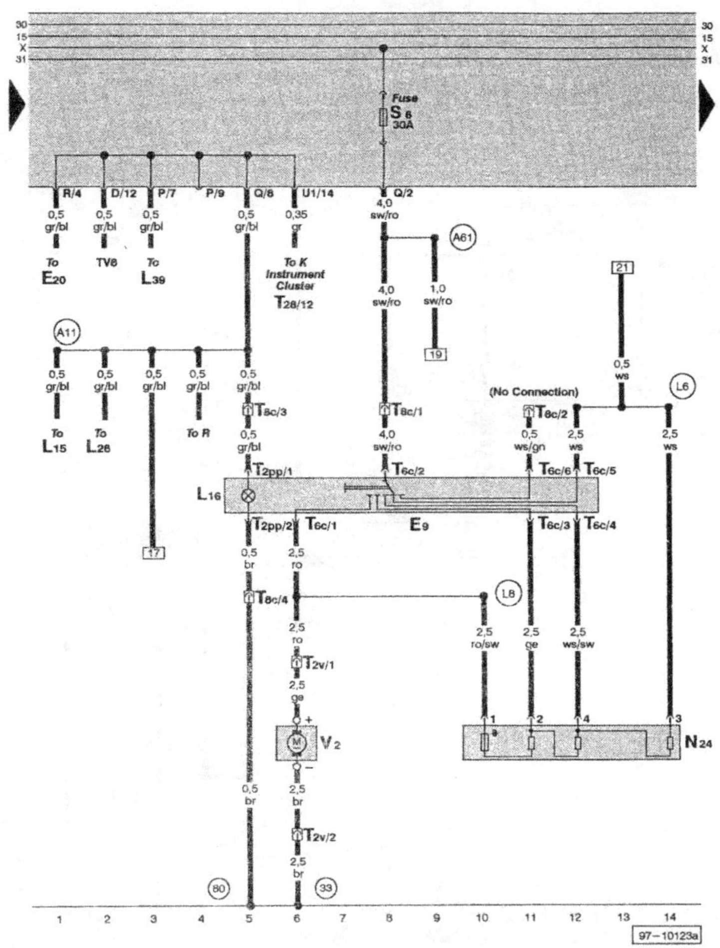

WIRING COLOR CODE		
ws	=	white
sw	=	black
ro	=	red
br	=	brown
gn	=	green
bl	=	blue
gr	=	grey
li	=	lilac
ge	=	yellow

E9 – Fresh Air Blower Switch
E20 – Instrument Panel Light Dimmer Switch
K – Instrument Cluster
L15 – Ashtray Light
L16 – Fresh Air Control Lever Light
L28 – Cigarette Lighter Light
L39 – Rear Window Defogger Switch Light
N24 – Fresh Air Blower Series Resistance with Fuse
R – Radio
V2 – Fresh Air Blower
T2pp – Double Connector, on Fresh Air Blower Switch
T2v – Double Connector, below right side of dash at blower motor
T6c – 6–Pin Connector, on Fresh Air Blower Switch
T8c – 8–Pin Connector, behind lower center of dash panel
T28 – 28–Pin Connector, on Instrument Cluster

TV8 – Terminal 58b Wire Connector, top of fuse/relay panel
(33) – Ground connection, behind instrument panel, right
(80) – Ground connection –1–, in instrument panel wiring harness
(A11) – Plus connection (58b), in instrument panel wiring harness
(A61) – Plus connection (X), in instrument panel wiring harness
(L6) – Wire connection –2–, in evaporator housing wiring harness
(L8) – Wire connection –4–, in evaporator housing wiring harness

1993-1994

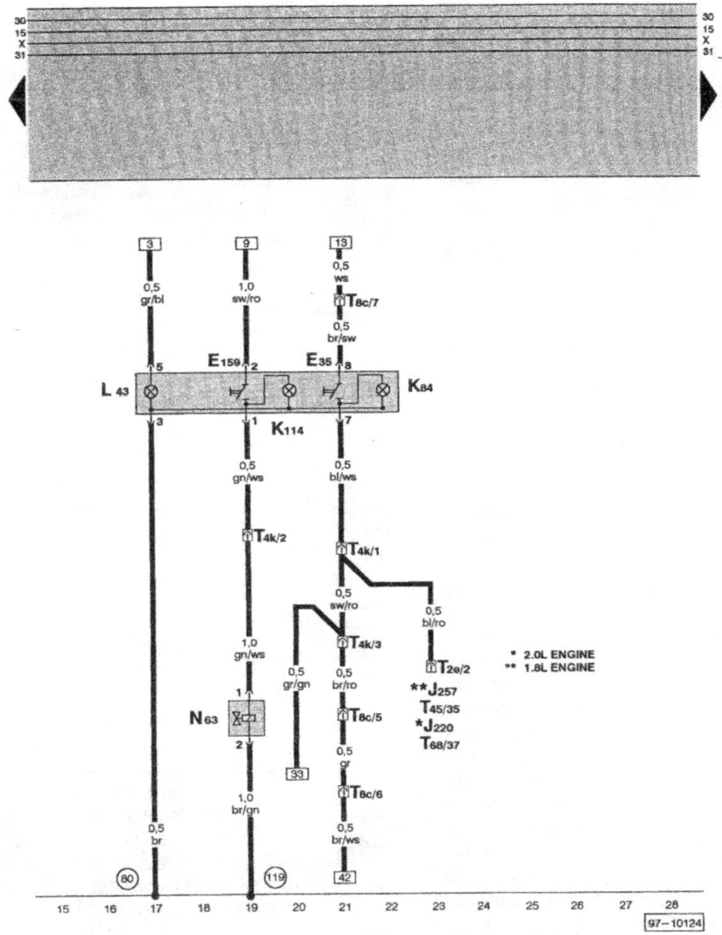

WIRING COLOR CODE		
ws	=	white
sw	=	black
ro	=	red
br	=	brown
gn	=	green
bl	=	blue
gr	=	grey
li	=	lilac
ge	=	yellow

E35 – A/C Switch
E159 – Fresh Air/Recirculating Flap Switch
J220 – Motronic Engine Control Module (ECM)
J257 – Mono–Motronic Engine Control Module (ECM)
K84 – A/C Indicator Light
K114 – Fresh Air and Recirculation Air Mode Indicator Light
L43 – A/C Switch Illumination Light
N63 – Fresh Air/Recirculating Flap Two–Way Valve
T2e – Double Connector, behind fuse/relay panel, A/C Connection

T4k – 4–Pin Connector, behind fuse/relay panel
T8c – 8–Pin Connector, behind lower center of dash panel
T45 – 45–Pin Connector, on Mono–Motronic Engine Control Module (ECM)
T68 – 68–Pin Connector, under right side of hood, on Motronic Engine Control Module
(80) – Ground connection –1–, in instrument panel wiring harness
(119) – Ground connection –1–, in headlight wiring harness

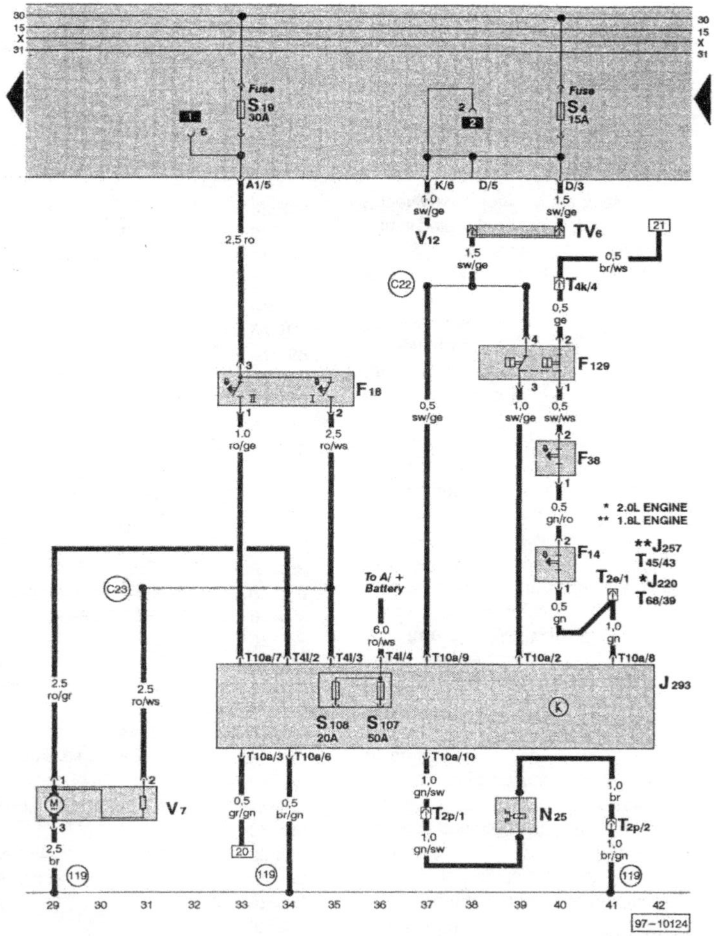

WIRING COLOR CODE		
ws	=	white
sw	=	black
ro	=	red
br	=	brown
gn	=	green
bl	=	blue
gr	=	grey
li	=	lilac
ge	=	yellow

A – Battery
F14 – Engine Coolant Temperature (ECT) Warning Switch
F18 – Coolant Fan Control (FC) Thermal Switch
F38 – Ambient Temperature Switch
F129 – A/C Pressure Switch
J220 – Motronic Engine Control Module (ECM)
J257 – Mono–Motronic Engine Control Module (ECM)
J293 – Coolant Fan Control (FC) Control Module
N25 – A/C Clutch
V7 – Coolant Fan
V12 – Rear Window Wiper Motor
S107 – Coolant Fan Second and Third Speed Fuse
S108 – Coolant Fan First Speed and A/C Clutch Fuse
T2e – Double Connector, behind fuse/relay panel, A/C connection
T2p – Double Connector, on A/C compressor

T4k – 4–Pin Connector, behind fuse/relay panel
T4l – 4–Pin Connector, on Coolant Fan Control (FC) Control Module
T10a – 10–Pin Connector, on Coolant Fan Control (FC) Control Module
T45 – 45–Pin Connector, on Mono–Motronic Engine Control Module (ECM)
T68 – 68–Pin Connector, under right side of hood, on Motronic Engine Control Module (ECM)
TV6 – Terminal X Wire Connector, top of fuse/relay panel
(119) – Ground connection –1–, in headlight wiring harness
(C22) – Wire connection (X), in headlight/air conditioner wiring harness
(C23) – Wire connection (blower, 1st speed), in headlight/air conditioner wiring harness

Golf, Jetta, Cabrio — 2.0L gas engine (ABA)

Golf — 1.8L gas engine (ACC), Canada only

Golf, Jetta — 1.9L Eco diesel engine (AAZ), Canada only

Golf, Jetta — 2.8L gas engine (AAA)

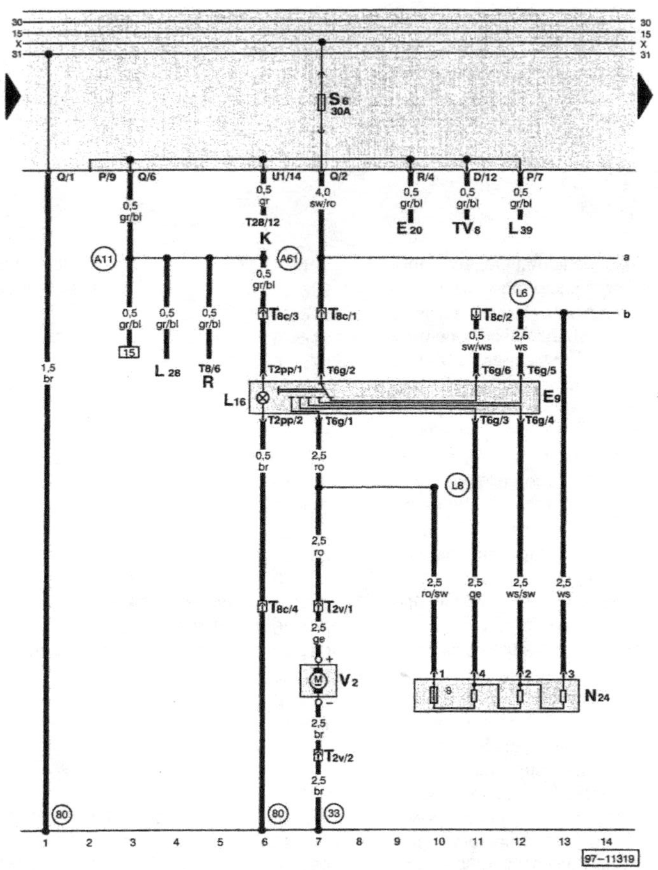

WIRING COLOR CODE

ws	=	white
sw	=	black
ro	=	red
br	=	brown
gn	=	green
bl	=	blue
gr	=	grey
li	=	lilac
ge	=	yellow

- E9 – Fresh Air Blower Switch
- E20 – Instrument Panel Light Dimmer Switch
- K – Instrument Cluster
- L16 – Fresh Air Control Level Light
- L28 – Cigarette Lighter Light
- L39 – Rear Window Defogger Switch Light
- N24 – Fresh Air Blower Series Resistance with Fuse
- R – Radio
- T2pp – Double Connector, on E9
- T2v – Double Connector, behind instrument panel, right
- T6g – 6-Pin Connector, on E9
- T8 – 8-Pin Connector, on radio
- T8c – 8-Pin Connector, behind instrument panel, center
- T28 – 28-Pin Connector, on instrument cluster
- TV8 – Terminal 58b Wire Connector, above fuse/relay panel
- V2 – Fresh Air Blower
- (33) – Ground connection, behind instrument panel, right
- (80) – Ground connection –1–, in instrument panel wiring harness
- (A11) – Plus connection (58b), in instrument panel wiring harness
- (A61) – Plus connection (X), in instrument cluster wiring harness
- (L6) – Wire connection –2–, in evaporator housing wiring harness
- (L8) – Wire connection –4–, in evaporator housing wiring harness

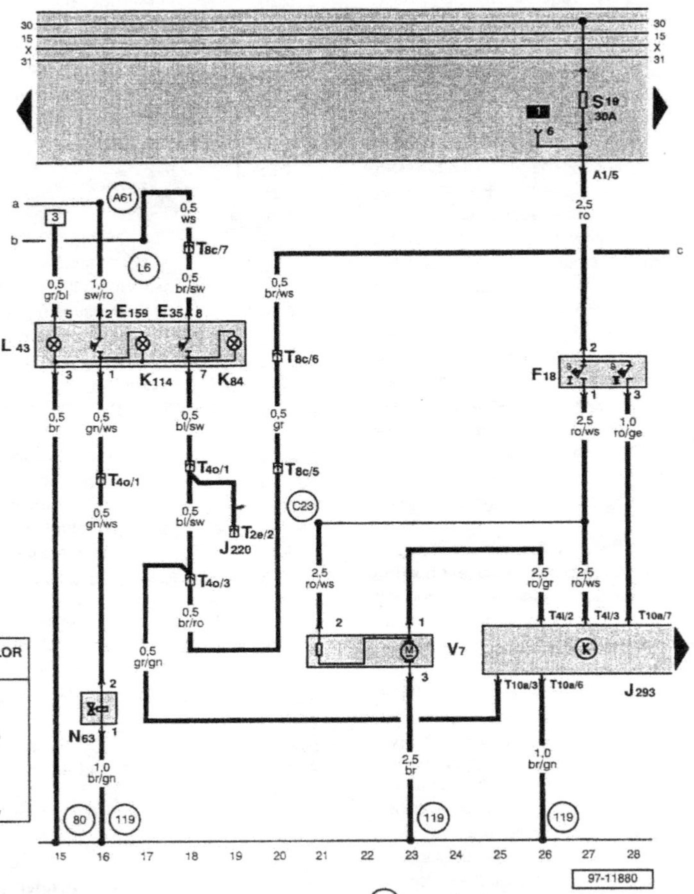

WIRING COLOR CODE

ws	=	white
sw	=	black
ro	=	red
br	=	brown
gn	=	green
bl	=	blue
gr	=	grey
li	=	lilac
ge	=	yellow

- E35 – A/C Switch
- E159 – Fresh Air/Recirculation Flap Switch
- F18 – Coolant Fan Control (FC) Thermal Switch
- J220 – Motronic Engine Control Module (ECM)
- J293 – Coolant Fan Control (FC) Control Module, in left side of engine compartment
- K84 – A/C Indicator Light
- K114 – Fresh Air and Recirculating Air Mode Indicator Light
- L43 – A/C Switch Illumination Light
- N63 – Fresh Air/Recirculating Flap Two-Way Valve (00605), in rear of engine compartment
- T2e – Double Connector, behind fuse/relay panel
- T4l – 4-Pin Connector, on J293
- T4o – 4-Pin Connector, behind fuse/relay panel
- T8c – 8-Pin Connector, behind instrument panel, center
- T10a – 10-Pin Connector, on J293
- V7 – Coolant Fan
- (80) – Ground connection –1–, in instrument panel wiring harness
- (119) – Ground connection –1–, in headlight wiring harness
- (A61) – Plus connection (X), in instrument cluster wiring harness
- (C23) – Wire connection (blower, 1st speed) in headlight/air conditioner wiring harness
- (L6) – Wire connection –2–, in evaporator housing wiring harness

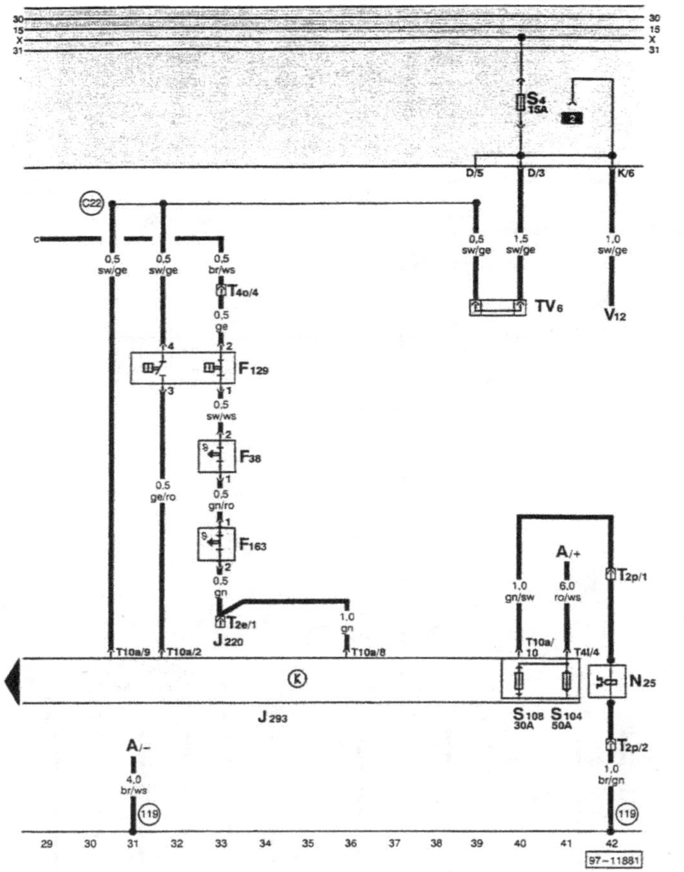

WIRING COLOR CODE		
ws	=	white
sw	=	black
ro	=	red
br	=	brown
gn	=	green
bl	=	blue
gr	=	grey
li	=	lilac
ge	=	yellow

97–11881

A – Battery
F38 – Ambient Temperature Switch, near horn
F87 – After-Run Coolant Fan Control (FC) Thermal Switch
F129 – A/C Pressure Switch, in right side of engine compartment
F163 – A/C Cut-Out Thermal Switch, in center of engine compartment
F165 – Third Speed Coolant Fan Control (FC) Thermal Switch, in center of engine compartment
J220 – Motronic Engine Control Module (ECM)
J293 – Coolant Fan Control (FC) Control Module
N25 – A/C Clutch
S104 – Coolant Fan Second and Third Speed Fuse
S108 – Coolant Fan First Speed and A/C Clutch Fuse

T2p – Double Connector, near compressor
T2e – Double Connector, behind fuse/relay panel (this connector is identified as T2e in the Main Wiring Diagram)
T4l – 4-Pin Connector, on J293
T4o – 4-Pin Connector, behind fuse/relay panel
T10a – 10-Pin Connector, on J293
TV6 – Terminal X Wire Connector, above fuse/relay panel
V12 – Rear Window Wiper Motor (not applicable to Cabrio vehicles)
(119) – Ground connection –1–, in headlight wiring harness
(C22) – Wire connection (X), in headlight/air conditioning wiring harness

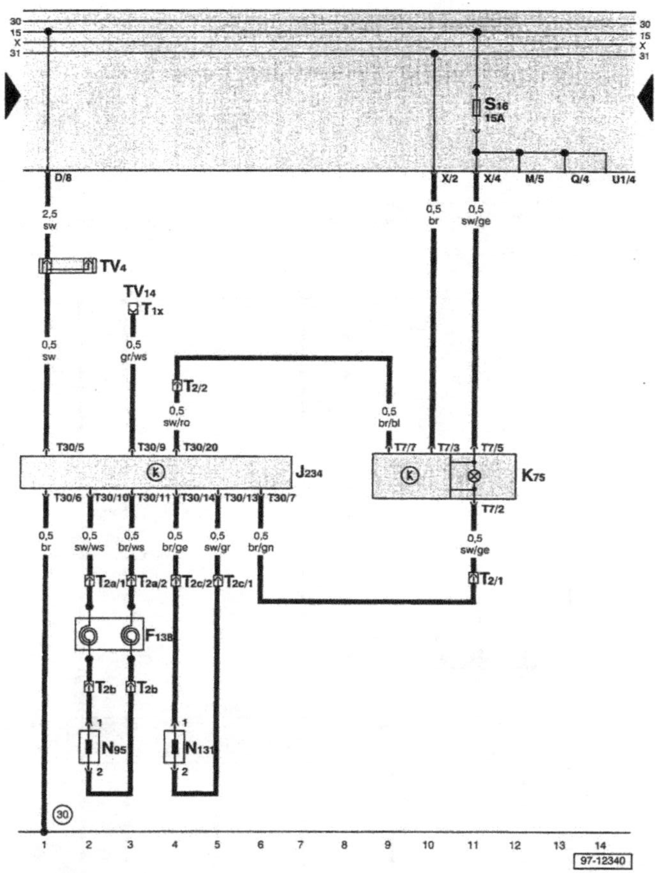

WIRING COLOR CODE		
ws	=	white
sw	=	black
ro	=	red
br	=	brown
gn	=	green
bl	=	blue
gr	=	grey
li	=	lilac
ge	=	yellow

97-12340

F138 – Airbag Spiral Spring/Return Spring With Slop Ring
J234 – Airbag Control Module
K75 – Airbag Malfunction Indicator Lamp (MIL)
N95 – Driver's Side Airbag Igniter (00588)
N131 – Passenger's Side Airbag Igniter 1 (00589)
T2a – Double Connector, behind instrument panel, center
T2b – Double Connector, behind steering column switch cover
T2c – Double Connector, behind Airbag-unit

T2 – Double Connector, behind Airbag-unit
T7 – 7-Pin Connector, near Airbag-unit
T30 – 30-Pin Connector on Airbag Control Module
TV4 – Terminal 15 Wire Connector, above fuse/relay panel
TV14 – Data Link Connector (DLC) wire Connector, above fuse/relay panel
(30) – Ground connection –1–, beside fuse/relay panel

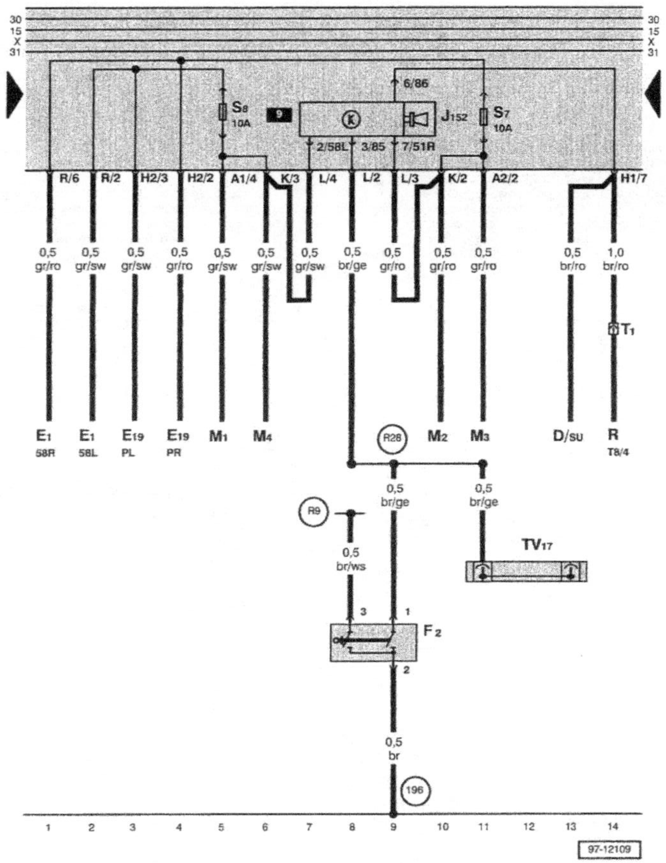

WIRING COLOR CODE

ws	=	white
sw	=	black
ro	=	red
br	=	brown
gn	=	green
bl	=	blue
gr	=	grey
li	=	lilac
ge	=	yellow

D – Ignition/Starter Switch
E1 – Light switch
E19 – Park Light Switch
F2 – Door contact Switch - Driver's side (01169)
J152 – Parking Light Warning Buzzer
M1 – Left Parking Light
M2 – Right Tail Light
M3 – Right Parking Light
M4 – Left Tail Light
R – Radio
T1 – Single Connector
T8 – 8-Pin Connector
TV17 – Door Contact Switch Wire Connector

(196) – Ground connection –1–, in rear wiring harness

(R9) – Wire Connection (door contact switch), in rear door wiring harness

(R28) – Wire Connection –2– (door contact switch), in rear wiring harness

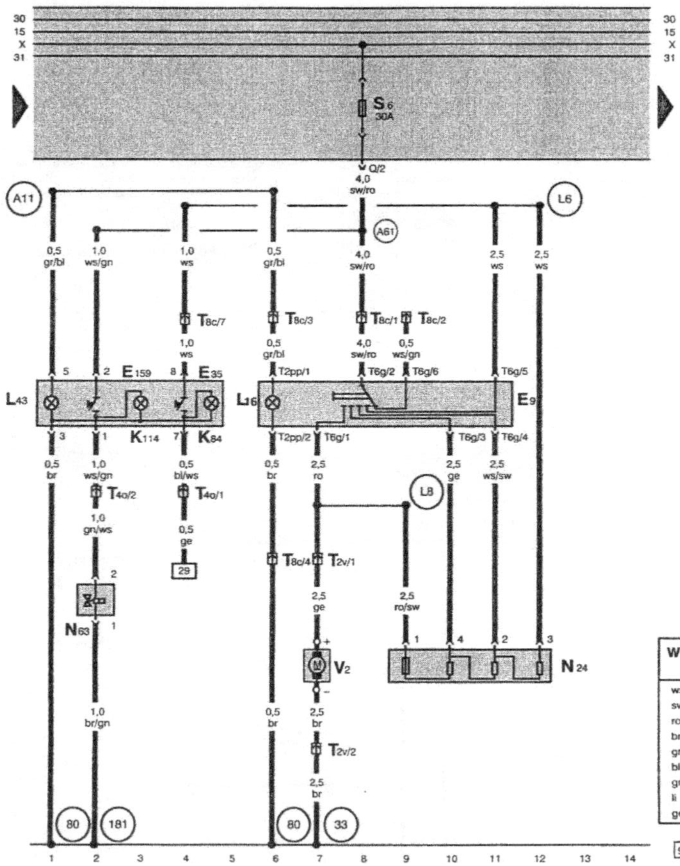

WIRING COLOR CODE

ws	=	white
sw	=	black
ro	=	red
br	=	brown
gn	=	green
bl	=	blue
gr	=	grey
li	=	lilac
ge	=	yellow

E9 – Fresh Air Blower Switch
E35 – A/C Switch
E159 – Fresh Air/Recirculating Flap Switch
K84 – A/C Indicator Light
K114 – Fresh Air and Recirculating Air Mode Indicator Light
L16 – Fresh Air Control Lever Light
L43 – A/C Switch Illumination Light
N24 – Fresh Air Blower Series Resistance with Fuse
N63 – Fresh Air/Recirculating Flap Two-Way Valve
T8c – 8-Pin Connector behind lower center of dash panel
T6g – 6-Pin Connector, on E9
T2pp – Double Connector, on E9
T2v – Double Connector below right side of dash at blower motor
T40 – 40-Pin Connector, behind fuse/relay panel
V2 – Fresh Air Blower

(33) – Ground connection, behind instrument panel, right

(80) – Ground connection –1–, in instrument panel wiring harness

(A61) – Plus connection (X), in instrument cluster wiring harness

(181) – Ground connection (for A/C), in engine compartment, left wiring harness

(A11) – Plus connection (58b), in instrument panel wiring harness

(L6) – Wire connection –2–, in evaporator housing wiring harness

(L8) – Wire connection –4–, in evaporator housing wiring harness

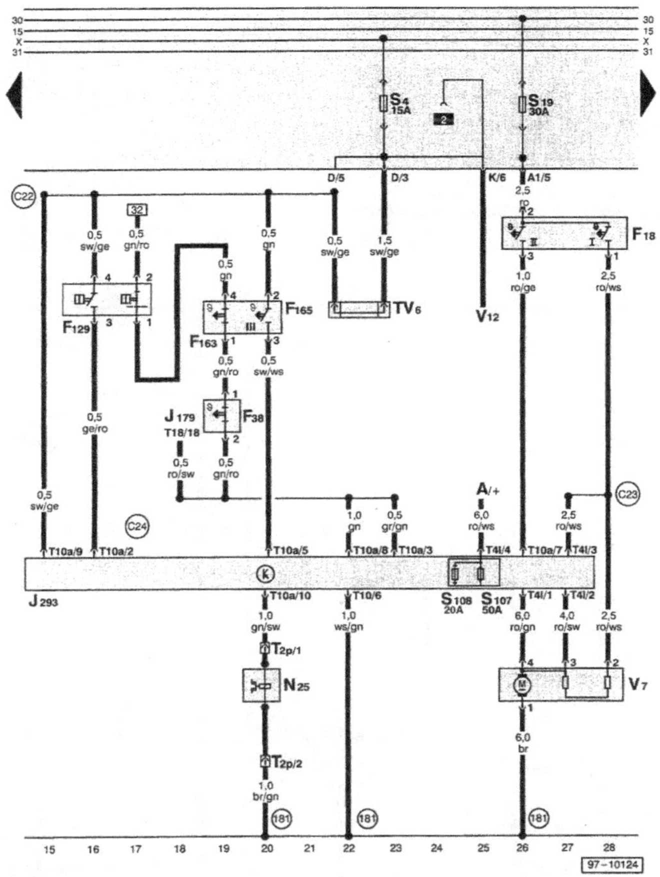

WIRING COLOR CODE

ws	=	white
sw	=	black
ro	=	red
br	=	brown
gn	=	green
bl	=	blue
gr	=	grey
li	=	lilac
ge	=	yellow

A – Battery
F18 – Coolant Fan Control (FC) Thermal Switch
F38 – Ambient Temperature Switch
F129 – A/C Pressure Switch
F163 – A/C Cut-Out Thermal Siwtch, in center of engine compartment
F165 – Third Speed Coolant Fan Control (FC) Thermal Switch, in center of engine compartment
J179 – Automatic Glow Time Control Module
J293 – Coolant Fan Control (FC) Module
N25 – A/C Clutch
S107 – Coolant Fan Second and Third Speed Fuse
S108 – Coolant Fan First Speed and A/C Clutch Fuse
T2p – Double Connector, on compressor
T4l – 4-Pin Connector, left side of engine, on Coolant Fan Control (FC) Module
T10 – 10-Pin Connector, left side of engine, on Coolant Fan Control (FC) Module
T10a – 10-Pin Connector, left side of engine, on Coolant Fan Control (FC) Module
T18 – 18-Pin Connector, on Automatic Glow Time Control Module
TV6 – Terminal X Wire Connector, above fuse/relay panel
V7 – Coolant Fan
V12 – Rear Window Wiper Motor
(181) – Ground connection (for A/C), in engine compartment wiring harness
(C22) – Wire connection (X), in headlight/air conditioner wiring harness
(C23) – Wire connection (blower, 1st speed), in headlight/air conditioner wiring harness
(C24) – Wire connection (blower, 2nd speed), in headlight/air conditioner wiring harness

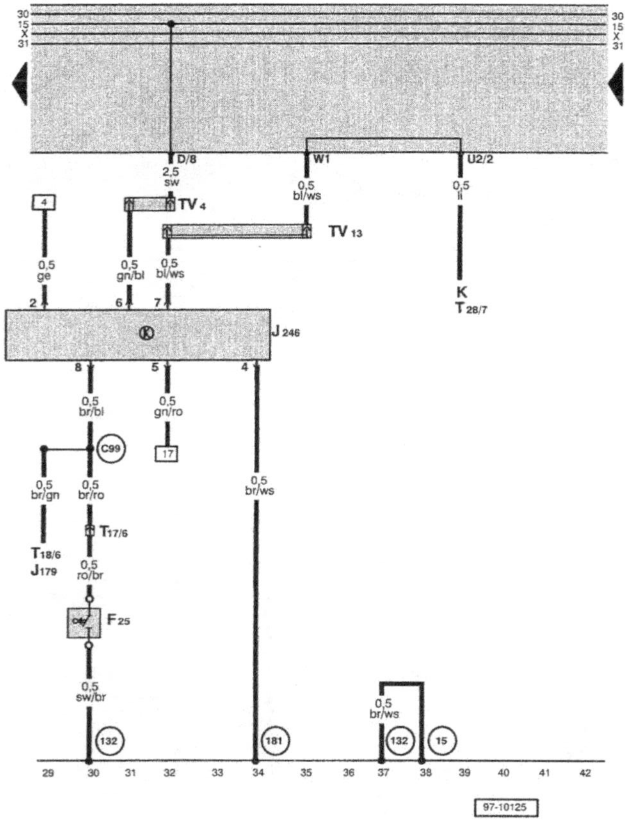

WIRING COLOR CODE

ws	=	white
sw	=	black
ro	=	red
br	=	brown
gn	=	green
bl	=	blue
gr	=	grey
li	=	lilac
ge	=	yellow

F25 – Wide Open Throttle (WOT) Switch
J179 – Automatic Glow Time Control Module
J246 – A/C Clutch Cut Off Relay
K – Instrument Cluster
T17 – 17-Pin Connector, on engine
T18 – 18-Pin Connector, on Automatic Glow Time Control Module
T28 – 28-Pin Connector, on instrument cluster
TV13 – Connector, vehicle speed signal
TV4 – Terminal 15 Wire Connector
(15) – Ground connection, on cylinder head
(132) – Ground connection –3–, in engine compartment wiring harness
(181) – Ground connection (for A/C), in Engine Compartment Left Wiring Harness
(C99) – Wire connection, in headlight/A/C harness

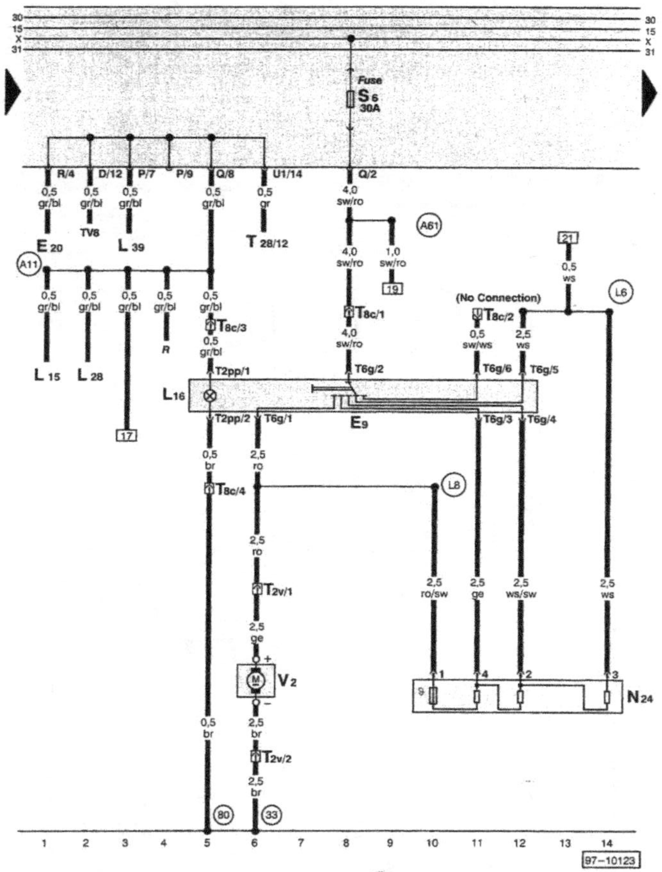

WIRING COLOR CODE		
ws	=	white
sw	=	black
ro	=	red
br	=	brown
gn	=	green
bl	=	blue
gr	=	grey
li	=	lilac
ge	=	yellow

E9 – Fresh Air Blower Switch
E20 – Instrument Panel Light Dimmer Switch
K – Instrument Cluster
L15 – Ashtray Light
L16 – Fresh Air Control Level Light
L28 – Cigarette Lighter Light
L39 – Rear Window Defogger Switch Light
N24 – Fresh Air Blower Series Resistance with Fuse
R – Radio
T2pp – Double Connector, on E9
T2v – Double Connector, behind instrument panel, right
T6g – 6-Pin Connector, on E9
T8c – 8-Pin Connector, behind instrument panel, center
T28 – 28-Pin Connector, on instrument cluster
TV8 – Terminal 58b Wire Connector, above fuse/relay panel
V2 – Fresh Air Blower

(33) – Ground connection, behind instrument panel, right

(80) – Ground connection –1–, in instrument panel wiring harness

(A11) – Plus connection (58b), in instrument panel wiring harness

(A61) – Plus connection (X), in instrument cluster wiring harness

(L6) – Wire connection –2–, in evaporator housing wiring harness

(L8) – Wire connection –4–, in evaporator housing wiring harness

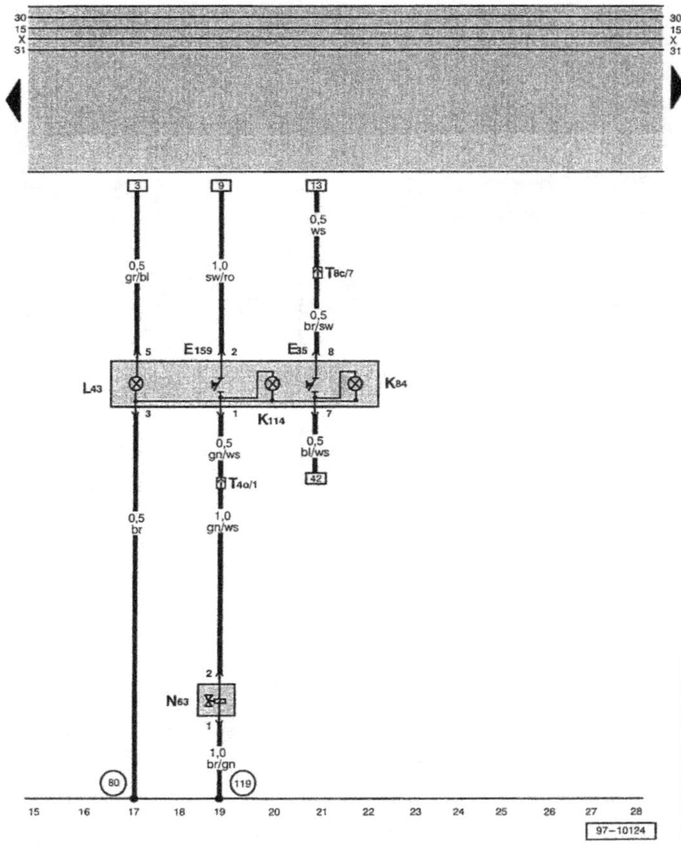

WIRING COLOR CODE		
ws	=	white
sw	=	black
ro	=	red
br	=	brown
gn	=	green
bl	=	blue
gr	=	grey
li	=	lilac
ge	=	yellow

E35 – A/C Switch
E159 – Free Air/Recirculation Flap Switch
K84 – A/C Indicator Light
K114 – Fresh Air and Recirculating Air Mode Indicator Light
L43 – A/C Switch Illumination Light
N63 – Fresh Air/Recirculating Flap Two-Way Valve (00605), in rear of engine compartment
T8c – 8-Pin Connector, behind instrument panel, center
T4o – 4-Pin Connector, Terminal 58/D2 Wire Connector

(80) – Ground connection –1–, in instrument panel wiring harness

(119) – Ground connection –1–, in headlight wiring harness

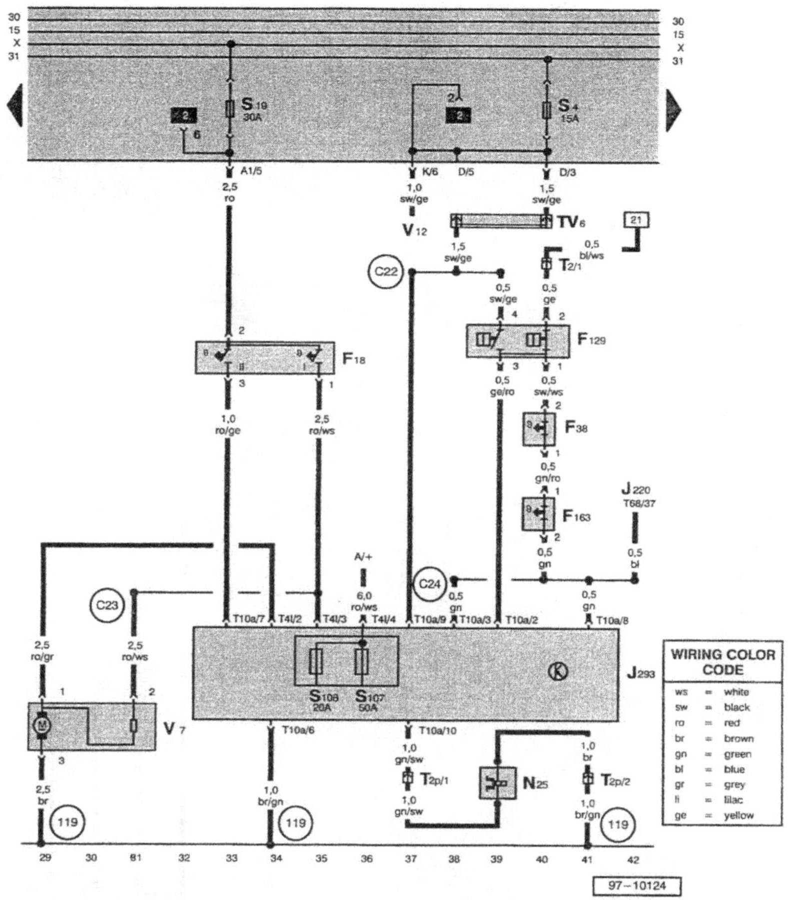

A – Battery
F18 – Coolant Fan Control (FC) Thermal Switch
F38 – Ambient Temperature Switch
F129 – A/C Pressure Switch
F163 – A/C Cut-out Thermal Switch
J220 – Montronic Engine Control Module (ECM)
J293 – Coolant Fan Control (FC) Control Module
N25 – A/C Clutch
S107 – Coolant Fan Second and Third Speed Fuse
S108 – Coolant Fan First Speed and A/C Clutch Fuse
T2p – Double Connector, near compressor
T2 – 2–Pin Connector, behind fuse/relay panel
T4l – 4–Pin Connector, on Coolant Fan Control (FC) Control Module
T10a – 10–Pin Connector, on Coolant Fan Control (FC) Control Module
T68 – 68–Pin Connector, under right side of hood, on Motronic Engine Control Module (ECM)

TV6 – Terminal X Wire Connector, above fuse/relay panel
V7 – Coolant Fan
V12 – Rear Window Wiper Motor
(119) – Ground connection –1–, in headlight wiring harness
(C22) – Wire connection (X), in headlight/air conditioning wiring harness
(C23) – Wire connection (blower, 2nd speed), in headlight/air conditioning wiring harness
(C24) – Wire connection (blower, 2nd speed), in headlight/air conditioning wiring harness

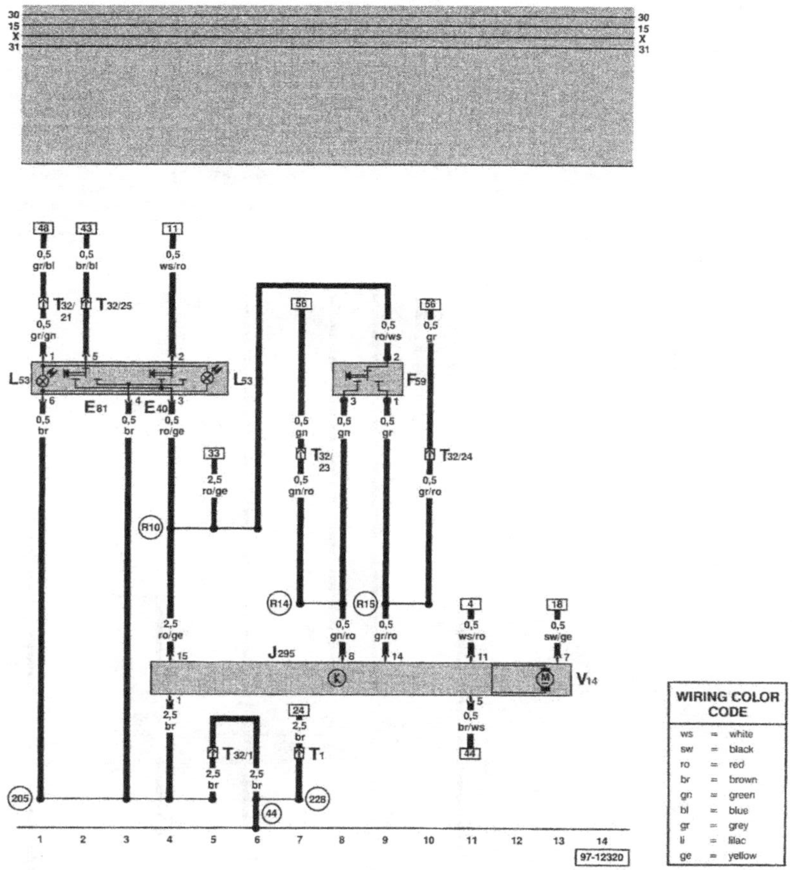

E40 – Left Front Window Switch
E81 – Right Front Window Switch (In RF Door)
F59 – Driver's Door Central Locking System Switch
J295 – Left Front Power Window Control Module
L53 – Power Window Switch Light
T1 – Single Connector, on left A-Pillar, lower part
T32 – 32-Pin Connector, on left A-Pillar, lower part
V14 – Left Window Motor
(44) – Ground Connection, on left A-Pillar, lower part

(205) – Ground Connection, in driver's door wiring harness
(228) – Wire Connection, in door ground wiring harness
(R10) – Plus connection –1– (30), in driver's door wiring harness
(R14) – Wire Connection –1– (open), in driver's door wiring harness
(R15) – Plus Connection –2– (closed), in driver's door wiring harness

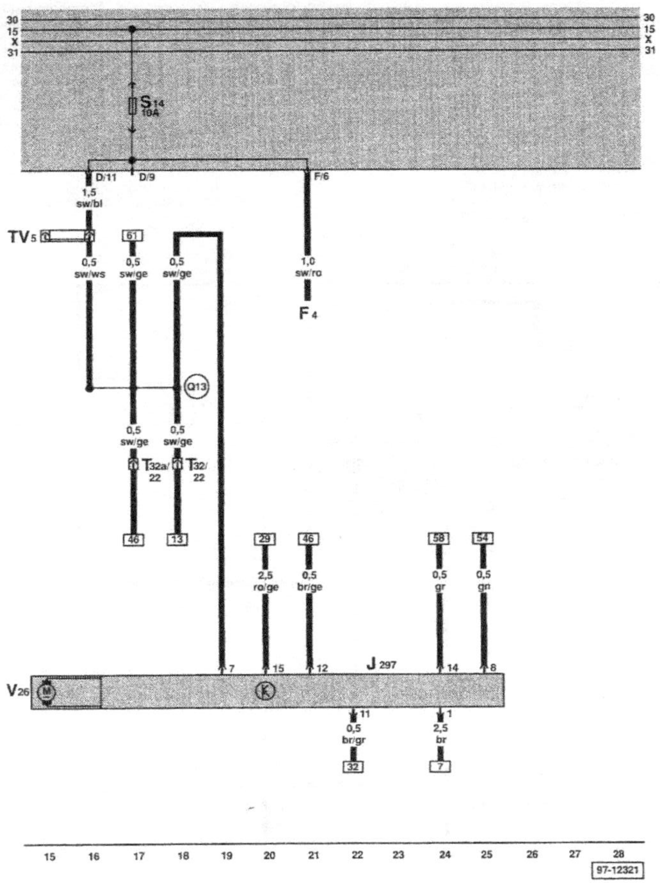

WIRING COLOR CODE		
ws	=	white
sw	=	black
ro	=	red
br	=	brown
gn	=	green
bl	=	blue
gr	=	grey
li	=	lilac
ge	=	yellow

- F4 – Back-Up Light Switch
- J297 – Left Rear Power Window Control Module
- T32 – 32-Pin Connector, on right A-Pillar, lower part
- TV5 – Terminal 15a Wire Connector above fuse/relay panel
- V26 – Left Rear Door Window Motor
- (Q13) – Plus Connection (75as/15), in power window/central locking system and door contact switch wiring harness

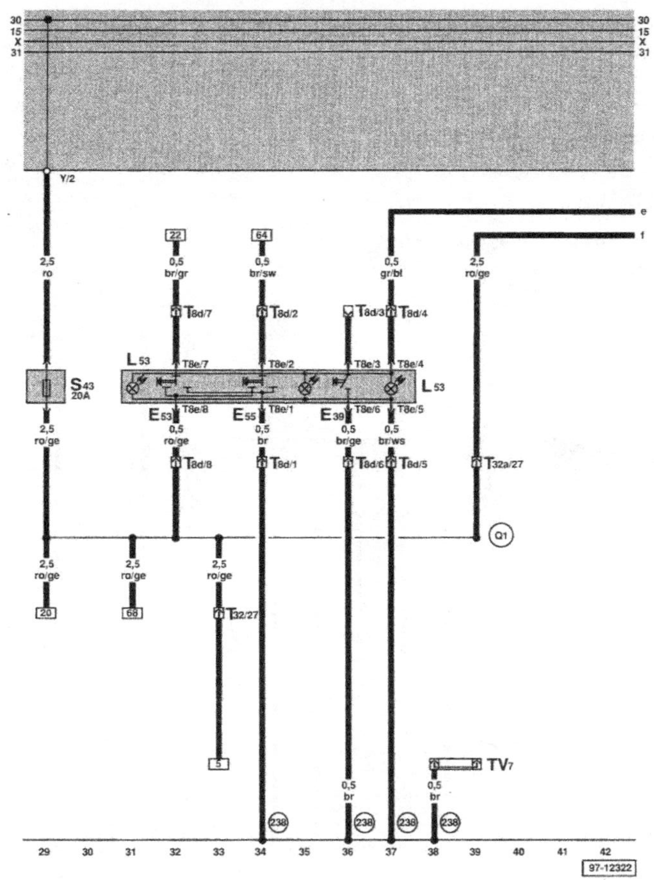

WIRING COLOR CODE		
ws	=	white
sw	=	black
ro	=	red
br	=	brown
gn	=	green
bl	=	blue
gr	=	grey
li	=	lilac
ge	=	yellow

- E39 – Window Lockout Switch
- E53 – Left Rear Window Switch, (in console)
- E55 – Right Rear Window Switch, (in console)
- L53 – Power Window Switch Light
- S43 – Power Window Circuit Breaker above fuse/relay panel
- T8d – 8-Pin Connector
- T8e – 8-Pin Connector, behind instrument panel, center
- T32 – 32-Pin Connector, on left A-Pillar, lower part
- T32a – 32-Pin Connector, on right A-Pillar, lower part
- TV7 – Terminal 31 (GND) Wire Connector above fuse/relay panel
- (238) – Ground connection, in wiring harness interior
- (Q1) – Plus connection (30), in power window wiring harness

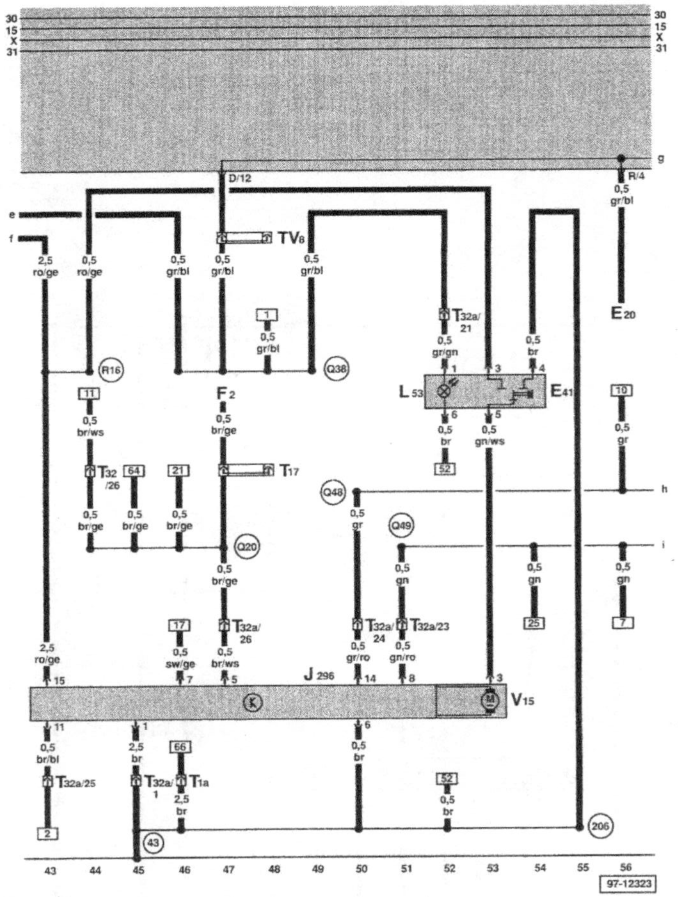

WIRING COLOR CODE		
ws	=	white
sw	=	black
ro	=	red
br	=	brown
gn	=	green
bl	=	blue
gr	=	grey
li	=	lilac
ge	=	yellow

E20 – Instrument Panel Light Dimmer Switch
E41 – Right Front Window Switch
J296 – Right Front Power Window Control Module
L53 – Power Window Switch Light
T1 – Single Connector, on right A-Pillar, lower part
T32 – 32-Pin Connector, on right A-Pillar, lower part
TV8 – Terminal 58b Wire Connector, above fuse/relay panel
TV17 – Door Contact Switch Wire Connector, above fuse/relay panel
V15 – Right Window Motor
(43) – Ground connection, on right A-Pillar, lower part
(206) – Ground connection, in passenger's door wiring harness
(Q20) – Wire connection –1– (door contact switch), in power window wiring harness
(Q38) – Plus connection (58b) –1–, in power window wiring harness
(Q48) – Connector (open), in wiring harness interior
(Q49) – Connector (closed), in wiring harness interior
(R16) – Plus connection –1– (30), in passenger's door wiring harness

1995

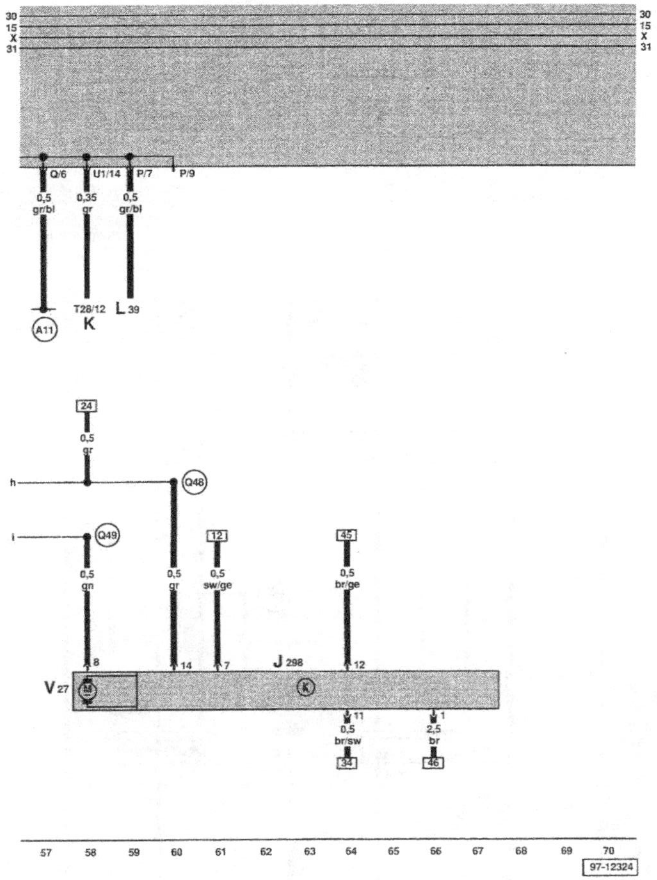

WIRING COLOR CODE		
ws	=	white
sw	=	black
ro	=	red
br	=	brown
gn	=	green
bl	=	blue
gr	=	grey
li	=	lilac
ge	=	yellow

J298 – Right Rear Power Window Control Module
K – Instrument Cluster
L39 – Rear Window Defogger Switch Light
T28 – 28-Pin Connector on instrument cluster
V27 – Right Rear Door Window Motor
(A11) – Plus connection (58b), in instrument panel wiring harness
(Q48) – Connector (open), in wiring harness interior
(Q49) – Connector (closed), in wiring harness interior

1995

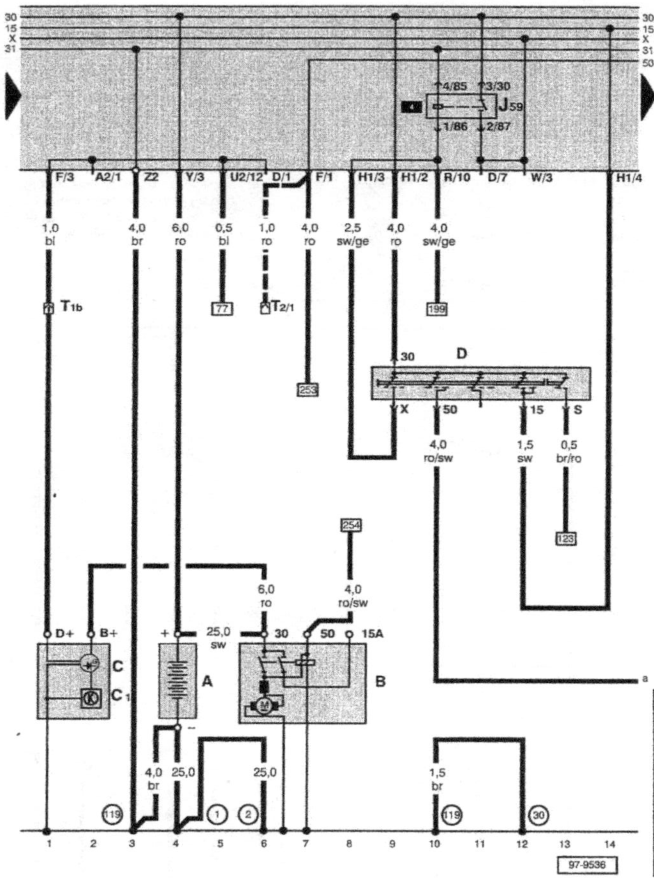

WIRING COLOR CODE		
ws	=	white
sw	=	black
ro	=	red
br	=	brown
gn	=	green
bl	=	blue
gr	=	grey
li	=	lilac
ge	=	yellow

97-9536

- A – Battery
- B – Starter
- C – Generator (Gen)
- C1 – Voltage Regulator (VR)
- D – Ignition/Starter Switch
- J59 – Load Reduction Relay
- T1b – Single Connector, near battery
- T2 – Double Connector, behind fuse/relay panel, only on automatic transmission
- (1) – Ground strap, battery to body
- (2) – Ground strap, transmission to body
- (30) – Ground connection –1–, beside fuse/relay panel
- (119) – Ground connection –1–, in headlight wiring harness
- – – – Automatic transmission only

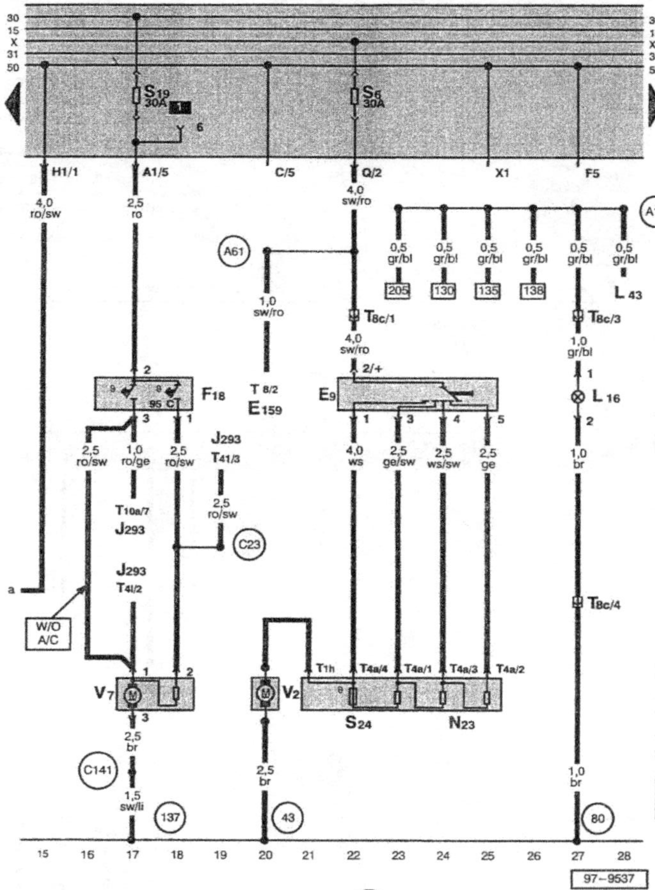

WIRING COLOR CODE		
ws	=	white
sw	=	black
ro	=	red
br	=	brown
gn	=	green
bl	=	blue
gr	=	grey
li	=	lilac
ge	=	yellow

97–9537

- E9 – Fresh Air Blower Switch
- E159 – Free Air/Recirculation Flap Switch
- F18 – Coolant Fan Control (FC) Thermal Switch
- J293 – Coolant Fan Control (FC) Module
- L16 – Fresh Air Control Lever Light
- N23 – Fresh Air Blower Series Resistance
- S24 – Heater, A/C Clutch Fuse
- T1h – Single Connector, behind right side of dash
- T4a – 4-Pin Connector, behind right side of dash
- T4l – 4-Pin Connector, on Coolant Fan Control (FC) Control Module
- T8 – 8-Pin Connector, on Free Air/Recirculation Flap Switch
- T8c – 8-Pin Connector, behind lower center of dash panel
- T10a – 10-Pin Connector, on Coolant Fan Control (FC) Module
- V2 – Fresh Air Blower
- V7 – Coolant Fan
- (43) – Ground connection, on right A-pillar, lower part
- (80) – Ground connection –1–, in instrument panel wiring harness
- (137) – Ground connection, Motronic MFI wiring harness
- (A11) – Plus connection (58b), in instrument panel wiring harness
- (A61) – Plus connection (X), in instrument cluster wiring harness
- (C23) – Wire connection, in Motronic MFI wiring harness
- (C141) – Wire connection, in Motronic MFI wiring harness

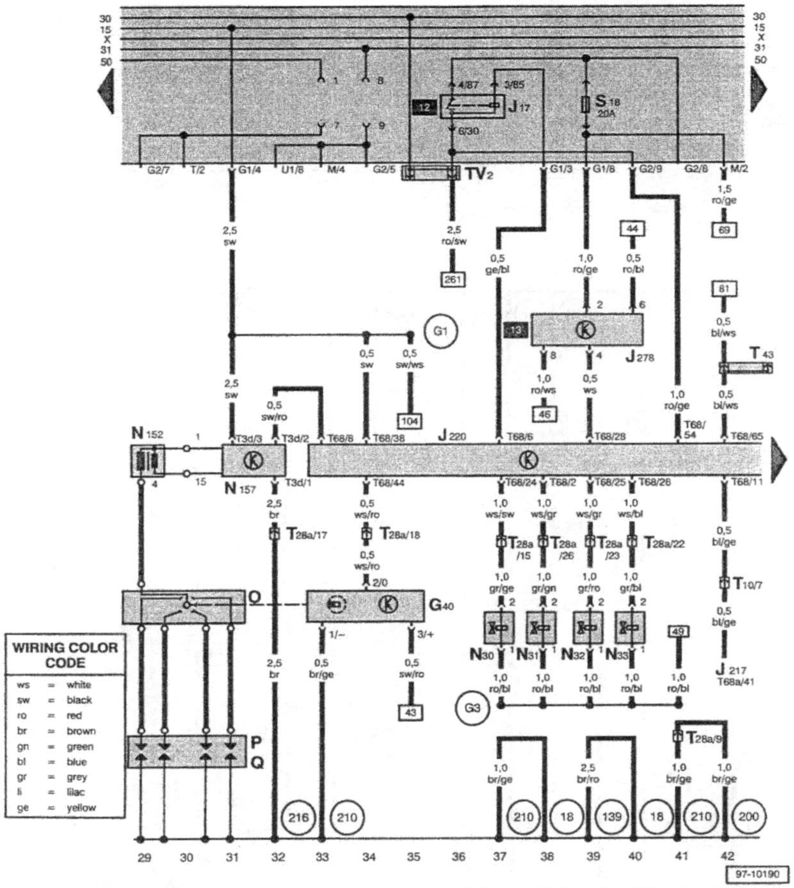

G40 – Camshaft Position (CMP) Sensor
J17 – Fuel Pump (FP) Relay
J217 – Transmission Control Module (TCM)
J220 – Motronic Engine Control Module (ECM)
J278 – Heated Oxygen Sensor (HO2S) Relay
N30 – Cylinder 1 Fuel Injector
N31 – Cylinder 2 Fuel Injector
N32 – Cylinder 3 Fuel Injector
N33 – Cylinder 4 Fuel Injector
N152 – Ignition Coil
N157 – Ignition Coil Power Output Stage
O – Ignition Distributor
P – Spark Plug Connector
Q – Spark Plugs
T10 – 10-Pin Connector, behind fuse/relay panel
T3d – 3-Pin Connector, left rear of engine compartment
T28a – 28-Pin Connector, left side of engine, near distributor
T68 – 68-Pin Connector, under right side of hood, on Motronic Engine Control Module (ECM)
T68a – 68-Pin Connector, on Transmission Control Module (TCM)
TV2 – Wire connector (Red), terminal 30 (B+)
TV43 – 43-Pin Connector, Vehicle Speed Signal
(18) – Ground connection, on engine block
(200) – Ground connection, (shielding) in eng. comp. wiring harness
(210) – Ground connection –1–, in Motronic Multiport Fuel Injection (MFI) wiring harness
(216) – Ground connection –2–, in Motronic Multiport Fuel Injection (MFI) wiring harness
(G1) – Plus connection, in Multiport Fuel Injection (MFI) harness
(G3) – Plus connection, in injector wiring harness
(139) – Ground connection (sensor ground), in Motronic Multiport Fuel Injection (MFI) wiring harness

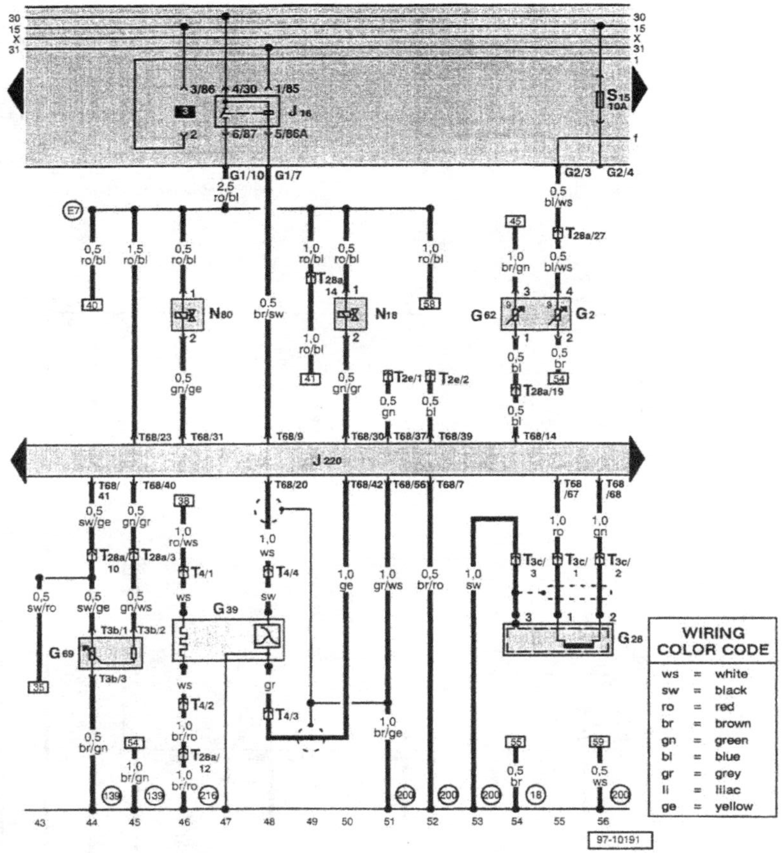

G2 – Engine Coolant Temperature (ECT) Sensor
G28 – Engine Speed (RPM) Sensor
G39 – Heated Oxygen Sensor (HO2S)
G62 – Engine Coolant Temperature (ECT) Sensor
G69 – Throttle Position (TP) Sensor
J16 – Power Supply Relay
J220 – Motronic Engine Control Module (ECM)
N18 – EGR Vacuum Regulator Solenoid Valve
N80 – Evaporative Emission (EVAP) Canister Purge Regulator Valve
T2e – Double Connector, behind fuse/relay panel, A/C connection
T3b – 3-Pin Connector, on throttle position sensor
T3c – 3-Pin Connector, left front of engine
T4 – 4-Pin Connector, bottom right rear of engine compartment
T28a – 28-Pin Connector, left side of engine, near distributor
T68 – 68-Pin Connector, under right side of hood, on Motronic Engine Control Module (ECM)
(18) – Ground connection, on engine block
(139) – Ground connection (sensor ground), in Motronic Multiport Fuel Injection (MFI) wiring harness
(200) – Ground connection (shielding), in engine compartment wiring harness
(216) – Ground connection –2–, in Motronic Multiport Fuel Injection (MFI) wiring harness
(E7) – Plus connection (87a), in Multiport Fuel Injection (MFI) wiring harness

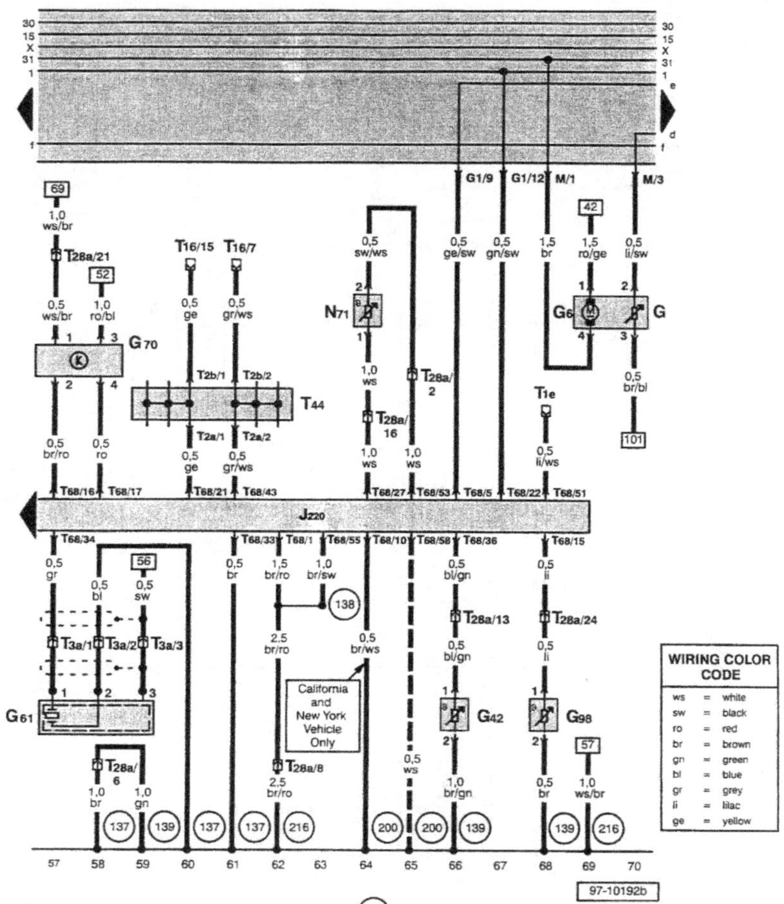

WIRING COLOR CODE

ws	=	white
sw	=	black
ro	=	red
br	=	brown
gn	=	green
bl	=	blue
gr	=	grey
li	=	lilac
ge	=	yellow

- G – Fuel Level Sensor
- G6 – Fuel Pump (FP)
- G42 – Intake Air Temperature (IAT) Sensor
- G61 – Knock Sensor (KS) 1
- G70 – Mass Air Flow (MAF) Sensor
- G98 – EGR Temperature Sensor
- J220 – Motronic Engine Control Module (ECM)
- N71 – Idle Air Control (IAC) Valve
- T1e – Single Connector, behind fuse/relay panel
- T2a – Double Connector, on junction connector (T44) above fuse/relay panel
- T2b – Double Connector, on junction connector (T44) above fuse/relay panel
- T3a – 3-Pin Connector, on engine
- T28a – 28-Pin Connector, left side of engine, near distributor
- T44 – 44-Pin Connector, Junction Box for on board diagnosis, above fuse/relay panel
- T68 – 68-Pin Connector, under right side of hood, on Motronic Engine Control Module (ECM)

- (137) – Ground connection (end stage), in Motronic Multiport Fuel Injection (MFI) wiring harness
- (138) – Ground connection (control module), in Motronic Multiport Fuel Injection (MFI) wiring harness
- (139) – Ground connection (sensor ground), in Motronic Multiport Fuel Injection (MFI) wiring harness
- (200) – Ground connection, (shielding) in engine compartment wiring harness
- (216) – Ground connection –2–, in Motronic Multiport Fuel Injection (MFI) wiring harness
- ▬ ▬ Automatic transmission only

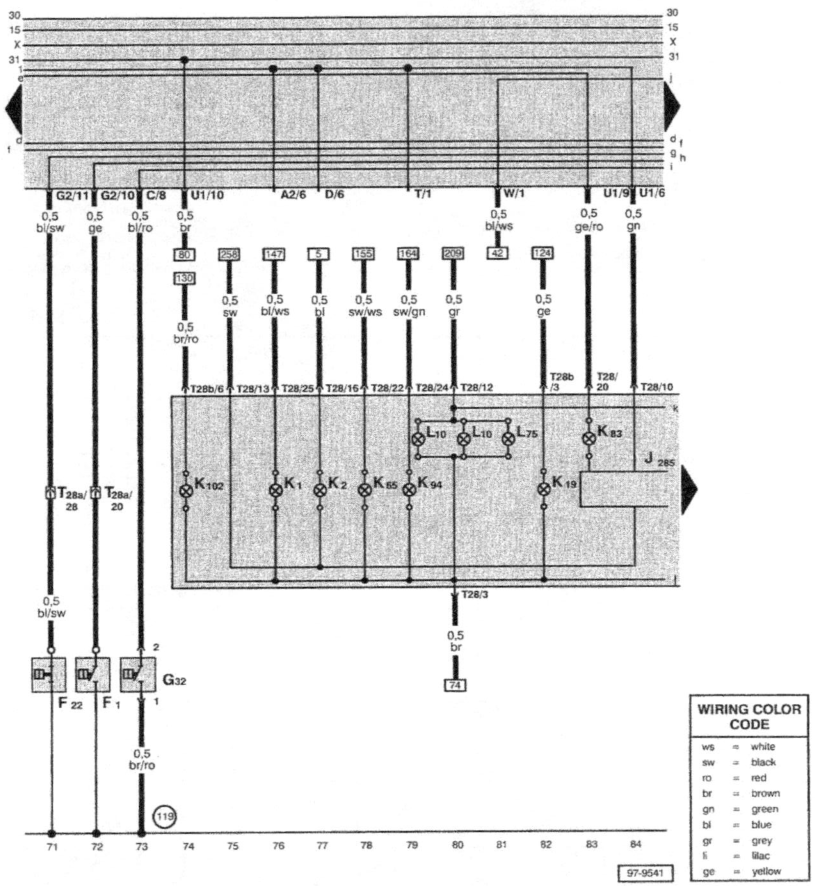

WIRING COLOR CODE

ws	=	white
sw	=	black
ro	=	red
br	=	brown
gn	=	green
bl	=	blue
gr	=	grey
li	=	lilac
ge	=	yellow

- F1 – Oil Pressure Switch
- F22 – 0.3 Bar Oil Pressure Switch
- G32 – Engine Coolant Level (ECL) Sensor
- J285 – Control Module With Display Unit In Instrument Cluster
- K1 – Headlight High Beam Indicator Light
- K2 – Generator (GEN) Warning Light
- K19 – Seat Belt Warning Light
- K65 – Left Turn Signal Indicator Light
- K83 – Multi-Function Indicator Lamp (MIL)
- K94 – Right Turn Signal Indicator Light
- K102 – Rear Hatch/Trunk Lid Ajar Warning Light
- L10 – Instrument Cluster Light
- L75 – Digital Display Light
- T28 – 28-Pin Connector, instrument cluster
- T28a – 28-Pin Connector, left side of engine, near distributor
- T28b – 28-Pin Connector, on instrument cluster

- (119) – Ground connection –1–, in headlight wiring harness

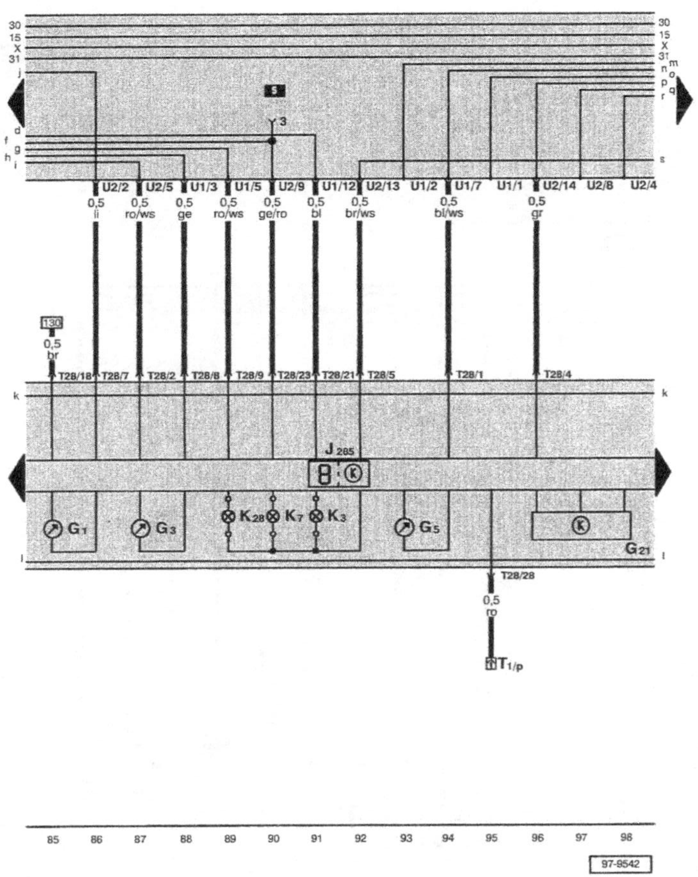

WIRING COLOR CODE		
ws	=	white
sw	=	black
ro	=	red
br	=	brown
gn	=	green
bl	=	blue
gr	=	grey
li	=	lilac
ge	=	yellow

G1 – Fuel Gauge
G3 – Engine Coolant Temperature (ECT) Gauge
G5 – Tachometer
G21 – Speedometer
J285 – Control Module With Display Unit In Instrument Cluster
K3 – Oil Pressure Warning Light
K7 – Brake and Parking Brake Warning Light
K28 – Engine Coolant Level/Temperature (ECL/ECT) Warning Light
T1p – Single Connector, behind Fuse/Relay Panel
T28 – 28-Pin Connector, instrument cluster

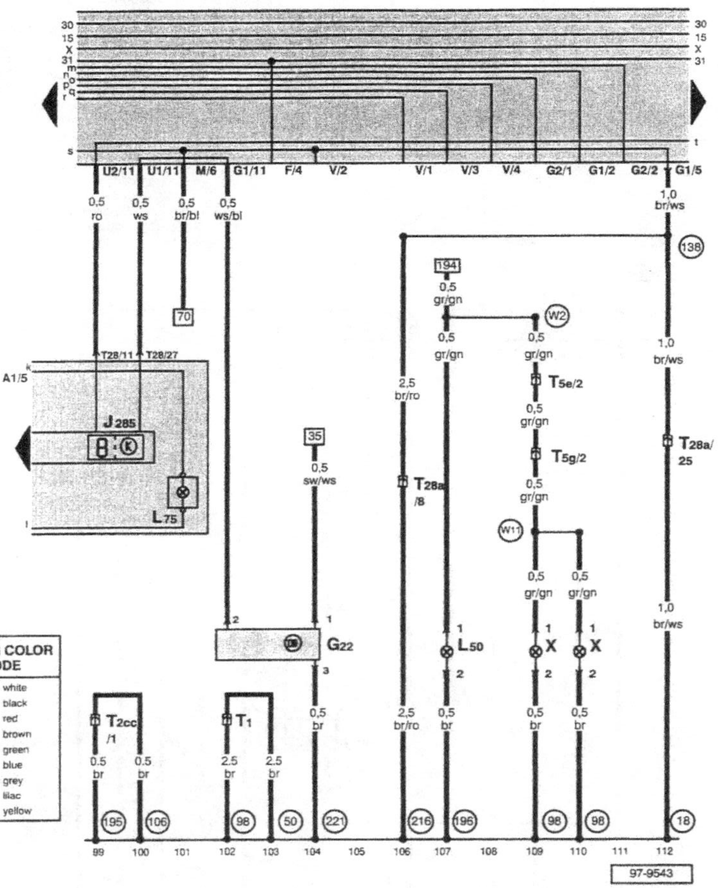

WIRING COLOR CODE		
ws	=	white
sw	=	black
ro	=	red
br	=	brown
gn	=	green
bl	=	blue
gr	=	grey
li	=	lilac
ge	=	yellow

G22 – Vehicle Speed Sensor (VSS), Speedometer
J285 – Control Module With Display Unit in Instrument Cluster
L50 – Center Rear Ashtray Light
L75 – Digital Display Light
T1 – Single Connector, in Luggage compartment lid
T2cc – Double Connector, on lower right A–pillar
T5e – 5–Pin Connector, in luggage compartment, left
T5g – 5–Pin Connector, in luggage compartment lid
T28 – 28–Pin Connector, instrument cluster
T28a – 28–Pin Connector, left side of engine, near distributor
X – License Plate Light

(18) – Ground connection, on engine block

(50) – Ground Connection, in luggage compartment,left

(98) – Ground connection, in rear lid wiring harness

(106) – Ground connection, in alarm system wiring harness

(138) – Ground connection (Control Module) in Motronic Multiport Fuel Injection (MFI) wiring harness

(195) – Ground connection, in rear door contact switch wiring harness

(196) – Ground connection –1–, rear wiring harness

(216) – Ground connection –2–, in Motronic Multiport Fuel Injection (MFI) wiring harness

(221) – Ground connection (engine Ground), in engine compartment wiring harness

(W2) – Plus connection (58), in rear wiring harness

(W11) – Wiring connection (58), in rear lid wiring harness

Door switches, interior light, seat belt warning system relay

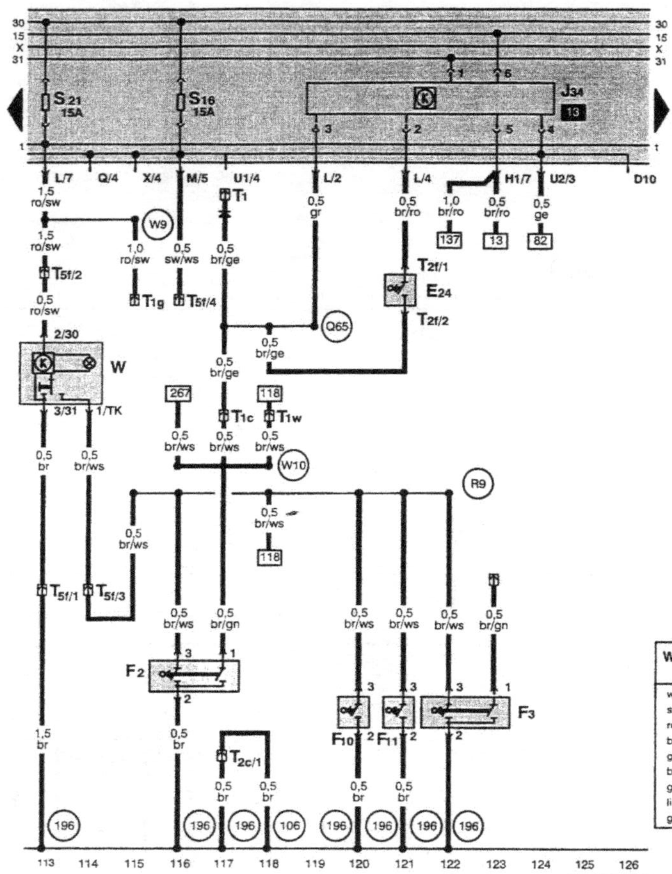

WIRING COLOR CODE

ws	=	white
sw	=	black
ro	=	red
br	=	brown
gn	=	green
bl	=	blue
gr	=	grey
li	=	lilac
ge	=	yellow

- E24 – Left Seat Belt Switch
- F2 – Left Front Door Contact Switch
- F3 – Right Front Door Contact Switch
- F10 – Left Rear Door Contact Switch
- F11 – Right Rear Door Contact Switch
- J34 – Seat Belt Warning System Relay
- T1c – Single Connector, behind fuse/relay panel
- T1g – Single Connector, behind fuse/relay panel
- T1y – Single Connector, behind fuse/relay panel
- T2c – Double Connector, on lower right A-pillar
- T2cc – Double Connector, on lower right A-pillar
- T2f – Double Connector, behind rear of center console
- T5f – 5-Pin Connector, above fuse/relay panel
- W – Interior Light, Front

- (106) – Ground connection, in alarm system wiring harness
- (Q65) – Wire connection, in seat belt warning system wiring harness
- (R9) – Wire connection (door contact switch), in rear wiring harness
- (W9) – Plus connection (30), in rear wiring harness
- (W10) – Wire connection (31TK), in rear wiring harness
- (196) – Ground connection –1–, in rear wiring harness

Luggage compartment release solenoid, luggage compartment light, cigarette lighter

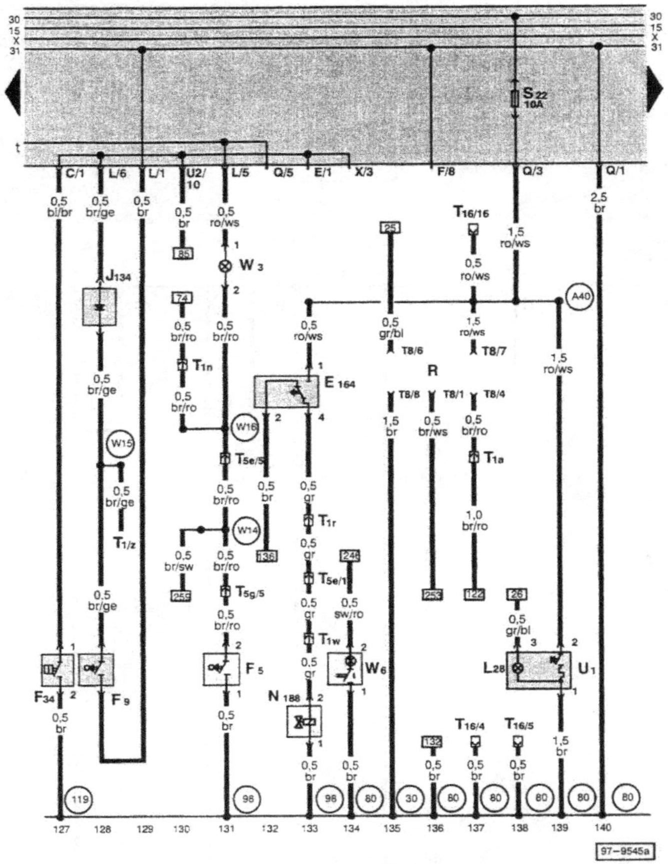

WIRING COLOR CODE

ws	=	white
sw	=	black
ro	=	red
br	=	brown
gn	=	green
bl	=	blue
gr	=	grey
li	=	lilac
ge	=	yellow

- E164 – Luggage Compartment Release Switch
- F5 – Luggage Compartment Light Switch
- F9 – Parking Brake Warning Light Switch
- F34 – Brake Fluid Level Warning Switch
- J134 – Daytime Running Lights Diode
- L28 – Cigarette Lighter Light
- N188 – Luggage Compartment Release Solenoid
- R – Radio
- T1a – Single Connector, behind fuse/relay panel
- T1n – Single Connector, behind fuse/relay panel
- T1r – Single Connector, behind fuse/relay panel
- T1w – Single Connector, on luggage compartment lid
- Tlz – Single Connector, behind Fuse/Relay Panel
- T5g – 5-Pin Connector, in luggage compartment lid
- T8 – 8-Pin Connector, on radio
- T16 – Data Link Connector (DLC), behind center of dash panel
- T5e – 5-Pin Connector, in luggage compartment, left
- U1 – Cigarette Lighter
- W3 – Luggage Compartment Light
- W6 – Glove Compartment Light

- (30) – Ground connection –1–, beside fuse/relay panel
- (80) – Ground connection –1–, in instrument panel wiring harness
- (98) – Ground connection, in rear lid wiring harness
- (119) – Ground connection –1–, in headlight wiring harness
- (A40) – Plus connection (30), in instrument panel wiring harness
- (W14) – Wire connection (58R), in rear wiring harness
- (W15) – Wire connection (HB), in rear wiring harness
- (W16) – Wire connection (31s), in rear warning harness

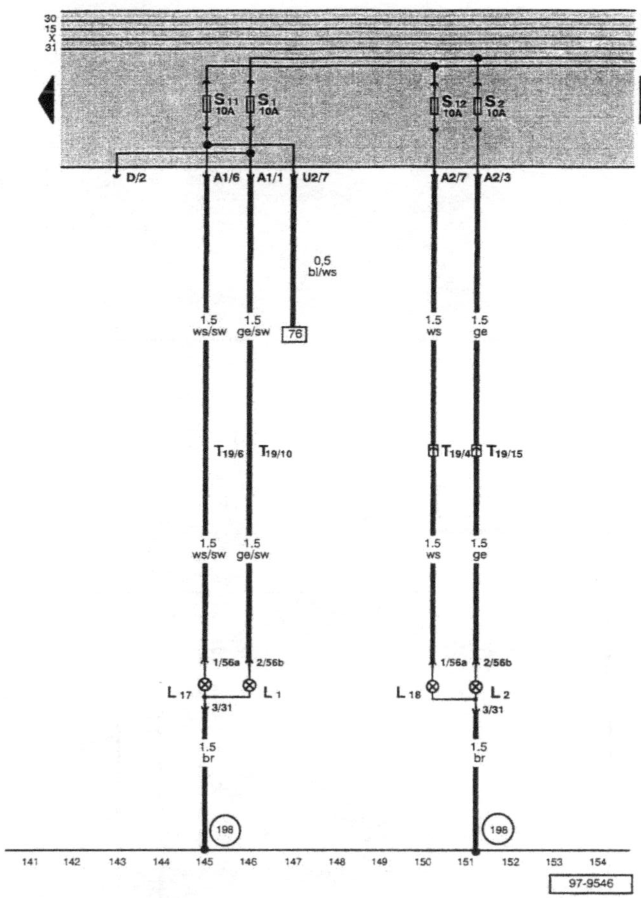

WIRING COLOR CODE		
ws	=	white
sw	=	black
ro	=	red
br	=	brown
gn	=	green
bl	=	blue
gr	=	grey
li	=	lilac
ge	=	yellow

L1 – Left Headlight
L2 – Right Headlight
L17 – Left Sealed Beam High Beam Headlight
L18 – Right Sealed Beam High Beam Headlight
T19 – 19-Pin Connector, behind radiator grill, center
(198) – Ground connection, in front light wiring harness

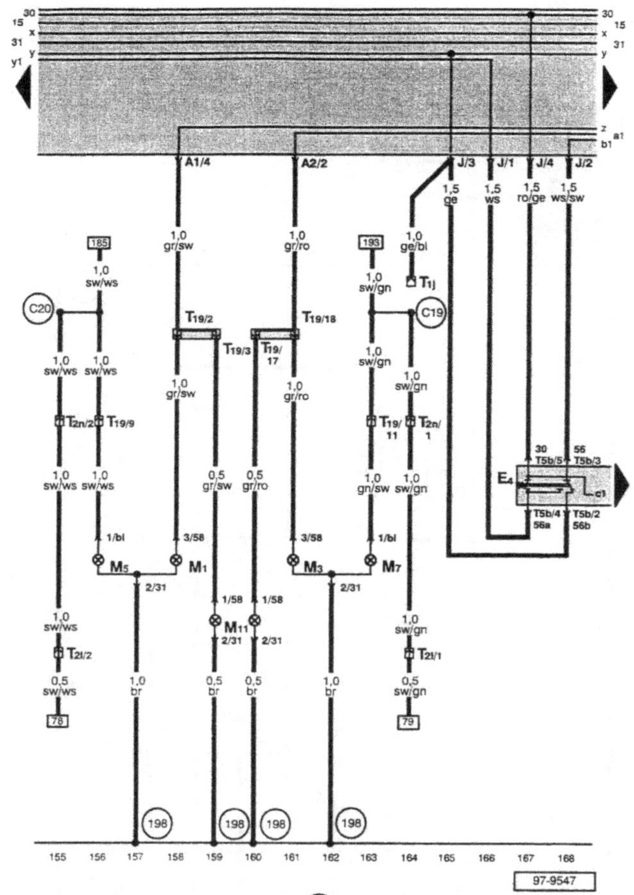

WIRING COLOR CODE		
ws	=	white
sw	=	black
ro	=	red
br	=	brown
gn	=	green
bl	=	blue
gr	=	grey
li	=	lilac
ge	=	yellow

E4 – Headlight Dimmer/Flasher Switch
M1 – Left Parking Light
M3 – Right Parking Light
M5 – Left Front Turn Signal Light
M7 – Right Front Turn Signal Light
M11 – Front Side Marker Lights
T1j – Single Connector, behind fuse/relay panel
T2l – Double Connector, behind fuse/relay panel
T2n – Double Connector, behind fuse/relay panel
T5b – 5-Pin Connector, behind steering column switch trim
T19 – 19-Pin Connector, behind radiator grill, center
(198) – Ground connection, in front light wiring harness

(C19) – Wire connection (right turn signal), in headlight wiring harness
(C20) – Wire connection (left turn signal), in headlight wiring harness
(Q54) – Wire connection (left turn signal), in alarm system wiring harness
(Q55) – Wire connection (right turn signal), in alarm system wiring harness

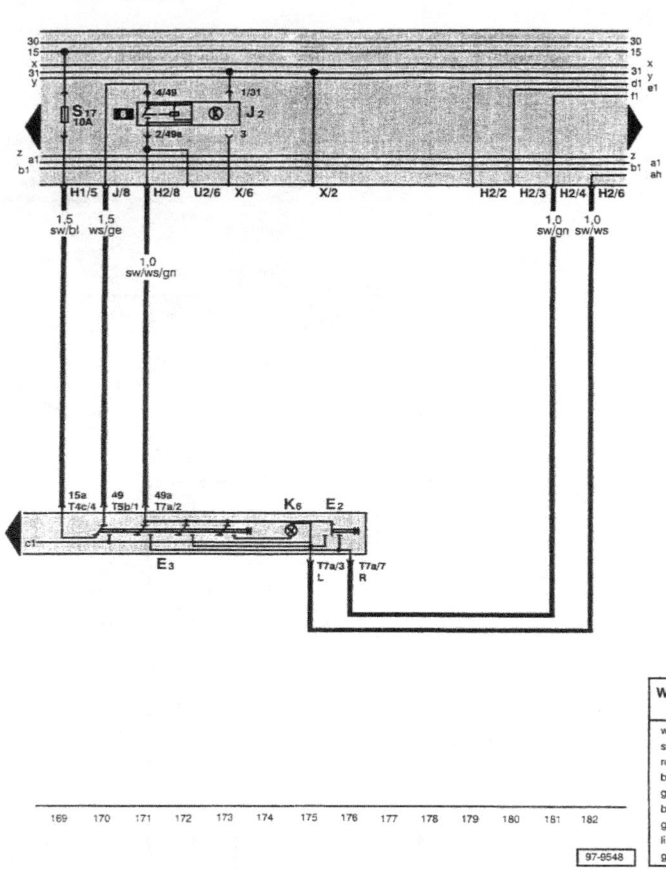

WIRING COLOR CODE

ws	=	white
sw	=	black
ro	=	red
br	=	brown
gn	=	green
bl	=	blue
gr	=	grey
li	=	lilac
ge	=	yellow

E2 – Turn Signal Switch
E3 – Emergency Flasher Switch
J2 – Emergency Flasher Relay
K6 – Emergency Flasher Warning Light
T4c – 4-Pin Connector, behind steering column switch trim
T5b – 5-Pin Connector, behind steering column switch trim
T7a – 7-Pin Connector, behind steering column switch trim

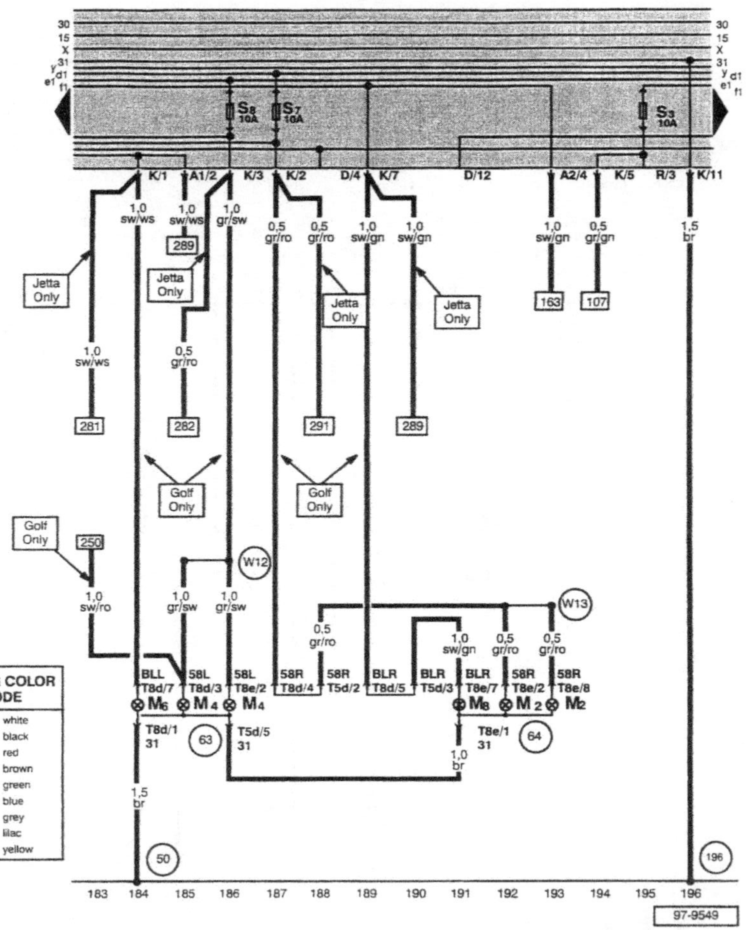

WIRING COLOR CODE

ws	=	white
sw	=	black
ro	=	red
br	=	brown
gn	=	green
bl	=	blue
gr	=	grey
li	=	lilac
ge	=	yellow

M2 – Right Tail Light
M4 – Left Tail Light
M6 – Left Rear Turn Signal Light
M8 – Right Rear Turn Signal Light
T5d – 5-Pin Connector, on tail light, left
T8d – 8-Pin Connector, on tail light, left
T8e – 8-Pin Connector, on tail light right

(50) – Ground connection, in luggage compartment, left

(63) – Ground Connection, on left tail light abulb holder

(64) – Ground connection, on right tail light bulb holder

(196) – Ground Connection –1–, in rear wiring harness

(W12) – Wire connection (dwa), in rear wiring harness, left

(W13) – Wire connection (58L), in rear wiring harness

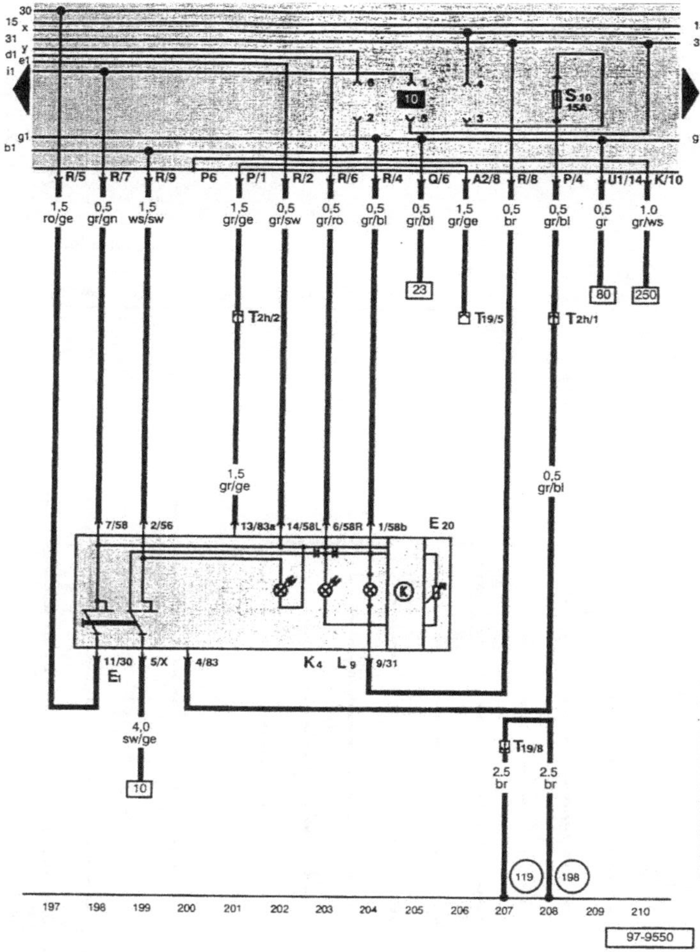

WIRING COLOR CODE		
ws	=	white
sw	=	black
ro	=	red
br	=	brown
gn	=	green
bl	=	blue
gr	=	grey
li	=	lilac
ge	=	yellow

- E1 – Light Switch
- E20 – Instrument Panel Light Dimmer Switch
- K4 – Park Light Indicator Light
- L9 – Headlight Switch Light
- T2h – 2-Pin Connector, behind fuse/relay panel
- T19 – 19-Pin Connector, behind radiator grill, center
- (119) – Ground Connection –1–, in headlight wiring harness
- (198) – Ground Connection, in front light wiring harness

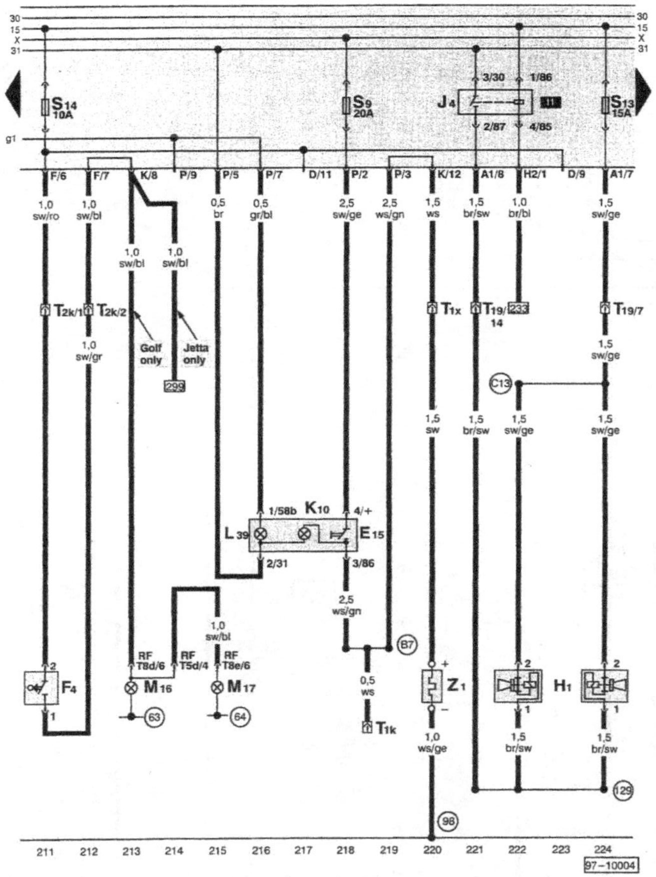

WIRING COLOR CODE		
ws	=	white
sw	=	black
ro	=	red
br	=	brown
gn	=	green
bl	=	blue
gr	=	grey
li	=	lilac
ge	=	yellow

- E15 – Rear Window Defogger Switch
- F4 – Back-Up Light Switch
- H1 – Dual Horns
- J4 – Dual Horn Relay
- K10 – Rear Window Defogger Indicator Light
- L39 – Rear Window Defogger Switch Light
- M16 – Left Back-Up Light
- M17 – Right Back-Up Light
- T1k – Single Connector behind fuse/relay panel
- T1x – Single Connector in luggage compartment, left
- T2k – Double Connector behind fuse/relay panel
- T5d – 5-Pin Connector on tail light, left
- T8d – 8-Pin Connector on tail light, left
- T8e – 8-Pin Connector on tail light right
- T19 – 19-Pin Connector behind radiator grill, center
- Z1 – Rear Window Defogger (Heat Element)
- (63) – Ground connection, on left tail light bulb holder
- (64) – Ground connection, on right tail light bulb holder
- (98) – Ground connection, in rear lid wiring harness
- (129) – Ground connection, in dual tone horn wiring harness
- (B7) – Wire connection (86), in rear window defogger wiring harness
- (C13) – Plus connection, in dual tone horn wiring harness

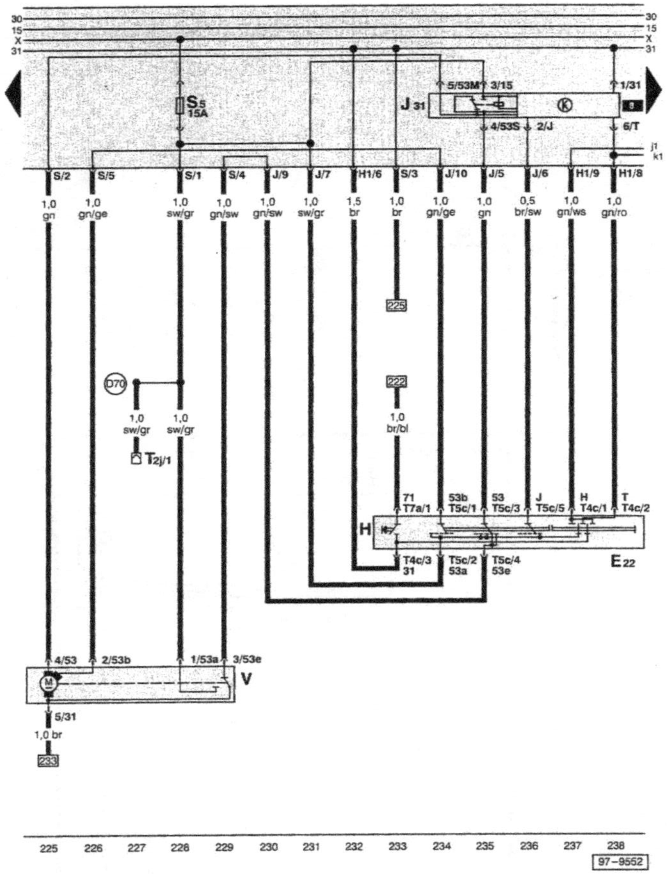

WIRING COLOR CODE

ws	=	white
sw	=	black
ro	=	red
br	=	brown
gn	=	green
bl	=	blue
gr	=	grey
li	=	lilac
ge	=	yellow

E22 – Windshield Wiper/Washer Switch
H – Horn Button
J31 – Washer/Wiper Intermittent Relay
T2j – Double Connector behind fuse/relay panel
T4c – 4-Pin Connector behind steering column switch trim
T5c – 5-Pin Connector behind steering column switch trim
T7a – 7-Pin Connector behind steering column switch trim
V – Windshield Wiper Motor
(D70) – Wire connection (53a), in engine compartment wiring harness

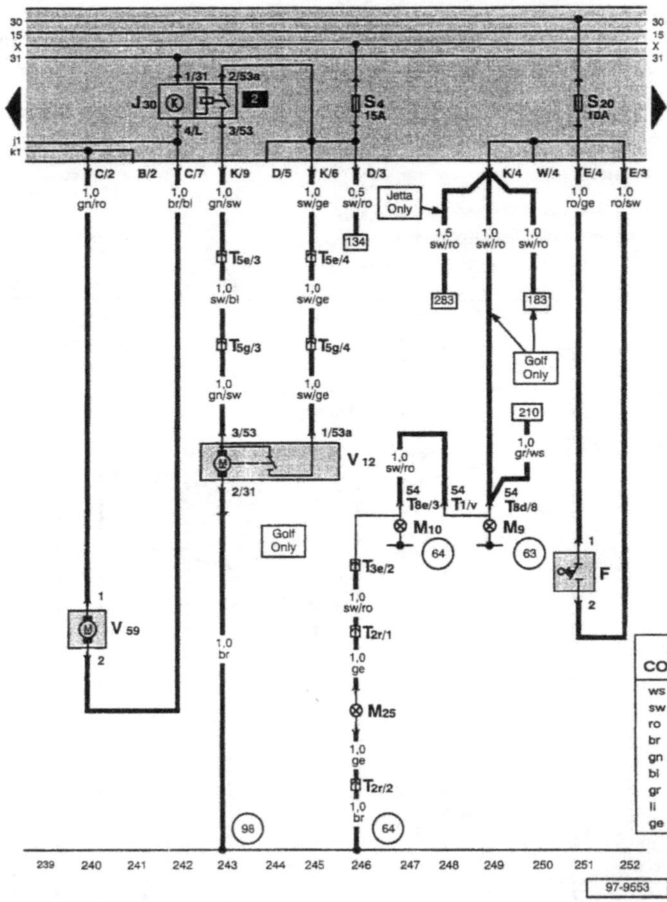

WIRING COLOR CODE

ws	=	white
sw	=	black
ro	=	red
br	=	brown
gn	=	green
bl	=	blue
gr	=	grey
li	=	lilac
ge	=	yellow

F – Brake Light Switch
J30 – Rear Window Wiper/Washer Relay
M9 – Left Brake Light
M10 – Right Brake Light
M25 – High-Mount Brake Light
V12 – Rear Window Wiper Motor
V59 – Windshield and Rear Window Washer Pump
T1v – Single Connector on tail light, left
T2r – Double Connector behind right grommet, on luggage compartment lid
T3e – 3-Pin Connector on tail light, right
T5e – 5-Pin Connector in luggage compartment, left
T5g – 5-Pin Connector in luggage compartment lid
T8d – 8-Pin Connector on tail light, left
T8e – 8-Pin Connector on tail light right
(63) – Ground connection, on left tail light bulb holder
(64) – Ground connection, on right tail light bulb holder
(98) – Ground connection, in rear lid wiring harness

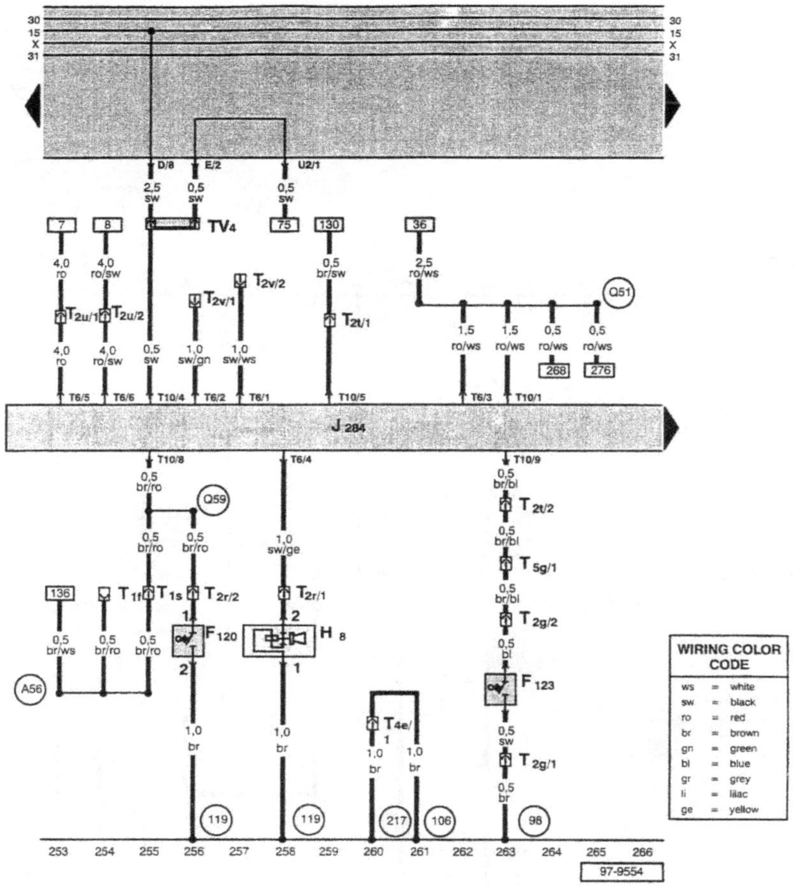

F120 – Hood Alarm Switch
F123 – Trunk Lid Alarm Switch
H8 – Alarm Horn
J284 – Alarm System Control Module
T1f – Single Connector behind dash panel, center, behind radio
T1s – Single Connector behind fuse/relay panel
T2g – Double Connectoron luggage compartment lid
T2t – Double Connector on luggage compartment left
T2u – Double Connector behind fuse/relay panel
T2r – 2-Pin Connector behind fuse/relay panel
T5g – 5-Pin Connector in luggage compartment lid
T6 – 6-Pin Connector behind dash panel, left, on Alarm System Control Module
T10 – 10-Pin Connector behind dash panel, left, on Alarm System Control Module
TV4 – Terminal 15 Wire Connector
T2v –
(98) – Ground connection, in rear lid wiring harness

(106) – Ground connection, in alarm system wiring harness
(119) – Ground connection -1-, in headlight wiring harness
(217) – Ground connection, in alarm horn wiring harness
(A56) – Plus connection -2-, (30) instrument panel wiring harness
(Q51) – Plus connection (30), in alarm system wiring harness
(Q59) – Wire connection (Sensor) in alarm system wiring harness

1995

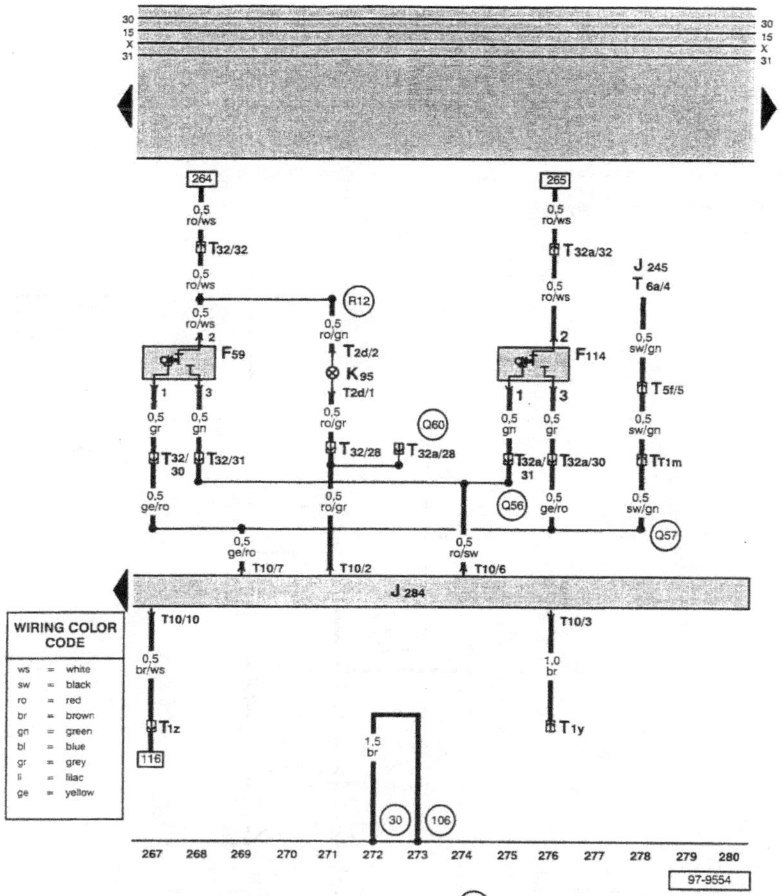

F59 – Driver's Door Central Locking System Door Switch
F114 – Passenger's Door Central Locking System Door Switch
J245 – Power Sunroof Control Module
J284 – Alarm System Control Module
K95 – Alarm System Indicator Light
T1m – Single Connector behind fuse/relay panel
T1y – Single connector in terminal 31 (end) wire connector
T1z – Single connector in headlight wiring harness
T2d – Double Connector in left front door
T5f – 5-Pin Connector above fuse/relay panel
T6a – 6-Pin Connector behind interior light, on Power Sunroof Control Module
T10 – 10-Pin Connector behind dash panel, left, on Alarm System Control Module
T32 – 32-Pin Connector on A-pillar, lower left
T32a – 32-Pin Connector on A-pillar, lower right
30 – Ground connection -1-, beside fuse/relay panel

(106) – Ground connection, in alarm system wiring harness
(Q56) – Wire connection (open), in alarm system wiring harness
(Q57) – Wire connection (closed), in alarm system wiring harness
(R12) – Plus connection -3- (30), in driver's door wiring harness
(Q60) – Wire connection, in alarm system wiring harness

1995

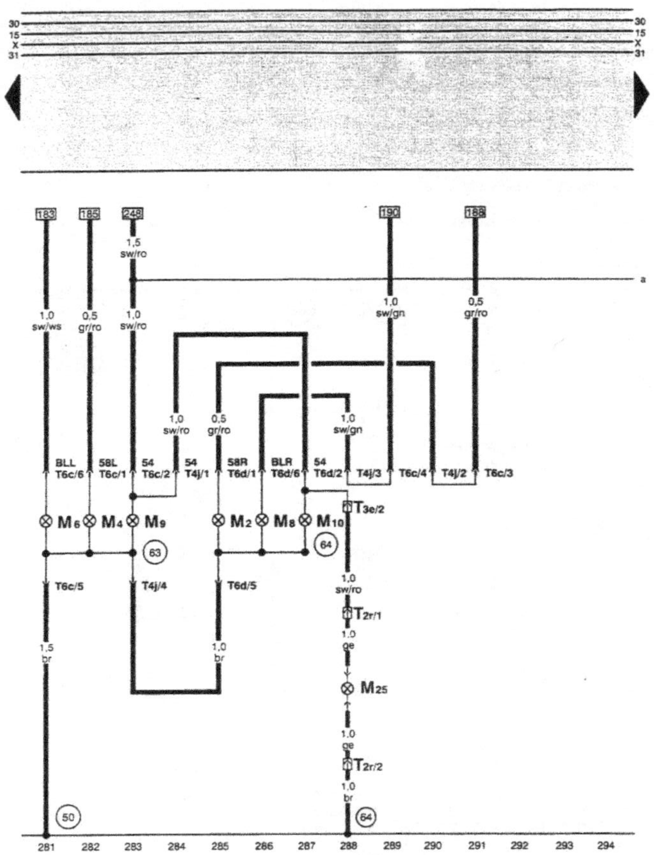

WIRING COLOR CODE

ws	=	white
sw	=	black
ro	=	red
br	=	brown
gn	=	green
bl	=	blue
gr	=	grey
li	=	lilac
ge	=	yellow

- M2 – Right Tail Light
- M4 – Left Tail Light
- M6 – Left Rear Turn Signal Light
- M8 – Right Rear Turn Signal Light
- M9 – Left Brake Light
- M10 – Right Brake Light
- M25 – High Mount Brake Light
- T2r – Double Connector behind right grommet, on luggage compartment lid
- T3e – 3-Pin Connector on tail light, right
- T4j – 4-Pin Connector on tail light, left
- T6c – 6-Pin Connector on tail light, left
- T6d – 6-Pin Connector on tail light, right
- (50) – Ground connection, in luggage compartment lid
- (63) – Ground connection, on left tail light bulb holder
- (64) – Ground connection, on right tail light bulb holder

WIRING COLOR CODE

ws	=	white
sw	=	black
ro	=	red
br	=	brown
gn	=	green
bl	=	blue
gr	=	grey
li	=	lilac
ge	=	yellow

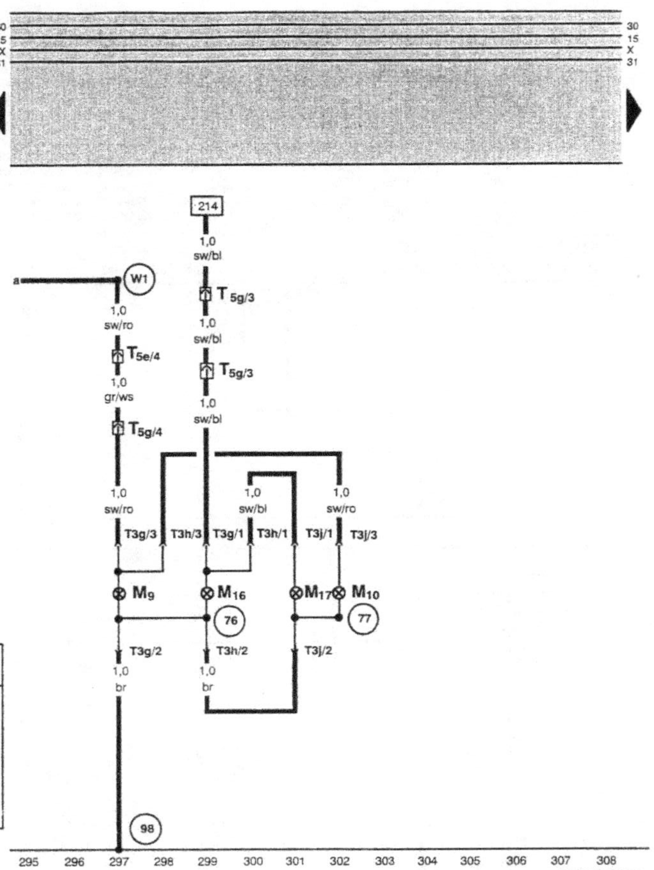

- M9 – Left Brake Light
- M10 – Right Brake Light
- M16 – Left Back-up Light
- M17 – Right Back-up Light
- T3g – 3-Pin Connector on back-up/brake light, left
- T3h – 3-Pin Connector on back-up/brake light, left
- T3j – 3-Pin Connector on back-up/brake light, right
- T5e – 5-Pin Connector on luggage compartment lid
- T5g – 5-Pin Connector on luggage compartment lid
- (W1) – Plus connection, (54), in rear wiring harness
- (76) – Ground connection, on left back-up/brake light bulb holder
- (77) – Ground connection, on right back-up/brake light bulb holder
- (98) – Ground connection, in rear lid wiring harness

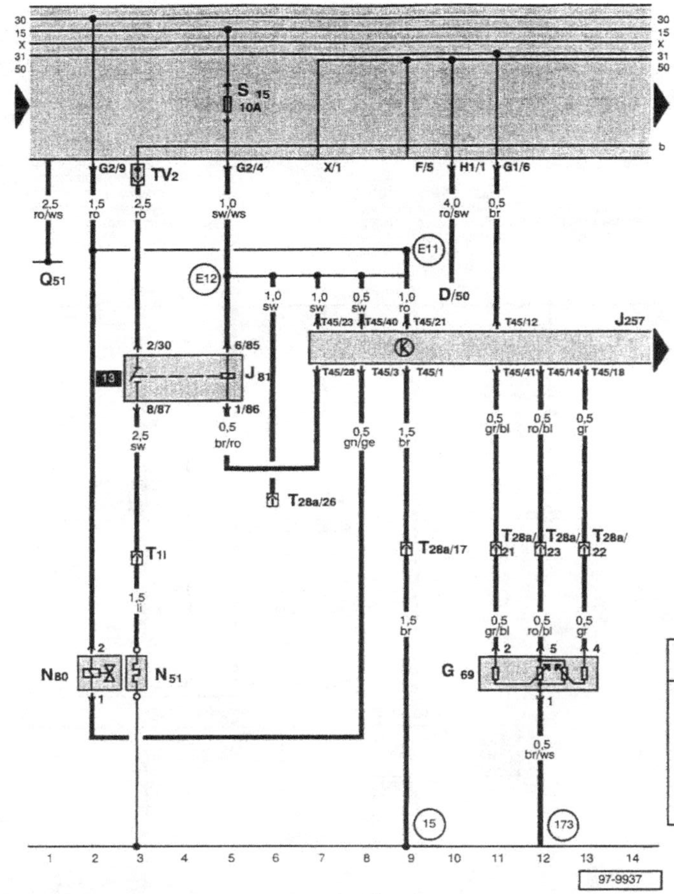

WIRING COLOR CODE		
ws	=	white
sw	=	black
ro	=	red
br	=	brown
gn	=	green
bl	=	blue
gr	=	grey
li	=	lilac
ge	=	yellow

- D – Ignition/Starter Switch
- G69 – Throttle Position (TP) Sensor
- J81 – Early Fuel Evaporation (EFE) Relay
- J257 – Mono–Motronic Engine Control Module (ECM)
- N51 – Early Fuel Evaporation (EFE) Heating Element
- N80 – Evaporative Emission (EVAP) Canister Purge Regulator Valve
- Q51 – Plus connection (30), in alarm system wiring harness
- T1i – Single Connector, left rear of engine
- T28a – 28-Pin Connector, on engine
- T45 – 45-Pin Connector, on Mono–Motronic Engine Control Module (ECM)
- TV2 – Terminal 30 (B+) Wire Connector
- (15) – Ground connection, on cylinder head
- (173) – Ground connection -1-, in Mono-Motronic Throttle Body Fuel injection (TBI) wiring harness
- (E11) – Plus connection (30), in Mono-Motronic Throttle Body Fuel Injection (TBI) wiring harness
- (E12) – Plus connection -1- (15), in Mono-Motronic Throttle Body Fuel Injection (TBI) wiring harness

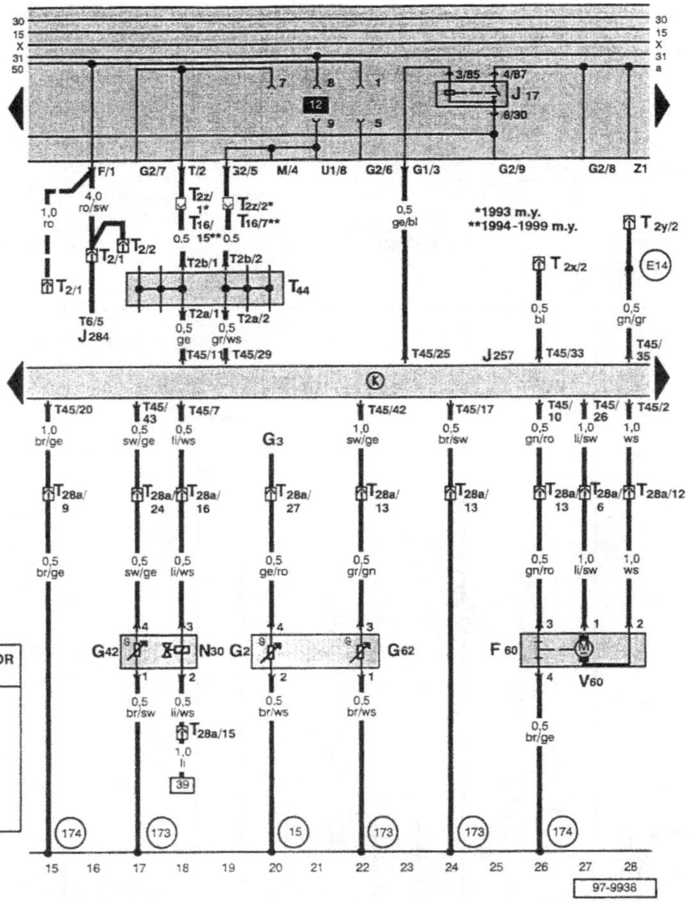

WIRING COLOR CODE		
ws	=	white
sw	=	black
ro	=	red
br	=	brown
gn	=	green
bl	=	blue
gr	=	grey
li	=	lilac
ge	=	yellow

- F60 – Closed Throttle Position (CTP) Switch
- G2 – Engine Coolant Temperature (ECT) Sensor
- G3 – Engine Coolant Temperature ECT Gauge
- G42 – Intake Air Temperature (IAT) Sensor
- G62 – Engine Coolant Temperature (ECT) Sensor
- J17 – Fuel Pump (FP) Relay
- J257 – Mono-Motronic Engine Control Module (ECM)
- J284 – Alarm System Control Module
- N30 – Cylinder 1 Fuel Injector
- T2 – Double Connector, behind fuse/relay panel, only on automatic transmission, on J226 (Park/Neutral PNP Relay)
- T2a – Double Connector, on junction connector (T44) above fuse/relay panel
- T2b – Double Connector, on junction connector (T44) above fuse/relay panel
- T2e – Double Connector, behind fuse/relay panel, A/C connection
- T2x – Double Connector, behind dash panel
- T2y – Double Connector, on A/C Cut out Thermal Switch
- T6 – 6-Pin Connector, behind dash panel, left, on Alarm System Control Module
- T28a – 28-Pin Connector, left side of engine, near distributor
- T44 – 44-Pin Connector, Junction Box for on board diagnosis, above fuse/relay panel
- T45 – 45-Pin Connector, on Mono-Motronic Engine Control Module (ECM)
- V60 – Throttle Position (TP) Actuator
- (15) – Ground connection, on cylinder head
- (173) – Ground connection -1-, in Mono-Motronic Throttle Body Fuel Injection (TBI) wiring harness
- (174) – Ground connection -2-, in Mono-Motronic Throttle Body Fuel Injection (TBI) wiring harness
- – – – Automatic transmission only

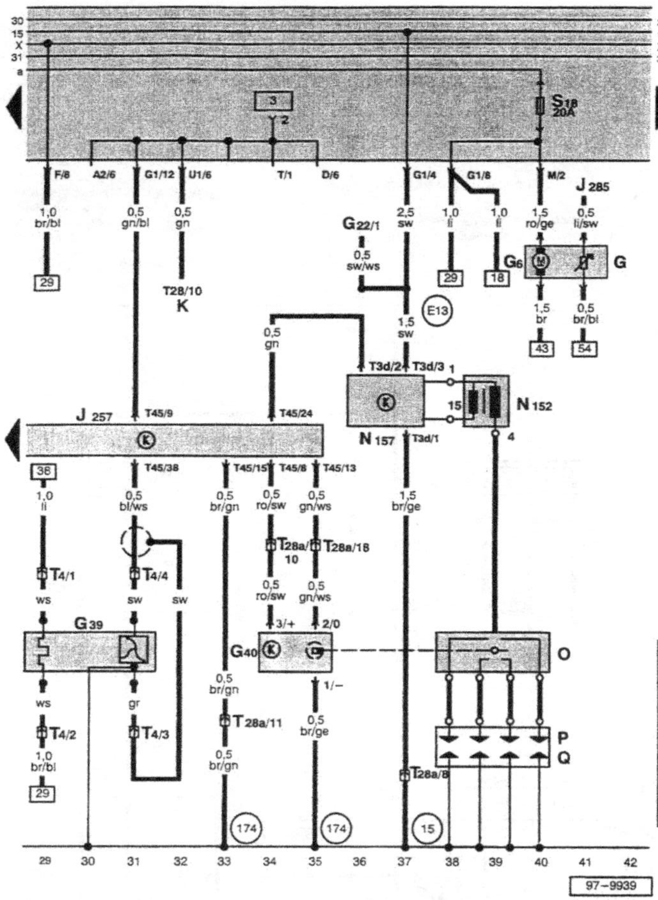

WIRING COLOR CODE

ws	=	white
sw	=	black
ro	=	red
br	=	brown
gn	=	green
bl	=	blue
gr	=	grey
li	=	lilac
ge	=	yellow

- G6 – Fuel Pump (FP)
- G22 – Speedometer Vehicle Speed Sensor (VSS)
- G39 – Heated Oxygen Sensor (HO2S)
- G40 – Camshaft Position (CMP) Sensor
- J257 – Mono-Motronic Engine Control Module
- J285 – Control Module with Display Unit in Instrument Cluster
- K – Instrument Cluster
- N152 – Ignition Coil
- N157 – Ignition Coil Power Output Stage
- O – Ignition Distributor
- P – Spark Plug Connector
- Q – Spark Plugs
- T3d – 3-Pin Connector, left rear of engine compartment
- T4 – 4-Pin Connector, bottom right rear of engine compartment
- T28 – 28-Pin Connector, instrument cluster
- T28a – 28-Pin Connector, left side of engine, near distributor
- T45 – 45-Pin Connector, on Mono-Motronic Engine Control Module (ECM)
- (15) – Ground connection, on cylinder head
- (174) – Ground connection –2–, in Mono-Motronic Throttle Body Fuel Injection (TBI) wiring harness
- (E13) – Plus connection –2– (15) in Mono-Motronic Throttle Body Fuel Injection (TBI) wiring harness

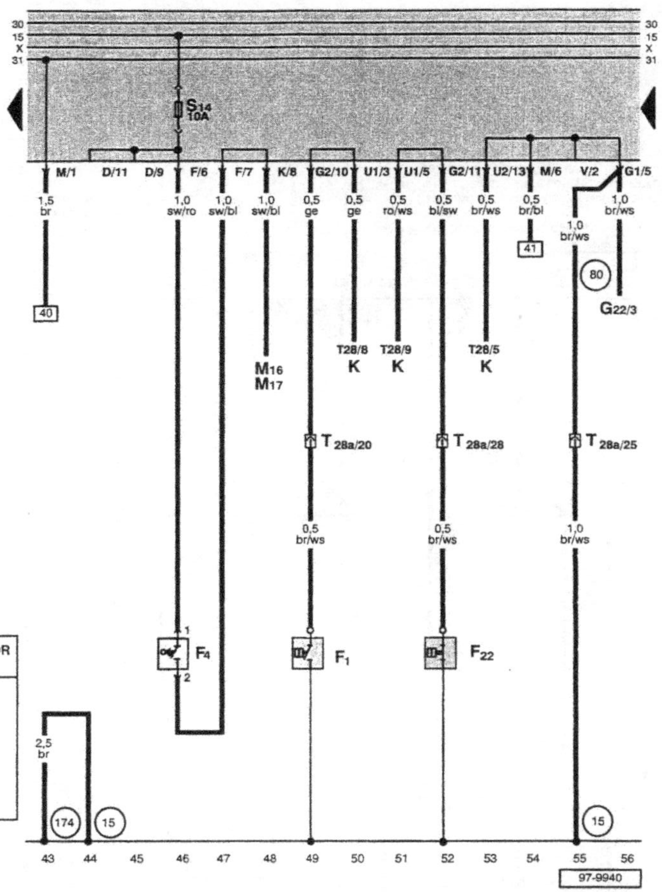

WIRING COLOR CODE

ws	=	white
sw	=	black
ro	=	red
br	=	brown
gn	=	green
bl	=	blue
gr	=	grey
li	=	lilac
ge	=	yellow

- F1 – Oil Pressure Switch
- F4 – Back-Up Light Switch
- F22 – 0.3 Bar Oil Pressure Switch
- G22 – Speedometer Vehicle Speed Sensor (VSS)
- K – Instrument Cluster
- M16 – Left Back-Up Light
- M17 – Right Back-Up Light
- T28 – 28-Pin Connector, instrument cluster
- T28a – 28-Pin Connector, left side of engine, near distributor
- (15) – Ground connector, on cylinder head
- (80) – Ground connection -1- in instrument panel wiring harness
- (174) – Ground connection -2-, in Mono-Motronic Throttle Body Fuel Injection (TBI) wiring harness

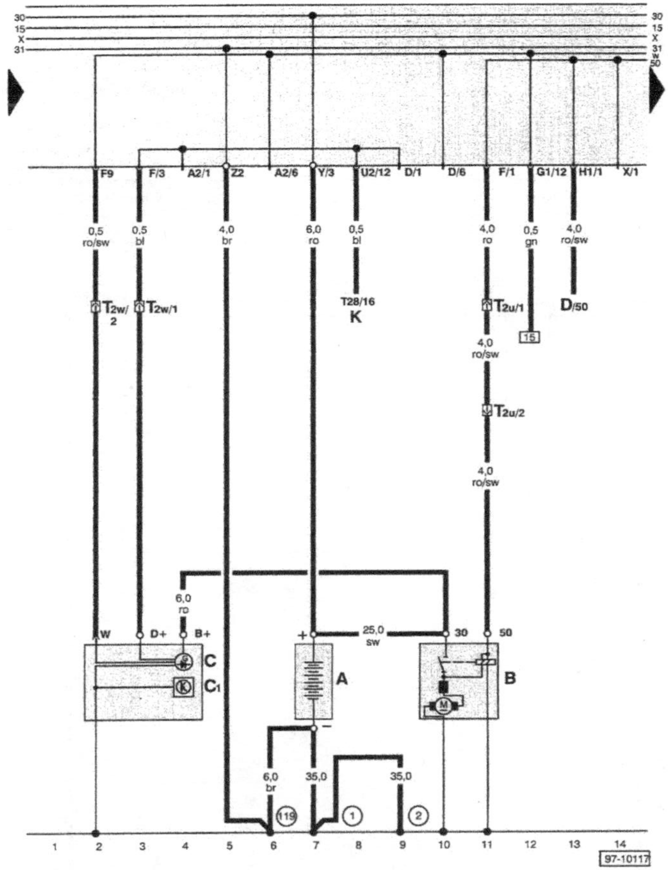

WIRING COLOR CODE		
ws	=	white
sw	=	black
ro	=	red
br	=	brown
gn	=	green
bl	=	blue
gr	=	grey
li	=	lilac
ge	=	yellow

A – Battery
B – Starter
C – Generator (Gen)
C1 – Voltage Regulator (VR)
D – Ignition/Starter Switch
K – Instrument Cluster
T2u – Double Connector, behind fuse/relay panel
T2w – Double Connector, near battery
T28 – 28-Pin Connector, instrument cluster

(1) – Ground strap, battery to body

(2) – Ground strap, transmission to body

(119) – Ground connection –1–, in headlight wiring harness

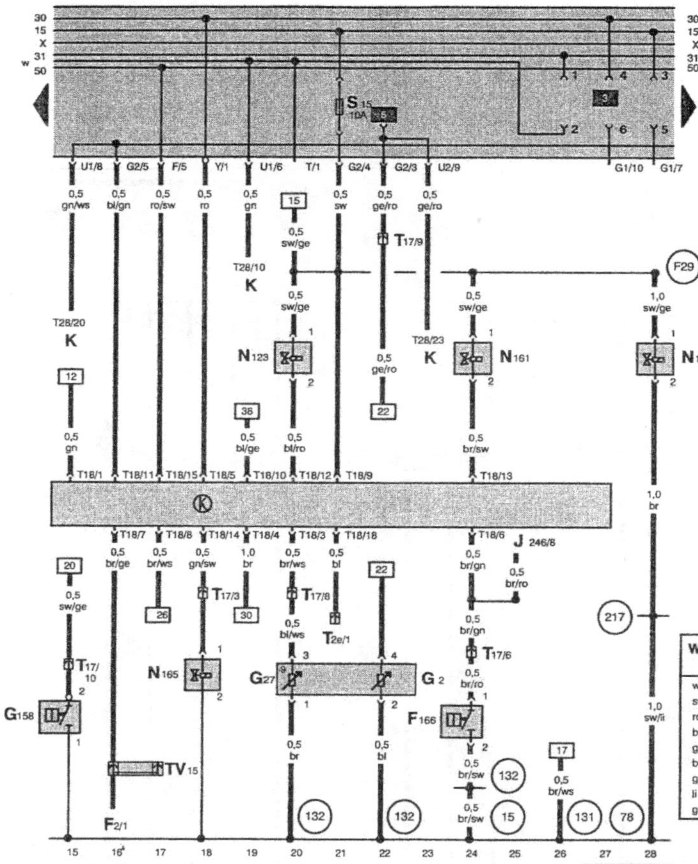

WIRING COLOR CODE		
ws	=	white
sw	=	black
ro	=	red
br	=	brown
gn	=	green
bl	=	blue
gr	=	grey
li	=	lilac
ge	=	yellow

F2 – Left Front Door Contact Switch
F166 – EGR Part Throttle Switch
G2 – Engine Coolant Temperature (ECT) Sensor
G27 – Engine Temperature Sensor
J179 – Automatic Glow Time Control Module
J246 – A/C Clutch Cut-off Relay
K – Instrument Cluster
N109 – Fuel Cut-off Valve
N123 – Idle Air Control (IAC) Solenoid
N161 – Two-way valve for EGR
N165 – Injection Start Positioner
G158 – Ignition Start Sensor
T2e – Double Connector, behind fuse/relay panel, A/C connection
T17 – 17-Pin Connector, on engine
T18 – 1-Pin Connector, on Automatic Glow Time Control Module
T28 – 28-Pin Connector, instrument cluster
TV15 – Wire Connector, door contact switch

(15) – Ground connection, on cylinder head

(78) – Ground connection, in engine compartment

(131) – Ground connection –2–, in engine compartment wiring harness

(132) – Ground connection –3–, in engine compartment wiring harness

(217) – Ground connection –1–, in glow plug

(C24) – Wire connection (blower, 2nd speed), in headlight/air conditioner wiring harness

(C99) – Wire connection in headlight/air conditioner wiring harness

(F29) – Wire connection (15a), in glow plug system wiring harness

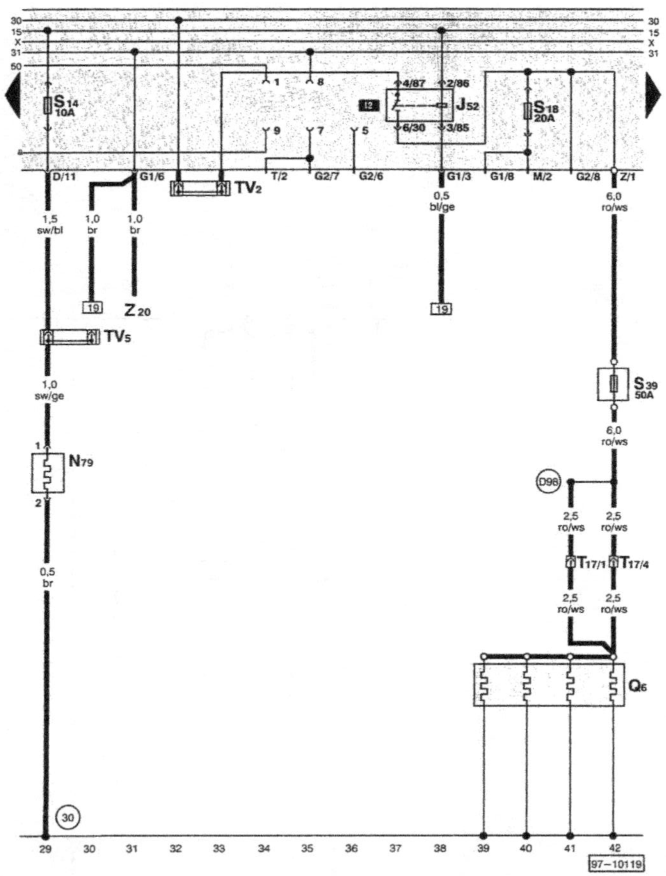

WIRING COLOR CODE		
ws	=	white
sw	=	black
ro	=	red
br	=	brown
gn	=	green
bl	=	blue
gr	=	grey
li	=	lilac
ge	=	yellow

J52 – Glow Plug Relay
N79 – Positive Crankcase Ventilation (PCV) Heating Element
Q6 – Engine Glow Plug
S39 – Engine Glow Plug Fuse Strip, 50A
T17 – 17-Pin Connector, on engine
TV2 – Terminal 30 (B+) Wire Connector
TV5 – Terminal 15a Wire Connector
Z20 – Left Washer Nozzle Heater

(30) – Ground connection –1–, beside fuse/relay panel

(D98) – Wire connection (glow plugs), in engine compartment wiring harness, right

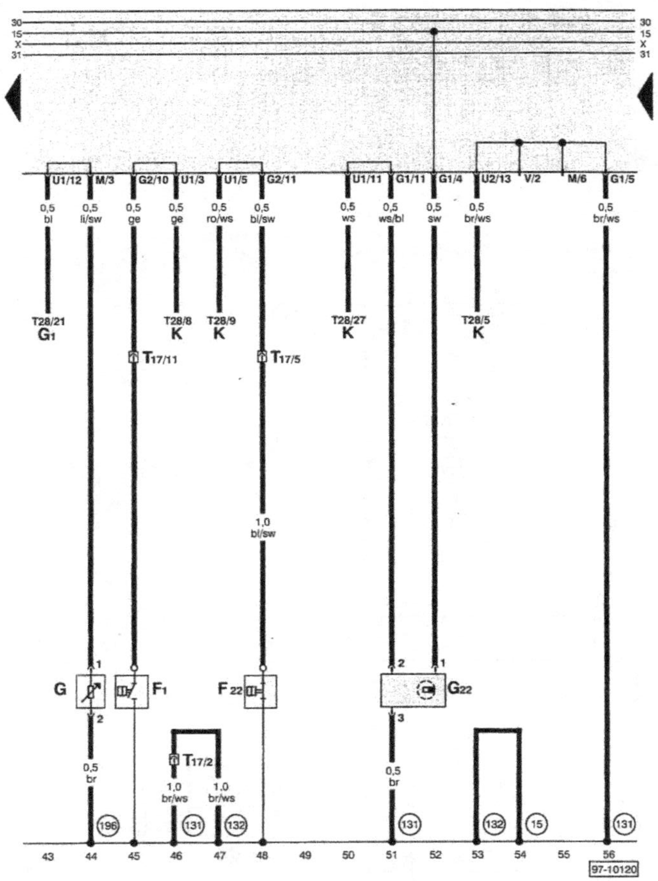

WIRING COLOR CODE		
ws	=	white
sw	=	black
ro	=	red
br	=	brown
gn	=	green
bl	=	blue
gr	=	grey
li	=	lilac
ge	=	yellow

F1 – Oil Pressure Switch
F22 – 0.3 Bar Oil Pressure Switch
G – Fuel Level Sensor
G1 – Fuel Gauge
G22 – Speedometer Vehicle Speed Sensor (VSS)
K – Instrument Cluster
T17 – 17-Pin Connector, on engine
T28 – 28-Pin Connector, instrument cluster

(15) – Ground connection, on cylinder head

(131) – Ground connection –2–, in engine compartment wiring harness

(132) – Ground connection –3–, in engine compartment wiring harness

(196) – Ground connection –1–, in rear wiring harness

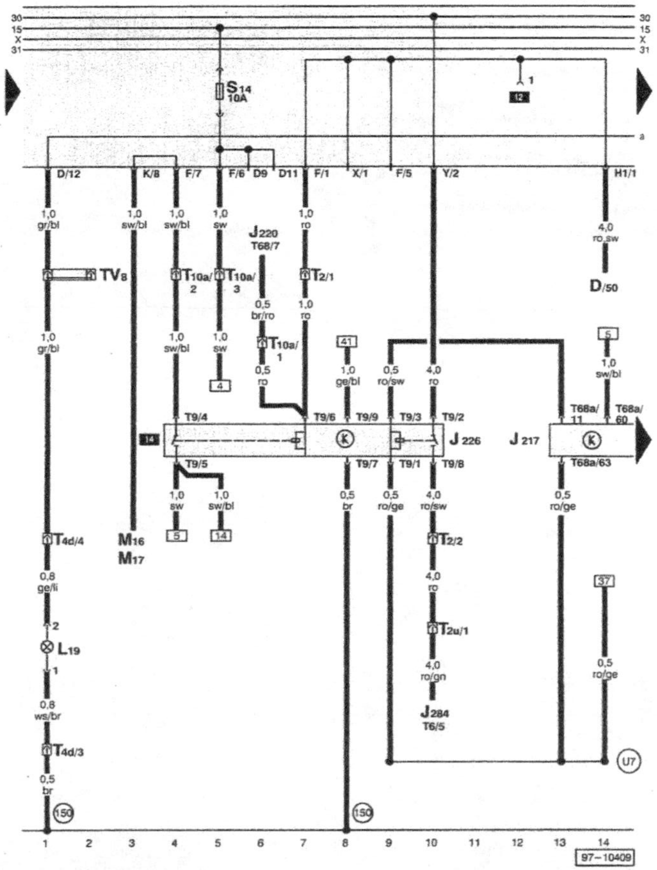

WIRING COLOR CODE		
ws	=	white
sw	=	black
ro	=	red
br	=	brown
gn	=	green
bl	=	blue
gr	=	grey
li	=	lilac
ge	=	yellow

D – Ignition/Starter Switch
J217 – Transmission Control Module (TCM)
J220 – Motronic Engine Control Module (ECM)
J226 – Park/Neutral Position (PNP) Relay
J284 – Alarm System Control Module
L19 – Automatic Transmission Console Light
M16 – Left Back-Up Light
M17 – Right Back-Up Light
T2 – Double Connector, behind fuse/relay panel, only on automatic transmission
T2u – Double Connector, behind fuse/relay panel
T4d – 4-Pin Connector, below center console
T6 – 6-Pin Connector, behind dash panel, left, on Alarm System Control Module
T10a – 10-Pin Connector, behind fuse/relay panel
T9 – 9-Pin Connector, on Park/Neutral Position (PNP) Relay
T68 – 68-Pin Connector, under right side of hood, on Motronic Engine Control Module (ECM)
T68a – 68-Pin Connector, on Transmission Control Module (TCM)
TV8 – Terminal 58b wire connector
(150) – Ground connection, in automatic transmission wiring harness
(U7) – Wire connection (G51), in automatic transmission wiring harness

Automatic transmission

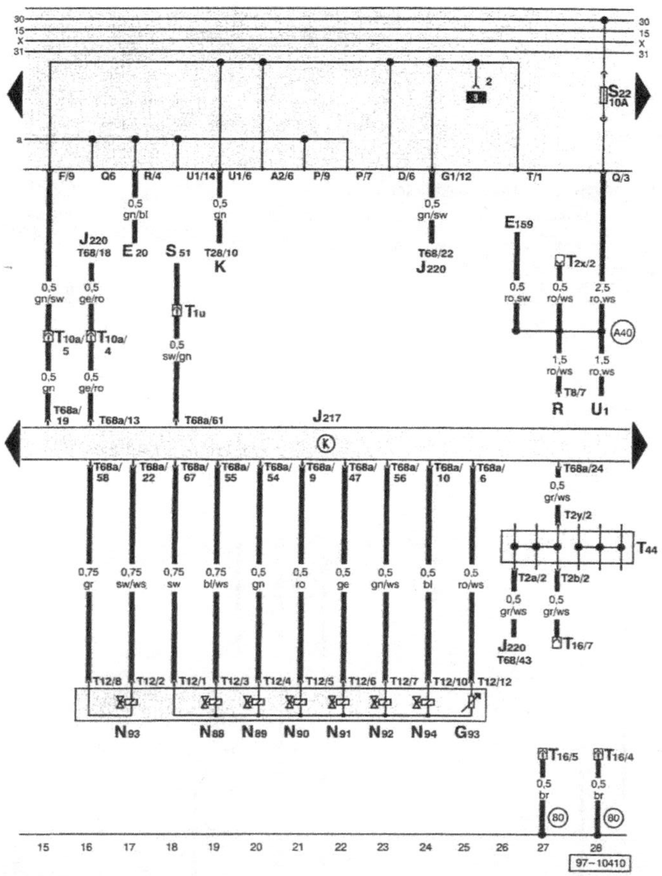

WIRING COLOR CODE		
ws	=	white
sw	=	black
ro	=	red
br	=	brown
gn	=	green
bl	=	blue
gr	=	grey
li	=	lilac
ge	=	yellow

E20 – Instrument Panel Light Dimmer Switch
E159 – Fresh Air/Recirculating Flap Switch
G93 – Transmission Fluid Temperature Sensor
J217 – Transmission Control Module (TCM)
J220 – Motronic Engine Control Module (ECM)
K – Instrument Cluster
N88 – Solenoid Valve 1
N89 – Solenoid Valve 2
N90 – Solenoid Valve 3
N91 – Solenoid Valve 4
N92 – Solenoid Valve 5
N93 – Solenoid Valve 6
N94 – Solenoid Valve 7
R – Radio
S51 – Fuse
T2a – Double Connector, on junction connector (T44) above fuse/relay panel
T2b – Double Connector, on junction connector (T44) above fuse/relay panel
T2y – Double Connector, on junction connector (T44) above fuse/relay panel
T8 – 8-Pin Connector, on radio
T10a – 10-Pin Connector, behind fuse/relay panel
T12 – 12-Pin Connector, on transmission
T16 – 16-Pin Connector, Data Link Connector, behind center of dash panel
T28 – 28-Pin Connector, instrument cluster
T1u – Single Connector, behind fuse/relay panel
T44 – 44-Pin Connector Junction Box for on board diagnosis, above fuse/relay panel
T68 – 68-Pin Connector, under right side of hood, on Motronic Engine Control Module (ECM)
T68a – 68-Pin Connector, below right front seat, on Transmission Control Module (TCM)
U1 – Cigarette Lighter
(80) – Ground connection – 1 –, in instrument panel wiring harness
(A40) – Plus connection (30), in instrument panel wiring harness

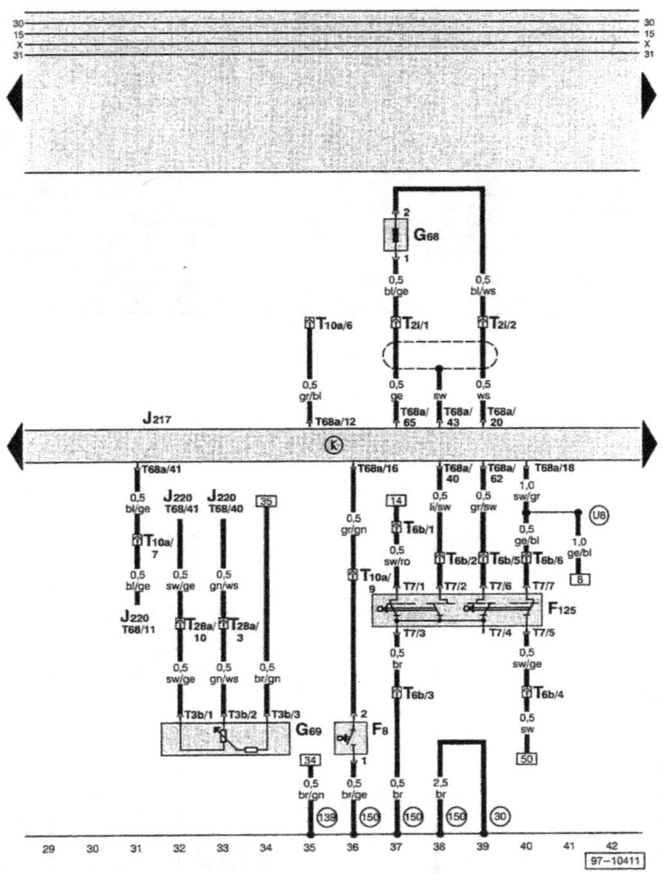

WIRING COLOR CODE

ws	=	white
sw	=	black
ro	=	red
br	=	brown
gn	=	green
bl	=	blue
gr	=	grey
li	=	lilac
ge	=	yellow

- F8 – Kick Down Switch
- F125 – Multi-Function Transmission Range (TR) Switch
- G68 – Vehicle Speed Sensor (VSS)
- G69 – Throttle Position (TP) Sensor
- J217 – Transmission Control Module (TCM)
- J220 – Motronic Engine Control Module (ECM)
- T1d – Single Connector, behind fuse/relay panel
- T2i – Double Connector, below battery
- T3b – 3-Pin Connector, on throttle position sensor
- T4b – 4-Pin Connector, on throttle position sensor
- T6b – 6-Pin Connector, below battery
- T7 – 7-Pin Connector, on transmission
- T10a – 10-Pin Connector, behind fuse/relay panel
- T28a – 28-Pin Connector, left side of engine, near distributor
- T68 – 68-Pin Connector, under right side of hood, on Motronic Engine Control Module (ECM)
- T68a – 68-Pin Connector, below right front seat, on Transmission Control Module (TCM)
- (30) – Ground connection –1–, beside fuse/relay panel
- (139) – Ground connection (sensor ground), in Motronic Multiport Fuel Injection (MFI) wiring harness
- (150) – Ground connection, in automatic transmission wiring harness
- (U8) – Wire connection (657), in automatic transmission wiring harness

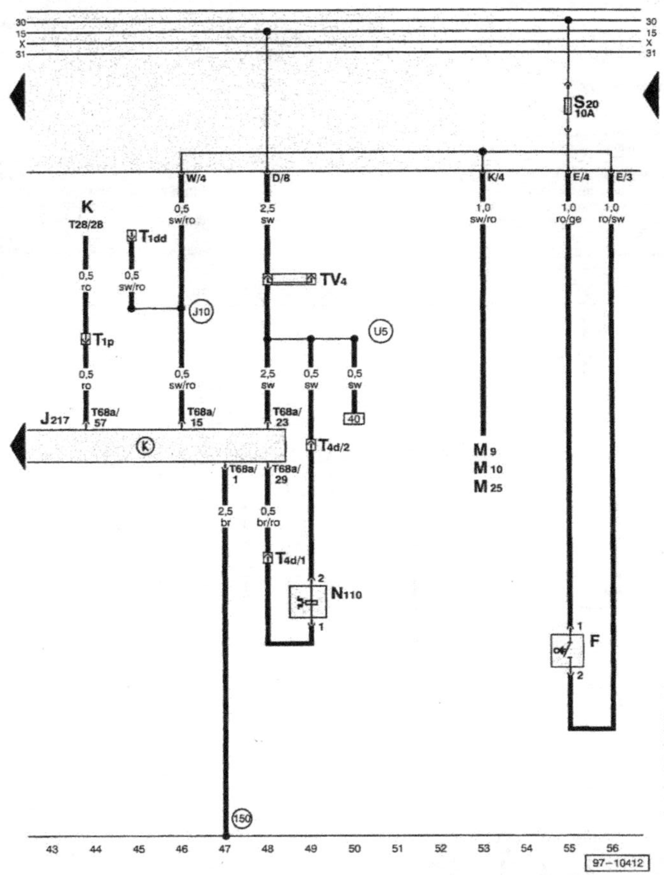

WIRING COLOR CODE

ws	=	white
sw	=	black
ro	=	red
br	=	brown
gn	=	green
bl	=	blue
gr	=	grey
li	=	lilac
ge	=	yellow

- F – Brake Light Switch
- J217 – Transmission Control Module (TCM)
- K – Instrument Cluster
- M9 – Left Brake Light
- M10 – Right Brake Light
- M25 – High-Mount Brake Light
- N110 – Shift Lock Solenoid
- T1p – Single Connector, behind fuse/relay panel
- T1dd – Single Connector, behind fuse/relay panel
- T4d – 4-Pin Connector, below center console
- T28 – 28-Pin Connector, instrument cluster
- T68a – 68-Pin Connector, below right front seat, on Transmission Control Module (TCM)
- TV4 – Terminal 15 Wire Connector
- (150) – Ground connection, in automatic transmission wiring harness
- (U5) – Wire connection –2– (15), in automatic transmission wiring harness
- (J10) – Wire connection (54), in ABS wiring harness

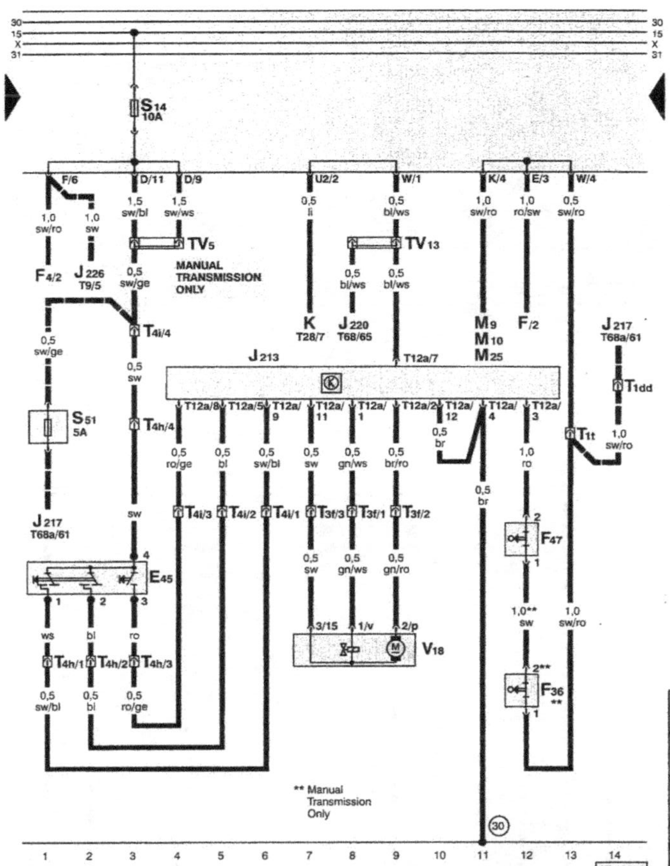

WIRING COLOR CODE		
ws	=	white
sw	=	black
ro	=	red
br	=	brown
gn	=	green
bl	=	blue
gr	=	grey
li	=	lilac
ge	=	yellow

97-10078

E45 – Cruise Control Switch
F – Brake Light Switch
F4 – Back-Up Light Switch
F36 – Cruise Control Clutch Pedal Position (CCP) Switch
F47 – Cruise Control Brake Pedal Position (BPP) Switch
J213 – Cruise Control, Control Module
J217 – Transmission Control Module (TCM)
J220 – Motronic Engine Control Module (ECM)
J226 – Park/Neutral Position (PNP) Relay
K – Instrument Cluster
M9 – Left Brake Light
M10 – Right Brake Light
M25 – High-Mount Brake Light
S51 – Fuse
T3f – 3-Pin Connector, below left side of dash
T1t – Single Connector, behind fuse/relay panel
T1dd – Single Connector, behind fuse/relay panel

T4h – 4-Pin Connector, behind steering column switch trim
T4i – 4-Pin Connector, behind fuse/relay panel
T9 – 9-Pin Connector, on Park/Neutral Position (PNP) Relay
T12a – 12-Pin Connector, on Cruise Control Module
T28 – 28-Pin Connector, instrument cluster
T68a – 68-Pin Connector, below right front seat, on Transmission Control Module (TCM)
TV13 – Connector, vehicle speed signal
T68 – 68-Pin Connector, under right side of hood, on Motronic Engine Control Module (ECM)
TV5 – Terminal 15a Wire Connector
V18 – Cruise Control Vacuum Pump
(30) – Ground connection –1–, beside fuse/relay panel
– – – Automatic transmission only

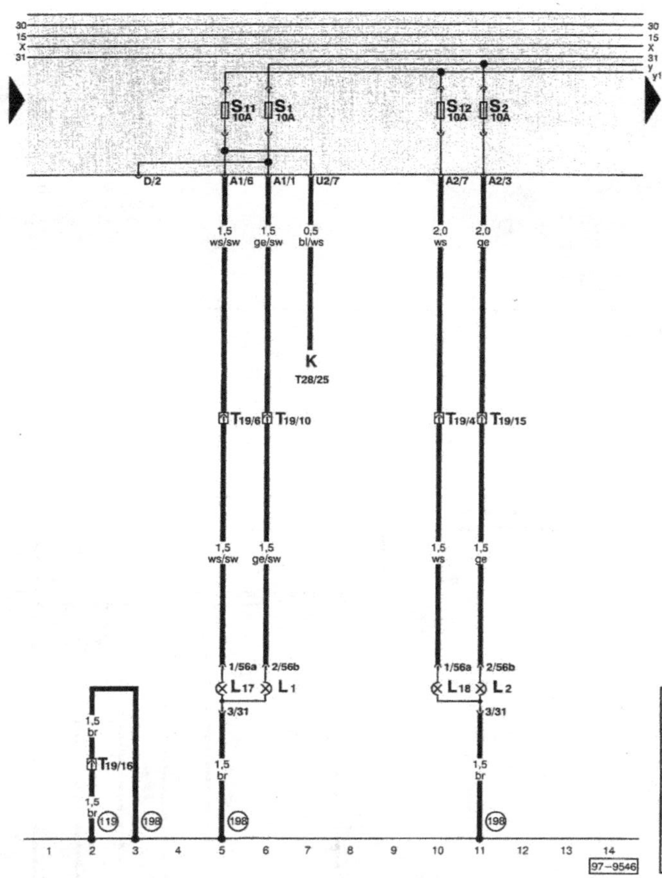

WIRING COLOR CODE		
ws	=	white
sw	=	black
ro	=	red
br	=	brown
gn	=	green
bl	=	blue
gr	=	grey
li	=	lilac
ge	=	yellow

97-9546

K – Instrument Cluster
L1 – Left Headlight
L2 – Right Headlight
L17 – Left Sealed Beam High Headlight
L18 – Right Sealed Beam High Headlight
T19 – 19-Pin Connector, behind radiator grill, center
T28 – 28-Pin Connector, instrument cluster
(119) – Ground connection –1–, in headlight wiring harness
(198) – Ground connection, in front light wiring harness

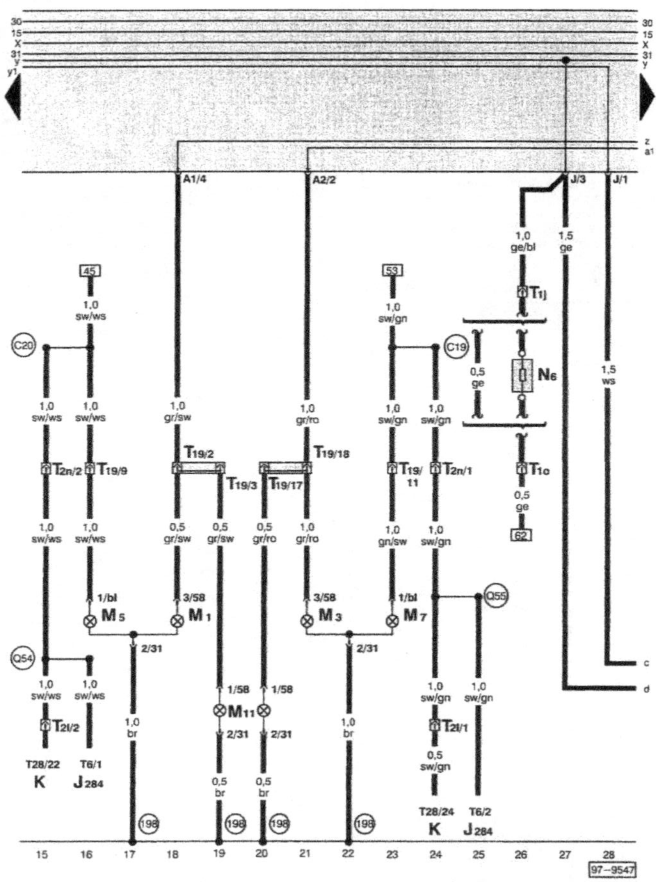

WIRING COLOR CODE		
ws	=	white
sw	=	black
ro	=	red
br	=	brown
gn	=	green
bl	=	blue
gr	=	grey
li	=	lilac
ge	=	yellow

J284 – Alarm System Control Module
K – Instrument Cluster
M1 – Left Parking Light
M3 – Right Parking Light
M5 – Left Front Turn Signal Light
M7 – Right Front Turn Signal Light
M11 – Front Side Marker Lights
N6 – Series Resistance
T1j – Single Connector, behind fuse/relay panel
T1o – Single Connector, behind fuse/relay panel
T2l – Double Connector, behind fuse/relay panel
T2n – Double Connector, behind fuse/relay panel
T6 – 6-Pin Connector, behind dash panel, left, on Alarm System Control Module
T19 – 19-Pin Connector, behind radiator grill, center
T28 – 28-Pin Connector, instrument cluster

(198) – Ground connection, in front light wiring harness

(C19) – Wire connection (right turn signal), in headlight wiring harness

(C20) – Wire connection (left turn signal), in headlight wiring harness

(Q54) – Wire connection (left turn signal), in alarm system wiring harness

(Q55) – Wire connection (right turn signal), in alarm system wiring harness

1995

Daytime running lights
(except GTI)

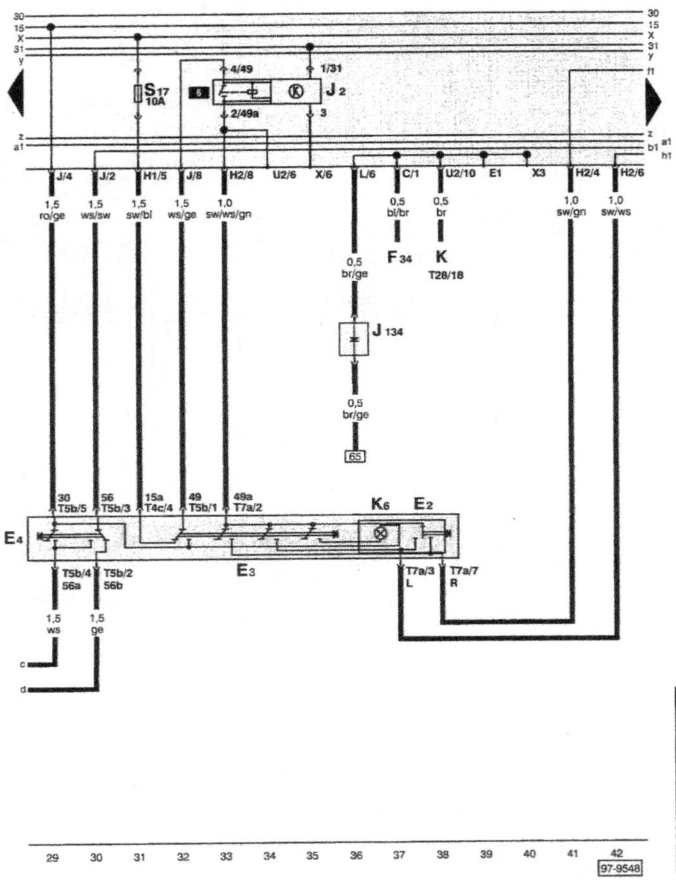

WIRING COLOR CODE		
ws	=	white
sw	=	black
ro	=	red
br	=	brown
gn	=	green
bl	=	blue
gr	=	grey
li	=	lilac
ge	=	yellow

E2 – Turn Signal Switch
E3 – Emergency Flasher Switch
E4 – Headlight Dimmer/Flasher Switch
F34 – Brake Fluid Level Warning Switch
J2 – Emergency Flasher Relay
J134 – Daytime Running Light Diode
K – Instrument Cluster
K6 – Emergency Flasher Warning Light
T4c – 4-Pin Connector, behind steering column switch trim
T5b – 5-Pin Connector, behind steering column switch trim
T7a – 7-Pin Connector, behind steering column switch trim
T28 – 28-Pin Connector, instrument cluster

1995

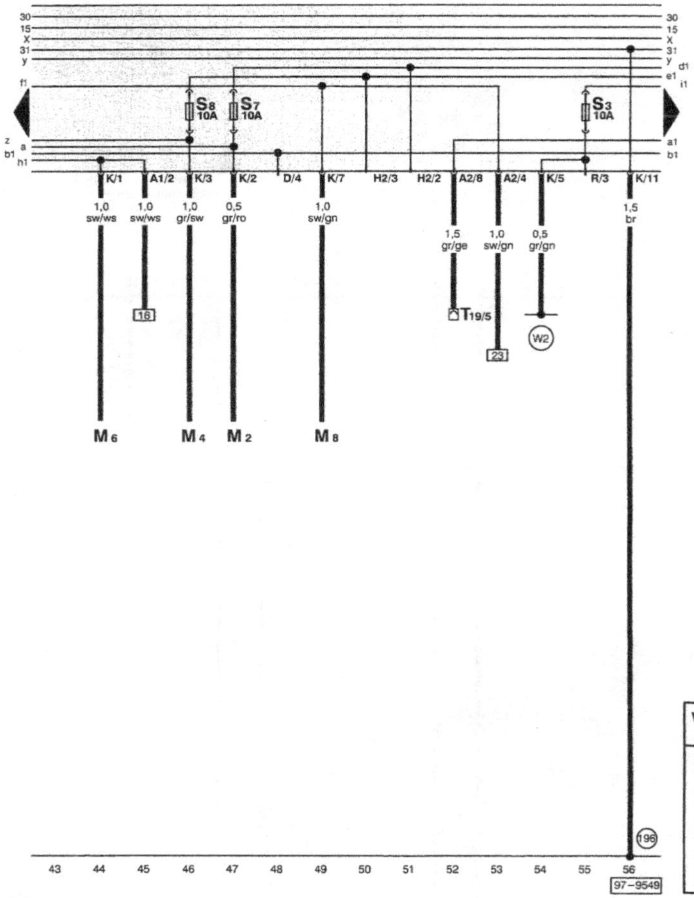

WIRING COLOR CODE

ws	=	white
sw	=	black
ro	=	red
br	=	brown
gn	=	green
bl	=	blue
gr	=	grey
li	=	lilac
ge	=	yellow

M2 – Right Tail Light
M4 – Left Tail Light
M6 – Left Rear Turn Signal Light
M8 – Right Rear Turn Signal Light
T19 – 19-Pin Connector, behind radiator grill, center
(196) – Ground connection –1–, in rear wiring harness
(W2) – Plus connection (58), in rear wiring harness

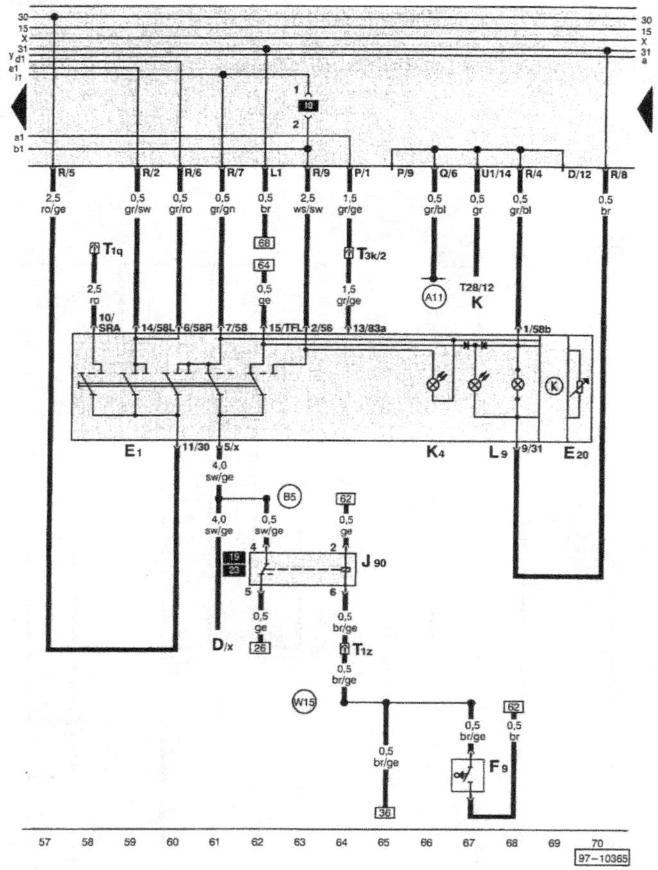

WIRING COLOR CODE

ws	=	white
sw	=	black
ro	=	red
br	=	brown
gn	=	green
bl	=	blue
gr	=	grey
li	=	lilac
ge	=	yellow

D – Ignition/Starter Switch
E1 – Light Switch
E20 – Instrument Panel Light Dimmer Switch
F9 – Parking Brake Warning Light Switch
J90 – Daytime Running Lights Switch On Relay
K – Instrument Cluster
K4 – Park Light Indicator Light
L9 – Headlight Switch Light
T1q – Single Connector, behind fuse/relay panel
T1z – Single Connector, behind fuse/relay panel
T3k – 3-Pin Connector, behind fuse/relay panel
T28 – 28-Pin Connector, instrument cluster
(A11) – Plus connection (58b), in instrument panel wiring harness
(B5) – Wire connection (X), in headlight switch wiring harness
(W15) – Wire connection (HB) in rear wiring harness

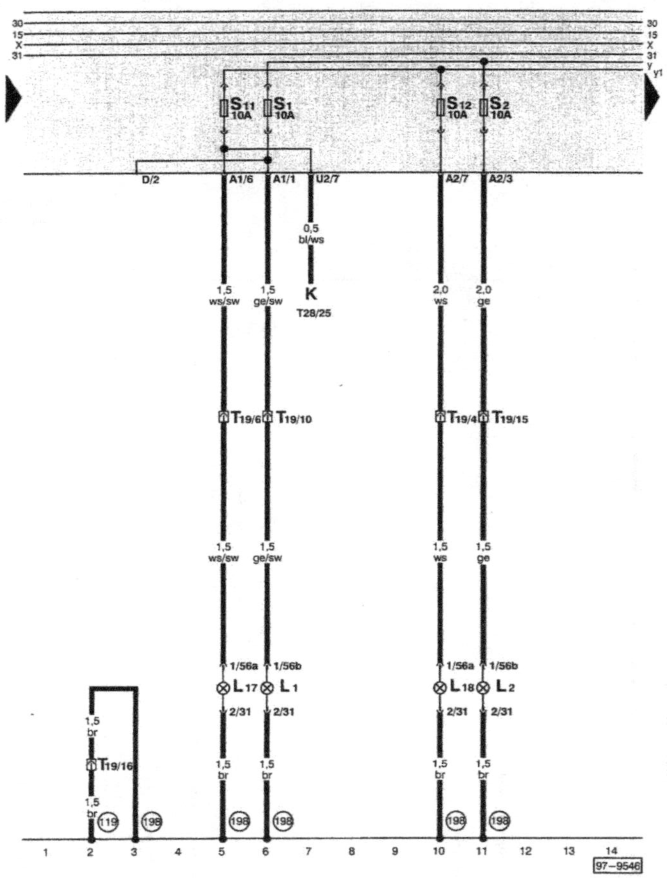

WIRING COLOR CODE

ws	=	white
sw	=	black
ro	=	red
br	=	brown
gn	=	green
bl	=	blue
gr	=	grey
li	=	lilac
ge	=	yellow

- K – Instrument Cluster
- L1 – Left Headlight
- L2 – Right Headlight
- L17 – Left Sealed Beam High Beam Headlight
- L18 – Right Sealed Beam High Beam Headlight
- T19 – 19-Pin Connector, behind radiator grill, center
- T28 – 28-Pin Connector, instrument cluster
- (119) – Ground connection –1–, in headlight wiring harness
- (198) – Ground connection, in front light wiring harness

1995

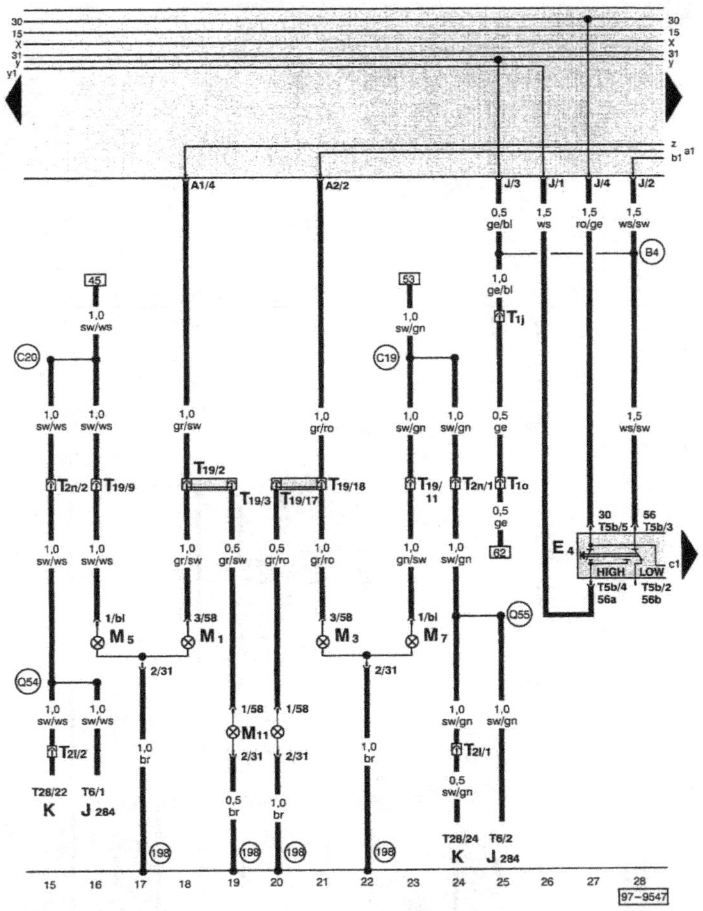

WIRING COLOR CODE

ws	=	white
sw	=	black
ro	=	red
br	=	brown
gn	=	green
bl	=	blue
gr	=	grey
li	=	lilac
ge	=	yellow

- E4 – Headlight Dimmer/Flasher Switch
- J284 – Alarm System Control Module
- K – Instrument Cluster
- M1 – Left Parking Light
- M3 – Right Parking Light
- M5 – Left Front Turn Signal Light
- M7 – Right Front Turn Signal Light
- M11 – Front Side Marker Lights
- T1j – Single Connector, behind fuse/relay panel
- T1o – Single Connector, behind fuse/relay panel
- T2l – Double Connector, behind fuse/relay panel
- T2n – Double Connector, behind fuse/relay panel
- T5b – 5-Pin Connector, behind steering column switch trim
- T6 – 6-Pin Connector, behind dash panel, left, on Alarm System Control Module
- T19 – 19-Pin Connector, behind radiator grill, center
- T28 – 28-Pin Connector, instrument cluster

- (198) – Ground connection, in front light wiring harness
- (B4) – Plus connection, (58e) in light switch wiring harness
- (C19) – Wire connection (right turn signal), in headlight wiring harness
- (C20) – Wire connection (left turn signal), in headlight wiring harness
- (Q54) – Wire connection (left turn signal), in alarm system wiring harness
- (Q55) – Wire connection (right turn signal), in alarm system wiring harness

1995

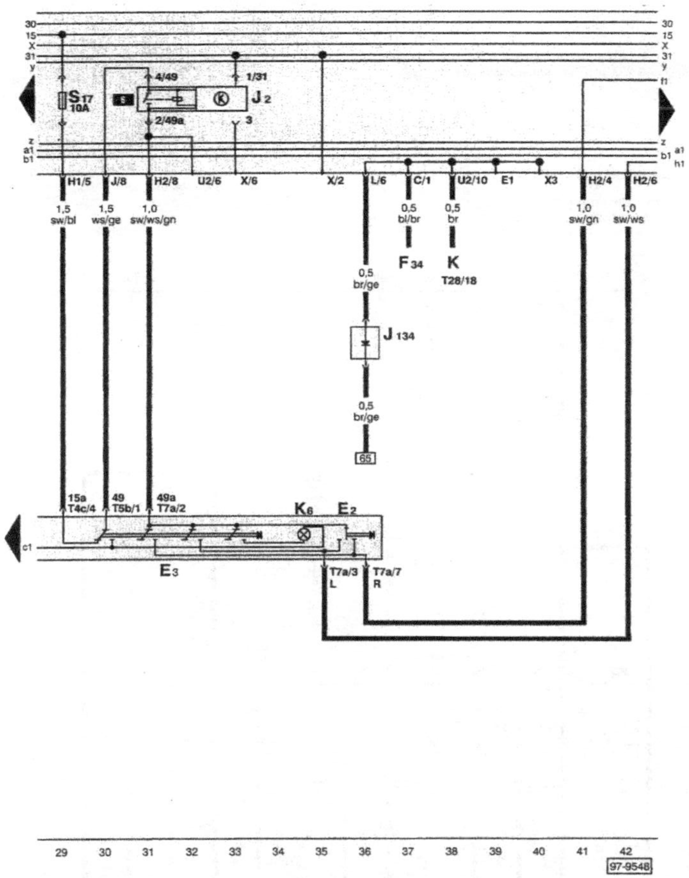

WIRING COLOR CODE

ws	=	white
sw	=	black
ro	=	red
br	=	brown
gn	=	green
bl	=	blue
gr	=	grey
li	=	lilac
ge	=	yellow

E2 – Turn Signal Switch
E3 – Emergency Flasher Switch
F34 – Brake Fluid Level Warning Switch
J2 – Emergency Flasher Relay
J134 – Daytime Running Light Diode
K – Instrument Cluster
K6 – Emergency Flasher Warning Light
T4c – 4-Pin Connector, behind steering column switch trim
T5b – 5-Pin Connector, behind steering column switch trim
T7a – 7-Pin Connector, behind steering column switch trim
T28 – 28-Pin Connector, instrument cluster

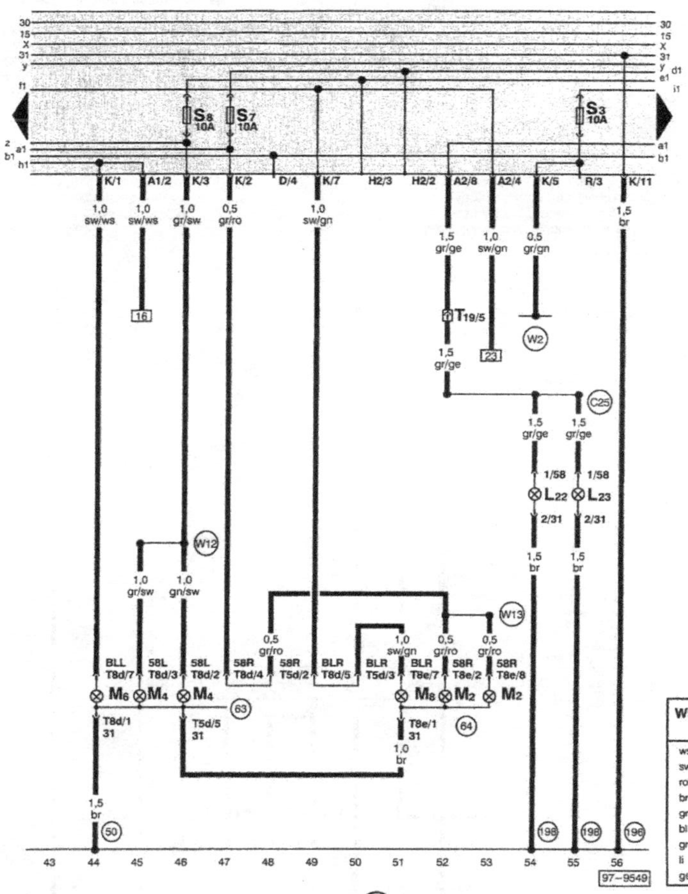

WIRING COLOR CODE

ws	=	white
sw	=	black
ro	=	red
br	=	brown
gn	=	green
bl	=	blue
gr	=	grey
li	=	lilac
ge	=	yellow

L22 – Left Front Fog Light
L23 – Right Front Fog Light
M2 – Right Tail Light
M4 – Left Tail Light
M6 – Left Rear Turn Signal Light
M8 – Right Rear Turn Signal Light
T5d – 5-Pin Connector, on tail light, left
T8d – 8-Pin Connector, on tail light, left
T8e – 8-Pin Connector, on tail light, right
T19 – 19-Pin Connector, behind radiator grill, center

(50) – Ground connection, in luggage compartment, left

(63) – Ground connection, on left tail light bulb holder

(64) – Ground connection, on right tail light bulb holder

(196) – Ground connection –1–, in rear wiring harness

(198) – Ground connection, in front light wiring harness

(C25) – Wire connection (fog lights), in front left wiring harness

(W2) – Plus connection (58), in rear wiring harness

(W12) – Wire connection (DWA), in rear wiring harness

(W13) – Wire connection (58L), in rear wiring harness

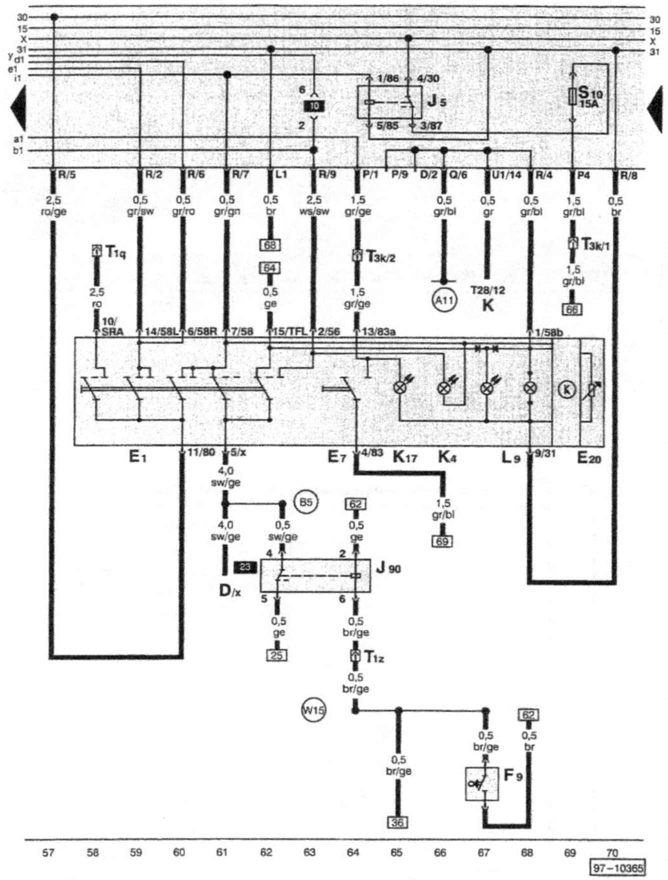

WIRING COLOR CODE		
ws	=	white
sw	=	black
ro	=	red
br	=	brown
gn	=	green
bl	=	blue
gr	=	grey
li	=	lilac
ge	=	yellow

D – Ignition/Starter Switch
E1 – Light Switch
E7 – Fog Light Switch
E20 – Instrument Panel Light Dimmer Switch
F9 – Parking Brake Warning Light Switch
J5 – Fog Light Relay
J90 – Daytime Running Lights Switch-on Relay
K – Instrument Cluster
K4 – Park Light Indicator Light
K17 – Fog Light Indicator Light
L9 – Headlight Switch Light
T1q – Single Connector, behind fuse/relay panel
T1z – Single Connector, behind fuse/relay panel
T3k – Double Connector, behind fuse/relay panel
T28 – 28-Pin Connector, instrument cluster
(A11) – Plus connection (58b), in instrument panel wiring harness
(B5) – Wire connection (X), in light switch wiring harness
(W15) – Wire connection (HB), in rear wiring harness

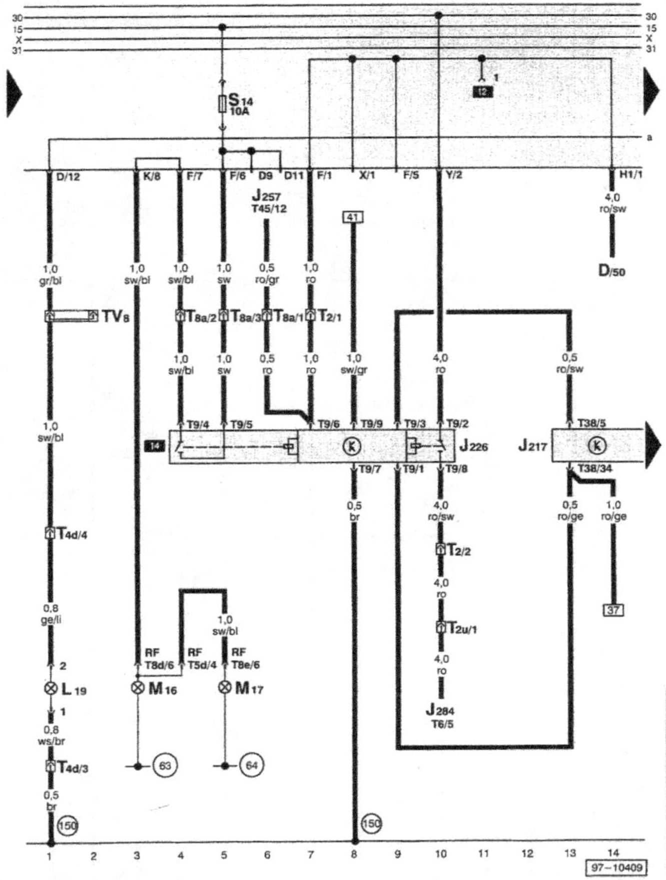

WIRING COLOR CODE		
ws	=	white
sw	=	black
ro	=	red
br	=	brown
gn	=	green
bl	=	blue
gr	=	grey
li	=	lilac
ge	=	yellow

D – Ignition/Starter Switch
J217 – Transmission Control Module (TCM)
J226 – Park/Neutral Position (PNP) Relay
J257 – Mono-Motronic Engine Control Module (ECM)
J284 – Alarm System Control Module
L19 – Automatic Transmission Console Light
M16 – Left Back-Up Light
M17 – Right Back-Up Light
T2 – Double Connector, behind fuse/relay panel, only on automatic transmission
T2u – Double Connector, behind fuse/relay panel
T4d – 4-Pin Connector, below center console
T5d – 5-Pin Connector, on tail light, left
T6 – 6-Pin Connector, behind dash panel, left, on Alarm System Control Module
T8a – 8-Pin Connector, behind fuse/relay panel
T8d – 8-Pin Connector, on tail light, left
T8e – 8-Pin Connector, on tail light, right
T9 – 9-Pin Connector, on Park/Neutral Position (PNP) Relay
T38 – 38-Pin Connector, below right front seat, on Transmission Control Module (TCM)
T45 – 45-Pin Connector, on Mono-Motronic Engine Control Module (ECM)
TV8 – Connector, terminal 58b
(63) – Ground connection, on left tail light bulb holder
(64) – Ground connection, on right tail light bulb holder
(150) – Ground connection, in automatic transmission (Digital AGH) wiring harness

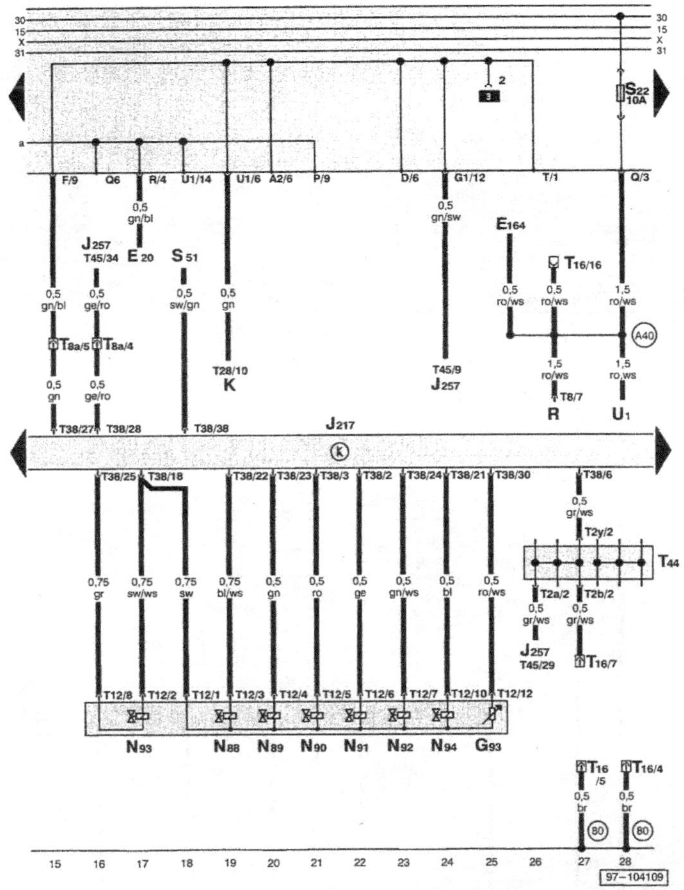

WIRING COLOR CODE		
ws	=	white
sw	=	black
ro	=	red
br	=	brown
gn	=	green
bl	=	blue
gr	=	grey
li	=	lilac
ge	=	yellow

E20 – Instrument Panel Light Dimmer Switch
E164 – Luggage Compartment Release Switch
G93 – Transmission Fluid Temperature Sensor
J217 – Transmission Control Module (TCM)
J257 – Mono-Motronic Engine Control Module (ECM)
K – Instrument Cluster
N88 – Solenoid Valve 1
N89 – Solenoid Valve 2
N90 – Solenoid Valve 3
N91 – Solenoid Valve 4
N92 – Solenoid Valve 5
N93 – Solenoid Valve 6
N94 – Solenoid Valve 7
R – Radio
S51 – Fuse
T2a – Double Connector, on junction connector (T44) above fuse/relay panel
T2b – Double Connector, on junction connector (T44) above fuse/relay panel
T2y – Double Connector, on junction connector (T44) above fuse/relay panel
T8 – 8-Pin Connector, on radio
T8a – 8-Pin Connector, behind fuse/relay panel
T12 – 12-Pin Connector, on transmission
T16 – 16-Pin Connector, Data Link Connector (DLC), behind center of dash panel
T28 – 28-Pin Connector, instrument cluster
T38 – 38-Pin Connector, below right front seat, on Transmission Control Module (TCM)
T44 – Junction Box for on board diagnosis, above fuse/relay panel
T45 – 45-Pin Connector, on Mono-Motronic Engine Control Module (ECM)
U1 – Cigarette Lighter
(80) – Ground connection –1–, in instrument panel wiring harness
(A40) – Plus connection (30), in instrument panel wiring harness

1995

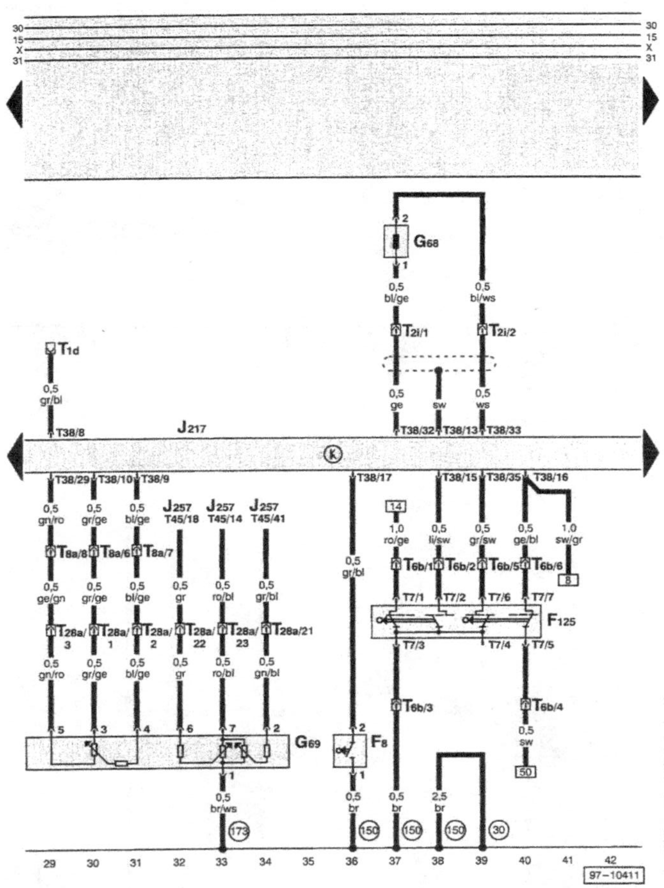

WIRING COLOR CODE		
ws	=	white
sw	=	black
ro	=	red
br	=	brown
gn	=	green
bl	=	blue
gr	=	grey
li	=	lilac
ge	=	yellow

F8 – Kick Down Switch
F125 – Multi-Function Transmission Range (TR) Switch
G68 – Vehicle Speed Sensor (VSS)
G69 – Throttle Position (TP) Sensor
J217 – Transmission Control Module (TCM)
J257 – Mono-Motronic Engine Control Module (ECM)
T1d – Single Connector, behind fuse/relay panel
T2i – Double Connector, below battery
T6b – 6-Pin Connector, below battery
T7 – 7-Pin Connector, on transmission
T8a – 8-Pin Connector, behind fuse/relay panel
T28a – 28-Pin Connector, left side of engine, near distributor
T38 – 38-Pin Connector, below right front seat, on Transmission Control Module (TCM)
T45 – 45-Pin Connector, on Mono-Motronic Engine Control Module (ECM)
(30) – Ground connection –1–, beside fuse/relay panel
(150) – Ground connection, in automatic transmission wiring harness
(173) – Ground connection –1–, in Mono-Motronic Throttle Body Fuel Injection (TBI) wiring harness

1995

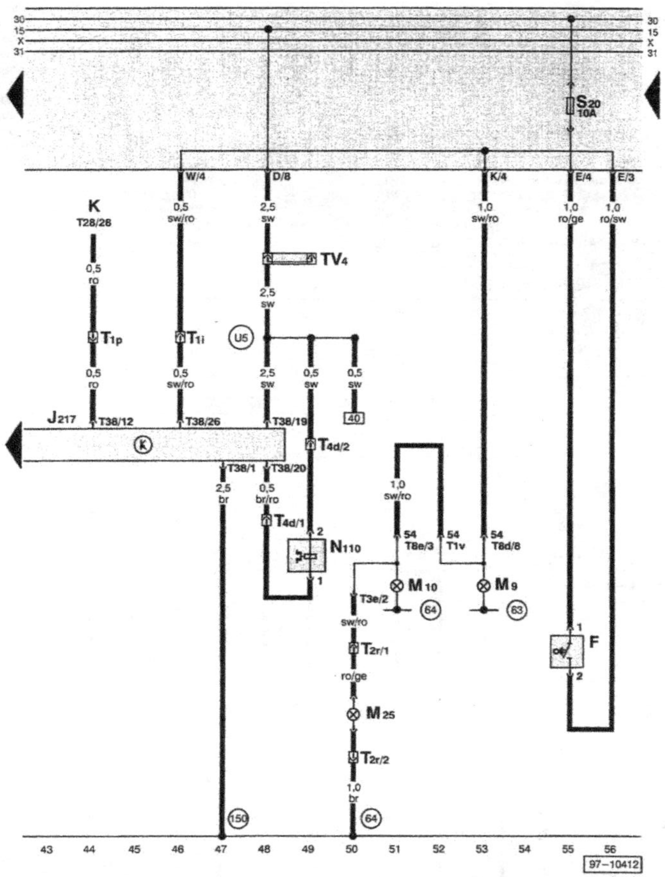

WIRING COLOR CODE

ws	=	white
sw	=	black
ro	=	red
br	=	brown
gn	=	green
bl	=	blue
gr	=	grey
li	=	lilac
ge	=	yellow

F – Brake Light Switch
J217 – Transmission Control Module (TCM)
K – Instrument Cluster
M9 – Left Brake Light
M10 – Right Brake Light
M25 – High-Mount Brake Light
N110 – Shift Lock Solenoid
T1i – Single Connector, behind fuse/relay panel
T1p – Single Connector, behind fuse/relay panel
T1v – Single Connector, on tail light, left
T2r – Double Connector, behind right grommet, on luggage compartment lid
T3e – 3-Pin Connector, on tail light, right
T4d – 4-Pin Connector, below center console
T8d – 8-Pin Connector, on tail light, left
T8e – 8-Pin Connector, on tail light right
T28 – 28-Pin Connector, instrument cluster
T38 – 38-Pin Connector, below right front seat, on Transmission Control Module (TCM)
TV4 – Terminal 15 Wire Connector

(63) – Ground connection, on left tail light bulb holder

(64) – Ground connection, on right tail light bulb holder

(150) – Ground connection, in automatic transmission wiring harness

(U5) – Wire connection –2– (15), in automatic transmission wiring harness

1995

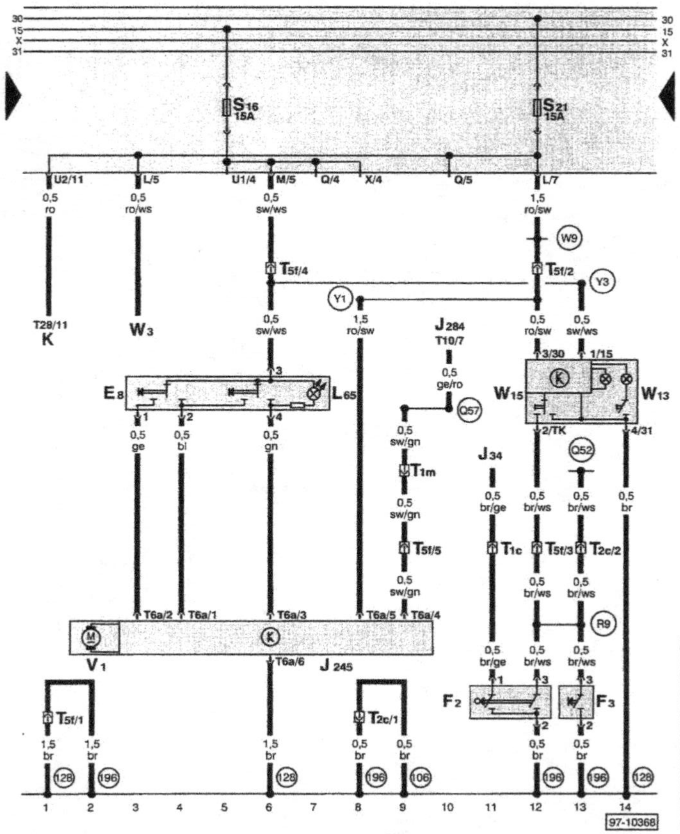

WIRING COLOR CODE

ws	=	white
sw	=	black
ro	=	red
br	=	brown
gn	=	green
bl	=	blue
gr	=	grey
li	=	lilac
ge	=	yellow

E8 – Sunroof Switch
F2 – Left Front Door Contact Switch
F3 – Right Front Door Contact Switch
J34 – Seat Belt Warning System Relay
J245 – Power Sunroof Control Module
J284 – Alarm System Control Module
K – Instrument Cluster
L65 – Sunroof Switch Light
T1c – Single Connector, behind fuse/relay panel
T1m – Single Connector, behind fuse/relay panel
T2c – Double Connector, on lower right A-pillar
T5f – 5-Pin Connector, above fuse/relay panel
T6a – 6-Pin Connector, behind interior light, on Power Sunroof Control Module
T10 – 10-Pin Connector, behind dash panel, left, on Alarm System Control Module
T28 – 28-Pin Connector, instrument cluster
V1 – Sunroof Motor
W3 – Luggage Compartment Light
W13 – Right Front Map/Reading Light
W15 – Interior Light With Delay Switch

(106) – Ground connection, in alarm system wiring harness

(128) – Ground connection –1–, in interior light wiring harness

(196) – Ground connection –1–, in rear wiring harness

(Q52) – Wire connection (door contact switch), in alarm system wiring harness

(Q57) – Wire connection (closed), in alarm system wiring harness

(R9) – Wire connection (door contact switch), in rear wiring harness

(W9) – Plus connection (30), in rear wiring harness

(Y1) – Plus connection (30), in interior light wiring harness

(Y3) – Plus connection (15) in interior light/sunroof wiring harness

1995

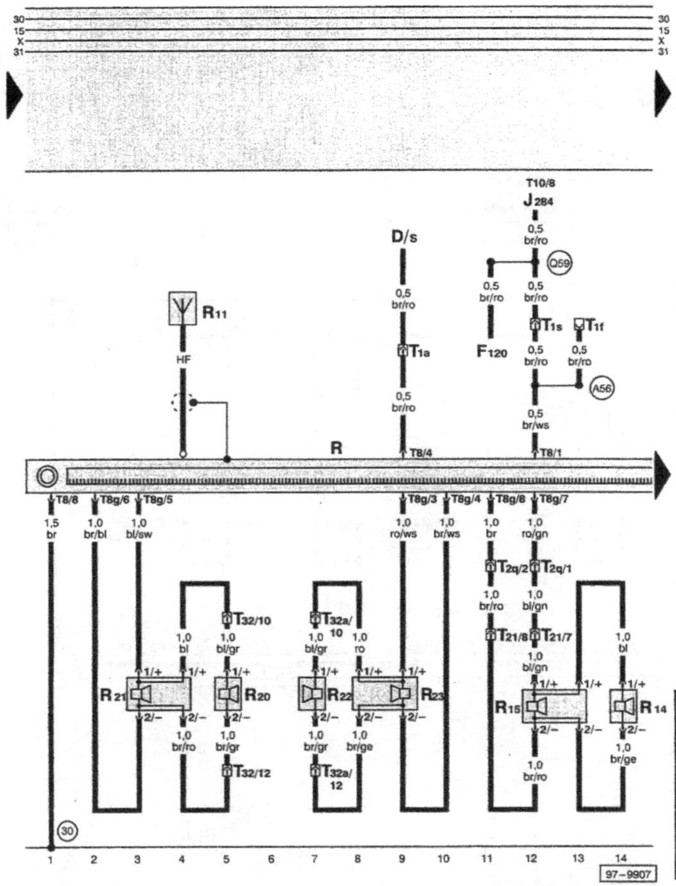

WIRING COLOR CODE		
ws	=	white
sw	=	black
ro	=	red
br	=	brown
gn	=	green
bl	=	blue
gr	=	grey
li	=	lilac
ge	=	yellow

D – Ignition/Starter Switch
F120 – Hood Alarm Switch
J284 – Alarm System Control Module
R – Radio
R11 – Antenna
R14 – Left Rear Tweeter
R15 – Left Rear Woofer
R20 – Left Front Tweeter
R21 – Left Front Woofer
R22 – Right Front Tweeter
R23 – Right Front Woofer,
T1a – Single Connector, behind fuse/relay panel
T1f – Single Connector, behind dash panel, center, behind radio
T1s – Single Connector, behind fuse/relay panel

T2q – Double Connector, on lower left A-pillar
T8 – 8-Pin Connector, on radio
T8g – 8-Pin Connector, on radio
T10 – 10-Pin Connector, behind dash panel, left, on Alarm System Control Module
T21 – 21-Pin Connector, on left B-pillar
T32 – 32-Pin Connector, on A-pillar, lower left
T32a – 32-Pin Connector, on A-pillar, lower right

(30) – Ground connection –1–, beside fuse/relay panel

(A56) – Wire connection –2–, (30) in instrument panel wiring harness

(Q59) – Wire connection (sensor), in alarm system wiring harness

1995

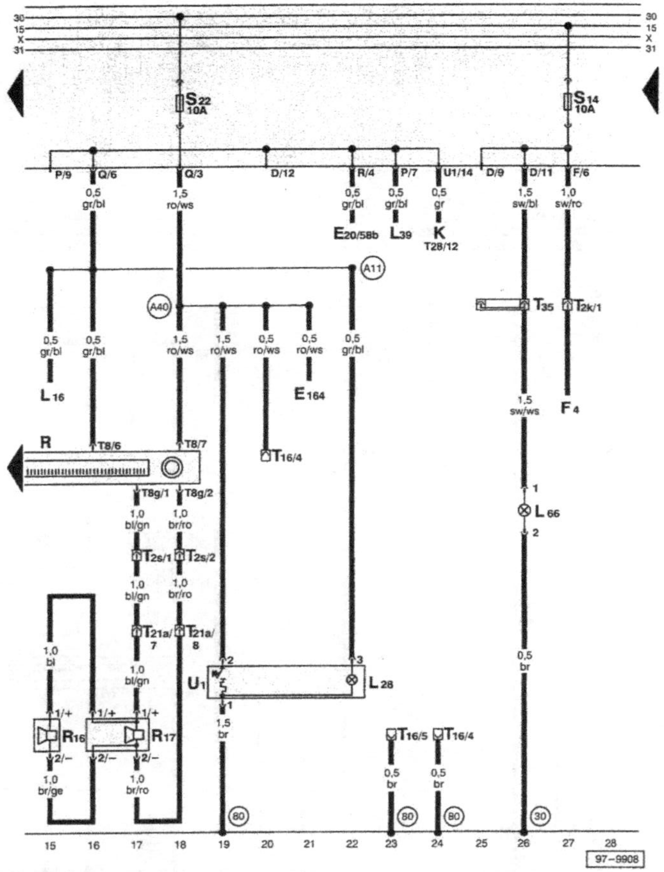

WIRING COLOR CODE		
ws	=	white
sw	=	black
ro	=	red
br	=	brown
gn	=	green
bl	=	blue
gr	=	grey
li	=	lilac
ge	=	yellow

E20 – Instrument Panel Light Dimmer Switch
E164 – Luggage Compartment Release Switch
F4 – Back-Up Light Switch
K – Instrument Cluster
L16 – Fresh Air Control Lever Light
L28 – Cigarette Lighter Light
L39 – Rear Window Defogger Switch Light
L66 – Cassette Storage Light
R – Radio
R16 – Right Rear Tweeter
R17 – Right Rear Woofer
T2k – Double Connector, behind fuse/relay panel
T2s – Double Connector, on lower right A-pillar
T8 – 8-Pin Connector, on radio
T8g – 8-Pin Connector, on radio

T16 – 6-Pin Double Connector Data Link Connector (DLC), behind center of dash panel
T21a – 21-Pin Connecter, on right B-pillar
T28 – 28-Pin Connector, instrument cluster
T35 – 35-Pin Connector, terminal 15
U1 – Cigarette Lighter

(30) – Ground connection –1–, beside fuse/relay panel

(80) – Ground connection –1–, in instrument panel wiring harness

(A11) – Plus connection (58b), in instrument panel wiring harness

(A40) – Plus connection (30), in instrument panel wiring harness

1995

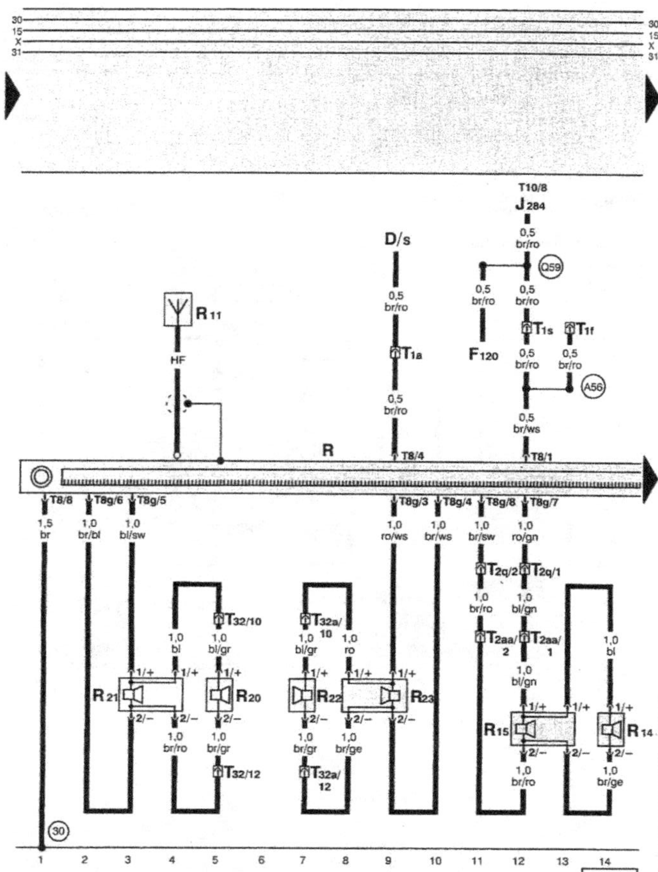

WIRING COLOR CODE		
ws	=	white
sw	=	black
ro	=	red
br	=	brown
gn	=	green
bl	=	blue
gr	=	grey
li	=	lilac
ge	=	yellow

D – Ignition/Starter Switch
F120 – Hood Alarm Switch
J284 – Alarm System Control Module
R – Radio
R11 – Antenna
R14 – Left Rear Tweeter
R15 – Left Rear Woofer
R20 – Left Front Tweeter
R21 – Left Front Woofer
R22 – Right Front Tweeter
R23 – Right Front Woofer
T1a – Single Connector, behind fuse/relay panel
T1f – Single Connector, behind dash panel, center, behind radio
T1s – Single Connector, behind fuse/relay panel
T2aa – Double Connector, on left B-pillar
T2q – Double Connector, on lower left A-pillar
T8 – 8-Pin Connector, on radio
T8g – 8-Pin Connector, on radio
T10 – 10-Pin Connector, behind dash panel, left, on Alarm System Control Module
T32 – 32-Pin Connector, on A-pillar, lower left
T32a – 32-Pin Connector, on A-pillar, lower right
(30) – Ground connection –1–, beside fuse/relay panel
(A56) – Wire connection –2– (30), in instrument panel wiring harness
(Q59) – Wire connection (sensor) in alarm system wiring harness

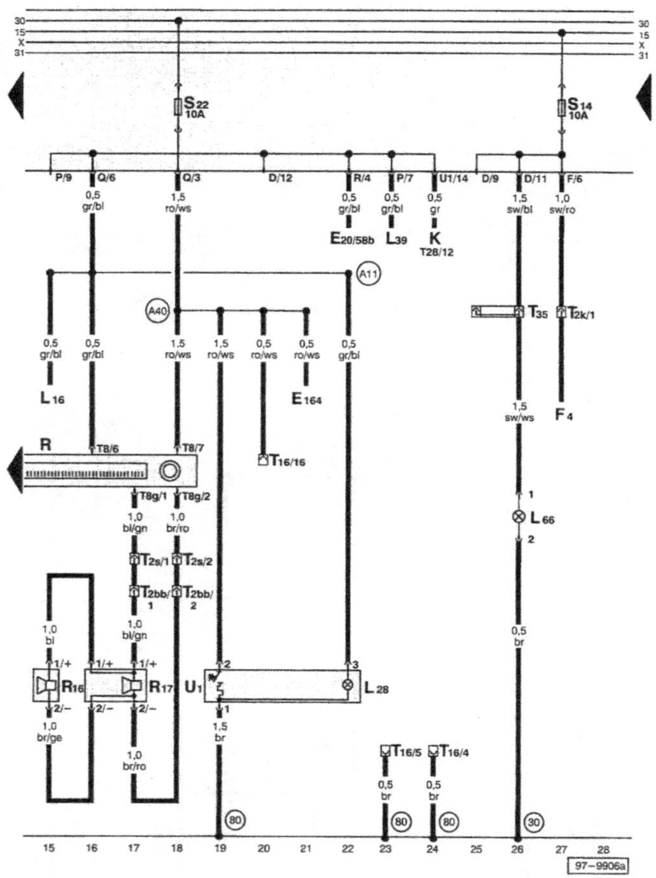

WIRING COLOR CODE		
ws	=	white
sw	=	black
ro	=	red
br	=	brown
gn	=	green
bl	=	blue
gr	=	grey
li	=	lilac
ge	=	yellow

E20 – Instrument Panel Light Dimmer Switch
E164 – Luggage Compartment Release Switch
F4 – Back-Up Light Switch
K – Instrument Cluster
L16 – Fresh Air Control Lever Light
L28 – Cigarette Lighter Light
L39 – Rear Window Defogger Switch Light
L66 – Cassette Storage Light
R – Radio
R16 – Right Rear Tweeter
R17 – Right Rear Woofer
T16 – 16-Pin Connector, Data Link Connector (DLC), behind center of dash panel
T2bb – Double Connector, on right B-pillar
T2k – Double Connector, behind fuse/relay panel
T2s – Double Connector, on lower right A-pillar
T8 – 8-Pin Connector, on radio
T8g – 8-Pin Connector, on radio
T28 – 28-Pin Connector, instrument cluster
T35 – 35-Pin Connector, terminal 15
U1 – Cigarette Lighter
(30) – Ground connection –1–, beside fuse/relay panel
(80) – Ground connection –1–, in instrument panel wiring harness
(A11) – Plus connection (58b), in instrument panel wiring harness
(A40) – Plus connection (30), in instrument panel wiring harness

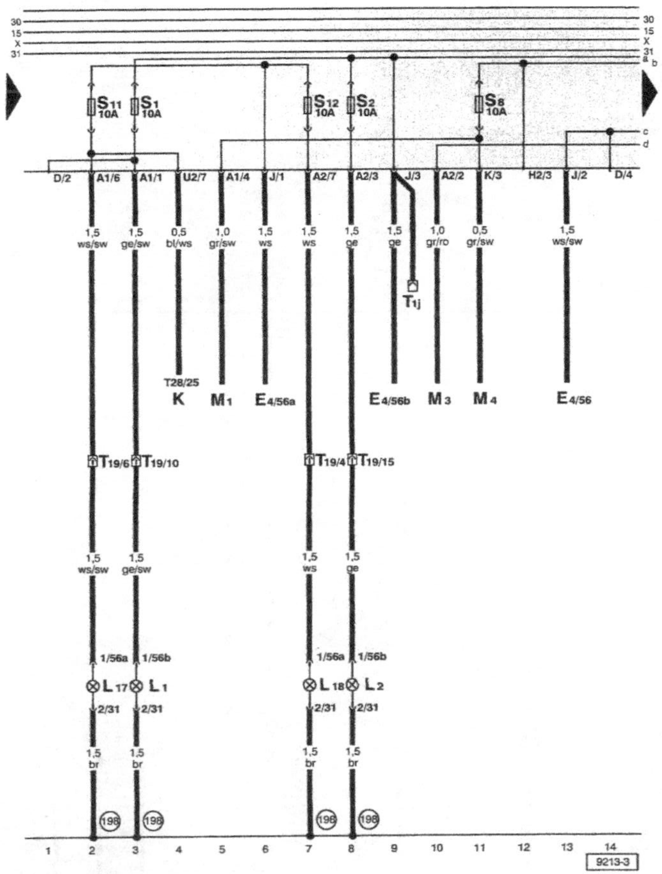

WIRING COLOR CODE		
ws	=	white
sw	=	black
ro	=	red
br	=	brown
gn	=	green
bl	=	blue
gr	=	grey
li	=	lilac
ge	=	yellow

- E4 – Headlight Dimmer/Flasher Switch
- K – Instrument Cluster
- L1 – Left Headlight
- L2 – Right Headlight
- L17 – Left Sealed Beam High Beam Headlight
- L18 – Right Sealed Beam High Beam Headlight
- M1 – Left Parking Light
- M3 – Right Parking Light
- M4 – Left Tail Light
- T1j – Single Connector, behind fuse/relay panel
- T19 – 19-Pin Connector, behind radiator grill, center
- T28 – 28-Pin Connector, instrument cluster
- (198) – Ground connection, in front light wiring harness

Headlights and fog lights
(GTI, USA only)

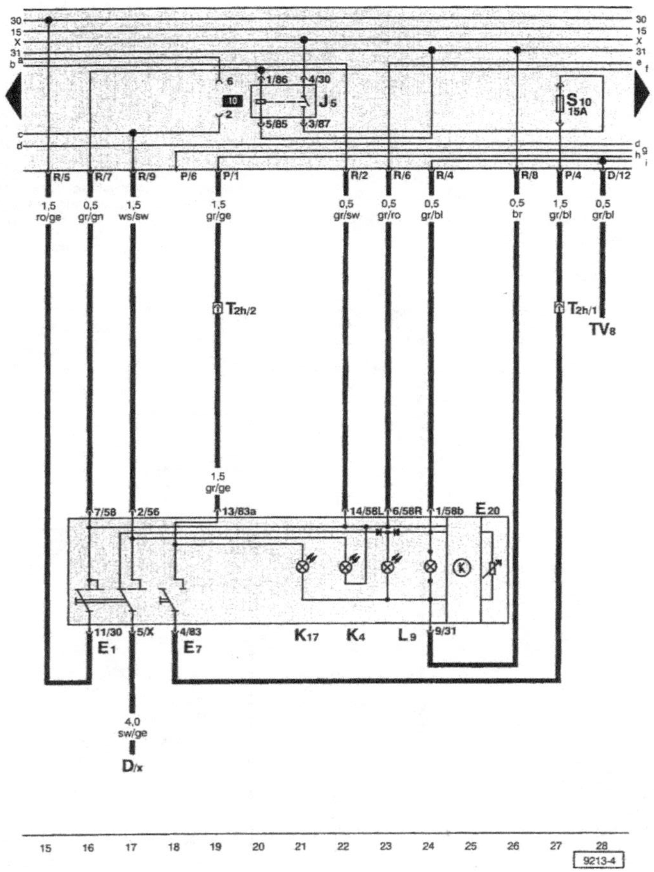

WIRING COLOR CODE		
ws	=	white
sw	=	black
ro	=	red
br	=	brown
gn	=	green
bl	=	blue
gr	=	grey
li	=	lilac
ge	=	yellow

- D – Ignition/Starter Switch
- E1 – Light Switch
- E7 – Fog Light Switch
- E20 – Instrument Panel Light Dimmer Switch
- J5 – Fog Light Relay
- K4 – Park Light Indicator Light
- K17 – Fog Light Indicator Light
- L9 – Headlight Switch Light
- T2h – 2-Pin Connector, behind fuse/relay panel
- TV8 – Terminal 58b Wire Connector

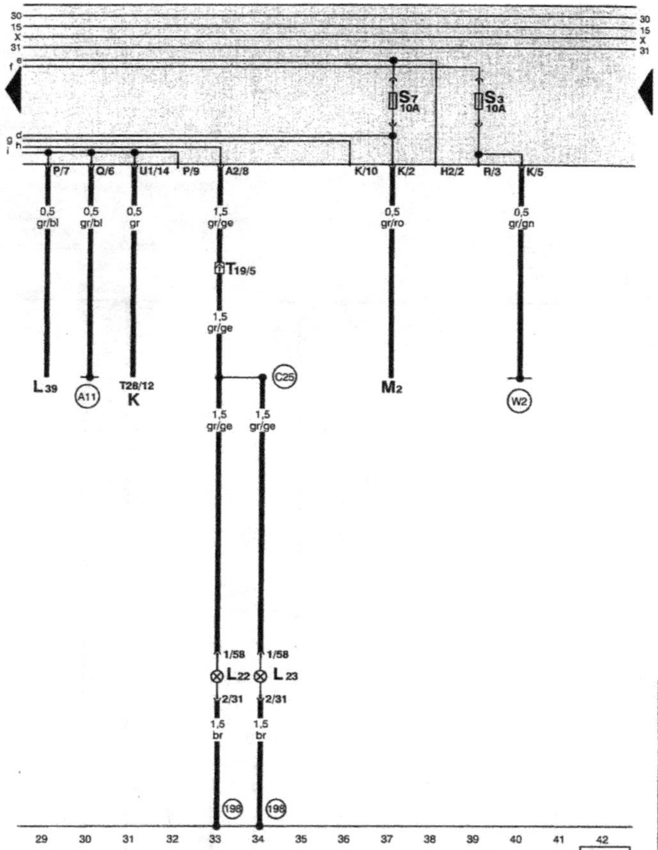

WIRING COLOR CODE

ws	=	white
sw	=	black
ro	=	red
br	=	brown
gn	=	green
bl	=	blue
gr	=	grey
li	=	lilac
ge	=	yellow

- K – Instrument Cluster
- L22 – Left Front Fog Light
- L23 – Right Front Fog Light
- L39 – Rear Window Defogger Switch Light
- M2 – Right Tail Light
- T19 – 19-Pin Connector, behind radiator grill, center
- T28 – 28-Pin Connector, instrument cluster

- (198) – Ground connection, in front light wiring harness
- (A11) – Plus connection (58b), in instrument panel wiring harness
- (C25) – Wire connection (fog lights), in front left wiring harness
- (W2) – Plus connection (58), in rear wiring harness

Heated washer nozzles

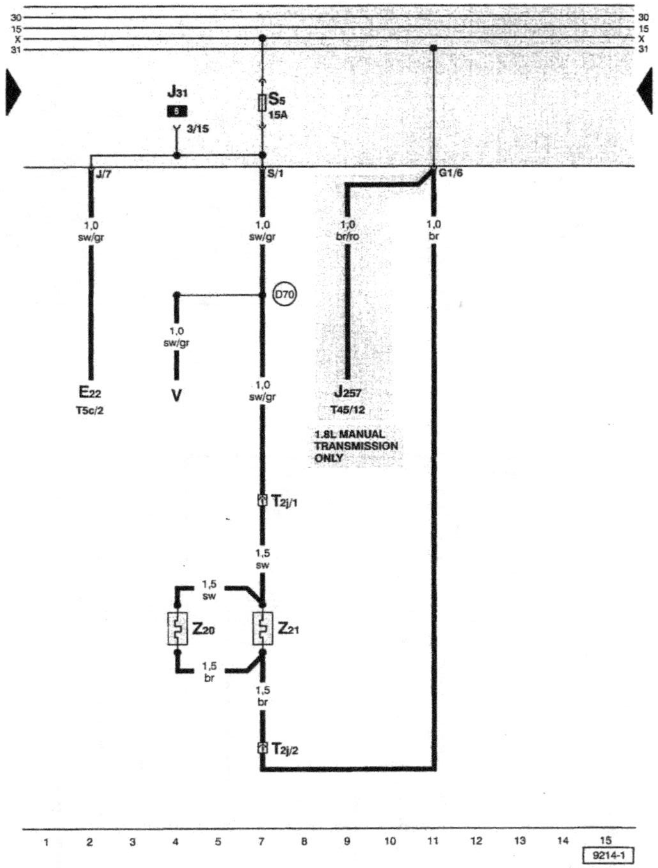

WIRING COLOR CODE

ws	=	white
sw	=	black
ro	=	red
br	=	brown
gn	=	green
bl	=	blue
gr	=	grey
li	=	lilac
ge	=	yellow

- E22 – Windshield Wiper Intermittent Switch
- J31 – Washer/Wiper Intermittent Relay
- J257 – Mono-Motronic Engine Control Module (ECM)
- T2j – Double Connector, behind fuse/relay panel
- T5c – 5-Pin Connector, behind steering column switch trim
- T45 – 45-Pin Connector, on Mono-Motronic Engine Control Module (ECM)
- V – Windshield Wiper Motor
- Z20 – Washer Nozzle Heater, Left
- Z21 – Washer Nozzle Heater, Right
- (D70) – Wire connection (53a), in engine compartment wiring harness

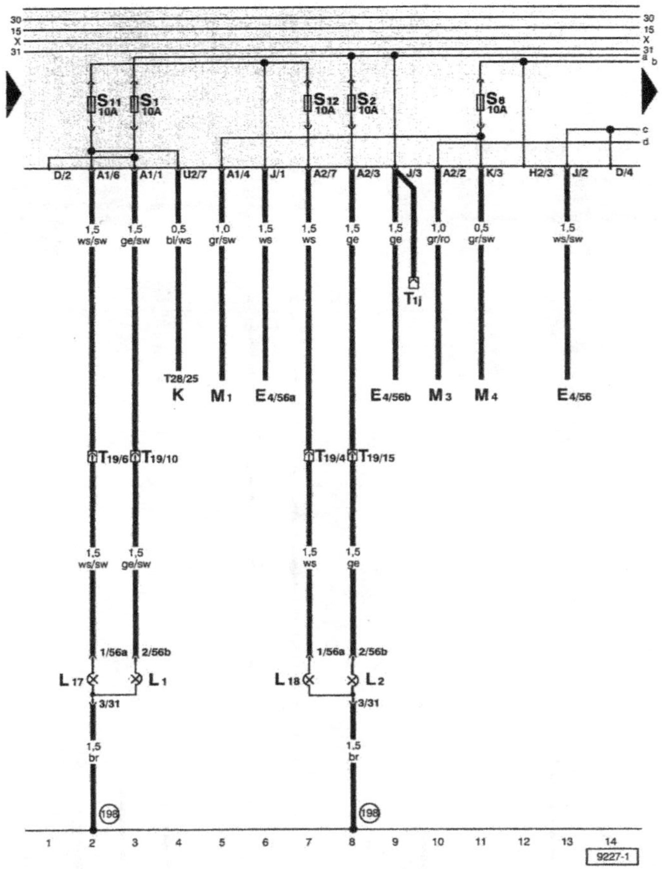

WIRING COLOR CODE		
ws	=	white
sw	=	black
ro	=	red
br	=	brown
gn	=	green
bl	=	blue
gr	=	grey
li	=	lilac
ge	=	yellow

E4 – Headlight Dimmer/Flasher Switch
K – Instrument Cluster
L1 – Left Headlight
L2 – Right Headlight
L17 – Left Sealed Beam High Beam Headlight
L18 – Right Sealed Beam High Beam Headlight
M1 – Left Parking Light
M3 – Right Parking Light
M4 – Left Tail Light
T1j – Single Connector, behind fuse/relay panel
T19 – 19-Pin Connector, behind radiator grill, center
T28 – 28-Pin Connector, on instrument cluster
(198) – Ground connection, in front light wiring harness

1995

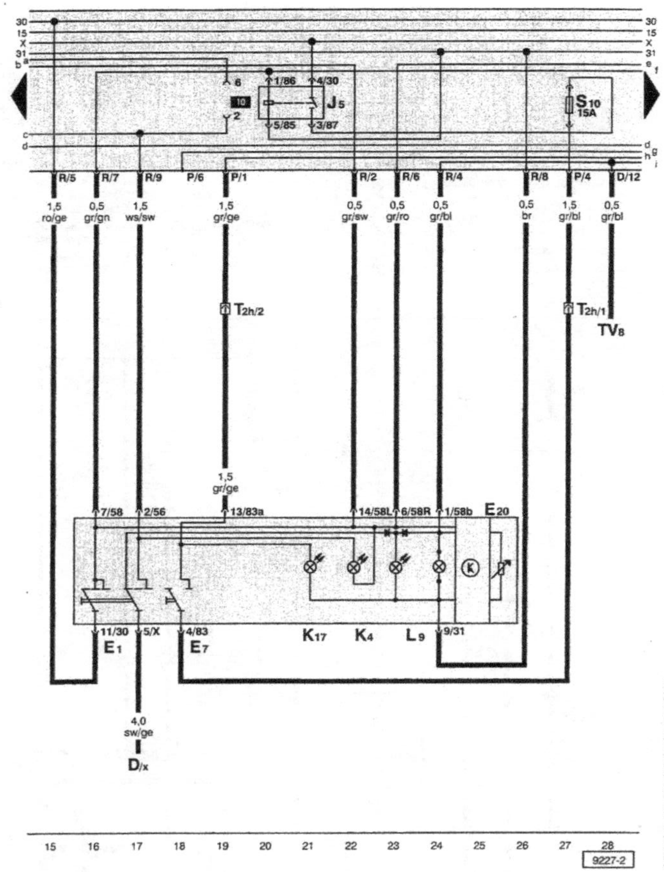

WIRING COLOR CODE		
ws	=	white
sw	=	black
ro	=	red
br	=	brown
gn	=	green
bl	=	blue
gr	=	grey
li	=	lilac
ge	=	yellow

D – Igntion/Starter Switch
E1 – Light Switch
E7 – Fog Light Switch
E20 – Instrument Panel Light Dimmer Switch
J5 – Fog Light Relay
K4 – Park Light Indicator Light
K17 – Fog Light Indicator Light
L9 – Headlight Switch Light
T2h – 2-Pin Connector, behind fuse/relay panel
TV8 – 8-Pin Connector, terminal 58b

1995

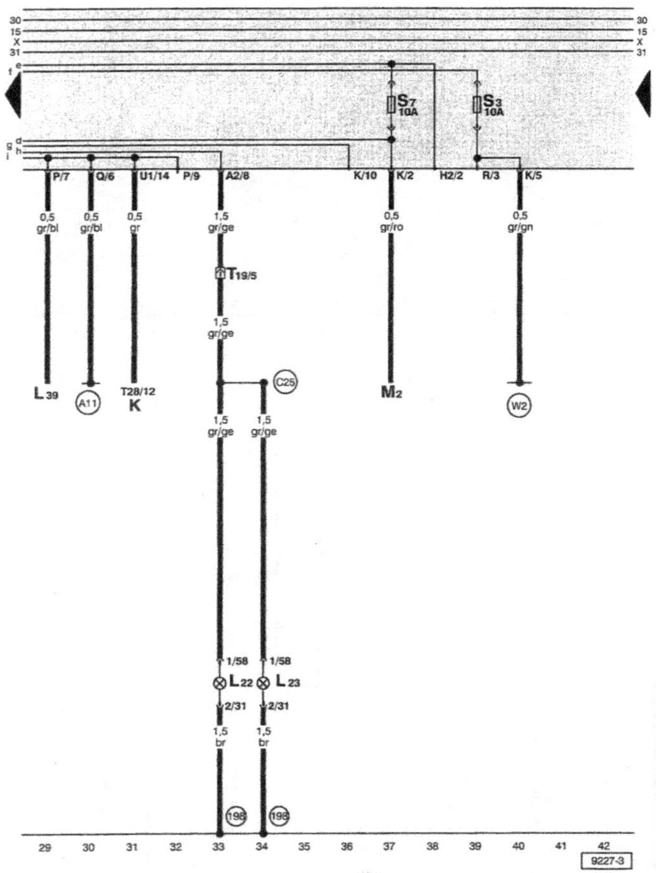

WIRING COLOR CODE		
ws	=	white
sw	=	black
ro	=	red
br	=	brown
gn	=	green
bl	=	blue
gr	=	grey
li	=	lilac
ge	=	yellow

K – Instrument Cluster
L22 – Left Front Fog Light
L23 – Right Front Fog Light
L39 – Rear Window Defogger Switch Light
M2 – Right Tail Light
T19 – 19-Pin Connector, behind radiator grill, center
T28 28-Pin Connector, on instrument cluster

(198) – Ground connection, in front light wiring harness

(A11) – Plus connection (58b), in instrument panel wiring harness

(C25) – Wire connection (fog lights), in front left wiring harness

(W2) – Plus connection (58), in rear wiring harness

1995

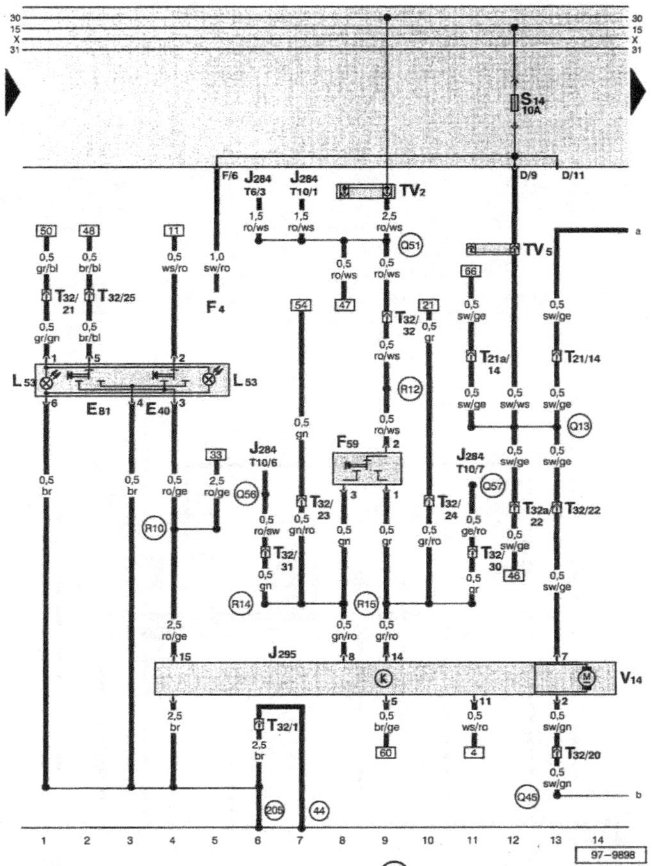

WIRING COLOR CODE		
ws	=	white
sw	=	black
ro	=	red
br	=	brown
gn	=	green
bl	=	blue
gr	=	grey
li	=	lilac
ge	=	yellow

E40 – Left Front Window Switch
E81 – Right Front Window Switch (In Driver's Door)
F4 – Back-Up Light Switch
F59 – Driver's Door Central Locking System Door Switch
J284 – Alarm System Control Module
J295 – Left Front Power Window Control Module
L53 – Power Window Switch Light
T6 – 6-Pin Connector, behind dash panel, left, on Alarm System Control Module
T10 – 10-Pin Connector, behind dash panel, left, on Alarm System Control Module
T21 – 21-Pin Connector, on left B-pillar
T21a – 21-Pin Connecter, on right B-pillar
T32 – 32-Pin Connector, on A-pillar, lower left
T32a – 32-Pin Connector, on A-pillar, lower right
TV2 – Terminal 30 (B+) Wire Connector
TV5 – Terminal 15a Wire Connector
V14 – Left Window Motor

(44) – Ground connection, on left A-pillar, lower part

(205) – Ground connection, in driver's door wiring harness

(Q45) – Plus connection (75as/15), in power window/central locking system and door contact switch wiring harness

(Q45) – Wire connection (comfort), in power window wiring harness

(Q51) – Plus connection (30), in alarm system wiring harness

(Q56) – Wire connection, (open), in alarm system wiring harness

(Q57) – Wire connection, (closed), in alarm system wiring harness

(R10) – Plus connection –1– (30), in driver's door wiring harness

(R12) – Plus connection –3– (30), in driver's door wiring harness

(R14) – Wire connection –1– (open), in driver's door wiring harness

(R15) – Wire connection –2– (closed), in driver's door wiring harness

1995

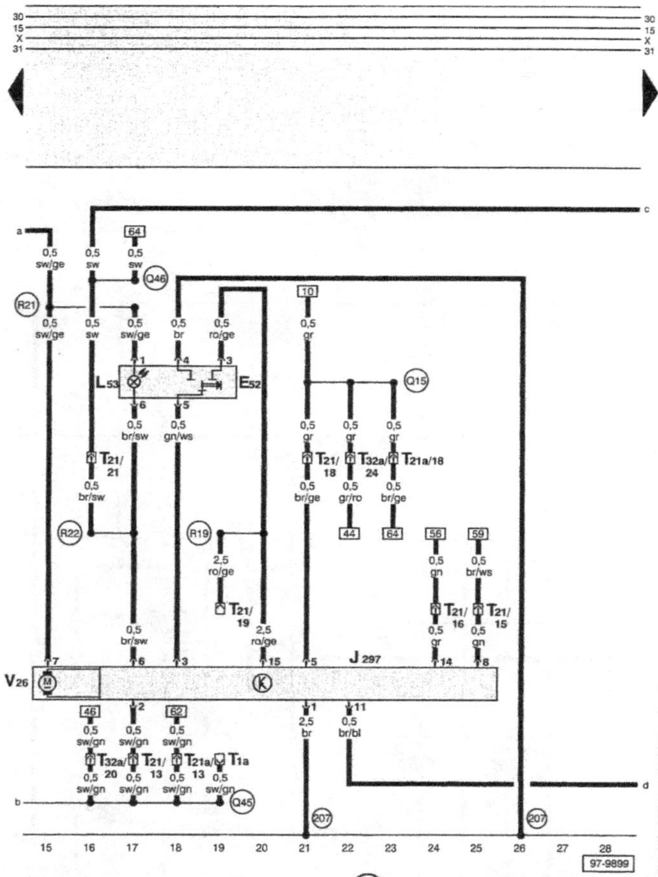

WIRING COLOR CODE

ws	=	white
sw	=	black
ro	=	red
br	=	brown
gn	=	green
bl	=	blue
gr	=	grey
li	=	lilac
ge	=	yellow

- E52 – Left Rear Window Switch, (In LR Door)
- J297 – Left Rear Power Window Control Module
- L53 – Power Window Switch Light
- T1a – Single Connector, behind fuse/relay panel
- T21 – 21-Pin Connector, on left B-pillar
- T21a – 21-Pin Connecter, on right B-pillar
- T32a – 32-Pin Connector, on A-pillar, lower right
- V26 – Left Rear Door Window Motor
- (207) – Ground connection, in left rear door wiring harness
- (Q15) – Wire connection, in power window wiring harness (Left up)
- (Q45) – Wire connection (comfort), in power window wiring harness
- (Q46) – Wire connection (child lock) in power window wiring harness
- (R19) – Plus connection –1– (30), in left rear door wiring harness
- (R21) – Plus connection –3– (15), in left rear door wiring harness
- (R22) – Wire connection (child lock), in left rear door wiring harness

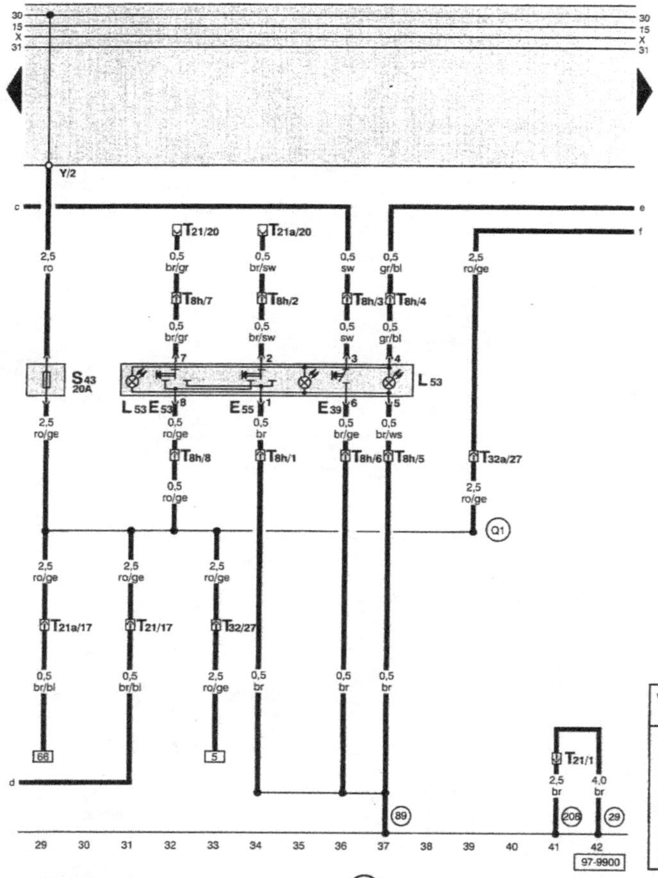

WIRING COLOR CODE

ws	=	white
sw	=	black
ro	=	red
br	=	brown
gn	=	green
bl	=	blue
gr	=	grey
li	=	lilac
ge	=	yellow

- E39 – Window Lockout Switch
- E53 – Left Rear Window Switch (In Console)
- E55 – Right Rear Window Switch (In Console)
- L53 – Power Window Switch Light
- S43 – Power Windows Circuit Breaker
- T8h – 8-Pin Connector, behind center of dash panel
- T21 – 21-Pin Connector, on left B-pillar
- T21a – 21-Pin Connecter, on right B-pillar
- T32 – 32-Pin Connector, on A-pillar, lower left
- T32a – 32-Pin Connector, on A-pillar, lower right
- (29) – Ground connection, near B-pillar, right
- (89) – Ground connection, in power window wiring harness
- (208) – Ground connection, in right rear door wiring harness
- (Q1) – Plus connection (30) in power window wiring harness

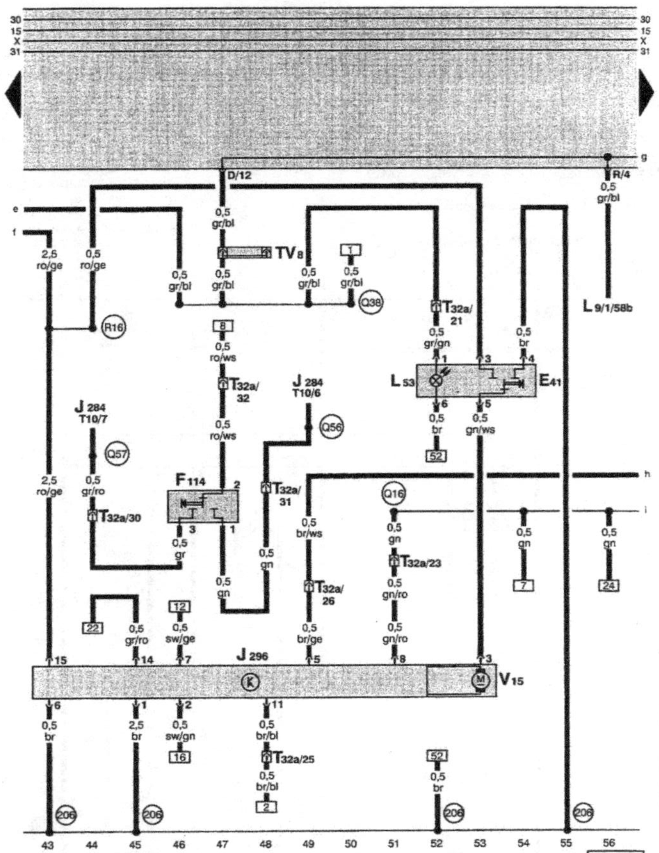

WIRING COLOR CODE		
ws	=	white
sw	=	black
ro	=	red
br	=	brown
gn	=	green
bl	=	blue
gr	=	grey
li	=	lilac
ge	=	yellow

E41 – Right Front Window Switch
F114 – Passenger's Door Central Locking System Door Switch
J284 – Alarm System Control Module
J296 – Right Front Power Window Control Module
L9 – Headlight Switch Light
L53 – Power Window Switch Light
T10 – 10-Pin Connector, behind dash panel, left, on Alarm System Control Module
T32a – 32-Pin Connector, on A-pillar, lower right
TV8 – Terminal 58b Wire Connector
V15 – Window Motor, Right Front Door

(206) – Ground connection, in passenger's door wiring harness
(Q16) – Wire connection, in power window wiring harness (left down)
(Q38) – Plus connection (58b) –1–, in power window wiring harness
(Q56) – Wire connection (open), in alarm system wiring harness
(Q57) – Wire connection (closed), in alarm system wiring harness
(R16) – Plus connection –1– (30), in passenger's door wiring harness

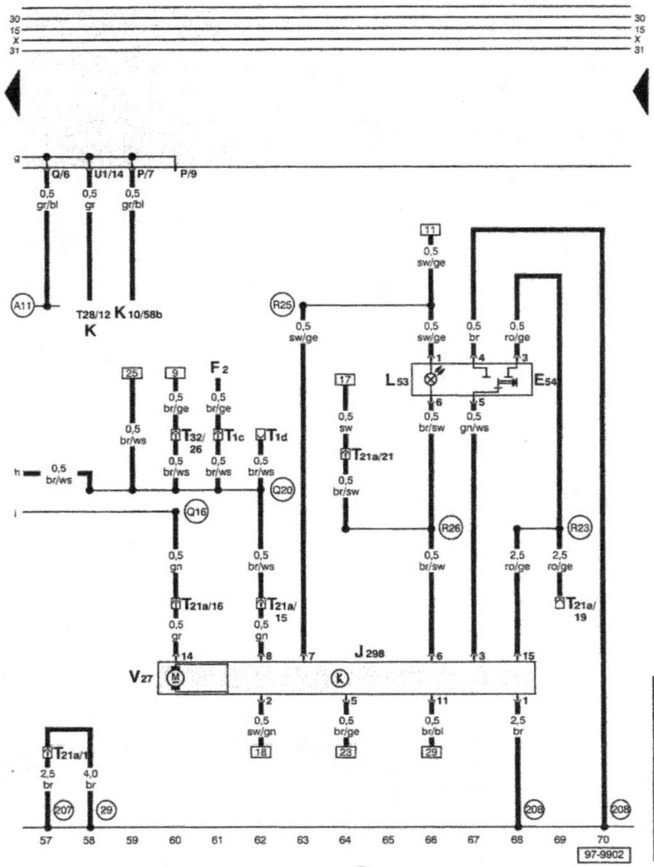

WIRING COLOR CODE		
ws	=	white
sw	=	black
ro	=	red
br	=	brown
gn	=	green
bl	=	blue
gr	=	grey
li	=	lilac
ge	=	yellow

E54 – Right Rear Window Switch, (In RR Door)
F2 – Door Contact Switch, Left Front
J298 – Right Rear Power Window Control Module
K – Instrument Cluster
K10 – Rear Window Defogger Indicator Light
L53 – Power Window Switch Light
T1c – Single Connector, behind fuse/relay panel
T1d – Single Connector, behind fuse/relay panel
T21a – 21-Pin Connecter, on right B-pillar
T28 – 28-Pin Connector, on instrument cluster
T32 – 32-Pin Connector, on A-pillar, lower left
V27 – Right Rear Door Window Motor

(29) – Ground connection, near B-pillar, right

(207) – Ground connection, in left rear door wiring harness

(208) – Ground connection, in right rear door wiring harness
(A11) – Plus connection (58b), in instrument panel wiring harness
(Q16) – Wire connection, in power window wiring harness (left down)
(Q20) – Wire connection –1– (door contact switch), in power window wiring harness
(R23) – Plus connection –1– (30), in right rear door wiring harness
(R25) – Plus connection –3– (15), in right rear door wiring harness
(R26) – Wire connection (child lock), in right rear door wiring harness

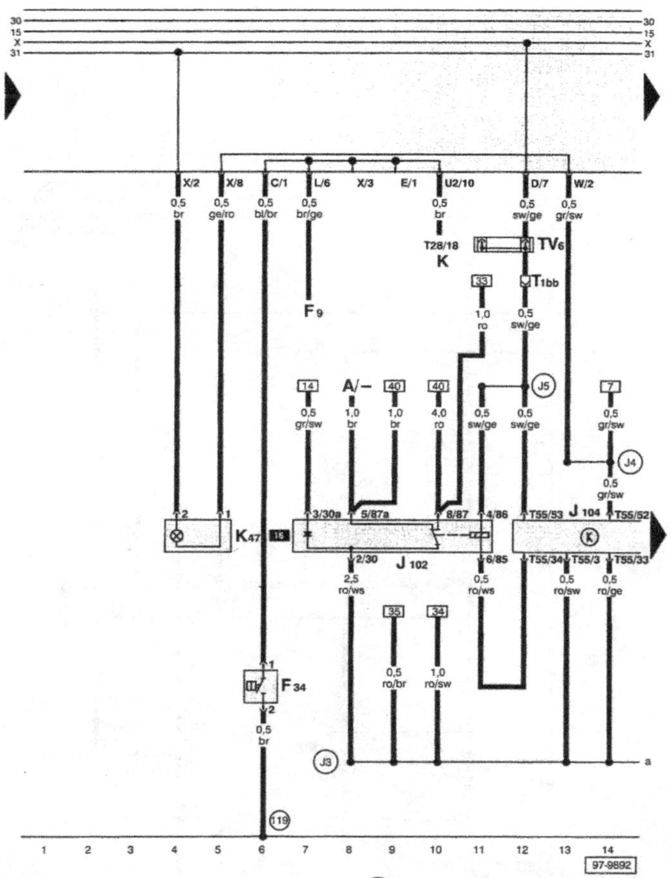

WIRING COLOR CODE

ws	=	white
sw	=	black
ro	=	red
br	=	brown
gn	=	green
bl	=	blue
gr	=	grey
li	=	lilac
ge	=	yellow

A – Battery
F9 – Parking Brake Warning Light Switch
F34 – Brake Fluid Level Warning Switch
J102 – ABS Relay
J104 – ABS (w/EDL) Control Module
K – Instrument Cluster
K47 – ABS Warning Light
T1bb – Single Connector, behind fuse/relay panel
T28 – 28-Pin Connector, on instrument cluster
T55 – 55-Pin Connector, on ABS Control Module
TV6 – Terminal X Wire Connector

(119) – Ground connection –1–, in headlight wiring harness

(J3) –Wire connection –1–, in ABS wiring harness

(J4) – Wire connection (OBD), in ABS wiring harness

(J5) – Wire connection –2–, in ABS wiring harness

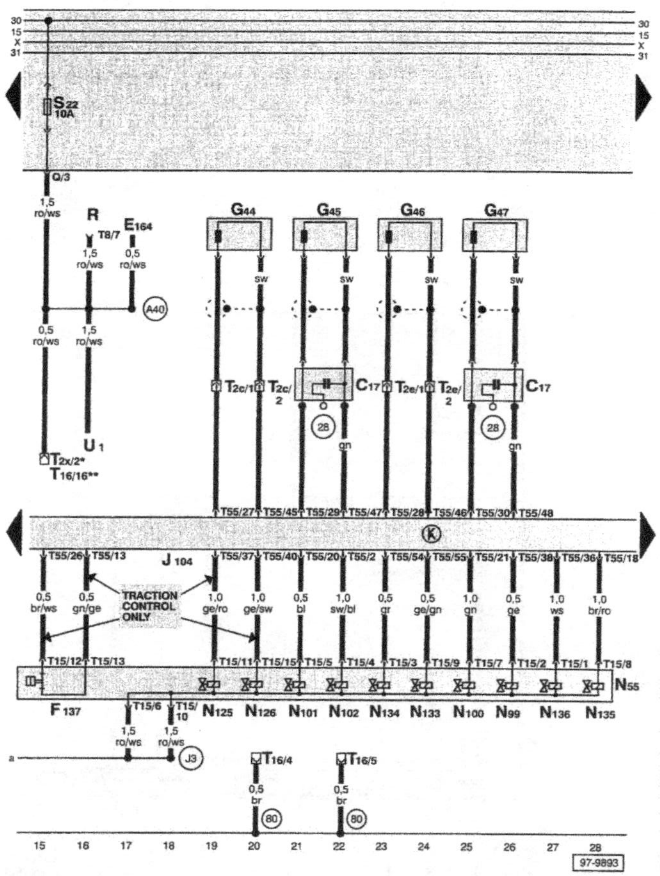

WIRING COLOR CODE

ws	=	white
sw	=	black
ro	=	red
br	=	brown
gn	=	green
bl	=	blue
gr	=	grey
li	=	lilac
ge	=	yellow

C17 – ABS Speed Sensor Capacitor
E164 – Luggage Compartment Release Switch
F137 – ABS Pressure Control Switch
G44 – Right Rear ABS Wheel Speed Sensor
G45 – Right Front ABS Wheel Speed Sensor
G46 – Left Rear ABS Wheel Speed Sensor
G47 – Left Front ABS Wheel Speed Sensor
J104 – ABS (w/EDL) Control Module
N55 – ABS Hydraulic Unit
N99 – Right Front ABS Inlet Valve
N100 – Right Front ABS Outlet Valve
N101 – Left Front ABS Inlet Valve
N102 – Left Front ABS Outlet Valve
N125 – Differential Lock Valve 1
N126 – Differential Lock Valve 2
N133 – Right Rear ABS Inlet Valve
N134 – Left Rear ABS Inlet Valve
N135 – Right Rear ABS Outlet Valve
N136 – Left Rear ABS Outlet Valve
R – Radio
T2c – 2-Pin Connector, near Right Rear ABS Wheel Speed Sensor
T2e – 2-Pin Connector, near Left Rear ABS Wheel Speed Sensor
T8 – 8-Pin Connector, on radio
T15 – 15-Pin Connector, on ABS Hydraulic Unit
T55 – 55-Pin Connector, on ABS Control Module
U1 – Cigarette Lighter

(28) – Ground connection, on firewall

(80) – Ground connection –1–, in instrument panel wiring harness

(A40) – Plus connection (30), in instrument panel wiring harness

(J3) – Wire connection –1–, in ABS wiring harness

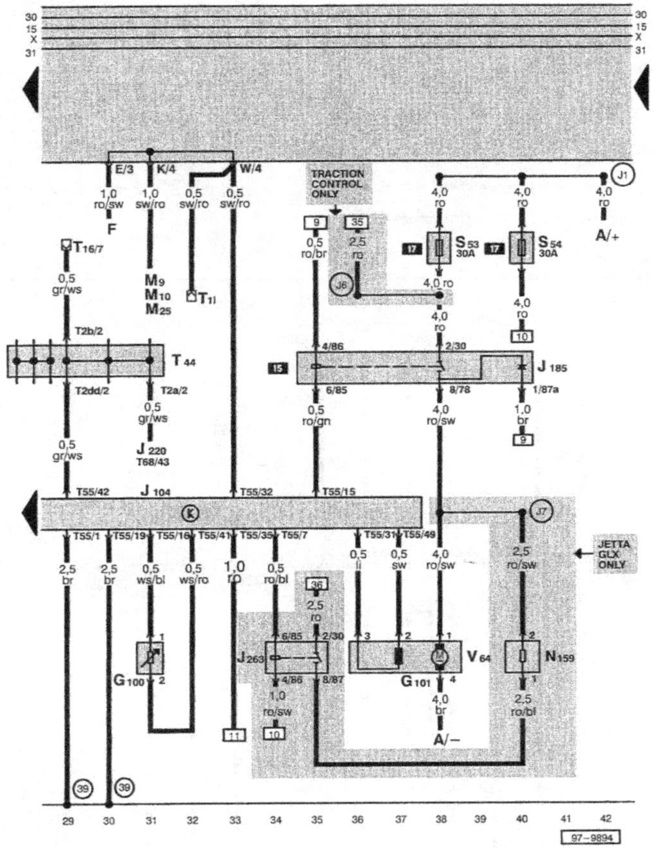

WIRING COLOR CODE		
ws	=	white
sw	=	black
ro	=	red
br	=	brown
gn	=	green
bl	=	blue
gr	=	grey
li	=	lilac
ge	=	yellow

A – Battery
F – Brake Light Switch
G100 – Brake Pedal Position Sensor
G101 – Hydraulic Pump Sensor
J104 – ABS (w/EDL) Control Module
J185 – ABS Hydraulic Pump Relay
J220 – Motronic Engine Control Module (ECM)
J263 – Electronic Differential Lock Cut-Off Relay
M9 – Left Brake Light
M10 – Right Brake Light
M25 – High-Mount Brake Light
N159 – Electronic Differential Lock Series Resistance
S53 – ABS Hydraulic Pump Fuse
S54 – ABS Valves Fuse
T1i – Single Connector, behind fuse/relay panel
T2a – Double Connector, on junction connector (T44) above fuse/relay panel
T2b – Double Connector, on junction connector (T44) above fuse/relay panel
T2dd – Double Connector, on junction connector (T44) behind fuse/relay panel
T44 – 44-Pin Connector, Junction Box for on board diagnosis, above fuse/relay panel
T55 – 55-Pin Connector, on ABS Control Module
T68 – 68-Pin Connector, right side of engine compartment, on Motronic Engine Control Module (ECM)
V64 – ABS Hydraulic Pump Fuse

(39) – Ground connection, below rear seat bench, left
(J1) – Plus connection (30) in ABS wiring harness
(J6) – Plus connection –2– (30), in ABS wiring harness
(J7) – Wire connection –3–, in ABS wiring harness

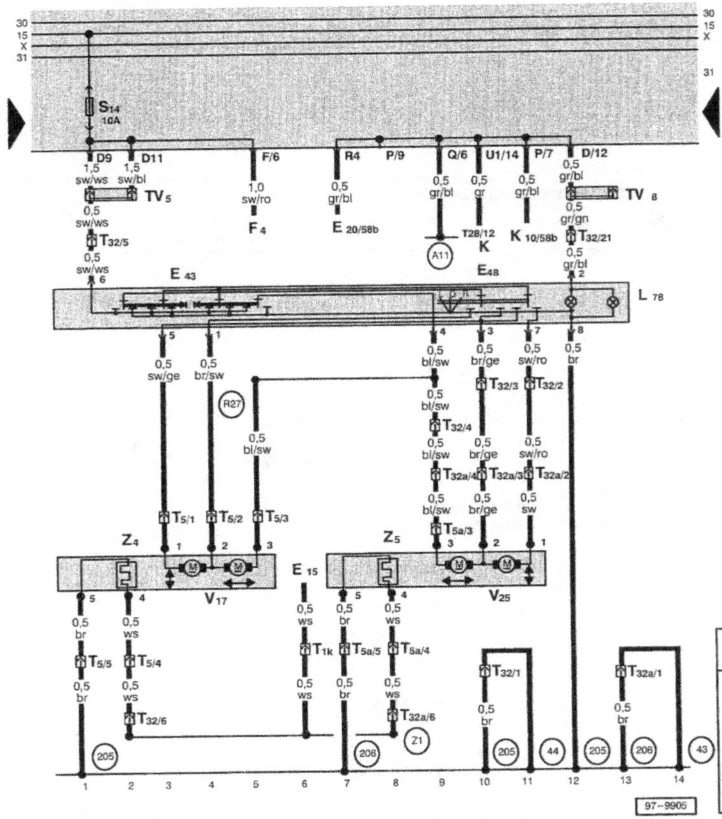

WIRING COLOR CODE		
ws	=	white
sw	=	black
ro	=	red
br	=	brown
gn	=	green
bl	=	blue
gr	=	grey
li	=	lilac
ge	=	yellow

E15 – Rear Window Defogger Switch
E20 – Instrument Panel Light Dimmer Switch
E43 – Mirror Adjustment Switch
E48 – Mirror Select Switch
F4 – Back-Up Light Switch
K – Instrument Cluster
K10 – Rear Window Defogger Indicator Light
L78 – Mirror Adjusting Switch Light
V17 – Driver's Side Mirror Adjustment Motor
V25 – Passenger's Side Heated Mirror Adjustment Motor
Z4 – Driver's Side Heated Mirror
Z5 – Passenger's Side Heated Mirror
T1k – Single Connector, behind fuse/relay panel
T5 – 5-Pin Connector, in driver's door
T5a – 5-Pin Connector, in passenger's door
T28 – 28-Pin Connector, on instrument cluster
T32 – 32-Pin Connector, on A-pillar, lower left
T32a – 32-Pin Connector, on A-pillar, lower right
TV5 – Terminal 15a Wire Connector
TV8 – Terminal 58b Wire Connector

(43) – Ground connection, on right A-pillar, lower part
(44) – Ground connection, on left A-pillar, lower part
(205) – Ground connection, in driver's door wiring harness
(206) – Ground connection, in passenger's door wiring harness
(A11) – Plus connection (58b), in instrument panel wiring harness
(R27) – Wire connection (power mirror), in driver's door wiring harness
(Z1) – Wire connection –1–, in power heated mirror wiring harness

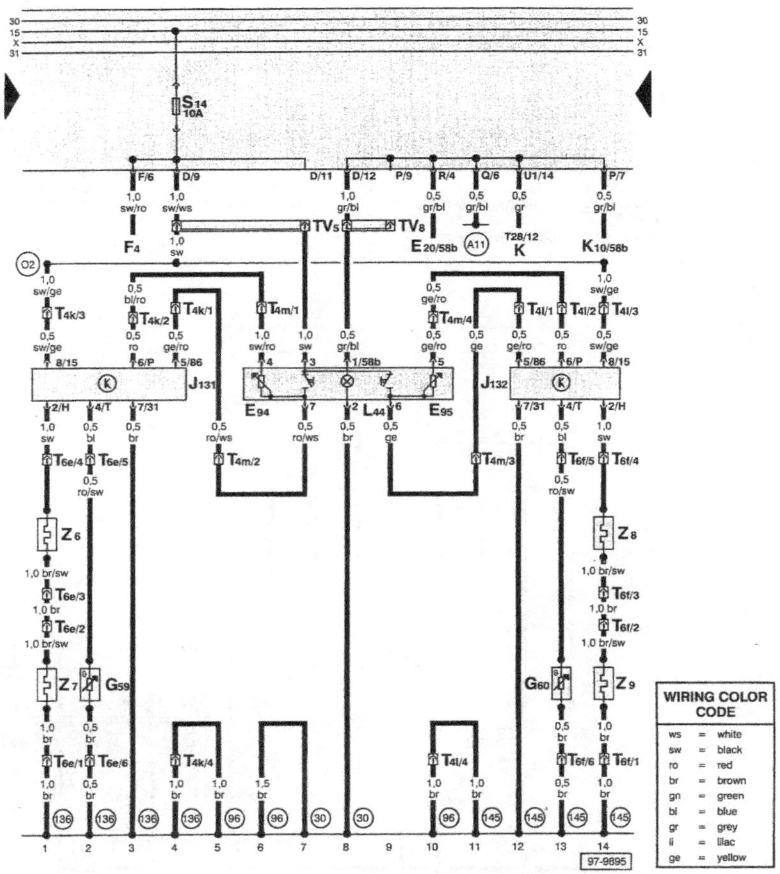

E20 – Instrument Panel Light Dimmer Switch
E94 – Driver's Seat Heat Regulating Switch
E95 – Passenger's Seat Heat Regulating Switch
F4 – Back-Up Light Switch
G59 – Driver's Seat Heater Temperature Sensor
G60 – Passenger's Seat Heater Temperature Sensor
J131 – Driver's Seat Heater Control Module
J132 – Passenger's Seat Heater Control Module
K – Instrument Cluster
K10 – Rear Window Defogger Indicator Light
L44 – Seat Heater Switch Illumination Light
T4k – 4-Pin Connector
T4l – 4-Pin Connector
T4m – 4-Pin Connector
T6e – 6-Pin Connector
T6f – 6-Pin Connector
T28 – 28-Pin Connector, on instrument cluster
TV5 – Terminal 15a Wire Connector
TV8 – Terminal 58b Wire Connector
Z6 – Driver's Seat Heat Element
Z7 – Driver's Backrest Heat Element
Z8 – Passenger's Seat Heat Element
Z9 – Passenger's Backrest Heat Element

(30) – Ground connection –1–, beside fuse/relay panel
(96) – Ground connection –1–, in heated seats wiring harness
(136) – Ground connection –2–, in heated seats wiring harness
(145) – Ground connection –3–, in heated seats wiring harness
(A11) – Plus connection (58b), in instrument panel wiring harness
(O2) – Plus connection (15) in heated seats wiring harness

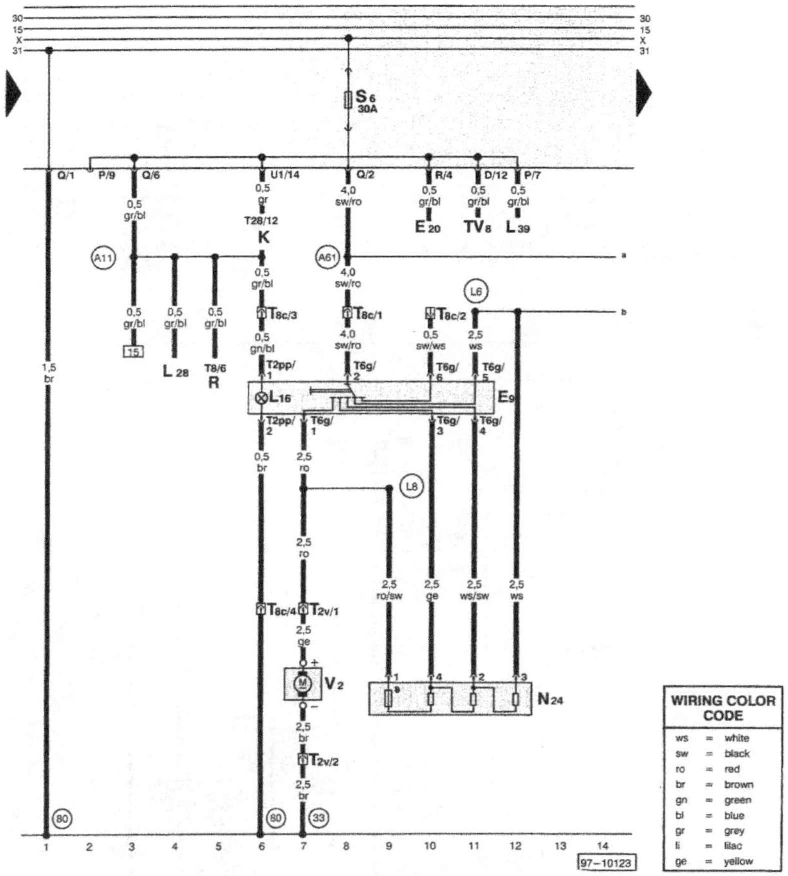

E9 – Fresh Air Blower Switch
E20 – Instrument Panel Light Dimmer Switch
K – Instrument Cluster
L16 – Fresh Air Control Level Light
L28 – Cigarette Lighter Light
L39 – Rear Window Defogger Switch Light
N24 – Fresh Air Blower Series Resistance with Fuse
R – Radio
T2pp – Double Connector, on E9
T2v – Double Connector, behind instrument panel, right
T6g – 6-Pin Connector, on E9
T8 – 8-Pin Connector, on radio
T8c – 8-Pin Connector, behind instrument panel, center
T28 – 28-Pin Connector, on instrument cluster
TV8 – Terminal 58b Wire Connector, above fuse/relay panel
V2 – Fresh Air Blower

(33) – Ground connection, behind instrument panel, right
(80) – Ground connection –1–, in instrument panel wiring harness
(A11) – Plus connection (58b), in instrument panel wiring harness
(A61) – Plus connection (X), in instrument cluster wiring harness
(L6) – Wire connection –2–, in evaporator housing wiring harness
(L8) – Wire connection –4–, in evaporator housing wiring harness

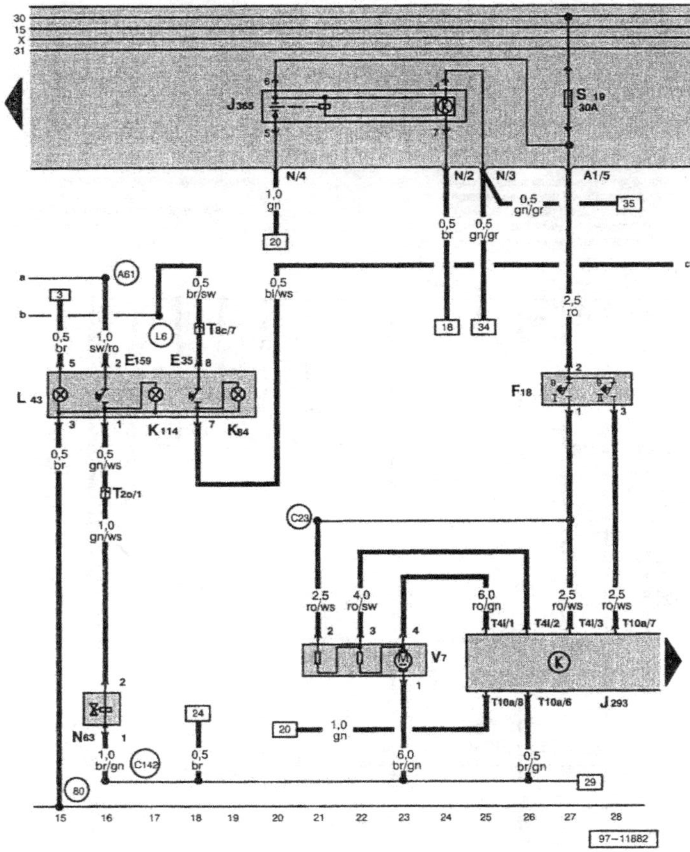

WIRING COLOR CODE		
ws	=	white
sw	=	black
ro	=	red
br	=	brown
gn	=	green
bl	=	blue
gr	=	grey
li	=	lilac
ge	=	yellow

E35 – A/C Switch
E159 – Fresh Air/Recirculation Flap Switch
F18 – Coolant Fan Control (FC) Thermal Switch
J220 – Motronic Engine Control Module (ECM)
J293 – Coolant Fan Control (FC) Control Module, in left side of engine compartment
J365 – Relay for A/C shutoff
K84 – A/C Indicator Light
K114 – Fresh Air and Recirculating Air Mode Indicator Light
L43 – A/C Switch Illumination Light
N63 – Fresh Air/Recirculating Flap Two-Way Valve (00605), in rear of engine compartment
T2e – Double Connector, behind fuse/relay panel
T2o – 4-Pin Connector, behind fuse/relay panel
T4l – 2-Pin Connector, on J293
T8c – 8-Pin Connector, behind instrument panel, center
T10a – 10-Pin Connector, on J293

V7 – Coolant Fan
(80) – Ground connection –1–, in instrument panel wiring harness
(119) – Ground connection –1–, in headlight wiring harness
(A61) – Plus connection (X), in instrument cluster wiring harness
(C23) – Wire connection (blower, 1st speed) in headlight/air conditioner wiring harness
(C142) – Wire connection, in Motronic MFI wiring harness
(L6) – Wire connection –2–, in evaporator housing wiring harness

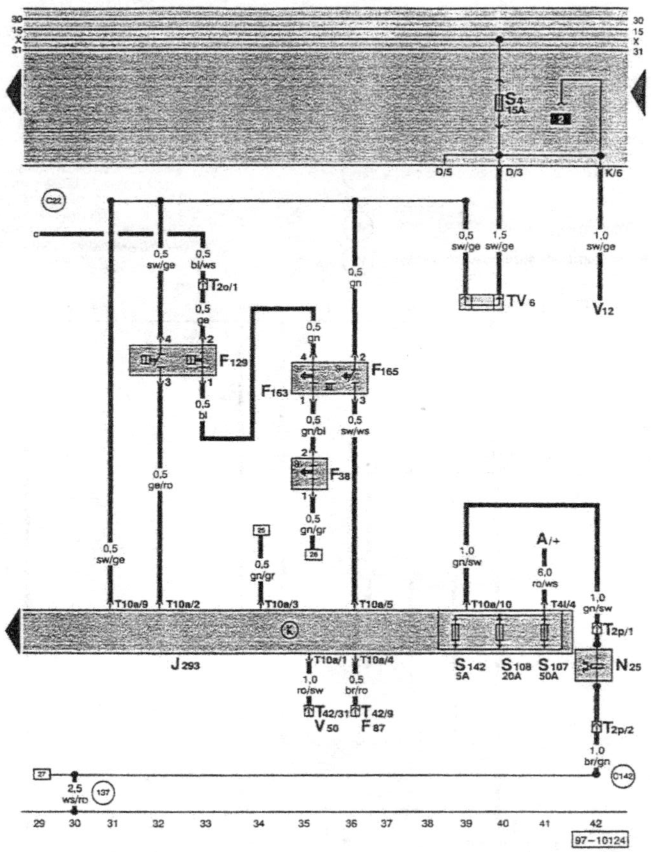

WIRING COLOR CODE		
ws	=	white
sw	=	black
ro	=	red
br	=	brown
gn	=	green
bl	=	blue
gr	=	grey
li	=	lilac
ge	=	yellow

A – Battery
F38 – Ambient Temperature Switch, near horn
F87 – After-Run Coolant Fan Control (FC) Thermal Switch
F129 – A/C Pressure Switch, in right side of engine compartment
F163 – A/C Cut-Out Thermal Switch, in center of engine compartment
F165 – Third Speed Coolant Fan Control (FC) Thermal Switch, in center of engine compartment
J220 – Motronic Engine Control Module (ECM)
J293 – Coolant FC (Fan Control) Control Module
N25 – A/C Clutch
S107 – Coolant Fan Second and Third Speed Fuse
S108 – Coolant Fan First Speed and A/C Clutch Fuse
S142 – Control Module Fuse for Coolant Fan
T2e – Double Connector, behind fuse/relay panel

T2o – 2-Pin Connector, behind fuse/relay panel
T2p – Double Connector, near compressor
T2qq – Double Connector, in engine compartment, left
T10a – 10-Pin Connector, on Coolant Fan (FC) Control Module
T4l – 4-Pin Connector, on Coolant Fan (FC) Control Module
TV6 – Terminal X Wire Connector, above fuse/relay panel
V12 – Rear Window Wiper Motor (not applicable to Jetta III)
V50 – Coolant Pump
(119) – Ground connection –1–, in headlight wiring harness
(C22) – Wire connection (X), in headlight/air conditioning wiring harness
(C142) – Wire Connection, in Motronic MFI wiring harness

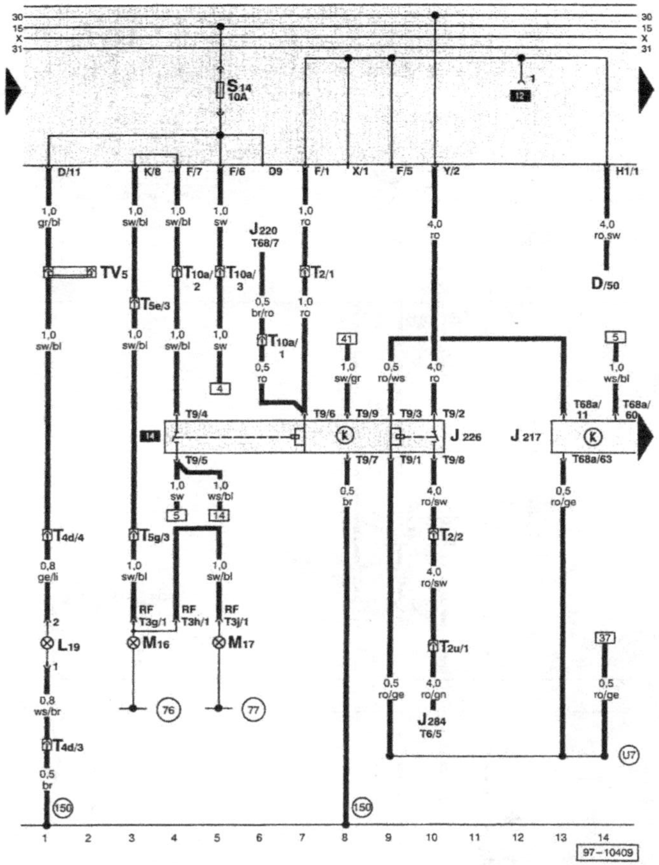

WIRING COLOR CODE

ws	=	white
sw	=	black
ro	=	red
br	=	brown
gn	=	green
bl	=	blue
gr	=	grey
li	=	lilac
ge	=	yellow

D – Ignition/Starter Switch
J217 – Transmission Control Module (TCM)
J220 – Motronic Engine Control Module (ECM)
J226 – Park/Neutral Position (PNP) Relay
J284 – Alarm System Control Module
L19 – Automatic Transmission Console Light
M16 – Left Back-Up Light
M17 – Right Back-Up Light
T2 – Double Connector, behind fuse/relay panel, only on automatic transmission
T2u – Double Connector, behind fuse/relay panel
T3g – 3-Pin Connector, on back-up/brake light, left
T3h – 3-Pin Connector, on back-up/brake light, left
T3j – 3-Pin Connector, on back-up/brake light, right
T4d – 4-Pin Connector, below center console
T5e – 5-Pin Connector, in luggage compartment, left
T5g – 5-Pin Connector, in luggage compartment lid
T6 – 6-Pin Connector, behind dash panel, left, on Alarm System Control Module
T9 – 9-Pin Connector, on Park/Neutral Position (PNP) Relay
T10a – 10-Pin Connector, behind fuse/relay panel
T68 – 68-Pin Connector, under right side of bonnet, on Motronic Engine Control Module (ECM)
T68a – 68-Pin Connector, below right front seat, on Transmission Control Module (TCM)
TV5 – Connector, terminal 15a
(76) – Ground connection, on left back-up/brake light bulb holder
(77) – Ground connection, on right back-up/brake light bulb holder
(150) – Ground connection, in automatic transmission wiring harness
(U7) – Wire connection (651), in automatic transmission wiring harness

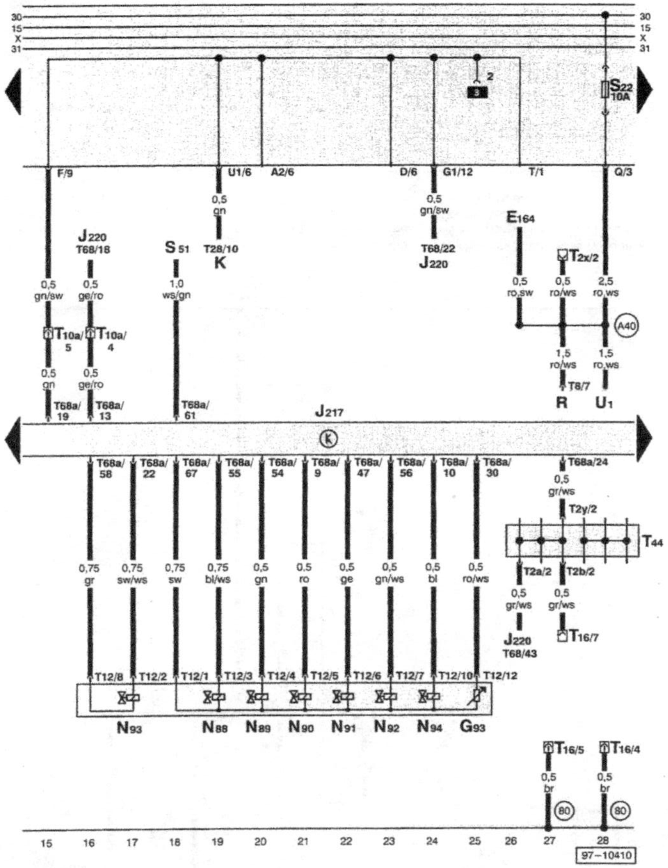

WIRING COLOR CODE

ws	=	white
sw	=	black
ro	=	red
br	=	brown
gn	=	green
bl	=	blue
gr	=	grey
li	=	lilac
ge	=	yellow

E164 – Luggage Compartment Release Switch
G93 – Transmission Fluid Temperature Sensor
J217 – Transmission Control Module (TCM)
J220 – Motronic Engine Control Module (ECM)
K – Instrument Cluster
N88 – Solenoid Valve 1
N89 – Solenoid Valve 2
N90 – Solenoid Valve 3
N91 – Solenoid Valve 4
N92 – Solenoid Valve 5
N93 – Solenoid Valve 6
N94 – Solenoid Valve 7
R – Radio
S51 – Fuse
T2a – Double Connector, on junction connector (T44) above fuse/relay panel
T2b – Double Connector, on junction connector (T44) above fuse/relay panel
T2y – Double Connector, on junction connector (T44) above fuse/relay panel
T8 – 8-Pin Connector, on radio
T10a – 8-Pin Connector, behind fuse/relay panel
T12 – 12-Pin Connector, on transmission
T16 – 16-Pin Connector, Data Link Connector (DLC), behind center of dash panel
T28 – 28-Pin Connector, on instrument cluster
T44 – 44-Pin Connector, Junction Box for on board diagnosis, above fuse/relay panel
T68 – 68-Pin Connector, under right side of engine compartment, on Motronic Engine Control Module (ECM)
T68a – 68-Pin Connector, below right front seat, on Transmission Control Module (TCM)
U1 – Cigarette Lighter
(80) – Ground connection – 1 –, in instrument panel wiring harness
(A40) – Plus connection (30), in instrument panel wiring harness

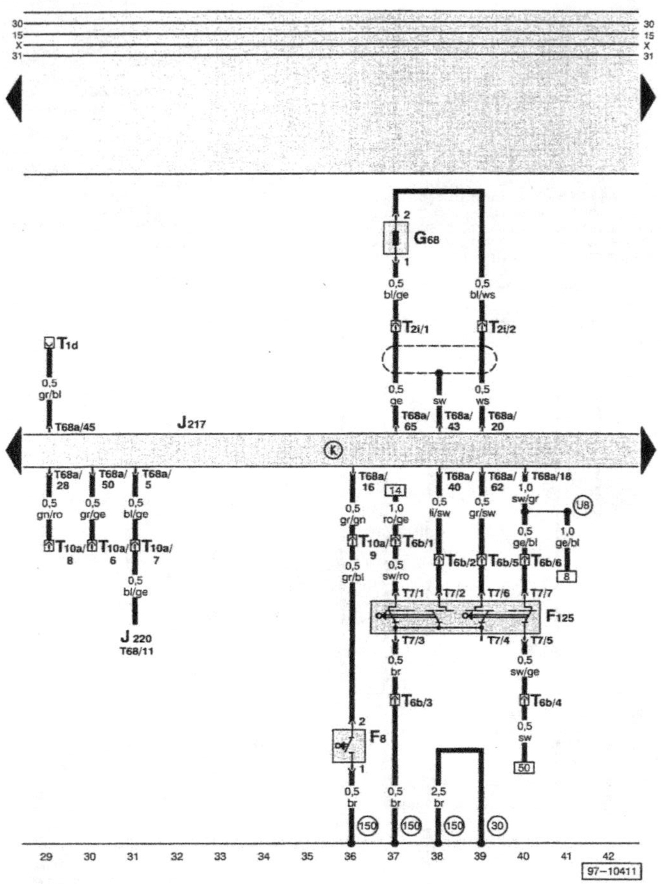

WIRING COLOR CODE		
ws	=	white
sw	=	black
ro	=	red
br	=	brown
gn	=	green
bl	=	blue
gr	=	grey
li	=	lilac
ge	=	yellow

F8 – Kick Down Switch
F125 – Multi-Function Transmission Range (TR) Switch
G68 – Vehicle Speed Sensor (VSS)
J217 – Transmission Control Module (TCM)
J220 – Motronic Engine Control Module (ECM)
T1d – Single Connector, behind fuse/relay panel
T2i – Double Connector, below battery
T6b – 6-Pin Connector, below battery
T7 – 7-Pin Connector, on transmission
T10a – 8-Pin Connector, behind fuse/relay panel
T68 – 68-Pin Connector, under right side of engine compartment, on Motronic Engine Control Module (ECM)

T68a – 68-Pin Connector, below right front seat, on Transmission Control Module (TCM)
(30) – Ground connection –1–, beside fuse/relay panel
(150) – Ground connection, in automatic transmission wiring harness
(U8) – Wire connection (657), in automatic transmission wiring

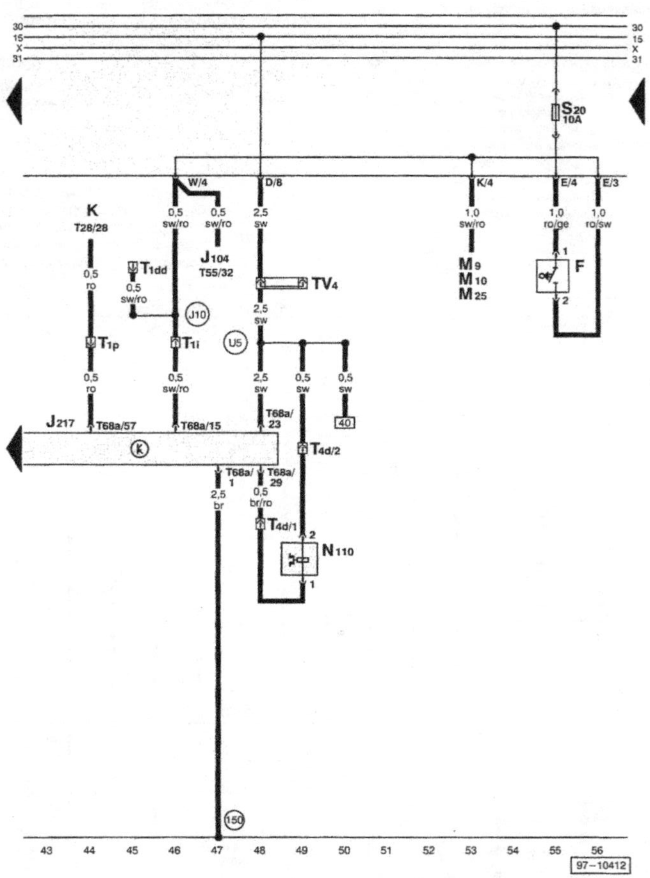

WIRING COLOR CODE		
ws	=	white
sw	=	black
ro	=	red
br	=	brown
gn	=	green
bl	=	blue
gr	=	grey
li	=	lilac
ge	=	yellow

F – Brake Light Switch
J104 – ABS Control Module (w/EDL)
J217 – Transmission Control Module (TCM)
K – Instrument Cluster
M9 – Left Brake Light
M10 – Right Brake Light
M25 – High-Mount Brake Light
N110 – Shift Lock Solenoid
T1dd – Single Connector, behind fuse/relay panel
T1p – Single Connector, behind fuse/relay panel
T4d – 4-Pin Connector, below center console
T28 – 28-Pin Connector, on instrument cluster

T55 – 55-Pin Connector, on ABS Control Module
T68a – 68-Pin Connector, below right front seat, on Transmission Control Module (TCM)
TV4 – Terminal 15 Wire Connector
(150) – Ground connection, in automatic transmission wiring harness
(J10) – Wire connection (54), in ABS wiring harness
(U5) – Wire connection –2– (15), in automatic transmission wiring harness

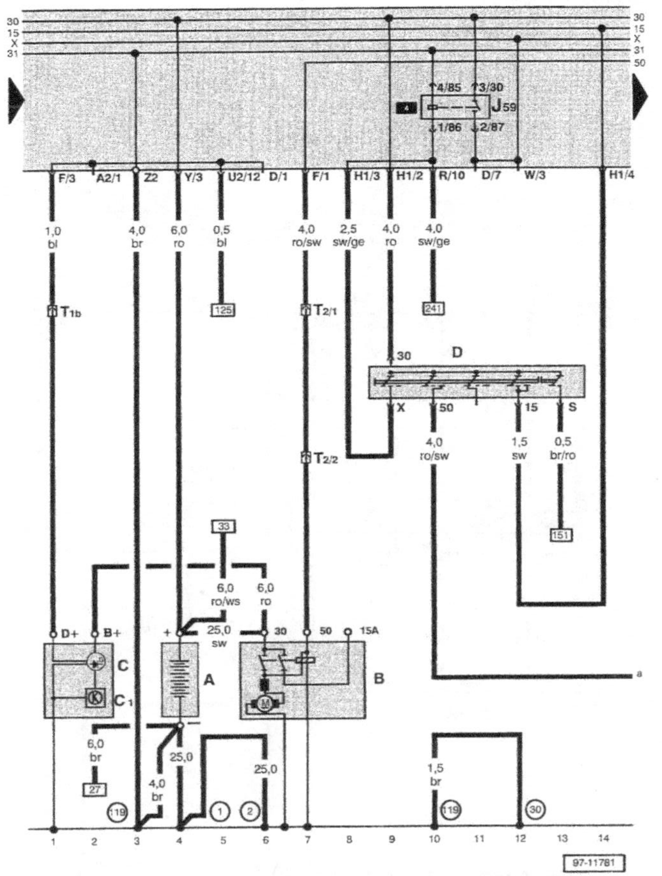

WIRING COLOR CODE

ws	=	white
sw	=	black
ro	=	red
br	=	brown
gn	=	green
bl	=	blue
gr	=	grey
li	=	lilac
ge	=	yellow

97-11781

A – Battery
B – Starter
C – Generator (GEN)
C1 – Voltage Regulator (VR)
D – Ignition/Starter Switch
J59 – Load Reduction Relay
T1b – Single Connector, near battery
T2 – Double Connector, behind fuse/relay panel

(1) – Ground strap, battery to body

(2) – Ground strap, transmission to body

(30) – Ground connection –1–, beside fuse/relay panel

(119) – Ground connection –1–, in headlight wiring harness

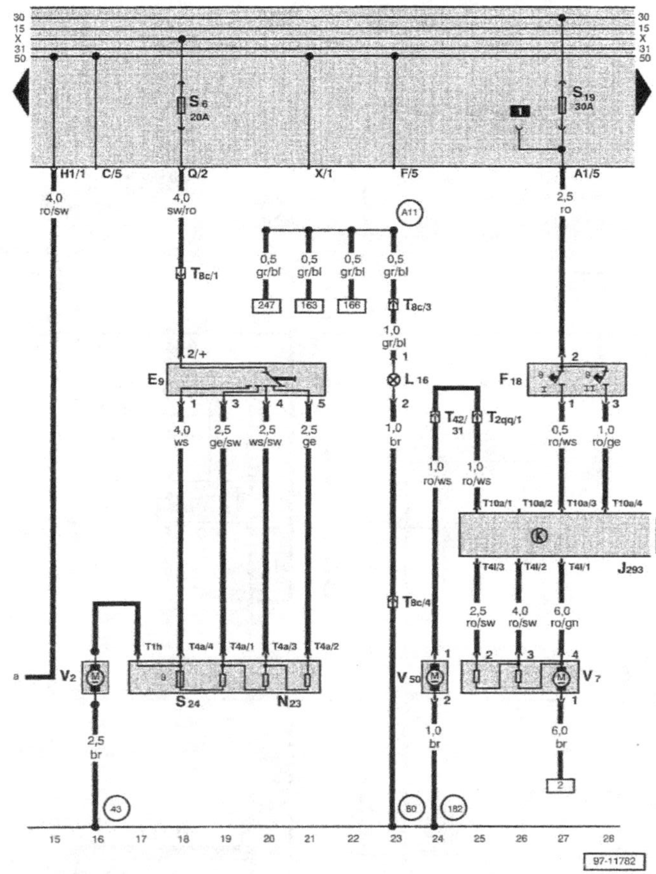

WIRING COLOR CODE

ws	=	white
sw	=	black
ro	=	red
br	=	brown
gn	=	green
bl	=	blue
gr	=	grey
li	=	lilac
ge	=	yellow

97-11782

E9 – Fresh Air Blower Switch
F18 – Coolant Fan Control (FC) Thermal Switch
J293 – Coolant Fan Control (FC) Control Module
L16 – Fresh Air Control Lever Light
N23 – Fresh Air Blower Series Resistance
S24 – Heater, A/C Clutch Fuse (5A)
T1h – Single Connector, on N23
T2qq – Double Connector, behind fuse/relay panel
T4l – 4-Pin Connector, on J293
T4a – 4-Pin Connector, on N23
T8c – 8-Pin Connector, behind instrument panel, center
T10a – 10-Pin Connector, on J293
T42 – 42-Pin Connector, on engine
V2 – Fresh Air Blower (01273-4124)

V7 – Coolant Fan (01018-3433)
V50 – Coolant Pump (00739)

(43) – Ground connection, on right A-pillar, lower part

(80) – Ground connection –1–, in instrument panel wiring harness

(182) – Ground connection –1–, in engine compartment wiring harness (6 cylinder)

(A11) – Plus connection (58b), in instrument panel wiring harness

Engine control module, ignition coil, coolant fan control module

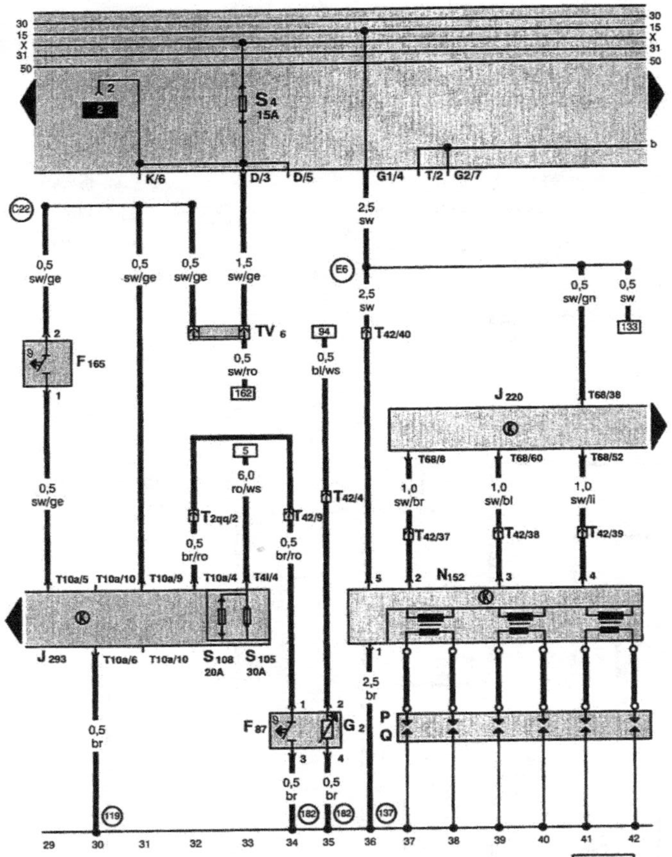

WIRING COLOR CODE		
ws	=	white
sw	=	black
ro	=	red
br	=	brown
gn	=	green
bl	=	blue
gr	=	grey
li	=	lilac
ge	=	yellow

F87 – After-Run Coolant Fan Control (FC) Thermal Switch
F165 – Third Speed Coolant Fan Control (FC) Thermal Switch
G2 – Engine Coolant Temperature (ECT) Sensor (01039)
J220 – Motronic Engine Control Module (ECM)
J293 – Coolant Fan Control (FC) Control Module
N152 – Ignition Coil
P – Spark Plug Connector
Q – Spark Plugs
S105 – First Speed Coolant Fan Fuse
S108 – Coolant Fan First Speed and A/C Clutch Fuse
T2qq – Double Connector, behind fuse/relay panel
T4l – 4-Pin Connector, on J293
T10a – 10-Pin Connector, on J293
T42 – 42-Pin Connector, one engine
T68 – 68-Pin Connector, on J220
TV6 – Terminal X Wire Connector
(119) – Ground connection –1–, in headlight wiring harness
(137) – Ground connection (end stage), in Motronic Multiport Fuel Injection (MFI) wiring harness
(182) – Ground connection –1–, in engine compartment wiring harness (6 cylinder)
(C22) – Wire connection (X), in headlight/air conditioner wiring harness
(E6) – Plug connection –1– (15), in Motronic Multiport Fuel Injection (MFI) wiring harness

Engine control module, heated oxygen sensor, knock sensors

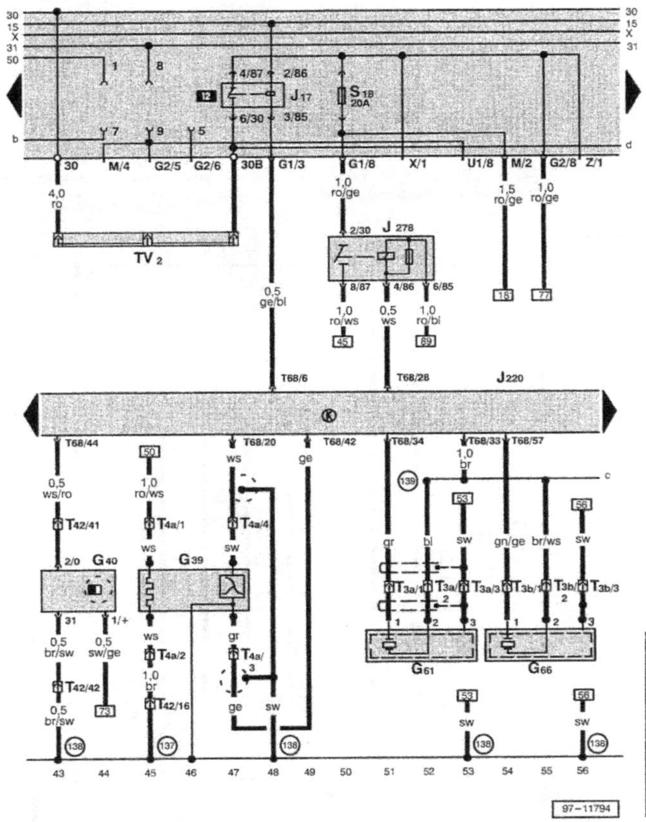

WIRING COLOR CODE		
ws	=	white
sw	=	black
ro	=	red
br	=	brown
gn	=	green
bl	=	blue
gr	=	grey
li	=	lilac
ge	=	yellow

G39 – Heated Oxygen Sensor (HO2S) (00525)
G40 – Camshaft Position (CMP) Sensor (00515)
G61 – Knock Sensor (KS) 1 (00524)
G66 – Knock Sensor (KS) 2 (00549)
J17 – Fuel Pump (FP) Relay (01259)
J220 – Motronic Engine Control Module (ECM)
J278 – Heated Oxygen Sensor (HO2S) Relay (00640)
T3a – 3-Pin Connector, near intake manifold
T3b – 3-Pin Connector, near intake manifold
T4a – 4-Pin Connector, near intake manifold
T42 – 42-Pin Connector, on engine
T68 – 68-Pin Connector, on J220
TV2 – Terminal 30 (B+) Wire Connector
(137) – Ground connection (end stage), in Motronic Multiport Fuel Injection (MFI) wiring harness
(138) – Ground connection (control module), in Motronic Multiport Fuel Injection (MFI) wiring harness
(139) – Ground Connection (sensor ground), in Motronic Multiport Fuel Injection (MFI) wiring harness

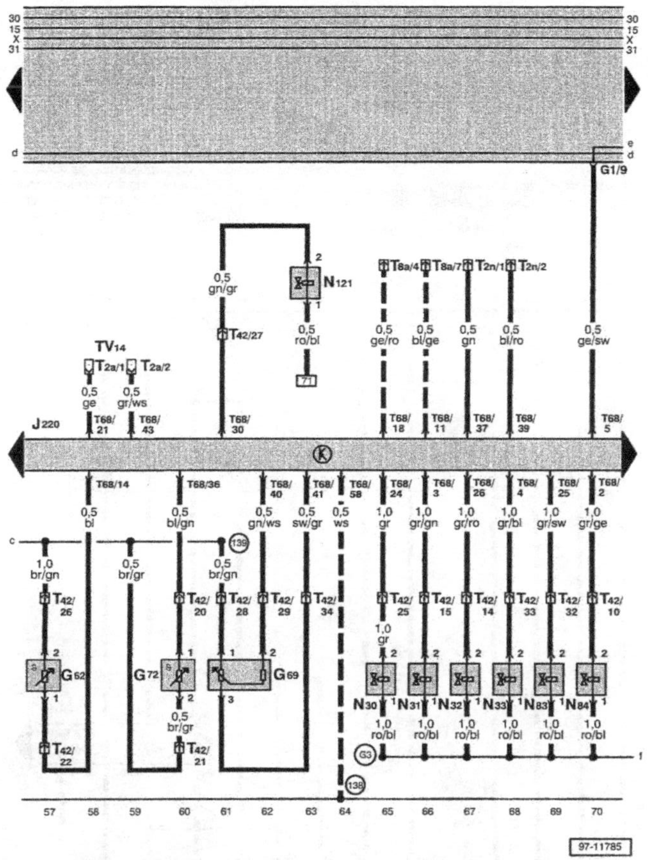

WIRING COLOR CODE		
ws	=	white
sw	=	black
ro	=	red
br	=	brown
gn	=	green
bl	=	blue
gr	=	grey
li	=	lilac
ge	=	yellow

G62 – Engine Coolant Temperature (ECT) Sensor (00522)
G69 – Throttle Position (TP) Sensor (00518)
G72 – Intake Air Temperature (IAT) Sensor (00527)
J220 – Motronic Engine Control Module (ECM)
N30 – Cylinder 1 Fuel Injector (01249)
N31 – Cylinder 2 Fuel Injector (01250)
N32 – Cylinder 3 Fuel Injector (01251)
N33 – Cylinder 4 Fuel Injector (01252)
N83 – Cylinder 5 Fuel Injector (01253)
N84 – Cylinder 6 Fuel Injector (01254)
N121 – EGR Vacuum Regulator Valve
T2a – Double Connector, behind fuse/relay panel
T2n – Double Connector, behind fuse/relay panel
T8a – 8-Pin Connector, behind fuse/relay panel

T42 – 42-Pin Connector, on engine
T68 – 68-Pin Connector, on J220
TV14 – Data Link Connector (DLC), wire connector
(138) – Ground connection (control module), in Motronic Multiport Fuel Injection (MFI) wiring harness
(139) – Ground connection (sensor ground), in Motronic Multiport Fuel Injection (MFI) wiring harness
(G3) – Plus connection in fuel injector wiring harness
– – – Automatic transmission only
* Connector T2a/1 connected to Data Link Connector T16/15
**Connector T2a/2 connected to Data Link Connector T16/7

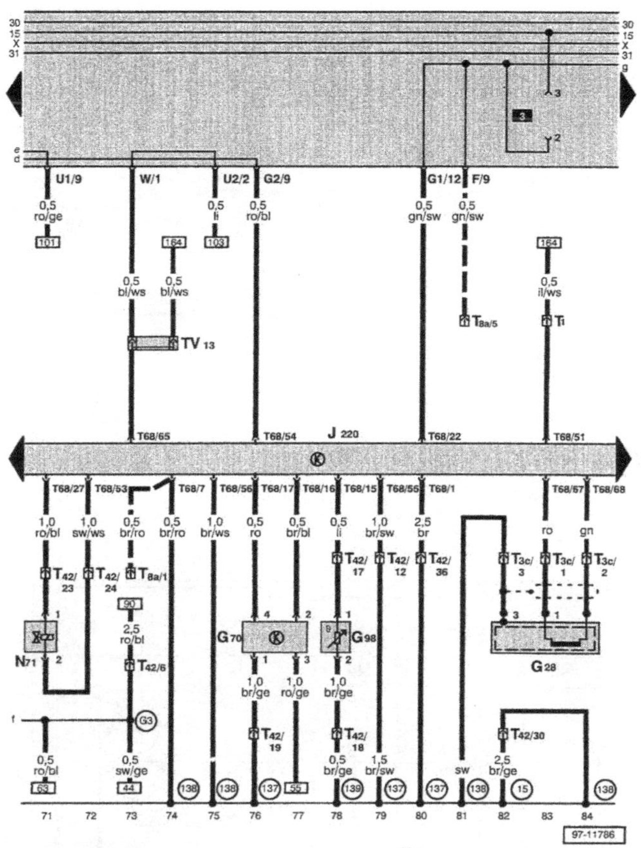

WIRING COLOR CODE		
ws	=	white
sw	=	black
ro	=	red
br	=	brown
gn	=	green
bl	=	blue
gr	=	grey
li	=	lilac
ge	=	yellow

G28 – Engine Speed (RPM) Sensor (00513)
G70 – Mass Air Flow (MAF) Sensor (00553)
G98 – EGR Temperature Sensor (00585)
J220 – Motronic Engine Control Module (ECM)
N71 – Idle Air Control (IAC) Valve (01257)
T1 – Single Connector, behind fuse/relay panel
T3c – 3-Pin Connector, near starter
T8a – 8-Pin Connector, behind fuse/relay panel
T42 – 42-Pin Connector, on engine
T68 – 68-Pin Connector, on J220
TV13 – Vehicle Speed Signal Wire Connector
(15) – Ground connection, on cylinder head

(137) – Ground connection (end stage), in Motronic Multiport Fuel Injection (MFI) wiring harness
(138) – Ground connection (control module), in Motronic Multiport Fuel Injection (MFI) wiring harness
(139) – Ground connection (sensor ground), in Motronic Multiport Fuel Injection (MFI) wiring harness
(G3) – Plus connection, in fuel injector wiring harness
– – – Automatic transmission only

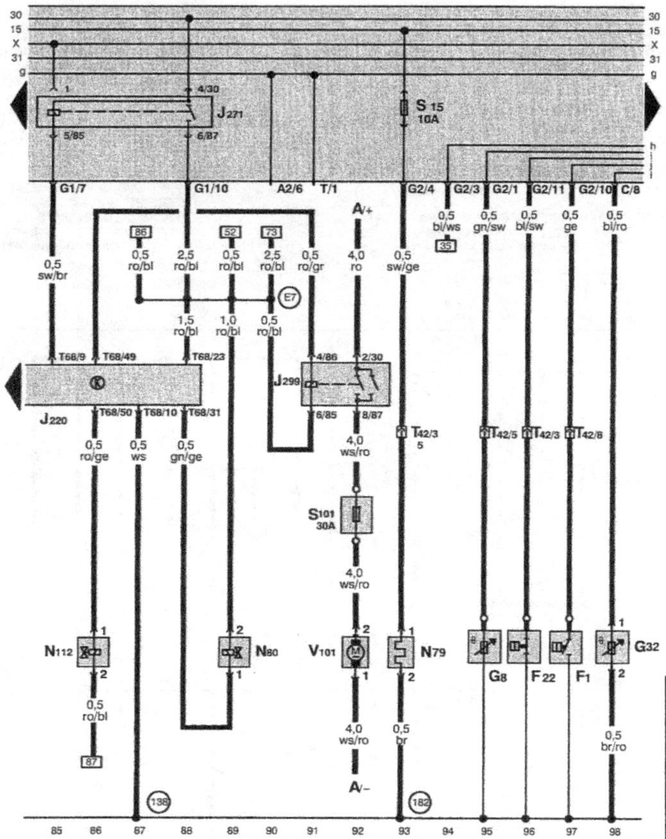

WIRING COLOR CODE		
ws	=	white
sw	=	black
ro	=	red
br	=	brown
gn	=	green
bl	=	blue
gr	=	grey
li	=	lilac
ge	=	yellow

A – Battery
F1 – Engine Oil Pressure Switch
F22 – 0.3 Bar Engine Oil Pressure Switch (00772)
G8 – Engine Oil Temperature Sensor
G32 – Engine Coolant Level (ECL) Sensor
J271 – Motronic Engine Control Module (ECM) Power Supply Relay
J299 – Secondary Air Injection (AIR) Pump Relay (01264)
N79 – Positive Crankcase Ventilation (PCV) Heating Element
N80 – Evaporative Emission (EVAP) Canister Purge Regulator Valve (01247)
N112 – Secondary Air Injection (AIR) Solenoid Valve (01235)
S101 – A/C Climatronic Control Module Fuse
T42 – 42-Pin Connector, on engine
T68 – 68-Pin Connector, on J220
V101 – Secondary Air Injection (AIR) Pump Motor
(138) – Ground connection (Control Module), in Motronic Multiport Fuel Injection (MFI) wiring harness
(182) – Ground connection –1–, in engine compartment wiring harness (6 cylinder)
(E7) – Ground connection (87a), in Motronic Multiport Fuel Injection (MFI) wiring harness

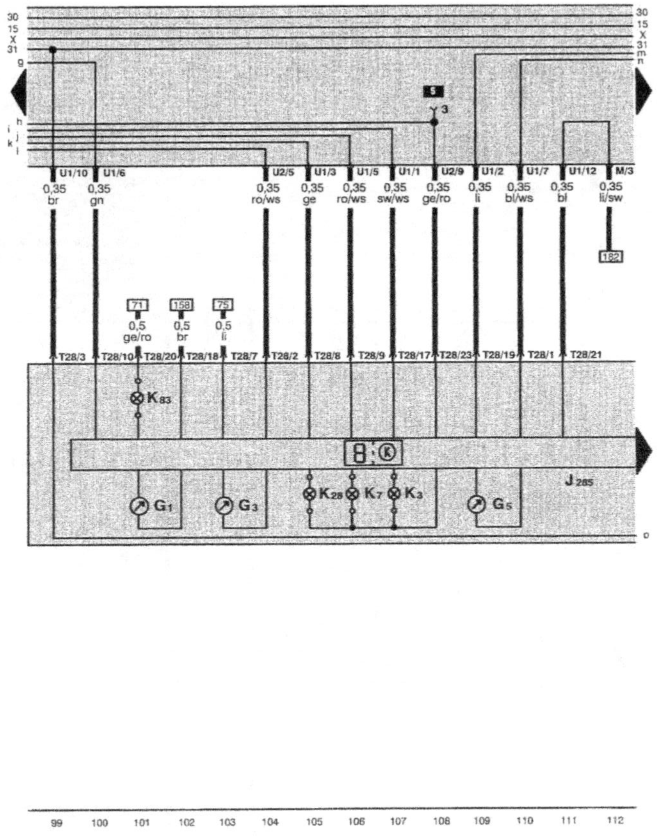

WIRING COLOR CODE		
ws	=	white
sw	=	black
ro	=	red
br	=	brown
gn	=	green
bl	=	blue
gr	=	grey
li	=	lilac
ge	=	yellow

G1 – Fuel Gauge (00664)
G3 – Engine Coolant Temperature (ECT) Gauge
G5 – Tachometer (00662)
J285 – Control Module With Display Unit In Instrument Cluster
K3 – Oil Pressure Warning Light (00755)
K7 – Brake and Parking Brake Warning Light (01047)
K28 – Engine Coolant Level/Temperature (ECL/ECT) Warning Light
K83 – Multi-Function Indicator (MIL)
T28 – 28-Pin Connector, on instrument cluster

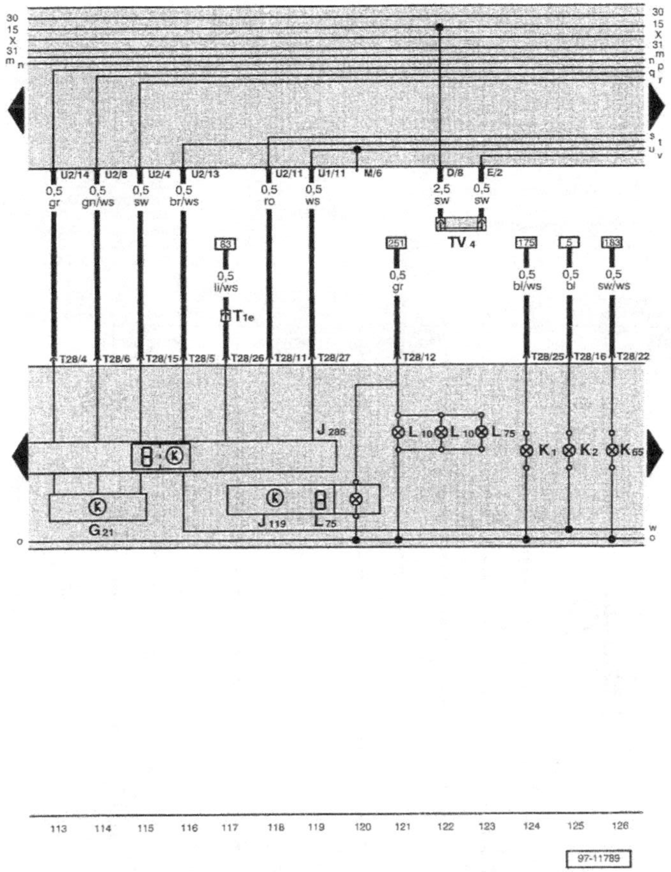

WIRING COLOR CODE		
ws	=	white
sw	=	black
ro	=	red
br	=	brown
gn	=	green
bl	=	blue
gr	=	grey
li	=	lilac
ge	=	yellow

G21 – Speedometer (00663)
J119 – Multi-Function Indicator
J285 – Control Module With Display Unit In Instrument Cluster
K1 – Headlight High Beam Indicator Light
K2 – Generator (GEN) Warning Light
K65 – Left Turn Signal Indicator Light

L10 – Instrument Cluster Light
L75 – Digital Display Light
T1e – Single Connector, behind fuse/relay panel
T28 – 28-Pin Connector, on instrument cluster
TV4 – Terminal 15 Wire Connector

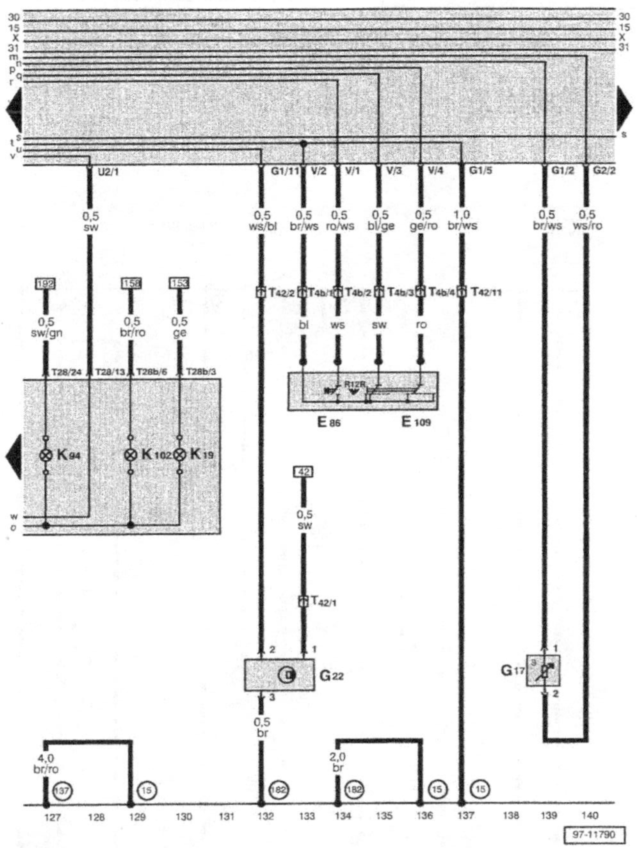

WIRING COLOR CODE		
ws	=	white
sw	=	black
ro	=	red
br	=	brown
gn	=	green
bl	=	blue
gr	=	grey
li	=	lilac
ge	=	yellow

E86 – Multi-Function Indicator Mode Select Switch
E109 – Multi-Function Indicator Memory Switch
G17 – Outside Air Temperature Sensor (00779-3133)
G22 – Speedometer Vehicle Speed Sensor (VSS) (01086), on transmission
K19 – Seat Belt Warning Light (00665)
K94 – Right Turn Signal Indicator Light
K102 – Rear Hatch/Trunk Lid Ajar Warning Light
T4b – 4-Pin Connector, behind steering column switch cover
T28b – 28-Pin Connector, on instrument cluster
T28 – 28-Pin Connector, on instrument cluster

T42 – 42-Pin Connector, on engine
(15) – Ground connection, on cylinder head
(137) – Ground connection (end stage), in Motronic Multiport Fuel Injection (MFI) wiring harness
(182) – Ground connection –1–, in engine compartment wiring harness (6 cylinder)

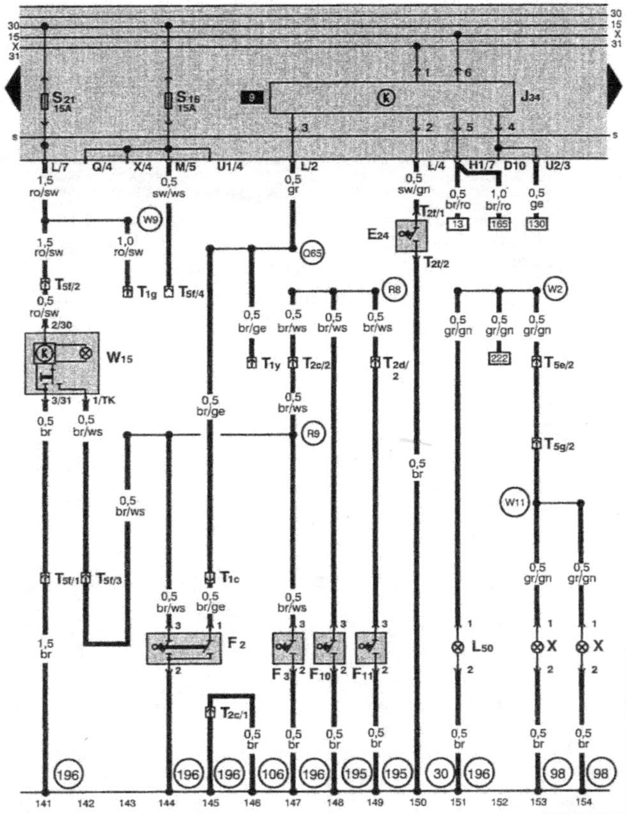

WIRING COLOR CODE

ws	=	white
sw	=	black
ro	=	red
br	=	brown
gn	=	green
bl	=	blue
gr	=	grey
li	=	lilac
ge	=	yellow

E24 – Left Seat Belt Switch (00591)
F2 – Left Front Door Contact Switch
F3 – Right Front Door Contact Switch
F10 – Left Rear Door Contact Switch
F11 – Right Rear Door Contact Switch
J34 – Seat Belt Warning System Relay
L50 – Center Rear Ashtray Light
T1c – Single Connector, behind fuse/relay panel
T1g – Single Connector, behind fuse/relay panel
T1y – Single Connector, behind fuse/relay panel
T2c – Double Connector, on right A-pillar, lower part
T2d – Double Connector, on left A-pillar, lower part
T2f – Double Connector, behind console
T5e – 5-Pin Connector, in luggage compartment, left rear
T5f – 5-Pin Connector, behind fuse/relay panel
T5g – 5-Pin Connector, on rear apron
W15 – Interior Light with Delay Switch
X – License Plate Light
(30) – Ground connection –1–, beside fuse/relay panel
(98) – Ground connection, in rear lid wiring harness
(106) – Ground connection, in alarm system wiring harness
(195) – Ground connection, in rear door contact switch wiring harness
(196) – Ground connection –1–, in rear wiring harness
(Q65) – Wire connection, in seat belt warning system wiring harness
(R8) – Wire connection (door contact switch), in rear door contact switch wiring harness
(R9) – Wire connection (door contact switch), in rear door wiring harness
(W2) – Plus connection (58), in rear wiring harness
(W9) – Plus connection (30), in rear wiring harness
(W11) – Wire connection (59), in rear lid wiring harness

1995

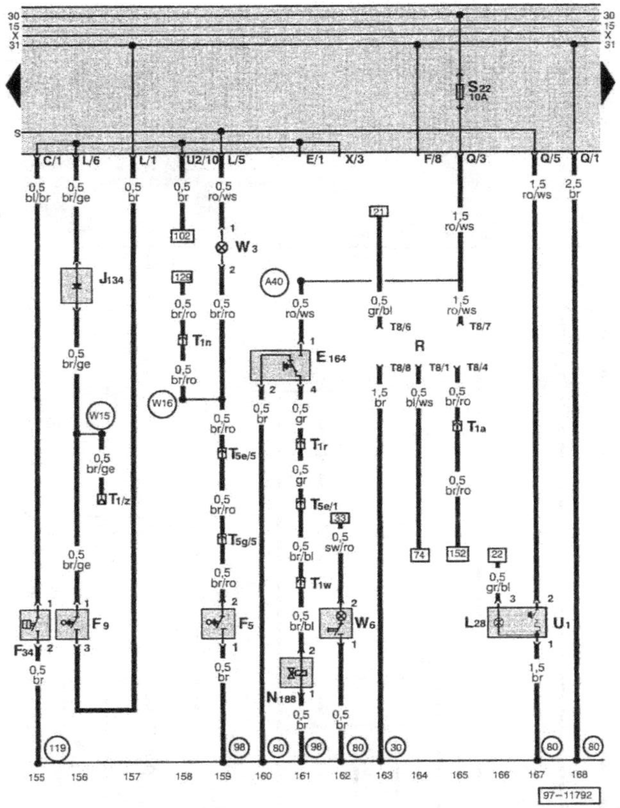

WIRING COLOR CODE

ws	=	white
sw	=	black
ro	=	red
br	=	brown
gn	=	green
bl	=	blue
gr	=	grey
li	=	lilac
ge	=	yellow

E164 – Luggage Compartment Release Switch
F5 – Luggage Compartment Light Switch
F9 – Parking Brake Warning Light Switch (01114)
F34 – Brake Fluid Level Warning Switch
J134 – Daytime Running Lights Diode
L28 – Cigarette Lighter Light
N188 – Luggage Compartment Release Solenoid
R – Radio
T1a – Single Connector, behind fuse/relay panel
T1n – Single Connector, behind fuse/relay panel
T1r – Single Connector, behind fuse/relay panel
T1w – Single Connector, on rear apron
T1z – Single Connector, behind Fuse/Relay Panel
T5e – 5-Pin Connector, in luggage compartment, left rear
T5g – 5-Pin Connector, on rear apron
T8 – 8-Pin Connector, on radio

U1 – Cigarette Lighter
W3 – Luggage Compartment Light
W6 – Glove Compartment Light (Not applicable on vehicles with passenger side air bag)
(30) – Ground connection –1–, beside fuse/relay panel
(80) – Ground connection –1–, in instrument panel wiring harness
(98) – Ground connection, in rear lid wiring harness
(119) – Ground connection –1–, in headlight wiring harness
(A40) – Plus connection (30), in instrument panel wiring harness
(W15) – Wire connection (HB), in rear wiring harness
(W16) – Wire connection, in rear warning harness

1995

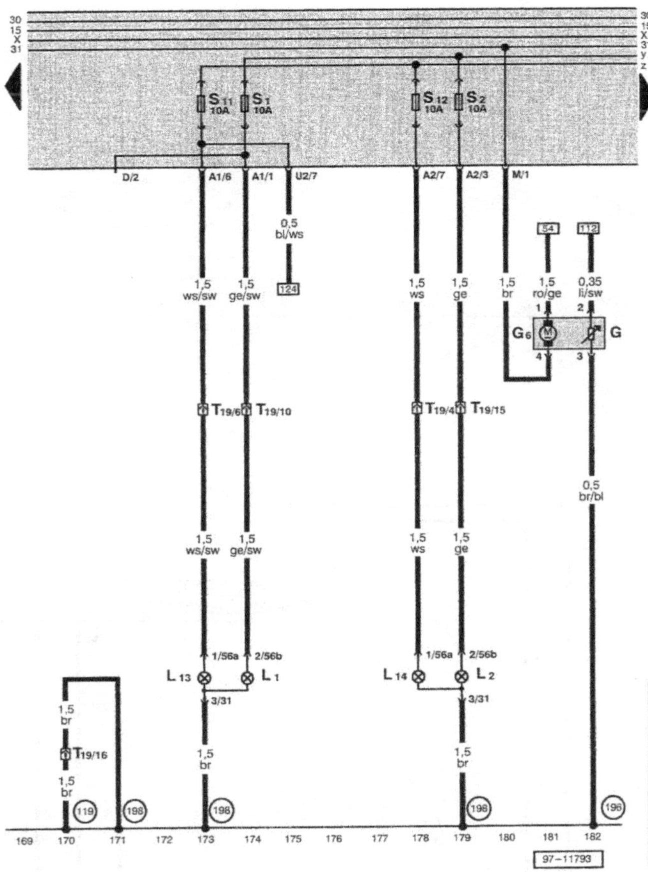

WIRING COLOR CODE		
ws	=	white
sw	=	black
ro	=	red
br	=	brown
gn	=	green
bl	=	blue
gr	=	grey
li	=	lilac
ge	=	yellow

- G – Fuel Level Sensor
- G6 – Fuel Pump (FP)
- L1 – Left Headlight
- L2 – Right Headlight
- L13 – Left High Beam Headlight
- L14 – Right High Beam Headlight
- T19 – 19-Pin Connector, behind radiator grill, center
- (119) – Ground connection –1–, in headlight wiring harness
- (196) – Ground connection –1–, in rear wiring harness
- (198) – Ground connection, in front light wiring harness

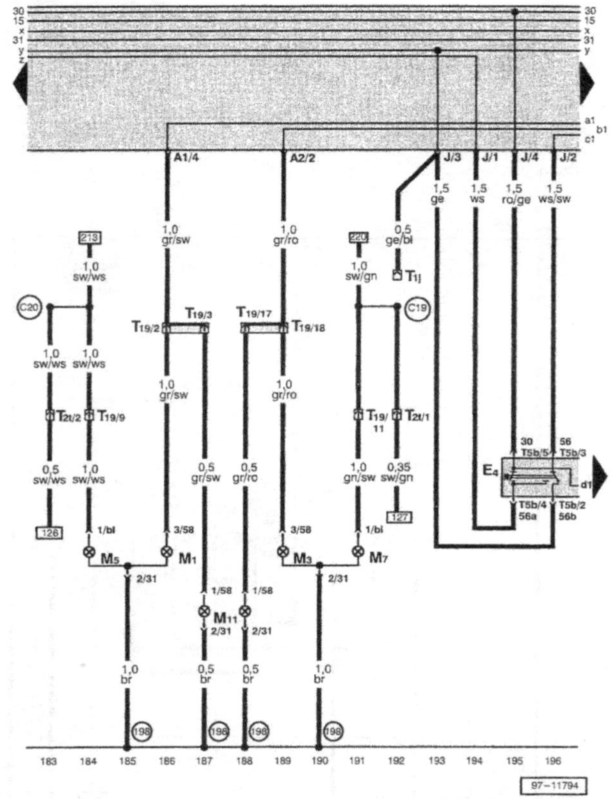

WIRING COLOR CODE		
ws	=	white
sw	=	black
ro	=	red
br	=	brown
gn	=	green
bl	=	blue
gr	=	grey
li	=	lilac
ge	=	yellow

- E4 – Headlight Dimmer/Flasher Switch
- M1 – Left Parking Light
- M3 – Right Parking Light
- M5 – Left Front Turn Signal Light
- M7 – Right Front Turn Signal Light
- M11 – Front Side Marker Lights
- T1j – Single Connector, behind fuse/relay panel
- T2t – Double Connector, behind fuse/relay panel
- T5b – 5-Pin Connector, behind steering column switch trim
- T19 – 19-Pin Connector, behind radiator grill, right
- (198) – Ground connection, in front light wiring harness
- (C19) – Wire connection (right turn signal), in headlight wiring harness
- (C20) – Wire connection (left turn signal), in headlight wiring harness

Turn signal switch, emergency flasher switch/relay

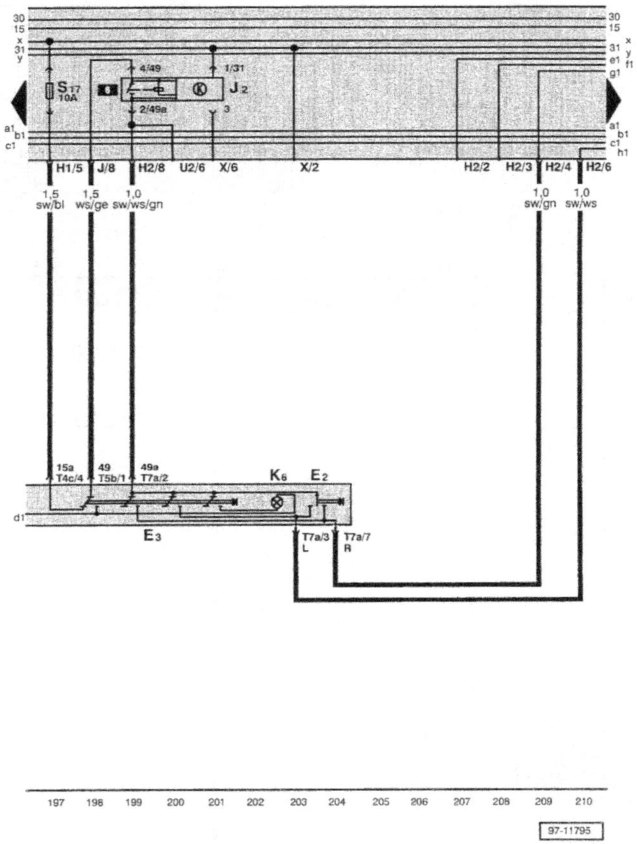

WIRING COLOR CODE		
ws	=	white
sw	=	black
ro	=	red
br	=	brown
gn	=	green
bl	=	blue
gr	=	grey
li	=	lilac
ge	=	yellow

- E2 – Turn Signal Switch
- E3 – Emergency Flasher Switch
- J2 – Emergency Flasher Relay
- K6 – Emergency Flasher Warning Light
- T4c – 4-Pin Connector, behind steering column switch cover
- T5b – 5-Pin Connector, behind steering column switch cover
- T7a – 7-Pin Connector, behind steering column switch cover

Taillights, brake lights (Golf only)

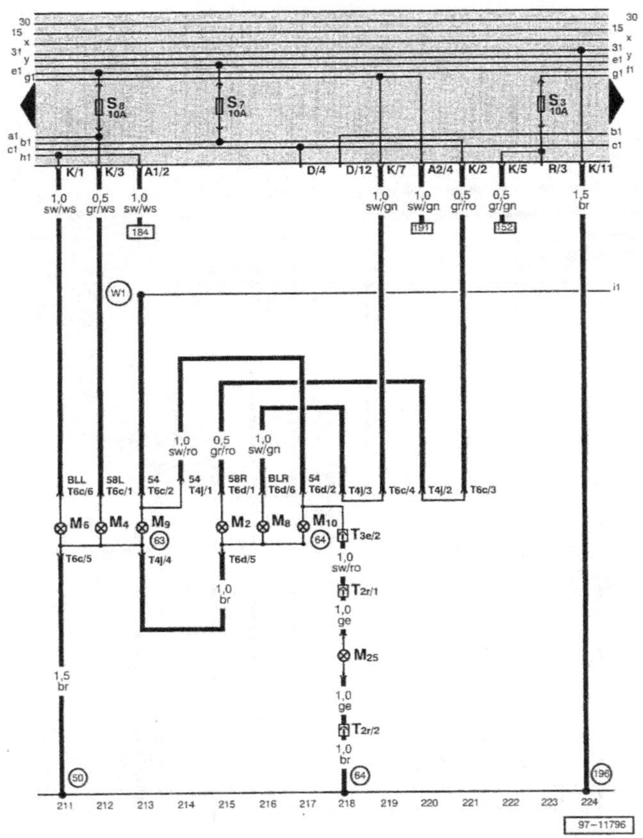

WIRING COLOR CODE		
ws	=	white
sw	=	black
ro	=	red
br	=	brown
gn	=	green
bl	=	blue
gr	=	grey
li	=	lilac
ge	=	yellow

- M2 – Right Tail light
- M4 – Left Tail light
- M6 – Left Rear Turn Signal Light
- M8 – Right Rear Turn Signal Light
- M9 – Left Brake Light
- M10 – Right Brake Light
- M25 – High-Mount Brake Light
- T2r – Double Connector, in luggage compartment, right rear
- T3e – 3-Pin Connector, on right tail light
- T4j – 4-Pin Connector, on left tail light
- T6c – 6-Pin Connector, on left tail light
- T6d – 6-Pin Connector, on right tail light

- (50) – Ground connection, in luggage compartment, left
- (63) – Ground connection, on left tail light bulb holder
- (64) – Ground connection, on right tail light bulb holder
- (196) – Ground connection –1–, in rear wiring harness
- (W1) – Plus connection (54), in rear wiring harness

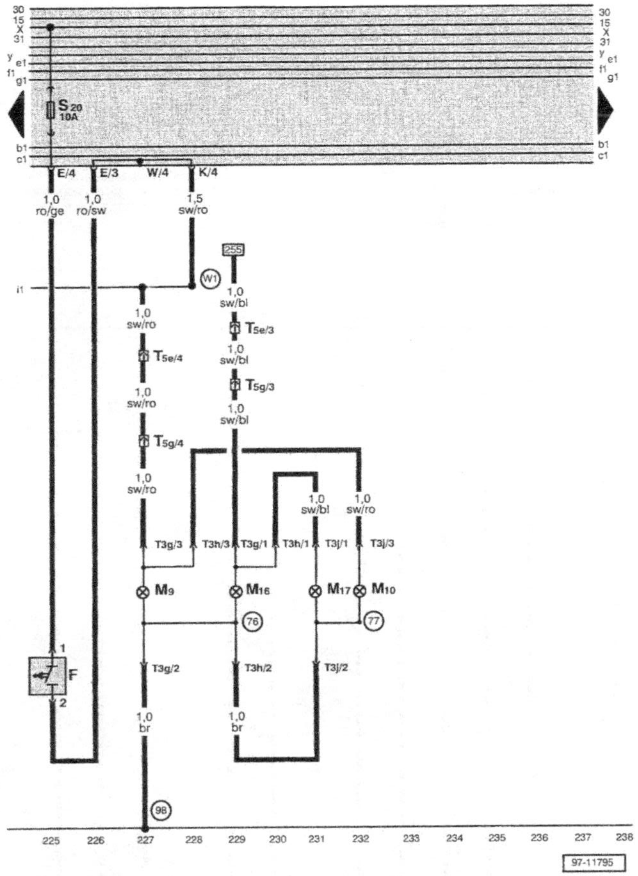

WIRING COLOR CODE		
ws	=	white
sw	=	black
ro	=	red
br	=	brown
gn	=	green
bl	=	blue
gr	=	grey
li	=	lilac
ge	=	yellow

F – Brake Light Switch (00526-2131)
M9 – Left Brake Light
M10 – Right Brake Light
M16 – Left Back-Up Light
M17 – Right Back-Up Light
T3j – 3-Pin Connector, on right tail light
T3h – 3-Pin Connector on left tail light
T3g – 3-Pin Connector, on left tail light
T5e – 5-Pin Connector, on rear apron
T5g – 5-Pin Connector, on rear apron

(76) – Ground connection, on left tail light bulb holder
(77) – Ground connection, on right tail light bulb holder
(98) – Ground connection, in rear lid wiring harness
(W1) – Plug connection (54), in rear wiring harness

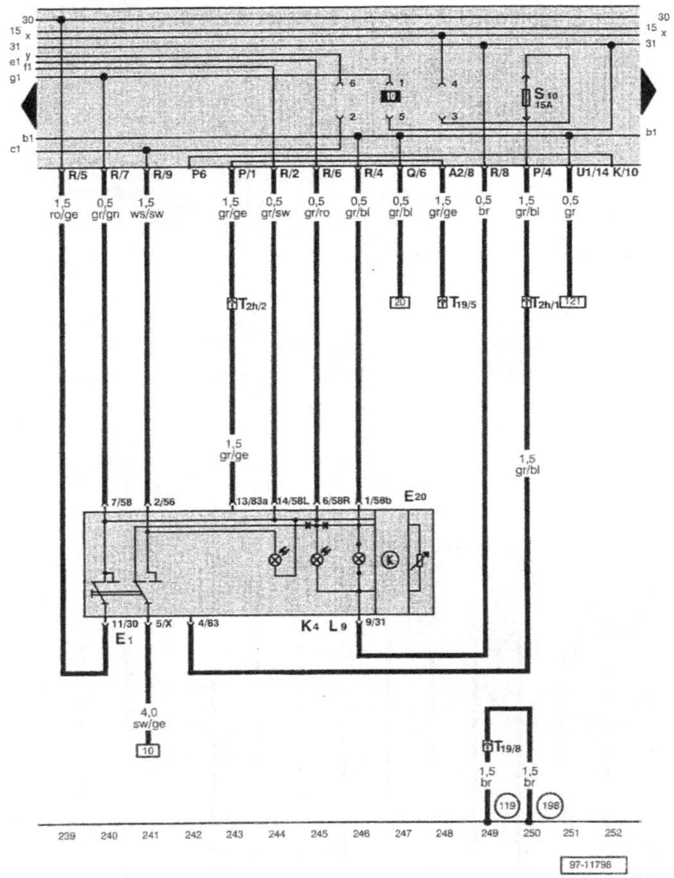

WIRING COLOR CODE		
ws	=	white
sw	=	black
ro	=	red
br	=	brown
gn	=	green
bl	=	blue
gr	=	grey
li	=	lilac
ge	=	yellow

E1 – Light Switch
E20 – Instrument Panel Light Dimmer Switch
K4 – Park Light Indicator Light
L9 – Headlight Switch Light
T2h – 2-Pin Connector, behind fuse/relay panel

T19 – 19-Pin Connector, behind radiator grill, center
(119) – Ground connection –1–, in headlight wiring harness
(198) – Ground connection, in front light wiring harness

Horns, rear window defogger

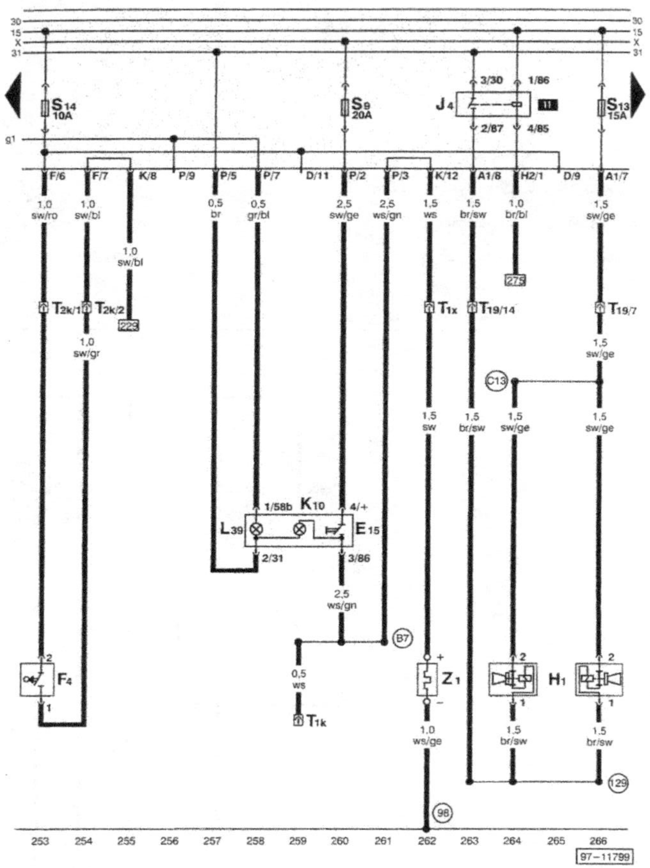

WIRING COLOR CODE		
ws	=	white
sw	=	black
ro	=	red
br	=	brown
gn	=	green
bl	=	blue
gr	=	grey
li	=	lilac
ge	=	yellow

E15 – Rear Window Defogger Switch
F4 – Back-Up Light Switch
H1 – Dual Horns
J4 – Dual Horn Relay
K10 – Rear Window Defogger Indicator Light
L39 – Rear Window Defogger Switch Light
T1k – Single Connector, behind fuse/relay panel
T1x – Single Connector, in luggage compartment, left
T2k – Double Connector, behind fuse/relay panel

T19 – 19-Pin Connector, behind radiator grill, right
Z1 – Rear Window Defogger/Heat Element

(98) – Ground connection, in rear lid wiring harness

(129) – Ground connection, in dual-tone horn wiring harness

(B7) – Wire connection (86), in rear window defogger wiring harness

(C13) – Plus connection, in dual tone horn wiring harness

Windshield wiper motor, windshield wiper/washer switch

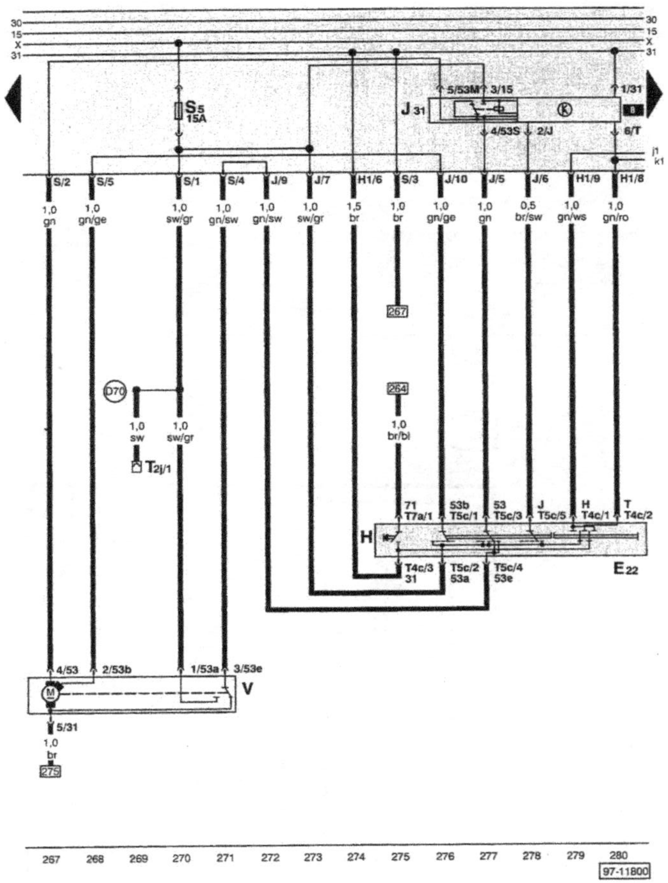

WIRING COLOR CODE		
ws	=	white
sw	=	black
ro	=	red
br	=	brown
gn	=	green
bl	=	blue
gr	=	grey
li	=	lilac
ge	=	yellow

E22 – Windshield Wiper/Washer Switch
H – Horn Button
J31 – Washer/Wiper Intermittent Relay
T2j – Double Connector, behind fuse/relay panel
T4c – 4-Pin Connector, on E22
T5c – 5-Pin Connector, behind steering column switch cover
T7a – 7-Pin Connector, behind steering column switch cover
V – Windshield Wiper Motor

(D70) – Wire connection (53a), in engine compartment wiring harness

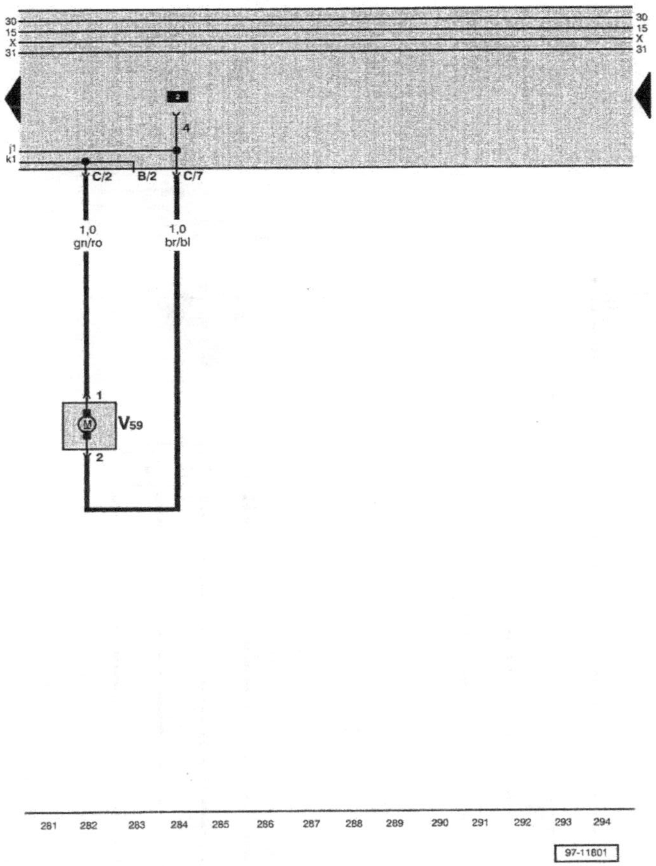

WIRING COLOR CODE

ws	=	white
sw	=	black
ro	=	red
br	=	brown
gn	=	green
bl	=	blue
gr	=	grey
li	=	lilac
ge	=	yellow

V59 – Windshield and Rear Window Washer Pump

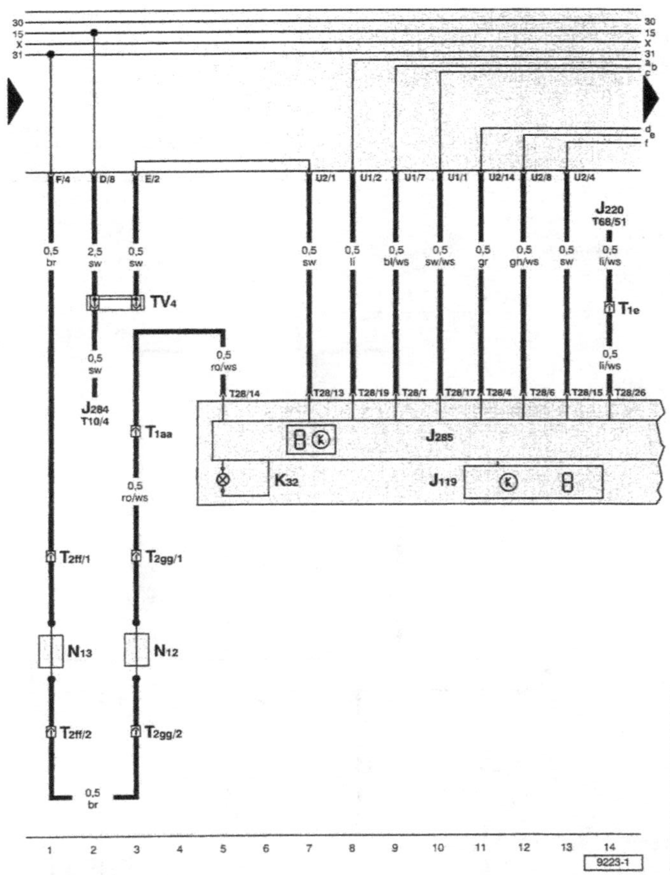

WIRING COLOR CODE

ws	=	white
sw	=	black
ro	=	red
br	=	brown
gn	=	green
bl	=	blue
gr	=	grey
li	=	lilac
ge	=	yellow

J119 – Multi-Function Indicator
J220 – Motronic Engine Control Module (ECM)
J284 – Alarm System Control Module
J285 – Control Module With Display Unit In Instrument Cluster
K32 – Brake Pad Wear Indicator Light
N12 – Right Brake Pad Wear Indicator (Sheer Element)
N13 – Left Brake Pad Wear Indicator (Sheer Element)
T1aa – Single Connector, behind fuse/relay panel
T1e – Single Connector, behind fuse/relay panel
T2ff – Double Connector, behind fuse/relay panel
T2gg – Double Connector, engine compartment, right
T10 – 10-Pin Connector, behind dash panel, left, on Alarm System Control Module
T28 – 28-Pin Connector, on instrument cluster
T68 – 68-Pin Connector, under right side of hood, on Motronic Engine Control Module (ECM)
TV4 – Terminal 15 Wire Connector

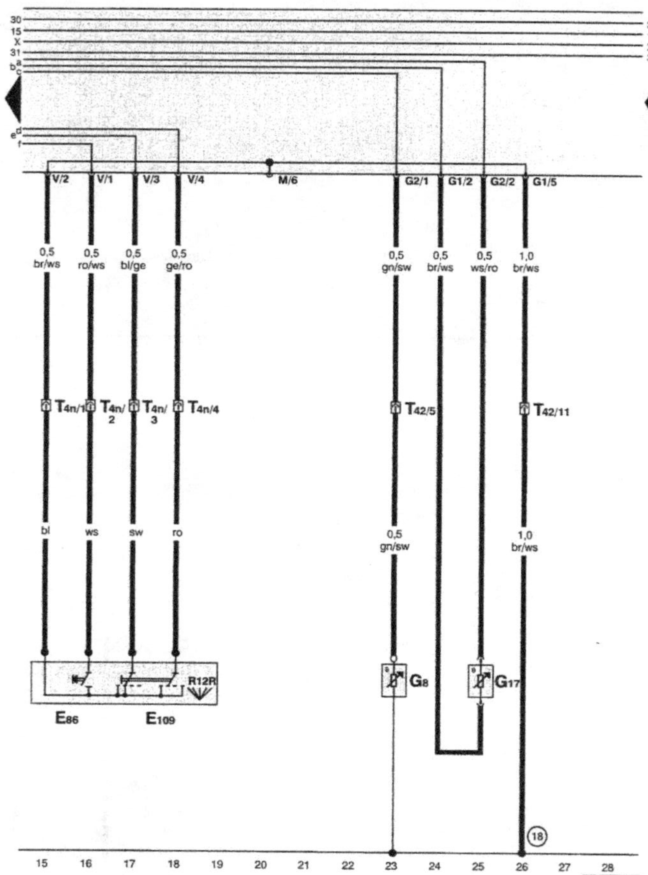

WIRING COLOR CODE		
ws	=	white
sw	=	black
ro	=	red
br	=	brown
gn	=	green
bl	=	blue
gr	=	grey
li	=	lilac
ge	=	yellow

E86 – Multi-Function Indicator Mode Select Switch
E109 – Multi-Function Indicator Memory Switch
G8 – Oil Temperature Sensor
G17 – Outside Air Temperature Sensor
T4n – 4-Pin Connector
T42 – 42-Pin Connector, left side of engine
(18) – Ground connection, on engine block

1995

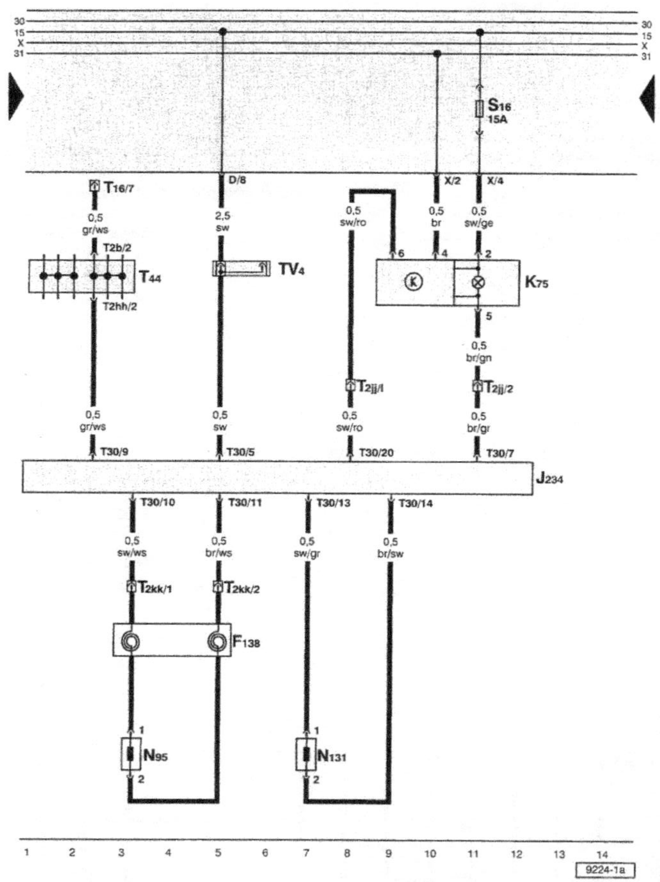

WIRING COLOR CODE		
ws	=	white
sw	=	black
ro	=	red
br	=	brown
gn	=	green
bl	=	blue
gr	=	grey
li	=	lilac
ge	=	yellow

F138 – Airbag Spiral Spring
J234 – Airbag Control Module
K75 – Airbag Malfunction Indicator Lamp (MIL)
N95 – Driver's Side Airbag Ignitor
N131 – Passenger Side Airbag Ignitor 1
T2b – Double Connector, on junction connector (T44) above fuse/relay panel
T2hh – Double Connector, on junction connector (T44) above fuse/relay panel
T2jj – Double Connector, behind fuse/relay panel
T2kk – Double Connector, behind fuse/relay panel
T16 – Double Connector, Data Link Connector (DLC), behind center of dash panel
T30 – 30-Pin Connector, on Airbag Control Module
T44 – Junction Box for on Board Diagnosis, above fuse/relay panel
TV4 – Terminal 15 Wire Connector

1995

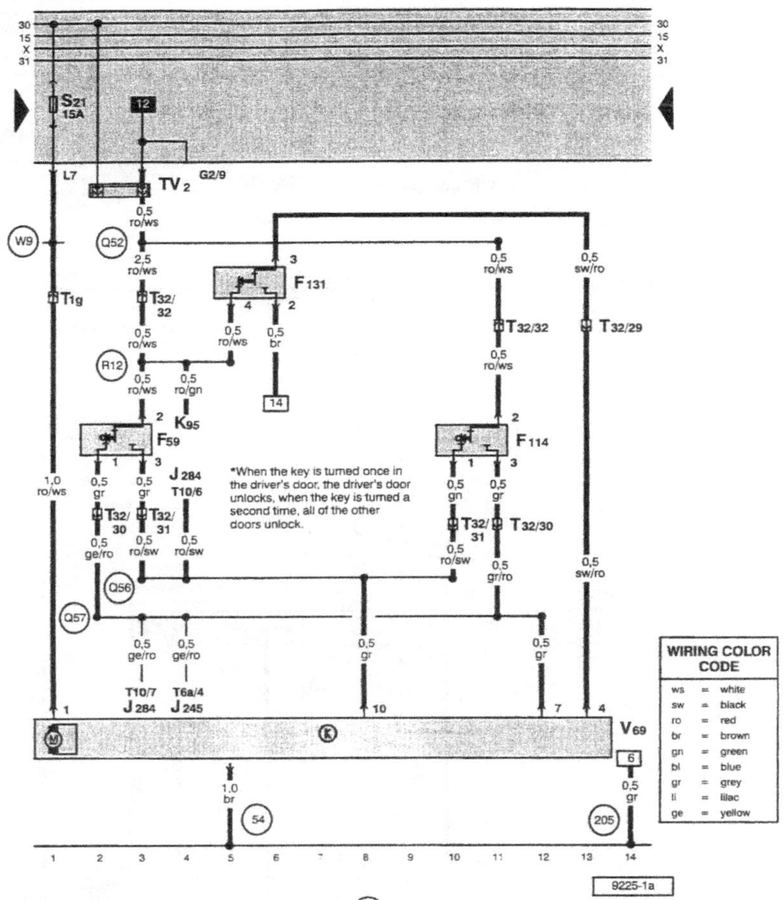

F59 – Driver's Door Central Locking System Door Switch
F114 – Passenger's Door Central Locking System Door Switch
F131 – Left Front Central Locking System Activator/Switch
J245 – Power Sunroof Control Module
J284 – Alarm System Control Module
K95 – Alarm System Indicator Light
T1g – Single Connector, behind fuse/relay panel
T6a – 6-Pin Connector, behind interior light, on Power Sunroof Control Module
T10 – 10-Pin Connector, behind dash panel, left, on Alarm System Control Module
T32 – 32-Pin Connector, on A-pillar, lower left
32a – 32-Pin Connector, on A–pillar, lower right
TV2 – Terminal 30 (B+) Wire Connector
V69 – Central Locking System Pump and Control Module
(54) – Ground connection, on rear apron

(205) – Ground connection, in driver's door wiring harness
(Q51) – Plus connection (30), in alarm system wiring harness
(Q56) – Wire connection (open), in alarm system wiring harness
(Q57) – Wire connection (closed), in alarm system wiring harness
(R12) – Plus connection -3- (30), in driver's door wiring harness
(W9) – Plus connection (30), in rear wiring harness

1995

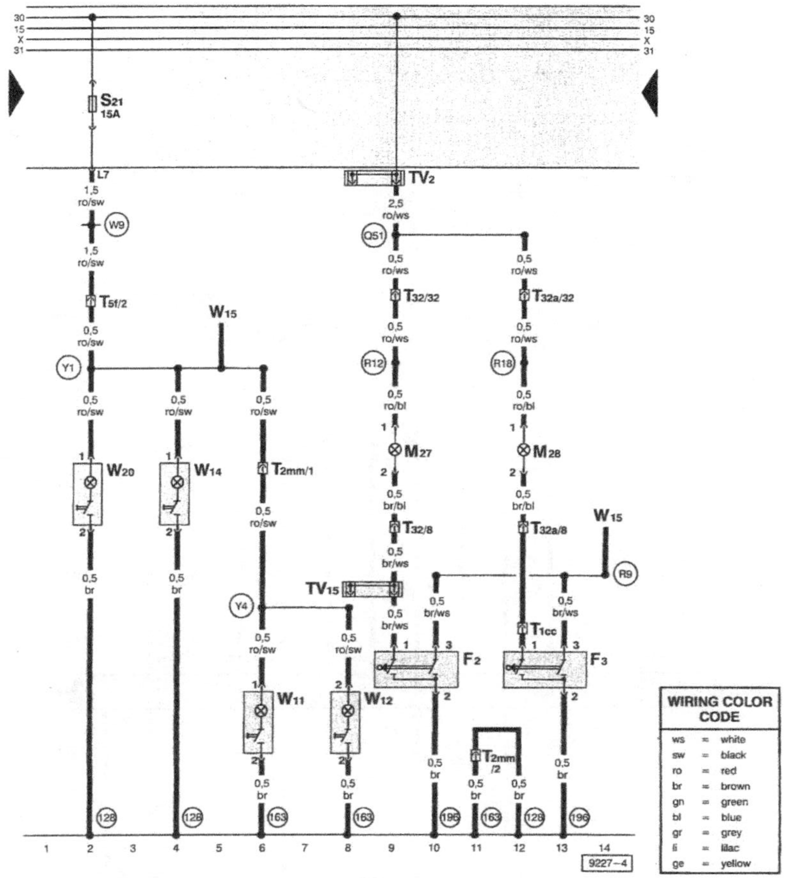

F2 – Left Front Door Contact Switch
F3 – Right Front Door Contact Switch
M27 – Left Door Warning Light
M28 – Right Door Warning Light
T1cc – Single Connector, behind fuse/relay panel
T2mm – Double Connector
T5f – 5-Pin Connector, above fuse/relay panel
T32 – 32-Pin Connector, on A-pillar, lower left
T32a – 32-Pin Connector, on A-pillar, lower right
TV2 – Terminal 30 Wire Connector
TV15 – Wire connector, door contact switch
W11 – Left Rear Reading Light
W12 – Right Rear Reading Light
W14 – Right Make-up Mirror Light
W15 – Interior Light with Delay Switch
W20 – Left Make-up Mirror Light
(128) – Ground connection –1–, in interior light wiring harness
(163) – Ground connection –2–, in interior light wiring harness

(196) – Ground connection –1–, in rear wiring harness
(Q51) – Plus connection (30), in alarm system wiring harness
(R9) – Wire connection (Door Contact Switch), in rear wiring harness
(R12) – Plus connection –3– (30), in driver's door wiring harness
(R18) – Plus connection –3– (30), in passenger's door wiring harness
(W9) – Plus connection (30), in rear wiring harness
(Y1) – Plus connection (30), in interior light wiring harness
(Y4) – Plus connection (30), in reading light wiring harness

1995

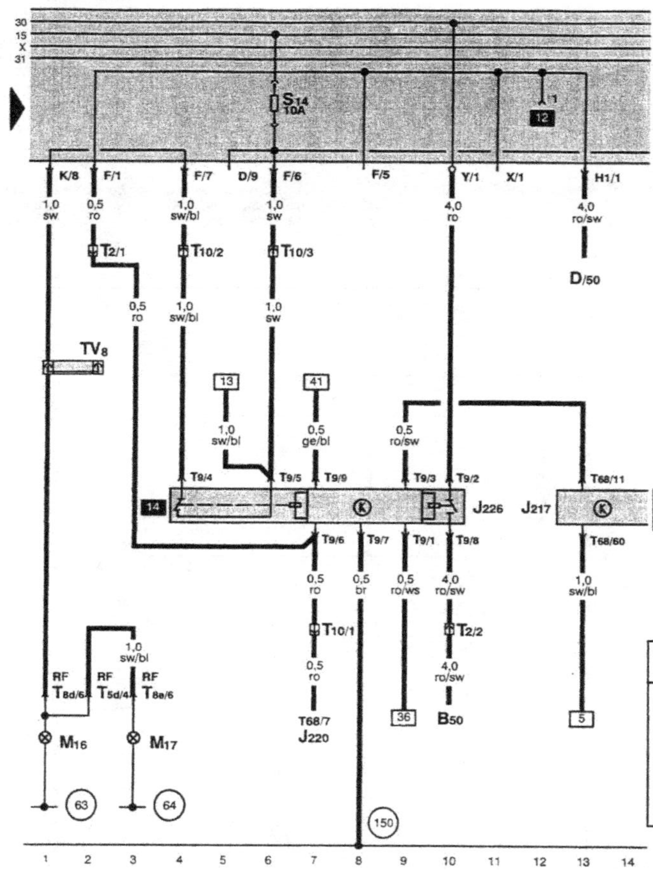

WIRING COLOR CODE

ws	=	white
sw	=	black
ro	=	red
br	=	brown
gn	=	green
bl	=	blue
gr	=	grey
li	=	lilac
ge	=	yellow

- B – Starter
- D – Ignition/Starter Switch
- E17 – Park/Neutral Position (PNP) Switch
- J217 – Transmission Control Module (TCM), below rear seat bench
- J220 – Motronic Engine Control Module (ECM), in plenum chamber, right
- J226 – Park/Neutral Position (PNP) Relay, on auxiliary relay panel
- M16 – Left Back-Up Light
- M17 – Right Back-Up Light
- T2 – Double Connector, near starter
- T5d – 5-Pin Connector, on taillight, left
- T6a – 6-Pin Connector
- T8c1 – 8-Pin Connector on taillight, left
- T8e – 8-Pin Connector, on taillight, right
- T9 – 9-Pin Connector
- T10 – 10-Pin Connector, behind fuse/relay panel
- T68 – 68-Pin Connector
- TV8 – Terminal 586 wire connector, above fuse/relay panel

- (63) – Ground connection, on left taillight bulb holder
- (64) – Ground connection, on right taillight bulb
- (150) – Ground connection, in automatic transmission wiring harness

1995

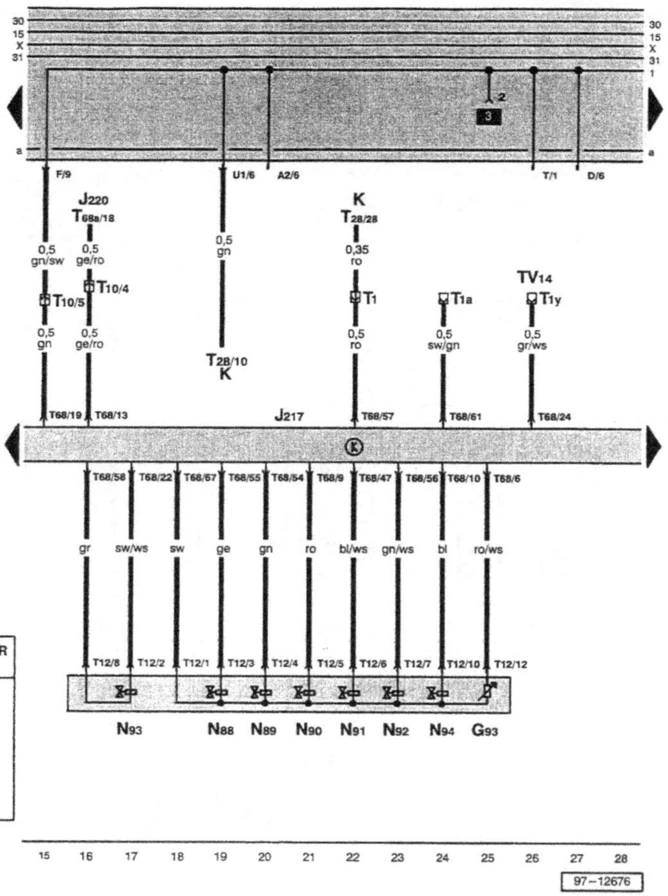

WIRING COLOR CODE

ws	=	white
sw	=	black
ro	=	red
br	=	brown
gn	=	green
bl	=	blue
gr	=	grey
li	=	lilac
ge	=	yellow

- G93 – Transmission Fluid Temperature Sensor (00300–1333)
- J217 – Transmission Control Module (TCM), below rear seat bench
- J220 – Motronic Engine Control Module (ECM)
- K – Instrument Cluster
- N88 – Solenoid Valve 1 (00258–1113)
- N89 – Solenoid Valve 2 (00260–1121)
- N90 – Solenoid Valve 3 (00262–1123)
- N91 – Solenoid Valve 4 (00264–1131)
- N92 – Solenoid Valve 5 (00266–1133)
- N93 – Solenoid Valve 6 (00268–1141)
- N94 – Solenoid Valve 7 (00270–1143)
- T1 – Single Connector, behind fuse/relay panel
- T1a – Single Connector, behind fuse/relay panel
- T1y – Single Connector, behind fuse/relay panel
- T10 – 10-Pin Connector, behind fuse/relay panel
- T10a – 10-Pin Connector
- T28 – 28-Pin Connector, on instrument cluster
- T68 – 68-Pin Connector
- T68a – 68-Pin Connector
- TV14 – Data Link Connector (DLC) Wire Connector, above fuse/relay panel

1995

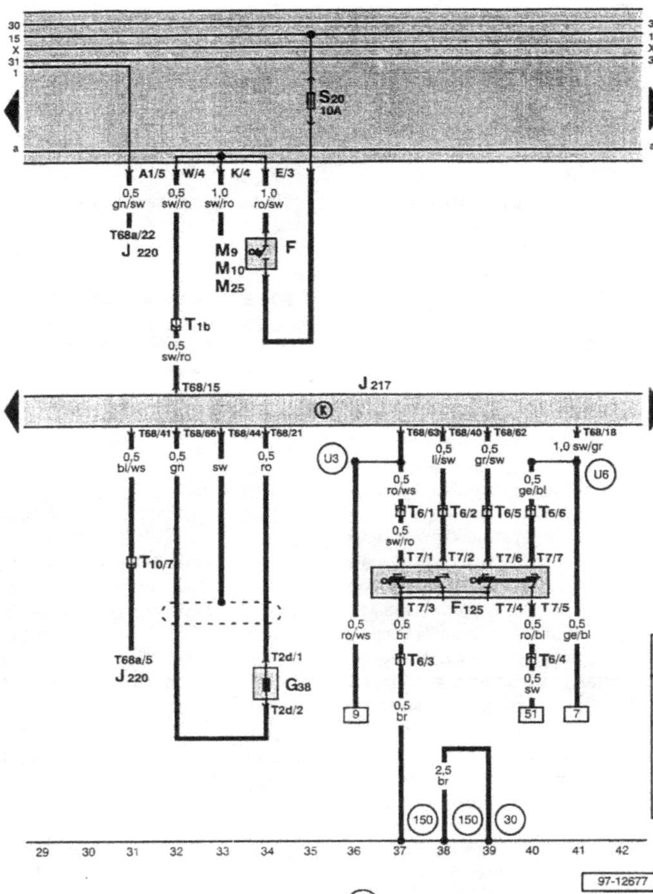

WIRING COLOR CODE

ws	=	white
sw	=	black
ro	=	red
br	=	brown
gn	=	green
bl	=	blue
gr	=	grey
li	=	lilac
ge	=	yellow

F – Brake Light Switch (00526-2131)
F125 – Multi-Function Transmission Range (TR) Switch (00293-1314)
G38 – Transmission Vehicle Speed Sensor (VSS) (00297-1324)
J217 – Transmission Control Module (TCM) below rear seat bench
J220 – Motronic Engine Control Module (ECM) in plenum chamber, right
M9 – Left Brake Light
M10 – Right Brake Light
M25 – High-Mount Brake Light
T1b – Single Connector behind fuse/relay panel
T2d – Double connector
T6 – 6-Pin Connector
T7 – 7-Pin Connector
T10 – 10-Pin Connector behind fuse/relay panel
T68 – 68-Pin Connector
T68a – 68-Pin Connector

(30) – Ground connection, -1-, beside fuse/relay panel
(150) – Ground connection, in automatic transmission
(U3) – Wire connection -1-, in automatic transmission
(U6) – Wire connection -2-, in automatic transmission wiring harness

1995

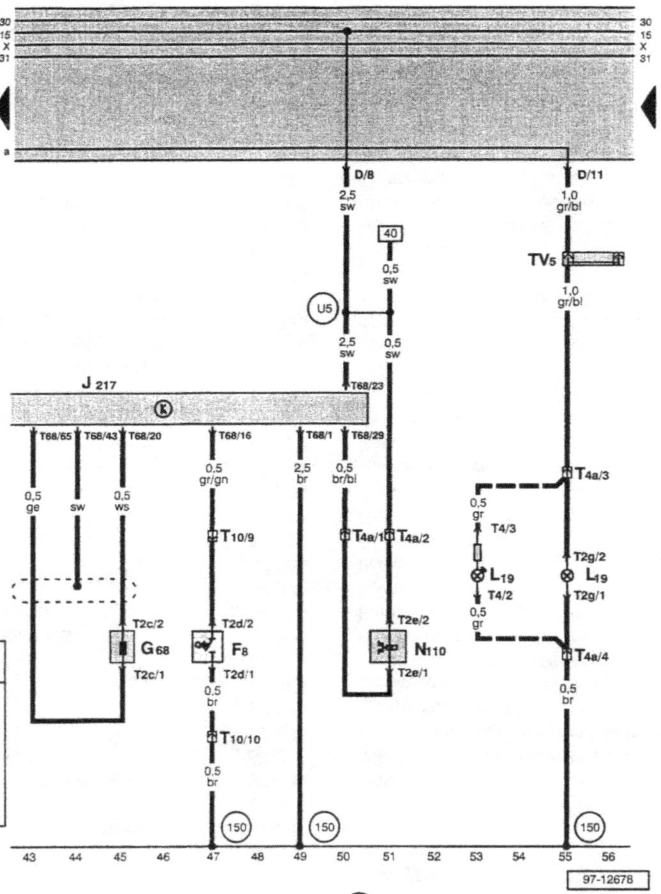

WIRING COLOR CODE

ws	=	white
sw	=	black
ro	=	red
br	=	brown
gn	=	green
bl	=	blue
gr	=	grey
li	=	lilac
ge	=	yellow

F8 – Kick Down Switch
G68 – Vehicle Speed Sensor (VSS) (00281–1231)
J217 – Transmission Control Module (TCM) Below rear seat bench
L19 – Automatic Transmission Console Light
N110 – Shift Lock Solenoid (01236–4314)
T2c – Double Connector
T2d – Double Connector
T2e – Double Connector
T2g – Double Connector
T4 – 4-Pin Connector
T4a – 4-Pin Connector Below gearshift lever cover
T10 – 10-Pin Connector behind fuse/relay panel
T68 – 68-Pin Connector
TV5 – Terminal 15a Wire Connector behind fuse/relay panel

(150) – Ground connection, in automatic transmission wiring harness
(U5) – Wire connection -2- (15), in automatic transmission wiring harness

1995

Golf, Jetta — 2.0L gas engine (ABA), OBD II (up to June 1996)

*applicable to m.y. 1996, from m.y. 1995

Golf, Jetta — 2.0L gas engine (ABA), OBD II (from June 1996)

*applicable to m.y. 1996-1999, from m.y. 1995

Cabrio — 2.0L gas engine (ABA), OBD II (1996-1998 m.y.)

*applicable to m.y. 1996-1999, from m.y. 1995

for 1999 m.y. and later Cabrio, see p. 97-416

Golf — 1.8L gas engine (ACC), Canada only

*applicable to m.y. 1996-1997, from m.y. 1995

Golf, Jetta — 1.9L Eco diesel engine (AAZ), Canada only

*applicable to m.y. 1996-1998, from m.y. 1995

Golf, Jetta — 2.8L gas engine (AAA), OBD II

*applicable to m.y. 1996-1999, from m.y. 1995

Golf, Jetta—1.9L TDI diesel engine (AHU)

*applicable to m.y. 1997, from m.y. 1995

Golf, Jetta—1.9L TDI diesel engine (AHU)

*applicable to m.y. 1998-1999, from m.y. 1995

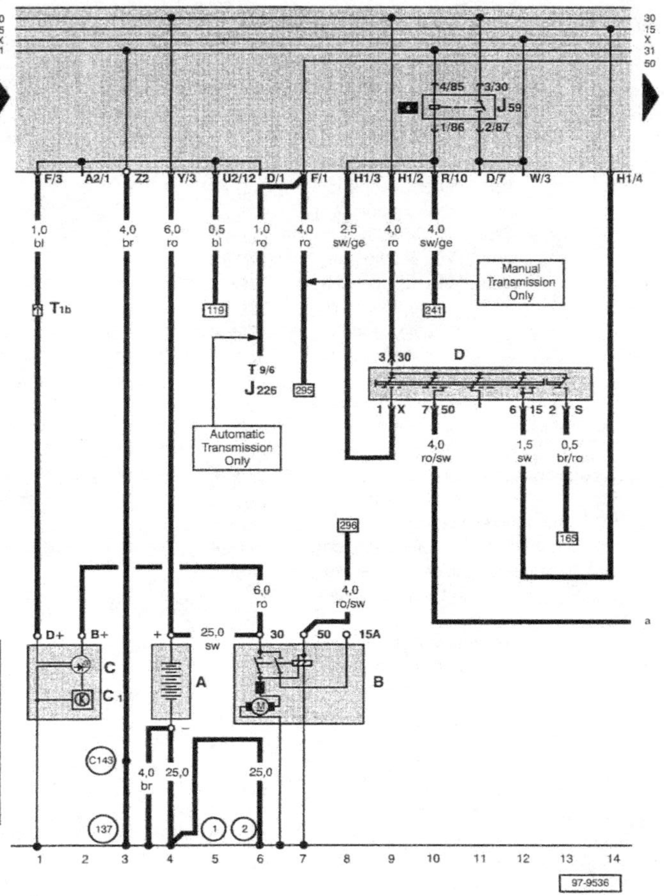

WIRING COLOR CODE		
ws	=	white
sw	=	black
ro	=	red
br	=	brown
gn	=	green
bl	=	blue
gr	=	grey
li	=	lilac
ge	=	yellow

A – Battery
B – Starter
C – Generator (Gen)
C1 – Voltage Regulator (VR)
D – Ignition/Starter Switch
J59 – Load Reduction Relay
J226 – Park/Neutral Position (PNP) Relay
T1b – Single Connector, near battery
T2 – Double Connector, behind fuse/relay panel, only on automatic transmission

(1) – Ground strap, battery to body
(2) – Ground strap, transmission to body
(137) – Ground connection, in Motronic MFI wiring harness
(C143) – Wire connection, in Motronic MFI wiring harness

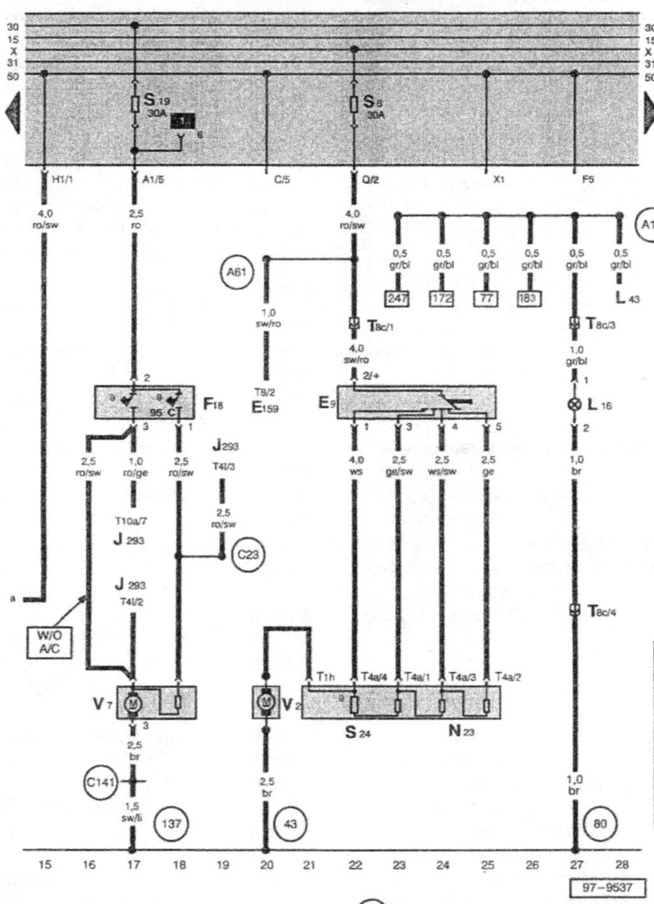

WIRING COLOR CODE		
ws	=	white
sw	=	black
ro	=	red
br	=	brown
gn	=	green
bl	=	blue
gr	=	grey
li	=	lilac
ge	=	yellow

E9 – Fresh Air Blower Switch
E159 – Free Air/Recirculation Flap Switch
F18 – Coolant Fan Control (FC) Thermal Switch
J293 – Coolant Fan Control (FC) Module
L16 – Fresh Air Control Lever Light
L43 – A/C Switch Illumination Light
N23 – Fresh Air Blower Series Resistance
S24 – Heater, A/C Clutch Fuse
T1h – Single Connector, behind right side of dash
T4a – 4-Pin Connector, behind right side of dash
T4l – 4-Pin Connector, on Coolant Fan Control (FC) Control Module
T8 – 8-Pin Connector, on Free Air/Recirculation Flap Switch
T8c – 8-Pin Connector, behind lower center of dash panel
T10a – 10-Pin Connector, on Coolant Fan Control (FC) Module
V2 – Fresh Air Blower
V7 – Coolant Fan

(43) – Ground connection, on right A-pillar, lower part
(80) – Ground connection –1–, in instrument panel wiring harness
(137) – Ground connection, Motronic MFI wiring harness
(A11) – Plus connection (58b), in instrument panel wiring harness
(A61) – Plus connection (X), in instrument cluster wiring harness
(C23) – Wire connection, in Motronic MFI wiring harness
(C141) – Wire connection, in Motronic MFI wiring harness

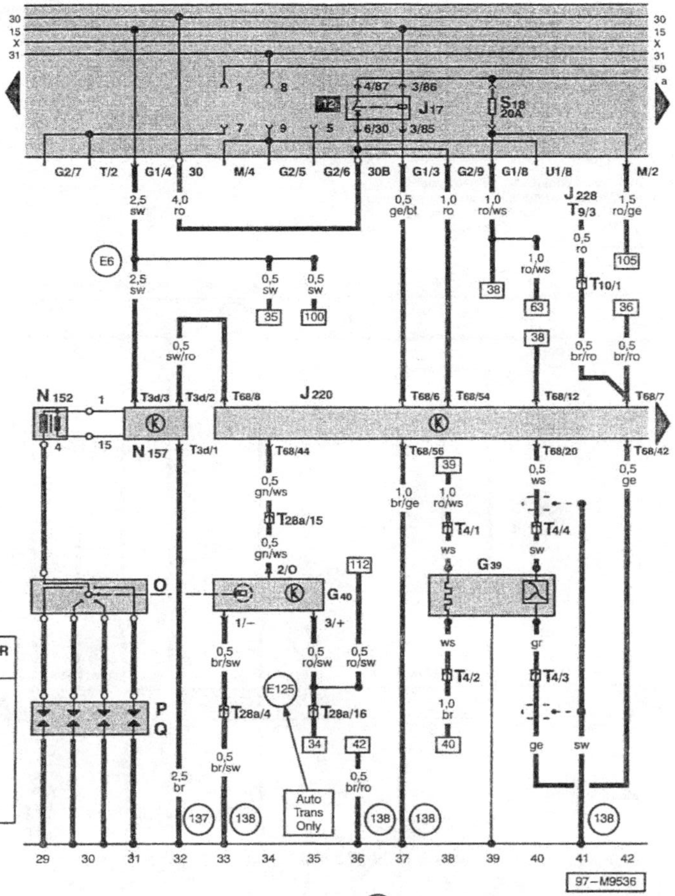

WIRING COLOR CODE

ws	=	white
sw	=	black
ro	=	red
br	=	brown
gn	=	green
bl	=	blue
gr	=	grey
li	=	lilac
ge	=	yellow

G6 – Fuel Pump
G39 – Heated Oxygen Sensor (HO2S)
G40 – Camshaft Position (CMP) Sensor
J17 – Fuel Pump Relay
J220 – Motronic Engine Control Module (ECM)
J228 – Right Seat Belt Control Module
N152 – Ignition Coil
N157 – Ignition Coil Power Output Stage
O – Ignition Distributor
P – Spark Plug Connectors
Q – Spark Plugs
S18 – Fuse
T4 – 4-Pin Connector, in engine compartment
T9 – 9-Pin Connector, in engine compartment
T10 – 10-Pin Connector, in engine compartment
T28a – 28-Pin Connector, left side of engine, near distributor
T28c – 28-Pin Connector, in engine compartment
T68 – 68-Pin Connector, under right side of hood, on Motronic Engine Control Module (ECM)

(137) – Ground connection (end stage), in Motronic Multiport Fuel Injection (MFI) wiring harness

(138) – Ground connection (control module), in Motronic Multiport Fuel Injection (MFI) wiring harness

(E6) – Wire connection, in Motronic Multiport Fuel Injection (MFI) wiring harness

(E125) Wire connection, in engine compartment wiring harness

1996-1999

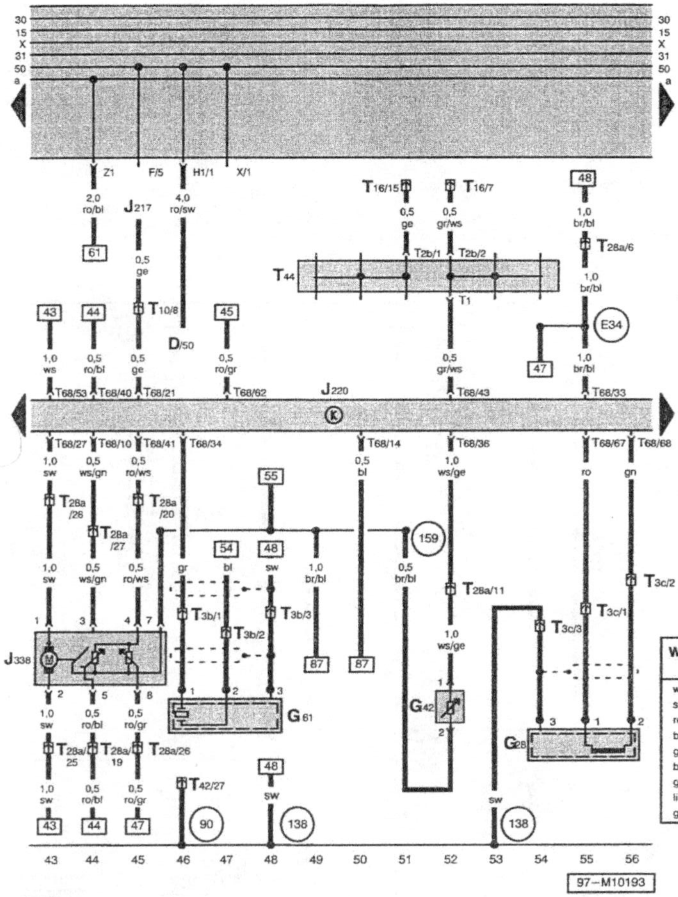

WIRING COLOR CODE

ws	=	white
sw	=	black
ro	=	red
br	=	brown
gn	=	green
bl	=	blue
gr	=	grey
li	=	lilac
ge	=	yellow

D – Ignition/Starter Switch
G28 – Engine Speed (RPM) Sensor
G42 – Intake Air Temperature (IAT) Sensor
G61 – Knock Sensor (KS) 1
J217 – Transmission Control Module (TCM)
J220 – Motronic Engine Control Module (ECM)
J338 – Throttle Valve Control Module
T1 – Single Connector, above fuse/relay panel
T2b – Double Connector, on junction connector (T44) above fuse relay/panel
T3b – 3-Pin Connector, on engine
T3c – 3-Pin Connector, left front of engine
T10 – 10-Pin Connector, in engine compartment
T16 – Data Link Connector (DLC), behind center of dash
T28a – 28-Pin Connector, left side of engine, near distributor
T42 – 42-Pin Connector, left side of engine
T44 – Junction Box for On Board Diagnosis, above fuse/relay panel
T68 – 68-Pin Connector, under right side of hood, on Motronic Engine Control Module (ECM)

T2z – Double connector, above fuse relay/panel

(90) – Wiring connection, in engine harness

(138) – Ground connection (control module), in Motronic Multiport Fuel Injection (MFI) wiring harness

(159) – Wiring connection, in engine harness

(E34) – Wiring connection, in engine harness

1996-1999

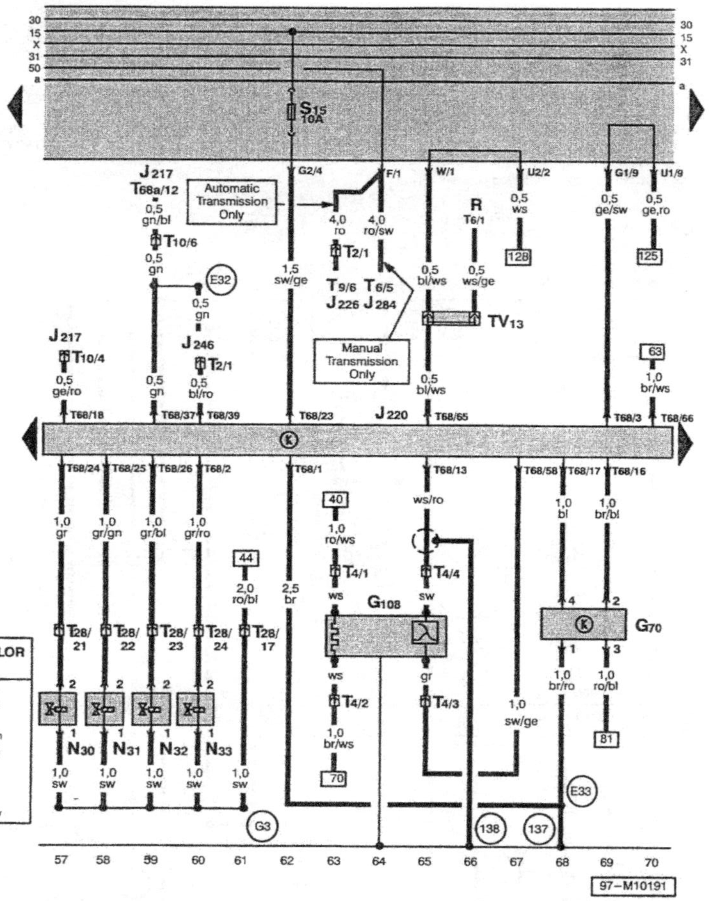

WIRING COLOR CODE

ws	=	white
sw	=	black
ro	=	red
br	=	brown
gn	=	green
bl	=	blue
gr	=	grey
li	=	lilac
ge	=	yellow

G70 – Mass Air Flow (MAF) Sensor
G108 – Heated Oxygen Sensor (HO2S) 2
J217 – Transmission Control Module (TCM)
J220 – Motronic Engine Control Module (ECM)
J226 – Park/Neutral Position (PNP) Relay
J246 – A/C Clutch Cut-Off Relay
J284 – Alarm System Control Module
N30 – Cylinder 1 Fuel Injector
N31 – Cylinder 2 Fuel Injector
N32 – Cylinder 3 Fuel Injector
N33 – Cylinder 4 Fuel Injector
R – Radio
T2 – Double Connector, behind fuse/relay panel, only on automatic transmission
T4 – 4-Pin Connector, bottom right rear of engine compartment
T6 – 6-Pin Connector, in engine compartment
T9 – 9-Pin Connector, in engine compartment
T10 – 10-Pin Connector, in engine compartment
T28 – 28-Pin Connector, instrument cluster

T68 – 68-Pin Connector, under right side of hood, on Motronic Multiport Fuel Injection (MFI) wiring harness
T68a – 68-Pin Connector, on Transmission Control Module (TCM)
TV13 – Vehicle Speed Signal Wire Connector
(137) – Ground connection (end stage), in Motronic Multiport Fuel Injection (MFI) wiring harness
(138) – Ground connection (control module), in Motronic Multiport Fuel Injection (MFI) wiring harness
(E33) – Ground connection, in Motronic Multiport Fuel Injection (MFI) wiring harness
(G3) – Plus connection, in injector wiring harness

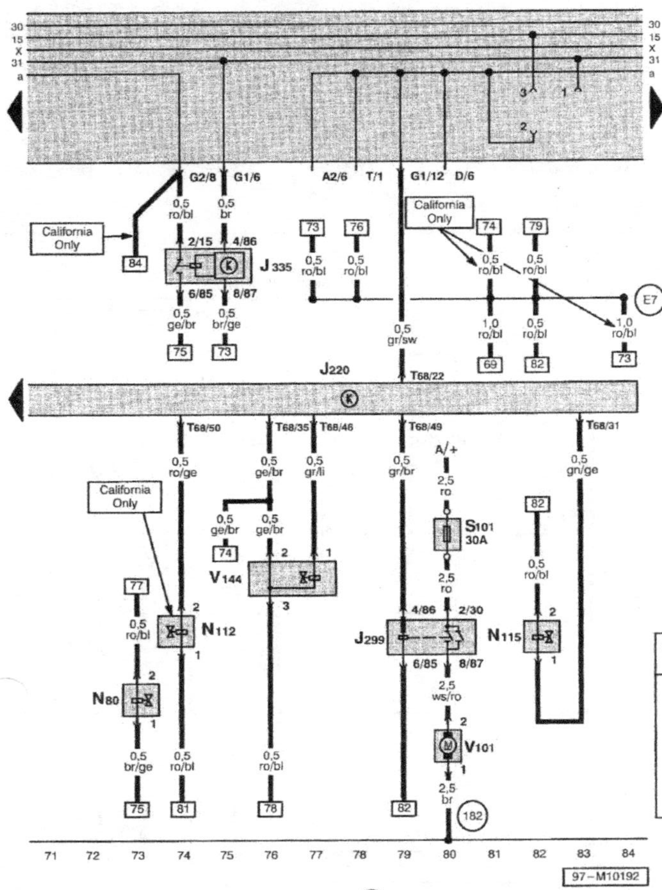

WIRING COLOR CODE

ws	=	white
sw	=	black
ro	=	red
br	=	brown
gn	=	green
bl	=	blue
gr	=	grey
li	=	lilac
ge	=	yellow

J220 – Motronic Engine Control Module (ECM)
J229 – First or Third Speed Coolant Fan Control (FC) Cut-off Relay
J335 – Fuel Cut-off Valve Shut-off Relay
N80 – Evaporative Emission (EVAP) Canister Purge Regulator Valve
N112 – Secondary Air Injection (AIR) Solenoid Valve
N115 – Evaporative Emission (EVAP) Canister Purge Solenoid Valve
P – Spark Plug Connector
T68 – 68-Pin Connector, under right side of hood, on Motronic Engine Control Module (ECM)
V101 – Secondary Air Injection (AIR) Pump Motor
V144 – Leak Detection Pump
V299 – Secondary Air Injection (AIR) Pump Relay

(E7) – Plus connection (87a) in Multiport Fuel Injection (MFI) wiring harness
(182) – Ground connection, in engine compartment wiring harness

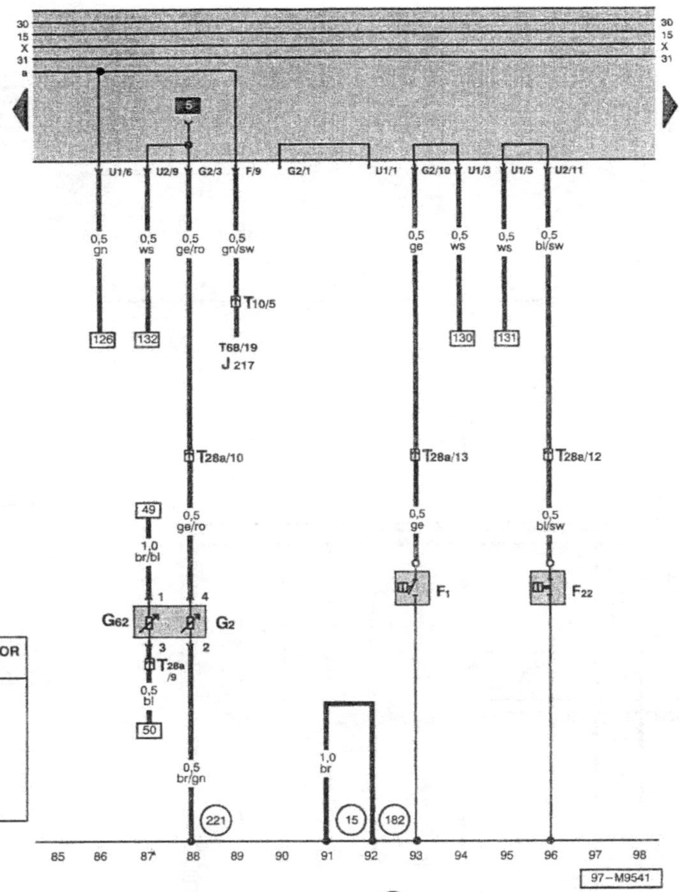

WIRING COLOR CODE

ws	=	white
sw	=	black
ro	=	red
br	=	brown
gn	=	green
bl	=	blue
gr	=	grey
li	=	lilac
ge	=	yellow

F1 – Oil Pressure Switch
F22 – 0.3 Bar Oil Pressure Switch
G2 – Engine Coolant Temperature (ECT) Sensor
G3 – Engine Coolant Temperature (ECT) Gauge
G62 – Engine Coolant Temperature (ECT) Sensor
J217 – Transmission Control Module
K – Instrument Cluster
T10 – 10-Pin Connector, behind fuse relay panel
T28 – 28-Pin Connector, left side of engine, near distributor
T28a – 28-Pin Connector, under right side of hood, on Motronic Engine Control Module (ECM)

(15) – Ground connection, in Motronic Multiport Fuel Injection (MFI) wiring harness

(182) – Ground connection, in Motronic Multiport Fuel Injection (MIF) wiring harness

(221) – Ground connection (engine ground), in engine compartment wiring harness

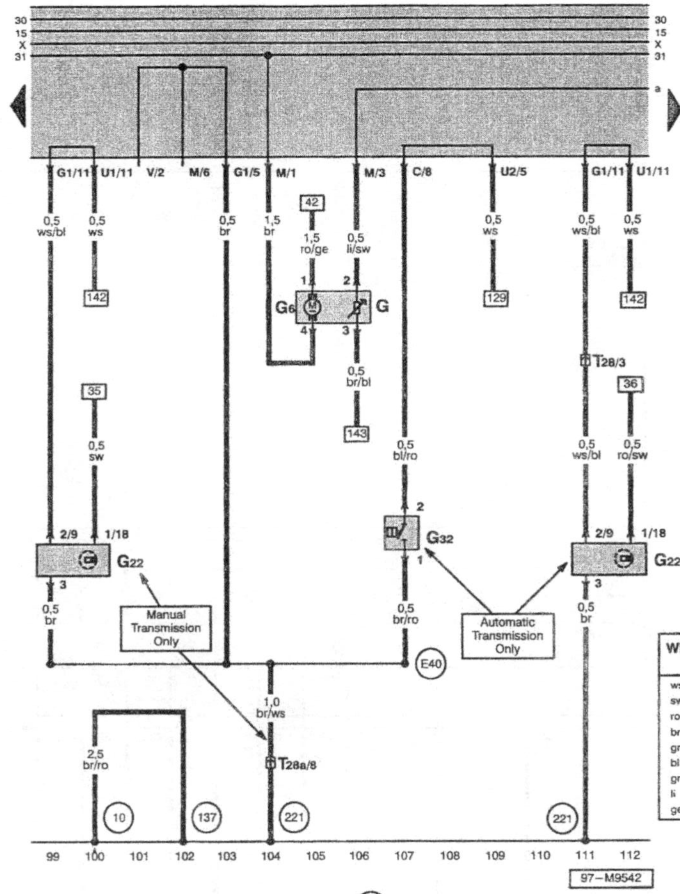

WIRING COLOR CODE

ws	=	white
sw	=	black
ro	=	red
br	=	brown
gn	=	green
bl	=	blue
gr	=	grey
li	=	lilac
ge	=	yellow

G – Fuel Level Sensor
G6 – Fuel Pump (FP)
G22 – Speedometer Vehicle Speed Sensor (VSS)
G32 – Engine Coolant Level (ECL) Sensor
K – Instrument Cluster

(10) – Ground connection, in Motronic Multiport Fuel Injection (MFI) wiring harness

(137) – Ground connection (end stake), in Motronic Multiport Fuel Injection (MFI) wiring harness

(221) – Ground connection (engine ground), in engine compartment wiring harness

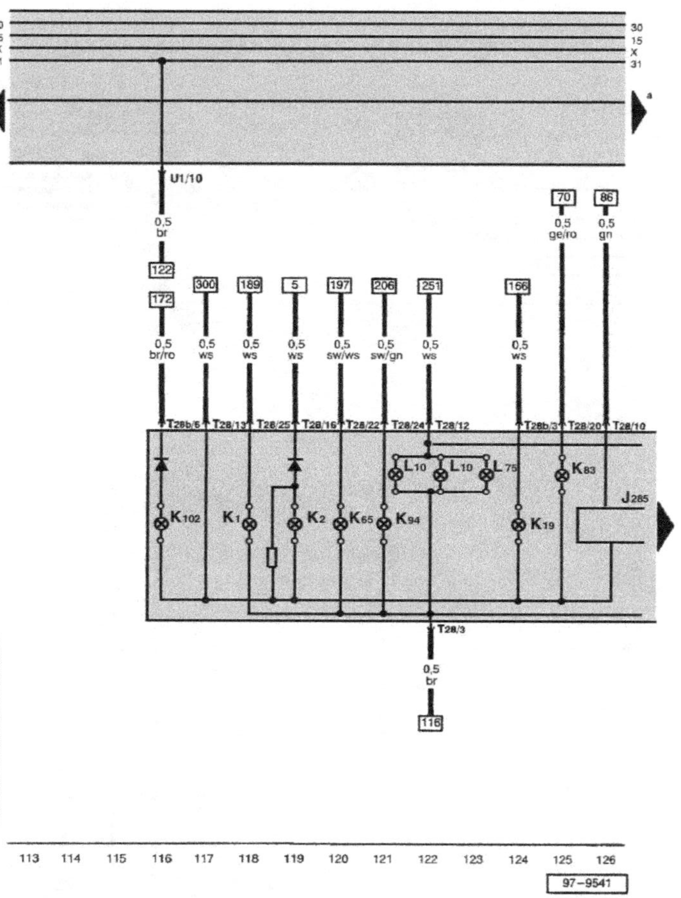

WIRING COLOR CODE

ws	=	white
sw	=	black
ro	=	red
br	=	brown
gn	=	green
bl	=	blue
gr	=	grey
li	=	lilac
ge	=	yellow

J285 – Control Module With Display Unit in Instrument Cluster
K1 – Headlight High Beam Indicator Light
K2 – Generator (GEN) Warning Light
K19 – Seat Belt Warning Light
K65 – Left Turn Signal Indicator Light
K83 – Multi-Function Indicator Lamp (MIL)
K94 – Right Turn Signal Indicator Light
K102 – Rear Hatch/Trunk Lid Ajar Warning Light
L10 – Instrument Cluster Light
L75 – Digital Display Light
T28 – 28-Pin Connector, instrument cluster
T28b – 28-Pin Connector, on instrument cluster

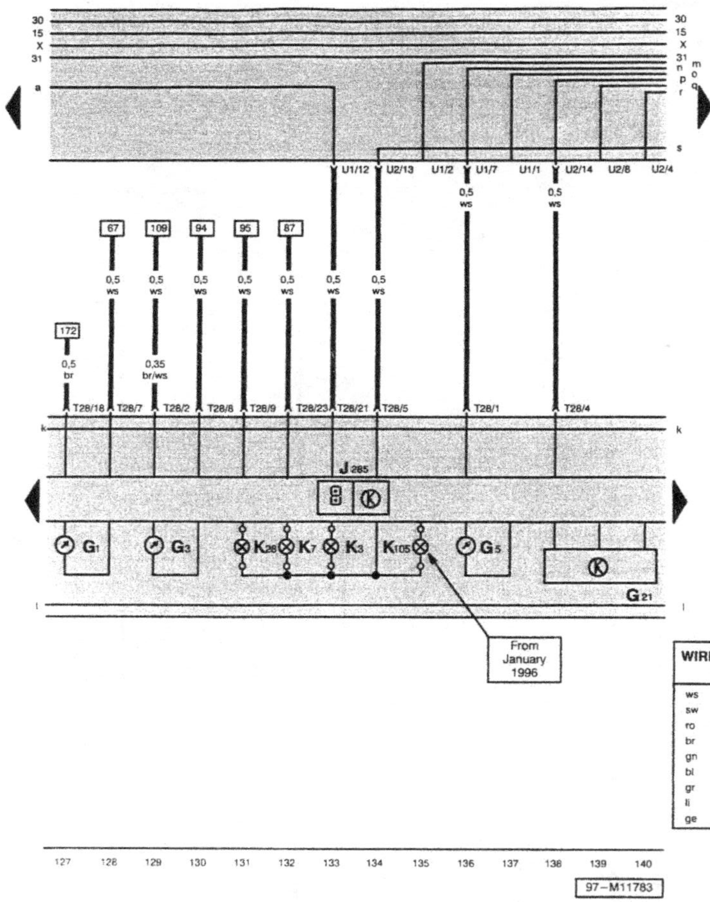

WIRING COLOR CODE

ws	=	white
sw	=	black
ro	=	red
br	=	brown
gn	=	green
bl	=	blue
gr	=	grey
li	=	lilac
ge	=	yellow

G1 – Fuel Gauge
G3 – Engine Coolant Temperature (ECT) Gauge
G5 – Tachometer
G21 – Speedometer
J285 – Control Module With Display Unit In Instrument Cluster
K3 – Oil Pressure Warning Light
K7 – Brake and Parking Brake Warning Light
K28 – Engine Coolant Level/Temperature (ECL/ECT) Warning Light
K105 – Low Fuel Level Warning Light
T28 – 28-Pin Connector, Instrument Cluster

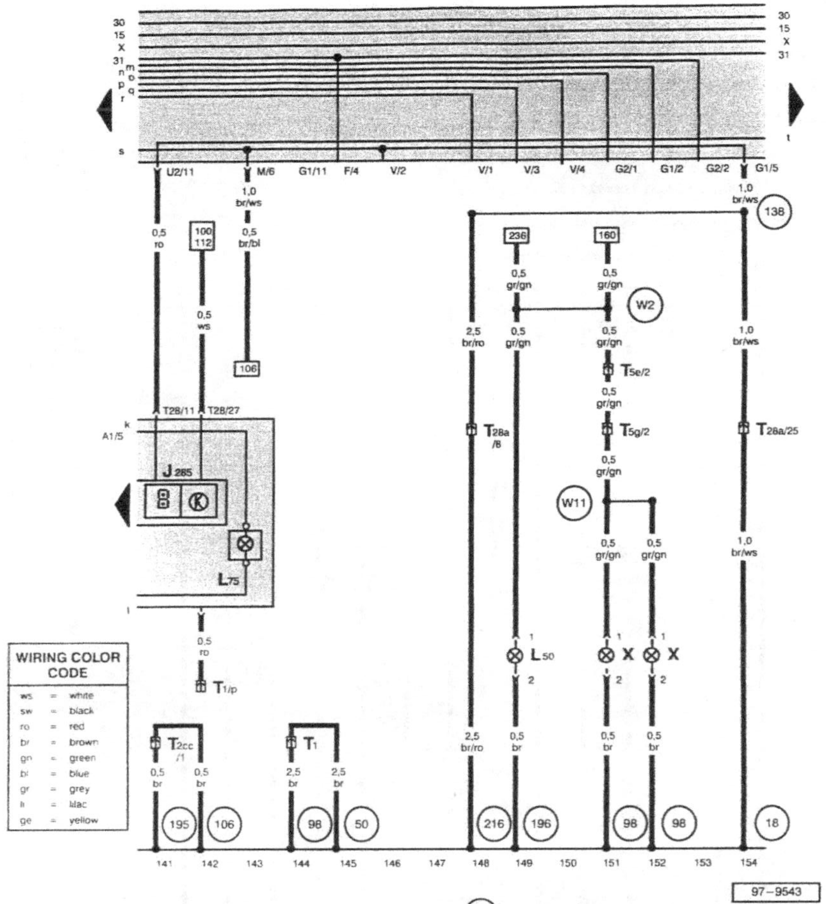

- J285 – Control Module With Display Unit in Instrument Cluster
- L50 – Center Rear Ashtray Light
- L75 – Digital Display Light
- T1 – Single Connector, in luggage compartment lid
- T1p – Single Connector, behind Fuse/Relay Panel
- T2cc – Double Connector, on lower right A-pillar
- T5e – 5-Pin Connector, in luggage compartment, left
- T5g – 5-Pin Connector, in luggage compartment, lid
- T28 – 28-Pin Connector, instrument cluster
- T28a – 28-Pin Connector, left side of engine, near distributor
- X – License Plate Light
- (18) – Ground connection, on engine block
- (50) – Ground connection, in luggage compartment, left
- (98) – Ground connection, in rear wiring harness
- (106) – Ground connectionm, in alarm system wiring harness
- (138) – Ground connection (Control Module) in Motronic Multiport Fuel Injection (MFI) wiring harness
- (195) – Ground connection, in rear door contact switch wiring harness
- (196) – Ground connection –1–, rear wiring harness
- (216) – Ground connection –2–, in Motronic Multiport Fuel Injection (MFI) wiring harness
- (W2) – Plus connection (58), in rear wiring harness
- (W11) – Wiring connection (58), in rear lid wiring harness

1996-1999

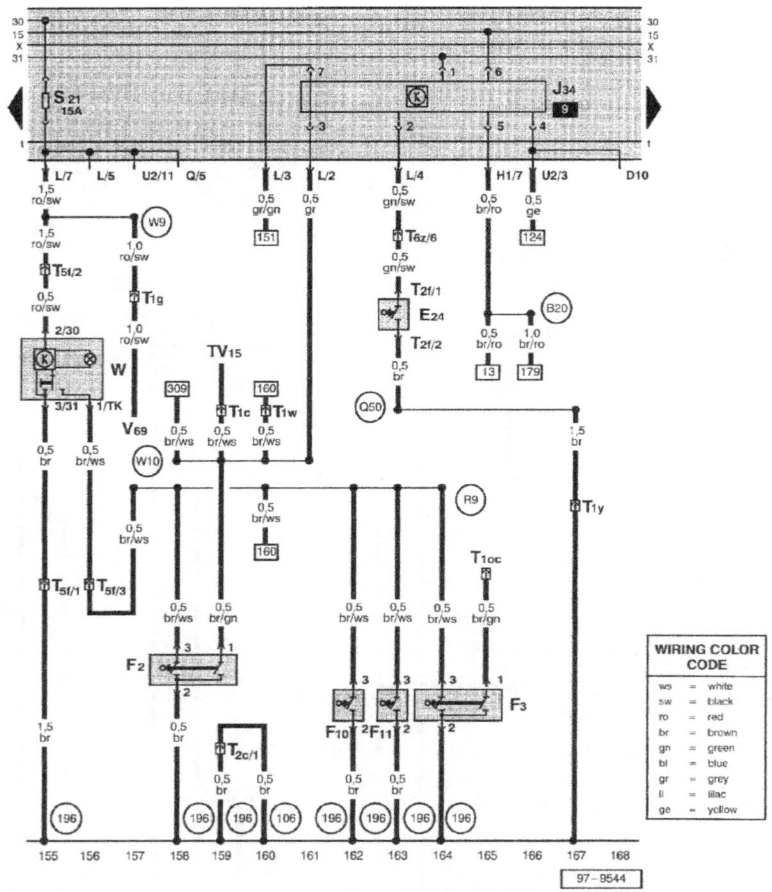

- E24 – Left Seat Belt Switch
- F2 – Left Front Door Contact Switch
- F3 – Right Front Door Contact Switch
- F10 – Left Rear Door Contact Switch
- F11 – Right Rear Door Contact Switch
- J34 – Seat Belt Warning System Relay
- M28 – Right Door Warning Light
- T1 – Single Connector, behind fuse/relay panel
- T1c – Single Connector, behind fuse/relay panel
- T1cc – Single Connector, behind fuse/relay panel
- T1dd – Single Connector, behind fuse/relay panel
- T1ee – Single Connector, behind fuse/relay panel
- T1ff – Single Connector, behind fuse/relay panel
- T1g – Single Connector, behind fuse/relay panel
- T1w – Single Connector, behind fuse/relay panel
- T1y – Single Connector, behind fuse/relay panel
- T2c – Double Connector, on lower right A-pillar
- T2cc – Double Connector, on lower right A-pillar
- T2f – Double Connector, behind rear of center console
- T5f – 5-Pin Connector, above fuse/relay panel
- T6z – 6-Pin Connector, behind dash panel
- TV15 – Wire Connector, door contact switch
- V69 – Central Locking System Pump and Control Module
- W – Interior Light, Front
- (106) – Ground connection, in alarm system wiring harness
- (B20) – Wire connection, in light switch wiring harness
- (Q50) – Wire connection, in alarm system wiring harness
- (R9) – Wire connection (door contact switch), in rear wiring harness
- (W9) – Plus connection (30), in rear wiring harness
- (W10) – Wire connection (31TK), in rear wiring harness
- (196) – Ground connection –1–, in rear wiring harness

1996-1999

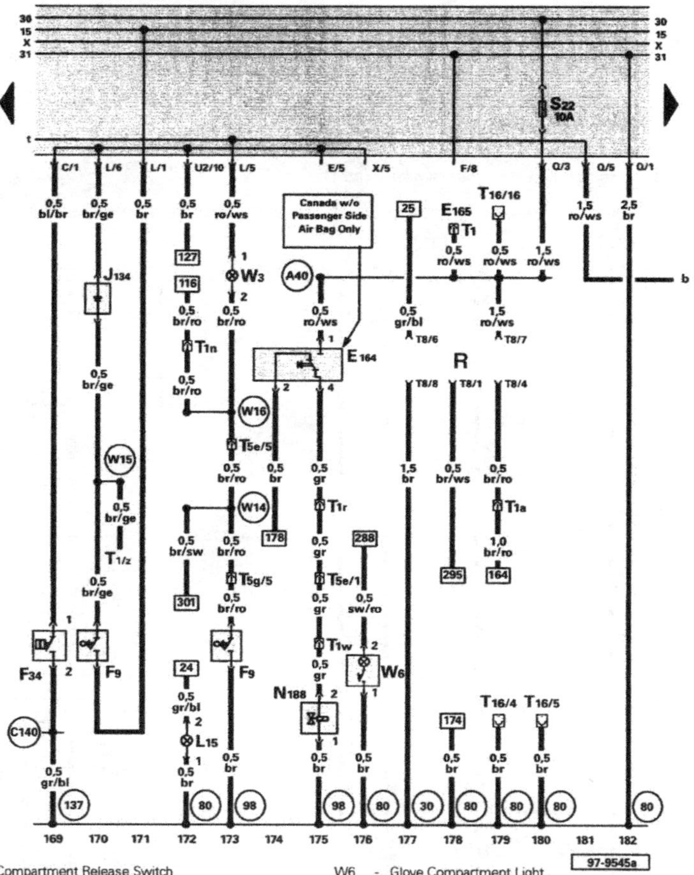

WIRING COLOR CODE		
ws	=	white
sw	=	black
ro	=	red
br	=	brown
gn	=	green
bl	=	blue
gr	=	grey
li	=	lilac
ge	=	yellow

E164 - Luggage Compartment Release Switch
E165 - Trunk Lid Release Switch
F5 - Luggage Compartment Light Switch
F9 - Parking Brake Warning Light Switch
F34 - Brake Fluid Level Warning Switch
J134 - Daytime Running Lights Diode
L15 - Ashtray Light
N188 - Luggage Compartment Release Solenoid
R - Radio
T1 - Single Connector, behind fuse/relay panel
T1a - Single Connector, behind fuse/relay panel
T1n - Single Connector, behind fuse/relay panel
T1r - Single Connector, behind fuse/relay panel
T1w - Single Connector, on luggage compartment lid
T1z - Single Connector, behind fuse/relay panel
T5e - 5-Pin Connector, in luggage compartment, left
T5g - 5-Pin Connector, in luggage compartment lid
T8 - 8-Pin Connector,on radio
T16 - Data link Connector (DLC), behind center of dash panel
W3 - Luggage compartment Light

W6 - Glove Compartment Light
(30) - Ground connection, -1-, beside fuse/relay panel
(80) - Ground connection -1-, in instrument panel wiring harness
(98) - Ground connection, in rear lid wiring harness
(137) - Ground connection , in Motronic MFI wiring harness
(A40) - Plus connection -1- (30), in instrument cluster wiring harness
(C140) - Wire connection, in Motronic MFI wiring harness
(W14) - Wire connection (58R), in rear wiring harness
(W15) - .Wire connection (Hb), in rear wiring harness
(W16) - .Wire connection (31s), in rear wiring harness

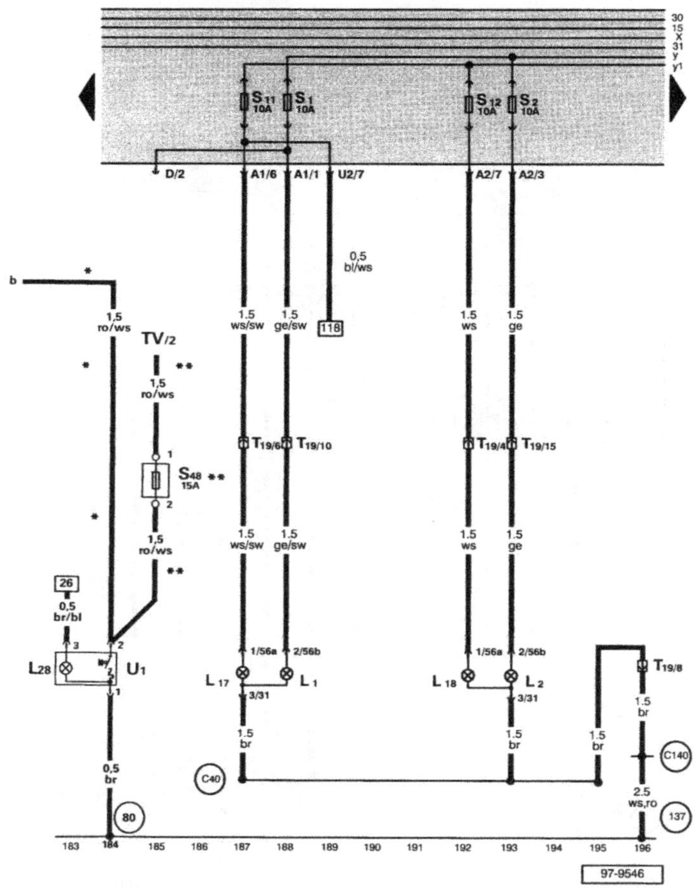

WIRING COLOR CODE		
ws	=	white
sw	=	black
ro	=	red
br	=	brown
gn	=	green
bl	=	blue
gr	=	grey
li	=	lilac
ge	=	yellow

L1 - Left Headlight
L2 - Right Headlight
L17 - Left High Beam Headlight
L18 - Right High Beam Headlight
L28 - Cigarette Light Light
S48 - Cigarette Lighter Fuse (behind fuse/relay panel)**
T19 - 19-Pin Connector, behind radiator grille
TV2 - Terminal 30 (B+) Wire Connector
U1 - Cigarette Lighter
(80) - Ground connection -1-, in instrument panel wiring harness
(137) - Ground connection , in Motronic MFI wiring harness
(C40) - Wire connection, in front light wiring harness
(C140) - Wire connection, in Motronic MFI wiring harness

* - Up to 10.96
** - From 10.96

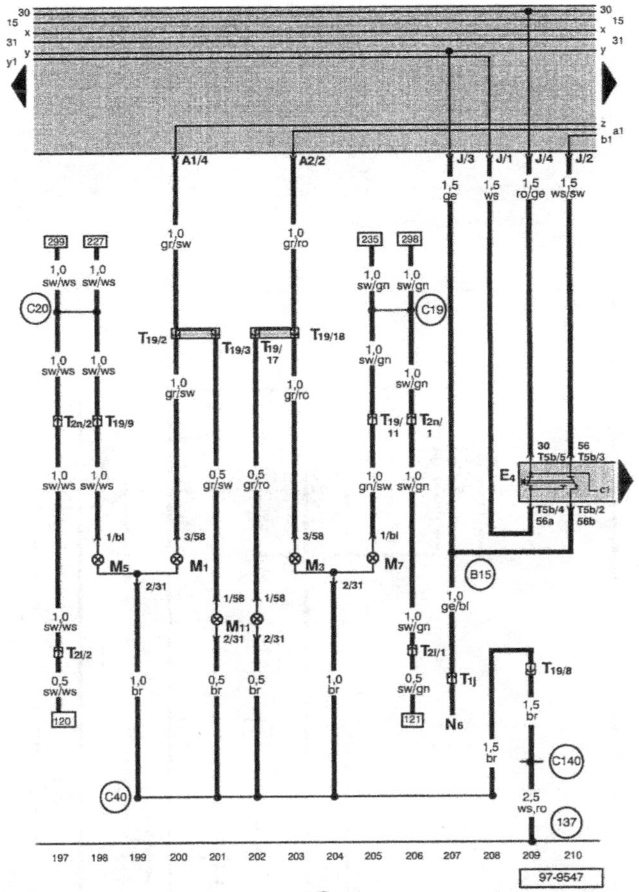

WIRING COLOR CODE		
ws	=	white
sw	=	black
ro	=	red
br	=	brown
gn	=	green
bl	=	blue
gr	=	grey
li	=	lilac
ge	=	yellow

E4 – Headlight Dimmer/Flasher Switch
M1 – Left Parking Light
M3 – Right Parking Light
M5 – Left Front Turn Signal Light
M7 – Right Front Turn Signal Light
M11 – Front Side Marker Lights
T1j – Single Connector, behind fuse/relay panel
T2l – Double Connector, behind fuse/relay panel
T2n – Double Connector, behind fuse/relay panel
T5b – 5-Pin Connector, behind steering column switch trim
T19 – 19-Pin Connector, behind radiator grill, center

(B15) – Wire connection, in steering column switch wiring harness

(C19) – Wire connection (right turn signal), in headlight wiring harness

(C20) – Wire connection (left turn signal), in headlight wiring harness

(C40) – Wire connection, in front light wiring harness

(C140) – Wire connection, in Motronic MFI wiring harness

(Q54) – Wire connection (left turn signal), in alarm system wiring harness

(Q55) – Wire connection (right turn signal), in alarm system wiring harness

(137) – Ground connection, in Motronic MFI wiring harness

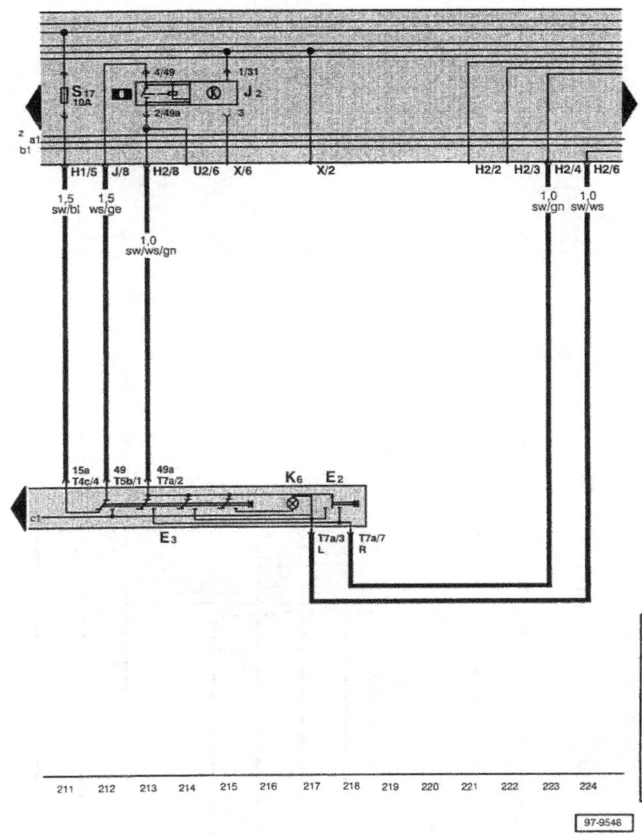

WIRING COLOR CODE		
ws	=	white
sw	=	black
ro	=	red
br	=	brown
gn	=	green
bl	=	blue
gr	=	grey
li	=	lilac
ge	=	yellow

E2 – Turn Signal Switch
E3 – Emergency Flasher Switch
J2 – Emergency Flasher Relay
K6 – Emergency Flasher Warning Light
T4c – 4-Pin Connector, behind steering column switch trim
T5b – 5-Pin Connector, behind steering column switch trim
T7a – 7-Pin Connector, behind steering column switch trim

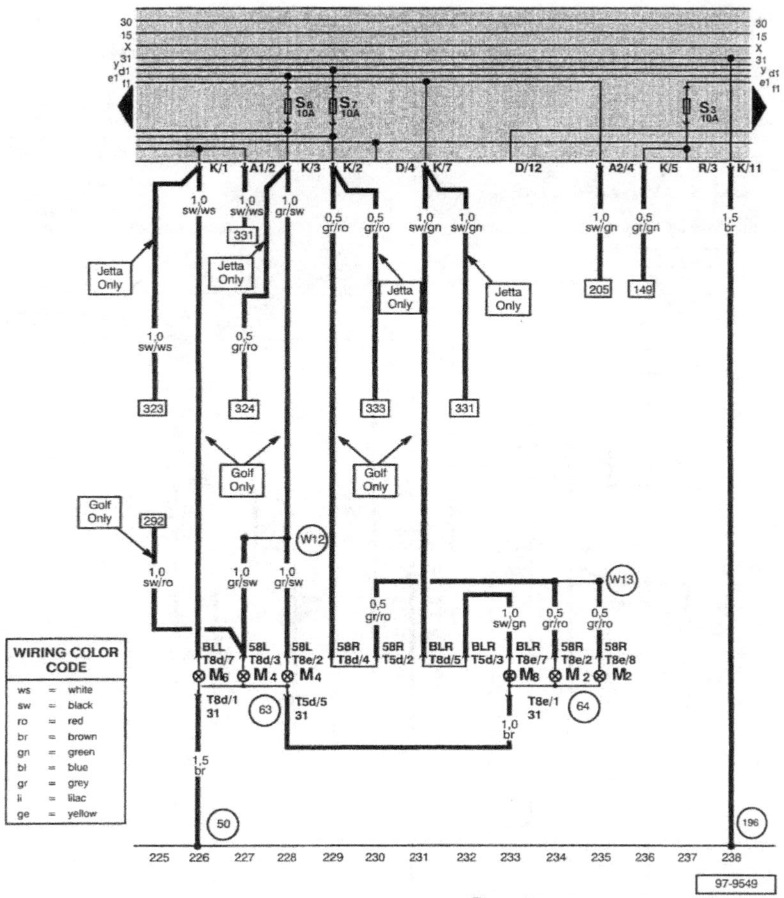

M2 – Right Tail Light
M4 – Left Tail Light
M6 – Left Rear Turn Signal Light
M8 – Right Rear Turn Signal Light
T5d – 5-Pin Connector, on tail light, left
T8d – 8-Pin Connector, on tail light, left
T8e – 8-Pin Connector, on tail light right

(50) – Ground connection, in luggage compartment, left

(63) – Ground Connection, on left tail light abulb holder

(64) – Ground connection, on right tail light bulb holder

(196) – Ground Connection –1–, in rear wiring harness

(W12) – Wire connection (dwa), in rear wiring harness, left

(W13) – Wire connection (58L), in rear wiring harness

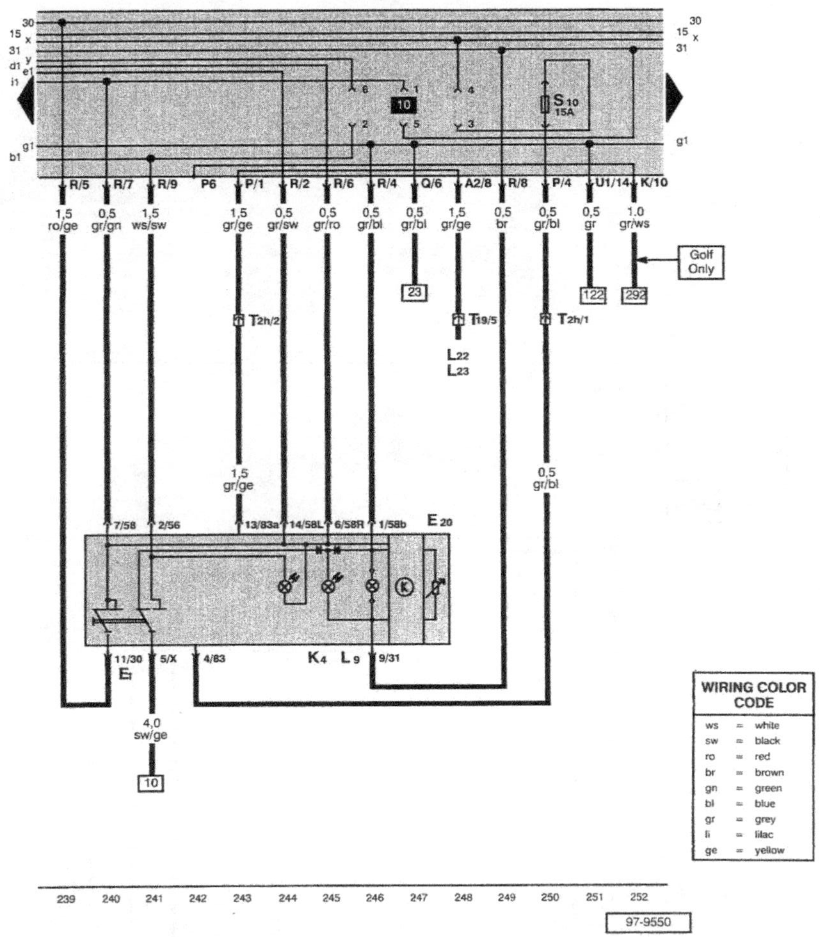

E1 – Light Switch
E20 – Instrument Panel Light Dimmer Switch
K4 – Park Light Indicator Light
L22 – Left Front Fog Light
L23 – Right Front Fog Light
L9 – Headlight Switch Light
T2h – 2-Pin Connector, behind fuse/relay panel
T19 – 19-Pin Connector, behind radiator grill, center

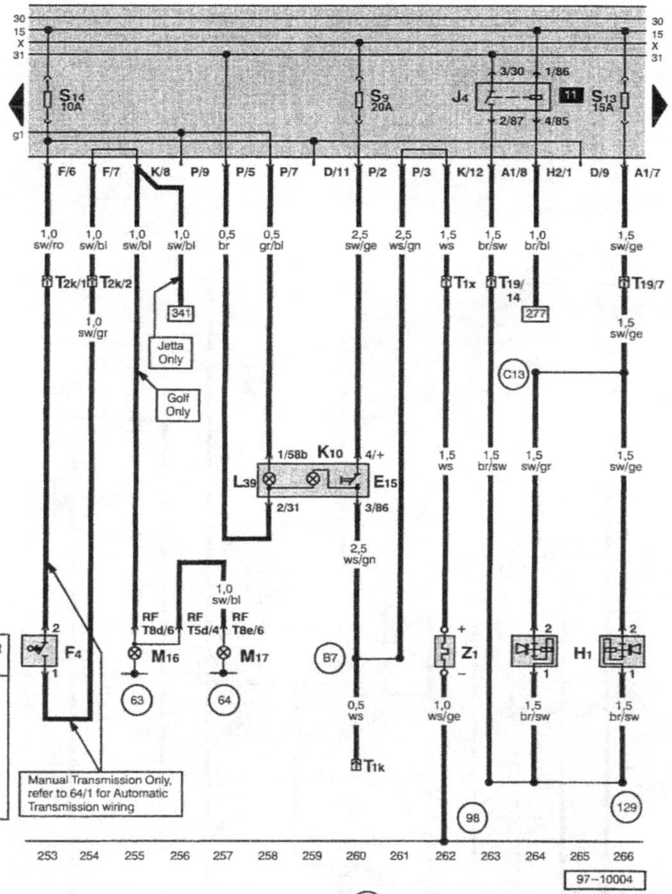

WIRING COLOR CODE

ws	=	white
sw	=	black
ro	=	red
br	=	brown
gn	=	green
bl	=	blue
gr	=	grey
li	=	lilac
ge	=	yellow

E15 – Rear Window Defogger Switch
F4 – Back-Up Light Switch
H1 – Dual Horns
J4 – Dual Horn Relay
K10 – Rear Window Defogger Indicator Light
L39 – Rear Window Defogger Switch Light
M16 – Left Back-Up Light
M17 – Right Back-Up Light
T1k – Single Connector, behind fuse/relay panel
T1x – Single Connector, in luggage compartment, left
T2k – Double Connector, behind fuse/relay panel
T5d – 5-Pin Connector, on tail light, left
T8d – 8-Pin Connector, on tail light, left
T8e – 8-Pin Connector, on tail light, right
T19 – 19-Pin Connector, behind radiator grill, center
Z1 – Rear Window Defogger (Heat Element)

(63) – Ground connection, on left tail light bulb holder
(64) – Ground connection, on right tail light bulb holder
(98) – Ground connection, in rear lid wiring harness
(129) – Ground connection, in dual tone horn wiring harness
(B7) – Wire connection (86), in rear window defogger wiring harness
(C13) – Plus connection, in dual tone horn wiring harness

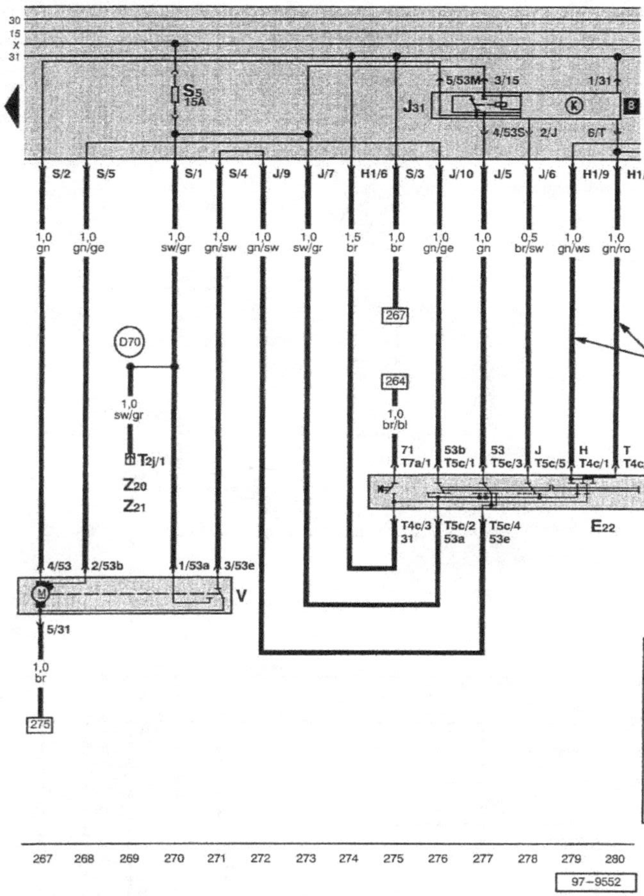

WIRING COLOR CODE

ws	=	white
sw	=	black
ro	=	red
br	=	brown
gn	=	green
bl	=	blue
gr	=	grey
li	=	lilac
ge	=	yellow

E22 – Windshield Wiper/Washer Switch
H – Horn Button
J31 – Washer/Wiper Intermittent Relay
T2j – Double Connector, behind fuse/relay panel
T4c – 4-Pin Connector, behind steering column switch trim
T5c – 5-Pin Connector, behind steering column switch trim
T7a – 7-Pin Connector, behind steering column switch trim
V – Windshield Wiper Motor
Z20 – Left Washer Nozzle Heater
Z21 – Right Washer Nozzle Heater
(D70) – Wire connection (53a), in engine compartment wiring harness

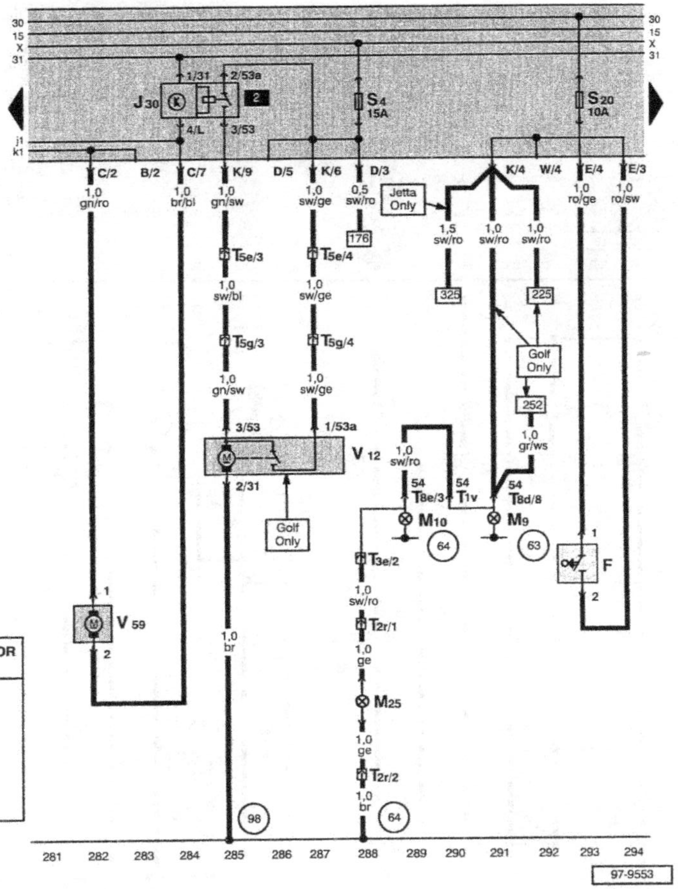

WIRING COLOR CODE

ws	=	white
sw	=	black
ro	=	red
br	=	brown
gn	=	green
bl	=	blue
gr	=	grey
li	=	lilac
ge	=	yellow

F – Brake Light Switch
J30 – Rear Window Wiper/Washer Relay
M9 – Left Brake Light
M10 – Right Brake Light
M25 – High-Mount Brake Light
V12 – Rear Window Wiper Motor
V59 – Windshield and Rear Window Washer Pump
T1v – Single Connector on tail light, left
T2r – Double Connector behind right grommet, on luggage compartment lid
T3e – 3-Pin Connector on tail light, right
T5e – 5-Pin Connector in luggage compartment, left
T5g – 5-Pin Connector in luggage compartment lid
T8d – 8-Pin Connector on tail light, left
T8e – 8-Pin Connector on tail light right

(63) – Ground connection, on left tail light bulb holder
(64) – Ground connection, on right tail light bulb holder
(98) – Ground connection, in rear lid wiring harness

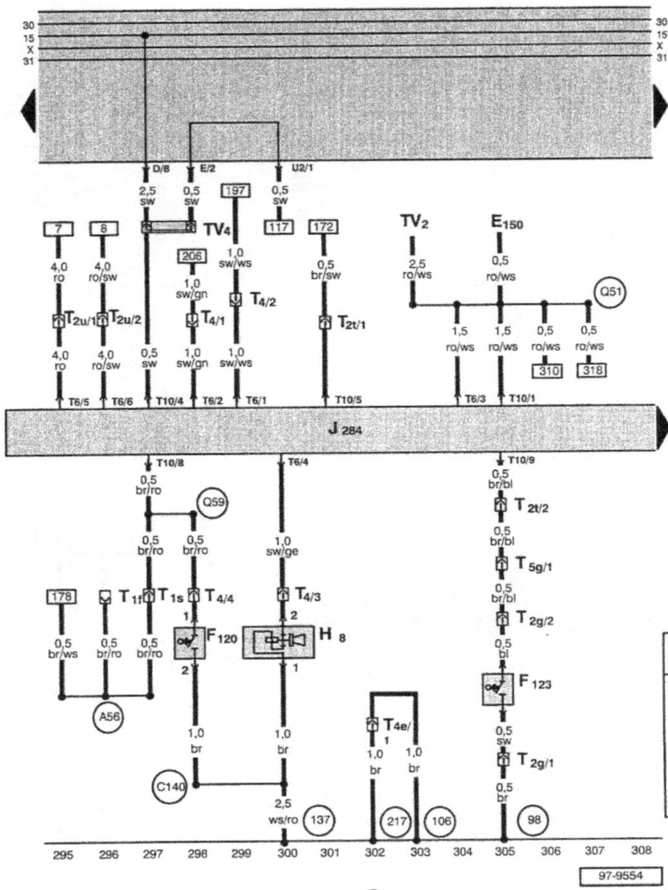

WIRING COLOR CODE

ws	=	white
sw	=	black
ro	=	red
br	=	brown
gn	=	green
bl	=	blue
gr	=	grey
li	=	lilac
ge	=	yellow

E150 – Door Lock - Unlock Switch
F120 – Hood Alarm Switch
F123 – Trunk Lid Alarm Switch
H8 – Alarm Horn
J284 – Alarm System Control Module
T1f – Single Connector, behind dash panel, center, behind radio
T1s – Single Connector, behind fuse/relay panel
T2g – Double Connector, on luggage compartment lid
T2t – Double Connector, on luggage compartment left
T2u – Double Connector, behind fuse/relay panel
T4 – 4-Pin Connector, behind fuse/relay panel
T5g – 5-Pin Connector, in luggage compartment lid
T6 – 6-Pin Connector, behind dash panel, left, on Alarm System Control Module
T10 – 10-Pin Connector, behind dash panel, left, on Alarm System Control Module
TV2 – Wire connector, terminal 30 (B+)
TV4 – Terminal 15 Wire Connector

(98) – Ground connection, in rear lid wiring harness
(106) – Ground connection, in alarm system wiring harness
(137) – Ground connection, in Motronic MFI wiring harness
(217) – Ground connection, in alarm horn wiring harness
(A56) – Plus connection -2-, (30) instrument panel wiring harness
(C140) – Wire connection, in Motronic MFI wiring harness
(Q51) – Plus connection (30), in alarm system wiring harness
(Q59) – Wire connection (Sensor) in alarm system wiring harness

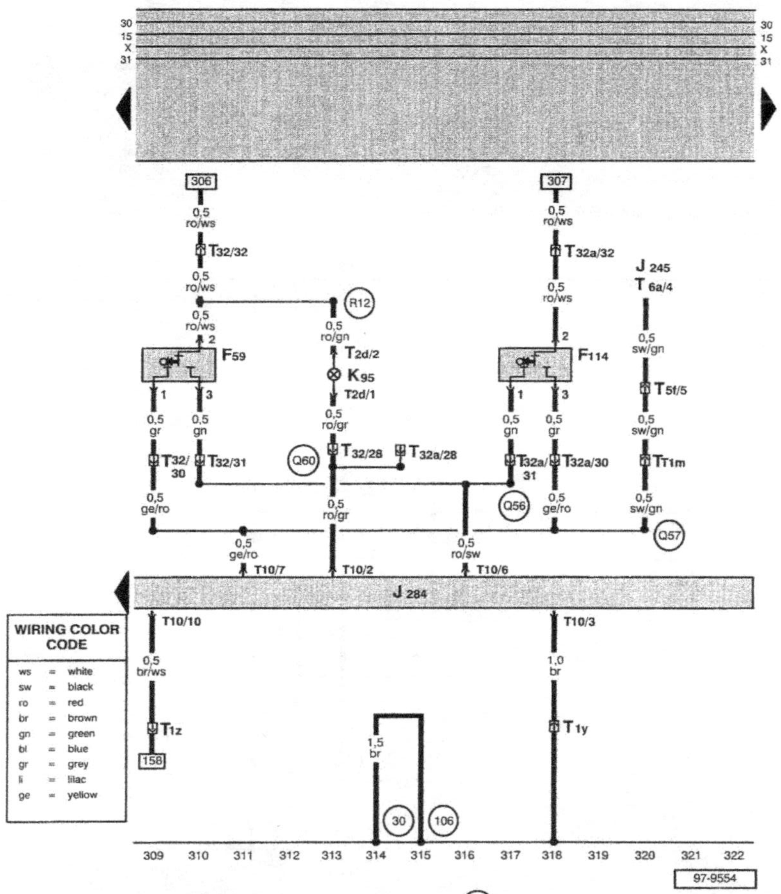

F59 – Driver's Door Central Locking System Door Switch
F114 – Passenger's Door Central Locking System Door Switch
J245 – Power Sunroof Control Module
J284 – Alarm System Control Module
K95 – Alarm System Indicator Light
T1m – Single Connector behind fuse/relay panel
T1y – Single connector in terminal 31 (end) wire connector
T1z – Single connector in headlight wiring harness
T2d – Double Connector in left front door
T5f – 5-Pin Connector above fuse/relay panel
T6a – 6-Pin Connector behind interior light, on Power Sunroof Control Module
T10 – 10-Pin Connector behind dash panel, left, on Alarm System Control Module
T32 – 32-Pin Connector on A-pillar, lower left
T32a – 32-Pin Connector on A-pillar, lower right

30 – Ground connection -1-, beside fuse/relay panel

106 – Ground connection, in alarm system wiring harness
Q56 – Wire connection (open), in alarm system wiring harness
Q57 – Wire connection (closed), in alarm system wiring harness
R12 – Plus connection -3- (30), in driver's door wiring harness
Q60 – Wire connection, in alarm system wiring harness

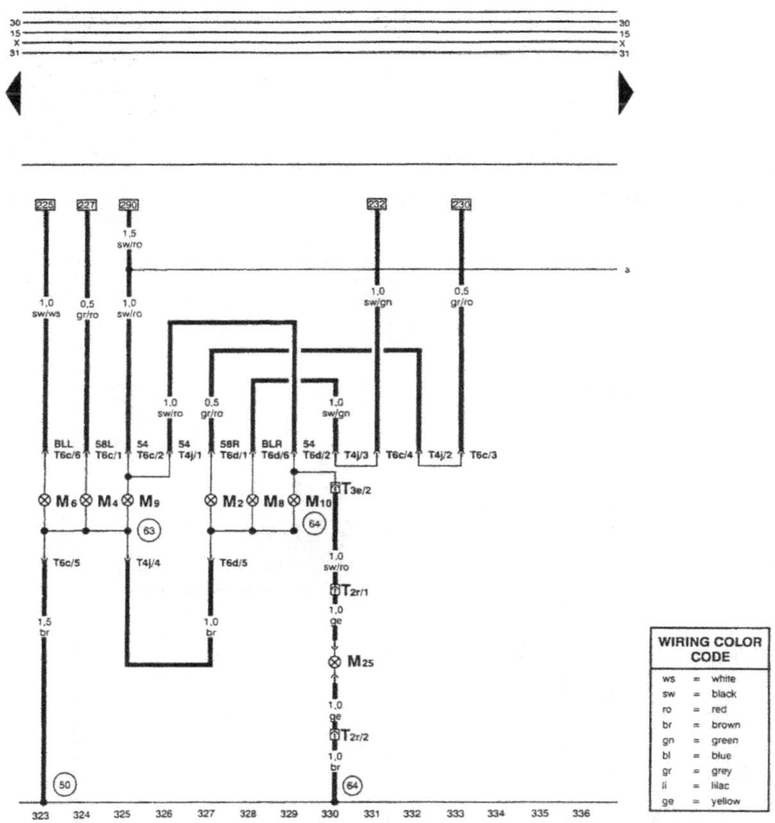

M2 – Right Tail Light
M4 – Left Tail Light
M6 – Left Rear Turn Signal Light
M8 – Right Rear Turn Signal Light
M9 – Left Brake Light
M10 – Right Brake Light
M25 – High Mount Brake Light
T2r – Double Connector behind right grommet, on luggage compartment lid
T3e – 3-Pin Connector on tail light, right
T4j – 4-Pin Connector on tail light, left
T6c – 6-Pin Connector on tail light, left
T6d – 6-Pin Connector on tail light, right

50 – Ground connection, in luggage compartment lid
63 – Ground connection, on left tail light bulb holder
64 – Ground connection, on right tail light bulb holder

Exterior lights (Jetta only)

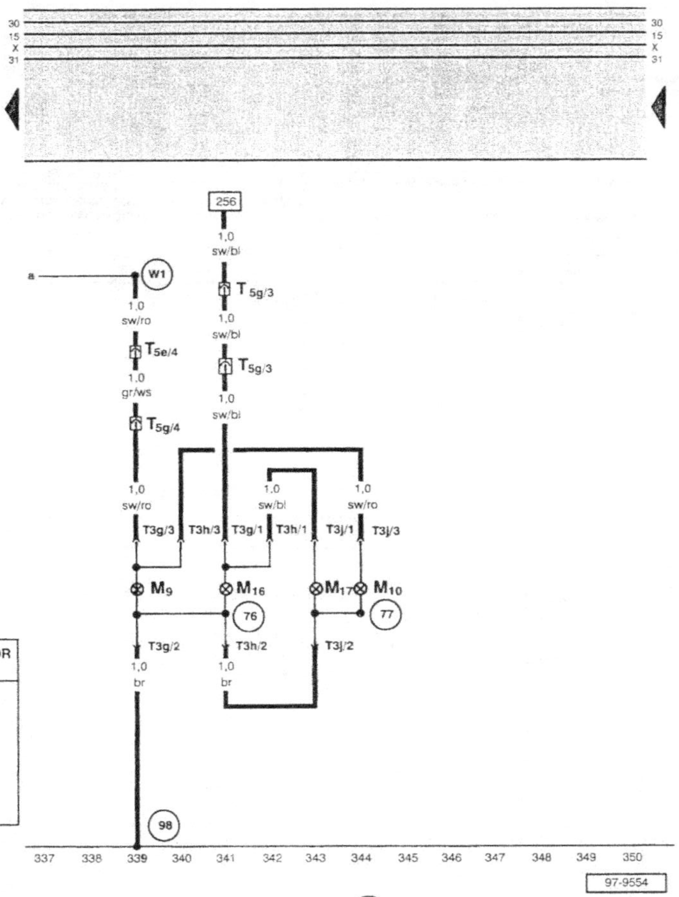

WIRING COLOR CODE		
ws	=	white
sw	=	black
ro	=	red
br	=	brown
gn	=	green
bl	=	blue
gr	=	grey
li	=	lilac
ge	=	yellow

M9 – Left Brake Light
M10 – Right Brake Light
M16 – Left Back-up Light
M17 – Right Back-up Light
T3g – 3-Pin Connector on back-up/brake light, left
T3h – 3-Pin Connector on back-up/brake light, left
T3j – 3-Pin Connector on back-up/brake light, right
T5e – 5-Pin Connector on luggage compartment lid
T5g – 5-Pin Connector on luggage compartment lid
(W1) – Plus connection, (54), in rear wiring harness

(76) – Ground connection, on left back-up/brake light bulb holder
(77) – Ground connection, on right back-up/brake light bulb holder
(98) – Ground connection, in rear lid wiring harness

Engine electronics (Cabrio)

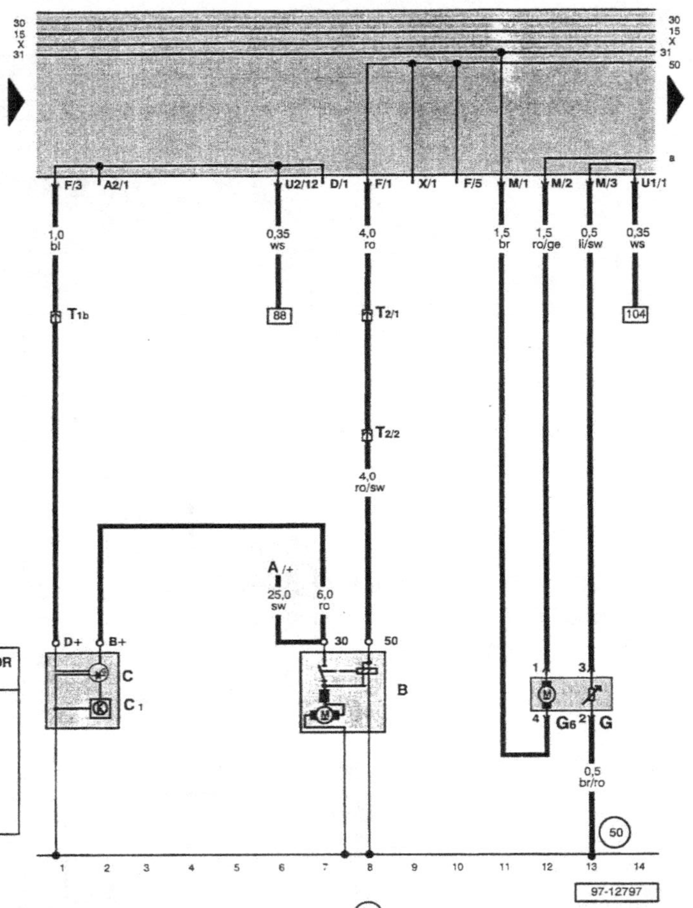

WIRING COLOR CODE		
ws	=	white
sw	=	black
ro	=	red
br	=	brown
gn	=	green
bl	=	blue
gr	=	grey
li	=	lilac
ge	=	yellow

A – Battery
B – Starter
C – Generator (Gen)
C1 – Voltage Regulator (VR)
G – Fuel Level Sensor
G6 – Fuel Pump (FP)
T1b – Single Connectonear battery
T2 – Double Connectorbehind fuse relay panel

(50) – Ground connection, in luggage compartment, left

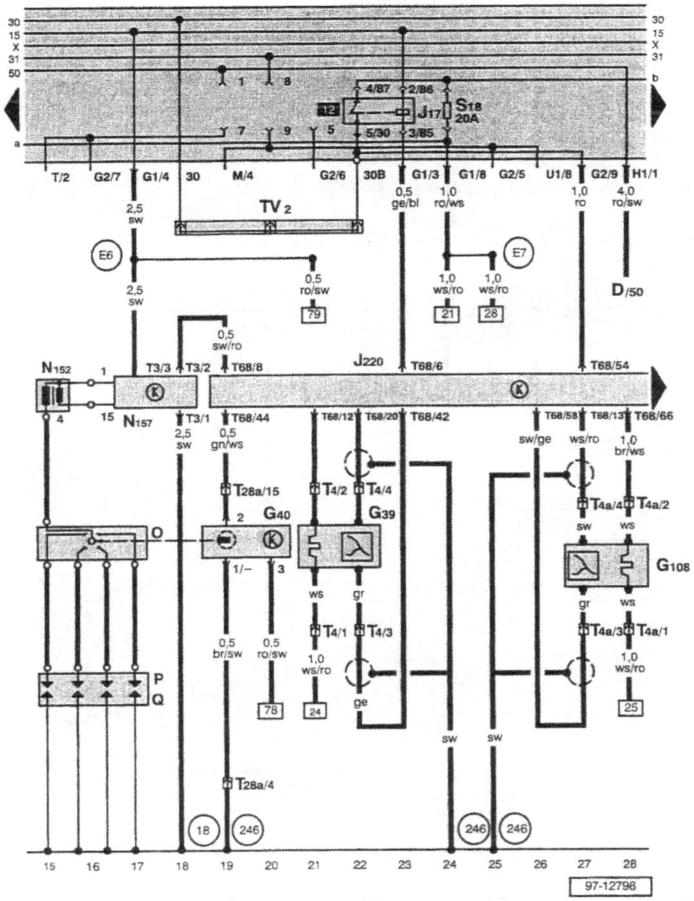

D – Ignition Switch
G39 – Heated Oxygen Sensor (HO2S)
G40 – Camshaft Position (CMP) Sensor
G108 – Heated Oxygen Sensor (HO2S) 2
J17 – Fuel Pump (FP) Relay
J220 – Motronic Engine Control Module (ECM)
N152 – Ignition Coil
N157 – Ignition Coil Power Output Stage
O – Ignition Distributor
P – Spark Plug Connectors
Q – Spark Plugs
T3 – 3-Pin Connector near cylinder head cover
T4 – 4-Pin Connector in engine wiring harness
T4a – 4-Pin Connector in engine wiring harness
T 28a – 28-Pin Connector on engine
T 68 – 68-Pin Connector, under right side of hood, on Motronic Engine Control Module (ECM)
TV2 – Terminal 30 (B+) Wire Connector

(18) – Ground Connection, on emgine block

(246) – Ground Connection (control module), in wiring harness

(E6) – Wiring connection in engine harness

(E7) – Plus connection (87a) in Multiport Fuel Injection (MFI) wiring harness

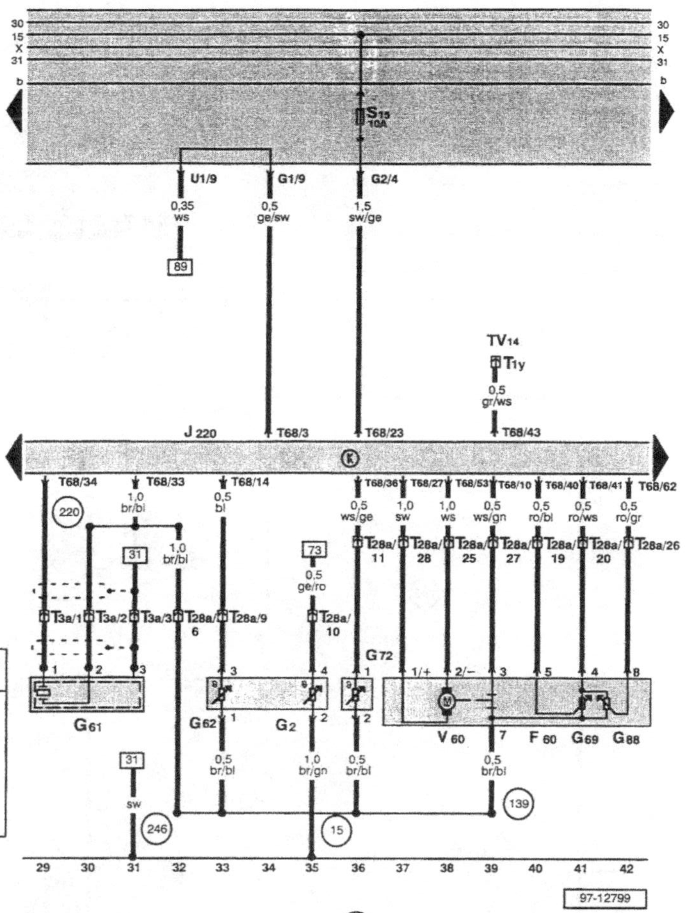

WIRING COLOR CODE

ws	=	white
sw	=	black
ro	=	rec
br	=	brown
gn	=	green
bl	=	blue
gr	=	grey
li	=	lilac
ge	=	yellow

F60 – Closed Throttle Position (CTP) Switch
G2 – Engine Coolant Temperature (ECT) Sensor
G61 – Knock Sensor (KS) 1
G62 – Engine Coolant Temperature (ECT) Sensor
G69 – Throttle Position (TP) Sensor
G72 – Intake Air Temperature (IAT) Sensor
G88 – Throttle Position (TP) Sensor
J220 – Motronic Engine Control Module (ECM)
T1 – Single Connector above fuse/relay panel
T2 – 3-Pin Connector near cylinder head cover
T28 – 28-Pin Connector on engine
T68 – 68-Pin Connector, on Motronic Engine Control Module (ECM)
TV14 – Data Link Connector (DLC) wire Connector above fuse/relay panel
V60 – Throttle Position (TP) Actuator

(15) – Ground connection, on cylinder head

(139) – Ground connection (sensor Ground), in Motronic Multiport Fuel Injection (MFI) wiring harness

(220) – Ground connection (sensor ground), in engine compartment wiring harness

(246) – Ground connection (control module), in wiring harness engine

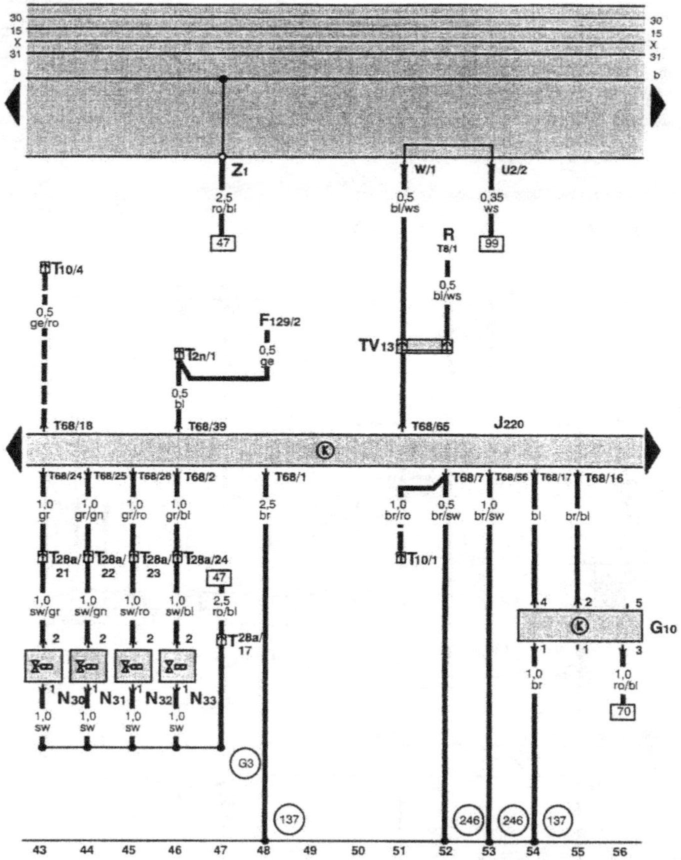

97-12800

F129 – A/C Pressure Switch
G10 – Engine Oil Pressure Switch
J220 – Motronic Engine Control Module (ECM)
N30 – Cylinder 1 Fuel Injector
N31 – Cylinder 2 Fuel Injector
N32 – Cylinder 3 Fuel Injector
N33 – Cylinder 4 Fuel Injector
R – Radio
T2n – Double Connector behind fuse/relay panel
T10 – 10-Pin Connector behind fuse/relay panel
T28 – 28-Pin Connector on engine
T68 – 68-Pin Connector, on Motronic Engine Control Module (ECM)
TV13 – Vehicle Speed Signal Wire Connector above fuse/relay panel

(137) – Ground connection (end stage), in Motronic Multiport Fue Injection (MFI) wiring harness

(246) – Ground connection (control module), in wiring harness engine

(G3) – Plus connection in injector wiring harness

WIRING COLOR CODE		
ws	=	white
sw	=	black
ro	=	red
br	=	brown
gn	=	green
bl	=	blue
gr	=	grey
li	=	lilac
ge	=	yellow

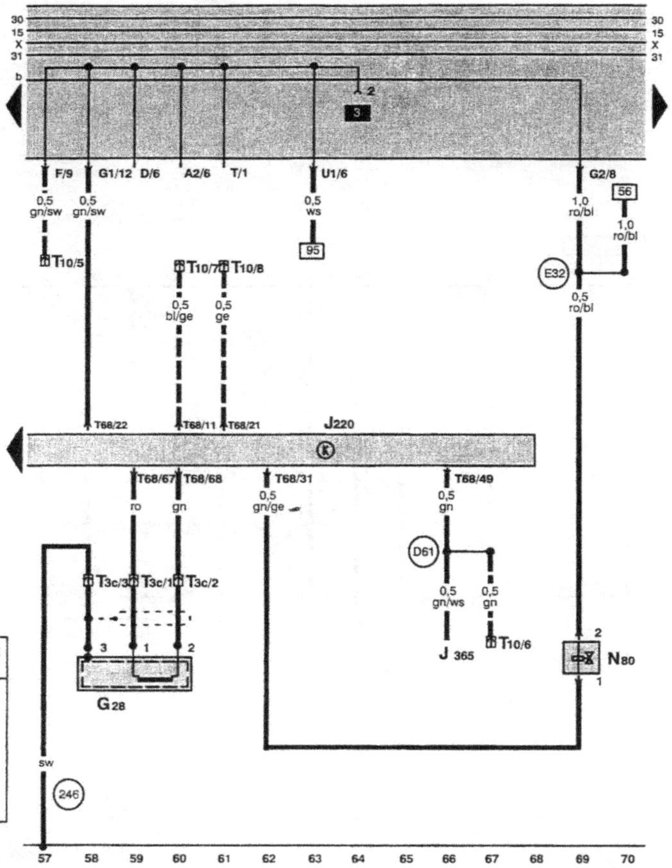

97-12801

G28 – Engine Speed (RPM) Sensor
J220 – Motronic Engine Control Module (ECM)
J365 – Relay for A/C shutoff
N80 – Evaporative Emission (EVAP) Canister Purge Regulator Valve
T3 – 3-Pin Connector on engine
T10 – 10-Pin Connector behind fuse/relay panel
T68 – 68-Pin Connector, on Motronic Engine Control Module (ECM)

(246) – Ground connection (control module), in wiring harness engine

(D61) – Wire connection (A/C), in engine compartment wiring harness

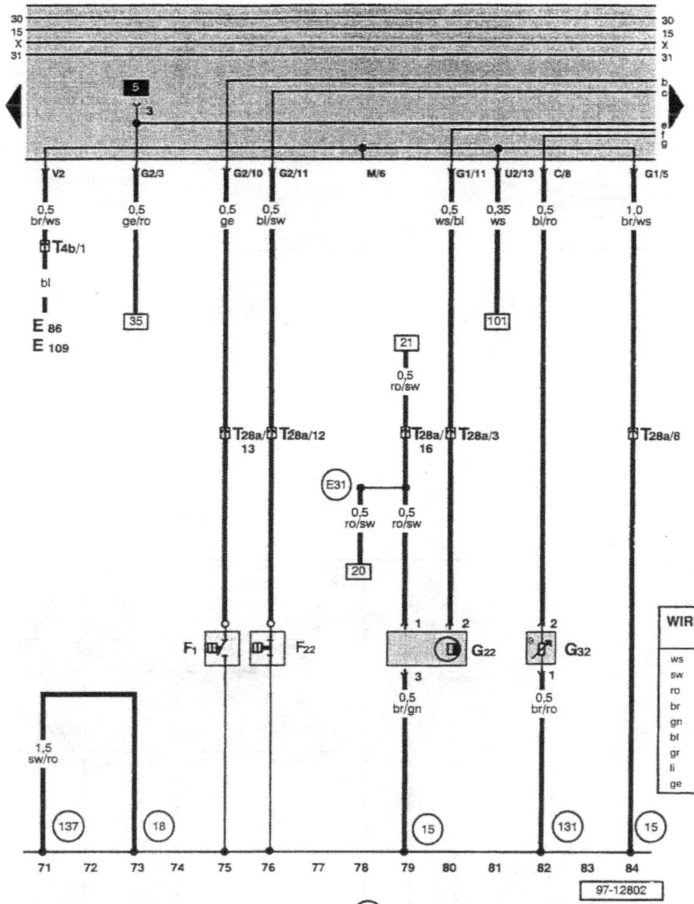

WIRING COLOR CODE

ws	=	white
sw	=	black
ro	=	red
br	=	brown
gn	=	green
bl	=	blue
gr	=	grey
li	=	lilac
ge	=	yellow

E86 – Multi-Function Indicator Mode Select Switch
E109 – Multi-Function Indicator Memory Switch
F1 – Oil Pressure Switch
F22 – 0.3 bar Oil Pressure Switch
G22 – Speedometer Vehicle Speed Sensor (VSS)
G32 – Engine Coolant Level (ECL) Sensor
T4 – 4-Pin Connector, behind steering column switch cover
T28 – 28-Pin Connector, on engine

(15) – Ground connection, on cylinder head

(18) – Ground connection, on engine block

(127) – Ground connection, -1-, in A/C compressor wiring harness

(131) – Ground connection -2-, in engine compartment wiring harness

(E31) – Connector (15), in wiring harness engine

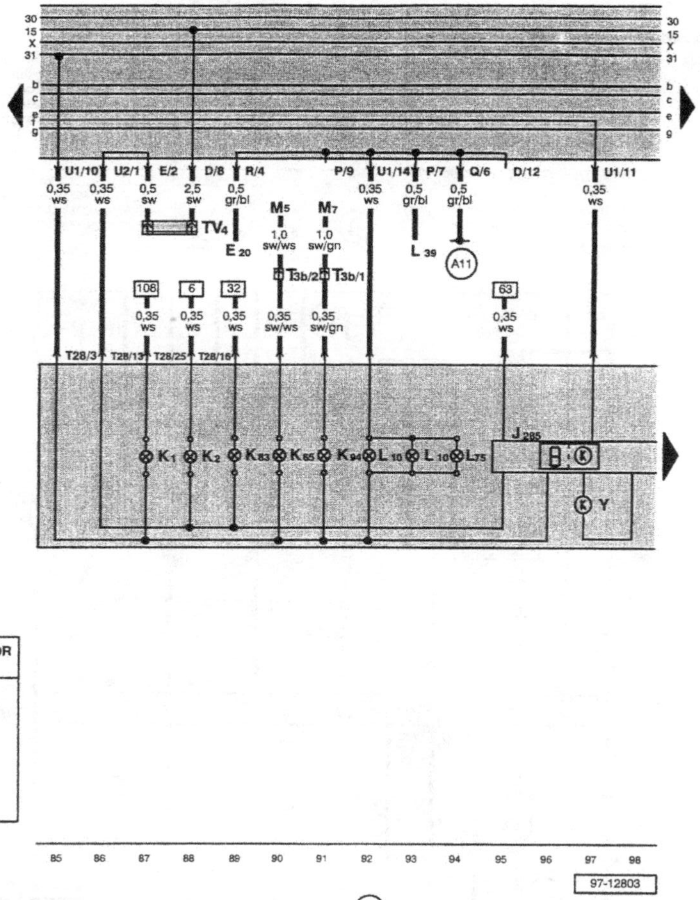

WIRING COLOR CODE

ws	=	white
sw	=	black
ro	=	red
br	=	brown
gn	=	green
bl	=	blue
gr	=	grey
li	=	lilac
ge	=	yellow

J285 – Control Module With Display Unit In Instrument Cluster
K1 – Headlight High Beam Indicator Light
K2 – Generator (GEN) Warning Light
K65 – Left Turn Signal Indicator Light
K83 – Malfunction Indicator Lamp (MIL)
K94 – Right Turn Signal Indicator Light
L10 – Instrument Cluster Light
L39 – Rear Window Defogger Switch Light
L75 – Digital Display Light
M5 – Left Front Turn Signal Light
M7 – Right Front Turn Signal Light
T3 – 3-Pin Connector, behind fuse/relay panel
T28 – 28-Pin Connector, on instrument cluster
TV4 – Terminal 15 Wire Connector, behind fuse/relay panel
Y – Analog Clock

(A11) – Plus connection (58b) instrument panel wiring harness

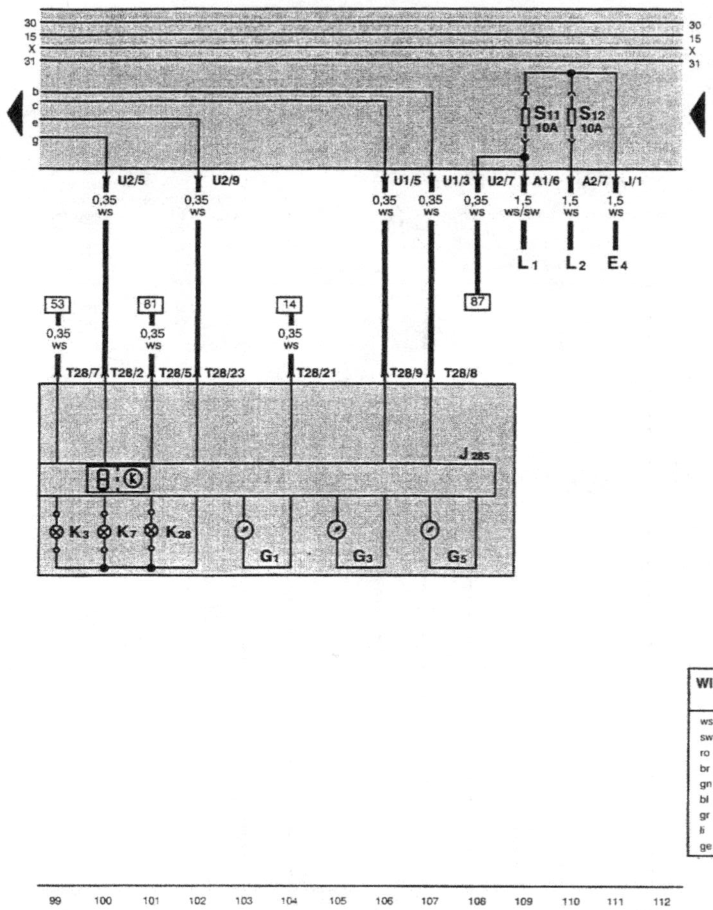

WIRING COLOR CODE		
ws	=	white
sw	=	black
ro	=	red
br	=	brown
gn	=	green
bl	=	blue
gr	=	grey
li	=	lilac
ge	=	yellow

- E4 – Headlight Dimmer/Flasher Switch
- G1 – Fuel Gauge
- G3 – Engine Coolant Temperature (ECT) Gauge
- G5 – Tachometer
- K3 – Oil Pressure Warning Light
- K7 – Brake and Parking Brake Warning Light
- K28 – Engine Coolant Level/Temperature (ECL/ECT) Warning Light
- L1 – Left Headlight
- L2 – Right Headlight
- T28 – 28-Pin Connector, on instrument cluster

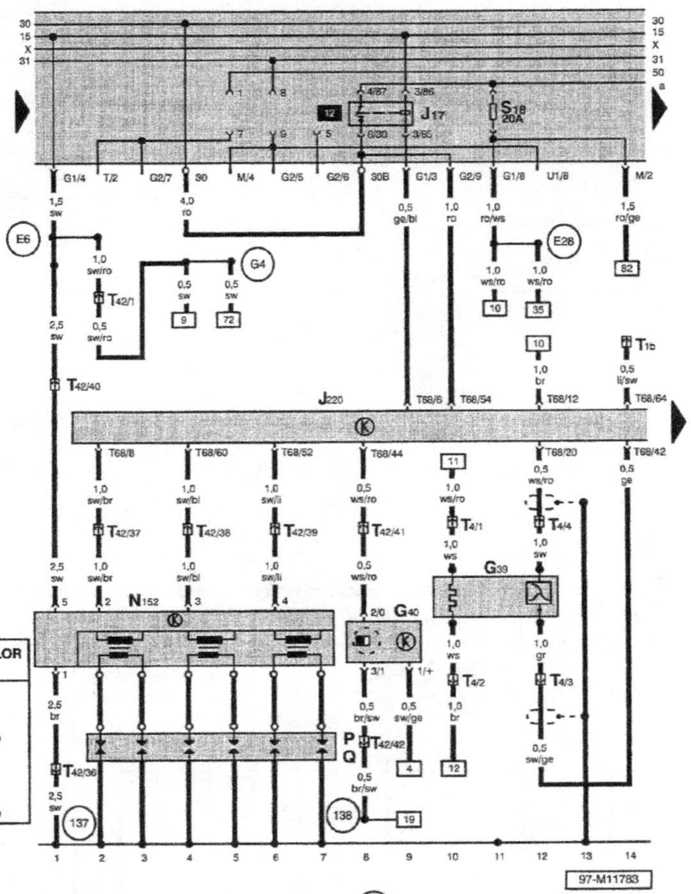

WIRING COLOR CODE		
ws	=	white
sw	=	black
ro	=	red
br	=	brown
gn	=	green
bl	=	blue
gr	=	grey
li	=	lilac
ge	=	yellow

- G39 – Heated Oxygen Sensor (HO2S) 1
- G40 – Camshaft Position (CMP) Sensor
- J17 – Fuel Pump (FP) Relay
- J220 – Motronic Engine Control Module (ECM)
- N152 – Ignition Coil
- P – Spark Plug Connector
- Q – Spark Plug
- T1b – 1-Pin Connector
- T4 – 4-Pin Connector, bottom right rear of engine compartment
- T42 – 42-Pin Connector, left side of engine
- T68 – 68-Pin Connector, under right side of hood, on Motronic Engine Control Module (ECM)

- (E6) – Plus connection (15), in Motronic Multiport Fuel Injection (MFI) wiring harness
- (G4) – Wire connection, in Injector wiring harness
- (E28) – Wire connection (Lambda probe heater), in wiring harness Motronic
- (137) – Ground connection (end stage), in Motronic Multiport Fuel Injection (MFI) wiring harness
- (138) – Ground connection (control module), in Motronic Multiport Fuel Injection (MFI) wiring harness

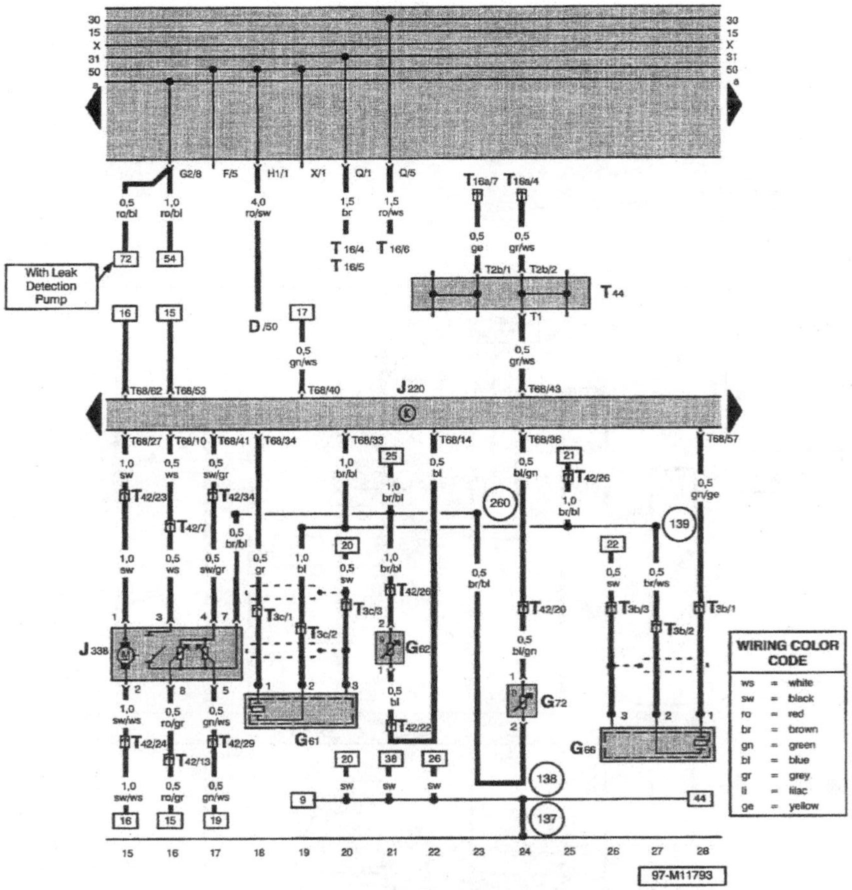

D – Ignition Switch
G61 – Knock Sensor (KS) 1
G66 – Knock Sensor (KS) 2
G62 – Engine Coolant Temperature (ECT) Sensor
G72 – Intake Air Temperature (IAT) Sensor
J217 – Transmission Control Module (TCM)
J220 – Motronic Engine Control Module (ECM)
J338 – Throttle Valve Control Module
T1 – Single Connector, above fuse/relay panel
T2a – Double Connector, on junction connector (T44) above fuse relay/panel
T2b – Double Connector, on junction connector (T44) above fuse relay/panel
T3a – 3-Pin Connector, on engine
T3b – 3-Pin Connector, in engine harness
T3c – 3-Pin Connector, left front of engine
T16 – 16-Pin Connector, in engine harness
T16a – Data Link Connector (DLC), behind center of dash
T28 – 28-Pin Connector, instrument cluster
T28a – 28-Pin Connector, left side of engine, near distributor
T42 – 42-Pin Connector, left side of engine
T44 – Junction Box for On-Board Diagnosis, above fuse/relay panel
T68 – 68-Pin Connector, under right side of hood, on Motronic Engine Control Module (ECM)
(137) – Ground connection (end stage), in Multiport Fuel Injection (MFI) wiring harness
(138) – Ground connection (control module), in Motronic Multiport Fuel Injection (MFI) wiring harness
(139) – Ground connection (sensor ground), in Motronic Multiport Fuel Injection (MFI) wiring harness
(260) – Ground connection (sensor ground), injector wiring harness

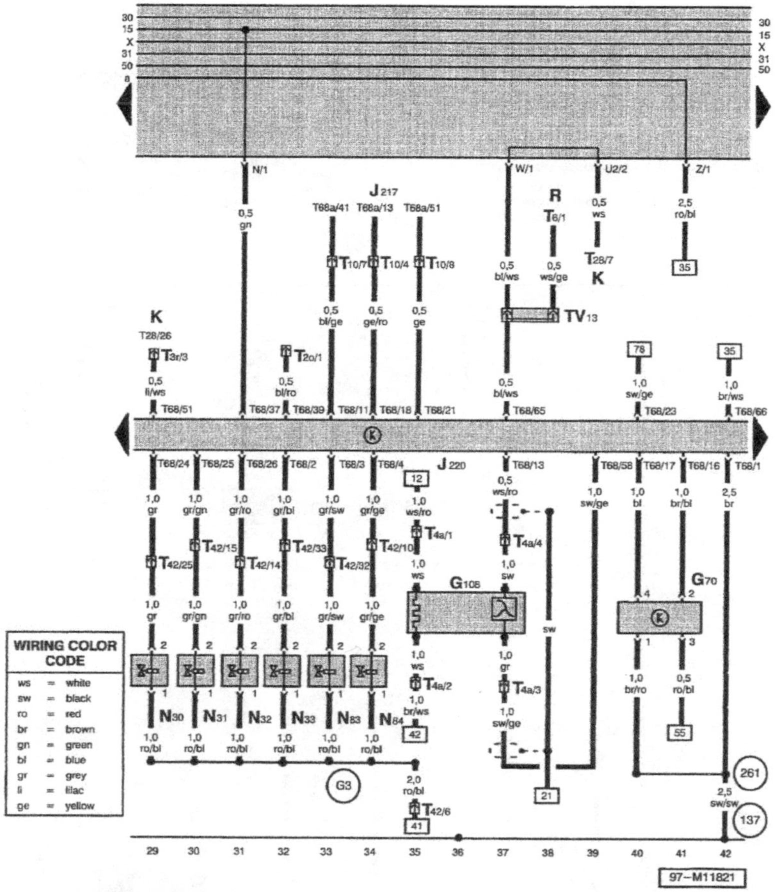

F194 – Clutch Pedal Position Switch
G70 – Mass Air Flow (MAF) Sensor
G108 – Heated Oxygen Sensor (HO2S) 2
J217 – Transmission Control Module (TCM)
J220 – Motronic Engine Control Module (ECM)
K – Instrument Cluster
N30 – Cylinder 1 Fuel Injector
N31 – Cylinder 2 Fuel Injector
N32 – Cylinder 3 Fuel Injector
N33 – Cylinder 4 Fuel Injector
N83 – Cylinder 5 Fuel Injector
N84 – Cylinder 6 Fuel Injector
TV13 – Vehicle Speed Signal Wire Connector
R – Radio
T2o – Double Connector, behind fuse/relay panel
T3r – 3-Pin Connector, in engine compartment
T4a – 4-Pin Connector, bottom right rear of engine compartment
T6 – 6-Pin Connector, in engine compartment
T10 – 10-Pin Connector, in engine compartment
T28 – 28-Pin Connector, instrument cluster
T42 – 42-Pin Connector, left side of engine
T68 – 68-Pin Connector, under right side of hood, on Motronic Multiport Fuel Injection (MFI) wiring harness
T68a – 68-Pin Connector, on Transmission Control Module (TCM)
(137) – Ground connection (end stage), in Motronic Multiport Fuel Injection (MFI) wiring harness
(261) – Ground connection, in Motronic Multiport Fuel Injection (MFI) wiring harness
(G3) – Plus connection, in injector wiring harness

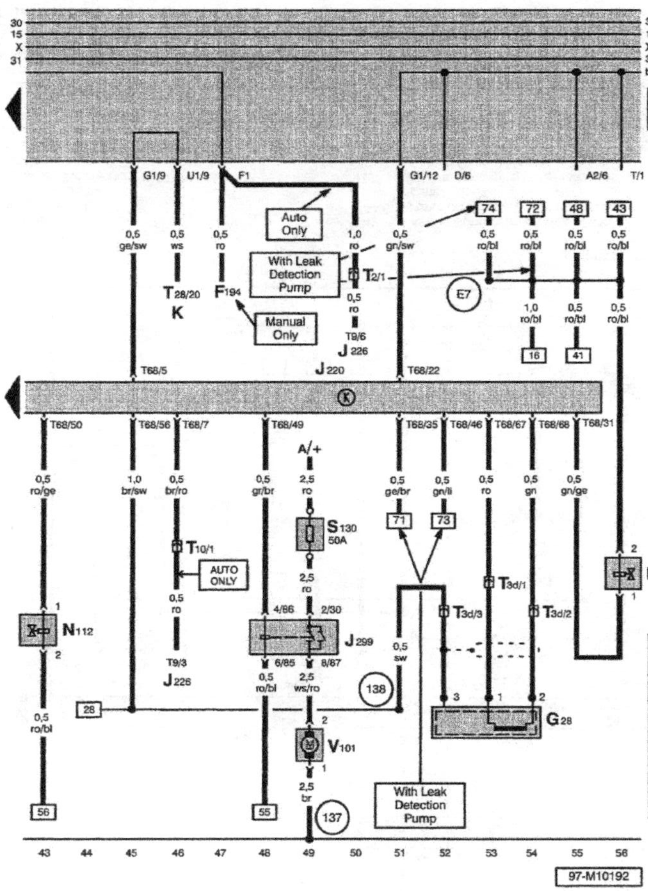

WIRING COLOR CODE		
ws	=	white
sw	=	black
ro	=	red
br	=	brown
gn	=	green
bl	=	blue
gr	=	grey
li	=	lilac
ge	=	yellow

A – Battery
F194 – Clutch Pedal Position Switch
G28 – Engine Speed (RPM) Sensor
J220 – Motronic Engine Control Module (ECM)
J226 – Park/Neutral Position (PNP) Relay
J299 – Secondary Air Injection (AIR) Pump Relay
K – Instrument Cluster
N80 – Evaporative Emission (EVAP) Canister Purge Regulator Valve
N112 – Secondary Air Injection (AIR) Solenoid Valve
S130 – Fuse for Secondary Air Pump
T2 – 2-Pin Connector
T3d – 3-Pin Connector, in engine compartment
T10 – 10-Pin Connector, in engine compartment
T28 – 28-Pin Connector, Instrument Cluster
T68 – 68-Pin Connector, under right side of hood, on Motronic Engine Control Module (ECM)
V101 – Secondary Air Injection (AIR) Pump Motor
(E7) – Plus connection (87a) in Multiport Fuel Injection (MFI) wiring harness
(137) – Ground connection (end stage), in Multiport Fuel Injection (MFI) wiring harness
(138) – Ground connection (control module), in Motronic Multiport Fuel Injection (MFI) wiring harness

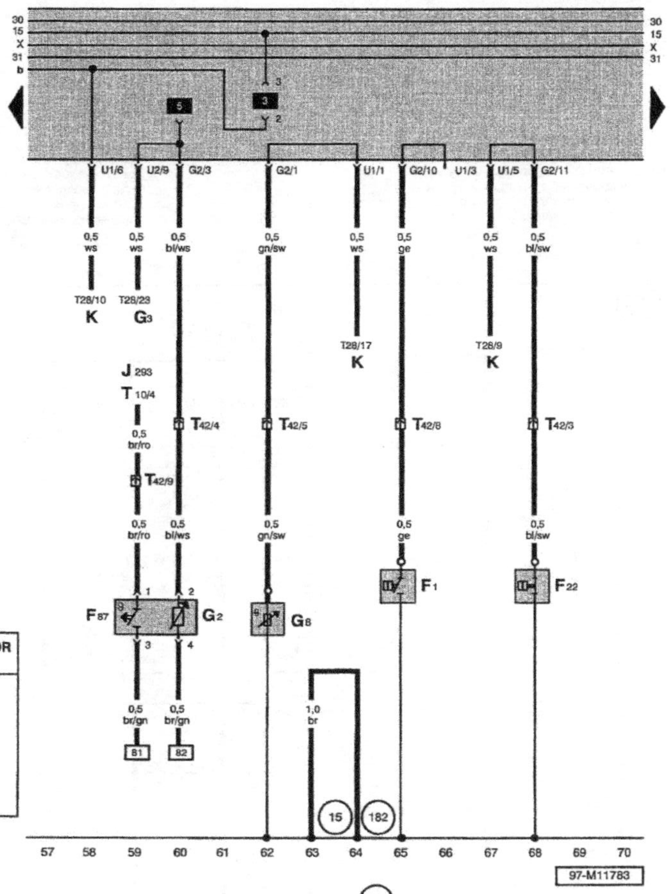

WIRING COLOR CODE		
ws	=	white
sw	=	black
ro	=	red
br	=	brown
gn	=	green
bl	=	blue
gr	=	grey
li	=	lilac
ge	=	yellow

F1 – Oil Pressure Switch
F22 – 0.3 Bar Oil Pressure Switch
F87 – After-Run Coolant Fan Control (FC) Thermal Switch
G2 – Engine Coolant Temperature (ECT) Sensor
G3 – Engine Coolant Temperature (ECT) Gauge
G8 – Engine Oil Temperature Sensor
J293 – Coolant FC (Fan Control) Control Module
K – Instrument Cluster
T10 – 10-Pin Connector, in engine compartment
T28 – 28-Pin Connector, left side of engine, near distributor
T42 – 42-Pin Connector, left side of engine
(15) – Ground connection, in Motronic Multiport Fuel Injection (MFI) wiring harness
(182) – Ground connection, in Motronic Multiport Fuel Injection (MIF) wiring harness

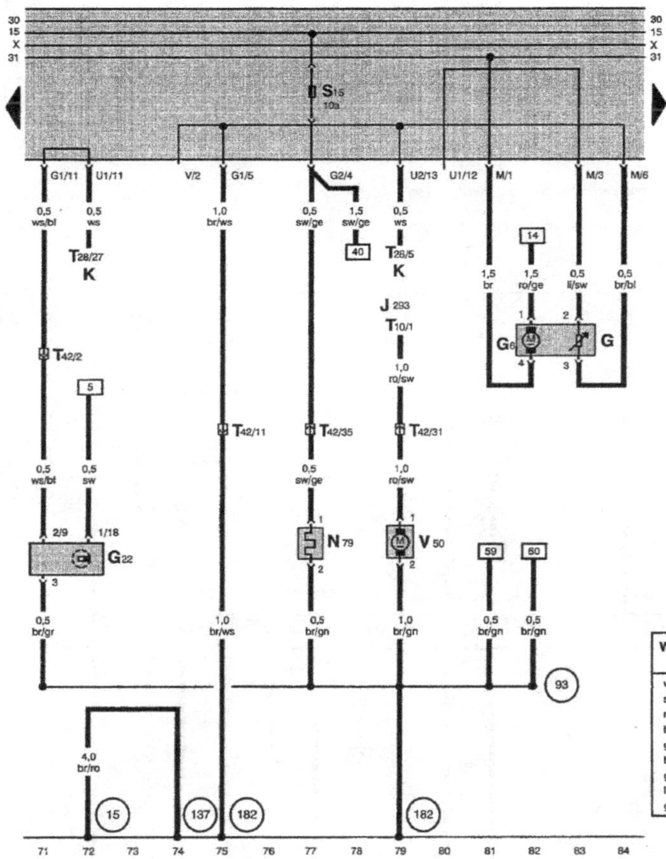

97-M11893

WIRING COLOR CODE		
ws	=	white
sw	=	black
ro	=	red
br	=	brown
gn	=	green
bl	=	blue
gr	=	grey
li	=	lilac
ge	=	yellow

G – Fuel Level Sensor
G6 – Fuel Pump (FP)
G22 – Speedometer Vehicle Speed Sensor (VSS)
J293 – Coolant FC (Fan Control) Control Module
K – Instrument Cluster
N79 – Positive Crankcase Ventilation (PCV) Heating Element
T10 – 10-Pin Connector, in engine compartment
T28 – 28-Pin Connector, instrument cluster
T42 – 42-Pin Connector, left side of engine
V50 – Coolant Pump

(15) – Ground connection, in Motronic Multiport Fuel Injection (MFI) wiring harness

(93) – Ground connection, engine block, in ignition system wiring harness

(137) – Ground connection (end stake), in Motronic Multiport Fuel Injection (MFI) wiring harness

(182) – Ground connection, in Motronic Multiport Fuel Injection (MFI) wiring harness

Engine electronics

WIRING COLOR CODE		
ws	=	white
sw	=	black
ro	=	red
br	=	brown
gn	=	green
bl	=	blue
gr	=	grey
li	=	lilac
ge	=	yellow

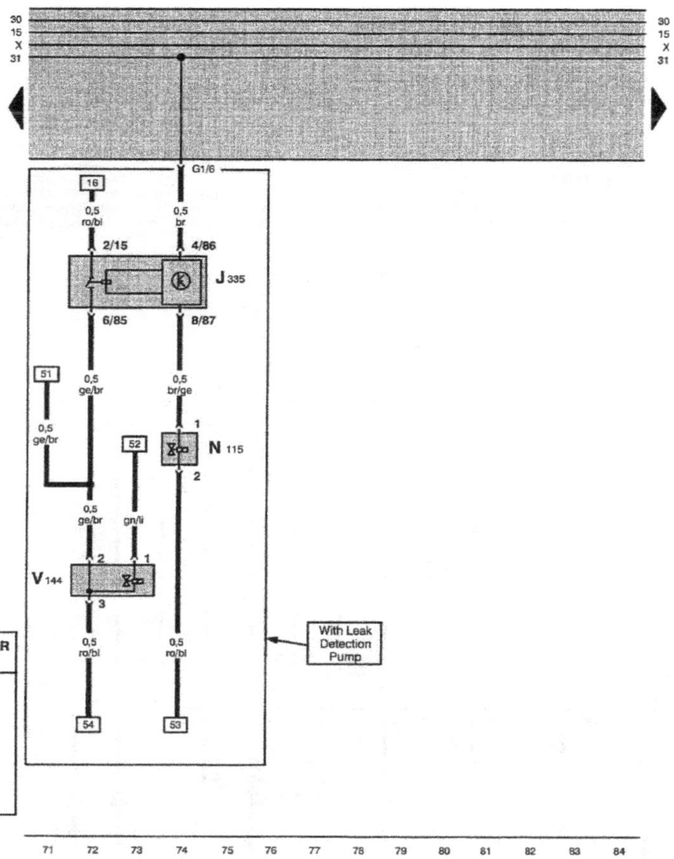

97-12804

J335 – Fuel Cut-off Valve Shut-off Relay
N115 – Evaporative Emission (EVAP) Canister Pure Regulator
V144 – Leak Detection Pump (LDP)

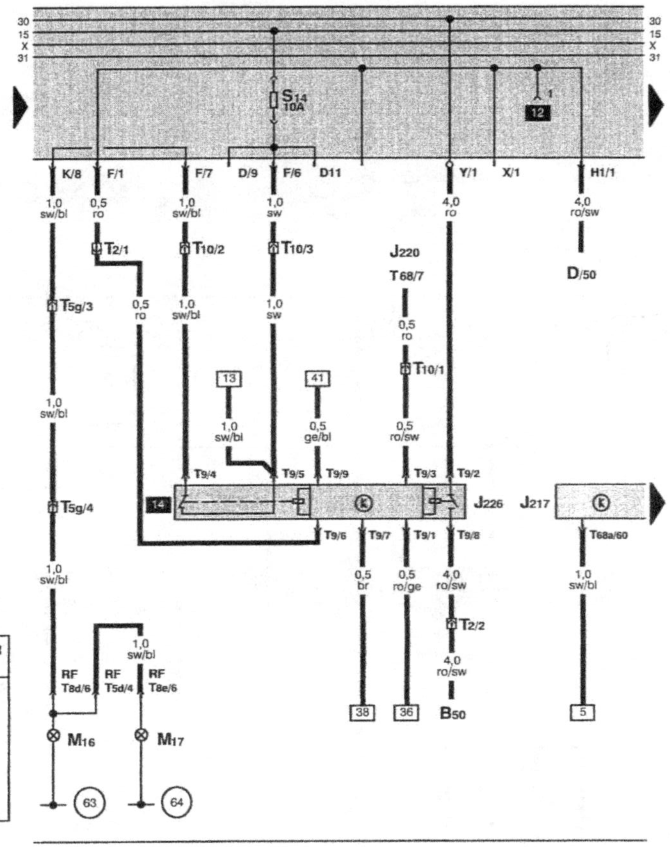

WIRING COLOR CODE		
ws	=	white
sw	=	black
ro	=	red
br	=	brown
gn	=	green
bl	=	blue
gr	=	grey
li	=	lilac
ge	=	yellow

B – Starter
D – Ignition/Starter Switch
E17 – Park/Neutral Position (PNP) Switch
J217 – Transmission Control Module (TCM), below rear seat bench
J220 – Motronic Engine Control Module (ECM), in plenum chamber, right
J226 – Park/Neutral Position (PNP) Relay, on auxiliary relay panel
M16 – Left Back-Up Light
M17 – Right Back-Up Light
T2 – Double Connector, near starter
T5d – 5-Pin Connector, on taillight, left
T5g – 5-Pin Connector, on luggage compartment lid
T8c1 – 8-Pin Connector, on taillight, left
T8d – 8-Pin Connector, on taillight, left
T8e – 8-Pin Connector, on taillight, right
T9 – 9-Pin Connector, on PNP relay
T10 – 10-Pin Connector, behind fuse/relay panel
T68 – 68-Pin Connector, on Motronic Engine Control Module (ECM)
TV8 – Terminal 58b wire connector, above fuse/relay panel

(63) – Ground connection, on left taillight bulb holder
(64) – Ground connection, on right taillight bulb

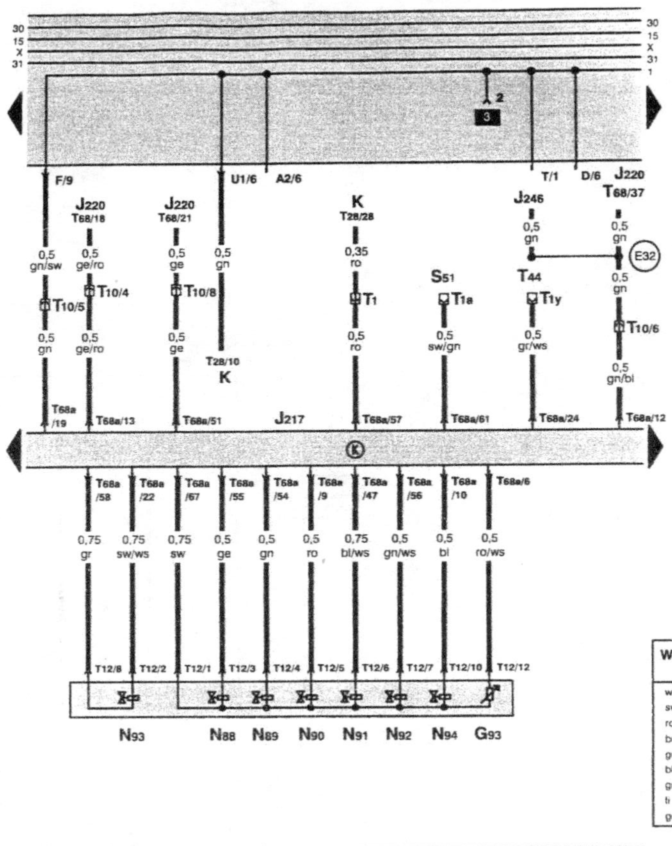

WIRING COLOR CODE		
ws	=	white
sw	=	black
ro	=	red
br	=	brown
gn	=	green
bl	=	blue
gr	=	grey
li	=	lilac
ge	=	yellow

G93 – Transmission Fluid Temperature Sensor (00300–1333)
J217 – Transmission Control Module (TCM), below rear seat bench
J220 – Motronic Engine Control Module (ECM)
J246 – A/C Clutch Cutoff Relay
K – Instrument Cluster
N88 – Solenoid Valve 1 (00258–1113)
N89 – Solenoid Valve 2 (00260–1121)
N90 – Solenoid Valve 3 (00262–1123)
N91 – Solenoid Valve 4 (00264–1131)
N92 – Solenoid Valve 5 (00266–1133)
N93 – Solenoid Valve 6 (00268–1141)
N94 – Solenoid Valve 7 (00270–1143)
S51 – Fuse

T1 – Single Connector, behind fuse/relay panel
T1a – Single Connector, behind fuse/relay panel
T1y – Single Connector, behind fuse/relay panel
T10 – 10-Pin Connector, behind fuse/relay panel
T28 – 28-Pin Connector, on instrument cluster
T44 – Junction Box for On Board Diagnosis, above fuse/relay panel
T68 – 68-Pin Connector, on Motronic Engine Control Module (ECM)
T68a – 68-Pin Connector, on Transmission Control Module (TCM)
(E32) – Wire connection, in Motronic MFI wiring harness

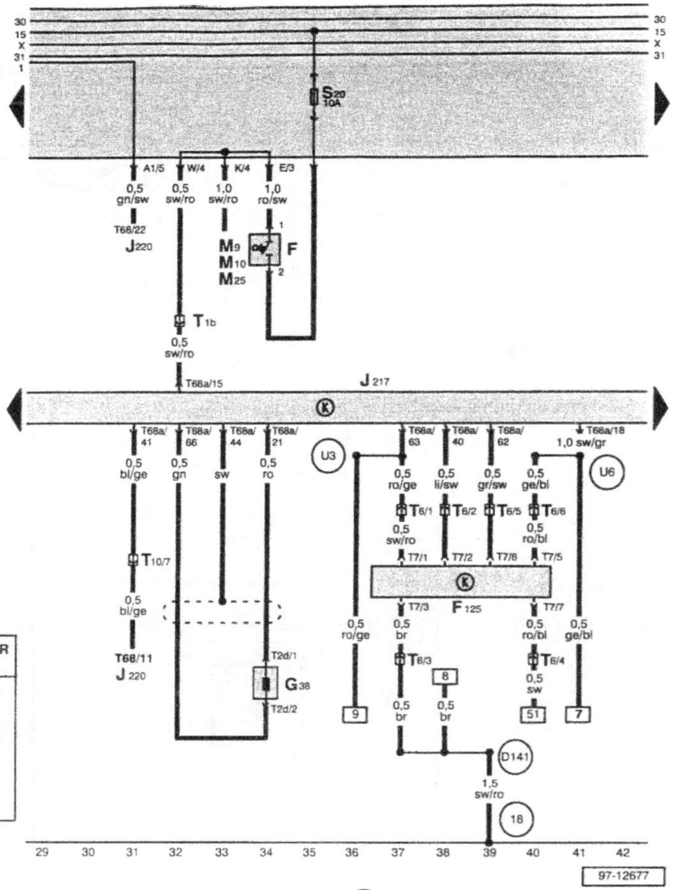

WIRING COLOR CODE

ws	=	white
sw	=	black
ro	=	red
br	=	brown
gn	=	green
bl	=	blue
gr	=	grey
li	=	lilac
ge	=	yellow

- F – Brake Light Switch
- F125 – Multi-Function Transmission Range (TR) Switch
- G38 – Transmission Vehicle Speed Sensor (VSS)
- J217 – Transmission Control Module (TCM) below rear seat bench
- J220 – Motronic Engine Control Module (ECM), in plenum chamber, right
- M9 – Left Brake Light
- M10 – Right Brake Light
- M25 – High-Mount Brake Light
- T1b – Single Connector, behind fuse/relay panel
- T2d – Double connector, on Vehicle Speed Sensor (VSS)
- T6 – 6-Pin Connector
- T7 – 7-Pin Connector, on Multi-Function Transmission Range Sensor
- T10 – 10-Pin Connector, behind fuse/relay panel
- T68 – 68-Pin Connector, on Motronic Engine Control Module (ECM)
- T68a – 68-Pin Connector, on Transmission Control Module (TCM)

- (18) – Ground connection, on engine block
- (D141) – Wire connection, in automatic transmission wiring harness
- (U3) – Wire connection -1-, in automatic transmission
- (U6) – Wire connection -2-, in automatic transmission wiring harness

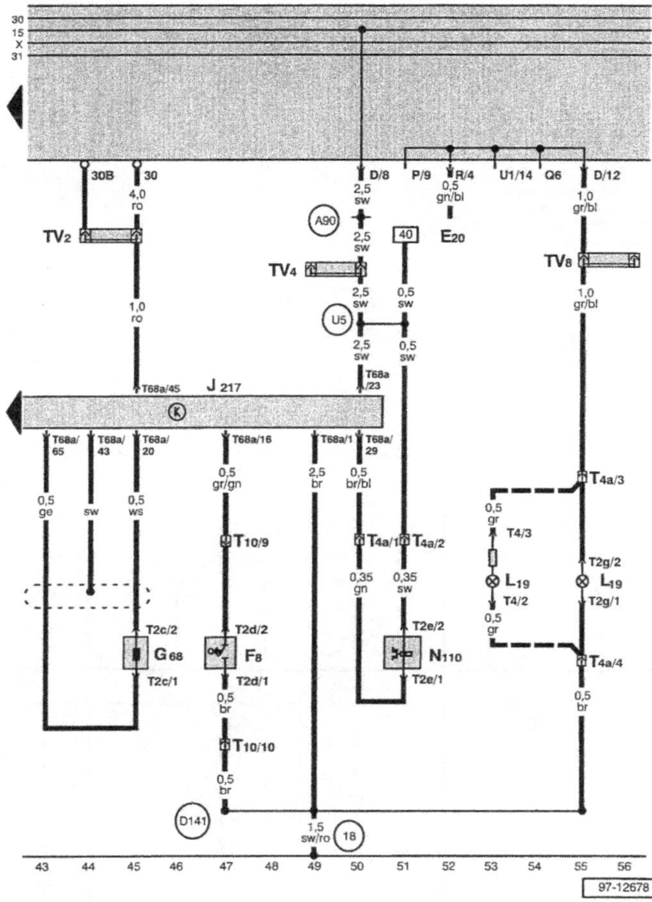

WIRING COLOR CODE

ws	=	white
sw	=	black
ro	=	red
br	=	brown
gn	=	green
bl	=	blue
gr	=	grey
li	=	lilac
ge	=	yellow

- E20 – Instrument Panel Light Dimmer Switch
- F8 – Kick Down Switch
- G68 – Vehicle Speed Sensor (VSS)
- J217 – Transmission Control Module (TCM) Below rear seat bench
- L19 – Automatic Transmission Console Light
- N110 – Shift Lock Solenoid
- T2c – Double Connector, on Vehicle Speed Sensor
- T2d – Double Connector, on Kick Down Switch
- T2e – Double Connector, on Shift Lock Solenoid
- T2g – Double Connector, on Automatic Transmission Console Light
- T4 – 4-Pin Connector, on Automatic Transmission Console Light
- T4a – 4-Pin Connector, below gearshift lever cover
- T10 – 10-Pin Connector, behind fuse/relay panel
- T68 – 68-Pin Connector
- TV2 – Terminal 30 (B+) Wire Connector (Red), behind fuse/relay panel
- TV4 – Terminal 15 Wire Connector (Black), behind fuse/relay pane
- TV8 – Terminal 58b Wire Connector (Blue), behind fuse/relay pane

- (18) – Ground connection, on engine block
- (A90) – Wire connection, in voltage distribution harness
- (D141) – Wire connection, in automatic transmission wiring harness
- (U5) – Wire connection -2- (15), in automatic transmission wiring harness

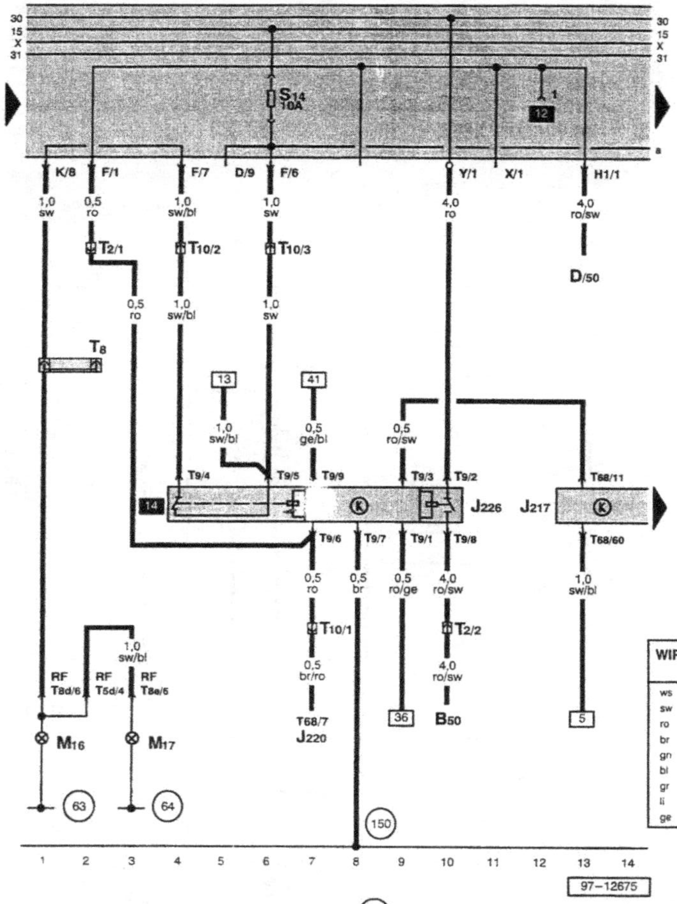

WIRING COLOR CODE		
ws	=	white
sw	=	black
ro	=	red
br	=	brown
gn	=	green
bl	=	blue
gr	=	grey
li	=	lilac
ge	=	yellow

B – Starter
D – Ignition/Starter Switch
E17 – Park/Neutral Position (PNP) Switch
J217 – Transmission Control Module (TCM), below rear seat bench
J220 – Motronic Engine Control Module (ECM), in plenum chamber, right
J226 – Park/Neutral Position (PNP) Relay, on auxiliary relay panel
M16 – Left Back-Up Light
M17 – Right Back-Up Light
T2 – Double Connector, near starter
T5d – 5-Pin Connector, on taillight, left
T6a – 6-Pin Connector
T8c1 – 8-Pin Connector on taillight, left
T8e – 8-Pin Connector, on taillight, right
T9 – 9-Pin Connector
T10 – 10-Pin Connector, behind fuse/relay panel
T68 – 68-Pin Connector, on Transmission Control Module (TCM)
TV8 – Terminal 586 wire connector, above fuse/relay panel

(63) – Ground connection, on left taillight bulb holder
(64) – Ground connection, on right taillight bulb
(150) – Ground connection, in automatic transmission wiring harness

1996-1999

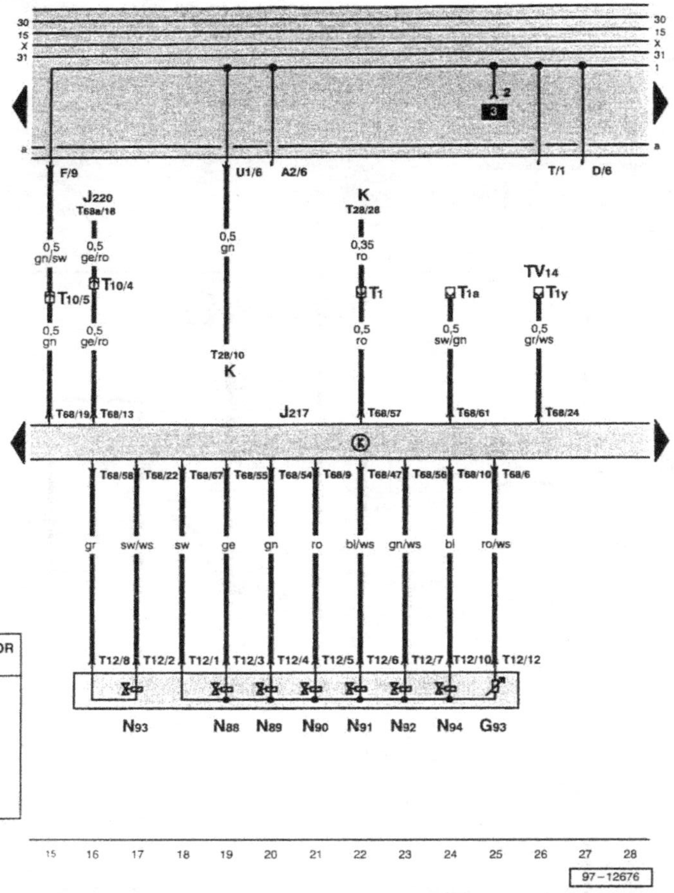

WIRING COLOR CODE		
ws	=	white
sw	=	black
ro	=	red
br	=	brown
gn	=	green
bl	=	blue
gr	=	grey
li	=	lilac
ge	=	yellow

G93 – Transmission Fluid Temperature Sensor (00300–1333)
J217 – Transmission Control Module (TCM), below rear seat bench
J220 – Motronic Engine Control Module (ECM)
K – Instrument Cluster
N88 – Solenoid Valve 1 (00258–1113)
N89 – Solenoid Valve 2 (00260–1121)
N90 – Solenoid Valve 3 (00262–1123)
N91 – Solenoid Valve 4 (00264–1131)
N92 – Solenoid Valve 5 (00266–1133)
N93 – Solenoid Valve 6 (00268–1141)
N94 – Solenoid Valve 7 (00270–1143)

T1 – Single Connector, behind fuse/relay panel
T1a – Single Connector, behind fuse/relay panel
T1y – Single Connector, behind fuse/relay panel
T10 – 10-Pin Connector, behind fuse/relay panel
T10a – 10-Pin Connector
T28 – 28-Pin Connector, on instrument cluster
T68 – 68-Pin Connector, on Transmission Control Module (TCM)
T68a – 68-Pin Connector, on Motronic Engine Control Module (ECM)
TV14 – Data Link Connector (DLC) Wire Connector, above fuse/relay panel

1996-1999

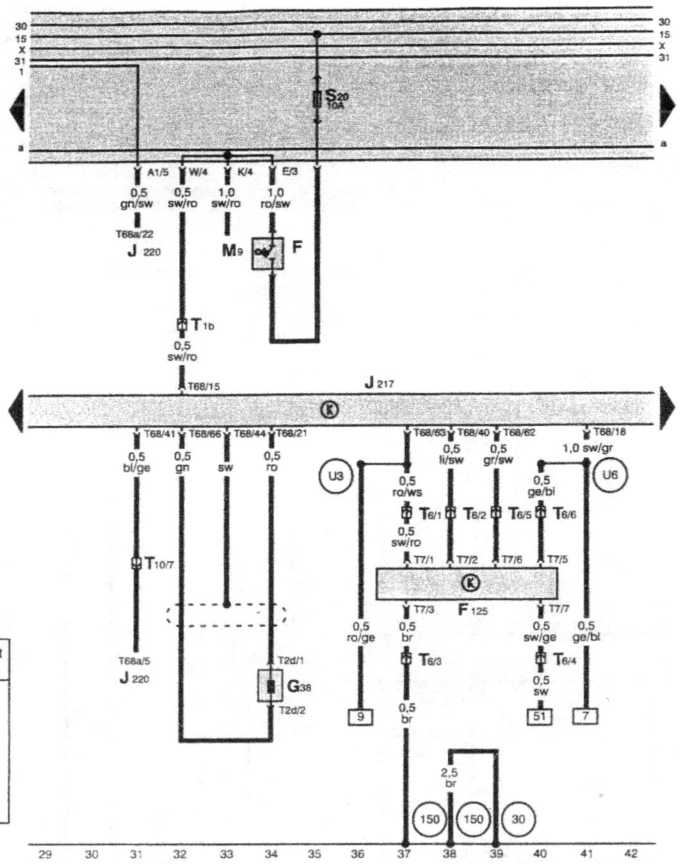

WIRING COLOR CODE		
ws	=	white
sw	=	black
ro	=	red
br	=	brown
gn	=	green
bl	=	blue
gr	=	grey
li	=	lilac
ge	=	yellow

F – Brake Light Switch
F125 – Multi-Function Transmission Range (TR) Switch
G38 – Transmission Vehicle Speed Sensor (VSS) (00297-1324)
J217 – Transmission Control Module (TCM) below rear seat bench
J220 – Motronic Engine Control Module (ECM), in plenum chamber, right
M9 – Left Brake Light
M10 – Right Brake Light
M25 – High-Mount Brake Light
T1b – Single Connector, behind fuse/relay panel
T2d – Double connector
T6 – 6-Pin Connector
T7 – 7-Pin Connector
T10 – 10-Pin Connector, behind fuse/relay panel
T68 – 68-Pin Connector, on Transmission Control Module (TCM)
T68a – 68-Pin Connector, on Motronic Engine Control Module (ECM)

(30) – Ground connection, -1-, beside fuse/relay panel
(150) – Ground connection, in automatic transmission
(U3) – Wire connection -1-, in automatic transmission
(U6) – Wire connection -2-, in automatic transmission wiring harness

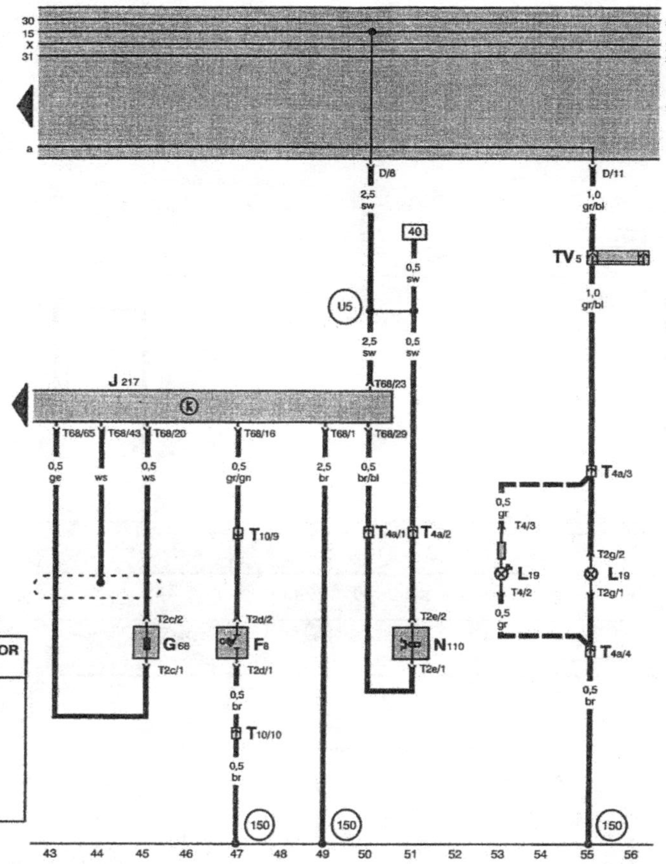

WIRING COLOR CODE		
ws	=	white
sw	=	black
ro	=	red
br	=	brown
gn	=	green
bl	=	blue
gr	=	grey
li	=	lilac
ge	=	yellow

F8 – Kick Down Switch
G68 – Vehicle Speed Sensor (VSS)
J217 – Transmission Control Module (TCM), below rear seat bench
L19 – Automatic Transmission Console Light
N110 – Shift Lock Solenoid
T2c – Double Connector, on VSS
T2d – Double Connector, on Kick Down Switch
T2e – Double Connector, on Shift Lock Solenoid
T2g – Double Connector, on Automatic Transmission Console Light
T4 – 4-Pin Connector, on Automatic Transmission Console Light
T4a – 4-Pin Connector, below gearshift lever cover
T10 – 10-Pin Connector, behind fuse/relay panel
T68 – 68-Pin Connector, on Transmission Control Module (TCM)
TV5 – Terminal 15a Wire Connector, behind fuse/relay panel

(150) – Ground connection, in automatic transmission wiring harness
(U5) – Wire connection -2- (15), in automatic transmission wiring harness

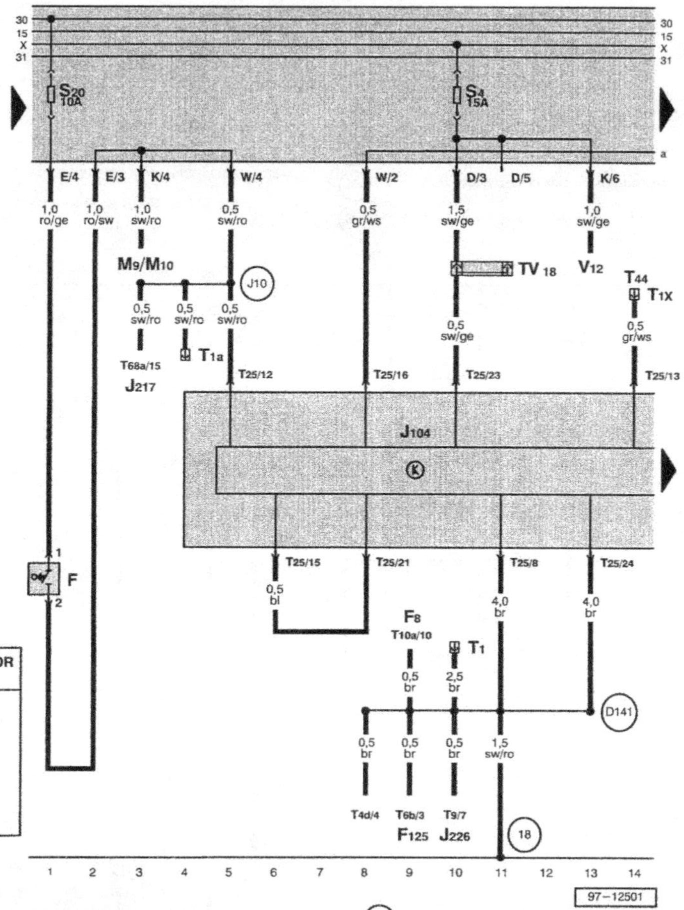

WIRING COLOR CODE		
ws	=	white
sw	=	black
ro	=	red
br	=	brown
gn	=	green
bl	=	blue
gr	=	grey
li	=	lilac
ge	=	yellow

- F – Brake Light Switch (00526–2131)
- F36 – Clutch Vacuum Vent Valve Switch (00650)
- F125 – Multi-Function Transmission Range (TR) Switch
- J104 – ABS Control Module (w/EDL), in engine compartment, left
- J217 – Transmission Control Module (TCM), below rear seat bench
- J220 – Motronic Engine Control Module (ECM)
- J226 – Park/Neutral Position (PNP) Relay
- M9 – Left Brake Light
- M10 – Right Brake Light
- T1 – Single Connector, above fuse/relay panel
- T1a – Single Connector, above fuse/relay panel
- T25 – 25-Pin Connector, on ABS Control Module
- T4d – 4-Pin Connector, below center console
- T6b – 6-Pin Connector, below battery
- T10a – 10-Pin Connector, behind fuse/relay panel
- T44 – Junction Box for On Board Diagnosis, above fuse/relay panel
- TV18 – Distributor for terminal Xa (yellow), above fuse/relay panel
- V12 – Rear Window Wiper Motor

- (18) – Ground connection, on engine block
- (D141) – Wire connection, in transmission wiring harness
- (J10) – Wire connection (54), in ABS wiring harness

Anti-lock brake system (ABS)

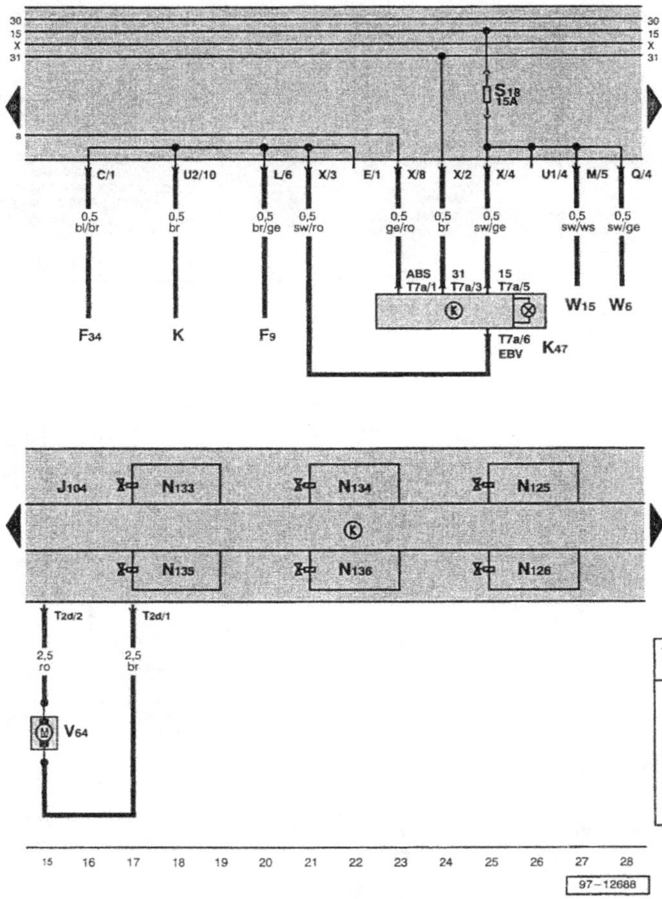

WIRING COLOR CODE		
ws	=	white
sw	=	black
ro	=	red
br	=	brown
gn	=	green
bl	=	blue
gr	=	grey
li	=	lilac
ge	=	yellow

- F9 – Parking Brake Warning Light Switch (01114)
- F34 – Brake Fluid Level Warning Switch
- J104 – ABS Control Module (w/EDL), in engine compartment, left
- K – Instrument Cluster
- K47 – ABS Warning Light
- N125 – Differential Lock Valve 1
- N126 – Differential Lock Valve 2
- N133 – Right Rear ABS Inlet Valve (00273–1211)
- N134 – Left Rear ABS Inlet Valve (00274–1212)
- N135 – Right Rear ABS Outlet Valve (00275–1213)
- N136 – Left Rear ABS Outlet Valve (00276–1214)
- T2d – 2-Pin Connector, on ABS control module
- T7a – 7-Pin Connector, on ABS warning light
- W6 – Glove Compartment Light
- W15 – Interior Light with Delay Switch
- V64 – ABS Hydraulic Pump

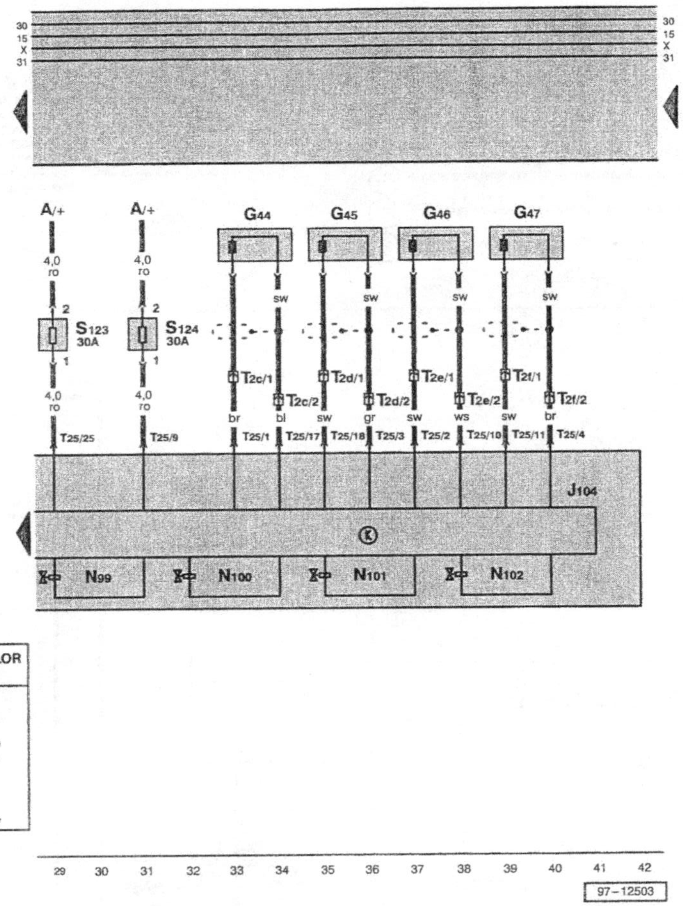

WIRING COLOR CODE		
ws	=	white
sw	=	black
ro	=	red
br	=	brown
gn	=	green
bl	=	blue
gr	=	grey
li	=	lilac
ge	=	yellow

A – Battery
G44 – Right Rear ABS Wheel Speed Sensor (00287–1243)
G45 – Right Front ABS Wheel Speed Sensor (00285–1241)
G46 – Left Rear ABS Wheel Speed Sensor (00290–1311)
G47 – Left Front ABS Wheel Speed Sensor (00283–1233)
J104 – ABS Control Module (w/EDL), in engine compartment, left
N99 – Right Front ABS Inlet Valve (00259–1114)
N100 – Right Front ABS Outlet Valve (00267–1134)
N101 – Left Front ABS Inlet Valve (00257–1112)
N102 – Left Front ABS Outlet Valve (00265–1132)
S123 – Fuse 1 for control module – ABS in engine compartment, left
S124 – Fuse 2 for control module – ABS in engine compartment, left
T2c – Double Connector, below rear seat bench
T2d – Double Connector, behind suspension strut, right
T2e – Double Connector, below rear seat bench
T2f – Double Connector, behind suspension strut, left
T25 – 25-Pin Connector, on ABS Control Module

THIS PAGE INTENTIONALLY LEFT

BLANK

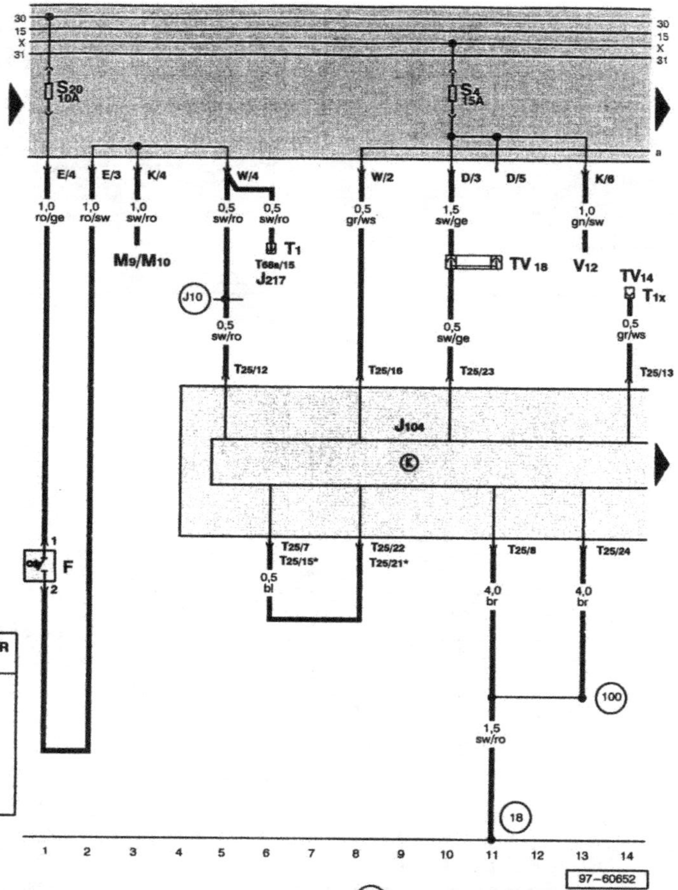

WIRING COLOR CODE		
ws	=	white
sw	=	black
ro	=	red
br	=	brown
gn	=	green
bl	=	blue
gr	=	grey
li	=	lilac
ge	=	yellow

F – Brake Light Switch
J104 – ABS Control Module (w/EDL), in engine compartment, left
J217 – Transmission Control Module (TCM)
M9 – Left Brake Light
M10 – Right Brake Light
T1 – Single Connector, above fuse/relay panel
T25 – 25-Pin Connector
T68a – 68-Pin Connector
TV14 – Data Link Connector (DLC) Wire Connector, above fuse/relay panel
TV18 – Distributor for terminal Xa, above fuse/relay panel
V12 – Rear Window Wiper Motor

(18) – Ground connection, on engine block

(100) – Ground connection –1–, in ABS wiring harness

(J10) – Wire connection (54), in ABS wiring harness

* applies only vehicles before m.y. 96

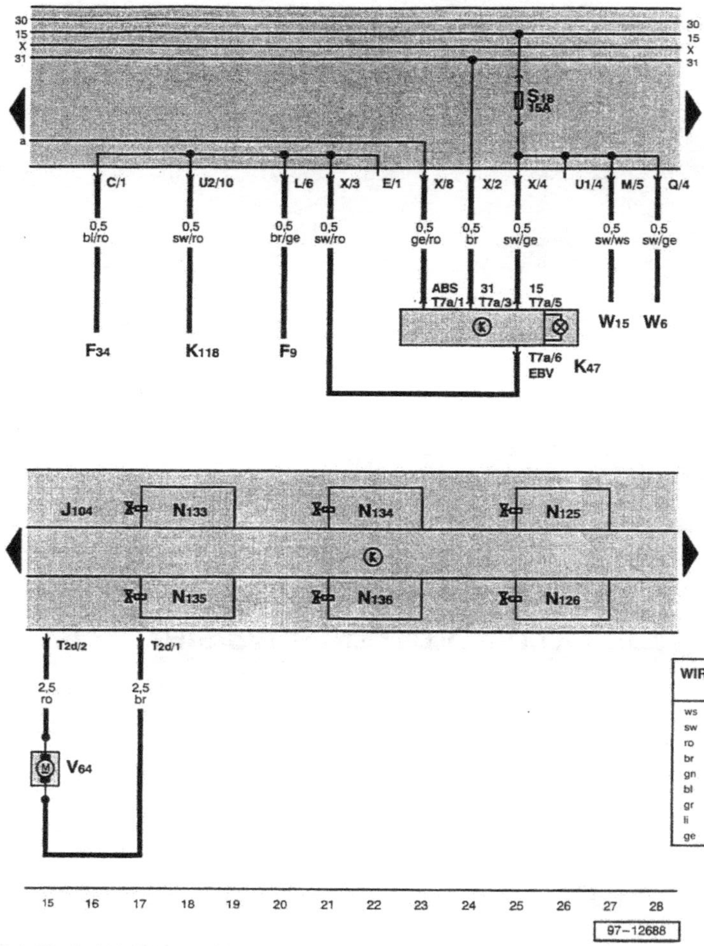

WIRING COLOR CODE		
ws	=	white
sw	=	black
ro	=	red
br	=	brown
gn	=	green
bl	=	blue
gr	=	grey
li	=	lilac
ge	=	yellow

F9 – Parking Brake Warning Light Switch (01114)
F34 – Brake Fluid Level Warning Switch
J104 – ABS Control Module (w/EDL), in engine compartment, left
K – Instrument Cluster
K47 – ABS Warning Light
N125 – Differential Lock Valve 1
N126 – Differential Lock Valve 2
N133 – Right Rear ABS Inlet Valve (00273–1211)
N134 – Left Rear ABS Inlet Valve (00274–1212)
N135 – Right Rear ABS Outlet Valve (00275–1213)
N136 – Left Rear ABS Outlet Valve (00276–1214)
T7a – 7-Pin Connector
W6 – Glove Compartment Light
W15 – Interior Light with Delay Switch
V64 – ABS Hydraulic Pump

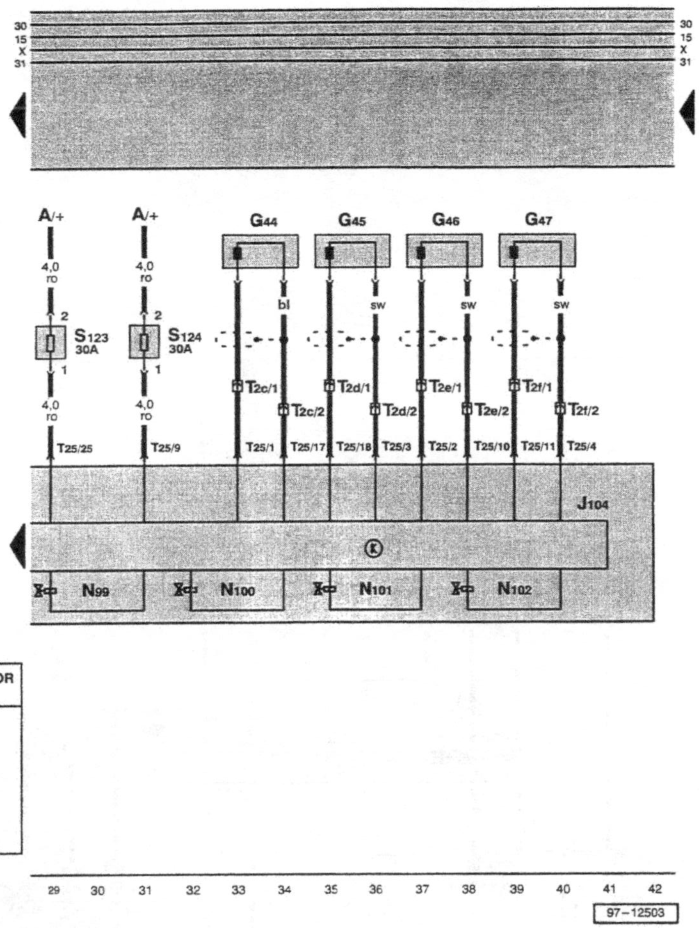

WIRING COLOR CODE		
ws	=	white
sw	=	black
ro	=	red
br	=	brown
gn	=	green
bl	=	blue
gr	=	grey
li	=	lilac
ge	=	yellow

- A – Battery
- G44 – Right Rear ABS Wheel Speed Sensor (00287–1243)
- G45 – Right Front ABS Wheel Speed Sensor (00285–1241)
- G46 – Left Rear ABS Wheel Speed Sensor (00290–1311)
- G47 – Left Front ABS Wheel Speed Sensor (00283–1233)
- J104 – ABS Control Module (w/EDL) , in engine compartment, left
- N99 – Right Front ABS Inlet Valve (00259–1114)
- N100 – Right Front ABS Outlet Valve (00267–1134)
- N101 – Left Front ABS Inlet Valve (00257–1112)
- N102 – Left Front ABS Outlet Valve (00265–1132)
- S123 – Fuse 1 for control module – ABS in engine compartment, left
- S124 – Fuse 2 for control module – ABS in engine compartment, left
- T2 – Double Connector, below rear seat bench
- T2 – Double Connector, behind suspension strut, right
- T2 – Double Connector, below rear seat bench
- T2 – Double Connector, behind suspension strut, left
- T25 – 25-Pin Connector, on ABS Control Module

THIS PAGE INTENTIONALLY LEFT

BLANK

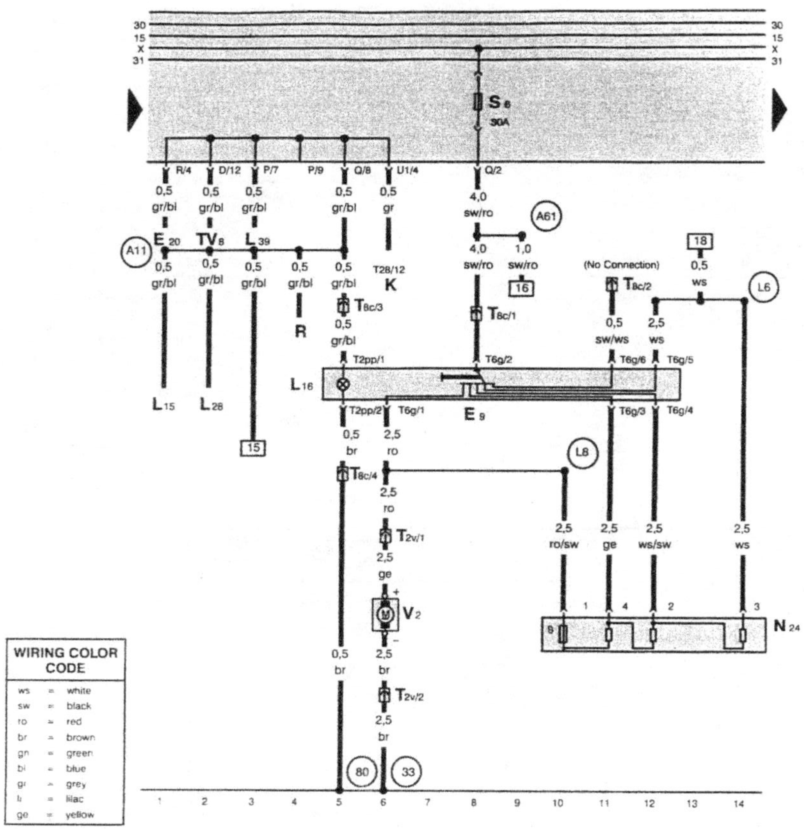

WIRING COLOR CODE		
ws	=	white
sw	=	black
ro	=	red
br	=	brown
gn	=	green
bl	=	blue
gr	=	grey
li	=	lilac
ge	=	yellow

- E9 – Fresh Air Blower Switch
- E20 – Instrument Panel Light Dimmer Switch
- K – Instrument Cluster
- L15 – Ashtray Light
- L16 – Fresh Air Control Level Light
- L28 – Cigarette Lighter Light
- L39 – Rear Window Defogger Switch Light
- N24 – Fresh Air Blower Series Resistance with Fuse
- R – Radio
- T2pp – Double Connector, on E9
- T2v – Double Connector, behind instrument panel, right
- T6g – 6-Pin Connector, on E9
- T8c – 8-Pin Connector, behind instrument panel, center
- T28 – 28-Pin Connector, on instrument cluster
- TV8 – Terminal 58b Wire Connector, above fuse/relay panel
- V2 – Fresh Air Blower

- (33) – Ground connection, behind instrument panel, right
- (80) – Ground connection -1-, in instrument panel wiring harness
- (A11) – Plus connection (58b), in instrument panel wiring harness
- (A61) – Plus connection (X), in instrument cluster wiring harness
- (L6) – Wire Connection -2-, in evaporator housing wiring harness
- (L8) – Wire connection -4-, in evaporator housing wiring harness

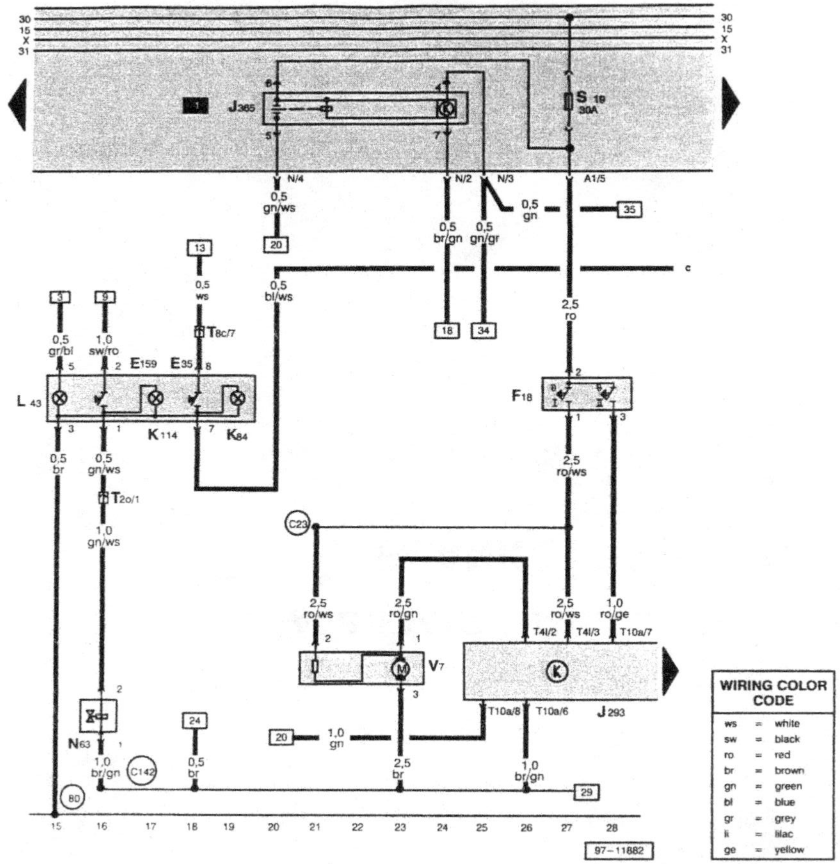

WIRING COLOR CODE		
ws	=	white
sw	=	black
ro	=	red
br	=	brown
gn	=	green
bl	=	blue
gr	=	grey
li	=	lilac
ge	=	yellow

- E35 – A/C Switch
- E159 – Fresh Air/Recirculation Flap Switch
- F18 – Coolant Fan Control (FC) Thermal Switch
- J293 – Coolant Fan Control (FC) Control Module, in left side of engine compartment
- J365 – Relay for A/C shutoff
- K84 – A/C Indicator Light
- K114 – Fresh Air and Recirculating Air Mode Indicator Light
- L43 – A/C Switch Illumination Light
- N63 – Fresh Air/Recirculating Flap Two-Way Valve (00605), in rear of engine compartment
- T2e – Double Connector, behind fuse/relay panel
- T2o – 4-Pin Connector, behind fuse/relay panel
- T4i – 2-Pin Connector, on J293
- T8c – 8-Pin Connector, behind instrument panel, center
- T10a – 10-Pin Connector, on J293
- V7 – Coolant Fan
- (80) – Ground connection –1–, in instrument panel wiring harness
- (C23) – Wire connection (blower, 1st speed) in headlight/air conditioner wiring harness
- (C142) – Wire connection, in Motronic MFI wiring harness

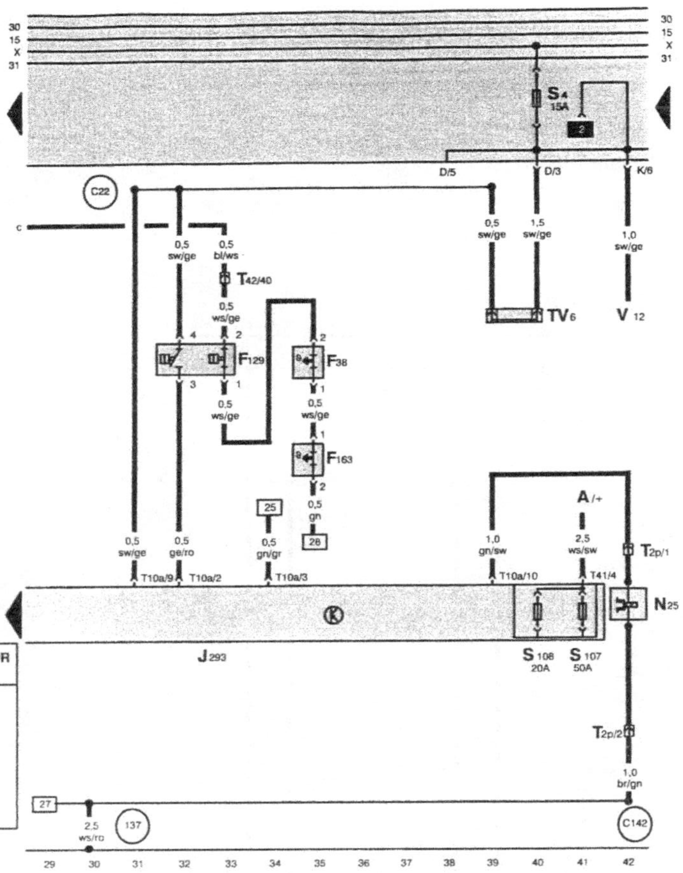

WIRING COLOR CODE		
ws	=	white
sw	=	black
ro	=	red
br	=	brown
gn	=	green
bl	=	blue
gr	=	grey
li	=	lilac
ge	=	yellow

A – Battery
F38 – Ambient Temperature Switch, near horn
F129 – A/C Pressure Switch, in right side of engine compartment
F163 – A/C Cut–Out Thermal Switch, in center of engine compartment
J293 – Coolant FC (Fan Control) Control Module
N25 – A/C Clutch
S107 – Coolant Fan Second and Third Speed Fuse
S108 – Coolant Fan First Speed and A/C Clutch Fuse
T2o – 2-Pin Connector, behind fuse/relay panel
T2p – Double Connector, near compressor

T10a – 10-Pin Connector, on Coolant Fan (FC) Control Module
T4l – 4-Pin Connector, on Coolant Fan (FC) Control Module
TV6 – Terminal X Wire Connector, above fuse/relay panel
V12 – Rear Window Wiper Motor (not applicable to JettaIII)

(137) – Ground connection, in motronic MFI wiring harness

(C22) – Wire connection (X), in headlight/air conditioning wiring harness

(C142) – Wire Connection, in Motronic MFI wiring harness

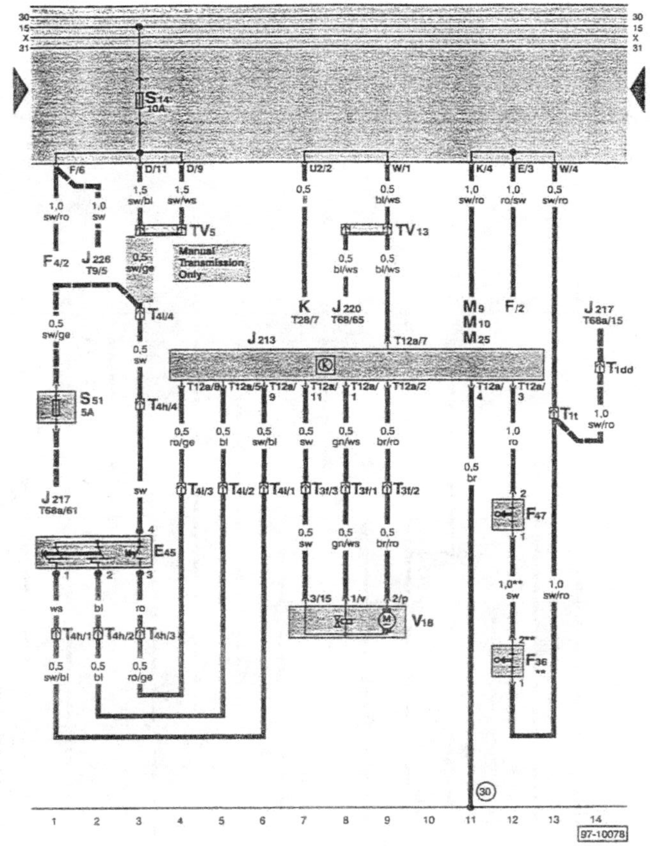

WIRING COLOR CODE		
ws	=	white
sw	=	black
ro	=	red
br	=	brown
gn	=	green
bl	=	blue
gr	=	grey
li	=	lilac
ge	=	yellow

E45 – Cruise Control Switch
F – Brake Light Switch
F4 – Back-Up Light Switch
F36 – Cruise Control Clutch Pedal Position (CCP) Switch
F47 – Cruise Control Brake Pedal Position (BPP) Switch
J213 – Cruise Control, Control Module
J217 – Transmission Control Module (TCM)
J220 – Motronic Engine Control Module (ECM)
J226 – Park/Neutral Position (PNP) Relay
K – Instrument Cluster
M9 – Left Brake Light
M10 – Right Brake Light
M25 – High-Mount Brake Light
S51 – Fuse
T3f – 3-Pin Connector, below left side of dash
T1t – Single Connector, behind fuse/relay panel
T1dd – Single Connector, behind fuse/relay panel

T4h – 4-Pin Connector, behind steering column switch trim
T4i – 4-Pin Connector, behind fuse/relay panel
T9 – 9-Pin Connector, on Park/Neutral Position (PNP) Relay
T12a – 12-Pin Connector, on Cruise Control Module
T28 – 28-Pin Connector, instrument cluster
T68a – 68-Pin Connector, below right front seat, on Transmission Control Module (TCM)
TV13 – Connector, vehicle speed signal
T68 – 68-Pin Connector, under right side of hood, on Motronic Engine Control Module (ECM)
TV5 – Terminal 15a Wire Connector
V18 – Cruise Control Vacuum Pump

(30) – Ground connection –1–, beside fuse/relay panel

– – – – Automatic transmission only

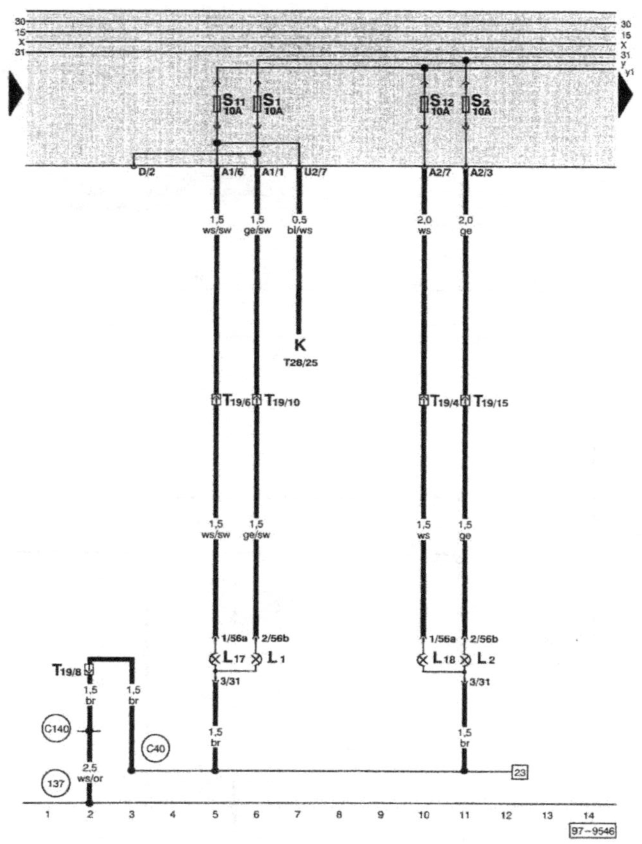

WIRING COLOR CODE

ws	=	white
sw	=	black
ro	=	red
br	=	brown
gn	=	green
bl	=	blue
gr	=	grey
li	=	lilac
ge	=	yellow

K – Instrument Cluster
L1 – Left Headlight
L2 – Right Headlight
L17 – Left Sealed Beam High Headlight
L18 – Right Sealed Beam High Headlight
T19 – 19-Pin Connector, behind radiator grill, center
T28 – 28-Pin Connector, instrument cluster

(132) – Ground connection, in Motronic MFI wiring harness
(C40) – Wire connection, in front light wiring harness
(C140) – Wire connection, in Motronic MFI wiring harness

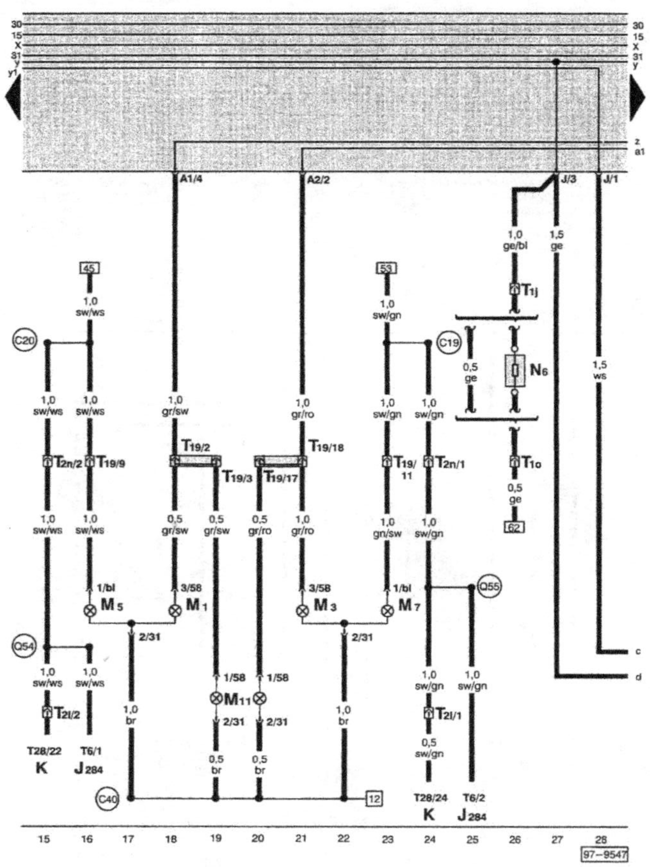

WIRING COLOR CODE

ws	=	white
sw	=	black
ro	=	red
br	=	brown
gn	=	green
bl	=	blue
gr	=	grey
li	=	lilac
ge	=	yellow

J284 – Alarm System Control Module
K – Instrument Cluster
M1 – Left Parking Light
M3 – Right Parking Light
M5 – Left Front Turn Signal Light
M7 – Right Front Turn Signal Light
M11 – Front Side Marker Lights
N6 – Series Resistance
T1j – Single Connector, behind fuse/relay panel
T1o – Single Connector, behind fuse/relay panel
T2l – Double Connector, behind fuse/relay panel
T2n – Double Connector, behind fuse/relay panel
T6 – 6-Pin Connector, behind dash panel, left, on Alarm System Control Module
T19 – 19-Pin Connector, behind radiator grill, center
T28 – 28-Pin Connector, instrument cluster

(C19) – Wire connection (right turn signal), in headlight wiring harness
(C20) – Wire connection (left turn signal), in headlight wiring harness
(C40) – Wire connection, in front light wiring harness
(Q54) – Wire connection (left turn signal), in alarm system wiring harness
(Q55) – Wire connection (right turn signal), in alarm system wiring harness

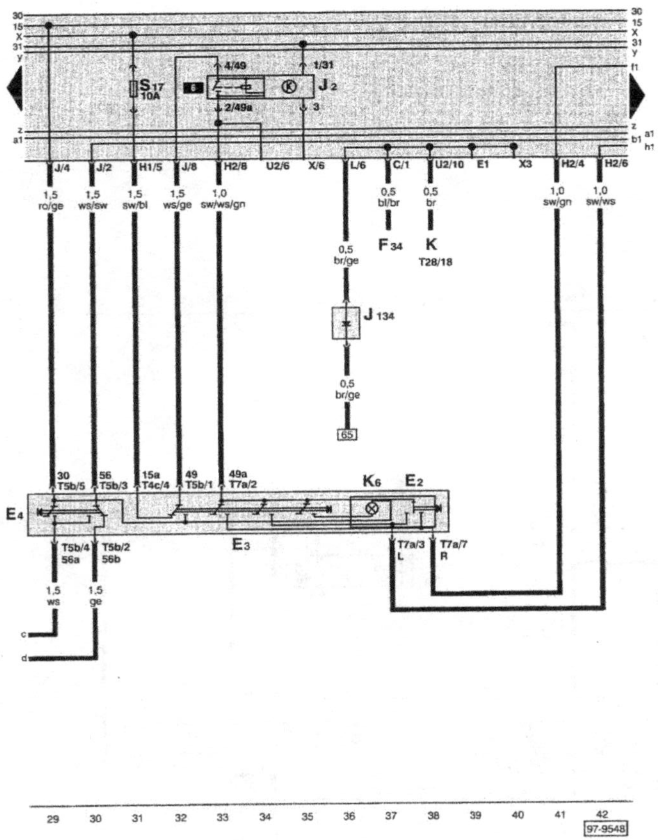

WIRING COLOR CODE		
ws	=	white
sw	=	black
ro	=	red
br	=	brown
gn	=	green
bl	=	blue
gr	=	grey
li	=	lilac
ge	=	yellow

- E2 – Turn Signal Switch
- E3 – Emergency Flasher Switch
- E4 – Headlight Dimmer/Flasher Switch
- F34 – Brake Fluid Level Warning Switch
- J2 – Emergency Flasher Relay
- J134 – Daytime Running Light Diode
- K – Instrument Cluster
- K6 – Emergency Flasher Warning Light
- T4c – 4-Pin Connector, behind steering column switch trim
- T5b – 5-Pin Connector, behind steering column switch trim
- T7a – 7-Pin Connector, behind steering column switch trim
- T28 – 28-Pin Connector, instrument cluster

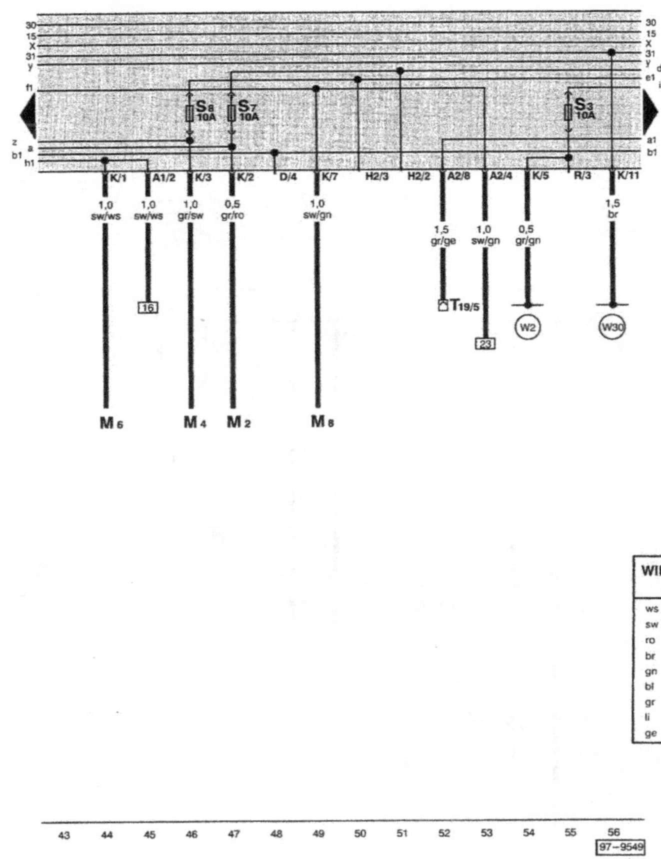

WIRING COLOR CODE		
ws	=	white
sw	=	black
ro	=	red
br	=	brown
gn	=	green
bl	=	blue
gr	=	grey
li	=	lilac
ge	=	yellow

- M2 – Right Tail Light
- M4 – Left Tail Light
- M6 – Left Rear Turn Signal Light
- M8 – Right Rear Turn Signal Light
- T19 – 19-Pin Connector, behind radiator grill, center
- (W2) – Plus connection (58), in rear wiring harness
- (W30) – Wire connection, in rear wiring harness

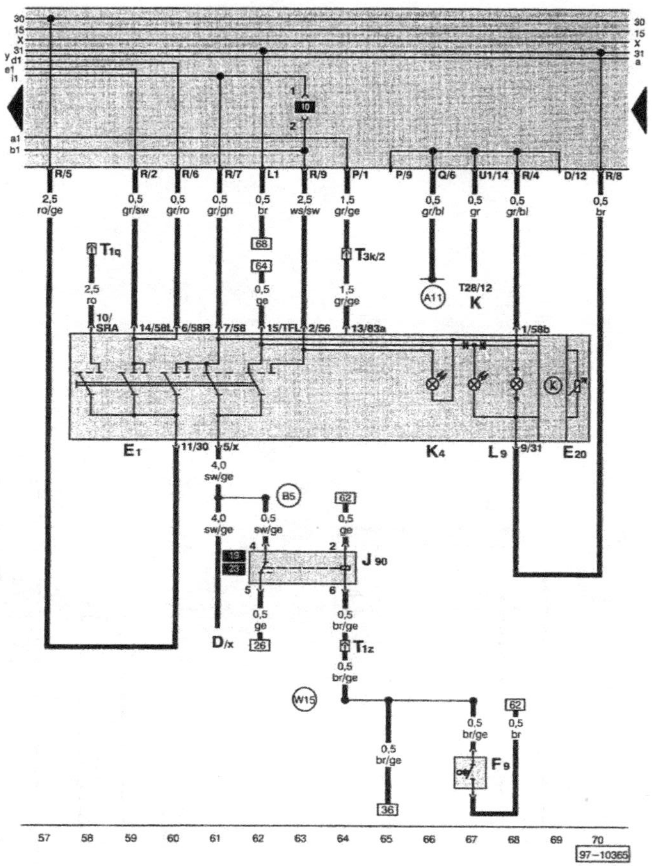

WIRING COLOR CODE

ws	=	white
sw	=	black
ro	=	red
br	=	brown
gn	=	green
bl	=	blue
gr	=	grey
li	=	lilac
ge	=	yellow

D – Ignition/Starter Switch
E1 – Light Switch
E20 – Instrument Panel Light Dimmer Switch
F9 – Parking Brake Warning Light Switch
J90 – Daytime Running Lights Switch On Relay
K – Instrument Cluster
K4 – Park Light Indicator Light
L9 – Headlight Switch Light
T1q – Single Connector, behind fuse/relay panel
T1z – Single Connector, behind fuse/relay panel
T3k – 3-Pin Connector, behind fuse/relay panel
T28 – 28-Pin Connector, instrument cluster

(A11) – Plus connection (58b), in instrument panel wiring harness

(B5) – Wire connection (X), in headlight switch wiring harness

(W15) – Wire connection (HB) in rear wiring harness

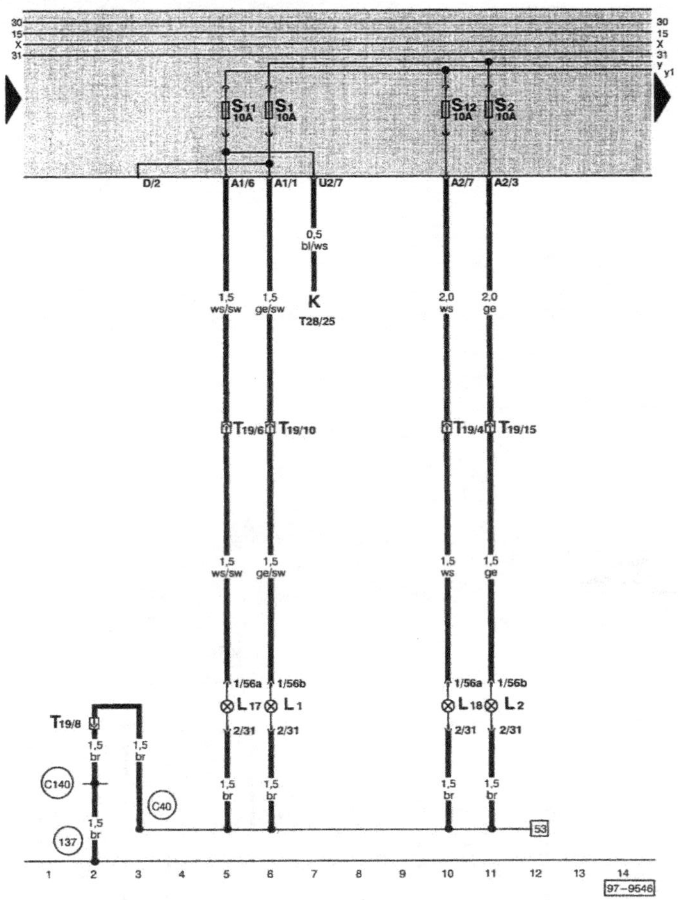

WIRING COLOR CODE

ws	=	white
sw	=	black
ro	=	red
br	=	brown
gn	=	green
bl	=	blue
gr	=	grey
li	=	lilac
ge	=	yellow

K – Instrument Cluster
L1 – Left Headlight
L2 – Right Headlight
L17 – Left Sealed Beam High Beam Headlight
L18 – Right Sealed Beam High Beam Headlight
T19 – 19-Pin Connector, behind radiator grill, center
T28 – 28-Pin Connector, instrument cluster

(C40) – Wire connection, in front light wiring harness

(C140) – Wire connection, in Motronic MFI wiring harness

(137) – Ground connection, in Motronic MFI wiring harness

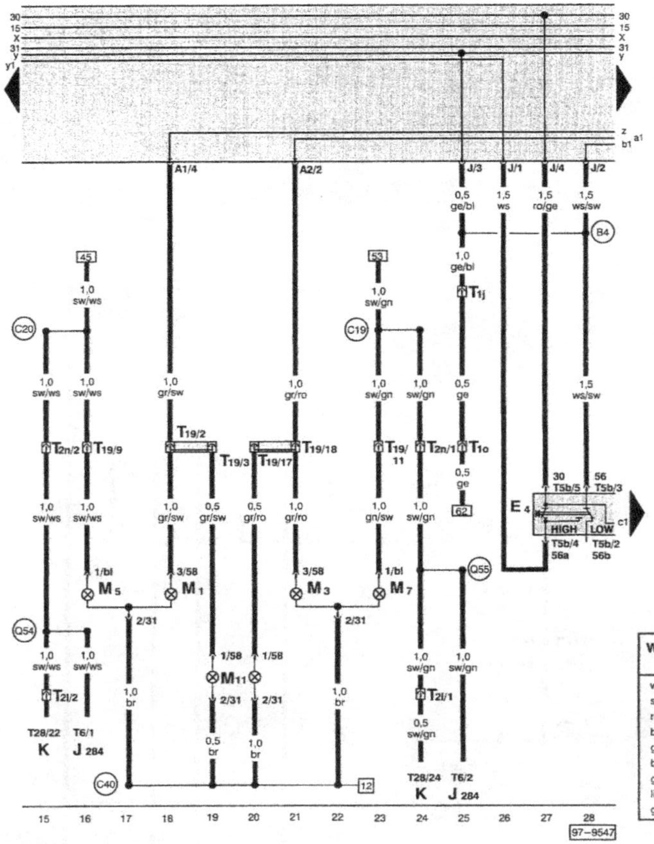

WIRING COLOR CODE		
ws	=	white
sw	=	black
ro	=	red
br	=	brown
gn	=	green
bl	=	blue
gr	=	grey
li	=	lilac
ge	=	yellow

- E4 – Headlight Dimmer/Flasher Switch
- J284 – Alarm System Control Module
- K – Instrument Cluster
- M1 – Left Parking Light
- M3 – Right Parking Light
- M5 – Left Front Turn Signal Light
- M7 – Right Front Turn Signal Light
- M11 – Front Side Marker Lights
- T1j – Single Connector, behind fuse/relay panel
- T1o – Single Connector, behind fuse/relay panel
- T2l – Double Connector, behind fuse/relay panel
- T2n – Double Connector, behind fuse/relay panel
- T5b – 5-Pin Connector, behind steering column switch trim
- T6 – 6-Pin Connector, behind dash panel, left, on Alarm System Control Module
- T19 – 19-Pin Connector, behind radiator grill, center
- T28 – 28-Pin Connector, instrument cluster
- (B4) – Plus connection, (58e) in light switch wiring harness

- (C19) – Wire connection (right turn signal), in headlight wiring harness
- (C20) – Wire connection (left turn signal), in headlight wiring harness
- (C40) – Wire connection, in front light wiring harness
- (Q54) – Wire connection (left turn signal), in alarm system wiring harness
- (Q55) – Wire connection (right turn signal), in alarm system wiring harness

1996-1999

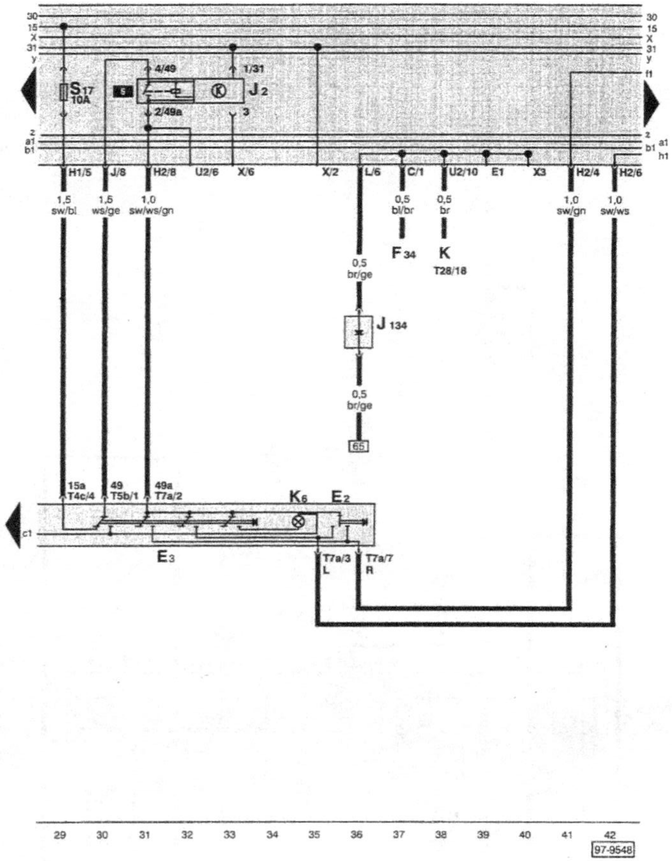

WIRING COLOR CODE		
ws	=	white
sw	=	black
ro	=	red
br	=	brown
gn	=	green
bl	=	blue
gr	=	grey
li	=	lilac
ge	=	yellow

- E2 – Turn Signal Switch
- E3 – Emergency Flasher Switch
- F34 – Brake Fluid Level Warning Switch
- J2 – Emergency Flasher Relay
- J134 – Daytime Running Light Diode
- K – Instrument Cluster
- K6 – Emergency Flasher Warning Light
- T4c – 4-Pin Connector, behind steering column switch trim
- T5b – 5-Pin Connector, behind steering column switch trim
- T7a – 7-Pin Connector, behind steering column switch trim
- T28 – 28-Pin Connector, instrument cluster

1996-1999

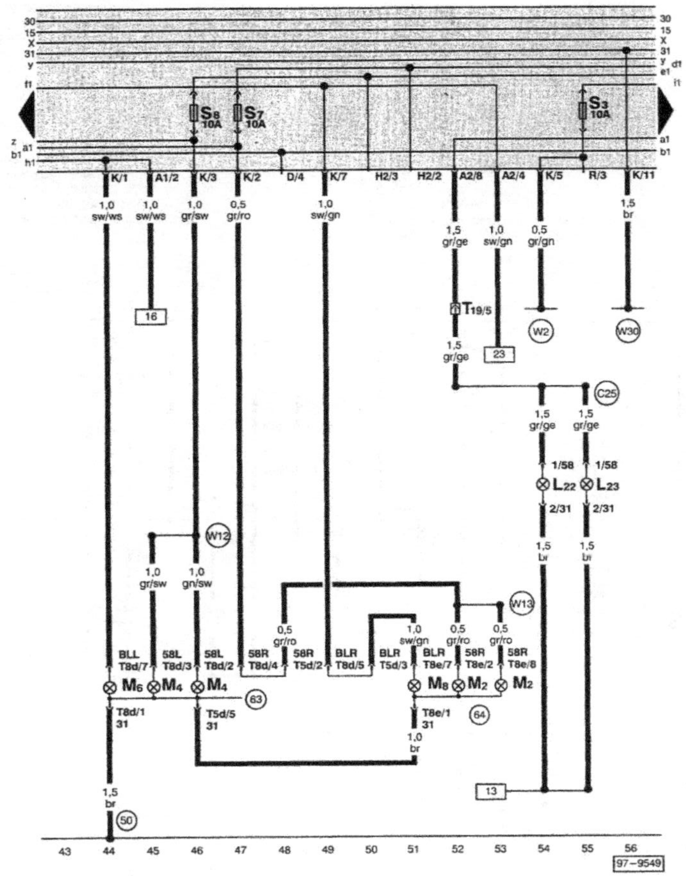

WIRING COLOR CODE

ws	=	white
sw	=	black
ro	=	red
br	=	brown
gn	=	green
bl	=	blue
gr	=	grey
li	=	lilac
ge	=	yellow

L22 – Left Front Fog Light
L23 – Right Front Fog Light
M2 – Right Tail Light
M4 – Left Tail Light
M6 – Left Rear Turn Signal Light
M8 – Right Rear Turn Signal Light
T5d – 5-Pin Connector, on tail light, left
T8d – 8-Pin Connector, on tail light, left
T8e – 8-Pin Connector, on tail light, right
T19 – 19-Pin Connector, behind radiator grill, center
(50) – Ground connection, in luggage compartment, left
(63) – Ground connection, on left tail light bulb holder
(64) – Ground connection, on right tail light bulb holder
(198) – Ground connection, in front light wiring harness
(C25) – Wire connection (fog lights), in front left wiring harness
(W2) – Plus connection (58), in rear wiring harness
(W12) – Wire connection (DWA), in rear wiring harness
(W13) – Wire connection (58L), in rear wiring harness
(W30) – Wire connection, in rear wiring harness

1996-1999

Daytime running lights
(GTI)

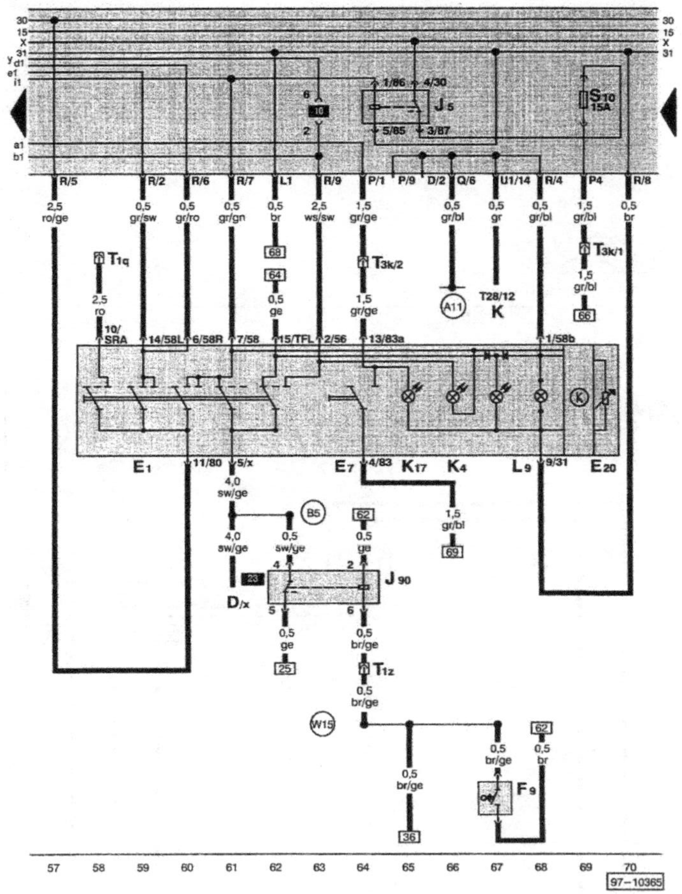

WIRING COLOR CODE

ws	=	white
sw	=	black
ro	=	red
br	=	brown
gn	=	green
bl	=	blue
gr	=	grey
li	=	lilac
ge	=	yellow

D – Ignition/Starter Switch
E1 – Light Switch
E7 – Fog Light Switch
E20 – Instrument Panel Light Dimmer Switch
F9 – Parking Brake Warning Light Switch
J5 – Fog Light Relay
J90 – Daytime Running Lights Switch-on Relay
K – Instrument Cluster
K4 – Park Light Indicator Light
K17 – Fog Light Indicator Light
L9 – Headlight Switch Light
T1q – Single Connector, behind fuse/relay panel
T1z – Single Connector, behind fuse/relay panel
T3k – Double Connector, behind fuse/relay panel
T28 – 28-Pin Connector, instrument cluster
(A11) – Plus connection (58b), in instrument panel wiring harness
(B5) – Wire connection (X), in light switch wiring harness
(W15) – Wire connection (HB), in rear wiring harness

1996-1999

Power sunroof, interior light with delay and reading light (1996)

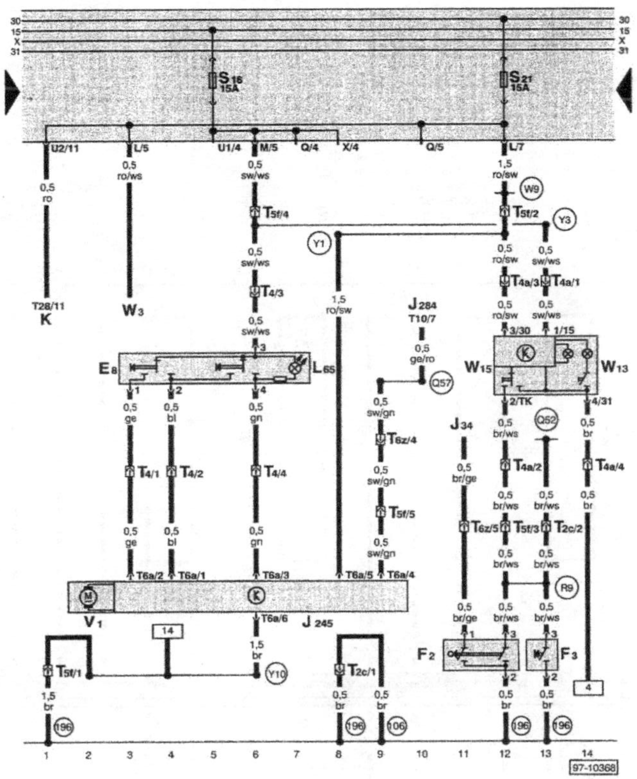

WIRING COLOR CODE		
ws	=	white
sw	=	black
ro	=	red
br	=	brown
gn	=	green
bl	=	blue
gr	=	grey
li	=	lilac
ge	=	yellow

E8 – Sunroof Switch
F2 – Left Front Door Contact Switch
F3 – Right Front Door Contact Switch
J34 – Seat Belt Warning System Relay
J245 – Power Sunroof Control Module
J284 – Alarm System Control Module
K – Instrument Cluster
L65 – Sunroof Switch Light
T1c – Single Connector, behind fuse/relay panel
T1m – Single Connector, behind fuse/relay panel
T2c – Double Connector, on lower right A-pillar
T4 – 4-Pin Connector
T4a – 4-Pin Connector
T5f – 5-Pin Connector, above fuse/relay panel
T6a – 6-Pin Connector, behind interior light, on Power Sunroof Control Module
T6z – 6-Pin Connector, behind dash panel
T10 – 10-Pin Connector, behind dash panel, left, on Alarm System Control Module
T28 – 28-Pin Connector, instrument cluster
V1 – Sunroof Motor
W3 – Luggage Compartment Light
W13 – Right Front Map/Reading Light
W15 – Interior Light With Delay Switch
(106) – Ground connection, in alarm system wiring harness
(196) – Wire connection, in rear wiring harness
(Q52) – Wire connection (door contact switch), in alarm system wiring harness
(Q57) – Wire connection (closed), in alarm system wiring harness
(R9) – Wire connection (door contact switch), in rear wiring harness
(W9) – Plus connection (30), in rear wiring harness
(Y1) – Plus connection (30), in interior light wiring harness
(Y3) – Plus connection (15) in interior light/sunroof wiring harness
(Y10) – Wire connection, in interior light wiring harness

THIS PAGE INTENTIONALLY LEFT

BLANK

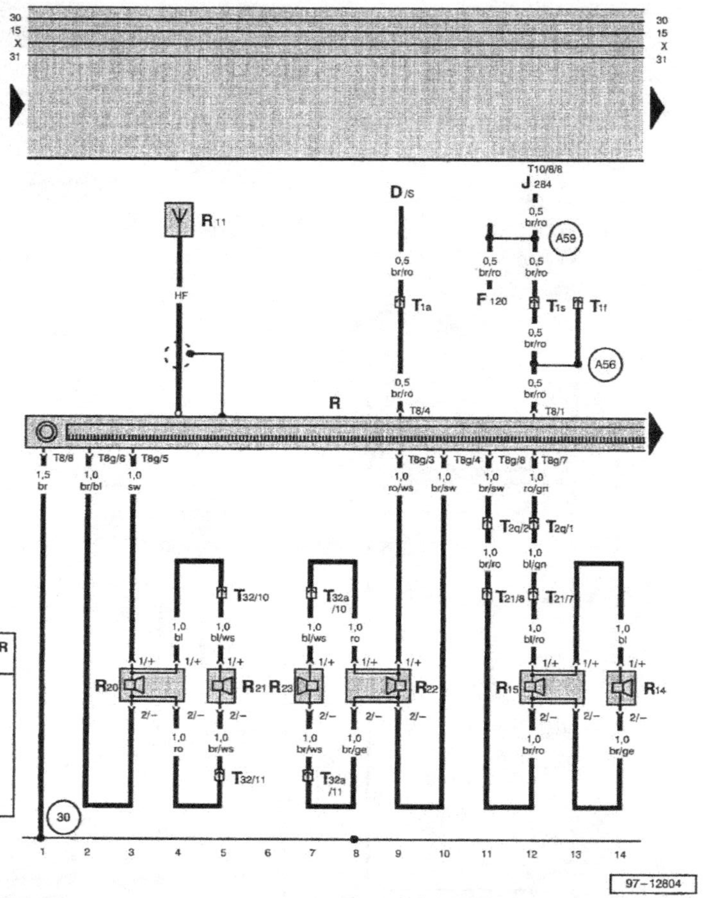

WIRING COLOR CODE

ws	=	white
sw	=	black
ro	=	red
br	=	brown
gn	=	green
bl	=	blue
gr	=	grey
li	=	lilac
ge	=	yellow

D – Ignition/Starter Switch
F120 – Hood Alarm Switch
J284 – Alarm System Control Module
R – Radio
R11 – Antenna
R14 – Left Reat Tweeter
R15 – Left Rear Woofer
R20 – Left Front Tweeter
R21 – Left Front Woofer
R22 – Right Front Tweeter
R23 – Right Front Woofer,
T1a – Signal Connector, behind fuse/relay panel
T1f – Signal Connector, behind dash panel, center, behind radio
T1s – Signal Connector, behind fuse/relay panel
T2q – Double Connector, on lower left A-pillar
T8 – 8-Pin Connector, on radio (Black)
T8g – 8-Pin Connector, on radio (Brown)
T21 – 21-Pin Connector, on left B-pillar
T32 – 32-Pin Connector, on A-pillar, lower left
T32a – 32-Pin Connector, on A-pillar, lower right
(30) – Ground connection –1–, beside fuse/relay panel
(A56) – Wire connection –2–, (30) in instrument panel wiring harness
(A59) – Wire connection (sensor), in alarm system wiring harness

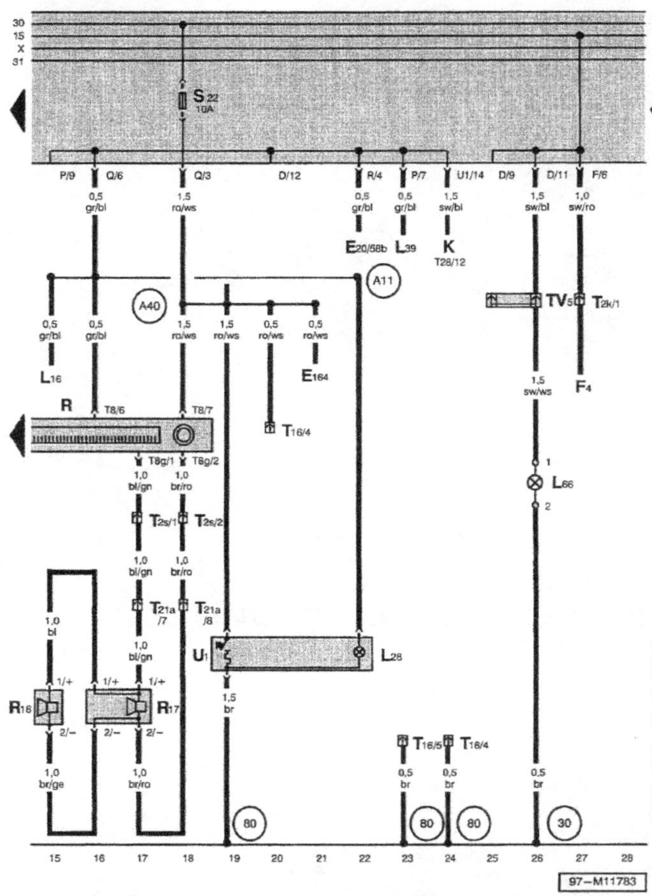

WIRING COLOR CODE

ws	=	white
sw	=	black
ro	=	red
br	=	brown
gn	=	green
bl	=	blue
gr	=	grey
li	=	lilac
ge	=	yellow

E20 – Instrument Panel Light Dimmer Switch
E164 – Luggage Compartment Release Switch
F4 – Back-Up Light Switch
K – Instrument Cluster
L16 – Fresh Air Control Lever Light
L28 – cigarette Lighter Light
L39 – Rear Window Defogger Switch Light
L66 – Cassette Storage Light
R – Radio
R16 – Right Rear Tweeter
R17 – Right Rear Woofer
T2k – Double Connector, behind fuse/relay panel
T2s – Double Connector, on lower right A-pillar
T8 – 8-Pin Connector, on radio
T8g – 8-Pin Connector, on radio
T16 – Data Link Connector (DLC), behind center of dash panel
T21a – 21-Pin Connector, on right B-pillar
T28 – 28-Pin Connector, instrument cluster
TV5 – Terminal 15a Wire Connector (Green)
U1 – Cigarette Lighter
(30) – Ground connection –1–, beside fuse/relay panel
(80) – Ground connection –1–, in instrument panel wiring harness
(A11) – Plus connection (58b), in instrument panel wiring harness
(A40) – Plus connection (30), in instrument panel wiring harness

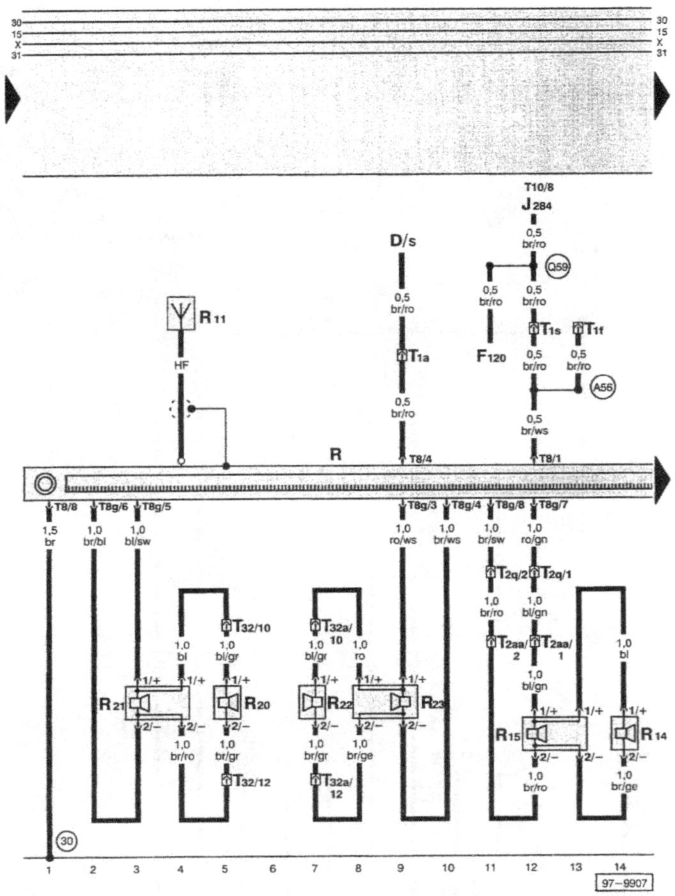

WIRING COLOR CODE

ws	=	white
sw	=	black
ro	=	red
br	=	brown
gn	=	green
bl	=	blue
gr	=	grey
li	=	lilac
ge	=	yellow

- D – Ignition/Starter Switch
- F120 – Hood Alarm Switch
- J284 – Alarm System Control Module
- R – Radio
- R11 – Antenna
- R14 – Left Rear Tweeter
- R15 – Left Rear Woofer
- R20 – Left Front Tweeter
- R21 – Left Front Woofer
- R22 – Right Front Tweeter
- R23 – Right Front Woofer
- T1a – Single Connector, behind fuse/relay panel
- T1f – Single Connector, behind dash panel, center, behind radio
- T1s – Single Connector, behind fuse/relay panel
- T2aa – Double Connector, on left B-pillar
- T2q – Double Connector, on lower left A-pillar
- T8 – 8-Pin Connector, on radio
- T8g – 8-Pin Connector, on radio
- T10 – 10-Pin Connector, behind dash panel, left, on Alarm System Control Module
- T32 – 32-Pin Connector, on A-pillar, lower left
- T32a – 32-Pin Connector, on A-pillar, lower right
- (30) – Ground connection –1–, beside fuse/relay panel
- (A56) – Wire connection –2– (30), in instrument panel wiring harness
- (Q59) – Wire connection (sensor) in alarm system wiring harness

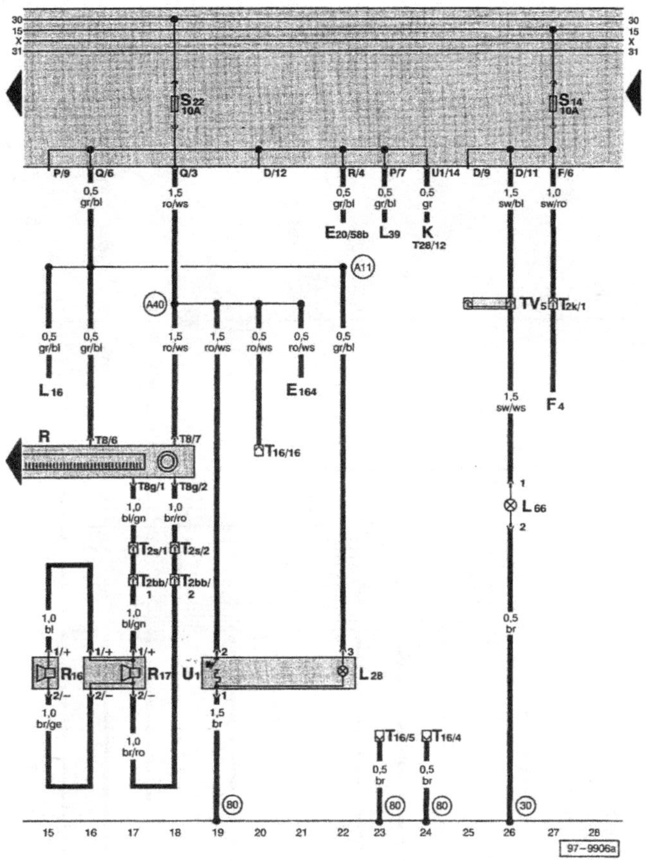

WIRING COLOR CODE

ws	=	white
sw	=	black
ro	=	red
br	=	brown
gn	=	green
bl	=	blue
gr	=	grey
li	=	lilac
ge	=	yellow

- E20 – Instrument Panel Light Dimmer Switch
- E164 – Luggage Compartment Release Switch
- F4 – Back-Up Light Switch
- K – Instrument Cluster
- L16 – Fresh Air Control Lever Light
- L28 – Cigarette Lighter Light
- L39 – Rear Window Defogger Switch Light
- L66 – Cassette Storage Light
- R – Radio
- R16 – Right Rear Tweeter
- R17 – Right Rear Woofer
- T16 – 16-Pin Connector, Data Link Connector (DLC), behind center of dash panel
- T2bb – Double Connector, on right B-pillar
- T2k – Double Connector, behind fuse/relay panel
- T2s – Double Connector, on lower right A-pillar
- T8 – 8-Pin Connector, on radio
- T8g – 8-Pin Connector, on radio
- T28 – 28-Pin Connector, instrument cluster
- TV5 – Terminal 15a Wire Connector (Green)
- U1 – Cigarette Lighter
- (30) – Ground connection –1–, beside fuse/relay panel
- (80) – Ground connection –1–, in instrument panel wiring harness
- (A11) – Plus connection (58b), in instrument panel wiring harness
- (A40) – Plus connection (30), in instrument panel wiring harness

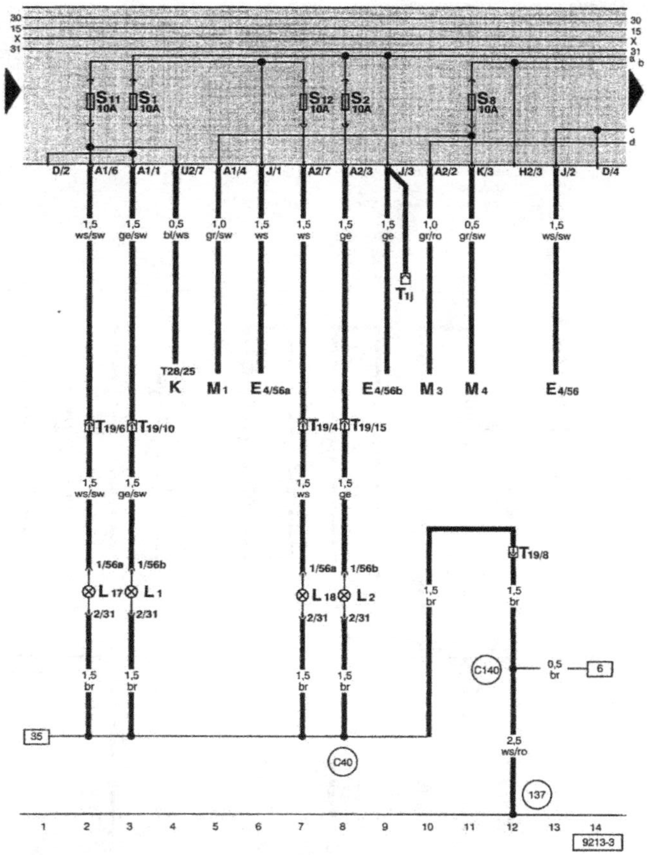

WIRING COLOR CODE		
ws	=	white
sw	=	black
ro	=	red
br	=	brown
gn	=	green
bl	=	blue
gr	=	grey
li	=	lilac
ge	=	yellow

E4 – Headlight Dimmer/Flasher Switch
K – Instrument Cluster
L1 – Left Headlight
L2 – Right Headlight
L17 – Left Sealed Beam High Beam Headlight
L18 – Right Sealed Beam High Beam Headlight
M1 – Left Parking Light
M3 – Right Parking Light
M4 – Left Tail Light
T1j – Single Connector, behind fuse/relay panel
T19 – 19-Pin Connector, behind radiator grill, center
T28 – 28-Pin Connector, instrument cluster
C40 – Wire connection, in front lighting wiring harness
C140 – Wire connection, in Motronic MFI wiring harness
137 – Ground connection, in Motronic MFI wiring harness

1996-1999

Headlights and fog lights
(GTI, USA only)

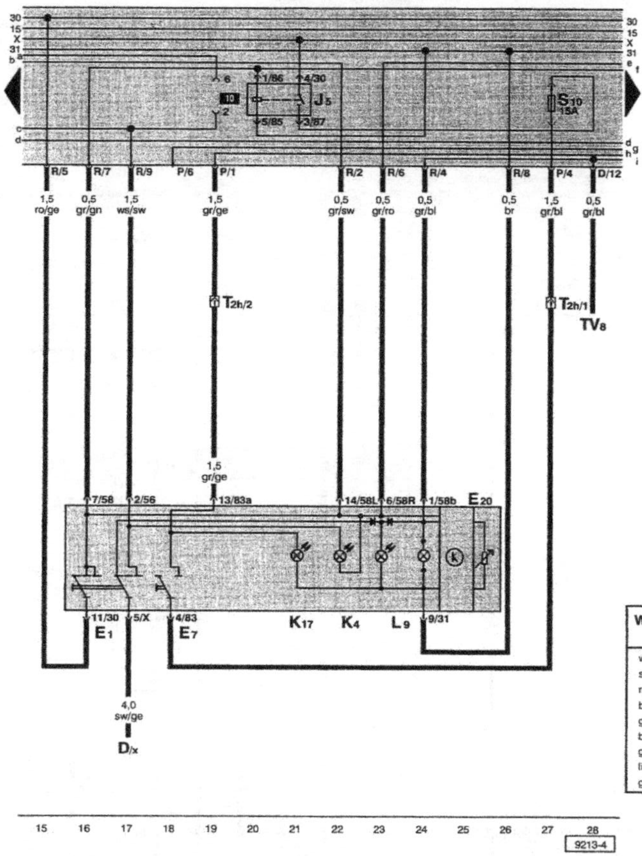

WIRING COLOR CODE		
ws	=	white
sw	=	black
ro	=	red
br	=	brown
gn	=	green
bl	=	blue
gr	=	grey
li	=	lilac
ge	=	yellow

D – Ignition Switch
E1 – Light Switch
E7 – Fog Light Switch
E20 – Instrument Panel Light Dimmer Switch
J5 – Fog Light Relay
K4 – Park Light Indicator Light
K17 – Fog Light Indicator Light
L9 – Headlight Switch Light
T2h – 2-Pin Connector, behind fuse/relay panel
TV8 – Terminal 58b Wire Connector

1996-1999

Headlights and fog lights
(GTI, USA only)

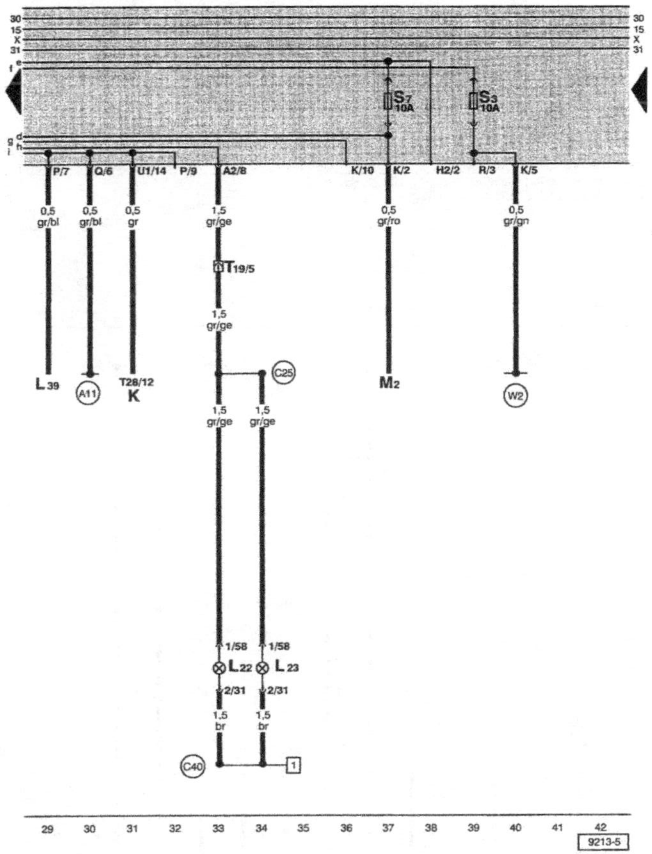

WIRING COLOR CODE		
ws	=	white
sw	=	black
ro	=	red
br	=	brown
gn	=	green
bl	=	blue
gr	=	grey
li	=	lilac
ge	=	yellow

K – Instrument Cluster
L22 – Left Front Fog Light
L23 – Right Front Fog Light
L39 – Rear Window Defogger Switch Light
M2 – Right Tail Light
T19 – 19-Pin Connector, behind radiator grill, center
T28 – 28-Pin Connector, instrument cluster

(A11) – Plus connection (58b), in instrument panel wiring harness

(C25) – Wire connection (fog lights), in front left wiring harness

(C40) – Wire connection, in front lighting wiring harness

(W2) – Plus connection (58), in rear wiring harness

Heated washer nozzles

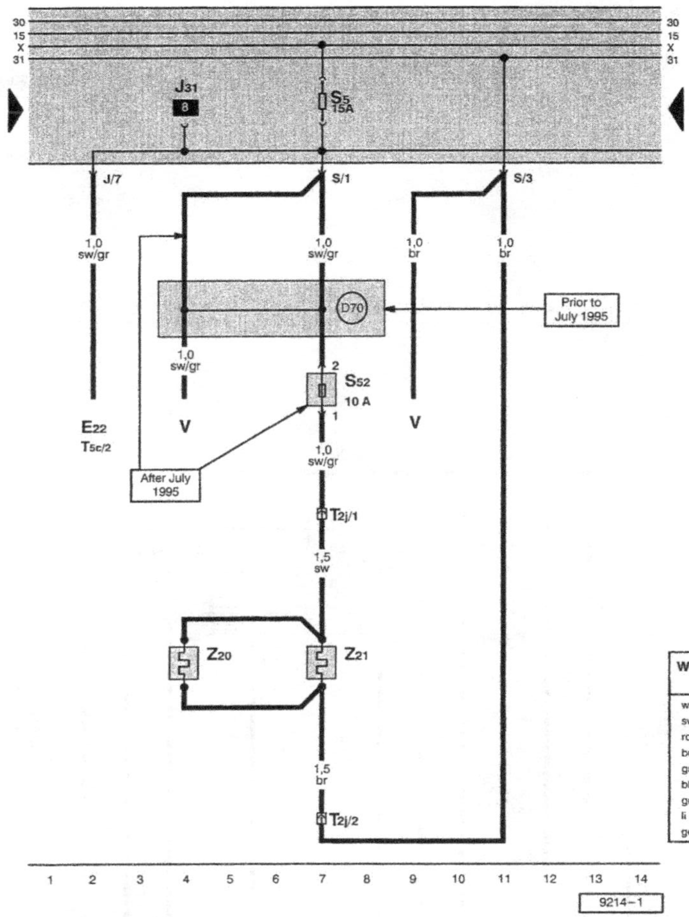

WIRING COLOR CODE		
ws	=	white
sw	=	black
ro	=	red
br	=	brown
gn	=	green
bl	=	blue
gr	=	grey
li	=	lilac
ge	=	yellow

E22 – Windshield Wiper Intermittent Switch
J31 – Washer/Wiper Intermittent Relay
S52 – Fuse
T2j – Double Connector, behind fuse/relay panel
T5c – 5-Pin Connector, behind steering column switch trim
V – Windshield Wiper Motor
Z20 – Washer Nozzle Heater, Left
Z21 – Washer Nozzle Heater, Right

(D70) – Wire connection (53a), in engine compartment wiring harness

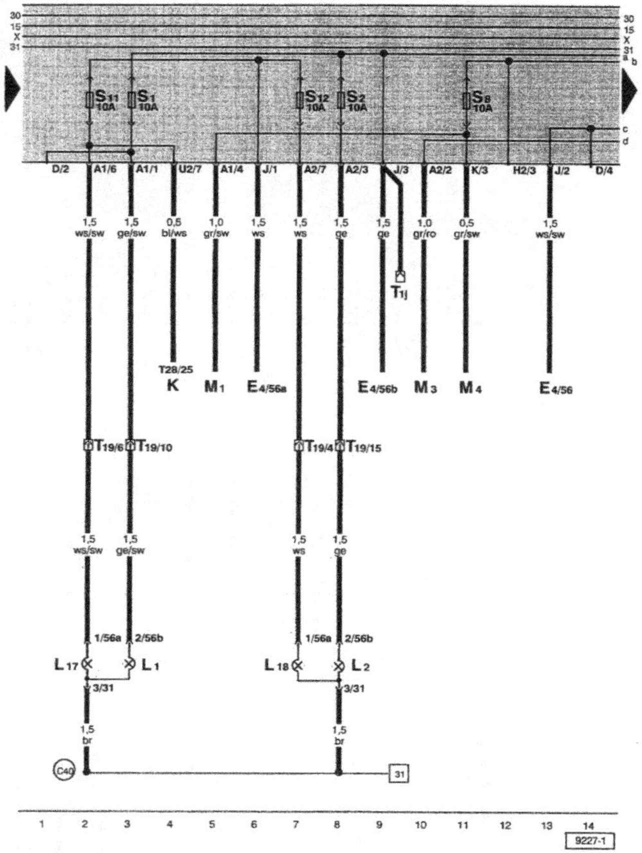

WIRING COLOR CODE		
ws	=	white
sw	=	black
ro	=	red
br	=	brown
gn	=	green
bl	=	blue
gr	=	grey
li	=	lilac
ge	=	yellow

- E4 – Headlight Dimmer/Flasher Switch
- K – Instrument Cluster
- L1 – Left Headlight
- L2 – Right Headlight
- L17 – Left Sealed Beam High Beam Headlight
- L18 – Right Sealed Beam High Beam Headlight
- M1 – Left Parking Light
- M3 – Right Parking Light
- M4 – Left Tail Light
- T1j – Single Connector, behind fuse/relay panel
- T19 – 19-Pin Connector, behind radiator grill, center
- T28 – 28-Pin Connector, on instrument cluster
- (198) – Ground connection, in front light wiring harness

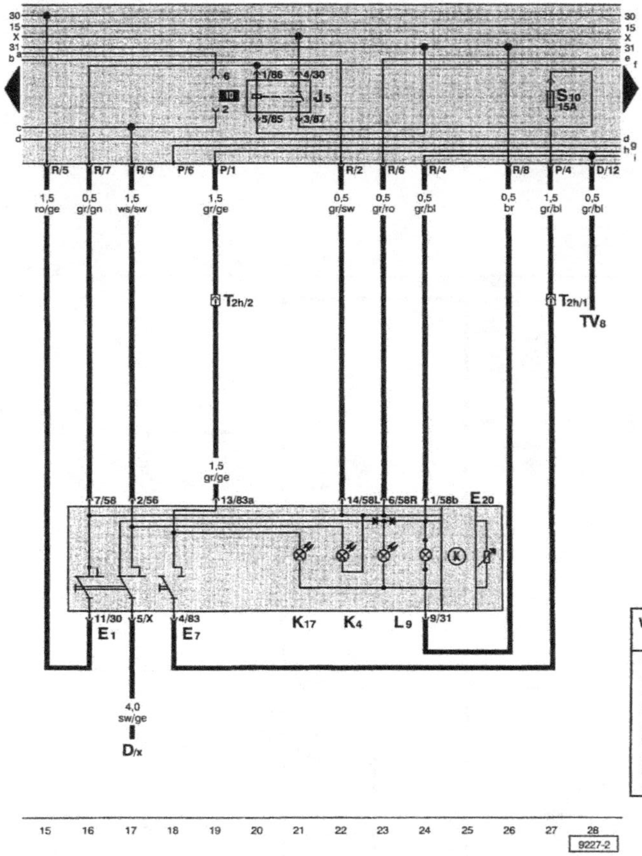

WIRING COLOR CODE		
ws	=	white
sw	=	black
ro	=	red
br	=	brown
gn	=	green
bl	=	blue
gr	=	grey
li	=	lilac
ge	=	yellow

- D – Igntion/Starter Switch
- E1 – Light Switch
- E7 – Fog Light Switch
- E20 – Instrument Panel Light Dimmer Switch
- J5 – Fog Light Relay
- K4 – Park Light Indicator Light
- K17 – Fog Light Indicator Light
- L9 – Headlight Switch Light
- T2h – 2-Pin Connector, behind fuse/relay panel
- TV8 – 8-Pin Connector, terminal 58b

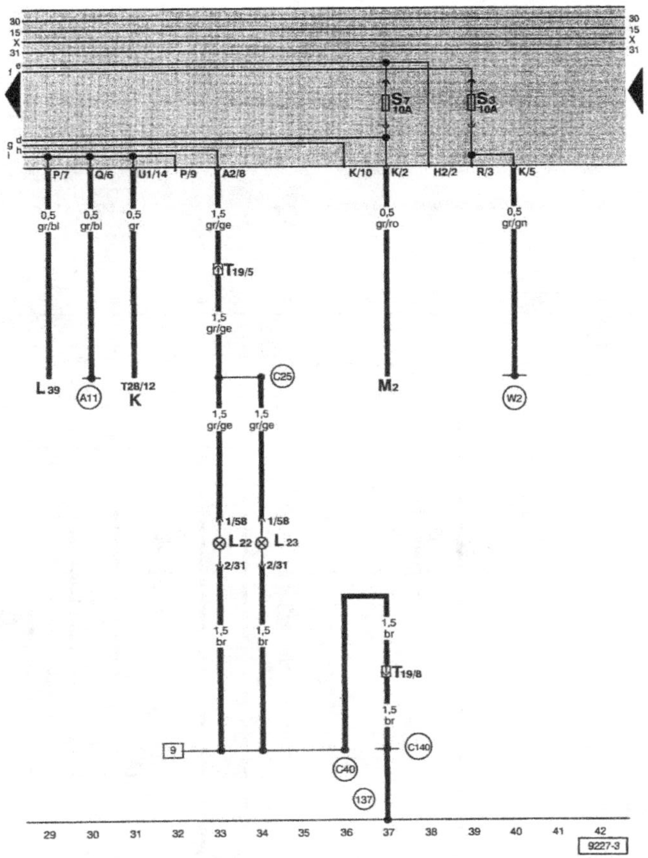

WIRING COLOR CODE		
ws	=	white
sw	=	black
ro	=	red
br	=	brown
gn	=	green
bl	=	blue
gr	=	grey
li	=	lilac
ge	=	yellow

K – Instrument Cluster
L22 – Left Front Fog Light
L23 – Right Front Fog Light
L39 – Rear Window Defogger Switch Light
M2 – Right Tail Light
T19 – 19-Pin Connector, behind radiator grill, center
T28 – 28-Pin Connector, on instrument cluster
(137) – Ground connection, in Motronic MFI wiring harness
(A11) – Plus connection (58b), in instrument panel wiring harness
(C25) – Wire connection (fog lights), in front left wiring harness
(C40) – Wire connection, in front light wiring harness
(C140) – Wire connection, in Motronic MFI wiring harness
(W2) – Plus connection (58), in rear wiring harness

1996-1999

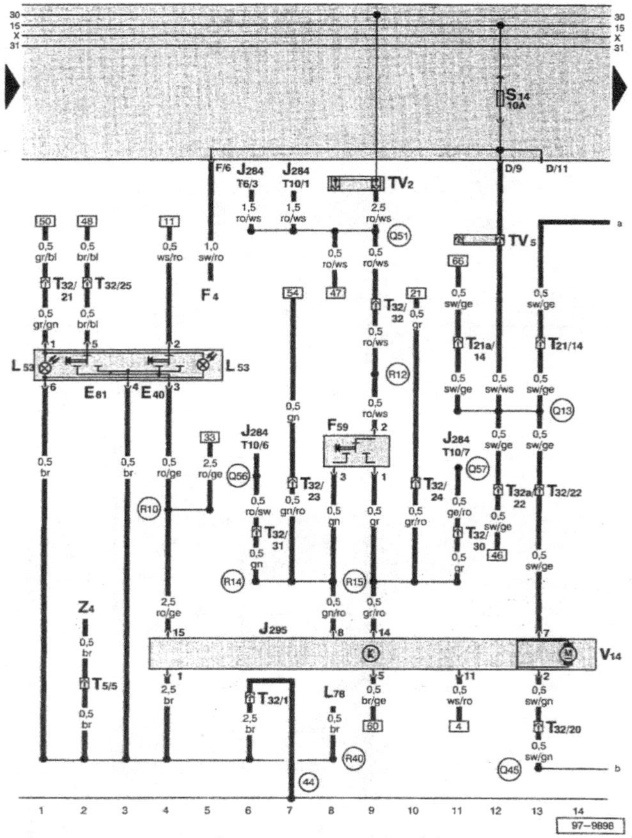

WIRING COLOR CODE		
ws	=	white
sw	=	black
ro	=	red
br	=	brown
gn	=	green
bl	=	blue
gr	=	grey
li	=	lilac
ge	=	yellow

E40 – Left Front Window Switch
E81 – Right Front Window Switch (In Driver's Door)
F4 – Back-Up Light Switch
F59 – Driver's Door Central Locking System Door Switch
J284 – Alarm System Control Module
J295 – Left Front Power Window Control Module
L53 – Power Window Switch Light
L78 – Mirror Adjusting Switch Light
T6 – 6-Pin Connector, behind dash panel, left, on Alarm System Control Module
T10 – 10-Pin Connector, behind dash panel, left, on Alarm System Control Module
T21 – 21-Pin Connector, on left B-pillar
T21a – 21-Pin Connector, on right B-pillar
T32 – 32-Pin Connector, on A-pillar, lower left
T32a – 32-Pin Connector, on A-pillar, lower right
TV2 – Terminal 30 (B+) Wire Connector
TV5 – Terminal 15a Wire Connector
V14 – Left Window Motor
Z4 – Driver's Side Heated Mirror
(44) – Ground connection, on left A-pillar, lower part
(Q45) – Wire connection (comfort), in power window wiring harness
(Q51) – Plus connection (30), in alarm system wiring harness
(Q56) – Wire connection, (open), in alarm system wiring harness
(Q57) – Wire connection, (closed), in alarm system wiring harness
(R10) – Plus connection –1– (30), in driver's door wiring harness
(R12) – Plus connection –3– (30), in driver's door wiring harness
(R14) – Wire connection –1– (open), in driver's door wiring harness
(R15) – Wire connection –2– (closed), in driver's door wiring harness
(R40) – Wire connection, in driver's door wiring harness

1996-1999

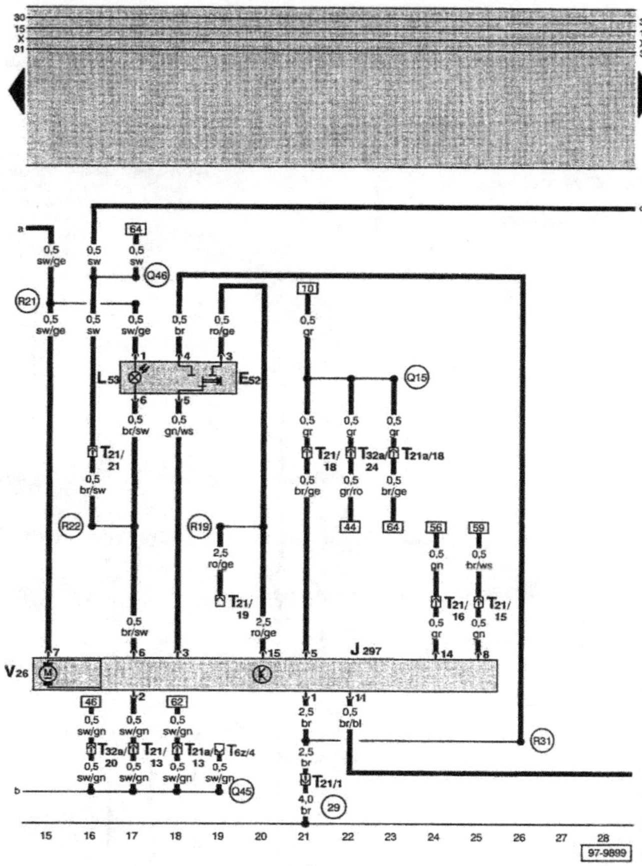

WIRING COLOR CODE		
ws	=	white
sw	=	black
ro	=	red
br	=	brown
gn	=	green
bl	=	blue
gr	=	grey
li	=	lilac
ge	=	yellow

E52 – Left Rear Window Switch, (In LR Door)
J297 – Left Rear Power Window Control Module
L53 – Power Window Switch Light
T1a – Single Connector, behind fuse/relay panel
T6z – 6-Pin Connector, behind dash panel
T21 – 21-Pin Connector, on left B-pillar
T21a – 21-Pin Connecter, on right B-pillar
T32a – 32-Pin Connector, on A-pillar, lower right
V26 – Left Rear Door Window Motor

(29) – Ground connection, near B-pillar, left

(Q15) – Wire connection, in power window wiring harness (Left up)

(Q45) – Wire connection (comfort), in power window wiring harness

(Q46) – Wire connection (child lock) in power window wiring harness

(R19) – Plus connection –1– (30), in left rear door wiring harness

(R21) – Plus connection –3– (15), in left rear door wiring harness

(R22) – Wire connection (child lock), in left rear door wiring harness

(R31) – Wire connection, in left rear door wiring harness

WIRING COLOR CODE		
ws	=	white
sw	=	black
ro	=	red
br	=	brown
gn	=	green
bl	=	blue
gr	=	grey
li	=	lilac
ge	=	yellow

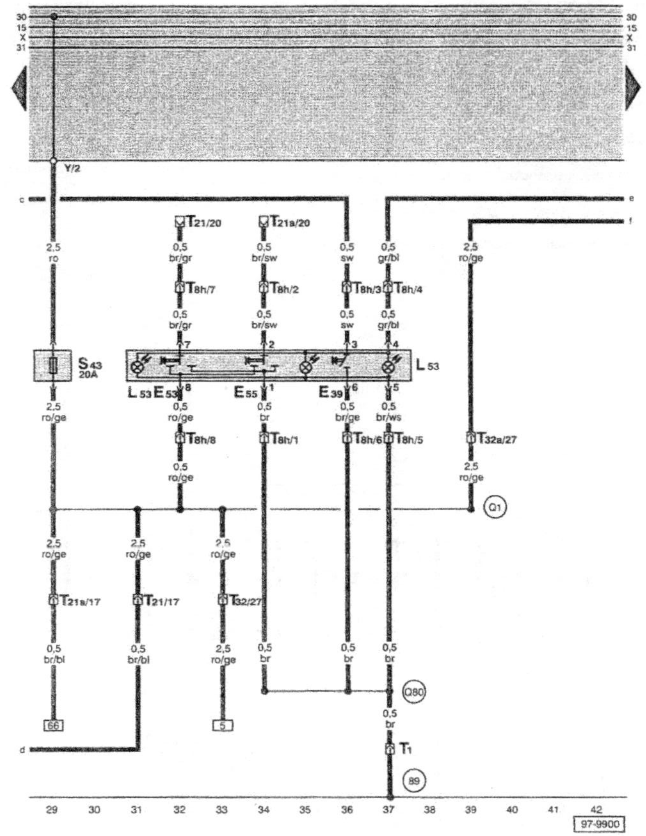

E39 – Window Lockout Switch
E53 – Left Rear Window Switch (In Console)
E55 – Right Rear Window Switch (In Console)
L53 – Power Window Switch Light
S43 – Power Windows Circuit Breaker
T1 – Single Connector
T8h – 8-Pin Connector, behind center of dash panel
T21 – 21-Pin Connector, on left B-pillar
T21a – 21-Pin Connecter, on right B-pillar
T32 – 32-Pin Connector, on A-pillar, lower left
T32a – 32-Pin Connector, on A-pillar, lower right

(89) – Ground connection

(Q1) – Plus connection (30) in power window wiring harness

(Q80) – Wire connection, in window lift wiring harness

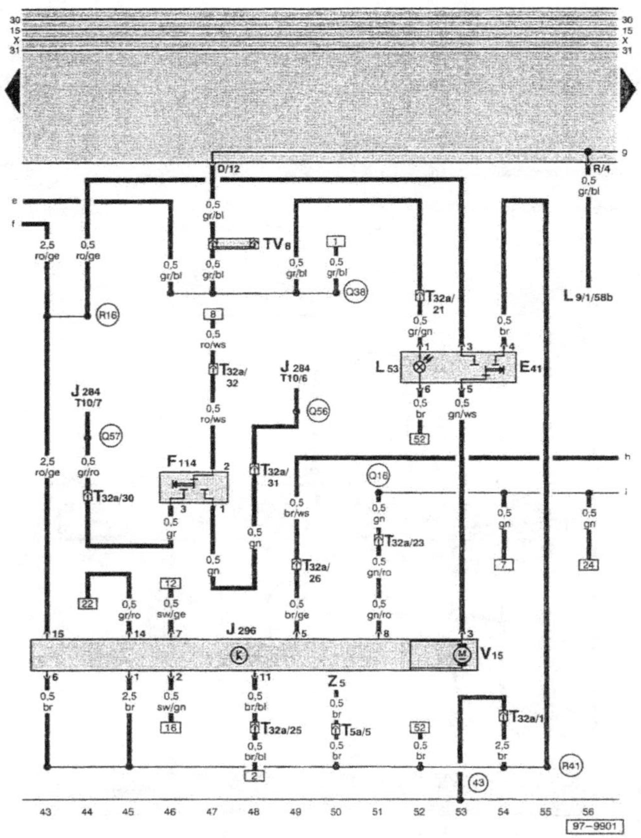

WIRING COLOR CODE		
ws	=	white
sw	=	black
ro	=	red
br	=	brown
gn	=	green
bl	=	blue
gr	=	grey
li	=	lilac
ge	=	yellow

E41 – Right Front Window Switch
F114 – Passenger's Door Central Locking System Door Switch
J284 – Alarm System Control Module
J296 – Right Front Power Window Control Module
L9 – Headlight Switch Light
L53 – Power Window Switch Light
T5a – 5-Pin Connector, in passenger's door
T10 – 10-Pin Connector, behind dash panel, left, on Alarm System Control Module
T32a – 32-Pin Connector, on A-pillar, lower right
TV8 – Terminal 58b Wire Connector
V15 – Window Motor, Right Front Door
Z5 – Passenger Side Heated Mirror

(43) – Ground connection, on right A-Pillar, lower part
(Q16) – Wire connection, in power window wiring harness (left down)
(Q38) – Plus connection (58b) –1–, in power window wiring harness
(Q56) – Wire connection (open), in alarm system wiring harness
(Q57) – Wire connection (closed), in alarm system wiring harness
(R16) – Plus connection –1– (30), in passenger's door wiring harness
(R41) – Wire Connection, in passenger's door wiring harness

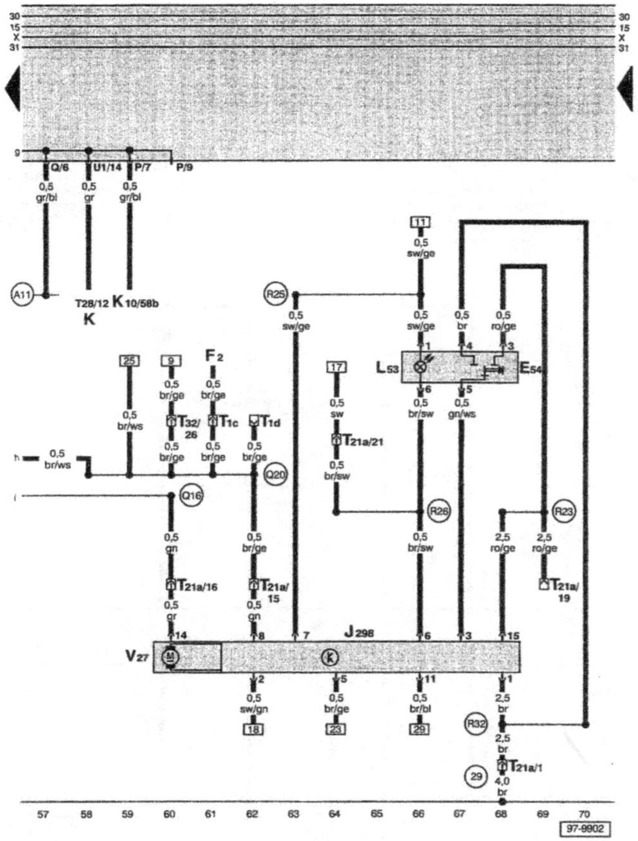

WIRING COLOR CODE		
ws	=	white
sw	=	black
ro	=	red
br	=	brown
gn	=	green
bl	=	blue
gr	=	grey
li	=	lilac
ge	=	yellow

E54 – Right Rear Window Switch, (In RR Door)
F2 – Door Contact Switch, Left Front
J298 – Right Rear Power Window Control Module
K – Instrument Cluster
K10 – Rear Window Defogger Indicator Light
L53 – Power Window Switch Light
T1c – Single Connector, behind fuse/relay panel
T1d – Single Connector, behind fuse/relay panel
T21a – 21-Pin Connecter, on right B-pillar
T28 – 28-Pin Connector, on instrument cluster
T32 – 32-Pin Connector, on A-pillar, lower left
V27 – Right Rear Door Window Motor
(29) – Ground connection, near B-pillar, right
(207) – Ground connection, in left rear door wiring harness

(208) – Ground connection, in right rear door wiring harness
(A11) – Plus connection (58b), in instrument panel wiring harness
(Q16) – Wire connection, in power window wiring harness (left down)
(Q20) – Wire connection –1– (door contact switch), in power window wiring harness
(R23) – Plus connection –1– (30), in right rear door wiring harness
(R25) – Plus connection –3– (15), in right rear door wiring harness
(R26) – Wire connection (child lock), in right rear door wiring harness
(R32) – Wire connection, in right rear door wiring harness

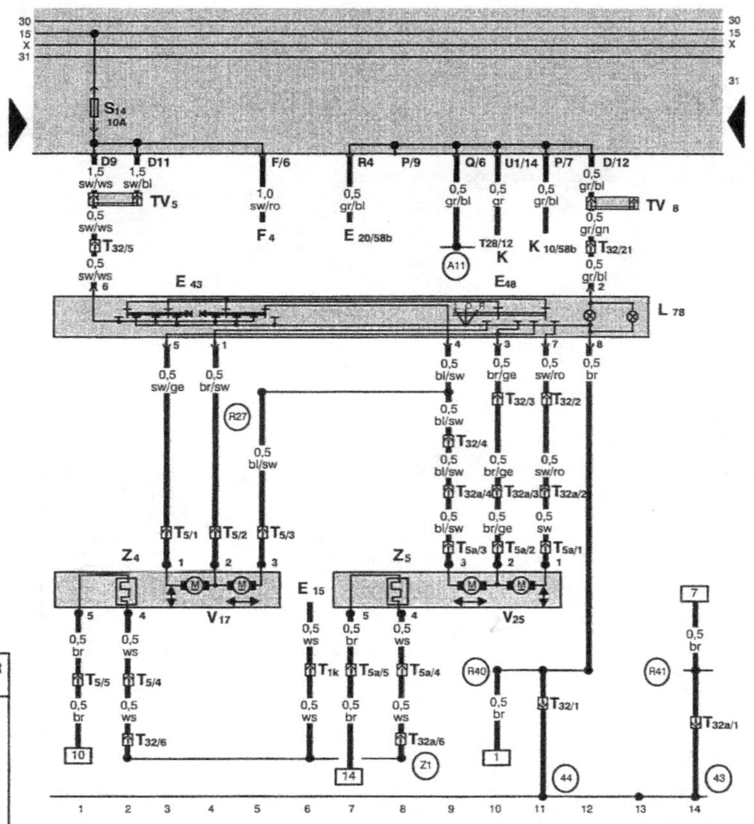

WIRING COLOR CODE		
ws	=	white
sw	=	black
ro	=	red
br	=	brown
gn	=	green
bl	=	blue
gr	=	grey
li	=	lilac
ge	=	yellow

E15 – Rear Window Defogger Switch
E20 – Instrument Panel Light Dimmer Switch
E43 – Mirror Adjustment Switch
E48 – Mirror Select Switch
F4 – Back-Up Light Switch
K – Instrument Cluster
K10 – Rear Window Defogger Indicator Light
L78 – Mirror Adjusting Switch Light
V17 – Driver's Side Mirror Adjustment Motor
V25 – Passenger's Side Heated Mirror Adjustment Motor
Z4 – Driver's Side Heated Mirror
Z5 – Passenger's Side Heated Mirror
T1k – Single Connector, behind fuse/relay panel
T5 – 5-Pin Connector, in driver's door
T5a – 5-Pin Connector, in passenger's door
T28 – 28-Pin Connector, on instrument cluster
T32 – 32-Pin Connector, on A-pillar, lower left
T32a – 32-Pin Connector, on A-pillar, lower right
TV5 – Terminal 15a Wire Connector
TV8 – Terminal 58b Wire Connector

(43) – Ground connection, on right A-pillar, lower part
(44) – Ground connection, on left A-pillar, lower part
(A11) – Plus connection (58b), in instrument panel wiring harness
(R27) – Wire connection (power mirror), in driver's door wiring harness
(R40) – Wire Connection, in driver's door wiring harness
(R41) – Wire Connection, in passenger's door wiring harness
(Z1) – Wire connection –1–, in power heated mirror wiring harness

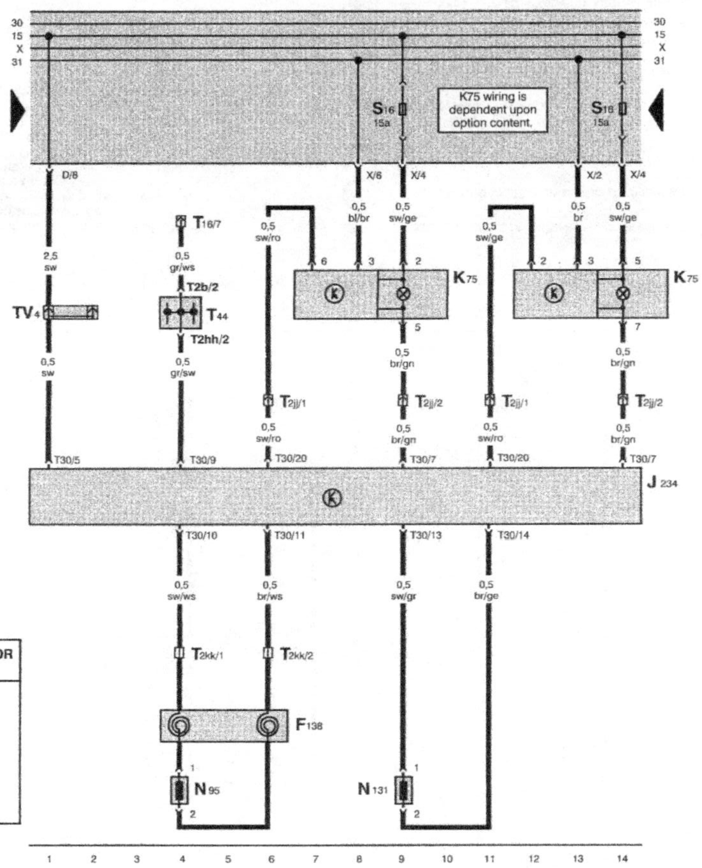

WIRING COLOR CODE		
ws	=	white
sw	=	black
ro	=	red
br	=	brown
gn	=	green
bl	=	blue
gr	=	grey
li	=	lilac
ge	=	yellow

F138 – Airbag Spiral Spring
J234 – Airbag Control Module
K75 – Airbag Malfunction Indicator Light (MIL), in Instrument Cluster
N95 – Driver's Side Airbag Igniter
N131 – Passenger's Side Airbag Igniter 1
T2b – Double Connector, on junction connector (T44) above fuse/relay panel
T2hh – Double Connector, on junction connector (T44) above fuse/relay panel
T2jj – Double Connector, behind fuse/relay panel
T2kk – Double Connector, behind fuse/relay panel
T16 – Data Link Connector (DLC), behind center of dash
T30 – 30-Pin Connector, on Airbag Control Module
T44 – Junction Block for On-board Diagnosis, next to fuse/relay panel
TV4 – Terminal 15 Wire Connector (Black), next to fuse/relay panel

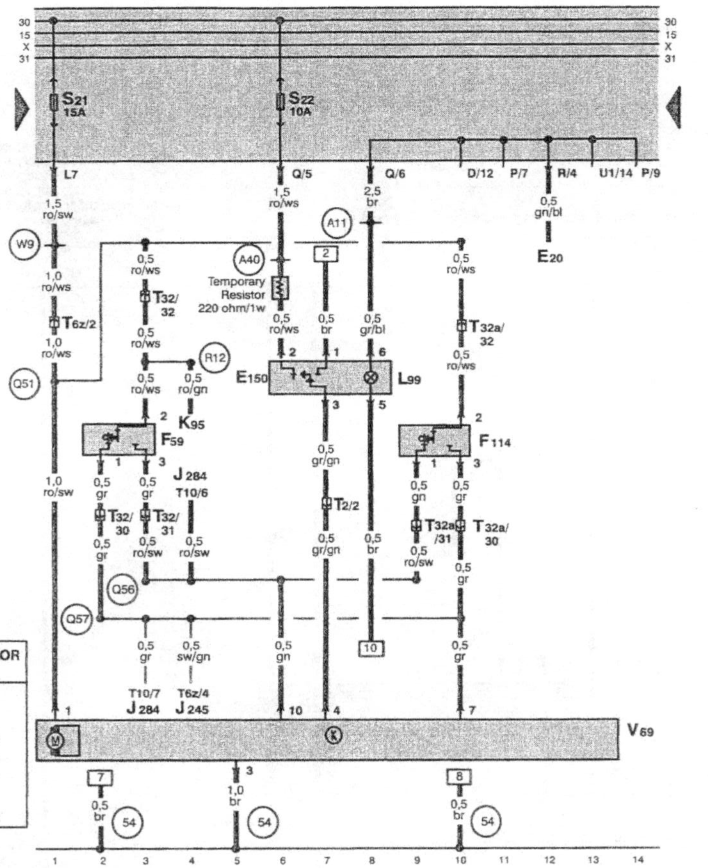

WIRING COLOR CODE		
ws	=	white
sw	=	black
ro	=	red
br	=	brown
gn	=	green
bl	=	blue
gr	=	grey
li	=	lilac
ge	=	yellow

E20 – Instrument Panel Light Dimmer Switch
E150 – Door Lock - Unlock Switch
F59 – Driver's Door Central Locking System Door Switch
F114 – Passenger's Door Central Locking System Door Switch
J245 – Power Sunroof Control Module
J284 – Alarm System Control Module
K95 – Alarm System Indicator Light
L99 – Lighting Switch for Interior Lock
T2 – Double Connector, near Door Lock - Unlock Switch
T6z – 6-Pin Connector, behind dash panel
T10 – 10-Pin Connector, behind dash panel, left, on Alarm System Control Module
T32 – 32-Pin Connector, on A-pillar, lower left
T32a – 32-Pin Connector, on A-pillar, lower right
V69 – Central Locking System Pump and Control Module

(54) – Ground connection, on rear apron
(A11) – Plus connection (58b), in instrument panel wiring harness
(A40) – Plus connection (30), in instrument panel wiring harness
(Q51) – Plus connection (30), in alarm system wiring harness
(Q56) – Wire connection (open), in alarm system wiring harness
(Q57) – Wire connection (closed), in alarm system wiring harness
(R12) – Plus connection -3- (30), in driver's door wiring harness
(W9) – Plus connection (30), in rear wiring harness

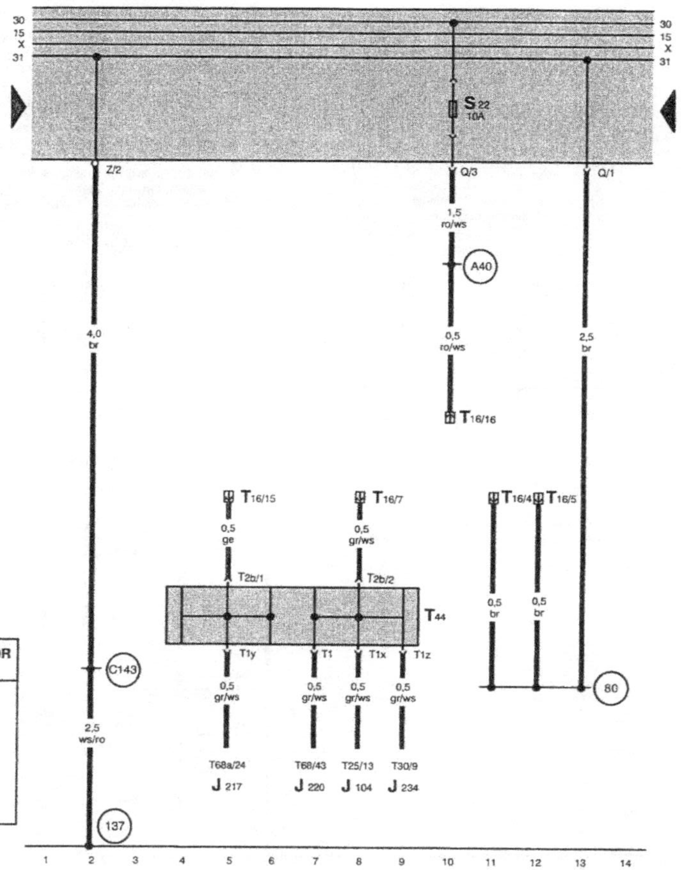

WIRING COLOR CODE		
ws	=	white
sw	=	black
ro	=	red
br	=	brown
gn	=	green
bl	=	blue
gr	=	grey
li	=	lilac
ge	=	yellow

J104 – ABS Control Module (w/EDL), in engine compartment, left
J217 – Transmission Control Module (TCM), below rear seat bench
J220 – Motronic Engine Control Module (ECM)
J234 – Airbag Control Module
T1 – Single Connector, above fuse/relay panel
T1x – Single Connector, above fuse/relay panel
T1y – Single Connector, above fuse/relay panel
T1z – Single Connector, above fuse/relay panel
T2b – Double Connector, on junction connector (T44) above fuse relay/panel
T16 – Data Link Connector (DLC), behind center of dash
T44 – Junction Box for On Board Diagnosis, above fuse/relay panel

(80) – Ground connection –1–, in instrument panel wiring harness
(137) – Ground connection (end stage), in Motronic Multiport Fuel Injection (MFI) wiring harness
(A40) – Plus connection –1– (30), in instrument cluster wiring harness
(C143) – Wire connection, in Motronic Multiport Fuel Injection (MFI) wiring harness

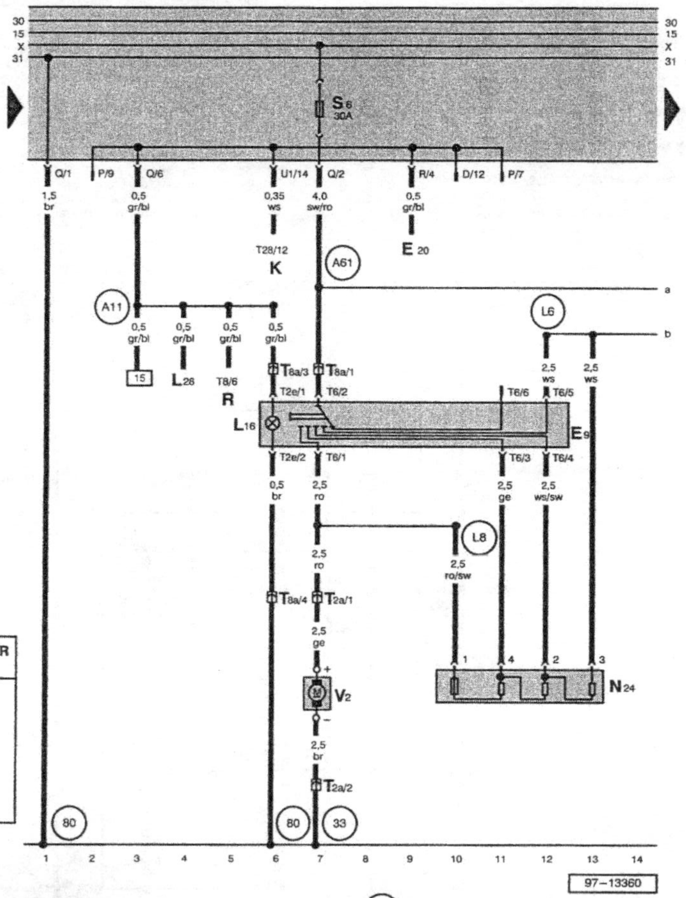

WIRING COLOR CODE		
ws	=	white
sw	=	black
ro	=	red
br	=	brown
gn	=	green
bl	=	blue
gr	=	grey
li	=	lilac
ge	=	yellow

E9 – Fresh Air Blower Switch
E20 – Instrument Panel Light Dimmer Switch
K – Instrument Cluster
L16 – Fresh Air Control Lever Light
L28 – Cigarette Lighter Light
N24 – Fresh Air Blower Series Resistance With Fuse
R – Radio
T2a – Double Connector, behind instrument panel, right
T2e – Double Connector
T6 – 6-Pin Connector
T8 – 8-Pin Connector, on radio
T8a – 8-Pin Connector, behind instrument panel, center
T28 – 28-Pin Connector, on instrument cluster
V2 – Fresh Air Blower (01273–4124)

(33) – Ground connection, behind instrument panel, right

(80) – Ground connection –1–, in instrument panel wiring harness

(A11) – Plus connection (58b), in instrument panel wiring harness

(A61) – Plus connection (X), in instrument cluster wiring harness

(L6) – Wire connection –2–, in evaporator housing wiring harness

(L8) – Wire connection –4–, in evaporator housing wiring harness

1996-1999

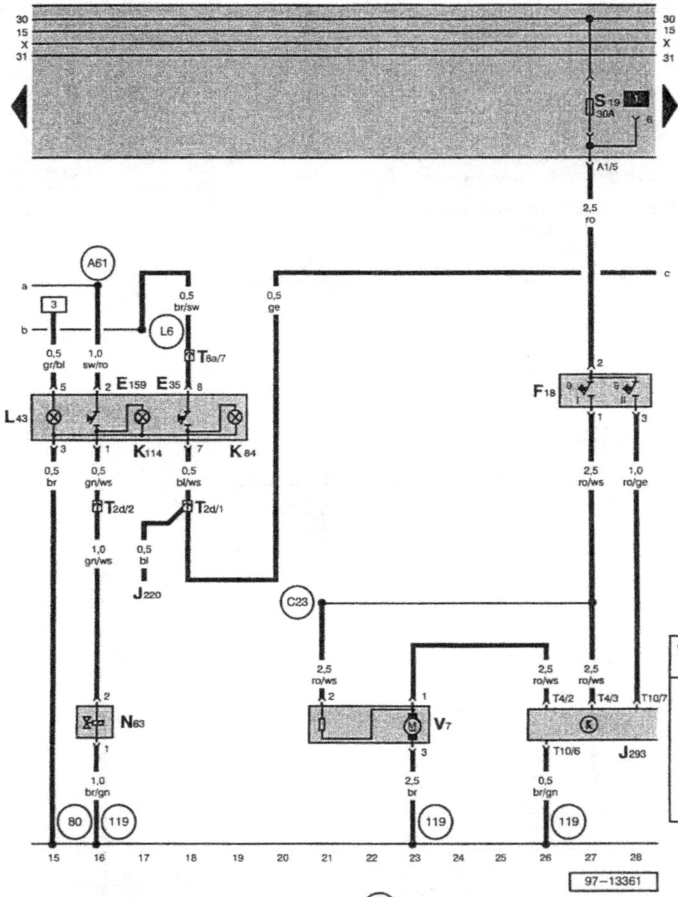

WIRING COLOR CODE		
ws	=	white
sw	=	black
ro	=	red
br	=	brown
gn	=	green
bl	=	blue
gr	=	grey
li	=	lilac
ge	=	yellow

E35 – A/C Switch
E159 – Fresh Air/Recirculating Flap Switch
F18 – Coolant Fan Control (FC) Thermal Switch
J220 – Motronic Engine Control Module (ECM), in plenum chamber, right
J293 – Coolant FC (Fan Control) Control Module, in engine compartment, left
K84 – A/C Indicator Light
K114 – Fresh Air and Recirculating Air Mode Indicator Light
L43 – A/C Switch Illumination Light
N63 – Fresh Air/Recirculating Flap Two-Way Valve (00605), in engine compartment, rear
T2 – Double Connector, behind fuse/relay panel
T4 – 4-Pin Connector
T8 – 8-Pin Connector, behind instrument panel, center
V7 – Coolant Fan (01018–3433)

(80) – Ground connection –1–, in instrument panel wiring harness

(119) – Ground connection –1–, in headlight wiring harness

(A61) – Plus connection (X), in instrument cluster wiring harness

(C23) – Wire connection (blower, 1st speed), in headlight/air conditioner wiring harness

(L6) – Wire connection –2–, in evaporator housing wiring harness

1996-1999

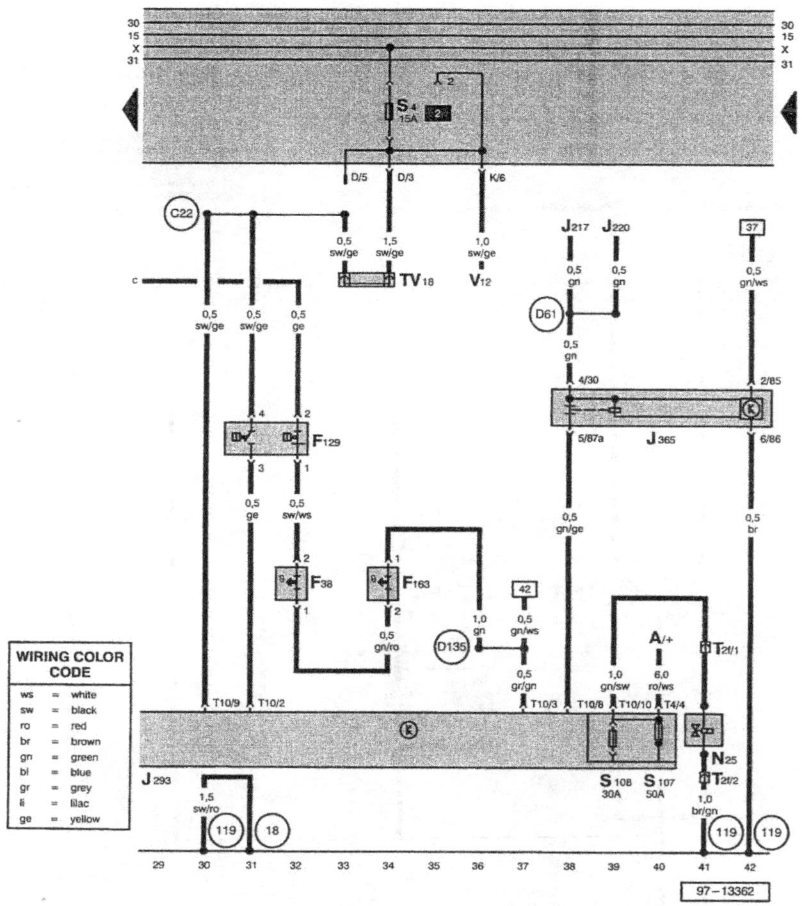

A – Battery
F38 – Ambient Temperature Switch, near dual tone horns
F129 – A/C Pressure Switch (00792–3224), in engine compartment, right
F163 – A/C Cut-Out Thermal Switch, in engine compartment, center
J217 – Transmission Control Module (TCM), below rear seat bench
J220 – Motronic Engine Control Module (ECM), in plenum chamber, right
J293 – Coolant FC (Fan Control) Control Module, in engine compartment, left
J365 – Relay for A/C shutoff, above fuse/relay panel
N25 – A/C Clutch (01270–4121)
S107 – Coolant Fan Second and Third Speed Fuse
S108 – Coolant Fan First Speed and A/C Clutch Fuse
T2f – Double Connector, near compressor
T4 – 4-Pin Connector
T10 – 10-Pin Connector
TV18 – Distributor for terminal Xa, above fuse/relay panel

V12 – Rear Window Wiper Motor

(18) – Ground connection, on engine block

(119) – Ground connection –1–, in headlight wiring harness

(C22) – Wire connection (x), in headlight/air conditioner wiring harness

(D61) – Wire connection (A/C), in engine compartment wiring harness

(D135) – Connector (fan stage I), in wiring harness engine-headlamp

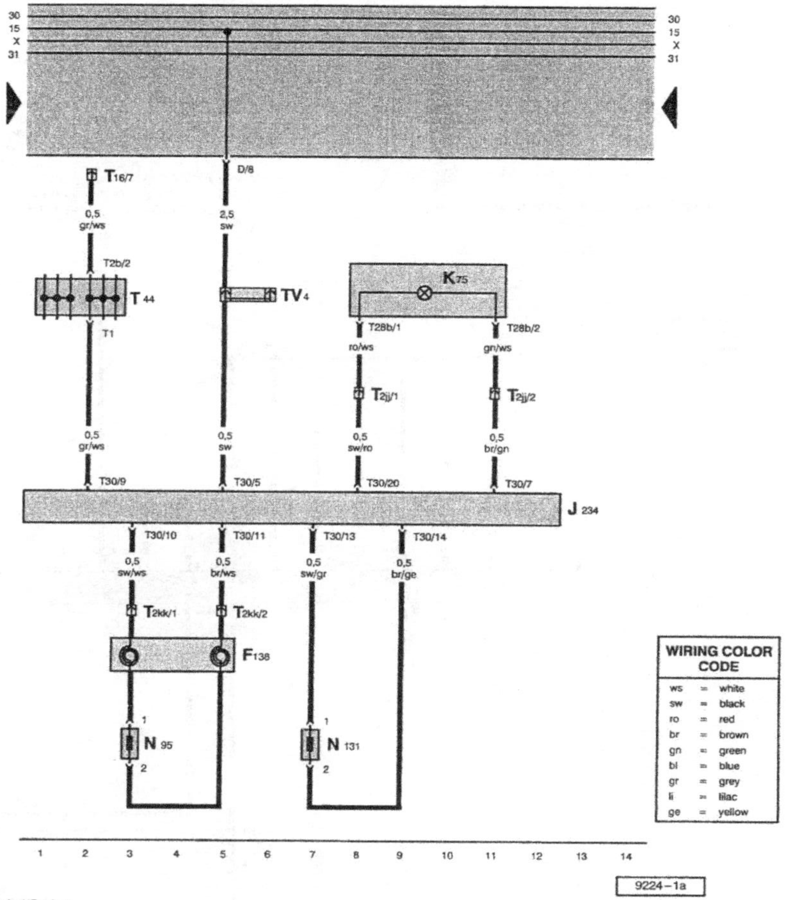

F138 – Airbag Spiral Spring
J234 – Airbag Control Module
K75 – Airbag Malfunction Indicator Lamp (MIL)
N95 – Driver's Side Airbag Ignitor
N131 – Passenger Side Airbag Ignitor 1
T1 – Single Connector, on junction connector (T44) above fuse/relay panel
T2b – Double Connector, on junction connector (T44) above fuse/relay panel
T16 – Data Link Connector (DLC), behind center of dash panel
T2jj – Double Connector, behind fuse/relay panel
T2kk – Double Connector, behind fuse/relay panel
T28b – 28-Pin Connector, on instrument cluster
T30 – 30-Pin Connector, on Airbag Control Module
T44 – Junction Box for on Board Diagnosis, above fuse/relay panel
TV4 – Terminal 15 Wire Connector

Central locking system (Cabrio only)

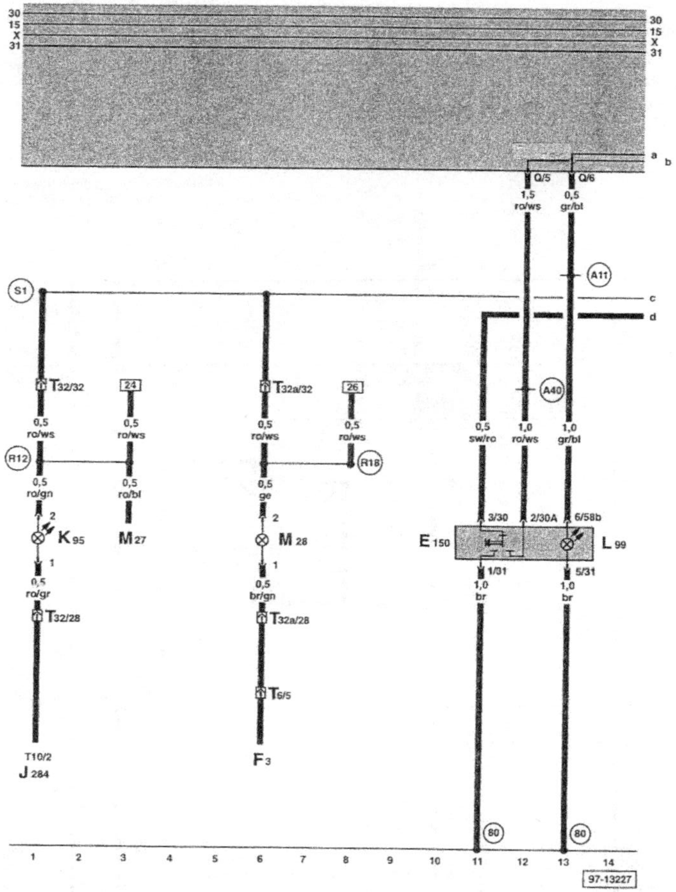

WIRING COLOR CODE		
ws	=	white
sw	=	black
ro	=	red
br	=	brown
gn	=	green
bl	=	blue
gr	=	grey
li	=	lilac
ge	=	yellow

E150 – Door Lock-Unlock Switch
F3 – Door Contact Switch-Passenger Side
J284 – Alarm System Control Module
K95 – Alarm System Indicator Light
L99 – Door Lock-Unlock Switch Illumination Light
M27 – Right Door Warning Light
M28 – Right Door Warning Light
T6 – 6-Pin Connector, behind fuse/relay panel
T10 – 10-Pin Connector
T32 – 32-Pin Connector, on B-pillar, right

(80) – Ground connection –1–, in instrument panel wiring harness

(A11) – Plus connection (58b), in instrument panel wiring harness

(A40) – Plus connection –1– (30), in instrument cluster wiring harness

(R12) – Plus connection –3– (30), in driver's door wiring harness

(R18) – Plus connection –3– (30), in passenger's door wiring harness

(S1) – Plus connection (30) –1–, in central locking system wiring harness

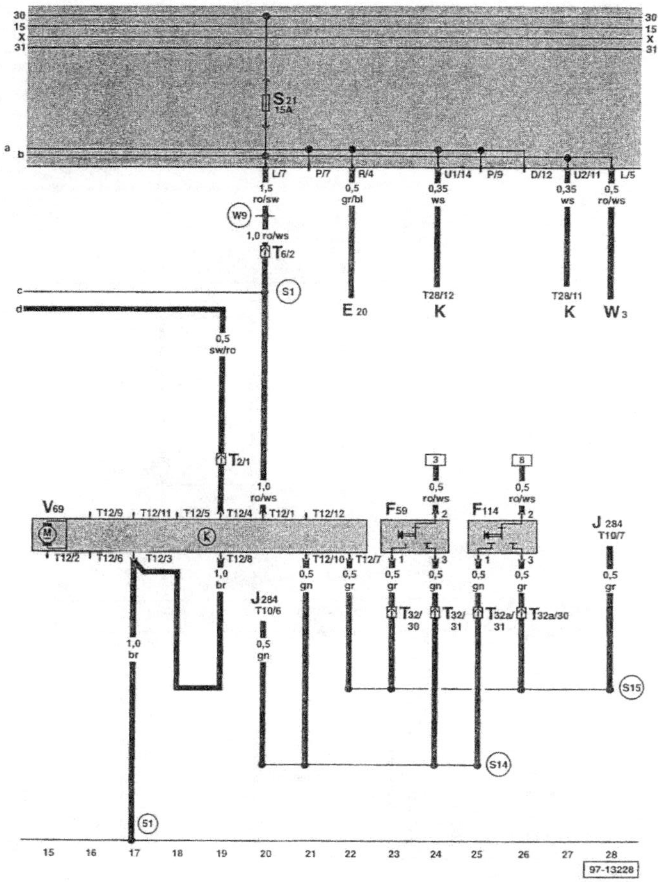

WIRING COLOR CODE		
ws	=	white
sw	=	black
ro	=	red
br	=	brown
gn	=	green
bl	=	blue
gr	=	grey
li	=	lilac
ge	=	yellow

E20 – Instrument Panel Light Dimmer Switch
F59 – Driver's Door Central Locking System Switch
F114 – Passenger's Door Central Locking System Switch
J284 – Alarm System Control Module
K – Instrument Cluster
T2 – Double Connector, behind fuse/relay panel
T6 – 6-Pin Connector, behind fuse/relay panel
T10 – 10-Pin Connector
T12 – 12-Pin Connector
T28 – 28-Pin Connector, on instrument cluster
T32 – 32-Pin Connector, on B-pillar, right
T32 – 32-Pin Connector, on B-pillar, left
V69 – Central Locking System Pump and Control Module
W3 – Luggage Compartment Light

(51) – Ground connection, in luggage compartment, right

(S1) – Plus connection (30) –1–, in central locking system wiring harness

(S14) – Wire connection (open), in central locking system wiring harness

(S15) – Wire connection (closed), in central locking system wiring harness

(W9) – Plus connection (30), in rear wiring harness

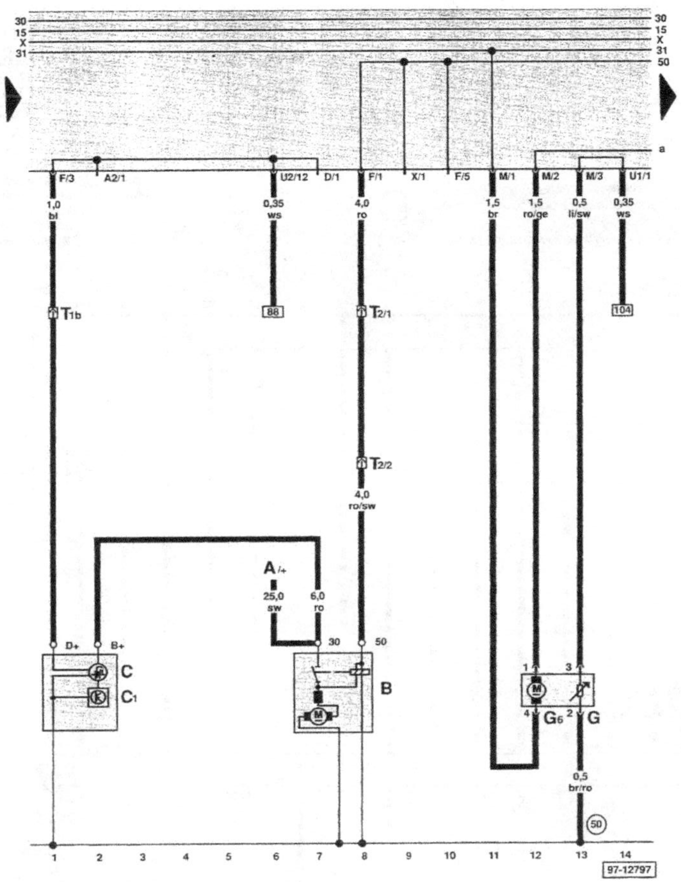

WIRING COLOR CODE		
ws	=	white
sw	=	black
ro	=	red
br	=	brown
gn	=	green
bl	=	blue
gr	=	grey
li	=	lilac
ge	=	yellow

A – Battery
B – Starter
C – Generator (GEN)
C1 – Voltage Regulator (VR)
G – Fuel Level Sensor
G6 – Fuel Pump (FP)
T1 – Single Connector, near battery
T2 – Double Connector, behind fuse/relay panel

50 – Ground connection, in luggage compartment, left

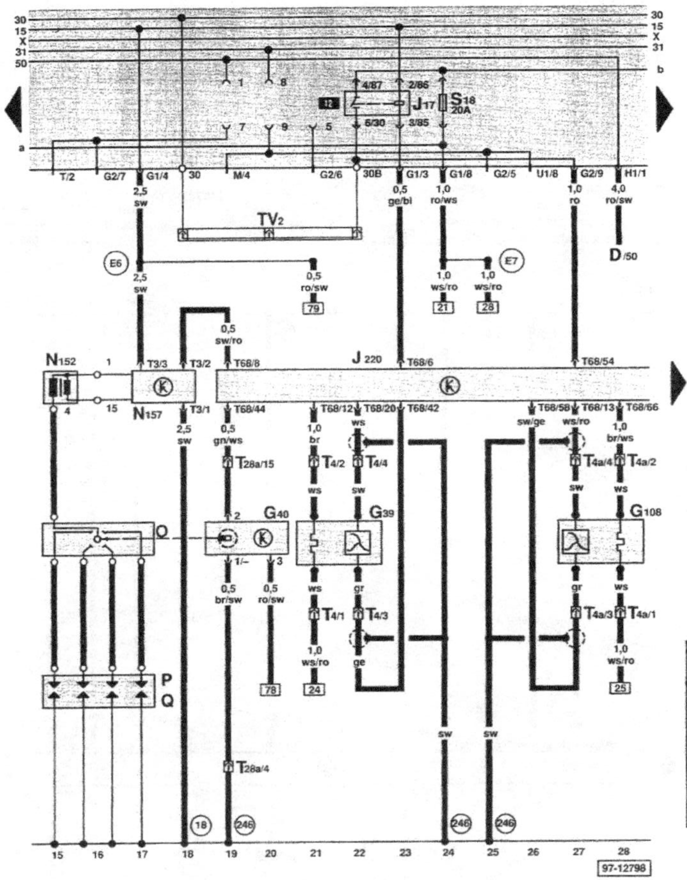

WIRING COLOR CODE		
ws	=	white
sw	=	black
ro	=	red
br	=	brown
gn	=	green
bl	=	blue
gr	=	grey
li	=	lilac
ge	=	yellow

D – Ignition/Starter Switch
G39 – Heated Oxygen Sensor (HO2S)
G40 – Camshaft Position (CMP) Sensor
G108 – Heated Oxygen Sensor (HO2S) 2
J17 – Fuel Pump (FP) Relay
J220 – Motronic Engine Control Module (ECM)
N152 – Ignition Coil
N157 – Ignition Coil Power Output Stage
O – Ignition Distributor
P – Spark Plug Connectors
Q – Spark Plugs
T3 – 3-Pin Connector
T4 – 4-Pin Connector, near intake manifold
T4 – 4-Pin Connector
T28 – 28-Pin Connector, on engine
T68 – 68-Pin Connector
TV2 – Terminal 30 (B+) wire connector, above fuse/relay panel

(18) – Ground connection, on engine block

(246) – Ground connection (control module), in wiring harness engine

(E6) – Plus connection –1– (15), in motronic Multiport Fuel Injection (MFI) wiring harness

(E7) – Plus connection (87a), in Motronic Multiport Fuel Injection (MFI) wiring harness

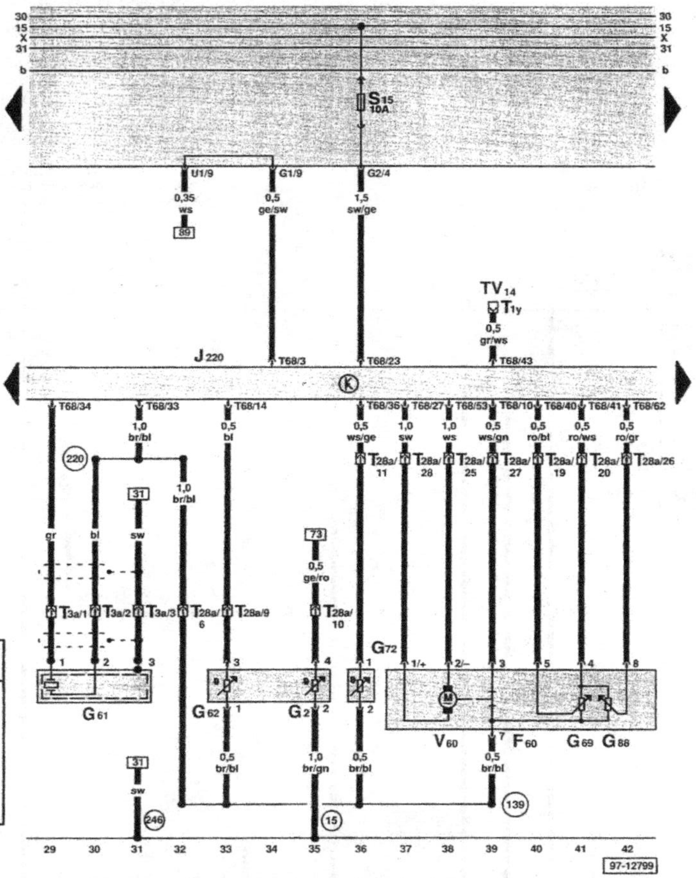

WIRING COLOR CODE		
ws	=	white
sw	=	black
ro	=	red
br	=	brown
gn	=	green
bl	=	blue
gr	=	grey
li	=	lilac
ge	=	yellow

F60 – Closed Throttle Position (CTP) Switch
G2 – Engine Coolant Temperature (ECT) Sensor
G61 – Knock Sensor (KS) 1
G62 – Engine Coolant Temperature (ECT) Sensor
G69 – Throttle Position (TP) Sensor
G72 – Intake Air Temperature (IAT) Sensor
G88 – Throttle Position (TP) Sensor
J220 – Motronic Engine Control Module (ECM)
T1 – Single Connector, above fuse/relay panel
T3 – 3-Pin Connector, near cylinder head cover
T28 – 28-Pin Connector, on engine
T68 – 68-Pin Connector
TV14 – Data Link Connector (DLC) wire Connector, above fuse/relay panel
V60 – Throttle Position (TP) Actuator

(15) – Ground connection, on cylinder head

(139) – Ground connection (sensor ground), in Motronic Multiport Fuel Injection (MFI) wiring harness

(220) – Ground connection (sensor ground), in engine compartment wiring harness

(246) – Ground connection (control module), in wiring harness engine

Engine electronics (Cabrio only)

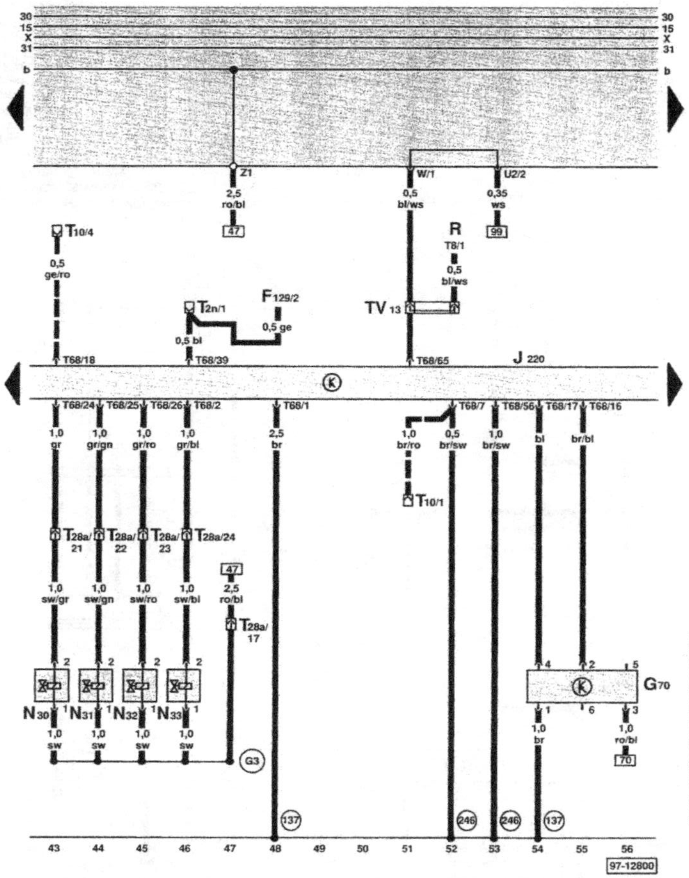

WIRING COLOR CODE		
ws	=	white
sw	=	black
ro	=	red
br	=	brown
gn	=	green
bl	=	blue
gr	=	grey
li	=	lilac
ge	=	yellow

F129 – A/C Pressure Switch
G3 – Engine Coolant Temperature (ECT) Gauge
G70 – Mass Air Flow (MAF) Sensor
J220 – Motronic Engine Control Module (ECM)
N30 – Cylinder 1 Fuel Injector
N31 – Cylinder 2 Fuel Injector
N32 – Cylinder 3 Fuel Injector
N33 – Cylinder 4 Fuel Injector
R – Radio
T2 – Double Connector, behind fuse/relay panel
T8 – 8-Pin Connector, on radio
T10 – 10-Pin Connector, behind fuse/relay panel
T28 – 28-Pin Connector, on engine
T68 – 68-Pin Connector
TV13 – Vehicle Speed Signal Wire Connector, above fuse/relay panel

(137) – Ground connection (end stage), in Motronic Multiport Fuel Injection (MFI) wiring harness

(246) – Ground connection (control module), in wiring harness engine

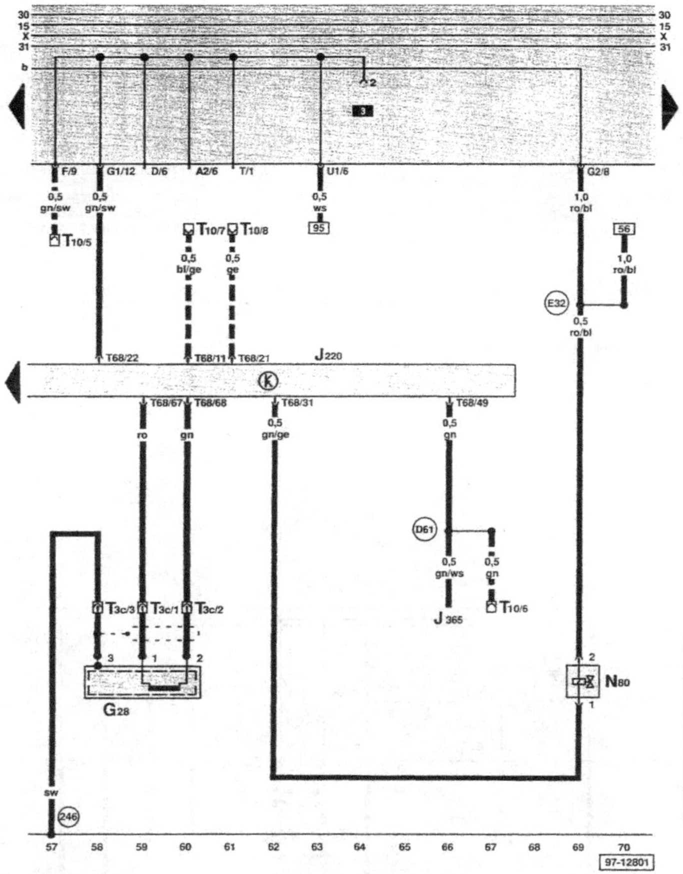

WIRING COLOR CODE		
ws	=	white
sw	=	black
ro	=	red
br	=	brown
gn	=	green
bl	=	blue
gr	=	grey
li	=	lilac
ge	=	yellow

G28 – Engine Speed (RPM) Sensor
J220 – Motronic Engine Control Module (ECM)
J365 – Relay for A/C Shutoff
N80 – Evaporative Emission (EVAP) Canister Purge Regulator Valve
T3 – 3-Pin Connector, on engine
T10 – 10-Pin Connector, behind fuse/relay panel
T68 – 68-Pin Connector

(246) – Ground connection (control module), in wiring harness engine
(D61) – Wire connection (A/C), in engine compartment wiring harness
(E32) – Wire connection

Engine electronics (Cabrio only)

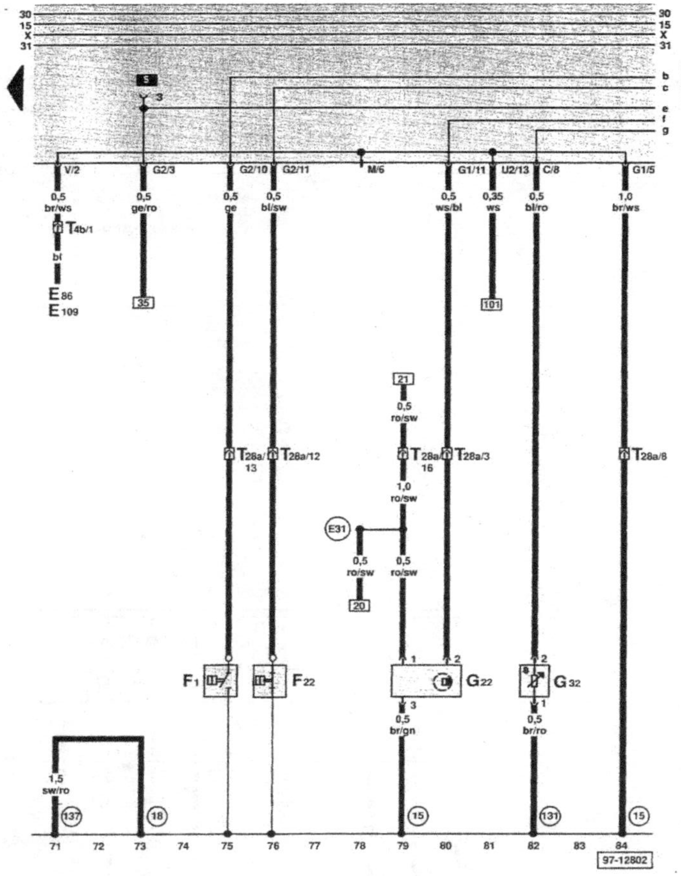

WIRING COLOR CODE		
ws	=	white
sw	=	black
ro	=	red
br	=	brown
gn	=	green
bl	=	blue
gr	=	grey
li	=	lilac
ge	=	yellow

E86 – Multi-Function Indicator Mode Select Switch
E109 – Multi-Function Indicator Memory Switch
F1 – Oil Pressure Switch
F22 – 0.3bar Oil Pressure Switch
G22 – Speedometer Vehicle Speed Sensor (VSS)
G32 – Engine Coolant Level (ECL) Sensor
T4 – 4-Pin Connector, behind steering column switch cover
T28 – 28-Pin Connector, on engine

(15) – Ground connection, on cylinder head
(18) – Ground connection, on engine block
(127) – Ground connection, –1–, in A/C compressor wiring harness
(131) – Ground connection –2–, in engine compartment wiring harness
(E31) – Connector (15), in wiring harness engine

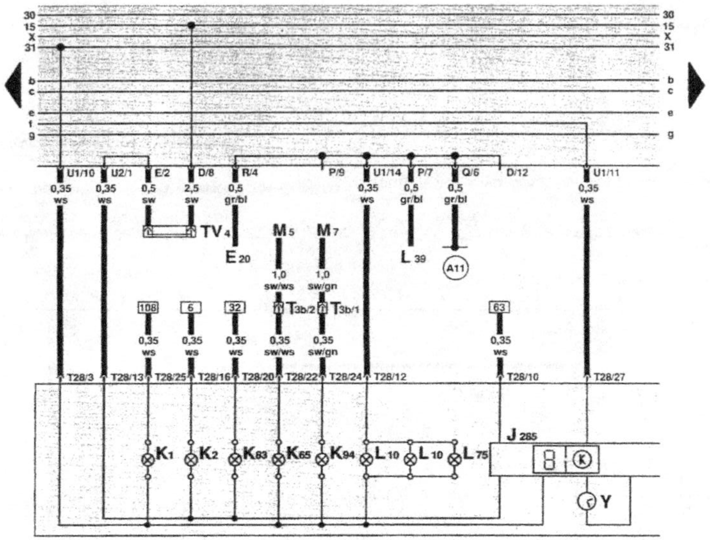

WIRING COLOR CODE		
ws	=	white
sw	=	black
ro	=	red
br	=	brown
gn	=	green
bl	=	blue
gr	=	grey
li	=	lilac
ge	=	yellow

J285 – Control Module with Display Unit in Instrument Cluster
K1 – Headlight High Beam Indicator Light
K2 – Generator (GEN) Warning Light
K65 – Left Turn Signal Indicator Light
K83 – Malfunction Indicator Lamp (MIL)
K94 – Right Turn Signal Indicator Light
L10 – Instrument Cluster Light
L39 – Rear Window Defogger Switch Light
L75 – Digital Display Light
M5 – Left Front Turn Signal Light
M7 – Right Front Turn Signal Light
T3 – 3-Pin Connector, behind fuse/relay panel
T28 – 28-Pin Connector, on instrument cluster
TV4 – Terminal 15 Wire Connector, behind fuse/relay panel
Y – Analog Clock
(A11) – Plus connection (58b) in instrument panel wiring harness

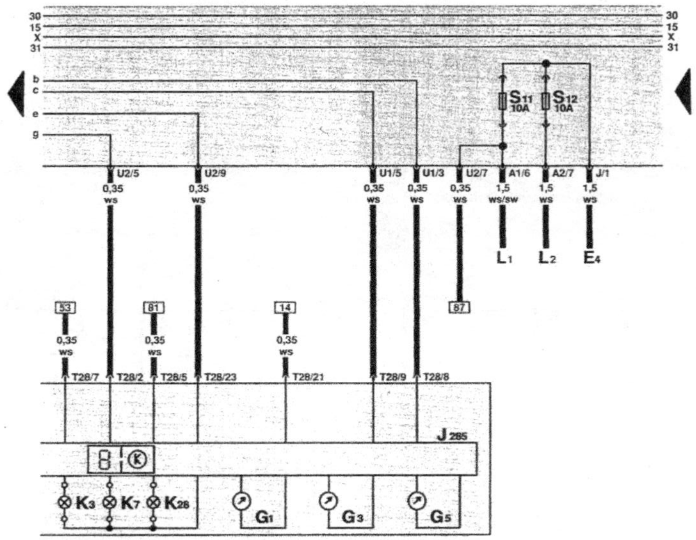

WIRING COLOR CODE		
ws	=	white
sw	=	black
ro	=	red
br	=	brown
gn	=	green
bl	=	blue
gr	=	grey
li	=	lilac
ge	=	yellow

E4 – Headlight Dimmer/Flasher Switch
G1 – Fuel Gauge
G3 – Engine Coolant Temperature (ECT) Gauge
G5 – Tachometer
K3 – Oil Pressure Warning Light
K7 – Brake and Parking Brake Warning Light
K28 – Engine Coolant Level/Temperature (ECL/ECT) Warning Light
L1 – Left Headlight
L2 – Right Headlight
T28 – 28-Pin Connector, on instrument cluster

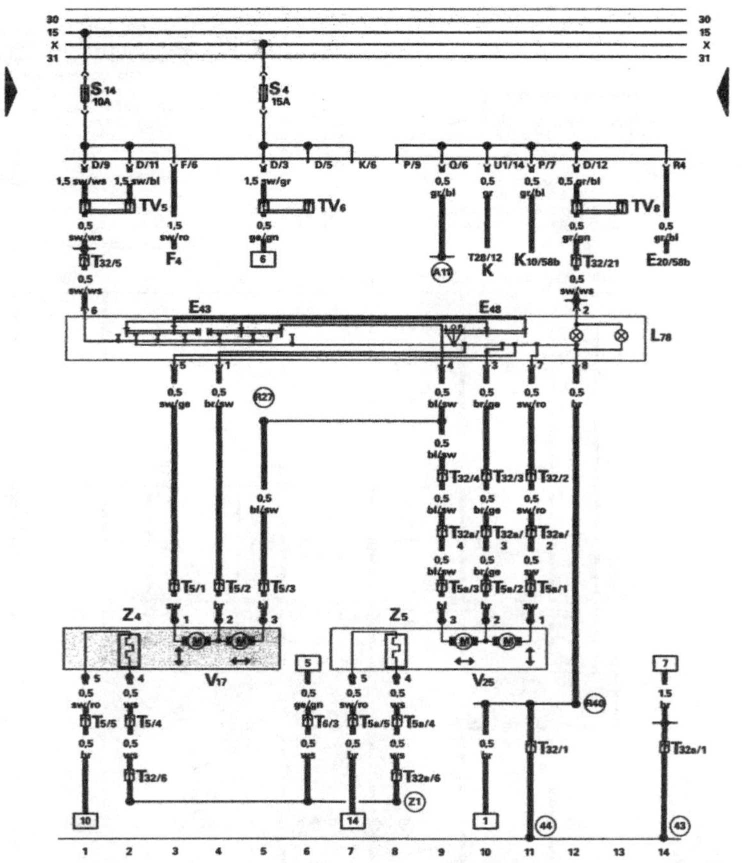

E20	-	Instrument Panel Light Dimmer Switch
E43	-	Mirror Adjustment Switch
E48	-	Mirror Selector Switch
F4	-	Back-Up Light Switch
K	-	Instrument Cluster
K10	-	Indicator light for heated rear windshield
L78	-	Mirror Adjusting Switch Light
V17	-	Driver's Side Mirror Adjustment Motor
V25	-	Passenger's Side Mirror Adjustment Motor
Z4	-	Heated outside mirror, driver side
Z5	-	Heated outside mirror, passenger side
T1k	-	Single Connector, behind fuse/relay panel
T5	-	5-Pin Connector, in driver's door
T5a	-	5-Pin Connector, in passenger's door
T6	-	6-Pin Connector
T28	-	28-Pin Connector, on instrument cluster
T32	-	32-Pin Connector, on A-pillar, lower left
T32a	-	32-Pin Connector, on A-pillar, lower right
TV5	-	Terminal 15a Wire Connector (green)
TV6	-	Terminal X Wire Connector (yellow)
TV8	-	Terminal 58b Wire Connector (grey)
(43)	-	Ground connection, on right A-pillar, lower part
(44)	-	Ground connection, on left A-pillar, lower part
(A11)	-	plus connection (58b) in instrument panel wiring harness
(R27)	-	wire connection (power mirror), in driver's door wiring harness
(R40)	-	Connection 1 (window lifter) in door wiring harness, right rear
(Z1)	-	Wire connection -1-, in power heated mirror wiring harness

ws = white
sw = black
ro = red
br = brown
gn = green
bl = blue
gr = grey
li = violet
ge = yellow

97-60737

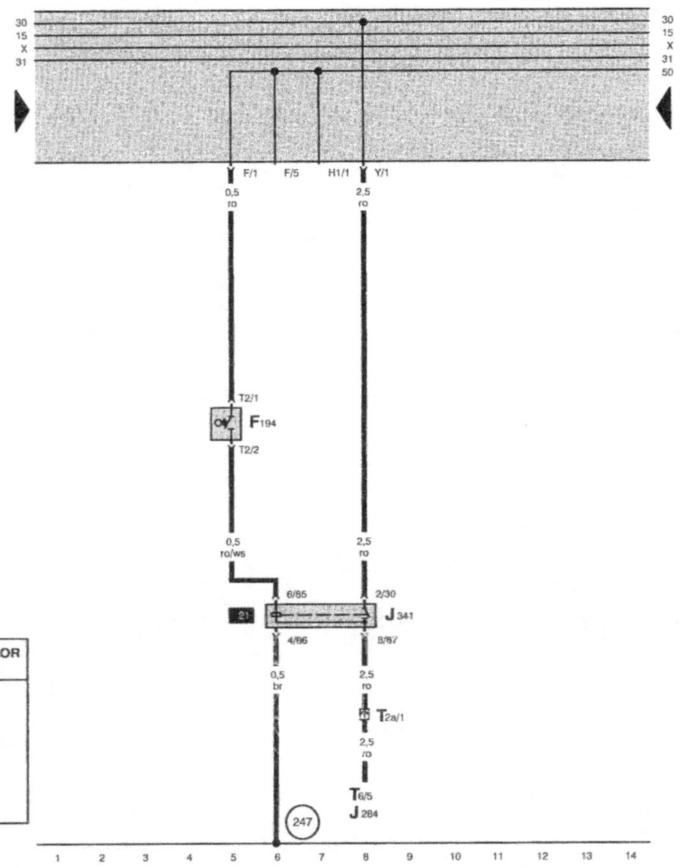

WIRING COLOR CODE

ws	=	white
sw	=	black
ro	=	red
br	=	brown
gn	=	green
bl	=	blue
gr	=	grey
li	=	lilac
ge	=	yellow

97-12804

F194	– Clutch Pedal Position Switch
J284	– Alarm System Control module
J341	– Anti-theft Engine Disable Relay 1
T2	– Double Connector, on Park/Neutral Position (PNP) Switch
T2a	– Double Connector, behind Fuse/Relay Panel
T6	– 6-Pin Connector, on Alarm System Control Module
(247)	– Ground connection (anti-theft immobilizer sensor), in wiring harness fittings

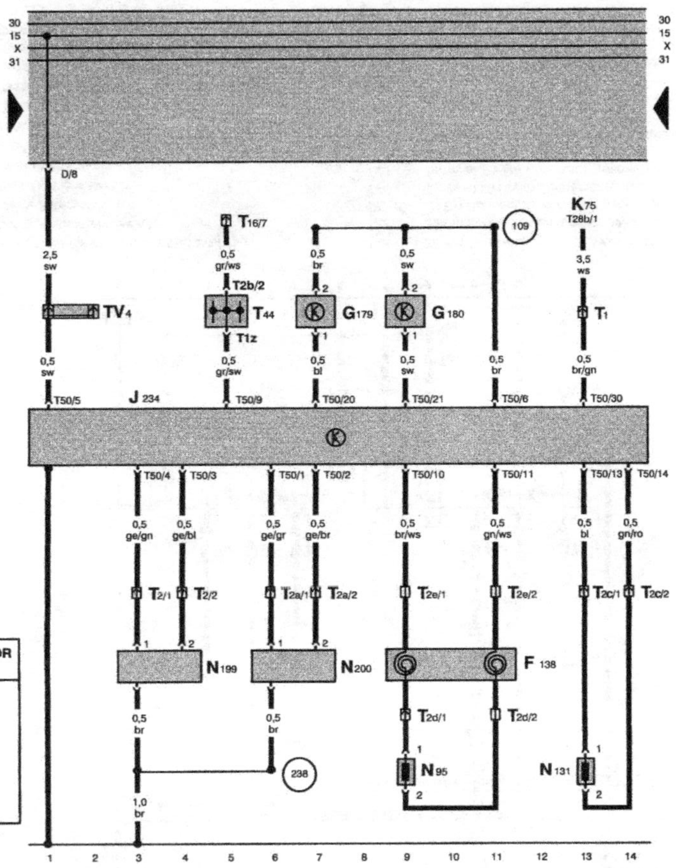

WIRING COLOR CODE		
ws	=	white
sw	=	black
ro	=	red
br	=	brown
gn	=	green
bl	=	blue
gr	=	grey
li	=	lilac
ge	=	yellow

F138 – Airbag Spiral Spring /Return Spring with Slip Ring
G179 – Crash Sensor for Side Airbag, Driver's Side
G180 – Crash Sensor for Side Airbag, Passenger's Side
J234 – Airbag Control Module
K75 – Airbag Malfunction Indicator Light (MIL), in Instrument Cluster
N95 – Driver's Side Airbag Igniter
N131 – Passenger's Side Airbag Igniter 1
N199 – Igniter for Side Airbag, Driver's Side
N200 – Igniter for Side Airbag, Passenger's Side
T1 – Single Connector, behind instrument panel, center
T1z – Single Connector, next to fuse/relay panel
T2 – Double Connector, on airbag wiring harness
T2a – Double Connector, on airbag wiring harness
T2b – Double Connector, next to fuse/relay panel
T2c – Double Connector, on airbag wiring harness
T2d – Double Connector, on airbag wiring harness
T2e – Double Connector, on airbag wiring harness
T16 – Data Link Connector, behind center of dash
T44 – Junction Block for On-board Diagnosis, next to fuse/relay panel
T50 – 50-Pin Connector, on Airbag Control Module
TV4 – Terminal 15 Wire Connector (Black), next to fuse/relay panel
(109) – Ground connection, in airbag wiring harness
(238) – Ground connection, in airbag wiring harness

THIS PAGE INTENTIONALLY LEFT

BLANK

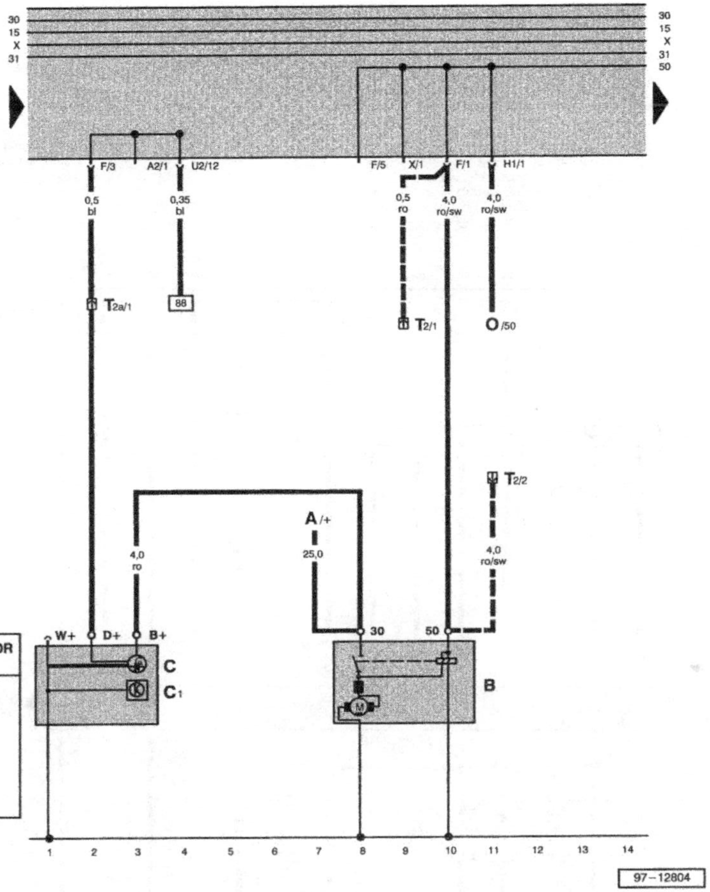

WIRING COLOR CODE		
ws	=	white
sw	=	black
ro	=	red
br	=	brown
gn	=	green
bl	=	blue
gr	=	grey
li	=	lilac
ge	=	yellow

A – Battery
B – Starter
C – Generator (GEN)
C1 – Voltage Regulator (VR)
O – Ignition Distributor
T2 – Double Connector
T2a – Double Connector

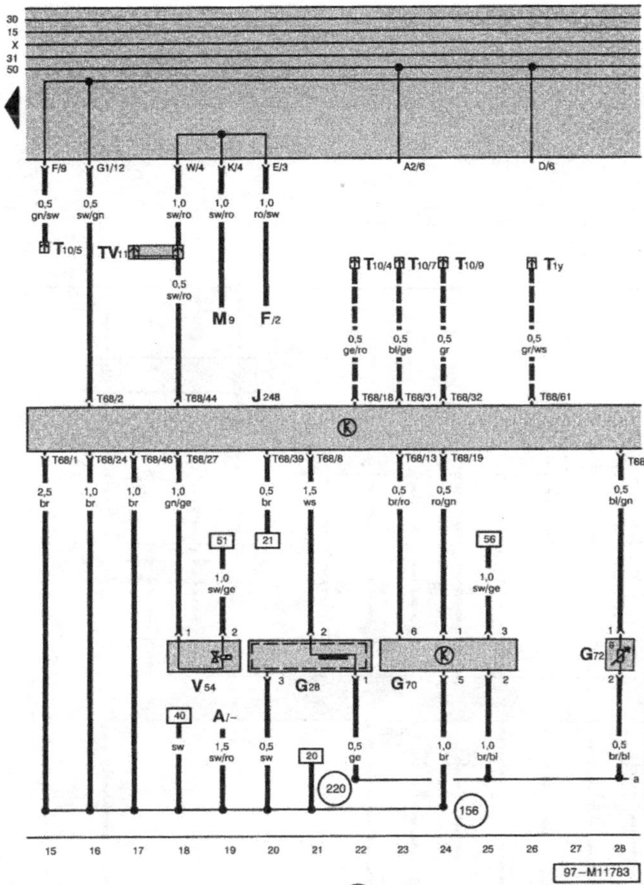

WIRING COLOR CODE		
ws	=	white
sw	=	black
ro	=	red
br	=	brown
gn	=	green
bl	=	blue
gr	=	grey
li	=	lilac
ge	=	yellow

A – Battery
F – Brake Light Switch
G28 – Engine Speed (RPM) Sensor
G70 – Mass Air Flow (MAF) Sensor
G72 – Intake Air Temperature (IAT) Sensor
J248 – Diesel Direct Fuel Injection (DFI) Engine Control Module (ECM)
M9 – Left Brake Light
T1y – Single Connector
T10 – 10-Pin Connector
T68 – 68-Pin Connector
TV11 – Terminal 54 Wire Connector
V54 – Metering Pump

(156) – Ground connection, in Diesel Direct Fuel Injection (DFI) wiring harness

(220) – Ground connection (sensor ground), in engine compartment wiring harness

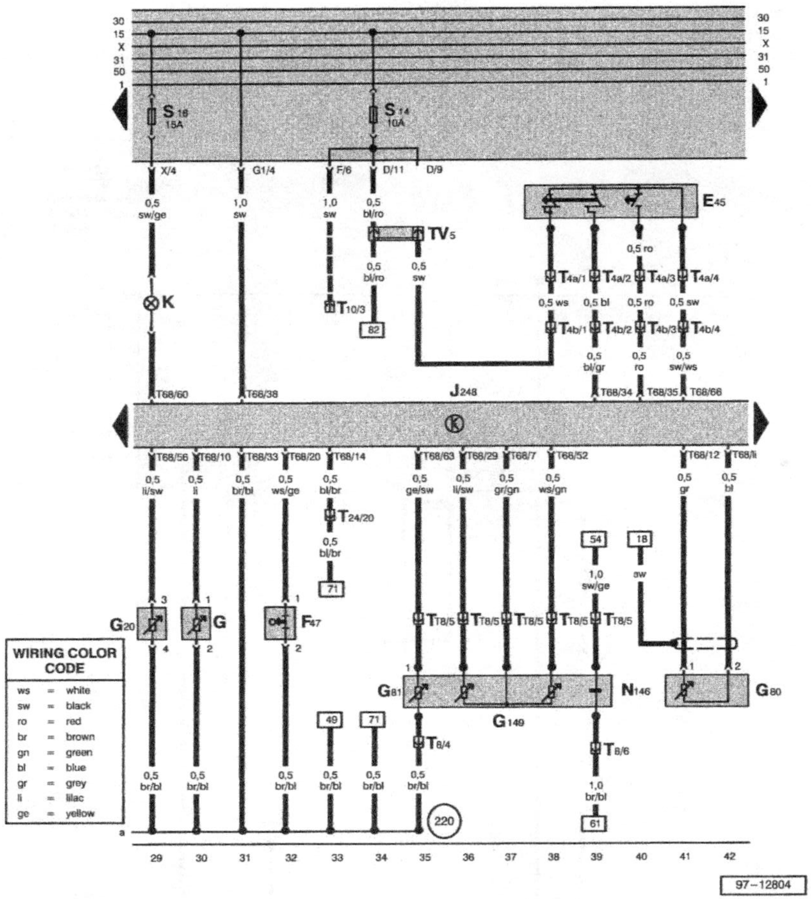

E45 – Cruise Control Switch
F47 – Brake Switch
G – Sender for Fuel Gauge
G20 – Temperature Sensor I for Catalyst
G80 – Needle Lift Sensor
G81 – Fuel Temperature Sensor
G149 – Modulating Piston Displacement Sensor
J248 – Diesel Direct Fuel Injection (DFI) Engine Control Module (ECM)
K – Instrument Cluster
N146 – Quantity Adjuster

T4a – 4-Pin Connector
T4b – 4-Pin Connector
T8 – 8-Pin Connector
T10 – 10-Pin Connector
T24 – 24-Pin Connector, in engine compartment
T68 – 68-Pin Connector, on Diesel DFI ECM
TV5 – Terminal 15a Wire Connector
(220) – Ground connection (sensor ground), in engine compartment wiring harness

Engine electronics

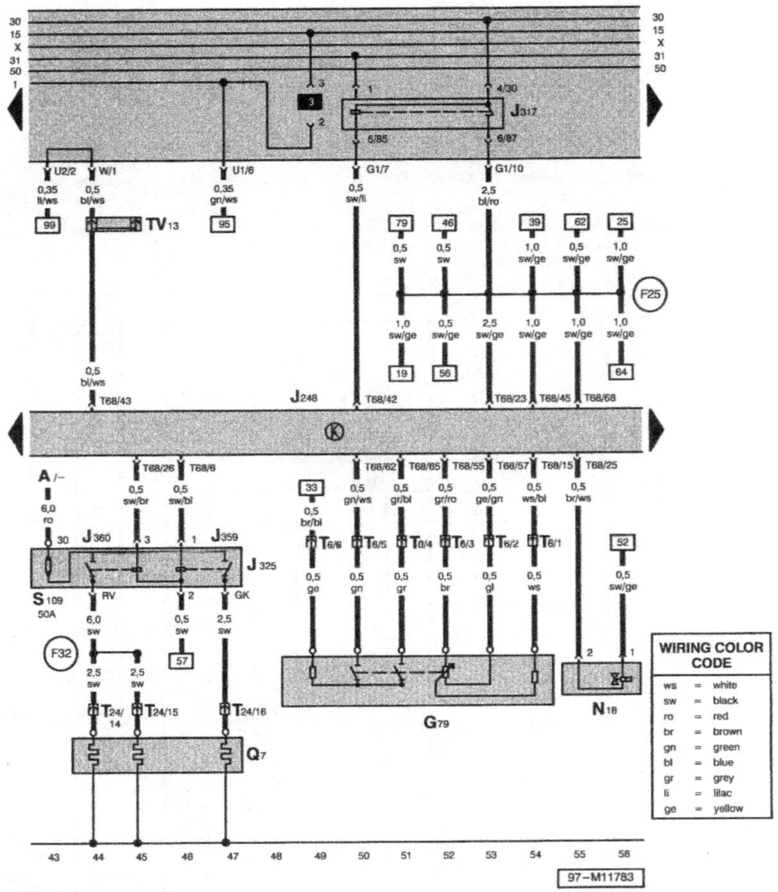

A – Battery
G79 – Throttle Position (TP) Sensor
J248 – Diesel Direct Fuel Injection (DFI) Engine Control Module (ECM)
J317 – Power Supply (Terminal 30, B+) Relay
J325 – Coolant System Glow Plug Relay, on bulkhead center
J359 – Low Heater Output Relay
J360 – High Heater Output Relay
N18 – EGR Vacuum Regulator Solenoid Valve
Q7 – Coolant System Glow Plugs
S109 – Coolant System Glow Plug Fuse
T6 – 6-Pin Connector, in footwell, front right
T24 – 24-Pin Connector, on engine

T68 – 68-Pin Connector, on engine control module
TV13 – Vehicle Speed Signal Wire Connector
(F25) – Wire connection -1-, in Diesel Direct Fuel Injection (DFI) system wiring harness
(F32) – Wire Connection (coolant system glow plugs), in Diesel direct injection system wiring harness

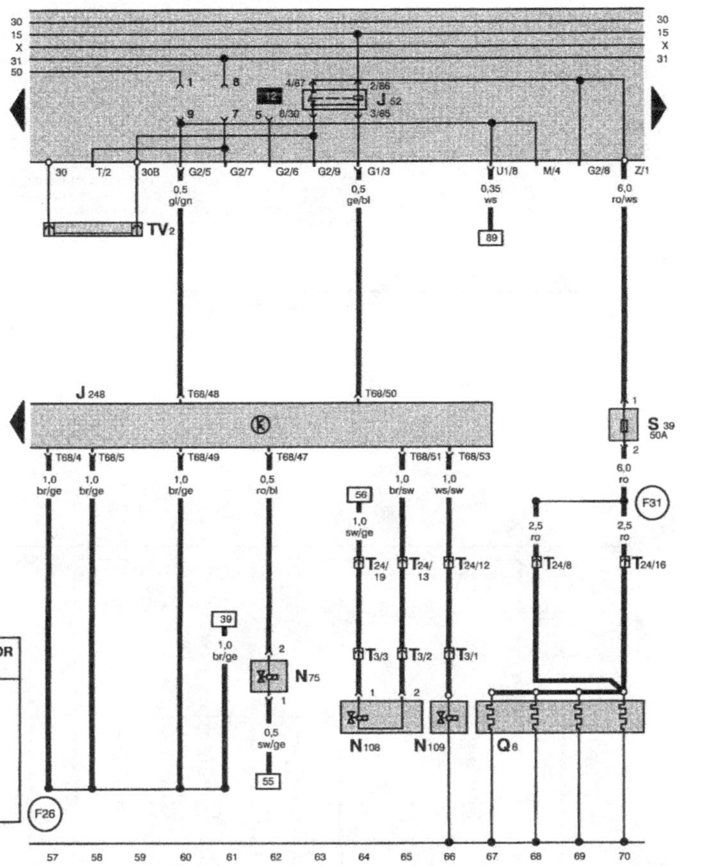

97–12804

WIRING COLOR CODE		
ws	=	white
sw	=	black
ro	=	red
br	=	brown
gn	=	green
bl	=	blue
gr	=	grey
li	=	lilac
ge	=	yellow

J52 – Glow Plug Relay
J248 – Diesel Direct Fuel Injection (DFI) Engine Control Module (ECM)
N75 – Wastegate Bypass Regulator Valve
N108 – Cold Start Injector
N109 – Fuel Cut-off Valve
Q6 – Engine Glow Plugs
S39 – Engine Glow Plug Fuse Strip
T3 – 3-Pin Connector
T24 – 24-Pin Connector, in engine compartment
T68 – 68-Pin Connector, on Diesel DFI ECM
TV2 – Terminal 30 (B+) Wire Connector

(F26) – Wire connection -2-, in Diesel Direct Fuel Injection (DFI) system wiring harness

(F31) – Wire connection (engine glow plugs), in Diesel Direct Fuel Injection (DFI) system wiring harness

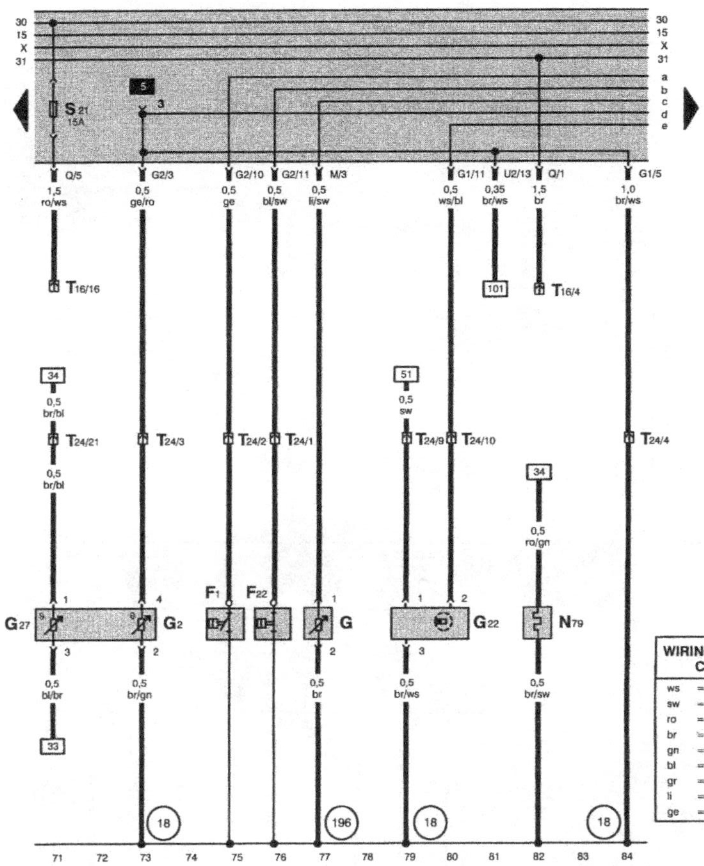

97–M11783

WIRING COLOR CODE		
ws	=	white
sw	=	black
ro	=	red
br	=	brown
gn	=	green
bl	=	blue
gr	=	grey
li	=	lilac
ge	=	yellow

F1 – Oil Pressure Switch
F22 – 0.3bar Oil Pressure Switch
G – Sender for Fuel Gauge
G2 – Engine Coolant Temperature (ECT) Sensor
G22 – Speedometer Vehicle Speed Sensor (VSS)
G27 – Sensor for Engine Temperature
N79 – Positive Crankcase Ventilation (PCV) Heating Element
T16 – Data Link Connector (DLC), behind center of dash
T24 – 24-Pin Connector, in engine compartment

(18) – Ground connection, on engine block

(196) – Ground connection -3-, in wiring harness, rear

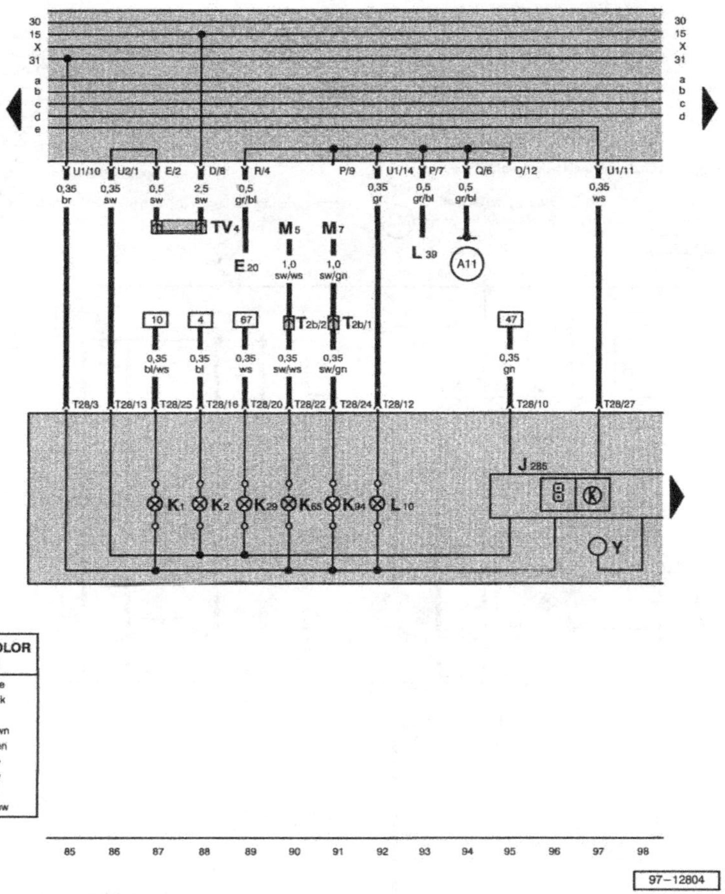

WIRING COLOR CODE

ws	=	white
sw	=	black
ro	=	red
br	=	brown
gn	=	green
bl	=	blue
gr	=	grey
li	=	lilac
ge	=	yellow

E20 – Instrument Panel Light Dimmer Switch
J285 – Control Module with Indicator Unit Instrument Panel Insert
K1 – Headlight High Beam Indicator Light
K2 – Generator (GEN) Warning Light
K29 – Glow Plug Indicator Light
K65 – Left Turn Signal Indicator Light
K94 – Right Turn Signal Indicator Light
L10 – Instrumant Cluster Light
L39 – Rear Window Defogger Switch Light
M5 – Left Front Turn Signal Light
M7 – Right Front Turn Signal Light
T2b – Double Connector
T28 – 28-Pin Connector, in engine compartment
TV4 – Terminal 14 Wire Connector
Y – Analog Clock
(A11) – Plus connection (58b), in instrument panel wiring harness

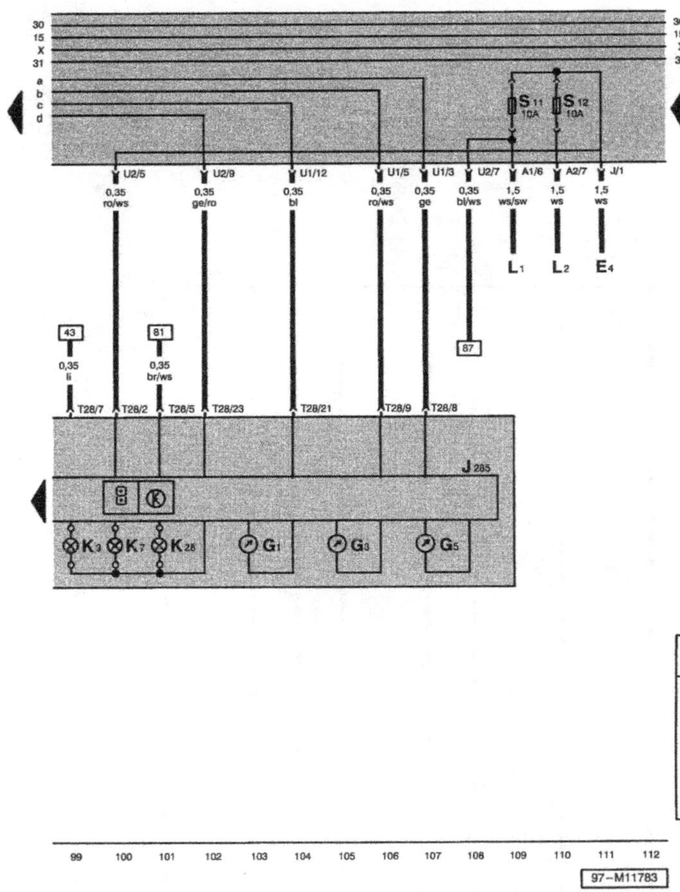

WIRING COLOR CODE

ws	=	white
sw	=	black
ro	=	red
br	=	brown
gn	=	green
bl	=	blue
gr	=	grey
li	=	lilac
ge	=	yellow

E4 – Headlight Dimmer/Flasher Switch
G1 – Fuel Gauge
G3 – Engine Coolant Temperature (ECT) Gauge
G5 – Tachometer
J285 – Control Module with Indicator Unit in Instrument Panel Insert
K3 – Oil Pressure Warning Light
K7 – Brake and Parking Brake Warning Light
K28 – Engine Coolant Level/Temperature Warning Light
L1 – Left Headlight
L2 – Right Headlight
T28 – 28-Pin Connector, instrument cluster

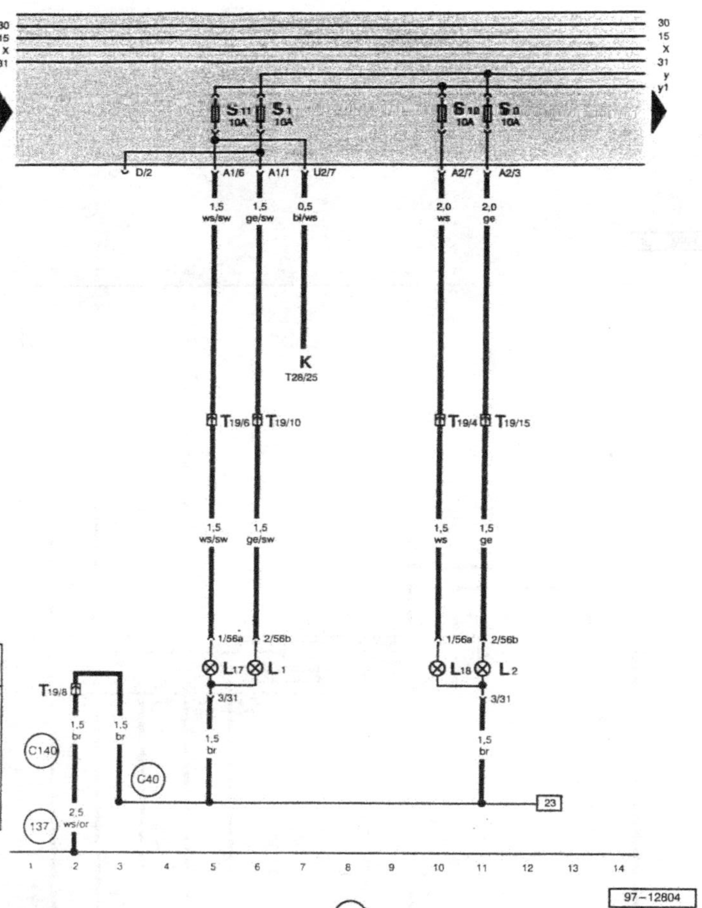

WIRING COLOR CODE		
ws	=	white
sw	=	black
ro	=	red
br	=	brown
gn	=	green
bl	=	blue
gr	=	grey
li	=	lilac
ge	=	yellow

K – Instrument Cluster
L1 – Left Headlight
L2 – Right Headlight
L17 – Left Sealed Beam High Headlight
L18 – Right Sealed Beam High Headlight
L19 – 19-Pin Connector, behind radiator grill, center
T28 – 28-Pin Connector, instrument cluster

(132) – Ground connection, in Motronic MFI wiring harness
(C40) – Wire connection, in front light wiring harness
(C104) – Wire connection, in Motronic MFI wiring harness

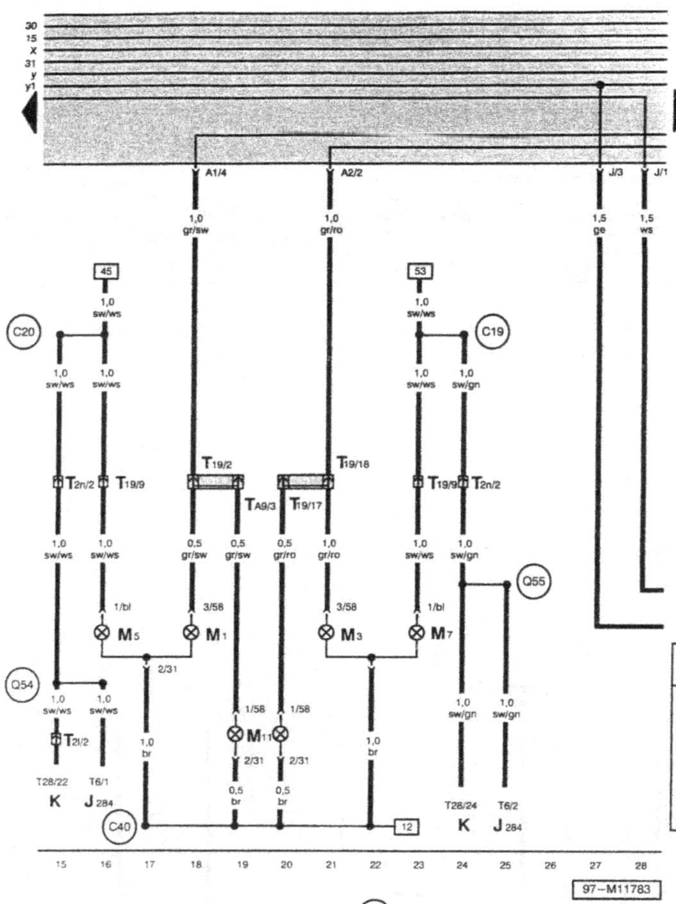

WIRING COLOR CODE		
ws	=	white
sw	=	black
ro	=	red
br	=	brown
gn	=	green
bl	=	blue
gr	=	grey
li	=	lilac
ge	=	yellow

J284 – Alarm System Control Module
K – Instrument Cluster
M1 – Left Parking Light
M3 – Right Parking Light
M5 – Left Front Turn Signal Light
M7 – Right Front Turn Signal Light
M11 – Front Side Marker Lights
T1o – Single Connector, behind fuse/relay panel
T2l – Double Connector, behind fuse/relay panel
T2n – Double Connector, behind fuse/relay panel
T6 – 6-Pin Connector, behind radiator grill, center
T19 – 19-Pin Connector, behind radiator grill, center
T28 – 28-Pin Connector, instrument cluster

(C19) – Wire connection (right turn signal), in headlight wire harness
(C20) – Wire connection (left turn signal), in headlight wiring harness
(C40) – Wire connection, in front light wiring harness
(Q54) – Wire connection (left turn signal), in alarm system wiring harness
(Q55) – Wire connection (right turn signal), in alarm system wiring harness

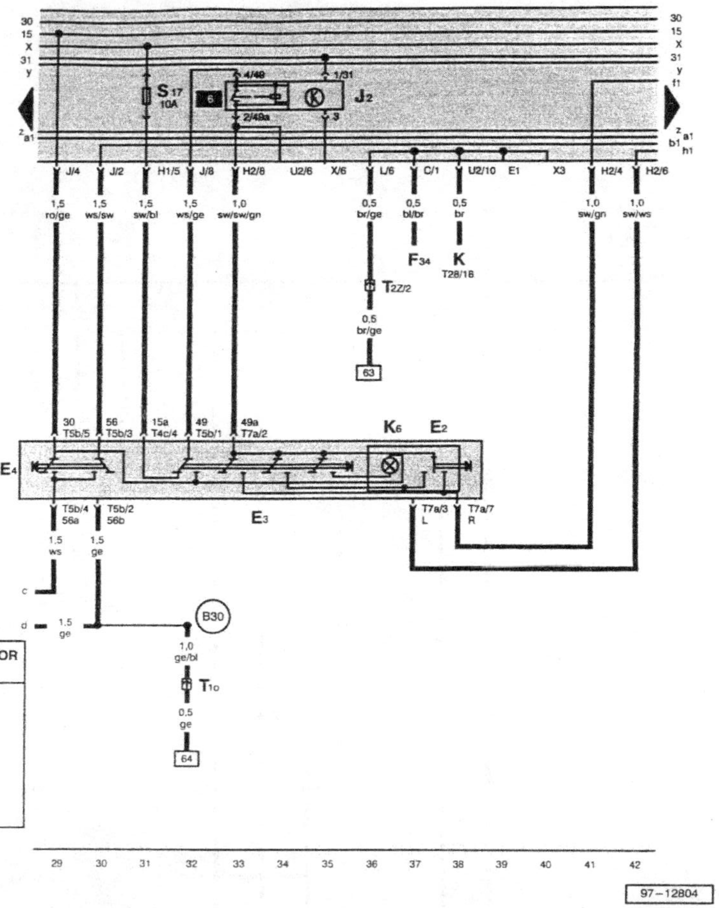

WIRING COLOR CODE

ws	=	white
sw	=	black
ro	=	red
br	=	brown
gn	=	green
bl	=	blue
gr	=	grey
li	=	lilac
ge	=	yellow

E2 – Trun Signal Switch
E3 – Emergency Flasher Switch
E4 – Headlight Dimmer/Flasher Switch
F34 – Brake Fluid Level Warning Switch
J2 – Emergency Flasher Relay
K – Instrument Cluster
K6 – Emergency Flasher Warning Light
T2z – Double Connector, behind fuse/relay panel
T4c – 4-Pin Connector, behind steering column switch trim
T5b – 5-Pin Connector, behind steering column switch trim
T7a – 7-Pin Connector, behind steering column switch trim
T10 – Single Connector, behind fuse/relay panel
T28 – 28-Pin Connector, instrument cluster
(B30) – Wire Connection, in light switch wiring harness

Daytime running lights (from January 1996)

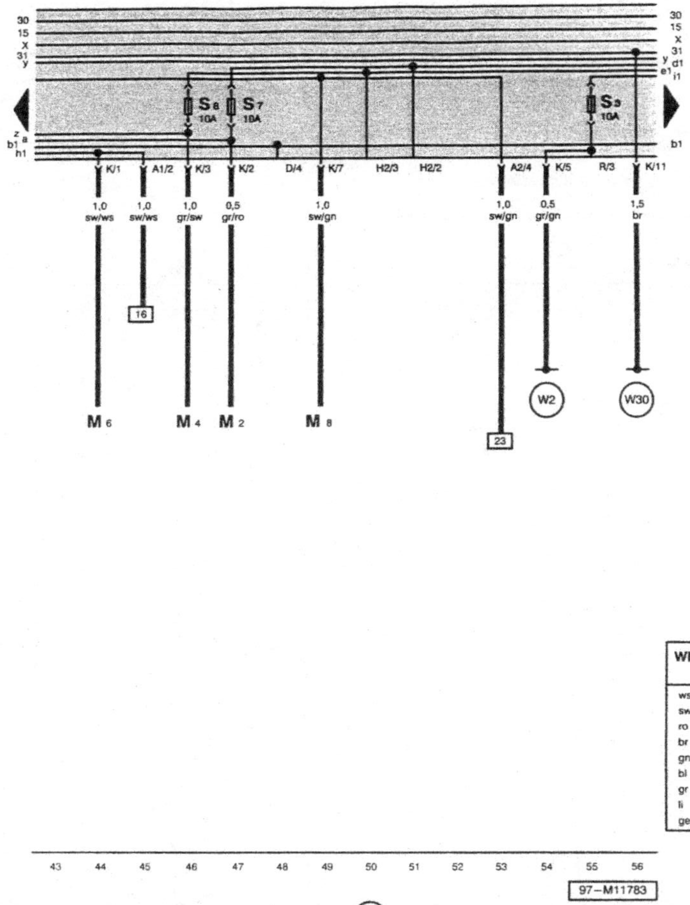

WIRING COLOR CODE

ws	=	white
sw	=	black
ro	=	red
br	=	brown
gn	=	green
bl	=	blue
gr	=	grey
li	=	lilac
ge	=	yellow

M2 – Right Tail Light
M4 – Left Tail Light
M6 – Left Rear Turn Signal Light
M8 – Right Rear Turn Signal Light

(W2) – Plus Connection (58), in rear wiring harness
(W30) – Wire connection, in rear wiring harness

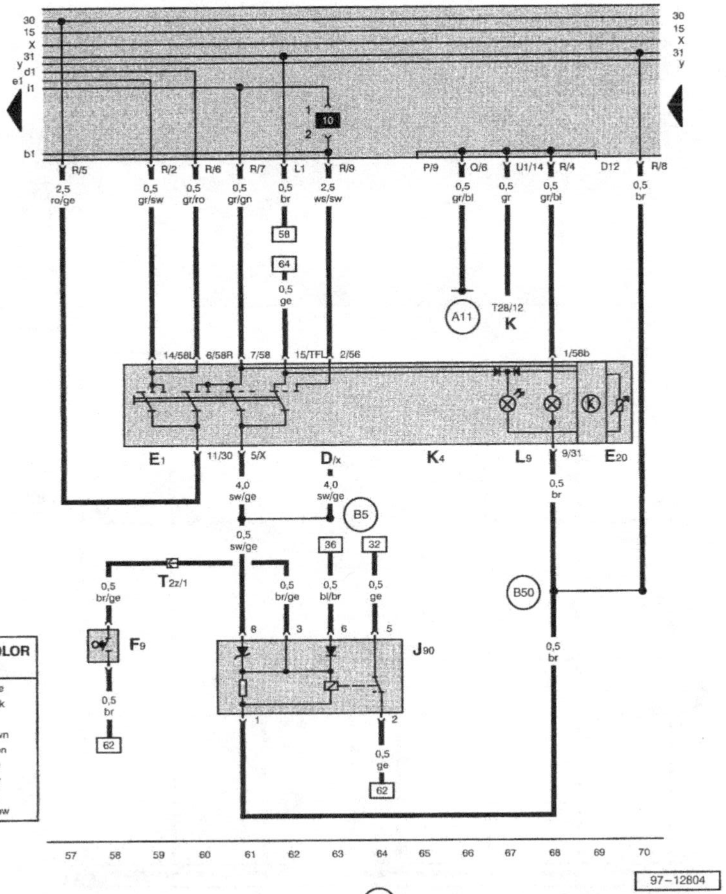

WIRING COLOR CODE		
ws	=	white
sw	=	black
ro	=	red
br	=	brown
gn	=	green
bl	=	blue
gr	=	grey
li	=	lilac
ge	=	yellow

D – Ignition/Starter Switch
E1 – Light Switch
E20 – Instrument Panel Light Dimmer Switch
F9 – Parking Brake Warning Light Switch
J90 – Daytime Running Lights Switch On Relay
K – Instrument Cluster
K4 – Park Light Indicator Light
L9 – Headlight Switch Light
T2z – Double Connector, behind fuse/relay panel
T28 – 28-Pin Connector, Instrument Cluster

A11 – Plus connection (58b), in instrument panel wiring harness
B5 – Wire connection (X), in headlight switch wiring harness
B50 – Wire connection (31), in headlight switch wiring harness

THIS PAGE INTENTIONALLY LEFT

BLANK

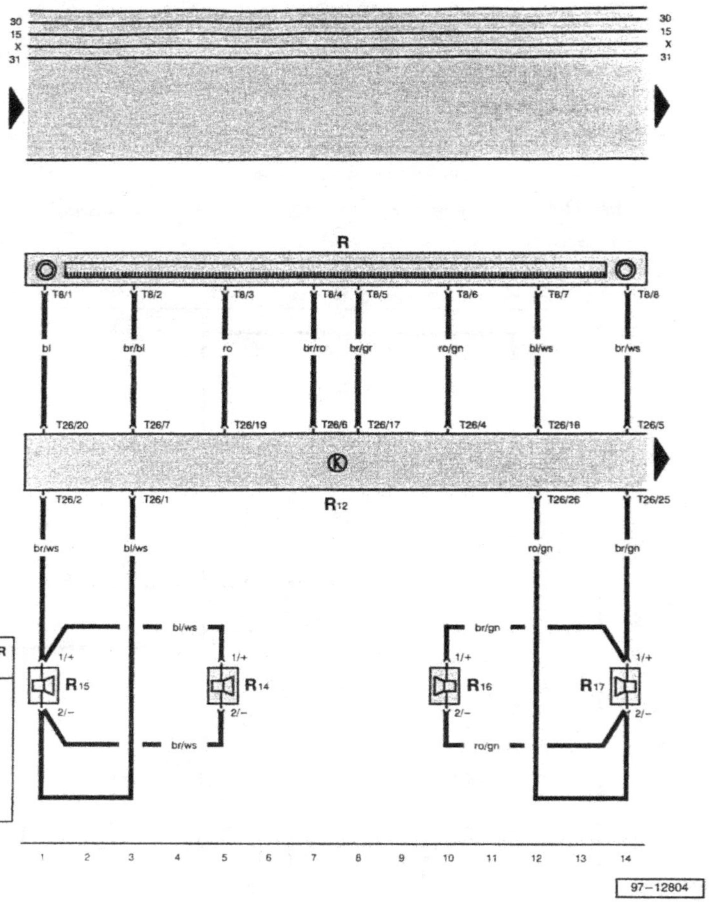

WIRING COLOR CODE

ws	=	white
sw	=	black
ro	=	red
br	=	brown
gn	=	green
bl	=	blue
gr	=	grey
li	=	lilac
ge	=	yellow

R – Radio
R12 – Amplifier
R14 – Left Rear Tweeter
R15 – Left Rear Woofer
R16 – Right Rear Tweeter
R17 – Right Rear Woofer
T8 – 8-Pin Connector, on Radio
T26 – 26-Pin Connector, on Amplifier

Stereo radio, Bose speakers (from January 1997)

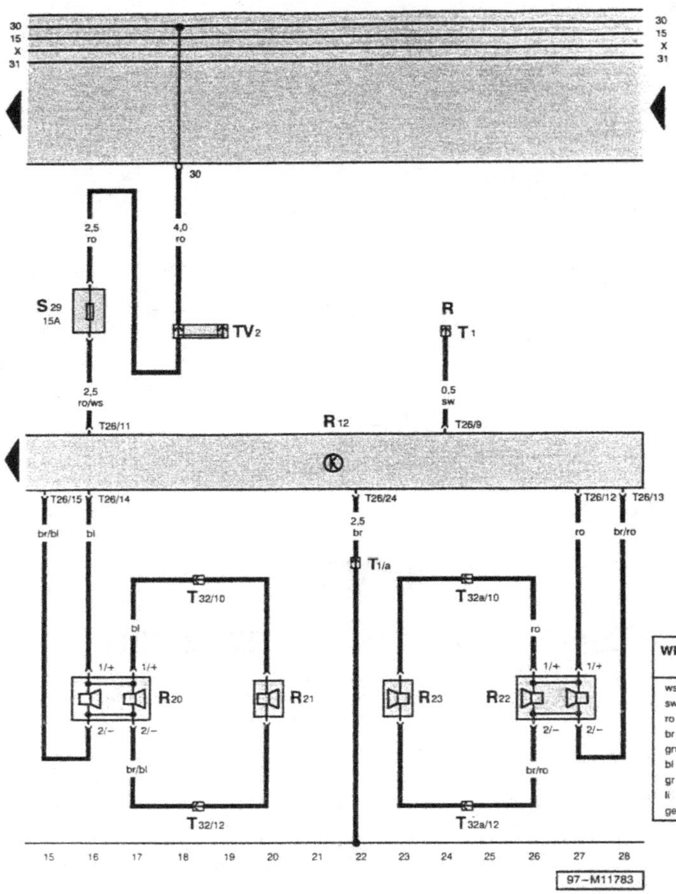

WIRING COLOR CODE

ws	=	white
sw	=	black
ro	=	red
br	=	brown
gn	=	green
bl	=	blue
gr	=	grey
li	=	lilac
ge	=	yellow

R – Radio
R12 – Amplifier
R20 – Left Front Tweeters
R21 – Left Front Woofer
R22 – Right Front Tweeters
R23 – Right Front Woofer
S29 – Radio fuse (in-line), above fuse/relay panel
T1 – Single Connector, to Radio
T1a – Single Connector
T2 – Double Connector
T8c – 8-Pin Connector
T26 – 26-Pin Connector, on Amplifier
T32 – 32-Pin Connector, on A-pillar, lower left
T32a – 32-Pin Connector, on A-pillar, lower right
TV2 – Terminal 30(B+) Wire Connector

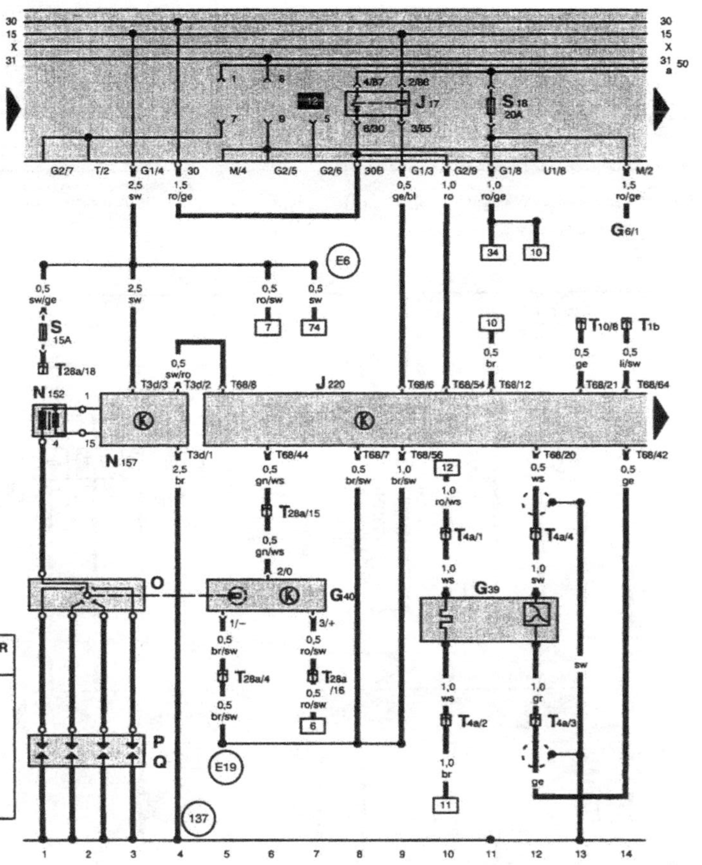

97–12804

WIRING COLOR CODE

ws	=	white
sw	=	black
ro	=	red
br	=	brown
gn	=	green
bl	=	blue
gr	=	grey
li	=	lilac
ge	=	yellow

- G6 – Fuel Pump (FP)
- G39 – Heated Oxygen Sensor (HO2S) 1
- G40 – Camshaft Position (CMP) Sensor
- J17 – Fuel Pump (FP) Relay
- J220 – Motronic Engine Control Module (ECM)
- N152 – Ignition Coil
- N157 – Ignition Coil Power Output Stage
- O – Ignition Distributor
- P – Spark Plug Connectors
- Q – Spark Plugs
- T1b – 1-Pin Connector
- T3d – 3-Pin Connector, on Ignition Coil Power Output Stage
- T4a – 4-Pin Connector
- T10 – 10-Pin Connector
- T28a – 28-Pin Connector, in engine compartment
- T68 – 68-Pin Connector, on Motronic ECM

- (137) – Ground connection (end stage), in Motronic MFI wiring harness
- (E6) – Plus connection -1- (15), in Motronic MFI wiring harness
- (E19) – Wire Connection– –, in Motronic Engine Control Module (EMC) wiring harness

Engine electronics (from June 1996)

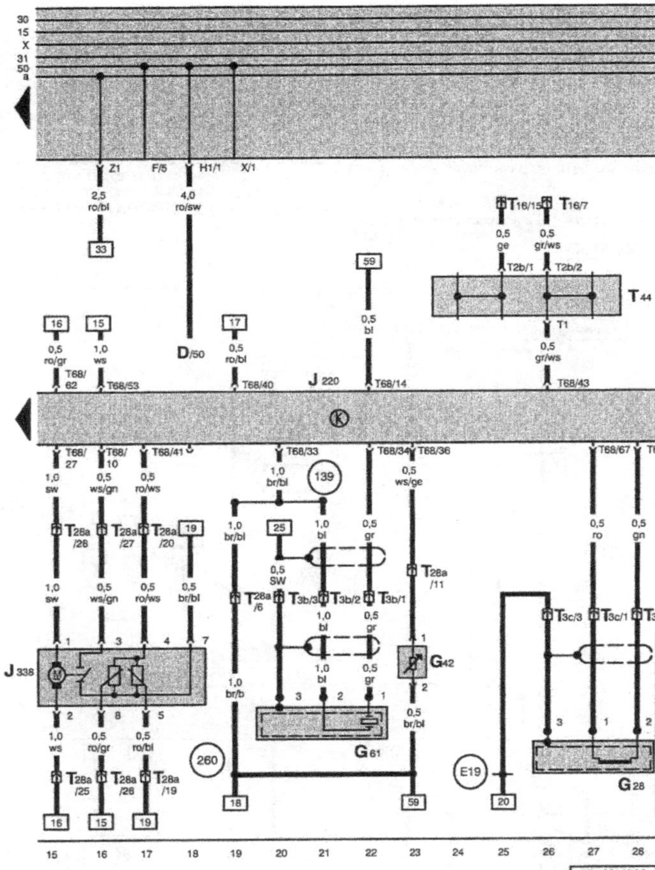

97–M11783

WIRING COLOR CODE

ws	=	white
sw	=	black
ro	=	red
br	=	brown
gn	=	green
bl	=	blue
gr	=	grey
li	=	lilac
ge	=	yellow

- D – Ignition/Starter Switch
- G28 – Engine Speed (RPM) Sensor
- G42 – Intake Air Temperature (IAT)Sensor
- G61 – Knock Sensor (KS) 1
- J220 – Motronic Engine Control Module (ECM)
- J338 – Throttle Valve Control Module
- T2b – Double Connector, above fuse/relay panel
- T3b – 3-Pin Connector
- T3c – 3-Pin Connector
- T16 – Data Link Connector (DLC), behind center of dash
- T28a – 28-Pin Connector, in engine compartment
- T42 – 42-Pin Connector
- T44 – Junction Box for On-Board Diagnosis, above fuse/relay panel
- T68 – 68-Pin Connector, on Motronic ECM

- (138) – Ground connection (control module), in Motronic MFI wiring harness
- (139) – Ground connection (sensor ground), in Motronic Multiport Fuel Injection (MFI) wiring harness
- (260) – Ground connection (sensor ground), in injector wiring harness
- (E19) – Wire Connection – – in Motronic Engine Control Module (ECM) wiring harness

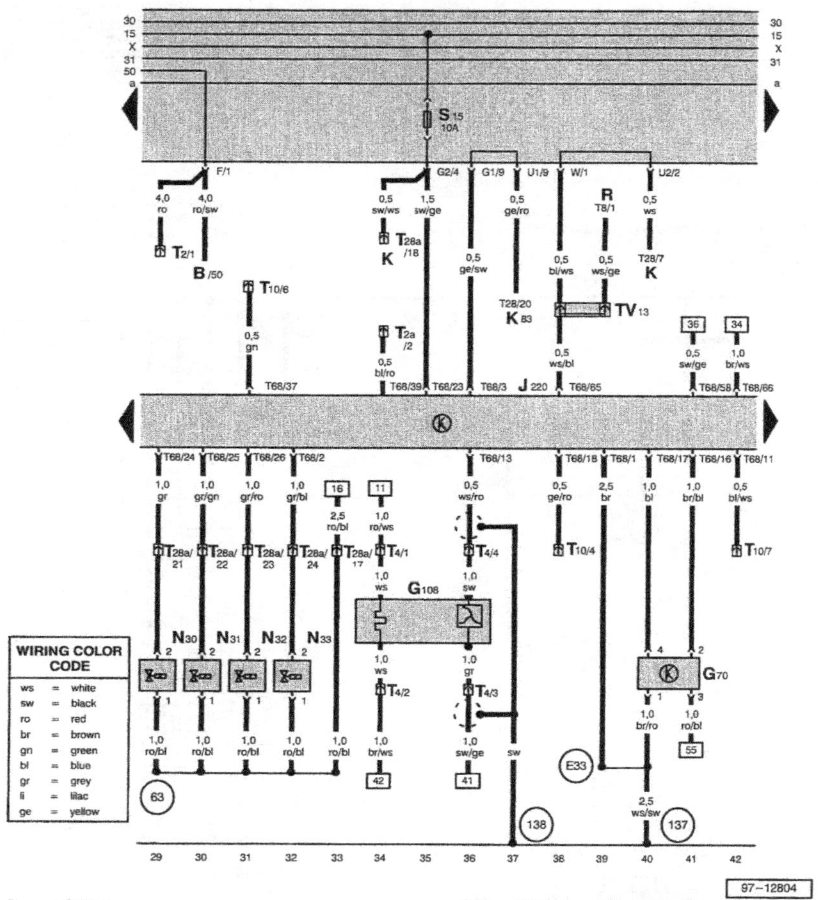

B – Starter
G70 – Mass Air Flow (MAF) Sensor
G108 – Heated Oxygen Sensor (HO2S)
J220 – Motronic Engine Control Module (ECM)
K – Instrument Cluster
K83 – Multi-Function Indicator Lamp (MIL)
N30 – Cylinder 1 Fuel Injector
N31 – Cylinder 2 Fuel Injector
N32 – Cylinder 3 Fuel Injector
N33 – Cylinder 4 Fuel Injector
R – Radio
T2 – Double Connector
T2a – Double Connector
T4 – 4-Pin Connector
T8 – 8-Pin Connector
T10 – 10-Pin Connector
T28 – 28-Pin Connector, instrument cluster
T28a – 28-Pin Connector, in engine compartment
T68 – 68-Pin Connector, on Motronic ECM

TV13 – Vehicle Speed Signal Wire Connector
(63) – Ground connection, on left taillight bulb holder
(137) – Ground connection (end stage), in Motronic MFI wiring harness
(138) – Ground connection (sensor ground), in Motronic MFI wiring harness
(E33) – Ground connection, in Motronic MFI wiring harness

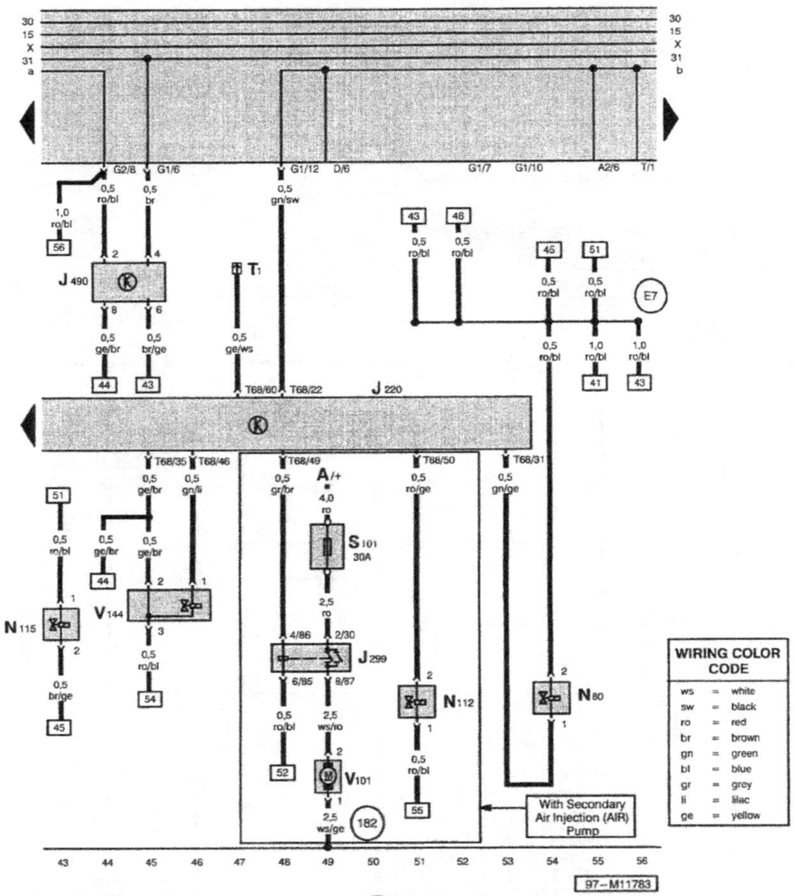

A – Battery
J220 – Motronic Engine Control Module (ECM)
J299 – Secondary Air Injection (AIR) Pump Relay
J490 – Fuel System Diagnostic Control Module
N80 – Evaporative Emission (EVAP) Canister Purge Regulator Valve
N112 – Secondary Air Injection (AIR) Solenoid Valve
N115 – Evaporative Emission (EVAP) Canister Purge Solenoid Valve

S101 – A/C-Climatronic Control Module Fuse
T68 – 68-Pin Connector, on Motronic ECM
V101 – Secondary Air Injection (AIR) Pump Motor
V144 – Leak Detection Pump (LDP)
T1 – 1-pin Connector, engine compartment wiring harness

(182) – Ground connection -1-, in engine compartment wiring harness
(E7) – Plus Connection (87a), in Motronic MFI wiring harness

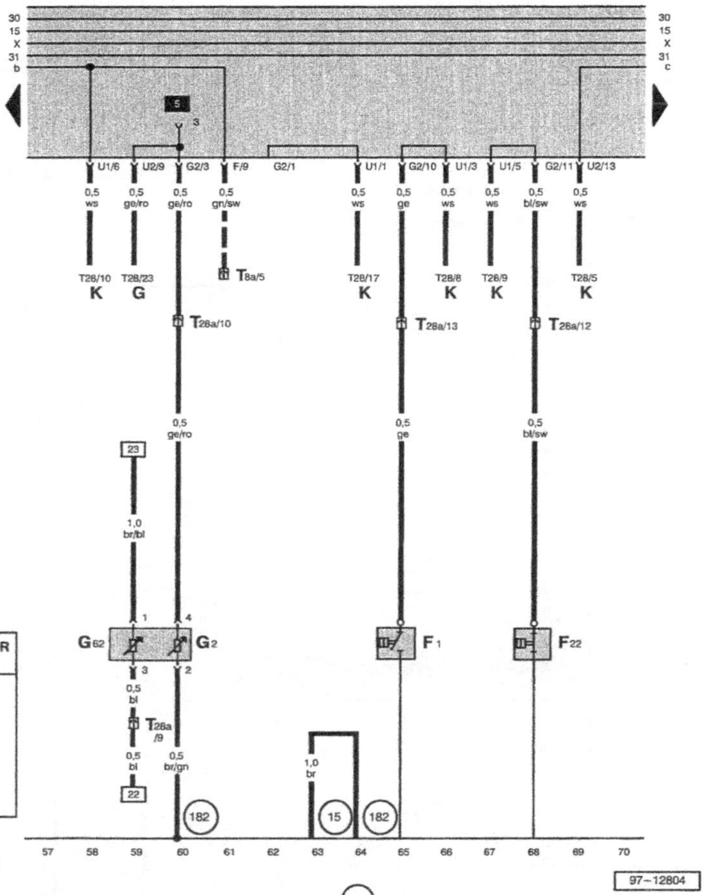

WIRING COLOR CODE		
ws	=	white
sw	=	black
ro	=	red
br	=	brown
gn	=	green
bl	=	blue
gr	=	grey
li	=	lilac
ge	=	yellow

F1 – Oil Pressure Switch
F22 – 0.3 bar Oil Pressure Switch
G – Sender for Fuel Gauge
G2 – Engine Coolant Temperature (ECT) Sensor
G62 – Engine Coolant Temperature (ECT) Sensor
K – Instrument Cluster
T8a – 8-Pin Connector
T28 – 28-Pin Connector, on Instrument Cluster
T28a – 28-Pin Connector, in Engine Compartment

(15) – Ground connection, on cylinder head

(182) – Ground connection -1-, in engine compartment wiring harness

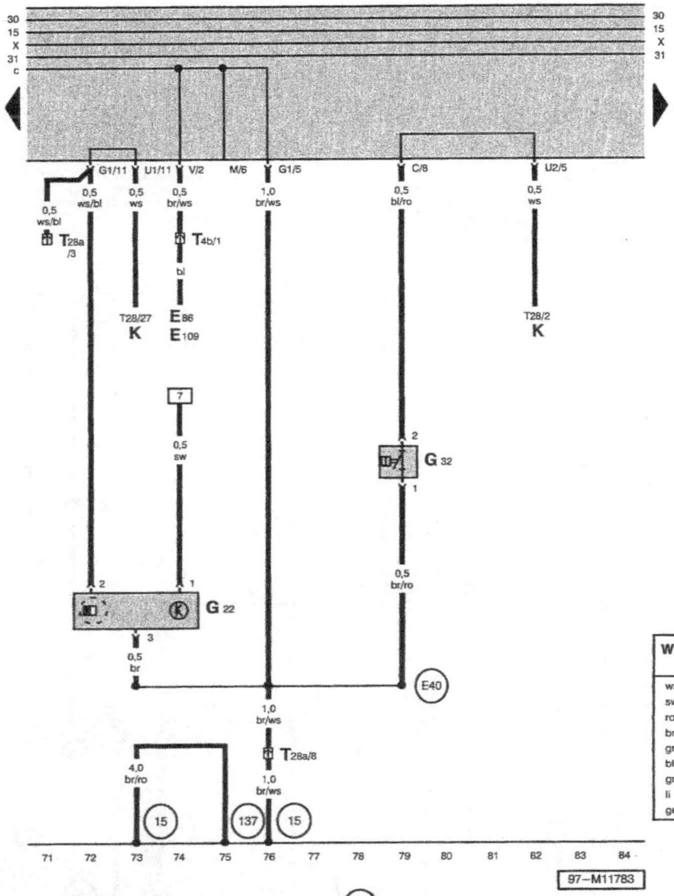

WIRING COLOR CODE		
ws	=	white
sw	=	black
ro	=	red
br	=	brown
gn	=	green
bl	=	blue
gr	=	grey
li	=	lilac
ge	=	yellow

E86 – Multi-Function Indicator Mode Select Switch
E109 – Multi-Function Indicator Memory Switch
G22 – Speedometer Vehicle Speed Sensor (VSS)
G32 – Engine Coolant Level (ECL) Sensor
K – Instrument Cluster
T4b – 4-Pin Connector
T28 – 28-Pin Connector, on Instrument Cluster
T28a – 28-Pin Connector, in engine compartment

(15) – Ground connection, on cylinder head

(137) – Ground connection (end stage), in Motronic MFI wiring harness

(E40) – Wire connection, in Motronic MFI Wiring harness

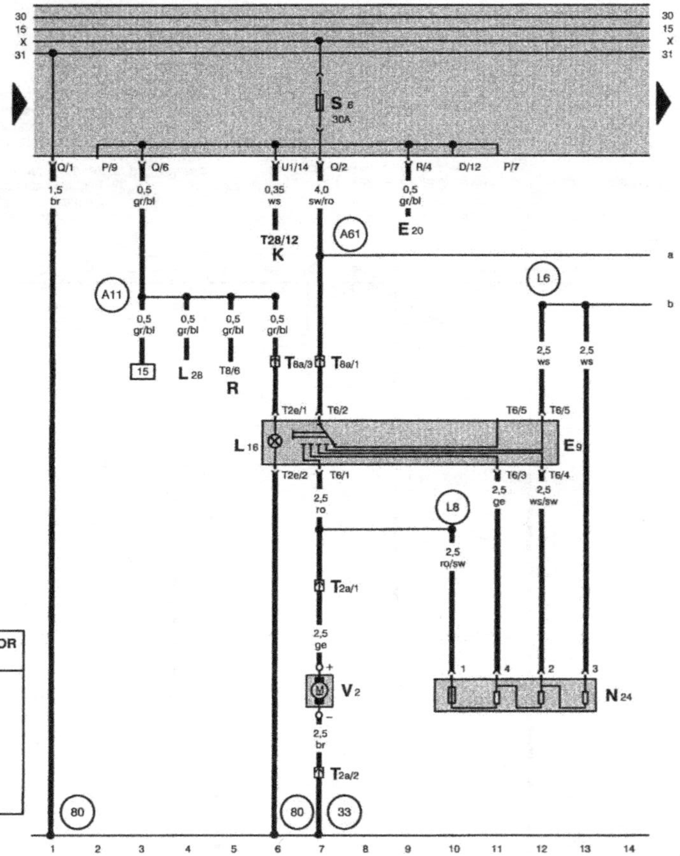

WIRING COLOR CODE		
ws	=	white
sw	=	black
ro	=	red
br	=	brown
gn	=	green
bl	=	blue
gr	=	grey
li	=	lilac
ge	=	yellow

E9 – Fresh Air Blower Switch
E20 – Instrument Panel Light Dimmer Switch
K – Instrument Cluster
L16 – Fresh Air Control Lever Light
L28 – Cigarette Lighter Light
N24 – Fresh Air Blower Series Rsistance with Fuse
R – Radio
T2a – Double Connector, behind dash panel, right
T2e – Double Connector, on Fresh Air Control Lever Light
T6 – 6-Pin Connector, on Fresh Air Blower Switch
T8 – 8-Pin Connector, on Radio
T8a – 8-Pin Connector, behind dash panel, center
T28 – 28-Pin Connector, on Instrument Cluster
V2 – Fresh Air Blower.

(33) – Ground connection, behind instrument panel, right
(80) – Ground connection -1-, in instrument panel wiring harness
(A11) – Plus connection (58b), in instrument panel wiring harness
(A61) – Plus connection (X), in instrument cluster wiring harness
(L6) – Wire connection -2-, in evaporator housing wiring harness
(L8) – Wire connection -4-, in evaporator housing wiring harness

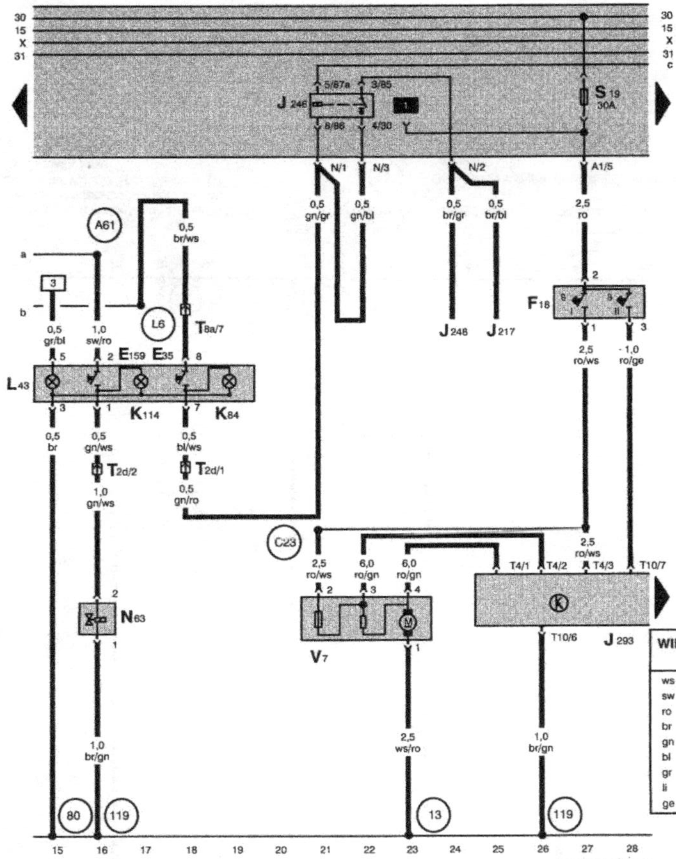

WIRING COLOR CODE		
ws	=	white
sw	=	black
ro	=	red
br	=	brown
gn	=	green
bl	=	blue
gr	=	grey
li	=	lilac
ge	=	yellow

E35 – A/C Switch
E159 – Fresh Air/Recirculating Flap Switch
F18 – Coolant Fan Control (FC) Thermal Switch
J217 – Transmission Control Module (TCM)
J246 – A/C Clutch Cut-off Relay
J248 – Diesel Direct Fuel Injection (DFI) Engine Control Module (ECM)
J293 – Coolant Fan Control (FC) Control Module
K84 – A/C Indicator Light
K114 – Fresh Air and Recirculating Air Mode Light
L43 – A/C Switch Illumination Light
N63 – Fresh Air/Recirculating Flap Two-Way Valve
T2d – Double Connector, behind relay plate
T4 – 4-Pin Connector, on Coolant Fan Control Unit
T8a – 8-Pin Connector, behind dash panel, center
T10 – 10-Pin Connector, on Coolant Fan Control Unit
V7 – Coolant fan

(80) – Ground Connection -1-, in instrument panel wiring harness
(119) – Ground connection -1-, in headlight wiring harness
(A61) – Plus connection (X), in instrument cluster wiring harness
(C23) – Wire connection (blower, 1st speed), in headlight/air conditioner wiring harness
(L6) – Wire connection -2-, in evaporator housing wiring harness

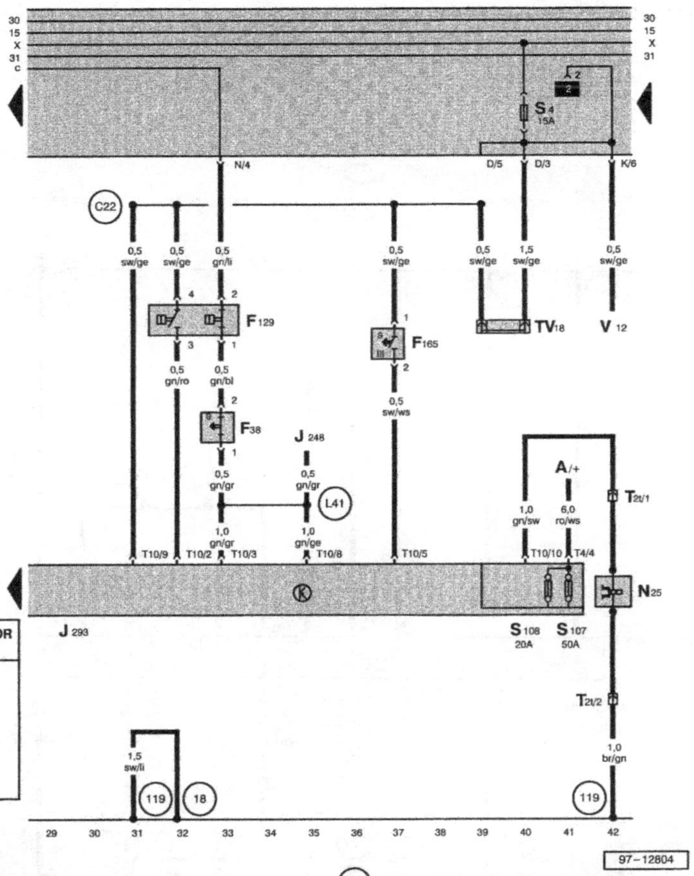

WIRING COLOR CODE

ws	=	white
sw	=	black
ro	=	red
br	=	brown
gn	=	green
bl	=	blue
gr	=	grey
li	=	lilac
ge	=	yellow

- A – Battery
- F38 – Ambient Temperature Switch
- F129 – A/C Pressure Switch
- F165 – Third Speed Coolant Fan Control (FC) Thermal Switch
- J248 – Diesel Direct Fuel Injection (DFI) Engine Control Module (ECM)
- J293 – Coolant Fan Control (FC) Control Module
- N25 – A/C Clutch
- S107 – Coolant Fan Second and Third Speed Fuse
- S108 – Coolant Fan First Speed and A/C Clutch Fuse
- T2t – Double Connector, near compressor
- T4 – 4-Pin Connector, on Coolant Fan Control Unit
- T10 – 10-Pin Connector, on Coolant Fan Control Unit
- TV18 – Distributor for Terminal Xa
- V12 – Rear Window Wiper Motor

- (18) – Ground connection, on engine block
- (119) – Ground connection -1-, in headlight wiring harness
- (C22) – Wire Connection (X), in headlight/air conditioner wiring harness
- (L41) – Wire connection (A/C on), in A/C system–coolant fan wiring harness

THIS PAGE INTENTIONALLY LEFT

BLANK

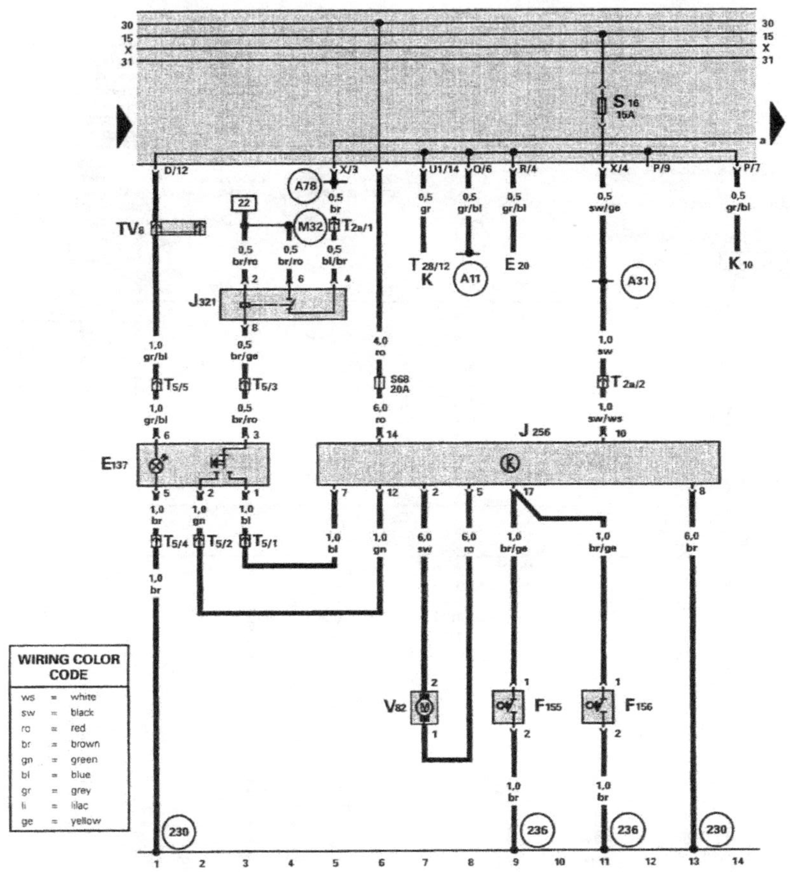

E20 – Instrument Panel Light Dimmer Switch
E137 – Convertible Top Operation Switch
F155 – Left Contact Switch (Cover Installed)
F156 – Right Contact Switch (Cover Installed)
J256 – Convertible Top Control Module
J321 – Convertible Top Hydraulic Pump Relay
K – Instrument Cluster
K10 – Rear Window Defogger Indicator Light
S68 – Convertible Top Circuit Breaker
T2a – 2-Pin Connector
T5 – 5-Pin Connector
TV8 – Adapter for Terminal 58B
T28 – 28-Pin Connector
V82 – Convertible Top Motor

(A11) – On-Board Charging Unit
(230) – Ground Connection – convertible top wiring harness
(236) – Ground Connection – convertible top wiring harness
(A31) – Plus Connection (15), in control light wiring harness
(A78) – Wire Connection (su), in control light wiring harness
(M32) – Wire Connection (su), in convertible top wiring harness

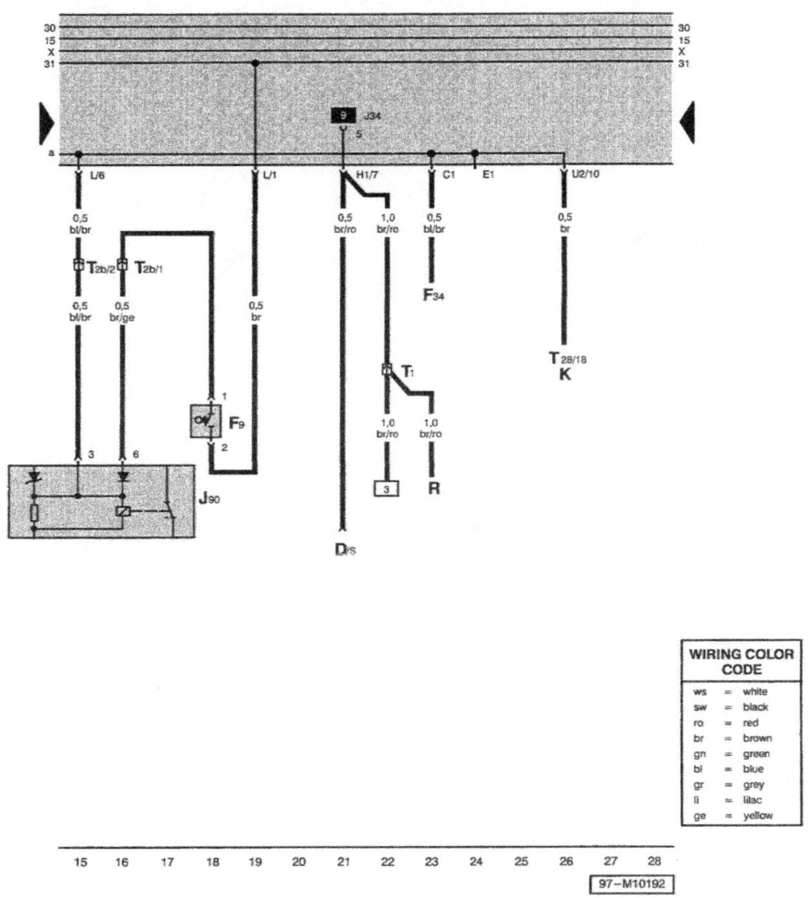

D – Ignition/Starter Switch
F9 – Parking Brake Warning Light Switch
F34 – Brake Fluid Level Warning Switch
J34 – Seat Belt Warning System Relay
J90 – Daytime Running Lights Switch ON Relay
K – Instrument Cluster
R – Radio
T1 – 1-Pin Connector
T2b – 2-Pin Connector
T28 – 28-Pin Connector

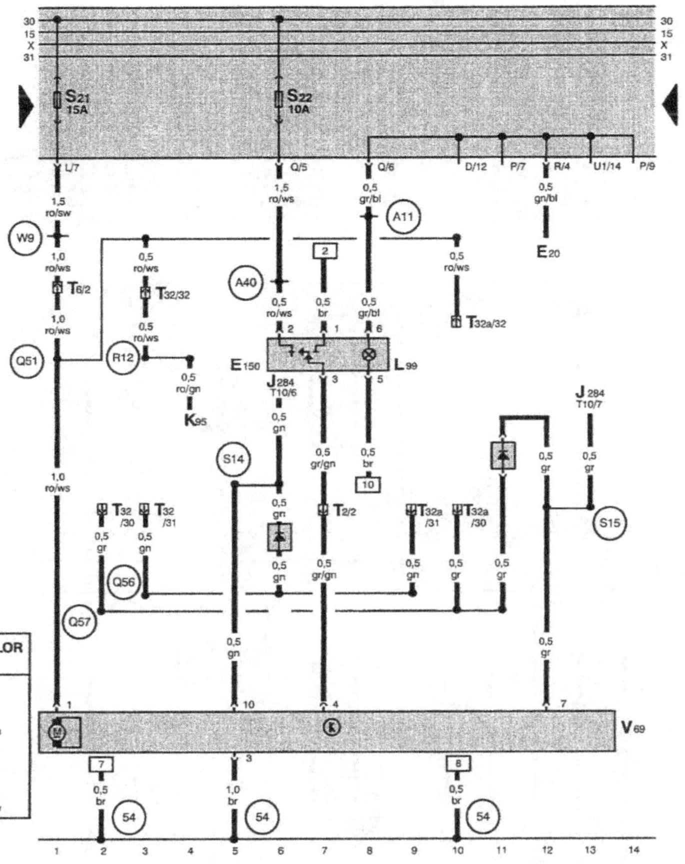

WIRING COLOR CODE		
ws	=	white
sw	=	black
ro	=	red
br	=	brown
gn	=	green
bl	=	blue
gr	=	grey
li	=	lilac
ge	=	yellow

E20 – Instrument Panel Light Dimmer Switch
E150 – Door Lock - Unlock Switch
F59 – Driver's Door Central Locking System Door Switch
F114 – Passenger's Door Central Locking System Door Switch
J284 – Alarm System Control Module
L99 – Lighting Switch for Interior Lock
K95 – Alarm System Indicator Light
T2 – Double Connector, near Door Lock - Unlock Switch
T6 – 6-Pin Connector
T10 – 10-Pin Connector, behind dash panel, left, on Alarm System Control Module
T32 – 32-Pin Connector, on A-pillar, lower left
T32a – 32-Pin Connector, on A-pillar, lower right
V69 – Central Locking System Pump and Control Module
(A11) – Plus connection (58b) in instrument panel wiring harness
(A40) – Plus connection -1- (30), in instrument cluster wiring harness
(54) – Ground connection, on rear apron
(Q51) – Plus connection (30), in alarm system wiring harness
(Q56) – Wire connection (open), in alarm system wiring harness
(Q57) – Wire connection (closed), in alarm system wiring harness
(R12) – Plus connection -3- (30), in driver's door wiring harness
(S14) – Wire connection (Open), in central locking system wiring harness
(S15) – Wire connection (Closed), in central locking system wiring harness
(W9) – Plus connection (30), in rear wiring harness

THIS PAGE INTENTIONALLY LEFT

BLANK

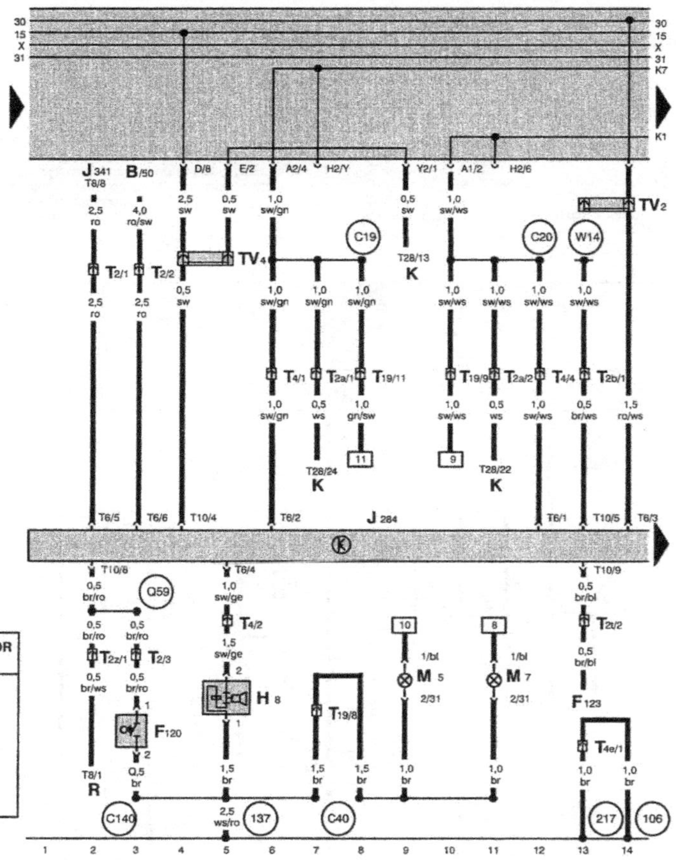

WIRING COLOR CODE

ws	=	white
sw	=	black
ro	=	red
br	=	brown
gn	=	green
bl	=	blue
gr	=	grey
li	=	lilac
ge	=	yellow

B – Starter
E19 – Connector –1– (right turn signal), in wire harness headlamp
F120 – Hood Alarm Switch
F123 – Truck Lid Alarm Switch
H8 – Alarm Horn
J284 – Alarm System Control Module
J341 – Anti-Theft Engine Disable Relay I
K – Instrument Cluster
M5 – Left Front Turn Signal Right
M7 – Right Front Turn Signal Left
R – Radio
T2 – 2-Pin Connector
T2a – 2-Pin Connector
T2b – 2-Pin Connector
T2t – 2-Pin Connector
T2z – 2-Pin Connector
T4 – 4-Pin Connector
T6 – 6-Pin Connector
T8/8 – 8-Pin Connector
T10 – 10-Pin Connector
T19 – 19-Pin Connector
T28 – 28-Pin Connector
TV2 – Wire Connector, terminal (B+)
TV4 – Terminal 15 Wire Connector
V69 – Central Locking System Pump and Control Module
C19 – Wire connection, -1- right turn signal, in wire harness headlamp
C20 – Wire connection (left turn signal), in headlight wiring harness
C40 – Wire connection, in front light wiring harness
C140 – Wire connection, in Motronic MFI wiring harness
Q59 – Wire connection, (sensor) in alarm system wiring harness
W14 – Wire connection (58R), in rear wiring harness
106 – Ground connection, in alarm system wiring harness
137 – Ground connection, in Motronic MFI wiring harness
217 – Ground connection, in alarm horn wiring harness

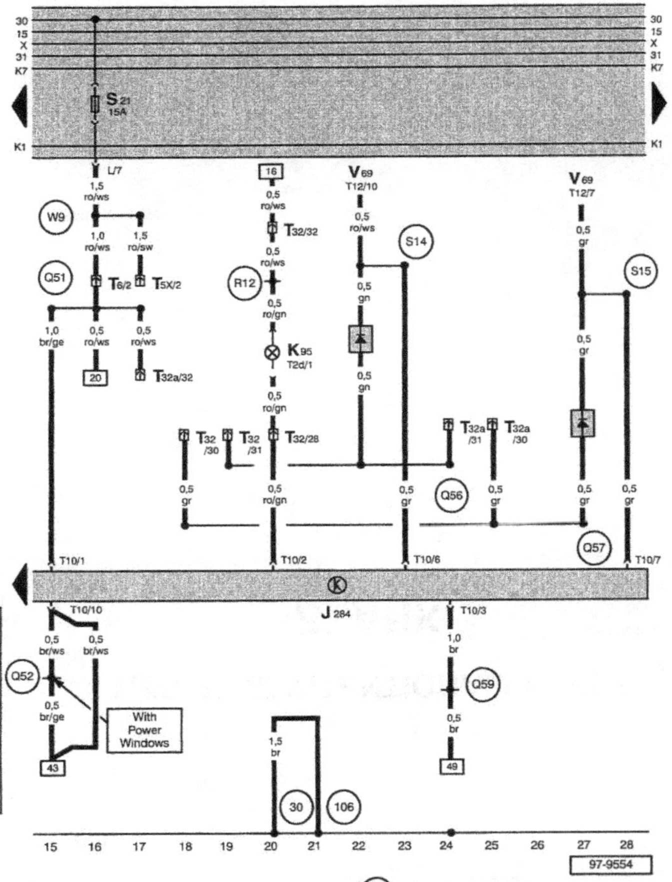

WIRING COLOR CODE

ws	=	white
sw	=	black
ro	=	red
br	=	brown
gn	=	green
bl	=	blue
gr	=	grey
li	=	lilac
ge	=	yellow

97-9554

J284 – Alarm System Control Module
K95 – Alarm System Indicator Light
T1y – Single connector in terminal 31 (end) wire connector
T1z – Single connector in headlight wiring harness
T2d – Double Connector in left front door
T5x – 5-Pin Connector
T10 – 10-Pin Connector behind dash panel, left, on Alarm System Control Module
T32 – 32-Pin Connector on A-pillar, lower left
T32a – 32-Pin Connector on A-pillar, lower right
V69 – Central locking pump and control module
30 – Ground connection -1-, beside fuse/relay panel
106 – Ground connection, in alarm system wiring harness
Q51 – Plus connection (30), in alarm system wiring harness
Q52 – Wire connection (door contact switch), in alarm system wiring harness
Q56 – Wire connection (open), in alarm system wiring harness
Q57 – Wire connection (closed), in alarm system wiring harness
Q59 – Wire connection (31) in alarm system wiring harness
Q60 – Wire connection, in alarm system wiring harness
R12 – Plus connection -3- (30), in driver's door wiring harness
S14 – Wire connection (open), in central locking system wiring harness
S15 – Wire connection (closed), in central locking system wiring harness
W9 – Plus connection (30), in rear wiring harness

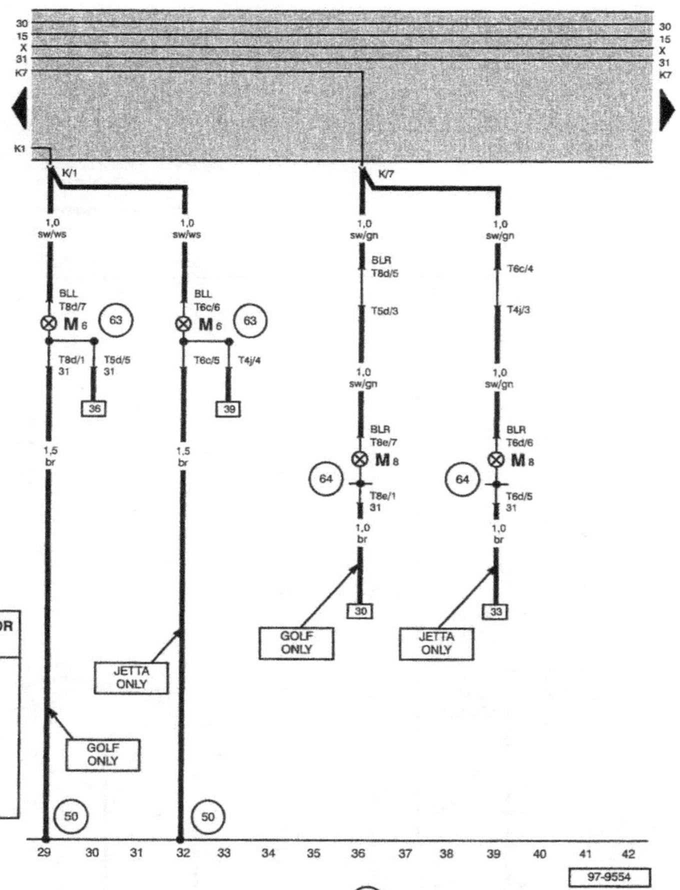

WIRING COLOR CODE

ws	=	white
sw	=	black
ro	=	red
br	=	brown
gn	=	green
bl	=	blue
gr	=	grey
li	=	lilac
ge	=	yellow

M6 – Left Rear Turn Signal Light
M8 – Right Rear Turn Signal Light
T5d – 5-Pin Connector, on tail light, left
T4j – 4-Pin Connector on tail light, left
T6c – 6-Pin Connector on tail light, left
T6d – 6-Pin Connector on tail light, right
T8d – 8-Pin Connector, on tail light, left
T8e – 8-Pin Connector, on tail light, right

(50) – Ground connection, in luggage compartment lid
(63) – Ground connection, on left tail light bulb holder
(64) – Ground connection, on right tail light bulb holder

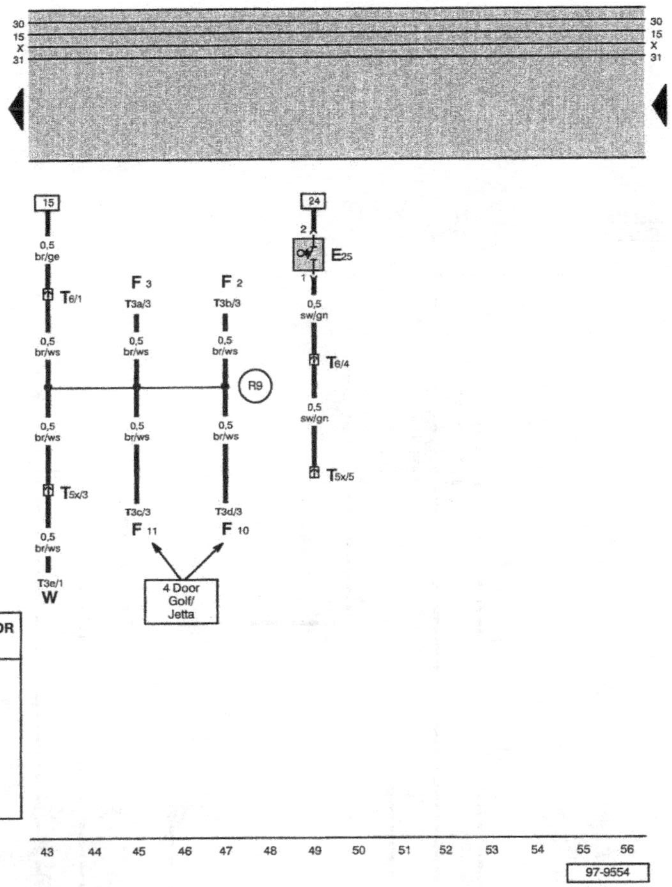

WIRING COLOR CODE

ws	=	white
sw	=	black
ro	=	red
br	=	brown
gn	=	green
bl	=	blue
gr	=	grey
li	=	lilac
ge	=	yellow

E25 – Right Seat Belt Switch
F2 – Left Front Door Contact Switch
F3 – Right Front Door Contact Switch
F10 – Left Rear Door Contact Switch
F11 – Right Rear Door Contact Switch
W – Interior Light, Front
T3a – 3-Pin Connector
T3b – 3-Pin Connector
T3c – 3-Pin Connector
T3d – 3-Pin Connector
T5x – 5-Pin Connector
T6 – 6-Pin Connector
(R9) – Wire connection (door contact switch), in rear wiring harness

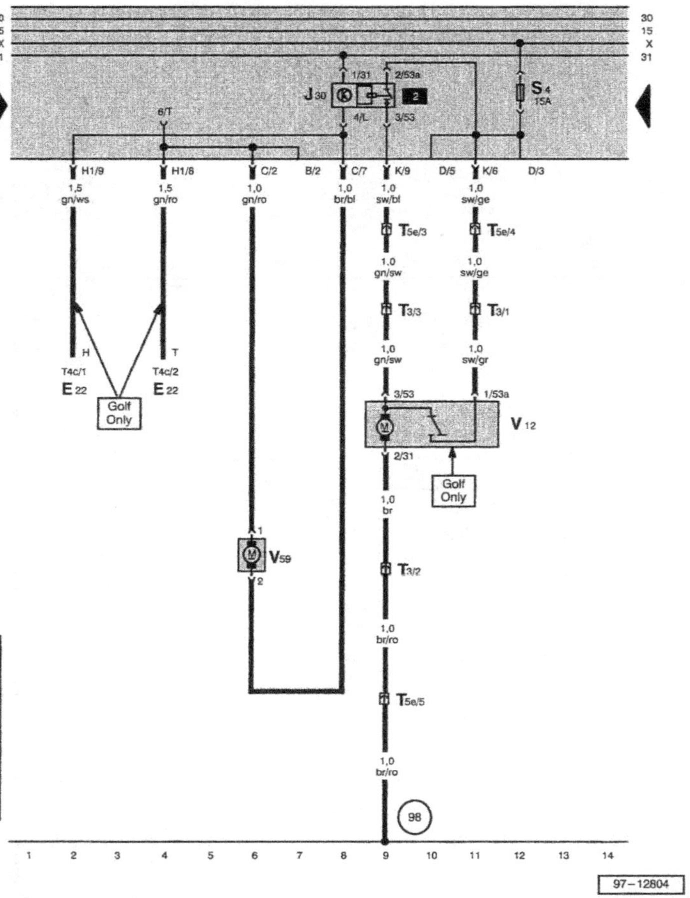

WIRING COLOR CODE		
ws	=	white
sw	=	black
ro	=	red
br	=	brown
gn	=	green
bl	=	blue
gr	=	grey
li	=	lilac
ge	=	yellow

E22 – Rear window Wiper/Washer switch
J30 – Rear Window Wiper/Washer Relay
V12 – Rear Window Wiper Motor
V59 – Windshield and Rear Window Washer Pump
T3 – 3-Pin Connector in Luggage compartment lid
T5e – 5-Pin Connector in Luggage compartment, left
T4c – 4-Pin Connector, behind steering column switch trim
(98) – Ground connection, in rear lid wiring harness

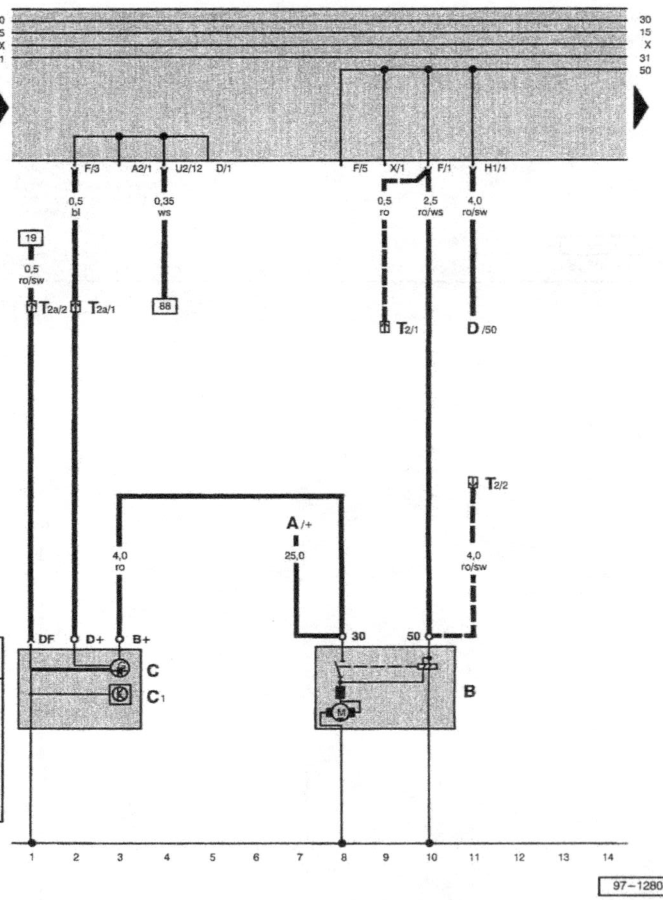

WIRING COLOR CODE		
ws	=	white
sw	=	black
ro	=	red
br	=	brown
gn	=	green
bl	=	blue
gr	=	grey
li	=	lilac
ge	=	yellow

A – Battery
B – Starter
C – Generator (GEN)
C1 – Voltage Regulator (VR)
D – Ignition/Starter switch
T2 – Double Connector
T2a – Double Connector
– – – Automatic transmission only

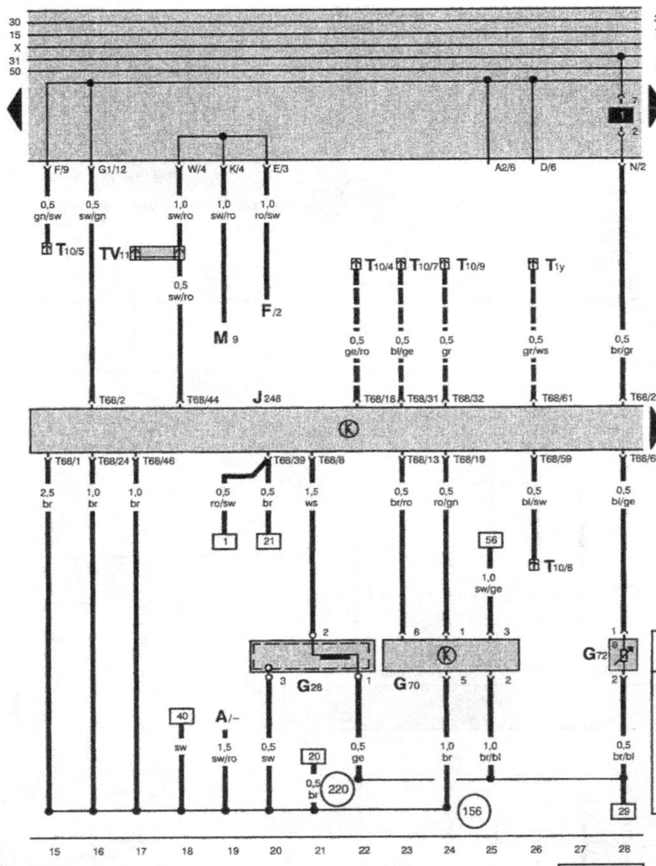

WIRING COLOR CODE		
ws	=	white
sw	=	black
ro	=	red
br	=	brown
gn	=	green
bl	=	blue
gr	=	grey
li	=	lilac
ge	=	yellow

97–M11783

- A – Battery
- F – Brake Light Switch
- G28 – Engine Speed (RPM) Sensor
- G70 – Mass Air Flow (MAF) Sensor
- G72 – Intake Air Temperature (IAT) Sensor
- J248 – Diesel Direct Fuel Injection (DFI) Engine Control Module (ECM)
- M9 – Left Brake Light
- T1y – Single Connector
- T10 – 10-Pin Connector
- T18 – 18-Pin Connector
- T2c – 2-Pin Connector
- T2a – 2-Pin Connector
- T4c – 4-Pin Connector
- T68 – 68-Pin Connector
- TV11 – Terminal 54 Wire Connector

(156) – Ground connection, in Diesel Direct Fuel Injection (DFI) wiring harness

(220) – Ground connection (sensor ground), in engine compartment wiring harness

– – – Automatic transmission only

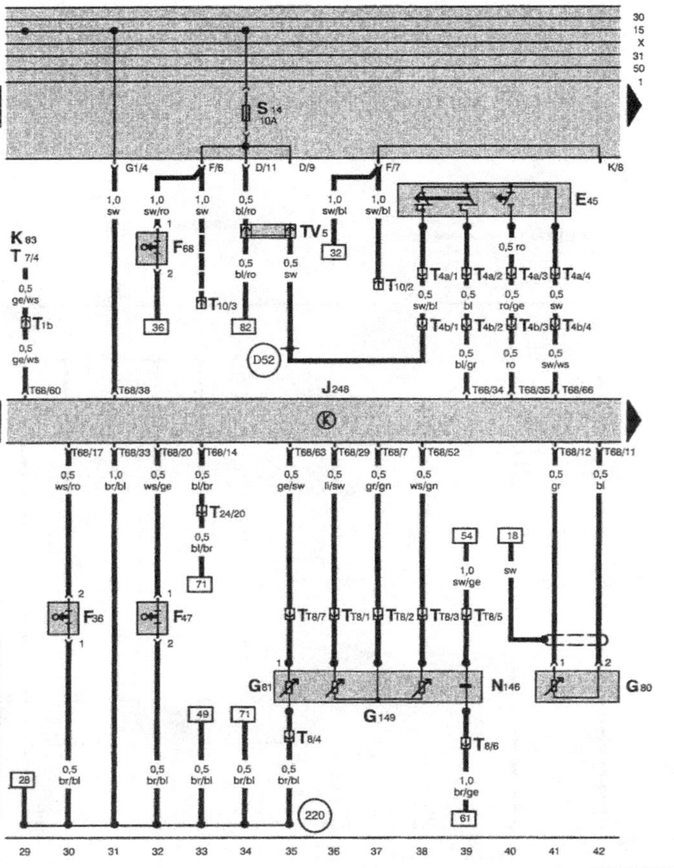

WIRING COLOR CODE		
ws	=	white
sw	=	black
ro	=	red
br	=	brown
gn	=	green
bl	=	blue
gr	=	grey
li	=	lilac
ge	=	yellow

97–12804

- E45 – Cruise Control Switch
- F36 – Clutch Vacuum Vent Valve Switch
- F47 – Brake Switch
- F68 – Transmission Range Sensor
- G80 – Needle Lift Sensor
- G81 – Fuel Temperature Sensor
- G149 – Modulating Piston Displacement Sensor
- J248 – Diesel Direct Fuel Injection (DFI) Engine Control Module (ECM)
- N146 – Quantity Adjuster
- **K83 – Malfunction Indicator Light (MIL)**
- T1b – 1-Pin Connector, in engine compartment
- T4a – 4-Pin Connector, behind dash panel
- T4b – 4-Pin Connector, in engine compartment
- T8 – 8-Pin Connector, in engine compartment
- T10 – 10-Pin Connector
- T24 – 24-Pin Connector, in engine compartment
- T68 – 68-Pin Connector, on Diesel DFI ECM
- TV5 – Terminal 15a Wire Connector

(220) – Ground connection (sensor ground), in engine compartment wiring harness

(D52) – Plus connection (15a), in engine compartment

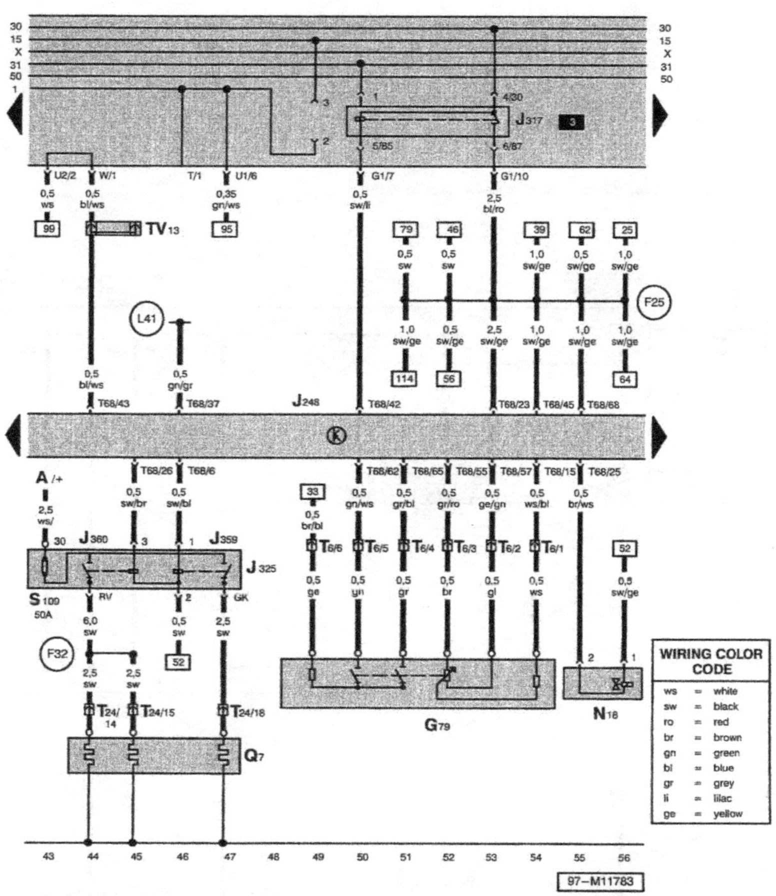

WIRING COLOR CODE		
ws	=	white
sw	=	black
ro	=	red
br	=	brown
gn	=	green
bl	=	blue
gr	=	grey
li	=	lilac
ge	=	yellow

A – Battery
G79 – Throttle Position (TP) Sensor
J248 – Diesel Direct Fuel Injection (DFI) Engine Control Module (ECM)
J317 – Power Supply (Terminal 30, B+) Relay
J325 – Coolant System Glow Plug Relay, on bulkhead center
J359 – Low Heater Output Relay
J360 – High Heater Output Relay
N18 – EGR Vacuum Regulator Solenoid Valve
Q7 – Coolant System Glow Plugs
S109 – Coolant System Glow Plug Fuse
T6 – 6-Pin Connector, in footwell, front right
T9 – 9-Pin Connector,
T24 – 24-Pin Connector, on engine
T68 – 68-Pin Connector, on engine control module
TV13 – Vehicle Speed Signal Wire Connector
(F25) – Wire connection -1-, in Diesel Direct Fuel Injection (DFI) system wiring harness
(F32) – Wire Connection (coolant system glow plugs), in Diesel direct injection system wiring harness
(L41) – Wire Connection (A/C ON), in A/C system coolant fan wiring harness

Engine electronics (from September 1997)

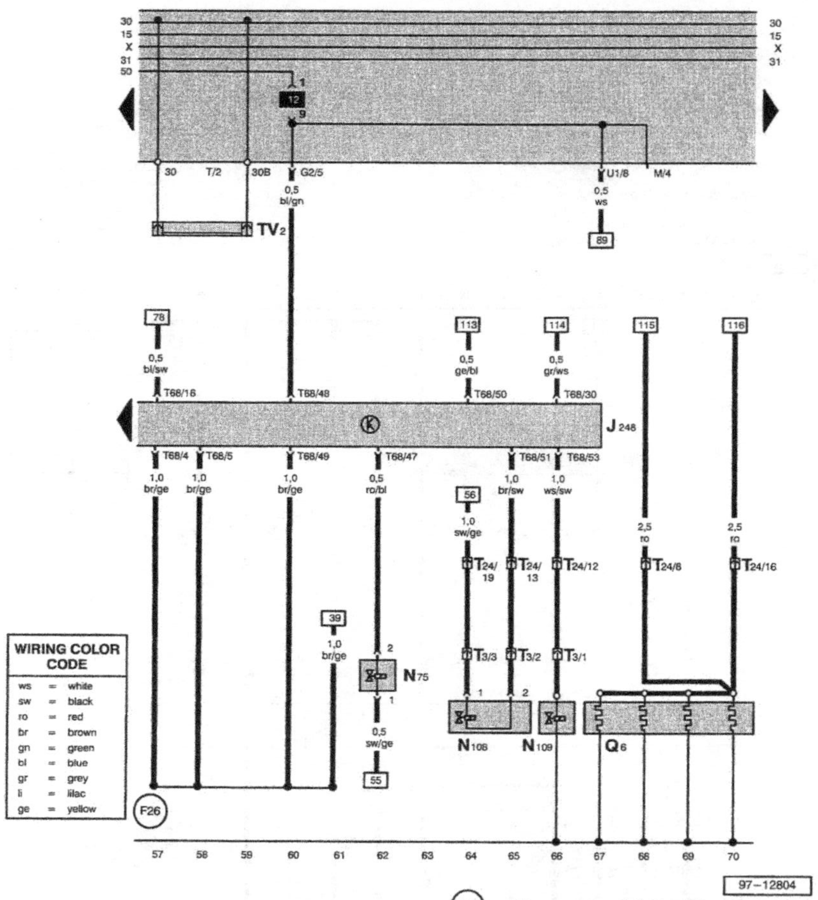

WIRING COLOR CODE		
ws	=	white
sw	=	black
ro	=	red
br	=	brown
gn	=	green
bl	=	blue
gr	=	grey
li	=	lilac
ge	=	yellow

J248 – Diesel Direct Fuel Injection (DFI) Engine Control Module (ECM)
N75 – Wastegate Bypass Regulator Valve
N108 – Cold Start Injector
N109 – Fuel Cut-off Valve
Q6 – Engine Glow Plugs
T3 – 3-Pin Connector, in engine compartment
T24 – 24-Pin Connector, in engine compartment
T68 – 68-Pin Connector, on Diesel DFI ECM
TV2 – Terminal 30 (B+) Wire Connector
(F26) – Wire connection -2-, in Diesel Direct Fuel Injection (DFI) system wiring harness

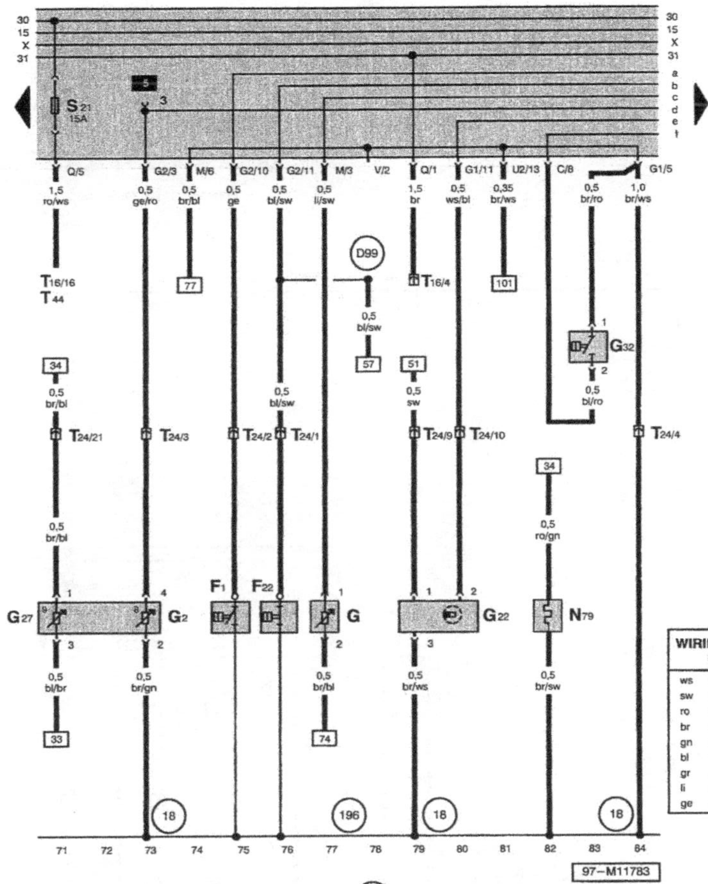

WIRING COLOR CODE

ws	=	white
sw	=	black
ro	=	red
br	=	brown
gn	=	green
bl	=	blue
gr	=	grey
li	=	lilac
ge	=	yellow

F1 – Oil Pressure Switch
F22 – 0.3bar Oil Pressure Switch
G – Sender for Fuel Gauge
G2 – Engine Coolant Temperature (ECT) Sensor
G22 – Speedometer Vehicle Speed Sensor (VSS)
G27 – Sensor for Engine Temperature
G32 – Engine Coolant Level (ECL) Sensor
N79 – Positive Crankcase Ventilation (PCV) Heating Element
T16 – Data Link Connector (DLC), behind center of dash
T24 – 24-Pin Connector, in engine compartment
T44 – Junction Box for-On-Board Diagnosis above fuse/ relay panel

(18) – Ground connection, on engine block

(196) – Ground connection -3-, in wiring harness, rear

(D99) –Wire connection (Engine Coolant Temperature Sensor), in engine compartment wiring harness

Engine electronics (from September 1997)

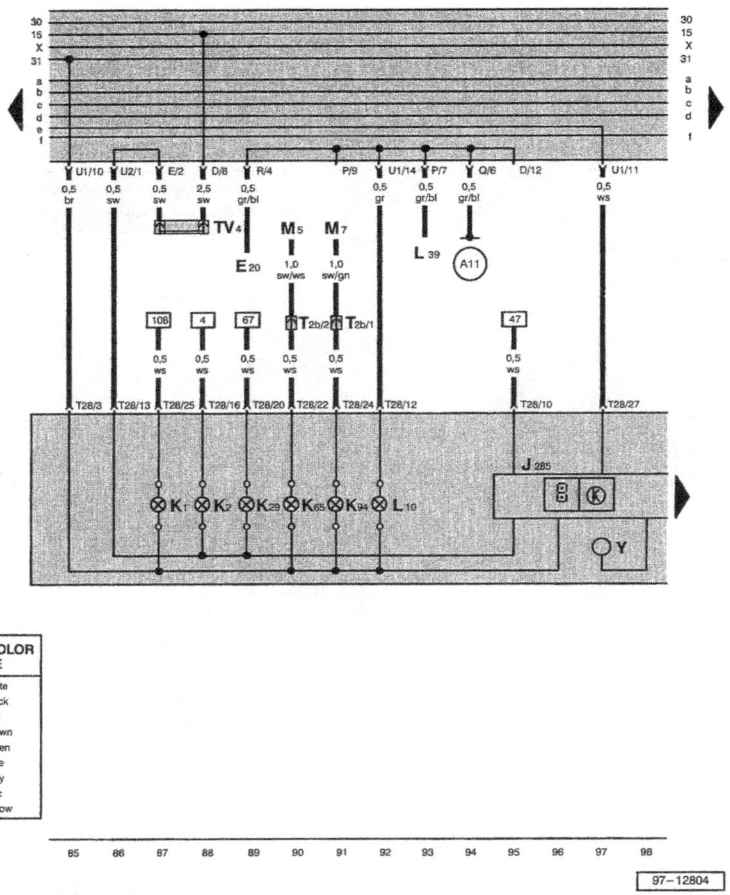

WIRING COLOR CODE

ws	=	white
sw	=	black
ro	=	red
br	=	brown
gn	=	green
bl	=	blue
gr	=	grey
li	=	lilac
ge	=	yellow

E20 – Instrument Panel Light Dimmer Switch
J285 – Control Module with Indicator Unit Instrument Panel Insert
K1 – Headlight High Beam Indicator Light
K2 – Generator (GEN) Warning Light
K29 – Glow Plug Indicator Light
K65 – Left Turn Signal Indicator Light
K94 – Right Turn Signal Indicator Light
L10 – Instrument Cluster Light
L39 – Rear Window Defogger Switch Light
M5 – Left Front Turn Signal Light
M7 – Right Front Turn Signal Light
T2b – Double Connector
T28 – 28-Pin Connector, in engine compartment
TV4 – Terminal 14 Wire Connector
Y – Analog Clock

(A11) – Plus connection (58b), in instrument panel wiring harness

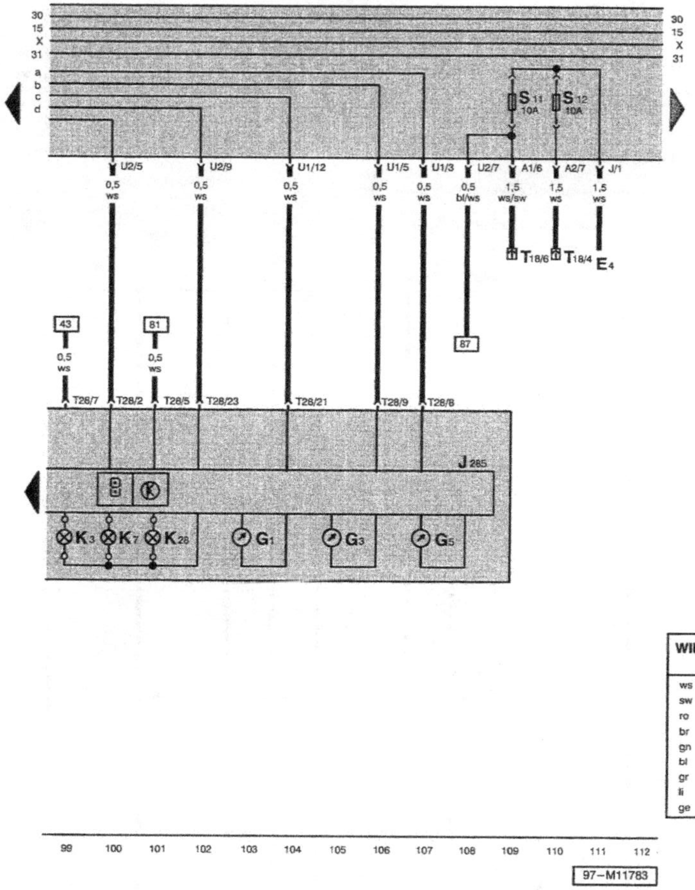

WIRING COLOR CODE		
ws	=	white
sw	=	black
ro	=	red
br	=	brown
gn	=	green
bl	=	blue
gr	=	grey
li	=	lilac
ge	=	yellow

E4 – Headlight Dimmer/Flasher Switch
G1 – Fuel Gauge
G3 – Engine Coolant Temperature (ECT) Gauge
G5 – Tachometer
J285 – Control Module with Indicator Unit in Instrument Panel Insert
K3 – Oil Pressure Warning Light
K7 – Brake and Parking Brake Warning Light
K28 – Engine Coolant Level/Temperature Warning Light
L1 – Left Headlight
L2 – Right Headlight
T28 – 28-Pin Connector, instrument cluster

Engine electronics (from September 1997)

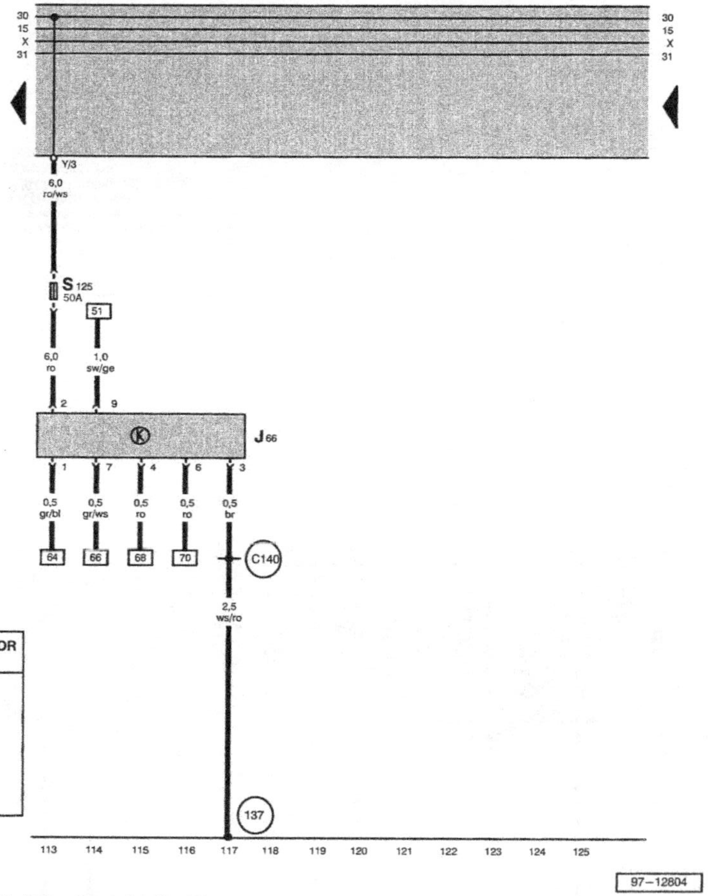

WIRING COLOR CODE		
ws	=	white
sw	=	black
ro	=	red
br	=	brown
gn	=	green
bl	=	blue
gr	=	grey
li	=	lilac
ge	=	yellow

J66 – Glow Plug Relay, relay panel position #16
S125 – Fuse for glow plugs, located in relay panel position #22
C140 – Wire connection, in Motronic MFI wiring harness
137 – Ground connection, in Motronic MFI wiring harness

Cabrio — 2.0L gas engine (ABA)

*applicable to 1999-2002 m.y., from previous m.y.

Service Information

Repair Manual Technical Bulletin

Cabrio 1999 ➤

Bulletin-No.: 4

New information in this Technical Bulletin supplements and updates previously published Wiring Diagram, Troubleshooting and Component Location information.

The following have been integrated into the Electrical Wiring Diagram section:

- Replace index pages 3 to 5 with new pages supplied.
- Insert wiwirng diagrams 19 - 24.
- Remove and replace wiring diagrams No. 5/1, 5/3, 5/5, 14/3, 5/1, 15/3 & 15/5.
- Remove and replace complete wiring diagrams No. 1, 2, 3, 4, 6, 7, 8, 9, 10, 11, 13, 16, 17 & 18.

This cover sheet is part of the Technical Bulletin. Please file in front of the Index.

i ***Wiring harness repairs must only be carried out using wiring harness repair kit VAS 1978.***

VMM-100
Printed in USA W42.USA.5403.04.21 **Edition** 09.01

Cabrio — Wiring diagram — No. 1/1

Immobilizer

from September 1999

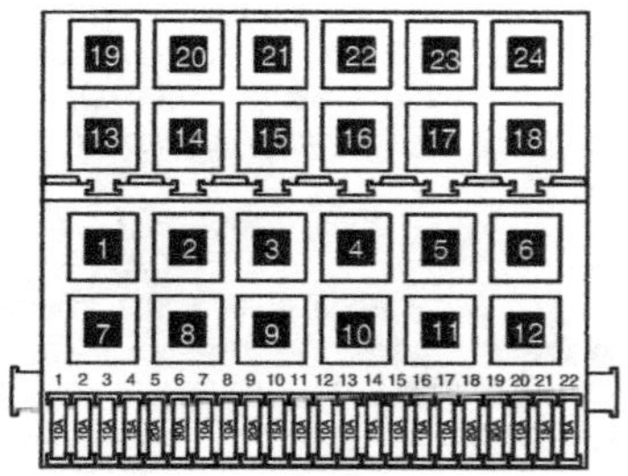

97-61067

Note:
Number in parentheses indicates production control number stamped on relay housing.

Fuse colors

30 A - green
25 A - white
20 A - yellow
15 A - blue
10 A - red
7,5 A - brown
5 A - beige
3 A - violet

Edition 09.01
USA.5403.04.21

No. 1/2 — Wiring diagram — Cabrio

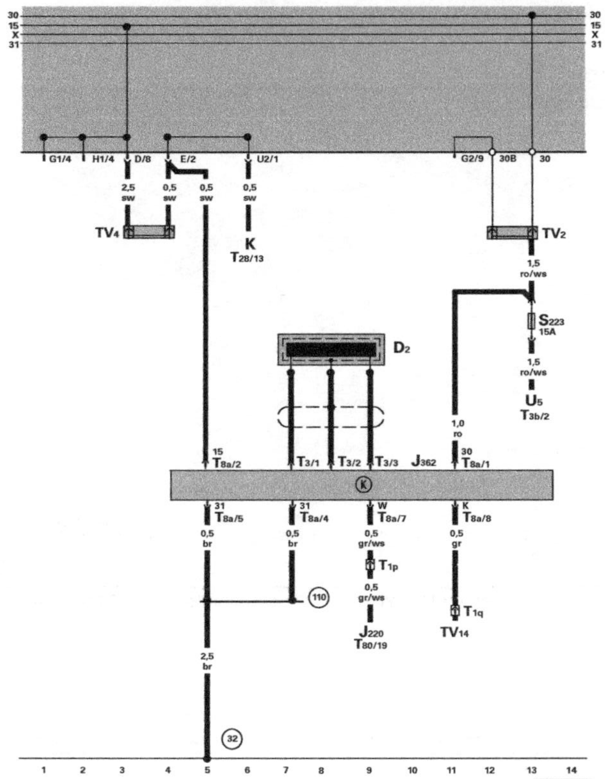

ws = white
sw = black
ro = red
br = brown
gn = green
bl = blue
gr = grey
li = violet
ge = yellow

97-61535

- D2 - Induction coil of antitheft immobilizer
- J220 - Motronic Engine Control Module (ECM)
- J362 - Control module for anti-theft immobilizer
- K - Instrument cluster
- S223 - Fuse in fuse holder
- T1p - Single Connector, brown, behind fuse/ relay panel
- T1q - Single Connector, brown, behind fuse/ relay panel
- T3 - 3-Pin Connector
- T3b - 3-Pin Connector, white
- T8a - 8-Pin Connector, black
- T28 - 28-Pin Connector, white, on instrument cluster
- T80 - 80-Pin Connector
- TV2 - Terminal 30 (B+) wire connector, red
- TV4 - Terminal 15 wire connector, black
- TV14 - Data Link Connector (DLC) Wire Connector, black, above fuse relay panel
- U5 - 12 V Socket
- (32) - Ground connection, behind instrument panel, left
- (110) - Ground connection -2-, in instrument panel wiring harness

Induction coil of antitheft immobilizer, control module for antitheft immobilizer

Edition 09.01
USA.5403.04.21

Cabrio — Wiring diagram — No. 2/1

Alarm system

from August 1998

Relay locations

Position of relays and multi-pin connector wiring, see also "Component Locations" section.

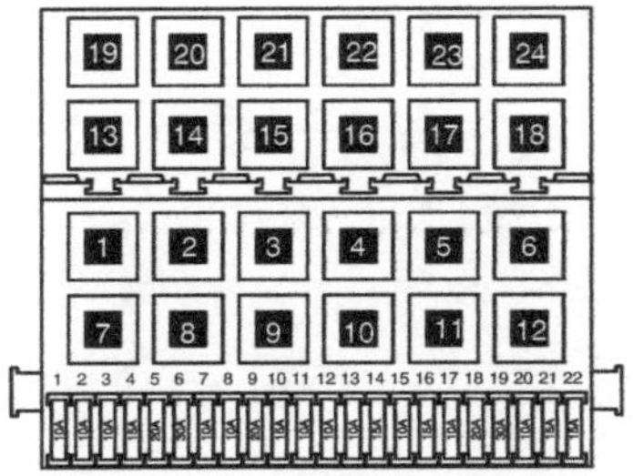

97-61067

Note:
Number in parentheses indicates production control number stamped on relay housing.

Fuse colors

30 A - green
25 A - white
20 A - yellow
15 A - blue
10 A - red
7,5 A - brown
5 A - beige
3 A - violet

Edition 09.01
USA.5403.04.21

No. 2/2 — Wiring diagram — Cabrio

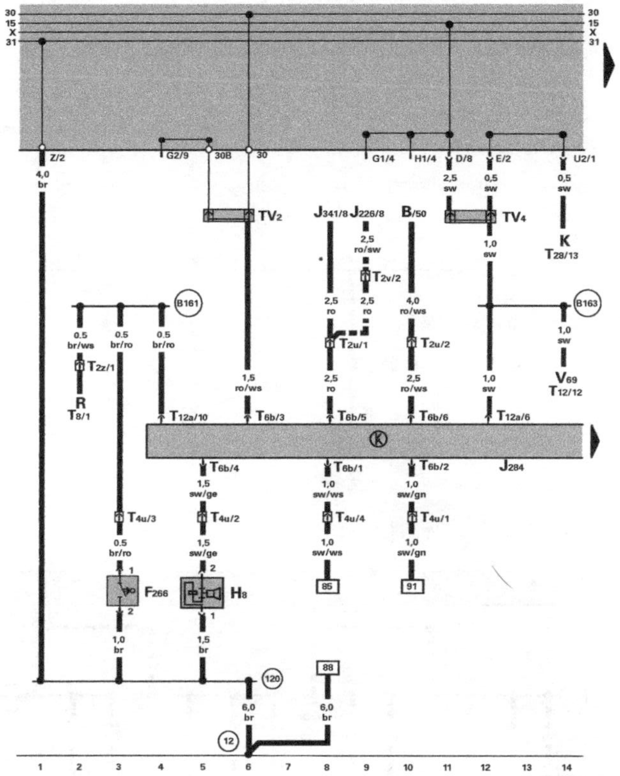

ws = white
sw = black
ro = red
br = brown
gn = green
bl = blue
gr = grey
li = violet
ge = yellow

B - Starter
F266 - Front Hood Switch
H8 - Alarm horn
J226 - Park Neutral Position (PNP) Relay
J284 - Alarm system control module
J341 - Anti-Theft Engine Disable Relay I (53)
K - Instrument Cluster
R - Radio
T2u - 2-Pin Connector
T2v - 2-Pin Connector
T2z - 2-Pin Connector
T4u - 4-Pin Connector
T6b - 6-Pin Connector, black
T8 - 8-Pin Connector, in Radio
T12 - 12-Pin Connector, near right taillight bulb holde
T12a - 12-Pin Connector, black
T28 - 28-Pin Connector, white
TV2 - Terminal 30 (B+) wire connector, red
TV4 - Terminal 15 wire connector, black
V69 - Central Looking System Pump and control Module
(12) - Ground connection, in engine compartment, left
(120) - Ground connection -2-, in head light wiring harness
(B161) - Connector (anti-theft warning system), in wiring harness interior
(B163) - Plus connector -1- (15) in wiring harness interior
* - Manual transmission only
--- - Automatic transmission only

Alarm system control module, front hood switch, alarm horn

Edition 09.01
USA.5403.04.21

Cabrio — Wiring diagram — No. 2/3

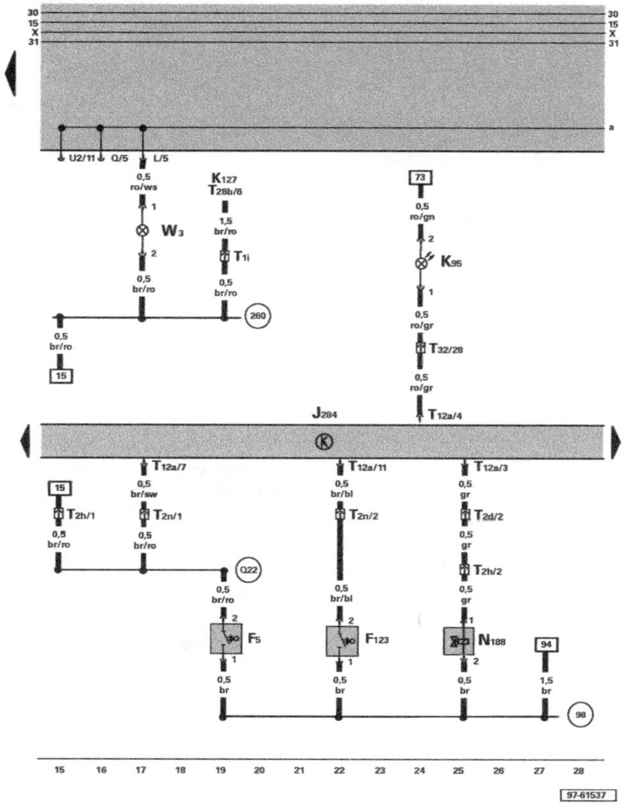

ws = white
sw = black
ro = red
br = brown
gn = green
bl = blue
gr = grey
li = violet
ge = yellow

F5 - Luggage Compartment Light Switch
F123 - Trunk Lid Alarm Switch
J284 - Alarm system control module
K95 - Alarm System Indicator Light
K127 - Indicator light for open trunk
N188 - Luggage Compartment Release Solenoid
T1i - Single Connector, white, behind fuse/relay panel
T2d - 2-Pin Connector, green, behind fuse/relay panel
T2h - 2-Pin Connector, yellow, near left taillight bulb holde
T2n - 2-Pin Connector, black, near left taillight bulb holde
T12a - 12-Pin Connector, black
T28b - 28-Pin Connector, red
T32 - 32-Pin Connector, white, on left A-pillar
W3 - Luggage Compartment Light
(98) - Ground connection, in rear lid wiring harness
(260) - Ground Connection, in wiring harness taillights
(Q22) - Wire connection -1-, in rear lid wiring harness

Edition 09.01
USA.5403.04.21

Alarm system control module, luggage compartment light switch, trunk lock alarm/central locking switch

No. 2/4 — Wiring diagram — Cabrio

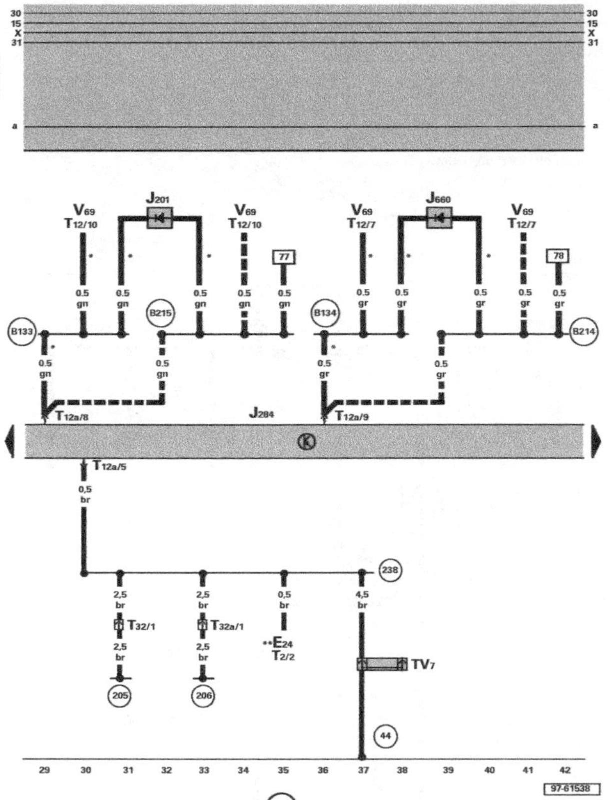

ws = white
sw = black
ro = red
br = brown
gn = green
bl = blue
gr = grey
li = violet
ge = yellow

E24 - Left Seat Belt Switch
J201 - Protective diode 1
J284 - Alarm System Control Module
J660 - Protective diode 2
T2 - 2-Pin Connector
T12 - 12-Pin Connector, near right taillight bulb holde
T12a - 12-Pin Connector, black
T32 - 32-Pin Connector, white, on left A-pillar
T32a - 32-Pin Connector, white, on right A-pillar
TV7 - Terminal 31 (GND) wire connector, brown, beside fuse /relay panel
V69 - Central Looking System Pump and control Module
(44) - Ground connection, on left A-pillar, lower part
(205) - Ground connection, in driver's door wiring harness
(206) - Ground connection, in passenger's door wiring harness
(238) - Ground connection -1-, in wiring harness interior
(B133) - Connector (central locking open), in wiring harness interior
(B134) - Connector (central locking closed), in wiring harness interior
(B214) - Connection 1 (central locking) in passenger compartment wiring harness
(B215) - Connection 2 (central locking) in passenger compartment wiring harness
* - With power windows
---- - Without power windows
** - Up to 10.00 production

Alarm system control module

Edition 09.01
USA.5403.04.21

Cabrio — Wiring diagram — No. 2/5

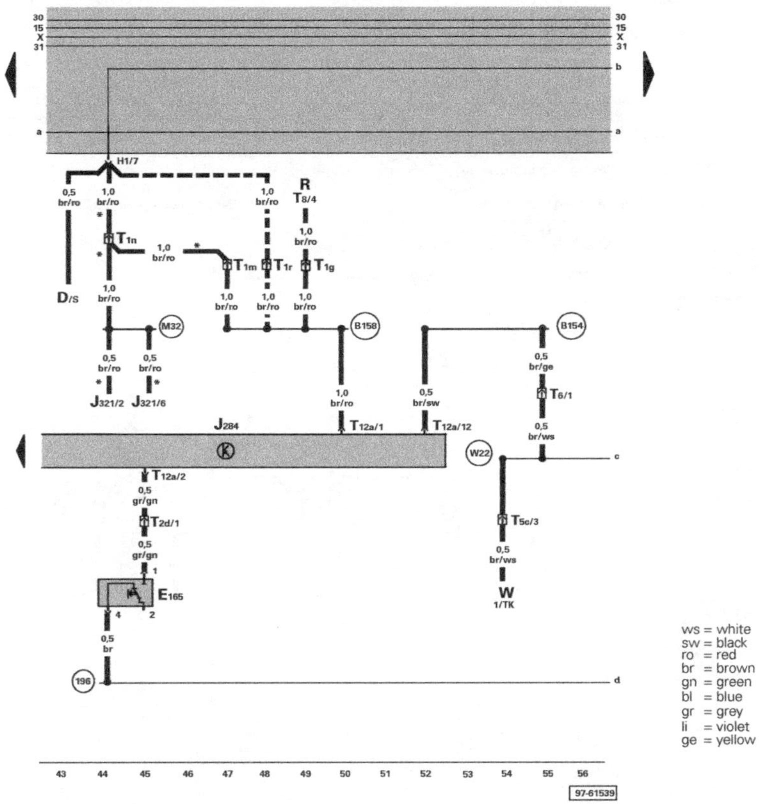

D - Ignition/Starter Switch
E165 - Trunk Lid Release Switch
J284 - Alarm system control module
J321 - Convertible Top Hydraulic Pump Relay (53)
R - Radio
T1g - 1-Pin Connector, red, behind fuse relay panel
T1m - 1-Pin Connector, red, behind fuse relay panel
T1n - 1-Pin Connector, red, behind fuse relay panel
T1r - 1-Pin Connector, red, behind fuse relay panel
T2d - 2-Pin Connector, green, behind fuse/relay panel
T5c - 5-Pin Connector, black
T6 - 6-Pin Connector, black
T8 - 8-Pin Connector, black, in radio
T12a - 12-Pin Connector
W - Front Interior Light

(196) - Ground connection -3-, in wiring harness, rear
(B154) - Connector -1- (TK), in wiring harness interior
(B158) - Connector (S) in wiring harness interior
(M32) - Wire Connection (su),in convertible top wiring harness
(W22) - Connector (door contact switch) in wiring harness, rear

* - Power convertible top
--- - Manual convertible top

Alarm system control module, trunk lid release switch

No. 2/6 — Wiring diagram — Cabrio

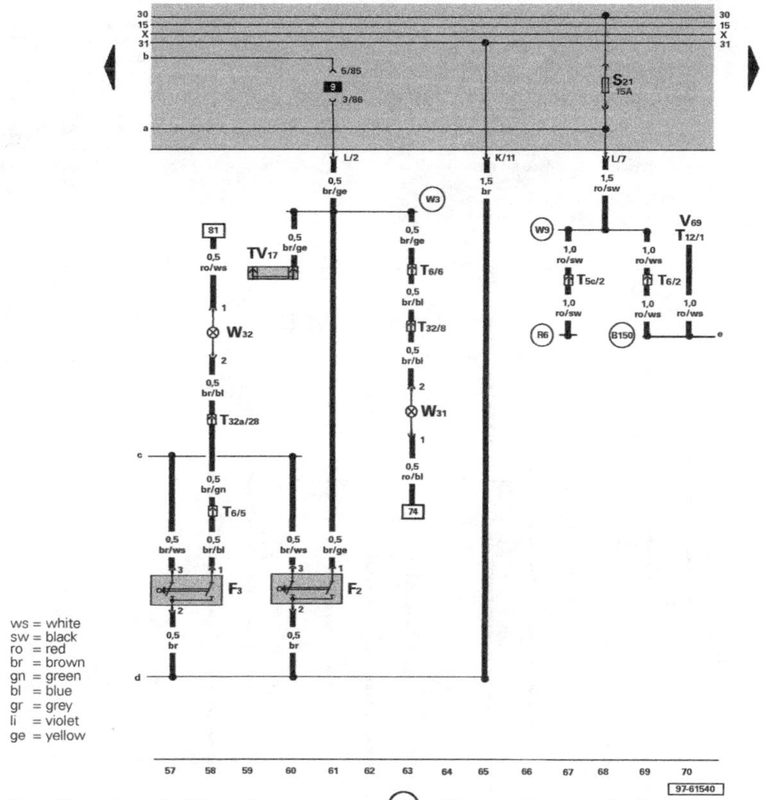

F2 - Door contact switch-Driver's side
F3 - Door contact switch-Passenger's side
T5c - 5-Pin Connector, black
T6 - 6-Pin Connector, black
T12 - 12-Pin Connector, near right taillight bulb holde
T32 - 32-Pin Connector, white, on left A-pillar
T32a - 32-Pin Connector, white, on right A-pillar
TV17 - Door Contact Switch Wire Connector, white
V69 - Central Looking System Pump and control Module
W31 - Entry light front, left
W32 - Entry light front, right

(B150) - Plus connector -2- (30a), in wiring harness interior
(R6) - Plus connection -1-, in interior light wiring harness

(W3) - Wire connection, in rear wiring harness
(W9) - Plus connection -1- (30), in rear wiring harness

Door contact switches, entry ligths front

Cabrio — Wiring diagram — No. 2/7

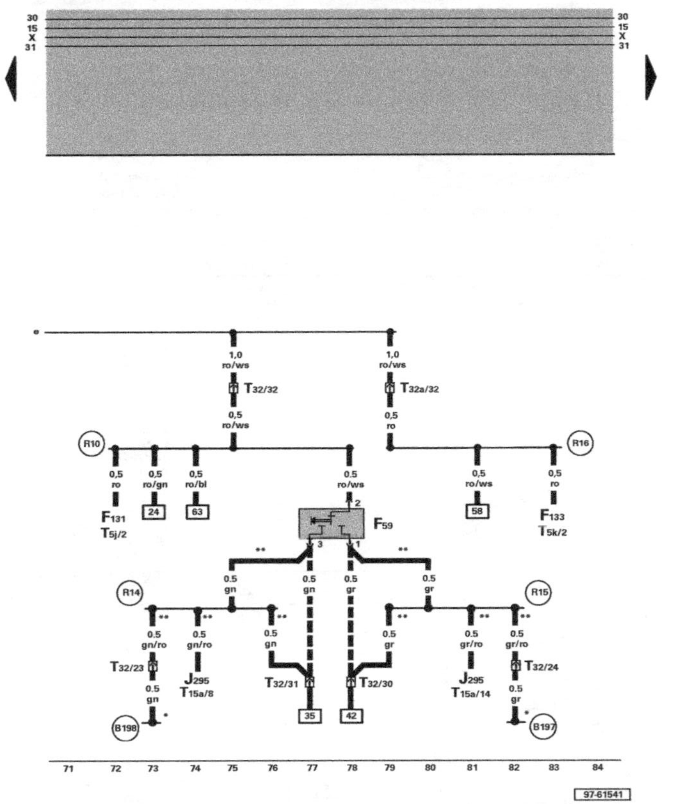

ws = white
sw = black
ro = red
br = brown
gn = green
bl = blue
gr = grey
li = violet
ge = yellow

F59 - Driver's Door Central Locking System Switch
F131 - Left Front Central Locking System Activator/Switch
F133 - Right Front Central Locking System Activator/Switch
J295 - Left Front Power Window Control Module
T5j - 5-Pin Connector, beige, on left door
T5k - 5-Pin Connector, beige, on right door
T15a - 15-Pin Connector
T32 - 32-Pin Connector, white, on left A-pillar
T32a - 32-Pin Connector, white, on right A-pillar
(B197) - Connection (close window) in passenger compartment wiring harness
(B198) - Connection (open window) in passenger compartment wiring harness

(R10) - Plus connection -1- (30), in driver's door wiring harness
(R14) - Wire connection -1- (open), in driver's door wiring harness
(R15) - Plus connection -2- (closed), in driver's door wiring harness
(R16) - Plus connection -1- (30), in passenger's door wiring harness

* - With power windows
** - Driver & passenger side power mirrors, heatable
\- - - - Driver & passenger side remote mirrors, heatable

Edition 09.01
USA.5403.04.21

Driver's door central locking system switch

No. 2/8 — Wiring diagram — Cabrio

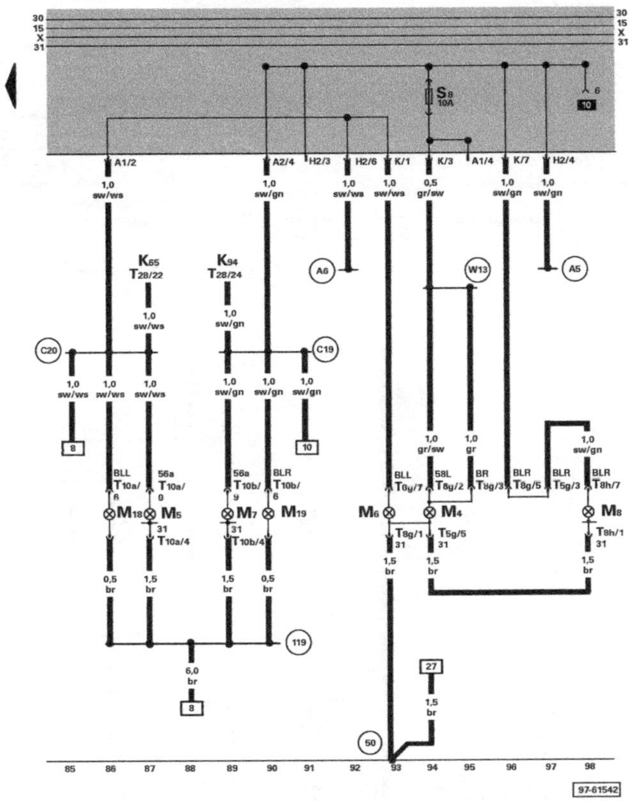

ws = white
sw = black
ro = red
br = brown
gn = green
bl = blue
gr = grey
li = violet
ge = yellow

K65 - Left Turn Signal Indicator Light
K94 - Right Turn Signal Indicator Light
M4 - Left Tail Light
M5 - Left Front Turn Signal Light
M6 - Left Rear Turn Signal Light
M7 - Right Front Turn Signal Light
M8 - Right Rear Turn Signal Light
M18 - Left, Side Turn Signal Light
M19 - Right, Side turn signal light
T5g - 5-Pin Connector, on taillight, left
T8g - 8-Pin Connector, on taillight, left
T8h - 8-Pin Connector, on taillight, right
T10a - 10-Pin Connector
T10b - 10-Pin Connector
T28 - 28-Pin Connector, white

(50) - Ground connection, in luggage compartment, left
(119) - Ground connection -1-, in headlight wiring harness
(A5) - Plus connection (right turn signal), in instrument panel wiring harness
(A6) - Plus connection (left turn signal), in instrument panel wiring harness
(C19) - Connector -1- (right turn signal), in wiring harness headlamp
(C20) - Connection (left turn signal), in headlight wiring harness
(W13) - Wire connection (58L),in rear wiring harness

Turn signal lights, front /rear turn signal lights, side marker lights

Edition 09.01
USA.5403.04.21

Cabrio — Wiring diagram — No. 3/1

Airbag systems

from August 1998

Relay locations

Position of relays and multi-pin connector wiring, see also "Component Locations" section.

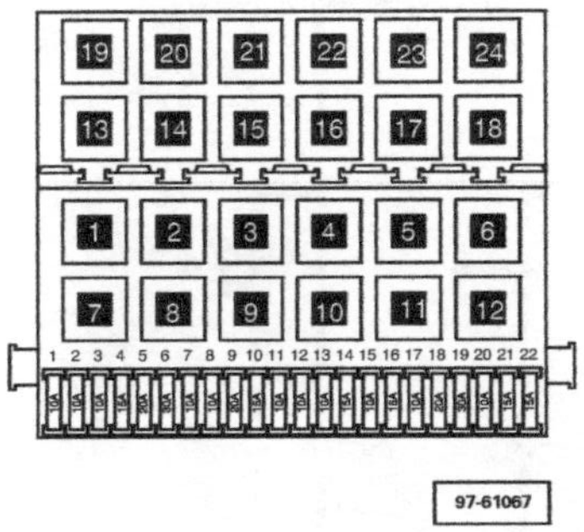

Note:
Number in parentheses indicates production control number stamped on relay housing.

Fuse colors

30 A - green
25 A - white
20 A - yellow
15 A - blue
10 A - red
7,5 A - brown
5 A - beige
3 A - violet

Edition 09.01
USA.5403.04.21

Airbag system (from August 1998)

No. 3/2 — Wiring diagram — Cabrio

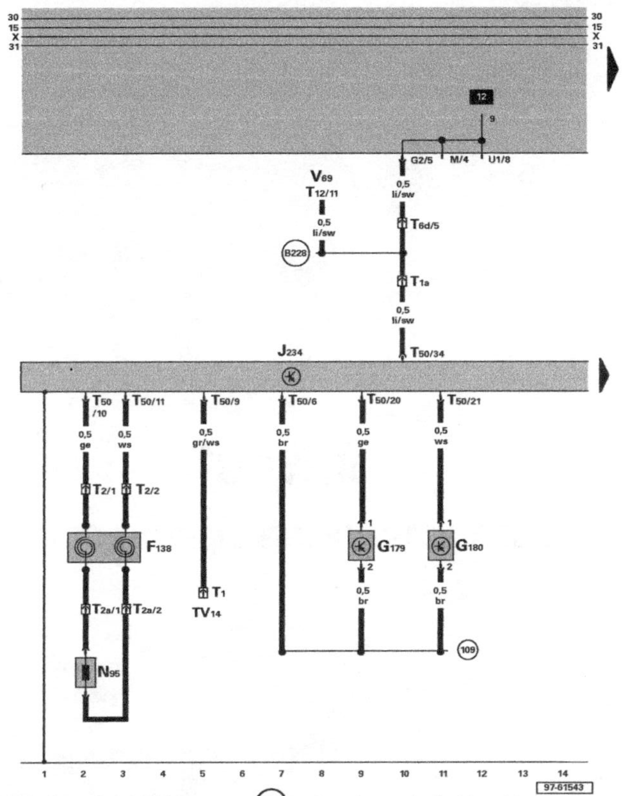

ws = white
sw = black
ro = red
br = brown
gn = green
bl = blue
gr = grey
li = violet
ge = yellow

F138 - Airbag Spiral Spring/Return Spring With Slip Ring
G179 - Crash sensor for side airbag, driver's side
G180 Crash sensor for side airbag, passenger side
J234 - Airbag Control Module
N95 - Driver's Side Airbag Igniter
T1 - Single connector, red
T1a - Single connector, behind fuse/relay panel
T2 - 2-Pin Connector
T2a - 2-Pin Connector
T6d - 6-Pin Connector, red
T12 - 12-Pin Connector
TV14 - Data Link Connector (DLC) wire Connector, black, above fuse relay panel
T50 - 50-Pin Connector
V69 - Central Locking System Pump and Control Module

(109) - Ground connection, in airbag wiring harness
(B228) - Connection (Crash-Signal) (interior wiring harness)

Airbag control module, airbag spiral spring, driver's side airbag igniter, crash sensors for side airbags

Edition 09.01
USA.5403.04.21

Cabrio — Wiring diagram — No. 3/3

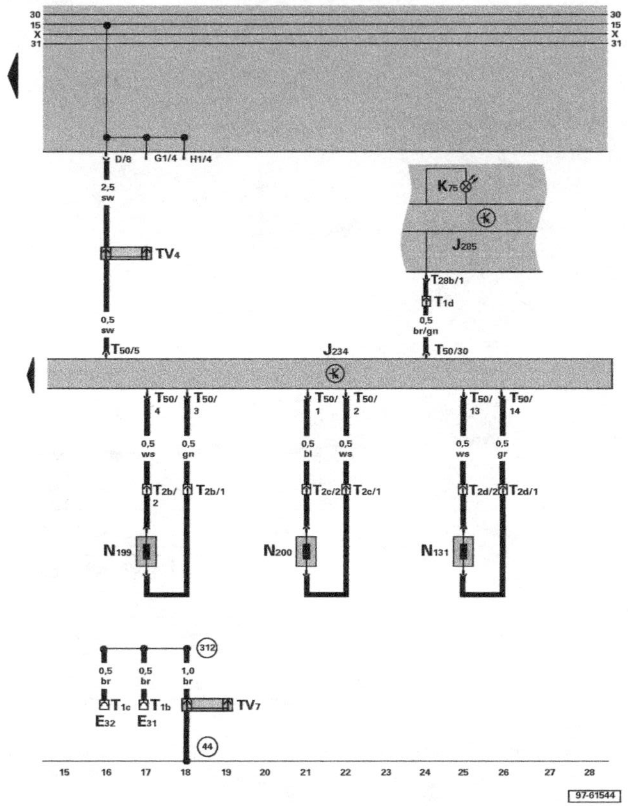

ws = white
sw = black
ro = red
br = brown
gn = green
bl = blue
gr = grey
li = violet
ge = yellow

E31 - Driver's Seat Contact Strip (Seat Belt Warning system)
E32 - Passenger's Seat Contact Strip (Seat Belt Warning system)
J234 - Airbag Control Module
J285 - Control module with indicator unit in instrument panel insert
K75 - Airbag Malfunction Indicator Lamp (MIL)
N131 - Passenger's Side Airbag Igniter 1
N199 - Igniter for side airbag, driver's side
N200 - Igniter for side airbag, passenger side
T1b - Single connector, yellow
T1c - Single connector, yellow
T1d - Single connector, brown
T2b - 2-Pin Connector
T2c - 2-Pin Connector, yellow
T2d - 2-Pin Connector, yellow
T28b - 28-Pin Connector, on instrument cluster, red
T50 - 50-Pin Connector
TV4 - Terminal 15 Wire Connector, black
TV7 - Terminal 31 (GND) wire connector, brown, beside fuse /relay panel
(44) - Ground connection,on left A-pillar, lower part
(312) - Ground connection -2-, in airbag wiring harness

Edition 09.01
USA.5403.04.21

Airbag control module, passenger's side airbag igniter, airbag malfuction indicator lamp (MIL)

THIS PAGE INTENTIONALLY LEFT

BLANK

Cabrio — Wiring diagram — No. 4/1

Alarm system

from September 1999

Relay locations

Position of relays and multi-pin connector wiring, see also "Component Locations" section.

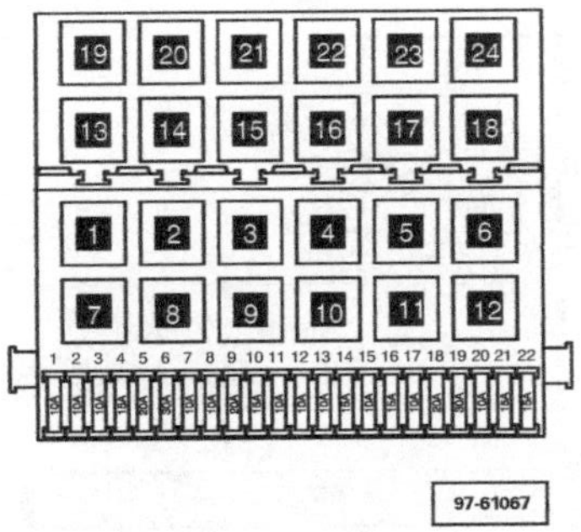

Note:

Number in parentheses indicates production control number stamped on relay housing.

Fuse colors

30 A - green
25 A - white
20 A - yellow
15 A - blue
10 A - red
7,5 A - brown
5 A - beige
3 A - violet

No. 4/2 — Wiring diagram — Cabrio

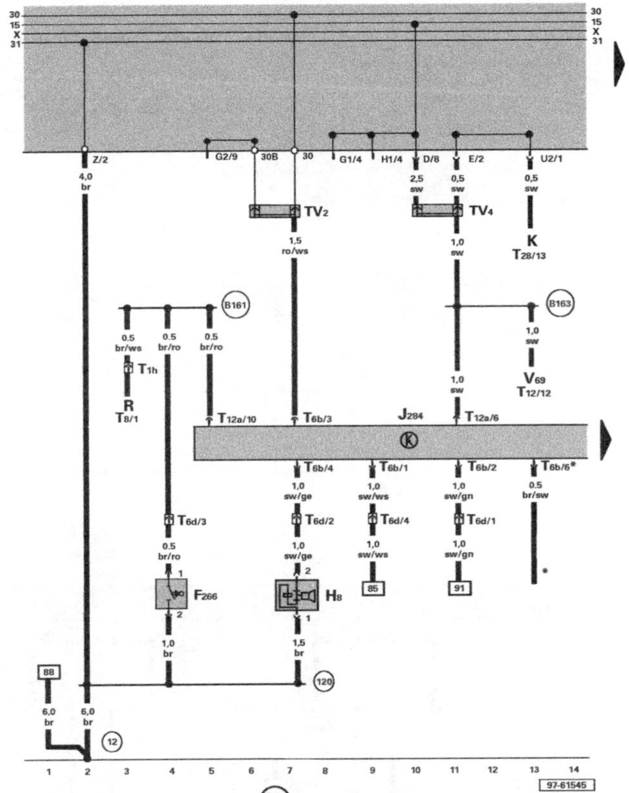

ws = white
sw = black
ro = red
br = brown
gn = green
bl = blue
gr = grey
li = violet
ge = yellow

F266 - Front Hood Switch
H8 - Alarm horn
J284 - Alarm system control module
K - Instrument Cluster
R - Radio
T1h - Single Connector; yellow, behind fuse/relay panel
T6b - 6-Pin Connector, black
T6d - 6-Pin Connector, red, behind fuse relay panel
T8 - 8-Pin Connector, black, in Radio
T12 - 12-Pin Connector, near right taillight bulb holde
T12a - 12-Pin Connector, black
T28 - 28-Pin Connector, white
TV2 - Terminal 30 (B+) Wire Connector
TV4 - Terminal 15 Wire Connector, black
V69 - Central Looking System Pump and control Module

(12) - Ground connection, in engine compartment, left
(120) - Ground connection -2-, in head light wiring harness
(B161) - Connector (anti-theft warning system), in wiring harness interior
(B163) - Plus connector -1- (15) in wiring harness interior

* - from 08.00 production (external antenna for radio control alarm)

Alarm system control module, front hood switch, alarm horn

Cabrio — Wiring diagram — No. 4/3

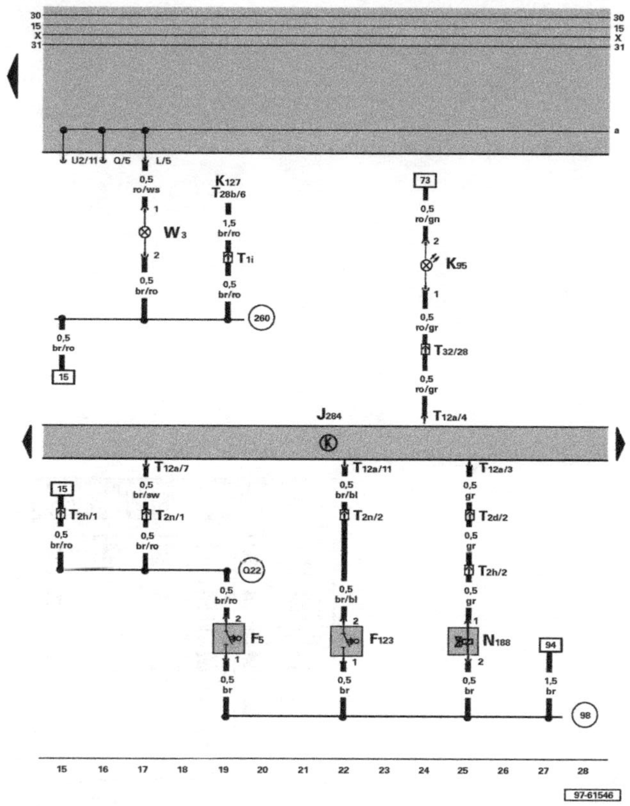

ws = white
sw = black
ro = red
br = brown
gn = green
bl = blue
gr = grey
li = violet
ge = yellow

F5 - Luggage Compartment Light Switch
F123 - Trunk Lid Alarm Switch
J284 - Alarm system control module
K95 - Alarm System Indicator Light
K127 - Indicator light for open trunk
N188 - Luggage Compartment Release Solenoid
T1i - Single Connector, white, behind fuse/relay panel
T2d - 2-Pin Connector, green, behind fuse/relay panel
T2h - 2-Pin Connector, yellow, near left taillight bulb holde
T2n - 2-Pin Connector, black, near left taillight bulb holde
T12a - 12-Pin Connector
T28b - 28-Pin Connector, red
T32 - 32-Pin Connector, white, on left A-pillar
W3 - Luggage Compartment Light
(98) - Ground connection, in rear lid wiring harness
(260) - Ground Connection, in wiring harness taillights
(Q22) - Wire connection -1-, in rear lid wiring harness

Alarm system control module, luggage compartment light switch, trunk lock alarm/central locking switch

No. 4/4 — Wiring diagram — Cabrio

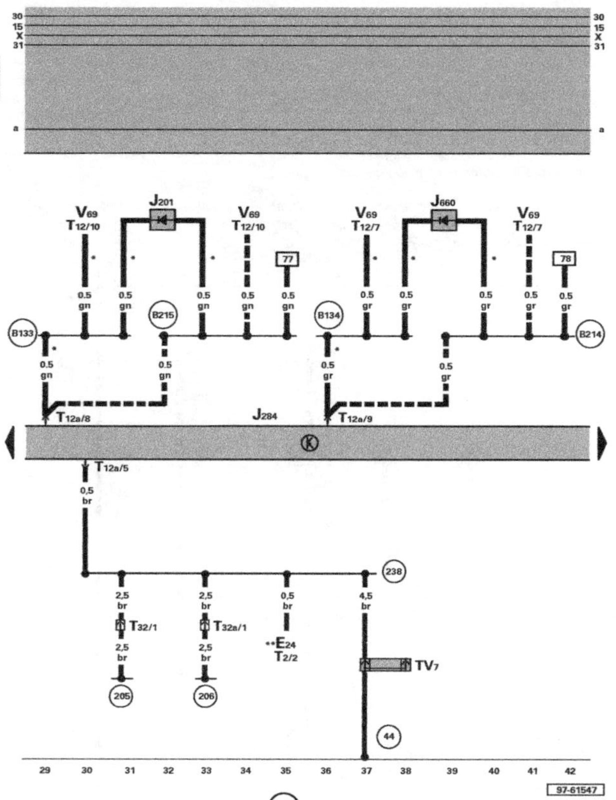

ws = white
sw = black
ro = red
br = brown
gn = green
bl = blue
gr = grey
li = violet
ge = yellow

E24 - Left Seat Belt Switch
J201 - Protective diode 1
J284 - Alarm System Control Module
J660 - Protective diode 2
T2 - 2-Pin Connector
T12 - 12-Pin Connector, near right taillight bulb holde
T12a - 12-Pin Connector, black
T32 - 32-Pin Connector, white, on left A-pillar
T32a - 32-Pin Connector, white, on right A-pillar
TV7 - Terminal 31 (GND) wire connector, brown, beside fuse /relay panel
V69 - Central Looking System Pump and control Module
(44) - Ground connection, on left A-pillar, lower part
(205) - Ground connection, in driver's door wiring harness
(206) - Ground connection, in passenger's door wiring harness
(238) - Ground connection -1-, in wiring harness interior
(B133) - Connector (central locking open), in wiring harness interior
(B134) - Connector (central locking closed), in wiring harness interior
(B214) - Connection 1 (central locking) in passenger compartment wiring harness
(B215) - Connection 2 (central locking) in passenger compartment wiring harness

* - With power windows
---- - Without power windows
** - Up to 10.00 production

Alarm system control module

Cabrio — Wiring diagram — No. 4/5

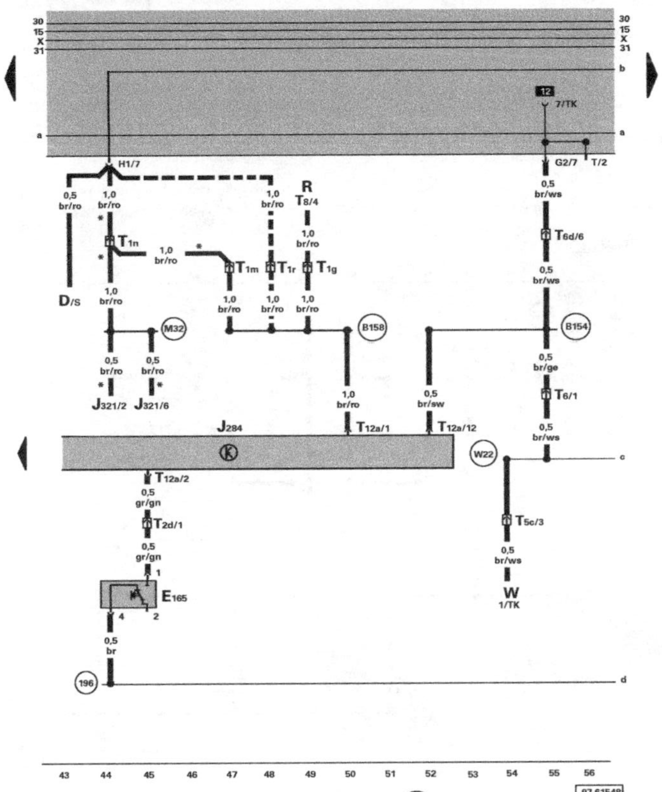

ws = white
sw = black
ro = red
br = brown
gn = green
bl = blue
gr = grey
li = violet
ge = yellow

D - Ignition/Starter Switch
E165 - Trunk Lid Release Switch
J284 - Alarm system control module
J321 - Convertible Top Hydraulic Pump Relay (53)
R - Radio
T1g - 1-Pin Connector, red, behind fuse relay panel
T1m - 1-Pin Connector, red, behind fuse relay panel
T1n - 1-Pin Connector, red, behind fuse relay panel
T1r - 1-Pin Connector, red, behind fuse relay panel
T2d - 2-Pin Connector, green, behind fuse/relay panel
T5c - 5-Pin Connector
T6 - 6-Pin Connector, black
T6d - 6-Pin Connector, red, behind fuse relay panel
T8 - 8-Pin Connector, black, in radio
T12a - 12-Pin Connector
W - Front Interior Light

(196) - Ground connection -3-, in wiring harness, rear
(B154) - Connector -1- (TK), in wiring harness interior
(B158) - Connector (S) in wiring harness interior
(M32) - Wire Connection (su),in convertible top wiring harness
(W22) - Connector (door contact switch) in wiring harness, rear

* - Power convertible top
--- - Manual convertible top

Edition 09.01
USA.5403.04.21

Alarm system control module, trunk lid release switch

No. 4/6 — Wiring diagram — Cabrio

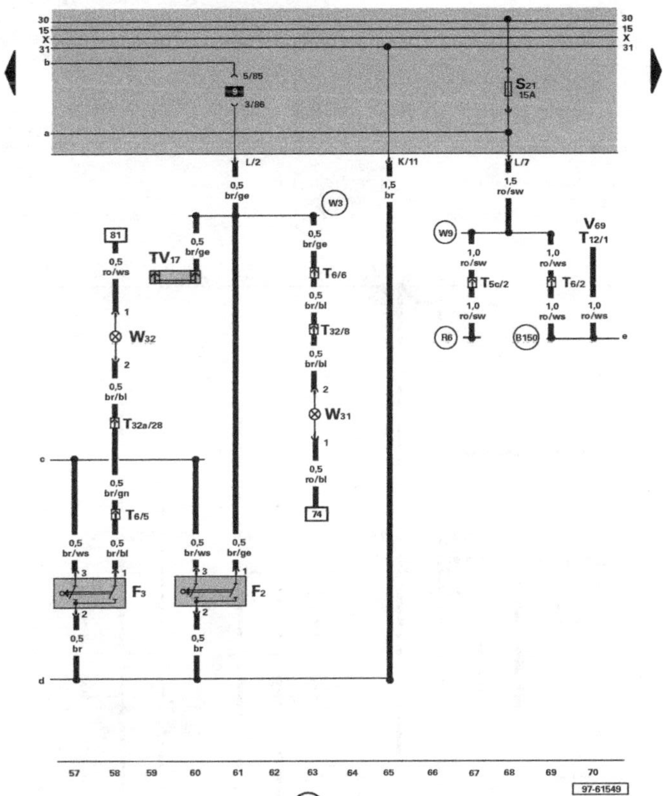

ws = white
sw = black
ro = red
br = brown
gn = green
bl = blue
gr = grey
li = violet
ge = yellow

F2 - Door contact switch-Driver's side
F3 - Door contact switch-Passenger's side
T5c - 5-Pin Connector
T6 - 6-Pin Connector, black
T12 - 12-Pin Connector, near right taillight bulb holde
T32 - 32-Pin Connector, white, on left A-pillar
T32a - 32-Pin Connector, white, on right A-pillar
TV17 - Door Contact Switch Wire Connector, white
V69 - Central Looking System Pump and control Module
W31 - Entry light front, left
W32 - Entry light front, right

(B150) - Plus connector -2- (30a), in wiring harness interior
(R6) - Plus connection -1-, in interior light wiring harness

(W3) - Wire connection, in rear wiring harness
(W9) - Plus connection -1- (30), in rear wiring harness

Door contact switches, entry ligths front

Edition 09.01
USA.5403.04.21

Cabrio — Wiring diagram — No. 4/7

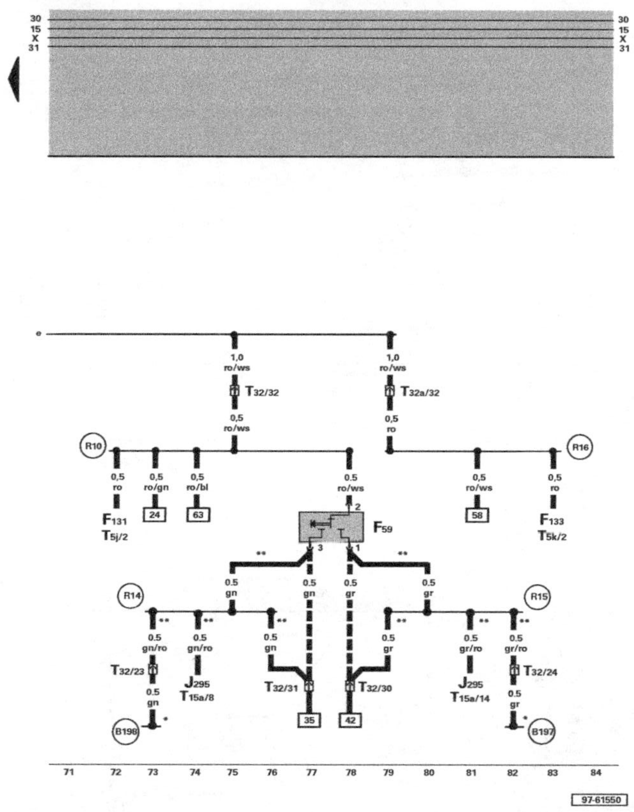

ws = white
sw = black
ro = red
br = brown
gn = green
bl = blue
gr = grey
li = violet
ge = yellow

F59 - Driver's Door Central Locking System Switch
F131 - Left Front Central Locking System Activator/Switch
F133 - Right Front Central Locking System Activator/Switch
J295 - Left Front Power Window Control Module
T5j - 5-Pin Connector, beige, on left door
T5k - 5-Pin Connector, beige, on right door
T15a - 15-Pin Connector
T32 - 32-Pin Connector, white, on left A-pillar
T32a - 32-Pin Connector, white, on right A-pillar
B197 - Connection (close window) in passenger compartment wiring harness
B198 - Connection (open window) in passenger compartment wiring harness

R10 - Plus connection -1- (30), in driver's door wiring harness
R14 - Wire connection -1- (open), in driver's door wiring harness
R15 - Plus connection -2- (closed), in driver's door wiring harness
R16 - Plus connection -1- (30), in passenger's door wiring harness

* - With power windows
** - Driver & passenger side power mirrors, heatable
--- - Driver & passenger side remote mirrors, heatable

Edition 09.01
USA.5403.04.21

Driver's door central locking system switch

Alarm system (from September 1999)

No. 4/8 — Wiring diagram — Cabrio

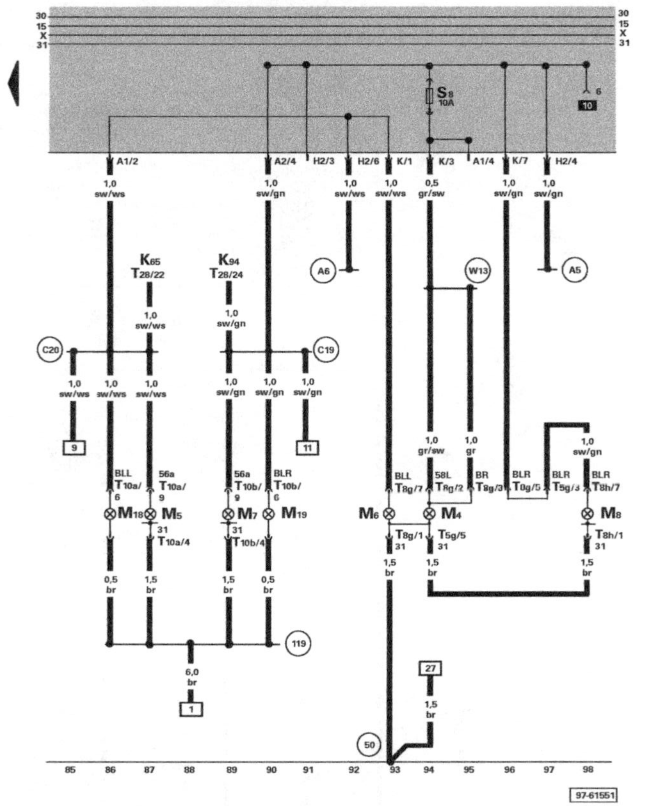

ws = white
sw = black
ro = red
br = brown
gn = green
bl = blue
gr = grey
li = violet
ge = yellow

K65 - Left Turn Signal Indicator Light
K94 - Right Turn Signal Indicator Light
M4 - Left Tail Light
M5 - Left Front Turn Signal Light
M6 - Left Rear Turn Signal Light
M7 - Right Front Turn Signal Light
M8 - Right Rear Turn Signal Light
M18 - Left, Side Turn Signal Light
M19 - Right, Side turn signal light
T5g - 5-Pin Connector, on taillight, left
T8g - 8-Pin Connector, on taillight, left
T8h - 8-Pin Connector, on taillight, right
T10a - 10-Pin Connector
T10b - 10-Pin Connector
T28 - 28-Pin Connector, white

50 - Ground connection, in luggage compartment, left
119 - Ground connection -1-, in headlight wiring harness
A5 - Plus connection (right turn signal), in instrument panel wiring harness
A6 - Plus connection (left turn signal), in instrument panel wiring harness
C19 - Connector -1- (right turn signal), in wiring harness headlamp
C20 - Connection (left turn signal), in headlight wiring harness
W13 - Wire connection (58L),in rear wiring harness

Turn signal lights, front /rear turn signal lights, side marker lights

Edition 09.01
USA.5403.04.21

Cabrio Wiring diagram **No. 5/1**

2.0L - Engine - Motronic Multiport Fuel Injection (MFI)/85 kW, engine code ABA

from September 1999

Relay locations

Position of relays and multi-pin connector wiring, see also "Component Locations" section.

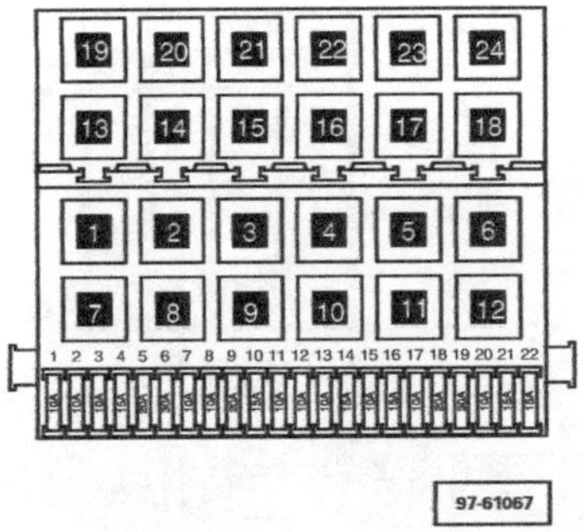

9 Radio/Parking Light Warning Buzzer (188)

12 Fuel Pump (FP) Relay (409)

Note: Number in parentheses indicates production control number stamped on relay housing.

Fuse colors

30 A - green
25 A - white
20 A - yellow
15 A - blue
10 A - red
7,5 A - brown
5 A - beige
3 A - violet

Edition 09.01
USA.5403.04.21

2.0L engine (code: ABA) fuel injection (from September 1999)

No. 5/2 Wiring diagram **Cabrio**

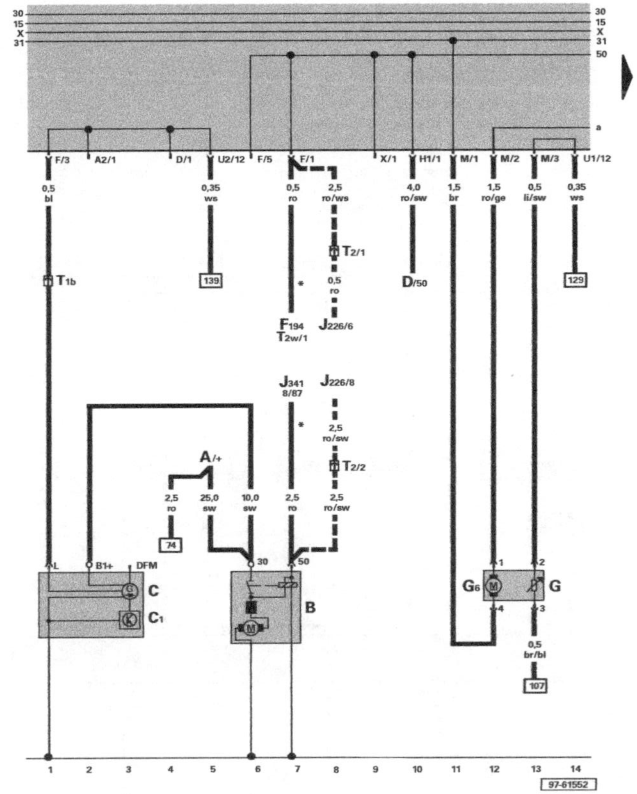

ws = white
sw = black
ro = red
br = brown
gn = green
bl = blue
gr = grey
li = violet
ge = yellow

A - Battery
B - Starter
C - Generator (GEN)
C1 - Voltage Regulator (VR)
D - Ignition/Starter Switch
F194 - Clutch Pedal Position (CPP) Switch, green
G - Sender for fuel gauge
G6 - Fuel Pump (FP)
J226 - Park/Neutral Position (PNP) Relay
J341 - Anti-Theft Engine Disable Relay I (53)
T1b - Single Connector, black, near starter
T2 - 2-Pin Connector, black, behind fuse relay panel
T2w - 2-Pin Connector, green, on clutch pedal bracket

* - Manual transmission only
--- - Automatic transmission only

Battery, starter, generator (GEN), voltage regulator (VR), fuel pump (FP), sender for fuel gauge

Edition 09.01
USA.5403.04.21

Cabrio — Wiring diagram — No. 5/3

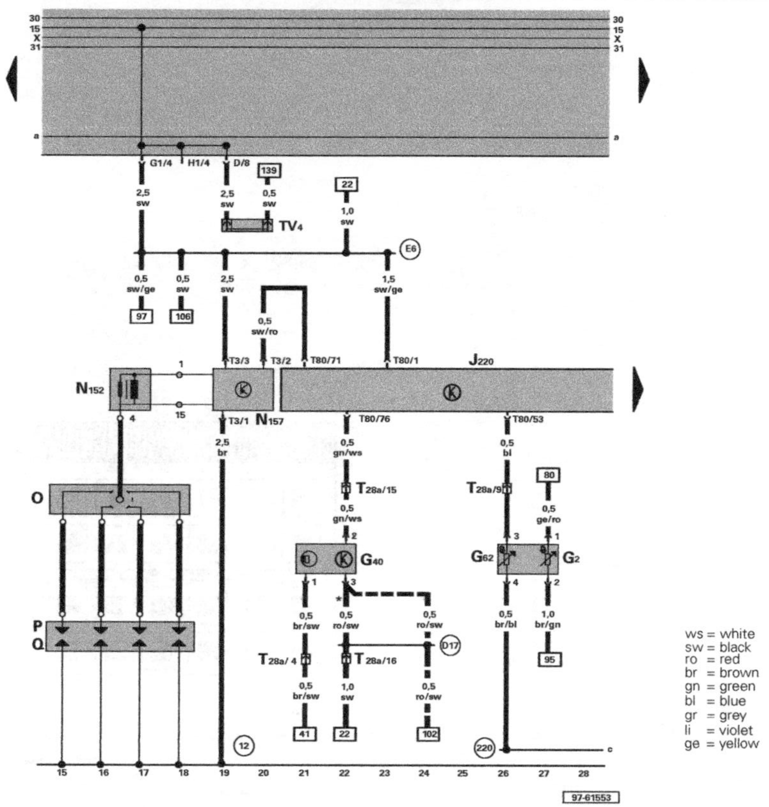

ws = white
sw = black
ro = red
br = brown
gn = green
bl = blue
gr = grey
li = violet
ge = yellow

- G2 - Engine Coolant Temperature (ECT) Sensor
- G40 - Camshaft Position (CMP) Sensor
- G62 - Engine Coolant Temperature (ECT) Sensor
- J220 - Motronic Engine Control Module (ECM)
- N152 - Ignition Coil
- N157 - Ignition Coil Power Output Stage
- O - Ignition Distributor
- P - Spark Plug Connectors
- Q - Spark Plugs
- T3 - 3-Pin connector
- T28a - 28-Pin connector, in engine, left
- T80 - Connector, 80 point
- TV4 - Terminal 15 wire connector, black
- (12) - Ground connection, in engine compartment, left
- (220) - Ground connection (sensor ground), in engine compartment wiring harness
- (D17) - Plus connection -1- (15), in engine compartment wiring harness
- (E6) - Plus connection -1- (15), in Motronic Multiport Fuel Injection (MFI)
- * - Manual transmission only
- --- - Automatic transmission only

Edition 09.01
USA.5403.04.21

Motronic ECM, engine coolant temperature (ECT) sensor, camshaft position (CMP) sensor, ignition coil, ignition power output stage, spark plugs

No. 5/4 — Wiring Diagram — Cabrio

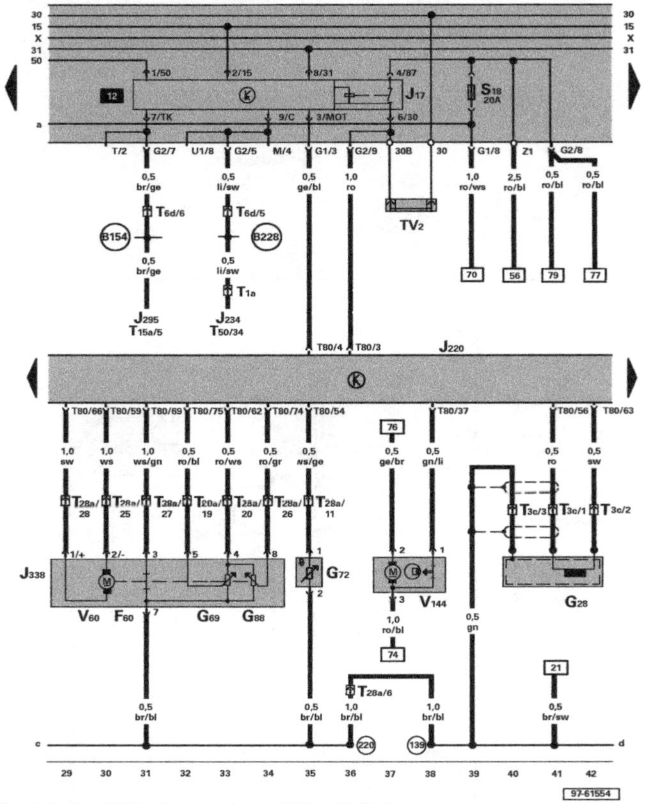

ws = white
sw = black
ro = red
br = brown
gn = green
bl = blue
gr = grey
li = violet
ge = yellow

- F60 - Closed Throttle Position (CTP) Switch
- G28 - Engine Speed (RPM) Sensor
- G69 - Throttle Position (TP) Sensor
- G72 - Intake Air Temperature (IAT) Sensor
- G88 - Throttle Position (TP) Sensor
- J17 - Fuel Pump (FP) Relay
- J220 - Motronic Engine Control Module (ECM)
- J234 - Airbag Control Module
- J295 - Left Front Power Window Control Module
- J338 - Throttle Valve Control Module
- T1a - Single Connector, behind relay panel
- T3c - 3-Pin Connector, in engine
- T6d - 6 Pin Connector, behind relay panel
- T15a - 15 Pin Connector
- T28a - 28-Pin Connector, in engine, left
- T50 - 50-Pin Connector
- T80 - 80-Pin Connector
- TV2 - Terminal 30 (B+) Wire Connector
- V60 - Throttle Position (TP) Actuator
- V144 - Leak Detection Pump (LDP)
- (139) - Ground connection (sensor Ground),in Motronic Multiport Fuel Injection (MFI) wiring harness
- (220) - Ground connection (sensor ground), in engine compartment wiring harness
- (B154) - Connector -1- (TK), in wiring harness interior
- (B228) - Connection (Crash-Signal) (interior wiring harness)

Motronic ECM, throttle valve control module, intake air temperature (IAT) sensor, engine speed (RPM) sensor, fuel pump (FP) relay

Edition 09.01
USA.5403.04.21

Cabrio — Wiring Diagram — No. 5/5

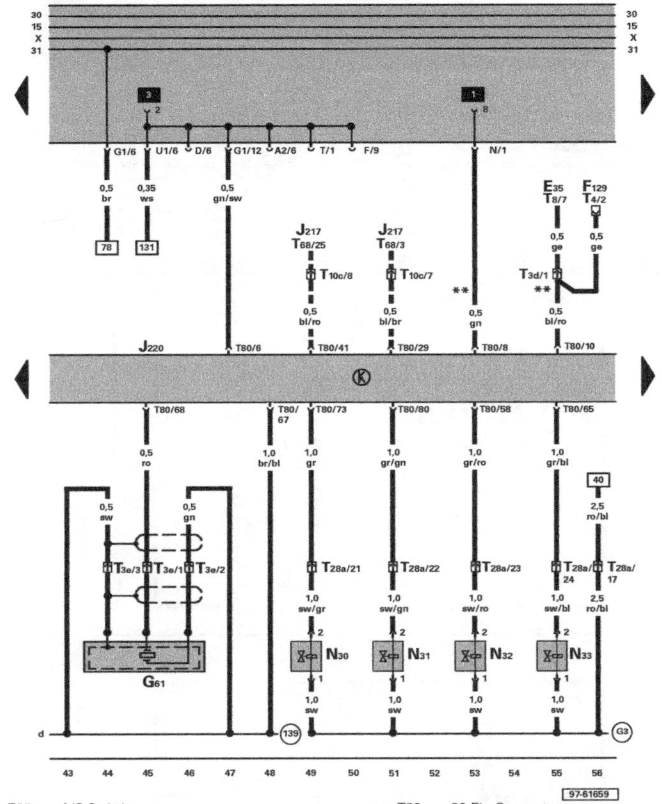

ws = white
sw = black
ro = red
br = brown
gn = green
bl = blue
gr = grey
li = violet
ge = yellow

- E35 - A/C Switch
- F129 - A/C Pressure Switch
- G61 - Knock Sensor (KS) 1
- J217 - Transmission Control Module (TCM)
- J220 - Motronic Engine Control Module (ECM)
- N30 - Cylinder 1 Fuel Injector
- N31 - Cylinder 2 Fuel Injector
- N32 - Cylinder 3 Fuel Injector
- N33 - Cylinder 4 Fuel Injector
- T3d - 3-Pin Connector, black, behind relay panel
- T3e - 3-Pin Connector, in engine compartment
- T4 - 4-Pin Connector, in engine compartment, rigth
- T8 - 8-Pin Connector
- T10c - 10-Pin Connector, black, behind relay panel
- T28a - 28-Pin Connector, in engine, left
- T68 - 68-Pin Connector
- T80 - 80-Pin Connector
- (139) - Ground connection (sensor Ground), in Motronic Multiport Fuel Injection (MFI), wiring harness
- (G3) - Plus connection, in injector wiring harness
- ** - A/C Connector
- --- - Automatic transmission only

Edition 09.01 USA.5403.04.21 — **Motronic engine control module (ECM), knock sensor (KS) 1, fuel injectors**

No. 5/6 — Wiring Diagram — Cabrio

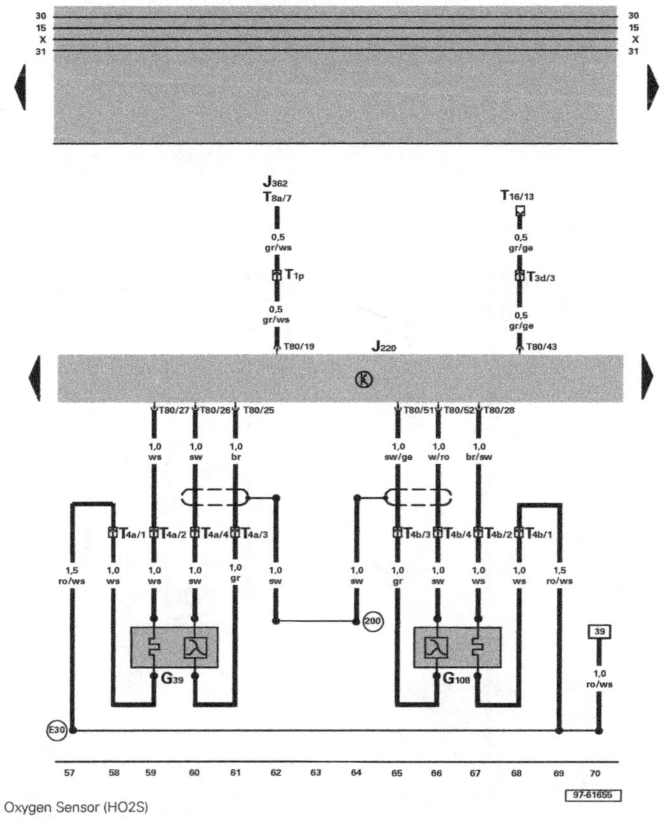

ws = white
sw = black
ro = red
br = brown
gn = green
bl = blue
gr = grey
li = violet
ge = yellow

- G39 - Heated Oxygen Sensor (HO2S)
- G108 - Heated Oxygen Sensor (HO2S) 2
- J220 - Motronic Engine Control Module (ECM)
- J362 - Control module for anti-theft immobilizer
- T1p - Single Connector, brown, behind fuse/ relay panel
- T3d - 3-Pin Connector, black, behind relay panel
- T4a - 4-Pin Connector, in engine compartment
- T4b - 4-Pin Connector, in engine compartment
- T8a - 8-Pin Connector, black
- T16 - 16-Pin Connector, Data Link Connector (DLC)
- T80 - 80-Pin Connector
- (200) - Ground connection (shielding), in engine compartment wiring harness
- (E30) - Connector (87a), in wiring harness engine

Motronic engine control module (ECM), heated oxygen sensors (HO2S), evaporative emission canister purge regulator valve — **Edition 09.01** USA.5403.04.21

Cabrio — Wiring Diagram — No. 5/7

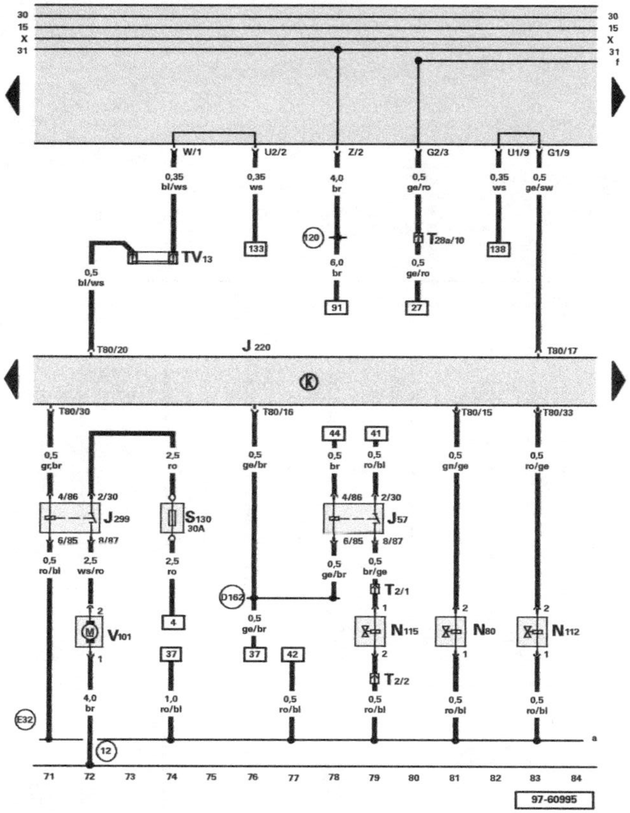

ws = white
sw = black
ro = red
br = brown
gn = green
bl = blue
gr = grey
li = violet
ge = yellow

- J220 - Motronic Engine Control Module (ECM)
- J299 - Secondary Air Injection (AIR) Pump Relay
- J57 - Vacuum Pump Relay
- N 80 - Evaporative Emission (EVAP) Canister Purge Regulator valve
- N112 - Secondary Air Injection (AIR) Solenoid Valve
- N115 - Evaporative Emission (EVAP) Canister Purge Solenoid valve
- S130 - Fuse for secondary air pump
- T2 - 2-Pin Connector, behind panel relay
- T28a - 28-Pin Connector, in engine, left
- T80 - Connector, 80 point
- TV13 - Vehicle Speed Signal Wire Connector
- V101 - Secondary Air Injection (AIR) Pump Motor
- (12) - Ground connection, in engine compartment, left
- (120) - Ground connection -2-, in headligth wiring harness
- (D162) - Connection for Leak Detection Pump (LDP),engine compartment wiring harness
- (E32) - Connector (87), in wiring harness Motronic

Edition 10.00
W42.USA.5403.02.21

Motronic (ECM), secondary air injection (AIR) pump motor, Evaporative Emission (EVAP) canister purge regulator valve

No. 5/8 — Wiring Diagram — Cabrio

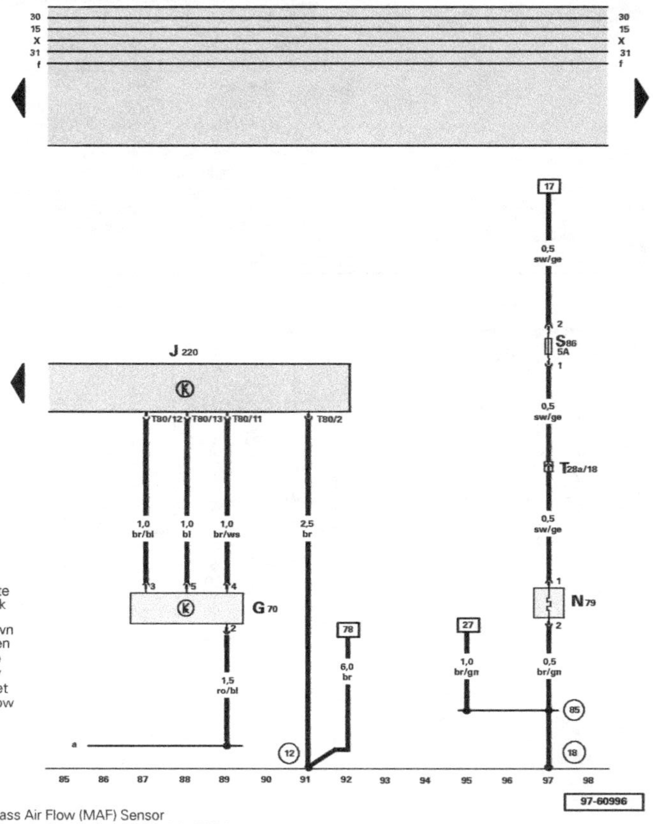

ws = white
sw = black
ro = red
br = brown
gn = green
bl = blue
gr = grey
li = violet
ge = yellow

- G70 - Mass Air Flow (MAF) Sensor
- J220 - Motronic Engine Control Module (ECM)
- N79 - Positive Crankcase Ventilation (PCV) Heating Element
- S86 - Door Lock Heating Element Circuit Breaker, above relay panel
- T28a - 28-Pin Connector, in engine, left
- T80 - Connector, 80 point
- (12) - Ground connection, in engine compartment, left
- (18) - Ground connection, on engine block
- (85) - Ground connection -1-, in engine compartment wiring harness

Motronic (ECM), mass air flow (MAF) sensor, positive crankcase ventilation (PVC) heating element, secondary air injection (AIR) solenoid valve

Edition 10.00
W42.USA.5403.02.21

Cabrio — Wiring Diagram — No. 5/9

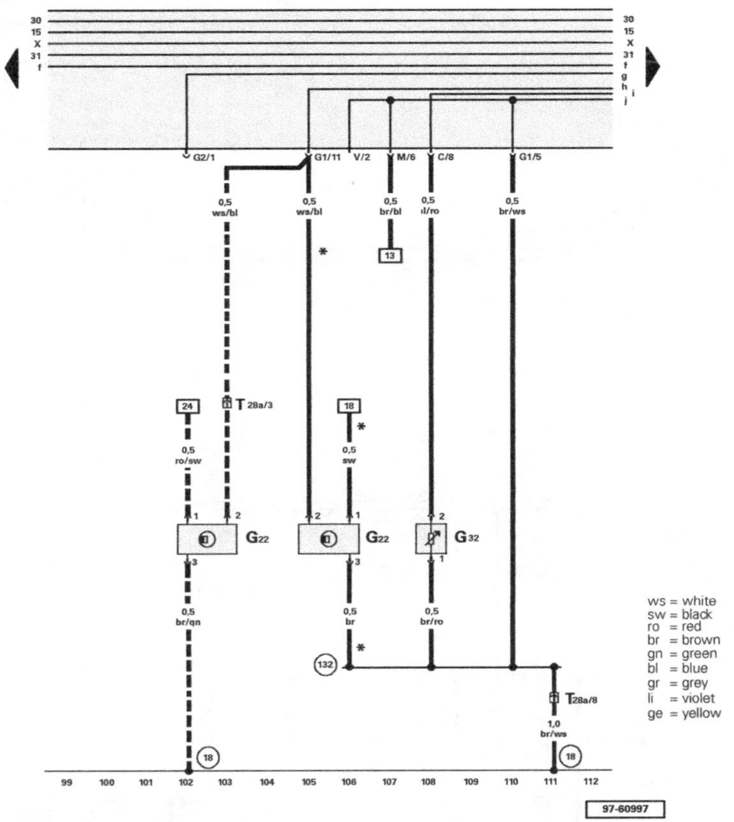

ws = white
sw = black
ro = red
br = brown
gn = green
bl = blue
gr = grey
li = violet
ge = yellow

G22 - Speedometer Vehicle Speed Sensor (VSS)
G32 - Engine Coolant Level (ECL) Sensor
T28a - 28-Pin Connector, in engine, left

(18) - Ground connection, on engine block

(132) - Ground connection -3-, in engine compartment wiring harness

* - Manual transmission only

--- - Automatic transmission only

Speedometer vehicle speed sensor(VSS), engine coolant level (ECL) sensor

2.0L engine (code: ABA) fuel injection (from September 1999)

No. 5/10 — Wiring Diagram — Cabrio

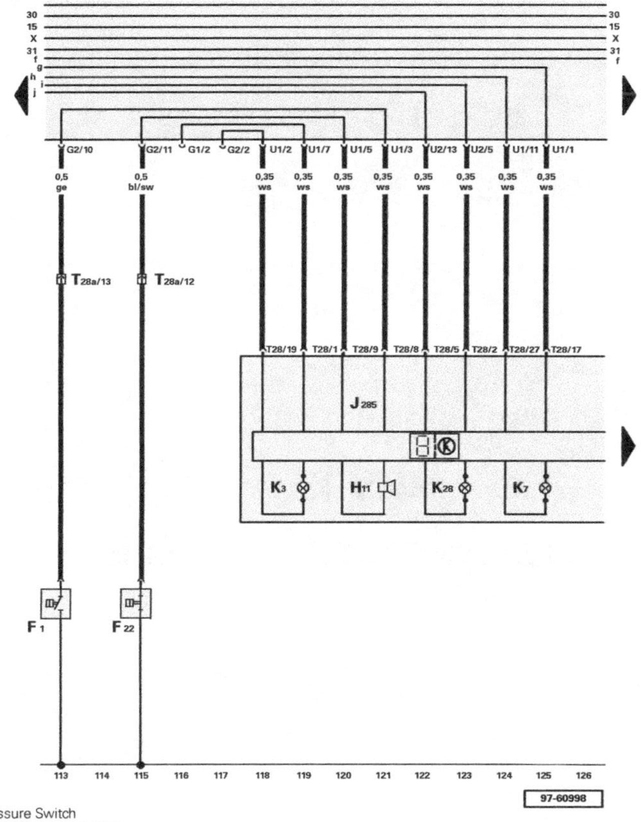

ws = white
sw = black
ro = red
br = brown
gn = green
bl = blue
gr = grey
li = violet
ge = yellow

F1 - Oil Pressure Switch
F22 - 0.3bar Oil Pressure Switch
H11 - Oil Pressure Warning Buzzer
J285 - Control module with indicator unit in instrument panel insert
K3 - Oil Pressure Warning Light
K7 - Brake and Parking Brake Warning Light
K28 - Engine Coolant Level/Temperature (ECL/ECT) Warning Light
T28 - 28-Pin Connector
T28a - 28-Pin Connector, in engine, left

Instrument cluster, oil pressure switch, 0.3 bar oil pressure switch, oil pressure warning buzzer, brake and parking brake warning light

Cabrio Wiring Diagram **No. 5/11**

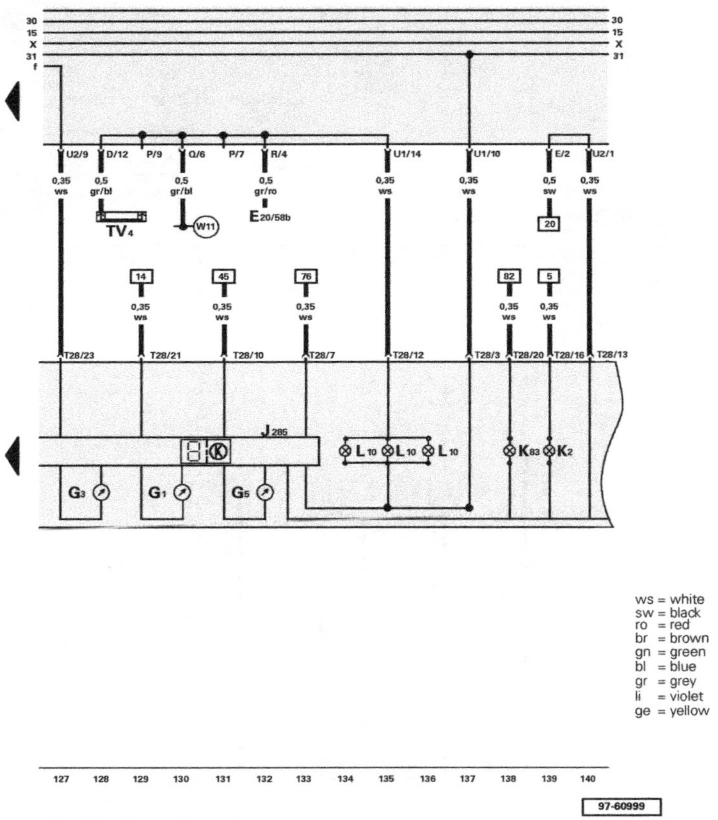

ws = white
sw = black
ro = red
br = brown
gn = green
bl = blue
gr = grey
li = violet
ge = yellow

E20 - Instrument Panel Light Dimmer Switch
G1 - Fuel Gauge
G3 - Engine Coolant Temperature (ECT) Gauge
G5 - Tachometer
J285 - Control module with indicator unit in instrument panel insert
K2 - Generator (GEN) Warning Light
K83 - Malfunction Indicator Lamp (MIL)
L10 - Instrument Cluster Light
T28 - 28-Pin Connector
TV4 - Terminal 15 Wire Connector
(W11) - Plus connection (58), in rear lid wiring harness

Edition 10.00
W42.USA.5403.02.21

Instrument cluster, fuel gauge, engine coolant temperature (ECT) gauge, tachometer, generator (GEN) warning light, malfunction indicator lamp (MIL)

THIS PAGE INTENTIONALLY LEFT

BLANK

Cabrio — Wiring diagram — No. 6/1

Automatic transmission, 4 speed automatic

from September 1999

2.0L-Engine - Motronic Multiport fuel Injection (MFI) /85kW, code ABA

Relay locations

Position of relays and multi-pin connector wiring, see also "Component Locations" section.

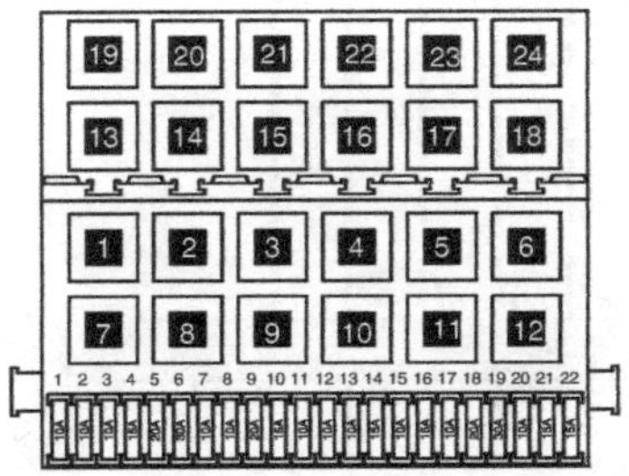

97-61067

- **1** - A/C Relay (147)
- **12** - Fuel Pump Relay (409)
- **14** - Park/Neutral Position (PNP) Relay (175)

Note:
Number in parentheses indicates production control number stamped on relay housing.

Fuse colors

30 A - green
25 A - white
20 A - yellow
15 A - blue
10 A - red
7,5 A - brown
5 A - beige
3 A - violet

Edition 09.01
USA.5403.04.21

Automatic transmission, 4-speed (from September 1999)

No. 6/2 — Wiring diagram — Cabrio

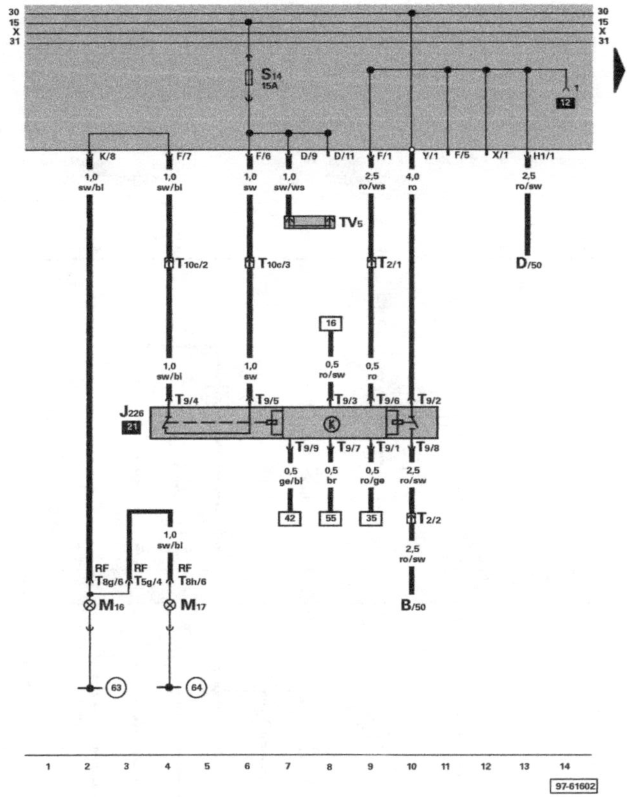

ws = white
sw = black
ro = red
br = brown
gn = green
bl = blue
gr = grey
li = violet
ge = yellow

- B - Starter
- D - Ignition/Starter Switch
- J226 - Park Neutral Position (PNP) Relay (175)
- M16 - Left Back-Up Light
- M17 - Right Back-Up Light
- T2 - 2-Pin Connector,black, behind fuse/relay panel
- T5g - 5-Pin Connector, on taillight, left
- T8g - 8-Pin Connector, on taillight, left
- T8h - 8-Pin Connector, on taillight, right
- T9 - 9-Pin Connector
- T10c - 10-Pin Connector, black
- TV5 - Terminal 15a wire connector, green
- (63) - Ground connection, on left taillight bulb holder
- (64) - Ground connection, on right taillight bulb holder

Transmission control module (TCM), park/neutral position (PNP) relay, starter, ignition/starter switch, back-up lights

Edition 09.01
USA.5403.04.21

Cabrio — Wiring diagram — No. 6/3

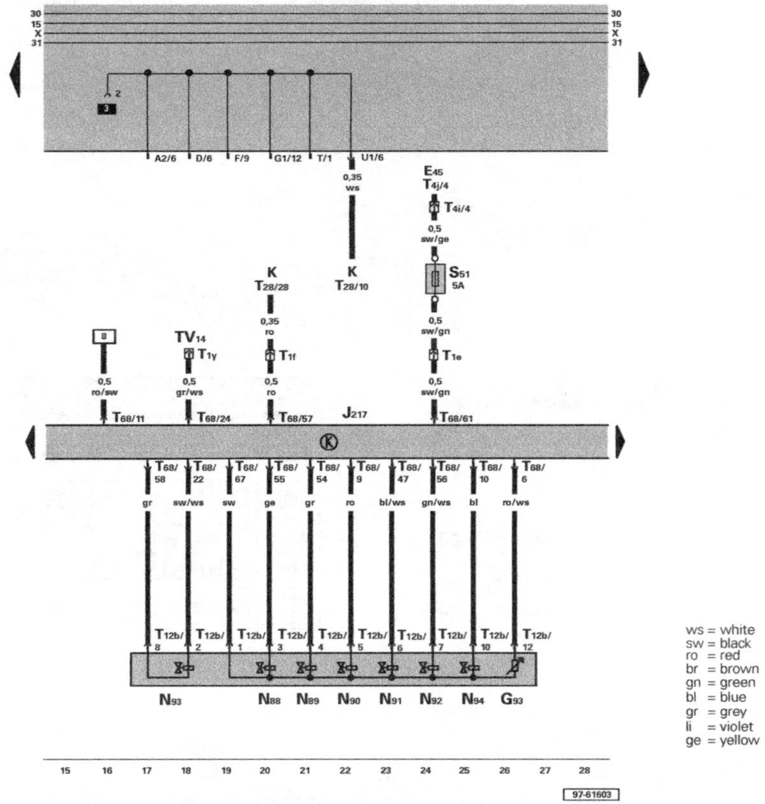

ws = white
sw = black
ro = red
br = brown
gn = green
bl = blue
gr = grey
li = violet
ge = yellow

E45 - Cruise Control Switch
G93 - Transmission Fluid Temperature Sensor
J217 - Transmission Control Module (TCM), beneath rear seat, left side
K - Instrument Cluster
N88 - Solenoid Valve 1
N89 - Solenoid Valve 2
N90 - Solenoid Valve 3
N91 - Solenoid Valve 4
N92 - Solenoid Valve 5
N93 - Solenoid Valve 6
N94 - Solenoid Valve 7
S51 - Fuse (Only Automatic Transmission)
T1e - Single Connector, black, behind fuse/relay panel
T1f - Single Connector, blue, behind fuse/relay panel
T1y - Single Connector, yellow, at TV14
T4i - 4-Pin Connector, black, behind fuse/relay panel
T4j - 4-Pin Connector, black, behind fuse/relay panel
T12b - 12-Pin Connector, top of transmission
T28 - 28-Pin Connector, white
T68 - 68-Pin Connector, on Transmission Control Module, black
TV14 - Data Link Connector (DLC) wire Connector, black, above fuse relay panel

Transmission control module (TCM), transmission fluid temperature sensor, solenoid valves

No. 6/4 — Wiring diagram — Cabrio

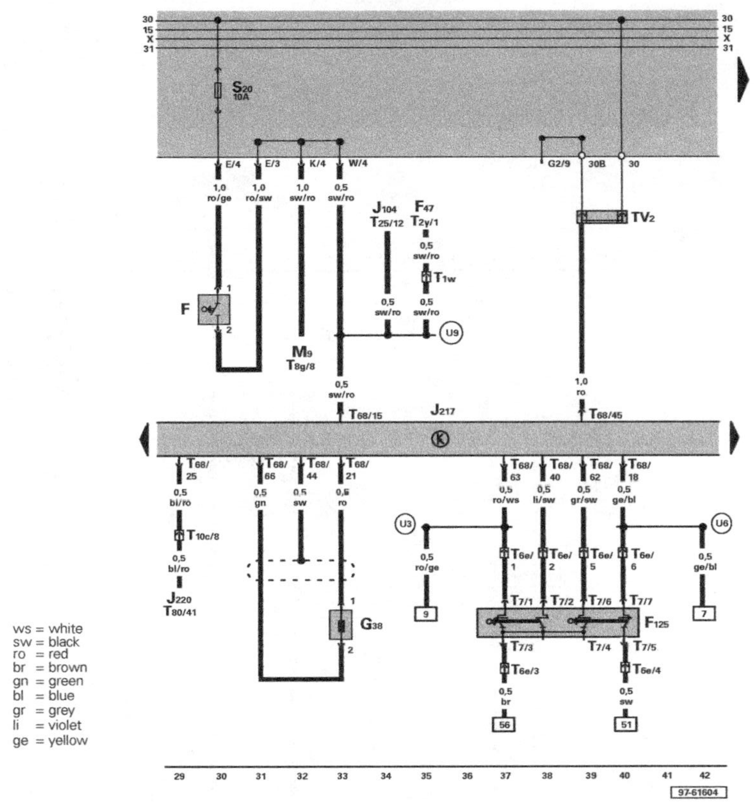

ws = white
sw = black
ro = red
br = brown
gn = green
bl = blue
gr = grey
li = violet
ge = yellow

F - Brake Light Switch
F47 - Brake Vacuum Vent Valve Switch for Cruise Control/Diesel Fuel Injection
F125 - Multi-Function Transmission Range (TR) Switch
G38 - Transmission Vehicle Speed Sensor (VSS)
J104 - ABS Control Module (w/EDL)
J217 - Transmission Control Module (TCM), beneath rear seat, left side
J220 - Motronic Engine Control Module (ECM)
M9 - Left Brake Light
T1w - Single Connector, green, behind fuse/relay panel
T2y - 2-Pin Connector, blue, on brake pedal bracket
T6e - 6-Pin Connector, black, top of transmission
T7 - 7-Pin Connector, top of transmission
T8g - 8-Pin Connector, on taillight, left
T10c - 10-Pin Connector, black
T25 - 25-Pin Connector
T68 - 68-Pin Connector, on Transmission Control Module
T80 - 80-Pin Connector, on engine Control Module
TV2 - Terminal 30 (B+) Wire Connector, red
(U3) - Wire connection -1-, in automatic transmission wiring harness
(U6) - Wire connection -2-, in automatic transmission wiring harness
(U9) - Connection (54) (in automatic transmission wiring harness)

Transmission control module, multifuction transmission range (TR) switch, transmisssion vehicle speed sensor (VSS)

Cabrio

Wiring diagram

No. 6/5

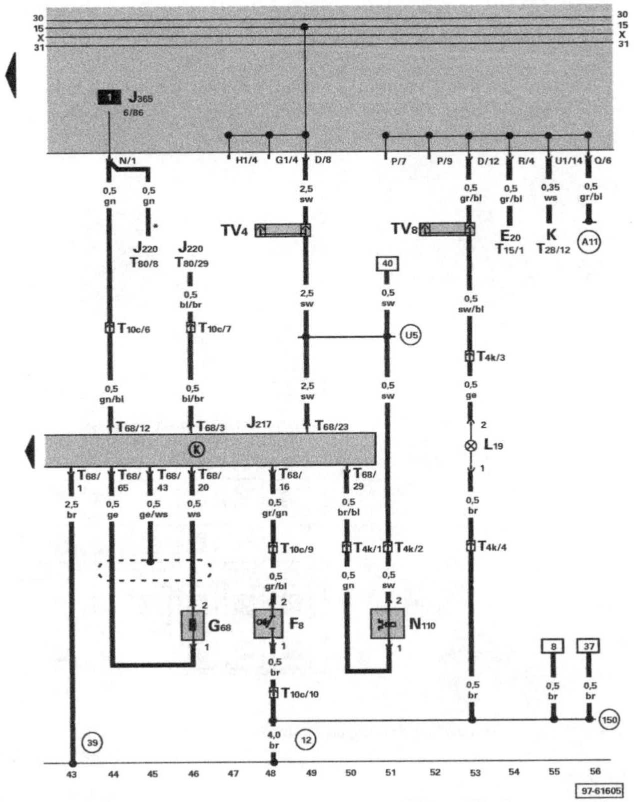

ws = white
sw = black
ro = red
br = brown
gn = green
bl = blue
gr = grey
li = violet
ge = yellow

E20 - Instrument Panel Light Dimmer Switch
F8 - Kick Down Switch
G68 - Vehicle Speed Sensor (VSS)
J217 - Transmission Control Module (TCM), beneath rear seat, left side
J220 - Motronic Engine Control Module (ECM)
J365 - Relay for A/C shutoff
K - Instrument Cluster
L19 - Automatic Transmission Console Light
N110 - Shift Lock Solenoid
T4k - 4-Pin Connector, black, below gearshift lever cover
T10c - 10-Pin Connector, black
T15 - 15-Pin Connector, behind fuse relay panel
T28 - 28-Pin Connector, white
T68 - 68-Pin Connector, on Transmission Control Module, black
T80 - 80-Pin Connector, on Motronic Engine Control Module (ECM)
TV4 - Terminal 15 Wire Connector, black
TV8 - Terminal 58b Wire Connector, grey
(12) - Ground connection, in engine compartment, left
(39) - Ground connection, below rear seat bench, left
(150) - Ground connection, in automatic transmission wiring harness
(A11) - Plus connection (58b), in instrument panel wiring harness
(U5) - Wire connection -2- (15), in automatic transmission wiring harness
* - Manual transmission only

Transmission control module (TCM), kick down switch, vehicle speed sensor (VSS), automatic transmission console ligth, shift lock solenoid

THIS PAGE INTENTIONALLY LEFT

BLANK

Cabrio — Wiring diagram — No. 7/1

Central locking system

from September 1999

Relay locations

Position of relays and multi-pin connector wiring, see also "Component Locations" section.

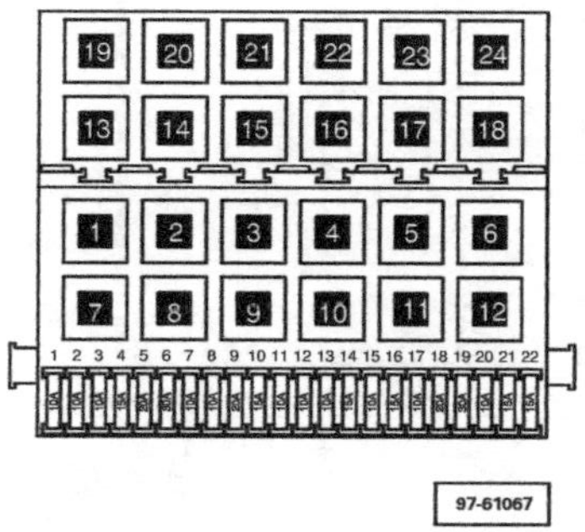

Note:
Number in parentheses indicates production control number stamped on relay housing.

Fuse colors

- 30 A - green
- 25 A - white
- 20 A - yellow
- 15 A - blue
- 10 A - red
- 7,5 A - brown
- 5 A - beige
- 3 A - violet

No. 7/2 — Wiring diagram — Cabrio

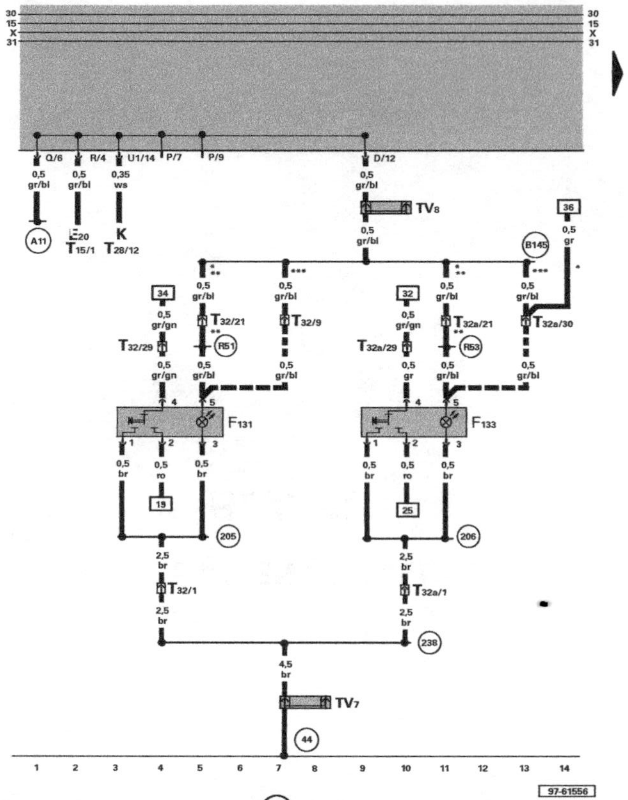

- ws = white
- sw = black
- ro = red
- br = brown
- gn = green
- bl = blue
- gr = grey
- li = violet
- ge = yellow

- E20 - Instrument Panel Light Dimmer Switch
- F131 - Left Front Central Locking System Activator/Switch
- F133 - Right Front Central Locking System Activator/Switch
- K - Instrument Cluster
- T15 - 15-Pin Connector
- T28 - 28-Pin Connector, white
- T32 - 32-Pin Connector, on left A-pillar
- T32a - 32-Pin Connector, on right A-pillar
- TV7 - Terminal 31 (GND) wire connector, brown
- TV8 - Terminal 58b wire connector, grey
- (44) - Ground connection, on left A-pillar, lower part
- (205) - Ground connection, in driver's door wiring harness
- (206) - Ground connection, in passenger's door wiring harness
- (238) - Ground connection -1-, in wiring harness interior
- (A11) - Plus connection (58b) in instrument panel wiring harness
- (B145) - Plus connector (58b), in wiring harness interior
- (R51) - Connection (58b) (driver's side door wiring harness)
- (R53) - Connection (58s) (in right front door wiring harness)
- * - With power windows
- ** - Outside power mirrors, heatable
- *** - Without power windows
- --- - Outside remote mirrors, heatable

Left/right front central locking system activator/switch

Cabrio — Wiring diagram — No. 7/3

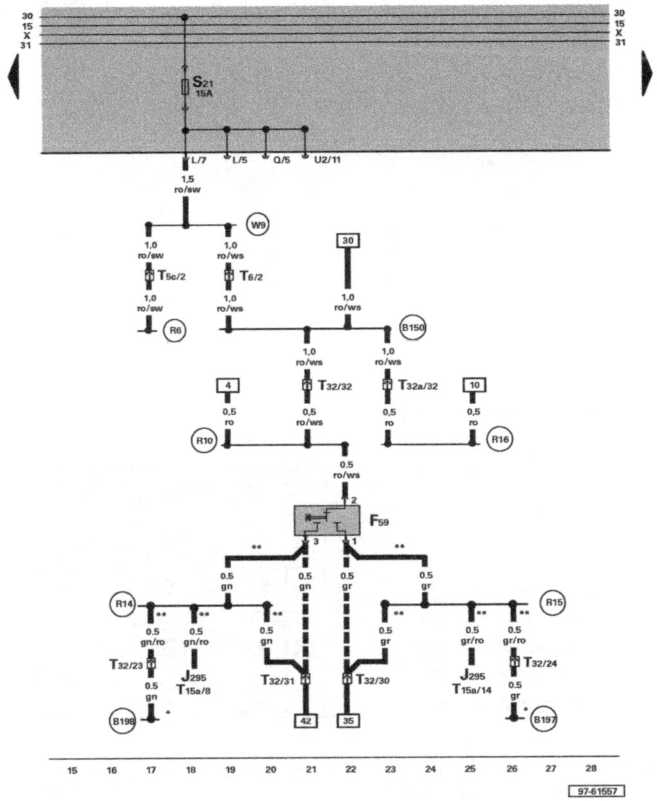

ws = white
sw = black
ro = red
br = brown
gn = green
bl = blue
gr = grey
li = violet
ge = yellow

- F59 - Driver's Door Central Locking System Switch
- J295 - Left Front Power Window Control Module
- T5c - 5-Pin Connector
- T6 - 6-Pin Connector, black
- T15a - 15-Pin Connector
- T32 - 32-Pin Connector, white, on left A-pillar
- T32a - 32-Pin Connector, white, on right A-pillar
- (B150) - Plus connector -2- (30a), in wiring harness interior
- (B197) - Connection (close window) in passenger compartment wiring harness
- (B198) - Connection (open window) in passenger compartment wiring harness
- (R6) - Plus connection -1-, in interior light wiring harness
- (R10) - Plus connection -1- (30), in driver's door wiring harness
- (R14) - Wire connection -1- (open), in driver's door wiring harness
- (R15) - Plus connection -2- (closed), in driver's door wiring harness
- (R16) - Plus connection -1- (30), in passenger's door wiring harness
- (W9) - Plus connection -1- (30), in rear wiring harness
- * - With power windows
- ** - Driver & passenger side power mirrors, heatable
- --- - Driver & passenger side remote mirrors, heatable

Edition 09.01
USA.5403.04.21

Driver's door central locking system switch

Central locking system (from September 1999)

No. 7/4 — Wiring diagram — Cabrio

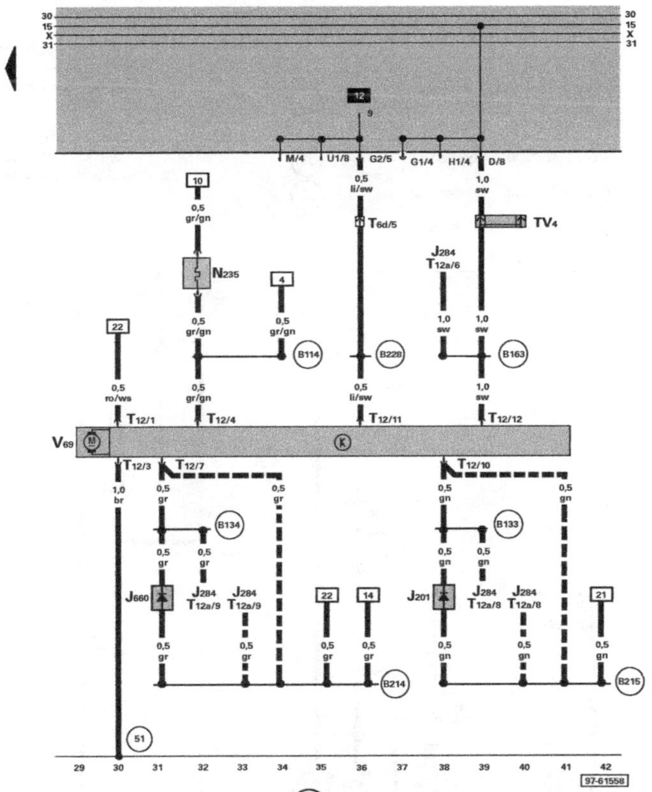

ws = white
sw = black
ro = red
br = brown
gn = green
bl = blue
gr = grey
li = violet
ge = yellow

- J201 - Protective diode 1
- J284 - Alarm System Control Module
- J660 - Protective diode 2
- N235 - Current limiting series resistor
- T6d - 6-Pin Connector, red, behind fuse relay panel
- T12 - 12-Pin Connector, in luggage compartment, right
- T12a - 12-Pin Connector, black
- TV4 - Terminal 15 wire connector, black
- V69 - Central Locking System Pump and Control Module
- (51) - Ground connection, in luggage compartment, right
- (B114) - Connector (lock Switch) in wiring harness interior
- (B133) - Connector (central locking open), in wiring harness interior
- (B134) - Connector (central locking closed), in wiring harness interior
- (B163) - Plus connector -1- (15) in wiring harness interior
- (B214) - Connection 1 (central locking) in passenger compartment wiring harness
- (B215) - Connection 2 (central locking) in passenger compartment wiring harness
- (B228) - Connection (Crash-Signal) (interior wiring harness)
- --- - Without power windows

Central Locking System Pump and Control Module

Edition 09.01
USA.5403.04.21

Cabrio | Wiring diagram | No. 8/1

Cabrio - Standard Equipment

from September 1999

Relay locations
Position of relays and multi-pin connector wiring, see also "Component Locations" section.

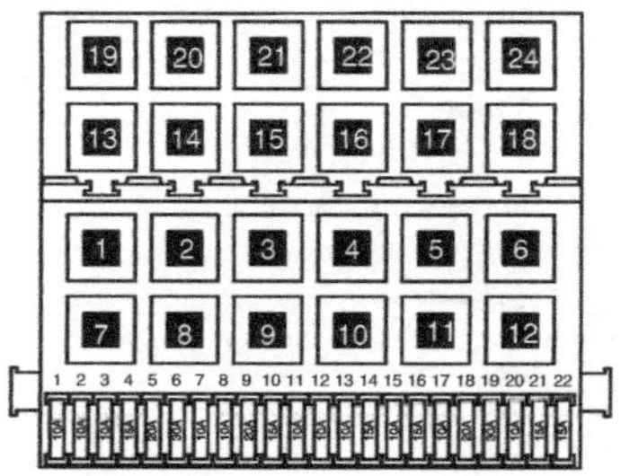

97-61067

- **4** - Load Reduction Relay (18)
- **6** - Emergency Flasher Relay (191)
- **8** - Wiper/Washer Intermitent Relay (196)
- **9** - Radio/Parking Light Warning Buzzer (188)
- **11** - Dual Tone Horn Relay (53)

Note:
Number in parentheses indicates production control number stamped on relay housing.

Fuse colors

30 A - green
25 A - white
20 A - yellow
15 A - blue
10 A - red
7,5 A - brown
5 A - beige
3 A - violet

Cabrio standard equipment (from September 1999)

No. 8/2 | Wiring diagram | Cabrio

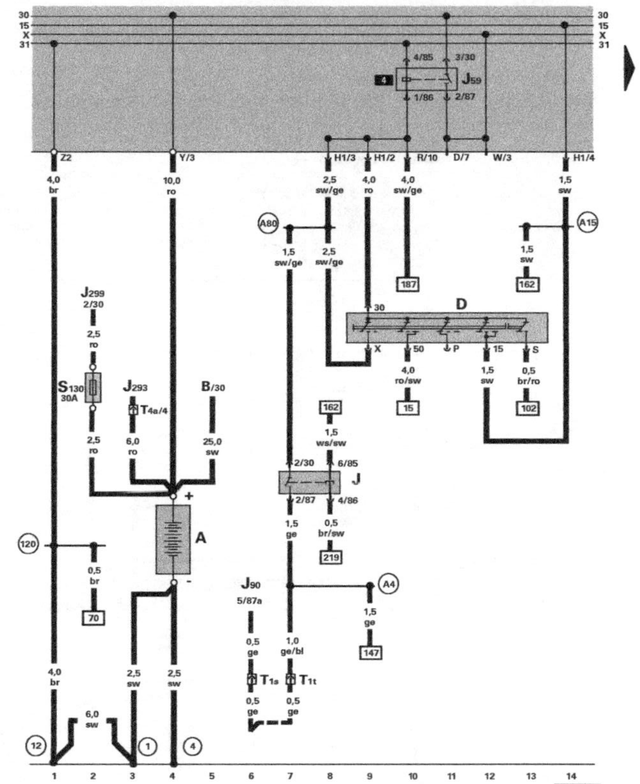

97-61559

ws = white
sw = black
ro = red
br = brown
gn = green
bl = blue
gr = grey
li = violet
ge = yellow

A - Battery
B - Starter
D - Ignition/Starter Switch
J - Headlight dimmer/flasher Relay (53)
J59 - Load Reduction Relay (18)
J90 - Daytime Running Lights Switch ON Relay (173)
J293 - Coolant FC (Fan Control) Control Module
J299 - Secondary Air Injection (AIR) Pump Relay
S130 - Fuse for Secondary Air Injection (AIR) Pump
T1s - 1-Pin Connector, yellow, behind fuse relay panel
T1t - 1-Pin Connector, yellow, behind fuse relay panel
T4a - 4-Pin Connector

(1) - Ground strap, battery to body
(4) - Ground strap, battery to body and transmission

(12) - Ground connection, in engine compartment, left
(120) - Ground connection -2-, in head light wiring harness
(A4) - Plus connection (58b), in instrument panel wiring harness
(A15) - Plus connection (15), in instrument panel wiring harness
(A80) - Connector -1- (X), in instrument panel wiring harness

--- - resistive wiring harness

Battery, starter, ignition/starter switch, load reduction relay, headlight dimmer/flasher relay

Cabrio — Wiring diagram — No. 8/3

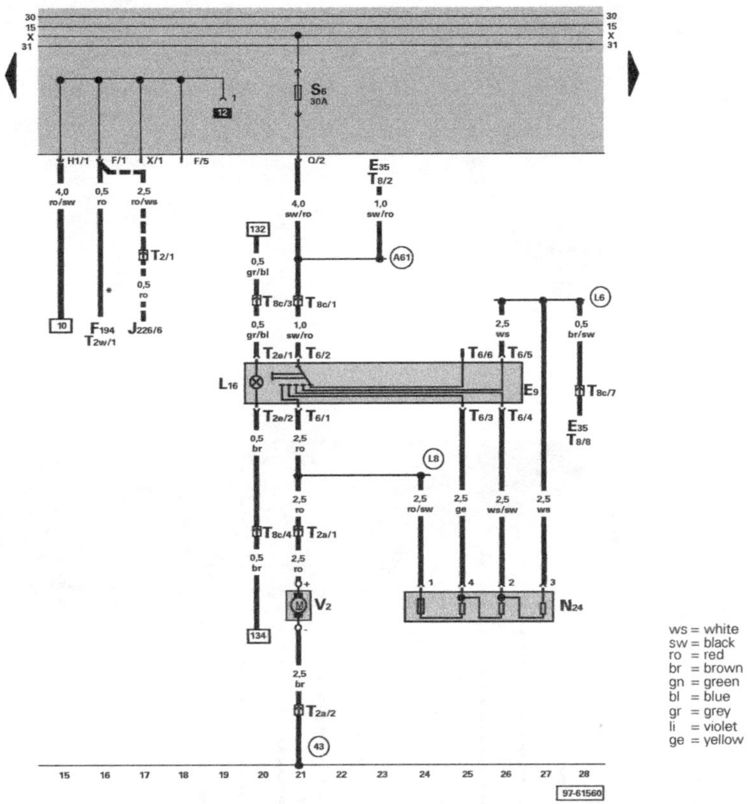

ws = white
sw = black
ro = red
br = brown
gn = green
bl = blue
gr = grey
li = violet
ge = yellow

E9 - Fresh Air Blower Switch
E35 - A/C switch
F194 - Clutch Pedal Position (CPP) Switch
J226 - Park Neutral Position (PNP) Relay
L16 - Fresh Air Control Lever Light
N24 - Fresh Air Blower Series Resistance With Fuse
T2 - 2-Pin Connector, black, behind fuse relay panel
T2a - 2-Pin Connector, red, behind instrument panel, right side
T2e - 2-Pin Connector
T2w - 2-Pin Connector, green, on clutch pedal bracket
T6 - 6-Pin Connector
T8 - 8-Pin Connector, black
T8c - 8-Pin Connector
V2 - Fresh Air Blower

(43) - Ground connection, on right A-pillar, lower part
(A61) - Plus connection (X) in instrument cluster wiring harness
(L6) - Wire connection -2-, in evaporator housing wiring harness
(L8) - Wire connection -4-, in evaporator housing wiring harness

* - Manual transmission only
--- - Automatic transmission only

Fresh air blower switch, fresh air control lever light, fresh air blower/ series resistance with fuse

No. 8/4 — Wiring diagram — Cabrio

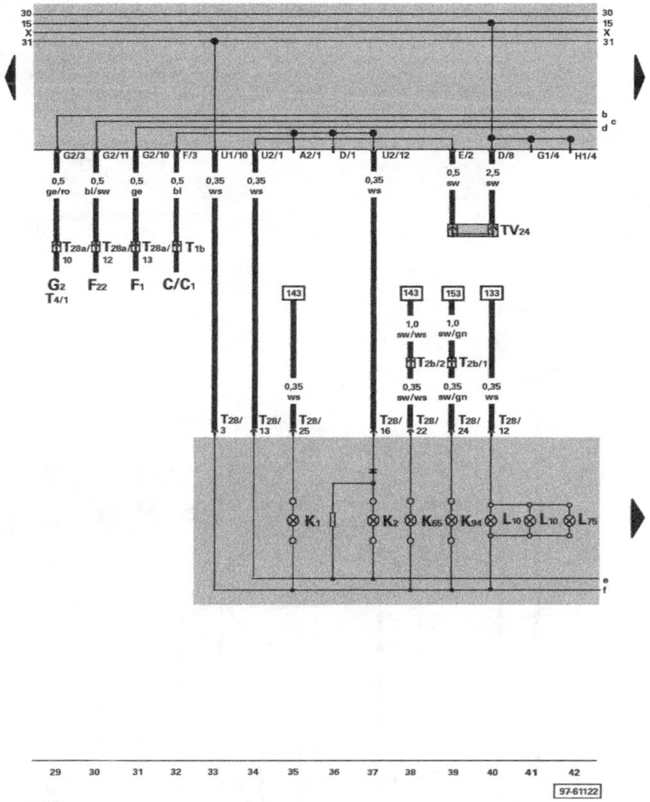

ws = white
sw= black
ro = red
br = brown
gn = green
bl = blue
gr = grey
li = violet
ge = yellow

C - Generator (GEN)
C1 - Voltage Regulator (VR)
F1 - Oil Pressure Switch
F22 - 0.3bar Oil Pressure Switch
G2 - Engine Coolant Temperature (ECT) Sensor
K1 - Headlight High Beam Indicator Light
K2 - Generator (GEN) Warning Light
K65 - Left Turn Signal Indicator Light
K94 - Right Turn Signal Indicator Light
L10 - Instrument Cluster Light
L75 - Digital Display Light
T1b - Single Connector
T2b - 2-Pin Connector, black, behind fuse/relay panel
T4 - 4-Pin Connector, in engine compartment
T28 - 28-Pin Connector, white
T28a - 28-Pin Connector, in engine, left
TV24 - Juction box, black

Generator (GEN), voltage regulator (VR), oil pressure switch, 0.3 bar oil pressure switch, Instrument cluster, indicator and warning lights, instrument illumination

Cabrio — Wiring diagram — No. 8/5

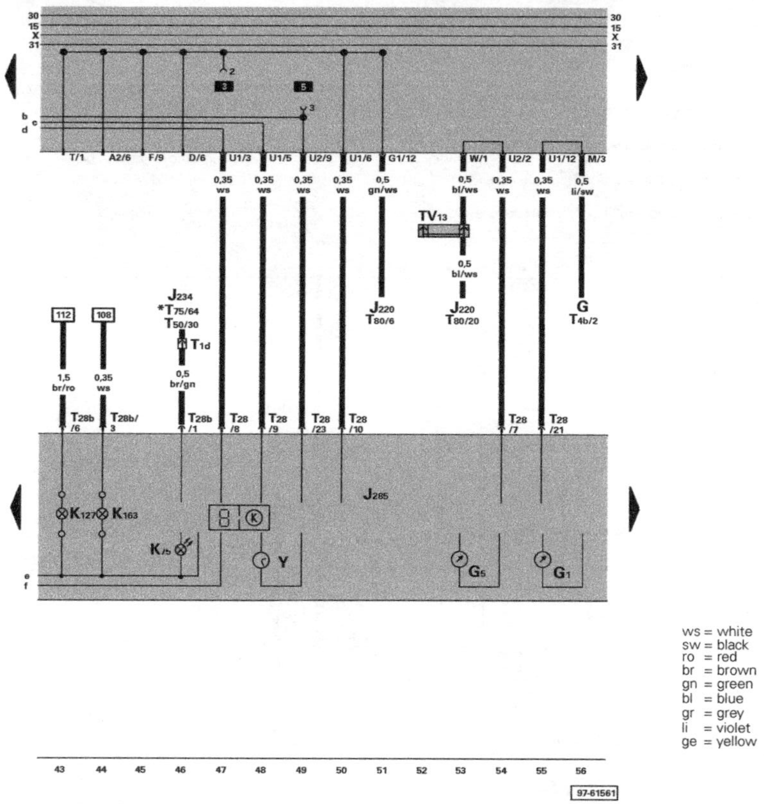

ws = white
sw = black
ro = red
br = brown
gn = green
bl = blue
gr = grey
li = violet
ge = yellow

- G - Sender for fuel gauge
- G1 - Fuel gauge
- G5 - Tachometer
- J220 - Motronic Engine Control Module (ECM)
- J234 - Airbag Control Module
- J285 - Control module with indicator unit in instrument panel insert
- K75 - Airbag Malfunction Indicator Lamp (MIL)
- K127 - Indicator light for open trunk
- K163 - Seat Belt Switch Warning Lamp
- T1d - Single connector, brown
- T4b - 4-Pin Connector
- T28 - 28-Pin Connector, white
- T28b - 28-Pin Connector, red
- T50 - 50-Pin Connector
- T75 - 75-Pin Connector
- T80 - 80-Pin Connector
- TV13 - Vehicle Speed Signal Wire Connector, blue
- Y - Analog Clock
- * - from October 2000

Edition 09.01
USA.5403.04.21

Tachometer, Instrument cluster, sender for fuel gauge, fuel gauge, Indicator light for open trunk, airbag malfunction indicator lamp (MIL), clock

No. 8/6 — Wiring diagram — Cabrio

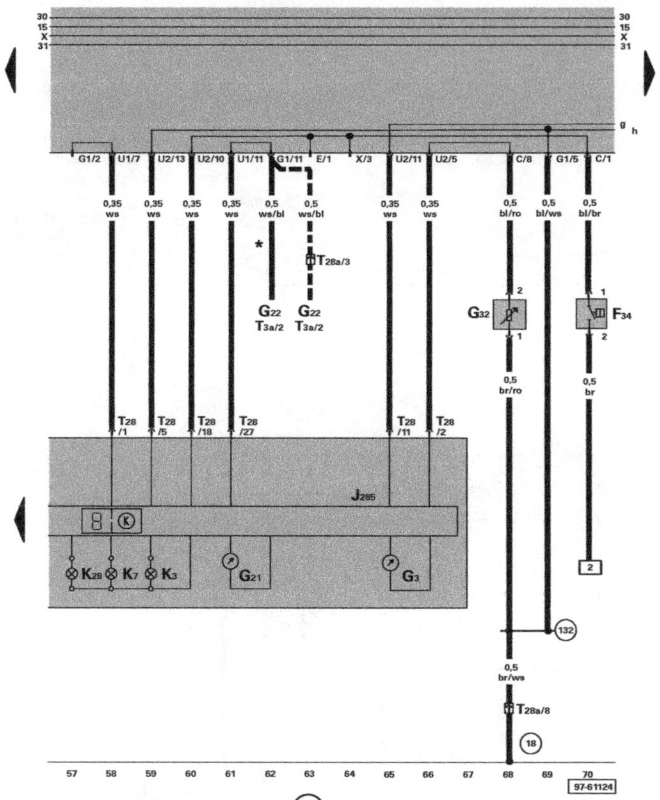

ws = white
sw = black
ro = red
br = brown
gn = green
bl = blue
gr = grey
li = violet
ge = yellow

- F34 - Brake Fluid Level Warning Switch
- G3 - Engine Coolant Temperature (ECT) Gauge
- G21 - Speedometer
- G22 - Speedometer Vehicle Speed Sensor (VSS)
- G32 - Engine Coolant Level (ECL) Sensor
- J285 - Control module with indicator unit in instrument panel insert
- K3 - Oil Pressure Warning Light
- K7 - Brake and Parking Brake Warning Light
- K28 - Engine Coolant Level/Temperature (ECL/ECT) Warning Light
- T3a - 3-Pin Connector
- T28 - 28-Pin Connector, white
- T28a - 28-Pin Connector, in engine, left
- (18) - Ground connection, on engine block
- (132) - Ground connection -3-, in engine compartment wiring harness
- * - Manual transmission only
- --- - Automatic transmission only

Instrument cluster, engine coolant temperature (ECT) gauge/sensor, speedomter, indicator and warning lights

Edition 09.01
USA.5403.04.21

Cabrio — Wiring diagram — No. 8/7

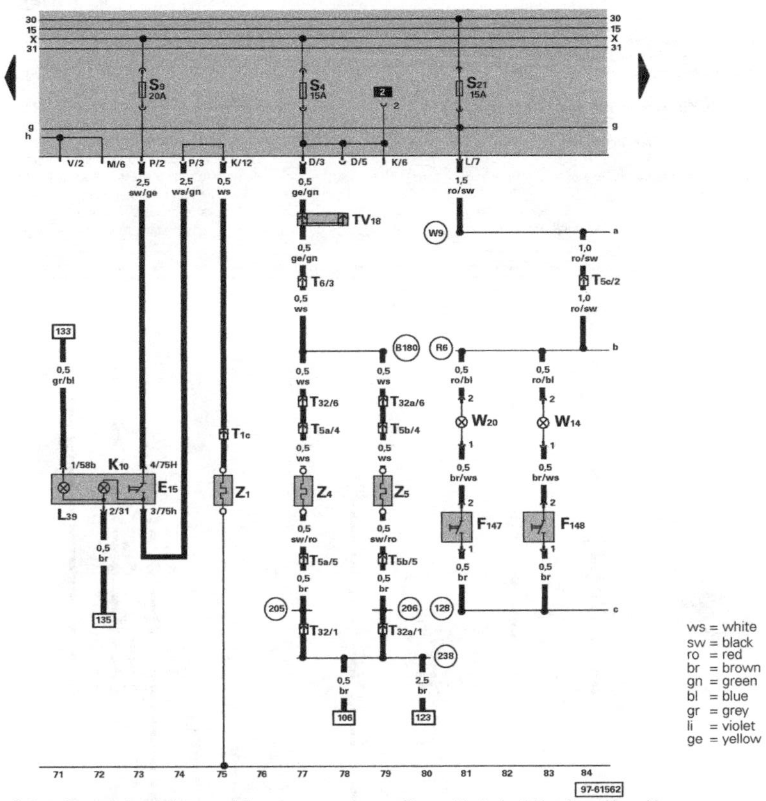

ws = white
sw = black
ro = red
br = brown
gn = green
bl = blue
gr = grey
li = violet
ge = yellow

E15 - Rear window defogger switch
F147 - Left Make-Up Mirror Light Switch
F148 - Right Make-Up Mirror Light Switch
K10 - Indicator light for heated rear windshield
L39 - Rear Window Defogger Switch Light
T1c - Single Connector, white, near left taillight bulb holde
T5a - 5-Pin Connector, on left door
T5b - 5-Pin Connector, on right door
T5c - 5-Pin Connector, black
T6 - 6-Pin Connector, black
T32 - 32-Pin Connector, white, on left A-pillar
T32a - 32-Pin Connector, white, on right A-pillar
TV18 - Distributor for terminal Xa, yellow
W14 - Right Make-up Mirror Light
W20 - Left Make-up Mirror Light
Z1 - Heated rear window
Z4 - Heated outside mirror, driver side
Z5 - Heated outside mirror, passenger side

(128) - Ground connection -1-, in interior light wiring harness
(205) - Ground connection, in driver's door wiring harness
(206) - Ground connection, in passenger's door wiring harness harness
(238) - Ground connection -1-, in wiring harness interior
(B180) - Connection (heated rear window) in passenger compartment wiring harness
(R6) - Plus connection -1-, in interior light wiring harness
(W9) - Plus connection -1- (30), in rear wiring harness

Edition 09.01
USA.5403.04.21

Rear window defogger switch, heated rear window, heated outside mirrors, make-up mirrors

No. 8/8 — Wiring diagram — Cabrio

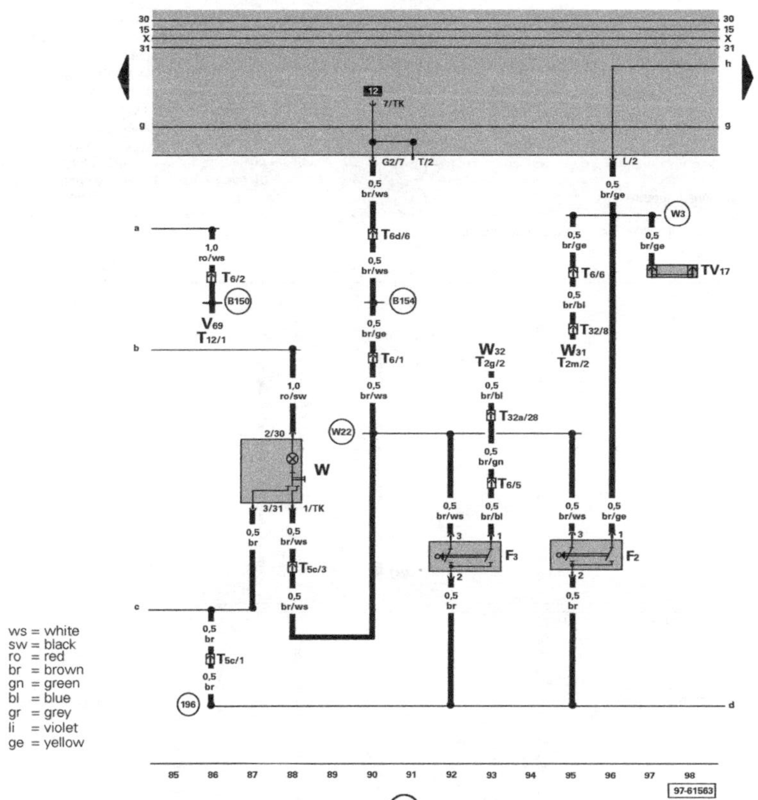

ws = white
sw = black
ro = red
br = brown
gn = green
bl = blue
gr = grey
li = violet
ge = yellow

F2 - Door contact switch-Driver's side
F3 - Door contact switch-Passenger's side
T2g - 2-Pin Connector, green, on right door
T2m - 2-Pin Connector, black, on left door
T32 - 32-Pin Connector, white, on left A-pillar
T32a - 32-Pin Connector, white, on right A-pillar
T5c - 5-Pin Connector, black
T6 - 6-Pin Connector, black
T6d - 6-Pin Connector, red
T12 - 12-Pin Connector
TV17 - Door Contact Switch Wire Connector, white
V69 - Central Looking System Pump and control Module
W - Front Interior Light
W31 - Entry light front, left
W32 - Entry light front, right

(196) - Ground connection -3-, in wiring harness, rear
(B150) - Plus connector -2- (30a), in wiring harness interior
(B154) - Connector -1- (TK), in wiring harness interior
(W3) - Wire connection, in rear wiring harness
(W22) - Connector (door contact switch) in wiring harness, rear

Door contact switches, front interior light

Edition 09.01
USA.5403.04.21

Cabrio — Wiring diagram — No. 8/9

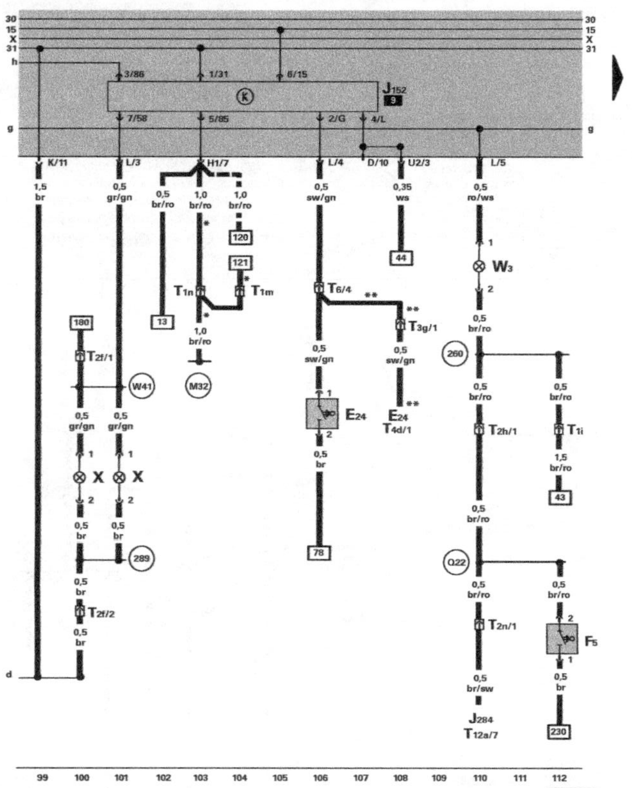

ws = white
sw = black
ro = red
br = brown
gn = green
bl = blue
gr = grey
li = violet
ge = yellow

97-61564

- E24 - Left Seat Belt Switch
- F5 - Luggage Compartment Light Switch
- J152 - Radio/Parking Light Warning Buzzer (188)
- J284 - Alarm system control module
- T1i - Single Connector, white
- T1m - 1-Pin Connector, red, behind fuse relay panel
- T1n - 1-Pin Connector, red, behind fuse relay panel
- T2f - 2-Pin Connector, grey, near left taillight bulb holde
- T2h - 2 Pin Connector, yellow, near left taillight bulb holde
- T2n - 2-Pin Connector, black, near left taillight bulb holde
- T3g - 3-Pin Connector, black, behind fuse/relay panel
- T4d - 4-Pin Connector, yellow, bellow driver's seat
- T6 - 6-Pin Connector, black, behind fuse/relay panel
- T12a - 12-Pin Connector
- W3 - Luggage Compartment Light
- X - License Plate Light
- (260) - Ground Connection, in wiring harness taillights
- (289) - Ground connection in plug wiring harness License Plate Light, rear
- (Q22) - Wire connection -1-, in rear lid wiring harness
- (W41) - Plus Connection (58) (license plate light wiring harness
- (M32) - Wire Connection (su),in convertible top wiring harness
- * - Power convertible top
- --- - Manual convertible top
- ** - From October 2000 production

Edition 09.01
USA.5403.04.21

Luggage compartment light/switch, radio/parking light warning buzzer, license plate light

1999-2002

Cabrio standard equipment (from September 1999)

No. 8/10 — Wiring diagram — Cabrio

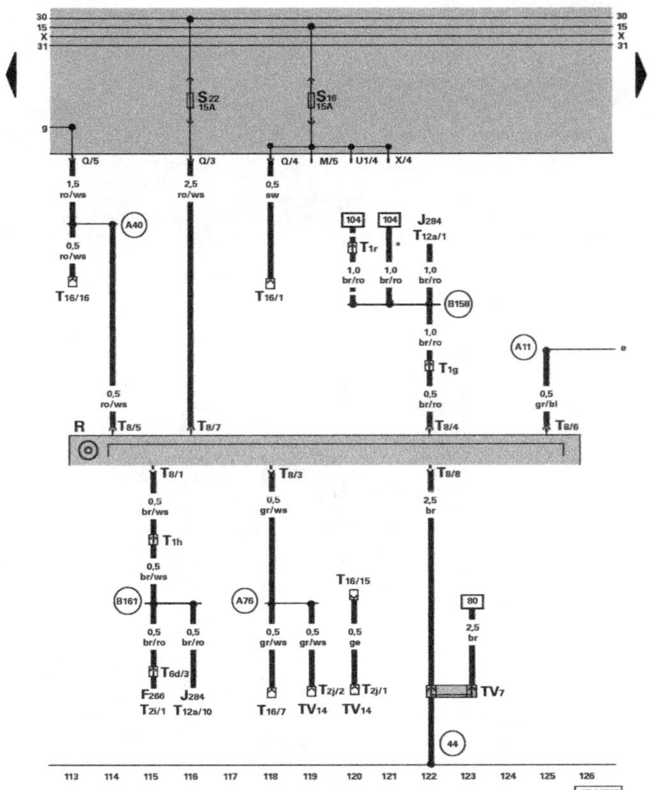

ws = white
sw = black
ro = red
br = brown
gn = green
bl = blue
gr = grey
li = violet
ge = yellow

97-61565

- F266 - Front Hood Switch
- J284 - Alarm system control module
- R - Radio
- T1g - Single Connector, red, behind fuse/relay panel
- T1h - Single Connector, yellow, behind fuse/relay panel
- T1r - Single Connector, red, behind fuse/relay panel
- T2i - 2 Pin Connector
- T2j - 2-Pin Connector, grey
- T8 - 8-Pin Connector, in Radio
- T6d - 6-Pin Connector, red
- T12a - 12-Pin Connector, black
- T16 - 16-Pin Connector, Data Link Connector (DLC)
- TV7 - Terminal 31 (GND) Wire Connector, brown, beside fuse /relay panel, left
- TV14 - Data Link Connector (DLC) Wire Connector, black, above fuse relay panel
- (44) - Ground connection, on left A-pillar, lower part
- (A11) - Plus connection (58b) in instrument panel wiring harness
- (A40) - Plus connection -1- (30), in instrument cluster wiring harness
- (A76) - Connector (k diagnosis wire) in instrument panel wiring harness
- (B158) - Connector (S) in wiring harness interior
- (B161) - Connector (anti-theft warning system), in wiring harness interior
- * - Power convertible top
- --- - Manual convertible top

Radio connection, data link connector (DLC)

Edition 09.01
USA.5403.04.21

1999-2002

Cabrio — Wiring diagram — No. 8/11

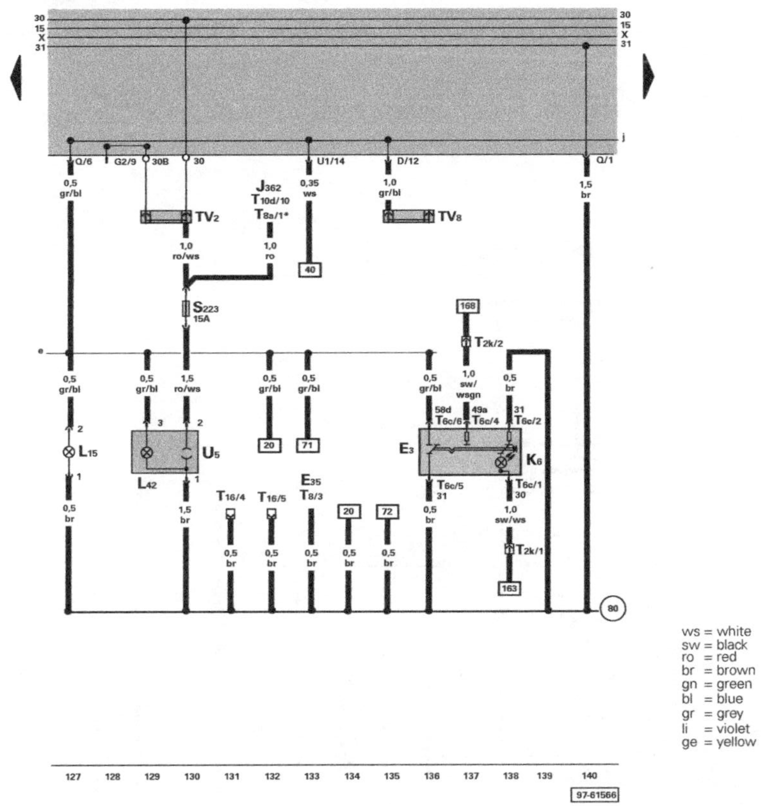

ws = white
sw = black
ro = red
br = brown
gn = green
bl = blue
gr = grey
li = violet
ge = yellow

- E3 - Emergency Flasher Switch
- E35 - A/C Switch
- J362 - Control module for anti-theft immobilizer
- K6 - Indicator light for emergency flasher system
- L15 - Ashtray Light
- L42 - Socket Light Bulb
- S223 - Fuse 23 in fuse holder, behind fuse/relay panel
- T2k - 2-Pin Connector, brown, behind fuse/relay panel
- T6c - 6-Pin Connector
- T8 - 8-Pin Connector
- T8a - 8-Pin Connector, black
- T10d - 10-Pin Connector, black
- T16 - 16-Pin Connector, Data Link Connector (DLC)
- TV2 - Terminal 30 (B+) Wire Connector, red
- TV8 - Terminal 58b Wire Connector, grey
- U5 - 12 V Socket /Cigarette Lighter (optional)
- (80) - Ground connection -1-, in instrument panel wiring harness
- * - Up to 02.00 production

Emergency flasher switch, 12 V socket, socket light bulb, ashtray light, data link connector (DLC)

No. 8/12 — Wiring diagram — Cabrio

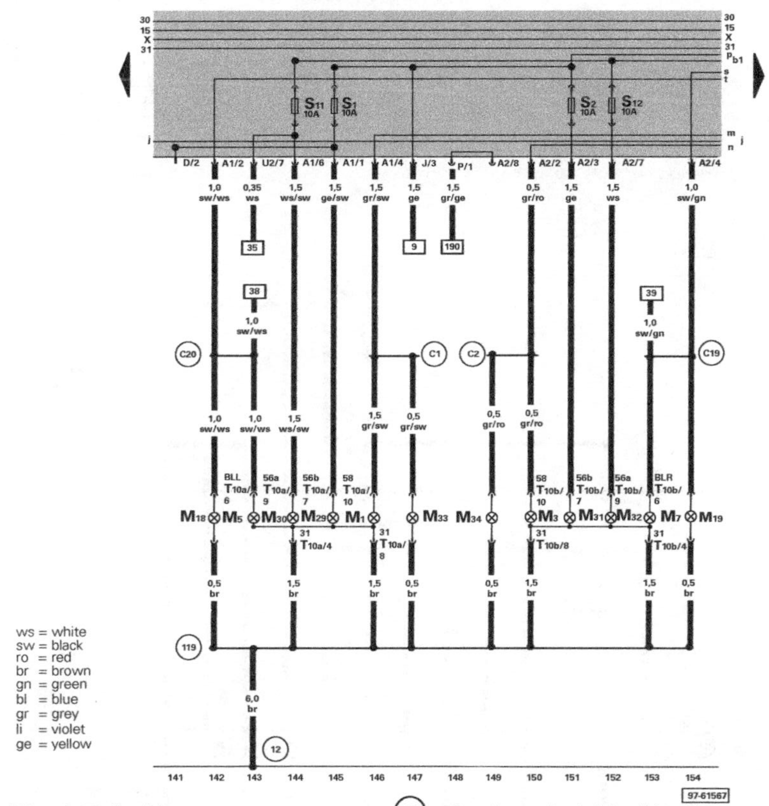

ws = white
sw = black
ro = red
br = brown
gn = green
bl = blue
gr = grey
li = violet
ge = yellow

- M1 - Left Parking Light
- M3 - Right Parking Light
- M5 - Left Front Turn Signal Light
- M7 - Right Front Turn Signal Light
- M18 - Left, Side Turn Signal Light
- M19 - Right, Side turn signal light
- M29 - Left Low Beam Headlight
- M30 - Left High Beam Headlight
- M31 - Right Low Beam Headlight
- M32 - Right High Beam Headlight
- M33 - Light for Side marker front left
- M34 - Light for Side marker front right
- T10a - 10-Pin Connector
- T10b - 10-Pin Connector
- (12) - Ground connection, in engine compartment, left
- (119) - Ground connection -1-, in headlight wiring harness
- (C1) - Plus connection (58L), in left front wiring harness
- (C2) - Plus connection (58R), in left front wiring harness
- (C19) - Connector -1- (right turn signal), in wiring harness headlamp
- (C20) - Connection (left turn signal), in headlight wiring harness

Headlights, parking lights, turn signal lights, side marker lights

Cabrio — Wiring diagram — No. 8/13

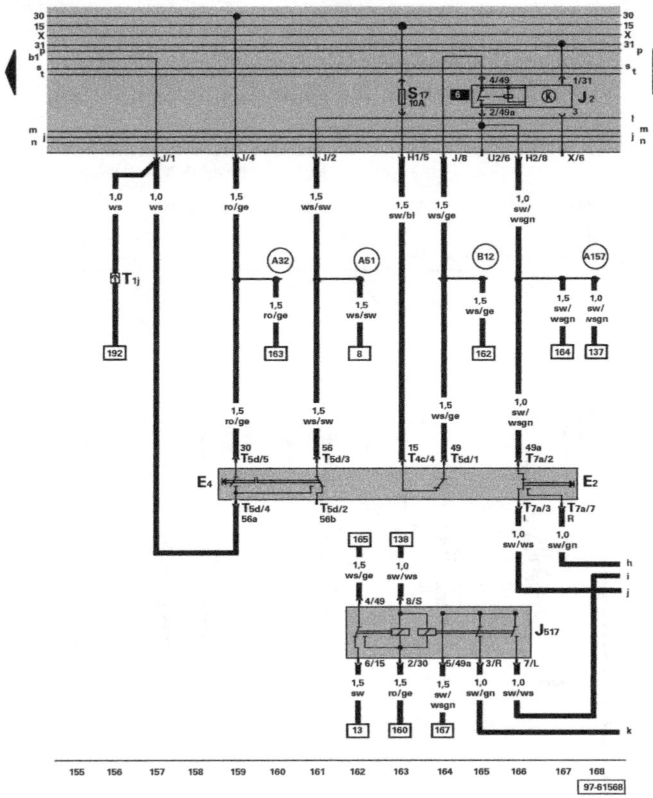

ws = white
sw = black
ro = red
br = brown
gn = green
bl = blue
gr = grey
li = violet
ge = yellow

E2 - Turn signal switch
E4 - Headlight Dimmer/Flasher Switch
J2 - Emergency Flasher Relay (191)
J517 - Auxiliary Emergency Flasher Relay (413)
T1j - Single Connector, yellow
T4c - 4-Pin Connector, red
T5d - 5-Pin Connector, black
T7a - 7-Pin Connector

(A32) - Plus connection (30), in instrument panel wiring harness
(A51) - Wire connection (56), in instrument panel wiring harness
(A157) - Connection (49a) (in instrument panel wiring harness
(B12) - Connector (49) in wiring harness steering column switch

Edition 09.01 USA.5403.04.21 — **Turn signal switch, headlight dimmer/flasher switch, turn signal relay, emergency flasher relay**

No. 8/14 — Wiring diagram — Cabrio

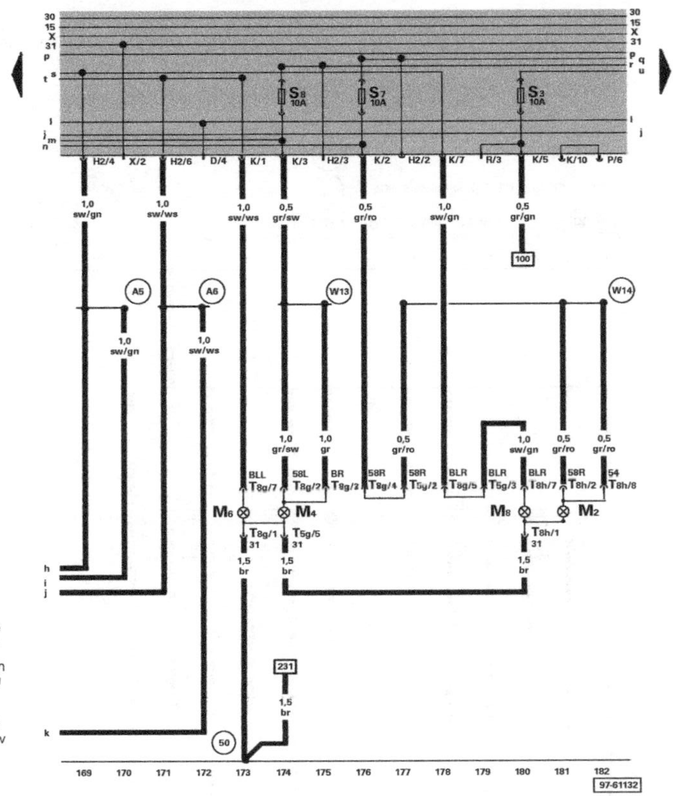

ws = white
sw = black
ro = red
br = brown
gn = green
bl = blue
gr = grey
li = violet
ge = yellow

M2 - Right Tail Light
M4 - Left Tail Light
M6 - Left Rear Turn Signal Light
M8 - Right Rear Turn Signal Light
T5g - 5-Pin Connector, on taillight, left
T8g - 8-Pin Connector, on taillight, left
T8h - 8-Pin Connector, on taillight, right

(50) - Ground connection, in luggage compartment, left
(A5) - Plus connection (right turn signal), in instrument panel wiring harness
(A6) - Plus connection (left turn signal), in instrument panel wiring harness

(W13) - Wire connection (58L), in rear wiring harness
(W14) - Wire connection (58R), in rear wiring harness

Right and left tail light, right and left rear turn signal light — **Edition 09.01** USA.5403.04.21

Cabrio — Wiring diagram — No. 8/15

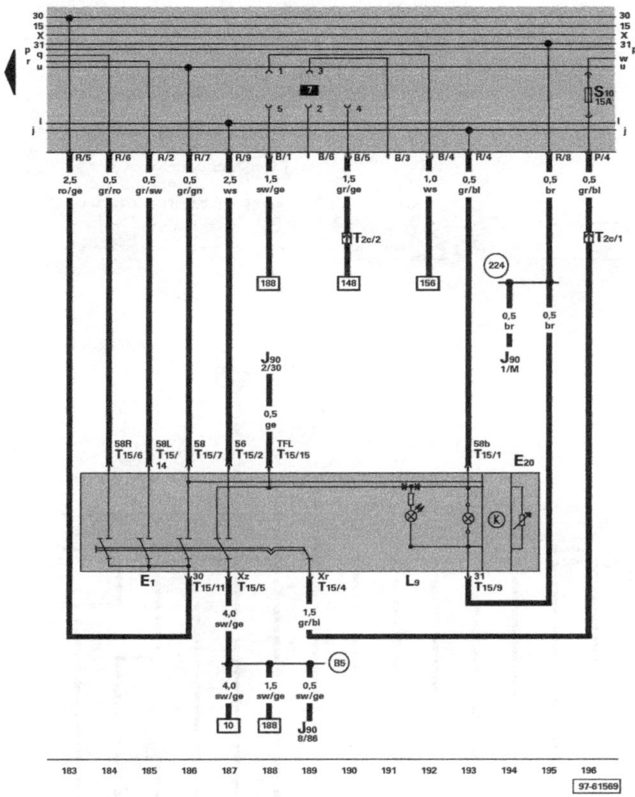

ws = white
sw = black
ro = red
br = brown
gn = green
bl = blue
gr = grey
li = violet
ge = yellow

E1 - Light switch
E20 - Instrument Panel Light Dimmer Switch
J90 - Daytime Running Lights Switch-On Relay (173)
L9 - Headlight Switch Light
T2c - 2-Pin Connector, black, behind fuse/relay panel
T15 - 15-Pin Connector

(B5) - Wire connection (X), in light switch wiring harness
(224) - Ground connection, in light switch wiring harness

Light switch, instrument panel light dimmer switch

Cabrio standard equipment (from September 1999)

No. 8/16 — Wiring diagram — Cabrio

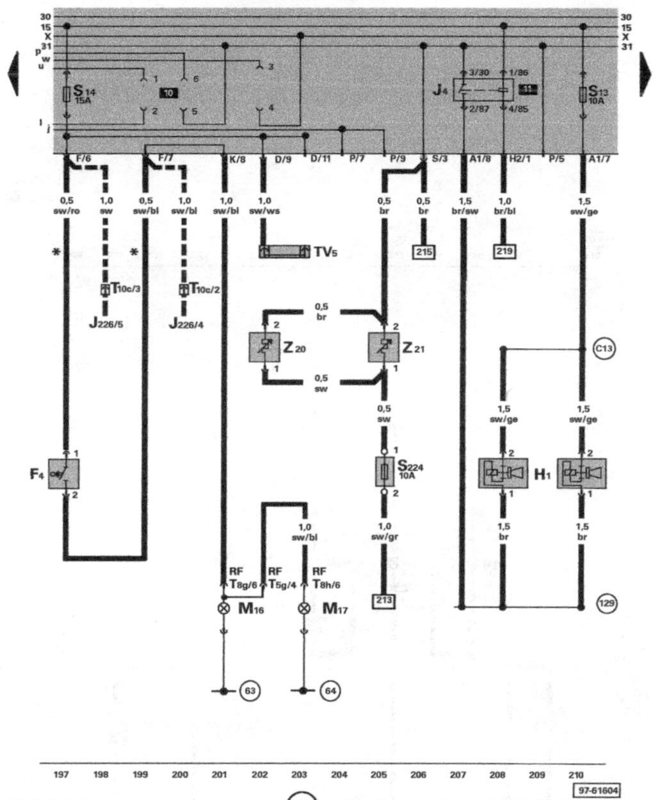

ws = white
sw = black
ro = red
br = brown
gn = green
bl = blue
gr = grey
li = violet
ge = yellow

F4 - Back-Up Light Switch
H1 - Horn/dual tone horn
J226 - Park Neutral Position (PNP) Relay
J4 - Dual tone horn relay (53)
M16 - Reversing light bulb left
M17 - Reversing light bulb right
T5g - 5-Pin Connector, on taillight, left
T8g - 8-Pin Connector, on taillight, left
T8h - 8-Pin Connector, on taillight, right
T10c - 10-Pin Connector
TV5 - Terminal 15a Wire Connector, green
S224 - Fuse in fuse holder
Z20 - Left Washer Nozzle Heater
Z21 - Right Washer Nozzle Heater

(63) - Ground connection, on left taillight bulb holder
(64) - Ground connection, on right taillight bulb holder
(129) - Ground connection, in dual-tone horn wiring harness
(C13) - Plus connection, in dual tone horn wiring harness

* - Manual transmission only
--- - Automatic transmission only

Back up light switch, left and right reversing light bulb, horn/dual tone horn/relay, washer nozzles heater

Cabrio — Wiring diagram — No. 8/17

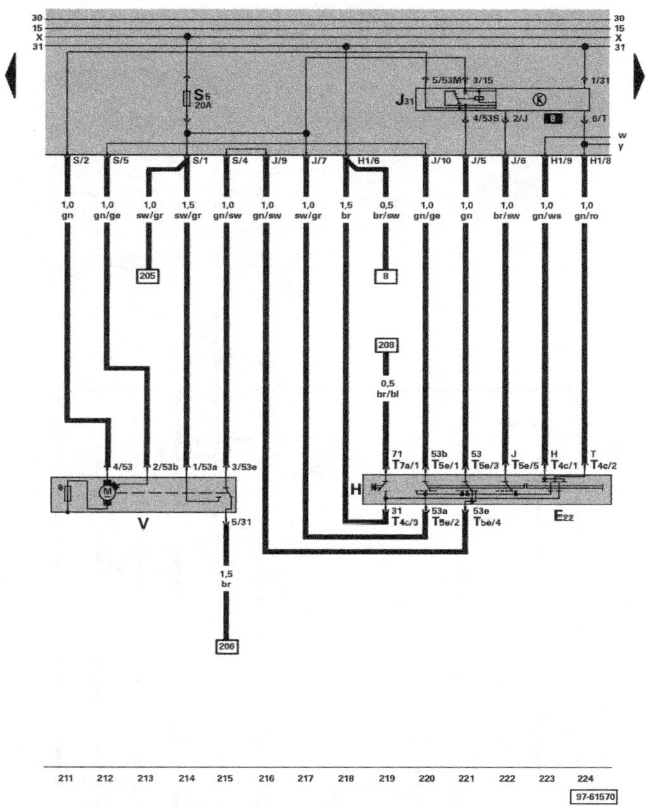

ws = white
sw = black
ro = red
br = brown
gn = green
bl = blue
gr = grey
li = violet
ge = yellow

E22 - Windshield Wiper/Washer Switch
H - Signal horn activation
J31 - Wiper/Washer Intermittent Relay (196)
T4c - 4-Pin Connector, red
T5e - 5-Pin Connector, black
T7a - 7-Pin Connector
V - Windshield Wiper Motor

Windshield wiper/washer switch, signal horn activation, wiper/washer intermittent relay

No. 8/18 — Wiring diagram — Cabrio

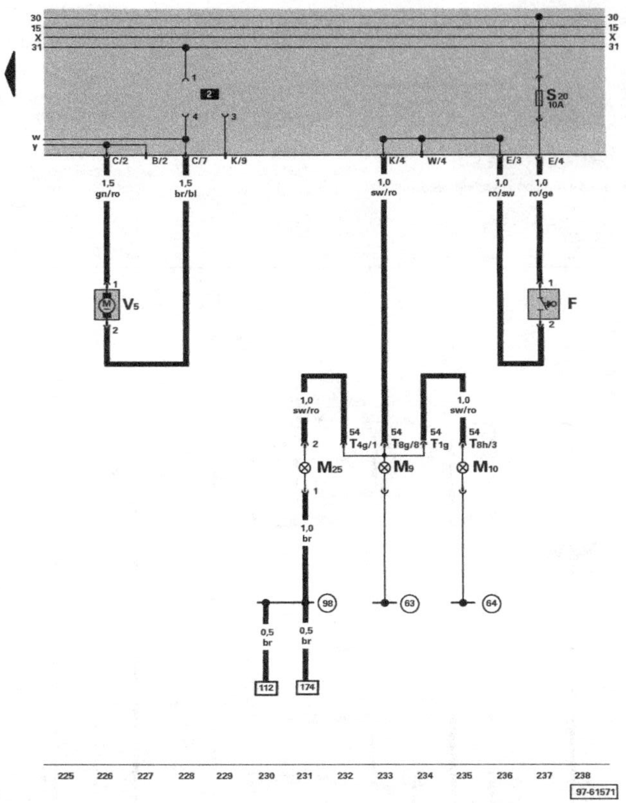

ws = white
sw = black
ro = red
br = brown
gn = green
bl = blue
gr = grey
li = violet
ge = yellow

F - Brake Light Switch
M9 - Left Brake Light
M10 - Right Brake Light
M25 - High-mount Brake Light
T1g - Single Connector, on taillight, left
T4g - 4-Pin Connector, on taillight, left
T8g - 8-Pin Connector, on taillight, left
T8h - 8-Pin Connector, on taillight, right
V5 - Windshield Washer Pump

(63) - Ground connection, on left taillight bulb holder
(64) - Ground connection, on right taillight bulb holder
(98) - Ground connection, in rear lid wiring harness

Brake light switch, brake lights, high mount brake light, windshield washer pump

Cabrio — Wiring diagram — No. 9/1

Power windows

from August 1998

Relay locations

Position of relays and multi-pin connector wiring, see also "Component Locations" section.

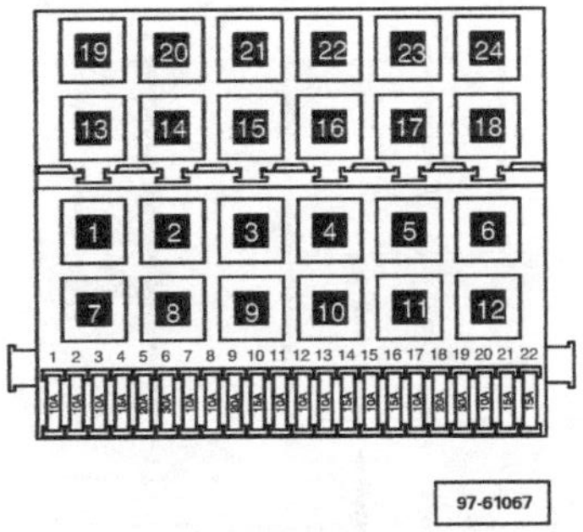

Note:
Number in parentheses indicates production control number stamped on relay housing.

Fuse colors

30 A - green
25 A - white
20 A - yellow
15 A - blue
10 A - red
7,5 A - brown
5 A - beige
3 A - violet

Edition 09.01
USA.5403.04.21

No. 9/2 — Wiring diagram — Cabrio

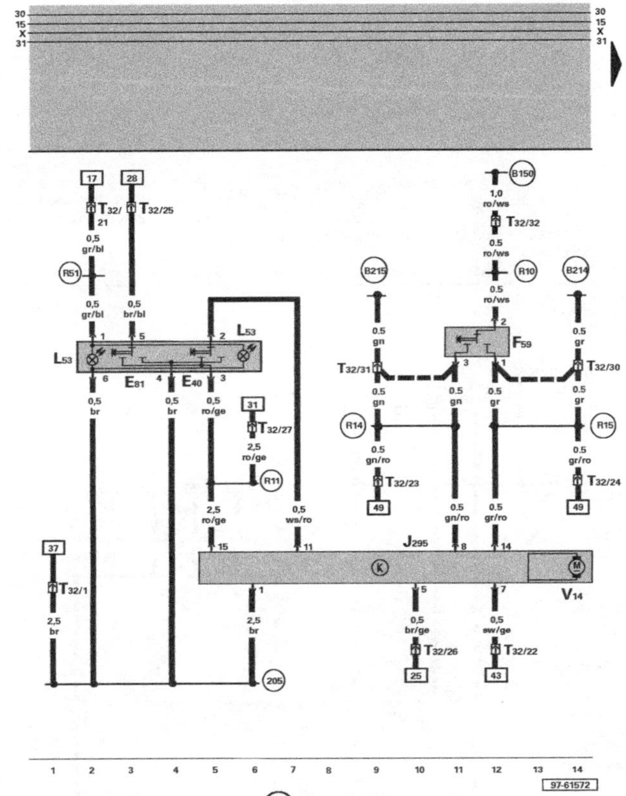

ws = white
sw = black
ro = red
br = brown
gn = green
bl = blue
gr = grey
li = violet
ge = yellow

E40 - Switch for window regulator front left
E81 - Switch for window regulator front right, driver
F59 - Driver's Door Central Locking System Switch
J295 - Left Front Power Window Control Module
L53 - Power Window Switch Light
T32 - 32-Pin Connector, on left A-pillar
V14 - Left Window Motor

(205) - Ground connection, in driver's door wiring harness
(B150) - Plus connector -2- (30a), in wiring harness interior
(B214) - Connection 1 (central locking) in passenger compartment wiring harness
(B215) - Connection 2 (central locking) in passenger compartment wiring harness

(R10) - Plus connection -1- (30), in driver's door wiring harness
(R11) - Plus connection -2- (30), in driver's door wiring harness
(R14) - Wire connection -1- (open), in driver's door wiring harness
(R15) - Plus connection -2- (closed), in driver's door wiring harness
(R51) - Connection (58b) (driver's side door wiring harness)

--- - Driver & passenger side remote mirrors, heatable

Control module with motor for power window LF, switch for central locking (driver side)

Edition 09.01
USA.5403.04.21

Cabrio — Wiring diagram — No. 9/3

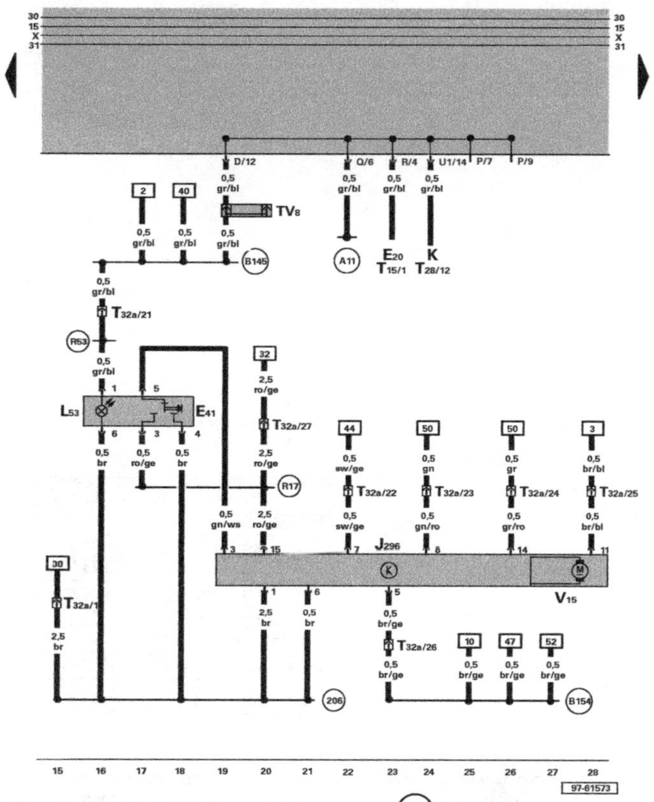

ws = white
sw = black
ro = red
br = brown
gn = green
bl = blue
gr = grey
li = violet
ge = yellow

- E20 - Instrument Panel Light Dimmer Switch
- E41 - Right Front Window Switch
- J296 - Right Front Power Window Control Module
- K - Instrument Cluster
- L53 - Power Window Switch Light
- T15 - 15-Pin Connector
- T28 - 28-Pin Connector, white
- T32a - 32-Pin Connector, on right A-pillar
- TV8 - Terminal 58b wire connector, grey
- V15 - Right Window Motor
- (206) - Ground connection, in passenger's door wiring harness
- (A11) - Plus connection (58b) in instrument panel wiring harness
- (B145) - Plus connector (58b), in wiring harness interior
- (B154) - Connector -1- (TK), in wiring harness interior
- (R17) - Plus connection -2- (30), in passenger's door wiring harness
- (R53) - Connection (58s) (in right front door wiring harness)

Edition 09.01
USA.5403.04.21

Control module with motor for power window RF, switch for power window RF

1999-2002

No. 9/4 — Wiring diagram — Cabrio

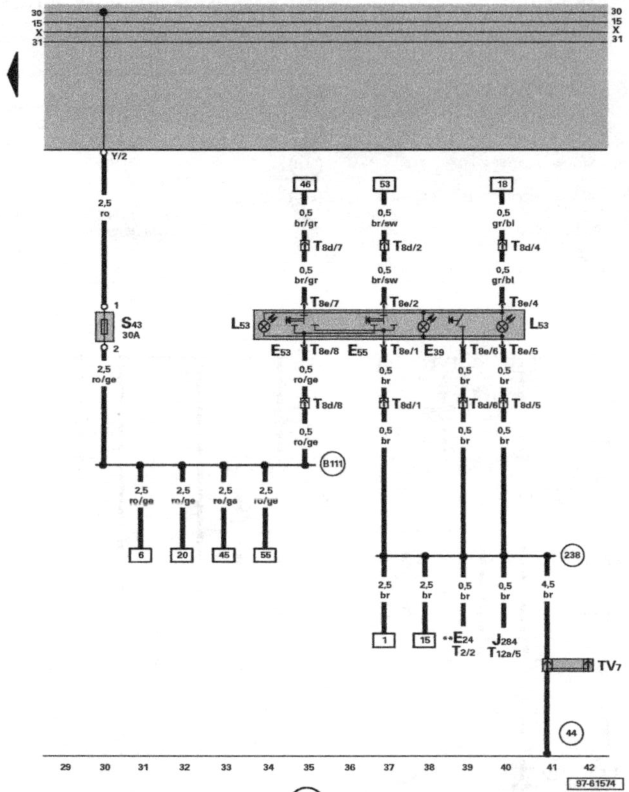

ws = white
sw = black
ro = red
br = brown
gn = green
bl = blue
gr = grey
li = violet
ge = yellow

- E24 - Left Seat Belt Switch
- E39 - Safety switch for window regulator rear
- E53 - Switch for window regulator rear left, driver
- E55 - Switch for window regulator rear right, driver
- J284 - Alarm System Control Module
- L53 - Power Window Switch Light
- S43 - Power Window Circuit Breaker
- T2 - 2-Pin Connector
- T8d - 8-Pin Connector, behind middle switchboard
- T8e - 8-Pin Connector
- T12a - 12-Pin Connector, black
- TV7 - Terminal 31 (GND) wire connector, brown, beside fuse /relay panel
- (44) - Ground connection, on left A-pillar, lower part
- (238) - Ground connection -1-, in wiring harness interior
- (B111) - Plus connector -1- (30a), in wiring harness interior
- ** - Up to 10.00 production

Switch for rear windows

Edition 09.01
USA.5403.04.21

1999-2002

Cabrio — Wiring diagram — No. 9/5

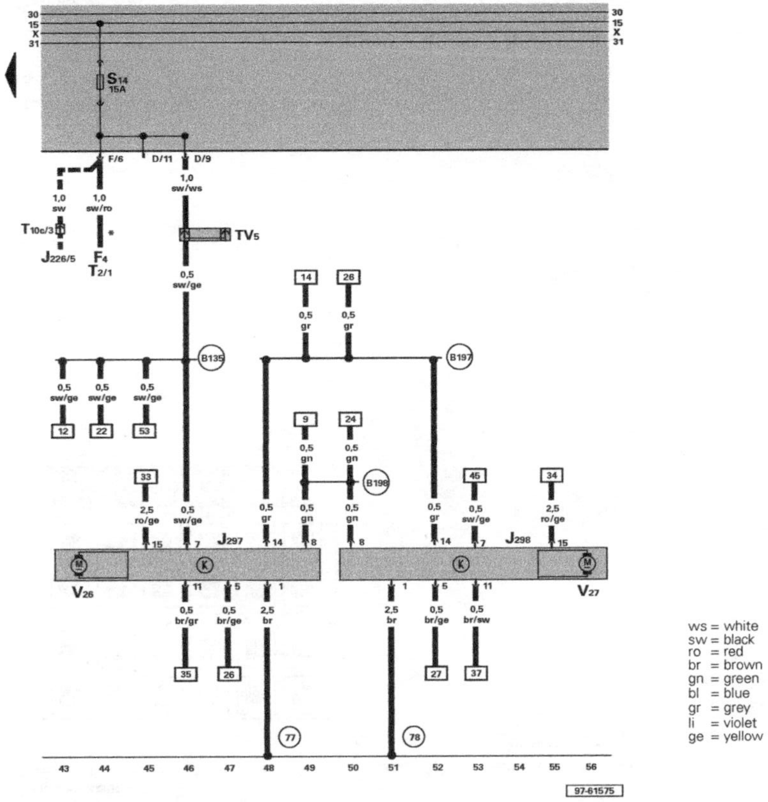

ws = white
sw = black
ro = red
br = brown
gn = green
bl = blue
gr = grey
li = violet
ge = yellow

F4 - Back-Up Light Switch
J226 - Park Neutral Position (PNP) Relay
J297 - Left Rear Power Window Control Module
J298 - Right Rear Power Window Control Module
T2 - 2-Pin Connector
T10c - 10-Pin Connector, black
TV5 - Terminal 15a Wire Connector, green
V26 - Motor for window regulator, rear, left
V27 - Motor for window regulator, rear, right

(77) - Ground Connection (lower left B-pillar)

(78) - Ground Connection (lower right B-pillar)

(B135) - Plus connector -1- (15a) in wiring harness interior

(B197) - Connection (close window) in passenger compartment wiring harness

(B198) - Connection (open window) in passenger compartment wiring harness

* - Manual transmission only

--- - Automatic transmission only

Edition 09.01
USA.5403.04.21

Control module with motor for power window, LR/RR

THIS PAGE INTENTIONALLY LEFT

BLANK

Cabrio — Wiring diagram — No. 10/1

Power Convertible Top

from August 1998

Relay locations

Position of relays and multi-pin connector wiring, see also "Component Locations" section.

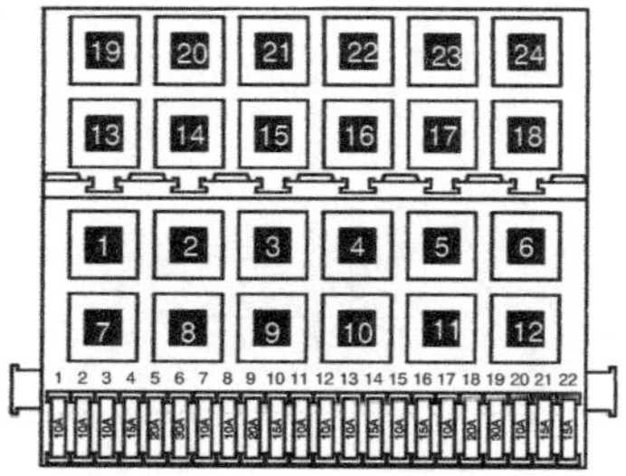

97-61067

Note:
Number in parentheses indicates production control number stamped on relay housing.

Fuse colors

30 A - green
25 A - white
20 A - yellow
15 A - blue
10 A - red
7,5 A - brown
5 A - beige
3 A - violet

Edition 09.01
USA.5403.04.21

No. 10/2 — Wiring diagram — Cabrio

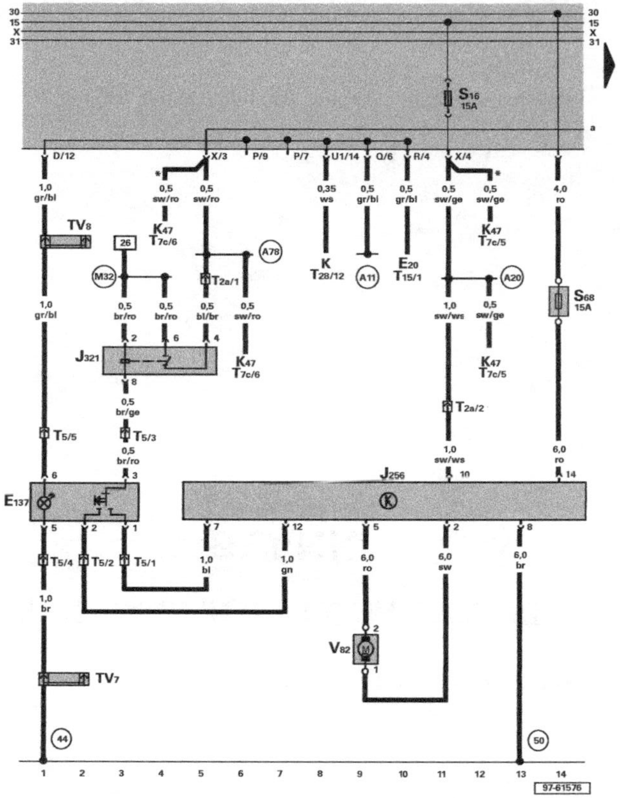

ws = white
sw = black
ro = red
br = brown
gn = green
bl = blue
gr = grey
li = violet
ge = yellow

E20 - Instrument Panel Light Dimmer Switch
E137 - Convertible Top Operation Switch
J256 - Convertible Top Control Module control (154)
J321 - Convertible Top Hydraulic Pump Relay (53)
K - Instrument Cluster
K47 - ABS Warning Light
S68 - Convertible Top Circuit Breaker
T2a - 2-Pin Connector, green, behind fuse relay panel
T5 - 5-Pin Connector, black
T7c - 7-Pin Connector, black
T15 - 15-Pin Connector, behind fuse relay panel
T28 - 28-Pin Connector, white
TV7 - Terminal 31 (GND) wire connector, brown, beside fuse /relay panel
TV8 - Terminal 58b wire connector, grey
V82 - Convertible Top Motor

(44) - Ground connection, on left A-pillar lower part
(50) - Ground connection, in luggage compartment, left
(A11) - Plus connection (58b), in instrument panel wiring harness
(A20) - Wire connection (15a), in instrument panel wiring harness
(A78) - Connector (SU), in instrument panel wiring harness
(M32) - Wire Connection (su),in convertible top wiring harness
* - Manual convertible top

Convertible top operation switch, convertible top control module, convertible top hydraulic pump relay, convertible top circuit breaker

Edition 09.01
USA.5403.04.21

Cabrio — Wiring diagram — No. 10/3

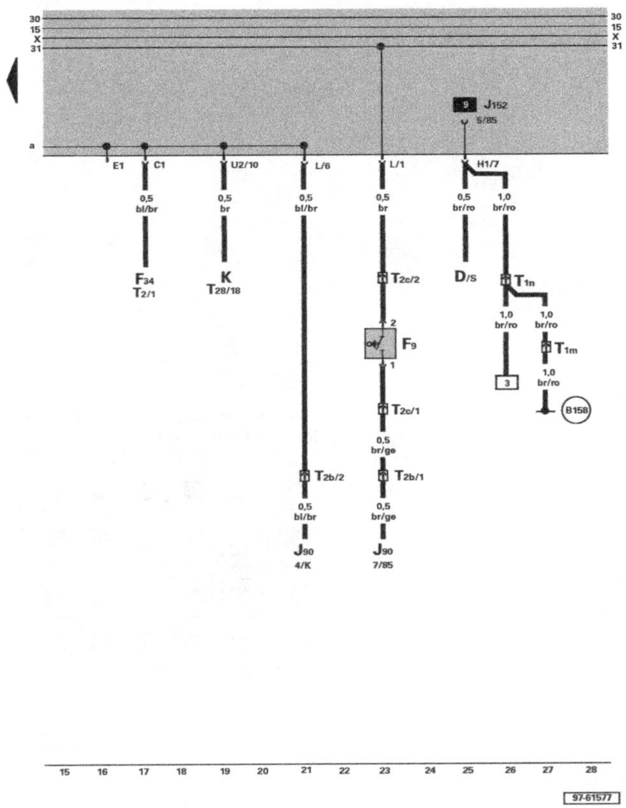

ws = white
sw = black
ro = red
br = brown
gn = green
bl = blue
gr = grey
li = violet
ge = yellow

D - Ignition/Starter Switch
F9 - Parking Brake Warning Light Switch
F34 - Brake Fluid Level Warning Switch
J90 - Daytime Running Lights Switch ON Relay (173)
J152 - Radio/parking Light Warning Buzzer (188)
K - Instrument Cluster
T1m - 1-Pin Connector, red, behind fuse relay panel
T1n - 1-Pin Connector, red, behind fuse relay panel
T2 - 2-Pin Connector, behind fuse relay panel
T2b - 2-Pin Connector, red, behind fuse relay panel
T2c - 2-Pin Connector, black, on Parking Brake
T28 - 28-Pin Connector, white

(B158) - Connector (S) in wiring harness interior

THIS PAGE INTENTIONALLY LEFT

BLANK

Cabrio — Wiring diagram — No. 11/1

Stereo radio

from September 1999

Relay locations

Position of relays and multi-pin connector wiring, see also "Component Locations" section.

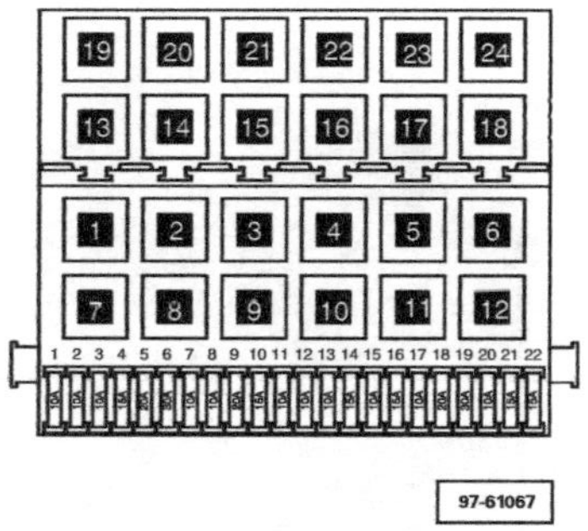

Note:
Number in parentheses indicates production control number stamped on relay housing.

Fuse colors

- 30 A - green
- 25 A - white
- 20 A - yellow
- 15 A - blue
- 10 A - red
- 7,5 A - brown
- 5 A - beige
- 3 A - violet

Edition 09.01
USA.5403.04.21

No. 11/2 — Wiring diagram — Cabrio

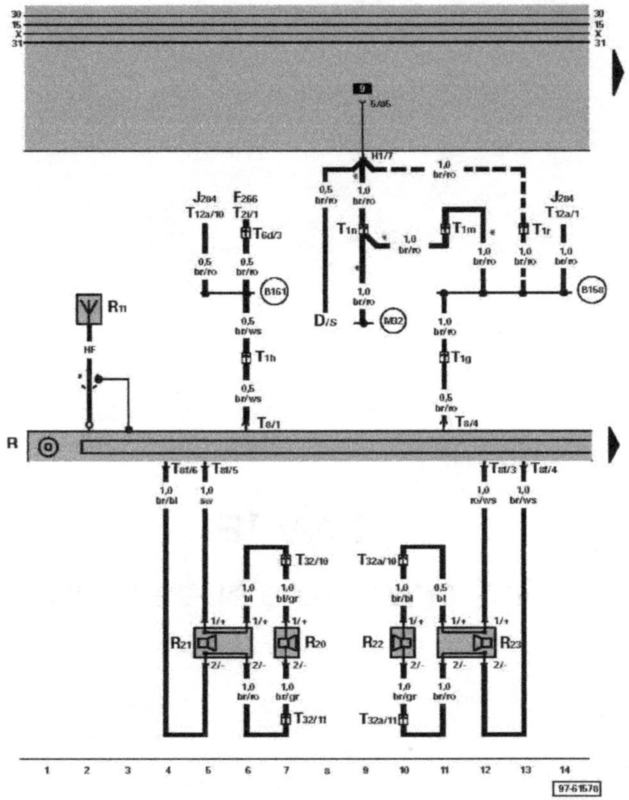

ws = white
sw = black
ro = red
br = brown
gn = green
bl = blue
gr = grey
li = violet
ge = yellow

D	-	Ignition/Starter Switch
F266	-	Front Hood Switch
J284	-	Alarm system control module
R	-	Radio
R11	-	Antenna
R20	-	Treble speaker, left front
R21	-	Bass speaker, left front
R22	-	Treble speaker, right front
R23	-	Bass speaker, right front
T1g	-	Single Connector, red, behind fuse/relay panel
T1h	-	Single Connector; yellow, behind fuse/relay panel
T1m	-	1-Pin Connector, red, behind fuse relay panel
T1n	-	1-Pin Connector, red, behind fuse relay panel
T1r	-	1-Pin Connector, red, behind fuse relay panel
T2i	-	2-Pin Connector
T6d	-	6-Pin Connector, red
T8	-	8-Pin Connector, in Radio, black
T8f	-	8-Pin Connector, brown
T12a	-	12-Pin Connector, black
T32	-	32-Pin Connector, white, on left A-pillar
T32a	-	32-Pin Connector, white, on right A-pillar
B158	-	Connector (S) in wiring harness interior
B161	-	Connector (anti-theft warning system), in wiring harness interior
M32	-	Wire Connection (su),in convertible top wiring harness
*	-	Power convertible top
---	-	Manual convertible top

Stereo radio

Edition 09.01
USA.5403.04.21

Cabrio

Wiring diagram

No. 11/3

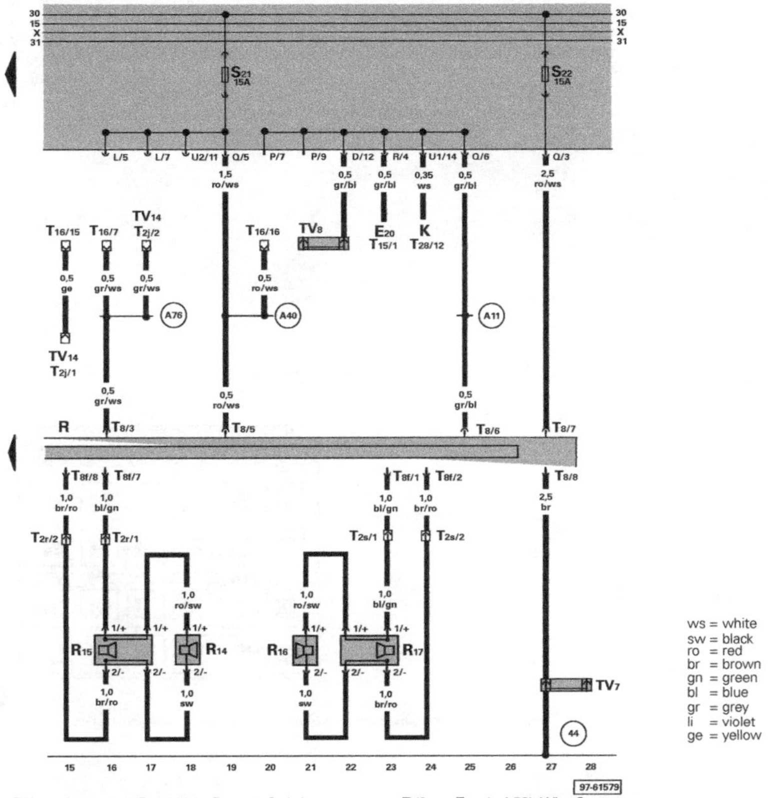

ws = white
sw = black
ro = red
br = brown
gn = green
bl = blue
gr = grey
li = violet
ge = yellow

- E20 - Instrument Panel Light Dimmer Switch
- K - Instrument Cluster
- R - Radio
- R14 - Treble speaker, left rear
- R15 - Bass speaker, left rear
- R16 - Treble speaker, right rear
- R17 - Bass speaker, right rear
- T2j - 2-Pin Connector, grey
- T2r - 2-Pin Connector, black
- T2s - 2-Pin Connector, black
- T8 - 8-Pin Connector, in Radio, black
- T8f - 8-Pin Connector, brown
- T15 - 15-Pin Connector
- T16 - 16-Pin Connector, Data Link Connector (DLC)
- T28 - 28-Pin Connector, white
- TV7 - Terminal 31 (GND) wire connector, brown, beside fuse /relay panel
- TV8 - Terminal 58b Wire Connector, grey
- TV14 - Data Link Connector (DLC) wire Connector, black, above fuse relay panel
- (44) - Ground connection, on left A-pillar, lower part
- (A11) - Plus connection (58b) in instrument panel wiring harness
- (A40) - Plus connection -1- (30), in instrument cluster wiring harness
- (A76) - Connector (k diagnosis wire) in instrument panel wiring harness

THIS PAGE INTENTIONALLY LEFT

BLANK

Cabrio — Wiring Diagram — No. 12/1

Heated seats

from August 1998

Relay locations

Position of relays and multi-pin connector wiring, see also "Component Locations" section.

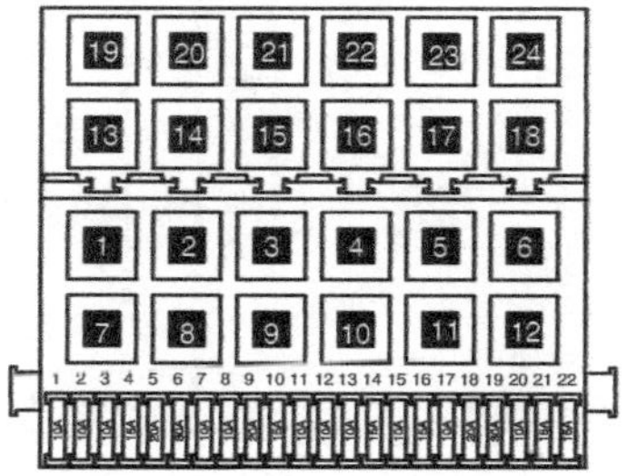

Fuse colors

30 A - green
25 A - white
20 A - yellow
15 A - blue
10 A - red
7,5 A - brown
5 A - beige
3 A - violet

VMM-067

Edition 04.01
USA.5403.03.21

No.12/2 — Wiring diagram — Cabrio

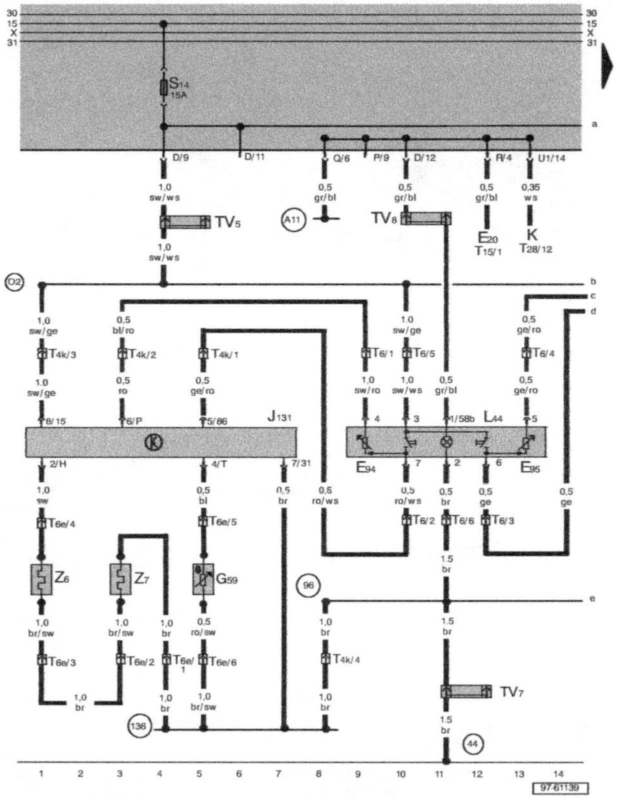

ws = white
sw = black
ro = red
br = brown
gn = green
bl = blue
gr = grey
li = violet
ge = yellow

E20 - Instrument Panel Light Dimmer Switch
E94 - Adjuster for heated driver's seat
E95 - Adjuster for heated passenger seat
G59 - Driver's Heated Seat Temperature Sensor
J131 - Driver's Heated Seat Control Module
K - Instrument Cluster
L44 - Heated Seat Switch Illumination Light
T4k - 4-Pin Connector
T6 - 6-Pin Connector
T6e - 6-Pin Connector
T15 - 15-Pin Connector
T28 - 28-Pin Connector
TV5 - Terminal 15a Wire Connector, green
TV7 - Terminal 31 (GND) wire connector, brown, beside fuse /relay panel
TV8 - Terminal 58b Wire Connector, grey
Z6 - Driver's Seat Heating Element
Z7 - Driver's Backrest Heating Element
(44) - Ground connection, on left A-pillar lower part
(96) - Ground connection -1-, in heated seats wiring harness
(136) - Ground connection -2-, in heated seats wiring harness
(A11) - Plus connection (58b), in instrument panel wiring harness
(O2) - Plus connection (15), in heated seats wiring harness

Heated seats

Edition 04.01
USA.5403.03.21

Cabrio — Wiring diagram — No. 12/3

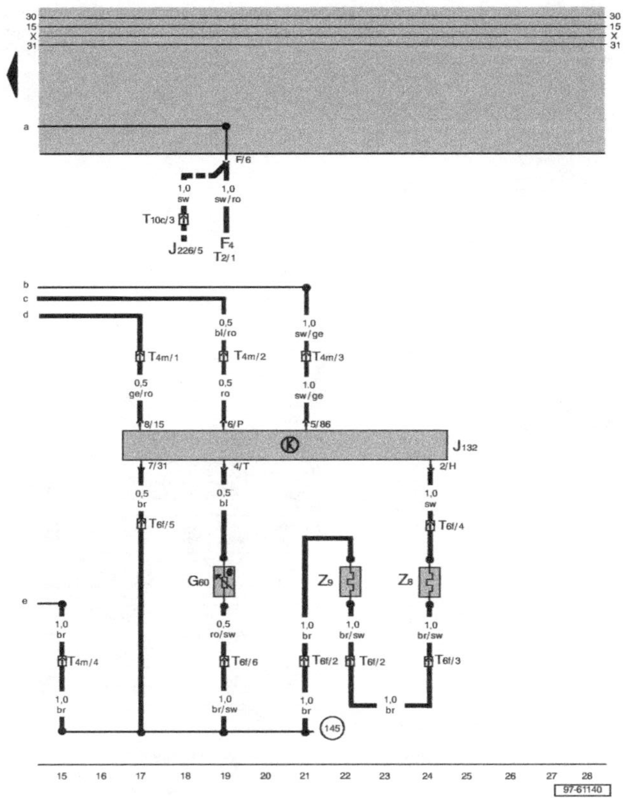

ws = white
sw = black
ro = red
br = brown
gn = green
bl = blue
gr = grey
li = violet
ge = yellow

- F4 - Back-Up Light Switch
- G60 - Passenger's Heated Seat Temperature Sensor
- J132 - Passenger's Heated Seat Control Module
- J226 - Park / Neutral Position (PNP) relay
- T2 - 2-Pin Connector
- T4m - 4-Pin Connector
- T6f - 6-Pin Connector
- T10c - 10-Pin Connector
- Z8 - Passenger's Seat Heating Element
- Z9 - Passenger's Backrest Heating Elements
- (145) - Ground connection -3-, in heated seats wiring harness
- - - - Automatic transmission only

Edition 04.01
USA.5403.03.21

Heated seats

THIS PAGE INTENTIONALLY LEFT BLANK

Cabrio — Wiring diagram — No. 13/1

Cruise Control

from August 1998

Relay locations
Position of relays and multi-pin connector wiring, see also "Component Locations" section.

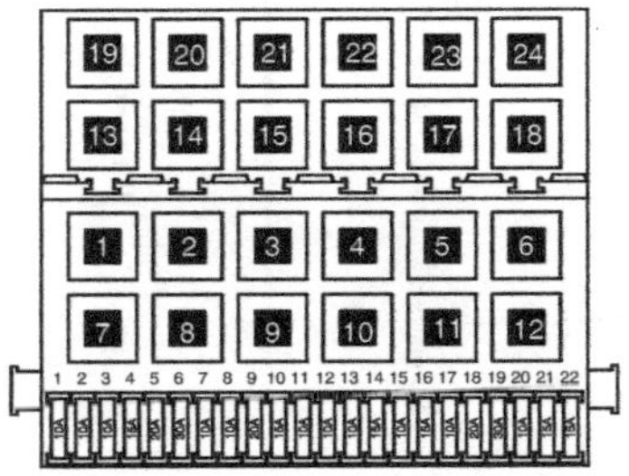

97-61067

Note:
Number in parentheses indicates production control number stamped on relay housing.

Fuse colors
30 A - green
25 A - white
20 A - yellow
15 A - blue
10 A - red
7,5 A - brown
5 A - beige
3 A - violet

Edition 09.01
USA.5403.04.21

No. 13/2 — Wiring diagram — Cabrio

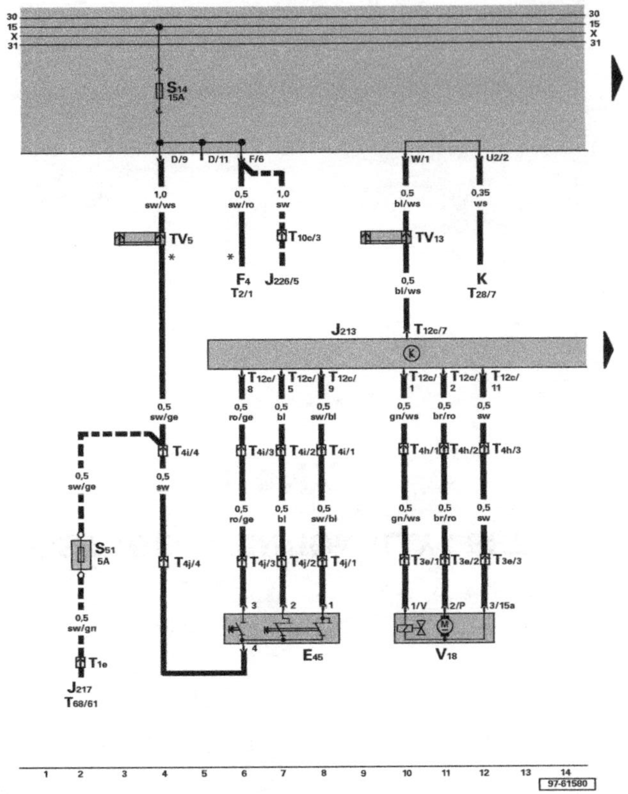

ws = white
sw = black
ro = red
br = brown
gn = green
bl = blue
gr = grey
li = violet
ge = yellow

97-61580

E45 - Cruise Control Switch
F4 - Back-Up Light Switch
J213 - Cruise Control, Control Module
J217 - Transmission Control Module (TCM)
J226 - Park / Neutral Position (PNP) relay
K - Instrument Cluster
S51 - Fuse (Only Automatic Transmission)
T1e - Single Connector, black, behind fuse/relay panel
T2 - 2-Pin Connector
T3e - 3-Pin Connector, black, behind fuse/relay panel
T4h - 4-Pin Connector, behind fuse/relay panel
T4i - 4-Pin Connector, black
T4j - 4-Pin Connector, black
T10c - 10-Pin Connector, black
T12c - 12-Pin Conenctor, black
T28 - 28-Pin Connector, white
T68 - 68-Pin Connector, on Transmission Control Module, black
TV5 - Terminal 15a wire connector, green
TV13 - Vehicle Speed Signal Wire Connector, blue
V18 - Cruise Control Vacuum Pump

* - Only for vehicles with standard transmission
\- - - - Only for vehicles with automatic transmission

Cruise control control module, cruise control switch

Edition 09.01
USA.5403.04.21

Cabrio — Wiring diagram — No. 13/3

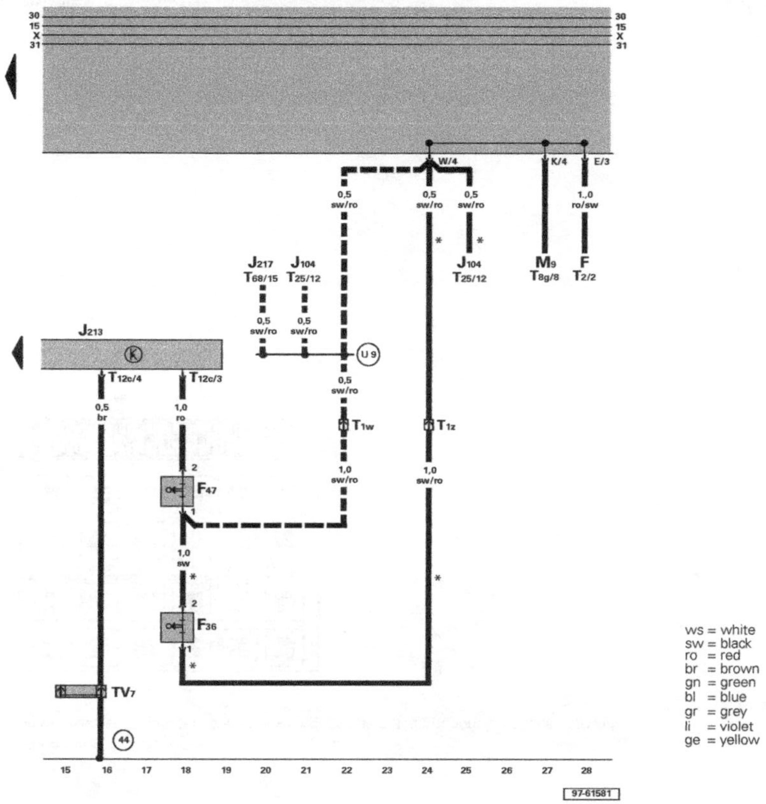

ws = white
sw = black
ro = red
br = brown
gn = green
bl = blue
gr = grey
li = violet
ge = yellow

F - Brake Light Switch
F36 - Clutch Vacuum Vent Valve Switch
F47 - Brake Vacuum Vent Valve Switch for Cruise Control/Diesel Fuel Injection
J104 - ABS Control Module (w/EDL)
J213 - Cruise Control, Control Module
J217 - Transmission Control Module
M9 - Left Brake Light
T1w - Single Connector,green, behind fuse/relay panel
T1z - Single Connector,green, behind fuse/relay panel
T2 - 2-Pin Connector
T8g - 8-Pin Connector, on taillight, left
T12c - 12-Pin Conenctor, black
T25 - 25-Pin Connector, black
T68 - 68-Pin Connector, on Transmission Control Module

TV7 - Terminal 31 (GND) wire connector, brown, beside fuse /relay panel
(44) - Ground connection, on left A-pillar lower part
(U9) - Connection (54) (in automatic transmission wiring harness)

* - Only for vehicles with standard transmission
- - - - Only for vehicles with automatic transmission

Edition 09.01
USA.5403.04.21

Cruise control control module, clutch vacuum vent valve switch, brake pedal cruise control switch

THIS PAGE INTENTIONALLY LEFT

BLANK

Cabrio — Wiring Diagram — No. 14/1

Air conditioning
from September 1999

Relay locations
Position of relays and multi-pin connector wiring, see also "Component Locations" section.

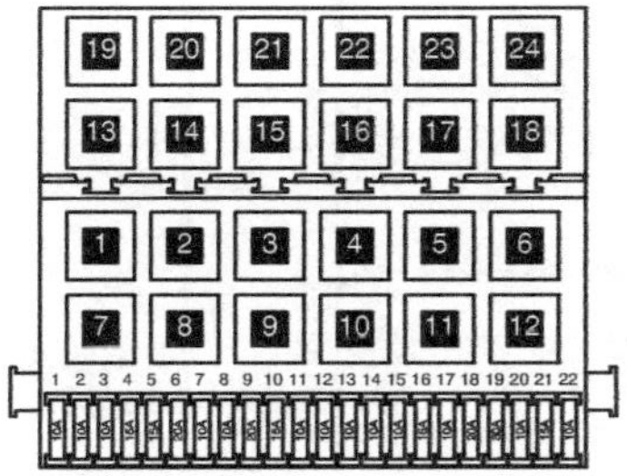

97-3000

1 - A/C Relay (147)

Fuse colors
30A – green
25A – white
20A – yellow
15A – blue
10A – red

Edition 12/99
W42. USA.5403.01.21

Air conditioning (from September 1999)

No. 14/2 — Wiring Diagram — Cabrio

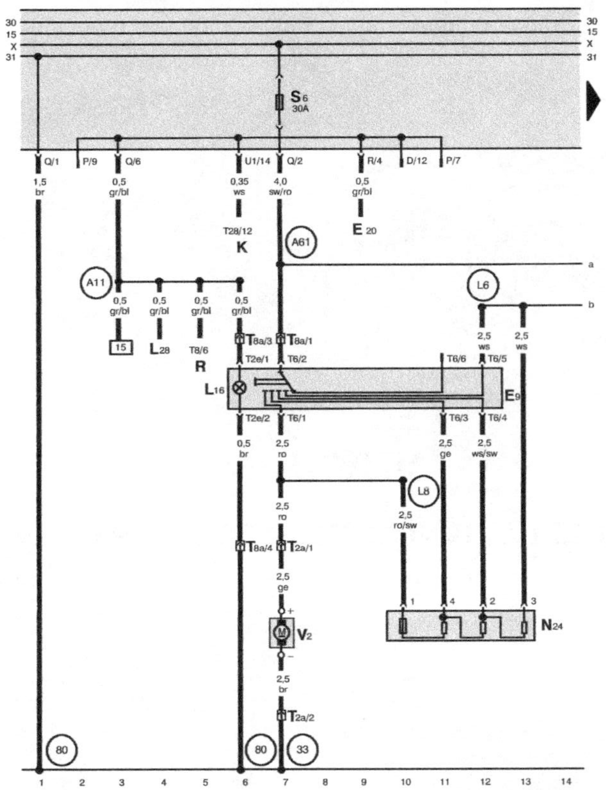

97–60734

ws = white
sw = black
ro = red
br = brown
gn = green
bl = blue
gr = grey
li = violet
ge = yellow

E9 – Fresh Air Blower Switch
E20 – Instrument Panel Light Dimmer Switch
K – Instrument Cluster
L16 – Fresh Air Control Lever Light
L28 – Cigarette Lighter Light
N24 – Fresh Air Blower Series Resistance With Fuse
R – Radio
T2a – Double Connector, behind instrument panel, right
T2e – Double Connector
T6 – 6-Pin Connector
T8 – 8-Pin Connector, on radio
T8a – 8-Pin Connector, behind instrument panel, center
T28 – 28-Pin Connector, on instrument cluster
V2 – Fresh Air Blower (01273–4124)

(33) – Ground connection, behind instrument panel, right
(80) – Ground connection –1–, in instrument panel wiring harness
(A11) – Plus connection (58b), in instrument panel wiring harness
(A61) – Plus connection (X), in instrument cluster wiring harness
(L6) – Wire connection –2–, in evaporator housing wiring harness
(L8) – Wire connection –4–, in evaporator housing wiring harness

Air Conditioning **Edition 12/99**
W42 .USA.5403.01.21

Cabrio — Wiring diagram — No. 14/3

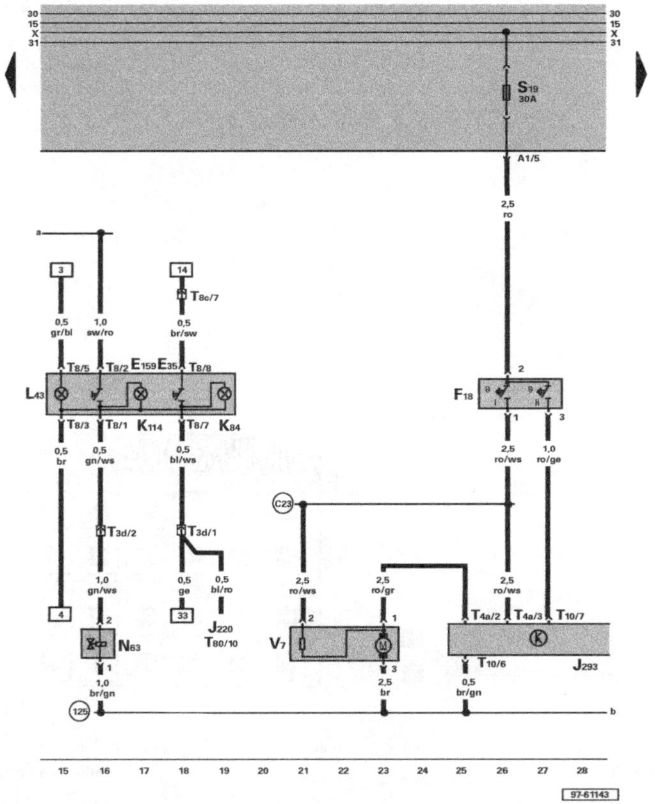

ws = white
sw = black
ro = red
br = brown
gn = green
bl = blue
gr = grey
li = violet
ge = yellow

E35 - A/C Switch
E159 - Fresh Air/Recirculating Flap Switch
F18 - Coolant Fan Control (FC) Thermal Switch
J220 - Motronic Engine Control Module (ECM), in plenum chamber, right
J293 - Coolant FC (Fan Control) Control Module,in engine compartment, left
K84 - A/C Indicator Light
K114 - Fresh Air and Recirculating Air Mode Indicator Light
L43 - A/C Switch Illumination Light
N63 - Fresh Air/Recirculating Flap Two-Way Valve, in engine compartment, rear
T3d - 3-Pin Connector, behind fuse/relay panel
T4a - 4-Pin Connector
T8 - 8-Pin Connector
T8c - 8-Pin Connector, behind instrument panel, center
T10 - 10-Pin Connector
T80 - 80-Pin Connector
V7 - Coolant fan

(125) - Ground connection -3-, in headlight wiring harness
(C23) - Wire connection (blower, 1st speed), in headlight/air conditioner wiring harness

A/C switch, fresh air/recirculating flap switch, coolant fan control (FC) thermal switch, fresh air and recirculating air mode indicator light

No. 14/4 — Wiring diagram — Cabrio

ws = white
sw = black
ro = red
br = brown
gn = green
bl = blue
gr = grey
li = violet
ge = yellow

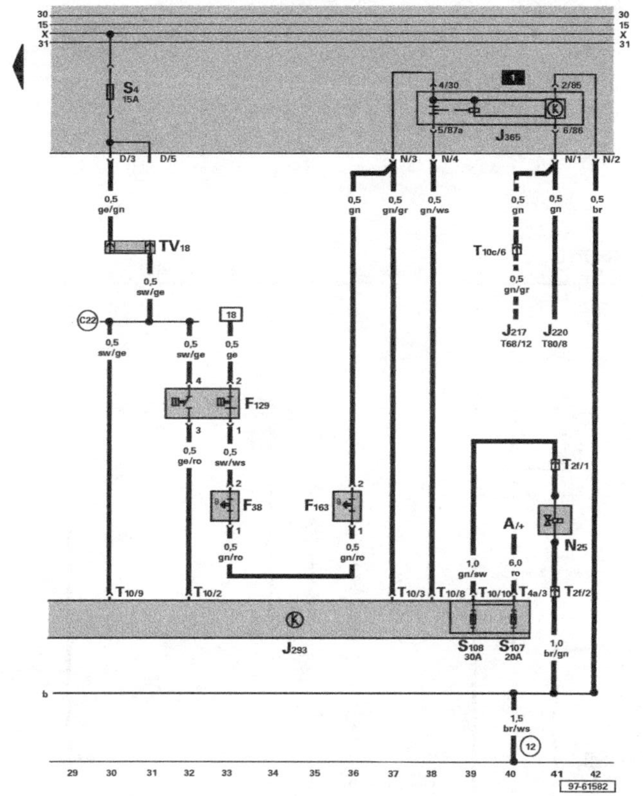

A - Battery
F38 - Ambient Temperature Switch, near dual tone horns
F129 - A/C Pressure Switch, in engine compartment, right
F163 - A/C Cut-Out Thermal Switch, in engine compartment, center
J217 - Transmission Control Module (TCM), below rear seat bench
J220 - Motronic Engine Control Module (ECM), in plenum chamber, right
J293 - Coolant FC (Fan Control) Control Module, in engine compartment, left
J365 - Relay for A/C shutoff, above fuse/relay panel (147)
N25 - A/C Clutch
S107 - Coolant Fan Second and Third Speed Fuse
S108 - Coolant Fan First Speed and A/C Clutch Fuse
T2f - Double Connector, near compressor
T4a - 4-Pin Connector
T10 - 10-Pin Connector
T10c - 10-Pin Connector
T68 - 68-Pin Connector
T80 - 80-Pin Connector
TV18 - Distributor for terminal Xa, geld, above fuse/relay panel

(12) - Ground connection,in engine compartment, left
(C22) - Wire connection (x), in headlight/air conditioner wiring harness

Ambient temperature switch, A/C pressure switch, A/C cut-out thermal switch, coolant FC (fan control) control module, relay for A/C shutoff, A/C clutch

Cabrio — Wiring diagram — No. 15/1

2.0L - Engine - Motronic Multipoint Fuel Injection (MFI) /85 kW, engine code ABA

from June 2000

Relay locations

Position of relays and multi-pin connector wiring, see also "Component Locations" section.

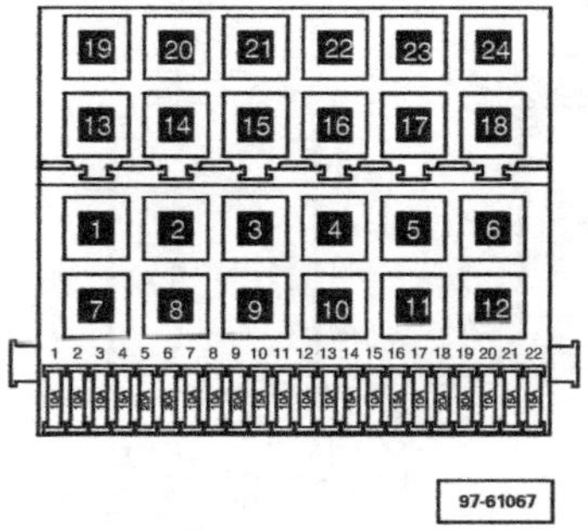

9 Radio/Parking Light Warning Buzzer (188)

12 Fuel Pump (FP) Relay (409)

Note: Number in parentheses indicates production control number stamped on relay housing.

Fuse colors

30 A - green
25 A - white
20 A - yellow
15 A - blue
10 A - red
7,5 A - brown
5 A - beige
3 A - violet

No. 15/2 — Wiring diagram — Cabrio

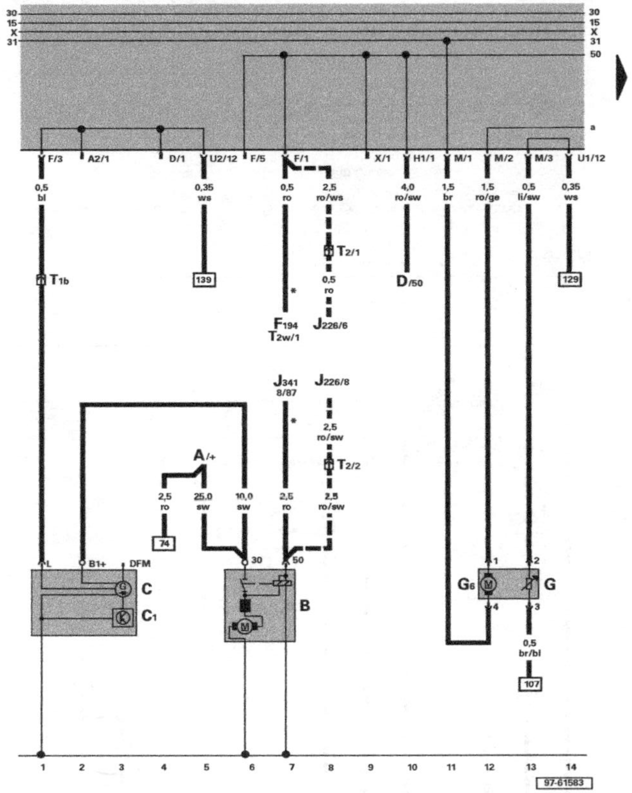

ws = white
sw = black
ro = red
br = brown
gn = green
bl = blue
gr = grey
li = violet
ge = yellow

A - Battery
B - Starter
C - Generator (GEN)
C1 - Voltage Regulator (VR)
D - Ignition/Starter Switch
F194 - Clutch Pedal Position (CPP) Switch, green
G - Sender for fuel gauge
G6 - Fuel Pump (FP)
J226 - Park/Neutral Position (PNP) Relay
J341 - Anti-Theft Engine Disable Relay I (53)
T1b - Single Connector, black, near starter
T2 - 2-Pin Connector, black, behind fuse relay panel
T2w - 2-Pin Connector, green, on clutch pedal bracket

* - Manual transmission only
--- - Automatic transmission only

Battery, starter, generator (GEN), voltage regulator (VR), fuel pump (FP), sender for fuel gauge

Cabrio — Wiring diagram — No. 15/3

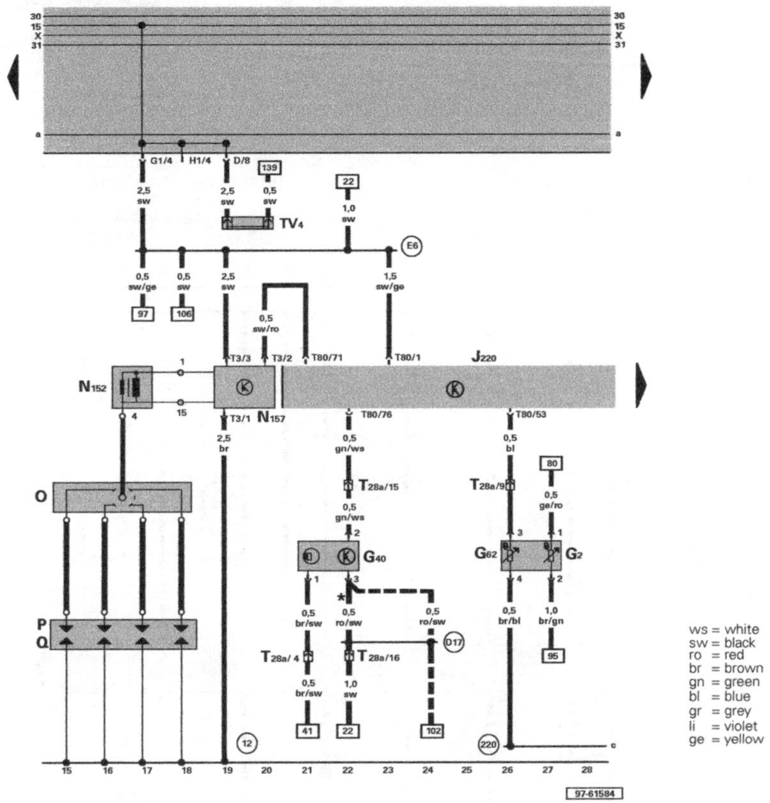

ws = white
sw = black
ro = red
br = brown
gn = green
bl = blue
gr = grey
li = violet
ge = yellow

- G2 - Engine Coolant Temperature (ECT) Sensor
- G40 - Camshaft Position (CMP) Sensor
- G62 - Engine Coolant Temperature (ECT) Sensor
- J220 - Motronic Engine Control Module (ECM)
- N152 - Ignition Coil
- N157 - Ignition Coil Power Output Stage
- O - Ignition Distributor
- P - Spark Plug Connectors
- Q - Spark Plugs
- T3 - 3-Pin connector
- T28a - 28-Pin connector, in engine, left
- T80 - Connector, 80 point
- TV4 - Terminal 15 Wire Connector, black
- (12) - Ground connection, in engine compartment, left
- (220) - Ground connection (sensor ground), in engine compartment wiring harness
- (D17) - Plus connection -1- (15), in engine compartment wiring harness
- (E6) - Plus connection -1- (15), in Motronic Multiport Fuel Injection (MFI)
- * - Manual transmission only
- --- - Automatic transmission only

Edition 09.01 USA.5403.04.21

Motronic ECM, engine coolant temperature (ECT) sensor, camshaft position (CMP) sensor, ignition coil, ignition power output stage, spark plugs

No. 15/4 — Wiring Diagram — Cabrio

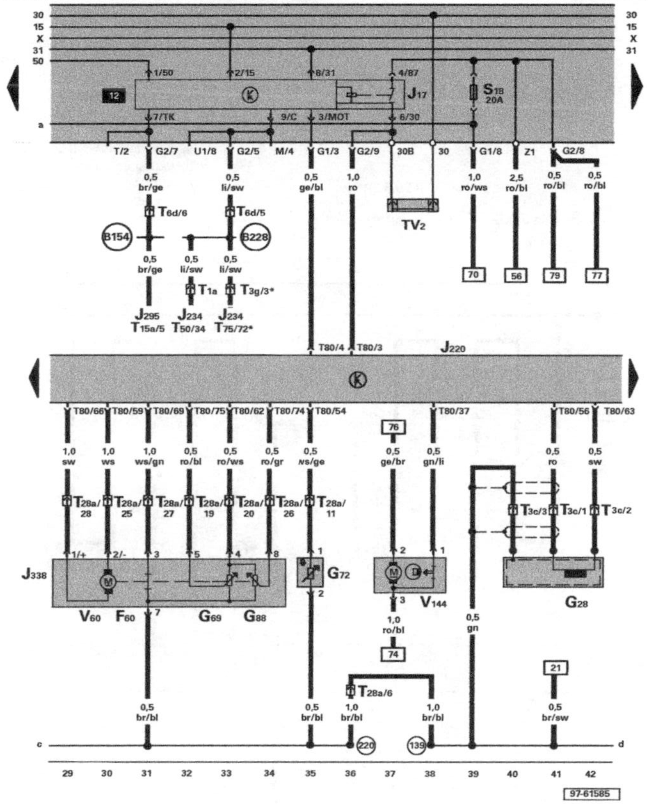

ws = white
sw = black
ro = red
br = brown
gn = green
bl = blue
gr = grey
li = violet
ge = yellow

- F60 - Closed Throttle Position (CTP) Switch
- G28 - Engine Speed (RPM) Sensor
- G69 - Throttle Position (TP) Sensor
- G72 - Intake Air Temperature (IAT) Sensor
- G88 - Throttle Position (TP) Sensor
- J17 - Fuel Pump (FP) Relay
- J220 - Motronic Engine Control Module (ECM)
- J234 - Airbag Control Module
- J295 - Left Front Power Window Control Module
- J338 - Throttle Valve Control Module
- T1a - Single Connector, behind relay panel
- T3c - 3-Pin Connector, in engine
- T3g - 3-Pin Connector, behind fuse relay panel
- T6d - 6 Pin Connector, behind fuse relay panel
- T15a - 15 Pin Connector
- T28a - 28-Pin Connector, in engine, left
- T50 - 50-Pin Connector
- T80 - 80-Pin Connector
- TV2 - Terminal 30 (B+) wire connector
- V60 - Throttle Position (TP) Actuator
- V144 - Leak Detection Pump (LDP)
- (139) - Ground connection (sensor Ground),in Motronic Multiport Fuel Injection (MFI) wiring harness
- (220) - Ground connection (sensor ground), in engine compartment wiring harness
- (B154) - Connector -1- (TK), in wiring harness interior
- (B228) - Connection (Crash-Signal) (interior wiring harness)
- * - from 10.00 production

Motronic ECM, throttle valve control module, intake air temperature (IAT) sensor, engine speed (RPM) sensor, fuel pump (FP) relay

Edition 09.01 USA.5403.04.21

Cabrio — Wiring Diagram — No. 15/5

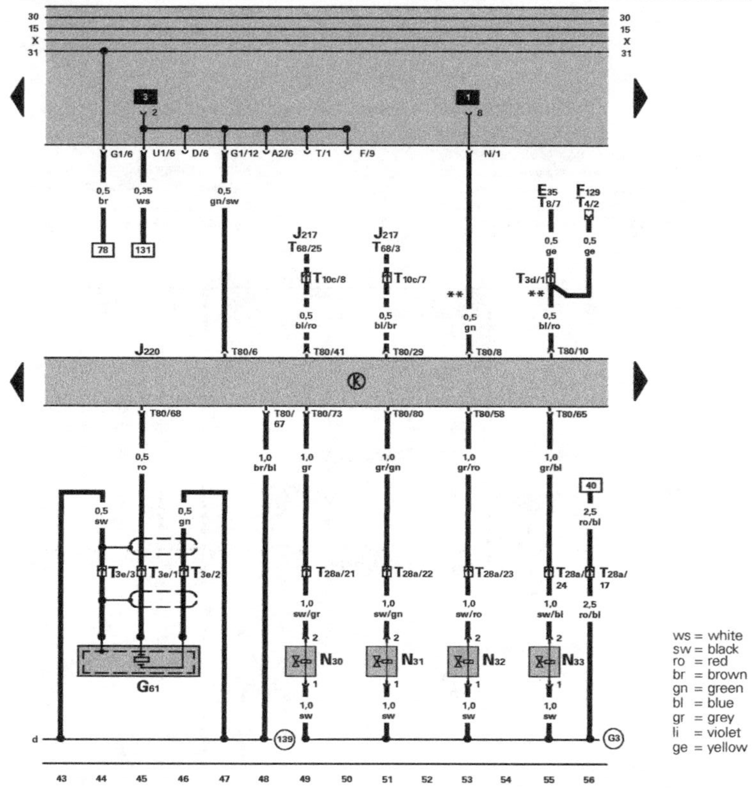

ws = white
sw = black
ro = red
br = brown
gn = green
bl = blue
gr = grey
li = violet
ge = yellow

E35 - A/C Switch
F129 - A/C Pressure Switch
G61 - Knock Sensor (KS) 1
J217 - Transmission Control Module (TCM)
J220 - Motronic Engine Control Module (ECM)
N30 - Cylinder 1 Fuel Injector
N31 - Cylinder 2 Fuel Injector
N32 - Cylinder 3 Fuel Injector
N33 - Cylinder 4 Fuel Injector
T3d - 3-Pin Connector, black, behind relay panel
T3e - 3-Pin Connector, in engine compartment
T4 - 4-Pin Connector, in engine compartment, right
T8 - 8-Pin Connector
T10c - 10-Pin Connector, black, behind relay panel
T28a - 28-Pin Connector, in engine, left
T68 - 68-Pin Connector
T80 - 80-Pin Connector
(139) - Ground connection (sensor Ground), in Motronic Multiport Fuel Injection (MFI), wiring harness
(G3) - Plus connection, in injector wiring harness

** - A/C Connector
--- - Automatic transmission only

Motronic engine control module (ECM), knock sensor (KS) 1, fuel injectors

No.15/6 — Wiring Diagram — Cabrio

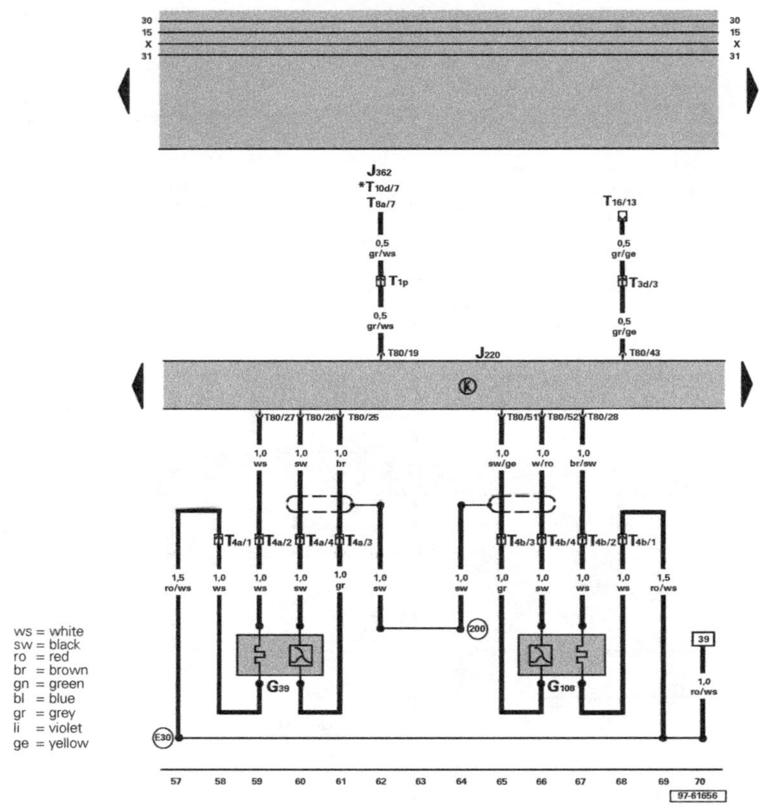

ws = white
sw = black
ro = red
br = brown
gn = green
bl = blue
gr = grey
li = violet
ge = yellow

G39 - Heated Oxygen Sensor (HO2S)
G108 - Heated Oxygen Sensor (HO2S) 2
J220 - Motronic Engine Control Module (ECM)
J362 - Control module for anti-theft immobilizer
T1p - Single Connector, brown, behind fuse/ relay panel
T3d - 3-Pin Connector, black, behind relay panel
T4a - 4-Pin Connector, in engine compartment
T4b - 4-Pin Connector, in engine compartment
T8a - 8-Pin Connector, black
T10d - 10-Pin Connector, black
T16 - 16-Pin Connector, Data Link Connector (DLC)
T80 - 80-Pin Connector
(200) - Ground connection (shielding), in engine compartment wiring harness
(E30) - Connector (87a), in wiring harness engine
* - from 08.00 production

Motronic engine control module (ECM), heated oxygen sensors (HO2S), evaporative emission canister purge regulator valve

Cabrio — Wiring Diagram — No. 15/7

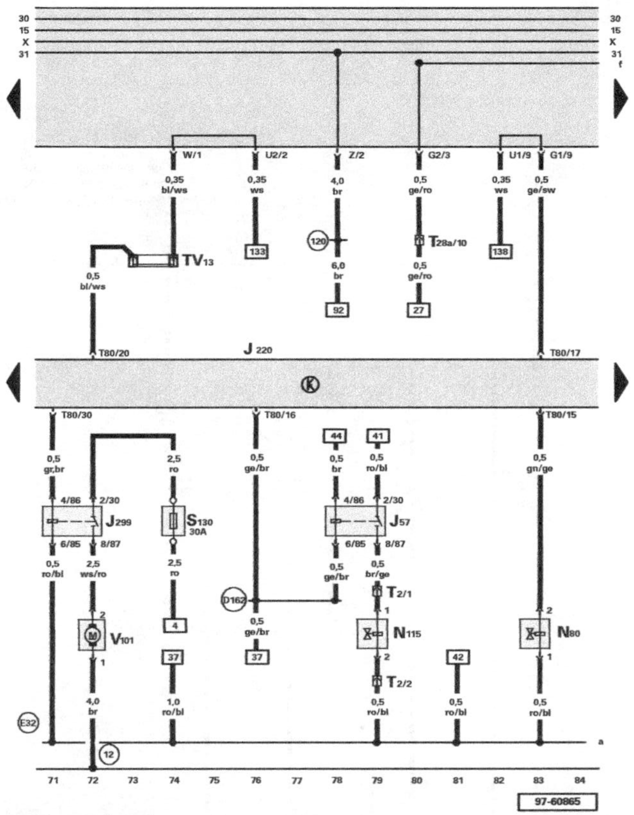

ws = white
sw = black
ro = red
br = brown
gn = green
bl = blue
gr = grey
li = violet
ge = yellow

J220 - Motronic Engine Control Module (ECM)
J299 - Secondary Air Injection (AIR) Pump Relay
J57 - Vacuum Pump Relay
N 80 - Evaporative Emission (EVAP) Canister Purge Regulator valve
N115 - Evaporative Emission (EVAP) Canister Purge Solenoid valve
S130 - Fuse for secondary air pump
T2 - 2-Pin Connector, behind panel relay
T28a - 28-Pin Connector, in engine, left
T80 - Connector, 80 point
TV13 - Vehicle Speed Signal Wire Connector
V101 - Secondary Air Injection (AIR) Pump Motor
(12) - Ground connection, in engine compartment, left
(120) - Ground connection -2- , in headligth wiring harness
(D162) - Connection for Leak Detection Pump (LDP),engine compartment wiring harness
(E32) - Connector (87), in wiring harness Motronic

Edition 10.00 W42.USA.5403.02.21

Motronic (ECM), secondary air injection (AIR) pump motor, Evaporative Emission (EVAP) canister purge regulator valve

2.0L engine (code: ABA) fuel injection (from June 2000)

No. 15/8 — Wiring Diagram — Cabrio

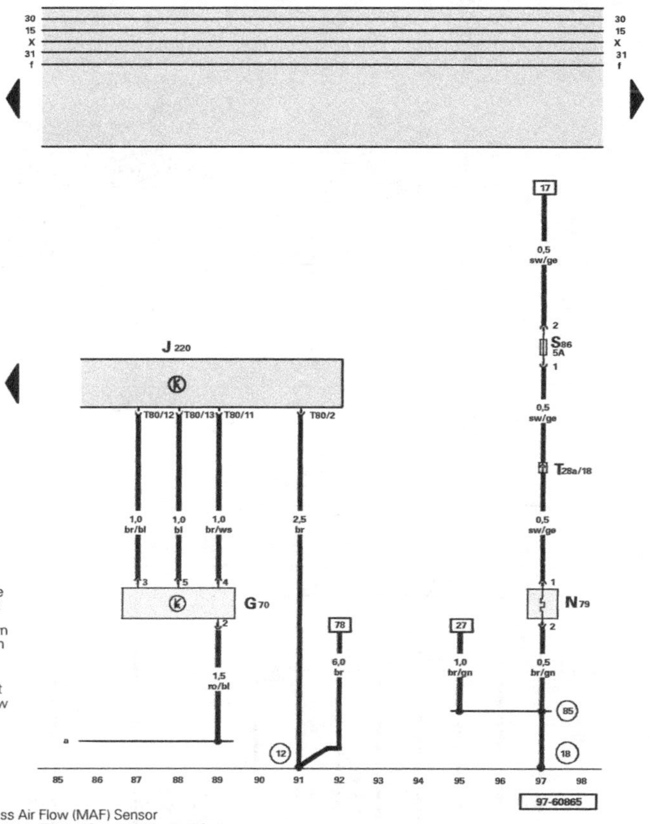

ws = white
sw = black
ro = red
br = brown
gn = green
bl = blue
gr = grey
li = violet
ge = yellow

G70 - Mass Air Flow (MAF) Sensor
J220 - Motronic Engine Control Module (ECM)
N79 - Positive Crankcase Ventilation (PCV) Heating Element
S86 - Door Lock Heating Element Circuit Breaker, above relay panel
T28a - 28-Pin Connector, in engine, left
T80 - Connector, 80 point
(12) - Ground connection, in engine compartment, left
(18) - Ground connection, on engine block
(85) - Ground connection -1-, in engine compartment wiring harness

Motronic (ECM), mass air flow (MAF) sensor, positive crankcase ventilation (PVC) heating element

Edition 10.00 W42.USA.5403.02.21

2.0L engine (code: ABA) fuel injection (from June 2000)

Cabrio — Wiring Diagram — No.15/9

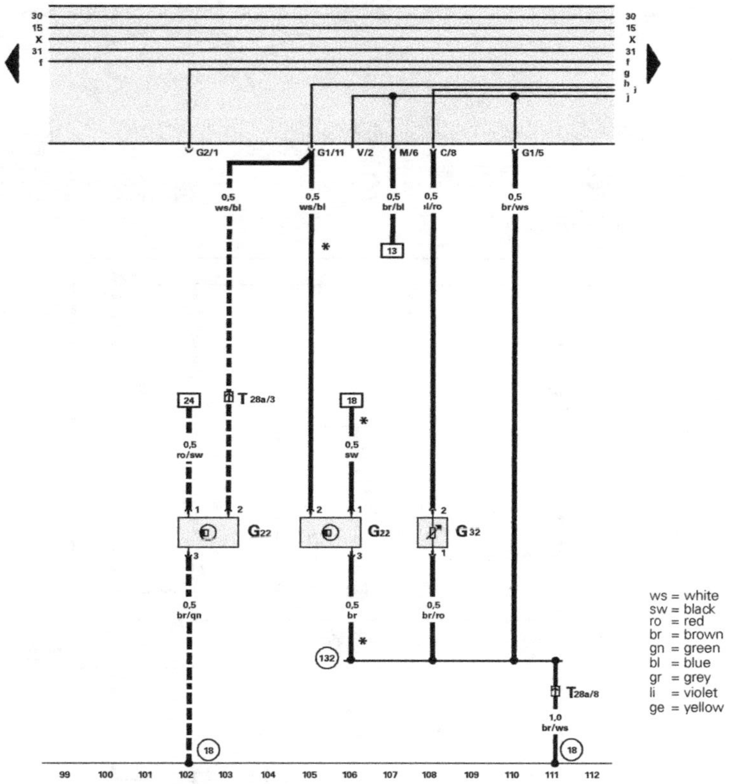

ws = white
sw = black
ro = red
br = brown
gn = green
bl = blue
gr = grey
li = violet
ge = yellow

G22 - Speedometer Vehicle Speed Sensor (VSS)
G32 - Engine Coolant Level (ECL) Sensor
T28a - 28-Pin Connector, in engine, left

(18) - Ground connection, on engine block

(132) - Ground connection -3-, in engine compartment wiring harness

* - Manual transmission only

--- - Automatic transmission only

Edition 10.00
W42.USA.5403.02.21

Speedometer vehicle speed sensor(VSS), engine coolant level (ECL) sensor

No.15/10 — Wiring Diagram — Cabrio

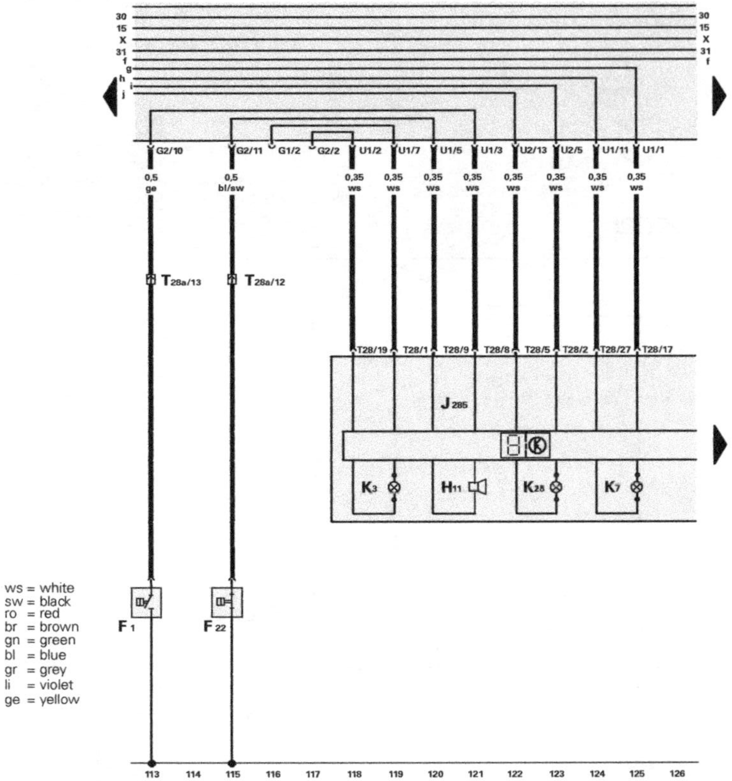

ws = white
sw = black
ro = red
br = brown
gn = green
bl = blue
gr = grey
li = violet
ge = yellow

F1 - Oil Pressure Switch
F22 - O.3bar Oil Pressure Switch
H11 - Oil Pressure Warning Buzzer
J285 - Control module with indicator unit in instrument panel insert
K3 - Oil Pressure Warning Light
K7 - Brake and Parking Brake Warning Light
K28 - Engine Coolant Level/Temperature (ECL/ECT) Warning Light
T28 - 28-Pin Connector
T28a - 28-Pin Connector, in engine, left

Instrument cluster, oil pressure switch, 0.3 bar oil pressure switch, oil pressure warning buzzer, brake and parking brake warning light

Edition 10.00
W42.USA.5403.02.21

Cabrio — Wiring Diagram — No.15/11

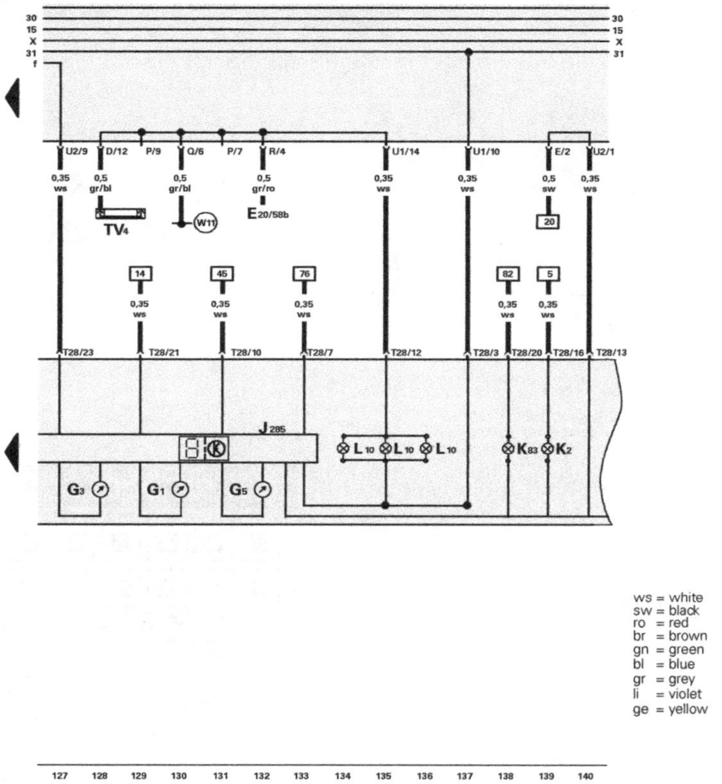

ws = white
sw = black
ro = red
br = brown
gn = green
bl = blue
gr = grey
li = violet
ge = yellow

E20 - Instrument Panel Light Dimmer Switch
G1 - Fuel Gauge
G3 - Engine Coolant Temperature (ECT) Gauge
G5 - Tachometer
J285 - Control module with indicator unit in instrument panel insert
K2 - Generator (GEN) Warning Light
K83 - Malfunction Indicator Lamp (MIL)
L10 - Instrument Cluster Light
T28 - 28-Pin Connector
TV4 - Terminal 15 Wire Connector
(W11) - Plus connection (58), in rear lid wiring harness

THIS PAGE INTENTIONALLY LEFT

BLANK

Cabrio — Wiring diagram — No. 16/1

Immobilizer

from August 2000

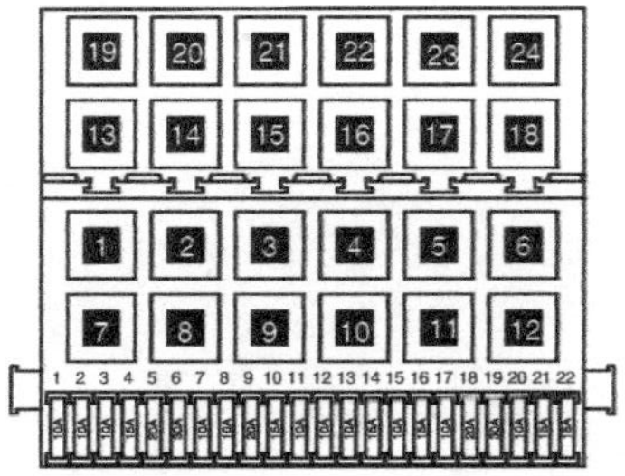

Note:
Number in parentheses indicates production control number stamped on relay housing.

Fuse colors

30 A - green
25 A - white
20 A - yellow
15 A - blue
10 A - red
7,5 A - brown
5 A - beige
3 A - violet

No. 16/2 — Wiring diagram — Cabrio

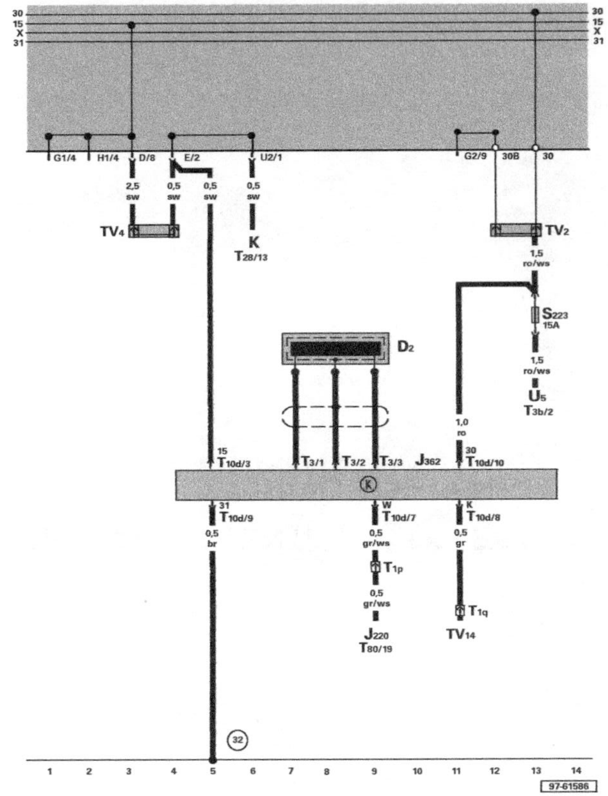

ws = white
sw = black
ro = red
br = brown
gn = green
bl = blue
gr = grey
li = violet
ge = yellow

D2 - Induction coil of antitheft immobilizer
J220 - Motronic Engine Control Module (ECM)
J362 - Control module for anti-theft immobilizer
K - Instrument Cluster
S223 - Fuse in fuse holder
T1p - Single Connector, brown, behind fuse/ relay panel
T1q - Single Connector, brown, behind fuse/ relay panel
T3 - 3-Pin Connector
T3b - 3-Pin Connector, white
T10d - 10-Pin Connector, black
T28 - 28-Pin Connector, white, on instrument cluster
T80 - 80-Pin Connector
TV2 - Terminal 30 (B+) Wire Connector, red
TV4 - Terminal 15 Wire Connector, black
TV14 - Data Link Connector (DLC) Wire Connector, black, above fuse relay panel
U5 - 12 V Socket
(32) - Ground connection, behind instrument panel, left

Induction coil of antitheft immobilizer, control module for antitheft immobilizer

Cabrio — Wiring diagram — No. 17/1

Airbag systems

from October 2000

Relay locations

Position of relays and multi-pin connector wiring, see also "Component Locations" section.

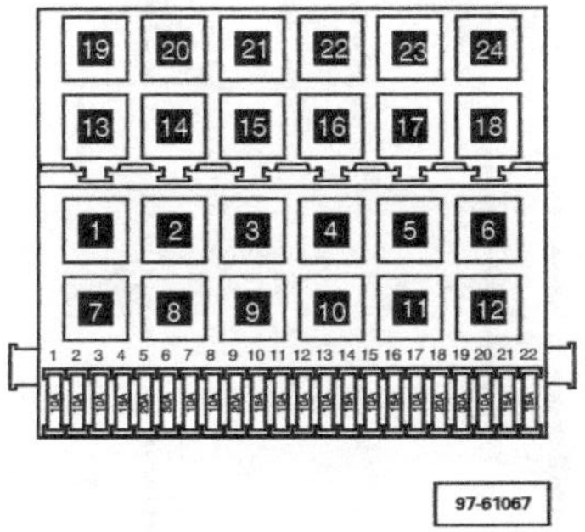

Note:
Number in parentheses indicates production control number stamped on relay housing.

Fuse colors

30 A - green
25 A - white
20 A - yellow
15 A - blue
10 A - red
7,5 A - brown
5 A - beige
3 A - violet

Edition 09.01
USA.5403.04.21

No. 17/2 — Wiring diagram — Cabrio

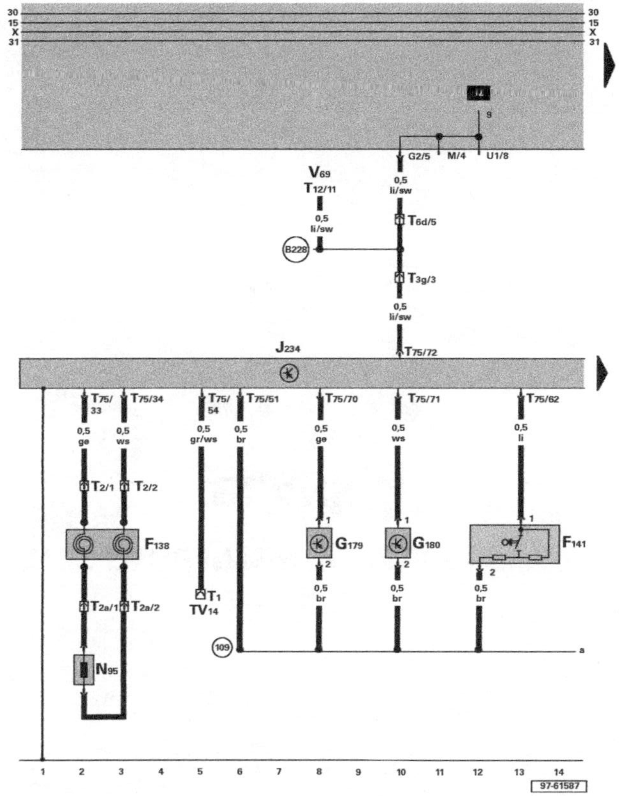

ws = white
sw = black
ro = red
br = brown
gn = green
bl = blue
gr = grey
li = violet
ge = yellow

F138 - Airbag Spiral Spring/Return Spring With Slip Ring
F141 - Right Front Seatbelt Microswitch
G179 - Crash sensor for side airbag, driver's side
G180 Crash sensor for side airbag, passenger side
J234 - Airbag Control Module
N95 - Driver's Side Airbag Igniter
T1 - Single connector, red
T2 - 2-Pin Connector
T2a - 2-Pin Connector
T3g - 3-Pin Connector, black, behind fuse relay panel
T6d - 6-Pin Connector, red
T12 - 12-Pin Connector
TV14 - Data Link Connector (DLC) wire Connector, black, above fuse relay panel
T75 - 75-Pin Connector

V69 - Central Locking System Pump and Control Module
(109) - Ground connection, in airbag wiring harness
(B228) - Connection (Crash-Signal) (interior wiring harness)

Airbag control module, right seatbelt microswitch, airbag spiral spring, driver's side airbag igniter, crash sensors for side airbags

Edition 09.01
USA.5403.04.21

Cabrio Wiring diagram **No. 17/3**

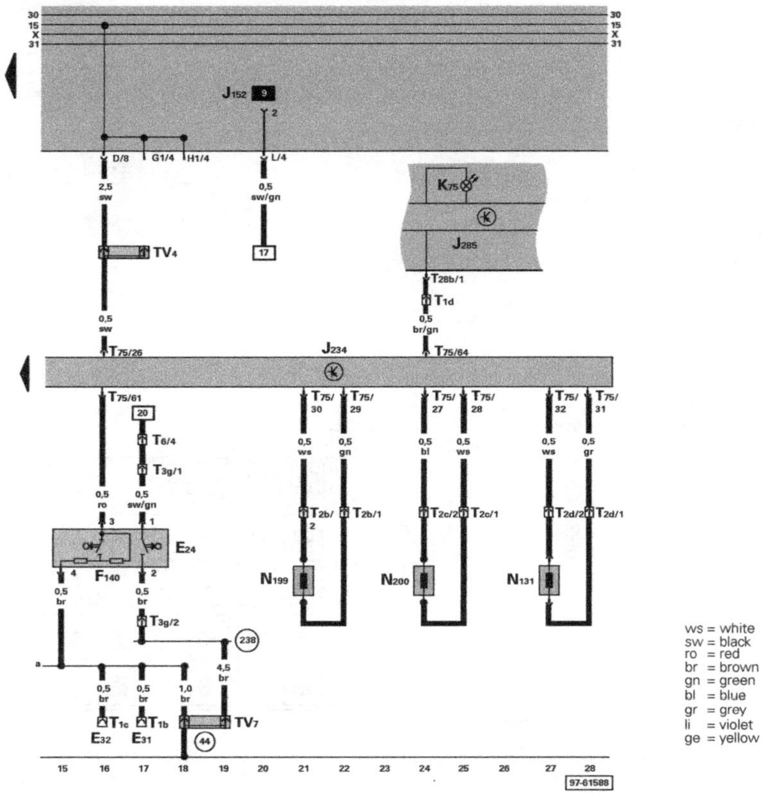

ws = white
sw = black
ro = red
br = brown
gn = green
bl = blue
gr = grey
li = violet
ge = yellow

E24 - Left Seat Belt Switch
F140 - Left Front Seatbelt Microswitch
E31 - Driver's Seat Contact Strip (Seat Belt Warning system)
E32 - Passenger 's Seat Contact Strip (Seat Belt Warning system)
J152 - Radio/Parking Light Warning Buzzer
J234 - Airbag Control Module
J285 - Control module with indicator unit in instrument panel insert
K75 - Airbag Malfunction Indicator Lamp (MIL)
N131 - Passenger's Side Airbag Igniter 1
N199 - Igniter for side airbag, driver's side
N200 - Igniter for side airbag, passenger side
T1b - Single connector, yellow
T1c - Single connector, yellow
T1d - Single connector, brown
T2b - 2-Pin Connector
T2c - 2-Pin Connector, yellow
T2d - 2-Pin Connector, yellow
T3g - 3-Pin Connector, black, behind fuse relay panel
T6 - 6-Pin Connector, black
T28b - 28-Pin Connector, on instrument cluster, red
T75 - 75-Pin Connector
TV4 - Terminal 15 Wire Connector, black
TV7 - Terminal 31 (GND) wire connector, brown, beside fuse /relay panel
(44) - Ground connection,on left A-pillar, lower part
(238) - Ground connection -1-, in wiring harness interior

Edition 09.01
USA.5403.04.21

Airbag control module, passenger's side airbag igniter, left seatbelt microswitch, airbag malfuction indicator lamp (MIL)

THIS PAGE INTENTIONALLY LEFT

BLANK

Cabrio — Wiring diagram — No. 18/1

Fog lights (Cabrio GLX only)

from September 1999

Relay locations

Position of relays and multi-pin connector wiring, see also "Component Locations" section.

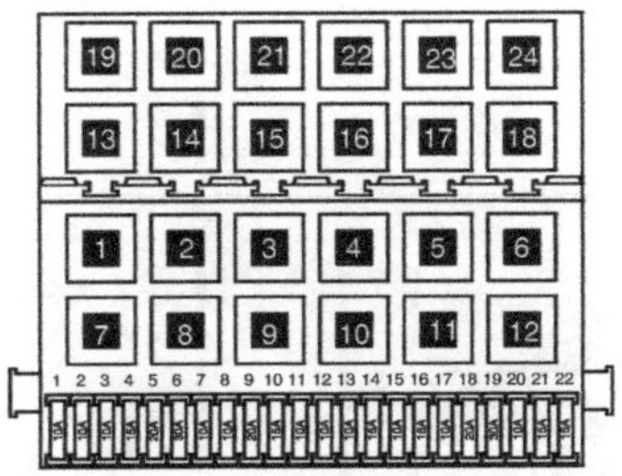

- 7 - Fog Light Relay (96)
- 10 - Metalic bridge

Note:
Number in parentheses indicates production control number stamped on relay housing.

Fuse colors

30 A - green
25 A - white
20 A - yellow
15 A - blue
10 A - red
7,5 A - brown
5 A - beige
3 A - violet

Edition 09.01
USA.5403.04.21

Fog lights

No. 18/2 — Wiring diagram — Cabrio

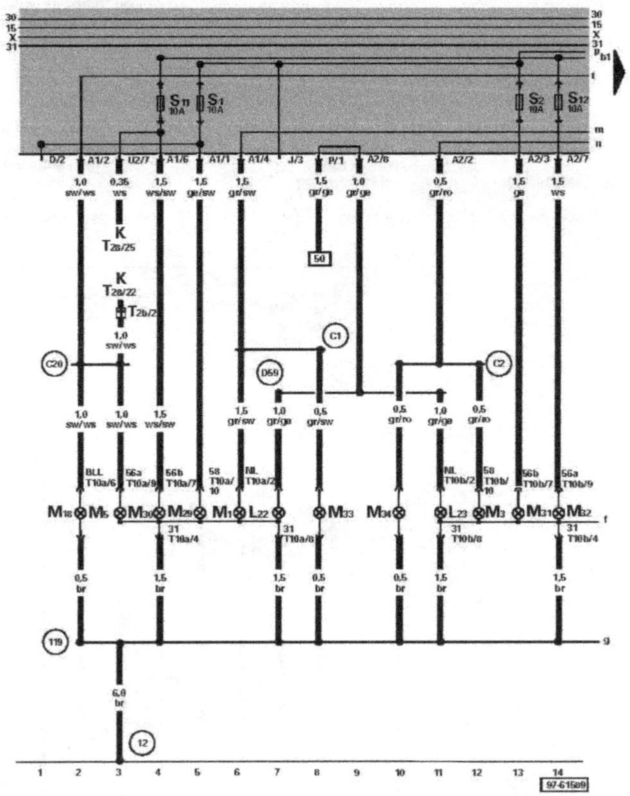

ws = white
sw = black
ro = red
br = brown
gn = green
bl = blue
gr = grey
li = violet
ge = yellow

K - Instrument Cluster
L 22 - Left Front Fog Light
L 23 - Right Front Fog Light
M1 - Left Parking Light
M3 - Right Parking Light
M5 - Left Front Turn Signal Light
M18 - Left, Side Turn Signal Light
M29 - Left Low Beam Headlight
M30 - Left High Beam Headlight
M31 - Right Low Beam Headlight
M32 - Right High Beam Headlight
M33 - Light for Side marker front left
M34 - Light for Side marker front right
T2b - 2-Pin Connector, black, behind fuse/relay panel
T10a - 10-Pin Connector
T10b - 10-Pin Connector
T28 - 28-Pin Connector, white, on instrument cluster
(12) - Ground connection, in engine compartment, left
(119) - Ground connection -1-, in headlight wiring harness
(C1) - Plus connection (58L), in left front wiring harness
(C2) - Plus connection (58R), in left front wiring harness
(C20) - Connection (left turn signal), in headlight wiring harness
(D59) - Wire connection (Fog light), in engine compartment wiring harness

Headlights, fog lights, parking lights, turn signal lights, side marker lights

Edition 09.01
USA.5403.04.21

Cabrio — Wiring diagram — No. 18/3

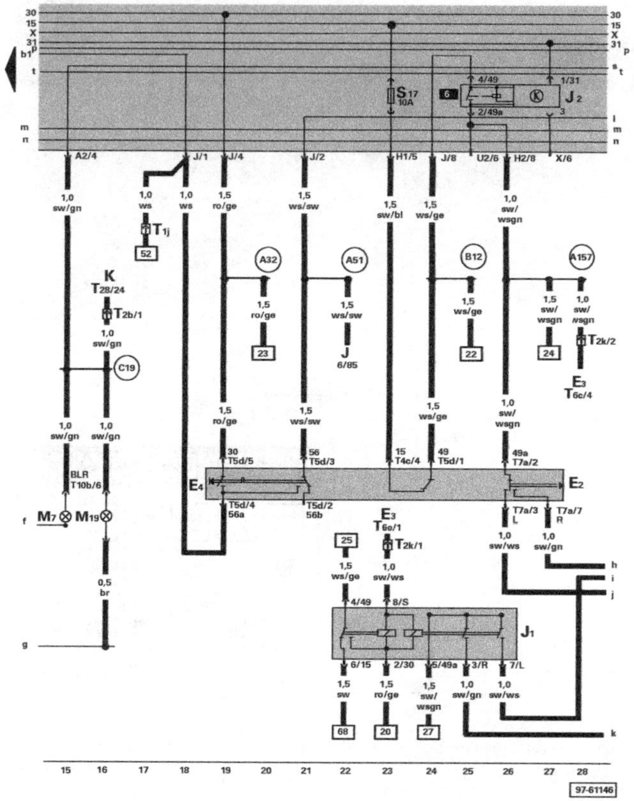

ws = white
sw = black
ro = red
br = brown
gn = green
bl = blue
gr = grey
li = violet
ge = yellow

- E3 - Emergency Flasher Switch
- E2 - Turn signal switch
- E4 - Headlight Dimmer/Flasher Switch
- J - Headlight Dimmer/Flasher Relay (53)
- J1 - Turn Signal Relay (413)
- J2 - Emergency Flasher Relay (191)
- K - Instrument Cluster
- T1j - Single Connector, yellow
- T2b - 2-Pin Connector, black, behind fuse/relay panel
- T2k - 2-Pin Connector, brown, behind fuse/relay panel
- T4c - 4-Pin Connector, red
- T5d - 5-Pin Connector, black
- T6c - 6-Pin Connector
- T7a - 7-Pin Connector
- T10b - 10-Pin Connector
- T28 - 28-Pin Connector, white, on instrument cluster
- M7 - Right Front Turn Signal Light
- M19 - Right, Side Turn Signal Light
- (A32) - Plus connection (30), in instrument panel wiring harness
- (A51) - Wire connection (56), in instrument panel wiring harness
- (A157) - Connection (49a) (in instrument panel wiring harness)
- (B12) - Connector (49) in wiring harness steering column switch
- (C19) - Connector -1- (right turn signal), in wiring harness headlamp

Edition 09.01
USA.5403.04.21

Turn signal switch, headlight dimmer/flasher switch, turn signal relay, emergency flasher relay, right front/right side turn signal light

No. 18/4 — Wiring diagram — Cabrio

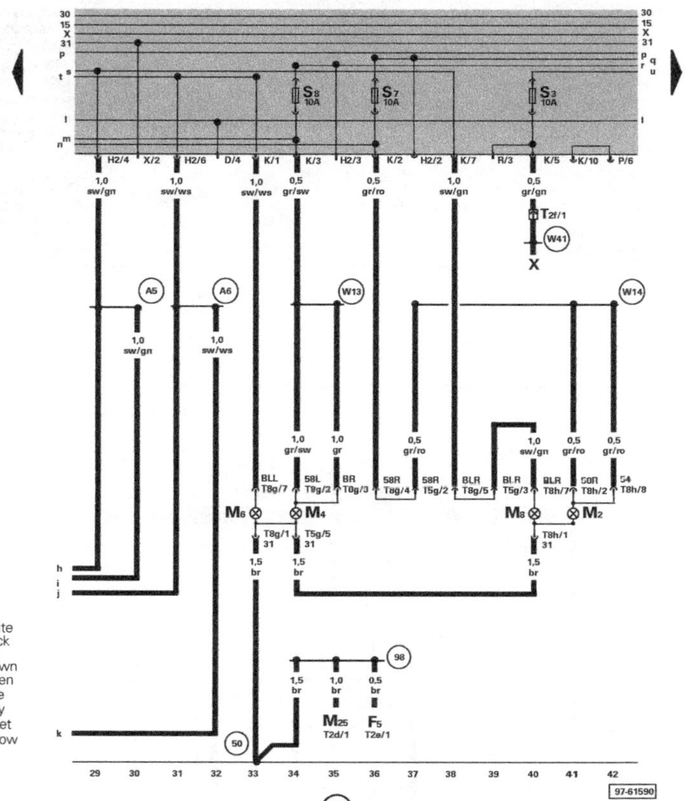

ws = white
sw = black
ro = red
br = brown
gn = green
bl = blue
gr = grey
li = violet
ge = yellow

- F5 - Luggage Compartment Light Switch
- M2 - Right Tail Light
- M4 - Left Tail Light
- M6 - Left Rear Turn Signal Light
- M8 - Right Rear Turn Signal Light
- M25 - High-mount Brake Light
- T2d - 2-Pin Connector
- T2e - 2-Pin Connector
- T2f - 2-Pin Connectorr, grey, near left taillight bulb holde
- T5g - 5-Pin Connector, on taillight, left
- T8g - 8-Pin Connector, on taillight, left
- T8h - 8-Pin Connector, on taillight, right
- X - License Plate Light
- (50) - Ground connection, in luggage compartment, left
- (98) - Ground connection, in rear lid wiring harness
- (A5) - plus connection (right turn signal), in instrument panel wiring harness
- (A6) - plus connection (left turn signal), in instrument panel wiring harness
- (W13) - Wire connection (58L),in rear wiring harness
- (W14) - Wire connection (58R), in rear wiring harness
- (W41) - Plus Connection (58) (license plate light wiring harness

Tail lights, rear turn signal lights

Edition 09.01
USA.5403.04.21

Cabrio — Wiring diagram — No. 18/5

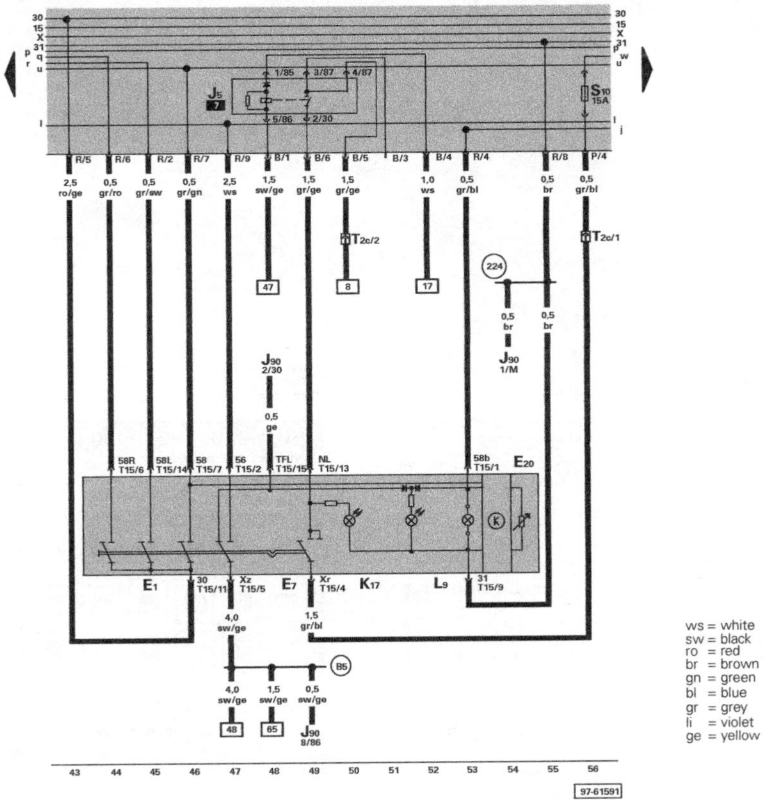

ws = white
sw = black
ro = red
br = brown
gn = green
bl = blue
gr = grey
li = violet
ge = yellow

E1 - Light switch
E7 - Fog Light Switch
E20 - Instrument Panel Light Dimmer Switch
J5 - Fog Light Relay (96)
J90 - Daytime Running Lights Switch-On Relay (173)
K17 - Fog Light Indicator Light
L9 - Headlight Switch Light
T2c - 2-Pin Connector, black, behind fuse/relay panel

(B5) - Wire connection (X), in light switch wiring harness
(224) - Ground connection, in light switch wiring harness

No. 18/6 — Wiring diagram — Cabrio

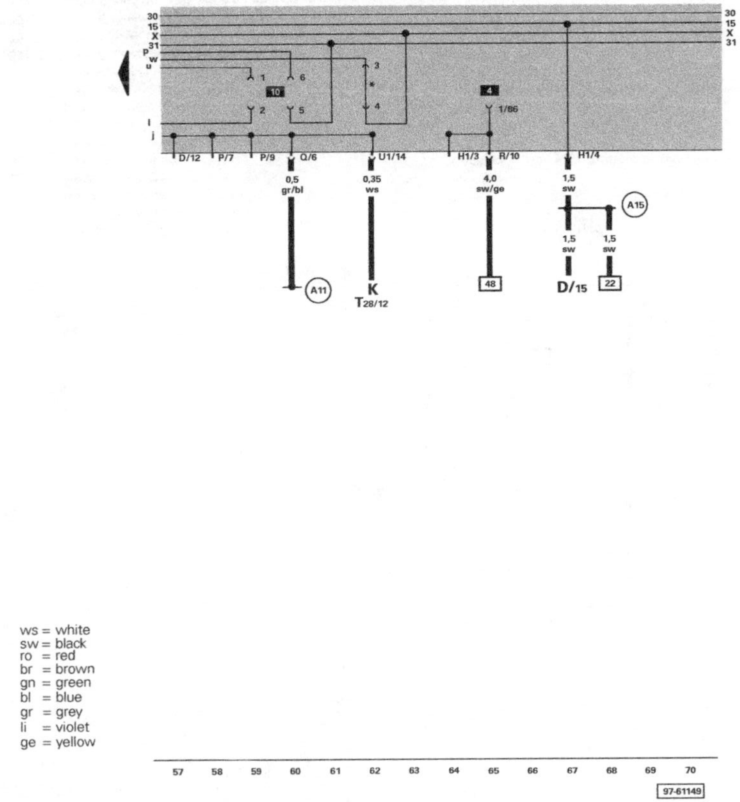

ws = white
sw = black
ro = red
br = brown
gn = green
bl = blue
gr = grey
li = violet
ge = yellow

D - Ignition/Starter Switch
K - Instrument Cluster
T28 - 28-Pin Connector, white, on instrument cluster

(A11) - Plus connection (58b) in instrument panel wiring harness
(A15) - Plus connection (15) in instrument panel wiring harness

* - Metalic bridge (only for vehicles with fog lights) (Cabrio GLX)

Cabrio — Wiring diagram — No. 19/1

Daytime running lights

from September 1999

Relay locations

Position of relays and multi-pin connector wiring, see also "Component Locations" section.

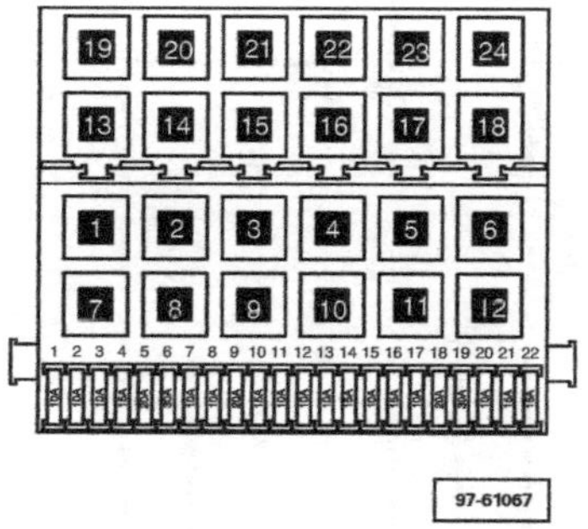

- **4** - Load reduction relay (18)
- **7** - Fog light relay (96)
- **18** - Daytime running lights switch-no relay (173)

Note:
Number in parentheses indicates production control number stamped on relay housing.

Fuse colors

- 30 A - green
- 25 A - white
- 20 A - yellow
- 15 A - blue
- 10 A - red
- 7,5 A - brown
- 5 A - beige
- 3 A - violet

Daytime running lights (from September 1999)

No. 19/2 — Wiring diagram — Cabrio

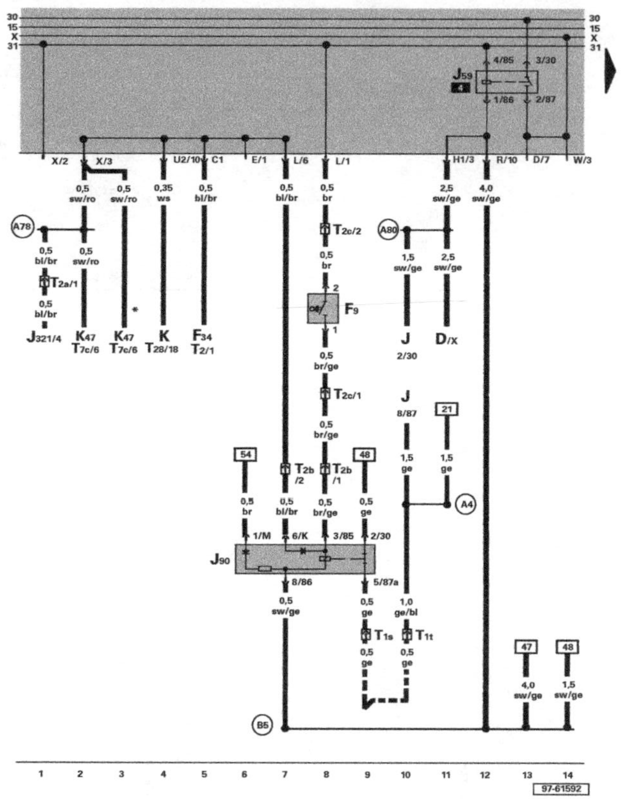

- ws = white
- sw = black
- ro = red
- br = brown
- gn = green
- bl = blue
- gr = grey
- li = violet
- ge = yellow

- D - Ignition/Starter Switch
- F9 - Parking Brake Warning Light Switch
- F34 - Brake Fluid Level Warning Switch
- J - Headlight dimmer/flasher Relay (53)
- J59 - Load Reduction Relay (18)
- J90 - Daytime Running Lights Switch ON Relay (173)
- J321 - Convertible Top Hydraulic Pump Relay (53)
- K - Instrument Cluster
- K47 - ABS Warning Light
- T1s - 1-Pin Connector, yellow, behind fuse relay panel
- T1t - 1-Pin Connector, yellow, behind fuse relay panel
- T2 - 2-Pin Connector, behind fuse relay panel
- T2a - 2-Pin Connector, green, behind fuse relay panel
- T2b - 2-Pin Connector, red, behind fuse relay panel
- T2c - 2-Pin Connector, black, on Parking Brake
- T7c - 7-Pin Connector, black
- T28 - 28-Pin Connector, white
- A4 - Plus connection (58b), in instrument panel wiring harness
- A78 - Connector (SU), in instrument panel wiring harness
- A80 - Connector -1- (X) , in instrument panel wiring harness
- B5 - Wire connection (X), in light switch wiring harness
- * - manual convertible top
- - - - - resistive wiring harness

Daytime running lights switch ON relay, parking brake warning light switch

Cabrio — Wiring diagram — No. 19/3

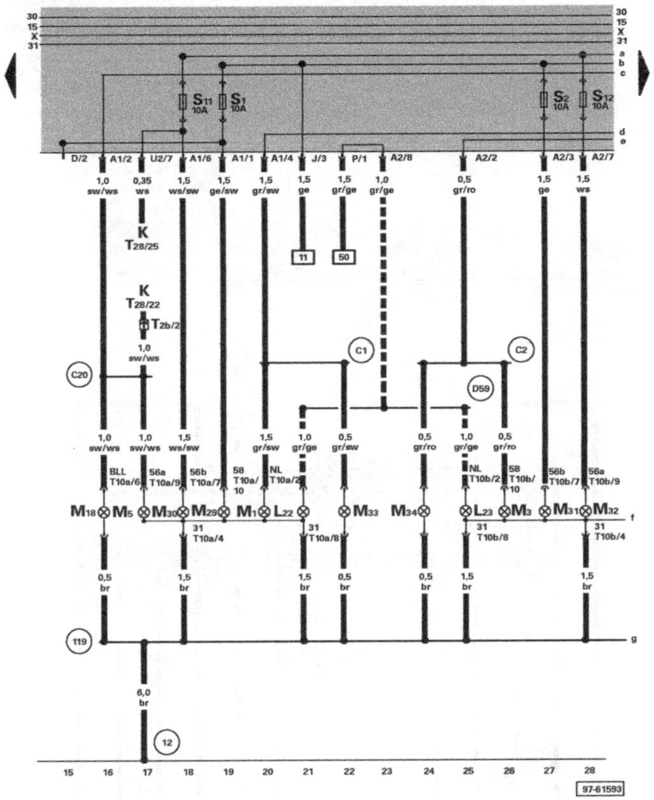

ws = white
sw = black
ro = red
br = brown
gn = green
bl = blue
gr = grey
li = violet
ge = yellow

- K - Instrument Cluster
- L 22 - Left Front Fog Light
- L 23 - Right Front Fog Light
- M1 - Left Parking Light
- M3 - Right Parking Light
- M5 - Left Front Turn Signal Light
- M18 - Left, Side Turn Signal Light
- M29 - Left Low Beam Headlight
- M30 - Left High Beam Headlight
- M31 - Right Low Beam Headlight
- M32 - Right High Beam Headlight
- M33 - Light for Side marker front left
- M34 - Light for Side marker front right
- T2b - 2-Pin Connector, black, behind fuse/relay panel
- T10a - 10-Pin Connector
- T10b - 10-Pin Connector
- T28 - 28-Pin Connector, white, on instrument cluster
- (12) - Ground connection, in engine compartment, left
- (119) - Ground connection -1-, in headlight wiring harness
- (C1) - Plus connection (58L), in left front wiring harness
- (C2) - Plus connection (58R), in left front wiring harness
- (C20) - Connection (left turn signal), in headlight wiring harness
- (D59) - Wire connection (Fog light), in engine compartment wiring harness
- --- - Only for vehicles with fog lights (Cabrio GLX)

Edition 09.01
USA.5403.04.21

Headlights, fog lights, parking lights, turn signal lights, side marker lights

1999-2002

No. 19/4 — Wiring diagram — Cabrio

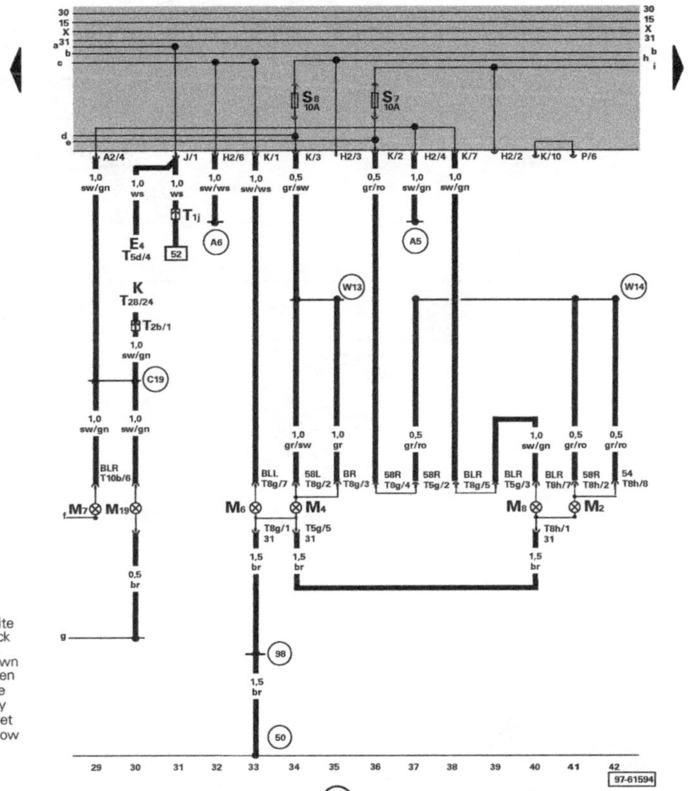

ws = white
sw = black
ro = red
br = brown
gn = green
bl = blue
gr = grey
li = violet
ge = yellow

- E4 - Headlight Dimmer/Flasher Switch
- K - Instrument Cluster
- M7 - Right Front Turn Signal Light
- M19 - Right, Side Turn Signal Light
- M2 - Right Tail Light
- M4 - Left Tail Light
- M6 - Left Rear Turn Signal Light
- M8 - Right Rear Turn Signal Light
- T1j - Single Connector, yellow
- T5d - 5-Pin Connector, black
- T5g - 5-Pin Connector, on taillight, left
- T8g - 8-Pin Connector, on taillight, left
- T8h - 8-Pin Connector, on taillight, right
- T28 - 28-Pin Connector, white, on instrument cluster
- (50) - Ground connection, in luggage compartment, left
- (98) - Ground connection, in rear lid wiring harness
- (A5) - plus connection (right turn signal), in instrument panel wiring harness
- (A6) - plus connection (left turn signal), in instrument panel wiring harness
- (C19) - Connector -1- (right turn signal), in wiring harness headlamp
- (W13) - Wire connection (58L),in rear wiring harness
- (W14) - Wire connection (58R), in rear wiring harness

Tail lights, rear turn signal lights

Edition 09.01
USA.5403.04.21

1999-2002

Cabrio — Wiring diagram — No. 19/5

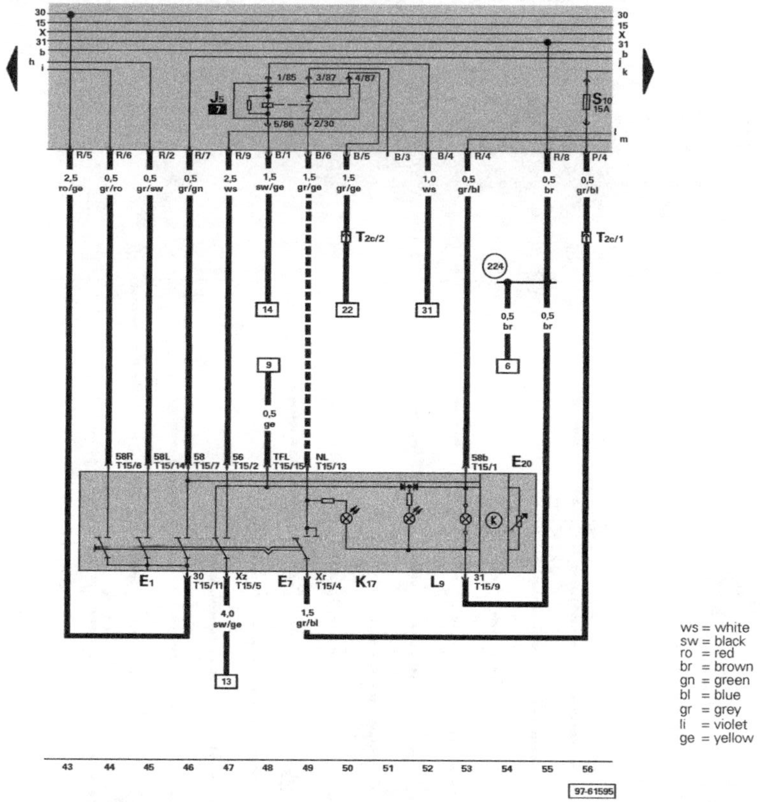

ws = white
sw = black
ro = red
br = brown
gn = green
bl = blue
gr = grey
li = violet
ge = yellow

E1 - Light switch
E7 - Fog Light Switch
E20 - Instrument Panel Light Dimmer Switch
J5 - Fog Light Relay (96)
K17 - Fog Light Indicator Light
L9 - Headlight Switch Light
T2c - 2-Pin Connector, black, behind fuse/relay panel

(224) - Ground connection, in light switch wiring harness

--- Only for vehicles with fog lights (Cabrio GLX)

No. 19/6 — Wiring diagram — Cabrio

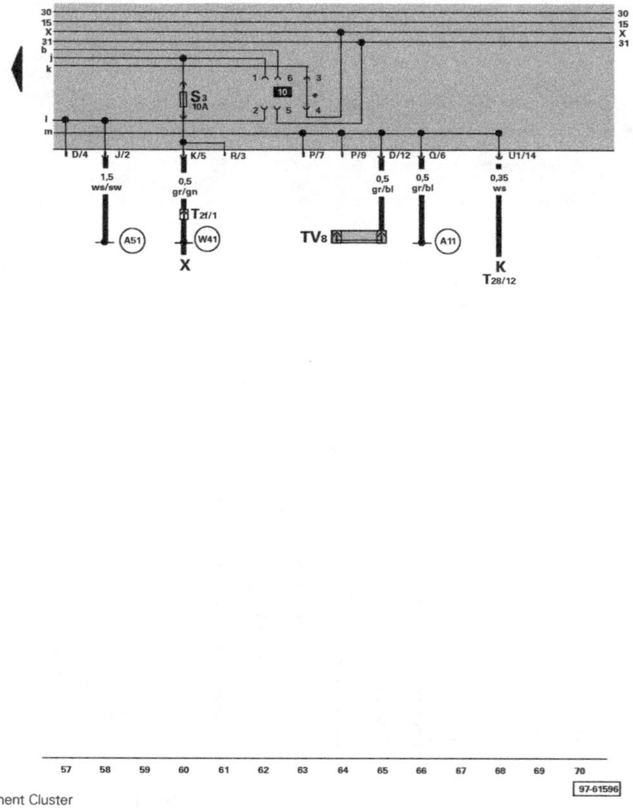

ws = white
sw = black
ro = red
br = brown
gn = green
bl = blue
gr = grey
li = violet
ge = yellow

K - Instrument Cluster
T2f - 2-Pin Connectorr, grey, near left taillight bulb holde
T28 - 28-Pin Connector, white, on instrument cluster
TV8 - Terminal 58b Wire Connector, grey
X - License Plate Light

(A11) - Plus connection (58b) in instrument panel wiring harness
(A51) - Wire connection (56), in instrument panel wiring harness
(W41) - Plus Connection (58) (license plate light wiring harness

* - Metalic bridge (only for vehicles with fog lights) (Cabrio GLX)

Cabrio — Wiring diagram — No. 20/1

Anti-lock Brake System (ABS)

from September 1999

Relay locations

Position of relays and multi-pin connector wiring, see also "Component Locations" section.

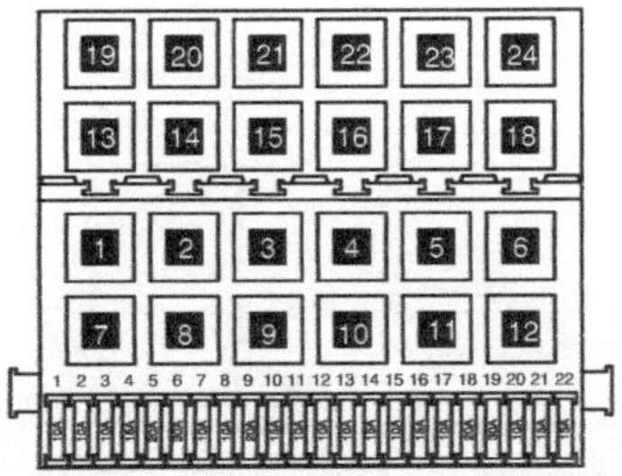

97-61067

Note:
Number in parentheses indicates production control number stamped on relay housing.

Fuse colors

30 A - green
25 A - white
20 A - yellow
15 A - blue
10 A - red
7,5 A - brown
5 A - beige
3 A - violet

Edition 09.01
USA.5403.04.21

No. 20/2 — Wiring diagram — Cabrio

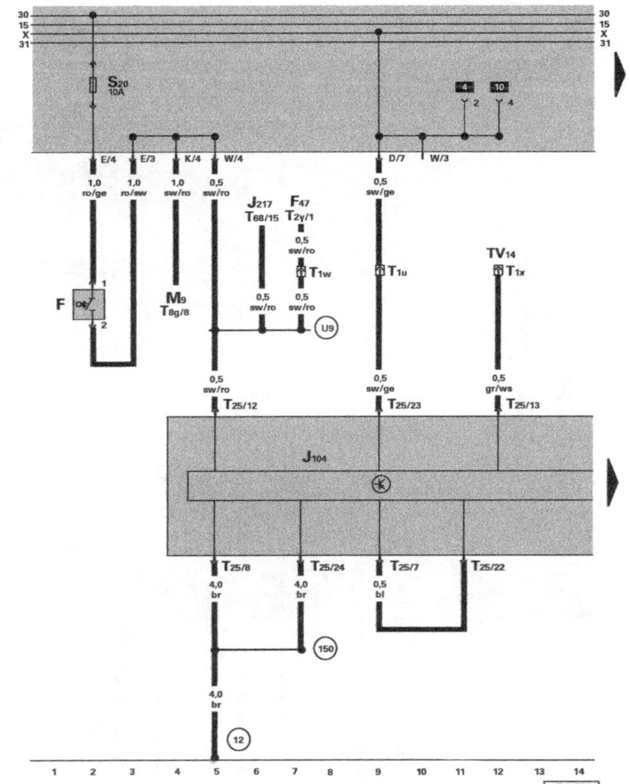

97-61597

ws = white
sw = black
ro = red
br = brown
gn = green
bl = blue
gr = grey
li = violet
ge = yellow

- F - Brake Light Switch
- F47 - Brake Pedal Switch (cruise control/Diesel Direct Fuel Injection)
- J104 - ABS Control Module (w/EDL)
- J217 - Transmission Control Module (TCM)
- M9 - Left Brake Light
- T1u - Single Connector, grey, behind fuse/relay panel
- T1w - Single Connector, green, behind fuse/relay panel
- T1x - Single Connector, grey, behind fuse/relay panel
- T2y - 2-Pin Connector, blue, on brake pedal bracket
- T8g - 8-Pin Connector, on taillight, left
- T25 - 25-Pin Connector
- T68 - 68-Pin Connector, on Transmission Control Module
- TV14 - Data Link Connector (DLC) wire Connector, black, above fuse relay panel

- (12) - Ground connection, in engine compartment, left
- (150) - Ground connection, in automatic transmission wiring harness
- (U9) - Connection (54) (in automatic transmission wiring harness)

Anti-lock brake system (ABS)

Edition 09.01
USA.5403.04.21

Anti-lock Brake System (ABS) (from September 1999)

Cabrio — Wiring diagram — No. 20/3

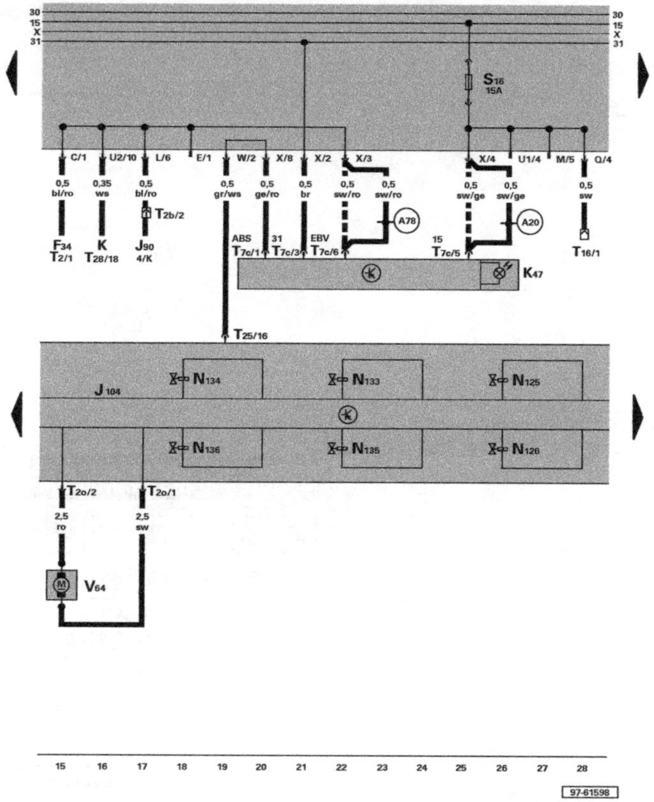

ws = white
sw = black
ro = red
br = brown
gn = green
bl = blue
gr = grey
li = violet
ge = yellow

F34 - Brake Fluid Level Warning Switch
J90 - Daytime Running Lights Switch ON Relay (173)
J104 - ABS Control Module (w/EDL)
K - Instrument Cluster
K47 - ABS Warning Light
N125 - Differential lock valve 1
N126 - Differential Lock Valve 2
N133 - Right Rear ABS Inlet Valve
N134 - Left Rear ABS Inlet Valve
N135 - Right Rear ABS Outlet Valve
N136 - Left Rear ABS Outlet Valve
T2 - 2-Pin Connector, black, in brake master cylinder fluid reservoir
T2b - 2-Pin Connector, red, behind fuse relay panel
T2o - 2-Pin Connector, black, in brake master cylinder
T7c - 7-Pin Connector
T16 - 16-Pin Connector, Data Link Connector (DLC)
T25 - 25-Pin Connector
T28 - 28-Pin Connector, white
V64 - ABS Hydraulic Pump
(A78) - Connector (SU), in instrument panel wiring harness
(A20) - Wire connection (15a), in instrument panel wiring harness
- - - - Manual convertible top

No. 20/4 — Wiring diagram — Cabrio

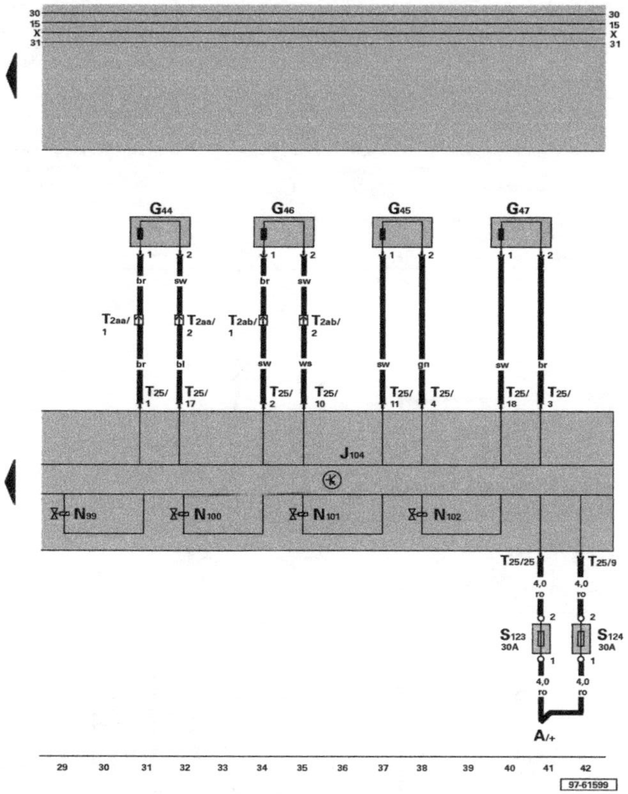

ws = white
sw = black
ro = red
br = brown
gn = green
bl = blue
gr = grey
li = violet
ge = yellow

A - Battery
G44 - Right Rear ABS Wheel Speed Sensor
G45 - Right Front ABS Wheel Speed Sensor
G46 - Left Rear ABS Wheel Speed Sensor
G47 - Left Front ABS Wheel Speed Sensor
J104 - ABS Control Module (w/EDL)
N99 - Right Front ABS Inlet Valve
N100 - Right Front ABS Outlet Valve
N101 - Left Front ABS Inlet Valve
N102 - Left Front ABS Outlet Valve
S123 - Fuse 1 for control module - ABS
S124 - Fuse 2 for control module - ABS
T25 - 25-Pin Connector
T2aa - 2-Pin Connector, black, beneath rear seat
T2ab - 2-Pin Connector, black, beneath rear seat

Cabrio — Wiring diagram — No. 21/1

Clutch interlock (manual transmission)

from September 1999

Relay locations
Position of relays and multi-pin connector wiring, see also "Component Locations" section.

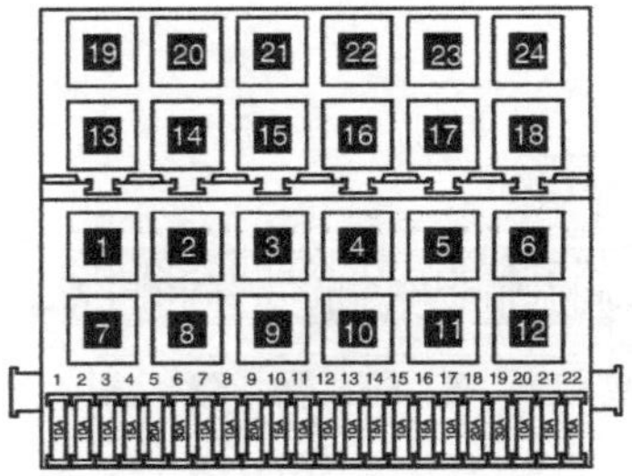

97-61067

22 - Anti-theft diseable relay (53)

Note:
Number in parentheses indicates production control number stamped on relay housing.

Fuse colors

30 A - green
25 A - white
20 A - yellow
15 A - blue
10 A - red
7,5 A - brown
5 A - beige
3 A - violet

No. 21/2 — Wiring diagram — Cabrio

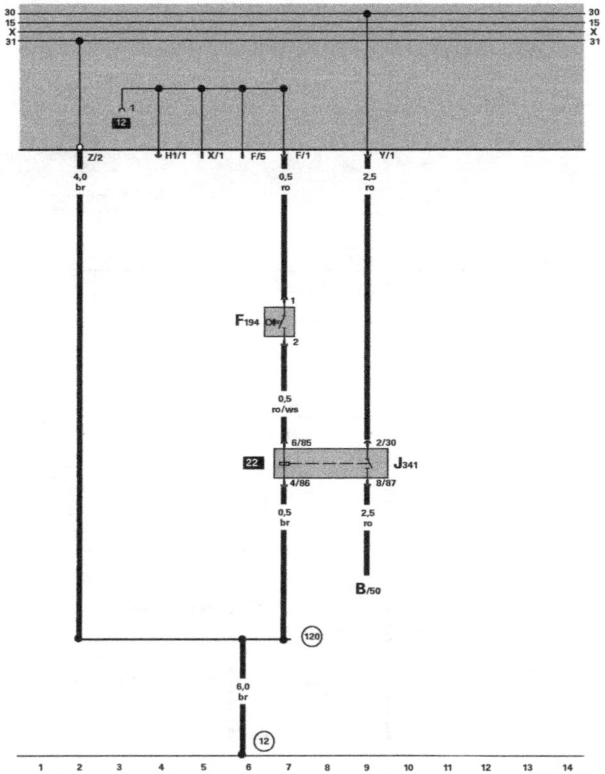

97-61600

ws = white
sw = black
ro = red
br = brown
gn = green
bl = blue
gr = grey
li = violet
ge = yellow

B - Starter
F194 - Clutch Pedal Position (CPP) Switch, green
J341 - Anti-Theft Engine Disable Relay I (53)

(12) - Ground connection, in engine compartment, left
(120) - Ground connection -2-, in head light wiring harness

Cabrio — Wiring diagram — No. 22/1

Heated washer nozzles

from September 1999

Relay locations

Position of relays and multi-pin connector wiring, see also "Component Locations" section.

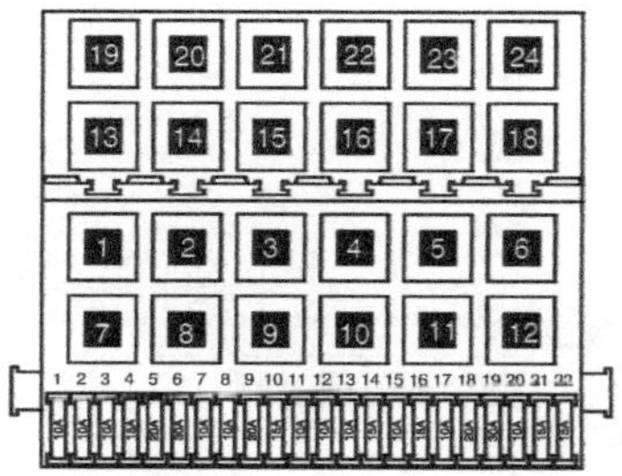

97-61067

8 - Wiper/Washer Intermitent Relay (196)

Note:
Number in parentheses indicates production control number stamped on relay housing.

Fuse colors

30 A - green
25 A - white
20 A - yellow
15 A - blue
10 A - red
7,5 A - brown
5 A - beige
3 A - violet

No. 22/2 — Wiring diagram — Cabrio

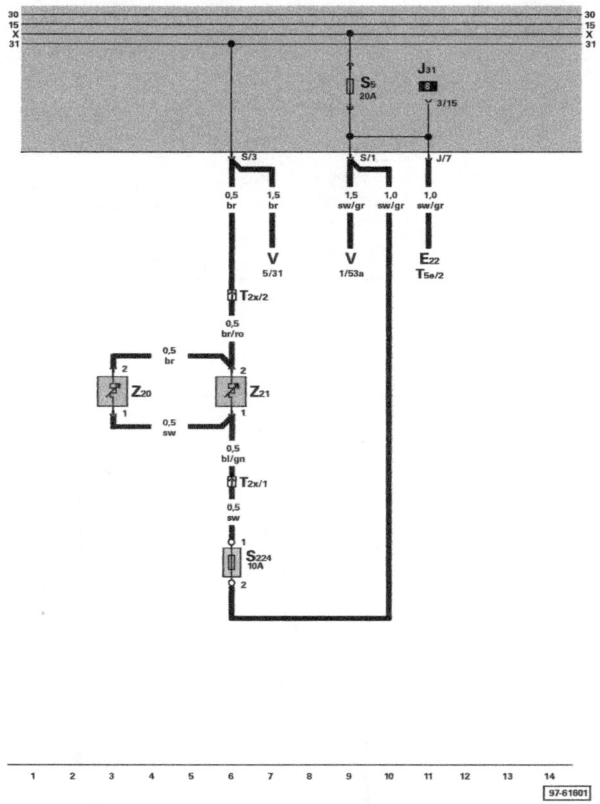

ws = white
sw = black
ro = red
br = brown
gn = green
bl = blue
gr = grey
li = violet
ge = yellow

97-61601

E22 - Windshield Wiper/Washer Switch
J31 - Wiper/Washer Intermittent Relay (196)
S224 - Fuse in fuse holder
T2x - 2-Pin Connector, black, in fresh air plenum, left side
T5e - 5-Pin Connector, black, in steering column
Z20 - Left Washer Nozzle Heater
Z21 - Right Washer Nozzle Heater
V - Windshield Wiper Motor

Cabrio — Wiring diagram — No. 23/1

Data Link Connector (DLC)

from September 1999

Relay locations

Position of relays and multi-pin connector wiring, see also "Component Locations" section.

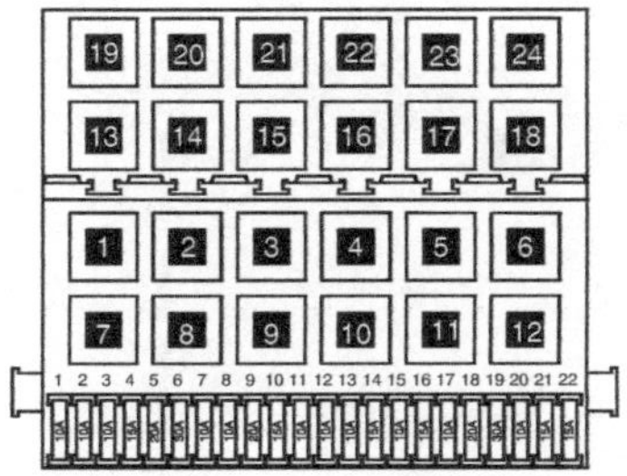

97-61067

Note:
Number in parentheses indicates production control number stamped on relay housing.

Fuse colors

30 A - green
25 A - white
20 A - yellow
15 A - blue
10 A - red
7,5 A - brown
5 A - beige
3 A - violet

Edition 09.01
USA.5403.04.21

Data Link Connector (DLC) (from September 1999)

No. 23/2 — Wiring diagram — Cabrio

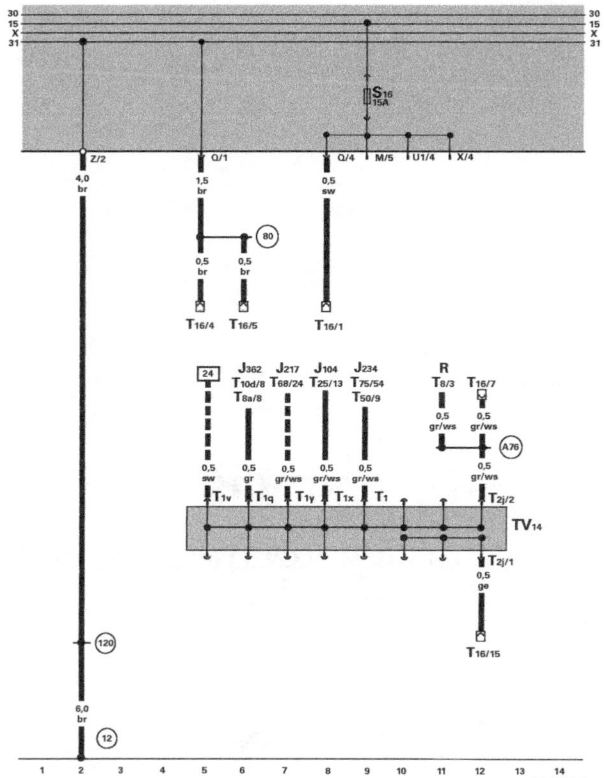

97-61602

ws = white
sw = black
ro = red
br = brown
gn = green
bl = blue
gr = grey
li = violet
ge = yellow

J104 - ABS Control Module (w/EDL)
J217 - Transmission Control Module (TCM)
J234 - Airbag Control Module
J362 - Control module for anti-theft immobilizer
R - Radio
T1 - Single connector, red
T1q - Single Connector, brown, behind fuse/ relay panel
T1v - Single Connector, grey, behind fuse/relay panel
T1x - Single Connector, grey, behind fuse/relay panel
T1y - Single Connector,yellow, behind fuse/relay panel
T2j - 2-Pin Connector, grey
T8 - 8-Pin Connector, in Radio
T8a - 8-Pin Connector, black
T10d - 10-Pin Connector, black (from 08.00 production)
T16 - 16-Pin Connector, Data Link Connector (DLC)
T25 - 25-Pin Connector
T50 - 50-Pin Connector
T75 - 75-Pin Connector (from 10.00 production)
T68 - 68-Pin Connector
TV14 - Data Link Connector (DLC) Wire Connector, black, above fuse relay panel
(12) - Ground connection, in engine compartment, left
(80) - Ground connection -1-, in instrument panel wiring harness
(120) - Ground connection -2-, in head light wiring harness
(A76) - Connector (k diagnosis wire) in instrument panel wiring harness
--- - Automatic transmission only

Data Link Connector (DLC)

Edition 09.01
USA.5403.04.21

Cabrio — Wiring diagram — No. 23/3

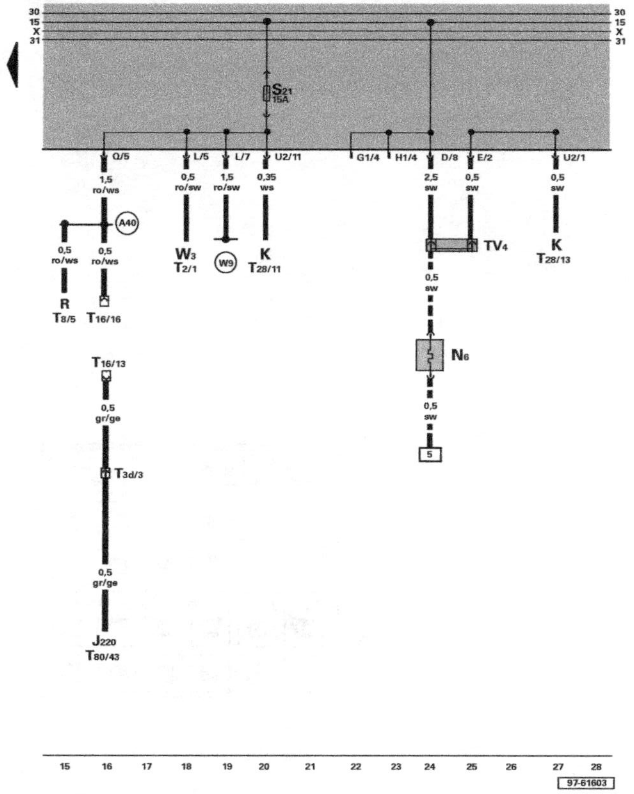

ws = white
sw = black
ro = red
br = brown
gn = green
bl = blue
gr = grey
li = violet
ge = yellow

J220 - Motronic Engine Control Module (ECM)
K - Instrument Cluster
N6 - Series Resistance
R - Radio
T2 - 2-Pin Connector, black
T3d - 3-Pin Connector, black, behind fuse/relay panel
T8 - 8-Pin Connector, black, in Radio
T16 - 16-Pin Connector, Data Link Connector (DLC)
T28 - 28-Pin Connector, white
T80 - 80-Pin Connector
TV4 - Terminal 15 Wire Connector, black
W3 - Luggage compartment Light

(A40) - Plus connection -1- (30), in instrument cluster wiring harness

(W9) - Plus connection 1 (30), in rear wiring harness

- - - - only automatic transmission

THIS PAGE INTENTIONALLY LEFT

BLANK

Cabrio — Wiring diagram — No. 24/1

Heated power mirrors

from September 1999

Relay locations

Position of relays and multi-pin connector wiring, see also "Component Locations" section.

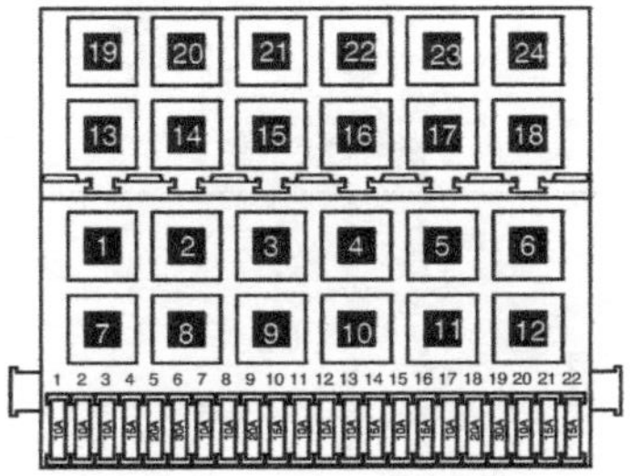

Note:
Number in parentheses indicates production control number stamped on relay housing.

Fuse colors

30 A - green
25 A - white
20 A - yellow
15 A - blue
10 A - red
7,5 A - brown
5 A - beige
3 A - violet

Edition 09.01
USA.5403.04.21

No. 24/2 — Wiring diagram — Cabrio

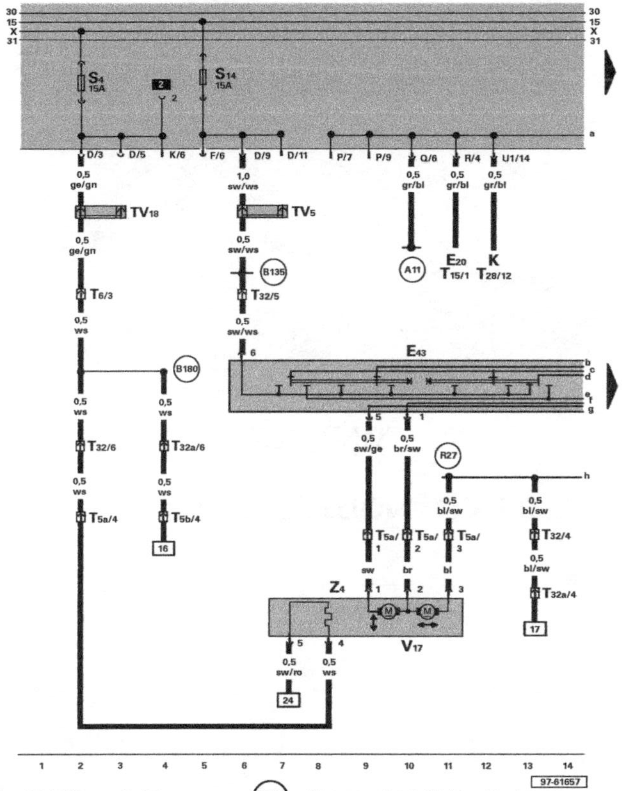

ws = white
sw = black
ro = red
br = brown
gn = green
bl = blue
gr = grey
li = violet
ge = yellow

E20 - Instrument Panel Light Dimmer Switch
E43 - Mirror Adjustment Switch
K - Instrument Cluster
T5a - 5-Pin Connector, on left door
T5b - 5-Pin Connector, on right door
T6 - 6-Pin Connector, black
T15 - 15-Pin Connector
T28 - 28-Pin Connector, white
T32 - 32 Pin Connector, white, on left A-pillar
T32a - 32-Pin Connector, white, on right A-pillar
TV5 - Terminal 15a wire connector, green
TV18 - Distributor for terminal Xa, yellow
V17 - Driver's Side Mirror Adjustment Motor
Z4 - Heated outside mirror, driver side

(A11) - Plus connection (58b) in instrument panel wiring harness

(B135) - Plus connector -1- (15a) in wiring harness interior
(B180) - Connection (heated rear window) in passenger compartment wiring harness
(R27) - Wire connection (power mirror), in driver's door wiring harness

Heated outside mirrors

Edition 09.01
USA.5403.04.21

Cabrio — Wiring diagram — No. 24/3

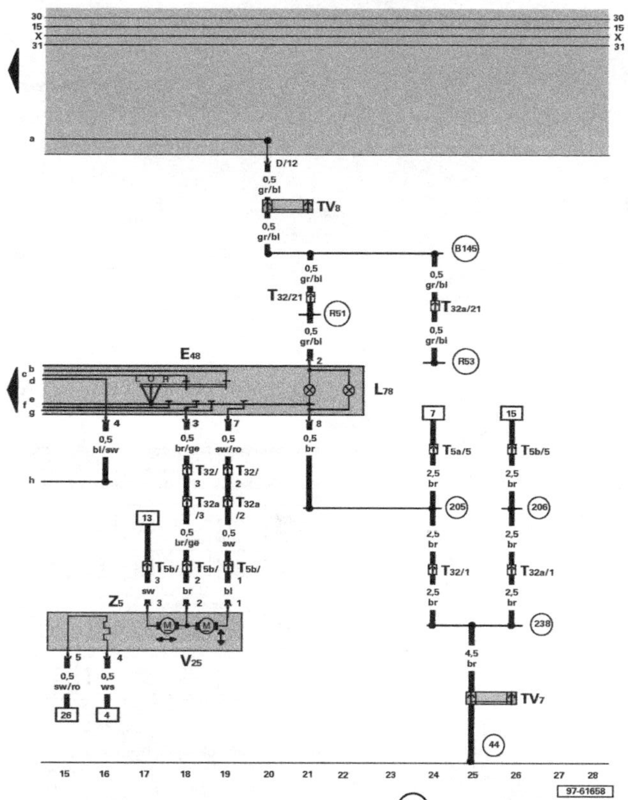

ws = white
sw = black
ro = red
br = brown
gn = green
bl = blue
gr = grey
li = violet
ge = yellow

- E48 - Mirror Selector Switch
- L78 - Mirror Adjusting Switch Light
- T5a - 5-Pin Connector, on left door
- T5b - 5-Pin Connector, on right door
- T32 - 32-Pin Connector, on A-pillar, lower left
- T32a - 32-Pin Connector, on A-pillar, lower right
- TV7 - Terminal 31 (GND) Wire Connector
- TV8 - Terminal 58b wire connector, grey
- V25 - Passengers Side Mirror Adjustment Motor
- Z5 - Heated outside mirror, passenger side

- (44) - Ground connection, on left A-pillar, lower part
- (205) - Ground connection, in driver's door wiring harness
- (206) - Ground connection, in passenger's door wiring harness harness
- (238) - Ground connection -1-, in wiring harness interior
- (B145) - Plus connector (58b), in wiring harness interior
- (R51) - Connection (58b) (driver's side door wiring harness)
- (R53) - Connection (58s) (in right front door wiring harness)

THIS PAGE INTENTIONALLY LEFT

BLANK

ST Scan Tool

GENERAL

Vehicles covered by this manual have On-Board Diagnostic capabilities (OBD). Generally, OBD is the ability of a system to monitor operation and identify and store information. This can include faults, operating data, and the ability to over-ride normal operation. The number of systems varies with each vehicle but can be determined by connecting a scan tool. For a complete list of all systems, see **VAG 1551/1552 Operating Flow Chart** on the last page of this section.

On-Board Diagnostic I (OBD I)

All vehicles through model year 1995 covered by this manual are equipped with engine management diagnostic systems known generally as On-Board Diagnostics I (OBD I). These early systems can be accessed by blink code or by a scan tool. Information on accessing these systems can be found in **24a Fuel Injection-Motronic** and **24b Fuel Injection-Mono-Motronic**.

On-Board Diagnostic II (OBD II)

All vehicles from model year 1996 covered by this manual are equipped with the Government required diagnostic system known as On-Board Diagnostics II (OBD II). This system monitors operation and function of all engine management system activity and automatic transmission operation to insure compliance with specified emission levels.

Vehicle emission levels are constantly monitored by the OBD II system and malfunctions are recognized and recorded. A Malfunction Indicator Light (MIL) in the instrument cluster alerts the driver to the fault and the need to have the system checked for fault codes. These codes follow a standard format and are known as Diagnostic Trouble Codes (DTCs).

DTCs are assigned two codes. The first code is referred to as a P-code and follows a structure required by law and defined by the Society of Automotive Engineers (SAE). This standard uses a letter to designate the system and four numbers to further identify and detail the malfunction as listed below.

First digit structure is as follows:

- Pxxxx for powertrain
- Bxxxx for body
- Cxxxx for chassis
- Uxxxx for future systems

Second digit structure is:

- P0xxx Government required codes
- P1xxx Manufacturer codes for additional emission system function; not required but reported to the government
- P3xxx Additional supplementary codes for expanded engine management supplied by the manufacturer and reported to the government

Third digit structure is:

- Px1xx measurement of air and fuel
- Px2xx measurement of air and fuel
- Px3xx ignition system
- Px4xx additional emission control
- Px5xx speed and idle regulation
- Px6xx computer and output signals
- Px7xx transmission
- Px8xx transmission
- Px9xx control modules, input and output signals

The fourth and fifth digits designate the individual components and systems.

The second code is a numerical code assigned by the factory. Each P-code has a corresponding factory code (VAG code) that follows manufacturer's definitions.

DTCs and data can be retrieved with Volkswagen Factory Scan Tools such as the VAG 1551, VAG 1552, or diagnostic computers VAS 5051 or VAS 5052. See Fig. 1. Several aftermarket scan tools and computer programs are also very capable of accurately retrieving this same information in the factory mode. The factory mode also allows the scan tool to be used for other system diagnostic functions and information retrieval.

Some DTC information can also be retrieved in a generic mode. The generic mode is not as complete as the factory mode, but allows inexpensive commercially available scan tools to be used simply to read DTCs. Generic scan tool mode does not have the capability to retrieve the detailed information of a manufacturer-specific scan tool.

All information in both modes is retrieved through the Data Link Connector (DLC) located in the center console above the lighter. See Fig. 2.

0024478

Fig. 2. Data Link Connector (**arrow**) for OBD II vehicles. This style of DLC pre-dates the OBD II requirement with installation of this type beginning in 1994. Access is attained by removing ashtray and sliding cover to the left.

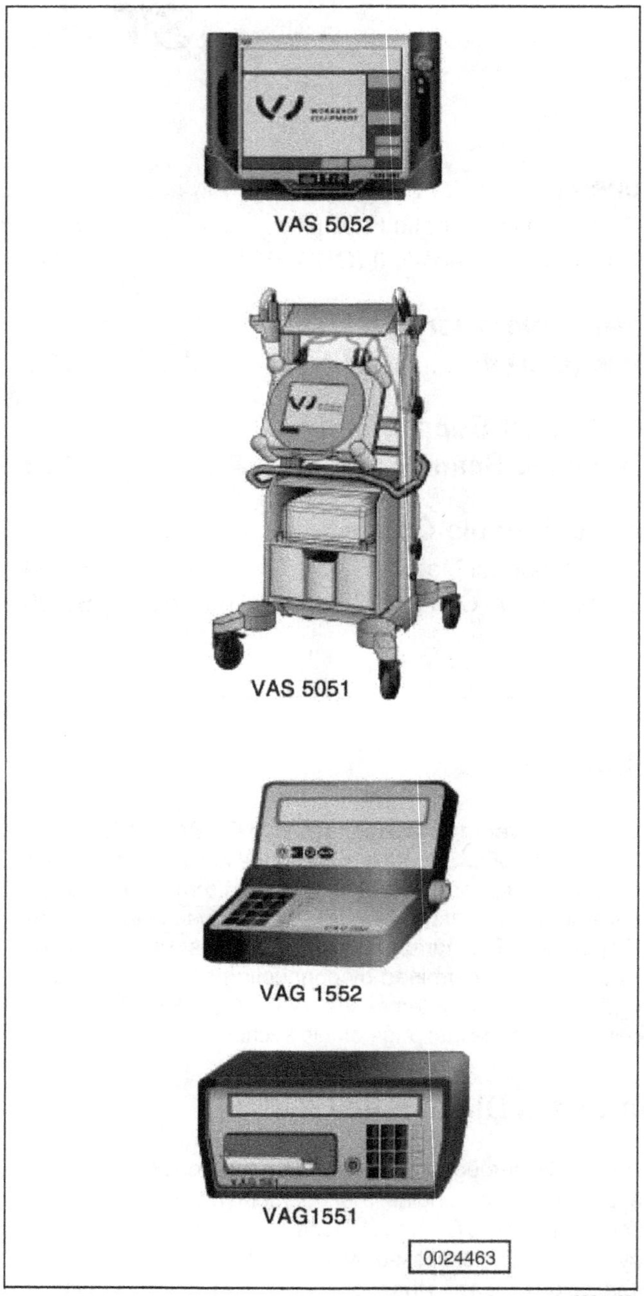

0024463

Fig. 1. Volkswagen Factory Scan Tools and Diagnostic Computers. (Not to shown to scale)

SCAN TOOL POWERTRAIN DTC TABLE (OBD II)

NOTE —
The following table contains a listing of OBD II Diagnostic Trouble Codes (DTC) available at the time of publication. This P-Code table is applicable to powertrains (engines and transmissions) installed in Volkswagen and Audi vehicles. Not all of the codes apply to the vehicles covered by this manual.

Scan Tool Diagnostic Trouble Codes - Powertrain

P-code	VAG code	Description
P0030	16414	O2 Sensor Heater Contr. Circ.(Bank1(1)Sensor 1)
P0031	16415	O2 Sensor Heater Contr. Circ.(Bank1(1)Sensor 1) Low
P0032	16417	O2 Sensor Heater Contr. Circ.(Bank1(1)Sensor 1) High
P0036	16420	O2 Sensor Heater Contr. Circ.(Bank1(1)Sensor 2)
P0037	16421	O2 Sensor Heater Contr. Circ.(Bank1(1)Sensor 2) Low
P0038	16422	O2 Sensor Heater Contr. Circ.(Bank1(1)Sensor 2) High
P0100	16484	Mass or Volume Air Flow Circuit
P0101	16485	Mass or Volume Air Flow Circ Range/Performance
P0102	16486	Mass or Volume Air Flow Circ Low Input
P0103	16487	Mass or Volume Air Flow Circ High Input
P0105	16489	Manifold Abs.Pressure or Bar.Pressure Voltage supply
P0106	16490	Manifold Abs.Pressure or Bar.Pressure Range/Performance
P0107	16491	Manifold Abs.Pressure or Bar.Pressure Low Input
P0108	16492	Manifold Abs.Pressure or Bar.Pressure High Input
P0112	16496	Intake Air Temp.Circ Low Input
P0113	16497	Intake Air Temp.Circ High Input
P0116	16500	Engine Coolant Temp.Circ Range/Performance
P0117	16501	Engine Coolant Temp.Circ Low Input
P0118	16502	Engine Coolant Temp.Circ High Input
P0120	16504	Throttle/Pedal Pos.Sensor A Circ Malfunction
P0121	16505	Throttle/Pedal Pos.Sensor A Circ Range/Performance
P0122	16506	Throttle/Pedal Pos.Sensor A Circ Low Input
P0123	16507	Throttle/Pedal Pos.Sensor A Circ High Input
P0125	16509	Insufficient Coolant Temp.for Closed Loop Fuel Control
P0128	16512	Coolant Thermostat/Valve Temperature below control range
P0130	16514	O2 Sensor Circ.,Bank1-Sensor1 Malfunction
P0131	16515	O2 Sensor Circ.,Bank1-Sensor1 Low Voltage
P0132	16516	O2 Sensor Circ.,Bank1-Sensor1 High Voltage

Scan Tool Diagnostic Trouble Codes - Powertrain

P-code	VAG code	Description
P0133	16517	O2 Sensor Circ.,Bank1-Sensor1 Slow Response
P0134	16518	O2 Sensor Circ.,Bank1-Sensor1 No Activity Detected
P0135	16519	O2 Sensor Heater Circ.,Bank1-Sensor1 Malfunction
P0136	16520	O2 Sensor Circ.,Bank1-Sensor2 Malfunction
P0137	16521	O2 Sensor Circ.,Bank1-Sensor2 Low Voltage
P0138	16522	O2 Sensor Circ.,Bank1-Sensor2 High Voltage
P0139	16523	O2 Sensor Circ.,Bank1-Sensor2 Slow Response
P0140	16524	O2 Sensor Circ.,Bank1-Sensor2 No Activity Detected
P0141	16525	O2 Sensor Heater Circ.,Bank1-Sensor2 Malfunction
P0170	16554	Fuel Trim,Bank1 Malfunction
P0171	16555	Fuel Trim,Bank1 System too Lean
P0172	16556	Fuel Trim,Bank1 System too Rich
P0182	16566	Fuel temperature sender-G81 Short to ground
P0183	16567	Fuel temperature sender-G81 Interruption/Short to B+
P0201	16585	Cyl.1, Injector Circuit Fault in electrical circuit
P0202	16586	Cyl.2, Injector Circuit Fault in electrical circuit
P0203	16587	Cyl.3, Injector Circuit Fault in electrical circuit
P0204	16588	Cyl.4, Injector Circuit Fault in electrical circuit
P0205	16589	Cyl.5 Injector Circuit Fault in electrical circuit
P0206	16590	Cyl.6 Injector Circuit Fault in electrical circuit
P0215	16599	Engine Shut-Off Solenoid Malfunction
P0216	16600	Injector/Injection Timing Control Malfunction
P0219	16603	Engine Overspeed Condition
P0221	16605	Throttle Pos. Sensor -B- Circuit Range/Performance
P0222	16606	Throttle Pos. Sensor -B- Circuit Low Input
P0223	16607	Throttle Pos. Sensor -B- Circuit High Input
P0225	16609	Throttle Pos. Sensor -C- Circuit Voltage supply
P0226	16610	Throttle Pos. Sensor -C- Circuit Range/Performance
P0227	16611	Throttle Pos. Sensor -C- Circuit Low Input
P0228	16612	Throttle Pos. Sensor -C- Circuit High Input
P0230	16614	Fuel Pump Primary Circuit Fault in electrical circuit
P0234	16618	Turbocharger Overboost Condition Control limit exceeded
P0235	16619	Turbocharger Boost Sensor (A) Circ Control limit not reached
P0236	16620	Turbocharger Boost Sensor (A) Circ Range/Performance
P0237	16621	Turbocharger Boost Sensor (A) Circ Low Input

SCAN TOOL POWERTRAIN DTC TABLE (OBD II)

Scan Tool Diagnostic Trouble Codes - Powertrain

P-code	VAG code	Description
P0238	16622	Turbocharger Boost Sensor (A) Circ High Input
P0243	16627	Turbocharger Wastegate Solenoid (A) Open/Short Circuit to Ground
P0245	16629	Turbocharger Wastegate Solenoid (A) Low Input/Short to ground
P0246	16630	Turbocharger Wastegate Solenoid (A) High Input/Short to B+
P0252	16636	Injection Pump Metering Control (A) Range/Performance
P0261	16645	Cyl.1 Injector Circuit Low Input/Short to ground
P0262	16646	Cyl.1 Injector Circuit High Input/Short to B+
P0264	16648	Cyl.2 Injector Circuit Low Input/Short to ground
P0265	16649	Cyl.2 Injector Circuit High Input/Short to B+
P0267	16651	Cyl.3 Injector Circuit Low Input/Short to ground
P0268	16652	Cyl.3 Injector Circuit High Input/Short to B+
P0270	16654	Cyl.4 Injector Circuit Low Input/Short to ground
P0271	16655	Cyl.4 Injector Circuit High Input/Short to B+
P0273	16657	Cyl.5 Injector Circuit Low Input/Short to ground
P0274	16658	Cyl.5 Injector Circuit High Input/Short to B+
P0276	16660	Cyl.6 Injector Circuit Low Input/Short to ground
P0277	16661	Cyl.6 Injector Circuit High Input/Short to B+
P0300	16684	Random/Multiple Cylinder Misfire Detected
P0301	16685	Cyl.1 Misfire Detected
P0302	16686	Cyl.2 Misfire Detected
P0303	16687	Cyl.3 Misfire Detected
P0304	16688	Cyl.4 Misfire Detected
P0305	16689	Cyl.5 Misfire Detected
P0306	16690	Cyl.6 Misfire Detected
P0313	16697	Misfire Detected Low Fuel Level
P0314	16698	Single Cylinder Misfire
P0321	16705	Ign./Distributor Eng.Speed Inp.Circ Range/Performance
P0322	16706	Ign./Distributor Eng.Speed Inp.Circ No Signal
P0325	16709	Knock Sensor 1 Circuit Electrical Fault in Circuit
P0326	16710	Knock Sensor 1 Circuit Range/Performance
P0327	16711	Knock Sensor 1 Circ Low Input
P0328	16712	Knock Sensor 1 Circ High Input
P0332	16716	Knock Sensor 2 Circ Low Input
P0333	16717	Knock Sensor 2 Circ High Input
P0335	16719	Crankshaft Pos. Sensor (A) Circ Malfunction

Scan Tool Diagnostic Trouble Codes - Powertrain

P-code	VAG code	Description
P0336	16720	Crankshaft Pos. Sensor (A) Circ Range/Performance/Missing tooth
P0337	16721	Crankshaft Pos.Sensor (A) Circ Low Input
P0340	16724	Camshaft Pos. Sensor (A) Circ Incorrect allocation
P0341	16725	Camshaft Pos.Sensor Circ Range/Performance
P0342	16726	Camshaft Pos.Sensor Circ Low Input
P0343	16727	Camshaft Pos.Sensor Circ High Input
P0351	16735	Ignition Coil (A) Cyl.1 Prim./Sec. Circ Malfunction
P0352	16736	Ignition Coil (B) Cyl.2 Prim./Sec. Circ Malfunction
P0353	16737	Ignition Coil (C) Cyl.3 Prim./Sec. Circ Malfunction
P0354	16738	Ignition Coil (D) Cyl.4 Prim./Sec. Circ Malfunction
P0355	16739	Ignition Coil (E) Cyl.5 Prim./Sec. Circ Malfunction
P0356	16740	Ignition Coil (F) Cyl.6 Prim./Sec. Circ Malfunction
P0380	16764	Glow Plug/Heater Circuit (A) Electrical Fault in Circuit
P0400	16784	Exhaust Gas Recirc.Flow Malfunction
P0401	16785	Exhaust Gas Recirc.Flow Insufficient Detected
P0402	16786	Exhaust Gas Recirc.Flow Excessive Detected
P0403	16787	Exhaust Gas Recirc. Contr. Circ Malfunction
P0404	16788	Exhaust Gas Recirc. Contr. Circ Range/Performance
P0405	16789	Exhaust Gas Recirc. Sensor (A) Circ Low Input
P0406	16790	Exhaust Gas Recirc. Sensor (A) Circ High Input
P0407	16791	Exhaust Gas Recirc. Sensor (B) Circ Low Input
P0408	16792	Exhaust Gas Recirc. Sensor (B) Circ High Input
P0410	16794	Sec.Air Inj.Sys Malfunction
P0411	16795	Sec.Air Inj.Sys. Incorrect Flow Detected
P0412	16796	Sec.Air Inj.Sys.Switching Valve A Circ Malfunction
P0418	16802	Sec. Air Inj. Sys. Relay (A) Contr. Circ Malfunction
P0420	16804	Catalyst System,Bank1 Efficiency Below Threshold
P0422	16806	Main Catalyst,Bank1 Below Threshold
P0427	16811	Catalyst Temperature Sensor, Bank 1 Low Input/Short to ground
P0428	16812	Catalyst Temperature Sensor, Bank 1 High Input/Open/Short Circuit to B+
P0440	16824	EVAP Emission Contr.Sys. Malfunction
P0441	16825	EVAP Emission Contr.Sys.Incorrect Purge Flow
P0442	16826	EVAP Emission Contr.Sys.(Small Leak) Leak Detected
P0443	16827	EVAP Emiss. Contr. Sys. Purge Valve Circ Electrical Fault in Circuit
P0452	16836	EVAP Emission Contr.Sys.Press.Sensor Low Input

SCAN TOOL POWERTRAIN DTC TABLE (OBD II)

Scan Tool Diagnostic Trouble Codes - Powertrain

P-code	VAG code	Description
P0453	16837	EVAP Emission Contr.Sys.Press.Sensor High Input
P0455	16839	EVAP Emission Contr.Sys.(Gross Leak) Leak Detected
P0461	16845	Fuel Level Sensor Circ Range/Performance
P0462	16846	Fuel Level Sensor Circuit Low Input
P0463	16847	Fuel Level Sensor Circuit High Input
P0501	16885	Vehicle Speed Sensor Range/Performance
P0503	16887	Vehicle Speed Sensor Intermittent/Erratic/High Input
P0505	16889	Idle Control System Malfunction
P0506	16890	Idle Control System RPM Lower than Expected
P0507	16891	Idle Control System Higher than Expected
P0510	16894	Closed Throttle Pos.Switch Malfunction
P0531	16915	A/C Refrigerant Pressure Sensor Circuit Range/Performance
P0532	16916	A/C Refrigerant Pressure Sensor Circuit Low Input
P0533	16917	A/C Refrigerant Pressure Sensor Circuit High Input
P0551	16935	Power Steering Pressure Sensor Circuit Range/Performance
P0560	16944	System Voltage Malfunction
P0562	16946	System Voltage Low Voltage
P0563	16947	System Voltage High Voltage
P0568	16952	Cruise Control Set Signal Incorrect Signal
P0571	16955	Cruise/Brake Switch (A) Circ Malfunction
P0600	16984	Serial Comm. Link (Data Bus) Message Missing
P0601	16985	Internal Contr.Module Memory Check Sum Error
P0602	16986	Control Module Programming Error/Malfunction
P0603	16987	Internal Contr.Module (KAM) Error
P0604	16988	Internal Contr.Module Random Access Memory (RAM) Error
P0605	16989	Internal Contr.Module ROM Test Error
P0606	16990	ECM/PCM Processor
P0642	17026	Knock Control Control Module Malfunction
P0645	17029	A/C Clutch Relay Control Circuit
P0650	17034	MIL Control Circuit Electrical Fault in Circuit
P0654	17038	Engine RPM Output Circuit Electrical Fault in Circuit
P0656	17040	Fuel Level Output Circuit Electrical Fault in Circuit
P0700	17084	Transm.Contr.System Malfunction
P0702	17086	Transm.Contr.System Electrical
P0703	17087	Torque Converter/Brake Switch B Circ Malfunction

Scan Tool Diagnostic Trouble Codes - Powertrain

P-code	VAG code	Description
P0705	17089	Transm.Range Sensor Circ.(PRNDL Inp.) Malfunction
P0706	17090	Transm.Range Sensor Circ Range/Performance
P0707	17091	Transm.Range Sensor Circ Low Input
P0708	17092	Transm.Range Sensor Circ High Input
P0710	17094	Transm.Fluid Temp.Sensor Circ. Malfunction
P0711	17095	Transm.Fluid Temp.Sensor Circ. Range/Performance
P0712	17096	Transm.Fluid Temp.Sensor Circ. Low Input
P0713	17097	Transm.Fluid Temp.Sensor Circ. High Input
P0715	17099	Input Turbine/Speed Sensor Circ. Malfunction
P0716	17100	Input Turbine/Speed Sensor Circ. Range/Performance
P0717	17101	Input Turbine/Speed Sensor Circ. No Signal
P0721	17105	Output Speed Sensor Circ Range/Performance
P0722	17106	Output Speed Sensor Circ No Signal
P0725	17109	Engine Speed Inp.Circ. Malfunction
P0726	17110	Engine Speed Inp.Circ. Range/Performance
P0727	17111	Engine Speed Inp.Circ. No Signal
P0730	17114	Gear Incorrect Ratio
P0731	17115	Gear 1 Incorrect Ratio
P0732	17116	Gear 2 Incorrect Ratio
P0733	17117	Gear 3 Incorrect Ratio
P0734	17118	Gear 4 Incorrect Ratio
P0735	17119	Gear 5 Incorrect Ratio
P0740	17124	Torque Converter Clutch Circ Malfunction
P0741	17125	Torque Converter Clutch Circ Performance or Stuck Off
P0748	17132	Pressure Contr.Solenoid Electrical
P0750	17134	Shift Solenoid A malfunction
P0751	17135	Shift Solenoid A Performance or Stuck Off
P0752	17136	Shift Solenoid A Stuck On
P0753	17137	Shift Solenoid A Electrical
P0756	17140	Shift Solenoid B Performance or Stuck Off
P0757	17141	Shift Solenoid B Stuck On
P0758	17142	Shift Solenoid B Electrical
P0761	17145	Shift Solenoid C Performance or Stuck Off
P0762	17146	Shift Solenoid C Stuck On
P0763	17147	Shift Solenoid C Electrical

SCAN TOOL POWERTRAIN DTC TABLE (OBD II)

Scan Tool Diagnostic Trouble Codes - Powertrain

P-code	VAG code	Description
P0768	17152	Shift Solenoid D Electrical
P0773	17157	Shift Solenoid E Electrical
P0790	17174	Normal/Performance Switch Circ Malfunction
P1101	17509	O2 Sensor Circ.,Bank1-Sensor1Voltage too Low/Air Leak
P1102	17510	O2 Sensor Heating Circ.,Bank1-Sensor1 Short to B+
P1103	17511	O2 Sensor Heating Circ.,Bank1-Sensor1 Output too Low
P1104	17512	Bank1-Sensor2 Voltage too Low/Air Leak
P1105	17513	O2 Sensor Heating Circ.,Bank1-Sensor2 Short to B+
P1111	17519	O2 Control (Bank 1) System too lean
P1112	17520	O2 Control (Bank 1) System too rich
P1113	17521	Bank1-Sensor1 Internal Resistance too High
P1114	17522	Bank1-Sensor2 Internal Resistant too High
P1115	17523	O2 Sensor Heater Circ.,Bank1-Sensor1 Short to Ground
P1116	17524	O2 Sensor Heater Circ.,Bank1-Sensor1 Open
P1117	17525	O2 Sensor Heater Circ.,Bank1-Sensor2 Short to Ground
P1118	17526	O2 Sensor Heater Circ.,Bank1-Sensor2 Open
P1123	17531	Long Term Fuel Trim Add.Air.,Bank1 System too Rich
P1124	17532	Long Term Fuel Trim Add.Air.,Bank1 System too Lean
P1127	17535	Long Term Fuel Trim mult.,Bank1 System too Rich
P1128	17536	Long Term Fuel Trim mult.,Bank1 System too Lean
P1131	17539	Bank2-Sensor1 Internal Resistance too High
P1132	17540	O2 Sensor Heating Circ.,Bank1+2-Sensor1 Short to B+
P1133	17541	O2 Sensor Heating Circ.,Bank1+2-Sensor1 Electrical Malfunction
P1134	17542	O2 Sensor Heating Circ.,Bank1+2-Sensor2 Short to B+
P1135	17543	O2 Sensor Heating Circ.,Bank1+2-Sensor2 Electrical Malfunction
P1136	17544	Long Term Fuel Trim Add.Fuel,Bank1 System too Lean
P1137	17545	Long Term Fuel Trim Add.Fuel,Bank1 System too Rich
P1141	17549	Load Calculation Cross Check Range/Performance
P1142	17550	Load Calculation Cross Check Lower Limit Exceeded
P1143	17551	Load Calculation Cross Check Upper Limit Exceeded
P1144	17552	Mass or Volume Air Flow Circ Open/Short to Ground
P1145	17553	Mass or Volume Air Flow Circ Short to B+
P1146	17554	Mass or Volume Air Flow Circ Supply Malfunction
P1149	17557	O2 Control (Bank 1) Out of range
P1151	17559	Bank1, Long Term Fuel Trim, Range 1 Leanness Lower Limit Exceeded

Scan Tool Diagnostic Trouble Codes - Powertrain

P-code	VAG code	Description
P1152	17560	Bank1, Long Term Fuel Trim, Range 2 Leanness Lower Limit Exceeded
P1154	17562	Manifold Switch Over Malfunction
P1155	17563	Manifold Abs.Pressure Sensor Circ. Short to B+
P1156	17564	Manifold Abs.Pressure Sensor Circ. Open/Short to Ground
P1157	17565	Manifold Abs.Pressure Sensor Circ. Power Supply Malfunction
P1158	17566	Manifold Abs.Pressure Sensor Circ. Range/Performance
P1160	17568	Manifold Temp.Sensor Circ. Short to Ground
P1161	17569	Manifold Temp.Sensor Circ. Open/Short to B+
P1162	17570	Fuel Temp.Sensor Circ. Short to Ground
P1163	17571	Fuel Temp.Sensor Circ. Open/Short to B+
P1164	17572	Fuel Temperature Sensor Range/Performance/Incorrect Signal
P1165	17573	Bank1, Long Term Fuel Trim, Range 1 Rich Limit Exceeded
P1166	17574	Bank1, Long Term Fuel Trim, Range 2 Rich Limit Exceeded
P1171	17579	Throttle Actuation Potentiometer Sign.2 Range/Performance
P1172	17580	Throttle Actuation Potentiometer Sign.2 Signal too Low
P1173	17581	Throttle Actuation Potentiometer Sign.2 Signal too High
P1174	17582	Fuel Trim, Bank 1 Different injection times
P1176	17584	O2 Correction Behind Catalyst,B1 Limit Attained
P1177	17585	O2 Correction Behind Catalyst,B2 Limit Attained
P1178	17586	Linear 02 Sensor / Pump Current Open Circuit
P1179	17587	Linear 02 Sensor / Pump Current Short to ground
P1180	17588	Linear 02 Sensor / Pump Current Short to B+
P1181	17589	Linear 02 Sensor / Reference Voltage Open Circuit
P1182	17590	Linear 02 Sensor / Reference Voltage Short to ground
P1183	17591	Linear 02 Sensor / Reference Voltage Short to B+
P1184	17592	Linear 02 Sensor / Common Ground Wire Open Circuit
P1185	17593	Linear 02 Sensor / Common Ground Wire Short to ground
P1186	17594	Linear 02 Sensor / Common Ground Wire Short to B+
P1187	17595	Linear 02 Sensor / Compens. Resistor Open Circuit
P1188	17596	Linear 02 Sensor / Compens. Resistor Short to ground
P1189	17597	Linear 02 Sensor / Compens. Resistor Short to B+
P1190	17598	Linear 02 Sensor / Reference Voltage Incorrect Signal
P1196	17604	O2 Sensor Heater Circ.,Bank1-Sensor1 Electrical Malfunction
P1198	17606	O2 Sensor Heater Circ.,Bank1-Sensor2 Electrical Malfunction
P1201	17609	Cyl.1-Fuel Inj.Circ. Electrical Malfunction

SCAN TOOL POWERTRAIN DTC TABLE (OBD II)

Scan Tool Diagnostic Trouble Codes - Powertrain

P-code	VAG code	Description
P1202	17610	Cyl.2-Fuel Inj.Circ. Electrical Malfunction
P1203	17611	Cyl.3-Fuel Inj.Circ. Electrical Malfunction
P1204	17612	Cyl.4-Fuel Inj.Circ. Electrical Malfunction
P1205	17613	Cyl.5-Fuel Inj.Circ. Electrical Malfunction
P1206	17614	Cyl.6-Fuel Inj.Circ. Electrical Malfunction
P1209	17617	Intake valves for cylinder shut-off Short circuit to ground
P1210	17618	Intake valves for cylinder shut-off Short to B+
P1211	17619	Intake valves for cylinder shut-off Open circuit
P1213	17621	Cyl.1-Fuel Inj.Circ. Short to B+
P1214	17622	Cyl.2-Fuel Inj.Circ. Short to B+
P1215	17623	Cyl.3-Fuel Inj.Circ. Short to B+
P1216	17624	Cyl.4-Fuel Inj.Circ. Short to B+
P1217	17625	Cyl.5-Fuel Inj.Circ. Short to B+
P1218	17626	Cyl.6-Fuel Inj.Circ. Short to B+
P1221	17629	Cylinder shut-off exhaust valves Short circuit to ground
P1222	17630	Cylinder shut-off exhaust valves Short to B+
P1223	17631	Cylinder shut-off exhaust valves Open circuit
P1225	17633	Cyl.1-Fuel Inj.Circ. Short to Ground
P1226	17634	Cyl.2-Fuel Inj.Circ. Short to Ground
P1227	17635	Cyl.3-Fuel Inj.Circ. Short to Ground
P1228	17636	Cyl.4-Fuel Inj.Circ. Short to Ground
P1229	17637	Cyl.5-Fuel Inj.Circ. Short to Ground
P1230	17638	Cyl.6-Fuel Inj.Circ. Short to Ground
P1237	17645	Cyl.1-Fuel Inj.Circ. Open Circ.
P1238	17646	Cyl.2-Fuel Inj.Circ. Open Circ.
P1239	17647	Cyl.3-Fuel Inj.Circ. Open Circ.
P1240	17648	Cyl.4-Fuel Inj.Circ. Open Circ.
P1241	17649	Cyl.5-Fuel Inj.Circ. Open Circ.
P1242	17650	Cyl.6-Fuel Inj.Circ. Open Circ.
P1245	17653	Needle Lift Sensor Circ. Short to Ground
P1246	17654	Needle Lift Sensor Circ. Range/Performance
P1247	17655	Needle Lift Sensor Circ. Open/Short to B+
P1248	17656	Injection Start Control Deviation
P1249	17657	Fuel consumption signal Electrical Fault in Circuit
P1250	17658	Fuel Level Too Low

Scan Tool Diagnostic Trouble Codes - Powertrain

P-code	VAG code	Description
P1251	17659	Start of Injection Solenoid Circ Short to B+
P1252	17660	Start of Injection Solenoid Circ Open/Short to Ground
P1253	17661	Fuel consumption signal Short to ground
P1254	17662	Fuel consumption signal Short to B+
P1255	17663	Engine Coolant Temp.Circ Short to Ground
P1256	17664	Engine Coolant Temp.Circ Open/Short to B+
P1257	17665	Engine Coolant System Valve Open
P1258	17666	Engine Coolant System Valve Short to B+
P1259	17667	Engine Coolant System Valve Short to Ground
P1280	17688	Fuel Inj.Air Contr.Valve Circ. Flow too Low
P1283	17691	Fuel Inj.Air Contr.Valve Circ. Electrical Malfunction
P1284	17692	Fuel Inj.Air Contr.Valve Circ. Open
P1285	17693	Fuel Inj.Air Contr.Valve Circ. Short to Ground
P1286	17694	Fuel Inj.Air Contr.Valve Circ. Short to B+
P1287	17695	Turbocharger bypass valve open
P1288	17696	Turbocharger bypass valve short to B+
P1289	17697	Turbocharger bypass valve short to ground
P1296	17704	Cooling system malfunction
P1297	17705	Connection turbocharger - throttle valve pressure hose
P1300	17708	Misfire detected Reason: Fuel level too low
P1319	17721	Knock Sensor 1 Circ. Short to Ground
P1320	17728	Knock Sensor 2 Circ. Short to Ground
P1325	17733	Cyl.1-Knock Contr. Limit Attained
P1326	17734	Cyl.2-Knock Contr. Limit Attained
P1327	17735	Cyl.3-Knock Contr. Limit Attained
P1328	17736	Cyl.4-Knock Contr. Limit Attained
P1329	17737	Cyl.5-Knock Contr. Limit Attained
P1330	17738	Cyl.6-Knock Contr. Limit Attained
P1335	17743	Engine Torque Monitoring 2 Control Limit Exceeded
P1336	17744	Engine Torque Monitoring Adaptation at limit
P1337	17745	Camshaft Pos.Sensor,Bank1 Short to Ground
P1338	17746	Camshaft Pos.Sensor,Bank1 Open Circ./Short to B+
P1339	17747	Crankshaft Pos./Engine Speed Sensor Cross Connected
P1340	17748	Crankshaft-/Camshaft Pos.Sens.Signals Out of Sequence
P1341	17749	Ignition Coil Power Output Stage 1 Short to Ground

SCAN TOOL POWERTRAIN DTC TABLE (OBD II)

Scan Tool Diagnostic Trouble Codes - Powertrain

P-code	VAG code	Description
P1342	17750	Ignition Coil Power Output Stage 1 Short to B+
P1343	17751	Ignition Coil Power Output Stage 2 Short to Ground
P1344	17752	Ignition Coil Power Output Stage 2 Short to B+
P1345	17753	Ignition Coil Power Output Stage 3 Short to Ground
P1346	17754	Ignition Coil Power Output Stage 3 Short to B+
P1348	17756	Ignition Coil Power Output Stage 1 Open Circuit
P1349	17757	Ignition Coil Power Output Stage 2 Open Circuit
P1350	17758	Ignition Coil Power Output Stage 3 Open Circuit
P1354	17762	Modulation Piston Displ.Sensor Circ. Malfunction
P1355	17763	Cyl. 1, ignition circuit Open Circuit
P1356	17764	Cyl. 1, ignition circuit Short to B+
P1357	17765	Cyl. 1, ignition circuit Short to ground
P1358	17766	Cyl. 2, ignition circuit Open Circuit
P1359	17767	Cyl. 2, ignition circuit Short Circuit to B+
P1360	17768	Cyl. 2, ignition circuit Short Circuit to Ground
P1361	17769	Cyl. 3, ignition circuit Open Circuit
P1362	17770	Cyl. 3, ignition circuit Short Circuit to B+
P1363	17771	Cyl. 3, ignition circuit Short Circuit to ground
P1364	17772	Cyl. 4 ignition circuit Open Circuit
P1365	17773	Cyl. 4 ignition circuit Short circuit to B+
P1366	17774	Cyl. 4 ignition circuit Short circuit to ground
P1367	17775	Cyl. 5, ignition circuit Open Circuit
P1368	17776	Cyl. 5, ignition circuit Short Circuit to B+
P1369	17777	Cyl. 5, ignition circuit short to ground
P1370	17778	Cyl. 6, ignition circuit Open Circuit
P1371	17779	Cyl. 6, ignition circuit Short Circuit to B+
P1372	17780	Cyl. 6, ignition circuit short to ground
P1386	17794	Internal Control Module Knock Control Circ.Error
P1387	17795	Internal Contr. Module altitude sensor error
P1388	17796	Internal Contr. Module drive by wire error
P1393	17801	Ignition Coil Power Output Stage 1 Electrical Malfunction
P1394	17802	Ignition Coil Power Output Stage 2 Electrical Malfunction
P1395	17803	Ignition Coil Power Output Stage 3 Electrical Malfunction
P1396	17804	Engine Speed Sensor Missing Tooth
P1397	17805	Engine speed wheel Adaptation limit reached

Scan Tool Diagnostic Trouble Codes - Powertrain

P-code	VAG code	Description
P1398	17806	Engine RPM signal, TD Short to ground
P1399	17807	Engine RPM signal, TD Short Circuit to B+
P1400	17808	EGR Valve Circ Electrical Malfunction
P1401	17809	EGR Valve Circ Short to Ground
P1402	17810	EGR Valve Circ Short to B+
P1403	17811	EGR Flow Deviation
P1404	17812	EGR Flow Basic Setting not carried out
P1406	17814	EGR Temp.Sensor Range/Performance
P1407	17815	EGR Temp.Sensor Signal too Low
P1408	17816	EGR Temp.Sensor Signal too High
P1409	17817	Tank Ventilation Valve Circ. Electrical Malfunction
P1410	17818	Tank Ventilation Valve Circ. Short to B+
P1411	17819	Sec.Air Inj.Sys.,Bank2 Flow too Flow
P1412	17820	EGR Different.Pressure Sensor Signal too Low
P1413	17821	EGR Different.Pressure Sensor Signal too High
P1417	17825	Fuel Level Sensor Circ Signal too Low
P1418	17826	Fuel Level Sensor Circ Signal too High
P1420	17828	Sec.Air Inj.Valve Circ Electrical Malfunction
P1421	17829	Sec.Air Inj.Valve Circ Short to Ground
P1422	17830	Sec.Air Inj.Sys.Contr.Valve Circ Short to B+
P1423	17831	Sec.Air Inj.Sys.,Bank1 Flow too Low
P1424	17832	Sec.Air Inj.Sys.,Bank1 Leak Detected
P1425	17833	Tank Vent.Valve Short to Ground
P1426	17834	Tank Vent.Valve Open
P1432	17840	Sec.Air Inj.Valve Open
P1433	17841	Sec.Air Inj.Sys.Pump Relay Circ. open
P1434	17842	Sec.Air Inj.Sys.Pump Relay Circ. Short to B+
P1435	17843	Sec.Air Inj.Sys.Pump Relay Circ. Short to ground
P1436	17844	Sec.Air Inj.Sys.Pump Relay Circ. Electrical Malfunction
P1440	17848	EGR Valve Power Stage Open
P1441	17849	EGR Valve Circ Open/Short to Ground
P1442	17850	EGR Valve Position Sensor Signal too high
P1443	17851	EGR Valve Position Sensor Signal too low
P1444	17852	EGR Valve Position Sensor range/performance
P1445	17853	Catalyst Temp.Sensor 2 Circ. Range/Performance

SCAN TOOL POWERTRAIN DTC TABLE (OBD II)

Scan Tool Diagnostic Trouble Codes - Powertrain

P-code	VAG code	Description
P1446	17854	Catalyst Temp.Circ Short to Ground
P1447	17855	Catalyst Temp.Circ Open/Short to B+
P1448	17856	Catalyst Temp.Sensor 2 Circ. Short to Ground
P1449	17857	Catalyst Temp.Sensor 2 Circ. Open/Short to B+
P1450	17858	Sec.Air Inj.Sys.Circ Short to B+
P1451	17859	Sec.Air Inj.Sys.Circ Short to Ground
P1452	17860	Sec.Air Inj.Sys. Open Circ.
P1453	17861	Exhaust gas temperature sensor 1 open/short to B+
P1454	17862	Exhaust gas temperature sensor short 1 to ground
P1455	17863	Exhaust gas temperature sensor 1 range/performance
P1456	17864	Exhaust gas temperature control bank 1 limit attained
P1461	17869	Exhaust gas temperature control bank 1 Range/Performance
P1465	17873	Additive Pump Short Circuit to B+
P1466	17874	Additive Pump Open/Short to Ground
P1467	17875	EVAP Canister Purge Solenoid Valve Short Circuit to B+
P1468	17876	EVAP Canister Purge Solenoid Valve Short Circuit to Ground
P1469	17877	EVAP Canister Purge Solenoid Valve Open Circuit
P1470	17878	EVAP Emission Contr.LDP Circ Electrical Malfunction
P1471	17879	EVAP Emission Contr.LDP Circ Short to B+
P1472	17880	EVAP Emission Contr.LDP Circ Short to Ground
P1473	17881	EVAP Emission Contr.LDP Circ Open Circ.
P1474	17882	EVAP Canister Purge Solenoid Valve electrical malfunction
P1475	17883	EVAP Emission Contr.LDP Circ Malfunction/Signal Circ.Open
P1476	17884	EVAP Emission Contr.LDP Circ Malfunction/Insufficient Vacuum
P1477	17885	EVAP Emission Contr.LDP Circ Malfunction
P1478	17886	EVAP Emission Contr.LDP Circ Clamped Tube Detected
P1500	17908	Fuel Pump Relay Circ. Electrical Malfunction
P1501	17909	Fuel Pump Relay Circ. Short to Ground
P1502	17910	Fuel Pump Relay Circ. Short to B+
P1503	17911	Load signal from Alternator Term. DF Range/performance/Incorrect Signal
P1504	17912	Intake Air Sys.Bypass Leak Detected
P1505	17913	Closed Throttle Pos. Does Not Close/Open Circ
P1506	17914	Closed Throttle Pos.Switch Does Not Open/Short to Ground
P1507	17915	Idle Sys.Learned Value Lower Limit Attained
P1508	17916	Idle Sys.Learned Value Upper Limit Attained

Scan Tool Diagnostic Trouble Codes - Powertrain

P-code	VAG code	Description
P1509	17917	Idle Air Control Circ. Electrical Malfunction
P1510	17918	Idle Air Control Circ. Short to B+
P1511	17919	Intake Manifold Changeover Valve circuit electrical malfunction
P1512	17920	Intake Manifold Changeover Valve circuit Short to B+
P1513	17921	Intake Manifold Changeover Valve2 circuit Short to B+
P1514	17922	Intake Manifold Changeover Valve2 circuit Short to ground
P1515	17923	Intake Manifold Changeover Valve circuit Short to Ground
P1516	17924	Intake Manifold Changeover Valve circuit Open
P1517	17925	Main Relay Circ. Electrical Malfunction
P1518	17926	Main Relay Circ. Short to B+
P1519	17927	Intake Camshaft Contr.,Bank1 Malfunction
P1520	17928	Intake Manifold Changeover Valve2 circuit Open
P1521	17929	Intake Manifold Changeover Valve2 circuit electrical malfunction
P1523	17931	Crash Signal from Airbag Control Unit range/performance
P1525	17933	Intake Camshaft Contr.Circ.,Bank1 Electrical Malfunction
P1526	17934	Intake Camshaft Contr.Circ.,Bank1 Short to B+
P1527	17935	Intake Camshaft Contr.Circ.,Bank1 Short to Ground
P1528	17936	Intake Camshaft Contr.Circ.,Bank1 Open
P1529	17937	Camshaft Control Circuit Short to B+
P1530	17938	Camshaft Control Circuit Short to ground
P1531	17939	Camshaft Control Circuit open
P1537	17945	Engine Shutoff Solenoid Malfunction
P1538	17946	Engine Shutoff Solenoid Open/Short to Ground
P1539	17947	Clutch Vacuum Vent Valve Switch Incorrect signal
P1540	17948	Vehicle Speed Sensor High Input
P1541	17949	Fuel Pump Relay Circ Open
P1542	17950	Throttle Actuation Potentiometer Range/Performance
P1543	17951	Throttle Actuation Potentiometer Signal too Low
P1544	17952	Throttle Actuation Potentiometer Signal too High
P1545	17953	Throttle Pos.Contr Malfunction
P1546	17954	Boost Pressure Contr.Valve Short to B+
P1547	17955	Boost Pressure Contr.Valve Short to Ground
P1548	17956	Boost Pressure Contr.Valve Open
P1549	17957	Boost Pressure Contr.Valve Short to Ground
P1550	17958	Charge Pressure Deviation

SCAN TOOL POWERTRAIN DTC TABLE (OBD II)

Scan Tool Diagnostic Trouble Codes - Powertrain

P-code	VAG code	Description
P1551	17959	Barometric Pressure Sensor Circ. Short to B+
P1552	17960	Barometric Pressure Sensor Circ. Open/Short to Ground
P1553	17961	Barometric/manifold pressure signal ratio out of range
P1554	17962	Idle Speed Contr.Throttle Pos. Basic Setting Conditions not met
P1555	17963	Charge Pressure Upper Limit exceeded
P1556	17964	Charge Pressure Contr. Negative Deviation
P1557	17965	Charge Pressure Contr. Positive Deviation
P1558	17966	Throttle Actuator Electrical Malfunction
P1559	17967	Idle Speed Contr.Throttle Pos. Adaptation Malfunction
P1560	17968	Maximum Engine Speed Exceeded
P1561	17969	Quantity Adjuster Deviation
P1562	17970	Quantity Adjuster Upper Limit Attained
P1563	17971	Quantity Adjuster Lower Limit Attained
P1564	17972	Idle Speed Contr.Throttle Pos. Low Voltage During Adaptation
P1565	17973	Idle Speed Control Throttle Position lower limit not attained
P1566	17974	Load signal from A/C compressor range/performance
P1567	17975	Load signal from A/C compressor no signal
P1568	17976	Idle Speed Contr.Throttle Pos. mechanical Malfunction
P1569	17977	Cruise control switch Incorrect signal
P1570	17978	Contr.Module Locked
P1579	17987	Idle Speed Contr.Throttle Pos. Adaptation not started
P1580	17988	Throttle Actuator B1 Malfunction
P1581	17989	Idle Speed Contr.Throttle Pos. Basic Setting Not Carried Out
P1582	17990	Idle Adaptation at Limit
P1600	18008	Power Supply (B+) Terminal 15 Low Voltage
P1602	18010	Power Supply (B+) Terminal 30 Low Voltage
P1603	18011	Internal Control Module Malfunction
P1604	18012	Internal Control Module Driver Error
P1605	18013	Rough Road/Acceleration Sensor Electrical Malfunction
P1606	18014	Rough Road Spec Engine Torque ABS-ECU Electrical Malfunction
P1607	18015	Vehicle speed signal Error message from instrument cluster
P1608	18016	Steering angle signal Error message from steering angle sensor
P1609	18017	Crash shut-down activated
P1611	18019	MIL Call-up Circ./Transm.Contr.Module Short to Ground
P1612	18020	Electronic Control Module Incorrect Coding

Scan Tool Diagnostic Trouble Codes - Powertrain

P-code	VAG code	Description
P1613	18021	MIL Call-up Circ Open/Short to B+
P1614	18022	MIL Call-up Circ./Transm.Contr.Module Range/Performance
P1615	18023	Engine Oil Temperature Sensor Circuit range/performance
P1616	18024	Glow Plug/Heater Indicator Circ. Short to B+
P1617	18025	Glow Plug/Heater Indicator Circ. Open/Short to Ground
P1618	18026	Glow Plug/Heater Relay Circ. Short to B+
P1619	18027	Glow Plug/Heater Relay Circ. Open/Short to Ground
P1620	18028	Engine coolant temperature signal open/short to B+
P1621	18029	Engine coolant temperature signal short to ground
P1622	18030	Engine coolant temperature signal range/performance
P1623	18031	Data Bus Powertrain No Communication
P1624	18032	MIL Request Sign.active
P1625	18033	Data-Bus Powertrain Unplausible Message from Transm.Contr.
P1626	18034	Data-Bus Powertrain Missing Message from Transm.Contr.
P1627	18035	Data-Bus Powertrain missing message from fuel injection pump
P1628	18036	Data-Bus Powertrain missing message from steering sensor
P1629	18037	Data-Bus Powertrain missing message from distance control
P1630	18038	Accelera.Pedal Pos.Sensor 1 Signal too Low
P1631	18039	Accelera.Pedal Pos.Sensor 1 Signal too High
P1632	18040	Accelera.Pedal Pos.Sensor 1 Power Supply Malfunction
P1633	18041	Accelera.Pedal Pos.Sensor 2 Signal too Low
P1634	18042	Accelera.Pedal Pos.Sensor 2 Signal too High
P1635	18043	Data Bus Powertrain missing message f.air condition control
P1636	18044	Data Bus Powertrain missing message from Airbag control
P1637	18045	Data Bus Powertrain missing message f.central electr.control
P1638	18046	Data Bus Powertrain missing message from clutch control
P1639	18047	Accelera.Pedal Pos.Sensor 1+2 Range/Performance
P1640	18048	Internal Contr.Module (EEPROM) Error
P1641	18049	Please check DTC Memory of Air Condition ECU
P1642	18050	Please check DTC Memory of Airbag ECU
P1643	18051	Please check DTC Memory of central electric ECU
P1644	18052	Please check DTC Memory of clutch ECU
P1647	18055	Please check coding of ECUs in Data Bus Powertrain
P1648	18056	Data Bus Powertrain Malfunction
P1649	18057	Data Bus Powertrain Missing message from ABS Control Module

SCAN TOOL POWERTRAIN DTC TABLE (OBD II)

Scan Tool Diagnostic Trouble Codes - Powertrain

P-code	VAG code	Description
P1650	18058	Data Bus Powertrain Missing message fr.instrument panel ECU
P1651	18059	Data Bus Powertrain missing messages
P1652	18060	Please check DTC Memory of transmission ECU
P1653	18061	Please check DTC Memory of ABS Control Module
P1654	18062	Please check DTC Memory of control panel ECU
P1655	18063	Please check DTC Memory of ADR Control Module
P1656	18064	A/C clutch relay circuit short to ground
P1657	18065	A/C clutch relay circuit short to B+
P1658	18066	Data Bus Powertrain Incorrect signal from ADR Control Module
P1676	18084	Drive by Wire-MIL Circ. Electrical Malfunction
P1677	18085	Drive by Wire-MIL Circ. Short to B+
P1678	18086	Drive by Wire-MIL Circ. Short to Ground
P1679	18087	Drive by Wire-MIL Circ. Open
P1681	18089	Contr.Unit Programming, Programming not Finished
P1684	18092	Contr.Unit Programming Communication Error
P1686	18094	Contr.Unit Error Programming Error
P1690	18098	Malfunction Indication Light Malfunction
P1691	18099	Malfunction Indication Light Open
P1692	18100	Malfunction Indication Light Short to Ground
P1693	18101	Malfunction Indication Light Short to B+
P1694	18102	Malfunction Indication Light Open/Short to Ground
P1704	18112	Kick Down Switch Malfunction
P1705	18113	Gear/Ratio Monitoring Adaptation limit reached
P1711	18119	Wheel Speed Signal 1 Range/Performance
P1716	18124	Wheel Speed Signal 2 Range/Performance
P1721	18129	Wheel Speed Signal 3 Range/Performance
P1723	18131	Starter Interlock Circ. Open
P1724	18132	Starter Interlock Circ. Short to Ground
P1726	18134	Wheel Speed Signal 4 Range/Performance
P1728	18136	Different Wheel Speed Signals Range/Performance
P1729	18137	Starter Interlock Circ. Short to B+
P1740	18148	Clutch temperature control
P1741	18149	Clutch pressure adaptation at limit
P1742	18150	Clutch torque adaptation at limit
P1743	18151	Clutch slip control signal too high

Scan Tool Diagnostic Trouble Codes - Powertrain

P-code	VAG code	Description
P1745	18153	Transm.Contr.Unit Relay Short to B+
P1746	18154	Transm.Contr.Unit Relay Malfunction
P1747	18155	Transm.Contr.Unit Relay Open/Short to Ground
P1748	18156	Transm.Contr.Unit Self-Check
P1749	18157	Transm.Contr.Unit Incorrect Coded
P1750	18158	Power Supply Voltage Low Voltage
P1751	18159	Power Supply Voltage High Voltage
P1752	18160	Power Supply Malfunction
P1760	18168	Shift Lock Malfunction
P1761	18169	Shift Lock Short to Ground
P1762	18170	Shift Lock Short to B+
P1763	18171	Shift Lock Open
P1764	18172	Transmission temperature control
P1766	18174	Throttle Angle Signal Stuck Off
P1767	18175	Throttle Angle Signal Stuck On
P1770	18178	Load Signal Range/Performance
P1771	18179	Load Signal Stuck Off
P1772	18180	Load Signal Stuck On
P1773	18181	Hydraulic Pressure Sensor 1 Too High
P1774	18182	Hydraulic Pressure Sensor 1 Too Low
P1775	18183	Hydraulic Pressure Sensor 1 adaptation at limit
P1776	18184	Hydraulic Pressure Sensor 1 range/performance
P1778	18186	Solenoid EV7 Electrical Malfunction
P1781	18189	Engine Torque Reduction Open/Short to Ground
P1782	18190	Engine Torque Reduction Short to B+
P1784	18192	Shift up/down Wire Open/Short to Ground
P1785	18193	Shift up/down Wire Short to B+
P1786	18194	Reversing Light Circ. Open
P1787	18195	Reversing Light Circ. Short to Ground
P1788	18196	Reversing Light Circ. Short to B+
P1789	18197	Idle Speed Intervention Circ. Error Message from Engine Contr.
P1790	18198	Transmission Range Display Circ. Open
P1791	18199	Transmission Range Display Circ. Short to Ground
P1792	18200	Transmission Range Display Circ. Short to B+
P1793	18201	Output Speed Sensor 2 Circ. No Signal

SCAN TOOL POWERTRAIN DTC TABLE (OBD II)

Scan Tool Diagnostic Trouble Codes - Powertrain

P-code	VAG code	Description
P1795	18203	Vehicle Speed Signal Circ. Open
P1796	18204	Vehicle Speed Signal Circ. Short to Ground
P1797	18205	Vehicle Speed Signal Circ. Short to B+
P1798	18206	Output Speed Sensor 2 Circ. Range/Performance
P1799	18207	Output Speed Sensor 2 Circ. Rpm too High
P1813	18221	Pressure Contr.Solenoid 1 Electrical
P1814	18222	Pressure Contr.Solenoid 1 Open/Short to Ground
P1815	18223	Pressure Contr.Solenoid 1 Short to B+
P1818	18226	Pressure Contr.Solenoid 2 Electrical
P1819	18227	Pressure Contr.Solenoid 2 Open/Short to Ground
P1820	18228	Pressure Contr.Solenoid 2 Short to B+
P1823	18231	Pressure Contr.Solenoid 3 Electrical
P1824	18232	Pressure Contr.Solenoid 3 Open/Short to Ground
P1825	18233	Pressure Contr.Solenoid 3 Short to B+
P1828	18236	Pressure Contr.Solenoid 4 Electrical
P1829	18237	Pressure Contr.Solenoid 4 Open/Short to Ground
P1830	18238	Pressure Contr.Solenoid 4 Short to B+
P1834	18242	Pressure Contr.Solenoid 5 Open/Short to Ground
P1835	18243	Pressure Contr.Solenoid 5 Short to B+
P1841	18249	Engine/Transmission Control Modules Versions do not match
P1842	18250	Please check DTC Memory of instrument panel ECU
P1843	18251	Please check DTC Memory of ADR Control Module
P1844	18252	Please check DTC Memory of central electric control ECU
P1847	18255	Please check DTC Memory of brake system ECU
P1848	18256	Please check DTC Memory of engine ECU
P1849	18257	Please check DTC Memory of transmission ECU
P1850	18258	Data-Bus Powertrain Missing Message from Engine Contr.
P1851	18259	Data-Bus Powertrain Missing Message from Brake Contr.
P1852	18260	Data-Bus Powertrain Unplausible Message from Engine Contr.
P1853	18261	Data-Bus Powertrain Unplausible Message from Brake Contr.
P1854	18262	Data-Bus Powertrain Hardware Defective
P1855	18263	Data-Bus Powertrain Software version Contr.
P1856	18264	Throttle/Pedal Pos.Sensor A Circ. Error Message from Engine Contr.
P1857	18265	Load Signal Error Message from Engine Contr.
P1858	18266	Engine Speed Input Circ. Error Message from Engine Contr.

Scan Tool Diagnostic Trouble Codes - Powertrain

P-code	VAG code	Description
P1859	18267	Brake Switch Circ. Error Message from Engine Contr.
P1860	18268	Kick Down Switch Error Message from Engine Contr.
P1861	18269	Throttle Position (TP) sensor Error Message from ECM
P1862	18270	Data Bus Powertrain Missing message from instr. panel ECU
P1863	18271	Data Bus Powertrain Missing Message from St. Angle Sensor
P1864	18272	Data Bus Powertrain Missing message from ADR control module
P1865	18273	Data Bus Powertrain Missing message from central electronics
P1866	18274	Data Bus Powertrain Missing messages

AFTERMARKET SUPPLIERS OF DIAGNOSTIC SCAN TOOLS

> ***NOTE —***
> *The following suppliers offer diagnostic scan tools or software that can be used to retrieve scan tool codes as listed in the table above.*

Assenmacher Specialty Tools, Inc.
6440 Odell Place
Boulder, CO 80301
(800)525-2943
www.asttool.com

Baum Tools Unlimited, Inc.
PO Box 5867
Sarasota, FL 34277
(800)848-6657
www.baumtools.com

Equipment Solutions
P.O. Box 1450
Kenosha, WI 53141-1450
(800)892-9650

Mac Tools
4635 Hilton Corporate Drive
Columbus, OH 43232
(800)622-8655
www.mactools.com

Ross-Tech
888 Sumneytown Pike
Lansdale, PA 19446
215-361-8942
www.ross-tech.com

Samstag Sales
115 Main St. N., Suite 216
Carthage, TN 37030
www.samstagsales.com

Snap-On Technologies, Inc.
2801 80th St.
Kenosha, WI 53141-1410
(262)656-5200
www.snapon.com

Zelenda Automotive, Inc.
66-02 Austin St.
Forest Hills, NY 11374
(888)892-8384
www.zelenda.com

CONTROL MODULE CODING

Control modules are produced in a few major variations and electronically matched to the exact system needs through a process known as coding. Coding is automatically displayed when the control module responds to the scan tool inquiry and can also be called up through Function 01 of the Function Menu of the VAG 1551/1552 diagnostic scan tool. The coding process requires a properly set-up scan tool or scan tool computer program and is accessed through Function 07.

Most codes are read from a table. Some 5 digit codes require assembly from the tables in this section. Careful attention must be paid to assemble accurate codes. Failure to code, or insertion of an incorrect code can cause any or all of the following conditions:

- Driveability problems
- Increased fuel consumption
- Increased emissions
- Storage of non-existent or erroneous DTCs resulting in Malfunction Indicator Light (MIL) illumination.
- Necessary functions not being performed (oxygen sensor regulation, activation of the EVAP canister system, non-functioning of anti-slip regulation (ASR) etc.).
- Necessary functions being performed incorrectly (mileage recorded instead of kilometers, warning light failure, etc.).
- Reduction in component service life.

NOTE —
- *New or replacement control units will usually require coding of the control module before they will function properly.*
- *After a new code is entered it will usually be necessary to switch the ignition off and back on for the new code to be recorded.*

Some control modules do not require coding; they will work correctly with all equipment versions as-is. Other control modules are coded by jumpers within the wiring harnesses and do not require coding with a scan tool.

Vehicle system coding overview

System	Scan tool address	Abbreviation	Coding required	Basic setting required
Engine Control Module	01	ECM	Some	Some
Transmission Control Module	02	TCM	No	Yes
Anti-Lock Braking Systems and Electronic Differential Lock	03	ABS ABS/EDL	Some	No
Airbag	15	n/a	Most	No
Instrument cluster	17	n/a	No	No
Radio	56	n/a	Some	No (requires resetting anti-theft code)

Engine Control Module (ECM) Coding Tables

The following tables list coding information for systems that require it. If your system is not listed, coding is either not required or the ECU is correctly coded as it is received from the supplier.

Engine Control Module Coding Variations, 1Z, AHU (1.9L TDI)

Vehicle Type	Equipment Level	Code
Golf, Jetta	Automatic Transmission	00001
	Manual Transmission	00002
Notes: Only the 2 codes listed above are recognized as valid. Access through Address Word 01, Function 07.		

Engine Control Module Coding Variations, ABA (2.0L 1993-1995)

Vehicle Type	Equipment Level	Code
Golf, Jetta, Cabrio	Manual Transmission	1
	Automatic Transmission	2
	Manual Transmission, with EGR	3
	Automatic Transmission, with EGR	4
Notes: Some vehicles with automatic transmissions may have been intentionally mis-coded as manual transmission vehicles to improve idle quality and to eliminate hesitation when accelerating from idle. In all cases, ECM coding is accomplished through the use of special coding wires in the ECM wiring harness; these systems cannot be coded with a scan tool.		

Engine Control Module Coding Variations, ABA (2.0L 1996-1999)

Vehicle Type	Equipment Level	Code
Golf, Jetta	Manual Transmission, Tier 1	00000
	Automatic Transmission, Tier 1	00001
	Manual Transmission, TLEV	00004
	Automatic Transmission, TLEV	00005
Cabrio	Manual Transmission	00002
	Automatic Transmission	00003
Notes: Only the codes listed above are recognized as valid. Enabling the coding process for these systems requires an extra step. First access the system through Address Word 01, Function 11 Log-In Procedure. Enter code 01283 in all cases. Return to Function Menu, enter Function 07 and then enter appropriate code.		

Engine Control Module Coding Variations, ABA (2.0L Cabrio 2000➤)		
Vehicle Type	**Equipment Level**	**Code**
Cabrio	Manual Transmission, TLEV	00004
	Automatic Transmission, TLEV	00005
Notes: Only the codes listed above are recognized as valid. Enabling the coding process for these systems requires an extra step. First access the system through Address Word 01, Function 11 Log-In Procedure. Enter code 01283 in all cases. Return to Function Menu, enter Function 07 and then enter appropriate code.		

Engine Control Module Coding Variations, AAA (2.8L VR6 1994-1995)		
Vehicle Type	**Equipment Level**	**Code**
Golf, Jetta	Manual Transmission	1
	Automatic Transmission	2
	Manual Transmission, with EGR	3
	Automatic Transmission, with EGR	4
Notes: ECM coding is accomplished through the use of special coding wires in the ECM wiring harness; these systems cannot be coded with a scan tool.		

Engine Control Module Coding Variations, AAA (2.8L VR6 1996-1999)		
Vehicle Type	**Equipment Level**	**Code**
Golf, Jetta	Manual Transmission	00000
	Automatic Transmission	00001
Notes: Only the codes listed above are recognized as valid. Enabling the coding process for these systems requires an extra step. First access the system through Address Word 01, Function 11 Log-In Procedure. Enter code 01283 in all cases. Return to Function Menu, enter Function 07 and then enter appropriate code.		

Transmission Control Module (TCM) Coding

Transmission Control Modules (TCM) for the 096 and 01M automatic transmissions do not require coding at this time. They will function properly when the Engine Control Module is properly coded (where required). Coding information, if displayed in the appropriate area of the scan tool, will usually be 00000.

Anti-Lock Brake System (ABS) and Anti-Lock Brake System with Electronic Differential Lock (ABS/EDL) Control Module Coding Table

Two different types of ABS and ABS/EDL systems are installed in the vehicles covered by this manual. Early vehicles use the Teves 04 system with a separate control module. This system does not require coding. Later vehicles use the Teves 20 GI system with integral ECU. This system requires coding upon replacement of the ECU according to the following table.

ABS/EDL Control Module Coding Variations, Teves 20 GI

Engine Code	Equipment Level	Code
2.0L (ABA) ➤1995	ABS	03604
2.8L (AAA) ➤1995	ABS/EDL	03604
4-cylinder vehicles 1996 ➤ engine codes ABA, ACC, AAZ, 1Z, AHU,	ABS	03704
6-cylinder vehicles 1996 ➤ engine code AAA	ABS/EDL	03704
Notes: Only the codes listed above are recognized as valid. Access through Address Word 03. If ABS ECU has been incorrectly coded, the ABS warning light (K47) will be illuminated continuously and the brake system warning light (K118) will illuminate at approximately 6 mph (10 kph) or when an engine speed of 2000 rpm has been exceeded once. DTCs will be stored.		

Airbag Control Module Coding Table

Airbag control units supplied as replacement parts for most early airbag-equipped vehicles should be properly coded for the application. If the scan tool displays a coding of 00000 and the ECU part number is NOT listed in the table, consult an Authorized Volkswagen Dealer for the correct airbag ECU.

Airbag-equipped vehicles with the ECU part numbers listed below require coding upon installation. An index code as the last digit of the part number determines the ECU code.

Airbag System Electronic Control Unit Coding Variations			
Vehicle Equipment Level	**ECU Part Number**	**Index Code (part number suffix)**	**ECU Code**
Driver's/Passenger's airbag, (VR6 and Cabrio only) (USA)	701 909 601	D	00068
Driver's/Passenger's airbag, (USA)	6N0 909 603	D	00068
Driver's/Passenger's airbag	6N0 909 603	E	00069
Driver's/Passenger's airbag, (USA)	1J0 909 603	T	00084
Driver's/Passenger's and side airbag	1J0 909 603	AQ	16721
Driver's/Passenger's airbag	1J0 909 608	A	00065
Driver's/Passenger's airbag, (USA)	1J0 909 603	BC	16963
Driver's/Passenger's and side airbag, with seat belt detection for switch-over activation limit, USA (Mexico production)	1J0 909 605	0H	12360
Driver's/Passenger's and side airbag, (USA)	1J0 909 608	J	00074
Driver's/Passenger's airbag	1J0 909 608	K	00075
Driver's/Passenger's and side airbag	1J0 909 608	BA	16961
Driver's and side airbags	1J0 909 608	BB	16962
Driver's and side airbags	1J0 909 608	BC	16963
Driver's/Passenger's and side airbag, (USA)	1J0 909 608	BF	16966
Driver's/Passenger's and side airbag, (USA)	1J0 909 608	ZF	23110
Notes: Airbag ECU will only recognize the values listed above as valid. Access through Address Word 15, Function 07.			

Sound System Coding Table

A new radio series was installed as a running change beginning with late October, 1997 production as follows: Jetta VIN WM142005, Golf/Cabrio VIN: WM146415 (VINs approximate). Radio descriptions "Deluxe" and "Premium" are no longer used. The new radio series is referred to as "Sound System". Radios of this series require coding to match the vehicle characteristics and the installed equipment.

Radio Head Unit Coding Variations[5], Sound System

Designations	Code Assembly Position[3,4]	
	1 + 2	**3, 4, 5**
Sound Field Tuning[1]		
Golf, GTI, Jetta	00	
Cabrio	07	
Radio System Equipment Configuration[2]		
Without CD changer, with standard mast antenna		400
Without CD changer, with amplified antenna		401
With CD changer, with standard mast antenna		402
With CD changer, with amplified antenna		403

Notes:
1. Sound field tuning provides various frequency response characteristics that are matched to the vehicle interior volume.
2. Sound system can have rear-mounted CD changer and/or amplified antenna.
3. Code will be 5 digits. For example, a VR6 Jetta with an amplified roof antenna and CD changer is coded 00403.
4. Only select from the listed codes.
5. Access through Address Word 56.

BASIC SETTINGS

Throttle valve control module, adapting

The Engine Control Module (ECM) and the throttle valve control Module share data concerning engine operation. Under certain circumstances, data that is shared may be lost. The procedure for "teaching" or "reteaching" the ECM the various positions of the throttle valve control module is known as adaptation. This position data is stored in the ECM and remains stored. Adaptation data influences engine operation and is set at the time that the vehicle is new. In some circumstances, adaptation data can be lost and must be restored using the VAG 1551/1552, VAS 5051/5052 or equivalent, or an appropriate scan tool program to insure proper engine operation. OBD I engines (1993-1995) do not have a throttle valve control module and, therefore, do not require this procedure.

Engine codes:
ABA (1996-1999)
AAA (1996-1999)
ABA (2000-2002)

The following circumstances will cause the throttle valve control module adaptation to be lost. It should be reset after completion of any of the following repairs:

- Throttle valve control unit removing and installing (replacement with new or used).
- Replacing or coding the Engine Control Module (ECM).
- The voltage supply is interrupted.

The following conditions must be observed prior to starting this procedure:

- No DTCs stored in memory.
- Battery voltage at least 11.5 volts
- All electrical consumers switched off.
- Throttle valve at idle position (correct cable adjustment).
- Throttle not operated during procedure.
- Throttle valve must not be damaged, dirty or contaminated with carbon build up.
- Cruise control (if installed) must be functional.

Adaptation can be initiated or restored using a suitable scan tool. Follow the scan tool manufacturer's instructions and/or use the procedure outlined below. Shown here is the display from the VAG 1551. Others scan tool displays are similar.

1. Connect scan tool to DLC and switch on ignition, but **do not** start engine and **do not** depress accelerator pedal.

2. Advance scan tool to address word menu and enter **01** for engine electronics.

Rapid Data Transfer **HELP**

3. Advance scan tool to function menu and enter **04** for basic setting.

Rapid Data Transfer
HELP

4. Advance scan tool to display group menu and enter **098** (1996-1999) or **060** (2000-2002). Depress the **Q** button to confirm the display.

5. Advance scan tool until system confirms that it is in basic setting mode with four display fields as shown. Display field four should indicate that adaptation is running.

System in basic setting 098			**-->**
XXX	**XXX%**	**XXX%**	**ADP runs**

6. During this time, the ECM switches off the voltage to the throttle valve control module to take a measurement and then switches back on to take additional measurements. Voltage will again be switched off allowing a spring to pull the throttle back to a previously measured position.

7. At the conclusion of this procedure, the ECM will confirm proper stored values from the throttle position sensors in the throttle valve control module with the value in the fourth display field as shown. Display field four should indicate that adaptation is OK.

System in basic setting 098			**-->**
XXX	**XXX**	**XXX**	**ADP OK**

8. Wait at least 30 seconds and advance scan tool with **-->** button to function menu and enter **06** to end output.

9. Switch off ignition to store values in ECM.

10. Remove scan tool.

NOTE —
If display reads ADP ERROR, test conditions may not have been met or procedure was interrupted while running. Confirm proper test conditions and rerun procedure.

Transmission control module, adapting

The Transmission Control Module (TCM) and the Engine Control Module (ECM) share data concerning engine and transmission operation. Data that the TCM may not always precisely "know" is throttle range and full throttle position. The acquisition of this data is known as basic setting. The basic setting influences automatic transmission shifting and is set at the time that the vehicle is new.

The basic setting will be lost and should be initiated after completion of the following repairs:

- Engine removing and installing (replacement).
- Transmission removing and installing.
- Replacing or coding the Engine Control Module (ECM).
- Replacing the Transmission Control Module (TCM).
- Removing, installing, adjusting, and/or replacing electronic throttle components at pedal or throttle housing.
- If the battery is disconnected or runs down.
- If the TCM is disconnected.

In these circumstances, the basic settings must be restored to insure proper automatic transmission operation.

Basic settings, initiating

Basic settings can be initiated or restored using a suitable scan tool. Follow the scan tool manufacturer's instructions and/or use the procedure outlined below. In either case, the kickdown switch on the accelerator cable (where applicable) must function properly. Shown here is the display from the VAG 1551. Others scan tool displays are similar.

1. Connect scan tool to DLC and switch **on** ignition, but **do not** start engine.

2. Advance scan tool to address word menu and enter **02** for transmission electronics.

```
Rapid Data Transfer
HELP
```

3. Advance scan tool to function menu and enter **04** for basic setting.

```
Rapid Data Transfer
HELP
```

4. Advance scan tool to display group menu and enter **000.**

```
Basic setting
HELP
```

5. Advance scan tool until system confirms that it is in basic setting mode as shown.

```
System in basic setting        -->
                               000
```

6. Push accelerator pedal all the way to the floor (past kickdown) and hold it there for a minimum of 3 seconds.

NOTE —
Most scan tools and scan tool programs will not confirm completion of basic settings by any on-screen display.

7. Release accelerator pedal.

8. Exit program and disconnect scan tool.

9. Switch off the ignition.

Procedure following interruption of voltage supply

NOTE —
If the Voltage supply is interrupted, the Engine Control Module loses all the "learned" and/or "basic" values. As a result, severe driveability problems may occur depending on vehicle mileage.

Special tools, testers, measuring instruments and auxiliary items required:

- VAG 1551/1552 or VAS 5051/5052 Scan Tool, with appropriate adaptor cable or
- Equivalent scan tool or scan tool computer program.

Perform the following operations after "Voltage supply open circuit" display appears in scan tool display.

1. Switch ON ignition for at least 10 seconds.

2. Switch OFF ignition.

3. Adapt the Engine Control Module to the Throttle Valve Control Module, see **Scan Tool** section at the back of manual.

4. For vehicles equipped with automatic transmissions, the kick down shift point must be adapted (adapt Transmission Control Module (TCM) to ECM), see **Scan Tool** section at the back of manual.

5. Test drive until any driveability complaints are eliminated.
   During the road test the following operating conditions must be fulfilled:

   - The Engine Coolant Temperature must exceed 80 °C.
   - Once the required temperature is reached, the following operating conditions must be attained several times.
   - Closed Throttle position
   - Part Throttle position
   - Wide Open Throttle position
   - Overrun
   - With Wide Open Throttle engine speed must exceed 3500 rpm.

6. Completely open and close all power windows. Enter anti-theft radio code and check function of radio, tape player and CD changer. Reset clock and check function of all other electrical consumers.

V.A.G. 1551/1552 OPERATING OVERVIEW FLOW CHART

MODE MENU

1 - Rapid Data =>
2 - Blink Code
3 - Self Test
4 - Dealership Code
5 - LT - Diagnosis

ADDRESS WORDS MENU (V9)

01 - Engine Electronics =>
11 - Engine Electronics II
41 - Diesel Pump Electronics
51 - Electric Drive
61 - Battery Control
71 - Battery Charger
02 - Transmission Electronics
12 - Clutch Electronics
22 - 4 Wheel Drive Electronics
03 - Brake Electronics
13 - Distance Control
33 - OBD II
14 - Suspension Electronics
24 - Anti - Slip Control
34 - Level Control
44 - Steering Assist
54 - Rear Spoiler
15 - Airbag
25 - Anti-Theft Immobilizer Sensor
35 - Central Locking
45 - Interior Monitor
55 - Headlamp Vert. Aim Control
65 - Tire Pressure Monitoring
75 - Emergency Call Module
16 - Steering Wheel Electronics
26 - Automatic Roof Control
36 - Seat Adjustment Driver's Side
46 - Comfort System (AC CM)
56 - Radio
66 - Seat & Mirror Adjustment
76 - Parking Aid
17 - Instrument Cluster
37 - Navigation System
47 - Sound System
57 - Airbag (Cabriolet only)
67 - Language Control
08 - A/C & Heating Electronics
18 - Auxiliary Heater
09 - Electronic Control Module
19 - Diagnostic Interface for Data Bus
29 - Left Light Control
39 - Right Light Control
49 - Auto Light Switch
00 - Automatic Test Sequence

FUNCTIONS MENU

01 - Check Control Module Versions
02 - Check DTC Memory
03 - Output Diagnostic Test Mode
04 - Basic Setting
    Display Group xxx =>
05 - Erase DTC Memory
06 - End Output
07 - Code Control Module
08 - Read Measuring Value Block
    Display Group xxx =>
09 - Read Individual Measuring Values
    Channel xx
10 - Adaptation
    Channel xx
11 - Log-In Procedure
15 - Readiness Code

DISPLAY GROUP FORMAT

Display Group 000 (only)
xx xx xx xx xx xx xx xx xx xx

Display Group 001 (and higher)
xxxxx xxxxx xxxxx xxxxx

V.A.G. 1551/1552 OPERATING OVERVIEW FLOW CHART

The New

Golf III

GTI

Jetta III

Service Training

Self-Study Program 405

CAUTION

This training module is included as background information and is based on introductory and preliminary information. It applies to only some of the vehicles covered by this manual and is intended to serve as a product familiarization guide. Note that this information is subject to change and should be used as reference only. Always refer to the latest Volkswagen factory service information for the most up to date service, repair and training information.

Table of Contents

Introduction

The new Golf III and Jetta III models are the third generation of A class vehicles (A3) from Volkswagen. These vehicles set high standards for safety, performance, and driving comfort.

Strength

Both models feature a body structure that has approximately a 30% increase in torsional rigidity from the previous Golf and Jetta models. This results in advances in ride and handling qualities. Special attention was given to the placement and use of sound dampening materials. This, combined with the increased body strength, provides very low vehicle noise levels for the occupants.

Power

Both vehicles feature new engines with more power. These new engines are long stroke designs which provide generous levels of torque with good fuel efficiency. The latest fully adaptive engine management systems are used and require practically no adjustments.

German Engineering

Many features of the previous Golf and Jetta, such as the torsion beam rear axle with track correction, MacPherson strut front suspension with asymmetrical load compliant bushings, and high capacity blend air heating and air conditioning systems, have been retained and improved upon for these new vehicles.

Safety

The new A3 models have also been designed with the latest safety developments. Reinforcements are incorporated in the doors and upper and lower door sills for side impact protection. A special crossmember is located under the dashboard and connects the "A" pillars together to complete the safety cell for the occupants.

The front long members of the body feature a new construction method which allow metals of different thickness to be joined together to provide progressive crushability during a frontal impact. The new A3 models also pass the 35 mph frontal barrier crash test defined in the NHTSA new car assessment program, which is over and above the current government standards.

Introduction

Model Line

Golf III

All Golf III models have a substantial list of standard equipment such as an anti-theft alarm system, power steering, service reminder indicator, eight speaker radio prep, rear spoiler and power remote trunk release.

CL (Canadian model only)

This model is powered by a 1.8 liter 90 hp engine with Mono-Motronic engine management. It comes equipped with 13-inch wheels and 175/70 all-season tires in addition to most of the standard equipment listed above. Options include A/C, sunroof and 4-speed automatic transmission.

GL

This model is powered by a new 2.0 liter 115 hp engine with cross flow cylinder head. It rides on 14-inch wheels with 185/60 all-season tires (195/60 on vehicles with automatic transmission). A central locking system is standard Equipment. Options also include A/C, power glass sunroof and 4-speed automatic transmission.

GTI

This model begins with 1994 model year. It is equipped with the 2.8 liter VR6 engine. Equipment includes ABS with traction control, 15 inch wheels and sports suspension. Interior features include sports seats and height adjustable steering column.

Model Line

Jetta III

All Jetta III models have power steering, a power remote trunk release, a central locking system, anti-theft alarm system and a service reminder indicator.

GL

The GL is powered by the new 2.0 liter 115 hp engine. This model features a height adjustable steering wheel/column, 185/60 14-inch all-season tires (195/60 on vehicles with automatics), eight speaker radio prep, rear seat headrests, velour seat trim, and lockable folding rear seat.

GLS

This model becomes available in 1994 model year and replaces the previous Jetta Carat. It features a higher level of standard equipment such as comfort seats with lumbar adjustments, 60/40 split fold and lockable rear seats, A/C, courtesy lights in the front door pockets, cruise control, power windows, and power mirrors. This model is also powered by the 2.0 liter 115 hp cross flow engine.

In Canada, a new 1.9 liter 75 hp ECOdiesel is available as an option on the Golf III GL, and Jetta III GL and GLS. Other options in the U.S. and Canada are 4-speed auto transmission and power glass sunroof.

GLX

The GLX becomes available later in model year 1994 and replaces the previous Jetta GLI model. It is equipped with the same convenience equipment as the GLS model (power windows, power mirrors and cruise control). The interior features partial leather sports seats and a 3-spoke leather-wrapped sports steering wheel.

This sports/performance model of the A3 Sedan is powered by the new 2.8 liter VR6 engine. It rides on 205/50 performance tires with sports suspension. ABS with traction control, BBS® wheels, Multi-function Indicator, heated windshield washer nozzles and power glass sunroof are standard equipment on this model.

Engine

1.8 Liter
(Canada only)

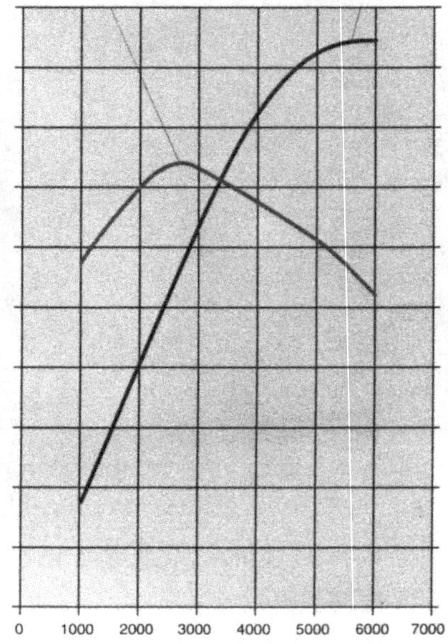

The Golf CL model (Canadian market only) is equipped with a new 1.8 liter 90 hp engine. This engine features a 10:1 compression ratio and is designed to provide good mid-range torque.

This 1.8 liter engine uses a new Monomotronic Engine Management system.

Displacement	1781 cc
Bore	81 mm
Stroke	86.4 mm
Compression ratio	10:1
Horsepower (SAE net)	90 hp @ 5,500 rpm
Torque	106 lbs./ft. @ 2,500 rpm
Engine Code	ACC

Diesel Engine

1.9 liter ECOdiesel
(Canada only)

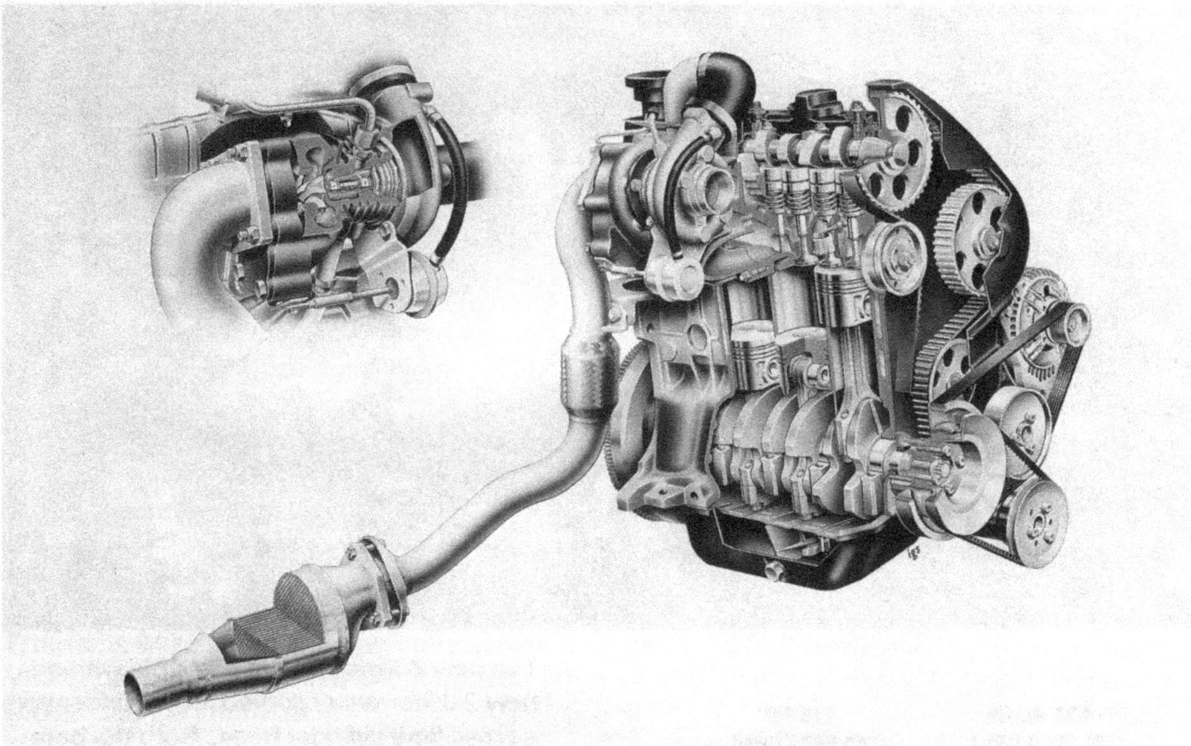

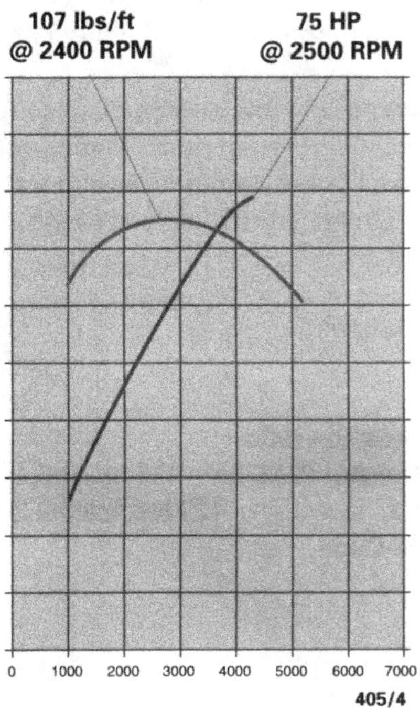

The Golf III GL and Jetta III GL and GLS models sold in Canada are available with a new optional 1.9 liter ECOdiesel engine. The engine is similar to the 1.6 liter ECOdiesel in that it uses a turbocharger to supply extra air for reduced particulate emissions together with a one-way catalytic converter to reduce HC emissions.

Displacement	1896 cc
Bore	79.5 mm
Stroke	95.5 mm
Compression ratio	22.5:1
Horsepower (SAE net)	75 hp @ 4,200 rpm
Torque	107 lbs./ft. @ 2,400 rpm
Engine Code	AAZ

Engine

2.0 Liter Four Cylinder Engine

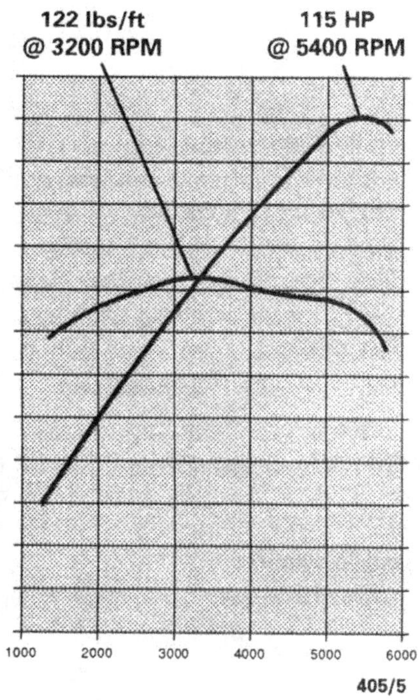

The new A3 models are available with a new 2.0 liter water cooled four cylinder with a cross flow cylinder head. Both the bore and the stroke have been increased when compared to the previous 1.8 liter engine.

The increase in engine size results in a generous amount of performance, and due to a cross flow cylinder head and long piston stroke, a broad torque range is available.

Displacement	1984 cc
Bore	82.5 mm
Stroke	92.8 mm
Compression ratio	10:1
Horsepower (SAE net)	115 hp @ 5,400 rpm
Torque	122 lbs./ft. @ 3,200 rpm
Engine Code	ABA

Engine

VR6 Six Cylinder Engine

The 2.8 liter VR6 engine is installed in the GTI and GLX models that begin in the 1994 model year.

The engine features a cast-iron cylinder block, one light alloy cross flow cylinder head with two valves per cylinder operated by a chain-driven overhead camshafts.

All fuel and ignition requirements of the VR6 engine are controlled by the Motronic® Engine Management system.

The engine features the latest revisions, such as new, distributorless ignition and secondary air injection.

For more information on the VR6 engine, refer to the Service Training Self-Study Programs #402, titled The New Volkswagen VR6 Engine, and program #406, titled 1993 Model Change Information.

VR6 Horsepower/Torque Specifications

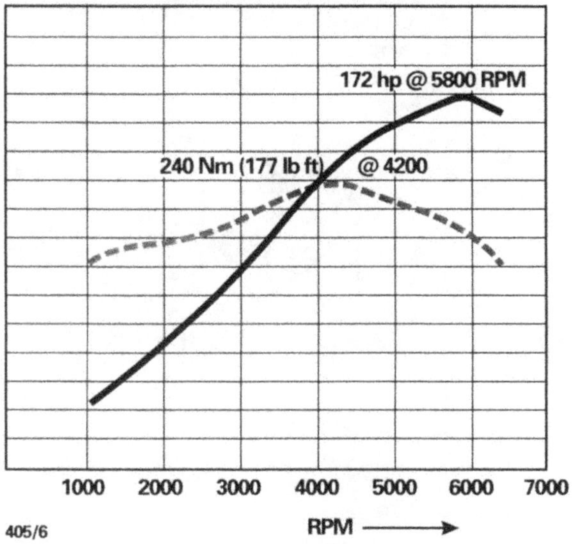

Displacement	2792 cc
Bore	81 mm
Stroke	90.3 mm
Compression ratio	10:1
Horsepower (SAE net)	172 hp @ 5,800 rpm
Torque	177 lbs./ft. @ 4,200 rpm
Engine Code	AAA

Fuel Supply

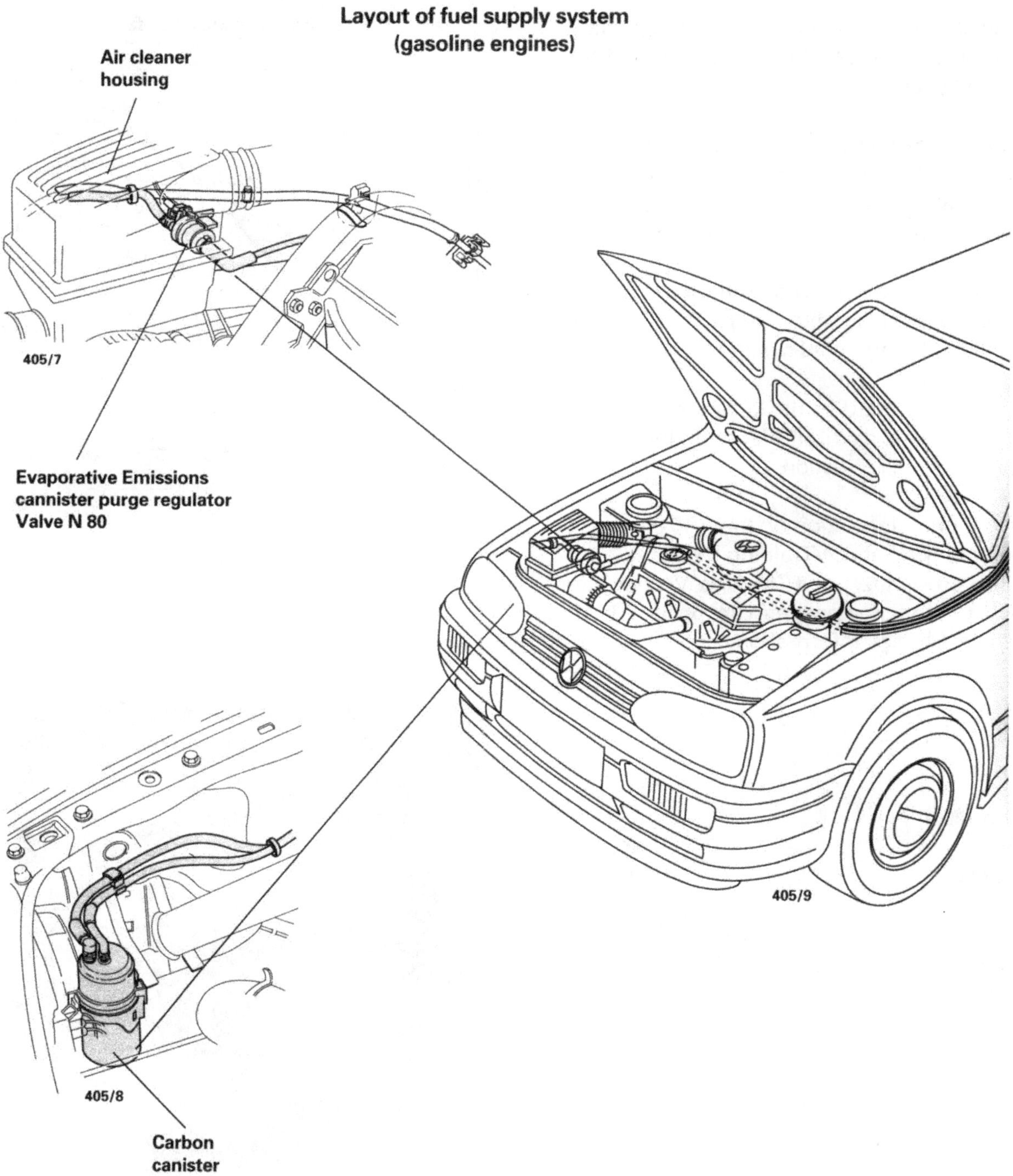
Layout of fuel supply system
(gasoline engines)
Air cleaner
housing
405/7
Evaporative Emissions
cannister purge regulator
Valve N 80
405/9
405/8
Carbon
canister

Fuel Supply

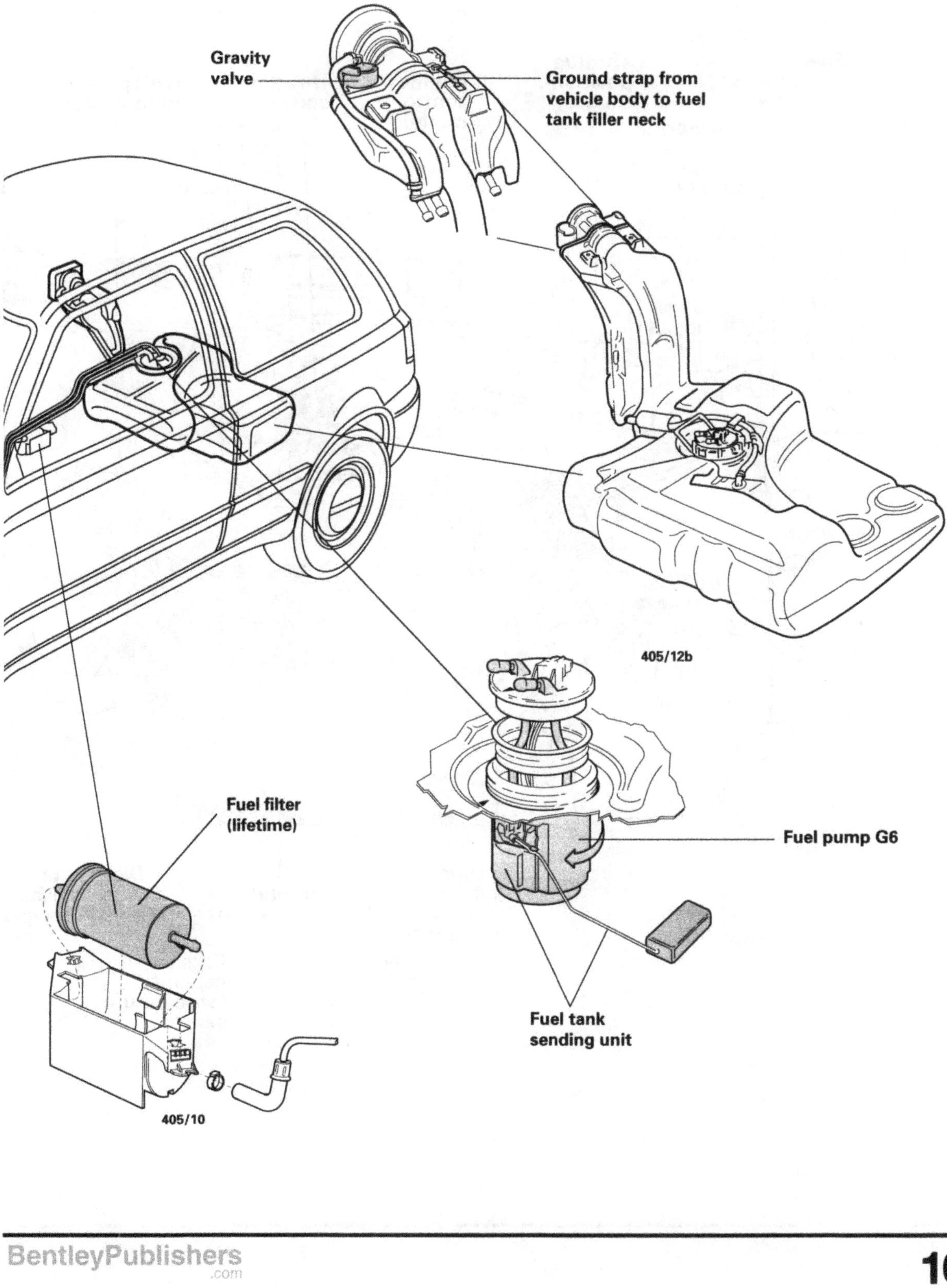
Gravity valve
Ground strap from vehicle body to fuel tank filler neck
405/12b
Fuel filter (lifetime)
Fuel pump G6
Fuel tank sending unit
405/10

Engine

1.8 liter Mono-Motronic
(Canada only)

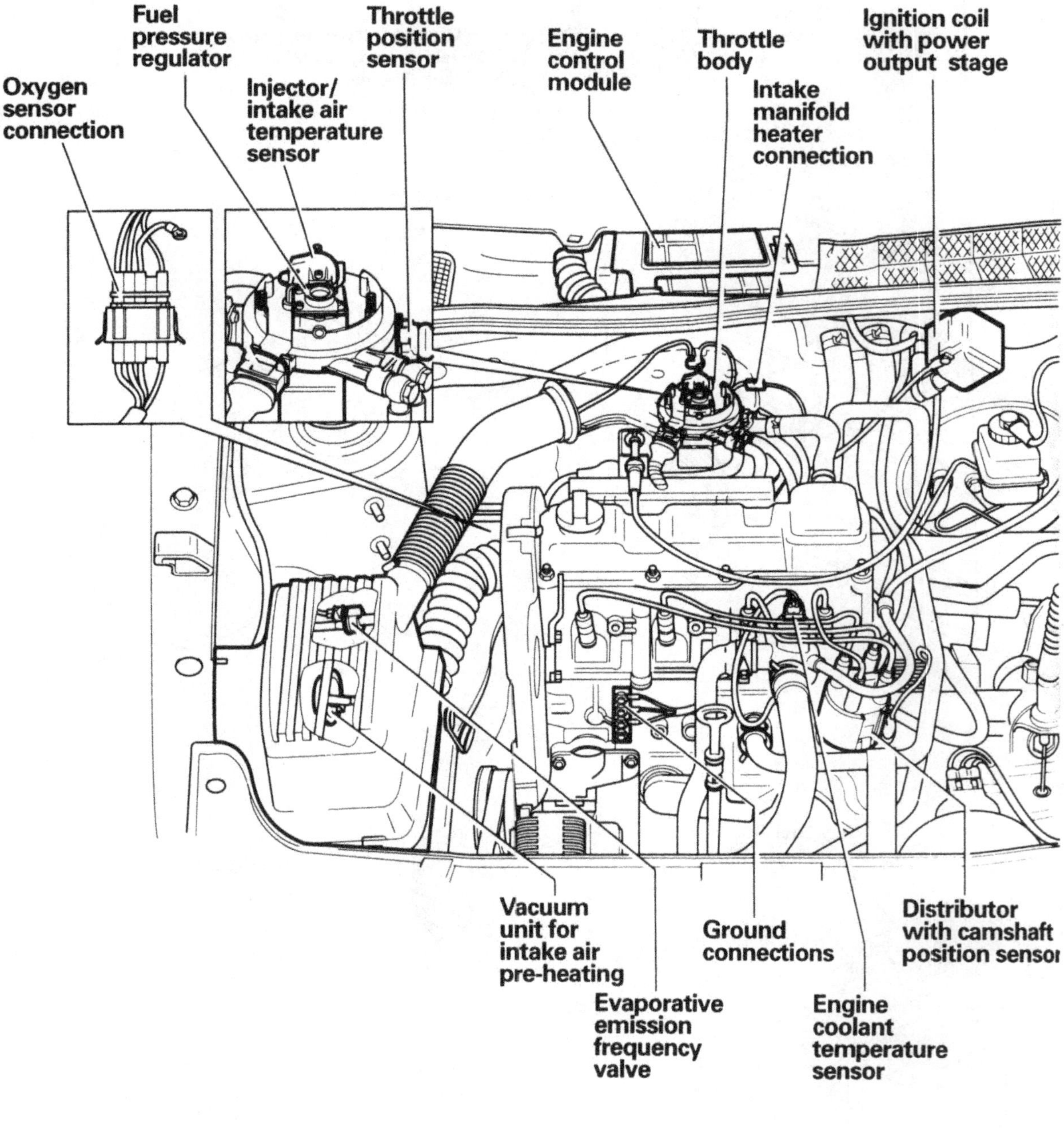

Mono-Motronic

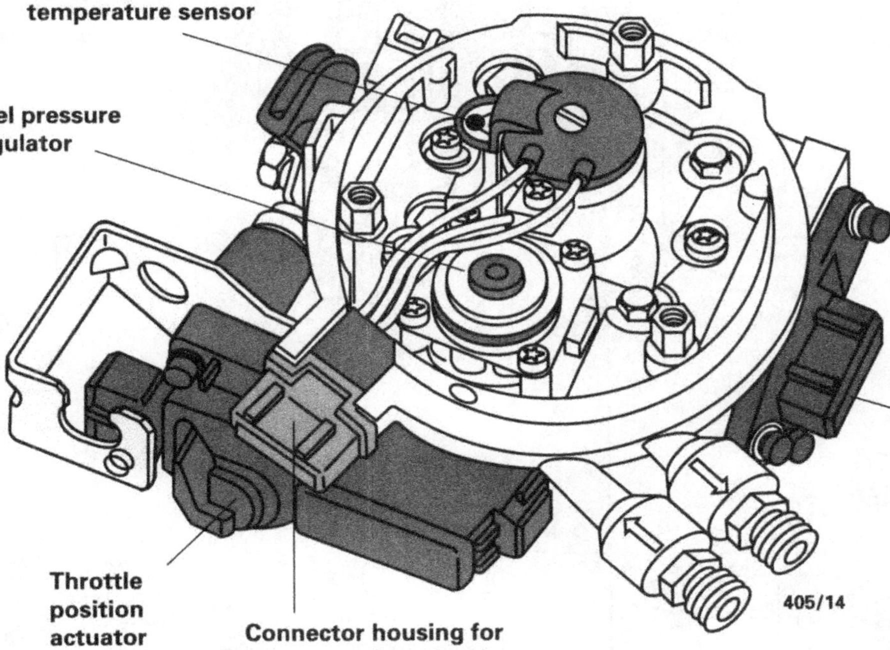

The Mono-Motronic system is a complete engine management system which controls:

- Ignition timing
- Idle stabilization via throttle position and digital idle stabilization
- Fuel mixture
- Fuel tank ventilation

The system is operated by an electronic control module. A Throttle body unit is attached to the intake manifold. This unit contains:

- Injector with intake air temperature sensor
- Fuel pressure regulator
- Throttle valve position sensor
- Throttle position actuator

The Throttle position actuator is a stepper motor, which is attached to the Throttle body unit. It is operated by a pulsed voltage from the control module. The motor will open and close the throttle within a specified range to control the idle speed. This unit works together with the DIS function in the control module, which adjusts the ignition timing at idle to stabilize the engine speed.

The Engine control module receives information on engine temperature from a coolant temperature sensor located in the cylinder head. The camshaft position sensor is a Hall sender in the ignition distributor that sends one signal for each cylinder. This is used to determine engine speed and crankshaft position.

Mono-Motronic

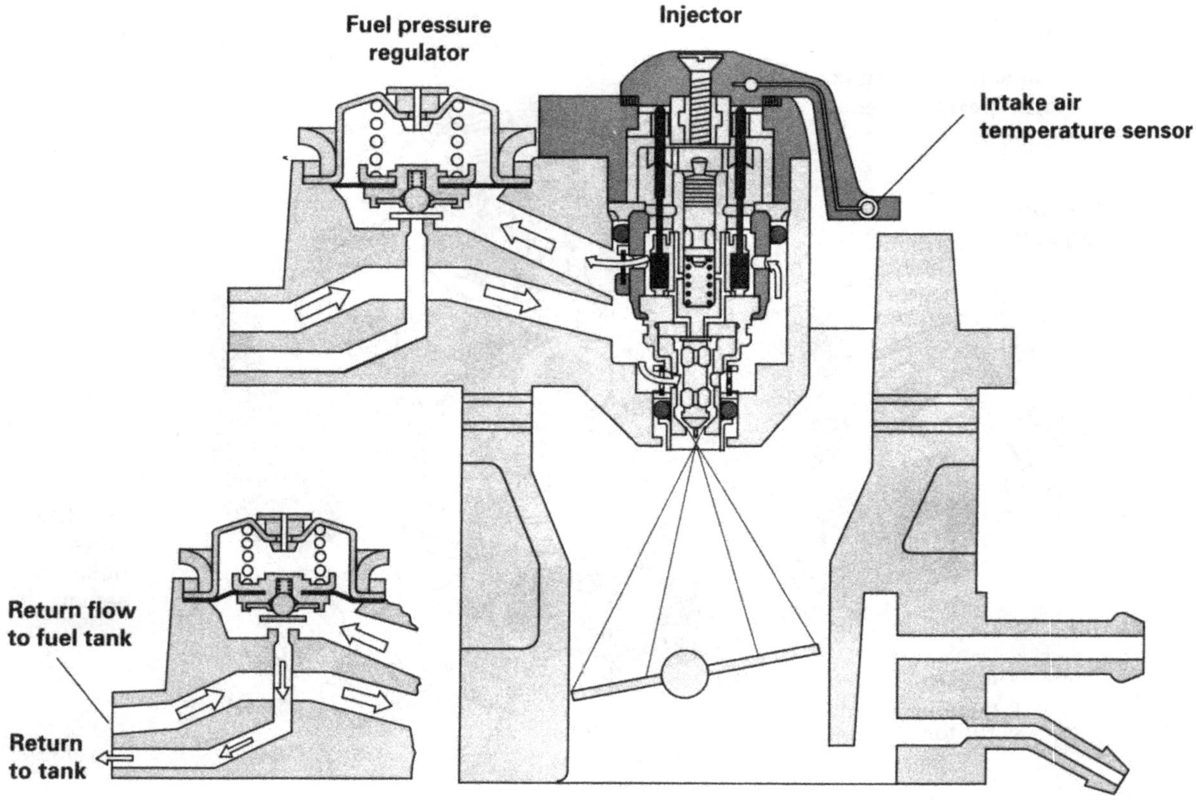

405/15

The Engine control module regulates the fuel mixture by determining the opening time of the central injector. The throttle position potentiometer sends a signal to the control module that is used to determine engine load.

The Mono-Motronic Engine control module has a permanent diagnostic trouble code memory. The following functions can be performed with the 1551 scan tool:

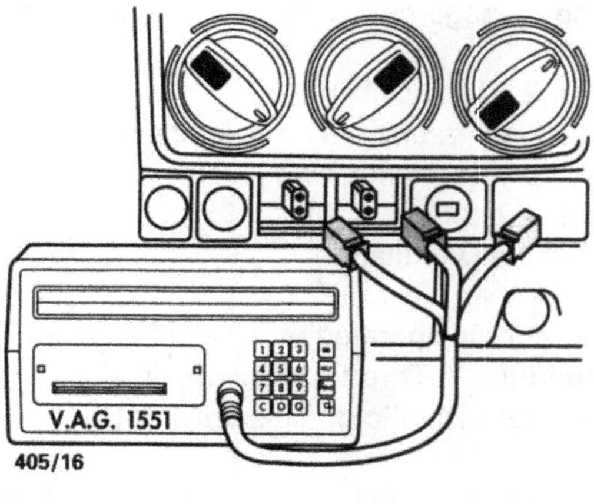

405/16

- 02 Read fault memory
- 03 Output check diagnosis
- 04 Start basic settings
- 05 Erase fault memory
- 06 End output
- 08 Read measuring block value

For more information, refer to the repair manual.

Diesel Engine

Injection Pump

The 1.9 liter ECOdiesel engine uses an injection pump with boost pressure enrichment. This unit will slightly enrich the mixture at high turbocharger boost conditions to increase engine power.

Load dependent timing control is also incorporated in the injection pump. This retards the injection timing during periods of low engine load to reduce engine noise. To improve exhaust emissions, an Injection Start Positioner, N 165, is used to shut this load dependent timing feature off at higher engine speeds.

When the A/C is switched on, the idling speed is increased by a vacuum unit on the injection pump. The vacuum to this servo is controlled by Idle Air Control Solenoid N123, located in the engine compartment on the right front shock absorber tower.

The solenoid valve opens when the A/C is switched on. Vacuum from the engine vacuum pump will then be applied to the vacuum unit, and the throttle lever will be opened slightly.

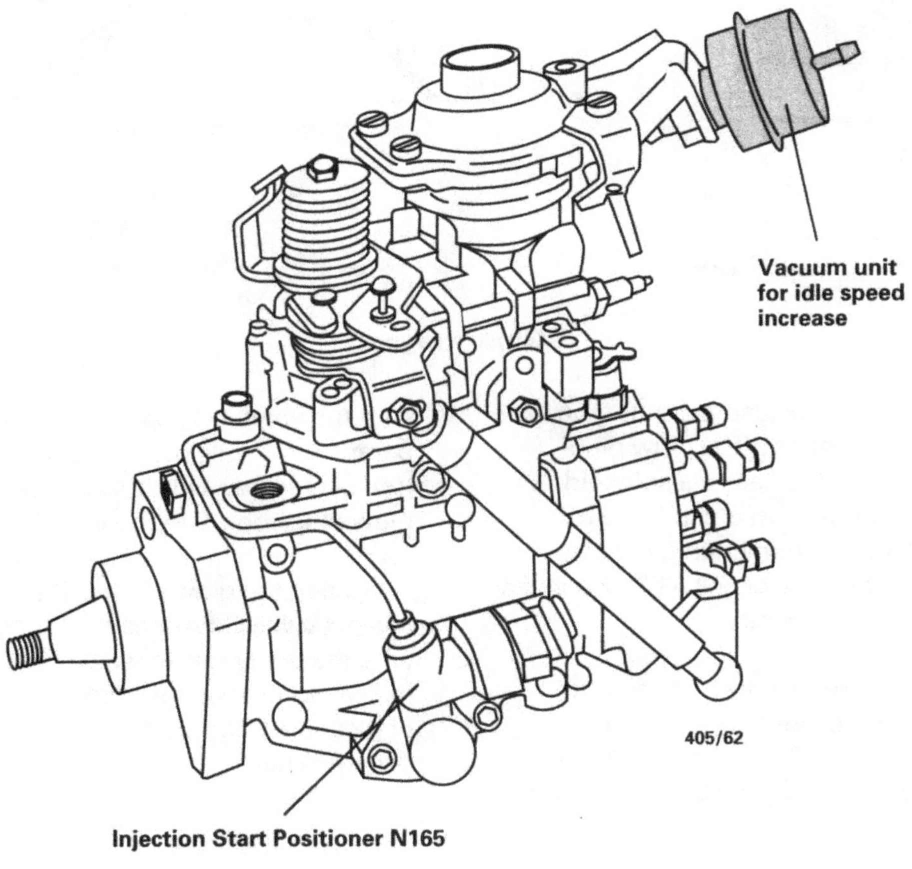

Diesel Engine

Injectors

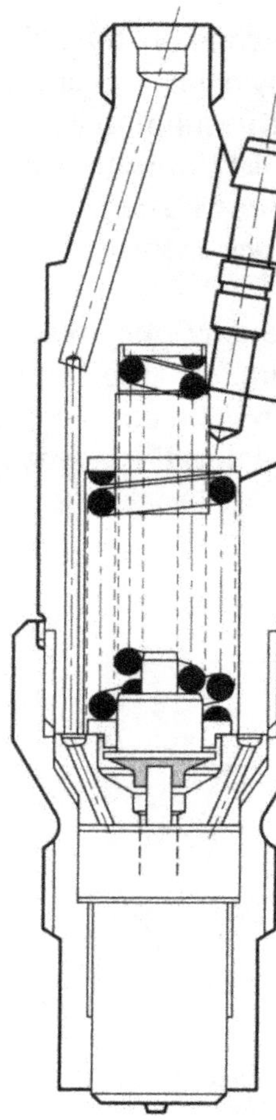

The 1.9 liter ECOdiesel uses a new double spring diesel fuel injector. The injector uses two springs of different strength to control the injector needle. This design results in the injector needle lift being smaller but the opening time being longer. The fuel is then injected over a longer period of time which provides a "softer" combustion and lower engine combustion noise. This also helps to reduce NOx exhaust emissions.

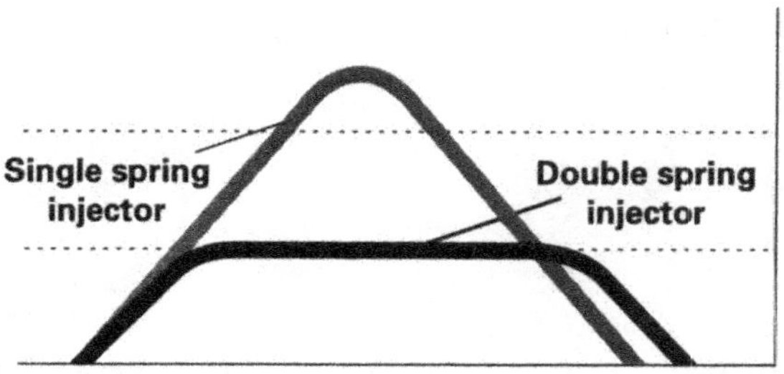

Glow Plugs

The 1.9 liter ECOdiesel also features a new quick glow plug system. The glow plug system, idle speed increase via solenoid N123, and the shut-off solenoid N165 for load dependent injection timing, are controlled by a single control unit J179, located above the fuse/relay panel.

The glow plugs are switched on after the door is closed when getting into the car with the engine temperature below 148°F (65°C). The amount of glow time needed to start the engine is based on engine temperature and will increase as the temperature decreases. This is calculated by the glow plug control module J179.

Activating the glow plugs with the door contact switch shortens the amount of glow time the driver has to wait. The glow plug indicator light will go out after the amount of glow time needed to start the engine has been reached.

Diesel Engine

Wiring Diagram

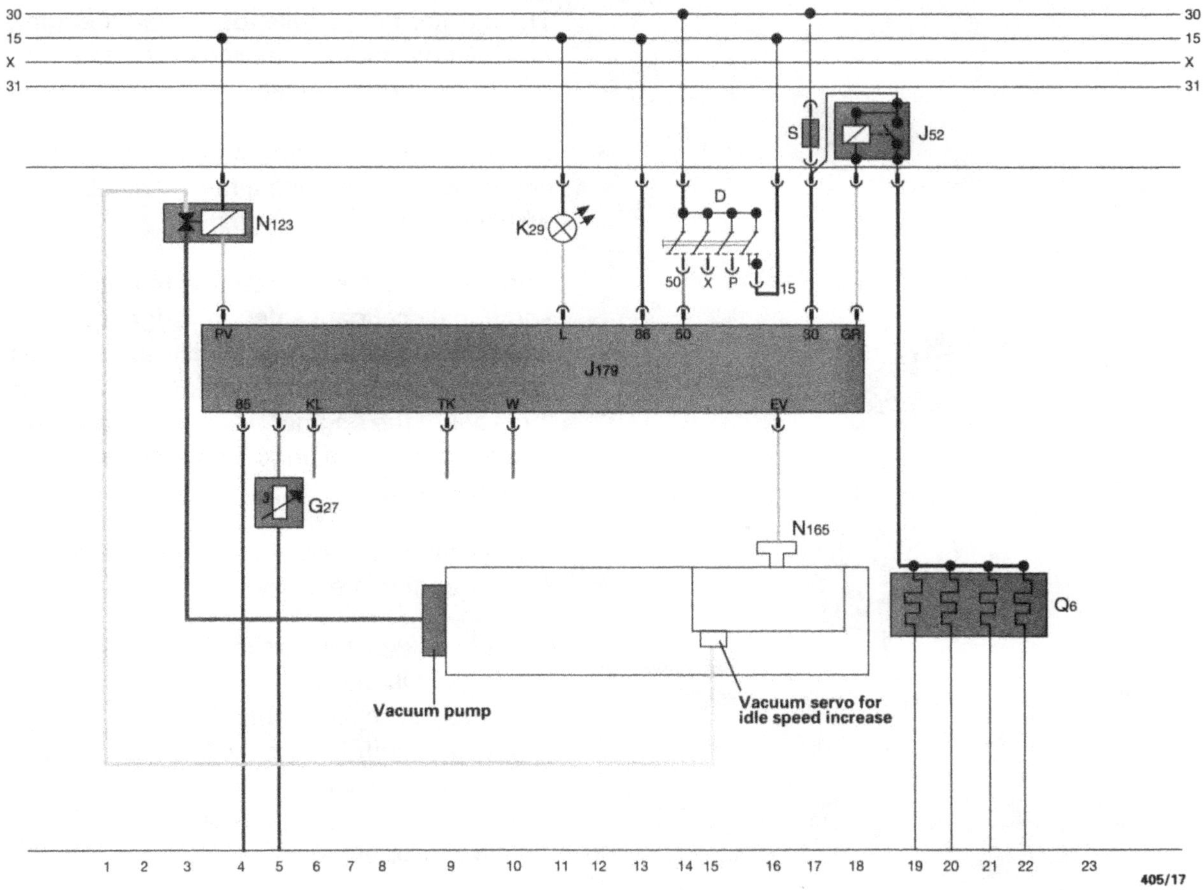

Wiring diagram legend

D	Ignition switch
J179	Automatic glow time control module
J52	Relay for glow plugs
Q6	Glow plugs
N123	Idle speed control solenoid
N165	Injection start positioner solenoid
K29	Glow plug indicator light
G27	Engine temperature sensor

Control module inputs

85	Ground
T	Engine temperature
KL	A/C on signal
50	Starter signal
TK	Signal from driver's door
W	Speed signal from alternator

2.0 Liter Four Cylinder Engine

Overview

The 2.0 liter long stroke four cylinder engine has been developed specifically for the North American market.

Some of the development objectives of this engine were to provide a fuel efficient engine with a broad torque range.

Another important objective is to provide additional occupant safety through impact absorption during a frontal collision. To help achieve this, the intake manifold is moved to the front of the engine. This provides more space between the engine and passenger compartment.

Engine changes when compared to the 1.8 liter engine are as follows:

- Increased piston stroke (6.4 mm)
- Larger cylinder bore (1.5 mm)
- Cross flow cylinder head
- Intake manifold with long intake runners
- Cylinder block height increased 16.5 mm
- Connecting rod lengthened from 144 mm to 159 mm

Increasing the height of the cylinder block and lengthening the connecting rods reduces the second order engine vibrations by reducing the operating angles of the connecting rods. This results in the increased smoothness through lower engine vibrations.

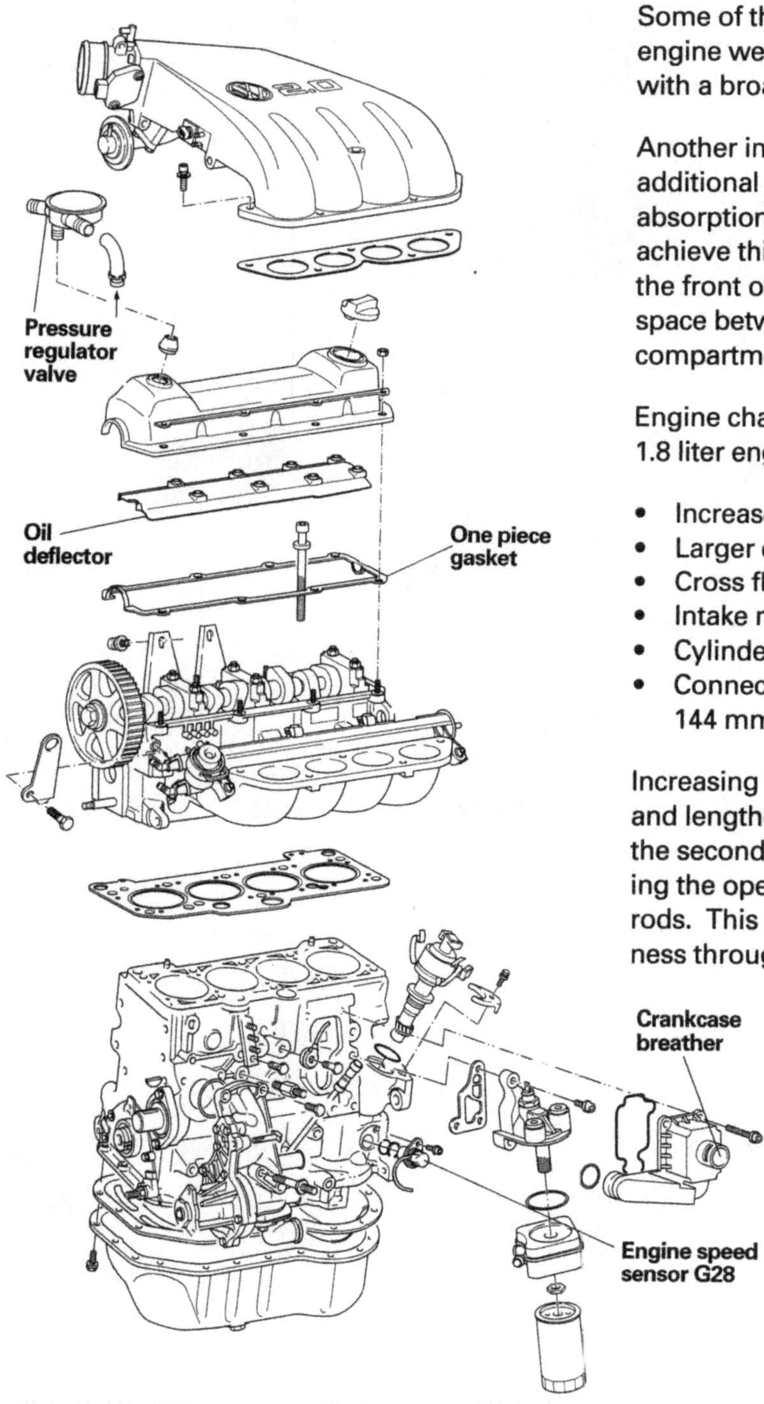

2.0 liter Four Cylinder Engine

Intake Manifold

The intake manifold is a two piece design. To remove the fuel rail, fuel injectors and cam cover, the manifold must be separated.

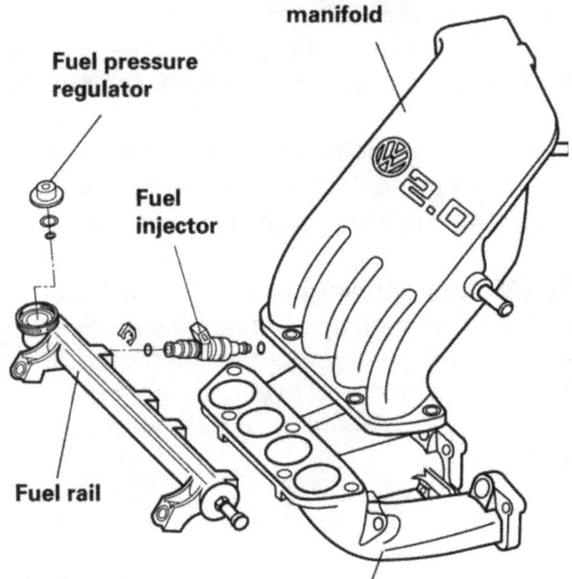

405/20

Drive Belts

A conventional "V" belt drives both the water pump and the power steering pump. This belt is adjusted by moving the power steering pump.

A poly-ribbed belt with automatic belt tensioning roller is used with this engine. The belt drives the alternator and A/C compressor. If the belt needs to be removed for any reason, special tool 3299 can be used to loosen the tensioner. Mark the running direction of the belt prior to removal.

For vehicles without A/C, see page 65.

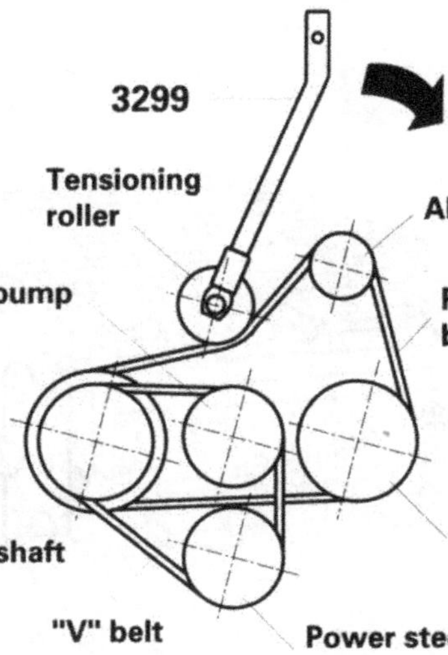

405/21

Motronic Engine Management

Overview

The Motronic Engine Management system used on the new 2.0 liter four cylinder engine for A3 models is similar to the Motronic system used on the VR6 engine.

The Motronic Engine Management system combines the ignition and fuel injection system and controls both electronically.

The Motronic Engine control module (J220) has a digital microcomputer. Maps are stored in the control modules memory. These maps are used to control ignition timing, knock regulation, injection timing and quantity.

The advantages of the Motronic system are:

- Optimum fuel economy due to low fuel consumption through adaptation of all operating conditions.
- Emission control with adaptation of injection rate and ignition timing.
- Reduced maintenance and ease of servicing to improved diagnostics.

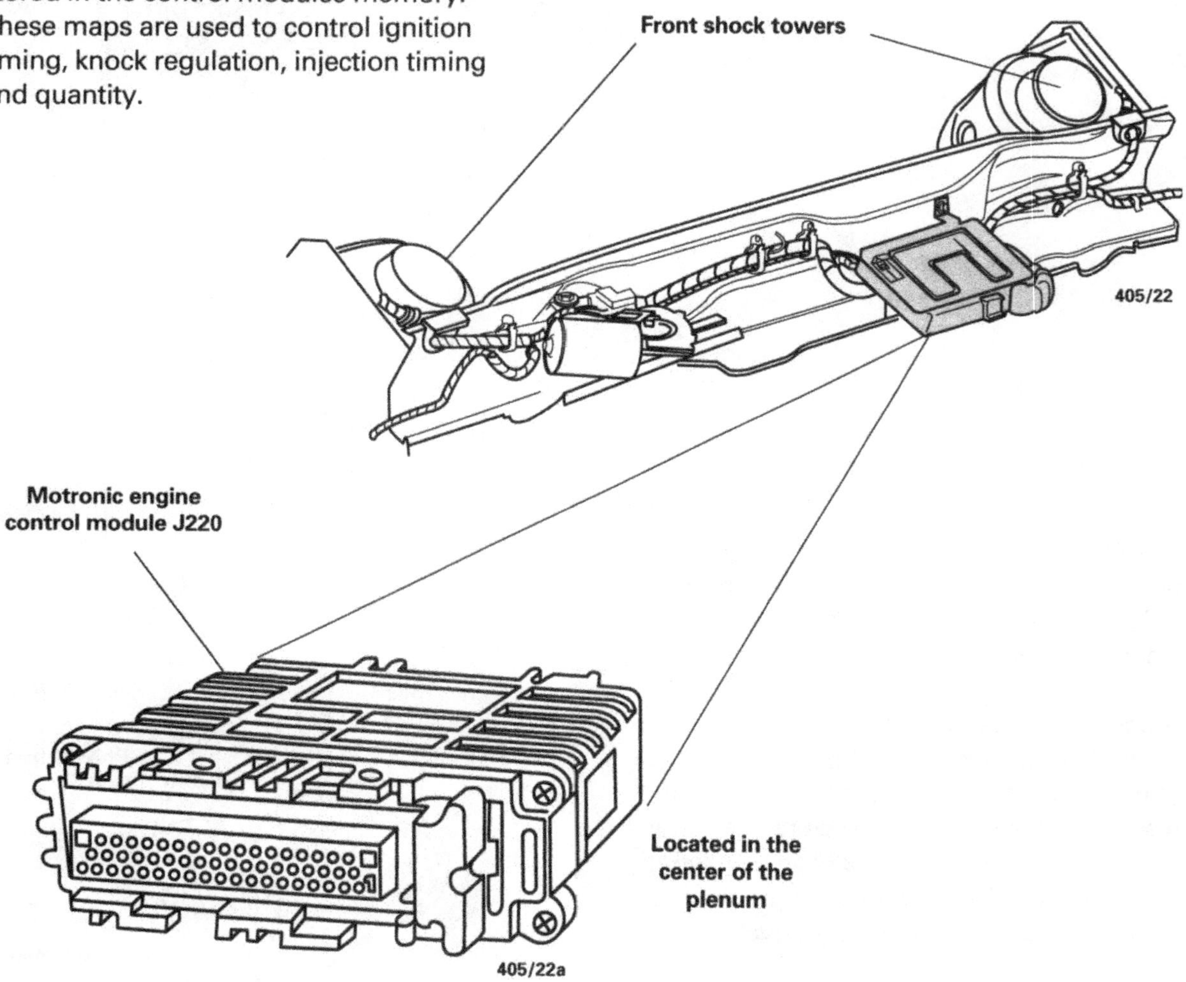

Motronic Engine Management

Overview

This system is adaptive. This means that the idle speed, ignition timing, injection timing and quantity will adapt to different engine operating conditions such as altitude changes and even small vacuum leaks. The engine control module (ECM) receives inputs from the sensors and determines the optimum settings. There are no basic engine settings or adjustments.

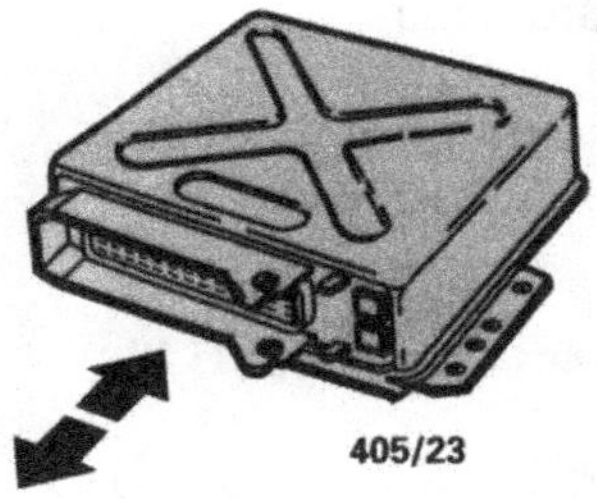

The Motronic Engine Management system for the A3 models has permanent Diagnostic Trouble Code memory (DTC memory). The data link connectors for the Scan tool V.A.G. 1551 are located in the center of the dash below the heating, ventilation and A/C controls.

These data link connectors are also used to interface with other control modules.

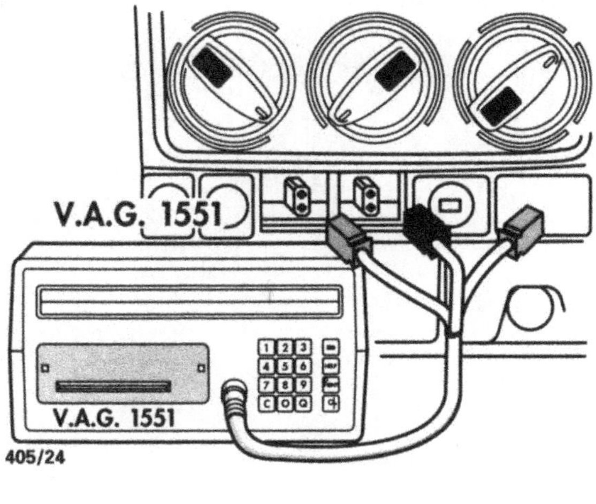

Motronic Engine Management

Component Location

1 Evaporate emissions canister purge Regulator N80

2 EGR vacuum regulator solenoid valve

3 Mass air flow sensor G70

4 Throttle position sensor G69

5 Engine control module

6 4-pin connector for heated oxygen sensor G39

7 Intake air temperature sensor G72

8 Idle air control valve N71

9 EGR temperature sensor G98

10 EGR valve

11 CO tap tube

12 Ignition coil N152

13 Vehicle speed sensor G22

14 Engine coolant temperature sensor G62

15 Central plug connection

16 3-pin connector for knock sensor (blue)

17 3-pin connector for engine speed sensor (white)

18 Engine speed sensor G28

19 Ignition distributor with camshaft position sensor G40

20 Fuel pressure measuring point

21 Fuel rail for injectors

22 Knock sensor G61

23 Ground connection

24 Fuel pressure regulator

25 Temperature regulator for intake air preheating

26 Carbon canister (below air cleaner housing)

Motronic Engine Management

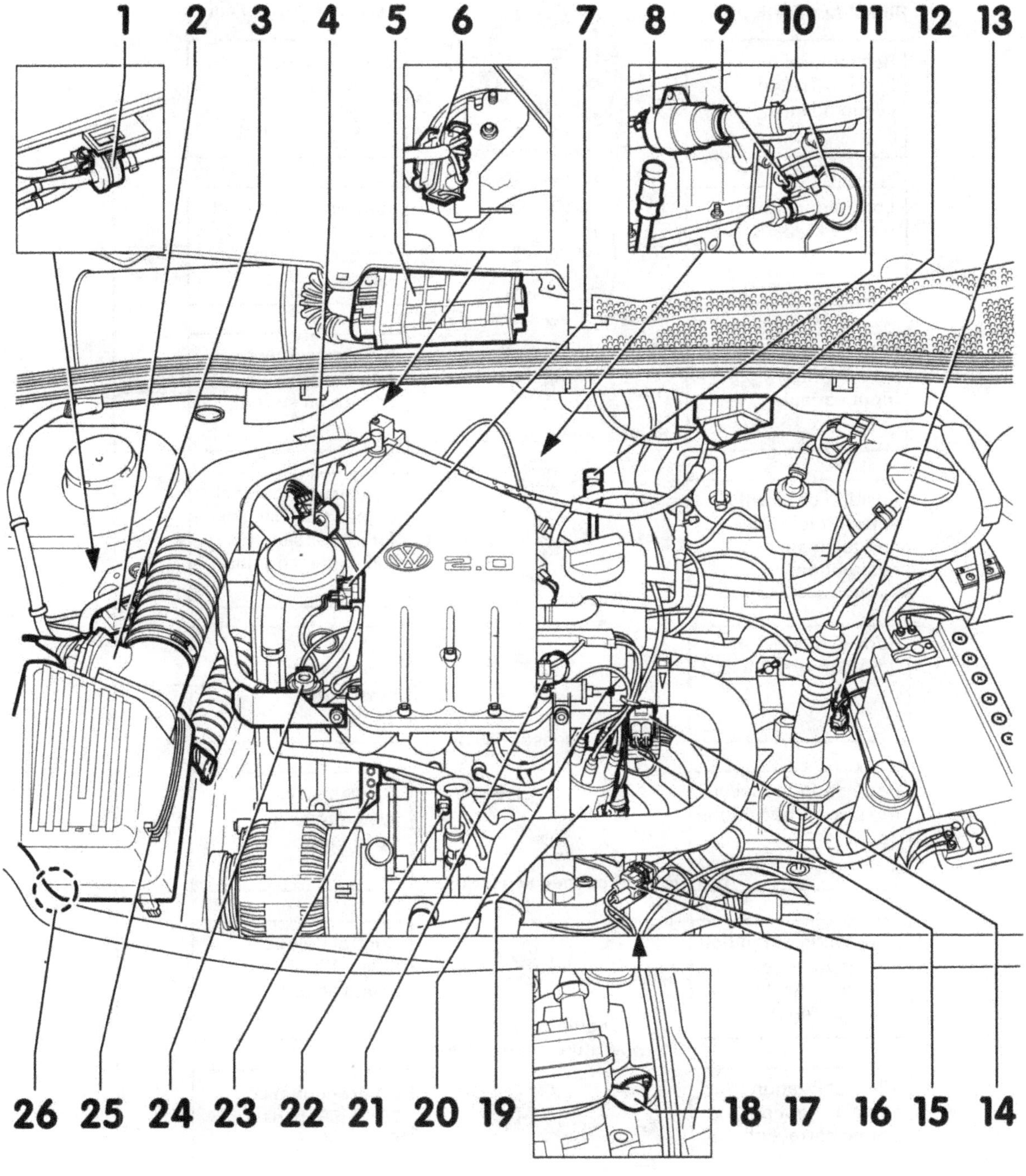

Motronic Engine Management

Signal Overview

Signal Application	Signal Overview	Measures or Effects in Case of Signal Failure
Recognition of ignition TDC in cylinder No. 1.	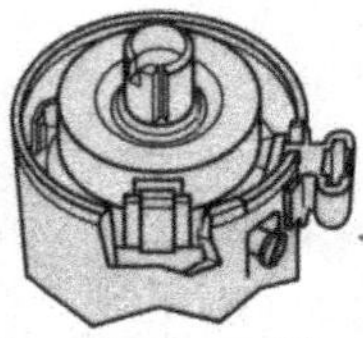 Camshaft position sensor (G40)	No sequential injection. No knock control.
Load signal (main input variable).	 Mass air flow sensor (G70)	Replacement signal from throttle position sensor and speed sensor.
Engine speed sensor (main input variable).	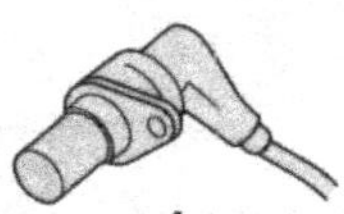 Engine speed sensor (G28)	No replacement function/engine cannot be started.
Position of throttle valve for idle speed and full load recognition.	 Throttle position sensor (G69)	Replacement signal comes from sensor for idle speed and full load recognition.
Control of ignition timing during knock.	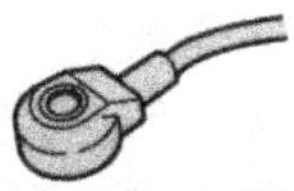 Knock sensor (G61)	Ignition timing retarded 15°. Power loss.
Regulation of fuel/air mixture composition.	 Heated oxygen sensor (G39)	Oxygen sensor system goes to open loop.
Adaptation of ignition timing and injection duration. Activation of knock regulation.	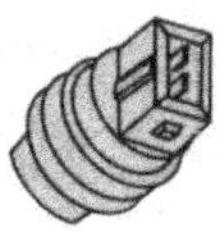 Engine coolant temperature sender (G62)	Control unit uses 80° C (176° F) as fixed value.
Idle stabilization control. Ignition timing correction.	 Intake air temperature sensor (G42)	Control unit uses 20° C (68° F) as fixed value.
Battery voltage. Air conditioner On.A/C compressor clutch engaged.Vehicle speed.		

Motronic Engine Management

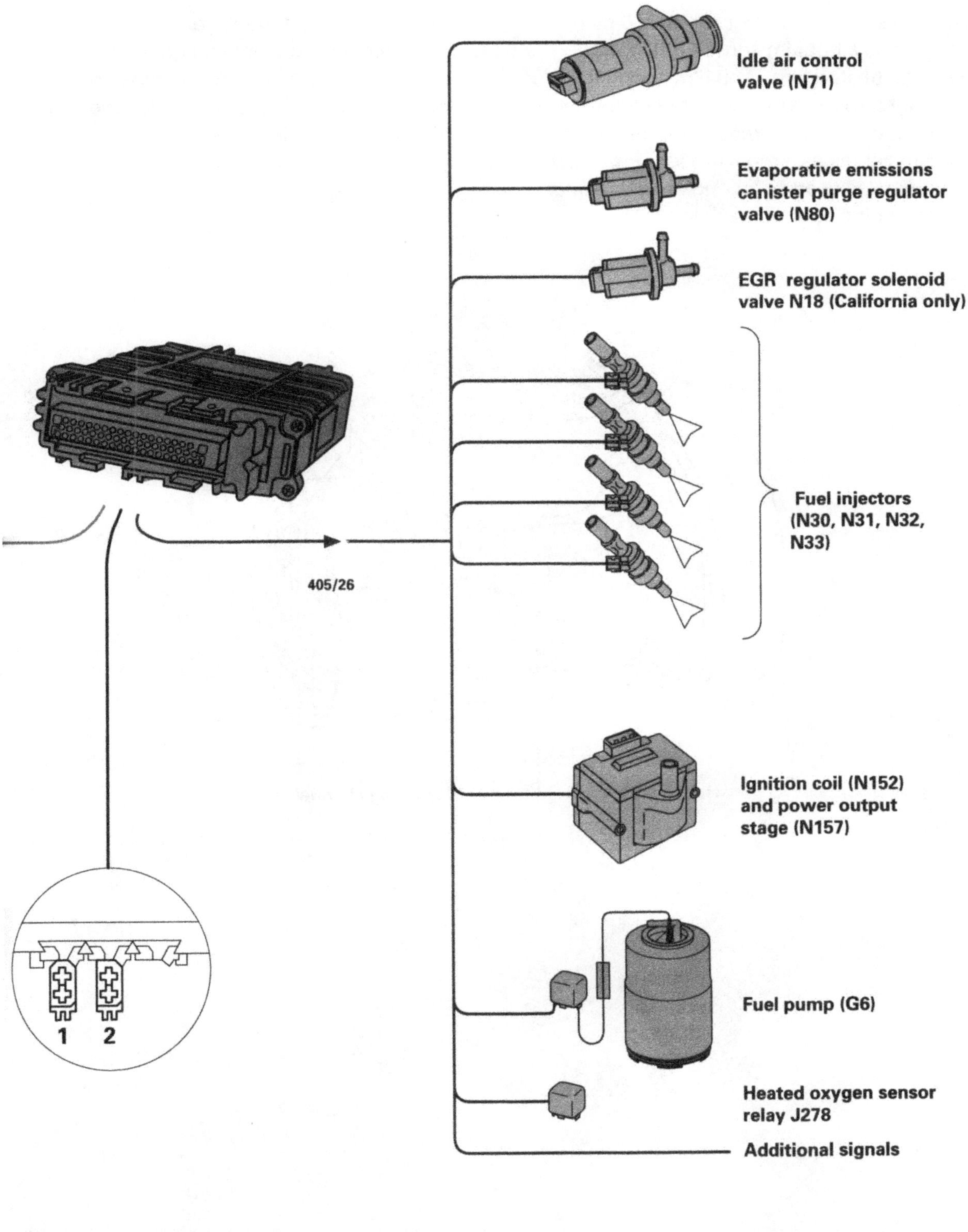

405/26

Motronic Engine Management

Mass Air Flow Sensor G70

The mass air flow sensor for the 2.0 liter engine has a hot film sensor. Intake air passes a heated film. The film is kept at a constant temperature above the intake air temperature. The current necessary to heat the film serves as a measure for air density. This signal is used as an engine load signal by the engine control module.

This sensor is different from the hot wire mass air flow sensor used on the VR6 engine. The hot film sensor does not require a "burn off" signal to clean the sensor after the vehicle is shut-off.

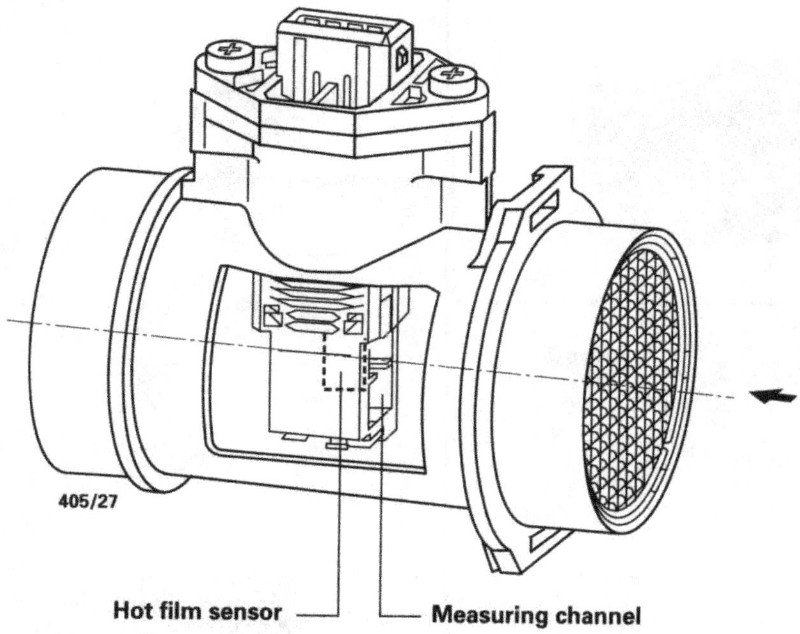

Motronic Engine Management

2.0 Liter Engine
Motronic 1551 Functions

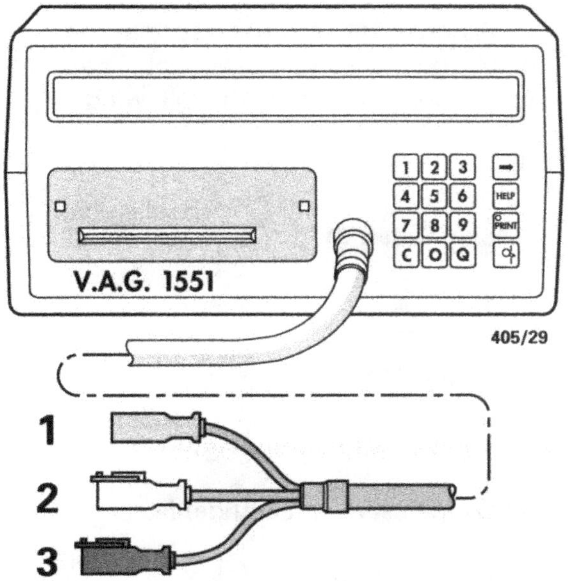

01 Check Control module versions

02 Check diagnostic trouble code memory

03 Output diagnostic test mode

04 Start basic setting

05 Erase diagnostic trouble code memory

06 End output

08 Read measuring value block

Function 02 - Check diagnostic trouble code memory

- Drive vehicle for 10 minutes before checking fault memory. During the road test, do the following to ensure that the necessary information is received by the control module:
  - Engine coolant temperature over 80°C for at least three minutes
  - Accelerate with wide open throttle with engine speed over 4600 RPM
  - Accelerate with half throttle in 4th gear from 30 mph for at least four seconds

Function 03 - Output diagnostic test mode

- The following components are activated by the control unit:
  - Injector #1 (N30)
  - Injector #3 (N32)
  - Injector #4 (N33)
  - Injector #2 (N31)
  - Idle air control valve (N71)
  - Evaporative emissions canister purge regulator valve (N80)
  - Relay for OXS heater (J278)
  - EGR vacuum regulator solenoid valve (N18) if equipped
  - Valve for intake manifold changeover (N156) (not used)
- The injectors are activated in the sequence of cylinder firing order. To activate the injectors, open and close the throttle. The injector will click 5 times and then the output will switch to the next injector.

Motronic Engine Management

Function 04 - Start basic setting

The Motronic system is adaptive and will automatically compensate for changes in engine operating conditions. This adaptation is done by the Motronic ECM which learns new operating values and ranges, and stores them in its memory.

If any Motronic system components are replaced, or the battery is disconnected, the function 04 "basic engine setting" should be performed using the 1551 scan tool. This helps provide the conditions for the Motronic ECM to learn any new operating conditions.

Refer to the repair manual for more information on this procedure.

Function 08 - Read measuring value block

In function 08, various operating values and conditions of the Motronic system can be displayed on the 1551 scan tool. These values are displayed in the following display groups.

Display Group 01

1 2 3 4

1 Engine speed

2 Engine coolant temperature

3 Oxygen sensor (Lambda) factor

4 Ignition timing

Display Group 02

1 2 3 4

1 Engine speed

2 Injection duration

3 Battery voltage

4 EGR temperature

Display Group 03

1 2 3 4

1 Engine speed

2 Engine load

3 Throttle angle

4 Intake air temperature

Motronic Engine Management

Function 08 (cont'd)

Display Group 04

1 2 3 4

1 Engine speed

2 Engine load

3 Vehicle speed

4 Switched inputs

00001 Fuel deceleration shut-off
00010 Closed throttle position
00100 Part load
01000 Wide open throttle
10000 Acceleration enrichment

Display Group 05

1 2 3 4

1 Engine speed

2 Idle air control adaptation

3 Idle air control duty cycle

4 Switched inputs

10 00 A/C compressor on
01 00 A/C on
00 01 Transmission in park or neutral (auto trans)

Display Group 06

1 2 3 4

1 Oxygen sensor (Lambda) factor

2 Oxygen sensor (Lambda) adaptation at idle speed

3 Not used

4 Oxygen sensor (Lambda) adaptation at part load

Transmission Range

020 5-speed Manual Transmission

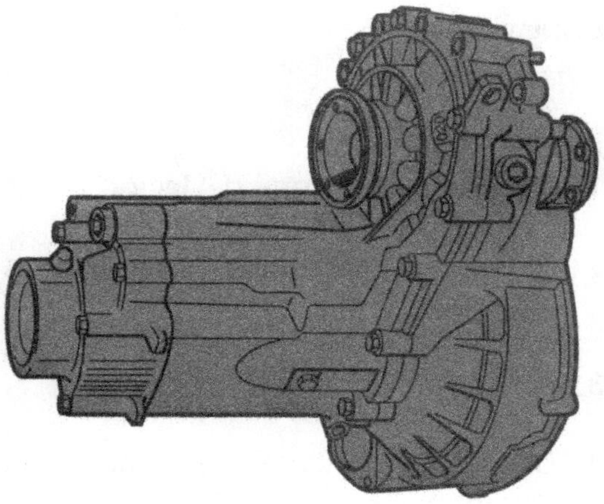

The 020 transmission is installed on all A3 models except the GTI and Jetta III GLX.

The type of shift linkage used for this transmission is rod style.

02A 5-speed Manual Transmission

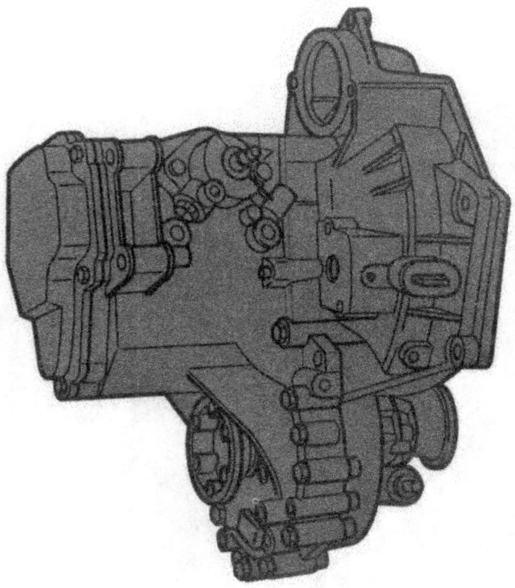

The 02A transmission is used on the GTI and Jetta III GLX (VR6 engine).

Cable operated shift linkage is used with this transmission.

Transmission Range

Phase One 096 Four-Speed Automatic Transmission

The new phase I 096 4-speed automatic transmission is available as an option on all models except GTI. This new transmission features the following changes when compared to the previous 4-speed automatic transmission:

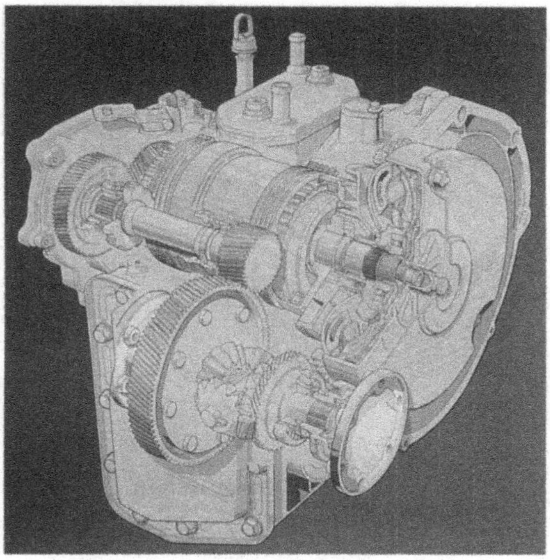

- Modified K1 and K3 clutches which provide quicker application and release for smoother shifts.
- New planetary gear set with revised gear ratios.
- Elimination of Economy/Sport switch.
- Revised shift points programmed into the transmission control module.

The previous version of the 4-speed automatic transmission was designed to shift from 3rd gear to 4th gear in several steps. This new Phase One 096 four speed transmission can now shift directly between 3rd and 4th gear. This results in smoother shifting characteristics.

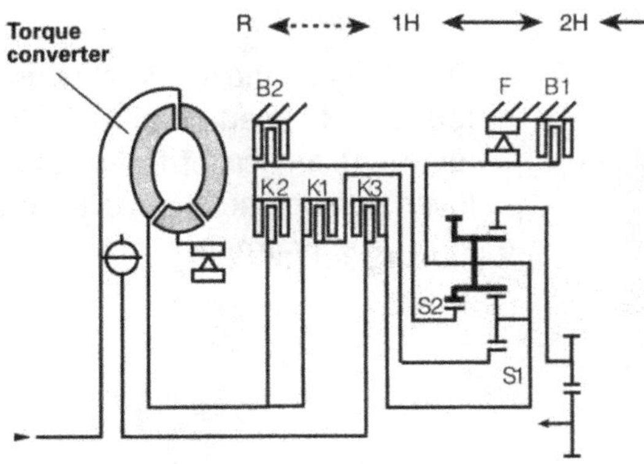

Gears	Clutches/Brakes B1	B2	K1	K2	K3	F	Gear Ratios
R	●			●			-2.88
1			●			●	2.71
2		●	●				1.44
3			●	●			1.00
4		●			●		0.74

405/135

Transmission

020 5-speed Transmission

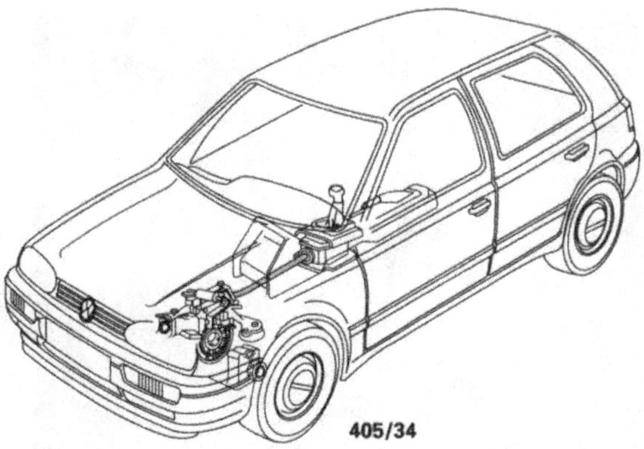

405/34

The shift linkage for the 020 transmission is a rod type. This linkage is similar to the shift linkage used on the previous Golf and Jetta models.

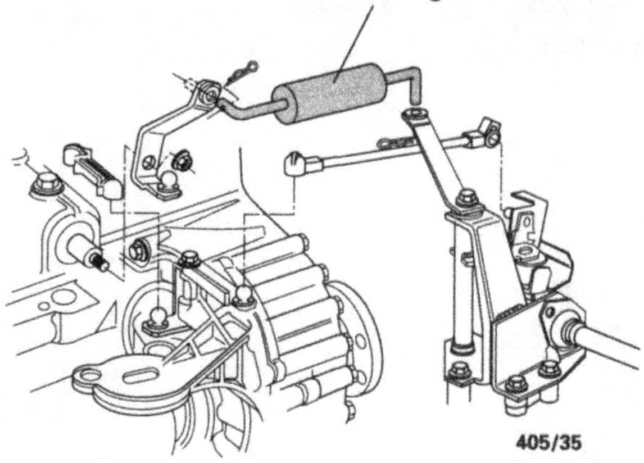

405/35

A balance weight is installed on the connecting link. The weight is used to help reduce engine noise and vibration that is transmitted into the interior of the vehicle through the shift linkage.

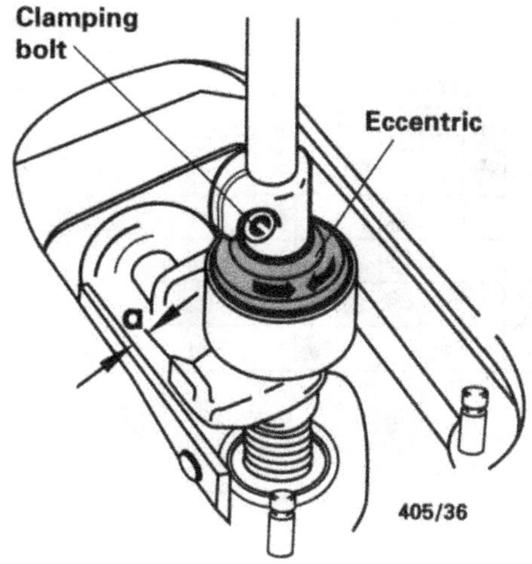

405/36

The shift linkage is adjusted using special tool #3104. In addition to the linkage adjustment, an eccentric, located at the shift lever, can be used to fine tune the shift linkage adjustment.

Transmission

02A 5-speed Transmission

For vehicles equipped with the VR6 engine, the 02A 5-speed manual transmission is used. This is the same transmission that is currently being used in the Passat and Corrado models.

As with the Passat and Corrado, gear selection is done via two cables.

One cable is used to select the proper shift fork, and the other cable moves the shift fork forward and backwards.

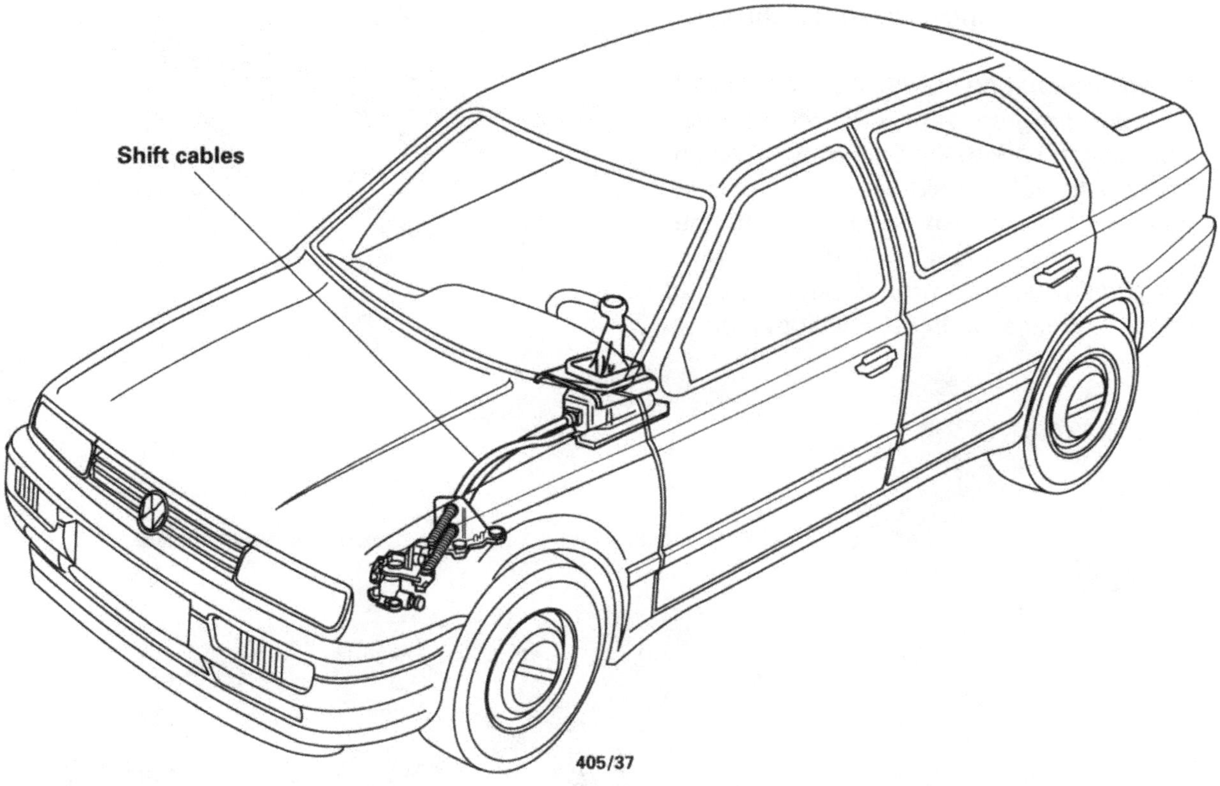

Transmission

Phase One 096 Four-Speed Automatic Transmission

An Economy/Sport switch is not used with the new phase I 096 4-speed automatic transmission. The transmission control module now selects the shift program automatically, adapting to different driving styles.

Two shift programs, Economy and Sport, are stored in the transmission control module memory. The TCM will automatically select one of the two based on the opening and closing speed of the throttle.

When accelerating quickly, the TCM selects the Sport program. The TCM will automatically switch back to the Economy program after the throttle is held below the full load range for a predetermined amount of time.

The transmission control module is located under the rear seat on the left side of the vehicle.

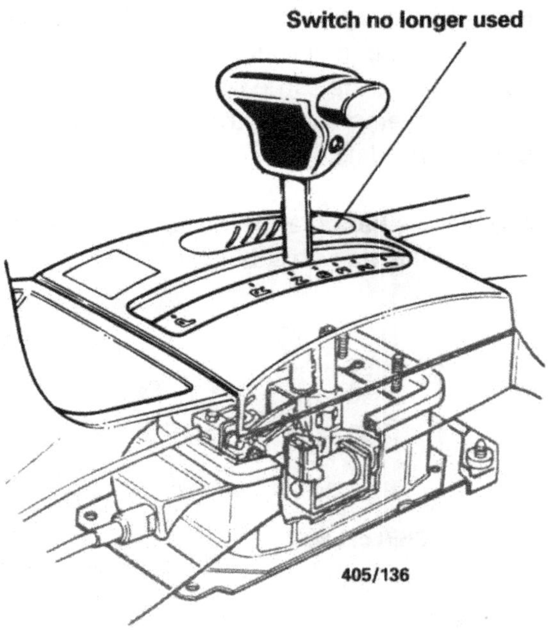

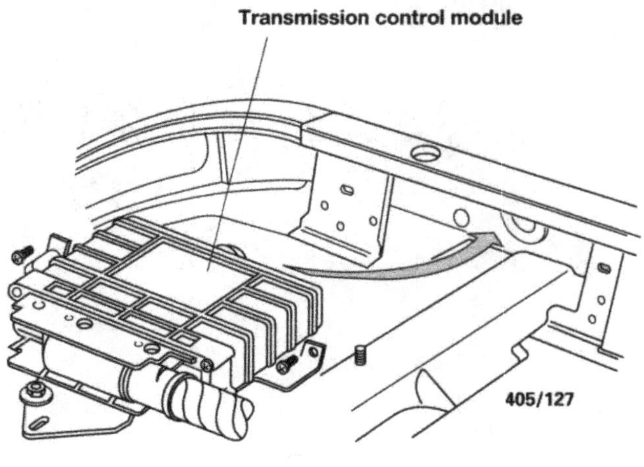

Automatic Shift-Lock III

Overview

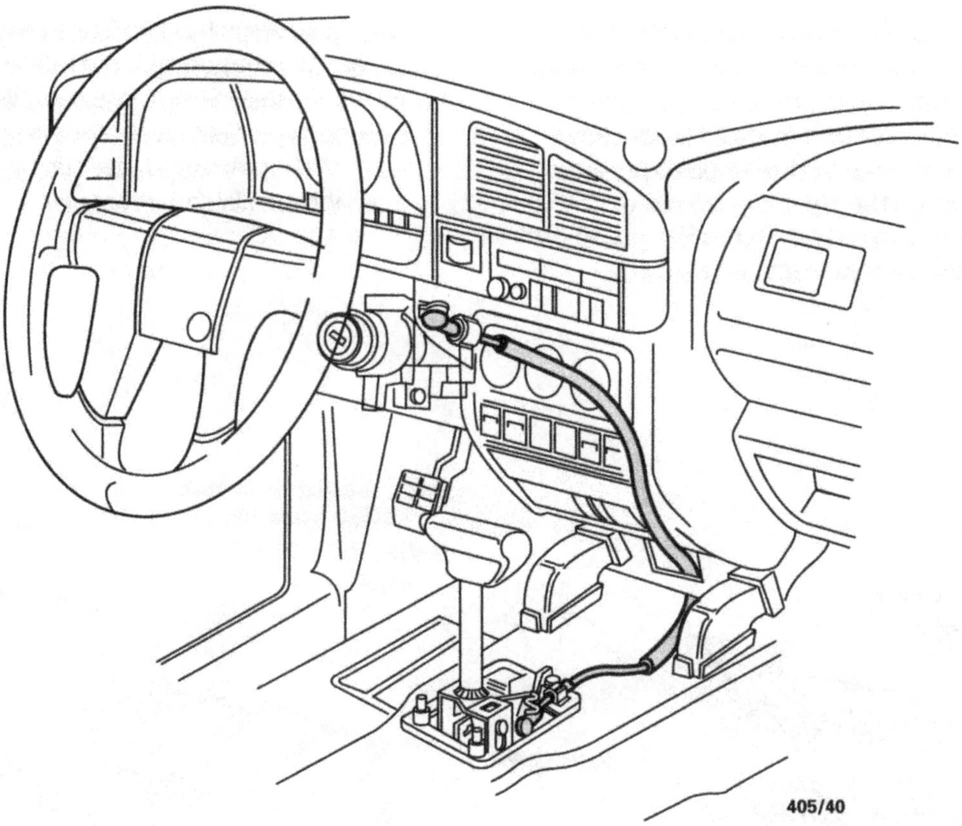

405/40

A new Shift-Lock III system is used on vehicles equipped with automatic transmissions. With this system, the ignition key can not be removed from the ignition switch until the shift lever is placed in the "Park" position. This helps ensure that the vehicle is not left with the transmission out of "Park."

The Shift-Lock III system is completely mechanical. The system is operated by a Bowden cable that is connected to a new ignition lock assembly and the shift lever assembly.

Once the ignition key is removed, the shift lever will be locked in the Park position. The selector lever can not be moved until the ignition key is inserted, the ignition is switched on and the brake pedal is depressed. This provides additional security against vehicle theft.

Automatic Shift-Lock III

Operation
(Ignition switched on)

When the ignition is switched ON, a bell crank assembly on the ignition switch pulls the shift lock cable. This cable pulls the stop lever away from the shift lever allowing, the shift lever button to be depressed and enabling the shift lever to move. Once the lever is moved out of the Park position, the ignition key can not be removed.

What prevents the key from being removed is the "S" shaped clip. The clip is designed to lock the stop lever in place when the selector lever is in any other position than Park. When the stop lever is locked, the shift lock cable can not move.

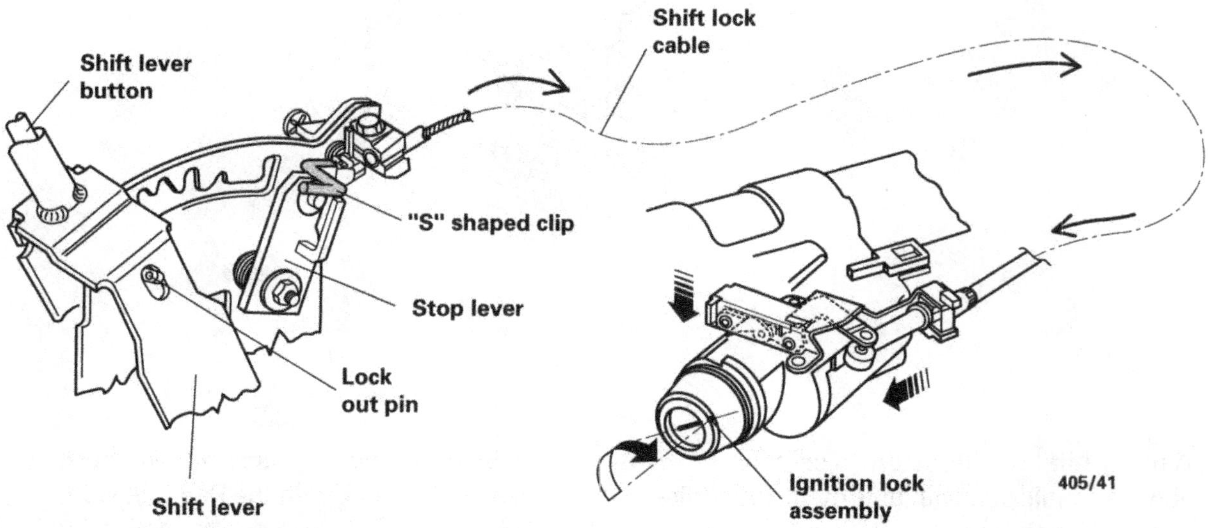

Automatic Shift-Lock III

Operation (Ignition switched off)

When the shift lever is returned to the Park position, and the ignition key is switched OFF, the shift lock cable pushes the stop lever into the shift lever. This does two things. The shift lever button can not be depressed to move the shift lever and the stop lever is moved to its most rearward position so that the ignition key can be removed.

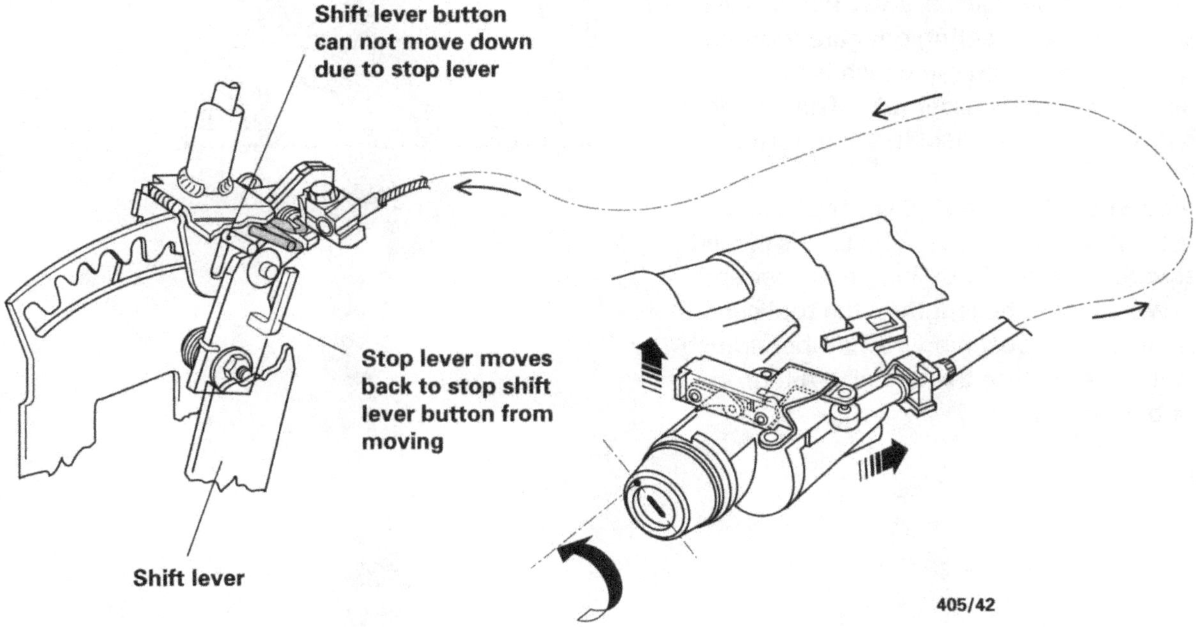

Suspension

Overview

The suspension for the A3 models is similar to the previous Golfs and Jettas. The suspension components have been further developed to improve the ride comfort and control.

The wheelbase remains at 2,475 mm (97.4 in.). The front and the rear track have been increased by approximately two inches. The increase in track width improves directional stability as well as ride comfort and handling.

Front Suspension

The front suspension utilizes the strut type design. This type of design takes less space when compared to a double wishbone design.

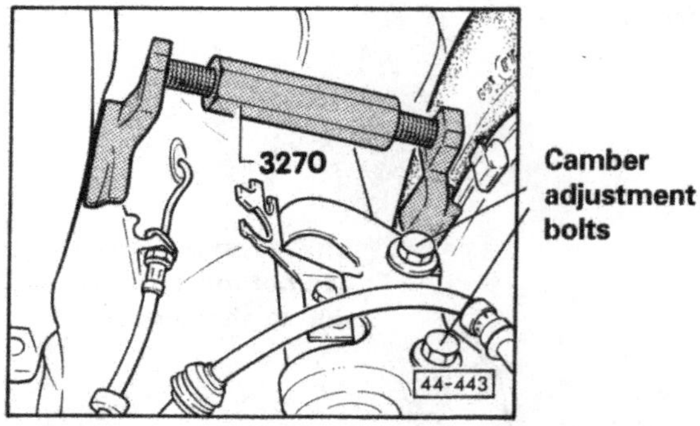

The lower control arms, sway bar and the rack and pinion steering gear are mounted on a separate subframe which is then bolted to the body assembly. This provides rigidity and helps to isolate road noise.

A new special tool, #3270 is used to adjust the front wheel camber. The tool is placed between the front body long member and the wheel rim. The length of the tool can be adjusted after loosening the camber adjustment bolts to place the wheel in the proper camber position.

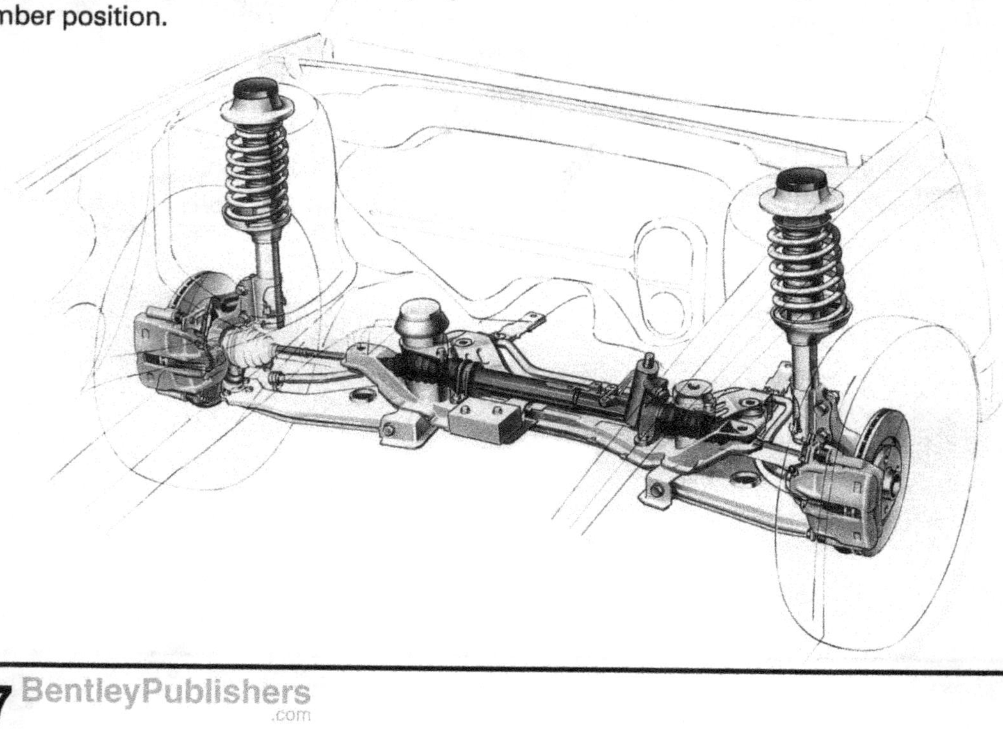

Suspension

Rear Suspension

A "V" profile independent torsion beam rear axle with integral trailing arms and sway bar is used on both the A3 models. The advantages of using this design is that the rear wheel camber and toe remain virtually unaffected by suspension movement.

The torsion beam rear axle is mounted to the vehicle with track correcting bushings. These bushings are asymmetrically shaped and minimize any steering effect from rear wheels during cornering.

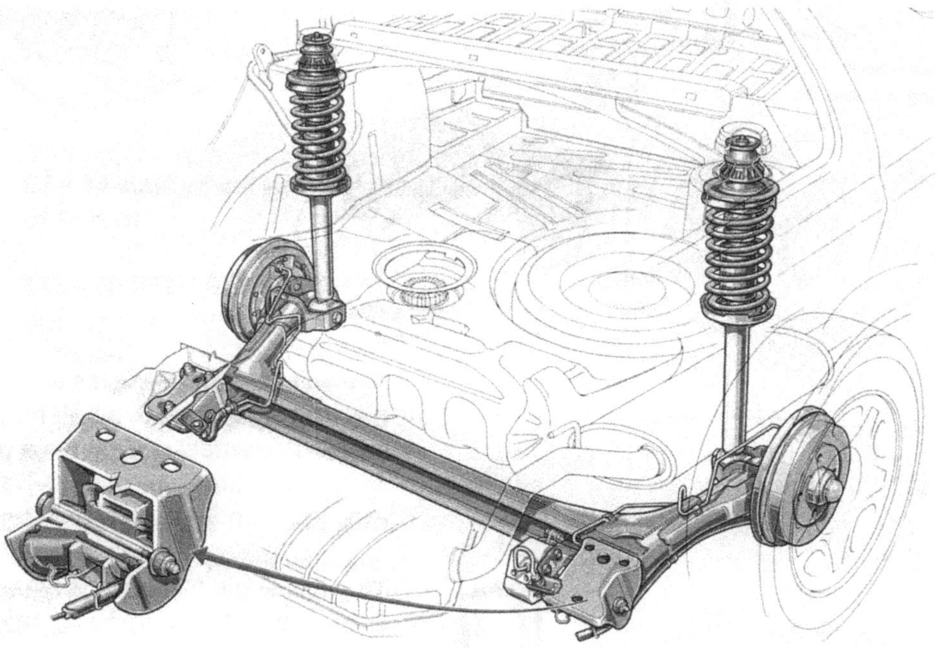

Suspension

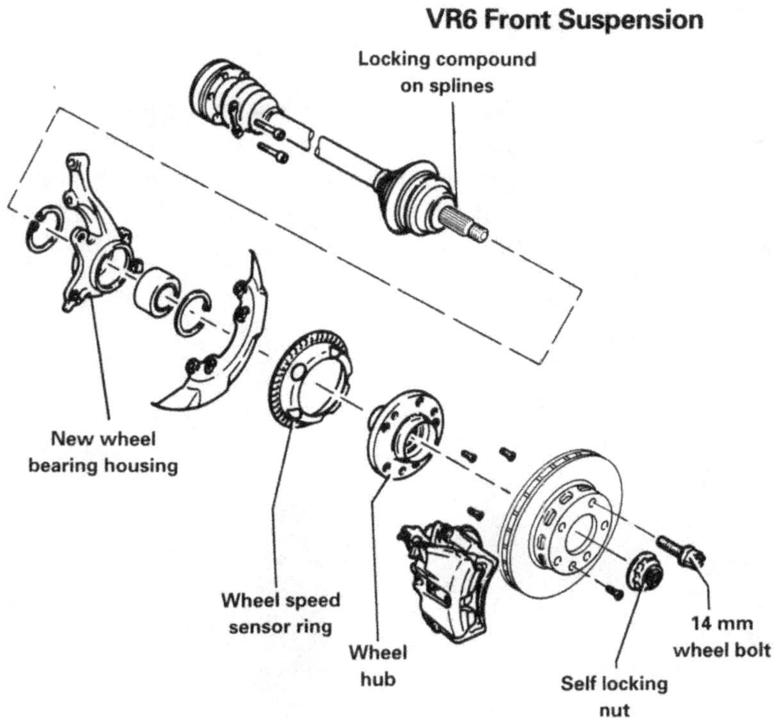

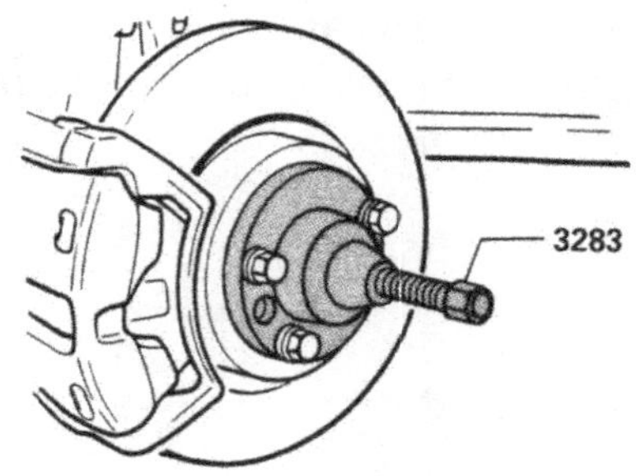

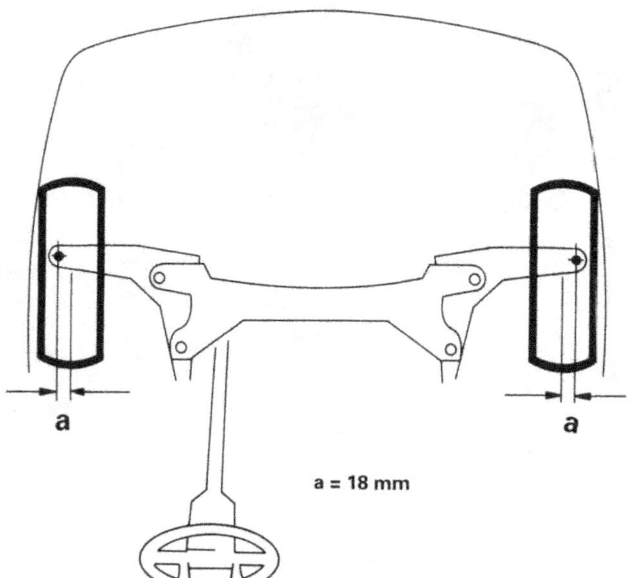

The suspension on vehicles equipped with the VR6 engine have five bolt hubs and wheels. The wheel bolts have a larger diameter and are also longer. The wheel offset has been increased to (43 mm).

The caster angle has been increased by approximately two degrees. The front lower control arms and wheel bearing housings are also different. These suspension changes help provide an increased negative steering radius. This helps reduce torque steer and provides good directional control.

The front CV joint splines are secured to the inner splines with a locking compound. The outer CV joint must be pressed out of the wheel hub using special tool 3283 when making repairs. New locking compound must be used on the splines and a new self-locking nut lock must be used during installation.

Suspension

Wheel and Tires

There are several wheels offered for the A3 models.

Three different wheel offsets are used. This is to ensure that the different tire and wheel combinations are flush with the vehicle body contours.

CL (Canada only)

Wheel size: 5.5 x 13 steel wheel with an offset of 38 mm

Tire size: 175 x 70 SR13

GL
GLS

Wheel size: 6J x 14 steel wheel with an offse of 45 mm

Tire size: 185 x 60 R14H (All Season) for manual transmission equipped vehicles
195 x 60 R14H (All Season) for models with automatic transmission

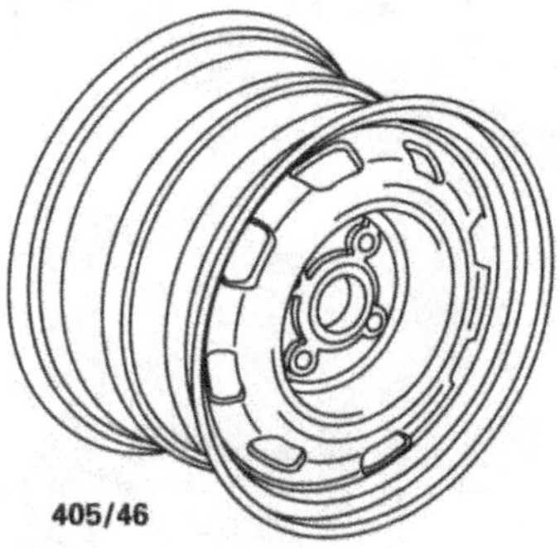

405/46

GTI

Wheel size: 6.5J x 15 alloy wheel with an offset of 43 mm

Tire size: 205 x 50 R15V (All Season)

GLX

Wheel size: 6.5J x 15 alloy wheel with an offset of 43 mm

Tire size: 205 x 50 R15V (All Season)

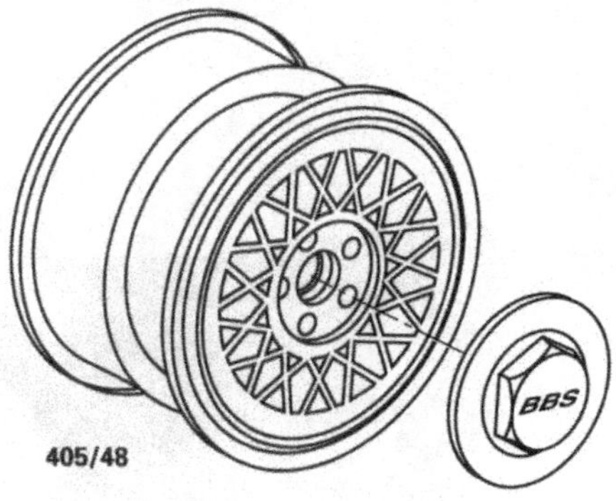

405/48

Brakes

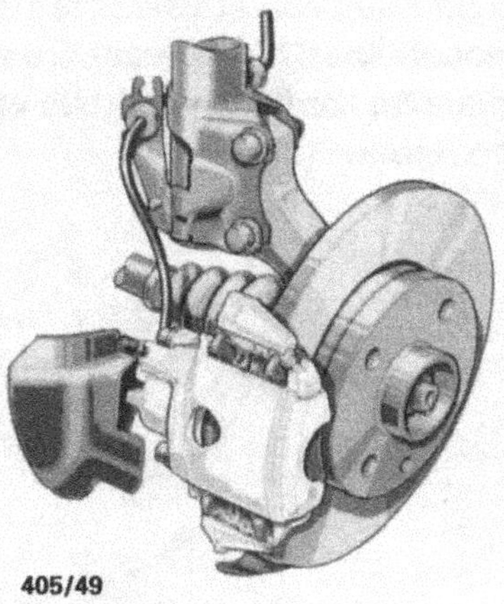
405/49

The A3 models are equipped with several different brake systems. All models have a dual diagonal hydraulic brake system.

The front brakes for models equipped with the 2.0 liter 4-cylinder engine are a single piston floating caliper design. The diameter of the disc is 256 mm.

The CL and all models with diesel engine (Canadian model) are equipped with 239 mm solid brake rotors. An air duct for brake cooling is attached to each front caliper.

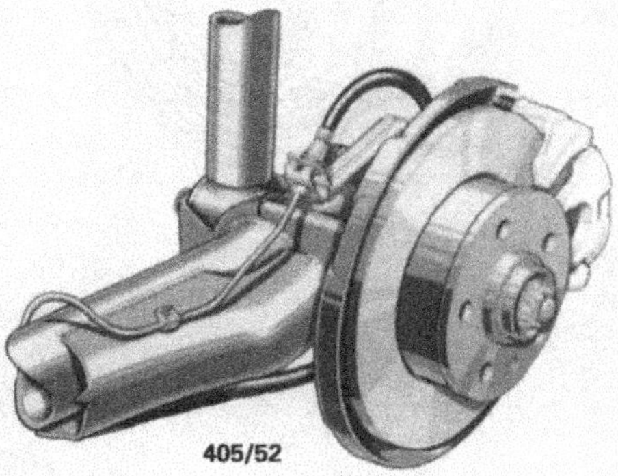
405/52

Some models are equipped with rear disc brakes. The rear discs are 226 mm in diameter. A load dependent brake pressure regulator is also used.

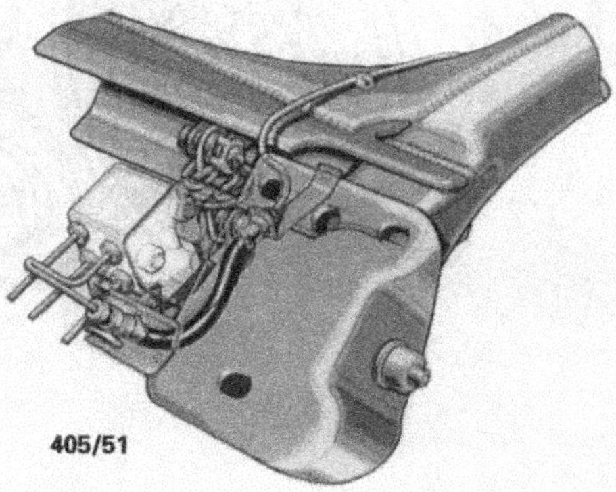
405/51

The load dependent brake pressure regulator is located on the left side of the rear axle. The regulator controls the amount of brake pressure to the rear wheels.

Brakes

On some models, the rear brakes are drum type with self-adjusting shoes. The diameter of each drum is 200 mm and the width is 40 mm.

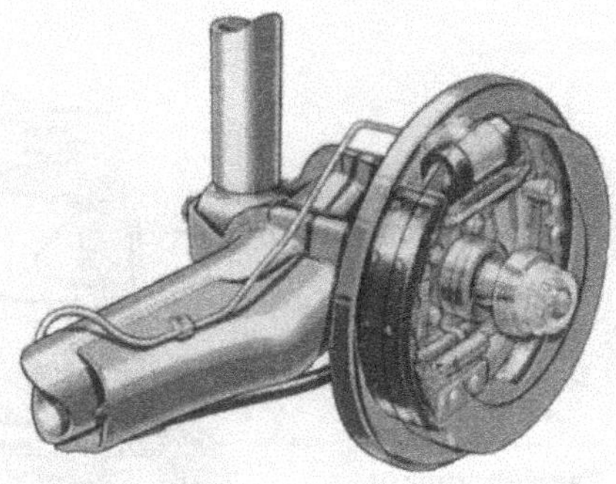

405/50

VR6 brakes

Vehicles equipped with the VR6 engine have larger front brakes to compensate for the additional power output. The larger front disc brakes, Teves Anti-lock brake system and traction control are standard equipment.

The internally ventilated front discs are 280 mm in diameter. The front calipers are a single piston floating caliper design.

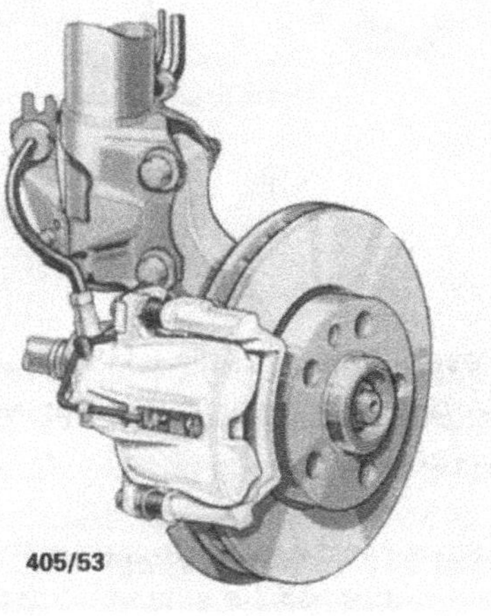

405/53

The rear brakes are disc. The diameter of the rear discs are 226 mm. Both front and rear wheel hubs are a five bolt pattern.

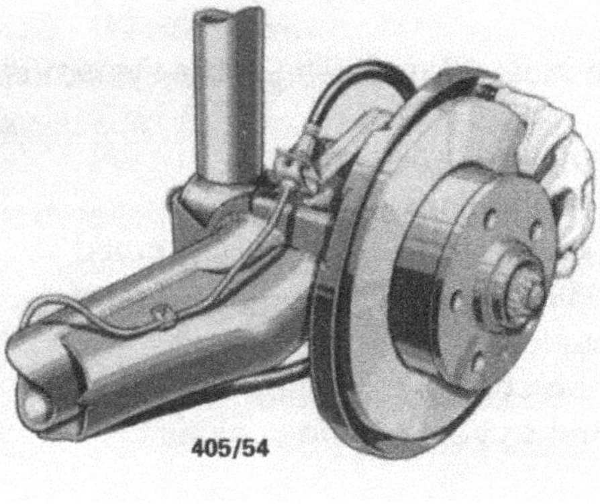

405/54

ABS

Anti-lock Brake System

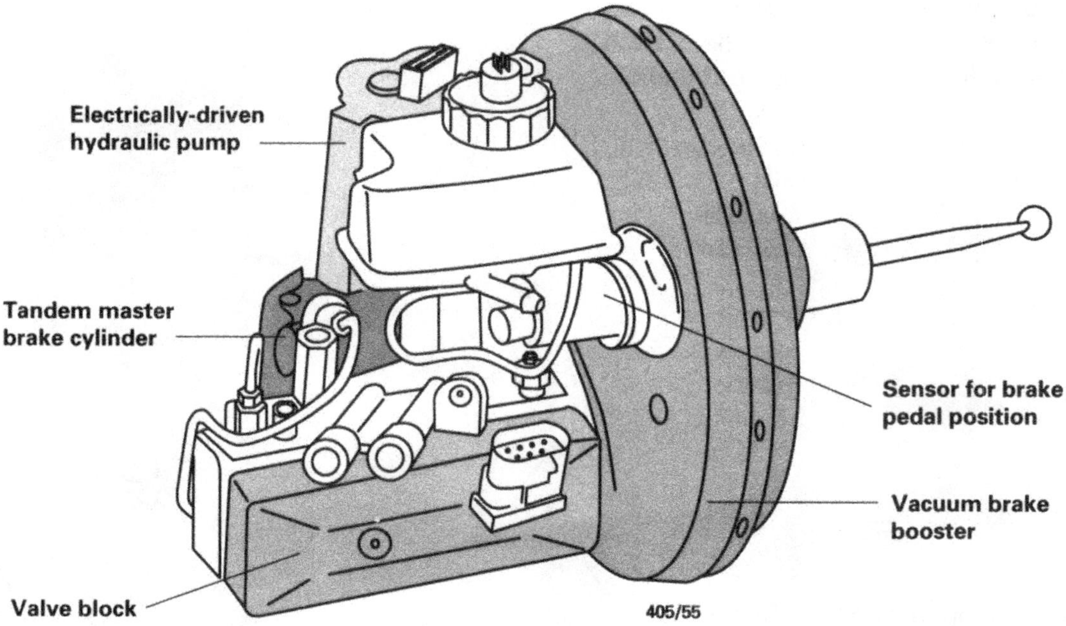

The Teves® MK IV anti-lock brake system is standard equipment on models with the VR6 engine.

The Teves MK IV system is a further development of the MK II system used on other Volkswagen models.

The main components of the system are:

- Vacuum brake booster
- The master cylinder
- Separate hydraulic control unit
- Wheel speed sensors (one for each wheel)
- ABS control module
- Brake pedal position sensor

As with the MK II system, the MK IV system is a three channel system which permits individual control of the front wheels and a joint "select low" control of the rear wheels.

The most obvious difference between the two systems is the addition of a vacuum brake booster on the MK IV system. The advantage of using a vacuum booster is that if the hydraulic pump should fail, the brakes can still be operated conventionally with power assistance from the vacuum booster.

ABS

System Layout

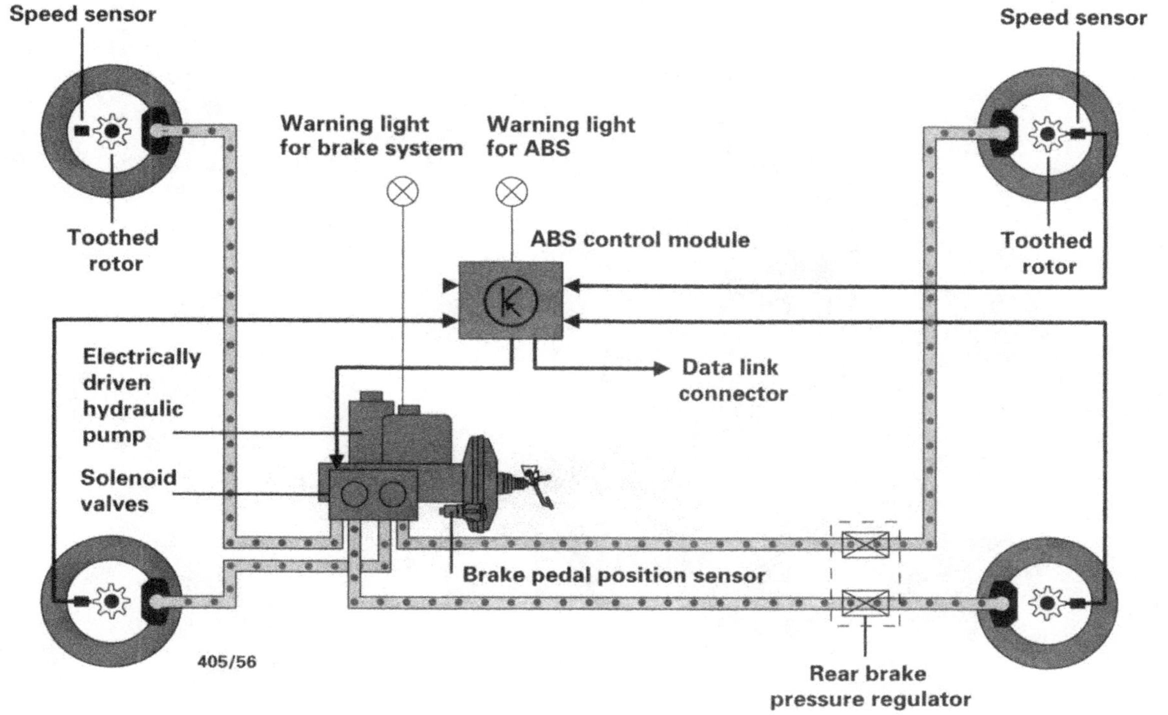

Another new feature of the MK IV system, when compared to the MK II system, is that it utilizes dual diagonal brake circuits.

The self-diagnosis function has been expanded to include the use of the V.A.G. 1551 scan tool. Faults that occur within the anti-lock brake system can be read using the 1551.

On vehicles equipped with the VR6 engines, a low speed traction control system (EDL) is included with the anti-lock brake system.

The traction control works in conjunction with the ABS system by alternately applying the front brakes to control wheel spin during low speed acceleration. By controlling wheel spin when the vehicle is starting to move from a stop, more power can be transmitted from the vehicle to the road surface. This increases traction and vehicle control when starting off in slippery conditions.

More detailed information on these systems can be found in other Service Training Self-study programs.

Steering

Steering

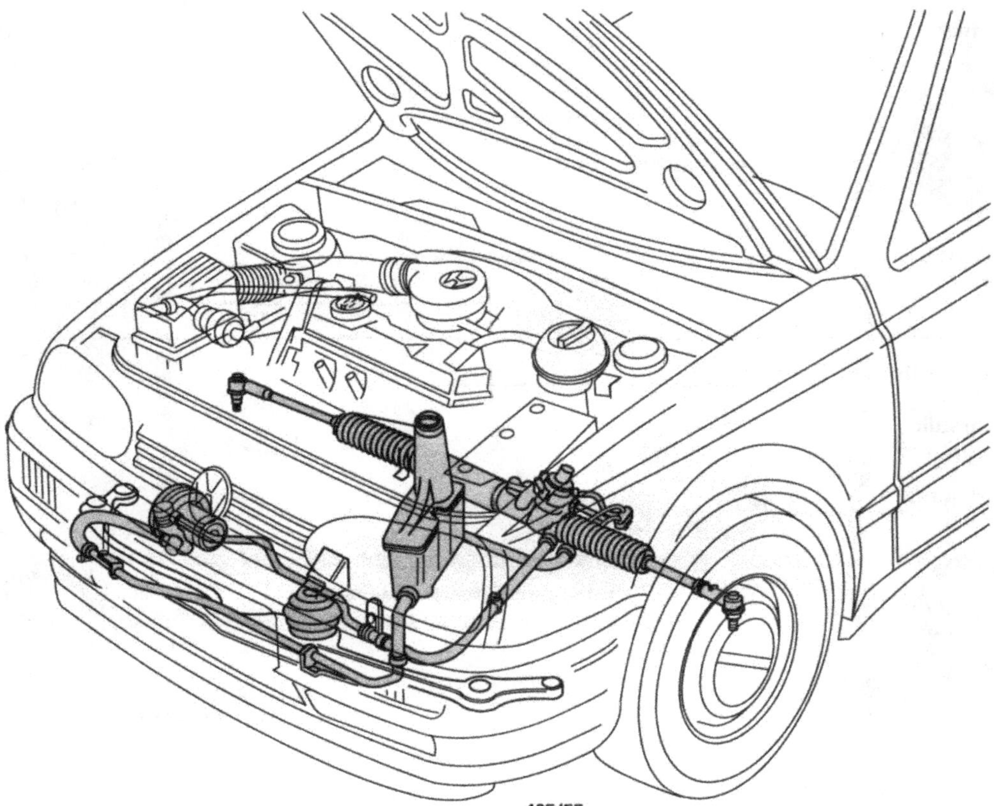

405/57

Power assisted rack and pinion steering is standard on all A3 models. The power steering pump is a vane type pump. The hydraulic (power steering fluid) oil is G.002.00 which is a mineral based oil.

The power steering belt is adjusted using a toothed rack mechanism. This type of adjustment is the same as the alternator belt adjustment on previous Golf and Jetta models.

The reservoir for the power steering pump has been designed with a longer filler neck for easy access. The fill capacity of the system is 0.7 - 0.9 liter (0.75 - 0.95 qt.).

Steering

Steering column

The steering column is also new. The steering wheel and steering column are not only functional components but are also essential safety components.

Normal condition

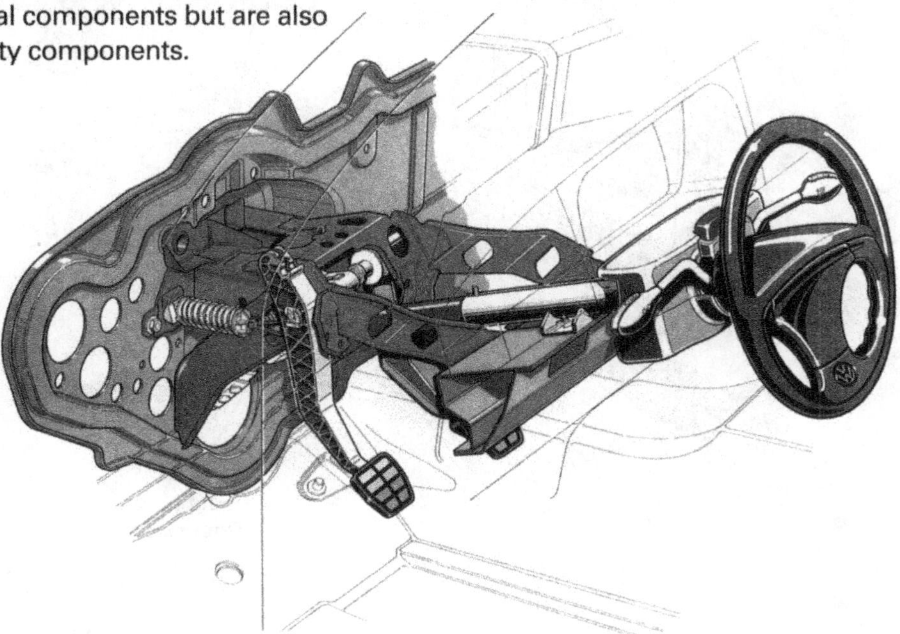

The foamed steering wheel is energy absorbing and acts as a crumple zone. The telescopic steering column can now retract as much as 170 mm, previously 110 mm.

After Severe Impact

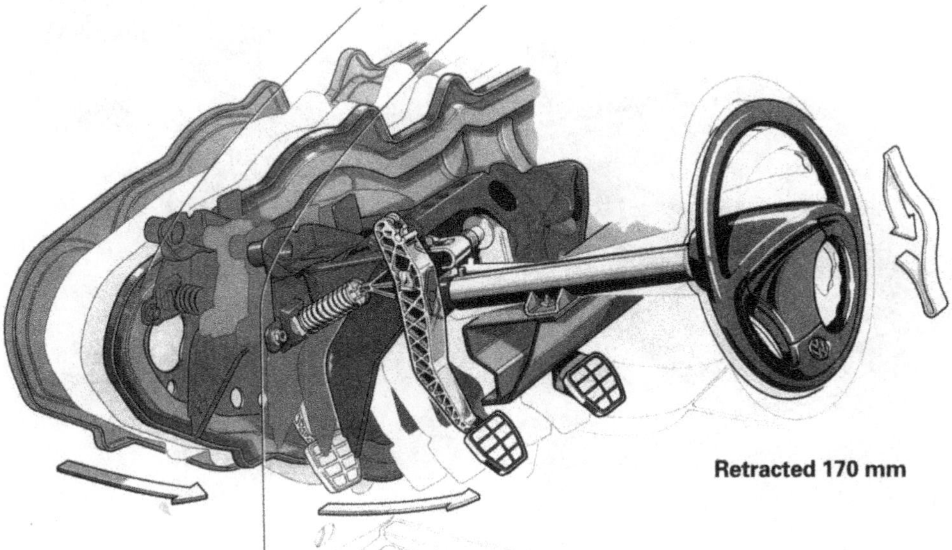

Body Assembly

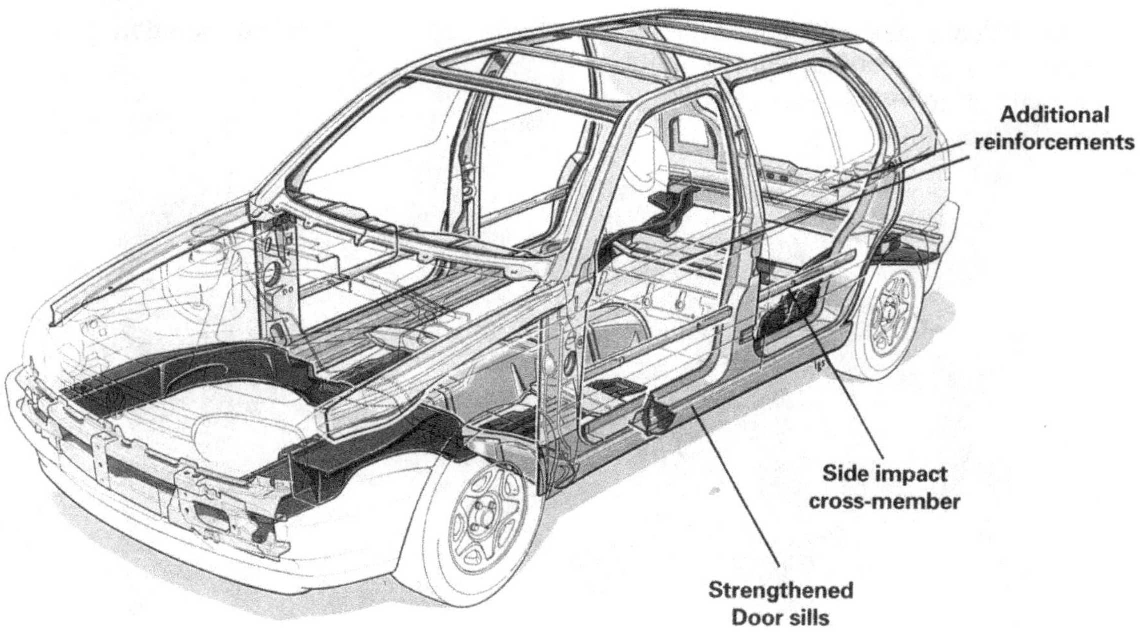

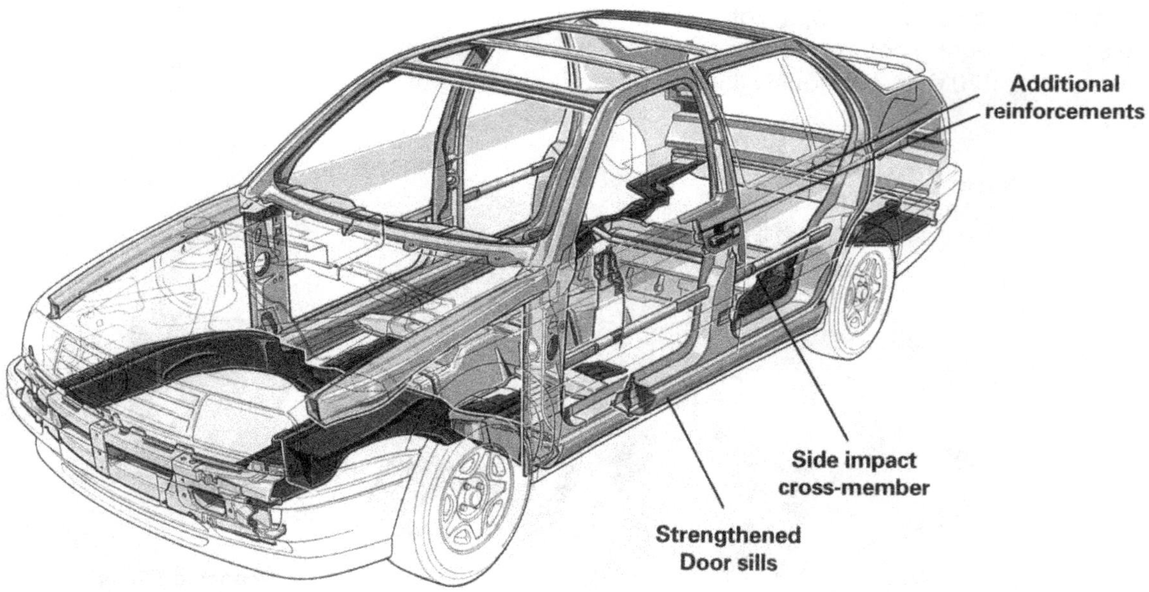

Body Assembly

The new A3 body assembly was designed to provide a very rigid body structure. The torsional stiffness has been increased by approximately 30% from the previous models. This increased strength helps provide lower vehicle interior noise levels, improved ride and handling qualities, and good occupant safety in the event of an accident.

The door sills and front and rear long members/side rails are connected together by several crossbeams that are welded in the floor assembly. A very rigid cross-member is bolted between the two front "A" pillars. The dashboard and steering column are attached to this cross-member (see page 49).

The front long members are made with a new construction process called mash seam welding. The members consist of three sections that are each a different thickness and welded together (thinnest at front, thicker at rear). This design provides a progressive deformation in a frontal collision for good energy absorption.

The body assemblies are designed to form a very rigid passenger cell around the interior compartment. The lower door sills have been significantly changed in profile and the metal thickness increased to provide good lateral protection. Side impact cross-members are bolted into each door assembly. Additional reinforcements are also welded in the top of each door to form profiled stiffening sections for additional side impact protection.

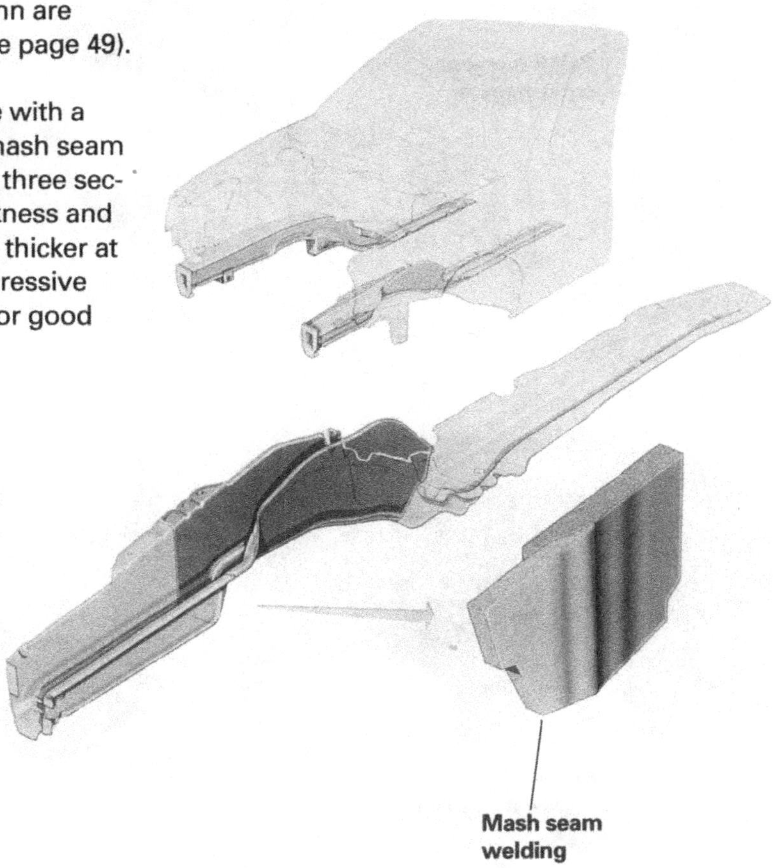

Body Assembly

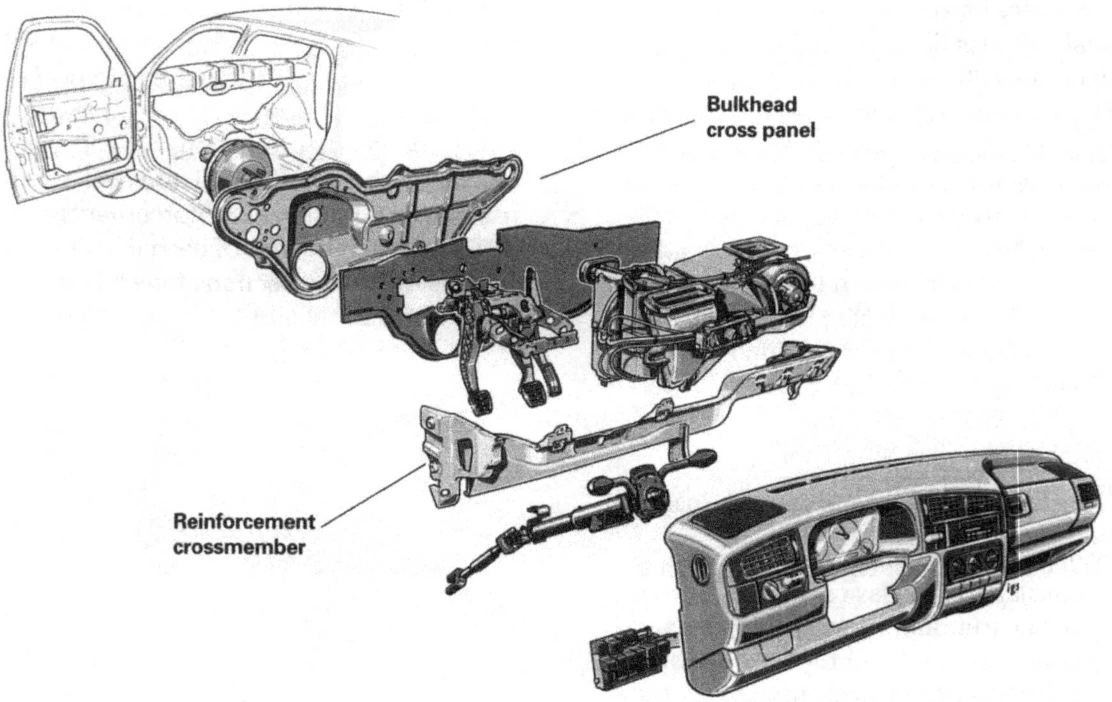

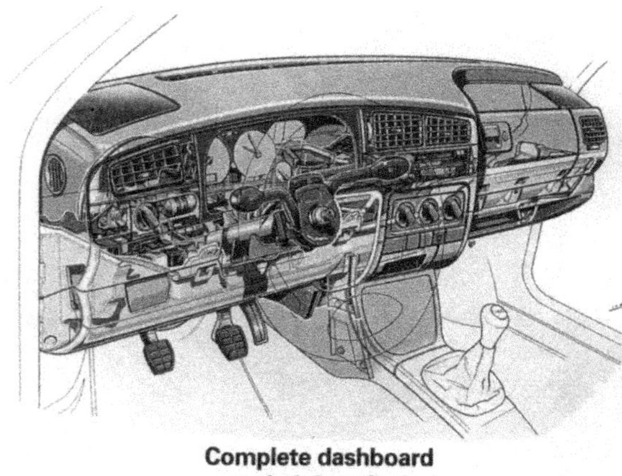

Complete dashboard
cockpit installed

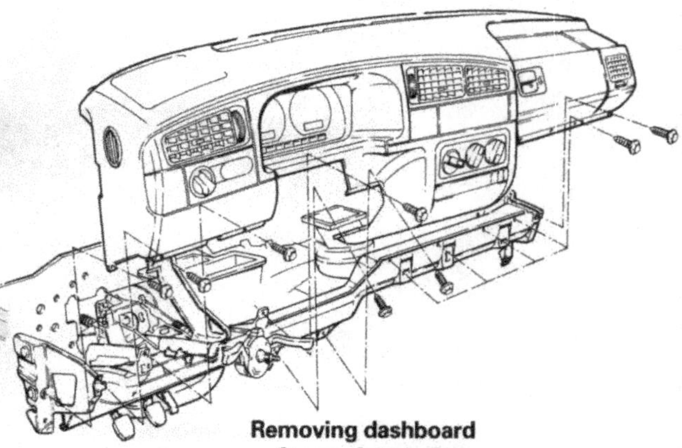

Removing dashboard
for service work

Body Assembly

Modular assembly is used extensively in the new A3 models. The dashboard unit, together with the heating and air conditioning unit, steering column, and pedal cluster with master cylinder, are assembled as a module outside the vehicle. These components are attached to the reinforcement cross-member and a bulkhead cross panel. The entire assembly is then installed in the vehicle.

This permits many of the more difficult assembly procedures on these units to be performed outside the vehicle where access to the components is easier. This modular assembly procedure is also used on the front end, door assemblies and sunroof. Newly designed central electrical connectors are used for most of these modular assemblies (See page 79).

For service work, the dashboard can be removed from the reinforcement cross-member by unbolting it. It is not necessary to remove all of the other components.

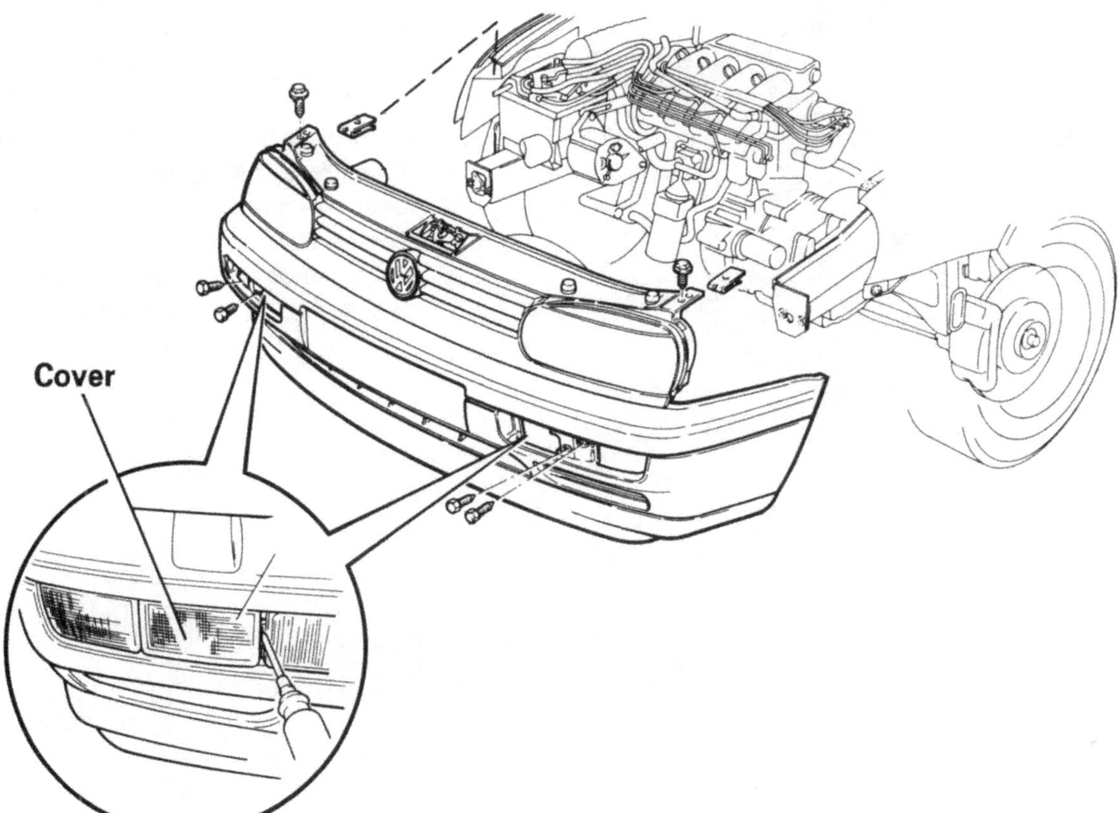

Bumpers

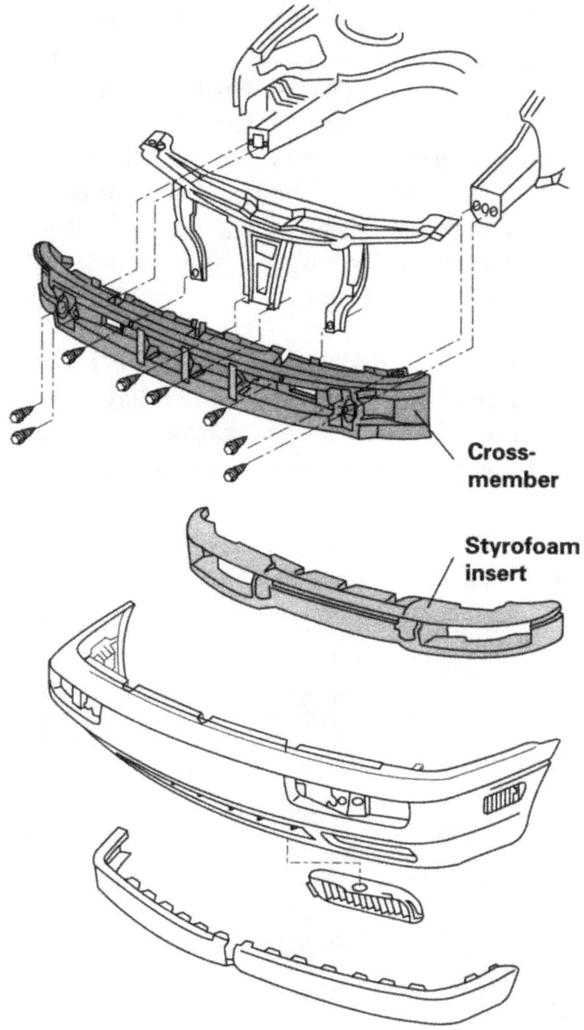

405/67

Bumpers

The bumper assemblies on the A3 models are a new design. A bumper cross-member is bolted directly to the body structure instead of using brackets. A molded styrofoam insert is installed over this cross-member. The body colored bumper covers are then bolted to the bumper cross-members with the styrofoam insert sandwiched in-between.

The bumper covers are specially shaped from special plastics to optimize their ability to absorb minor impacts. The styrofoam insert will compress and absorb major impacts. The bumper cover can then be easily removed for replacement of the styrofoam insert. This design spreads the bumper loads over a larger area and saves weight over other bumper designs.

A tow hook, which is included in the vehicle tow kit, can be threaded into the front bumper cross-member. The clear plastic cover can be snapped out in the center of the right directional light/fog light assembly for access to the front tow hook installation.

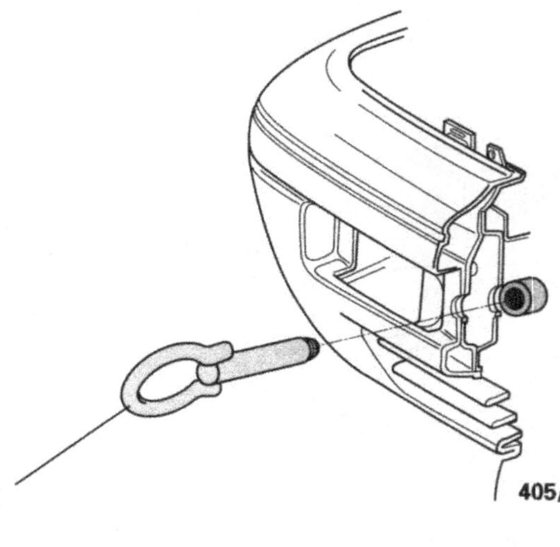

405/68

Doors

The door lock assemblies are new. The lock assemblies are a wedge type latch and are now located inside the door. This design lowers the likelihood of the door jamming shut in the event of a severe collision in which the door is deformed, and also provides a more positive alignment of the door lock.

The locking plate striker pin installed on the "B" pillar can be positioned to properly adjust it with the door and lock assembly position. The door-to-body position can be adjusted, if necessary, at the door hinge attachment bolts via elongated holes.

Each door lock has an adjustment to eliminate play in the linkage between the door lock and release handle. The adjustment is performed by tightening an adjustment screw to a specified torque after the lock is installed. Right side locks have a left hand threaded adjustment screw. Left hand locks have a right hand thread. Refer to the repair information for the specified torque and adjustment procedure.

Doors

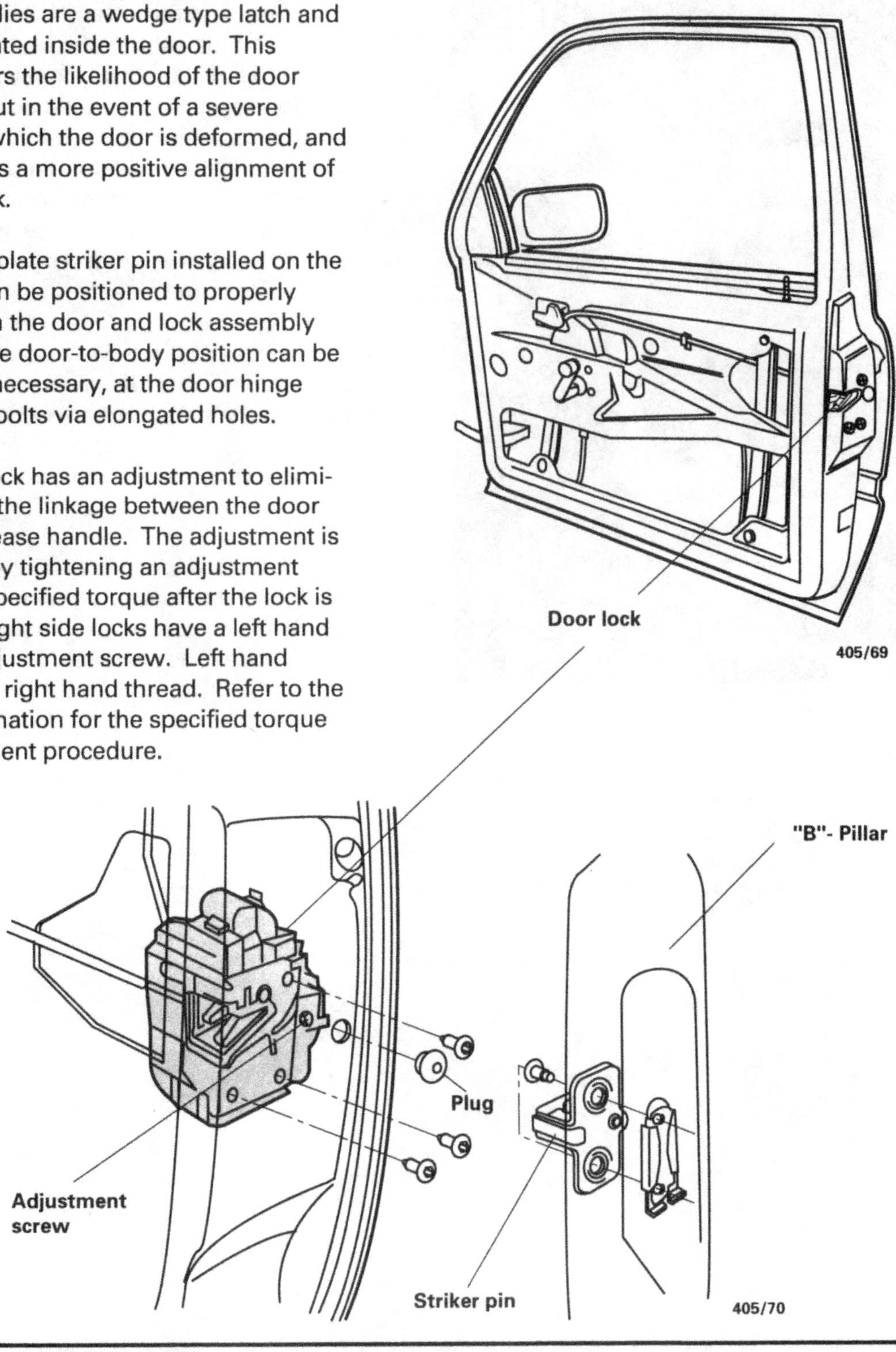

Seats

Front Seat

A3 models have multi-density foam seat cushions. The thickness of the foam is greater on the sides of the cushion than in the center. This provides good seating comfort together with good lateral support. The foam cushion is supported by a steel seat pan. The seat pan has a ramp design at the front to prevent the occupant from "submarining" down into the seat and under the seat belt during heavy breaking or a frontal impact.

The driver's seat has a new parallel height adjustment. This maintains the correct angle of the seat cushion as the height is adjusted. Some models will also have this height adjustment for the front passenger seat.

On models equipped with heated front seats, the seat heater control relay is located on a bracket which is riveted to the under side of the seat pan.

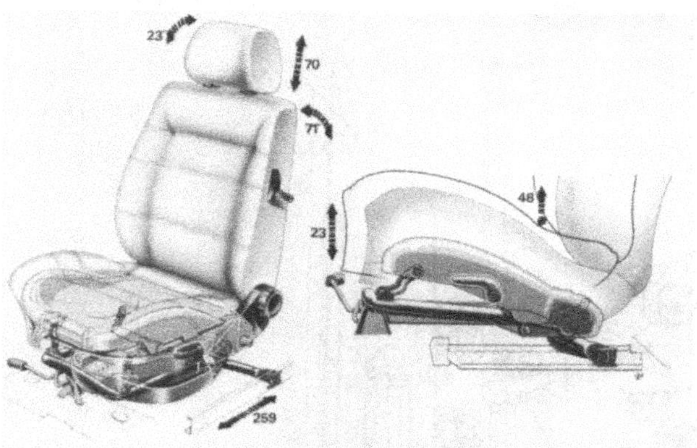

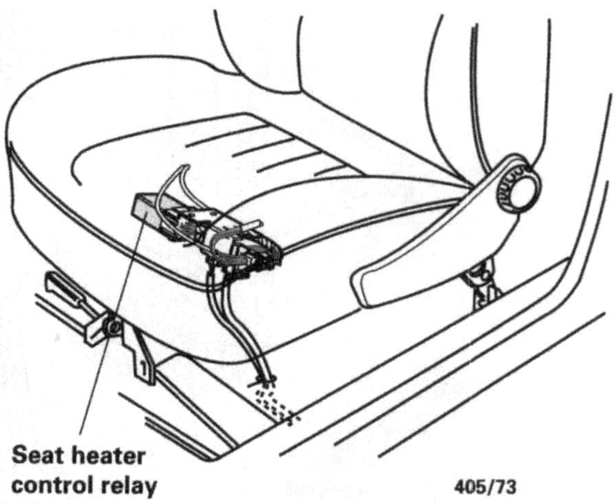

Seats

Rear Seat

The rear seats are a flat folding design for increased storage area. The seat back is made from sheet metal for added security.

Jetta III GLS and GLX models are equipped with a 60/40 split folding rear seat.

All Jetta III models are equipped with a locking feature for the rear seat backs. The seat backs can be locked with the vehicle master ignition key. Lock cylinders are installed next to the seat back release levers.

When the key is turned in the lock position, a locking lever prevents the seat back release lever from moving back to unlatch the seat.

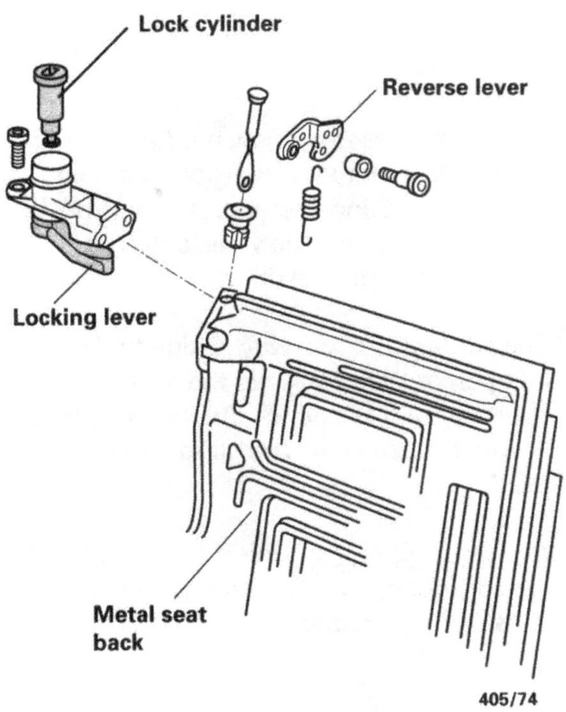

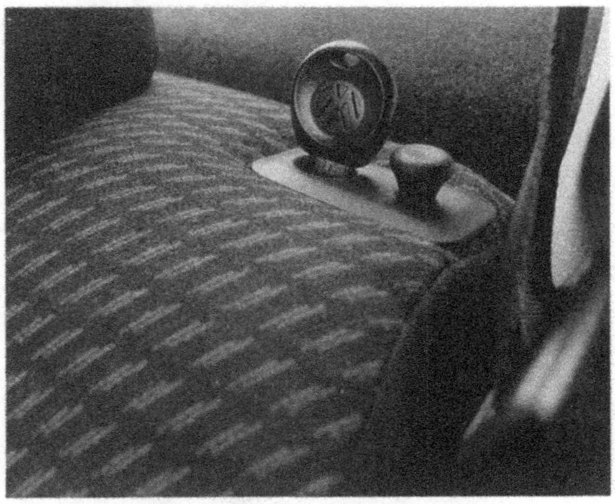

Sunroof

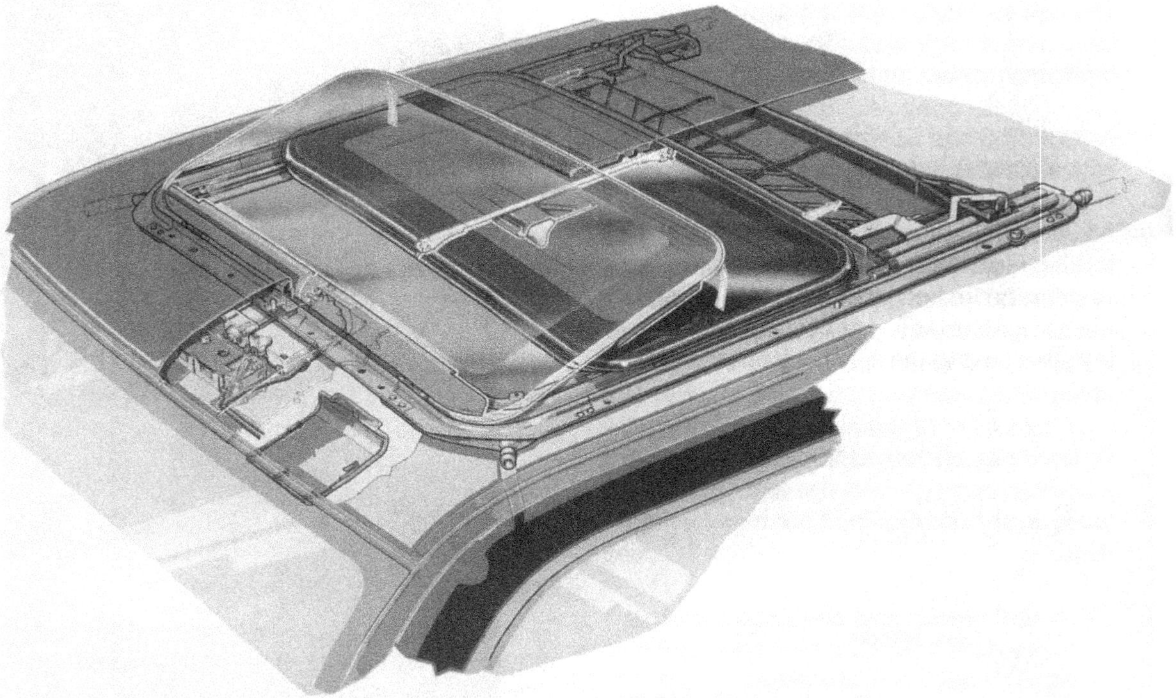

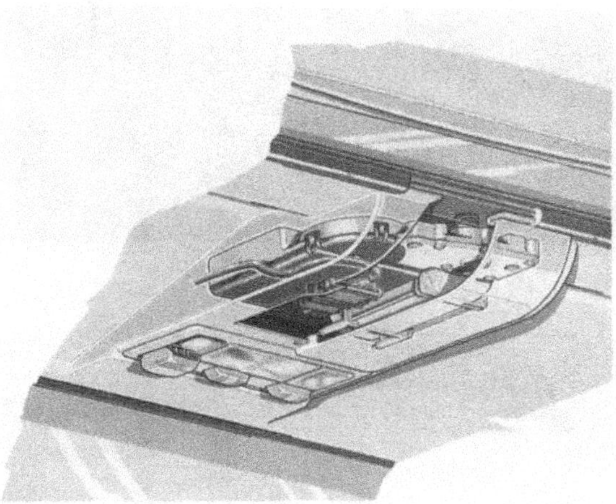

The A3 models use a new modular power sunroof similar to the sunroof introduced on the Passat and Corrado. The sunroof is assembled outside the vehicle then installed as a complete unit.

The sunroof is a two way design that can tilt open or slide back into the roof. The sunroof is a glass panel. An interior sliding shade that can cover the glass panel is also used.

An electronic control unit is combined with the sunroof motor. This control unit regulates the motor operation.

Sunroof

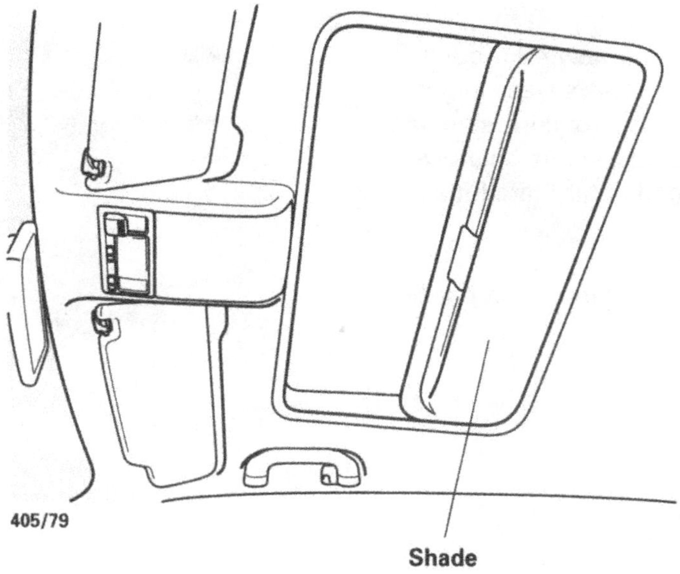

The sunroof motor is operated by a new "logic" switch. This switch opens and closes the roof in imitation of the roof movement. Moving the switch backwards opens the roof, forward closes it. Pushing up on the rear of the switch tilts the roof up at the rear. Pulling down will lower the roof closed.

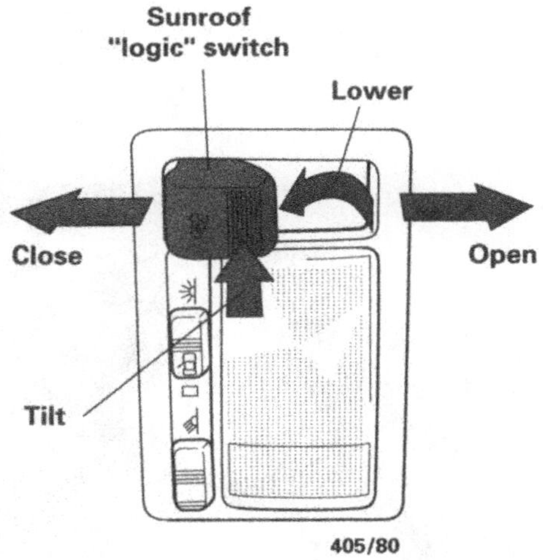

Included also in the sunroof operation are the following features:

- If the shade is closed when tilting the roof, the shade will automatically open partially for ventilation.
- When locking the car from either front door, an open sunroof can also be closed by holding the key in the locked position.

Plastics

Many components of the new A3 models are made from plastic. This saves weight and reduces the possibility of corrosion. All of the major plastic components on these vehicles are now coded to facilitate future recycling.

405/127

The label which is molded into each plastic component specifies:

- Name of manufacturer
- Part number
- Material code

Example: > PP /EPDM - T 20

PP = Polypropylene
EPDM = Ethylene propylene rubber
T 20 = Talc reinforced

405/128

Windows

The windshield and rear window are bonded in position. This contributes to the strength of the body structure and provides a finish more flush with the body surface. On two door models the rear side windows are also bonded in place.

The window glass is supplied with a sealing lip that is bonded to the glass. The PUR adhesive that bonds the glass to the body is applied directly to this sealing lip.

Eccentrics are used to adjust the windshield so that it fits properly within the window opening when it is installed. The eccentrics should be used only to lower the window in the opening.

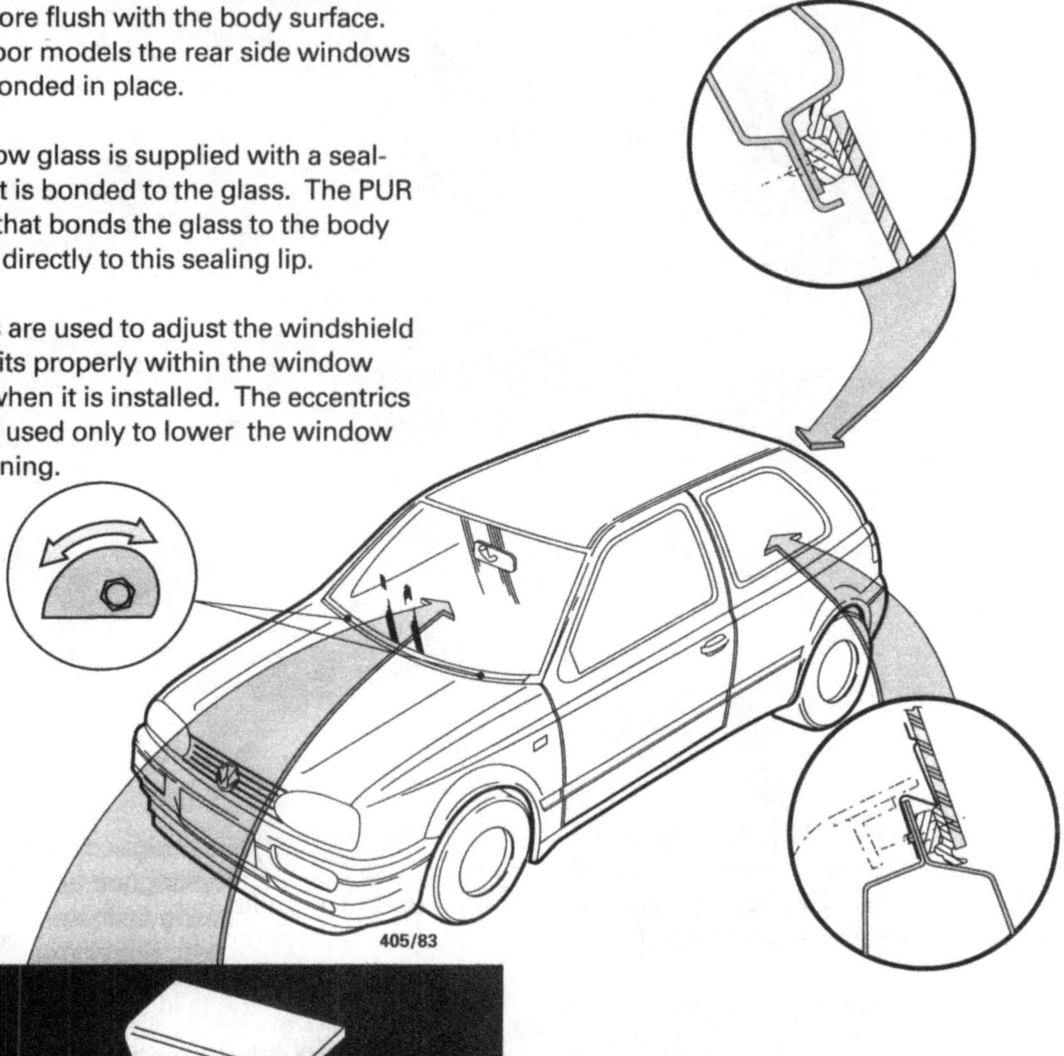

405/83

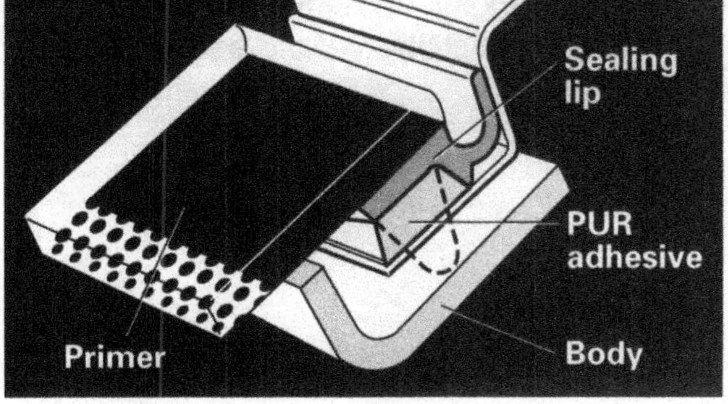

405/84

Heating/Ventilation

Air Flow

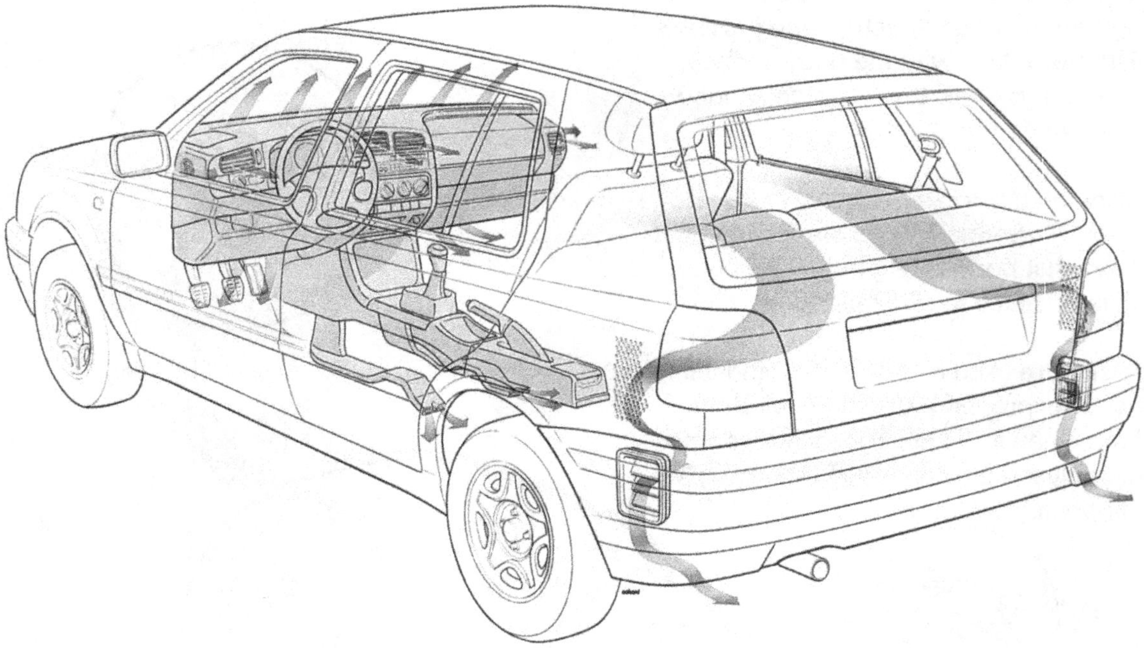

The air flow in the A3 models has been increased by 25% from the previous Golf and Jetta models. The heating ducts to the rear seats have also been enlarged.

The interior air is extracted through vents that are located in the rear of the vehicle, behind the bumper covers. This is a relatively neutral pressure zone, which results in the flow of air being quiet and less affected by changes in vehicle speed.

The duct work is designed to provide two layers of air flow for good occupant comfort. The air flowing from the center dashboard vents moves at a higher speed than air from the foot well and rear heating vents. This results in an upper layer of air that is slightly cooler than the lower air layer around the feet and legs.

Heating/Ventilation

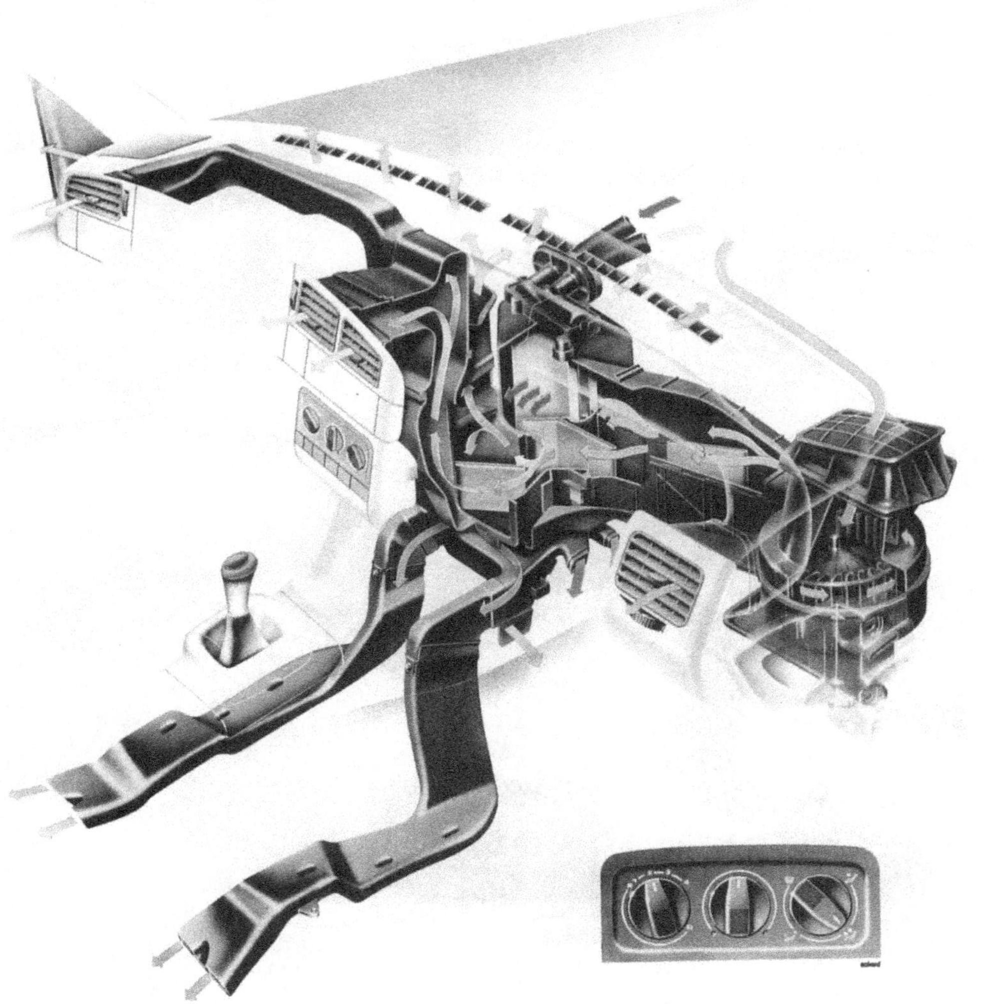

The heating and ventilation system is a blend type system. A heater on/off valve is not used. Coolant flows unrestricted through the heat exchanger. Interior temperature is regulated by a blend door which controls the amount of air that is passed over the heat exchanger. This blend door is operated by a cable connected to the temperature control dial.

The ventilation controls have been revised for ease of operation. Three round dials are now used to control blower speed, air temperature and air distribution.

The fresh air blower has four speeds. The series resistance for the blower is located in the blower motor housing.

Air Conditioning

A/C System

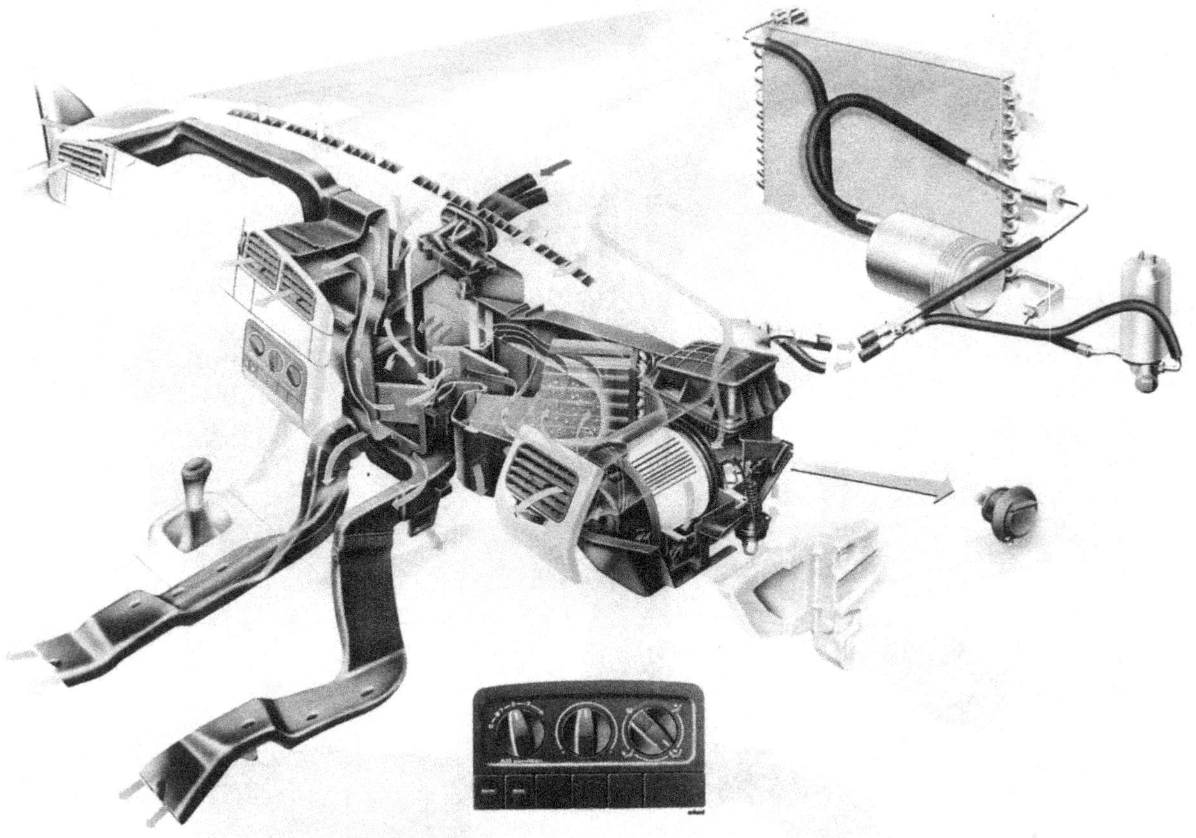

The air conditioning system also uses the air blend principle. Fresh air is first drawn in and passed through the evaporator to be cooled and dehumidified when the A/C compressor is switched on. A portion of this air can then be directed over the heat exchanger by the temperature regulation door to maintain the desired temperature.

Vehicles with A/C have two additional buttons on the ventilation controls. The A/C button turns the A/C compressor on and off. A second button is used to switch closed the air recirculation door for maximum cooling.

The series resistance for the blower is located in the A/C Evaporator case.

The air circulation door is also closed when the blower switch is turned off.

Air Conditioning

A/C System

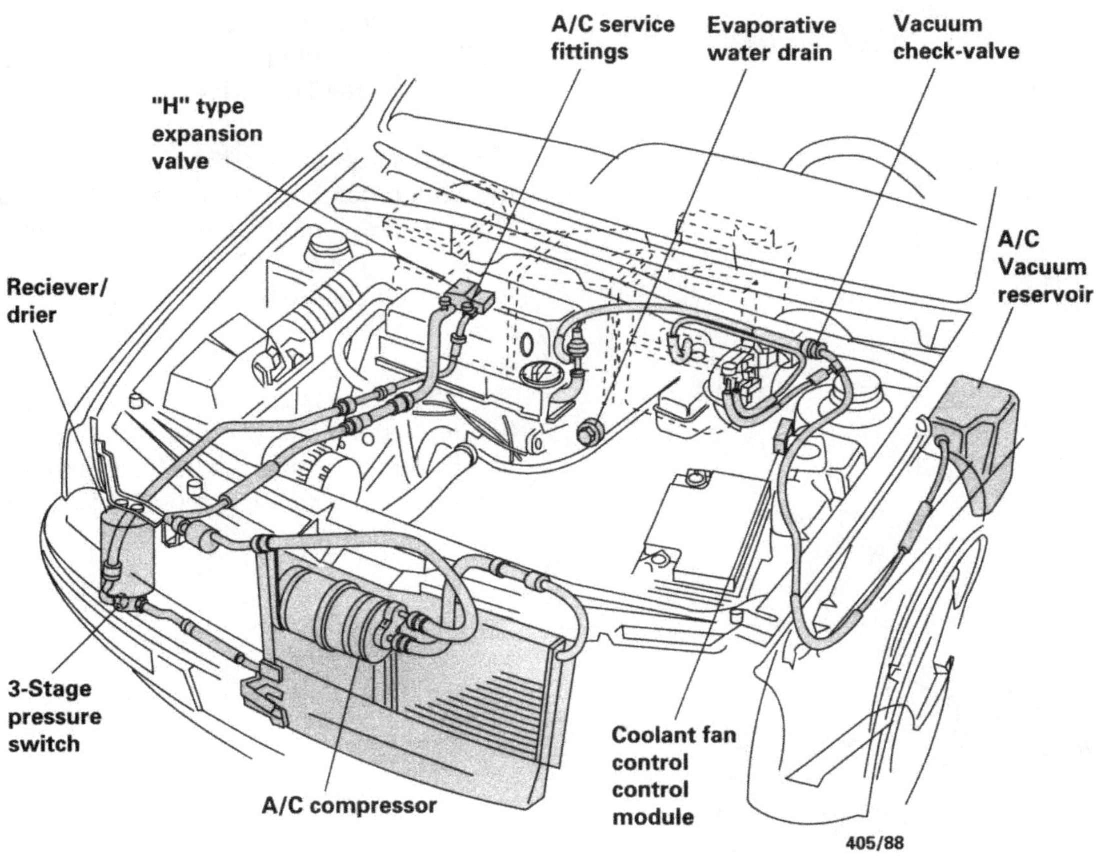

The A/C refrigerant system on the new A3 models uses a variable displacement A/C compressor. The A/C compressor no longer cycles on and off when the A/C is switched on. Instead, the displacement of the A/C compressor will increase or decrease to match the cooling load on the A/C system.

This eliminates the surging caused by a cycling A/C compressor clutch and increases the thermal efficiency of the system. For more information on this compressor, see Self-Study program #404, 1992 Model Change Information.

The A3 models use the new R-134a A/C refrigerant. All of the system's components are new and made from mostly different materials, in order to be compatible with the R-134a. These components must not be interchanged with other systems which use R-12 or other refrigerants.

Air Conditioning

R-134a Introduction

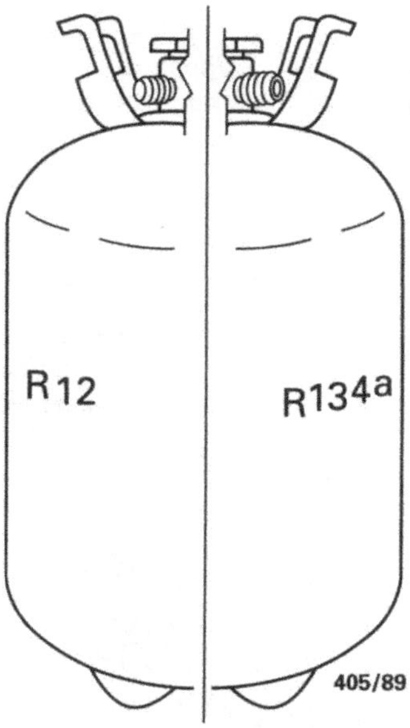

405/89

Due to the harmful effect that R-12 refrigerant has on the Earth's ozone layer, R-12 refrigerant is being phased out and being replaced by R-134a refrigerant.

The benefit of using R-134a (hydrofluorocarbon) refrigerant is that it is chlorine free. Chlorinated fluorocarbons (CFCs), present in R-12, destroy the ozone in the stratosphere through chemical reaction.

The A3 models factory-installed air conditioning system uses the new R-134a refrigerant. The A/C components for this system differ from a system using R-12 refrigerant.

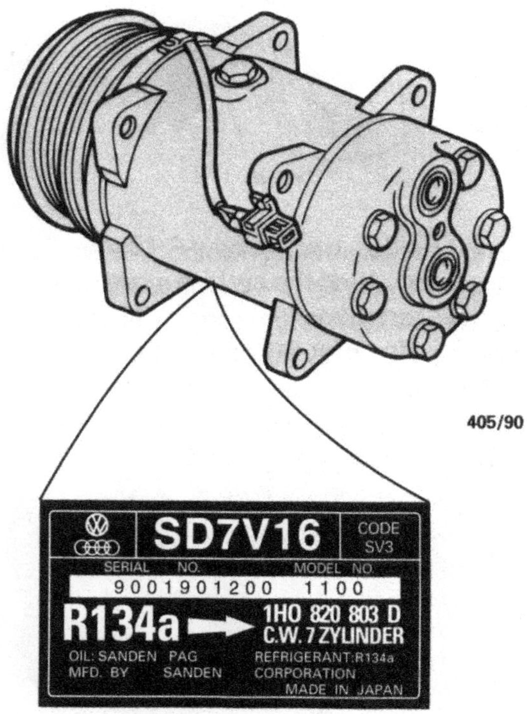

405/90

Green label designating R134a and PAG oil

Although the R-134a A/C system and the R-12 A/C system are very similar in appearance; DON'T let this fool you. There are differences in the refrigerant, lubrication, components and servicing.

A/C components used in R-134a systems can be identified by green markings. For example, the A/C hoses have a green stripe on the hose; the identification label for the A/C compressor is green.

If you are working with an R-134a system, and a green marking is not visible, make sure you check the part number prior to installation.

Air Conditioning

R-134a Introduction

A/C systems using R-12 use a mineral-oil-based lubricant. The refrigerant oil used in R-134a A/C systems is a synthetic oil (polyalkyleneglycol) otherwise known as PAG oil.

The two refrigerants and refrigerant oils must not be mixed under any circumstances or component failure will result. There are different kinds of PAG oils. Make sure that when you service an A/C system with R-134a, use the correct oil for the system you are working on.

405/133

The equipment needed to service an R-134a system is also different. This is necessary since the lubricants and the two refrigerants are not compatible.

The equipment designed to service R-134a systems have different thread configurations. The fittings are designed to prevent connection of R-12 tools and equipment to R-134a components.

For more information on Volkswagen A/C systems using the new R-134a refrigerant, refer to the Service Training Self-study program #403 titled "A/C Systems with R-134a, Part number WSP-521-403-00."

405/92a

High Pressure Service Fitting

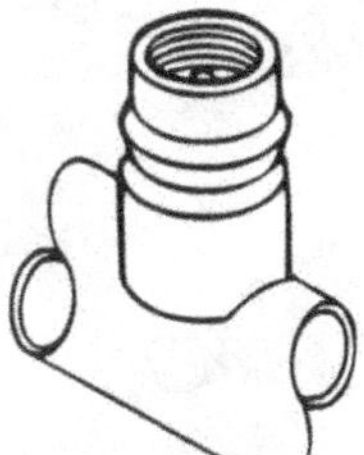

405/93a

Low Pressure Service Fitting

Generator

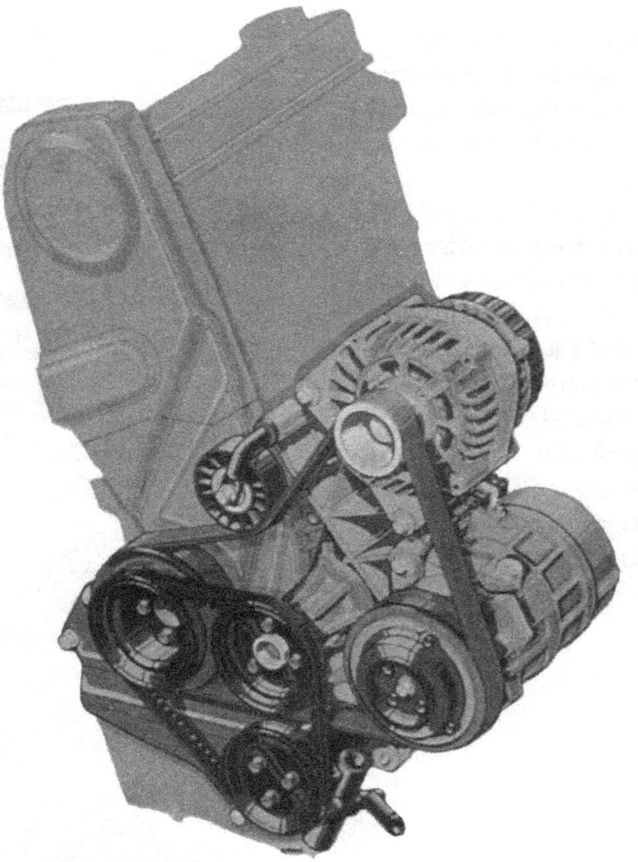

Tensioner for vehicles without A/C

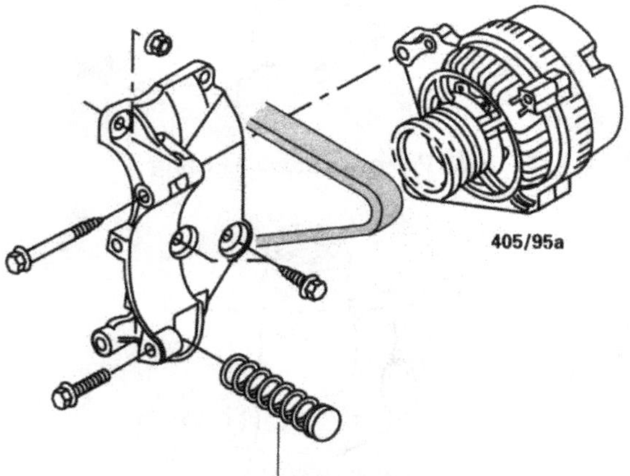

The generator and A/C compressor are now driven by a poly-ribbed belt. The belt is tensioned by an automatic tensioner that requires no adjustment.

An additional V belt is used to drive the water pump and power steering pump.

On vehicles without A/C, the alternator is spring loaded and tensions the poly-ribbed belt. A separate belt tensioner is not used.

Fuse/Relay Panel

The A3 models use the central electric III system that was introduced on the Corrado and Passat. This system is a modular design in which the harnesses are individualized. Each harness is used for only one or two vehicle systems, thus the number of wires in each harness is reduced considerably.

This simplifies the design and production of the wiring harnesses, and can reduce service diagnostic time since fewer wires are in each harness.

The fuse/relay panel is located below the left side of the dashboard. The wiring harness connectors are attached to the back of the panel. A sliding lock is located between the two rows of harness connectors. The lock will only slide and lock into place when all the connectors are fully seated.

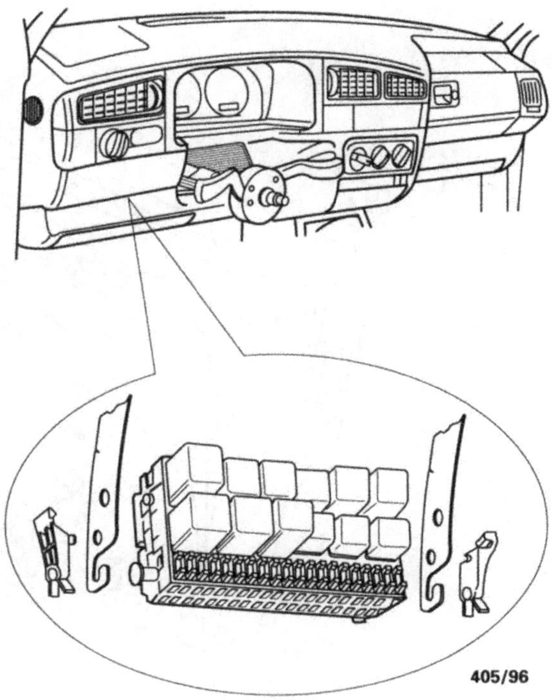

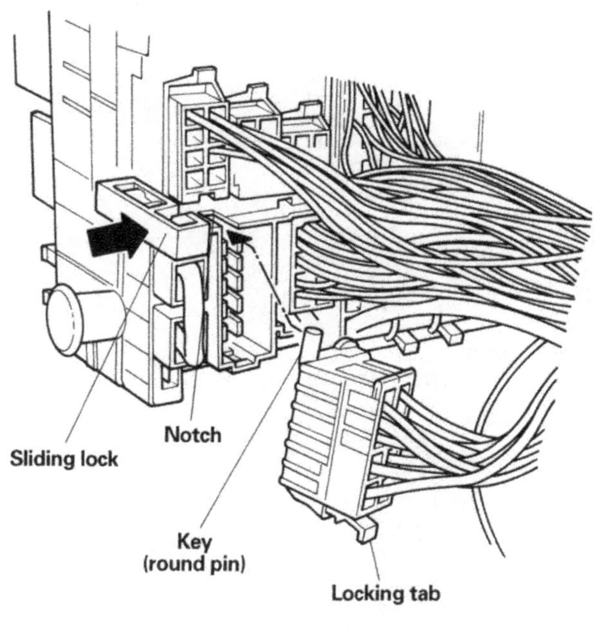

Instrument Cluster

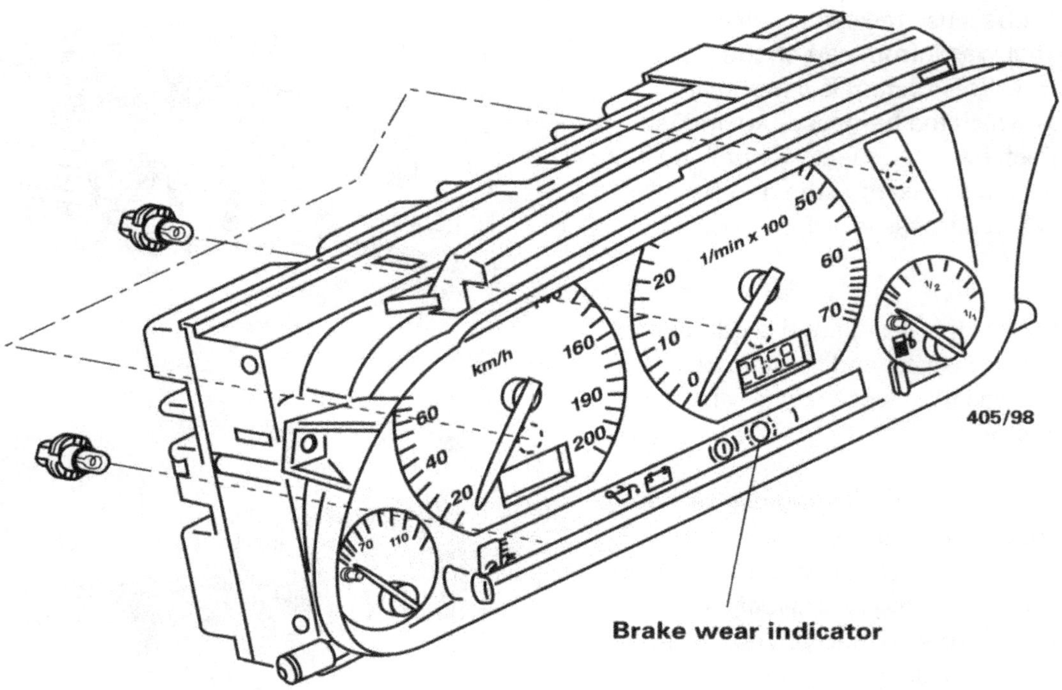

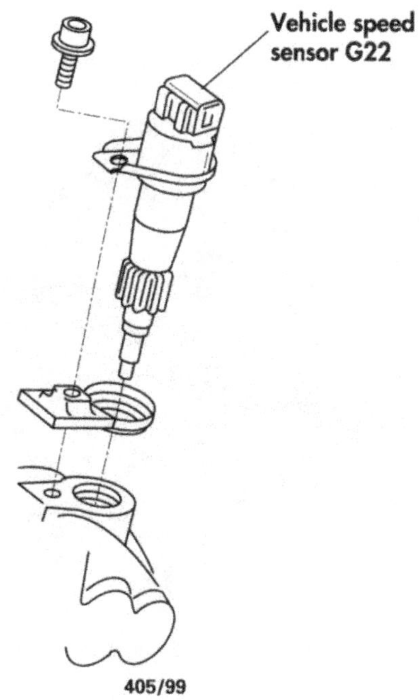

The instrument cluster has many new features. An electronic speedometer with a digital odometer display is used on all models. The speedometer receives an electronic signal from a Hall sender on the transmission.

Jetta III GLX models will include the Multi-function Indicator system. Vehicles with the VR6 engine also have a new front brake pad wear indicator.

The only repairs allowed to the instrument cluster are the replacement of warning light bulbs and the illumination light bulbs. No other repairs are possible. Remanufactured instrument cluster assemblies are available through the Parts department.

Electrical System

A service reminder indicator is standard equipment on all models. The indicator tracks vehicle mileage and time and will indicate when a service is due. The indicator appears in the odometer display window below the speedometer. The service indicator will only be shown for three seconds after the ignition is switched on (engine not running).

One of the following displays will appear, depending on the type of service due.

After a service is performed, the indicator display should be reset. To reset the display:

- Switch the ignition on then push and hold the odometer reset button.
- Switch the ignition off and release reset button.
- Push the odometer reset button until the display you wish to reset appears.
- Push the clock set button, and the service indicator display is reset.

The service indicator display has a "hard" memory and will not be erased if the battery is disconnected.

After a service has been performed, any previous service displays must also be reset. For example, when resetting INO1, the OEL display must also be reset.

When installing a new instrument cluster, the new service reminder must be programmed to come on at the correct time. Refer to the Repair Manual for information on this procedure.

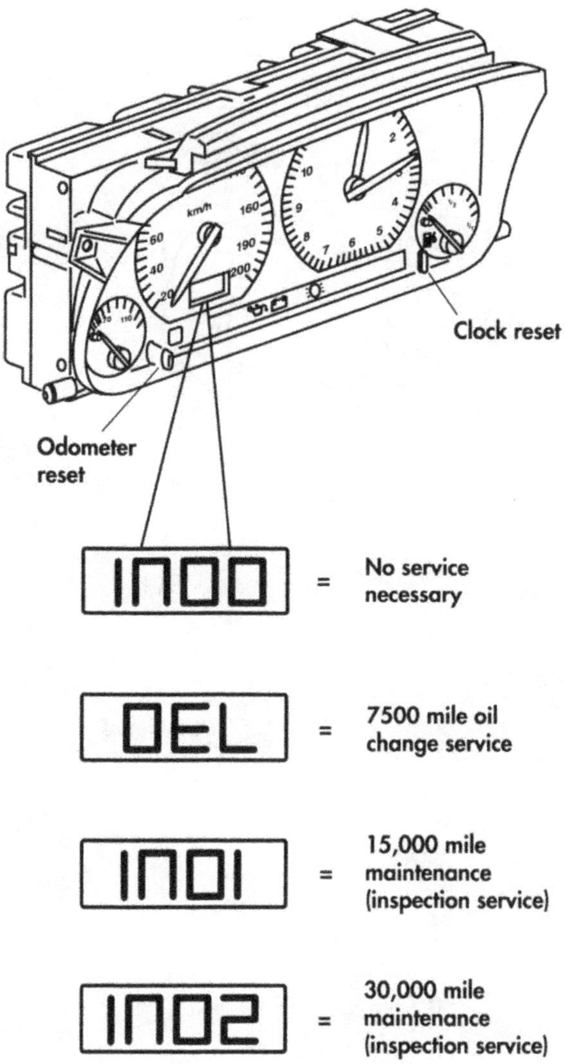

405/100

Radio

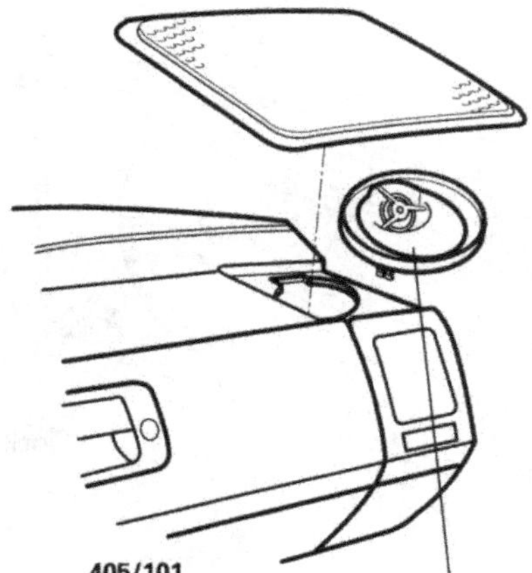

405/101

The A3 models are equipped with an 8 speaker sound system. These speakers do not use separate amplifiers like some Volkswagen "active" speaker systems. The speakers are powered directly by amplifiers located within the radio.

Full range speakers are located in the front and rear doors (on two door models the rear speakers are located in the rear quarter interior trim panel).

High frequency directional tweeters are located above the rear speakers and also in the dashboard.

Directional tweeters

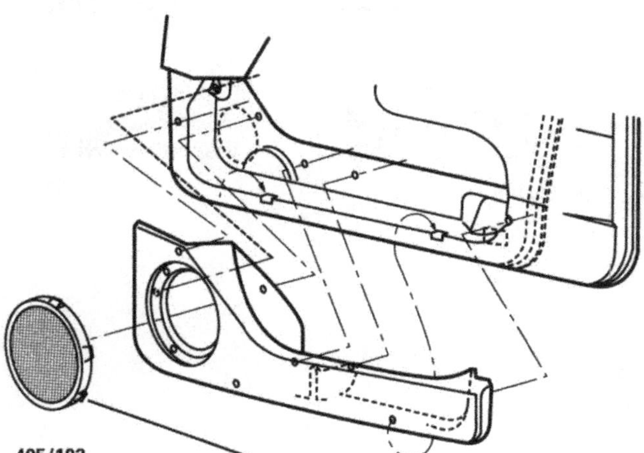

405/102

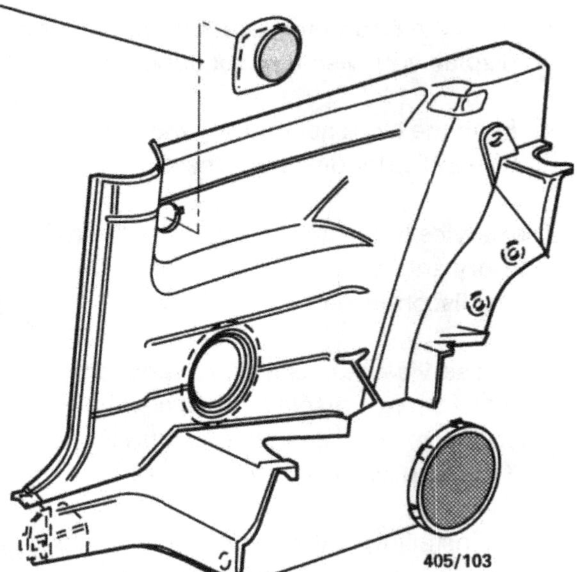

405/103

Full range speakers

Radio

Two new radios are available. The VW Deluxe radio provides 40 watts of power (2 x 20W), and an external jack for portable CD players. The VW Premium radio has 80 watts (4 x 20W), tape, scan, CD changer connectability and other features. Both AM/FM stereo cassette radios have the ignition switch power shut-off feature and are security coded with a blinking LED like other Volkswagen radios.

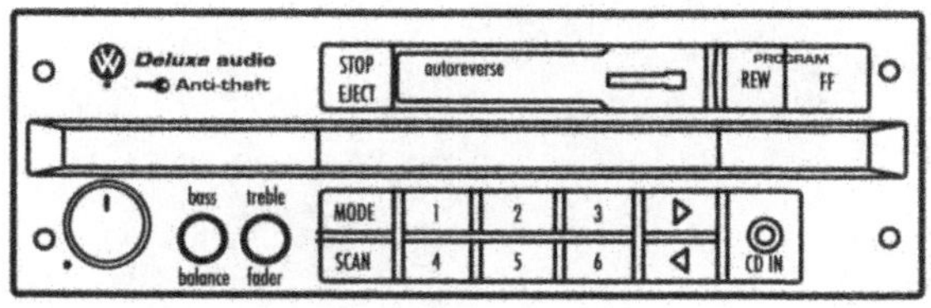

405/104

The radios are "hard" programmed with a security code as they are manufactured. Once this security code is entered into the radio (during the PDI) it must be entered again to activate the radio after the power has been disconnected from it.

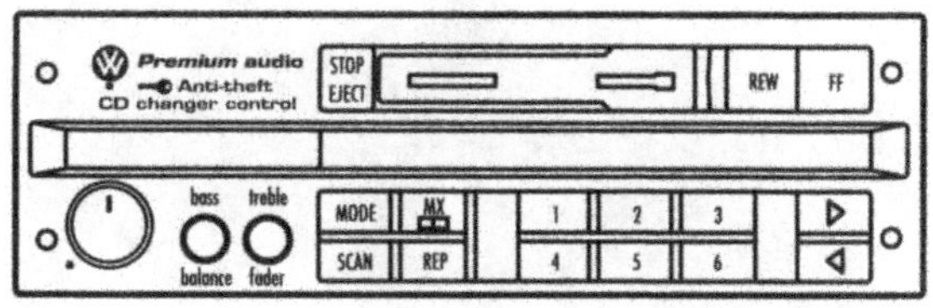

405/105

The coding process to enter the code is activated by holding the Mode and Scan buttons depressed. The radio code can then be entered by using the first four radio station program buttons. In addition to this security feature, these radios are also connected to the vehicles alarm system (see alarm system).

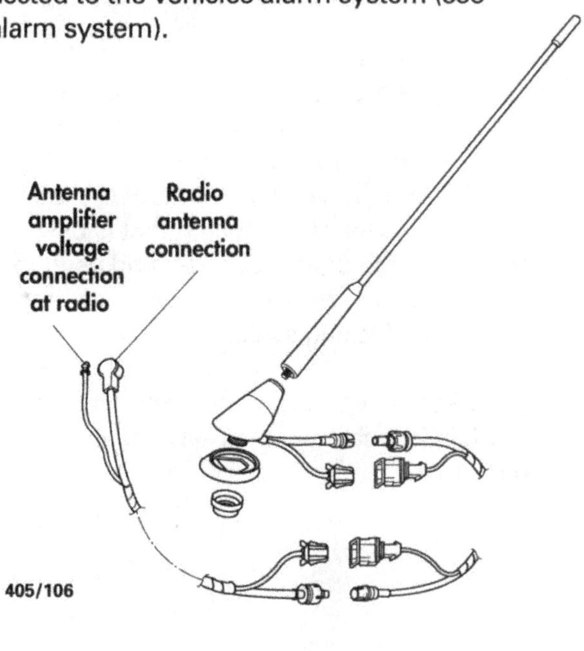

405/106

Two radio antennas are used. GTI and GLX models use the roof mounted antenna that includes an amplifier. All other models have collapsible antenna mounted in the left front fender.

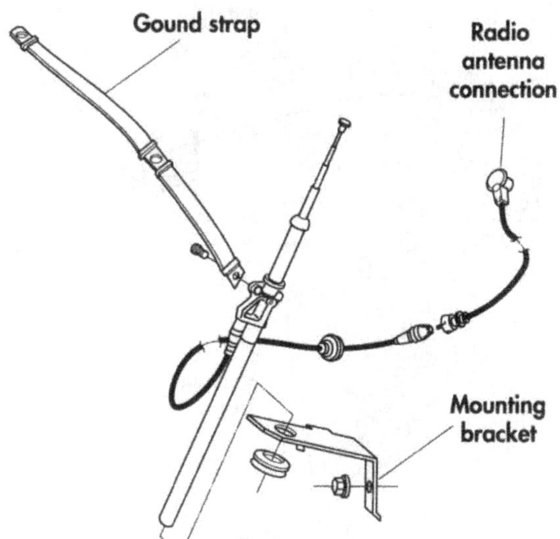

405/107

Electrical System

Central Locking

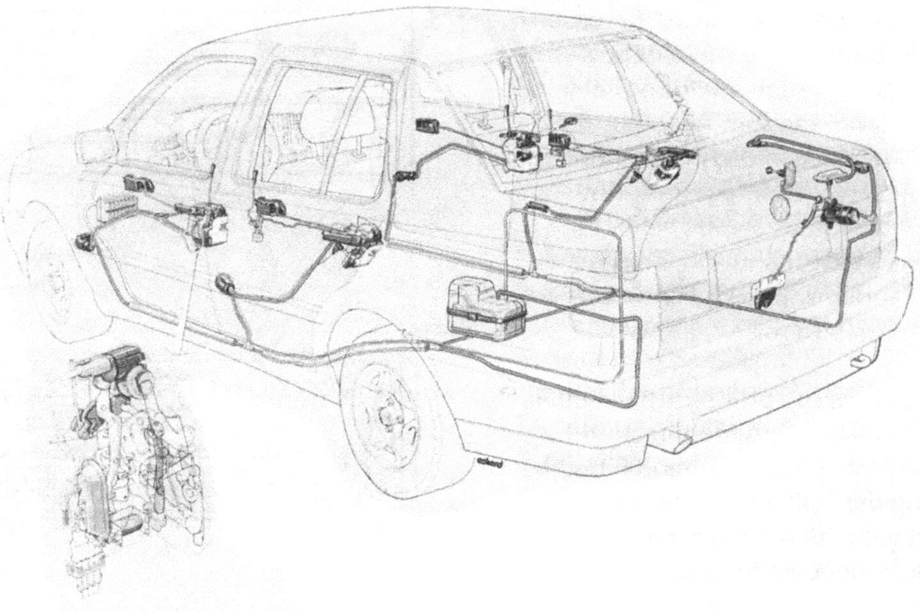

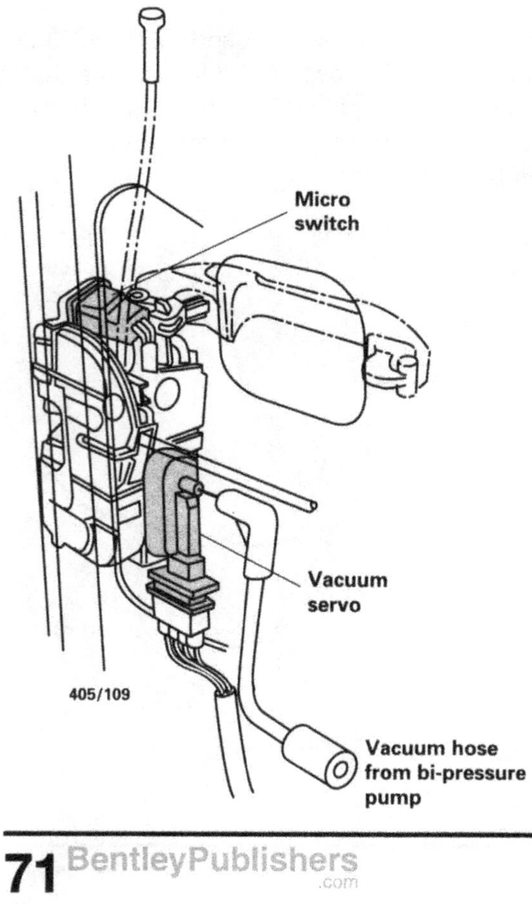

All models are equipped with central locking for the doors, rear lid and fuel filler door. The system has new lock assemblies. The vacuum servo is now part of the latch assembly. The micro-switch that is operated by the door lock cylinder and used to switch the alarm on and off also activates the central locking system.

The bi-pressure pump and systems control module is located in the rear luggage compartment.

Electrical System

Central Locking

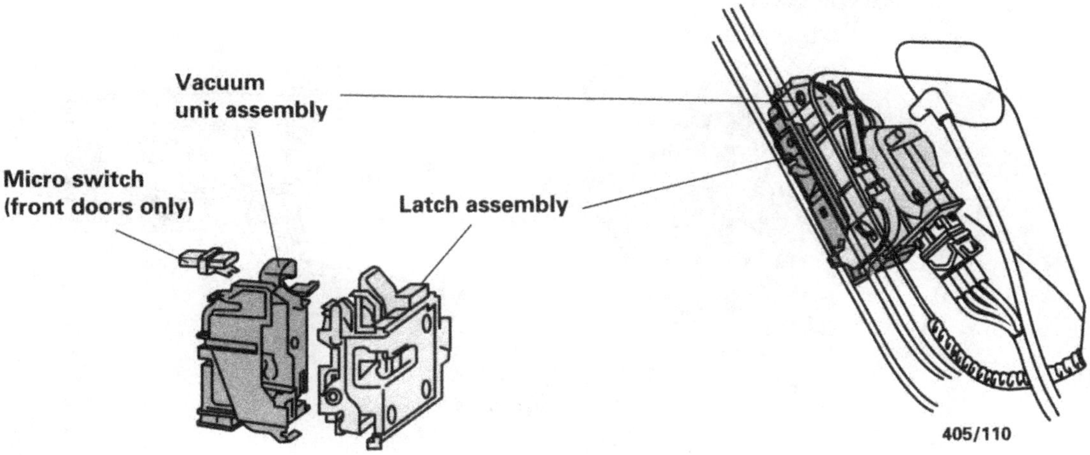

405/110

The micro-switch for the alarm and central locking system can be replaced separately from the door latch assembly.

Separate vacuum servos are used for the trunk lid and fuel filler door.

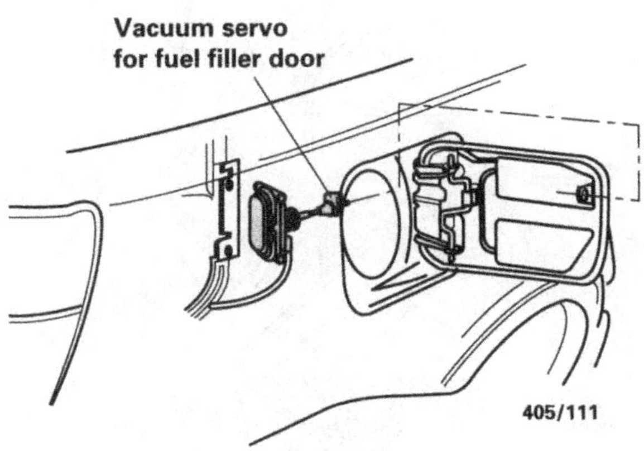

405/111

Cruise Control

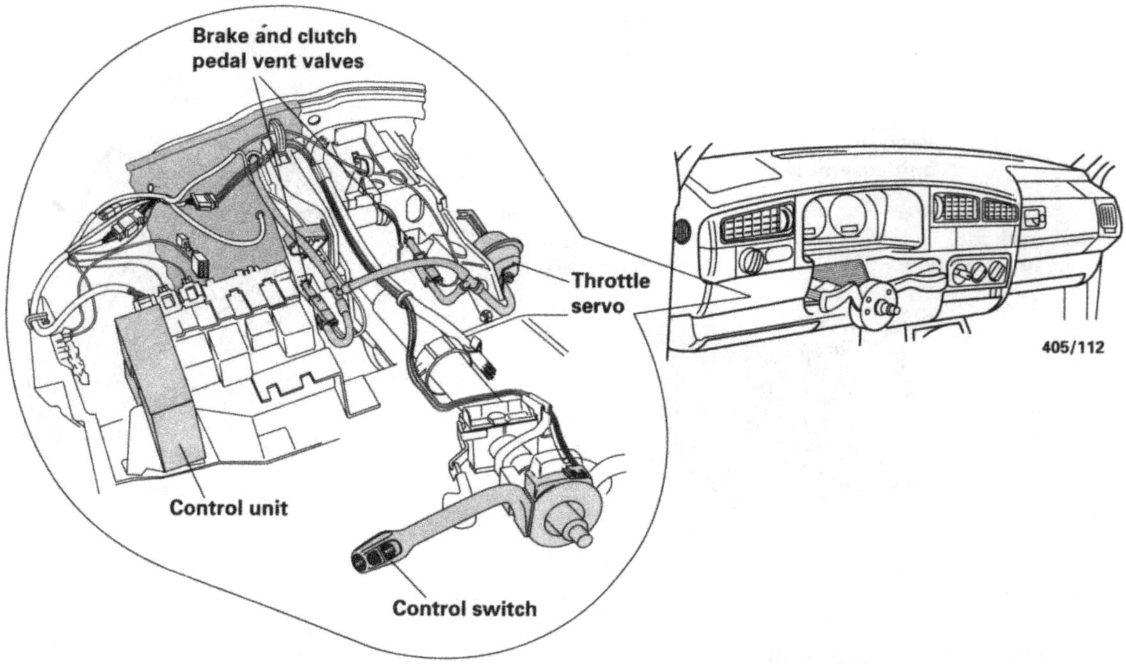

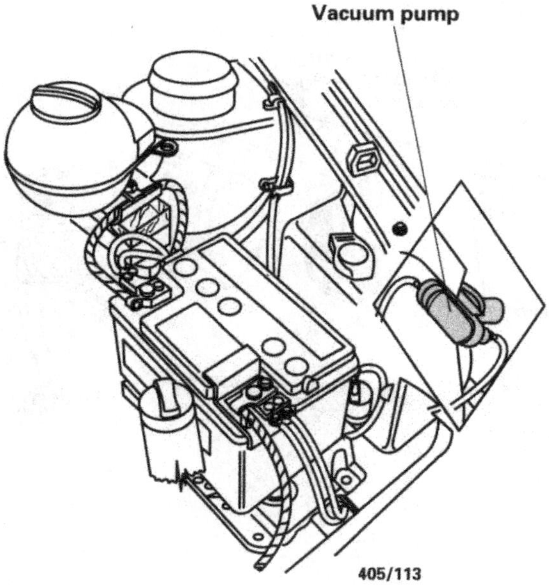

Cruise control will be available on Jetta III GLS and GLX models. The system is similar to other Volkswagen models. The vacuum servo for the throttle is located on the pedal assembly. A electrical/mechanical valve is used on the clutch and brake pedal. These switches vent the vacuum to the throttle servo when the brake or clutch pedal is depressed. They also open an electrical ground signal to the control module which switches the system off.

The control module is located behind the top left of the dashboard, next to the alarm control module. The vacuum pump is located inside the top of the left front fender, ahead of the wheel liner.

On vehicles with automatic transmissions, the cruise control module receives vehicle voltage from the transmission control module. The cruise control module will not receive voltage until the vehicle speed has exceeded approximately 20 mph (35 kmh).

Remote Trunk Release

All U.S. models of the A3 are equipped with a power remote trunk release. A release switch is located in the left side of the glove box.

The switch operates a release motor that is located in the rear lid. The motor is activated when the release button is pushed. The motor operates a lever which is attached to the motor. The lever pushes the trunk latch rod in the open or unlatch direction.

An indicator light is also included in the instrument cluster to show when the trunk is open.

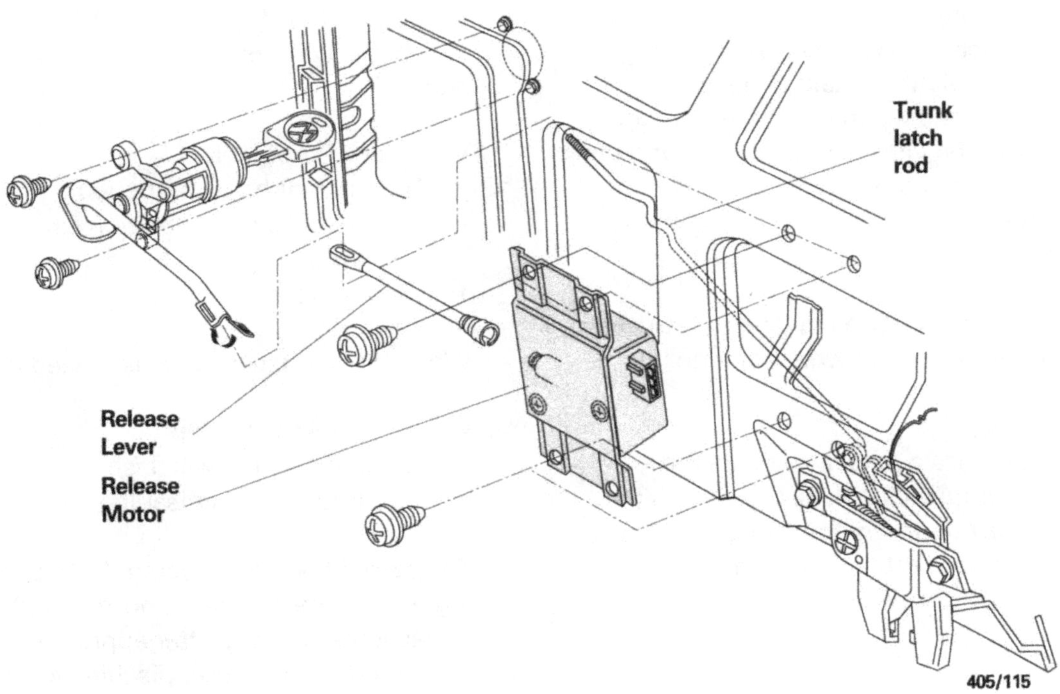

Alarm System

All A3 models are equipped with the Volkswagen alarm system. The alarm system is armed when either front door is locked with the vehicle ignition key, and the trunk lid, hood and all doors are closed. The system is similar to that installed on the Passat and Corrado.

The alarm will not arm if a door, the trunk lid, or hood is opened when the door is locked.

The system has a defeat switch in the luggage compartment lock that allows the luggage compartment to be opened using the key, without activating the alarm.

When the alarm is armed, it will become active if:

- A door is opened
- The trunk or hood is opened
- Terminal #15 of the ignition circuit is energized
- The radio is removed

When the alarm system is activated the:

- Alarm horn will beep
- Directional lights will flash
- The starter will be disabled

To prevent the alarm system from becoming a nuisance, the alarm horn and directional lights will stop after approximately 30 seconds. The starter disable will, however, remain active.

The alarm control module is located behind the upper left side of the dashboard, next to the left side dashboard fresh air vent.

Alarm System

An LED is located next to the driver's door lock knob. The LED will flash at 1-second intervals after the doors are locked to indicate that the system will soon be armed. After 30 seconds, the LED will begin to flash faster (50 ms on, 950 ms off) to indicate that the system is armed.

To prevent battery drain, the LED will stop flashing after 24 hours. The alarm system will, however, remain armed.

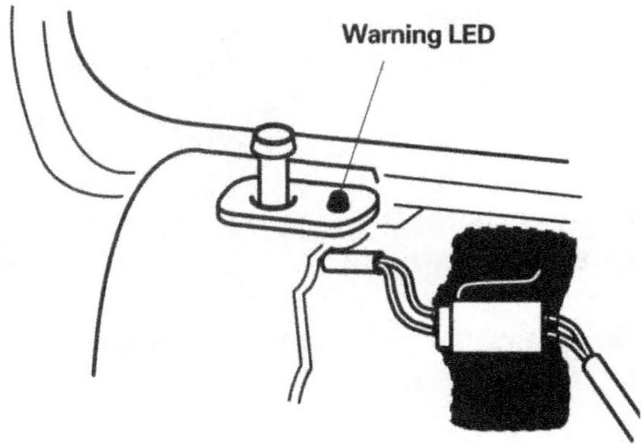

405/17

The switch used to turn the alarm system on and off is not incorporated in the door handle like other Volkswagen models. Instead, a separate micro-switch is located on top of the door latch assembly. This switch is operated by a cam which is turned by a shaft attached to the door lock cylinder. The switch can be replaced separately from the door latch assembly.

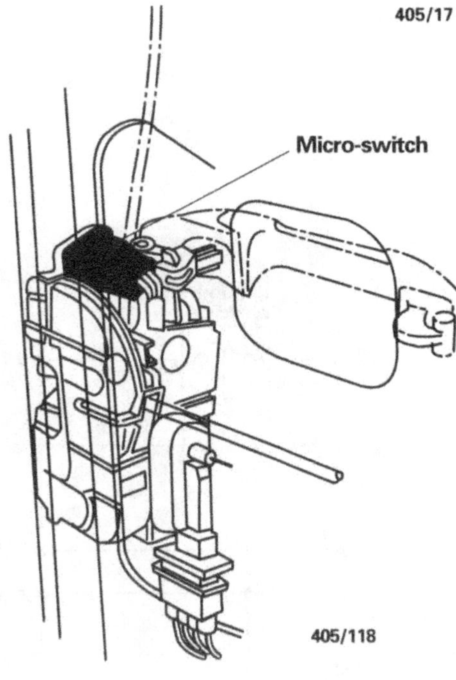

405/118

The radio is also connected to the alarm system through a wire in one of the radio multi-plug connectors. This wire is connected to a switch in the left radio retention tab. If the retention tab is depressed to remove the radio when the alarm is armed, a ground circuit is completed. This will activate the alarm.

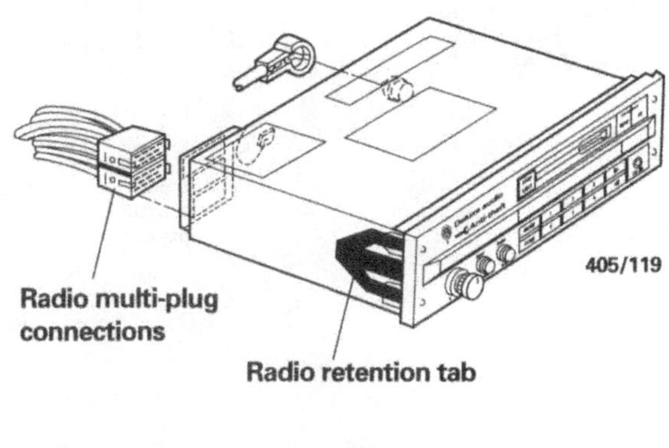

405/119

Alarm System

The alarm control module includes three mechanical relays. These relays operate the following components:

- Starter motor
- Alarm horn
- Emergency flashers

No repairs can be made to the alarm control module.

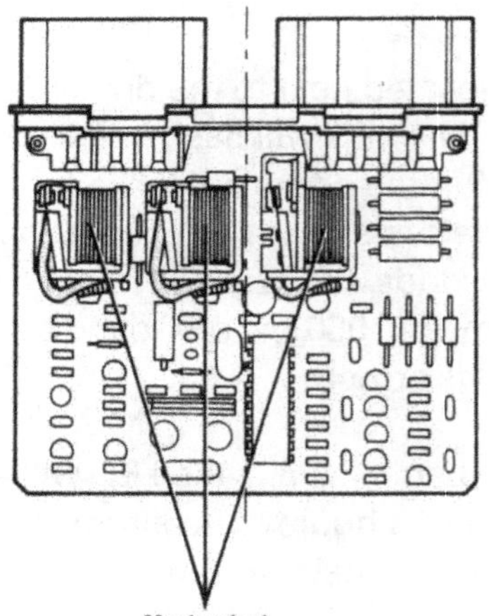

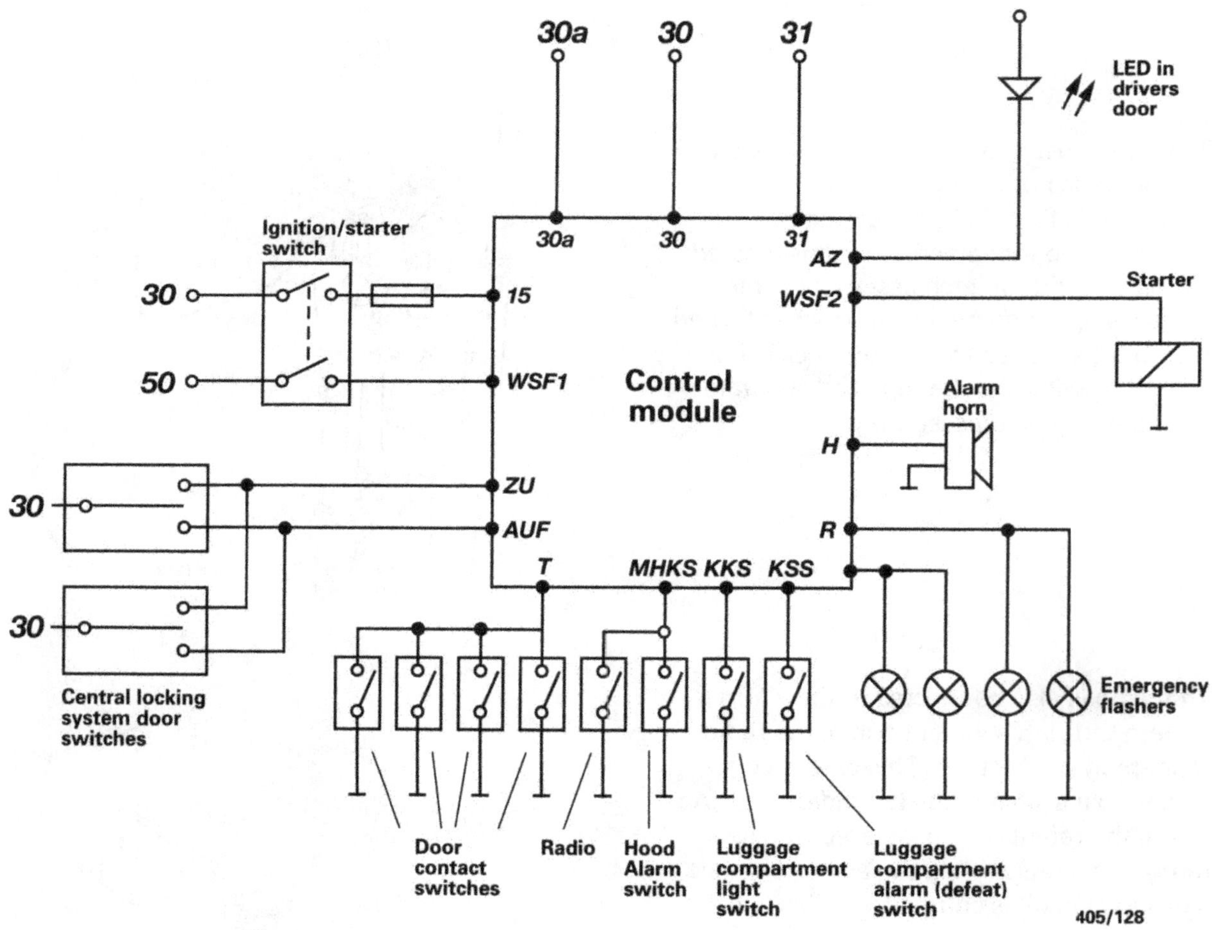

Fog Lights

GTI and Jetta III GLX models are equipped with a new fog light design. These fog lights are ellipsoid reflector lights. The fog lights have an ellipsoid shaped (similar to an egg shape) reflector to direct the light from the bulb through a lens. The lens works as a projector lens. An imaging screen is placed between the lens and reflector to give the light beam a precisely defined cut-off area.

This design results in:

- A very uniform and sharp light beam
- Extremely low glare

The fog light bulbs can be replaced separately, and the fog lights are aimed with adjusting screws, like other light assemblies.

The fog light switch is integral with the headlight switch. With the headlights turned on, the fog lights can be switched on by pulling the switch knob out.

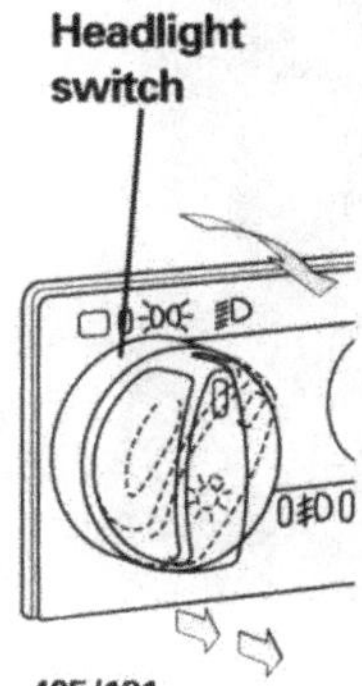

405/121

405/122

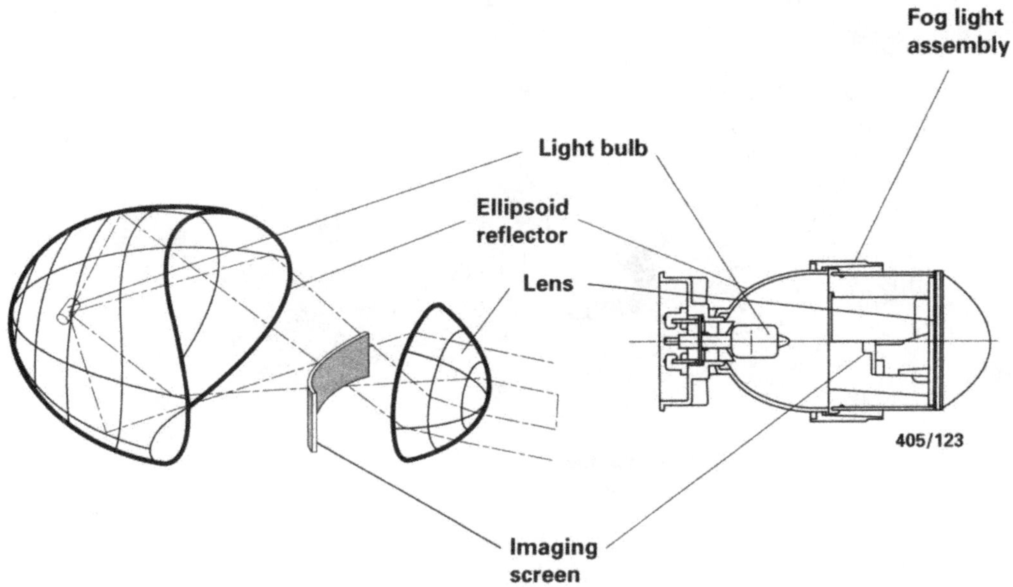

405/123

Electrical System

Central Connectors

Centralized electrical connectors are used to simplify assembly. These connectors will also aid in service procedures. Instead of several smaller electrical connectors, one large central connector is used for each of the door assemblies, the engine, and the front end of the vehicle.

This makes it possible to install the wiring and assemble these components away from the vehicle in a modular fashion, before installation. In service, for example, an entire door assembly can be removed from the vehicle by simply unbolting the hinges and door strap, then disconnecting this one electrical connector.

The connectors are installed and removed by turning the body of the connector housing. The connector is properly installed when all three recessed portions on the connector body are aligned.

The bi-pressure hoses for the central locking servos are also routed through these connectors which are installed in the door assemblies.

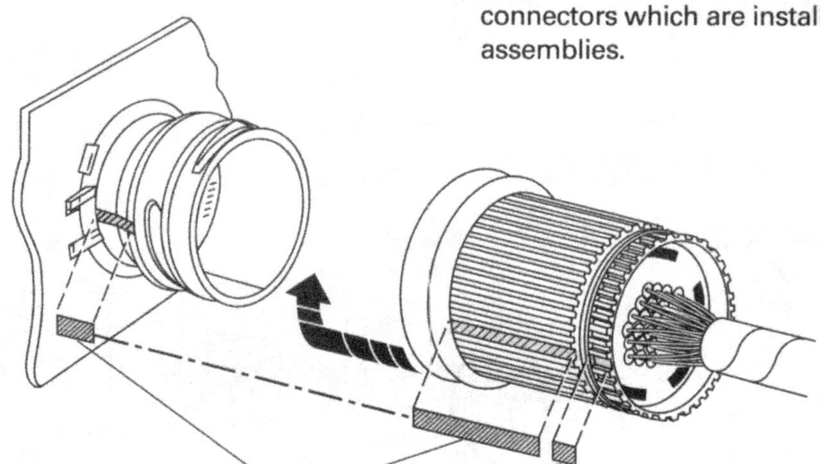

Electrical System

Wiring

Many of the vehicle wiring harnesses for the doors and the rear of the vehicle are routed below the door sill panels. Special cable holders are clipped to the body under the sill panel trim.

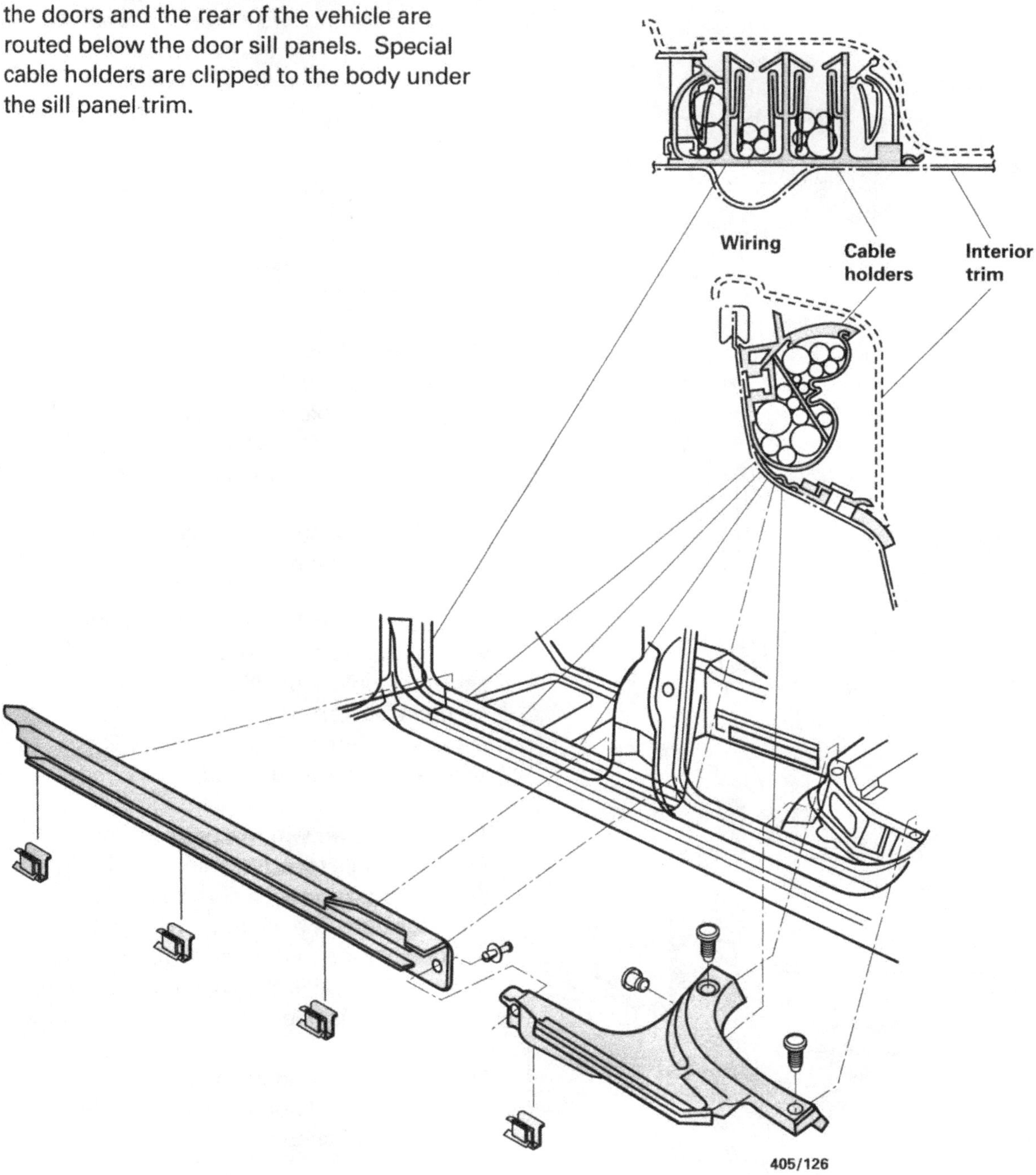

Power Windows

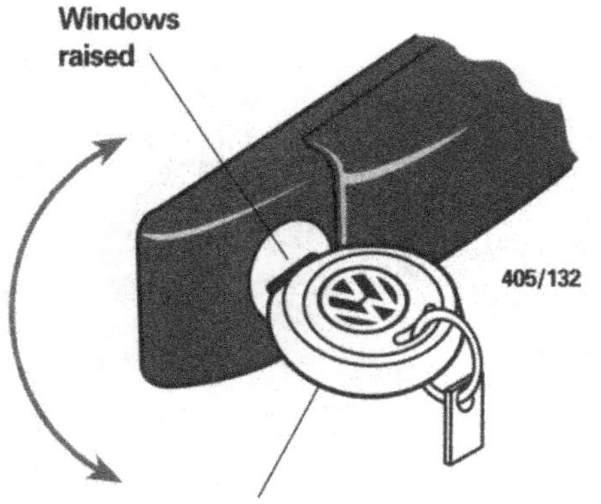

Jetta III GLS and GLX models equipped with power windows feature a new system that can operate all four windows in parallel. The power windows can also be raised and lowered by using the key in either front door lock.

By holding the key in the locked position for more than 1 second, all windows that are opened will be raised to the closed position at the same time.

If the vehicle is equipped with a sunroof, it will close after all the windows are closed. This prevents excessive drain on the battery. This convenience feature for the sunroof only works in the close direction.

The windows and sunroof will stop as soon as the key is released.

As a ventilation aid in hot weather, all four windows can be lowered. By holding the key in the unlock position for more than 2 seconds, all four windows will be lowered at the same time. The windows will stop as soon as the key is released.

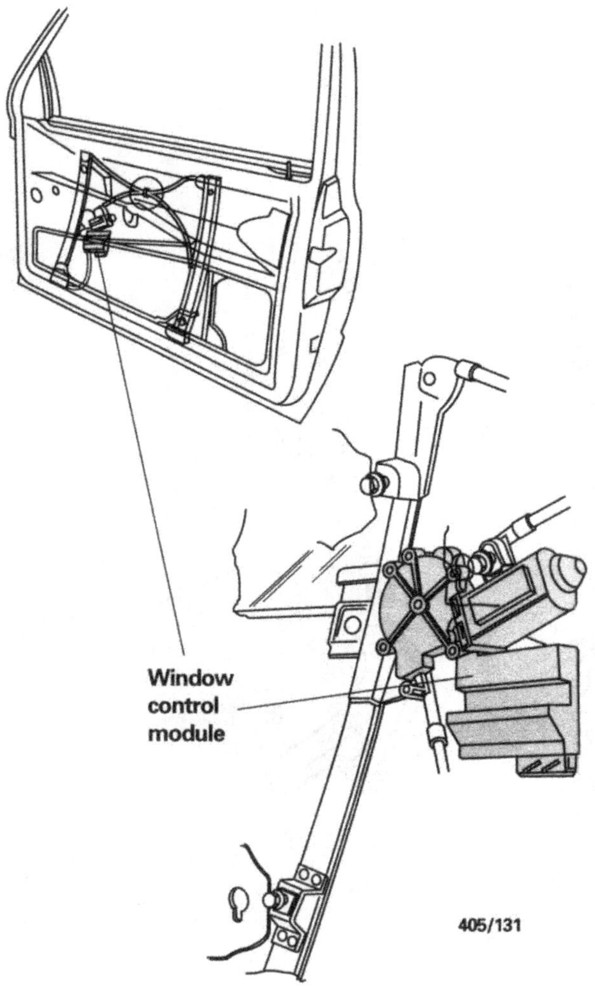

This new power window system also features single touch opening for all windows and single touch closing for the driver's door. The windows can continue to operate for approximately 10 minutes after the ignition is switched off or until the driver's door is opened.

The single touch closing feature for the driver's door will not operate when the ignition is switched off.

Power Windows

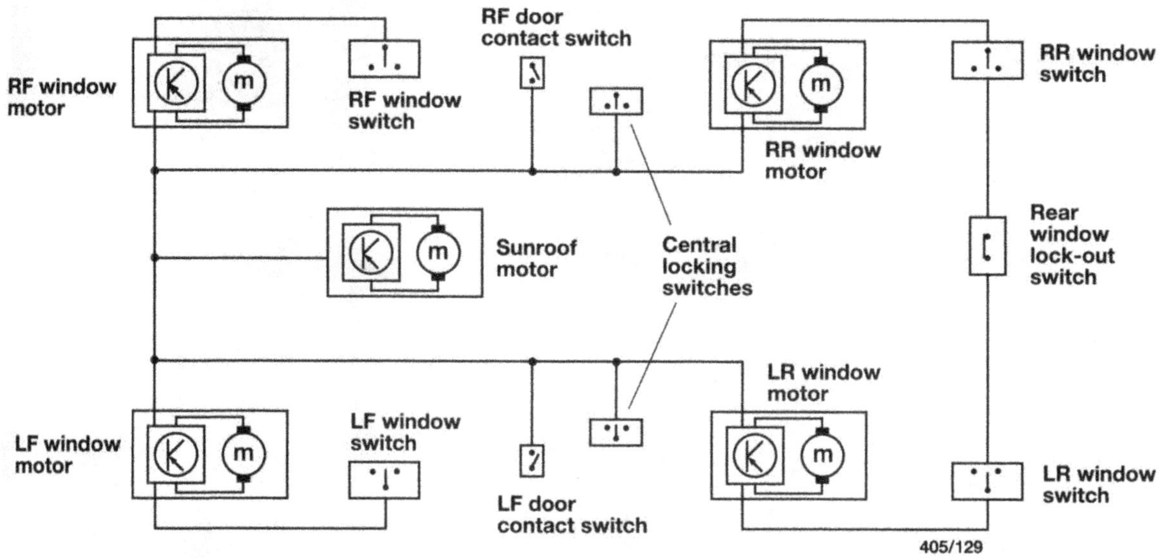

The window motors are operated by control modules that are attached to each window regulator motor. The window motor control modules are electrically connected in parallel with each other and the central locking switches in the front doors.

Each window control module contains a pair of relays, various ICs and a window motor position sensor. The window motor and control module are supplied as a single unit.

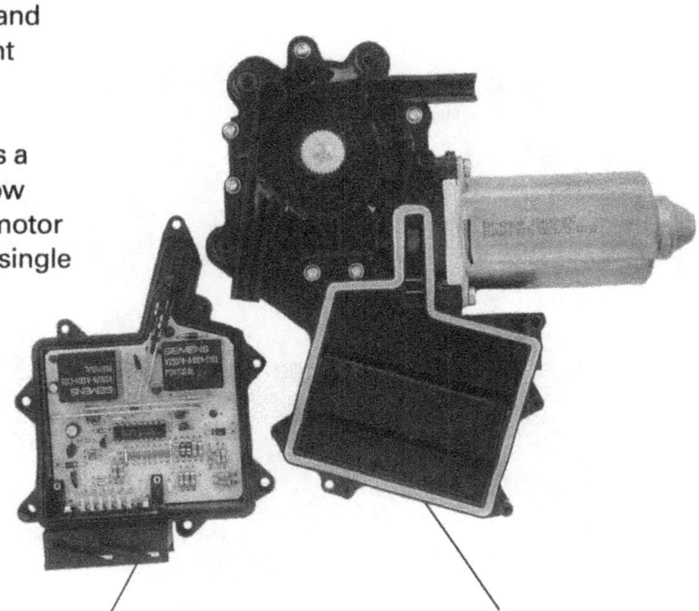

Notes

CONTROLS AND EQUIPMENT

General illustration

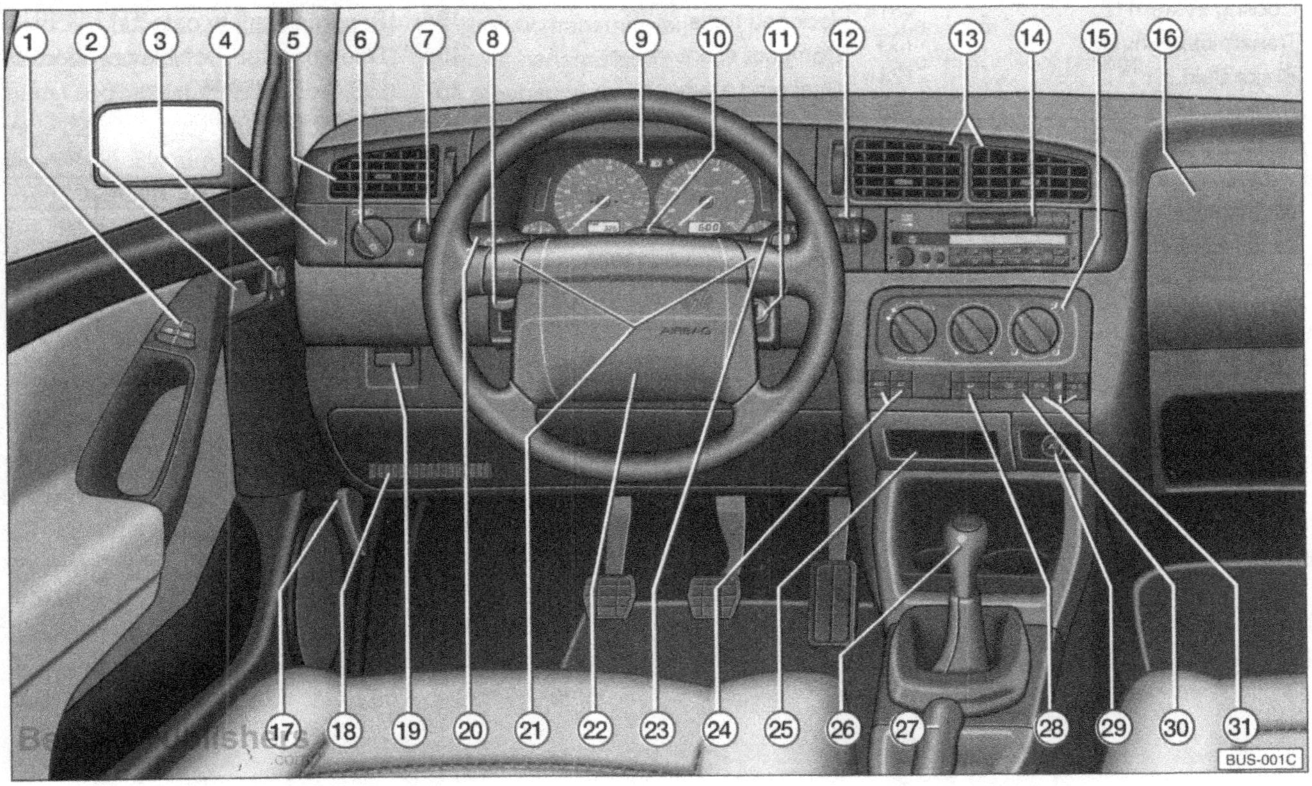

1 - Power window switches
2 - Door handle
3 - Adjuster control for outside mirrors
4 - ABS indicator light
5 - Air vent
6 - Switches
7 - Instrument cluster illumination
8 - Height adjuster for steering wheel
9 - Instrument cluster: Instruments Warning and Indicator lights
10 - Emergency flasher switch
11 - Steering lock / ignition / starter switch
12 - Switches
13 - Air vents
14 - Radio[1]
15 - Climate controls
16 - Passenger's side airbag
17 - Release lever for engine hood
18 - Fuse panel behind knee bar
19 - Cold start assist handle* / Idle speed boost* (Diesel models only) .... Not available
20 - Turn signals/headlight dimmer switch lever with knobs for cruise control
21 - Horn (works only when the ignition is switched on)

[1] A separate brochure is provided for your factory-installed radio. If you replace your radio, please be sure to read the notes in the chapter "Do-it-yourself Service", page 168.

22 - Driver's side air bag
23 - Windshield wiper / washer lever with knobs for multi-function indicator
24 - Switches
25 - Ashtray
26 - Selector lever (Automatic transmission) Gearshift lever (manual transmission)
27 - Parking brake lever
28 - Central locking switch.
29 - Cigarette lighter
30 - Rear window defogger switch
31 - Power window switches

Notes

■ Some features mentioned are standard equipment on some models only and options on others.

■ For technical reasons, the arrangement of the switches and indicator/warning lights may differ. The governing factor is the symbol on the switch itself.

> **WARNING**
>
> Your common sense, good judgement, and general alertness are crucial to safe and successful service work. Before attempting any work on your Volkswagen, read the warnings and cautions on page vii and the copyright page at the front of the manual. Review these warnings and cautions each time you prepare to work on your Volkswagen. Please also read any warnings and cautions that accompany the procedures in the manual.

> **WARNING**
>
> Your common sense, good judgement, and general alertness are crucial to safe and successful service work. Before attempting any work on your Volkswagen, read the warnings and cautions on page vii and the copyright page at the front of the manual. Review these warnings and cautions each time you prepare to work on your Volkswagen. Please also read any warnings and cautions that accompany the procedures in the manual.

WARNING

Your common sense, good judgement, and general alertness are crucial to safe and successful service work. Before attempting any work on your Volkswagen, read the warnings and cautions on page vii and the copyright page at the front of the manual. Review these warnings and cautions each time you prepare to work on your Volkswagen. Please also read any warnings and cautions that accompany the procedures in the manual.